2006 The ARRL Handbook

For Radio Communications

Editor
R. Dean Straw, N6BV

Contributing Editors
Steven R. Ford, WB8IMY
Charles L. Hutchinson, K8CH
Rick Lindquist, N1RL
Larry D. Wolfgang, WR1B

Editorial Assistants
Helen W. Dalton, KB1HLF
Maty Weinberg, KB1EIB

Technical Consultants
Michael E. Gruber, W1MG
Edward F. Hare, Jr., W1RFI
Zachary H.J. Lau, W1VT
Michael D. Tracy, KC1SX

Cover Design
Sue Fagan
Bob Inderbitzen, NQ1R

Production
Michelle Bloom, WB1ENT
Michael Daniels
Jodi Morin, KA1JPA
David F. Pingree, N1NAS
Joe Shea

CD-ROM Production
Dan Wolfgang

Proofreader
Kathy Ford

Additional Contributors to the 2006 Edition:
Wayde S. Bartholomew, K3MF
David J. Benson, K1SWL
Albert C. Buxton, W8NX
L. B. Cebik, W4RNL
John J. Champa, K8OCL
Paul M. Danzer, N1II
Donald R. Greenbaum, N1DG
John C. Hennessee, N1KB
George L. Heron, N2APB
Howard S. Huntington, K9KM
Sylvia K. Hutchinson, K8SYL
Richard M. Jansson, WD4FAB
Shawn A. Reed, N1HOQ
Douglas T. Smith, KF6DX
Frederick J. Telewski, WA7TZY
James L. Tonne, WB6BLD
Edward E. Wetherhold, W3NQN
Rosalie A. White, K1STO

Cover Info:
New high-power HF amplifier by Jerry Pittenger, K8RA, using new 3CX1500D7 tube.

Eighty-Third Edition

Published by:
ARRL—the national association for Amateur Radio
Newington, CT 06111 USA

Printed in the USA

ISBN: 0-87259-948-5 Softcover
ISBN: 0-87259-949-3 Hardcover

Eighty-Third Edition

Contents

Chapter 4 — Electrical Fundamentals

Chapter 5 — Electrical Signals and Components

Chapter 6 — Real-World Component Characteristics

Chapter 7 — Component Data and References

Chapter 8 — Circuit Construction

Chapter 9 — Modes and Modulation Sources

Chapter 10 — Oscillators and Synthesizers

Chapter 11 — Mixers, Modulators and Demodulators

Chapter 12 — RF and AF Filters

Chapter 13 — EMI/Direction Finding

Chapter 14 — Receivers and Transmitters

Chapter 15 — Transceivers, Transverters and Repeaters

Chapter 16 — DSP and Software Radio Design

Chapter 17 — Power Supplies

Chapter 18 — RF Power Amplifiers

Chapter 19 — Station Layout and Accessories

Chapter 20 — Propagation of RF Signals

Chapter 21 — Transmission Lines

Chapter 22 — Antennas

Foreword

The enormously successful 2005 Edition of *The ARRL Handbook* has been updated in this 2006 Edition. Thanks to the sharp eyes of a number of readers, the minor errors in the massive 2005 Edition rewrite have been fixed.

A brand-new, high-power HF linear amplifier project by Jerry Pittenger, K8RA, has been added to this 2006 Edition. Feast your eyes on Jerry's beautiful workmanship building a full-featured amplifier using the new 3CX1500D7 power triode. This rugged new Eimac tube has a 50-W grid dissipation, making it capable of withstanding almost any condition of drive or tuning.

Once again, we are including the fully searchable CD-ROM containing all of the almost 1200 pages of the printed book. Readers have been very enthusiastic about the "instant search" capabilities they have when they load the *Handbook* on their hard drives.

Whether you want to tackle projects using surface-mount components or learn how a computer can enhance many aspects and areas of your Amateur Radio pursuits, this *Handbook* has something for everyone.

So, whether you prefer to open this comprehensive reference book on your workbench, in your favorite reading chair, or using the CD drive of your computer, enjoy all that this 2006 Edition *Handbook* has to offer.

David Summer, K1ZZ
Executive Vice President
Newington, Connecticut
September 2005

The Amateur's Code

The Radio Amateur is:

CONSIDERATE...never knowingly operates in such a way as to lessen the pleasure of others.

LOYAL...offers loyalty, encouragement and support to other amateurs, local clubs, and the American Radio Relay League, through which Amateur Radio in the United States is represented nationally and internationally.

PROGRESSIVE...with knowledge abreast of science, a well-built and efficient station and operation above reproach.

FRIENDLY...slow and patient operating when requested; friendly advice and counsel to the beginner; kindly assistance, cooperation and consideration for the interests of others. These are the hallmarks of the amateur spirit.

BALANCED...radio is an avocation, never interfering with duties owed to family, job, school or community.

PATRIOTIC...station and skill always ready for service to country and community.

—The original Amateur's Code was written by Paul M. Segal, W9EEA, in 1928.

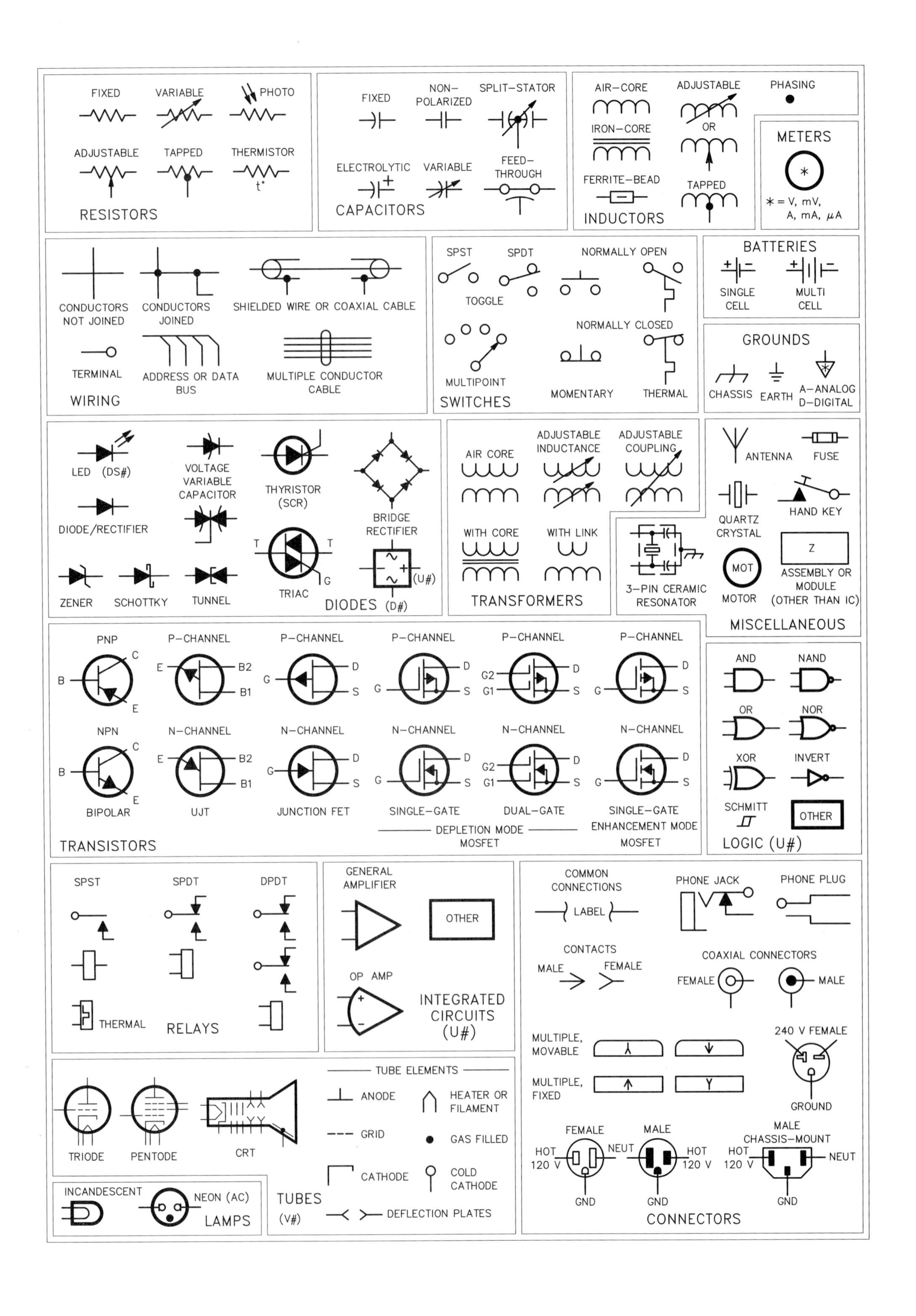

FIXED
VARIABLE
PHOTO
ADJUSTABLE
TAPPED
THERMISTOR
t°
RESISTORS
FIXED
NON-POLARIZED
SPLIT-STATOR
ELECTROLYTIC
VARIABLE
FEED-THROUGH
CAPACITORS
AIR-CORE
IRON-CORE
FERRITE-BEAD
ADJUSTABLE
OR
TAPPED
INDUCTORS
PHASING
METERS
∗ = V, mV, A, mA, μA
CONDUCTORS NOT JOINED
CONDUCTORS JOINED
SHIELDED WIRE OR COAXIAL CABLE
TERMINAL
ADDRESS OR DATA BUS
MULTIPLE CONDUCTOR CABLE
WIRING
SPST
SPDT
TOGGLE
NORMALLY OPEN
NORMALLY CLOSED
MULTIPOINT
MOMENTARY
THERMAL
SWITCHES
BATTERIES
SINGLE CELL
MULTI CELL
GROUNDS
CHASSIS
EARTH
A-ANALOG
D-DIGITAL
LED (DS#)
VOLTAGE VARIABLE CAPACITOR
THYRISTOR (SCR)
BRIDGE RECTIFIER
DIODE/RECTIFIER
T
T
G
TRIAC
(U#)
ZENER
SCHOTTKY
TUNNEL
DIODES (D#)
AIR CORE
ADJUSTABLE INDUCTANCE
ADJUSTABLE COUPLING
WITH CORE
WITH LINK
TRANSFORMERS
3-PIN CERAMIC RESONATOR
ANTENNA
FUSE
QUARTZ CRYSTAL
HAND KEY
MOT
MOTOR
Z
ASSEMBLY OR MODULE (OTHER THAN IC)
MISCELLANEOUS
PNP
C
B
E
P-CHANNEL
E
B2
B1
G
D
S
G2
G1
NPN
N-CHANNEL
BIPOLAR
UJT
JUNCTION FET
SINGLE-GATE
DUAL-GATE
DEPLETION MODE MOSFET
SINGLE-GATE ENHANCEMENT MODE MOSFET
TRANSISTORS
AND
NAND
OR
NOR
XOR
INVERT
SCHMITT
OTHER
LOGIC (U#)
SPST
SPDT
DPDT
THERMAL
RELAYS
GENERAL AMPLIFIER
OTHER
OP AMP
+
-
INTEGRATED CIRCUITS (U#)
COMMON CONNECTIONS
LABEL
PHONE JACK
PHONE PLUG
CONTACTS
MALE
FEMALE
COAXIAL CONNECTORS
FEMALE
MALE
MULTIPLE, MOVABLE
MULTIPLE, FIXED
240 V FEMALE
GROUND
FEMALE
MALE
MALE CHASSIS-MOUNT
HOT 120 V
NEUT
HOT 120 V
HOT 120 V
NEUT
GND
GND
GND
CONNECTORS
TRIODE
PENTODE
CRT
TUBE ELEMENTS
ANODE
HEATER OR FILAMENT
GRID
GAS FILLED
CATHODE
COLD CATHODE
DEFLECTION PLATES
TUBES (V#)
INCANDESCENT
NEON (AC)
LAMPS

The ARRL—At Your Service

ARRL Headquarters is open from 8 AM to 5 PM Eastern Time, Monday through Friday, except holidays. Call **toll free** to join the ARRL or order ARRL products: **1-888-277-5289** (US), M-F only, 8 AM to 8 PM Eastern Time.

If you have a question, try one of these Headquarters departments . . .

	Telephone	*Electronic Mail*
Joining ARRL	860-594-0338	**membership@arrl.org**
***QST* Delivery**	860-594-0338	**circulation@arrl.org**
Permission Requests	860-594-0229	**permission@arrl.org**
Publication Orders	860-594-0355	**pubsales@arrl.org**
Amateur Radio News	860-594-0222	**n1rl@arrl.org**
Regulatory Info	860-594-0236	**reginfo@arrl.org**
Exams	860-594-0300	**vec@arrl.org**
Educational Materials	860-594-0230	**ead@arrl.org**
CCE/EmComm Courses	860-594-0340	**dmiller@arrl.org**
Contests	860-594-0232	**contests@arrl.org**
Technical Questions	860-594-0214	**tis@arrl.org**
Awards/VUCC	860-594-0288	**awards@arrl.org**
Development Office	860-594-0397	**mhobart@arrl.org**
DXCC	860-594-0234	**dxcc@arrl.org**
Advertising	860-594-0207	**ads@arrl.org**
Media Relations	860-594-0328	**newsmedia@arrl.org**
QSL Service	860-594-0274	**buro@arrl.org**
Scholarships	860-594-0397	**foundation@arrl.org**
Emergency Comm	860-594-0265	**emergency@arrl.org**
Clubs	860-594-0292	**clubs@arrl.org**
Hamfests	860-594-0262	**hamfests@arrl.org**

You can send e-mail to any ARRL Headquarters employee if you know his or her name or call sign. The second half of every Headquarters e-mail address is **@arrl.org**. To create the first half, simply use the person's call sign. If you don't know their call sign, use the first letter of their first name, followed by their complete last name. For example, to send a message to John Hennessee, N1KB, Regulatory Information Specialist, you could address it to **jhennessee@arrl.org** or **N1KB@arrl.org**.

If all else fails, send e-mail to **hq@arrl.org** and it will be routed to the right people or departments.

Technical Information Service

The ARRL answers questions of a technical nature for ARRL members and nonmembers alike through the Technical Information Service. Questions may be submitted via e-mail (**tis@arrl.org**); Phone (860-594-0214); Fax (860-594-0259); or mail (TIS at ARRL, 225 Main Street, Newington, CT 06111). The TIS also maintains a home page on **www.arrl.org/tis**. See the **Component Data and References** chapter (page 7.11) of this *Handbook* for more details. Also, please note that the "Technical Information Server" or *Info Server* service previously available via e-mail has been discontinued.

ARRL ON THE WORLD WIDE WEB

You'll find ARRL at: **www.arrl.org/**

At the ARRL Web page you'll find the latest W1AW bulletins, a hamfest calendar, exam schedules, an on-line ARRL Publications Catalog and much more. We're always adding new features, so check it often!

Members-Only Web Features

As an ARRL member you enjoy exclusive access to our Members-Only Web features. Just point your browser to **www.arrl.org/members/** and you'll open the door to benefits that you won't find anywhere else.

• *QST* Product Review Archive. Get copies of *QST* product reviews from 1980 to the present.

• *QST/QEX* searchable index (find that article you were looking for!)

• Previews of contest results. See them here before they appear in *QST!*

• Access to your information in the ARRL membership database. Enter corrections or updates on line!

Stopping by for a visit?

We offer tours of Headquarters and W1AW at 9, 10 and 11 AM, and at 1, 2 and 3 PM, Monday to Friday (except holidays). Special tour times may be arranged in advance. Bring your license and you can operate W1AW anytime between 10 AM and noon, and 1 to 3:45 PM!

Would you like to write for *QST*?

We're always looking for new material of interest to hams. Send a self-addressed, stamped envelope (2 units of postage) and ask for a copy of the *Author's Guide*. (It's also available via the ARRL Web page at **www.arrl.org/qst/aguide/**.)

Press Releases and New Products/Books

Send your press releases and new book announcements to the attention of the ***QST* Editor** (e-mail **qst@arrl.org**). New product announcements should be sent to the Product Review Editor (e-mail **reviews@arrl.org**).

ARRL Amateur Radio News on the Web

The primary focus of the ARRL Web site, **www.arrl.org**, is Amateur Radio news and general-interest features and columns—available to all. This is the Amateur Radio community's most comprehensive and immediate source for news and information on issues of importance and interest to radio amateurs.

The ARRL Letter

The ARRL Letter has become the League's flagship Amateur Radio news medium. The *Letter* is a weekly news summary for those who want to be on top of what's happening in the world of Amateur Radio. ARRL members can request Friday e-mail delivery via their Member Data Page. It's also available to all on the Web, **www.arrl.org/arrlletter**. *The ARRL Letter* is published 50 times a year.

ARRL Audio News

Another way to keep up with fast-moving events in the ham community is to listen to the *ARRL Audio News*. It's as close as your telephone at 860-594-0384, or on the Web at **www.arrl.org/arrlletter/audio/**.

Interested in Becoming a Ham?

Just pick up the telephone and call toll free 1-800-326-3942, or send e-mail to **newham@arrl.org**. We'll provide helpful advice on obtaining your Amateur Radio license, and we'll be happy to send you our informative Prospective Ham Package.

Chapter 1

What is Amateur (Ham) Radio?

You are reading this book because you are an interesting person whose curiosity is piqued by unusual things. Or you enjoy talking to other people, or you want to understand a little about electronics — our world is full of electronic tools. You may want to know enough about electronics to tackle building your own electronics project. You may be a user of a personal radio station, purely for noncommercial purposes, with other radio hobbyists. We call that ham radio or Amateur Radio. We call ourselves Amateur Radio operators, ham radio operators or just plain "hams."

You know that ham radio operators communicate with other hams — on a distant continent or around the block — or from an orbiting space station! Some talk

Fig 1.1 — Hams are always willing to help others who are excited about becoming ham radio operators. You will find more information in this chapter about how to locate ham radio operators in your local area.

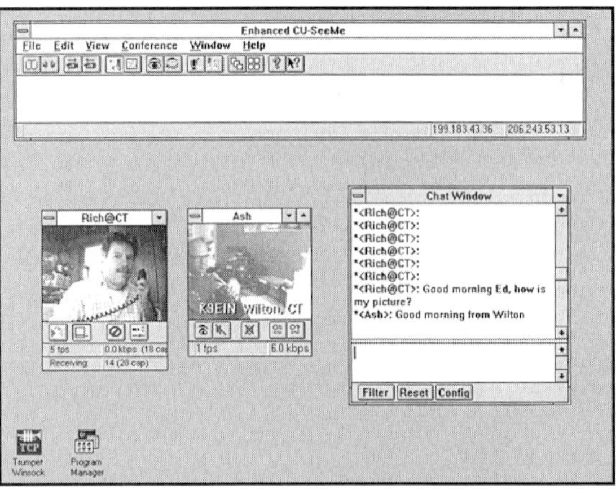

Fig 1.2 — Computers are an integral part of ham radio, and the Internet plays a large role. Rich Roznoy, K1OF and Ed Ashway, K3EIN, use their PCs to hold a video conference. Audio transmission was by way of their VHF transmitters. ***(Photo courtesy of Rich Roznoy, K1OF)***

Fig 1.3 — Brian Wood, WØDZ, like many hams, enjoys restoring antique radios.

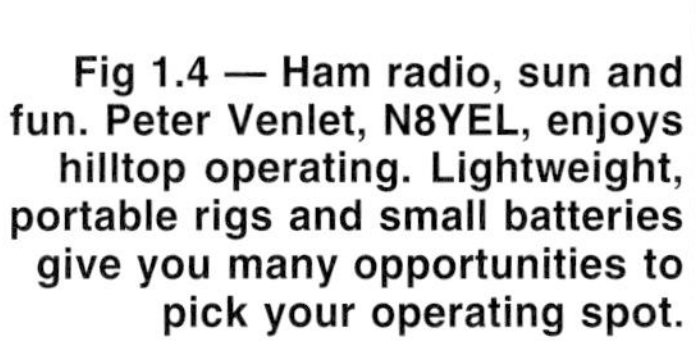

Fig 1.4 — Ham radio, sun and fun. Peter Venlet, N8YEL, enjoys hilltop operating. Lightweight, portable rigs and small batteries give you many opportunities to pick your operating spot.

via computers and ham radio; others prefer to use regular voice communication. Still others enjoy using one of the oldest forms of radio communication — Morse code — even though it is not a requirement to know Morse in order to earn a ham license. Some hams help save people's lives by handling emergency communication following a natural disaster or other emergency. Some become close friends with the people they talk to on the other side of the globe — then make it a point to travel and meet one or more of them in person. Some can take a bag full of electrical parts and turn it into a radio.

This chapter, by Rosalie White, K1STO, covers the basics — what hams do, and how they do it.

HOBBY OF DIVERSITIES

If you wrote down a list of all of the unusual, interesting things you can do as an Amateur Radio operator, you would fill a few pages. What types of people will you meet as a ham? If you walk down a city street, you'll pass men and women, girls and boys, and people of all ages, ethnic backgrounds and physical abilities. Any of them might be a ham you will meet tonight on your radio. They're office workers and students, nurses and mail carriers, engineers and truck drivers, housewives and bankers.

If you drive your car on the interstate this weekend, you'll see people on their way to a state park, a Scout camp, a convention, an airport or a computer show. The young couple going to the park to hike for the day have their hand-held ham radio transceivers in their backpacks. When they stop on a scenic hilltop for a rest, they'll pull out their radios and see how far away they can communicate with the radio's 3 watts of power. And, the radios will be handy just in case they break down on the road or lose the hiking trail.

The father and son on their way to Scout camp will soon be canoeing with their Scout troop. After setting up camp, they'll get out a portable radio, throw a wire antenna over a branch, and get on the air. Aside from the enjoyment of talking with other hams from their campsite, their radios give them the security of having reliable communication with the outside world, in case of emergency.

The family driving to the ham radio convention will spend the day talking with their ham friends, including two they've never met but know quite well from talking to them on the air every week. They will also look at new and used radio equipment, listen to a speaker talk about the latest ways computers can be used to operate on the Amateur Radio bands, and enjoy a banquet talk by a NASA astronaut who is also a ham radio operator.

The couple on the way to the airport to take a pleasure flight in their small plane has their hand-held radios in their flight bags. Once they're airborne, they'll contact hams on the ground all along their flight path. Up at 5,000 feet, they can receive and transmit over much greater distances than they can from the ground. Hams enjoy the novelty of talking to other hams in a plane. The radios are an ideal means of backup communication, too.

The two friends on their way to the computer show are discussing the best interface to use between their computers and their radios for several of the operating modes that use computers. They're looking forward to seeing a number of their ham friends who are into computers.

What other exciting things will you do on the Amateur Radio bands? You might catch yourself excitedly calling (along with 50 other hams) a Russian cosmonaut in space or a sailor on the Coast Guard's tall ship *Eagle*. You could be linked via packet radio with an Alaskan sled-dog driver, a rock star, a US legislator, a major league baseball player, a ham operating the Amateur Radio station aboard the ocean liner *Queen Mary*, an active-duty soldier, a king or someone who is building the same power supply that you are from a design in this *ARRL Handbook*.

On the other hand, a relaxing evening at

Fig 1.5 — A ham's operating area is called *the shack*. It may be a corner of a room, a basement area or in this case part of a former battleship. The Mobile (AL) Amateur Radio Club installed a temporary station, W4IAX, on this imposing structure.

All Types of Physical Abilities

People who aren't as mobile as they'd like find the world of Amateur Radio a rewarding place to make friends — around the block or the globe. Many hams with and without certain physical abilities belong to Courage Handi-Hams, an international organization of radio amateurs. Courage Handi-Ham members live in all states and many countries, and they are ready to help however they can. The Handi-Ham System provides members with study materials and aids for physical disabilities. Local Handi-Hams will assist you with studies. Once you receive your license, the Handi-Ham System may lend you basic radio equipment to get you on the air.

handihams

home could find you in a friendly radio conversation with a ham in Frankfort, Kentucky, or Frankfurt, Germany. Unlike any other hobby, Amateur Radio knows no country boundaries and brings the world together as good friends.

Although talking with astronauts isn't exactly an everyday ham radio occurrence, more and more hams are doing just that, as many NASA astronauts are ham radio operators.

Fig 1.6 — Bill Carter, KG4FXG, helps young Andrea Hartlage, KG4IUM, work her way through her first CW contact.

YOUR LICENSE

You must have a license granted by the FCC to operate an Amateur Radio Station in the United States, in any of its territories and possessions, or from any vessel or aircraft registered in the United States. The FCC sets no minimum or maximum age requirement for obtaining a license, nor does it require that an applicant be a US citizen. However, US citizens are required to pass an FCC license exam to obtain a US Amateur Radio license.

An Amateur Radio license incorporates two kinds of authorization. For the individual, the license grants operator privileges in one class. The class of license the operator has earned determines these privileges. The license also authorizes the licensee to operate transmitting equipment physically present at the licensed station location. Therefore, the full license includes a class-specific *Operator license* and a *Station license*.

Fig 1.7 — The gang's all here. Who said ham radio is a solo activity? Standing are KC9XT, N9LVL, KB9ATR, WZ9M, N9IOX and N9LBT. Seated are Chris Kratzer, W9XD and KB9GRP. *(Photo courtesy of Mike McCauley, KB9GNU)*

The Technician License: Your Path to Amateur Radio Fun!

Most people start in Amateur Radio with a Technician Class license. They hold that license for a while to learn the ways of the hobby. Then they move to the next class of ham license. The Technician license presents an excellent way for beginners to start enjoying the fun and excitement of Amateur Radio. The only requirement for the Technician license is that you pass a single 35-question written exam. The exam covers Federal Communications Commission (FCC) rules and regulations that govern the airways, courteous operating procedures and techniques, and some basic electronics. There is no Morse code requirement for a Technician license.

Technician license frequency privileges begin at 50 MHz and extend through the very high frequency (VHF) and ultra high frequency (UHF) ranges, and into the microwave region. All of these frequency bands give Technicians plenty of room to explore. They aren't restricted to only certain operating modes, either. They can use any communication methods allowed to hams.

Most new hams will operate first on the popular 2-meter band. With plenty of *repeaters* across the country, the FM voice signals from their low-power hand-held and mobile radios reach many other hams. Technicians also communicate through *satellites* and *packet radio* networks. They use *single-sideband (SSB)* voice and *Morse code (CW)*.

Technician licensees can gain additional operating privileges on the amateur high frequency (HF) bands by passing a 5-wpm Morse code test. Passing another 35-question written exam completes the upgrade to a General class license. Generals enjoy worldwide communication using SSB and CW as well as *slow-scan television (SSTV)* and a variety of *digital communication* modes.

One more written exam, this one containing 50 questions, will take you to the top of the Amateur Radio license ladder — the Amateur Extra class license. Amateur Extra class licensees enjoy full amateur privileges on all bands. The exam may be challenging, but many hams find it to be well worth the effort!

WHAT'S IN A CALLSIGN?

When you earn your Amateur Radio license, you receive a unique *call sign*. Many hams are known by their call signs more than by their names! All US hams get a call sign, a set of letters and numbers, assigned to them by the Federal Communications Commission (FCC). No one else *owns* your call sign — it's unique. Your Amateur Radio license with your unique call sign gives you permission to operate your Amateur Radio station on the air. US call signs begin with W, K, N or A, with some combination of letters and numbers that follow. When a US call sign begins with A there are always two letters for the

US Amateur Bands

ARRL *The national association for AMATEUR RADIO*

160 METERS

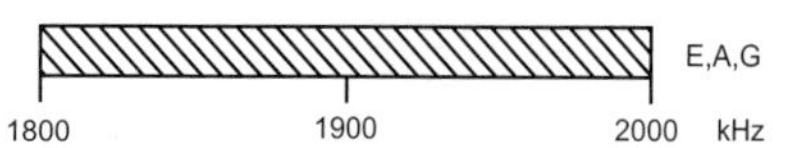

Amateur stations operating at 1900-2000 kHz must not cause harmful interference to the radiolocation service and are afforded no protection from radiolocation operations.

80 METERS

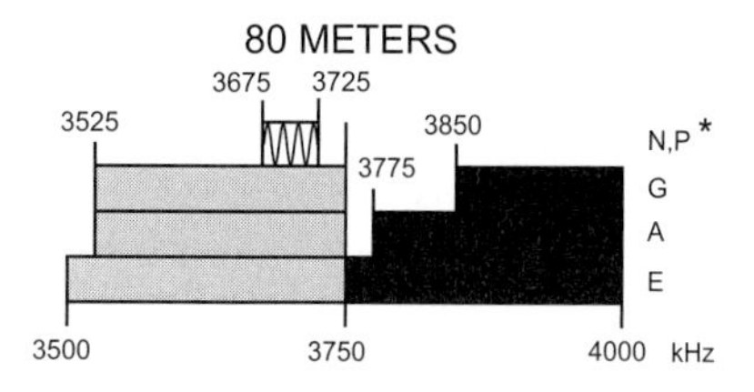

60 METERS

General, Advanced, and Amateur Extra licensees may use the following five channels on a secondary basis with a maximum effective radiated power of 50 W PEP relative to a half wave dipole. Only upper sideband suppressed carrier voice transmissions may be used. The frequencies are 5330.5, 5346.5, 5366.5, 5371.5 and 5403.5 kHz. The occupied bandwidth is limited to 2.8 kHz centered on 5332, 5348, 5368, 5373, and 5405 kHz respectively.

40 METERS

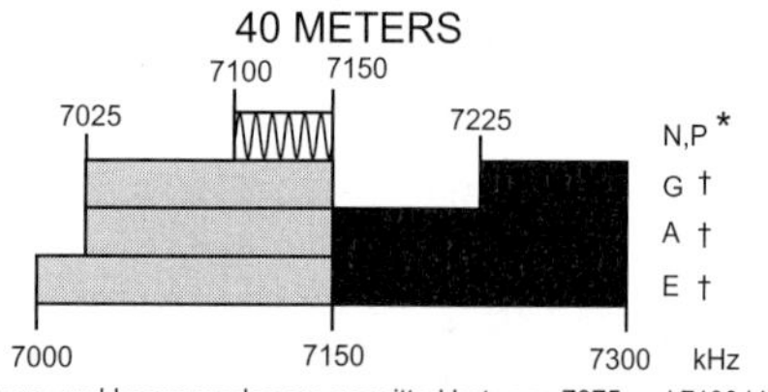

† Phone and Image modes are permitted between 7075 and 7100 kHz for FCC licensed stations in ITU Regions 1 and 3 and by FCC licensed stations in ITU Region 2 West of 130 degrees West longitude or South of 20 degrees North latitude. See Sections 97.305(c) and 97.307(f)(11). Novice and Technician Plus licensees outside ITU Region 2 may use CW only between 7050 and 7075 kHz. See Section 97.301(e). These exemptions do not apply to stations in the continental US.

30 METERS

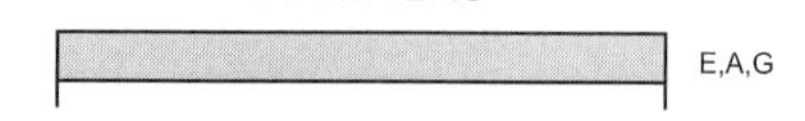

Maximum power on 30 meters is 200 watts PEP output.
Amateurs must avoid interference to the fixed service outside the US.

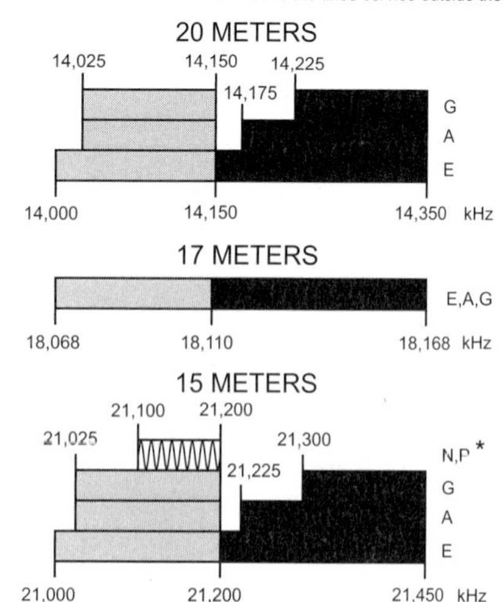

Novice, Advanced and Technician Plus Allocations

New Novice, Advanced and Technician Plus licenses are no longer being issued, but *existing* Novice, Technician Plus and Advanced class licenses are unchanged. Amateurs can continue to renew these licenses. Technicians who pass the 5 wpm Morse code exam *after* that date have Technician Plus privileges, although their license says Technician. They must retain the 5 wpm Certificate of Successful Completion of Examination (CSCE) as proof. The CSCE is valid indefinitely for operating authorization, but is valid only for 365 days for upgrade credit.

12 METERS

E,A,G

24,890 24,930 24,990 kHz

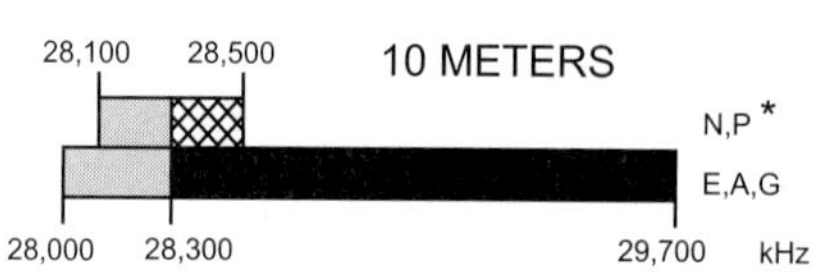

Novices and Technician Plus Licensees are limited to 200 watts PEP output on 10 meters.

US AMATEUR POWER LIMITS

At all times, transmitter power should be kept down to that necessary to carry out the desired communications. Power is rated in watts PEP output. Unless otherwise stated, the maximum power output is 1500 W. Power for all license classes is limited to 200 W in the 10,100-10,150 kHz band and in all Novice subbands below 28,100 kHz. Novices and Technicians are restricted to 200 W in the 28,100-28,500 kHz subbands. In addition, Novices are restricted to 25 W in the 222-225 MHz band and 5 W in the 1270-1295 MHz subband.

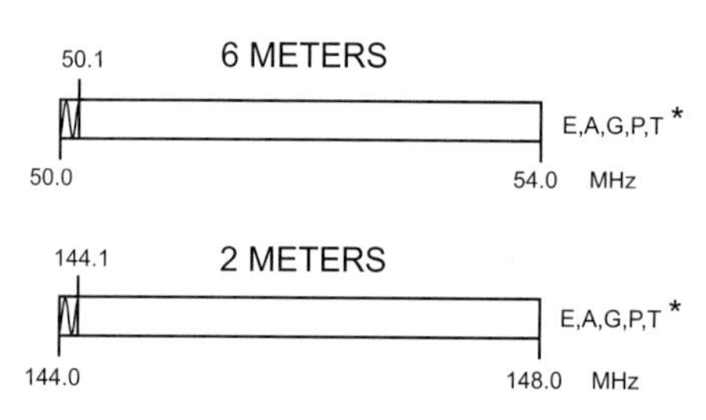

1.25 METERS ***

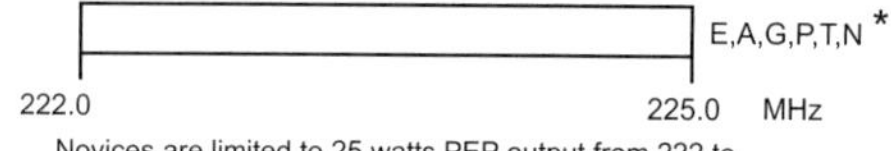

Novices are limited to 25 watts PEP output from 222 to 225 MHz.

70 CENTIMETERS **

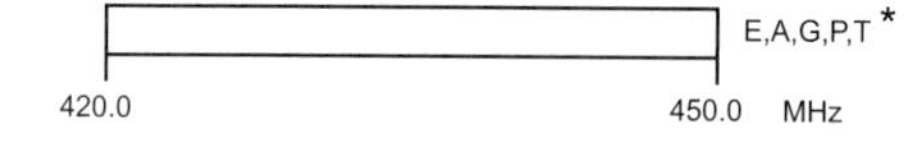

33 CENTIMETERS **

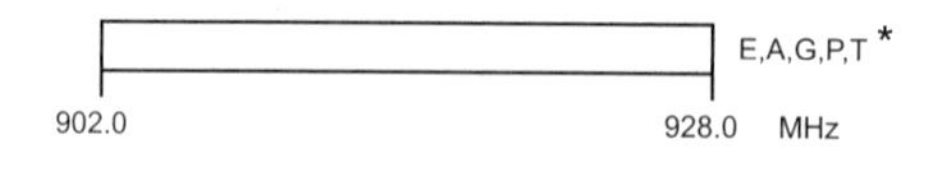

23 CENTIMETERS **

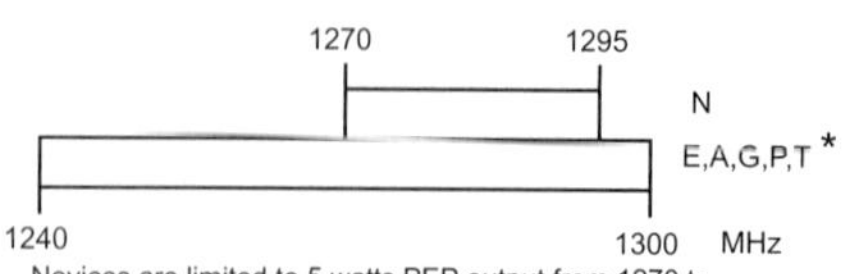

Novices are limited to 5 watts PEP output from 1270 to 1295 MHz.

KEY

- = CW, RTTY and data
- = CW, RTTY, data, MCW, test, phone and image
- = CW, phone and image
- = CW and SSB phone
- = CW, RTTY, data, phone, and image
- = CW only

E = AMATEUR EXTRA
A = ADVANCED
G = GENERAL
P = TECHNICIAN PLUS
T = TECHNICIAN
N = NOVICE

*Technicians who have passed the 5 wpm Morse code exam are indicated as "P".

**Geographical and power restrictions apply to all bands with frequencies above 420 MHz. See *The ARRL FCC Rule Book* for more information about your area.

***219-220 MHz allocated to amateurs on a secondary basis for fixed digital message forwarding systems only and can be operated by all licensees except Novices.

All licensees except Novices are authorized all modes on the following frequencies:

2300-2310 MHz
2390-2450 MHz
3300-3500 MHz
5650-5925 MHz
10.0-10.5 GHz
24.0-24.25 GHz
47.0-47.2 GHz
75.5-76.0, 77.0-81.0 GHz
119.98-120.02 GHz
142-149 GHz
241-250 GHz
All above 300 GHz

ARRL We're At Your Service

ARRL Headquarters	860-594-0200 (Fax 860-594-0259)	hq@arrl.org
Publication Orders	Toll-Free 1-888-277-5289 (860-594-0355)	orders@arrl.org
Membership/Circulation Desk	Toll-Free 1-888-277-5289 (860-594-0338)	membership@arrl.org
Getting Started in Amateur Radio	Toll-Free 1-800-326-3942 (860-594-0355)	newham@arrl.org
Exams	860-594-0300	vec@arrl.org
ARRL on the World Wide Web	www.arrl.org/	

Field Day: Ham Radio Alfresco

Most hams enjoy operating during Field Day. On the fourth full weekend in June, US and Canadian amateurs plus some in other parts of the world, take their radios into the great outdoors to operate away from power mains. They test how to set up efficient temporary stations to operate under emergency conditions and contact as many other stations as possible. But Field Day is, above all, fun! Hams enjoy working with a group to compete against perhaps 100,000 other hams who are also out in the elements!

Many different Field Days are possible — from a camper, cabin or tent, maybe a view of the mountains, which is what the W7AV team did.

prefix — AA through AL. The number in the middle of the call sign indicates the station location within the country when the license was first issued. In the US, call signs that contain a 9 indicate that the ham lived in the Midwest when the license was issued, and call signs with the number 1 indicate a New England location. However, with the creation of the FCC's vanity callsign program, which permits certain FCC licensees to select their own callsign from a list of unassigned call signs based on the license class of the holder, one can no longer assume that a specific number in a US amateur call sign denotes that licensee's actual physical location. You can tell what country issued a ham's call sign by the prefix — the letters before the number.

HAM RADIO ACTION

Amateur Radio and public service go together. On a warm early-summer weekend, hams can be found directing radio communication in the aftermath of a train derailment — a simulated one, that is — to prepare for a real emergency. Others provide radio communication for a walk-a-thon. Still others hone their communication skills by setting up a station outdoors, away from electrical power. This largest public-service-related ham activity is called *Field Day*.

Biking and Cruising

Hams even operate their radios while riding on bicycle treks. They easily carry their lightweight hand-held radios in their packs, and can pull them out quickly, if needed. Or they might pull along a small trailer with sleeping bag, food and water and a small ham radio transceiver and wire antenna that they can set up in the evenings.

Fig 1.8 — A radio set up can be simple or complex.

Other hams enjoy going to far-flung, exotic parts of the world. They get to go to new places, operate the radios, and meet new friends. Hams call these trips *DXpeditions*. Many hams enjoy operating from a sparsley populated country to provide rare ham contacts for their excited fellow hams around the world.

Nets: Scheduled Get-Togethers

If you'd enjoy finding other hams with interests like yours (such as chess, gardening, rock climbing, railroads, computer programming or teaching), you'll soon learn about *nets*. A net forms when hams with similar interests get together on the air on a regular schedule. You can find your special interest — from the Armenian Amateur Radio and Traffic Net to ARES nets — listed in *The ARRL Net Directory* on the *ARRLWeb*.

Awards and Contests: Competitive Fun

If you're competitive by nature, you'll want to explore ham radio awards and contests. These activities recognize your ability to contact many hams under specific guidelines. In the ARRL DX Contest, for example, you'll try to contact as many DX (foreign) stations as possible over a weekend. During such a weekend, experienced hams with top-notch stations easily contact more than 100 different countries!

Awards you can earn include *Worked All States*, earned by communicating with a ham in every US state, *Worked All VE*, earned by contacting hams in every Canadian province, and *DX Century Club*, for working stations in 100 or more different countries.

In an outdoor orienteering competition, "fox-hunters" track and locate hidden transmitters by car or on foot. This activity, also called direction finding, has a serious side: Skills learned in tracking the fox are useful when there's a suspected pirate (unlicensed ham station) in the area.

QRP: Talk to the World with 5 Watts of Power

For a new challenge, try operating *QRP* — using low power. Some hams enjoy operating with only 1 watt or less. It's a challenge, but with decent antennas and skillful operating, QRP enthusiasts are heard around the world. The best reason for operating QRP is that equipment is lightweight, inexpensive and easy to build. Some hams use nothing but "homebrew" equipment.

Computers and Ham Radio

If computers are your favorite aspect of today's technology, you'll soon discover that you can connect your computer to Amateur Radio equipment and operate using such digital modes as *packet radio*, RTTY, *PSK31* and High-Speed Multimedia (*HSMM*). HSMM is a popular multimedia mode used on the ultra-high-

Fig 1.9 — Ham radio operators are well-known for providing emergency communication. Famous newscaster Walter Cronkite, KB2GSD, honored hams by narrating a show on hams saving the day with communications when all else failed.

frequency (UHF) bands for the simultaneous transmission of voice, video, text and data. With packet, you can leave messages that other packet enthusiasts will pick up and answer later. PSK31 is a popular digital mode used on the amateur high-frequency (HF) bands. It is used with common computers and sound cards, allowing you to talk great distances with a modest ham station. Computers can help you practice taking amateur license examinations or improve your Morse code abilities. ARRL, and others, offer software especially designed to help you pass amateur radio exams. You'll discover software for many ham applications, from keeping track of the stations you've contacted to designing a great antenna. See also the chapter on **Web, Wi-Fi, Wireless and PC Technology** in this *Handbook* for additional information.

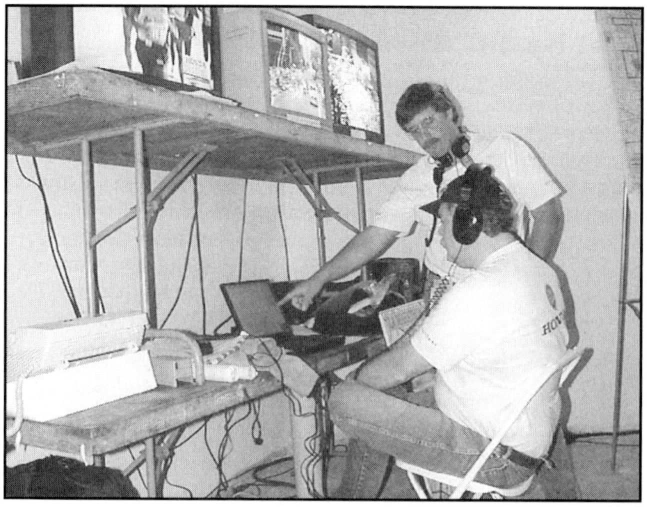

Fig 1.10 — John Tallon, N6OMB (standing) and Randy Hammock, KC6HUR, discuss operation of the APRS net in the LA Police Department command center during the 1998 LA Marathon. *(KN6F Photo)*

Enhancing Radio Signals

Radio signals normally travel in straight lines, which limits their range. But hams have found some ingenious ways of extending the distance and improving the quality of the signals they transmit. High-frequency (HF) radio waves can be refracted or bent by a layer of the atmosphere called the ionosphere. Signals are returned to Earth, often after several "hops." This *ionospheric propagation* of radio signals allows worldwide communication on the HF bands. Hams have also learned how to bounce signals off the moon, airplanes and even meteor trails! Repeaters, located on hilltops or tall buildings, strengthen signals and retransmit them. This provides communication over distances much farther than would be possible without repeaters.

Helping out in Emergencies

When commercial communication services are disrupted by power failures or during natural disasters such as earthquakes and hurricanes, Amateur Radio operators are often first at the scene. Battery-powered equipment allows hams to provide essential communication even when power is knocked out. If need be, hams can make and install antennas on the spot from whatever materials and supports they find available. Many hams join their local Amateur Radio Emergency Service (ARES)® group sponsored by ARRL, and regularly practice their emergency communication skills.

Working with emergency personnel such as police and fire departments, the Red Cross and medical personnel, ham volunteers provide communication. Hams can handle communication between agencies whose normal radios are incompatible with one another, for example. The ability of radio amateurs to help the public in emergencies is one of the reasons Amateur Radio has prospered since the early days of the 20th century.

Community Events

To keep their emergency-preparedness skills honed, and to help their community, hams enjoy assisting with communication to aid the public at any number of activities. Hams volunteer to provide communication for marathons, bike races, parades and other community events. In fact, it's rare to see a large community event that doesn't make use of public-spirited ham radio operators.

Build It Yourself!

Another favorite activity hams enjoy is building their own radio equipment. Many amateurs proudly stay at the forefront of technology, continually being challenged to keep up with advances that could be applied to the hobby. They have an incessant curiosity and an eagerness to try new techniques. They try to find ways to allow the radio frequency bands to support more users, since some portions of the

Fig 1.11 — The art of *homebrewing*, or building your own equipment, is thriving. This small receiver contains two transistors and one integrated circuit. It is assembled without special tools in one or two evenings. The thrill of operating a radio you built lasts a long time! See Chapters 14 and 15 for some receivers and transceivers you can build.

Hams at the Forefront

Over the years, the military and the electronics industry have often drawn on ham ingenuity to improve designs or solve problems. Hams provided the keystone for the development of modern military communication equipment, for example. In the 1950s, the Air Force needed to convert its long-range communication from Morse code to voice — and jet bombers had no room for skilled radio operators. At the time, hams communicated by voice at great distance with both home-built and commercial single-sideband (SSB) equipment. Air Force Generals LeMay and Griswold, both radio amateurs, hatched an experiment that used ham equipment at the Strategic Air Command headquarters and an airplane flying around the world. They found that the equipment would need only slight modification to meet Air Force needs. By using ham radio technology, two generals saved the government millions of dollars in research costs.

OSCARs: The Ham Satellites

You will be thrilled hearing your own radio signal returned from space by an orbiting "repeater in the sky" — a ham radio satellite. Hams regularly use Amateur Radio satellites, called OSCARs (for *O*rbiting *S*atellites *C*arrying *A*mateur *R*adio). Your VHF and UHF ham signals normally won't travel much beyond the horizon. But if you route signals through a satellite, you can make global radio contacts on VHF and UHF. OSCAR-16 was built and launched in the '90s. It allows messages from Earth to be stored and forwarded back down to Earth when the spacecraft is within range of the designated station. Hams from around the globe worked together to design, build and test OSCAR 40, launched in 2000. It provided hams with communication on several radio frequencies up through the microwave bands.

OSCAR-E — the "Echo Project" — was built and launched in June 2004. The satellite has two UHF transmitters for running simultaneous operation, four VHF receivers and a multiband, multimode receiver for the 10-m, 2-m, 70-cm and 23-cm bands. Hams can use FM voice and various digital modes — including PSK31 on a 10-meter SSB uplink. Does this sound beyond your skills? It shouldn't. All it takes is a Technician license to enjoy satellites.

Fig 1.13 — Most communication with unmanned satellites uses VHF, UHF or microwave frequencies, and many hams build their own antennas for this pursuit.

Fig 1.12 — Amateur Radio operators on the International Space Station (ISS) enjoy using its on-board ham station during their off-duty hours. They talk to friends, school students and hams around the world by voice and packet radio. Many astronauts are ham radio operators now that they are assigned long-duration ISS stints.

Fig 1.14 — Astronaut Susan Helms, KC7NHZ, made several dozen Field Day contacts operating from the ISS as NA1SS. *(NASA Photo)*

ham bands are very popular and can be crowded.

The projects you'll find in this book provide a wide variety of equipment and accessories that make ham radio more convenient and enjoyable. Many manufacturers provide parts kits and etched circuit boards to make building even easier.

Hams in Space

In 1983, the first ham/astronaut made history by communicating from the space shuttle *Columbia* with ground-based hams. On that mission, NASA Payload Specialist Owen Garriott, whose Amateur Radio call sign is W5LFL, took along a hand-held amateur transceiver and placed a specially designed antenna in an orbiter window. It was the first time ham radio operators throughout the world were to experience the thrill of working an astronaut aboard an orbiting spacecraft. In 1985, Mission Specialist Tony England, WØORE, transmitted slow-scan television via Amateur Radio while orbiting the Earth in the shuttle *Challenger*. He named the payload *SAREX*, for *S*huttle *A*mateur *R*adio *EX*periment.

From then on, NASA routinely scheduled SAREX missions. In 1991, all five members of mission STS-37 had earned a ham radio license! NASA promotes ham activity because of its proven educational and PR value. It could be used for backup communication. In 1995, five shuttle crews requested that NASA install the SAREX payload on their flights. A ham station was one of the first "extra" items installed in the International Space Station (ISS).

Plans began in 1996 to have Amateur Radio on the ISS, in a program called ARISS. During crew members' on-board leisure time, the ham astronauts sometimes talk to hams, school students and friends. Because ISS missions are long-duration stints, many astronauts have decided to earn their ham licenses.

GETTING STARTED

Now that you know some things hams do, you may be asking, "Okay, how do I get started?" The first step is to earn a license.

Fig 1.15 — The St Xavier High School Amateur Radio Club assembled for this photo just before a big weekend contest. Their enthusiastic efforts led them to a high score for their type of station.

Fig 1.16 — There were no computers, satellites or TV when Clarice, W7FTX, was first licensed in 1935. One thing has not changed — the friendship of her fellow hams.

Most people start with the Technician license, which has a 35-question written exam. Exams are given by local ham Volunteer Examiners (VEs). Many clubs sponsor exam sessions regularly, so you shouldn't have to travel far. Exams are given on weekends or evenings. Exam questions are taken from a large pool of questions for each license class. The complete question pools for the three licenses are published in study guides or can be found on the Internet. (See the URL for *ARRLWeb* in the Resources section at the end of this chapter.)

Study Guides

You can prepare for the exams in a class or on your own. Help is available at every step. The ARRL publishes study materials for all classes of licenses. Contact ARRL's New Ham Desk (the phone number and address are at the end of this chapter) for more information on getting started: nearby ham clubs, instructors who have registered with ARRL, and local VEs. You'll find a lot of information on the ARRL Web site, such as frequencies hams can use, popular operating activities and the latest versions of ARRL study guides.

ARRL's beginner manual, *Now You're Talking!*, includes the complete, up-to-date question pool with the correct answers and clear explanations. You'll also find tips for how to choose your equipment and put together your ham radio station, simple and inexpensive antennas and much more.

Now You're Talking! assumes no prior electronics background. Children as young as five years old have passed ham radio exams! You could choose *ARRL's Tech Q & A manual,* instead, if you already have some electronics background or just want brief explanations to help you understand the correct answers. Every question in the Technician question pool is included to help you prepare.

When you are ready to upgrade to a General class license, *The ARRL General Class License Manual* or *ARRL's General Q & A* will help you prepare. *The ARRL Extra Class License Manual* will guide your study efforts for the Amateur Extra class license. ARRL's *Your Introduction to Morse Code* will teach you the Morse code and prepare you for the 5-wpm exam. *ARRL's Technician Video Course* and *ARRL's General Video Course* can bring our expert instructors to your TV. Even newer is ARRL's online Technician license course—learn what you need to know to earn your license by studying at your computer when it's convenient for you!

Mentors, Sometimes Called Elmers

Ham radio operators often learn how to get on the air from a mentor. An experienced ham, called an "Elmer," teaches newcomers about Amateur Radio on a one-to-one basis. Many ham clubs have special mentor programs. Elmering was first documented in ARRL's monthly magazine, *QST*, in a story meant to be a public thank you from an appreciative student whose mentor's name was Elmer. Elmers are there for you as you study, buy your first radio and set up your station. They are proud to help you with your first on-the-air contacts.

Putting Together a Station

As with any other hobby, you can enjoy ham radio no matter how small or large

Fig 1.17 — Many hams operate radios while on the road.

your budget. You can start with a hand-held transceiver that fits in your pocket or purse, and take it along when you hike, canoe or aviate. Or you can fill your "radio shack" with the latest and fanciest radios technology offers and money can buy, and talk to people in all corners of the world. You can put up a simple, inexpensive wire antenna between two trees in your backyard, or install a tower with antennas on top. Either way you'll talk to the world.

Accessories and equipment for your ham radio station come in all price ranges. *QST* contains display advertisements for new ham gear plus classified ads for previously owned items.

Used Versus New

Hams are continually upgrading their stations, so you can always find a ready supply of good previously-owned Amateur Radio gear. You can find new hand-held transceivers and used HF radios for less than $300. Many hams start with a radio that costs between $300 and $600. Antennas and other gear can add appreciably to the cost, but less-expensive alternatives, such as putting together your own antenna or low-power transceiver, are available.

HAMS AS WORLD CITIZENS

When you become an Amateur Radio operator, you become a "world citizen" — you join people who have earned the privilege of talking to other hams around the corner or around the world. Hams have a long tradition of spreading international goodwill. One way hams do this is to assist with getting needed medical advice or medicine to developing countries. Another is by learning about the lives and cultures of those they contact. But you should avoid talking about sensitive political or ethical issues.

Although English is the standard language on the ham bands, English-speakers will make a good impression on hams in foreign countries if they can speak a little of the other person's language — even if it's as simple as *danke* or *sayonara*.

International Amateur Radio

Hams in other countries have formed national organizations, just as US hams organized the ARRL — the national association for Amateur Radio. These sister societies work to have a united voice in international radio affairs, such as when governments together decide how radio frequencies will be divided among users. The International Amateur Radio Union (IARU), composed of about 160 national Amateur Radio societies, works to help protect the amateur frequencies.

THE ADMINISTRATORS: ITU AND FCC

Our world has a limited spectrum of radio frequencies. These frequencies must be shared by many competing radio services: aeronautical, marine, land mobile, to name a few.

The International Telecommunication Union (ITU), an agency of the United Nations, allocates frequencies among the services. With its long tradition of public service and technological savvy, ham radio enjoys the use of many different frequency bands.

In the US, a government agency, the Federal Communications Commission (FCC), regulates the radio services, including Amateur Radio. The section of the FCC Rules that deals with Amateur Radio is Part 97. Hams are expected to know the important sections of Part 97, as serious violations (such as causing malicious interference or operating without the appropriate license) can lead to fines and even imprisonment! Aside from writing and enforcing the rules governing Amateur Radio, the FCC also assigns call signs and issues licenses.

THE ARRL

Since it was founded in 1914, the ARRL — the national association for Amateur Radio — has grown and evolved along with Amateur Radio. The ARRL Headquarters building and Maxim Memorial Station, W1AW, are in Newington, Connecticut, near Hartford. Through its dedicated volunteers and a professional staff, the ARRL promotes the advancement of the amateur service in the US and around the world.

The ARRL is a nonprofit, educational and scientific organization dedicated to the promotion and protection of the many privileges that ham radio operators enjoy. Of, by and for the radio amateur, ARRL numbers 150,000 members — the vast majority of active amateurs in North America. Licensed hams become Full Members, while unlicensed persons become Associate Members with all membership privileges except for voting in ARRL elections. Anyone with a good interest in Amateur Radio belongs in the ARRL.

The ARRL volunteer corps is called the Field Organization. Working at the state and local level, these volunteers work on ARRL goals to further Amateur Radio. They organize emergency communication in times of disaster and work with agencies such as Red Cross and Citizen Corps. Other

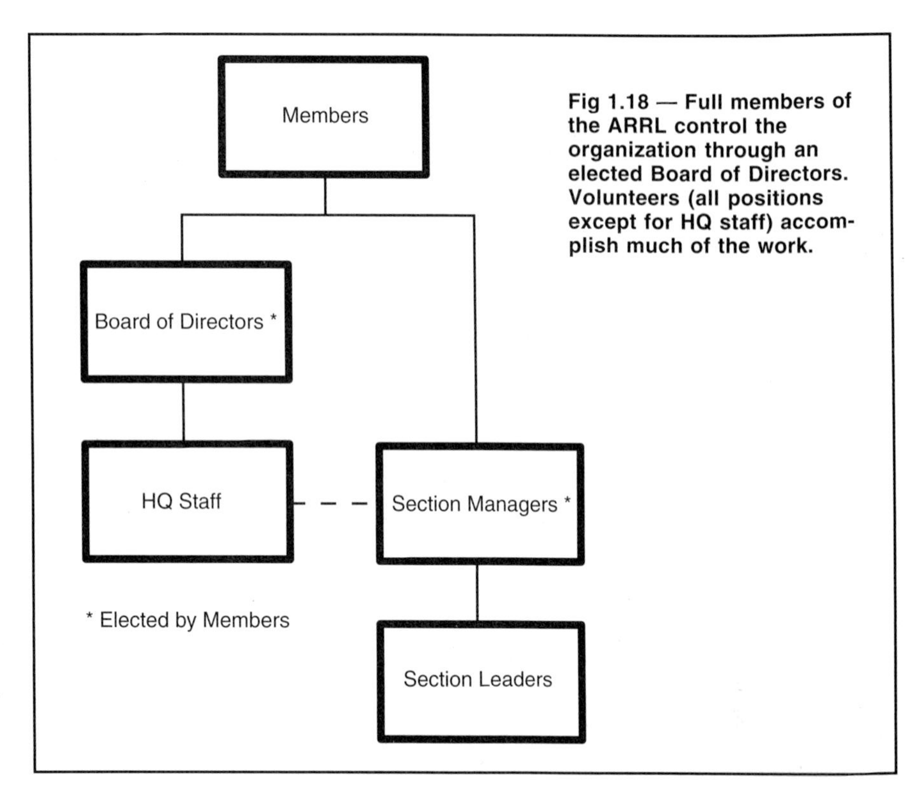

Fig 1.18 — Full members of the ARRL control the organization through an elected Board of Directors. Volunteers (all positions except for HQ staff) accomplish much of the work.

Fig 1.19 — W1AW, the station operated by the ARRL in Newington, Connecticut, is known around the world as the home of ham radio. W1AW memorializes Hiram Percy Maxim, one of the founders of the ARRL. Visitors are welcome and often operate the station. *(Photo courtesy of W2ABE)*

Fig 1.20 — Many hams enjoy talking to far away stations, and learn geography in the process. Here Bill, KB7JAH, and William, KB7JAG, chat with a ham in Kiribati. Don't know where Kiribati is? (Hint: Part of the island is the first in the world to greet each new day.)

Fig 1.21 — All-ham families are not unusual. Class instructor Robert Lavin, K6BOB, and VE Jonathan Fleischer, AC6GW, handed graduation certificates to the Harrises: Wesley, KG6QHT, age 7; Desiree Domingo Foraste, KD6LEW; Wesley's twin Ryan, KG6QHU, and David, KD6LET.

Fig 1.22 — Three generations of Andrew Maroneys visit W1AW. From left to right, Andrew IV, W2AJM, Andrew III, WA2QAX, and Andrew Jr, W2SON. While at Headquarters, Andrew Jr turned in QSL cards for his DXCC Honor Roll verification, bringing his total to 325.

volunteers keep state and local government officials abreast of the good that hams do at the state and local level.

Membership Services

When you join ARRL, you add your voice to those who are most involved with ham radio. The most prominent benefit of ARRL membership is *QST*, the premiere Amateur Radio magazine. *QST* has stories you'll want to read, articles on projects to build, announcements on upcoming activities, equipment reviews, reports on the role hams play in emergencies, and much more.

But being an ARRL member is far more than a subscription to *QST*. The ARRL represents your interests to the FCC and Congress, sponsors operating events, and offers membership services at a personal level. A few are:

- Low-cost equipment insurance
- the Volunteer Examiner program
- the Technical Information Service (which gets you answers to questions about any technical subject in Amateur Radio)
- the QSL bureau (which lets you exchange postcards with hams in foreign countries as a confirmation of your contacts with them)

Schoolteachers and Volunteer Instructors

ARRL Field & Educational Services (F&ES) provides teachers with aids for using Amateur Radio in schools, and even has equipment grants. Hundreds of teachers have found Amateur Radio an ideal way to provide hands-on, intercurricular learning, while enticing students to become interested in science and technology. F&ES also has materials, such as online newsletters and lesson plans for hams that wish to teach Amateur Radio licensing classes.

WELCOME!

For answers to any questions you may have about Amateur Radio, write or call ARRL Headquarters. See Resources, at the end of this chapter, for information.

Governing Regulations

International and national radio regulations govern the operational and technical standards of all radio stations. The International Telecommunication Union (ITU) governs telecommunication on the international level and broadly defines radio services through the international Radio Regulations. In the United States, the agency responsible for nongovernmental and nonmilitary stations is the Federal Communications Commission (FCC). Title 47 of the *US Code of Federal Regulations* governs telecommunication. Different rule Parts of Title 47 govern the various radio services in the US. The Amateur Radio Service is governed by Part 97. Some other Parts are described in the sidebar "Other FCC Rule 'Parts'." *The ARRL RFI Book* contains a detailed chapter on these FCC Rule parts, which affect Amateur Radio directly and indirectly.

Experimentation has been the backbone of Amateur Radio for almost a century and the amateur rules provide a framework within which amateurs have wide latitude to experiment in accordance with the basis and purpose of the service. The rules should be viewed as vehicles to promote healthy activity and growth, not as constraints that lead to stagnation. A brief overview of Amateur Radio regulations follows with special emphasis on technical standards.

Other FCC Rule "Parts"

Part 97 is just a small piece of the overall regulatory picture. An up-to-date copy of Part 97 can be found on the Web at **www.arrl.org/FandES/field/regulations/news/part97** and in *The ARRL FCC Rule Book*, which also contains hundreds of pages of interpretive materials. The *US Code of Federal Regulations*, Title 47, consists of telecommunication rules numbered as Parts 0 through 300. These Parts contain specific rules for the many telecommunication services the FCC administers. Individuals may purchase or obtain from the Web (**wireless.fcc.gov/rules.html**) a specific rule Part for a particular service from the Superintendent of Documents, US Government Printing Office. Here is a list of FCC Parts amateurs may find of interest:

Part	
0	Commission organization
1	Practice and procedure
2	Frequency allocation and radio treaty matters; general rules and regulations.
15	Low-power radio-frequency transmitting devices
17	Construction, marking and lighting of antenna structures
18	Industrial, scientific and medical equipment
73	Radio broadcast services
76	Cable Television Service
90	Private Land Mobile Radio Service
95	Personal radio services, including CB and GMRS
97	Amateur Radio Service

BASIS AND PURPOSE OF THE AMATEUR RADIO SERVICE

There's much more in the regulatory scheme than Part 97. The basis for the FCC regulations is found in treaties, international agreements and statutes that provide for the allocation of frequencies and place conditions on how the frequencies are to be used. For example, Article 25 of the international *Radio Regulations* limits the types of international communication amateur stations may transmit and mandates that the technical qualifications of amateur operators be verified.

It's the FCC's responsibility to see that amateurs are able to operate their stations in a manner consistent with the basis and purpose of the amateur rules. The FCC must also ensure that amateurs have the knowledge and ability to operate powerful and potentially dangerous equipment safely without causing harmful interference to others. A review of each of the five basic purposes of the Amateur Radio Service, as they appear in Part 97, follows:

Recognition and enhancement of the value of the amateur service to the public as a voluntary noncommercial communication service, particularly with respect to providing emergency communication [97.1(a)].

Probably the best known aspect of Amateur Radio to the general public is its ability to provide emergency communication, such as following the World Trade Center attack by terrorists in 2001. One of the most important aspects of the service is its noncommercial nature. Amateurs are prohibited from receiving any form of payment for operating their stations.

Continuation and extension of the amateur's proven ability to contribute to the advancement of the radio art [97.1(b)].

For nearly a century, hams have carried on a tradition of learning by doing, and since the beginning have remained at the forefront of technology. Through experimenting and building, hams have pioneered advances, such as techniques for single-sideband transmissions, and are currently engaged in the development of new digital schemes which continue to improve the efficiency of such communication. Hams' practical experience has led to technical refinements and cost reductions beneficial to the commercial radio industry. In addition, amateurs have designed and built a series of sophisticated satellites at a fraction of the cost of their commercial equivalents.

Encouragement and improvement of the Amateur Radio Service through rules which provide for advancing skills in both the communication and technical phases of the art [97.1(c)].

Amateurs have always been experimenters and that's what sets the Amateur Service apart from other services. The cost to the government for licensing and enforcement is minimal when compared to the benefit the public receives. Hams have contributed greatly to the development of computer communication techniques. The FCC and industry have also credited the amateur community with the development of Low-Earth-Orbit (LEO) satellite technology. The same can be said for a number of digital modes that have arrived on the scene in the 1990s, such as PacTOR, CLOVER and PSK31.

Expansion of the existing reservoir within the Amateur Radio Service of trained operators, technicians and electronic experts [97.1(d)].

Amateurs learn by doing. While all amateurs may not be able to troubleshoot and repair a transceiver, all amateurs have some degree of technical competence.

Continuation and extension of the amateur's unique ability to enhance international goodwill [97.1(e)].

Amateur Radio is one of the few truly international hobbies. It is up to amateurs to maintain high standards and to represent the US as its ambassadors, because, in a sense, all US amateurs serve that function.

A QUICK JOURNEY THROUGH PART 97

The Amateur Radio Service rules, Part 97, are organized in six major subparts: General Provisions, Station Operation Standards, Special Operations, Technical Standards, Providing Emergency Communication and Qualifying Examination Systems. A brief discussion of the highlights of each subpart follows:

General Provisions

Subpart A covers the basics that apply to all facets of Amateur Radio. The "Basis and Purpose" of Amateur Radio, discussed above, is found at the beginning of Part 97 [97.1]. Definitions of key terms used throughout Part 97 form the foundation of Part 97 [97.3].

The remainder of the subpart is devoted to Federal restrictions on amateur installations, which include a mention of FCC standards for RF exposure. The ARRL publication *RF Exposure and You* details these RF exposure requirements.

Station Operation Standards

Subpart B, "Station Operation Standards," concerns the basic operating practices that apply to all types of operation. Amateurs must operate their stations in accordance with good engineering and amateur practice [97.101(a)]. Part 97 doesn't always tell amateurs specifically how to operate their stations, particularly concerning technical issues, but the FCC provides broad guidelines. The use of good engineering and amateur practice means, for example, that amateurs shouldn't operate a station with a distorted signal and that amateurs shouldn't operate on a busy band like 20 m just to talk to a ham across town. Also, amateurs must share the frequencies with others — no one ham or group has any special claim to any frequency [97.101(b)]. The station licensee is always responsible for the proper operation of an amateur station, except where the control operator is someone other than the station licensee, in which case both share responsibility equally [97.103(a)]. This subpart also contains regulations for the reciprocal operating authority.

The requirements for control operators, station control and reciprocal operating authority are also addressed in Subpart B. Each station must have a control point [97.109(a)]. A control operator must always be at the control point, except in a few cases where the transmitter is controlled automatically [97.109(b), (c), (d) and (e)]. The purpose of the Amateur Radio Service is to communicate with other amateurs [97.111]. Certain one-way transmissions are allowed. Amateurs can send a one-way transmission to:

- make adjustments to equipment for test purposes
- call CQ
- remotely control devices
- communicate information in emergencies
- send code practice and information bulletins of interest to amateurs [97.111(b)]

Broadcasting to the public is strictly prohibited [97.113(b)]. The section on prohibited transmissions states that amateurs cannot: be paid for operating a station, make transmissions on behalf of an employer, transmit music (unless otherwise allowed in the rules), transmit obscenity, use amateur stations for news-gathering purposes or transmit false signals and ciphers. The FCC has relaxed the previously restrictive business rules to encourage public service and personal communication [97.113].

Station identification is addressed in this subpart. The purpose of station identification is to make the source of its transmissions known to those receiving them, including FCC monitors. The rules cover identification requirements for the various operating modes. Section 97.119 details the station-identification requirements. Amateurs must transmit their call sign at the end of the communication and every 10 minutes during communication. CW and phone may be used to identify an amateur station. RTTY and data (using a specified digital code) may be used when all or part of the communication are transmitted using such an emission. Images (Amateur Television, for example) may be used to identify when all or part of the transmission is in that mode. A final section addresses restricted operation and sets forth the conditions that must exist in an interference case involving a neighbor's TV or radio before the Commission can impose "quiet hours" — hours of the day when a particular amateur may not operate an amateur transmitter [97.121(a)]. Imposition of quiet hours by the FCC is rare, however.

Special Operations

Subpart C, "Special Operations," addresses specialized activities of Amateur Radio including the various types of stations an amateur may operate. This subpart gives specific guidelines concerning repeaters, beacons, space stations, Earth stations, message forwarding systems, and telecommand (remote control) stations. These rules are of particular interest to the technically minded amateur. An amateur may send ancillary functions (user functions) of a repeater on the input of the repeater — to turn on and off an autopatch, for example. However, the primary control links used to turn the repeater on and off, for example, may be transmitted only above 222.150 MHz since such one-way transmissions are auxiliary transmissions. Every repeater trustee/licensee and user should understand the rules for repeaters and auxiliary links. An important regulatory approach to solving interference problems between repeaters is addressed in that section: Repeater station licensees are equally responsible for resolving an interference problem, unless one of the repeaters has been approved for operation by the recognized repeater coordinator for the area and the other has not. In that case, the owner of the uncoordinated repeater has primary responsibility to resolve the problem [97.205(c)]. The control operator of a repeater that inadvertently retransmits communication in violation of the rules is not held accountable [97.205(g)]. The originator and first forwarding station of a message transmitted through a message forwarding system are held accountable for any violations of Part 97. Other forwarding stations are not held accountable [97.219]. For a detailed explanation, see *The ARRL FCC Rule Book.*

Technical Standards

The word *standard* means consistency and order — and this is what the technical standards in Subpart D are all about. The FCC outlines the specific frequency bands available to US amateurs [97.301] as well as the sharing agreements [97.303]. (This chapter includes a table of frequencies allocated to the Amateur Radio Service.) The Commission made these standards a basic framework so all types of amateur operation may peacefully coexist with other radio occupants in the spectrum neighborhood. Emission standards for RTTY, data and spread spectrum are discussed, as are standards for the Certification of RF power amplifiers. FCC Certification is not needed for most amateur equipment. This gives amateurs the freedom to experiment without being bound by specific equipment standards.

Providing Emergency Communication

Subpart E, "Providing Emergency Communication," addresses disaster communication, stations in distress, communication for the safety of life and protection of property and the Radio Amateur Civil Emergency Service (RACES).

Qualifying Examination Systems

The final subpart of the rules, Subpart F, deals with the examination system and covers exam requirements and elements and standards. In 1983, the Commission delegated much of the exam administration program to amateurs themselves. The rules provide for checks and balances on volunteer examiners (VEs), who administer exams at the local and regional levels. These checks and balances protect against fraud and provide integrity for the exam process.

The classes of Amateur Radio license are Novice, Technician, Technician Plus, General, Advanced, and Amateur Extra. Since April 15, 2000, the FCC has issued only *new* Technician, General and Amateur Extra licenses. (A Technician with credit for 5 wpm has the same privileges as the "old Technician Plus.") Those who hold any of the six classes of license may have their license renewed at the same license class, however.

EMISSION STANDARDS, BANDWIDTH, POWER AND EXTERNAL POWER AMPLIFIERS

Like most of Part 97, the technical standards exist to promote operating techniques that make efficient use of the spectrum and minimize interference. The standards in Part 97 identify problems that must be solved. Section 97.307 spells out the standards FCC expects amateur signals to meet. It states, in part: "No amateur station transmission shall occupy more bandwidth than necessary for the information rate and emission type being transmitted, in accordance with good amateur practice" [97.307(a)]. Simply stated, don't transmit a wide signal when a narrow one will do. Specific bandwidth limits are given for RTTY and data emissions. Specific bandwidth limits are not given for other modes of operation, but amateurs must still observe good engineering and operator practice.

The rules state: "Emissions resulting from modulation must be confined to the band or segment available to the control operator" [97.307(b)]. Every modulated signal produces sidebands. Amateurs must not operate so close to the band edge that the sidebands extend out of the subband, even if the frequency readout says that the carrier is inside the band. Further: "Emissions outside the necessary bandwidth must not cause splatter or key-click interference to operations on adjacent frequencies" [97.307(b)]. The rules simply codify good operating practice. Key clicks or over-processed voice signals shouldn't cause interference up and down the band.

Spurious emissions

Spurious emissions include harmonic emissions, parasitic emissions, intermodulation products and frequency conversion products, but do not include splatter [97.307(c)]. Definitions for *necessary bandwidth* and *out-of-band emission* appear in the **Glossary** at the end of this chapter. Also see **Fig 1.23**.

Emission standards

The FCC is very specific concerning spurious emission standards for emissions below 30 MHz and emissions from 30-225 MHz. The following requirements apply only to amateur transmitters or external HF amplifiers **below 30 MHz** [97.307(d)]:

If the amateur transmitter was installed (FCC terminology for put into operation) ***after*** **January 1, 2003**, the mean power of any spurious emission from the station transmitter or external RF power amplifier on a frequency below 30 MHz must be **at least 43 dB below the mean power of the fundamental emission.** This applies to *all* power levels, including QRP transmitters.

For transmitters installed ***on or before*** **January 1, 2003**, the mean power of any spurious emission from a station transmitter or external RF power amplifier transmitting on a frequency below 30 MHz must **not exceed 50 mW** and must be **at least 40 dB below the mean power of the fundamental emission**. For a transmitter of mean ***power less than 5 W*** installed ***on or before*** **January 1, 2003**, the attenuation must be ***at least 30 dB***. A transmitter built before April 15, 1977, or first marketed before January 1, 1978, is exempt from this requirement.

The following spurious emission standards apply **between 30 and 225 MHz** [97.307(e)]:

- In transmitters with **25 W or less mean output power**, spurs must be **at least 40 dB below the mean power of the fundamental** emission and never greater than 25 µW (microwatts), but need not be reduced further than 10 µW. This means that the spurs from a 25-W transmitter must be at least 60 dB down to meet the 25-µW restriction.
- In transmitters with more than 25 W mean output power, spurious emissions must be at least 60 dB below the mean power of the fundamental emission.

The situation for transmitters operating between 30 and 225 MHz is more complex. The combination of the requirement that spurious emissions be less than 25 µW and the stipulation that they don't need to be reduced below 10 µW makes the requirements vary significantly with power level. This ranges from 0 dB suppression required for a transmitter whose power is 10 µW to 60 dB of suppression required for power levels above 25 W. The requirements for transmitter operation between 30 and 225 MHz are shown graphically in **Fig 1.24**. There are no absolute limits for

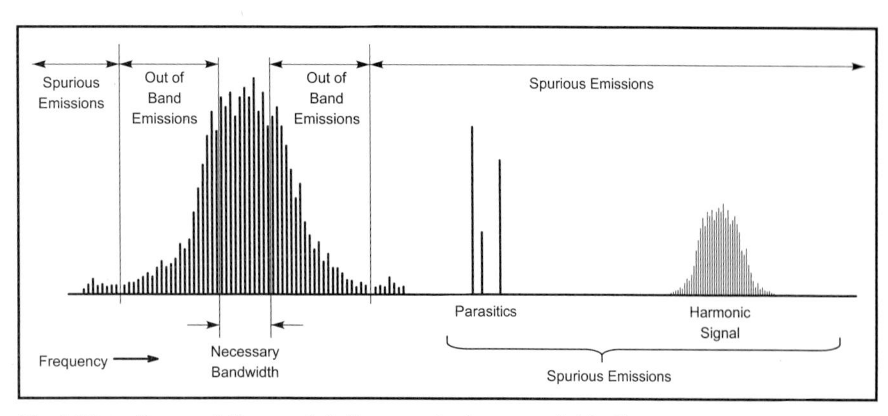

Fig 1.23 — Some of the modulation products are outside the necessary bandwidth. These are out-of-band emissions, but they are not considered spurious emissions. On the other hand, these out-of-band emissions must not interfere with other stations [97.307(b)]. The harmonics and parasitics shown in this figure are spurious emissions, and they must be reduced to the levels specified in Part 97. The FCC states that all spurious emissions must be reduced "to the greatest extent practicable" [97.307(c)]. Further, if any spurious emission, including chassis or power-line radiation, causes harmful interference to the reception of another radio station, the licensee of the interfering amateur station is required to take steps to eliminate the interference.

Emission Designators

ITU system of designating emission, modulation and transmission characteristics employed appears in the International Radio Regulations at **life.itu.int/radioclub/rr/aps01.htm**. The same text appears in FCC rules. The cite is 47 CFR, Section 2.201 — Emission, modulation and transmission characteristics.

transmitters operating above 225 MHz, although the requirements for good engineering practice would still apply.

Transmitter Power Standards

Amateurs shall not use more power than necessary to carry out the desired communication [97.313(a)]. Don't use 700 W when 10 m is wide open, for example. No station may use more than 1.5 kW peak envelope power [97.313(b)] and no station may use more than 200 W in the 30-m band. Novices and Technicians with 5 WPM credit are limited to 200 W in their HF segments [97.313(c)]. Amateurs may use no more than 50 W in the 70-cm band near certain military installations [97.313(f) and (g)].

The FCC has chosen and published the following standards of measurement: (1) Read an in-line peak-reading RF wattmeter that is properly matched; and (2) calculate the power using the peak RF voltage as indicated by an oscilloscope or other peak-reading device. Multiply the peak RF voltage by 0.707, square the result and divide by the load resistance. The SWR must be 1:1 for calculation accuracy.

The FCC requires that you meet the power output regulations, but does not require that you make such measurements or possess measurement equipment. The methods listed simply indicate how the Commission would measure your transmitter's output during a station inspection.

As a practical matter, most hams don't have to worry about special equipment to check their transmitter's output because they never approach the 1500-W PEP output limit. Many common amplifiers aren't capable of generating this much power. However, if you do have a capable amplifier and do operate close to the limit, you should be prepared to measure your output along the lines detailed above.

External RF Power Amplifiers: Certification and Standards

In 1978, the FCC banned the manufacture and marketing of any external RF power amplifier or amplifier kit capable of operation on any frequency below 144 MHz, unless the FCC has issued a grant of type acceptance (now called FCC Certification) for that model amplifier. The FCC also banned the manufacture and marketing of HF amplifiers that were capable of operation on 10 m to stem the flow of amplifiers being distributed for illegal use in and around frequencies used by CB operators.

Amateurs may still use amplifiers capable of operation on 10 m. While the rules may make it difficult to buy a new amplifier capable of operation on 10 m, the FCC allows amateurs to modify an amplifier to restore or include 10-m capability. An amateur may modify no more than one unit of the same model amplifier in any year without FCC Certification [97.315(a)].

Of course, amateurs are permitted to build amplifiers, convert equipment from any other radio service for this use or to buy used amplifiers. When converting equipment from other services, it must meet all technical standards outlined in Part 97, and it can no longer be used in the service for which it was intended since the Certification would have been voided. Non-amateurs are specifically prohibited from building or modifying amplifiers capable of operation below 144 MHz without FCC Certification [97.315(a)]. All external amplifiers and amplifier kits capable of operation below 144 MHz must be FCC Certificated in order to be marketed [97.315(b)]. A number of amplifiers, manufactured prior to the April 28, 1978 cutoff were issued a waiver of the new regulations [97.315(b)(2)]. Amateurs may buy or sell an amplifier that has either been FCC Certificated, granted a waiver or modified so that the Certification is no longer valid. There are restrictions that would be valid regardless of whether the amplifier was capable of operation below 144 MHz. Some amplifiers marketed before April 28, 1978, are covered under the waiver if they are the same model that was granted a waiver [97.315(b)(2)]. An individual amateur may sell his amplifier regardless of grants or waivers, provided that he sells it only to another amateur operator [97.315(b)(4)]. Amateurs may also sell a

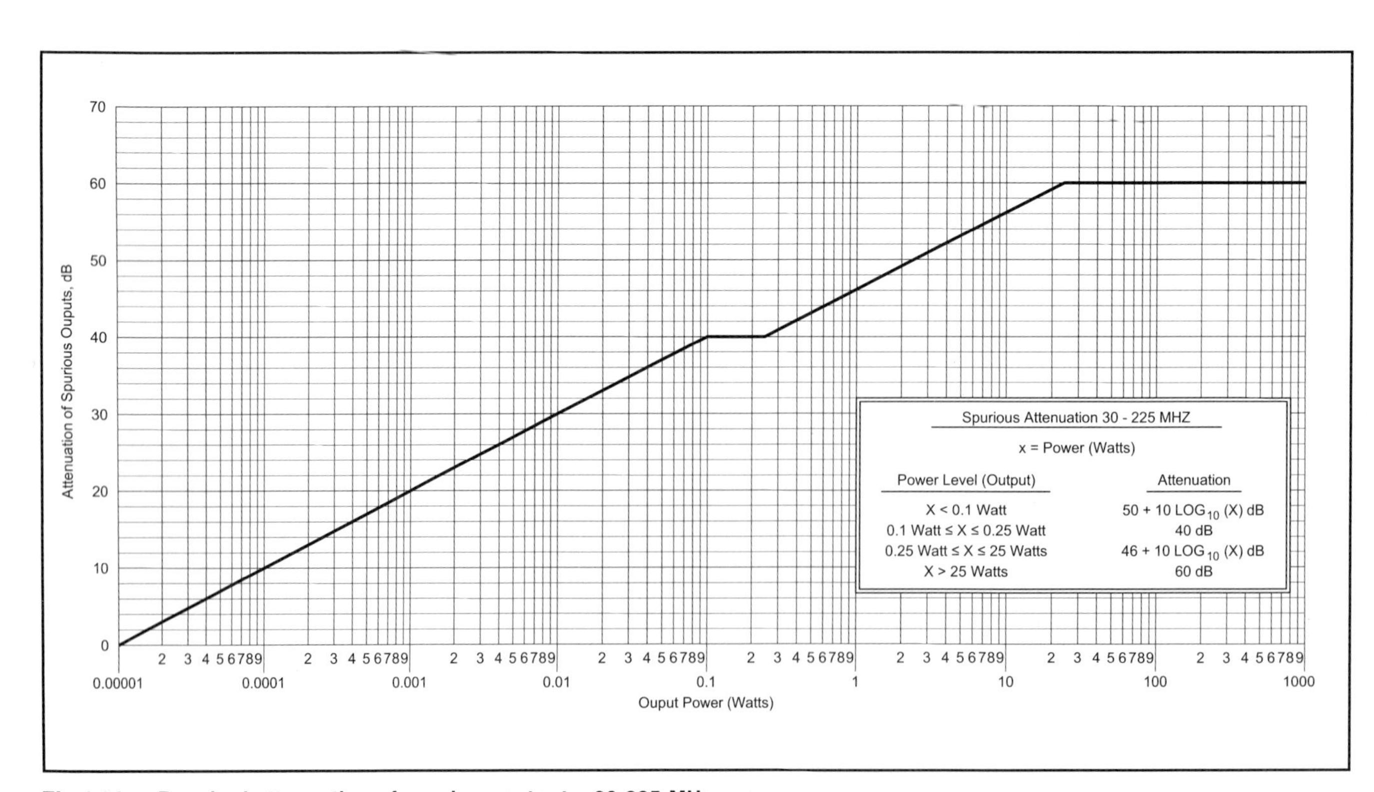

Fig 1.24 — Required attenuation of spurious outputs, 30-225 MHz.

used amplifier to a *bona fide* amateur equipment dealer. The dealer could sell those amplifiers only to other hams [97.315(b)(5)].

In some cases, the FCC will deny Certification. Some features that may cause a denial are (1) any accessible wiring which, when altered, would permit operation in a manner contrary to FCC rules; (2) circuit boards or similar circuitry to facilitate the addition of components to change the amplifier's operation characteristics in a manner contrary to FCC rules; (3) for operation or modification of the amplifier in a manner contrary to FCC rules; (4) any internal or external controls or adjustments to facilitate operation of the amplifier in a manner contrary to FCC rules; (5) any internal RF sensing circuitry or any external switch, the purpose of which is to place the amplifier in the transmit mode; (6) the incorporation of more gain than is necessary to operate in the amateur service.

CONCLUSION

A common thread in Amateur Radio's history has been a dynamic regulatory environment that has nurtured technological growth and diversity. This thread continues to sew together the elements of Amateur Radio today and prepare it for tomorrow's challenges.

RESOURCES

ARRL — the national association for Amateur Radio
225 Main St
Newington, CT 06111-1494
860-594-0200
Fax: 860-594-0259
e-mail: hq@arrl.org
Prospective hams call 1-800-32 NEW HAM (1-800-326-3942)
ARRLWeb: **www.arrl.org**
Membership organization of US ham radio operators and those interested in ham radio. Publishes study guides for all Amateur Radio license classes, a monthly journal, *QST*, and many books on Amateur Radio and electronics.

AMSAT NA (The Radio Amateur Satellite Corporation, Inc)
PO Box 27
Washington, DC 20044
301-589-6062
www.amsat.org
Membership organization for those interested in Amateur Radio satellites. Publishes *The AMSAT Journal*, monthly.

Courage Handi-Ham System
3915 Golden Valley Rd
Golden Valley, MN 55422
763-520-0511
www.handiham.org
Provides assistance to persons with disabilities who want to earn a ham radio license or set up a station.

Now You're Talking! All You Need for Your First Amateur Radio License
(Newington, CT: ARRL)
www.arrl.org/catalog/
Complete introduction to ham radio, including the exam question pool, complete explanations of the subjects on the exams. Tips on buying equipment, setting up a station and more.

The ARRL's Tech Q & A
(Newington, CT: ARRL)
www.arrl.org/catalog/
All the questions on the Technician exam, with correct answers highlighted and explained in plain English. Many helpful diagrams.

Your Introduction to Morse Code
(Newington, CT: ARRL)
www.arrl.org/catalog/
A set of audio CDs (or cassette tapes) that make learning Morse code fun. Teaches all letters, numbers and other required characters, and provides practice text.

Ham University
(Available from ARRL)
www.arrl.org/catalog/
Windows-based PC software. Learn Morse code to pass your 5 words-per-minute exam. Ham University includes the questions for all written exams so you can quiz yourself.

Glossary

Note: Words in ***boldface italics*** have separate entries in the Glossary.

ADV (Amateur digital video) — A mode of operation in which ***Amateur Radio*** operators exchange video motion images using their personal computers.

AM (Amplitude modulation) — The oldest voice operating mode still found on the amateur bands. The more common voice mode, ***SSB***, is actually a narrower-bandwidth variation of AM.

Amateur Radio — A radiocommunication service for the purpose of self-training, intercommunication and technical investigations carried out by amateurs, that is, duly authorized persons interested in radio technique solely with a personal aim and without pecuniary interest. (*Pecuniary* means payment of any type, whether money or goods.) Also called ***ham radio***.

Amateur Radio operator — A person holding a license to operate a ham radio station.

Amateur Radio station — A station licensed in the amateur service, including necessary equipment.

Amateur Service — A radio communication service for the purpose of self-training, intercommunication and technical investigations carried out by amateurs, that is, duly authorized persons interested in radio technique solely with a personal aim and without pecuniary interest.

AMSAT (Radio Amateur Satellite Corporation) — An international membership organization that designs, builds and promotes the use of Amateur Radio satellites.

ARES® — An ARRL program specializing in emergency communication.

ARISS — An acronym for Amateur Radio on the International Space Station.

ARRL — The membership organization for Amateur Radio operators in the US.

ATV (Amateur television) — A mode of operation that amateur radio operators use to exchange pictures from their ham stations.

Auxiliary station — An amateur station, other than in a message-forwarding system, transmitting communication point-to-point within a system of cooperating amateur stations.

Band — A range of frequencies. Hams are authorized to transmit on many different bands.

Bandwidth — The width of a frequency band outside of which the mean power of the transmitted signal is attenuated at least 26 dB below the mean power of the transmitted signal within the band.

Beacon — An amateur station transmitting communication for the purposes of observation of propagation and reception or other related experimental activities.

Beam antenna — A type of ham radio antenna that can be pointed in any direction.

Broadcasting — Transmissions intended for reception by the general public, either direct or relayed.

Call sign — A series of unique letters and numbers assigned to a person who has earned an Amateur Radio license.

Carrier power — The average power supplied to the antenna transmission line by a transmitter during one RF cycle taken under the condition of no modulation.

Certification — An equipment authorization granted by the FCC. It is used to ensure that equipment will function properly in the service for which it has been accepted. Most amateur equipment does not require FCC Certification, although HF power amplifiers and amplifier kits do. Part 15 Rules require FCC Certification for all receivers operating anywhere between 30 and 960 MHz. Amateur transmitters may not be legally used in any other service that requires FCC equipment authorization. For example, it is illegal to use a modified amateur transmitter in the police, fire or business radio services.

Contact — A two-way communication between Amateur Radio operators.

Contest — An Amateur Radio activity in which hams and their stations compete to contact the most stations within a designated time period.

Control operator — An amateur operator designated by the licensee of a station to be responsible for the transmissions from that station to assure compliance with the FCC Rules.

Control point — The location at which the control operator function is performed.

Courage Handi-Ham System — Membership organization for ham radio enthusiasts with various physical disabilities and abilities.

CW — Abbreviation for *continuous wave;* another name for ***Morse code*** telegraphy by radio. Also, International Morse code telegraphy emissions having designators with A, C, H, J or R as the first symbol; 1 as the second symbol; A or B as the third symbol; and emissions J2A and J2B.

Data — Telemetry, telecommand and computer communication emissions having designators with A, C, D, F, G, H, J or R as the first symbol; 1 as the second symbol; D as the third symbol; and emission J2D. Only a digital code of a type specifically authorized in this Part may be transmitted.

Digital communication — Computer-based communication modes such as PSK31, ***packet radio*** and ***HSMM***.

Dipole antenna — A wire antenna often used on the high-frequency amateur bands.

DSP (Digital signal processing) — A newer technology that allows software to replace electronic circuitry.

DX — A ham radio abbreviation for *distance* or *foreign countries*.

DXCC — A popular ARRL award earned for contacting Amateur Radio operators in 100 different countries.

DX PacketCluster — A method of informing hams, via their computers, about the activities of stations operating from unusual locations.

DXpedition — A trip to an unusual location, such as an uninhabited island or other geographical or political entity which has few, if any, Amateur Radio operators, where hams operate while visiting. DXpeditions provide sought-after contacts for hams who are anxious to have a radio contact with someone in a rare location.

Elmer — A traditional term for someone who enjoys helping newcomers get started in ham radio. A mentor.

Emergency communication — Amateur Radio communication that take place during a situation where there is danger to lives or property.

External RF power amplifier — A device capable of increasing power output when used in conjunction with, but not an integral part of, a transmitter.

External RF power amplifier kit — A number of electronic parts, which, when assembled, is an external RF power amplifier, even if additional parts are required to complete assembly.

FCC (Federal Communications Commission) — The government agency that regulates Amateur Radio in the US.

Field & Educational Services (F&ES) — Staff at ARRL Headquarters that helps newcomers get started in ham radio and supports hams who help newcomers.

Field Day — A popular Amateur Radio activity during which hams set up radio stations outdoors and away from electrical service to simulate emergencies.

Field Organization — A cadre of ARRL volunteers who perform various services for the Amateur Radio community at the local and state level.

FM (Frequency modulation) — An operating *mode* commonly used on ham radio *repeaters*.

Fox hunt — A competitive Amateur Radio activity in which hams track down a transmitted signal.

Frequency coordinator — An entity, recognized in a local or regional area by amateur operators whose stations are eligible to be auxiliary or repeater stations, that recommends transmit/receive channels and associated operating and technical parameters for such stations in order to avoid or minimize potential interference.

FSTV (Fast-scan television) — A mode of operation that Amateur Radio operators can use to exchange live TV images from their stations.

Ham band — A range of frequencies on which ham communication are authorized.

Ham radio — Another name for *Amateur Radio*.

Ham radio operator — An *Amateur Radio operator* holding a written authorization to operate a ham station.

Harmful interference — Interference which endangers the functioning of a radionavigation service or of other safety services or seriously degrades, obstructs or repeatedly interrupts a radiocommunication service — including ham radio — operating in accordance with the international Radio Regulations.

HF (High frequency) — The radio frequencies from 3 to 30 MHz.

HSMM (High Speed Multimedia) — A digital radio communication technique using spread spectrum modes primarily on UHF to simultaneously send and receive video, voice, text, and data.

IARU (International Amateur Radio Union) — The international organization made up of national Amateur Radio organizations such as the ARRL.

Image — Facsimile and television emissions having designators with A, C, D, F, G, H, J or R as the first symbol; 1, 2 or 3 as the second symbol; C or F as the third symbol; and emissions having B as the first symbol; 7, 8 or 9 as the second symbol; W as the third symbol.

Information bulletin — A message directed only to amateur operators consisting solely of subject matter of direct interest to the amateur service.

International Morse code — A dot-dash code as defined in International Telegraph and Telephone Consultative Committee (CCITT) Recommendation F.1 (1984), Division B, I. Morse Code.

ITU (International Telecommunication Union) — An agency of the United Nations that allocates the radio spectrum among the various radio services.

Key clicks — Undesired switching transients beyond the necessary bandwidth of a Morse code transmission caused by improperly shaped modulation envelopes.

Mean power — The average power supplied to an antenna transmission line during an interval of time sufficiently long compared with the lowest frequency encountered in the modulation taken under normal operating conditions.

Mode — A type of ham radio communication; examples are *frequency modulation (FM voice), slow-scan television (SSTV) and SSB (single sideband voice).*

Morse code — A popular communication mode transmitted by on/off keying of a radio-frequency signal. Hams use the *international Morse code.*

Necessary bandwidth — The width of the transmitted frequency band which is just sufficient to ensure the transmission of information at the rate and with the quality required under specified conditions.

Net — An on-the-air meeting of hams at a set time, day and radio frequency.

Network — A system of interconnected radios to allow more than one station access to shared resources.

Out-of-band emission (splatter) — An emission on a frequency immediately outside the necessary bandwidth caused by overmodulation on peaks (excluding spurious emissions).

Packet radio — A computer-to-computer radio communication mode in which information is broken into short bursts. The bursts (packets) contain addressing and error-detection information.

PEP (peak envelope power) — The average power supplied to the antenna transmission line by a transmitter during one RF cycle at the crest of the modulation envelope taken under normal operating conditions.

Phone — Emissions carrying speech or other sound information having designators with A, C, D, F, G, H, J or R as the first symbol; 1, 2 or 3 as the second symbol; E as the third symbol. Also speech emissions having B as the first symbol; 7, 8 or 9 as the second symbol; E as the third symbol.

Power — Power is expressed in three ways: (1) Peak envelope power (PEP); (2) Mean power; and (3) Carrier power.

Public service — Activities involving Amateur Radio that hams perform to benefit their communities.

Pulse — Emissions having designators with K, L, M, P, Q, V or W as the first symbol; 0, 1, 2, 3, 7, 8, 9 or X as the second symbol; A, B, C, D, E, F, N, W or X as the third symbol.

QRP — An abbreviation for low power.

QSL bureau — A system for sending *QSL cards* to and from ham radio operators.

QSL cards — Cards that serve to confirm communication between two hams.

QST — The premiere ham radio monthly magazine, published by the *ARRL*. *QST* means "calling all radio amateurs."

RACES (Radio Amateur Civil Emergency Service) — A radio service that uses amateur stations for civil defense communication during periods of local, regional or national civil emergencies.

Radio Regulations — The latest ITU *Radio Regulations*.

RF (Radio frequencies) — The range of frequencies that can travel through space in the form of electromagnetic radiation.

RIC (Radio interface card) — A PCMCIA device with an antenna port used in *HSMM* radio to allow a personal computer to control a radio transceiver.

RLAN (Radio Local Area Network) — This is similar to a wireless LAN, except that hams replace the small antennas with larger outdoor antennas and use the equipment to cover several miles or more.

RMAN (Radio Metropolitan Area Network) — This is an *HSMM* radio technique, usually longer range or higher data rates, to interconnect amateur radio RLANs.

Radio (*or* Ham) shack — The room where Amateur Radio operators keep their station.

Radiotelegraphy — See *Morse code*.

Receiver — A device that converts radio signals into a form that can be heard.

Remote control — The use of a control operator who indirectly manipulates the operating adjustments in the station through a control link to achieve compliance with the FCC Rules.

Repeater — An amateur station, usually located on a mountaintop, hilltop or tall building, that receives and simultaneously retransmits the signals of other stations on a different channel or channels for greater range.

RTTY — Narrow-band direct-printing telegraphy emissions having designators with A, C, D, F, G, H, J or R as the first symbol; 1 as the second symbol; B as the third symbol; and emission J2B.

SAREX (*S*pace *A*mateur *R*adio *Ex*periment) — Amateur Radio equipment flown in space and operated by astronauts who are licensed Amateur Radio operators.

Space station — An amateur station located more than 50 km above the Earth's surface.

Splatter — See ***Out-of-band emission***.

Spread Spectrum — A technology, originated during World War II, which distributes or spreads a radio signal over a broad frequency range. This spreading prevents narrow band signals and noise sources from interfering with the spread spectrum signal. The spread spectrum signal is heard as noise to the traditional narrow band receiver. Also, emissions using bandwidth-expansion modulation emissions having designators with A, C, D, F, G, H, J or R as the first symbol; X as the second symbol; X as the third symbol.

Spurious emission — An emission, on frequencies outside the necessary bandwidth of a transmission, the level of which may be reduced without affecting the information being transmitted. They include harmonic emissions, intermodulation products and frequency conversion products, but exclude out-of-band emissions.

SSB (Single sideband) — A common ***mode*** of voice operation on the amateur bands.

SSTV (Slow-scan television) — A ***mode*** of operation in which ham radio operators exchange still pictures from their stations.

SWL (Shortwave listener) — A person who enjoys listening to shortwave radio broadcasts or Amateur Radio conversations.

Telecommand — A one-way transmission to initiate, modify, or terminate functions of a device at a distance.

Telecommand station — An amateur station that transmits communication to initiate, modify, or terminate functions of a space station.

Telemetry — A one-way transmission of measurements at a distance from the measuring instrument.

Test — Emissions containing no information having the designators with N as the third symbol. Test does not include pulse emissions with no information or modulation unless pulse emissions are also authorized in the frequency band.

TIS (Technical Information Service) — A service of the ***ARRL*** that helps hams solve technical problems.

Transceiver — A radio transmitter and receiver combined in one unit.

Transmitter — A device that produces radio-frequency signals.

UHF (Ultra-high frequencies) — The radio frequencies from 300 to 3000 MHz.

VE (Volunteer Examiners) — Amateur Radio operators who give Amateur Radio licensing examinations.

VHF (Very-high frequencies) — The radio frequencies from 30 to 300 MHz.

WAS (Worked All States) — An ***ARRL*** award that is earned when an Amateur Radio operator talks to and exchanges QSL cards with a ham in each of the 50 states in the US.

WAVE (Worked All VE) — An award that is earned when a ham talks to and exchanges QSL cards with a ham in each Canadian province.

Wavelength — A means of designating a frequency ***band,*** such as the 80-meter band.

Work — To contact another ham.

Chapter 2

Activities in Amateur Radio

One of the best things about this hobby we call Amateur Radio is its *flexibility*. In other words, Amateur Radio can be whatever *you* want it to be. Whether you are looking for relaxation, excitement, or a way to stretch your mental (and physical) horizons, Amateur Radio can provide it. This chapter was written by Larry Kollar, KC4WZK, with some new material by John Champa, K8OCL and Shawn Reed, N1HOQ. Let's take a brief tour through the following topic areas:

Awards — the individual and competitive pursuits that make up the tradition we call "paper chasing."

Contests — the challenge of on-the-air competition.

Nets — both traffic nets, where amateurs pass messages on behalf of hams and non-hams, and the casual nets, where groups of people with common interests often meet on the air to swap equipment, anecdotes and information.

Ragchewing — meeting new friends on the air.

Amateur Radio Education — Educating current and future hams brings in new blood (and revitalizes old blood!); educating our neighbors about ham radio is good for public relations and awareness.

ARRL Field Organization — Amateur Radio in general, and the ARRL in particular, depend on the volunteer spirit. As part of the Field Organization, you can exercise your administrative, speaking and diplomatic skills in service of the amateur community.

Emergency Communications — When disaster strikes, hams often have the only reliable means to communicate with the outside world. Practice and preparation are key to fulfilling this mission.

DF (Direction Finding) — If you've ever wanted to know where a transmitter (hidden or otherwise) is located, you'll find DFing is an enjoyable and useful skill.

HSMM (High Speed Multimedia) — Making contacts using video, voice, text, and data simultaneously on the ham radio version of a wireless Internet called the Hinternet.

Satellite Operation — You may be surprised to learn that hams have their own communications satellites! Satellite operation can be great fun and a technical challenge for those who want to operate on the "final frontier."

Repeaters — Using and operating repeaters is one of the most popular activities for both new and old hams.

Image Communications — Although it's fun to talk to other amateurs, it's even more fun to *see* them.

Digital Communications — Use your computer to communicate with stations around your town or around the world.

VHF, UHF and Microwave Weak-Signal Operating — Explore the challenging, quirky and surprising world above 50 MHz.

EME (Earth-Moon-Earth), Meteor Scatter and Aurora — Making contacts by bouncing your signals off the moon, the fiery trails of meteors and auroras.

AWARDS

Winning awards, or "paper chasing," is a time-honored amateur tradition. For those who enjoy individual pursuits or friendly competition, the ARRL and other organizations offer awards ranging from the coveted to the humorous.

DX Awards

The two most popular DX awards are DXCC (DX Century Club), sponsored by the ARRL and WAC (Worked All Continents), sponsored by the International Amateur Radio Union (IARU). The WAC award is quite simple: all you have to do is work one station on each of six continents. The DXCC is more challenging: you must work at least one station in each of 100 countries!

Fig 2.1 — One of the most prized awards in Amateur Radio: the DX Century Club.

How-to's of DXCC — Direct QSLs and DX Bureaus

Since DX stations are often inundated with QSL cards (and QSL requests) from US hams, it is financially impossible for most of them to pay for the return postage. Hams have hit upon several ways to lighten the load on popular DX stations.

The fastest, but most expensive, way to get QSL cards is the *direct* approach. You send your QSL card, with one or two International Reply Coupons (IRCs) or one or two dollars and a self-addressed airmail envelope to the DX station. International Reply Coupons are available from your local post office and can be used nearly anywhere in the world for return postage. Some DX hams prefer that you send one or two "green stamps" (dollar bills) because they can be used to defray posting, printing and other expenses. However, it is illegal in some countries to possess foreign currency. If you're not sure, ask the DX station or check DX bulletins available on the DX Cluster System, accessible by either packet radio or Telnet.

Many DX hams have recruited *QSL managers*, hams who handle the QSL chores of one or more DX stations. QSL managers are convenient for everyone. The DX station need only send batches of blank cards and a copy of the logs; hams wanting that station's card need only send a First Class stamp for US return postage and can expect a prompt reply. (In the case of QSL managers located outside the United States, you must still send IRCs (or dollars) and a self-addressed return envelope.)

The easiest (and slowest) way to send and receive large batches of QSL cards is through the incoming and outgoing QSL bureaus. The outgoing bureau is available to ARRL members. The incoming bureaus are available to all amateurs. Bureau instructions and addresses are printed periodically in *QST;* they appear in the *ARRL Operating Manual*, and they are available from ARRL Headquarters for an SASE.

Alternatively, you can submit your QSO log electronically to ARRL's Logbook of The World. All submissions are free; you only pay when you "redeem" your QSO credits for an award, such as DXCC. Once you are signed up as a Logbook user, you can submit new contact records whenever you wish. Your contacts will be matched against the logs of other Logbook users. Whenever a match occurs, you receive instant credit for the contact. You can learn more about Logbook of The World by visiting its Web site at **www.arrl.org/lotw/**.

DXpeditions

What does the avid DXer who has worked them all (or almost all of them) do for an encore? Answer: *become* the DX! DXpeditions journey to countries with few or no hams, often making thousands of contacts in the space of a few days.

In 1991, Albania opened its borders and legalized Amateur Radio for the first time in many years. To train the first new generation of Albanian hams and to relieve the pileups that were sure to happen, a contingent of European and American hams organized a DXpedition to Albania. The DXpedition made over 10,000 contacts and changed Albania from one of the rarest and most-desired countries to an "easy one."

In March/April, 2004, one of the largest DXpeditions ever organized completed 153,113 QSOs during a 25-day period from tiny Rodriguez Island in the Indian Ocean. With only one resident amateur, Rodriguez is quite difficult to work for most hams, and it counts as a separate country for the DXCC award. The 3B9C DXpedition team worked amateurs worldwide on all current HF bands, including contacts on the 6-meter band. 3B9C even worked stations via EME (earth-moon-earth) and satellite communications on the 70-cm (432 MHz) UHF band.

DX Nets

The beginning DXer can get a good jump on DXCC by frequenting DX nets. On DX nets, a net control station keeps track of which DX stations have checked into the net. He or she then allows a small group of operators (usually 10) to check in and work one of the DX stations. This permits weaker stations to be heard instead of being buried in a pileup. Since the net control station does not tolerate net members making contacts out-of-turn, beginning operators have a better chance of snagging a new country. Nets and frequencies on which they operate vary. For the latest information on DX nets, check with local DXers and DX bulletins.

Efficient DX Operation

The best DXers will tell you the best equipment you have is "the equipment between your ears." Good operators can make contacts with modest power. The details of efficient DX operating cannot be covered in this brief space.

WAS (Worked All States)

The WAS certificate is awarded to amateurs who have QSL cards from at least one operator in each of the 50 United States. Chasing WAS is often a casual affair, although there are also nets dedicated to operators who are looking for particular states.

Endorsements

The initial DXCC or WAS award does not mean the end. There are over 300 DXCC countries. As you reach certain levels in your country count, you qualify for endorsements. Endorsements arrive in the form of stickers that you attach to your DXCC certificate.

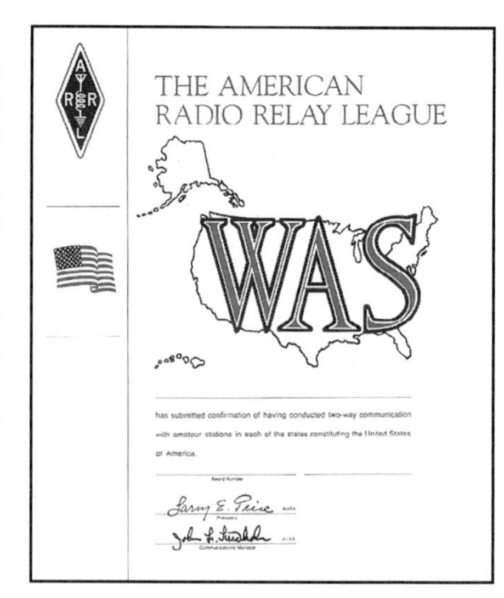

Fig 2.2 — Work one station in each of the 50 states and you're eligible for the ARRL's Worked All States (WAS) award.

Both WAS and DXCC offer endorsements for single-band or single-mode operation. For example, if you work all 50 United States on the 15-m band, your certificate has an endorsement for 15 m. The most difficult endorsement is the 5-band (5B) endorsement. Rare indeed is the operator who can display a 5BDXCC certificate!

CONTESTS

Some people enjoy the thrill of competition, and Amateur Radio provides challenges at all levels in the form of operating contests. Besides the competitive outlet, contests have provided many hams with a means to hone their operating skills under less-than-optimum conditions. On the VHF and higher bands, contests are one way to stimulate activity on little-used segments of the amateur spectrum.

This section briefly discusses a few ARRL-sponsored contests. The Contest Corral section of *QST* provides up-to-date information on these and other contests. The ARRL also publishes the *National Contest Journal* (*NCJ*), which is good reading for any serious (or semi-serious) contester.

Field Day

Every year on the fourth full weekend in June, thousands of hams take to the hills, forests, campsites and parking lots to participate in Field Day. The object of Field Day is not only to make contacts, but also to make contacts under conditions that simulate the aftermath of a disaster. Most stations are set up outdoors and use

Fig 2.3 — Elaine Larson, KD6DUT, takes a turn at logging as Fred Martin, KI6YN, works the paddles during the Conejo Valley Amateur Radio Club's Field Day operation.

Keeping a Log Book

At one time, keeping a log of your contacts was an FCC requirement. The FCC has dropped this requirement in recent years, but many amateurs, both new and old, still keep logs.

Why Keep a Log?

If keeping a log is optional, why do it? Some of the more important reasons for keeping a log include:

Legal protection — If you can show a complete log of your activity, it can help you deal with interference complaints. Good recordkeeping can help you protect yourself if you are ever accused of intentional interference, or have a problem with unauthorized use of your call sign.

Awards tracking — A log helps you keep track of contacts required for DXCC, WAS, or other awards. Keeping a log lets you quickly see how well you are progressing toward your goal.

An operating diary — A log book is a good place for recording general information about your station. You may be able to tell just how well that new antenna is working compared to the old one by comparing recent QSOs with older contacts. The log book is also a logical place to record new acquisitions (complete with serial numbers in case your gear is ever stolen). You can also record other events, such as the names and calls of visiting operators, license upgrades, or contests, in your log.

Paper and Computer Logs

Many hams, even those with computers, choose to keep their logs on paper. Paper logs still offer several advantages (such as flexibility) and do not require power. Paper logs also survive hard-drive crashes!

Preprinted log sheets are available, or you can create your own. Computers with word processing and publishing software let you create customized log sheets in no time.

On the other hand, computer logs offer many advantages to the serious contester or DXer. For example, the computer can search a log and instantly tell you whether you need a particular station for DXCC. Contesters use computer logs in place of *dupe sheets* to weed out duplicate contacts before they happen, saving valuable time. Computer logs can also tell you at a glance how far along you are toward certain awards.

Computer logging programs are available from commercial vendors. Some programs may be available as shareware (you can download it from a website and pay for the program if you like the way it works). If you can program your computer, you can also create your own custom logging program, and then give it to your friends or even sell it!

emergency power sources.

Many clubs and individuals have built elaborate Field Day equipment, and that is all to the best — if a real disaster were to strike, those stations could be set up quickly, wherever needed, and need not depend on potentially unreliable commercial power!

Other Contests

Other popular contests include:

QSO Parties. These are fairly relaxed contests — good for beginners. There are many state QSO parties, and others for special interests, such as the QRP ARCI Spring QSO Party.

Sweepstakes. This is a high-energy contest that brings thousands of operators out of the woodwork each year.

Various DX contests. DX contests offer good opportunities for amateurs to pursue their DXCC award contacts. A good operator can work over 100 countries in a weekend!

VHF, UHF and microwave contests. These contests are designed to stimulate activity on the weak-signal portions of our highest-frequency bands. The ARRL VHF/UHF contests are held in January, June and September. There is also a contest for 10-GHz operators, and another one for EME (moonbounce) enthusiasts.

Each issue of *QST* lists the contests to be held during the next two months.

NETS

A net is simply a group of hams that meet on a particular frequency at a particular time. Nets come in three classes: public service, traffic and special interest.

Public Service/Traffic Nets

Public service and traffic nets are part of a tradition that dates back almost to the dawn of Amateur Radio. The ARRL, in fact, was formed to coordinate and promote the formation of traffic nets. In those early days, nets were needed to communicate over distances longer than a few miles. (Thus the word "Relay" in "American Radio Relay League.")

Public service and traffic nets benefit hams and non-hams alike. Any noncommercial message — birthday and holiday greetings, personal information or a friendly hello — may be sent anywhere in the US and to foreign countries that have third-party agreements with the United States. Many missionaries in South America, for example, keep in touch with stateside families and sponsors via Amateur Radio.

The ARRL National Traffic System (NTS) oversees many of the existing traffic nets. Most nets are local or regional. They use many modes, from slow-speed CW nets in the Novice HF bands, to FM repeater nets on 2 m.

Since the amateur packet-radio network now covers much of the US and the world, many messages travel over packet links. Amateurs use the packet radio network not only for personal or third-party traffic, but also for lively conferences, discussions and for trading equipment.

HF and Repeater Nets

HF nets usually cover a region, although some span the entire country. This has obvious advantages for amateurs sending traffic over long distances. Repeater nets usually cover only a local area, but some linked repeater nets can cover several states.

Both types of nets work together to speed traffic to its destination. For example, think of the HF nets as a "trunk" or highway that carries traffic quickly and reliably toward its *approximate* destination. From there, the local and regional nets take over and pass the traffic directly to the city or town. Finally, a local amateur delivers the message to the recipient.

Routine traffic handling keeps the National Traffic System (NTS) prepared for emergencies. In the wake of Hurricane Andrew in 1992, hams carried thousands of messages in and out of the stricken south Florida region. The work that hams do during crisis situations ensures good relations with neighbors and local governments.

Fig 2.4 — The ARRL VEC processes 30,000 applications annually.

Other Nets

Many nets exist for hams with common interests inside and outside of Amateur Radio. Some examples include computers, owners of Collins radio equipment, religious groups and scattered friends and families. Most nets meet on the 80- and 20-m phone bands, where propagation is fairly predictable and there are no shortwave broadcast stations to dodge.

RAGCHEWING

Ragchewing is the fine art of the long contact. Old friends often get together on the air to catch up on current events. Family members use ham radio to keep in touch. And, of course, new acquaintances get to know each other!

In many cases, friends scattered across the country get together to create ragchewing nets. These nets are very informal and may not make much sense to the outsider listening in.

AMATEUR RADIO EDUCATION

Elmering (helping new and prospective operators) is a traditional amateur activity. Much of an amateurs' educational efforts go toward licensing (original and upgrading), but there are other opportunities for education, including public relations.

License Classes

Anyone can set up license classes. Many Amateur Radio clubs hold periodic classes, usually for the Novice and Technician elements with CW practice sessions. The ARRL supports Registered Amateur Radio Instructors, but registration is not necessary to conduct a class.

If you are looking for a class to attend, and do not have an "Elmer" to answer your questions, write ARRL Field & Educational Services for a list and schedule of classes in your area. If you want to become an instructor, you can request the same list of classes from Field & Educational Services — most classes will welcome another helping hand.

Volunteer Examiners (VEs)

To become a VE, you must hold a General or higher amateur license and be certified by one of the VE Coordinators (VECs). The ARRL supports the largest VE program in the nation; other organizations run VE programs on a national or regional basis. General and Advanced licensees on a VE team must be supervised by at least one Extra Class licensee.

School Presentations

Amateur Radio complements any school program. Schoolchildren suddenly find that Amateur Radio gives them a chance to apply their studies immediately. The math and science used in Amateur Radio applies equally to the classroom. Even geography takes on a new meaning when a student works a new country!

Unfortunately, many schools do not have an active Amateur Radio presence — and that is why local volunteers are important. An HF or satellite station, or even a 2-m hand-held transceiver tuned to the local repeater, can prove an exciting and educational experience for both the volunteer and the students.

Fig 2.5 — Dry run just before an International Space Station pass. Keilah Meuser is practicing with others looking on.

Thanks to NASA's ARISS (Amateur Radio aboard International Space Station) program, amateurs all over the nation have put schoolchildren in direct contact with astronauts. Who knows how many future scientists received their inspiration while sitting behind an amateur's microphone?

ARRL FIELD ORGANIZATION

ARRL members elect the Board of Directors and the Section Managers. Each Section Manager appoints volunteers to posts that promote Amateur Radio within that Section. (The United States is divided into 15 ARRL *Divisions*. These Divisions are further broken down into 69 *Sections*.) A few of the posts include:

Assistant Section Managers — ASMs are appointed as necessary by the SM to assist the SM in responding to membership needs within the Section.

Official Observers (OO) / Amateur Auxiliary — Official Observers are authorized by the FCC to monitor the amateur bands for rules discrepancies or violations. The Amateur Auxiliary is administered by Section Managers and OO Coordinators, with support from ARRL Headquarters.

Technical Coordinators (TC) and ***Technical Specialists (TS)*** — Technical Coordinators and Technical Specialists assist hams with technical questions and interference problems. They also represent the ARRL at technical symposiums, serve on cable TV advisory committees and advise municipal governments on technical matters.

EMERGENCY COMMUNICATIONS

The FCC Rules list emergency communications as one of the purposes of the

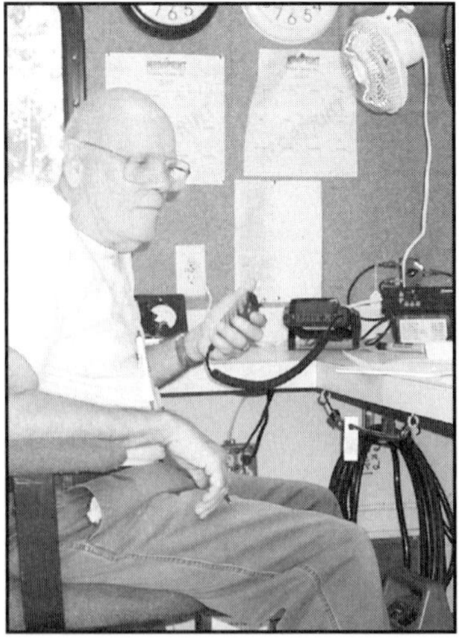

Fig 2.6 — Ham Radio on the scene after terrorists attack New York City in September 2001. On the left, in the American Red Cross radio room in Brooklyn, Mark Dieterich, N2PGD (standing), and Simone Lambert, KA1YVF, check the volunteer shift schedule and handle schedule management from the World Trade Center Disaster Relief Communications Website. On the right, Ed Cravey, KF4HPY, at the controls of the Chattahoochee Baptist Association's W4CBA mobile unit in Edison, New Jersey. The W4CBA station used a local 2-Meter repeater to communicate with deployed kitchens and showers in the old Brooklyn Navy Yard and near Ground Zero in Manhattan.

Amateur Radio Service — and in reality, the ability to provide emergency communications justifies Amateur Radio's existence. The FCC has recognized Amateur Radio as being among the most reliable means of medium- and long-distance communication in disaster areas. The terrorist attacks on the United States September 11, 2001 launched the Amateur Radio community into a real-life test of individual and collective communications skills.

Amateur Radio operators have a long tradition of operating from backup power sources. Through events such as Field Day, hams have cultivated the ability to set up communication posts wherever they are needed. Moreover, Amateur Radio can provide computer networks (with over-the-air links where needed) and provide other services such as video (ATV) and store-and-forward satellite links that no other service can deploy on a wide scale. One can argue, therefore, that widespread technology makes Amateur Radio even more crucial in a disaster situation.

If you are interested in participating in this important public service, you should contact your local EC (Emergency Coordinator). Plan to participate in preparedness nets and a yearly SET (Simulated Emergency Test).

ARES AND RACES

The Amateur Radio Emergency Service (ARES) and the Radio Amateur Civil Emergency Service (RACES) are the umbrella organizations of Amateur Radio emergency communications. The ARES is sponsored by ARRL, although ARRL membership is not required for ARES participation, and handles many different kinds of public-service activities. On the other hand, RACES is administered by the Federal Emergency Management Agency (FEMA) and operates only for civil preparedness and in times of civil emergency. RACES is activated at the request of a local, state or federal official.

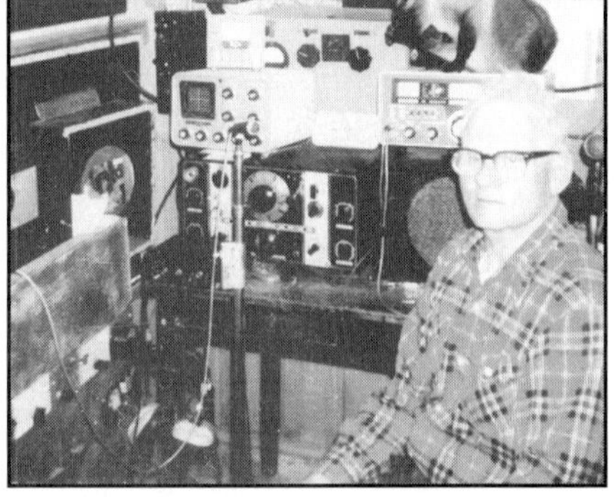

Fig 2.7 — Shown here in his Worcester, MA ham shack, the original W1LC — Herman R. Sanborn (SK). In June 1953, Sanborn provided emergency communications for the American Red Cross and others in the aftermath of the deadly Worcester Tornado. Sanborn's station was also on the air to assist during two other historic New England weather events — the Flood of 1936 and the Hurricane of '38. The cat atop the radio gear is *Scorpio*. (*photo courtesy of Nancy Riik*)

Amateurs serious about emergency communication may carry dual RACES/ARES membership. FCC rules make it possible for ARES and RACES to use many of the same frequencies, so that an ARES group also enrolled in RACES can work in either organization as required by the situation.

MILITARY AFFILIATE RADIO SERVICE (MARS)

MARS is administered by the US armed forces, and exists for the purpose of transmitting communications between those serving in the armed forces and their families. This service has existed in one form or another since 1925.

There are three branches of MARS: Army MARS, Navy/Marine Corps MARS and Air Force MARS. Each branch has its own requirements for membership, although all three branches require members to hold a valid US Amateur Radio license and to be 18 years of age or older (amateurs from 14 to 18 years of age may join with the signature of a parent or legal guardian).[1]

MARS operation takes place on frequencies adjoining the amateur bands and usually consists of nets. Nets are usually scheduled to handle traffic or to handle administrative tasks. Various MARS branches may also maintain repeaters or packet systems.

MARS demonstrated its importance during the 1991 Desert Storm conflict, when MARS members handled thousands of messages between the forces on the front lines and their friends and families at home. While MARS usually handles routine traffic, the organization is set up to handle official and emergency traffic if needed.

DIRECTION FINDING (DF)

If you've ever wanted to learn a skill that's both fun *and* useful, then you'll enjoy direction finding, or DFing. DFing is the art of locating a signal or noise source by tracking it with portable receivers and directional antennas. Direction finding is not only fun, it has a practical side as well. Hams have been instrumental in hunting down signals from aircraft ELTs (emergency locator transmitters), saving lives and property in the process.

We will just scratch the surface of DF activities in this section. There is much more in the **EMI/Direction Finding** chapter.

Fox Hunting

Fox hunting, also called *T-hunting* or sometimes *bunny hunting*, is ham radio's answer to hide-and-seek. One player is designated the fox; he or she hides a transmitter, and the other player attempts to find it. Rules change from place to place, but the fox must generally locate the transmitter within certain boundaries and transmit at specific intervals.

Fox hunts vary around the world. American fox hunts often employ teams of fox hunters cruising in their cars over a wide area. European and other fox hunters employ a smaller area and conduct fox hunts on foot. *Radiosport* competitions are usually European style.

Locating Interference

Imagine trying to check into your favorite repeater or HF net one day, only to find reception totally destroyed by noise or a rogue signal. If you can track down the interference, then you can figure out how to eliminate it.

Finding interference sources, accidental or otherwise, has both direct and indirect benefits. Touch lamps are a notorious noise source, especially on 80 m. If you can find one, the owner is legally obligated to eliminate the interference. Even better, if you can show your neighbors that something other than your station is interfering with their TV reception, you might gain an ally next time you petition the local government to let you have a higher tower!

Fig 2.8 — Dave Pingree, N1NAS, hunts down a transmitter on 2-m FM. (*photo by Kirk Kleinschmidt, NT0Z*)

Fig 2.9 — Dick Esneault, W4IJC, a member of the original Project OSCAR team, looks over a model of the Phase 3D satellite. The body of the actual satellite is well over 7 feet wide. (*photo courtesy AMSAT-NA*)

SPECIALIZED COMMUNICATIONS

Satellite Operation

Amateur Radio has maintained a presence in space since 1961, with the launch of OSCAR 1 (OSCAR is an acronym for Orbiting Satellite Carrying Amateur Radio). Since then, amateurs have launched over three dozen satellites, with over 20 still in orbit today.

Amateurs have pioneered several developments in the satellite industry, including low-orbit communication "birds" and PACSATs — orbiting packet bulletin board systems. Operating awards are available from ARRL and other organizations specifically for satellite operation, such as WAS (Worked All States), WAZ (Worked All Zones), DXCC and many more.

Satellite operation does not have to be complex and difficult to learn. When someone mentions satellite operation, many people conjure up an image of large dishes and incredibly complex equipment. Actually, you can probably work several OSCARs with the equipment you have in your shack right now!

The entire collection of OSCARs — and their operating modes — can be broken down into four basic categories:

Voice/CW (Analog)

Analog satellites range from the low-orbit FO (Fuji OSCARs) built in Japan, and AO (AMSAT OSCAR) birds built by a co-op of many nations, to the high orbit Phase 3 satellites such as OSCAR 40 and soon to be launched AMSAT EAGLE and PHASE 3-E. Operating on analog satellites is much like operating on HF — you'll find lots of SSB and CW contacts, with some RTTY, Hellschreiber, even SSTV and other digital modes thrown in. You may even work a vintage satellite (AO-7), launched in November of 1974!

Fig 2.10 — A portable, all purpose, mode V/U/S earth station. This station works LEOs and HEOs (high earth orbiters) like AO-40, and future birds like Eagle. If you live in an area of dense tree growth that impedes your antenna performance, you might consider portable operation at a location with a very low horizon. (*photo courtesy of Shawn Reed, N1HOQ*).

Packet (Digital)

The digital satellites are orbiting packet mailboxes and/or APRS digipeaters. APRS stands for Automatic Position Reporting System. Store and forward packet "mail" can be sent and received, and position reports can be sent to be plotted on maps in APRS software. Real-time "keyboarding" QSOs are also seen digipeating through these birds. Several digital satellites carry an experiment called RUDAK, a versatile system that allows experimentation with packet, analog and crossband FM modes. Many of these satellites carry systems that allow switching between digital and analog operation.

FM Repeater Satellites

Some of the digital satellites are fixed on, or can be switched to FM "bent pipe" repeater mode. This allows one FM channel of communication very similar to terrestrial FM repeaters. These birds are often the first satellite experience for many amateurs, and because of the relatively modest equipment and skill level required to operate them, they are referred to as "Easy Sats". Most of these LEO, low-earth orbiters, can be operated with as little as a dual-band handheld transceiver and handheld beam antenna! They are popular with portable and mobile operators, award chasers and "Grid Expeditioners", who seek contacts from different maidenhead grid squares around the country and the world. Some of these birds include AO-27 (AMRAD OSCAR), SO-41 and SO-50 (SAUDI OSCARs) built by the Kingdom of Saudi Arabia. The FM repeater mode will be available on AMSAT-ECHO, which may be launched and in orbit by late 2004, VUSAT, being built by AMSAT India, and others.

ARISS

The multi-national team that developed the International Space Station recognizes the value of Amateur Radio in space. Carrying forward the tradition of the SAREX (Shuttle Amateur Radio EXperiment) and Amateur Radio aboard the former MIR space station, a permanent ham shack now resides on the new space station. The ARISS (Amateur Radio aboard International Space Station) program allows for digital and analog/voice communication with the expedition crew aboard ISS and other amateurs via an on board packet mailbox and APRS digipeater. Scheduled contacts are frequently made with demonstration stations in school classrooms around the world. Questions and answers are exchanged between astronauts and schoolchildren via FM voice on VHF.

Nearly all the astro/cosmonauts assigned to ISS are licensed amateurs in their home countries, so your chances of making a contact with space travelers is quite good! ARISS and the former SAREX and MIR projects have proven their worth in education and goodwill, over and over. The Amateur equipment also provides a backup for the astronauts in case of normal communications failure.

Repeaters

Many amateurs make their first contacts on repeaters. Repeaters carry the vast majority of VHF/UHF traffic, making local mobile communication possible for many hams.

Hams in different regions have different opinions on repeater usage. In some areas, hams use repeaters only for brief contacts, while those in other areas encourage socializing and ragchewing. All repeater users give priority to mobile emergency communications.

The best way to learn the customs of a particular repeater is to listen for a while before transmitting. This avoids the misunderstandings and embarrassment that can occur when a newcomer jumps in. For example, in some repeater systems it is assumed that the word "break" indicates an urgent or emergency situation. Other systems recognize "break" as a simple request to join or interrupt a conversation in progress. Neither usage is more "correct," but you can imagine what might happen to a traveling ham who was unaware of the local customs!

Most repeaters are *open*, meaning that any amateur may use the repeater. Other repeaters are *closed*, meaning that usage is restricted to members. Many repeaters have an *autopatch* capability that allows amateurs to make telephone calls. However, most autopatches are closed, even on otherwise open repeaters. The *ARRL Repeater Directory* shows repeater locations, frequencies, capabilities and whether the repeater is open or closed.

Most repeaters are maintained by clubs and other local organizations. If you use a particular repeater frequently, you should join and support the repeater organization. Some hams set up their own repeaters as a service to the community.

Image Communications

Several communications modes allow amateurs to exchange still or moving images over the air. Advances in technology in the last few years have brought the price of image transmission equipment within reach of the average ham's budget. This has caused a surge of interest in image communication.

ADV

Amateur Digital Video (ADV) is the transmission of full-motion digital video on the Amateur Radio microwave bands. The output of a conventional, full motion, video source, such as a camcorder, web cam, or digital camera, is digitized by PC-based communications software called a CODEC (coder-decoder). The processed video is then transmitted on a amateur microwave band. This process is sometimes called videoconferencing or streaming video.

The image quality of ADV is often lower than that of ATV, but the equipment cost is significantly less. Digital video can be transmitted and received on some microwave ham bands by using inexpensive WiFi equipment, also known as IEEE 802.11 WLAN. This equipment comes with a rubber duck antenna, which can be replaced with an outdoor, high-gain directive antenna array. A useful range of several miles can be easily covered in most terrain. This increasing popular approach is part of new amateur technology called HSMM (high speed multimedia) radio.

Why multimedia? Along with your PC, HSSM technology, and inexpensive microwave equipment, you can share high-speed Internet access, send text files, high resolution still images, or instant messages all at the same time. Some hams play interactive electronic games or use HSMM to control another ham's entire radio station so they can split the cost of the equipment, or take advantage of a better radio site, such as a higher altitude, lower noise remote location. For more information on HSMM radio, see the Spread Spectrum and Multimedia section of Chapter 9, Modes and Modulation Sources.

ATV

Amateur TV is full-motion video over the air. (It is sometimes referred to as *fast scan*, or *FSTV*.) ATV signals use the same format as broadcast (and cable) TV. Watching an ATV transmission is the same as watching your own television. With ATV, however, you can turn a small

Fig 2.11 — Give him the specifications and Sam, K6LVM, can show you the radiation pattern of your antenna — via ATV! (*photo by Tom O'Hara, W6ORG*)

space in your home into your own television studio. Amateur communication takes on an exciting, new dimension when you can actually *see* the person you're communicating with!

The costs of ATV equipment have declined steadily over the years. The popularity of the camcorder has also played a significant role. (The family camcorder can do double duty as a station camera!) It is now possible to assemble a versatile station for well under $1000. Amateur groups in many areas have set up ATV repeaters, allowing lower-powered stations to communicate over a fairly wide area. If you're fortunate enough to live within range of an ATV repeater, you won't need complicated antenna arrays or high power.

If you can erect high-gain directional antennas for your ATV station, you can try your hand at DXing. When the bands are open, it's not uncommon to enjoy conversations with stations several hundred miles away. In addition to your directional antennas, you must run moderate power levels to work ATV DX. Most DXers use at least 50 W or more.

Since this is a wide-bandwidth mode, operation is limited to the UHF bands (70 cm and higher). The *ARRL Repeater Directory* and the *ARRL Operating Manual* list band plans. The *Repeater Directory* includes lists of ATV repeaters. The **Modes and Modulation Sources** chapter provides details on setting up an ATV station with dedicated or converted video gear.

SSTV

SSTV, or slow-scan TV, is a narrow-bandwidth image mode. Instead of full-motion video at roughly 24 frames per second, SSTV pictures are transmitted at 8, 16 or 32 seconds per frame. In the beginning, SSTV was strictly a black-and-white mode. The influx of computers (and digital interfaces) has spawned color SSTV modes. Since SSTV is a narrow-band mode, it is popular on HF. Some experimenters run SSTV on satellites as well.

An SSTV signal is generated by breaking an image into individual *pixels*, or dots. Each color or shade is represented by a different audio tone. This tone is fed into the audio input of an SSB transmitter, converting the tones into RF. On the receive end, the audio tones are regenerated and fed into a dedicated SSTV converter or into a simple computer interface to regenerate the picture. For more information about SSTV, see the **Modes and Modulation Sources** chapter.

Fig 2.12 — An SSTV image as seen on a standard TV set using a digital scan converter.

Fax

Fax, or *facsimile* transmission, is one of the original image communication modes. Fax was once unavailable to amateurs due to FCC regulations, but is now a legal communication mode on most HF and higher bands.

Amateur Radio fax works much like old analog fax systems: an image is scanned from paper and converted into a series of tones representing white or black portions of a page. Amateurs are working on standards for the use of digital fax machines over radio as well.

Uses for amateur fax are as limitless as

QSLing

A QSL card (or just "QSL") is an Amateur Radio tradition. QSL cards are nearly as old as Amateur Radio itself, and the practice has spread so that short-wave listeners (SWLs) can get cards from shortwave and AM broadcast stations.

Most amateurs have printed QSL cards. QSL card printers usually have several standard layouts from which to choose. Some offer customized designs at extra cost. If you are just starting out, or anticipate changing your call sign (just think, you could get a call like "KC4WZK"), you may want to purchase a pack of "generic" QSL cards available from many ham stores and mail-order outlets.

YANKEE CLIPPER

CONTEST CLUB

MASSACHUSETTS

K5MA

Barnstable County FOC 1256

FN41QO Life Member ARRL

Confirming QSO with:

Day/Month/Year	UTC	MHz	2-way	RST
			CW SSB	

PSE QSL TNX Contest: 73,

Jan Carman
P. O. Box 930
West Falmouth, MA
02574 USA

K5MA's QSL is an example of a properly formatted card. Notice how all of the information is on one side of the card.

Filling Out Your Cards

QSL cards must have certain information for them to be usable for award qualification. At a minimum, the card must have:

- Your call sign, street address, city, state or province and country. This information should be preprinted on one side of your QSL card.
- The call of the station worked.
- The date and time (in UTC) of the contact.
- The signal report.
- The band and mode used for the contact.

Awards for VHF and UHF operations may also require the grid locator (or "grid square") in which your station is located. Current practice is to include your 6-digit grid square on your QSL card even if you have no plans to operate VHF and UHF, since some HF competitions and awards require your grid square designator.

Many hams provide additional information on their QSL cards such as the equipment and antennas used during the contact, power levels, former calls and friendly comments.

Sending and Receiving Domestic QSLs

Although most QSL cards can be sent as post cards within the United States, usually saving some postage costs, post card style QSL cards often arrive with multiple cancellations and other unintended markings that can obscure or obliterate the printed and written information. It is best to send all QSL cards in a protective envelope. Back when postage was cheap, you could send out 100 post cards for a few dollars and domestic stations would send QSLs as a matter of course. Currently, if you really need a particular QSL, it is best to send a self-addressed stamped envelope along with your card.

QSLing for DX stations is somewhat more involved and is discussed elsewhere in this chapter.

Fig 2.13 — Dave Patterson, WB8ISZ, checks into his local packet bulletin board as his cat Sam looks on. *(photo by WB8IMY)*

Fig 2.14 — John Shew, N4QQ (at a portable station set up by KG5OG) makes an EME CW contact with VE3ONT. *(photo by WB8IMY)*

your imagination. Suppose you were having trouble with the design of your new home-brewed widget. You could fax a copy of the schematic to a sympathetic ham, who could mark in some changes and fax it back to you. And how about faxing QSL cards? No hunting for stamps or waiting for the mail to arrive!

Digital Communications

Digital communications predate the personal computer by many years. In fact, some amateurs consider CW to be a digital mode in which the amateur's brain handles the encoding and decoding of information. For the purposes of this *Handbook*, however, we consider digital modes to be those traditionally encoded and decoded by electronic means. Currently, the use of PC digital sound cards to encode and decode digital communication modes has become the ham radio standard. Common sound card-based digital modes in use today include RTTY, PSK31, Packet Radio, and the vhf/uhf weak-signal mode software package WSJT, created by Joe Taylor, K1JT.

Packet Radio

Packet radio is much less popular in today's ham radio world than it was in the 1990s. Packet radio's most important applications include networking and unattended operation. The two most common uses are the worldwide DX Cluster network and regional or local general use networks. Do you need to give some information to an absent friend? Send an electronic mail message (or *e-mail* in networking parlance). Would you like to see what 20-meter DX stations in Asia have been worked or heard recently by east-coast USA stations? Log onto your local DX Cluster node and find out. Is your friend out of range of your 2-meter packet radio? Send your message through the packet network.

In packet radio, transmitted data is broken into "packets" of data by a *TNC* (terminal node controller). Before sending these packets over the air, the TNC calculates each packet's checksum and makes sure the frequency is clear. On the receive end, a TNC checks packets for accuracy and requests retransmission of bad packets to ensure error-free communication.

Packet radio works best on frequencies that are relatively uncrowded. On busy frequencies (or LANs), it is possible for two stations to begin transmitting at once, garbling both packets (this is called a *collision*). Another common problem is the *hidden transmitter*, which happens when one of two stations (that are out of range of each other) is in contact with a third station within range of *both* (see **Fig 2.15**). Collisions can easily occur at the third station since neither of the other two stations can hear each other and thus may transmit simultaneously.

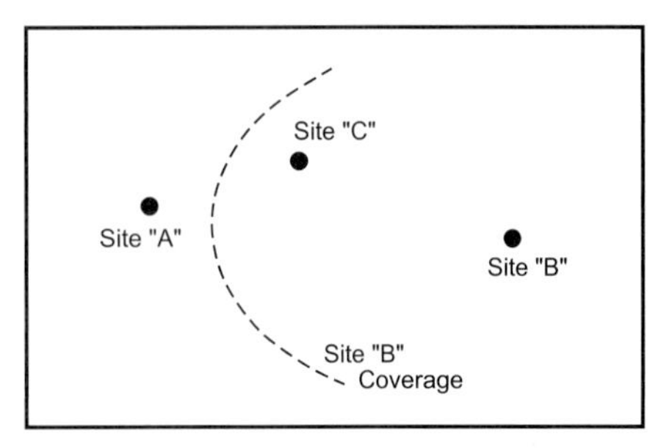

Fig 2.15 — The "hidden transmitter" problem. Site B has established contact with site C, but cannot hear site A. However, site C can hear *both stations*. If site B transmits while site A is transmitting, or vice versa, a packet collision occurs.

APRS

APRS (*A*utomatic *P*osition *R*eporting *S*ystem) uses the unconnected packet radio mode to graphically indicate the position of moving and stationary objects on maps displayed on a computer monitor. Unconnected packets are used to permit all stations to receive each transmitted APRS packet on a one-to-all basis rather than the one-to-one basis required by connected packets.

APRS is used for tracking stations or objects in motion or in fixed positions. Weather-monitoring equipment can be interfaced to an APRS station to disseminate real-time weather information.

Like standard packet-radio transmissions, APRS data are relayed through digipeaters. Unlike standard packet radio, APRS stations use generic digipeater paths so that no prior knowledge of the network is needed. In addition, the Internet is an integral part of the system that is used for collecting and disseminating current APRS data in real time.

Virtually all VHF APRS activity occurs on 2 meters, specifically on 144.39 MHz, which is recognized as *the* APRS operating channel in the United States and Canada. On UHF, you'll find the activity on 445.925 MHz.

Many groups and individuals that participate in public service and disaster communications find APRS a useful tool. Others find it interesting to view real-time weather reports from around their area.

RTTY

RTTY is the original data communication mode, and it remains in active use today. While RTTY does not support the features of the newer data modes, such as frequency sharing or error correction, RTTY is better suited for "roundtable" QSOs with several stations. It also is the most popular mode for worldwide digital contest events.

RTTY was originally designed for use with mechanical teleprinters, predating personal computers by several decades. Amateurs first put RTTY on the air using surplus teletypewriters (TTYs) and home-brewed vacuum-tube-based interfaces. Today, of course, RTTY uses computers, computer sound cards, or dedicated controllers, many of which also support other digital modes such as CW, PSK31 and packet.

Other HF Digital Modes

One of the most common HF digital modes for general domestic and DX contacts is PSK31. This mode is particularly effective for low power (QRP) communications. PSK31 uses phase-shift keying techniques most often generated by PC-based software operating through a common PC sound card. There are many PC software (terminal) programs available that support PSK31, most of which are available as 'freeware.'

There are a number of HF digital modes based on an old system called Hellschreiber, which uses facsimile technology. These modes can all be generated by PC-based software using the PC sound card to handle the processing and generate the audio signal for your HF transmitter. One of the Hellschreiber modes, called PSK Hell, is quite popular for HF keyboard-to-keyboard communications. Another variant called Feld-Hell is often heard on the HF bands. Finally, there are two freeware programs written by IZ8BLY, MFSK8 and MFSK16, which transmit at higher data rates and show improved immunity to HF propagation variations.

Microwave and VHF/UHF Weak-Signal Operating

Hams use many modes and techniques to extend the range of line-of-sight signals. Those who explore the potential of VHF/UHF communications are often known as *weak-signal* operators. Weak-signal enthusiasts probe the limits of propagation. Their goal is to discover just how far they can communicate.

They use directional antennas (beams or parabolic dishes) and very sensitive receivers. In some instances, they employ considerable output power, too. As a result of their efforts, distance records are

Amateur Voice Over the Internet: VoIP

Fueled partly by the improvements in Internet bandwidth, hams have been using the Internet as a means to link stations over great distances. This typically involves passing voice signals through a technique known as *VoIP*, or Voice Over Internet Protocol.

Some FM repeater systems now include Internet VoIP linking so that they can share signals with other repeaters and individuals throughout the world. Hams who cannot otherwise set up HF stations are using VoIP links to enjoy long-distance communication.

One of the most successful amateur VoIP systems was (and still is) the Internet Radio Linking Project, or *IRLP*. *IRLP* experimenting began in the fall of 1997 with use of the Internet to link several Amateur systems in Canada, with the first full-time hookup running coast-to-coast between Vancouver and Saint John, New Brunswick. After a few false starts, Dave Cameron, VE7LTD, settled on the open-source operating system called Linux as the platform for the next generation of *IRLP*. Since then, *IRLP* has flourished into a reliable, worldwide network, with more than 1200 repeaters and simplex stations in early 2004.

A few other enterprising hams continued experimenting with *Windows*-based systems, on which the original *Internet Phone* experiments had been conducted. In 2001, Graeme Barnes, MØCSH, developed a program called *iLINK*, which had two compelling features: it allowed access directly from a desktop computer, and it worked remarkably well over conventional dial-up Internet connections. This helped put Internet linking in the hands of thousands of hams who already had *Windows*-based computers and dial-up Internet access, and gave rise to a whole new classification of Internet-linked Amateur "stations": licensed amateurs who were connecting themselves to distant repeaters using only a computer.

The doors opened even wider with *eQSO*, developed by Paul Davies, MØZPD. Like *iLINK*, *eQSO* runs on conventional *Windows* PCs, and allows direct PC-based access to remote links. It specifically permits access by

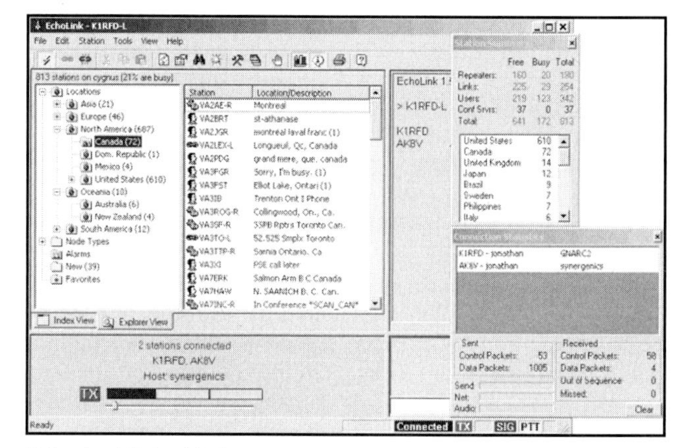

A typical EchoLink screen.

non-hams, either in a listen-only mode, or in off-air conferences. *eQSO* remains popular today.

Yaesu (Vertex Standard) entered the world of Amateur Internet linking with a hardware-software product called *WIRES-II*, in 2002. In conjunction with a set of Internet-based servers operated by Yaesu, the product functions similarly to some of the other linking systems, providing a "plug-and-play" solution for hams wishing to set up a link on a *Windows*-based computer.

Following the success of *iLINK*, my own contribution (in mid-2002) was a compatible software package called *EchoLink*. The program was originally designed to offer an alternative look-and-feel and feature set for *iLINK* users, but quickly evolved into a complete system of its own, due to its rapid (and unexpected) rate of acceptance. By early 2004, *EchoLink* had been installed and registered by more than 110,000 licensed hams in 147 countries, and typically carried more than 1,400 repeaters and simplex links and about 700 PC-based users at any given time.—*Jonathan Taylor, K1RFD*

broken almost yearly! On 2 m, for example, conversations between stations hundreds and even thousands of miles apart are not uncommon. The distances decrease as frequencies increase, but communications have spanned several hundred miles even at microwave frequencies.

One of the more recent developments in VHF/UHF communications is a suite of computer programs using state of the art digital techniques. WSJT by Joe Taylor, K1JT, contains four separate modules optimized for meteor scatter, EME and extreme troposcatter, meteor scatter optimized for the 50 MHz band, and a special program for measuring the strength of your own echoes off the moon. The programs run on Microsoft Windows based computers equipped with a sound card. In addition to the four communication modes, WSJT offers a 'measure' mode for testing sun noise and an EME Calculator to help you estimate the strength of your own and other stations' echoes from the moon. For further information on the WSJT suite of programs, visit K1JT's Web site at **pulsar.princeton.edu/~joe/K1JT/index.htm**.

EME

EME (Earth-Moon-Earth) communication, also known as "moonbounce," continues to fascinate many amateurs. The concept is simple: use the moon as a passive reflector for VHF and UHF signals. With a total path length of about 500,000 miles, EME is the ultimate DX.

Amateur involvement in moonbounce grew out of experiments by the military after World War II. While the first amateur signals reflected from the moon were received in 1953, it took until 1960 for the first two-way amateur EME contacts to take place. Using surplus parabolic dish antennas and high-power klystron amplifiers, the Eimac Radio Club, W6HB, and the Rhododendron Swamp VHF Society, W1BU, achieved the first EME QSO in July 1960 on 1296 MHz. Since then, EME activity has proliferated onto most VHF and higher amateur bands.

Advances in low-noise semiconductors and Yagi arrays in the 1970s and 1980s have put EME within the grasp of most serious VHF and UHF operators. Further advances in technology will bring forth sophisticated receivers with digital signal processing (DSP), such as the WSJT programs discussed above, that may make EME affordable to most amateurs.

EME activity is primarily a CW mode. However, improvements in station equipment now allow the best-equipped stations to make SSB contacts under the right conditions. Regardless of the transmission mode, successful EME operating requires:

- Power output as close to the legal limit as possible.
- A good-sized antenna array. Arrays of 8, 16, or more Yagis are common on the VHF frequencies, while large parabolic dish antennas are common on UHF and microwave frequencies.
- Accurate azimuth and elevation.
- Minimal transmission line losses.
- The best possible receiving equipment, generally a receiver with a low system noise figure and a low-noise preamplifier mounted at the antenna.

The ARRL sponsors EME contests to stimulate activity. Given the marginal nature of most EME contacts, EME contests designate a "liaison frequency" on HF where EME participants can schedule contacts. Contest weekends give smaller stations the opportunity to make many contacts with stations of all sizes. See the **Space Communications** chapter for more about EME.

Meteor Scatter

As a meteor enters the Earth's atmosphere, it vaporizes into an ionized trail of matter. Such trails are often strong enough to reflect VHF radio signals for several seconds. During meteor showers, the ionized region becomes large enough (and lasts long enough) to sustain short QSOs.

Amateurs experimenting with meteor scatter propagation use high power (100 W or more) and beam antennas with an elevation rotor (to point the beam upward at the incoming meteors). Most contacts are made using CW, as voice modes experience distortion and fading. Reflected CW signals often have a rough note.

The *ARRL UHF/Microwave Experimenter's Handbook* contains detailed information about the techniques and equipment used for meteor scatter. Also, the WSJT software modules, FSK441 and JT6M support meteor scatter communications.

Auroral Propagation

During intense solar storms, the Earth's magnetic field around the poles can become heavily charged with ions. In higher latitudes, this often produces a spectacular phenomenon called the *aurora borealis* (or northern lights) in the Northern Hemisphere and the *aurora australis* (or southern lights) in the Southern Hemisphere. The ionization is often intense enough to reflect VHF radio signals. Many amateurs experiment with aurora contacts on 10, 6 and 2 m. Aurora contacts are often possible even when the aurora is not visible.

Equipment used to make aurora contacts is similar to that used for meteor-scatter contacts: high power, directional antennas and CW. Antenna pointing is less critical, however, since the antenna need only be aimed at the aurora curtain. Reflected CW signals often have a rough buzz-saw-like note and can also be Doppler-shifted. Reflected aurora SSB signals are difficult to understand, but careful listening will often produce understandable voice on the 10- and 6-meter bands. SSB aurora signals are extremely difficult to understand at 144 MHz and higher frequencies, which is why CW is so popular for the aurora mode.

The *ARRL UHF/Microwave Experimenter's Handbook* contains detailed information about the techniques and equipment used for auroral propagation.

Notes

[1]You can find more information online about the three branches of the Military Affiliate Radio Service at their respective Web sites:

public.afca.af.mil/LIBRARY/MARS1.HTM
(Air Force MARS)

navymars.org/
(Navy-Marine Corps MARS)

www.asc.army.mil/mars/default.htm
(Army MARS)

Chapter 3

Safety

This chapter was written by James N. Woods, W7PUP, and includes additional contributors as well. This chapter will focus on how to avoid potential hazards as we explore Amateur Radio and its many facets. We need to learn as much as possible about what could *go wrong* so we can avoid factors that might result in accidents. Amateur Radio activities are not inherently hazardous, but like many things in modern life, it pays to be informed. Stated another way, while we long to be creative and innovative, there is still the need to act responsibly. Safety begins with our attitude. Make it a habit to plan work carefully. Don't be the one to say, "I didn't think it could happen to me."

Having a good attitude about safety is not enough, however. We must be knowledgeable about common safety guidelines and follow them faithfully. Safety guidelines cannot possibly cover all situations, but if we approach each task with a measure of common sense, we should be able to work safely.

This chapter will address some of the most popular ham radio activities: building and erecting antennas, constructing radio equipment, and the testing and troubleshooting of our radios. Safety associated with emergency disaster operations are covered best by the agencies and organizations affected.

Although the RF, ac and dc voltages in most amateur stations pose a potentially grave threat to life and limb, common sense and knowledge of good safety practices will help us avoid accidents. Building and operating an Amateur Radio station can be, and is for almost all amateurs, a perfectly safe pastime. Carelessness can lead to severe injury, or even death, however. The ideas presented here are only guidelines; it would be impossible to cover all safety precautions. *Remember: There is no substitute for common sense.*

Fires in well-designed electronic equipment are not common but are known to occur. Proper use of a suitable fire extinguisher can make the difference between a small fire with limited damage and loss of an entire home. Make sure you know the limitations of your extinguisher and the importance of reporting the fire to your local fire department immediately.

Several types of extinguishers are suitable for electrical fires. The multipurpose dry chemical or "ABC" type units are relatively inexpensive and contain a solid powder that is nonconductive. Avoid buying the smallest size; a 5-pound capacity will meet most requirements in the home. ABC extinguishers are also the best choice for kitchen fires (the most common location of home fires). One disadvantage of this type is the residue left behind that might cause corrosion in electrical connectors. Another type of fire extinguisher suitable for energized electrical equipment is the carbon dioxide unit. CO_2 extinguishers require the user to be much closer to the fire, are heavy and difficult to handle, and are relatively expensive. For obvious reasons, water extinguishers are not suitable for fires in or near electronic equipment.

Involve your family in Amateur Radio. Having other people close by is always beneficial in the event that you need immediate assistance. Take the valuable step of showing family members how to turn off the electrical power to your equipment safely. Additionally, cardiopulmonary resuscitation (CPR) training can save lives in the event of electrical shock. Classes are offered in most communities. Take the time to plan with your family members exactly what action should be taken in the event of an emergency, such as electrical shock, equipment fire or power outage. Practice your plan!

Antenna and Tower Safety

Since antennas are generally outdoors, they are affected by such potentially hazardous weather as wind, ice and lightning. Learning about the potential hazards of towers and antennas and how to do antenna work safely will pay dividends.

ARRL Technical Advisor Paul Krugh, N2NS, reminds us to remember that putting up a tower has a set of responsibilities associated with it. *Any heavy, large and permanent structure that fails or collapses can potentially hurt or even kill somebody.* The complete installation *must* comply with all applicable structural and building codes. Professional engineers design towers to withstand code loadings — that is, dead weight, wind and ice loadings that are applicable to the environment at your particular location. The latest revision of the EIA-222 standard is the document from which professional engineers work to ensure that their tower designs are structurally safe. For further information, contact the Electronic Industries Alliance (EIA) in Arlington, CA.

To ensure structural safety and integrity, you must demonstrate that your tower has been designed by a qualified engineer to withstand EIA-222 loadings at your specific geographic area. Further, the tower, foundation, guys and anchors must be installed (and maintained) according to any drawings, instructions and specifications supplied by the professional engineer. Remember: A properly designed, installed and maintained tower should be as safe as a building or a bridge!

It is not feasible to discuss each type of antenna and tower in detail, so this section will include only highlights. For a full understanding of the specific hardware you will be working with, consult the manufacturer or supplier. You should discuss your antenna plans with a qualified engineer. The ARRL Volunteer Consulting Engineer program can steer you to a knowledgeable engineer.

In addition, your town or city will probably require that you obtain a building permit to erect a tower or antenna. This is their way to help ensure that the installation follows good practices and that the installation is safe. Wise amateurs realize that an independent review of drawings and site inspections are beneficial and can result in fewer problems in the future.

Towers must have a properly engineered support, both for the tower sections themselves as well as guy wire attachments. Sometimes towers are braced to buildings for added support. The Antenna Supports chapter of *The ARRL Antenna Book* covers this subject in greater detail. Towers are available commercially in both guyed and self-supporting styles, and constructed of both steel and aluminum materials. Masts may be wood or metal. One popular and inexpensive mast used to support small antennas is the tubular mast often sold for TV antenna use. These come in telescoping sections, in heights from 20 to 50 ft.

Aluminum extension ladders are sometimes used for temporary antenna supports, such as at Field Day sites. One problem with this approach is the difficulty in holding down the bottom section while "walking up" the ladder. Do *not* try to erect this type of support alone.

Trees are sometimes pressed into service for holding one end of a wire antenna. When using slingshots or arrows to string up the antenna, be sure no one is in range before you launch.

FACTORS TO CONSIDER WHEN SELECTING A TOWER

- Towers have design load limitations. Make very sure the tower you consider has the capacity to safely handle the antenna(s) you intend to install in the kind of environment that is applicable to your QTH.
- The antenna must be located in such a position that *it cannot possibly tangle with power lines, both during normal operation or if the structure should fall.*
- Sufficient yard space must be available to position a guyed tower properly. A rule of thumb is that the guy anchors should be between 60% and 80% of the tower height in distance from the base of the tower.
- Provisions must be made to keep children from climbing the support.
- Always write to the manufacturer of the tower before purchasing and ask for installation specifications, including guying data.
- Soil conditions at the tower site should be investigated. The footings need to be designed around actual soil conditions, particularly on a rocky site.

TOWER TIPS

- Beware of used towers. Have them professionally inspected and contact the manufacturer for installation criteria.
- Always follow manufacturer's instructions, using only parts that are designed for the model you have.
- Never rush into projects. Consult the most experienced amateurs in your community for assistance, especially if you are new to tower installation.
- Check with your local building officials.
- Liability may be increased with a tower installation. Check with your insurer to ensure your coverage is adequate.
- Consider your neighbors about any hazards your antennas may present to them.
- Don't let your installation become an "attractive nuisance." Take steps to install barriers so your tower cannot easily be climbed by others, particularly adventurous children.
- Use only the highest quality materials in your system.
- Make sure you have all the tools needed before starting. Some specialized tools (such as a gin pole) may be required.
- Never erect an antenna, tower or rotor during an electrical storm or rainstorm, or when lightning is a possibility.
- The assembly crew as well as those climbing the tower during erection must wear hard hats and use appropriate personal protective equipment including gloves, boots, climbing belt or harness. Don't forget that lifelines are needed when the belt is unattached from the tower while moving.
- Be careful not to over-stress the tower when it is being assembled. The tower manufacturer can offer suggestions that will avoid jeopardizing the tower.
- Install guy wires using the proper tools. Care should be exercised especially when handling loose, un-terminated, and sharp guy wire ends! Avoid wrapping guy wire around your hands to pull it into place, and instead use sufficient length to easily attach it to the anchors. Use tower-rated turnbuckles or similar devices to adjust tension evenly around the tower.
- Assign someone in the erection crew to monitor the use of safety equipment.
- After the tower is installed, keep the installation safe. Inspection and maintenance recommended by the tower's manufacturer should be carefully followed.
- If making attachments to houses or installations on roofs, have a qualified person determine that the method is adequate and the loading conditions are satisfactory.
- Avoid metal ladders if there are any utility lines in the vicinity. Assume that any line is energized — including cable television and telephone lines.

POWER LINES

Hundreds of people have been killed or seriously injured when attempting to install or dismantle antennas. In virtually all cases, the victim was aware of the hazards, including the potential for serious electrical shock, but did not take the necessary steps to eliminate the risks. Never install antennas, towers and masts near power lines. How far away is considered safe? Towers and masts should be installed twice the height of the installation away from power lines. Every electrical wire must be considered dangerous. If the installation should contact power lines, you or those around you could be killed! If you have any questions about power lines, contact your electrical utility, city inspector or a qualified professional.

If, for some reason your tower or antenna structure begins to fall, get away from it immediately! If it contacts energized lines, it can become a lethal hazard if you are touching any part of the conductive structure. If a coworker becomes energized, **do not touch the person!** The safest practice is to keep all others clear of the area, call 911, and just wait for the power company and rescue team to arrive and assist the victim. At some greater risk, a well-insulated pole such as fiberglass or PVC pipe — as long as possible for safety — can be utilized in an attempt to dislodge the live wire or collapsed metal structure from the victim (with moisture, etc., wood can be a *poor insulator* — especially at high voltages!). If the victim can be well cleared of the hazard and is not breathing, immediately start CPR procedures and seek emergency assistance. **Remember, use caution and understand that during such an accident, the live conductor or live antenna structure can further move (lurch) suddenly and without warning. One accident is bad enough — there is no need to have two victims! It is best to just seek qualified emergency help if you are unsure of the situation-specific hazards.**

Further information about tower safety appears in *The ARRL Antenna Book.*

Electrical Wiring Around the Shack

The standard power available from commercial mains in the United States for residential service is 120/240-V ac. The "primary" voltages that feed transformers in our neighborhoods may range from 2000 to about 10,000 V. Generally, the responsibility for maintaining the power distribution system belongs to a utility company, electric cooperative or city. The "ownership" of conductors usually transfers from the electric utility supplier to the homeowner where the power connects to the meter or weatherhead. If you are unsure where the division of responsibility falls in your community, a call to your electrical utility will provide the answer. **Fig 3.1** shows the typical division of responsibility between the utility company and the homeowner.

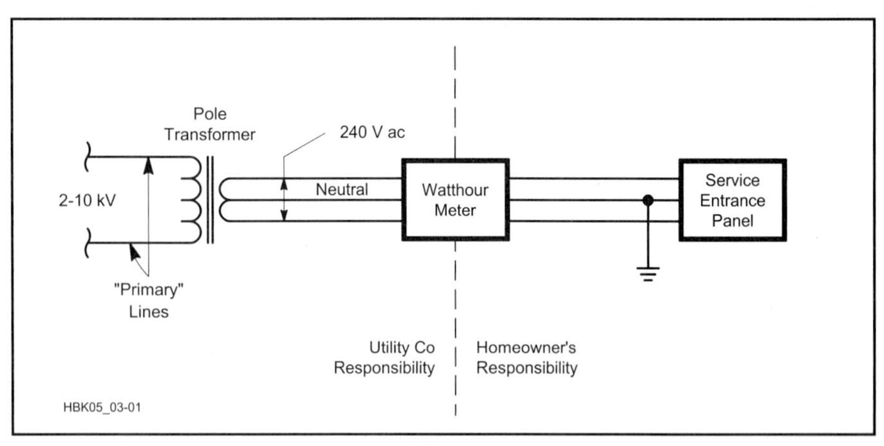

Fig 3.1 — Typical division of responsibility for maintenance of electrical power conductors and equipment. The meter is supplied by the utility company.

There are two facets to success with electrical power: safety and performance. Since we are not professionals, we need to pursue safety first and consult professionals for alternative solutions if performance is unacceptable.

STATION CONCERNS

The primary electrical power supplied to your radio equipment should be controlled by one master switch so that it is easy to kill the power in an emergency. One convenient means is a switched outlet strip, as used for computer equipment. The strip should be listed by a nationally recognized testing laboratory such as Underwriters Lab and incorporate a circuit breaker. See "What Does UL Listing Mean?" and "How Safe are Outlet Strips?" for warnings about poor quality products. It is poor practice to "daisy-chain" several power strips. If you need more outlets than are available on a strip, have additional convenience outlets installed.

Before adding equipment to your home, be sure that it does not overload the circuit. National and local codes set permissible branch capacities according to a rather complex process. Here's a safe rule of thumb: consider adding a new circuit if the total load is more than 80% of the circuit breaker or fuse rating. (This assumes that the fuse or breaker is correct. If you have any doubts, have an electrician check it.)

What Does UL Listing Mean?

CAUTION: Listing *does not* mean what most consumers expect it to mean! More often than not the listing *does not* relate to the performance of the listed product. The listing simply indicates that a sample of the device meets certain manufacturers' construction criteria. Similar devices from the same or different manufacturers may differ significantly in overall construction and performance even though all are investigated and listed against the same UL product category

Fig 3.2 — If the switch box feeding power to your shack is equipped with a lock-out hole, use it. With a lock through the hole on the box, the power cannot be accidentally turned back on. *(Photo courtesy of American ED-CO)*

Do It Yourself Wiring?

Amateurs sometimes "rewire" parts of their homes to accommodate their hobby. Most local codes *do* allow for modification of wiring (by building owners), so long as the electrical codes are met. Generally, the building owner must obtain an electrical permit before beginning changes or additions to permanent wiring. Some jobs may require drawings of planned work. Often the permit fee pays for an inspector to review the work. Considering the risk of injury or fire if critical mistakes are left uncorrected, a permit and inspection are well worth the effort. *Don't take chances* — seek assistance from the building officials or an experienced electrician if you have *any* questions or doubts about proper wiring techniques.

Ordinary 120-V circuits are the most common source of fatal electrical accidents. Never use bare wire for exposed circuits or open-chassis construction with exposed connections! Remember that high-current, low-voltage power sources can be just as dangerous as high-voltage sources.

Never work on electrical wiring with the conductors energized! Switch off the circuit breaker or remove the fuse and take positive steps to ensure that others do not restore the power while you are working. (**Fig 3.2** illustrates one way to ensure that power will be off until you want it turned on.) Check the circuit with an ac voltmeter to be sure that it is "dead" *each time you begin work.* Before restoring power, check your work with an ohm meter: There should be good continuity between the neutral conductor (white wire, "silver" screw) and the grounding conductor (green or bare wire, green screw). An ohmmeter should indicate a closed circuit between the conductors.

There should be no continuity between the hot conductor (black wire, "brass" screw) and the grounding conductor or the neutral conductor. An ohmmeter should indicate an *open* circuit between the hot wire and either of the other two conductors.

A commercially available plug-in tester is the best way to test regular three-wire receptacles.

NATIONAL ELECTRICAL CODE

Fortunately, much has been learned about how to harness electrical energy safely. This collective experience has been codified into the *National Electrical Code*, or *NEC*. The *Code* details safety requirements for many kinds of electrical installations. Compliance with the *NEC* provides an installation that is *essentially* free from hazard, but not necessarily efficient, convenient or adequate for good service (paraphrased from NEC Article 90-1a and b). For example, the *NEC* requirements discussed here are *not* adequate for lightning protection and high transient voltage events. Look at "Lightning/Transient Protection" for more information. While the *NEC* is national in nature and sees wide application, it is not universal.

Local building authorities set the codes for their area of jurisdiction. They often incorporate the *NEC* in some form, while considering local issues. For example, Washington State specifically exempts telephone, telegraph, radio and television wires and equipment from conformance to electrical codes, rules and regulations. However, some local jurisdictions (city, county and so on) do impose a higher level of installation criteria, including some of the requirements exempted by the state.

Code interpretation is a complex subject, and untrained individuals should steer clear of the *NEC* itself. The *NEC* is not written to be understood by do-it-yourselfers. Therefore, the best sources of information about code compliance and acceptable practices are local building officials, engineers and practicing electricians. With that said, let's look at a few *NEC* requirements for radio installations.

Antenna conductors — Transmitting antennas using hard-drawn copper wire: #14 for unsupported spans less than 150 ft, and #10 for longer spans. Copper-clad steel, bronze or other high-strength conductors must be #14 for spans less than 150 ft and #12 for longer spans. Open-wire transmission line conductors must be at least as large as those specified for antennas.

Lead-ins — There are several *NEC* requirements for antenna lead-in conductors. For transmitting stations, their size must be equal to or greater than that of the antenna. Lead-ins attached to buildings must be firmly mounted at least 3 inches clear of the surface of the building on nonabsorbent insulators. Lead-in conductors must enter through rigid, noncombustible, nonabsorbent insulating tubes or bushings, through an opening provided for the purpose that provides a clearance of at least 2 inches; or through a drilled windowpane. All lead-in conductors to transmitting equipment must be arranged so that accidental contact is difficult.

Lightning arrestors — Transmitting stations are required to have a means of draining static charges from the antenna system. An antenna discharge unit (lightning arrestor) must be installed on each lead-in conductor that is not protected by a permanently and effectively grounded metallic shield, unless the antenna itself is permanently and effectively grounded. (The code exception for shielded lead-ins does *not* apply to coax, but to shields such as thin-wall conduit. Coaxial braid is neither "adequate" nor "effectively grounded" for lightning protection purposes.) An acceptable alternative to lightning arrestor installation is a switch (capable of withstanding many kilovolts) that connects the lead-in to ground when the transmitter is not in use.

Ground Conductors

Grounding conductors may be made from copper, aluminum, copper-clad steel, bronze or similar erosion-resistant materials. Insulation is not required. *[Lightning and high-voltage transient events may require much larger conductors. —Ed.]* The "protective grounding conductor" (main conductor running to the ground rod) must be as large as the antenna lead-in, but not smaller than #10. The "operating grounding conductor" (to bond equipment chassis together) must be at least #14. There is a "unified" grounding electrode require-

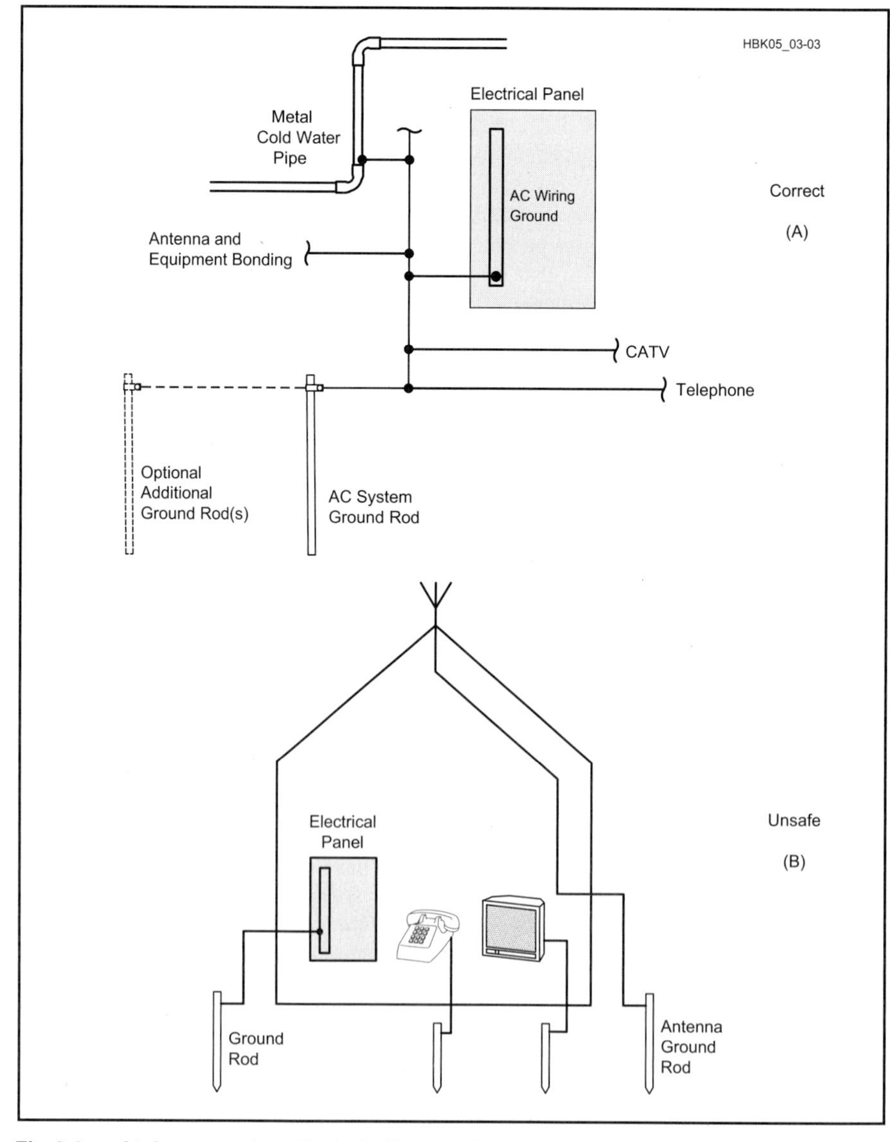

Fig 3.3 — At A, proper bonding of all grounds to electrical service panel. Installation shown at B is unsafe — the separate grounds are not bonded. This could result in a serious accident or electrical fire.

ment — it is necessary to bond *all* ground rods to the electric service entrance ground. All utilities, antennas and any separate grounding rods used must be bonded together. **Fig 3.3** shows correct (A) and incorrect (B) ways to bond ground rods. **Fig 3.4** demonstrates the importance of correctly bonding ground rods. (Note: The *NEC* requirements do not address effective RF grounds. See the **EMI/Direction Finding** chapter of this book for information about RF grounding practices.)

Additionally, the *Code* covers some information on safety inside the station. All conductors inside the building must be at least 4 inches away from conductors of any lighting or signaling circuit except when they are separated from other conductors by conduit or insulator. Transmitters must be enclosed in metal cabinets, and the cabinets must be grounded. All metal handles and controls accessible by the operator must be grounded. Access doors must be fitted with interlocks that will automatically disconnect all voltages above 350 when the door is opened.

How Safe are Outlet Strips?

CAUTION: The switch in outlet strips is generally *not* rated for repetitive *load break* duty. Early failure and fire hazard may result from using these devices to switch loads. Misapplications are common (another bit of bad technique that has evolved from the use of personal computers), and manufacturers are all too willing to accommodate the market with marginal products that are "cheap."

Nonindicating and poorly designed surge protection also add to the safety hazard of using power strips. Marginally rated MOVs often fail in a manner that could cause a fire hazard, especially in outlet strips that have nonmetallic enclosures.

A lockable disconnect switch or circuit breaker, as shown in Fig 3.2, is a better and safer station master switch.

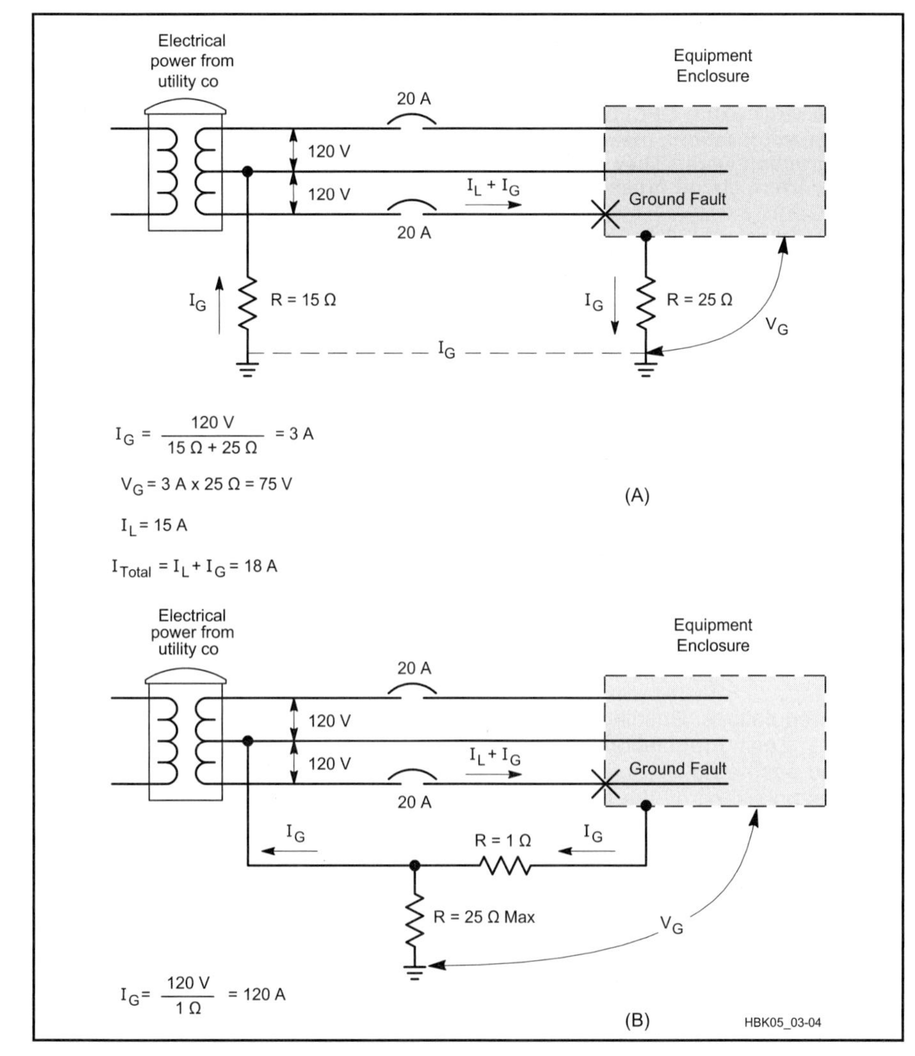

Fig 3.4 — These drawings show the importance of properly bonded ground rods. In the system shown in A, the 20-A breaker will not trip. In the system in B, the 20-A circuit breaker trips instantly. There is an equipment internal short to ground — the ground rod is properly bonded back to the power system ground. Of course, the main protection should be in a circuit ground wire in the equipment power cord itself!

Ground-Fault Circuit Interrupters

GFCIs are devices that can be used with common 120-V circuits to reduce the chance of electrocution when the path of current flow leaves the branch circuit (say, through a person's body to another branch or ground). The *NEC* requires GFCI outlets in all wet or potentially wet locations, such as: bathrooms, kitchens, and any outdoor outlet with ground-level access, garages and unfinished basements. Any area with bare concrete floors or concrete masonry walls should be GFCI equipped. GFCIs are available as portable units, duplex outlets and as individual circuit breakers. Some early units may have been sensitive to RF radiation but this problem appears to have been solved. Ham radio shacks in potentially wet areas (basements, out buildings) should be GFCI equipped. **Fig 3.5** is a simplified diagram of a GFCI.

LIGHTNING/TRANSIENT PROTECTION

Nearly everyone recognizes the need to protect themselves from lightning. From miles away, the sight and sound of lightning boldly illustrates its destructive potential. Many people don't realize that destructive transients from lightning and other events can reach electronic equipment from many sources, such as outside antennas, power, telephone and cable TV installations. Many hams don't realize that the standard protection scheme of several decades, a ground rod and simple "lightning arrestor" is *not* adequate.

Lightning and transient high-voltage protection follows a familiar communications scenario: identify the unwanted signal, isolate it and dissipate it. The difference here is that the unwanted signal is many megavolts at possibly 200,000 A. What can we do?

Hams *cannot* expect to design or install effective lightning protection systems, but reasonably complete protection from lightning is available in systems designed by lightning protection professionals. Hams *can* easily follow some general guidelines that will protect their stations against high-voltage events that are induced by nearby lightning strikes or that arrive via utility lines. Let's talk about where to find professionals first, and then consider construction guidelines.

Professional Help

Start with your local government. Find out what building codes apply in your area and have someone explain the regulations about antenna installation and safety. For more help, look in your telephone yellow pages for professional engineers, lightning protection suppliers and contractors.

Companies that sell lightning-protection products may offer considerable help to apply their products to specific installations. One such source is PolyPhaser Corporation. Look under "Ground References," later in this chapter, for a partial list of PolyPhaser's publications.

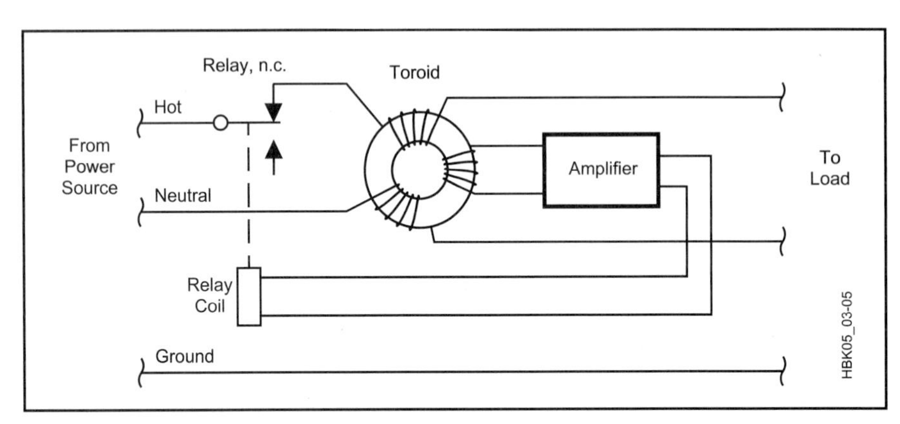

Fig 3.5 — Simplified diagram of a 120-V ac ground fault circuit interrupter (GFCI). When a stray current flows from the load (or outlet) side to ground, the toroidal current becomes unbalanced allowing detection, amplification and relay actuation to immediately cut off power to the load (and to the stray path!) GFCI units require a manual reset after tripping. GFCI's are required in wet locations (near kitchen sinks, in garages, in outdoor circuits and for construction work.) They are available as portable units or combined with over-current circuit breakers for installation in entrance panels.

About the National Electrical Code

Exactly how does the National Electrical Code become a requirement? How is it enforced?

Cities and other political subdivisions have the responsibility to act for the public safety and welfare. To address safety and fire hazards in buildings, regulations are adopted by local laws and ordinances usually including some form of permit and accompanying inspections. Because the technology for the development of general construction, mechanical and electrical codes is beyond most city building departments, model codes are incorporated by reference. There are several general building code models used in the US: Uniform, BOCA and Southern Building Codes are those most commonly adopted. For electrical issues, the *National Electrical Code* is in effect in virtually every community. City building officials will serve as "the authority having jurisdiction" and interpret the provisions of the *Code* as they apply it to specific cases.

Building codes differ from planning or zoning regulations: Building codes are directed only at safety, fire and health issues. Zoning regulations often are aimed at preservation of property values and aesthetics.

The *NEC* is part of a series of reference codes published by the National Fire Protection Association, a nonprofit organization. Published codes are regularly kept up-to-date and are developed by a series of technical committees whose makeup represents a wide consensus of opinion. The *NEC* is updated every three years.

Do I have to update my electrical wiring as code requirements are updated or changed?

Generally, no. Codes are typically applied for new construction and for renovating existing structures. Room additions, for example, might not directly trigger upgrades in the existing service panel unless the panel was determined to be inadequate. However, the wiring of the new addition would be expected to meet current codes. Prudent homeowners, however, may want to add safety features for their own value. Many homeowners, for example, have added GFCI protection to bathroom and outdoor convenience outlets.

Construction Guidelines

Ground rods — Ground rods should be either solid copper, copper-clad steel, hot-dipped galvanized steel or stainless steel. They should be at least 8 ft long by $^1/_2$ inch in diameter ($^5/_8$ inch diameter for iron or steel).

Bonding Conductors — Copper strapping (or *flashing*) comes in a number of sizes; use $1^1/_2$ inches wide and 0.051 inches thick as a *minimum* for ground connections. Copper strap is a better lightning and RF ground than wire because straps have less inductance than wires. On the other hand, straps are more expensive than wire and more difficult to find.

Use bare copper for buried ground wires. (There are some exceptions; seek an expert's advice if your soil is corrosive.) Exposed runs above ground that are subject to physical damage may require additional protection (a conduit) to meet code requirements. Wire size depends on the application, but never use anything smaller than #6 AWG for bonding conductors. Local lightning-protection experts or building inspectors can recommend sizes for each application.

Tower and Antennas

Because a tower is usually the highest metal object on the property, it is the most likely strike target. Proper tower grounding is essential to lightning protection. The goal is to establish short multiple paths to the Earth so that the strike energy is divided and dissipated.

Connect each tower leg and each fan of metal guy wires to a separate ground rod. Space rods at least 6 ft apart. Bond the leg ground rods together with a #6 AWG or larger copper bonding conductor (form a ring around the tower base, see **Fig 3.6**). Connect a continuous bonding conductor between the tower ring ground and the entrance panel. Make all connections with fittings approved for grounding applications. ***Do not use solder for these connections.*** Solder will be destroyed in the heat of a lightning strike.

Unless the tower is also a shunt-fed antenna, use grounded metal guys. For crank-up or telescoping towers, connect the sections with strap jumpers. Because galvanized steel (which has a zinc coating) reacts with copper when combined with moisture, use stainless steel hardware between the galvanized metal and the

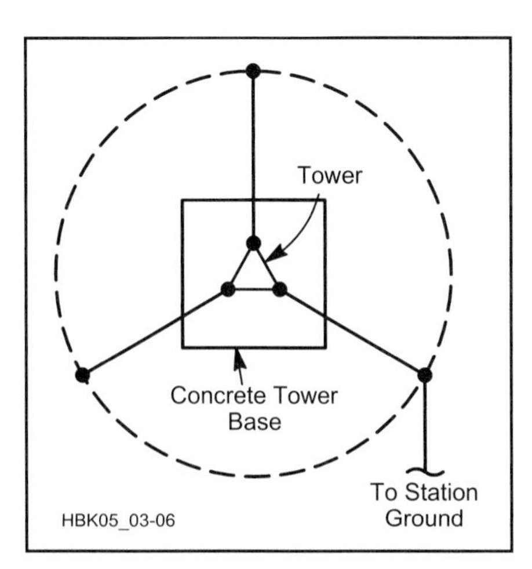

Fig 3.6 — Schematic of a properly grounded tower. A bonding conductor connects each tower leg to a ground rod and a buried (1 ft deep) bare, tinned copper ring (dashed line), which is also connected to the station ground and then to the ac safety ground. Locate ground rods on the ring, as close as possible to their respective tower legs. All connectors should be compatible with the tower and conductor materials to prevent corrosion. See text for conductor sizes and details of lightning and voltage transient protection.

copper grounding materials.

To prevent strike energy from entering a shack via the feed line, ground the feed line *outside* the home. Ground the coax shield *to the tower* at the antenna and the base to keep the tower and line at the same potential. Several companies offer grounding blocks that make this job easy.

All grounding media at the home must be bonded together. This includes lightning-protection conductors, electrical service, telephone, antenna system grounds and underground metal pipes. Any ground rods used for lightning protection or entrance-panel grounding must be spaced at least 6 ft from each other and the electrical service or other utility grounds and then bonded to the ac system ground as required by the *NEC*.

A Radio Entrance Panel

We want to control the flow of the energy in a strike. Eliminate any possible paths for surges to enter the building. This involves routing the feed lines, rotator control cables, and so on at least 6 ft away from other nearby grounded metal objects.

Every conductor that enters the structure should have its own surge suppressor including antenna system control lines at the Radio Entrance Panel and other services where they connect to the ac system ground. They are available from a number of manufacturers, including ICE and PolyPhaser.

Both balanced line and coax arrestors should be mounted to a secure ground connection on the *outside* of the building. The easiest way to do this is to install a large metal enclosure as a bulkhead and ground block. This bulkhead serves as the last line of lightning defense, so it's critical that it be installed properly. You can home-brew a bulkhead panel from 1/8-inch copper sheet, bent into a box shape. Position the bulkhead on the building exterior, 4 to 6 inches (minimum) away from nearby combustible materials. Install a separate ground rod for this panel and connect it to the bulkhead with a short, direct connection. Bond this ground rod to the rest of the ground system. Mount all protective devices, switches and relay disconnects on the outside face wall of the bulkhead.

Lightning Arrestors

Feed line lightning arrestors are available for both coax cable and balanced line. Most of the balanced line arrestors use a simple spark gap arrangement, but a balanced line *impulse* suppresser is available from Industrial Communication Engineers, Ltd (ICE), Indianapolis, IN.

Coaxial Cable Arrestors — DC blocking arrestors have a fixed frequency range. They present a high-impedance to lightning (less than 1 MHz) while offering a low impedance to RF.

DC continuity arrestors (gas tubes and spark gaps) can be used over a wider frequency range than those that block dc. Where the coax carries supply voltages to remote devices (such as a mast-mounted preamp or remote coax switch), dc-continuous arrestors *must* be used.

GROUNDS

As hams we are concerned with three kinds of ground, which are easily confused because we call each of them "ground." The first is the power line ground, which is required by building codes to ensure the safety of life and property surrounding electrical systems. The *NEC* requires that all grounds be *bonded* together; this is a very important safety feature as well as an *NEC* requirement. Ground systems to prevent shock hazards are generally referred to as the *dc ground* by amateurs, although *safety ground* is a more appropriate term.

The previous section discussed some of the features of a lightning protection grounding system. Additional information on lightning, surge and EMI grounding can be found in *The ARRL Antenna Book*. The *National Electrical Code* requires lightning protection ground rods to be separate from the power line safety grounding electrodes. As discussed later, however, all grounding systems must eventually be bonded together.

An effective safety ground system is necessary for every amateur station. It provides a common reference potential for all parts of the ac system and reduces the possibility of electrical shock by ensuring that all exposed conductors remain at that (low) potential. Three-wire electrical systems effectively ground our equipment for dc and low frequencies. Unfortunately, an effective ground conductor at 60 Hz (5,000,000 m wavelength) may be an excellent antenna for a 20 m signal.

When stray RF causes interference or other problems, we need another kind of ground — a low-impedance path for RF to reach the earth or some other "ground" that dissipates, rather than radiates, the RF energy. Let's call this an *RF ground.*

In most stations, dc ground and RF ground are provided by the same system. If you install ground rods, however, bond them to each other and to the safety ground at the electrical service entrance. In older houses, water lines are sometimes used for the service entrance panel ground. It is a good idea to check that the pipes are electrically continuous from the panel to earth. (Consider that Teflon tape is often used to seal pipe joints in modern repairs.)

For decades, amateurs have been advised to bond all equipment cabinets to an RF ground located near the station. That's a good idea, but it's not easily achieved. "Near" in this use is 10 ft or less for HF operation, even less for higher frequen-

Suppliers of Lightning Protection Equipment

For current vendor contact information, use your favorite Internet search tool.

- Alpha Delta Communications: Coax lightning arrestors, coax switches with surge protectors.
- The Wireman: copper wire up to #4 AWG, 2-inch flat copper strap, 8-ft copper clad ground rods and 1 × 1/4-inch buss bar.
- Industrial Communication Engineers, Ltd (ICE): Coax lightning arrestors.
- PolyPhaser Corporation: Many lightning protection products for feed lines, towers, equipment, and so on.
- Rohn: Copper strap and other tower grounding products.
- Zero Surge Inc: Power line surge protector.

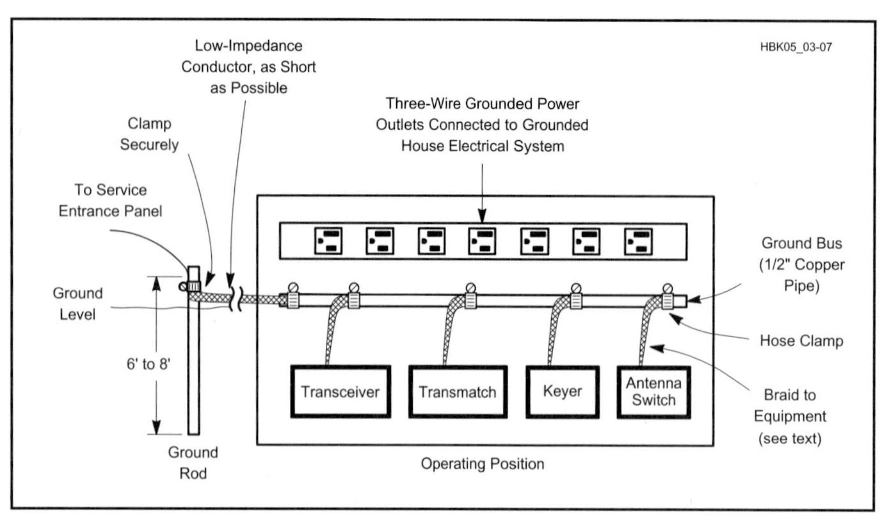

Fig 3.7 — An effective station ground bonds the chassis of all equipment together with low-impedance conductors and ties into a good earth ground. Note that the ground bus is in turn bonded to the service entrance panel. This connection should be made by a licensed electrician with #6 AWG (minimum size) copper wire.

cies. At some stations, it is very difficult to produce an effective RF ground. When levels of unwanted RF are low, an RF ground may not be needed. (See the **EMI/ Direction Finding** chapter for more about RF grounds and interference.) Some think that RF grounds should be isolated from the safety ground system — *that is not true! All grounds, including safety, RF, lightning protection and commercial communications, must be bonded together in order to protect life and property.*

The first step in building an RF ground system is to bond together the chassis of all equipment in your station. Choose conductors large enough to provide a low impedance path. The *NEC* requires that grounding conductors be as large as the largest conductor in the primary power circuit (#14 for a 15-A circuit, #12 for 20 A). Copper strap, sold as "flashing copper," is excellent for this application. Coax braid is a popular choice; but it is not a good ground conductor unless tinned, and then it's no longer very flexible. It is best to use commercially made copper braid ground strap that is tinned and ampacity rated — wider straps make better RF grounds. Avoid solid conductors; they tend to break.

Grounding straps can be run from equipment chassis to equipment chassis, but a more convenient approach is illustrated in **Fig 3.7**. In this installation, a $^1/_2$-inch-diameter copper water pipe runs the entire length of the operating bench. A wide copper ground braid runs from each piece of equipment to a stainless-steel clamp on the pipe.

After the equipment is bonded to a common ground bus, the ground bus must be wired to a good earth ground. This run should be made with a heavy conductor (copper braid is a good choice again) and should be as short and direct as possible. The earth ground usually takes one of two forms.

In most cases, the best approach is to drive one or more ground rods into the earth at the point where the conductor from the station ground bus leaves the house. The best ground rods to use are those available from an electrical supply house. These rods are generally 8 ft long and made from steel with a heavy copper plating. Do not depend on shorter, thinly plated rods sold by some home electronics suppliers, as they can quickly rust and soon become worthless.

Once the ground rod is installed, clamp the conductor from the station ground bus to it with a clamp that can be tightened securely and will not rust. Copper-plated clamps made specially for this purpose (and matching the rods) are available from electrical supply houses. Multiple ground rods reduce the electrical resistance and improve the effectiveness of the ground system.

Building cold water supply systems were used as station grounds in years past. Connection was made via a low-impedance conductor from the station ground bus to a convenient cold water pipe, preferably somewhere near the point where the main water supply enters the house. (Hot water lines are unsuitable for grounding conductors.) Increased used of plastic plumbing both inside and outside houses is reducing the availability of this option. If you do use the cold water line, ensure that it has a good electrical connection to the earth and attach it *outside* the structure to reduce EMI. As with ground rods, ensure that the water line is also bonded to the service entrance panel.

For some installations, especially those located above the first floor, a conventional ground system such as that just described will make a fine dc ground but will not provide the necessary low-impedance path to ground for RF. The length of the conductor between the ground bus and the ultimate ground point becomes a problem. For example, the ground wire may be about $^1/_4$ wavelength (or an odd multiple of $^1/_4$ wavelength) long on some amateur band. A $^1/_4$-wavelength wire acts as an impedance inverter from one end to the other. Since the grounded end is at a very low impedance, the equipment end will be at a high impedance. The likely result is RF hot spots around the station while the transmitter is in operation. In this case, this ground system may be worse (from an RF viewpoint) than no ground at all.

Ground References

Federal Information Processing Standards (FIPS) publication 94: *Guideline on Electrical Power for ADP Installations.* FIPS are available from the National Technical Information Service.

IAEI: *Soares' Book on Grounding*, available from International Association of Electrical Inspectors (IAEI).

IEEE Std 1100: Powering and Grounding Sensitive Electronics Equipment.

PolyPhaser: The Grounds for Lightning and EMP Protection. PolyPhaser's quarterly newsletter, Striking News, contains articles on Amateur Radio station lightning protection in the February and May 1994 issues. Complimentary copies of these issues are available from PolyPhaser.

AN EARTH-CONTINUITY TESTER

This project was first published by the Radio Society of Great Britain (RSGB) and is reproduced here by permission. This simple ground continuity tester determines the quality of the ground (earth) connection in an Amateur Radio station installation. In the context of this discussion, the terms *ground* and *earth* are synonymous.

When using mains-powered electrical equipment, a good-quality protective earth system is very important for safety. Good earth connections are additionally important for radio operation, both for protection against lightning strikes and also for the greater effectiveness of antennas that use earth as one half of a dipole. In situations where the earth path is a *functional* earth as opposed to a *protective* earth, a simple low-voltage, low-current continuity tester or resistance meter is usually sufficient for checking earthing resistance, but for a proper test of a protective earth a high-current tester is needed, as shown in **Fig 3.8**. This is because a deteriorating earth connection in the form of a stranded wire where many of the strands are broken will still show a low resistance to a low-current tester but, in a fault situation when the earth path needs to pass a high current to ground and thus trigger a protective device, the high current causes the remaining strands to *burn out*, i.e., go open-circuit, before the protective device has time to operate. The protection is then nonexistent!

SAFETY STANDARDS

Recognizing this situation, the British and European safety standards for electrical safety, for example BS EN 60335-1 for household equipment, demand that the resistance of the protective earth path between an exposed metal part and the protective earth pin is less than 0.1 Ω.

The equipment needed for checking to this standard is specialized and expensive, but this simple project provides a low-cost alternative that will check resistance at 2-3 A if good-quality batteries are used. To simplify use, the circuit gives a pass/fail indication instead of a resistance value.

WHEATSTONE BRIDGE

The circuit can be considered in three parts; TEST, DETECTOR and OUTPUT INDICATOR. See the diagram in **Fig 3.9** and the components list in **Table 3.1**. The TEST part of the circuit is based on a Wheatstone bridge, where the earth resistance path forms one of the resistance arms. See **Fig 3.10** for the principle behind a Wheatstone bridge. As a consequence of the values of resistance chosen (the test leads are assumed to have a resistance of 0.1 Ω), if the earth resistance is less than 0.1 Ω, the voltage between the midpoints of the two halves of the Wheatstone bridge will be positive, and if it is less than 0.1 Ω, it will be negative. This is fed to the detector part of the circuit. The DETECTOR is an op-amp wired as a comparator. Connected in this way, it has such a high gain that its output is roughly equal to either the positive or negative supply rail voltage, depending on whether the PD between its non-inverting and inverting inputs is positive or negative. It doesn't matter whether the PD is large or small — the output will always be at either extreme. This means there will always be a definite pass or fail indication from the detector, no matter how large or small the output from the Wheatstone bridge. This is important, as it means correct operation of the circuit doesn't depend on the voltage of the high-current battery, particularly as it is a chemical type whose output voltage can fall dramatically when a high current is being drawn. The pass/fail voltage V_{pf} from the detector then passes to the output indicator circuit.

Table 3.1
Components List

Resistors

R1, 2	0.1 Ω, 2.5 W
R3, 4	10k Ω
R6, 7	24k Ω
R5, 8	150 Ω
R9, 10	330 Ω

All resistors metal oxide 0.4 W 1%, except R1 & R2

Semiconductors

U1	LM324
D1, 2	TLY114A yellow, or TLR114A red and TLG114A green
Q1	BC179 (general-purpose pnp)
Q2	BC109C (general-purpose npn)

Additional items

B1	1 × AA Duracell
B2	PP3, 9V
S1	Double pole, momentary on, or push-to-make
S2	SPST

Battery clips/holders
Stripboard
Plastic case*
2 × 4mm plugs & sockets*
2 × alligator clips

*Only required if you are building the project in a case.

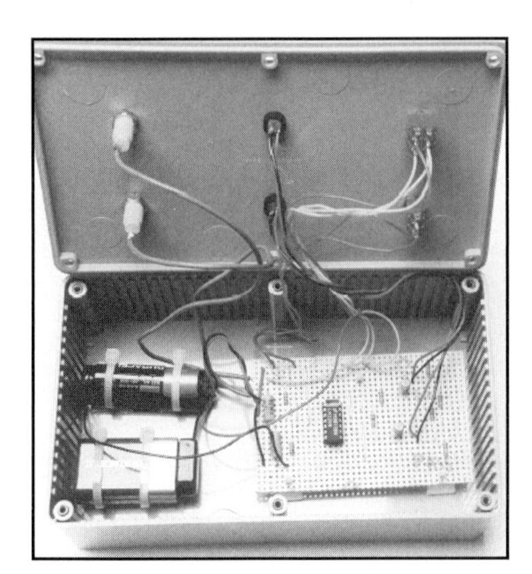

Fig 3.8 — Inside a completed tester. Note that in an enclosed project, the LEDs are brought out to the front panel.

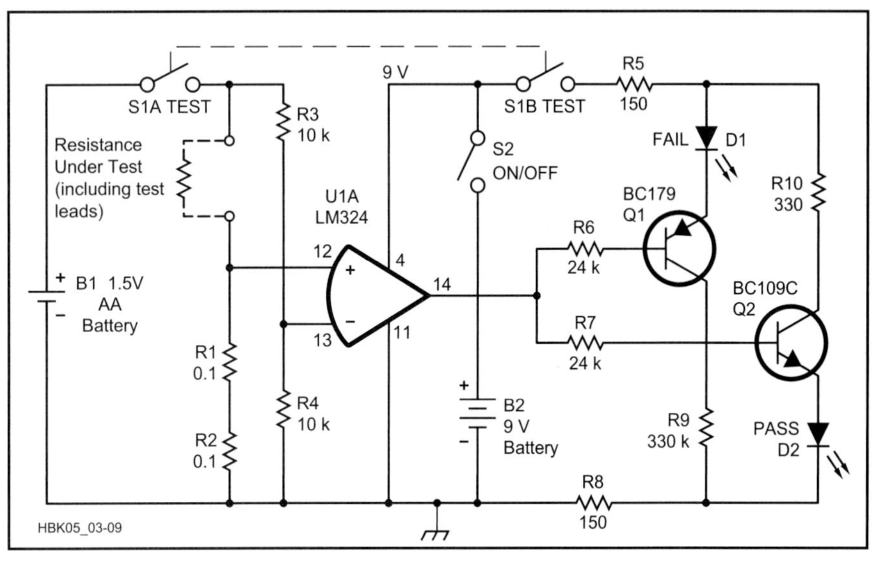

Fig 3.9 — Circuit diagram of the earth-continuity tester.

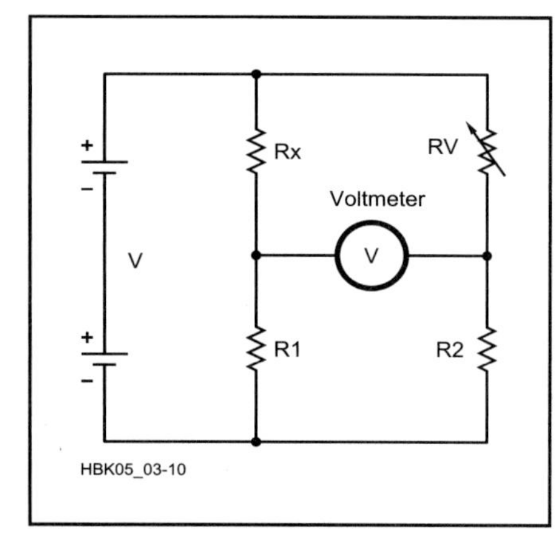

Fig 3.10 — In a conventional Wheatstone bridge circuit, the value of an unknown resistance (Rx) is determined by adjusting a variable resistor (RV) with a calibrated scale until the reading on the voltmeter is zero. At that point (Rx/R1) = (RV/R2) so the value of Rx is then given by Rx = (R1 × RV)/R2. The advantage of this method is that, at the balance point, no current flows through the voltmeter, so the resistance of it doesn't affect the measurement. This is a sensitive method for detecting small changes in resistance, as a small change causes a large meter reading.

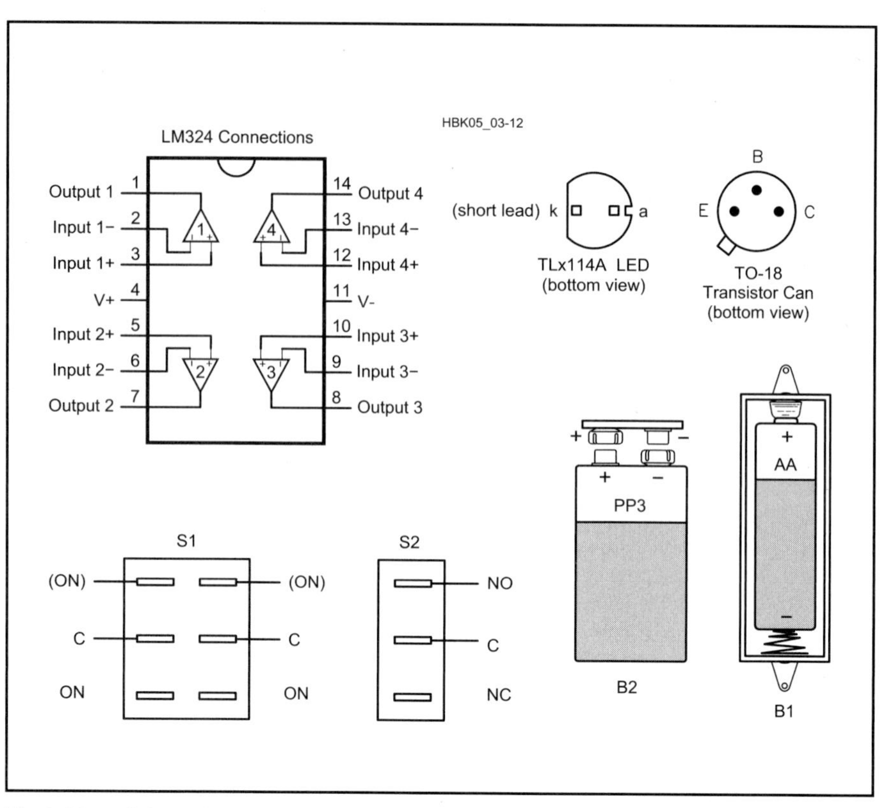

Fig 3.12 — Orientation (and pin-outs) of the batteries, U1, LEDs and transistors.

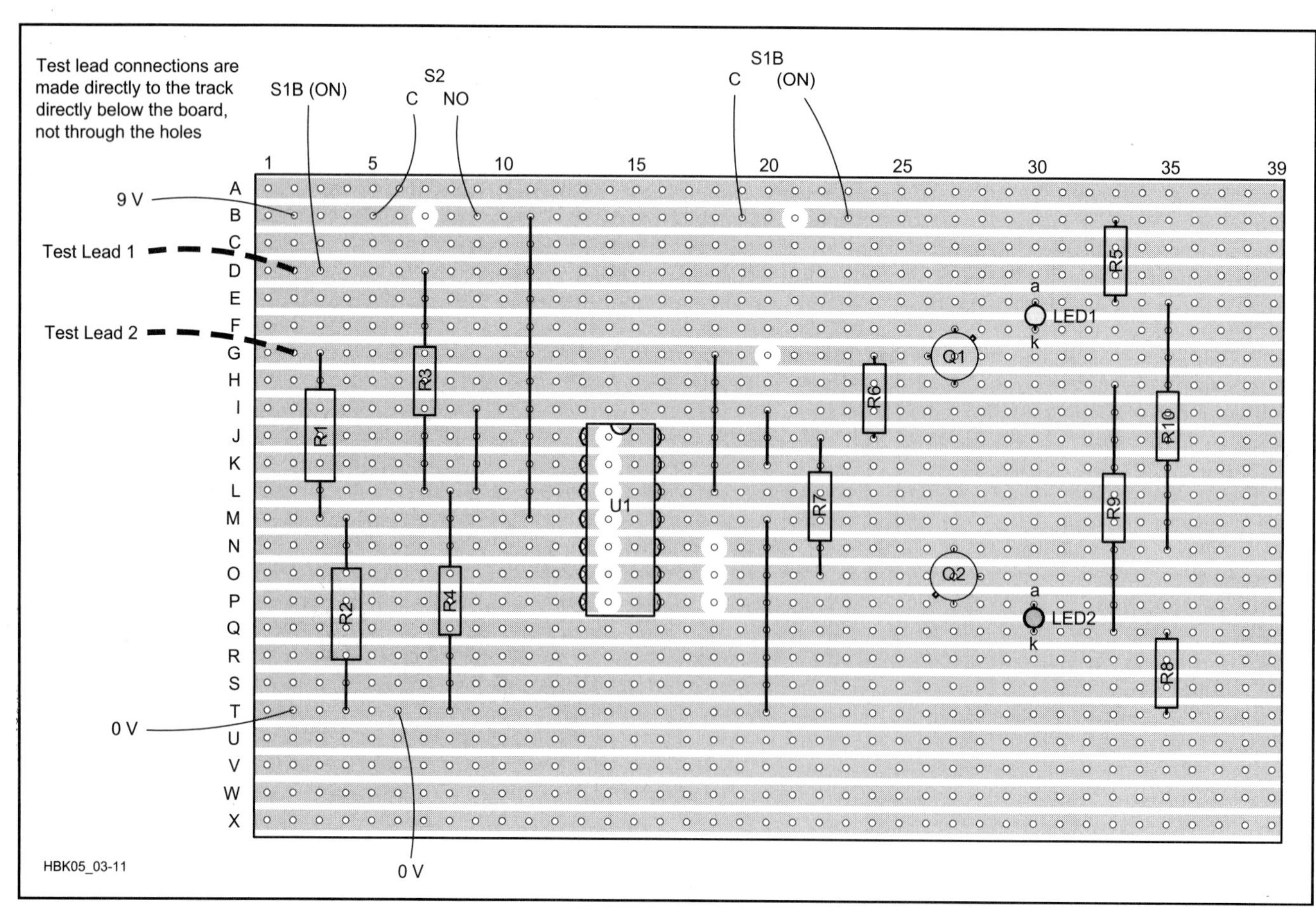

Fig 3.11 — Layout of the circuit on stripboard.

The OUTPUT INDICATOR circuit consists of two LEDs, driven by transistors to provide sufficient current, which indicate either a pass or a fail for an earth path resistance of less than or more than 0.1 Ω. Q2 is an npn transistor that switches on when its input is high, while Q1 is a pnp type that switches on when its input is low. A separate supply voltage is needed for the op-amp and LED circuit, since the test battery voltage will drop under a heavy load current.

Because the output of the op-amp does not swing completely to the positive and negative supply rails, measures need to be taken to ensure that the LED driver transistors switch off correctly.

Construction

A suitable stripboard layout is shown in **Fig 3.11**, and **Fig 3.12** shows how to identify and orientate several of the components. Use thick wire for the test leads!

How To Use It

Using flying leads with suitable connectors, eg, alligator clips, connect the circuit to each end of the earth path to be tested. This would usually be the mains plug earth pin and any metal part meant to be earthed. Then press the test button. Release the test switch as soon as a pass/fail indicator lights (certainly within 5 to 10 seconds, to lengthen battery life and prevent possible overheating of R1 and R2).

Safety Notice

The project described here may be used to test the resistance of appliance earth connections, but it is *not* intended to conform to any *legal* requirements for the testing of electrical safety. The RSGB, ARRL and the author accept no responsibility for any accident or injury caused by its use. *Never* use on mains equipment plugged into the mains — the connection to the mains plug earth pin mentioned in the previous paragraph implies that the plug is free. — *Ed*

STATION POWER

Amateur Radio stations generally require a 120-V ac power source. (In residential systems voltages from 110 V through 125 V are considered equivalent, as are those from 220 V through 250 V.) 120-V ac is converted to the proper ac or dc levels required for the station equipment. Power supplies should accommodate the measured voltage range at each station. (The measured voltage usually varies by hour, day, season and location.) Power supply theory is covered in the **Power Supplies** chapter. If your station is located in a room with electrical outlets, you're in luck. If your station is located in the basement, an attic or other area without a convenient 120-V source, you may need to have a new line run to your operating position.

Stations with high-power amplifiers should have a 240-V ac power source in addition to the 120-V supply. Some amplifiers may be powered from 120 V, but they require current levels that may exceed the limits of standard house wiring. For safety, and for the best possible voltage regulation in the equipment, it is advisable to install a separate 240 or 120-V line with an appropriate current rating if you use an amplifier.

The usual line running to baseboard outlets is rated at 15 A, although 20-A outlets may be installed in newer houses. This may or may not be enough current to power your station. To determine how much current your station requires, check the ratings for each piece of gear. Usually, the manufacturer will specify the required current at 120 V; if the power consumption is rated in watts, divide that rating by 120 V to get amperes. If the total current required for your station is near 12 (0.8 × 15 = 12 A), you need to install another circuit. Keep in mind that other rooms may be powered from the same branch of the electrical system, so the power consumption of any equipment connected to other outlets on the branch must be taken into account. Whenever possible, power your station from a separate, heavy-duty line run directly to the distribution panel through a disconnect switch or circuit breaker that can be locked in the off position.

If you decide to install a separate heavy-duty 120-V line or a 240-V line, consult the power company for local requirements. In some areas, a licensed electrician must perform this work. Others may require a special building permit. Even if you are allowed to do the work yourself, it might need inspection by a licensed electrician. Go through the system and get the necessary permits and inspections! Faulty wiring can destroy your possessions and take away your loved ones. Many fire insurance policies are void if there is unapproved wiring in the structure.

If you decide to do the job yourself, work closely with local building officials. Most home-improvement centers sell books to guide do-it-yourself wiring projects. If you have any doubts about doing the work yourself, get a licensed electrician to do the installation.

Three-Wire 120-V Power Cords

Most metal-cased electrical tools and appliances are equipped with three-conductor power cords. Two of the conductors carry power to the device, while the third conductor is connected to the case or frame. **Fig 3.13** shows two commonly used connectors.

When both plug and receptacle are prop-

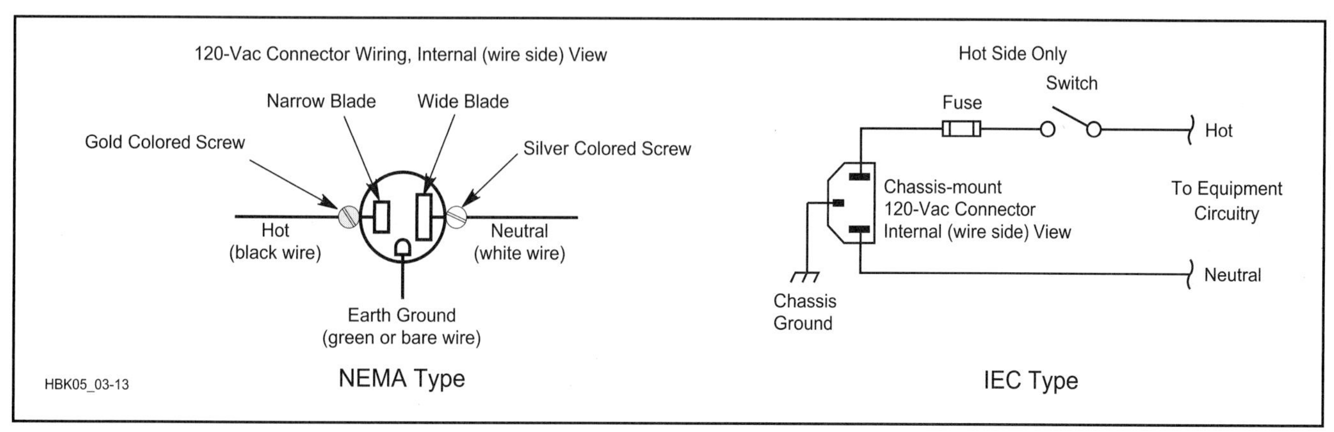

Fig 3.13 — 120-V ac chassis connector wiring.

FCC RF-Exposure Regulations

FCC regulations control the amount of RF exposure that can result from your station's operation (§§97.13, 97.503, 1.1307 (b)(c)(d), 1.1310 and 2.1093). The regulations set limits on the maximum permissible exposure (MPE) allowed from operation of transmitters in all radio services. They also require that certain types of stations be evaluated to determine if they are in compliance with the MPEs specified in the rules. The FCC has also required that five questions on RF environmental safety practices be added to Novice, Technician and General license examinations.

These rules went into effect on January 1, 1998 for new stations or stations that file a 610 application with the FCC. Other existing stations have until September 1, 2000 to be in compliance with the rules.

THE RULES

Maximum Permissible Exposure (MPE)

All radio stations regulated by the FCC must comply with the requirements for MPEs, even QRP stations running only a few watts or less. The MPEs vary with frequency, as shown in **Table A**. MPE limits are specified in maximum electric and magnetic fields for frequencies below 30 MHz, in power density for frequencies above 300 MHz and all three ways for frequencies from 30 to 300 MHz. For compliance purposes, all of these limits must be considered *separately*. If any one is exceeded, the station is not in compliance.

The regulations control human exposure to RF fields, not the strength of RF fields. There is no limit to how strong a field can be as long as no one is being exposed to it, although FCC regulations require that amateurs use the minimum necessary power at all times (§97.311 [a]).

Environments

The FCC has defined two exposure environments — *controlled* and *uncontrolled*. A controlled environment is one in which the people who are being exposed are aware of that exposure and can take steps to minimize that exposure, if appropriate. In an uncontrolled environment, the people being exposed are not normally aware of the exposure. The uncontrolled environment limits are more stringent than the controlled environment limits.

Although the controlled environment is usually intended as an occupational environment, the FCC has determined that it generally applies to amateur operators and members of their immediate households. In most cases, controlled-environment limits can be applied to your home and property to which you can control physical access. The uncontrolled environment is intended for areas that are accessible by the general public, such as your neighbors' properties.

The MPE levels are based on average exposure. An averaging time of 6 minutes is used for controlled exposure; an averaging period of 30 minutes is used for uncontrolled exposure.

Station Evaluations

The FCC requires that certain amateur stations be evaluated for compliance with the MPEs. Although an amateur can have someone else do the evaluation, it

Table A — (From §1.1310) Limits for Maximum Permissible Exposure (MPE)

(A) Limits for Occupational/Controlled Exposure

Frequency Range (MHz)	*Electric Field Strength (V/m)*	*Magnetic Field Strength (A/m)*	*Power Density (mW/cm²)*	*Averaging Time (minutes)*
0.3-3.0	614	1.63	(100)*	6
3.0-30	1842/f	4.89/f	$(900/f^2)$*	6
30-300	61.4	0.163	1.0	6
300-1500	—	—	f/300	6
1500-100,000	—	—	5	6

f = frequency in MHz
* = Plane-wave equivalent power density (see Note 1).

(B) Limits for General Population/Uncontrolled Exposure

Frequency Range (MHz)	*Electric Field Strength (V/m)*	*Magnetic Field Strength (A/m)*	*Power Density (mW/cm²)*	*Averaging Time (minutes)*
0.3-1.34	614	1.63	(100)*	30
1.34-30	824/f	2.19/f	$(180/f^2)$*	30
30-300	27.5	0.073	0.2	30
300-1500	—	—	f/1500	30
1500-100,000	—	—	1.0	30

f = frequency in MHz
* = Plane-wave equivalent power density (see Note 1).

Note 1: This means the equivalent far-field strength that would have the E or H-field component calculated or measured. It does not apply well in the near field of an antenna. The equivalent far-field power density can be found in the near or far field regions from the relationships: $P_d = |E_{total}|^2 / 3770$ mW/cm² or from $P_d = H_{total}|^2 \times 37.7$ mW/cm².

is not difficult for hams to evaluate their own stations. The ARRL book *RF Exposure and You* contains extensive information about the regulations and a large chapter of tables that show compliance distances for specific antennas and power levels. Generally, hams will use these tables to evaluate their stations. Some of these tables have been included in the FCC's information — *OET Bulletin 65* and its *Supplement B*. If hams choose, however, they can do more extensive calculations, use a computer to model their antenna and exposure, or make actual measurements.

Categorical Exemptions

Some types of amateur stations do not need to be evaluated, but these stations must still comply with the MPE limits. The station licensee remains responsible for ensuring that the station meets these requirements.

The FCC has exempted these stations from the evaluation requirement because their output power, operating mode and frequency are such that they are presumed to be in compliance with the rules.

Stations using power equal to or less than the levels in **Table B** do not have to be evaluated. For the 100-W HF ham station, for example, an evaluation would be required *only* on 12 and 10 meters.

Hand-held radios and vehicle-mounted mobile radios that operate using a push-to-talk (PTT) button are also categorically exempt from performing the routine evaluation. Repeater stations that use less than 500 W ERP or those with antennas not mounted on buildings, if the antenna is at least 10 meters off the ground, also do not need to be evaluated.

Correcting Problems

Most hams are already in compliance with the MPE requirements. Some amateurs, especially those using indoor antennas or high-power, high-duty-cycle modes such as a RTTY bulletin station and specialized stations for moonbounce operations and the like may need to make adjustments to their station or operation to be in compliance.

The FCC permits amateurs considerable flexibility in complying with these regulations. As an example, hams can adjust their operating frequency, mode or power to comply with the MPE limits. They can also adjust their operating habits or control the direction their antenna is pointing.

Table B — Power Thresholds for Routine Evaluation of Amateur Radio Stations

Wavelength Band	*Evaluation Required if Power* (watts) Exceeds:*
MF	
160 m	500
HF	
80 m	500
75 m	500
40 m	500
30 m	425
20 m	225
17 m	125
15 m	100
12 m	75
10 m	50
VHF (all bands)	50
UHF	
70 cm	70
33 cm	150
23 cm	200
13 cm	250
SHF (all bands)	250
EHF (all bands)	250
Repeater stations (all bands)	*non-building-mounted antennas*: height above ground level to lowest point of antenna < 10 m *and* power > 500 W ERP *building-mounted antennas*: power > 500 W ERP

*Transmitter power = Peak-envelope power input to antenna. For repeater stations ***only,*** power exclusion based on ERP (effective radiated power).

More Information

This discussion offers only an overview of this topic; additional information can be found in *RF Exposure and You* and on *ARRLWeb* at **www.arrl.org/news/rfsafety/**. *ARRLWeb* has links to the FCC Web site, with *OET Bulletin 65* and *Supplement B* and links to software that hams can use to evaluate their stations.

erly wired, the three-contact polarized plug connects the equipment to the system ground. This grounds the chassis or frame of the appliance and prevents the possibility of electrical shock to the user. Most commercially manufactured test equipment and ac-operated amateur equipment is supplied with these three-wire cords. Unfortunately, the ground wire is sometimes improperly installed. Before connecting any new equipment, check for continuity from case to ground pin with an ohmmeter. If there is no continuity, have the equipment repaired before use. Use such equipment only with properly installed three-wire outlets. If your house does not have such outlets, consult with an electrician or local building officials to learn about safe alternatives.

Equipment with plastic cases is often "double insulated" and fed with a two-wire cord. Such equipment is safe because both conductors are completely insulated from the user. Nonetheless, there is still a hazard if, say, a double insulated drill were used to drill an improperly grounded case of a transmitter that was still plugged in. Remember, all insulation is prey to age, damage and wear that may erode its initial protection.

Safe Homebrewing

Since Amateur Radio began, building equipment in home workshops has been a major part of an amateur's activity. In fact, in the early days, building equipment with your hands was the *only* option available. While times and interests change, home construction of radio equipment and related accessories remains very popular and enjoyable. Building your own gear need not be hazardous if you become familiar with the hazards, learn how to perform the necessary functions and follow some basic safe practices including the ones listed below.

*Consider your State of Mind...*when working on projects or troubleshooting (especially where high voltage is present). Some activities require a lot of concentration. As we grow older, this may be a challenge for some of us. Put another way, if we aren't able to be highly alert, we should put off doing hazardous work until we are better able to focus on the hazards.

Read instructions carefully...and follow them. The manufacturers of tools are the most knowledgeable about how to use their products safely. Tap their knowledge by carefully reading all operating instructions and warnings. Avoiding injuries with power tools requires safe tool design as well as proper operation by the user. Keep the instructions in a place where you can refer to them in the future.

Keep your tools in good condition. Always take care of your investment. Store tools in a way to prevent damage or use by untrained persons (young children, for example). Keep the cutting edges of saws, chisels and drill bits sharp. Protect metal surfaces from corrosion. Frequently inspect the cords and plugs of electrical equipment and make any necessary repairs. If you find that your power cord is becoming frayed, do not delay its repair. Often the best solution is to buy a replacement cord with a molded connector already attached.

Protect yourself. Use of drills, saws, grinders and other wood- or metal-working equipment can release small fragments that could cause serious eye damage. Always wear safety glasses or goggles when doing work that might present a flying object hazard. If you use hammers, wirecutters, chisels and other hand tools, you will also need the protection that safety eyewear offers. Dress appropriately — loose clothing (or even hair) can be caught in exposed rotating equipment such as drill presses.

Take your time. If you hurry, not only will you make more mistakes and possibly spoil the appearance of your new equipment, you won't have time to think things through. Always plan ahead. Do not work with shop tools if you can't concentrate on what you are doing.

Know what to do in an emergency. Despite your best efforts to be careful, accidents may still occur from time to time. Ensure that everyone in your household knows basic first aid procedures and understands how to summon help in an emergency. They should also know where to find and how to safely shut down electrical power in your shack and shop. Keep your shop neat and orderly, with everything in its place. Do not store an excessive amount of flammable materials. Keep clutter off the floor so no one will trip or lose their footing. Exemplary housekeeping is contagious — set a good example for everyone!

Soldering. Soldering requires a certain degree of practice and, of course, the right tools. What potential hazards are involved?

- Since the solder used for virtually all electronic components is a lead-tin alloy, the first thing in most people's mind is lead, a well-known health hazard. There are two primary ways lead might enter our bodies when soldering: we could breathe lead fumes into our lungs or we could ingest (swallow) lead or lead-contaminated food. Inhalation of lead fumes is extremely unlikely because the temperatures ordinarily used in electronic soldering are far below those needed to vaporize lead. But since lead is soft and we may tend to handle it with our fingers, contaminating our food is a real possibility. For this reason, wash your hands carefully after any soldering (or touching of solder connections).
- Generally, solder used for electronic components contains a flux, often a rosin material. When heated the flux flows freely and emits a vapor in the form of a light gray smoke-like plume. This flux vapor, which often contains aldehydes, is a strong irritant and can cause potentially serious problems to persons who may have respiratory sensitivity conditions including those who suffer from asthma. In most cases it is relatively easy to use a small fan to move the flux vapor away from your eyes and face. Open a window, if there is one, to provide additional air exchange. In extreme cases use an organic vapor cartridge respirator.
- Although it is fairly obvious, be careful when soldering not to burn yourself. A soldering iron stand is helpful.
- Solvents are often used to remove excess flux after the parts have cooled to room temperature. Minimize skin contact with solvents by wearing molded gloves that are resistant to the solvent.

RF Burns!

There's a lot of talk about hazards of RF radiation, but most people don't think about RF burns. Happily, most ham shacks offer little exposure to RF current. Transmitters are enclosed, coaxial cable is the most common feed line, and antennas are located well out of reach.

Some people have experienced a mild tingling on their lips while operating with a metal microphone — a gentle reminder of "RF in the shack." When first licensed in 1963, I learned a stronger lesson. Lightbulbs were often used as dummy loads then: they give a nice visual indication of output power, but provide a poor load for the transmitter (not 50 Ω). Also, you can work a lot of people on such a "dummy" antenna. (Don't try this with a modern solid-state transmitter; the mismatch could be fatal to the radio!)

While tuning my Viking Adventurer one day, I bumped the lit bulb and it fell off the table. I prevented a broken lightbulb by catching it — with my finger across the cable ends that were soldered to the bulb. 50-W of RF went through my finger tip and cauterized a path about $^3/_{16} \times {}^1/_8$ inch. It was an extremely painful burn; I would rather have broken the bulb. To avoid RF burns, insulate or enclose any exposed RF conductors and keep your antennas out of reach. Ground mounted vertical antennas that carry more than a few watts should be enclosed by an insulator such as a PVC pipe slipped over the radiator or an 8-ft-high fence around the antenna base.

— Bob Schetgen, KU7G, QEX *Managing Editor*

RF Radiation and Electromagnetic Field Safety

Amateur Radio is basically a safe activity. In recent years, however, there has been considerable discussion and concern about the possible hazards of electromagnetic radiation (EMR), including both RF energy and power-frequency (50-60 Hz) electromagnetic (EM) fields. FCC regulations set limits on the maximum permissible exposure (MPE) allowed from the operation of radio transmitters. These regulations do not take the place of RF-safety practices, however. This section deals with the topic of RF safety.

This section was prepared by members of the ARRL RF Safety Committee and coordinated by Dr. Robert E. Gold, WBØKIZ. It summarizes what is now known and offers safety precautions based on the research to date.

All life on Earth has adapted to survive in an environment of weak, natural, low-frequency electromagnetic fields (in addition to the Earth's static geomagnetic field). Natural low-frequency EM fields come from two main sources: the sun, and thunderstorm activity. But in the last 100 years, man-made fields at much higher intensities and with a very different spectral distribution have altered this natural EM background in ways that are not yet fully understood. Researchers continue to look at the effects of RF exposure over a wide range of frequencies and levels.

Both RF and 60-Hz fields are classified as *nonionizing radiation,* because the frequency is too low for there to be enough photon energy to ionize atoms. (*Ionizing radiation*, such as X-rays, gamma rays and even some ultraviolet radiation has enough energy to knock electrons loose from their atoms. When this happens, positive and negative ions are formed.) Still, at sufficiently high power densities, EMR poses certain health hazards. It has been known since the early days of radio that RF energy can cause injuries by heating body tissue. (Anyone who has ever touched an improperly grounded radio chassis or energized antenna and received an *RF burn* will agree that this type of injury can be quite painful.) In extreme cases, RF-induced heating in the eye can result in cataract formation, and can even cause blindness. Excessive RF heating of the reproductive organs can cause sterility. Other health problems also can result from RF heating. These heat-related health hazards are called *thermal effects*. A microwave oven is a positive application of this thermal effect.

There also have been observations of changes in physiological function in the presence of RF energy levels that are too low to cause heating. These functions return to normal when the field is removed. Although research is ongoing, no harmful health consequences have been linked to these changes.

In addition to the ongoing research, much else has been done to address this issue. For example, FCC regulations set limits on exposure from radio transmitters. The Institute of Electrical and Electronics Engineers, the American National Standards Institute and the National Council for Radiation Protection and Measurement, among others, have recommended voluntary guidelines to limit human exposure to RF energy. The ARRL has established the RF Safety Committee, consisting of concerned medical doctors and scientists, serving voluntarily to monitor scientific research in the fields and to recommend safe practices for radio amateurs.

THERMAL EFFECTS OF RF ENERGY

Body tissues that are subjected to *very high* levels of RF energy may suffer serious heat damage. These effects depend on the frequency of the energy, the power density of the RF field that strikes the body and factors such as the polarization of the wave.

At frequencies near the body's natural resonant frequency, RF energy is absorbed more efficiently, and an increase in heating occurs. In adults, this frequency usually is about 35 MHz if the person is grounded, and about 70 MHz if insulated from the ground. Individual body parts may be resonant at different frequencies. The adult head, for example, is resonant around 400 MHz, while a baby's smaller head resonates near 700 MHz. Body size thus determines the frequency at which most RF energy is absorbed. As the frequency is moved farther from resonance, less RF heating generally occurs. *Specific absorption rate (SAR)* is a term that describes the rate at which RF energy is absorbed in tissue.

Maximum permissible exposure (MPE) limits are based on whole-body SAR values, with additional safety factors included as part of the standards and regulations. This helps explain why these safe exposure limits vary with frequency. The MPE limits define the maximum electric and magnetic field strengths or the plane-wave equivalent power densities associated with these fields, that a person may be exposed to without harmful effect — and with an acceptable safety factor. The regulations assume that a person exposed to a specified (safe) MPE level also will experience a safe SAR.

Nevertheless, thermal effects of RF energy should not be a major concern for most radio amateurs, because of the power levels we normally use and the intermittent nature of most amateur transmissions. Amateurs spend more time listening than transmitting, and many amateur transmissions such as CW and SSB use low-duty-cycle modes. (With FM or RTTY, though, the RF is present continuously at its maximum level during each transmission.) In any event, it is rare for radio amateurs to be subjected to RF fields strong enough to produce thermal effects, unless they are close to an energized antenna or unshielded power amplifier. Specific suggestions for avoiding excessive exposure are offered later in this chapter.

ATHERMAL EFFECTS OF EMR

Research about possible health effects resulting from exposure to the lower level energy fields, the athermal effects, has been of two basic types: epidemiological research and laboratory research.

Scientists conduct laboratory research into biological mechanisms by which EMR may affect animals including humans. Epidemiologists look at the health patterns of large groups of people using statistical methods. These epidemiological studies have been inconclusive. By their basic design, these studies do not demonstrate cause and effect, nor do they postulate mechanisms of disease. Instead, epidemiologists look for associations between an environmental factor and an observed pattern of illness. For example, in the earliest research on malaria, epidemiologists observed the association between populations with high prevalence of the disease and the proximity of mosquito infested swamplands. It was left to the biological and medical scientists to isolate the organism causing malaria in the blood of those with the disease, and identify the same organisms in the mosquito population.

In the case of athermal effects, some studies have identified a weak association between exposure to EMF at home or at work and various malignant conditions including leukemia and brain cancer. A larger number of equally well-designed and performed studies, however, have found no association. A risk ratio of between 1.5 and 2.0 has been observed in positive studies (the number of observed cases of malignancy being 1.5 to 2.0 times the "expected" number in the population). Epidemiologists generally regard a risk ratio of 4.0 or greater to be indicative of a strong association between the cause and effect under study. For example, men who smoke one pack of cigarettes per day increase their risk for lung cancer tenfold compared to nonsmokers, and two packs per day increases the risk to more than 25 times the nonsmokers' risk.

Epidemiological research by itself is rarely conclusive, however. Epidemiology only identifies health patterns in groups — it does not ordinarily determine their cause. And there are often confounding factors: Most of us are exposed to many different environmental hazards that may affect our health in various ways. Moreover, not all studies of persons likely to be exposed to high levels of EMR have yielded the same results.

There also has been considerable laboratory research about the biological effects of EMR in recent years. For example, some separate studies have indicated that even fairly low levels of EMR might alter the human body's circadian rhythms, affect the manner in which T lymphocytes function in the immune system and alter the nature of the electrical and chemical signals communicated through the cell membrane and between cells, among other things. Although these studies are intriguing, they do not demonstrate any effect of these low-level fields on the overall organism.

Much of this research has focused on low-frequency magnetic fields, or on RF fields that are keyed, pulsed or modulated at a low audio frequency (often below 100 Hz). Several studies suggested that humans and animals could adapt to the presence of a steady RF carrier more readily than to an intermittent, keyed or modulated energy source.

The results of studies in this area, plus speculations concerning the effect of various types of modulation, were and have remained somewhat controversial. None of the research to date has demonstrated that low-level EMR causes adverse health effects.

Given the fact that there is a great deal of ongoing research to examine the health consequences of exposure to EMF, the American Physical Society (a national group of highly respected scientists) issued a statement in May 1995 based on its review of available data pertaining to the possible connections of cancer to 60-Hz EMF exposure. This report is exhaustive and should be reviewed by anyone with a serious interest in the field. Among its general conclusions were the following:

1. The scientific literature and the reports of reviews by other panels show no consistent, significant link between cancer and power line fields.

2. No plausible biophysical mechanisms for the systematic initiation or promotion of cancer by these extremely weak 60-Hz fields have been identified.

3. While it is impossible to prove that no deleterious health effects occur from exposure to any environmental factor, it is necessary to demonstrate a consistent, significant, and causal relationship before one can conclude that such effects do occur.

In a report dated October 31, 1996, a committee of the National Research Council of the National Academy of Sciences has concluded that no clear, convincing evidence exists to show that residential exposures to electric and magnetic fields (EMFs) are a threat to human health.

A National Cancer Institute epidemiological study of residential exposure to magnetic fields and acute lymphoblastic leukemia in children was published in the *New England Journal of Medicine* in July 1997. The exhaustive, seven-year study concludes that if there is any link at all, it is far too weak to be concerned about.

Readers may want to follow this topic as further studies are reported. Amateurs should be aware that exposure to RF and ELF (60 Hz) electromagnetic fields at all power levels and frequencies has not been fully studied under all circumstances. "Prudent avoidance" of any avoidable EMR is always a good idea. Prudent avoidance doesn't mean that amateurs should be fearful of using their equipment. Most amateur operations are well within the MPE limits. If any risk does exist, it will almost surely fall well down on the list of causes that may be harmful to your health (on the other end of the list from your automobile). It does mean, however, that hams should be aware of the potential for exposure from their stations, and take whatever reasonable steps they can take to minimize their own exposure and the exposure of those around them.

Safe Exposure Levels

How much EM energy is safe? Scientists and regulators have devoted a great deal of effort to deciding upon safe RF-exposure limits. This is a very complex problem, involving difficult public health and economic considerations. The recommended safe levels have been revised downward several times over the years — and not all scientific bodies agree on this question even today. An Institute of Electrical and Electronics Engineers (IEEE) standard for recommended EM exposure limits was published in 1991 (see Bibliography). It replaced a 1982 American National Standards Institute (ANSI) standard. In the new standard, most of the permitted exposure levels were revised downward (made more stringent), to better reflect the current research. The new IEEE standard was adopted by ANSI in 1992.

The IEEE standard recommends frequency-dependent and time-dependent maximum permissible exposure levels. Unlike earlier versions of the standard, the 1991 standard recommends different RF exposure limits in *controlled environments* (that is, where energy levels can be accurately determined and everyone on the premises is aware of the presence of EM fields) and in *uncontrolled environments* (where energy levels are not known or where people may not be aware of the presence of EM fields). FCC regulations also include controlled/occupational and uncontrolled/general population exposure environments.

The graph in **Fig 3.14** depicts the 1991 IEEE standard. It is necessarily a complex graph, because the standards differ not only for controlled and uncontrolled environments but also for electric (E) fields and magnetic (H) fields. Basically, the lowest E-field exposure limits occur at frequencies between 30 and 300 MHz. The lowest H-field exposure levels occur at 100-300 MHz. The ANSI standard sets the maximum E-field limits between 30 and 300 MHz at a power density of 1 mW/cm^2 (61.4 V/m) in controlled environments — but at one-fifth that level (0.2 mW/cm^2 or 27.5 V/m) in uncontrolled environments. The H-field limit drops to 1 mW/cm^2 (0.163 A/m) at 100-300 MHz in controlled environments and 0.2 mW/cm^2 (0.0728 A/m) in uncontrolled environments. Higher power densities are permitted at frequencies below 30 MHz (below 100 MHz for H fields) and above 300 MHz, based on the concept that the body will not be resonant at those frequencies and will therefore absorb less energy.

In general, the 1991 IEEE standard requires averaging the power level over time periods ranging from 6 to 30 minutes for power-density calculations, depending on the frequency and other variables. The ANSI exposure limits for uncontrolled environments are lower than those for controlled environments, but to compensate for that the standard allows exposure levels in those environments to be averaged over much longer time periods (generally 30 minutes). This long averaging time means that an intermittently operating RF source (such as an Amateur Radio transmitter) will show a much lower power density than a continuous-duty station — for a given power level and antenna configuration.

Time averaging is based on the concept that the human body can withstand a greater rate of body heating (and thus, a higher level of RF energy) for a short

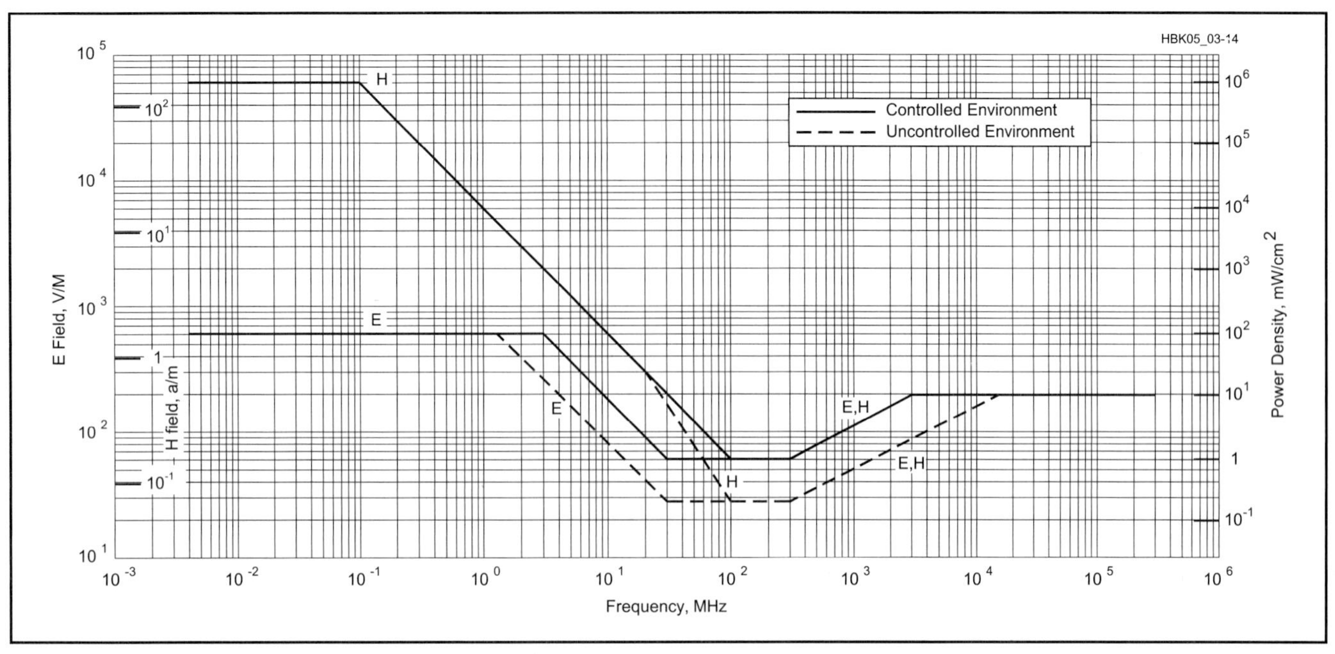

Fig 3.14 — 1991 RF protection guidelines for body exposure of humans. It is known officially as the "IEEE Standard for Safety Levels with Respect to Human Exposure to Radio Frequency Electromagnetic Fields, 3 kHz to 300 GHz."

time than for a longer period. Time averaging may not be appropriate, however, when considering nonthermal effects of RF energy.

The IEEE standard excludes any transmitter with an output below 7 W because such low-power transmitters would not be able to produce significant whole-body heating. (Recent studies show that hand-held transceivers often produce power densities in excess of the IEEE standard within the head.)

There is disagreement within the scientific community about these RF exposure guidelines. The IEEE standard is still intended primarily to deal with thermal effects, not exposure to energy at lower levels. A small but significant number of researchers now believe athermal effects also should be taken into consideration. Several European countries and localities in the United States have adopted stricter standards than the recently updated IEEE standard.

Another national body in the United States, the National Council for Radiation Protection and Measurement (NCRP), also has adopted recommended exposure guidelines. NCRP urges a limit of 0.2 mW/cm^2 for nonoccupational exposure in the 30-300 MHz range. The NCRP guideline differs from IEEE in two notable ways: It takes into account the effects of modulation on an RF carrier, and it does not exempt transmitters with outputs below 7 W.

The FCC MPE regulations are based on parts of the 1992 IEEE/ANSI standard and recommendations of the National Council for Radiation Protection and Measurement (NCRP). The MPE limits under the regulations are slightly different than the IEEE/ANSI limits. Note that the MPE levels apply to the FCC rules put into effect for radio amateurs on January 1, 1998. These MPE requirements do not reflect and include all the assumptions and exclusions of the IEEE/ANSI standard.

Cardiac Pacemakers and RF Safety

It is a widely held belief that cardiac pacemakers may be adversely affected in their function by exposure to electromagnetic fields. Amateurs with pacemakers may ask whether their operating might endanger themselves or visitors to their shacks who have a pacemaker. Because of this, and similar concerns regarding other sources of electromagnetic fields, pacemaker manufacturers apply design methods that for the most part shield the pacemaker circuitry from even relatively high EM field strengths.

It is recommended that any amateur who has a pacemaker, or is being considered for one, discuss this matter with his or her physician. The physician will probably put the amateur into contact with the technical representative of the pacemaker manufacturer. These representatives are generally excellent resources, and may have data from laboratory or "in the field" studies with specific model pacemakers.

One study examined the function of a modern (dual chamber) pacemaker in and around an Amateur Radio station. The pacemaker generator has circuits that receive and process electrical signals produced by the heart, and also generate electrical signals that stimulate (pace) the heart. In one series of experiments, the pacemaker was connected to a heart simulator. The system was placed on top of the cabinet of a 1-kW HF linear amplifier during SSB and CW operation. In another test, the system was placed in close proximity to several 1 to 5-W 2-meter hand-held transceivers. The test pacemaker was connected to the heart simulator in a third test, and then placed on the ground 9 meters below and 5 meters in front of a three-element Yagi HF antenna. No interference with pacemaker function was observed in these experiments.

Although the possibility of interference cannot be entirely ruled out by these few observations, these tests represent more severe exposure to EM fields than would ordinarily be encountered by an amateur — with an average amount of common sense. Of course prudence dictates that amateurs with pacemakers, who use hand-held VHF transceivers, keep the antenna as far as possible from the site of the implanted pacemaker generator. They also should use the lowest transmitter output required for adequate communication. For high power HF transmission, the antenna should be as far as possible from the operating position, and all equipment should be properly grounded.

LOW-FREQUENCY FIELDS

Although the FCC doesn't regulate 60-Hz fields, some recent concern about EMR has focused on low-frequency energy rather than RF. Amateur Radio equipment can be a significant source of low-frequency magnetic fields, although there are many other sources of this kind of energy in the typical home. Magnetic fields can be measured relatively accurately with inexpensive 60-Hz meters that are made by several manufacturers.

Table 3.2 shows typical magnetic field intensities of Amateur Radio equipment and various household items. Because these fields dissipate rapidly with distance, "prudent avoidance" would mean staying perhaps 12 to 18 inches away from most Amateur Radio equipment (and 24 inches from power supplies with 1-kW RF amplifiers).

DETERMINING RF POWER DENSITY

Unfortunately, determining the power density of the RF fields generated by an amateur station is not as simple as measuring low-frequency magnetic fields. Although sophisticated instruments can be used to measure RF power densities quite accurately, they are costly and require frequent recalibration. Most amateurs don't have access to such equipment, and the inexpensive field-strength meters that we do have are not suitable for measuring RF power density.

Table 3.3 shows a sampling of measurements made at Amateur Radio stations by the Federal Communications Commission and the Environmental Protection Agency in 1990. As this table indicates, a good antenna well removed from inhabited areas poses no hazard under any of the IEEE/ANSI guidelines. However, the FCC/EPA survey also indicates that amateurs must be careful about using indoor or attic-mounted antennas, mobile antennas, low directional arrays or any other antenna that is close to inhabited areas, especially when moderate to high power is used.

Ideally, before using any antenna that is in close proximity to an inhabited area, you should measure the RF power density. If that is not feasible, the next best option is make the installation as safe as possible by observing the safety suggestions listed in **Table 3.4**.

It also is possible, of course, to calculate the probable power density near an antenna using simple equations. Such calculations have many pitfalls. For one, most of the situations where the power density would be high enough to be of concern are in the near field. In the near field, ground interactions and other variables produce power densities that cannot be determined by simple arithmetic. In the far field, conditions become easier to predict with simple calculations.

The boundary between the near field and the far field depends on the wavelength of the transmitted signal and the physical size and configuration of the antenna. The boundary between the near field and the far field of an antenna can be as much as several wavelengths from the antenna.

Computer antenna-modeling programs are another approach you can use. *MININEC* or other codes derived from *NEC* (Numerical Electromagnetics Code) are suitable for estimating RF magnetic and electric fields around amateur antenna systems.

These models have limitations. Ground interactions must be considered in estimating near-field power densities, and the "correct ground" must be modeled. Computer modeling is generally not sophisticated enough to predict "hot spots" in the near field — places where the field intensity may be far higher than would be expected, due to reflections from nearby objects. In addition, "nearby objects" often change or vary with weather or the season, so the model so laboriously crafted may not be representative of the actual situation, by the time it is running on the computer.

Intensely elevated but localized fields often can be detected by professional measuring instruments. These "hot spots" are often found near wiring in the shack, and metal objects such as antenna masts or equipment cabinets. But even with the best instrumentation, these measurements also may be misleading in the near field.

One need not make precise measurements or model the exact antenna system, however, to develop some idea of the relative fields around an antenna. Computer modeling using close approximations of the geometry and power input of the antenna will generally suffice. Those who are familiar with *MININEC* can estimate their power densities by computer modeling, and those who have access to profes-

Table 3.2

Typical 60-Hz Magnetic Fields Near Amateur Radio Equipment and AC-Powered Household Appliances

Values are in milligauss.

Item	*Field*	*Distance*
Electric blanket	30-90	Surface
Microwave oven	10-100	Surface
	1-10	12"
IBM personal computer	5-10	Atop monitor
	0-1	15" from screen
Electric drill	500-2000	At handle
Hair dryer	200-2000	At handle
HF transceiver	10-100	Atop cabinet
	1-5	15" from front
1-kW RF amplifier	80-1000	Atop cabinet
	1-25	15" from front

(Source: measurements made by members of the ARRL RF Safety Committee)

Table 3.3

Typical RF Field Strengths Near Amateur Radio Antennas

A sampling of values as measured by the Federal Communications Commission and Environmental Protection Agency, 1990

Antenna Type	*Freq (MHz)*	*Power (W)*	*E Field (V/m)*	*Location*
Dipole in attic	14.15	100	7-100	In home
Discone in attic	146.5	250	10-27	In home
Half sloper	21.5	1000	50	1 m from base
Dipole at 7-13 ft	7.14	120	8-150	1-2 m from earth
Vertical	3.8	800	180	0.5 m from base
5-element Yagi at 60 ft	21.2	1000	10-20	In shack
			14	12 m from base
3-element Yagi at 25 ft	28.5	425	8-12	12 m from base
Inverted V at 22-46 ft	7.23	1400	5-27	Below antenna
Vertical on roof	14.11	140	6-9	In house
			35-100	At antenna tuner
Whip on auto roof	146.5	100	22-75	2 m antenna
			15-30	In vehicle
			90	Rear seat
5-element Yagi at 20 ft	50.1	500	37-50	10 m antenna

Table 3.4
RF Awareness Guidelines

These guidelines were developed by the ARRL RF Safety Committee, based on the FCC/EPA measurements of Table 3.2 and other data.

- Although antennas on towers (well away from people) pose no exposure problem, make certain that the RF radiation is confined to the antennas' radiating elements themselves. Provide a single, good station ground (earth), and eliminate radiation from transmission lines. Use good coaxial cable or other feed line properly. Avoid serious imbalance in your antenna system and feed line. For high-powered installations, avoid end-fed antennas that come directly into the transmitter area near the operator.
- No person should ever be near any transmitting antenna while it is in use. This is especially true for mobile or ground-mounted vertical antennas. Avoid transmitting with more than 25 W in a VHF mobile installation unless it is possible to first measure the RF fields inside the vehicle. At the 1-kW level, both HF and VHF directional antennas should be at least 35 ft above inhabited areas. Avoid using indoor and attic-mounted antennas if at all possible. If open-wire feeders are used, ensure that it is not possible for people (or animals) to come into accidental contact with the feed line.
- Don't operate high-power amplifiers with the covers removed, especially at VHF/UHF.
- In the UHF/SHF region, never look into the open end of an activated length of waveguide or microwave feed-horn antenna or point it toward anyone. (If you do, you may be exposing your eyes to more than the maximum permissible exposure level of RF radiation.) Never point a high-gain, narrow-bandwidth antenna (a paraboloid, for instance) toward people. Use caution in aiming an EME (moonbounce) array toward the horizon; EME arrays may deliver an effective radiated power of 250,000 W or more.
- With hand-held transceivers, keep the antenna away from your head and use the lowest power possible to maintain communications. Use a separate microphone and hold the rig as far away from you as possible. This will reduce your exposure to the RF energy.
- Don't work on antennas that have RF power applied.
- Don't stand or sit close to a power supply or linear amplifier when the ac power is turned on. Stay at least 24 inches away from power transformers, electrical fans and other sources of high-level 60-Hz magnetic fields.

sional power-density meters can make useful measurements.

While our primary concern is ordinarily the intensity of the signal radiated by an antenna, we also should remember that there are other potential energy sources to be considered. You also can be exposed to RF radiation directly from a power amplifier if it is operated without proper shielding. Transmission lines also may radiate a significant amount of energy under some conditions. Poor microwave waveguide joints or improperly assembled connectors are another source of incidental radiation.

FURTHER RF EXPOSURE SUGGESTIONS

Potential exposure situations should be taken seriously. Based on the FCC/EPA measurements and other data, the "RF awareness" guidelines of Table 3.4 were developed by the ARRL RF Safety Committee. A longer version of these guidelines, along with a complete list of references, appeared in a *QST* article by Ivan Shulman, MD, WC2S ("Is Amateur Radio Hazardous to Our Health?" *QST*, Oct 1989, pp 31-34).

In addition, the ARRL has published a book, *RF Exposure and You*, that is helping hams comply with the FCC's RF-exposure regulations. The ARRL also maintains an RF-exposure news page on its Web site. See **www.arrl.org/news/rfsafety**. This site contains reprints of selected *QST* articles on RF exposure and links to the FCC and other useful sites.

Other Hazards in the Ham Shack

CHEMICALS

We can't seem to live without the use of chemicals, even in the electronics age. A number of substances are used everyday by amateurs without causing ill effects. A sensible approach is to become knowledgeable of the hazards associated with the chemicals we use in our shack and then treat them with respect.

A few key suggestions:

- Read the information that accompanies the chemical and follow the manufacturer's recommended safety practices. If you would like more information than is printed on the label, ask for a material safety data sheet.
- Store chemicals properly away from sunlight and sources of heat. Provide security so they won't fall off the shelf. Secure them so that children and untrained persons will not gain access.
- Always keep containers labeled so there is no confusion about the contents. Use the container in which the chemical was purchased.
- Handle chemicals carefully to avoid spills.
- Clean up any spills or leaks promptly but don't overexpose yourself in the process. Never dispose of chemicals in household sinks or drains. Instead, contact your local waste plant operator or fire department to determine the proper disposal procedures for your area. Many communities have household hazardous waste collection programs. Of course, the best solution is to only buy the amount of chemical that you will need, and use it all if possible. Always label any waste chemicals, especially if they are no longer in their original containers. Oil-filled capacitors and transformers were once commonly filled with oil containing PCB's. Never dispose of any such items that may contain PCB's in landfills.
- Always use recommended personal protective equipment (such as gloves, face shield, splash goggles and aprons).
- If corrosives (acids or caustics) are splashed on you *immediately* rinse with cold water for a minimum of 15 minutes to flush the skin thoroughly. If splashed in the eyes, direct a gentle stream of cold water into the eyes for at least 15 minutes. Gently lift the eyelids so trapped liquids can be flushed completely. Start flushing before removing contaminated clothing. Seek professional medical assistance. It is unwise to work alone since people splashed with chemicals need

the calm influence of another person.

- Food and chemicals don't mix. Keep food, drinks and cigarettes *away* from areas where chemicals are used and don't bring your chemicals to places where you eat.

Table 3.5 summarizes the uses and hazards of chemicals used in the ham shack. It includes preventive measures that can minimize risk.

ERGONOMICS

Ergonomics is a term that loosely means "fitting the work to the person." If tools and equipment are designed about what people can accommodate, the results will be much more satisfactory. For example, in the 1930s research was done in telephone equipment manufacturing plants because use of long-nosed pliers for wiring switchboards required considerable force at the end of the hand's range of motion. A simple tool redesign resolved this issue. Considerable attention has been focused on ergonomics in recent years because we have come to realize that long periods of time spent in unnatural positions can lead to repetitive-motion illness. Much of this attention has been focused on people whose job tasks have required them to operate video display terminals (VDTs). While most Amateur Radio operators do not devote as much time to their hobby as they might in a full-time job, it does make sense to consider comfort and flexibility when choosing furniture and arranging it in the shack or workshop. Adjustable height chairs are available with air cylinders to serve as a shock absorber.

Table 3.5
Properties and Hazards of Chemicals often used in the Shack or Workshop

Generic Chemical Name	*Purpose or Use*	*Hazards*	*Ways to Minimize Risks*
Lead-tin solder	Bonding electrical components	• Lead exposure (mostly from hand contact)	• Always wash hands after soldering or touching solder.
		• Flux exposure (inhalation)	• Use good ventilation.
Isopropyl alcohol	Flux remover	• Dermatitis (skin rash)	• Wear molded gloves suitable for solvents.
		• Vapor inhalation	• Use good ventilation and avoid aerosol generation.
		• Fire hazard	• Use good ventilation, limit use to small amounts, keep ignition sources away, dispose of rags only in tightly sealed metal cans.
Freons	Circuit cooling and	• Vapor inhalation	• Use adequate ventilation.
	general solvent	• Dermatitis	• Wear molded gloves suitable for solvents.
Phenols and	Enameled wire/ methylene chloride	• Strong skin corrosive paint stripper	• Avoid skin contact; wear suitable molded gloves; use adequate ventilation.
Beryllium oxide	Ceramic insulator which can conduct heat well	• Toxic when in fine dust form and inhaled	• Avoid grinding, sawing or reducing to dust form.
Beryllium metal	Lightweight metal, alloyed with copper.	• Same as beryllium oxide	• Avoid grinding, sawing, welding, or often reducing to dust. Contact supplier for special procedures.
Various paints	Finishing	• Exposures to solvents	• Adequate ventilation; use respirator when spraying.
		• Exposures to sensitizers (especially urethane paint)	• Adequate ventilation and use respirator. Contact supplier for more info.
		• Exposure to toxic metals (lead, cadmium, chrome, and so on) in pigments	• Adequate ventilation and use respirator. Contact supplier for more info.
		• Fire hazard (especially when spray painting)	• Adequate ventilation; control of residues; eliminate ignition sources.
Ferric chloride	Printed circuit board etchant	• Skin and eye contact	• Use suitable containers; wear splash goggles and molded gloves suitable for acids.
Ammonium persulphate and mercuric chloride	Printed circuit board etchants	• Skin and eye contact	• Use suitable containers; wear splash goggles and molded gloves suitable for acids.
Epoxy resins	General purpose cement or paint	• Dermatitis and possible sensitizer	• Avoid skin contact. Mix only amount needed.
Sulfuric acid	Electrolyte in lead-acid batteries	•Strong corrosive when on skin or eyes. •Will release hydrogen when charging (fire, explosion hazard).	•Always wear splash goggles and molded plastic gloves (PVC) when handling. Keep ignition sources away from battery when charging. Provide adequate ventilation.

Footrests might come in handy if the chair is so high that your feet cannot support your lower leg weight. The height of tables and keyboards often is not adjustable.

Placement of VDT screens should take into consideration the reflected light coming from windows. It is always wise to build into your sitting sessions time to walk around and stimulate blood circulation. Your muscles are less likely to stiffen, while the flexibility in your joints can be enhanced by moving around.

Selection of hand tools is another area where there are choices to make that may affect how comfortable you will be while working in your shack. Look for screwdrivers with pliable grips. Take into account how heavy things are before picking them up — your back will thank you.

ENERGIZED CIRCUITS

Working with energized circuits can be very hazardous since our senses cannot directly detect dangerous voltages. The first thing we should ask ourselves when faced with troubleshooting, aligning or other "live" procedures is, "Is there a way to reduce the hazard of electrical shock?" Here are some ways of doing just that.

1. If at all possible, troubleshoot with an ohmmeter. With a reliable schematic diagram and careful consideration of how various circuit conditions may reflect resistance readings, it will often be unnecessary to do live testing.

2. Keep a fair distance from energized circuits. What is considered "good practice" in terms of distance? The *National Electrical Code* specifies minimum working space about electric equipment in Sections 110-16 and 110-34, depending on the voltage level. The principle here is that a person doing live work needs adequate space so they are not forced to be dangerously close to energized equipment.

3. If you need to measure the voltage of a circuit, install the voltmeter with the power safely off, back up, and only then energize the circuit. Remove the power before disconnecting the meter.

4. If you are building equipment that has hinged or easily removable covers that could expose someone to an energized circuit, install interlock switches that safely remove power in the event that the enclosure was opened with the power still on. Interlock switches are generally not used if tools are required to open the enclosure.

5. Never assume that a circuit is at zero potential even if the power is switched off and the power cable disconnected. Capacitors can retain a charge for a considerable period of time. Bleeder resistors should be installed, but don't assume they have bled off the voltage. Instead, after power is removed and disconnected use a "shorting stick" to ground all exposed conductors and ensure that voltage is not present. Avoid using screwdrivers, as this brings the amateur too close to the circuit and could ruin the screwdriver's blade.

6. If you must hold a probe to take a measurement, always keep one hand in your pocket. As mentioned in the sidebar on the effects of high voltages, the worst path current could take through your body is from hand to hand since the flow would pass through the chest cavity.

7. Make sure someone is in the room with you and that they know how to remove the power safely. If they grab you with the power still on they will be shocked as well.

8. Test equipment probes and their leads must be in very good condition and rated for the conditions they will encounter.

9. Be wary of the hazards of "floating" (ungrounded) test equipment. A number of options are available to avoid this hazard. Contact your test equipment manufacturer for suggested procedures.

10. Ground-fault circuit interrupters can offer additional protection for stray currents that flow through the ground on 120-V circuits. Know their limitations. They cannot offer protection for the plate supply voltages in linear amplifiers, for example.

11. Older radio equipment containing ac/dc power supplies have their own hazards. If working on these live, use an isolation transformer, as the chassis may be connected directly to the hot or neutral power conductor.

12. Be aware of electrolytic capacitors that might fail if used outside their intended applications.

13. Replace fuses only with those having proper ratings.

SUMMARY

The ideas presented in this chapter are intended to reinforce the concept that ham radio, like many other activities in

High-Voltage Hazards

What happens when someone receives an electrical shock?

Electrocutions (fatal electric shocks) usually are caused by the heart ceasing to beat in its normal rhythm. This condition, called ventricular fibrillation, causes the heart muscles to quiver and stop working in a coordinated pattern, in turn preventing the heart from pumping blood.

The current flow that results in ventricular fibrillation varies between individuals but may be in the range of 100 mA to 500 mA. At higher current levels the heart may have less tendency to fibrillate but serious damage would be expected. Studies have shown 60-Hz alternating current to be more hazardous than dc currents. Emphasis is placed on application of cardiopulmonary resuscitation (CPR), as this technique can provide mechanical flow of some blood until paramedics can "restart" the heart's normal beating pattern. Defibrillators actually apply a carefully controlled dc voltage to "shock" the heart back into a normal heartbeat. It doesn't always work but it's the best procedure available.

What are the most important factors associated with severe shocks?

You may have heard that the current that flows through the body is the most important factor, and this is generally true. The path that current takes through the body affects the outcome to a large degree. While simple application of Ohm's Law tells us that the higher the voltage applied with a fixed resistance, the greater the current that will flow. Most electrical shocks involve skin contact. Skin, with its layer of dead cells and often fatty tissues, is a fair insulator. Nonetheless, as voltage increases the skin will reach a point where it breaks down. Then the lowered resistance of deeper tissues allows a greater current to flow. This is why electrical codes refer to the term "high voltage" as a voltage above 600 V.

How little a voltage can be lethal?

This depends entirely on the resistance of the two contact points in the circuit, the internal resistance of the body, and the path the current travels through the body. Historically, reports of fatal shocks suggest that as little as 24 V *could* be fatal under extremely adverse conditions. To add some perspective, one standard used to prevent serious electrical shock in hospital operating rooms limits leakage flow from electronic instruments to only 50 µA due to the use of electrical devices and related conductors inside the patient's body.

modern life, does have certain risks. But by understanding the hazards and how to deal effectively with them, the risk can be minimized. Common-sense measures can go a long way to help us prevent accidents. Traditionally, amateurs are inventors, and experimenting is a major part of our nature. But reckless chance-taking is never wise, especially when our health and well-being is involved. A healthy attitude toward doing things the right way will help us meet our goals and expectations.

BIBLIOGRAPHY

Source material and more extended discussion of topics covered in this chapter can be found in the references given below.

Lightning Protection Code, NFPA 780,National Fire Protection Association, Quincy, MA, 1992.

National Electrical Code, NFPA 70, National Fire Protection Association, Quincy, MA, 1993. *National Electrical Code* and *NEC* are registered trademarks of the National Fire Protection Association, Inc, Quincy, MA 02269.

R. P. Haviland, "Amateur Use of Telescoping Masts," *QST*, May 1994, pp 41-45.

For more information about soldering hazards, symptoms, and protection, see "Making Soldering Safer," by Bryan P. Bergeron, MD, NU1N (Mar 1991 *QST*, pp 28-30) and "More on Safer Soldering," by Gary E. Myers, K9CZB (Aug 1991 *QST*, p 42).

Chapter 4

Electrical Fundamentals

DC Circuits and Resistance Glossary

Alternating current — A flow of charged particles through a conductor, first in one direction, then in the other direction.

Ampere — A measure of flow of charged particles per unit time. One ampere represents one coulomb of charge flowing past a point in one second.

Atom — The smallest particle of matter that makes up an element. Consists of protons and neutrons in the central area called the nucleus, with electrons surrounding this central region.

Coulomb — A unit of measure of a quantity of electrically charged particles. One coulomb is equal to 6.25×10^{18} electrons.

Direct current — A flow of charged particles through a conductor in one direction only.

EMF — Electromotive Force is the term used to define the force of attraction between two points of different charge potential. Also called voltage.

Energy — Capability of doing work. It is usually measured in electrical terms as the number of watts of power consumed during a specific period of time, such as watt-seconds or kilowatt-hours.

Joule — Measure of a quantity of energy. One joule is defined as one newton (a measure of force) acting over a distance of one meter.

Ohm — Unit of resistance. One ohm is defined as the resistance that will allow one ampere of current when one volt of EMF is impressed across the resistance.

Power — Power is the rate at which work is done. One watt of power is equal to one volt of EMF, causing a current of one ampere through a resistor.

Volt — A measure of electromotive force.

Introduction

The DC Circuits and Resistance section of this chapter was written by Roger Taylor, K9ALD.

The atom is the primary building block of the universe. The main parts of the atom include protons, electrons and neutrons. Protons have a positive electrical charge, electrons a negative charge and neutrons have no electrical charge. All atoms are electrically neutral, so they have the same number of electrons as protons. If an atom loses electrons, so it has more protons than electrons, it has a net positive charge. If an atom gains electrons, so it has more electrons than protons, it has a negative charge. Particles with a positive or negative charge are called ions. Free electrons are also called ions, because they have a negative charge.

When there are a surplus number of positive ions in one location and a surplus number of negative ions (or electrons) in another location, there is an attractive force between the two collections of particles. That force tries to pull the collections together. This attraction is called electromotive force, or EMF.

If there is no path (conductor) to allow electric charge to flow between the two locations, the charges cannot move together and neutralize one another. If a conductor is provided, then electric current (usually electrons) will flow through the conductor.

Electrons move from the negative to the positive side of the voltage, or EMF source. *Conventional current* has the opposite direction, from positive to negative. This comes from an arbitrary decision made by Benjamin Franklin in the 18th century. The conventional current direction is important in establishing the proper polarity sign for many electronics calculations. Conventional current is used in much of the technical literature. The arrows in semiconductor schematic symbols point in the direction of conventional current, for example.

To measure the quantities of charge, current and force, certain definitions have been adopted. Charge is measured in *coulombs*. One coulomb is equal to 6.25×10^{18} electrons (or protons). Charge flow is measured in *amperes*. One ampere represents one coulomb of charge flowing past a point in one second. Electromotive force is measured in *volts*. One volt is defined as the potential force (electrical) between two points for which one ampere of current will do one *joule* (measure of energy) of work flowing from one point to another. (A joule of work per second represents a power of one watt.)

Voltage can be generated in a variety of ways. Chemicals with certain characteristics can be combined to form a battery. Mechanical motion such as friction (static electricity, lightning) and rotating conductors in a magnetic field (generators) can also produce voltage.

Any conductor between points at different voltages will allow current to pass between the points. No conductor is perfect or lossless, however, at least not at normal temperatures. Charged particles such as electrons resist being moved and it requires energy to move them. The amount of resistance to current is measured in *ohms*.

OHM'S LAW

One ohm is defined as the amount of resistance that allows one ampere of current to flow between two points that have a potential difference of one volt. Thus, we get Ohm's Law, which is:

$$R = \frac{E}{I} \quad (1)$$

where:
R = resistance in ohms,
E = potential or EMF in volts and
I = current in amperes.

Transposing the equation gives the other common expressions of Ohm's Law as:

$$E = I \times R \quad (2)$$

and

$$I = \frac{E}{R} \quad (3)$$

All three forms of the equation are used often in radio work. You must remember that the quantities are in volts, ohms and amperes; other units cannot be used in the equations without first being converted. For example, if the current is in milliamperes you must first change it to the equivalent fraction of an ampere before substituting the value into the equations.

The following examples illustrate the use of Ohm's Law. The current through a 20000-Ω resistance is 150 mA. See **Fig 4.1**. What is the voltage? To find voltage, use equation 2 (E = I × R). Convert the current from milliamperes to amperes. Divide by 1000 mA / A (or multiply by 10^{-3} A / mA) to make this conversion. (Notice the conversion factor of 1000 does not limit the number of significant figures in the calculated answer.)

$$I = \frac{150\,\text{mA}}{1000\,\frac{\text{mA}}{\text{A}}} = 0.150\,\text{A}$$

Then:

$$E = 0.150\,\text{A} \times 20000\,\Omega = 3000\,\text{V}$$

When 150 V is applied to a circuit, the current is measured at 2.5 A. What is the resistance of the circuit? In this case R is the unknown, so we will use equation 1:

$$R = \frac{E}{I} = \frac{150\,\text{V}}{2.5\,\text{A}} = 60\,\Omega$$

No conversion was necessary because the voltage and current were given in volts and amperes.

How much current will flow if 250 V is applied to a 5000-Ω resistor? Since I is unknown,

$$I = \frac{E}{R} = \frac{250\,\text{V}}{5000\,\Omega} = 0.05\,\text{A}$$

It is more convenient to express the current in mA, and 0.05 A × 1000 mA / A = 50 mA.

RESISTANCE AND CONDUCTANCE

Suppose we have two conductors of the same size and shape, but of different materials. The amount of current that will flow when a given EMF is applied will vary with the resistance of the material. The lower the resistance, the greater the current for a given EMF. The *resistivity* of a material is the resistance, in ohms, of a cube of the material measuring one centimeter on each edge. One of the best conductors is copper, and in making resistance calculations it is frequently convenient to compare the resistance of the material under consideration with that of a copper conductor of the same size and shape. **Table 4.1** gives the ratio of the resistivity of various conductors to the resistivity of copper.

The longer the physical path, the higher the resistance of that conductor. For direct current and low-frequency alternating currents (up to a few thousand hertz) the resistance is inversely proportional to the cross-sectional area of the path the current must travel; that is, given two conductors of the same material and having the same length, but differing in cross-sectional area, the one with the larger area will have the lower resistance.

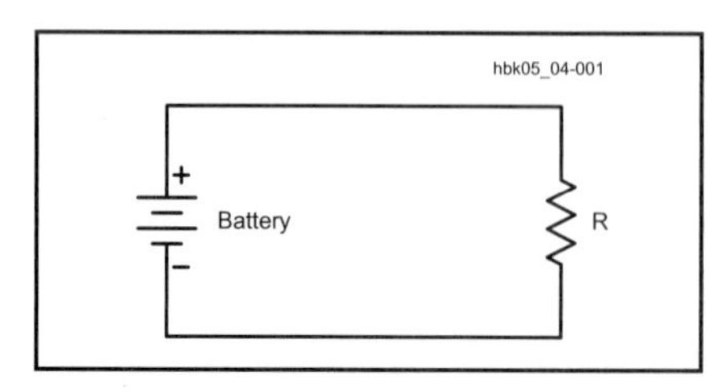

Fig 4.1 — A simple circuit consisting of a battery and a resistor.

Table 4.1
Relative Resistivity of Metals

Material	*Resistivity Compared to Copper*
Aluminum (pure)	1.60
Brass	3.7-4.90
Cadmium	4.40
Chromium	1.80
Copper (hard-drawn)	1.03
Copper (annealed)	1.00
Gold	1.40
Iron (pure)	5.68
Lead	12.80
Nickel	5.10
Phosphor bronze	2.8-5.40
Silver	0.94
Steel	7.6-12.70
Tin	6.70
Zinc	3.40

RESISTANCE OF WIRES

The problem of determining the resistance of a round wire of given diameter and length—or its converse, finding a suitable size and length of wire to provide a desired amount of resistance—can easily be solved with the help of the copper wire table given in the **Component Data and References** chapter. This table gives the resistance, in ohms per 1000 ft, of each standard wire size. For example, suppose you need a resistance of 3.5 Ω, and some #28 wire is on hand. The wire table in the **Component Data and References** chapter shows that #28 wire has a resistance of 66.17 Ω / 1000 ft. Since the desired resistance is 3.5 Ω, the required wire length is:

$$\text{Length} = \frac{R_{DESIRED}}{\frac{R_{WIRE}}{1000\,\text{ft}}} = \frac{3.5\,\Omega}{\frac{66.17\,\Omega}{1000\,\text{ft}}}$$

$$= \frac{3.5\,\Omega \times 1000\,\text{ft}}{66.17\,\Omega} = 53\,\text{ft} \quad (4)$$

As another example, suppose that the resistance of wire in a circuit must not exceed 0.05 Ω and that the length of wire required for making the connections totals 14 ft. Then:

$$\frac{R_{WIRE}}{1000\,\text{ft}} < \frac{R_{MAXIMUM}}{\text{Length}} = \frac{0.05\,\Omega}{14.0\,\text{ft}} \quad (5)$$

$$= 3.57 \times 10^{-3}\,\frac{\Omega}{\text{ft}} \times \frac{1000\,\text{ft}}{1000\,\text{ft}}$$

$$\frac{R_{WIRE}}{1000\,\text{ft}} < \frac{3.57\,\Omega}{1000\,\text{ft}}$$

Find the value of R_{WIRE} / 1000 ft that is less than the calculated value. The wire table shows that #15 is the smallest size having a resistance less than this value. (The resistance of #15 wire is given as 3.1810 Ω / 1000 ft.) Select any wire size larger than this for the connections in your circuit, to ensure that the total wire resistance will be less than 0.05 Ω.

When the wire in question is not made of copper, the resistance values in the wire table should be multiplied by the ratios shown in Table 4.1 to obtain the resulting resistance. If the wire in the first example were made from nickel instead of copper, the length required for 3.5 Ω would be:

$$\text{Length} = \frac{R_{DESIRED}}{\frac{R_{WIRE}}{1000\,\text{ft}}} \qquad (6)$$

$$= \frac{3.5\,\Omega}{\frac{66.17\,\Omega}{1000\,\text{ft}} \times 5.1}$$

$$= \frac{3.5\,\Omega \times 1000\,\text{ft}}{66.17\,\Omega \times 5.1}$$

$$\text{Length} = \frac{3500\ \text{ft}}{337.5} = 10.37\,\text{ft}$$

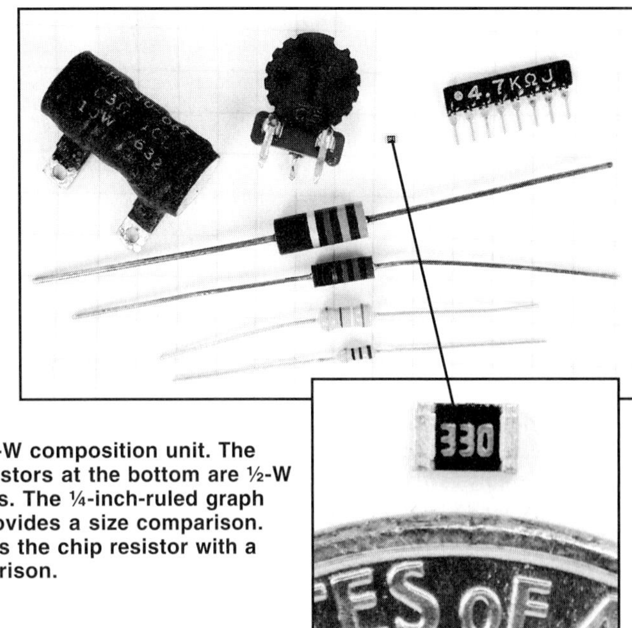

Fig 4.2 — Examples of various resistors. At the top left is a small 10-W wirewound resistor. A single in-line package (SIP) of resistors is at the top right. At the top center is a small PC-board-mount variable resistor. A tiny surface-mount (chip) resistor is also shown at the top. Below the variable resistor is a 1-W carbon compo-sition resistor and then a ½-W composition unit. The dog-bone-shaped resistors at the bottom are ½-W and ¼-W film resistors. The ¼-inch-ruled graph paper background provides a size comparison. The inset photo shows the chip resistor with a penny for size comparison.

TEMPERATURE EFFECTS

The resistance of a conductor changes with its temperature. The resistance of practically every metallic conductor increases with increasing temperature. Carbon, however, acts in the opposite way; its resistance decreases when its temperature rises. It is seldom necessary to consider temperature in making resistance calculations for amateur work. The temperature effect is important when it is necessary to maintain a constant resistance under all conditions, however. Special materials that have little or no change in resistance over a wide temperature range are used in that case.

RESISTORS

A package of material exhibiting a certain amount of resistance, made up into a single unit is called a resistor. Different resistors having the same resistance value may be considerably different in physical size and construction (see **Fig 4.2**). Current through a resistance causes the conductor to become heated; the higher the resistance and the larger the current, the greater the amount of heat developed. Resistors intended for carrying large currents must be physically large so the heat can be radiated quickly to the surrounding air. If the resistor does not dissipate the heat quickly, it may get hot enough to melt or burn.

The amount of heat a resistor can safely dissipate depends on the material, surface area and design. Typical carbon resistors used in amateur electronics (1/8 to 2-W resistors) depend primarily on the surface area of the case, with some heat also being carried off through the connecting leads. Wirewound resistors are usually used for higher power levels. Some have finned cases for better convection cooling and/or metal cases for better conductive cooling.

In some circuits, the resistor value may be critical. In this case, precision resistors are used. These are typically wirewound, or carbon-film devices whose values are carefully controlled during manufacture. In addition, special material or construction techniques may be used to provide temperature compensation, so the value does not change (or changes in a precise manner) as the resistor temperature changes. There is more information about the electrical characteristics of real resistors in the **Real-World Component Characteristics** chapter.

CONDUCTANCE

The reciprocal of resistance (1/R) is *conductance*. It is usually represented by the symbol G. A circuit having high conductance has low resistance, and vice versa. In radio work, the term is used chiefly in connection with electron-tube and field-effect transistor characteristics. The unit of conductance is the siemens, abbreviated S. A resistance of 1 Ω has a conductance of 1 S, a resistance of 1000 Ω has a conductance of 0.001 S, and so on. A unit frequently used in connection with electron devices is the µS or one millionth of a siemens. It is the conductance of a 1-MΩ resistance.

Series and Parallel Resistances

Very few actual electric circuits are as simple as Fig 4.1. Commonly, resistances are found connected in a variety of ways. The two fundamental methods of connecting resistances are shown in **Fig 4.3**. In part A, the current flows from the source of EMF (in the direction shown by the arrow) down through the first resistance, R1, then through the second, R2 and then back to the source. These resistors are connected in series. The current everywhere in the circuit has the same value.

In part B, the current flows to the common connection point at the top of the two resistors and then divides, one part of it flowing through R1 and the other through R2. At the lower connection point these two currents again combine; the total is the same as the current into the upper common connection. In this case, the two resistors are connected in parallel.

RESISTORS IN PARALLEL

In a circuit with resistances in parallel, the total resistance is less than that of the lowest resistance value present. This is because the total current is always greater than the current in any individual resistor. The formula for finding the total resistance of resistances in parallel is:

$$R = \frac{1}{\frac{1}{R1}+\frac{1}{R2}+\frac{1}{R3}+\frac{1}{R4}+...} \quad (7)$$

where the dots indicate that any number of resistors can be combined by the same method. For only two resistances in parallel (a very common case) the formula becomes:

$$R = \frac{R1 \times R2}{R1 + R2} \quad (8)$$

Example: If a 500-Ω resistor is connected in parallel with one of 1200 Ω, what is the total resistance?

$$R = \frac{R1 \times R2}{R1+R2} = \frac{500\,\Omega \times 1200\,\Omega}{500\,\Omega + 1200\,\Omega}$$

$$R = \frac{600000\,\Omega^2}{1700\,\Omega} = 353\,\Omega$$

KIRCHHOFF'S FIRST LAW (KIRCHHOFF'S CURRENT LAW)

Suppose three resistors (5.00 kΩ, 20.0 kΩ and 8.00 kΩ) are connected in parallel as shown in **Fig 4.4**. The same EMF, 250 V, is applied to all three resistors. The current in each can be found from

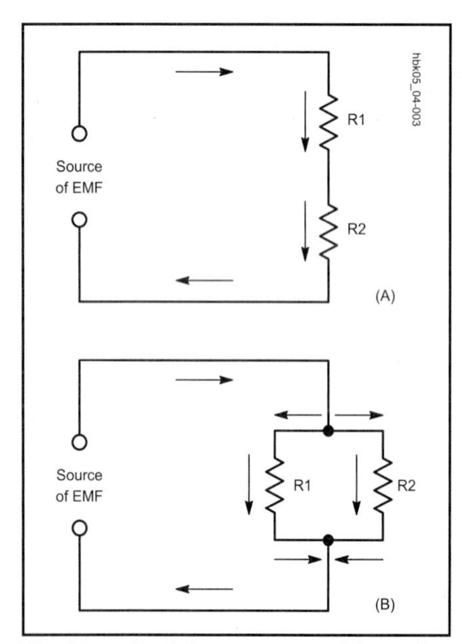

Fig 4.3 — Resistors connected in series at A, and in parallel at B.

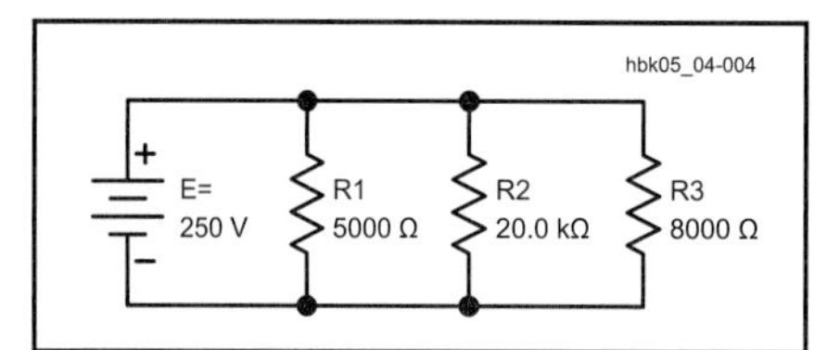

Fig 4.4 — An example of resistors in parallel. See text for calculations.

Ohm's Law, as shown below. The current through R1 is I1, I2 is the current through R2 and I3 is the current through R3.

For convenience, we can use resistance in kΩ, which gives current in milliamperes.

$$I1 = \frac{E}{R1} = \frac{250\,V}{5.00\,k\Omega} = 50.0\,mA$$

$$I2 = \frac{E}{R2} = \frac{250\,V}{20.0\,k\Omega} = 12.5\,mA$$

$$I3 = \frac{E}{R3} = \frac{250\,V}{8.00\,k\Omega} = 31.2\,mA$$

Notice that the branch currents are inversely proportional to the resistances. The 20000-Ω resistor has a value four times larger than the 5000-Ω resistor, and has a current one quarter as large. If a resistor has a value twice as large as another, it will have half as much current through it when they are connected in parallel.

The total circuit current is:

$$I_{TOTAL} = I1 + I2 + I3 \quad (9)$$

$$I_{TOTAL} = 50.0\,mA + 12.5\,mA + 31.2\,mA$$

$$I_{TOTAL} = 93.7\,mA$$

This example illustrates Kirchhoff's Current Law: The current flowing into a node or branching point is equal to the sum of the individual currents leaving the node or branching point. The total resistance of the circuit is therefore:

$$R = \frac{E}{I} = \frac{250\,V}{93.7\,mA} = 2.67\,k\Omega$$

You can verify this calculation by combining the three resistor values in parallel, using equation 7.

RESISTORS IN SERIES

When a circuit has a number of resistances connected in series, the total resistance of the circuit is the sum of the individual resistances. If these are numbered R1, R2, R3 and so on, then:

$$R_{TOTAL} = R1 + R2 + R3 + R4... \quad (10)$$

where the dots indicate that as many resistors as necessary may be added.

Example: Suppose that three resistors are

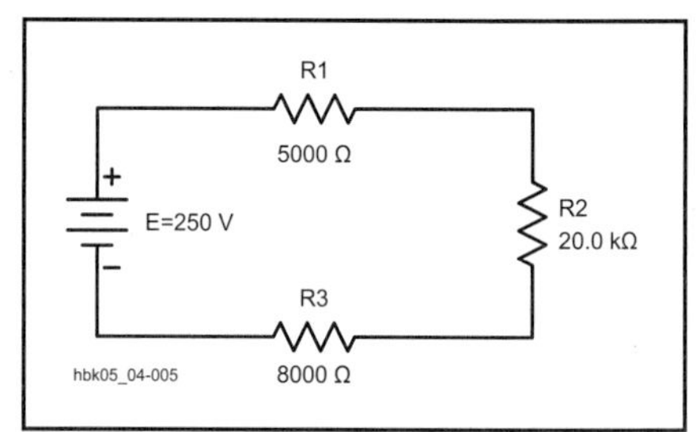

Fig 4.5 — An example of resistors in series. See text for calculations.

connected to a source of EMF as shown in **Fig 4.5**. The EMF is 250 V, R1 is 5.00 kΩ, R2 is 20.0 kΩ and R3 is 8.00 kΩ. The total resistance is then

$$R_{TOTAL} = R1 + R2 + R3$$

$$R = 5.00\,k\Omega + 20.0\,k\Omega + 8.00\,k\Omega$$

$$R = 33.0\,k\Omega.$$

The current in the circuit is then

$$I = \frac{E}{R} = \frac{250\,V}{33.0\,k\Omega} = 7.58\,mA$$

(We need not carry calculations beyond three significant figures; often, two will suffice because the accuracy of measurements is seldom better than a few percent.)

KIRCHHOFF'S SECOND LAW (KIRCHHOFF'S VOLTAGE LAW)

Ohm's Law applies in any portion of a circuit as well as to the circuit as a whole. Although the current is the same in all three of the resistances in the example of Fig 4.5, the total voltage divides between them. The voltage appearing across each resistor (the voltage drop) can be found from Ohm's Law.

Example: If the voltage across R1 is called E1, that across R2 is called E2 and that across R3 is called E3, then

$$E = IR1 = 0.00758\,A \times 5000\,\Omega = 37.9\,V$$

$$E = IR2 = 0.00758\,A \times 20000\,\Omega = 152\,V$$

$$E = IR3 = 0.00758\,A \times 8000\,\Omega = 60.6\,V$$

Notice here that the voltage drop across each resistor is directly proportional to the resistance. The 20000-Ω resistor value is four times larger than the 5000-Ω resistor, and the voltage drop across the 20000-Ω resistor is four times larger. A resistor that has a value twice as large as another will have twice the voltage drop across it when they are connected in series.

Kirchhoff's Voltage Law accurately describes the situation in the circuit: The sum of the voltages in a closed current loop is zero. The resistors are power sinks, while the battery is a power source. It is common to assign a + sign to power sources and a

– sign to power sinks. This means the voltages across the resistors have the opposite sign from the battery voltage. Adding all the voltages yields zero. In the case of a single voltage source, algebraic manipulation implies that the sum of the individual voltage drops in the circuit must be equal to the applied voltage.

$$E_{TOTAL} = E1 + E2 + E3 \quad (11)$$

$$E_{TOTAL} = 37.9\,V + 152\,V + 60.6\,V$$

$$E_{TOTAL} = 250\,V$$

(Remember the significant figures rule for addition.)

In problems such as this, when the current is small enough to be expressed in milliamperes, considerable time and trouble can be saved if the resistance is expressed in kilohms rather than in ohms. When the resistance in kilohms is substituted directly in Ohm's Law, the current will be milliamperes, if the EMF is in volts.

RESISTORS IN SERIES-PARALLEL

A circuit may have resistances both in parallel and in series, as shown in **Fig 4.6A**. The method for analyzing such a circuit is as follows: Consider R2 and R3 to be the equivalent of a single resistor, R_{EQ} whose value is equal to R2 and R3 in parallel.

$$R_{EQ} = \frac{R2 \times R3}{R2 + R3} = \frac{20000\,\Omega \times 8000\,\Omega}{20000\,\Omega + 8000\,\Omega}$$

$$= \frac{1.60 \times 10^8\,\Omega^2}{28000\,\Omega}$$

$$R_{EQ} = 5710\,\Omega = 5.71\,k\Omega$$

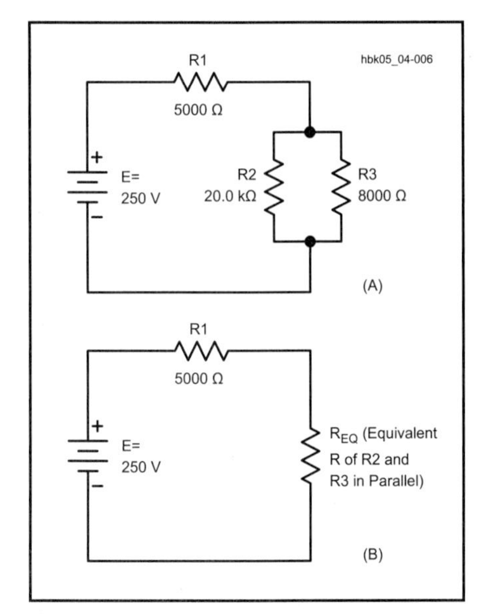

Fig 4.6 — At A, an example of resistors in series-parallel. The equivalent circuit is shown at B. See text for calculations.

This resistance in series with R1 forms a simple series circuit, as shown in Fig 4.6B. The total resistance in the circuit is:

$$R_{TOTAL} = R1 + R_{EQ} = 5.00\,k\Omega + 5.71\,k\Omega$$

$$R_{TOTAL} = 10.71\,k\Omega$$

The current is:

$$I = \frac{E}{R} = \frac{250\,V}{10.71\,k\Omega} = 23.3\,mA$$

The voltage drops across R1 and R_{EQ} are:

$$E1 = I \times R1 = 23.3\,mA \times 5.00\,k\Omega = 117\,V$$

$$E2 = I \times R_{EQ} = 23.3\,mA \times 5.71\,k\Omega = 133\,V$$

with sufficient accuracy. These two voltage drops total 250 V, as described by Kirchhoff's Current Law. E2 appears across both R2 and R3 so,

$$I2 = \frac{E2}{R2} = \frac{133\,V}{20.0\,k\Omega} = 6.65\,mA$$

$$I3 = \frac{E3}{R3} = \frac{133\,V}{8.00\,k\Omega} = 16.6\,mA$$

where:
I2 = current through R2 and
I3 = current through R3.

The sum of I2 and I3 is equal to 23.3 mA, conforming to Kirchhoff's Voltage Law.

THEVENIN'S THEOREM

Thevenin's Theorem is a useful tool for simplifying electrical networks. Thevenin's Theorem states that any two-terminal network of resistors and voltage or current sources can be replaced by a single voltage source and a series resistor. Such a transformation can simplify the calculation of current through a parallel branch. Thevenin's Theorem can be readily applied to the circuit of Fig 4.6A, to find the current through R3.

In this example, R1 and R2 form a voltage divider circuit, with R3 as the load (**Fig 4.7A**). The current drawn by the load (R3) is simply the voltage across R3, divided by its resistance. Unfortunately, the value of R2 affects the voltage across R3, just as the presence of R3 affects the potential appearing across R2. Some means of separating the two is needed; hence the Thevenin-equivalent circuit.

The voltage of the Thevenin-equivalent battery is the open-circuit voltage, measured when there is no current from either terminal A or B. Without a load connected between A and B, the total current through the circuit is (from Ohm's Law):

$$I = \frac{E}{R1 + R2} \quad (12)$$

and the voltage between terminals A and B (E_{AB}) is:

$$E_{AB} = I \times R2 \quad (13)$$

By substituting the first equation into the second, we can find a simplified expression for E_{AB}:

$$E_{AB} = \frac{R2}{R1 + R2} \times E \quad (14)$$

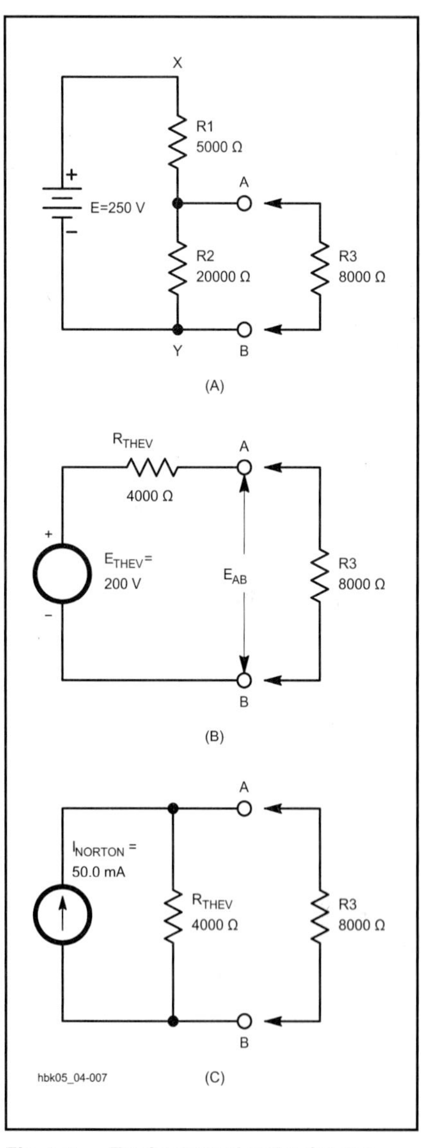

Fig 4.7 — Equivalent circuits for the circuit shown in Fig 4.6. A shows the load resistor (R3) looking into the circuit. B shows the Thevenin-equivalent circuit, with a resistor and a voltage source in series. C shows the Norton-equivalent circuit, with a resistor and current source in parallel.

Using the values in our example, this becomes:

$$E_{AB} = \frac{20.0\,k\Omega}{25.0\,k\Omega} \times 250\,V = 200\,V$$

when nothing is connected to terminals A or B. With no current drawn, E is equal to E_{AB}.

The Thevenin-equivalent resistance is the total resistance between terminals A and B. The ideal voltage source, by definition, has zero internal resistance. Assuming the battery to be a close approximation of an ideal source, put a short between points X and Y in the circuit of Fig 4.7A. R1 and R2 are then effectively placed in parallel, as viewed from terminals A and B. The Thevenin-equivalent resistance is then:

$$R_{THEV} = \frac{R1 \times R2}{R1 + R2} \qquad (15)$$

$$R_{THEV} = \frac{5000\,\Omega \times 20000\,\Omega}{5000\,\Omega + 20000\,\Omega}$$

$$R_{THEV} = \frac{1.00 \times 10^8\,\Omega^2}{25000\,\Omega} = 4000\,\Omega$$

This gives the Thevenin-equivalent circuit as shown in Fig 4.7B. The circuits of Figures 4.7A and 4.7B are equivalent as far as R3 is concerned.

Once R3 is connected to terminals A and B, there will be current through R_{THEV}, causing a voltage drop across R_{THEV} and reducing E_{AB}. The current through R3 is equal to

$$I3 = \frac{E_{THEV}}{R_{TOTAL}} = \frac{E_{THEV}}{R_{THEV} + R3} \qquad (16)$$

Substituting the values from our example:

$$I3 = \frac{200\,V}{4000\,\Omega + 8000\,\Omega} = 16.7\,mA$$

This agrees with the value calculated earlier.

NORTON'S THEOREM

Norton's Theorem is another tool for analyzing electrical networks. Norton's Theorem states that any two-terminal network of resistors and current or voltage sources can be replaced by a single current source and a parallel resistor. Norton's Theorem is to current sources what Thevenin's Theorem is to voltage sources. In fact, the Thevenin resistance calculated previously is also used as the Norton equivalent resistance.

The circuit just analyzed by means of Thevenin's Theorem can be analyzed just as easily by Norton's Theorem. The equivalent Norton circuit is shown in Fig 4.7C. The current I_{SC} of the equivalent current source is the short-circuit current through terminals A and B. In the case of the voltage divider shown in Fig 4.7A, the short-circuit current is:

$$I_{SC} = \frac{E}{R1} \qquad (17)$$

Substituting the values from our example, we have:

$$I_{SC} = \frac{E}{R1} = \frac{250\,V}{5000\,\Omega} = 50.0\,mA$$

The resulting Norton-equivalent circuit consists of a 50.0-mA current source placed in parallel with a 4000-Ω resistor. When R3 is connected to terminals A and B, one-third of the supply current flows through R3 and the remainder through R_{THEV}. This gives a current through R3 of 16.7 mA, again agreeing with previous conclusions.

A Norton-equivalent circuit can be transformed into a Thevenin-equivalent circuit and vice versa. The equivalent resistor stays the same in both cases; it is placed in series with the voltage source in the case of a Thevenin-equivalent circuit and in parallel with the current source in the case of a Norton-equivalent circuit. The voltage for a Thevenin-equivalent source is equal to the no-load voltage appearing across the resistor in the Norton-equivalent circuit. The current for a Norton-equivalent source is equal to the short-circuit current provided by the Thevenin source.

Power and Energy

Regardless of how voltage is generated, energy must be supplied if current is drawn from the voltage source. The energy supplied may be in the form of chemical energy or mechanical energy. This energy is measured in joules. One joule is defined from classical physics as the amount of energy or work done when a force of one newton (a measure of force) is applied to an object that is moved one meter in the direction of the force.

Power is another important concept. In the USA, power is often measured in horsepower in mechanical systems. We use the metric power unit of watts in electrical systems, however. In metric countries, mechanical power is usually expressed in watts also. One watt is defined as the use (or generation) of one joule of energy per second. One watt is also defined as one volt of potential pushing one ampere of current through a resistance. Thus,

$$P = I \times E \qquad (18)$$

where:
- P = power in watts
- I = current in amperes
- E = EMF in volts.

When current flows through a resistance, the electrical energy is turned into heat. Common fractional and multiple units for power are the milliwatt (one thousandth of a watt) and the kilowatt (1000 W).

Example: The plate voltage on a transmitting vacuum tube is 2000 V and the plate current is 350 mA. (The current must be changed to amperes before substitution in the formula, and so is 0.350 A.) Then:

$$P = I \times E = 2000\ V \times 0.350\ A = 700\ W$$

By substituting the Ohm's Law equivalent for E and I, the following formulas are obtained for power:

$$P = \frac{E^2}{R} \qquad (19)$$

and

$$P = I^2 \times R \qquad (20)$$

These formulas are useful in power calculations when the resistance and either the current or voltage (but not both) are known.

Example: How much power will be converted to heat in a 4000-Ω resistor if the potential applied to it is 200 V? From equation 19,

$$P = \frac{E^2}{R}$$

$$= \frac{40000\ V^2}{4000\ \Omega} = 10.0\ W$$

As another example, suppose a current

of 20 mA flows through a 300-Ω resistor. Then:

$$P = I^2 \times R = 0.020^2\ A^2 \times 300\ \Omega$$

$$P = 0.00040\ A^2 \times 300\ \Omega$$

$$P = 0.12\ W$$

Note that the current was changed from milliamperes to amperes before substitution in the formula.

Electrical power in a resistance is turned into heat. The greater the power, the more rapidly the heat is generated. Resistors for radio work are made in many sizes, the smallest being rated to dissipate (or carry safely) about $^1/_{16}$ W. The largest resistors commonly used in amateur equipment will dissipate about 100 W. Large resistors, such as those used in dummy-load antennas, are often cooled with oil to increase their power-handling capability.

If you want to express power in horsepower instead of watts, the following relationship holds:

$$1\ \text{horsepower} = 746\ W \quad (21)$$

This formula assumes lossless transformation; practical efficiency is taken up shortly. This formula is especially useful if you are working with a system that converts electrical energy into mechanical energy, and vice versa, since mechanical power is often expressed in horsepower, in the US.

This discussion relates to direct current in resistive circuits. See the **AC Theory and Reactive Components** section of this chapter for a discussion about power in ac circuits, including reactive circuits.

The Ohm's Law and Power Circle

During the first semester of my *Electrical Power Technology* program, one of the first challenges issued by our dedicated instructor—Roger Crerie—to his new freshman students was to identify and develop 12 equations or formulas that could be used to determine voltage, current, resistance and power. Ohm's Law is expressed as

$$R = \frac{E}{I}$$

and it provided three of these equation forms while the basic equation relating power to current and voltage (P = I×E) accounted for another three. With six known equations, it was just a matter of applying mathematical substitution for his students to develop the remaining six. Together, these 12 equations compose the *circle* or *wheel* of voltage (E), current (I), resistance (R) and power (P) shown in **Fig A**. Just as Roger's previous students had learned at the Worcester Industrial Technical Institute (Worcester, Massachusetts), our Class of '82 now held the basic electrical formulas needed to proceed in our studies or professions. As can be seen in Fig A, we can determine any one of these four electrical quantities by knowing the value of any two others. You may want to keep this page bookmarked for your reference. You'll probably be using many of these formulas as the years go by—this has certainly been my experience—*Dana G. Reed, W1LC, ARRL Handbook Editor*

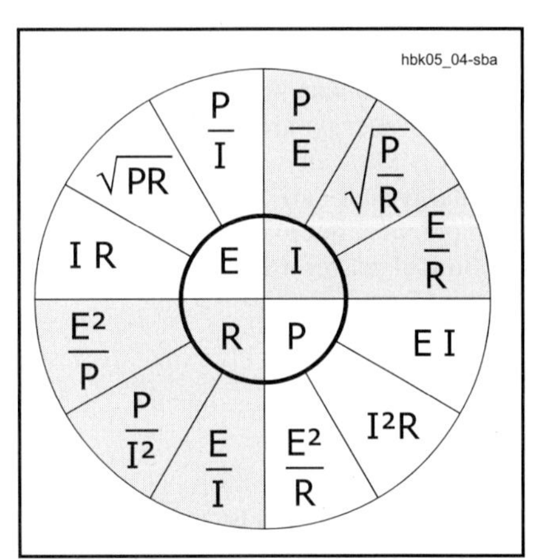

Fig A—Electrical formulas.

GENERALIZED DEFINITION OF RESISTANCE

Electrical energy is not always turned into heat. The energy used in running a motor, for example, is converted to mechanical motion. The energy supplied to a radio transmitter is largely converted into radio waves. Energy applied to a loudspeaker is changed into sound waves. In each case, the energy is converted to other forms and can be completely accounted for. None of the energy *just disappears*! This is a statement of the Law of Conservation of Energy. When a device converts energy from one form to another, we often say it *dissipates* the energy, or power. (Power is energy divided by time.) Of course the device doesn't really "use up" the energy, or make it disappear, it just converts it to another form. Proper operation of electrical devices often requires that the power must be supplied at a specific ratio of voltage to current. These features are characteristics of resistance, so it can be said that any device that "dissipates power" has a definite value of resistance.

This concept of resistance as something that absorbs power at a definite voltage-to-current ratio is very useful; it permits substituting a simple resistance for the load or power-consuming part of the device receiving power, often with considerable simplification of calculations. Of course, every electrical device has some resistance of its own in the more narrow sense, so a part of the energy supplied to it is converted to heat in that resistance even though the major part of the energy may be converted to another form.

EFFICIENCY

In devices such as motors and vacuum tubes, the objective is to convert the supplied energy (or power) into some form other than heat. Therefore, power converted to heat is considered to be a loss, because it is not useful power. The efficiency of a device is the useful power output (in its converted form) divided by the power input to the device. In a vacuum-tube transmitter, for example, the objective is to convert power from a dc source into ac power at some radio frequency. The ratio of the RF power output to the dc input is the efficiency of the tube. That is:

$$\text{Eff} = \frac{P_O}{P_I} \quad (22)$$

where:

Eff = efficiency (as a decimal)

P_O = power output (W)

P_I = power input (W).

Example: If the dc input to the tube is 100 W, and the RF power output is 60 W, the efficiency is:

$$\text{Eff} = \frac{P_O}{P_I} = \frac{60\ W}{100\ W} = 0.6$$

Efficiency is usually expressed as a percentage — that is, it tells what percent of the input power will be available as useful output. To calculate percent efficiency, just multiply the value from equation 22 by 100%. The efficiency in the example above is 60%.

Suppose a mobile transmitter has an RF power output of 100 W with 52% efficiency at 13.8 V. The vehicle's alternator system charges the battery at a 5.0-A rate at this voltage. Assuming an alternator efficiency of 68%, how much horsepower must the engine produce to operate the transmitter and charge the battery? Solution: To charge the battery, the alternator must produce 13.8 V × 5.0 A = 69 W. The transmitter dc input power is 100 W / 0.52 = 190 W. Therefore, the total electrical power required from the alternator is 190 + 69 = 260 W. The engine load then is:

$$P_I = \frac{P_O}{Eff} = \frac{260\ W}{0.68} = 380\ W$$

We can convert this to horsepower using the formula given earlier to convert between horsepower and watts:

$$380\ W \times \frac{1\ horsepower}{746\ W} = 0.51\ horsepower$$

ENERGY

When you buy electricity from a power company, you pay for electrical energy, not power. What you pay for is the *work* that electricity does for you, not the rate at which that work is done. Work is equal to power multiplied by time. The common unit for measuring electrical energy is the watt-hour, which means that a power of 1 W has been used for one hour. That is:

$$W\ hr = P\ T$$

where:
W hr = energy in watt-hours
P = power in watts
T = time in hours.

Actually, the watt-hour is a fairly small energy unit, so the power company bills you for kilowatt-hours of energy used. Another energy unit that is sometimes useful is the watt-second (joule).

Energy units are seldom used in amateur practice, but it is obvious that a small amount of power used for a long time can eventually result in a power bill that is just as large as if a large amount of power had been used for a very short time.

One practical application of energy units is to estimate how long a radio (such as a hand-held unit) will operate from a certain battery. For example, suppose a fully charged battery stores 900 mA hr of energy, and a radio draws 30 mA on receive. You might guess that the radio will receive 30 hrs with this battery, assuming 100% efficiency. You shouldn't expect to get the full 900 mA hr out of the battery, and you will probably spend some of the time transmitting, which will also reduce the time the battery will last. The **Real-World Component Characteristics** and **Power Supplies** chapters include additional information about batteries and their charge/discharge cycles.

Circuits and Components

SERIES AND PARALLEL CIRCUITS

Passive components (resistors for dc circuits) can be used to make voltage and current dividers and limiters to obtain a desired value. For instance, in **Fig 4.8A**, two resistors are connected in series to provide a voltage divider. As long as the device connected at point A has a much higher resistance than the resistors in the divider, the voltage will be approximately the ratio of the resistances. Thus, if E = 10 V, R1 = 5 Ω and R2 = 5 Ω, the voltage at point A will be 5 V measured on a high-impedance voltmeter. A good rule of thumb is that the load at point A should be at least ten times the value of the highest resistor in the divider to get reasonably close to the voltage you want. As the load resistance gets closer to the value of the divider, the current drawn by the load affects the division and causes changes from the desired value. If you need precise voltage division from fixed resistors and know the value of the load resistance, you can use Kirchhoff's Laws and Thevenin's Theorem (explained earlier) to calculate exact values.

Similarly, resistors can be used, as shown in Fig 4.8B, to make current dividers. Suppose you had two LEDs (light emitting diodes) and wanted one to glow twice as brightly as the other. You could use one resistor with twice the value of the other for the dimmer LED. Thus, approximately two-thirds of the current would flow through one LED and one-third through the other (neglecting any effect of the 0.7-V drop across the diode).

Resistors can also be used to limit the current through a device from a fixed voltage source. A typical example is shown in Fig 4.8C. Here a high-voltage source feeds a battery in a battery charger. This is typical of nickel cadmium chargers. The high resistor value limits the current that can possibly flow through the battery to a value that is low enough so it will not damage the battery.

SWITCHES

Switches are used to start or stop a signal (current) flowing in a particular circuit. Most switches are mechanical devices, although the same effect may be achieved with solid-state devices. Relays are switches that are controlled by another electrical signal rather than manual or mechanical means.

Switches come in many different forms and a wide variety of ratings. The most important ratings are the voltage and current handling capabilities. The voltage rating usually includes both the breakdown rating and the interrupt rating. Normally, the interrupt rating is the lower value, and therefore the one given on (for) the switch. The current rating includes both the current carrying capacity and the interrupt capability.

Most power switches are rated for alternating current use. Because ac voltage goes through zero with each cycle, switches can successfully interrupt much more alternating current than direct current without arcing. A switch that has a 10-A ac current rating may arc and dam-

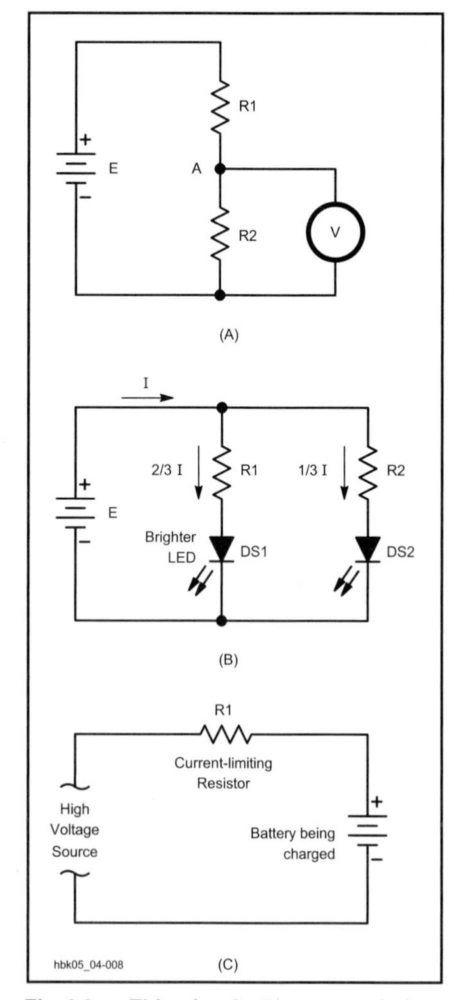

Fig 4.8 — This circuit shows a resistive voltage divider at A, a resistive current divider at B, and a current-limiting resistor at C.

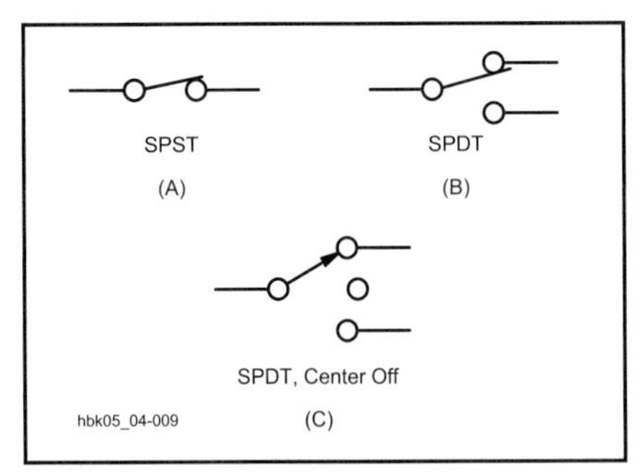

Fig 4.9 — Schematic diagrams of various types of switches. A is an SPST, B is an SPDT, and C is an SPDT switch with a center-off position.

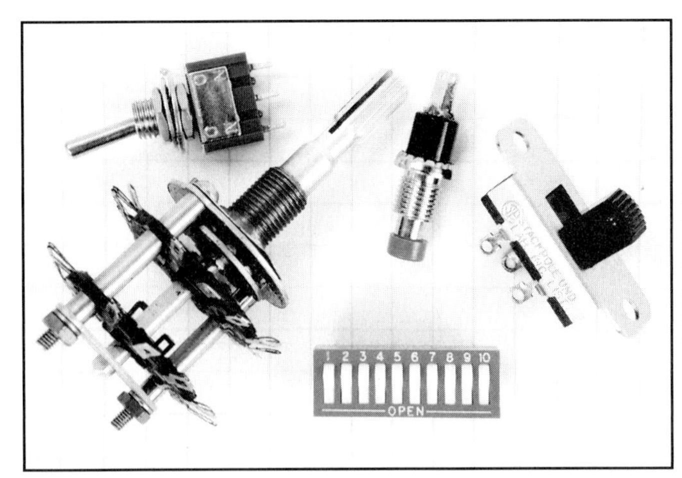

Fig 4.10 — This photo shows examples of various styles of switches. The ¼-inch-ruled graph paper background provides for size comparison.

age the contacts if used to turn off more than an ampere or two of dc.

Switches are normally designated by the number of *poles* (circuits controlled) and *positions* (circuit path choices). The simplest switch is the on-off switch, which is a single-pole, single-throw (SPST) switch as shown in **Fig 4.9A**. The off position does not direct the current to another circuit. The next step would be to change the current path to another path. This would be a single-pole, double-throw (SPDT) switch as shown in Fig 4.9B. Adding an off position would give a single-pole, double-throw, center-off switch as shown in Fig 4.9C.

Several such switches can be "ganged" to the same mechanical activator to provide double pole, triple pole or even more, separate control paths all activated at once. Switches can be activated in a variety of ways. The most common methods include lever, push button and rotary switches. Samples of these are shown in **Fig 4.10**. Most switches stay in the position set, but some are spring loaded so they only stay in the desired position while held there. These are called momentary switches.

Switches typically found in the home are usually rated for 125 V ac and 15 to 20 A. Switches in cars are usually rated for 12 V dc and several amperes. The breakdown voltage rating of a switch, which is usually higher than the interrupt rating, primarily depends on the insulating material surrounding the contacts and the separation between the contacts. Plastic or phenolic material normally provides both structural support and insulation. Ceramic material may be used to provide better insulation, particularly in rotary (wafer) switches.

The current carrying capacity of the switch depends on the contact material and size and on the pressure between the contacts. It is primarily determined from the allowable contact temperature rise. On larger ac switches, or most dc switches, the interrupt capability may be lower than the current carrying value.

Rotary/wafer switches can provide very complex switching patterns. Several poles (separate circuits) can be included on each wafer. Many wafers may be stacked on the same shaft. Not only may many different circuits be controlled at once, but by wiring different poles/positions on different wafers together, a high degree of circuit switching logic can be developed. Such switches can select different paths as they are turned and can also "short" together successive contacts to connect numbers of components or paths. They can also be designed to either break one contact before making another, or to short two contacts together before disconnecting the first one (make before break) to eliminate arcing or perform certain logic functions.

In choosing a switch for a particular task, consideration should be given to function, voltage and current ratings, ease of use, availability and cost. If a switch is to be operated frequently, a slightly higher cost for a better-quality switch is usually less costly over the long run. If signal noise or contact corrosion is a potential problem, (usually in low-current signal applications) it is best to get gold plated contacts. Gold does not oxidize or corrode, thus providing surer contact, which can be particularly important at very low signal levels. Gold plating will not hold up under high-current-interrupt applications, however.

FUSES

Fuses self-destruct to protect circuit wiring or equipment. The fuse element that melts is a carefully shaped piece of soft metal, usually mounted in a cartridge of some kind. The element is designed to safely carry a given amount of current and to melt at a current value that is a certain percentage over the rated value. The melting value depends on the type of material, the shape of the element and the heat dissipation capability of the cartridge and holder, among other factors. Some fuses (Slo-blo) are designed to carry an overload for a short period of time. They typically are used in motor starting and power-supply circuits that have a large inrush current when first started. Other fuses are designed to blow very quickly to protect delicate instruments and solid-state circuits. A replacement fuse should have the same current rating and the same characteristics as the fuse it replaces. **Fig 4.11** shows a variety of fuse types and sizes.

The most important fuse rating is the nominal current rating that it will safely carry. Next most important are the timing characteristics, or how quickly it opens under a given current overload. A fuse also has a voltage rating, both a value in volts and whether it is expected to be used in ac or dc circuits. While you should never substitute a fuse with a higher current rating than the one it replaces, you can use a fuse with a higher voltage rating. There is no danger in replacing a 12-V, 2-A fuse with a 250-V, 2-A unit.

Fuses fail for several reasons. The most obvious reason is that a problem develops in the circuit, which causes too much current to flow. In this case, the circuit problem needs to be fixed. A fuse may just fail eventually, particularly when cycled on and off near its current rating. A kind of metal fatigue sets in, and eventually the fuse goes. A fuse can also blow because of a momentary power surge, or even turning something on and off several times quickly when there is a large inrush current. In these cases it is only necessary to replace the fuse with the same type and value. Never substitute a fuse with a larger current rating. You may cause permanent

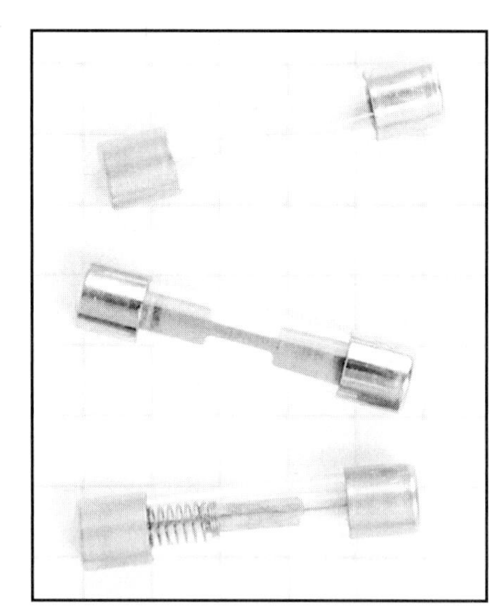

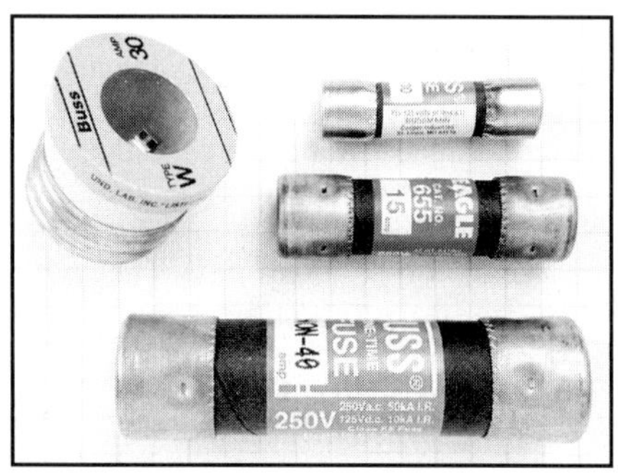

Fig 4.11 — These photos show examples of various styles of fuses. The ¼-inch-ruled graph paper background provides a size comparison.

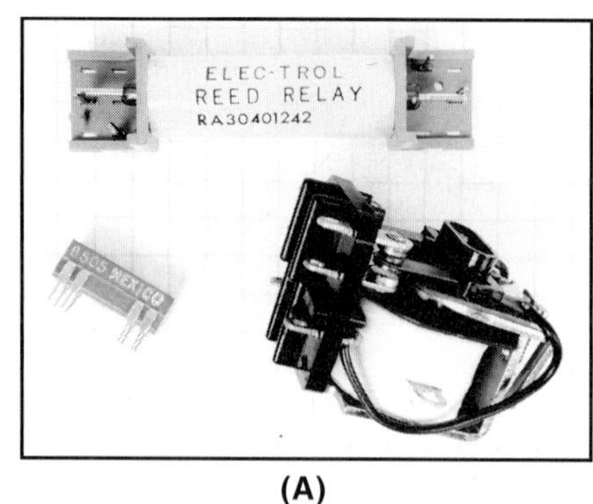

(A)

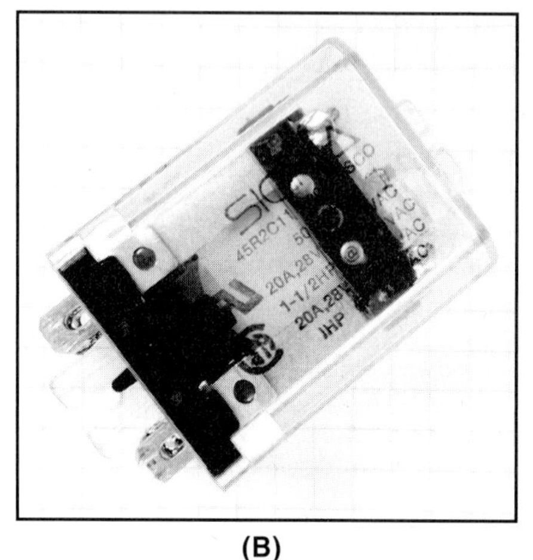

(B)

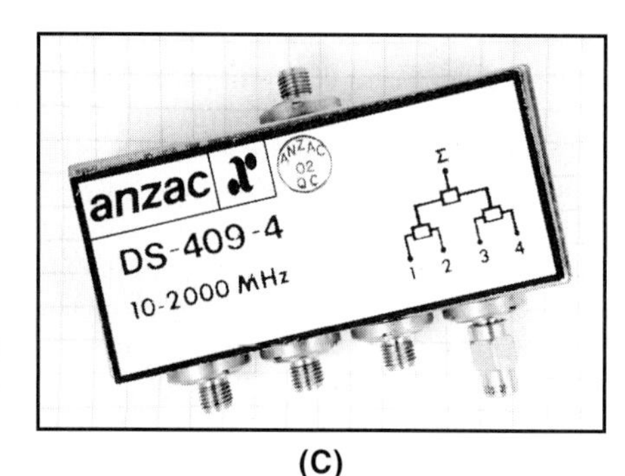

(C)

Fig 4.12 — These photos show examples of various styles and sizes of relays. Photo A shows a large reed relay, and a small reed relay in a package the size of a DIP IC. The contacts and coil can clearly be seen in the open-frame relay. Photo B shows a relay inside a plastic case. Photo C shows a four-position relay with SMA coaxial connectors. The ¼-inch-ruled graph paper background provides a size comparison.

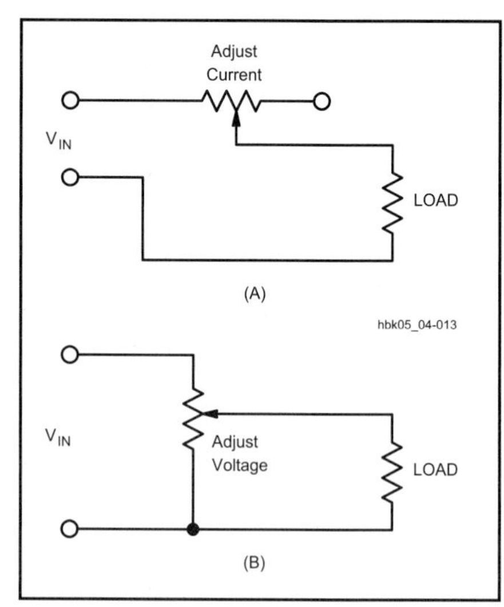

Fig 4.13 — Various uses of potentiometers.

damage (maybe even a fire) to the wiring or circuit elements if/when there is an internal problem in the equipment.

RELAYS

Relays are switches that are driven by an electrical signal, usually through a magnetic coil. An armature that moves when current is applied pushes the switch contacts together, or pulls them apart. Many such contacts can be connected to the same armature, allowing many circuits to be controlled by a single signal. Usually, relays have only two positions (opening some contacts and closing others) although there are special cases.

Like switches, relays have specific voltage and current ratings for the contacts. These may be far different from the voltage and current of the coil that drives the relay. That means a small signal voltage might control very large values of voltage and/or current. Relay contacts (and housings) may be designed for ac, dc or RF signals. The control voltages are usually 12 V dc or 125 V ac for most amateur applications, but the coils may be designed to be "current sensing" and operate when the current through the coil exceeds a specific value. **Fig 4.12** shows some typical relays found in amateur equipment. Relays with 24- and 28-V coils are also common.

Coaxial relays are specially designed to handle RF signals and to maintain a characteristic impedance to match certain values of coaxial-cable impedance. They typically are used to switch an antenna between a receiver and transmitter or between a linear amplifier and a transceiver.

POTENTIOMETERS

Potentiometer is a big name for a variable resistor. They are commonly used as volume controls on radios, televisions and stereos. A typical potentiometer is a circular pattern of resistive material, usually a carbon compound, that has a wiper on a shaft moving across the material. For higher power applications, the resistive material may be wire, wound around a core. As the wiper moves along the material, more resistance is introduced between the wiper and one of the fixed contacts on the material. A potentiometer may be used primarily to control current, voltage or resistance in a circuit. **Fig 4.13** shows several circuits to demonstrate various uses. **Fig 4.14** shows several different types of potentiometers.

Typical specifications for a potentiometer include maximum resistance, power dissipation, voltage and current ratings, number of turns (or degrees) the shaft can rotate, type and size of shaft, mounting arrangements and resistance "taper."

Not all potentiometers have a *linear* taper. That is, the resistance may not be the same for a given number of degrees of shaft rotation along different portions of the resistive material. A typical use of a potentiometer with a nonlinear taper is as

Fig 4.14 — This photo shows examples of different styles of potentiometers. The ¼ -inch-ruled graph paper background provides a size comparison.

a volume control. Since the human ear has a logarithmic response to sound, a volume control may actually change the volume (resistance) much more near one end of the potentiometer than the other (for a given amount of rotation) so that the "perceived" change in volume is about the same for a similar change in the control. This is commonly called an "audio taper" as the change in resistance per degree of rotation attempts to match the response of the human ear. The taper can be designed to match almost any desired control function for a given application. Linear and audio tapers are the most common.

AC Theory and Reactance Glossary

Admittance (Y) — The reciprocal of impedance, measured in siemens (S).

Capacitance (C) — The ability to store electrical energy in an electrostatic field, measured in farads (F). A device with capacitance is a capacitor.

Conductance (G)— The reciprocal of resistance, measured in siemens (S).

Current (I) — The rate of electron flow through a conductor, measured in amperes (A).

Flux density (B) — The number of magnetic-force lines per unit area, measured in gauss.

Frequency (f) — The rate of change of an ac voltage or current, measured in cycles per second, or hertz (Hz).

Impedance (Z) — The complex combination of resistance and reactance, measured in ohms (Ω).

Inductance (L) — The ability to store electrical energy in a magnetic field, measured in henrys (H). A device, such as a coil, with inductance is an inductor.

Peak (voltage or current) — The maximum value relative to zero that an ac voltage or current attains during any cycle.

Peak-to-peak (voltage or current) — The value of the total swing of an ac voltage or current from its peak negative value to its peak positive value, ordinarily twice the value of the peak voltage or current.

Period (T) — The duration of one ac voltage or current cycle, measured in seconds (s).

Permeability (μ) — The ratio of the magnetic flux density of an iron, ferrite, or similar core in an electromagnet compared to the magnetic flux density of an air core, when the current through the electromagnet is held constant.

Power (P) — The rate of electrical-energy use, measured in watts (W).

Q (quality factor) — The ratio of energy stored in a reactive component (capacitor or inductor) to the energy dissipated, equal to the reactance divided by the resistance.

Reactance (X) — Opposition to alternating current by storage in an electrical field (by a capacitor) or in a magnetic field (by an inductor), measured in ohms (Ω).

Resistance (R) — Opposition to current by conversion into other forms of energy, such as heat, measured in ohms (Ω).

Resonance — Ordinarily, the condition in an ac circuit containing both capacitive and inductive reactance in which the reactances are equal.

RMS (voltage or current) — Literally, "root mean square," the square root of the average of the squares of the instantaneous values for one cycle of a waveform. A dc voltage or current that will produce the same heating effect as the waveform. For a sine wave, the RMS value is equal to 0.707 times the peak value of ac voltage or current.

Susceptance (B) — The reciprocal of reactance, measured in siemens (S).

Time constant (τ) — The time required for the voltage in an RC circuit or the current in an RL circuit to rise from zero to approximately 63.2% of its maximum value or to fall from its maximum value 63.2% toward zero.

Toroid — Literally, any donut-shaped solid; most commonly referring to ferrite or powdered-iron cores supporting inductors and transformers.

Transducer — Any device that converts one form of energy to another; for example an antenna, which converts electrical energy to electromagnetic energy or a speaker, which converts electrical energy to sonic energy.

Transformer — A device consisting of at least two coupled inductors capable of transferring energy through mutual inductance.

Voltage (E) — Electromotive force or electrical pressure, measured in volts (V).

AC Theory and Reactive Components

AC IN CIRCUITS

A circuit is a complete conductive route for electrons to follow from a source, through a load and back to the source. If the source permits the electrons to flow in only one direction, the current is *dc* or *direct current*. If the source permits the current periodically to change direction, the current is *ac* or *alternating current*. **Fig 4.15** illustrates the two types of circuits. Drawing A shows the source as a battery, a typical dc source. Drawing B shows a more abstract source symbol to indicate ac. In an ac circuit, not only does the current change direction periodically; the voltage also periodically reverses. The rate of reversal may range from a few times per second to many billions per second.

Graphs of current or voltage, such as Fig 4.15, begin with a horizontal axis that represents time. The vertical axis represents the amplitude of the current or the voltage, whichever is graphed. Distance above the zero line means a greater positive amplitude; distance below the zero line means a greater negative amplitude. Positive and negative simply designate the opposing directions in which current may flow in an alternating current circuit or the opposing directions of force of an ac voltage.

If the current and voltage never change direction, then from one perspective, we have a dc circuit, even if the level of dc constantly changes. **Fig 4.16** shows a current that is always positive with respect to 0. It varies periodically in amplitude, however. Whatever the shape of the variations, the current can be called *pulsating dc*. If the current periodically reaches 0, it can be called *intermittent dc*. From another perspective, we may look at intermittent and pulsating dc as a combination of an ac and a dc current. Special circuits can separate the two currents into ac and dc components for separate analysis or use. There are also circuits that combine ac and dc currents and voltages for many purposes.

We can combine ac and dc voltages and currents. Different ac voltages and currents also form combinations. Such combinations will result in complex waveforms. A *waveform* is the pattern of amplitudes reached by the voltage or current as measured over time. **Fig 4.17** shows two ac waveforms fairly close in frequency, and their resultant combination. **Fig 4.18** shows two ac waveforms dissimilar in both frequency and wavelength, along with the resultant combined waveform. Note the similarities (and the differences) between the resultant waveform in Fig 4.18 and the combined ac-dc waveform in Fig 4.16.

Alternating currents may take on many useful wave shapes. **Fig 4.19** shows a few that are commonly used in practical circuits and in test equipment. The square

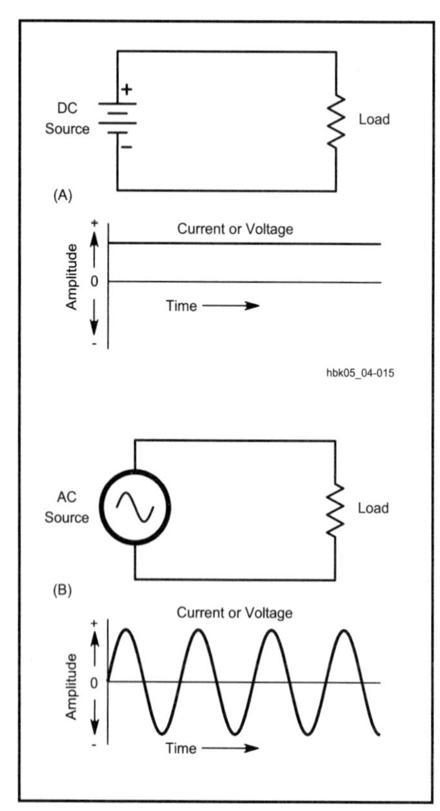

Fig 4.15 — Basic circuits for direct and alternating currents. With each circuit is a graph of the current, constant for the dc circuit, but periodically changing direction in the ac circuit.

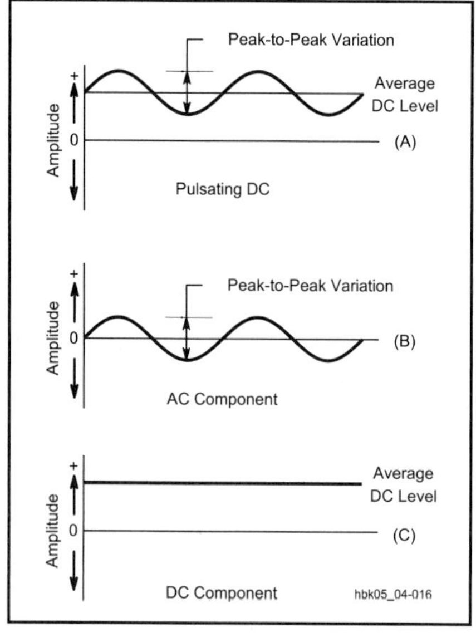

Fig 4.16 — A pulsating dc current (A) and its resolution into an ac component (B) and a dc component (C).

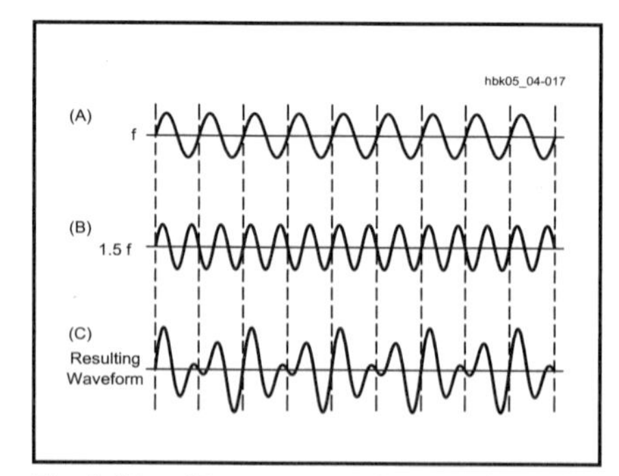

Fig 4.17 — Two ac waveforms of similar frequencies (f1 = 1.5 f2) and amplitudes form a composite wave. Note the points where the positive peaks of the two waves combine to create high composite peaks: this is the phenomenon of beats. The beat note frequency is 1.5f – f = 0.5f and is visible in the drawing.

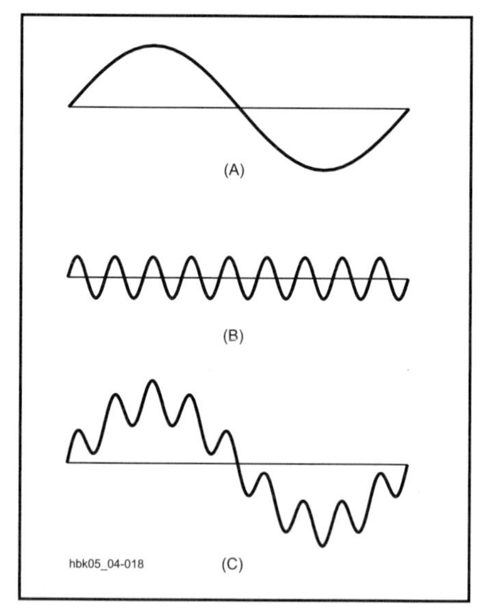

Fig 4.18 — Two ac waveforms of widely different frequencies and amplitudes form a composite wave in which one wave appears to ride upon the other.

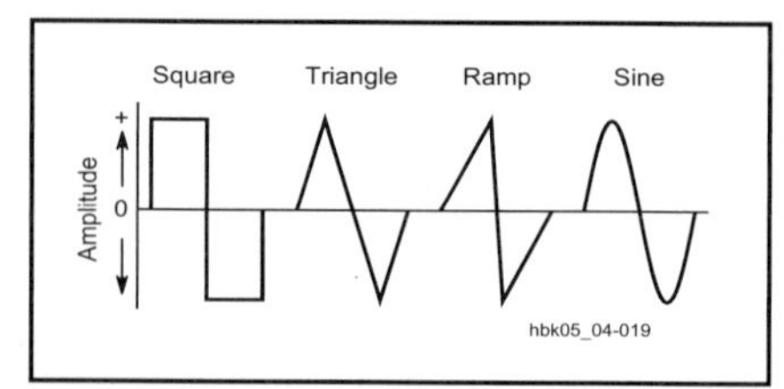

Fig 4.19 — Some common ac waveforms: square, triangle, ramp and sine.

wave is vital to digital electronics. The triangular and ramp waves — sometimes called "sawtooth" waves — are especially useful in timing circuits. The sine wave is both mathematically and practically the foundation of all other forms of ac; the other forms can usually be reduced to (and even constructed from) a particular collection of sine waves.

There are numerous ways to generate alternating currents: with an ac power generator (an *alternator*), with a transducer (for example, a microphone) or with an electronic circuit (for example, an RF oscillator). The basis of the sine wave is circular motion, which underlies the most usual methods of generating alternating current. The circular motion of the ac generator may be physical or mechanical, as in an alternator. Currents in the resonant circuit of an oscillator may also produce sine waves without mechanical motion.

Fig 4.20 demonstrates the relationship of the current (and voltage) amplitude to relative positions of a circular rotation through one complete revolution of 360°. Note that the current is zero at point 1. It rises to its maximum value at a point 90° from point 1, which is point 3. At a point 180° from point 1, which is point 4, the current level falls back to zero. Then the current begins to rise again. The direction of the current after point 4 and prior to its return to point 1, however, is opposite the direction of current from point 1 to point 4. Point 2 illustrates one of the innumerable intermediate values of current throughout the cycle.

Tracing the rise and fall of current over a linear time line produces the curve accompanying the circle in Fig 4.20. The curve is *sinusoidal* or a *sine wave*. The amplitude of the current varies as the sine of the angle made by the circular movement with respect to the zero point. The sine of 90° is 1, and 90° is also the point of maximum current (along with 270°). The sine of 45° (point 2) is 0.707, and the value of current at the 45° point of rotation is 0.707 times the maximum current. Similar considerations apply to the variation of ac voltage over time.

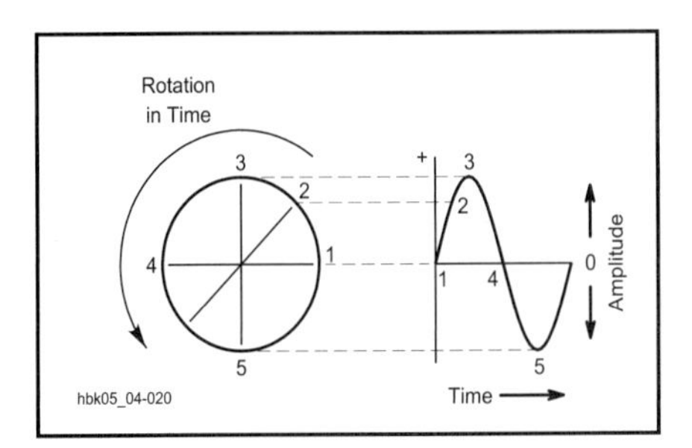

Fig 4.20 — The relationship of circular motion and the resultant graph of ac current or voltage. The curve is sinusoidal, a sine wave.

FREQUENCY AND PERIOD

With a continuously rotating generator, alternating current will pass through many equal cycles over time. Select an arbitrary point on any one cycle and use it as a marker. For this example, the positive peak will work as an unambiguous marker. The number of times per second that the current (or voltage) reaches this positive peak in any one second is called the *frequency* of the ac. In other words, frequency expresses the *rate* at which current (or voltage) cycles occur. The unit of frequency is *cycles per second*, or *hertz*—abbreviated Hz (after the 19th century radio-phenomena pioneer, Heinrich Hertz).

The length of any cycle in units of time is the *period* of the cycle, as measured from and to equivalent points on succeeding cycles. Mathematically, the period is simply the inverse of the frequency. That is,

$$\text{Frequency (f) in Hz} = \frac{1}{\text{Period (T) in seconds}} \tag{23}$$

and

$$\text{Period (T) in seconds} = \frac{1}{\text{Frequency (f) in Hz}} \tag{24}$$

Example: What is the period of a 400-hertz ac current?

$$T = \frac{1}{f} = \frac{1}{400\ \text{Hz}} = 0.00250\ \text{s} = 2.5\ \text{ms}$$

The frequency of alternating currents used in Amateur Radio circuits varies from a few hertz, or cycles per second, to thousands of millions of hertz. Likewise, the period of alternating currents amateurs use ranges from significant fractions of a second down to nanoseconds or smaller. In order to express units of frequency, time and almost everything else in electronics compactly, a standard system of prefixes is used. In magnitudes of 1000 or 10^3, frequency is measurable in hertz, in kilohertz (1000 hertz or kHz), in megahertz (1 million hertz or MHz), gigahertz (1 billion hertz or GHz) and even in tera-hertz (1 trillion hertz or THz). For units smaller than one, as in the measurement of period, the basic unit seconds can become milliseconds (1 thousandth of a second or ms), microseconds (1 millionth of a second or µs), nanoseconds (1 billionth of a second or ns) and picoseconds (1 trillionth of a second or ps).

The uses of ac in Amateur Radio circuits are many and varied. Most can be cataloged by reference to ac frequency ranges used in circuits. For example, ac power used in the home, office and factory is ordinarily 60 Hz in the United States and Canada. In Great Britain and much of Europe, ac power is 50 Hz. For special purposes, ac power has been generated up to about 400 Hz.

Sonic and ultrasonic applications of ac run from about 20 Hz up to several MHz. Audio work makes use of the lower end of the sonic spectrum, with communications audio focusing on the range from about 300 to 3000 Hz. High-fidelity audio uses ac circuits capable of handling 20 Hz to at least 20 kHz. Ultrasonics — used in medicine and industry — makes use of ac circuits above 20 kHz.

Amateur Radio circuits include both power- and sonic-frequency-range circuits. Radio communication and other electronics work, however, require ac circuits capable of operation with frequencies up to the gigahertz range. Some of the applications include signal sources for transmitters (and for circuits inside receivers); industrial induction heating; diathermy; microwaves for cooking, radar and communication; remote control of appliances, lighting, model planes and boats and other equipment; and radio direction finding and guidance.

AC IN CIRCUITS AND TRANSDUCED ENERGY

Alternating currents are often loosely classified as audio frequency (AF) and radio frequency (RF). Although these designations are handy, they actually represent something other than the electrical energy of ac circuits: They designate special forms of energy that we find useful.

Audio or *sonic* energy is the energy imparted by the mechanical movement of a medium, which can be air, metal, water or even the human body. Sound that humans can hear normally requires the movement of air between 20 Hz and 20 kHz, although the human ear loses its ability to detect the extremes of this range as we age. Some animals, such as elephants, can apparently detect air vibrations well below 20 Hz, while others, such as dogs and cats, can detect air vibrations well above 20 kHz.

Electrical circuits do not directly produce air vibrations. Sound production requires a *transducer*, a device to transform one form of energy into another form of energy; in this case electrical energy into sonic energy. The speaker and the microphone are the most common audio transducers. There are numerous ultrasonic transducers for various applications.

Likewise, converting electrical energy

into radio signals also requires a transducer, usually called an *antenna*. In contrast to RF alternating currents in circuits, RF *energy* is a form of electromagnetic energy. The frequencies of electromagnetic energy run from 3 kHz to above 10^{12} GHz. They include radio, infrared, visible light, ultraviolet and a number of energy forms of greatest interest to physicists and astronomers. **Table 4.2** provides a brief glimpse at the total spectrum of electromagnetic energy.

All electromagnetic energy has one thing in common: it travels, or *propagates,* at the speed of light. This speed is approximately 300000000 (or 3.00×10^8) meters per second in a vacuum. Electromagnetic-energy waves have a length uniquely associated with each possible frequency. The wavelength (λ) is simply the speed of propagation divided by the frequency (f) in hertz.

$$f\,(\text{Hz}) = \frac{3.00 \times 10^8 \left(\frac{\text{m}}{\text{s}}\right)}{\lambda(\text{m})} \qquad (25)$$

and

$$\lambda(\text{m}) = \frac{3.00 \times 10^8 \left(\frac{\text{m}}{\text{s}}\right)}{f\,(\text{Hz})} \qquad (26)$$

Example: What is the frequency of an 80.0-m RF wave?

$$f\,(\text{Hz}) = \frac{3.00 \times 10^8 \left(\frac{\text{m}}{\text{s}}\right)}{\lambda(\text{m})}$$

$$= \frac{3.00 \times 10^8 \left(\frac{\text{m}}{\text{s}}\right)}{80.0\ \text{m}}$$

$$f\,(\text{Hz}) = 3.75 \times 10^6\ \text{Hz}$$

We could use a similar equation to calculate the wavelength of a sound wave in air, but we would have to use the speed of sound instead of the speed of light in the numerator of the equation. The speed of propagation of the mechanical movement of air that we call sound varies considerably with air temperature and altitude. The speed of sound at sea level is about 331 m/s at 0°C and 344 m/s at 20°C.

To calculate the frequency of an electromagnetic wave directly in kilohertz, change the speed constant to 300,000 (3.00×10^5) km/s.

$$f\,(\text{kHz}) = \frac{3.00 \times 10^5 \left(\frac{\text{km}}{\text{s}}\right)}{\lambda(\text{m})} \qquad (27)$$

and

$$\lambda\,(\text{m}) = \frac{3.00 \times 10^5 \left(\frac{\text{km}}{\text{s}}\right)}{f\,(\text{kHz})} \qquad (28)$$

For frequencies in megahertz, use:

$$f\,(\text{MHz}) = \frac{300 \left(\frac{\text{Mm}}{\text{s}}\right)}{\lambda(\text{m})} \qquad (29)$$

and

$$\lambda(\text{m}) = \frac{300 \left(\frac{\text{Mm}}{\text{s}}\right)}{f\,(\text{MHz})} \qquad (30)$$

You would normally just drop the units that go with the speed of light constant to make the equation look simpler.

Example: What is the wavelength of an RF wave whose frequency is 4.0 MHz?

$$\lambda\,(\text{m}) = \frac{300}{f\,(\text{MHz})} = \frac{300}{4.0\ \text{MHz}} = 75\ \text{m}$$

At higher frequencies, circuit elements act like transducers. This property can be put to use, but it can also cause problems for some circuit operations. Therefore, wavelength calculations are of some importance in designing ac circuits for those frequencies.

Within the part of the electromagnetic-energy spectrum of most interest to radio applications, frequencies have been classified into groups and given names. **Table 4.3** provides a reference list of these classifications. To a significant degree, the frequencies within each group exhibit similar properties. For example, HF or high frequencies, from 3 to 30 MHz, all exhibit *skip* or ionospheric refraction that permits regular long-range radio communications. This property also applies occasionally both to MF (medium frequencies) and to VHF (very high frequencies).

Despite the close relationship between RF electromagnetic energy and RF ac circuits, it remains important to distinguish the two. To the ac circuit producing or amplifying a 15-kHz alternating current, the ultimate transformation and use of the electrical energy may make no difference to the circuit's operation. By choosing the right transducer, one can produce either an audio tone or a radio signal — or both. Such was the accidental fate of many horizontal oscillators and amplifiers in early television sets; they found ways to vibrate parts audibly and to radiate electromagnetic energy.

PHASE

When tracing a sine-wave curve of an ac voltage or current, the horizontal axis represents time. We call this the *time domain* of the sine wave. Events to the right take place later; events to the left occur earlier. Although time is measurable in parts of a second, it is more convenient

Table 4.2
Key Regions of the Electromagnetic Energy Spectrum

Region Name	*Frequency Range*
Radio frequencies	3.0×10^3 Hz to 3.0×10^{11} Hz
Infrared	3.0×10^{11} Hz to 4.3×10^{14} Hz
Visible light	4.3×10^{14} Hz to 7.5×10^{14} Hz
Ultraviolet	7.5×10^{14} Hz to 6.0×10^{16} Hz
X-rays	6.0×10^{16} Hz to 3.0×10^{19} Hz
Gamma rays	3.0×10^{19} Hz to 5.0×10^{20} Hz
Cosmic rays	5.0×10^{20} Hz to 8.0×10^{21} Hz

Table 4.3
Classification of the Radio Frequency Spectrum

Abbreviation	*Classification*	*Frequency Range*
VLF	Very low frequencies	3 to 30 kHz
LF	Low frequencies	30 to 300 kHz
MF	Medium frequencies	300 to 3000 kHz
HF	High frequencies	3 to 30 MHz
VHF	Very high frequencies	30 to 300 MHz
UHF	Ultrahigh frequencies	300 to 3000 MHz
SHF	Superhigh frequencies	3 to 30 GHz
EHF	Extremely high frequencies	30 to 300 GHz

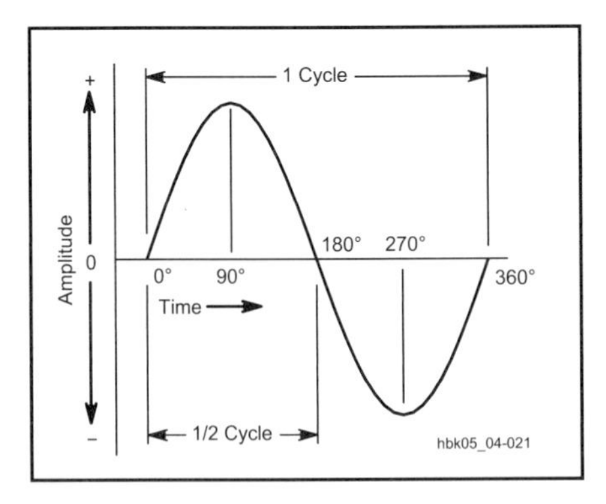

Fig 4.21 — An ac cycle is divided into 360° that are used as a measure of time or phase.

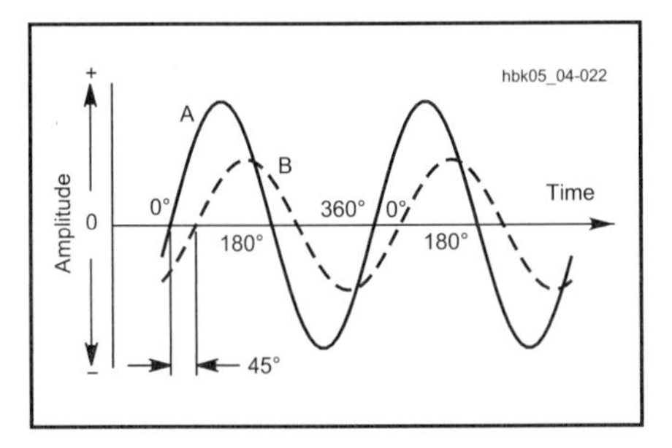

Fig 4.22 — When two waves of the same frequency start their cycles at slightly different times, the time difference or phase difference is measured in degrees. In this drawing, wave B starts 45° (one-eighth cycle) later than wave A, and so lags 45° behind A.

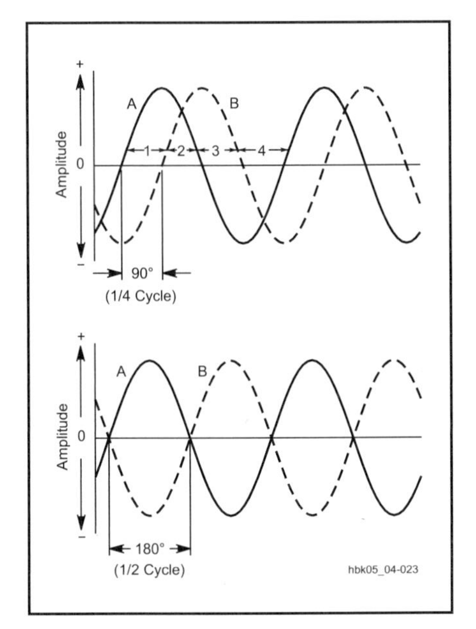

Fig 4.23 — Two important special cases of phase difference: In the upper drawing, the phase difference between A and B is 90°; in the lower drawing, the phase difference is 180°.

to treat each cycle as a complete time unit that we divide into 360°. The conventional starting point for counting degrees is the zero point as the voltage or current begins the positive half cycle. The essential elements of an ac cycle appear in **Fig 4.21**.

The advantage of treating the ac cycle in this way is that many calculations and measurements can be taken and recorded in a manner that is independent of frequency. The positive peak voltage or current occurs at 90° along the cycle. Relative to the starting point, 90° is the *phase* of the ac at that point. Thus, a complete description of an ac voltage or current involves reference to three properties: frequency, amplitude and phase.

Phase relationships also permit the comparison of two ac voltages or currents at the same frequency, as **Fig 4.22** demonstrates. Since B crosses the zero point in the positive direction after A has already done so, there is a *phase difference* between the two waves. In the example, B *lags* A by 45°, or A *leads* B by 45°. If A and B occur in the same circuit, their composite waveform will also be a sine wave at an intermediate phase angle relative to each. Adding any number of sine waves of the same frequency always results in a sine wave at that frequency.

Fig 4.22 might equally apply to a voltage and a current measured in the same ac circuit. Either A or B might represent the voltage; that is, in some instances voltage will lead the current and in others voltage will lag the current.

Two important special cases appear in **Fig 4.23**. In Part A, line B lags 90° behind line A. Its cycle begins exactly one quarter cycle later than the A cycle. When one wave is passing through zero, the other just reaches its maximum value.

In Part B, lines A and B are 180° *out of phase*. In this case, it does not matter which one is considered to lead or lag. Line B is always positive while line A is negative, and vice versa. If the two waveforms are of two voltages or two currents in the same circuit and if they have the same amplitude, they will cancel each other completely.

MEASURING AC VOLTAGE, CURRENT AND POWER

Measuring the voltage or current in a dc circuit is straightforward, as **Fig 4.24A** demonstrates. Since the current flows in only one direction, for a resistive load, the voltage and current have constant values until the circuit components change.

Fig 4.24B illustrates a perplexing problem encountered when measuring voltages and currents in ac circuits. The current and voltage continuously change direction and value. Which values are meaningful? In fact, several values of constant sine-wave voltage and current in ac circuits are important to differing applications and concerns.

Instantaneous Voltage and Current

Fig 4.25 shows a sine wave of some arbitrary frequency and amplitude with respect to either voltage or current. The instantaneous voltage (or current) at point A on the curve is a function of three factors: the maximum value of voltage (or current) along the curve (point B), the frequency of the wave, and the time elapsed in seconds or fractions of a second. Thus,

$$E_{inst} = E_{max} \sin(2\pi ft)\,\theta \quad (31)$$

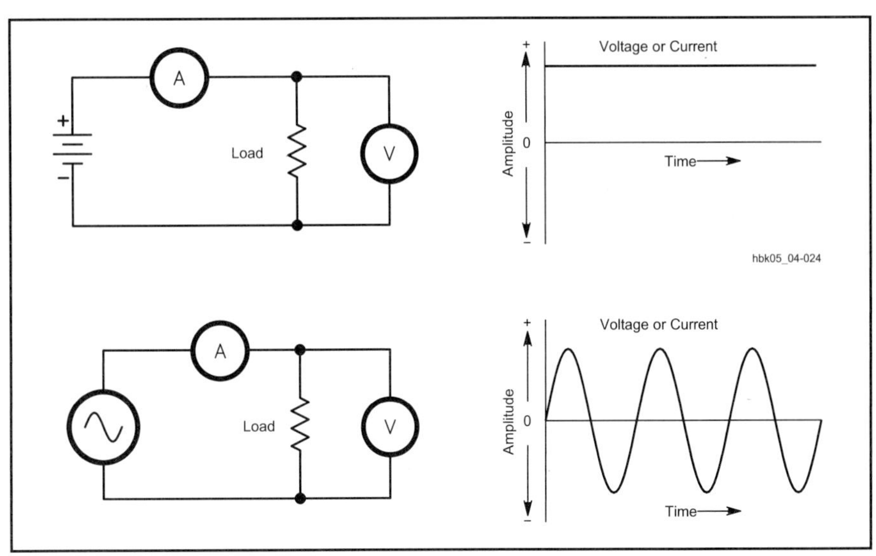

Fig 4.24 — Voltage and current measurements in dc and ac circuits.

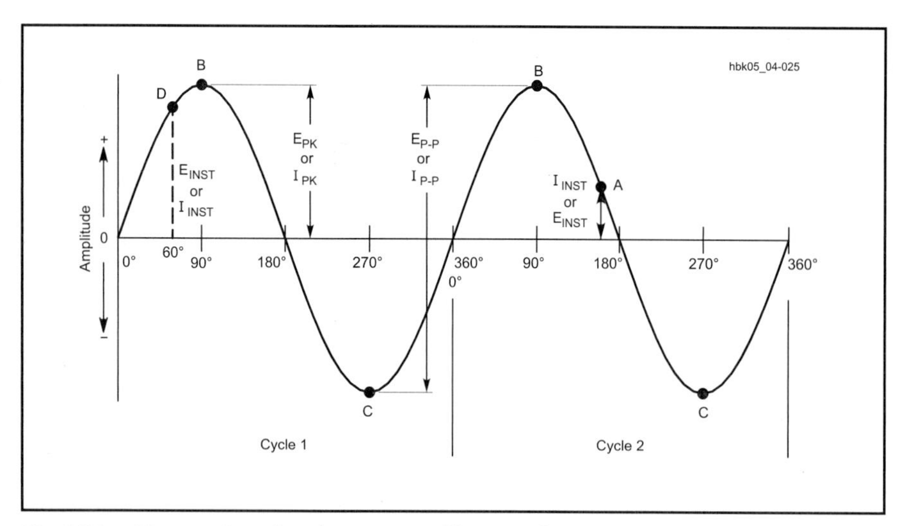

Fig 4.25 — Two cycles of a sine wave to illustrate instantaneous, peak, and peak-to-peak ac voltage and current values.

Considering just one sine wave, independent of frequency, the instantaneous value of voltage (or current) becomes

$$E_{inst} = E_{max} \sin \theta \tag{32}$$

where θ is the angle in degrees through which the voltage has moved over time after the beginning of the cycle.

Example: What is the instantaneous value of voltage at point D in Fig 4.25, if the maximum voltage value is 120 V and the angular travel is 60.0°?

$$E_{inst} = 120 \text{ V} \times \sin 60.0°$$
$$= 120 \times 0.866 = 104 \text{ V}$$

Peak and Peak-to-Peak Voltage and Current

The most important instantaneous voltages and currents are the maximum or peak values reached on each positive and negative half cycle of the sine wave. In Fig 4.25, points B and C represent the positive and negative peaks of voltage or current. Peak (pk) values are especially important with respect to component ratings, which the voltage or current in a circuit must not exceed without danger of component failure.

The peak power in an ac circuit is simply the product of the peak voltage and the peak current, or

$$P_{pk} = E_{pk} \times I_{pk} \tag{33}$$

The span from points B to C in Fig 4.25 represents the largest voltage or current swing of the sine wave. Designated the *peak-to-peak* (P-P) voltage (or current), this span is equal to twice the peak value of the voltage (or current). Thus,

$$E_{P-P} = 2E_{pk} \tag{34}$$

Amplifying devices often specify their input limits in terms of peak-to-peak voltages. Operational amplifiers, which have almost unlimited gain potential, often require input-level limiting to prevent the output signals from distorting if they exceed the peak-to-peak output rating of the devices.

RMS Voltages and Currents

The *root mean square* or *RMS* values of voltage and current are the most common values encountered in electronics. Sometimes called the *effective* values of ac voltage and current, they are based upon equating the values of ac and dc power required to heat a resistive element to exactly the same temperature. The peak ac power required for this condition is twice the dc power needed. Therefore, the average ac power equivalent to a corresponding average dc power is half the peak ac power.

$$P_{ave} = \frac{P_{pk}}{2} \tag{35}$$

Since a circuit with a constant resistance is linear — that is, raising or lowering the voltage will raise or lower the current proportionally — the voltage and current values needed to arrive at average ac power are related to their peak values by the factor.

$$E_{RMS} = \frac{E_{pk}}{\sqrt{2}} = \frac{E_{pk}}{1.414} = E_{pk} \times 0.707 \tag{36}$$

$$I_{RMS} = \frac{I_{pk}}{\sqrt{2}} = \frac{I_{pk}}{1.414} = I_{pk} \times 0.707 \tag{37}$$

In the time domain of a sine wave, the RMS values of voltage and current occur at the 45°, 135°, 225° and 315° points along the cycle shown in **Fig 4.26**. (The sine of 45° is approximately 0.707.) The absolute instantaneous value of voltage or current is greater than the RMS value for half the cycle and less than the RMS value for half the cycle.

The RMS values of voltage and current get their name from the means used to derive their value relative to peak voltage and current. Square the individual values of all the instantaneous values of voltage or current in a single cycle of ac. Take the average of these squares and then find the square root of the average. This *root mean square* procedure produces the RMS value of voltage or current.

If the RMS voltage is the peak voltage divided by the $\sqrt{2}$, then the peak voltage must be the RMS voltage multiplied by

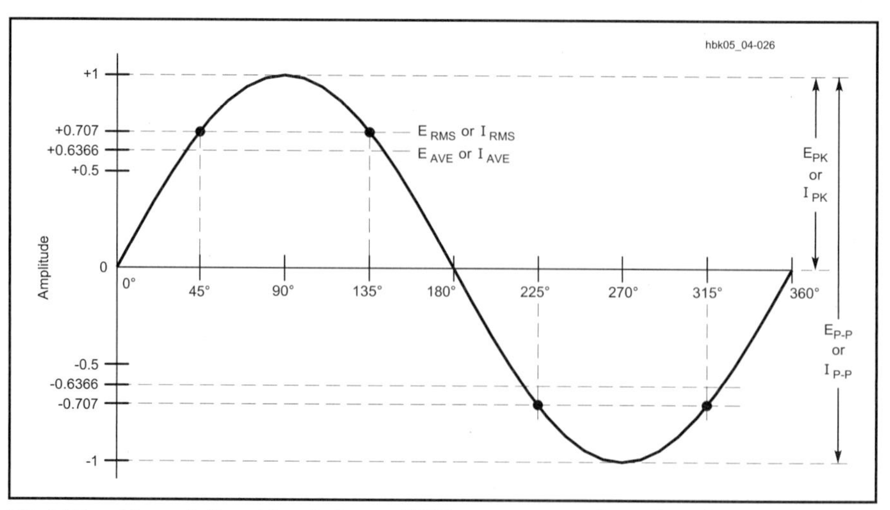

Fig 4.26 — The relationships between RMS, average, peak, and peak-to-peak values of ac voltage and current.

Table 4.4
Conversion Factors for AC Voltage or Current

From	*To*	*Multiply By*
Peak	Peak-to-Peak	2
Peak-to-Peak	Peak	0.5
Peak	RMS	$1 / \sqrt{2}$ or 0.707
RMS	Peak	$\sqrt{2}$ or 1.414
Peak-to-Peak	RMS	$1 / (2 \times \sqrt{2})$ or 0.35355
RMS	Peak-to-Peak	$2 \times \sqrt{2}$ or 2.828
Peak	Average	$2 / \pi$ or 0.6366
Average	Peak	$\pi / 2$ or 1.5708
RMS	Average	$(2 \times \sqrt{2}) / \pi$ or 0.90
Average	RMS	$\pi / (2 \times \sqrt{2})$ or 1.11

Note: These conversion factors apply only to continuous pure sine waves.

the $\sqrt{2}$, or

$$Epk = ERMS \times 1.414 \quad (38)$$

$$I_{pk} = I_{RMS} \times 1.414 \quad (39)$$

Since circuit specifications will most commonly list only RMS voltage and current values, these relationships are important in finding the peak voltages or currents that will stress components.

Example: What is the peak voltage on a capacitor if the RMS voltage of a sinusoidal waveform signal across it is 300 V ac?

$$E_{pk} = 300\ V \times 1.414 = 424\ V$$

The capacitor must be able to withstand this higher voltage, plus a safety margin. The capacitor must also be rated for ac use. A capacitor rated for 1 kV dc may explode if used in this application. In power supplies that convert ac to dc and use capacitive input filters, the output voltage will approach the peak value of the ac voltage rather than the RMS value.

Example: What is the peak voltage and the peak-to-peak voltage at the usual household ac outlet, if the RMS voltage is 120 V?

$$E_{pk} = 120\ V \times 1.414 = 170\ V$$

$$E_{p-p} = 2 \times 170\ V = 340\ V$$

Unless otherwise specified, unlabeled ac voltage and current values found in most electronics literature are normally RMS values.

Average Values of Voltage and Current

Certain kinds of circuits respond to the *average* value of an ac waveform. Among these circuits are electrodynamic meter movements and power supplies that convert ac to dc and use heavily inductive ("choke") input filters, both of which use the pulsating dc output of a full-wave rectifier. The average value of each ac half cycle is the *mean* of all the instantaneous values in that half cycle. Related to the peak values of voltage and current, average values are $2 / \pi$ (or 0.6366) times the peak value.

$$E_{ave} = 0.6366\, E_{pk} \quad (40)$$

$$I_{ave} = 0.6366\, I_{pk} \quad (41)$$

For convenience, **Table 4.4** summarizes the relationships between all of the common ac values. All of these relationships apply only to pure sine waves.

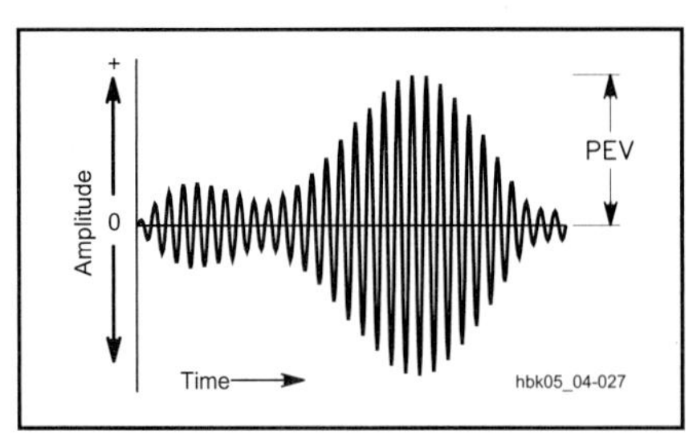

Fig 4.27 — The peak envelope voltage (PEV) for a composite waveform.

Complex Waves and Peak-Envelope Values

Complex waves, as shown earlier in Fig 4.18, differ from pure sine waves. The amplitude of the peak voltage may vary significantly from one cycle to the next. Therefore, other amplitude measures are required, especially for accurate measurement of voltage and power with single sideband (SSB) waveforms. **Fig 4.27** illustrates a multitone composite waveform with an RF ac waveform as the basis.

The RF ac waveform has a frequency many times that of the audio-frequency ac waveform with which it is usually combined in SSB operations. Therefore, the resultant waveform appears as an amplitude envelope superimposed upon the RF waveform. The *peak envelope voltage* (PEV), then, is the maximum or peak value of voltage achieved.

Peak envelope voltage permits the calculation of *peak envelope power* (PEP). The Federal Communications Commission (FCC) uses the concept of peak envelope power to set the maximum power standards for amateur transmitters. PEP is the *average* power supplied to the antenna transmission line by a transmitter during one RF cycle at the crest of the modulation envelope, taken under normal operating conditions. Since calculation of PEP requires the average power of the cycle, multiply the PEV by 0.707 to obtain the RMS value. Then calculate power by using the square of the voltage divided by the load resistance.

$$PEP = \frac{(PEV \times 0.707)^2}{R} \quad (42)$$

Capacitance and Capacitors

Without the ability to store electrical energy, radio would not be possible. One may build and hold an electrical charge in an *electrostatic field*. This phenomenon is called *capacitance*, and the devices that exhibit capacitance are called *capacitors*. See Chapter 6 for more information on practical capacitor applications and problems. **Fig 4.28** shows several schematic symbols for capacitors. Part A shows a fixed capacitor; one that has a single value of capacitance. Part B shows variable capacitors; these are adjustable over a range of values. Ordinarily, the straight line in each symbol connects to a positive voltage, while the curved line goes to a negative voltage or to ground. Some capacitor designs require rigorous adherence to polarity markings; other designs are symmetrical and non-polarized.

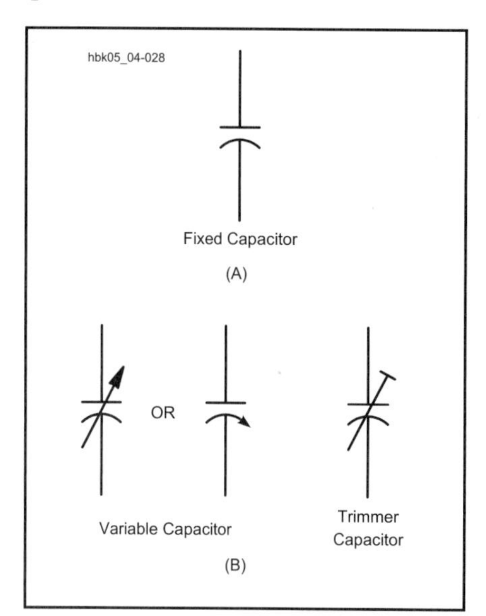

Fig 4.28 — Schematic symbol for a fixed capacitor is shown at A. The symbols for a variable capacitor are shown at B.

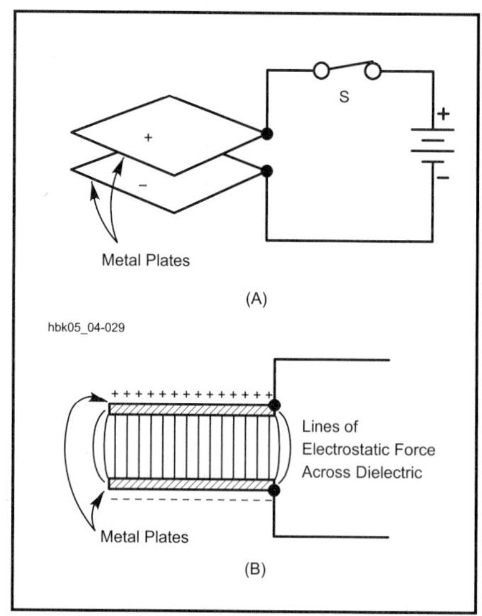

Fig 4.29 — A simple capacitor showing the basic charging arrangement at A, and the retention of the charge due to the electrostatic field at B.

CHARGE AND ELECTROSTATIC ENERGY STORAGE

Suppose two flat metal plates are placed close to each other (but not touching) and are connected to a battery through a switch, as illustrated in **Fig 4.29A**. At the instant the switch is closed, electrons are attracted from the upper plate to the positive terminal of the battery, and the same number are repelled into the lower plate from the negative battery terminal. Enough electrons move into one plate and out of the other to make the voltage between the plates the same as the battery voltage.

If the switch is opened after the plates have been charged in this way, the top plate is left with a deficiency of electrons and the bottom plate with an excess. Since there is no current path between the two, the plates remain charged despite the fact that the battery no longer is connected. The charge remains due to the electrostatic field between the plates. The large number of opposite charges exert an attractive force across the small distance between plates, as illustrated in Fig 4.29B.

If a wire is touched between the two plates (short-circuiting them), the excess electrons on the bottom plate flow through the wire to the upper plate, restoring electrical neutrality. The plates are discharged.

These two plates represent an electrical capacitor, a device possessing the property of storing electrical energy in the electric field between its plates. During the time the electrons are moving — that is, while the capacitor is being charged or discharged — a current flows in the circuit even though the circuit apparently is broken by the gap between the capacitor

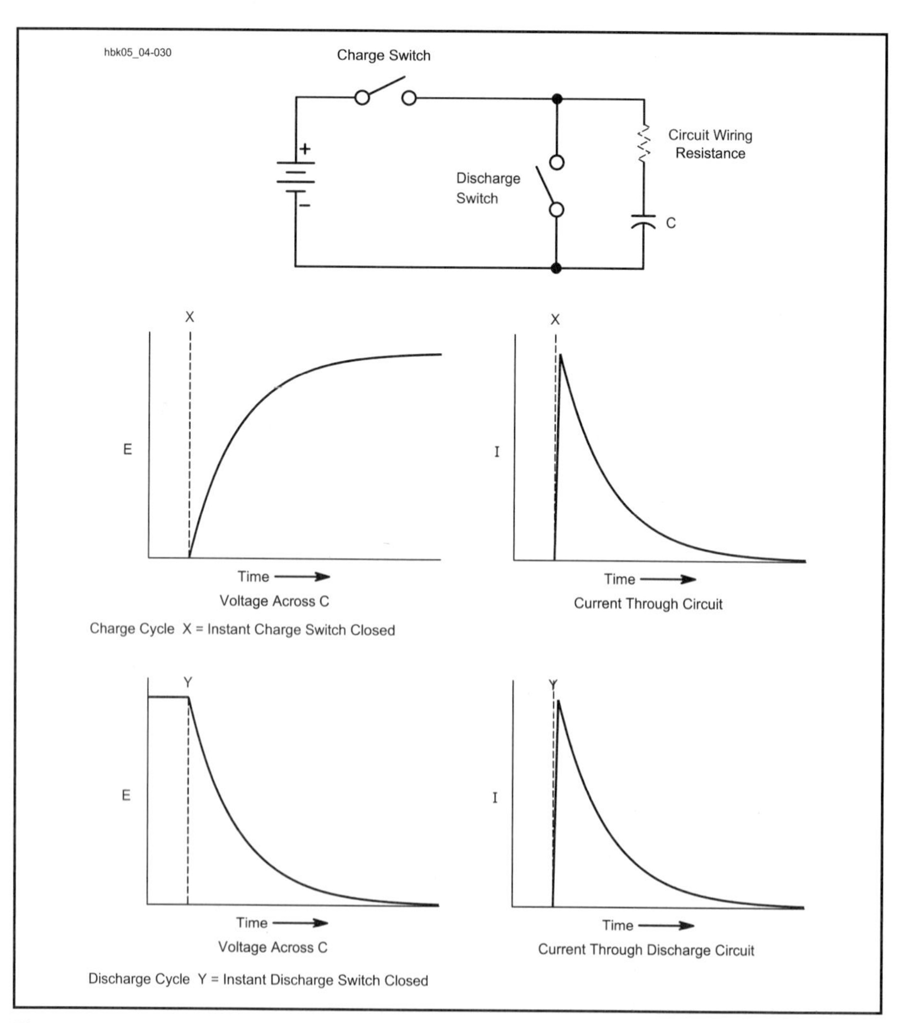

Fig 4.30 — The flow of current during the charge and discharge of a capacitor. The charge graphs assume that the charge switch is closed and the discharge switch is open. The discharge graphs assume just the opposite.

plates. The current flows only during the time of charge and discharge, however, and this time is usually very short. There can be no continuous flow of direct current through a capacitor.

Fig 4.30 demonstrates the voltage and current in the circuit, first, at the moment the switch is closed to charge the capacitor and, second, at the moment the shorting switch is closed to discharge the unit. Note that the periods of charge and discharge are very short, but that they are not zero. This finite charging and discharging time can be lengthened and will prove useful later in timing circuits.

Although dc cannot pass through a capacitor, alternating current can. As fast as one plate is charged positively by the positive excursion of the alternating current, the other plate is being charged negatively. Positive charges flowing into one plate causes a current to flow out of the other plate during one half of the cycle, resulting in a negative charge on that plate. The reverse occurs during the second half of the cycle.

The charge or quantity of electricity that can be held on the capacitor plates is proportional to the applied voltage and to the capacitance of the capacitor:

$$Q = CE \qquad (43)$$

where:
Q = charge in coulombs,
C = capacitance in farads, and
E = electrical potential in volts.

The energy stored in a capacitor is also a function of electrical potential and capacitance:

$$W = \frac{E^2 C}{2} \qquad (44)$$

where:
W = energy in joules (watt-seconds),
E = electrical potential in volts (some texts use V instead of E), and
C = capacitance in farads.

The numerator of this expression can be derived easily from the definitions for charge, capacitance, current, power and energy. The denominator is not so obvious, however. It arises because the voltage across a capacitor is not constant, but is a function of time. The average voltage over the time interval determines the energy stored. The time dependence of the capacitor voltage is a very useful property; see the section on time constants.

UNITS OF CAPACITANCE AND CAPACITOR CONSTRUCTION

A capacitor consists, fundamentally, of two plates separated by an insulator or *dielectric*. The *larger* the plate area and the *smaller* the spacing between the plates, the *greater* the capacitance. The capacitance also depends on the kind of insulating material between the plates; it is smallest with air insulation or a vacuum. Substituting other insulating materials for air may greatly increase the capacitance.

The ratio of the capacitance with a material other than a vacuum or air between the plates to the capacitance of the same capacitor with air insulation is called the dielectric constant, or K, of that particular insulating material. The dielectric constants of a number of materials commonly used as dielectrics in capacitors are given in **Table 4.5**. For example, if a sheet of polystyrene is substituted for air between the plates of a capacitor, the capacitance will be 2.6 times greater.

Table 4.5
Relative Dielectric Constants of Common Capacitor Dielectric Materials

Material	*Dielectric Constant (k)*	*(O)rganic or (I)norganic*
Vacuum	1 (by definition)	I
Air	1.0006	I
Ruby mica	6.5 - 8.7	I
Glass (flint)	10	I
Barium titanate (class I)	5 - 450	I
Barium titanate (class II)	200 - 12000	I
Kraft paper	≈ 2.6	O
Mineral Oil	≈ 2.23	O
Castor Oil	≈ 4.7	O
Halowax	≈ 5.2	O
Chlorinated diphenyl	≈ 5.3	O
Polyisobutylene	≈ 2.2	O
Polytetrafluoroethylene	≈ 2.1	O
Polyethylene terephthalate	≈ 3	O
Polystyrene	≈ 2.6	O
Polycarbonate	≈ 3.1	O
Aluminum oxide	≈ 8.4	I
Tantalum pentoxide	≈ 28	I
Niobium oxide	≈ 40	I
Titanium dioxide	≈ 80	I

(Adapted from: Charles A. Harper, *Handbook of Components for Electronics*, p 8-7.)

The basic unit of capacitance, the ability to store electrical energy in an electrostatic field, is the farad. This unit is generally too large for practical radio work, however. Capacitance is usually measured in microfarads (abbreviated µF), nanofarads (abbreviated nF) or picofarads (pF). The microfarad is one millionth of a farad (10^{-6} F), the nanofarad is one thousandth of a microfarad (10^{-9} F) and the picofarad is one millionth of a microfarad (10^{-12} F).

In practice, capacitors often have more than two plates, the alternate plates being connected to form two sets, as shown in **Fig 4.31**. This practice makes it possible to obtain a fairly large capacitance in a small space, since several plates of smaller individual area can be stacked to form the equivalent of a single large plate of the same total area. Also, all plates except the two on the ends are exposed to plates of the other group on both sides, and so are twice as effective in increasing the capacitance.

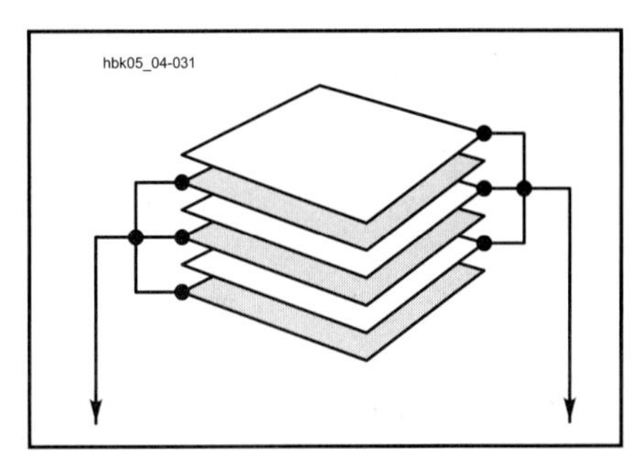

Fig 4.31 — A multiple-plate capacitor. Alternate plates are connected to each other.

The formula for calculating capacitance from these physical properties is:

$$C = \frac{0.2248\, K\, A\, (n-1)}{d} \qquad (45)$$

where:
C = capacitance in pF,
K = dielectric constant of material between plates,
A = area of one side of one plate in square inches,

d = separation of plate surfaces in inches, and

n = number of plates.

If the area (A) is in square centimeters and the separation (d) is in centimeters, then the formula for capacitance becomes

$$C = \frac{0.0885\,K\,A\,(n-1)}{d} \qquad (46)$$

If the plates in one group do not have the same area as the plates in the other, use the area of the smaller plates.

Example: What is the capacitance of 2 copper plates, each 1.50 square inches in area, separated by a distance of 0.00500 inch, if the dielectric is air?

$$C = \frac{0.2248\,K\,A\,(n-1)}{d}$$

$$C = \frac{0.2248 \times 1 \times 1.50\,(2-1)}{0.00500}$$

$$C = 67.4\text{ pF}$$

KINDS OF CAPACITORS AND THEIR USES

The capacitors used in radio work differ considerably in physical size, construction and capacitance. Representative kinds are shown in **Fig 4.32**. In variable capacitors, which are almost always constructed with air for the dielectric, one set of plates is made movable with respect to the other set so the capacitance can be varied. Fixed capacitors — those having a single, nonadjustable value of capacitance — can also be made with metal plates and with air as the dielectric.

Fixed capacitors are usually constructed from plates of metal foil with a thin solid or liquid dielectric sandwiched between, so a relatively large capacitance can be obtained in a small unit. The solid dielectrics commonly used are mica, paper and special ceramics. An example of a liquid dielectric is mineral oil. Electrolytic capacitors use aluminum-foil plates with a semiliquid conducting chemical compound between them. The actual dielectric is a very thin film of insulating material that forms on one set of plates through electrochemical action when a dc voltage is applied to the capacitor. The capacitance obtained with a given plate area in an electrolytic capacitor is very large compared to capacitors having other dielectrics, because the film is so thin — much less than any thickness practical with a solid dielectric.

The use of electrolytic and oil-filled capacitors is confined to power-supply filtering and audio-bypass applications because their dielectrics have high losses at higher frequencies. Mica and ceramic capacitors are used throughout the frequency range from audio to several hundred megahertz.

New dielectric materials appear from time to time and represent improvements in capacitor performance. Silvered-mica capacitors, formed by spraying thin coats of silver on each side of the mica insulating sheet, improved the stability of mica capacitors in circuits sensitive to temperature changes. Polystyrene and other synthetic dielectrics, along with tantalum electrolytics, have permitted the size of capacitors to shrink per unit of capacitance.

VOLTAGE RATINGS AND BREAKDOWN

When high voltage is applied to the plates of a capacitor, considerable force is exerted on the electrons and nuclei of the dielectric. The dielectric is an insulator; its electrons do not become detached from

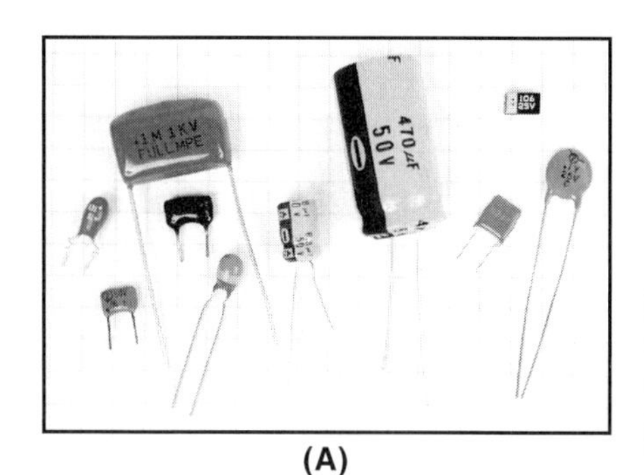

(A)

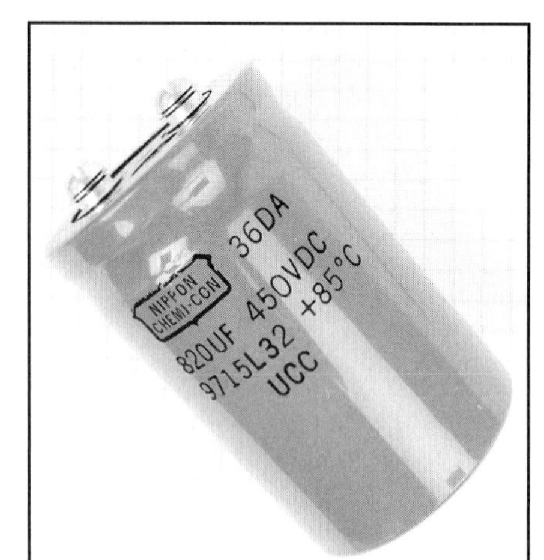

(B)

(C)

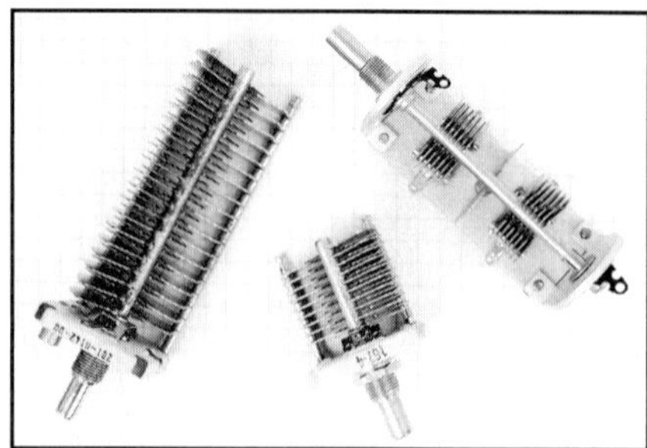

(D)

(E)

Fig 4.32 — Fixed-value capacitors are shown in parts A and B. Aluminum electrolytic capacitors are pictured near the center of photo A. The small tear-drop units to the left of center are tantalum electrolytic capacitors. The rectangular units are silvered-mica, polystyrene film and monolithic ceramic. At the right edge is a disc-ceramic capacitor and near the top right corner is a surface-mount capacitor. B shows a large "computer-grade" electrolytic. These have very low equivalent series resistance (ESR) and are often used as filter capacitors in switch-mode power supplies, and in series-strings for high-voltage supplies of RF power amplifiers. Parts C and D show a variety of variable capacitors, including air variable capacitors and mica compression units. Part E shows a vacuum variable capacitor such as is sometimes used in high-power amplifier circuits. The 1/4-inch-ruled graph paper backgrounds provide size comparisons.

atoms the way they do in conductors. If the force is great enough, however, the dielectric will break down. Failed dielectrics usually puncture and offer a low-resistance current path between the two plates.

The *breakdown voltage* a dielectric can withstand depends on the chemical composition and thickness of the dielectric. Breakdown voltage is not directly proportional to the thickness; doubling the thickness does not quite double the breakdown voltage. Gas dielectrics also break down, as evidenced by a spark or arc between the plates. Spark voltages are generally given with the units *kilovolts per centimeter*. For air, the spark voltage or V_s may range from more than 120 kV/cm for gaps as narrow as 0.006 cm down to 28 kV/cm for gaps as wide as 10 cm. In addition, a large number of variables enter into the actual breakdown voltage in a real situation. Among the variables are the electrode shape, the gap distance, the air pressure or density, the voltage, impurities in the air (or any other dielectric material) and the nature of the external circuit (with air, for instance, the humidity affects conduction on the surface of the capacitor plate).

Dielectric breakdown occurs at a lower voltage between pointed or sharp-edged surfaces than between rounded and polished surfaces. Consequently, the breakdown voltage between metal plates of any given spacing in air can be increased by buffing the edges of the plates. With most gas dielectrics such as air, once the voltage is removed, the arc ceases and the capacitor is ready for use again. If the plates are damaged so they are no longer smooth and polished, they may have to be polished or the capacitor replaced. In contrast, solid dielectrics are permanently damaged by dielectric breakdown, and often will totally short out and melt or explode.

A thick dielectric must be used to withstand high voltages. Since the capacitance is inversely proportional to dielectric thickness (plate spacing) for a given plate area, a high-voltage capacitor must have more plate area than a low-voltage one of the same capacitance. High-voltage, high-capacitance capacitors are therefore physically large.

Dielectric strength is specified in terms of a dielectric withstanding voltage (DWV), given in volts per mil (0.001 inch) at a specified temperature. Taking into account the design temperature range of a capacitor and a safety margin, manufacturers specify *dc working voltage* (dcwv) to express the maximum safe limits of dc voltage across a capacitor to prevent dielectric breakdown.

It is not safe to connect capacitors across an ac power line unless they are rated for such use. Capacitors with dc ratings may short the line. Several manufacturers make capacitors specifically rated for use across the ac power line.

For use with other ac signals, the peak value of ac voltage should not exceed the dc working voltage, unless otherwise specified in component ratings. In other words, the RMS value of ac should be 0.707 times the dcwv value or lower. With many types of capacitors, further derating is required as the operating frequency increases. An additional safety margin is good practice.

Any two surfaces having different electrical potentials, and which are close enough to exhibit a significant electrostatic field, constitute a capacitor. The arrangement of circuit components and leads sometimes results in the creation of unintended capacitors. This is called *stray capacitance*: It often results in the passage of signals in ways that disrupt the normal operation of a circuit. Good design minimizes stray capacitance.

Stray capacitance may have a greater affect in a high-impedance circuit because the capacitive reactance may be a greater percentage of the circuit impedance. Also, because stray capacitance often appears in parallel with the circuit, the stray capacitor may bypass more of the desired signal at higher frequencies. Stray capacitance can often adversely affect sensitive circuits.

For further information about the physical and electrical characteristics of various types of capacitors in actual use, see the **Real-World Component Characteristics** chapter.

CAPACITORS IN SERIES AND PARALLEL

When a number of capacitors are connected in parallel, as in **Fig 4.33A**, the total capacitance of the group is equal to the sum of the individual capacitances:

$$C_{total} = C1 + C2 + C3 + C4 + ... + C_n \quad (47)$$

When two or more capacitors are connected in series, as in Fig 4.33B, the total capacitance is less than that of the smallest capacitor in the group. The rule for finding the capacitance of a number of series-connected capacitors is the same as that for finding the resistance of a number of parallel-connected resistors.

$$C_{total} = \frac{1}{\frac{1}{C1} + \frac{1}{C2} + \frac{1}{C3} + ... + \frac{1}{C_n}} \quad (48)$$

For only two capacitors in series, the formula becomes:

$$C_{total} = \frac{C1 \times C2}{C1 + C2} \quad (49)$$

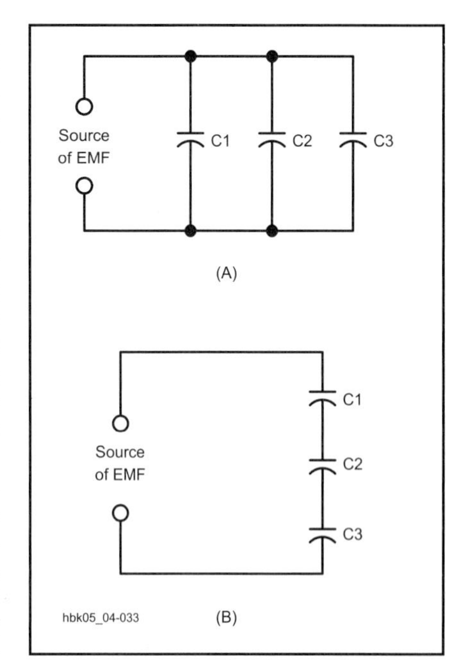

Fig 4.33 — Capacitors in parallel are shown at A, and in series at B.

The same units must be used throughout; that is, all capacitances must be expressed in either µF, nF or pF. Different units cannot be used in the same equation.

Capacitors are usually connected in parallel to obtain a larger total capacitance than is available in one unit. The largest voltage that can be applied safely to a parallel-connected group of capacitors is the voltage that can be applied safely to the one having the lowest voltage rating.

When capacitors are connected in series, the applied voltage is divided between them according to Kirchhoff's Voltage Law. The situation is much the same as when resistors are in series and there is a

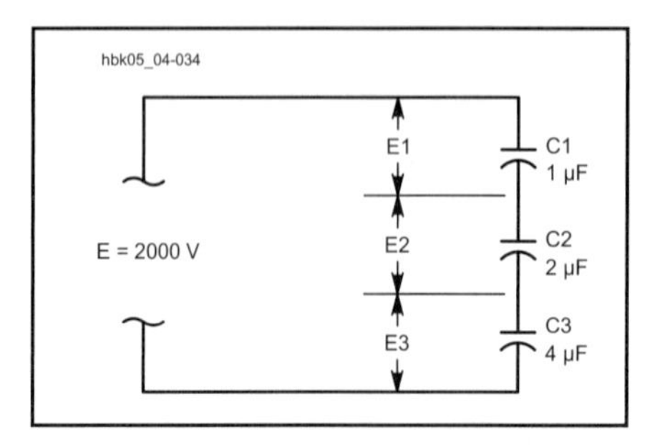

Fig 4.34 — An example of capacitors connected in series. The text shows how to find the voltage drops, E1 through E3.

voltage drop across each. The voltage that appears across each series-connected capacitor is inversely proportional to its capacitance, as compared with the capacitance of the whole group. (This assumes ideal capacitors.)

Example: Three capacitors having capacitances of 1, 2 and 4 μF, respectively, are connected in series as shown in **Fig 4.34**. The voltage across the entire series is 2000 V. What is the total capacitance? (Since this is a calculation using theoretical values to illustrate a technique, we will not follow the rules of significant figures for the calculations.)

$$C_{total} = \frac{1}{\frac{1}{C1} + \frac{1}{C2} + \frac{1}{C3}}$$

$$= \frac{1}{\frac{1}{1\,\mu F} + \frac{1}{2\,\mu F} + \frac{1}{4\,\mu F}}$$

$$C_{total} = \frac{1}{\frac{7}{4\,\mu F}} = \frac{4\,\mu F}{7} = 0.5714\,\mu F$$

The voltage across each capacitor is proportional to the total capacitance divided by the capacitance of the capacitor in question. So the voltage across C1 is:

$$E1 = \frac{0.5714\,\mu F}{1\,\mu F} \times 2000\text{ V} = 1143\text{ V}$$

Similarly, the voltages across C2 and C3 are:

$$E2 = \frac{0.5714\,\mu F}{2\,\mu F} \times 2000\text{ V} = 571\text{ V}$$

and

$$E3 = \frac{0.5714\,\mu F}{4\,\mu F} \times 2000\text{ V} = 286\text{ V}$$

The sum of these three voltages equals 2000 V, the applied voltage.

Capacitors may be connected in series to enable the group to withstand a larger voltage than any individual capacitor is rated to withstand. The trade-off is a decrease in the total capacitance. As shown by the previous example, the applied voltage does not divide equally between the capacitors except when all the capacitances are precisely the same. Use care to ensure that the voltage rating of any capacitor in the group is not exceeded. If you use capacitors in series to withstand a higher voltage, you should also connect an "equalizing resistor" across each capacitor. Use resistors with about 100 Ω per volt of supply voltage, and be sure they have sufficient power-handling capability for the circuit. With real capacitors, the leakage resistance of the capacitors may have more effect on the voltage division than does the capacitance. A capacitor with a high parallel resistance will have the highest voltage across it. Adding equalizing resistors reduces this effect.

RC TIME CONSTANT

Connecting a dc voltage source directly to the terminals of a capacitor charges the capacitor to the full source voltage almost instantaneously. Any resistance added to the circuit as in **Fig 4.35A** limits the current, lengthening the time required for the voltage between the capacitor plates to build up to the source-voltage value. During this charging period, the current flowing from the source into the capacitor gradually decreases from its initial value. The increasing voltage stored in the capacitor's electric field offers increasing opposition to the steady source voltage.

While it is being charged, the voltage between the capacitor terminals is an exponential function of time, and is given by:

$$V(t) = E\left(1 - e^{-\frac{t}{RC}}\right) \qquad (50)$$

where:

V(t) = capacitor voltage in volts at time t;
E = potential of charging source in volts;
t = time in seconds after initiation of charging current;
e = natural logarithmic base = 2.718;
R = circuit resistance in ohms; and
C = capacitance in farads.

Theoretically, the charging process is never really finished, but eventually the charging current drops to an unmeasurable value. For many purposes, it is convenient to let t = RC. Under this condition, the above equation becomes:

$$V(RC) = E(1 - e^{-1}) \approx 0.632\text{ E} \qquad (51)$$

The product of R in ohms times C in farads is called the *time constant* of the circuit and is the time in seconds required to charge the capacitor to 63.2% of the supply voltage. (The lower-case Greek letter tau [τ] is often used to represent the time constant in electronics circuits.) After two time constants (t = 2τ) the capacitor charges another 63.2% of the difference between the capacitor voltage at one time constant and the supply voltage, for a total charge of 86.5%. After three time constants the capacitor reaches 95% of the supply voltage, and so on, as illustrated in the curve of **Fig 4.36A**. After 5 RC time periods, a capacitor is considered fully charged, having reached 99.24% of the source voltage.

If a charged capacitor is discharged through a resistor, as indicated in Fig 4.35B, the same time constant applies for the decay of the capacitor voltage. A direct short circuit applied between the capacitor terminals would discharge the capacitor almost instantly. The resistor, R, limits the current, so the capacitor voltage decreases only as rapidly as the capacitor can discharge itself through R. A capacitor discharging through a resistance

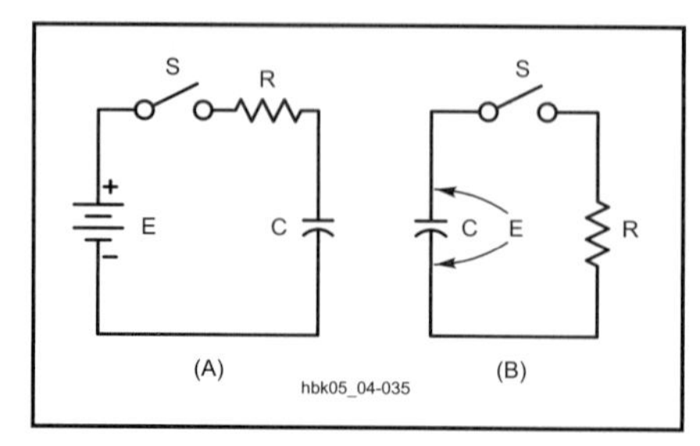

Fig 4.35 — An illustration of the time constant in an RC circuit.

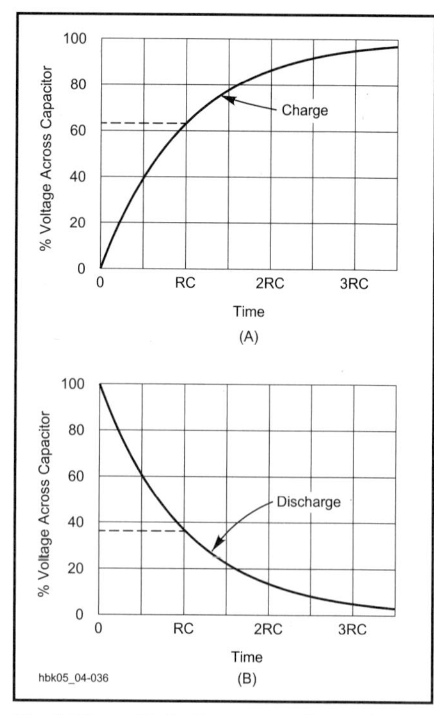

Fig 4.36 — At A, the curve shows how the voltage across a capacitor rises, with time, when charged through a resistor. The curve at B shows the way in which the voltage decreases across a capacitor when discharging through the same resistance. For practical purposes, a capacitor may be considered charged or discharged after 5 RC periods.

exhibits the same time-constant characteristics (calculated in the same way as above) as a charging capacitor. The voltage, as a function of time while the capacitor is being discharged, is given by:

$$V(t) = E\left[e^{-\frac{t}{RC}}\right] \qquad (52)$$

where t = time in seconds after initiation of discharge.

Again, by letting t = RC, the time constant of a discharging capacitor represents a decrease in the voltage across the capacitor of about 63.2%. After 5 time-constant periods, the capacitor is considered fully discharged, since the voltage has dropped to less than 1% of the full-charge voltage.

Time constant calculations have many uses in radio work. The following examples are all derived from practical-circuit applications.

Example 1: A 100-µF capacitor in a high-voltage power supply is shunted by a 100-kΩ resistor. What is the minimum time before the capacitor may be considered fully discharged? Since full discharge is approximately 5 RC periods,

$$t = 5 \times RC = 5 \times 100 \times 10^{3}\ \Omega \times 100 \times 10^{-6}\ F$$

$$= 50000 \times 10^{-3}\ \text{seconds}$$

$$t = 50.0\ s$$

Note: Although waiting almost a minute for the capacitor to discharge seems safe in this high-voltage circuit, never rely solely on capacitor-discharging resistors (often called *bleeder resistors*). Be certain the power source is removed and the capacitors are totally discharged before touching any circuit components.

Example 2: Smooth CW keying without clicks requires approximately 5 ms (0.005 s) of delay in both the make and break edges of the waveform, relative to full charging and discharging of a capacitor in the circuit. What typical values might a builder choose for an RC delay circuit in a keyed voltage line? Since full charge and discharge require 5 RC periods,

$$RC = \frac{t}{5} = \frac{0.005\ s}{5} = 0.001\ s$$

Any combination of resistor and capacitor whose values, when multiplied together, equal 0.001 would do the job. A typical capacitor might be 0.05 µF. In that case, the necessary resistor would be:

$$R = \frac{0.001\ s}{0.05 \times 10^{-6}\ F}$$

$$= 0.02 \times 10^{6}\ \Omega = 20000\ \Omega = 20\ k\Omega$$

In practice, a builder would likely either experiment with values or use a variable resistor. The final value would be selected after monitoring the waveform on an oscilloscope.

Example 3: Many modern integrated circuit (IC) devices use RC circuits to control their timing. To match their internal circuitry, they may use a specified threshold voltage as the trigger level. For example, a certain IC uses a trigger level of 0.667 of the supply voltage. What value of capacitor and resistor would be required for a 4.5-second timing period?

First we will solve equation 50 for the time constant, RC. The threshold voltage is 0.667 times the supply voltage, so we use this value for V(t).

$$V(t) = E\left[1 - e^{-\frac{t}{RC}}\right]$$

$$0.667\ E = E\left[1 - e^{-\frac{t}{RC}}\right]$$

$$e^{-\frac{t}{RC}} = 1 - 0.667$$

$$\ln\left[e^{-\frac{t}{RC}}\right] = \ln(0.333)$$

$$-\frac{t}{RC} = -1.10$$

We want to find a capacitor and resistor combination that will produce a 4.5 s timing period, so we substitute that value for t.

$$RC = \frac{4.5\ s}{1.10} = 4.1\ s$$

If we select a value of 10 µF, we can solve for R.

$$R = \frac{4.1\ s}{10 \times 10^{-6}\ F} = 0.41 \times 10^{6}\ \Omega = 410\ k\Omega$$

A 1% tolerance resistor and capacitor will give good precision. You could also use a variable resistor and an accurate method to measure the time to set the circuit to a 4.5 s period.

As the examples suggest, RC circuits have numerous applications in electronics. The number of applications is growing steadily, especially with the introduction of integrated circuits controlled by part or all of a capacitor charge or discharge cycle.

ALTERNATING CURRENT IN CAPACITANCE

Everything said about capacitance and capacitors in a dc circuit applies to capacitance in an ac circuit with one major exception. Whereas a capacitor in a dc circuit will appear as an open circuit except for the brief charge and discharge periods, the same capacitor in an ac circuit will both pass and limit current. A capacitor in an ac circuit does not handle electrical energy like a resistor, however. Instead of converting the energy to heat and dissipating it, capacitors store electrical energy and return it to the circuit.

In **Fig 4.37** a sine-wave ac voltage having a maximum value of 100 is applied to a ca-

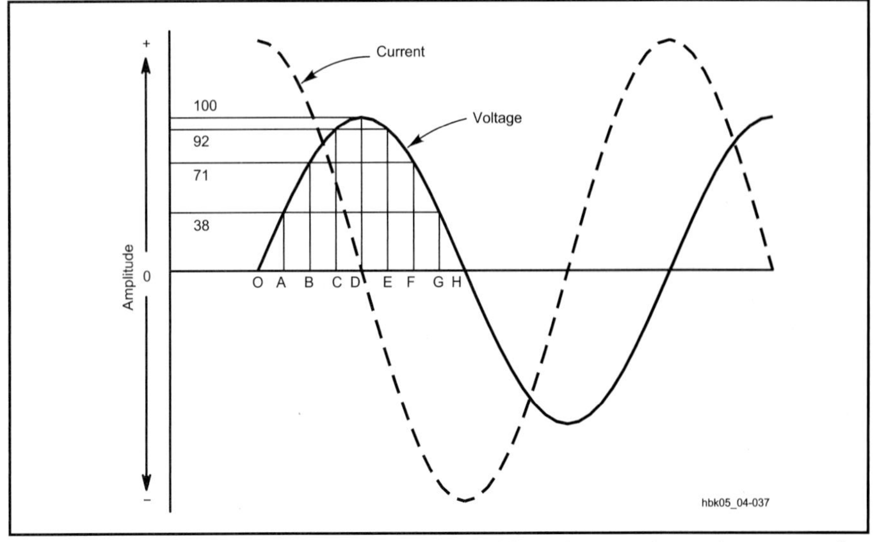

Fig 4.37 — Voltage and current phase relationships when an alternating current is applied to a capacitor.

pacitor. In the period OA, the applied voltage increases from 0 to 38; at the end of this period the capacitor is charged to that voltage. In interval AB the voltage increases to 71; that is, 33 V additional. During this interval a smaller quantity of charge has been added than in OA, because the voltage rise during interval AB is smaller. Consequently the average current during interval AB is smaller than during OA. In the third interval, BC, the voltage rises from 71 to 92, an increase of 21 V. This is less than the voltage increase during AB, so the quantity of electricity added is less; in other words, the average current during interval BC is still smaller. In the fourth interval, CD, the voltage increases only 8 V; the charge added is smaller than in any preceding interval and therefore the current also is smaller.

By dividing the first quarter cycle into a very large number of intervals, it could be shown that the current charging the capacitor has the shape of a sine wave, just as the applied voltage does. The current is largest at the beginning of the cycle and becomes zero at the maximum value of the voltage, so there is a phase difference of 90° between the voltage and the current. During the first quarter cycle the current is flowing in the normal direction through the circuit, since the capacitor is being charged. Hence the current is positive, as indicated by the dashed line in Fig 4.37.

In the second quarter cycle — that is, in the time from D to H — the voltage applied to the capacitor decreases. During this time the capacitor loses its charge. Applying the same reasoning, it is evident that the current is small in interval DE and continues to increase during each succeeding interval. The current is flowing against the applied voltage, however, because the capacitor is discharging into the circuit. The current flows in the negative direction during this quarter cycle.

The third and fourth quarter cycles repeat the events of the first and second, respectively, with this difference: the polarity of the applied voltage has reversed, and the current changes to correspond. In other words, an alternating current flows in the circuit because of the alternate charging and discharging of the capacitance. As shown in Fig 4.37, the current starts its cycle 90° before the voltage, so the current in a capacitor *leads* the applied voltage by 90°. You might find it helpful to remember the word "ICE" as a mnemonic because the current (I) in a capacitor (C) comes before voltage (E). We can also turn this statement around, to say the voltage in a capacitor *lags* the current by 90°.

CAPACITIVE REACTANCE

The quantity of electric charge that can be placed on a capacitor is proportional to the applied voltage and the capacitance. This amount of charge moves back and forth in the circuit once each cycle; hence, the rate of movement of charge (the current) is proportional to voltage, capacitance and frequency. When the effects of capacitance and frequency are considered together, they form a quantity that plays a part similar to that of resistance in Ohm's Law. This quantity is called *reactance*. The unit for reactance is the ohm, just as in the case of resistance. The formula for calculating the reactance of a capacitor at a given frequency is:

$$X_C = \frac{1}{2 \pi f C} \quad (53)$$

where:

X_C = capacitive reactance in ohms,
f = frequency in hertz,
C = capacitance in farads
π = 3.1416

Note: In many references and texts, the symbol ω is used to represent 2 π f. In such references, equation 53 would read

$$X_C = \frac{1}{\omega C}$$

Although the unit of reactance is the ohm, there is no power dissipated in reactance. The energy stored in the capacitor during one portion of the cycle is simply returned to the circuit in the next.

The fundamental units for frequency and capacitance (hertz and farads) are too cumbersome for practical use in radio circuits. If the capacitance is specified in microfarads (µF) and the frequency is in megahertz (MHz), however, the reactance calculated from the previous formula retains the unit ohms.

Example: What is the reactance of a capacitor of 470 pF (0.000470 µF) at a frequency of 7.15 MHz?

$$X_C = \frac{1}{2 \pi f C}$$

$$= \frac{1}{2 \pi \times 7.15 \text{ MHz} \times 0.000470 \text{ µF}}$$

$$= \frac{1 \Omega}{0.0211} = 47.4 \Omega$$

Example: What is the reactance of the same capacitor, 470 pF (0.000470 µF), at a frequency of 14.29 MHz?

$$X_C = \frac{1}{2 \pi f C}$$

$$= \frac{1}{2 \pi \times 14.30 \text{ MHz} \times 0.000470 \text{ µF}}$$

$$= \frac{1 \Omega}{0.0422} = 23.7 \Omega$$

The rate of change of voltage in a sine wave increases directly with the frequency. Therefore, the current into the capacitor also increases directly with frequency. Since, for a given voltage, an increase in current is equivalent to a decrease in reactance, the reactance of any capacitor decreases proportionally as the frequency increases. **Fig 4.38** traces the decrease in reactance of an arbitrary-value capacitor with respect to increasing frequency. The only limitation on the application of the graph is the physical make-up of the capacitor, which may favor low-frequency uses or high-frequency applications.

Among other things, reactance is a measure of the ability of a capacitor to limit the flow of ac in a circuit. For some purposes it is important to know the ability of a capacitor to pass current. This ability is called *susceptance*, and it corresponds to conductance in resistive circuit elements. In an ideal capacitor with no resistive losses — that is, no energy lost as heat — susceptance is simply the reciprocal of reactance. Hence,

$$B = \frac{1}{X_C} \quad (54)$$

where:

X_C is the reactance, and
B is the susceptance.

The unit of susceptance (and conductance and admittance) is the *siemens* (abbreviated S). In literature only a few years old, the term *mho* is also sometimes given as the unit of susceptance (as well as of conductance and admittance). The role of reactance and susceptance in current and other Ohm's Law calculations will appear in a later section of this chapter.

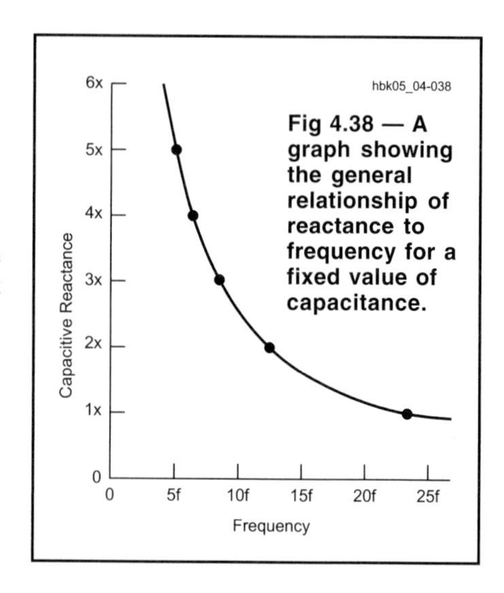

Fig 4.38 — A graph showing the general relationship of reactance to frequency for a fixed value of capacitance.

Inductance and Inductors

A second way to store electrical energy is in a *magnetic field*. This phenomenon is called *inductance*, and the devices that exhibit inductance are called *inductors*. Inductance depends upon some basic underlying magnetic properties. See Chapter 6 for more information on practical inductor applications and problems.

MAGNETISM

Magnetic Fields, Flux and Flux Density

Magnetic fields are closed fields that surround a magnet, as illustrated in **Fig 4.39**. The field consists of lines of magnetic force or *flux*. It exhibits polarity, which is conventionally indicated as north-seeking and south-seeking poles, or *north* and *south poles* for short. Magnetic flux is measured in the SI unit of the weber, which is a volt second (Wb = Vs). In the *centimeter gram second* (*cgs*) metric system units, we measure magnetic flux in maxwells (1Mx = 10^{-8} Wb).

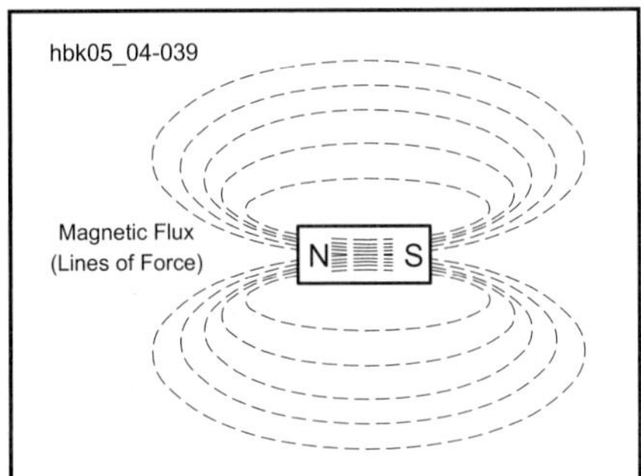

Fig 4.39 — The magnetic field and poles of a permanent magnet. The magnetic field direction is from the north to the south pole.

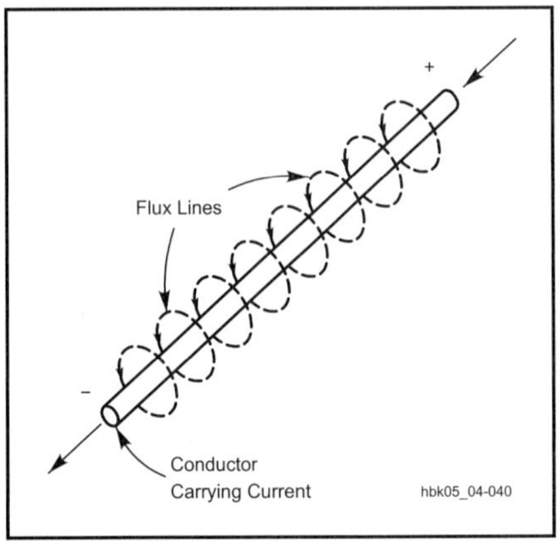

The field intensity, known as the *flux density*, decreases with the square of the distance from the source. Flux density (*B*) is represented in gauss (G), where one gauss is equivalent to one line of force per square centimeter of area across the field (G = Mx / cm2). The gauss is a *cgs* unit. In SI units, flux density is represented by the tesla (T), which is one weber per square meter (T = Wb/m^2).

Magnetic fields exist around two types of materials. First, certain ferromagnetic materials contain molecules aligned so as to produce a magnetic field. Lodestone, Alnico and other materials with high *retentivity* form *permanent magnets* because they retain their magnetic properties for long periods. Other materials, such as soft iron, yield temporary magnets that lose their magnetic properties rapidly.

The second type of magnetic material is an electrical conductor with a current through it. As shown in **Fig 4.40**, moving electrons are surrounded by a closed magnetic field lying in planes perpendicular to their motion. The needle of a compass placed near a wire carrying direct current will be deflected by the magnetic field around the wire. This phenomenon is one aspect of a two-way relationship: a moving magnetic field whose lines cut across a wire will induce an electrical current in the wire, and an electrical current will produce a magnetic field.

If the wire is coiled into a solenoid, the magnetic field greatly intensifies as the individual flux lines add together. **Fig 4.41** illustrates the principle by showing a coil section. Note that the resulting *electromagnet* has magnetic properties identical in principle to those of a permanent magnet, including poles and lines of force or flux. The strength of the magnetic field depends on several factors: the number of turns of the coil, the magnetic properties of the materials surrounding the coil (both inside and out), the length of the magnetic path and the amplitude of the current.

The magnetizing or *magnetomotive force* that produces a flux or total magnetic field is measured in gilberts (Gb). The force in gilberts equals 0.4π (approximately 1.257) times the number of turns in the coil times the current in amperes. (The SI unit of magnetomotive force is the ampere turn, abbreviated A, just like the ampere.) The magnetic field strength, H, measured in oersteds (Oe) produced by any particular magnetomotive force (measured in gilberts) is given by:

$$H = \frac{0.4\,\pi N I}{\ell} \qquad (55)$$

where:

H = magnetic field strength in oersteds,
N = number of turns,
I = dc current in amperes,
π = 3.1416, and
ℓ = mean magnetic path length in centimeters.

The gilbert and oersted are *cgs* units. These are given here because most amateur calculations will use these units. You may also see the preferred SI units in some literature. The SI unit of magnetic field strength is the ampere (turn) per meter.

A force is required to produce a given magnetic field strength. This implies that there is a resistance, called *reluctance*, to be overcome.

Fig 4.40 — The magnetic field around a conductor carrying an electrical current. If the thumb of your right hand points in the direction of the conventional current (plus to minus), your fingers curl in the direction of the magnetic field around the wire.

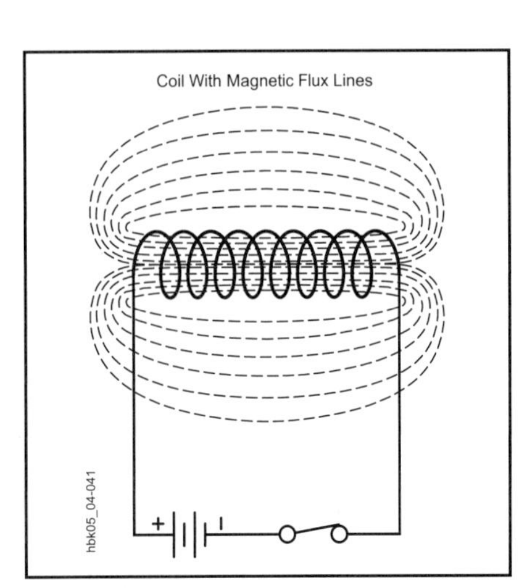

Fig 4.41 — Cross section of an inductor showing its flux lines and overall magnetic field.

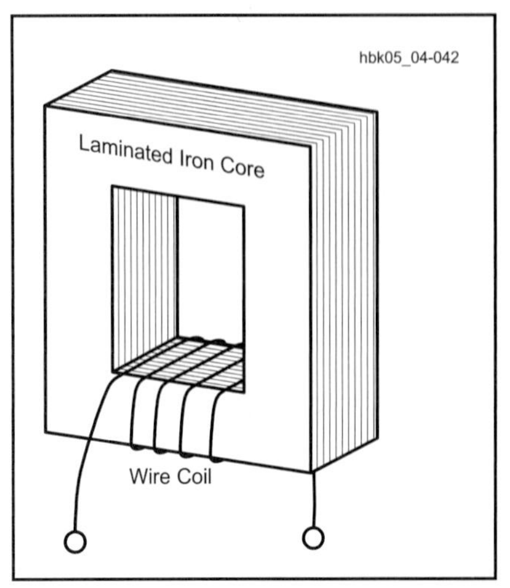

Fig 4.42 — A coil of wire wound around a laminated iron core.

Table 4.6
Properties of Some High-Permeability Materials

Material	*Approximate Percent Composition*					*Maximum Permeability*
	Fe	Ni	Co	Mo	Other	
Iron	99.91	—	—	—	—	5000
Purified Iron	99.95	—	—	—	—	180000
4% silicon-iron	96	—	—	—	4 Si	7000
45 Permalloy	54.7	45	—	—	0.3 Mn	25000
Hipernik	50	50	—	—	—	70000
78 Permalloy	21.2	78.5	—	—	0.3 Mn	100000
4-79 Permalloy	16.7	79	—	—	0.3 Mn	100000
Supermalloy	15.7	79	—	5	0.3 Mn	800000
Permendur	49.7	—	50	—	0.3 Mn	5000
2V Permendur	49	—	49	—	2 V	4500
Hiperco	64	—	34	—	2 Cr	10000
2-81 Permalloy*	17	81	—	2	—	130
Carbonyl iron*	99.9	—	—	—	—	132
Ferroxcube III**	($MnFe_2O_4$ + $ZnFe_2O_4$)				1500	

Note: all materials in sheet form except * (insulated powder) and ** (sintered powder).
(Reference: L. Ridenour, ed., *Modern Physics for the Engineer*, p 119.)

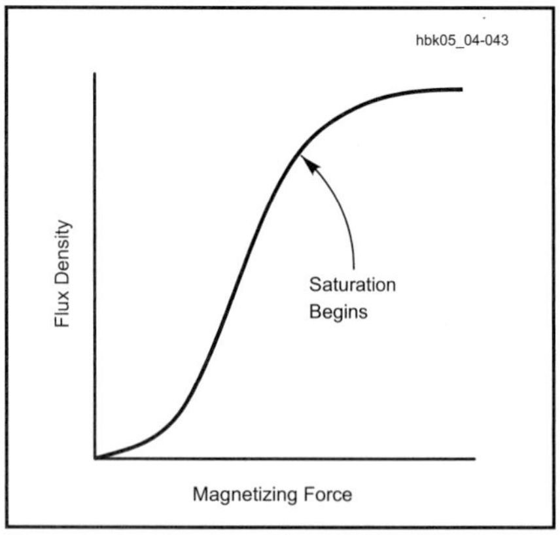

Fig 4.43 — A typical permeability curve for a magnetic core, showing the point where saturation begins.

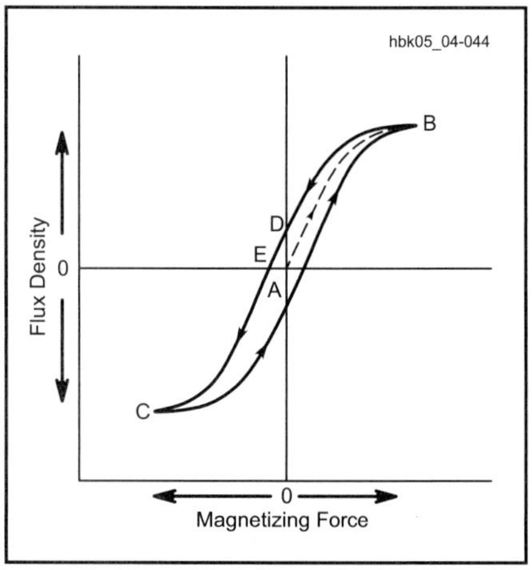

Fig 4.44 — A typical hysteresis curve for a magnetic core, showing the additional energy needed to overcome residual flux.

Core Properties: Permeability, Saturation, Reluctance, Hysteresis

The nature of the material within the coil of an electromagnet, where the lines of force are most concentrated, has the greatest effect upon the magnetic field established by the coil. All materials are compared to air. The ratio of flux density produced by a given material compared to the flux density produce by an air core is the *permeability* of the material. Suppose the coil in **Fig 4.42** is wound on an iron core having a cross-sectional area of 2 square inches. When a certain current is sent through the coil, it is found that there are 80000 lines of force in the core. Since the area is 2 square inches, the magnetic flux density is 40000 lines per square inch. Now suppose that the iron core is removed and the same current is maintained in the coil. Also suppose the flux density without the iron core is found to be 50 lines per square inch. The ratio of these flux densities, iron core to air, is 40000 / 50 or 800, the core's permeability.

Permeabilities as high as 10^6 have been attained. The three most common types of materials used in magnetic cores are these:

A. stacks of laminated steel sheets (for power and audio applications);

B. various ferrite compounds (for cores shaped as rods, toroids, beads and numerous other forms); and

C. powdered iron (shaped as slugs, toroids and other forms for RF inductors).

Brass has a permeability less than 1. A brass core inserted into a coil will decrease the inductance compared to an air core.

The permeability of silicon-steel power-transformer cores approaches 5000 in high-quality units. Powdered-iron cores used in RF tuned circuits range in permeability from 3 to about 35, while ferrites of nickel-zinc and manganese-zinc range from 20 to 15000. **Table 4.6** lists some common magnetic materials, their composition and their permeabilities. Core materials are often frequency sensitive, exhibiting excessive losses outside the frequency band of intended use.

As a measure of the ease with which a magnetic field may be established in a material as compared with air, permeability (μ) corresponds roughly to electrical conductivity. Permeability is given as:

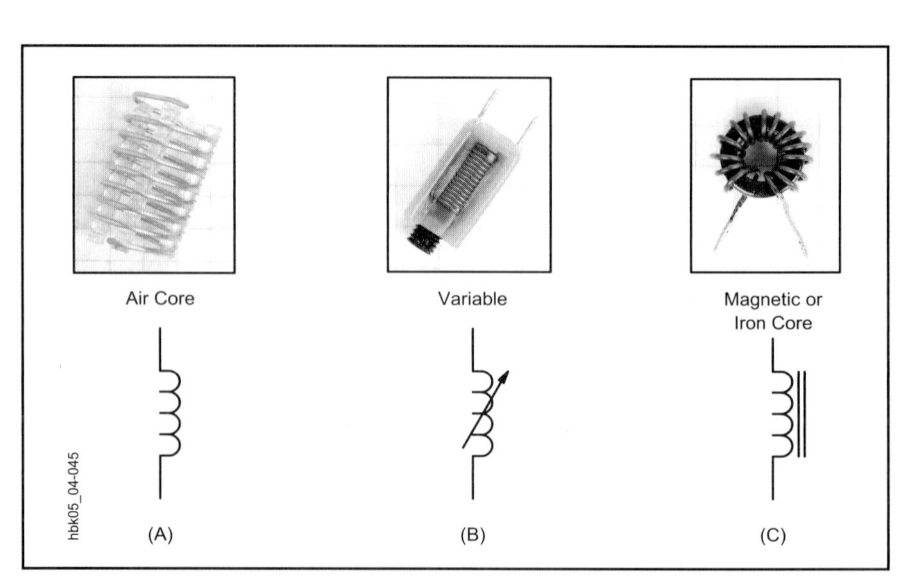

Fig 4.45 — Photos and schematic symbols for representative inductors. A, an air-core inductor; B, a variable inductor with a nonmagnetic slug and C, an inductor with a toroidal magnetic core. The 1/4-inch-ruled graph paper background provides a size comparison.

$$\mu = \frac{B}{H} \qquad (56)$$

where:

B is the flux density in gauss, and

H is the magnetic field strength in oersteds.

Unlike electrical conductivity, which is independent of other electrical parameters, the permeability of a magnetic material varies with the flux density. At low flux densities (or with an air core), increasing the current through the coil will cause a proportionate increase in flux. But at very high flux densities, increasing the current beyond a certain point may cause no appreciable change in the flux. At this point, the core is said to be *saturated*. Saturation causes a rapid decrease in permeability, because it decreases the ratio of flux lines to those obtainable with the same current using an air core. **Fig 4.43** displays a typical permeability curve, showing the region of saturation. The saturation point varies with the makeup of different magnetic materials. Air and other nonmagnetic materials do not saturate and have a permeability of one. *Reluctance*, which is the reciprocal of permeability and corresponds roughly to resistance in an electrical circuit, is also one for air and other nonmagnetic cores.

The retentivity of magnetic core materials creates another potential set of losses caused by *hysteresis*. **Fig 4.44** illustrates the change of flux density (*B*) with a changing magnetizing force (*H*). From starting point A, with no residual flux, the flux reaches point B at the maximum magnetizing force. As the force decreases, so too does the flux, but it does not reach zero simultaneously with the force at point D. As the force continues in the opposite direction, it brings the flux density to point C. As the force decreases to zero, the flux once more lags behind. In effect, a reverse force is necessary to overcome the residual magnetism retained by the core material, a *coercive force*. The result is a power loss to the magnetic circuit, which appears as heat in the core material. Air cores are immune to hysteresis effects and losses.

INDUCTANCE AND DIRECT CURRENT

In an electrical circuit, any element having a magnetic field is called an *inductor*. **Fig 4.45** shows schematic-diagram symbols and photographs of a few representative inductors: an air-core inductor, a slug-tuned variable inductor with a nonmagnetic core and an inductor with a magnetic (iron) core.

The transfer of energy to the magnetic field of an inductor represents work performed by the source of the voltage. Power is required for doing work, and since power is equal to current multiplied by voltage, there must be a voltage drop in the circuit while energy is being stored in the field. This voltage drop, exclusive of any voltage drop caused by resistance in the circuit, is the result of an opposing voltage induced in the circuit while the field is building up to its final value. Once the field becomes constant, the *induced voltage* or back-voltage disappears, because no further energy is being stored. The induced voltage opposes the voltage of the source and tends to prevent the current from rising rapidly when the circuit is closed. **Fig 4.46A** illustrates the situation of energizing an inductor or magnetic circuit, showing the relative amplitudes of induced voltage and the delayed rise in current to its full value.

The amplitude of the induced voltage is proportional to the rate at which the current changes (and consequently, the rate at which the magnetic field changes) and to a constant associated with the circuit itself: the *inductance* (or *self-inductance*) of the circuit. Inductance depends on the physical configuration of the inductor. Coiling a conductor increases its inductance. In effect, the growing (or shrinking) magnetic field of each turn produces magnetic lines of force that — in their expansion (or contraction) — cut across the other turns of the coil, inducing a voltage in every other turn. The mutuality of the effect multiplies the ability of the coiled conductor to store electrical energy.

A coil of many turns will have more inductance than one of few turns, if both coils are otherwise physically similar. Furthermore, if an inductor is placed around a magnetic core, its inductance will increase in proportion to the permeability of that core, if the circuit current is below the point at which the core saturates.

The polarity of an induced voltage is always such as to oppose any change in the circuit current. This means that when the current in the circuit is increasing, work is being done against the induced voltage by storing energy in the magnetic field. Likewise, if the current in the circuit tends to decrease, the stored energy of the field returns to the circuit, and adds to the energy being supplied by the voltage

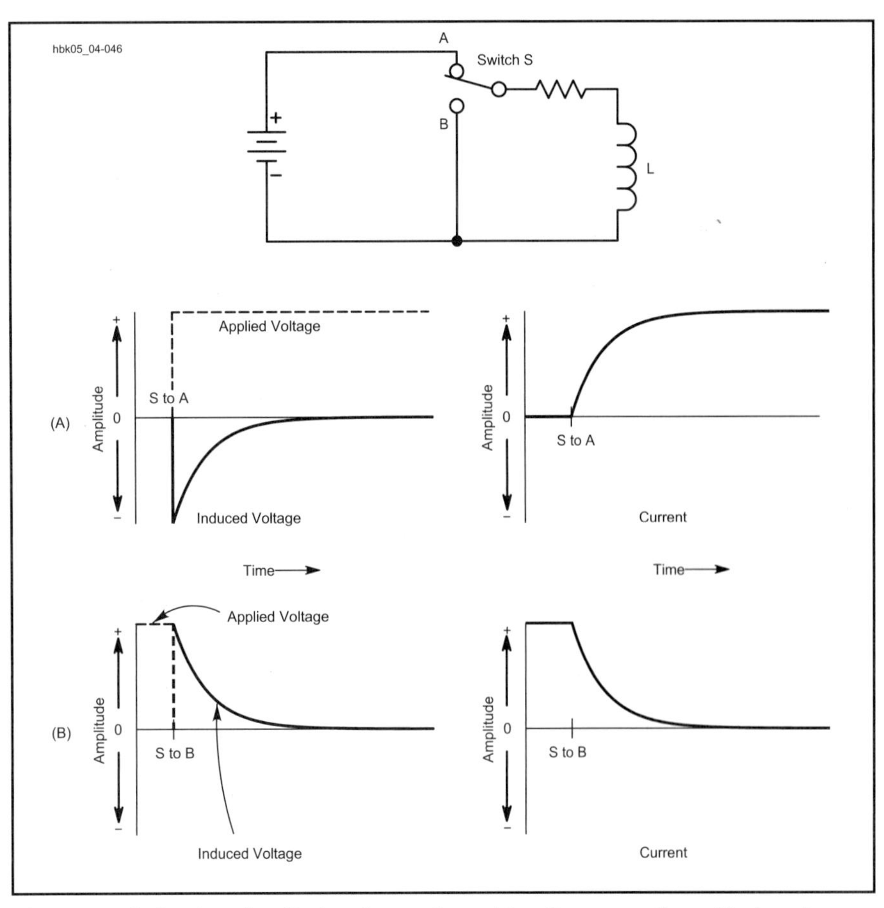

Fig 4.46 — Inductive circuit showing and graphing the generation of induced voltage and the rise of current in an inductor at A, and the decay of current as power is removed and the coil shorted at B.

source. Inductors try to maintain a constant current through the circuit. This phenomenon tends to keep the current flowing even though the applied voltage may be decreasing or be removed entirely. Fig 4.46B illustrates the decreasing but continuing flow of current caused by the induced voltage after the source voltage is removed from the circuit.

The energy stored in the magnetic field of an inductor is given by the formula:

$$W = \frac{I^2 L}{2} \qquad (57)$$

where:

W = energy in joules,
I = current in amperes, and
L = inductance in henrys.

This formula corresponds to the energy-storage formula for capacitors: energy storage is a function of current squared over time. As with capacitors, the time dependence of inductor current is a significant property; see the section on time constants.

The basic unit of inductance is the *henry* (abbreviated H), which equals an induced voltage of one volt when the inducing current is varying at a rate of one ampere per second. In various aspects of radio work, inductors may take values ranging from a fraction of a nanohenry (nH) through millihenrys (mH) up to about 20 H.

MUTUAL INDUCTANCE AND MAGNETIC COUPLING

Mutual Inductance

When two coils are arranged with their axes on the same line, as shown in **Fig 4.47**, current sent through coil 1 creates a magnetic field that cuts coil 2. Consequently, a voltage will be induced in coil 2 whenever the field strength of coil 1 is changing. This induced voltage is similar to the voltage of self-induction, but since it appears in the second coil because of current flowing in the first, it is a mutual effect and results from the *mutual inductance* between the two coils.

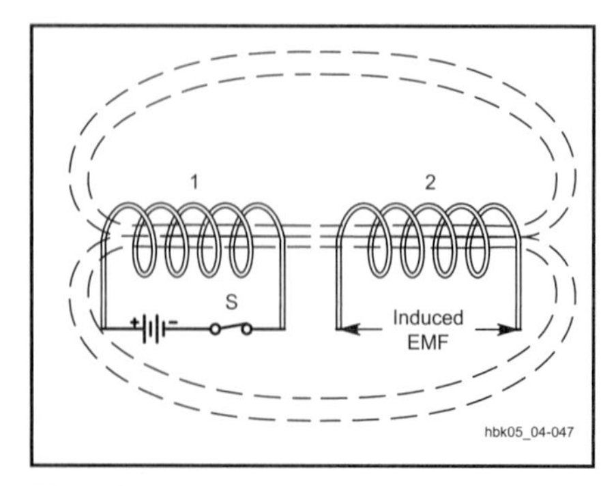

Fig 4.47 — Mutual inductance: When S is closed, current flows through coil number 1, setting up a magnetic field that induces a voltage in the turns of coil number 2.

When all the flux set up by one coil cuts all the turns of the other coil, the mutual inductance has its maximum possible value. If only a small part of the flux set up by one coil cuts the turns of the other, the mutual inductance is relatively small. Two coils having mutual inductance are said to be *coupled.*

The ratio of actual mutual inductance to the maximum possible value that could theoretically be obtained with two given coils is called the *coefficient of coupling* between the coils. It is frequently expressed as a percentage. Coils that have nearly the maximum possible mutual inductance (coefficient = 1 or 100%) are said to be closely, or tightly, coupled. If the mutual inductance is relatively small the coils are said to be loosely coupled. The degree of coupling depends upon the physical spacing between the coils and how they are placed with respect to each other. Maximum coupling exists when they have a common axis and are as close together as possible (for example, one wound over the other). The coupling is least when the coils are far apart or are placed so their axes are at right angles.

The maximum possible coefficient of coupling is closely approached when the two coils are wound on a closed iron core. The coefficient with air-core coils may run as high as 0.6 or 0.7 if one coil is wound over the other, but will be much less if the two coils are separated. Although unity coupling is suggested by Fig 4.47, such coupling is possible only when the coils are wound on a closed magnetic core.

Unwanted Couplings: Spikes, Lightning and Other Pulses

Every conductor passing current has a magnetic field associated with it — and therefore inductance — even though the conductor is not formed into a coil. The inductance of a short length of straight wire is small, but it may not be negligible. If the current through it changes rapidly, the induced voltage may be appreciable. This is the case in even a few inches of wire with an alternating current having a frequency on the order of 100 MHz or higher. At much lower frequencies or at dc, the inductance of the same wire might be ignored because the induced voltage would seemingly be negligible.

There are many phenomena, however, both natural and man-made, which create sufficiently strong magnetic fields to induce voltages into straight wires. Many of them are brief but intense pulses of energy that act like the turning on of the switch in a circuit containing self-inductance. Because the fields created grow to very high levels rapidly, they cut across wires leading into and out of — and wires wholly within — electronic equipment, inducing unwanted voltages by mutual coupling.

Short-duration, high-level voltage spikes occur on ac and dc power lines. Because the field intensity is great, these spikes may induce voltages upon conducting elements in sensitive circuits, disrupting them and even injuring components. Lightning in the vicinity of the equipment can induce voltages on power lines and other conductive paths (even ground conductors) that lead to the equipment location. Lightning that seems a safe distance away can induce large spikes on power lines that ultimately lead to the equipment. Closer at hand, heavy equipment with electrical motors can induce significant spikes into power lines within the equipment location. Even though the power lines are straight, the powerful magnetic field of a spike source can induce damaging voltages on equipment left "plugged in" during electrical storms or during the operation of heavy equipment that inadequately filters its spikes.

Parallel-wire cables linking elements of electronic equipment consist of long wires in close proximity to each other. Signal pulses can couple both magnetically and capacitively from one wire to another. Since the magnetic field of a changing current decreases as the square of distance, separating the signal-carrying lines diminishes inductive coupling. Placing a grounded wire between signal-carrying lines reduces capacitive coupling. Unless they are well-shielded and filtered, however, the lines are still susceptible to the inductive coupling of pulses from other sources.

INDUCTORS IN RADIO WORK

Various facets of radio work make use of inductors ranging from the tiny up to the massive. Small values of inductance, as illustrated by **Fig 4.48A**, serve mostly in RF circuits. They may be self-supporting air-core or air-wound coils or the winding may be supported by nonmagnetic strips or a form. Phenolic, certain plastics and ceramics are the most common coil forms for air-core inductors. These inductors range in value from a few hundred µH for medium- and high-frequency circuits down to tenths of a µH for VHF and UHF work. The smallest values of inductance in radio work result from component leads. For VHF work and higher frequencies, component lead

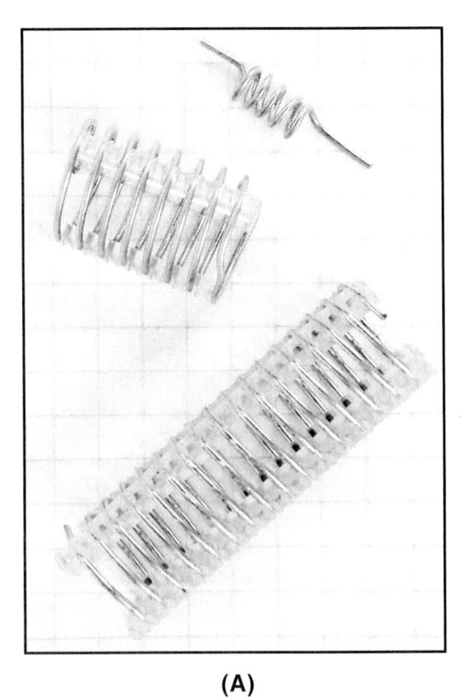

(A)

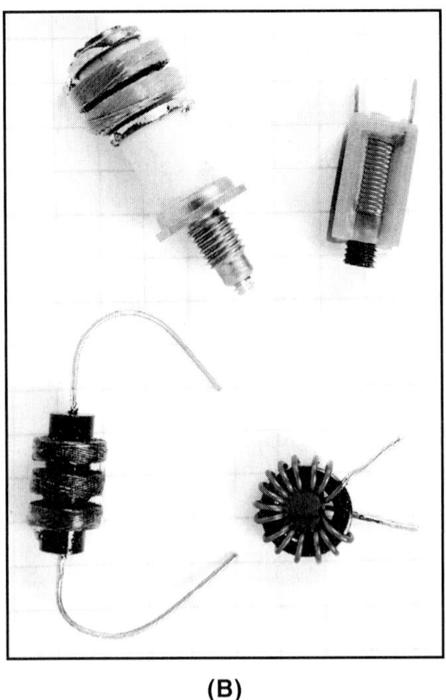

(B)

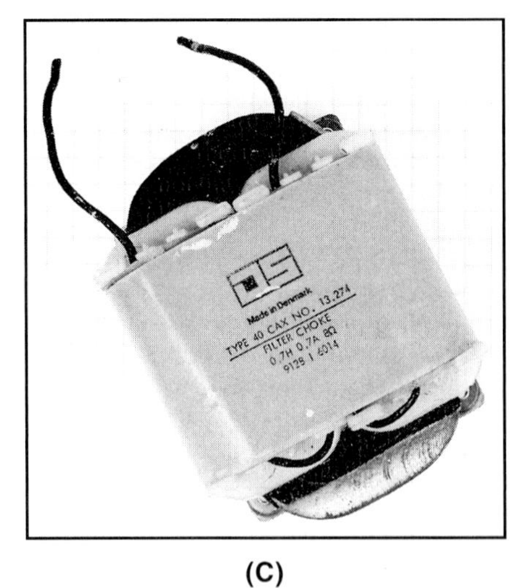

(C)

Fig 4.48 — Part A shows small-value air-wound inductors. Part B shows some inductors with values in the range of a few millihenrys and C shows a large inductor such as might be used in audio circuits or as power-supply chokes. The ¼-inch-ruled graph paper background provides a size comparison.

length is often critical. Circuits may fail to operate properly because leads are a little too short or too long.

It is possible to make these solenoid coils variable by inserting a slug in the center of the coil. (Slug-tuned coils normally have a ceramic, plastic or phenolic insulating form between the conductive slug and the coil winding.) If the slug material is magnetic, such as powdered iron, the inductance increases as the slug is centered along the length of the coil. If the slug is brass or some other conductive but nonmagnetic material, centering the slug will reduce the coil's inductance. This effect stems from the fact that brass has low electrical resistance and acts as an effective short-circuited one-turn secondary for the coil. (See more on transformer effects later in this chapter.)

An alternative to air-core inductors for RF work are toroidal coils wound on cores composed of powdered iron mixed with a binder to hold the material together. The availability of many types and sizes of powdered-iron cores has made these inductors popular for low-power fixed-value service. The toroidal shape concentrates the inductor's field tightly about the coil, eliminating the need in many cases for other forms of shielding to limit the interaction of the inductor's magnetic field with the fields of other inductors.

Fig 4.48B shows samples of inductors in the millihenry range. Among these inductors are multisection RF chokes designed to keep RF currents from passing beyond them to other parts of circuits. Low-frequency radio work may also use inductors in this range of values, sometimes wound with *litz* wire. Litz wire is a special version of stranded wire, with each strand insulated from the others. For audio filters, toroidal coils with values below 100 mH are useful. Resembling powdered-iron-core RF toroids, these coils are wound on ferrite or molybdenum-permalloy cores having much higher permeabilities.

Audio and power-supply inductors appear in Fig 4.48C. Lower values of these iron-core coils, in the range of a few henrys, are useful as audio-frequency chokes. Larger values up to about 20 H may be found in power supplies, as choke filters, to suppress 120-Hz ripple. Although some of these inductors are open frame, most have iron covers to confine the powerful magnetic fields they produce.

INDUCTANCES IN SERIES AND PARALLEL

When two or more inductors are con-

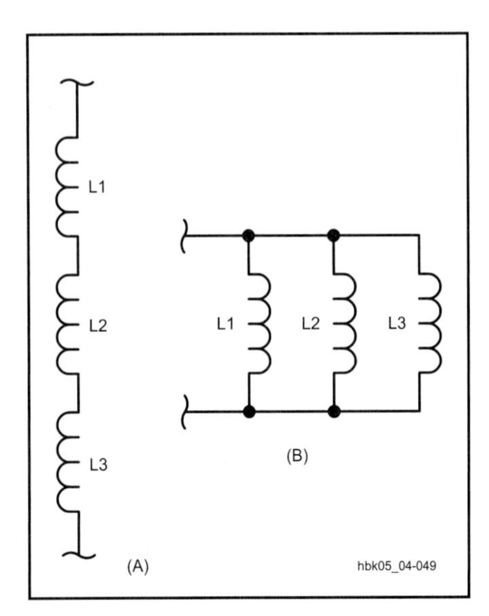

Fig 4.49 — Part A shows inductances in series, and Part B shows inductances in parallel.

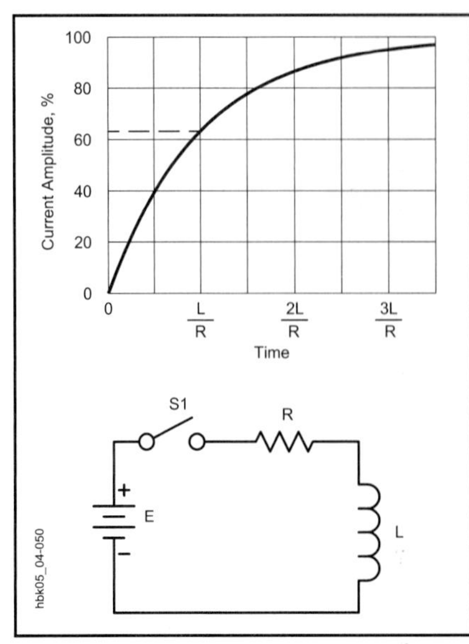

Fig 4.50 — Time constant of an RL circuit being energized.

nected in series (**Fig 4.49A**), the total inductance is equal to the sum of the individual inductances, provided that the coils are sufficiently separated so that coils are not in the magnetic field of one another. That is:

$$L_{total} = L1 + L2 + L3 ... + L_n \qquad (58)$$

If inductors are connected in parallel (Fig 4.49B), and if the coils are separated sufficiently, the total inductance is given by:

$$L_{total} = \frac{1}{\frac{1}{L1} + \frac{1}{L2} + \frac{1}{L3} + ... + \frac{1}{L_n}} \qquad (59)$$

For only two inductors in parallel, the formula becomes:

$$L_{total} = \frac{L1 \times L2}{L1 + L2} \qquad (60)$$

Thus, the rules for combining inductances in series and parallel are the same as those for resistances, assuming that the coils are far enough apart so that each is unaffected by another's magnetic field. When this is not so, the formulas given above will not yield correct results.

RL TIME CONSTANT

A comparable situation to an RC circuit exists when resistance and inductance are connected in series. In **Fig 4.50**, first consider L to have no resistance and also consider that R is zero. Closing S1 sends a current through the circuit. The instantaneous transition from no current to a finite value, however small, represents a rapid change in current, and a reverse voltage is developed by the self-inductance of L. The value of reverse voltage is almost equal and opposite to the applied voltage. The resulting initial current is very small.

The reverse voltage depends on the change in the value of the current and would cease to offer opposition if the current did not continue to increase. With no resistance in the circuit (which, by Ohm's Law, would lead to an infinitely large current), the current would increase forever, always growing just fast enough to keep the self-induced voltage equal to the applied voltage.

When resistance in the circuit limits the current, Ohm's Law defines the value that the current can reach. The reverse voltage generated in L must only equal the difference between E and the drop across R, because the difference is the voltage actually applied to L. This difference becomes smaller as the current approaches the final Ohm's Law value. Theoretically, the reverse voltage never quite disappears, and so the current never quite reaches the Ohm's Law value. In practical terms, the differences become unmeasurable after a time.

The current at any time after the switch in Fig 4.50 has been closed, can be found from:

$$I(t) = \frac{E\left[1 - e^{-\frac{tR}{L}}\right]}{R} \qquad (61)$$

where:
- $I(t)$ = current in amperes at time t,
- E = power supply potential in volts,
- t = time in seconds after initiation of current,
- e = natural logarithmic base = 2.718,
- R = circuit resistance in ohms, and
- L= inductance in henrys.

The time in seconds required for the current to build up to 63.2% of the maximum value is called the time constant, and is equal to L / R, where L is in henrys and R is in ohms. After each time interval equal to this constant, the circuit conducts an additional 63.2% of the remaining current. This behavior is graphed in Fig 4.50. As is the case with capacitors, after 5 time constants the current is considered to have reached its maximum value. As with capacitors, we often use the lowercase Greek tau (τ) to represent the time constant.

Example: If a circuit has an inductor of 5.0 mH in series with a resistor of 10 Ω, how long will it take for the current in the circuit to reach full value after power is applied? Since achieving maximum current takes approximately five time constants,

$$t = 5\,L / R = (5 \times 5.0 \times 10^{-3}\ H) / 10\,\Omega$$

$$= 2.5 \times 10^{-3} \text{ seconds or } 2.5 \text{ ms}$$

Note that if the inductance is increased to 5.0 H, the required time increases by a factor of 1000 to 2.5 seconds. Since the circuit resistance didn't change, the final current is the same for both cases in this example. Increasing inductance increases the time required to reach full current.

Zero resistance would prevent the circuit from ever achieving full current. All inductive circuits have some resistance, however, if only the resistance of the wire making up the inductor.

An inductor cannot be discharged in the simple circuit of Fig 4.50 because the magnetic field collapses as soon as the current ceases. Opening S1 does not leave the inductor charged in the way that a capacitor would remain charged. The energy stored in the magnetic field returns instantly to the circuit when S1 is opened. The rapid collapse of the field causes a very large voltage to be induced in the coil. Usually the induced voltage is many times larger than the applied voltage, because the induced voltage is proportional to the rate at which the field changes. The common result of opening the switch in such a circuit is that a spark or arc forms at the switch contacts during the instant the switch opens. When the inductance is large and the current in the circuit is high, large amounts of energy are released in a very short time. It is not at all unusual for the switch contacts to burn or melt under such circumstances. The spark or arc at the opened switch can be reduced or suppressed by connecting a suitable capacitor and resistor in series across the contacts. Such an RC combination is called a *snubber network*.

Transistor switches connected to and controlling coils, such as relay solenoids, also require protection. In most cases, a small power diode connected in reverse across the relay coil will prevent field-collapse currents from harming the transistor.

If the excitation is removed without breaking the circuit, as theoretically diagrammed in **Fig 4.51**, the current will decay according to the formula:

$$I(t) = \frac{E}{R}\left[e^{\frac{-tR}{L}}\right] \qquad (62)$$

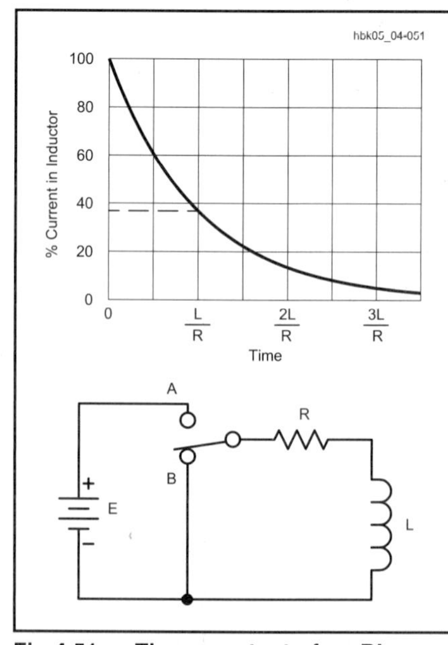

Fig 4.51 — Time constant of an RL circuit being deenergized. This is a theoretical model only, since a mechanical switch cannot change state instantaneously.

where t = time in seconds after removal of the source voltage.

After one time constant the current will lose 63.2% of its steady-state value. (It will decay to 36.8% of the steady-state value.) The graph in Fig 4.51 shows the current-decay waveform to be identical to the voltage-discharge waveform of a capacitor. Be careful about applying the terms *charge* and *discharge* to an inductive circuit, however. These terms refer to energy storage in an electric field. An inductor stores energy in a magnetic field.

ALTERNATING CURRENT IN INDUCTORS

When an alternating voltage is applied to an ideal inductance (one with no resistance — all practical inductors have some resistance), the current is 90° out of phase with the applied voltage. In this case the current *lags* 90° behind the voltage, the opposite of the capacitor current-voltage relationship, as shown in **Fig 4.52**. (Here again, we can also say the voltage across an inductor *leads* the current by 90°.)

If you have difficulty remembering the phase relationships between voltage and current with inductors and capacitors, you may find it helpful to think of the phrase, "ELI the ICE man." This will remind you that voltage across an inductor leads the current through it, because the E comes before the I, with an L between them, as you read from left to right. (The letter L represents inductance.) It will also help you remember the capacitor conditions because I comes before E with a C between them.

Interpreting Fig 4.52 begins with understanding that the primary cause for current lag in an inductor is the reverse voltage generated in the inductance. The amplitude of the reverse voltage is proportional to the rate at which the current changes. In time segment OA, when the applied voltage is at its positive maximum, the reverse or induced voltage is also maximum, allowing the least current to flow. The rate at which the current is changing is the highest, a 38% change in the time period OA. In the segment AB, the current changes by only 33%, yielding a reduced level of induced voltage, which is in step with the decrease in the applied voltage. The process continues in time segments BC and CD, the latter producing only an 8% rise in current as the applied and induced voltage approach zero.

In segment DE, the applied voltage changes direction. The induced voltage also changes direction, which returns current to the circuit from storage in the magnetic field. The direction of this current is now opposite to the applied voltage, which sustains the current in the positive direction. As the applied voltage continues to increase negatively, the current — although positive — decreases in value, reaching zero as the applied voltage reaches its negative maximum. The negative half-cycle continues just as did the positive half-cycle.

Compare Fig 4.52 with Fig 4.37. Whereas in a pure capacitive circuit, the current *leads* the voltage by 90°, in a pure inductive circuit, the current *lags* the voltage by 90°. These phenomena are especially important in circuits that combine inductors and capacitors.

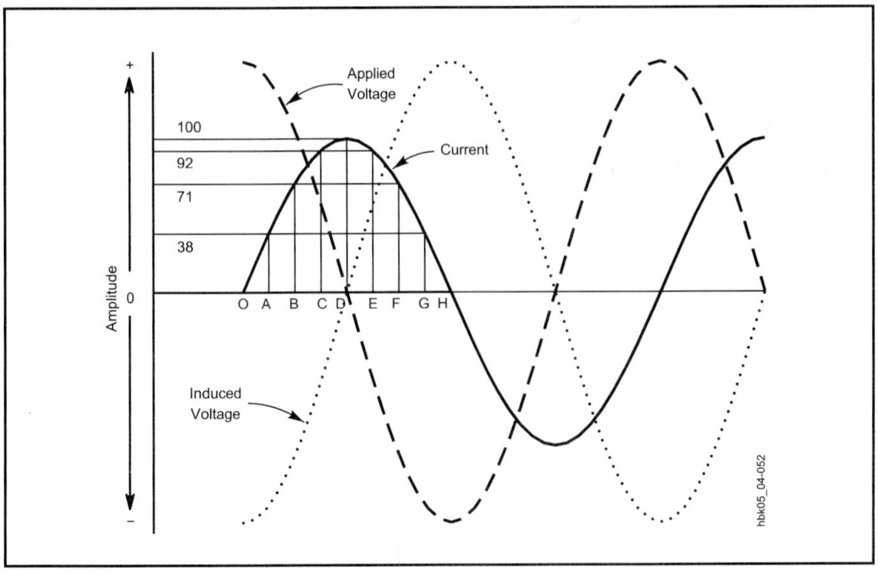

Fig 4.52 — Phase relationships between voltage and current when an alternating current is applied to an inductance.

INDUCTIVE REACTANCE

The amplitude of alternating current in an inductor is inversely proportional to the applied frequency. Since the reverse voltage is directly proportional to inductance for a given rate of current change, the current is inversely proportional to inductance for a given applied voltage and frequency.

The combined effect of inductance and frequency is called inductive reactance, which — like capacitive reactance — is expressed in ohms. The formula for inductive reactance is:

$$X_L = 2 \pi f L \qquad (63)$$

where:

X_L = inductive reactance,
f = frequency in hertz,
L = inductance in henrys, and
π = 3.1416.
(If $\omega = 2 \pi f$, then $X_L = \omega L$.)

Example: What is the reactance of a coil having an inductance of 8.00 H at a frequency of 120 Hz?

$$X_L = 2 \pi f L$$
$$= 6.2832 \times 120 \text{ Hz} \times 8.00 \text{ H}$$
$$= 6030 \, \Omega$$

In RF circuits the inductance values are usually small and the frequencies are large. When the inductance is expressed in millihenrys and the frequency in kilohertz, the conversion factors for the two units cancel, and the formula for reactance may be used without first converting to fundamental units. Similarly, no conversion is necessary if the inductance is expressed in microhenrys and the frequency in megahertz.

Example: What is the reactance of a 15.0-microhenry coil at a frequency of 14.0 MHz?

$$X_L = 2 \pi f L$$
$$= 6.2832 \times 14.0 \text{ MHz} \times 15.0 \, \mu\text{H}$$
$$= 1320 \, \Omega$$

The resistance of the wire used to wind the coil has no effect on the reactance, but simply acts as a separate resistor connected in series with the coil.

Example: What is the reactance of the same coil at a frequency of 7.0 MHz?

$$X_L = 2 \pi f L$$
$$= 6.2832 \times 7.0 \text{ MHz} \times 15.0 \, \mu\text{H}$$
$$= 660 \, \Omega$$

Comparing the two examples suggests correctly that inductive reactance varies directly with frequency. The rate of change of the current varies directly with the frequency, and this rate of change also determines the amplitude of the induced or reverse voltage. Hence, the opposition to the flow of current increases proportionally to frequency. This opposition is called *inductive reactance*. The direct relationship between frequency and reactance in inductors, combined with the inverse relationship between reactance and frequency in the case of capacitors, will be of fundamental importance in creating resonant circuits.

As a measure of the ability of an inductor to limit the flow of ac in a circuit, inductive reactance is similar to capacitive reactance in having a corresponding *susceptance*, or ability to pass ac current in a circuit. In an ideal inductor with no resistive losses — that is, no energy lost as heat — susceptance is simply the reciprocal of reactance.

$$B = \frac{1}{X_L} \qquad (64)$$

where:

X_L = reactance, and

B = susceptance.

The unit of susceptance for both inductors and capacitors is the *siemens*, abbreviated S.

Quality Factor, or Q of Components

Components that store energy, like capacitors and inductors, may be compared in terms of quality or Q. The Q of any such component is the ratio of its ability to store energy to the sum total of all energy losses within the component. In practical terms, this ratio reduces to the formula:

$$Q = \frac{X}{R} \qquad (65)$$

where:

Q = figure of merit or quality (no units),

X = X_L (inductive reactance) for inductors and X_C (capacitive reactance) for capacitors (in ohms), and

R = the sum of all resistances associated with the energy losses in the component (in ohms).

The Q of cxapacitors is ordinarily high. Good quality ceramic capacitors and mica capacitors may have Q values of 1200 or more. Small ceramic trimmer capacitors may have Q values too small to ignore in some applications. Microwave capacitors can have poor Q values; 10 or less at 10 GHz and higher frequencies.

Inductors are subject to many types of electrical energy losses, however, such as wire resistance, core losses and skin effect. All electrical conductors have some resistance through which electrical energy is lost as heat. Moreover, inductor wire must be sized to handle the anticipated current through the coil. Wire conductors suffer additional ac losses because alternating current tends to flow on the conductor surface. As the frequency increases, the current is confined to a thinner layer of the conductor surface. This property is called *skin effect*. If the inductor's core is a conductive material, such as iron, ferrite, or brass, the core will introduce additional losses of energy. The specific details of these losses are discussed in connection with each type of core material.

The sum of all core losses may be depicted by showing a resistor in series with the inductor (as in Figs 4.50 and 4.51), although there is no separate component represented by the resistor symbol. As a result of inherent energy losses, inductor Q rarely, if ever, approaches capacitor Q in a circuit where both components work together. Although many circuits call for the highest Q inductor obtainable, other circuits may call for a specific Q, even a very low one.

AC Component Summary

Component	*Resistor*	*Capacitor*	*Inductor*
Basic Unit	ohm (Ω)	farad (F)	henry (H)
Units Commonly Used		microfarads (µF) picofarads (pF)	millihenrys (mH) microhenrys (µH)
Quantity Stored (Does not want to change in circuit)	(None)	Voltage	Current
Combining components in series	R1 + R2	C1 × C2 / C1 + C2	L1 + L2
Combining components in parallel	R1 × R2 / R1 + R2	C1 + C2	L1 × L2 / L1 + L2
Time constant	(None)	RC	L/R
Voltage-Current Phase	In phase	Current leads voltage Voltage lags current	Voltage leads current Current lags voltage
Resistance or Reactance	Resistance	$X_C = 1 / 2\pi fC$	$X_L = 2\pi fL$
Change with increasing frequency	No	Reactance decreases	Reactance increases
Q of circuit	Not defined	X_C / R	X_L / R

Calculating Practical Inductors

Although builders and experimenters rarely construct their own capacitors, inductor fabrication is common. In fact, it is often necessary, since commercially available units may be unavailable or expensive. Even if available, they may consist of coil stock to be trimmed to the required value. Core materials and wire for winding both solenoid and toroidal inductors are readily available. The following information includes fundamental formulas and design examples for calculating practical inductors, along with additional data on the theoretical limits in the use of some materials.

AIR-CORE INDUCTORS

Many circuits require air-core inductors using just one layer of wire. The approximate inductance of a single-layer air-core coil may be calculated from the simplified formula:

$$L\,(\mu H) = \frac{d^2\, n^2}{18d + 40\ell} \quad (66)$$

where:

L = inductance in microhenrys,
d = coil diameter in inches (from wire center to wire center),
ℓ = coil length in inches, and
n = number of turns.

The notation is explained in **Fig 4.53**. This formula is a close approximation for coils having a length equal to or greater than 0.4 d. (Note: Inductance varies as the square of the turns. If the number of turns is doubled, the inductance is quadrupled. This relationship is inherent in the equation, but is often overlooked. For example, if you want to double the inductance, put on additional turns equal to 1.4 times the original number of turns, or 40% more turns.)

Example: What is the inductance of a coil if the coil has 48 turns wound at 32 turns per inch and a diameter of $^3/_4$ inch? In this case, d = 0.75, ℓ = $^{48}/_{32}$ = 1.5 and n = 48.

$$L = \frac{0.75^2 \times 48^2}{(18 \times 0.75) + (40 \times 1.5)}$$

$$= \frac{1300}{74} = 18\,\mu H$$

To calculate the number of turns of a single-layer coil for a required value of inductance, the formula becomes:

$$n = \frac{\sqrt{L\,(18\,d + 40\,\ell)}}{d} \quad (67)$$

Example: Suppose an inductance of 10.0 µH is required. The form on which the coil is to be wound has a diameter of one inch and is long enough to accommodate a coil of $1^1/_4$ inches. Then d = 1.00 inch, ℓ = 1.25 inches and L = 10.0. Substituting:

$$n = \frac{\sqrt{10.0\,[(18 \times 1.00) + (40 \times 1.25)]}}{1}$$

$$= \sqrt{680} = 26.1 \text{ turns}$$

A 26-turn coil would be close enough in practical work. Since the coil will be 1.25 inches long, the number of turns per inch will be 26.1 / 1.25 = 20.9. Consulting the wire table in the **Component Data and References** chapter, we find that #17 enameled wire (or anything smaller) can be used. The proper inductance is obtained by winding the required number of turns on the form and then adjusting the spacing between the turns to make a uniformly spaced coil 1.25 inches long.

Most inductance formulas lose accuracy when applied to small coils (such as are used in VHF work and in low-pass filters built for reducing harmonic interference to televisions) because the conductor thickness is no longer negligible in comparison with the size of the coil. **Fig 4.54** shows the measured inductance of VHF coils and may be used as a basis for circuit design. Two curves are given; curve A is for coils wound to an inside diameter of $^1/_2$ inch; curve B is for coils of $^3/_4$-inch inside diameter. In both curves, the wire size is #12, and the winding pitch is eight turns to the inch ($^1/_8$-inch center-to-center turn spacing). The inductance values given include leads $^1/_2$-inch long.

Machine-wound coils with the preset diameters and turns per inch are available in many radio stores, under the trade names of B&W Miniductor, Airdux and Polycoil. The **Component Data and References** chapter provides information on using such coil stock to simplify the process of designing high-quality inductors for most HF applications. Forming a wire into a solenoid increases its inductance, and also introduces distributed capacitance. Since each turn is at a slightly different ac potential, each pair of turns effectively forms a parasitic capacitor. See the **Real-World Component Characteristics** chapter for information on the effects of these complications to the "ideal" inductors under discussion in this chapter. Moreover, the Q of air-core inductors is, in part, a function of the coil shape, specifically its ratio of length to diameter. Q tends to be highest when these dimensions are nearly equal. With wire properly sized to the current carried by the coil, and with high-caliber construction, air-core inductors can achieve Qs above 200. Air-core inductors with Qs as high as 400 are possible.

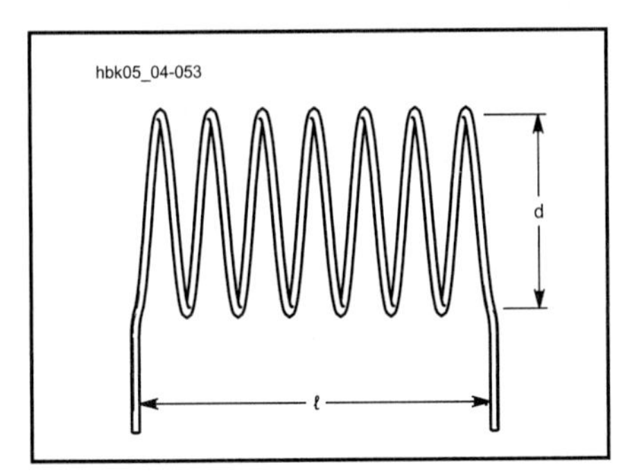

Fig 4.53 — Coil dimensions used in the inductance formula for air-core inductors.

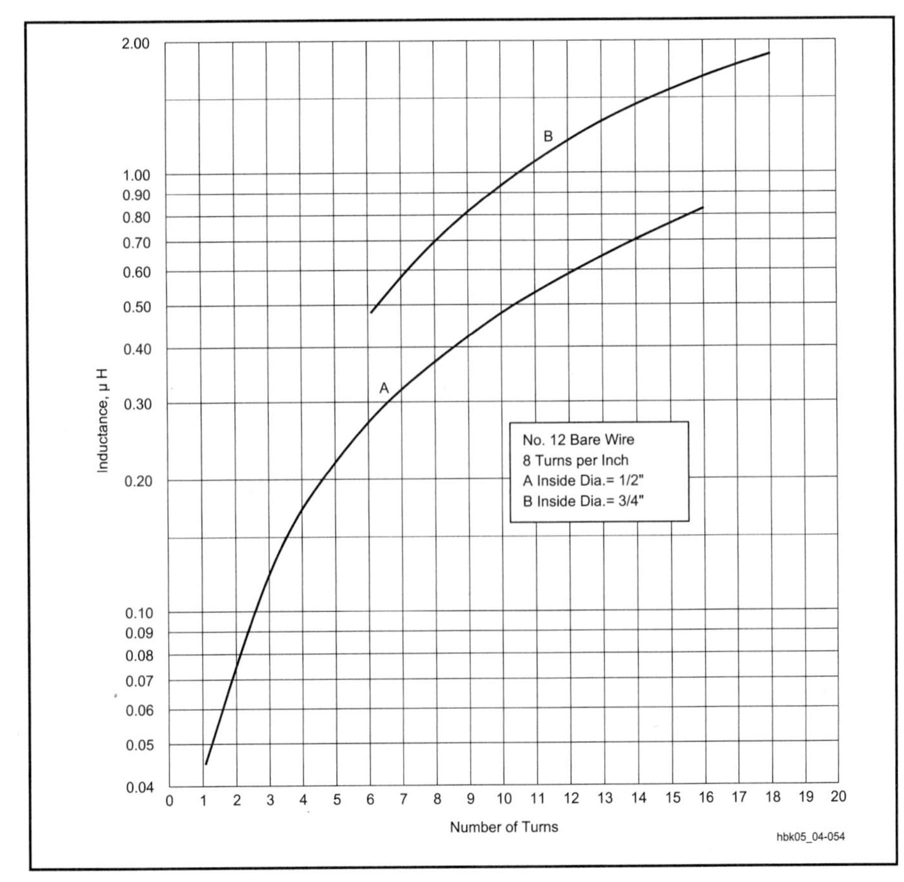

Fig 4.54 — Measured inductance of coils wound with #12 bare wire, eight turns to the inch. The values include half-inch leads.

STRAIGHT-WIRE INDUCTANCE

At low frequencies the inductance of a straight, round, nonmagnetic wire in free space is given by:

$$L = 0.00508\, b \left\{ \left[\ln\left(\frac{2b}{a}\right) \right] - 0.75 \right\} \quad (68)$$

where:

L = inductance in μH,
a = wire radius in inches,
b = wire length in inches, and
ln = natural logarithm = 2.303 × common logarithm (base 10).

If the dimensions are expressed in millimeters instead of inches, the equation may still be used, except replace the 0.00508 value with 0.0002.

Skin effect reduces the inductance at VHF and above. As the frequency approaches infinity, the 0.75 constant within the brackets approaches unity. As a practical matter, skin effect will not reduce the inductance by more than a few percent.

Example: What is the inductance of a wire that is 0.1575 inch in diameter and 3.9370 inches long? For the calculations, a = 0.0787 inch (radius) and b = 3.9370 inch.

$$L = 0.00508\, b \left\{ \left[\ln\left(\frac{2b}{a}\right) \right] - 0.75 \right\}$$

$$= 0.00508\,(3.9370) \times \left\{ \left[\ln\left(\frac{2 \times 3.9370}{0.0787}\right) \right] - 0.75 \right\}$$

$$L = 0.0200\,[\ln(100) - 0.75]$$
$$= 0.0200\,(4.60 - 0.75)$$
$$= 0.0200 \times 3.85 = 0.077\ \mu H$$

Fig 4.55 is a graph of the inductance for wires of various radii as a function of length.

A VHF or UHF tank circuit can be fabricated from a wire parallel to a ground plane, with one end grounded. A formula for the inductance of such an arrangement is given in **Fig 4.56**.

Example: What is the inductance of a wire 3.9370 inches long and 0.0787 inch in radius, suspended 1.5748 inch above a ground plane? (The inductance is measured between the free end and the ground plane, and the formula includes the inductance of the 1.5748-inch grounding link.) To demonstrate the use of the formula in Fig 4.56, begin by evaluating these quantities:

$$b + \sqrt{b^2 + a^2} = 3.9370 + \sqrt{3.9370^2 + 0.0787^2}$$
$$= 3.9370 + 3.94 = 7.88$$

$$b + \sqrt{b^2 + 4\left(h^2\right)}$$
$$= 3.9370 + \sqrt{3.9370^2 + 4\left(1.5748^2\right)}$$
$$= 3.9370 + \sqrt{15.500 + 4\,(2.4800)}$$
$$= 3.9370 + \sqrt{15.500 + 9.9200}$$
$$= 3.9370 + 5.0418 = 8.9788$$

$$\frac{2h}{a} = \frac{2 \times 1.5748}{0.0787} = 40.0$$

$$\frac{b}{4} = \frac{3.9370}{4} = 0.98425$$

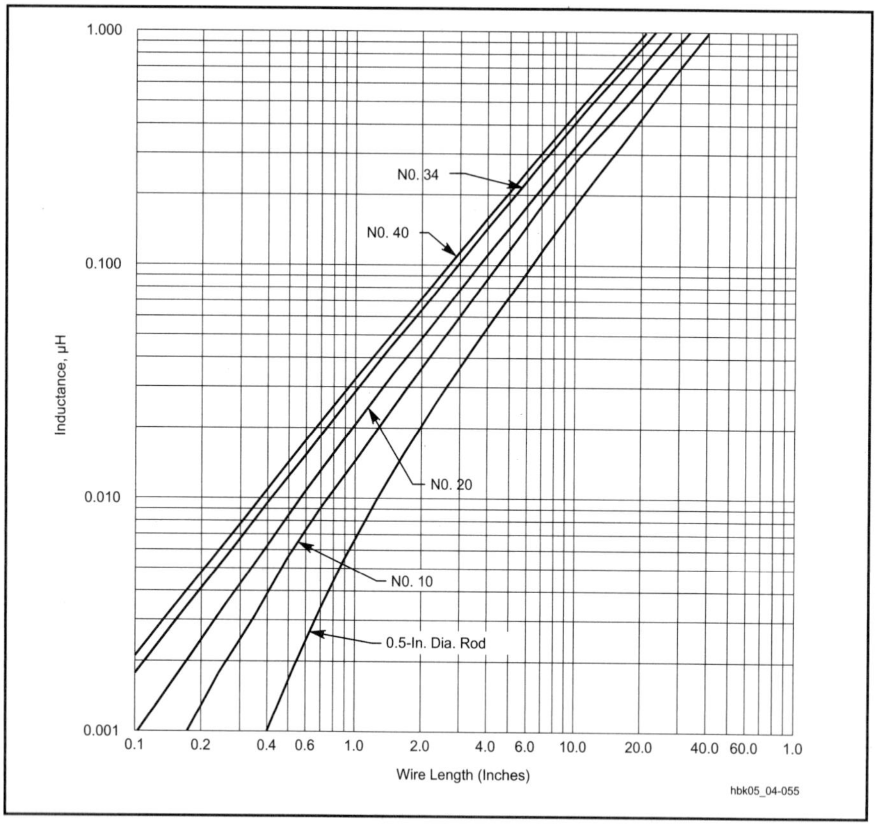

Fig 4.55 — Inductance of various conductor sizes as straight wires.

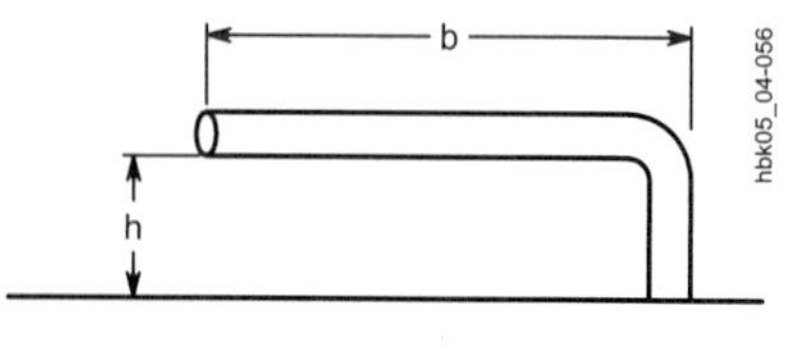

$$L = 0.0117\, b \left\{ \log_{10}\left[\frac{2h}{a} \left(\frac{b + \sqrt{b^2 + a^2}}{b + \sqrt{b^2 + 4h^2}} \right) \right] \right\} + 0.00508 \left(\sqrt{b^2 + 4h^2} - \sqrt{b^2 + a^2} + \frac{b}{4} - 2h + a \right)$$

where
L=inductance in μH
a=wire radius in inches
b=wire length parallel to ground plane in inches
h= wire height above ground plane in inches

Fig 4.56 — Equation for determining the inductance of a wire parallel to a ground plane, with one end grounded. If the dimensions are in millimeters, the numerical coefficients become 0.0004605 for the first term and 0.0002 for the second term.

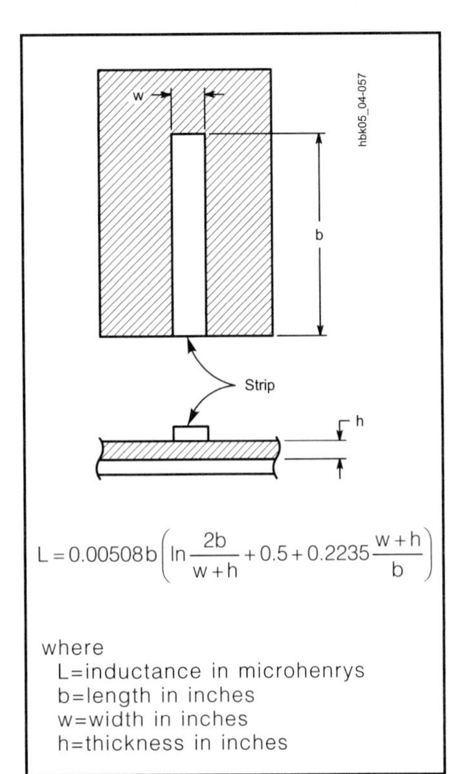

Fig 4.57 — Equation for determining the inductance of a flat strip inductor.

Substituting these values into the formula yields:

$$L = 0.0117 \times 3.9370 \left\{ \log_{10} \left[40.0 \times \left(\frac{7.88}{8.9788} \right) \right] \right\}$$

+ 0.00508 × (5.0418 – 3.94 + 0.98425 – 3.1496 + 0.0787)

$L = 0.0662 \mu H$

Another conductor configuration that is frequently used is a flat strip over a ground plane. This arrangement has lower skin-effect loss at high frequencies than round wire because it has a higher surface-area to volume ratio. The inductance of such a strip can be found from the formula in **Fig 4.57**. For a large collection of formulas useful in constructing air-core inductors of many configurations, see the "Circuit Elements" section in Terman's *Radio Engineers' Handbook* or the "Transmission Media" chapter of *The ARRL UHF/Microwave Experimenter's Manual.*

IRON-CORE INDUCTORS

If the permeability of an iron core in an inductor is 800, then the inductance of any given air-wound coil is increased 800 times by inserting the iron core. The inductance will be proportional to the magnetic flux through the coil, other things being equal. The inductance of an iron-core inductor is highly dependent on

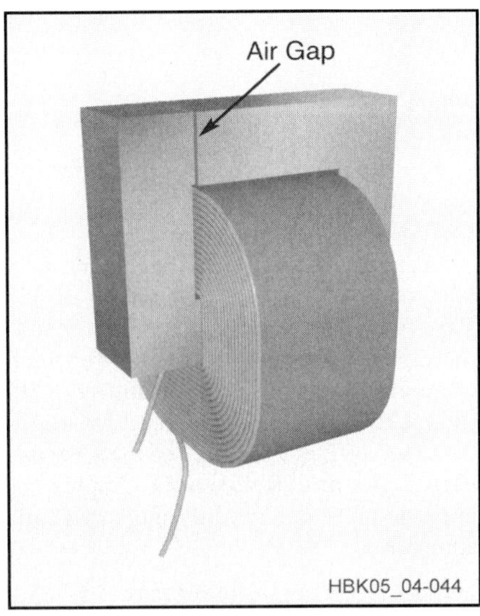

Fig 4.58 — Typical construction of an iron-core inductor. The small air gap prevents magnetic saturation of the iron and thus maintains the inductance at high currents.

the current flowing in the coil, in contrast to an air-core coil, where the inductance is independent of current because air does not saturate.

Iron-core coils such as the one sketched in **Fig 4.58** are used chiefly in power-supply equipment. They usually have direct current flowing through the winding, and any variation in inductance with current is usually undesirable. Inductance variations may be overcome by keeping the flux density below the saturation point of the iron. Opening the core so there is a small air gap, indicated by the dashed lines in Fig 4.58, will achieve this goal. The reluctance or magnetic resistance introduced by such a gap is very large compared with that of the iron, even though the gap is only a small fraction of an inch. Therefore, the gap — rather than the iron — controls the flux density. Air gaps in iron cores reduce the inductance, but they hold the value practically constant regardless of the current magnitude.

When alternating current flows through a coil wound on an iron core, a voltage is induced. Since iron is a conductor, a current also flows in the core. Such currents are called *eddy currents*. Eddy currents represent lost power because they flow through the resistance of the iron and generate heat. Losses caused by eddy currents can be reduced by laminating the core (cutting the core into thin strips). These strips or laminations are then insulated from each other by painting them with some insulating material such as varnish or shellac. These losses add to hysteresis losses, which are also significant in iron-core inductors.

Eddy-current and hysteresis losses in iron increase rapidly as the frequency of the alternating current increases. For this reason, ordinary iron cores can be used only at power-line and audio frequencies — up to approximately 15000 Hz. Even then, a very good grade of iron or steel is necessary for the core to perform well at the higher audio frequencies. Laminated iron cores become completely useless at radio frequencies.

SLUG-TUNED INDUCTORS

For RF work, the losses in iron cores can be reduced to a more useful level by grinding the iron into a powder and then mixing it with a "binder" of insulating material in such a way that the individual iron particles are insulated from each other. Using this approach, cores can be made that function satisfactorily even into the VHF range.

Because a large part of the magnetic path is through a nonmagnetic material (the "binder"), the permeability of the iron is low compared with the values obtained at power-line frequencies. The core is usually shaped in the form of a slug or cylinder for fit inside the insulating form on which the coil is wound. Despite the fact that the major portion of the magnetic path for the flux is in air, the slug is quite effective in increasing the coil inductance. By pushing (or screwing) the slug in and out of the coil, the inductance can be varied over a considerable range.

POWDERED-IRON TOROIDAL INDUCTORS

For fixed-value inductors intended for use at HF and VHF, the powdered-iron toroidal core has become almost the standard core and material in low power circuits. **Fig 4.59** shows the general outlines of a toroidal coil on a magnetic core. Manufacturers offer a wide variety of core materials, or mixes, to provide units that will perform over a desired frequency range with a reasonable permeability. Initial permeabilities for powdered-iron cores fall in the range of 3 to 35 for various mixes. In addition, core sizes are available in the range of 0.125-inch outside diameter (OD) up to 1.06-inch OD, with larger sizes to 5-inch OD available in certain mixes. The range of sizes permits the builder to construct single-layer inductors for almost any value using wire sized to meet the circuit current demands. While powdered-iron toroids are often painted various colors, you must know the manufacturer to identify the mix. There seems to be no set standard between manufac-

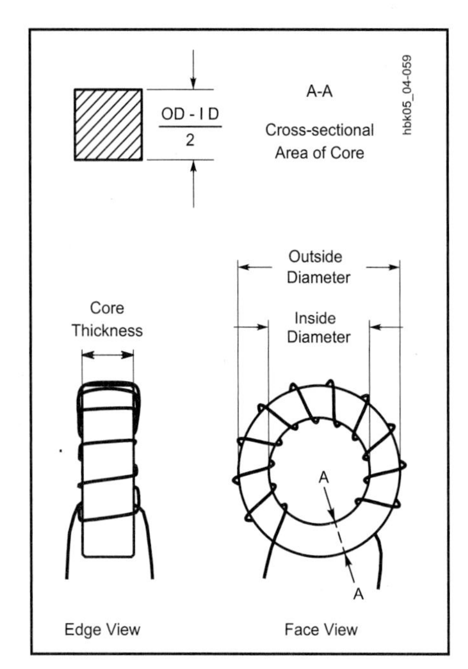

Fig 4.59 — A typical toroidal inductor wound on a powdered-iron or ferrite core. Some key physical dimensions are noted. Equally important are the core material, its permeability, its intended range of operating frequencies, and its A_L value. This is an 11-turn toroid.

turers. Iron-powder toroids usually have rounded edges.

The use of powdered iron in a binder reduces core losses usually associated with iron, while the permeability of the core permits a reduction in the wire length and associated resistance in forming a coil of a given inductance. Therefore, powdered-iron-core toroidal inductors can achieve Qs well above 100, often approaching or exceeding 200 within the frequency range specified for a given core. Moreover, these coils are considered self-shielding since most of the flux lines are within the core, a fact that simplifies circuit design and construction.

Each powdered-iron core has a value of A_L determined and published by the core manufacturer. For powdered-iron cores, A_L represents the *inductance index*, that is, the inductance in µH per 100 turns of wire on the core, arranged in a single layer. The builder must select a core size capable of holding the calculated number of turns, of the required wire size, for the desired inductance. Otherwise, the coil calculation is straightforward. To calculate the inductance of a powdered-iron toroidal coil, when the number of turns and the core material are known, use the formula:

$$L = \frac{A_L \times N^2}{10000} \tag{69}$$

where:

L = the inductance in µH,

A_L = the inductance index in µH per 100 turns, and

N = the number of turns.

Example: What is the inductance of a 60-turn coil on a core with an A_L of 55? This A_L value was selected from manufacturer's information about a 0.8-inch OD core with an initial permeability of 10. This particular core is intended for use in the range of 2 to 30 MHz. See the **Component Data and References** chapter for more detailed data on the range of available cores.

$$L = \frac{A_L \times N^2}{10000} = \frac{55 \times 60^2}{10000}$$

$$= \frac{198000}{10000} = 19.8\ \mu H$$

To calculate the number of turns needed for a particular inductance, use the formula:

$$N = 100\sqrt{\frac{L}{A_L}} \tag{70}$$

Example: How many turns are needed for a 12.0-µH coil if the A_L for the selected core is 49?

$$N = 100\sqrt{\frac{L}{A_L}} = 100\sqrt{\frac{12.0}{49}}$$

$$= 100\sqrt{0.245} = 100 \times 0.495 = 49.5 \text{ turns}$$

If the value is critical, experimenting with 49-turn and 50-turn coils is in order, especially since core characteristics may vary slightly from batch to batch. Count turns by each pass of the wire through the center of the core. (A straight wire through a toroidal core amounts to a one-turn coil.) Fine adjustment of the inductance may be possible by spreading or squeezing inductor turns.

The power-handling ability of toroidal cores depends on many variables, which include the cross-sectional area through the core, the core material, the numbers of turns in the coil, the applied voltage and the operating frequency. Although powdered-iron cores can withstand dc flux densities up to 5000 gauss without saturating, ac flux densities from sine waves above certain limits can overheat cores. Manufacturers provide guideline limits for ac flux densities to avoid overheating. The limits range from 150 gauss at 1 MHz to 30 gauss at 28 MHz, although the curve is not linear. To calculate the maximum anticipated flux density for a particular coil, use the formula:

$$B_{max} = \frac{E_{RMS} \times 10^8}{4.44 \times A_e \times N \times f} \tag{71}$$

where:

B_{max} = the maximum flux density in gauss,

E_{RMS} = the voltage across the coil,

A_e = the cross-sectional area of the core in square centimeters,

N = the number of turns in the coil, and

f = the operating frequency in Hz.

Example: What is the maximum ac flux density for a coil of 15 turns if the frequency is 7.0 MHz, the RMS voltage is 25 V and the cross-sectional area of the core is 0.133 cm^2?

$$B_{max} = \frac{E_{RMS} \times 10^8}{4.44 \times A_e \times N \times f}$$

$$= \frac{25 \times 10^8}{4.44 \times 0.133 \times 15 \times 7.0 \times 10^6}$$

$$= \frac{25 \times 10^8}{62 \times 10^6} = 40 \text{ gauss}$$

Since the recommended limit for cores operated at 7 MHz is 57 gauss, this coil is well within guidelines.

FERRITE TOROIDAL INDUCTORS

Although nearly identical in general appearance to powdered-iron cores, ferrite cores differ in a number of important characteristics. They are often unpainted, unlike powdered-iron toroids. Ferrite toroids often have sharp edges, while powdered-iron toroids usually have rounded edges. Composed of nickel-zinc ferrites for lower permeability ranges and of manganese-zinc ferrites for higher per-meabilities, these cores span the permeability range from 20 to above 10000. Nickel-zinc cores with permeabilities from 20 to 800 are useful in high-Q applications, but function more commonly in amateur applications as RF chokes. They are also useful in wide-band transformers (discussed later in this chapter).

Because of their higher permeabilities, the formulas for calculating inductance and turns require slight modification. Manufacturers list ferrite A_L values in mH per 1000 turns. Thus, to calculate inductance, the formula is

$$L = \frac{A_L \times N^2}{1000000} \tag{72}$$

where:

L = the inductance in mH,

A_L = the inductance index in mH per 1000 turns, and

N = the number of turns.

Example: What is the inductance of a 60-turn coil on a core with an A_L of 523? (See the **Component Data and References** chapter for more detailed data on the range of available cores.)

$$L = \frac{A_L \times N^2}{1000000} = \frac{523 \times 60^2}{1000000}$$

$$= \frac{1.88 \times 10^6}{1 \times 10^6} = 1.88 \text{ mH}$$

To calculate the number of turns needed for a particular inductance, use the formula:

$$N = 1000 \sqrt{\frac{L}{A_L}} \qquad (73)$$

Example: How many turns are needed for a 1.2-mH coil if the A_L for the selected core is 150?

$$N = 1000 \sqrt{\frac{L}{A_L}} = 1000 \sqrt{\frac{1.2}{150}}$$

$$= 1000 \sqrt{0.008} = 1000 \times 0.089 = 89 \text{ turns}$$

For inductors carrying both dc and ac currents, the upper saturation limit for most ferrites is a flux density of 2000 gauss, with power calculations identical to those used for powdered-iron cores. For detailed information on available cores and their characteristics, see *Iron-Powder and Ferrite Coil Forms*, a combination catalog and information book from Amidon Associates, Inc.

Ohm's Law for Reactance

Only ac circuits containing capacitance or inductance (or both) have reactance. Despite the fact that the voltage in such circuits is 90° out of phase with the current, circuit reactance does limit current in a manner that corresponds to resistance. Therefore, the Ohm's Law equations relating voltage, current and resistance apply to purely reactive circuits:

$$E = I\,X \qquad (74)$$

$$I = \frac{E}{X} \qquad (75)$$

$$X = \frac{E}{I} \qquad (76)$$

where:

E = ac voltage in RMS,

I = ac current in amperes, and

X = inductive or capacitive reactance.

Example: What is the voltage across a capacitor of 200 pF at 7.15 MHz, if the current through the capacitor is 50 mA?

Since the reactance of the capacitor is a function of both frequency and capacitance, first calculate the reactance:

$$X_C = \frac{1}{2\,\pi\,f\,C}$$

$$= \frac{1}{2 \times 3.1416 \times 7.15 \times 10^6 \text{ Hz} \times 200 \times 10^{-12} \text{ F}}$$

$$= \frac{10^6\ \Omega}{8980}\ 111\,\Omega$$

Next, use Ohm's Law:

$$E = I \times X_C = 0.050 \text{ A} \times 111\,\Omega = 5.6 \text{ V}$$

Example: What is the current through an 8.00-H inductor at 120 Hz, if 420 V is applied?

$$X_L = 2\,\pi\,f\,L = 2 \times 3.1416 \times 120 \text{ Hz} \times 8.00 \text{ H} = 6030\,\Omega$$

Fig 4.60 charts the reactances of capacitors from 1 pF to 100 µF, and the reactances of inductors from 0.1 µH to 10 H, for frequencies between 100 Hz and 100 MHz. Approximate values of reactance can be read or interpolated from the chart. The formulas will produce more exact values, however.

Although both inductive and capacitive reactance limit current, the two types of reactance differ. With capacitive reactance, the current *leads* the voltage by 90°, whereas with inductive reactance, the current *lags* the voltage by 90°. The convention for charting the two types of reactance appears in **Fig 4.61**. On this graph, inductive reactance is plotted along the +90° vertical line, while capacitive reactance is plotted along the –90° vertical line. This convention of assigning a positive value to inductive reactance and a negative value to capacitive reactance results from the mathematics involved in impedance calculations.

REACTANCES IN SERIES AND PARALLEL

If a circuit contains two reactances of the same type, whether in series or in parallel, the resultant reactance can be determined by applying the same rules as for resistances in series and in parallel. Series reactance is given by the formula

$$X_{total} = X1 + X2 + X3 + \ldots + X_n \qquad (77)$$

Example: Two noninteracting inductances are in series. Each has a value of 4.0 µH, and the operating frequency is 3.8 MHz. What is the resulting reactance?

The reactance of each inductor is:

$$X_L = 2\,\pi\,f\,L = 2 \times 3.1416 \times 3.8 \times 10^6 \text{ Hz} \times 4 \times 10^{-6} \text{ H} = 96\,\Omega$$

$$X_{total} = X1 + X2 = 96\,\Omega + 96\,\Omega = 192\,\Omega$$

We might also calculate the total reactance by first adding the inductances:

$$L_{total} = L1 + L2 = 4.0\,\mu\text{H} + 4.0\,\mu\text{H} = 8.0\,\mu\text{H}$$

$$X_{total} = 2\,\pi\,f\,L = 2 \times 3.1416 \times 3.8 \times 10^6 \text{ Hz} \times 8.0 \times 10^{-6} \text{ H}$$

$$X_{total} = 191\,\Omega$$

(The fact that the last digit differs by one illustrates the uncertainty of the calculation caused by the uncertainty of the measured values in the problem, and differences caused by rounding off the calculated values. This also shows why it is important to follow the rules for significant figures.

Example: Two noninteracting capacitors are in series. One has a value of 10.0 pF, the other of 20.0 pF. What is the resulting reactance in a circuit operating at 28.0 MHz?

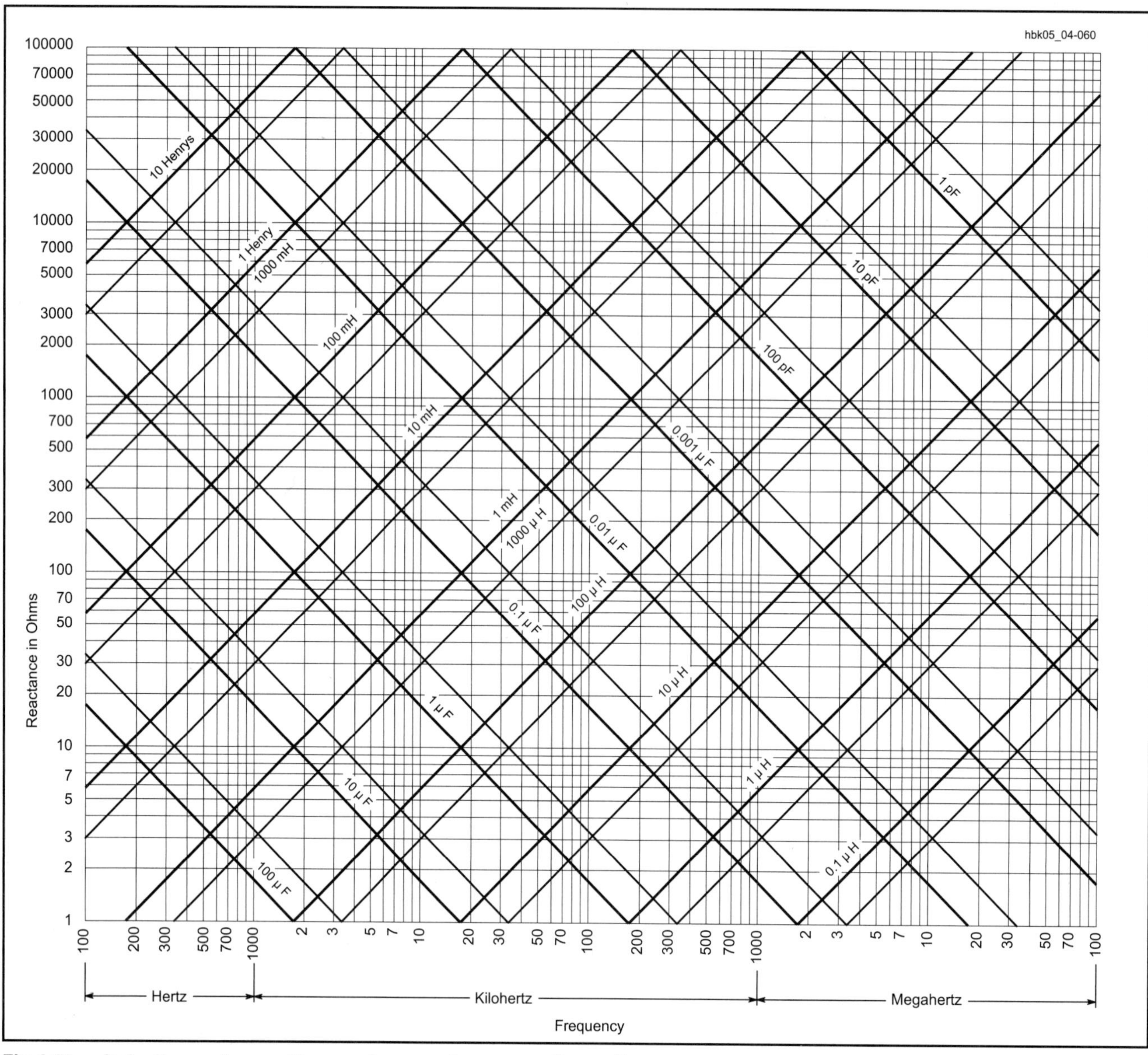

Fig 4.60 — Inductive and capacitive reactance vs frequency. Heavy lines represent multiples of 10, intermediate lines multiples of 5. For example, the light line between 10 µH and 100 µH represents 50 µH; the light line between 0.1 µF and 1 µF represents 0.5µF, and so on. Other values can be extrapolated from the chart. For example, the reactance of 10 H at 60 Hz can be found by taking the reactance of 10 H at 600 Hz and dividing by 10 for the 10 times decrease in frequency.

$$X_{C1} = \frac{1}{2 \pi f C}$$

$$= \frac{1}{2 \times 3.1416 \times 28.0 \times 10^6 \text{ Hz} \times 10.0 \times 10^{-12} \text{ F}}$$

$$= \frac{10^6 \ \Omega}{1760} = 568 \ \Omega$$

$$X_{C2} = \frac{1}{2 \pi f C}$$

$$= \frac{1}{2 \times 3.1416 \times 28.0 \times 10^6 \text{ Hz} \times 20.0 \times 10^{-12} \text{ F}}$$

$$= \frac{10^6 \ \Omega}{3520} = 284 \ \Omega$$

$$X_{total} = X_{C1} + X_{C2} = 568 \ \Omega + 284 \ \Omega = 852 \ \Omega$$

Alternatively, for series capacitors, the total capacitance is 6.67×10^{-12} F or 6.67 pF. Then:

$$X_{total} = \frac{1}{2 \pi f C}$$

$$= \frac{1}{2 \times 3.1416 \times 28.0 \times 10^6 \text{ Hz} \times 6.67 \times 10^{-12} \text{ F}}$$

$$= \frac{10^6 \ \Omega}{1170} = 855 \ \Omega$$

(Within the uncertainty of the measured values and the rounding of values in the calculations, this is the same result as we obtained with the first method.)

This example serves to remind us that *series capacitance* is not calculated in the manner used by other series resistance and inductance, but *series capacitive reactance* does follow the simple addition formula.

For reactances of the same type in parallel, the general formula is:

$$X_{total} = \frac{1}{\frac{1}{X1} + \frac{1}{X2} + \frac{1}{X3} + \ldots + \frac{1}{X_n}} \quad (78)$$

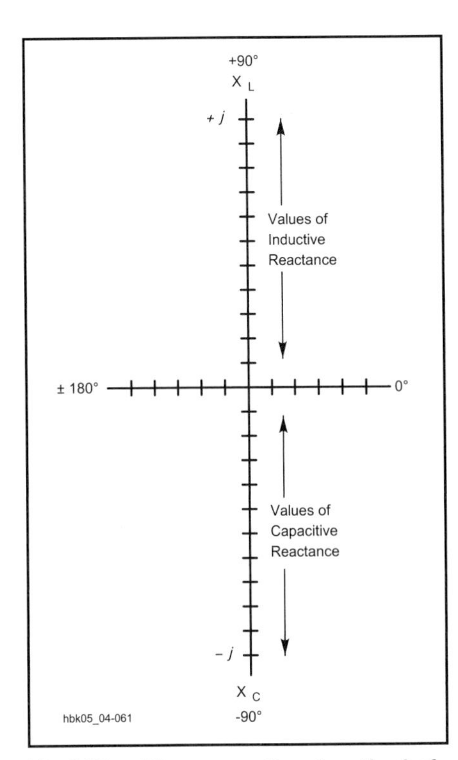

Fig 4.61 — The conventional method of plotting reactances on the vertical axis of a graph, using the upward or "plus" direction for inductive reactance and the downward or "minus" direction for capacitive reactance. The horizontal axis will be used for resistance in later examples.

or, for exactly two reactances in parallel

$$X_{total} = \frac{X1 \times X2}{X1 + X2} \qquad (79)$$

Example: Place the capacitors in the last example (10.0 pF and 20.0 pF) in parallel in the 28.0 MHz circuit. What is the resultant reactance?

$$X_{total} = \frac{X1 \times X2}{X1 + X2}$$

$$= \frac{568\,\Omega \times 284\,\Omega}{568\,\Omega + 284\,\Omega} = 189\,\Omega$$

Alternatively, two capacitors in parallel add their capacitances.

$$C_{total} = C_1 + C_2 = 10.0\text{ pF} + 20.0\text{ pF} = 30\text{ pF}$$

$$X_C = \frac{1}{2\,\pi\,f\,C}$$

$$= \frac{1}{2 \times 3.1416 \times 28.0 \times 10^6\text{ Hz} \times 30 \times 10^{-12}\text{ F}}$$

$$= \frac{10^6\,\Omega}{5280} = 189\,\Omega$$

Example: Place the series inductors above (4.0 µH each) in parallel in a 3.8-MHz circuit. What is the resultant reactance?

$$X_{total} = \frac{X_{L1} \times X_{L2}}{X_{L1} + X_{L2}}$$

$$= \frac{96\,\Omega \times 96\,\Omega}{96\,\Omega + 96\,\Omega} = 48\,\Omega$$

Of course, equal reactances (or resistances) in parallel yield a reactance that is the value of one of them divided by the number (n) of equal reactances, or:

$$X_{total} = \frac{X}{n} = \frac{96\,\Omega}{2} = 48\,\Omega$$

All of these calculations apply only to reactances of the same type; that is, all capacitive or all inductive. Mixing types of reactances requires a different approach.

UNLIKE REACTANCES IN SERIES

When combining unlike reactances — that is, combinations of inductive and capacitive reactance — in series, it is necessary to take into account that the voltage-to-current phase relationships differ for the different types of reactance. **Fig 4.62** shows a series circuit with both types of reactance. Since the reactances are in series, the current must be the same in both. The voltage across each circuit element differs in phase, however. The voltage E_L *leads* the current by 90°, and the voltage E_C *lags* the current by 90°. Therefore, E_L and E_C have opposite polarities and cancel each other in whole or in part. The dotted line in Fig 4.62 approximates the resulting voltage E, which is the *difference* between E_L and E_C.

Since, for a constant current, the reactance is directly proportional to the voltage, the net reactance must be the difference between the inductive and the capacitive reactances, or:

$$X_{total} = X_L - X_C \qquad (80)$$

For this and subsequent calculations in which there is a mixture of inductive and capacitive reactance, use the absolute value of each reactance. The convention of recording inductive reactances as positive and capacitive reactances as negative is built into the mathematical operators in the formulas.

Example: Using Fig 4.62 as a visual aid, let X_C = 20.0 Ω and X_L = 80.0 Ω. What is the resulting reactance?

$$X_{total} = X_L - X_C$$

$$= 80.0\,\Omega - 20.0\,\Omega = +60.0\,\Omega$$

Since the result is a positive value, reactance is inductive. Had the result been a negative number, the reactance would have been capacitive.

When reactance types are mixed in a series circuit, the resulting reactance is always smaller than the larger of the two reactances. Likewise, the resulting voltage across the series combination of reactances is always smaller than the larger of the two voltages across individual reactances.

Every series circuit of mixed reactance types with more than two circuit elements can be reduced to the type of circuit covered here. If the circuit has more than one capacitor or more than one inductor in the overall series string, first use the formulas given earlier to determine the total series inductance alone and the total series capacitance alone (or their respective reactances). Then combine the resulting single capacitive reactance and single inductive reactance as shown in this section.

UNLIKE REACTANCES IN PARALLEL

The situation of parallel reactances of mixed type appears in **Fig 4.63**. Since the elements are in parallel, the voltage is common to both reactive components. The current through the capacitor, I_C, *leads* the voltage by 90°, and the current through the inductor, I_L, *lags* the voltage by 90°. The two currents are 180° out of phase and thus cancel each other in whole or in part. The total current is the difference between the individual currents, as indicated by the dotted line in Fig 4.63.

Since reactance is the ratio of voltage to current, the total reactance in the circuit is:

$$X_{total} = \frac{E}{I_L - I_C} \qquad (81)$$

In the drawing, I_C is larger than I_L, and the resulting differential current retains the phase of I_C. Therefore, the overall reactance, X_{total}, is capacitive in this case. The total reactance of the circuit will be smaller than the larger of the individual reactances, because the total current is smaller than the larger of the two individual currents.

In parallel circuits of this type, reactance and current are inversely proportional to each other for a constant voltage. Therefore, to calculate the total reactance directly from the individual reactances, use the formula:

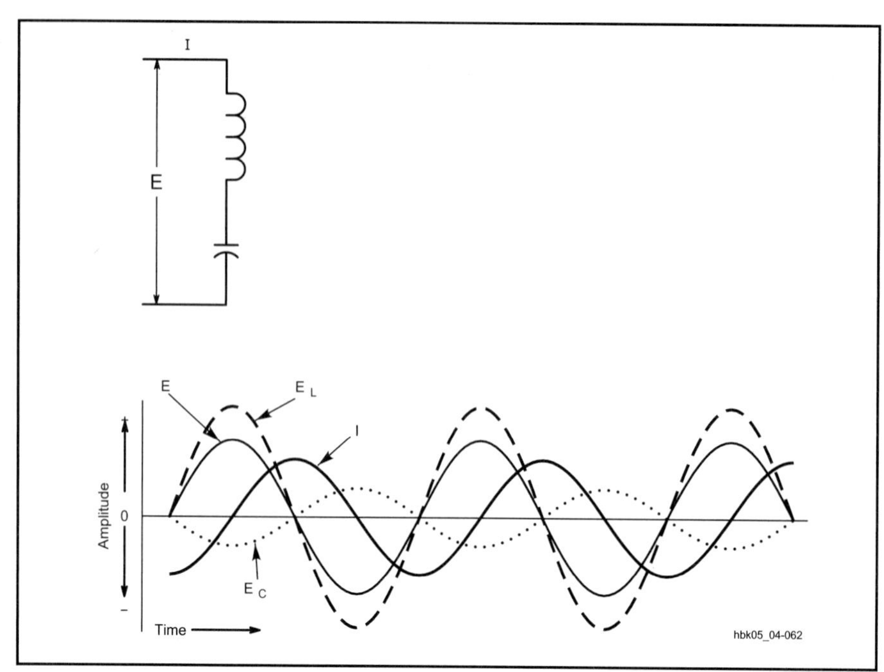

Fig 4.62 — A series circuit containing both inductive and capacitive components, together with representative voltage and current relationships.

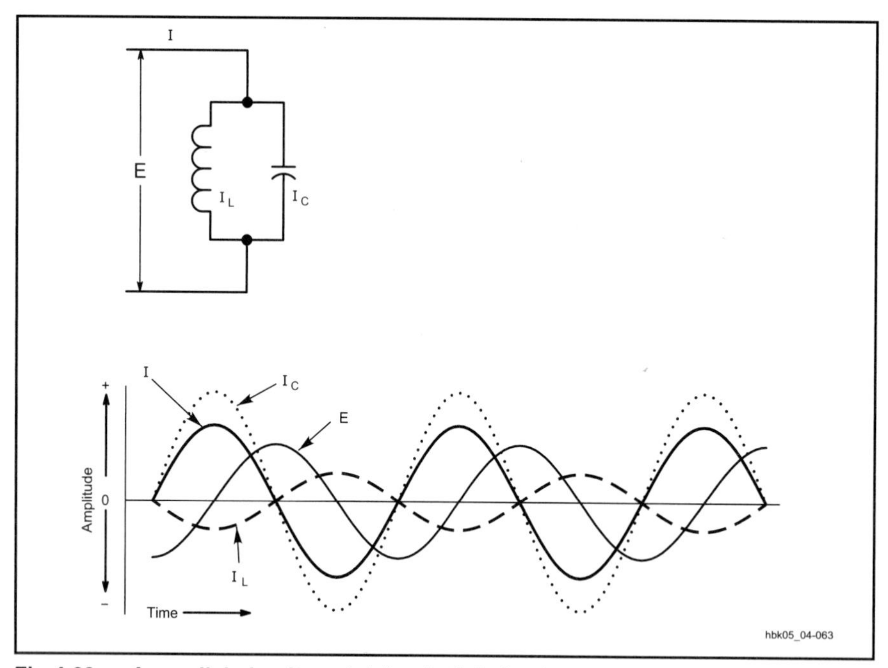

Fig 4.63 — A parallel circuit containing both inductive and capacitive components, together with representative voltage and current relationships.

$$X_{total} = \frac{-X_L \times X_C}{X_L - X_C} \quad (82)$$

As with the series formula for mixed reactances, use the absolute values of the reactances, since the minus signs in the formula take into account the convention of treating capacitive reactances as negative numbers. If the solution yields a negative number, the resulting reactance is capacitive, and if the solution is positive, then the reactance is inductive.

Example: Using Fig 4.63 as a visual aid, place a capacitive reactance of 10.0 Ω in parallel with an inductive reactance of 40.0 Ω. What is the resulting reactance?

$$X_{total} = \frac{-X_L \times X_C}{X_L - X_C}$$

$$= \frac{-40.0\ \Omega \times 10.0\ \Omega}{40.0\ \Omega - 10.0\ \Omega}$$

$$= \frac{-400\ \Omega}{30.0\ \Omega} = -13.3\ \Omega$$

The reactance is capacitive, as indicated by the negative solution. Moreover, the resultant reactance is always smaller than the larger of the two individual reactances.

As with the case of series reactances, if each leg of a parallel circuit contains more than one reactance, first simplify each leg to a single reactance. If the reactances are of the same type in each leg, the series reactance formulas for reactances of the same type will apply. If the reactances are of different types, then use the formulas shown above for mixed series reactances to simplify the leg to a single value and type of reactance.

APPROACHING RESONANCE

When two unlike reactances have the same numerical value, any series or parallel circuit in which they occur is said to be *resonant*. For any given inductance or capacitance, it is theoretically possible to find a value of the opposite reactance type to produce a resonant circuit for any desired frequency.

When a series circuit like the one shown in Fig 4.62 is resonant, the voltages E_C and E_L are equal and cancel; their sum is zero. Since the reactance of the circuit is proportional to the sum of these voltages, the total reactance also goes to zero. Theoretically, the current, as shown in **Fig 4.64**, can rise without limit. In fact, it is limited only by power losses in the components and other resistances that would be in a real circuit of this type. As the frequency of operation moves slightly off resonance, the reactance climbs rapidly and then begins to level off. Similarly, the current drops rapidly off resonance and then levels.

In a parallel-resonant circuit of the type in Fig 4.63, the current I_L and I_C are equal and cancel to zero. Since the reactance is inversely proportional to the current, as the current approaches zero, the reactance rises without limit. As with series circuits, component power losses and other resistances in the circuit limit the current drop to some point above zero. **Fig 4.65** shows the theoretical current curve near and at resonance for a purely reactive parallel-resonant circuit. Note that in both Fig 4.64 and Fig 4.65, the departure of current from the resonance value is close to, but not quite, symmetrical above and below the resonant frequency.

Example: What is the reactance of a series L-C circuit consisting of a 56.04-pF capacitor and an 8.967-µH inductor at 7.00, 7.10 and 7.20 MHz? Using the formulas from earlier in this chapter, we calculate a table of values:

Frequency (MHz)	X_L (Ω)	X_C (Ω)	X_{total} (Ω)
7.000	394.4	405.7	–11.3
7.100	400.0	400.0	0
7.200	405.7	394.4	11.3

The exercise shows the manner in which the reactance rises rapidly as the frequency moves above and below resonance. Note that in a series-resonant circuit, the reactance at frequencies below resonance is capacitive, and above resonance, it is inductive. **Fig 4.66** displays this fact graphically. In a parallel-resonant circuit, where the reactance increases without limit at resonance, the opposite condition exists: above resonance, the reactance is capacitive and below resonance it is inductive, as shown in **Fig 4.67**. Of course, all graphs and calculations in this section are theoretical and presume a purely reactive circuit. Real circuits are never purely reactive; they contain some resistance that modifies their performance considerably. Real resonant circuits will be discussed later in this chapter.

REACTIVE POWER

Although purely reactive circuits, whether simple or complex, show a measurable ac voltage and current, we cannot simply multiply the two together to arrive at power. Power is the rate at which energy is consumed by a circuit, and purely reactive circuits do not consume power. The charge placed on a capacitor during part of an ac cycle is returned to the circuit during the next part of a cycle. Likewise, the energy stored in the magnetic field of an inductor returns to the circuit as the field collapses later in the ac cycle. A reactive circuit simply cycles and recycles energy into and out of the reactive components. If a purely reactive circuit were possible in reality, it would consume no power at all.

In reactive circuits, circulation of energy accounts for seemingly odd phenomena. For example, in a series circuit with capacitance and inductance, the voltages across the components may exceed the supply voltage. That condition can exist because, while energy is being stored by the inductor, the capacitor is returning energy to the circuit from its previously charged state, and vice versa. In a parallel circuit with inductive and capacitive branches, the current circulating through the components may exceed the current drawn from the source. Again, the phenomenon occurs because the inductor's collapsing field supplies current to the capacitor, and the discharging capacitor provides current to the inductor.

To distinguish between the non-dissipated power in a purely reactive circuit and the dissipated power of a resistive circuit, the unit of reactive power is called the *volt-ampere reactive*, or VAR. The term watt is not used; sometimes reactive power is called wattless power. Formulas similar to those for resistive power are used to calculate VAR:

$$VAR = I \times E \quad (83)$$

$$VAR = I^2 \times X \quad (84)$$

$$VAR = \frac{E^2}{X} \quad (85)$$

These formulas have only limited use in radio work.

REACTANCE AND COMPLEX WAVEFORMS

All of the formulas and relationships shown in this section apply to alternating current in the form of regular sine waves. Complex wave shapes complicate the reactive situation considerably. A complex or nonsinusoidal wave can be resolved into a fundamental frequency and a series of harmonic frequencies whose amplitudes depend on the original wave shape. When such a complex wave — or

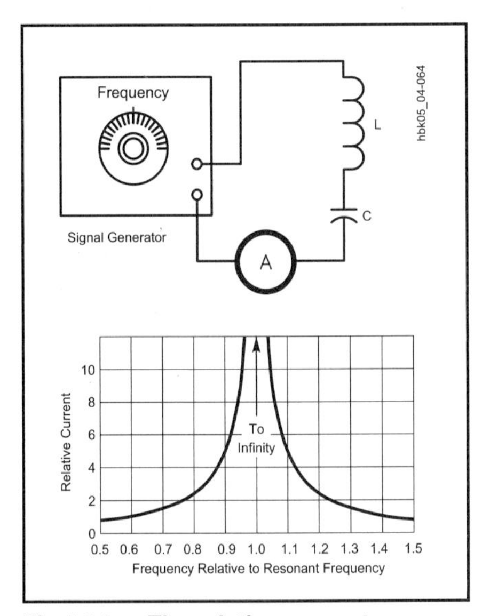

Fig 4.64 — The relative generator current with a fixed voltage in a series circuit containing inductive and capacitive reactances as the frequency approaches and departs from resonance.

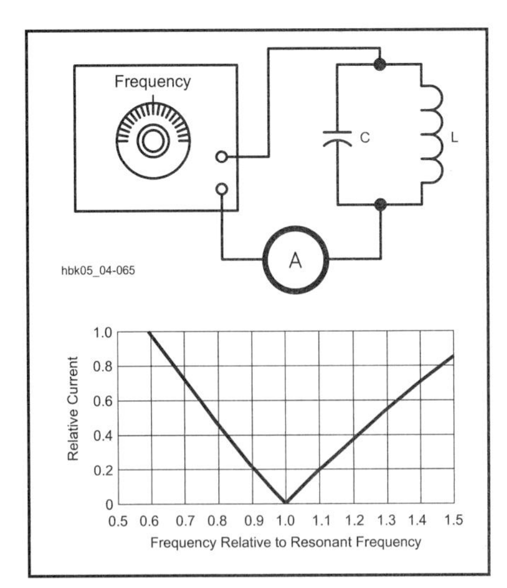

Fig 4.65 — The relative generator current with a fixed voltage in a parallel circuit containing inductive and capacitive reactances as the frequency approaches and departs from resonance. (The circulating current through the parallel inductor and capacitor is a maximum at resonance.)

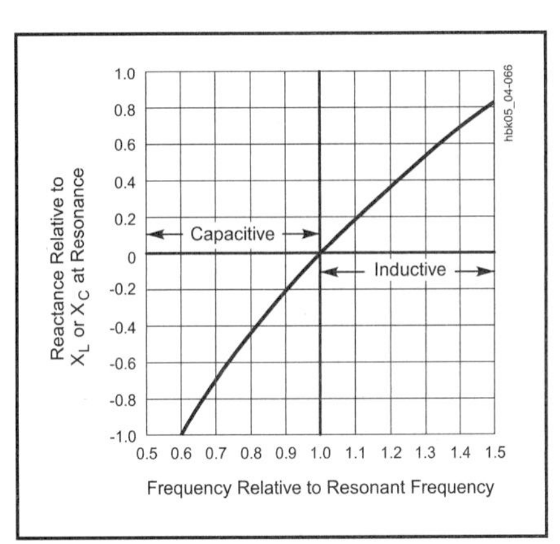

Fig 4.66 — The transition from capacitive to inductive reactance in a series-resonant circuit as the frequency passes resonance.

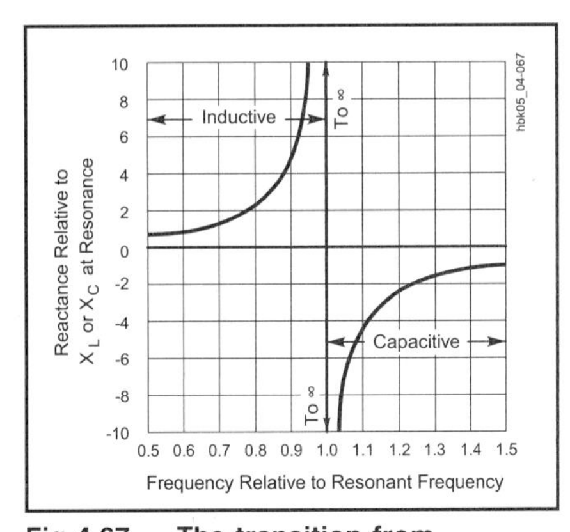

Fig 4.67 — The transition from inductive to capacitive reactance in a parallel-resonant circuit as the frequency passes resonance.

collection of sine waves — is applied to a reactive circuit, the current through the circuit will not have the same wave shape as the applied voltage. The difference results because the reactance of an inductor and capacitor depend in part on the applied frequency.

For the second-harmonic component of the complex wave, the reactance of the inductor is twice and the reactance of the capacitor is half their respective values at the fundamental frequency. A third-harmonic component produces inductive reactances that are triple and capacitive reactances that are one-third those at the fundamental frequency. Thus, the overall circuit reactance is different for each harmonic component.

The frequency sensitivity of a reactive circuit to various components of a complex wave shape creates both difficulties and opportunities. On the one hand, calculating the circuit reactance in the presence of highly variable as well as complex waveforms, such as speech, is difficult at best. On the other hand, the frequency sensitivity of reactive components and circuits lays the foundation for filtering, that is, for separating signals of different frequencies and passing them into different circuits. For example, suppose a coil is in the series path of a signal and a capacitor is connected from the signal line to ground, as represented in **Fig 4.68**. The reactance of the coil to the second harmonic of the signal will be twice that at the fundamental frequency and oppose more effectively the flow of harmonic current. Likewise, the reactance of the capacitor to the harmonic will be half that to the fundamental, allowing the harmonic an easier current path away from the signal line toward ground. See the **RF and AF Filters** chapter for detailed information on filter theory and construction.

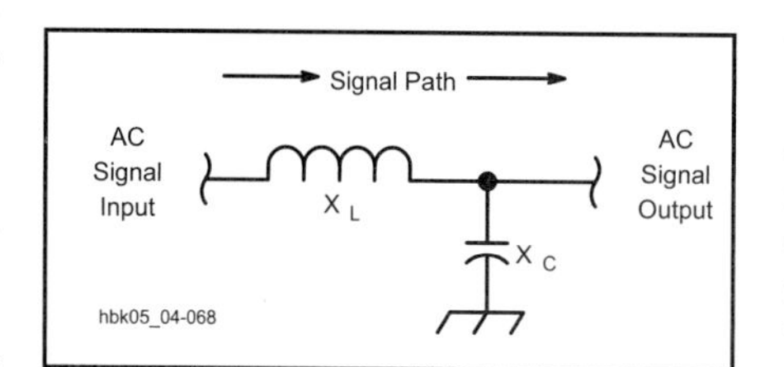

Fig 4.68 — A signal path with a series inductor and a shunt capacitor. The circuit presents different reactances to an ac signal and to its harmonics.

Impedance

When a circuit contains both resistance and reactance, the combined opposition to current is called *impedance*. Symbolized by the letter Z, impedance is a more general term than either resistance or reactance. Frequently, the term is used even for circuits containing only resistance or reactance. Qualifications such as "resistive impedance" are sometimes added to indicate that a circuit has only resistance, however.

The reactance and resistance comprising an impedance may be connected either in series or in parallel, as shown in **Fig 4.69**. In these circuits, the reactance is shown as a box to indicate that it may be either inductive or capacitive. In the series circuit at A, the current is the same in both elements, with (generally) different voltages appearing across the resistance and reactance. In the parallel circuit at B, the same voltage is applied to both elements, but different currents may flow in the two branches.

In a resistance, the current is in phase with the applied voltage, while in a reactance it is 90° out of phase with the voltage. Thus, the phase relationship between current and voltage in the circuit as a whole may be anything between zero and 90°, depending on the relative amounts of resistance and reactance.

As shown in Fig 4.61 in the preceding section, reactance is graphed on the vertical (Y) axis to record the phase difference between the voltage and the current. **Fig 4.70** adds resistance to the graph. Since the voltage is in phase with the current, resistance is recorded on the horizontal axis, using the positive or right side of the scale.

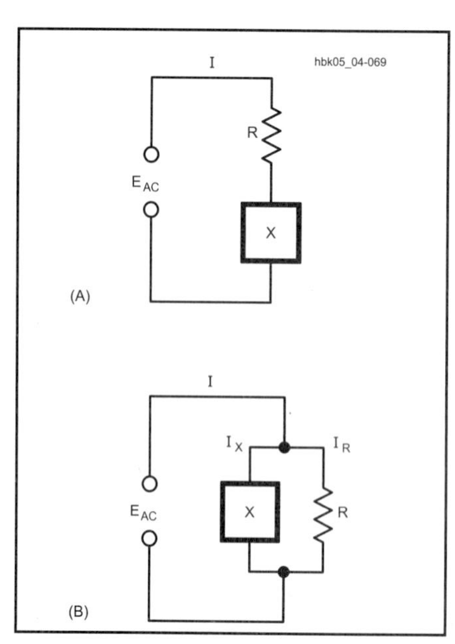

Fig 4.69 — Series and parallel circuits containing resistance and reactance.

CALCULATING Z FROM R AND X IN SERIES CIRCUITS

Impedance is the complex combination of resistance and reactance. Since there is a 90° phase difference between resistance and reactance (whether inductive or capacitive), simply adding the two values will not yield what actually happens in a

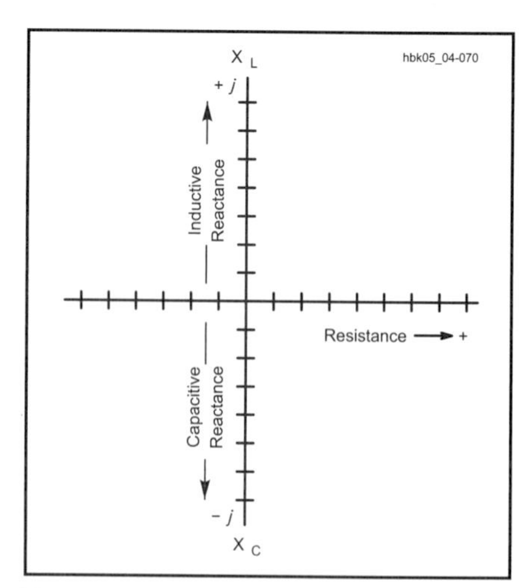

Fig 4.70 — The conventional method of charting impedances on a graph, using the vertical axis for reactance (the upward or "plus" direction for inductive reactance and the downward or "minus" direction for capacitive reactance), and using the horizontal axis for resistance.

circuit. Therefore, expressions like "Z = R ± X" can be misleading, because they suggest simple addition. As a result, impedance is often expressed "Z = R ± *j*X."

In pure mathematics, "*i*" indicates an imaginary number. Because i represents current in electronics, we use the letter "*j*" for the same mathematical operator, although there is nothing imaginary about what it represents in electronics. With respect to resistance and reactance, the letter *j* is normally assigned to those figures on the vertical scale, 90° out of phase with the horizontal scale. The actual function of *j* is to indicate that calculating impedance from resistance and reactance requires *vector addition*. In vector addition, the result of combining two values at a 90° phase difference results in a new quantity for the combination, and also in a new combined phase angle relative to the base line.

Consider **Fig 4.71**, a series circuit consisting of an inductive reactance and a resistance. As given, the inductive reactance is 100 Ω and the resistance is 50 Ω. Using *rectangular coordinates*, the impedance becomes

$$Z = R + jX \quad (86)$$

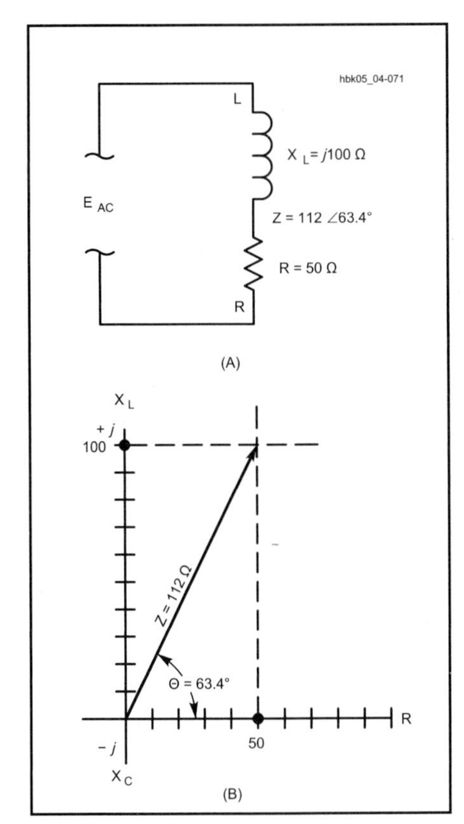

Fig 4.71 — A series circuit consisting of an inductive reactance of 100 Ω and a resistance of 50 Ω. At B, the graph plots the resistance, reactance, and impedance.

where:

Z = the impedance in ohms,

R = the resistance in ohms, and

X = the reactance in ohms.

In the present example,

$$Z = 50 + j100\ \Omega$$

As the graph shows, the combined opposition to current (or impedance) is represented by a line triangulating the two given values. The graph will provide an estimate of the value. A more exact way to calculate the resultant impedance involves the formula for right triangles, where the square of the hypotenuse equals the sum of the squares of the two sides. Since impedance is the hypotenuse:

$$Z = \sqrt{R^2 + X^2} \quad (87)$$

In this example:

$$Z = \sqrt{(50\ \Omega)^2 + (100\ \Omega)^2}$$

$$= \sqrt{2500\ \Omega^2 + 10000\ \Omega^2}$$

$$= \sqrt{12500\ \Omega^2} = 112\ \Omega$$

The impedance that results from combining 50 Ω of resistance with 100 Ω of inductive reactance is 112 Ω. The phase angle of the resultant is neither 0° nor +90°. Instead, it lies somewhere between the two. Let θ be the angle between the horizontal axis and the line representing the impedance. From trigonometry, the tangent of the angle is the side opposite the angle divided by the side adjacent to the angle, or

$$\tan\theta = \frac{X}{R} \quad (88)$$

where:

X = the reactance, and

R = the resistance.

Find the angle by taking the inverse tangent, or arctan:

$$\theta = \arctan\frac{X}{R} \quad (89)$$

In the example shown in Fig 6.57,

$$\theta = \arctan\frac{100\ \Omega}{50\ \Omega} = \arctan 2.0 = 63.4°$$

Combining the resultant impedance with the angle provides the impedance in *polar coordinate* form:

$$Z\angle\theta \quad (90)$$

Using the information just calculated, the impedance is:

$$Z = 112\ \Omega \angle 63.4°$$

The expressions R ± *j*X and Z ∠θ both provide the same information, but in two different forms. The procedure just given permits conversion from rectangular coordinates into polar coordinates. The reverse procedure is also important. **Fig 4.72** shows an impedance composed of a capacitive reactance and a resistance. Since capacitive reactance appears as a negative value, the impedance will be at a negative phase angle, in this case, 12.0 Ω at a phase angle of – 42.0° or Z = 12.0 Ω ∠– 42.0°.

Think of the impedance as forming a triangle with the values of X and R from the rectangular coordinates. The reactance axis forms the side opposite the angle θ.

$$\sin\theta = \frac{\text{side opposite}}{\text{hypotenuse}} = \frac{X}{Z} \quad (91)$$

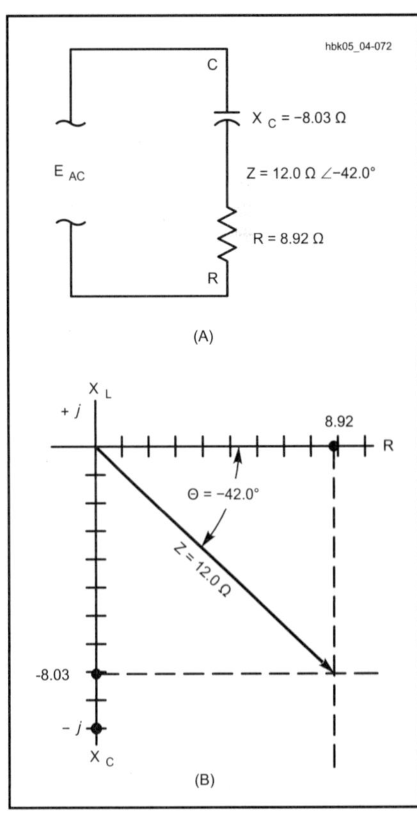

Fig 4.72 — A series circuit consisting of a capacitive reactance and a resistance: the impedance is given as 12.0 Ω at a phase angle θ of -42 degrees. At B, the graph plots the resistance, reactance, and impedance.

Solving this equation for reactance, we have:

$$X = Z \times \sin\theta \text{ (ohms)} \quad (92)$$

Likewise, the resistance forms the side adjacent to the angle.

$$\cos\theta = \frac{\text{side adjacent}}{\text{hypotenuse}} = \frac{R}{Z}$$

Solving for resistance, we have:

$$R = Z \times \cos\theta \text{ (ohms)} \quad (93)$$

Then from our example:

$$X = 12.0\,\Omega \times \sin(-42°)$$
$$= 12.0\,\Omega \times -0.669 = -8.03\,\Omega$$
$$R = 12.0\,\Omega \times \cos(-42.0°)$$
$$= 12.0\,\Omega \times 0.743 = 8.92\,\Omega$$

Since X is a negative value, it plots on the lower vertical axis, as shown in Fig 4.72, indicating capacitive reactance. In rectangular form, Z = 8.92 Ω – *j*8.03 Ω.

In performing impedance and related calculations with complex circuits, rectangular coordinates are most useful when formulas require the addition or subtraction of values. Polar notation is most useful for multiplying and dividing complex numbers.

All of the examples shown so far in this section have presumed values of reactance that contribute to the circuit impedance. Reactance is a function of frequency, however, and many impedance calculations may begin with a value of capacitance or inductance and an operating frequency. In terms of these values, the series impedance formula (Eq 87) becomes two formulas:

$$Z = \sqrt{R^2 + (2\pi f L)^2} \quad (94)$$

$$Z = \sqrt{R^2 + \left(\frac{1}{2\pi f C}\right)^2} \quad (95)$$

Example: What is the impedance of a circuit like Fig 4.71 with a resistance of 100 Ω and a 7.00-µH inductor operating at a frequency of 7.00 MHz? Using equation 94,

$$Z = \sqrt{R^2 + (2\pi f L)^2}$$

$$= \sqrt{(100\,\Omega)^2 + \left(2\pi \times 7.00 \times 10^{-6}\,\text{H} \times 7.00 \times 10^{6}\,\text{Hz}\right)^2}$$

$$Z = \sqrt{10{,}000\,\Omega^2 + (308\,\Omega)^2}$$

$$= \sqrt{10{,}000\,\Omega^2 + 94{,}900\,\Omega^2}$$

$$= \sqrt{104900\,\Omega^2} = 323.9\,\Omega$$

Since 308 Ω is the value of inductive reactance of the 7.00-µH coil at 7.00 MHz, the phase angle calculation proceeds as given in the earlier example (equation 89):

$$\theta = \arctan\frac{X}{R} = \arctan\left(\frac{308.0\,\Omega}{100.0\,\Omega}\right)$$

$$= \arctan(3.08) = 72.0°$$

Since the reactance is inductive, the phase angle is positive.

CALCULATING Z FROM R AND X IN PARALLEL CIRCUITS

In a parallel circuit containing reactance and resistance, such as shown in **Fig 4.73**, calculation of the resultant impedance from the values of R and X does not proceed by direct triangulation. The general formula for such parallel circuits is:

$$Z = \frac{RX}{\sqrt{R^2 + X^2}} \quad (96)$$

where the formula uses the absolute (unsigned) reactance value. The phase angle for the parallel circuit is given by:

$$\theta = \arctan\left(\frac{R}{X}\right) \quad (97)$$

If the parallel reactance is capacitive, then θ is a negative angle, and if the parallel reactance is inductive, then θ is a positive angle.

Example: An inductor with a reactance of 30.0 Ω is in parallel with a resistor of 40.0 Ω. What is the resulting impedance and phase angle?

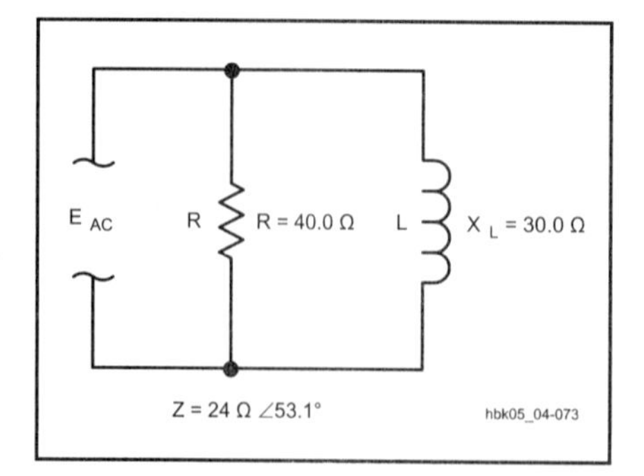

Fig 4.73 — A parallel circuit containing an inductive reactance of 30.0 Ω and a resistor of 40.0 Ω. No graph is given, since parallel impedances do not triangulate in the simple way of series impedances.

$$Z = \frac{RX}{\sqrt{R^2 + X^2}} = \frac{30.0\,\Omega \times 40.0\,\Omega}{\sqrt{(30.0\,\Omega)^2 + (40.0\,\Omega)^2}}$$

$$= \frac{1200\,\Omega^2}{\sqrt{900\,\Omega^2 + 1600\,\Omega^2}} = \frac{1200\,\Omega^2}{\sqrt{2500\,\Omega^2}}$$

$$Z = \frac{1200\,\Omega^2}{50.0\,\Omega} = 24.0\,\Omega$$

$$\theta = \arctan\left(\frac{R}{X}\right) = \arctan\left(\frac{40.0\,\Omega}{30.0\,\Omega}\right)$$

$$\theta = \arctan(1.33) = 53.1°$$

Since the parallel reactance is inductive, the resultant angle is positive.

Example: A capacitor with a reactance of 16.0 Ω is in parallel with a resistor of 12.0 Ω. What is the resulting impedance and phase angle?

$$Z = \frac{RX}{\sqrt{R^2 + X^2}} = \frac{16.0\,\Omega \times 12.0\,\Omega}{\sqrt{(16.0\,\Omega)^2 + (12.0\,\Omega)^2}}$$

$$= \frac{192\,\Omega^2}{\sqrt{256\,\Omega^2 + 144\,\Omega^2}} = \frac{192\,\Omega^2}{\sqrt{400\,\Omega^2}}$$

$$Z = \frac{192\,\Omega^2}{20.0\,\Omega} = 9.60\,\Omega$$

$$\theta = \arctan\left(\frac{R}{X}\right) = \arctan\left(\frac{12.0\,\Omega}{16.0\,\Omega}\right)$$

$$\theta = \arctan(0.750) = -36.9°$$

Because the parallel reactance is capacitive, the resultant phase angle is negative.

ADMITTANCE

Just as the inverse of resistance is conductance (*G*) and the inverse of reactance is susceptance (*B*), so too impedance has an inverse: admittance (*Y*), measured in siemens (S). Thus,

$$Y = \frac{1}{Z} \quad (98)$$

Since resistance, reactance and impedance are inversely proportional to the current (Z = E / I), conductance, susceptance and admittance are directly proportional to current. That is,

$$Y = \frac{1}{E} \quad (99)$$

One handy use for admittance is in simplifying parallel circuit impedance calculations. A parallel combination of reactance and resistance reduces to a vector addition of susceptance and conductance, if admittance is the desired outcome. In other words, for parallel circuits:

$$Y = \sqrt{G^2 + B^2} \quad (100)$$

where:
Y = admittance,
G = conductance or 1 / R, and
B = susceptance or 1 / X.

Example: An inductor with a reactance of 30.0 Ω is in parallel with a resistor of 40.0 Ω. What is the resulting impedance and phase angle? The susceptance is 1 / 30.0 Ω = 0.0333 S and the conductance is 1 / 40.0 Ω = 0.0250 S.

$$Y = \sqrt{(0.0333\ S)^2 + (0.0250\ S)^2}$$

$$= \sqrt{0.00173\ S^2} = 0.0417\ S$$

$$Z = \frac{1}{Y} = \frac{1}{0.0417\ S} = 24.0\ \Omega$$

The phase angle in terms of conductance and susceptance is:

$$\theta = \arctan\left(\frac{B}{G}\right) \quad (101)$$

In this example,

$$\theta = \arctan\left(\frac{0.0333\ S}{0.0250\ S}\right) = \arctan(1.33) = 53.1°$$

Again, since the reactive component is inductive, the phase angle is positive. For a capacitively reactive parallel circuit, the phase angle would have been negative. Compare these results with the direct calculation earlier in the section.

Conversion from resistance, reactance and impedance to conductance, susceptance and admittance is perhaps most useful in complex-parallel-circuit calculations. Many advanced facets of active-circuit analysis will demand familiarity both with the concepts and with the calculation strategies introduced here, however.

More than Two Elements in Series or Parallel

When a circuit contains several resistances or several reactances in series, simplify the circuit before attempting to calculate the impedance. Resistances in series add, just as in a purely resistive circuit. Series reactances of the same kind — that is, all capacitive or all inductive — also add, just as in a purely reactive circuit. The goal is to produce a single value of resistance and a single value of reactance for the impedance calculation.

Fig 4.74 illustrates a more difficult case in which a circuit contains two different reactive elements in series, along with a further series resistance. The series combination of X_C and X_L reduce to a single value using the same rules of combination discussed in the section on purely reactive components. As Fig 4.74B demonstrates, the resultant reactance is the difference between the two series reactances.

For parallel circuits with multiple resistances or multiple reactances of the same type, use the rules of parallel combination to reduce the resistive and reactive components to single elements. Where two or more reactive components of different types appear in the same circuit, they can be combined using formulas shown earlier for pure reactances. As **Fig 4.75** suggests, however, they can also be combined as susceptances. Parallel susceptances of different types add, with attention to their differing signs. The resulting single susceptance can then be combined with the conductance to arrive at the overall circuit admittance. The inverse of the admittance is the final circuit impedance.

Equivalent Series and Parallel Circuits

The two circuits shown in Fig 4.69 are equivalent if the same current flows when a given voltage of the same frequency is applied, and if the phase angle between voltage and current is the same in both cases. It is possible, in fact, to transform any given series circuit into an equivalent parallel circuit, and vice versa.

A series RX circuit can be converted into its parallel equivalent by means of the formulas:

$$R_P = \frac{R_S^2 + X_S^2}{R_S} \quad (102)$$

$$X_P = \frac{R_S^2 + X_S^2}{X_S} \quad (103)$$

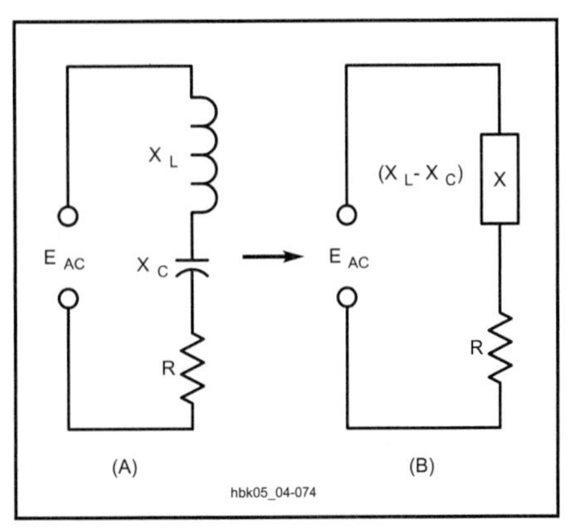

Fig 4.74 — A series impedance containing mixed capacitive and inductive reactances can be reduced to a single reactance plus resistance by combining the reactances algebraically.

where the subscripts P and S represent the parallel- and series-equivalent values, respectively. If the parallel values are known, the equivalent series circuit can be found from:

$$R_S = \frac{R_P X_P^2}{R_P^2 + X_P^2} \quad (104)$$

and

$$X_S = \frac{R_P^2 X_P}{R_P^2 + X_P^2} \quad (105)$$

Example: Let the series circuit in Fig 4.69 have a series reactance of –50.0 Ω (indicating a capacitive reactance) and a resistance of 50.0 Ω. What are the values of the equivalent parallel circuit?

$$R_P = \frac{R_S^2 + X_S^2}{R_S} = \frac{(50.0\ \Omega)^2 + (-50.0\ \Omega)^2}{50.0\ \Omega}$$

$$= \frac{2500\ \Omega^2 + 2500\ \Omega^2}{50.0\ \Omega} = \frac{5000\ \Omega^2}{50.0\ \Omega} = 100\ \Omega$$

$$X_P = \frac{R_S^2 + X_S^2}{X_S} = \frac{(50.0\ \Omega)^2 + (-50.0\ \Omega)^2}{-50.0\ \Omega}$$

$$= \frac{2500\ \Omega^2 + 2500\ \Omega^2}{-50.0\ \Omega} = \frac{5000\ \Omega^2}{-50.0\ \Omega}$$

$$= -100\ \Omega$$

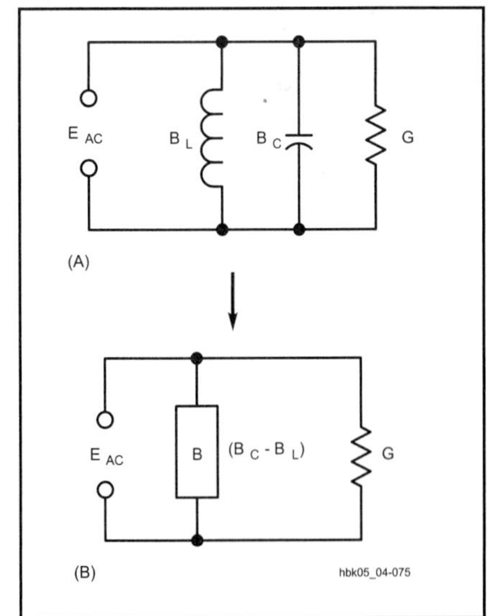

Fig 4.75 — A parallel impedance containing mixed capacitive and inductive reactances can be reduced to a single reactance plus resistance using formulas shown earlier in the chapter. By converting reactances to susceptances, as shown in A, you can combine the susceptances algebraically into a single susceptance, as shown in B.

The parallel circuit in Fig 4.69 calls for a capacitive reactance of 100 Ω and a resistance of 100 Ω to be equivalent to the series circuit.

OHM'S LAW FOR IMPEDANCE

Ohm's Law applies to circuits containing impedance just as readily as to circuits having resistance or reactance only. The formulas are:

$$E = IZ \qquad (106)$$

$$I = \frac{E}{Z} \qquad (107)$$

$$Z = \frac{E}{I} \qquad (108)$$

where:
E = voltage in volts,
I = current in amperes, and
Z = impedance in ohms.

Fig 4.76 shows a simple circuit consisting of a resistance of 75.0 Ω and a reactance of 100 Ω in series. From the series-impedance formula previously given, the impedance is

$$Z = \sqrt{R^2 + X_L^2} = \sqrt{(75.0\ \Omega)^2 + (100\ \Omega)^2}$$

$$= \sqrt{5630\ \Omega^2 + 10000\ \Omega^2} = \sqrt{15600\ \Omega^2}$$

$$= 125\ \Omega$$

If the applied voltage is 250 V, then

$$I = \frac{E}{Z} = \frac{250\ V}{125\ \Omega} = 2.00\ A$$

This current flows through both the resistance and reactance, so the voltage drops are:

$$E_R = IR = 2.00 A \times 75.0 \Omega = 150 V$$

$$E_{XL} = IX_L = 2.00 A \times 100 \Omega = 200 V$$

The simple arithmetical sum of these two drops, 350 V, is greater than the applied voltage because the two voltages are 90° out of phase. Their actual resultant, when phase is taken into account, is:

$$E = \sqrt{(150\ V)^2 + (200\ V)^2}$$

$$= \sqrt{22500\ V^2 + 40000\ V^2} = \sqrt{62500\ V^2}$$

$$= 250\ V$$

POWER FACTOR

In the circuit of Fig 4.76, an applied voltage of 250 V results in a current of 2.00 A, giving an apparent power of 250 V × 2.00 A = 500 W. Only the resistance actually consumes power, however. The power in the resistance is:

$$P = I^2 R = (2.00\ A)^2 \times 75.0\ V = 300\ W$$

The ratio of the consumed power to the apparent power is called the power factor of the circuit.

$$PF = \frac{P_{consumed}}{P_{apparent}} = \frac{R}{Z} \qquad (109)$$

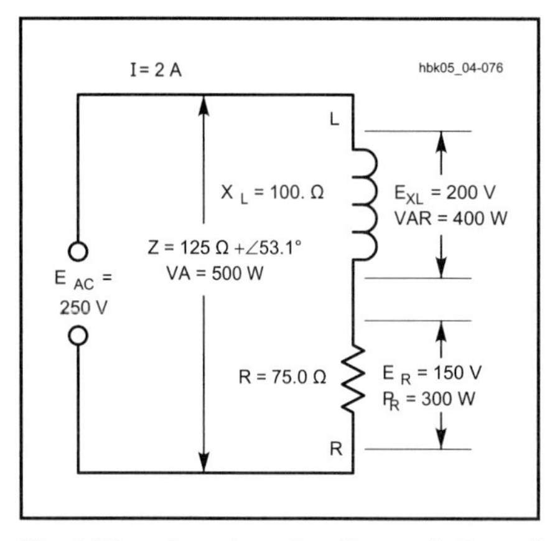

Fig 4.76 — A series circuit consisting of an inductive reactance of 100 Ω and a resistance of 75.0 Ω. Also shown is the applied voltage, voltage drops across the circuit elements, and the current.

In this example the power factor would be 300 W / 500 W = 0.600. Power factor is frequently expressed as a percentage; in this case, 60%. An equivalent definition of power factor is:

$$PF = \cos\theta$$

where θ is the phase angle. Since the phase angle equals:

$$\theta = \arctan\left(\frac{X}{R}\right) = \arctan\left(\frac{100\ \Omega}{75.0\ \Omega}\right)$$

$$= \arctan(1.33) = 53.1°$$

Then the power factor is:

$$PF = \cos 53.1° = 0.600$$

as the earlier calculation confirms.

Real, or dissipated, power is measured in watts. Apparent power, to distinguish it from real power, is measured in volt-amperes (VA). It is simply the product of the voltage across and the current through an overall impedance. It has no direct relationship to the power actually dissipated unless the power factor of the circuit is known. The power factor of a purely resistive circuit is 100% or 1, while the power factor of a pure reactance is zero. In this illustration, the reactive power is:

$$VAR = I^2 X_L = (2.00\ A)^2 \times 100\ \Omega = 400\ VA$$

Since power factor is always rendered as a positive number, the value must be followed by the words "leading" or "lagging" to identify the phase of the voltage with respect to the current. Specifying the numerical power factor is not always sufficient. For example, many dc-to-ac power inverters can safely operate loads having a large net reactance of one sign but only a small reactance of the opposite sign. Hence, the final calculation of the power factor in this example yields the value 0.600, leading.

Resonant Circuits

A circuit containing both an inductor and a capacitor — and therefore, both induc-tive and capacitive reactance — is often called a *tuned circuit*. There is a particular frequency at which the inductive and capacitive reactances are the same, that is, $X_L = X_C$. For most purposes, this is the *resonant frequency* of the circuit. (Special considerations apply to parallel circuits; they will emerge in the section devoted to such circuits.) At the resonant frequency — or at resonance, for short:

$$X_L = 2\pi f L = X_C = \frac{1}{2\pi f C}$$

By solving for f, we can find the resonant frequency of any combination of inductor and capacitor from the formula:

$$f = \frac{1}{2\pi\sqrt{LC}} \qquad (110)$$

where:
f = frequency in hertz (Hz),
L = inductance in henrys (H),
C = capacitance in farads (F), and
π = 3.1416.

For most high-frequency (HF) radio work, smaller units of inductance and capacitance and larger units of frequency are more convenient. The basic formula becomes:

$$f = \frac{10^3}{2\pi\sqrt{LC}} \qquad (111)$$

where:
f = frequency in megahertz (MHz),
L = inductance in microhenrys (μH),
C = capacitance in picofarads (pF), and
π = 3.1416.

Example: What is the resonant frequency of a circuit containing an inductor of 5.0 μH and a capacitor of 35 pF?

$$f = \frac{10^3}{2\pi\sqrt{LC}} = \frac{10^3}{6.2832 \times \sqrt{5.0 \times 35}}$$

$$= \frac{10^3}{83} = 12 \text{ MHz}$$

To find the matching component (inductor or capacitor) when the frequency and one component is known (capacitor or inductor) for general HF work, use the formula:

$$f^2 = \frac{1}{4\pi^2 LC} \qquad (112)$$

where F, L and C are in basic units. For HF work in terms of MHz, μH and pF, the basic relationship rearranges to these handy formulas:

$$L = \frac{25330}{f^2 C} \qquad (113)$$

$$C = \frac{25330}{f^2 L} \qquad (114)$$

where:
f = frequency in MHz,
L = inductance in μH, and
C = capacitance in pF.

Example: What value of capacitance is needed to create a resonant circuit at 21.1 MHz, if the inductor is 2.00 μH?

$$C = \frac{25330}{f^2 L} = \frac{25330}{(21.1^2 \times 2.00)}$$

$$= \frac{25330}{890} = 28.5 \text{ pF}$$

For most radio work, these formulas will permit calculations of frequency and component values well within the limits of component tolerances. Resonant circuits have other properties of importance, however, in addition to the resonant frequency. These include impedance, voltage drop across components in series-resonant circuits, circulating current in parallel-resonant circuits, and bandwidth. These properties determine such factors as the selectivity of a tuned circuit and the component ratings for circuits handling considerable power. Although the basic determination of the tuned-circuit resonant frequency ignored any resistance in the circuit, that resistance will play a vital role in the circuit's other characteristics.

SERIES-RESONANT CIRCUITS

Fig 4.77 presents a basic schematic diagram of a *series-resonant circuit*. Although most schematic diagrams of radio circuits would show only the inductor and the capacitor, resistance is always present in such circuits. The most notable resistance is associated with losses in the inductor at HF; resistive losses in the capacitor are low enough at those frequencies to be ignored. The current meter shown in the circuit is a reminder that in series circuits, the same current flows through all elements.

At resonance, the reactance of the capacitor cancels the reactance of the inductor. The voltage and current are in phase with each other, and the impedance of the circuit is determined solely by the resistance. The actual current through the circuit at resonance, and for frequencies near resonance, is determined by the formula:

$$I = \frac{E}{Z} = \frac{E}{\sqrt{R^2 + \left[2\pi f L - \frac{1}{(2\pi f C)}\right]^2}} \qquad (115)$$

where all values are in basic units.

At resonance, the reactive factor in the formula is zero. As the frequency is shifted above or below the resonant frequency without altering component values, however, the reactive factor becomes significant, and the value of the current becomes smaller than at resonance. At frequencies far from resonance, the reactive components become dominant, and the resistance no longer significantly affects the current amplitude.

The exact curve created by recording the current as the frequency changes depends on the ratio of reactance to resistance. When the reactance of either the coil or capacitor is of the same order of magnitude as the resistance, the current decreases rather slowly as the frequency is moved in either direction away from resonance. Such a curve is said to be *broad*. Conversely, when the reactance is considerably larger than the resistance, the current decreases rapidly as the frequency moves away from resonance, and the circuit is said to be *sharp*. A sharp circuit will respond a great deal more readily to the resonant frequency than to frequencies quite close to resonance; a broad circuit will respond almost equally well to a

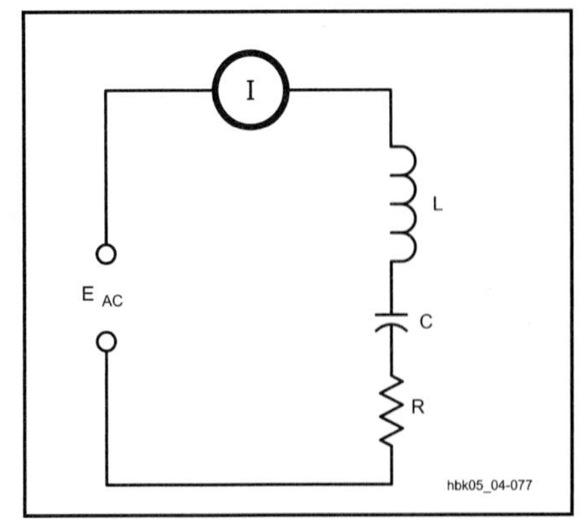

Fig 4.77 — A series circuit containing L, C, and R is resonant at the applied frequency when the reactance of C is equal to the reactance of L. The I in the circle is the schematic symbol for an ammeter.

group or band of frequencies centered around the resonant frequency.

Both types of resonance curves are useful. A sharp circuit gives good selectivity — the ability to respond strongly (in terms of current amplitude) at one desired frequency and to discriminate against others. A broad circuit is used when the apparatus must give about the same response over a band of frequencies, rather than at a single frequency alone.

Fig 4.78 presents a family of curves, showing the decrease in current as the frequency deviates from resonance. In each case, the reactance is assumed to be 1000 Ω. The maximum current, shown as a relative value on the graph, occurs with the lowest resistance, while the lowest peak current occurs with the highest resistance. Equally important, the rate at which the current decreases from its maximum value also changes with the ratio of reactance to resistance. It decreases most rapidly when the ratio is high and most slowly when the ratio is low.

Q

As noted in earlier sections of this chapter, the ratio of reactance or stored energy to resistance or consumed energy is Q. Since both terms of the ratio are measured in ohms, Q has no units and is variously known as the *quality factor*, the *figure of merit* or the *multiplying factor*. Since the resistive losses of the coil dominate the energy consumption in HF series-resonant circuits, the inductor Q largely determines the resonant-circuit Q. Since this value of Q is independent of any external load to which the circuit might transfer power, it is called the *unloaded Q* or Q_U of the circuit.

Example: What is the unloaded Q of a series-resonant circuit with a loss resistance of 5 Ω and inductive and capacitive components having a reactance of 500 Ω each? With a reactance of 50 Ω each?

$$Q_{U1} = \frac{X1}{R} = \frac{500\,\Omega}{5\,\Omega} = 100$$

$$Q_{U2} = \frac{X2}{R} = \frac{50\,\Omega}{5\,\Omega} = 10$$

Bandwidth

Fig 4.79 is an alternative way of drawing the family of curves that relate current to frequency for a series-resonant circuit. By assuming that the peak current of each curve is the same, the rate of change of current for various values of Q_U and the associated ratios of reactance to resistance are more easily compared. From the curves, it is evident that the lower Q_U circuits pass frequencies over a greater *bandwidth* of frequencies than the circuits with a higher Q_U. For the purpose of comparing tuned circuits, bandwidth is often defined as the frequency spread between the two frequencies at which the current amplitude decreases to 0.707 (or $1/\sqrt{2}$) times the maximum value. Since the power consumed by the resistance, R, is proportional to the square of the current, the power at these points is half the maximum power at resonance, assuming that R is constant for the calculations. The half-power, or –3 dB, points are marked on Fig 4.79.

For Q values of 10 or greater, the curves shown in Fig 4.79 are approximately symmetrical. On this assumption, bandwidth (BW) can be easily calculated:

$$BW = \frac{f}{Q_U} \qquad (116)$$

where BW and f are in the same units, that is, in Hz, kHz or MHz.

Example: What is the bandwidth of a series-resonant circuit operating at 14 MHz with a Q_U of 100?

$$BW = \frac{f}{Q_U} = \frac{14\text{ MHz}}{100} = 0.14\text{ MHz} = 140\text{ kHz}$$

The relationship between Q_U, f and BW provides a means of determining the value of circuit Q when inductor losses may be difficult to measure. By constructing the series-resonant circuit and measuring the current as the frequency varies above and below resonance, the half-power points can be determined. Then:

$$Q_U = \frac{f}{BW} \qquad (117)$$

Example: What is the Q_U of a series-resonant circuit operating at 3.75 MHz, if the bandwidth is 375 kHz?

$$Q_U = \frac{f}{BW} = \frac{3.75\text{ MHz}}{0.375\text{ MHz}} = 10.0$$

Table 4.7 provides some simple formulas for estimating the maximum current and phase angle for various bandwidths, if both f and Q_U are known.

Voltage Drop Across Components

The voltage drop across the coil and across the capacitor in a series-resonant circuit are each proportional to the reactance of the component for a given current (since E = I X). These voltages may be many times the source voltage for a high-Q circuit. In fact, at resonance, the voltage drop is:

$$E_X = Q_U E \qquad (118)$$

where:

E_X = the voltage across the reactive component,

Q_U = the circuit unloaded Q, and

E = the source voltage.

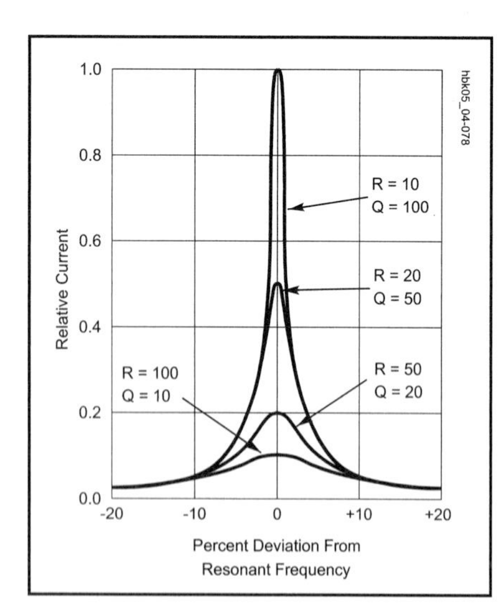

Fig 4.78 — Current in series-resonant circuits with various values of series resistance and Q. The current values are relative to an arbitrary maximum of 1.0. The reactance for all curves is 1000 Ω. Note that the current is hardly affected by the resistance in the circuit at frequencies more than 10% away from the resonant frequency.

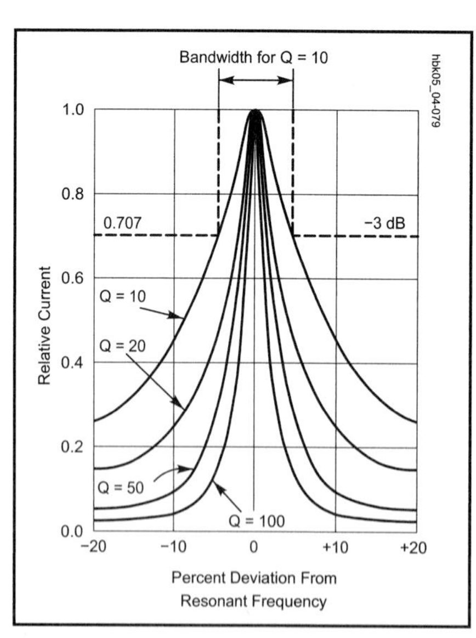

Fig 4.79 — Current in series-resonant circuits having different values of Q_u. The current at resonance is set at the same level for all curves in order to show the rate of change of decrease in current for each value of Q_u. The half-power points are shown to indicate relative bandwidth of the response for each curve. The bandwidth is indicated for a circuit with a Q_u of 10.

Table 4.7
The Selectivity of Resonant Circuits

Approximate percentage of current at resonance[1] or of impedance at resonance[2]	*Bandwidth (between half-power or –3 dB points on response curve)*	*Series circuit current phase angle (degrees)*
95	f / 3Q	18.5
90	f / 2Q	26.5
70.7	f / Q	45
44.7	2f / Q	63.5
24.2	4f / Q	76
12.4	8f / Q	83

[1]For a series resonant circuit
[2]For a parallel resonant circuit

(Note that the voltage drop across the inductor is the vector sum of the voltages across the resistance and the reactance; however, for Qs greater than 10, the error created by using equation 96 is not ordinarily significant.) Since the calculated value of E_X is the RMS voltage, the peak voltage will be higher by a factor of 1.414. Antenna couplers and other high-Q circuits handling significant power may experience arcing from high values of E_X, even though the source voltage to the circuit is well within component ratings.

Capacitor Losses

Although capacitor energy losses tend to be insignificant compared to inductor losses up to about 30 MHz, the losses may affect circuit Q in the VHF range. Leakage resistance, principally in the solid dielectric that forms the insulating support for the capacitor plates, is not exactly like the wire resistance losses in a coil. Instead of forming a series resistance, capacitor leakage usually forms a parallel resistance with the capacitive reactance. If the leakage resistance of a capacitor is significant enough to affect the Q of a series-resonant circuit, the parallel resistance must be converted to an equivalent series resistance before adding it to the inductor's resistance.

$$R_S = \frac{{X_C}^2}{R_P} = \frac{1}{R_p \times (2 \pi f C)^2} \qquad (119)$$

Example: A 10.0 pF capacitor has a leakage resistance of 10000 Ω at 50.0 MHz. What is the equivalent series resistance?

$$R_S = \frac{1}{R_p \times (2 \pi f C)^2}$$

$$= \frac{1}{1.00 \times 10^4 \times \left(6.283 \times 50.0 \times 10^6 \times 10.0 \times 10^{-12}\right)^2}$$

$$R_S = \frac{1}{1.00 \times 10^4 \times 9.87 \times 10^{-6}}$$

$$= \frac{1}{0.0987} = 10.1\,\Omega$$

In calculating the impedance, current and bandwidth for a series-resonant circuit in which this capacitor might be used, the series-equivalent resistance of the unit is added to the loss resistance of the coil. Since inductor losses tend to increase with frequency because of skin effect, the combined losses in the capacitor and the inductor can seriously reduce circuit Q, without special component- and circuit-construction techniques.

PARALLEL-RESONANT CIRCUITS

Although series-resonant circuits are common, the vast majority of resonant circuits used in radio work are *parallel-resonant circuits*. **Fig 4.80** represents a typical

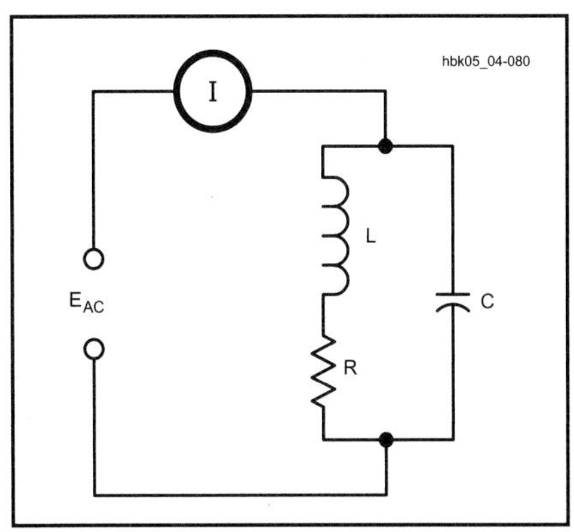

Fig 4.80 — A typical parallel-resonant circuit, with the resistance shown in series with the inductive leg of the circuit. Below a Q_U of 10, resonance definitions may lead to three separate frequencies which converge at higher Q_U levels. See text.

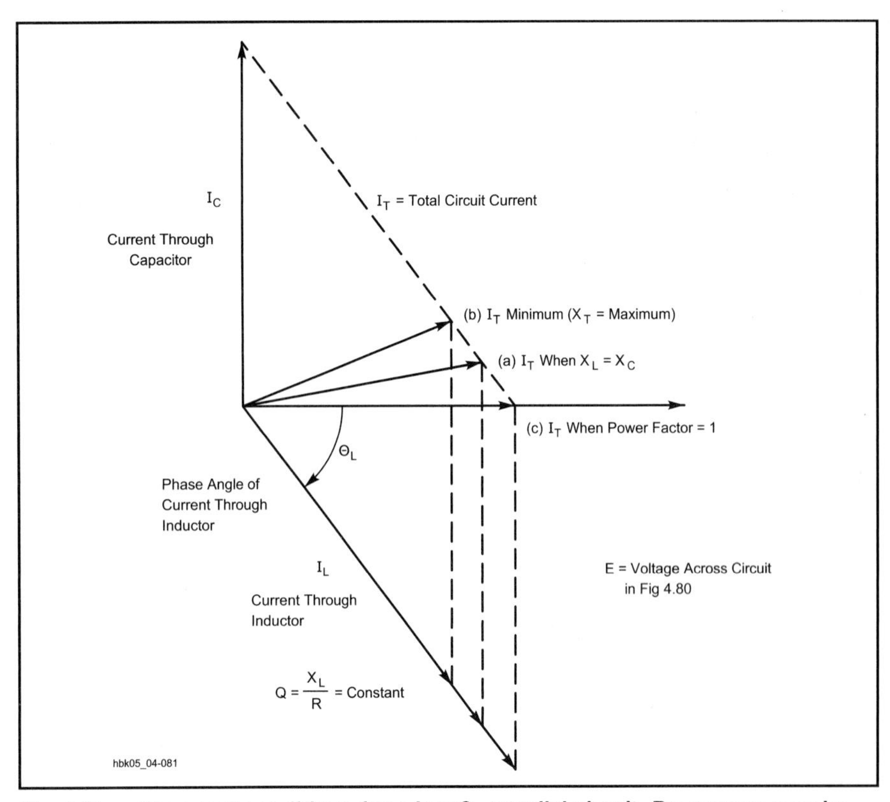

Fig 4.81 — Resonant conditions for a low-Q_U parallel circuit. Resonance may be defined as (a) $X_L = X_C$, (b) minimum current flow and maximum impedance or (c) voltage and current in phase with each other. With the circuit of Fig 4.80 and a Q_U of less than 10, these three definitions may represent three distinct frequencies.

HF parallel-resonant circuit. As is the case for series-resonant circuits, the inductor is the chief source of resistive losses, and these losses appear in series with the coil. Because current through parallel-resonant circuits is lowest at resonance, and impedance is highest, they are sometimes called *antiresonant* circuits. Likewise, the names *acceptor* and *rejector* are occasionally applied to series- and parallel-resonant circuits, respectively.

Because the conditions in the two legs of the parallel circuit in Fig 4.80 are not the same — the resistance is in only one of the legs — all of the conditions by which series resonance is determined do not occur simultaneously in a parallel-resonant circuit. **Fig 4.81** graphically illustrates the situation by showing the currents through the two components. When the inductive and capacitive reactances are identical, the condition defined for series resonance is met as shown in line (A). The impedance of the inductive leg is composed of both XL and R, which yields an impedance that is greater than XC and that is not 180° out of phase with XC. The resultant current is greater than its minimum possible value and not in phase with the voltage.

By altering the value of the inductor slightly (and holding the Q constant), a new frequency can be obtained at which the current reaches its minimum. When parallel circuits are tuned using a current meter as an indicator, this point (B) is ordinarily used as an indication of resonance. The current "dip" indicates a condition of maximum impedance and is sometimes called the *antiresonant* point or *maximum impedance resonance* to distinguish it from the condition where $X_C = X_L$. Maximum impedance is achieved by vector addition of X_C, X_L and R, however, and the result is a current somewhat out of phase with the voltage.

Point (C) on the curve represents the *unity-power-factor* resonant point. Adjusting the inductor value and hence its reactance (while holding Q constant) produces a new resonant frequency at which the resultant current is in phase with the voltage. The inductor's new value of reactance is the value required for a parallel-equivalent inductor and its parallel-equivalent resistor (calculated according to the formulas in the last section) to just cancel the capacitive reactance. The value of the parallel-equivalent inductor is always smaller than the actual inductor in series with the resistor and has a proportionally smaller reactance. (The parallel-equivalent resistor, conversely, will always be larger than the coil-loss resistor shown in series with the inductor.) The result is a resonant frequency slightly different from the one for minimum current and the one for $X_L = X_C$.

The points shown in the graph in Fig 4.81 represent only one of many possible situations, and the relative positions of the three resonant points do not hold for all possible cases. Moreover, specific circuit designs can draw some of the resonant points together, for example, compensating for the resistance of the coil by retuning the capacitor. The differences among these resonances are significant for circuit Qs below 10, where the inductor's series resistance is a significant percentage of the reactance. Above a Q of 10, the three points converge to within a percent of the frequency and can be ignored for practical calculations. Tuning for minimum current will not introduce a sufficiently large phase angle between voltage and current to create circuit difficulties.

Parallel Circuits of Moderate to High Q

The resonant frequencies defined above converge in parallel-resonant circuits with Qs higher than about 10. Therefore, a single set of formulas will sufficiently approximate circuit performance for accurate predictions. Indeed, above a Q of 10, the performance of a parallel circuit appears in many ways to be simply the inverse of the performance of a series-resonant circuit using the same components.

Accurate analysis of a parallel-resonant circuit requires the substitution of a parallel-equivalent resistor for the actual inductor-loss series resistor, as shown in **Fig 4.82**. Sometimes called the *dynamic resistance* of the parallel-resonant circuit, the parallel-equivalent resistor value will increase with circuit Q, that is, as the series resistance value decreases. To calculate the approximate parallel-equivalent resistance, use the formula:

$$R_P = \frac{{X_L}^2}{R_S} = \frac{(2\pi f L)^2}{R_S} = Q_U X_L \qquad (120)$$

Example: What is the parallel-equivalent resistance for a coil with an inductive reactance of 350 Ω and a series resistance of 5.0 Ω at resonance?

$$R_P = \frac{{X_L}^2}{R_S} = \frac{(350\ \Omega)^2}{5.0\ \Omega}$$

$$= \frac{122{,}500\ \Omega^2}{5.0\ \Omega} = 24{,}500\ \Omega$$

Since the coil Q_U remains the inductor's reactance divided by its series resistance, the coil Q_U is 70. Multiplying Q_U by the reactance also provides the approximate parallel-equivalent resistance of the coil series resistance.

At resonance, where $X_L = X_C$, R_P defines the impedance of the parallel-resonant circuit. The reactances just equal each other, leaving the voltage and current in phase with each other. In other words, the circuit shows only the parallel resistance. Therefore, equation 120 can be rewritten as:

$$Z = \frac{{X_L}^2}{R_S} = \frac{(2\pi f L)^2}{R_S} = Q_U X_L \qquad (121)$$

In this example, the circuit impedance at resonance is 24,500 Ω.

At frequencies below resonance the current through the inductor is larger than that through the capacitor, because the reactance of the coil is smaller and that of the capacitor is larger than at resonance. There is only partial cancellation of the two reactive currents, and the line current therefore is larger than the current taken by the resistance alone. At frequencies above resonance the situation is reversed and more current flows through the capacitor than through the inductor, so the line current again increases. The current at resonance, being determined wholly by R_P, will be small if R_P is large, and large if R_P is small. **Fig 4.83** illustrates the relative current flows through a parallel-tuned circuit as the frequency is moved from below resonance to above resonance. The base line represents the minimum current level for the particular circuit. The actual current at any frequency off resonance is simply the vector sum of the currents through the parallel equivalent resistance and through the reactive components.

To obtain the impedance of a parallel-tuned circuit either at or off the resonant

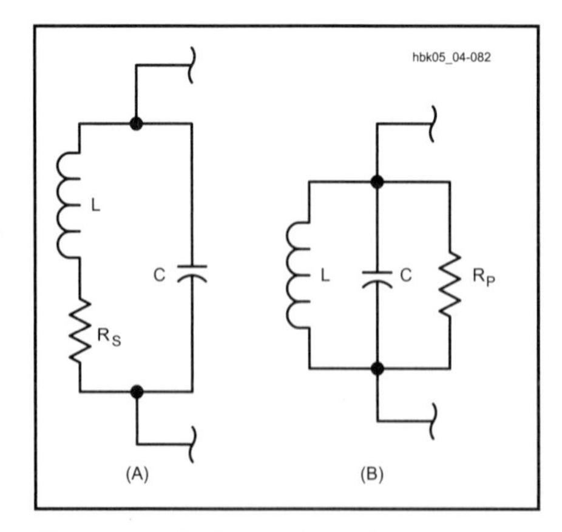

Fig 4.82 — Series and parallel equivalents when both circuits are resonant. The series resistance, R_S in A, is replaced by the parallel resistance, R_P in B, and vice versa. $R_P = {X_L}^2 / R_S$.

frequency, apply the general formula:

$$Z = \frac{Z_C Z_L}{Z_S} \qquad (122)$$

where:

Z = overall circuit impedance

Z_C = impedance of the capacitive leg (usually, the reactance of the capacitor),

Z_L = impedance of the inductive leg (the vector sum of the coil's reactance and resistance), and

Z_S = series impedance of the capacitor-inductor combination as derived from the denominator of equation 115.

After using vector calculations to obtain Z_L and Z_S, converting all the values to polar form — as described earlier in this chapter — will ease the final calculation. Of course, each impedance may be derived from the resistance and the application of the basic reactance formulas on the values of the inductor and capacitor at the frequency of interest.

Since the current rises off resonance, the parallel-resonant-circuit impedance must fall. It also becomes complex, resulting in an ever greater phase difference between the voltage and the current. The rate at which the impedance falls is a function of Q_U. **Fig 4.84** presents a family of curves showing the impedance drop from resonance for circuit Qs ranging from 10 to 100. The curve family for parallel-circuit impedance is essentially the same as the curve family for series-circuit current.

As with series tuned circuits, the higher the Q of a parallel-tuned circuit, the sharper the response peak. Likewise, the lower the Q, the wider the band of frequencies to which the circuit responds. Using the half-power (–3 dB) points as a comparative measure of circuit performance, equations 116 and 117 apply equally to parallel-tuned circuits. That is, $BW = f / Q_U$ and $Q_U = f / BW$, where the resonant frequency and the bandwidth are in the same units. As a handy reminder, **Table 4.8** summarizes the performance of parallel-resonant circuits at high and low Qs and above and below resonant frequency.

It is possible to use either series or parallel-resonant circuits do the same work in many circuits, thus giving the designer considerable flexibility. **Fig 4.85** illustrates this general principle by showing a series-resonant circuit in the signal path and a parallel-resonant circuit shunted from the signal path to ground. Assume both circuits are resonant at the same frequency, f, and have the same Q. The series tuned circuit at A has its lowest impedance at f, permitting the maximum possible current to flow along the signal path. At all other frequencies, the impedance is greater and the current at those frequencies is less. The circuit passes the desired signal and tends to impede signals at undesired frequencies. The parallel circuit at B provides the highest impedance at resonance, f, making the signal path the lowest impedance

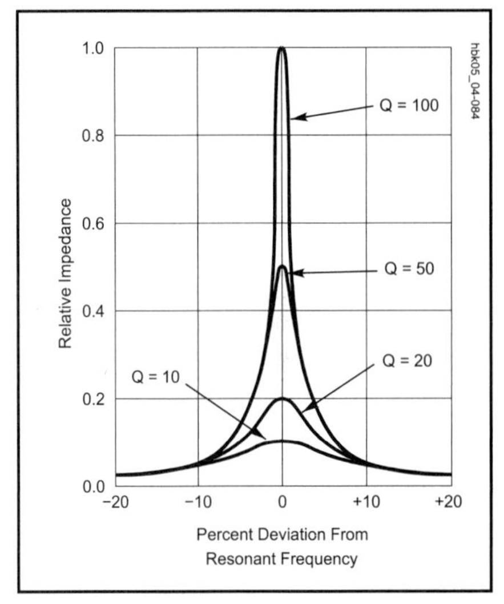

Fig 4.84 — Relative impedance of parallel-resonant circuits with different values of Q_U. The curves are similar to the series-resonant circuit current level curves of Fig 4.78. The effect of Q_U on impedance is most pronounced within 10% of the resonance frequency.

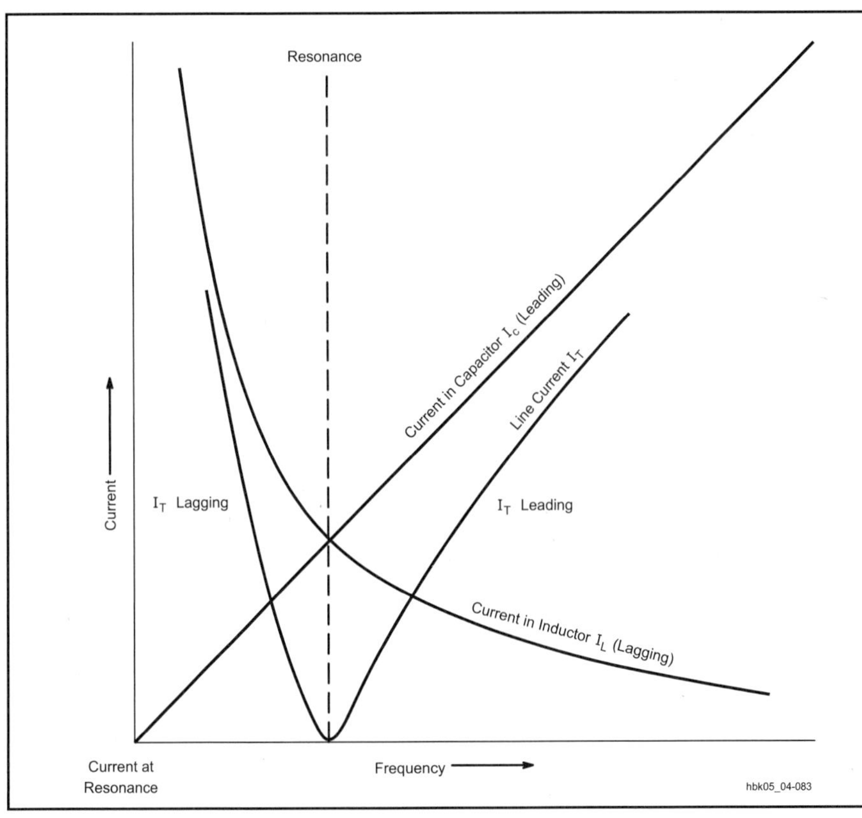

Fig 4.83 — The currents in a parallel-resonant circuit as the frequency moves through resonance. Below resonance, the current lags the voltage; above resonance the current leads the voltage. The base line represents the current level at resonance, which depends on the impedance of the circuit at that frequency.

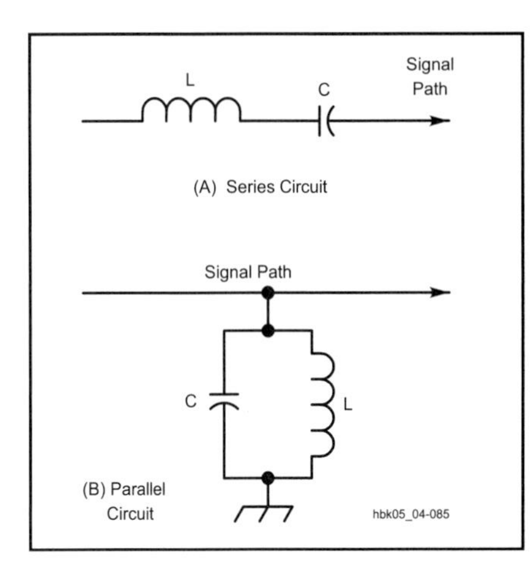

Fig 4.85 — Series- and parallel-resonant circuits configured to perform the same theoretical task: passing signals in a narrow band of frequencies along the signal path. A real design example would consider many other factors.

Table 4.8
The Performance of Parallel-Resonant Circuits

A. High and Low Q Circuits (in relative terms)

Characteristic	High Q Circuit	Low Q Circuit
Selectivity	high	low
Bandwidth	narrow	wide
Impedance	high	low
Line current	low	high
Circulating current	high	low

B. Off-Resonance Performance for Constant Values of Inductance and Capacitance

Characteristic	Above Resonance	Below Resonance
Inductive reactance	increases	decreases
Capacitive reactance	decreases	increases
Circuit resistance	unchanged*	unchanged*
Circuit impedance	decreases	decreases
Line current	increases	increases
Circulating current	decreases	decreases
Circuit behavior	capacitive	inductive

*This is true for frequencies near resonance. At distant frequencies, skin effect may alter the resistive losses of the inductor.

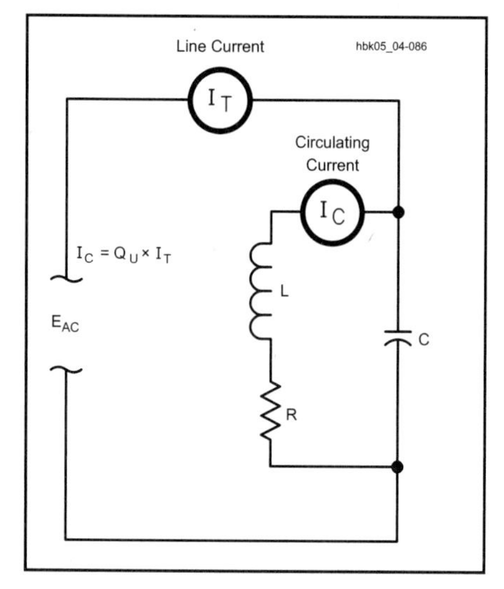

Fig 4.86 — A parallel-resonant circuit redrawn to illustrate both the line current and the circulating current.

path for the signal. At frequencies off resonance, the parallel-resonant circuit presents a lower impedance, thus presenting signals with a path to ground and away from the signal path. In theory, the effects will be the same relative to a signal current on the signal path. In actual circuit design exercises, of course, many other variables will enter the design picture to make one circuit preferable to the other.

Circulating Current

In a parallel-resonant circuit, the source voltage is the same for all the circuit elements. The current in each element, however, is a function of the element's reactance. **Fig 4.86** redraws the parallel-tuned circuit to indicate the line current and the current circulating between the coil and the capacitor. The current drawn from the source may be low, because the overall circuit impedance is high. The current through the individual elements may be high, however, because there is little resistive loss as the current circulates through the inductor and capacitor. For parallel-resonant circuits with an unloaded Q of 10 or greater, this *circulating current* is approximately:

$$I_C = Q_U I_T \tag{123}$$

where:

I_C = circulating current in A, mA or μA,
Q_U = unloaded circuit Q, and
I_T = line current in the same units as I_C.

Example: A parallel-resonant circuit permits an ac or RF line current of 30 mA and has a Q of 100. What is the circulating current through the elements?

$$I_X = Q_U I = 100 \times 30 \text{ mA} = 3000 \text{ mA} = 3 \text{ A}$$

Circulating currents in high-Q parallel-tuned circuits can reach a level that causes component heating and power loss. Therefore, components should be rated for the anticipated circulating currents, and not just the line current.

The Q of Loaded Circuits

In many resonant-circuit applications, the only power lost is that dissipated in the resistance of the circuit itself. At frequencies below 30 MHz, most of this resistance is in the coil. Within limits, increasing the number of turns in the coil increases the reactance faster than it raises the resistance, so coils for circuits in which the Q must be high are made with relatively large inductances for the frequency.

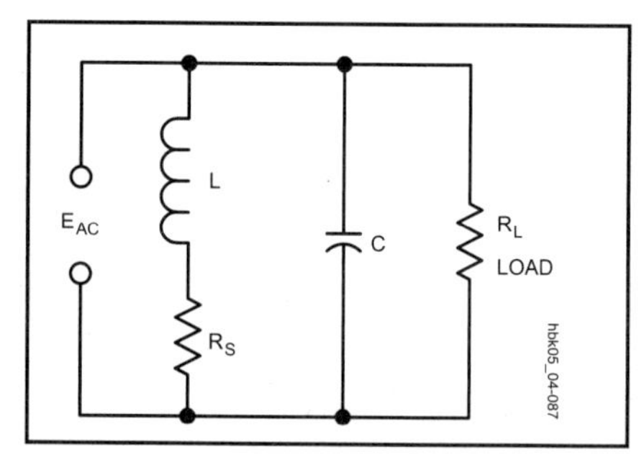

Fig 4.87 — A loaded parallel-resonant circuit, showing both the inductor-loss resistance and the load, R_L. If smaller than the inductor resistance, R_L will control the loaded Q of the circuit (Q_L).

When the circuit delivers energy to a load (as in the case of the resonant circuits used in transmitters), the energy consumed in the circuit itself is usually negligible compared with that consumed by the load. The equivalent of such a circuit is shown in **Fig 4.87**, where the parallel resistor, R_L, represents the load to which power is delivered. If the power dissipated in the load is at least 10 times as great as the power lost in the inductor and capacitor, the parallel impedance of the resonant circuit itself will be so high compared with the resistance of the load that for all practical purposes the impedance of the combined circuit is equal to the load impedance. Under these conditions, the load resistance replaces the circuit impedance in calculating Q. The Q of a parallel-resonant circuit loaded by a resistive impedance is:

$$Q_L = \frac{R_L}{X} \tag{124}$$

where:

Q_L = circuit loaded Q,
R_L = parallel load resistance in ohms, and
X = reactance in ohms of either the inductor or the capacitor.

Example: A resistive load of 3000 Ω is connected across a resonant circuit in which the inductive and capacitive reactances are each 250 Ω. What is the circuit Q?

$$Q_L = \frac{R_L}{X} = \frac{3000\,\Omega}{250\,\Omega} = 12$$

The effective Q of a circuit loaded by a parallel resistance increases when the reactances are decreased. A circuit loaded with a relatively low resistance (a few thousand ohms) must have low-reactance elements (large capacitance and small inductance) to have reasonably high Q. Many power-handling circuits, such as the output networks of transmitters, are designed by first choosing a loaded Q for the circuit and then determining component values. See the **RF Power Amplifiers** chapter for more details.

Parallel load resistors are sometimes added to parallel-resonant circuits to lower the circuit Q and increase the circuit bandwidth. By using a high-Q circuit and adding a parallel resistor, designers can tailor the circuit response to their needs. Since the parallel resistor consumes power, such techniques ordinarily apply to receiver and similar low-power circuits, however.

Example: Specifications call for a parallel-resonant circuit with a bandwidth of 400 kHz at 14.0 MHz. The circuit at hand has a Q_U of 70.0 and its components have reactances of 350 Ω each. What is the parallel load resistor that will increase the bandwidth to the specified value? The bandwidth of the existing circuit is:

$$BW = \frac{f}{Q_U} = \frac{14.0\ \text{MHz}}{70.0} = 0.200\ \text{MHz}$$

$$= 200\ \text{kHz}$$

The desired bandwidth, 400 kHz, requires a circuit with a Q of:

$$Q = \frac{f}{BW} = \frac{14.0\ \text{MHz}}{0.400\ \text{MHz}} = 35.0$$

Since the desired Q is half the original value, halving the resonant impedance or parallel-resistance value of the circuit is in order. The present impedance of the circuit is:

$$Z = Q_U\ X_L = 70.0 \times 350\ \Omega = 24500\ \Omega$$

The desired impedance is:

$$Z = Q_U\ X_L = 35.0 \times 350\ \Omega$$

$$= 12250\ \Omega = 12.25\ \text{k}\Omega$$

or half the present impedance.

A parallel resistor of 24500 Ω, or the nearest lower value (to guarantee sufficient bandwidth), will produce the required reduction in Q and bandwidth increase. Although this example simplifies the situation encountered in real design cases by ignoring such factors as the shape of the band-pass curve, it illustrates the interaction of the ingredients that determine the performance of parallel-resonant circuits.

Impedance Transformation

An important application of the parallel-resonant circuit is as an impedance matching device in the output circuit of an RF power amplifier. There is an optimum value of load resistance for each type of tube or transistor and each set of required operating conditions. The resistance of the load to which the active device delivers power may be considerably lower than the value required for proper device operation, or the load impedance may be considerably higher than the amplifier output impedance.

To transform the actual load resistance to the desired value, the load may be tapped across part of the coil, as shown in **Fig 4.88**. This is equivalent to connecting a higher value of load resistance across the whole circuit, and is similar in principle to impedance transformation with an iron-core transformer (described in the next section of this chapter). In high-frequency resonant circuits, the impedance ratio does not vary exactly as the square of the turns ratio, because all the magnetic flux lines do not cut every turn of the coil. A desired impedance ratio usually must be obtained by experimental adjustment.

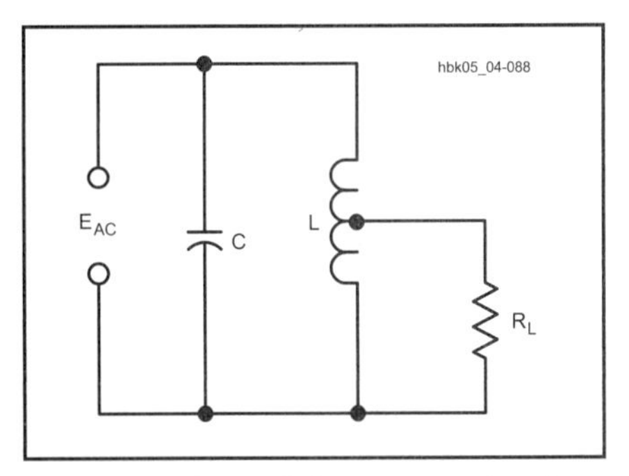

Fig 4.88 — A parallel-resonant circuit with a tapped coil to effect an impedance match. Although the impedance presented by the entire circuit is very high, the impedance "seen" by the load, R_L, is lower.

When the load resistance has a very low value (say below 100 Ω) it may be connected in series in the resonant circuit (R_S in Fig 4.82A, for example), in which case it is transformed to an equivalent parallel impedance as previously described. If the Q is at least 10, the equivalent parallel impedance is:

$$Z_R = \frac{X^2}{R_L} \qquad (125)$$

where:

Z_R = resistive parallel impedance at resonance,

X = reactance (in ohms) of either the coil or the capacitor, and

R_L = load resistance inserted in series.

If the Q is lower than 10, the reactance will have to be adjusted somewhat — for the reasons given in the discussion of low-Q circuits — to obtain a resistive impedance of the desired value.

Networks like the one in Fig 4.88 have some serious disadvantages for some applications. For instance, the common connection between the input and the output provides no dc isolation. Also, the common ground is sometimes troublesome with regard to ground-loop currents. Consequently, a network with only mutual magnetic coupling is often preferable. With the advent of ferrites, constructing impedance transformers that are both broadband and permit operation well up into the VHF portion of the spectrum has become relatively easy. The basic principles of broadband impedance transformers appear in the following section.

Transformers

When the ac source current flows through every turn of an inductor, the generation of a counter-voltage and the storage of energy during each half cycle is said to be by virtue of *self-inductance*. If another inductor — not connected to the source of the original current — is positioned so the expanding and contracting magnetic field of the first inductor cuts across its turns, a current will be induced into the second coil. A load such as a resistor may be connected across the second coil to consume the energy transferred magnetically from the first inductor. This phenomenon is called *mutual inductance*.

Two inductors positioned so that the magnetic field of one (the *primary* inductor) induces a current in the other (the *secondary* inductor) are *coupled*. **Fig 4.89** illustrates a pair of coupled inductors, showing an ac energy source connected to one and a load connected to the other. If the coils are wound tightly on an iron core so that nearly all the lines of force or magnetic flux from the first coil link with the turns of the second coil, the pair is said to be tightly coupled. Coils with air cores separated by a distance would be loosely coupled. The signal source for the primary inductor may be household ac power lines, audio or other waveforms at lower frequencies, or RF currents. The load may be a device needing power, a speaker converting electrical energy into sonic energy, an antenna using RF energy for communications or a particular circuit set up to process a signal from a preceding circuit. The uses of magnetically coupled energy in electronics are innumerable.

Mutual inductance (M) between coils is measured in henrys. Two coils have a mutual inductance of 1 H under the following conditions: as the primary inductor current changes at a rate of 1 A/s, the voltage across the secondary inductor is 1 V. The level of mutual inductance varies with many factors: the size and shape of the inductors, their relative positions and distance from each other, and the permeability of the inductor core material and of the space between them.

If the self-inductance values of two coils are known, it is possible to derive the mutual inductance by way of a simple experiment schematically represented in **Fig 4.90**. Without altering the physical setting or position of two coils, measure the inductance of the series-connected coils with their windings complementing each other and again with their windings opposing each other. Since, for the two coils, $L_C = L1 + L2 + 2M$, in the complementary case, and $L_O = L1 + L2 - 2M$ for the opposing case,

$$M = \frac{L_C - L_O}{4} \qquad (126)$$

The ratio of magnetic flux set up by the secondary coil to the flux set up by the primary coil is a measure of the extent to which two coils are coupled, compared to the maximum possible coupling between them. This ratio is the *coefficient of coupling* (k) and is always less than 1. If k were to equal 1, the two coils would have the maximum possible mutual coupling. Thus:

$$M = k\sqrt{L1\,L2} \qquad (127)$$

where:

M = mutual inductance in henrys,
L1 and L2 = individual coupled inductors, each in henrys, and
k = the coefficient of coupling.

Using the experiment above, it is possible to solve equation 127 for k with reasonable accuracy.

Any two coils having mutual inductance comprise a *transformer* having a *primary winding* or inductor and a *secondary winding* or inductor. **Fig 4.91** provides a pictorial representation of a typical iron-core transformer, along with the schematic symbols for both iron-core and air-core

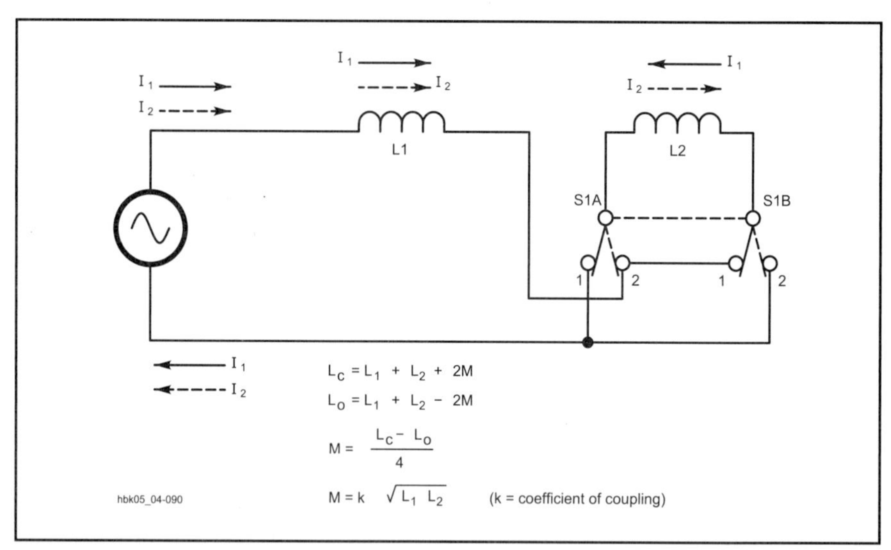

Fig 4.90 — An experimental setup for determining mutual inductance. Measure the inductance with the switch in each position and use the formula in the text to determine the mutual inductance.

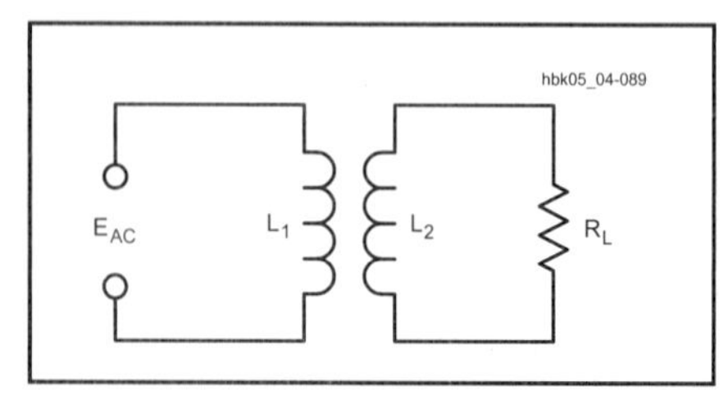

Fig 4.89 — A basic transformer: two inductors — one connected to an ac energy source, the other to a load — with coupled magnetic fields.

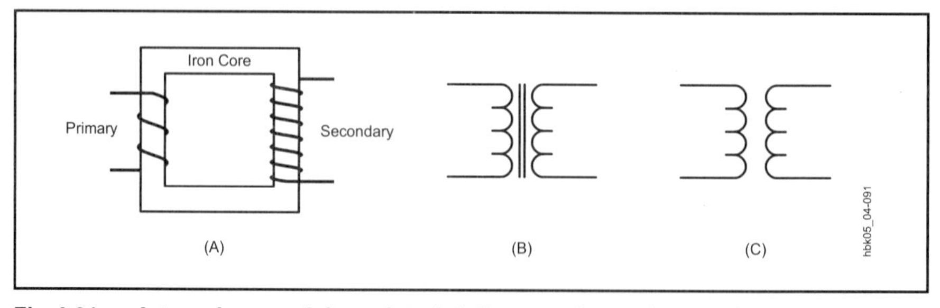

Fig 4.91 — A transformer. A is a pictorial diagram. Power is transferred from the primary coil to the secondary by means of the magnetic field. B is a schematic diagram of an iron-core transformer, and C is an air-core transformer.

transformers. Conventionally, the term *transformer* is most commonly applied to coupled inductors having a magnetic core material, while coupled air-wound inductors are not called by that name. They are still transformers, however.

We normally think of transformers as ac devices, since mutual inductance only occurs when magnetic fields are expanding or contracting. A transformer connected to a dc source will exhibit mutual inductance only at the instants of closing and opening the primary circuit, or on the rising and falling edges of dc pulses, because only then does the primary winding have a changing field. The principle uses of transformers are three: to physically isolate the primary circuit from the secondary circuit, to transform voltages and currents from one level to another, and to transform circuit impedances from one level to another. These functions are not mutually exclusive and have many variations.

IRON-CORE TRANSFORMERS

The primary and secondary coils of a transformer may be wound on a core of magnetic material. The permeability of the magnetic material increases the inductance of the coils so a relatively small number of turns may be used to induce a given voltage value with a small current. A closed core having a continuous magnetic path, such as that shown in Fig 4.91, also tends to ensure that practically all of the field set up by the current in the primary coil will cut the turns of the secondary coil. For power transformers and impedance-matching transformers used in audio work, cores of iron strips are most common and generally very efficient.

The following principles presume a coefficient of coupling (k) of 1, that is, a perfect transformer. The value k = 1 indicates that all the turns of both coils link with all the magnetic flux lines, so that the voltage induced per turn is the same with both coils. This condition makes the induced voltage independent of the inductance of the primary and secondary inductors. Iron-core transformers for low frequencies most closely approach this ideal condition. **Fig 4.92** illustrates the conditions for transformer action.

Voltage Ratio

For a given varying magnetic field, the voltage induced in a coil within the field is proportional to the number of turns in the coil. When the two coils of a transformer are in the same field (which is the case when both are wound on the same closed core), it follows that the induced voltages will be proportional to the number of turns in each coil. In the primary, the induced voltage practically equals, and opposes, the applied voltage, as described earlier. Hence:

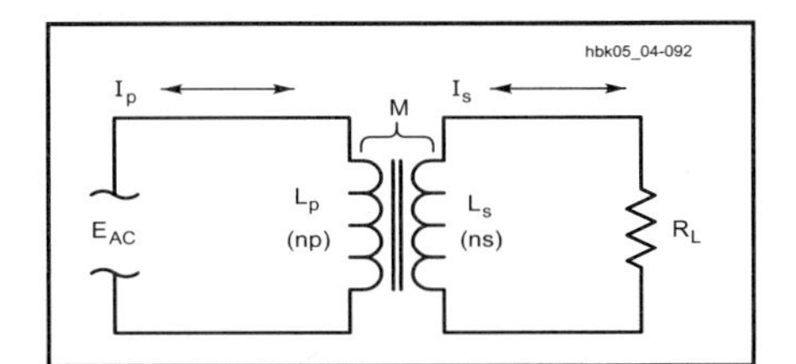

Fig 4.92 — The conditions for transformer action: two coils that exhibit mutual inductance, an ac power source, and a load. The magnetic field set up by the energy in the primary circuit transfers energy to the secondary for use by the load, resulting in a secondary voltage and current.

$$E_S = E_P \left(\frac{N_S}{N_P} \right) \qquad (128)$$

where:

E_S = secondary voltage,

E_P = primary applied voltage,

N_S = number of turns on secondary,

and

N_P = number of turns on primary.

Example: A transformer has a primary of 400 turns and a secondary of 2800 turns, and a voltage of 120 V is applied to the primary. What voltage appears across the secondary winding?

$$E_S = 120\text{ V} \left(\frac{2800}{400} \right) = 120\text{ V} \times 7 = 840\text{ V}$$

(Notice that the number of turns is taken as a known value rather than a measured quantity, so they do not limit the significant figures in the calculation.) Also, if 840 V is applied to the 2800-turn winding (which then becomes the primary), the output voltage from the 400-turn winding will be 120 V.

Either winding of a transformer can be used as the primary, provided the winding has enough turns (enough inductance) to induce a voltage equal to the applied voltage without requiring an excessive current. The windings must also have insulation with a voltage rating sufficient for the voltage present.

Current or Ampere-Turns Ratio

The current in the primary when no current is taken from the secondary is called the *magnetizing current* of the transformer. An ideal transformer, with no internal losses, would consume no power, since the current through the primary inductor would be 90° out of phase with the voltage. In any properly designed transformer, the power consumed by the transformer when the secondary is open (not delivering power) is only the amount necessary to overcome the losses in the iron core and in the resistance of the wire with which the primary is wound.

When power is taken from the secondary winding by a load, the secondary current sets up a magnetic field that opposes the field set up by the primary current. For the induced voltage in the primary to equal the applied voltage, the original field must be maintained. The primary must draw enough additional current to set up a field exactly equal and opposite to the field set up by the secondary current.

In practical transformer calculations it may be assumed that the entire primary current is caused by the secondary load. This is justifiable because the magnetizing current should be very small in comparison with the primary load current at rated power output.

If the magnetic fields set up by the primary and secondary currents are to be equal, the primary current multiplied by the primary turns must equal the secondary current multiplied by the secondary turns.

$$I_P = I_S \left(\frac{N_S}{N_P} \right) \qquad (129)$$

where:

I_P = primary current,

I_S = secondary current,

N_P = number of turns on primary, and

N_S = number of turns on secondary.

Example: Suppose the secondary of the transformer in the previous example is delivering a current of 0.20 A to a load. What will be the primary current?

$$I_P = 0.20\text{ A} \times \left(\frac{2800}{400} \right) = 0.20\text{ A} \times 7 = 1.4\text{ A}$$

Although the secondary voltage is higher than the primary voltage, the secondary current is lower than the primary current, and by the same ratio. The secondary current in an ideal transformer is 180° out of phase with the primary current, since the field in the secondary just offsets the field in the primary. The phase relationship between the currents in the windings holds true no matter what the phase difference between the current and the voltage of the secondary. In fact, the phase difference, if any, between voltage and current in the secondary winding will be reflected back to the primary as an identical phase difference.

Power Ratio

A transformer cannot create power; it can only transfer it and change the voltage level. Hence, the power taken from the secondary cannot exceed that taken by the primary from the applied voltage source. There is always some power loss in the resistance of the coils and in the iron core, so in all practical cases the power taken from the source will exceed that taken from the secondary.

$$P_O = n\,P_I \qquad (130)$$

where:

P_O = power output from secondary,
P_I = power input to primary, and
n = efficiency factor.

The efficiency, n, is always less than 1. It is usually expressed as a percentage: if n is 0.65, for instance, the efficiency is 65%.

Example: A transformer has an efficiency of 85.0% at its full-load output of 150 W. What is the power input to the primary at full secondary load?

$$P_I = \frac{P_O}{n} = \frac{150\text{ W}}{0.850} = 176\text{ W}$$

A transformer is usually designed to have the highest efficiency at the power output for which it is rated. The efficiency decreases with either lower or higher outputs. On the other hand, the losses in the transformer are relatively small at low output but increase as more power is taken. The amount of power that the transformer can handle is determined by its own losses, because these losses heat the wire and core. There is a limit to the temperature rise that can be tolerated, because too high a temperature can either melt the wire or cause the insulation to break down. A transformer can be operated at reduced output, even though the efficiency is low, because the actual loss will be low under such conditions. The full-load efficiency of small power transformers such as are used in radio receivers and transmitters usually lies between about 60 and 90%, depending on the size and design.

IMPEDANCE RATIO

In an ideal transformer — one without losses or leakage reactance — the following relationship is true:

$$Z_P = Z_S\left(\frac{N_P}{N_S}\right)^2 \qquad (131)$$

where:

Z_P = impedance looking into the primary terminals from the power source,
Z_S = impedance of load connected to secondary, and
N_P, N_S = turns ratio, primary to secondary.

A load of any given impedance connected to the transformer secondary will be transformed to a different value looking into the primary from the power source. The impedance transformation is proportional to the square of the primary-to-secondary turns ratio.

Example: A transformer has a primary-to-secondary turns ratio of 0.6 (the primary has six-tenths as many turns as the secondary) and a load of 3000 Ω is connected to the secondary. What is the impedance at the primary of the transformer?

$$Z_P = 3000\,\Omega \times (0.6)^2 = 3000\,\Omega \times 0.36$$

$$Z_P = 1080\,\Omega$$

By choosing the proper turns ratio, the impedance of a fixed load can be transformed to any desired value, within practical limits. If transformer losses can be neglected, the transformed (reflected) impedance has the same phase angle as the actual load impedance. Thus, if the load is a pure resistance, the load presented by the primary to the power source will also be a pure resistance. If the load impedance is complex, that is, if the load current and voltage are out of phase with each other, then the primary voltage and current will show the same phase angle.

Many devices or circuits require a specific value of load resistance (or impedance) for optimum operation. The impedance of the actual load that is to dissipate the power may differ widely from the impedance of the source device or circuit, so a transformer is used to change the actual load into an impedance of the desired value. This is called impedance matching.

$$\frac{N_P}{N_S} = \sqrt{\frac{Z_P}{Z_S}} \qquad (132)$$

where:

N_P / N_S = required turns ratio, primary to secondary,
Z_P = primary impedance required, and
Z_S = impedance of load connected to secondary.

Example: A transistor audio amplifier requires a load of 150 Ω for optimum performance, and is to be connected to a loudspeaker having an impedance of 4.0 Ω. What is the turns ratio, primary to secondary, required in the coupling transformer?

$$\frac{N_P}{N_S} = \sqrt{\frac{Z_P}{Z_S}} = \frac{N_P}{N_S}\sqrt{\frac{150\,\Omega}{4.0\,\Omega}} = \sqrt{38} = 6.2$$

The primary therefore must have 6.2 times as many turns as the secondary.

These relationships may be used in practical work even though they are based on an ideal transformer. Aside from the normal design requirements of reasonably low internal losses and low leakage reactance, the only other requirement is that the primary have enough inductance to operate with low magnetizing current at the voltage applied to the primary.

The primary terminal impedance of an iron-core transformer is determined wholly by the load connected to the secondary and by the turns ratio. If the characteristics of the transformer have an appreciable effect on the impedance presented to the power source, the transformer is either poorly designed or is not suited to the voltage and frequency at which it is being used. Most transformers will operate quite well at voltages from slightly above to well below the design figure.

Transformer Losses

In practice, none of the formulas given so far provides truly exact results, although they afford reasonable approximations. Transformers in reality are not simply two coupled inductors, but a network of resistances and reactances, most of which appear in **Fig 4.93**. Since only the terminals numbered 1 through 4 are accessible to the user, transformer ratings and specifications take into account the additional losses created by these complexities.

In a practical transformer not all of the magnetic flux is common to both windings, although in well designed transformers the amount of flux that cuts one coil and not the other is only a small percentage of the total flux. This *leakage flux* causes a voltage of self-induction. Consequently, there are small amounts of leakage inductance associated with both windings of the transformer. Leakage inductance acts in exactly the same way as an equivalent amount of ordinary inductance inserted in series with the circuit. It has, therefore, a certain reactance, depending on the amount of leakage inductance and the frequency. This reactance is called *leakage reactance*, shown as X_{L1} and X_{L2} in Fig 4.93.

Current flowing through the leakage reactance causes a voltage drop. This voltage drop increases with increasing current; hence, it increases as more power is taken from the secondary. Thus, the greater the secondary current, the smaller the secondary terminal voltage becomes.

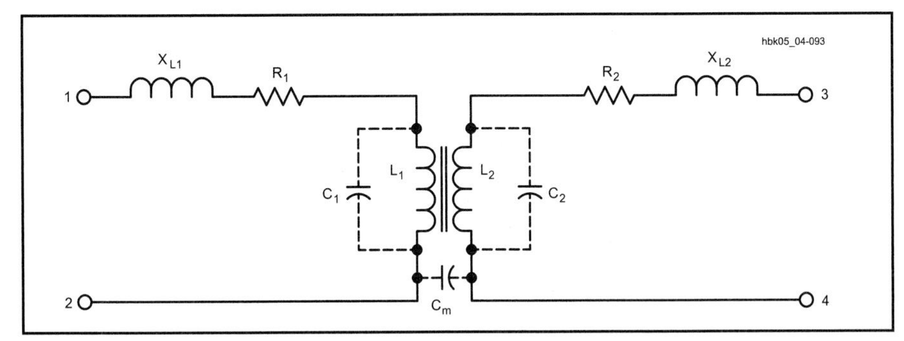

Fig 4.93 — A transformer as a network of resistances, inductances and capacitances. Only L1 and L2 contribute to the transfer of energy.

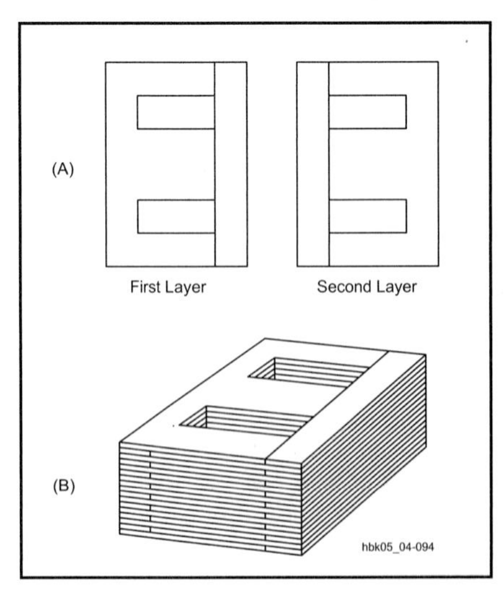

Fig 4.94 — A typical transformer iron core. The E and I pieces alternate direction in successive layers to improve the magnetic path while attenuating eddy currents in the core.

The resistances of the transformer windings, R1 and R2, also cause voltage drops when there is current. Although these voltage drops are not in phase with those caused by leakage reactance, together they result in a lower secondary voltage under load than is indicated by the transformer turns ratio.

Thus, the voltage regulation in a real transformer is not perfect. At ac line frequencies (50 or 60 Hz), the voltage at the secondary, with a reasonably well-designed transformer, should not drop more than about 10% from open-circuit conditions to full load. The voltage drop may be considerably more than this in a transformer operating at voice and music frequencies, because the leakage reactance increases directly with the frequency.

In addition to wire resistances and leakage reactances, certain stray capacitances occur in transformers. An electric field exists between any two points having a different voltage. When current flows through a coil, each turn has a slightly different voltage than its adjacent turns, creating a capacitance between turns. This *distributed capacitance* appears in Fig 4.93 as C1 and C2. Another capacitance, C_M, appears between the two windings for the same reason. Moreover, transformer windings can exhibit capacitance relative to nearby metal, for example, the chassis, the shield and even the core.

Although these stray capacitances are of little concern with power and audio transformers, they become important as the frequency increases. In transformers for RF use, the stray capacitance can resonate with either the leakage reactance or, at lower frequencies, with the winding reactances, L1 or L2, especially under very light or zero loads. In the frequency region around resonance, transformers no longer exhibit the properties formulated above or the impedance properties to be described below.

Iron-core transformers also experience losses within the core itself. *Hysteresis losses* include the energy required to overcome the retentivity of the core's magnetic material. Circulating currents through the core's resistance are *eddy currents*, which form part of the total core losses. These losses, which add to the required magnetizing current, are equivalent to adding a resistance in parallel with L1 in Fig 4.93.

Core Construction

Audio and power transformers usually employ one or another grade of silicon steel as the core material. With permeabilities of 5000 or greater, these cores saturate at flux densities approaching 10^5 lines per square inch of cross section. The cores consist of thin insulated laminations to break up potential eddy current paths.

Each core layer consists of an E and an I piece butted together, as represented in **Fig 4.94**. The butt point leaves a small gap. Since the pieces in adjacent layers have a continuous magnetic path, however, the flux density per unit of applied magnetic force is increased and flux leakage reduced.

Two core shapes are in common use, as shown in **Fig 4.95**. In the shell type, both windings are placed on the inner leg, while in the core type the primary and secondary windings may be placed on separate legs, if desired. This is sometimes done when it is necessary to minimize capacitive effects between the primary and secondary, or when one of the windings must operate at very high voltage.

The number of turns required in the primary for a given applied voltage is determined by the size, shape and type of core material used, as well as the frequency. The number of turns required is inversely proportional to the cross-sectional area of the core. As a rough indication, windings of small power transformers frequently have about six to eight turns per volt on a core of 1-square-inch cross section and have a magnetic path 10 or 12 inches in length. A longer path or smaller cross sec-

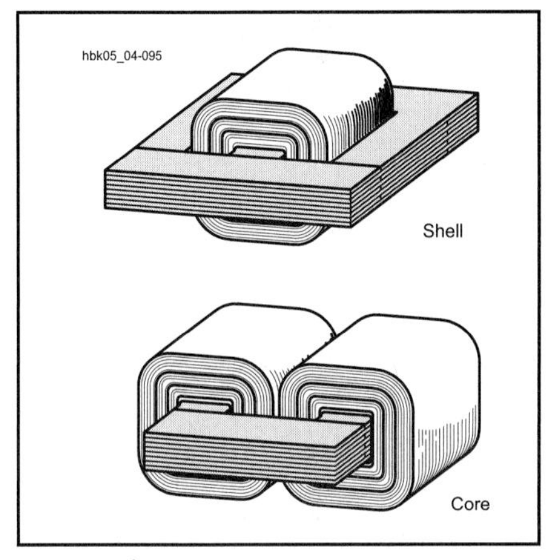

Fig 4.95 — Two common transformer constructions: shell and core.

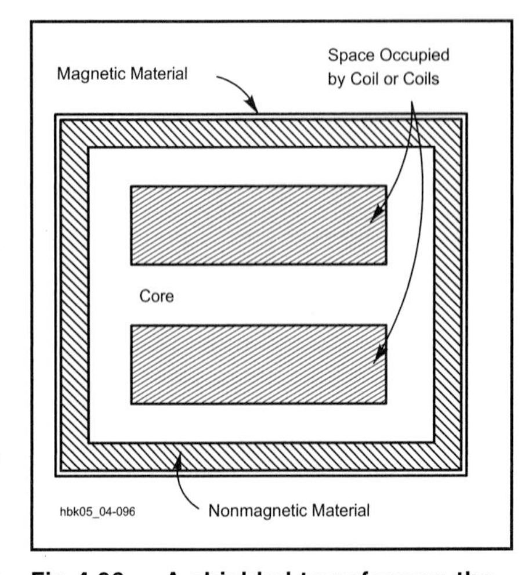

Fig 4.96 — A shielded transformer: the core plus an outer shield of magnetic material contain nearly all of the magnetic field.

tion requires more turns per volt, and vice versa.

In most transformers the coils are wound in layers, with a thin sheet of treated-paper insulation between each layer. Thicker insulation is used between adjacent coils and between the first coil and the core.

Shielding

Because magnetic lines of force are continuous and closed upon themselves, shielding requires a path for the lines of force of the leakage flux. The high-permeability of iron cores tends to concentrate the field, but additional shielding is often needed. As depicted in **Fig 4.96**, enclosing the transformer in a good magnetic material can restrict virtually all of the magnetic field in the outer case. The nonmagnetic material between the case and the core creates a region of high reluctance, attenuating the field before it reaches the case.

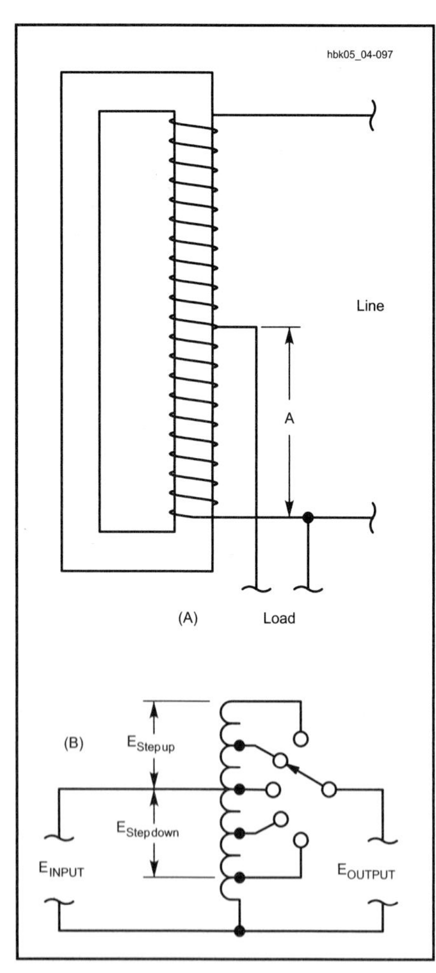

Fig 4.97 — The autotransformer is based on the transformer, but uses only one winding. The pictorial diagram at A shows the typical construction of an autotransformer. The schematic diagram at B demonstrates the use of an autotransformer to step up or step down ac voltage, usually to compensate for excessive or deficient line voltage.

AUTOTRANSFORMERS

The transformer principle can be used with only one winding instead of two, as shown in **Fig 4.97A**. The principles that relate voltage, current and impedance to the turns ratio also apply equally well. A one-winding transformer is called an *autotransformer*. The current in the common section (A) of the winding is the difference between the line (primary) and the load (secondary) currents, since these currents are out of phase. Hence, if the line and load currents are nearly equal, the common section of the winding may be wound with comparatively small wire. The line and load currents will be equal only when the primary (line) and secondary (load) voltages are not very different.

Autotransformers are used chiefly for boosting or reducing the power-line voltage by relatively small amounts. Fig 4.97B illustrates the principle schematically with a switched, stepped autotransformer. Continuously variable autotransformers are commercially available under a variety of trade names; Variac and Powerstat are typical examples.

Technically, tapped air-core inductors, such as the one in the network in Fig 4.88 at the close of the discussion of resonant circuits, are also autotransformers. The voltage from the tap to the bottom of the coil is less than the voltage across the entire coil. Likewise, the impedance of the tapped part of the winding is less than the impedance of the entire winding. Because leakage reactances are great and the co-efficient of coupling is quite low, the relationships true of a perfect transformer grow quite unreliable in predicting the exact values. For this reason, tapped induc-tors are rarely referred to as transformers. The stepped-down situation in Fig 4.88 is better approximated — at or close to resonance — by the formula

$$R_P = \frac{R_L \, X_{COM}^{\;2}}{X_L} \qquad (133)$$

where:

R_P = tuned-circuit parallel-resonant impedance,

R_L = load resistance tapped across part of the coil,

X_{COM} = reactance of the portion of the coil common to both the resonant circuit and the load tap, and

X_L = reactance of the entire coil.

The result is approximate and applies only to circuits with a Q of 10 or greater.

AIR-CORE RF TRANSFORMERS

Air-core transformers often function as mutually coupled inductors for RF applications. They consist of a primary winding and a secondary winding in close proximity. Leakage reactances are ordinarily high, however, and the coefficient of coupling between the primary and secondary windings is low. Consequently, unlike transformers having a magnetic core, the turns ratio does not have as much significance. Instead, the voltage induced in the secondary depends on the mutual inductance.

Nonresonant RF Transformers

In a very basic transformer circuit operating at radio frequencies, such as in **Fig 4.98A**, the source voltage is applied to L1. R_S is the series resistance inherent in the source. By virtue of the mutual inductance, M, a voltage is induced in L2. A current flows in the secondary circuit through the reactance of L2 and the load resistance of R_L. Let X_{L2} be the reactance of L2 independent of L1, that is, independent of the effects of mutual inductance. The impedance of the secondary circuit is then:

$$Z_S = \sqrt{R_L^{\,2} + X_{L2}^{\,2}} \qquad (134)$$

where:

Z_S = the impedance of the secondary circuit in ohms,

R_L = the load resistance in ohms, and

X_{L2} = the reactance of the secondary inductance in ohms.

The effect of Z_S upon the primary circuit is the same as a coupled impedance in series with L1. Fig 4.98B displays the coupled impedance (Z_P) in a dashed enclosure to indicate that it is not a new physical component. It has the same absolute value of phase angle as in the secondary impedance, but the sign of the reactance is reversed; it appears as a capacitive reactance. The value of Z_P is:

$$Z_P = \frac{(2 \pi f M)^2}{Z_S} \qquad (135)$$

where:

Z_P = the impedance introduced into the primary,

Z_S = the impedance of the secondary circuit in ohms, and

$2 \pi f M$ = the mutual reactance between the reactances of the primary and secondary coils (also designated as X_M).

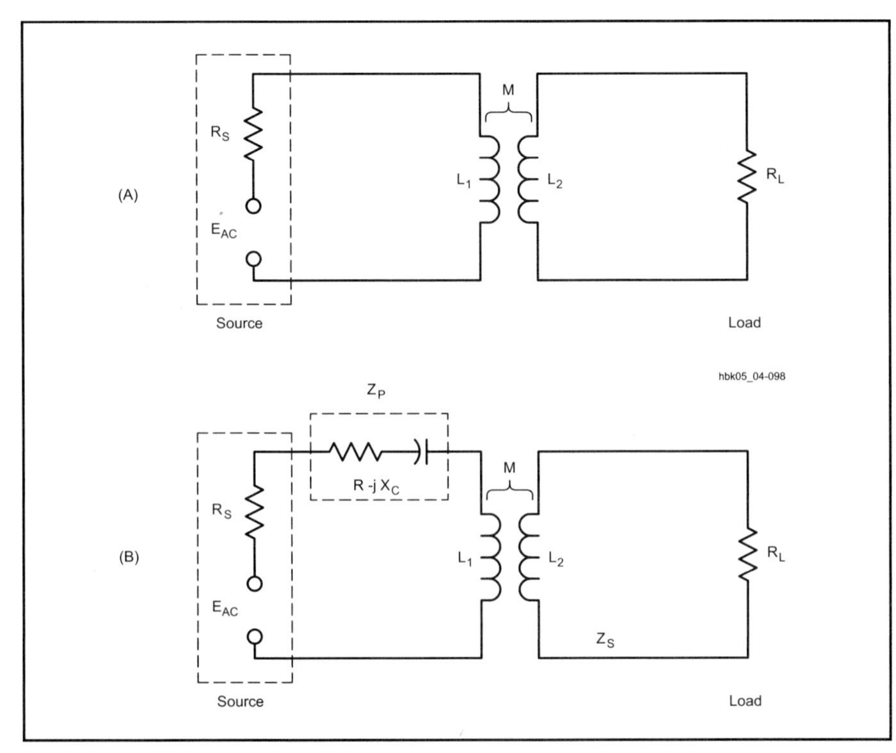

Fig 4.98 — The coupling of a complex impedance back into the primary circuit of a transformer composed of nonresonant air-core inductors.

Resonant RF Transformers

The use of at least one resonant circuit in place of a pair of simple reactances eliminates the reactance from the transformed impedance in the primary. For loaded or operating Qs of at least 10, the resistances of individual components is negligible. **Fig 4.99** represents just one of many configurations in which at least one of the inductors is in a resonant circuit. The reactance coupled into the primary circuit is cancelled if the circuit is tuned to resonance while the load is connected. If the reactance of the load capacitance, C_L is at least 10 times any stray capacitance in the circuit, as is the case for low impedance loads, the value of resistance coupled to the primary is

$$R1 = \frac{{X_M}^2 R_L}{X2^2 + {R_L}^2} \tag{136}$$

where:

R1 = series resistance coupled into the primary circuit,

X_M = mutual reactance,

R_L = load resistance, and

X2 = reactance of the secondary inductance.

The parallel impedance of the resonant circuit is just R1 transformed from a series to a parallel value by the usual formula, $R_P = X^2 / R1$.

The higher the loaded or operating Q of the circuit, the smaller the mutual inductance required for the same power transfer. If both the primary and secondary circuits consist of resonant circuits, they can be more loosely coupled than with a single tuned circuit for the same power transfer. At the usual loaded Q of 10 or greater, these circuits are quite selective, and consequently narrowband.

Although coupling networks have to a large measure replaced RF transformer coupling that uses air-core transformers, these circuits are still useful in antenna tuning units and other circuits. For RF work, powdered-iron toroidal cores have generally replaced air-core inductors for almost all applications except where the circuit handles very high power or the coil must be very temperature stable. Slug-tuned solenoid coils for low-power circuits offer the ability to tune the circuit precisely to resonance. For either type of core, reasonably accurate calculation of impedance transformation is possible. It is often easier to experiment to find the correct values for maximum power transfer, however. For further information on coupled circuits, see the section on Tuned (Resonant) Networks in the **Receivers and Transmitters,** chapter.

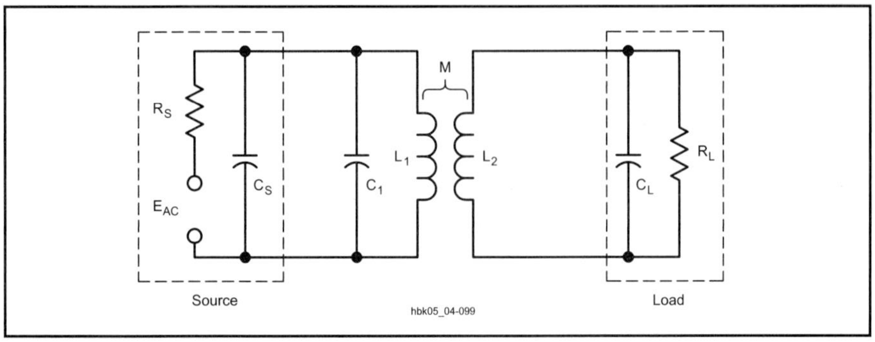

BROADBAND FERRITE RF TRANSFORMERS

The design concepts and general theory of ideal transformers presented earlier in this chapter apply also to transformers wound on ferromagnetic-core materials (ferrite and powdered iron). As is the case with stacked cores made of laminations in the classic I and E shapes, the core material has a specific permeability factor that determines the inductance of the windings versus the number of wire turns used.

Toroidal cores are useful from a few hundred hertz well into the UHF spectrum. The principal advantage of this type of core is the self-shielding characteristic. Another feature is the compactness of a transformer or inductor. Therefore, toroidal-core transformers are excellent for use not only in dc-to-dc converters, where tape-wound steel cores are employed, but at frequencies up to at least 1000 MHz with the selection of the proper core material for the range of operating frequencies. Toroidal cores are available from microminiature sizes up to several inches in diameter. The latter can be used, as one example, to build a 20-kW balun for use in antenna systems.

One of the most common ferromagnetic transformers used in Amateur Radio work

Fig 4.99 — An air-core transformer circuit consisting of a resonant primary circuit and an untuned secondary. R_S and C_S are functions of the source, while R_L and C_L are functions of the load circuit.

is the *conventional broadband transformer*. Broadband transformers with losses of less than 1 dB are employed in circuits that must have a uniform response over a substantial frequency range, such as a 2- to 30-MHz broadband amplifier. In applications of this sort, the reactance of the windings should be at least four times the impedance that the winding is designed to look into at the lowest design frequency.

Example: What should be the winding reactances of a transformer that has a 300-Ω primary and a 50-Ω secondary load? Relative to the 50-Ω secondary load:

$$X_S = 4\,Z_S = 4 \times 50\,\Omega = 200\,\Omega$$

The primary winding reactance (XP) is:

$$X_P = 4\,Z_P = 4 \times 300\,\Omega = 1200\,\Omega$$

The core-material permeability plays a vital role in designing a good broadband transformer. The effective permeability of the core must be high enough to provide ample winding reactance at the low end of the operating range. As the operating frequency is increased, the effects of the core tend to disappear until there are scarcely any core effects at the upper limit of the operating range. The limiting factors for high frequency response are distributed capacity and leakage inductance due to uncoupled flux. A high-permeability core minimizes the number of turns needed for a given reactance and therefore also minimizes the distributed capacitance at high frequencies.

Ferrite cores with a permeability of 850 are common choices for transformers used between 2 and 30 MHz. Lower frequency ranges, for example, 1 kHz to 1 MHz, may require cores with permeabilities up to 2000. Permeabilities from 40 to 125 are useful for VHF transformers. Conventional broadband transformers require resistive loads. Loads with reactive components should use appropriate networks to cancel the reactance.

Conventional transformers are wound in the same manner as a power transformer. Each winding is made from a separate length of wire, with one winding placed over the previous one with suitable insulation between. Unlike some transmission-line transformer designs, conventional broadband transformers provide dc isolation between the primary and secondary circuits. The high voltages encountered in high-impedance-ratio step-up transformers may require that the core be wrapped with glass electrical tape before adding the windings (as an additional protection from arcing and voltage breakdown), especially with ferrite cores that tend to have rougher edges. In addition, high voltage applications should also use wire with high-voltage insulation and a high temperature rating.

Fig 4.100 illustrates one method of transformer construction using a single toroid as the core. The primary of a step-down impedance transformer is wound to occupy the entire core, with the secondary wound over the primary. The first step in planning the winding is to select a core of the desired permeability. Convert the required reactances determined earlier into inductance values for the lowest frequency of use. To find the number of turns for each winding, use the A_L value for the selected core and equation 73 from the section on ferrite toroidal inductors earlier in this chapter. Be certain the core can handle the power by calculating the maximum flux using equation 71, given earlier in the chapter, and comparing the result with the manufacturer's guidelines.

Example: Design a small broadband transformer having an impedance ratio of 16:1 for a frequency range of 2.0 to 20.0 MHz to match the output of a small-signal stage (impedance ≈ 500 Ω) to the input (impedance ≈ 32 Ω) of an amplifier.

1. Since the impedance of the smaller winding should be at least 4 times the lower impedance to be matched at the lowest frequency, $X_S = 4 \times 32\,\Omega = 128\,\Omega$.

2. The inductance of the secondary winding should be

$$L_S = X_S / 2\pi f = 128 / \left(6.2832 \times 2.0 \times 10^6\ \text{Hz}\right) = 0.0101\ \text{mH}$$

3. Select a suitable core. For this low-power application, a 3/8-inch ferrite core with permeability of 850 is suitable. The core has an value of 420. Calculate the number of turns for the secondary.

$$N_S = 1000\sqrt{\frac{L}{A_L}} = 1000\sqrt{\frac{0.010}{420}}$$

4. A 5-turn secondary winding should suffice. The primary winding derives from the impedance ratio:

$$NP = N_S\sqrt{\frac{Z_P}{Z_S}} = 5\sqrt{\frac{16}{1}}$$

$$= 5 \times 4 = 20\ \text{turns}$$

This low power application will not approach the maximum flux density limits for the core, and #28 enamel wire should both fit the core and handle the currents involved.

A second style of broadband transformer

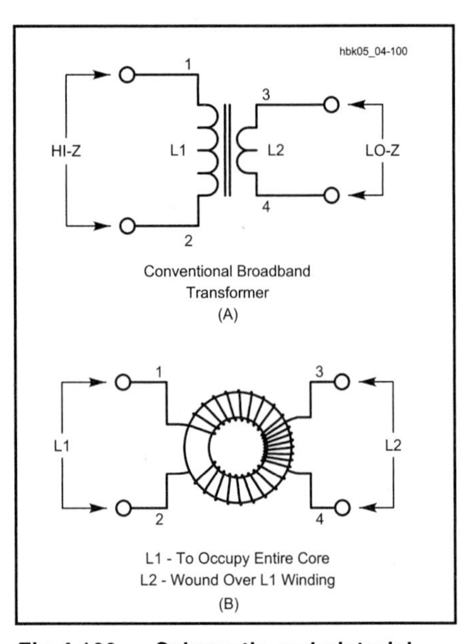

Fig 4.100 — Schematic and pictorial representation of a conventional broadband transformer wound on a ferrite toroidal core. The secondary winding (L2) is wound over the primary winding (L1).

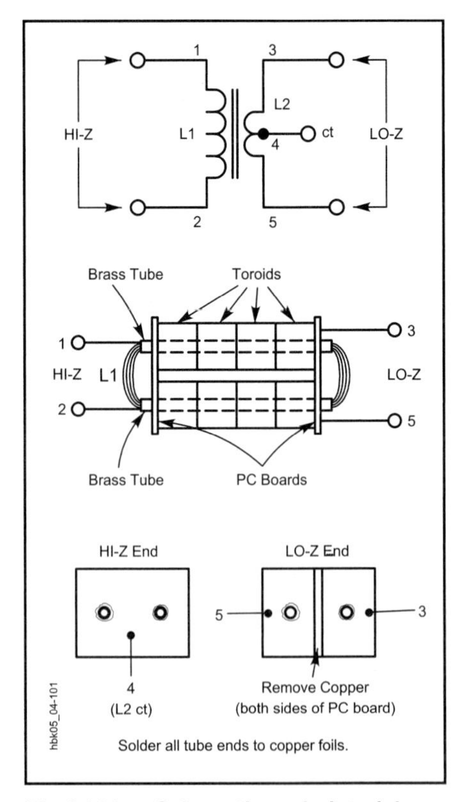

Fig 4.101 — Schematic and pictorial representation of a "binocular" style of conventional broadband transformer. This style is used frequently at the input and output ports of transistor RF amplifiers. It consists of two rows of high-permeability toroidal cores, with the winding passed through the center holes of the resulting stacks.

construction appears in **Fig 4.101**. The key elements in this transformer are the stacks of ferrite cores aligned with tubes soldered to pc-board end plates. This style of transformer is suited to high power applications, for example, at the input and output ports of transistor RF power amplifiers. Low-power versions of this transformer can be wound on "binocular" cores having pairs of parallel holes through them.

For further information on conventional transformer matching using ferromagnetic materials, see the Matching Networks section in the **RF Power Amplifiers** chapter. Refer to the **Component Data and References** chapter for more detailed information on available ferrite cores. A standard reference on conventional broadband transformers using ferro-magnetic materials is *Ferromagnetic Core Design and Applications Handbook* by Doug DeMaw, W1FB, published by Prentice Hall.

TRANSMISSION-LINE TRANSFORMERS

Conventional transformers use flux linkages to deliver energy to the output circuit. *Transmission line transformers* use transmission line modes of energy transfer between the input and the output terminals of the devices. Although toroidal versions of these transformers physically resemble toroidal conventional broadband transformers, the principles of operation differ significantly. Stray inductances and interwinding capacitances form part of the characteristic impedance of the transmission line, largely eliminating resonances that limit high frequency response. The limiting factors for transmission line transformers include line length, deviations in the constructed line from the design value of characteristic impedance, and parasitic capacitances and inductances that are independent of the characteristic impedance of the line.

The losses in conventional transformers depend on current and include wire, eddy-current and hysteresis losses. In contrast, transmission line transformers exhibit voltage-dependent losses, which make higher impedances and higher VSWR values limiting factors in design. Within design limits, the cancellation of flux in the cores of transmission line transformers permits very high efficiencies across their passbands. Losses may be lower than 0.1 dB with the proper core choice.

Transmission-line transformers can be configured for several modes of operation, but the chief amateur use is in *baluns* (*bal*anced-to-*un*balanced transformers) and in *ununs* (*un*balanced-to-*un*balanced transformers). The basic principle behind a balun appears in **Fig 4.102**, a representation of the classic Guanella 1:1 balun. The input and output impedances are the same, but the output is balanced about a real or virtual center point (terminal 5). If the characteristic impedance of the transmission line forming the inductors with numbered terminals equals the load impedance, then E2 will equal E1. With respect to terminal 5, the voltage at terminal 4 is E1 / 2, while the voltage at terminal 2 is –E1 / 2, resulting in a balanced output.

The small losses in properly designed baluns of this order stem from the potential gradient that exists along the length of transmission line forming the transformer. The value of this potential is –E1 / 2, and it forms a dielectric loss that can't be eliminated. Although the loss is very small in well-constructed 1:1 baluns at low impedances, the losses climb as impedances climb (as in 4:1 baluns) and as the VSWR climbs. Both conditions yield higher voltage gradients.

The inductors in the transmission-line transformer are equivalent to — and may be — coiled transmission line with a characteristic impedance equal to the load. They form a choke isolating the input from the output and attenuating undesirable currents, such as antenna current, from the remainder of the transmission line to the energy source. The result is a *current* or *choke* balun. Such baluns may take many forms: coiled transmission line, ferrite beads placed over a length of transmission line, windings on linear ferrite cores or windings on ferrite toroids.

Reconfiguring the windings of Fig 4.102 can alter the transformer operation. For example, if terminal 2 is connected to terminal 3, a positive potential gradient appears across the lengths of line, resulting in a terminal 4 potential of 2 E1 with respect to ground. If the load is disconnected from terminal 2 and reconnected to ground, 2 E1 appears across the load — instead of ±E1 / 2. The product of this experiment is a 4:1 impedance ratio, forming an unun. The bootstrapping effect of the new connection is applicable to many other design configurations involving multiple windings to achieve custom impedance ratios from 1:1 up to 9:1.

Balun and unun construction for the impedances of most concern to amateurs requires careful selection of the feed line used to wind the balun. Building transmission line transformers on ferrite toroids may require careful attention to wire size and spacing to approximate a 50-Ω line. Wrapping wire with polyimide tape (one or two coatings, depending upon the wire size) and then glass taping the wires together periodically produces a reasonable 50-Ω transmission line. Ferrite cores in the permeability range of 125 to 250 are generally optimal for transformer windings, with 1.25-inch cores suitable to 300-Ω power levels and 2.4-inch cores usable to the 5 kW level. Special designs may alter the power-handling capabilities of the core sizes. For the 1:1 balun shown in Fig 4.102, 10 bifilar turns (#16 wire for the smaller core and #12 wire for the larger, both Thermaleze wire) yields a transformer operable from 160 to 10 m.

Transmission-line transformers have their most obvious application to antennas, since they isolate the antenna currents from the feed line, especially where a coaxial feed line is not exactly perpendicular to the antenna. The balun prevents antenna currents from flowing on the outer surface of the coax shielding, back to the trans-mitting equipment. Such currents would distort the antenna radiation pattern. Appropriately designed baluns can also transform impedance values at the same time. For example, one might use a 4:1 balun to match a 12.5-Ω Yagi antenna impedance to a 50-Ω feed line. A 4:1 balun

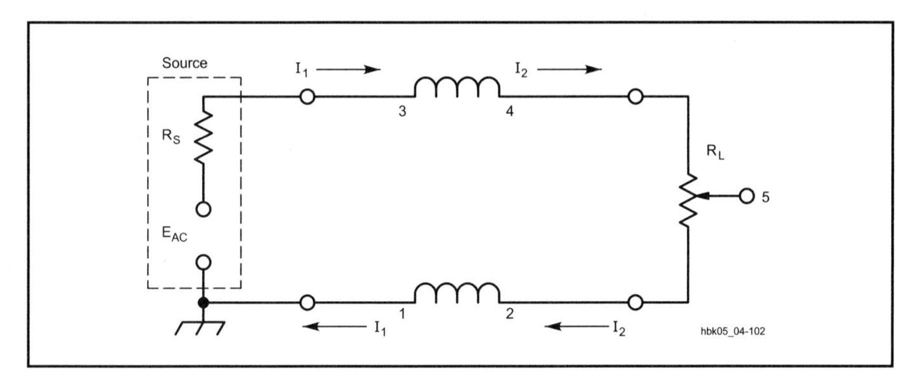

Fig 4.102 — Schematic representation of the basic Guanella "choke" balun or 1:1 transmission line transformer. The inductors are a length of two-wire transmission line. R_S is the source impedance and R_L is the load impedance.

might also be used to match a 75-Ω TV antenna to 300-Ω feed line.

Interstage coupling within solid-state transmitters represents another potential for transmission-line transformers. Broadband coupling between low-impedance, but mismatched stages can benefit from the low losses of transmission-line transformers. Depending upon the losses that can be tolerated and the bandwidth needed, it is often a matter of designer choice between a transmission-line transformer and a conventional broadband transformer as the coupling device.

For further information on transmission-line transformers and their applications, see the **RF Power Amplifiers** chapter. Another reference on the subject is *Transmission Line Transformers*, by Jerry Sevick, W2FMI, published by Noble Publishing.

Chapter 5

Electrical Signals and Components

Analog Glossary

Active Region — The region in the characteristic curve of an analog device in which the signal is amplified linearly.

Amplification — The process of increasing the size of a signal. Also called gain.

Analog signal — A signal, usually electrical, that can have any amplitude (voltage or current) value and exists at any point in time.

Anode — The element of an analog device that accepts electrons.

Base — The middle layer of a bipolar transistor, often the input.

Biasing — The addition of a dc voltage or current to a signal at the input of an analog device, which changes the signal's position on the characteristic curve.

Bipolar Transistor — An analog device made by sandwiching a layer of doped semiconductor between two layers of the opposite type: PNP or NPN.

Buffer — An analog stage that prevents loading of one analog stage by another.

Cascade — Placing one analog stage after another to combine their effects on the signal.

Cathode — The element of an analog device that emits electrons.

Characteristic Curve — A plot of the relative responses of two or three analog-device parameters, usually output with respect to input.

Clamping — A nonlinearity in amplification where the signal can be made no larger.

Collector — One of the outer layers of a bipolar transistor, often the output.

Compensation — The process of counteracting the effects of signals that are inadvertently fed back from the output to the input of an analog system. The process increases stability and prevents oscillation.

Cutoff Region — The region in the characteristic curve of an analog device in which there is no current through the device. Also called the OFF region.

Diode — A two-element vacuum tube or semiconductor with only a cathode and an anode (or plate).

Drain — The connection at one end of a field-effect-transistor channel, often the output.

Electron — A subatomic particle that has a negative charge and is the basis of electrical current.

Emitter — One of the outer layers of a bipolar transistor, often the reference.

Field-Effect Transistor (FET) — An analog device with a semiconductor channel whose width can be modified by an electric field. Also called a unipolar transistor.

Gain — see ***Amplification***.

Gain-Bandwidth Product — The interrelationship between amplification and frequency that defines the limits of the ability of a device to act as a linear amplifier. In many amplifiers, gain times bandwidth is approximately constant.

Gate — The connection at the control point of a field-effect transistor, often the input.

Grid — The vacuum-tube element that controls the electron flow from cathode to plate. Additional grids in some tubes perform other control functions to improve performance.

Hole — A positively charged "particle" that results when an electron is removed from an atom in a semiconductor crystal structure.

Integrated Circuit (IC) — A semiconductor device in which many components, such as diodes, bipolar transistors, field-effect transistors, resistors and capacitors are fabricated to make an entire circuit.

Junction FET (JFET) — A field-effect transistor that forms its electric field across a PN junction.

Linearity — The property found in nature and most analog electrical circuits that governs the processing and combination of signals by treating all signal levels the same way.

Load Line — A line drawn through a family of characteristic curves that shows the operating points of an analog device for a given output load impedance.

Loading — The condition that occurs when a cascaded analog stage modifies the operation of the previous stage.

Metal-Oxide Semiconductor (MOSFET) — A field-effect transistor that forms its electric field through an insulating oxide layer.

N-Type Impurity — A doping atom with an excess of electrons that is added to semiconductor material to give it a net negative charge.

Noise — Any unwanted signal.

Noise Figure (NF) — A measure of the noise added to a signal by an analog processing stage.

Operational Amplifier (op amp) — An integrated circuit that contains a symmetrical circuit of transistors and resistors with highly improved characteristics over other forms of analog amplifiers.

Oscillator — An unstable analog system, which causes the output signal to vary spontaneously.

P-Type Impurity — A doping atom with an excess of holes that is added to semiconductor material to give it a net positive charge.

Peak Inverse Voltage (PIV) — The highest voltage that can be tolerated by a reverse biased PN junction before current is conducted.

Pentode — A five-element vacuum tube with a cathode, a control grid, a screen grid, a suppressor grid, and a plate.

Plate — See anode, usually used with vacuum tubes.

PN Junction — The region that occurs when P-type semiconductor material is placed in contact with N-type semiconductor material.

Saturation Region — The region in the characteristic curve of an analog device in which the output signal can be made no larger. See ***Clamping***.

Semiconductor — An elemental material whose current conductance can be controlled.

Signal-To-Noise Ratio (SNR) — The ratio of the strength of the desired signal to that of the unwanted signal (noise).

Slew Rate — The maximum rate at which a signal may change levels and still be accurately amplified in a particular device.

Source — The connection at one end of the channel of a field-effect transistor, often the reference.

Superposition — The natural process of adding two or more signals together and having each signal retain its unique identity.

Tetrode — A four-element vacuum tube with a cathode, a control grid, a screen grid, and a plate.

Triode — A three-element vacuum tube with a cathode, a grid, and a plate.

Unipolar Transistor — see ***Field-Effect Transistor (FET)***.

Zener Diode — A PN-junction diode with a controlled peak inverse voltage so that it will start conducting current at a preset reverse voltage.

Introduction

This section, written by Greg Lapin, N9GL, treats analog signal processing in two major parts. Analog signals behave in certain well-defined ways regardless of the specific hardware used to implement the processing. Signal processing involves various electronic stages to perform func-tions such as amplifying, filtering, modulation and demodulation. A piece of electronic equipment, such as a radio, cascades a number of these circuits. How these stages interact with each other and how they affect the signal individually and in tandem is the subject of the first part of this chapter.

Implementing analog signal processing functions involves several types of active components. An active electronic component is one that requires a power source to function, and is distinguished in this way from passive components (such as resistors, capacitors and inductors) that are described in the **Real-World Component Characteristics** chapter. The second part of this chapter describes the various technologies that implement active devices. Vacuum tubes, bipolar semiconductors, field-effect semiconductors and integrated semiconductor circuitry comprise a wide spectrum of active devices used in analog signal processing. Several different devices can perform the same function. The second part of the chapter describes the physical basis of each device. Understanding the specific characteristics of each device allows you to make educated decisions about which device would be best for a particular purpose when designing analog circuitry, or understanding why an existing circuit was designed in a particular way.

Analog Signal Processing

LINEARITY

The term, *analog signal*, refers to the continuously variable voltage of which all radio and audio signals are made. Some signals are man-made and others occur naturally. In nature, analog signals behave according to laws that make radio communication possible. These same laws can be put to use in electronic instruments to allow us to manipulate signals in a variety of ways.

The premier properties of signals in nature are *superposition* and *scaling*. Superposition is the property by which signals combine. If two signals are placed together, whether in a circuit, in a piece of wire, or even in air, they become one combined signal that is the sum of the individual signals. This is to say that at any one point in time, the voltage of the combined signal is the sum of the voltages of the two original signals at the same time. In a linear system any number of signals will add in this way to give a single combined signal.

One of the more important features of superposition, for the purposes of signal processing, is that signals that have been combined can be separated into their original components. This is what allows signals that have been contaminated with noise to be separated from the noise, for example.

Amplification and attenuation scale signals to be larger and smaller, respectively. The operation of scaling is the same as multiplying the signal at each point in time by a constant value; if the constant is greater than one then the signal is amplified, if less than one then the signal is attenuated.

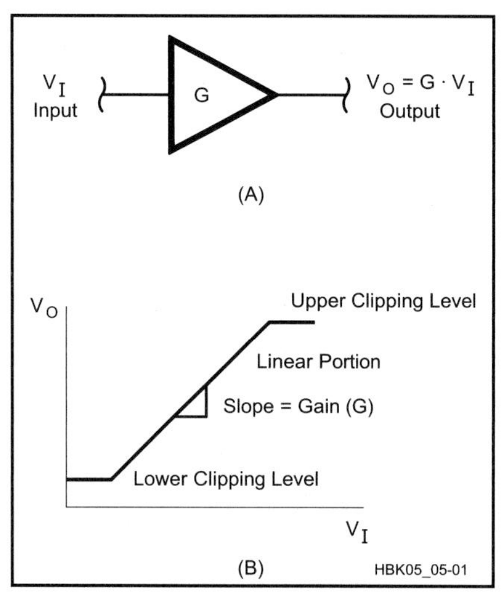

Fig 5.1 — Generic amplifier. (A) Symbol. For the linear amplifier, gain is the constant value, G, and the output voltage is equal to the input voltage times G; (B) Transfer function, input voltage along the x-axis is converted to the output voltage along the y-axis. The linear portion of the response is where the plot is diagonal; its slope is equal to the gain, G. Above and below this range are the clipping limits, where the response is not linear and the output signal is clipped.

Linear Operations

Any operation that modifies a signal and obeys the rules of superposition and scaling is a *linear operation*. The most basic linear operation occurs in an amplifier, a circuit that increases the amplitude of a signal. Schematically, a generic amplifier is signified by a triangular symbol, its input along the left face and its output at the point on the right (see **Fig 5.1**). The linear amplifier multiplies every value of a signal by a constant value. Amplifier gain is often expressed as a multiplication factor (x 5, for example).

$$\text{Gain} = \frac{V_o}{V_i} \qquad (1)$$

where V_o is the output voltage from an amplifier when an input voltage, V_i, is applied.

Ideal linear amplifiers have the same gain for all parts of a signal. Thus, a gain of 10 changes 10 V to 100 V, 1 V to 10 V and –1 V to –10 V. Amplifiers are limited by their dynamic range and frequency response, however. An amplifier can only produce output levels that are within the range of its power supply. The power-supply voltages are also called the *rails* of an amplifier. As the amplified output approaches one of the rails, the output will not go beyond a given voltage that is near the rail. The output is limited at the *clipping level* of an amplifier. When an amplifier tries to amplify a signal to be larger than this value, the output remains at this level; this is called output *clipping*. Clipping is a nonlinear effect; an amplifier is considered linear only between its clipping levels. See Fig 5.1.

Another limitation of an amplifier is its frequency response. Signals within a range of frequencies are amplified consistently but outside that range the amplification changes. At higher frequencies an amplifier acts as a low-pass filter, decreasing amplification with increasing frequency. For lower frequencies, amplifiers are of two kinds: dc and ac coupled.

A dc-coupled amplifier equally amplifies signals with frequencies down to dc. An ac-coupled amplifier acts as a high-pass filter, decreasing amplification as the frequency decreases toward dc.

The combination of gain and frequency limitations is often expressed as a *gain-bandwidth product*. At high gains many amplifiers work properly only over a small range of frequencies. In many amplifiers, gain times bandwidth is approximately constant. As gain increases, bandwidth decreases, and vice versa. Another similar descriptor is called *slew rate*. This term describes the maximum rate at which a signal can change levels and still be accurately amplified in a particular device. There is a direct correlation between the signal-level rate of change and the frequency content of that signal.

Feedback and Oscillation

The stability of an amplifier refers to its ability to provide gain to a signal without tending to oscillate. For example, an amplifier just on the verge of oscillating is not generally considered to be "stable." If the output of an amplifier is fed back to the input, the feedback can affect the amplifier stability. If the amplified output is added to the input, the output of the sum will be larger. This larger output, in turn, is also fed back. As this process continues, the amplifier output will continue to rise until the amplifier cannot go any higher (clamps). Such *positive feedback* increases the amplifier gain, and is called *regeneration*.

Most practical amplifiers have intrinsic feedback that is unavoidable. To improve the stability of an amplifier, *negative feedback* can be added to counteract any unwanted positive feedback. Negative feedback is often combined with a phase-shift *compensation* network to improve the amplifier stability.

Although negative feedback reduces amplifier or stage gain, the advantages of *stable* gain, freedom from unwanted oscillations, and the reduction of distortion are often key design objectives and advantages of using negative feedback.

The design of feedback networks depends on the desired result. For amplifiers, which should not oscillate, the feedback network is customized to give the desired frequency response without loss of stability. For oscillators, the feedback network is designed to create a steady oscillation at the desired frequency.

Filtering

A filter is a common linear stage in radio equipment. Filters are characterized by their ability to selectively attenuate certain frequencies (stop band) while passing or amplifying others (pass band). Passive filters are described in the **RF and AF Filters** chapter. Filters can also be designed using active devices. All practical amplifiers are low-pass filters or band-pass filters, because the gain decreases as the frequency increases beyond their gain-bandwidth products.

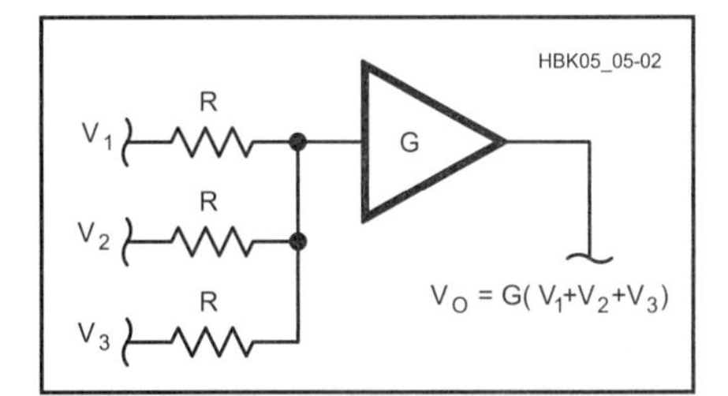

Fig 5.2 — Summing amplifier. The output voltage is equal to the sum of the input voltages times the amplifier gain, G. As long as the resistance values, R, are equal and the amplifier input impedance is much higher, the actual value of R does not affect the output signal.

Summing Amplifiers

In a linear system, nature does most of the work for us when it comes to adding signals; placing two signals together naturally causes them to add. When processing signals, we would like to control the summing operation so the signals do not distort. If two signals come from separate stages and they are connected, the stages may interact, causing both stages to distort their signals. Summing amplifiers generally use a resistor in series with each stage, so the resistors connect to the common input of the following stage. **Fig 5.2** illustrates the resistors connecting to a summing amplifier. Ideally, any time we wanted to combine signals (for example, combining an audio signal with a PL tone in a 2-m FM transmitter prior to modulating the RF signal) we could use a summing amplifier.

Buffering

It is often necessary to isolate the stages of an analog circuit. This isolation reduces the loading, coupling and feedback between stages. An intervening stage, called a *buffer*, is often used for this purpose. A buffer is a linear circuit that is a type of amplifier. It is often necessary to change the characteristic impedance of a circuit

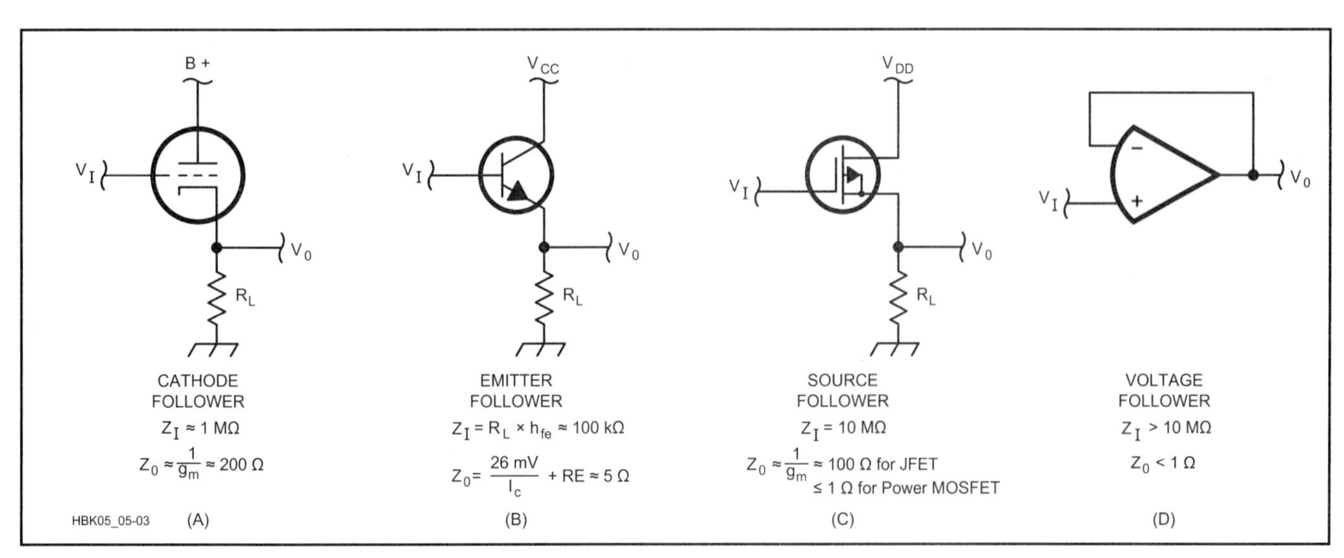

Fig 5.3 — Common buffer stages and some typical input (Z_I) and output (Z_O) impedances. (A) Cathode follower, made with triode tube; (B) Emitter follower, made with NPN bipolar transistor; (C) Source follower, made with FET; and (D) Voltage follower, made with operational amplifier. All of these buffers are terminated with a load resistance, R_L, and have an output voltage that is approximately equal to the input voltage (gain ≈ 1).

between stages. Buffers can have high values of amplification but this is unusual. A buffer performs impedance transformations most efficiently when it has a low or unity gain. **Fig 5.3** shows common forms of buffers with low-impedance outputs: the cathode follower using a triode tube, the emitter follower using a bipolar transistor, the source follower using a field-effect transistor and the voltage follower, using an operational amplifier.

In some circuits, notably power amplifiers, the desired goal is to deliver a maximum amount of power to the output device (such as a speaker or an antenna). Matching the amplifier output impedance to the output-device impedance provides maximum power transfer. A buffer amplifier may be just the circuit for this type of application. Such amplifier circuits must be carefully designed to avoid distortion.

Amplitude Modulation/ Demodulation

Voice signals are transmitted over the air by amplitude modulating them on higher frequency carrier signals (see the **Mixers, Modulators and Demodulators** chapter). The process of amplitude modulation can be mathematically described as the multiplication (product) of the voice signal and the carrier signal. Multiplication is a linear process since amplitude modulating the sum of two audio signals produces a signal that is identical to the sum of amplitude modulating each audio signal individually. When two equal-strength SSB signals are transmitted on the same frequency, the observer hears both of the voices simultaneously. Another aspect of the linear behavior of amplitude modulation is that amplitude-modulated signals can be demodulated to be exactly in their original form. Amplitude demodulation is the converse of amplitude modulation, and is represented as a division operation.

In the linear model of amplitude modulation, the signal to be modulated (such as the audio signal in an AM transmitter) is shifted in frequency by multiplying it with the carrier. The modulated waveform is considered to be a linear function of the signal. The carrier is considered to be part of a time-varying linear system and not a second signal.

A curious trait of amplitude modulation is that it can be performed nonlinearly. Each nonlinear form of amplitude modulation generates the desired linear product term in addition to other unwanted terms that must be removed. Accurate analog multipliers and dividers are difficult and expensive to fabricate. Two common nonlinear amplitude-modulating schemes are much simpler to implement but have disadvantages as well.

Power-law modulators generate many frequencies in addition to the desired ones. These unwanted frequencies, often called *intermodulation products*, steal energy from the desired *first order product*. The unwanted signals must be filtered out. The inefficiency of this process makes this type of modulator good only for low-level modulation, with additional amplification required for the modulated signal. A *square-law modulator* can be implemented with a single FET, biased in its saturation region, as the only active component.

Switching modulators are more efficient and provide high-level modulation. A single active device acts as a switch to turn the signal on and off at the carrier frequency. Both the signal and the carrier must be amplified to relatively high levels prior to this form of modulation. The modulated carrier must be filtered by a tank circuit to remove unwanted frequency components generated by the switching artifacts.

Nonlinear demodulation of an amplitude-modulated signal can be realized with a single diode. The diode rectifies the signal (a nonlinear process) and then the nonlinear products are filtered out before the desired signal is recovered.

NONLINEAR OPERATORS

All signal processing doesn't have to be linear. Any time that we treat various signal levels differently, the operation is called *nonlinear*. This is not to say that all signals must be treated the same for a circuit to be linear. High-frequency signals are attenuated in a low-pass filter while low-frequency signals are not, yet the filter can be linear. The distinction is that all voltages of the high-frequency signal are attenuated by the same amount, thus satisfying one of the linearity conditions. What if we do not want to treat all voltage levels the same way? This is commonly desired in analog signal processing for clipping, rectification, compression, modulation and switching.

Clipping and Rectification

Clipping is the process of limiting the range of signal voltages passing through a circuit (in other words, *clipping* those voltages outside the desired range from the signals). There are a number of reasons why we would like to do this. Clipping generally refers to the process of limiting the positive and negative peaks of a signal. We might use this technique to avoid overdriving an amplifier, for example. Another kind of clipping results in rectification. The rectifier clips off all voltages of one polarity (positive or negative) and allows only the other polarity through, thus changing ac to pulsating dc (see the **Power Supplies** chapter). Another use of clipping is when only one signal polarity is allowed to drive an amplifier input; a clipping stage precedes the amplifier to ensure this.

Logarithmic Amplification

It is sometimes desirable to amplify a signal logarithmically, which means amplifying low levels more than high levels. This type of amplification is often called *signal compression*. Speech compression is sometimes used in audio amplifiers that feed modulators. The voice signal is compressed into a small range of amplitudes, allowing more voice energy to be transmitted without overmodulation (see the **Modes and Modulation Sources** chapter).

ANALOG BUILDING BLOCKS

Many kinds of electronic equipment are developed by combining basic analog signal processing circuits called "building blocks." This section describes several of these building blocks and how they are combined to perform complex functions. Although not all basic electronic functions are discussed here, the characteristics of combining them can be applied generally.

An analog building block can contain any number of discrete components. Since our main concern is the effect that circuitry has on a signal, we often describe the building block by its actions rather than its specific components. For this reason, an analog building block is often referred to as a *two-port network* or a *black box*. Two basic properties of analog networks are of principal concern: the effect that the network has on an analog signal and the interaction that the network has with the circuitry surrounding it. The two network ports are the input and output connections. The signal is fed into the input port, is modified inside the network and then exits from the output port.

An analog network modifies a signal in a specific way that can be described mathematically. The output is related to the input by a *transfer function*. The mathematical operation that combines a signal with a transfer function is pictured symbolically in **Fig 5.4**. The output signal, w(t), has a

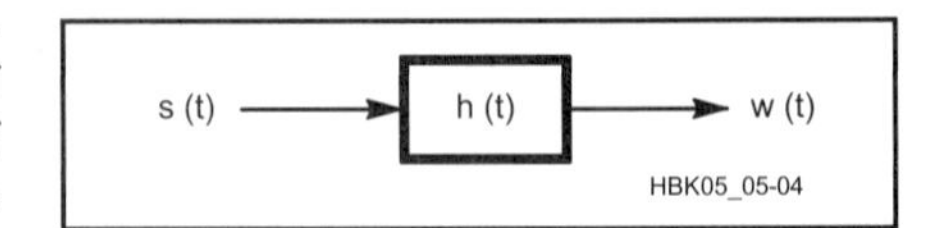

Fig 5.4 — Linear function block. The output signal, w(t) is produced by the action of the transfer function, h(t) on the input signal s(t).

value that changes with time. The output signal is created by the action of an analog transfer function, h(t), on the input signal, g(t).

While it is not necessary to understand transfer functions mathematically to work with analog circuits, it is useful to realize that they describe how a signal interacts with other signals in an electronic system. In general, the output signal of an analog system depends not only on the input signal at the same time, but also on past values of the input signal. This is a very important concept and is the basis of such essential functions as analog filtering.

Cascading Stages

If an analog circuit can be described with a transfer function, a combination of analog circuits can also be described similarly. This description of the combined circuits depends upon the relationship between the transfer functions of the parts and that of the combined circuits. In many cases this relationship allows us to predict the behavior of large and complex circuits from what we know about the parts of which they are made. This aids in the design and analysis of analog circuits.

When two analog circuits are cascaded (the output signal of one stage becomes the input signal to the next stage) their transfer functions are combined. The mechanism of the combination depends on the interaction between the stages. The ideal case is when there is no interaction between stages. In other words, the action of the first stage is unchanged, regardless of whether or not the second stage follows it. Just as the signal entering the first stage is modified by the action of the first transfer function, the ideal cascading of analog circuits results in changes produced only by the individual transfer functions. For any number of stages that are cascaded, the combination of their transfer functions results in a new transfer function. The signal that enters the circuit is changed by the composite transfer function, to produce the signal that exits the cascaded circuits.

While each stage in a series may use feedback within itself, feedback around more than one stage may create a function — and resultant performance — different from any of the included stages (oscillation or negative feedback).

Cascaded Buffers

Buffer stages that are made with single active devices can be more effective if cascaded. Two types of such buffers are in common use. The *Darlington pair* is a cascade of two common-collector transistors as shown in **Fig 5.5**. (The various amplifier configurations will be described later in this chapter.) The input impedance of the Darlington pair is equal to the load impedance times the current gain, h_{FE}. The current gain of the Darlington pair is the product of the current gains for the two transistors.

$$Z_I = Z_{LOAD} \times h_{FE1} \times h_{FE2} \tag{2}$$

For example, if a typical bipolar transistor has h_{FE} = 100 and a circuit has a Z_{LOAD} = 15 kΩ, a pair of these transistors in the Darlington-pair configuration would have:

$$Z_I = 15\,k\Omega \times 100 \times 100 = 150\,M\Omega$$

The shunt capacitance at the input of real transistors can lower the actual impedance as the frequency increases.

A common-emitter amplifier followed by a common-base amplifier is called a *cascode buffer* (see **Fig 5.6**). Cascodes are also made with FETs by following a common-source amplifier by a common-gate configuration. The input impedance and current gain of the cascode are approximately the same as those of the first stage. The output impedance is much higher than that of a single stage. Cascode amplifiers have excellent input/output isolation (very low unwanted feedback) and this can provide high gain with good stability. An example of a cascode buffer made with bipolar transistors has moderate input impedance, Z_I = 1 kΩ, high current gain, h_{FE} = 50 and high output impedance, Z_O = 1 MΩ. There is very little reverse internal feedback in the cascode design, making it very stable, and the amplifier design has little effect on external tuning components. Cascode circuits are often used in tuned amplifier designs for these reasons.

Interstage Loading and Impedance Matching

If the transfer function of a stage changes when it is cascaded with another stage, we say that the second stage has *loaded* the first stage. This often occurs when an appreciable amount of current passes from one stage to the next.

Every two-port network can be further defined by its input and output impedance. The input impedance is the opposition to current, as a function of frequency, seen when looking into the input port of the network. Likewise, the output impedance is similarly defined when looking back into a network through its output port. Interstage loading is related to the relative output impedance of a stage and the input impedance of the stage that is cascaded after it.

In some applications the goal is to transfer a maximum amount of power. In an RF amplifier, the impedance at the input of the transmission line feeding an antenna is transformed by means of a matching network to produce the resistance the amplifier needs in order to efficiently produce RF power.

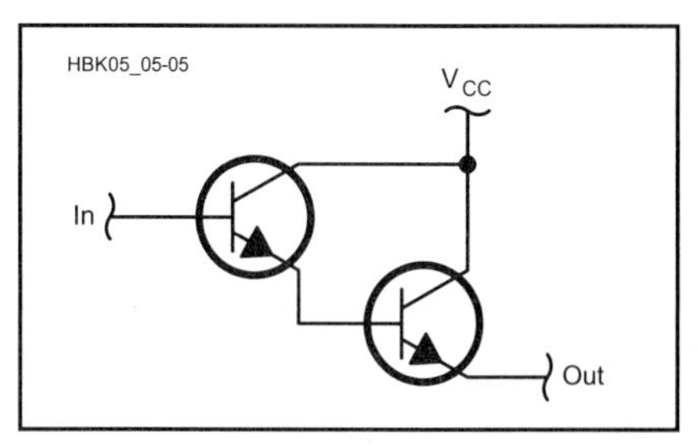

Fig 5.5 — Darlington pair made with two emitter followers. Input impedance, Z_I, is far higher than for a single transistor and output impedance, Z_O, is nearly the same as for a single transistor. DC biasing has been omitted for simplicity.

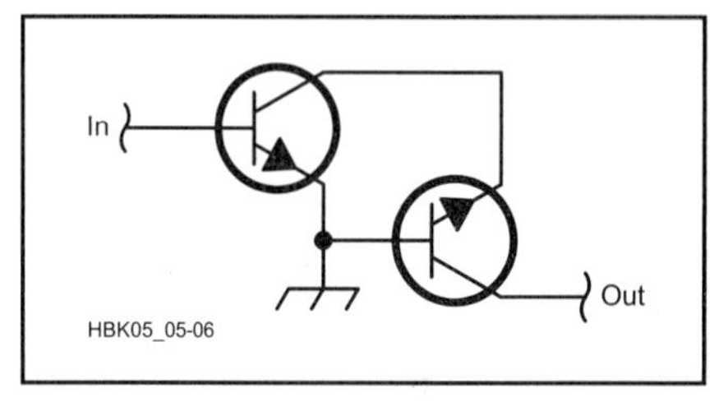

Fig 5.6 — Cascode pair made with two NPN bipolar transistors has a medium input impedance and high output impedance. DC biasing has been omitted for simplicity.

In contrast, it is the goal of most analog signal processing circuitry to modify a signal rather than to deliver large amounts of energy. Thus, an impedance-matched condition may not be what is desired. Instead, current between stages can be minimized by using mismatched impedances. Ideally, if the output impedance of a network approaches zero ohms and the input impedance of the following stage is very high, very little current will pass between the stages, and interstage loading will be negligible.

Noise

Generally we are only interested in specific man-made signals. Nature allows many signals to combine, however, so the desired signal becomes combined with many other unwanted signals, both man-made and naturally occurring. The broadest definition of noise is any signal that is not the one in which we are interested. One of the goals of signal processing is to separate desired signals from noise.

One form of noise that occurs naturally and must be dealt with in low-level processing circuits is called *thermal noise*, or *Johnson noise*. Thermal noise is produced by random motion of free electrons in conductors and semiconductors. This motion increases as temperature increases, hence the name. This kind of noise is present at all frequencies and is proportional to temperature. Naturally occurring noise can be reduced either by decreasing the bandwidth or by reducing the temperature in the system. Thermal noise voltage and current vary with the circuit impedance, according to Ohm's Law. Low-noise-amplifier-design techniques are based on these relationships.

Analog signal processing stages are characterized in part by the noise they add to a signal. A distinction is made between enhancing existing noise (such as amplifying it) and adding new noise. The noise added by analog signal processing is commonly quantified by the *noise factor, f*. Noise factor is the ratio of the total output noise power (thermal noise plus noise added by the stage) to the input noise power when the termination is at the standard temperature of 290 K (17°C). When the noise factor is expressed in dB, we often call it *noise figure, NF*. NF is calculated as:

$$NF = 10 \log \frac{P_{NO}}{A\, P_{N\ TH}} \tag{3}$$

where:

P_{NO} = total noise output power,

A = amplification gain, and

$P_{N\ TH}$ = input thermal noise power.

The noise factor can also be calculated as the difference between the input and output signal-to-noise ratios (SNR), with SNR expressed in dB.

In a system of many cascaded signal processing stages, each stage affects the noise of the system. The noise factor of the first stage dominates the noise factor of the entire system. Designers try to optimize system noise factor by using a first stage with a minimum possible noise factor and maximum possible gain. A circuit that overloads is often as useless as one that generates too much noise. See the **Receivers and Transmitters** chapter for more information about circuit noise.

Analog Devices

There are several different kinds of components that can be used to build circuits for analog signal processing. The same processing can be performed with vacuum tubes, bipolar semiconductors, field-effect semiconductors or integrated circuitry, each with its own advantages and disadvantages.

TERMINOLOGY

A similar terminology is used for most active electronic devices. The letter V stands for voltages and I for currents. Voltages generally have two subscripts indicating the terminals between which the voltage is measured (V_{BE} is the voltage between the base and the emitter of a bipolar transistor). Currents have a single subscript indicating the terminal into which the current flows (I_P is the current into the plate of a vacuum tube). If the current flows out of the device, it is generally indicated with a negative sign. Power supply voltages have two subscripts that are the same, indicating the terminal to which the voltage is applied (V_{DD} is the power supply voltage applied to the drain of a field-effect transistor). A transfer characteristic is a ratio of an output parameter to an input parameter, such as output current divided by input current. Transfer characteristics are represented with letters, such as h, s, y or z. Resistance is designated with the letter r, and impedance with the letter Z. For example, r_{DS} is resistance between drain and source of an FET and Z_i is input impedance. In some designators, values differ for dc and ac signals. This is indicated by using capital letters in the subscripts for dc and lowercase subscripts for ac. For example, the common-emitter dc current gain for a bipolar transistor is designated as h_{FE}, and h_{fe} is the ac current gain. Qualifiers are sometimes added to the subscripts to indicate certain operating modes of the device. SS for saturation, BR for breakdown, ON and OFF are all commonly used.

The abbreviations for tubes existed before these standards were adopted so some tube-performance descriptors are different. For example, B+ is usually used for the plate bias voltage. Since integrated circuits are collections of semiconductor components, the abbreviations for the type of semiconductor used also apply to the integrated circuit. V_{CC} is a power supply voltage for an integrated circuit made with bipolar transistor technology.

Amplifier Types

Amplifier configurations are described by the *common* part of the device. The word "common" is used to describe the connection of a lead directly to a reference. The most common reference is ground, but positive and negative power sources are also valid references. The type of reference used depends on the type of device (vacuum tube, transistor [NPN or PNP], FET [P-channel or N-channel]), which lead is common and the range of signal levels. Once a common lead is chosen, the other two leads are used for signal input and output. Based on the biasing conditions, there is only one way to select these leads. Thus, there are three possible amplifier configurations for each type of three-lead device.

The operation of an amplifier is specified by its gain. A gain in this sense is defined as the change (Δ) in the output parameter divided by the corresponding change in the input parameter. If a particular device measures its input and output as currents, the gain is called a current gain. If the input and output are voltages, the amplifier is defined by its voltage gain. If the input is a voltage and the output is a current, the ratio is called the *transconductance*.

Characteristic Curves

Analog devices are described most completely with their *characteristic curves*. Almost all devices of concern are nonlinear over a wide range of operating parameters. We are often interested in using a device only in the region that approximates a linear response. The characteristic curve is a plot of the interrelationships between two or three variables. The vertical (y) axis parameter is the output, or result of the device being operated with an input parameter on the horizontal (x) axis. Often the output is the result of two input values. The first input parameter is represented along the x axis and the second input parameter by several curves, each for a different value. For example, a vacuum tube characteristic curve may have the plate current along the y axis, the grid voltage along the x axis and several curves, each representing a different value of the plate bias voltage (see **Fig 5.7**).

The parameters plotted in the characteristic curve depend on how the device will be used. The common amplifier configuration defines the input and output leads, and their relationship is diagrammed by the curves. Device parameters are usually derived from the characteristic curve. To calculate a gain,

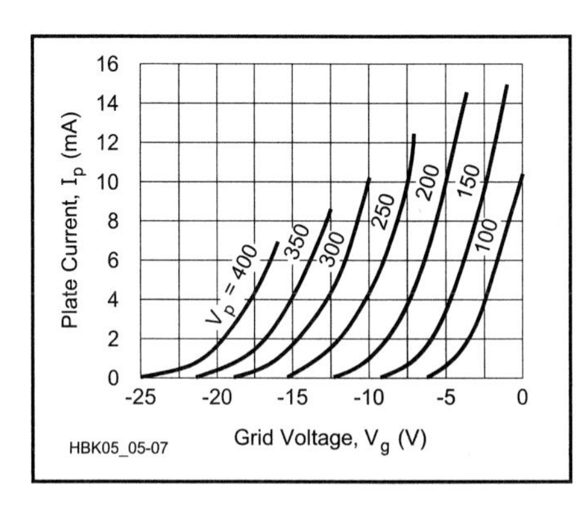

Fig 5.7 — Tube characteristic curve. Input signal is the grid voltage, V_g, along the x-axis, and the output signal is the plate current, I_p, along the y-axis. Different curves are plotted for various values of plate bias voltage, V_p (also called B+).

the operating region of the curve is specified, usually a straight portion of the curve if linear operation is desired. Two points along that portion of the curve are selected, each defined by its location along the x and y axes. If the two points are defined by (x_1,y_1) and (x_2,y_2), the slope, m, of the curve, which can be a gain, a resistance or a conductance, is calculated as:

$$m = \frac{\Delta y}{\Delta x} = \frac{y_1 - y_2}{x_1 - x_2} \qquad (4)$$

A characteristic curve that plots device output voltage and current along the x and y axes permits the inclusion of an additional curve. The *load line* is a straight line with a slope that is equal to the load impedance. The intersections between the load line and the characteristic curves indicate the operating points for that circuit. Load lines are only applicable to output characteristic plots; they cannot be used with input or transfer (input versus output) characteristic curves.

BIASING

The operation of an analog signal-processing device is greatly affected by which portion of the characteristic curve is used to do the processing. As an example, consider the vacuum tube characteristic curves in **Fig 5.8** and **Fig 5.9**.

The relationship between the input and the output of a tube amplifier is illustrated in Fig 5.8. The input signal (a sine wave in this example) is plotted in the vertical direction and below the graph. For a grid bias level of –5 V, the sine wave causes the grid voltage, V_g, to deviate between –3 and –7 V. These values correspond to a range of plate currents, I_p, between 1.4 and 2.6 mA. With a plate bias of 200 V and a load resistance, R_p, of 50 kΩ, the corresponding change in plate voltage, V_p, is between 70 and 130 V. Thus, this triode amplifier configuration changes a range of 4 V at the input to 60 V at the output. Also there is a change of output-signal voltage polarity; this amplifier both amplifies the signal magnitude 15 times and shifts the phase of the signal by 180°.

In the previous example the signal was biased so that it fell on a linear (straight) portion of the characteristic curve. If a different bias voltage is selected so that the signal does not fall on a linear portion of the curve, the output signal will be a distorted version of the input signal. This is illustrated in Fig 5.9. The input signal is amplified within a curved region of the characteristic curve. The positive part of the signal is amplified more than the negative part of the signal. Proper biasing is crucial to ensure amplifier linearity.

Input biasing serves to modify the relative level (dc offset) of the input signal so that it falls on the desired portion of the characteristic curve. Devices that perform signal processing (vacuum tubes, diodes, bipolar transistors, field-effect transistors and operational amplifiers) usually require appropriate input signal biasing.

Manufacturers' Data Sheets

Manufacturer's data sheets list device characteristics, along with the specifics of the part type (polarity, semiconductor type), identification of the pins, and the typical use (such as small signal, RF, switching or power amplifier). The pin identification is important because, although common package pinouts are normally used, there are exceptions. Manufacturers may differ slightly in the values reported, but certain basic parameters are listed. Different batches of the same devices are rarely identical, so manufacturers specify the guaranteed limits for the parameters of their device. There are usually three columns of values listed in the data sheet. For each parameter, the columns may list the guaranteed minimum value, the guaranteed maximum value and/or the typical value.

Another section of the data sheet lists ABSOLUTE MAXIMUM RATINGS, beyond which device damage may result. For example, the parameters listed in the ABSOLUTE MAXIMUM RATINGS section for a solid-state device are typically voltages, continuous currents, total device power dissipation (P_D) and operating- and storage-temperature ranges.

Rather than plotting the characteristic curves for each device, the manufacturer often selects key operating parameters that describe the device operation for the configurations and parameter ranges that are most commonly used. For example, a bipolar transistor data sheet might include an OPERATING PARAMETERS section. Parameters are listed in an OFF CHARACTERISTICS subsection and an ON CHARACTERISTICS subsection that describe the conduction properties of the device for dc voltages. The SMALL-SIGNAL CHARACTERISTICS section

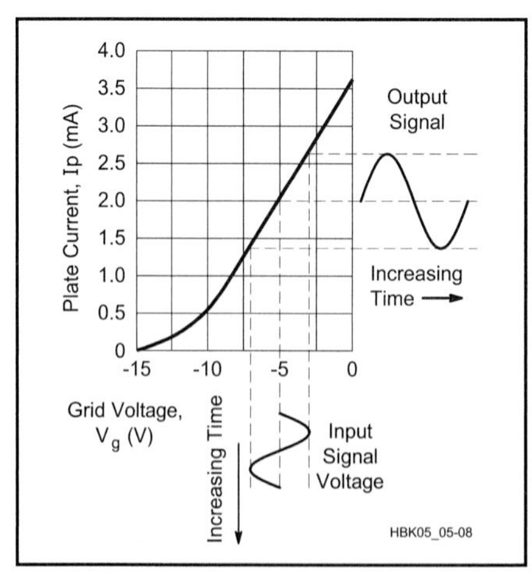

Fig 5.8 — Determination of output signal (to the right of the plot) for a given input signal (below the plot, turned on its side) with a tube characteristic curve plotted for a given plate bias. Note that the grid bias voltage, –5 V, causes the entire range of the input signal to be mapped onto the linear (diagonal straight line) portion of the characteristic curve. The output signal has the same shape as the input signal except that it is larger in amplitude.

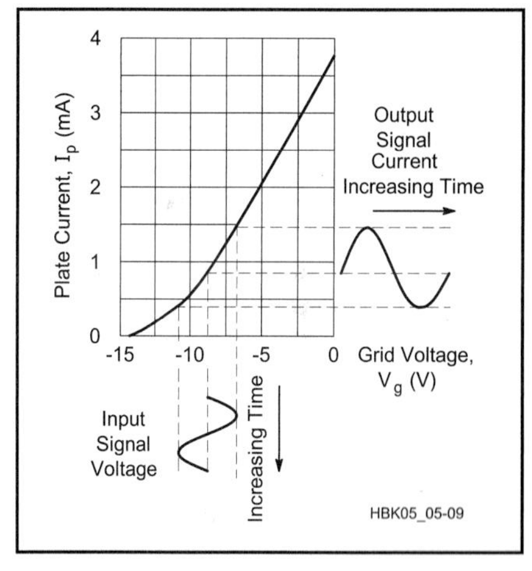

Fig 5.9 — Same characteristic curve and input signal as in Fig 5.8 except the grid bias voltage is now about –8.75 V. The input signal falls on the curved (non-linear) portion of the plot and causes distortion in the output signal. Note how the upper portion of the output sine wave was amplified more than the lower portion.

often contains the guaranteed minimum Gain-Bandwidth Product (f_T), the guaranteed maximum output capacitance, the guaranteed maximum input capacitance and the guaranteed range of the transfer parameters applicable to a given device. Finally, the SWITCHING CHARACTERISTICS section lists absolute maximum ratings for Delay Time (t_d), Rise Time (t_r), Storage Time (t_s) and Fall Time (t_f). Other types of devices list characteristics important to operation of that specific device.

When selecting equivalent parts for replacement of specified devices, the data sheet provides the necessary information to tell if a given part will perform the functions of another. Lists of equivalencies generally only specify devices that have nearly identical parameters. There are usually a large number of additional devices that can be chosen as replacements. Knowledge of the circuit requirements adds even more to the list of possible replacements. The device parameters should be compared individually to make sure that the replacement part meets or exceeds the parameter values of the original part required by the circuit. Be aware that in some applications a far superior part may fail as a replacement, however. A transistor with too much gain could easily oscillate if there were insufficient negative feedback to ensure stability.

VACUUM TUBES

Current is generally described as the flow of electrons through a conductor, such as metal. The vacuum tube controls the flow of electrons in a vacuum, which is analogous to a faucet that adjusts the flow of a fluid. The British commonly refer to vacuum tubes as *valves*. Although the physics of the operation of vacuum tubes varies greatly from that of semiconductors, there are many similarities in the way that they behave in analog circuits.

Thermionic Theory

Metals are elements that are characterized by their large number of free electrons. Individual atoms do not hold onto all of their electrons very tightly, and it is relatively easy to dislodge them. This property makes metals good conductors of electricity. Under electrical pressure (voltage), electrons collide with metal atoms, dislodging an equal number of free electrons from the metal. These collide with adjoining metal atoms to continue the process, resulting in a flow of electrons.

It is also possible to cause the free electrons to be emitted into space if enough energy is added to them. Heat is one way of adding energy to metal atoms, and the resulting flow of electrons into space is called *thermionic emission*. It is important to remember that the metal atoms don't permanently lose electrons; the emitted electrons are replaced by others that come from an electrical connection to the heated metal. Thus, an electron that flows into the heated metal collides with and is captured by a metal atom, knocking loose a highly energized electron that is emitted into space.

In a vacuum, there are no other atoms with which the emitted electron can collide, so it follows a straight path until it collides with another atom. A *vacuum tube* has nearly all of the air evacuated from it, so the emitted electrons proceed unhindered to another piece of metal, where they continue to move as part of the electrical current.

Components of a Vacuum Tube

A basic vacuum tube contains at least two parts: a *cathode* and a *plate*. The electrons are emitted from the *cathode*. The cathode can either be heated directly by passing a large dc current through it, or it can be located adjacent to a heating element. Although ac currents can also be used to directly heat cathodes, if any of the ac voltage mixes with the signal, ac hum will be introduced into the output. If the ac heater supply voltage can be obtained from a center tapped transformer, and the center tap is connected to the signal ground, hum can be minimized. Cathodes are made of substances that have the highest emission of electrons for the lowest temperatures and voltages. Tungsten, thoriated-tungsten and oxide-coated metals are commonly used.

Every vacuum tube needs a receptor for the emitted electrons. After moving though the vacuum, the electrons are absorbed by the *plate*. Since the plate receives electrons, it is also called the *anode*. Each electron has a negative charge, so a positively biased plate will attract the emitted electrons to it, and a current will result. For every electron that is accepted by the plate, another electron flows into the cathode; the plate and cathode currents must be the same. As the plate voltage is increased, there is a larger electrical field attracting electrons, causing more of them to be emitted from the cathode. This increases the current through the tube. This relationship continues until a limit is reached where further increases to the electrical field do not cause any more electrons to be emitted. This is the *saturation point* of the vacuum tube.

A vacuum tube that contains only a cathode and a plate is called a *diode tube* (di- for two components). See **Fig 5.10**. The diode tube is similar to a semiconductor diode since it allows current to pass in only one direction; it is used as a rectifier. When the plate voltage becomes

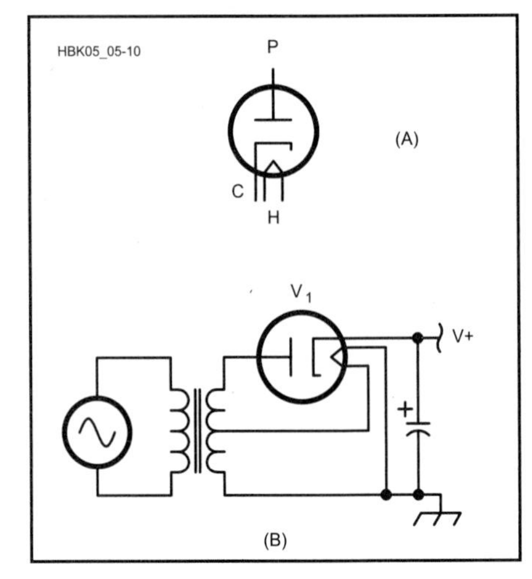

Fig 5.10 — Vacuum tube diode. (A) Schematic symbol detailing heater (H), cathode (C) and plate (P). (B) Power supply circuit using diode as a half wave rectifier.

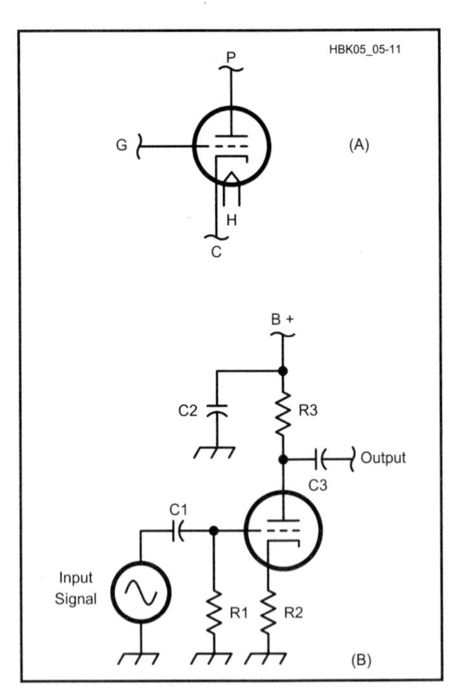

Fig 5.11 — Vacuum tube triode. (A) Schematic symbol detailing heater (H), cathode (C), grid (G) and plate (P). (B) Audio amplifier circuit using a triode. C1 and C3 are dc blocking capacitors for the input and output signals to isolate the grid and plate bias voltages. C2 is a bypass filter capacitor to decrease noise in the plate bias voltage, B+. R1 is the grid bias resistor, R2 is the cathode bias resistor and R3 is the plate bias resistor. Note that although the cathode and grid bias voltages are positive with respect to ground, they are still negative with respect to the plate.

negative, the electrical field that is set up repels electrons, preventing them from being emitted from the cathode.

To amplify signals, a vacuum tube must also contain a control *grid*. This name comes from its physical construction. The grid is a mesh of wires located between the cathode and the plate. Electrons from the cathode pass between the grid wires on their way to the plate. The electrical field that is set up by the voltage on these wires affects the electron flow from cathode to plate. A negative grid voltage sets up an electrical field that repels electrons, decreasing emission from the cathode because of the higher energy needed for the electrons to escape from their atoms into the vacuum. A positive grid voltage will have the opposite effect. Since the plate voltage is always positive, however, grid voltages are usually negative. The more negative the grid, the less effective the electrical field from the plate will be at attracting electrons from the cathode.

Vacuum tubes containing a cathode, a grid and a plate are called *triode* tubes (tri- for three components). See **Fig 5.11**. They are generally used as amplifiers, particularly at frequencies in the HF range and below. Characteristic curves for triodes normally relate grid bias voltage and plate bias voltage to plate current for the triode (Fig 5.7). There are three descriptors of a tube's performance that can be derived from the characteristic curves. The *plate resistance*, r_P, describes the resistance to the flow of electrons from cathode to plate. The r_P is calculated by selecting a vertical line in the characteristic curve and dividing the change in plate-to-cathode voltage (ΔV_P) of two of the lines by the corresponding change in plate current (ΔI_P).

$$r_P = \frac{\Delta V_P}{\Delta I_P} \qquad (5)$$

The ratio of change in plate voltage (ΔV_P) to the change in grid-to-cathode voltage (ΔV_g) for a given plate current is the *amplification factor* (μ). Amplification factor is calculated by selecting a horizontal line in the characteristic curve and dividing the difference in plate voltage of two of the lines by the difference in grid voltages that corresponds to the same points.

$$\mu = \frac{\Delta V_P}{\Delta V_g} \qquad (6)$$

Triode amplification factors range from 10 to about 100.

The plate current flows to the plate bias supply, so the output from a triode amplifier is often expressed as the voltage that is developed as this current passes through a load resistor. The value of the load resistance affects the tube amplification, as illustrated by the dynamic characteristic curves in **Fig 5.12**, so the tube μ does not fully describe its action as an amplifier. *Grid-plate transconductance* (g_m) takes into account the change of amplification due to load resistance. The slope of the lines in the characteristic curve represents g_m. (Since the various lines are nearly parallel in the linear operating region, they have about the same slope.)

$$g_m = \frac{\Delta I_P}{\Delta V_g} \qquad (7)$$

This ratio represents a conductance, which is measured in siemens. Triodes have g_m values that range from about 1000 to several thousand microsiemens, the higher values indicating greater possible amplification.

The input impedance of a vacuum tube amplifier is directly related to the grid current. Grid current varies with grid voltage, increasing as the voltage becomes more positive. The normal operation uses a negative grid-bias voltage, and the input impedance can be in the megohm range for very negative grid bias values. This is, however, limited by the desired operating point on the characteristic curve as illustrated in Figs 5.8 and 5.9. The output impedance of the amplifier is a function of the plate resistance, r_P, in parallel with the output capacitance. Typical output impedance is on the order of hundreds of ohms.

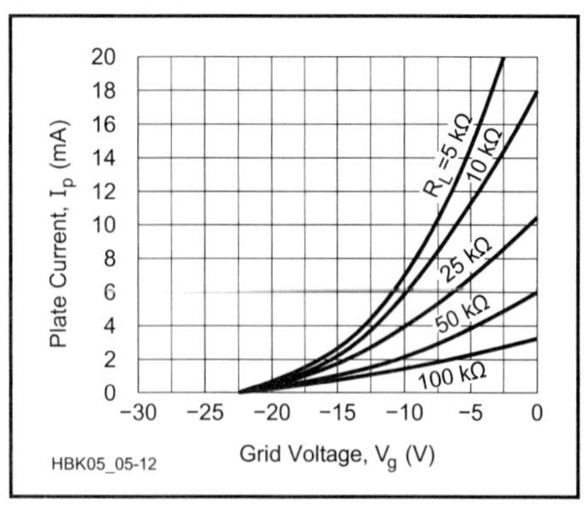

Fig 5.12 — Vacuum tube dynamic characteristic curve. This corresponds to the V_P = 300 line in Fig 5.7 with different values of load resistance. This shows how the tube will behave when cascaded to circuits with different input impedances.

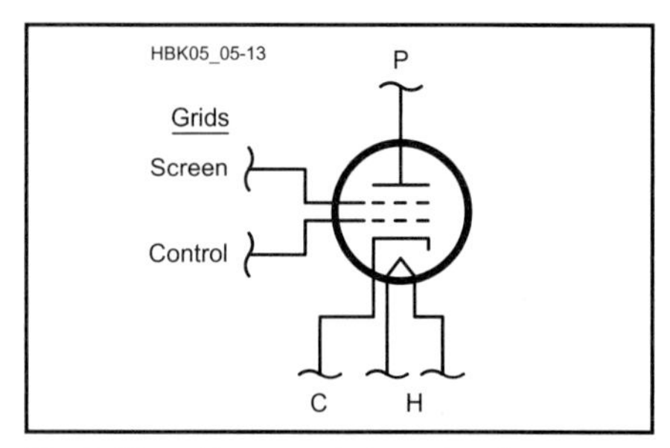

Fig 5.13 — Vacuum tube tetrode. Schematic symbol detailing heater (H), cathode (C), the two grids: control and screen, and the plate (P).

The physical configuration of the components within the vacuum tube appears as conductors that are separated by an insulator (in this case, the vacuum). This description is very similar to that of a capacitor. The capacitance between the cathode and grid, between the grid and plate, and between the cathode and plate can be large enough to affect the operation of the amplifier at high frequencies. These capacitances, which are usually on the order of a few picofarads, can limit the frequency response of a vacuum tube amplifier and can also provide signal feedback paths that may lead to unwanted oscillation. Neutralizing circuits are sometimes used to counteract the effects of internal capacitances and to prevent oscillations.

The grid-to-plate capacitance is the chief source of unwanted signal feedback. A special form of vacuum tube has been developed to deal with the grid-to-plate capacitance. A second grid, called a *screen grid*, is inserted between the original grid (now called a *control grid*) and the plate. The additional tube component leads to the name for this new tube — *tetrode* (tetra- for four components). See **Fig 5.13**. The screen grid reduces the capacitance between the control grid and the plate, but it also reduces the electrical field from the plate that attracts electrons from the cathode. Like the control grid, the screen grid is made of a wire mesh and electrons pass through the spaces between the wires to get to the plate. The bias of the screen grid is positive with respect to the cathode, in order to enhance the attraction of electrons from the cathode. The electrons accelerate toward the screen grid and most of them pass through the spaces and continue to accelerate until they reach the plate. The presence of the screen grid adversely affects the overall efficiency of the tube, since some of the electrons strike the grid wires. A bypass capacitor with a low reactance at the frequency being amplified by the vacuum tube is generally connected between the screen grid and the cathode.

A special form of tetrode concentrates the electrons flowing between the cathode and the plate into a tight beam. The decreased electron-beam area increases the

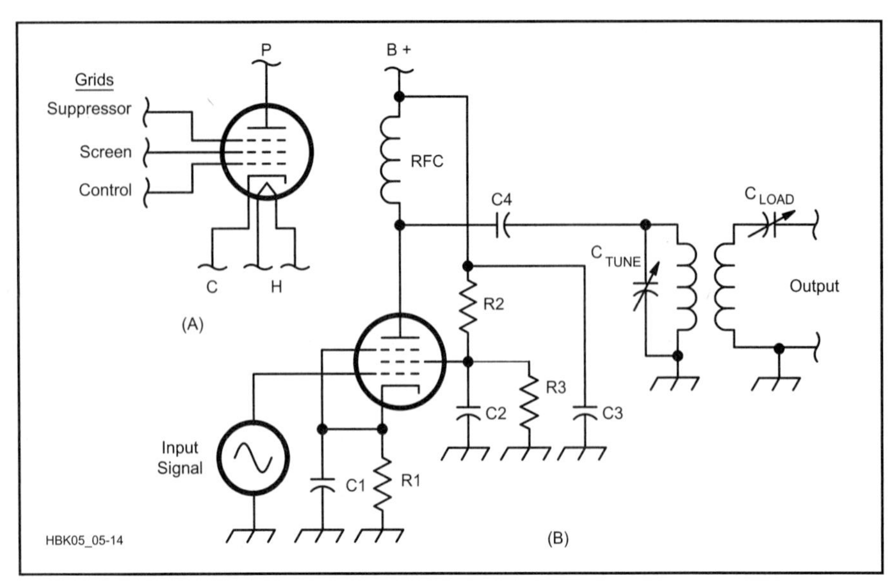

Fig 5.14 — Vacuum tube pentode. (A) Schematic symbol detailing heater (H), cathode (C), the three grids: control, screen and suppressor, and plate (P). (B) RF amplifier circuit using a pentode. C1, C2 and C3 are bypass (filter) capacitors and C4 is a dc blocking capacitor to isolate the plate voltage from the tank circuit components. R1 is the cathode bias resistor. R2 and R3 comprise a screen-dropping voltage divider. The plate tank circuit is tuned to the desired frequency bandpass. As is common, the heater circuit is not shown.

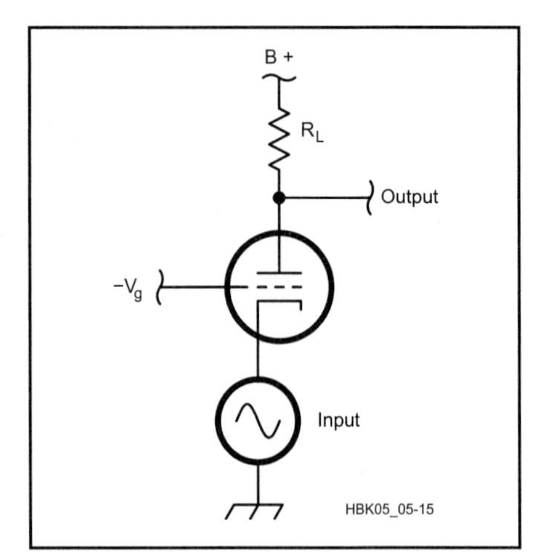

Fig 5.15 — Grounded grid amplifier schematic. The input signal is connected to the cathode, the grid is biased to the appropriate operating point by a dc bias voltage, $-V_G$, and the output voltage is obtained by the voltage drop through R_L that is developed by the plate current, I_P.

efficiency of the tube. *Beam tetrodes* permit higher plate currents with lower plate voltages and large power outputs with smaller grid driving power. RF power amplifiers are usually made with this type of vacuum tube.

Another unwanted effect in vacuum tubes is the emission of electrons from the plate. The electrons flowing within the tube have so much energy that they are capable of dislodging electrons from the metal atoms in the plate. These *secondary emission* electrons are repelled back to the plate by the negative bias of the grid in a triode and are of no concern. In the tetrode, the screen grid is positively biased and attracts the secondary emission electrons, causing a reverse current from the plate to the screen grid.

A third grid, called the *suppressor grid*, can be added between the screen grid and the plate. This overcomes the effects of secondary emission in tetrodes. A vacuum tube with three grids is called a *pentode* (penta for five components). See **Fig 5.14**. The suppressor grid is negatively biased with respect to the screen grid and the plate. In some tube designs it is internally connected to the cathode. The suppressor grid repels the secondary emission electrons back to the plate.

As the number of grids is increased between the cathode and the plate, the effect of the electrical field from the positive plate voltage at the cathode is decreased. This limits the number of electrons that can be emitted from the cathode and the characteristic curves tend to flatten out as the grid bias becomes less negative. This flattening is another nonlinearity of the tube as an amplifier, since the response saturates at a given plate current and will go no higher. Tube saturation can be used advantageously in some circuits if a constant current source is desired, since the current does not change within the saturation region regardless of changes in plate voltage.

Types of Vacuum Tube Amplifiers

The descriptions of vacuum tube amplifiers up to this point have been for only one configuration, the common cathode, where the cathode is connected to the signal reference point, the grid is the input and the plate is the output. Although this is the most common configuration of the vacuum tube as an amplifier, other configurations exist. If the signal is introduced into the cathode and the grid is at a reference level (still negatively biased but with no ac component), with the output at the plate, the amplifier is called a *grounded-grid* (**Fig 5.15**). This amplifier is characterized by a very low input impedance, on the order of a few hundred ohms, and a low output impedance, that is mainly determined by the plate resistance of the tube.

The third configuration is called the *cathode follower* (**Fig 5.16**). The plate is the common element, the grid is the input and the cathode is the output. This type of amplifier is often used as a buffer stage due to its high input impedance, similar to that of the common cathode amplifier, and its very low output impedance. The output impedance (Z_O) can be calculated from the tube characteristics as:

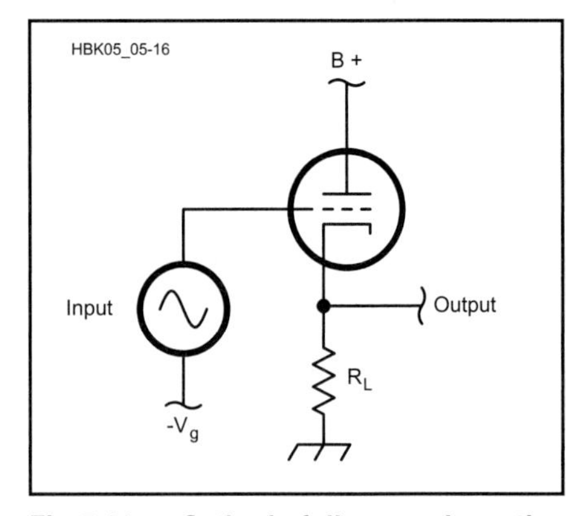

Fig 5.16 — Cathode follower schematic. The input signal is biased by $-V_G$ and fed into the grid. The plate bias, B+ is fed directly into the plate terminal. The output is derived by the cathode current (which is equal to the plate current, I_P) dropping the voltage through the load resistor, R_L.

$$Z_o = \frac{r_P}{1+\mu} \quad (8)$$

where:

r_P = tube plate resistance

μ = tube amplification factor.

For a close approximation, we can simplify this equation as:

$$Z_o \approx \frac{r_P}{\mu} = \frac{1}{g_m}$$

Other Types of Tubes

Vacuum tube identifiers do not generally indicate the tube type. The format is typically a number, one or two letters and a number (such as 6AU6 or 12AT7). The first number in the identifier indicates the heater voltage (usually either 6 or 12 V). The last number often indicates the number of elements, including the heater. Some tubes also have an additional letter following the identifier (usually A or B) that indicates a revision of the tube design that represents an improvement in its operating parameters. There are also tubes that do not follow this naming convention, many of which are power amplifiers or military-type tubes (such as 6146 and 811).

To reduce stray reactances, some tubes do not have the plate connection in the tube base, where all the other connections are located. Rather, a connection is made at the top of the tube through a metallic cap. This requires an additional connector for the plate circuitry.

Tubes may share components in a single envelope to reduce size and incidental power requirements. A very common example of this is the dual triode tube (such as 12AT7 or 12AU7) that contains a single heater circuit and two complete triode tubes in the same device. Other configurations of multiple devices contained in a single vacuum tube also exist. The 6GW8 and 6EA8 tubes each contain both a triode and a pentode. The 6BN8 contains three distinct devices, one triode and two diodes.

Most common vacuum tubes are encased in glass. It is also possible to encase them in metal or ceramic materials to attain higher tube power and smaller size. Since heat dissipation from the plate is one of the major limiting factors for vacuum tube power amplifiers, the alternate materials remove heat more efficiently. These tubes can be cooled by convection, with the casing connected to a large heat sink, or with water flowing past the tube for hydraulic cooling.

A variation of the vacuum tube that is widely used in oscilloscopes and television monitors is the *cathode ray tube (CRT)*, diagrammed in **Fig 5.17**. The CRT has a cathode and grid much like a triode tube. The plate, usually referred to as the *anode* in this device, is designed to accelerate the electrons to very high velocities, with anode voltages that can be as high as tens of thousands of volts. The anode of the CRT differs from the plates of other vacuum tubes, since it is designed as a set of plates that are parallel to the electron beam. The anode voltage accelerates the electrons but does not absorb them. The electron beam passes by the anode and continues to the face of the tube. The cathode, grid and anode are all located in the neck of the CRT and are collectively referred to as the *electron gun.*

The electron beam is deflected from its path by either magnetic deflectors that surround the yoke of the tube or by electrostatic deflection plates that are built into the tube neck just beyond the electron gun. A CRT typically has two sets of deflectors: vertical and horizontal. When a potential is applied to a set of deflectors, the passing electron beam is bent, altering its path. In an oscilloscope, the time base typically drives the horizontal deflectors and the input signal drives the vertical deflectors, although in many oscilloscopes it is possible to connect another input signal to the horizontal deflectors to obtain an X-Y, or vector, display. In televisions and some computer monitors the deflectors typically are driven by a raster generator. The horizontal deflectors are driven by a sawtooth pattern that causes the beam to move repeatedly from left to right and then retrace quickly to the left. The vertical deflectors are driven by a slower sawtooth pattern that causes the beam to move repeatedly from top to bottom and then retrace quickly to the top. The relative timing of the two sawtooth patterns is such that the beam scans from left to right, retraces to the left and then begins the next horizontal trace just below the previous one.

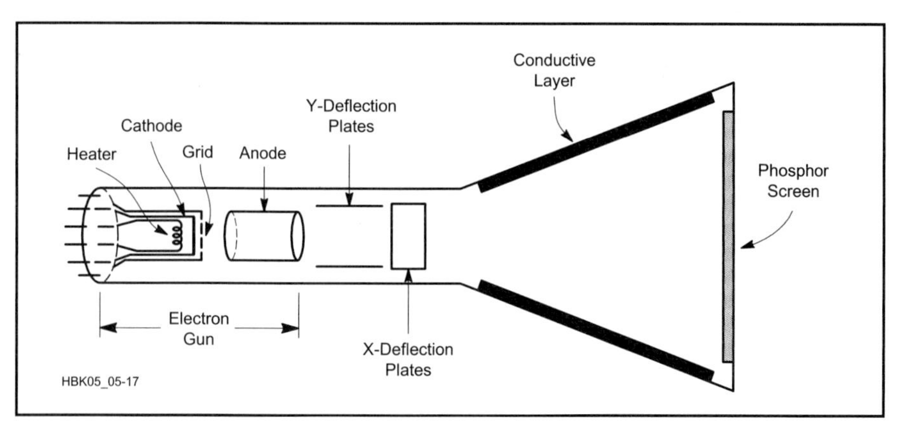

Fig 5.17 — Cross section of CRT. The electron gun generates a stream of electrons and is made up of a heater, cathode, grid and anode (plate). The electron beam passes by two pairs of deflection plates that deviate the path of the beam in the vertical (y) direction and then the horizontal (x) direction. The deflected electron beam strikes a phosphor screen and causes it to glow at that spot. Any electrons that bounce off the screen are absorbed by the conductive layer along the sides of the tube, preventing spurious luminescence.

Beyond the deflectors, the CRT flares out. The front face is coated with a phosphorescent material that glows when struck by the electron beam. To prevent spurious phosphorescence, a conductive layer along the sides of the tube absorbs any electrons that reflect off the glass.

Vector displays have better resolution than raster scanning. The trace lines are clearer, which is the reason oscilloscope displays use this technique. It is faster to fill the screen using raster scanning, however. This is why TVs use raster scanning.

Some CRT tubes are designed with multiple electron beams. The beams are sometimes generated by different electron guns that are placed next to each other in the neck of the tube. They can also be generated by splitting the output of a single electron gun into two or more beams. Very high quality oscilloscopes use two electron beams to trace two input channels rather than the more common method of alternating a single beam between the two inputs. Color television tubes use three electron beams for the three primary colors (red, green and blue). Each beam is focused on only one of these colored phosphors, which are interleaved on the face of the tube. A metal shadow mask keeps the colors separate as the beams scan across the tube.

A variation of the CRT is the *vidicon tube*. The vidicon is used in many video cameras and operates in a similar fashion to the CRT. The vidicon absorbs light from the surroundings, which charges the plate at the location of the light. This charge causes the cathode-to-plate current to increase when the raster scan points the electron beam at that location. The current increase is converted to a voltage that is proportional to the amount of light absorbed. This results in an electrical signal that represents the pattern of a visual image.

Standard vacuum tubes work well for frequencies up to hundreds of megahertz. At frequencies higher than this, the amount of time that it takes for the electrons to move between the cathode and the plate becomes a limiting factor. There are several special tubes designed to work at microwave frequencies. The *klystron* tube uses the principle of velocity modulation of the electrons to avoid transit time limitations. The beam of electrons travels down a metal drift tube that has interaction gaps along its sides. RF voltages are applied to the gaps and the electric fields that they generate accelerate or decelerate the passing electrons. The relative positions of the electrons shift due to their changing velocities causing the electron density of the beam to vary. The modulation of the electron density is used to perform amplification or oscillation. Klystron tubes tend to be relatively large, with lengths ranging from 10 cm to 2 m and weights ranging from as little as 150 g to over 100 kg. Unfortunately, klystrons have relatively narrow bandwidths, and are not retunable by amateurs for operation on different frequencies.

The *magnetron* tube is an efficient oscillator for microwave frequencies. Magnetrons are most commonly found in microwave ovens and high-powered radar equipment. The anode of a magnetron is made up of a number of coupled resonant cavities that surround the cathode. The magnetic field causes the electrons to rotate around the cathode and the energy that they give off as they approach the anode adds to the RF electric field. The RF power is obtained from the anode through a vacuum window. Magnetrons are self-oscillating with the frequency determined by the construction of their anodes; however, they can be tuned by coupling either inductance or capacitance to the resonant anode. The range of frequencies depends on how fast the tuning must be accomplished. The tube may be tuned slowly over a range of approximately 10% of the center frequency. If faster tuning is necessary, such as is required for frequency modulation, the range decreases to about 5%.

A third type of tube capable of operating in the microwave range is the *traveling wave tube*. For wide band amplifiers in the microwave range this is the tube of choice. Either permanent magnets or electromagnets are used to focus the beam of electrons that emerges from an electron gun similar to the one described for the CRT tube. The electron beam passes through a helical *slow-wave structure*, in which electrons are accelerated or decelerated, providing density modulation due to the applied RF signal, similar to that in the klystron. The modulated electron beam induces voltages in the helix that provides an amplified tube output whose gain is proportional to the length of the slow-wave structure. After the RF energy is extracted from the electron beam by the helix, the electrons are collected and recycled to the cathode. Traveling wave tubes can often be operated outside their designed frequencies by carefully optimizing the beam voltage.

PHYSICAL ELECTRONICS OF SEMICONDUCTORS

Every atom of matter consists of, among other things, an equal number of protons and electrons. These two subatomic particles must match in number to neutralize the electric charge: one positive charge for a proton and one negative charge for an electron.

Electrons orbit the nucleus, which contains the protons, at different energy levels. The binding of the electrons to the nucleus determines how an atom will behave electrically. Loosely bound electrons are easily liberated from their nuclei; atoms with this property are called *conductors*. In contrast, tightly bound electrons require considerable energy to be dislodged from their atoms; these atoms are called *insulators*. In between these two extremes is a class of elements called *semiconductors*, or partial conductors. As energy is added to a semiconductor atom, electrons are more easily freed. This property leads to many potential applications for this type of material.

In a conductor, such as a metal, the outer, or *valence*, electrons of each atom are shared with the adjacent atoms so there are many electrons that can move about freely between atoms. The moving free electrons are the constituents of electrical current. In a good conductor, the concentration of these free electrons is very high, on the order of 10^{22} electrons/cm^3. In an insulator, nearly all the electrons are tightly held by their atoms; the concentration of free electrons is very small, on the order of 10 electrons/cm^3.

Semiconductor atoms (germanium — Ge and silicon — Si) share their valence electrons in a chemical bond that holds adjacent atoms together. The electrons are not free to leave their atom in order to move into the sphere of the adjacent atom, as in a conductor. They can be shared by the adjacent atom, however. The sharing of electrons means that the adjacent atoms are attracted to each other, forming a bond that gives the semiconductor its physical structure.

When energy is added to a semiconductor lattice, generally in the form of heat, some electrons are liberated from their bonds and move freely throughout the structure. The bond that loses an electron is then unbalanced and the space that the electron came from is referred to as a *hole*. Electrons from adjacent bonds can leave their positions and fill the holes, thus creating new holes in the adjacent bonds. Two opposite movements can be said to occur: negatively charged electrons move from bond to bond in one direction and positively charged holes move from bond to bond in the opposite direction. Both of these movements represent forms of electrical current, but this is very different from the current in a conductor. While the conductor has *free electrons* that flow regardless of the crystalline structure, the current in a semiconductor is constrained to move only along the crystalline lattice between adjacent bonds.

Crystals formed from pure semiconductor atoms (Ge or Si) are called *intrinsic* semiconductors. In these materials the number of free electrons is equal to the number of holes. Each atom has four valence electrons that form bonds with adjacent atoms. Impurities can be added to the semiconductor material to enhance the formation of electrons or holes. These are *extrinsic* semiconductors. There are two types of impurities that can be added: one kind with five valence electrons *donates* free electrons to the crystalline structure; this is called an *N-type* impurity, for the negative charge that it adds. Some examples are antimony (Sb), phosphorus (P) and arsenic (As). N-type extrinsic semiconductors have more electrons and fewer holes than intrinsic semiconductors. Impurities with three valence electrons accept free electrons from the lattice, adding holes to the overall structure. These are called P-type impurities, for the net positive charge; some examples are boron (B), gallium (Ga) and indium (In).

Intrinsic semiconductor material can be formed by combining equal amounts of N-type and P-type impurity materials. Some examples of this include gallium-arsenide (GaAs), gallium-phosphate (GaP) and indium-phosphide (InP). To make an N-type compound semiconductor, a slightly higher amount of N-type material is used in the mixture. A P-type compound semiconductor has a little more P-type material in the mixture.

The conductivity of an extrinsic semiconductor depends on the charge density (in other words, the concentration of free electrons in N-type, and holes in P-type, semiconductor material). As the energy in the semiconductor increases, the charge density also increases. This is the basis of how all semiconductor devices operate: the major difference is the way in which the energy level is increased. Variations are: The *transistor*, where conductivity is altered by injecting current into the device via a wire; the *thermistor*, where the level of heat in the device is detected by its conductivity, and the *photoconductor*, where light energy that is absorbed by the semiconductor material increases the conductivity.

The PN Semiconductor Junction

If a piece of N-type semiconductor material is placed against a piece of P-type semiconductor material, the location at which they join is called a *PN semiconductor junction*. The junction has characteristics that make it possible to develop diodes and transistors. The action of the junction is best described by a diode operating as a rectifier. Initially, when the two types of semiconductor material are placed in contact, each type of material will have only its majority carriers: P-type will have only holes and N-type will have only free electrons. The net positive charge of the P-type material attracts free electrons from across the junction and the opposite is true in the N-type material. These attractions lead to diffusion of some of the majority carriers across the junction, which neutralize the carriers immediately on the other side. The region close to the junction is then *depleted* of carriers, and, as such, is named the *depletion region* (or the *space-charge region* or the *transition region*). The width of the depletion region is very small, on the order of 0.5 µm.

If the N-type material is placed at a more negative voltage than the P-type material, current will pass through the junction because electrons are attracted from the lower potential to the higher potential and holes are attracted in the opposite direction. When the polarity is reversed, current does not flow because the electrons that are trying to enter the N-type material are repelled, as are the holes trying to enter the P-type material. This unidirectional current is what allows a semiconductor diode to act as rectifier.

Diodes are commonly made of silicon or germanium. Although they act similarly, they have slightly different characteristics. The *junction threshold voltage*, or *junction barrier voltage,* is the forward bias voltage at which current begins to pass through the device. This voltage is different for the two kinds of diodes. In the diode response curve of **Fig 5.18**, this value corresponds to the voltage at which the positive portion of the curve begins to rise sharply from the x axis. Most silicon diodes have a junction threshold voltage of about 0.7 V, while the value for germanium diodes typically is 0.3 V. The reverse biased leakage current is much lower for silicon diodes than for germanium diodes. The forward resistance of a diode is typically very low and varies with the amount of forward current.

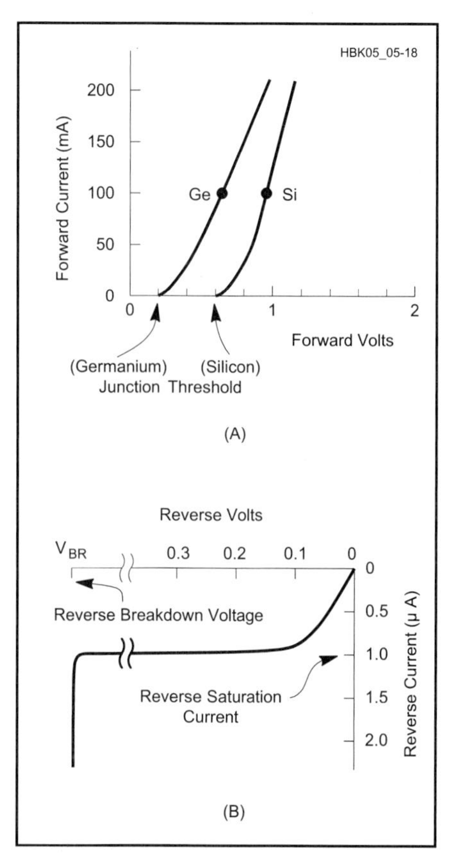

Fig 5.18 — Semiconductor diode (PN junction) response curve. (A) Forward biased (anode voltage higher than cathode) response for Germanium (Ge) and Silicon (Si) devices. Each curve breaks away from the x-axis at its junction threshold voltage. The slope of each curve is its forward resistance. (B) Reverse biased response. Very small reverse current increases until it reaches the reverse saturation current (I_0). The reverse current increases suddenly and drastically when the reverse voltage reaches the reverse breakdown voltage, V_{BR}.

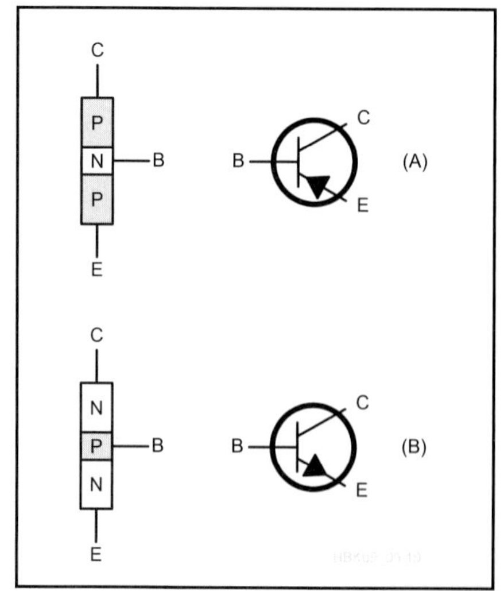

Fig 5.19 — Bipolar transistors. (A) A layer of N-type semiconductor sandwiched between two layers of P-type semiconductor makes a PNP device. The schematic symbol has three leads: collector (C), base (B) and emitter (E), with the arrow pointing in toward the base. (B) A layer of P-type semiconductor sandwiched between two layers of N-type semiconductor makes an NPN device. The schematic symbol has three leads: collector (C), base (B) and emitter (E), with the arrow pointing out away from the base.

Multiple Junctions

A bipolar transistor is formed when two PN junctions are placed next to each other. If N-type material is surrounded by P-type material, the result is a PNP transistor. Alternatively, if P-type material is in the middle of two layers of N-type material, the NPN transistor is formed (**Fig 5.19**).

Physically, we can think of the transistor as two PN junctions back-to-back, such as two diodes connected at their *anodes* (the positive terminal) for an NPN transistor or two diodes connected at their *cathodes* (the negative terminal) for a PNP transistor. The connection point is the base of the transistor. (You can't actually *make* a transistor this way.) A transistor conducts when the base-emitter junction is forward biased and the base-collector is reverse biased. Under these conditions, the emitter region emits majority carriers into the base region, where they are minority carriers because the materials of the emitter and base regions have opposite polarity. The excess minority carriers in the base are attracted across the base-collector junction, where they are collected and are once again considered majority carriers. The flow of majority carriers from emitter to collector can be modified by the application of a bias current to the base terminal. If the bias current has the same polarity as the base material (for example holes flowing into a P-type base) the emitter-collector current increases. A transistor allows a small base current to control a much larger collector current.

As in a semiconductor diode, the forward biased base-emitter junction has a threshold voltage (V_{BE}) that must be exceeded before the emitter current increases.

PNPN Diode

If four alternate layers of P-type and N-type material are placed together, a PNPN (usually pronounced like *pinpin*) diode with three junctions is obtained (see **Fig 5.20**). This device, when the anode is at a higher potential than the cathode, has its first and third junctions forward biased and its center junction reverse biased. In this state, there is little current, just as in the reverse biased diode. As the forward bias voltage is increased, the current through the device increases slowly until the *breakover (or firing) voltage*, V_{BO}, is reached and the flow of current abruptly increases. The PNPN diode is often considered to be a switch that is off below V_{BO} and on above it.

Bilateral Diode Switch

A semiconductor device similar to two PNPN diodes facing in opposite directions and attached in parallel is the *bilateral diode switch* or *diac*. This device has the characteristic curve of the PNPN diode for both positive and negative bias voltages. Its construction, schematic symbol and characteristic curve are shown in **Fig 5.21**.

Silicon Controlled Rectifier

Another device with four alternate layers of P-type and N-type semiconductor is the *silicon controlled rectifier (SCR)*, or *thyristor*. In addition to the connections to the outer two layers, two other terminals can be brought out for the inner two layers. The connection to the P-type material near the cathode is called the *cathode gate* and the N-type material near the anode is called the *anode gate*. In nearly all commercially available SCRs, only the cathode gate is connected (**Fig 5.22**).

Like the PNPN diode switch, the SCR is used to abruptly start conducting when the voltage exceeds a given level. By biasing the gate terminal appropriately, the breakover voltage can be adjusted. The SCR is highly efficient and is used in power control applications. SCRs are available that can handle currents of greater than 100 A and voltage differentials of greater than 1000 V, yet can be switched with gate currents of less than 50 mA.

Triac

A five-layered semiconductor whose operation is similar to a bidirectional SCR is the *triac* (**Fig 5.23**). This is also similar to a bidirectional diode switch with a bias control gate. The gate terminal of the triac can control both positive and negative breakover voltages and the devices can pass both polarities of voltage.

SCRs and triacs are often used to modify ac power sources. A sine wave with a given RMS value can be switched on and off at preset points during the cycle to decrease the RMS voltage. When conduction is delayed until after the peak (as **Fig 5.24** shows) the peak-to-peak voltage is reduced. If conduction starts before the

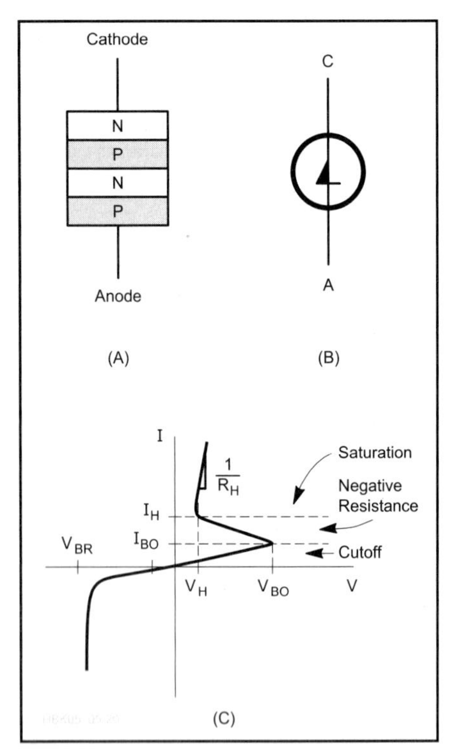

Fig 5.20 — PNPN diode. (A) Alternating layers of P-type and N-type semiconductor. (B) Schematic symbol with cathode (C) and anode (A) leads. (C) Voltage-current response curve. Reverse biased response is the same as normal PN junction diodes. Forward biased response acts as a hysteresis switch. Resistance is very high until the bias voltage reaches V_{BO} and exceeds the cutoff current, I_{BO}. The device exhibits a negative resistance when the current increases as the bias voltage decreases until a voltage of V_H and saturation current of I_H is reached. After this, the resistance is very low, with large increases in current for small voltage increases.

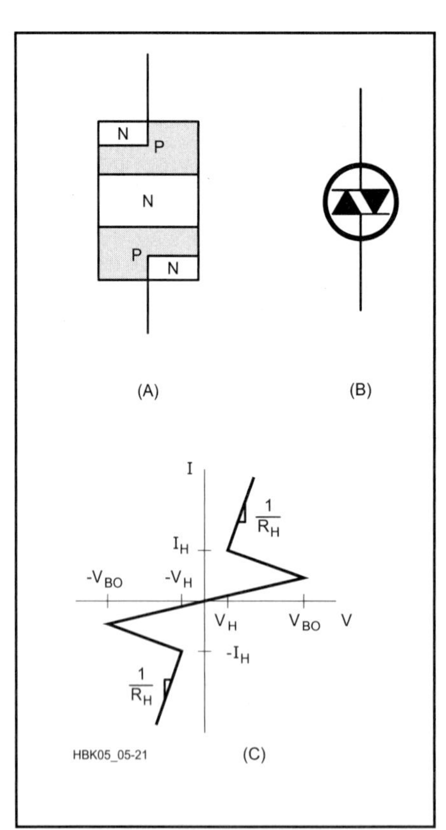

Fig 5.21 — Bilateral switch. (A) Alternating layers of P-type and N-type semiconductor. (B) Schematic symbol. (C) Voltage-current response curve. The right-hand side of the curve is identical to the PNPN diode response in Fig 5.20. The device responds identically for both forward and reverse bias so the left-hand side of the curve is symmetrical with the right-hand side.

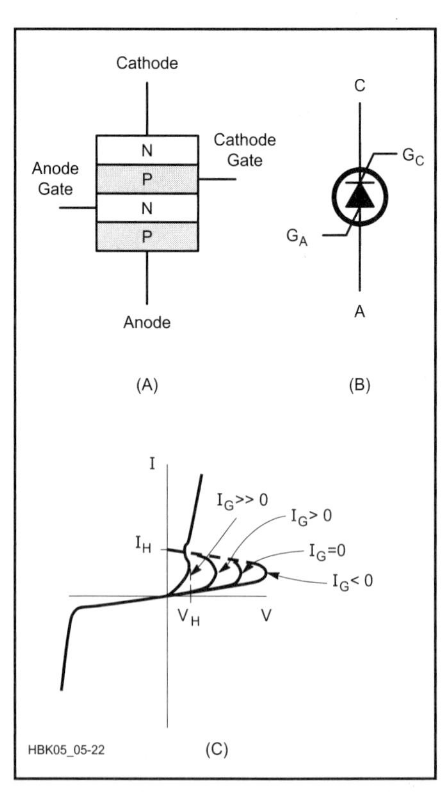

Fig 5.22 — SCR. (A) Alternating layers of P-type and N-type semiconductor. This is similar to a PNPN diode with gate terminals attached to the interior layers. (B) Schematic symbol with anode (A), cathode (C), anode gate (G_A) and cathode gate (G_C). Many devices are constructed without G_A. (C) Voltage-current response curve with different responses for various gate currents. $I_G = 0$ has the same response as the PNPN diode.

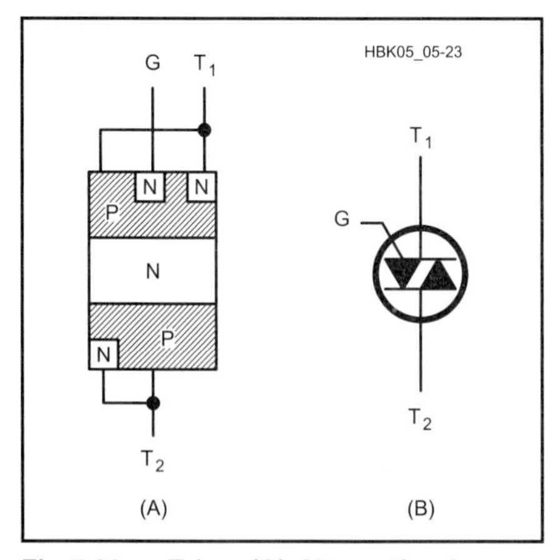

Fig 5.23 — Triac. (A) Alternating layers of P-type and N-type semiconductor. This behaves as two SCR devices facing in opposite directions with the anode of one connected to the cathode of the other and the cathode gates connected together. (B) Schematic symbol.

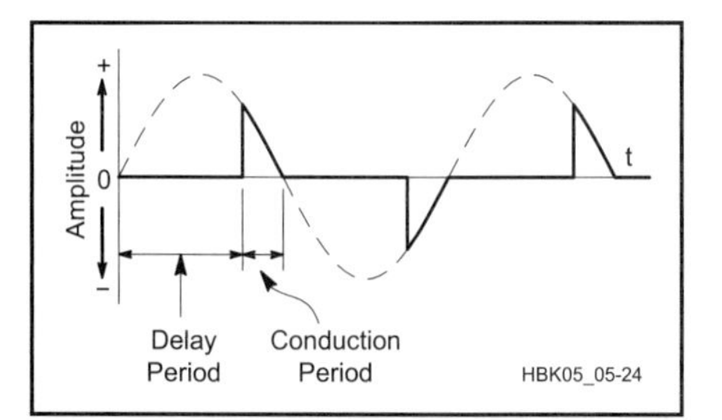

Fig 5.24 — Triac operation on sine wave. The dashed line is the original sine wave and the solid line is the portion that conducts through the triac. The relative delay and conduction period times are controlled by the amount or timing of gate current, I_G. The response of an SCR is the same as this for positive voltages (above the x-axis) and with no conduction for negative voltages.

peak, the RMS voltage is reduced, but the peak-to-peak value remains the same. This method is used to operate light dimmers and 240 V ac to 120 V ac converters. The sharp switching transients created when these devices switch are common sources of RF interference. SCRs are used as "crowbars" in power supply circuits, to short the output to ground and blow a fuse when an overvoltage condition exists.

FIELD-EFFECT TRANSISTORS

The *field-effect transistor (FET)* controls the current between two points but does so differently than the bipolar transistor. The FET operates by the effects of an electric field on the flow of electrons through a single type of semiconductor material. This is why the FET is sometimes called a *unipolar* transistor. Also, unlike bipolar semiconductors that can be arranged in many configurations to provide diodes, transistors, photoelectric devices, temperature sensitive devices and so on, the field effect is usually only used to make transistors, although FETs are also available as special-purpose diodes, for use as constant current sources.

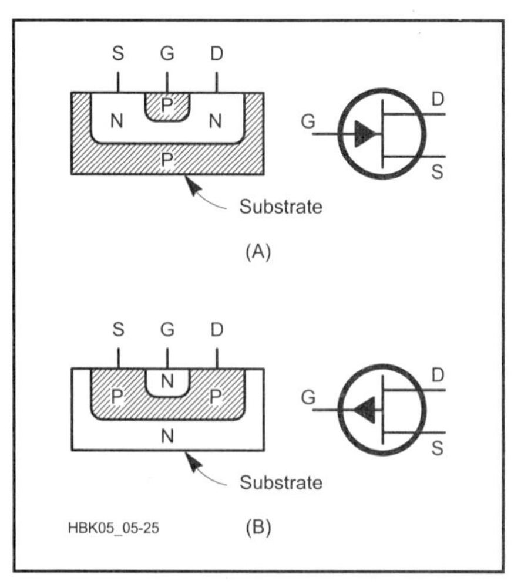

Fig 5.25 — JFET devices with terminals labeled: source (S), gate (G) and drain (D). A) Pictorial of N-type channel embedded in P-type substrate and schematic symbol. B) P-channel embedded in N-type substrate and schematic symbol.

Current moves within the FET in a channel, from the source connection to the drain connection. A gate terminal generates an electric field that controls the current (see **Fig 5.25**). The channel is made of either N-type or P-type semiconductor material; an FET is specified as either an N-channel or P-channel device. Majority carriers flow from source to drain. In N-channel devices, electrons flow so the drain potential must be higher than that of the source ($V_{DS} > 0$). In P-channel devices, the flow of holes requires that $V_{DS} < 0$. The polarity of the electric field that controls current in the channel is determined by the majority carriers of the channel, ordinarily positive for P-channel FETs and negative for N-channel FETs.

Variations of FET technology are based on different ways of generating the electric field. In all of these, however, electrons at the gate are used only for their charge in order to create an electric field around the channel, and there is a minimal flow of electrons through the gate. This leads to a very high dc input resistance in devices that use FETs for their input circuitry. There may be quite a bit of capacitance between the gate and the other FET terminals, however. The input impedance may be quite low at RF.

The current through an FET only has to pass through a single type of semiconductor material. There is very little resistance in the absence of an electric field (no bias voltage). The drain-source resistance ($r_{DS\ ON}$) is between a few hundred ohms to less than an ohm. The output impedance of devices made with FETs is generally quite low. If a gate bias voltage is added to operate the transistor near cutoff, the circuit output impedance may be much higher.

FET devices are constructed on a *substrate* of doped semiconductor material. The channel is formed within the substrate and has the opposite polarity (a P-channel FET has N-type substrate). Most FETs are constructed with silicon. In order to achieve a higher gain-bandwidth product, other materials have been used. Gallium Arsenide (GaAs) has electron mobility and drift velocities that are far higher than the standard doped silicon. Amplifiers designed with *GaAs FET* devices have much higher frequency response and lower noise factor at VHF and UHF than those made with standard FETs.

JFET

There are two basic types of FET. In the *junction FET (JFET)*, the gate material is made of the opposite polarity semiconductor to the channel material (for a P-channel FET the gate is made of N-type semiconductor material). The gate-channel junction is similar to a diode's PN junction. As with the diode, current is high if the junction is forward biased and is extremely small when the junction is reverse biased. The latter case is the way that JFETs are used, since any current in the gate is undesirable. The magnitude of the reverse bias at the junction is proportional to the size of the electric field that "pinches" the channel. Thus, the current in the channel is reduced for higher reverse gate bias voltage.

Because the gate-channel junction in a JFET is similar to a bipolar junction diode, this junction must never be forward biased; otherwise large currents will pass through the gate and into the channel. For an N-channel JFET, the gate must always be at a lower potential than the source ($V_{GS} < 0$). The channel is as fully open as it can get when the gate and source voltages are equal ($V_{GS} = 0$). The prohibited condition is when $V_{GS} > 0$. For P-channel JFETs these conditions are reversed (in normal operation $V_{GS} > 0$ and the prohibited condition is when $V_{GS} < 0$).

MOSFET

Placing an insulating layer between the gate and the channel allows for a wider range of control (gate) voltages and further decreases the gate current (and thus increases the device input resistance). The insulator is typically made of an oxide (such as silicon dioxide, SiO_2). This type of device is called a *metal-oxide-semiconductor FET (MOSFET)* or *insulated-gate FET (IGFET)*. The substrate is often connected to the source internally. The insulated gate is on the opposite side of the channel from the substrate (see **Fig 5.26**). The bias voltage on the gate terminal either attracts or repels the majority carriers of the substrate across the PN junction with the channel. This narrows (depletes) or widens (enhances) the channel, respectively, as V_{GS} changes polarity. For N-channel MOSFETs, positive gate voltages with respect to the substrate and the source ($V_{GS} > 0$) repel holes from the channel into the substrate, thereby widening the channel and decreasing channel resistance. Conversely, $V_{GS} < 0$ causes holes to be attracted from the substrate, narrowing the channel and increasing the channel resistance. Once again, the polarities discussed in this example are reversed for P-channel devices. The common abbreviation for an N-channel MOSFET is *NMOS*, and for a P-channel MOSFET, *PMOS*.

Because of the insulating layer next to the gate, input resistance of a MOSFET is usually greater than 10^{12} Ω (a million megohms). Since MOSFETs can both deplete the channel, like the JFET, and also enhance it, the construction of MOSFET devices differs based on the channel size in the resting state, $V_{GS} = 0$. A *depletion mode* device (also called a *normally on MOSFET*) has a channel in resting state that gets smaller as a reverse bias is applied; this device conducts current with no bias applied (see Fig 5.26 A and B). An *enhancement mode* device (also called a *normally off MOSFET*) is built without a channel and does not conduct current when $V_{GS} = 0$; increasing forward bias forms a channel that conducts current (see Fig 5.26 C and D).

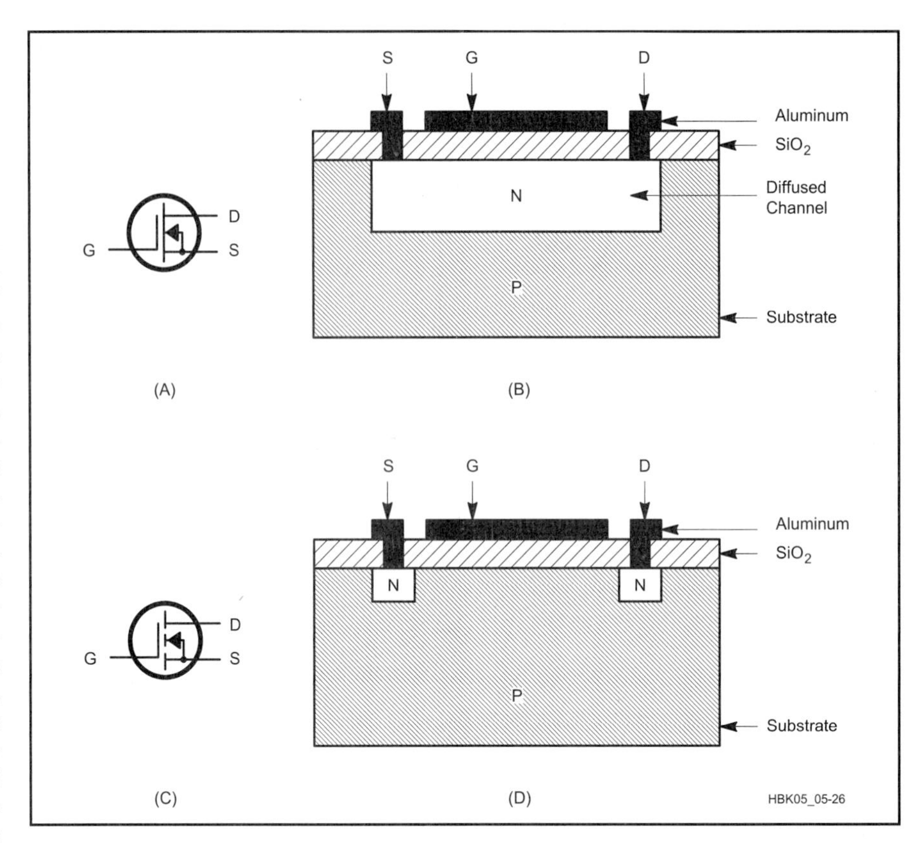

Fig 5.26 — MOSFET devices with terminals labeled: source (S), gate (G) and drain (D). N-channel devices are pictured. P-channel devices have the arrows reversed in the schematic symbols and the opposite type semiconductor material for each of the layers. (A) N-channel depletion mode device schematic symbol and (B) pictorial of P-type substrate, diffused N-type channel, SiO_2 insulating layer and aluminum gate region and source and drain connections. The substrate is connected to the source internally. A negative gate potential narrows the channel. (C) N-channel enhancement mode device schematic and (D) pictorial of P-type substrate, N-type source and drain wells, SiO_2 insulating layer and aluminum gate region and source and drain connections. Positive gate potential forms a channel between the two N-type wells.

Semiconductor Temperature Effects

The number of excess holes and electrons is increased as the temperature of a semiconductor increases. Since the conductivity of a semiconductor is related to the number of excess carriers, this also increases with temperature. With respect to resistance, semiconductors have a negative temperature coefficient. The resistance of silicon *decreases* by about 8% / °C and by about 6% / °C for germanium. Semiconductor temperature properties are the opposite of most metals, which *increase* their resistance by about 0.4% / °C. These opposing temperature characteristics permit the design of circuits with opposite temperature coefficients that cancel each other out, making a temperature insensitive circuit. Left alone, the semiconductor can experience an effect called *thermal runaway* as the current causes an increase in temperature. The increased temperature decreases resistance and may lead to a further increase in current (depending on the circuit) that leads to an additional temperature increase. This sequence of events can continue until the semiconductor destroys itself.

Semiconductor Failure

There are several common failure modes for semiconductors that are related to heat. The semiconductor material is connected to the outside world through metallic leads. The point at which the metal and the semiconductor are connected is one common place for the semiconductor device to fail. As the device heats up and cools down, the materials expand and contract. The rate of expansion and contraction of semiconductor material is different from that of metal. Over many cycles of heating and cooling the bond between the semiconductor and the metal can break. Some experts have suggested that the lifetime of semiconductor equipment can be extended by leaving the devices powered on all the time. While this would decrease the type of failure just described, inadequate cooling can lead to another type of semiconductor failure.

Impurities are introduced into intrinsic semiconductors by diffusion, the same physical property that lets you smell cookies baking from several rooms away. Smells diffuse through air much faster than molecules diffuse through solids. Once the impurities diffuse into the semiconductor, they tend to stay in place. Rates of diffusion are proportional to temperature, and semiconductors are doped with impurities at high temperature to save time. Once the doped semiconductor material is cooled, the rate of diffusion of the impurities is so low that they are essentially immobile for many years to come.

A common failure mode of semiconductors is due to the heat generated during semiconductor use. If the temperatures at the junctions rise to high enough levels for long enough periods of time, the impurities start to diffuse across the PN junctions. When enough of these atoms get across the junction, it stops functioning properly and the semiconductor device fails.

Thermistors

A *thermistor* is an intrinsic (no N or P doping) semiconductor metal-oxide compound device that has a large negative temperature coefficient (NTC) of resistance. Thermally generated free electrons and holes become available as current carriers in thermistors. Metal oxides, such as nickel-oxide (NiO), dimanganese-trioxide (Mn_2O_3) and cobalt-trioxide (Co_2O_3), are chosen for their stable electrical properties. Silicon and Germanium are not used as thermistors because their temperature properties are very sensitive to impurities.

A related device, the *sensistor*, uses large amounts of doping materials to achieve a large positive temperature coefficient (PTC) of resistance, usually over some restricted temperature range.

Practical Semiconductors

SEMICONDUCTOR DIODES

Although many types of semiconductor diodes are available, there are not many differences between them. The diode is made of a single PN junction that affects current differently depending on its direction. This leads to a large number of applications in electronic circuitry.

The diode symbol is shown in **Fig 5.27**. Current passes most easily from anode to cathode, in the direction of the arrow. This is often referred to as the *forward* direction and the opposite is the *reverse* direction. Remember that *current* refers to the flow of electricity from higher to lower potentials and is in the opposite direction to the flow of electrons (current moves from anode to cathode and electrons flow from cathode to anode, as based on the definitions of the words, *anode* and *cathode*). The anode of a semiconductor junction diode is made of P-type material and the cathode is made of N-type material, as indicated in Fig 5.27. Most diodes are marked with a band on the cathode end (Fig 5.27). The ideal diode would have zero resistance in the forward direction and infinite resistance in the reverse direction. This is not the case for actual devices, which behave as shown in the plot of a diode response in Fig 5.18. Note that the scales of the two parts of the graph are drastically different. The inverse of the slope of the line (the change in voltage between two points on a straight portion of the line divided by the corresponding change in current) on the upper right is the resistance of the diode in the forward direction. The range of voltages is small and the range of currents is large since the forward resistance is very small (in this example, about 2 Ω). The lower left portion of the curve illustrates a much higher resistance that increases from tens of kilohms to thousands of megohms as the reverse voltage gets larger, and then decreases to near zero (a nearly vertical line) very suddenly at the peak inverse voltage (PIV = 100 V in this example).

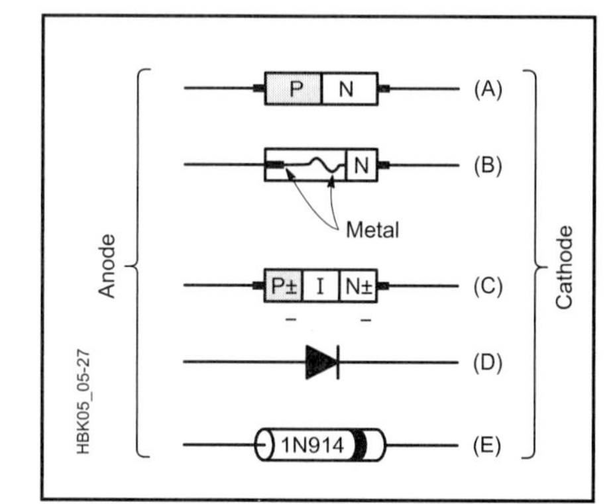

Fig 5.27 — Practical semiconductor diodes. All devices are aligned with anode on the left and cathode on the right. (A) Standard PN junction diode. (B) Point-contact or "cat's whisker" diode. (C) PIN diode formed with heavily doped P-type (P+), undoped (intrinsic) and heavily doped N-type (N+) semiconductor material. (D) Diode schematic symbol. (E) Diode package with marking stripe on the cathode end.

There are five major characteristics that distinguish standard junction diodes from one another: the PIV, the current or power handling capacity, the response speed, reverse leakage current and the junction barrier voltage. Each of these characteristics can be manipulated during manufacture to produce special purpose diodes.

The most common application of a diode is to perform rectification; that is, allowing positive voltages to pass and stopping negative voltages. Rectification is used in power supplies that convert ac to dc and in amplitude demodulation. The most important diode parameters to consider for power rectification are the PIV and current ratings. The peak negative voltages that are stopped by the diode must be smaller in magnitude than the PIV and the peak current through the diode when it is forward biased must be less than the maximum amount for which the device was designed. Exceeding the current rating in a diode will cause excessive heating (based on $P = I \times V_F$) that leads to PN junction failure as described earlier.

Fast Diodes

The speed of a diode affects the frequencies on which it can act. The diode response in Fig 5.18 is a steady state response, showing how that diode will act at dc. As the frequency increases, the diode may not be able to keep up with the changing polarity of the signal and its response will not be as expected. Diode speed mainly depends on charge storage in the depletion region. Under reverse bias, excess charges move away from the junction, forming a larger space-charge region that is the equivalent of a dielectric. The diode thus exhibits capacitance, which is inversely proportional to the width of the dielectric and directly proportional to the cross-sectional surface area of the junction.

One way to decrease charge storage time in the depletion region is to form a metal-semiconductor junction. This can be accomplished with a point-contact diode, where a thin piece of aluminum wire, often called a *whisker*, is placed in contact with one face of a piece of lightly doped N-type material. In fact, the original diodes used for detecting radio signals ("cat's whisker diodes") were made this way. A more recent improvement to this technology, the *hot-carrier diode*, is like a point-contact diode with more ideal characteristics attained by using more efficient metals, such as platinum and gold, that act to lower forward resistance and increase PIV. This type of contact is known as a *Schottky barrier*, and diodes made this way are called *Schottky diodes*.

The PIN diode, shown in Fig 5.27C is a *slow response* diode that is capable of passing microwave signals when it is forward biased. This device is constructed with a layer of intrinsic (undoped) semiconductor placed between very highly doped P-type and N-type material (called P^+-type and N^+-type material to indicate the high level of doping), creating a PIN junction. These devices provide very effective switches for RF signals and are often used in TR switches in transceivers. PIN diodes have longer than normal carrier lifetimes, resulting in a slow switching process that causes them to act more like resistors than diodes at high radio frequencies.

Varactors

If the PN junction capacitance is controlled rather than reduced, a diode can be made to act as a variable capacitor. As the reverse bias voltage on a diode increases, the width of the junction increases, which decreases its capacitance. A *varactor* is a diode whose junction is specially formulated to have a relatively large range of capacitance values for a modest range of reverse bias voltages (**Fig 5.28**). Although special forms of varactors are available from manufacturers, other types of diodes may be used as inexpensive varactor diodes, but the relationship between reverse voltage and capacitance is not always reliable. When designing with varactor diodes, the reverse bias voltage must be absolutely free of noise since any variations in the bias voltage will cause changes in capacitance. Unwanted frequency shifts or instability will result if the reverse bias voltage is noisy. It is possible to frequency modulate a signal by adding the audio signal to the reverse bias on a varactor diode used in the carrier oscillator.

Zener Diodes

When the PIV of a reverse biased diode is exceeded, the diode begins to conduct current as it does when it is forward biased. This current does not destroy the diode if it is limited to less than the device's maximum allowable value. When the PIV is controlled during manufacture to be at desired levels, the device is called a *Zener diode*. Zener diodes (named after the American physicist Clarence Zener) provide accurate voltage references and

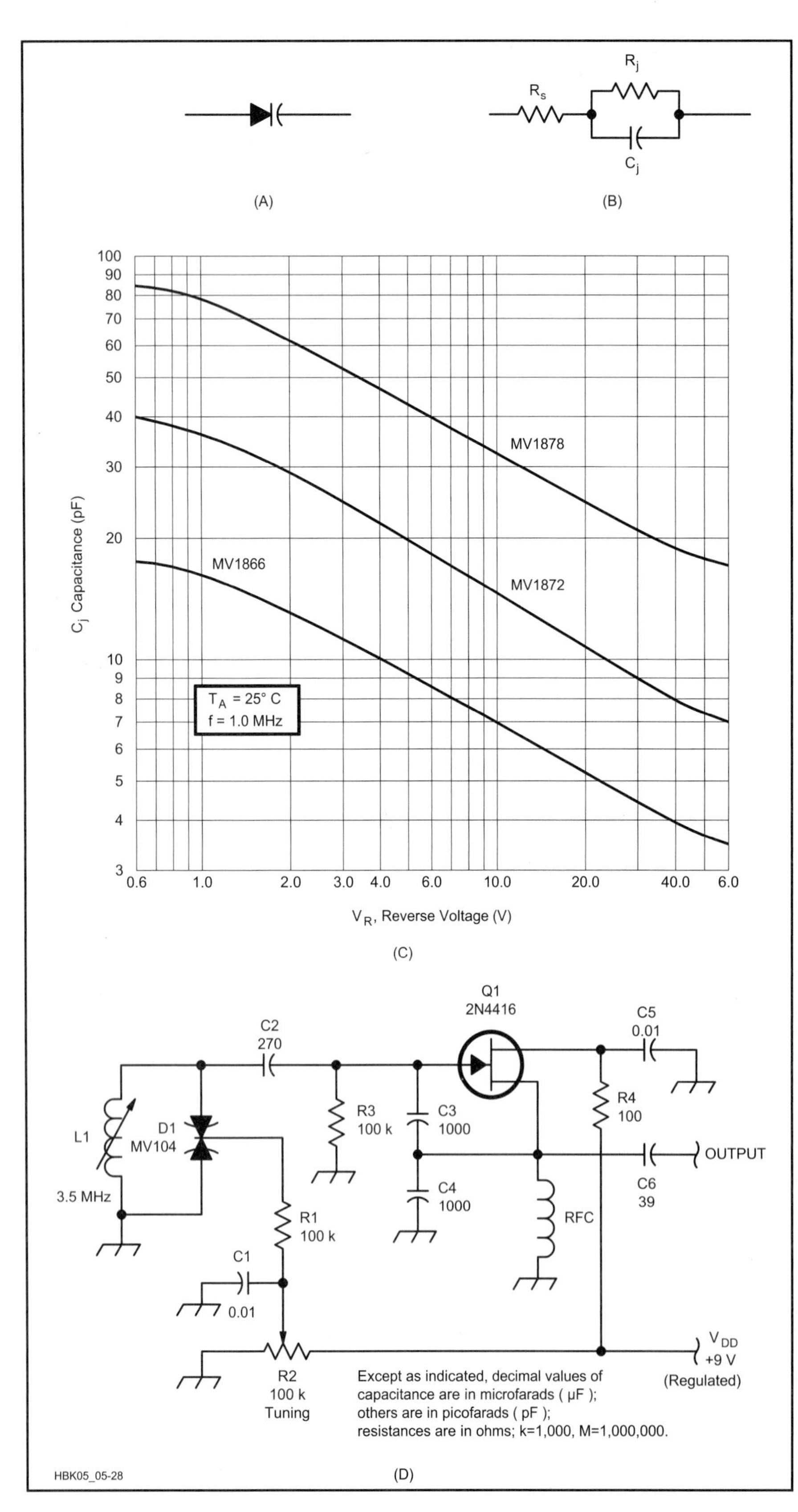

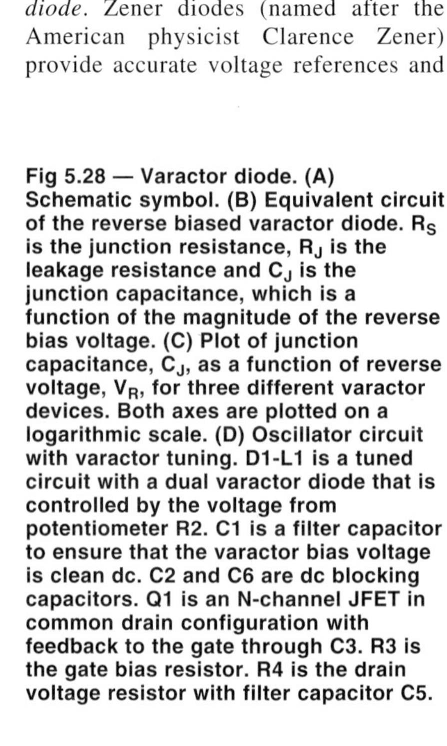

Fig 5.28 — Varactor diode. (A) Schematic symbol. (B) Equivalent circuit of the reverse biased varactor diode. R_S is the junction resistance, R_J is the leakage resistance and C_J is the junction capacitance, which is a function of the magnitude of the reverse bias voltage. (C) Plot of junction capacitance, C_J, as a function of reverse voltage, V_R, for three different varactor devices. Both axes are plotted on a logarithmic scale. (D) Oscillator circuit with varactor tuning. D1-L1 is a tuned circuit with a dual varactor diode that is controlled by the voltage from potentiometer R2. C1 is a filter capacitor to ensure that the varactor bias voltage is clean dc. C2 and C6 are dc blocking capacitors. Q1 is an N-channel JFET in common drain configuration with feedback to the gate through C3. R3 is the gate bias resistor. R4 is the drain voltage resistor with filter capacitor C5.

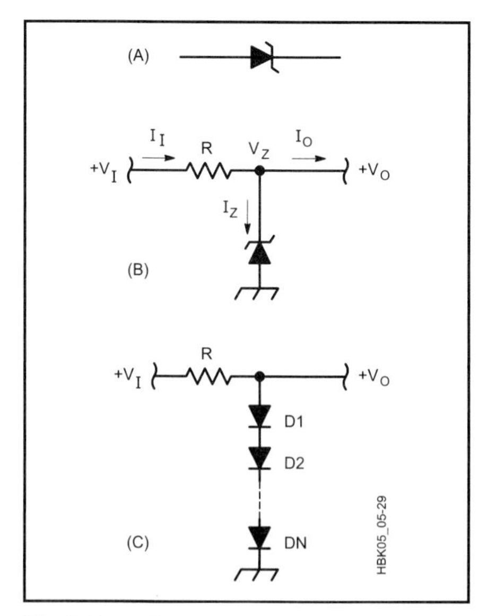

Fig 5.29 — Zener diode. (A) Schematic symbol. (B) Basic voltage regulating circuit. V_Z is the Zener reverse breakdown voltage. The Zener diode draws more current until $V_I - I_I R = V_Z$. The circuit design should select R so that when the maximum current is drawn, $R < (V_I - V_Z) / I_O$. The diode should be capable of passing the same current when there is no output current drawn. (C) For small voltages, several forward biased diodes can be used in place of Zener diodes. Each diode will drop the voltage by about 0.7 V for silicon or 0.3 V for germanium.

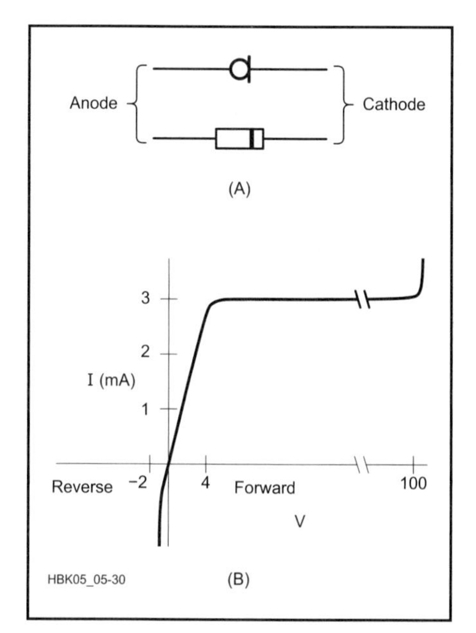

Fig 5.30 — Current regulator diode. (A) Schematic symbol and package with line marking cathode end. (B) Diode characteristic curve (1N5283 device). When forward bias voltage exceeds about 4 V the current passing through the device is held constant regardless of the voltage across the device.

are often used for this purpose in power supply regulators.

When the reverse breakdown voltage is exceeded, the reverse voltage drop across the Zener diode remains constant. With an appropriate current limiting resistor in series with it, the Zener diode provides an accurate voltage reference (**Fig 5.29**). Zener diodes are rated by their reverse-breakdown voltage and their power-handling capacity. The power is a product of the current passing through the reverse-biased Zener diode "in breakdown" (that is, in the breakdown mode of operation) and the breakdown voltage. Since the same current must always pass though the resistor to drop the source voltage down to the reference voltage, with that current divided between the Zener diode and the load, this type of power source is very wasteful of current. The Zener diode does make an excellent and efficient voltage reference in a larger voltage regulating circuit where the load current is provided from another device whose voltage is set by the reference. (See the **Power Supplies** chapter for more information about using Zener diodes as voltage regulators.) The major sources of error in Zener-diode derived voltages are the variation with load current and the variation due to heat. Temperature compensated Zener diodes are available with temperature coefficients as low as 0.0005 % / °C. If this is unacceptable, voltage reference integrated circuits based on Zener diodes have been developed that include additional circuitry to counteract temperature effects.

Constant Current Diodes

A form of diode, called a *field-effect regulator diode*, provides a constant current over a wide range of forward biased voltages. The schematic symbol and characteristic curve for this type of device are shown in **Fig 5.30**. Constant current diodes are very useful in any application where a constant current is desired. Some part numbers are 1N5283 through 1N5314.

Common Diode Applications

Standard semiconductor diodes have many uses in analog circuitry. Several examples of diode circuits are shown in **Fig 5.31**. Rectification has already been described. There are three basic forms of rectification using semiconductor diodes: half wave (1 diode), full-wave center-tapped (2 diodes) and full-wave bridge (4 diodes). These are more fully described in the **Power Supplies** chapter.

Diodes are commonly used to protect circuits. In battery powered devices a forward biased series diode is often used to protect the circuitry from the user inadvertently inserting the batteries backwards. Likewise, when a circuit is powered from an external dc source, a diode is often placed in series with the power connector in the device to prevent incorrectly wired power supplies from destroying the equipment. Diodes are commonly used to protect analog meters from both reverse voltage and over voltage conditions that would destroy the delicate needle movement.

Zener diodes are sometimes used to protect low-current (a few amps) circuits from over-voltage conditions. A reverse biased Zener diode connected between the positive power lead and ground will conduct excessive current if its breakdown voltage is exceeded. Used in conjunction with a fuse in series with the power lead, the Zener diode will cause the fuse to blow when an over-voltage condition exists.

Very high, short-duration voltage spikes can destroy certain semiconductors, particularly MOS devices. Standard Zener diodes can't handle the high pulse powers found in these voltage spikes. Special Zener diodes are designed for this purpose, such as the *mosorb*. (General Semiconductor Industries, Inc calls these devices *TransZorbs*.) A reverse biased TransZorb with a low-value series resistor can decrease the voltage reaching the sensitive device. Since the polarity of the spike can be positive, negative, or both, over voltage transient suppressor circuits can be designed with two devices wired back-to-back. They protect a circuit over a range of voltages rather than just suppressing positive peaks.

Diodes can be used to clip signals, similar to rectification. If the signal is appropriately biased it can be clipped at any level. Two Zener diodes placed back-to-back can be used to clip both the positive and negative peaks of a signal. Such an arrangement is used to convert a sine wave to an approximate square wave.

Care must be taken when using Zener diodes to process signals. The Zener diode is a relatively noisy device and can add excessive noise to the signals if it operates in breakdown. The Zener diode is often specified for at intentionally generate noise, such as the noise bridge (see the **Test Procedures** chapter). The reverse biased Zener diode in breakdown generates wide band (nearly white) noise levels as high as $2000\ \mu V / \sqrt{Hz}$. The noise voltage is determined by multiplying this value by the square root of the circuit bandwidth in Hz.

Diodes are used as switches for ac coupled signals when a dc bias voltage can be added to the signal to permit or inhibit

Fig 5.31 — Diode circuits. (A) Half wave rectifier circuit. Only when the ac voltage is positive does current pass through the diode. Current flows only during half of the cycle. (B) Full-wave center-tapped rectifier circuit. Center-tap on the transformer secondary is grounded and the two ends of the secondary are 180° out of phase. During the first half of the cycle the upper diode conducts and during the second half of the cycle the lower diode conducts. There is conduction during the full cycle with only positive voltages appearing at the output. (C) Full-wave bridge rectifier circuit. In each half of the cycle two diodes conduct. (D) Polarity protection for external power connection. J1 is the connector that power is applied to. If polarity is correct, the diode will conduct and if reversed the diode will block current, protecting the circuit that is being powered. (E) Over-voltage protection circuit. If excessive voltage is applied to J1, D1 will conduct current until fuse, F1, is blown. (F) Bipolar voltage clipping circuit. In the positive portion of the cycle, D2 is forward biased, but no current is shunted to ground because D1 is reverse biased. D1 starts to conduct when the voltage exceeds the Zener breakdown voltage, and the positive peak is clipped. When the negative portion of the cycle is reached, D1 is forward biased, but no current is shunted to ground because D2 is reverse biased. When the voltage exceeds the Zener breakdown voltage of D2, it also begins to conduct, and the negative peak is clipped. (G) Diode switch. The signal is ac coupled to the diode by C1 at the input and C2 at the output. R2 provides a reference for the bias voltage. When switch S1 is in the ON position, a positive dc voltage is added to the signal so it is forward biased and is passed through the diode. When S1 is in the OFF position, the negative dc voltage added to the signal reverse biases the diode, and the signal does not get through.

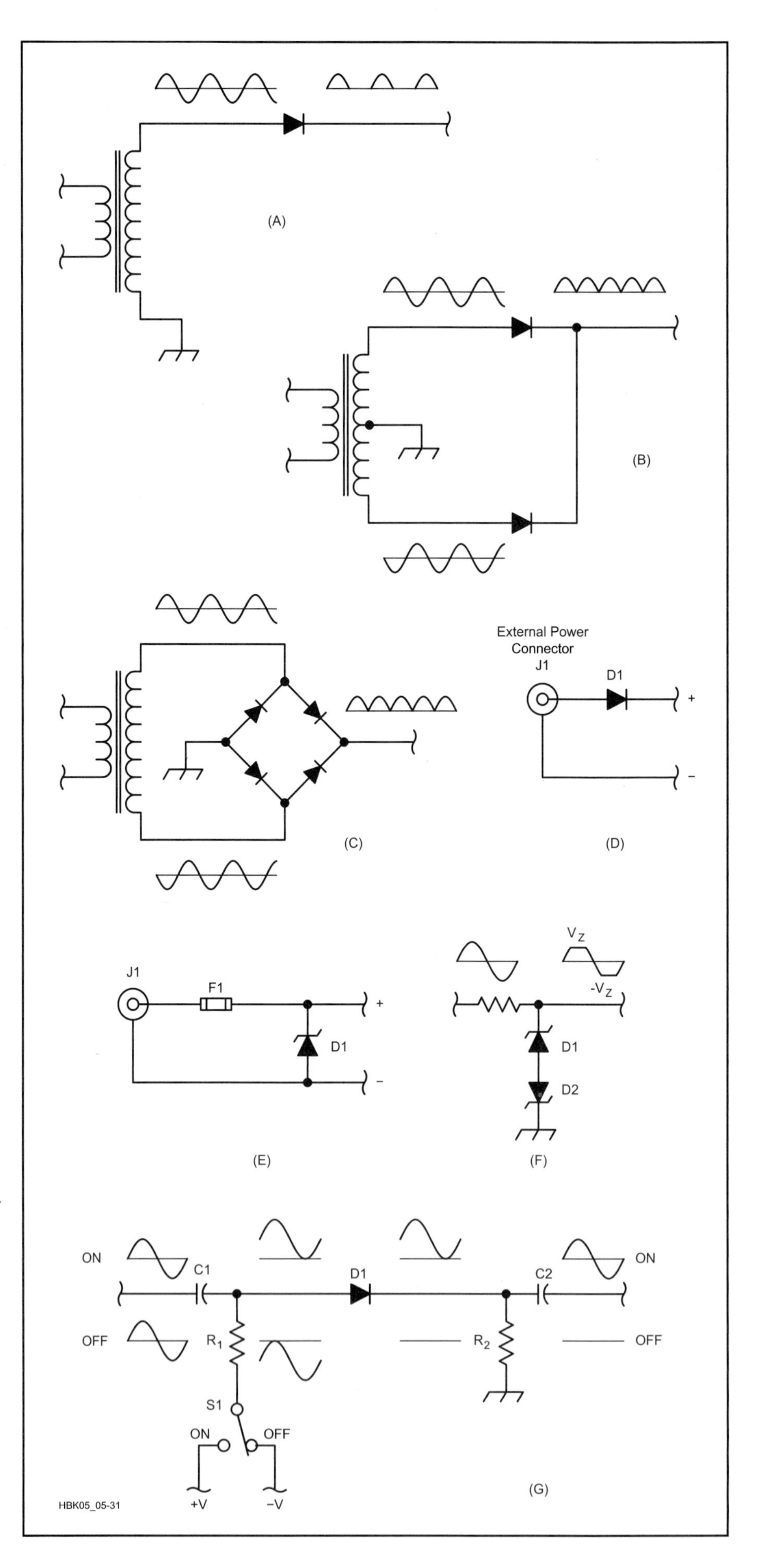

the signal from passing through the diode. In this case the bias voltage must be added to the ac signal and be of sufficient magnitude so that the entire envelope of the ac signal is above or below the junction barrier voltage, with respect to the cathode, to pass through the diode or inhibit the signal. Special forms of diodes, such as the PIN diode described earlier, which are capable of passing higher frequencies, are used to switch RF signals.

BIPOLAR TRANSISTORS

The bipolar transistor is a *current-controlled device*. The current between the emitter and the collector is governed by the current that enters the base. The convention when discussing transistor opera-

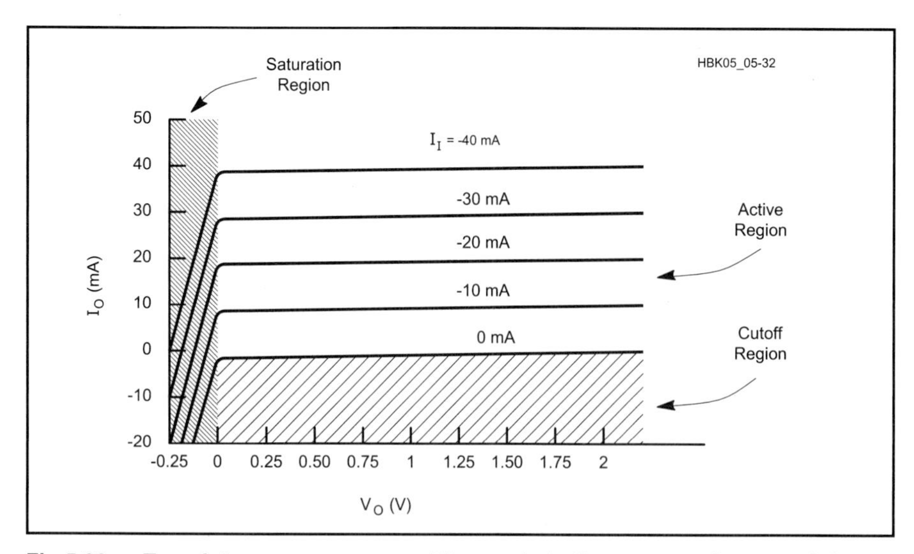

Fig 5.32 — Transistor response curve. The x-axis is the output voltage, and the y-axis is the output current. Different curves are plotted for various values of input current. The three regions of the transistor are its cutoff region, where no current flows in any terminal, its active region, where the output current is nearly independent of the output voltage and there is a linear relationship between the input current and the output current, and the saturation region, where the output current has large changes for small changes in output voltage.

tion is that the three currents into the device are positive (I_c into the collector, I_b into the base and I_e into the emitter). Kirchhoff's current law applies to transistors just as it does to passive electrical networks: the total current entering the device must be zero. Thus, the relationship between the currents into a transistor can be generalized as

$$0 = I_c + I_b + I_e \tag{9}$$

which can be rearranged as necessary. For example, if we are interested in the emitter current,

$$I_e = -(I_c + I_b) \tag{10}$$

The back-to-back diode model is appropriate for visualization of transistor construction. In actual transistors, however, the relative sizes of the collector, base and emitter regions differ. A common transistor configuration that spans a distance of 3 mm between the collector and emitter contacts typically has a base region that is only 25 µm across.

Current conduction between collector and emitter is described by regions in the common-base response curves of the transistor device (see **Fig 5.32**). The transistor is in its *active region* when the base-collector junction is reverse biased and the base-emitter junction is forward biased. The slope of the output current (I_O) versus the output voltage (V_O) is virtually flat, indicating that the output current is nearly independent of the output voltage. The slight slope that does exist is due to base-width modulation (known as the "Early effect"). Under these conditions, there is a linear relationship between the input current (I_I) and I_O. When both the junctions in the transistor are forward biased, the transistor is said to be in its *saturation region*. In this region, V_O is nearly zero and large changes in I_O occur for very small changes in V_O. The *cutoff region* occurs when both junctions in the transistor are reverse biased. Under this condition, there is very little current in the output, only the nanoamperes or microamperes that result from the very small leakage across the input-to-output junction. These descriptions of junction conditions are the basis for the use of transistors. Various configurations of the transistor in circuitry make use of the properties of the junctions to serve different purposes in analog signal processing.

In the common base configuration, where the input is at the emitter and the output is at the collector, the current gain is defined as

$$\alpha = -\frac{\Delta I_C}{\Delta I_E} = 1 \tag{11}$$

In the common emitter configuration, with the input at the base and the output at the collector, the current gain is

$$\beta = \frac{\Delta I_C}{\Delta I_B} \tag{12}$$

and the relationship between α and β is defined as

$$\alpha = \frac{\beta}{1+\beta} \tag{13}$$

Since the common-emitter configuration is the most used transistor-amplifier configuration, another designation for β is often used: h_{FE}, the forward dc current gain. (The "h" refers to "h parameters," a set of parameters for describing a two-port network.) The symbol, h_{fe}, is used for the forward current gain of ac signals. Other transistor transfer function relationships that are measured are h_{ie}, the input impedance, h_{oe}, the output admittance (reciprocal of impedance) and h_{re}, the voltage feedback ratio.

The behavior of a transistor can be defined in many ways, depending on which type of amplifier it is wired to be. A complete description of a transistor must include characteristic curves for each configuration. Typically, two sets of characteristic curves are presented: one describing the input behavior and the other describing the output behavior in each amplifier configuration. Different transistor amplifier configurations have different gains, input and output impedances. At low frequencies, where parasitic capacitances aren't a factor, the common emitter configuration has a high current gain (about – 50, with the negative sign indicating a 180° phase shift), medium to high input impedance (about 50 kΩ) and a medium to low output impedance (about 1 kΩ). The common collector has a high current gain (about 50), a high input impedance (about 150 kΩ) and a low output impedance (about 80 Ω). The common base amplifier has a low current gain (about 1), a low input impedance (about 25 Ω) and a very high output impedance (about 2 MΩ). Depending on the intended use of the transistor amplifier in an analog circuit, one configuration will be more appropriate than others. Once the common lead of the transistor amplifier configuration is chosen, the input and output impedance are functions of the device bias levels and circuit loading (**Fig 5.33**). The actual input and output impedances of a transistor amplifier are highly dependent on the input, biasing and load resistors that are used in the circuit.

A typical general-purpose bipolar-transistor data sheet lists important device specifications. Parameters listed in the ABSOLUTE MAXIMUM RATINGS section are the three junction voltages (V_{CEO}, V_{CBO} and V_{EBO}), the continuous collector current (I_C), the total device power dissipation (P_D) and the operating and storage temperature range. In the OPERATING PARAMETERS section, the three guaranteed minimum junction breakdown voltages

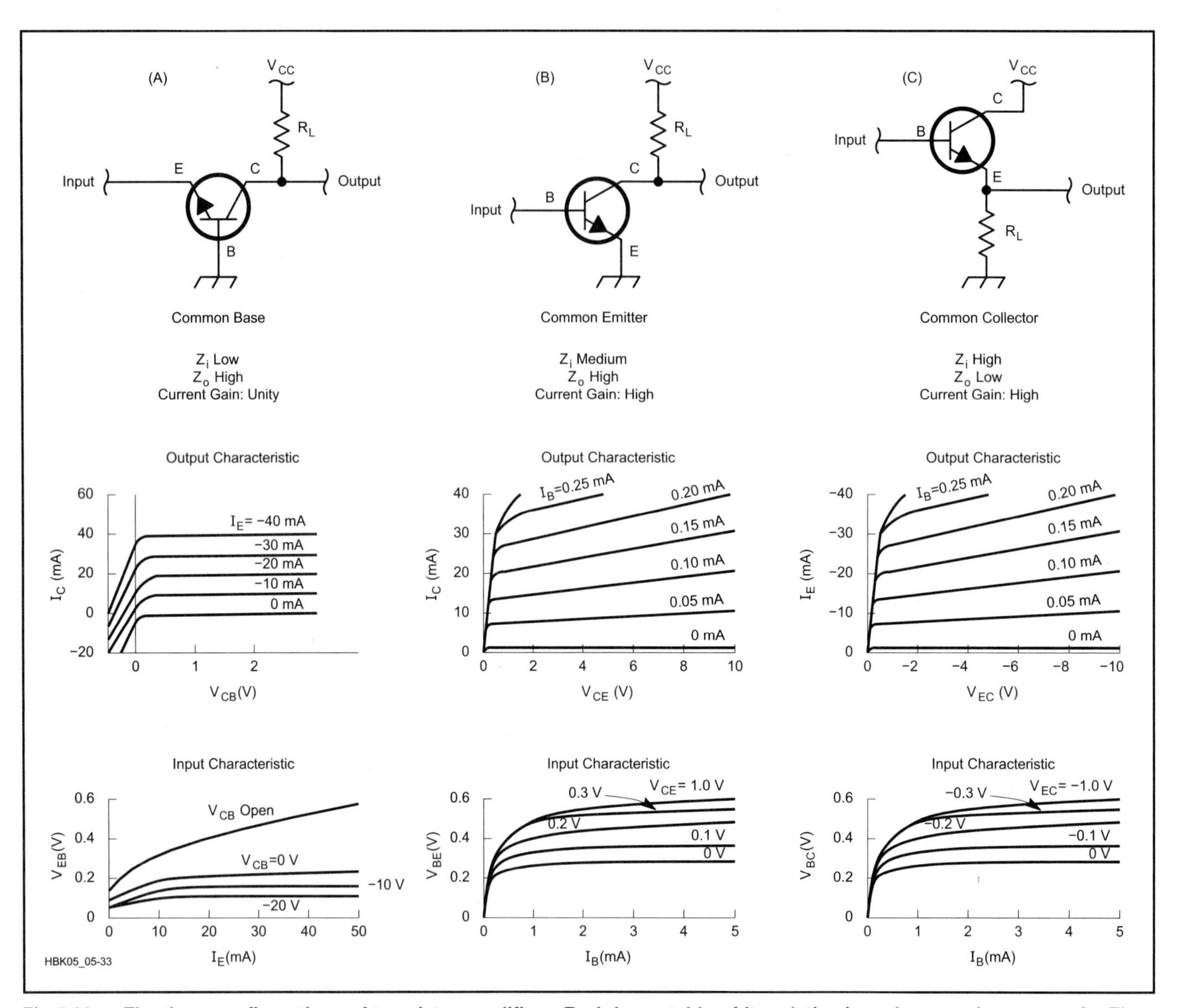

Fig 5.33 — The three configurations of transistor amplifiers. Each has a table of its relative impedance and current gain. The output characteristic curve is plotted for each, with the output voltage along the x-axis, the output current along the y-axis and various curves plotted for different values of input current. The input characteristic curve is plotted for each configuration with input current along the x-axis, input voltage along the y-axis and various curves plotted for different values of output voltage. (A) Common base configuration with input terminal at the emitter and output terminal at the collector. (B) Common emitter configuration with input terminal at the base and output terminal at the collector. (C) Common collector with input terminal at the base and output terminal at the emitter.

are listed $V_{(BR)CEO}$, $V_{(BR)CBO}$ and $V_{(BR)EBO}$ — along with the two guaranteed maximum collector cutoff currents — I_{CEO} and I_{CBO} — under OFF CHARACTERISTICS. Under ON CHARACTERISTICS are the guaranteed minimum dc current gain (h_{FE}), guaranteed maximum collector-emitter saturation voltage — $V_{CE(SAT)}$ — and the guaranteed maximum base-emitter on voltage — $V_{BE(ON)}$. The next section is SMALL-SIGNAL CHARACTERISTICS, where the guaranteed minimum current gain-bandwidth product — f_T, the guaranteed maximum output capacitance — C_{obo}, the guaranteed maximum input capacitance — C_{ibo}, the guaranteed range of input impedance — h_{ie}, the small-signal current gain — h_{fe}, the guaranteed maximum voltage feedback ratio — h_{re} and output admittance — h_{oe} are listed. Finally, the SWITCHING CHARACTERISTICS section lists absolute maximum ratings for delay time — t_d, rise time — t_r, storage time — t_s and fall time — t_f.

Transistor Biasing

Biasing in a transistor adds or subtracts a fixed amount of current from the signal at the input port. This differs from vacuum tube, FET and operational amplifier biasing where a bias *voltage* is added to the input signal. Fixed bias is the simplest form, as shown in **Fig 5.34A**. The operating point is determined by the intersection between the characteristic curves, the load line and the quiescent current bias line (Fig 5.34B). The problem with fixing the bias current is that if the transistor parameters drift due to heat, the operating point will change. The operating point can be stabilized by self biasing, also called emitter biasing, as pictured in Fig 5.34C. If I_C increases due to temperature changes, the current in R_E increases. The larger current through R_E increases the voltage drop across that resistor, causing a decrease in the base current, I_B. This, in turn, leads to a decreasing I_C, minimizing its variation

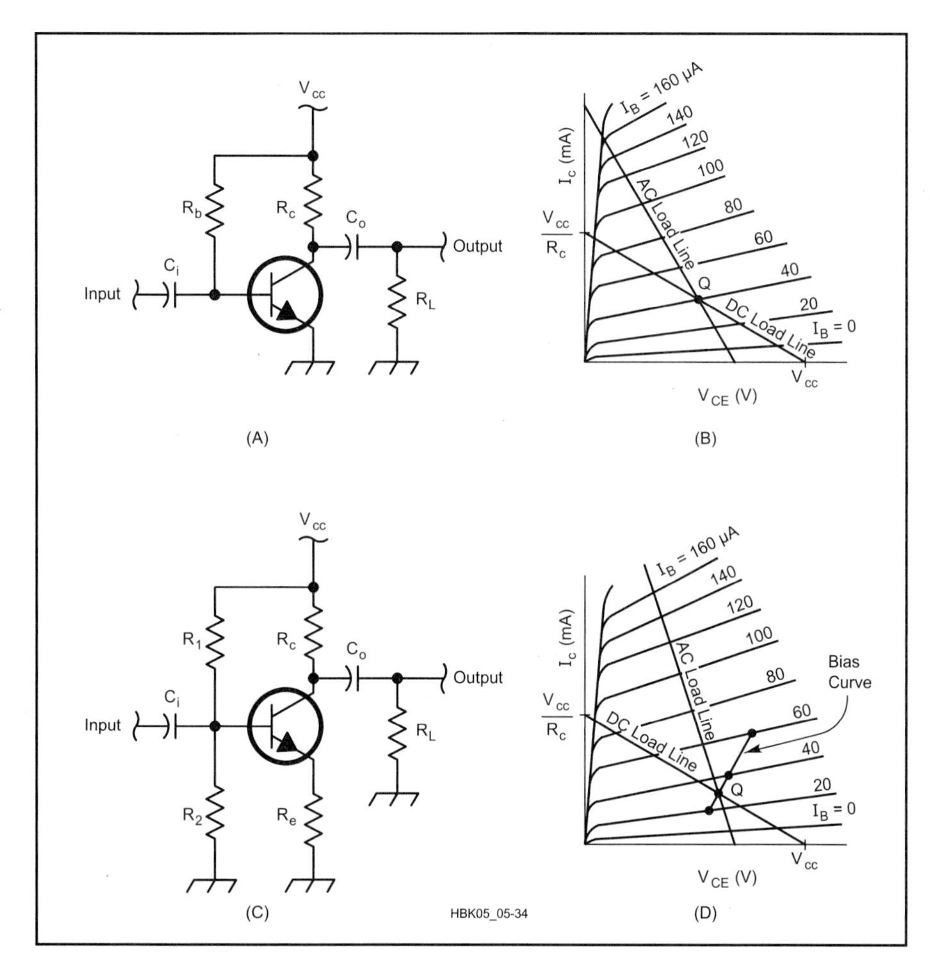

Fig 5.34 — Transistor biasing circuits. (A) Fixed bias. Input signal is ac coupled through C_i. The output has a voltage that is equal to $V_{CC} - I_C \times R_C$. This signal is ac coupled to the load, R_L, through C_O. For dc signals, the entire output voltage is based on the value of R_C. For ac signals, the output voltage is based on the value of R_C in parallel with R_L. (B) Characteristic curve for the transistor amplifier pictured in (A). The slope of the dc load line is equal to $-1 / R_C$. For ac signals, the slope of the ac load line is equal to $-1 / (R_C \,||\, R_L)$. The quiescent operating point, Q, is based on the base bias current with no input signal applied and the point where this characteristic line crosses the dc load line. The ac load line must also pass through point Q. (C) Self-bias. Similar to fixed bias circuit with the base bias resistor split into two: R1 connected to V_{CC} and R2 connected to ground. Also an emitter bias resistor, R_E, is included to compensate for changing device characteristics. (D) This is similar to the characteristic curve plotted in (B) but with an additional "bias curve" that shows how the base bias current varies as the device characteristics change with temperature. The operating point, Q, moves along this line and the load lines continue to intersect it as it changes.

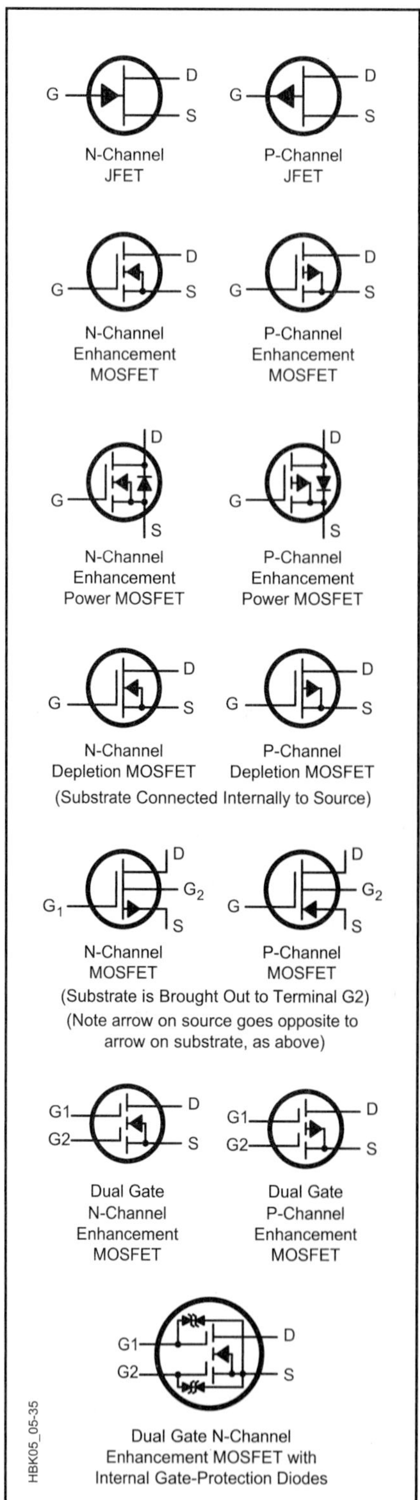

Fig 5.35—FET schematic symbols.

due to heat. The operating point for this type of biasing is plotted in Fig 5.34D.

FIELD-EFFECT TRANSISTORS

FET devices are more closely related to vacuum tubes than are bipolar transistors. Both the vacuum tube and the FET are controlled by the voltage level of the input rather than the input current, as in the bipolar transistor. FETs have three basic terminals, the gate, the source and the drain. These are related to both vacuum tube and bipolar transistor terminals: the gate to the grid and the base, the source to the cathode and the emitter, and the drain to the plate and the collector. Different forms of FET devices are pictured in **Fig 5.35**.

The characteristic curves for FETs are similar to those of vacuum tubes. The two most useful relationships are called the transconductance and output curves (**Fig 5.36**). Transconductance curves give the drain current, I_D, due to different gate-source voltage differences, V_{GS}, for various drain-source voltages, V_{DS}. The same parameters are interrelated in a different way in the output curve. For different values of V_{GS}, I_D is plotted against V_{DS}. In both of these representations, the device output is the drain current and these curves describe the FET in the common-source configuration. The action of the FET channel is so nearly ideal that, as long as the JFET gate does not become forward biased, the drain and source currents are virtually identical. For JFETs the gate leakage current, I_G, is a function of V_{GS} and this is often expressed with an input curve (**Fig 5.37**). The point at which there

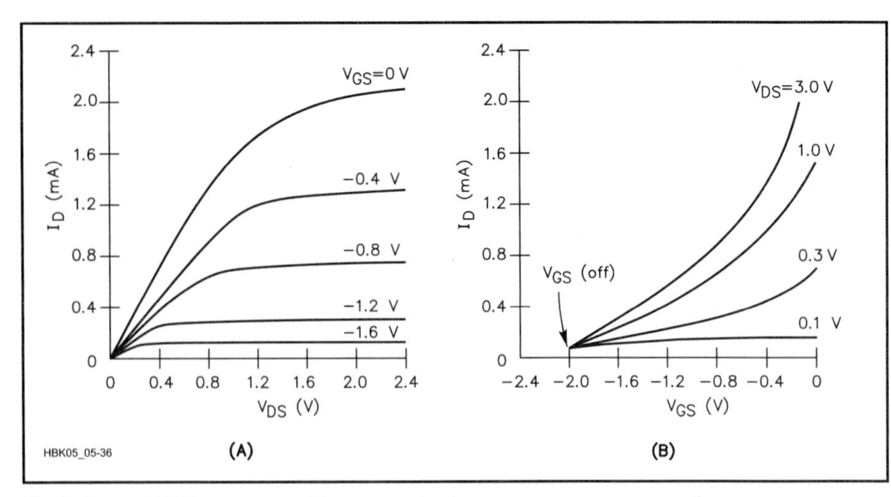

Fig 5.36 — JFET output and transconductance response curves for common source amplifier configuration. (A) Output voltage (V_{DS}) on the x-axis versus output current (I_D) on the y-axis, with different curves plotted for various values of input voltage (V_{GS}). (B) Transconductance curve has the same three variables rearranged, V_{GS} on the x-axis, I_D on the y-axis and curves plotted for different values of V_{DS}.

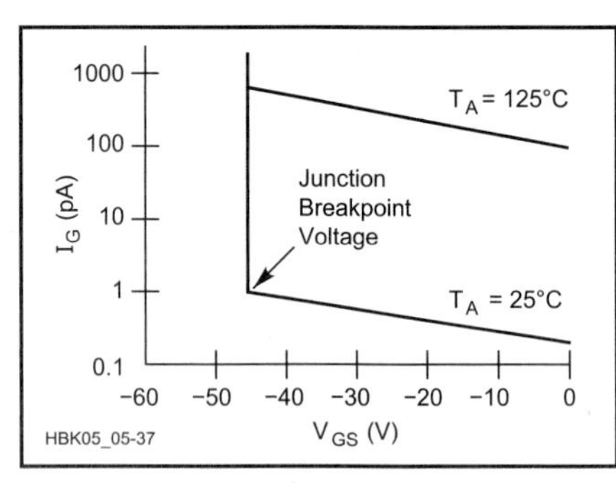

Fig 5.37 — JFET input leakage curves for common source amplifier configuration. Input voltage (V_{GS}) on the x-axis versus input current (I_G) on the y-axis, with two curves plotted for different operating temperatures, 25°C and 125°C. Input current increases greatly when the gate voltage exceeds the junction breakpoint voltage.

is a great increase in I_G is called the *junction break point voltage*. The insulated gates in MOSFET devices do away with any appreciable gate leakage current. MOSFETs do not need input and reverse transconductance curves. Their output curves (**Fig 5.38**) are similar to those of the JFET.

The parameters used to describe a FET's performance are also similar to those of vacuum tubes. The dc channel resistance, r_{DS}, is specified in data sheets to be less than a maximum value when the device is biased on ($r_{DS(on)}$). For ac signals, $r_{ds(on)}$ is not necessarily the same as $r_{DS(on)}$, but it is not very different as long as the frequency is not so high that capacitive reactance becomes significant. The common source forward transconductance, g_{fs}, is obtained as the slope of one of the lines in the forward transconductance curve,

$$g_{fs} = \frac{\Delta I_D}{\Delta V_{GS}} \qquad (14)$$

When the gate voltage is maximum ($V_{GS} = 0$ for a JFET), $r_{DS(on)}$ is minimum. This describes the effectiveness of the device as an analog switch.

A typical FET data sheet gives ABSOLUTE MAXIMUM RATINGS for V_{DS}, V_{DG}, V_{GS} and I_D, along with the usual device dissipation (P_D) and storage temperature range. The OFF CHARACTERISTICS listed are the gate-source breakdown voltage, $V_{GS(BR)}$, the reverse gate current, I_{GSS} and the gate-source cutoff voltage, $V_{GS(OFF)}$. The ON CHARACTERISTIC is the zero-gate-voltage drain current (I_{DSS}). The SMALL SIGNAL CHARACTERISTICS include the forward transfer admittance, y_{fs}, the output admittance, y_{os}, the static drain-source on resistance, $r_{ds(on)}$ and various capacitances such as input capacitance, C_{iss}, reverse transfer capacitance, C_{rss}, the drain-substrate capacitance, $C_{d(sub)}$. FUNCTIONAL CHARACTERISTICS include the noise figure, NF and the common source power gain G_{ps}.

The relatively flat regions in the MOSFET output curves are often used to provide a constant current source. As is plotted in these curves, the drain current,

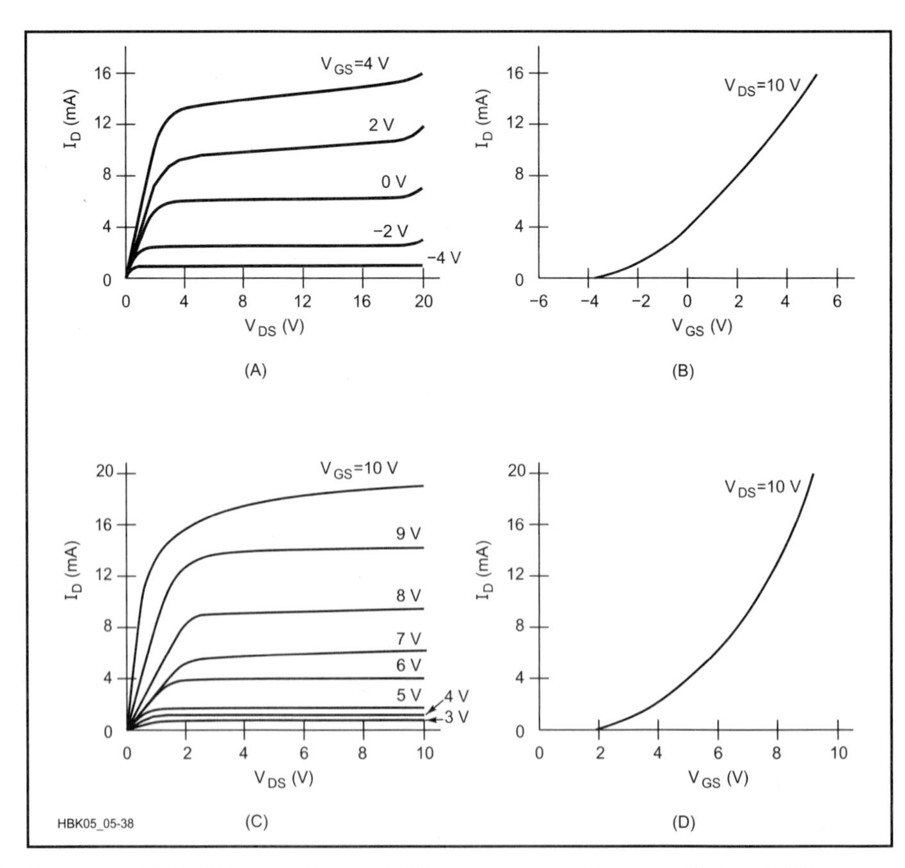

Fig 5.38 — MOSFET output [(A) and (C)] and transconductance [(B) and (D)] response curves. Plots (A) and (B) are for an N-channel depletion mode device. Note that V_{GS} varies from negative to positive values. Plots (C) and (D) are for an N-channel enhancement mode device. V_{GS} has only positive values.

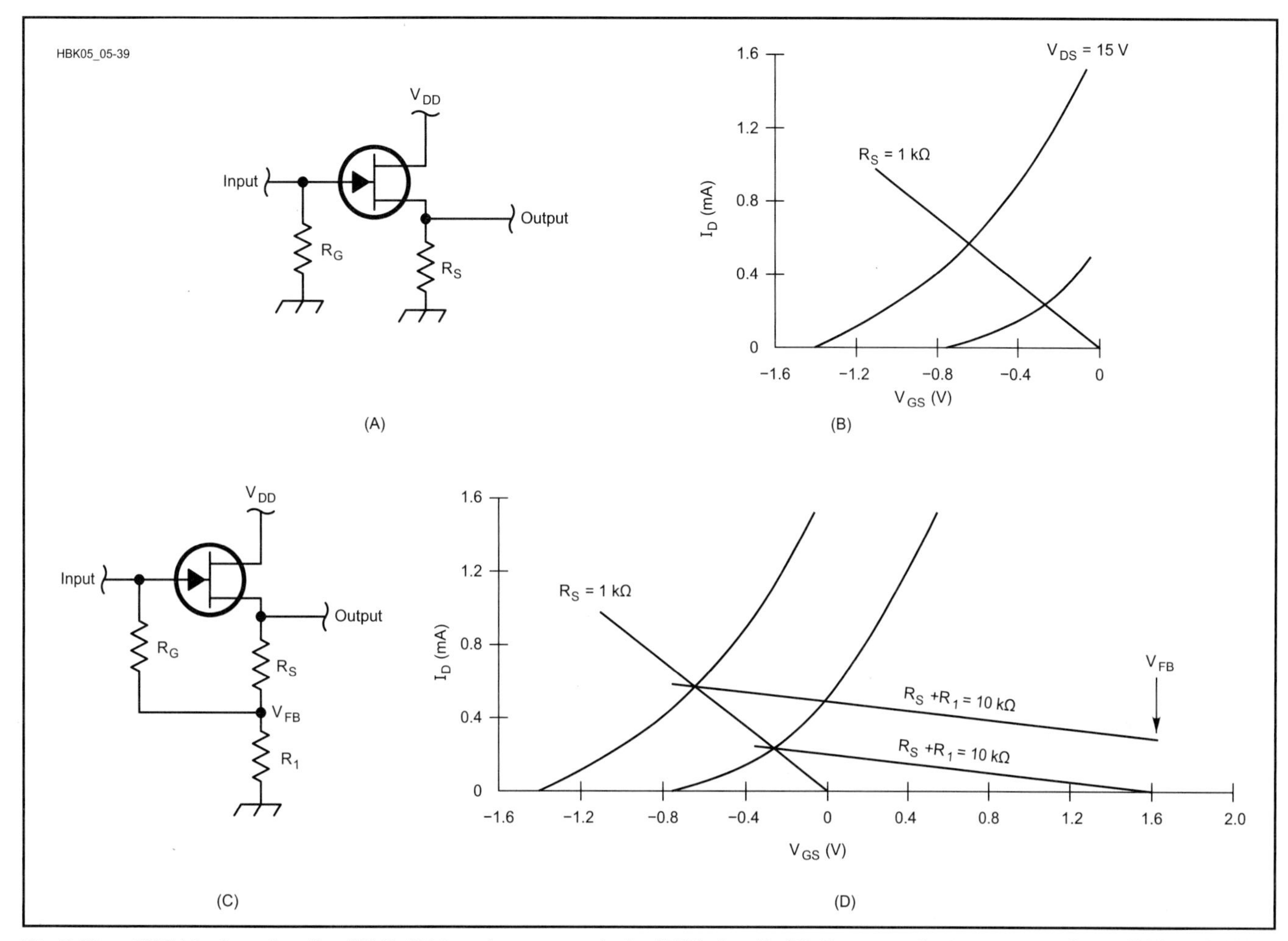

Fig 5.39 — FET biasing circuits. (A) Self-biased common drain JFET circuit. (B) Transconductance curve for self biased JFET in (A). Gate bias is determined by current through R_S. Load line has a slope of $-1 / R_S$, and gate bias voltage can vary between where the load line crosses the characteristic curves. (C) Feedback bias common drain JFET circuit.

I_D, changes very little as the drain-source voltage, V_{DS}, varies in this portion of the curve. Thus, for a fixed gate-source voltage, V_{GS}, the drain current can be considered to be constant over a wide range of drain-source voltages.

Multiple gate MOSFETs are also available (MFE130, MPF201, MPF211, MPF521). Due to the insulating layer, the two gates are isolated from each other and allow two signals to control the channel simultaneously with virtually no loading of one signal by the other. A common application of this type of device is an automatic gain control (AGC) amplifier. The signal is applied to one gate and a rectified, low-pass filtered form of the output (the AGC voltage) is fed back to the other gate. Another common application is for mixers.

FET Biasing

There are two ways to bias an FET, with and without feedback. Source self biasing for an N-channel JFET is pictured in **Fig 5.39A**. In this common-drain amplifier circuit, bias level is determined by the current through R_S, since I_G is very small and there is essentially no voltage drop across R_G. The characteristic curve for this configuration is plotted in Fig 5.39B. The operating points of the amplifier are where the load line intersects the curves. An example of feedback biasing is shown in Fig 5.39C. R_1 is generally much larger than R_S and the load line is determined by the sum of these resistors, as shown in Fig 5.39D. Feedback biasing increases the input impedance of the amplifier, but is rarely required, since input resistance (R_G) can be made very large.

MOSFET Gate Protection

The MOSFET is constructed with a very thin layer of SiO_2 for the gate insulator. This layer is extremely thin in order to improve the gain of the device but this makes it susceptible to damage from high voltage levels. If enough charge accumulates on the gate terminal, it can punch through the gate insulator and destroy it. The insulation of the gate terminal is so good that virtually none of this potential is eased by leakage of the charge into the device. While this condition makes for nearly ideal input impedance (approaching infinity), it puts the device at risk of destruction from even such seemingly innocuous electrical sources as static electricity in the air.

Some MOSFET devices contain an internal Zener diode with its cathode connected to the gate and its anode to the substrate. If the voltage at the gate rises to a damaging level the Zener junction breaks down and bleeds the excess charges off to the substrate. When voltages are within normal operating limits the Zener has little effect on the signal at the gate, although it may decrease the input impedance of the MOSFET. This solution will not work for all MOSFETs. The Zener diode must always be reverse biased to be effective. In the enhancement-mode MOSFET, $V_{GS} > 0$ for all valid uses of the part. In depletion mode devices V_{GS} can be both positive and negative; when negative, a protection Zener diode would be

Transistor Amplifier Design — a Practical Approach

The design of a transistorized amplifier is a straightforward process. Just as you don't need a degree in mechanical engineering to drive an automobile, neither do you need detailed knowledge of semiconductor physics in order to design a transistor amplifier with predictable and repeatable properties.

This sidebar will describe how to design a small-signal "Class A" transistor amplifier, following procedures detailed in one of the best books on the subject — *Solid State Design for the Radio Amateur*, by Wes Hayward, W7ZOI, and Doug DeMaw, W1FB. For many years, both hams and professional engineers have used this classic ARRL book to design untold numbers of working amplifiers.

How Much Gain?

One of the simple, yet profound, observations made in *Solid State Design for the Radio Amateur* is that a designer should *not* attempt to extract every last bit of gain from a single amplifier stage. Trying to do so virtually guarantees that the circuit will be "touchy" — it may end up being more oscillator than amplifier! While engineers might debate the exact number, modern semiconductor circuits are inexpensive enough that you should try for no more than 25 dB of gain in a single stage.

For example, if you are designing a high-gain amplifier system to follow a direct-conversion receiver mixer, you will need a total of about 100 dB of audio amplification. We would recommend a conservative approach where you use four stages, each with 25 dB of gain. You might risk oscillation and instability by using only two stages, with 50 dB gain each. The component cost will not be greatly different between these approaches, but the headaches and lack of reproducibility of the "simpler" two-stage design will very likely far outweigh any small cost advantages!

Biasing the Transistor Amplifier

The first step in amplifier design is to *bias* the transistor properly. A small-signal linear amplifier is biased properly when there is current at all times. Once you have biased the stage, you can then use several simple rules of thumb to determine all the major properties of the resulting amplifier.

Solid State Design for the Radio Amateur introduces several elegant transistor models. We won't get into that much detail here, except to say that the most fundamental property of a transistor is this: When there is current in the base-emitter junction, a larger current will flow in the collector-emitter junction. When the base-emitter junction is thus *forward biased*, the voltage across the base and emitter leads of a silicon transistor will be relatively constant, at 0.7 V. For most modern transistors, the dc current in the collector-emitter junction will be at least 50 to 100 times greater than the base-emitter current. This dc current gain is called the transistor's *Beta* (β).

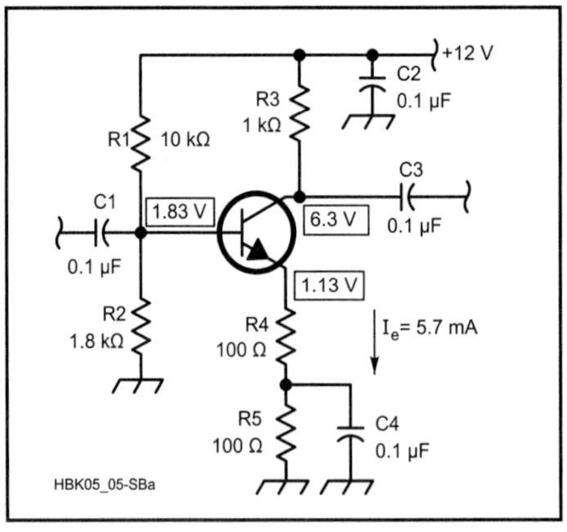

+12 V
C2
0.1 µF
R3
1 kΩ
R1
10 kΩ
C3
C1
1.83 V
6.3 V
0.1 µF
0.1 µF
1.13 V
R2
1.8 kΩ
R4
100 Ω
I_e= 5.7 mA
R5
100 Ω
C4
0.1 µF
HBK05_05-SBa

Fig A — Example of a simple low-frequency capacitively coupled transistorized small-signal amplifier. The voltages shown are the preliminary values desired for a collector current of 5 mA. The ac voltage gain is the ratio of the collector load resistor, R3, divided by the unbypassed portion of the emitter resistor, R4.

See **Fig A**, which shows a simple capacitively coupled low-frequency amplifier suitable for use at 1 MHz. Resistors R1 and R2 form a voltage divider feeding the base of the transistor. The amount of current in the resistive voltage divider is purposely made large enough so the base current is small in comparison, thus creating a "stiff" voltage supply for the base. As stated above, the voltage at the emitter will be 0.7 V less than the base voltage for this NPN transistor. The emitter voltage V_E appears across the series combination of R4 and R5. Note that R5 is bypassed by capacitor C4 for ac current.

By Ohm's Law, the emitter current is equal to the emitter voltage V_E divided by the sum of R4 plus R5. Now, the emitter current is made up of both the base-emitter and the collector-emitter current, but since the base current is much smaller than the collector current, the amount of collector current is essentially equal to the emitter current, at V_E / (R4 + R5).

Our design process starts by specifying the amount of current we want to flow in the collector, with the dc collector voltage equal to half the supply voltage. For good bias stability with temperature variation, the total emitter resistor should be at least 100 Ω for a small-signal amplifier. Let's choose a collector current of 5 mA, and use a total emitter resistance of 200 Ω, with R4 = R5 = 100 Ω each. The voltage across 200 Ω for 5 mA of current is 1.0 V. This means that the voltage at the base must be 1.0 V + 0.7 V = 1.7 V, provided by the voltage divider R1 and R2.

The dc base current requirements for a collector current of 5 mA is approximately 5 mA / 50 = 0.1 mA if the transistor's dc Beta is at least 50, a safe assumption for modern transistors. To provide a "stiff" base voltage, we want the current through the voltage divider to be about five to ten times greater than the base current. For convenience then, we choose the current through R1 to be 1 mA. This is a convenient current value, because the math is simplified — we don't have to worry about decimal points for current or resistance: 1 mA × 1.8 kΩ = 1.8 V. This is very close to the 1.7 V we are seeking. We thus choose a standard value of 1.8 kΩ for R2. The voltage drop across R1 is 12 V – 1.8 V = 10.2 V. With 1 mA in R1, the necessary value is 10.2 kΩ, and we choose the closest standard value, 10 kΩ.

Let's now look at what is happening in the collector part of the circuit. The collector resistor R3 is 1 kΩ, and the 5 mA of collector current creates a 5 V drop across R3. This means that the collector dc voltage must be 12 V – 5 V = 7 V. The dc power dissipated in the transistor will be essentially all in the collector-emitter junction, and will be the collector-emitter voltage (7 V – 1 V = 6 V) times the collector current of 5 mA = 0.030 W, or 30 mW. This dissipation is well within the 0.5 W rating typical of small-signal transistors.

Now, let's calculate more accurately the result from using standard values for R1 and R2. The actual base voltage will be 12 V × [1.8 kΩ / (1.8 kΩ + 10 kΩ)] = 1.83 V, rather than 1.7 V. The resulting emitter voltage is 1.83 V – 0.7 V = 1.13 V, resulting in 1.13 V / 200 Ω = 5.7 mA of collector current, rather than our desired 5 mA. We are close enough — we have finished designing the bias circuitry!

Performance: Voltage Gain

Now we can analyze how our little amplifier will work. The use of the unbypassed emitter resistor R4 results in *emitter degeneration* — a fancy word describing a form of negative feedback. The bottom line for us is that we can use several handy rules of thumb. The first is for the ac voltage gain of an amplifier: A_V = R3 / R4, where A_V is shorthand for *voltage gain.* The ac voltage gain of such an amplifier is simply the ratio of the collector load resistor and the unbypassed emitter resistor. In this case, the gain is 1000 / 100 = 10, which is 20 dB of voltage gain. This expression for gain is true virtually without regard for the exact kind of transistor used in the circuit, provided that we design for moderate gain in a single stage, as we have done.

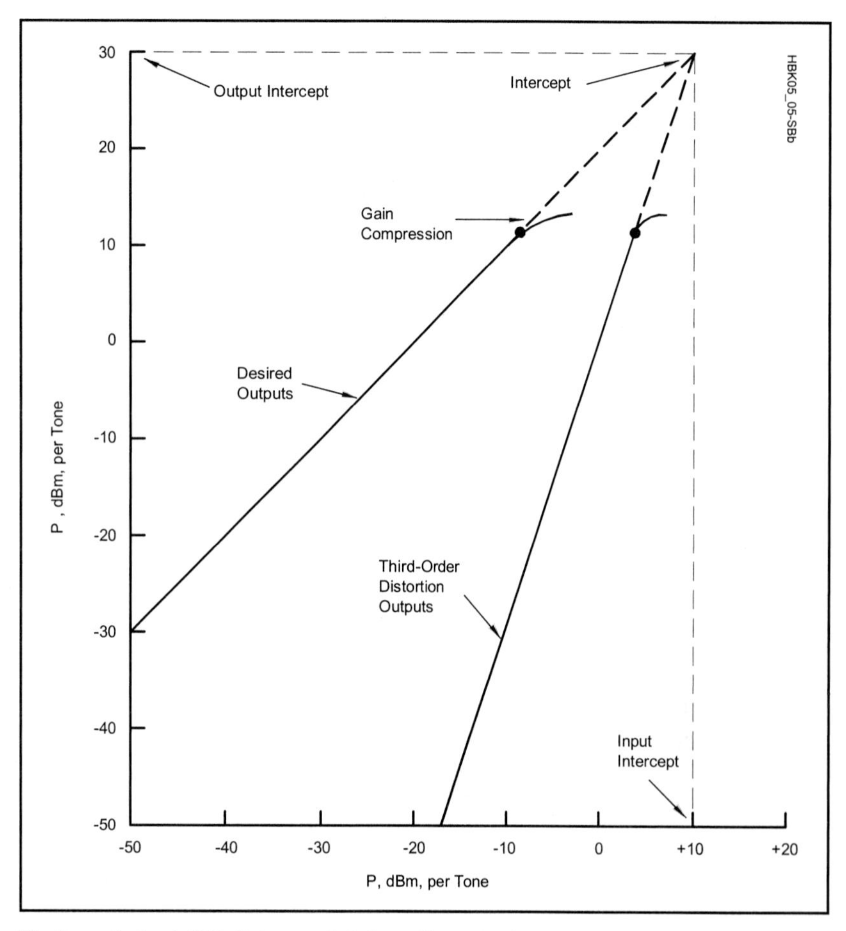

Fig B — Output IMD (intermodulation distortion) as a function of input. In the region below the 1-dB compression point, a decrease in input level of 10 dB results in a drop of IMD products by 30 dB below the level of each output tone in a two-tone signal.

Performance: Input Resistance

Another useful rule of thumb stemming from use of an unbypassed emitter resistor is the expression for the ac input resistance: R_{IN} = Beta × R4. If the ac Beta at low frequencies is about 50, then the input resistance of the transistor is 50 × 100 Ω = 5000 Ω. The actual input resistance includes the shunt resistance of voltage divider R2 and R1, about 1.5 kΩ. Thus the biasing resistive voltage divider essentially sets the input resistance of the amplifier.

Performance: Overload

We can accurately predict how this amplifier will perform. If we were to supply a peak positive 1 V signal to the base, the voltage at the collector will try to fall by the voltage gain of 10. However, since the dc voltage at the collector is only 7 V, it is clear that the collector voltage cannot fall 10 V. In theory, the collector voltage could fall as low as the 1.13 V dc level at the emitter. This amplifier will "run out of voltage" at a negative collector voltage swing of about 6.3 V – 1.13 V = 5.17 V, when the input voltage is 5.17 divided by the gain of 10 = 0.517 V.

When a negative-going ac voltage is supplied to the base, the collector current falls, and the collector voltage will rise by the voltage gain of 10. The maximum amount of voltage possible is the 12 V supply voltage, where the transistor is cut off with no collector current. The maximum positive collector swing is from the standing collector dc voltage to the supply voltage: 12 V – 6.3 V = 5.7 V positive swing. This occurs with a peak negative input voltage of 5.7 V / 10 = 0.57 V. Our amplifier will overload rather symmetrically on both negative and positive peaks. This is no accident — we biased it to have a collector voltage halfway between ground and the supply voltage.

When the amplifier "runs out of output voltage" in either direction, another useful rule of thumb is that this is the *1 dB compression point.* This is where the amplifier just begins to depart from linearity, where it can no longer provide any more output for further input. For our amplifier, this is with a peak-to-peak output swing of approximately 5.1 V × 2 = 10.2 V, or 3.6 V rms. The output power developed in output resistor R3 is $(3.6)^2$ / 1000 = 0.013 W = 13 mW, which is +11.1 dBm (referenced to 1 mW on 50 Ω).

At the 1 dB compression point, the third-order *IMD* (intermodulation distortion) will be roughly 25 dB below the level of each tone. **Fig B** shows a graph of output versus input levels for both the desired signal and for third-order IMD products. The rule of thumb for IMD is that if the input level is decreased by 10 dB, the IMD will decrease by 30 dB. Thus, if input is restricted to be 10 dB below the 1 dB compression point, the IMD will be 25 dB + 30 dB = 55 dB below each output tone.

With very simple math we have thus designed and characterized a simple amplifier. This amplifier will be stable for both dc and ac under almost any thermal and environmental conditions conceivable. That wasn't too difficult, was it? — *R. Dean Straw, N6BV, ARRL Senior Assistant Technical Editor*

forward biased and the MOSFET would not work properly. In some depletion mode MOSFET devices, back-to-back Zener diodes are used to protect the gate.

MOSFET devices are at greatest risk of damage from static electricity when they are out of circuit. Even though static electricity is capable of delivering little current, it can generate thousands of volts. When storing MOSFETs, the leads should be placed into conductive foam. When working with MOSFETs, it is a good idea to minimize static by wearing a grounded wrist strap and working on a grounded table. A humidifier may help to decrease the static electricity in the air. Before inserting a MOSFET into a circuit board it helps to first touch the device leads with your hand and then touch the circuit board. This serves to equalize the excess charge so that when the device is inserted into the circuit board little charge will flow into the gate terminal.

OPTICAL SEMICONDUCTORS

In addition to electrical energy and heat energy, light energy also affects the behavior of semiconductor materials. If a device is made to allow light to fall on the surface of the semiconductor material, the light energy will break covalent bonds and increase the number of electron-hole pairs, decreasing the resistance of the material.

Photoconductors

In commercial *photoconductors* (also called *photoresistors*) the resistance can change by as much as several kilohms for a light intensity change of 100 ft-candles. The most common material used in photoconductors is cadmium sulfide (CdS), with a resistance range of more than 2 MΩ in total darkness to less than 10 Ω in bright light. Other materials used in photoconductors respond best at specific colors. Lead sulfide (PbS) is most sensitive to infrared light and selenium (Se) works best in the blue end of the visible spectrum.

A similar effect is used in some diodes and transistors so that their operation can be controlled by light instead of electrical current biasing. These devices are called *photodiodes* and *phototransistors*. The flow of minority carriers across the reverse biased PN junction is increased by light falling on the doped semiconductor material. In the dark, the junction acts the same as any reverse biased PN junction, with a very low current (on the order of 10 µA) that is nearly independent of reverse voltage. The presence of light not only increases the current but also provides a resistance-like relationship (reverse current increases as reverse voltage increases). See **Fig 5.40** for the characteristic response of a photodiode. Even with no reverse voltage applied, the presence of light causes a small reverse current, as indicated by the points at which the lines in Fig 5.40 intersect the left side of the graph. Photoconductors and photodiodes are generally used to produce light-related analog signals that require further processing. The phototransistor can often be used to serve both purposes, acting as an amplifier whose gain varies with the amount of light present. It is also more sensitive to light than the other devices. Phototransistors have lots of gain, but photodiodes normally have less noise, so they make sensitive detectors.

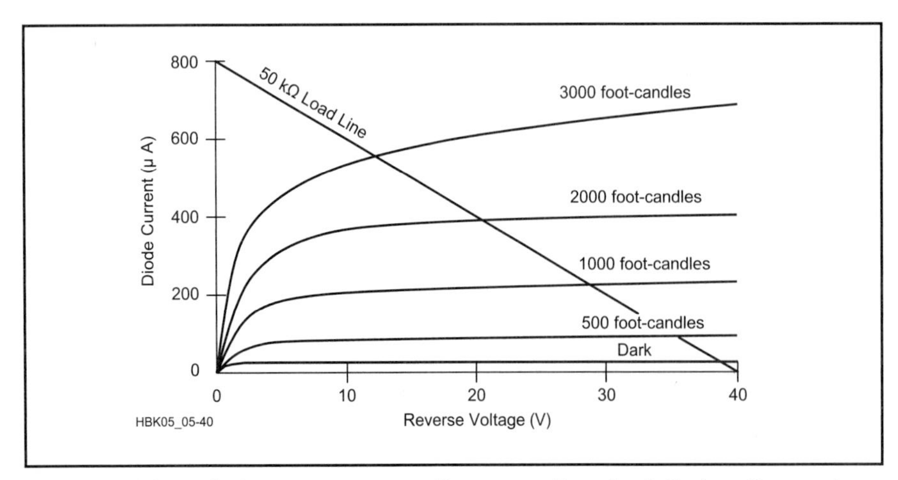

Fig 5.40 — Photodiode response curve. Reverse voltage is plotted on the x-axis and current through diode is plotted on the y-axis. Various response lines are plotted for different illumination. Except for the zero illumination line, the response does not pass through the origin since there is current generated at the PN junction by the light energy. A load line is shown for a 50 kΩ resistor in series with the photodiode.

Photovoltaic Effect

When illuminated, the reverse biased photodiode has a reverse current due to excess minority carriers. As the reverse voltage is reduced, the potential barrier to the forward flow of majority carriers is also reduced. Since light energy leads to the generation of both majority and minority carriers, when the resistance to the flow of majority carriers is decreased these carriers form a forward current. The voltage at which the forward current equals the reverse current is called the *photovoltaic potential* of the junction. If the illuminated PN junction is not connected to a load, a voltage equal to the photovoltaic potential can be measured across it. Devices that use light from the sun to produce electricity in this way are called *solar cells* or *solar batteries*. Common operating characteristics of silicon photovoltaic cells are an open circuit voltage of about 0.6 V and a conversion efficiency of about 10 to 15%.

Light Emitting Diodes

In the photodiode, energy from light falling on the semiconductor material is absorbed to make additional electron-hole pairs. When the electrons and holes recombine, the same amount of energy is given off. In normal diodes the energy from recombination of carriers is given off as heat. In certain forms of semiconductor material, the recombination energy is given off as light with a mechanism called *electroluminescence*. Unlike the incandescent light bulb, electroluminescence is a cold light source that typically operates with low voltages and currents (such as 1.5 V and 10 mA). Devices made for this purpose are called *light emitting diodes (LEDs)*. They have the advantages of low power requirements, fast switching times (on the order of 10 ns) and narrow spectra (relatively pure color). The LED emits light when it is forward biased and excess carriers are present. As the carriers recombine, light is produced with a color that depends on the properties of the semiconductor material used. Gallium arsenide (GaAs) generates light in the infrared region, gallium phosphide (GaP) gives off red light when doped with oxygen or green light when doped with nitrogen. Orange light is attained with a mixture of GaAs and GaP (GaAsP). Silicon doped with carbon gives off yellow light but does not produce much illumination. Other colors are also possible with different types and concentrations of dopants but usually have lower illumination efficiencies.

The LED is very simple to use. It is connected across a voltage source with a series resistor that limits the current to the desired level for the amount of light to be generated. The cathode lead is connected to the lower potential, and is usually specially marked (flattening of the lead near the package, a dot of paint next to the lead, and a flat portion of the round device located next to the lead are all common methods).

Optoisolators

An interesting combination of optoelectronic components proves very useful in many analog signal processing applications. An *optoisolator* consists of an LED optically coupled to a phototransistor, usually in an enclosed package. The optoisolator, as its name suggests, isolates different circuits from each other. Typically, isolation resistance is on the order of 10^{11} Ω and isolation capacitance is less than 1 pF. Maximum voltage isolation varies from 1,000 to 10,000 V ac. The most common optoisolators are available in 6-pin DIP packages.

Optoisolators are used for voltage level shifting and signal isolation. The isolation has two purposes: to protect circuitry from excessive voltage spikes and to isolate noisy circuitry from noise sensitive circuitry. A disadvantage of an optoisolator is that it adds a finite amount of noise and is not appropriate for use in many applications with low-level signals. Optoisolators also cannot transfer signals with high power levels. The power rating of the LED in a 4N25 device is 120 mW. Optoisolators have a limited frequency response due to the high capacitance of the LED. A typical bandwidth for the 4N25 series is 300 kHz.

As an example of voltage level shifting, the input to an optoisolator can be derived from a tube amplifier that has a signal varying between 0 and 150 V by using a series current limiting resistor. In order to drive a semiconductor circuit that operates in the –1 to 0 V range, the output of the optoisolator can be biased to operate in that range. This conversion of voltage levels, without a common ground connection between the circuits, is not easily performed in any other way.

A 1000 V spike that is high enough to destroy a semiconductor circuit will only saturate the LED in the optoisolator and will not propagate to the next stage. The worst that will happen is the LED will be destroyed, but very often it is capable of surviving even very high voltage spikes.

Optoisolators are also useful for isolating different ground systems. The input and output signals are totally isolated from each other, even with respect to the references for each signal. A common application for optoisolation is when a computer is used to control radio equipment. The computer signal, and even its ground reference, typically contains considerable wide band noise due to the digital circuitry. The best way to keep this noise out of the radio is to isolate both the signal and its reference; this is easily done with an optoisolator.

The design of circuits with optoisolators is not different from the design of circuits with LEDs and with transistors. The LED is forward biased and usually driven with a series current limiting resistor whose value is set so that the forward current will be less than the maximum value for the device (such as 60 mA in a 4N25). Signals must be appropriately dc shifted so that the LED is always forward biased. The phototransistor typically has all three leads available for connection. The base lead is used for biasing, since the signal is usually derived from the optics, and the collector and emitter leads are used as they would be in any transistor amplifier circuit.

Fiber Optics

An interesting variation of the optoisolator is the *fiber-optic* connection. Like the optoisolator, the signal is introduced to an LED device that modulates light. The signal is recovered by a photodetecting device (photoresistor, photodiode, or phototransistor). Instead of locating the input and output devices next to each other, the light is transmitted in a fiber optic cable, an extruded glass fiber that efficiently carries light over long distances and around fairly sharp bends. The fiber optic cable isolates the two circuits and provides an interesting transmission line. Fiber optics generally have far less loss than coaxial cable transmission lines. They do not leak RF energy, nor do they pick up electrical noise. Fiber optic cables are virtually immune from electromagnetic interference! Special forms of LEDs and phototransistors are available with the appropriate optical couplers for connecting to fiber optic cables. These devices are typically designed for higher frequency operation with bandwidths in the tens and hundreds of megahertz.

LINEAR INTEGRATED CIRCUITS

If you look into a transistor, the actual size of the semiconductor is quite small compared to the size of the packaging. For most semiconductors, the packaging takes considerably more space than the actual semiconductor device. Thus, an obvious way to reduce the physical size of circuitry is to combine more of the circuit inside a single package.

Hybrid Integrated Circuits

It is easy to imagine placing several small semiconductor chips in the same package. This is known as *hybrid circuitry*, a technology in which several semiconductor chips are placed in the same package and miniature wires are connected between them to make complete circuits.

Hybrid circuits miniaturize analog electronic circuits by replacing much of the packaging that is inherent in discrete electronics. The term *discrete* refers to the use of individual components to make a circuit, each in its own package. One application that still exists for hybrid circuitry is microwave amplifiers. The components of the amplifier are placed in a standard TO-39 package that is only 1 cm in diameter. The small dimensions of these circuits permit operation at VHF. For example, the Motorola MWA5157 can provide over 23 dB of gain at 1 GHz.

Both discrete and hybrid circuitry require that connections be made between the leads of the components. This takes space, is relatively expensive to construct and is the source of most failures in electronic circuitry. If multiple components could be placed on a single piece of semiconductor with the connections between them as part of the semiconductor chip, these three disadvantages would be overcome.

Monolithic Integrated Circuits

In order to build entire circuits on a single piece of semiconductor, it must be possible to fabricate other devices, such as resistors and capacitors, as well as transistors and diodes. The entire circuit is combined into a single unit, or chip, that is called a *monolithic integrated circuit.*

An integrated circuit (IC) is fabricated in layers. An example of a semiconductor circuit schematic and its implementation in an IC is pictured in **Fig 5.41**. The base layer of the circuit, the *substrate*, is made of P-type semiconductor material. Although less common, the polarity of the substrate can also be N-type material. Since the mobility of electrons is about three times higher than that of holes, bipolar transistors made with N-type collectors and FETs made with N-type channels are capable of higher speeds and power handling. Thus, P-type substrates are far more common. For devices with N-type substrates, all polarities in the ensuing discussion would be reversed. Other substrates have been used, one of the most successful of which is the silicon-on-sapphire (SOS) construction that has been used to increase the bandwidth of integrated circuitry. Its relatively high manufacturing cost has impeded its use, however.

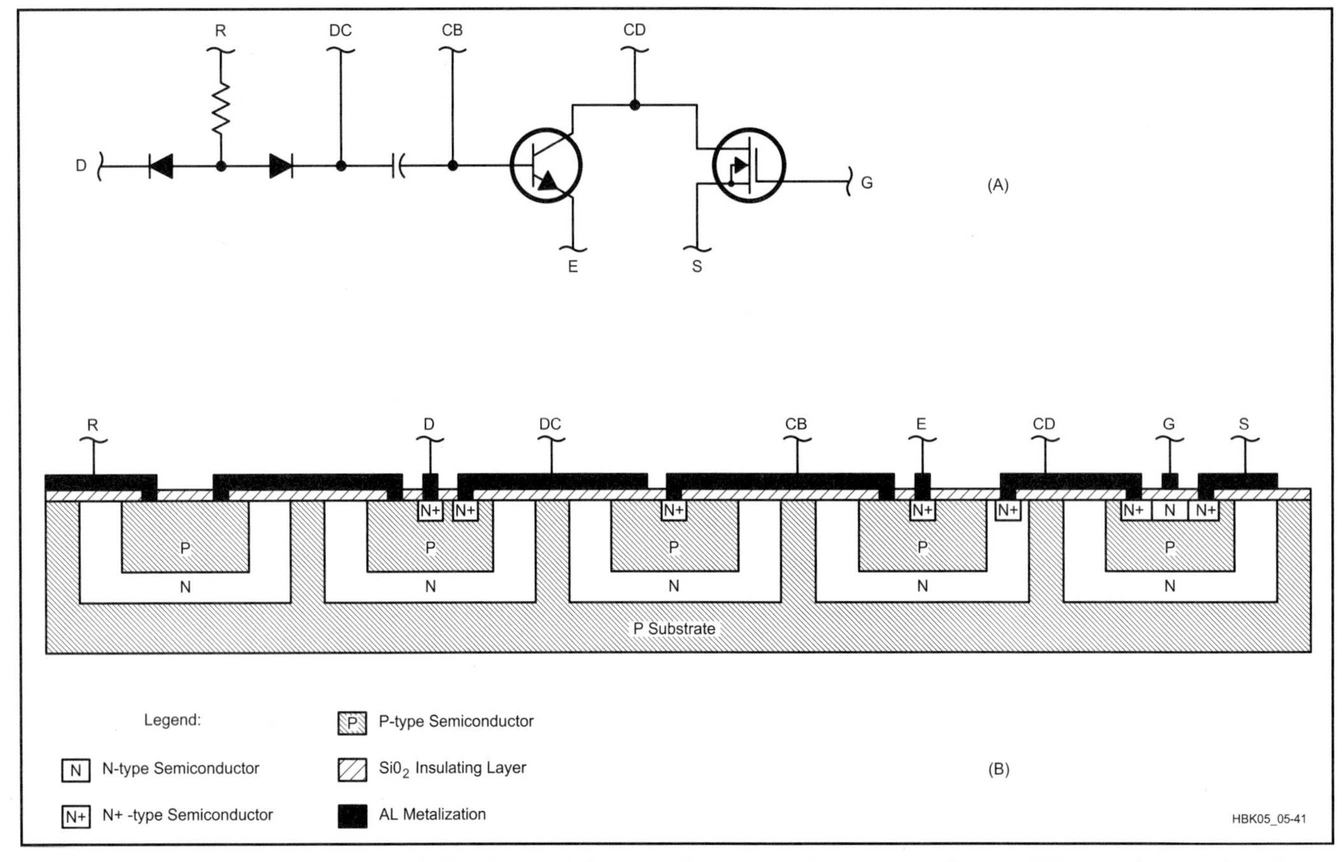

Fig 5.41 — Integrated circuit layout. (A) Circuit containing two diodes, a resistor, a capacitor, an NPN transistor and an N-channel MOSFET. Labeled leads are D for diode, R for resistor, DC for diode-capacitor, E for emitter, S for source, CD for collector-drain and G for gate. (B) Integrated circuit that is identical to circuit in (A). Same leads are labeled for comparison. Circuit is built on a P-type semiconductor substrate with N-type wells diffused into it. An insulating layer of SiO_2 is above the semiconductor and is etched away where aluminum metal contacts are made with the semiconductor. Most metal-to-semiconductor contacts are made with heavily doped N-type material (N^+-type semiconductor).

On top of the P-type substrate is a thin layer of N-type material in which the active and passive components are built. Impurities are diffused into this layer to form the appropriate component at each location. To prevent random diffusion of impurities into the N-layer, its upper surface must be protected. This is done by covering the N-layer with a layer of silicon dioxide (SiO_2). Wherever diffusion of impurities is desired, the SiO_2 is etched away. The precision of placing the components on the semiconductor material depends mainly on the fineness of the etching. The fourth layer of an IC is made of metal (usually aluminum) and is used to make the interconnections between the components.

Different components are made in a single piece of semiconductor material by first diffusing a high concentration of acceptor impurities into the layer of N-type material. This process creates P-type semiconductor — often referred to as P^+-type semiconductor because of its high concentration of acceptor atoms — that isolates regions of N-type material. Each of these regions is then further processed to form single components. A component is produced by the diffusion of a lesser concentration of acceptor atoms into the middle of each isolation region. This results in an N-type *isolation well* that contains P-type material, is surrounded on its sides by P^+-type material and has P-type material (substrate) below it. The cross sectional view in Fig 5.41B illustrates the various layers. Connections to the metal layer are often made by diffusing high concentrations of donor atoms into small regions of the N-type well and the P-type material in the well. The material in these small regions is N^+-type and facilitates electron flow between the metal contact and the semiconductor. In some configurations, it is necessary to connect the metal directly to the P-type material in the well.

An isolation well can be made into a resistor by making two contacts into the P-type semiconductor in the well. Resistance is inversely proportional to the cross-sectional area of the well. An alternate type of resistor that can be integrated in a semiconductor circuit is a *thin film resistor*, where a metallic film is deposited on the SiO_2 layer, masked on its upper surface by more SiO_2 and then etched to make the desired geometry, thus adjusting the resistance.

There are two ways to form capacitors in a semiconductor. One is to make use of the PN junction between the N-type well and the P-type material that fills it. Much like a varactor diode, when this junction is reverse biased a capacitance results. Since a bias voltage is required, this type of capacitor is polarized, like an electrolytic capacitor. Nonpolarized capacitors can also be formed in an integrated circuit by using thin film technology. In this case, a very high concentration of donor ions is diffused into the well, creating an N^+-type region. A thin metallic film is deposited over the SiO_2 layer covering the well and the capacitance is created between the metallic film and the well. The value of the capacitance is adjusted by varying the thickness of the SiO_2 layer and the cross-sectional size of the well. This type of thin film capacitor is also known as a metal oxide semiconductor (MOS) capacitor.

Unlike resistors and capacitors, it is very difficult to create inductors in integrated circuits. Generally, RF circuits that need inductance require external inductors to be connected to the IC. In some cases, particularly at lower frequencies, the behavior of an inductor can be mimicked by an amplifier circuit. In many cases the appropriate design of IC amplifiers can obviate the need for external inductors.

Transistors are created in integrated circuitry in much the same way that they are fabricated in their discrete forms. The NPN transistor is the easiest to make since the wall of the well, made of N-type semiconductor, forms the collector, the P-type material in the well forms the base and a small region of N^+-type material formed in the center of the well becomes the emitter. A PNP transistor is made by diffusing donor ions into the P-type semiconductor in the well to make a pattern with P-type material in the center (emitter) surrounded by a ring of N-type material that connects all the way down to the well material (base), and this is surrounded by another ring of P-type material (collector). This configuration results in a large base width separating the emitter and collector, causing these devices to have much lower current gain than the NPN form. This is one reason why integrated circuitry is designed to use many more NPN transistors than PNP transistors.

The simplest form of diode is generated by connecting to an N^+-type connection point in the well for the cathode and to the P-type well material for the anode. Diodes are often converted from NPN transistor configurations. Integrated circuit diodes made this way can either short the collector to the base or leave the collector unconnected. The base contact is the anode and the emitter contact is the cathode.

FETs can also be fabricated in IC form. Due to its many functional advantages, the MOSFET is the most common form used for digital ICs. MOSFETs are made in a semiconductor chip much the same way as MOS capacitors, described earlier. In addition to the signal processing advantages offered by MOSFETs over other transistors, the MOSFET device can be fabricated in 5% of the physical space required for bipolar transistors. MOSFET ICs can contain 20 times more circuitry than bipolar ICs with the same chip size. Just as discrete MOSFETs are at risk of gate destruction, IC chips made with MOSFET devices have a similar risk. They should be treated with the same care to protect them from static electricity as discrete MOSFETs. Integrated circuits need not be made exclusively with MOSFETs or bipolar transistors. It is common to find IC chips designed with both technologies, taking advantage of the strengths of each.

Complementary Metal Oxide Semiconductors

Power dissipation in a circuit can be reduced to very small levels (on the order of a few nW) by using the MOSFET devices in complementary pairs (CMOS). Each amplifier is constructed of a series circuit of MOSFET devices, as in **Fig 5.42**. The gates are tied together for the input signal, as are the drains for the output signal. In saturation and cutoff, only one of the devices conducts. The current drawn by the circuit under no load is equal to the OFF leakage current of either device and the voltage drop across the pair is equal to V_{DD}, so the steady state power used by the circuit is always equal to $V_{DD} \times I_{D(OFF)}$. For ac signals, power consumption is proportional to frequency.

CMOS circuitry could be built with discrete components; however, the number of extra parts and the need for the complementary components to be matched has made it an unusual design technique. Although CMOS is most commonly used in digital integrated circuitry, its low power consumption has been put to advantage by several manufacturers of analog ICs.

Integrated Circuit Advantages

There are many advantages of monolithic integrated circuitry over similar circuitry implemented with discrete components. The integration of the interconnections is one that has already been mentioned. This procedure alone serves to greatly decrease the physical size of the circuit and to improve its reliability. In fact, in one study performed on failures of electronic circuitry, it was found that the failure rate is not necessarily related to the complexity of the circuit, as had been previously thought, but is more closely a function of the number of interconnections between packages. Thus, the more circuitry that can be integrated onto a single piece of semiconductor material, the more reliable the circuit should be.

The amount of circuitry that can be placed onto a single semiconductor chip is a function of two factors: the size of the chip and how closely the various components are spaced. A revolution in IC manufacture occurred when semiconductor material was created in the laboratory rather than found in nature. The man-made semiconductor wafers are more pure and allow for larger wafer sizes. This, along with the steady improvement of the etching resolution on the chips, has caused an exponential increase over the past two decades in the amount of circuitry that can be placed in a single IC package. Currently, it is not unusual to find chips with more than one million transistors on them.

Decreased circuit size and improved reliability are only two of the advantages of monolithic integrated circuitry. The uncertainty of the exact behavior of the integrated components is the same as it is for discrete components, as discussed earlier. The relative properties of the devices on a single chip are very predictable, how-

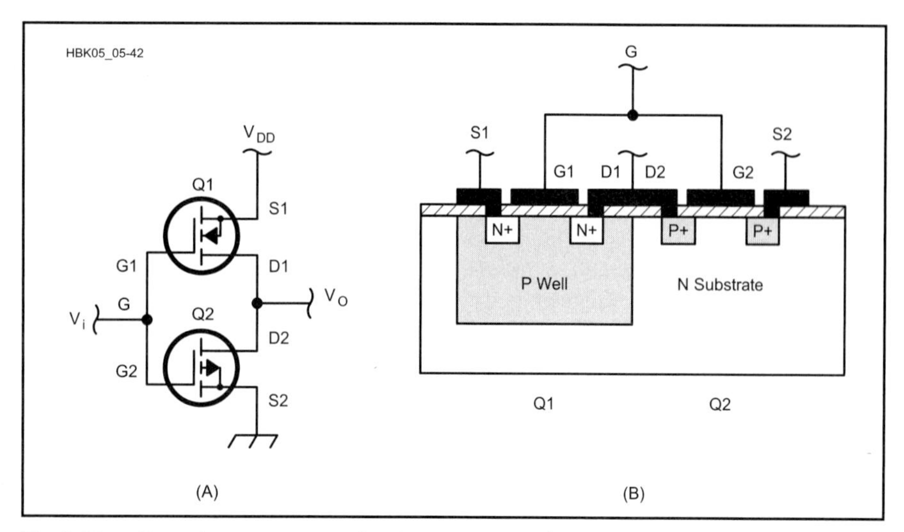

Fig 5.42 — Complementary metal oxide semiconductor (CMOS). (A) CMOS device is made from a pair of enhancement mode MOS transistors. The upper is an N-channel device, and the lower is a P-channel device. When one transistor is biased on, the other is biased off; therefore, there is minimal current from V_{DD} to ground. (B) Implementation of a CMOS pair as an integrated circuit.

ever. Since adjacent components on a semiconductor chip are made simultaneously (the entire N-type layer is grown at once, a single diffusion pass isolates all the wells and another pass fills them), the characteristics of identically formed components on a single chip of silicon should be identical. Even if the exact characteristics of the components are unknown, very often in analog circuit design the major concern is how components interact. For instance, push-pull amplifiers require perfectly matched transistors, and the gain of many amplifier configurations is governed by the ratio between two resistors and not their absolute values of resistance.

Integrated circuits often have an advantage over discrete circuits in their temperature behavior. The variation of performance of the components on an integrated circuit due to heat is no better than that of discrete components. While a discrete circuit may be exposed to a wide range of temperature changes, the entire semiconductor chip generally changes temperature by the same amount; there are fewer "hot spots" and "cold spots." Thus, integrated circuits can be designed to better compensate for temperature changes.

A designer of analog devices implemented with integrated circuitry has more freedom to include additional components that could improve the stability and performance of the implementation. The inclusion of components that could cause a prohibitive increase in the size, cost or complexity of a discrete circuit would have very little effect on any of these factors in an integrated circuit.

Once an integrated circuit is designed and laid out, the cost of making copies of it is very small, often only pennies per chip. Integrated circuitry is responsible for the incredible increase in performance with a corresponding decrease in price of electronics over the last 20 years. While this trend is most obvious in digital computers, analog circuitry has also benefited from this technology.

The advent of integrated circuitry has also improved the design of high frequency circuitry. One problem in the design and layout of RF equipment is the radiation and reception of spurious signals. As frequencies increase and wavelengths approach the dimensions of the wires in a circuit board, the interconnections act as efficient antennas. The dimensions of the circuitry within an IC are orders of magnitude smaller than in discrete circuitry, thus greatly decreasing this problem and permitting the processing of much higher frequencies with fewer problems of interstage interference. Another related advantage of the smaller interconnections in an IC is the lower inherent inductance of the wires, and lower stray capacitance between components and traces.

Integrated Circuit Disadvantages

Despite the many advantages of integrated circuitry, disadvantages also exist. ICs have not replaced discrete components, even tubes, in some applications. There are some tasks that ICs cannot perform, even though the list of these continues to decrease over time as IC technology improves.

Although the high concentration of components on an IC chip is considered to be an advantage of that technology, it also leads to a major limitation. Heat generated in the individual components on the IC chip is often difficult to dissipate. Since there are so many heat generating components so close together, the heat can build up and destroy the circuitry. It is this limitation that currently causes many power amplifiers to be designed with discrete components.

Integrated circuits, despite their short interconnection lengths and lower stray inductance, do not have as high a frequency response as similar circuits built with appropriate discrete components. (There are exceptions to this generalization, of course. Monolithic microwave integrated circuits — MMICs — are available for operation on frequencies up through 10 GHz.) The physical architecture of an integrated circuit is the cause of this limitation. Since the substrate and the walls of the isolation wells are made of opposite types of semiconductor material, the PN junction between them must be reverse biased to prevent current from passing into the substrate. Like any other reverse biased PN junction, a capacitance is created at the junction and this limits the frequency response of the devices on the IC. This situation has improved over the years as isolation wells have gotten smaller, thus decreasing the capacitance between the well and the substrate, and techniques have been developed to decrease the PN junction capacitance at the substrate. One such technique has been to create an N^+-type layer between the well and the substrate, which decreases the capacitance of the PN junction as seen by the well. As an example, in the 1970s the LM324 operational amplifier IC package was developed by National Semiconductor and claimed a gain-bandwidth product of 1 MHz. In the 1990s the HFA1102 operational amplifier IC, developed by Harris Semiconductor, was introduced with a gain-bandwidth product of 600 MHz.

A major impediment to the introduction of new integrated circuits, particularly with special applications, is the very high cost of development of new designs. The masking cost alone for a designed and tested integrated circuit can exceed $100,000. Adding the design, layout and debugging costs motivates IC manufacturers to produce devices that will be widely used so that they can recoup the development costs by volume of sales. While a particular application would benefit from customization of circuitry on an IC, the popularity of that application may not be wide enough to compel an IC manufacturer to develop that design. A designer who wishes to use IC chips must often settle for circuits that do not behave exactly as desired for the specific application. This trade-off between the advantages afforded by the use of integrated circuitry and the loss of performance if the available IC products do not exactly meet the desired specifications must be considered by equipment designers. It often leads to the use of discrete circuitry in sensitive applications. Once again, the improvements afforded by technology have mitigated this problem somewhat. The design and layout of ICs has been made more affordable by computer-based aids. Interaction between the computer aided design (CAD) software and modern chip masking hardware has also decreased the masking costs. As these development costs decrease, we are seeing an increase in the number of specialty chips that are being marketed and also of small companies that are created to fill the needs of the niche markets.

Common Types of Linear Integrated Circuits

The three main advantages of designing a circuit into an IC are to take advantage of the matched characteristics of like components, to make highly complex circuitry more economical, and to miniaturize the circuit. As a particular technology becomes popular, a rash of integrated circuitry is developed to service that technology. A recent example is the cellular telephone industry. Cellular phones have become so pervasive that IC manufacturers have developed a large number of devices targeted toward this technology. Space limitations prohibit a comprehensive listing of all analog special function ICs but a sampling of those that are more useful in the radio field is presented.

Component Arrays

The most basic form of linear integrated circuit is the component array. The most common of these are the resistor, diode and transistor arrays. Though capacitor arrays are also possible, they are used less often. Component arrays usually provide space saving but this is not the major advantage of these devices. They are the least densely packed of the integrated circuits, limited mainly by the number of off-chip connections needed. While it may be possible to place over a million transistors on a single semiconductor chip, individual access to these would require a total of three million pins and this is beyond the limits of practicability. More commonly, resistor and diode arrays contain from five to 16 individual devices and transistor arrays contain from three to six individual transistors. The advantage of these arrays is the very close matching of component values within the array. In a circuit that needs matched components, the component array is often a good method of obtaining this feature. The components within an array can be internally combined for special functions, such as termination resistors, diode bridges and Darlington pair transistors. A nearly infinite number of possibilities exists for these combinations of components and many of these are available in arrays.

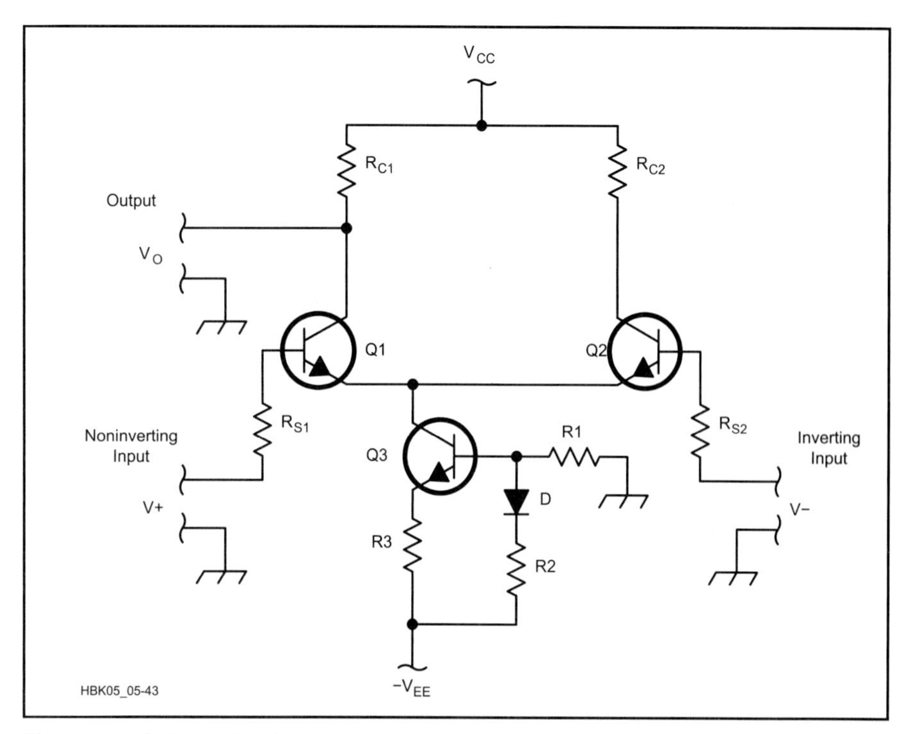

Fig 5.43 — Schematic of the components that make up an operational amplifier. Q1 and Q2 are matched emitter-coupled amplifiers. Q3 provides a constant current source. The symmetry of this device makes the matching of the components critical to its operation. This is why this circuit is usually implemented only in integrated circuitry. This simple op amp design has a large dc offset voltage at the output. Most practical designs include a level-shifting circuit, so the output voltage can exist near ground potential.

Multivibrators

A *multivibrator* is a circuit that oscillates, usually with a square wave output in the audio frequency range. The frequency of oscillation is accurately controlled with the addition of appropriate values of external resistance and capacitance. The most common multivibrator in use today is the 555 (NE555 by Signetics [now Philips] or LM555 by National Semiconductor). This very simple eight-pin DIP device has a frequency range from less than one hertz to several hundred kilohertz. Such a device can also be used in *monostable* operation, where an input pulse generates an output pulse of a different duration, or in *astable* operation, where the device freely oscillates. Some other applications of a multivibrator are as a frequency divider, a delay line, a pulse width modulator and a pulse position modulator.

Operational Amplifiers

An *operational amplifier*, or *op amp*, is one of the most useful linear devices that has been developed in integrated circuitry. While it is possible to build an op amp with discrete components, the symmetry of this circuit requires a close match of many components and is more effective, and much easier, to implement in integrated circuitry. **Fig 5.43** shows a basic op-amp circuit. The op amp approaches a perfect analog circuit building block.

Ideally, an op amp has an infinite input impedance (Z_i), a zero output impedance (Z_o) and an open loop voltage gain (A_v) of infinity. Obviously, practical op amps do not meet these specifications, but they do come closer than most other types of amplifiers. An older op amp that is based on bipolar transistor technology, the LM324, has the following characteristics: guaranteed minimum CMRR of 65 dB, guaranteed minimum A_v of 25000, an input bias current (related to Z_i) guaranteed to be below 250 nA (2.5×10^{-7} A), output current capability (which determines Z_o) guaranteed to be above 10 mA and a gain-bandwidth product of 1 MHz. The TL084, which is a pin compatible replacement for the LM324 but is made with both JFET and bipolar transistors, has a guaranteed minimum CMRR of 80 dB, an input bias current guaranteed to be below 200 pA (2.0×10^{-10} A, almost 1000 times smaller than the LM324) and a gain-bandwidth product of 3 MHz. Philips has recently introduced the LMC6001 op amp with an input bias current of 25 fA (2.5×10^{-14} A, almost 10,000 times smaller than the TL084). This is equivalent to 156 electrons entering the device every millisecond and corresponds to nearly infinite input impedance. Op amps can be customized to perform a large variety of functions by the addition of external components.

The typical op amp has three signal terminals (see **Fig 5.44**). There are two input terminals, the noninverting terminal marked with a + sign and the inverting terminal marked with a – sign. The output of the amplifier has a single terminal and

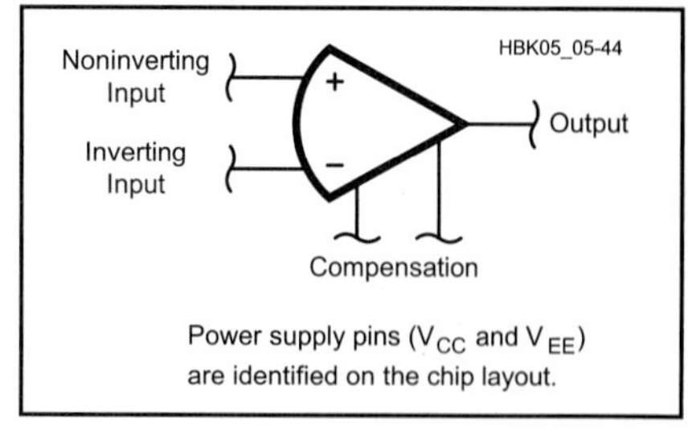

Fig 5.44 — Operational amplifier schematic symbol. The terminal marked with a + sign is the noninverting input. The terminal marked with a – sign is the inverting input. The output is to the right. On some op amps, external compensation is needed and leads are provided, pictured here below the device. Usually, the power supply leads are not shown on the op amp itself but are specified in the data sheet.

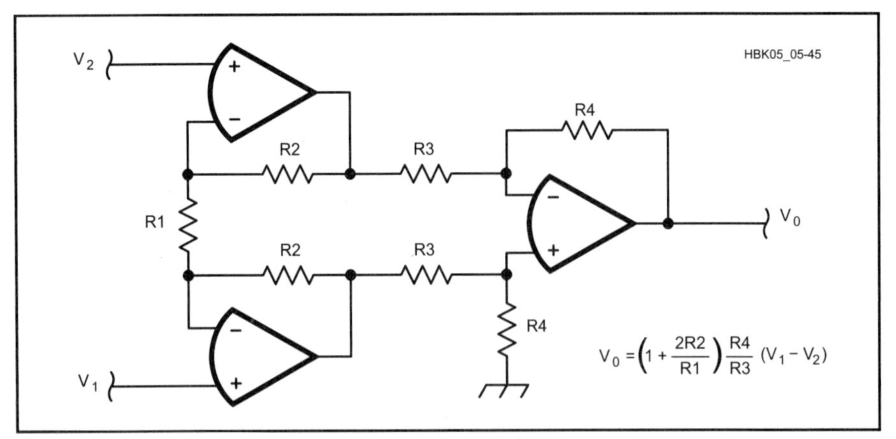

Fig 5.45 — Operational amplifiers arranged as an instrumentation amplifier. The balanced and cascaded series of op amps work together to perform differential amplification with good common-mode rejection and very high input impedance (no load resistor required) on both the inverting (V_1) and noninverting (V_2) inputs.

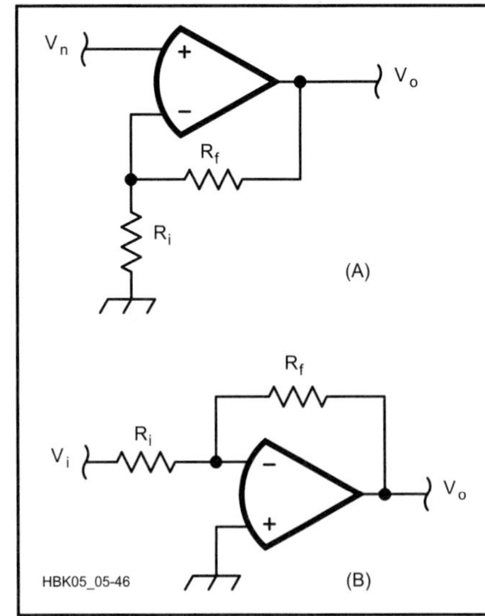

Fig 5.46 — Operational amplifier circuits. (A) Noninverting configuration. (B) Inverting configuration.

all signal levels within the op amp float, which means they are not tied to a specific reference. Rather, the reference of the input signals becomes the reference for the output signal. In many circuits this reference level is ground. Older operational amplifiers have an additional two connections for *compensation*. To keep the amplifier from going into oscillation at very high gains (increase its stability) it is often necessary to place a capacitor across the compensation terminals. This also decreases the frequency response of the op amp. Most modern op amps are internally compensated and do not have separate pins to add compensation capacitance. Additional compensation can be attained by connecting a capacitor between the op amp output and the inverting input.

One of the major advantages of using an op amp is its very high common mode rejection ratio (CMRR). Since there are two input terminals to an op amp, anything that is common to both terminals will be subtracted from the signal during amplification. The CMRR is a measure of the effectiveness of this removal. High CMRR results from the symmetry between the circuit halves. The rejection of power-supply noise is also an important parameter of an op amp. This is attained similarly, since the power supply is connected equally to both symmetrical halves of the op amp circuit. Thus, the power supply rejection ratio (PSRR) is similar to the CMRR and is often specified on the device data sheets.

Just as the symmetry of the transistors making up an op amp leads to a device with high values of Z_i, A_v and CMRR and a low value of Z_o, a symmetric combination of op amps is used to further improve these parameters. This circuit, shown in **Fig 5.45**, is called an instrumentation amplifier.

The op amp is capable of amplifying signals to levels limited mainly by the power supplies. Two power supplies are required, thus defining the range of signal voltages that can be processed. In most op amps the signal levels that can be handled are less than the power supply limits (rails), usually one or two diode drops (0.7 V or 1.4 V) away from each rail. Thus, if an op amp has 15 V connected as its upper rail (usually denoted V^+) and ground connected as its lower rail (V^-), input signals can be amplified to be as high as 13.6 V and as low as 1.4 V in most amplifiers. Any values that would be amplified beyond those limits are clamped (output voltages that should be 1.4 V or less appear as 1.4 V and those that should be 13.6 V or more appear as 13.6 V). This clamping action is illustrated in Fig 5.1. Op amps have been developed to handle signals all the way out to the power supply rails (for example, the MAX406, from Maxim Integrated Products).

If a signal is connected to the input terminals of an op amp, it will be amplified as much as the device is able (up to A_v), and will probably grow so large that it clamps, as described above. Even if such large gains are desired, A_v varies from one device to the next and cannot be guaranteed. In most applications the op amp gain is limited to a more reasonable value and this is usually realized by providing a negative feedback path from the output terminal to the inverting input terminal. The *closed loop gain* of an op amp depends solely on the values of the passive components used to form the loop (usually resistors and, for frequency-selective circuits, capacitors). Some examples of different circuit configurations that manipulate the loop gain follow.

The op amp is often used as either an inverting or a noninverting amplifier. Accurate amplification can be achieved with just two resistors: the feedback resistor, R_f, and the input resistor, R_i (see **Fig 5.46**). If connected in the noninverting configuration, the input signal is connected to the noninverting terminal. The feedback resistor is connected between the output and the inverting terminal. The inverting terminal is connected to R_i, which is connected to ground. The gain of this configuration is:

$$\frac{V_o}{V_n} = \left(1 + \frac{R_f}{R_i}\right) \tag{15}$$

where:
- V_o is the output voltage, and
- V_n is the input voltage to the noninverting terminal.

In the inverting configuration, the input signal (V_i) is connected through R_i to the inverting terminal. The feedback resistor is again connected between the inverting terminal and the output. The noninverting terminal can be connected to ground or to a dc-offset voltage. The gain of this circuit is:

$$\frac{V_o}{V_i} = -\frac{R_f}{R_i} \tag{16}$$

where V_i represents the voltage input to R_i.

The negative sign in equation 16 indicates that the signal is inverted. For ac signals, inversion represents a 180° phase shift. The gain of the noninverting op amp can vary from a minimum of × 1 to the maximum of which the device is capable. The gain of the inverting op amp configuration can vary from a minimum of × 0 (gains from × 0 to × 1 attenuate the signal while gains of × 1 and higher amplify the signal) to the maximum of which the device is capable, as indicated by A_v for dc signals, or the gain-bandwidth product for ac signals. Both parameters are usually specified in the manufacturer's data sheet.

A voltage follower is a type of op amp that is commonly used as a buffer stage. The voltage follower has the input connected directly to the noninverting terminal and the output connected directly to the inverting terminal (**Fig 5.47**). This configuration has unity gain and provides the maximum possible input impedance and the minimum possible output impedance of which the device is capable.

A *differential amplifier* is a special application of an operational amplifier (see **Fig 5.48**). It amplifies the difference between two analog signals and is very useful to cancel noise under certain conditions. For instance, if an analog signal and a reference signal travel over the same cable they may pick up noise, and it is likely that both signals will have the same amount of noise. When the differential amplifier subtracts them, the signal will be unchanged but the noise will be completely removed, within the limits of the CMRR. The equation for differential amplifier operation is

$$V_o = \frac{R_f}{R_i}\left[\frac{1}{\frac{R_n}{R_g}+1}\left(\frac{R_i}{R_f}+1\right)V_n - V_i\right] \quad (17)$$

which, if the ratios

$$\frac{R_i}{R_f} \text{ and } \frac{R_n}{R_g}$$

are equal, simplifies to

$$V_o = \frac{R_f}{R_i}\left(V_n - V_i\right) \quad (18)$$

Note that the differential amplifier response is identical to the inverting op amp response (equation 16) if the voltage source to the noninverting terminal is equal to zero. If the voltage source to the inverting terminal (V_i) is set to zero, the analysis is a little more complicated but it is possible to derive the noninverting op amp response (equation 15) from the differential amplifier response by taking into account the influence of R_n and R_g.

DC offset is an important consideration in op amps for two reasons. Actual op amps have a slight mismatch between the inverting and noninverting terminals that can become a substantial dc offset in the output, depending on the amplifier gain. The op amp output must not be too close to the clamping limits or distortion will occur. Introduction of a small dc correction voltage to the noninverting terminal is sometimes used to apply an offset voltage that counteracts the internal mismatch and centers the signal in the rail-to-rail range.

The high input impedance of an op amp makes it ideal for use as a *summing amplifier*. In either the inverting or noninverting configuration, the single input signal can be replaced by multiple input signals that are connected together through series resistors, as shown in **Fig 5.49**. For the inverting summing amplifier, the gain of each input signal can be calculated individually using equation 16 and, because of the superposition property, the output becomes the sum of each input signal multiplied by its gain. In the noninverting configuration, the output is the gain times the weighted sum of the m different input signals:

$$V_n = V_{n1}\frac{R_{p1}}{R_1+R_{p1}} + V_{n2}\frac{R_{p2}}{R_2+R_{p2}} + \ldots + V_{nm}\frac{R_{pm}}{R_m+R_{pm}} \quad (19)$$

where R_{pm} is the parallel resistance of all m resistors excluding R_m. For example, with three signals being summed, R_{p1} is the parallel combination of R_2 and R_3.

Other combinations of summing and difference amplification can be realized with a single op amp. The analyses of such circuits use the standard op amp equations coupled with the principle of superposition.

A *voltage comparator* is another special form of an operational amplifier. It takes in two analog signals and provides a binary output that is true if the voltage of one signal is bigger than that of the other, and false if not. A standard operational amplifier can be made to act as a comparator by connecting the two voltages to the noninverting and inverting inputs with no input or feedback resistors. If the voltage of the noninverting input is higher than that of the inverting input, the output voltage will be clamped to the positive clamping limit. If the inverting input is at a higher potential than the noninverting input, the output voltage will be clamped to the negative clamping limit (although this is not necessarily a negative voltage, depending on the value of the lower rail). Some applications of a voltage comparator are a zero crossing detector, a signal squarer (which turns other cyclical wave

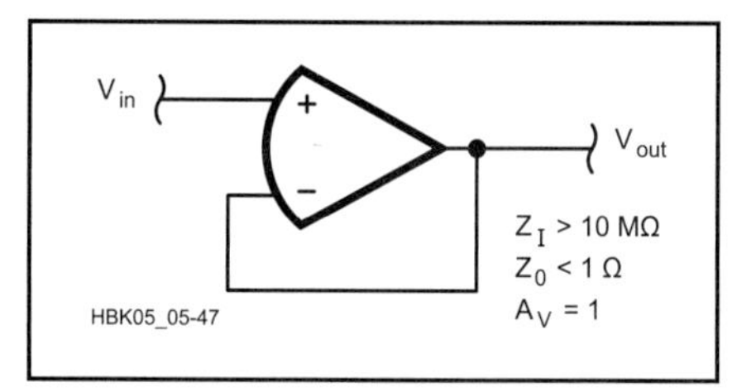

Fig 5.47 — Voltage follower. This operational amplifier circuit makes a nearly ideal buffer with a voltage gain of about one, and with extremely high input impedance and extremely low output impedance.

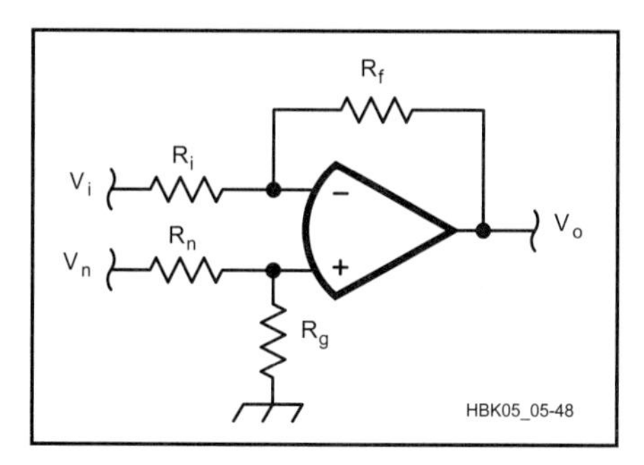

Fig 5.48 — Differential amplifier. This operational amplifier circuit amplifies the difference between the two input signals.

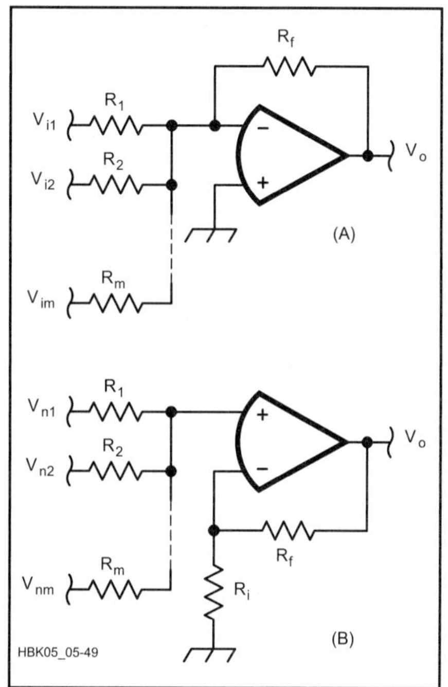

Fig 5.49 — Summing operational amplifier circuits. (A) Inverting configuration. (B) Noninverting configuration.

forms into square waves) and a peak detector.

Charge Coupled Devices

As the speed of integrated circuitry increases, it becomes possible to process some of the signals digitally while other processing occurs in analog form, all of this on the same IC chip. Such a chip is often called a *mixed modality* or *hybrid* chip (not to be confused with the hybrid circuitry discussed earlier). An example of this is the *charge-coupled device (CCD)*. Pure digital analysis of signals requires digitization in two domains, namely the time sampling of a signal into individual packets and the amplitude sampling of each time packet into digital levels. CCDs perform time sampling but the time packets remain in analog form; they can take on any voltage value rather than a fixed number of discrete values. The CCD is often used to produce a delay filter. While most analog filters introduce some phase shift or delay into the signal, the relationship between the phase shift and the frequency is not always linear; different frequencies are delayed by different amounts of time. The goal of an ideal delay filter is to delay all parts of the signal by the same time. The CCD is used to realize this by sampling the signal, shifting the time packets through a series of capacitors and then reconstructing the continuous signal at the other end. The rate of shifting the time packets and the number of stages determines the amount of the delay. When originally introduced in the late 1970s, CCDs were described as bucket brigade devices (after the old fire fighting technique), where the buckets filled with signal packets are passed along the line until they are dumped at the end and recombined into an analog signal. These devices are simply constructed in an IC where each bucket is a MOS capacitor that is surrounded by two MOSFETs. When the transistors are biased to conduct, the charge moves from one bucket to the next and, while biased off, the charges are held in their capacitors. Very accurate filters, called *switched capacitor filters*, can be made with CCDs (see the **RF and AF Filters** chapter).

A special form of CCD has also become quite popular in recent years, replacing the vidicon in modern camera circuitry. A two dimensional array of CCD elements has been developed with light sensitive semiconductor material; the charge that enters the capacitors is proportional to the amount of light incident on that location of the chip. The charges are held in their array of capacitors until shifted out, one horizontal line at a time, in a raster format. The CCD array mimics the operation of the vidicon camera and has many advantages. CCD response linearity across the field is superior to that of the vidicon. Very bright light at one location saturates the CCD elements only at that location rather than the blooming effect in vidicons where bright light spreads radially from the original location. CCD imaging elements do not suffer from image retention, which is another disadvantage of vidicon tubes.

Balanced Mixers

The *balanced mixer* is a device with many applications in modern radio transceivers (see the **Mixers, Modulators and Demodulators** chapter). Audio signals can be modulated onto a carrier or demodulated from the carrier with a balanced mixer. RF signals can be downconverted to intermediate frequency (IF) or IF can be upconverted to RF with a balanced mixer. This device is made with a bridge of four matched Schottky diodes and the necessary transformers packaged in a small metal, plastic or ceramic container. The consequence of unmatched diodes is poor isolation between the local oscillator (LO) and the two signals. IC mixers often use a "Gilbert cell" to provide LO isolation as high as –30 dB at 500 MHz. The isolation improves with decreasing frequency.

Receiver Subsystems

High performance ICs have been designed that make up complete receivers with the addition of only a few external components. Two examples that are very similar are the Motorola MC3363 and the Philips NE627. Both of these chips have all the active RF stages necessary for a double conversion FM receiver. The MC3363 has an internal local oscillator (LO) with varactor diodes that can generate frequencies up to 200 MHz, although the rest of the circuit is capable of operating at frequencies up to 450 MHz with an external oscillator. The RF amplifier has a low noise factor and gives this chip a 0.3 μV sensitivity. The intermediate frequency stages contain limiter amplifiers and quadrature detection. The necessary circuitry to implement receiver squelch and zero crossing detection of FSK modulation is also present. The circuit also contains received signal strength ("S-meter") circuitry (RSSI). The input and output of each stage are also brought out of the chip for versatility. The audio signal out of this chip must be appropriately amplified to drive a low-impedance speaker. This chip can be driven with a dc power source from 2 to 7 V and it draws only 3 mA with a 2 V supply.

The Philips NE627 is a newer chip than the MC3363 and has better performance characteristics even though it has essentially the same architecture. Its LO can generate frequencies up to 150 MHz and external oscillator frequencies up to 1 GHz can be used. The chip has a 4.6 dB noise figure and 0.22 μV sensitivity. The circuit can be powered with a dc voltage between 4.5 and 8 V and it draws between 5.1 and 6.7 mA. This chip is also ESD hardened so it resists damage from electrostatic discharges, such as from nearby lightning strikes.

The various stages in the receiver subsystem ICs are made available by connections on the package. There are two reasons that this is done. Filtering that is added between stages can be performed more effectively with inductors and crystal or ceramic filters, which are difficult to fabricate in integrated circuitry, so the output of one stage can be filtered externally before being fed to the next stage. It also adds to the versatility of the device. Filter frequencies can be customized for different intermediate frequencies. Stages can be used individually as well, so these devices can be made to perform direct conversion or single conversion reception or other forms of demodulation instead of FM.

Older integrated circuits that are subsets of the receiver subsystems are popular. The NE602 contains one double balanced mixer and a local oscillator, along with voltage regulation and buffering (**Fig 5.50**). It contains almost everything required to construct a direct conversion receiver. Its small size, an 8-pin DIP, makes it more desirable for this purpose than using part of an MC3363, which is in a 24-pin DIP and is more expensive. The NE604 contains the IF amplifiers and quadrature detector that, together with two NE602s and an RF amplifier, could almost duplicate the functions of the MC3363 or the NE367.

Transmitter Subsystems

Single chips are available to implement FM transmitters. One implementation is the Motorola MC2831A. This chip contains a mike preamplifier with limiting, a tone generator for CTCSS or AFSK, and a frequency modulator. It has an internal voltage controlled oscillator that can be controlled with a crystal or an LC circuit. This chip also contains circuitry to check the power supply voltage and produce a warning if it falls too low. Together with an FM receiver IC, an entire transceiver can be fabricated with very few parts.

Monolithic Microwave Integrated Circuit

A class of bipolar IC that is capable of higher frequency responses is the *mono-*

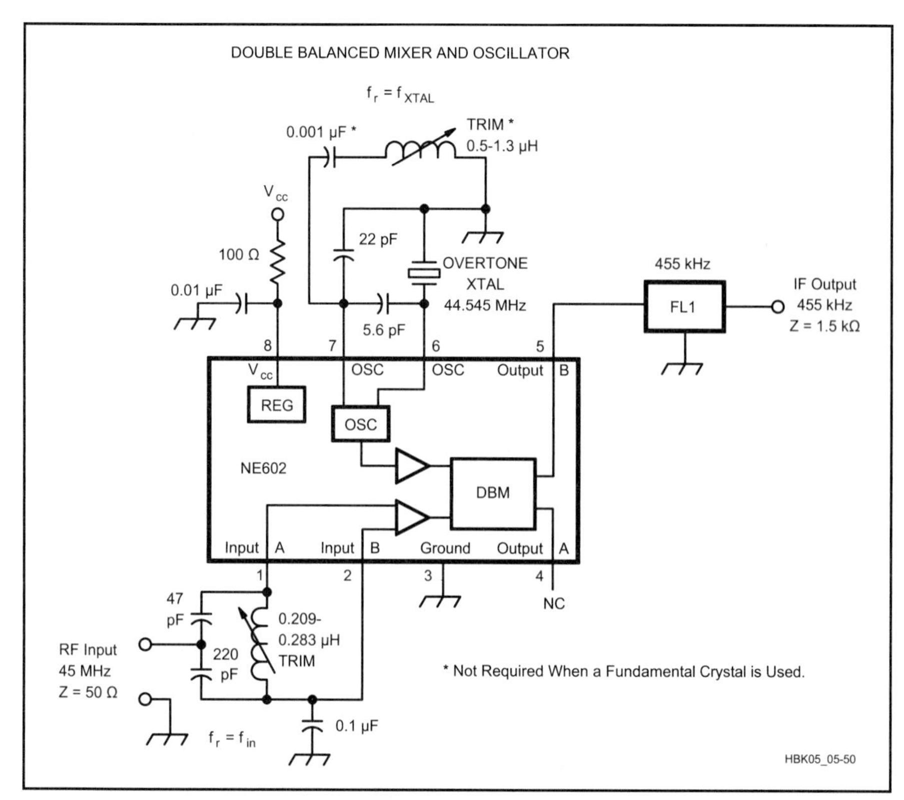

Fig 5.50 — The NE602 functional block diagram in circuit. This device contains a doubly balanced mixer, a local oscillator, buffers and a voltage regulator. This application uses the NE602 to convert an RF signal in a receiver to IF.

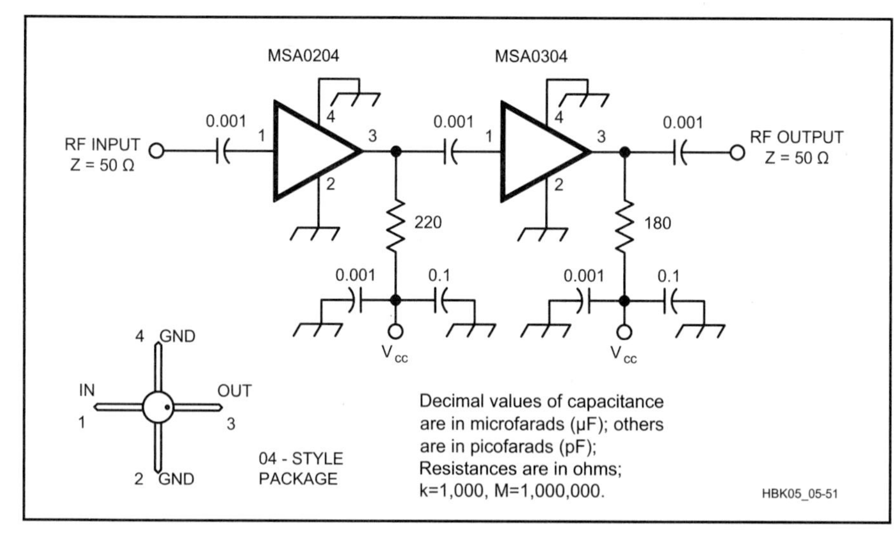

Fig 5.51 — The MSA0204 and MSA0304 MMICs in circuit. Both amplifiers have both input and output impedance of 50 Ω and a bandwidth of more than 2.5 GHz.

lithic microwave integrated circuit (MMIC). There is no formal definition of when an IC amplifier becomes an MMIC and, as the performance of IC devices improves, particularly MOS based devices, the distinction is becoming blurred. MMIC devices typically have predefined operating characteristics and require few external components. An example of an MMIC is a fixed gain amplifier, the MSA0204 (**Fig 5.51**), which can deliver 12 dB of gain up to 1 GHz. More modern MMIC devices are being developed with bandwidths in the tens of GHz.

Comparison of Analog Signal Processing Components

Analog signal processing deals with changing a signal to a desired form. Vacuum tubes, bipolar transistors, field-effect transistors and integrated circuitry perform similar functions, each with specific advantages and disadvantages. These are summarized here.

Of the four component types, vacuum tubes are physically the largest and require the most operating power. They have more limited life spans, usually because the heater filament burns out just as a light bulb does. Regardless of its use, a vacuum tube always generates heat. Miniaturization is difficult with vacuum tubes both because of their size and because of the need for air space around them for cooling. Vacuum tubes do have advantages, however. They are electrically robust. You need not be as concerned about static charges destroying vacuum tubes. A transmitter with vacuum tube finals usually has a variable matching network built in, and can be loaded into a higher SWR than one with semiconductor finals. Tubes are generally able to withstand the high voltages generated by reflections under high SWR conditions. They are not as easily damaged by short-term overloads or the electromagnetic pulses generated by lightning. The relatively high plate voltages mean that the plate current is lower for a given power output; thus power supplies do not need as high a current handling capability. Vacuum tubes are capable of considerable heat dissipation and many high power applications still use them. Special forms of vacuum tubes are also still used. Most video displays use CRTs, and microwave transmitting tubes are still common.

Bipolar transistors have many advantages over vacuum tubes. When treated properly they can have virtually unlimited life spans. They are relatively small and, if they do not handle high currents, do not generate much heat, improving miniaturization. They make excellent high-frequency amplifiers. Compared to MOSFET devices they are less susceptible to damage from electrostatic discharge. RF amplifiers designed with bipolar transistors in their finals generally include circuitry to protect the transistors from the high voltages generated by reflections under high SWR conditions. Lightning strikes in the area (not direct hits) have been known to destroy all kinds of semiconductors, including bipolar transistors. Semiconductors have replaced almost all small-signal applications of tubes.

There are many performance advantages to FET devices, particularly MOSFETs. The extremely low gate currents allow the design of analog stages with nearly infinite input resistance. Signal distortion due to loading is minimized in this way. As these characteristics are improved by technology, we are seeing an increase in FET design at the expense of bipolar transistors.

The current trend in electronics is portability. Transceivers are decreasing in size and in their power requirements. Integrated circuitry has played a large part in this trend. Extremely large circuits have been designed with microscopic proportions. It is more feasible to use MOSFETs within an IC chip than as discrete components since the devices at risk are usually those that are connected to the outside world. It is not necessary to use electrostatic discharge protection circuitry on the gate of every MOSFET in an IC; only the ones that connect to the pins on the chip need this protection. This arrangement both improves the performance of the internal MOSFETs and decreases the circuit size even further. Semiconductors are slowly replacing the last tube applications. CCD chips have been so successful in video cameras that it is difficult to find an application for vidicon tubes. The liquid crystal displays (LCDs) in laptop computers have given considerable competition to the CRT tube.

An important consideration in the use of analog components is the future availability of parts. At an ever increasing rate, as new components are developed to replace older technology, the older components are discontinued by the manufacturers and become unavailable for future use. This tends to be a fairly long term process but it is not unusual for a manufacturer to stop offering a component when demand for it falls. This has become evident with vacuum tubes, which are becoming more difficult to find and more expensive as fewer manufacturers produce them.

The major disadvantages of IC technology have been power handling capability, frequency response and non-customized circuitry. These characteristics have improved at an amazing pace over recent years; it is a process that feeds itself. As ICs are improved they are used to make more powerful tools (such as computers and electronic test equipment) that are used in the design of further IC improvements. Entire transceivers are designed with just a few IC chips and the appropriate transistors for power amplification. The quiescent current draw of these devices has been reduced to the microampere level so they can operate effectively from small battery packs. The improved noise performance of circuitry has also decreased the need for high transmitter power, further decreasing the current requirements for these devices. If this trend continues, we should eventually see a near total switch to IC components with few discrete semiconductors and no vacuum tubes.

Digital Fundamentals

Radio Amateurs have been involved with digital technology since the first spark transmitters, a form of pulse-coded transmission, were connected to an "aerial." Modern digital technology use by Radio Amateurs probably arrived first in automatic keyers, where hams learned about flip-flops and gates to replace their semi-automatic mechanical keyers (bugs). Amateur use of digital technology echoes public use of these new abilities, starting with using the first home computers for calculations and later as Teletype and data terminals.

The first PC hobbyists, on the west coast, worked on the idea that technology, and in particular software, should be free and available to all. It is not very surprising to learn that many of these PC pioneers were also Radio Amateurs, who were accustomed to seeing new technical ideas freely distributed in the pages of *QST*

The remainder of this chapter, written by Christine Montgomery, KGØGN, with additional material by Paul Danzer, N1II, presents digital-theory fundamentals and some applications of that theory in Amateur Radio. The fundamentals introduce digital mathematics, including number systems, logic devices and simple digital circuits. Next, the implementation of these simple circuits is explored in integrated circuits, their families and interfacing. Integrated circuits continue with memory chips and microprocessors, culminating in a synthesis of these components in the modern digital computer. Where possible, this section mentions Amateur Radio applications associated with the technologies being discussed, as well as pointers to other chapters that discuss such applications in greater depth.

DIGITAL VS. ANALOG

An essential first step in understanding digital theory is to understand the difference between a *digital* and an *analog* signal. An analog value, a real number, has no end; for example, the number 1/3 is 0.333... where the 3 can be repeated forever, or 3/4 equals 0.7500... with infinite repeated 0s. A digital approximation of an analog number breaks the real number line into discrete steps, for example, the integers. This process of approximating a value with discrete steps either truncates or rounds an analog value to some number of decimal places. For example, rounding 1/3 to an integer gives 0 and rounding 3/4 gives 1.

For a simple physical example, look at your wristwatch. A watch with a face — with the hands of the watch rotating in a continuous, smooth motion — is an analog display. Here, the displayed time has a *continuous* range of values, such as from 12:00 exactly to 12:00 and 1/3 second or any values in between. In contrast, a watch with a digital display is limited to *discrete* states. Here the displayed time jumps from 12:00 and 0 seconds to 12:00 and 1 second, without showing the time in between. (A watch with a second hand that jerks from one second to another could also fit the digital analogy.)

In the digital watch example, time is represented by ten distinct states (0, 1, 2, 3, 4, 5, 6, 7, 8 and 9). Digital electronic signals, however, will usually be much more limited in the number of states allowed.

The digital system used is also called the *binary* system, since only two values are allowed. By using coding, as discussed in the following pages, these two binary values can represent any number of real values. **Fig 5.52** illustrates the contrast of an analog signal (in this case a sine wave) and its digital approximation. Three positive and three negative values are shown as an approximation to the sine wave, but any number of coded value steps can be used as an approximation.

While the focus in this chapter will be on digital theory, many circuits and systems involve *both* digital and analog components. Often, a designer may choose between using digital technology, analog technology or a combination.

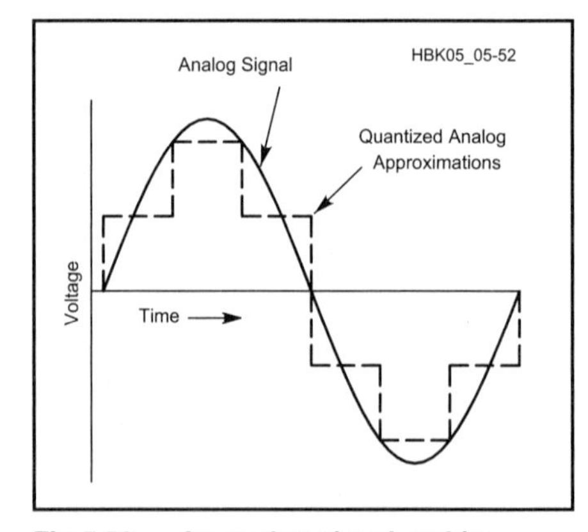

Fig 5.52 — An analog signal and its analog approximation. Note that the analog waveform has continuously varying voltage while the approximated waveform is composed of discrete steps.

Number Systems

In order to understand digital electronics, you must first understand the digital numbering system. Any number system has two distinct characteristics: a set of *symbols* (digits or numerals) and a *base* or radix. A *number* is a collection of these digits, where the left-most digit is the *most significant digit (MSD)* and the right-most digit is the *least significant digit (LSD)*. The value of this number is a weighted sum of its digits. The *weights* are determined by the system's base and the digit's position relative to the decimal point.

While these definitions may seem strange with all the technical terms, they will be more familiar when seen in a decimal system example. This is the "traditional" number system with which we are all familiar.

DECIMAL

The decimal system is a base-10 system, with ten symbols: {0, 1, 2, 3, 4, 5, 6, 7, 8, 9}. To count, we start at 0, and then work our way up to the highest single value allowed — 9. Therefore we count 0, 1, 2,3, 4, 5, 6, 7, 8, 9.

Consider 3 digits, represented by XXX. We start at 000, and then fill up the first (least significant) column, on the right:

XXX
000
001
002
003
...
009

We now reset the first column to the lowest possible value, 0, and increase the second column by 1.

010
011
012
013
...
019

We have again filled up the first column, so again reset it to 0, and increase the second column by one.

020
021
022
023
...
029

We repeat this process, until we hit 099. At this point the second column is filled, so we reset the first two columns to 00 and increase the third column by 1, giving us 100. This is how our familiar decimal or 10-digit number system works; number systems working on other bases work the same way.

Each column in a number has a property called weight. As an example, look at the decimal number, 548. The digits are 5, 4, 8, where 5 is the most significant digit since it is positioned to the far left and 8 is the least significant digit since it is positioned to the far right. The value of this number is a weighted sum of its digits, as shown in **Table 5.1**.

The weight of a position is the system's base raised to a power. In this case, for a decimal system the base is 10, so each position is weighted by 10^P with the power determined by the position relative to the decimal. For example, digit 8, immediately to the left of the decimal, is at position 0; therefore, its weight factor is $10^0 = 1$. Similarly, digit 5 is 2 positions to the left of the decimal and has a weight factor $10^2 = 100$. The value of the number is the sum of each digit times its weight.

Table 5.1
Decimal Numbers
Example: 548

Digit = 5; Weight = 10; Position = 2

548						
	=	$5(10^2)$	+	$4(10^1)$	+	$8(10^0)$
	=	5(100)	+	4(10)	+	8(1)
	=	500	+	40	+	8
	=	5		4		8
		MSD				LSD

BINARY

Binary is a *base-2* number system and therefore limited to two symbols: {0, 1}. The weight factors are now powers of 2, like 2^0, 2^1 and 2^2. For example, the decimal number, 163 and its equivalent binary number, 10100011, are shown in **Table 5.2**.

The digits of a binary number are now *bits* (short for binary digit). The MSD is the *most significant bit (MSB)* and the LSD is the *least significant bit (LSB)*. Four bits make a *nibble* and two nibbles, or eight bits, make a *byte*. A *word* can consist of two or four or more bytes. These groupings are useful when converting to hexadecimal notation, which is explained later.

Counting in binary follows the same pattern illustrated for decimal. Consider the three digit binary number XXX. First fill up the right-hand column.

XXX
000
001

The column has been filed, and much quicker then with decimal, since there are only two values instead of 10. But just like decimal, now reset the right-hand column to 0, increase the next column by 1, and continue.

XXX
000
001
010 ←
011

Table 5.2
Decimal and Binary Number Equivalents

163	=	128	+ 0	+ 32	+ 0	+ 0	+0	+ 2+	1	decimal
	=	1(128)	+ 0(64)	+ 1(32)	+ 0(16)	+ 0(8)	+0(4)	+ 1(2)	+1(1)	
	=	$1(2^7)$	+ $0(2^6)$	+ $1(2^5)$	+ $0(2^4)$	+ $0(2^3)$	$+0(2^2)$	+ $1(2^1)$	$+1(2^0)$	
10100011	=	1	0	1	0	0	0	1	1	binary
		MSB							LSB	
		Nibble				Nibble				
		Byte = 8 digits								

Now the first two columns are full, so reset both back to 0 and increase the next column by 1 and continue:

XXX
000
001
010
011
100←
101
110
111 and so on.

Examination of the set of binary numbers from 0 to 15 shows some important characteristics:

Binary Value	*Decimal Value*
0000	0
0001	1
0010	2
0011	3
0100	4
0101	5
0110	6
0111	7
1000	8
1001	9
1010	10
1011	11
1100	12
1101	13
1110	14
1111	15

Notice each column starts with 0. The first (right-most) column alternates; every other value is a 0 and a 1. The second column alternates every two values, that is there are two 0's followed by two 1's. The third column has groups of four 0's and four 1's, and the fourth column has groups of eight 0's and eight 1's. Thus you can make up a binary counting table by simply following this pattern.

HEXADECIMAL

The hexadecimal, or hex, *base-16* number system is widely used in personal computers for its ease in conversion to and from binary numbers and the fact that it is somewhat more human-friendly than long strings of 1's and 0's. A base-16 number requires 16 symbols. Since our normal mathematical number, as set up in the decimal system, has only 10 digits (0 through 9), a set of additional new symbols is required. Hex uses both numbers and characters in its set of sixteen symbols: {0,1,2,3,4,5,6,7,8,9,A,B,C,D,E,F}. Here, the letters A to F have the decimal equivalents of 10 to 15 respectively: A=10, B=11, C=12, D=13, E=14 and F=15. Again, the weights are powers of the base, such as 16^0, 16^1 and 16^2.

The four-bit binary listing in the previous paragraph shows that the individual 16 hex digits can be represented by a four-bit binary number. Four binary digits are called a *nibble*.

Since a byte is equal to eight binary digits, two hex digits provide a byte — the equivalent of 8 binary digits. Conversion from binary to hex is therefore simplified. Take a binary number, divide it into groups of four binary digits starting from the right, and convert each of the four binary digits to an individual value.

Conversion from hex to binary is equally convenient; replace each hex digit with its four-bit binary equivalent. As an example, the decimal number 163 is shown in Table 5.2 as binary 10100011. Divide the binary number in groups of four, so 1010 is equivalent to decimal 10 or "A" hex, and 0011 is equivalent to decimal 3, thus decimal number 163 is equivalent to hex A3.

BINARY CODED DECIMAL (BCD)

Scientists have experimented with many devices out of a desire for fast computations. Initially, analog computers were developed and used for many applications, especially military applications. It was not unusual to see analog computers aboard US navy ships as recently as the mid-1960s, where they were used to direct naval gunfire.

Analog computers have a very large disadvantage; they could not be readily reprogrammed. They did have a very great advantage; their output was more human readable than digital computers. Very few humans can get used to either binary or hexadecimal read-outs!

The binary number system representation is the most appropriate form for fast internal computations since there is a direct mathematical relationship for every bit in the number. To interface with a user — who usually wants to see I/O in terms of decimal numbers — other codes are more useful. The *Binary Coded Decimal (BCD)* system is the simplest and most widely used form for inputs and outputs of user-oriented digital systems.

In the Binary Coded Decimal (BCD) system, each decimal digit is expressed as a corresponding 4-bit binary number. In other words, the decimal digits 0 to 9 are encoded as the bit strings 0000 to 1001. To make the number easier to read, a space is left between each 4-bit group. For example, the decimal number 163 is equivalent to the BCD number 0001 0110 0011, as shown in **Table 5.3**.

A generic code could use any n-bit string to represent a piece of information. BCD uses 4 bits because that is the minimum needed to represent a 9. All four bits are always written; even a decimal 0 is written as 0000 in BCD.

The important difference between BCD and the previous number systems is that, starting with decimal 10, BCD loses the standard mathematical relationship of a weighted sum. BCD is simply a cut-off hexadecimal. Instead of using the 4-bit code strings 1010 to 1111 for decimal 10 to 15, BCD uses 0001 0000 to 0001 0101. There are other n-bit decimal codes in use and, even for specifically 4 bits, there are millions of combinations to represent the decimal digits 0-9. BCD is the simplest way to convert between decimal and a binary code; thus it is the ideal form for I/O interfacing. The binary number system, since it maintains the mathematical relationship between bits, is the ideal form for the computer's internal computations.

CONVERSION TECHNIQUES

An easy way to convert a number from decimal to another number system is to do repeated division, recording the remainders in a tower just to the right. The converted number, then, is the remainders, reading up the tower. This technique is illustrated in **Table 5.4** for hexadecimal and binary conversions of the decimal number 163.

For example, to convert decimal 163 to hex, repeated divisions by 16 are performed. The first division gives 163/16 = 10 remainder 3. The remainder 3 is written in a column to the right. The second division gives 10/16 = 0 remainder 10. Since 10 decimal = A hex, A is written in the remainder column to the right. This division gave a divisor of 0 so the process is complete. Reading up the remainders column, the result is A3. The most common mistake in this technique is to forget that the Most Significant Digit ends up at the bottom.

Another technique that should be

Table 5.3
Binary Coded Decimal Number Conversion

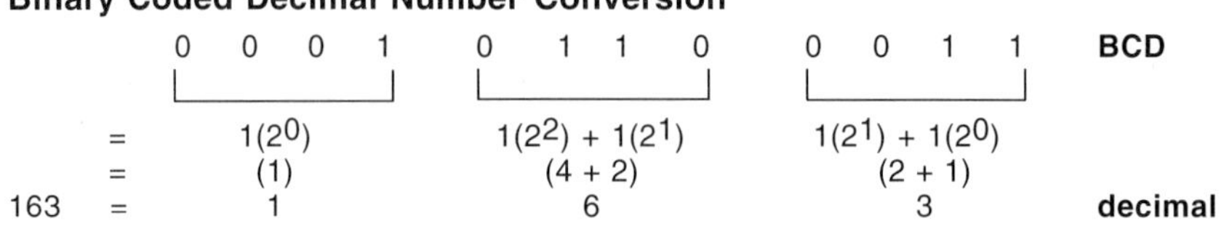

	0 0 0 1	0 1 1 0	0 0 1 1	**BCD**
=	$1(2^0)$	$1(2^2) + 1(2^1)$	$1(2^1) + 1(2^0)$	
=	(1)	(4 + 2)	(2 + 1)	
163 =	1	6	3	**decimal**

briefly mentioned can be even easier: get a calculator with a binary and/or hex mode option. One warning for this technique: this chapter doesn't discuss negative binary numbers. If your calculator does not give you the answer you expected, it may have interpreted the number as negative. This would happen when the number's binary form has a 1 in its MSB, such as the highest (leftmost) bit for the binary mode's default size. To avoid learning about negative binary numbers, always use a leading 0 when you enter a number in binary or hex into your calculator.

Table 5.4
Number System Conversions

Hex		*Remainder*	
16	|163		
	|10	3	LSB
	|0	A	MSB
A3 hex			

Binary		*Remainder*	
2	|163		
	|81	1	LSB
	|40	1	
	|20	0	
	|10	0	
	|5	0	
	|2	1	
	|1	0	
	|0	1	MSB
1010 0011 binary			

Physical Representation of Binary States

STATE LEVELS

Most digital systems use the binary number system because many simple physical systems are most easily described by two state levels (0 and 1). For example, the two states may represent "on" and "off," a punched hole or the absence of a hole in paper tape or a card, or a "mark" and "space" in a communications transmission. In electronic systems, state levels are physically represented by voltages. A typical choice is

state 0 = 0 V
state 1 = 5 V

Since it is unrealistic to obtain these exact voltage values, a more practical choice is a range of values, such as

state 0 = 0.0 to 0.4 V
state 1 = 2.4 to 5.0 V

Fig 5.53 illustrates this representation of states by voltage levels. The undefined region between the two binary states is also known as the *transition region* or *noise margin*.

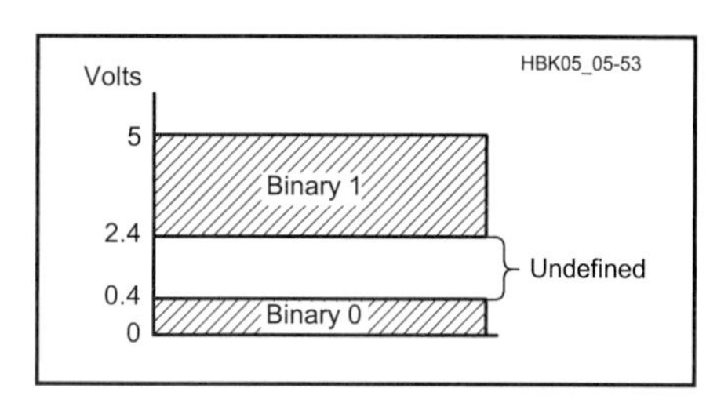

Fig 5.53 — Representation of binary states 1 and 0 by a selected range of voltage levels.

Transition Time

The gap in **Fig 5.53**, between binary 0 and binary 1, shows that a change in state does not occur instantly. There is a *transition time* between states. This transition time is a result of the time it takes to charge or discharge the stray capacitance in wires and other components because voltage cannot change instantaneously across a capacitor. (Stray inductance in the wires also has an effect because the current through an inductor can't change instantaneously.) The transition from a 0 to a 1 state is called the *rise time*, and is usually specified as the time for the pulse to rise from 10% of its final value to 90% of its final value. Similarly, the transition from a 1 to a 0 state is called the *fall time,* with a similar 10% t0 90% definition. Note that these times need not be the same. **Fig 5.54A** shows an ideal signal, or *pulse*, with zero-time switching. Fig 5.54B shows a typical pulse, as it changes between states in a smooth curve.

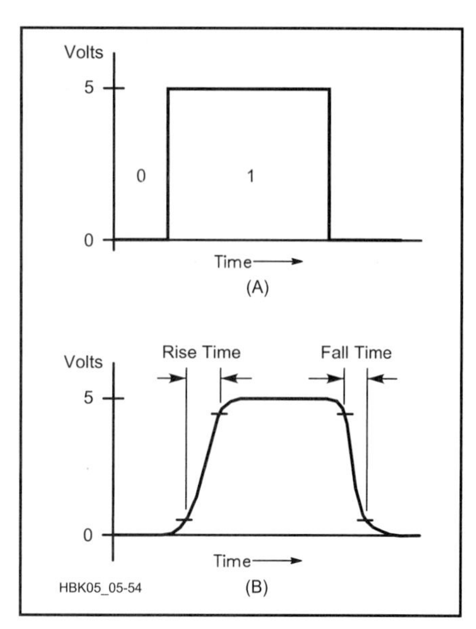

Fig 5.54 — (A) An ideal digital pulse and (B) a typical actual pulse, showing the gradual transition between states.

Rise and fall times vary with the logic family used and the location in a circuit. Typical values of transition time are in the microsecond to nanosecond range. In a circuit, distributed inductances and capacitances in wires or PC-board traces may cause rise and fall times to increase as the pulse moves away from the source.

Propagation Delay

Rise and fall times only describe a relationship within a pulse. For a circuit, a pulse input into the circuit must propagate through the circuit; in other words it must pass through each component in the circuit until eventually it arrives at the circuit output. The time delay between providing an input to a circuit and seeing a response at the output is the *propagation delay* and is illustrated by **Fig 5.55**.

For modern switching logic, typical propagation delay values are in the 1 to 15 nanosecond range. (It is useful to remem-

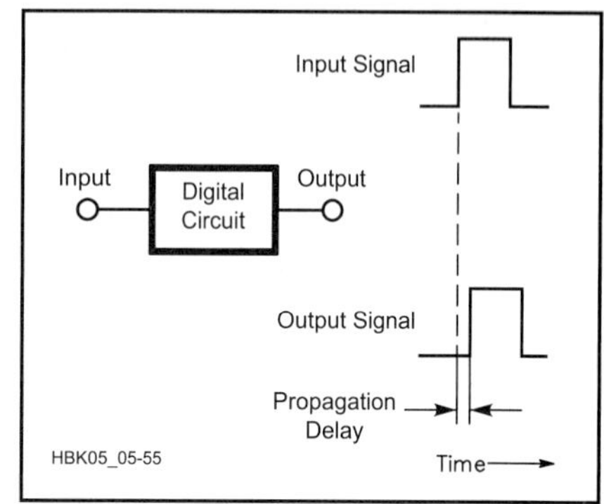

Fig 5.55 — Propagation delay in a digital circuit.

ber that the propagation delay along a wire or printed-circuit-board trace is about 1.0 to 1.5 ns per inch.) Propagation delay is the result of cumulative transition times as well as transistor switching delays, reactive element charging times and the time for signals to travel through wires. In complex circuits, different propagation delays through different paths can cause problems when pulses must arrive somewhere at exactly the same time.

The effect of these delays on digital devices can be seen by looking at the speed of the digital pulses. Most digital devices and all PCs use *clock pulses,* which are one of the more prominently advertised feature of new PCs. A PC having a 2-GHz clock rate translates into each clock pulse, if it is a symmetrical square wave, being at a logic 1 and a logic 0 state for 2.5 ns as a 1 and 2.5 ns as a zero. Therefore if two pulses are supposed to arrive at a logic circuit at the same time, or very close to the same time, the path length for the two signals cannot be any different than two to three inches. This can be a very significant design problem for high-speed logic designs.

Combinational Logic

Having defined a way to use voltage levels to physically represent digital numbers, we can apply digital signal theory to design useful circuits. Digital circuits combine binary inputs to produce a desired binary output or combination of outputs. This simple combination of 0s and 1s can become very powerful, implementing everything from simple switches to powerful computers.

A digital circuit falls into one of two types: combinational logic or sequential logic. In a *combinational logic* circuit, the output depends only on the *present inputs*. (If we ignore propagation delay.) In contrast, in a *sequential logic* circuit, the output depends on the present inputs, the *previous sequence of inputs* and often a clock signal. The next section discusses combinational logic circuits. Later, we will build sequential logic circuits from the basics established here.

BOOLEAN ALGEBRA AND THE BASIC LOGICAL OPERATORS

Combinational circuits are composed of logic gates, which perform binary operations. Logic gates manipulate binary numbers, so you need an understanding of the algebra of binary numbers to understand how logic gates operate. *Boolean algebra* is the mathematical system to describe and design binary digital circuits. It is named after George Boole, the mathematician who developed the system. Standard algebra has a set of basic operations: addition, subtraction, multiplication and division. Similarly, Boolean algebra has a set of basic operations, called *logical operations*: NOT, AND and OR.

The function of these operators can be described by either a Boolean equation or a truth table. A Boolean *equation* describes an operator's function by representing the inputs and the operations performed on them. An equation is of the form "B = A," while an *expression* is of the form "A." In an assignment equation, the inputs and operations appear on the right and the result, or output, is assigned to the variable on the left.

A *truth table* describes an operator's function by listing all possible inputs and the corresponding outputs. Truth tables are sometimes written with Ts and Fs (for true and false) or with their respective equivalents, 1s and 0s. In company databooks (catalogs of logic devices a company manufactures), truth tables are usually written with Hs and Ls (for high and low). In the figures, 1 will mean high and 0 will mean low. This representation is called positive logic. The meaning of different logic types and why they are useful is discussed in a later section.

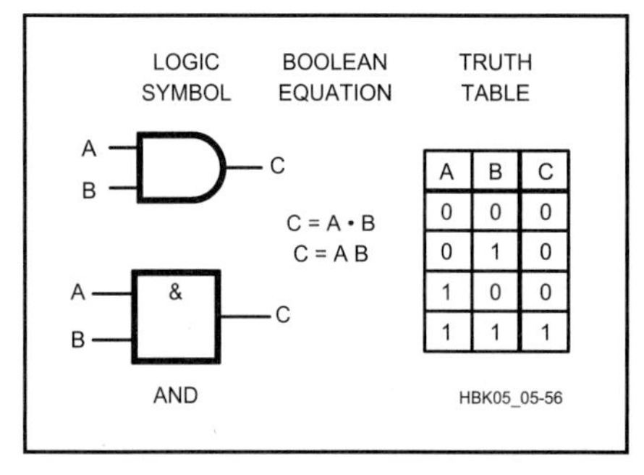

Fig 5.56 — Two-input AND gate.

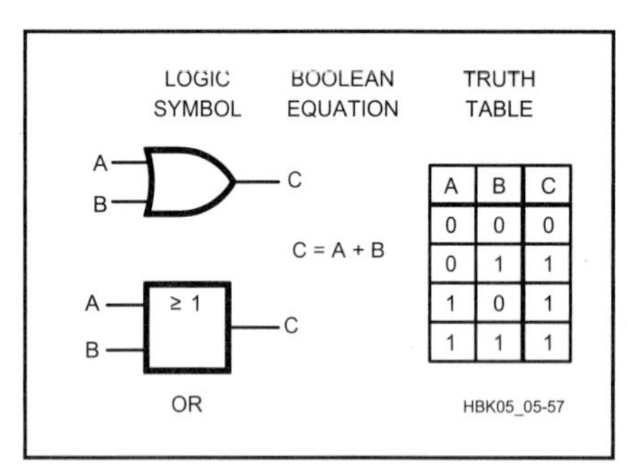

Fig 5.57 — Two-input OR gate.

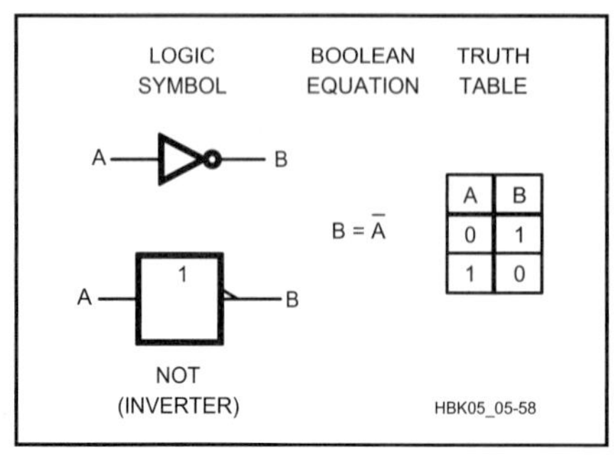

Fig 5.58 — Inverter.

Each Boolean operator also has two circuit symbols associated with it. The traditional symbol — used by ARRL and other US publications — appears on top in each of the figures; for example, the triangle and bubble for the NOT function in **Fig 5.58**. In the traditional symbols, a small circle, or *bubble*, always represents "NOT." (This *bubble* is called a state indicator.) Appearing just below the traditional symbol is the newer ANSI/IEEE Standard symbol. This symbol is always a square box with notations inside it. In these newer symbols, a small flag represents "NOT." The new notation is an attempt to replace the detailed logic drawing of a complex function with a simpler block symbol.

COMMON GATES

Figs **5.56**, **5.57** and 5.58 show the truth tables, Boolean algebra equations and circuit symbols for the three basic Boolean operations: AND, OR and NOT, respectively. All combinational logic functions, no matter how complex, can be described in terms of these three operators. Each truth table can be converted into words. The truth table for the two-input AND gate can be expressed as "the output C is a 1 only when the inputs are both 1's." This can be seen by examining the output column "C" — it remains at a 0 and becomes a 1 only when the input column "A" and the input column "B" are both 1's — the last line of the table.

The NOT operation is also called *inversion, negation* or *complement*. The circuit that implements this function is called an *inverter* or *inverting buffer*. The most common notation for NOT is a bar over a variable or expression. For example, NOT A is denoted $\overline{A}$. This is read as either "Not A" or as "A bar." A less common notation is to denote Not A by A′, which is read as "A prime."

While the inverting buffer and the noninverting buffer covered later have only one input and output, many combinational logic elements can have multiple inputs. When a combinational logic element has two or more inputs and one output, it is called a *gate*. (The term "gate" has many different but specific technical uses. For a clarification of the many definitions of gate, see the section on Synchronicity and Control Signals, later in this chapter.) For simplicity, the figures and truth tables for multiple-input elements will show the operations for only two inputs, the minimum number.

The output of an AND function is 1 only if *all* of the inputs are 1. Therefore, if *any* of the inputs are 0, then the output is 0. The notation for an AND is either a dot (•) between the inputs, as in C = A•B, or nothing between the inputs, as in C = AB. Read these equations as "C equals A AND B."

The OR gate detects if one or more inputs are 1. In other words, if *any* of the inputs are 1, then the output of the OR gate is 1. Since this includes the case where more than one input may be 1, the OR operation is also known as an INCLUSIVE OR. The OR operation detects if *at least one* input is 1. Only if all the inputs are 0, then the output is 0. The notation for an OR is a plus sign (+) between the inputs, as in C = A + B. Read this equation as "C equals A OR B."

Additional Gates

More complex logical functions are derived from combinations of the basic logical operators. These operations — NAND, NOR, XOR and the noninverter or buffer — are illustrated in **Figs 5.59** through **5.62**, respectively. As before, each is described by a truth table, Boolean algebra equation and circuit symbols. Also as before, except for the noninverter, each could have more inputs than the two illustrated.

The NAND gate (short for NOT AND) is equivalent to an AND gate followed by a NOT gate. Thus, its output is the complement of the AND output: The output is a 0 only if all the inputs are 1. If any of the inputs is 0, then the output is a 1.

The NOR gate (short for NOT OR) is equivalent to an OR gate followed by a NOT gate. Thus, its output is the complement of the OR output: If any of the inputs are 1, then the output is a 0. Only if all the inputs are 0, then the output is a 1.

The operations so far enable a designer to determine two general cases: (1) if *all* inputs have a desired state or (2) if *at least one* input has a desired state. The XOR and XNOR gates enable a designer to determine if *one and only one* input of a desired state is present.

The XOR gate (read as EXCLUSIVE OR) has an output of 1 if one and only one of the inputs is a 1 state. The output is 0 otherwise. The symbol for XOR is ⊕. This is easy to remember if you think of the "+" OR symbol enclosed in an "O" for *only one*.

The XOR gate is also known as a "half adder," because in binary arithmetic it does everything but the "carry" operation. The following examples show the possible binary additions for a two-input XOR.

0	0	1	1
0	1	0	1
0	1	1	0

The XNOR gate (read as EXCLUSIVE NOR) is the complement of the XOR gate. The output is 0 if one and only one of the inputs is a 1. The output is 1 either if all inputs are 0 or more than one input is 1.

Buffers

A *noninverter*, also known as a *buffer*, *amplifier* or *driver*, at first glance does not seem to do anything. It simply receives an input and produces the same output. In reality, it is changing other properties of the signal in a useful fashion, such as amplifying the current level. The practical uses of a noninverter include (A) providing sufficient current to drive a number of gates, (B) interfacing between two logic families, (C) obtaining a desired pulse rise time and (D) providing a slight delay to make pulses arrive at the proper time.

Tri-State Gates

Under normal circumstances, a logic element can drive or feed several other logic elements. A typical AND gate might be able to drive or feed 10 other gates. This is known as *fan-out*. However, only one gate output can be connected to a single wire. If you have two possible driving sources to feed one particular wire, some logic network that probably includes a number OR gates must be used. The symbol and truth table for a tri-state gate is in Fig **5.63**.

In each PC data is routed internally on a set of wires called *buses*. A bus consists of a set of wires, and many input logic elements are connected to *listen* on the bus. However the data on the bus can come from many circuits or drivers. To eliminate the need for the network of OR gates to drive each bus wire, as set of gates known as *tri-state* gates are used.

A tri-state gate can be any of the common gates previously described, but with one additional control lead. When this lead is enabled (it can be designed to allow

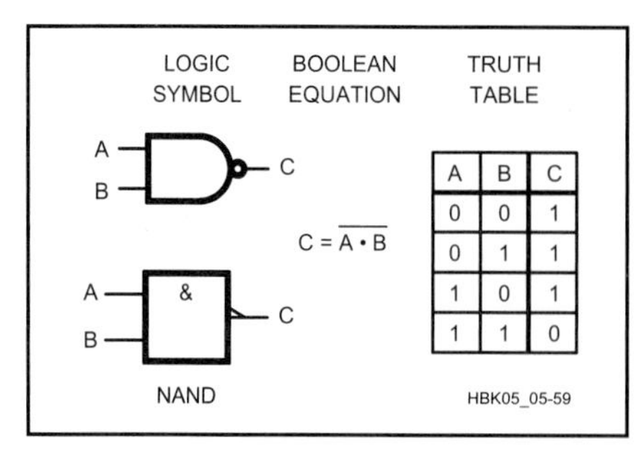

Fig 5.59 — Two-input NAND gate.

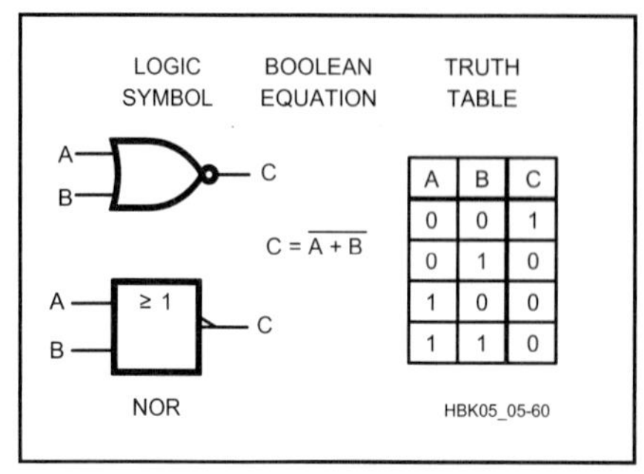

Fig 5.60 — Two-input NOR gate.

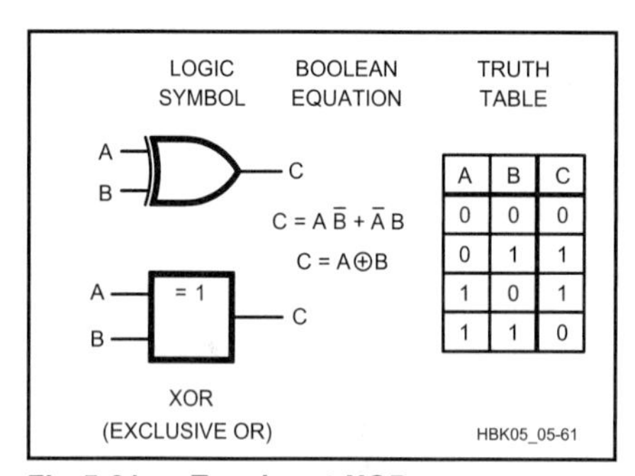

Fig 5.61 — Two-input XOR gate.

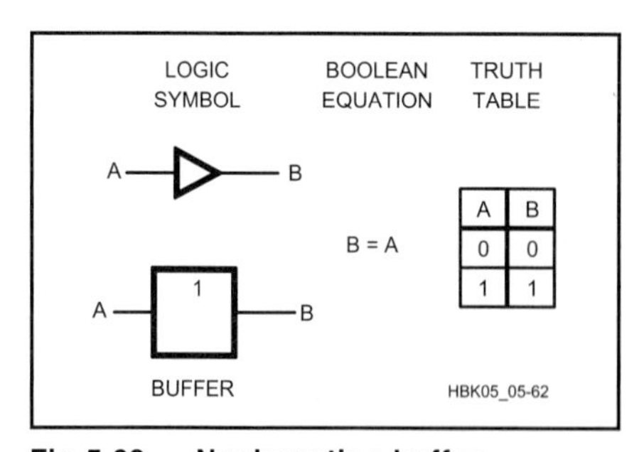

Fig 5.62 — Noninverting buffer.

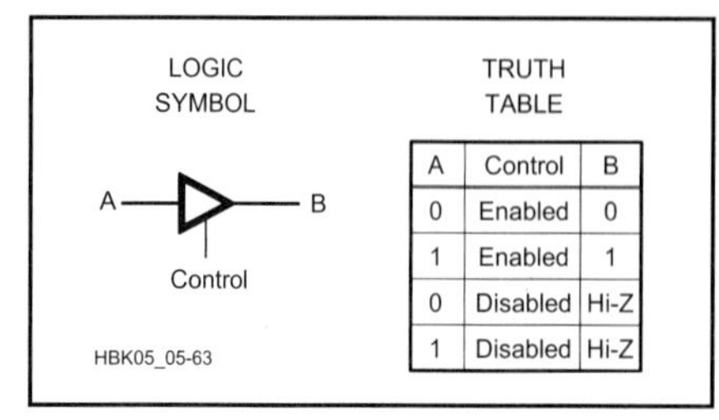

Fig 5.63 — Tri-State Gate.

either a 0 or a 1 to enable it) the gate operates normally, according to the truth table for that type of gate. However, when the gate is not enabled, the output goes to a high impedance, and so far as the output wire is concerned, the gate does not exist.

Each device that has to send data down a bus wire is connected to the bus wire through a tri-state gate. However, as long as only one device, through its tri-state gate, is enabled, it is as though all the other connected tri-state gates do not exist.

BOOLEAN THEOREMS

The analysis of a circuit starts with a logic diagram and then derives a circuit description. In digital circuits, this description is in the form of a truth table or logical equation. The *synthesis*, or design, of a circuit goes in the reverse: starting with an informal description, determining an equation or truth table and then expanding the truth table to components that will implement the desired response. In both of these processes, we need to either simplify or expand a complex logical equation.

To manipulate an equation, we use mathematical *theorems*. Theorems are statements that have been proven to be true. The theorems of Boolean algebra are very similar to those of standard algebra, such as commutivity and associativity. Proofs of the Boolean algebra theorems can be found in an introductory digital design textbook.

Table 5.5

Boolean Algebra Single Variable Theorems

Identities:	$A \bullet 1 = A$	$A + 0 = A$
Null elements:	$A \bullet 0 = 0$	$A + 1 = 1$
Idempotence:	$A \bullet A = A$	$A + A = A$
Complements:	$A \bullet \overline{A} = 0$	$A + \overline{A} = 1$
Involution:	$\overline{(A)} = A$	

BASIC THEOREMS

Table 5.5 lists the theorems for a single variable and **Table 5.6** lists the theorems for two or more variables. These tables illustrate the *principle of duality* exhibited by the Boolean theorems: Each theorem has a dual in which, after swapping all ANDs with ORs and all 1s with 0s, the statement is still true.

Table 5.6

Boolean Algebra Multivariable Theorems

Commutativity:	$A \bullet B = B \bullet A$ $A + B = B + A$
Associativity:	$(A \bullet B) \bullet C = A \bullet (B \bullet C)$ $(A + B) + C = A + (B + C)$
Distributivity:	$(A + B) \bullet (A + C) = A + B \bullet C$ $A \bullet B + A \bullet C = A \bullet (B + C)$
Covering:	$A \bullet (A + B) = A$ $A + A \bullet B = A$
Combining:	$(A + B) \bullet (A + \overline{B}) = A$ $A \bullet B + A \bullet B = A$
Consensus:	$A \bullet B + \overline{A} \bullet C + B \bullet C = A \bullet B + \overline{A} \bullet C$ $(A + B) \bullet (\overline{A} + C) \bullet (B + C) = (A + B) \bullet (\overline{A} + C)$ $A + \overline{A}B = A + B$

The tables also illustrate the *precedence* of the Boolean operations: the order in which operations are performed when not specified by parenthesis. From highest to lowest, the precedence is NOT, AND then OR. For example, the distributive law includes the expression "A + B•C." This is equivalent to "A + (B•C)." The parenthesis around (B•C) can be left out since an AND operation has higher priority than an OR operation. Precedence for Boolean algebra is similar to the convention of standard algebra: raising to a power, then multiplication, then addition.

DeMorgan's Theorem

One of the most useful theorems in Boolean algebra is DeMorgan's Theorem: $\overline{A \bullet B} = \overline{A} + \overline{B}$ and its dual $\overline{A + B} = \overline{A} \bullet \overline{B}$. The truth table in **Table 5.7** proves these statements. DeMorgan's Theorem provides a way to simplify the complement of a large expression. It also enables a designer to interchange a number of equivalent gates, as shown by **Fig 5.64**.

The equivalent gates show that the duality principle works with symbols the same as it does for Boolean equations: just swap ANDs with ORs and switch the bubbles. For example, the NAND gate — an AND gate followed by an inverter bubble — becomes an OR gate preceded by two inverter bubbles. DeMorgan's Theorem is important because it means any logical function can be implemented using either inverters and AND gates or inverters and OR gates. Also, the ability to change placement of the bubbles using DeMorgan's Theorem is useful in dealing with mixed logic, to be discussed next.

POSITIVE AND NEGATIVE LOGIC

The truth tables shown in the figures in this chapter are drawn for positive logic. In *positive logic*, or *high true*, a higher voltage means true (logic 1) while a lower

Table 5.7

DeMorgan's Theorem

(A) $\overline{A \bullet B} = \overline{A} + \overline{B}$

(B) $\overline{A + B} = \overline{A} \bullet \overline{B}$

(C)

(1)	(2)	(3)	(4)	(5)	(6)	(7)	(8)	(9)	(10)
A	B	$\overline{A}$	$\overline{B}$	$A \bullet B$	$\overline{A \bullet B}$	$A + B$	$\overline{A + B}$	$\overline{A} \bullet \overline{B}$	$\overline{A} + \overline{B}$
0	0	1	1	0	1	0	1	1	1
0	1	1	0	0	1	1	0	0	1
1	0	0	1	0	1	1	0	0	1
1	1	0	0	1	0	1	0	0	0

(A) and (B) are statements of DeMorgan's Theorem. The truth table at (C) is proof of these statements: (A) is proven by the equivalence of columns 6 and 10 and (B) by columns 8 and 9.

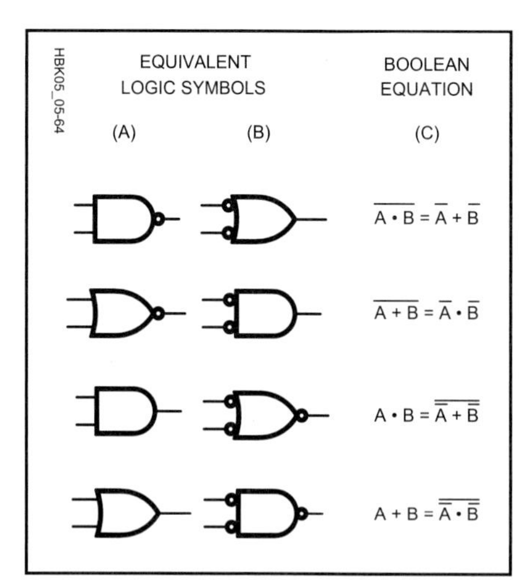

Fig 5.64 — Equivalent gates from DeMorgan's Theorem: Each gate in column A is equivalent to the opposite gate in column B. The Boolean equations in column C formally state the equivalences.

voltage means false (logic 0). This is also referred to as *active high*: a signal performs a named action or denotes a condition when it is "high" or 1. In *negative logic*, or *low true*, a lower voltage means true (1) and a higher voltage means false (0). An *active low* signal performs an action or denotes a condition when it is "low" or 0.

In both logic types, true = 1 and false = 0; but whether true means high or low differs. Company databooks are drawn for general truth tables: an "H" for high and an "L" for low. (Some tables also have an "X" for a "don't care" state.) The function of the table can differ depending on whether it is interpreted for positive logic or negative logic.

Device data sheets often show positive logic convention, or same is assumed. However, a signal into an IC is represented with a bar above it, indicating that the "enable" on that wire is active low — it does ***not*** *mean negative logic (0 volts = a logical 1) is used! Similarly a bubble on the input of a logic element also usually means active low. These can be sources of confusion.*

Fig 5.65 shows how a general truth table differs when interpreted for different logic types. The same truth table gives two equivalent gates: positive logic gives the function of a NAND gate while negative logic gives the function of a NOR gate.

(A)

A	B	C
L	L	H
L	H	H
H	L	H
H	H	L

(B) True = 1 = High

A	B	C
0	0	1
0	1	1
1	0	1
1	1	0

(C) True = 1 = Low

A	B	C
1	1	0
1	0	0
0	1	0
0	0	1

HBK05_05-65

Fig 5.65 — (A) A general truth table, (B) a truth table and NAND symbol for positive logic and (C) a truth table and NOR symbol for negative logic.

Note that these gates correspond to the equivalent gates from DeMorgan's Theorem. A bubble on an input or output terminal indicates an active low device. The absence of bubbles indicates an active high device.

Like the bubbles, signal names can be used to indicate logic states. These names can aid the understanding of a circuit by indicating control of an action (GO, /ENABLE) or detection of a condition (READY, /ERROR). The action or condition occurs when the signal is in its active state. When a signal is in its active state, it is called *asserted*; a signal not in its active state is called *negated* or *deasserted*. A prefix can easily indicate a signal's active state: active low signals are preceded by a "/," like /READY, while active high signals have no prefix. Standard practice is that the signal name and input pin match (have the same active level). For example, an input with a bubble (active low) may be called /READY while an input with no bubble (active high) is called READY. Output signal names should always match the device output pin.

In this chapter, positive logic is used unless indicated otherwise. Although using mixed logic can be confusing, it does have some advantages. Mixed logic combined with DeMorgan's Theorem can promote more effective use of available gates. Also, well-chosen signal names and placement of bubbles can promote more understandable logic diagrams.

Sequential Logic

The previous section discussed combinational logic, whose outputs depend only on the present inputs. In contrast, in *sequential logic* circuits, the new output depends not only on the present inputs but also on the present outputs. The present outputs depended on the previous inputs and outputs and those earlier outputs depended on even earlier inputs and outputs and so on. Thus, the present outputs depend on the previous *sequence of inputs* and the system has *memory*. Having the outputs become part of the new inputs is known as *feedback*.

This section first introduces a number of terms necessary to understand sequential logic: types of synchronicity, types of control signals and ways to illustrate circuit function. Numerous sequential logic circuits are then introduced. These circuits provide an overview of the basic sequential circuits that are commercially available. Depending on your approach to learning, you may choose to either (1) read the material in the order presented, definitions then examples, or (2) start with the example circuits, which begin with the flip-flop, referring back to the definitions as needed.

SYNCHRONICITY AND CONTROL SIGNALS

When a combinational circuit is given a set of inputs, the outputs take on the expected values after a propagation delay during which the inputs travel through the circuit to the output. In a sequential circuit, however, the travel through the circuit is more complicated. After application of the first inputs and one propagation delay, the outputs take on the resulting state; but then the outputs start trickling back through and, after a second propagation delay, new outputs appear. The same happens after a third propagation delay. With propagation delays in the nanosecond range, this cycle around the circuit is rapidly and continually generating new outputs. A user needs to know when the outputs are valid.

There are two types of sequential circuits: synchronous circuits and asynchronous circuits, which are analyzed differently for valid outputs. In *asynchronous* operation, the outputs respond to the inputs immediately after the propagation delay. To work properly, this type of circuit must eventually reach a *stable* state: the inputs and the fed back outputs result in the new outputs staying the same. When the nonfeedback inputs are changed, the feedback cycle needs to eventually reach a new stable state. Generally, the output of this type of logic is not valid until the last input has changed, and enough time has elapsed for all propagation delays to have occurred.

In *synchronous* operation, the outputs change state only at specific times. These times are determined by the presence of a particular input signal: a clock, toggle, latch or enable. Synchronicity is important because it ensures proper timing: all the inputs are present where needed when the control signal causes a change of state.

Some authors vary the meanings slightly for the different control signals. The following is a brief illustration of common uses, as well as showing uses for noun, verb and adjective. *Enabling* a circuit generally means the control signal goes to its asserted level, allowing the circuit to change state. *Latch* implies memory: (noun) a circuit that stores a bit of information or (verb) to hold at the same output state. *Gate* has many meanings, some unrelated to synchronous control: (A) a signal used to trigger the passage of other signals through a circuit (for example, "A gate circuit passes a signal only when a gating pulse is present."), (B) any logic circuit with two or more inputs and one output (used earlier in this chapter) or (C) one of the electrodes of an FET (as described in the analog portion of this chapter). To *toggle* means a signal changes state, from 1 to 0 or vice versa. A *clock* signal is one that toggles at a regular rate.

Clock control is the most common method, so it has some additional terms, illustrated by **Fig 5.66**. The *clock period* is the time between successive transitions in the same direction; the *clock frequency* is the reciprocal of the period. A *pulse* or *clock tick* is the first edge in a clock period, or sometimes the period itself or the first half of the period. The *duty cycle* is the percentage of time that the clock signal is at its asserted level. A common application of the use of clock pulses is to limit the input to a logic circuit such that the circuit is only enabled on one clock phase; that is the inputs occur before the clock changes to a logic 1, The outputs are sampled only after this point; perhaps when the clock next changes back to a logic 0.

The reaction of a synchronous circuit to its control signal is *static* or *dynamic*. Static, *gated* or *level-triggered* control allows the circuit to change state whenever the control signal is at its active or asserted level. Dynamic, or *edge-triggered*, control allows the circuit to change state only when the control signal *changes* from unasserted to asserted. By convention, a control signal is active high if state changes occur when the signal is high or at the rising edge and active low in the opposite case. Thus, for positive logic, the convention is enable = 1 or enable goes from 0 to 1. This transition from 0 to 1 is called

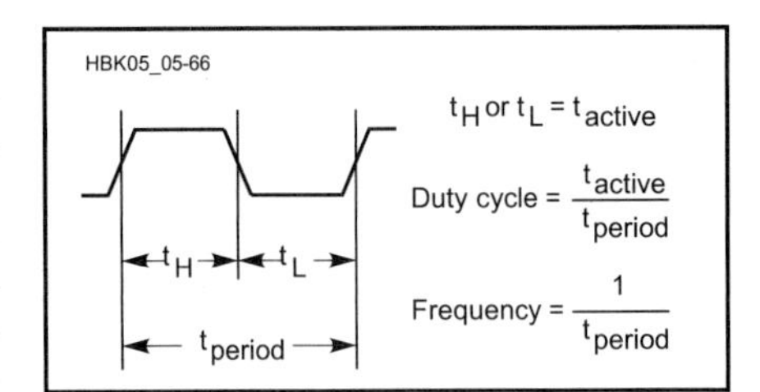

Fig 5.66 — Clock signal terms. The duty cycle would be t_H / t_{PERIOD} for an active high signal and t_L / t_{PERIOD} for an active low signal.

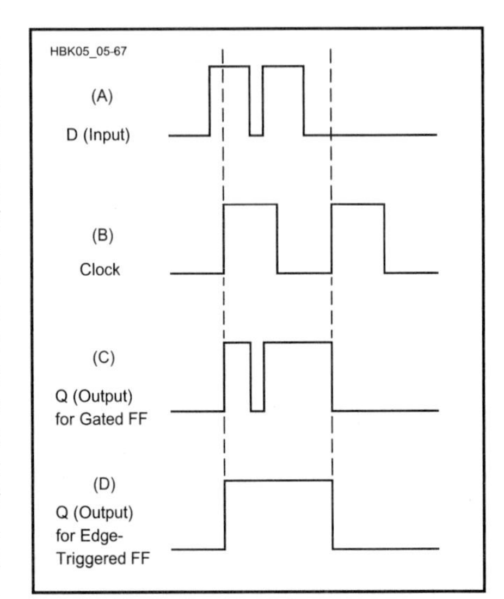

Fig 5.67 — Level-triggered vs edge-triggered for a D flip-flop: (A) input D, (B) clock input, (C) output Q for level-triggered: circuit responds whenever clock is 1. (D) output Q for edge-triggered: circuit responds only at rising edge of clock. Notice that the short negative pulse on the input D is not reproduced by the edge-triggered flip-flop.

positive edge-triggered and is indicated by a small triangle inside the circuit box. A circuit responding to the opposite transition, from 1 to 0, is called *negative edge-triggered*, indicated by a bubble with the triangle. Whether a circuit is level-triggered or edge-triggered can affect its output, as shown by **Fig 5.67**. Input D includes a very brief pulse, called a *glitch*, which may be caused by noise. The differing results at the output illustrate how noise can cause errors.

CIRCUIT DESIGN — FLIP-FLOPS

Flip-flops are the basic building blocks of sequential circuits. A *flip-flop* is a device with two stable states: the *set* state (1) or the *reset* state (0). (The reset state is also called the *cleared* state.) The flip-flop can be placed in one or the other of the two states by applying the appropriate input. (Since a common use of flip-flops is to store one bit of information, some use the term *latch* interchangeably with flip-flop. A set of latches, or flip-flops holding an n-bit number is called a register.) While gates have special symbols, the schematic symbol for most components is a rectangular box with the circuit name or abbreviation, the signal names and assertion bubbles. For flip-flops, the circuit name is usually omitted since the signal names are enough to indicate a flip-flop and its type. The four basic types of flip-flops are the S-R, D, T and J-K. The most common flip-flops available to Amateurs today are the J-K and D- flip-flops; the others can be synthesized by utilizing these two varieties.

Triggering a Flip-Flop

Although the S-R (Set-reset) flip flop is no longer generally available or used, it does provide insight in basic flip-flop operations and triggering. In **Fig 5.68** the symbol for an S-R flip flop and its truth table are accompanied by a logic implementation, using NAND gates. As the truth table shows, this basic implementation requires a positive or logic 1 input on the set input to put the flip-flop in the Q or set state. Remove the input, and the flip flop stays in the Q state, which is what is expected of a flip-flop. Not until the S input receives a logic 1 input does the flip-

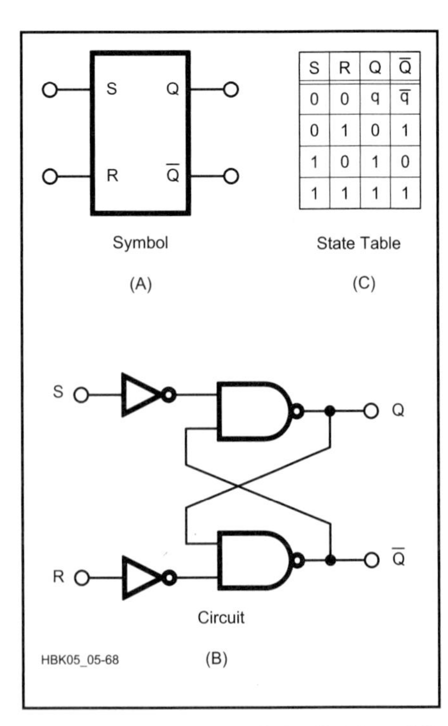

Fig 5.68 — Unclocked S-R Flip-Flop. (A) schematic symbol. (B) circuit diagram. (C) state table or truth table.

flop change state and go to the reset or Q=0 state.

Note that the input can be a short pulse or a level; as long as it is there for some minimum duration, the flip-flop will respond. By contrast the clocked S-R flip-flop in **Fig 5.69** requires both a positive level to be present at either the S or R inputs and a positive clock pulse; the clock pulse is ANDed with the S or R input to trigger the flip–flop. In this case the flip-flop shown is implemented with a set of NOR gates.

A final triggering method is edge triggering. Here, instead of using the clock pulse as shown in the timing diagram of Fig 5.69, just the edge of the clock pulse is used. This will be discussed more in the section on J-K flip-flops. The edge-triggered flip-flop helps solves a problem with noise. Edge-triggering minimizes the time during which a circuit responds to its inputs: the chance of a glitch occurring during the nanosecond transition of a clock pulse is remote. A side benefit of edge-triggering is that only one new output is produced per clock period. Edge-triggering is denoted by a small rising-edge or falling-edge symbol in the clock column of the flip-flop's truth table. It can also appear, instead of the clock triangle, inside the schematic symbol.

Although this description uses positive levels and positive clock pulses, other implementations of these flip-flops can be made with negative levels, negative clock pulses or combinations of positive and negative levels and pulses. This will again be illustrated in the section on J-K flip-flops.

Master/Slave Flip-Flop

One major problem with the simple flip-flop shown up to now is the question of when is there a valid output. Suppose a flip-flop receives input that causes it to change state; at the same time the output of this flip-flop is being sampled to control some other logic element. There is a real risk here that the output will be sampled just as it is changing and thus the validity of the output is questionable.

A solution to this problem is a circuit that samples and stores its inputs before changing its outputs. Such a circuit is built by placing two flip-flops in series; both flip-flops are triggered by a common clock but an inverter on the second flip-flop's clock input causes it to be asserted only when the first flip-flop is not asserted. The action for a given clock pulse is as follows: The first, or master, flip-flop can change only when the clock is high, sampling and storing the inputs. The second, or slave, flip-flop gets its input from the master and changes when the clock is low. Hence, when the clock is 1, the input is sampled; then when the clock becomes 0, the output is generated. Note that a bubble may appear on the schematic symbol's clock input, reminding us that the output appears when the clock is asserted low. This is conventional for TTL-style J-K flip-flops, but it can be different for CMOS devices.

The master/slave method isolates output changes from input changes, eliminating the problem of series-fed circuits. It also ensures only one new output per clock period, since the slave flip-flop responds to only the single sampled input. A problem can still occur, however, because the master flip-flop can change more than once while it is asserted; thus, there is the potential for the master to sample at the wrong time. There is also the potential that either flip-flop can be affected by noise.

A master-slave, S-R clocked input flip-flop synthesized from NAND gates, Fig 5.69B, is accompanied by its logic symbol, Fig 5.69A. From the logic symbols you can tell that the output changes on a negative-going clock edge.

G3A and G3B form the master set-rest flip-flop, and G4A and G4B the slave flip flop. The input signals S and R are controlled by the positive going edge of the clock through gates G1A and G1B. G2A and G2B control the inputs into the slave flip-flop; these inputs are the outputs of the master flip-flop. Note G5 inverts the clock; thus while the positive-going edge places new data into the master flip-flop, the other edge of the clock transfers the output of the master into the slave on the following negative clock edge.

Table 5.8 provides a summary of the two basic flip-flops currently readily available, the D flip-flop and the J-K flip-flop. The S-R (Set-Reset) used for illustration previously and the T or toggle flip-flop can be synthesized by using the J-K or D flip-flops. The T flip-flop is primarily used to count. With proper input, it changes state for each input pulse – alternating between Q = 1 and Q = 0 each time a clock pulse appears.

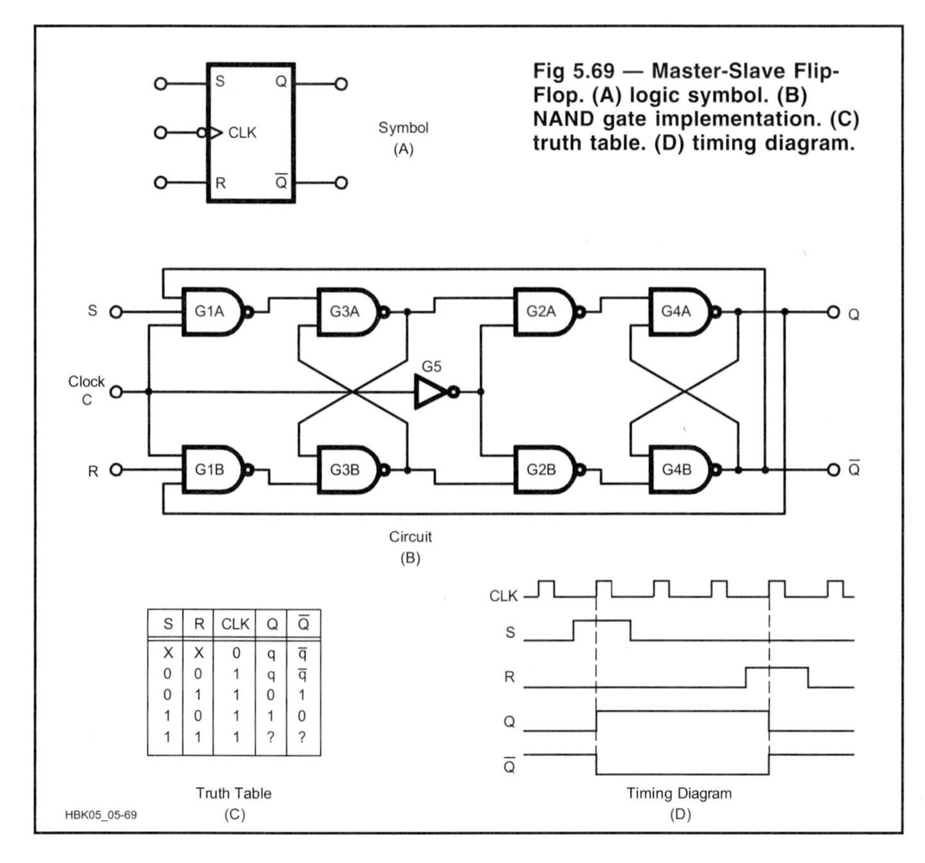

S	R	CLK	Q	Q̄
X	X	0	q	q̄
0	0	1	q	q̄
0	1	1	0	1
1	0	1	1	0
1	1	1	?	?

Fig 5.69 — Master-Slave Flip-Flop. (A) logic symbol. (B) NAND gate implementation. (C) truth table. (D) timing diagram.

D Flip-Flop

In a D (data) flip-flop, the *data* input is transferred to the outputs when the flip-flop is enabled. The logic level at input D is transferred to Q when the clock is positive; the Q output retains this logic level until the next positive clock pulse (see **Fig 5.70**). The truth table summarizes this operation. If D = 1 the next clock pulse makes Q = 1. If D = 0, the next clock pulse makes Q = 0. A D flip-flop is useful to store one bit of information. A collection of D flip-flops forms a register.

Table 5.8

Summary of Standard Flip-Flops

q = current state Q = next state X = don't care

Flip-Flop Type	Symbol	Truth Table	Characteristic Equation	Excitation Table
D	D Q > Q̄	D CLK | Q X ┘ | q 0 ┘ | 0 1 ┘ | 1	Q = D • CLK	q Q | D CLK 0 0 | X ┘ 0 1 | 1 ┘ 1 0 | 0 ┘ 1 1 | X ┘
J K	J Q > K Q̄	J K CLK | Q X X ┘ | q 0 0 ┘ | q 0 1 ┘ | 0 1 0 ┘ | 1 1 1 ┘ | t	$Q = (J \cdot \bar{q} + \bar{K} \cdot q)$ CLK Positive Edge Clock	q Q | J K CLK 0 0 | 0 X ┘ 0 1 | 1 X ┘ 1 0 | X 1 ┘ 1 1 | X 0 ┘
J K	J Q o> K Q̄	J K CLK | Q 0 0 ┐ | 0 0 1 ┐ | 0 1 0 ┐ | 1 1 1 ┐ | t	$Q = (J \cdot \bar{q} + \bar{K} \cdot q)$ CLK Negative Edge Clock	q Q | J K CLK 0 0 | 0 X ┐ 0 1 | 1 X ┐ 1 0 | X 1 ┐ 1 1 | X 0 ┐

HBK05_05-Tab08

t: If J = K, the clock toggles the flip-flop

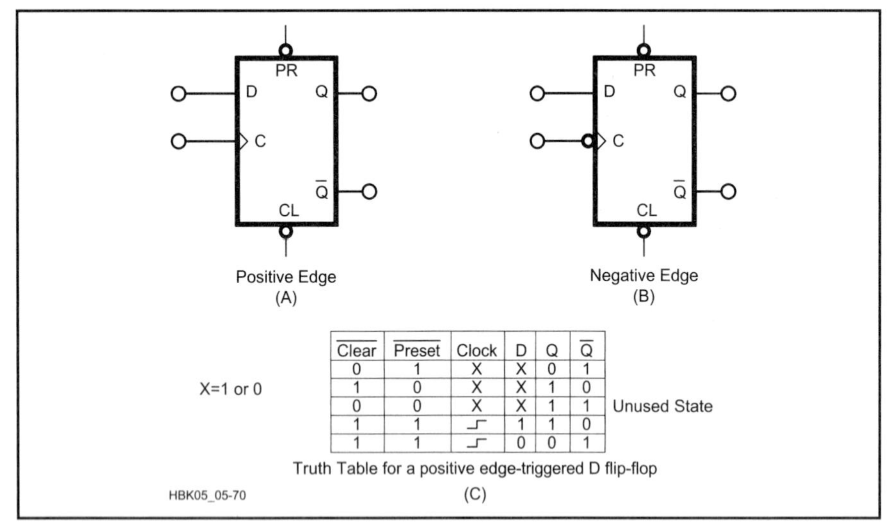

Clear	Preset	Clock	D	Q	Q̄	
0	1	X	X	0	1	
1	0	X	X	1	0	
0	0	X	X	1	1	Unused State
1	1	_┌	1	1	0	
1	1	_┌	0	0	1	

Fig 5.70 — (A & B) The D flip-flop. (C) A truth table for the positive edge-triggered D flip-flop.

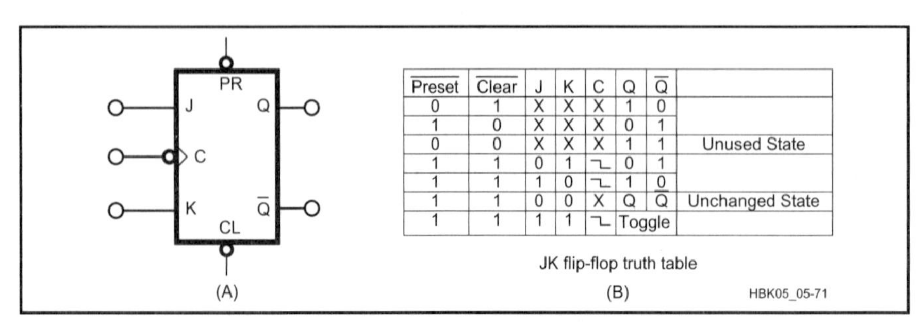

Preset	Clear	J	K	C	Q	Q̄	
0	1	X	X	X	1	0	
1	0	X	X	X	0	1	
0	0	X	X	X	1	1	Unused State
1	1	0	1	┐_	0	1	
1	1	1	0	┐_	1	0	
1	1	0	0	X	Q	Q̄	Unchanged State
1	1	1	1	┐_	Toggle		

Fig 5.71 — (A) JK flip-flop. (B) JK flip-flop truth table.

J-K Flip-Flop

The most readily available flip-flop today is the J-K flip-flop, shown schematically in **Fig 5.71A**. It has five inputs, and the unit shown uses both positive active (the J and K inputs) and negative active inputs (notes the "bubbles" on the C or clock, PR or preset and CL or clear inputs). With these inputs almost any other type of flip-flop may be synthesized.

The truth table of Fig 5.71B provides an explanation. Lines 1 and 2 show the preset and clear inputs and their use. These are active low, meaning that when one (and only one) of them goes to a logic 0, the flip-flop responds, just as if it was a S-R or set-reset flip-flop. Make PR a logic 0, and leave CL a logic 1, and the flip-flop goes into the Q = 1 state (line 1). Do the reverse (line 2) – PR = 1, CL = 0 and the flip-flop goes into a Q' = 1 state. When these two inputs are used, J, K and C are marked as X or don't care, because the PR and CL inputs override them. Line 3 corresponds to the unused state of the R-S flip-flop.

Lines 4 and 5 show that if J = 1 and K = 0, the next clock transition from high to low sets Q = 1 and Q' = 0. Alternately, J = 0 and K = 1 sets Q = 0 and Q' = 1. Therefore if a signal is applied to J, and the inverted signal sent to K, the J-K flip-flop will mimic a D flip-flop, echoing its input.

The most unique feature of the J-K flip-flop is line 7. If both J and K are connected to a 1, then each clock 1 to 0 transition will flip or toggle the flop-flop. Thus the J-K flip-flop can be used as a T flip-flop, as in a ripple counter (see the following COUNTERS section.)

Summary

Only the D and J-K flip-flops are generally available as commercial integrated circuit chips. Since memory and temporary storage are so often desirable, the D flip-flop is manufactured as the simplest way to provide memory. When more functionality is needed, the J-K flip-flop is available. The J-K flip-flop (using its PR and CL inputs) can substitute for an S-R flip-flop and a T flip-flop can be created from either the D or J-K flip-flop.

GROUPS OF FLIP-FLOPS

Counters

Groups of flip-flops can be combined to make counters. Intuitively, a counter is a circuit that starts at state 0 and sequences up through states 1, 2, 3, to m, where m is the maximum number of states available. From state m, the next state will return the counter to 0. This describes the most common counter: the *n-bit binary counter*, with n outputs corresponding to 2^n = m states.

Such a counter can be made from n flip-flops, as shown in **Fig 5.72**. This figure shows implementations for each of the types of synchronicity. Both circuits pass the data count from stage to stage. In the asynchronous counter, Fig 5.72A, the clock is also passed from stage to stage and the circuit is called *ripple* or *ripple-carry*.

The J-K flip-flop truth table shows that with PR (Preset) and CL (Clear) both positive, and therefore not effecting the operation, the flip-flop will toggle if J and K are tied to a logic 1. In Fig 5.72A the first stage has its J and K inputs permanently tied to a logic 1, and each succeeding stage has its J and K inputs tied to Q of the proceeding stage. This provides a direct ripple counter implementation.

Design of a synchronous counter is bit more involved. It consists of determining, for a particular count, the conditions that will make the next stage change at the same clock edge when all the stages are changing.

To illustrate this, notice the binary counting table of Fig 5.72. The right-hand column represents the lowest stage of the counter. It alternates between 1 and 0 on every line. Thus, for the first stage the J and K inputs are tied to logic 1. This provides the alternation required by the counting table.

The middle column or second stage of the counter changes state right after the lower stage is a one (lines C, E and G). Thus if the Q output of the lowest stage is tied to the J and K inputs of the second stage, each time the output of the lowest stage is a 1 the second stage toggles on the next clock pulse.

Finally, the third column (third stage) toggles when both the first stage and the second stage are both 1s (line D). Thus by ANDing the Q outputs of the first two stages, and then connecting them to the J and K inputs of the third stage, the third stage will toggle whenever the first two stages are 1s.

There are formal methods for determining the wiring of synchronous counters. The illustration above is one manual method that may be used to design a counter of this type. The advantage of the synchronous counter is that at any instant, except during clock pulse transition, all counter stage outputs *are correct* and delay due to propagation through the flip flops is not a problem.

In the synchronous counter, Fig 5.72B, each stage is controlled by a common clock signal.

There are numerous variations on this first example of a counter. Most counters have the ability to *clear* the count to 0. Some counters can also *preset* to a desired count. The clear and preset control inputs are often asynchronous — they change the output state without being clocked. Counters may either count up (increment) or down (decrement). *Up/down* counters can be controlled to count in either direction. Counters can have sequences other than the standard numbers, for example a BCD counter.

Counters are also not restricted to changing state on every clock cycle. An n-bit counter that changes state only after m clock pulses is called a *divider* or *divide-by-m* counter. There are still 2^n = m states; however, the output after p clock pulses is now p / m. Combining different divide-by-m counters can result in almost any desired count. For example, a base 12 counter can be made from a divide-by-2 and a divide-by-6 counter; a base 10 (decade) counter consists of a divide-by-2 and a BCD divide-by-5 counter.

The outputs of these counters are binary. To produce output in decimal form, the output of a counter would be provided to a binary-to-decimal decoder chip and/or an LED display.

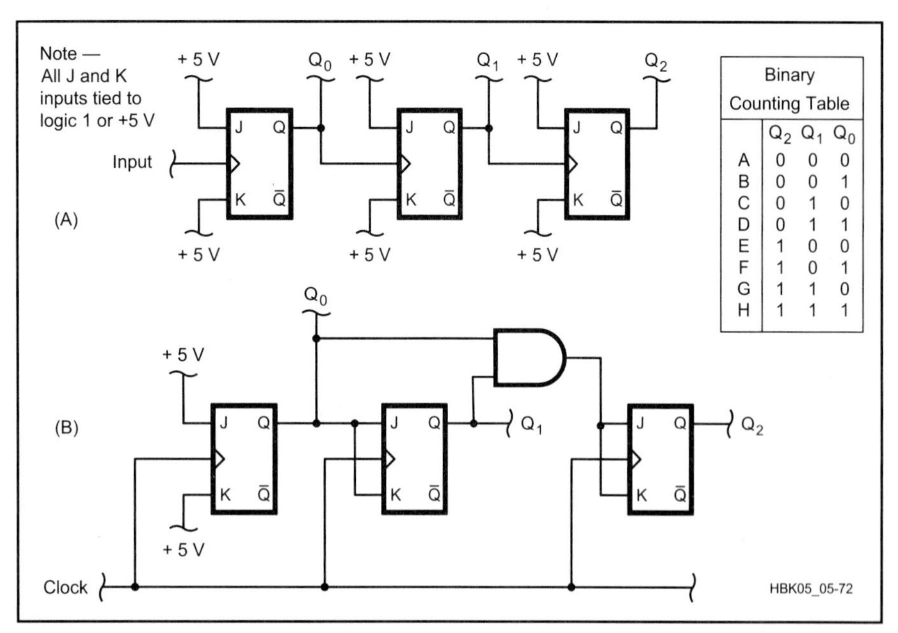

Fig 5.72 — Three-bit binary counter using J-K flip-flops: (A) asynchronous or ripple, (B) synchronous.

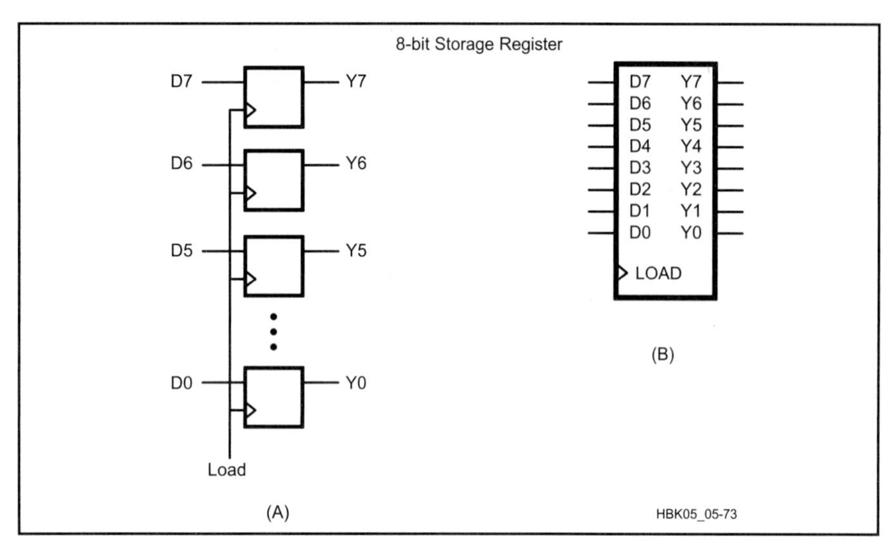

Fig 5.73 — An eight-bit storage register: (A) circuit and (B) schematic symbol.

Registers

Groups of flip-flops can be combined to make registers, usually implemented with D flip-flops. A *register* stores n bits of information, delivering that information in response to a clock pulse. Registers usually have asynchronous *set* to 1 and *clear* to 0 capabilities.

Storage Register

A storage register simply stores temporary information, for example, incoming information or intermediate results. The size is related to the basic size of information handled by a computer: 8 flip-flops for an 8-bit or *byte register* or 16 bits for a *word register*. **Fig 5.73** shows a typical circuit and schematic symbols for an 8-bit storage register. In (C), although the bits are passed on 8 separate lines (from 8 flip-flops), a slash and number, "/8," is used to simplify the symbol. Storage registers are important to computer architecture; this topic is discussed in depth later in the chapter.

Shift Register

Shift registers also store information and provide it in response to a clock signal, but they handle their information differently: When a clock pulse occurs, instead of each flip-flop passing its result to the output, the flip-flops pass their data to each other, up and down the row. For example, in up mode, each flip-flop receives the output of the preceding flip-flop. A data bit starting in flip-flop D0 in a left shifter would move to D1, then D2 and so on until it is shifted out of the register. If a 0 was input to the least significant bit, D0, on each clock pulse then, when the last data bit has been shifted out, the register contains all 0s.

Shift registers can be left shifters, right shifters or controlled to shift in either direction. The most general form, a *universal shift register*, has two control inputs for four states: Hold, Shift right, Shift left and Load. Most also have asynchronous inputs for preset, clear and parallel load. The primary use of shift registers is to convert parallel information to serial or vice versa. This is useful in interfacing between devices, and is discussed in detail in the Interfacing section.

Additional uses for a shift register are to (1) delay or synchronize data, (2) multiply or divide a number by a factor 2^n or (3) provide random data. Data can be delayed simply by taking advantage of the Hold feature of the register control inputs. Multiplication and division with shift registers is best explained by example: Suppose a 4-bit shift register currently has the value 1000 = 8. A right shift results in the new parallel output 0100 = 4 = 8 / 2. A second right shift results in 0010 = 2 = (8 / 2) / 2. Together the 2 right shifts performed a division by 2^2. In general, shifting right n times is equivalent to dividing by 2^n. Similarly, shifting left multiplies by 2^n. This can be useful to compiler writers to make a computer program run faster. Random data is provided via a ring counter. A *ring counter* is a shift register with its output fed back to its input. At each clock pulse, the register is shifted up or down and some of the flip-flops feedback to other flip-flops, generating a random binary number. Shift registers with several feedback paths can be used as a *pseudorandom number generator*, where the sequence of bits output by the generator meets one or more mathematical criteria for randomness.

MULTIVIBRATORS

Multivibrators are a general type of circuit with three varieties: bistable, monostable and astable. The only truly digital multivibrator is bistable, having two stable states. The flip-flop is a *bistable multivibrator*: both of its two states are stable; it can be triggered from one stable state to the other by an external signal. The other two varieties of multivibrators are partly analog circuits and partly digital. While their output is one or more pulses, the internal operation is strictly analog.

Monostable Multivibrator

A *monostable* or *one-shot* multivibrator has one energy-storing element in its feedback paths, resulting in one stable and one quasi-stable state. It can be switched, or *triggered*, to its quasi-stable state; then returns to the stable state after a time delay. Thus, when triggered, the one-shot multivibrator puts out a pulse of some duration, T.

A very common integrated circuit used for non-precision generation of a signal pulse is the 555 timer IC. **Fig 5.74** shows a 555 connected as a one-shot multivibrator. The one-shot is activated by a negative-going pulse between the trigger input and ground. The trigger pulse causes the output (Q) to go positive and capacitor C to charge through resistor R. When the voltage across C reaches two-thirds of V_{CC}, the capacitor is quickly discharged to ground and the output returns to 0. The output remains at logic 1 for a time determined by T = 1.1 RC, where R is the resistance in ohms and C is the capacitance in farads.

A very common, but again, non-precision application of this circuit is the generation a delayed pulse. If there is a requirement to generate a 50-microsecond pulse, but delayed from a trigger by 20 ms, two 555s might be used. The first 555, configured as an astable multivibrator, generates the 20-ms pulse, and the trailing edge of the 20-ms pulse is used to trigger a second 555 that in turn generates the 10 microsecond pulse.

Astable Multivibrator

An *astable* or *free-running* multivibrator has two energy-storing elements in its feedback paths, resulting in two quasi-stable states. It continuously switches between these two states without external excitation. Thus, the astable multivibrator puts out a sequence of pulses. By properly selecting circuit components, these pulses can be of a desired frequency and width.

Fig 5.75 shows a 555 timer IC connected as an astable multivibrator. The

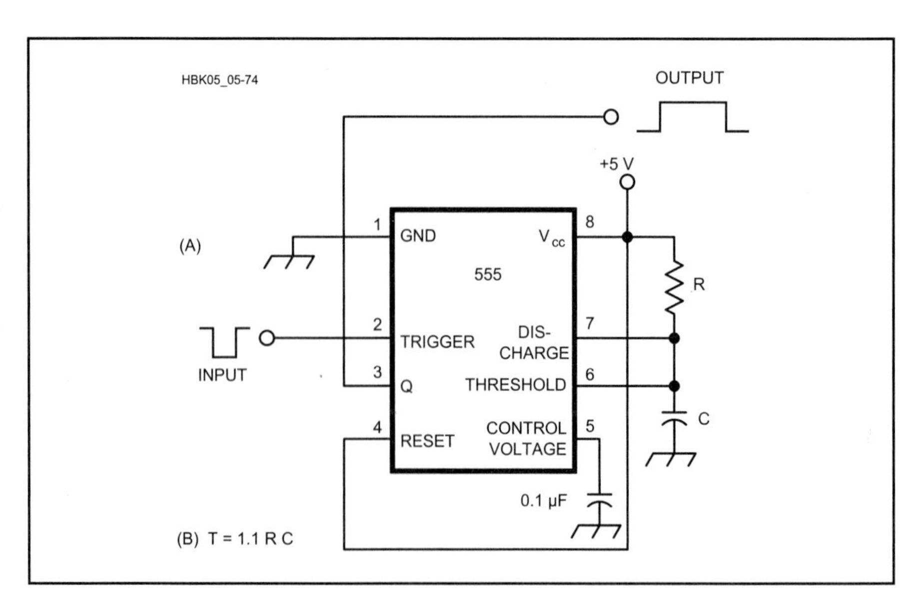

Fig 5.74 — (A) A 555 timer connected as a monostable multivibrator. (B) The equation to calculate values for R in ohms and C in farads, where T is the pulse duration in seconds.

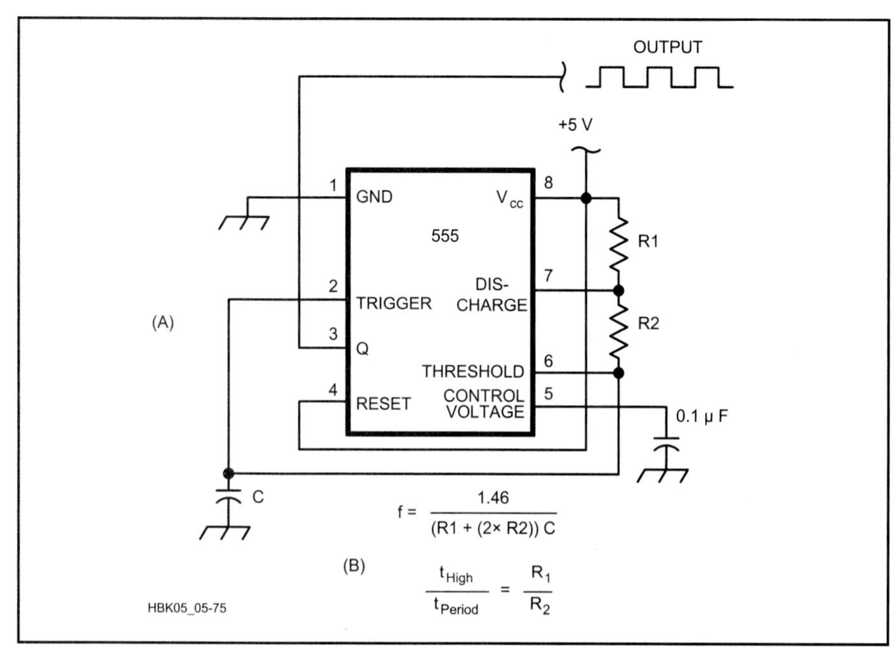

$$f = \frac{1.46}{(R1 + (2 \times R2))\,C}$$

$$\frac{t_{High}}{t_{Period}} = \frac{R_1}{R_2}$$

Fig 5.75 — (A) A 555 timer connected as an astable multivibrator. (B) The equations to calculate values for R1, R2 in ohms and C in farads, where f is the clock frequency in Hertz.

capacitor C charges to two-thirds V_{CC} through R1 and R2 and discharges to one-third V_{CC} through R2. The ratio R1 : R2 sets the asserted high duty cycle of the pulse: t_{HIGH} / t_{PERIOD}. The output frequency is determined by:

$$f = \frac{1.46}{(R1 + 2\,R2)C}$$

where:

R1 and R2 are in ohms,
C is in farads and
f is in hertz.

It may be difficult to produce a 50% duty cycle due to manufacturing tolerance for the resistors R1 and R2. One way to ensure a 50% duty cycle is to run the astable multivibrator at 2f and then divide by 2 with a toggle flip-flop.

Astable multivibrators, and the 555 integrated circuit in particular, are very often used to generate clock pulses. Although this is a very inexpensive and minimum hardware approach, the penalty is stability with temperature. Since the frequency and the pulse dimensions are set by resistors and capacitors, drift with temperature and to some extent aging of components will result in changes with time. However, this is no different than the problem faced by designers of L-C controlled VFOs.

SUMMARY

Digital logic plays an increasingly important role in Amateur Radio. Most of this logic is binary and can be described and designed using Boolean algebra. Using the NOT, AND and OR gates of combinational logic, designers can build sequential logic circuits that have memory and feedback. The simplest sequential logic circuit is called a flip-flop. By using control inputs, a flip-flop can latch a data value, retaining one bit of information and acting as memory. Combinations of flip-flops can form useful circuits such as counters, storage registers and shift registers. While this section discussed discrete logic, synthesized with available integrated circuits, most new commercial designs begin with a gate/flip-flop design and then use automated tools to build the entire system or major parts of it with programmable logic devices (PLDs) or equivalent custom integrated circuits.

Digital Integrated Circuits

Integrated circuits (ICs) are the cornerstone of digital logic devices. Modern technology has enabled electronics to become miniature in size and less expensive. Today's complex digital equipment would be impossible with vacuum tubes or even with discrete transistors.

An IC is a miniature electronic module of components and conductors manufactured as a single unit. All you see is a ceramic or black plastic package and the silver-colored pins sticking out. Inside the package is a piece of material, usually silicon, created (fabricated) in such a way that it conducts an electric current to perform logic functions, such as a gate, flip-flop or decoder.

As each generation of ICs surpassed the previous one, they became classified according to the number of gates on a single chip. These classifications are roughly defined as:

Small-scale integration (SSI):
10 or fewer gates on a chip.
Medium-scale integration (MSI):
10-100 gates.
Large-scale integration (LSI):
100-1000 gates.
Very-large-scale integration (VLSI):
1000 or more gates.

This chapter will primarily deal with SSI ICs, the basic digital building blocks. Microprocessors, memory chips and programmable logic devices are discussed later in the Computer Hardware section of this chapter.

The previous section discussed the design of a digital circuit. To build that circuit, the designer must choose between IC chips available in various logic families. Each family and subfamily has its own desirable characteristics. This section reviews the primary IC logic families of interest to radio amateurs. The designer may also be challenged to interface between different logic families or between a logic device and a peripheral device. The former is discussed at the end of this section; the latter with Computer Hardware, later in the chapter.

COMPARING LOGIC FAMILIES

When selecting devices for a circuit, a designer is faced with choosing between many families and subfamilies of logic ICs. The determination of which logic subfamily is right for a specific application is based upon several desirable characteristics: logic speed, power consumption, fan-out, noise immunity and cost. From a practical view-point, the primary integrated circuit families available from most suppliers today are the TTL and CMOS ICs. Within these families, there are tradeoffs that can be made with respect to individual circuit capabilities, especially in the areas of speed and power consumption. Except under the most demanding circumstances, normal commercial grade temperature rating will do for amateur service. However, at a premium price and with perhaps some problems in availability, military temperature grade equivalent circuits can be selected.

Fan-out

A gate output can supply only a limited amount of current. Therefore, a single output can only drive a limited number of inputs. The measure of driving ability is called fan-out, expressed as the number of inputs (of the same subfamily) that can be driven by a single output. If a logic family that is otherwise desirable does not have sufficient fan-out, consider using noninverting buffers to increase fan-out, as shown by **Fig 5.76**.

Noise Immunity

The noise margin was illustrated in Fig 5.53. The choice of voltage levels for the binary states determines the noise margin. If the gap is too small, a spurious signal can too easily produce the wrong state. Too large a gap, however, produces longer, slower transitions and thus decreased switching speeds.

Circuit impedance also plays a part in noise immunity, particularly if the noise is from external sources such as radio transmitters. At low impedances, more energy is needed to change a given voltage level than at higher impedances.

BIPOLAR LOGIC FAMILIES

Two broad categories of digital logic ICs are *bipolar* and *metal-oxide semiconductor* (MOS). Numerous manufacturing techniques have been developed to fabricate each type. Each surviving, commercially available family has its particular advantages and disadvantages and has found its own special niche in the market.

Bipolar semiconductor ICs usually employ NPN junction transistors. (Bipolar ICs can be manufactured using PNP transistors, but NPN transistors make faster circuits.) While early bipolar logic was faster and had higher power consumption than MOS logic, these distinctions have blurred as manufacturing technology has developed. There are several families of bipolar logic devices, and within some of these families there are subfamilies. The most-used digital logic family is Transistor-Transistor Logic (TTL). Another bipolar logic family, Emitter Coupled Logic (ECL), has exceptionally high speed but high power consumption.

Transistor-Transistor Logic (TTL)

The TTL family has seen widespread acceptance because it is fast and has good noise immunity. It is by far the most commonly used logic family. TTL levels were shown earlier in Fig 5.53: An input voltage between 0.0-0.4 V will represent LOW and an input voltage between 2.4-5.0 V will represent HIGH.

TTL Subfamilies

The original standard TTL is infrequently used today. In the standard TTL circuit, the transistors saturate, reducing the operating speed. TTL variations cure this by clamping the transistors with Schottky diodes to prevent saturation, or by using a dopant in the chip fabrication to reduce transistor recovery time. Schottky-clamped TTL is the faster of these two manufacturing processes.

TTL IC identification numbers begin with either 54 or 74. The 54 prefix denotes a military temperature range of –55 to 125°C, while 74 indicates a commercial temperature range of 0 to 70°C. The next letters, in the middle of the TTL device number, indicate the TTL subfamily. Following the subfamily designation is a 2, 3 or 4-digit device-identification number. For example, a 7400 is a standard TTL NAND gate and a 74LS00 is a low-power Schottky NAND gate. (The NAND gate is the workhorse TTL chip. Recall, from Fig 5.68, the alternative implementation of the S-R flip-flop.) The following TTL subfamilies are available:

	74xx	standard TTL
H	74Hxx	High-speed
L	74Lxx	Low-power
S	74Sxx	Schottky
LS	74LSxx	Low-power Schottky
AS	74ASxx	Advanced Schottky
ALS	74ALSxx	Advanced Low-power Schottky

Each subfamily is a compromise between speed and power consumption. Because the speed-power product is approximately constant, less power consumption results in less speed and vice versa. For the amateur, an additional consideration to the speed-versus-power trade-off is the cost trade-off. The advanced Schottky devices offer both increased speed and reduced power consumption but at a higher cost.

In addition to the above power/speed/cost trade-offs, each TTL subfamily has particular characteristics that can make it suitable or unsuitable for a specific design. **Table 5.9** shows some of these parameters. The actual parameter values may vary slightly from manufacturer to manu-

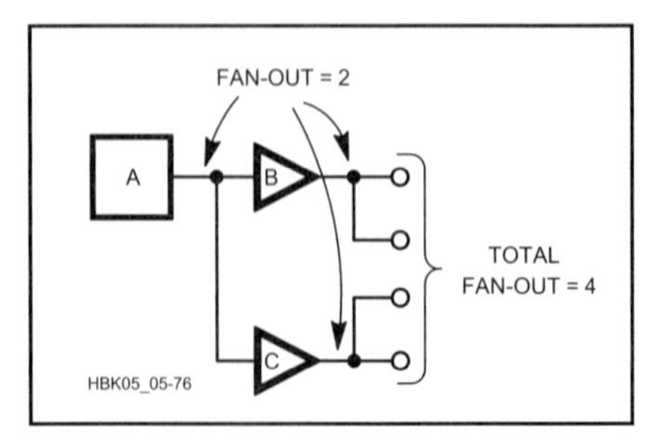

Fig 5.76 — Nonverting buffers used to increase fan-out: Gate A (fan-out = 2) is connected to two buffers, B and C, each with a fan-out of 2. Result is a total fan-out of 4.

Table 5.9

TTL and CMOS Subfamily Performance Characteristics

TTL Family	*Propagation Delay (ns)*	*Per Gate Power Consumption (mW)*	*Speed Power Product (pico-joules)*
Standard	9	10	90
L	33	1	33
H	6	22	132
S	3	20	60
LS	9	2	18
AS	1.6	20	32
ALS	5	1.3	6.5

CMOS Family Operating with $4.5 < V_{CC} < 5.5$ V		*f=100 kHz*	*f=1 MHz*	*f=10 MHz*	*f=100 kHz*	*f=1 MHz*	*f=10 MHz*
HC	18	0.0625	0.6025	6.0025	1.1	10.8	108
HCT	18	0.0625	0.6025	6.0025	1.1	10.8	108
AC	5.25	0.080	0.755	7.505	0.4	3.9	39
ACT	4.75	0.080	0.755	7.505	0.4	3.6	36

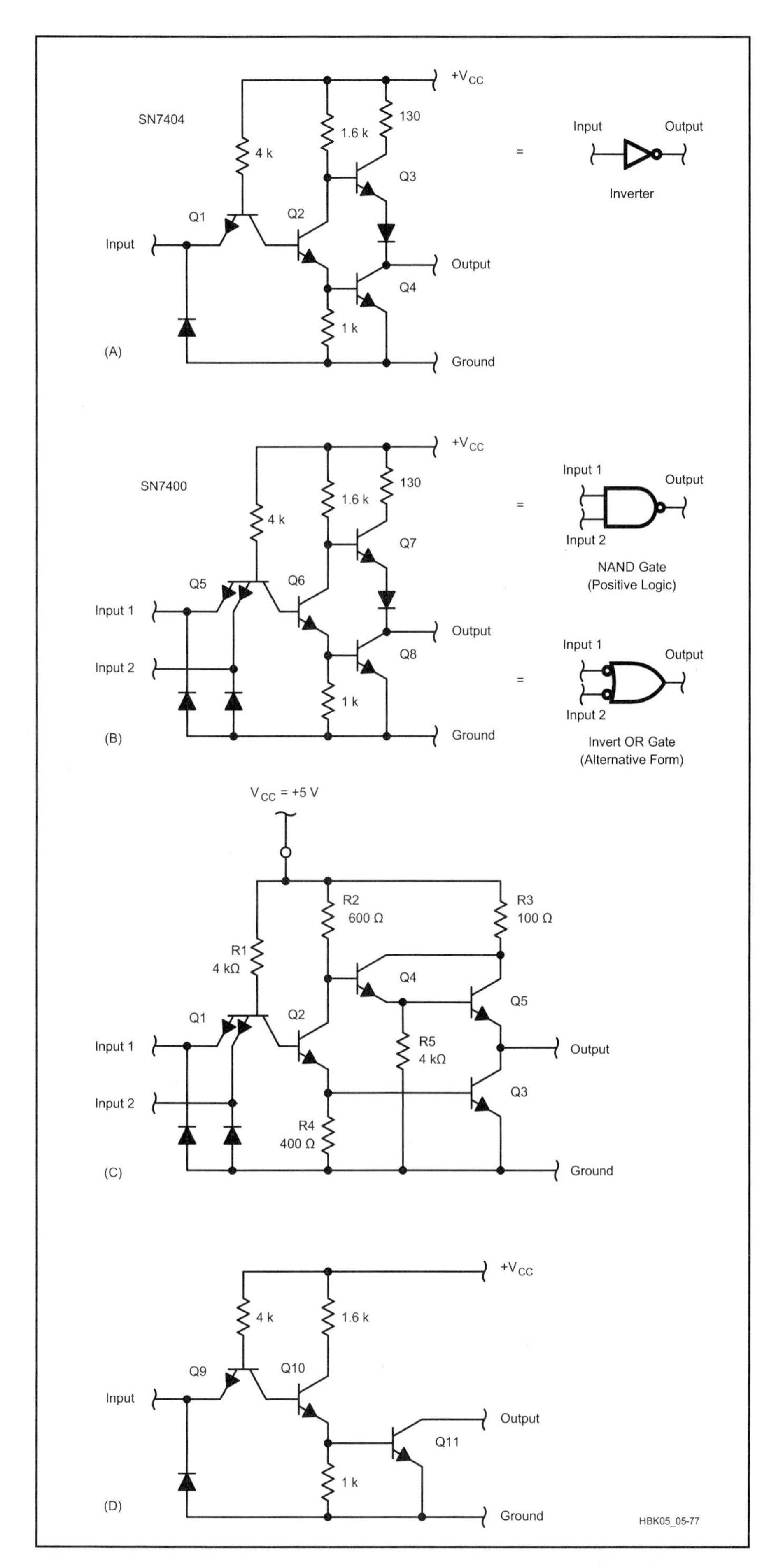

Fig 5.77 — Example TTL circuits and their equivalent logic symbols: (A) an inverter and (B) a NAND gate, both with totem-pole outputs. (C) A NAND gate with a Darlington output. (D) A NAND gate with an open-collector output. (Indicated resistor values are typical. Identification of transistors is for text reference only. These are not discrete components but parts of the silicon die.)

facturer, so always consult the manufacturers' data books for complete information.

TTL Circuits

Fig 5.77A shows the schematic representation of a TTL hex inverter. A 7404 chip contains four of these inverters. When the input is low, Q1 is ON, conducting current from base to emitter through the input lead and into ground. Thus, a low TTL input device must be prepared to sink current from the input. Since Q1 is saturated, Q2 is OFF because there is not enough voltage at its base. Similarly, Q4 is also OFF. With Q2 and Q4 OFF, Q3 will be ON and pull the output high, about one volt below V_{CC}. When the input is high, an unusual situation occurs: Q1 is operating in the inverse mode, with current flowing from base to collector. This current causes Q2 to be ON, which causes Q4 to be ON. With Q2 and Q4 ON, there is not enough current left for Q3, so Q3 is OFF. Q4 is pulling the output low.

By replacing Q1 with a multiple-emitter transistor, as is done with the two-input Q5 in Fig 5.77B, the inverter circuit becomes a NAND gate. Commercially available TTL NAND gates have as many as 13 inputs, the limiting factor being the number of input pins on the standard 16-pin chip. The operation of this multiple-input NAND circuit is the same as described for the inverter, the difference being that any one of the emitter inputs being low will conduct current through the emitter, leading to the conditions described above to produce a high at the output. Similarly, all inputs must be high to produce the low output.

In the TTL circuit of Fig 5.77A, transistors Q3 and Q4 are arranged in a *totem-pole* configuration. This configuration gives the output circuit a low source impedance, allowing the gate to source (supply) or sink substantial output current. The 130-Ω resistor between the collector of Q3 and $+V_{CC}$ limits the current through Q3.

When a TTL gate changes state, the amount of current that it draws changes rapidly. These changes in current, called switching transients, appear on the power supply line and can cause false triggering of other devices. For this reason, the power bus should be adequately decoupled. For proper decoupling, connect a 0.01 to 0.1 µF capacitor from V_{CC} to ground near each device to minimize the transient currents caused by device switching and magnetic coupling. These capacitors must be low-inductance, high-frequency RF capacitors (disk-ceramic capacitors are preferred). In addition, a large-value (50 to 100 µF) capacitor should be connected from V_{CC} to ground somewhere on the board to accommodate the continually changing I_{CC} requirements of the total V_{CC} bus line. These are generally low-inductance tantalum capacitors rather than rolled-foil Mylar or aluminum-electrolytic capacitors.

Darlington and Open-Collector Outputs

Fig 5.77C and D show variations from the totem-pole configuration. They are the Darlington transistor pair and the open-collector configuration respectively.

The Darlington pair configuration replaces the single transistor Q4 with two transistors, Q4 and Q5. The effect is to provide more current-sourcing capability in the high state. This has two benefits: (1) the rise time is decreased and (2) the fan-out is increased.

Transistor(s) on the output in both the totem-pole and Darlington configurations provide active pull-up. Omitting the transistor(s) and providing an external resistor for passive pull-up gives the open-collector configuration. This configuration, unfortunately, results in slower rise time, since a relatively large external resistor must be used. The technique has some very useful applications, however: driving other devices, performing wired logic, busing and interfacing between logic devices.

Devices that need other than a 5-V supply can be driven with the open-collector output by substituting the device for the external resistor. Example devices include light-emitting diodes (LEDs), relays and solenoids. Inductive devices like relay coils and solenoids need a "flyback" protection diode across the coil. You must pay attention to the current ratings of open-collector outputs in such applications. You may need a switching transistor to drive some relays or other high-current loads.

Open-collector outputs can perform wired logic, rather than gated IC logic, by wire-ANDing the outputs. This can save the designer an AND gate, potentially simplifying the design. Wire-ANDed outputs are several open-collector outputs connected to a single external pull-up resistor. The overall output, then, will only be high when all pull-down transistors are OFF (all connected outputs are high), effectively performing an AND of the connected outputs. If any of the connected outputs are low, the output after the external resistor will be low. **Fig 5.78** illustrates the wire-ANDing of open-collector outputs.

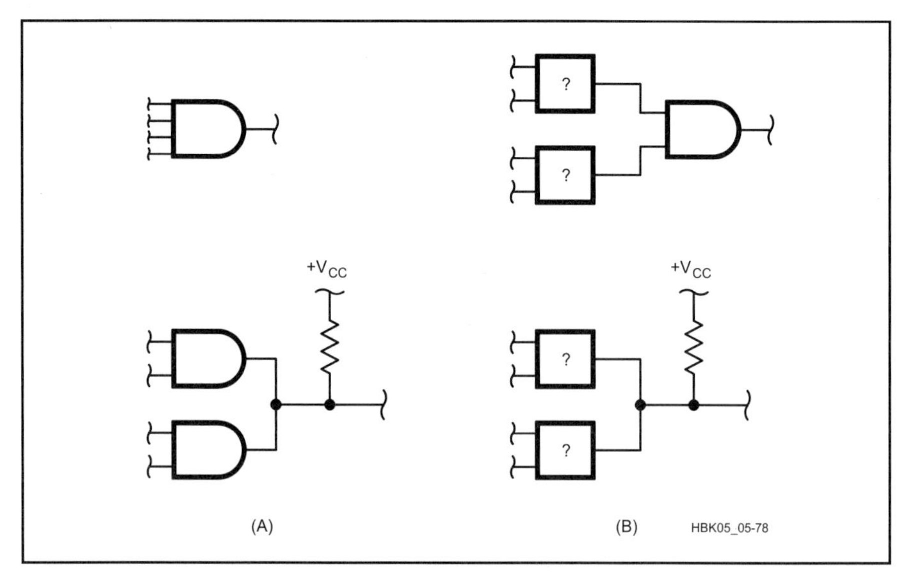

Fig 5.78 — The outputs of two open-collector-output AND gates are shorted together (wire ANDed) to produce an output the same as would be obtained from a 4-input AND gate.

The wire-ANDed concept can be applied to several devices sharing a common bus. At any time, all but one device has a high-impedance (off) output. The remaining device, enabled with control circuitry, drives the bus output.

Open-collector outputs are also useful for interfacing TTL gates to gates from other logic families. TTL outputs have a minimum high level of 2.4 V and a maximum low level of 0.4 V. When driving non-TTL circuits, a pull-up resistor (typically 2.2 kΩ) connected to the positive supply can raise the high level to 5 V. If a higher output voltage is needed, a pull-up resistor on an open-collector output can be connected to a positive supply greater than 5 V, so long as the chip output voltage and current maximums are not exceeded.

Three-State Outputs

While open-collector outputs can perform bus sharing, a more popular method is three-state output, or tristate, devices. The three states are low, high and high impedance, also called Hi-Z or *floating*. An output in the high-impedance state behaves as if it is disconnected from the circuit, except for possibly a small leakage current. Three-state devices have an additional disable input. When enable is low, the device provides high and low outputs just as it would normally; when enable is high, the device goes into its high-impedance state.

A bus is a common set of wires, usually used for data transfer. A three-state bus has several three-state outputs wired together. With control circuitry, all devices on the bus but one have outputs in the high-impedance state. The remaining device is enabled, driving the bus with high and low outputs. Care should be taken to ensure only one of the output devices can be enabled at any time, since simultaneously connected high and low outputs may result in an incorrect logic voltage. (The condition when more than one driver is enabled at the same time is called *bus contention.*) Also, the large current drain from V_{CC} to ground through the high driver to the low driver can potentially damage the circuit or produce noise pulses that can affect overall system behavior.

Unused TTL Inputs

A design may result in the need for an n-input gate when only an n + m input gate is available. In this case, the recommended solution for extraneous inputs is to give the extra inputs a constant value that won't affect the output. A low input is easily provided by connecting the input to ground. A high input can be provided with either an inverter whose input is ground or with a pull-up resistor. The pull-up resistor is preferred rather than a direct connection to power because the resistor limits the current, thus protecting the circuit from transient voltages. Usually, a 1-kΩ to 5-kΩ resistor is used; a single

1-kΩ resistor can handle up to 10 inputs.

It's important to properly handle all inputs. Design analysis would show that an unconnected, or floating TTL input is usually high but can easily be changed low by only a small amount of capacitively-coupled noise.

METAL-OXIDE SEMICONDUCTOR (MOS) LOGIC FAMILIES

While bipolar devices use junction transistors, MOS devices use field effect transistors (FETs). MOS is characterized by simple device structure, small size (high density) and ease of fabrication. MOS circuits use the NOR gate as the workhorse chip rather than the NAND. MOS families are used extensively in digital watches, calculators and VLSI circuits such as microprocessors and memories.

P-Channel MOS (PMOS)

The first MOS devices to be fabricated were PMOS, conducting electrical current by the flow of positive charges (holes). PMOS power consumption is much lower than that of bipolar logic, but its operating speed is also lower. The only extensive use of PMOS is in calculators and watches, where low speed is acceptable and low power consumption and low cost are desirable.

N-Channel MOS (NMOS)

With improved fabrication technology, NMOS became feasible and provided improved performance and TTL compatibility. The speed of NMOS is at least twice that of PMOS, since electrons rather than holes carry the current. NMOS also has greater gain than PMOS and supports greater packaging density through the use of smaller transistors.

Complementary MOS (CMOS)

CMOS combines both P-channel and N-channel devices on the same substrate to achieve high noise immunity and low power consumption: less than 1 mW per gate and negligible power during standby. This accounts for the widespread use of CMOS in battery-operated equipment. The high impedance of CMOS gates makes them susceptible to electromagnetic interference, however, particularly if long traces are involved. Consider a trace ¼-wavelength long between input and output. The output is a low-impedance point, hence the trace is effectively grounded at this point. You can get high RF potentials ¼-wavelength away, which disturbs circuit operation.

A notable feature of CMOS devices is that the logic levels swing to within a few millivolts of the supply voltages. The input-switching threshold is approximately one half the supply voltage ($V_{DD} - V_{SS}$). This characteristic contributes to high noise immunity on the input signal or power supply lines. CMOS input-current drive requirements are minuscule, so the fan-out is great, at least in low-speed systems. For high-speed systems, the input capacitance increases the dynamic power dissipation and limits the fan-out.

CMOS Subfamilies

There are a number of CMOS subfamilies available. Like TTL, the original CMOS has largely been replaced by later subfamilies using improved technologies. The original family, called the 4000-series, has numbers beginning with 40 or 45 followed by two or three numbers to indicate the specific device. 4000B is second generation CMOS. When introduced, this family offered low power consumption but was fairly slow and not easy to interface with TTL.

Later CMOS subfamilies provided improved performance and TTL compatibility. For simplicity, the later subfamilies were given numbers similar to the TTL numbering system, with the same leading numbers, 54 or 74, followed by 1 to 3 letters indicating the subfamily and as many as 5 numbers indicating the specific device. The subfamily letters usually include a "C" to distinguish them as CMOS.

The following CMOS device families are available:

4000	4071B	standard CMOS
C	74Cxx	CMOS versions of TTL

Devices in the 74C subfamily are pin and functional equivalents of many of the most popular parts in the 7400 TTL family. It may be possible to replace all TTL ICs in a particular circuit with 74C-series CMOS, but this family should not be mixed with TTL in a circuit without careful design considerations. Devices in the C series are typically 50% faster than the 4000 series.

HC	74HCxx	High-speed CMOS

Devices in the 74HC subfamily have speed and drive capabilities similar to Low-power Schottky (LS) TTL but with better noise immunity and greatly reduced power consumption. High-speed refers to faster than the previous CMOS family, the 4000-series.

HCT	74HCTxx	High-Speed CMOS, TTL compatible

Devices in this subfamily were designed to interface TTL to CMOS systems. The HCT inputs recognize TTL levels, while the outputs are CMOS compatible.

AC	74ACxxxxx	Advanced CMOS

Devices in this family have reduced propagation delays, increased drive capabilities and can operate at higher speeds than standard CMOS. They are comparable to Advanced Low-power Schottky (ALS) TTL devices.

ACT	74ACTxxxxx	Advanced CMOS, TTL compatible

This subfamily combines the improved performance of the AC series with TTL-compatible inputs.

As with TTL, each CMOS subfamily has characteristics that make it suitable or unsuitable for a particular design. You should consult the manufacturer's data books for complete information on each subfamily being considered.

CMOS Circuits

A simplified diagram of a CMOS logic inverter is shown in **Fig 5.79**. When the

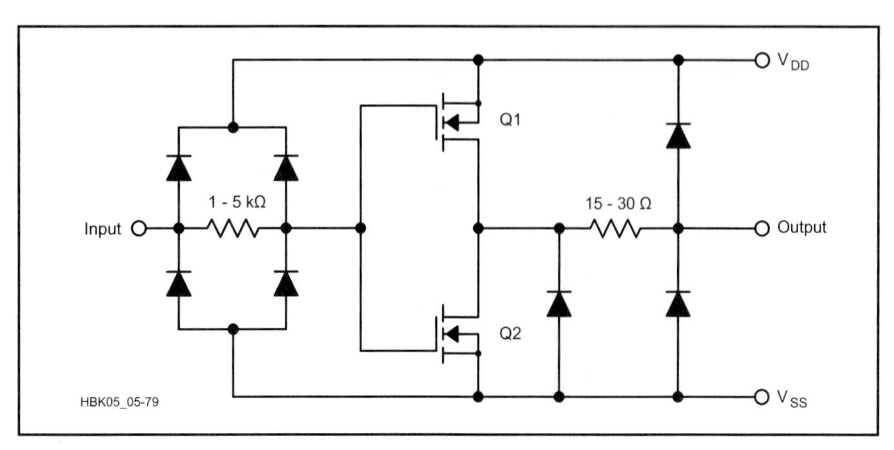

Fig 5.79 — Internal structure of a CMOS inverter.

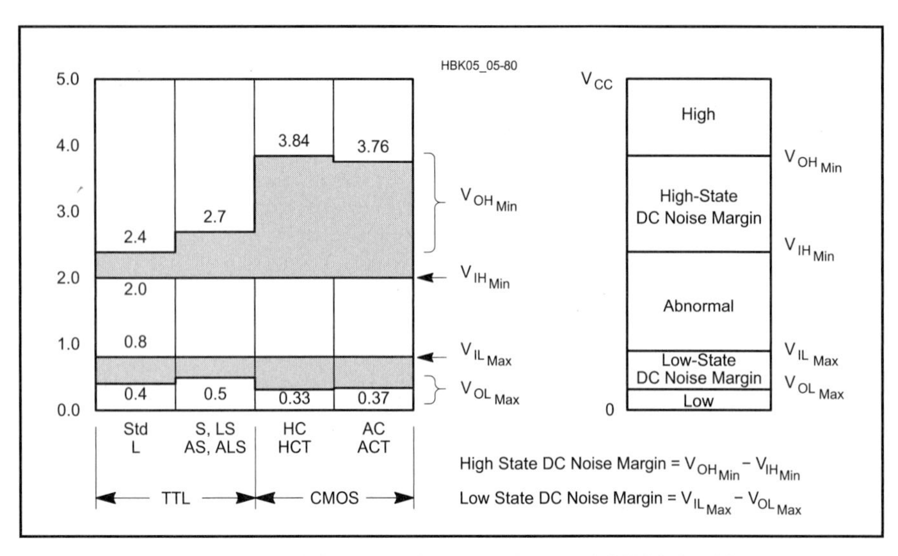

Fig 5.80 — Differences in logic levels for some TTL and CMOS families.

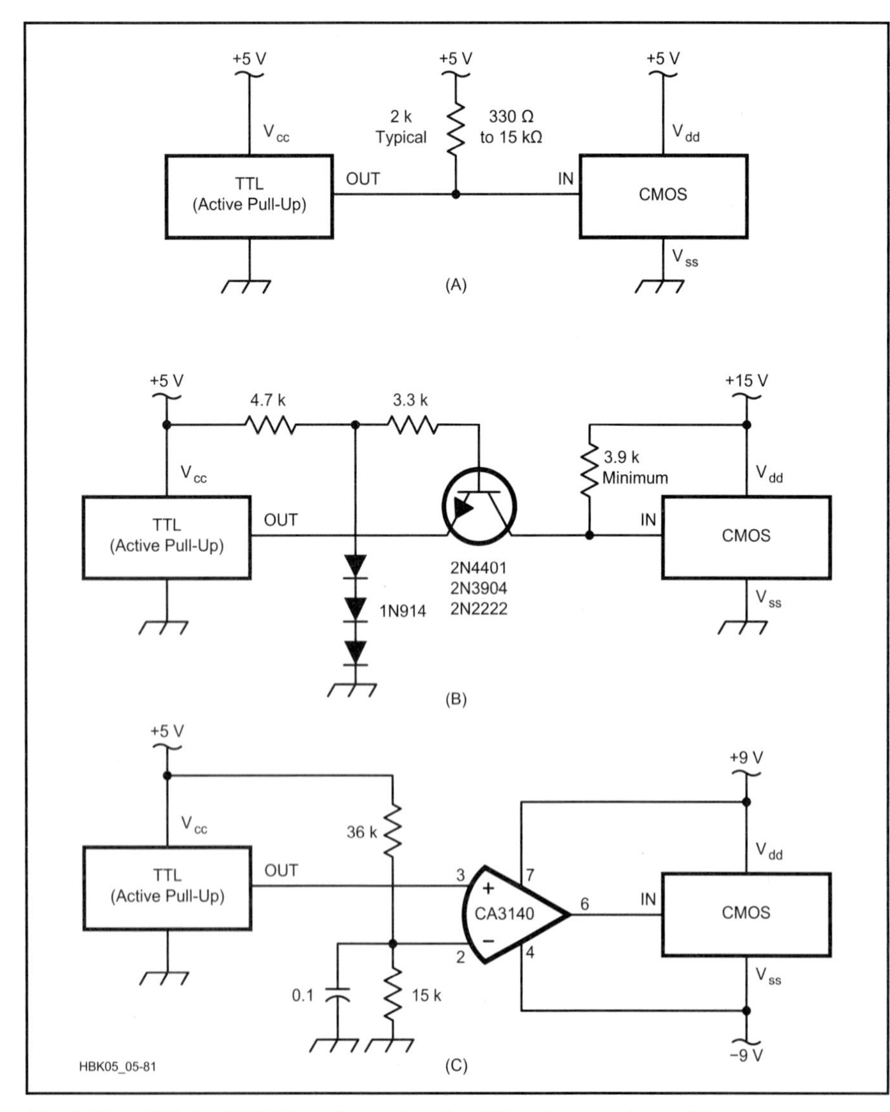

Fig 5.81 — TTL to CMOS interface circuits: (A) pull-up resistor, (B) common-base level shifter and (C) op amp configured as a comparator.

input is low, the resistance of Q2 is low so a high current flows from V_{CC}. Since Q1's resistance is high, the high current flows to the output. When the input is high, the opposite occurs: Q2's resistance is low, Q1's is high and the output is low. The diodes are to protect the circuit against static charges.

Special Considerations

Some of the diodes in the input- and output-protection circuits are an inherent part of the manufacturing process. Even with the protection circuits, however, CMOS ICs are susceptible to damage from static charges. To protect against damage from static, the pins should not be inserted in Styrofoam as is sometimes done with other components. Instead, a spongy conductive material is available for this purpose. Before removing a CMOS IC from its protective material, make certain that your body is grounded. Touching nearly any large metal object before handling the ICs is probably adequate to drain any static charge off your body. Some people prefer to touch a grounded metal object or to use a conductive bracelet connected to the ground terminal of a three-wire ac outlet through a 10-MΩ resistor. Since wall outlets aren't always wired properly, you should measure the voltage between the ground terminal and any metal objects you might touch. Connecting yourself to ground through a 1-MΩ to 10-MΩ resistor will limit any current that might flow through your body.

All CMOS inputs should be tied to an input signal. A positive supply voltage or ground is suitable if a constant input is desired. Undetermined CMOS inputs, even on unused gates, may cause gate outputs to oscillate. Oscillating gates draw high current, overheat and self destruct.

The low power consumption of CMOS ICs made them attractive for satellite applications, but standard CMOS devices proved to be sensitive to low levels of radiation — cosmic rays, gamma rays and X rays. Later, radiation-hardened CMOS ICs, able to tolerate 10^6 rads, made them suitable for space applications. (A rad is a unit of measurement for absorbed doses of ionizing radiation, equivalent to 10^{-2} joules per kilogram.)

SUMMARY

There are many types of logic ICs, each with its own advantages and disadvantages. Regardless of the application, consult up-to-date literature when designing logic circuits. IC databooks and application notes are usually available from IC manufacturers and distributors. Also, just about all of this information is available on the Internet. By using a search engine and entering a few key word specifications, you will locate application notes, tutorials and a host of other information.

INTERFACING LOGIC FAMILIES

Each semiconductor logic family has its own advantages in particular applications. When a design mixes ICs from different logic families, the designer must account for the differing voltage and current requirements each logic family recognizes. The designer must ensure the appropriate interface exists between the point at which one logic family ends and another begins. A knowledge of the specific input/output (I/O) characteristics of each device is necessary, and a knowledge of the general internal structure is desirable to ensure reliable digital interfaces. Typical internal structures have been illustrated for each common logic family. **Fig 5.80** illustrates the logic level changes for different TTL and CMOS families. Databooks should be consulted for manufacturer's specifications.

Often more than one conversion scheme is possible, depending on whether the designer wishes to optimize power consumption or speed. Usually one quality must be traded off for the other. The following section discusses some specific logic conversions. Where an electrical connection between two logic systems isn't possible, an optoisolator can sometimes be used.

TTL Driving CMOS

TTL and low-power TTL can drive 74C series CMOS directly over the commercial temperature range without an external pull-up resistor. However, they cannot drive 4000-series CMOS directly, and for HC-series devices, a pull-up resistor is recommended. The pull-up resistor, connected between the output of the TTL gate and V_{CC} as shown in **Fig 5.81A**, ensures proper operation and enough noise margin by making the high output equal to V_{DD}. Since the low output voltage will also be affected, the resistor value must be chosen with both desired high and low voltage ranges in mind. Resistor values in the range 1.5 kΩ to 4.7 kΩ should be suitable for all TTL families under worst conditions. A larger resistance reduces the maximum possible speed of the CMOS gate; a lower resistance generates a more favorable RC product but at the expense of increased power dissipation.

HCT-series and ACT-series CMOS devices were specifically designed to interface non-CMOS devices to a CMOS system. An HCT device acts as a simple buffer between the non-CMOS (usually TTL) and CMOS device and may be com-

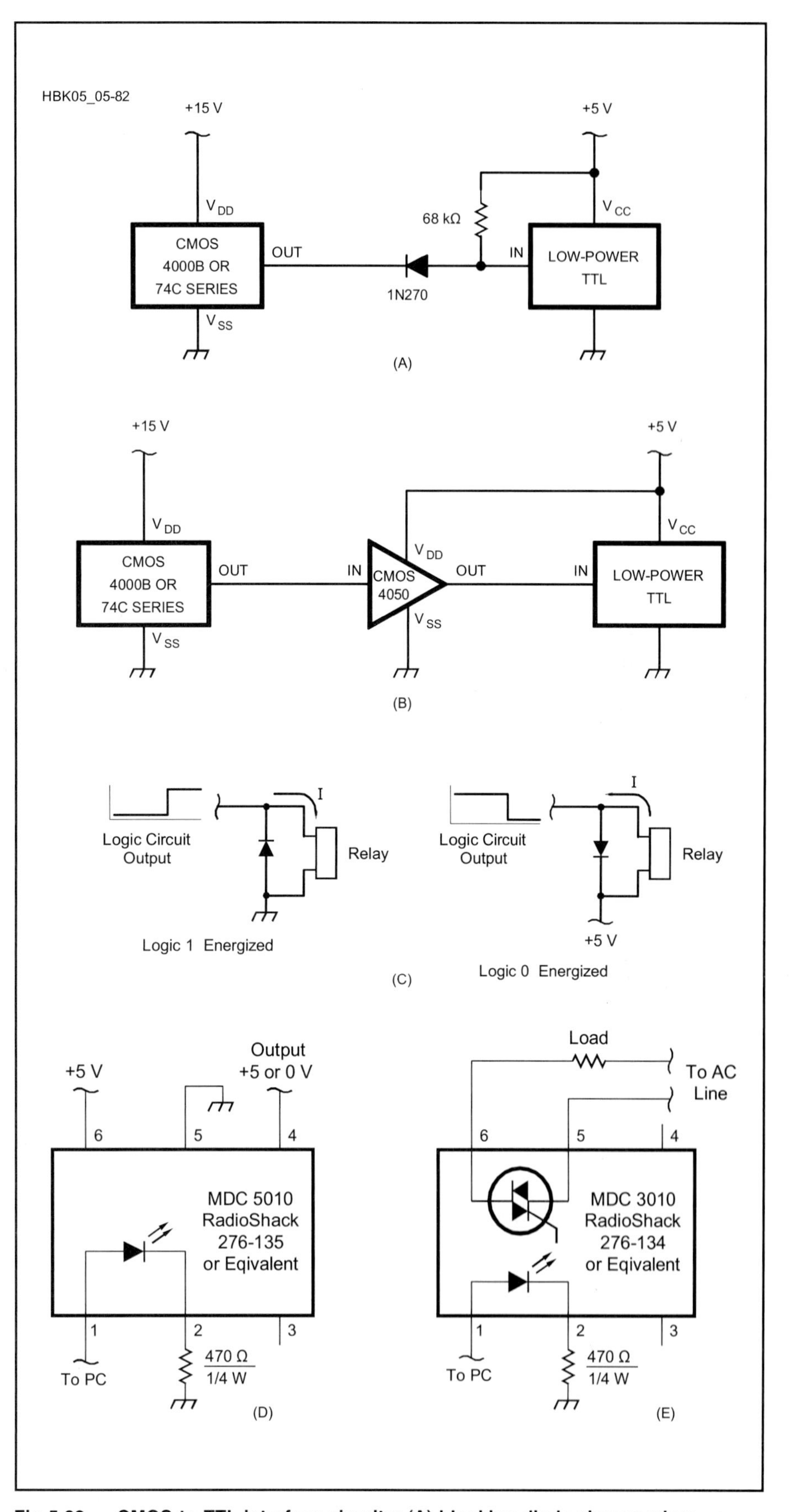

Fig 5.82 — CMOS to TTL interface circuits: (A) blocking diode chosen when different supply voltages are used. The diode is not necessary if both devices operate with a +5 V supply. (B) CMOS noninverting buffer IC. (C) Relay interface for total isolation. (D) An electro-optical coupler can be used in place of a relay. (E) An electro-optical coupler can also drive a triac-based unit for switching ac.

bined with a logic function if a suitable HCT device is available.

When the CMOS device is operating from a power supply other than +5 V, the TTL interface is more complex. One fairly simple technique uses a TTL open-collector output connected to the CMOS input, with a pull-up resistor from the CMOS input to the CMOS power supply. Another method, shown in Fig 5.81B, is a common-base level shifter. The level shifter translates a TTL output signal to a +15 V CMOS signal while preserving the full noise immunity of both gates. An excellent converter from TTL to CMOS using dual power supplies is to configure an operational amplifier as a comparator, as shown in Fig 5.81C. An FET op amp is shown because its output voltage can usually swing closer to the rails (+ and – supply voltages) than a bipolar device.

CMOS Driving TTL

Certain CMOS devices can drive TTL loads directly. The output voltages of CMOS are compatible with the input requirements of TTL, but the input-current requirement of TTL limits the number of TTL loads that a CMOS device can drive from a single output (the fan-out).

Interfacing CMOS to TTL is a bit more complicated when the CMOS is operating at a voltage other than +5 V. One technique is shown in **Fig 5.82A**. The diode blocks the high voltage from the CMOS gate when it is in the high output state. A germanium diode is used because its lower forward-voltage drop provides higher noise immunity for the TTL device in the low state. The 68-kΩ resistor pulls the input high when the diode is back biased.

There are two CMOS devices specifically designed to interface CMOS to TTL when TTL is using a lower supply voltage. The CD4050 is a noninverting buffer that allows its input high voltage to exceed the supply voltage. This capability allows the CD4050 to be connected directly between the CMOS and TTL devices, as shown in Fig 5.82B. The CD4049 is an inverting buffer that has the same capabilities as the CD4050.

Real-World Interfacing

Quite often logic circuits must either drive or be driven from non-logic sources. A very common requirement is sensing the presence or absence of a high (as compared to +5 volts) voltage or perhaps turning on or off a 120-VAC motor, such as an antenna rotor. A similar problem occurs when two different units in the shack must be interfaced since induced AC voltages or ground loops can cause problems with the desired signals.

A slow speed but safe way to interface such circuits is to use a relay. This provides absolute isolation between the logic circuits and the load. Fig 5.82C shows the correct way to provide this connection. The relay coil is selected to draw less than the available current from the driving logic circuit. The diode, most often a 1N914 or equivalent switching diode, prevents the inductive load from back-biasing the logic circuit and possibly destroying it.

Electro-optical couplers can also be used for this circuit interfacing. Fig 5.82D uses one to interface two sets of logic circuits, and Fig 5.82E interfaces with the AC line.

MSI, LSI VLSI Circuits and Controllers

In addition to using the basic logic elements discussed in the previous sections, there are many integrated circuits available for Amateur Radio applications that include the equivalent of dozens, hundreds or perhaps even thousands of gates. Often, it is no harder to use one of these units than it is to use a few gates and flip-flops.

The analog to digital (A/D) and digital-to-analog ((D/A) converters discussed in the section following (on the parallel port) are examples. Hybrid (containing both analog and digital functions) integrated circuits provide other opportunities for builders. As additional examples, the LM3914 is a LED driver that takes an analog signal in and turns on one or more LEDs, depending on the analog voltage. A self-contained signal generator, the MAX038, can accept either switch closures or digital control signals and generates selectable sin, square or triangle waves with variable frequency.

At the other end of the spectrum are microcontrollers, which can be considered *PCs on a chip.* These include a reduced version of an arithmetic-logic unit, as described in the next section as well as interfacing and some data and program storage. However, these units require a special purpose programmer to load and store the programs, and programming them can become quite involved.

A larger, but perhaps friendlier microcontroller is the various versions (and clones) of the *BASIC Stamp* (*BASIC Stamp* is a registered trademark of Parallax, Inc.) Several varieties are available, with varying capabilities. They are programmed in a version of the *BASIC* language using a PC, and, then, the program is downloaded from the PC into the BASIC Stamp.

Computer Hardware

So far, this chapter segment has discussed digital logic, the implementation of that logic with integrated circuits, interfacing IC logic families and the use of memory to store information used by the ICs. The synthesis of all this technology is the microcomputer — combining a microprocessor IC, memory, peripheral devices, and user interface into the modern personal computer. A computer has both physical components (hardware) and a collection of programs (software) to tell it what to do. This section (by Bob Wolbert, K6XX with additional material by Paul Danzer, N1II) will focus on the physical components of the computer: its internal physical components, their interaction, and peripheral I/O devices that communicate with other systems and the operator.

Material relating to computers in general, and PCs in particular, tends to become obsolete very quickly. For that reason, a quick look at the ever-growing PC-related section at your local bookstore will show very few books on the shelf older than perhaps 12 to 18 months. While the basic technology of electronics as applied to PCs does not change, the standard, performance and configurations change monthly, weekly and perhaps even daily. Thus, this section will provide only an overview of PC hardware and technology.

A WORD ABOUT AVAILABILITY

While many of an application's programs such as RTTY, SSTV, PSK AMTOR (and several dozen others) work best with fast, new PC hardware, many Amateur applications using the hardware as a controller do not require very much capability. Since many used computers are available at prices ranging from nothing (just haul it away) to perhaps $100, these units are worth considering for single, special applications. Want a modulation monitor or a recording voltmeter? How about basing the design on an old PC? Application of the material in this chapter to interfacing with an old PC can provide a host of ham-shack capabilities.

WHAT IS A COMPUTER?

The strictest definition of the term "computer" includes special purpose digital systems optimized for a particular task. For example, a modern synthesized transceiver, with its memory, I/O, serial control, DSP, etc. meets the definition of a computing device. Many of the concepts discussed in this section apply equally well to your HF rig as well as to your PC; however, our definition of a computer will be restricted to a general purpose machine whose task is quickly and easily changed by loading or changing software. If its task cannot be readily modified — to compute a spreadsheet or compose e-mail, for example — we will exclude that system from our discussion. The personal computer (PC) will be emphasized due to its ubiquitous nature.

The three major divisions of a PC are its hardware, its software, and its firmware. See **Fig 5.83**. The hardware includes the *central processing unit* (CPU) and input/output (I/O) devices. Software refers to the programs that are loaded into the computer to configure it for the task at hand. Firmware, also called microcode or BIOS (Basic Input/Output System), is a hybrid of both hardware and software that is used to perform specific tasks. The microcode is the basis of the microprocessor's command set that tells it how to fetch data and add numbers. For example, the BIOS is firmware generally used to start-up (boot) the system.

COMPUTER ARCHITECTURE

Unlike many present textbooks, where computer architecture is narrowly defined as only including those attributes of the system that interest programmers, our discussion deals with the structural organization and hardware design of the digital computer system. All modern computer systems consist of three basic sections: the CPU, memory, and peripherals for interfacing with the operator and the real world. The architecture of a computer is the arrangement of these two specific internal subsystems, the CPU and the bus. The CPU (central processing unit), called a *microprocessor* in personal computers, is an IC consisting of three major parts: a control unit, an arithmetic logic unit (ALU), and temporary storage registers. The *bus* is a set of wires carrying address, data and control information, which interconnects all of the subsystems. Virtually all computers are designed based on the basic "Von Neumann" architecture shown in **Fig 5.84**.

The microprocessor, memory chips and other circuitry are all part of the system's hardware, the physical components of a system. The computer case, the nuts and bolts and physical parts are other parts of the hardware. A computer also includes software, a collection of programs or sequence of instructions to perform a specified task. The design of computers is so complex, however, that it is nearly impossible to design an original architecture without any bugs. Thus many designers use microprocessors that include *microcode* or microinstructions, which are instructions in the control unit of a microprocessor. This hybrid between hardware and software is called firmware. Firmware also includes software stored in ROM or EPROM rather than being stored on magnetic disk or tape.

Computer designers make decisions on hardware, software and firmware based on cost versus performance. Today's computer market includes a wide range of systems, from high-performance super-computers costing millions of dollars, to the personal microcomputer, with prices in the high hundreds to a few thousands new and ranging from free on up for older used models.

THE CENTRAL PROCESSING UNIT

The central processing unit is usually a single microprocessor chip, although its subsystems can be on more than one chip. The CPU at least includes a control unit, timing circuitry, an arithmetic logic unit (ALU) and registers for temporary storage. Modern microprocessors have tens of millions of transistors and are designed in modules.

Control Unit

The control unit directs the operation of the computer, managing the interaction between subunits. It takes instructions from the memory and executes them, performing tasks such as accessing data in memory, calling on the ALU or performing I/O. Control is one of the most difficult parts to design; thus it is the most likely source of bugs in designing an original architecture.

Microprocessors consist of both hardwired control and micro-programmed control. In both cases, the designer determines a sequence of states through which the computer cycles, each with inputs to examine and outputs to activate other CPU sub-

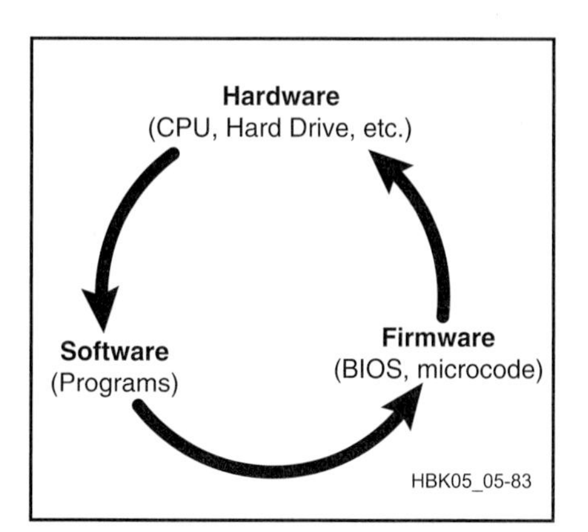

Fig 5.83 — Hardware, software and firmware comprise a computer.

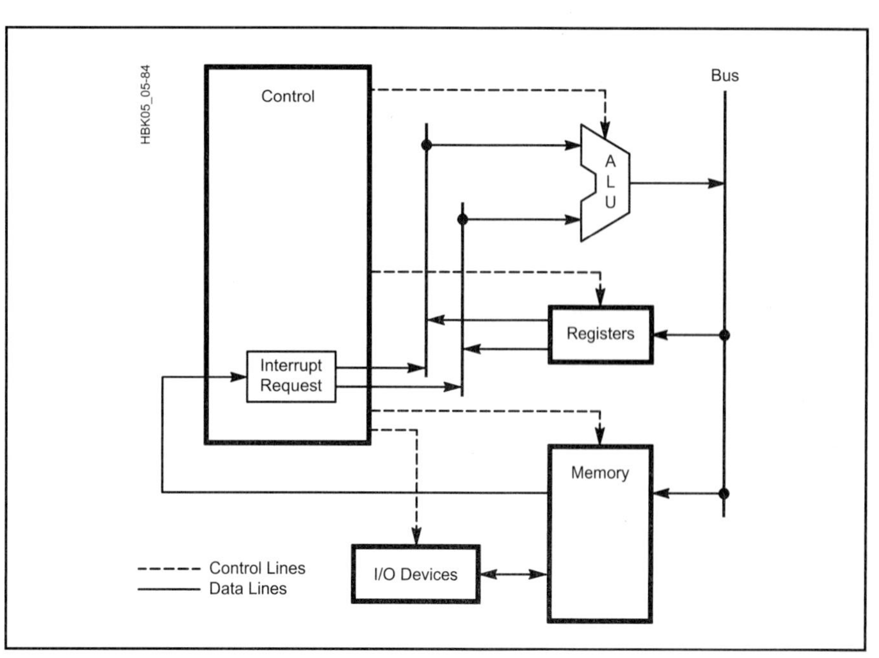

Fig 5.84 — Example of a basic computer architecture.

systems (including activating itself, indicating which state to do next). For example, the sequence usually starts with "Fetch the next instruction from memory," with control outputs to activate memory for a read, a program counter to send the address to be fetched and an instruction register to receive the memory contents. Hardwired control is completely via circuitry, usually with a programmed logic array. Microprogrammed control uses a microprocessor with a modifiable control memory, containing microcode or micro-instructions. An advantage of microprogrammed control is flexibility; the code can be changed without changing the hardware, making it easier to correct design errors.

Timing

Usually, an oscillator controlled by a quartz crystal generates the microcomputer's clock signal. The output of this clock goes to the microprocessor and to other ICs. The clock synchronizes the microcomputer subunits. For example, each of the microinstructions is designed to take only one clock cycle to execute, so any components triggered by a microinstruction's control outputs should finish their actions by the end of the clock cycle. The exception to this is memory, which may take multiple clock cycles to finish, so the control unit repeats in its same state until memory says it's done. Since the clock rate effectively controls the rate at which instructions are executed, the clock frequency is one way to measure the speed of a computer. Clock frequency, however, cannot be the only criteria considered because the actions performed during a clock cycle vary for different designs, particularly processors with superscalar or pipelined designs capable of computing multiple items simultaneously.

Arithmetic Logic Unit

The *arithmetic logic unit* (ALU) performs logical operations such as AND, OR and SHIFT and two number arithmetic operations such as addition, subtraction, multiplication and division. The ALU depends on the control unit to tell it which operation to perform and also to trigger other devices (memory, registers and I/O) to supply its input data and to send out its results to the appropriate place.

The ALU often only performs simple operations. Complex operations, such as multiplication, division and operations involving decimal numbers, are performed by dedicated hardware, called floating-point processors, or *coprocessors*.

Registers

Microprocessor chips have some internal memory locations that are used by the control unit and ALU. Because they are inside the microprocessor IC, these registers can be accessed more quickly than main memory locations. Special purpose registers or *dedicated registers* are purely internal, have predefined uses and cannot be directly accessed by programs. *Genera-purpose registers* hold data and addresses in use by programs and can be directly accessed, although usually only by assembly level programs.

The dedicated registers include the instruction register, program counter, effective address register and status register. The first step to execute an instruction is to fetch it from memory and put it in the *instruction register* (IR). The *program counter* (PC) is then incremented to contain the address of the next instruction to be fetched. An instruction may change the program counter as a result of a conditional branch (if-then), loop, subroutine call or other nonlinear execution. If data from memory is needed by an instruction, the address of the data is calculated and fetched with the *effective address register* (EAR). The *status register* (SR) keeps track of various conditions in the computer. For example, it tells the control unit when the keyboard has been typed on so the control unit knows to get input. It also notices if something goes wrong during an instruction execution, for example an attempted divide by 0, and tells the control unit to halt the program or fix the error. Certain bits in the status register are known as the *condition codes*, flags set by each instruction. These flags tell information about the result of the latest instruction, such as if the result was negative, positive or zero and if an arithmetic overflow or a carry error occurred. The flags can then be used by a conditional branch to decide if that branch should be chosen.

MEMORY

Computers and other digital circuits rely on stored information, either data to be acted upon or instructions to direct circuit actions. This information is stored in memory devices, in binary form. Computers use four main types of memory, as shown in **Fig 5.85**.

Accessing a Memory Item

Memory devices consist of a large number of memory cells each capable of remembering one bit of binary information. The information in memory is stored in digital form with collections of bits, called words, representing numbers and symbols. The most common symbol set is the American National Standard Code for Information Interchange (ASCII). Words in memory, just like the letters in this sentence, are stored one after the other. They are accessed by their location or address. The number of bits in each word, equal to the number of memory cells per memory location, is constant within a memory device but can vary for different devices. Common memory devices have word sizes of 8, 16 and 32 bits.

Addresses and Chip Size

An *address* is the identifier, or name, given to a particular location in memory. Since this address is expressed as a binary number, the number of unique addresses available in a particular memory chip is determined by the number of bits to express the address. For example, a memory chip with 8 bit addresses has $2^8 = 256$ memory locations. These locations are accessed as the addresses 00000000 through 11111111, 0 through 255 decimal or 00 through FF hex. (For ease of notation, programmers and circuit designers use hexadecimal (base 16) notation to avoid long strings of 1s and 0s.) The memory chip size can be expressed as M × N, where M is the number of unique addresses, or memory locations and N is the word size, or number of bits per memory location. Memory chips come in a variety of sizes and can be arranged, together with control circuitry and decoders, to meet a designer's needs.

Memory chips, no matter how large or small, have several things in common. Each chip has address, data and control lines. A memory chip must have enough

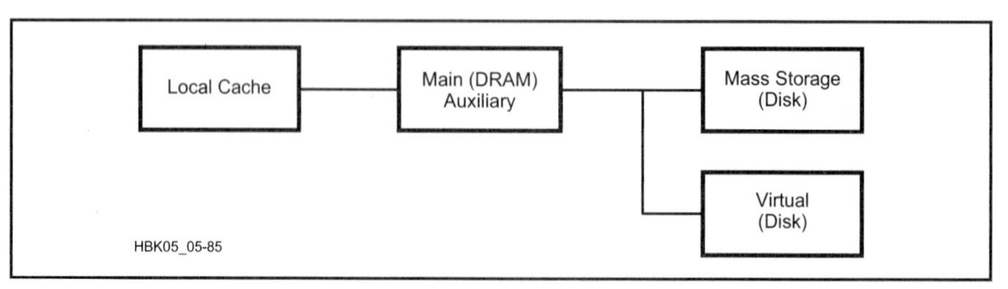

Fig 5.85 — Computer memory types.

address lines to uniquely address each of its words and as many data lines as there are bits per word. Memory size is usually specified in kilobytes (kbyte). This is usually abbreviated as K. Notice this is not quite the same as the metric prefix kilo, because it represents 1024, rather than 1000. When checking memory size on a PC, using various programs and tools, it is very common to see slightly different numbers as a result of these checks and tools. These differences are usually due to some processes giving results in increments of 1024 and some is actual increments of 1000.

Memory Types

The concepts described above are applied to several types of random-access, semiconductor memory. Semiconductor memories are categorized by the ease and speed with which they can be accessed and their ability to "remember" in the absence of power.

SAM versus RAM

One way to categorize memory is by which memory cells can be accessed at a given instant. *Sequential-access memory* (SAM) must be accessed by stepping past each memory location until the desired location is reached. Magnetic tapes implement SAM. To reach information in the middle of the tape, the tape head must pass over all of the information on the beginning of the tape. Two special types of SAM are the queue and the push-down stack. In a *queue*, also called a first-in, first-out (FIFO) memory, locations must be read in the order that they were written. The queue is a first-come, first-served device, like a line at a ticket window. The push-down stack is also called last-in, first-out (LIFO) memory. In LIFO memory, the location written most recently is the next location read. LIFO can be visualized as a stack, always adding to and removing from the top of the stack. *Random-access memory* (RAM) allows any memory cell to be accessed at any instant, with no time wasted stepping past the beginning parts of the data. Random-access memory is like a bookcase; any book can be pulled out at any time.

It is usually faster to access a desired word in RAM than in SAM. Also, all words in RAM have the same access time, while each word in a SAM has a different access time based on its position. Generally, the semiconductor memory devices internal to computers are random-access memories. Magnetic devices, such as tapes and disks, have at least some sequential access characteristics. We will leave tapes and disks for a later section and concentrate here on random-access, solid-state memories.

Random Access Memory

Most RAM chips are *volatile*, meaning that stored information is lost if power is removed. RAM is either static or dynamic. *Dynamic RAM* (DRAM) stores a bit of information as the presence or absence of charge. This charge, since it is stored in a capacitor, slowly leaks away and must be refreshed periodically. Memory refresh typically occurs every few milliseconds and is usually performed by a dynamic RAM controller chip. *Static RAM* (SRAM) stores a bit of information in a flip-flop. Since the bit will retain its value until either power is removed or another bit replaces it, refresh is not necessary.

Both types of RAM have their advantages and disadvantages. The advantage of DRAM is increased density and ease of manufacture, making them significantly less expensive. SRAMs, however, have much faster access times. Most general purpose computers use DRAMs, since large memory size and low cost are the major objectives. Where the amount of memory required doesn't justify the use of DRAM, and the faster access time is important, SRAMs are common, for example, in embedded systems (telephones, toasters), battery powered devices, and for cache memories. Cost, power consumption and access time, provided in manufacturers' data sheets, are factors to consider in selecting the best RAM for a given application.

New acronyms for various new memory types appear to be invented daily. The only solution to this problem is to do an internet-based search on an unfamiliar acronym; any result brought up by the search engine over 6-months old is almost certain to be obsolete!

Read-Only Memory

Read-only memory (ROM) is nonvolatile; its contents are not lost when power is removed from the memory. Despite its name, all ROMs can be written or programmed at least once. The earliest ROM designs were "written" by clipping a diode between the memory bit and power supply wherever a 0 was desired. Modern MOS ROMs use a transistor instead of a diode. Mask ROMs are programmed by having ones and zeros etched into their semiconductors at manufacturing time, according to a pattern of connections and non-connections provided in a mask. Since the programming of a mask ROM must be done by the manufacturer, adding expense and time delays, this type of ROM is primarily used only in high volume applications.

For low-volume applications, the programmable ROM (PROM) is the most effective choice since the data can be written after manufacture. A PROM is manufactured with all its diodes or transistors connected. A PROM programmer device then burns away undesired connections. This type of PROM can be written only once.

Two types of PROMs that can be erased and reprogrammed are EPROMs and EEPROMs. The transistors in UV erasable PROMs (EPROMs) have a floating gate surrounded by an insulating material. When programming with a bit value, a high voltage creates a negative charge on the floating gate. Exposure to ultraviolet light erases the negative charge. Similarly, electrically erasable PROMs (EEPROMs) erase their floating-gate values by applying a voltage of the opposite polarity.

Besides being nonvolatile, PROMs are also distinguished from RAMs by their read and write times. RAM read and write times are nearly equal, in the nanosecond range. Naturally, since PROMs are only written to infrequently, they can have slow write times (in the millisecond range). Their read times, however, are near those of RAM. Two factors make it hard to write to PROMs: (1) PROMs must be erased before they can be reprogrammed and (2) PROMs often require a programming voltage higher than their operating voltage.

ROMs are practical only for storing data or programs that do not change frequently and must survive when power is removed from the memory. The BIOS program that starts a computer when it is first switched on or the memory that holds the call sign in a repeater IDer are prime candidates for ROM.

Nonvolatile RAM

For some situations, the ideal memory would be as nonvolatile as ROM but as easy to write to as RAM. The primary example is data that must not be allowed to perish despite a power failure. Low-power RAMs can be used in such applications if they are supplied with NiCd or lithium cells for backup power. A more elegant and durable solution is nonvolatile RAM (NVRAM), which includes both RAM and ROM. The standard volatile RAM, called shadow RAM, is backed up by nonvolatile EEPROM. When the RECALL control is asserted, such as when power is first applied, the contents of the ROM are copied into the RAM. During normal operation, the system reads and writes to the RAM. When the STORE control is triggered, such as by a power failure or before turning off the system, the entire contents of the RAM are copied into the ROM for nonvolatile storage. In the event of primary power failure, to successfully save the RAM data, some power must be maintained until the memory store is complete, generally about 20 ms.

Cache versus Main Memory

Memory is in high demand for many applications. To balance the trade-off of speed versus cost, most computers use a larger, slower, but cheaper main memory in conjunction with a smaller, faster, but more expensive cache memory. As you run a computer program, it accesses memory frequently. When it needs an item, a piece of data or the next part of the program to execute, it first looks in the cache. If the item is not found in the cache, it is copied to the cache from the main memory. As you run a computer program, it often repeats certain parts of the program and repeatedly uses pieces of data. Since this information has been copied to the high-speed cache, your computer game or other application can run faster. Information used less often or not being used at all (programs not currently being run) can stay in the slower main memory.

A "cache" is a place to store treasure; the treasure, the information you are using frequently, can be accessed quickly because it is in the high-speed cache. The use of cache versus main memory is managed by a computer's CPU so it is transparent to the user. The improvement in program execution time is similar to accessing a floppy disk versus the computer's internal memory.

I/O TRANSFERS

No computer will perform useful work without some means of communicating with the real world. Its input and output system allow the computer to react to and affect the outside world. The ability to interact with their environment is a primary reason why computers are so useful and cost-effective. Often, I/O is provided by a user, and a great deal of effort goes towards making computers user-friendly. Alongside the drive for user-friendly computers is the drive for automation. Data are acquired and operations are performed automatically, such as the packet bulletin board automatically forwarding a message.

BUS STRUCTURE: LOCAL BUSSES

Tying the blocks together is the data bus, the main information corridor inside the computer. The bus carries signals to and from various components, such as the CPU, the keyboard, mass storage, and communications ports. The first IBM PC and compatibles used the 62-pin, 8-bit, 8-MHz ISA bus, which was revised to the AT-bus, a 98-pin, 16-bit wide 8-MHz bus. Other pre-PCI bus architectures include Apple's NuBus, the Extended Industry Standard Architecture (EISA), and the VESA Local Bus.

The slow ISA bus was a bottleneck to system performance, so a separate bus was implemented between the microprocessor and main memory. This bus was called a *local bus*, as it was local to the CPU and memory only. Eventually, graphics and hard disk drive speeds increased to the point where they could ride on the local bus as well, without impacting memory access performance. Present computer systems are built around the Peripheral Components Interface (PCI-X) bus, a 64-bit wide system running at 133 MHz or higher.

The need to remain compatible with older PC architectures and conventions has hampered bus development and changes. As a result, instead of the invention of a host of new bus structures, most new PCs now include external bus attachments, such as USB. This permits attachment of more and faster devices than the older bus structures would allow.

PERIPHERALS

Peripherals work with the CPU and memory to provide additional capabilities. One of the most common examples is communication with a user via input devices and output devices. Peripherals may be divided into three groups: bidirectional (input/output) devices, input devices, and output devices. Bidirectional devices allow data storage and communications with the outside world. Input devices provide the computer both data to work on and programs to tell it what to do. Output devices present the results of computer operations to the user or another system and may even control an external system. Both input and output combine to provide user-friendly interaction. Most of these devices have adapted to certain standards and use readily available connectors and cables, enabling easy incorporation into a system. Knowledge of how external memory devices work is more useful and will be discussed in more detail.

Bidirectional Input/Output Devices

Mass storage and communications devices provide data input as well as output. Perhaps the most important peripheral in the computer system is the mass storage unit, such as the *disk drive*. Another important I/O unit is the *modem* (a contraction of MODulator/DEModulator), which allows easy communication between computers across standard telephone lines. A third is the *local area network* (LAN) card, which provides high speed communications between nearby computers.

Hard Disk Drives

An electromechanical hybrid, the hard disk drive provides the largest capacity at the lowest cost per byte of any random-access storage media. Hard drives are an essential part of present computer systems. Their key features include:

- Low cost per byte of storage.
- Large capacity available.
- Random access to data.
- Non-volatile magnetic storage.

Hard drives consist of three main units, the *head/disk assembly* (HDA), the *read/write channel*, and the *controller*. The HDA comprises the mechanical portion of the assembly, with one or more aluminum disks mounted on a spindle, which is rotated by a brushless dc motor, generally between 3600 and 7200 rpm. Read/write heads are mounted on an actuator arm that sweeps across the disk surfaces. The read/write amplifier is affixed to the actuator by a flexible cable to provide the lowest possible noise pickup. Read/write heads do not touch the disk surface; instead, air flowing over the rapidly spinning disk causes them to fly slightly above the disk. "Slightly" is no exaggeration — typical flying heights of drives made in the year 2000 are approximately two millionths of an inch (about 500 nanometers). Due to these extremely tight tolerances, the entire HDA is enclosed in a sealed aluminum casting that prevents contamination. As shown in **Fig 5.86**, debris such as a hair or a dust particle tower over the flying heads.

The drive controller performs data caching and communication with the host bus. Common hard drive busses include the EIDE (Extended Integrated Device Electronics), the (similar) ATA (AT-Attachment), and SCSI (small computer systems interface). External hard drives, interfacing with USB ports, are now both readily available and provide portability from PC to PC. In fact, some newer PCs now will boot from an external drive. Instead of carrying a PC from location to location, just unplug these thumb-sized drives and plug them back in at the new location.

Data is stored on both surfaces of each disk in concentric arcs called sectors, as shown in **Fig 5.87**. All sectors on one surface a given distance from the disk edge constitute a track. A cylinder is the collection of all similar tracks on all surfaces of the drive. When a new drive is installed, most newer PCs and their BIOS will rec-

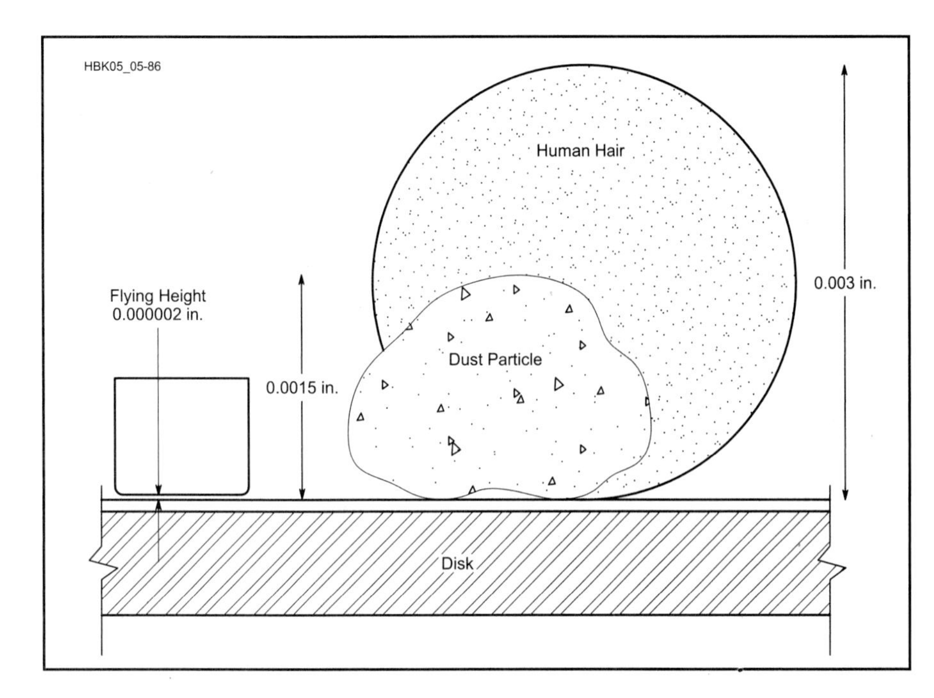

Fig 5.86 — Hard disk drive head flying height compared to common debris size.

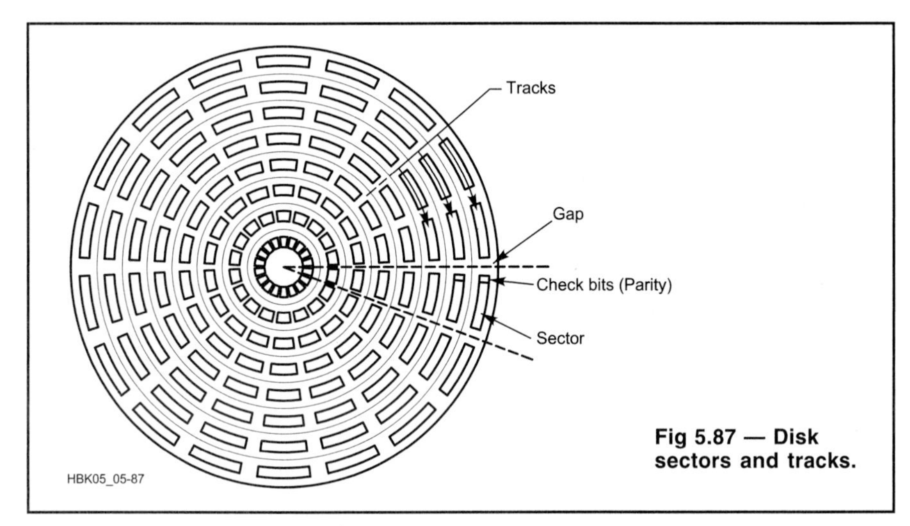

Fig 5.87 — Disk sectors and tracks.

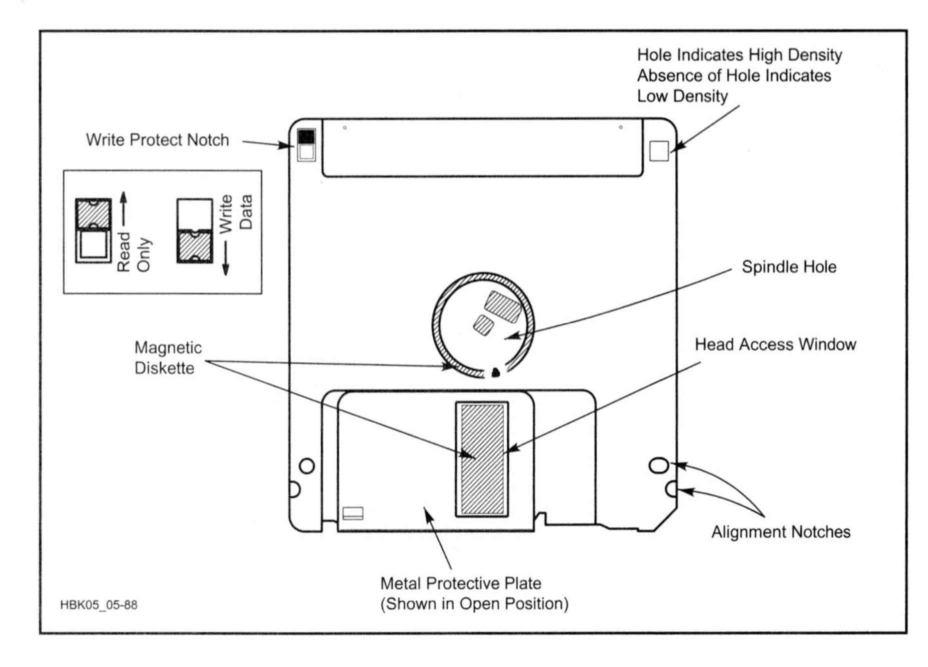

Fig 5.88 — Standard 3.5-in. floppy disk.

ognize the new drive and set the BIOS directly for the new drive.

Occasionally, when installing a hard drive in an older PC, the motherboard must be informed of the number of disk heads and sectors available. Often, the number entered is not the actual number of physical heads and sectors, as intelligent controllers map data requests to the proper physical location without burdening the CPU with the fine details.

Important hard disk drive parameters include bus type (for compatibility with your system), storage capacity, average and worst-case seek time, and size of onboard cache. Seek times are dominated by mechanical considerations, such as the time the read head arms require to settle from track to track or from one edge to the other, and by the rotational speed of the disk. Better overall performance is expected with larger on-board cache memory. The best drives have high capacity, fast seek times, and large caches.

Floppy Disk Drive

The most common storage peripheral in the PC is the 3.5-in., 1.44 MB floppy disk drive. The floppy drive operates similarly to its higher speed/higher capacity brother, but its media is removable and it does not offer much on-board intelligence. Floppy disks enclose the magnetic-media platter in a protective casing, as shown in **Fig 5.88**, so the disk can be carried around. The floppy disk can be inserted into a disk drive and the read/write head automatically extended. When the recording is complete, the read/write head is automatically retracted before the disk is ejected from the drive.

Most standard PCs contain a cable to connect a single floppy. This cable may be replaced with one that allows connection of two floppies. Some PCs have a cable that permits the connection of two floppies. The first drive, drive A, attaches to the *end* of the standard "twisted" control cable; if only one drive is used, leave the middle connector free. The twist in the cable causes the floppy controller to recognize the end drive as **drive A:** while the second drive — if you connect one — registers as **drive B:**, and is before the twist.

Other Magnetic Disk Drives

Disk drives featuring moderate storage capacity and performance with removable media are available, such as the Iomega Zip and Jazz, and the Imation SuperDrive. These drives use the same basic techniques as the hard drives, but with somewhat looser mechanical tolerances that allow an unsealed disk chamber and more flexible media. These drives are available with EIDE, SCSI, parallel port, USBs, and PC-Card interfaces for internal or external use. Previously very widely used for both back-up and transferring data from PC to PC, these proprietary drives are now competing with very inexpensive hard drives and USB connected hard drives. Adding a second hard drive and cloning the first hard drive (Drive C:) to the second is a popular back-up technique. Using a USB-connected hard drive for data transfer also is very convenient and popular.

Optical Storage

The audio compact disc evolved into the CD-ROM offering over 600 MB of storage on a very low-cost, single-sided platter. Initially a read-only media, writeable and re-writable discs are used as removable mass storage. While media cost is low, the low performance and restricted number of write sessions, even on the rewritable discs, prevent these technologies from competing with the magnetic hard drive except in archival or other applications where removability is important. Another transplant from the consumer entertainment industry, the DVD is another important optical storage format. Similar in appearance to CD-ROM discs, DVDs offer up to 8.5GB of storage capability per disc.

Both the CD-ROM and the DVD use a laser and an optical system to detect the surface deformities that represent data. Unlike the hard disk and floppy drives, which employ concentric tracks, the optical drives use a spiral pattern that works out from the center. Also, more elaborate data handling is necessary since the raw data error rate is significantly higher than that from magnetic drives.

All drives and disks eventually fail, and the data on the disk can be lost. Therefore, it is prudent to make backup copies of your disks, stored in a clean, dry, cool place.

Tape

Tape is one of the more inexpensive options for auxiliary memory. Tape access time is slow, since the data must be accessed sequentially, so tape is primarily used for backup copies of a system's hard drive. Tape is available in cassette form (common sizes are comparable to the cassettes for a portable tape player and VCR tapes) and on digital-audio tape (DAT). A single 4-mm-wide DAT cartridge, which fits in the palm of your hand, can hold over 2 gigabytes (GB) of data (1 GB = 1000 MB). Tape units are rarely used now in home PCs, but they are often seen at hamfests as obsolescent equipment.

Modems

Nearly all computers assembled today include a modem for connecting to the Internet via standard telephone lines. Besides connecting to an ISP or online service, this peripheral may call another modem-equipped PC or send standard facsimiles. The so-called 56-k modem uses V.90 protocol with its sophisticated DSP techniques providing echo cancellation and dynamic line equalization. This protocol uses the telephone line to gain every possible bit per second of data transfer rate. While the data rate never reaches 56 kbps, if the modem is less than four miles from the telephone central office, expect speeds of 40 kbps to 53 kbps. Achieving the 56 kbps rate would necessitate increasing transmission power above the –9 dBm limit and could cause excessive crosstalk between lines. The V.90 specification has an unusual characteristic in that the bit rate is non-symmetrical. The modem originating a call is limited to approximately 33 kbps while the answering modem runs up to 53 kbps. This compromise was made because most modem traffic is between a user and an ISP, where the user downloads much more data from the World Wide Web than he uploads.

Modems are available in three configurations, internal, external, and PC-Card. Internal modems plug directly into the computer motherboard, external devices attach to a serial port, and PC-Card (often called PCMCIA) modems plug into a PC-Card slot. The internal version is the least expensive and most common. PC-Card modems are popular with notebook computers due to their small size. External modems offer the advantage of providing immediate visual feedback of all data transfer activity. Additionally, external modems provide another level of surge protection to the computer system; if a destructive surge travels through the phone line it *might* be stopped outside the PC cabinet by the external modem.

Input Devices

The keyboard is probably the most familiar input device. A keyboard simply makes and breaks electrical contacts. The open or closed contacts are usually sensed by a microprocessor built into the circuit board under the keys. This microprocessor decodes the key closures and sends the appropriate ASCII code to the main computer unit. Keyboards will generate the entire 128-character ASCII set and often, with CONTROL and ALT (Alternate) keys, the 256-character extended ASCII set.

The mouse is a close second in familiarity to the keyboard. This pointing device controls the position of a cursor on the screen, and switches on the mouse make and break connections (clicking) to select and activate items (icons) on the screen. Touchpads, trackballs, and pen input on a sensitive screen are variations of the mouse and may offer a more natural, human-friendly interface to the computer. Voice recognition systems promise even easier data entry.

Digital cameras and streaming video cameras allow quick transfer of images into the computer realm.

Image scanners digitize photographs and older printed pages, allowing reuse of material without laboriously recreating the work. Scanners are available in four major configurations: handheld, sheet feed, flatbed, and drum. The handheld scanner is low cost and very portable. The sheet feed scanner is also small and is easier to use, since pages are better aligned. Flatbed scanners are an economical and a moderately high resolution means of entering data from book or other bound sources. Drum scanners provide the best resolution and color reproduction, but their high cost relegates them to professional graphics shops.

Output Devices

The most familiar output device is the computer screen, or monitor. The next most common output device is the printer, to produce paper hardcopy. Sound cards provide high-fidelity-stereo audio.

Monitors and most printers share a common display technique: images, such as characters and graphics, are formed by tiny dots, called *pixels* (picture elements). On screens, these are dots of light turned on and off. In printers, they are dots of ink or electrostatic toner imposed onto the paper. For color displays, pixels in red, green and blue (RGB) are spaced closely together and appear as colors to the human eye.

Video Displays

Video monitors are usually specialized high-resolution cathode-ray tube (CRT) displays, except in notebook computers, which use screens fabricated with liquid-crystal displays (LCDs). Most monitors employ *raster scanning* techniques to turn on the screen pixels, similar to that used by standard broadcast television receiv-

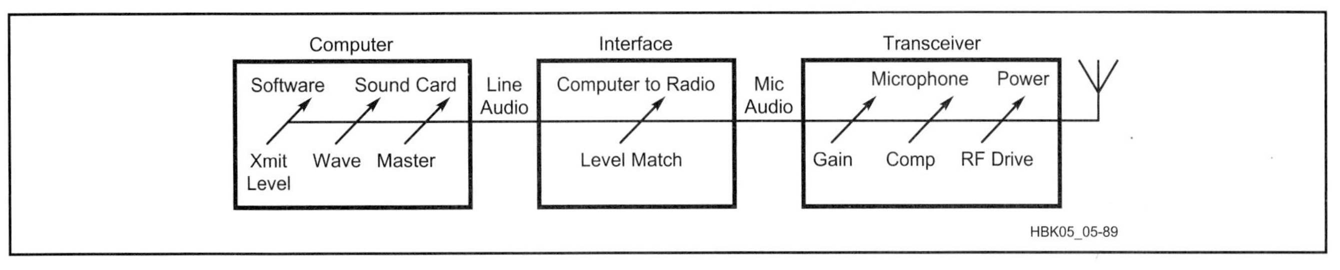

Fig 5.89 — An example of daisy-chained transmit level controls. Note that the controls are in series, and the resultant output is affected by each of the prior control settings.

ers. The electron beam paints the screen one row of pixels at a time, from left to right and top to bottom. Then, a vertical retrace brings the beam back to the top of the screen to begin again. Raster scanning signals every pixel on or off for each screen pass.

Printers

Printers suitable for hamshack use generally fall into two categories, inkjet and laser. The inkjet printer uses a controlled spray of liquid ink to produce images. Photographic-quality full-color prints are possible when the proper paper and ink is employed.

Laser printers produce exceptionally crisp text and graphics in black or a few colors at relatively high speed and low cost per page. While color laser printers do not (yet) produce the lifelike quality images of inkjets, they are not as fussy about the quality of the paper used, and its powdered *toner*, or "ink," does not dry up when stored in the printer over time as does the inkjet pigment. The use of dry paper is important, especially with many inkjet printers.

Today's state of the art in manufacturing and construction has made possible a host of very inexpensive inkjet printers, both black and white and color. There are several disadvantages to many of these units. First, and most noticeable, is the cost of replacement cartridges, especially color cartridges. It is not uncommon to pay under $100 for one of these printers and discover that two or three sets of cartridges, easily used up in a single year, cost the same as the printer. Some manufacturers have configured their printers so that when the ink goes below a certain level, the printer stops and no further operation is possible. Thus even if you wish to print in black and white, an exhausted color cartridge might prevent you from doing so. Finally, some printers are configured with a *smart chip* that prevents you from manually refilling a cartridge. The refilled cartridge may be full; the chip tells the printer it is empty.

Sound Cards

Your PC will produce and record full CD-quality audio when a suitable sound card and speaker system is deployed. Microphone and auxiliary inputs and line level outputs on the sound card let the PC serve as a contest voice keyer. When the proper software is used, it can also serve as a RTTY, CW, Pactor, PSK31, etc, terminal.

PCs running various versions of the Windows operating system do have a major annoyance in their use of soundcards. **Figure 5.89** shows a typical installation, where a PC with a sound card is the source of transmitted data. There may be as many as three level or gain controls on the PC software, perhaps one in the interfacing unit and effectively one to three in the transmitter. While all controls in the interface units and the transmitter tend to remain set, unless manually changed, that is not always true for the controls in the PC software. Some of these level set controls may revert back to a nominal value after re-booting the PC, and often several tries will be needed to set a system so that the constantly reverting levels do not have to be touched. For further information on this problem, see *QST*, Oct 2003, page 33, *The Ins and Outs of a Sound Card.*

COMMUNICATIONS: INTERNAL AND EXTERNAL INTERFACING

Designing an interface, or simply using an existing interface, to connect two devices involves a number of issues. For example, digital interfacing can be categorized as parallel or serial, internal or external and asynchronous or synchronous. Additional issues are the data rate, error detection methods and the signaling format or standards. The format can be especially important since many standards and conventions have developed that should be taken into consideration. This section focuses on some basic concepts of digital communications for interfacing between devices.

Parallel Versus Serial Signaling

To communicate a word to you across the room, you could hold up flash cards displaying the letters of the word. If you hold up four flash cards, each with a letter on it, all at once, then you are transmitting in parallel. If instead, you hold up each of the flashcards only one at a time, then you are transmitting in serial. *Parallel* means all the bits in a group are handled exactly at the same time. *Serial* means each of the bits is sent in turn over a single channel or wire, according to an agreed sequence. **Fig 5.90** gives a graphic illustration of parallel and serial signaling.

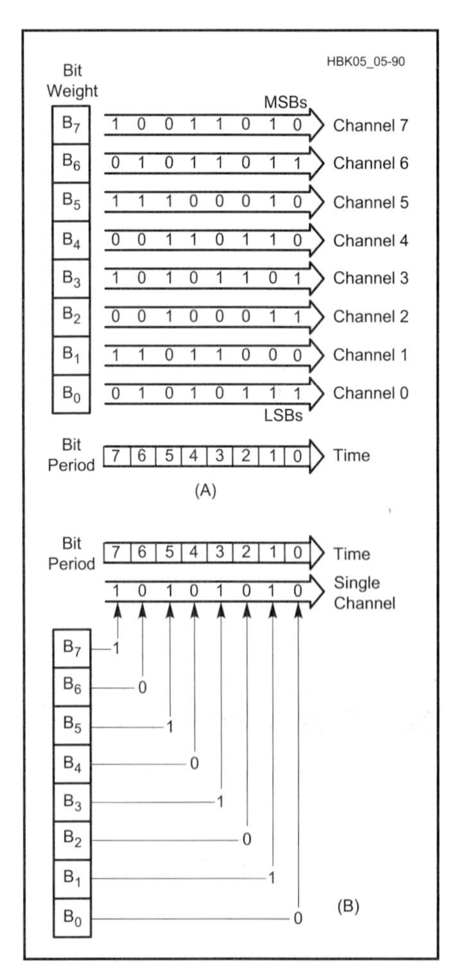

Fig 5.90 — Parallel (A) and serial (B) signaling. Parallel signaling in this example uses 8 channels and is capable of transferring 8 bits per bit period. Serial transfer only uses 1 channel and can send only 1 bit per bit period.

Interfacing To The Parallel Port

While there is a choice of ports on the PC for Amateur use and direct interfacing, the parallel port is probably the simplest. With eight data wires, several control wires and bidirectional capability, it offers a convenient way to get information in and out of a PC. The examples in the next two sections use an older software language, BASIC or GWBASIC to get information in and out. Newer languages can be used, however several varieties of BASIC are available on the internet at no cost, and the learning curve for someone who has never used a programming language is very short — usually a matter of a few minutes. The two examples that follow interface single chip analog to digital and digital to analog converters to the parallel port of a PC.

Single-Chip Dual-Channel A/D

In this analog world, often there is need to measure an analog voltage and convert it to a digital value for further processing in a PC. This single chip converter and accompanying software performs this task for two analog voltages.

Circuit Description

The circuit consists of a single-chip A/D converter, U2, and a DB-25 male plug (**Fig A**). Pins 2 and 3 are identical voltage inputs, with a range from 0 to slightly less than the supply voltage V_{CC} (+5 V). Rl, R2, C3 and C4 provide some input isolation and RF bypass. There are four signal leads on U2. DO is the converted data from the AID out to the computer; DI and CS are control signals from the computer, and CLK is a computer-generated clock signal sent to pin 7 of U2.

The +5-V supply is required. It may be obtained from a +12-V source and regulator Ul. Current drain is usually less than 20 mA, so any 5-V regulator may be used for Ul. The power supply ground, the circuit ground and the computer ground are all tied together. If you already have a source of regulated 5 V, Ul is not needed.

In this form the circuit will give you two identical dc voltmeters.

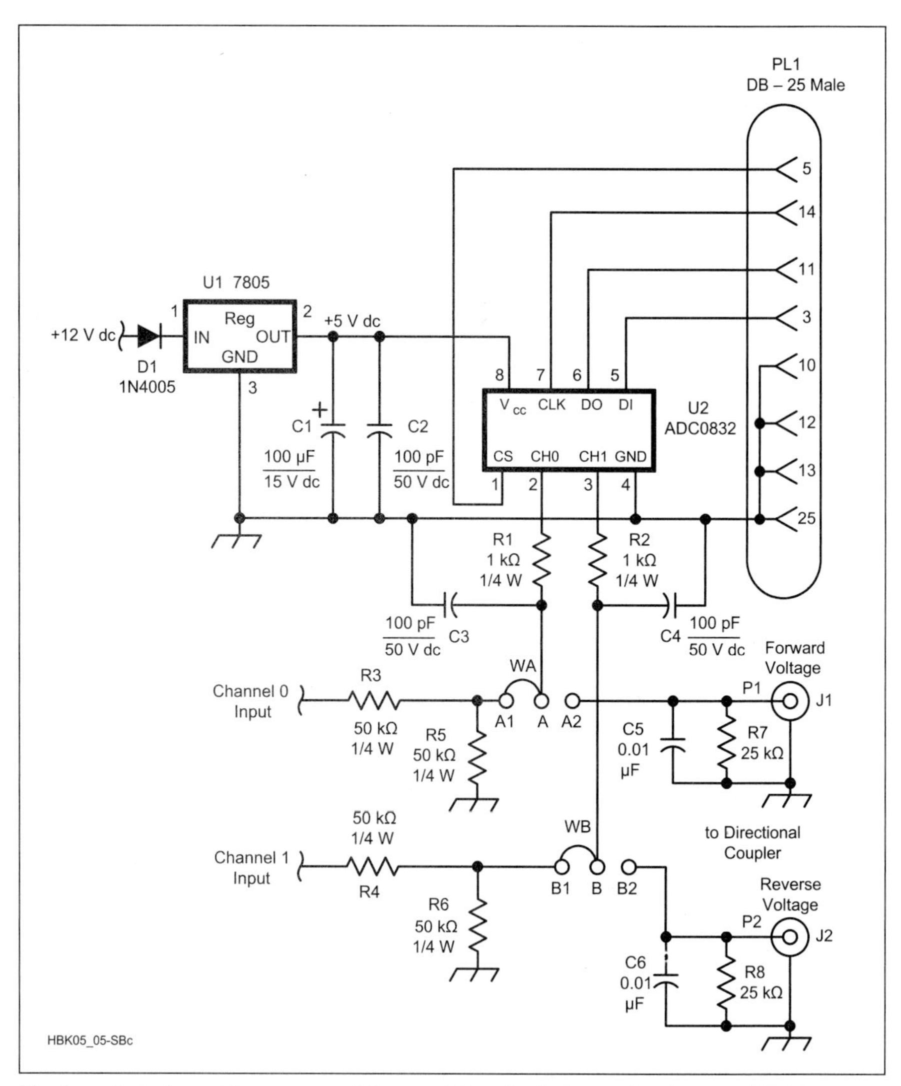

Fig A — Only two chips are used to provide a dual-channel voltmeter. PL1 is connected through a standard 25-pin cable to your computer printer port. U2 requires an 8-pin IC socket. All resistors are ¼ W. You can use the A/D as an SWR display by connecting it to a sensor such as the one shown in Chapter 19 of this *Handbook* (Tandem Match Wattmeter project). A few more resistors are all that are needed to change the voltmeter scale. The 50-kΩ resistors from 2:1 voltage dividers, extending the voltmeter scale on both channels to almost 10 V dc.

To extend their range, connect voltage dividers to the input points A and B. A typical 2:1 divider, using 50-kΩ resistors, is shown in the figure. Resistor accuracy is not important, since the circuit is calibrated in the accompanying software.

Software

The software, *A2D.BAS*, includes a voltmeter function and an SWR function. It is written in *GWBASIC* and saved as an ASCII file. Therefore, you can read it on any word processor, but if you modify it, make sure you re-save it as an ASCII file. It can be imported into *QBasic* and most other Basic dialects.

The program was written to be understandable rather than to be highly efficient. Each line of basic code has a comment or explanation. It can be modified for most PCs. The printer port used is LPT1, which is at a hex address of 378h. If you wish to use LPT2 (printer port 2), try changing the address to 278h. To find the addresses of your printer ports, run *FINDLPT.BAS*.

A2D.BAS was written to run on computers as slow as 4.7-MHz PC/XTs. If you get erratic results with a much faster computer, set line 1020(CD=1) to a higher value to increase the width of the computer-generated clock pulses.

The software is set up to act as an SWR meter. Connecting points A and B to the forward and reverse voltage points on any conventional SWR bridge will result in the program calculating the value of SWR.

Initially the software reads the value of voltage at point A into the computer, followed by the voltage at point B. It then prints these two values on the screen, and computes their sum and difference to derive the SWR. If you use the project as a voltmeter, simply ignore the SWR reading on the screen or suppress it by deleting lines 2150, 2160 and 2170. If the two voltages are very close to each other (within 1 mV), the program declares a bad reading for SWR.

Calibration

Lines 120 and 130 in the program independently set the calibration for the two voltage inputs. To calibrate a channel, apply a known voltage to input point A. Read the value on the PC screen. Now multiply the constant in line 120 by the correct value and divide the result by the value you previously saw on the screen. Enter this constant on line 120. Repeat the procedure for input point B and line 130.

D to A CONVERTERS — CONTROLLING ANALOG DEVICES

The complement to A/D converters is D/A (digital-to-analog) converters. Once there is a digital value in your PC, a D/A will provide an analog voltage proportional to the digital value. Normally the actual value is scaled. As an example, an 8-bit converter allows a maximum count of 255. If the converter is set up with a +5 Vdc reference voltage, a maximum value digital value of 255 would result in a D/A output value of 5-V. Lower digital inputs would give proportionally lower voltages.

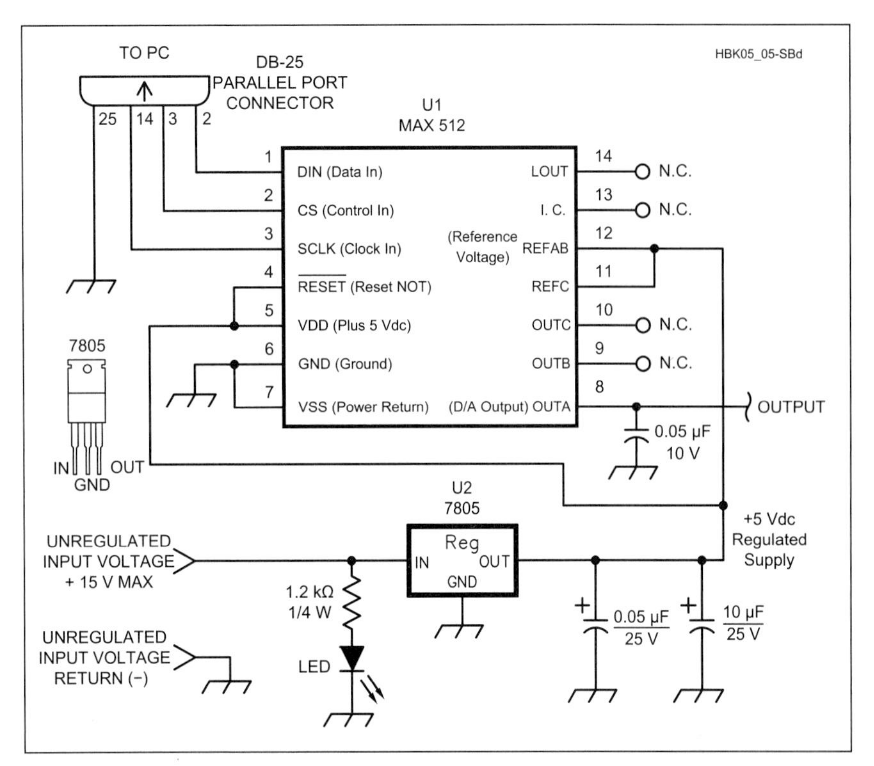

Fig B — Only three wires and a ground lead are needed to connect the converter to your PC.

Circuit Description

This project is the complement of the parallel port A/D converter described earlier. It takes a digital number from the computer, and converts it to a voltage from 0 to 5 V dc. Only one chip, the MAX 512, is required. It operates from a 5-V supply and is connected to the computer by a standard DB-25 parallel port connector. The chip may be ordered from Digi-Key, Allied Electronics and other ham suppliers as MAX512CPD-ND. The voltage regulator in **Fig B** provides the 5-volt source required to power the chip.

Software

The software needed to run the chip, *D2A.BAS*, can be downloaded from Internet sites. It is about 60-lines long, fully commented and written in *GWBASIC*, so it may be readily modified. The parallel port address is defined on line 105 as PORTO=&H378. Your computer may use a different address. To find the correct address, run *FINDLPT.BAS*.

The program takes the value AIN from the keyboard (line 230), converts it to a number between 0 and 255, and then sends it out as a serial word to the DIA chip. If you would like to use the project with another program, use your other program to set AIN to the value you want to generate, and then run this program as a subroutine.

At the end of the program is the clock pulse subroutine. In the event your computer is too fast for the converter chip, you can stretch the clock pulses by changing CD in line 5010 to a value greater than the default value of 1.

Applications

This circuit provides the capability of setting a voltage under computer control. It can be calibrated to match the power supply and the actual chip used. Tests with several chips showed an error of 25 mV or less over the range of 0 to 5 V dc output.

— *Paul Danzer, N1II*

Both parallel and serial signaling are appropriate for certain circumstances. Parallel signaling is faster, since all bits are transmitted simultaneously, but each bit needs its own conductor, which can be expensive. Parallel signaling is more likely to be used for internal communications. For spanning longer distances, such as to an external device, serial signaling is more appropriate. Each bit is sent in turn, so communication is slower, but it is also less expensive, since fewer channels are needed between the devices.

Most amateur digital communications use serial transmission to minimize cost and complexity. The number of channels needed for signaling also depends on the operational mode. One channel is required per bit for simplex (one-way, from sender to receiver only) and for half-duplex (two-way communication, but only one person can talk at a time), but two channels per bit are needed for full-duplex (simultaneous communications in both directions).

Parallel I/O Interfacing

Fig 5.91 shows an example of a parallel input/output chip. Typically, they have eight data lines and one or more handshaking lines. *Handshaking* involves a number of functions to coordinate the data transfer. For example, the READY line indicates that data are available on all 8 data lines. If only the READY line is used, however, the receiver may not be able to keep up with the data. Thus, the STROBE line is added so the receiver can determine when the transmitter is ready for the next character.

On standard PCs, a parallel port is available using a 25-pin DB-25 connector. This port was originally intended for use with a printer. Several versions of the parallel port exist, and late-model PCs feature a high speed, bidirectional parallel port that, in addition to high-speed printing, may also interface with mass storage devices, scanners, and other I/O devices. The most common parallel ports are:

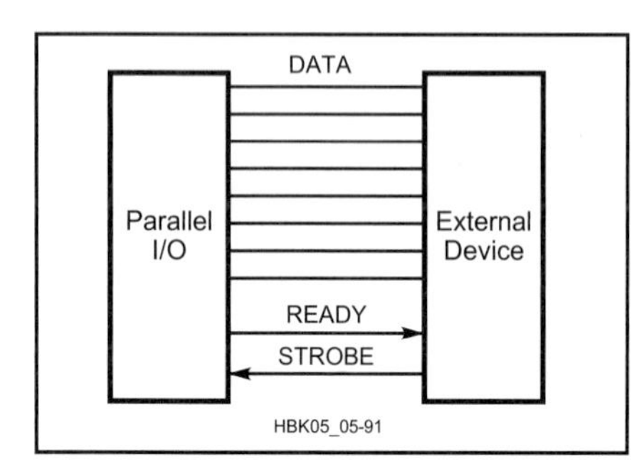

Fig 5.91 — Parallel interface with READY and STROBE handshaking lines.

- Printer Mode — The most basic, output mode only port.
- Standard & Bidirectional (SPP) — The low-speed bidirectional port
- Enhanced Parallel Port (EPP) — Uses local hardware handshaking and strobing to accomplish 500 kB/s to 2 MB/s transfer rates.
- Extended Capabilities Port (ECP) — Similar to EPP, except negotiates a reverse channel with the external peripheral and requires that peripheral controls handshaking. It is optimized for the Windows operating system and uses DMA channels, a FIFO buffer, and real time data compression of up to 64:1.

Most PCs offer a choice of port protocol in the BIOS setup. Unless you have a reason to do otherwise, select the "ECP and EPP 1.9 Mode" for maximum flexibility and performance.

Serial I/O Interfacing

Serial input/output interfacing is more complex than parallel, since the data must be transmitted based on an agreed sequence. For example, transmitting the 8 bits (b7, b6, . . . b0) of a word includes specifying whether the least significant bit, b0, or the most significant bit, b7, is sent first. Fortunately, a number of standards have been developed to define the agreed sequence, or encoding scheme. Use only IEEE specified cable to ensure availability of the correct number of conductors.

Data Rate

There are a number of limitations on how fast data can be transferred: (1) The sending equipment has an upper limit on how fast it can produce a continuous stream of data. (2) The receiving equipment has an upper limit on how fast it can accept and process data. (3) The signaling channel itself has a speed limit, often based on how fast data can be sent without errors. (4) Finally, standards and the need for compatibility with other equipment may have a strong influence on the data rate.

Two ways to express data transmission rates are *baud* and *bits per second (bps)*. These two terms are not interchangeable: Baud describes the signaling, or symbol, rate — a measure of how fast individual signal elements *could* be transmitted through a communications system. Specifically, the baud is defined as the reciprocal of the shortest element (in seconds) in the data- encoding scheme. For example, in a system where the shortest element is 1-ms long, the maximum signaling rate would be 1000 elements per second. (Note that, since baud is measured in elements per second, the term "baud rate" is incorrect since baud is already a measure of speed, or rate.) Continuous transmission is not required, because signaling speed is based only on the shortest signaling element.

Signaling rate in baud says nothing about actual information transfer rate. The maximum information transfer rate is defined as the number of equivalent binary digits transferred per second; this is measured in bits per second.

When binary data encoding is employed, each signaling element represents one bit. Complications arise when more sophisticated data encoding schemes are used. In a quadrature- phase-shift keying (QPSK) system, a phase transition of 90° represents a level shift. There are four possible states in a QPSK system; thus, two binary digits are required to represent the four possible states. If 1000 elements per second are transmitted in a quadriphase system where each element is represented by two bits, then the actual information rate is 2000 bps.

This scheme can be extended. It is possible to transmit three bits at a time using eight different phase angles (bps = 3 × baud). In addition, each angle can have more than one amplitude. A 9600 bps modem uses 12 phase angles, 4 of which have two amplitude values. This yields 16 distinct states, each represented by four binary digits. Using this technique, the information transfer rate is four times the signaling speed. This is what makes it possible to transfer data over a phone line at a rate that produces an unacceptable bandwidth using simpler binary encoding. This also makes it possible to transfer data at 2400 bps on 10 m, where FCC regulations allow only 1200-baud signals.

When are transmission speed in bauds and information rate in bps equal? Three conditions must be met: (1) binary encoding must be used, (2) all elements used to encode characters must be equal in width and (3) synchronous transmission at a constant rate must be employed. In all other cases, the two terms are not equivalent.

Within a given piece of equipment, it is desirable to use the highest possible data rate. When external devices are interfaced, it is normal practice to select the highest standard signaling rate at which both the sending and receiving equipment can operate.

Error Detection

Since data transfers are subject to errors, data transmission should include some method of detecting and correcting errors. Numerous techniques are available, each used depending on the specific circumstances, such as what types of errors are likely to be encountered. Some error detection techniques are discussed in the Modes and Modulation Sources chapter. One of the simplest and most common techniques, parity check, is discussed here.

Parity Check

Parity check provides adequate error detection for some data transfers. This method transmits a parity bit along with the data bits. In systems using odd parity, the parity bit is selected such that the number of 1 bits in the transmitted character (data bits plus parity bit) is odd. In even parity systems, the parity bit is chosen to give the character an even number of ones. For example, if the data 1101001 is to be transmitted, there are 4 (an even number) ones in the data. Thus, the parity bit should be set to 1 for odd parity (to give a total of 5 ones) or should be 0 for even parity (to maintain the even number, 4). When a character is received, the receiver checks parity by counting the ones in the character. If the parity is correct, the data is assumed to be correct. If the parity is wrong, an error has been detected.

Parity checking only detects a small fraction of possible errors. This can be intuitively understood by noting that a randomly chosen word has a 50% chance of having even parity and a 50% chance of having odd parity. Fortunately, on relatively error-free channels, single-bit errors are the most common and parity checking will always detect a single bit in error. However, an even number of errors will go undetected, whereas an odd number of errors will be detected. Parity checking is a simple error detection strategy. Because it is easy to implement, it is frequently used. Other more complex techniques are used in commercial data transmission services.

Standard Interface Busses

Signaling Levels

Inside equipment and for short runs of wire between equipment, the normal practice is to use neutral keying; that is, simply to key a voltage such as + 5 V on and off. In neutral keying, the off condition is considered to be 0 V. Over longer runs of wire, the line is viewed as a transmission line, with distributed inductance and capacitance. It takes longer to make the transition from 0 to 1 or vice versa because of the additional inductance and capacitance. This decreases the maximum speed at which data can be transferred on the wire and may also cause the 1s and 0s to be different lengths, called bias distortion. Also, longer lines are more likely to pick up noise, which can make it difficult for the receiver to decide exactly when the transition takes place. Because of these problems, bipolar keying is used on longer lines. Bipolar keying uses one polarity (for example +) for a logical 1 and the other (– in this example) for a 0. This means that the decision threshold at the receiver is 0 V. Any positive voltage is taken as a 1 and any negative voltage as a 0.

EIA-RS-232

The most common serial bus protocol, EIA-RS-232, addresses this issue (however, a Mark "1" is a negative voltage and a Space "0" is positive). Generally called RS-232, this protocol defines connectors and voltages between data terminal equipment (DTE) such as a PC, and data communications equipment (DCE), such as a modem or TNC. The connector is the DB-25, or the presently more popular DB-9 version. Signaling voltages are defined between + 3V and + 25V for logic "0" and between – 3V and – 25V for logic "1." Although the top data rate addressed in the specification is only 20 kbps, speeds of up to 115 kbps are commonly used. Communications distances of hundreds of meters are possible at reasonable data rates.

Since neutral keying is usually used inside equipment and bipolar keying for lines leaving equipment, signals must be converted between bipolar and neutral. Discrete level shifters or op amp circuits may perform this task, or low cost specialized IC line drivers and receivers are available.

RS-422

RS-422 is a serial protocol similar to RS-232, but employing fully differential data lines. Differential data offers the important advantage that common grounds between remote units are not necessary, and an important cause of ground loops (and their associated problems) is eliminated. Available on many Apple Macintosh computers, RS-422 systems may connect to standard RS-232 modems and TNCs by building a cable that makes the following translations:

RS-422 DTE	*RS-232 DCE*
RXD–	RXD
TXD–	TXD
RXD+	GND
TXD+	No Connection
GPi	CD

IrDA (Infrared Data Access)

Another high-speed serial protocol is the IrDA, which is a simple, short range wireless system using infrared LEDs and detectors. Data rates up to 3 MB/sec are possible between compatible units.

Universal Serial Bus (USB)

USB is a computer standard for an intelligent serial data transfer protocol. In addition to its higher speed than RS-232, USB offers reasonable power availability to its loads, or *functions*. Under certain circumstances, up to 127 hubs and functions may connect to a single computer. USBs requires that each function have onboard intelligence and that it negotiate with the host for power and bandwidth allocation, and has the major advantage of *hot-pluggablity* — the PC need not reboot when new functions are added. The USB connectors use four-conductor cable, with two bidirectional, differential data lines, power, and ground. Approximately 5 V at 100 mA is allowed per function, with up to 500 mA available if the host system has the capability. This means that relatively sophisticated devices, such as modems, small video cameras, or hand-held scanners may operate from the bus without additional power supplies. Preventing power back-flowing up from function to host is accomplished by configuring the connector shapes such that the host has a rectangular connector while that of functions are nearly square.

There are currently two USB standards. USB 1.1, somewhat obsolete but common in PCs just a few years old, is capable of 12 megabits per second (12 Mbps). USB 2.0 is the later standard and is rated up to 480 Mbps. Most USB 2.0 ports will allow the use of older USB 1.1 devices—that is they are backwards compatible. However maximum cable lengths and available power to devices may be affected.

IEEE-1394 (FireWire)

A very high speed serial protocol, IEEE-1394 (christened "FireWire" by its creator, Apple Computer), is capable of up to 400 Mbps of sustained transfer. It is ideal for high bandwidth systems, such as live video, external hard drives, or high-speed DVD player/recorders. Up to 63 devices may daisy-chain together at once via a standard six-wire cable. Unlike USB, 1394 is peer-to-peer, meaning any device may initiate a data transfer — the PC does not have to initiate a data transfer. Similar to USB, IEEE-1394 is hot-pluggable and provides power on the cable, but the voltage may vary from 7 V to almost 40 V, and may be sourced by any device. Allowable current drain per device may reach 1 A.

PC-Card (PCMCIA)

The PC-Card Standard is a collection of specifications for miniature plug-in peripherals. The most common is colloquially referred to as PCMCIA cards — 68-pin devices the size of thick credit cards that contain a modem, LAN, GPS receiver, USB port, FireWire port, high resolution video, an extra serial or parallel port, or memory storage expansion. The standard PC-Card allows up to 5 V at 1 A peak current, or 3.3 V at 1 A, depending upon configuration. Other voltages may be used if available from the host. Other portions of the specification define memory storage-only cards in even smaller footprints.

Small Computer System Interface (SCSI)

SCSI interfaces provide for up to 320 MB/s transmission rates with up to 15 devices. Used mostly with disk drives, the "skuzzy" bus also supports a very wide variety of high-speed peripherals. A wide variety of bus widths, speeds, and connectors exist. With the widespread use of USB, SCSI is no longer generally used except where large numbers of storage disks are needed — primarily in commercial installations.

10Base2, 10BaseT, 10Base4, 100BaseT

Common office/home networks use 10BaseN protocol. 10Base2 is generally recommended for Amateur Radio installations, since it uses shielded cable (RG-58, renamed "thin coax" in this application). Additionally, no separate hub is needed as the connected computers work on a peer-to-peer basis. A drawback of 10Base2 is its maximum data rate is limited to 10 Mbps. Also, shorter runs may occasionally require a simple extension in length of approximately 30-50 feet for proper operation. The other protocols use one or more hubs, RJ-45 connectors, and unshielded Category 5 cable. 100BaseT systems are rated to 100 Mbps.

Explanation of the *MBaseN* and *Category N* terminology can be found in many available networking books. Newer home networks are wireless; often, they are more trouble-free from high power Amateur Radio installations because the unshielded cable used for wired networks tends to pick up RF readily. Unfortunately, high power VHF and UHF installations may pose problems with wireless networks.

POWER SYSTEMS AND ATX

When the initial personal computers were designed, the most common logic family was TTL and CMOS that interfaced with TTL levels at 5 V. Disk drive and fan motors preferred +12 V. RS-232 demanded a higher voltage bipolar supply, so –12 V was added to complement the 12 V already used. The analog portion of the early modems required –5 V. Thus, the initial "silver box" PC supply provided +5 V at high current, +12 V at moderate current, and –5 V and –12 V at low current. Advances in semiconductor technology allowed shrinking transistor geometries. The smaller transistors were faster, but they had lower breakdown voltages, thus a new logic voltage of 3.3 V was introduced. Initially, computer manufacturers responded by placing IC regulators on the motherboard to power the 3.3-V circuits, but eventually the current demanded by these circuits exceeded that of the traditional +5V components and a new physical standard, called ATX, was introduced.

The ATX standard defines a layout physically different from older "AT-type" computers. The computer case, motherboard mounting holes, expansion slot location, and the power supply and its connector are all changed. ATX computers are recommended for the hamshack due to their better RFI control, resulting from careful mechanical design of connectors and consideration of card slot case penetration. ATX power supplies produce +3.3 V, +5 V, +12 V, –5 V and –12 V. They also provide an output voltage, even when the power supply is otherwise off, allowing *sleep mode* operation. Sleep mode retains the computer RAM contents and configuration so a reboot is not necessary each time the computer is used, and is especially critical to extending battery life in notebook PCs.

As semiconductor technology continued improving, the 3.3 V source became too high for the fine-geometry microprocessors, and an even lower voltage was required. At present, there is no standard for the next lower voltage, but devices are available that need anywhere between 1 V to 2.8 V for the microprocessor core (densest portion). Motherboard manufacturers have addressed this issue by again using on-board voltage regulators to drop an existing ATX standard voltage to the value needed by the CPU. Since there is no standard — in fact, the exact voltage preferred by a given processor family decreases as the manufacturing process evolves — motherboards provide means of selecting the matching voltage. Sometimes this process is automatic, as the on-board power supply communicates with the microprocessor before initializing and rises to the proper voltage, but other times, jumpers must be manually positioned *before* initial power is applied. Using a higher than recommended supply voltage causes excessive operating temperature and stresses the gates of the CMOS transistors, leading to reduced reliability and early (sometimes immediate) circuit death.

Power supplies removed from older computers are readily available, and there is a great temptation to reuse them for other amateur applications. Unfortunately, there is very little control on the manufacturers of these units, and quite often the expected *5-VDC, 10-amp* output is poorly regulated except at currents such as one or two amperes. Many of the newer units do not have an on-off switch, but depend on connecting one lead to ground for some period to turn the supply on and off. Even when off, a portion of the supply is on to provide control voltages. Many computer handbooks and texts contain tables of the connections to the various PC power supplies, complete with wire colors. However, there is no way to tell if compliance with the color codes has been followed. Hence, use of these supplies should be limited to those voltage sources that can be measured on a particular supply.

Powering Small Circuits

Small amounts of power can be 'stolen' from the serial port to power devices interfacing with a PC. The signals on the serial port, in accordance with the various versions of RS-232, generally range from +15 VDC to –15 VDC. By selecting several control signals and knowing that one of these signals will be at the positive voltage level, the power source in **Fig 5.92** can be used. The amount of current available will vary depending on the specific interface chip and chip supplier for the particular PC.

By reversing the diodes and filter capacitor a negative supply can be built. The regulator should be chosen for the negative or positive supply, as needed. Most PCs will not have any difficulty supplying 10 ma, or slightly more.

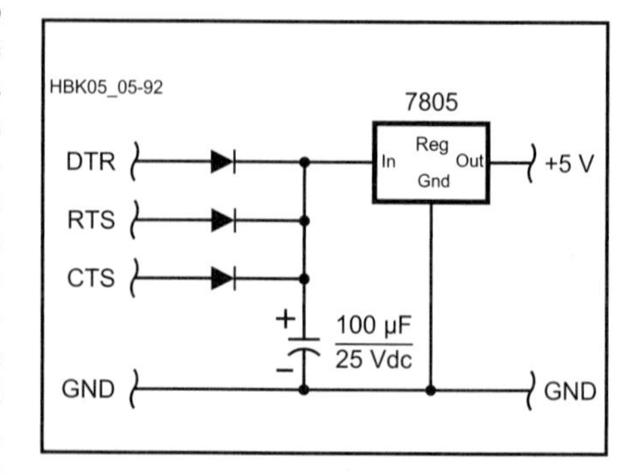

Fig 5.92 — Small amounts of power can be 'stolen' from the serial port with this circuit. See the Component Data and References chapter for the 7805 pin connections.

Power Quality

As operating voltages drop, power quality — the measure of voltage accuracy and transient response to changing load currents — becomes simultaneously more critical and more difficult. A 500-mV spike represents a 10% error in a 5-V supply, but the same 500-mV spike applied to a 2-V processor represents an overvoltage of 25%, grossly exceeding the maximum rated supply voltage for that controller. Further, if the spike is of the opposite polarity, it seriously reduces the noise margin of the logic-high levels, possibly corrupting data.

Providing clean power becomes more difficult when the effects of sleep mode are considered. Microprocessors reduce their power consumption when idle by slowing down internal clocks and other techniques, but when called back to duty, their response occurs in nanoseconds. The result is a huge change in supply current, from nearly zero to maximum current flow in those few nanoseconds. Large voltage spikes may result from this fast rise-time current step working against the inductance of the PC board power supply traces. Careful power supply design, especially during layout, and judicious use of low ESR, and low-inductance bypass capacitors mitigate the transients and keep the system reliable.

STANDARD COMPUTER CONNECTIONS

See the **Component Data and References** chapter for details on computer connector pinouts. You'll also find details on cables, such as a null modem cable.

Chapter 6

Real-World Component Characteristics

When is an inductor not an inductor? When it's a capacitor! This statement may seem odd, but it suggests the main message of this chapter. In the earlier chapter about **Electrical Fundamentals**, the basic components of electronic circuits were introduced. You saw that each has its own unique function to perform. For example, a capacitor stores energy in an electric field, a diode rectifies current and a battery provides voltage. All of these unique and different functions are necessary in order to build large circuits that perform useful tasks. The first part of this chapter, up to Low-Frequency Transistor Models, was written by Leonard Kay, K1NU.

As you may know from experience, these component pictures are *ideal*. That is, they are perfect mathematical pictures. An ideal component (or *element*) by definition behaves exactly like the mathematical equations that describe it, and *only* in that fashion. For example, an ideal capacitor passes a current that is equal to the capacitance C times the rate of change of the voltage across it. Period, end of sentence.

We call any other exhibited behavior either *nonideal*, *nonlinear* or *parasitic*. Nonideal behavior is a general term that covers any deviation from the theoretical picture. Ideal circuit elements are often linear: The graphs of their current versus voltage characteristics are straight lines when plotted on a suitable (Cartesian, rectangular) set of axes. We therefore call deviation from this behavior nonlinear: For example, as current through a resistor exceeds its power rating, the resistor heats up and its resistance changes. As a result, a graph of current versus voltage is no longer a straight line.

When a component begins to exhibit properties of a different component, as when a capacitor allows a dc current to pass through it, we call this behavior *parasitic*. Nonlinear and parasitic behavior are both examples of nonideal behavior.

Much to the bane of experimenters and design engineers, ideal components exist only in electronics textbooks and computer programs. Real components, the ones we use, only approximate ideal components (albeit very closely in most cases).

Real diodes store minuscule energy in electric fields (junction capacitance) and magnetic fields (lead inductance); *real* capacitors conduct some dc (modeled by a parallel resistance), and real battery voltage is not *perfectly* constant (it may even decrease nonlinearly during discharge).

Knowing to what extent and under what conditions real components cease to behave like their ideal counterparts, and what can be done to account for these behaviors, is the subject of component or circuit *modeling*. In this chapter, we will explore how and why the real components behave differently from ideal components, how we can account for those differences when analyzing circuits, how to select components to minimize, or exploit, nonideal behaviors and give a brief introduction to computer-aided circuit modeling.

Much of this chapter may seem intuitive. For example, you probably know that #20 hookup wire works just fine for wiring many experimental circuits. When testing a circuit after construction, you probably don't even think about the voltage drops across those pieces of wire (you inherently assume that they have zero resistance). But, connect that same #20 wire directly across a car battery—with no *other* series resistance—and suddenly, the resistance of the wire *does* matter, and it changes—as the wire heats, melts and breaks—from a very small value to infinity!

LUMPED VS DISTRIBUTED ELEMENTS

Most electronic circuits that we use everyday are inherently and math-ematically considered to be composed of *lumped elements*. That is, we assume each component acts at a single point in space, and the wires that connect these lumped elements are assumed to be *perfect conductors* (with zero resistance and insignificant length). This concept is illustrated in **Fig 6.1**. These assumptions are perfectly reasonable for many applications, but they have limits. Lumped element models break down when:

- Circuit impedance is so low that the small, but non-zero, resistance in the wires is important. (A significant portion of the circuit power may be lost to heat in the conductors.)
- Operating frequency (f_0) is high enough that the length of the connecting wires is a significant fraction (>0.1) of the wavelength. (Radiation from the conductors may be significant.)
- Transmission lines are used as conductors. (Their characteristic impedance is usually significant, and impedances connected to them are transformed as a function of the line length. See the **Transmission Lines** chapter for more information.)

Effects such as these are called *distributed*, and we talk of *distributed* elements or effects to contrast them to lumped elements.

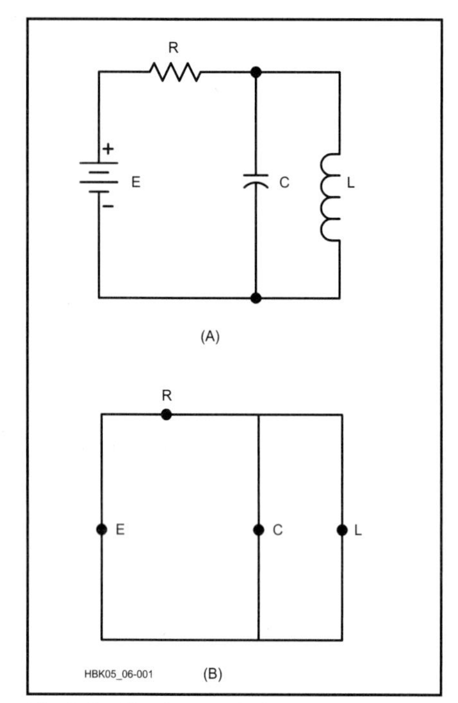

Fig 6.1—The lumped element concept. Ideally, the circuit at A is assumed to be as shown at B, where the components are isolated points connected by perfect conductors. Many components exhibit nonideal behavior when these assumptions no longer hold.

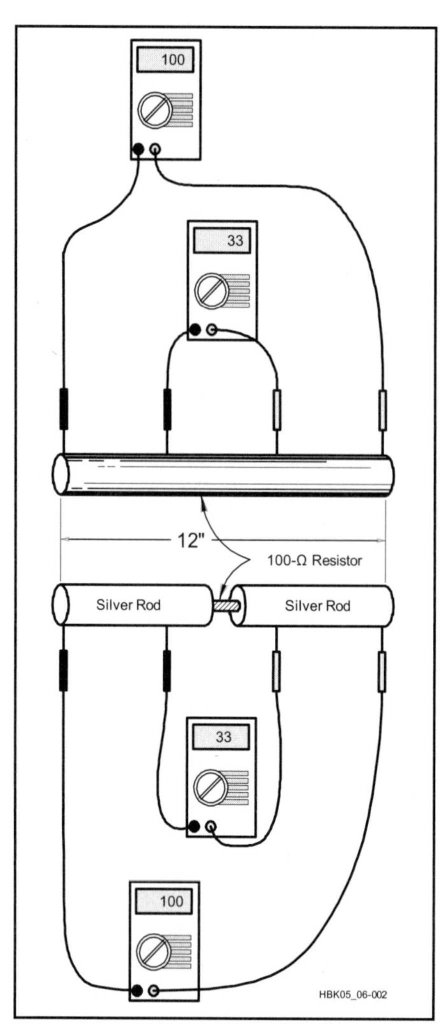

Fig 6.2—A, distributed and B, lumped resistances. See text for discussion.

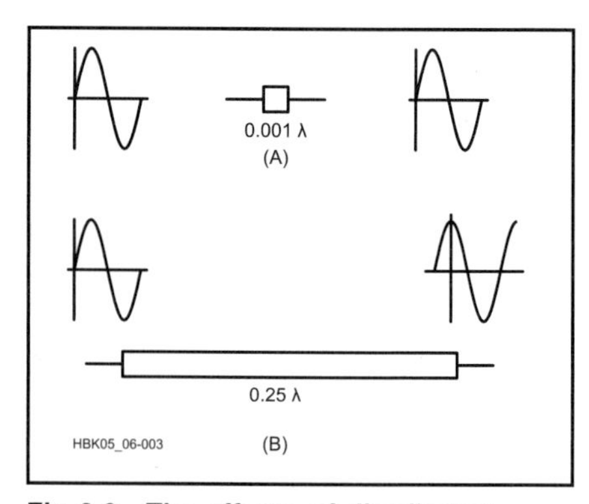

Fig 6.3—The effects of distributed resistance on the phase of a sinusoidal current. There is no phase delay between ends of a lumped element.

To illustrate the differences between lumped and distributed elements, consider the two resistors in **Fig 6.2**, which are both 12-inches long. The resistor at A is a uniform rod of carbon—a battery anode, for example. The second "resistor" B is made of two 6-inch pieces of silver rod (or other highly conductive material), with a small resistor soldered between them. Now imagine connecting the two probes of an ohmmeter to each of the two resistors, as in the figure. Starting with the probes at the far ends, as we slide the probes toward the center, the carbon rod will display a constantly decreasing resistance on the ohmmeter. This represents a *distributed* resistance. On the other hand, the ohmmeter connected to the other 12-inch "resistor" will display a constant resistance as long as one probe remains on each side of the small resistance. (Oh yes, as long as we neglect the resistance of the silver rods!) This represents a *lumped* resistance connected by perfect conductors.

Lumped elements also have the very desirable property that they introduce no phase shift resulting from propagation delay through the element. (Although combinations of lumped elements can produce phase shifts by virtue of their R, L and C properties.) Consider a lumped element that is carrying a sinusoidal current, as in **Fig 6.3A**. Since the element has negligible length, there is no phase difference in the current between the two sides of the element—*no matter how high the frequency*—precisely *because* the element length is negligible. If the physical length of the element were long, say 0.25 λ as shown in Fig 6.3B, the current phase would *not* be the same from end to end. In this instance, the current is delayed 90 electrical degrees. The amount of phase difference depends on the circuit's electrical length.

Because the relationship between the physical size of a circuit and the wavelength of an ac current present in the circuit will vary as the frequency of the ac signal varies, the ideas of lumped and distributed effects actually occupy two ends of a spectrum. At HF (30 MHz and below), where λ ≥10 m, the lumped element concept is almost always valid. In the UHF region and above, where λ ≥10 m and physical component size can represent a significant fraction of a wavelength, everything shows distributed effects to one degree or another. From roughly 30 to 300 MHz, problems are usually examined on a case-by-case basis.

Of course, if we could make resistors, capacitors, inductors and so on, very small, we could treat them as lumped elements at much higher frequencies.

Thanks to the advances constantly being made in microelectronic circuit fabrication, this is in fact possible. Commercial monolithic microwave integrated circuits (MMICs) in the early 1990s often use lumped elements that are valid up to 50 GHz. Since this frequency represents a wavelength of roughly 5 mm, this implies a component size of less than 0.5 mm.

LOW-FREQUENCY COMPONENT MODELS

Every circuit element behaves nonideally in some respect and under some conditions. It helps to know the most common types of nonideal behavior so we can design and build circuits that will perform as intended under expected ranges of operating conditions. In the sections below, we will discuss the basic components in order of increasing common nonideal behavior. Please note that much of this section applies only through HF; the peculiarities of VHF frequencies and above are discussed later in the chapter.

First, remember that some of the common limitations associated with real components are manufacturing concerns: tolerance and standard values. Tolerance is the measure of how much the actual value of a component may differ from its labeled value; it is usually expressed in percent. By convention, a "*higher*" (actually *closer*) tolerance component indicates a *lower* (lesser) percentage deviation and will usually cost more. For example, a 1-kΩ resistor with a 5% tolerance has an actual resistance of anywhere from 950 to 1050 Ω. When designing circuits be careful to include the effects of tolerance. More specific examples will be discussed below.

Keep standard values in mind because not every conceivable component value is available. For instance, if circuit calculations yield a 4,932-Ω resistor for a demanding application, there may be trouble. The nearest commonly available values are 4700 Ω and 5600 Ω, and they'll be rated at 5% tolerance!

These two constraints prevent us from building circuits that do precisely what we wish. In fact, the measured performance of any circuit is the summation of the tolerances and temperature characteristics of all the components in both the circuit under test and the test equipment itself. As a result, most circuits are designed to operate within tolerances such as "up to a certain power" or "within this frequency range." Some circuits have one or more adjustable components that can compensate for variations in others. These limitations are then further complicated by other problems.

RESISTORS

Resistors are made in several different ways: carbon composition, carbon film, metal film, and wire wound. Carbon composition resistors are simply small cylinders of carbon mixed with various binding agents. Carbon is technically a semiconductor and can be *doped* with various impurities to produce any desired resistance. Most common everyday $^1/_2$- and $^1/_4$-W resistors are of this sort. They are moderately stable from 0 to 60 °C (their resistance increases above and below this temperature range). They are not inductive, but they are relatively noisy, and have relatively wide tolerances.

The other resistors exploit the fact that resistance is proportional to the length of the resistor and inversely proportional to its cross-sectional area:

Wire-wound resistors are made from wire, which is cut to the proper length and wound on a coil form (usually ceramic). They are capable of handling high power; their values are very stable, and they are manufactured to close tolerances.

Metal-film resistors are made by depositing a thin film of aluminum, tungsten or other metal on an insulating substrate. Their resistances are controlled by careful adjustments of the width, length and depth of the film. As a result, they have very close tolerances. They are used extensively in surface-mount technology. As might be expected, their power handling capability is somewhat limited. They also produce very little electrical noise.

Carbon film resistors use a film of doped carbon instead of metal. They are not quite as stable as other film resistors and have wider tolerances than metal-film resistors, but they are still as good as (or better than) composition resistors.

Resistors behave much like their ideal through AF; lead inductance becomes a problem only at higher frequencies. The major departure from ideal behavior is their temperature coefficient (TC). The resistivity of most materials changes with temperature, and typical TC values for resistor materials are given in **Table 6.1**. TC values are usually expressed in parts-per-million (PPM) for each degree (centigrade) change from some nominal temperature, usually room temperature (77 °F/27 °C). A positive TC indicates an increase in resistance with increasing temperature while a negative TC indicates a decreasing resistance. For example, if a 1000-Ω resistor with a TC of +300 PPM/°C is heated to 50°C, the *change* in resistance is 300(50 – 27) = 6900 PPM, yielding a new resistance of

$$1000\left(1+\frac{6900}{1000000}\right)=1006.9\ \Omega$$

Carbon-film resistors are unique among the major resistor families because they alone have a negative temperature coefficient. They are often used to "offset" the thermal effects of the other components (see Thermal Considerations, below).

If the temperature increase is small (less than 30-40 °C), the resistance change with temperature is nondestructive—the resistor will return to normal when the temperature returns to its nominal value. Resistors that get too hot to touch, however, may be permanently damaged even if they appear normal. For this reason, be conservative when specifying power ratings for resistors. It's common to specify a resistor rated at 200% to 400% of the expected dissipation.

Table 6.1

Temperature Coefficients for Various Resistor Compositions

1 PPM = 1 part per million = 0.0001%

Type	*TC (PPM/°C)*
Wire wound	±(30 - 50)
Metal Film	±(100 - 200)
Carbon Film	+350 to –800
Carbon composition	±800

Wire-wound resistors are essentially inductors used as resistors. Their use is therefore limited to dc or low-frequency ac applications where their reactance is negligible. Remember that this inductance will also affect switching transient waveforms, even at dc, because the component will act as an RL circuit.

As a rough example, consider a 1-Ω, 5-W wire-wound resistor that is formed from #24 wire on a 0.5-inch diameter form. What is the approximate associated inductance? First, we calculate the length of wire using the wire tables:

$$L=\frac{R}{\Omega/\text{ ft for \#24 wire}}$$

$$=\frac{1}{25.7\ \Omega/1000\text{ ft}}=39\text{ ft}$$

This yields a total of (39 × 12 inches) / (0.5 π inch/turn) ≈ 298 turns, which further yields a coil length (for #24 wire close-wound at 46.9 turns per inch) of 6.3 inches, assuming a single-layer winding. Then, from the inductance formula for air coils in the **Electrical Fundamentals** chapter, calculate

$$L=\frac{(0.5)^2\times 298^2}{18\times 0.5+40\times 6.3}=85\ \mu\text{H}$$

Real wire-wound resistors have multiple windings layered over each other to minimize both size and parasitic inductance (by winding each layer in opposite directions, much of the inductance is canceled). If we assume a five-layer winding, the length is reduced to 1.8 inches and the inductance to approximately 17 µH. If we want the inductive reactance to stay below 10% of the resistor value, then this resistor cannot be used above f = 0.1 / (2π 17 µH) = 937 Hz, or roughly 1 kHz.

Another exception to the rule that resistors are, in general, fairly ideal has to do with skin effect at RF. This will be discussed later in the chapter. **Fig 6.4** shows some more accurate circuit models for resistors at low frequencies. For a treatment of pure resistance theory, look at the **Electrical Fundamentals** chapter.

VOLTAGE AND CURRENT SOURCES

An ideal voltage source maintains a constant voltage across its terminals no matter how much current is drawn. Consequently, it is capable of providing infinite power, for an infinite period of time. Similarly, an ideal current source provides a

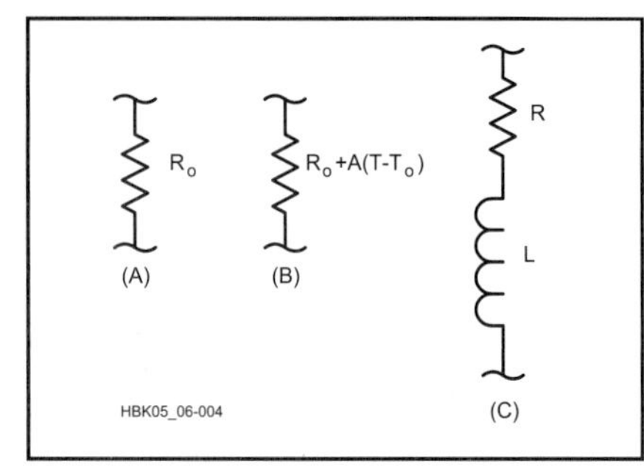

Fig 6.4—Circuit models for resistors. A is the ideal element. At B is the simple temperature-varying model for noninductive resistors. The wire-wound model with associated inductance is shown at C. For UHF or microwave designs, the model at C could be used with L representing lead inductance.

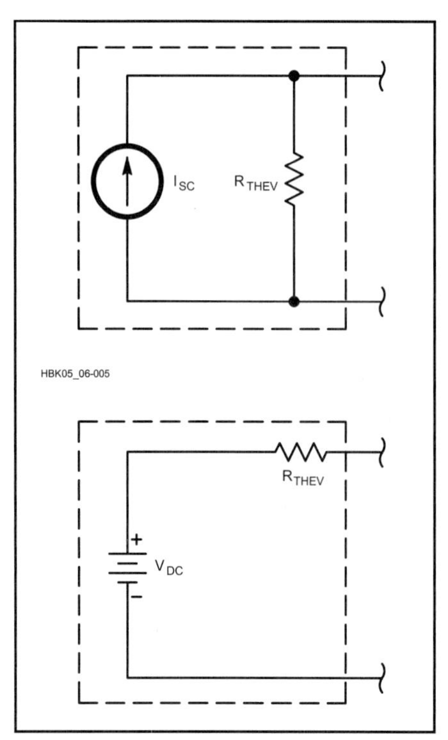

Fig 6.5—Norton and Thevenin equivalent circuits for real voltage and current sources.

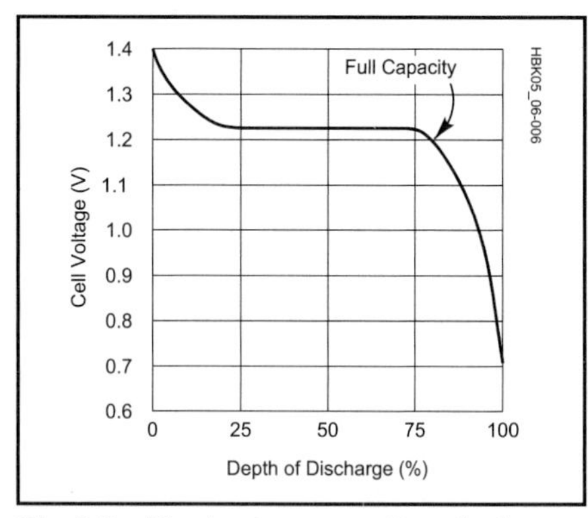

Fig 6.6—Discharge curve (terminal voltage vs percent capacity remaining) for a typical NiCd rechargeable battery. Note the dramatic voltage drop near the end.

constant current through its terminals no matter what voltage appears across it. It, too, can deliver an infinite amount of power for an infinite time.

Internal Resistance

As you may have learned through experience, *real* voltage and current sources—batteries and power supplies—do not meet these expectations. All real power sources have a finite *internal resistance* associated with them that limits the maximum power they can deliver.

We can model a real dc voltage source as a Thevenin-equivalent circuit of an *ideal* source in series with a resistance R_{thev} that is equal to the source's internal resistance. Similarly, we can model a real dc current source as a Norton-equivalent circuit: an ideal current source in *parallel* with R_{thev}. These two circuits shown in **Fig 6.5** are interchangeable through the relation

$$V_{OC} = I_{SC} \times R_{thev} \quad (1)$$

Using these more realistic models, the maximum current that a real voltage source can deliver is seen to be I_{sc} and the maximum voltage is V_{oc}.

We can model sinusoidal voltage or current sources in much the same way, keeping in mind that the internal *impedance*, Z_{thev}, for such a source may not be purely resistive, but may have a reactive component that varies with frequency.

Battery Capacity and Discharge Curves

Batteries have a finite energy capacity as well as an internal resistance. Because of the physical size of most batteries, a convenient unit for measuring capacity is the *milliampere hour* (mAh) or, for larger units, the *ampere hour* (Ah). Note that these are units of charge (coulombs per second × seconds) and thus they measure the total net charge that the battery holds on its plates when full.

A capacity of 1 mAh indicates that a battery, when fully charged, holds sufficient electrochemical energy to supply a steady current of 1 mA for 1 hour. If a battery discharges at a constant rate, you could estimate the useful time of a battery charge by the simple formula

$$\text{Time (hr)} = \frac{\text{capacity (mAh)}}{\text{current (mA)}} \quad (2)$$

In practice, however, the usable capacity of a rechargeable battery depends on the discharge *rate*, decreasing slightly as the current increases. For example, a 500 mAh nickel-cadmium (NiCd) battery may supply a current of 50 mA for almost 10 hours while maintaining a current of 500 mA for only 45 minutes.

The finite capacity of a battery does not *in itself* change the circuit model for a real source as shown in Fig 6.5. However, the chemistry of electrolytic cells, from which batteries are made, does introduce a minor, but significant, change. The voltage of a discharging battery does not stay constant, but slowly drops as time goes on. **Fig 6.6** illustrates the *discharge curve* for a typical NiCd rechargeable battery. For a given battery design, such curves are fairly reproducible and often used by battery-charging circuits to sense when a battery is fully charged or discharged.

CAPACITORS

The ideal capacitor is a pair of infinitely large parallel metal plates separated by an insulating or *dielectric* layer, ideally a vacuum (see **Fig 6.7**). For a discussion of pure capacitance see the **Electrical Fundamentals** chapter. For this case, the capacitance is given by

$$C = \frac{A\, \varepsilon_r\, \varepsilon_0}{d} \quad (3)$$

where

C = capacitance, in farads

A = area of plates, cm

d = spacing of the plates, cm

ε_r = dielectric constant of the insulating material

ε_0 = permittivity of free space, 8.85 × 10^{-14} F/cm.

If the plates are not infinite, the actual capacitance is somewhat higher due to *end effect*. This is the same phenomenon that causes a dipole to resonate at a lower frequency if you place large insulators (which add capacitance) on the ends.

If we built such a capacitor, we would find that (neglecting the effects of quantum mechanics) it would be a perfect open circuit at dc, and it could be operated at whatever voltage we desire without breakdown.

Leakage Conductance and Breakdown

If we use anything other than a vacuum for the insulating layer, even air, we introduce two problems. Because there are

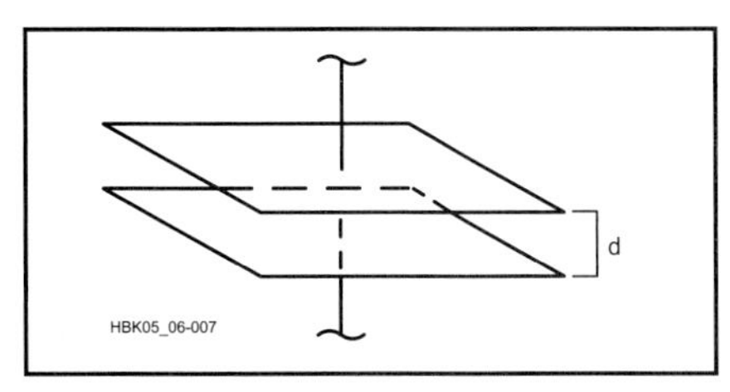

Fig 6.7—The ideal parallel-plate capacitor.

atoms between the plates, the capacitor will now be able to conduct a dc current. The magnitude of this *leakage current* will depend on the insulator quality, and the current is usually very small. Leakage current can be modeled by a resistance R_L in parallel with the capacitance (in the ideal case, this resistance is infinite).

In addition, when a high enough voltage is applied to the capacitor, the atoms of the dielectric will ionize due to the extremely high electric field and cause a large dc current to flow. This is *dielectric breakdown,* and it is destructive to the capacitor if the dielectric is ruined. To avoid dielectric breakdown, a capacitor has a *working voltage* rating, which represents the maximum voltage that can be permitted to develop across it.

Dielectrics

The leakage conductance and breakdown voltage characteristics of a capacitor are strongly dependent on the composition and quality of the dielectric. Various materials are used for different reasons such as availability, cost, and desired capacitance range. In rough order of "best" to "worst" they are:

Vacuum Both fixed and variable vacuum capacitors are available. They are rated by their maximum working voltages (3 to 60 kV) and currents. Losses are specified as negligible for most applications.

Air An *air-spaced* capacitor provides the best commonly available approximation to the ideal picture. Since $e_r = 1$ for air, air-dielectric capacitors are large when compared to those of the same value using other dielectrics. Their capacitance is very stable over a wide temperature range, leakage losses are low, and therefore a high Q can be obtained. They also can withstand high voltages. For these reasons (and ease of construction) most variable capacitors in tuning circuits are air-spaced.

Plastic film Capacitors with plastic film (polystyrene, polyethylene or Mylar) dielectrics are more expensive than paper capacitors, but have much lower leakage rates (even at high temperatures) and low TCs. Capacitance values are more stable than those of paper capacitors. In other respects, they have much the same characteristics as paper capacitors. Plastic-film variable capacitors are available.

Mica The capacitance of mica capacitors is very stable with respect to time, temperature and electrical stress. Leakage and losses are very low. Values range from 1 pF to 0.1 µF, with tolerances from 1 to 20%. High working voltages are possible, but they must be derated severely as operating frequency increases.

Silver mica capacitors are made by depositing a thin layer of silver on the mica dielectric. This makes the value even more stable, but it presents the possibility of silver migration through the dielectric. The migration problem worsens with increased dc voltage, temperature and humidity. Avoid using silver-mica capacitors under such conditions.

Ceramic There are two kinds of ceramic capacitors. Those with a low dielectric constant are relatively large, but very stable and nearly as good as mica capacitors at HF. High dielectric constant ceramic capacitors are physically small for their capacitance, but their value is not as stable. Their dielectric properties vary with temperature, applied voltage and operating frequency. They also exhibit piezoelectric behavior. Use them only in coupling and bypass roles. Tolerances are usually +100% and –20%. Ceramic capacitors are available in a wide range of values: 10 pF to 1 µF. Some variable units are available.

Electrolytic These capacitors have the space between their foil plates filled with a chemical paste. When voltage is applied, a chemical reaction forms a layer of insulating material on the foil.

Electrolytic capacitors are popular because they provide high capacitance values in small packages at a reasonable cost. Leakage is high, as is inductance, and they are polarized—there is a definite positive and negative plate, due to the chemical reaction that provides the dielectric. Internal inductance restricts aluminum-foil electrolytics to low-frequency applications. They are available with values from 1 to 500,000 µF.

Tantalum electrolytic capacitors perform better than aluminum units but their cost is higher. They are smaller, lighter and more stable, with less leakage and inductance than their aluminum counterparts. Reformation problems are less frequent, but working voltages are not as high as with aluminum units.

Electrolytics should not be used if the dc potential is well below the capacitor working voltage.

Paper Paper capacitors are inexpensive; capacitances from 500 pF to 50 µF are available. High working voltages are possible, but paper-dielectric capacitors have high leakage rates and tolerances are no better than 10 to 20%. Paper-dielectric capacitors are not polarized; however, the body of the capacitor is usually marked with a color band at one end. The band indicates the terminal that is connected to the outermost plate of the capacitor. This terminal should be connected to the side of the circuit at the lower potential as a safety precaution.

Loss Angle

For ac signals (even at low frequencies), capacitors exhibit an additional parasitic resistance that is due to the electromagnetic properties of dielectric materials. This resistance is often quantified in catalogs as *loss angle,* θ, because it represents the angle, in the complex impedance plane, between $Z_C = R + jX_C$ and X_C. This angle is usually quite small, and would be zero for an ideal capacitor.

Loss angle is normally specified as tan θ at a certain frequency, which is simply the ratio R/X_C. The loss angle of a given capacitor is relatively constant over frequency, which means the *effective series resistance* or *ESR* = (tan θ) / (2 π f C) goes down as frequency goes up. This resistance is placed in *series* with the capacitor because it came (mathematically) from the equation for Z_C above. It can always be converted into a parallel resistance if desired. To summarize, **Figs 6.8** and **6.9** show reasonable models for the capacitor that are good up to VHF.

Temperature Coefficients and Tolerances

As with resistors, capacitor values vary in production, and most capacitors have a tolerance rating either printed on them or listed on a data sheet. Capacitance varies with temperature, and this is important to consider when constructing a circuit that will carry high power levels or operate in a hot environment. Also, as just described, not all capacitors are available in all ranges of values due to inherent differences in material properties. Typical values, temperature coefficients and leakage conductances for several capacitor types are given in **Table 6.2.**

DIODES

An ideal diode acts as a rectifying switch—it is a short circuit when forward biased and an open circuit when reverse biased. Many circuit components can be completely described in terms of current-voltage (or *I-V*) characteristics, and to

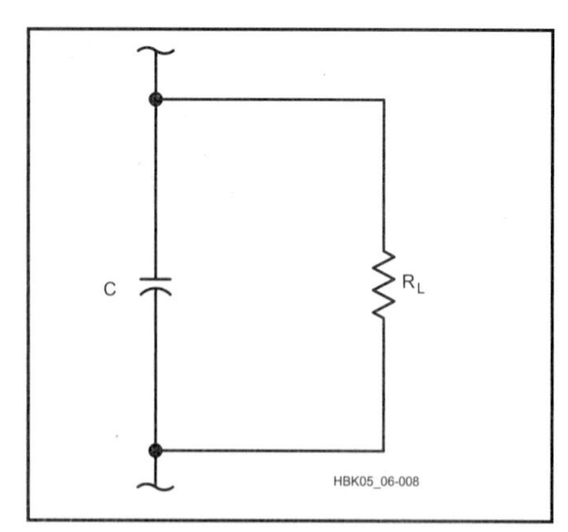

Fig 6.8—A simple capacitor model for frequencies well below self-resonance.

Table 6.2

Typical Temperature Coefficients and Leakage Conductances for Various Capacitor Constructions

Type	*TC @ 20°C (PPM/°C)*	*DC Leakage Conductance (Ω)*
Ceramic Disc	±300(NP0)	> 10 M
	+150/–1500(GP)	> 10 M
Mica	–20 to +100	> 100,000 M
Polyester	±500	> 10 M
Tantalum Electrolytic	±1500	> 10 MΩ
Small Al Electrolytic(≈ 100 µF)	–20,000	500 k - 1 M
Large Al Electrolytic(≈ 10 mF)	–100,000	10 k
Vacuum (glass)	+100	≈ ∞
Vacuum (ceramic)	+50	≈ ∞

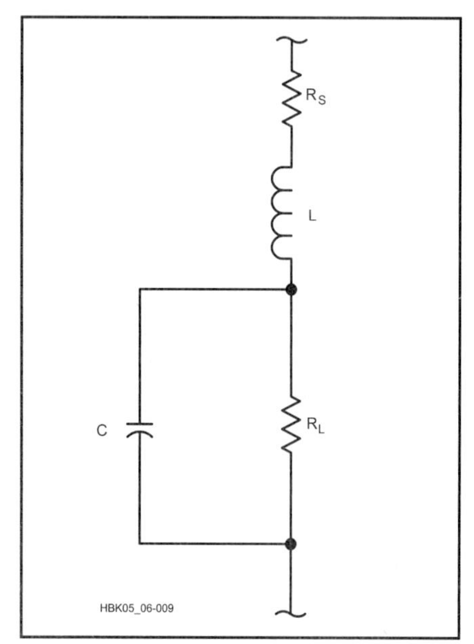

Fig 6.9—A capacitor model for VHF and above including series resistance and distributed inductance.

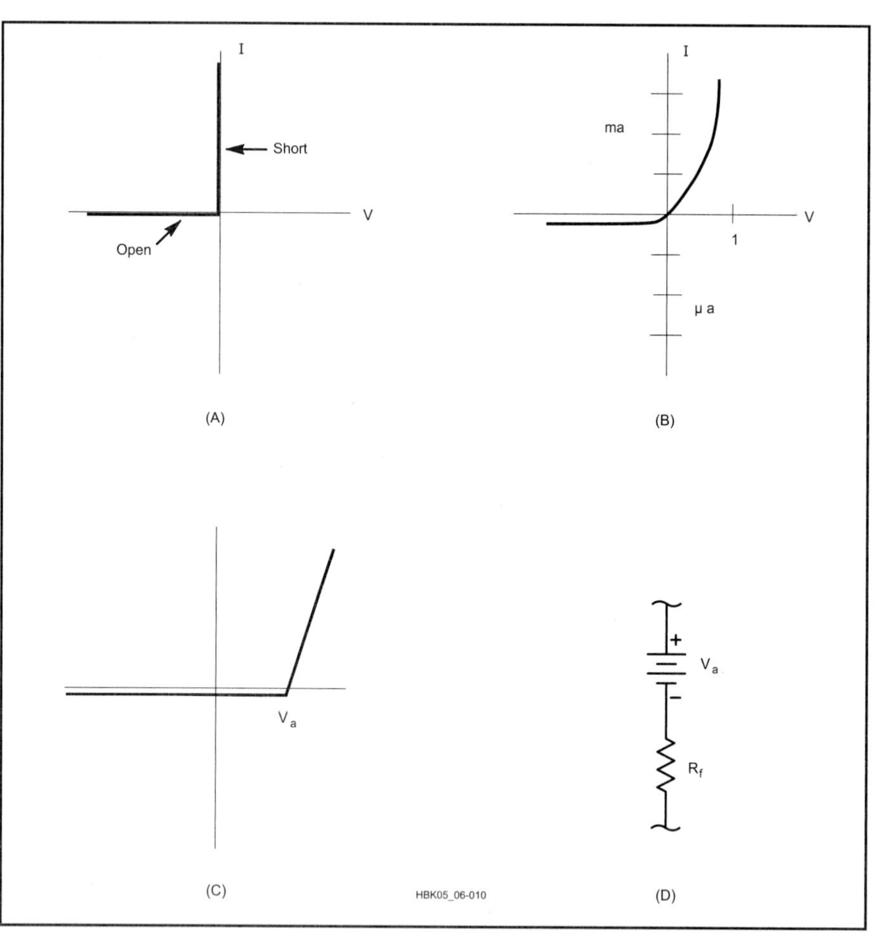

Fig 6.10—Circuit models for rectifying switches (diodes). A: I-V curve of the ideal rectifier. B: I-V curve of a typical semiconductor diode. Note the different scales for + and – current. C shows a simplified diode I-V curve for dc-circuit calculations. D is an equivalent circuit for C.

discuss the diode, this approach is especially helpful. **Fig 6.10A** shows the I-V curve for an ideal rectifier.

In contrast, the I-V curve for a semiconductor diode junction is given by the following equation (slightly simplified).

$$I = I_S \left(e^{\frac{V}{V_t}} \right) \quad (4)$$

where

I = diode current
V = diode voltage
I_s = reverse-bias saturation current
V_t = kT/q, the thermal equivalent of voltage (about 25 mV at room temperature).

This curve is shown in Fig 6.10B.

The obvious differences between Fig 6.10A and B are that the semiconductor diode has a finite *turn-on* voltage—it requires a small but nonzero forward bias voltage before it begins conducting. Furthermore, once conducting, the diode voltage continues to increase very slowly with increasing current, unlike a true short circuit. Finally, when the applied voltage is negative, the current is not exactly zero but very small (microamperes).

For bias (dc) circuit calculations, a useful model for the diode that takes these two effects into account is shown by the artificial I-V curve in Fig 6.10C. The small reverse bias current I_s is assumed to be completely negligible.

When converted into an equivalent circuit, the model in Fig 6.10C yields the picture in Fig 6.10D. The ideal voltage source

V_a represents the turn-on voltage and R_f represents the effective resistance caused by the small increase in diode voltage as the diode current increases. The turn-on voltage is material-dependent: approximately 0.3 V for germanium diodes and 0.7 for silicon. R_f is typically on the order of 10 Ω, but it can vary according to the specific component. R_f can often be completely neglected in comparison to the other resistances in the circuit. This very common simplification leaves only a pure voltage drop for the diode model.

Temperature Bias Dependence

The reverse saturation current I_s is not constant but is itself a complicated function of temperature. For silicon diodes (and transistors) near room temperature, I_s increases by a factor of 2 every 4.8 °C. This means that for every 4.8 °C rise in temperature, either the diode current doubles (if the voltage across it is constant), or if the current is held constant by other resistances in the circuit, the diode voltage will *decrease* by $V_t \times \ln 2$ = 18 mV. For germanium, the current doubles every 8 °C and for gallium arsenide (GaAs), 3.7 °C. This dependence is highly reproducible and may actually be exploited to produce temperature-measuring circuits.

While the change resulting from a rise of several degrees may be tolerable in a circuit design, that from 20 or 30 degrees may not. Therefore it's a good idea with diodes, just as with other components, to specify power ratings conservatively (2 to 4 times over) to prevent self-heating.

While component derating does reduce self-heating effects, circuits must be designed for the expected operating environment. For example, mobile radios may face temperatures from –20° to +140°F (–29° to 60°C).

Junction Capacitance

Immediately surrounding a PN junction is a *depletion layer*. This is an electrically charged region consisting primarily of ionized atoms with relatively few electrons and holes (see **Fig 6.11**). Outside the depletion layer are the remainder of the P and N regions, which do not contribute to diode operation but behave primarily as parasitic series resistances.

We can treat the depletion layer as a tiny capacitor consisting of two parallel plates. As the reverse bias applied to a diode changes, the width of the depletion layer, and therefore the capacitance, also changes. There is an additional *diffusion capacitance* that appears under forward bias due to electron and hole storage in the bulk regions, but we will not discuss this here because diodes are usually reverse biased when their capacitance is exploited. The diode junction capacitance under a reverse bias of V volts is given by

$$C_j = C_{j0} \Big/ \sqrt{V_{on} - V} \qquad (5)$$

where C_{j0} = measured capacitance with zero applied voltage.

Note that the quantity under the radical is a large *positive* quantity for reverse bias.

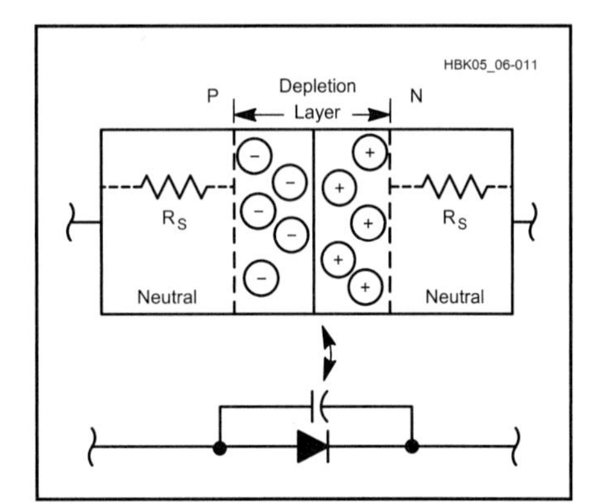

Fig 6.11—A more detailed picture of the PN junction, showing depletion layer, bulk regions and junction capacitance.

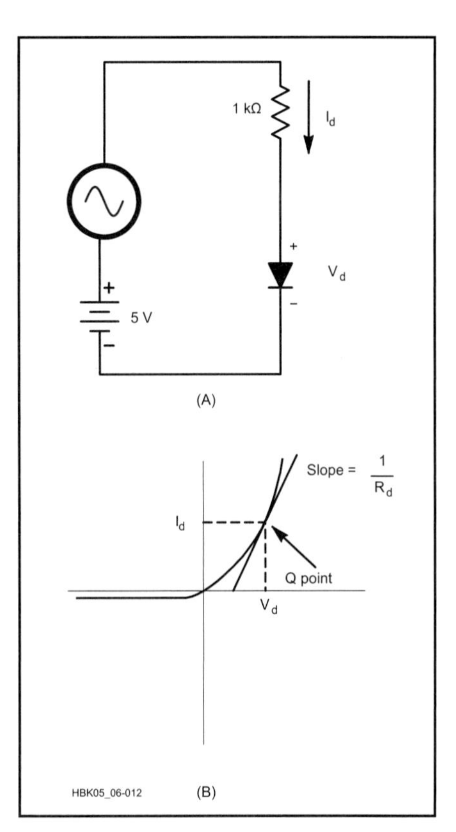

Fig 6.12—A simple resistor-diode circuit used to illustrate dynamic resistance. The ac input voltage "sees" a diode resistance whose value is the slope of the line at the Q-point, shown in B.

As seen from the equation, for large reverse biases C_j is inversely proportional to the square root of the voltage.

Junction capacitances are small, on the order of pF. They become important, however, for diode circuits at RF, as they can affect the resonant frequency. In fact, *varactors* are diodes used for just this purpose.

Reverse Breakdown and Zener Diodes

If a sufficiently large reverse bias is applied to a real diode, the internal electric field becomes so strong that electrons and holes are ripped from their atoms, and a large reverse current begins to flow. This is called *breakdown* or *avalanche*. Unless the current is so large that the diode fails from overheating, breakdown is not destructive and the diode will again behave normally when the bias is removed.

Once a diode breaks down, the voltage across it remains relatively constant regardless of the level of breakdown current. *Zener* diodes are diodes that are specially made to operate in this region with a very constant voltage, called the *Zener voltage*. They are used primarily as voltage regulators. When operating in this region, they can be modeled as a simple voltage source. Zener diodes are rated by their Zener voltage and power dissipation.

A Diode AC Model

Fig 6.12A shows a simple resistor-diode circuit to which is applied a dc bias voltage plus an ac signal. Assuming that the voltage drop across the diode is 0.6 V and R_f is negligible, we can calculate the bias current to be I » (5 – 0.6) / 1 kΩ = 4.4 mA. This point is marked on the diodes I-V curve in Fig 6.12B. If we draw a line tangent to this point, as shown, the slope of this line represents the *dynamic* resistance R_d of the diode seen by a small ac signal, which at room temperature can be approximated by

$$R_d = \frac{25}{I} \Omega \qquad (6)$$

where I is the diode current in mA. Note that this resistance changes with bias current and should not be confused with the dc forward resistance in the previous section, which has a similar value but represents a different concept. **Fig 6.13** shows a low-frequency ac model for the diode, including the dynamic resistance and junction capacitance.

Switching Time

If you change the polarity of a signal applied to the ideal switch whose I-V curve appears in Fig 6.10A, the switch turns on or off instantaneously. A real diode cannot do this, as a finite amount of time is required to move electrons and holes in or out of the diode as it changes states (effectively, the diode capacitances must be charged or discharged). As a result, diodes have a maximum useful frequency when used in switching applications. The operation of diode switching circuits can often be modeled by the picture in **Fig 6.14.** The approximate switching time (in seconds) for this circuit is given by

$$t_s = \tau_p \frac{\left(\frac{V_1}{R_1}\right)}{\left(\frac{V_2}{R_2}\right)} = \tau_p \frac{I_1}{I_2} \qquad (7)$$

where τ_p is the minority carrier lifetime of the diode (a material constant determined during manufacture, on the order of 1 ms). I_1 and I_2 are currents that flow during the switching process. The minimum time in which a diode can switch from one state to the other and back again is therefore 2 t_s, and thus the maximum usable switching frequency is f_{sw} (Hz) = $^1/_2$ t_s. It is usually a good idea to stay below this by a factor of two. Diode data sheets usually give typical switching times and show the circuit used to measure them.

Note that f_{sw} depends on the forward and reverse currents, determined by I_1 and I_2 (or equivalently V_1, V_2, R_1, and R_2). Within a reasonable range, the switching time can be reduced by manipulating these currents. Of course, the maximum power that other circuit elements can handle places an upper limit on switching currents.

Schottky Diodes

Schottky diodes are made from metal-semiconductor junctions rather than PN junctions. They store less charge internally and as a result, have shorter switching times and junction capacitances than standard PN junction diodes.

Their turn-on voltage is also less, typically 0.3 to 0.4 V. In most other respects they behave similarly to PN diodes.

INDUCTORS

Inductors are the problem children of the component world. (Fundamental inductance is discussed in the **Electrical Fundamentals** chapter). Besides being difficult to fabricate on integrated circuits, they are perhaps the most nonideal of real-world components. While the leakage conductance of a capacitor is usually negligible, the series resistance of an inductor often is not. This is basically because an inductor is made of a long piece of relatively thin wire wound into a small coil. As an example, consider a typical air-core tuning coil from a component catalog, with L = 33 mH and a minimum Q of 30 measured at 2.5 MHz. This would indicate a series resistance of $R_s = 2 \pi f L / Q =$ 17 Ω. This R_s could significantly alter the resonant frequency of a circuit. For frequencies up to HF this is the only significant nonlinearity, and a low-frequency circuit model for the inductor is shown in **Fig 6.15.** Many of the problems associated with inductors are actually due to core materials, and not the coil itself.

Magnetic Materials

Many discrete inductors and transformers are wound on a core of iron or other magnetic material. As discussed in the **Electrical Fundamentals** chapter, the inductance value of a coil is proportional to the density of magnetic field lines that pass through it. A piece of magnetic material placed inside a coil will con-centrate the field lines inside itself. This permits a higher number of field lines to exist inside a coil of a given cross-sectional area, thus allowing a much higher inductance than what would be possible with an air core. This is especially important for transformers that operate at low frequencies (such as 60 Hz) to ensure the reactance of the windings is high.

Core Saturation

Magnetic (more precisely, ferro-magnetic) materials exhibit two kinds of nonlinear behavior that are important to circuit design. The first is the phenomenon of *saturation*. A magnetic core increases the magnetic flux density of a coil because the current passing through the coil forces the atoms of the iron (or other material) to line up, just like many small compass needles, and the magnetic field that results from the atomic alignment is *much* larger than that produced by the current with no core. As coil current increases, more and more atoms line up. At some high current, all of the atoms will be aligned and the core is *saturated.* Any further increase in current can't increase the core alignment any further. Of course, added current generates its own magnetic flux, but it is very small compared to the magnetic field contributed by the aligned atoms in the magnetic core.

The important concept in terms of circuit design is that as long as the coil current remains below saturation, the inductance of the coil is essentially constant. **Fig 6.17** shows graphs of magnetic flux linkage (N f) and inductance (L) vs current (i) for a typical iron-core inductor. These quantities are related by the equation

$$N\phi = Li \qquad (8)$$

where
- N = number of turns,
- ϕ = flux density
- L = inductance
- i = current.

In the lower graph, a line drawn from any point on the curve to the (0,0) point will show the effective inductance, $L = N \phi / i$,

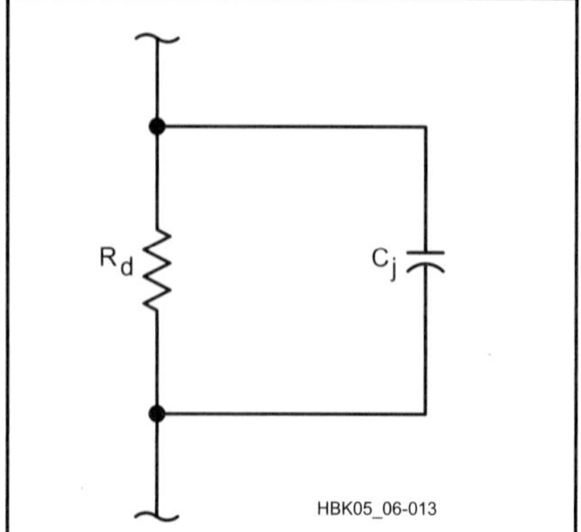

Fig 6.13—An ac model for diodes. R_d is the dynamic resistance and C_j is the junction capacitance.

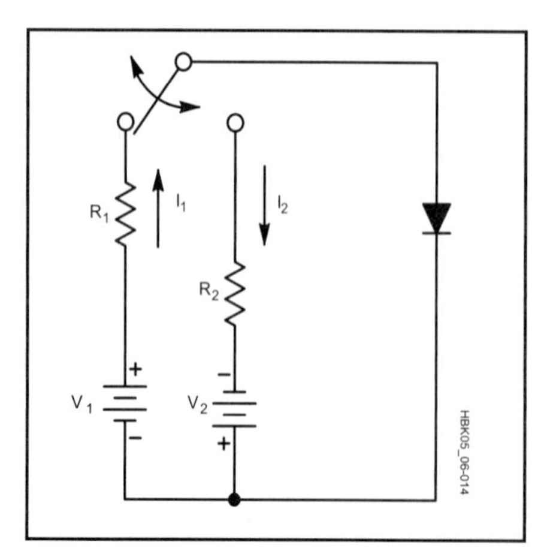

Fig 6.14—Circuit used for computation of diode switching time.

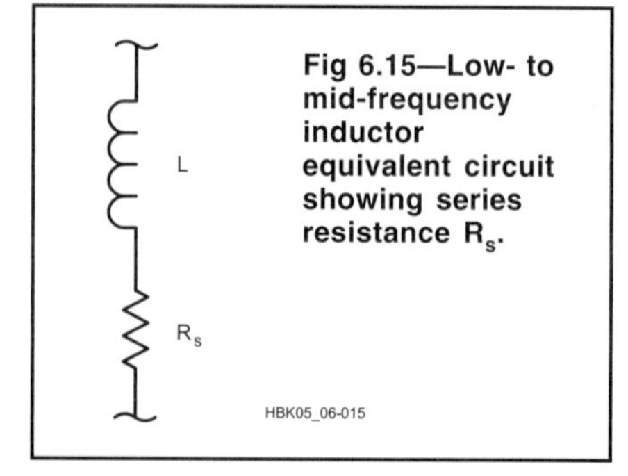

Fig 6.15—Low- to mid-frequency inductor equivalent circuit showing series resistance R_s.

at that current. These results are plotted on the upper graph.

Note that below saturation, the inductance is constant because both N ϕ and i are increasing at a steady rate. Once the saturation current is reached, the inductance decreases because N ϕ does not increase anymore (except for the tiny additional magnetic field the current itself provides). This may render some coils useless at VHF and higher frequencies. Air-coil inductors do not suffer from saturation.

One common method of increasing the saturation current level is to cut a small air gap in the core (see **Fig 6.16**). This gap forces the flux lines to travel through air for a short distance. Since the saturation flux linkage of the core is unchanged, this method works by requiring a higher current to achieve saturation. The price that is paid is a reduced inductance below saturation. Curves B in Fig 6.17 show the result of an air gap added to that inductor.

Manufacturer's data sheets for magnetic cores usually specify the saturation flux density. Saturation flux density, ϕ, in gauss can be calculated for ac and dc currents from the following equations:

$$\phi_{ac} = \frac{3.49\ V}{fNA} \quad (9)$$

$$\phi_{dc} = \frac{NIA_L}{10\ A} \quad (10)$$

where

V = RMS ac voltage

f = frequency, in MHz

N = number of turns

A = equivalent area of the magnetic path in square inches (from the data sheet)

I = dc current, in A

A_L = inductance index (also from the data sheet).

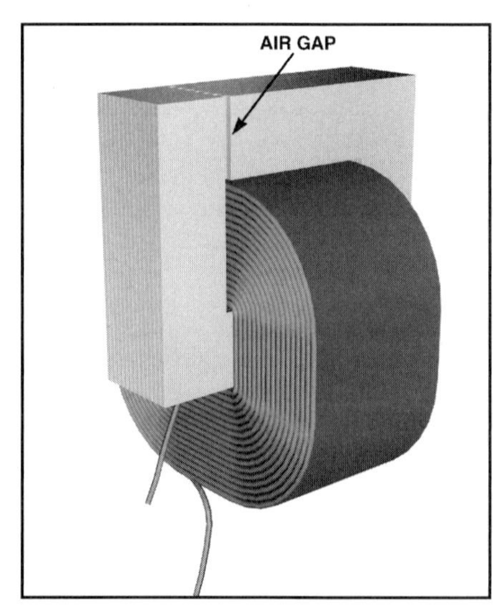

Fig 6.16—Typical construction of a magnetic-core inductor. The air gap greatly reduces core saturation at the expense of some inductance. The insulating laminations between the core layers help to minimize eddy currents.

Hysteresis

Consider **Fig 6.18**. If the current passed through a magnetic-core inductor is increased from zero (point a) to near the point of saturation (point b) and then decreased back to zero, we find that a magnetic field remains, because some of the core atoms retain their alignment. If we then increase the current in the opposite direction and again return to zero (through points c, d and e), the curve does *not* retrace itself. This is the property of *hysteresis*.

If a circuit carries a large ac current (that is, equal to or larger than saturation), the path shown in Fig 6.18 (from b to e and back again) will be retraced many times each second. Since the curve is nonlinear, this will introduce distortion in the resulting waveform. Where linear circuit operation is crucial, it is important to restrict the operation of magnetic-core inductors well below saturation.

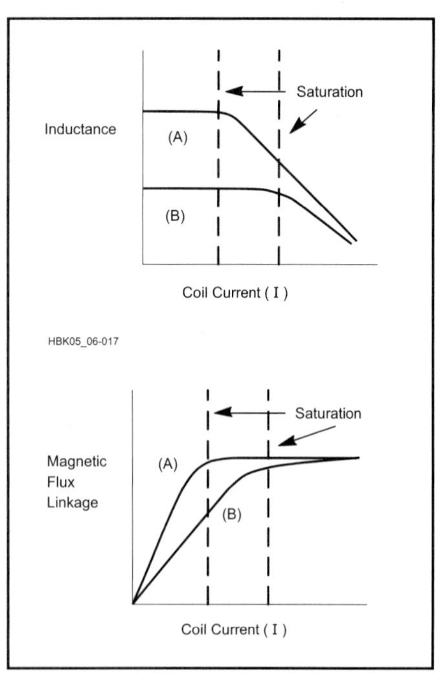

Fig 6.17—Magnetic flux linkage and inductance plotted versus coil current for (A) a typical iron-core inductor. As the flux linkage N ϕ in the coil saturates, the inductance begins to decrease since inductance = flux linkage / current. The curves marked B show the effect of adding an air gap to the core. The current handling capability has increased but at the expense of reduced inductance.

Eddy Currents

The changing magnetic field produced by an ac current generates a "back voltage" in the core as well as the coil itself. Since magnetic core material is usually conductive, this voltage causes a current to flow in the core. This eddy current serves no useful purpose and represents power lost to heat. Eddy currents can be substantially reduced by laminating the core—slicing the core into thin sheets and placing a suitable insulating material (such as varnish) between them (see Fig 6.16).

Transformers

In an ideal transformer (described in the **Electrical Fundamentals** chapter), *all* the power supplied to its primary terminals is available at the secondary terminals. An air-core transformer provides a very good approximation to the ideal case, the major loss being the series resistances of the windings. As you might guess, there are several ways that a magnetic-core transformer can lose power. For example, useless heat can be generated through hysteresis and eddy currents; power may be lost to harmonic generation through nonlinear saturation effects.

Another source of loss in transformers (or actually, any pair of mutual inductances) is *leakage reactance*. While the main purpose of a magnetic core is to concentrate the magnetic flux entirely within itself, in a real-life device there is always some small amount of flux that does not pass through both windings. This leakage reactance can be pictured as a small amount of self-inductance appearing on

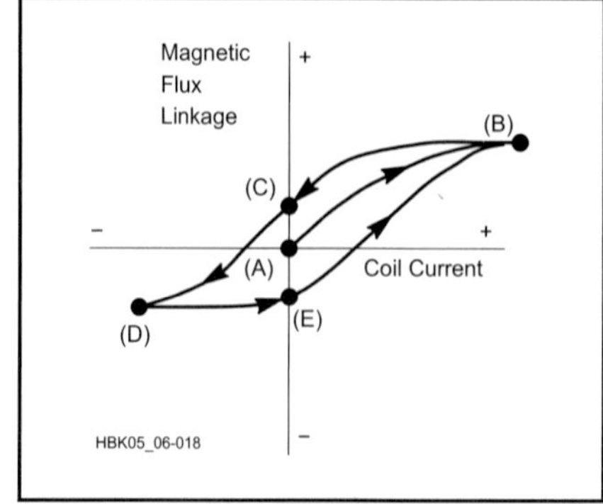

Fig 6.18—Hysteresis loop. A large current (larger than saturation current) passed through a magnetic-core coil will cause some permanent magnetization of the core. This results in a different path of flux linkage vs current to be traced as the current decreases.

each winding. **Fig 6.19** shows a fairly complete circuit model for a transformer at low to medium frequencies.

COMPONENTS AT RF

The models described in the previous section are good for dc, and ac up through AF and low RF. At HF and above (where we do much of our circuit design) several other considerations become very important, in some cases dominant, in our component models. To understand what happens to circuits at RF (a good cutoff is 30 MHz and above) we turn to a brief discussion of some electromagnetic and microwave theory concepts.

Parasitic Inductance

Maxwell's equations—the basic laws of electromagnetics that govern the propagation of electromagnetic waves and the operation of all electronic components—tell us that any wire carrying a current that changes with time (one example is a sine wave) develops a changing magnetic field around it. This changing magnetic field in turn induces an opposing voltage, or back EMF, on the wire. The back EMF is proportional to how fast the current changes (see **Fig 6.20**).

We exploit this phenomenon when we make an inductor. The reason we typically form inductors in the shape of a coil is to concentrate the magnetic field lines and thereby maximize the inductance for a given physical size. However, *all* wires carrying varying currents have these inductive properties. This includes the wires we use to connect our circuits, and even the *leads* of capacitors, resistors and so on.

The inductance of a straight, round, nonmagnetic wire in free space is given by:

$$L = 0.00508\,b\left[\ln\left(\frac{2b}{a}\right) - 0.75\right] \qquad (11)$$

where

L = inductance, in µH

a = wire radius, in inches

b = wire length, in inches

ln = natural logarithm ($2.303 \times \log_{10}$).

Skin effect (see below) changes this formula slightly at VHF and above. As the frequency approaches infinity, the constant 0.75 in the above equation approaches 1. This effect usually presents no more than a few percent change.

As an example, let's find the inductance of a #18 wire (diam = 0.0403 inch) that is 4 inches long (a typical wire in a circuit). Then a = 0.0201 and b = 4:

$$L = 0.00508\,(4)\left[\ln\left(\frac{8}{0.0201}\right) - 0.75\right]$$

$$= 0.0203[5.98 - 0.75] = 0.106\ \mu\text{H}$$

It is obvious that this *parasitic* inductance is usually very small. It becomes important *only at very high frequencies*; at AF or LF, the parasitic inductive reactance is practically zero. To use this example, the reactance of a 0.106 µH inductor even at 10 MHz is only 6.6 Ω. **Fig 6.21** shows a graph of the inductance for wires of various gauges (radii) as a function of length.

Any circuit component that has wires attached to it, or is fabricated from wire, will have a parasitic inductance associated with it. We can treat this parasitic inductance in component models by adding an inductor of appropriate value in *series* with the component (since the wire lengths are in series with the element). This (among other reasons) is why minimizing lead lengths and interconnecting wires becomes very important when designing circuits for VHF and above.

Parasitic Capacitance

Maxwell's equations also tell us that if the voltage between any two points changes with time, a *displacement* current is generated between these points. See **Fig 6.22.** This displacement current results from the propagation, at the speed of light, of the electromagnetic field between the two points and is not to be confused with conduction current, which is caused by the movement of electrons. This displacement current is directly proportional to how fast the voltage is changing.

A capacitor takes advantage of this consequence of the laws of electromagnetics. When a capacitor is connected to an ac voltage source, a steady ac current can flow because taken together, conduction current and displacement current "complete the loop" from the positive source terminal, across the plates of the capacitor, and back to the negative terminal.

In general, parasitic capacitance shows up *wherever* the voltage between two points is changing with time, because the laws of electromagnetics require a displacement current to flow. Since this phenomenon represents an *additional* current path from one point in space to another, we can add this parasitic capacitance to our component models by adding a capacitor of appropriate value in *parallel* with the component. These parasitic capacitances are typically less than 1 pF, so that below VHF they can be treated as open circuits (infinite resistances) and thus neglected.

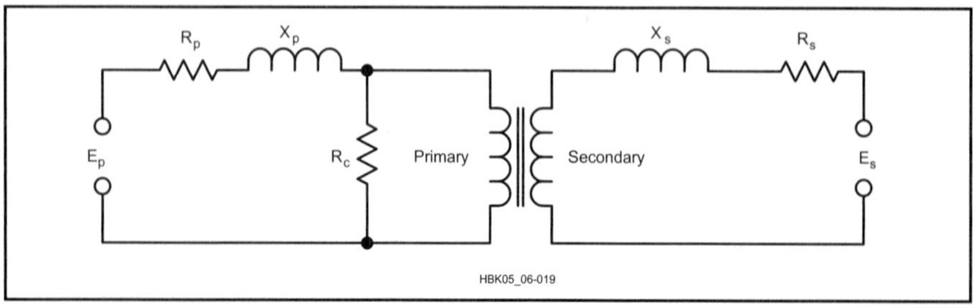

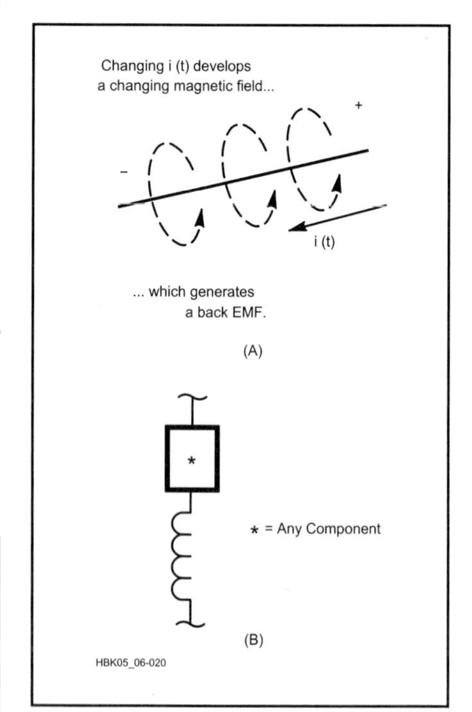

Fig 6.20—Inductive consequences of Maxwell's equations. At A, any wire carrying a changing current develops a voltage difference along it. This can be mathematically described as an effective inductance. B adds parasitic inductance to a generic component model.

Fig 6.19—An equivalent circuit for a transformer at low to medium frequencies. R_p and R_s represent the series resistances of the windings, X_p and X_s are the leakage inductances, and R_c represents the dissipative losses in the core due to eddy currents and so on.

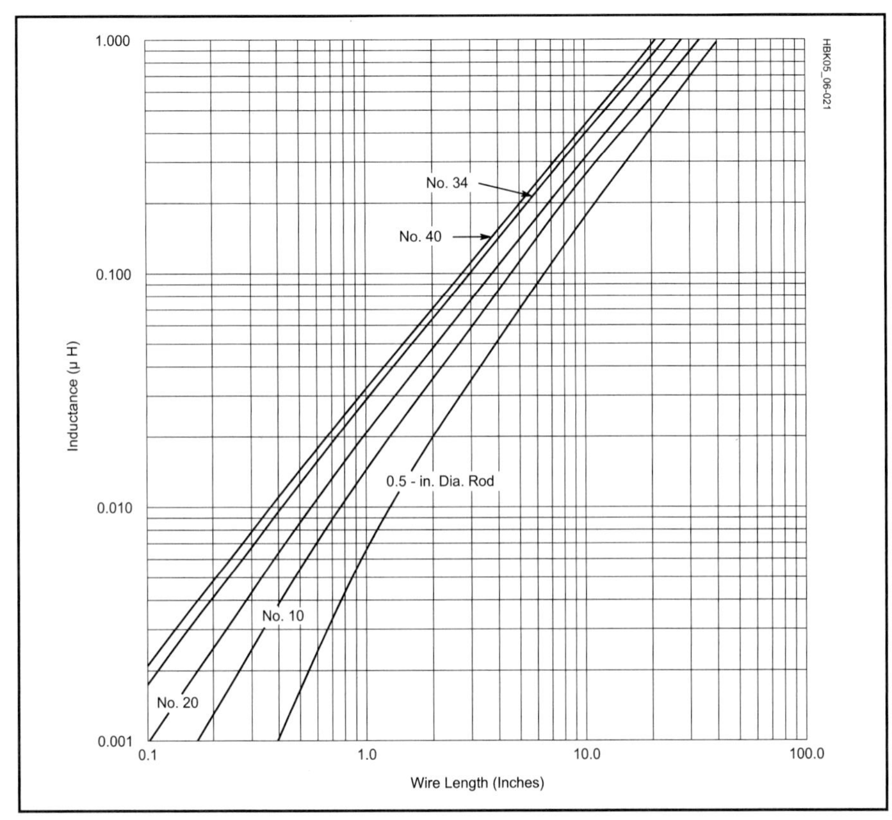

Fig 6.21—A plot of inductance vs length for straight conductors in several wire sizes.

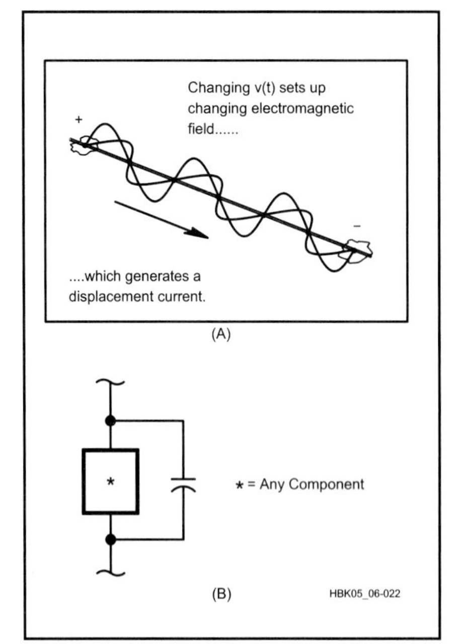

Fig 6.22—Capacitive consequences of Maxwell's equations. A: Any changing voltage between two points, for example along a bent wire, generates a displacement current running between them. This can be mathematically described as an effective capacitance. B adds parasitic capacitance to a generic component model.

Consider the inductor in **Fig 6.23**. If this coil has n turns, then the ac voltage between identical points of two neighboring turns is 1/n times the ac voltage across the entire coil. When this voltage changes due to an ac current passing through the coil, the effect is that of many small capacitors acting in parallel with the inductance of the coil. Thus, in addition to the capacitance resulting from the leads, inductors have higher parasitic capacitance due to their physical shape.

Package Capacitance

Another source of capacitance, also in the 1-pF range and therefore important only at VHF and above, is the packaging of the component itself. For example, a power transistor packaged in a TO-220 case (see **Fig 6.24**), often has either the emitter or collector connected to the metal tab itself. This introduces an extra *inter-electrode capacitance* across the junctions.

The copper traces on a PC board also present capacitance to the circuit it holds. Double-sided PC boards have a certain capacitance per square inch. It is possible to create capacitors by leaving unetched areas of copper on both sides of the board. The capacitance is not well controlled on inexpensive low-frequency boards, however. For this reason, the copper on one side of a double-sided board should be completely removed under frequency-determining circuits such as VFOs. Board capacitance is *exploited* to make microwave transmission lines (microstrip lines). The capacitance of boards for microwave use is better controlled than that of less expensive board material.

Stray capacitance (a general term used for any "extra" capacitance that exists due to physical construction) appears in any circuit where two metal surfaces exist at different voltages. Such effects can be modeled as an extra capacitor in parallel with the given points in the circuit. A rough value can be obtained with the parallel-plate formula given above.

Thus, similar to inductance, *any* circuit component that has wires attached to it, or is fabricated from wire, or is near or attached to metal, will have a parasitic capacitance associated with it, which again, becomes important only at RF.

A GENERAL MODEL

The parasitic problems due to component leads, packaging, leakage and so on are relatively common to all components. When working at frequencies where many or all of the parasitics become important, a complex but completely general model such as that in **Fig 6.25** can be used for just about any component, with the actual component placed in the box marked "*". Parasitic capacitance C_p and leakage conductance G_L appear in parallel across the device, while series resistance R_s and parasitic inductance L_p appear in series with it. Package capacitance C_{pkg} appears as an additional capacitance in parallel across the whole device. This maze of effects may seem overwhelming, but remember that it is very seldom necessary to consider all parasitics. In the Computer-Aided-Design section of this chapter, we will use the power of the computer to examine the combined effects of these multiple parasitics on circuit performance.

Self-Resonance

Because of the effects just discussed, a capacitor or inductor—all by itself—exhibits the properties of a resonant RLC circuit as we increase the applied frequency. **Fig 6.26** illustrates RF models for the capacitor and inductor, which are based on the general model in Fig 6.25, leaving out the packaging capacitance. Note the slight difference in configuration; the pairs C_p, R_p and L_s, R_s are in series in the capacitor but in parallel in the

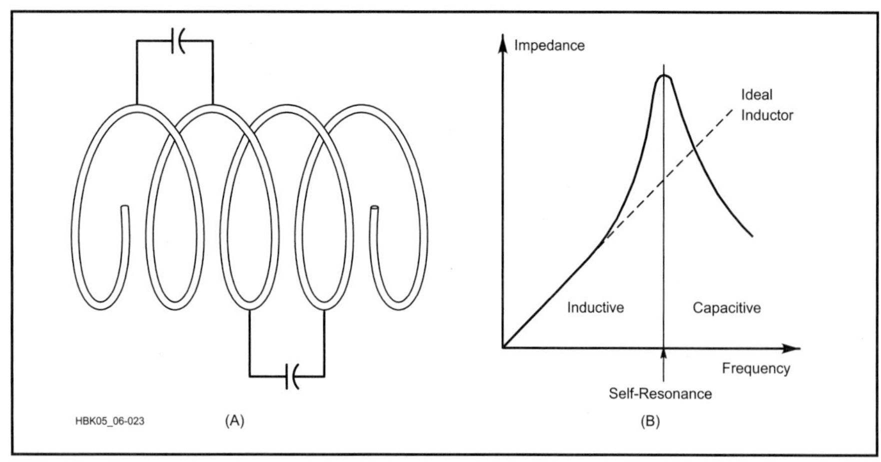

Fig 6.23—Coils exhibit distributed capacitance as explained in the text. The graph at B shows how distributed capacitance resonates with the inductance. Below resonance, the reactance is predominantly inductive which increases as frequency increases. However, above resonance, the reactance becomes predominantly capacitive which decreases as frequency increases.

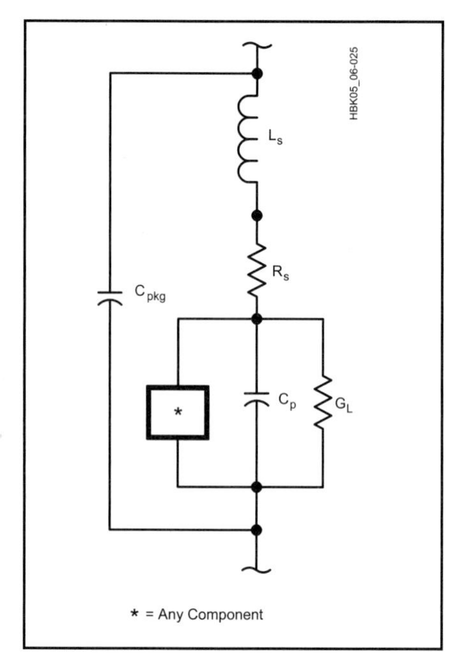

Fig 6.25—A general model for electrical components at VHF frequencies and above. The box marked "*" represents the component itself. See text for discussion.

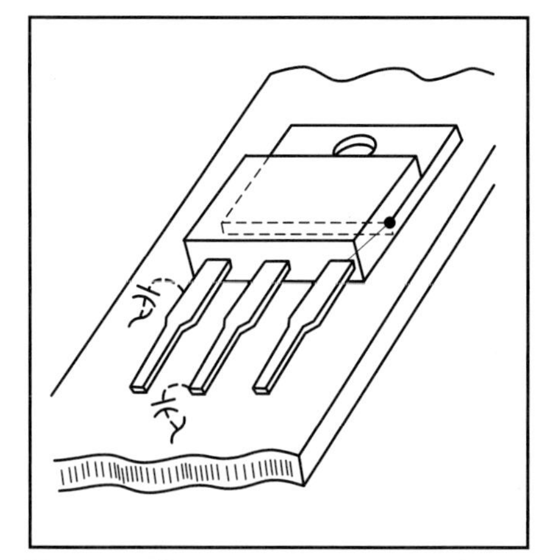
Fig 6.24—Unexpected stray capacitance. The mounting tab of TO-220 transistors is often connected to one of the device leads. Because one lead is connected to the chassis, small capacitances from the other lead to the chassis appear as additional package capacitance at the device. Similar capacitance can appear at any device with a conductive package.

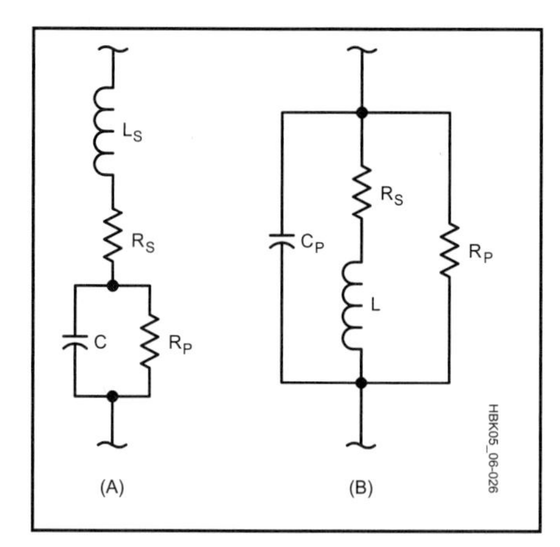

Fig 6.26—Capacitor (A) and inductor (B) models for RF frequencies.

inductor. This is because in the inductor, C_p and R_p are the parasitics, while in the capacitor, L_s and R_s are the added effects.

At some sufficiently high frequency, both inductors and capacitors become *self-resonant*. Just like a tuned circuit, above that frequency the capacitor will appear inductive, and the inductor will appear capacitive.

For an example, let's calculate the approximate self-resonant frequency of a 470-pF capacitor whose leads are made from #20 wire (diam 0.032 inch), with a total length of 1 inch. From the formula above, we calculate the approximate parasitic inductance

$$L(\mu H) = 0.00508(1)\left[\ln\left(\frac{2(1)}{(0.032/2)}\right) - 0.75\right]$$

$$= 0.021\,\mu H$$

and then the self-resonant frequency is roughly

$$f = \frac{1}{2\pi\sqrt{LC}} = 50.6\text{ MHz}$$

The purpose of making these calculations is to give you a rough feel for actual component values. They could be used as a rough design guideline, but should not be used quantitatively. Other factors such as lead orientation, shielding and so on, can alter the parasitic effects to a large extent. Large-value capacitors tend to have higher parasitic inductances (and therefore a lower self-resonant frequency) than small-value ones.

Self-resonance becomes critically important at VHF and UHF because the self-resonant frequency of many common components is at or below the frequency where the component will be used. In this case, either special techniques can be used to construct components to operate at these frequencies, by reducing the parasitic effects, or else the idea of lumped elements must be abandoned altogether in favor of microwave techniques such as striplines and waveguides.

Skin Effect

The resistance of a conductor to ac is different than its value for dc. A consequence of Maxwell's equations is that thick, near-perfect conductors (such as metals) conduct ac only to a certain depth that is proportional to the wavelength of the signal. This decreases the effective cross-section of the conductor at high frequencies and thus increases its resistance.

This resistance increase, called *skin effect*, is insignificant at low (audio) frequencies, but beginning around 1 MHz (depending on the size of the conductor) it is so pronounced that practically all the current flows in a very thin layer near the conductor's surface. For this reason, at RF a hollow tube and a solid rod of the same diameter and made of the same metal will have the same resistance. The depth of this skin layer decreases by a factor of 10 for every 100 × increase in frequency. Consequently, the RF resistance is often much higher than the dc resistance. Also, a thin highly conductive layer, such as silver plating, can lower resistance for UHF or microwaves, but does little to improve HF conductivity.

A rough estimate of the cutoff frequency where a nonferrous wire will begin to show skin effect can be calculated from

$$f = \frac{124}{d^2} \quad (12)$$

where

f = frequency, in MHz

d = diam, in mils (a mil is 0.001 inch).

Above this frequency, increase the resistance of the wire by 10 × for every 2 decades of frequency (roughly 3.2 × for every decade). For example, say we wish to find the RF resistance of a 2-inch length of #18 copper wire at 100 MHz. From the wire tables, we see that this wire has a dc resistance of (2 in.) (6.386 Ω/1000 ft) = 1.06 milliohms. From the above formula, the cutoff frequency is found to be $124 / 40.3^2$ = 76 kHz. Since 100 MHz is roughly three decades above this (100 kHz to 100 MHz), the RF resistance will be approximately (1.06 mΩ) (10 × 3.2) = 34 mΩ. Again, values calculated in this manner are approximate and should be used qualitatively—that is, when you want an answer to a question such as, "Can I neglect the RF resistance of this length of connecting wire at 100 MHz?"

For additional information, see *Reference Data for Engineers*, Howard W. Sams & Co, Indianapolis, IN 46268. Chapter 6 contains a discussion and several design charts.

Effects on Q

Recall from the **AC Theory** portion of Chapter 4 that circuit Q, a useful figure of merit for tuned RLC circuits, can be defined in several ways:

$$Q = \frac{X_L \text{ or } X_C \text{ (at resonance)}}{R} \quad (13)$$

$$= \frac{\text{energy stored per cycle}}{\text{energy dissipated per cycle}}$$

Q is also related to the bandwidth of a tuned circuit's response by

$$Q = \frac{f_0}{BW_{3\,dB}} \quad (14)$$

Parasitic inductance, capacitance and resistance can significantly alter the performance and characteristics of a tuned circuit if the design frequency is anywhere near the self-resonant frequencies of the components.

As an example, consider the resonant circuit of **Fig 6.27A**, which could represent the input tank circuit of an oscillator. Neglecting any parasitics, $f_0 = 1 / (2 \pi (LC)^{0.5})$ = 10.06 MHz. As in many real cases, assume the resistance arises entirely from the inductor series resistance. The data sheet for the inductor specified a minimum Q of 30, so assuming Q = 30 yields an R value of $X_L / Q = 2 \pi$ (10.06 MHz) (5 µH) / 30 = 10.5 Ω.

Next, let's include the parasitic inductance of the capacitor (Fig 6.27B). A reasonable assumption is that this capacitor has the same physical size as the example from the Parasitic Inductance discussion above, for which we calculated L_s = 0.106 µH. This would give the capacitor a self-resonant frequency of 434 MHz—well above our area of interest. However, the added parasitic inductance does account for an extra 0.106/5.00 = 2% inductance. Since this circuit is no longer strictly series or parallel, we must convert it to an equivalent form before calculating the new f_0.

An easier and faster way is to *simulate* the altered circuit by computer. This analysis was performed on a desktop computer using *SPICE*, a standard circuit simulation program; for more details, see the Computer-Aided Design section below. The voltage response of the circuit (given an input current of 1 mA) was calculated as a function of frequency for both cases, with and without parasitics. The results are shown in the plot in Fig 6.27C, where we can see that the parasitic circuit has an f_0 of 9.96 MHz (a shift of 1%) and a Q (measured from the –3 dB points) of 31.5. For comparison, the simulation of the unaltered circuit does in fact show f_0 = 10.06 MHz and Q = 30.

Inductor Coupling

Mutual inductance will also have an effect on the resonant frequency and Q of the involved circuits. For this reason, inductors in frequency-critical circuits should always be shielded, either by constructing compartments for each circuit block or through the use of "can"-mounted coils. Another helpful technique is to mount nearby coils with their axes perpendicular as in **Fig 6.28**. This will minimize coupling.

As an example, assume we build an oscillator circuit that has both input and output filters similar to the resonant circuit in Fig 6.27A. If we are careful to keep the two coils in these circuits uncoupled, the frequency response of either of the two circuits is that of the solid line in **Fig 6.29**, reproduced from Fig 6.27C.

If the two coils are coupled either through careless placement or improper shielding, the resonant frequency and Q will be affected. The dashed line in Fig 6.29 shows the frequency response that results from a coupling coefficient of k = 0.05, a reasonable value for air-wound inductors mounted perpendicularly in close proximity on a circuit chassis. Note the resonant frequency shifted from 10.06 to 9.82 MHz, or 2.4%. The Q has gone up slightly from 30.0 to 30.8 as a result of the slightly higher inductive reactance at the resonant frequency.

To summarize, even small parasitics can significantly affect frequency responses of RF circuits. Either take steps to minimize or eliminate them, or use simple circuit theory to predict and anticipate changes.

Dielectric Breakdown and Arcing

Anyone who has ever watched a capacitor burn out, or heard the hiss of an arc across the inductor of an antenna tuner, while loading the 2-m vertical on 160 m, or touched a doorknob on a cold winter day has seen the effects of dielectric breakdown. When the dielectric is gaseous, especially air, we often call this *arcing*.

In the ideal world, we could take any two conductors and put as large a voltage as we want across them, no matter how close together they are. In the real world, there is a voltage limit (*dielectric strength*, measured in kV/cm and determined by the insulator between the two conductors) above which the insulator will break down.

Because they are charged particles, the electrons in the atoms of a dielectric material feel an attractive force when placed in an electric field. If the field is sufficiently strong, the force will strip the electron from the atom. This electron is available to conduct current, and furthermore, it is traveling at an extremely high velocity. It is very likely that this electron will hit another atom, and free another electron. Before long, there are many stripped electrons producing a large current. When this happens, we say the dielectric has suffered *breakdown.*

If the dielectric is liquid or gas, it can heal when the applied voltage is removed. A solid dielectric, however, cannot repair itself. A good example of this is a CMOS integrated circuit. When exposed to the very high voltages associated with static electricity, the electric field across the very thin gate oxide layer exceeds the dielectric strength of silicon dioxide, and the device is permanently damaged.

Capacitors are, by nature, perhaps the component most often associated with dielectric failure. To prevent damage, the working voltage of a capacitor—and there

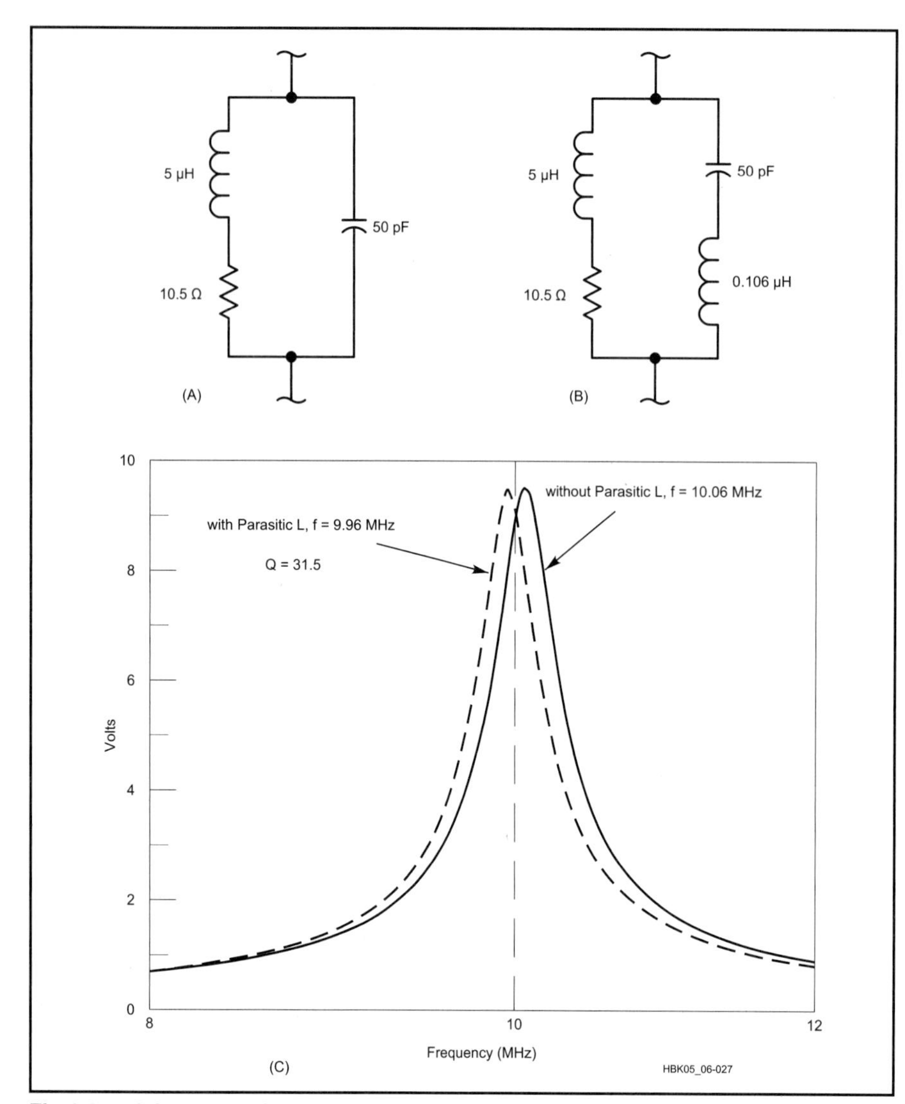

Fig 6.27—A is a tank circuit, neglecting parasitics. B same circuit including Lp on capacitor. C frequency response curves for A and B. The solid line represents the unaltered circuit (also see Fig 6.29) while the dashed line shows the effects of adding parasitic inductance.

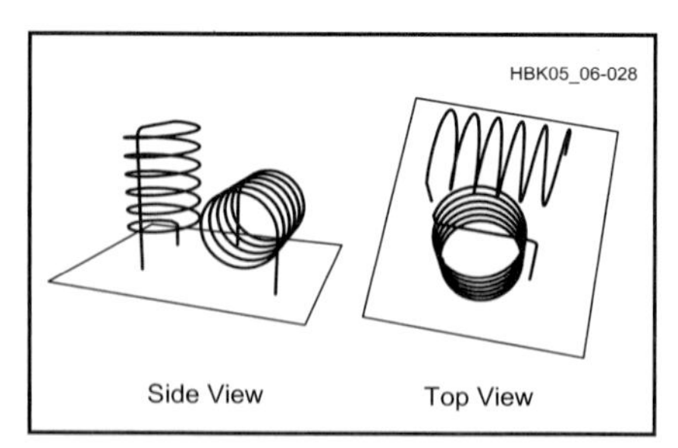

Fig 6.28—Unshielded coils in close proximity should be mounted perpendicular to each other to minimize coupling.

are separate dc and ac ratings—should ideally be 2 or 3 times the expected maximum voltage in the circuit.

Arcing is most often seen in RF circuits where the voltages are normally high, but it is possible anywhere two components at significantly different voltage levels are closely spaced.

The breakdown voltage of a dielectric layer depends on its composition and thickness (see **Table 6.3**). The variation with thickness is not linear; doubling the thickness does not quite double the breakdown voltage. Breakdown voltage is also a function of geometry: Because of electromagnetic considerations, the breakdown voltage between two conductors separated by a fixed distance is less if the surfaces are pointed or sharp-edged than if they are smooth or rounded. Therefore, a simple way to help prevent breakdown in many projects is to file and smooth the edges of conductors.

Radiative Losses and Coupling

Another consequence of Maxwell's equations states that any conductor placed in an electromagnetic field will have a current induced in it. We put this principle to good use when we make an antenna. The unwelcome side of this law of nature is the phrase "any conductor"; even conductors we don't intend as antennas will act this way.

Fortunately, the *efficiency* of such "antennas" varies with conductor length. They will be of importance only if their length is a significant fraction of a wavelength. When we make an antenna, we usually choose a length on the order of λ/2. Therefore, when we *don't* want an antenna, we should be sure that the conductor length is *much less* than λ/2, no more than 0.1 λ. This will ensure a very low-efficiency antenna. This is why 60-Hz power lines do not lose a significant fraction of the power they carry over long distances—at 60 Hz, 0.1 λ is about 300 miles!

In addition, we can use shielded cables. Such cables do allow some penetration of EM fields if the shield is not solid, but even 95% coverage is usually sufficient, especially if some sort of RF choke is used to reduce shield current.

Radiative losses and coupling can also be reduced by using twisted pairs of conductors—the fields tend to cancel. In some applications, such as audio cables, this may work better than shielding.

This argument also applies to large components, and remember that a component or long wire can *radiate* RF, as well as receive them. Critical stages such as tuned circuits should be placed in shielded compartments where possible. See the **EMI** and **Transmission Lines** chapters for more information.

REMEDIES FOR PARASITICS

The most common effect (always the most annoying) of parasitics is to influence the resonant frequency of a tuned circuit. This shift could cause an oscillator to fail, or more commonly, to cause a stable circuit to oscillate. It can also degrade filter performance (more on this later) and basically causing any number of frequency-related problems.

We can often reduce parasitic effects in discrete-component circuits by simply exploiting the models in Fig 6.26. Since parasitic inductance and loss resistance appear in series with a capacitor, we can reduce both by using several smaller capacitances in parallel, rather than of one large one.

An example is shown in the circuit block in **Fig 6.30**, which is representative of the input tank circuit used in many HF VFOs. C_{main}, C_{trim}, C1 and C3 act with L to set the oscillator frequency. Therefore, temperature effects are critical in these components. By using several capacitors in

Table 6.3
Dielectric Constants and Breakdown Voltages

Material	*Dielectric Constant**	*Puncture Voltage***
Aisimag 196	5.7	240
Bakelite	4.4-5.4	240
Bakelite, mica filled	4.7	325-375
Cellulose acetate	3.3-3.9	250-600
Fiber	5-7.5	150-180
Formica	4.6-4.9	450
Glass, window	7.6-8	200-250
Glass, Pyrex	4.8	335
Mica, ruby	5.4	3800-5600
Mycalex	7.4	250
Paper, Royalgrey	3.0	200
Plexiglas	2.8	990
Polyethylene	2.3	1200
Polystyrene	2.6	500-700
Porcelain	5.1-5.9	40-100
Quartz, fused	3.8	1000
Steatite, low loss	5.8	150-315
Teflon	2.1	1000-2000

*At 1 MHz
**In volts per mil (0.001 inch)

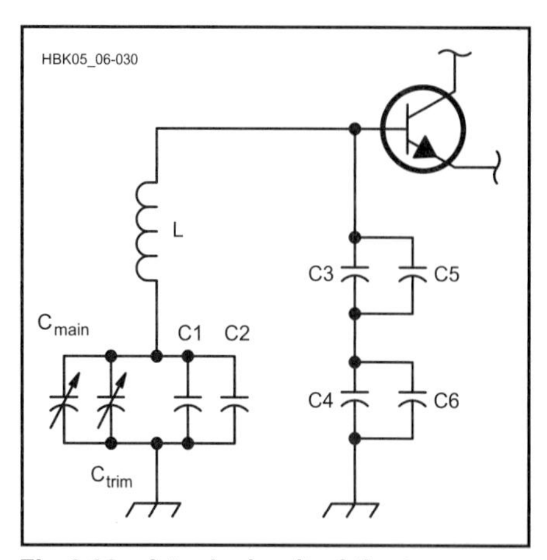

Fig 6.30—A tank circuit of the type commonly used in VFOs. Several capacitors are used in parallel to distribute the RF current, which reduces temperature effects.

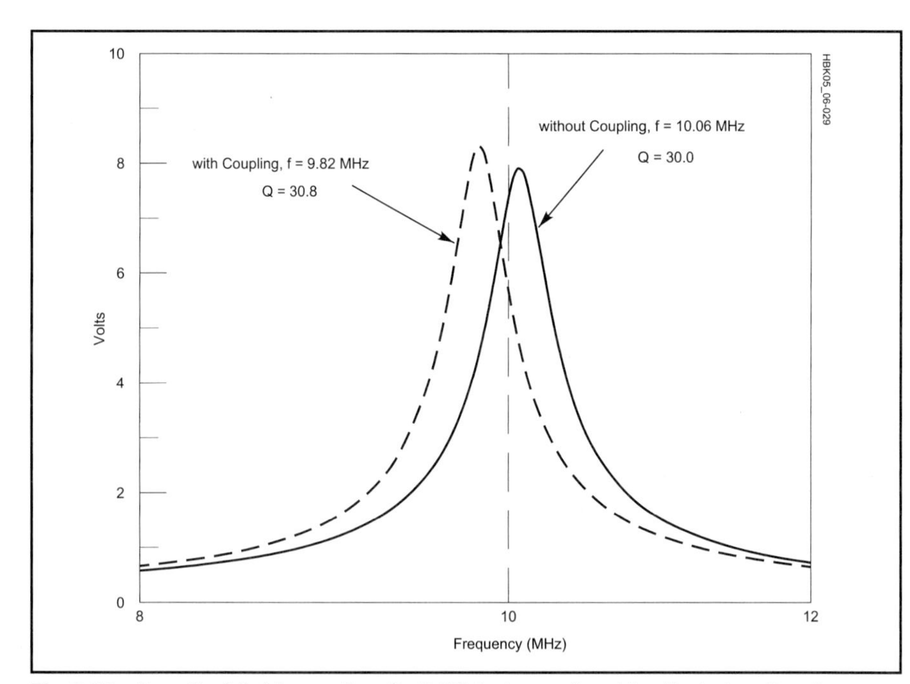

Fig 6.29—Result of light coupling (k=0.05) between two identical circuits of Fig 6.27A on their frequency responses.

parallel, the RF current (and resultant heat) is reduced in each component. Parallel combinations are used at the feedback capacitors for the same reason.

Another example of this technique is the common capacitor bypass arrangement shown in **Fig 6.31**. C1 provides a bypass path at audio frequencies, but it has a low self-resonant frequency due to its large capacitance. How low? Even if we assume the 0.02-µH value shown previously in this chapter, f_{sc} = 355 kHz. Adding C2 (with a smaller capacitance but much higher self-resonant frequency) in parallel provides bypass at high frequencies where C1 appears inductive.

Construction Techniques

These concepts also help explain why so many different components exist with similar values. As an example, assume you're working on a project that requires you to wind a 5 µH inductor. Looking at the coil inductance formula in the **AC Theory** section of Chapter 4, it comes to mind that many combinations of length and diameter could yield the desired inductance. If you happen to have both 0.5 and 1-inch coil forms, why should you select one over the other? To eliminate some other variables, let's make both coils 1 inch long, close-wound, and give them 1-inch leads on each end.

Let's calculate the number of turns required for each. On a 0.5-inch-diameter form:

$$n = \frac{\sqrt{5\left[(18 \times 0.5) + (40 \times 1)\right]}}{0.5} = 31.3 \text{ turns}$$

This means coil 1 will be made from #20 wire (29.9 turns per inch). Coil 2, on the 1-inch form, yields

$$n = \frac{\sqrt{5\left[(18 \times 1) + (40 \times 1)\right]}}{1} = 17.0 \text{ turns}$$

which requires #15 wire in order to be close-wound.

What are the series resistances associated with each? For coil 1, the total wire length is 2 inches + (31.3 × π × 0.5) = 51 inches, which at 10.1 Ω /1000 ft gives R_s = 0.043 Ω at dc. Coil 2 has a total wire length of 2 inches + (17.0 × π × 1) = 55 inches, which at 3.18 Ω /1000 ft gives a dc resistance of R_s = 0.015 Ω, or about 1/3 that of coil 1. Furthermore, at RF, coil 1 will begin to suffer from skin effect at a frequency about 3 times lower than coil 2 because of its smaller conductor diameter. Therefore, if Q were the sole consider-

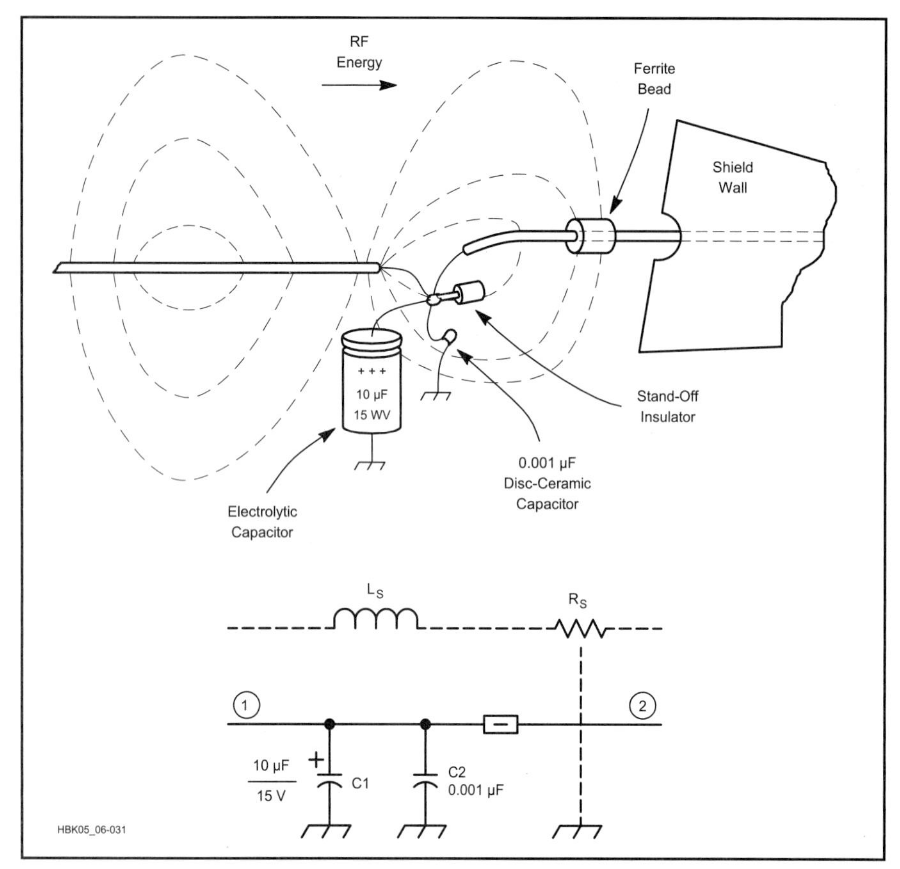

Fig 6.31—A typical method to provide bypassing at high frequencies when a large capacitor with a low-self-resonant frequency is required.

ation, it would be better to use the larger diameter coil.

Q is not the only concern, however. Such coils are often placed in shielded enclosures. Rule of thumb says the enclosure should be at least one coil diameter from the coil on all sides. That is, 3×3×2 inches for the large coil and 1.5×1.5×1.5 for the small coil, a volume difference of over 500%.

THERMAL CONSIDERATIONS

Any real energized circuit consumes electric power because any real circuit contains components that convert electricity into other forms of energy.[1] This *dissipated power* appears in many forms. For example, a loudspeaker converts electrical energy into the motion of air molecules we call sound. An antenna (or a light bulb) converts electricity into electromagnetic radiation. Charging a battery converts electrical energy into chemical energy (which is then converted back to electrical energy upon discharge). But the most common transformation by far is the conversion, through some form of *resistance*, of electricity into heat.

Sometimes the power lost to heat serves a useful purpose—toasters and hair dryers come to mind. But most of the time, this heat represents a power loss that is to be minimized wherever possible or at least taken into account. Since all real circuits contain resistance, even those circuits (such as a loudspeaker) whose primary purpose is to convert electricity to some *other* form of energy also convert some part of their input power to heat. Often, such losses are negligible, but sometimes they are not.

If unintended heat generation becomes significant, the involved components will get warm. Problems arise when the temperature increase affects circuit operation by either

- causing the component to fail, by explosion, melting, or other catastrophic event, or, more subtly,
- causing a slight change in the properties of the component, such as through a temperature coefficient (TC).

In the first case, we can design conservatively, ensuring that components are rated to safely handle two, three or more times the maximum power we expect them to dissipate. In the second case, we can specify components with low TCs, or we can design the circuit to minimize the effect of any one component. Occasionally we even exploit temperature effects (for example, using a resistor, capacitor or diode as a temperature sensor). Let's look more closely at the two main categories of thermal effects.

HEAT DISSIPATION

Not surprisingly, heat dissipation (more correctly, the efficient removal of generated heat) becomes important in medium- to high-power circuits: power supplies, transmitting circuits and so on. While these are not the only examples where elevated temperatures and related failures are of concern, the techniques we will discuss here are applicable to all circuits.

Thermal Resistance

The transfer of heat energy, and thus the change in temperature, between two ends of a block of material is governed by the following heat flow equation (see **Fig 6.32**):

$$P = \frac{k\,A}{L}\Delta T = \frac{\Delta T}{\theta} \qquad (15)$$

where

P = power (in the form of heat) conducted between the two points

k = *thermal conductivity*, measured in W/(m °C), of the material between the two points, which may be steel, silicon, copper, PC board material and so on

L = length of the block

A = area of the block

ΔT = *change* in temperature between the two points;

$\theta = \frac{L}{kA}$ is often called the *thermal resistance* and has units of °C/W.

Thermal conductivities of various common materials at room temperature are given in **Table 6.4.**

A very useful property of the above equation is that it is *exactly* analogous to Ohm's Law, and therefore the same principles and methods apply to heat flow problems as circuit problems. The following correspondences hold:

- Thermal conductivity W/(m °C) ↔ Electrical conductivity (S/m).
- Thermal resistance (°C/W) ↔ Electrical resistance (Ω).
- Thermal current (heat flow) (W) ↔ Electrical current (A).
- Thermal potential (T) ↔ Electrical potential (V).
- Heat source ↔ Current source.

For example, calculate the temperature of a 2-inch (0.05 m) long piece of #12 copper wire at the end that is being heated by a 25 W (input power) soldering iron, and whose other end is clamped to a large metal vise (assumed to be an infinite heat sink), if the ambient temperature is 25 °C (77 °F).

First, calculate the thermal resistance of the copper wire (diameter of #12 wire is 2.052 mm, cross-sectional area is 3.31×10^{-6} m²)

$$\theta = \frac{L}{k\,A} = \frac{(0.05\,\text{m})}{(390\,\text{W}/(\text{m}\,^\circ\text{C}))(3.31 \times 10^{-6}\,\text{m}^2)}$$

$$= 38.7\,^\circ\text{C/W}$$

Then, rearranging the heat flow equation above yields (after assuming the heat energy actually transferred to the wire is around 10 W)

$$\Delta T = P\,\theta = (10\,\text{W})(38.7\,^\circ\text{C}/\text{W}) = 387\,^\circ\text{C}$$

So the wire temperature at the hot end is 25 C + ΔT = 412°C (or 774°F). If this sounds a little high, remember that this is for the steady state condition, where you've been holding the iron to the wire for a long time.

From this example, you can see that things can get very hot even with application of moderate power levels. In the case of a soldering iron, that's good, but in the case of a 25-W power transistor in a transmitter output stage, it's bad. For this reason, circuits that generate sufficient heat to alter, not necessarily damage, the components must employ some method of cooling, either active or passive. Passive methods include heat sinks or careful component layout for good ventilation. Active methods include forced air (fans) or some sort of liquid cooling (in some high-power transmitters).

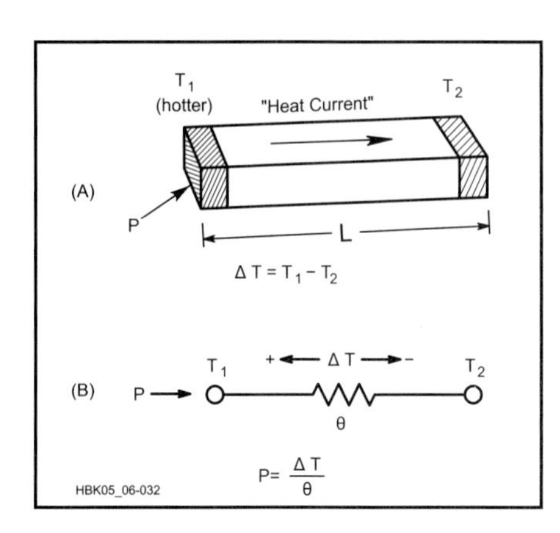

Fig 6.32—Physical and "circuit" models for the heat-flow equation.

Heat Sink Design and Use

The purpose of a heat sink is to provide a high-power component with a large surface area through which to dissipate heat. To use the models above, it provides a low thermal-resistance path to a cooler temperature, thus allowing the hot component to conduct a large "thermal current" away from itself.

Power supplies probably represent one of the most common high-power circuits amateurs are likely to encounter. Everyone has certainly noticed that power supplies get warm or even hot if not ventilated properly. Performing the thermal design for a properly cooled power supply is a very well-defined process and is a good illustration of heat-flow concepts.

A 28-V, 10-A power supply will be used for this design example. This material was originally prepared by ARRL Technical Advisor Dick Jansson, WD4FAB, and the steps described below were actually followed during the design of that supply.

An outline of the design procedure shows the logic applied:

1. Determine the expected power dissipation (P_{in}).
2. Identify the requirements for the dissipating elements (maximum component temperature).
3. Estimate heat-sink requirements.
4. Rework the electronic device (if necessary) to meet the thermal requirements.
5. Select the heat exchanger (from heat sink data sheets).

The first step is to estimate the filtered, unregulated supply voltage under full load. Since the transformer secondary output is 32 V ac (RMS) and feeds a full-wave bridge rectifier, let's estimate 40 V as the filtered dc output at a 10-A load.

Table 6.4
Thermal Conductivities of Various Materials

Gases at 0°C, Others at 25°C; from *Physics,* by Halliday and Resnick, 3rd Ed.

Material	k in units of $\left(\frac{W}{m\,^\circ C}\right)$
Aluminum	200
Brass	110
Copper	390
Lead	35
Silver	410
Steel	46
Silicon	150
Air	0.024
Glass	0.8
Wood	0.08

The next step is to determine our critical components and estimate their power dissipations. In a regulated power supply, the pass transistors are responsible for nearly all the power lost to heat. Under full load, and allowing for some small voltage drops in the power-transistor emitter circuitry, the output of the series pass transistors is about 29 V for a delivered 28 V under a 10-A load. With an unregulated input voltage of 40 V, the total energy heat dissipated in the pass transistors is (40 V – 29 V) × 10 A = 110 W. The heat sink for this power supply must be able to handle that amount of dissipation and still keep the transistor junctions below the specified safe operating temperature limits. It is a good rule of thumb to select a transistor that has a maximum power dissipation of twice the desired output power.

Now, consider the ratings of the pass transistors to be used. This supply calls for 2N3055s as pass transistors. The data sheet shows that a 2N3055 is rated for 15-A service and 115-W dissipation. But the design uses *four* in parallel. Why? Here we must look past the big, bold type at the top of the data sheet to such subtle characteristics as the junction-to-case thermal resistance, q_{jc}, and the maximum allowable junction temperature, T_j.

The 2N3055 data sheet shows θ_{jc} = 1.52°C/W, and a maximum allowable case (and junction) temperature of 220°C. While it seems that one 2N3055 could barely, on paper at least, handle the electrical requirements, at what temperature would it operate?

To answer that, we must model the entire "thermal circuit" of operation, starting with the transistor junction on one end and ending at some point with the ambient air. A reasonable model is shown in **Fig 6.33**. The ambient air is considered here as an infinite heat sink; that is, its temperature is assumed to be a constant 25°C (77°F). θ_{jc} is the thermal resistance from the transistor junction to its case. θ_{cs} is the resistance of the mounting interface between the transistor case and the heat sink. θ_{sa} is the thermal resistance between the heat sink and the ambient air. In this "circuit," the generation of heat (the "thermal current source") occurs in the transistor at P_{in}.

Proper mounting of most TO-3 package power transistors such as the 2N3055 requires that they have an electrical insulator between the transistor case and the heat sink. However, this electrical insulator must at the same time exhibit a low thermal resistance. To achieve a quality mounting, use thin polyimid or mica formed washers and a suitable thermal compound to exclude air from the intersti-

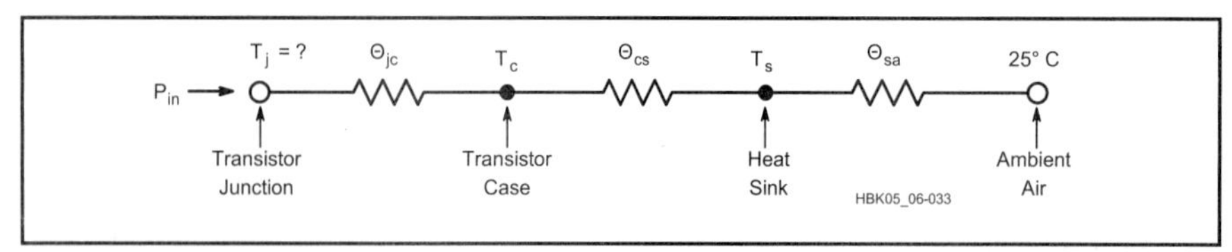

Fig 6.33—Resistive model of thermal conduction in a power transistor and associated heat sink. See text for calculations.

tial space. "Thermal greases" are commonly available for this function. Any silicone grease may be used, but filled silicone oils made specifically for this purpose are better.

Using such techniques, a conservatively high value for θ_{cs} is 0.50°C/W. Lower values are possible, but the techniques needed to achieve them are expensive and not generally available to the average amateur. Furthermore, this value of θ_{cs} is already much lower than θ_{jc}, which cannot be lowered without going to a somewhat more exotic pass transistor.

Finally, we need an estimate of θ_{sa}. **Fig 6.34** shows the relationship of heat-sink volume to thermal resistance for natural-convection cooling. This relationship presumes the use of suitably spaced fins (0.35 inch or greater) and provides a "rough order-of-magnitude" value for sizing a heat sink. For a first calculation, let's assume a heat sink of roughly 6×4×2 inch (48 cubic inches). From Fig 6.34, this yields a θ_{sa} of about 1°C/W.

Returning to Fig 6.33, we can now calculate the approximate temperature increase of a single 2N3055:

$$\delta T = P\,\theta_{total} = (110\,W) \times (1.52\,°C/W + 0.5\,°C/W + 1.0\,°C/W) = 332\,°C$$

Given the ambient temperature of 25°C, this puts the junction temperature T_j of the 2N3055 at 25 + 332 = 357°C! This is clearly too high, so let's work backward from the air end and calculate just how many transistors we need to handle the heat.

First, putting more 2N3055s in parallel means that we will have the thermal model illustrated in **Fig 6.35**, with several identical θ_{jc} and θ_{cs} in parallel, all funneled through the same θ_{as} (we have one heat sink).

Keeping in mind the physical size of the project, we could comfortably fit a heat sink of approximately 120 cubic inches (6×5×4 inches), well within the range of commercially available heat sinks. Furthermore, this application can use a heat sink where only "wire access" to the transistor connections is required. This allows the selection of a more efficient design. In contrast, RF designs require the transistor mounting surface to be completely exposed so that the PC board can be mounted close to the transistors to minimize parasitics. Looking at Fig 6.34, we see that a 120-cubic-inch heat sink yields a θ_{sa} of 0.55°C/W. This means that the temperature of the heat sink when dissipating 110 W will be 25°C + (110 W) (0.55°C/W) = 85.5°C.

Industrial experience has shown that silicon transistors suffer substantial failure when junctions are operated at highly elevated temperatures. Most commercial and military specifications will usually not permit design junction temperatures to exceed 125°C. To arrive at a safe figure for our maximum allowed T_j, we must consider the intended use of the power supply. If we are using it in a 100% duty-cycle transmitting application such as RTTY or FM, the circuit will be dissipating 110 W continuously. For a lighter duty-cycle load such as CW or SSB, the "key-down" temperature can be slightly higher as long as the average is less than 125°C. In this intermittent type of service, a good conservative figure to use is T_j = 150°C.

Given this scenario, the temperature rise across each transistor can be 150 – 85.5 = 64.5°C. Now, referencing Fig 6.35, remembering the total θ for each 2N3055 is 1.52 + 0.5 = 2.02°C/W, we can calculate the maximum power each 2N3055 can safely dissipate:

$$P = \frac{\delta T}{\theta} = \frac{64.5\,°C}{2.02\,°C/W} = 31.9\,W$$

Thus, for 110 W full load, we need four 2N3055s to meet the thermal requirements of the design. Now comes the big question: What is the "right" heat sink to use? We have already established its requirements: it must be capable of dissipating 110 W, and have a θ_{sa} of 0.55°C/W (see above).

A quick consultation with several manufacturer's catalogs reveals that Wakefield Thermal Solutions, Inc. model nos. 441 and 435 heat sinks meet the needs of this application.[2] A Thermalloy model no. 6441 is suitable as well. Data published in the catalogs of these manufacturers show that in natural-convection service, the expected temperature rise for 100 W dissipation would be just under 60°C, an almost perfect fit for this application. Moreover, the no. 441 heat sink can easily mount four TO-3-style 2N3055 transistors. See **Fig 6.36**. Remember: heat sinks should be mounted with the fins and transistor mounting area vertical to promote convection cooling.

The design procedure just described is applicable to any circuit where heat buildup is a potential problem. By using the thermal-resistance model, we can easily calcu-

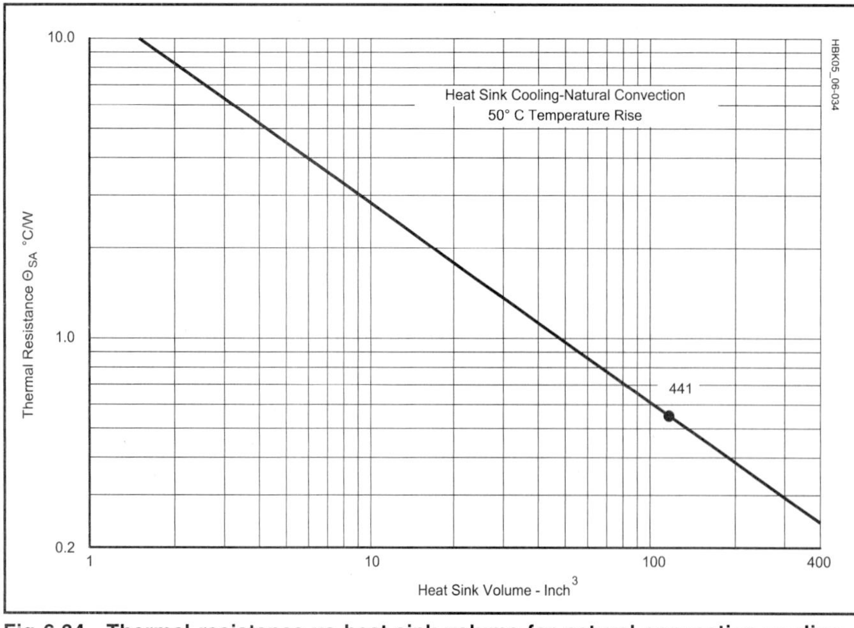

Fig 6.34—Thermal resistance vs heat-sink volume for natural convection cooling and 50°C temperature rise. The graph is based on engineering data from Wakefield Thermal Solutions, Inc.

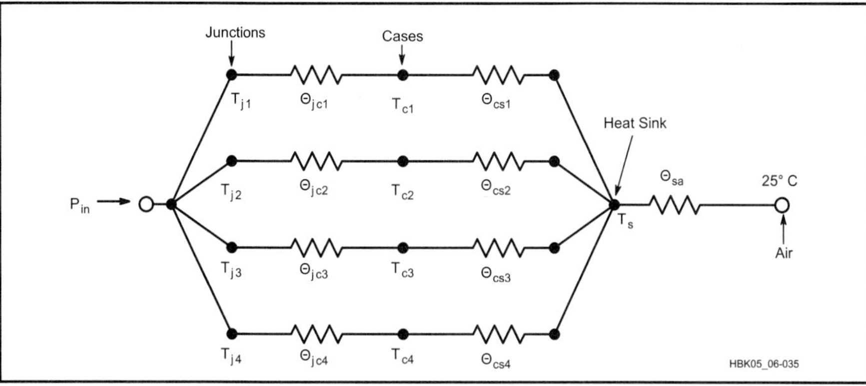

Fig 6.35—Thermal model for multiple power transistors mounted on a common heat sink.

late whether or not an external means of cooling is necessary, and if so, how to choose it. Aside from heat sinks, forced air cooling (fans) is another common method. In commercial transceivers, heat sinks with forced-air cooling are common.

Transistor Derating

Maximum ratings for power transistors are usually based on a case temperature of 25°C. These ratings will decrease with increasing operating temperature. Manufacturer's data sheets usually specify a *derating* figure or curve that indicates how the maximum ratings change per degree rise in temperature. If such information is not available (or even if it is!), it is a good rule of thumb to select a power transistor with a maximum power dissipation of at least twice the desired output power.

Rectifiers

Diodes are physically quite small, and they operate at high current densities. As a result their heat-handling capabilities are somewhat limited. Normally, this is not a problem in high-voltage, low-current supplies. The use of high-current (2 A or greater) rectifiers at or near their maximum ratings, however, requires some form of heat sinking. Frequently, mounting the rectifier on the main chassis (directly, or with thin mica insulating washers) will suffice. If the diode is insulated from the chassis, thin layers of silicone grease should be used to ensure good heat conduction. Large, high-current rectifiers often require special heat sinks to maintain a safe operating temperature. Forced-air cooling is sometimes used as a further aid.

Forced-Air Cooling

In Amateur Radio today, forced-air cooling is most commonly found in vacuum-tube circuits or in power supplies built in small enclosures, such as those in solid-state transceivers or computers. Fans or blowers are commonly specified in cubic feet per minute (CFM). While the nomenclature and specifications differ from those used for heat sinks, the idea remains the same: to offer a low thermal resistance between the inside of the enclosure and the (ambient) exterior.

For forced air cooling, we basically use the "one resistor" thermal model of Fig 6.32. The important quantity to be determined is heat generation, P_{in}. For a power supply, this can be easily estimated as the difference between the input power, measured at the transformer primary, and the output power at full load. For variable-voltage supplies, the worst-case output condition is minimum voltage with maximum current.

A discussion of forced-air tube cooling appears in the **RF Power Amplifiers** chapter.

TEMPERATURE STABILITY

Aside from catastrophic failure, temperature changes may also adversely affect circuits if the TCs of one or more components is too large. If the resultant change is not too critical, adequate temperature stability can often be achieved simply by using higher-precision components with low TCs (such as NP0/C0G capacitors or metal-film resistors). For applications where this is impractical or impossible (such as many solid-state circuits), we can minimize temperature sensitivity by *compensation* or *matching*—using temperature coefficients to our advantage.

Compensation is accomplished in one of two ways. If we wish to keep a certain circuit quantity constant, we can interconnect pairs of components that have equal but opposite TCs. For example, a resistor with a negative TC can be placed in series with a positive TC resistor to keep the total resistance constant. Conversely, if the important point is to keep the *difference* between two quantities constant, we can use components with the *same* TC so that the pair "tracks." That is, they both change by the same amount with temperature.

An example of this is a Zener reference circuit. Since the I-V equation for a diode is strongly affected by operating temperature, circuits that use diodes or transistors to generate stable reference voltages must use some form of temperature compensation. Since, for a constant current, a

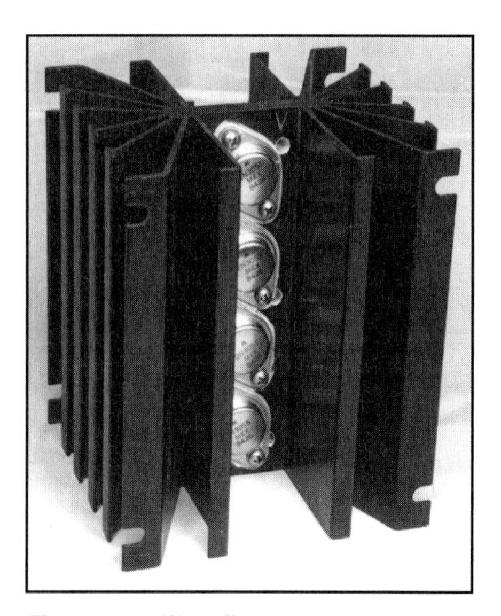
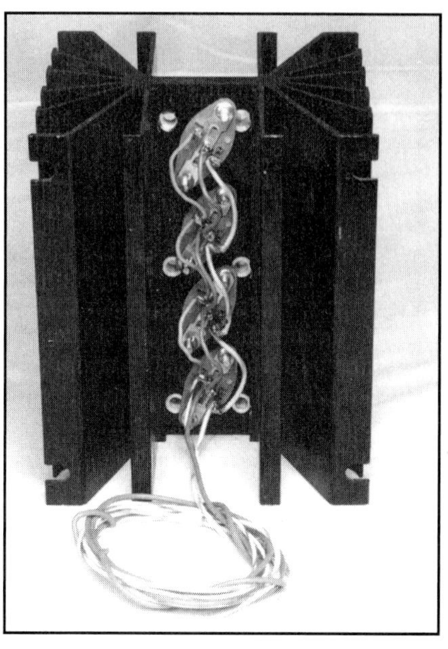

Fig 6.36—A Wakefield 441 heat sink with four 2N3055 transistors mounted.

The Thermistor in Homebrew Projects

Thermistors can be used to enhance project performance. Circuit temperature variations affect gain, distortion as well as control functions like receiver AGC or transmitter ALC. Thermistors can compensate for temperature changes. This greatly reduces the dangers of self destruction of overheated power transistors. It also can help reduce oscillator drift. In this sidebar, William Sabin, WØIYH describes how a simple temperature control circuit can be used to protect a MOSFET power amplifier or produce a temperature-stable heater to regulate the temperature of a VFO.

A thermistor is a small bit of intrinsic (no N or P doping) metal-oxide semiconductor compound material between two wire leads. As temperature increases, the number of liberated hole/electron pairs increases exponentially, causing the resistance to decrease exponentially. You can see this in the resistance equation:

$$R(T) = R(T0)e^{-\beta\left(\frac{1}{T0} - \frac{1}{T}\right)} \quad \text{(Eq 1)}$$

where T is some temperature in kelvins and T0 is a reference temperature, usually 298 K (25°C), at which the manufacturer specifies R(T0). The constant β is experimentally determined by measuring resistance at various temperatures and finding the value of β that best agrees with the measurements. A simple way to get an approximate value of β is to make two measurements, one at room temperature, say 25°C (298 K) and one at 100°C (373 K) in boiling water. Suppose the resistances are 10 kΩ and 938 Ω. Eq 1 is solved for β

$$\beta = \frac{\ln\left(\frac{R(T)}{R(T0)}\right)}{\frac{1}{T} - \frac{1}{T0}} = \frac{\ln\left(\frac{938}{1000}\right)}{\frac{1}{373} - \frac{1}{298}} = 3507 \quad \text{(Eq 2)}$$

For many electronics applications, the exact value of temperature is not as important as the ability to maintain that temperature. The following examples illustrate some thermistor circuit designs.

MOSFET Power Transistor Protector

It is very desirable to compensate the temperature sensitivity of power transistors. In bipolar (BJT) transistors, thermal runaway occurs because the dc current gain increases as the transistors get hotter. In MOSFET transistors, thermal runaway is less likely to occur. With excessive drain dissipation or inadequate cooling, however, the junction temperature may increase until its maximum allowable value is exceeded. You can place a thermistor on the heat sink close to the transistors so that the bias adjustment tracks the flange temperature. Another good idea is to attach a thermistor to the ceramic case of the FET with a small drop of epoxy. That way it will respond more quickly to a sudden temperature increase, possibly saving the transistor from destruction.

The circuit of **Fig A** uses a thermistor to detect a case temperature of about 93°C, with an ON/OFF operating range of about 0.3°C. A red LED warns of an overtemperature condition that requires attention. The LM339 comparator toggles when Rt = R5. Resistors R3, R4, R5 and the thermistor form a Wheatstone bridge. R3 and R4 are chosen to make the inputs to the LM339 about 0.5 V at the desired temperature. This greatly reduces self-heating of the thermistor, which could cause a substantial error in the circuit behavior.

The RadioShack Precision Thermistor RS 271-110 used in this circuit is rated at 10 kΩ ±1% at 25°C. It comes with a calibration chart from –50°C to +110°C that you can use to get an approximate resistance at any temperature in that range. For many temperature protection applications you don't have to know the exact temperature. You won't need an exact calibration, but you can verify that the thermistor is working properly by measuring its resistance at 20°C (68°F) and in boiling water (≈100°C).

As one example of how you can use this circuit, suppose you want to protect a power amplifier that uses a

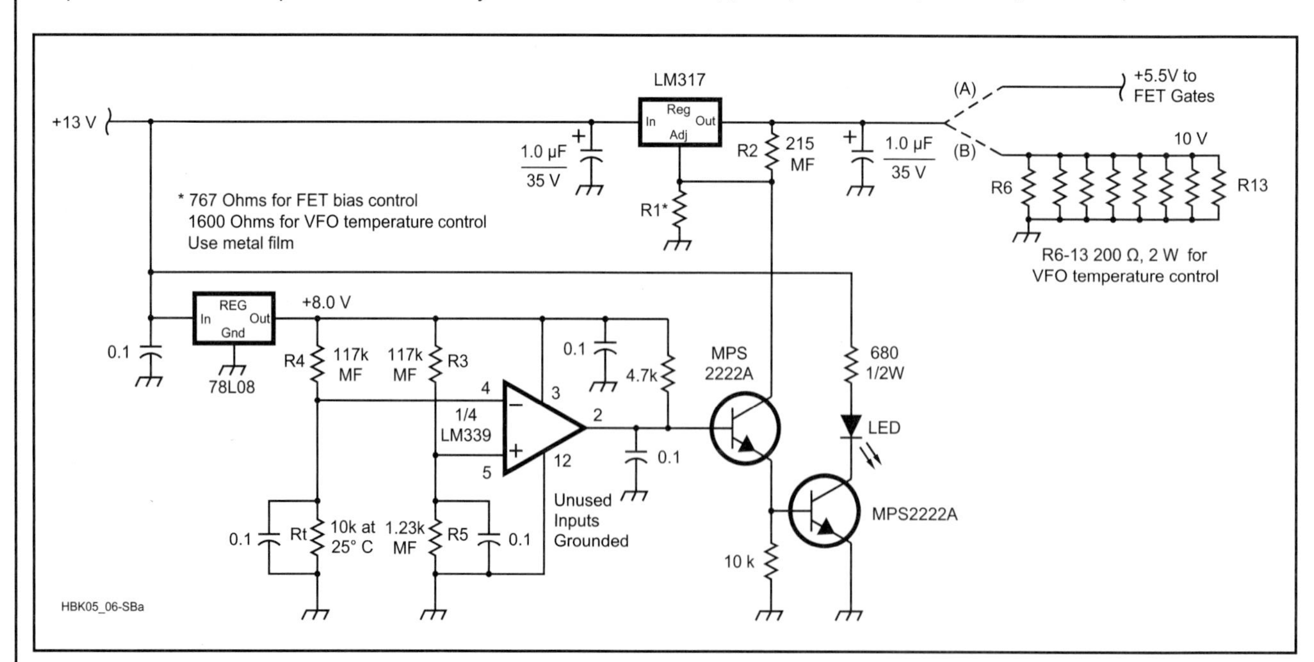

Fig A—Temperature controller for PA (A) or VFO (B).

Fig B—Temperature controlled VFO cabinet. 1/4-inch plexiglass on the outside and 1/4-inch styrofoam panels on the inside are used to improve temperature stability.

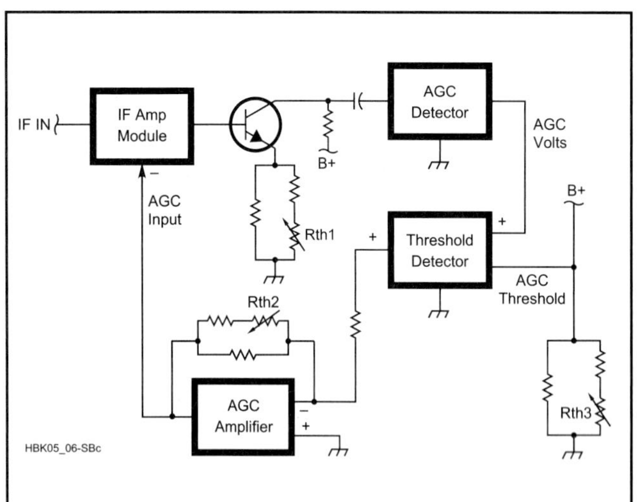

Fig C—Temperature compensation of IF amplifier and AGC circuitry.

pair of MRF150 MOSFETs. Use the following procedure to get the desired temperature control:

The MRF150 MOSFET has a maximum allowed junction temperature of 200°C. The thermal resistance θ_{JC} from junction to case is 0.6°C per watt. The maximum expected dissipation of the FET in normal operation is 110 W. Select a case temperature of 93°C. This makes the junction temperature 93°C + (0.6°C/W) × (110 W) =159°C, which is a safe 41°C below the maximum allowed temperature.

The FET has a rating of 300 W maximum dissipation at a case temperature of 25°C, derated at 1.71 W per °C. At 93°C case temperature the maximum allowed dissipation is 300 W – 1.71 W/°C × (93°C – 25°C) = 184 W. The safety margin *at that temperature* is 184 – 110 = 74 W.

A very simple way to determine the correct value of R5 is to put the thermistor in 93°C water (let it stabilize) and adjust R5 so that the circuit toggles. At 93°C, the measured value of the thermistor should be about 1230 Ω.

To use the circuit of Fig A to control the FET bias voltage, set R1 to be 767 Ω. This value, along with R2 adjusts the LM317 voltage regulator output to 5.5V for the FET gate bias. When the thermistor heats to the set point, the LM339 comparator toggles on. This brings the FET gate voltage to a low level and completely turns off the FETs until the temperature drops about 0.3°C. Use metal film resistors throughout the circuit.

Temperature controlled variable frequency oscillator (VFO)

The same circuit can be used to control the temperature of a VFO (variable frequency oscillator) with a few simple changes. Select R1 to be 1600 Ω (metal film) to adjust the LM317 for 10 V. Add resistors R6 through R13 as a heater element. Place the VFO in a thermally insulated enclosure such as shown in **Fig B**. The eight 200-Ω, 2-W, metal-oxide resistors at the output of the LM317 supply about 4 W to maintain a temperature of about 33°C inside the enclosure. The resistors are placed so that heat is distributed uniformly. Five are placed near the bottom and three near the top. The thermistor is mounted in the center of the box, close to the tuned circuit and in physical contact with the oscillator groundplane surface, using a small drop of epoxy.

Use a massive and well-insulated enclosure to slow the rate of temperature change. In one project, over a 0.1°C range, the frequency at 5.0 MHz varied up and down ±20 Hz or less, with a period of about five minutes. Superimposed is a very slow drift of average frequency that is due to settling down of components, including possibly the thermistor. These gradual changes became negligible after a few days of "burn-in." One problem that is virtually eliminated by the constant temperature is a small but significant "retrace" effect of the cores and capacitors, and perhaps also the thermistor, where a substantial temperature transient of some kind may take from minutes to hours to recover the previous L and C values.

For a much more detailed discussion of this material, see Sabin W.E., WØIYH "Thermistors in Homebrew Projects," *QEX* Nov/Dec 2000. This article is included (with all *QST*, *QEX* and *NCJ* articles from 2000) on the *2000 ARRL Periodicals* CD-ROM. (Order No. 8209.)

reverse-biased PN junction has a negative voltage TC while a forward-biased junction has a positive voltage TC, a good way to temperature-compensate a Zener reference diode is to place one or more forward-biased diodes in series with it.

RF Heating

RF current often causes component heating problems where the same level of dc current may not. An example is the tank circuit of an RF oscillator. If several small capacitors are connected in parallel to achieve a desired capacitance, skin effect will be reduced and the total surface area available for heat dissipation will be increased, thus significantly reducing the RF heating effects as compared to a single large capacitor. This technique can be applied to any similar situation; the general idea is to divide the heating among as many components as possible.

CAD TOOLS FOR CIRCUIT DESIGN

Hams today enjoy the easy availability of tremendous computing power to aid them in their hobby. A commercial application that computers perform very well is circuit simulation. Today anyone with access to a computer software bulletin board is likely to find shareware or other inexpensive software to perform such analysis. Indeed, ARRL offers *ARRL Radio Designer,* which provides excellent RF design capabilities at a reasonable price. *Radio Designer* is discussed in October 1994 *QST* and a subsequent *QST* column, "Exploring RF."

The advantages of computer circuit simulation over pencil-and-paper calculation are many. For example, while the analysis of a small circuit, say 5 to 10 components, may be recomputed without too much effort if the design changes, the same is not true for a 50 to 100-component circuit. Indeed, the larger circuit may require too many simplifying assumptions to calculate by hand at all. It is much faster to watch and adjust the response of a circuit on a computer screen than to breadboard the circuit in search of the same information.

All circuit simulators, including *SPICE,* work from a *netlist* that is simply a component-by-component list of the circuit that tells the program how components are interconnected. The program then uses standard techniques of numerical analysis and matrix mathematics to calculate the analysis you wish to perform. Common analyses include dc and ac bias and operating point calculations, frequency response curves, transient analysis (looking at waveforms in the circuit over a specified period of time), Fourier transforms, transfer functions, two-port parameter calculations, and pole-zero analysis. Many packages also perform sensitivity analyses: calculating what happens to the voltages and currents as you sweep the value(s) of an individual component (or group of components), and statistical analyses: determining the variational "window" of the system response that would result if all component values randomly varied by a given tolerance (also known as *Monte Carlo* analysis).

Beware!

Before we turn to a brief introduction to circuit simulation for the amateur, a *caveat* to the reader is definitely in order: However fast and powerful the computer may seem, keep in mind that it is only another tool in your workshop, just like your 'scope or soldering iron. Circuit simulations are only meaningful if you can interpret the results correctly, and a good initial circuit design based on real experience and common sense is mandatory. *SPICE* and other programs have no problem with a bench-top power supply delivering 10,000 V to a 5-W resistor because you made a mistake when specifying the circuit. Software also won't remind you that the resistor better be rated for 20 megawatts! Remember, the real power behind any simulation software is *your mind.*

SPICE

The electronics industry has been the main consumer of circuit simulation packages ever since the development of *SPICE* (or *S*imulation *P*rogram with *I*ntegrated *C*ircuit *E*mphasis) on mainframe systems in the mid 1970s. While many companies today produce other simulation software, *SPICE* remains the *de facto* industry standard. *SPICE* itself now exists in many "flavors", and specially tailored versions for desktop computers are sold by several companies, including *PSPICE* by Microsim Corp and *HSPICE* by Intusoft. In fact, an "evaluation" copy of *PSPICE* has been placed in the public domain by Microsim and is probably the easiest and most inexpensive way for amateurs to begin circuit simulation. It may be obtained (with a companion reference book) from technical bookstores, and also from many sites on the Internet.

A Basic Circuit

Consider the simple RC low-pass filter in **Fig 6.37A**. To simulate this circuit, we must first precisely describe it. Convention sets ground as *node* 0, and we label all other nodes (a node is a place where two or more components meet) with numbers, as shown in the figure. We give each element a unique name, and then prepare a *netlist,* listing each element with the numbers of its "+" and "–" nodes (*SPICE* assumes the ground node to be labeled 0), and its value, as shown at B in the figure.

The question we wish to answer by simulation is the following: what is the frequency response of this filter? Of course, for this simple problem we can calculate the answer by hand. The cutoff frequency of a simple RC filter is given by:

$$f_{co} = \frac{1}{2 \pi R C} \qquad (17)$$

where

f_{co} = cutoff frequency (–3 dB)
R = resistance, in ohms
C = capacitance, in farads.

In this case f_{co} is 1.59 kHz. Beyond this, with one storage element (the capacitor) we expect the output to decrease by 20 dB for each decade of frequency.

Fig 6.38 shows the graph of the ratio of output to input voltages V_{out}/V_{in}, in dB for that frequency range as calculated by *SPICE.* Note the 3-dB point is indeed 1.59 kHz, and above that frequency the response drops by 20-dB/decade.

At this point we could use this circuit to ask some "What if?" questions, but for a circuit this simple we could answer those questions with pencil and paper. Below is a more extensive design to show the power of simulation, where "What if?" questions and the interaction of numerous components are much more easily and clearly examined with a computer.

A Case Study: Amplifier Distortion

An application where computer circuit simulation does a much faster, easier, and more accurate job than pencil and paper is Fourier analysis. Recall that any periodic signal can be broken down into a sum of sine waves of different amplitudes whose frequencies are multiples of the fundamental frequency of the signal. Performing such an analysis allows us to identify how much signal power is contained in the fundamental frequency and its harmonics. A Fourier *transform* is the representation of a signal by its frequency components, and resembles what you would see if you fed the signal into a spectrum analyzer.

Consider the amplifier circuit in **Fig 6.39**, which is a simple common-emitter audio amplifier. The frequency response of its voltage gain, calculated from *SPICE,* is shown in **Fig 6.40**. Pay particular attention to the voltage gain at 1 kHz, which is 148, or 43.4 dB. If we place a small ac signal on the input as shown in Fig 6.39, the output will be a faithfully amplified sine wave. However, the amplifier is powered from a 9-V supply, so when the output reaches this level, the waveform will begin to show signs of clipping. For a 1-kHz signal, this will occur at roughly V_{in} = 4.5/148 = 0.03 V amplitude.

Total Harmonic Distortion is a common figure of merit for audio amplifiers, and is defined as

$$THD = \frac{P_H}{P_F} \times 100 \qquad (18)$$

where

THD = total harmonic distortion, in percent
P_H = total power in all harmonics above the fundamental
P_F = power in fundamental.

SPICE readily calculates the Fourier content of a signal and also THD. Assume that we will tolerate 5% THD for our design. What is the maximum swing we can allow the input signal?

Fig 6.41 shows the output waveforms that *SPICE* calculates for our circuit with input signals of 0.001, 0.003, 0.01, 0.03

and 0.1-V amplitude. Note that, as expected, the last two input voltages show definite signs of clipping. What is not so evident is that the smaller inputs have some distortion as well, due to the small but finite nonlinearity of the amplifier even at those voltages.

Fig 6.42 shows the relative amplitudes of the harmonic content (Fourier transform) of the Fig 6.41 waveforms, all normalized to fundamental = 100%. You can see how the higher harmonics grow in strength as the signal increases, especially when the clipping starts. The THD for each case is also shown, and from this simulation we see we must limit our input to somewhere between 0.003 and 0.01 V. We would pin this down more closely by running another simulation around this "window."

A caution is in order here. Circuit models have limits just as do circuits. The models apply only over a limited dynamic range. That's why we have both small- and large-signal models. One challenge of CAD is determining the limits of your models. There is more information about this in the sections on modeling transistors.

Conclusion

Start using your computer for more challenging tasks than QSL bookkeeping! Just like any other piece of software, you'll find yourself using *SPICE* to do things you never thought of trying before. It can even do digital circuits, transmission lines and antennas. Remember that a computer is only as smart as the person using it! Have fun!

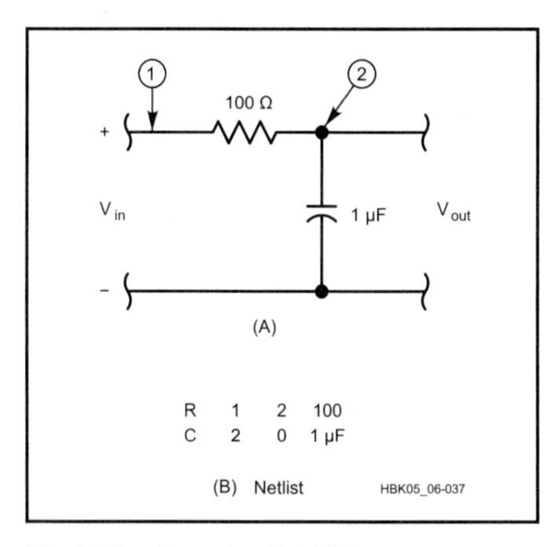

Fig 6.37—A, a simple RC low-pass filter.

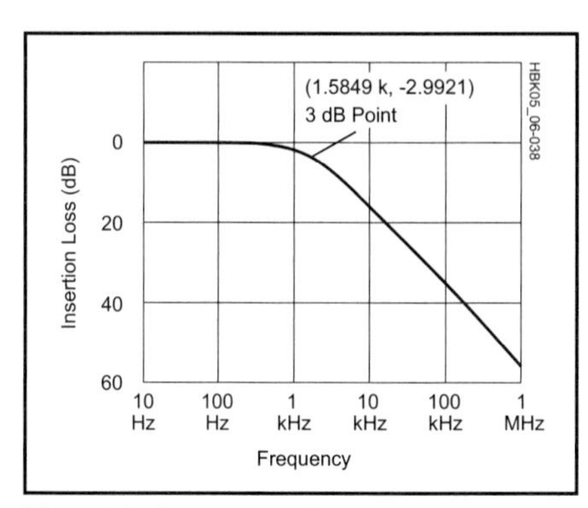

Fig 6.38—Low-pass filter output simulation.

References

J. Carr, *Secrets of RF Circuit Design*, Tab Books, 1991.

D. DeMaw, *Practical RF Design Manual*, Prentice-Hall, 1982.

R. Dorf, Ed., *The Electrical Engineering Handbook*, CRC Press and IEEE Press, 1993.

W. Hayward, *Introduction to RF Design*, ARRL, 1994.

P. Tuinenga, *SPICE: A Guide to Circuit Simulation and Analysis using PSPICE*, by Prentice-Hall, ISBN 0-13-747270-6. This is a wonderful PSPICE reference manual that comes with copy of the software. It is available at most university or technical bookstores.

A. Vladimirescu, *The SPICE Book*, Wiley & Sons, 1994.

D. Pederson and K. Mayaram, *Analog Integrated Circuits for Communication: Principles, Simulation and Design, 1991*, Kluwer Academic Publishers. Pederson is one of the inventors of SPICE. This book is about SPICE simulation.

J. White, Thermal Design of Transistor Circuits," April 1972 *QST*, pp 30-34.

LOW-FREQUENCY TRANSISTOR MODELS

The Fundamental Equations

Design models are based on the physics of the components we use. The complexity of models can vary widely though, and many times we can use very simple models to achieve our goals. Increasingly complex models are developed and used only when demanded by the circuit application.

In this discussion, we will focus on simple models for bipolar transistors (BJTs) and FETs. These models are reasonably accurate at low frequencies, and they are of some use at RF. For more sophisticated RF models, look to professional RF-design literature. This discussion is adapted from Wes Hayward's *Introduction to RF Design*. Derivations of the material shown here appear in that book, an excellent text for the beginning RF designer.

This discussion is centered on NPN BJTs and N-channel JFETs. The material here applies to PNPs and P-channel FETs when you simply change the bias polarities.

First, consider the bipolar transistor as a current controlled device. When the base current controls the collector current, this equation defines the transistor operation

$$I_c = \beta I_b \tag{19}$$

where β = common-emitter current gain.

When we consider a transistor as a voltage-controlled device, this equation describes it (emitter current, I_e, in terms of base-emitter voltage, V_{be}):

$$\begin{aligned} I_e &= I_{es}\left[\exp\left(qV/kT\right)-1\right] \\ &\approx I_{es}\exp\left(qV/kT\right) \end{aligned} \tag{20}$$

where
$V = V_{be}$
q = electronic charge
k = Boltzmann's constant
T = temperature in kelvins (K)
I_b = emitter saturation current, typically 1×10^{-13} A.

Both equations approximate models of more complex behavior. Equation 20 is a simplification of the first Ebers-Moll model (Ref 1). More sophisticated models for BJTs are described by Getreu in Ref 2. Equations 19 and 20 apply to both transistor dc biasing and signal design.

The operation of an N-channel JFET can be characterized by

$$I_D = I_{DSS}\left(1 - V_{sg}/V_p\right)^2 \tag{21}$$

where
I_{DSS} = drain saturation current
V_{sg} = the source-gate voltage
V_p = the pinch-off voltage.

This equation applies only so long as V_{sg} is between 0 and V_p. JFETs are seldom used with the gate-to-channel diode forward biased. Drain current, I_D, is 0 when V_{sg} exceeds V_p. This equation applies to both biasing and signal design.

Now that we have some basic equations, let's go on to some other areas:

- Small-signal amplifier design (and application limits).
- Large-signal amplifier design (distortion from nonlinearity).

BIPOLAR TRANSISTORS (SMALL SIGNALS)

Transistors are usually driven by both biasing and signal voltages. Small-signal models treat only the signal components. We will consider bias and nonlinear signal effects later.

A Basic Common-Emitter Model

Fig 6.43 shows a BJT amplifier. The circuit is adequately described by equation 19. A mathematical analysis of the control and output currents and voltages yields the small-signal common-emitter

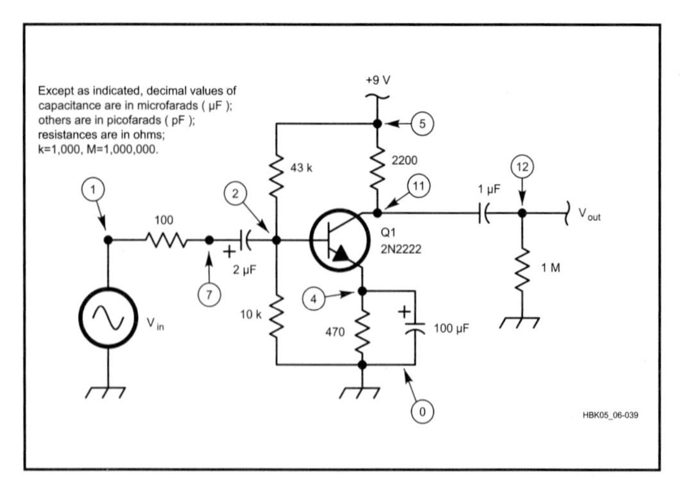

Fig 6.39—Common-emitter audio amplifier used in the text.

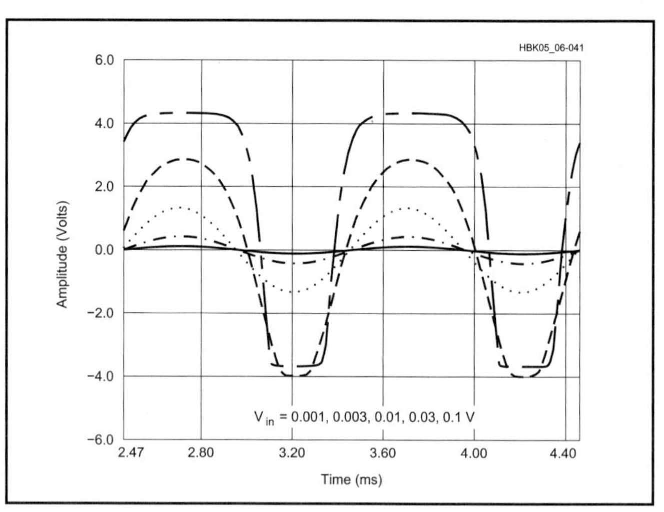

Fig 6.41—Output waveforms for 1-kHz inputs of different magnitudes. Note the clipping that begins to appear at approximately V_{in} = 0.03 V.

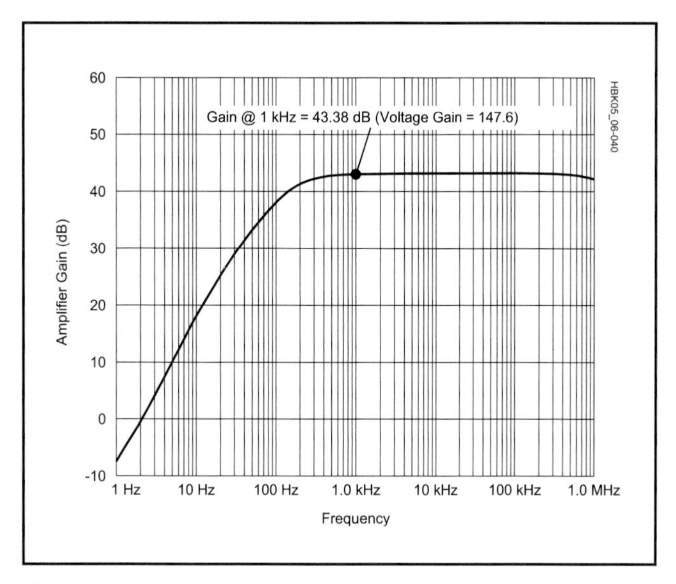

Fig 6.40—Frequency response of the voltage gain of the amplifier in Fig 6.39. The gain at 1 kHz is approximately 148 or 43 dB.

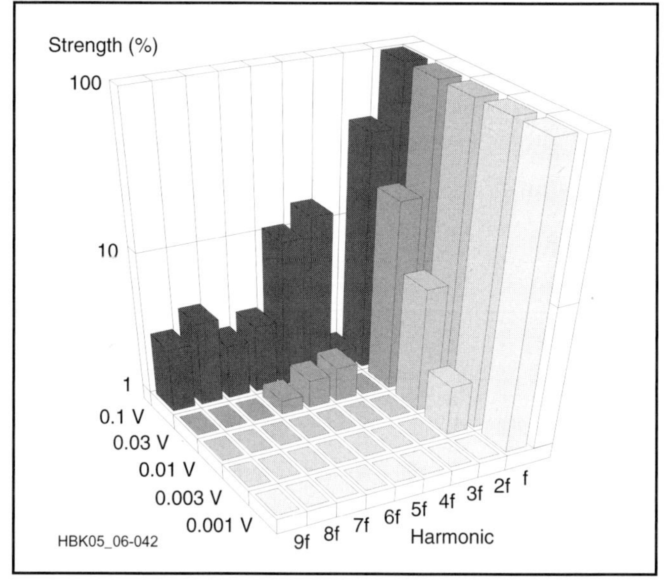

Fig 6.42—Fourier (harmonic) decomposition and Total Harmonic Distortion values for the waveforms in Fig 6.41.

amplifier model shown in **Fig 6.44**. This is the most common of all transistor small-signal models, a controlled current source with emitter resistance.

In order to use this model, however, we must have a value for r_e. That's no trouble, however, $r_e = kT / qI_0$, or $r_e = 26 / I_e$, where I_e is the dc bias current in milliamperes. This value applies at a typical ambient temperature of 300 K.

The device output resistance is infinite because it is a pure current source, which is a good approximation for most silicon transistors at low frequencies. In use, the collector lead would feed a load resistor, R_L. For this model, that resistance must be small enough so that the collector bias voltage is positive for the chosen bias current.

Gain vs Frequency

Fig 6.44 is a low frequency approximation. As signal frequency increases, however, current gain appears to decrease. The low-frequency current gain is β_0. β_0 is constant through the audio spectrum, but it eventually decreases, and at some high frequency it will drop by a factor of 2 for each doubling of signal frequency. A transistor's frequency vs current gain relationship is specified by its *gain-bandwidth product*, or F_T. F_T is the frequency at which the current gain is 1. Common transistors for lower RF applications might have $\beta_0 = 100$ and $F_T = 500$ MHz. The frequency at which current gain is β_0 is called F_b and related to F_T by $F_b = F_T / \beta_0$.

The frequency dependence of current gain is modeled by adding a capacitor across the base resistor of Fig 6.50A; **Fig 6.45**, the *hybrid-pi* model results. The capacitor reactance should equal the low-frequency input resistance, $(\beta + 1)r_e$, at F_b. This simulates a frequency-dependent current gain.

Three Simple Models

Even though transistor gain varies with frequency, the simple model is still useful under certain conditions. Calculations show that the simple model is valid, with $\beta = F_T / F$, for frequencies well above F_β.

APPENDIX: SPICE Input Files

CE Audio Amp (Case Study)

```
Amplifier with distortion test
*
VSIG  1 0  ac 0.01 SIN (0 0.001 1KHZ)
RIN   1 7  100
CIN   7 2  2UF
RB1   5 2  43K
RB2   2 0  10K
Q1    11 2 4  Q2N2222A
RC    5 11  2200
RE    4 0  470
CE    4 0  100UF
COUT  11 12   1UF
ROUT  12 0    1MEG
VBIAS 5 0  DC 9
*
..MODEL Q2N2222A NPN(Is=14.34f Xti=3 Eg=1.11
Vaf=74.03  Bf=255.9  Ne=1.307  +Ise=14.34f
Ikf=.2847 Xtb=1.5 Br=6.092 Nc=2 Isc=0 Ikr=0 Rc=1
+Cjc=7.306p Mjc=.3416 Vjc=.75 Fc=.5 Cje=22.01p
Mje=.377 Vje=.75 +Tr=46.91n Tf=411.1p Itf=.6
Vtf=1.7 Xtf=3 Rb=10)
*National    pid=19   case=TO18
*
..AC DEC 10 1HZ 1MEGHZ
..TRAN 1MS 10MS 2MS 20US
..FOUR 1KHZ V(12)
..PROBE v(12) v(1)
..END
```

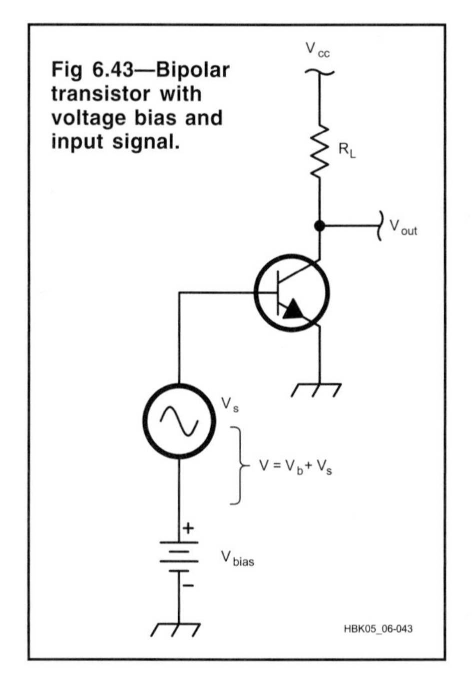

Fig 6.43—Bipolar transistor with voltage bias and input signal.

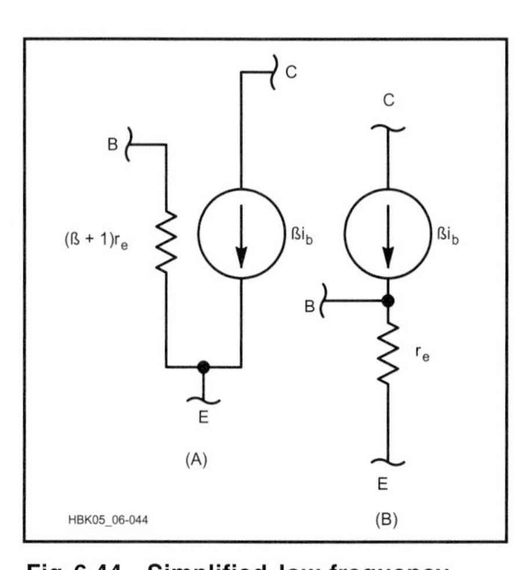

Fig 6.44—Simplified low-frequency model for the bipolar transistor, a "beta generator with emitter resistance." r_e=26 / I_e(mA dc).

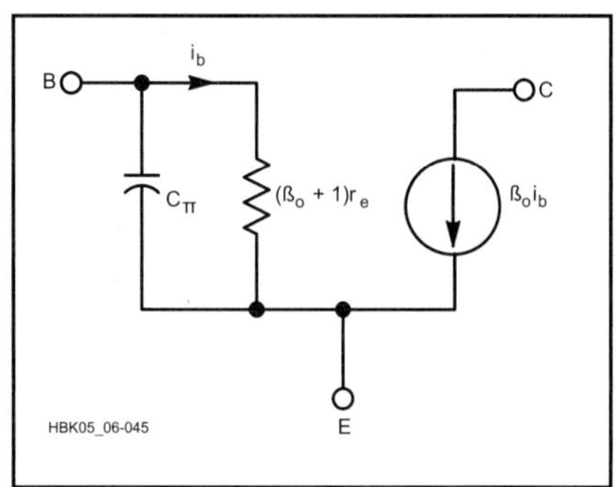

Fig 6.45—The hybrid-pi model for the bipolar transistor.

The approximation worsens, however, as the operating frequency (f_0) approaches F_T. **Fig 6.46** shows small-signal models for the three common amplifier configurations: common emitter (ce), common base (cb) and common collector (cc).

The common-collector amplifier, unlike the common-emitter or common-base configurations, has a finite output resistance. This resistance is calculated by short circuiting the input voltage source, V_s, and "driving" the output port with either a voltage or current source. The result is the equation for R_{out} that appears in the figure.

The common-collector example shows characteristics that are more typical of practical RF amplifiers than the idealized ce and cb amplifiers. Specifically, the input resistance is a function of both the device and the termination at the output. The output resistance is critically dependent upon the input driving source resistance.

These examples have used the simplest of models, the controlled current genera-

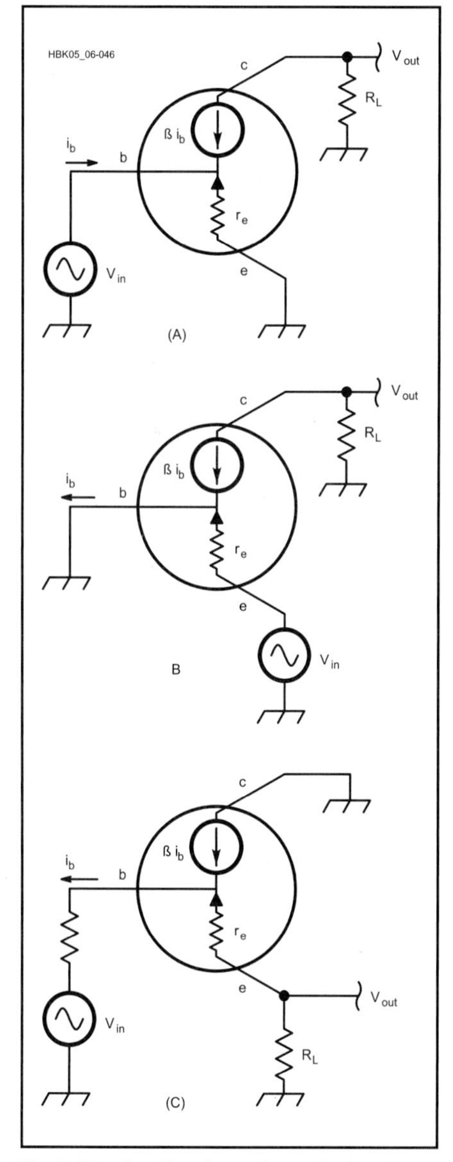

Fig 6.46—Application of small-signal models for analysis of (A) the ce amplifier, (B) the cb and (C) the cc bipolar transistor amplifiers.

tor with an emitter resistance, r_e. Better models are necessary to design at high frequencies. The simplified hybrid-pi of Fig 6.45 is often suitable.

Small-Signal Design at RF

Fig 6.47 shows a better small-signal model for RF design that expands on the hybrid-pi. Consider the physical aspects of a real transistor: There is some capacitance across each of the PN junctions (C_{cb} and C_p) and capacitance from collector to emitter (C_c). There are also capacitances between the device leads (C_e, C_b and C_c). There is a resistance in each current path, emitter to base and collector. From emitter to base, there is r_π from the hybrid-pi model and r´b, the "base spreading" resistance. From emitter to collector is R_o, the output resistance. The leads from the die to the circuit present three inductances.

Manual circuit analysis with this model doesn't look like much fun. It's best tackled with the aid of a computer and specialized software. Other methods are presented in Hayward's *Introduction to RF Design.*

Don't be intimidated by the complexity of the model, however. Surprisingly accurate results may be obtained, even at RF, from the simple models. Simple models also give a better "feel" for device characteristics that might be obscured by the mathematics of a more rigorous treatment. Use the simplest model that describes the important features of the device and circuit at hand.

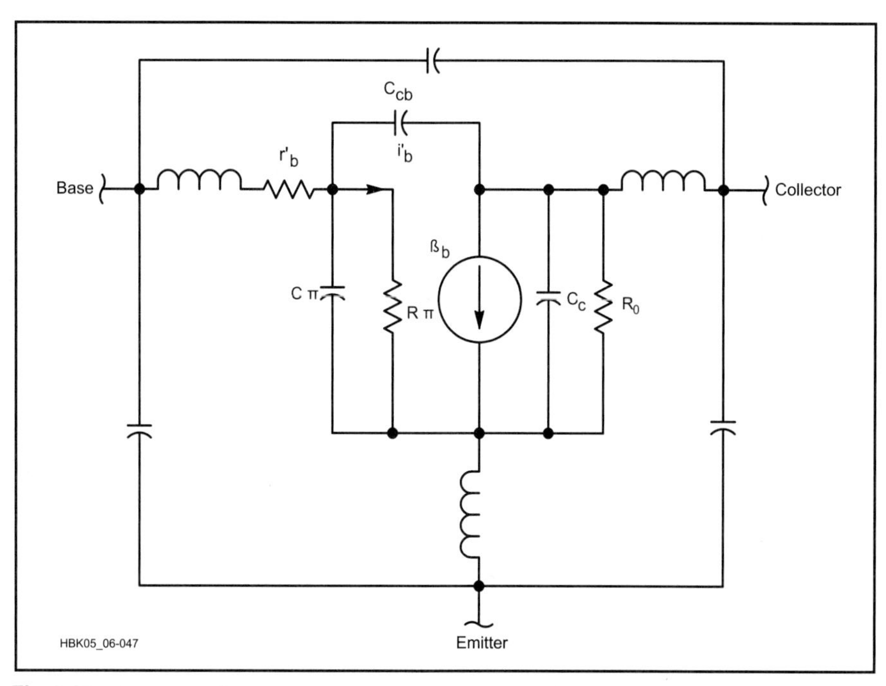

Fig 6.47—A more refined small-signal model for the bipolar transistor. Suitable for many applications near the transistor F_T.

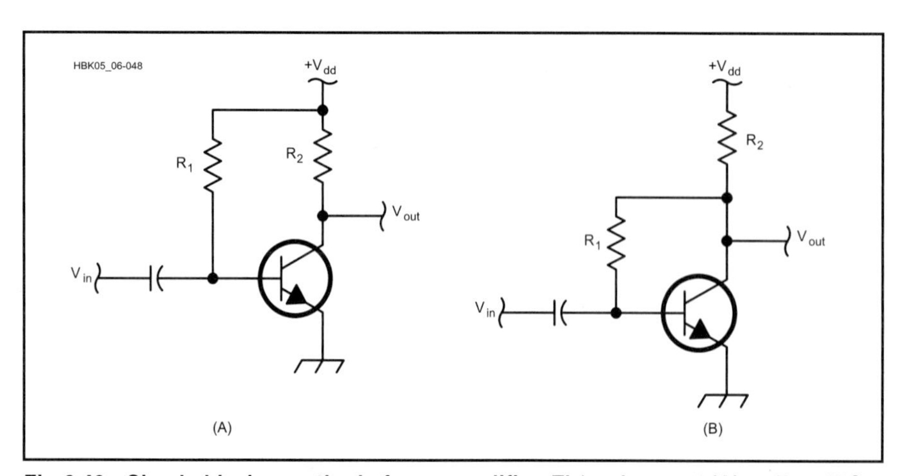

Fig 6.48—Simple biasing methods for ce amplifier. The scheme at (A) suffers if β is not well known. Negative feedback is used in (B).

Biasing Bipolar Transistors

Proper biasing of the bipolar transistor is more complicated than it might appear. The Ebers-Moll equation would suggest that a common-emitter amplifier could be built as shown in Fig 6.43, grounding the emitter and biasing the base with a constant voltage source. Further examination shows that this presents many problems. The worst is that constant-voltage bias ultimately leads to thermal *runaway*. Constant-voltage biasing applied to the base is almost never used.

Constant base-current biasing is shown in **Fig 6.48A**. This works reasonably well if the current gain is known, which is rarely true. A transistor with a typical β of 100 might actually have values ranging from 50 to 250. A slightly improved method is shown in Fig 6.48B, where the bias is derived from the collector. As current increases, collector voltage decreases, as does the bias current flowing through R_1. This ensures operation in the transistor active region.

The most common biasing method is shown in **Fig 6.49A**. The device model used, shown in Fig 6.49B, is based on the Ebers-Moll equation, which shows that virtually no transistor current flows until the base-emitter voltage reaches about 0.6. Then, current increases dramatically with small additional voltage change. The transistor is thus modeled as a current controlled generator with a battery in series with the base. The battery voltage is ΔV.

The circuit is analyzed with nodal equations. The collector resistance, R_5, is initially assumed to be zero. The analysis results in three equations for V_b, $V_c{}'$ and I_b:

$$V_b = \frac{V_{cc}\, R2R3 + \Delta V R2\left(\dfrac{R4+R1}{\beta+1}\right)}{R3R4 + R2R4 + R2R3 + R1R3 + \dfrac{R1R2}{\beta+1}} \tag{22}$$

$$V_C{}' = \frac{R1V_{cc}R4V_b + \beta R1R4\left(\dfrac{\Delta V - V_b}{R3(\beta+1)}\right)}{R1+R4} \tag{23}$$

$$I_b = \frac{V_b - \Delta V}{R3(\beta+1)} \tag{24}$$

The emitter current is then $I_b(\beta + 1)$. Once the circuit has been analyzed, R5 may be taken into account. The final collector voltage is

$$V_c = V_c{}' - \beta I_b\, R5 \tag{25}$$

The solution is valid so long as V_c exceeds V_b.

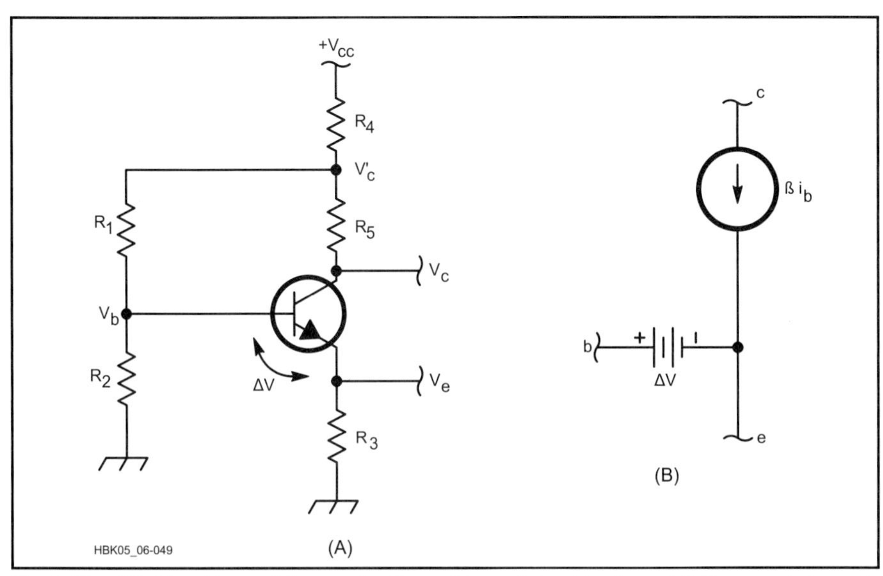

Fig 6.49—(A) Circuit used for evaluation of transistor biasing. (B) The model used for bias calculations.

Ready-Made Models

Many manufacturers provide computer models or data helpful when modeling their devices. The *ARRL Radio Designer* software is specifically designed for ham radio designs. Other commercial packages are:

1. Avantek, "S Parameter Performance and Data Library." Collection of transistor S Parameters. To be used in *SPICE* programs. Two disks.
2. Linear Technology Co, "*SPICE* Models for LTC Linear IC Products." A second disk contains "Noise Models" for *SPICE* that are used to predict noise performance using *SPICE.*
3. Burr-Brown, "*PSPICE* Macromodels" for Burr-Brown op amps.
4. Analog Devices Co, "*SPICE* Model Library" for Analog Devices Op amps and other linear ICs.
5. Texas Instruments, "Operational Amplifier, Comparator and Building Block Macromodels."
6. Motorola, "Discrete Data Disk."
7. Motorola, "Scattering Parameter Library," for EESOF *Touchstone.*
8. Motorola, "Scattering Parameter Plotting Utility," for EESOF *Touchstone.*
9. Motorola, "Impedance Matching Program," for EESOF *Touchstone.*
10. ARRL, "UHF/Microwave Experimenter's Manual," Software Disk.
11. Precision Monolithics, "*SPICE* Macro-Models," op amps.
12. Tektronix, "BJT *SPICE* Subcircuits."
13. Harris, "*SPICE* Macro-Models," op amps.

Analysis of these equations with a computer or hand-held programmable calculator shows that I_e is not a strong function of the transistor parameters, ΔV and β. In practice, the base biasing resistors, R1 and R2, should be chosen to draw a current that is much larger than I_b (to eliminate effects of β variation). V_b should be much larger than ΔV to reduce the effects of variations in ΔV.

Three additional biasing schemes are presented in **Fig 6.50**. All provide bias that is stable regardless of device parameter variations. A and B require a negative power supply. The circuit of Fig 6.50C uses a second, PNP, transistor for bias control. The PNP transistor may be replaced with an op amp if desired. All three circuits have the transistor emitter grounded directly. This is often of great importance in microwave amplifiers. These circuits may be analyzed using the simple model of Fig 6.49.

The biasing equations presented may be solved for the resistors in terms of desired operating conditions and device parameters. It is generally sufficient, however, to repetitively analyze the circuit, using standard resistor values.

The small-signal transconductance of a common-emitter amplifier was found in the previous section. If biased for constant current, the small-signal voltage gain will vary inversely with temperature. Gain may be stabilized against temperature variations with a biasing scheme that causes the bias current to vary in *proportion to absolute temperature.* Such methods, termed PTAT methods, are often used in modern integrated circuits and are finding increased application in circuits built from discrete components.

Large-Signal Operation

The models presented in previous sections have dealt with small signals applied to a bipolar transistor. While small-signal design is exceedingly powerful, it is not sufficient for many designs. Large signals must also be processed with transistors. Two significant questions must be considered with regard to transistor modeling. First, what is a reasonable limit to accurate application of small-signal methods? Second, what are the consequences of exceeding these limits?

The same analysis of the Ebers-Moll model yields an equation for collector current. The mathematics show that current will vary in a complicated way, for the sinusoidal signal voltage is embedded within an exponential function. Nonetheless, the output is a sinusoidal current if the signal voltage is sufficiently low.

The current of the equation may be studied by normalizing the current to its peak value. The result is relative current, I_r, which is plotted in **Fig 6.51** for V_p values of 1, 10, 30, 100, and 300 mV. The 1-mV case is very sinusoidal. Similarly, the 10-mV curve is generally sinusoidal with only minor distortions. The higher amplitude cases show increasing distortion.

Constant base-voltage biasing is unusual. More often, a transistor is biased to produce nearly constant emitter current. When such an amplifier is driven by a large input signal, the average bias voltage will adjust itself until the time average of the nonlinear current equals the previous constant bias current. Hence, it is vital to consider the average relative current of the waveforms of Fig 6.51. This is evaluated through calculus.

The average relative currents for the cases analyzed occur at the intersection of the curves with the dotted lines of Fig 6.51. For example, the dotted curve intersects the V_p = 300-mV waveform at an average relative current of 0.12. If an amplifier was biased to a constant current of 1 mA, but was driven with a 300-mV signal, the positive peak current would reach a value greater than the average by a factor of 1/(0.12). The average current would remain at 1 mA, but the positive peak would be 8 mA. The transistor would not conduct for most of the cycle.

The curves have presented data based upon the simplest of large-signal models, the Ebers-Moll equation. Still, the simple model has yielded considerable information. The analysis suggests that a reasonable upper limit for accurate small-signal analysis is a peak base signal of about 10 mV. The effect of emitter degeneration is also evident. Assume a transistor is biased for $r_e = 5\ \Omega$ and an external emitter resistor of 10 Ω is used. Only the r_e portion of the 15 Ω total is nonlinear. Hence, this amplifier would tolerate a 30-mV signal while still being well described with a small-signal analysis.

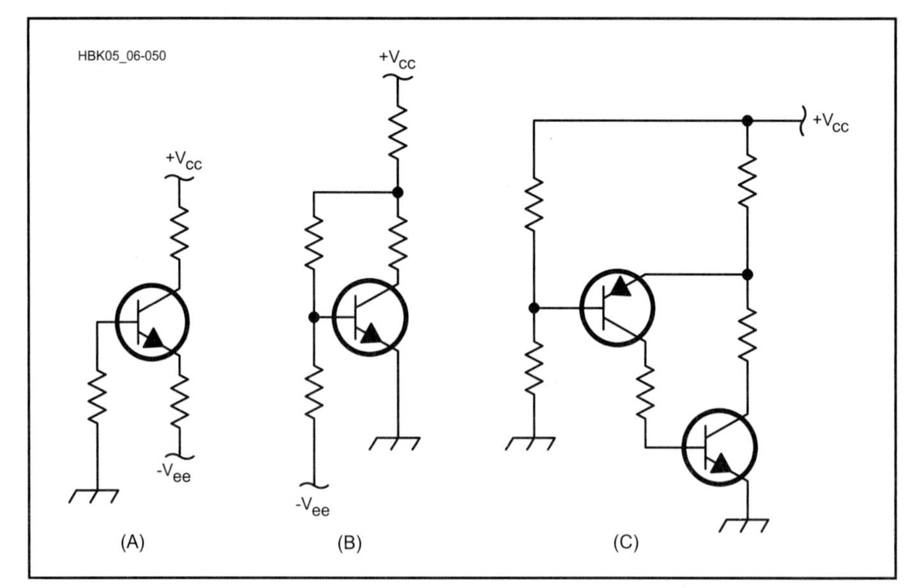

Fig 6.50—Alternative biasing methods. (A) and (B) use dual power supplies, (B) and (C) allow the emitter to be at ground while still providing temperature-stable operation.

FETS

An often used device in RF applications is the field-effect transistor (FET). There are many kinds: JFETs, MOSFETs and so on. Here we will discuss JFETs, with the understanding that other FETs are similar.

We viewed the bipolar transistor as controlled by either voltage or current. The JFET, however, is purely a voltage controlled element, at least at low frequencies. The input gate is usually a reverse biased diode junction with virtually no current flow. The drain current is related to the source-gate voltage by:

$$I_D = I_{DSS}\left(1 - \frac{V_{sg}}{V_p}\right)^2$$

for $0 \le V_{sg} \le V_p$; $I_D = 0$; and $V_{sg} > V_p$, (26)

where I_{DSS} is the drain saturation current and V_p is the pinch-off voltage. Operation is not defined when V_{sg} is less than zero because the gate diode is then forward biased. Equation 26 is a reasonable approximation as long as the drain bias voltage exceeds the magnitude of the pinch-off voltage.

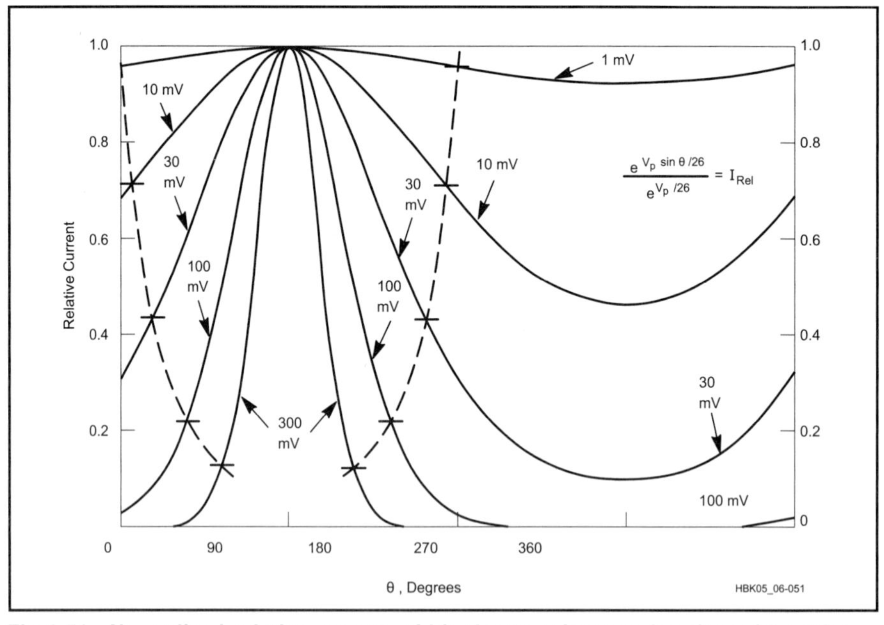

Fig 6.51—Normalized relative current of bipolar transistor under sinusoidal drive at the base.

Biasing FETs

Two virtually identical amplifiers using N-channel JFETs are shown in **Fig 6.52**. The two circuits illustrate the two popular methods for biasing the JFET. Fixed gate-voltage bias, Fig 6.52A, is feasible for JFETs because of their favorable temperature characteristics. As the temperature of the usual FET increases, current decreases, avoiding the thermal-runaway problem of bipolar transistors.

A known source resistor, R_s in Fig 6.52B, will lead to a known source voltage. This is obtained from a solution of equation 26 (see **Eq 27**).

$$V_{sg} = \frac{\left[\frac{1}{R_s I_{DSS}} + \frac{2}{V_p}\right] - \left[\left(\frac{1}{R_s I_{DSS}} + \frac{2}{V_p}\right)^2 - \left(\frac{2}{V_p}\right)^2\right]^{0.5}}{\frac{2}{V_p^{\,2}}} \quad (27)$$

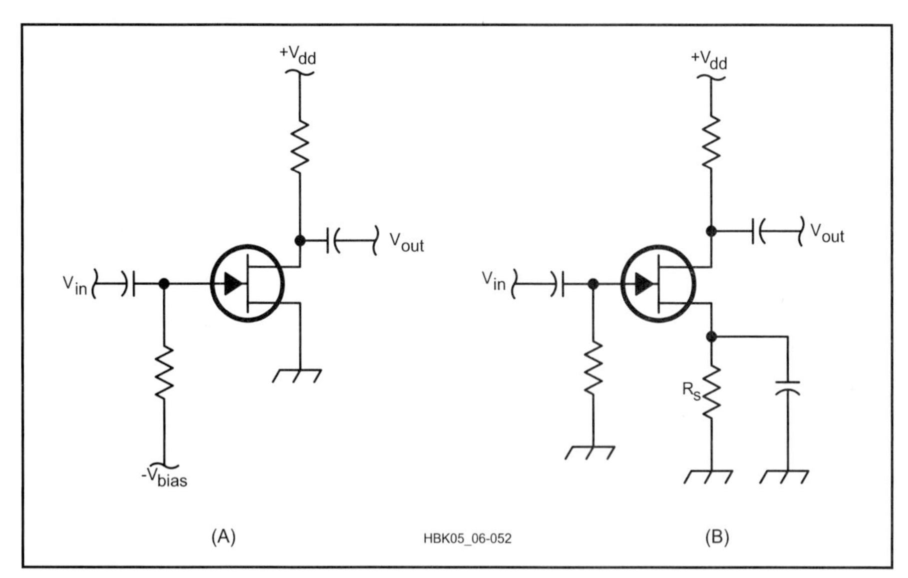

Fig 6.52—Biasing schemes for a common-source JFET amplifier. $-V_{bias}$ is normally adjusted to suit each device; there is a significant spread over a product run. Also note that some FETs can exhibit thermal runaway in some current/ temperature ranges.

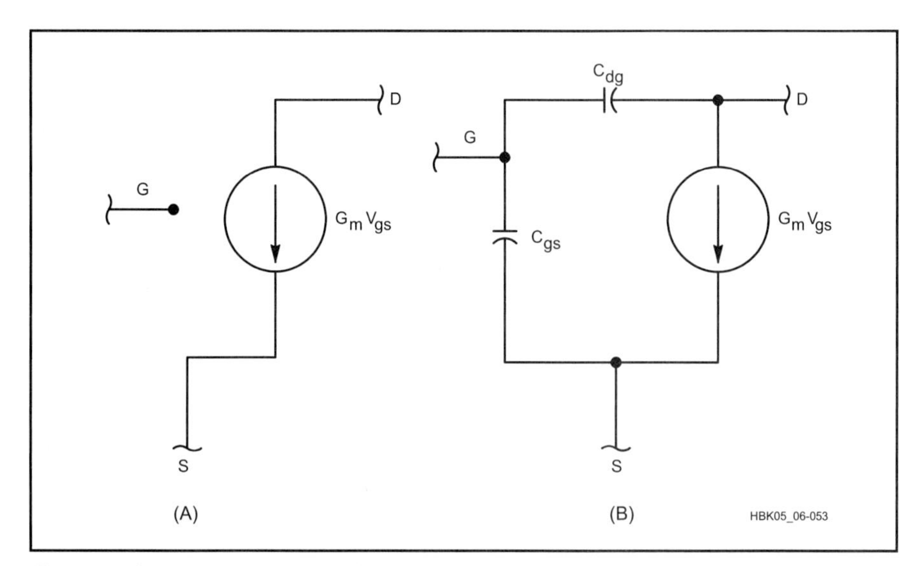

Fig 6.53—Small-signal models for the JFET. (A) is useful at low frequency, (B) is a modification to approximate high-frequency behavior.

The drain current is then obtained by direct substitution.

Alternatively, a desired drain current less than I_{DSS} may be achieved with a proper choice of source resistor

$$R_s = \frac{V_p\left(1 - \frac{I_D}{I_{DSS}}\right)^{0.5}}{I_D} \quad (28)$$

The small-signal transconductance of the JFET is obtained by differentiating equation 26

$$g_m = \frac{dI_D}{dV_{sg}} = \frac{-2\,I_{DSS}}{V_p}\left(1 - \frac{V_{sg}}{V_p}\right) \quad (29)$$

The minus sign indicates that the equation describes a common-gate configuration. The amplifiers of Fig 6.52 are both common-source types and are described by equation 29 except that g_m is now positive. Small-signal models for the JFET are shown in **Fig 6.53**. The simple model is that inferred from the equations while the model of Fig 6.53B contains capacitive elements that are effective in describing high-frequency behavior. Like the bipolar transistor, the JFET model will grow in complexity as more sophisticated applications are encountered.

Large-Signal Operation

Large-signal JFET operation is examined by normalizing the previous equation to $V_p = 1$ and $I_{DSS} = 1$ and injecting a sinusoidal signal. The circuit is shown in **Fig 6.54**. Also shown in the figure are examples for a variety of bias and sinusoid amplitude conditions. The main feature is the asymmetry of the curves. The positive portions of the oscillations are farther from the mean than are the negative excursions. This is especially dramatic when the bias, v_0, is large, which places the quiescent point close to pinch-off. With such bias and high-amplitude drive, conduction occurs only over a small fraction of the total input waveform period.

The average current for these operating conditions can be determined by calculus. The average current values obtained may be further normalized by dividing by the corresponding dc bias current, $I_0 = (1 - v_0)^2$. The results are shown in **Fig 6.55**. The curves show that the average current increases as the amplitude of the drive increases. This, again, is most pronounced when the FET is biased close to pinch-off.

Although practical for the JFET, constant-voltage operation in the previous curves is not common. Instead, a resistive bias is usually employed, Fig 6.52B. With this form of bias, the increased current from high signal drive will cause the voltage drop across the bias resistor to increase. This will then move the quiescent operating level closer to pinch-off, accompanied by a reduced small-signal trans-conductance. This behavior is vital in describing the limiting found in FET oscillators.

The limits on small-signal operation are not as well defined for a FET as they were for the bipolar transistor. Generally, a maximum voltage of 50 to 100 mV is allowed at the input (normalized to a 1-V pinch-off) without severe distortion. The voltages are much higher than they were for the bipolar transistor. However, the input resistance of the usual common source amplifier is so high and the corresponding transconductance low enough that the available gain is no greater than could be obtained with a bipolar transistor. The distortion is generally less with FETs, owing to the lack of high-order curvature in the defining equations.

Many of the standard circuits used with bipolar transistors are also practical with FETs. Noting that the transconductance of a bipolar transistor is $g_m = I_e$ (dc mA) / 26, the previous equations may be applied directly. The "emitter current" is chosen to correspond with the FET transconductance. A very large value is used for current gain. The same calculator or computer program is then used directly. In prac-

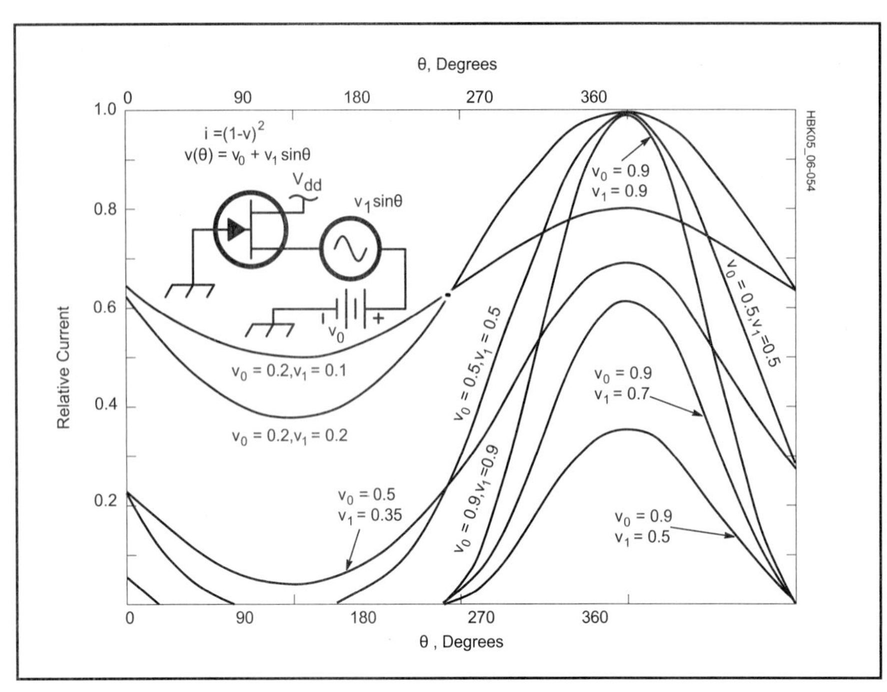

Fig 6.54—Relative normalized drain current for a JFET with constant voltage bias and sinusoidal signals. Relatively "clean" waveforms exist for low signals while large input amplitudes cause severe distortion.

tice, much higher terminating impedances are needed to obtain transducer gain values similar to those of bipolar transistors.

References

1. Ebers, J., and Moll, J., "Large-Signal Behavior of Junction Transistors," *Proceedings of the IRE*, 42, pp 1761-1772, December 1954.
2. Getreu, I., *Modeling the Bipolar Transistor*, Elsevier, New York, 1979. Also available from Tektronix, Inc, Beaverton, Oregon, in paperback form. Must be ordered as Part Number 062-2841-00.
3. Searle, C., Boothroyd, A., Angelo, E. Jr., Gray, P., and Pederson, D., *Elementary Circuit Properties of Transistors*, Semiconductor Electronics and Education Committee, Vol 3, John Wiley & Sons, New York, 1964.
4. Clarke, K. and Hess, D., *Communication Circuits: Analysis and Design*, Addison-Wesley, Reading, Massachusetts, 1971.
5. Middlebrook, R., *Design-Oriented Circuit Analysis and Measurement Techniques*. Copyrighted notes for a short course presented by Dr Middlebrook in 1978.

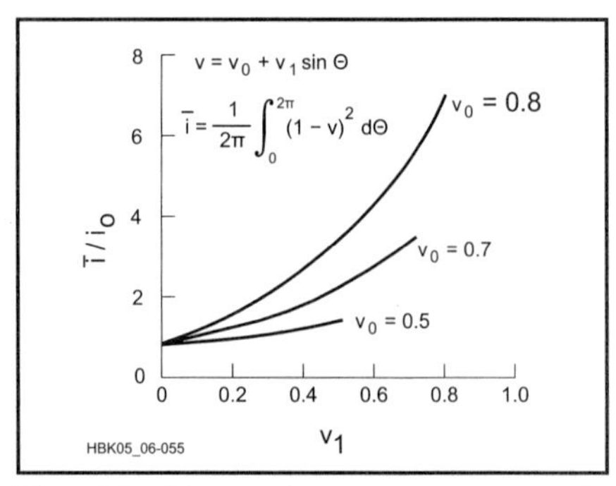

Fig 6.55—Change in average current of a JFET with increasing input signals. The average current with no input signals is I_0, while v_1 is the normalized drive amplitude, and v_0 is the bias voltage.

Suggested Additional Reading

Alley, C. and Atwood, K., *Electronic Engineering*, John Wiley & Sons, New York, 1962.

Terman, F., *Electronic and Radio Engineering*, McGraw-Hill, New York, 1955.

Gilbert, B., "A New Wide-Band Amplifier Technique," *IEEE Journal of Solid-State Circuits*, SC-3, 4, pp 353-365, December 1968.

Egenstafer, F., "Design Curves Simplify Amplifier Analysis," *Electronics*, pp 62-67, August 2, 1971.

Notes

[1]Superconducting circuits are the one exception; but they are outside the scope of this book.

[2]There are numerous manufacturers of excellent heat sinks. References to any one manufacturer are not intended to exclude the products of any others, nor to indicate any particular predisposition to one manufacturer. The catalogs and products referred to here are from: Wakefield Thermal Solutions, Inc., 33 Bridge Street, Pelham, NH 03076, Phone: (603) 635-2800.

Chapter 7

Component Data and References

Component Data

None of us has the time or space to collect all the literature available on the many different commercially available manufactured components. Even if we did, the task of keeping track of new and obsolete devices would surely be formidable. Fortunately, amateurs tend to use a limited number of component types. This section, by Douglas Heacock, AAØMS, provides information on the components most often used by the Amateur Radio experimenter.

COMPONENT VALUES

Throughout this Handbook, composition resistors and small-value capacitors are specified in terms of a system of "preferred values." This system allows manufacturers to supply these components in a standard set of values, which, when considered along with component tolerances, satisfy the vast majority of circuit requirements.

The preferred values are based on a roughly logarithmic scale of numbers between 1 and 10. One decade of these values for three common tolerance ratings is shown in **Table 7.1**.

Table 7.1 represents the two significant digits in a resistor or capacitor value. Multiply these numbers by multiples of ten to get other standard values. For example, 22 pF, 2.2 µF, 220 µF, and 2200 µF are all standard capacitance values, available in all three tolerances. Standard resistor values include 3.9 Ω, 390 Ω, 39000 Ω and 3.9 MΩ in ±5% and ±10% tolerances. All standard resistance values, from less than 1 Ω to about 5 MΩ are based on this table.

Each value is greater than the next smaller value by a multiplier factor that depends on the1 tolerance. For ±5% devices, each value is approximately 1.1 times the next lower one. For ±10% devices, the multiplier is 1.21, and for ±20% devices, the multiplier is 1.47. The resultant values are rounded to make up the series.

Tolerance refers to a range of acceptable values above and below the specified component value. For example, a 4700-Ω resistor rated for ±20% tolerance can have an actual value anywhere between 3760 Ω and 5640 Ω. You may always substitute a closer-tolerance device for one with a wider tolerance. For projects in this Handbook, assume a 10% tolerance if none is specified.

Table 7.1
Standard Values for Resistors and Capacitors

±5%	*±10%*	*±20%*
1.0	1.0	1.0
1.1		
1.2	1.2	
1.3		
1.5	1.5	1.5
1.6		
1.8	1.8	
2.0		
2.2	2.2	2.2
2.4		
2.7	2.7	
3.0		
3.3	3.3	3.3
3.6		
3.9	3.9	
4.3		
4.7	4.7	4.7
5.1		
5.6	5.6	
6.2		
6.8	6.8	6.8
7.5		
8.2	8.2	
9.1		
10.0	10.0	10.0

COMPONENT MARKINGS

The values, tolerances or types of most small components are typically marked with a color code or an alphanumeric code according to standards agreed upon by component manufacturers. The Electronic Industries Alliance (EIA) is a US agency that sets standards for electronic components, testing procedures, performance and device markings. The EIA cooperates with other standards agencies such as the International Electrotechnical Commission (IEC), a worldwide standards agency. You can often find published EIA standards in the engineering library of a college or university.

The standard EIA color code is used to identify a variety of electronic components. Most resistors are marked with color bands according to the code, shown in **Table 7.2**. Some types of capacitors and inductors are also marked using this color code.

Resistor Markings

Carbon-composition, carbon-film, and metal-film resistors are typically manufactured in roughly cylindrical cases with axial leads. They are marked with color bands as shown in **Fig 7.1A**. The first two bands represent the two significant digits of the component value, the third band represents the multiplier, and the fourth band (if there is one) represents the tolerance. Some units

Table 7.2
Resistor-Capacitor Color Codes

Color	*Significant Figure*	*Decimal Multiplier*	*Tolerance (%)*	*Voltage Rating**
Black	0	1	-	-
Brown	1	10	1*	100
Red	2	100	2*	200
Orange	3	1,000	3*	300
Yellow	4	10,000	4*	400
Green	5	100,000	5*	500
Blue	6	1,000,000	6*	600
Violet	7	10,000,000	7*	700
Gray	8	100,000,000	8*	800
White	9	1,000,000,000	9*	900
Gold	-	0.1	5	1000
Silver	-	0.01	10	2000
No color	-	-	20	500

*Applies to capacitors only

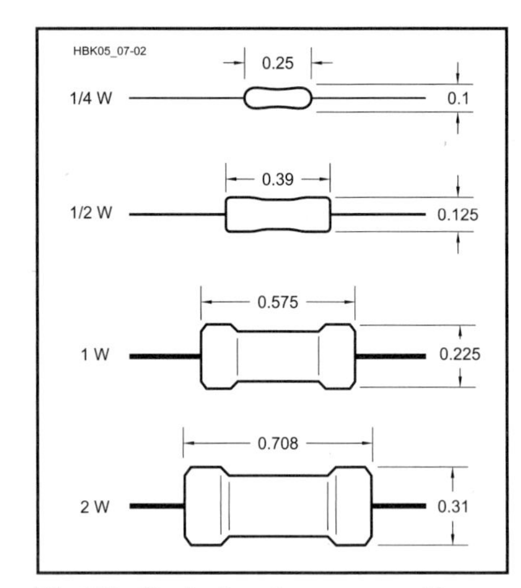

Fig 7.2—Typical carbon-composition resistor sizes.

are marked with a fifth band that represents the percentage of resistance change per 1000 hours of oper-ation: brown = 1%; red = 0.1%; orange = 0.01%; and yellow = 0.001%. Precision resistors (EIA Std RS-279,Fig 7.1B) and some mil-spec (MIL STD-1285A) resistors also use five color bands. On precision resistors, the first *three* bands are used for significant figures and the space between the fourth and fifth bands is wider than the others, to identify the tolerance band. On the military resistors, the fifth band indicates reliability information, such as failure rate.

For example, if a resistor of the type shown in Fig 7.1A is marked with A = red; B = red; C = orange; D = no color, the significant figures are 2 and 2, the multiplier is 1000, and the tolerance is ±20%. The device is a 22,000-Ω, ±20% unit.

Some resistors are made with radial leads (Fig 7.1C) and are marked with a color code in a slightly different scheme. For example, a resistor as shown in Fig 7.1C is marked as follows: A (body) = blue; B (end) = gray; C (dot) = red; D (end) = gold. The significant figures are 6 and 8, the multiplier is 100, and the tolerance is ±5%; 6800 Ω with ±5% tolerance.

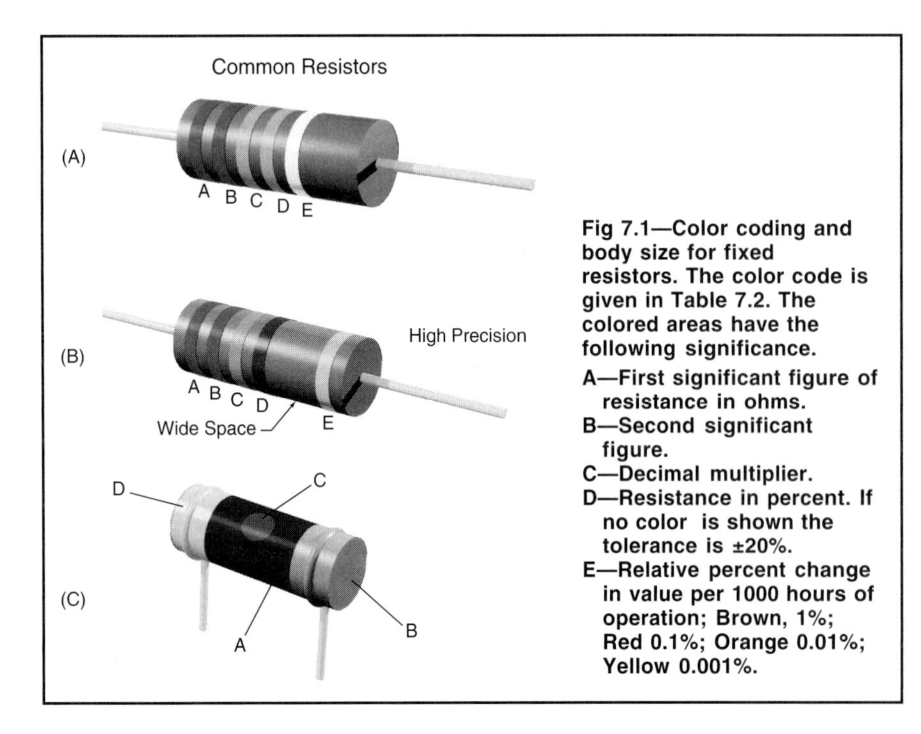

Fig 7.1—Color coding and body size for fixed resistors. The color code is given in Table 7.2. The colored areas have the following significance.
A—First significant figure of resistance in ohms.
B—Second significant figure.
C—Decimal multiplier.
D—Resistance in percent. If no color is shown the tolerance is ±20%.
E—Relative percent change in value per 1000 hours of operation; Brown, 1%; Red 0.1%; Orange 0.01%; Yellow 0.001%.

Resistor Power Ratings

Carbon-composition and metal-film resistors are available in standard power ratings of 1/10, 1/8, 1/4, 1/2, 1 and 2 W. The 1/10- and 1/8-W sizes are relatively expensive and difficult to purchase in small quantities. They are used only where miniaturization is essential. The 1/4, 1/2, 1, and 2-W composition resistor packages are drawn to scale in **Fig 7.2.** Metal-film resistors are typically slightly smaller than carbon-composition units of the same power rating. Film resistors can usually be identified by a glossy enamel coating and an hourglass profile. Carbon-film and metal-film are the most commonly available resistors today, having largely replaced the less-stable carbon-composition resistors.

Capacitor Markings

A variety of systems for capacitor markings are in use. Some use color bands, some use combinations of numbers and letters. Capacitors may be marked with their value, tolerance, temperature characteristics, voltage ratings or some subset of these specifications. **Fig 7.3** shows several popular capacitor marking systems.

In addition to the value, ceramic disk capacitors may be marked with an alphanumeric code signifying temperature characteristics**. Table 7.3** explains the EIA code for ceramic-disk capacitor temperature characteristics. The code is made up of one character from each column in the table. For example, a capacitor marked Z5U is suitable for use between +10 and +85°C, with a maximum change in capacitance of –56% or +22%.

Capacitors with highly predictable temperature coefficients of capacitance are sometimes used in oscillators that must be frequency stable with temperature. If an application called for a temperature coefficient of –750 ppm/°C (N750), a capacitor marked U2J would be suitable. The older industry code for these ratings is being replaced with the EIA code shown

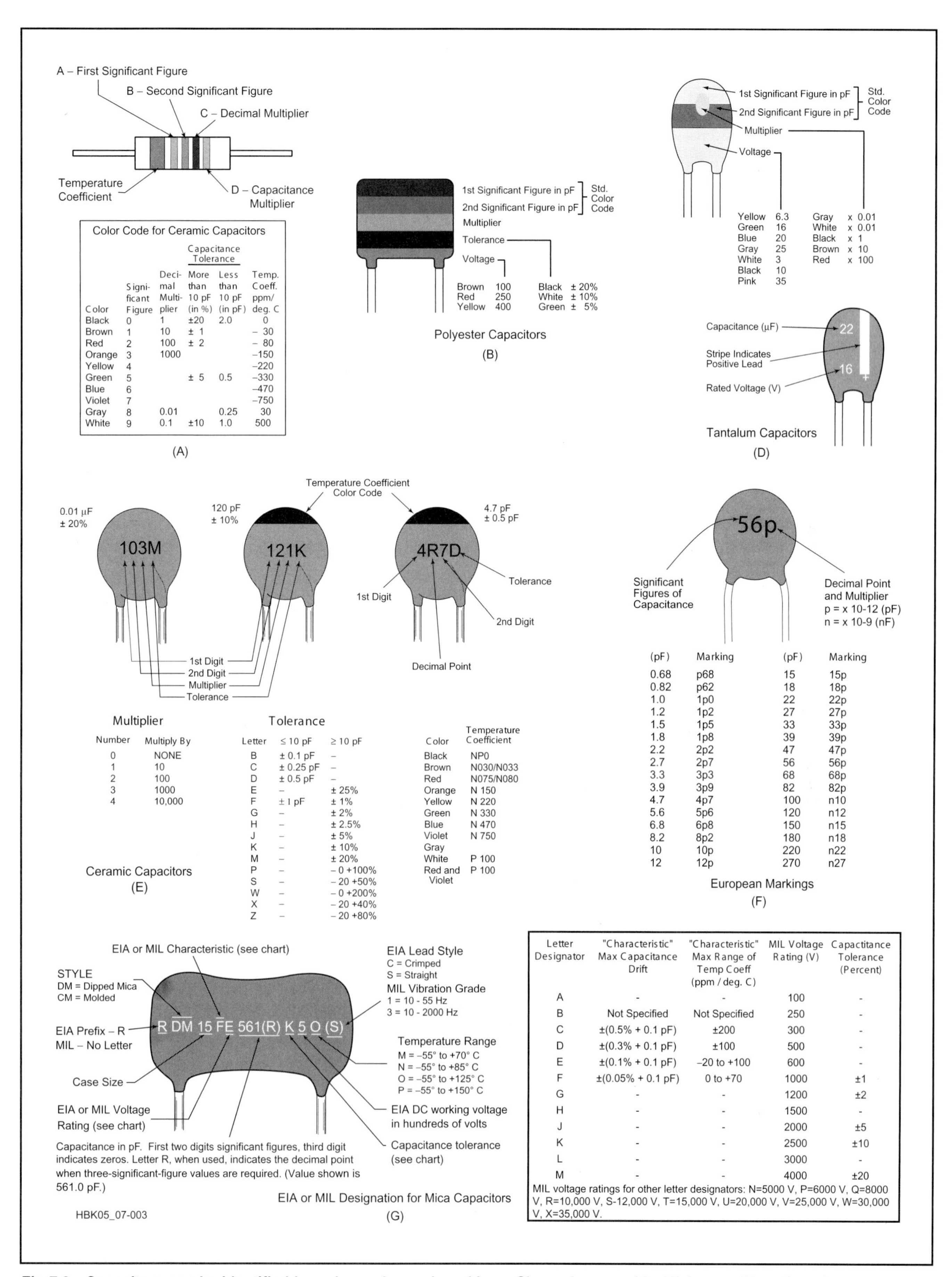

Color Code for Ceramic Capacitors

Color	Significant Figure	Decimal Multiplier	Capacitance Tolerance: More than 10 pF (in %)	Capacitance Tolerance: Less than 10 pF (in pF)	Temp. Coeff. ppm/deg. C
Black	0	1	±20	2.0	0
Brown	1	10	± 1		– 30
Red	2	100	± 2		– 80
Orange	3	1000			–150
Yellow	4				–220
Green	5		± 5	0.5	–330
Blue	6				–470
Violet	7				–750
Gray	8	0.01		0.25	30
White	9	0.1	±10	1.0	500

Multiplier

Number	Multiply By
0	NONE
1	10
2	100
3	1000
4	10,000

Tolerance

Letter	≤ 10 pF	≥ 10 pF
B	± 0.1 pF	-
C	± 0.25 pF	-
D	± 0.5 pF	-
E	-	± 25%
F	± 1 pF	± 1%
G	-	± 2%
H	-	± 2.5%
J	-	± 5%
K	-	± 10%
M	-	± 20%
P	-	– 0 +100%
S	-	– 20 +50%
W	-	– 0 +200%
X	-	– 20 +40%
Z	-	– 20 +80%

Color	Temperature Coefficient
Black	NP0
Brown	N030/N033
Red	N075/N080
Orange	N 150
Yellow	N 220
Green	N 330
Blue	N 470
Violet	N 750
Gray	
White	P 100
Red and Violet	P 100

(pF)	Marking	(pF)	Marking
0.68	p68	15	15p
0.82	p62	18	18p
1.0	1p0	22	22p
1.2	1p2	27	27p
1.5	1p5	33	33p
1.8	1p8	39	39p
2.2	2p2	47	47p
2.7	2p7	56	56p
3.3	3p3	68	68p
3.9	3p9	82	82p
4.7	4p7	100	n10
5.6	5p6	120	n12
6.8	6p8	150	n15
8.2	8p2	180	n18
10	10p	220	n22
12	12p	270	n27

Letter Designator	"Characteristic" Max Capacitance Drift	"Characteristic" Max Range of Temp Coeff (ppm / deg. C)	MIL Voltage Rating (V)	Capactitance Tolerance (Percent)
A	-	-	100	-
B	Not Specified	Not Specified	250	-
C	±(0.5% + 0.1 pF)	±200	300	-
D	±(0.3% + 0.1 pF)	±100	500	-
E	±(0.1% + 0.1 pF)	–20 to +100	600	-
F	±(0.05% + 0.1 pF)	0 to +70	1000	±1
G	-	-	1200	±2
H	-	-	1500	-
J	-	-	2000	±5
K	-	-	2500	±10
L	-	-	3000	-
M	-	-	4000	±20

MIL voltage ratings for other letter designators: N=5000 V, P=6000 V, Q=8000 V, R=10,000 V, S-12,000 V, T=15,000 V, U=20,000 V, V=25,000 V, W=30,000 V, X=35,000 V.

Fig 7.3—Capacitors can be identified by color codes and markings. Shown here are identifying markings found on many common capacitor types.

in **Table 7.4**. NPO (that is, N-P-zero) means "negative, positive, zero." It is a characteristic often specified for RF circuits requiring temperature stability, such as VFOs. A capacitor of the proper value marked C0G is a suitable replacement for an NP0 unit.

Some capacitors, such as dipped silver-mica units, have a letter designating the capacitance tolerance. These letters are deciphered in **Table 7.5**.

Surface-Mount Resistor and Capacitor Markings

Many different types of electronic components, both active and passive, are now available in surface-mount packages. These are commonly known as *chip* resistors and capacitors. The very small size of these components leaves little space for marking with conventional codes, so brief alphanumeric codes are used to convey the most information in the smallest possible space.

Surface-mount resistors are typically marked with a three- or four-digit value code and a character indicating tolerance. The nominal resistance, expressed in ohms, is identified by three digits for 2% (and greater) tolerance devices. The first two digits represent the significant figures; the last digit specifies the multiplier as the exponent of ten. (It may be easier to remember the multiplier as the number of zeros you must add to the significant figures.) For values less than 100 Ω, the letter R is substituted for one of the significant digits and represents a decimal point. Here are some examples:

Resistor Code	*Value*
101	10 and 1 zero = 100 Ω
224	22 and 4 zeros = 220,000 Ω
1R0	1.0 and no zeros = 1 Ω
22R	22.0 and no zeros = 22 Ω
R10	0.1 and no zeros = 0.1 Ω

If the tolerance of the unit is narrower than ±2%, the code used is a four-digit code where the first three digits are the significant figures and the last is the multiplier. The letter R is used in the same way to represent a decimal point. For example, 1001 indicates a 1000-Ω unit, and 22R0 indicates a 22-Ω unit. The tolerance rating for a surface-mount resistor is expressed with a single character at the end of the numeric value code in **Table 7.6**.

Surface-mount capacitors are marked with a two-character code consisting of a letter indicating the significant digits (see **Table 7.7**) and a number indicating the multiplier (see **Table 7.8**). The code represents the capacitance in picofarads. For example, a chip capacitor marked "A4" would have a capacitance of 10,000 pF, or 0.01 µF. A unit marked "N1" would be a 33-pF capacitor. If there is sufficient space on the device package, a tolerance code

Table 7.3

EIA Temperature Characteristic Codes for Ceramic Disc Capacitors

Minimum temperature	*Maximum temperature*	*Maximum capacitance change over temperature range*
X −55°C	2 +45°C	A ±1.0%
Y −30°C	4 +65°C	B ±1.5%
Z +10°C	5 +85°C	C ±2.2%
	6 +105°C	D ±3.3%
	7 +125°C	E ±4.7%
		F ±7.5%
		P ±10%
		R ±15%
		S ±22%
		T −33%, +22%
		U −56%, +22%
		V −82%, +22%

Table 7.4

EIA Capacitor Temperature-Coefficient Codes

Industry	*EIA*
NP0	C0G
N033	S1G
N075	U1G
N150	P2G
N220	R2G
Industry	*EIA*
N330	S2H
N470	U2J
N1500	P3K
N2200	R3L

Table 7.5

EIA Capacitor Tolerance Codes

Code	*Tolerance*
C	±1/4 pF
D	±1/2 pF
F	±1 pF or ±1%
G	±2 pF or ±2%
J	±5%
K	±10%
L	±15%
M	±20%
N	±30%
P or GMV*	−0%, +100%
W	−20%, +40%
Y	−20%, +50%
Z	−20%, +80%

*GMV = guaranteed minimum value.

Table 7.6

SMT Resistor Tolerance Codes

Letter	*Tolerance*
D	±0.5%
F	±1.0%
G	±2.0%
J	±5.0%

Table 7.7

SMT Capacitor Significant Figures Code

Character	*Significant Figures*	*Character*	*Significant Figures*
A	1.0	T	5.1
B	1.1	U	5.6
C	1.2	V	6.2
D	1.3	W	6.8
E	1.5	X	7.5
F	1.6	Y	8.2
G	1.8	Z	9.1
H	2.0	a	2.5
J	2.2	b	3.5
K	2.4	d	4.0
L	2.7	e	4.5
M	3.0	f	5.0
N	3.3	m	6.0
P	3.6	n	7.0
Q	3.9	t	8.0
R	4.3	y	9.0
S	4.7		

Table 7.8

SMT Capacitor Multiplier Codes

Numeric Character	*Decimal Multiplier*
0	1
1	10
2	100
3	1,000
4	10,000
5	100,000
6	1,000,000
7	10,000,000
8	100,000,000
9	0.1

Table 7.9
Powdered-Iron Toroidal Cores: Magnetic Properties

Inductance and Turns Formula

The turns required for a given inductance or inductance for a given number of turns can be calculated from:

$$N = 100\sqrt{\frac{L}{A_L}} \qquad L = A_L\left(\frac{N^2}{10{,}000}\right)$$

where N = number of turns; L = desired inductance (μH); A_L = inductance index (μH per 100 turns).*

AL Values

	Mix										
Size	*26***	*3*	*15*	*1*	*2*	*7*	*6*	*10*	*12*	*17*	*0*
T-12	na	60	50	48	20	18	17	12	7.5	7.5	3.0
T-16	145	61	55	44	22	na	19	13	8.0	8.0	3.0
T-20	180	76	65	52	27	24	22	16	10.0	10.0	3.5
T-25	235	100	85	70	34	29	27	19	12.0	12.0	4.5
T-30	325	140	93	85	43	37	36	25	16.0	16.0	6.0
T-37	275	120	90	80	40	32	30	25	15.0	15.0	4.9
T-44	360	180	160	105	52	46	42	33	18.5	18.5	6.5
T-50	320	175	135	100	49	43	40	31	18.0	18.0	6.4
T-68	420	195	180	115	57	52	47	32	21.0	21.0	7.5
T-80	450	180	170	115	55	50	45	32	22.0	22.0	8.5
T-94	590	248	200	160	84	na	70	58	32.0	na	10.6
T-106	900	450	345	325	135	133	116	na	na	na	19.0
T-130	785	350	250	200	110	103	96	na	na	na	15.0
T-157	870	420	360	320	140	na	115	na	na	na	na
T-184	1640	720	na	500	240	na	195	na	na	na	na
T-200	895	425	na	250	120	105	100	na	na	na	na

*The units of AL (μH per 100 turns) are an industry standard; however, to get a correct result use AL only in the formula above.
**Mix-26 is similar to the older Mix-41, but can provide an extended frequency range.

Magnetic Properties Iron Powder Cores

Mix	*Color*	*Material*	*μ*	*Temp stability (ppm/°C)*	*f (MHz)*	*Notes*
26	Yellow/white	Hydrogen reduced	75	825	dc - 1	Used for EMI filters and dc chokes
3	Gray	Carbonyl HP	35	370	0.05 - 0.50	Excellent stability, good Q for lower frequencies
15	Red/white	Carbonyl GS6	25	190	0.10 - 2	Excellent stability, good Q
1	Blue	Carbonyl C	20	280	0.50 - 5	Similar to Mix-3, but better stability
2	Red	Carbonyl E	10	95	2 - 30	High Q material
7	White	Carbonyl TH	9	30	3 - 35	Similar to Mix-2 and Mix-6, but better temperature stability
6	Yellow	Carbonyl SF	8	35	10 - 50	Very good Q and temp. stability for 20-50 MHz
10	Black	Powdered iron W	6	150	30 - 100	Good Q and stability for 40 - 100 MHz
12	Green/white	Synthetic oxide	4	170	50 - 200	Good Q, moderate temperature stability
17	Blue/yellow	Carbonyl	4	50	40 - 180	Similar to Mix-12, better temperature stability, Q drops about 10% above 50 MHz, 20% above 100 MHz
0	Tan	phenolic	1	0	100 - 300	Inductance may vary greatly with winding technique

Courtesy of Amidon Assoc and Micrometals
Note: Color codes hold only for cores manufactured by Micrometals, which makes the cores sold by most Amateur Radio distributors.

may be included (see Fig 7.3E for tolerance codes). Surface-mount capacitors can be very small; you may need a magnifying glass to read the markings.

INDUCTORS AND CORE MATERIALS

Inductors, both fixed and variable, are available in a wide variety of types and packages, and many offer few clues as to their values. Some coils and chokes are marked with the EIA color code shown in Table 7.2. See **Fig 7.4** for another marking system for tubular encapsulated RF chokes.

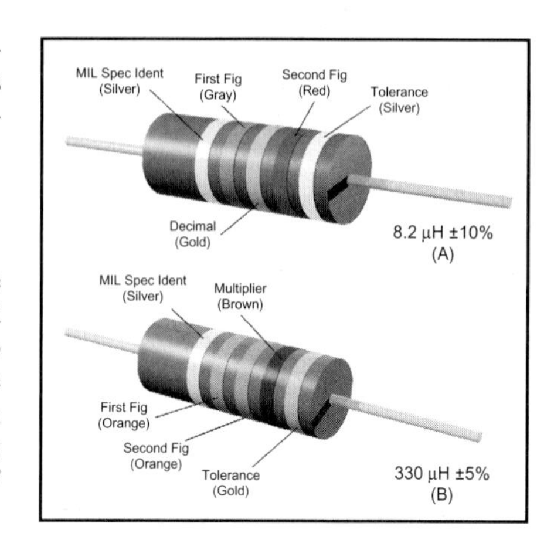

Fig 7.4—Color coding for tubular encapsulated RF chokes. At A, an example of the coding for an 8.2-μH choke is given. At B, the color bands for a 330-μH inductor are illustrated. The color code is given in Table 7.2.

Table 7.10
Powdered-Iron Toroidal Cores: Dimensions

Red E Cores—500 kHz to 30 MHz (μ = 10)

No.	*OD (in)*	*ID (in)*	*H (in)*
T-200-2	2.00	1.25	0.55
T-94-2	0.94	0.56	0.31
T-80-2	0.80	0.50	0.25
T-68-2	0.68	0.37	0.19
T-50-2	0.50	0.30	0.19
T-37-2	0.37	0.21	0.12
T-25-2	0.25	0.12	0.09
T-12-2	0.125	0.06	0.05

Black W Cores—30 MHz to 200 MHz (μ=6)

No.	*OD (In)*	*ID (In)*	*H (In)*
T-50-10	0.50	0.30	0.19
T-37-10	0.37	0.21	0.12
T-25-10	0.25	0.12	0.09
T-12-10	0.125	0.06	0.05

Yellow SF Cores—10 MHz to 90 MHz (μ=8)

No.	*OD (In)*	*ID (In)*	*H (In)*
T-94-6	0.94	0.56	0.31
T-80-6	0.80	0.50	0.25
T-68-6	0.68	0.37	0.19
T-50-6	0.50	0.30	0.19
T-26-6	0.25	0.12	0.09
T-12-6	0.125	0.06	0.05

Number of Turns vs Wire Size and Core Size

Approximate maximum number of turns—single layer wound—enameled wire.

Wire Size	*T-200*	*T-130*	*T-106*	*T-94*	*T-80*	*T-68*	*T-50*	*T-37*	*T-25*	*T-12*
10	33	20	12	12	10	6	4	1		
12	43	25	16	16	14	9	6	3		
14	54	32	21	21	18	13	8	5	1	
16	69	41	28	28	24	17	13	7	2	
18	88	53	37	37	32	23	18	10	4	1
20	111	67	47	47	41	29	23	14	6	1
22	140	86	60	60	53	38	30	19	9	2
24	177	109	77	77	67	49	39	25	13	4
26	223	137	97	97	85	63	50	33	17	7
28	281	173	123	123	108	80	64	42	23	9
30	355	217	154	154	136	101	81	54	29	13
32	439	272	194	194	171	127	103	68	38	17
34	557	346	247	247	218	162	132	88	49	23
36	683	424	304	304	268	199	162	108	62	30
38	875	544	389	389	344	256	209	140	80	39
40	1103	687	492	492	434	324	264	178	102	51

Actual number of turns may differ from above figures according to winding techniques, especially when using the larger size wires. Chart prepared by Michel J. Gordon, Jr, WB9FHC.
Courtesy of Amidon Assoc.

Most powdered-iron toroid cores that we amateurs use are manufactured by Micrometals, who uses paint to identify the material used in the core. The Micrometals color code is part of **Table 7.9**. **Table 7.10** gives the physical characteristics of powdered-iron toroids. Ferrite cores are not typically painted, so identification is more difficult. See **Table 7.11** for information about ferrite cores.

TRANSFORMERS

Many transformers, including power transformers, IF transformers and audio transformers, are made to be installed on PC boards, and have terminals designed for that purpose. Some transformers are manufactured with wire leads that are color-coded to identify each connection. When colored wire leads are present, the color codes in **Tables 7.12**, **7.13** and **7.14** usually apply.

In addition, many miniature IF transformers are tuned with slugs that are color-coded to signify their application. **Table 7.15** lists application vs slug color.

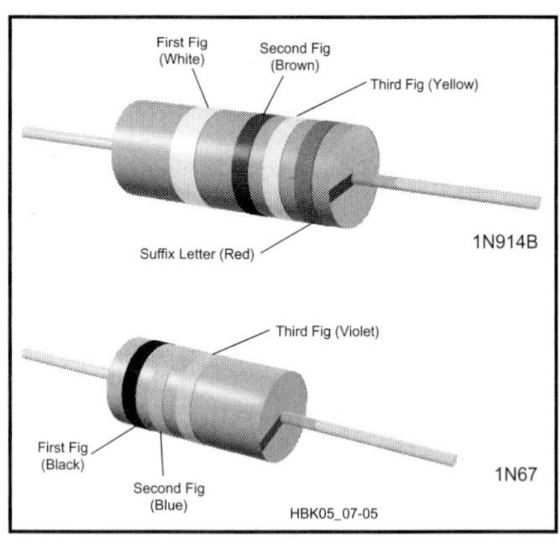

Fig 7.5—Color coding for semi-conductor diodes. At A, the cathode is identified by the double-width first band. At B, the bands are grouped toward the cathode. Two-figure designations are signified by a black first band. The color code is given in Table 7.2. The suffix-letter code is A—Brown, B—red, C—orange, D—yellow, E—green, F—blue. The 1N prefix is understood.

SEMICONDUCTORS

Most semiconductor devices are clearly marked with the part number and in some cases, a manufacturer's date code as well. Identification of semiconductors can be difficult, however, when the parts are "house-marked" (marked with codes used by an equipment manufacturer instead of the standard part numbers). In such cases, it is often possible to find the standard equivalent or a suitable replacement by using one of the semiconductor cross-reference directories available from various replacement-parts distributors. If you look up the house number and find the recommended replacement part, you can often find other standard parts that are replaced by that same part.

Diodes

Most diodes are marked with a part number and some means of identifying which lead is the cathode. Some diodes are marked with a color-band code (see **Fig 7.5**). Important diode parameters include maximum forward current, maximum peak inverse voltage (PIV) and the power-handling capacity.

Transistors

Some important parameters for transistor selection are voltage and current limits, power-handling capability, beta or

Table 7.11

Ferrite Toroids: A_L Chart (mH per 1000 turns) Enameled Wire

Core Size	*63/67-Mix* $\mu = 40$	*61-Mix* $m = 125$	*43-Mix* $\mu = 850$	*77 (72)-Mix* $\mu = 2000$	*J (75)-Mix* $\mu = 5000$
FT-23	7.9	24.8	188.0	396	980
FT-37	19.7	55.3	420.0	884	2196
FT-50	22.0	68.0	523.0	1100	2715
FT-82	22.4	73.3	557.0	1170	NA
FT-114	25.4	79.3	603.0	1270	3170

$$\text{Number of turns} = 1000\sqrt{\text{desired L (mH)} \div A_L \text{ value (above)}}$$

Ferrite Magnetic Properties

Property	*Unit*	*63/67-Mix*	*61-Mix*	*43-Mix*	*77 (72)-Mix*	*J (75)-Mix*
Initial perm.	(μ_i)	40	125	850	2000	5000
Max. perm.		125	450	3000	6000	8000
Saturation flux density @ 10 oer	Gauss	1850	2350	2750	4600	3900
Residual flux density	Gauss	750	1200	1200	1150	1250
Curie temp.	°C	450	350	130	200	140
Vol. resistivity	ohm/cm	1×10^8	1×10^8	1×10^5	1×10^2	5×10^2
Resonant circuit frequency	MHz	15-25	0.2-10	0.01-1	0.001-1	0.001-1
Specific gravity		4.7	4.7	4.5	4.8	4.8
Loss factor	$\frac{1}{\mu_i Q}$	110×10^{-6} @25 MHz	32×10^{-6} @2.5 MHz	120×10^{-6} @1 MHz	4.5×10^{-6} @0.1 MHz	15×10^{-6} @0.1 MHz
Coercive force	Oer	2.40	1.60	0.30	0.22	0.16
Temp. Coef. of initial perm.	%/°C (20°-70°)	0.10	0.15	1.0	0.60	0.90

Ferrite Toroids—Physical Properties

Core Size	*OD*	*ID*	*Height*	A_e	l_e	V_e	A_S	A_W
FT-23	0.230	0.120	0.060	0.00330	0.529	0.00174	0.1264	0.01121
FT-37	0.375	0.187	0.125	0.01175	0.846	0.00994	0.3860	0.02750
FT-50	0.500	0.281	0.188	0.02060	1.190	0.02450	0.7300	0.06200
FT-82	0.825	0.520	0.250	0.03810	2.070	0.07890	1.7000	0.21200
FT-114	1.142	0.750	0.295	0.05810	2.920	0.16950	2.9200	0.43900

OD—Outer diameter (inches)
ID—Inner diameter (inches)
Height (inches)
A_W—Total window area $(in)^2$
A_e—Effective magnetic cross-sectional area $(in)^2$
l_e—Effective magnetic path length (inches)
V_e—Effective magnetic volume $(in)^3$
A_S—Surface area exposed for cooling $(in)^2$
Courtesy of Amidon Assoc.

Table 7.12

Power-Transformer Wiring Color Codes

Non-tapped primary leads:	Black
Tapped primary leads:	Common: Black Tap: Black/yellow striped Finish: Black/red striped
High-voltage plate winding:	Red
Center tap:	Red/yellow striped
Rectifier filament winding:	Yellow
Center tap:	Yellow/blue striped
Filament winding 1:	Green
Center tap:	Green/yellow striped
Filament winding 2:	Brown
Center tap:	Brown/yellow striped
Filament winding 3:	Slate
Center tap:	Slate/yellow striped

Table 7.13

IF Transformer Wiring Color Codes

Plate lead:	Blue
B+ lead:	Red
Grid (or diode) lead:	Green
Grid (or diode) return:	Black

Note: If the secondary of the IF transformer is center-tapped, the second diode plate lead is green-and-black striped, and black is used for the center-tap lead.

Table 7.14

IF Transformer Slug Color Codes

Frequency	*Application*	*Slug color*
455 kHz	1st IF	Yellow
	2nd IF	White
	3rd IF	Black
	Osc tuning	Red
10.7 MHz	1st IF	Green
	2nd or 3rd IF	Orange, Brown or Black

Table 7.15

Audio Transformer Wiring Color Codes

Plate lead of primary	Blue
B+ lead (plain or center-tapped)	Red
Plate (start) lead on center-tapped primaries	Brown (or blue if polarity is not important)
Grid (finish) lead to secondary	Green
Grid return (plain or center tapped)	Black
Grid (start) lead on green center tapped secondaries	Yellow (or if polarity not important)

Note: These markings also apply to line-to-grid and tube-to-line transformers.

gain characteristics and useful frequency range. The case style may also be an issue; some transistors are available in several different case styles.

Integrated Circuits

Integrated circuits (ICs) come in a variety of packages, including transistor-like metal cans, dual and single in-line packages (DIPs and SIPs), flat-packs and surface-mount packages. Most are marked with a part number and a four-digit manufacturer's date code indicating the year (first two digits) and week (last two digits) that the component was made. ICs are frequently house-marked, and the cross-reference directories mentioned above can be helpful in identification and replacement.

Another very useful reference tool for

Table 7.16

Copper Wire Specifications

Bare and Enamel-Coated Wire

Wire Size (AWG)	Diam (Mils)	Area (CM[1])	*Enamel Wire Coating Turns / Linear inch[2]* Single	Heavy	Triple	Feet per Pound Bare	Ohms per 1000 ft 25° C	*Current Carrying Capacity Continuous Duty[3]* at 700 CM per Amp[4]	Open air	Conduit or bundles	Nearest British SWG No.
1	289.3	83694.49				3.948	0.1239	119.564			1
2	257.6	66357.76				4.978	0.1563	94.797			2
3	229.4	5267.36				6.277	0.1971	75.178			4
4	204.3	41738.49				7.918	0.2485	59.626			5
5	181.9	33087.61				9.98	0.3134	47.268			6
6	162.0	26244.00				12.59	0.3952	37.491			7
7	144.3	20822.49				15.87	0.4981	29.746			8
8	128.5	16512.25				20.01	0.6281	23.589			9
9	114.4	13087.36				25.24	0.7925	18.696			11
10	101.9	10383.61				31.82	0.9987	14.834			12
11	90.7	8226.49				40.16	1.2610	11.752			13
12	80.8	6528.64				50.61	1.5880	9.327			13
13	72.0	5184.00				63.73	2.0010	7.406			15
14	64.1	4108.81	15.2	14.8	14.5	80.39	2.5240	5.870	32	17	15
15	57.1	3260.41	17.0	16.6	16.2	101.32	3.1810	4.658			16
16	50.8	2580.64	19.1	18.6	18.1	128	4.0180	3.687	22	13	17
17	45.3	2052.09	21.4	20.7	20.2	161	5.0540	2.932			18
18	40.3	167.09	23.9	23.2	22.5	203.5	6.3860	2.320	16	10	19
19	35.9	1288.81	26.8	25.9	25.1	256.4	8.0460	1.841			20
20	32.0	107.00	29.9	28.9	27.9	322.7	10.1280	1.463	11	7.5	21
21	28.5	812.25	33.6	32.4	31.3	406.7	12.7700	1.160			22
22	25.3	640.09	37.6	36.2	34.7	516.3	16.2000	0.914		5	22
23	22.6	510.76	42.0	40.3	38.6	646.8	20.3000	0.730			24
24	20.1	404.01	46.9	45.0	42.9	817.7	25.6700	0.577			24
25	17.9	320.41	52.6	50.3	47.8	1031	32.3700	0.458			26
26	15.9	252.81	58.8	56.2	53.2	1307	41.0200	0.361			27
27	14.2	201.64	65.8	62.5	59.2	1639	51.4400	0.288			28
28	12.6	158.76	73.5	69.4	65.8	2081	65.3100	0.227			29
29	11.3	127.69	82.0	76.9	72.5	2587	81.2100	0.182			31
30	10.0	100.00	91.7	86.2	80.6	3306	103.7100	0.143			33
31	8.9	79.21	103.1	95.2		4170	130.9000	0.113			34
32	8.0	64.00	113.6	105.3		5163	162.0000	0.091			35
33	7.1	50.41	128.2	117.6		6553	205.7000	0.072			36
34	6.3	39.69	142.9	133.3		8326	261.3000	0.057			37
35	5.6	31.36	161.3	149.3		10537	330.7000	0.045			38
36	5.0	25.00	178.6	166.7		13212	414.8000	0.036			39
37	4.5	20.25	200.0	181.8		16319	512.1000	0.029			40
38	4.0	16.00	222.2	204.1		20644	648.2000	0.023			
39	3.5	12.25	256.4	232.6		26969	846.6000	0.018			
40	3.1	9.61	285.7	263.2		34364	1079.2000	0.014			
41	2.8	7.84	322.6	294.1		42123	1323.0000	0.011			
42	2.5	6.25	357.1	333.3		52854	1659.0000	0.009			
43	2.2	4.84	400.0	370.4		68259	2143.0000	0.007			
44	2.0	4.00	454.5	400.0		82645	2593.0000	0.006			
45	1.8	3.10	526.3	465.1		106600	3348.0000	0.004			
46	1.6	2.46	588.2	512.8		134000	4207.0000	0.004			

Teflon Coated, Stranded Wire

(As supplied by Belden Wire and Cable)

Size	Strands[5]	*Turns per Linear inch[2] UL Style No.* 1180	1213	1371
16	19×29	11.2		
18	19×30	12.7		
20	7×28	14.7	17.2	
20	19×32	14.7	17.2	
22	19×34	16.7	20.0	23.8
22	7×30	16.7	20.0	23.8
24	19×36	18.5	22.7	27.8
24	7×32		22.7	27.8
26	7×34		25.6	32.3
28	7×36		28.6	37.0
30	7×38		31.3	41.7
32	7×40			47.6

Notes

[1]A circular mil (CM) is a unit of area equal to that of a one-mil-diameter circle ($\pi/4$ square mils). The CM area of a wire is the square of the mil diameter.

[2]Figures given are approximate only; insulation thickness varies with manufacturer.

[3]Maximum wire temperature of 212°F (100°C) with a maximum ambient temperature of 13°F (57°C) as specified by the manufacturer. The *National Electrical Code* or local building codes may differ.

[4]700 CM per ampere is a satisfactory design figure for small transformers, but values from 500 to 1000 CM are commonly used. The *National Electrical Code* or local building codes may differ.

[5]Stranded wire construction is given as "count" × "strand size" (AWG).

Table 7.17

Color Code for Hookup Wire

Wire Color	*Type of Circuit*
Black	Grounds, grounded elements and returns
Brown	Heaters or filaments, off ground
Red	Power Supply B plus
Orange	Screen grids and base 2 of transistors
Yellow	Cathodes and transistor emitters
Green	Control grids, diode plates, and base 1 of transistors
Blue	Plates and transistor collectors
Violet	Power supply, minus leads
Gray	Ac power line leads
White	Bias supply, B or C minus, AGC

Note: Wires with tracers are coded in the same manner as solid-color wires, allowing additional circuit identification over solid-color wiring. The body of the wire is white and the color band spirals around the wire lead. When more than one color band is used, the widest band represents the first color.

Table 7.18

Aluminum Alloy Characteristics

Common Alloy Numbers

Type	*Characteristic*
2024	Good formability, high strength
5052	Excellent surface finish, excellent corrosion resistance, normally not heat treatable for high strength
6061	Good machinability, good weldability, can be brittle at high tempers
7075	Good formability, high strength

General Uses

Type	*Uses*
2024-T3	Chassis boxes, antennas, anything that will be bent or flexed repeatedly
7075-T3	
6061-T6	Mounting plates, welded assemblies or machined parts

Common Tempers

Type	*Characteristics*
T0	Special soft condition
T3	Hard
T6	Very hard, possibly brittle
TXXX	Three digit tempers—usually specialized high-strength heat treatments, similar to T6

working with ICs is *IC Master*, a master selection guide that organizes ICs by type, function and certain key parameters. A part number index is included, along with application notes and manufacturer's information for tens of thousands of IC devices. Some of the data from *IC Master* is also available on computer disk.

IC part numbers usually contain a few digits that identify the circuit die or function and several other letters and/or digits that identify the production process, manufacturer and package. For example, a '4066 IC contains four independent SPST switches. Harris (CD74HC4066, CD4066B and CD4066BE), National (MM74HC4066, CD4066BC and CD4066BM) and Panasonic (MN74HC4066 and MN4066B) all make similar devices (as do many other manufacturers) with slight differences. Among the numbers listed, "CD" (CMOS Digital), "MM" (MOS Monolithic), and "MN" indicate CMOS parts. The number "74" indicates a commercial quality product (for applications from 0°C to 70°C), which is pin compatible with the 74/54 TTL families. "HC" means high-speed CMOS family, which is as fast as the LS TTL family. The "B" suffix, as is CD4066B, indicates a buffered output. This is only a small example of the conventions used in IC part numbers. For more information look at data books from the various manufacturers.

When choosing ICs that are not exact replacements, several operating needs and performance aspects should be considered. First, the replacement power requirements must be met: Some ICs require 5 V dc, others 12 V and some need both positive and negative supplies. Current requirements vary among the various IC families, so be sure that sufficient current is available from the power supply. If a replacement IC uses much more current than the device it replaces, a heat sink or blower may be needed to keep it cool.

Next consider how the replacement interacts with its neighboring components. Input capacitance and "fanout" are critical factors in digital circuits. Increased input capacitance may overload the driving circuits. Overload slows circuit operation, which may prevent lines from reaching the "high" condition. Fanout tells how many inputs a device can drive. The fanout of a replacement should be equal to, or greater than, that required in the circuit. Operating speed and propagation delay are also significant. Choose a replacement IC that operates at or above the circuit clock speed. (Be careful: Increased speed can increase EMI and cause other problems.) Some circuits may not function if the propagation delay varies much from the specified part. Look at the **Electrical Signals and Components** chapter for details of how these operating characteristics relate to circuit performance.

Analog ICs have similar characteristics. Input and output capacities are often defined as how much current an analog IC can "sink" (accept at an input) or "source" (pass to a load). A replacement should be able to source or sink at least as much current as the device it replaces. Analog speed is sometimes listed as bandwidth (as in discrete-component circuits) or slew rate (common in op amps). Each of these quantities should meet or exceed that of the replaced component.

Some ICs are available in different operating temperature ranges. Op amps, for example, are commonly available in three standard ranges:

- Commercial: 0°C to 70°C
- Industrial: –25°C to 85°C
- Military: –55°C to 125°C

In some cases, part numbers reflect the temperature ratings. For example, an LM301A op amp is rated for the commercial temperature range; an LM201A op amp for the industrial range and an LM101A for the military range.

When necessary, you can add interface circuits or buffer amplifiers that improve the input and output capabilities of replacement ICs, but auxiliary circuits cannot improve basic device ratings, such as speed or bandwidth.

An excellent source of information on many common ICs is *The ARRL Electronics Data Book*, which contains detailed data for digital ICs (CMOS and TTL), op amps and other analog ICs.

Table 7.19
Crystal Holders

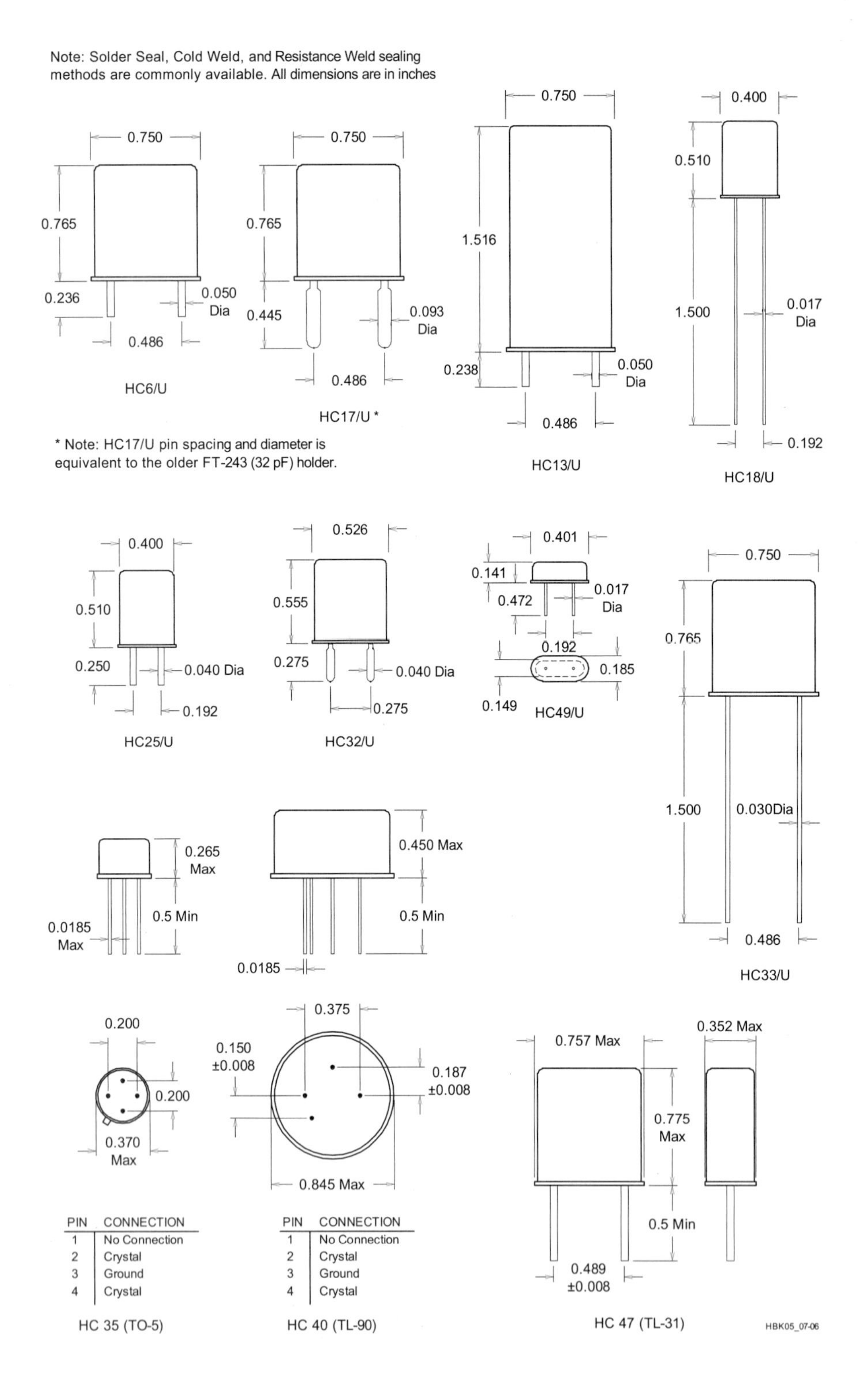

* Note: HC17/U pin spacing and diameter is equivalent to the older FT-243 (32 pF) holder.

OTHER SOURCES OF COMPONENT DATA

There are many sources you can consult for detailed component data. Many manufacturers publish data books for the components they make. Many distributors will include data sheets for parts you order if you ask for them. Parts catalogs themselves are often good sources of component data. The following list is representative of some of the data resources available from manufacturers and distributors.

Motorola Small-Signal Transistor Data
Motorola RF Device Data
Motorola Linear and Interface ICs
Signetics: General Purpose/Linear ICs
NTE Technical Manual and Cross Reference
TCE SK Replacement Technical Manual and Cross Reference
National Semiconductor:
Discrete Semiconductor Products Databook
CMOS Logic Databook
Linear Applications Handbook
Linear Application-Specific ICs Databook
Operational Amplifiers Databook

THE ARRL TECHNICAL INFORMATION SERVICE (TIS)

The ARRL answers questions of a technical nature for ARRL members and nonmembers alike through the Technical Information Service. Questions may be submitted via e-mail (**tis@arrl.org**); Fax (860-594-0259); or mail (TIS, ARRL, 225 Main St, Newington, CT 06111). The TIS also maintains a home page on ARRLWeb: **www.arrl.org/tis**. This site contains links to several technical areas.

TISfind — This search engine contains over 2000 providers of products, services and information of interest to radio amateurs. Before contacting TIS for the address of someone who can repair your radio, or sells antennas, or has old manuals or schematics, look in *TISfind*. Instructions and categories are on the *TISfind* page on ARRLWeb at: **www.arrl.org/tis/tisfind.html**.

SOURCING SUPPLIERS AND CONTACTS ON THE WEB

If you need to extend your search area beyond the TIS search engine described above, please refer to the Web section of the **Web, WiFi, Wireless and PC Technology** chapter in this *Handbook*. It contains information on using a search engine on the Web and tips that can help tailor your search.

Table 7.20
Miniature Lamp Guide

HBK05_07-07

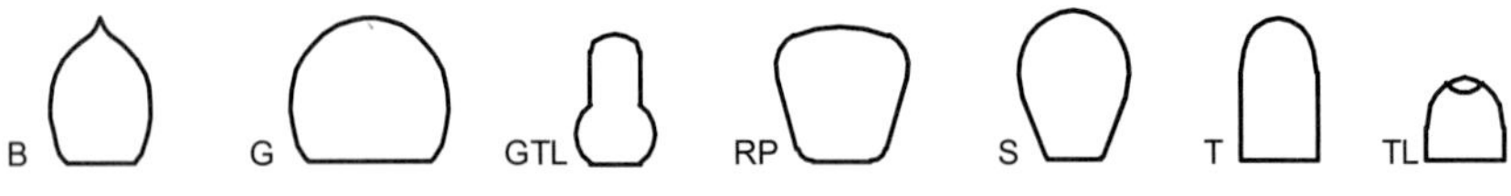

** Bulbs are described by a letter indicating shape and a number that is an approximation of diameter expressed in eighths of an inch. For example S - 8 is "S" shape, 8 eighths or 1 inch in diameter.

BULB STYLES

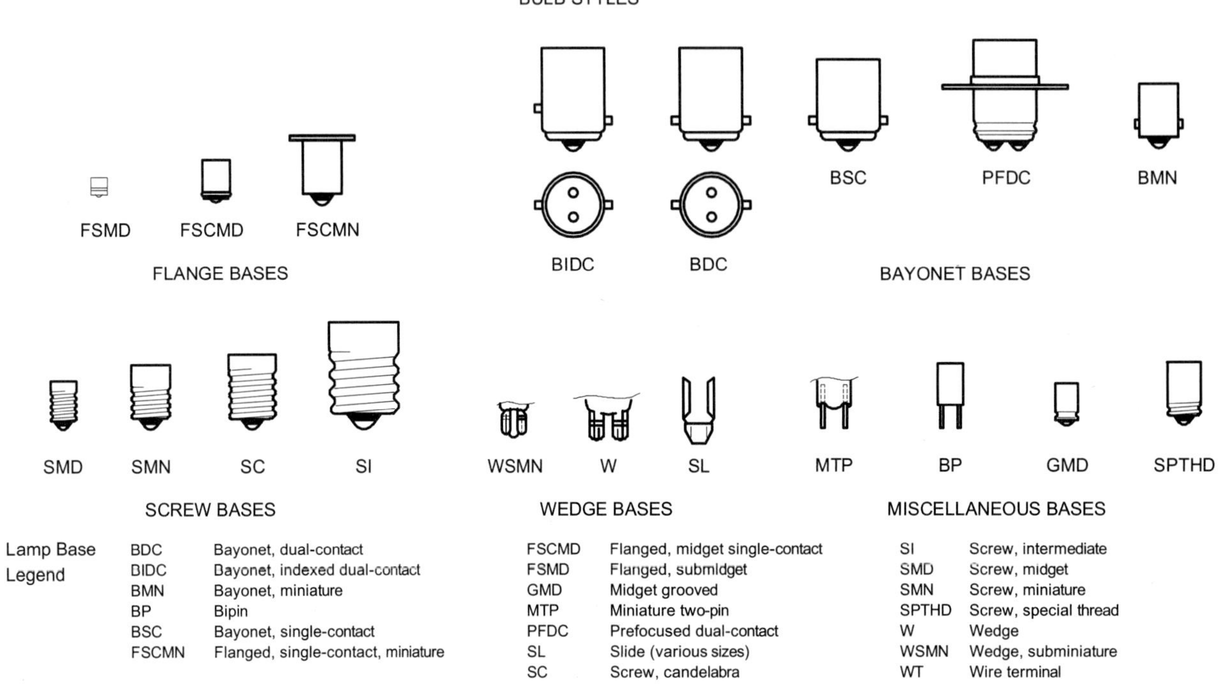

Type	*Bulb*	*Base*	*V*	*A*	*Life*
PR2	B-3½	FSCMN	2.38	0.500	15
PR3	B-3½	FSCMN	3.57	0.500	15
PR4	B-3½	FSCMN	2.33	0.270	10
PR6	B-3½	FSCMN	2.47	0.300	30
PR7	B-3½	FSCMN	3.70	0.300	30
PR12	B-3½	FSCMN	5.95	0.500	15
PR13	B-3½	FSCMN	4.75	0.500	15
10	G-3½	MTP	2.50	0.500	3K
12	G-3½	MTP	6.30	0.150	5K
13	G-3½	SMN	3.70	0.300	15
14	G-3½	SMN	2.47	0.300	15
19	G-3½	MTP	14.4	0.100	1K
27	G-4½	SMN	4.90	0.300	30
37	T-1¾	WSMN	14.00	0.090	1.5K
40	T-3¼	SMN	6.30	0.150	3K
43	T-3¼	BMN	2.50	0.500	3K
44	T-3¼	BMN	6.30	0.250	3K
45	T-3¼	BMN	3.20	0.350	3K
46	T-3¼	SMN	6.30	0.250	3K
47	T-3¼	BMN	6.30	0.150	3K
48	T-3¼	SMN	2.00	0.060	1K
49	T-3¼	BMN	2.00	0.060	1K
50	G-3½	SMN	7.50	0.220	1K
51	G-3½	BMN	7.50	0.220	1K
52	G-3½	SMN	14.40	0.100	1K
53	G-3½	BMN	14.40	0.120	1K
55	G-4½	BMN	7.00	0.410	500
57	G-4½	BMN	14.00	0.240	500
63	G-6	BSC	7.00	0.630	1K
73	T-1¾	WSMN	14.00	0.080	15K
74	T-1¾	WSMN	14.00	0.100	500
82	G-6	BDC	6.50	1.020	500
85	T-1¾	WSMN	28.00	0.040	7K
86	T-1¾	WSMN	6.30	0.200	20K
88	S-8	BDC	6.80	1.910	300
93	S-8	BSC	12.80	1.040	700
112	TL-3	SMN	1.20	0.220	5
130	G-3½	BMN	6.30	0.150	5K
131	G-3½	SMN	1.30	0.100	50
158	T-3¼	W	14.00	0.240	500
159	T-3¼	W	6.30	0.150	5K
161	T-3¼	W	14.00	0.190	4K
168	T-3¼	W	14.00	0.350	1.5K
219	G-3½	BMN	6.30	0.250	5K
222	TL-3	SMN	2.25	0.250	0.5
239	T-3¼	BMN	6.30	0.360	5K
240	T-3¼	BMN	6.30	0.360	5K
259	T-3¼	W	6.30	0.250	5K
268	T-1¾	FSCMD	2.50	0.350	10K
305	S-8	BSC	28.00	0.510	300
307	S-8	BSC	28.00	0.670	300
308	S-8	BDC	28.00	0.670	300
313	T-3¼	BMN	28.00	0.170	500
323	T-1¼	SPTHD	3.00	0.190	350
327	T-1¾	FSCMD	28.00	0.040	4K
327AS15	T-1¾	FSCMD	28.00	0.040	4K
328	T-1¾	FSCMD	6.00	0.200	1K
330	T-1¾	FSCMD	14.00	0.080	1.5K
331	T-1¾	FSCMD	1.35	0.060	3K
334	T-1¾	GMD	28.00	0.040	4K
335	T-1¾	SMD	28.00	0.040	4K
336	T-1¾	GMD	14.00	0.080	1.5K

Type	Bulb	Base	V	A	Life
337	T-1¾	GMD	6.00	0.200	1K
338	T-1¾	FSCMD	2.70	0.060	6K
342	T-1¾	SMD	6.00	0.040	10K
344	T-1¾	FSCMD	10.00	0.014	50K
345	T-1¾	FSCMD	6.00	0.040	10K
346	T-1¾	GMD	18.00	0.040	10K
349	T-1¾	FSCMD	6.30	0.200	5K
370	T-1¾	FSCMD	18.00	0.040	10K
373	T-1¾	SMD	14.00	0.080	1.5K
375	T-1¾	FSCMD	3.00	0.015	10K
376	T-1¾	FSCMD	28.00	0.060	25K
380	T-1¾	FSCMD	6.30	0.040	20K
381	T-1¾	FSCMD	6.30	0.200	20K
382	T-1¾	FSCMD	14.00	0.080	15K
385	T-1¾	FSCMD	28.00	0.040	10K
386	T-1¾	GMD	14.00	0.080	15K
387	T-1¾	FSCMD	28.00	0.040	7K
388	T-1¾	GMD	28.00	0.040	7K
397	T-1¾	GMD	10.00	0.040	5K
398	T-1¾	GMD	6.30	0.200	5K
399	T-1¾	SMD	28.00	0.040	7K
502	G-4½	SMN	5.10	0.150	100
555	T-3¼	W	6.30	0.250	3K
656	T-3¼	W	28.00	0.060	2.5K
680AS15	T-1	WT	5.00	0.060	60K
682AS15	T-1	FSMD	5.00	0.060	60K
683AS15	T-1	WT	5.00	0.060	25K
685AS15	T-1	FSMD	5.00	0.060	25K
715AS15	T-1	WT	5.00	0.115	40K
715AS25	T-1	WT	5.00	0.115	40K
718AS25	T-1	FSMD	5.00	0.115	40K
755	T-3¼	BMN	6.30	0.150	20K
756	T-3¼	BMN	14.00	0.080	15K
757	T-3¼	BMN	28.00	0.080	7.5K
1034	S-8	BIDC	14.00	0.590	5K
1073	S-8	BSC	12.80	1.800	200
1130	S-8	BDC	6.40	2.630	200
1133	RP-11	BSC	6.20	3.910	200
1141	S-8	BSC	12.80	1.440	1K
1143	RP-11	BSC	12.50	1.980	400
1184	RP-11	BDC	5.50	6.250	100
1251	G-6	BSC	28.00	0.230	2K
1445	G-3½	BMN	14.40	0.130	2K
1487	T-3¼	SMN	14.00	0.200	3K
1488	T-3¼	BMN	14.00	0.150	200
1490	T-3¼	BMN	3.20	0.160	3K
1493	S-8	BDC	6.50	2.750	100
1619	S-8	BSC	6.70	1.900	500
1630	S-8	PFDC	6.50	2.750	100
1691	S-8	BSC	28.00	0.610	1K
1705	T-1¾	WT	14.00	0.080	1.5K
1728	T-1¾	WT	1.35	0.060	3K
1730	T-1¾	WT	6.00	0.040	20K
1738	T-1¾	WT	2.70	0.060	6K
1762	T-1¾	WT	28.00	0.040	4K
1764	T-1¾	WT	28.00	0.040	4K
1767	T-1¾	SMD	2.50	0.200	500
1768	T-1¾	SMD	6.00	0.200	1K
1775	T-1¾	SMD	6.30	0.075	1K
1813	T-3¼	BMN	14.40	0.100	1K
1815	T-3¼	BMN	14.00	0.200	3K
1816	T-3¼	BMN	13.00	0.330	1K
1818	T-3¼	BMN	24.00	0.170	250
1819	T-3¼	BMN	28.00	0.040	2.5K
1820	T-3¼	BMN	28.00	0.100	1K
1821	T-3¼	SMN	28.00	0.170	500
1822	T-3¼	BMN	36.00	0.100	1K
1828	T-3¼	BMN	37.50	0.050	3K
1829	T-3¼	BMN	28.00	0.070	1K
1835	T-3¼	BMN	55.00	0.050	5K
1847	T-3¼	BMN	6.30	0.150	5K
1850	T-3¼	BMN	5.00	0.090	1.5K
1864	T-3¼	BMN	28.00	0.170	1.5K
1866	T-3¼	BMN	6.30	0.250	5K
1869	T-1¾	WT	10.00	0.014	50K
1891	T-3¼	BMN	14.00	0.240	500
1892	T-3¼	BMN	14.40	0.120	1K
1893	T-3¼	BMN	14.00	0.330	7.5K
1895	G-4½	BMN	14.00	0.270	2K
2102	T-1¾	WT	18.00	0.040	10K
2107	T-1¾	WT	10.00	0.040	5K
2158	T-1¾	WT	3.00	0.015	10K
2162	T-1¾	WT	14.00	0.100	10K
2169	T-1¾	WT	2.50	0.350	20K
2180	T-1¾	WT	6.30	0.040	20K
2181	T-1¾	WT	6.30	0.200	20K
2182	T-1¾	WT	14.00	0.080	40K
2187	T-1¾	WT	28.00	0.040	7K
2304	T-1¾	BP	3.00	0.300	1.5K
2307	T-1¾	BP	6.30	0.200	5K
2314	T-1¾	BP	28.00	0.050	1K
2316	T-1¾	BP	18.00	0.040	10K
2324	T-1¾	BP	28.00	0.040	4K
2335	T-1¾	BP	14.00	0.080	15K
2337	T-1¾	BP	6.30	0.200	20K
2342	T-1¾	BP	28.00	0.040	25K
3149	T-1¾	BP	5.00	0.060	5K
6803AS25	T-¾	WT	5.00	0.060	60K
6833AS15	T-¾	WT	5.00	0.060	25K
6838	T-1	WT	28.00	0.024	4K
6839	T-1	FSMD	28.00	0.024	4K
7001	T-1¾	BP	24.00	0.050	2K
7003	T-1¾	BP	24.00	0.050	2K
7153AS15	T-¾	WT	5.00	0.115	40K
7265	T-1	BP	5.00	0.060	5K
7327	T-1¾	BP	28.00	0.040	4K
7328	T-1¾	BP	6.00	0.200	1K
7330	T-1¾	BP	14.00	0.080	1.5K
7344	T-1¾	BP	10.00	0.014	50K
7349	T-1¾	BP	6.30	0.200	5K
7361	T-1¾	BP	5.00	0.060	25K
7362	T-1¾	BP	5.00	0.115	40K
7367	T-1¾	BP	10.00	0.040	5K
7370	T-1¾	BP	18.00	0.040	10K
7371	T-1¾	BP	12.00	0.040	10K
7373	T-1¾	BP	14.00	0.100	10K
7374	T-1¾	BP	28.00	0.040	10K
7375	T-1¾	BP	3.00	0.015	10K
7376	T-1¾	BP	28.00	0.065	10K
7377	T-1¾	BP	6.30	0.075	1K
7380	T-1¾	BP	6.30	0.040	30K
7381	T-1¾	BP	6.30	0.200	20K
7382	T-1¾	BP	14.00	0.080	15K
7387	T-1¾	BP	28.00	0.040	7K
7410	T-1¾	BP	14.00	0.080	15K
7839	T-1	BP	28.00	0.025	4K
7876	T-1¾	BP	28.00	0.060	25K
7931	T-1¾	BP	1.35	0.060	3K
7945	T-1¾	BP	6.00	0.040	20K
7968	T-1¾	BP	2.50	0.200	500
8099	T-1	BP	18.00	0.020	16K
8362	T-1¾	SMD	14.00	0.080	15K
8369	T-1¾	SMD	28.00	0.065	10K

(continued on next page)

Standard Line-Voltage Lamps

Type	*V*	*W*	*Bulb*	*Base*
10C7DC	115-125	10	C-7	BDC
3S6	120, 125	3	S-6	SC
6S6	30, 48, 115, 120, 125, 130, 135, 145, 155	6	S-6	SC
6S6/R	115-125	6	S-6 (red)	SC
6S6/W	115-125	6	S-6 (white)	SC
6T4-1/2	120, 130	6	T-4$\frac{1}{2}$	SC
7C7	115-125	7	C-7	SC
7C7/W	115-125	7	C-7 (white)	SC
10C7	115-125	10	C-7	SC
10S6	120	10	S-6	SC
10S6/10	220, 230, 250	10	S-6	SC
6S6DC	30, 120, 125, 145	6	S-6	BDC
10S6/10DC	230, 250	10	S-6	BDC
40S11 N	115-125	40	S-11	SI
120MB	120	3	T-2$\frac{1}{2}$	BMN
120MB/6	120	6	T-2$\frac{1}{2}$	BMN
120PSB	120	3	T-2	SL
120PS	120	3	T-2	WT
120PS/6	120	6	T-2$\frac{1}{2}$	WT

Indicator Lamps

Each has a T-2 bulb and a slide base.

Type	*V*	*A*	*Life*
6PSB	6.00	0.140	20K
12PSB	12.00	0.170	12K
24PSB	24.00	0.073	10K
28PSB	28.00	0.040	5K
48PSB	48.00	0.050	10K
60PSB	60.00	0.050	7.5K
120PSB	120.00	0.025	7.5K

Neon Glow Lamps

Operating circuit voltage 105-125

	Breakdown Voltage					
Type	*AC*	*DC*	*Bulb*	*Base*	*W*	*External Resistance*
NE-2	65	90	T-2	WT	$\frac{1}{12}$	150k
NE-2A	65	90	T-2	WT	$\frac{1}{15}$	100k
NE-2D	65	90	T-2	FSCMD	$\frac{1}{12}$	100k
NE-2E	65	90	T-2	WT	$\frac{1}{12}$	100k
NE-2H	95	135	T-2	WT	$\frac{1}{4}$	30k
NE-2J	95	135	T-2	FSCMD	$\frac{1}{4}$	30k
NE-2V	65	90	T-2	WT	$\frac{1}{2}$	100k
NE-45	65	90	T-4$\frac{1}{2}$	SC	$\frac{1}{4}$	None
NE-51	65	90	T-3$\frac{1}{4}$	BMN	$\frac{1}{25}$	220k
NE-51H	95	135	T-3$\frac{1}{4}$	BMN	$\frac{1}{7}$	47k
NE-84	95	135	T-2	SL	$\frac{1}{4}$	30k
NE-120PSB	95	95	T-2	SL	$\frac{1}{4}$	None

Table 7.21

Metal-Oxide Varistor (MOV) Transient Suppressors

Listed by voltage

Type No.	*ECG/NTE†† no.*	*V ac_{RMS}*	*Maximum Applied Voltage V ac_{Peak}*	*Maximum Energy (Joules)*	*Maximum Peak Current (A)*	*Maximum Power (W)*	*Maximum Varistor Voltage (V)*
V180ZA1	1V115	115	163	1.5	500	0.2	285
V180ZA10	2V115	115	163	10.0	2000	0.45	290
V130PA10A		130	184	10.0	4000	8.0	350
V130PA20A		130	184	20.0	4000	15.0	350
V130LA1	1V130	130	184	1.0	400	0.24	360
V130LA2	1V130	130	184	2.0	400	0.24	360
V130LA10A	2V130	130	184	10.0	2000	0.5	340
V130LA20A	524V13	130	184	20.0	4000	0.85	340
V150PA10A		150	212	10.0	4000	8.0	410
V150PA20A		150	212	20.0	4000	15.0	410
V150LA1	1V150	150	212	1.0	400	0.24	420
V150LA2	1V150	150	212	2.0	400	0.24	420
V150LA10A	524V15	150	212	10.0	2000	0.5	390
V150LA20A	524V15	150	212	20.0	4000	0.85	390
V250PA10A		250	354	10.0	4000	0.85	670
V250PA20A		250	354	20.0	4000	7.0	670
V250PA40A		250	354	40.0	4000	13.0	670
V250LA2	1V250	250	354	2.0	400	0.28	690
V250LA4	1V250	250	354	4.0	400	0.28	690
V250LA15A	2V250	250	354	15.0	2000	0.6	640
V250LA20A	2V250	250	354	20.0	2000	0.6	640
V250LA40A	524V25	250	354	40.0	4000	0.9	640

††ECG and NTE numbers for these parts are identical, except for the prefix. Add the "ECG" or "NTE" prefix to the numbers shown for the complete part number.

Table 7.22

Voltage-Variable Capacitance Diodes†

Listed numerically by device

Device	Nominal Capacitance pF ±10% @ V_R = 4.0 V f = 1.0 MHz	Capacitance Ratio 2-30 V Min.	Q @ 4.0 V 50 MHz Min.	Case Style
1N5441A	6.8	2.5	450	
1N5442A	8.2	2.5	450	
1N5443A	10	2.6	400	DO-7
1N5444A	12	2.6	400	
1N5445A	15	2.6	450	
1N5446A	18	2.6	350	
1N5447A	20	2.6	350	
1N5448A	22	2.6	350	DO-7
1N5449A	27	2.6	350	
1N5450A	33	2.6	350	
1N5451A	39	2.6	300	
1N5452A	47	2.6	250	
1N5453A	56	2.6	200	DO-7
1N5454A	68	2.7	175	
1N5455A	82	2.7	175	
1N5456A	100	2.7	175	
1N5461A	6.8	2.7	600	
1N5462A	8.2	2.8	600	
1N5463A	10	2.8	550	DO-7
1N5464A	12	2.8	550	
1N5465A	15	2.8	550	
1N5466A	18	2.8	500	
1N5467A	20	2.9	500	
1N5468A	22	2.9	500	DO-7
1N5469A	27	2.9	500	
1N5470A	33	2.9	500	
1N5471A	39	2.9	450	
1N5472A	47	2.9	400	
1N5473A	56	2.9	300	DO-7
1N5474A	68	2.9	250	
1N5475A	82	2.9	225	
1N5476A	100	2.9	200	
MV2101	6.8	2.5	450	TO-92
MV2102	8.2	2.5	450	
MV2103	10	2.0	400	
MV2104	12	2.5	400	
MV2105	15	2.5	400	
MV2106	18	2.5	350	TO-92
MV2107	22	2.5	350	
MV2108	27	2.5	300	
MV2109	33	2.5	200	
MV2110	39	2.5	150	
MV2111	47	2.5	150	TO-92
MV2112	56	2.6	150	
MV2113	68	2.6	150	
MV2114	82	2.6	100	
MV2115	100	2.6	100	

†For package shape, size and pin-connection information, see manufacturers' data sheets. Many retail suppliers offer data sheets to buyers free of charge on request. Data books are available from many manufacturers and retailers.

Table 7.23

Zener Diodes

Volts	Power (Watts) 0.25	0.4	0.5	1.0	1.5	5.0	10.0	50.0
1.8	1N4614							
2.0	1N4615							
2.2	1N4616							
2.4	1N4617	1N4370, A	1N4370,A 1N5221,B 1N5985,B					
2.5			1N5222B					
2.6	1N702,A							
2.7	1N4618	1N4371,A	1N4371,A 1N5223,B 1N5839, 1N5986					
2.8			1N5224B					
3.0	1N4619	1N4372,A	1N4372 1N5225,B 1N5987					
3.3	1N4620	1N746,A 1N764,A 1N5518	1N746,A 1N5226,B 1N5988	1N3821 1N4728,A	1N5913	1N5333,B		
3.6	1N4621	1N747,A 1N5519	1N747,A 1N5227,B 1N5989	1N3822 1N4729,A	1N5914	1N5334,B		
3.9	1N4622	1N748,A IN5520	1N748A 1N5228,B 1N5844, 1N5990	1N3823 1N4730,A	1N5915	1N5335,B	1N3993A	1N4549,B 1N4557,B
4.1	1N704,A							
4.3	1N4623	1N749,A 1N5521	1N749,A, 1N5229,B 1N5845 1N5991	1N3824 1N4731,A	1N5916	1N5336,B	1N3994,A	1N4550,B 1N4558,B

(continued on next page)

Volts	Power (Watts) 0.25	0.4	0.5	1.0	1.5	5.0	10.0	50.0
4.7	1N4624	1N750A 1N5522	1N750A 1N5230,B 1N5846, 1N5992	1N3825 1N4732,A	1N5917	1N5337,B	1N3995,A 1N4559,B	1N4551,B
5.1	1N4625 1N4689	1N751,A 1N5523	1N751,A, 1N5231,B 1N5847 1N5993	1N3826 1N4733	1N5918	1N5338,B	1N3996,A	1N4552,B 1N4560
5.6	1N708A 1N4626	1N752,A 1N5524	1N752,A 1N5232,B 1N5848, 1N5994	1N3827 1N4734,A	1N5919	1N5339,B	1N3997,A	2N4553,B 1N4561,B
5.8	1N706A	1N762						
6.0			1N5233B 1N5849			1N5340,B		
6.2	1N709, 1N4627 MZ605, MZ610 MZ620,MZ640	1N753,A	1N753,A 1N821,3,5,7,9;A	1N3828,A 1N5234,B, 1N5850 1N5995	1N5920	1N5341,B 1N4735,A	1N3998,A	1N4554,B 1N4562,B
6.4	1N4565-84,A							
6.8	1N4099	1N754,A 1N957,B 1N5526	1N754,A, 1N757,B 1N5235,B 1N5851 1N5996	1N3016,B 1N3829 1N4736,A	1N3785 1N5921	1N5342,B	1N2970,B 1N3999,A	1N2804B 1N3305B 1N4555, 1N4563
7.5	1N4100	1N755,A 1N958,B 1N5527	1N755A, 1N958,B 1N5236,B 1N5852 1N5997	1N3017,A,B 1N3830 1N4737,A	1N3786 1N5922	1N5343,B	1N2971,B 1N4000,A	1N2805,B 1N3306,B 1N4556, 1N4564
8.0	1N707A							
8.2	1N712A 1N4101	1N756,A 1N959,B 1N5528	1N756,A 1N959,B 1N5237,B 1N5853 1N5998	1N3018,B 1N4738,A	1N3787 1N5923	1N5344,B	1N2972,B	1N2806,B 1N3307,B
8.4		1N3154-57,A	1N3154,A 1N3155-57					
8.5	1N4775-84,A		1N5238,B 1N5854					
8.7	1N4102					1N5345,B		
8.8		1N764						
9.0		1N764A	1N935-9;A,B					
9.1	1N4103	1N757,A 1N960,B 1N5529	1N757,A, 1N960,B 1N5239,B, 1N5855 1N5999	1N3019,B 1N4739,A	1N3788 1N5924	1N5346,B	1N2973,B	1N2807,B 1N3308,B
10.0	1N4104	1N758,A 1N961,B 1N5530,B	1N758,A, 1N961,B 1N5240,B, 1N5856 1N6000	1N3020,B 1N4740,A	1N3789 1N5925	1N5347,B	1N2974,B	1N2808,B 1N3309,A,B
11.0	1N715,A 1N4105	1N962,B 1N5531	1N962,B 1N5241,B 1N5857, 1N6001	1N3021,B 1N4741,A	1N3790 1N5926	1N5348,B	1N2975,B	1N2809,B 1N3310,B
11.7	1N716,A 1N4106		1N941-4;A,B					
12.0		1N759,A 1N963,B 1N5532	1N759,A, 1N963,B 1N5242,B, 1N5858 1N6002	1N3022,B 1N4742,A	1N3791 1N5927	1N5349,B	1N2976,B	1N2810,B 1N3311,B
13.0	1N4107	1N964,B 1N5533	1N964,B 1N5243,B, 1N5859 1N6003	1N3023,B 1N4743,A	1N3792 1N5928	1N5350,B	1N2977,B	1N2811,B 1N3312,B
14.0	1N4108	1N5534	1N5244B 1N5860			1N5351,B	1N2978,B	1N2812,B 1N3313,B
15.0	1N4109	1N965,B 1N5535	1N965,B 1N5245,B, 1N5861 1N6004	1N3024,B 1N4744A	1N3793 1N5929	1N5352,B	1N2979,A,B	1N2813,A,B 1N3314,B
16.0	1N4110	1N966,B 1N5536	1N966,B, 1N5246,B 1N5862, 1N6005	1N3025,B 1N4745,A	1N3794 1N5930	1N5353,B	1N2980,B	1N2814,B 1N3315,B
17.0	1N4111	1N5537	1N5247,B 1N5863			1N5354,B	1N2981B	1N2815,B 1N3316,B
18.0	1N4112	1N967,B 1N5538	1N967,B 1N5248,B 1N5864, 1N6006	1N3026,B 1N4746,A	1N3795 1N5931	1N5355,B	1N2982,B	1N2816,B 1N3317,B
19.0	1N4113	1N5539	1N5249,B 1N5865			1N5356,B	1N2983,B	1N2817,B 1N3318,B
20.0	1N4114	1N968,B 1N5540	1N968,B 1N5250,B 1N5866, 1N6007	1N3027,B 1N4747,A	1N3796 1N5932,A,B	1N5357,B	1N2984,B	1N2818,B 1N3319,B
22.0	1N4115	1N969,B 1N5541	1N969,B 1N5241,B 1N5867, 1N6008	1N3028,B 1N4748,A	1N3797 1N5933	1N5358,B	1N2985,B	1N2819,B 1N3320,A,B

Volts	Power (Watts)							
	0.25	0.4	0.5	1.0	1.5	5.0	10.0	50.0
24.0	1N4116	1N5542 1N970B	1N970,B 1N5252,B, 1N5868 1N6009	1N3029,B 1N4749,A	1N3798 1N5934	1N5359,B	1N2986,B	1N2820,B 1N3321,B
25.0	1N4117	1N5543	1N5253,B 1N5869			1N5360,B	1N2987B	1N2821,B 1N3322,B
27.0	1N4118	1N971,B	1N971 1N5254,B, 1N5870 1N6010	1N3030,B 1N4750,A	1N3799 1N5935	1N5361,B	1N2988,B	1N2822B 1N3323,B
28.0	1N4119	1N5544	1N5255,B 1N5871			1N5362,B		
30.0	1N4120	1N972,B 1N5545	1N972,B 1N5256,B, 1N5872 1N6011	1N3031,B 1N4751,A	1N3800 1N5936	1N5363,B	1N2989,B	1N2823,B 1N3324,B
33.0	1N4121	1N973,B 1N5546	1N973,B 1N5257,B 1N5873, 1N6012	1N3032,B 1N4752,A	1N3801 1N5937	1N5364,B	1N2990,A,B	1N2824,B 1N3325,B
36.0	1N4122	1N974,B	1N974,B 1N5258,B 1N5874, 1N6013	1N3033,B 1N4753,A	1N3802 1N5938	1N5365,B	1N2991,B	1N2825,B 1N3326,B
39.0	1N4123	1N975,B	1N975,B, 1N5259,B 1N5875, 1N6014	1N3034,B 1N4754,A	1N3803 1N5939	1N5366,B	1N2992,B	1N2826,B 1N3327,B
43.0	1N4124	1N976,B	1N976,B 1N5260,B, 1N5876 1N6015	1N3035,B 1N4755,A	1N3804 1N5940	1N5367,B	1N2993,A,B	1N2827,B 1N3328,B
45.0							1N2994B	1N2828B 1N3329B
47.0	1N4125	1N977,B	1N977,B, 1N5261,B 1N5877, 1N6016	1N3036,B 1N4756,A	1N3805 1N5941	1N5368,B	1N2996,B	1N2829,B 1N3330,B
50.0								1N2830B 1N3331B
51.0	1N4126	1N978,B	1N978,B, 1N5262,A,B 1N5878, 1N6017	1N3037,B 1N4757,A	1N3806 1N5942	1N5369,B	1N2997,B	1N2831,B 1N3332,B
52.0							1N2998B	1N3333
56.0	1N4127	1N979,B	1N979 1N5263,B 1N6018	1N3038,B 1N4758,A	1N3807 1N5943	1N5370,B	1N2999,B	1N2822,B 1N3334,B
60.0	1N4128		1N5264,A,B			1N5371,B		
62.0	1N4129	1N980,B	1N980 1N5265,A,B, 1N6019	1N3039,B 1N4759,A	1N3808 1N5944	1N5372,B	1N3000,B	1N2833,B 1N3335,B
68.0	1N4130	1N981,B	1N981,B 1N5266,A,B, 1N6020	1N3040,A,B 1N4760,A	1N3809 1N5945	1N5373,B	1N3001,B	1N2834,B 1N3336,B
75.0	1N4131	1N982,B	1N982 1N5267,A,B, 1N6021	1N3041,B 1N4761,A	1N3810 1N5946	1N5374,B	1N3002,B	1N2835,B 1N3337,B
82.0	1N4132	1N983,B	1N983 1N5268,A,B, 1N6022	1N3042,B 1N4762,A	1N3811 1N5947	1N5375,B	1N3003,B	1N2836,B 1N3338,B
87.0	1N4133		1N5269,B			1N5376,B		
91.0	1N4134	1N984,B	1N984 1N5270,B, 1N6023	1N3043,B 1N4763,A	1N3812 1N5948	1N5377,B	1N3004,B	1N2837,B 1N3339,B
100.0	1N4135	1N985	1N985,B 1N5271,B, 1N6024	1N3044,A,B 1N4764,A	1N3813 1N5949	1N5378,B	1N3005,B	1N2838,B 1N3340,B
105.0							1N3006B	1N2839,B 1N3341,B
110.0		1N986	1N986 1N5272,B, 1N6025	1N3045,B 1M110ZS10	1N3814 1N5950	1N5379,B	1N3007A,B	1N2840,B 1N3342,B
120.0		1N987	1N987,B 1N5273,B, 1N6026	1N3046,B 1M120ZS10	1N3815 1N5951	1N5380,B	1N3008A,B	1N2841,B 1N3343,B
130.0		1N988	1N988,B 1N5274,B, 1N6027	1N3047,B 1M130ZS10	1N3816 1N5952	1N5381,B	1N3009,B	1N2842,B 1N3344,B
140.0			1N5275,B			1N5382B	1N3010B	1N3345B
150.0		1N989	1N989,B 1N5276,B, 1N6028	1N3048,B 1M150ZS10	1N3817 1N5953	1N5383,B	1N3011,B	1N2843,B 1N3346,B
160.0		1N990	1N990,B 1N5277,B, 1N6029	1N3048,B 1M160ZS10	1N3818 1N5954	1N5384,B	IN3012A,B	1N2844B 1N3347,B
170.0			1N5278,B	1M170ZS10		1N5385,B		
175.0							1N3013B	1N3348B
180.0			1N991,B, 1N5279,B 1N6030	1M180ZS10 1N3819	1N5955			1N3349,B
190.0			1N5280,B			1N5387,B		
200.0			1N992, 1N5281,B 1N6031	1N3051,B 1M200ZS10	1N3820 1N5956	1N5388B	1N3015,B	1N2846,B 1N3350,B

Table 7.24

Semiconductor Diode Specifications[†]

Listed numerically by device

Device	*Type*	*Material*	*Peak Inverse Voltage, PIV (V)*	*Average Rectified Current Forward (Reverse) $I_O(A)(I_R(A))$*	*Peak Surge Current, I_{FSM} 1 s @ 25°C (A)*	*Average Forward Voltage, V_F (V)*
1N34	Signal	Ge	60	8.5 m (15.0 μ)		1.0
1N34A	Signal	Ge	60	5.0 m (30.0 μ)		1.0
1N67A	Signal	Ge	100	4.0 m (5.0 μ)		1.0
1N191	Signal	Ge	90	15.0 m		1.0
1N270	Signal	Ge	80	0.2 (100 μ)		1.0
1N914	Fast Switch	Si	75	75.0 m (25.0 n)	0.5	1.0
1N1183	RFR	Si	50	40 (5 m)	800	1.1
1N1184	RFR	Si	100	40 (5 m)	800	1.1
1N2071	RFR	Si	600	0.75 (10.0 μ)		0.6
1N3666	Signal	Ge	80	0.2 (25.0 μ)		1.0
1N4001	RFR	Si	50	1.0 (0.03 m)		1.1
1N4002	RFR	Si	100	1.0 (0.03 m)		1.1
1N4003	RFR	Si	200	1.0 (0.03 m)		1.1
1N4004	RFR	Si	400	1.0 (0.03 m)		1.1
1N4005	RFR	Si	600	1.0 (0.03 m)		1.1
1N4006	RFR	Si	800	1.0 (0.03 m)		1.1
1N4007	RFR	Si	1000	1.0 (0.03 m)		1.1
1N4148	Signal	Si	75	10.0 m (25.0 n)		1.0
1N4149	Signal	Si	75	10.0 m (25.0 n)		1.0
1N4152	Fast Switch	Si	40	20.0 m (0.05 μ)		0.8
1N4445	Signal	Si	100	0.1 (50.0 n)		1.0
1N5400	RFR	Si	50	3.0 (500 μ)	200	
1N5401	RFR	Si	100	3.0 (500 μ)	200	
1N5402	RFR	Si	200	3.0 (500 μ)	200	
1N5403	RFR	Si	300	3.0 (500 μ)	200	
1N5404	RFR	Si	400	3.0 (500 μ)	200	
1N5405	RFR	Si	500	3.0 (500 μ)	200	
1N5406	RFR	Si	600	3.0 (500 μ)	200	
1N5408	RFR	Si	1000	3.0 (500 μ)	200	
1N5711	Schottky	Si	70	1 m (200 n)	15 m	0.41 @ 1 mA
1N5767	Signal	Si		0.1 (1.0 μ)		1.0
1N5817	Schottky	Si	20	1.0 (1 m)	25	0.75
1N5819	Schottky	Si	40	1.0 (1 m)	25	0.9
1N5821	Schottky	Si	30	3.0		
ECG5863	RFR	Si	600	6	150	0.9
1N6263	Schottky	Si	70	15 m	50 m	0.41 @ 1 mA
5082-2835	Schottky	Si	8	1 m (100 n)	10 m	0.34 @ 1 mA

Si = Silicon; Ge = Germanium; RFR = rectifier, fast recovery.

[†]For package shape, size and pin-connection information see manufacturers' data sheets. Many retail suppliers offer data sheets to buyers free of charge on request. Data books are available from many manufacturers and retailers.

Table 7.25

Suggested Small-Signal FETs

Device	*Type*	*Max Diss (mW)*	*Max V_{DS} (V)*[3]	*$V_{GS(off)}$ (V)*[3]	*Min gfs (µS)*	*Input C (pF)*	*Max ID (mA)*[1]	*f_{max} (MHz)*	*Noise Figure (typ)*	*Case*	*Base*	*Mfr*[2]	*Applications*
2N4416	N-JFET	300	30	–6	4500	4	–15	450	4 dB @400 MHz	TO-72	1	S, M	VHF/UHF amp, mix, osc
2N5484	N-JFET	310	25	–3	2500	5	30	200	4 dB @200 MHz	TO-92	2	M	VHF/UHF amp, mix, osc
2N5485	N-JFET	310	25	–4	3500	5	30	400	4 dB @400 MHz	TO-92	2	S	VHF/UHF amp, mix, osc
2N5486	N-JFET	360	25	–2	5500	5	15	400	4 dB @400 MHz	TO-92	2	M	VHF/UHF amp, mix, osc
3N200 NTE222 SK3065	N-dual-gate MOSFET	330	20	–6	10,000	4-8.5	50	500	4.5 dB @400 MHz	TO-72	3	R	VHF/UHF amp, mix, osc
3N202 NTE454 SK3991	N-dual-gate MOSFET	360	25	–5	8000	6	50	200	4.5 dB @200 MHz	TO-72	3	S	VHF amp, mixer
MPF102 ECG451 SK9164	N-JFET	310	25	–8	2000	4.5	20	200	4 dB @400 MHz	TO-92	2	N, M	HF/VHF amp, mix, osc
MPF106 2N5484	N-JFET	310	25	–6	2500	5	30	400	4 dB @200 MHz	TO-92	2	N, M	HF/VHF/UHF amp, mix, osc
40673 NTE222 SK3050	N-dual-gate MOSFET	330	20	–4	12,000	6	50	400	6 dB @200 MHz	TO-72	3	R	HF/VHF/UHF amp, mix, osc
U304	P-JFET	350	–30	+10	27		–50	—	—	TO-18	4	S	analog switch chopper
U310	N-JFET	500 300	30 30	–6	10,000	2.5	60	450	3.2 dB @450 MHz	TO-52	5	S	common-gate VHF/UHF amp,
U350	N-JFET Quad	1W	25	–6	9000	5	60	100	7 dB @100 MHz	TO-99	6	S	matched JFET doubly bal mix
U431	N-JFET Dual	300	25	–6	10,000	5	30	100	—	TO-99	7	S	matched JFET cascode amp and bal mix
2N5670	N-JFET	350	25	8	3000	7	20	400	2.5 dB @100 MHz	TO-92	2	M	VHF/UHF osc, mix, front-end amp
2N5668	N-JFET	350	25	4	1500	7	5	400	2.5 dB @100 MHz	TO-92	2	M	VHF/UHF osc, mix, front-end amp
2N5669	N-JFET	350	25	6	2000	7	10	400	2.5 dB @100 MHz	TO-92	2	M	VHF/UHF osc, mix, front-end amp
J308	N-JFET	350	25	6.5	8000	7.5	60	1000	1.5 dB @100 MHz	TO-92	2	M	VHF/UHF osc, mix, front-end amp
J309	N-JFET	350	25	4	10,000	7.5	30	1000	1.5 dB @100 MHz	TO-92	2	M	VHF/UHF osc, mix, front-end amp
J310	N-JFET	350	25	6.5	8000	7.5	60	1000	1.5 dB @100 MHz	TO-92	2	M	VHF/UHF osc, mix, front-end amp
NE32684A	HJ-FET	165	2.0	–0.8	45,000	—	30	20 GHz	0.5 dB @12 GHz	84A		NE	Low-noise amp

Notes:

[1]25°C.

[2]M = Motorola; N = National Semiconductor; NE=NEC; R = RCA; S = Siliconix.

[3]For package shape, size and pin-connection information, see manufacturers' data sheets. Many retail suppliers offer data sheets to buyers free of charge on request. Data books are available from many manufacturers and retailers.

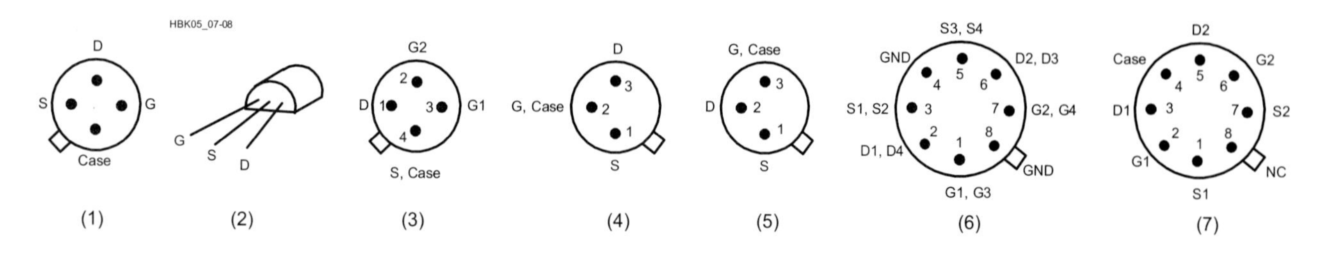

Table 7.26

Low-Noise Transistors

Device	*NF (dB)*	*F (MHz)*	*f_T (GHz)*	*I_C (mA)*	*Gain (dB)*	*F (MHz)*	*$V_{(BR)CEO}$ (V)*	*I_C (mA)*	*P_T (mW)*	*Case*
MRF904	1.5	450	4	15	16	450	15	30	200	TO-206AF
MRF571	1.5	1000	8	50	12	1000	10	70	1000	Macro-X
MRF2369	1.5	1000	6	40	12	1000	15	70	750	Macro-X
MPS911	1.7	500	7	30	16.5	500	12	40	625	TO-226AA
MRF581A	1.8	500	5	75	15.5	500	15	200	2500	Macro-X
BFR91	1.9	500	5	30	16	500	12	35	180	Macro-T
BFR96	2	500	4.5	50	14.5	500	15	100	500	Macro-T
MPS571	2	500	6	50	14	500	10	80	625	TO-226AA
MRF581	2	500	5	75	15.5	500	18	200	2500	Macro-X
MRF901	2	1000	4.5	15	12	1000	15	30	375	Macro-X
MRF941	2.1	2000	8	15	12.5	2000	10	15	400	Macro-X
MRF951	2.1	2000	7.5	30	12.5	2000	10	100	1000	Macro-X
BFR90	2.4	500	5	14	18	500	15	30	180	Macro-T
MPS901	2.4	900	4.5	15	12	900	15	30	300	TO-226AA
MRF1001A	2.5	300	3	90	13.5	300	20	200	3000	TO-205AD
2N5031	2.5	450	1.6	5	14	450	10	20	200	TO-206AF
MRF4239A	2.5	500	5	90	14	500	12	400	3000	TO-205AD
BFW92A	2.7	500	4.5	10	16	500	15	35	180	Macro-T
MRF521*	2.8	1000	4.2	–50	11	1000	–10	–70	750	Macro-X
2N5109	3	200	1.5	50	11	216	20	400	2500	TO-205AD
2N4957*	3	450	1.6	–2	12	450	–30	–30	200	TO-206AF
MM4049*	3	500	5	–20	11.5	500	–10	–30	200	TO-206AF
2N5943	3.4	200	1.5	50	11.4	200	30	400	3500	TO-205AD
MRF586	4	500	1.5	90	9	500	17	200	2500	TO-205AD
2N5179	4.5	200	1.4	10	15	200	12	50	200	TO-206AF
2N2857	4.5	450	1.6	8	12.5	450	15	40	200	TO-206AF
2N6304	4.5	450	1.8	10	15	450	15	50	200	TO-206AF
MPS536*	4.5	500	5	–20	4.5	500	–10	–30	625	TO-226AA
MRF536*	4.5	1000	6	–20	10	1000	–10	–30	300	Macro-X

*denotes a PNP device

Complementary devices

NPN	*PNP*
2N2857	2N4957
MRF904	MM4049
MRF571	MRF521

For package shape, size and pin-connection information, see manufacturers' data sheets. Many retail suppliers offer data sheets to buyers free of charge on request. Data books are available from many manufacturers and retailers.

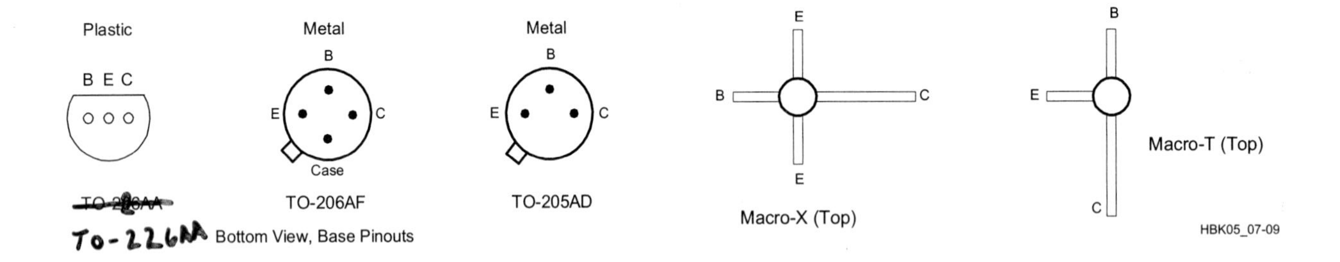

Table 7.27
Monolithic Amplifiers (50 Ω)
Mini-Circuits Labs MMICs

Device	Freq Range (MHz)	Gain (dB) at 1000 MHz	Output Level 1 dB Comp (dBm)	NF (dB)	I_{max} (mA)	P_{max} (mW)
ERA-1	dc - 8000	12.1	+12.0	4.3	75	330
ERA-2	dc - 6000	15.8	+13.0	4.0	75	330
ERA-3	dc - 3000	21.0	+12.5	3.5	75	330
ERA-4	dc - 4000	14.0	+17.3	4.2	120	650
ERA-5	dc - 4000	19.5	+18.4	4.3	120	650
ERA-6	dc - 4000	12.5	+17.9	4.5	12	650
GAL-1	dc - 8000	12.5	+12.2	4.5	55	225
GAL-2	dc - 8000	15.8	+12.9	4.6	55	225
GAL-3	dc - 8000	21.1	+12.5	3.5	55	225
GAL-4	dc - 8000	14.1	+17.5	4.0	85	475
GAL-5	dc - 8000	19.4	+18.0	3.5	85	475
GAL-6	dc - 8000	12.2	+18.2	4.5	85	475
GAL-21	dc - 8000	13.9	+12.6	4.0	55	225
GAL-33	dc - 8000	18.7	+13.4	3.9	55	265
GAL-51	dc - 8000	17.5	+18.0	3.5	85	475
HELA-10B	50 - 1000	12.0	+30.0	3.5	525	7150
HELA-10D	8 - 300	11.0	+30.0	3.5	525	7150
MAR-1	dc - 1000	15.5	+1.5	5.5	40	200
MAR-2	dc - 2000	12.0	+4.5	6.5	60	325
MAR-3	dc - 2000	12.0	+10.0	6.0	70	400
MAR-4	dc - 1000	8.0	+12.5	6.5	85	500
MAR-6	dc - 2000	16.0	+2.0	3.0	50	200
MAR-7	dc - 2000	12.5	+5.5	5.0	60	275
MAR-8	dc - 1000	22.5	+12.5	3.3	65	500
MAV-1	dc - 1000	15.0	+1.5	5.5	40	200
MAV-2	dc - 1500	11.0	+4.5	6.5	60	325
MAV-3	dc - 1500	11.0	+10.0	6.0	70	400
MAV-4	dc - 1000	7.5	+11.5	7.0	85	500
MAV-11	dc - 1000	10.5	+17.5	3.6	80	550
RAM-1	dc - 1000	15.5	+1.5	5.5	40	200
RAM-2	dc - 2000	11.8	+4.5	6.5	60	325
RAM-3	dc - 2000	12.0	+10.0	6.0	80	425
RAM-4	dc - 1000	8.0	+12.5	6.5	100	540
RAM-6	dc - 2000	16.0	+2.0	2.8	50	200
RAM-7	dc - 2000	12.5	+5.5	4.5	60	275
RAM-8	dc - 1000	23.0	+12.5	3.0	65	420
VAM-3	dc - 2000	11.0	+9.0	6.0	60	240
VAM-6	dc - 2000	15.0	+2.0	3.0	40	125
VAM-7	dc - 2000	12.0	+5.5	5.0	50	175
VNA-25	500 - 2500	18.0	+18.2	5.5	105	1000

Mini-circuits Labs Web site: **www.minicircuits.com/**.

Agilent MMICs

Device	Freq Range (MHz)	Typical Gain (dB)	Output Level 1 dB Comp (dBm)	NF (dB)	I_{max} (mA)	P_{max} (mW)
INA-01170	dc - 500	32.5	+11.0	2.0	50	400
INA-02184	dc - 1500	26.0	+11.0	2.0	50	400
INA-32063	dc - 2400	16.8	+3.6	4.4	25	75
INA-51063	dc - 2400	25.0	+1.0	2.5	14	170
MGA-725M4	100 - 6000	17.6	+13.1	1.2	80	250
MGA-86576	1.5 - 8000	23.0	+6.3	2.0	16	—
MSA-Olxx	dc - 1300	18.5	+1.5	5.5	40	200
MSA-02xx	dc - 2800	12.5	+4.5	6.5	60	325
MSA-03xx	dc - 2800	12.5	+10.0	6.0	80	425
MSA-04xx	dc - 4000	8.3	+11.5	7.0	85	500
MSA-05xx	dc- 2800	7.0	+19.0	6.5	135	1.5
MSA-06xx	dc - 800	19.5	+2.0	3.0	50	200
MSA-07xx	dc - 2500	13.0	+5.5	4.5	50	175
MSA-08xx	dc - 6000	32.5	+12.5	3.0	65	500
MSA-09xx	dc - 6000	7.2	+10.5	6.2	65	500
MSA-11xx	50-1300	12.0	+17.5	3.6	80	550

Agilent Web site: **www.agilent.com/Products/English/index.html**.

(continued on next page)

Motorola Hybrid Amplifiers (50 Ω)

Device type	*Freq Range (MHz)*	*Gain (dB) min/typ*	*Supply Voltage (V)*	*Output Level, 1 dB Comp (dBm)*	*NF at 250 MHz (dB)*
MWA110	0.1 - 400	13/14	2.9	–2.5	4.0
MWA120	0.1 - 400	13/14	5	+8.2	5.5
MWA130	0.1- 400	13/14	5.5	+18.0	7.0
MWA131	0.1 - 400	13/14	5.5	+20.0	5.0
MWA210	0.1- 600	9/10	1.75	+1.5	6.0
MWA220	0.1- 600	9/10	3.2	+10.5	6.5
MWA230	0.1- 600	9/10	4.4	+18.5	7.5
MWA310	0.1- 1000	7/8	1.6	+3.5	6.5
MWA320	0.1- 1000	7/8	2.9	+11.5	6.7
MWA330	0.1- 1000	na/6.2	4	+15.2	9.0

Motorola Web site: **merchant.hibbertco.com/servlet/MtrlDeactServlet**.

Table 7.28

General-Purpose Transistors

Listed numerically by device

Device	*Type*	V_{CEO} *Maximum Collector Emitter Voltage (V)*	V_{CBO} *Maximum Collector Base Voltage (V)*	V_{EBO} *Maximum Emitter Base Voltage (V)*	I_C *Maximum Collector Current (mA)*	P_O *Maximum Device Dissipation (W)*	**Minimum DC Current Gain** h_{FE} I_C = 0.1 mA	I_C = 150 mA	*Current-Gain Bandwidth Product* f_T *(MHz)*	*Noise Figure NF Maximum (dB)*	*Base*
2N918	NPN	15	30	3.0	50	0.2	20 (3 mA)	—	600	6.0	3
2N2102	NPN	65	120	7.0	1000	1.0	20	40	60	6.0	2
2N2218	NPN	30	60	5.0	800	0.8	20	40	250		2
2N2218A	NPN	40	75	6.0	800	0.8	20	40	250		2
2N2219	NPN	30	60	5.0	800	3.0	35	100	250		2
2N2219A	NPN	40	75	6.0	800	3.0	35	100	300	4.0	2
2N2222	NPN	30	60	5.0	800	1.2	35	100	250		2
2N2222A	NPN	40	75	6.0	800	1.2	35	100	200	4.0	2
2N2905	PNP	40	60	5.0	600	0.6	35	—	200		2
2N2905A	PNP	60	60	5.0	600	0.6	75	100	200		2
2N2907	PNP	40	60	5.0	600	0.4	35	—	200		2
2N2907A	PNP	60	60	5.0	600	0.4	75	100	200		2
2N3053	NPN	40	60	5.0	700	5.0	—	50	100		2
2N3053A	NPN	60	80	5.0	700	5.0	—	50	100		2
2N3563	NPN	15	30	2.0	50	0.6	20	—	800		1
2N3904	NPN	40	60	6.0	200	0.625	40	—	300	5.0	1
2N3906	PNP	40	40	5.0	200	0.625	60	—	250	4.0	1
2N4037	PNP	40	60	7.0	1000	5.0	—	50			2
2N4123	NPN	30	40	5.0	200	0.35	—	25 (50 mA)	250	6.0	1
2N4124	NPN	25	30	5.0	200	0.35	120 (2 mA)	60 (50 mA)	300	5.0	1
2N4125	PNP	30	30	4.0	200	0.625	50 (2 mA)	25 (50 mA)	200	5.0	1
2N4126	PNP	25	25	4.0	200	0.625	120 (2 mA)	60 (50 mA)	250	4.0	1
2N4401	NPN	40	60	6.0	600	0.625	20	100	250		1
2N4403	PNP	40	40	5.0	600	0.625	30	100	200		1
2N5320	NPN	75	100	7.0	2000	10.0	—	30 (I A)			2
2N5415	PNP	200	200	4.0	1000	10.0	—	30 (50 mA)	15		2
MM4003	PNP	250	250	4.0	500	1.0	20 (10 mA)	—			2
MPSA55	PNP	60	60	4.0	500	0.625	—	50 (0.1 A)	50		1
MPS6531	NPN	40	60	5.0	600	0.625	60 (10 mA)	90 (0.1 A)			1
MPS6547	NPN	25	35	3.0	50	0.625	20 (2 mA)	—	600		1

Test conditions: I_C = 20 mA dc; V_{CE} = 20 V; f = 100 MHz

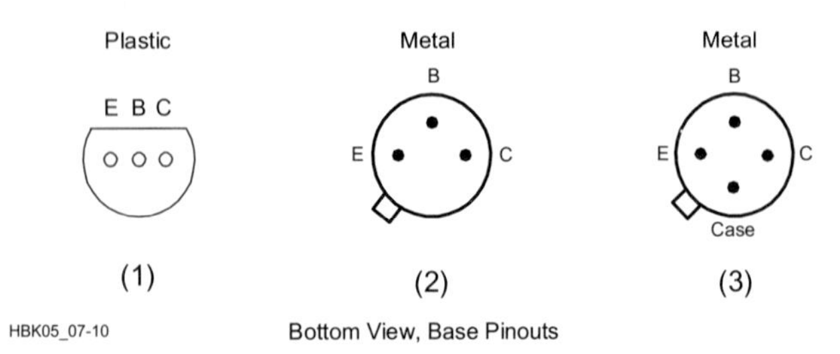

Bottom View, Base Pinouts

Table 7.29

RF Power Amplifier Modules

Listed by frequency

Device	*Supply (V)*	*Frequency Range (MHz)*	*Output Power (W)*	*Power Gain (dB)*	*Package*†	*Mfr/ Notes*
M57735	17	50-54	14	21	H3C	MI; SSB mobile
M57719N	17	142-163	14	18.4	H2	MI; FM mobile
S-AV17	16	144-148	60	21.7	5-53L	T, FM mobile
S-AV7	16	144-148	28	21.4	5-53H	T, FM mobile
MHW607-1	7.5	136-150	7	38.4	301K-02/3	MO; class C
BGY35	12.5	132-156	18	20.8	SOT132B	P
M67712	17	220-225	25	20	H3B	MI; SSB mobile
M57774	17	220-225	25	20	H2	MI; FM mobile
MHW720-1	12.5	400-440	20	21	700-04/1	MO; class C
MHW720-2	12.5	440-470	20	21	700-04/1	MO; class C
M57789	17	890-915	12	33.8	H3B	MI
MHW912	12.5	880-915	12	40.8	301R-01/1	MO; class AB
MHW820-3	12.5	870-950	18	17.1	301G-03/1	MO; class C

Manufacturer codes: MO = Motorola; MI = Mitsubishi; P = Philips; T = Toshiba.

†For package shape, size and pin-connection information, see manufacturers' data sheets. Many retail suppliers offer data sheets to buyers free of charge on request. Data books are available from many manufacturers and retailers.

Table 7.30

General Purpose Silicon Power Transistors

TO-220 Case, Pin 1=Base, Pin 2, Case = Collector; Pin 3 = Emitter

NPN	*PNP*	I_C *Max (A)*	V_{CEO} *Max (V)*	h_{FE} *Min*	F_T *(MHz)*	*Power Dissipation (W)*
D44C8		4	60	100/220	50	30
	D45C8	4	60	40/120	50	30
TIP29		1	40	15/75	3	30
	TIP30	1	40	15/75	3	30
TIP29A		1	50	15/75	3	30
	TIP30A	1	60	15/75	3	30
TIP29B		1	80	15/75	3	30
TIP29C		1	100	15/75	3	30
	TIP30C	1	100	15/75	3	30
TIP47		1	250	30/150	10	40
TIP48		1	300	30/150	10	40
TIP49		1	350	30/150	10	40
TIP50		1	400	30/150	10	40
TIP110 *		2	60	500	> 5	50
	TIP115 *	2	60	500	> 5	50
TIP116		2	80	500	25	50
TIP31		3	40	25	3	40
	TIP32	3	40	25	3	40
TIP31A		3	60	25	3	40
	TIP32A	3	60	25	3	40
TIP31B		3	80	25	3	40
	TIP32B	3	80	25	3	40
TIP31C		3	100	25	3	40
	TIP32C	3	100	25	3	40
2N6124		4	45	25/100	2.5	40
2N6122		4	60	25/100	2.5	40
MJE1300		4	300	6/30	4	60
TIP120 *		5	60	1000	> 5	65
	TIP125 *	5	60	1000	> 10	65
	TIP42	6	40	15/75	3	65
TIP41A		6	60	15/75	3	65
TIP41B		6	80	15/75	3	65
2N6290		7	50	30/150	4	40
	2N6109		50	30/150	4	40
2N6292		7	70	30/150	4	40
	2N6107	7	70	30/150	4	40
MJE3055T		10	50	20/70	2	75
	MJE2955T	10	60	20/70	2	75
2N6486		15	40	20/150	5	75
2N6488		15	80	20/150	5	75
TIP140 *		10	60	500	> 5	125
	TIP145 *	10	60	600	> 10	125

TO-204 Case (TO-3), Pin 1=Base, Pin 2 = Emitter, Case = Collector;

NPN	*PNP*	I_C *Max (A)*	V_{CEO} *Max (V)*	h_{FE} *Min*	F_T *(MHz)*	*Power Dissipation (W)*
2N3055A		15	60	20/70	0.8	115
2N3055		15	60	20/70	2.5	115
	MJ2955	15	60	20/70	2.5	115
2N6545		8	400	7/35	6	125
2N5039		20	75	20/100	—	140
2N3771		30	40	15	0.2	150
2N3789		10	60	15	4	150
2N3715		10	60	30	4	150
	2N3791	10	60	30	4	150
	2N5875	10	60	20/100	4	150
	2N3790	10	80	15	4	150
2N3716		10	80	30	4	150
	2N3792	10	80	30	4	150
2N3773		16	140	15/60	4	150
2N6284		20	100	750/18K	—	160
	2N6287	20	100	750/18K	—	160
2N5881		15	60	20/100	4	160
2N5880		15	80	20/100	4	160
2N6249		15	200	10/50	2.5	175
2N6250		15	275	8/50	2.5	175
2N6546		15	300	6/30	6-28	175
2N6251		15	350	6/50	2.5	175
2N5630		16	120	20/80	1	200
2N5301		30	40	15/60	2	200
2N5303		20	80	15/60	2	200
2N5885		25	60	20/100	4	200
2N5302		30	60	15/60	2	200
	2N4399	30	60	15/60	4	200
2N5886		25	80	20/100	4	200
	2N5884	25	80	20/100	4	200
MJ802		30	100	25/100	2	200
	MJ4502	30	100	25/100	2	200
MJ15003		20	140	25/150	2	250
	MJI5004	20	140	25/150	2	250
MJ15024		25	250	15/60	4	250

(shaded) = Complimentary pairs
* = Darlington transistor

Useful URLs for finding transistor/IC data sheets:

1. General-purpose URL: **members.nbci.com/cbradiomods/transistors/sigtransistors.html**.
2. General-purpose substitution URL: **www.nteinc.com/**.
3. Philips semiconductors: **www.semiconductors.philips.com/pack/discretes.html** and **www.semiconductors.philips.com/catalog**.
4. Mitsubishi: **www.mitsubishichips.com/data/datasheets/hf-optic/index.html**, then click "Si Modules."
5. Motorola: **design-net.com/redirect/books/index.html**. Look for archive section for older products.
6. STMicroelectronics (Thompson): **us.st.com/stonline/products/index.htm**.
7. Toshiba: **www.semicon.toshiba.co.jp/seek/us/td/16ktran/160021.htm**.

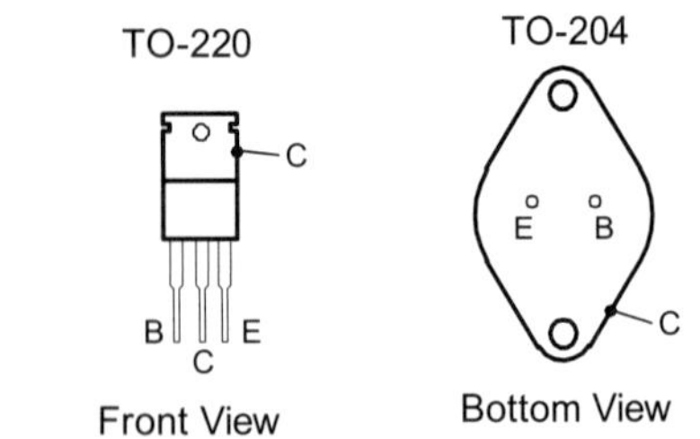

Table 7.31
RF Power Transistors

Device	Output Power (W)	Input Power (W)	Gain (dB)	Typ Supply Voltage (V)	Case	Mfr
1.5 to 30 MHz, HF SSB/CW						
2SC2086	0.3		13	1.2	TO-92	MI
BLV10	1		18	1.2	SOT123	PH
BLV11	2		18	12	SOT123	PH
MRF476	3	0.1	15	12.5-13.6	221A-04/1	MO
BLW87	6		18	1.2	SOT123	PH
2SC2166	6		13.8	1.2	TO-220	MI
BLW83	1.0		20	26	SOT123	PH
MRF475	1.2	1.2	10	12.5-13.6	221A-04/1	MO
MRF433	12.5	0.125	20	12.5-13.6	211-07/1	MO
2SC3133	1.3		14	1.2	TO-220	MI
MRF485	1.5	1.5	10	28	221 A-04/1	MO
2SC1969	1.6		1.2	1.2	TO-220	MI
BLW50F	1.6		19.5	45	SOT123	PH
MRF406	20	1.25	12	12.5-13.6	221-07/1	MO
SD1285	20	0.65	15	12.5	M113	ST
MRF426	25	0.16	22	28	211-07/1	MO
MRF427	25	0.4	18	50	211-11/1	MO
MRF477	40	1.25	15	12.5-13.6	211-11/1	MO
MRF466	40	1.25	15	28	211-07/1	MO
BLW96	50		1.9	40	SOT121	PH
2SC3241	75		12.3	12.5	T-45E	MI
SDI405	75	3.8	13	12.5	M174	ST
2SC2097	75		12.3	13.5	T-40E	MI
MRF464	80	2.53	10	28	211-11/1	MO
MRF421	100	10	10	12.5-13.6	211-11/1	MO
SD1487	100	7.9	11	12.5	M174	ST
2SC2904	100		11.5	12.5	T-40E	MI
SD1729	130	8.2	12	28	M174	ST
MRF422	150	15	10	28	211-11/1	MO
MRF428	150	7.5	13	50	211-11/1	MO
SD1726	150	6	14	50	M174	ST
PT9790	150	4.8	15	50	211-11/1	MO
MRF448	250	15.7	12	50	211-11/1	MO
MRF430	600	60	10	50	368-02/1	MO
50 MHz						
MRF475	4	0.4	10	12.5-13.6	*221A-04/1*	MO
MRF497	40	4	10	12.5-13.6	221A-04/2	MO
SDI446	70	7	10	12.5	M113	ST
MRF492	70	5.6	11	12.5-13.6	211-11/1	MO
SD1405	100	20	7	12.5	M174	ST
VHF to 175 MHz						
2N4427	0.7		8	7.5	TO-39	PH
2N3866	1		10	28	TO-39	PH
BFQ42	1.5		8.4	7.5	TO-39	PH
2SC2056	1.6		9	7.2	T-41	MI
2N3553	2.5	0.25	10	28	79-04/1	MO
BF043	3		9.4	7.5	TO-39	PH
SD1012	4	0.25	12	12.5	M135	ST
2SC2627	5		13	12.5	T-40	MI
2N5641	7	1	8.4	28	144B-05/1	MO
MRF340	8	0.4	13	28	221A-04/2	MO
BLW29	9		7.4	7.5	SOT120	PH
SD1143	10	1	10	12.5	M135	ST
2SC1729	1.4		10	13.5	T-31 E	MI
SD1014-02	15	3.5	6.3	12.5	M135	ST
BLVII	15		8	13.5	SOT123	PH
2N5642	20	3	8.2	28	145A-09/1	MO
MRF342	24	1.9	11	28	221A-04/2	MO
BLW87	25		6	13.5	SOT123	PH
2SC1946	28		6.7	13.5	T-31 E	MI
MRF314	30	3	10	28	211-07/1	MO
SD1018	40	14	4.5	12.5	M135	ST
2N5643	40	6.9	7.6	28	145A-09/1	MO
BLW40	40		10	12.5	SOT120	PH
MRF315	45	5.7	9	28	211-07/1	MO
PT9733	50	10	7	28	145A-09/1	MO
MRF344	60	15	6	28	221A-04/2	MO
2SC2694	70		6.7	12.5	T-40	MI
BLV75/12	75		6.5	12.5	SOT119	PH
MRF316	80	8	10	28	316-01/1	MO
SD1477	100	25	6	12.5	M111	ST
BLW78	100		6	28	SOT121	PH
MRF317	100	12.5	9	28	316-01/1	MO
TP9386	150	15	10	28	316-01/1	MO
220 MHz						
MRF207	1	0.15	8.2	12.5	79-04/1	MO
2N5109	2.5		11	12	TO-205AD	MO
MRF227	3	0.13	13.5	12.5	79-05/5	MO
MRF208	1.0	1	10	12.5	145A-09/1	MO
MRF226	1.3	1.6	9	12.5	145A-09/1	MO
2SC2133	30		8.2	28	T-40E	MI
2SC2134	60		7	28	T-40E	MI
2SC2609	100		6	28	T-40E	MI
UHF to 512 MHz						
2N4427	0.4		10	12.5	TO-39	PH
2SC3019	0.5		14	12.5	T-43	MI
MRF581	0.6	0.03	13	12.5	317-01/2	MO
2SC908	1		4	12.5	TO-39	MI
2N3866	1		10	28	TO-39	PH
2SC2131	1.4		6.7	13.5	TO-39	MI
BLX65E	2		9	12.5	TO-39	PH
BLW89	2		12	28	SOT122	PH
MRF586	2.5		16.5	1.5	79-04	MO
MRF630	3	0.33	9.5	12.5	79-05/5	MO
2SC3020	3	0.3	10	12.5	T-31 E	MI
BLW80	4		8	12.5	SOT122	PH
BLW90	4		11	12.5	SOT122	PH
MRF652	5	0.5	10	12.5	244-04/1	MO
MRF587	5		16.5	15	244A-01/1	MO
2SC3021	7	1.2	7.6	12.5	T-31 E	MI
BLW81	10		6	12.5	SOT122	PH
MRF653	10	2	7	12.5	244-04/1	MO
BLW91	10		9	28	SOT122	PH
MRF654	15	2.5	7.8	12.5	244-04/1	MO
2SC3022	18	6	4.7	12.5	T-31 E	MI
BLU20/12	20		6.5	12.5	SOT119	PH
BLX94A	25		6	28	SOT48/2	PH
2SC2695	28		4.9	13.5	T-31 E	MI
BLU30/12	30		6	12.5	SOT119	PH
BLU45/12	45		4.8	12.5	SOT119	PH
2SC2905	45		4.8	12.5	T-40E	MI
MRF650	50	15.8	5	12.5	316-01/1	MO
TP5051	50	6	9	24	333A-02/2	MO
BLU60/12	60		4.4	12.5	SOT119	PH
2SC3102	60	20	4.8	12.5	T-41 E	MI
BLU60/28	60		7	28	SOT119	PH
MRF658	65	25	4.15	12.5	316-01/1	MO
MRF338	80	15	7.3	28	333-04/1	MO
SD1464	100	28.2	5.5	28	M168	ST
UHF to 960 MHz						
MRF581	0.6	0.06	10	12.5	317-01/2	MO
MRF8372	0.75	0.11	8	12.5	751-04/1	MO
MRF557	1.5	0.23	8	12.5	317D-02/2	MO
BLV99	2		9	24	SOT172	PH
SD1420	2.1	0.27	9	24	M122	ST
MRF839	3	0.46	8	12.5	305A-01/1	MO
MRF896	3	0.3	10	24	305-01/1	MO
MRF891	5	0.63	9	24	319-06/2	MO

(continued on next page)

Device	Output Power (W)	Input Power (W)	Gain (dB)	Typ Supply Voltage (V)	Case	Mfr
2SC2932	6		7.8	12.5	T-31 B	MI
SD1398	6	0.6	10	24	M142	ST
2SC2933	14	3	6.7	12.5	T-31 B	MI
SD1400-03	14	1.6	9.5	24	M118	ST
MRF873	15	3	7	12.5	319-06/2	MO
SD1495-03	30	6	7	24	M142	ST
SD1424	30	5.3	7.5	24	M156	ST
MRF897	30	3	10	24	395B-01/1	MO
MRF847	45	16	4.5	12.5	319-06/1	MO
BLV101A	50		8.5	26	SOT273	PH
SD1496-03	55	10	7.4	24	M142	ST
MRF898	60	12	7	24	333A-02/1	MO
MRF880	90	12.7	8.5	26	375A-01/1	MO
MRF899	150	24	8	26	375A-01/1	MO

Manufacturer codes:
MI = Mitsubishi; MO = Motorola; PH Philips;
ST = STMicroelectronics

There is a bewildering variety of package types, sizes and pin-out connections. (For example, for the 137 different transistors in this table there are 54 different packages.) See the data sheets on each manufacturer's Web pages for details.

Mitsubishi: www.mitsubishichips.com/data/datasheets/hf-optic/index.html, then click "Si Modules." Scroll to section for "Si Discrete" and then choose frequency range and device.

Motorola: design-net.com/redirect/books/index.html. Type the part number in the search window at the upper left of the screen. If you receive a message that "No results were found for your search" the part you want is probably obsolete. Click on the text highlighted in red as "Motorola's SPS Literature Distribution Center Archive Site." In the Description box, type the part number you want, click the Search button and then click on the Document Number for the latest Revision level of that obsoleted part number.

Philips: www.semiconductors.philips.com/. Type the part number in the "search" box at the upper right corner of the screen. Click on the highlighted part number in the Description field and then click on Datasheet. Finally, view the PDF by clicking on the highlighted "Download" text or hold down the right-mouse button while clicking on "Download" to save the PDF to disk.

STMicroelectronics: us.st.com/stonline/discretes/index.shtml. Click on "Datasheets" at the top and then scroll down to the bottom of the listing to find "Radio Frequency, RF Power." Then specify the frequency range you want and either view the PDF directly or download the PDF by holding down the right-mouse button while clicking on the device's part number.

Table 7.32

RF Power Transistors Recommended for New Designs

Device	Output Power (W)	Type	Gain (dB)	Typ Supply Voltage (V)	Case	Mfr
1.5 to 30 MHz, HF SSB/CW						
MRF171A	30	MOS	20	28	211-07/2	MO
BLF145	30	MOS	24	28	SOT123A	PH
MRF148A	30	MOS	18	50	211-07/2	MO
SD2918	30	MOS	18	50	M113	ST
SD1405	75	BJT	13	12.5	M174	ST
SD1733	75	BJT	14	50	M135	ST
SD1487	100	BJT	11	12.5	M174	ST
SD1407	125	BJT	15	28	M174	ST
SD1729	130	BJT	12	28	M174	ST
BLF147	150	MOS	17	28	SOT121B	PH
BLF177	150	MOS	20	50	SOT121B	PH
BLF175	150	MOS	24	50	SOT123A	PH
SD1726	150	BJT	14	50	M174	ST
SD1727	150	BJT	14	50	M164	ST
MRF150	150	MOS	17	50	211-07/2	MO
SD1411	200	BJT	16	40	M153	ST
SD1730	220	BJT	12	28	M174	ST
SD1731	220	BJT	13	50	M174	ST
SD1728	250	BJT	14.5	50	M177	ST
SD2923	300	MOS	16	50	M177	ST
SD2933	300	MOS	18	50	M177	ST
MRF154	600	MOS	17	50	368-03/2	MO
50 to 175 MHz						
BLF202	2	MOS	10	12.5	SOT409A	PH
BLF242	5	MOS	13	28	SOT123A	PH
SD1274	30	BJT	10	13.6	M135	ST
BLF245	30	MOS	13	28	SOT123	PH
SD1275	40	BJT	9	13.6	M135	ST
BLF246B	60	MOS	14	28	SOT161A	PH
SD1477	100	BJT	6	12.5	M111	ST
SD1480	100	BJT	9.2	28	M111	ST
SD2921	150	MOS	12.5	50	M174	ST
MRF141	150	MOS	13	28	211-11/2	MO
MRF151	150	MOS	13	50	211-11/2	MO
SD2931	150	MOS	14	50	M174	ST
BLF248	300	MOS	10	28	SOT262	PH
SD2932	300	MOS	15	50	M244	ST
VHF to 220 MHz						
MRF134	5	MOS	10.6	28	211-07/2	MO
MRF136	15	MOS	16	28	211-07/2	MO
MRF173	80	MOS	13	28	211-11/2	MO
MRF174	125	MOS	11.8	28	211-11/2	MO
BLF278	250	MOS	14	50	SOT261A1	PH
VHF to 470 MHz						
BLT50	1.2	BJT	10	7.5	SOT223	PH
SD2900	5	MOS	13.5	28	M113	ST
SD1433	10	BJT	7	12.5	M122	ST
SD2902	15	MOS	12.5	28	M113	ST
SD2904	30	MOS	10	28	M113	ST
SD2903	30	MOS	13	28	M229	ST
SD1488	38	BJT	5.8	12.5	M111	ST
SD1434	45	BJT	5	12.5	M111	ST
MRF392	125	BJT	8	28	744A-01/1	MO
SD2921	150	MOS	12.5	50	M174	ST
VHF to 512 MHz						
BLF521	2	MOS	10	12.5	SOT172D	PH
MRF158	2	MOS	17.5	28	305A-01/2	MO
MRF160	4	MOS	17	28	249-06/3	MO
BLF542	5	MOS	13	28	SOT171A	PH
VLF544	20	MOS	11	28	SOT171A	PH
MRF166C	20	MOS	16	28	319-07/3	MO
MRF166W	40	MOS	14	28	412-01/1	MO
BLF546	80	MOS	11	28	SOT268A	PH
MRF393	100	BJT	7.5	28	744A-01/1	MO
MRF275L	100	MOS	8.8	28	333-04/2	MO
BLF548	150	MOS	10	28	SOT262A	PH
MRF275G	150	MOS	10	28	375-04/2	MO
UHF to 960 MHz						
BLT70	0.6	BJT	6	4.8	SOT223	PH
BLT80	0.6	BJT	6	7.5	SOT223	PH
BLT71/8	1.2	BJT	6	4.8	SOT223	PH
BLT81	1.2	BJT	6	7.5	SOT223	PH
BLF1043	10	MOS	16	26	SOT538A	PH
BLF1046	45	MOS	14	26	SOT467C	PH
BLF1047	70	MOS	14	26	SOT541A	PH
BLF1048	90	MOS	14	26	SOT502A	PH

Notes:

Manufacturer codes: MI = Mitsubishi; MO = Motorola; PH = Philips; ST = STMicroelectronics

There is a bewildering variety of package types, sizes and pin-out connections. (For example, for the 71 different transistors in this table there are 35 different packages.) See the data sheets on each manufacturer's Web pages for details.

Mitsubishi: **www.mitsubishichips.com/data/datasheets/hf-optic/index.html**, then click on "Si Modules." Scroll to section for "Si Discrete" and then choose frequency range and device.

Motorola: **design-net.com/redirect/books/index.html**. Type the part number in the search window at the upper left of the screen. If you receive a message that "No results were found for your search" the part you want is probably obsolete. Click on the text highlighted in red as "Motorola's SPS Literature Distribution Center Archive Site." In the Description box, type the part number you want, click the Search button and then click on the Document Number for the latest Revision level of that obsoleted part number.

Philips: **www.semiconductors.philips.com/**. Type the part number in the "search" box at the upper right corner of the screen. Click on the highlighted part number in the Description field and then click on Datasheet. Finally, view the PDF by clicking on the highlighted "Download" text or hold down the right-mouse button while clicking on "Download" to save the PDF to disk.

STMicroelectronics: **us.st.com/stonline/discretes/index.shtml**. Click on "Datasheets" at the top and then scroll down to the bottom of the listing to find "Radio Frequency, RF Power." Then specify the frequency range you want and either view the PDF directly or download the PDF by holding down the right-mouse button while clicking on the device's part number.

Table 7.33
Power FETs

Device	Type	VDSS min (V)	RDS(on) max (Ω)	ID max (A)	PD max (W)	Case†	Mfr
BS250P	P-channel	45	14	0.23	0.7	E-line	Z
IRFZ30	N-channel	50	0.050	30	75	TO-220	IR
MTP50N05E	N-channel	50	0.028	25	150	TO-220AB	M
IRFZ42	N-channel	50	0.035	50	150	TO-220	IR
2N7000	N-channel	60	5	0.20	0.4	E-line	Z
VN10LP	N-channel	60	7.5	0.27	0.625	E-line	Z
VN10KM	N-channel	60	5	0.3	1	TO-237	S
ZVN2106B	N-channel	60	2	1.2	5	TO-39	Z
IRF511	N-channel	60	0.6	2.5	20	TO-220AB	M
MTP2955E	P-channel	60	0.3	6	25	TO-220AB	M
IRF531	N-channel	60	0.180	14	75	TO-220AB	M
MTP23P06	P-channel	60	0.12	11.5	125	TO-220AB	M
IRFZ44	N-channel	60	0.028	50	150	TO-220	IR
IRF531	N-channel	80	0.160	14	79	TO-220	IR
ZVP3310A	P-channel	100	20	0.14	0.625	E-line	Z
ZVN2110B	N-channel	100	4	0.85	5	TO-39	Z
ZVP3310B	P-channel	100	20	0.3	5	TO-39	Z
IRF510	N-channel	100	0.6	2	20	TO-220AB	M
IRF520	N-channel	100	0.27	5	40	TO-220AB	M
IRF150	N-channel	100	0.055	40	150	TO-204AE	M
IRFP150	N-channel	100	0.055	40	180	TO-247	IR
ZVP1320A	P-channel	200	80	0.02	0.625	E-line	Z
ZVN0120B	N-channel	200	16	0.42	5	TO-39	Z
ZVP1320B	P-channel	200	80	0.1	5	TO-39	Z
IRF620	N-channel	200	0.800	5	40	TO-220AB	M
MTP6P20E	P-channel	200	1	3	75	TO-220AB	M
IRF220	N-channel	200	0.400	8	75	TO-220AB	M
IRF640	N-channel	200	0.18	10	125	TO-220AB	M

Manufacturers: IR = International Rectifier; M = Motorola; S = Siliconix; Z = Zetex.

†For package shape, size and pin-connection information, see manufacturers' data sheets. Many retail suppliers offer data sheets to buyers free of charge on request. Data books are available from many manufacturers and retailers.

Table 7.34
Logic IC Families

Type	***Propagation Delay for C_L = 50 pF*** (ns) Typ	Max	Max Clock Frequency (MHz)	Power Dissipation (CL = 0) @ 1 MHz (mW/gate)	Output Current @ 0.5 V max (mA)	Input Current (Max mA)	Threshold Voltage (V)	***Supply Voltage (V)*** Min	Typ	Max
CMOS										
74AC	3	5.1	125	0.5	24	0	V+/2	2	5 or 3.3	6
74ACT	3	5.1	125	0.5	24	0	1.4	4.5	5	5.5
74HC	9	18	30	0.5	8	0	V+/2	2	5	6
74HCT	9	18	30	0.5	8	0	1.4	4.5	5	5.5
4000B/74C (10 V)	30	60	5	1.2	1.3	0	V+/2	3	5 - 15	18
4000B/74C (5V)	50	90	2	3.3	0.5	0	V+/2	3	5 - 15	18
TTL										
74AS	2	4.5	105	8	20	0.5	1.5	4.5	5	5.5
74F	3.5	5	100	5.4	20	0.6	1.6	4.75	5	5.25
74ALS	4	11	34	1.3	8	0.1	1.4	4.5	5	5.5
74LS	10	15	25	2	8	0.4	1.1	4.75	5	5.25
ECL										
ECL III	1.0	1.5	500	60	—	—	−1.3	−5.19	−5.2	−5.21
ECL 100K	0.75	1.0	350	40	—	—	−1.32	−4.2	−4.5	−5.2
ECL100KH	1.0	1.5	250	25	—	—	−1.29	−4.9	−5.2	−5.5
ECL 10K	2.0	2.9	125	25	—	—	−1.3	−5.19	−5.2	−5.21
GaAs										
10G	0.3	0.32	2700	125	—	—	−1.3	−3.3	−3.4	−3.5
10G	0.3	0.32	2700	125	—	—	−1.3	−5.1	−5.2	−5.5

Source: Horowitz (W1HFA) and Hill, *The Art of Electronics—2nd edition,* page 570. © Cambridge University Press 1980, 1989. Reprinted with the permission of Cambridge University Press.

Table 7.35

Three-Terminal Voltage Regulators

Listed numerically by device

Device	*Description*	*Package*	*Voltage*	*Current (Amps)*
317	Adj Pos	TO-205	+1.2 to +37	0.5
317	Adj Pos	TO-204,TO-220	+1.2 to +37	1.5
317L	Low Current Adj Pos	TO-205,TO-92	+1.2 to +37	0.1
317M	Med Current Adj Pos	TO-220	+1.2 to +37	0.5
338	Adj Pos	TO-3	+1.2 to +32	5.0
350	High Current Adj Pos	TO-204,TO-220	+1.2 to +33	3.0
337	Adj Neg	TO-205	−1.2 to -37	0.5
337	Adj Neg	TO-204,TO-220	−1.2 to -37	1.5
337M	Med Current Adj Neg	TO-220	−1.2 to -37	0.5
309		TO-205	+5	0.2
309		TO-204	+5	1.0
323		TO-204,TO-220	+5	3.0
140-XX	Fixed Pos	TO-204,TO-220	Note 1	1.0
340-XX		TO-204,TO-220		1.0
78XX		TO-204,TO-220		1.0
78LXX		TO-205,TO-92		0.1
78MXX		TO-220		0.5

Device	*Description*	*Package*	*Voltage*	*Current (Amps)*
78TXX		TO-204		3.0
79XX	Fixed Neg	TO-204,TO-220	Note 1	1.0
79LXX		TO-205,TO-92		0.1
79MXX		TO-220		0.5

Note 1—XX indicates the regulated voltage; this value may be anywhere from 1.2 V to 35 V. A 7815 is a positive 15-V regulator, and a 7924 is a negative 24-V regulator.

The regulator package may be denoted by an additional suffix, according to the following:

Package	**Suffix**
TO-204 (TO-3)	K
TO-220	T
TO-205 (TO-39)	H, G
TO-92	P, Z

For example, a 7812K is a positive 12-V regulator in a TO-204 package. An LM340T-5 is a positive 5-V regulator in a TO-220 package. In addition, different manufacturers use different prefixes. An LM7805 is equivalent to a mA7805 or MC7805.

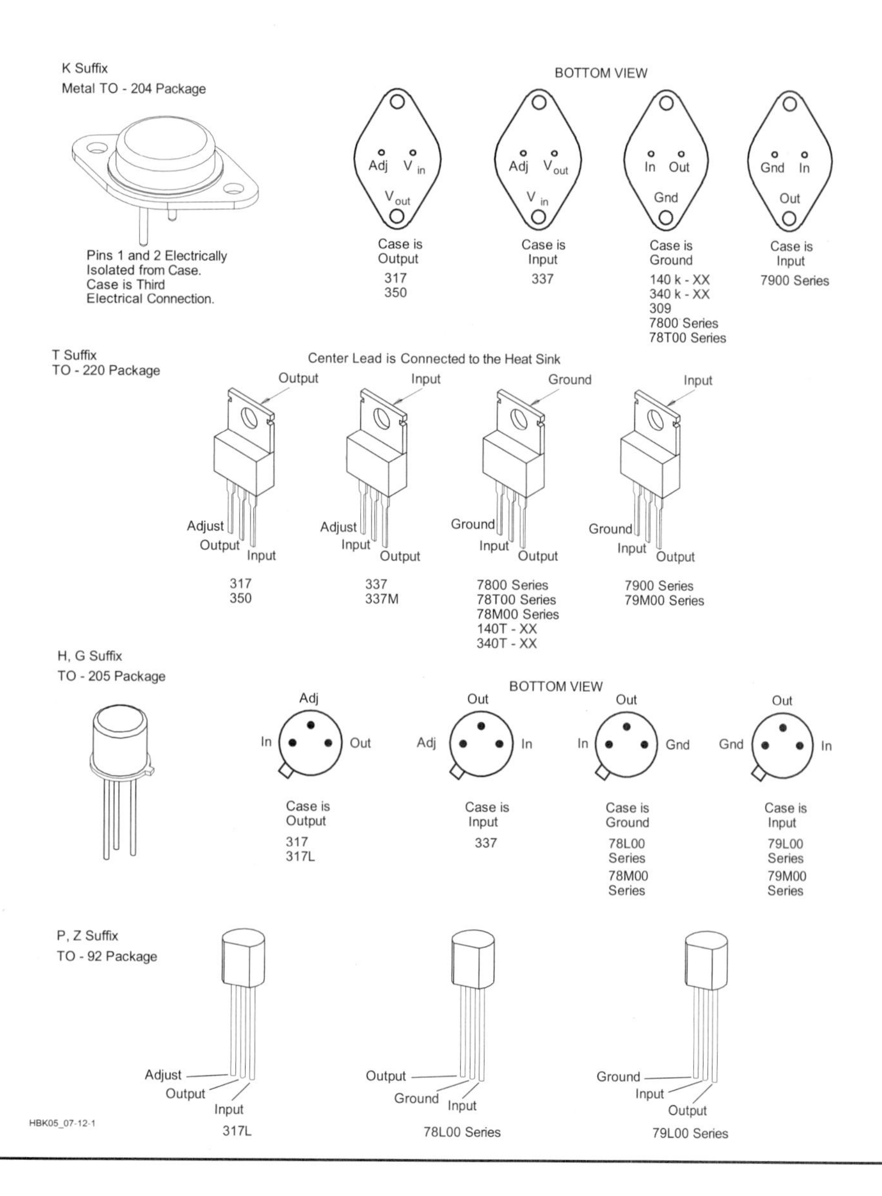

Table 7.36

Op Amp ICs

Listed by device number

Device	*Type*	*Freq Comp*	*Max Supply* (V)*	*Min Input Resistance (MΩ)*	*Max Offset Voltage (mV)*	*Min dc Open-Loop Gain (dB)*	*Min Output Current (mA)*	*Min Small-Signal Bandwidth (MHz)*	*Min Slew Rate (V/μs)*	*Notes*
101A	Bipolar	ext	44	1.5	3.0	79	15	1.0	0.5	General purpose
108	Bipolar	ext	40	30	2.0	100	5	1.0		
124	Bipolar	int	32		5.0	100	5	1.0		Quad op amp, low power
148	Bipolar	int	44	0.8	5.0	90	10	1.0	0.5	Quad 741
158	Bipolar	int	32		5.0	100	5	1.0		Dual op amp, low power
301	Bipolar	ext	36	0.5	7.5	88	5	1.0	10	Bandwidth extendable with external components
324	Bipolar	int	32		7.0	100	10	1.0		Quad op amp, single supply
347	BiFET	ext	36	106	5.0	100	30	4	13	Quad, high speed
351	BiFET	ext	36	106	5.0	100	20	4	13	
353	BiFET	ext	36	106	5.0	100	15	4	13	
355	BiFET	ext	44	106	10.0	100	25	2.5	5	
355B	BiFET	ext	44	106	5.0	100	25	2.5	5	
356A	BiFET	ext	36	106	2.0	100	25	4.5	12	
356B	BiFET	ext	44	106	5.0	100	25	5.0	12	
357	BiFET	ext	36	106	10.0	100	25	20.0	50	
357B	BiFET	ext	36	106	5.0	100	25	20.0	30	
358	Bipolar	int	32		7.0	100	10	1.0		Dual op amp, single supply
411	BiFET	ext	36	106	2.0	100	20	4.0	15	Low offset, low drift
709	Bipolar	ext	36	0.05	7.5	84	5	0.3	0.15	
741	Bipolar	int	36	0.3	6.0	88	5	0.4	0.2	
741S	Bipolar	int	36	0.3	6.0	86	5	1.0	3	Improved 741 for AF
1436	Bipolar	int	68	10	5.0	100	17	1.0	2.0	High-voltage
1437	Bipolar	ext	36	0.050	7.5	90		1.0	0.25	Matched, dual 1709
1439	Bipolar	ext	36	0.100	7.5	100		1.0	34	
1456	Bipolar	int	44	3.0	10.0	100	9.0	1.0	2.5	Dual 1741
1458	Bipolar	int	36	0.3	6.0	100	20.0	0.5	3.0	
1458S	Bipolar	int	36	0.3	6.0	86	5.0	0.5	3.0	Improved 1458 for AF
1709	Bipolar	ext	36	0.040	6.0	80	10.0	1.0		
1741	Bipolar	int	36	0.3	5.0	100	20.0	1.0	0.5	
1747	Bipolar	int	44	0.3	5.0	100	25.0	1.0	0.5	Dual 1741
1748	Bipolar	ext	44	0.3	6.0	100	25.0	1.0	0.8	Non-comp-ensated 1741
1776	Bipolar	int	36	50	5.0	110	5.0		0.35	Micro power, programmable
3140	BiFET	int	36	1.5 × 106	2.0	86	1	3.7	9	Strobable output
3403	Bipolar	int	36	0.3	10.0	80		1.0	0.6	Quad, low power
3405	Bipolar	ext	36		10.0	86	10	1.0	0.6	Dual op amp and dual comparator
3458	Bipolar	int	36	0.3	10.0	86	10	1.0	0.6	Dual, low power

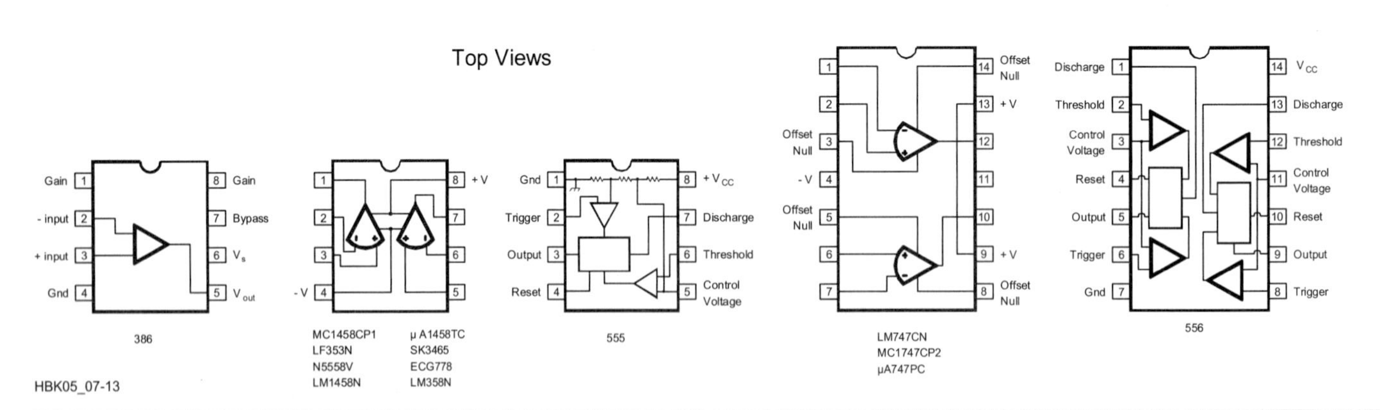

Device	Type	Freq Comp	Max Supply* (V)	Min Input Resistance (MΩ)	Max Offset Voltage (mV)	Min dc Open-Loop Gain (dB)	Min Output Current (mA)	Min Small-Signal Bandwidth (MHz)	Min Slew Rate (V/µs)	Notes
3476	Bipolar	int	36	5.0	6.0	92	12		0.8	
3900	Bipolar	int	32	1.0		65	0.5	4.0	0.5	Quad, Norton single supply
4558	Bipolar	int	44	0.3	5.0	88	10	2.5	1.0	Dual, wideband
4741	Bipolar	int	44	0.3	5.0	94	20	1.0	0.5	Quad 1741
5534	Bipolar	int	44	0.030	5.0	100	38	10.0	13	Low noise, can swing 20V P-P across 600
5556	Bipolar	int	36	1.0	12.0	88	5.0	0.5	1	Equivalent to 1456
5558	Bipolar	int	36	0.15	10.0	84	4.0	0.5	0.3	Dual, equivalent to 1458
34001	BiFET	int	44	106	2.0	94		4.0	13	JFET input
AD745	BiFET	int	±18	104	0.5	63	20	20	12.5	Ultra-low noise, high speed

LT1001 Precision op amp, low offset voltage (15 µV max), low drift (0.6 µV/°C max), low noise (0.3 µV p-p)
LT1007 Extremely low noise (0.06 µV p-p), very high gain (20 x 10^6 into 2 kΩ load)
LT1360 High speed, very high slew rate (800 V/µs), 50 MHz gain bandwidth, ±2.5 V to ±15 V supply range

Device	Type	Freq Comp	Max Supply* (V)	Min Input Resistance (MΩ)	Max Offset Voltage (mV)	Min dc Open-Loop Gain (dB)	Min Output Current (mA)	Min Small-Signal Bandwidth (MHz)	Min Slew Rate (V/µs)	Notes
NE5514	Bipolar	int	±16	100	1		10	3	0.6	
NE5532	Bipolar	int	±20	0.03	4	47	10	10	9	Low noise
OP-27A	Bipolar	ext	44	1.5	0.025	115		5.0	1.7	Ultra-low noise, high speed
OP-37A	Bipolar	ext	44	1.5	0.025	115		45.0	11.0	
TL-071	BiFET	int	36	10^6	6.0	91		4.0	13.0	Low noise
TL-081	BiFET	int	36	10^6	6.0	88		4.0	8.0	
TL-082	BiFET	int	36	10^6	15.0	99		4.0	8.0	Low noise
TL-084	BiFET	int	36	10^6	15.0	88		4.0	8.0	Quad, high-performance AF
TLC27M2	CMOS	int	18	10^6	10	44		0.6	0.6	Low noise
TLC27M4	CMOS	int	18	10^6	10	44		0.6	0.6	Low noise

*From –V to +V terminals

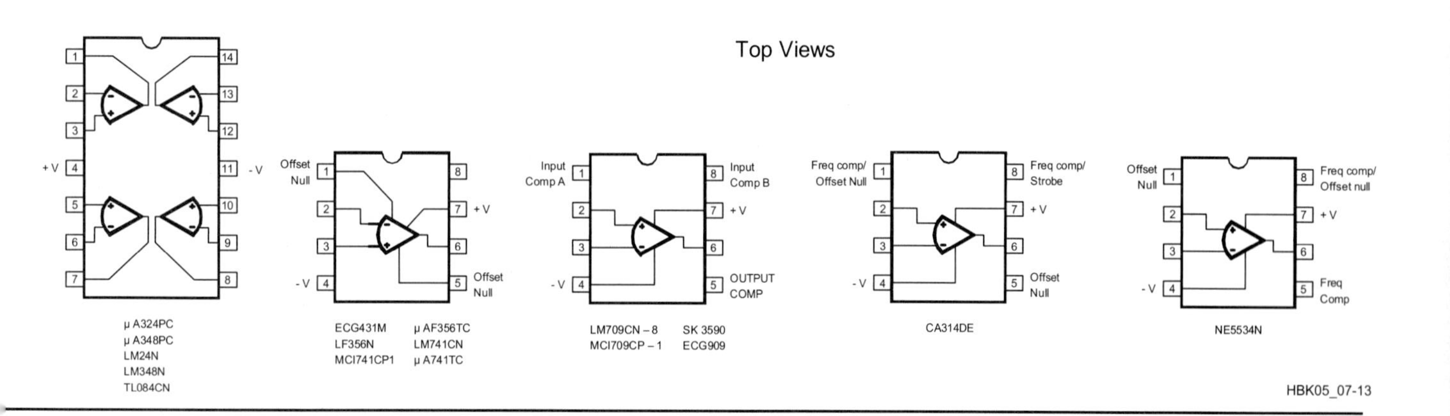

Table 7.37

Triode Transmitting Tubes

The full 1988 *Handbook* table of power tube specifications and base diagrams can be viewed in pdf format on the *ARRLWeb* at www.arrl.org/notes/1921/pwrtubes.pdf.

Type	*Power Diss.(W)*	*Plate (V)*	*Plate (mA)*	*Grid dc (mA)*	*Freq (MHz)*	*Ampl Factor*	*Fil (V)*	*Fil (A)*	C_{IN} *(pF)*	C_{GP} *(pF)*	C_{OUT} *(pF)*	*Base Diagram*	*Service Class*[1]	*Plate (V)*	*Grid (V)*	*Plate (mA)*	*Grid dc (mA)*	*Input (W)*	*P-P (kΩ)*	*Output (W)*
5675	5	165	30	8	3000	20	6.3	0.135	2.3	1.3	0.09	Fig 21	GG0	120	–8	25	4	¾	—	0.05
2C40	6.5	500	25	—	500	36	6.3	0.75	2.1	1.3	0.05	Fig 11	CT0	250	–5	20	0.3	¾	—	0.075
5893	8.0	400	40	13	1000	27	6.0	0.33	2.5	1.75	0.07	Fig 21	CT	350	–33	35	13	2.4	—	6.5
													CP	300	–45	30	12	2.0	—	6.5
2C43	12	500	40	—	1250	48	6.3	0.9	2.9	1.7	0.05	Fig 11	CTO	470	—	38[7]	—	¾	—	9[2]
811-A	65	1000	175	50	60	160	6.3	4.0	5.9	5.6	0.7	3G	CT	1500	–70	173	40	7.1	—	200
													CP	1250	–120	140	45	10.0	—	135
													B/CG	1250	0	21/175	28	12	—	165
													AB_1	1250	0	27/175	13	3.0	—	155
812-A	65	1500	175	35	60	29	6.3	4.0	5.4	5.5	0.77	3G	CT	1500	–120	173	30	6.5	—	190
													CP	1250	–115	140	35	7.6	—	130
													B[2]	1500	–48	28/310	270[4]	5.0	13.2	340
3CX100A5[6]	100	1000	125[5]	50	2500	100	6.0	1.05	7.0	2.15	0.035	—	AGG	800	–20	80	30	6	—	27
	70	600	100[5]										CP	600	–15	75	40	6	—	18
2C39	100	1000	60	40	500	100	6.3	1.1	6.5	1.95	0.03	—	G1C	600	–35	60	40	5.0	—	20
													CTO	900	–40	90	30	¾	—	40
													CP	600	–150	100[5]	50	¾	—	¾
AX9900, 5866	135	2500	200	40	150	25	6.3	5.4	5.8	5.5	0.1	Fig 3	CT	2500	–200	200	40	16	—	390
													CP	2000	–225	127	40	16	—	204
													B[2]	2500	–90	80/330	350[4]	14[3]	15.68	560
572B	160	2750	275	—	—	170	6.3	4.0	—	—	—	3G	CT	1650	–70	165	32	6	—	205
T160L													B/GG[2]	2400	–2.0	90/500	—	100	—	600
8873	200	2200	250	—	500	160	6.3	3.2	19.5	7.0	0.03	Fig 87	AB_2	2000	—	22/500	98[3]	27[3]	—	505
8875	300	2200	250	—	500	160	6.3	3.2	19.5	7.0	0.03	—	AB_2	2000	—	22/500	98[3]	27[3]	—	505
833A	350	3300	500	100	30	35	10	10	12.3	6.3	8.5	Fig 41	CTO	2250	–125	445	85	23	—	780
													CTO	3000	–160	335	70	20	—	800
													CP	2500	–300	335	75	30	—	635
	450[6]	4000[6]	500	100	20[6]	35	10	10	12.3	6.3	8.5	Fig 41	CP	3000	–240	335	70	26	—	800
													B[2]	3000	–70	100/750	400[4]	20[4]	9.5	1650
8874	400	2200	350	—	500	160	6.3	3.2	19.5	7.0	0.03	—	AB_2	2000	—	22/500	98[3]	27[3]	—	505
3-400Z	400	3000	400	—	110	200	5	14.5	7.4	4.1	0.07	Fig 3	B/GG	3000	0	100/333	120	32	—	655
3-500Z	500	4000	400	—	110	160	5	14.5	7.4	4.1	0.07	Fig 3	B/GG	3000	—	370	115	30	5	750
3-600Z	600	4000	425	—	110	165	5	15.0	7.8	4.6	0.08	Fig 3	B/GG	3000	—	400	118	33	—	810
													B/GG	3500	—	400	110	35	—	950
3CX800A7	800	2250	600	60	350	200	13.5	1.5	26	—	6.1	Fig 87	AB_2GG[7]	2200	–8.2	500	36	16	—	750
3-1000Z	1000	3000	800	—	110	200	7.5	21.3	17	6.9	0.12	Fig 3	B/GG	3000	0	180/670	300	65	—	1360
3CX1200A7	1200	5000	800	—	110	200	7.5	21.0	20	12	0.2	Fig 3	AB_2GG	3600	–10	700	230	85	—	1500
8877	1500	4000	1000	—	250	200	5.0	10	42	10	0.1	—	AB_2	2500	–8.2	1000	—	57	—	1520

Table 7.38

Tetrode and Pentode Transmitting Tubes

www.arrl.org/notes/1921/pwrtubes.pdf.

Type	Max. Plate Diss. (W)	Max. Plate Volts (V)	Max. Screen Diss. (W)	Max. Screen Volts (V)	Max. Freq. (MHz)	Filament Volts (V)	Amps (A)	C_{IN} (pF)	C_{GP} (pF)	C_{OUT} (pF)	Base	Serv. Class[1]	Plate (V)	Screen (V)	Grid (V)	Plate (mA)	Screen (mA)	Grid (mA)	P_{IN} (W)	P-P (kΩ)	P_{OUT} (W)
6146/	25	750	3	250	60	6.3	1.25	13	0.24	8.5	7CK	CT	500	170	–66	135	9	2.5	0.2	—	48
6146A												CT	700	160	–62	120	11	3.1	0.2	—	70
8032	25	750	3	250	60	12.6	0.585	13	0.24	8.5	7CK	CT[6]	400	190	–54	150	10.4	2.2	3.0	—	35
6883												CP	400	150	–87	112	7.8	3.4	0.4	—	32
												CP	600	150	–87	112	7.8	3.4	0.4	—	52
6159B/	25	750	3	250	60	26.5	0.3	13	0.24	8.5	7CK	AB_2[8]	600	190	–48	28/270	1.2/20	22	0.3	5	113
												AB_2[8]	750	165	–46	22/240	0.3/20	2.6[2]	0.4	7.4	131
												AB_1[8]	750	195	–50	23/220	1/26	100[3]	0	8	120
807, 807W	30	750	3.5	300	60	6.3	0.9	12	0.2	7	5AW	CT	750	250	–45	100	6	3.5	0.22	—	50
5933												CP	600	275	–90	100	6.5	4	0.4	—	42.5
												AB_1	750	300	–35	15/70	3/8	75[3]	0	—	72
1625	30	750	3.5	300	60	12.6	0.45	12	0.2	7	5AZ	B[5]	750	—	0	15/240	—	555[3]	5.3[2]	6.65	120
6146B/	35	750	3	250	60	6.3	1.125	13	0.22	8.5	7CK	CT	750	200	–77	160	10	2.7	0.3	—	85
8298A												CP	600	175	–92	140	9.5	3.4	0.5	—	62
												AB_1	750	200	–48	24/125	6.3	—	—	3.5	61
813	125	2500	20	800	30	10.0	5.0	16.3	0.25	14.0	5BA	CTO	1250	300	–75	180	35	12	1.7	—	170
												CTO	2250	400	–155	220	40	15	4	—	375
												AB1	2500	750	–95	25/145	27[2]	0	0	—	245
												AB_2[8]	2000	750	–90	40/315	1.5/58	230[3]	0.1[2]	16	455
												AB_2[8]	2500	750	–95	35/260	1.2/55	235[3]	0.35[2]	17	650
4CX250B	250	2000	12	400	175	6.0	2.9	18.5	0.04	4.7	—	CTO	2000	250	–90	250	25	27	2.8	—	410
												CP	1500	250	–100	200	25	17	2.1	—	250
												AB_1[8]	2000	350	–50	500	30	100	0	8.26	650
4-400A	400[4]	4000	35	600	110	5.0	14.5	12.5	0.12	4.7	5BK	CT/CP	4000	300	–170	270	22.5	10	10	—	720
												GG	2500	0	0	80/270[9]	55[9]	100[9]	39[9]	4.0	435
												AB_1	2500	750	–130	95/317	0/14	0	0	—	425
4CX400A	400	2500	8	400	500	6.3	3.2	24	0.08	7	See[11]	AB_2GD	2200	325	–30	100/270	22	2	9	—	405
												AB_2GD	2500	400	–35	100/400	18	1	13	—	610
4CX800A	800	2500	15	350	150	12.6	3.6	51	0.9	11	See[12]	AB_2GD	2200	350	–56	160/550	24	1	32	—	750
4-1000A	1000	6000	75	1000	—	7.5	21	27.2	0.24	7.6	—	CT	3000	500	–150	700	146	38	11	—	1430
8166												CP	3000	500	–200	600	145	36	12	—	1390
												AB_2	4000	500	–60	300/1200	0/95	—	11	7	3000
												GG	3000	0	0	100/700[9]	105[9]	170[9]	130[9]	2.5	1475
4CX1000A	1000	3000	12	400	110	6.0	9.0	81.5	0.01	11.8	—	AB_1[8]	2000	325	–55	500/2000	–4/60	—	—	2.5	2160
												AB_1[8]	2500	325	–55	500/2000	–4/60	—	—	3.1	2920
												AB_1[8]	3000	325	–55	500/1800	–4/60	—	—	3.85	3360
4CX1500B	1500	3000	12	400	110	6.0	10.0	81.5	0.02	11.8	—	AB_1	2750	225	–34	300/755	–14/60	0.95	1.5	1.9	1100
4CX1600B	1600	3300	20	350	250	12.6	4.4	86	0.15	12	See[13]	AB_2GD	2400	350	–53	500/1100	20	2	28	—	1600
												AB_2GD	2400	350	–70	200/870	48	2	83[10]	—	1500
												AB_2GD	3200	240	–57	200/740	21	1	33	—	1600

[1]Service Class Abbreviations:
AB_2GD=AB_2 linear with 50-Ω passive grid circuit.
B=Class-B push-pull
CP=Class-C plate-modulated phone
CT=Class-C telegraph
GG=Grounded-grid (grid and screen connected together)
[2]Maximum signal value
[3]Peak grid-grid volts
[4]Forced-air cooling required.
[5]Two tubes triode-connected, G2 to G1 through 20kΩ to G2.
[6]Typical operation at 175 MHz.
[7]±1.5 V.
[8]Values are for two tubes.
[9]Single tone.
[10]24-Ω cathode resistance.
[11]Base same as 4CX250B. Socket is Russian SK2A.
[12]Socket is Russian SK1A.
[13]Socket is Russian SK3A.

Table 7.39
TV Deflection Tubes

Type	Plate Diss. (W)	Screen Diss. (W)	Transcond. µMho	Heater 6.3 V (A)	C_{IN} (pF)	C_{GP} (pF)	C_{OUT} (pF)	Base	Class of Service	Plate Volt. (V)	Screen Volt. (V)	Grid Volt. (V)	Plate Curr. (A)	Screen Curr. (A)	Grid Curr. (A)	Drive Power (W)	Output Power (W)
6DQ5	24	3.2	10.5k	2.5	23	0.5	11	8JC	C	400	200	–40	100	12	1.5	0.1	25
6DQ6B	18	3.6	7.3k	1.2	15	0.5	7	6AM	C	400	200	–40	100	12	1.5	0.1	25
6FH6	17	3.6	6k	1.2	33	0.4	8	6AM	C	400	200	–40	100	12	1.5	0.1	25
6GC6	17.5	4.5	6.6k	1.2	15	0.55	7	8JX	C	400	200	–40	100	12	1.5	0.1	25
6GJ5	17.5	3.5	7.1k	1.2	15	0.26	6.5	9NM	C AB_1	500 500	200 200	–75 –43	180 85	15 4	5 —	0.43 —	63 35
6HF5	28	5.5	11.3k	2.25	24	0.56	10	12FB	C AB_1	500 500	140 140	–85 –46	232 133	12.5 4.5	77 —	0.76 —	8 58
6JB6	17.5	3.5	7.1k	1.2	15	0.2	6	9QL	C AB_1	500 500	200 200	–75 –42	180 85	13.3 4.2	5 —	0.43 —	63 35
6JE6	30	5	10.5k	2.5	24.3	—	14.5	9QL	C AB_1	450 450	150 150	–80 –35	202 98	20 4.5	8 —	0.75 —	63 38
6JM6	17.5	3.5	7.3k	1.2	16	0.6	7	12FJ	C AB_1	500 500	200 200	–75 –42	190 85	13.7 4.4	4 —	0.32 —	61 37
6JN6	17.5	3.5	7.3k	1.2	16	0.34	7	12FK	GC	800	0	–11	150	—	—	12.5	82
6JS6C	30	5.5	—	2.25	24	0.7	10	12FY	GC	800	0	–11	150	—	—	12.5	82
6KD6	33	5	14k	2.85	40	0.8	16	12GW	GC	800	0	–11	150	—	—	12.5	82
6LB6	30	5	13.4k	2.25	33	0.4	18	12GJ	GC	800	0	–11	150	—	—	12.5	82
6LG6	28	5	11.5k	2	25	0.8	13	12HL	GC	800	0	–11	150	—	—	12.5	82
6LQ6	30	5	9.6k	2.5	22	0.46	11	9QL	GC	800	0	–11	150	—	—	12.5	82
6MH6	38.5	7	14k	2.65	40	1.0	20	12GW	GC	800	0	–11	150	—	—	12.5	82

Table 7.40
EIA Vacuum-Tube Base Diagrams

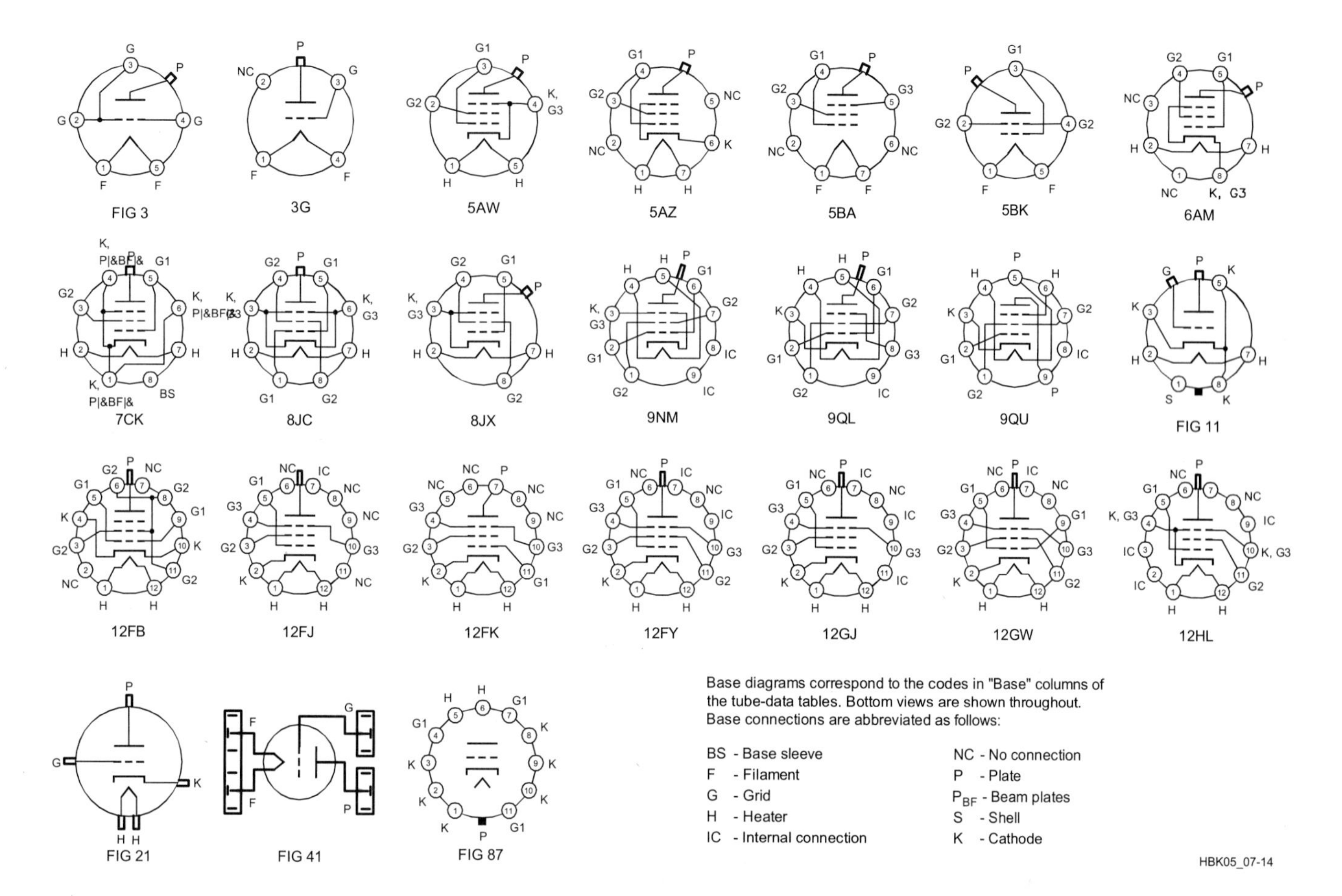

Alphabetical subscripts (D = diode, P = pentode, T = triode and HX = hexode) indicate structures in multistructure tubes. Subscript CT indicates filament or heater center tap.

Generally, when pin 1 of a metal-envelope tube (except all triodes) is shown connected to the envelope, pin 1 of a glass-envelope counterpart (suffix G or GT) is connected to an internal shield.

Table 7.41

Properties of Common Thermoplastics

Polyvinyl Chloride (PVC)

Advantages:
- Can be compounded with plasticizers, filters, stabilizers, lubricants and impact modifiers to produce a wide range of physical properties
- Can be pigmented to almost any color
- Rigid PVC has good corrosion and stain resistance, thermal & electrical insulation, and weatherability

Disadvantages:
- Base resin can be attacked by aromatic solvents, ketones, aldehydes, naphthalenes, and some chloride, acetate, and acrylate esters
- Should not be used above 140°

Applications:
- Conduit
- Conduit boxes
- Housings
- Pipe
- Wire and cable insulation

Polystyrene

Advantages:
- Low cost
- Moderate strength
- Electrical properties only slightly affected by temperature and humidity
- Sparkling clarity
- Impact strength is increased by blending with rubbers, such as polybutadiene

Disadvantages:
- Brittle
- Low heat resistance

Applications:
- Capacitors
- Light shields
- Knobs

Polyphenylene Sulfide (PPS)

Advantages:
- Excellent dimensional stability
- Strong
- High-temperature stability
- Chemical resistant
- Inherently completely flame retardant
- Completely transparent to microwave radiation

Applications:
- R3-R5 have various glass-fiber levels that are suitable for applications demanding high mechanical and impact strength as well as good dielectric properties
- R8 and R10 are suitable for high arc-resistance applications
- R9-901 is suitable for encapsulation of electronic devices

Polypropylene

Advantages:
- Low density
- Good balance of thermal, chemical, and electrical properties
- Moderate strength (increases significantly with glass-fiber reinforcement)

Disadvantages:
- Electrical properties affected to varying degrees by temperature (as temperature goes up, dielectric strength increases and volume resistivity decreases)
- Inherently unstable in presence of oxidative and UV radiation

Applications:
- Automotive battery cases
- Blower housings
- Fan blades
- Insulators
- Lamp housings
- Support for current-carrying electrical components
- TV yokes

Polyethylene (PE)

Advantages: Low Density PE
- Good toughness
- Excellent chemical resistance
- Excellent coefficient of friction
- Near zero moisture absorption
- Easy to process
- Relatively low heat resistance

Disadvantages:
- Susceptible to environmental and some chemical stress cracking
- Wetting agents (such as detergents) accelerate stress cracking

Advantages: High Density PE
- Same as above, plus increased rigidity and tensile strength

Advantages: Ultra-High Molecular Weight PE
- Outstanding abrasion resistance
- Low coefficient of friction
- High impact strength
- Excellent chemical resistance
- Material does not break in impact strength tests using standard notched specimens

Applications:
- Bearings
- Components requiring maximum abrasion resistance, impact strength, and low coefficient of friction

Phenolic

Advantages:
- Low cost
- Superior heat resistance
- High heat-deflection temperatures
- Good electrical properties
- Good flame resistance
- Excellent moldability
- Excellent dimensional stability
- Good water and chemical resistance

Applications:
- Commutators and housings for small motors
- Heavy duty electrical components
- Rotary-switch wafers
- Insulating spacers

Nylon

Advantages:
- Excellent fatigue resistance
- Low coefficient of friction
- Toughness as a function of degree of crystalinity
- Resists many fuels and chemicals
- Good creep- and cold-flow resistance as compared to less rigid thermoplastics
- Resists repeated impacts

Disadvantages:
- All nylons absorb moisture
- Nylons that have not been compounded with a UV stabilizer are sensitive to UV light, and thus not suitable for extended outdoor use

Applications:
- Bearings
- Housings and tubing
- Rope
- Wire coatings
- Wire connectors
- Wear plates

Table 7.42
Coaxial Cable End Connectors

UHF Connectors

Military No.	*Style*	*Cable RG- or Description*
PL-259	Str (m)	8, 9, 11, 13, 63, 87, 149, 213, 214, 216, 225
UG-111	Str (m)	59, 62, 71, 140, 210
SO-239	Pnl (f)	Std, mica/phenolic insulation
UG-266	Blkhd (f)	Rear mount, pressurized, copolymer of styrene ins.

Adapters

PL-258	Str (f/f)	Polystyrene ins.
UG-224,363	Blkhd (f/f)	Polystyrene ins.
UG-646	Ang (f/m)	Polystyrene ins.
M-359A	Ang (m/f)	Polystyrene ins.
M-358	T (f/m/f)	Polystyrene ins.

Reducers

UG-175		55, 58, 141, 142 (except 55A)
UG-176		59, 62, 71, 140, 210

Family Characteristics:

All are nonweatherproof and have a nonconstant impedance. Frequency range: 0-500 MHz. Maximum voltage rating: 500 V (peak).

N Connectors

Military No.	*Style*	*Cable RG-*	*Notes*
UG-21	Str (m)	8, 9, 213, 214	50 Ω
UG-94A	Str (m)	11, 13, 149, 216	70 Ω
UG-536	Str (m)	58, 141, 142	50 Ω
UG-603	Str (m)	59, 62, 71, 140, 210	50 Ω
UG-23, B-E	Str (f)	8, 9, 87, 213, 214, 225	50 Ω
UG-602	Str (f)	59, 62, 71, 140, 210	—
UG-228B, D, E	Pnl (f)	8, 9, 87, 213, 214, 225	—
UG-1052	Pnl (f)	58, 141, 142	50 Ω
UG-593	Pnl (f)	59, 62, 71, 140, 210	50 Ω
UG-160A, B, D	Blkhd (f)	8, 9, 87, 213, 214, 225	50 Ω
UG-556	Blkhd (f)	58, 141, 142	50 Ω
UG-58, A	Pnl (f)		50 Ω
UG-997A	Ang (f)		50 Ω 11/16″

Panel mount (f) with clearance above panel

M39012/04-	Blkhd (f)	Front mount hermetically sealed
UG-680	Blkhd (f)	Front mount pressurized

N Adapters

Military No.	*Style*	*Notes*
UG-29,A,B	Str (f/f)	50 Ω, TFE ins.
UG-57A.B	Str (m/m)	50 Ω, TFE ins.
UG-27A,B	Ang (f/m)	Mitre body
UG-212A	Ang (f/m)	Mitre body
UG-107A	T (f/m/f)	—
UG-28A	T (f/f/f)	—
UG-107B	T (f/m/f)	—

Family Characteristics:

N connectors with gaskets are weatherproof. RF leakage: –90 dB min @ 3 GHz. Temperature limits: TFE: –67° to 390°F (–55° to 199°C). Insertion loss 0.15 dB max @ 10 GHz. Copolymer of styrene: –67° to 185°F (–55° to 85°C). Frequency range: 0-11 GHz. Maximum voltage rating: 1500 V P-P. Dielectric withstanding voltage 2500 V RMS. SWR (MIL-C-39012 cable connectors) 1.3 max 0-11 GHz.

BNC Connectors

Military No.	*Style*	*Cable RG-*	*Notes*
UG-88C	Str (m)	55, 58, 141, 142, 223, 400	

Military No.	*Style*	*Cable RG-*	*Notes*
UG-959	Str (m)	8, 9	
UG-260,A	Str (m)	59, 62, 71, 140, 210	Rexolite ins.
UG-262	Pnl (f)	59, 62, 71, 140, 210	Rexolite ins.
UG-262A	Pnl (f)	59, 62, 71, 140, 210	nwx, Rexolite ins.
UG-291	Pnl (f)	55, 58, 141, 142, 223, 400	
UG-291A	Pnl (f)	55, 58, 141, 142, 223, 400	nwx
UG-624	Blkhd (f)	59, 62, 71, 140, 210	Front mount Rexolite ins.
UG-1094A	Blkhd		Standard
UG-625B	Receptacle		
UG-625			

BNC Adapters

Military No.	*Style*	*Notes*
UG-491,A	Str (m/m)	
UG-491B	Str (m/m)	Berylium, outer contact
UG-914	Str (f/f)	
UG-306	Ang (f/m)	
UG-306A,B	Ang (f/m)	Berylium outer contact
UG-414,A	Pnl (f/f)	# 3-56 tapped flange holes
UG-306	Ang (f/m)	
UG-306A,B	Ang (f/m)	Berylium outer contact
UG-274	T (f/m/f)	
UG-274A,B	T (f/m/f)	Berylium outer contact

Family Characteristics:

Z = 50 Ω. Frequency range: 0-4 GHz w/low reflection; usable to 11 GHz. Voltage rating: 500 V P-P. Dielectric withstanding voltage 500 V RMS. SWR: 1.3 max 0-4 GHz. RF leakage –55 dB min @ 3 GHz. Insertion loss: 0.2 dB max @ 3 GHz. Temperature limits: TFE: –67° to 390°F (–55° to 199°C); Rexolite insulators: –67° to 185°F (–55° to 85°C). "Nwx" = not weatherproof.

HN Connectors

Military No.	*Style*	*Cable RG-*	*Notes*
UG-59A	Str (m)	8, 9, 213, 214	
UG-1214	Str (f)	8, 9, 87, 213, 214, 225	Captivated contact
UG-60A	Str (f)	8, 9, 213, 214	Copolymer of styrene ins.
UG-1215	Pnl (f)	8, 9, 87, 213, 214, 225	Captivated contact
UG-560	Pnl (f)		
UG-496	Pnl (f)		
UG-212C	Ang (f/m)		Berylium outer contact

Family Characteristics:

Connector Styles: Str = straight; Pnl = panel; Ang = Angle; Blkhd = bulkhead. Z = 50 Ω. Frequency range = 0-4 GHz. Maximum voltage rating = 1500 V P-P. Dielectric withstanding voltage = 5000 V RMS SWR = 1.3. All HN series are weatherproof. Temperature limits: TFE: –67° to 390°F (–55° to 199°C); copolymer of styrene: –67° to 185°F (–55° to 85°C).

Cross-Family Adapters

Families	*Description*	*Military No.*
HN to BNC	HN-m/BNC-f	UG-309
N to BNC	N-m/BNC-f	UG-201,A
	N-f/BNC-m	UG-349,A
	N-m/BNC-m	UG-1034
N to UHF	N-m/UHF-f	UG-146
	N-f/UHF-m	UG-83,B
	N-m/UHF-m	UG-318
UHF to BNC	UHF-m/BNC-f	UG-273
	UHF-f/BNC-m	UG-255

Table 7.43

International System of Units (SI)—Metric Units

Prefix	*Symbol*	*Multiplication Factor*		
exe	E	10^{18}	=	1,000,000 000,000,000,000
peta	P	10^{15}	=	1,000 000,000,000,000
tera	T	10^{12}	=	1,000,000,000,000
giga	G	10^{9}	=	1,000,000,000
mega	M	10^{6}	=	1,000,000
kilo	k	10^{3}	=	1,000
hecto	h	10^{2}	=	100
deca	da	10^{1}	=	10
(unit)		10^{0}	=	1
deci	d	10^{-1}	=	0.1
centi	c	10^{-2}	=	0.01
milli	m	10^{-3}	=	0.001
micro	μ	10^{-6}	=	0.000001
nano	n	10^{-9}	=	0.000000001
pico	p	10^{-12}	=	0.000000000001
femto	f	10^{-15}	=	0.000000000000001
atto	a	10^{-18}	=	0.000000000000000001

Linear

1 meter (m) = 100 centimeters (cm) = 1000 millimeters (mm)

Area

1 $m^2 = 1 \times 10^4\ cm^2 = 1 \times 10^6\ mm^2$

Volume

1 $m^3 = 1 \times 10^6\ cm^3 = 1 \times 10^9\ mm^3$

1 liter (l) = 1000 $cm^3 = 1 \times 10^6\ mm^3$

Mass

1 kilogram (kg) = 1000 grams (g)
(Approximately the mass of 1 liter of water)
1 metric ton (or tonne) = 1000 kg

Table 7.44
US Customary Units

Linear Units

12 inches (in) = 1 foot (ft)
36 inches = 3 feet = 1 yard (yd)
1 rod = 5½ yards = 16½ feet
1 statute mile = 1760 yards = 5280 feet
1 nautical mile = 6076.11549 feet

Area

1 ft^2 = 144 in^2
1 yd^2 = 9 ft^2 = 1296 in^2
1 rod^2 = 30¼ yd^2
1 acre = 4840 yd^2 = 43,560 ft^2
1 acre = 160 rod^2
1 $mile^2$ = 640 acres

Volume

1 ft^3 = 1728 in^3
1 yd^3 = 27 ft^3

Liquid Volume Measure

1 fluid ounce (fl oz) = 8 fluid drams = 1.804 in
1 pint (pt) = 16 fl oz
1 quart (qt) = 2 pt = 32 fl oz = 57¾ in^3
1 gallon (gal) = 4 qt = 231 in^3
1 barrel = 31½ gal

Dry Volume Measure

1 quart (qt) = 2 pints (pt) = 67.2 in^3
1 peck = 8 qt
1 bushel = 4 pecks = 2150.42 in^3

Avoirdupois Weight

1 dram (dr) = 27.343 grains (gr) or (gr a)
1 ounce (oz) = 437.5 gr
1 pound (lb) = 16 oz = 7000 gr
1 short ton = 2000 lb, 1 long ton = 2240 lb

Troy Weight

1 grain troy (gr t) = 1 grain avoirdupois
1 pennyweight (dwt) or (pwt) = 24 gr t
1 ounce troy (oz t) = 480 grains
1 lb t = 12 oz t = 5760 grains

Apothecaries' Weight

1 grain apothecaries' (gr ap)
= 1 gr t = 1 gr
1 dram ap (dr ap) = 60 gr
1 oz ap = 1 oz t = 8 dr ap = 480 gr
1 lb ap = 1 lb t = 12 oz ap = 5760 gr

Conversion

Metric Unit = Metric Unit × US Unit

Metric Unit	Factor	US Unit
(Length)		
mm	25.4	inch
cm	2.54	inch
cm	30.48	foot
m	0.3048	foot
m	0.9144	yard
km	1.609	mile
km	1.852	nautical mile
(Area)		
mm^2	645.16	$inch^2$
cm^2	6.4516	in^2
cm^2	929.03	ft^2
m^2	0.0929	ft^2
cm^2	8361.3	yd^2
m^2	0.83613	yd^2
m^2	4047	acre
km^2	2.59	mi^2
(Mass)	**(Avoirdupois Weight)**	
grams	0.0648	grains
g	28.349	oz
g	453.59	lb
kg	0.45359	lb
tonne	0.907	short ton
tonne	1.016	long ton
(Volume)		
mm^3	16387.064	in^3
cm^3	16.387	in^3
m^3	0.028316	ft^3
m^3	0.764555	yd^3
ml	16.387	in^3
ml	29.57	fl oz
ml	473	pint
ml	946.333	quart
l	28.32	ft^3
l	0.9463	quart
l	3.785	gallon
l	1.101	dry quart
l	8.809	peck
l	35.238	bushel
(Mass)	**(Troy Weight)**	
g	31.103	oz t
g	373.248	lb t
(Mass)	**(Apothecaries' Weight)**	
g	3.387	dr ap
g	31.103	oz ap
g	373.248	lb ap

Multiply 

Metric Unit = Conversion Factor × US Customary Unit

← **Divide**

Metric Unit ÷ Conversion Factor = US Customary Unit

Table 7.45

Abbreviations List

A

a—atto (prefix for 10^{-18})
A—ampere (unit of electrical current)
ac—alternating current
ACC—Affiliated Club Coordinator
ACSSB—amplitude-compandored single sideband
A/D—analog-to-digital
ADC—analog-to-digital converter
AF—audio frequency
AFC—automatic frequency control
AFSK—audio frequency-shift keying
AGC—automatic gain control
Ah—ampere hour
ALC—automatic level control
AM—amplitude modulation
AMRAD—Amateur Radio Research and Development Corporation
AMSAT—Radio Amateur Satellite Corporation
AMTOR—Amateur Teleprinting Over Radio
ANT—antenna
ARA—Amateur Radio Association
ARC—Amateur Radio Club
ARES—Amateur Radio Emergency Service
ARQ—Automatic repeat request
ARRL—American Radio Relay League
ARS—Amateur Radio Society (station)
ASCII—American National Standard Code for Information Interchange
ATV—amateur television
AVC—automatic volume control
AWG—American wire gauge
az-el—azimuth-elevation

B

B—bel; blower; susceptance; flux density, (inductors)
balun—balanced to unbalanced (transformer)
BC—broadcast
BCD—binary coded decimal
BCI—broadcast interference
Bd—baud (bids in single-channel binary data transmission)
BER—bit error rate
BFO—beat-frequency oscillator
bit—binary digit
bit/s—bits per second
BM—Bulletin Manager
BPF—band-pass filter
BPL—Brass Pounders League
BPL—Broadband over Power Line
BT—battery
BW—bandwidth
Bytes—Bytes

C

c—centi (prefix for 10^{-2})
C—coulomb (quantity of electric charge); capacitor
CAC—Contest Advisory Committee
CATVI—cable television interference
CB—Citizens Band (radio)
CBBS—computer bulletin-board service
CBMS—computer-based message system
CCITT—International Telegraph and Telephone Consultative Committee
CCTV—closed-circuit television
CCW—coherent CW
ccw—counterclockwise
CD—civil defense
cm—centimeter
CMOS—complementary-symmetry metal-oxide semiconductor
coax—coaxial cable
COR—carrier-operated relay
CP—code proficiency (award)
CPU—central processing unit
CRT—cathode ray tube
CT—center tap
CTCSS—continuous tone-coded squelch system
cw—clockwise
CW—continuous wave

D

d—deci (prefix for 10^{-1})
D—diode
da—deca (prefix for 10)
D/A—digital-to-analog
DAC—digital-to-analog converter
dB—decibel (0.1 bel)
dBi—decibels above (or below) isotropic antenna
dBm—decibels above (or below) 1 milliwatt
DBM—double balanced mixer
dBV—decibels above/below 1 V (in video, relative to 1 V P-P)
dBW—decibels above/below 1 W
dc—direct current
D-C—direct conversion
DDS—direct digital synthesis
DEC—District Emergency Coordinator
deg—degree
DET—detector
DF—direction finding; direction finder
DIP—dual in-line package
DMM—digital multimeter
DPDT—double-pole double-throw (switch)
DPSK—differential phase-shift keying
DPST—double-pole single-throw (switch)
DS—direct sequence (spread spectrum); display
DSB—double sideband
DSP—digital signal processing
DTMF—dual-tone multifrequency
DVM—digital voltmeter
DX—long distance; duplex
DXAC—DX Advisory Committee
DXCC—DX Century Club

E

e—base of natural logarithms (2.71828)
E—voltage
EA—ARRL Educational Advisor
EC—Emergency Coordinator
ECL—emitter-coupled logic
EHF—extremely high frequency (30-300 GHz)
EIA—Electronic Industries Alliance
EIRP—effective isotropic radiated power
ELF—extremely low frequency
ELT—emergency locator transmitter
EMC—electromagnetic compatibility
EME—earth-moon-earth (moonbounce)
EMF—electromotive force
EMI—electromagnetic interference
EMP—electromagnetic pulse
EOC—emergency operations center
EPROM—erasable programmable read only memory

F

f—femto (prefix for 10^{-5}); frequency
F—farad (capacitance unit); fuse
fax—facsimile
FCC—Federal Communications Commission
FD—Field Day
FEMA—Federal Emergency Management Agency
FET—field-effect transistor
FFT—fast Fourier transform
FL—filter
FM—frequency modulation
FMTV—frequency-modulated television
FSK—frequency-shift keying
FSTV—fast-scan (real-time) television
ft—foot (unit of length)

G

g—gram (unit of mass)
G—giga (prefix for 10^{9}); conductance
GaAs—gallium arsenide
GB—gigabytes
GDO—grid- or gate-dip oscillator
GHz—gigahertz (10^{9} Hz)
GND—ground

H

h—hecto (prefix for 10^{2})
H—henry (unit of inductance)
HF—high frequency (3-30 MHz)
HFO—high-frequency oscillator; heterodyne frequency oscillator
HPF—highest probable frequency; high-pass filter
Hz—hertz (unit of frequency, 1 cycle/s)

I

I—current, indicating lamp
IARU—International Amateur Radio Union
IC—integrated circuit
ID—identification; inside diameter
IEEE—Institute of Electrical and Electronics Engineers
IF—intermediate frequency
IMD—intermodulation distortion

in.—inch (unit of length)
in./s—inch per second (unit of velocity)
I/O—input/output
IRC—international reply coupon
ISB—independent sideband
ITF—Interference Task Force
ITU—International Telecommunication Union
ITU-T—ITU Telecommunication Standardization Bureau

J-K

j—operator for complex notation, as for reactive component of an impedance (+*j* inductive; –*j* capacitive)
J—joule (kg m^2/s^2) (energy or work unit); jack
JFET—junction field-effect transistor
k—kilo (prefix for 10^3); Boltzmann's constant (1.38×10^{-23} J/K)
K—kelvin (used without degree symbol) absolute temperature scale; relay
kB—kilobytes
kBd—1000 bauds
kbit—1024 bits
kbit/s—1024 bits per second
kbyte—1024 bytes
kg—kilogram
kHz—kilohertz
km—kilometer
kV—kilovolt
kW—kilowatt
kΩ—kilohm

L

l—liter (liquid volume)
L—lambert; inductor
lb—pound (force unit)
LC—inductance-capacitance
LCD—liquid crystal display
LED—light-emitting diode
LF—low frequency (30-300 kHz)
LHC—left-hand circular (polarization)
LO—local oscillator; Leadership Official
LP—log periodic
LS—loudspeaker
lsb—least significant bit
LSB—lower sideband
LSI—large-scale integration
LUF—lowest usable frequency

M

m—meter (length); milli (prefix for 10^{-3})
M—mega (prefix for 10^6); meter (instrument)
mA—milliampere
mAh—milliampere hour
MB—megabytes
MCP—multimode communications processor
MDS—Multipoint Distribution Service; minimum discernible (or detectable) signal
MF—medium frequency (300-3000 kHz)
mH—millihenry
MHz—megahertz
mi—mile, statute (unit of length)
mi/h (MPH)—mile per hour
mi/s—mile per second
mic—microphone
min—minute (time)
MIX—mixer
mm—millimeter
MOD—modulator
modem—modulator/demodulator
MOS—metal-oxide semiconductor
MOSFET—metal-oxide semiconductor field-effect transistor
MS—meteor scatter
ms—millisecond
m/s—meters per second
msb—most-significant bit
MSI—medium-scale integration
MSK—minimum-shift keying
MSO—message storage operation
MUF—maximum usable frequency
mV—millivolt
mW—milliwatt
MΩ—megohm

N

n—nano (prefix for 10^{-9}); number of turns (inductors)
NBFM—narrow-band frequency modulation
NC—no connection; normally closed
NCS—net-control station; National Communications System
nF—nanofarad
NF—noise figure
nH—nanohenry
NiCd—nickel cadmium
NM—Net Manager
NMOS—N-channel metal-oxide silicon
NO—normally open
NPN—negative-positive-negative (transistor)
NPRM—Notice of Proposed Rule Making (FCC)
ns—nanosecond
NTIA—National Telecommunications and Information Administration
NTS—National Traffic System

O

OBS—Official Bulletin Station
OD—outside diameter
OES—Official Emergency Station
OO—Official Observer
op amp—operational amplifier
ORS—Official Relay Station
OSC—oscillator
OSCAR—Orbiting Satellite Carrying Amateur Radio
OTC—Old Timer's Club
oz—ounce ($^1/_{16}$ pound)

P

p—pico (prefix for 10^{-12})
P—power; plug
PA—power amplifier
PACTOR—digital mode combining aspects of packet and AMTOR
PAM—pulse-amplitude modulation
PBS—packet bulletin-board system
PC—printed circuit
PD—power dissipation
PEP—peak envelope power
PEV—peak envelope voltage
pF—picofarad
pH—picohenry
PIC—Public Information Coordinator
PIN—positive-intrinsic-negative (semiconductor)
PIO—Public Information Officer
PIV—peak inverse voltage
PLC—Power Line Carrier
PLL—phase-locked loop
PM—phase modulation
PMOS—P-channel (metal-oxide semiconductor)
PNP—positive negative positive (transistor)
pot—potentiometer
P-P—peak to peak
ppd—postpaid
PROM—programmable read-only memory
PSAC—Public Service Advisory Committee
PSHR—Public Service Honor Roll
PTO—permeability-tuned oscillator
PTT—push to talk

Q-R

Q—figure of merit (tuned circuit); transistor
QRP—low power (less than 5-W output)
R—resistor
RACES—Radio Amateur Civil Emergency Service
RAM—random-access memory
RC—resistance-capacitance
R/C—radio control
RCC—Rag Chewer's Club
RDF—radio direction finding
RF—radio frequency
RFC—radio-frequency choke
RFI—radio-frequency interference
RHC—right-hand circular (polarization)
RIT—receiver incremental tuning
RLC—resistance-inductance-capacitance
RM—rule making (number assigned to petition)
r/min (RPM)—revolutions per minute
rms—root mean square
ROM—read-only memory
r/s—revolutions per second
RS—Radio Sputnik (Russian ham satellite)
RST—readability-strength-tone (CW signal report)
RTTY—radioteletype
RX—receiver, receiving

S

s—second (time)
S—siemens (unit of conductance); switch

SASE—self-addressed stamped envelope
SCF—switched capacitor filter
SCR—silicon controlled rectifier
SEC—Section Emergency Coordinator
SET—Simulated Emergency Test
SGL—State Government Liaison
SHF—super-high frequency (3-30 GHz)
SM—Section Manager; silver mica (capacitor)
S/N—signal-to-noise ratio
SPDT—single-pole double-throw (switch)
SPST—single-pole single-throw (switch)
SS—ARRL Sweepstakes; spread spectrum
SSB—single sideband
SSC—Special Service Club
SSI—small-scale integration
SSTV—slow-scan television
STM—Section Traffic Manager
SX—simplex
sync—synchronous, synchronizing
SWL—shortwave listener
SWR—standing-wave ratio

T

T—tera (prefix for 10^{12}); transformer
TA—ARRL Technical Advisor
TC—Technical Coordinator
TCC—Transcontinental Corps (NTS)
TCP/IP—Transmission Control Protocol/ Internet Protocol
tfc—traffic
TNC—terminal node controller (packet radio)
TR—transmit/receive
TS—Technical Specialist
TTL—transistor-transistor logic
TTY—teletypewriter
TU—terminal unit
TV—television
TVI—television interference
TX—transmitter, transmitting

U

U—integrated circuit
UHF—ultra-high frequency (300 MHz to 3 GHz)
USB—upper sideband
UTC—Coordinated Universal Time (also abbreviated Z)
UV—ultraviolet

V

V—volt; vacuum tube
VCO—voltage-controlled oscillator
VCR—video cassette recorder
VDT—video-display terminal
VE—Volunteer Examiner
VEC—Volunteer Examiner Coordinator
VFO—variable-frequency oscillator
VHF—very-high frequency (30-300 MHz)
VLF—very-low frequency (3-30 kHz)
VLSI—very-large-scale integration
VMOS—V-topology metal-oxide-semiconductor
VOM—volt-ohmmeter
VOX—voice-operated switch
VR—voltage regulator
VSWR—voltage standing-wave ratio
VTVM—vacuum-tube voltmeter
VUCC—VHF/UHF Century Club
VXO—variable-frequency crystal oscillator

W

W—watt (kg m^2s^{-3}), unit of power
WAC—Worked All Continents
WAS—Worked All States
WBFM—wide-band frequency modulation
WEFAX—weather facsimile
Wh—watthour
WPM—words per minute
WRC—World Radiocommunication Conference
WVDC—working voltage, direct current

X

X—reactance
XCVR—transceiver
XFMR—transformer
XIT—transmitter incremental tuning
XO—crystal oscillator
XTAL—crystal
XVTR—transverter

Y-Z

Y—crystal; admittance
YIG—yttrium iron garnet
Z—impedance; also see UTC

Numbers/Symbols

5BDXCC—Five-Band DXCC
5BWAC—Five-Band WAC
5BWAS—Five-Band WAS
6BWAC—Six-Band WAC
°—degree (plane angle)
°C—degree Celsius (temperature)
°F—degree Fahrenheit (temperature)
α—(alpha) angles; coefficients, attenuation constant, absorption factor, area, common-base forward current-transfer ratio of a bipolar transistor
β—(beta) angles; coefficients, phase constant, current gain of common-emitter transistor amplifiers
γ—(gamma) specific gravity, angles, electrical conductivity, propagation constant
Γ—(gamma) complex propagation constant
δ—(delta) increment or decrement; density; angles
Δ—(delta) increment or decrement determinant, permittivity
ε—(epsilon) dielectric constant; permittivity; electric intensity
ζ—(zeta) coordinates; coefficients
η—(eta) intrinsic impedance; efficiency; surface charge density; hysteresis; coordinate
θ—(theta) angular phase displacement; time constant; reluctance; angles
ι—(iota) unit vector
Κ—(kappa) susceptibility; coupling coefficient
λ—(lambda) wavelength; attenuation constant
Λ—(lambda) permeance
μ—(mu) permeability; amplification factor; micro (prefix for 10^{-6})
μF—microfarad
μH—microhenry
μP—microprocessor
ξ—(xi) coordinates
π—(pi) ≈3.14159
ρ—(rho) resistivity; volume charge density; coordinates; reflection coefficient
σ—(sigma) surface charge density; complex propagation constant; electrical conductivity; leakage coefficient; deviation
Σ—(sigma) summation
τ—(tau) time constant; volume resistivity; time-phase displacement; transmission factor; density
ϕ—(phi) magnetic flux angles
Φ—(phi) summation
χ—(chi) electric susceptibility; angles
Ψ—(psi) dielectric flux; phase difference; coordinates; angles
ω—(omega) angular velocity 2 πF
Ω—(omega) resistance in ohms; solid angle

Table 7.46
Computer Connector Pinouts

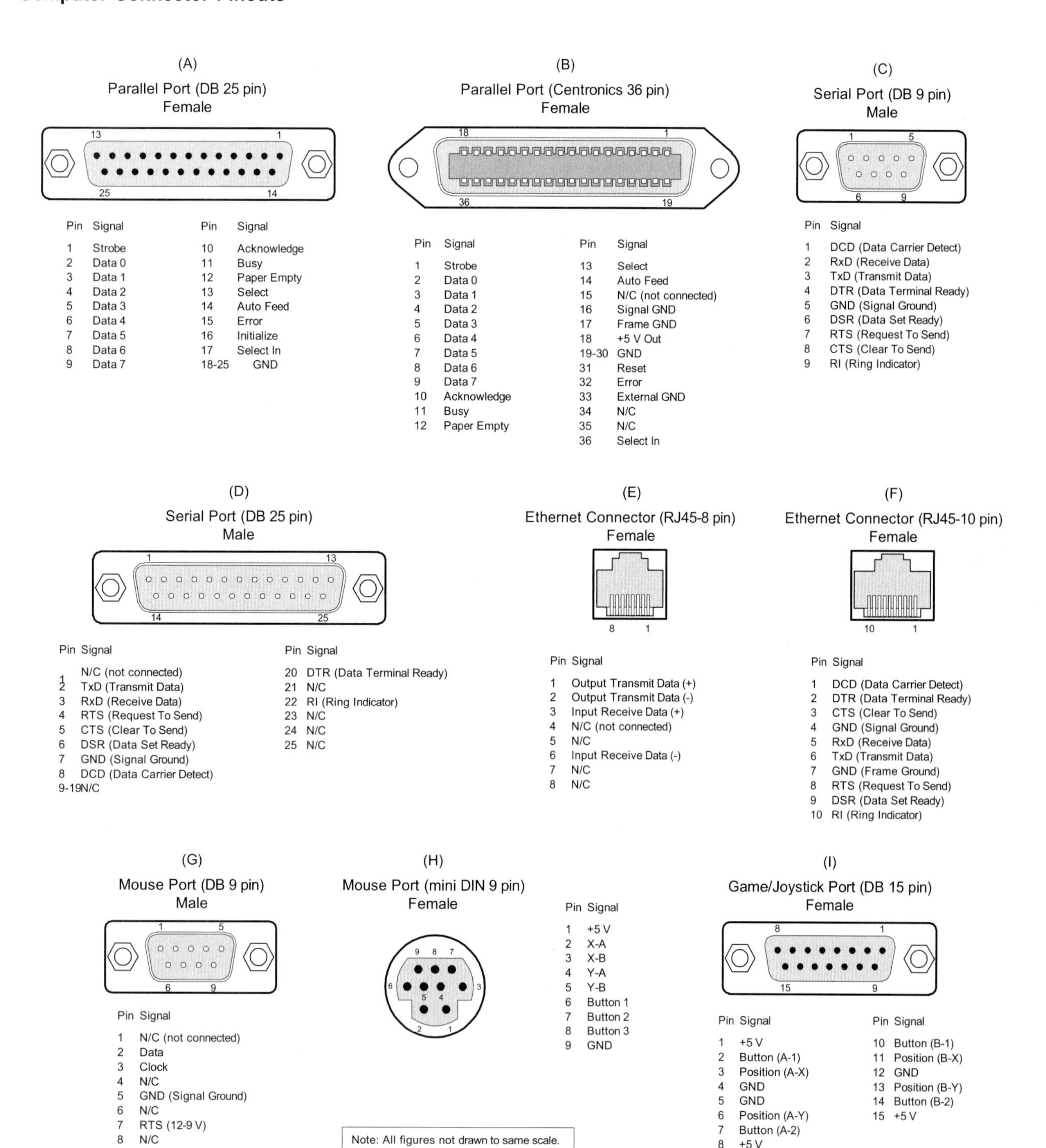

(A)
Parallel Port (DB 25 pin)
Female

Pin	Signal	Pin	Signal
1	Strobe	10	Acknowledge
2	Data 0	11	Busy
3	Data 1	12	Paper Empty
4	Data 2	13	Select
5	Data 3	14	Auto Feed
6	Data 4	15	Error
7	Data 5	16	Initialize
8	Data 6	17	Select In
9	Data 7	18-25	GND

(B)
Parallel Port (Centronics 36 pin)
Female

Pin	Signal	Pin	Signal
1	Strobe	13	Select
2	Data 0	14	Auto Feed
3	Data 1	15	N/C (not connected)
4	Data 2	16	Signal GND
5	Data 3	17	Frame GND
6	Data 4	18	+5 V Out
7	Data 5	19-30	GND
8	Data 6	31	Reset
9	Data 7	32	Error
10	Acknowledge	33	External GND
11	Busy	34	N/C
12	Paper Empty	35	N/C
		36	Select In

(C)
Serial Port (DB 9 pin)
Male

Pin	Signal
1	DCD (Data Carrier Detect)
2	RxD (Receive Data)
3	TxD (Transmit Data)
4	DTR (Data Terminal Ready)
5	GND (Signal Ground)
6	DSR (Data Set Ready)
7	RTS (Request To Send)
8	CTS (Clear To Send)
9	RI (Ring Indicator)

(D)
Serial Port (DB 25 pin)
Male

Pin	Signal	Pin	Signal
1	N/C (not connected)	20	DTR (Data Terminal Ready)
2	TxD (Transmit Data)	21	N/C
3	RxD (Receive Data)	22	RI (Ring Indicator)
4	RTS (Request To Send)	23	N/C
5	CTS (Clear To Send)	24	N/C
6	DSR (Data Set Ready)	25	N/C
7	GND (Signal Ground)		
8	DCD (Data Carrier Detect)		
9-19	N/C		

(E)
Ethernet Connector (RJ45-8 pin)
Female

Pin	Signal
1	Output Transmit Data (+)
2	Output Transmit Data (-)
3	Input Receive Data (+)
4	N/C (not connected)
5	N/C
6	Input Receive Data (-)
7	N/C
8	N/C

(F)
Ethernet Connector (RJ45-10 pin)
Female

Pin	Signal
1	DCD (Data Carrier Detect)
2	DTR (Data Terminal Ready)
3	CTS (Clear To Send)
4	GND (Signal Ground)
5	RxD (Receive Data)
6	TxD (Transmit Data)
7	GND (Frame Ground)
8	RTS (Request To Send)
9	DSR (Data Set Ready)
10	RI (Ring Indicator)

(G)
Mouse Port (DB 9 pin)
Male

Pin	Signal
1	N/C (not connected)
2	Data
3	Clock
4	N/C
5	GND (Signal Ground)
6	N/C
7	RTS (12-9 V)
8	N/C
9	N/C

(H)
Mouse Port (mini DIN 9 pin)
Female

Pin	Signal
1	+5 V
2	X-A
3	X-B
4	Y-A
5	Y-B
6	Button 1
7	Button 2
8	Button 3
9	GND

(I)
Game/Joystick Port (DB 15 pin)
Female

Pin	Signal	Pin	Signal
1	+5 V	10	Button (B-1)
2	Button (A-1)	11	Position (B-X)
3	Position (A-X)	12	GND
4	GND	13	Position (B-Y)
5	GND	14	Button (B-2)
6	Position (A-Y)	15	+5 V
7	Button (A-2)		
8	+5 V		
9	+5 V		

Note: All figures not drawn to same scale.

HBK05_07-15

Table 7.47

Voltage-Power Conversion Table

Based on a 50-ohm system

Voltage			Power	
RMS	*Peak-to-Peak*	*dBmV*	*Watts*	*dBm*
0.01 μV	0.0283 μV	−100	2×10^{-18}	−147.0
0.02 μV	0.0566 μV	−93.98	8×10^{-18}	−141.0
0.04 μV	0.113 μV	−87.96	32×10^{-18}	−134.9
0.08 μV	0.226 μV	−81.94	128×10^{-18}	−128.9
0.1 μV	0.283 μV	−80.0	200×10^{-18}	−127.0
0.2 μV	0.566 μV	−73.98	800×10^{-18}	−121.0
0.4 μV	1.131 μV	−67.96	3.2×10^{-15}	−114.9
0.8 μV	2.236 μV	−61.94	12.8×10^{-15}	−108.9
1.0 μV	2.828 μV	−60.0	20.0×10^{15}	−107.0
2.0 μV	5.657 μV	−53.98	80.0×10^{-15}	−101.0
4.0 μV	11.31 μV	−47.96	320.0×10^{-15}	−94.95
8.0 μV	22.63 μV	−41.94	1.28×10^{-12}	−88.93
10.0 μV	28.28 μV	−40.00	2.0×10^{-12}	− 86.99
20.0 μV	56.57 μV	−33.98	8.0×10^{-12}	−80.97
40.0 μV	113.1 μV	−27.96	32.0×10^{-12}	−74.95
80.0 μV	226.3 μV	−21.94	128.0×10^{-12}	−68.93
100.0 μV	282.8 μV	−20.0	200.0×10^{-12}	−66.99
200.0 μV	565.7 μV	−13.98	800.0×10^{-12}	−60.97
400.0 μV	1.131 mV	−7.959	3.2×10^{-9}	−54.95
800.0 μV	2.263 mV	−1.938	12.8×10^{-9}	−48.93
1.0 mV	2.828 mV	0.0	20.0×10^{-9}	−46.99
2.0 mV	5.657 mV	6.02	80.0×10^{-9}	−40.97
4.0 mV	11.31 mV	12.04	320×10^{-9}	−34.95
8.0 mV	22.63 mV	18.06	1.28 μW	−28.93
10.0 mV	28.28 mV	20.00	1 2.0 μW	−26.99
20.0 mV	56.57 mV	26.02	8.0 μW	−20.97
40.0 mV	113.1 mV	32.04	32.0 μW	−14.95
80.0 mV	226.3 mV	38.06	128.0 μW	−8.93
100.0 mV	282.8 mV	40.0	200.0 μW	−6.99
200.0 mV	565.7 mV	46.02	800.0 μW	−0.97
223.6 mV	632.4 mV	46.99	1.0 mW	0
400.0 mV	1.131 V	52.04	3.2 mW	5.05
800.0 mV	2.263 V	58.06	12.80 mW	11.07
1.0 V	2.828 V	60.0	20.0 mW	13.01
2.0 V	5.657 V	66.02	80.0 mW	19.03
4.0 V	11.31 V	72.04	320.0 mW	25.05
8.0 V	22.63 V	78.06	1.28 W	31.07
10.0 V	28.28 V	80.0	2.0 W	33.01
20.0 V	56.57 V	86.02	8.0 W	39.03
40.0 V	113.1 V	92.04	32.0 W	45.05
80.0 V	226.3 V	98.06	128.0 W	51.07
100.0 V	282.8 V	100.0	200.0 W	53.01
200.0 V	565.7 V	106.0	800.0 W	59.03
223.6 V	632.4 V	107.0	1,000.0 W	60.0
400.0 V	1,131.0 V	112.0	3,200.0 W	65.05
800.0 V	2,263.0 V	118.1	12,800.0 W	71.07
1000.0 V	2,828.0 V	120.0	20,000 W	73.01
2000.0 V	5,657.0 V	126.0	80,000 W	79.03
4000.0 V	11,310.0 V	132.0	320,000 W	85.05
8000.0 V	22,630.0 V	138.1	1.28 MW	91.07
10,000.0 V	28,280.0 V	140.0	2.0 MW	93.01

$$\text{Voltage}, V_{p-p} = V_{RMS} \times 2\sqrt{2}$$

$$\text{Voltage, dBmV} = 20 \times \log_{10}\left[\frac{V_{RMS}}{0.001V}\right] \text{ or } 20 \times \log_{10}\left[mV_{RMS}\right]$$

$$\text{Power, watts} = \left[\frac{V_{RMS}^{\;2}}{50\ \Omega}\right]$$

$$\text{Power, dBm} = 10 \times \log_{10}\left[\frac{\text{Power (watts)}}{0.001W}\right] \text{ or } 10 \times \log_{10}\left[mW_{RMS}\right]$$

Table 7.48
Large Machine-Wound Coil Specifications

Coil Dia, Inches	*Turns Per Inch*	*Inductance in μH*
1 1/4	4	2.75
	6	6.3
	8	11.2
	10	17.5
	16	42.5
1 1/2	4	3.9
	6	8.8
	8	15.6
	10	24.5
	16	63
1 3/4	4	5.2
	6	11.8
	8	21
	10	33
	16	85
2	4	6.6
	6	15
	8	26.5
	10	42
	16	108
2 1/2	4	10.2
	6	23
	8	41
	10	64
3	4	14
	6	31.5
	8	56
	10	89

Table 7.49
Inductance Factor for Large Machine-Wound Coils

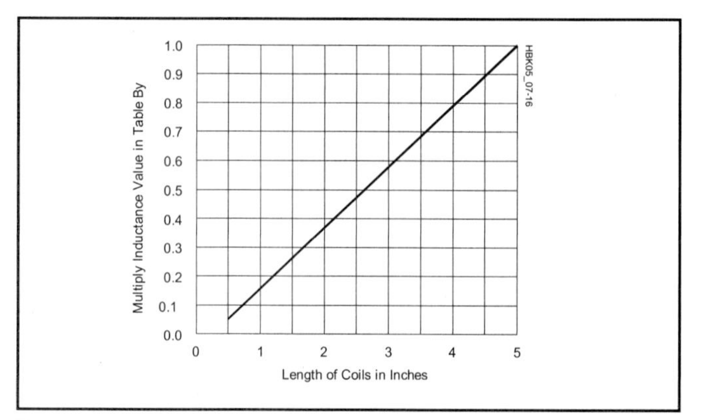

Factor to be applied to the inductance of large coils for coil lengths up to 5 inches.

Table 7.50
Small Machine-Wound Coil Specifications

Coil Dia, Inches	*Turns Per Inch*	*Inductance in μH*
1/2 (A)	4	0.18
	6	0.40
	8	0.72
	10	1.12
	16	2.8
	32	12
5/8 (A)	4	0.28
	6	0.62
	8	1.1
	10	1.7
	16	4.4
	32	18
3/4 (B)	4	0.6
	6	1.35
	8	2.4
	10	3.8
	16	9.9
	32	40
1 (B)	4	1.0
	6	2.3
	8	4.2
	10	6.6
	16	16.9
	32	68

Table 7.51
Inductance Factor for Small Machine-Wound Coils

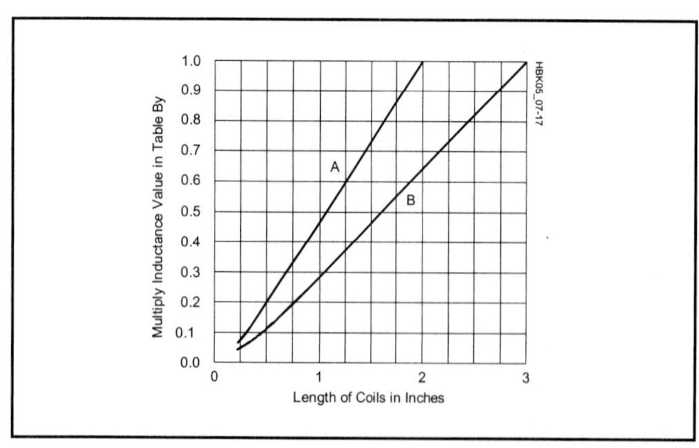

Factor to be applied to the inductance of small coils as a function of coil length. Use curve A for coils marked A, and curve B for coils marked B.

Table 7.52
Measured Inductance for #12 Wire Windings

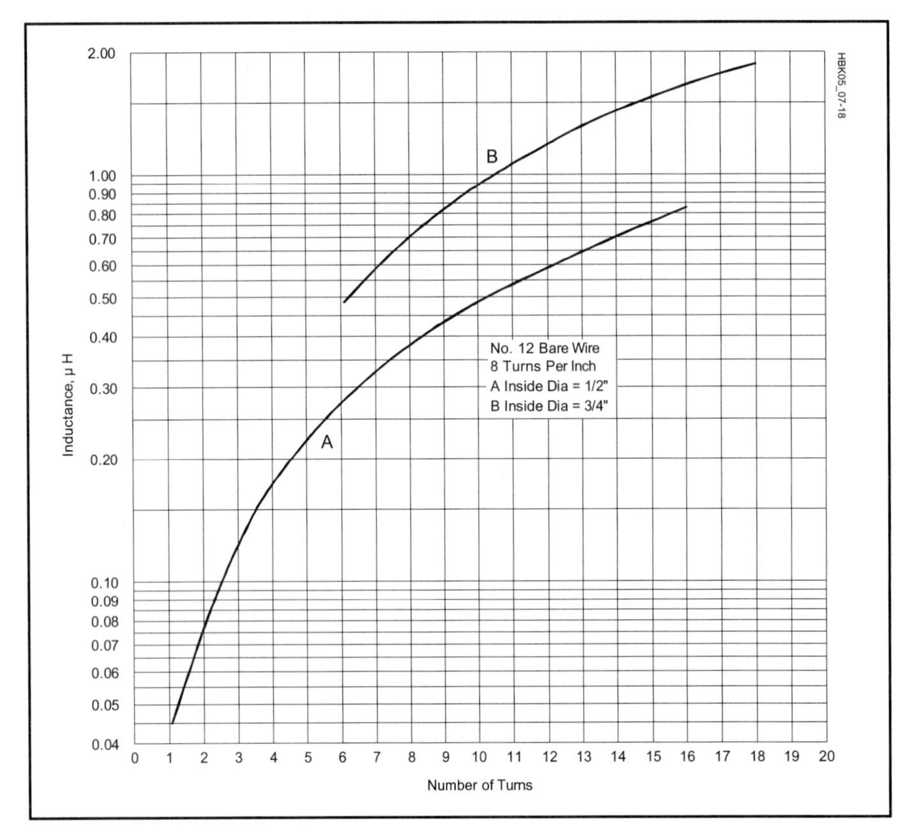

Values are for inductors with half-inch leads and wound with eight turns per inch.

Table 7.53
Relationship Between Noise Figure and Noise Temperature

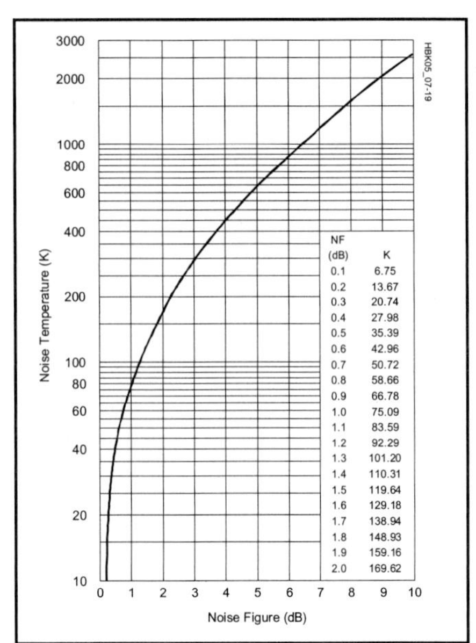

Table 7.54
Antenna Wire Strength

American Wire Gauge	***Recommended Tension[1] (pounds)***		***Weight (pounds per 1000 feet)***	
	Copper-clad steel[2]	*Hard-drawn copper*	*Copper-clad steel[2]*	*Hard-drawn copper*
4	495	214	115.8	126
6	310	130	72.9	79.5
8	195	84	45.5	50
10	120	52	28.8	31.4
12	75	32	18.1	19.8
14	50	20	11.4	12.4
16	31	13	7.1	7.8
18	19	8	4.5	4.9
20	12	5	2.8	3.1

[1]Approximately one-tenth the breaking load. Might be increased 50% if end supports are firm and there is no danger of ice loading.
[2]"Copperweld," 40% copper.

Table 7.55
Standard vs American Wire Gauge

SWG	*Diam (in.)*	*Nearest AWG*
12	0.104	10
14	0.08	12
16	0.064	14
18	0.048	16
20	0.036	19
22	0.028	21
24	0.022	23
26	0.018	25
28	0.0148	27
30	0.0124	28
32	0.0108	29
34	0.0092	31
36	0.0076	32
38	0.006	34
40	0.0048	36
42	0.004	38
44	0.0032	40
46	0.0024	—

Table 7.56
Pi-Network Resistive Attenuators (50 Ω)

dB Atten.	*R1 (Ohms)*	*R2 (Ohms)*
1.0	870	5.77
2.0	436	11.6
3.0	292	17.6
4.0	221	23.8
5.0	178	30.4
6.0	150	37.4
7.0	131	44.8
8.0	116	52.8
9.0	105	61.6
10.0	96.2	71.2
11.0	89.2	81.7
12.0	83.5	93.2
13.0	78.8	106
14.0	74.9	120
15.0	71.6	136
16.0	68.8	154
17.0	66.4	173
18.0	64.4	195
19.0	62.6	220
20.0	61.1	248
21.0	59.8	278
22.0	58.6	313
23.0	57.6	352
24.0	56.7	395
25.0	56.0	443
30.0	53.2	790
35.0	51.8	1405
40.0	51.0	2500
45.0	50.5	4446
50.0	50.3	7906
55.0	50.2	14,058
60.0	50.1	25,000

Note: A PC board kit for the Low-Power Step Attenuator (Sep 1982 *QST*) is available from FAR Circuits. Project details are in the Handbook **template package STEP ATTENUATOR.**

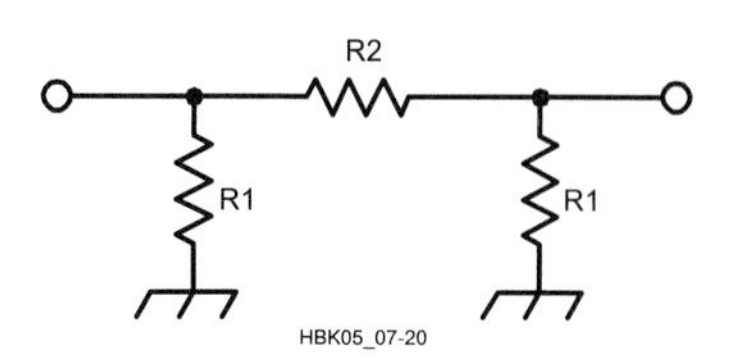

Table 7.57
T-Network Resistive Attenuators (50 Ω)

dB Atten.	*R1 (Ohms)*	*R2 (Ohms)*
1.0	2.88	433
2.0	5.73	215
3.0	8.55	142
4.0	11.3	105
5.0	14.0	82.2
6.0	16.6	66.9
7.0	19.1	55.8
8.0	21.5	47.3
9.0	23.8	40.6
10.0	26.0	35.1
11.0	28.0	30.6
12.0	30.0	26.8
13.0	31.7	23.5
14.0	33.3	20.8
15.0	35.0	18.4
16.0	36.3	16.2
17.0	37.6	14.4
18.0	38.8	12.8
19.0	40.0	11.4
20.0	41.0	10.0
21.0	41.8	9.0
22.0	42.6	8.0
23.0	43.4	7.1
24.0	44.0	6.3
25.0	44.7	5.6
30.0	47.0	3.2
35.0	48.2	1.8
40.0	49.0	1.0
45.0	49.4	0.56
50.0	49.7	0.32
55.0	49.8	0.18
60.0	49.9	0.10

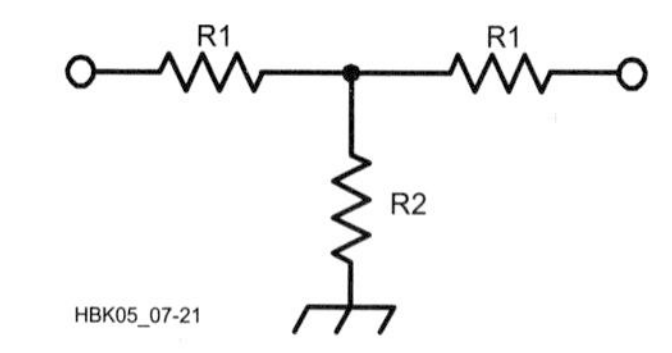

Table 7.58
Impedance of Various Two-Conductor Lines

	— *Twists per Inch* —				
Wire Size	*2.5*	*5*	*7.5*	*10*	*12.5*
no. 20	43	39	35		
no. 22	46	41	39	37	32
no. 24	60	45	44	43	41
no. 26	65	57	54	48	47
no. 28	74	53	51	49	47
no. 30			49	46	47

Measured in ohms at 14.0 MHz.

This illustrates the impedance of various two-conductor lines as a function of the wire size and number of twists per inch.

Table 7.59
Attenuation per Foot for Lines

Wire Size	Twists per Inch 2.5	5	7.5	10	12.5
no. 20	0.11	0.11	0.12		
no. 22	0.11	0.12	0.12	0.12	0.12
no. 24	0.11	0.12	0.12	0.13	0.13
no. 26	0.11	0.13	0.13	0.13	0.13
no. 28	0.11	0.13	0.13	0.16	0.16
no. 30			0.25	0.27	0.27

Measured in decibels at 14.0 MHz.

Attenuation in dB per foot for the same lines as shown above.

Table 7.60
Equivalent Values of Reflection Coefficient, Attenuation, SWR and Return Loss

Reflection Coefficient (%)	*Attenuation (dB)*	*Max SWR*	*Return Loss, dB*
1.000	0.000434	1.020	40.00
1.517	0.001000	1.031	36.38
2.000	0.001738	1.041	33.98
3.000	0.003910	1.062	30.46
4.000	0.006954	1.083	27.96
4.796	0.01000	1.101	26.38
5.000	0.01087	1.105	26.02
6.000	0.01566	1.128	24.44
7.000	0.02133	1.151	23.10
7.576	0.02500	1.164	22.41
8.000	0.02788	1.174	21.94
9.000	0.03532	1.198	20.92
10.000	0.04365	1.222	20.00
10.699	0.05000	1.240	19.41
11.000	0.05287	1.247	19.17
12.000	0.06299	1.273	18.42
13.085	0.07500	1.301	17.66
14.000	0.08597	1.326	17.08
15.000	0.09883	1.353	16.48
15.087	0.10000	1.355	16.43
16.000	0.1126	1.381	15.92
17.783	0.1396	1.433	15.00
18.000	0.1430	1.439	14.89
19.000	0.1597	1.469	14.42
20.000	0.1773	1.500	13.98
22.000	0.2155	1.564	13.15
23.652	0.2500	1.620	12.52
24.000	0.2577	1.632	12.40
25.000	0.2803	1.667	12.04
26.000	0.3040	1.703	11.70
27.000	0.3287	1.740	11.37
28.000	0.3546	1.778	11.06
30.000	0.4096	1.857	10.46
31.623	0.4576	1.925	10.00
32.977	0.5000	1.984	9.64
33.333	0.5115	2.000	9.54
34.000	0.5335	2.030	9.37
35.000	0.5675	2.077	9.12
36.000	0.6028	2.125	8.87
37.000	0.6394	2.175	8.64
38.000	0.6773	2.226	8.40
39.825	0.75000	2.324	8.00
40.000	0.7572	2.333	7.96
42.000	0.8428	2.448	7.54
42.857	0.8814	2.500	7.36
44.000	0.9345	2.571	7.13
45.351	1.0000	2.660	6.87
48.000	1.1374	2.846	6.38
50.000	1.2494	3.000	6.02
52.000	1.3692	3.167	5.68
54.042	1.5000	3.352	5.35
56.234	1.6509	3.570	5.00
58.000	1.7809	3.762	4.73
60.000	1.9382	4.000	4.44
60.749	2.0000	4.095	4.33
63.000	2.1961	4.405	4.01
66.156	2.5000	4.909	3.59
66.667	2.5528	5.000	3.52
70.627	3.0000	5.809	3.02
70.711	3.0103	5.829	3.01

$$\rho = \frac{SWR - 1}{SWR + 1}$$

where ρ = 0.01 × (reflection coefficient in %)

$$\rho = 10^{\frac{-RL}{20}}$$

where RL = return loss (dB)

$$\rho = \sqrt{1 - (0.1^{X})}$$

where X = A/10 and A = attenuation (dB)

$$SWR = \frac{1+\rho}{1-\rho}$$

Return loss (dB) = –8.68589 ln (ρ)
where ln is the natural log (log to the base e)

Attenuation (dB) = –4.34295 ln ($1-\rho^2$)
where ln is the natural log (log to the base e)

Table 7.61
Guy Wire Lengths to Avoid

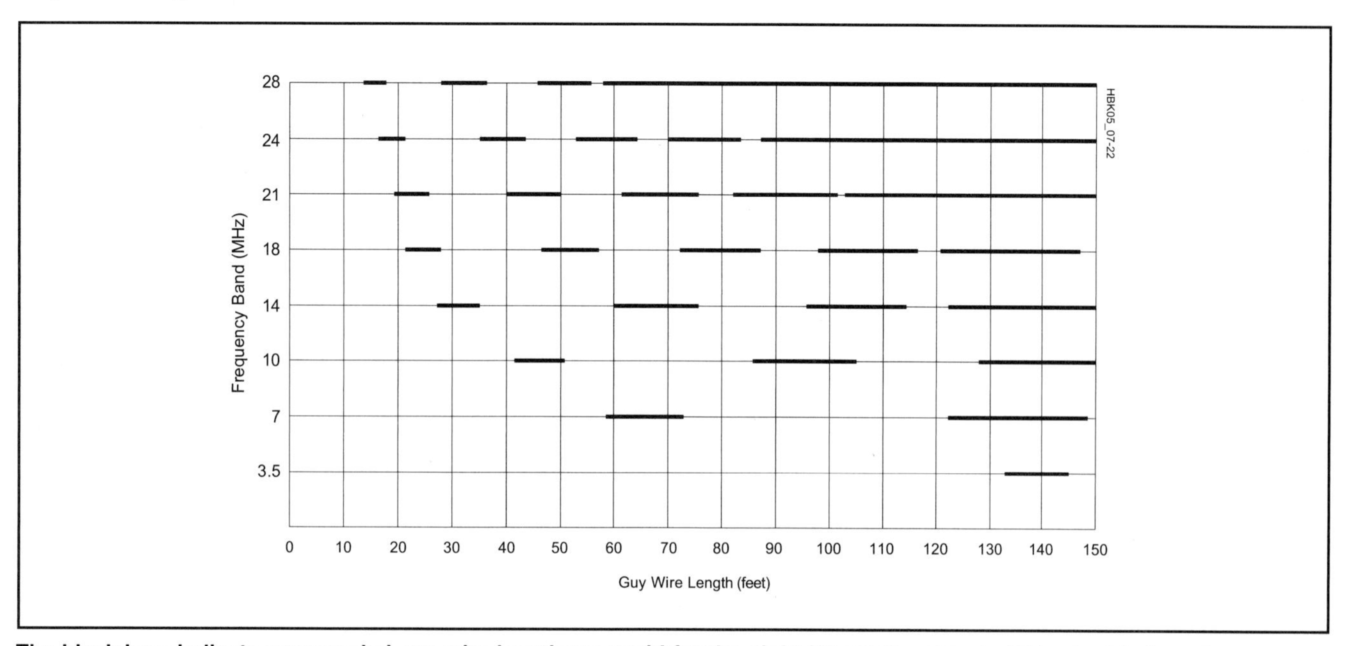

The black bars indicate ungrounded guy wire lengths to avoid for the eight HF amateur bands. This chart is based on resonance within 10% of any frequency in the band. Grounded wires will exhibit resonance at odd multiples of a quarter wavelength. ***(Jerry Hall, K1TD)***

Table 7.62

Morse Code Character Set[1]

Character	Sound	Code
A	didah	• –
B	dahdididit	– •••
C	dahdidahdit	– • – •
D	dahdidit	– ••
E	dit	•
F	dididahdit	•• – •
G	dahdahdit	– – •
H	dididit	••••
I	didit	••
J	didahdahdah	•– – –
K	dahdidah	– •–
L	didahdidit	• – ••
M	dahdah	– –
N	dahdit	– •
O	dahdahdah	– – –
P	didahdahdit	• – – •
Q	dahdahdidah	– – • –
R	didahdit	• – •
S	dididit	•••
T	dah	–
U	dididah	•• –
V	didididah	••• –
W	didahdah	•– –
X	dahdididah	– •• –
Y	dahdidahdah	– • – –
Z	dahdahdidit	– – ••
1	didahdahdahdah	• – – – –
2	dididahdahdah	•• – – –
3	didididahdah	••• – –
4	dididididah	•••• –
5	dididididit	•••••
6	dahdididit	– ••••
7	dahdahdididit	– – •••
8	dahdahdahdidit	– – – ••
9	dahdahdahdahdit	– – – – •
0	dahdahdahdahdah	– – – – –

Character	Sound	Code	Signal
Period [.]:	didahdidahdidah	• – • – • –	$\overline{AAA}$
Comma [,]:	dahdahdidididahdah	– – •• – –	$\overline{MIM}$
Question mark or request for repetition [?]:	dididahdahdidit	•• – – ••	$\overline{IMI}$
Error:	dididididididit	•••••••••	$\overline{HH}$
Hyphen or dash [–]:	dahdidididdah	– •••• –	$\overline{DU}$
Double dash [=]	dahdidididah	– ••• –	$\overline{BT}$
Colon [:]:	dahdahdahdididit	– – – •••	$\overline{OS}$
Semicolon [;]:	dahdidahdidahdit	– • – • – •	$\overline{KR}$
Left parenthesis [(]:	dahdidahdahdit	– • – – •	$\overline{KN}$
Right parenthesis [)]:	dahdidahdahddidah	– • – – • –	$\overline{KK}$
Fraction bar [/]:	dahdididahdit	– •• – •	$\overline{DN}$
Quotation marks ["]:	didahdididahdit	• – • • – •	$\overline{AF}$
Dollar sign [$]:	didididahdididah	••• – • •–	$\overline{SX}$
Apostrophe [']:	didahdahdahdahdit	• – – – – •	$\overline{WG}$
Paragraph [¶]:	didahdidahdidit	• – • – ••	$\overline{AL}$
Underline [_]:	dididahdahdidah	•• – – • –	$\overline{IQ}$
Starting signal:	dahdidahdidah	– • – •–	$\overline{KA}$
Wait:	didahdididit	• – •••	$\overline{AS}$
End of message or cross [+]:	didahdidahdit	• – • – •	$\overline{AR}$
Invitation to transmit [K]:	dahdidah	– • –	K
End of work:	didididahdidah	••• – • –	$\overline{SK}$
Understood:	didididahdit	••• – •	$\overline{SN}$

Notes:

1. Not all Morse characters shown are used in FCC code tests. License applicants are responsible for knowing, and may be tested on, the 26 letters, the numerals 0 to 9, the period, the comma, the question mark, $\overline{AR}$, $\overline{SK}$, $\overline{BT}$ and fraction bar [$\overline{DN}$].

2. The following letters are used in certain European languages which use the Latin alphabet:

Character	Sound	Code
Ä, Ą	didahdidah	• – • –
Á, Å, À, Â	didahdahdidah	• – – • –
Ç, Ć	dahdidahdidit	– • – ••
É, È, Ę	dididahdidit	•• – ••
È	didahdididah	• – ••–
Ê	dahdididahdit	– •• – •
Ö, Ô, Ó	dahdahdahdit	– – –•
Ñ	dahdahdidahdah	– – • – –
Ü	dididahdah	•• – –
Ź	dahdahdidit	– –••
Z	dahdahdididah	– –•• –
CH, Ş	dahdahdahdah	– – – –

3. Special Esperanto characters:

Character	Sound	Code
Ĉ	dahdidahdidit	– • – ••
Ŝ	didididahdit	••• – •
Ĵ	didahdahdahdit	• – – –•
Ĥ	dahdidahdahdit	–• – –•
Û	dididahdah	•• – –
Ĝ	dahdahdidahdit	– –• –•

4. Signals used in other radio services:

Signal	Sound	Code	Characters
Interrogatory	dididahdidah	•• – • –	$\overline{INT}$
Emergency silence	dididididahdah	•••• – –	$\overline{HM}$
Executive follows	dididahdididah	•• – •• –	$\overline{IX}$
Break–in signal	dahdahdahdahdah	– – – – –	$\overline{TTTTT}$
Emergency signal	didididahdahdahdididit	••• – – – •••	$\overline{SOS}$
Relay of distress	dahdididahdididahdidit	– •• – •• – ••	$\overline{DDD}$

Morse Abbreviated Numbers

Numeral		*Long Number*		*Abbreviated Number*	*Equivalent Character*
1	didahdahdahdah	• – – – –	didah	• –	A
2	dididahdahdah	•• – – –	dididah	•• –	U
3	didididahdah	••• – –	didididah	••• –	V
4	dididididah	•••• –	dididididah	•••• –	4
5	dididididit	•••••	dididididit	••••• or •	5 or E
6	dahdididdidit	– ••••	dahdididdidit	– ••••	6
7	dahdahdididit	– – •••	dahdididit	– •••	B
8	dahdahdahdidit	– – – ••	dahdidit	– ••	D
9	dahdahdahdahdit	– – – – •	dahdit	– •	N
0	dahdahdahdahdah	– – – – –	dah	–	T

Note: These abbreviated numbers are not legal for use in call signs. They should be used only where there is agreement between operators and when no confusion will result.

Table 7.63

Morse Abbeviated ("Cut") Numbers

Numeral		*Long Number*		*Abbreviated Number*	*Equivalent Character*
1	didahdahdahdah	• – – – –	didah	• –	A
2	dididahdahdah	• • – – –	divididah	• • –	U
3	didididahdah	• • • – –	didididah	• • • –	V
4	dididididah	• • • • –	dididididah	• • • • –	4
5	dididididit	• • • • •	dididididit	• • • • • or •	5 or E
6	dahdidididit	– • • • •	dahdidididit	– • • • •	6
7	dahdahdididit	– – • • •	dahdididit	– • • •	B
8	dahdahdahdidit	– – – • •	dahdidit	– • •	D
9	dahdahdahdahdit	– – – – •	dahdit	– •	N
0	dahdahdahdahdah	– – – – –	dah	–	T

Note: These abbreviated numbers are not legal for use in call signs. They should be used only where there is agreement between operators and when no confusion will result.

Table 7.64

The ASCII Coded Character Set

					6	0	0	0	0	1	1	1	1
Bit					**5**	**0**	**0**	**1**	**1**	**0**	**0**	**1**	**1**
Number					**4**	**0**	**1**	**0**	**1**	**0**	**1**	**0**	**1**
				Hex	***1st***	***0***	***1***	***2***	***3***	***4***	***5***	***6***	***7***
3	***2***	***1***	***0***	***2nd***									
0	0	0	0	0		NUL	DLE	SP	0	@	P	'	p
0	0	0	1	1		SOH	DC1	!	1	A	Q	a	q
0	0	1	0	2		STX	DC2	"	2	B	R	b	r
0	0	1	1	3		ETX	DC3	#	3	C	S	c	s
0	1	0	0	4		EOT	DC4	$	4	D	T	d	t
0	1	0	1	5		ENQ	NAK	%	5	E	U	e	u
0	1	1	0	6		ACK	SYN	&	6	F	V	f	v
0	1	1	1	7		BEL	ETB	'	7	G	W	g	w
1	0	0	0	8		BS	CAN	(	8	H	X	h	x
1	0	0	1	9		HT	EM	)	9	I	Y	i	y
1	0	1	0	A		LF	SUB	*	:	J	Z	j	z
1	0	1	1	B		VT	ESC	+	;	K	[	k	{
1	1	0	0	C		FF	FS	,	<	L	\	l	|
1	1	0	1	D		CR	GS	-	=	M	]	m	}
1	1	1	0	E		SO	RS	.	>	N	^	n	~
1	1	1	1	F		SI	US	/	?	O	—	o	DEL

Notes

1. "1" = mark, "0" = space.
2. Bit 6 is the most-significant bit (MSB). Bit 0 is the least-significant bit (LSB).

ACK	acknowledge
BEL	bell
BS	backspace
CAN	cancel
CR	carriage return
DC1	device control 1
DC2	device control 2
DC3	device control 3
DC4	device control 4
DEL	(delete)
DLE	data link escape
ENQ	enquiry
EM	end of medium
EOT	end of transmission
ESC	escape
ETB	end of block
ETX	end of text
FF	form feed
FS	file separator
GS	group separator
HT	horizontal tab
LF	line feed
NAK	negative acknowledge
NUL	null
RS	record separator
SI	shift in
SO	shift out
SOH	start of heading
SP	space
STX	start of text
SUB	substitute
SYN	synchronous idle
US	unit separator
VT	vertical tab

Table 7.65

Voluntary HF Band Plans

The following frequencies are generally recognized for certain modes or activities (all frequencies are in MHz).

Nothing in the rules recognizes a net's, group's or any individual's special privilege to any specific frequency. Section 97.101(b) of the Rules states that "Each station licensee and each control operator must cooperate in selecting transmitting channels and in making the most effective use of the amateur service frequencies. No frequency will be assigned for the exclusive use of any station." No one "owns" a frequency.

It's good practice—and plain old common sense—for any operator, regardless of mode, to check to see if the frequency is in use prior to engaging operation. If you are there first, other operators should make an effort to protect you from interference to the extent possible given that 100% inter-ference-free operation is an unrealistic expectation in today's congested bands.

Frequency	Use
1.800-1.810	Digital Modes
1.810	CW QRP
1.800-2.000	CW
1.843-2.000	SSB, SSTV and other wideband modes
1.910	SSB QRP
1.995-2.000	Experimental
1.999-2.000	Beacons
3.500-3.510	CW DX
3.590	RTTY DX
3.580-3.620	Data
3.620-3.635	Automatically controlled data stations
3.790-3.800	DX window
3.845	SSTV
3.885	AM calling frequency
3.985	QRP SSB calling frequency
7.040	RTTY DX
	QRP CW calling frequency
7.075-7.100	Phone in KH/KL/KP *only*
7.080-7.100	Data
7.100-7.105	Automatically controlled data stations
7.171	SSTV
7.290	AM calling frequency
10.106	QRP CW calling frequency
10.130-10.140	Data
10.140-10.150	Automatically controlled data stations
14.060	QRP CW calling frequency
14.070-14.095	Data
14.095-14.0995	Automatically controlled data stations
14.100	IBP/NCDXF beacons
14.1005-14.112	Automatically controlled data stations
14.230	SSTV
14.285	QRP SSB calling frequency
14.286	AM calling frequency
18.100-18.105	Data
18.105-18.110	Automatically controlled data stations
21.060	QRP CW calling frequency
21.070-21.090	Data
21.090-21.100	Automatically controlled data stations
21.340	SSTV
21.385	QRP SSB calling frequency
24.920-24.925	Data
24.925-24.930	Automatically controlled data stations
28.060	QRP CW calling frequency
28.070-28.120	Data
28.120-28.189	Automatically controlled data stations
28.190-28.225	Beacons
28.385	QRP SSB calling frequency
28.680	SSTV
29.000-29.200	AM
29.300-29.510	Satellite downlinks
29.520-29.580	Repeater inputs
29.600	FM simplex
29.620-29.680	Repeater outputs

Notes

ARRL band plans for frequencies above 28.300 MHz are shown in *The ARRL Repeater Directory* *The FCC Rule Book* *QST*

Table 7.66

VHF/UHF/EHF Calling Frequencies

Band (MHz)	*Calling Frequency*
50	50.125 SSB
	50.620 digital (packet)
	52.525 National FM simplex frequency
144	144.010 EME
	144.100, 144.110 CW
	144.200 SSB
	146.520 National FM simplex frequency
222	222.100 CW/SSB
	223.500 National FM simplex frequency
432	432.010 EME
	432.100 CW/SSB
	446.000 National FM simplex frequency
902	902.100 CW/SSB
	903.1 Alternate CW, SSB
	906.500 National FM simplex frequency
1296	1294.500 National FM simplex frequency
	1296.100 CW/SSB
2304	2304.4
	2305.2 FM simplex frequency
10000	10368.1 Narrow-band

VHF/UHF Activity Nights

Some areas do not have enough VHF/UHF activity to support contacts at all times. This schedule is intended to help VHF/UHF operators make contact. This is only a starting point; check with others in your area to see if local hams have a different schedule.

Band (MHz)	*Day*	*Local Time*
50	Sunday	6 PM
144	Monday	7 PM
222	Tuesday	8 PM
432	Wednesday	9 PM
902	Friday	9 PM
1296	Thursday	10 PM

Table 7.67
ITU Regions

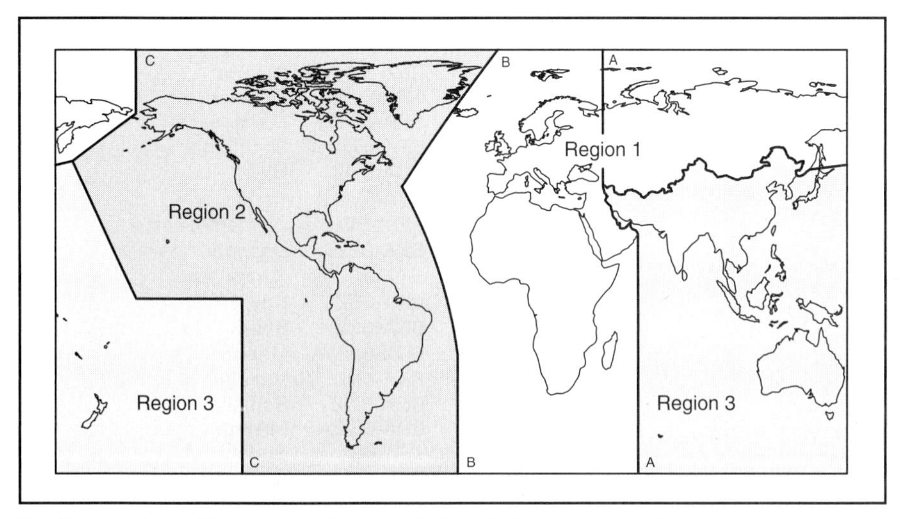

The International Telecommunication Union divides the world into three regions. Geographic details appear in *The FCC Rule Book*.

Table 7.68
Allocation of International Call Signs

Call Signs	Country
AAA-ALZ	United States of America
AMA-AOZ	Spain
APA-ASZ	Pakistan
ATA-AWZ	India
AXA-AXZ	Australia
AYA-AZZ	Argentina
A2A-A2Z	Botswana
A3A-A3Z	Tonga
A4A-A4Z	Oman
A5A-A5Z	Bhutan
A6A-A6Z	United Arab Emirates
A7A-A7Z	Qatar
A8A-A8Z	Liberia
A9A-A9Z	Bahrain
BAA-BZZ	China (People's Republic of)
CAA-CEZ	Chile
CFA-CKZ	Canada
CLA-CMZ	Cuba
CNA-CNZ	Morocco
COA-COZ	Cuba
CPA-CPZ	Bolivia
CQA-CUZ	Portugal
CVA-CXZ	Uruguay
CYA-CZZ	Canada
C2A-C2Z	Nauru
C3A-C3Z	Andorra
C4A-C4Z	Cyprus
C5A-C5Z	Gambia
C6A-C6Z	Bahamas
* C7A-C7Z	World Meteorological Organization
C8A-C9Z	Mozambique
DAA-DRZ	Germany
DSA-DTZ	Republic of Korea
DUA-DZZ	Philippines
D2A-D3Z	Angola
D4A-D4Z	Cape Verde
D5A-D5Z	Liberia
D6A-D6Z	Comoros
D7A-D9Z	Republic of Korea
EAA-EHZ	Spain
EIA-EJZ	Ireland
EKA-EKZ	Armenia
ELA-ELZ	Liberia
EMA-EOZ	Ukraine
EPA-EQZ	Iran
ERA-ERZ	Moldova
ESA-ESZ	Estonia
ETA-ETZ	Ethiopia
EUA-EWZ	Belarus
EXA-EXZ	Kyrgyzstan
EYA-EYZ	Tajikistan
EZA-EZZ	Turkmenistan
E2A-E2Z	Thailand
E3A-E3Z	Eritrea
†E4A-E4Z	Palestinian Authority
FAA-FZZ	France
GAA-GZZ	United Kingdom of Great Britain and Northern Ireland
HAA-HAZ	Hungary
HBA-HBZ	Switzerland
HCA-HDZ	Ecuador
HEA-HEZ	Switzerland
HFA-HFZ	Poland
HGA-HGZ	Hungary
HHA-HHZ	Haiti
HIA-HIZ	Dominican Republic
HJA-HKZ	Colombia
HLA-HLZ	Republic of Korea
HMA-HMZ	Democratic People's Republic of Korea
HNA-HNZ	Iraq
HOA-HPZ	Panama
HQA-HRZ	Honduras
HSA-HSZ	Thailand
HTA-HTZ	Nicaragua
HUA-HUZ	El Salvador
HVA-HVZ	Vatican City State
HWA-HYZ	France
HZA-HZZ	Saudi Arabia
H2A-H2Z	Cyprus
H3A-H3Z	Panama
H4A-H4Z	Solomon Islands
H6A-H7Z	Nicaragua
H8A-H9Z	Panama
IAA-IZZ	Italy
JAA-JSZ	Japan
JTA-JVZ	Mongolia
JWA-JXZ	Norway
JYA-JYZ	Jordan
JZA-JZZ	Indonesia
J2A-J2Z	Djibouti
J3A-J3Z	Grenada
J4A-J4Z	Greece
J5A-J5Z	Guinea-Bissau
J6A-J6Z	Saint Lucia
J7A-J7Z	Dominica
J8A-J8Z	St. Vincent and the Grenadines
KAA-KZZ	United States of America
LAA-LNZ	Norway
LOA-LWZ	Argentina
LXA-LXZ	Luxembourg
LYA-LYZ	Lithuania
LZA-LZZ	Bulgaria
L2A-L9Z	Argentina
MAA-MZZ	United Kingdom of Great Britain and Northern Ireland
NAA-NZZ	United States of America
OAA-OCZ	Peru
ODA-ODZ	Lebanon
OEA-OEZ	Austria
OFA-OJZ	Finland
OKA-OLZ	Czech Republic
OMA-OMZ	Slovak Republic
ONA-OTZ	Belgium
OUA-OZZ	Denmark
PAA-PIZ	Netherlands
PJA-PJZ	Netherlands Antilles
PKA-POZ	Indonesia
PPA-PYZ	Brazil
PZA-PZZ	Suriname
P2A-P2Z	Papua New Guinea

Series	Allocated to
P3A-P3Z	Cyprus
P4A-P4Z	Aruba
P5A-P9Z	Democratic People's Republic of Korea
RAA-RZZ	Russian Federation
SAA-SMZ	Sweden
SNA-SRZ	Poland
SSA-SSM	Egypt
SSN-STZ	Sudan
SUA-SUZ	Egypt
SVA-SZZ	Greece
S2A-S3Z	Bangladesh
S5A-S5Z	Slovenia
S6A-S6Z	Singapore
S7A-S7Z	Seychelles
S8A-S8Z	South Africa
S9A-S9Z	Sao Tome and Principe
TAA-TCZ	Turkey
TDA-TDZ	Guatemala
TEA-TEZ	Costa Rica
TFA-TFZ	Iceland
TGA-TGZ	Guatemala
THA-THZ	France
TIA-TIZ	Costa Rica
TJA-TJZ	Cameroon
TKA-TKZ	France
TLA-TLZ	Central Africa
TMA-TMZ	France
TNA-TNZ	Congo (Republic of the)
TOA-TQZ	France
TRA-TRZ	Gabon
TSA-TSZ	Tunisia
TTA-TTZ	Chad
TUA-TUZ	Ivory Coast
TVA-TXZ	France
TYA-TYZ	Benin
TZA-TZZ	Mali
T2A-T2Z	Tuvalu
T3A-T3Z	Kiribati
T4A-T4Z	Cuba
T5A-T5Z	Somalia
T6A-T6Z	Afghanistan
T7A-T7Z	San Marino
T8A-T8Z	Palau
T9A-T9Z	Bosnia and Herzegovina
UAA-UIZ	Russian Federation
UJA-UMZ	Uzbekistan
UNA-UQZ	Kazakhstan
URA-UZZ	Ukraine
VAA-VGZ	Canada
VHA-VNZ	Australia
VOA-VOZ	Canada
VPA-VQZ	United Kingdom of Great Britain and Northern Ireland
VRA-VRZ	China (People's Republic of)—Hong Kong
VSA-VSZ	United Kingdom of Great Britain and Northern Ireland
VTA-VWZ	India
VXA-VYZ	Canada
VZA-VZZ	Australia
V2A-V2Z	Antigua and Barbuda
V3A-V3Z	Belize
V4A-V4Z	Saint Kitts and Nevis
V5A-V5Z	Namibia
V6A-V6Z	Micronesia
V7A-V7Z	Marshall Islands
V8A-V8Z	Brunei
WAA-WZZ	United States of America
XAA-XIZ	Mexico
XJA-XOZ	Canada
XPA-XPZ	Denmark
XQA-XRZ	Chile
XSA-XSZ	China
XTA-XTZ	Burkina Faso
XUA-XUZ	Cambodia
XVA-XVZ	Viet Nam
XWA-XWZ	Laos
XXA-XXZ	Portugal
XYA-XZZ	Myanmar
YAA-YAZ	Afghanistan
YBA-YHZ	Indonesia
YIA-YIZ	Iraq
YJA-YJZ	Vanuatu
YKA-YKZ	Syria
YLA-YLZ	Latvia
YMA-YMZ	Turkey
YNA-YNZ	Nicaragua
YOA-YRZ	Romania
YSA-YSZ	El Salvador
YTA-YUZ	Yugoslavia
YVA-YYZ	Venezuela
YZA-YZZ	Yugoslavia
Y2A-Y9Z	Germany
ZAA-ZAZ	Albania
ZBA-ZJZ	United Kingdom of Great Britain and Northern Ireland
ZKA-ZMZ	New Zealand
ZNA-ZOZ	United Kingdom of Great Britain and Northern Ireland
ZPA-ZPZ	Paraguay
ZQA-ZQZ	United Kingdom of Great Britain and Northern Ireland
ZRA-ZUZ	South Africa
ZVA-ZZZ	Brazil
Z2A-Z2Z	Zimbabwe
Z3A-Z3Z	Macedonia (Former Yugoslav Republic)
2AA-2ZZ	United Kingdom of Great Britain and Northern Ireland
3AA-3AZ	Monaco
3BA-3BZ	Mauritius
3CA-3CZ	Equatorial Guinea
3DA-3DM	Swaziland
3DN-3DZ	Fiji
3EA-3FZ	Panama
3GA-3GZ	Chile
3HA-3UZ	China
3VA-3VZ	Tunisia
3WA-3WZ	Viet Nam
3XA-3XZ	Guinea
3YA-3YZ	Norway
3ZA-3ZZ	Poland
4AA-4CZ	Mexico
4DA-4IZ	Philippines
4JA-4KZ	Azerbaijani Republic
4LA-4LZ	Georgia
4MA-4MZ	Venezuela
4NA-4OZ	Yugoslavia
4PA-4SZ	Sri Lanka
4TA-4TZ	Peru
*4UA-4UZ	United Nations
4VA-4VZ	Haiti
*4WA-4WZ	United Nations
4XA-4XZ	Israel
*4YA-4YZ	International Civil Aviation Organization
4ZA-4ZZ	Israel
5AA-5AZ	Libya
5BA-5BZ	Cyprus
5CA-5GZ	Morocco
5HA-5IZ	Tanzania
5JA-5KZ	Colombia
5LA-5MZ	Liberia
5NA-5OZ	Nigeria
5PA-5QZ	Denmark
5RA-5SZ	Madagascar
5TA-5TZ	Mauritania
5UA-5UZ	Niger
5VA-5VZ	Togo
5WA-5WZ	Western Samoa
5XA-5XZ	Uganda
5YA-5ZZ	Kenya
6AA-6BZ	Egypt
6CA-6CZ	Syria
6DA-6JZ	Mexico
6KA-6NZ	Republic of Korea
6OA-6OZ	Somalia
6PA-6SZ	Pakistan
6TA-6UZ	Sudan
6VA-6WZ	Senegal
6XA-6XZ	Madagascar
6YA-6YZ	Jamaica
6ZA-6ZZ	Liberia
7AA-7IZ	Indonesia
7JA-7NZ	Japan
7OA-7OZ	Yemen
7PA-7PZ	Lesotho
7QA-7QZ	Malawi
7RA-7RZ	Algeria
7SA-7SZ	Sweden
7TA-7YZ	Algeria
7ZA-7ZZ	Saudi Arabia
8AA-8IZ	Indonesia
8JA-8NZ	Japan
8OA-8OZ	Botswana
8PA-8PZ	Barbados
8QA-8QZ	Maldives
8RA-8RZ	Guyana
8SA-8SZ	Sweden
8TA-8YZ	India
8ZA-8ZZ	Saudi Arabia
9AA-9AZ	Croatia
9BA-9DZ	Iran
9EA-9FZ	Ethiopia
9GA-9GZ	Ghana
9HA-9HZ	Malta
9IA-9JZ	Zambia
9KA-9KZ	Kuwait
9LA-9LZ	Sierra Leone
9MA-9MZ	Malaysia
9NA-9NZ	Nepal
9OA-9TZ	Democratic Republic of the Congo
9UA-9UZ	Burundi
9VA-9VZ	Singapore
9WA-9WZ	Malaysia
9XA-9XZ	Rwanda
9YZ-9ZZ	Trinidad and Tobago

Notes:

*Series allocated to an international organization

†In response to Resolution 99 (Minneapolis, 1998) of the Plenipotentiary Conference

Table 7.69

FCC-Allocated Prefixes for Areas Outside the Continental US

Prefix	*Location*
AH1, KH1, NH1, WH1	Baker, Howland Is
AH2, KH2, NH2, WH2	Guam
AH3, KH3, NH3, WH3	Johnston I
AH4, KH4, NH4, WH4	Midway I
AH5K, KH5K, NH5K, WH5K	Kingman Reef
AH5, KH5, NH5, WH5 (except K suffix)	Palmyra, Jarvis Is
AH6-7, KH6-7, NH6-7, WH6-7	Hawaii
AH7K, KH7K, NH7K, WH7K	Kure I
AH8, KH8, NH8, WH8	American Samoa
AH9, KH9, NH9, WH9	Wake, Wilkes, Peale Is
AHØ, KHØ, NHØ, WHØ	Northern Mariana Is
AL, KL, NL, WL	Alaska
KP1, NP1, WP1	Navassa
KP2, NP2, WP2	Virgin Is
KP3-4, NP3-4, WP3-4	Puerto Rico
KP5, NP5, WP5	Desecheo

Table 7.70

DX Operating Code

For W/VE Amateurs

Some DXers have caused considerable confusion and interference in their efforts to work DX stations. The points below, if observed by all W/VE amateurs, will help make DX more enjoyable for all.

1) *Call* DX only after he calls CQ, QRZ? or signs SK, or voice equivalents thereof. Make your calls short.

2) Do not call a DX station:

a) On the frequency of the station he is calling until you are sure the QSO is over (SK).

b) Because you hear someone else calling him.

c) When he signs KN, AR or CL.

d) Exactly on his frequency.

e) After he calls a directional CQ, unless of course you are in the right direction or area.

3) Keep within frequency band limits. Some DX stations can get away with working outside, but you cannot.

4) Observe calling instructions given by DX stations. Example: 15U means "call 15 kHz up from my frequency." 15D means down, etc.

5) Give honest reports. Many DX stations depend on W/VE reports for adjustment of station and equipment.

6) Keep your signal clean. Key clicks, ripple, feedback or splatter gives you a bad reputation and may get you a citation from the FCC.

7) *Listen* and call the station you want. Calling CQ DX is not the best assurance that the rare DX will reply.

8) When there are several W or VE stations waiting, avoid asking DX to "listen for a friend." Also avoid engaging him in a ragchew against his wishes.

For Overseas Amateurs

To all overseas amateur stations:
In their eagerness to work you, many W and VE amateurs resort to practices that cause confusion and QRM. Most of this is good-intentioned but ill-advised; some of it is intentional and selfish. The key to the cessation of unethical DX operating practices is in your hands. We believe that your adoption of certain operating habits will increase your enjoyment of Amateur Radio and that of amateurs on this side who are eager to work you. We recommend your adoption of the following principles:

1) Do not answer calls on your own frequency.

2) Answer calls from W/VE stations only when their signals are of good quality.

3) Refuse to answer calls from other stations when you are already in contact with someone, and do not acknowledge calls from amateurs who indicate they wish to be "next."

4) Give *everybody* a break. When many W/VE amateurs are patiently and quietly waiting to work you, avoid complying with requests to "listen for a friend."

5) Tell listeners where to call you by indicating how many kilohertz up (U) or down (D) from your frequency you are listening.

6) Use the ARRL-recommended ending signals, especially KN to indicate to impatient listeners the status of the QSO. KN means "Go ahead (specific station); all others keep out."

7) Let it be known that you avoid working amateurs who are constant violators of these principles.

Table 7.71

W1AW SCHEDULE

Pacific	Mtn	Central	East	Mon	Tue	Wed	Thu	Fri
6 AM	7 AM	8 AM	9 AM		Fast Code	Slow Code	Fast Code	Slow Code
7 AM - 1 PM	8 AM - 2 PM	9 AM - 3 PM	10 AM - 4 PM	Visiting Operator Time (12 PM - 1 PM closed for lunch)				
1 PM	2 PM	3 PM	4 PM	Fast Code	Slow Code	Fast Code	Slow Code	Fast Code
2 PM	3 PM	4 PM	5 PM	Code Bulletin				
3 PM	4 PM	5 PM	6 PM	Teleprinter Bulletin				
4 PM	5 PM	6 PM	7 PM	Slow Code	Fast Code	Slow Code	Fast Code	Slow Code
5 PM	6 PM	7 PM	8 PM	Code Bulletin				
6 PM	7 PM	8 PM	9 PM	Teleprinter Bulletin				
6:45 PM	7:45 PM	8:45 PM	9:45 PM	Voice Bulletin				
7 PM	8 PM	9 PM	10 PM	Fast Code	Slow Code	Fast Code	Slow Code	Fast Code
8 PM	9 PM	10 PM	11 PM	Code Bulletin				

W1AW's schedule is at the same local time throughout the year. The schedule according to your local time will change if your local time does not have seasonal adjustments that are made at the same time as North American time changes between standard time and daylight time. From the first Sunday in April to the last Sunday in October, UTC = Eastern Time + 4 hours. For the rest of the year, UTC = Eastern Time + 5 hours.

Morse code transmissions:

Frequencies are 1.818, 3.5815, 7.0475, 14.0475, 18.0975, 21.0675, 28.0675 and 147.555 MHz.
Slow Code = practice sent at 5, 71/2, 10, 13 and 15 wpm.
Fast Code = practice sent at 35, 30, 25, 20, 15, 13 and 10 wpm.
Code practice text is from the pages of *QST*. The source is given at the beginning of each practice session and alternate speeds within each session. For example, "Text is from June 2003 *QST*, pages 9 and 81," indicates that the plain text is from the article on page 9 and mixed number/letter groups are from page 81.
Code bulletins are sent at 18 wpm.
W1AW qualifying runs are sent on the same frequencies as the Morse code transmissions. West Coast qualifying runs are transmitted on approximately 3.590 MHz by K6YR. At the beginning of each code practice session, the schedule for the next qualifying run is presented. Underline one minute of the highest speed you copied, certify that your copy was made without aid, and send it to ARRL for grading. Please include your name, call sign (if any) and complete mailing address. The fee structure is $10 for a certificate and $7.50 for endorsements.

Teleprinter transmissions:

Frequencies are 3.625, 7.095, 14.095, 18.1025, 21.095, 28.095 and 147.555 MHz.
Bulletins are sent at 45.45-baud Baudot and 100-baud AMTOR, FEC Mode B. 110-baud ASCII will be sent only as time allows.
On Tuesdays and Fridays at 6:30 PM Eastern Time, Keplerian elements for many amateur -satellites are sent on the regular teleprinter frequencies.

Voice transmissions:

Frequencies are 1.855, 3.99, 7.29, 14.29, 18.16, 21.39, 28.59 and 147.555 MHz.

Miscellanea:

On Fridays, UTC, a DX bulletin replaces the regular bulletins.
W1AW is open to visitors from 10 AM until noon and from 1 PM until 3:45 PM on Monday through Friday. FCC-licensed amateurs may operate the station during that time. Be sure to bring your current FCC amateur license or a photocopy.
In a communication emergency, monitor W1AW for special bulletins as follows: voice on the hour, teleprinter at 15 minutes past the hour, and CW on the half hour.
Headquarters and W1AW are closed on New Year's Day, President's Day, Good Friday, Memorial Day, Independence Day, Labor Day, Thanksgiving and the following Friday, and Christmas Day and the following day.

Table 7.72

ARRL Procedural Signals (Prosigns)

In general, the CW prosigns are used on all data modes as well, although word abbreviations may be spelled out. That is, "CLEAR" might be used rather than "CL" on radioteletype. Additional radioteletype conventions appear at the end of the table.

Situation	*CW*	*Voice*
check for a clear frequency	QRL?	Is the frequency in use?
seek contact with any station	CQ	CQ
after call to specific named station or to indicate end of message	AR	over, end of message
invite any station to transmit	K	go
invite a specific named station to transmit	KN	go only
invite receiving station to transmit	BK	back to you
all received correctly	R	received
please stand by	AS	wait, stand by
end of contact (sent before call sign)	SK	clear
going off the air	CL	closing station

Additional RTTY prosigns

SK QRZ—Ending contact, but listening on frequency.
SK KN—Ending contact, but listening for one last transmission from the other station.
SK SZ—Signing off and listening on the frequency for any other calls.

Table 7.73

Q Signals

These Q signals most often need to be expressed with brevity and clarity in amateur work. (Q abbreviations take the form of questions only when each is sent followed by a question mark.)

QRA What is the name of your station? The name of your station is _______.

QRG Will you tell me my exact frequency (or that of ______)? Your exact frequency (or that of ______) is ______ kHz.

QRH Does my frequency vary? Your frequency varies.

QRI How is the tone of my transmission? The tone of your transmission is ______ (1. Good; 2. Variable; 3. Bad).

QRJ Are you receiving me badly? I cannot receive you. Your signals are too weak.

QRK What is the intelligibility of my signals (or those of ______)? The intelligibility of your signals (or those of ______) is ______ (1. Bad; 2. Poor; 3. Fair; 4. Good; 5. Excellent).

QRL Are you busy? I am busy (or I am busy with ______). Please do not interfere.

QRM Is my transmission being interfered with? Your transmission is being interfered with (1. Nil; 2. Slightly; 3. Moderately; 4. Severely; 5. Extremely.)

QRN Are you troubled by static? I am troubled by static ______ (1-5 as under QRM).

QRO Shall I increase power? Increase power.

QRP Shall I decrease power? Decrease power.

QRQ Shall I send faster? Send faster (______ WPM).

QRS Shall I send more slowly? Send more slowly (______ WPM).

QRT Shall I stop sending? Stop sending.

QRU Have you anything for me? I have nothing for you.

QRV Are you ready? I am ready.

QRW Shall I inform ______ that you are calling on ______ kHz? Please inform ______ that I am calling on ______ kHz.

QRX When will you call me again? I will call you again at ______ hours (on ______ kHz).

QRY What is my turn? Your turn is numbered ______

QRZ Who is calling me? You are being called by ______ (on ______ kHz).

QSA What is the strength of my signals (or those of _____)? The strength of your signals (or those of ______) is ______ (1. Scarcely perceptible; 2. Weak; 3. Fairly good; 4. Good; 5. Very good).

QSB Are my signals fading? Your signals are fading.

QSD Is my keying defective? Your keying is defective.

QSG Shall I send ______ messages at a time? Send ______ messages at a time.

QSK Can you hear me between your signals and if so can I break in on your transmission? I can hear you between my signals; break in on my transmission.

QSL Can you acknowledge receipt? I am acknowledging receipt.

QSM Shall I repeat the last message which I sent you, or some previous message? Repeat the last message which you sent me [or message(s) number(s) ______].

QSN Did you hear me (or ______) on ______ kHz? I did hear you (or ______) on ______ kHz.

QSO Can you communicate with ______ direct or by relay? I can communicate with ______ direct (or by relay through ______).

QSP Will you relay to ______? I will relay to ______

QST General call preceding a message addressed to all amateurs and ARRL members. This is in effect "CQ ARRL."

QSU Shall I send or reply on this frequency (or on ______ kHz)? Send or reply on this frequency (or ______ kHz).

QSV Shall I send a series of Vs on this frequency (or on ______ kHz)? Send a series of Vs on this frequency (or on ______ kHz).

QSW Will you send on this frequency (or on ______ kHz)? I am going to send on this frequency (or on _____ kHz).

QSX Will you listen to ______ on ______ kHz? I am listening to ______ on ______ kHz.

QSY Shall I change to transmission on another frequency? Change to transmission on another frequency (or on ______ kHz).

QSZ	Shall I send each word or group more than once? Send each word or group twice (or _____ times).
QTA	Shall I cancel message number _____? Cancel message number _____
QTB	Do you agree with my counting of words? I do not agree with your counting of words. I will repeat the first letter or digit of each word or group.
QTC	How many messages have you to send? I have ______ messages for you (or for _____).
QTH	What is your location? My location is _____
QTR	What is the correct time? The correct time is _____
QTV	Shall I stand guard for you? Stand guard for me.
QTX	Will you keep your station open for further communication with me? Keep your station open for me.
QUA	Have you news of _____? I have news of _____.

ARRL QN Signals

QNA*	Answer in prearranged order.
QNB	Act as relay between _____ and _____.
QNC	All net stations copy. I have a message for all net stations.
QND*	Net is Directed (Controlled by net control station.)
QNE*	Entire net stand by.
QNF	Net is Free (not controlled).
QNG	Take over as net control station
QNH	Your net frequency is High.
QNI	Net stations report in. I am reporting into the net. (Follow with a list of traffic or QRU.)
QNJ	Can you copy me?
QNK*	Transmit messages for _____ to _____.
QNL	Your net frequency is Low.
QNM*	You are QRMing the net. Stand by.
QNN	Net control station is _____. What station has net control?
QNO	Station is leaving the net.
QNP	Unable to copy you. Unable to copy _____.
QNQ*	Move frequency to _____ and wait for _____ to finish handling traffic. Then send him traffic for _____.
QNR*	Answer _____ and Receive traffic.
QNS	Following Stations are in the net.* (follow with list.) Request list of stations in the net.
QNT	I request permission to leave the net for _____ minutes.
QNU*	The net has traffic for *you*. Stand by.
QNV*	Establish contact with _____ on this frequency. If successful, move to _____ and send him traffic for _____.
QNW	How do I route messages for _____?
QNX	You are excused from the net.*
QNY*	Shift to another frequency (or to _____ kHz) to clear traffic with _____.
QNZ	Zero beat your signal with mine.

***For use only by the Net Control Station.**

Notes on Use of QN Signals

These QN signals are special ARRL signals for use in amateur CW nets *only*. They are not for use in casual amateur conversation. Other meanings that may be used in other services do not apply. Do not use QN signals on phone nets. *Say it with words.* QN signals need not be followed by a question mark, even though the meaning may be interrogatory.

Table 7.74
The RST System

Readability

1—Unreadable.
2—Barely readable, occasional words distinguishable.
3—Readable with considerable difficulty.
4—Readable with practically no difficulty.
5—Perfectly readable.

Signal Strength

1—Faint signals, barely perceptible.
2—Very weak signals.
3—Weak signals.
4—Fair signals.
5—Fairly good signals.
6—Good signals.
7—Moderately strong signals.
8—Strong signals.
9—Extremely strong signals.

Tone

1—Sixty-cycle ac or less, very rough and broad.
2—Very rough ac, very harsh and broad.
3—Rough ac tone, rectified but not filtered.
4—Rough note, some trace of filtering.
5—Filtered rectified ac but strongly ripple-modulated.
6—Filtered tone, definite trace of ripple modulation.
7—Near pure tone, trace of ripple modulation.
8—Near perfect tone, slight trace of modulation.
9—Perfect tone, no trace of ripple of modulation of any kind.

If the signal has the characteristic steadiness of crystal control, add the letter X to the RST report. If there is a chirp, add the letter C. Similarly for a click, add K. (See FCC Regulations §97.307, Emissions Standards.) The above reporting system is used on both CW and voice; leave out the "tone" report on voice.

Table 7.75

CW Abbreviations

Abbreviation	Meaning
AA	All after
AB	All before
AB	About
ADR	Address
AGN	Again
ANT	Antenna
BCI	Broadcast interference
BCL	Broadcast listener
BK	Break; break me; break in
BN	All between; been
BUG	Semi-automatic key
B4	Before
C	Yes
CFM	Confirm; I confirm
CK	Check
CL	I am closing my station; call
CLD-CLG	Called; calling
CQ	Calling any station
CUD	Could
CUL	See you later
CW	Continuous wave (i.e., radiotelegraph)
DE	From
DLD-DLVD	Delivered
DR	Dear
DX	Distance, foreign countries
ES	And, &
FB	Fine business, excellent
FM	Frequency modulation
GA	Go ahead (or resume sending)
GB	Good-by
GBA	Give better address
GE	Good evening
GG	Going
GM	Good morning
GN	Good night
GND	Ground
GUD	Good
HI	The telegraphic laugh; high
HR	Here, hear
HV	Have
HW	How
LID	A poor operator
MA, MILS	Milliamperes
MSG	Message; prefix to radiogram
N	No
NCS	Net control station
ND	Nothing doing
NIL	Nothing; I have nothing for you
NM	No more
NR	Number
NW	Now; I resume transmission
OB	Old boy
OC	Old chap
OM	Old man
OP-OPR	Operator
OT	Old timer; old top
PBL	Preamble
PSE	Please
PWR	Power
PX	Press
R	Received as transmitted; are
RCD	Received
RCVR (RX)	Receiver
REF	Refer to; referring to; reference
RFI	Radio Frequency Interference
RIG	Station equipment
RPT	Repeat; I repeat; report
RTTY	Radioteletype
RX	Receiver
SASE	Self-addressed, stamped envelope
SED	Said
SIG	Signature; signal
SINE	Operator's personal initials or nickname
SKED	Schedule
SRI	Sorry
SSB	Single sideband
SVC	Service; prefix to service message
T	Zero
TFC	Traffic
TMW	Tomorrow
TNX-TKS	Thanks
TT	That
TU	Thank you
TVI	Television interference
TX	Transmitter
TXT	Text
UR-URS	Your; you're; yours
VFO	Variable-frequency oscillator
VY	Very
WA	Word after
WB	Word before
WD-WDS	Word; words
WKD-WKG	Worked; working
WL	Well; will
WUD	Would
WX	Weather
XCVR	Transceiver
XMTR (TX)	Transmitter
XTAL	Crystal
XYL (YF)	Wife
YL	Young lady
73	Best regards
88	Love and Kisses

Although abbreviations help to cut down unnecessary transmission, make it a rule not to abbreviate unnecessarily when working an operator of unknown experience.

Table 7.76

ITU Recommended Phonetics

A — Alfa (**AL** FAH)
B — Bravo (**BRAH** VOH)
C — Charlie (**CHAR** LEE OR **SHAR** LEE)
D — Delta (**DELL** TAH)
E — Echo (**ECK** OH)
F — Foxtrot (**FOKS** TROT)
G — Golf (GOLF)
H — Hotel (HOH **TELL**)
I — India (**IN** DEE AH)
J — Juliet (**JEW** LEE ETT)
K — Kilo (**KEY** LOH)
L — Lima (**LEE** MAH)
M — Mike (MIKE)
N — November (NO **VEM** BER)
O — Oscar (**OSS** CAH)
P — Papa (PAH **PAH**)
Q — Quebec (KEH **BECK**)
R — Romeo (**ROW** ME OH)
S — Sierra (SEE *AIR* RAH)
T — Tango (**TANG** GO)
U — Uniform (**YOU** NEE FORM or **OO** NEE FORM)
V — Victor (**VIK** TAH)
W — Whiskey (**WISS** KEY)
X — X-Ray (**ECKS** RAY)
Y — Yankee (**YANG** KEY)
Z — Zulu (**ZOO** LOO)

Note: The **Boldfaced** syllables are emphasized. The pronunciations shown in the table were designed for speakers from all international languages. The pronunciations given for "Oscar" and "Victor" may seem awkward to English-speaking people in the U.S.

Table 7.77
ARRL Log

See inside front cover. (MODE) — Output in Watts. (POWER) — UTC recommended. (TIME) — RST. See back inside cover. (REPORT) — This column may also be used for contest-exchange info received. (TIME OFF)

FIXED				VARIABLE									
DATE	FREQ.	MODE	POWER	TIME	STATION WORKED	REPORT SENT	REPORT REC'D	TIME OFF	COMMENTS QTH	NAME	QSL VIA	QSL S	QSL R
28 JUL	146.52	FM	10	0430	WA1CCR				Wallingford	Eric	New converter works!		
3 OCT	7.0	CW	150	2319	WA6VEF	001	322	contra cos	CALIFORNIA QSO PARTY				
				22	N6OJ	002	157	Sono					
				24	K6NA	003	331	SD					
				31	N6OP/M	004	117	Calav					
9 OCT	28.6	SSB	1 KW	0301	JA1OCA	59	57		Tokyo	Isao	Buro	✓	
	21	CW		1545	EA9GD	559	579		Melilla	Jose	Box 348	✓	✓
				56	6OØDX	599	599		Somalia		I2YAE	✓	
5 NOV	3.810.2	SSB	150	0030	W9NA	59+	59+	0117	Wausau, WI	Reno			
9 Nov	21	CW	10	1642	G4BUE	339	449	1657		1 watt!			

The ARRL Log is adaptable for all types of operating—ragchewing, contesting, DXing. References are to pages in the ARRL Log.

Table 7.78
ARRL Operating Awards

Award	*Qualification*
Worked All States (WAS)	QSLs from all 50 US states
Worked All Continents (WAC)	QSLs from all six continents
DX Century Club (DXCC)	QSLs from at least 100 different countries
VHF/UHF Century Club (VUCC)	QSLs from many grid squares
A-1 Operator Club	Recommendation by two A-1 operators
Code Proficiency	One minute of perfect copy from W1AW qualifying run
ARRL Membership	ARRL membership for 25, 40, 50, 60 or 70 years

Table 7.79
ARRL Membership QSL Card

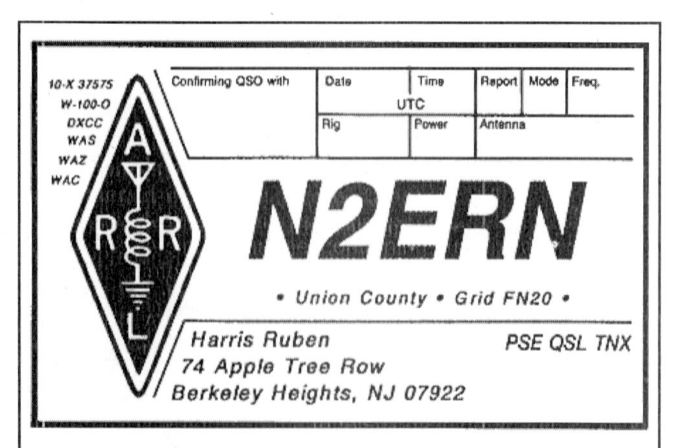

The ARRL membership QSL card. This example is from Harris Ruben, N2ERN, who designed the card. Your card would reflect your own call sign and address; awards and VUCC grid-square are optional. ARRL does not print or sell the cards. Inquire with printers who advertise in the *QST* Ham Ads.

Table 7.80
Mode Abbreviations for QSL Cards

Abbreviation	*Explanation*
CW	Telegraphy
DATA	Telemetry, telecommand and computer communications (includes packet radio)
IMAGE	Facsimile and television
MCW	Tone-modulated telegraphy
PHONE	Speech and other sound
PULSE	Modulated main carrier
RTTY	Direct-printing telegraphy (includes AMTOR)
SS	Spread Spectrum
TEST	Emissions containing no information

Note: For additional information on emission types refer to latest edition of *The FCC Rule Book*.

Table 7.81 US/Canada Map

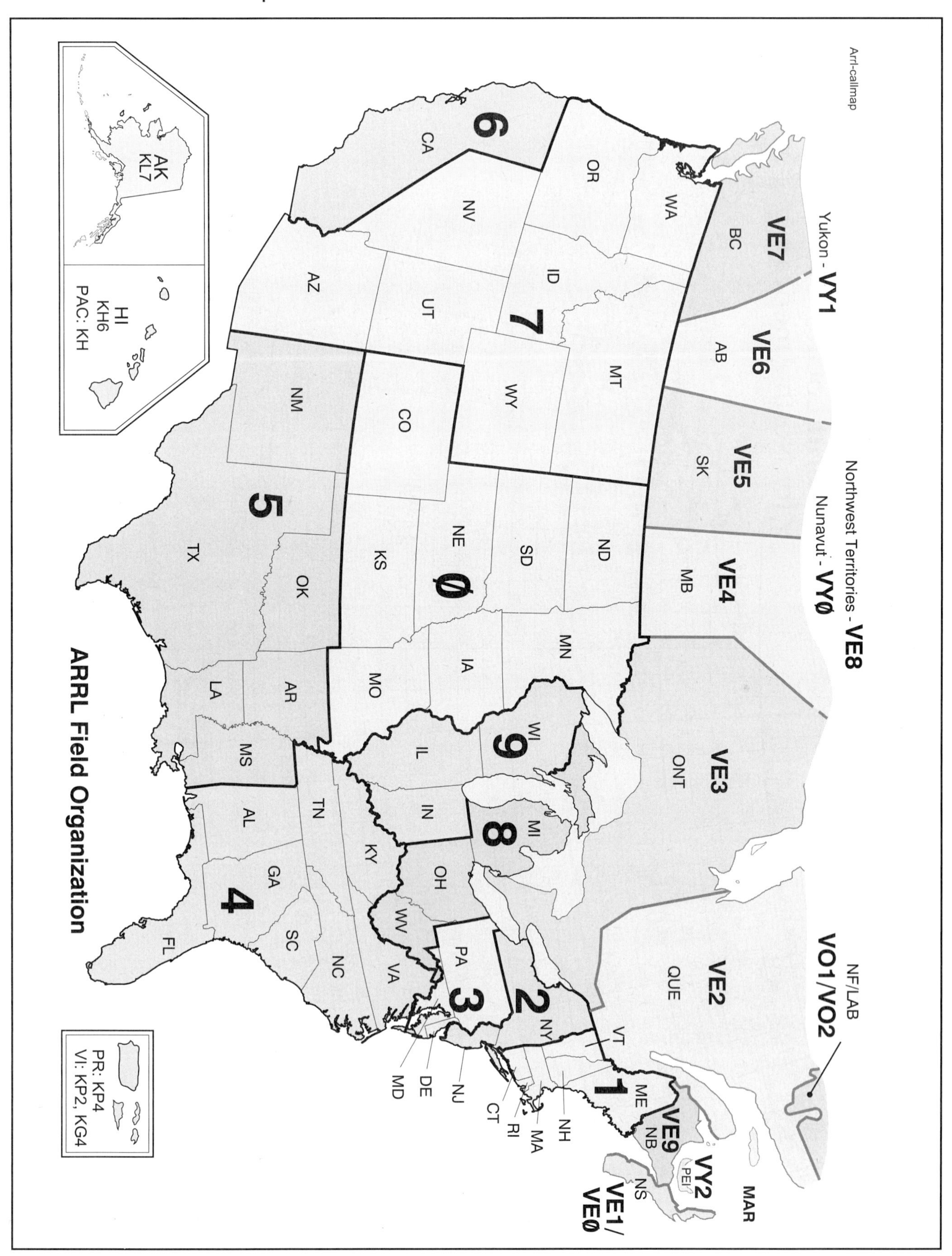

A map showing US states, Canadian provinces and ARRL/RAC Sections.

Table 7.82

ARRL Grid Locator Map for North America

This and a World Grid Locator Map are available from ARRL.

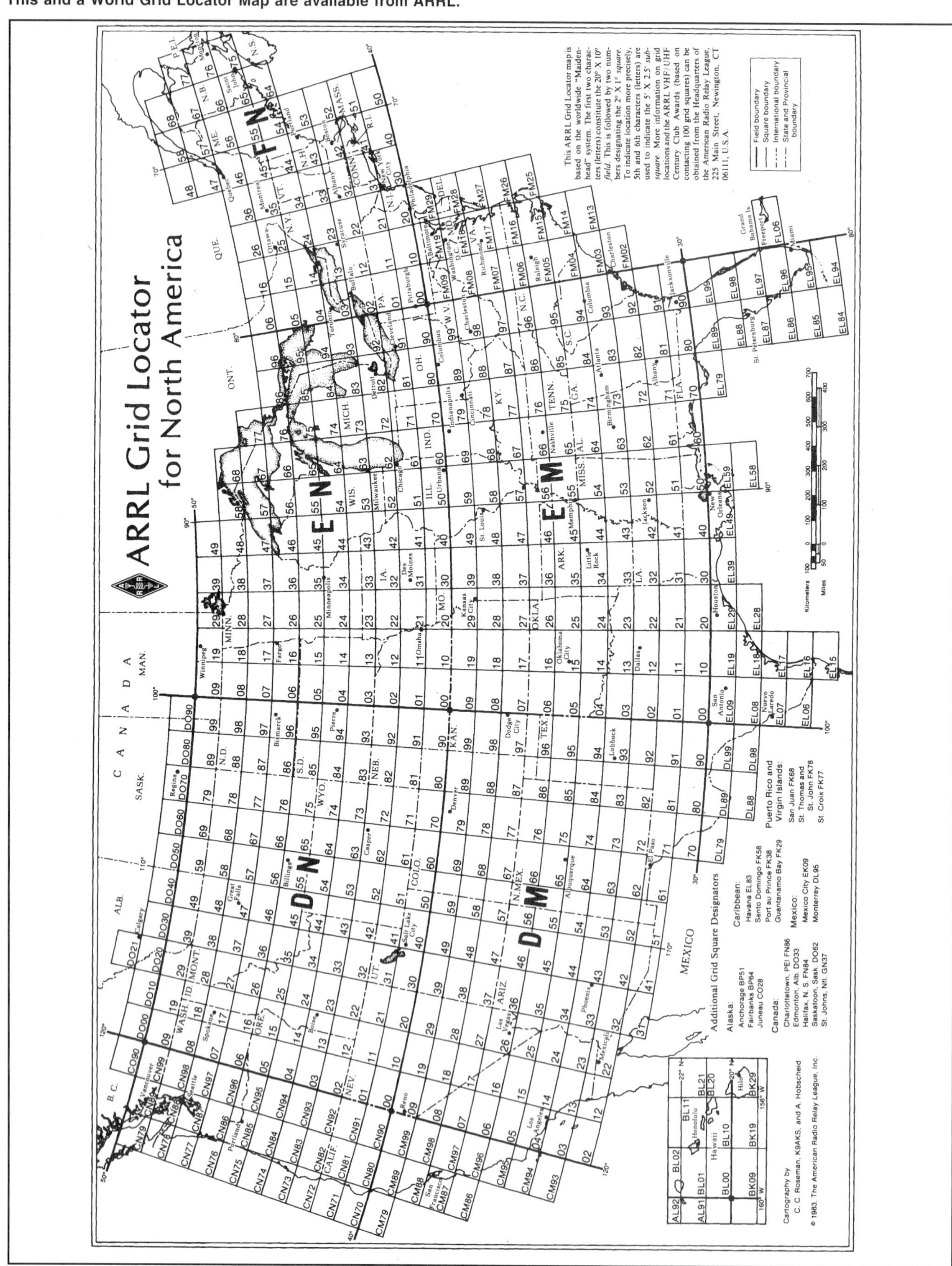

Table 7.83
Amateur Message Form

Every formal radiogram message originated and handled should contain the following component parts in the order given.

I PREAMBLE

- *a*. Number (begin with 1 each month or year)
- *b*. Precedence (R, W, P or EMERGENCY)
- *c*. Handling Instructions (optional, see text)
- *d*. Station of Origin (first amateur handler)
- *e*. Check (number of words/groups in text only)
- *f*. Place of Origin (not necessarily location of station of origin)
- *g*. Time Filed (optional with originating station)
- *h*. Date (must agree with date of time filed)

II ADDRESS (as complete as possible, include zip code and telephone number)

III TEXT (limit to 25 words of less, if possible)

IV SIGNATURE

CW MESSAGE EXAMPLE

I   NR 1 R HXG W1AW 8 NEWINGTON CT 1830Z JULY 1

a b c d e f g h

II   DONALD SMITH $\overline{AA}$
160 EAST SIXTH AVE $\overline{AA}$
NORTH RIVER CITY MO 00789 $\overline{AA}$
733 4868 $\overline{BT}$

III   HAPPY BIRTHDAY X SEE YOU SOON X LOVE $\overline{BT}$

IV   DIANA $\overline{AR}$

Note that X, when used in the text as punctuation, counts as a word.

CW: The prosign $\overline{AA}$ separates the parts of the message. $\overline{BT}$ separates the address from the text the text from the signature. $\overline{AR}$ marks the end of message; this is followed by B if there is another message to follow, by N if this is the only or last message. It is customary to copy the preamble, parts of the address, text and signature on separate lines.

RTTY: Same as CW procedure above, except (1) use extra space between parts of address, instead of $\overline{AA}$; (2) omit CW procedure sign $\overline{BT}$ to separate text from address and signature, using line spaces instead; (3) add a CFM line under the signature, consisting of all names, numerals and unusual words in the message in the order transmitted.

PACKET/AMTOR BBS: Same format as shown in the CW message example above, except that the $\overline{AA}$ and $\overline{AR}$ prosigns may be omitted. Most AMTOR and Packet BBS software in use today allow formal message traffic to be sent with the "ST" command. Always avoid the use of spectrum-wasting multiple line feeds and indentations.

PHONE: Use prowords instead of prosigns, but it is not necessary to name each part of the message as you send it. For example, the above message would be sent on phone as follows: "Number one routine HX Golf W1AW eight Newington Connecticut one eight three zero zulu July one Donald Smith *Figures* one six four East Sixth Avenue North River City Missouri zero zero seven eight nine *Telephone* seven three three four nine six eight *Break* Happy Birthday X-ray see you soon X-ray love *Break* Diana *End of Message Over*. "End of Message" is followed by "More" if there is another message to follow, "No More" if it is the only or last message. Speak clearly using VOX (or pause frequently on push-to-talk) so that the receiving station can get his fills. Spell phonetically all difficult or unusual words—do *not* spell out common words. Do not use CW abbreviations or Q-signals in phone traffic handling.

PRECEDENCES

The precedence will fill the message number. For example, on CW 207 R or 207 EMERGENCY. On phone, "Two Zero Seven Routine (or Emergency)."

EMERGENCY—Any message having life and death urgency to any person or group of persons, which is transmitted by Amateur Radio in the absence of regular commercial facilities. This includes official messages of welfare agencies during emergencies requesting supplies, materials or instructions vital to relief of stricken populace in emergency areas. During normal times, it will be *very rare*. On CW, RTTY and other digital modes this designation will always be spelled out. When in doubt, *do not* use it.

PRIORITY—Important messages having a specific time limit. Official messages not covered in the Emergency category. Press dispatches and other emergency-related traffic not of the utmost urgency. Notification of death or injury in a disaster area, personal or official. Use the abbreviation P on CW.

WELFARE—A message that is either (a) an inquiry as to the health and welfare of an individual in the disaster area (b) an advisory or reply from the disaster area that indicates that all is well should carry this precedence, which is abbreviated W on CW. These messages are handled *after* Emergency and Priority traffic but before Routine.

ROUTINE—Most traffic normal times will bear this designation. In disaster situations, traffic labeled Routine (R on CW) should be handled *last*, or not at all when circuits are busy with Emergency, Priority or Welfare traffic.

Handling Instructions (Optional)

HXA—(Followed by number.) Collect landline delivery authorized by addressee within ......miles. (If no number, authorization is unlimited.)
HXB—(Followed by number.) Cancel message if not delivered within ......hours of filing time; service originating station.
HXC—Report date and time of delivery (TOD) to originating station.
HXD—Report to originating station the identify of station from which received, plus date and time. Report identity of station to which relayed, plus date and time, or if delivered report date, time and method of delivery.
HXE—Delivering station get reply from addressee, originate message back.
HXF—(Followed by number.) Hold delivery until......(date).
HXG—Delivery by mail or landline toll call not required. If toll or other expense involved, cancel message and service originating station.

For further information on traffic handling, consult the *Public Service Communications Manual* or *The ARRL Operating Manual*, both published by the ARRL.

Table 7.84
A Simple NTS Formal Message

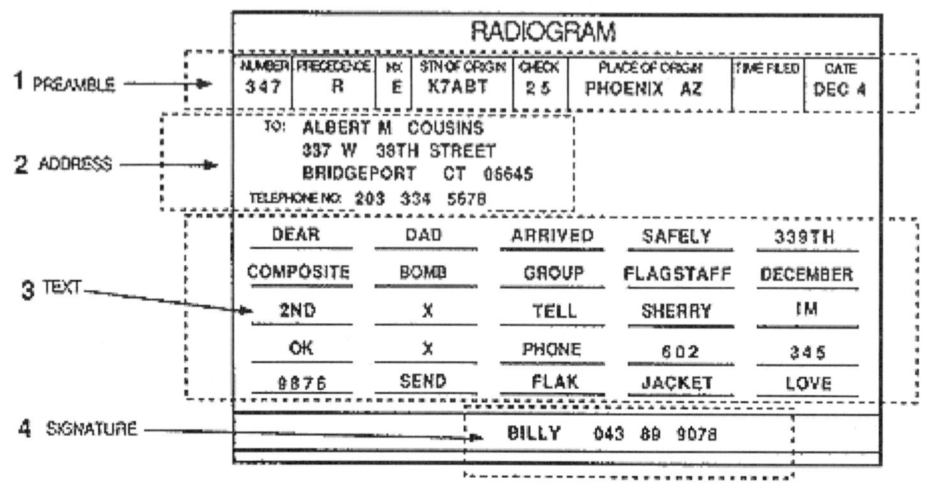

Table 7.85
Handling Instructions

HXA—(Followed by number.) Collect landline delivery authorized by addressee within _____ miles. (If no number, authorization is unlimited.)

HXB—(Followed by number.) Cancel messages if not delivered within ____ hours of filing time; service originating station.

HXC—Report date and time of delivery (TOD) to originating station.

HXD—Report to originating station the identity of station from which received, plus date and time. Report identity of station to which relayed, plus date and time, or if delivered report date, time and method of delivery.

HXE—Delivering station get reply from addressee, originate message back.

HXF—(Followed by number.) Hold delivery until ______ (date).

HXG—Delivery by mail or landline toll call not required. If toll or other expense involved, cancel message and service originating station.

An HX prosign (when used) will be inserted in the message preamble before the station of origin, thus: NR 207 R HXA50 W1AW 12...(etc). If more than one HX prosign is used they can be combined if no numbers are to be inserted; otherwise the HX should be repeated, thus: NR 207 R HXAC W1AW... (etc), but: NR 207 R HXA50 HXC W1AW...(etc). On phone, use phonetics for the letter or letters following the HX, to ensure accuracy.

Table 7.86

ARL Numbered Radiograms

The letters ARL are inserted in the preamble in the check and in the text before spelled out numbers, which represent texts from this list. Note that some ARL texts include insertion of numerals. *Example*: NR 1 R W1AW ARL 5 NEWINGTON CONN DEC 25 DONALD R SMITH AA 164 EAST SIXTH AVE AA NORTH RIVER CITY MO AA PHONE 733 3968 BT ARL FIFTY ARL SIXTY ONE BT DIANA AR.

Group One—For possible "Relief Emergency" Use

ONE	Everyone safe here. Please don't worry.
TWO	Coming home as soon as possible.
THREE	Am in _____ hospital. Receiving excellent care and recovering fine.
FOUR	Only slight property damage here. Do not be concerned about disaster reports.
FIVE	Am moving to new location. Send no further mail or communication. Will inform you of new address when relocated.
SIX	Will contact you as soon as possible.
SEVEN	Please reply by Amateur Radio through the amateur delivering this message. This is a free public service.
EIGHT	Need additional _______ mobile or portable equipment for immediate emergency use.
NINE	Additional _________ radio operators needed to assist with emergency at this location.
TEN	Please contact ______. Advise to standby and provide further emergency information, instructions or assistance.
ELEVEN	Establish Amateur Radio emergency communications with _____ on _____ MHz.
TWELVE	Anxious to hear from you. No word in some time. Please contact me as soon as possible.
THIRTEEN	Medical emergency situation exists here.
FOURTEEN	Situation here becoming critical. Losses and damage from _____ increasing.
FIFTEEN	Please advise your condition and what help is needed.
SIXTEEN	Property damage very severe in this area.
SEVENTEEN	REACT communications services also available. Establish REACT communications with _____ on channel _____.
EIGHTEEN	Please contact me as soon as possible at _____.
NINETEEN	Request health and welfare report on _______. (State name, address and telephone number.)
TWENTY	Temporarily stranded. Will need some assistance. Please contact me at _______.
TWENTY ONE	Search and Rescue assistance is needed by local authorities here. Advise availability.
TWENTY TWO	Need accurate information on the extent and type of conditions now existing at your location. Please furnish this information and reply without delay.
TWENTY THREE	Report at once the accessibility and best way to reach your location.
TWENTY FOUR	Evacuation of residents from this area urgently needed. Advise plans for help.
TWENTY FIVE	Furnish as soon as possible the weather conditions at your location.
TWENTY SIX	Help and care for evacuation of sick and injured from this location needed at once.

Emergency/priority messages originating from official sources must carry the signature of the originating official.

Group Two—Routine messages

FORTY SIX	Greetings on your birthday and best wishes for many more to come.
FIFTY	Greetings by Amateur Radio.
FIFTY ONE	Greetings by Amateur Radio. This message is sent as a free public service by ham radio operators here at ______. Am having a wonderful time.
FIFTY TWO	Really enjoyed being with you. Looking forward to getting together again.
FIFTY THREE	Received your _____. It's appreciated; many thanks.
FIFTY FOUR	Many thanks for your good wishes.
FIFTY FIVE	Good news is always welcome. Very delighted to hear about yours.
FIFTY SIX	Congratulations on your ______, a most worthy and deserved achievement.
FIFTY SEVEN	Wish we could be together.
FIFTY EIGHT	Have a wonderful time. Let us know when you return.
FIFTY NINE	Congratulations on the new arrival. Hope mother and child are well.
*SIXTY	Wishing you the best of everything on _______.
SIXTY ONE	Wishing you a very merry Christmas and a happy New Year.
*SIXTY TWO	Greetings and best wishes to you for a pleasant ________ holiday season.
SIXTY THREE	Victory or defeat, our best wishes are with you. Hope you win.
SIXTY FOUR	Arrived safely at ______.
SIXTY FIVE	Arriving _______ on _______. Please arrange to meet me there.
SIXTY SIX	DX QSLs are on hand for you at the ______ QSL Bureau. Send ______ self-addressed envelopes.
SIXTY SEVEN	Your message number ______ undeliverable because of ______. Please advise.
SIXTY EIGHT	Sorry to hear you are ill. Best wishes for a speedy recovery.
SIXTY NINE	Welcome to the ______. We are glad to have you with us and hope you will enjoy the fun and fellowship of the organization.

* Can be used for all holidays.

Note: ARL numbers should be spelled out at all times.

Table 7.87

How to be the Kind of Net Operator the Net Control Station (NCS) Loves

As a net operator, you have a duty to be self-disciplined. A net is only as good as its worst operator. You can be an exemplary net operator by following a few easy guidelines.

1) *Zero beat the NCS.* The NCS doesn't have time to chase all over the band for you. Make sure you're on frequency, and you will never be known at the annual net picnic as "old so-and-so who's always off frequency."

2) *Don't be late.* There's no such thing as "fashionably late" on a net. Liaison stations are on a tight timetable. Don't hold them up by checking in 10 minutes late with three pieces of traffic.

3) *Speak only when spoken to by the NCS.* Unless it is a bona fide emergency situation, you don't need to "help" the NCS unless asked. If you need to contact the NCS, make it brief. Resist the urge to help clear the frequency for the NCS or to "advise" the NCS. The NCS, not you, is boss.

4) Unless otherwise instructed by the NCS, *transmit only to the NCS.* Side comments to another station in the net are out of order.

5) *Stay until you are excused.* If the NCS calls you and you don't respond because you're getting a "cold one" from the fridge, the NCS may assume you've left the net, and net business may be stymied. If you need to leave the net prematurely, contact the NCS and simply ask to be excused (QNX PSE ON CW).

6) *Be brief when transmitting to the NCS.* A simple "yes" (C) or "no" (N) will usually suffice. Shaggy dog tales only waste valuable net time.

7) *Know how the net runs.* The NCS doesn't have time to explain procedure to you. After you have been on the net for a while, you should already know these things.

Table 7.88

Checking Your Message

Traffic handlers don't have to dine out to fight over the check! Even good ops find much confusion when counting up the text of a message. You can eliminate some of this confusion by remembering these basic rules:

1) Punctuation ("X-rays," "Querys") count separately as a word.

2) Mixed letter-number groups (1700Z, for instance) count as one word.

3) Initial or number groups count as one word if sent together, two if sent separately.

4) The signature does not count as part of the text, but any closing lines, such as "Love" or "Best wishes" do.

Here are some examples:

- Charles J McClain—3 words
- W B Stewart—3 words
- St Louis—2 words
- 3 PM—2 words
- SASE—1 word
- ARL FORTY SIX—3 words
- 2N1601—1 word
- Seventy-three—2 words
- 73—1 word

Telephone numbers count as 3 words (area code, prefix, number), and ZIP codes count as one, ZIP + 4 codes count as two words. Canadian postal codes count as two words (first three characters, last three characters.)

Although, it is improper to change the text of a message, you may change the check. Always do this by following the original check with a slash bar, then the corrected check. On phone, use the words "corrected to."

Table 7.89

Tips on Handling NTS Traffic by Packet Radio

Listing Messages

• After logging on to your local NTS-supported bulletin board, type the command LT, meaning List Traffic. The BBS will sort and display an index of all NTSXX traffic awaiting delivery.

Receiving Messages

• To take a message off the Bulletin Board for telephone delivery to the third party, or for relay to a NTS Local or Section Net, type the R command, meaning Read Traffic, and the message number. R 188 will cause the BBS to find the BBS message number 188. This RADIOGRAM will look like any other, with preamble, address, text and signature; only some additional packet-related message header information is added. This information includes the routing path of the message for auditing purposes; e.g., to discern any excessive delays in the system.

• After the message is saved to the printer or disk, the message should be KILLED by using the KT command, meaning Kill Traffic, and the message number. In the above case, at the BBS prompt, type KT 188. This prevents the message from being delivered twice. Some of the newer BBS software requires use of K rather than KT.

• At the time the message is killed, many BBSs will automatically send a message back to the station in the FROM field with information on who took the traffic, and when it was taken!

Delivering or Relaying A Message

• A downloaded RADIOGRAM should, of course, be handled expeditiously in the traditional way: telephone delivery, or relay to another net.

Sending Messages

• To send a RADIOGRAM, use the ST command meaning Send Traffic. The BBS will prompt you for the NTS routing (0611@NTSCT, for example), the message title which should contain the city in the address of the RADIOGRAM (QTC 1 Dayton), and the text of the message in RADIOGRAM format. The BBS, usually within the hour, will check its outgoing mailpouch, find the NTSCT message and automatically forward it to the next packet station in line to the NTSCT node. Note: Some states have more than one ARRL Section. If you do not know the destination ARRL Section ("Is San Angelo in the ARRL North, South or West Texas Section?"), then simply use the state designator NTSTX.

*Note: While NTS/packet radio message forwarding is evolving rapidly, there are still some gaps. When uploading an NTS message destined for a distant state, use handling instruction "HXC" to ask the delivering station to report back to you the date and time of delivery.

We Want You!

Local and Section BBSs need to be checked daily for NTS traffic. SYSOPs and STMs can't do it alone. They need your help to clear NTS RADIOGRAMs every day, seven days a week, for delivery and relay. If you are a traffic handler/packeteer, contact your Section Traffic Manager or Section Manager for information on existing NTS/packet procedures in your Section.

If you are a packeteer, and know nothing of NTS traffic handling, contact ARRL HQ, your Section Manager or Section Traffic Manager for information on how you can put your packet radio gear to use in serving the public in routine times, but especially in time of emergency!

And, if you enjoy phone/CW traffic handling, but aren't on packet yet, discover the incredible speed and accuracy of packet radio traffic handling. You probably already have a small computer and 2-meter rig; all you need is a packet radio "black box" to connect between your 2-meter rig and computer. For more information on packet radio, see *Practical Packet Radio*, published by the ARRL.

Chapter 8

Circuit Construction

Home construction of electronics projects can be a fun part of Amateur Radio. Some folks have said that hams don't build things nowadays; this just isn't so! An ARRL survey shows that 53% of active hams build some electronic projects. When you go to any ham flea market, you see row after row of dealers selling electronic components; people are leaving those tables with bags of parts. They must be doing something with them.

Even experienced constructors will find valuable tips in this chapter. It discusses tools and their uses, electronic construction techniques, tells how to turn a schematic into a working circuit and then summarizes common mechanical construction practices. This chapter was written by Ed Hare, W1RFI, and includes material contributions from Bruce Hale, KB1MW, Ian White, G3SEK, and Chuck Adams, K7QO.

SHOP SAFETY

All the fun of building a project will be gone if you get hurt. To make sure this doesn't happen, let's first review some safety rules.

- Read the manual! The manual tells all you need to know about the operation and safety features of the equipment you are using.
- Do not work when you are tired. You will be more likely to make a mistake or forget an important safety rule.
- Never disable any safety feature of any tool. If you do, sooner or later someone will make the mistake the safety feature was designed to prevent.
- Never fool around in the shop. Practical jokes and horseplay are in bad taste at social events; in a shop they are downright dangerous. A work area is a dangerous place at all times; even hand tools can hurt someone if they are misused.
- Keep your shop neat and organized. A messy shop is a dangerous shop. A knife left laying in a drawer can cut someone looking for another tool; a hammer left on top of a shelf can fall down at the worst possible moment; a sharp tool left on a chair can be a dangerous surprise for the weary constructor who sits down.
- Wear the proper safety equipment. Wear eye-protection goggles when working with chemicals or tools. Use earplugs or earphones when working near noise. If you are working with dangerous chemicals, wear the proper protective clothing.
- Make sure your shop is well ventilated. Paint, solvents, cleaners or other chemicals can create dangerous fumes. If you feel dizzy, get into fresh air immediately, and seek medical help if you do not recover quickly.
- Get medical help when necessary. Every workshop should contain a good first-aid kit. Keep an eye-wash kit near any dangerous chemicals or power tools that can create chips. If you become injured, apply first aid and then seek medical help if you are not sure that you are okay. Even a small burn or scratch on your eye can develop into a serious problem.
- Respect power tools. Power tools are not forgiving. A drill can go through your hand a lot easier than metal. A power saw can remove a finger with ease. Keep away from the business end of power tools. Tuck in your shirt, roll up your sleeves and remove your tie before using any power tool. If you have long hair, tie it back so it can't become entangled in power equipment.
- Don't work alone. Have someone nearby who can help if you get into trouble when working with dangerous equipment, chemicals or voltages.
- Think! Pay attention to what you are doing. No list of safety rules can cover all possibilities. Safety is always your responsibility. You must think about what you are doing, how it relates to the tools and the specific situation at hand.

TOOLS AND THEIR USES

All electronic construction makes use of tools, from mechanical tools for chassis fabrication to the soldering tools used for circuit assembly. A good understanding of tools and their uses will enable you to perform most construction tasks.

While sophisticated and expensive tools often work better or more quickly than simple hand tools, with proper use, simple hand tools can turn out a fine piece of equipment. **Table 8.1** lists tools indispensable for construction of electronic equipment. These tools can be used to perform nearly any construction task. Add tools to your collection from time to time, as finances permit.

Sources of Tools

Radio-supply houses, mail-order stores and most hardware stores carry the tools required to build or service Amateur Radio equipment. Bargains are available at ham flea markets or local neighborhood sales, but beware! Some flea-market bargains are really shoddy imports that won't work very well or last very long. Some used tools are offered for sale because the owner is not happy with their performance.

There is no substitute for quality! A high-quality tool, while a bit more expensive, will last a lifetime. Poor quality tools don't last long and often do a poor job even when brand new. You don't need to buy machinist-grade tools, but stay away from cheap tools; they are not the bargains they might appear to be.

Care of Tools

The proper care of tools is more than a matter of pride. Tools that have not been cared for properly will not last long or

Table 8.1

Recommended Tools and Materials

Simple Hand Tools

Screwdrivers

Slotted, 3-inch, 1/8-inch blade
Slotted, 8-inch, 1/8-inch blade
Slotted, 3-inch, 3/16-inch blade
Slotted, stubby, 1/4-inch blade
Slotted, 4-inch, 1/4-inch blade
Slotted, 6-inch, 5/16-inch blade
Phillips, 2 1/2-inch, #0 (pocket clip)
Phillips, 3-inch, #1
Phillips, stubby, #2
Phillips, 4-inch, #2
Phillips, 4-inch, #2
Long-shank screwdriver with holding clip on blade
Jeweler's set
Right-angle, slotted and Phillips

Pliers, Sockets and Wrenches

Long-nose pliers, 6- and 4-inch
Diagonal cutters, 6- and 4-inch
Channel-lock pliers, 6-inch
Slip-joint pliers
Locking pliers (Vise Grip or equivalent)
Socket nut-driver set, 1/16- to 1/2-inch
Set of socket wrenches for hex nuts
Allen (hex) wrench set
Wrench set
Adjustable wrenches, 6- and 10-inch
Tweezers, regular and reverse-action
Retrieval tool/parts holder, flexible claw
Retrieval tool, magnetic

Cutting and Grinding Tools

File set consisting of flat, round, half-round, and triangular. Large and miniature types recommended
Burnishing tool
Wire strippers
Wire crimper
Hemostat, straight
Scissors
Tin shears, 10-inch
Hacksaw and blades
Hand nibbling tool (for chassis-hole cutting)
Scratch awl or scriber (for marking metal)
Heavy-duty jackknife
Knife blade set (X-ACTO or equivalent)
Machine-screw taps, #4-40 through #10-32 thread
Socket punches, 1/2 in, 5/8 in, 3/4 in, 1 1/8 in, 1 1/4 in, and 1 1/2 in
Tapered reamer, T-handle, 1/2-inch maximum width
Deburring tool

Miscellaneous Hand Tools

Combination square, 12-inch, for layout work
Hammer, ball-peen, 12-oz head
Hammer, tack
Bench vise, 4-inch jaws or larger
Center punch
Plastic alignment tools
Mirror, inspection
Flashlight, penlight and standard
Magnifying glass
Ruler or tape measure
Dental pick
Calipers
Brush, wire
Brush, soft
Small paintbrush
IC-puller tool

Hand-Powered Tools

Hand drill, 1/4-inch chuck or larger
High-speed drill bits, #60 through 3/8-inch diameter

Power Tools

Motor-driven emery wheel for grinding
Electric drill, hand-held
Drill press
Miniature electric motor tool (Dremel or equivalent) and accessory drill press

Soldering Tools and Supplies

Soldering pencil, 30-W, 1/8-inch tip
Soldering iron, 200-W, 5/8-inch tip
Solder, 60/40, resin core
Soldering gun, with assorted tips
Desoldering tool
Desoldering wick

Safety

Safety glasses
Hearing protector, earphones or earplugs
Fire extinguisher
First-aid kit

Useful Materials

Medium-weight machine oil
Contact cleaner, liquid or spray can
Duco modeling cement or equivalent
Electrical tape, vinyl plastic
Sandpaper, assorted
Emery cloth
Steel wool, assorted
Cleaning pad, Scotchbrite or equivalent
Cleaners and degreasers
Contact lubricant
Sheet aluminum, solid and perforated, 16- or 18-gauge, for brackets and shielding.
Aluminum angle stock, 1/2 × 1/2-inch × 1/4-inch-diameter round brass or aluminum rod (for shaft extensions)
Machine screws: Round-head and flat head, with nuts to fit. Most useful sizes: 4-40, 6-32 and 8-32, in lengths from 1/4-inch to 1 1/2 inches. (Nickel-plated steel is satisfactory except in strong RF fields, where brass should be used.)
Bakelite, Lucite, polystyrene and copper-clad PC-board scraps.
Soldering lugs, panel bearings, rubber grommets, terminal-lug wiring strips, varnished-cambric insulating tubing, heat-shrinkable tubing
Shielded and unshielded wire
Tinned bare wire, #22, #14 and #12
Enameled wire, #20 through #30

work well. Dull or broken tools can be safety hazards. Tools that are in good condition do the work for you; tools that are misused or dull are difficult to use.

Store tools in a dry place. Tools do not fit in with most living-room decors, so they are often relegated to the basement or garage. Unfortunately, many basements or garages are not good places to store tools; dampness and dust are not good for tools. If your tools are stored in a damp place, use a dehumidifier. Sometimes you can minimize rust by keeping your tools lightly oiled, but this is a second-best solution. If you oil your tools, they may not rust, but you will end up covered in oil every time you use them. Wax or silicone spray is a better alternative.

Store tools neatly. A messy toolbox, with tools strewn about haphazardly, can be more than an inconvenience. You may waste a lot of time looking for the right tool and sharp edges can be dulled or nicked by tools banging into each other in the bottom of the box. As the old adage says, every tool should have a place, and every tool should be in its place. If you must search the workbench, garage, attic and car to find the right screwdriver, you'll spend more time looking for tools than building projects.

Sharpening

Many cutting tools can be sharpened. Send a tool that has been seriously dulled to a professional sharpening service. These services can resharpen saw blades, some files, drill bits and most cutting blades. Touch up the edge of cutting tools with a whetstone to extend the time between sharpenings.

Sharpen drill bits frequently to minimize the amount of material that must be removed each time. Frequent sharpening also makes it easier to maintain the critical surface angles required for best cutting with least wear. Most inexpensive drill-bit sharpeners available for shop use do a poor job, either from the poor quality of the sharpening tool or inexperience of the operator. Also, drills should be sharpened at different angles for different applications. Commercial sharpening services do a much better job.

Intended Purpose

Don't use tools for anything other than their intended purpose! If you use a pair of wire cutters to cut sheet metal, pliers as a vise or a screwdriver as a pry bar, you ruin a good tool. Although an experienced constructor can improvise with tools, most take pride in not abusing them.

Tool Descriptions and Uses

Specific applications for tools are discussed throughout this chapter. Hand tools are used for so many different applications that they are discussed first, followed by some tips for proper use of power tools.

Soldering Iron

Soldering is used in nearly every phase of electronic construction so you'll need soldering tools. A soldering tool must be hot enough to do the job and lightweight enough for agility and comfort. A 100-W soldering gun is overkill for printed-circuit work, for example. A temperature-controlled iron works well, although the cost is not justified for occasional projects. Get an iron with a small conical or chisel tip.

You may need an assortment of soldering irons to do a wide variety of soldering tasks. They range in size from a small 25-W iron for delicate printed-circuit work to larger 100 to 300-W sizes used to solder large surfaces. Several manufacturers also sell soldering guns. Small "pencil" butane torches are also available, with optional soldering-iron tips. A small butane torch is available from the Solder-It Company. This company also sells a soldering kit that contains paste solders (in syringes) for electronics, pot metal and plumbing.

Keep soldering tools in good condition by keeping the tips well tinned with solder. Do not run them at full temperature for long periods when not in use. After each period of use, remove the tip and clean off any scale that may have accumulated. Clean an oxidized tip by dipping the hot tip in sal ammoniac (ammonium chloride) and then wiping it clean with a rag. Sal ammoniac is somewhat corrosive, so if you don't wipe the tip thoroughly, it can contaminate electronic soldering.

If a copper tip becomes pitted, file it smooth and bright and then tin it immediately with solder. Modern soldering iron tips are nickel or iron clad and should not be filed.

The secret of good soldering is to use the right amount of heat. Many people who have not soldered before use too little heat, dabbing at the joint to be soldered and making little solder blobs that cause unintended short circuits.

Solders have different melting points, depending on the ratio of tin to lead. Tin melts at 450°F and lead at 621°F. Solder made from 63% tin and 37% lead melts at 361°F, the lowest melting point for a tin and lead mixture. Called 63-37 (or eutectic), this type of solder also provides the most rapid solid-to-liquid transition and the best stress resistance.

Solders made with different lead/tin ratios have a plastic state at some temperatures. If the solder is deformed while it is in the plastic state, the deformation remains when the solder freezes into the solid state. Any stress or motion applied to "plastic solder" causes a poor solder joint.

60-40 solder has the best wetting qualities. Wetting is the ability to spread rapidly and bond materials uniformly. 60-40 solder also has a low melting point. These factors make it the most commonly used solder in electronics.

Some connections that carry high current can't be made with ordinary tin-lead solder because the heat generated by the current would melt the solder. Automotive starter brushes and transmitter tank circuits are two examples. Silver-bearing solders have higher melting points, and so prevent this problem. High-temperature silver alloys become liquid in the 1100°F to 1200°F range, and a silver-manganese (85-15) alloy requires almost 1800°F.

Because silver dissolves easily in tin, tin bearing solders can leach silver plating from components. This problem can be greatly reduced by partially saturating the tin in the solder with silver or by eliminating the tin. Tin-silver or tin-lead-silver alloys become liquid at temperatures from 430°F for 96.5-3.5 (tin-silver), to 588°F for 1.0-97.5-1.5 (tin-lead-silver). A 15.0-80.0-5.0 alloy of lead-indium-silver melts at 314°F.

Never use acid-core solder for electrical work. It should be used only for plumbing or chassis work. For circuit construction, only use fluxes or solder-flux combinations that are labeled for electronic soldering.

The resin or the acid is a *flux*. Flux removes oxide by suspending it in solution and floating it to the top. Flux is not a cleaning agent! Always clean the work before soldering. Flux is not a part of a soldered connection—it merely aids the soldering process. After soldering, remove any remaining flux. Resin flux can be removed with isopropyl or denatured alcohol. A cotton swab is a good tool for applying the alcohol and scrubbing the excess flux away. Commercial flux-removal sprays are available at most electronic-part distributors.

The two key factors in quality soldering are time and temperature. Generally, rapid heating is desired, although most unsuccessful solder jobs fail because insufficient heat has been applied. Be careful; if heat is applied too long, the components or PC board can be damaged, the flux may be used up and surface oxidation can become a problem. The soldering-iron tip should be hot enough to readily melt the solder without burning, charring or discoloring components, PC boards or wires. Usually, a tip temperature about 100°F above the solder melting point is about right for mounting components on PC boards. Also, use solder that is sized appropriately for the job. As the cross section of the solder decreases, so does the amount of heat required to melt it. Diameters from 0.025 to 0.040 inches are good for nearly all circuit wiring.

Here's how to make a good solder joint. This description assumes that solder with a flux core is used to solder a typical PC board connection such as an IC pin.

- Prepare the joint. Clean all conductors thoroughly with fine steel wool or a plastic scrubbing pad. Do the circuit board at the beginning of assembly and individual parts such as resistors and capacitors immediately before soldering. Some parts (such as ICs and surface-mount components) cannot be easily cleaned; don't worry unless they're exceptionally dirty.
- Prepare the tool. It should be hot enough to melt solder applied to its tip quickly (half a second when dry, instantly when wet with solder). Apply a little solder directly to the tip so that the surface is shiny. This process is called "tinning" the tool. The solder coating helps conduct heat from the tip to the joint.
- Place the tip in contact with one side of the joint. If you can place the tip on the underside of the joint, do so. With the tool below the joint, convection helps transfer heat to the joint.
- Place the solder against the joint directly opposite the soldering tool. It should melt within a second for normal PC connections, within two seconds for most other connections. If it takes longer to melt, there is not enough heat for the job at hand.
- Keep the tool against the joint until the solder flows freely throughout the joint. When it flows freely, solder tends to form concave shapes between the conductors. With insufficient heat solder does not flow freely; it forms convex shapes—blobs. Once solder shape changes from convex to concave, remove the tool from the joint.
- Let the joint cool without movement at room temperature. It usually takes no more than a few seconds. If the joint is moved before it is cool, it may take on a dull, satin look that is characteristic of a "cold" solder joint. Reheat cold joints until the solder flows freely and hold them still until cool.
- When the iron is set aside, or if it loses its shiny appearance, wipe away any dirt

with a wet cloth or sponge. If it remains dull after cleaning, tin it again.

Overheating a transistor or diode while soldering can cause permanent damage. Use a small heat sink when you solder transistors, diodes or components with plastic parts that can melt. Grip the component lead with a pair of pliers up close to the unit so that the heat is conducted away (be careful—it is easy to damage delicate component leads). A small alligator clip also makes a good heat sink.

Mechanical stress can damage components, too. Mount components so there is no appreciable mechanical strain on the leads.

Soldering to the pins of coil forms or male cable plugs can be difficult. Use a suitable small twist drill to clean the inside of the pin and then tin it with resin-core solder. While it is still liquid, clear the surplus solder from each pin with a whipping motion or by blowing through the pin from the inside of the form or plug. Watch out for flying hot solder! Next, file the nickel plate from the pin tip. Then insert the wire and solder it. After soldering, remove excess solder with a file, if necessary.

When soldering to the pins of plastic coil forms, hold the pin to be soldered with a pair of heavy pliers to form a heat sink. Do not allow the pin to overheat; it will loosen and become misaligned.

In order to remove components, you need to learn the art of desoldering—removing solder from components and PC boards so they can be separated easily. Use commercially made wicking material (braid) to soak up excess solder from a joint. Another useful tool is an air-suction solder remover. Another method is to heat the joint and "flick" the wet solder off. (Watch out for solder splashes!)

Soldering equipment gets *hot*! Be careful. Treat a soldering burn as you would any other. Handling lead or breathing soldering fumes is also hazardous. Observe these precautions to protect yourself and others:

- Properly ventilate the work area. If you can smell fumes, you are breathing them.
- Wash your hands after soldering, especially before handling food.
- Minimize direct contact with flux and flux solvents.

For more information about soldering hazards and the ways to make soldering safer, see "Making Soldering Safer," by Brian P. Bergeron, MD, NU1N (Mar 1991 *QST*, pp 28-30) and "More on Safer Soldering," by Gary E. Meyers, K9CZB (Aug 1991 *QST*, p 42).

Screwdrivers

For construction or repair, you need to have an assortment of screwdrivers. Each blade size is designed to fit a specific range of screw-head sizes. Using the wrong size blade usually damages the blade, the screw head or both. You may also need stubby sizes to fit into tight spaces. Right-angle screwdrivers are inexpensive and can get into tight spaces that can't otherwise be reached.

Electric screwdrivers are relatively inexpensive. If you have a lot of screws to fasten, they can save a lot of time and effort. They come with a wide assortment of screwdriver and nut-driver bits. An electric drill can also function as an electric screwdriver, although it may be heavy and over-powered for some applications.

Keep screwdriver blades in good condition. If a blade becomes broken or worn out, replace the screwdriver. A screwdriver only costs a few dollars; do not use one that is not in perfect condition. Save old screwdrivers to use as pry bars and levers, but use only good ones on screws. Filing a worn blade seldom gives good results.

Pliers and Locking-Grip Pliers

Pliers and locking-grip pliers are used to hold or bend things. They are not wrenches! If pliers are used to remove a nut or bolt, the nut or the pliers is usually damaged. Pliers are not intended for heavy-duty applications. Use a metal brake to bend heavy metal; use a vise to hold a heavy component. To remove a nut, use a wrench or nut driver. There is one exception to this rule of thumb: To remove a nut that is stripped too badly for a wrench, use a pair of pliers, locking-grip pliers, or a diagonal cutter to bite into the nut and turn it a bit. If you do this, use an old tool or one dedicated to just this purpose; this technique is not good for the tool. If the pliers' jaws or teeth become worn, replace the tool.

Wire Cutters

Wire cutters are primarily used to cut wires or component leads. The choice of diagonal blades (sometimes called "dikes") or end-nip blades depends on the application. Diagonal blades are most often used to cut wires, while the end-nip blades are useful to cut off the ends of components that have been soldered into a printed-circuit board. Some delicate components can be damaged by cutting their leads with dikes. Scissors designed to cut wire can be used.

Wire strippers are handy, but you can usually strip wires using a diagonal cutter or a knife. This is not the only use for a knife, so keep an assortment handy.

Do not use wire cutters or strippers on anything other than wire! If you use a cutter to trim a protruding screw head, or cut

Table 8.2
Numbered Drill Sizes

No.	*Diameter (Mils)*	*Will Clear Screw*	*Drilled for Tapping from Steel or Brass*
1	228.0	12-24	—
2	221.0	—	—
3	213.0	—	14-24
4	209.0	12-20	—
5	205.0	—	—
6	204.0	—	—
7	201.0	—	—
8	199.0	—	—
9	196.0	—	—
10	193.5	—	—
11	**191.0**	**10-24** **10-32**	—
12	189.0	—	—
13	185.0	—	—
14	182.0	—	—
15	180.0	—	—
16	177.0	—	12-24
17	173.0	—	—
18	169.5	—	—
19	166.0	8-32	12-20
20	161.0	—	—
21	**159.0**	—	**10-32**
22	157.0	—	—
23	154.0	—	—
24	152.0	—	—
25	**149.5**	—	**10-24**
26	147.0	—	—
27	144.0	—	—
28	**140.0**	**6-32**	—
29	**136.0**	—	**8-32**
30	128.5	—	—
31	120.0	—	—
32	116.0	—	—
33	**113.0**	**4-40**	—
34	111.0	—	—
35	110.0	—	—
36	**106.5**	—	**6-32**
37	104.0	—	—
38	101.5	—	—
39	**099.5**	**3-48**	—
40	098.0	—	—
41	096.0	—	—
42	093.5	—	—
43	**089.0**	—	**4-40**
44	**086.0**	**2-56**	—
45	082.0	—	—
46	081.0	—	—
47	078.5	—	3-48
48	076.0	—	—
49	073.0	—	—
50	**070.0**	—	**2-56**
51	067.0	—	—
52	063.5	—	—
53	059.5	—	—
54	055.0	—	—

a hardened-steel spring, you will usually damage the blades.

Files

Files are used for a wide range of tasks. In addition to enlarging holes and slots, they are used to remove burrs, shape metal, wood or plastic and clean some surfaces in preparation for soldering. Files are especially prone to damage from rust and moisture. Keep them in a dry place. The cutting edge of the blades can also become clogged with the material you are removing. Use file brushes (also called file cards) to keep files clean. Most files cannot be sharpened easily, so when the teeth become worn, the file must be replaced. A worn file is sometimes worse than no file at all. At best, a worn file requires more effort.

Drill Bits

Drill bits are made from carbon steel, high-speed steel or carbide. Carbon steel is more common and is usually supplied unless a specific request is made for high-speed bits. Carbon-steel drill bits cost less than high-speed or carbide types; they are sufficient for most equipment construction work. Carbide drill bits last much longer under heavy use. One disadvantage of carbide bits is that they are brittle and break easily, especially if you are using a hand-held power drill.

Twist drills are available in a number of sizes. Those listed in bold type in **Table 8.2** are the most commonly used in construction of amateur equipment. You may not use all of the drills in a standard set, but it is nice to have a complete set on hand. You should also buy several spares of the more common sizes. Although Table 8.2 lists drills down to #54, the series extends to number #80.

Specialized Tools

Most constructors know how to use common tools, such as screwdrivers, wrenches and hammers. Let's discuss other tools that are not so common.

A hand nibbling tool is shown in **Fig 8.1**. Use this tool to remove small "nibbles" of metal. It is easy to use; position the tool where you want to remove metal and squeeze the handle. The tool takes a small bite out of the metal. When you use a nibbler, be careful that you don't remove too much metal, clip the edge of a component mounted to the sheet metal or grab a wire that is routed near the edge of a chassis. Fixing a broken wire is easy, but something to avoid if possible. It is easy to remove metal but nearly impossible to put it back. Do it right the first time!

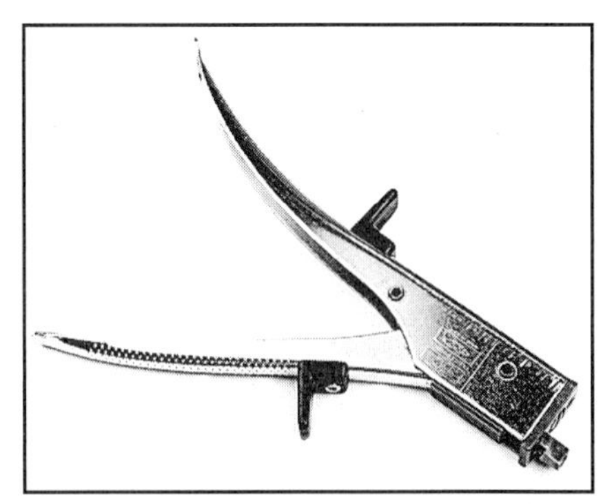

Fig 8.1—A nibbling tool is used to remove small sections of sheet metal.

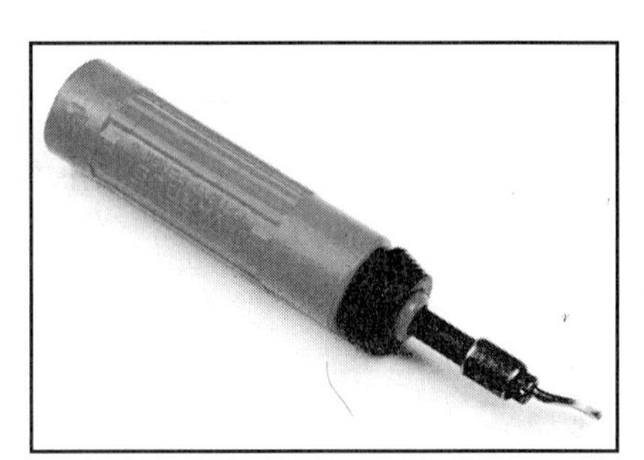

Fig 8.2—A deburring tool is used to remove the burrs left after drilling a hole.

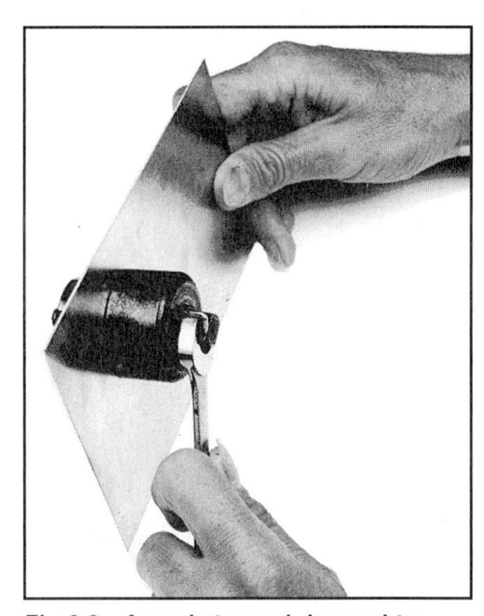

Fig 8.3—A socket punch is used to easily punch a hole in sheet metal.

Deburring Tool

A deburring tool is just the thing to remove the sharp edges left on a hole after most drilling or punching operations. See **Fig 8.2**. Position the tool over the hole and rotate it around the hole edge to remove burrs or rough edges. As an alternative, select a drill bit that is somewhat larger than the hole, position it over the hole, and spin it lightly to remove the burr.

Socket Punches

Greenlee is the most widely known of the socket-punch manufacturers. Most socket punches are round, but they do come in other shapes. To use one, drill a pilot hole large enough to clear the bolt that runs through the punch. Then, mount the punch as shown in **Fig 8.3**, with the cutter on one side of the sheet metal and the socket on the other. Tighten the nut with a wrench until the cutter cuts all the way through the sheet metal.

Useful Shop Materials

Small stocks of various materials are used when constructing electronics equipment. Most of these are available from hardware or radio-supply stores. A representative list is shown at the end of Table 8.1.

Small parts, such as machine screws, nuts, washers and soldering lugs can be economically purchased in large quantities (it doesn't pay to buy more than a lifetime supply). For items you don't use often, many radio-supply stores or hardware stores sell small quantities and assortments.

A DELUXE SOLDERING STATION

The simple tool shown in **Figs 8.4** through **8.6** can enhance the usefulness and life of a soldering iron as well as make electronic assembly more convenient. It includes a protective heat sink and a tip-cleaning sponge rigidly attached to a sturdy base for efficient one-handed operation.

Soldering-iron tips and heating elements last longer if operated at a reduced temperature when not being used. Temperature reduction is accomplished by half-wave rectification of the applied ac. D1 conducts during only one-half of the ac cycle. With current flowing only in one direction, only one electrode of the neon bulb glows. Closing S1 short-circuits the diode and applies full power to the soldering iron, igniting both bulb electrodes brightly.

The base for the unit is a 2 × 6 × 4-inch (HWD) aluminum chassis (Bud AC-431 or equivalent). A 30- or 40-W soldering iron fits neatly on the chassis top. The holder has two mounting holes in each foot. A sponge tray nests between the feet and the case. In this model, a sardine tin is used for the sponge tray.

The tray and iron holder are secured to the chassis by 6-32 × 1/2-inch pan-head machine screws and nuts, with flat washers under the screw heads (sponge tray) and lock washers under the nuts (chassis underside). One of these nuts fastens a six-lug tie point strip to the chassis bottom. Use the soldering-iron holder base as a template for drilling the chassis and sponge tray. The floor of the sponge tray must be sealed around the screw heads to prevent moisture from leaking into the electrical components below the chassis. RTV compound was used for this purpose in the unit pictured.

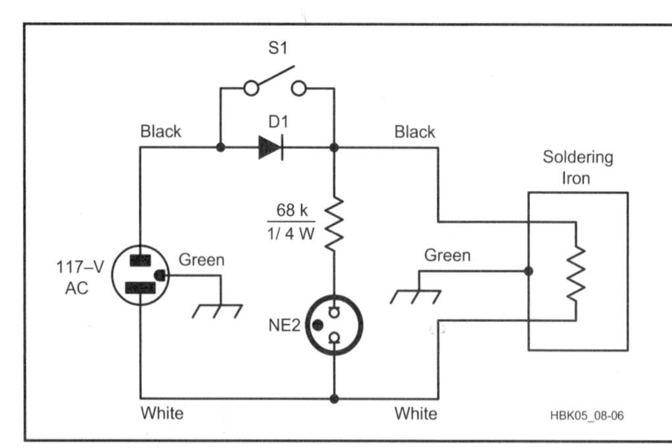

Fig 8.6—Schematic diagram of the soldering station. D1 is a silicon diode, 1-A, 400-PIV. S1 is a miniature SPST toggle switch rated 3 A at 125 V. This circuit is satisfactory for use with irons having power ratings up to 100 W.

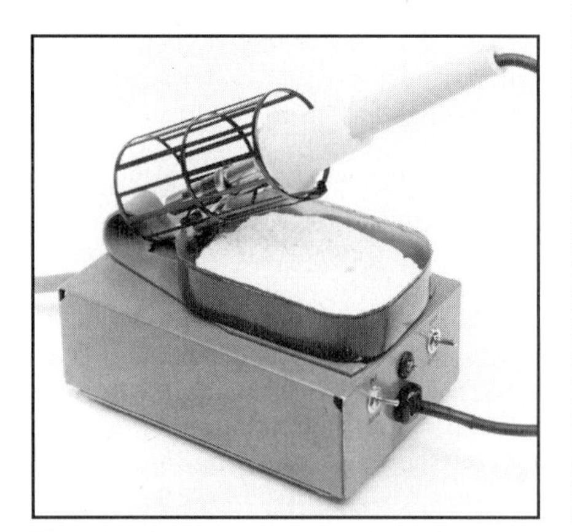

Fig 8.4—A compact assembly of commonly available items, this soldering station makes soldering easier. Miniature toggle switches are used because they are easy to operate.

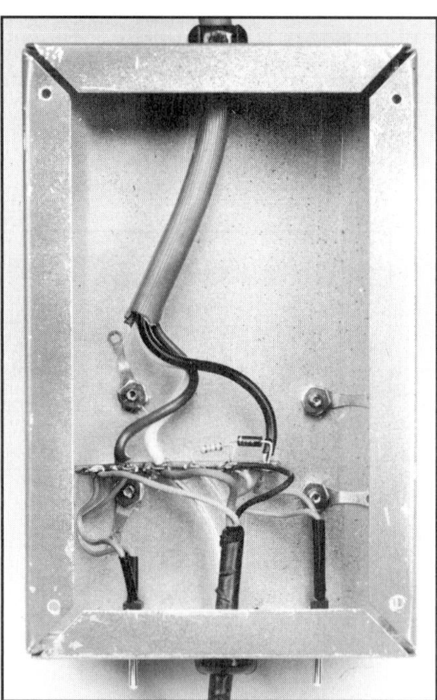

Fig 8.5—View of the soldering-station chassis underside with the bottom plate removed. #24 hookup wire is adequate for all connections. Make sure no possibility of a short circuit exists.

Notice that the soldering iron and the soldering station use separate ac line cords. This ensures that the cord of the soldering iron will be long enough to do useful work. Bushings are used to anchor both cords. If these aren't available, grommets and cable clamps work well. Knotting the cords inside the chassis is a simple technique that normally provides adequate strain relief.

The underchassis assembly is shown in Fig 8.5. The neon bulb is installed in a 3/16-inch-ID grommet. The leads are insulated with spaghetti insulation or heat-shrink tubing to prevent short circuits. If you mount the bulb in a fixture or socket, use a clear lens to ensure that the electrodes are distinctly visible. Install a cover on the bottom of the chassis to prevent accidental contact with the live ac wiring. Stick-on rubber feet prevent the bottom of the unit from scratching your work surface.

SOLDERING-IRON TEMPERATURE CONTROL

A temperature control gives greater flexibility than the simple control just described. An incandescent-light dimmer can be used to control the working temperature of the tip. **Fig 8.7** shows a temperature control built into an electrical box. A dimmer and a duplex outlet are mounted in the box; the wiring diagram is shown in **Fig 8.8.** Only one of the two ac outlets is controlled by the dimmer. A jumper on the duplex outlet connects the hot terminals of both outlets together. This jumper must be removed. The hot terminal is narrower than the neutral one and its connecting screw is usually brass. Neutral terminals remain interconnected.

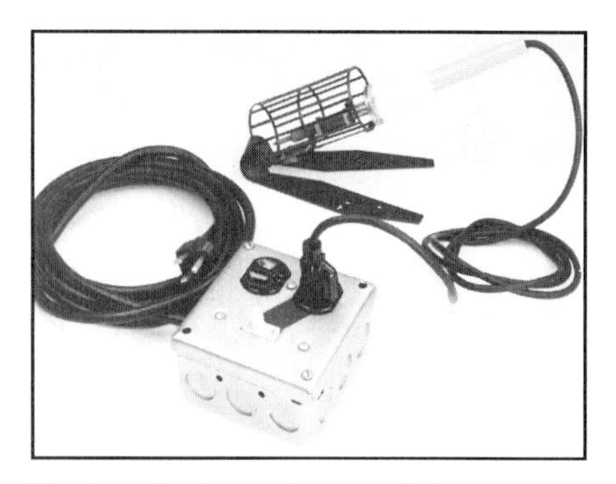

Fig 8.7—An incandescent-light dimmer controls soldering-iron tip temperature. Only one of the duplex outlets is connected through the dimmer.

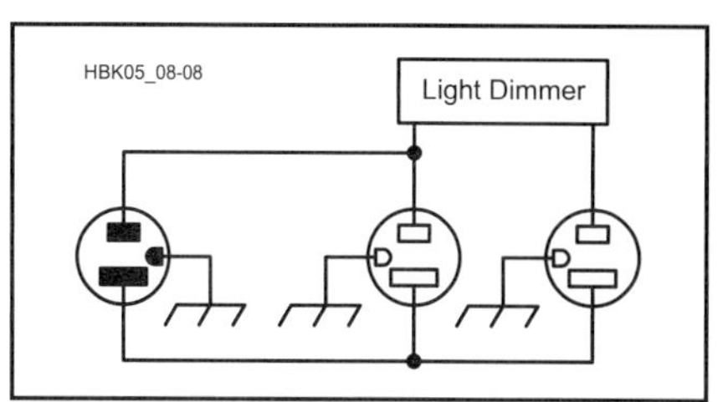

Fig 8.8—Schematic diagram of the soldering-iron temperature control.

The dimmer shown in Fig 8.7 can be purchased at any hardware or electrical-supply store. The knob is capable of fine control of the soldering temperature.

Electronic Circuits

Most of the construction projects undertaken by the average amateur involve electronic circuitry. The circuit is the "heart" of most amateur equipment. It might seem obvious, but in order for you to build it, the circuit must work! Don't always assume that a "cookbook" circuit that appears in an applications note or electronics magazine is flawless. These are sometimes design examples that have not always been thoroughly debugged. Many home-construction projects are "one-time" deals; the author has put one together and it worked. In some cases, component tolerances or minor layout changes might make it difficult to get a second unit working.

Protecting Components

You need to take steps to protect the electronic and mechanical components you use in circuit construction. Some components can be damaged by rough handling. Dropping a $^1/_4$-W resistor causes no harm, but dropping a vacuum tube or other delicate subassemblies usually causes damage.

Some components are easily damaged by heat. Some of the chemicals used to clean electronic components (such as flux removers, degreasers or control-lubrication sprays) can damage plastic. Check them for safety before you use them.

Electrostatic Discharge

Some components, especially high-impedance components such as FETs and CMOS gates, can be damaged by electrostatic discharge (ESD). Protect these parts from static charges. Most people are familiar with the static charge that builds up when one walks across a carpet then touches a metal object; the resultant spark can be quite lively. Walking across a carpet on a dry day can generate 35 kV! A worker sitting at a bench can generate voltages up to 6 kV, depending on conditions, such as when relative humidity is less than 20%.

You don't need this much voltage to damage a sensitive electronic component; damage can occur with as little as 30 V. The damage is not always catastrophic. A MOSFET can become noisy, or lose gain; an IC can suffer damage that causes early failure. To prevent this kind of damage, you need to take some precautions.

The energy from a spark can travel inside a piece of equipment to effect internal components. Protection of sensitive electronic components involves the prevention of static build-up together with the removal of any existing charges by dissipating any energy that does build up.

Several techniques can be used to minimize static build-up. First, remove any carpet in your work areas. You can replace it with special antistatic carpet, but this is expensive. It's less expensive to treat the carpet with antistatic spray, which is available from Chemtronics, GC Thorsen and other lines carried by electronics wholesalers.

Even the choice of clothing you wear can affect the amount of ESD. Polyester has a much greater ESD potential than cotton.

Many builders who have their workbench on a concrete floor use a rubber mat to minimize the risk of electric shocks from the ac line. Unfortunately, the rubber mat increases the risk of ESD. An antistatic rubber mat can serve both purposes.

Many components are shipped in antistatic packaging. Leave components in their conductive packaging. Other components, notably MOSFETs, are shipped with a small metal ring that temporarily shorts all of the leads together. Leave this ring in place until the device is fully installed in the circuit.

These precautions help reduce the build-up of electrostatic charges. Other techniques offer a slow discharge path for the charges or keep the components and the operator handling them at the same ground potential.

One of the best techniques is to connect the operator and the devices being handled to earth ground, or a common reference point. It is not a good idea to directly ground an operator working on electronic equipment, though; the risk of shock is too great. If the operator is grounded through a high-value resistor, ESD protection is still offered but there is no risk of shock.

The operator is usually grounded through a conductive wrist strap. 3M makes a grounding wrist band. This wrist band is equipped with a snap-on ground lead. A 1-MΩ resistor is built into the snap of the strap to protect the user should a live circuit be contacted. Build a similar resistor into any homemade ground strap.

The devices and equipment being handled are also grounded, by working on a charge-dissipating mat that is connected to ground. The mat should be an insulator that has been impregnated with a resistance material. Suitable mats and wrist straps are made by 3M, GC Electronics and others; they are available from most electronics supply houses. **Fig 8.9** shows a typical ESD-safe work station.

The work area should also be grounded, directly or through a conductive mat. Use a soldering iron with a grounded tip to solder sensitive components. Most irons that have three-wire power cords are properly grounded. When soldering static-sensitive devices, use two or three jumpers to ground you, the work and the iron. If the iron does not have a ground wire in the power cord, clip a jumper from the metal part of the iron near the handle to the metal box that houses the temperature control. Another jumper connects the box to the work. Finally, a jumper goes from the box to an elastic wrist band for static grounding.

Use antistatic bags to transport susceptible components or equipment. Keep your workbench free of objects such as paper, plastic and other static-generating items. Use conductive containers with a dissipative surface coating for equipment storage.

All of the antistatic products described above are available from Newark Electronics and other suppliers.

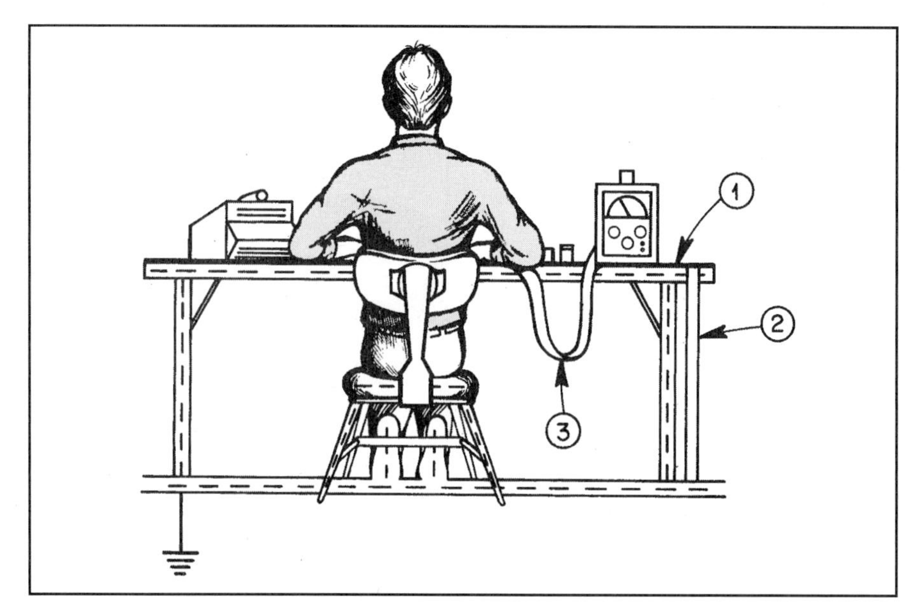

Fig 8.9—A work station that has been set up to minimize ESD features (1) a grounded dissipative work mat and (2) a wrist strap that (3) grounds the worker through high resistance.

Electronics Construction Techniques

Several different point-to-point wiring techniques or printed-circuit boards (PC boards) can be used to construct electronic circuits. Most circuit projects use a combination of techniques. The selection of techniques depends on many different factors and builder preferences.

The simple audio amplifier shown in **Fig 8.10** will be built using various point-to-point or PC-board techniques. This shows how the different construction methods are applied to a typical circuit.

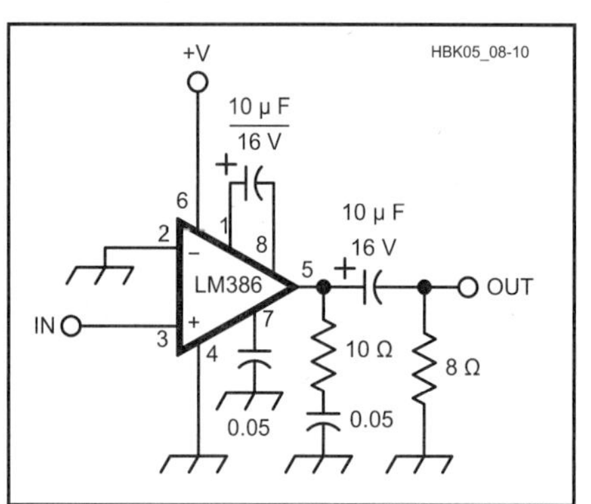

Fig 8.10—Schematic diagram of the audio amplifier used as a design example of various construction techniques.

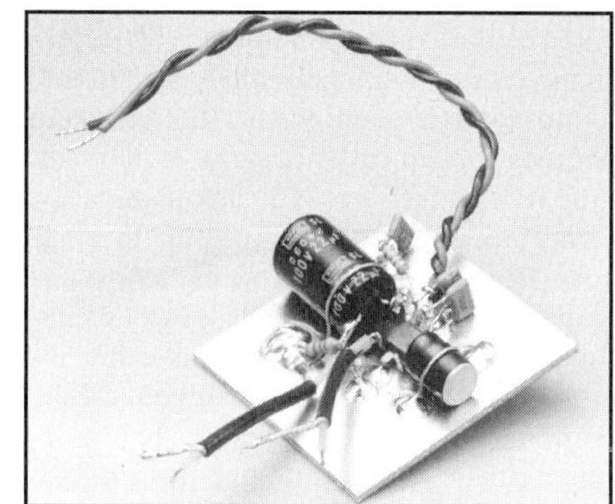
Fig 8.11—The example audio amplifier of Fig 8.10 built using ground-plane construction.

Point-to-Point Techniques

Point-to-point techniques include all circuit construction techniques that rely on tie points and wiring, or component leads, to build a circuit. This is the technique used in most home-brew construction projects. It is sometimes used in commercial construction, such as old vacuum-tube receivers and modern tube amplifiers.

Point-to-point is also used to connect the "off-board" components used in a printed-circuit project. It can be used to interconnect the various modules and printed-circuit boards used in more complex electronic systems. Most pieces of electronic equipment have at least some point-to-point wiring.

Ground-Plane Construction

A point-to-point construction technique that uses the leads of the components as tie points for electrical connections is known as "ground-plane construction," "dead-bug" or "ugly construction." (The term "ugly construction" was coined by Wes Hayward, W7ZOI.) "Dead-bug construction" gets its name from the appearance of an IC with its leads sticking up in the air. In most cases, this technique uses copper-clad circuit-board material as a foundation and ground plane on which to build a circuit using point-to-point wiring, so in this chapter it is called "ground-plane construction." An example is shown in **Fig 8.11**.

Ground-plane construction is quick and simple: You build the circuit on an unetched piece of copper-clad circuit board. Wherever a component connects to ground, you solder it to the copper board. Ungrounded connections between components are made point-to-point. Once you learn how to build with a ground-plane board, you can grab a piece of circuit board and start building any time you see an interesting circuit.

A PC board has strict size limits; the components must fit in the space allotted. Ground-plane construction is more flexible; it allows you to use the parts on hand. The circuit can be changed easily—a big help when you are experimenting. The greatest virtue of ground-plane construction is that it is fast.

Ground-plane construction is something like model building, connecting parts using solder almost—but not exactly—like glue. In ground-plane construction you build the circuit directly from the schematic, so it can help you get familiar with a circuit and how it works. You can build subsections of a large circuit on small ground-plane modules and string them together into a larger design.

Circuit connections are made directly, minimizing component lead length. Short lead lengths and a low-impedance ground conductor help prevent circuit instability.

There is usually less intercomponent capacitive coupling than would be found between PC-board traces, so it is often better than PC-board construction for RF, high-gain or sensitive circuits.

Use circuit components to support other circuit components. Start by mounting one component onto the ground plane, building from there. There is really only one two-handed technique to mount a component to the ground plane. Bend one of the component leads at a 90° angle, and then trim off the excess. Solder a blob of solder to the board surface, perhaps about 0.1 inch in diameter, leaving a small dome of solder. Using one hand, hold the component in place on top of the soldered spot and reheat the component and the solder. It should flow nicely, soldering the component securely. Remove the iron tip and hold the component perfectly still until the solder cools. You can then make connections to the first part.

Connections should be mechanically secure before soldering. Bend a small hook in the lead of a component, then "crimp" it to the next component(s). Do not rely only on the solder connections to provide mechanical strength; sooner or later one of these connections will fail, resulting in a dead circuit.

In most cases, each circuit has enough grounded components to support all of the components in the circuit. This is not always possible, however. In some circuits, high-value resistors can be used as standoff insulators. One resistor lead is soldered to the copper ground plane; the other lead is used as a circuit connection point. You can use 1/4- or 1/2-W resistors in values from 1 to 10 MΩ. Such high-value resistors permit almost no current to flow, and in low-impedance circuits they act more like insulators than resistors. As a rule of thumb, resistors used as standoff insulators should have a value that is at least 10 times the circuit impedance at that circuit point.

Fig 8.12A shows how to use the standoff technique to wire the circuit shown at Fig 8.12C. Fig 8.12B shows how the resistor leads are bent before the standoff component is soldered to the ground plane. Components E1 through E5 are resistors that are used as standoff insulators. They do not appear in the schematic diagram. The base circuitry at Q1 of Fig 8.12A has been stretched out to reduce clutter in the drawing. In a practical circuit, all of the signal leads should be kept as short as possible. E4 would, therefore, be placed much closer to Q1 than the drawing indicates.

No standoff posts are required near R1 and R2 of Fig 8.12. These two resistors serve two purposes: They are not only the normal circuit resistances, but function as standoff posts as well. Follow this practice wherever a capacitor or resistor can be employed in the dual role.

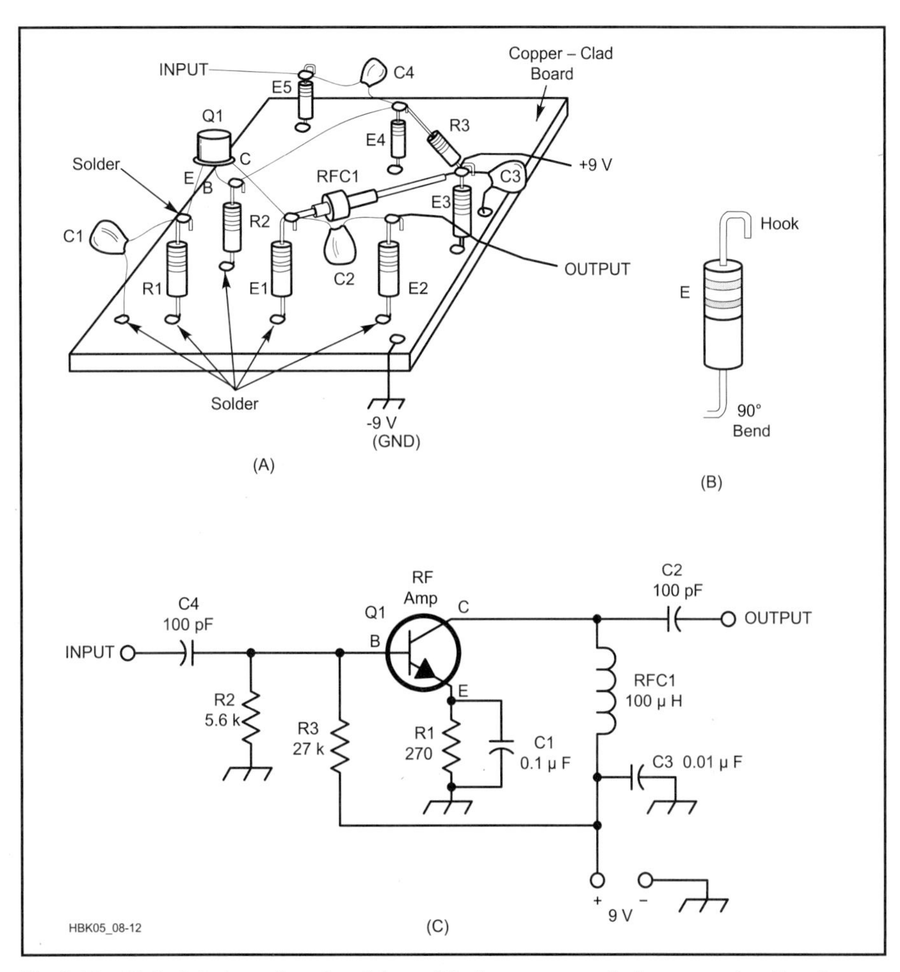

Fig 8.12—Pictorial view of a circuit board that uses ground-plane construction is shown at A. A close-up view of one of the standoff resistors is shown at B. Note how the leads are bent. The schematic diagram at C shows the circuit displayed at A.

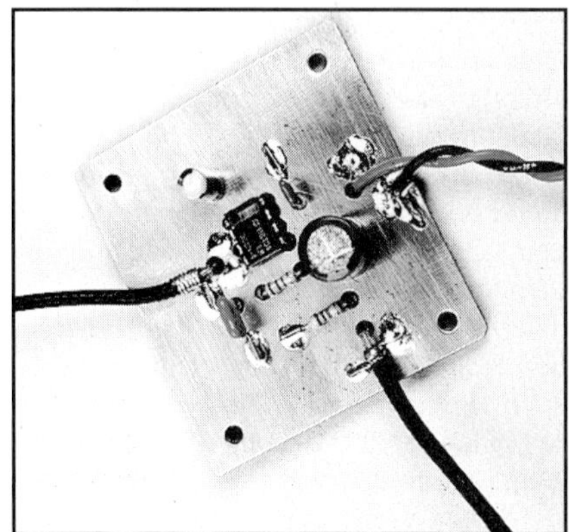

Fig 8.13—The audio amplifier built using wired-traces construction.

Wired Traces—the Lazy PC Board

If you already have a PC-board design, but don't want to copy the entire circuit—or you don't want to make a double-sided PC board—then the easiest construction technique is to use a bare board (or perfboard) and hard-wire the traces.

Drill the necessary holes in a piece of single-sided board, remove the copper ground plane from around the holes, and then wire up the back using component leads and bits of wire instead of etched traces (**Fig 8.13**).

To transfer an existing board layout, make a 1:1 photocopy and tape it to your piece of PC board. Prick through the holes with an automatic (one-handed) center punch or by firm pressure with a sharp scriber, remove the photocopy and drill all the holes. Holes for ground leads are optional—you generally get a better RF ground by bending the component lead flat to the board and soldering it down. Remove the copper around the rest of the holes by pressing a drill bit lightly against the hole and twisting it between your fingers. A drill press can also be used, but either way, don't remove too much board material. Then wire up the circuit beneath the board. The results look very neat and tidy—from the top, at least!

Circuits that contain components originally designed for PC-board mounting are good candidates for this technique. Wired traces would also be suitable for circuits involving multipin RF ICs, double-balanced mixers and similar components. To bypass the pins of these components to ground, connect a miniature ceramic capacitor on the bottom of the board directly from the bypassed pin to the ground plane.

A wired-trace board is fairly sturdy, even though many of the components are only held in by their bent leads and blobs of solder. A drop of cyanoacrylate "super glue" can hold down any larger components, components with fragile leads or any long leads or wires that might move.

Perforated Construction Board

A simple approach to circuit building uses a perforated board (perfboard). Perfboard is available with many different hole patterns. Choose the one that suits your needs. Perfboard is usually unclad, although it is made with pads that facilitate soldering.

Circuit construction on perforated board is easy. Start by placing the components loosely on the board and moving them around until a satisfactory layout is obtained. Most of the construction techniques described in this chapter can be applied to perfboard. The audio amplifier of Fig 8.10 is shown constructed with this technique in **Fig 8.14**.

Perfboard and accessories are widely available. Accessories include mounting hardware and a variety of connection terminals for solder and solderless construction.

Terminal and Wire

A perfboard is usually used for this technique (**Fig 8.15**). Push terminals are inserted into the hole in a perfboard. Components can then be easily soldered to the terminals. As an alternative, drill holes into a bare or copper-clad board wherever they are needed. The components are usually mounted on one side of the board and wires are soldered to the bottom of the board, acting as wired PC-board "traces." If a component has a reasonably rigid lead to which you can attach other components, use that instead of a push terminal, a modification of the ground-plane construction technique.

If you are using a bare board to provide a ground plane, drill holes for your terminals with a high-speed PC-board drill and drill press. Mark the position of the hole with a center punch to prevent the drill from skidding. The hole should provide a snug fit for the push terminal.

Mount RF components on top of the board, keeping the dc components and much of the interconnecting wiring underneath. Make dc feed-through connections with terminals having bypass capacitors on top of the board. Use small solder-in feedthrough capacitors for more critical applications.

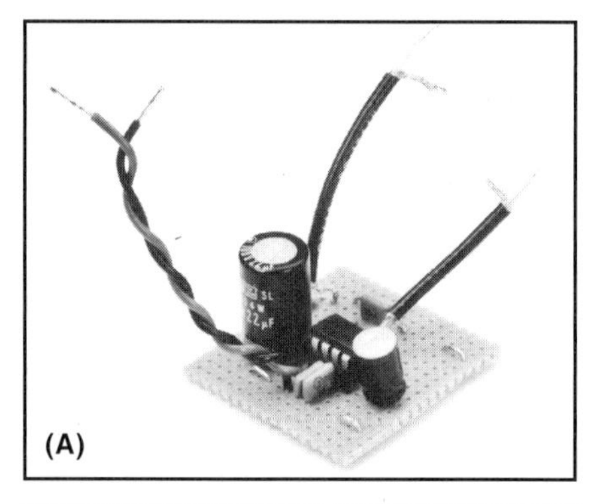

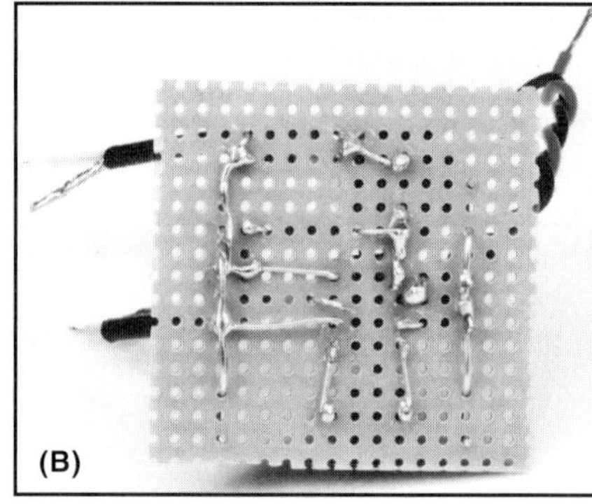

Fig 8.14—The audio amplifier built on perforated board. Top view at A; bottom view at B.

Solderless Prototype Board

One construction alternative that works well for audio and digital circuits is the solderless prototype board (protoboard), shown in **Fig 8.16**. It is usually not suitable for RF circuits.

A protoboard has rows of holes with spring-loaded metal strips inside the board. Circuit components and hookup wire are inserted into the holes, making contact with the metal strips. Components that are inserted into the same row are connected together. Component and interconnection changes are easy to make.

Protoboards have some minor disadvantages. The boards are not good for building RF circuits; the metal strips add too much stray capacitance to the circuit.

Fig 8.16—The audio amplifier built on a solderless prototyping board.

Fig 8.15—The audio amplifier built using terminal-and-wire construction.

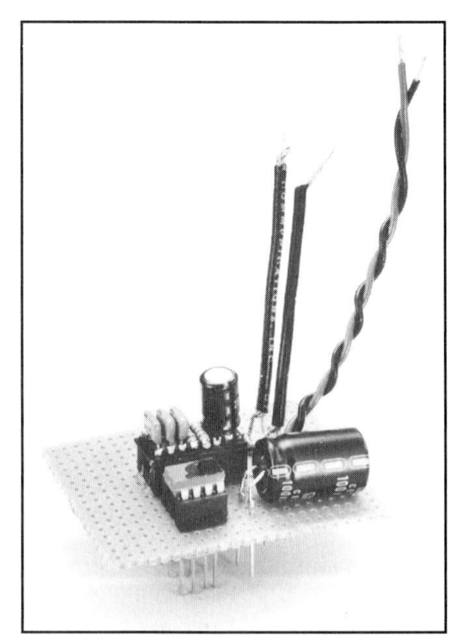

Fig 8.17—The audio amplifier built using wire-wrap techniques.

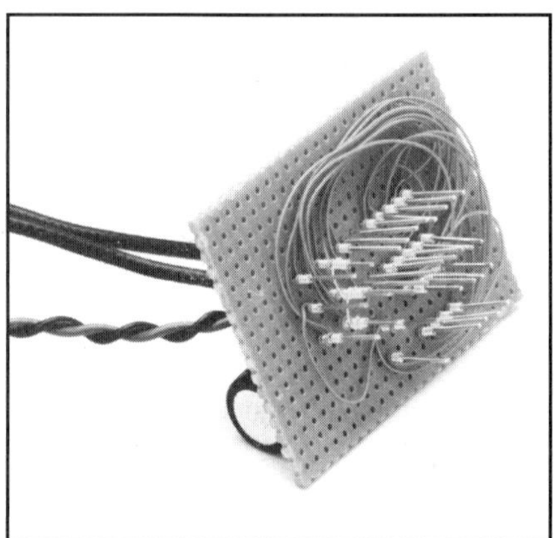

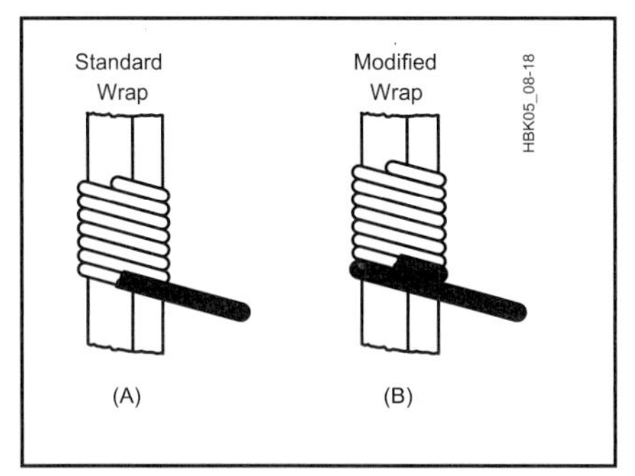

Fig 8.18—Wire-wrap connections. Standard wrap is shown at A; modified wrap at B.

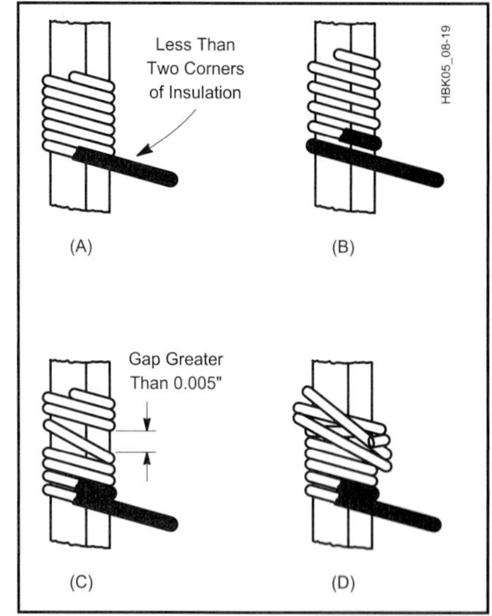

Fig 8.19—Improper wire-wrap connections. Insufficient insulation for modified wrap is shown at A; a spiral wrap at B, where there is too much space between turns; an open wrap at C, where one or more turns are improperly spaced and an overwrap at D, where the turns overlap on one or more turns.

Large component leads can deform the metal strips.

Wire Wrap

Wire-wrap techniques can be used to quickly construct a circuit without solder. Low- and medium-speed digital circuits are often assembled on a wire-wrap board. The technique is not limited to digital circuits, however. **Fig 8.17** shows the audio amplifier built using wire wrap. Circuit changes are easy to make, yet the method is suitable for permanent assemblies.

Wire wrap is done by wrapping a wire around a small square post to make each connection. A wrapping tool resembles a thick pencil. Electric wire-wrap guns are convenient when many connections must be made. The wire is almost always #30 wire with thin insulation. Two wire-wrap methods are used: the standard and the modified wrap (**Fig 8.18**). The modified wrap is more secure. The wrap-post terminals are square (wire wrap works only on posts with sharp corners). They should be long enough for at least two connections. Fig 8.18 and **Fig 8.19** show proper and improper wire-wrap techniques. Mount small components on an IC header plug. Insert the header into a wire-wrap IC socket as shown in Fig 8.17. The large capacitor in that figure has its leads soldered directly to wire-wrap posts.

Surface Mounting

Surface mounting is not new—it was an established ground-plane and professional technique for years before its appearance in consumer and amateur electronics. This technique is particularly suitable for PC-board construction, although it can be applied to many other construction techniques. Surface-mounted components take up very little space on a PC board.

Modern automated manufacturing techniques and surface-mount technology have evolved together; most modern ICs are being made specifically for this technique. Chip resistors and capacitors are common in UHF and microwave designs. Chip devices have low stray inductance and capacitance, making them excellent components to use in this frequency range. Other components, such as transistors and diode arrays are also available in this space-saving format.

Surface-mount techniques are not limited to "surface-mount" ICs, however. This technique can be used to mount standard resistors, capacitors or ICs. See the sidebar "Surface Mount Construction Techniques."

Removing SMT Components

The surface-mount ICs used in commercial equipment are not easy for experimenters to replace. They have tiny pins designed for precision PC boards. Sooner or later, you may need to replace one, though. If you do, don't try to get the old IC out in one piece! This will damage the IC beyond use anway, and will probably damage the PC board in the process.

Although it requires a delicate touch and small tools, it's possible to change a surface mount IC at home. To remove the old one, use small, sharp wire cutters to cut the IC pins flush with the IC. This usually leaves just enough old pin to grab with a tiny pair of needle-nose pliers or a hemostat. Heat the soldered connection with a small iron and use the pliers to gently pull the pin from the PC board. Solder in the new component using the techniques discussed in the sidebar.

Printed-Circuit Boards

Many builders prefer the neatness and miniaturization made possible by the use of etched printed-circuit boards (PC boards). Once designed, a PC board is easily duplicated, making PC boards ideal for group projects. To make a PC board, resist material is applied to a copper-clad bare PC board, which is then immersed into an acid etching bath to remove selected areas of copper. In a finished board, the conductive copper is formed into a pattern of conductors or "traces" that form the actual wiring of the circuit.

PC Board Stock

PC board stock consists of a sheet made from insulating material, usually glass epoxy or phenolic, coated with conduc-

Surface Mount Construction Techniques

"Oh no, this project uses SMT parts!" Some homebrewers recoil at the thought of assembling a kit that uses surface mount technology (SMT) components. They fear the parts are too small to see, handle, solder or debug when assembled. I had these same concerns until I tried it and found that it wasn't so difficult when using the right tools. Further, I discovered some benefits of using SMT parts that made my QRP projects smaller, lighter and more portable for optimized field use.

I've chosen two quite different projects to illustrate some successful SMT assembly techniques. One is a small DDS signal generator "daughtercard" kit that comes with an assortment of SMT capacitors, resistors and inductors, and an SOIC integrated circuit. The other example circuit is a small one-stage audio amplifier built "Manhattan-style"! Yes, you *can* homebrew using SMT parts—results can sometimes be even better than when using conventional leaded parts.

But first, here is some component history and what you need to do to get your work area ready for constructing an SMT project.

What is an SMT component?

Resistors and capacitors with axial or radial leads have been most common over the years. Same too for integrated circuits arranged in dual inline package (DIP) format with rows of leads separated by a generous 0.3 inch or so. This open-leaded component and easily accessed IC pins made for easy circuit board assembly back in the Heathkit days. Although these types of components are still available today, parts miniaturization has brought about more compact and less expensive products. Discrete components packaging has shrunk to 0.12 × 0.06 inches, as shown in the '1206' capacitor in **Fig A** compared to a penny. Even smaller packages are common today, requiring much less pc board area for the same equivalent circuits. Integrated circuit packaging has also been miniaturized to create 10 × 5 mm SOIC packages with lead separations of 0.025 inches. You need some extra skills beyond what was necessary when assembling that Heath SB-104 transceiver back in 1974!

Fig A—SMT components are small. Clockwise from left: MMIC RF amp, 1206 resistor, SOIC integrated circuit, 1206 capacitor and ferrite inductor.

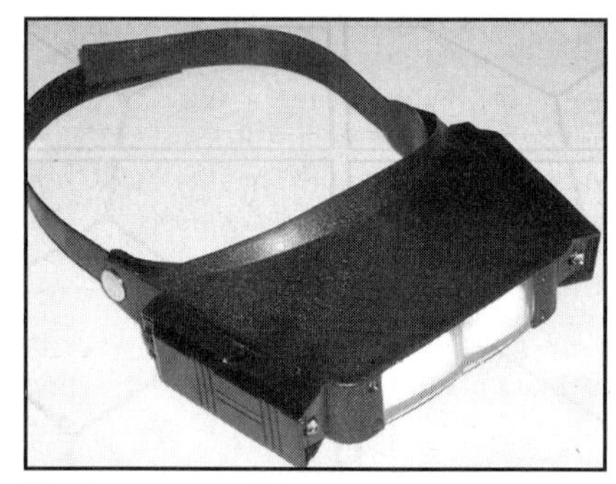

Fig B—A magnifying visor is great for close-up work on a circuit board. These headsets are often available for less than $10 at hamfests and some even come with superbright LEDs mounted on the side to illuminate the components being soldered.

Fig C—This DDS Daughtercard has all interconnections on the top side. Connections to the ground plane on the backside of the board are made by the use of "vias," wires through the PC board. Pin 28 of the SMT IC is shown being tack-soldered to hold it to the board, keeping all other pins carefully aligned on their pads. Then the other pins are carefully soldered, starting with pin 14 (opposite pin 28). Finally, pin 28 is reheated to ensure a good connection there. If you bridge solder across adjacent pads or pins, use solder wick or a vacuum solder sucker to draw off the excess solder.

Preparing for the job

The key to success with any construction project is selecting and using the proper tools. A magnifying lamp is essential for well-lighted, close-up work on the components. **Fig B** shows a convenient magnifying visor. Tweezers or fine-tipped pliers allow you to grab the small chip components with dexterity. Thinner solder (0.015 inches) than you might normally use is preferred because it melts more quickly and leaves a smaller amount of solder on the component lead. Use of a super fine-tipped soldering iron make soldering the leads of these small parts straightforward and easy. A clean work surface is of paramount importance because SMT components have a tendency to fly away even when held with the utmost care by tweezers—you'll have the best chance of recovering your wayward part if your table is clear. When the inevitable happens, you'll have lots of trouble finding it if the part falls onto a rug. It's best to have your work area in a non-carpeted room, for this reason as well as to protect static-sensitive parts.

Assembling SMT parts on a PC board

The first project example is the **DDS** Daughtercard—a small module that generates precision RF signals for a variety of projects. This kit has become immensely popular in homebrew circles and is supplied with the chip components contained in color-coded packaging that makes and easy job of identifying the little parts, a nice touch by a kit supplier.

Fig C shows the DDS PC board, a typical layout for SMT components. All traces are on one side, since the component leads are not "through-hole." The little square pads are the places where the 1206 package-style chips will eventually be soldered.

The trick to soldering surface-mount devices to PC boards is to (a) pre-solder ("tin") one of the pads on the board where the component will ultimately go by placing a small blob of solder there; (b) carefully hold the component in place with small needle-nose pliers or sharp tweezers on the tinned pad; (c) reheat the tinned pad and component to reflow the solder onto the component lead, thus temporarily holding the component in place; and (d) solder the other end of the component to its pad. Finally, check all connections to make sure there are no bridges or shorts. **Fig D** illustrates how to use this technique. **Fig E** shows the completed DDS board.

Homebrewing with SMT parts

The second project example is the **K8IQY Audio Amp**—a discrete component audio amplifier that is constructed "Manhattan-style." You glue little pads to the board wherever you need to attach component leads or wires. See **Fig F**.

Instead of using little squares or dots of pcb material for pads, you might decide to create isolated connection points by cutting an "island" in the copper using an end mill. No matter how the pads are created, SMT components may be easily soldered from pad-to-pad, or from pad-to-ground plane to build up the circuit. **Fig G** shows the completed board, combining SMT and leaded components.

Fig E—The fully-populated DDS Daughtercard PC board contains a mix of SMT and through-hole parts, showing how both packaging technologies can be used together.

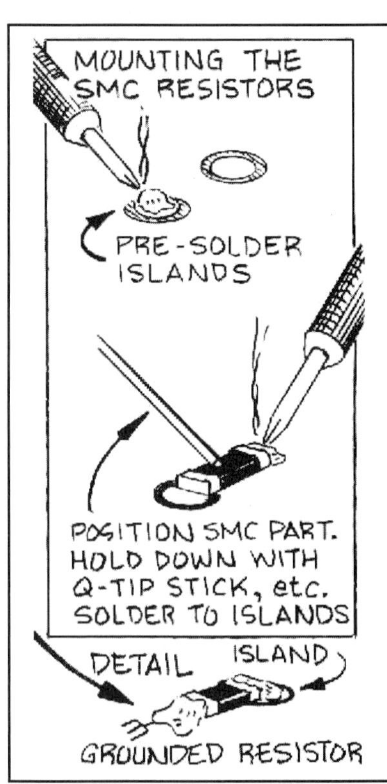

Fig D—Attaching an SMT part. Things are a lot easier attaching capacitors, resistors and other discrete components compared to multi-pin ICs. Carefully hold the component in place and properly aligned using needle-nose pliers or tweezers and then solder one end of the component. Then reheat the joint while gently pushing down on the component with the pliers or a Q-tip stick to ensure it is lying flat on the board. Finally, solder the other side of the component.

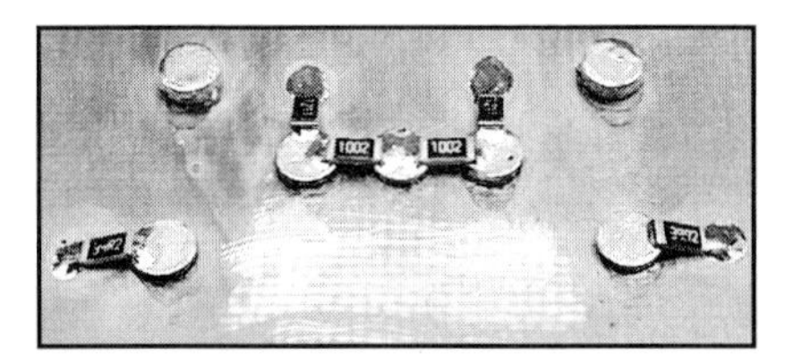

Fig F—SMT resistors soldered to base board of the Audio Amp in the beginning stages of assembly.

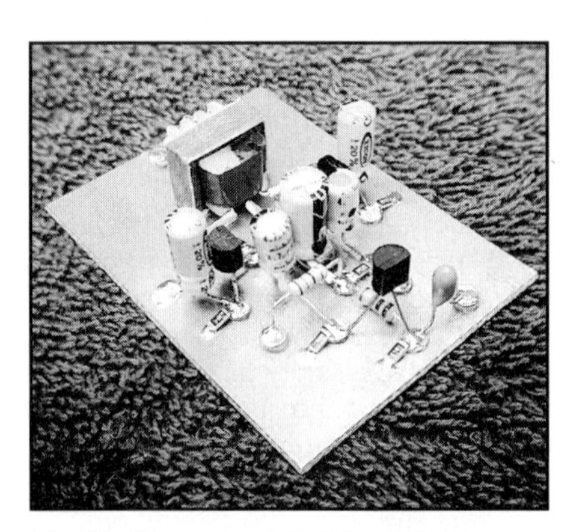

Fig G—The completed homebrew Audio Amp assembly shows simple, effective use of SMT components used together with conventional leaded components when constructed "Manhattan-style."

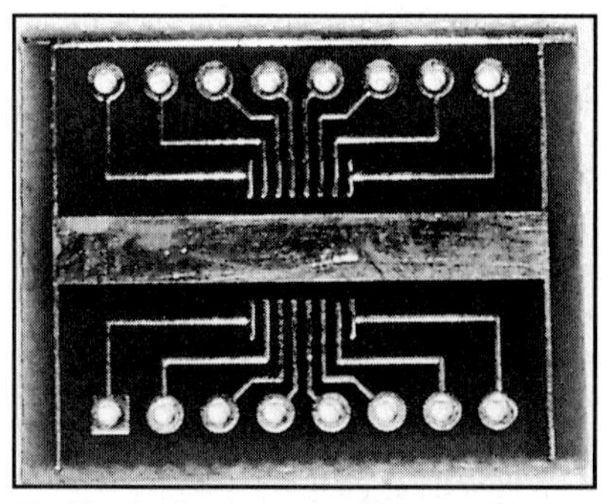

Fig H—Surface mount ICs can be mounted to general-purpose carrier boards, then attached as a submodule with wires to the base board of the homebrew project.

Homebrewing with SOIC-packaged integrated circuits is a little trickier and typically requires the use of an "SOIC carrier board" such as the one shown in **Fig H**, onto which you solder your surface mount integrated circuit. You can then wire the carrier board onto your homebrew project, copper-clad base board or whatever you're using to hold your other circuit components.

So start melting solder!

The techniques are easy, the SMT components are actually cheaper than conventional through-hole leaded components and you'll have a smaller, more portable project when you're done. Go for it!—*George Heron, N2APB.*

References

Full details on the DDS Daughtercard, the K8IQY Islander Audio Amp, and the Islander Pad Cutter may be found online at **www.njqrp.org/dds**, **www.njqrp.org/islanderamp**, and **www.njqrp.org/islanderpadcutter**, respectively. Also see: **www.arrl.org/tis/info/surface.html**.

tive copper. Copper-clad stock is manufactured with phenolic, FR-4 fiberglass and Teflon base materials in thicknesses up to 1/8 inch. The copper thickness varies. It is usually plated from 1 to 2 oz per square foot of bare stock.

Resists

Resist is a material that is applied to a PC board to prevent the acid etchant from eating away the copper on those areas of the board that are to be used as conductors. There are several different types of resist materials, both commercial and home brew. When resist is applied to those areas of the board that are to remain as copper traces, it "resists" the acid action of the etchant.

The PC board stock must be clean before any resist is applied. This is discussed later in the chapter. After you have applied resist, by whatever means, protect the board by handling it only at its edges. Do not let it get scraped. Etch the board as soon as possible, to minimize the likelihood of oxidation, moisture or oils contaminating the resist or bare board.

Tape

To make a single PC board, Scotch, adhesive or masking tape, securely applied, makes a good resist. (Don't use drafting tape; its glue may be too weak to hold in the etching bath.) Apply the tape to the entire board, transfer the circuit pattern by means of carbon paper, then cut out and remove the sections of tape where the copper is to be etched away. An X-Acto hobby knife is excellent for this purpose.

Resist Pens

Several electronics suppliers sell resist pens. Use a resist pen to draw PC-board artwork directly onto a bare board. Commercially available resist pens work well. Several types of permanent markers also function as resist, especially the "Sharpie" brand. They come in fine-point and regular sizes; keep two of each on hand.

Paint

Some paints are good resists. Exterior enamel works well. Nail polish is also good, although it tends to dry quickly so you must work fast. Paint the pattern onto the copper surface of the board to be etched. Use an artist's brush to duplicate the PC board pattern onto bare PC-board stock. Tape a piece of carbon paper to the PC-board stock. Tape the PC-board pattern to the carbon paper. Trace over the original layout with a ballpoint pen. The carbon paper transfers the outline of the pattern onto the bare board. Fill in the outline with the resist paint. After paint has been applied, allow it to dry thoroughly before etching.

Rub-On Transfer

Several companies (Kepro Circuit Systems, DATAK Corp, GC Electronics) produce rub-on transfer material that can also be used as resist. Patterns are made with various width traces and for most components, including ICs. As the name implies, the pad or trace is positioned on the bare board and rubbed to adhere to the board.

Etchant

Etchant is an acid solution that is designed to remove the unwanted copper areas on PC-board stock, leaving other areas to function as conductors. Almost any strong acid bath can serve as an etchant, but some acids are too strong to be safe for general use. Two different etchants are commonly used to fabricate prototype PC boards: ammonium persulphate and ferric chloride. The latter is the more common of the two.

Ferric chloride etchant is usually sold ready-mixed. It is made from one part ferric chloride crystals and two parts water, by volume. No catalyst is required.

Etchant solutions become exhausted as they are used. Keep a supply on hand. Dispose of the used solution safely; follow the instructions of your local environmental protection authority.

Most etchants work better if they are hot. A board that takes 45 minutes to etch at room temperature will take only a few minutes if the etchant is hot. Use a heat lamp to warm the etchant to the desired temperature. A darkroom thermometer is handy for monitoring the temperature of the bath.

Be careful! Do not heat your etchant above the recommended temperature, typically 160°F. If it gets too hot, it will probably damage the resist. Hot or boiling etchant is also a safety hazard.

Insert the board to be etched into the solution and agitate it continuously to keep fresh chemicals near the board surface. This speeds up the etching process. Normally, the circuit board should be placed in the bath with the copper side facing up.

After the etching process is completed, remove the board from the tray and wash it thoroughly with water. Use medium-grade steel wool to rub off the resist.

WARNING: Use a glass or other nonreactive container to hold etching chemicals. Most etchants will react with a metal container. Etchant is caustic and can burn eyes or skin easily. Use rubber gloves and wear old clothing, or a lab smock, when working with any chemicals. If you get some on your skin, wash it with soap and cold water. Wear safety goggles (the kind that fit snugly on your face) when working with any dangerous chemicals. Read the safety labels and follow them carefully. If you get etchant in your eyes, wash immediately with large amounts of cool water and seek immediate medical help. Even a small chemical burn on your eye can develop into a serious problem.

Planning and Layout

A PC board can be a real convenience. If you want to build a project and a ready-made PC board is available, you can assemble the project quickly and expect it to work. This is true because someone else has done most of the real work involved—designing the PC board layout and fixing any "bugs" caused by intertrace capacitive coupling, ground loops and similar problems. In most cases, if a ready-made board is not available, ground-plane construction is a lot less work than designing, debugging and then making a PC board.

A later section of this chapter explains how to turn a schematic into a working circuit. It is not as simple as laying out the PC board just like the circuit is drawn on the schematic. Read that section before you design a PC board.

Rough Layout

Start by drawing a rough scale pictorial diagram of the layout. Draw the interconnecting leads to represent the traces that are needed on the board. Rearrange the layout as necessary to find an arrangement that completes all of the circuit traces with a minimum number of jumper-wire connections. In some cases, however, it is not possible to complete a design without at least a few jumpers.

Layout

After you have completed a rough layout, redraw the physical layout on a grid. Graph paper works well for this. Most IC pins are on 0.1-inch centers. Use graph paper that has 10 lines per inch to draw artwork at 1:1 and estimate the distance halfway between lines for 0.05-inch spacing. Drafting templates are helpful in the layout stage. Local drafting-supply stores should be able to supply them. The templates usually come in either full-scale or twice normal size.

To lay out a double-sided board, ensure that the lines on both sides of the paper line up (hold the paper up to the light). You can then use each side of the paper for each side of the board.

When using graph paper for a PC-board layout, include bolt holes, notches for wires and other mechanical considerations. Fit the circuit into and around these, maintaining clearance between parts.

Most modern components have leads on 0.1-inch centers. The rows of dual-inline-package (DIP) IC pins are spaced 0.3 or 0.4 inch. Measure the spacing for other components. Transfer the dimensions to the graph paper. It is useful to draw a schematic symbol of the component onto the layout.

Most IC specification sheets show a top view of the pin locations. If you are designing the "foil" side of a PC board, be sure to invert the pin out.

Draw the traces and pads the way they will look. Using dots and lines is confusing. It's okay to connect more than one lead per pad, or run a lead through a pad, although using more than two creates a complicated layout. In that case, there may be problems with solder bridges that form short circuits. Traces can run under some components; it is possible to put two or three traces between 0.4-inch centers for a 1/4-W resistor, for example.

Leave power-supply and other dc paths for last. These can usually run just about anywhere, and jumper wires are fine for these noncritical paths.

Do not use traces less than 0.010 inch (10 mil) wide. If 1-oz stock is used, a 10-mil trace can safely carry up to 500 mA. To carry higher current, increase the width of the traces in proportion. (A trace should be 0.200 inch to carry 10 A, for example.) Allow 0.1 inch between traces for each kilovolt in the circuit.

When doing a double-sided board, use pads on both sides of the board to connect traces through the board. Home-brew PC boards do not use plated-through holes (a manufacturing technique that has copper and tin plating inside all of the holes to form electrical connections). Use a through hole and solder the associated component to both sides of the board. Make other through-hole connections with a small piece of bus wire providing the connection through the board; solder it on both sides. This serves the same purpose as the plated-through holes found in commercially manufactured boards.

After you have planned the physical design of the board, decide the best way to complete the design. For one or two simple boards, draw the design directly onto the board, using a resist pen, paint or rub-on resist materials. To transfer the design to the PC board, draw light, accurate pencil lines at 0.1- or 0.05-inch centers on the PC board. Draw both horizontal and vertical lines, forming a grid. You only need lines on one side. For single-sided boards, use this grid to transfer the layout directly onto the board surface. To make drilling easier, use a center punch to punch the centers of holes accurately. Do this before applying the resist so the grid is visible.

When drawing a pad with plenty of room around it, use a pad about 0.05 to 0.1 inch in diameter. For ICs, or other close quarters, make the pad as small as 0.03 inch or so. A "ring" that is too narrow invites soldering problems; the copper may delaminate from the heat. Pads need not be round. It's okay to shave one or more edges if necessary, to allow a trace to pass nearby.

Draw the traces next. A drafting triangle can help. It should be spaced about 0.1 inch above the table, to avoid smudging the artwork. Use a 9-inch or larger triangle, with a rubber grommet taped to each corner (to hold it off the table). Select a sturdy triangle that doesn't bend easily.

Align the triangle with the grid lines by eye and make straight, even traces similar to the layout drawing. The triangle can help with angled lines, too. Practice on a few pieces of scrap board.

Make sure that the resist adheres well to the PC board. Most problems can be seen by eye; there can be weak areas or bare spots. If necessary, touch up problems with additional resist. If the board is not clean the resist will not adhere properly. If necessary, remove the resist, clean the board and start from the beginning.

Discard troublesome pens. Resist pens dry out quickly. Keep a few on hand, switch back and forth and put the cap back on each for a bit to give the pen a chance to recover.

Once all of the artwork on the board is drawn, check it against the original artwork. It is easy to leave out a trace. It is not easy to put copper back after a board is etched. In a pinch, replace the missing trace with a small wire.

Applied resist takes about an hour to dry at room temperature. Fifteen minutes in a 200°F oven is also adequate.

Special techniques are used to make double-sided PC boards. See the section on double-sided boards for a description.

Making a PC Board

Several techniques can be used to make PC boards. They usually start with a PC-board "pattern" or artwork. All of the techniques have one thing in common: this pattern needs to be transferred to the copper surface of the PC board. Unwanted copper is then removed by chemical or mechanical means.

Most variations in PC-board manufacturing technique involve differences in resist or etchant materials or techniques.

Cut the Board to Size

No matter what technique you use, you should determine the required size of the PC board, and then cut the board to size. Trimming off excess PC-board material can be difficult after the components are installed.

Board Preparation

The bare (unetched) PC-board stock should be clean and dry before any resist is applied. (This is not necessary if you are using stock that has been treated with presensitized photoresist.) Wear rubber gloves when working with the stock to avoid getting fingerprints on the copper surface. Clean the board with soap and water, and then scrub the board with #000 steel wool. Rinse the board thoroughly then dry it with a clean, lint-free cloth. Keep the board clean and free of fingerprints or foreign substances throughout the entire manufacturing process.

No-Etch PC Boards

The simplest way to make PC boards is to mechanically remove the unwanted copper. Use a grinding tool, such as the Moto-Tool manufactured by the Dremel Company (available at most hardware or hobby stores). Another technique is to score the copper with a strong, sharp knife, then remove unwanted copper by heating it with a soldering iron and lifting it off with a knife while it is still hot. This technique requires some practice and is not very accurate. It often fails with thin traces, so use it only for simple designs.

Photographic Process

Many magazine articles feature printed-circuit layouts. Some of these patterns are difficult to duplicate accurately by hand. A photographic process is the most efficient way to transfer a layout from a magazine page to a circuit board.

The resist ink, tape or dry-transfer processes can be time consuming and tedious for very complex circuit boards. As an alternative, consider the photo process. Not only does the accuracy improve, you need not trace the circuit pattern yourself!

A copper board coated with a light-sensitive chemical is at the heart of the photographic process. In a sense, this board becomes your photographic film.

Make a contact print of the desired pattern by transferring the printed-circuit artwork to special copy film. This film is attached to the copper side of the board and both are exposed to intense light. The areas of the board that are exposed to the light—those areas not shielded by the black portions of the artwork—undergo a chemical

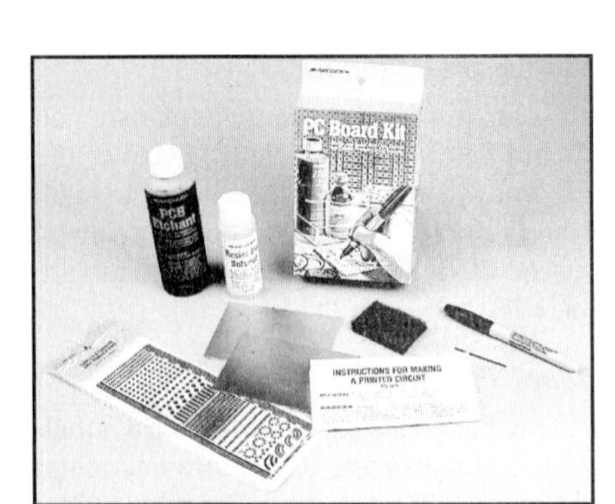

Fig 8.20—PC-board materials are available from several sources. This kit is from RadioShack.

change. This creates a transparent image of the artwork on the copper surface.

Develop the PC board, using techniques and chemicals specified by the manufacturer. After the board is developed, etch it to remove the copper from all areas of the board that were exposed to the light. The result is a PC board that looks like it was made in a factory.

Kepro sells materials and supplies for all types of PC-board manufacturing. RadioShack also sells PC-board materials. See **Fig 8.20.** If you're looking for printed-circuit board kits, chemicals, tools and other materials, contact Ocean State Electronics. They carry products by Kepro and the Meadowlake Corporation.

Iron-On Resist

One company that makes an iron-on resist is the Meadowlake Corporation. Their products make an artwork positive using a standard photocopier. A clothes iron transfers the printed resist pattern to the bare PC board.

Some experimenters have reported satisfactory results using standard photocopier paper or the output from a laser printer. Apparently the toner makes a reasonable resist. Note that the artwork for this method must be reversed with respect to a normal etching pattern because the print must be placed with the toner against the copper.

To transfer the resist pattern onto the board, place the pattern on the board (image side toward the copper), and then firmly press a hot iron onto the entire surface. Use plenty of heat and even pressure. This melts the resist, which then sticks to the bare PC board. This is not a perfect process; there will probably be bad areas on the resist. The amount of heat, the cleanliness of the bare board and the "skill" of the operator may affect the outcome.

The key to making high quality boards with the photocopy techniques is to be good at retouching the transferred resist.

Double-Sided PC Boards—by Hand!

Forget those nightmares about expensive photoresists that didn't work; forget that business of fifty bucks a board! You don't need computer-aided design to make a double-sided PC board; just improve on the basics, and keep it simple. Anyone can make low-cost double-sided boards with traces down to 0.020 inch, with perfect front-to-back hole registration.

To make a double-sided board, drill the holes before applying the resist artwork; that is the only way to assure good front-to-back registration. The artwork on both sides can then be properly positioned to the holes. PC-board drilling was discussed earlier in the text.

After you have drilled the board, clean its surface thoroughly. After that, wear clean rubber or cotton gloves to keep it clean. One fingerprint can really mess up the application of resist or the etchant.

Tape the board to your work surface, making sure it can't move around. Transfer the artwork from your layout grid to the PC board, drawing by hand with a resist pen.

Fig A—Make a permanent marker into a specialized PC-board drawing tool. Simply press the marker point into a drilled hole to form a modified point as shown. More pressure produces a wider shoulder that makes larger pads on the PC board.

Allot enough time to finish at least one side of the artwork in one sitting. Start with the pads. To make a handy pad-drawing tool, press the tip of a regular-size Sharpie into one of the drilled PC-board holes. This "smooshes" the tip into the shape of the hole, leaving a flat shoulder to draw the pad. See **Fig A**. The diameter of the pad is determined by how hard the pen is pressed; pressing too hard forms a pad that is way too large for most applications. Practice on scrap board first. Use this modified pen to fill in all the holes and draw the pads at the same time. Use an unmodified resist pen to draw all of the traces and to touch up any voids or weak areas in the pads. For the rest of the drawing, the procedure described for single-sided boards applies to double-sided boards, too.

After the resist is applied to the first side, carefully draw the second side. Inspect the board thoroughly; you may have scratched or smudged the first side while you were drawing the second.

Etching a double-sided board is not much different than etching a single-sided board, except that you must ensure that the etchant is able to reach both sides of the board. If you dunk the board in and out of the etchant solution, both sides are exposed to the etchant. If you use a tray, put some spacers on the bottom and rest the board on the spacers. (The spacers must be put on the board edges, not where you want to actually etch.) This ensures that etchant gets to both sides. If you use this method, turn the board over once or twice during the process.—*Dave Reynolds, KE7QF*

Or Photo-Etched

You can also make double-sided boards at home without drawing the layout by hand. This procedure can't produce results to match the finest professionally made double-sided boards, but it can make boards that are good enough for many moderately complex projects.

Start with the same sort of artwork used for single-sided boards, but leave a margin for taping at one edge. It is critical that the patterns for the two sides are accurately sized. The chief limiting factor in this technique is the requirement that matching pads on the two sides are positioned correctly. Not only must the two sides match each other, but they must also be the correct size for the parts in the project. Slight reproduction errors can accumulate to major problems in the length of a 40-pin DIP IC. One good tool to achieve this requirement is a photocopy machine that can make reductions and enlargements in 1% steps. Perform a few experiments to arrive at settings that yield accurately sized patterns.

Choose two holes at opposite corners of the etching patterns. Tape one of the two patterns to one side of the PC board. Choose some small wire and a drill bit that closely matches the wire diameter. For example, #20 enameled wire is a close match for a #62 or a #65 drill, depending on the thickness of the wire's enamel coating. Drill through the pattern and the board at the two chosen holes. Drill the chosen holes through the second pattern. Place two pieces of the wire through the PC board and slide the second pattern down these wire "pins" to locate the pattern on the board. Tape the second pattern in position and remove the pins. From this point on, expose and process each side of the board as if it were a single-sided board, but take care when exposing each side to keep the reverse side protected from light.—*Bob Schetgen, KU7G*

Fortunately, the problems are usually easy to retouch, if you have a bit of patience. A resist pen does a good job of reinforcing any spotty areas in large areas of copper.

Double-Sided PC Boards

All of the examples used to describe the above techniques were single-sided PC boards, with traces on one side of the board and either a bare board or a ground plane on the other side. PC boards can also have patterns etched onto both sides, or even have multiple layers. Most home-construction projects use single-sided boards, although some kit builders supply double-sided boards. Multilayer boards are rare in ham construction. One method for making double-sided boards is described in the sidebar, "Double-Sided PC Boards—by Hand!"

Tin Plating

Most commercial PC boards are tin plated, to make them easier to solder. Commercial tin-plating techniques require electroplating equipment not readily available to the home constructor. Immersion tin plating solutions can deposit a thin layer of tin onto a copper PC board. Using them is easy; put some of the solution into a plastic container and immerse the board in the solution for a few minutes. The chemical action of the tin-plating solution replaces some of the copper on the board with tin. The result looks nearly as good as a commercially made board. Agitate the board or solution from time to time. When the tinning is complete, take the board out of the solution and rinse it for five minutes under running water. If you don't remove all of the residue, solder may not adhere well to the surface. Kepro sells immersion tin plating solution.

Drilling a PC Board

After you make a PC board using one of the above techniques, you need to drill holes in the board for the components. Use a drill press, or at least improvise one. Boards can be drilled entirely "free hand" with a hand-held drill but the potential for error is great. A drill press or a small Moto-Tool in an accessory drill press makes the job a lot easier. A single-sided board should be drilled after it is etched; the easiest way to do a double-sided board is to do it before the resist is applied.

To drill in straight lines, build a small movable guide for the drill press so you can slide one edge of the board against it and line up all of the holes on one grid line at a time. See **Fig 8.21.** This is similar to the "rip fence" set up by most woodworkers to cut accurately and repeatably with a table saw.

The drill-bit sizes available in hardware stores are too big for PC boards. You can use high-speed steel bits, but glass epoxy stock tends to dull these after a few hundred holes. (When your drill bit becomes worn, it makes a little "hill" around each drilled hole, as the worn bit pushes and pulls the copper rather than drilling it.) A PC-board drill bit, available from many electronic suppliers, will last for thousands of holes! If you are doing a lot of boards, it is clearly worth the investment.

Small drill bits are usually ordered by number. Here are some useful numbers and their sizes:

Number	*Diameter*
68	0.0310"
65	0.0350"
62	0.0380"
60	0.0400"

Use high RPM and light pressure to make good holes. Count the holes on both the board and your layout drawing to ensure that none are missed. Use a larger-size drill bit, lightly spun between your fingers, to remove any burrs. Don't use too much pressure; remove only the burr.

"READY-MADE" PC BOARDS

Utility PC Boards

"Utility" PC boards are an alternative to custom-designed etched PC boards. They offer the flexibility of perforated board construction and the mechanical and electrical advantages of etched circuit connection pads. Utility PC boards can be used to build anything from simple passive filter circuits to computers.

Circuits can be built on boards on which the copper cladding has been divided into connection pads. Power supply voltages can be distributed on bus strips. Boards like those shown in **Fig 8.22** are commercially available.

An audio amplifier constructed on a utility PC board is shown in **Fig 8.23**. Component leads are inserted into the board and soldered to the etched pads. Wire jumpers connect the pads together to complete the circuit.

Utility boards with one or more etched plugs for use in computer-bus, interface and general purpose applications are widely available. Connectors, mounting hardware and other accessories are also available. Check with your parts supplier for details.

PC-Board Assembly Techniques

Once you have etched and drilled a PC board you are ready to use it in a project. Several tools come in handy: needle-nose pliers, diagonal cutters, pocket knife, wire strippers, clip leads and soldering iron.

Fig 8.21—This home-built drill fence makes it easy to drill PC-board holes in straight rows.

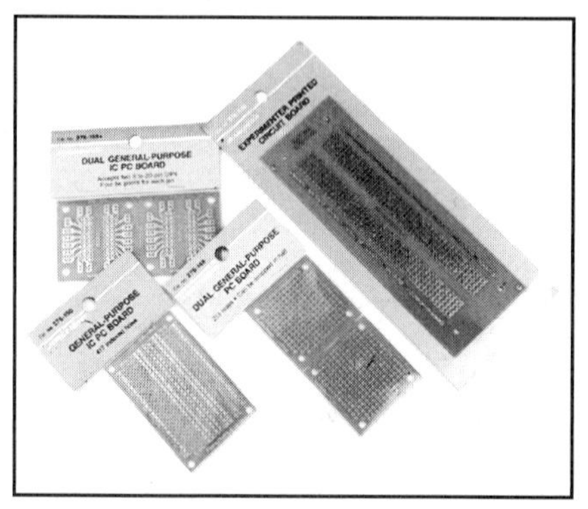

Fig 8.22—Utility PC boards like these are available from many suppliers.

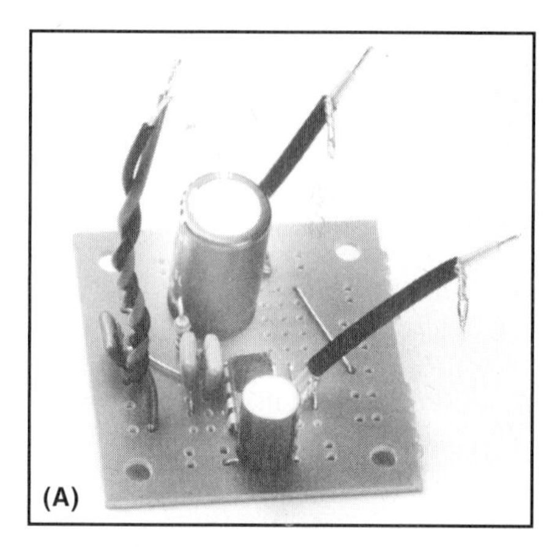

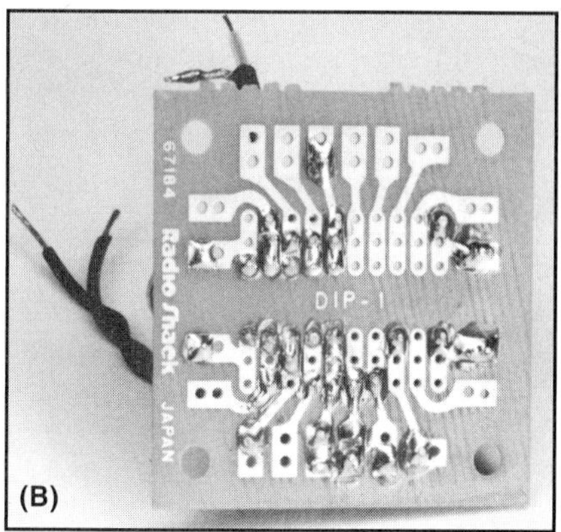

Fig 8.23—The audio amplifier built on a multipurpose PC breadboard. Top view at A; bottom view at B.

Cleanliness

Make sure your PC board and component leads are clean. Clean the entire PC board before assembly; clean each component before you install it. Corrosion looks dark instead of bright and shiny. Don't use sandpaper to clean your board. Use a piece of fine steel wool or a Scotchbrite cleaning pad to clean component leads or PC board before you solder them together.

Installing Components

In a construction project that uses a PC board, most of the components are installed on the board. Installing components is easy—stick the components in the right board holes, solder the leads, and cut off the extra lead length. Most construction projects have a parts-placement diagram that shows you where each component is installed.

Getting the components in the right holes is called "stuffing" the circuit board. Inserting and soldering one component at a time takes too long. Some people like to put the components in all at once, and then turn the board over and solder all the leads. If you bend the leads a bit (about 20°) from the bottom side after you push them through the board, the components are not likely to fall out when you turn the board over.

Start with the shortest components—horizontally mounted diodes and resistors. Larger components sometimes cover smaller components, so these smaller parts must be installed first. Use adhesive tape to temporarily hold difficult components in place while you solder.

PC-Board Soldering

To solder components to a PC board, bend the leads at a slight angle; apply the soldering iron to one side of the lead, and flow the solder in from the other side of the lead. See **Fig 8.24**. Too little heat causes a bad or "cold" solder joint; too much heat can damage the PC board. Practice a bit on some spare copper stock before you tackle your first PC board project. After the connection is soldered properly, clip the lead flush with the solder.

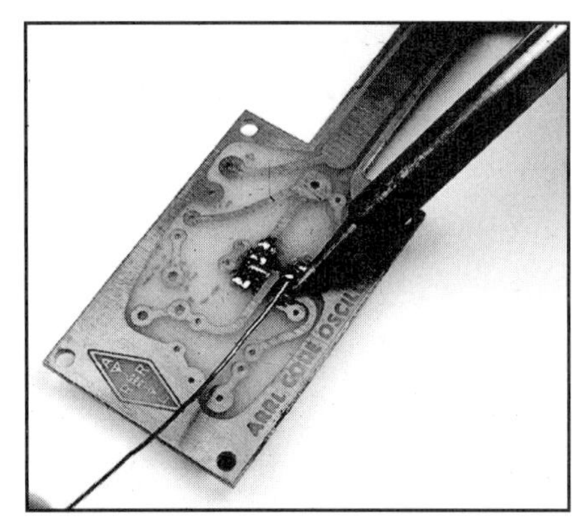

Fig 8.24—This is how to solder a component to a PC board. Make sure that the component is flush with the board on the other side.

Special Concerns

Make sure you have the components in the right holes before you solder them. Components that have polarity, such as diodes, ICs and some capacitors must be oriented as shown on the parts-placement diagram.

FROM SCHEMATIC TO WORKING CIRCUIT

Some people don't know how to turn a schematic into a working circuit. One thing is usually true—you can't build it the way it looks on the schematic. Many design and layout considerations that apply in the real world of practical electronics don't appear on a schematic.

PC Boards—Always the Best Choice?

PC boards are everywhere—in all kinds of consumer electronics, in most of your Amateur Radio equipment. They are also used in most kits and construction projects. A newcomer to electronics might think that there is some unwritten law against building equipment in any other way!

The misconception that everything needs to be built on a printed-circuit board is often a stumbling block to easy project construction. In fact, a PC board is probably the worst choice for a one-time project. In actuality, a moderately complex project (like a QRP transmitter) can be built in much less time using other techniques. The additional design, layout and manufacturing is usually much more work than it would take to build the project by hand.

So why does everyone use PC boards? The most important reason is that they are reproducible. They allow many units to be mass-produced with exactly the same layout, reducing the time and work of conventional wiring and minimizing the possibilities of wiring errors. If you can buy a ready-made PC board or kit for your project, it can save a lot of construction time.

Using a PC board usually makes project construction easier by minimizing the risk of wiring errors or other construction blunders. Inexperienced constructors usually feel more confident when construction has been simplified to the assembly of components onto a PC board. One of the best ways to get started with home construction (to some the best part of Amateur Radio) is to start by assembling a few kits using PC boards. Contact information for kit manufacturers can be found on *ARRLWeb* (**www.arrl.org/cgi-bin/tisfind?patt=amateur+radio+kit**).

One-Time Projects

Kits are fun, but another facet of electronics construction is building and developing your own circuits, starting from circuit diagrams. For one-time construction, PC boards are really not necessary. It takes time to lay out, drill and etch a PC board. Alterations are difficult to make if you change your ideas or make a mistake. Most important, PC boards aren't always the best technique for building RF circuits.

Layout

A circuit diagram is a poor guide toward a proper layout. Circuit diagrams are drawn to look attractive on paper. They follow drafting conventions that have very little to do with the way the circuit works. On a schematic, ground and supply voltage symbols are scattered all over the place. The first rule of RF layout is—*do not wire RF circuits as they are drawn!* How a circuit works in practice depends on the layout. Poor layout can ruin the performance of even a well-designed circuit.

How to Design a Good Circuit Layout

The easiest way to explain good layout practices is to take you through an example. **Fig 8.25** is the circuit diagram of a two-stage receiver IF amplifier using dual-gate MOSFETs. It is only a design example, so the values are only typical. To analyze which things are important to the layout of this circuit, ask these questions:

- Which are the RF components, and which are only involved with LF or dc?
- Which components are in the main RF signal path?
- Which components are in the ground return paths?

Use the answers to these questions to plan the layout. The RF components that are in the main RF signal path are usually the most critical. The AF or dc components can usually be placed anywhere. The components in the ground return path should be positioned so they are easily connected to the circuit ground. Answer the questions, apply the answers to the layout, and then follow these guidelines:

- Avoid laying out circuits so their inputs and outputs are close together. If a stage's output is too near a previous stage's input, the output signal can feedback into the input and cause problems.
- Keep component leads as short as prac-

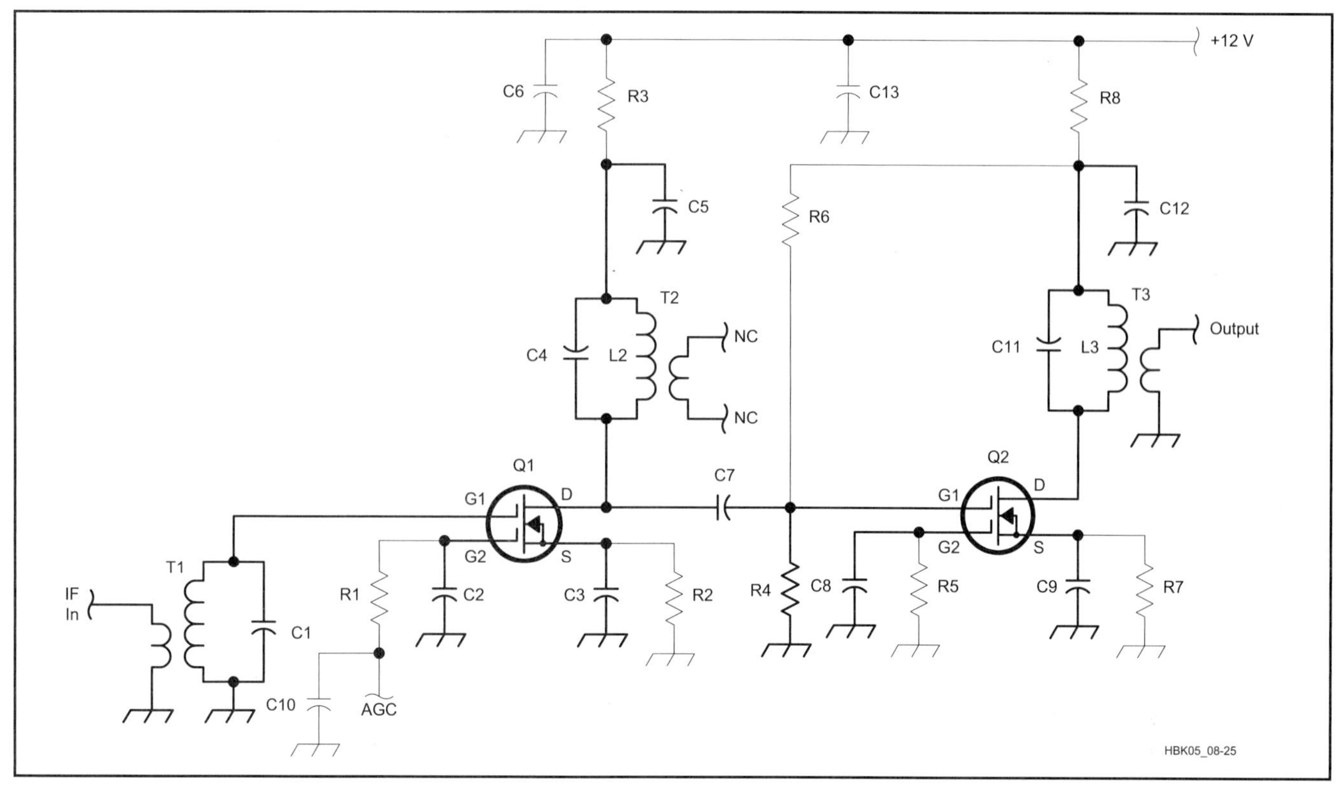

Fig 8.25—The IF amplifier used in the design example. C1, C4 and C11 are not specified because they are internal to the IF transformers.

tical. This doesn't necessarily mean as short as possible—just consider lead length as part of your design.

- Remember that metal transistor cases conduct, and that a transistor's metal case is usually connected to one of its leads. Prevent cases from touching ground or other components, unless called for in the design.

In our design example, the RF components are shown in heavy lines, though not all of these components are in the main RF signal path. The RF signal path consists of T1/C1, Q1, T2/C4, C7, Q2, T3/C11. These need to be positioned in almost a straight line, to avoid feedback from output to input. They form the backbone of the layout, as shown in **Fig 8.26A**.

The question about ground paths requires some further thought—what is really meant by "ground" and "ground-return paths"? Some points in the circuit need to be kept at RF ground potential. The best RF ground potential on a PC board is a copper ground plane covering one entire side. Points in the circuit that cannot be connected directly to ground for dc reasons must be bypassed ("decoupled") to ground by capacitors that provide ground-return paths for RF.

In Fig 8.26, the components in the ground-return paths are the RF bypass capacitors C2, C3, C5, C8, C9 and C12.

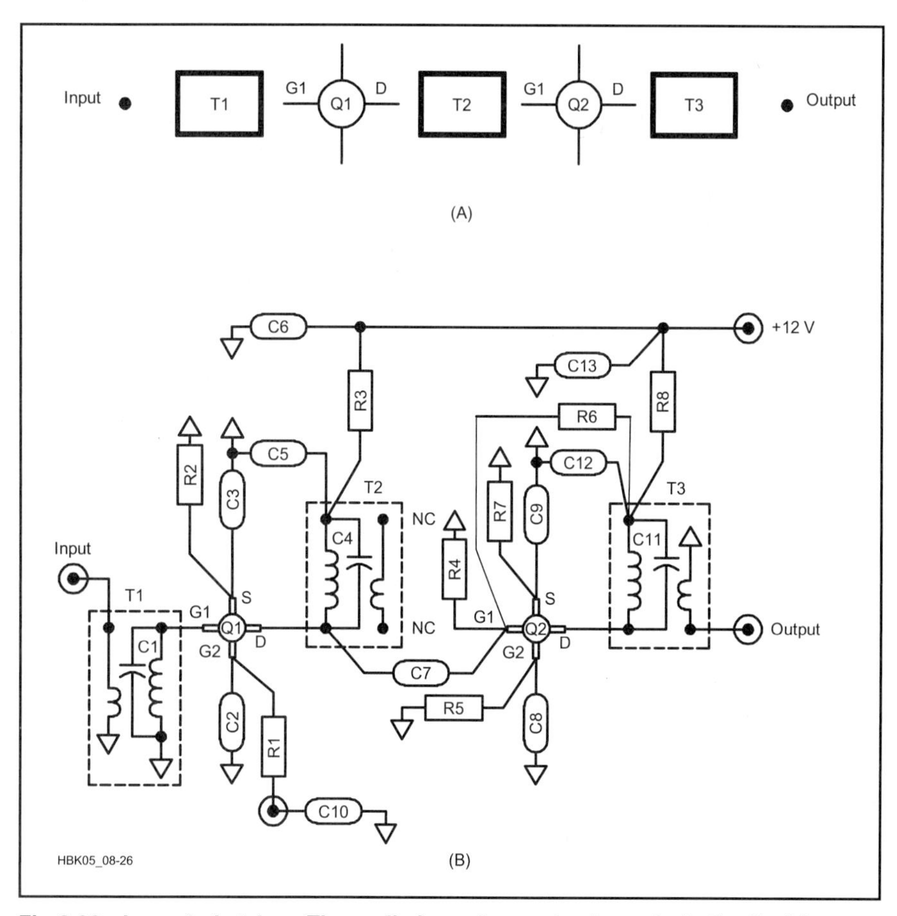

Fig 8.26—Layout sketches. The preliminary line-up is shown in A; the final layout in B.

R4 is primarily a dc biasing component, but it is also a ground return for RF so its location is important. The values of RF bypass capacitors are chosen to have a low reactance at the frequency in use; typical values would be 0.1 µF at LF, 0.01 µF at HF, and 0.001 µF or less at VHF. Not all capacitors are suitable for RF decoupling; the most common are disc ceramic capacitors. RF decoupling capacitors should always have short leads.

Almost every RF circuit has an input, an output and a common ground connection. Many circuits also have additional ground connections, both at the input side and at the output side. Maintain a low-impedance path between input and output ground connections. The input ground connections for Q1 are the grounded ends of C1 and the two windings of T1. (The two ends of an IF transformer winding are generally not interchangeable; one is designated as the "hot" end, and the other must be connected or bypassed to RF ground.) The capacitor that resonates with the adjustable coil is often mounted inside the can of the IF transformer, leaving only two component leads to be grounded as shown in Fig 8.26B.

The RF ground for Q1 is its source connection via C3. Since Q1 is in a plastic package that can be mounted in any orientation, you can make the common ground either above or below the signal path in Fig 8.26B. Although the circuit diagram shows the source at the bottom. The practical circuit works much better with the source at the top, because of the connections to T2.

It's a good idea to locate the hot end of the main winding close to the drain lead of the transistor package, so the other end is toward the top of Fig 8.26B. If the source of Q1 is also toward the top of the layout, there is a common ground point for C3 (the source bypass capacitor) and the output bypass capacitor C5. Gate 2 of Q1 can safely be bypassed toward the bottom of the layout.

C7 couples the signal from the output of Q1 to the input of Q2. The source of Q2 should be bypassed toward the top of the layout, in exactly the same way as the source of Q1. R4 is not critical, but it should be connected on the same side as the other components. Note how the pinout of T3 has placed the output connection as far as possible from the input. With this layout for the signal path and the critical RF components, the circuit has an excellent chance of working properly.

DC Components

The rest of the components carry dc, so their layout is much less critical. Even so, try to keep everything well separated from the main RF signal path. One good choice is to put the 12-V connections along the top of the layout, and the AGC connection at the bottom. The source bias resistors R2 and R7 can be placed alongside C3 and C9. The gate-2 bias resistors for Q2, R5 and R6 are not RF components so their locations aren't too critical. R7 has to cross the signal path in order to reach C12, however, and the best way to avoid signal pickup would be to mount R7 on the opposite side of the copper ground plane from the signal wiring. Generally speaking, 1/8-W or 1/4-W metal-film or carbon-film resistors are best for low-level RF circuits.

Actually, it is not quite accurate to say that resistors such as R3 and R8 are not "RF" components. They provide a high impedance to RF in the positive supply lead. Because of R8, for example, the RF signal in T2 is conducted to ground through C5 rather than ending up on the 12-V line, possibly causing unwanted RF feedback. Just to be sure, C6 bypasses R3 and C13 serves the same function for R8. Note that the gate-1 bias resistor R6 is connected to C12 rather than directly to the 12-V supply, to take advantage of the extra decoupling provided by R8 and C13.

If you build something, you want it to work the first time, so don't cut corners! Some commercial PC boards take liberties with layout, bypassing and decoupling. Don't assume that you can do the same. Don't try to eliminate "extra" decoupling components such as R3, C6, R8 and C13, even though they might not all be absolutely necessary. If other people's designs have left them out, put them in again. In the long run it's far easier to take a little more time and use a few extra components, to build in some insurance that your circuit will work. For a one-time project, the few extra parts won't hurt your pocket too badly; they may save untold hours in debugging time.

A real capacitor does not work well over a large frequency range. A 10-µF electrolytic capacitor cannot be used to bypass or decouple RF signals. A 0.1-µF capacitor will not bypass UHF or microwave signals. Choose component values to fit the range. The upper frequency limit is limited by the series inductance, L_S. In fact, at frequencies higher than the frequency at which the capacitor and its series inductance form a resonant circuit, the capacitor actually functions as an inductor. This is why it is a common practice to use two capacitors in parallel for bypassing, as shown in **Fig 8.27**. At first glance, this might appear to be unnecessary. However, the self-resonant frequency of C1 is usually 1 MHz or less; it cannot supply any bypassing above that frequency. However, C2 is able to bypass signals up into the lower VHF range.

Let's summarize how we got from Fig 8.24 to Fig 8.26B:

- Lay out the signal path in a straight line.
- By experimenting with the placement and orientation of the components in the RF signal path, group the RF ground connections for each stage close together, without mixing up the input and output grounds.
- Place the non-RF components well clear of the signal path, freely using decoupling components for extra measure.

Practical Construction Hints

Now it's time to actually construct a project. The layout concepts discussed earlier can be applied to nearly any construction technique. Although you'll eventually learn from your own experience, the following guidelines give a good start:

- Divide the unit into modules built into separate shielded enclosures—RF, IF, VFO, for example. Modular construction improves RF stability, and makes the individual modules easier to build and test. It also means that you can make major changes without rebuilding the whole unit. RF signals between the modules can usually be connected using small coaxial cable.
- Use a full copper ground plane. This is your largest single assurance of RF stability and good performance.
- Keep inputs and outputs well separated for each stage, and for the whole unit. If possible, lay out all stages in a straight line. If an RF signal path doubles back or recrosses itself it usually results in instability.
- Keep the stages at different frequencies well-separated to minimize interstage coupling and spurious signals.
- Use interstage shields where necessary, but don't rely on them to cure a bad layout.
- Make all connections to the ground plane short and direct. Locate the common ground for each stage between the input and the output ground. Single-point grounding may work for a single stage, but it is rarely effective in a complex RF system.
- Locate frequency-determining components away from heat sources and mount them so as to maximize mechanical strength.
- Avoid unwanted coupling between tuned circuits. Use shielded inductors or toroids rather than open coils. Keep the RF high-voltage points close to the ground plane. Orient air-wound coils at right angles to minimize mutual coupling.

- Use lots of extra RF bypassing, especially on dc supply lines.
- Try to keep RF and dc wiring on opposite sides of the board, so the dc wiring is well away from RF fields.
- Compact designs are convenient, but don't overdo it! If the guidelines cited above mean that a unit needs to be bigger, make it bigger.

Combination Techniques

You can use a mixture of construction techniques on the same board and in most cases you probably should. Even though you choose one style for most of the wiring, there will probably be places where other techniques would be better. If so, do whatever is best for that part of the circuit. The resulting hybrid may not be pretty (these techniques aren't called "ugly construction" for nothing), but it will work!

Mount dual-in-line package (DIP) ICs in an array of drilled holes, then connect them using wired traces as described earlier. It is okay to mount some of the components using a ground-plane method, push pins or even wire wrap. On any one board, you may use a combination of these techniques, drilling holes for some ICs, or gluing others upside down, then surface mounting some of the pins, and other techniques to connect the rest. These combination techniques are often found in a project that combines audio, RF and digital circuitry.

A Final Check

No matter what construction technique is chosen, do a final check before applying power to the circuit! Things do go wrong, and a careful inspection minimizes the risk of a project beginning and ending its life as a puff of smoke! Check wiring carefully. Make a photocopy of the schematic and mark each lead on the schematic with a red X when you've verified that it's connected to the right spot in the circuit.

Inspect solder connections. A bad solder joint is much easier to find before the PC board is mounted to a chassis. Look for any damage caused to the PC board by soldering. Look for solder "bridges" between adjacent circuit-board traces. Solder bridges (**Fig 8.28**) occur when solder accidentally connects two or more conductors that are supposed to be isolated. It is often difficult to distinguish a solder bridge from a conductive trace on a tin-plated board. If you find a bridge, remelt it and the adjacent trace or traces to allow the solder's surface tension to absorb it. Double check that each component is installed in the proper holes on the board and that the orientation is correct. Make sure that no component leads or transistor tabs are touching other components or PC board connections. Check the circuit voltages before installing ICs in their sockets. Ensure that the ICs are oriented properly and installed in the correct sockets.

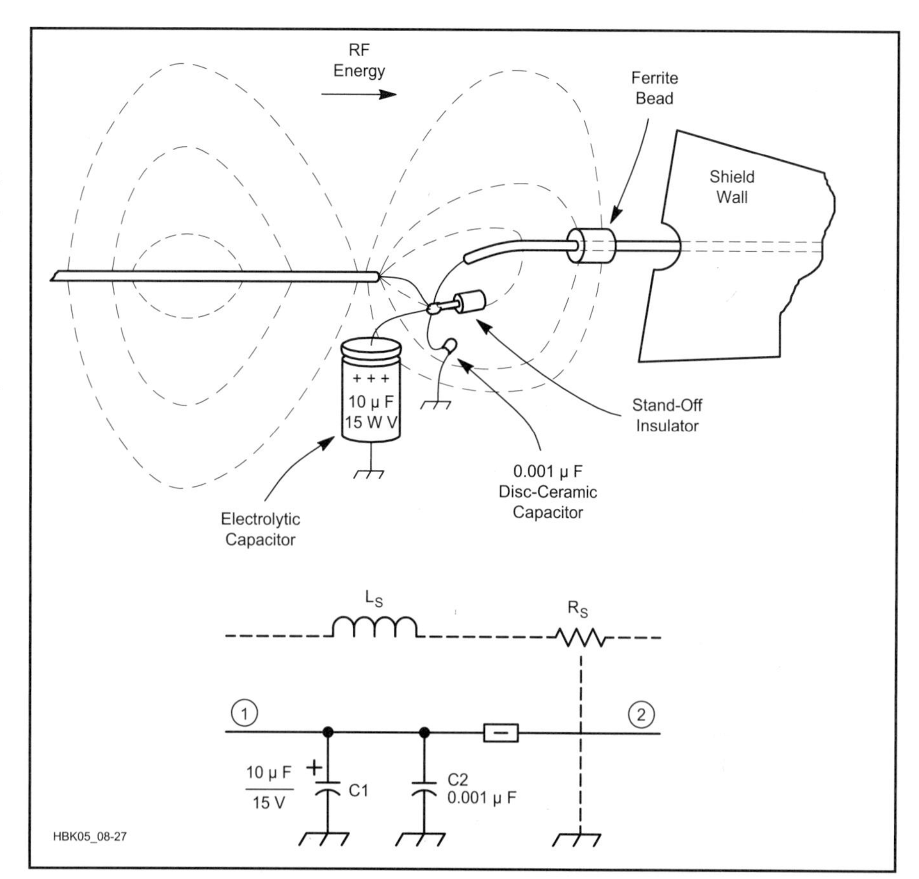

Fig 8.27—Two capacitors in parallel afford better bypassing across a wide frequency range.

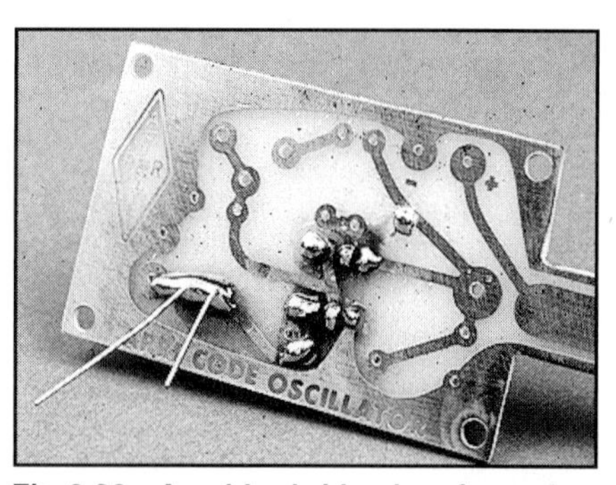

Fig 8.28—A solder bridge has formed a short circuit between PC board traces.

OTHER CONSTRUCTION TECHNIQUES

Wiring

Select the wire used in connecting amateur equipment by considering: the maximum current it must carry, the voltage its insulation must withstand and its use.

To minimize EMI, the power wiring of all transmitters should use shielded wire. Receiver and audio circuits may also require the use of shielded wire at some points for stability or the elimination of hum. Coaxial cable is recommended for all 50-Ω circuits. Use it for short runs of high-impedance audio wiring.

When choosing wire, consider how much current it will carry. Stranded wire is usually preferred over solid wire because stranded wire better withstands the inevitable bending that is part of building and troubleshooting a circuit. Solid wire is more rigid than stranded wire; use it where mechanical rigidity is needed or desired.

Wire with typical plastic insulation is good for voltages up to about 500 V. Use Teflon-insulated or other high-voltage wire for higher voltages. Teflon insulation does not melt when a soldering iron is applied. This makes it particularly helpful in tight places or large wiring harnesses. Although Teflon-insulated wire is more expensive, it is often available from industrial surplus houses. Inexpensive wire strippers make the removal of insulation from hookup wire an easy job. Solid wire is often used to wire HF circuits. Bare soft-drawn tinned wire, #22 to #12 (depending on mechanical requirements) is suitable. Avoid kinks by stretching a piece 10 or

15 ft long and then cutting it into short, convenient lengths. Run RF wiring directly from point to point with a minimum of sharp bends and keep the wire well-spaced from the chassis or other grounded metal surfaces. Where the wiring must pass through the chassis or a partition, cut a clearance hole and line it with a rubber grommet. If insulation is necessary, slip spaghetti insulation or heat-shrink tubing over the wire. For power-supply leads, bring the wire through the chassis via a feed-through capacitor.

In transmitters where the peak voltage does not exceed 500 V, shielded wire is satisfactory for power circuits. Shielded wire is not readily available for higher voltages—use point-to-point wiring instead. In the case of filament circuits carrying heavy current, it is necessary to use #10 or #12 bare or enameled wire. Slip the bare wire through spaghetti then cover it with copper braid pulled tightly over the spaghetti. Slide the shielding back over the insulation and flow solder into the end of the braid; the braid will stay in place, making it unnecessary to cut it back or

Microwave Construction Techniques

Microwave construction is becoming more popular, but at these frequencies the size of physical component leads and PC-board traces cannot be neglected. Microwave construction techniques either minimize these stray values or make them part of the circuit design.

Microwave construction does not always require tight tolerances and precision construction. A fair amount of error can often be tolerated if you are willing to tune your circuits, as you do at MF/HF. This usually requires the use of variable components that can be expensive and tricky to adjust.

Proper design and construction techniques, using high precision, can result in a "no-tune" microwave design. To build one of these no-tune projects, all you need do is buy the parts and install them on the board. The circuit tuning has been precisely controlled by the board and component dimensions so the project should work.

One tuning technique you can use with a microwave design, if you have the suitable test equipment, is to use bits of copper foil or EMI shielding tape as "stubs" to tune circuits. Solder these small bits of conductor into place at various points in the circuit to make reactances that can actually tune a circuit. After their position has been determined as part of the design, tuning is accomplished by removing or adding small amounts of conductor, or slightly changing the placement of the tuning stub. The size of the foil needed depends on your ability to determine changes in circuit performance, as well as the frequency of operation and the circuit board parameters. A precision setup that lets you see tiny changes allows you to use very small pieces of foil to get the best tuning possible.

From a mechanical accuracy point of view, the most tolerant type of construction is waveguide construction. Tuning is usually accomplished via one or more screws threaded into the waveguide. It becomes unwieldy to use waveguide on the amateur bands below 10 GHz because the dimensions get too large.

At 24 GHz and above, even waveguide becomes small and difficult to work with. At these frequencies, most readily available coax connectors work unreliably, so these higher bands are really a challenge. Special SMA connectors are available for use at 24 GHz.

Modular construction is a useful technique for microwave circuits. Often, circuits are tested by hooking their inputs and output to known 50-Ω sources and loads. Modules are typically kept small to prevent the chassis and PC board from acting as a waveguide, providing a feedback path between the input and output of a circuit, resulting in instability.

At microwave frequencies, the mechanical aspects and physical size of circuits become very much a part of the design. A few millimeters of conductor has significant reactance at these frequencies. This even affects VHF and HF designs! The traces and conductors used in an HF or VHF design resonate on microwave frequencies. If a high-performance FET has lots of gain in this region, a VHF preamplifier might also function as a 10-GHz oscillator if the circuit stray reactances were just right (or wrong!). You can prevent this by using shields between the input and output or by adding microwave absorptive material to the lid of the shielded module. (SHF Microwave sells absorptive materials.)

It is important to copy microwave circuits exactly, unless you really know what you are doing. "Improvements," such as better shielding or grounding can sometimes cause poor performance. It isn't usually attractive to substitute components, particularly with the active devices. It may look possible to substitute different grades of the same wafer, such as the ATF13135 and the ATF13335, but these are really the same transistor with different performance measurements. While two transistors may have exactly the same gain and noise figure at the desired operating frequency, often the impedances needed to maintain stability at other frequencies are be different. Thus, the "substitute" may oscillate, while the proper transistor would work just fine.

You can often substitute MMICs (monolithic microwave integrated circuits) for one another because they are designed to be stable and operate with the same input and output impedances (50 Ω).

The size of components used at microwaves can be critical—in some cases, a chip resistor 80 mils across is not a good substitute for one 60 mils across. Hopefully, the author of a construction project tells you which dimensions are critical, but you can't always count on this; the author may not know. It's not unusual for a person to spend years building just one prototype, so it's not surprising that the author might not have built a dozen different samples to try possible substitutions.

When using glass-epoxy PC board at microwave frequencies, the crucial board parameter is the thickness of the dielectric. It can vary quite a bit, in excess of 10%. This is not surprising; digital and lower-frequency analog circuits work just fine if the board is a little thinner or thicker than usual. Some of the board types used in microwave-circuit construction are a generic Teflon PC board, Duroid 5870 and 5880. These boards are available from Microwave Components of Michigan.

Proper connectors are a necessary expense at microwaves. At 10 GHz, the use of the proper connectors is essential for repeatable performance. Do not hook up microwave circuits with coax and pigtails. It might work but it probably can't be duplicated. SMA connectors are common because they are small and work well. SMA jacks are sometimes soldered in place, although 2-56 hardware is more common.—*Zack Lau, W1VT, ARRL Laboratory Engineer*

secure it in place. Clean the braid first so solder will take with a minimum of heat.

For receivers, RF wiring follows the methods described above. At RF, most of the current flows on the surface of the wire (a phenomenon called "skin effect"). Hollow tubing is just as good a conductor at RF as solid wire.

High-Voltage Techniques

High-voltage wiring requires special care. You need to use wire rated for the voltage it is carrying. Most standard hookup wire is inadequate. High-voltage wire is usually insulated with Teflon or special multilayer plastic. Some coaxial cable is rated at up to 3700 V.

Air is a great insulator, but high voltage can break down its resistance and form an arc. You need to leave ample room between any circuit carrying voltage and any nearby conductors. At dc, leave a gap of at least 0.1 inch per kilovolt. The actual breakdown voltage of air varies with the frequency of the signal, humidity and the shape of the conductors.

High voltage is also prone to corona discharge, a bleeding off of charge, primarily from sharp edges. For this reason, all connections need to be soldered, leaving only rounded surfaces on the soldered connection. It takes a little practice to get a "ball" of solder on each joint, but for voltages above 5 kV it is important.

Be careful working near high-voltage circuits! Most high-voltage power supplies can deliver a lethal shock.

Cable Lacing and Routing

Where power or control leads run together for more than a few inches, they present a better appearance when bound together in a single cable. Both plastic and waxed-linen lacing cords are available. You can also use a variety of plastic devices to bundle wires into cables and to clamp or secure them in place. Check with your local electronic parts supplier for items that are in stock.

To give a commercial look to the wiring of any unit, route any dc leads and shielded signal leads along the edge of the chassis. If this isn't possible, the cabled leads should then run parallel to an edge of the chassis. Further, the generous use of the tie points mounted parallel to an edge of the chassis, for the support of one or both ends of a resistor or fixed capacitor, adds to the appearance of the finished unit. In a similar manner, arrange the small components so that they are parallel to the panel or sides of the chassis.

Tie Points

When power leads have several branches in the chassis, it is convenient to use fiber-insulated multiple tie points as anchors for junction points. Strips of this kind are also useful as insulated supports for resistors, RF chokes and capacitors. Hold exposed points of high-voltage wiring to a minimum; otherwise, make them inaccessible to accidental contact.

Winding Coils

Winding coils seems so simple, yet many new constructors run into difficulty. Understanding the techniques prevents some of the frustration or construction errors associated with coil winding.

Close-wound coils are readily wound on the specified form by anchoring one end of the length of wire (in a vise or to a doorknob) and the other end to the coil form. Straighten any kinks in the wire and then pull to keep the wire under slight tension. Wind the coil to the required number of turns while walking toward the anchor, always maintaining a slight tension on the wire.

To space-wind the coil, wind the coil simultaneously with a suitable spacing medium (heavy thread, string or wire) in the manner described above. When the winding is complete, secure the end of the coil to the coil-form terminal and then carefully unwind the spacing material. If the coil is wound under suitable tension, the spacing material can be easily removed without disturbing the winding. Finish space-wound coils by judicious applications of Duco cement to hold the turns in place.

The "cold" end of a coil is the end at (or close to) chassis or ground potential. Wind coupling links on the cold end of a coil to minimize capacitive coupling.

Winding Toroidal Inductors

Toroidal inductors and transformers are specified for many projects in this *Handbook*. The advantages of these cores include compactness and a self-shielding property. **Figs 8.29** and **8.30** illustrate the proper way to wind and count turns on a toroidal core.

The task of winding a toroidal core, when more than just a few turns are required, can be greatly simplified by the use of a homemade bobbin upon which the wire is first wound. A simple yet effective bobbin can be fashioned from a wooden popsicle stick. Cut a "V" notch at each end and first wind the wire coil on the popsicle stick lengthwise through the notches. Once this is done, the wound bobbin can be easily passed through the toroid's inside diameter. While firmly grasping one of the wire ends against the toroidal core, the bobbin can be moved up, around, and through the toroidal core repeatedly until the wire has been completely transferred from the bobbin. The choice of bobbin used is somewhat dependent on the inside diameter of the toroid, the wire size, and the number of turns required.

When you wind a toroid inductor, count each pass of the wire through the toroid center as a turn. You can count the number of turns by counting the number of times the wire passes through the center of the core. See Fig 8.30A.

Multiwire Windings

A bifilar winding is one that has two identical lengths of wire, which when

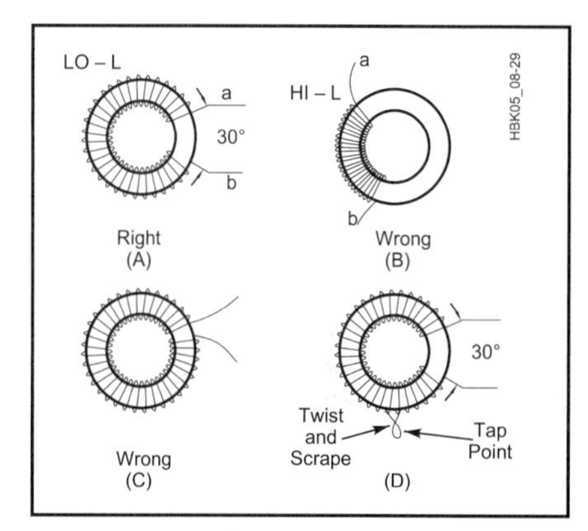

Fig 8.29—The maximum-Q method for winding a single-layer toroid is shown at A. A 30° gap is best. Methods at B and C have greater distributed capacitance. D shows how to place a tap on a toroidal coil winding.

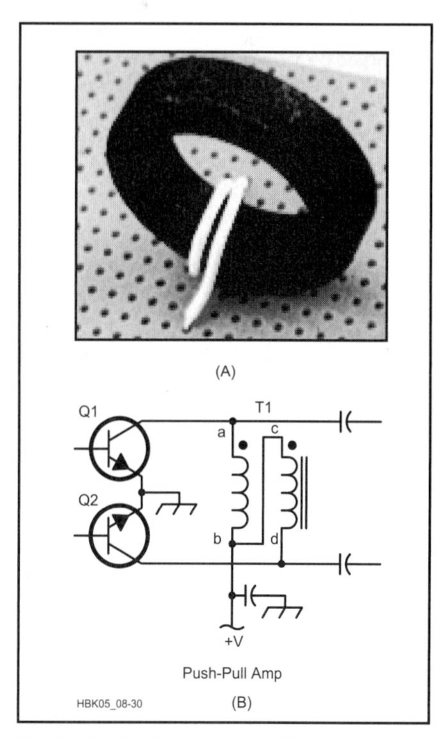

Fig 8.30—A shows a toroidal core with two turns of wire (see text). Large black dots, like those at T1 in B, indicate winding polarity (see text).

placed on the core result in the same number of turns for each wire. The two wires are wound on the core side by side at the same time, just as if a single winding were being applied. An easier and more popular method is to twist the two wires (8 to 15 turns per inch is adequate), then wind the twisted pair on the core. The wires can be twisted handily by placing one end of each in a bench vise. Tighten the remaining ends in the chuck of a small hand drill and turn the drill to twist the pair.

A trifilar winding has three wires, and a quadrifilar winding has four. The procedure for preparation and winding is otherwise the same as for a bifilar winding. **Fig 8.31** shows a bifilar toroid in schematic and pictorial form. The wires have been twisted together prior to placing them on the core. It is helpful, though by no means essential, to use wires of different color when multifilar-winding a core. It is more difficult to identify multiple windings on a core after it has been wound. Various colors of enamel insulation are available, but it is not easy for amateurs to find this wire locally or in small-quantity lots. This problem can be solved by taking lengths of wire (enameled magnet wire), cleaning the ends to remove dirt and

How to Buy Parts for Electronics Projects

The number one question received by the ARRL Technical Information Service starts out "Where can I buy..." It seems that one of the most perplexing problems faced by the would-be constructor is where to get parts. Sometimes you are lucky—the circuit author has made a kit available. But not every project has a ready-made kit. If you would like to expand your construction horizons, you can learn to be your own "purchasing manager." That means searching out parts sources and dealing with them in person.

In reality, it is not all that difficult to find most parts. Unfortunately, though, the days of the local electronic parts supplier seem to be gone. This is not surprising. Years ago an electronics supplier had to stock a relatively small number of electronic components—resistors, capacitors, tube sockets, a few relays and variable resistors. Technology has increased the number of components by a few orders of magnitude. Nowadays, the number of integrated circuits alone is enough to fill a multivolume book. No single electronics supplier could possibly stock them all. It has become a mail-order world; the electronics world is no exception.

Although it is no longer always possible to purchase all of your electronic needs from a local electronics supplier, the good news is that you don't have to! For a few dollars, mail-order companies are willing to supply whatever you need. You only need do two things to obtain nearly any electronic component—make a phone call and write a check.

Become an electronic catalog collector. Electronic suppliers also advertise in magazines that cater to ham-radio and electronics enthusiasts. If you are lucky, you may have a local source of electronics parts. Look in the Yellow Pages under "Electronic Equipment Suppliers" to find the local outlets. RadioShack is one local source found nearly everywhere. They carry an assortment of the more common electronic parts. You'll probably need to order from more than one mail-order company. (It's almost a corollary to Murphy's Law: No matter how wide a selection you find in one mail-order catalog, there's always at least one part you must buy somewhere else!)

While you're waiting for your catalogs, look at the parts list for the project you want to build. Unfortunately, you can't just photocopy the list and send it off to a mail-order company with a note that says "please send me these parts." You need to convert the part list into a part-order list that shows the order number and quantity required of each component. This may require a similar list for each parts supplier where one supplier does not have all the parts.

Check the type, tolerance, power rating and other key characteristics of the parts. Group the parts by those parameters before grouping them by value. If all of the circuit's components are already grouped by value on the parts list, you can just count the number of each value. Each time you add parts to the order list, check them off the published parts list. Sometimes the parts list does not include common components like resistors and capacitors. If this is the case, make a copy of the schematic and check off the parts as you build your shopping list.

Although you'll probably be able to order exactly the right number of each part for a project, buy a few extras of some parts for your junk box. It's always good to have a few extra parts on hand; you may break a component lead during assembly, or damage a solid-state component with too much heat or by wiring it in backwards. If you don't have extras, you'll need to order another part. Even if you don't need the extras for this project, they may come in handy, and you'll be encouraged to build another project! Pick up an extra toroid or two as well.

Now's the time to decide whether you're going to build your project with ground-plane construction or PC board. If you need a PC-board, FAR Circuits and others have them for many ARRL book and *QST* projects. If you're going to use ground-plane construction, buy a good-sized piece of single-sided copper-clad board—glass-epoxy board if you can. Phenolic board is inferior because it is brittle and deteriorates rapidly with soldering heat.

Don't forget an enclosure for your project. This is often overlooked in parts lists for most projects, because different builders like different enclosures. Make sure there's room in the box for all of the components used in your project. Some people like to cram projects in the smallest possible box, but miniaturization can be extremely frustrating if you're not good at it.

There are almost always a few items you can't get from one company and most have minimum orders. You may need to distribute your order between two or more companies to meet minimum-order requirements. Some companies put out beautiful catalogs, but their minimum order is $25 or they charge $5 for shipping if you place a small order.

If you order enough parts, you'll soon find out which companies you like to deal with and which have slow service. It is frustrating to receive most of an order, then wait months for the parts that are on back-order. If you don't want the company to back-order your parts, write clearly on the order form, "Do not back-order parts." They will then ship the parts they have and leave you to order the rest from somewhere else.

If you are in a hurry, call the company to inquire about the availability of the parts in your order. Some companies take credit-card orders over the telephone. Some companies hold orders a few weeks to allow personal checks to clear.

If you are familiar with the catalogs and policies of electronic-component suppliers, you will find that getting parts is not difficult. Concentrate on the fun part—building the circuit and getting it working.—*Bruce Hale, KB1MW*

grease, then spray painting them. Ordinary aerosol-can spray enamel works fine. Spray lacquer is not as satisfactory because it is brittle when dry and tends to flake off the wire.

The winding sense of a multifilar toroidal transformer is important in most circuits. Fig 8.30B illustrates this principle. The black dots (called phasing dots) at the top of the T1 windings indicate polarity. That is, points a and c are both start or finish ends of their respective windings. In this example, points a and d are of opposite phase (180° phase difference) to provide push-pull voltage output from Q1 and Q2.

After you wind a coil, scrape the insulation off the wire before you solder it into the circuit.

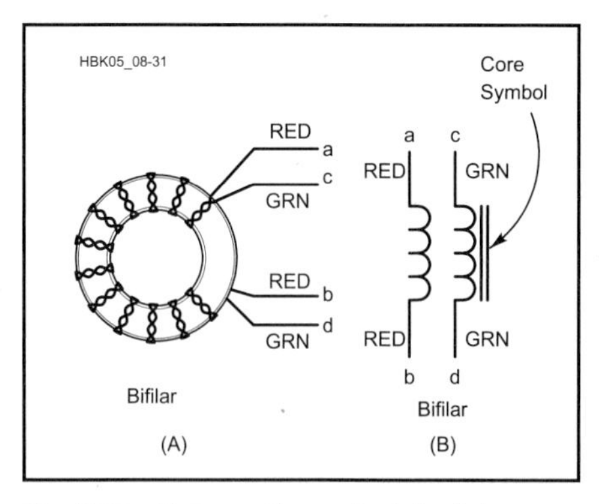

Fig 8.31—Schematic and pictorial presentation of a bifilar-wound toroidal transformer.

MECHANICAL FABRICATION

Buy or Build a Chassis?

Most projects end up in some sort of an enclosure, and most hams choose to purchase a ready-made chassis for small projects, but some projects require a custom enclosure. Even a ready-made chassis may require a fabricated sheet-metal shield or bracket, so it's good to learn something about sheet-metal and metal-fabrication techniques.

Most often, you can buy a suitable enclosure. These are sold by RadioShack and most electronics distributors.

Select an enclosure that has plenty of room. A removable cover or front panel can make any future troubleshooting or modifications easy. A project enclosure should be strong enough to hold all of the components without bending or sagging; it should also be strong enough to stand up to expected use and abuse.

Cutting and Bending Sheet Metal

Enclosures, mounting brackets and shields are usually made of sheet metal. Most sheet metal is sold in large sheets, 4 to 8 ft or larger. It must be cut to the size needed.

Most sheet metal is thin enough to cut with metal shears or a hacksaw. A jigsaw or bandsaw makes the task easier. If you use any kind of saw, select a blade that has teeth fine enough so that at least two teeth are in contact with the metal at all times.

If a metal sheet is too large to cut conveniently with a hacksaw, it can be scored and broken. Make scratches as deep as possible along the line of the cut on both sides of the sheet. Then, clamp it in a vise and work it back and forth until the sheet breaks at the line. Do not bend it too far before the break begins to weaken, or the edge of the sheet might bend. A pair of flat bars, slightly longer than the sheet being bent, make it easier to hold a sheet firmly in a vise. Use "C" clamps to keep the bars from spreading at the ends.

Smooth rough edges with a file or by sanding with a large piece of emery cloth or sandpaper wrapped around a flat block.

Finishing Aluminum

Give aluminum chassis, panels and parts a sheen finish by treating them in a caustic bath. Use a plastic container to hold the solution. Ordinary household lye can be dissolved in water to make a bath solution. Follow the directions on the container. A strong solution will do the job more rapidly.

Stir the solution with a stick of wood until the lye crystals are completely dissolved. If the lye solution gets on your skin, wash with plenty of water. If you get any in your eyes, immediately rinse with plenty of clean, room-temperature water and seek medical help. It can also damage your clothing, so wear something old. Prepare sufficient solution to cover the piece completely. When the aluminum is immersed, a very pronounced bubbling takes place. Provide ventilation to disperse the escaping gas. A half hour to two hours in the bath is sufficient, depending on the strength of the solution and the desired surface characteristics.

Chassis Working

With a few essential tools and proper procedure, building radio gear on a metal chassis is a relatively simple matter. Aluminum is better than steel, not only because it is a superior shielding material, but also because it is much easier to work and provides good chassis contact when used with secure fasteners.

Spend sufficient time planning a project to save trouble and energy later. The actual construction is much simpler when all details are worked out beforehand. Here we discuss a large chassis-and-cabinet project, such as a high-power amplifier. The techniques are applicable to small projects as well.

Cover the top of the chassis with a piece of wrapping paper or graph paper. Fold the edges down over the sides of the chassis and fasten them with adhesive tape. Place the front panel against the chassis front and draw a line there to indicate the chassis top edge.

Assemble the parts to be mounted on the chassis top and move them about to find a satisfactory arrangement. Consider that some will be mounted underneath the chassis and ensure that the two groups of components won't interfere with each other.

Place controls with shafts that extend through the cabinet first, and arrange them so that the knobs will form the desired pattern on the panel. Position the shafts perpendicular to the front chassis edge. Locate any partition shields and panel brackets next, then sockets and any other parts. Mark the mounting-hole centers of each part accurately on the paper. Watch out for capacitors with off-center shafts that do not line up with the mounting holes. Do not forget to mark the centers of socket holes and holes for wiring leads. Make the large center hole for a socket *before* the small mounting holes. Then use the socket itself as a template to mark the centers of the mounting holes. With all chassis holes marked, center-punch and drill each hole.

Next, mount on the chassis the capacitors and any other parts with shafts extending to the panel. Fasten the front panel to the chassis temporarily. Use a machinist's square to extend the line (vertical axis) of any control shaft to the chassis front and mark the location on the front panel at the chassis line. If the layout is complex, label each mark with an identifier. Also mark the back of the front panel with the locations of any holes in the chassis front that must go through the front panel. Remove the front panel.

PC-Board Materials

Much tedious sheet-metal work can be eliminated by fabricating chassis and enclosures from copper-clad printed-circuit board material. While it is manufactured in large sheets for industrial use, some hobby electronics stores and surplus outlets market usable scraps at reasonable prices. PC-board stock cuts easily with a small hacksaw. The nonmetallic base material isn't malleable, so it can't be bent. Corners are easily formed by holding two pieces at right angles and soldering the seam. This technique makes excellent RF-tight enclosures. If mechanical rigidity is required of a large copper-clad surface, solder stiffening ribs at right angles to the sheet.

Fig 8.32 shows the use of PC-board stock to make a project enclosure. This enclosure was made by cutting the pieces to size, then soldering them together. Start by laying the bottom piece on a workbench, then placing one of the sides in place at right angles. Tack-solder the second piece in two or three places, then start at one end and run a bead of solder down the entire seam. Use plenty of solder and plenty of heat. Continue with the rest of the pieces until all but the top cover is in place.

In most cases, it is better to drill all needed holes in advance. It can sometimes be difficult to drill holes after the enclosure is soldered together.

You can use this technique to build enclosures, subassemblies or shields. This technique is easy with practice; hone your skills on a few scrap pieces of PC-board stock.

Drilling Techniques

Before drilling holes in metal with a hand drill, indent the hole centers with a center punch. This prevents the drill bit from "walking" away from the center when starting the hole. Predrill holes greater than 1/2-inch in diameter with a smaller bit that is large enough to contain the flat spot at the large bit's tip. When the metal being drilled is thinner than the depth of the drill-bit tip, back up the metal with a wood block to smooth the drilling process.

The chuck on the common hand drill is limited to 3/8-inch bits. Some bits are much larger, with a 3/8-inch shank. If necessary, enlarge holes with a reamer or round file. For very large or odd-shaped holes, drill a series of closely spaced small holes just inside of the desired opening. Cut the metal remaining between the holes with a cold chisel and file or grind the hole to its finished shape. A nibbling tool also works well for such holes.

Use socket-hole punches to make socket holes and other large holes in an aluminum chassis. Drill a guide hole for the punch center bolt, assemble the punch with the bolt through the guide hole and tighten the bolt to cut the desired hole. Oil the threads of the bolt occasionally.

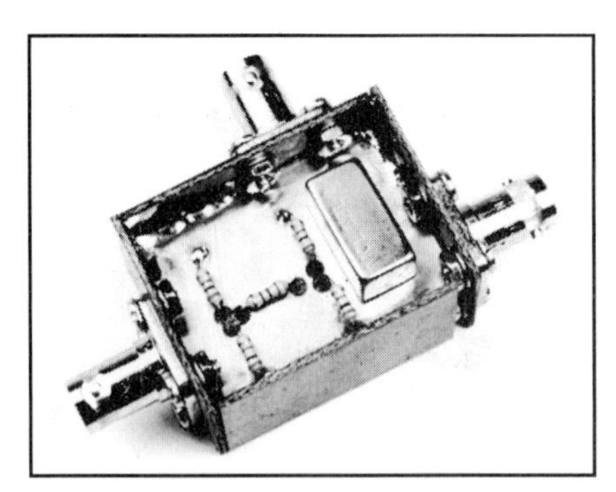

Fig 8.32—A box made entirely from PC-board stock.

Cut large circular holes in steel panels or chassis with an adjustable circle cutter ("flycutter"). Occasionally apply machine oil to the cutting groove to speed the job. Test the cutter's diameter setting by cutting a block of wood or scrap material first.

Remove burrs or rough edges that result from drilling or cutting with a burr-remover, round or half-round file, a sharp knife or chisel. Keep an old chisel sharpened and available for this purpose.

Rectangular Holes

Square or rectangular holes can be cut with a nibbling tool or a row of small holes as previously described. Large openings can be cut easily using socket-hole punches.

Construction Notes

If a control shaft must be extended or insulated, a flexible shaft coupling with adequate insulation should be used. Satisfactory support for the shaft extension, as well as electrical contact for safety, can be provided by means of a metal panel bushing made for the purpose. These can be obtained singly for use with existing shafts, or they can be bought with a captive extension shaft included. In either case the panel bushing gives a solid feel to the control. The use of fiber washers between ceramic insulation and metal brackets, screws or nuts will prevent the ceramic parts from breaking.

Painting

Painting is an art, but, like most arts, successful techniques are based on skills that can be learned. The surfaces to be painted must be clean to ensure that the paint will adhere properly. In most cases, you can wash the item to be painted with soap, water and a mild scrub brush, then rinse thoroughly. When it is dry, it is ready for painting. Avoid touching it with your bare hands after it has been cleaned. Your skin oils will interfere with paint adhesion. Wear rubber or clean cotton gloves.

Sheet metal can be prepared for painting by abrading the surface with medium-grade sandpaper, making certain the strokes are applied in the same direction (not circular or random). This process will create tiny grooves on the otherwise smooth surface. As a result, paint or lacquer will adhere well. On aluminum, one or two coats of zinc chromate primer applied before the finish paint will ensure good adhesion.

Keep work areas clean and the air free of dust. Any loose dirt or dust particles will probably find their way onto a freshly painted project. Even water-based paints produce some fumes, so properly ventilate work areas.

Select a paint suitable to the task. Some paints are best for metal, others for wood and so on. Some dry quickly, with no fumes; others dry slowly and need to be thoroughly ventilated. You may want to select a rust-preventative paint for metal surfaces that might be subjected to high moisture or salts.

Most metal surfaces are painted with some sort of spray, either from a spray gun or from spray cans of paint. Either way, follow the manufacturer's instructions for a high-quality job.

SUMMARY

If you're like most amateurs, once you've got the building bug, you won't let your soldering iron stay cold for long. Starting is the hardest part. Now, the next time you think about adding another project to your station, you'll know where to start.

Chapter 9

Modes and Systems

The various modes, modulation types and protocols we use partly reflect the different types of information we might wish to transmit, such as data, voice, image or even multimedia communications. Other considerations are the behavior of the radio link including fading, delay, Doppler frequency shift and distortion. We are also limited by regulatory restrictions such as bandwidth and following certain conventions or protocols. Some of the bands used by amateurs are wide and well behaved, such as VHF links over short paths. Others may be narrow, unstable and hostile to our signals, such as a long HF path through the auroral zones. Such conditions dictate which mode will be most successful.

ISSUES COMMON TO ALL TRANSMISSION MODES

Bandwidth is the amount of frequency spectrum that a signal occupies. There are narrow-band modes, such as CW and PSK31, and wideband modes, such as TV and spread spectrum. Not all modes are permitted on all amateur bands. Wideband modes can be used only where the total width of the amateur allocation is sufficient to contain the wide signal. In addition, voluntary agreements and regulatory restrictions keep some wideband modes out of certain bands or subbands so that one station's signal does not preclude operation by a large number of others using the narrower modes.

All users of the radio spectrum must comply with FCC bandwidth rules. The occupied bandwidth is determined not only by the mode being used, but by proper operation of that mode. Many of the permitted modes can become too wide when improperly adjusted. Perhaps the greatest source of conflict between ham operators is "splatter" or "key clicks" caused by overmodulated or otherwise improperly operated equipment, regardless of the mode being used. An amateur signal must be no wider than is necessary for good communication, and as clean as the "state of the art" will allow. Section 97.3 (a)(8) of the FCC rules defines occupied bandwidth as the point where spurious energy drops to 26 dB below the mean power of the transmitted signal.

Sensitivity refers to the relative ability of a mode to decode weak signals. Some modes are favored by DXers in that they have a greater ability to "get through" when the signals are very weak. For local communications, sensitivity may not be the major concern. *Fidelity* is not a major issue for most amateurs, although they rightly take pride in the clarity of their transmissions and some amateurs take audio quality quite seriously. *Intelligibility* is related to fidelity in a complex way and, sometimes, voice signals are modified in such a way as to make them more understandable, perhaps under difficult conditions, even though not as natural as they might otherwise be.

Quality is the corresponding term for images, and *accuracy* describes the degree to which a text mode reproduces the original message. *Robustness* or *reliability* refers to the ability of a mode to maintain continuous communication under difficult conditions. For example, a very robust signal is desired when controlling a model airplane. DXers are not overly concerned with reliability in that continuous contact is not needed. However, they do want a signal that gets through when needed to work a rare station.

Efficiency is the ability of a mode to get the signal through with minimum energy expended. Within the regulatory power limit, energy cost is not a major concern for most home stations. Thus, efficiency is a concern mainly to those on battery power—using handheld or portable stations. Emergency operators also need to consider using efficient modes. For radio services that use high power, such as shortwave broadcasters, efficiency is very important. QRP is a popular activity, where operators take pride in making contact with a very small amount of transmitter power (maximum miles per milliwatt!).

Stability is the ability to maintain the frequency of the transmission very precisely. Some modes require precise frequency control. Most modern equipment is very stable, but some vintage or homemade gear may be limited in frequency stability. Higher frequency work can put tight limits on frequency stability. Channel stability refers to both frequency, amplitude and phase variations of the transmission medium itself. The inherent instability of a radio channel may permit some modes but preclude others.

Noise immunity is the ability of a radio system to reject noise of various types that could otherwise destroy the meaning or impair the quality of the message. This is all-important in HF mobile operations and for those living in densely populated areas. Man-made electrical noise is an increasingly serious threat to ham operations and requires both regulatory and technical solutions.

Emission Classifications

Emissions are designated according to their classification and their necessary bandwidth. A minimum of three symbols is used to describe the basic characteristics of radio waves. Emissions are classified and symbolized according to the following characteristics:

I. First symbol—Type of modulation of the main carrier
II. Second symbol—Nature of signal(s) modulating the main carrier
III. Third symbol—Type of information to be transmitted
Note: A fourth and fifth symbol are provided for in the ITU Radio Regulations. Use of the fourth and fifth symbol is optional.
IV. Details of signal(s)
V. Nature of multiplexing

First symbol—type of modulation of the main carrier

(1) Emission of an unmodulated carrier .... N
(2) Emission in which the main carrier is amplitude-modulated (including cases where subcarriers are angle-modulated):
- Double sideband .... A
- Single sideband, full carrier .... H
- Single sideband, reduced or variable level carrier .... R
- Single sideband, suppressed carrier .... J
- Independent sidebands .... B
- Vestigial sideband C

(3) Emission in which the main carrier is angle-modulated:
- Frequency modulation .... F
- Phase modulation .... G

Note: Whenever frequency modulation (F) is indicated, phase modulation (G) is also acceptable.

(4) Emission in which the main carrier is amplitude and angle-modulated either simultaneously or in a pre-established sequence .... D
(5) Emission of pulses[1]
- Sequence of unmodulated pulses .... P
- A sequence of pulses:
- Modulated in amplitude .... K
- Modulated in width/duration L
- Modulated in position/phase .... M
- In which the carrier is angle-modulated during the period of the pulse .... Q
- Which is a combination of the foregoing or in produced by other means V

(6) Cases not covered above, in which an emission consists of the main carrier modulated, either simultaneously or in a pre-established sequence in a combination of two or more of the following modes: amplitude, angle, pulse .... W
(7) Cases not otherwise covered X

Second symbol—nature of signal(s) modulating the main carrier

(1) No modulating signal 0
(2) A single channel containing quantized or digital information without the use of a modulating subcarrier, excluding time-division multiplex 1
(3) A single channel containing quantized or digital information with the use of a modulating subcarrier, excluding time-division multiplex 2
(4) A single channel containing analog information 3
(5) Two or more channels containing quantized or digital information 7
(6) Two or more channels containing analog information 8
(7) Composite system with one or more channel containing quantized or digital information, together with one or more channels containing analog information 9
(8) Cases not otherwise covered X

Third symbol—type of information to be transmitted[2]

(1) No information transmitted .... N
(2) Telegraphy, for aural reception .... A
(3) Telegraphy, for automatic reception .... B
(4) Facsimile C
(5) Data transmission, telemetry, telecommand .... D
(6) Telephony (including sound broadcasting) .... E
(7) Television (video) .... F
(8) Combination of the above .... W
(9) Cases not otherwise covered X

Where the fourth or fifth symbol is used it shall be used as indicated below. Where the fourth or the fifth symbol is not used this should be indicated by a dash where each symbol would otherwise appear.

Fourth symbol—Details of signal(s)

(1) Two-condition code with elements of differing numbers and/or durations .... A
(2) Two-condition code with elements of the same number and duration without error-correction ...... B
(3) Two-condition code with elements of the same number and duration with error-correction C
(4) Four-condition code in which each condition represents a signal element (of one or more bits) .... D
(5) Multi-condition code in which each condition represents a signal element (of one or more bits) .... E
(6) Multi-condition code in which each condition or combination of conditions represents a character .... F
(7) Sound of broadcasting quality (monophonic) .... G
(8) Sound of broadcasting quality (stereophonic or quadraphonic) .... H
(9) Sound of commercial quality (excluding categories given in (10) and (11) below .... J
(10) Sound of commercial quality with frequency inversion or band-splitting .... K
(11) Sound of commercial quality with separate frequency-modulated signals to control the level of demodulated signal L
(12) Monochrome .... M
(13) Color .... N
(14) Combination of the above .... W
(15) Cases not otherwise covered X

Fifth symbol—Nature of multiplexing

(1) None .... N
(2) Code-division multiplex[3] C
(3) Frequency-division multiplex .... F
(4) Time-division multiplex .... T
(5) Combination of frequency-division and time-division multiplex .... W
(6) Other types of multiplexing X

[1] Emissions where the main carrier is directly modulated by a signal which has been coded into quantized form (eg, pulse code modulation) should be designated under (2) or (3).
[2] In this context the word "information" does not include information of a constant unvarying nature such as is provided by standard frequency emissions, continuous wave and pulse radars, etc.
[3] This includes bandwidth expansion techniques.

Emission, Modulation and Transmission Characteristics

Emission designators are generally expressed as characters representing the necessary bandwidth and emission classification symbols. *Necessary bandwidth* is expressed as a maximum of five numerals and one letter. The letter occupies the position of the decimal point and represents the unit of bandwidth, as follows: H = hertz, K = kilohertz, M = megahertz and G = gigahertz. For example, a bandwidth of 2.8 kHz is expressed as 2K8 or 2K80 and a bandwidth of 150 Hz is noted as 150H.

Emission classification symbols are (1) type of modulation of the main carrier, (2) nature of the signal(s) modulating the main carrier and (3) the information to be transmitted. They may be supplemented by (4) details of signal(s) and (5) nature of multiplexing, but the FCC does not require these. These designators are found in Appendix 1 of the ITU Radio Regulations, ITU-R Recommendation SM.1138 and in the FCC rules §2.201.

TRANSMISSION IMPAIRMENTS

In addition to attenuation of a signal from a transmitting station to a receiving station, the signal is subject to a variety of impairments. These include flat fading, frequency-selective fading, wave polarization rotation fading, Doppler shift, interference from other signals, atmospheric noise, galactic noise and manmade noise. Receiver thermal noise is not usually an issue at HF because external noise often dominates but can be a limiting factor at VHF and above.

Effect of the HF Path on Pulses

Digital signals, including Morse signals copied by ear, are transmissions in which the wave abruptly changes state. That means that something about them is varied in order to carry the digital information. Understanding what happens to pulses of all types when they travel via the ionosphere is important for knowing how to design a workable HF digital system. VHF and UHF signals generally use more benign paths but some of the same principles important in HF will also apply to them.

Since Morse CW using OOK (on-off keying) is the oldest as well as the simplest digital mode, it is useful to analyze the propagation effects on a simple Morse character, namely the letter "E." This consists of a single pulse, which can be distinguished from the letter "T" by its length, provided the keying speed is known beforehand. Using the ARRL-recommended values for shaping to prevent excessive key clicks, a single Morse "E" will appear as shown on the left in **Fig 9.1** as it leaves the transmitter. A ham living nearby would observe the pulse essentially unchanged and could comment usefully on the shape of the keying.

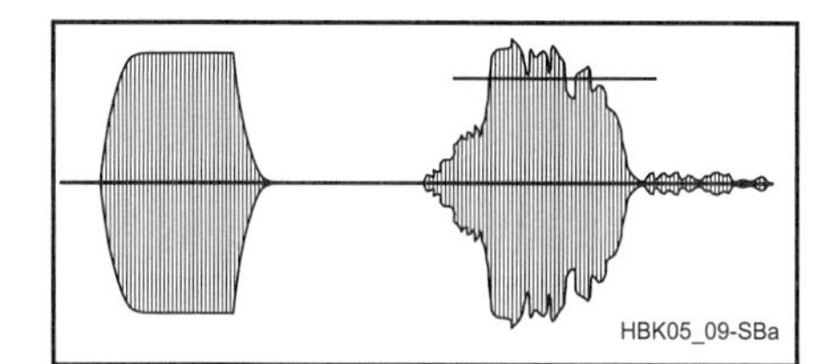

Fig 9.1—Morse "E" as transmitted (left) and received (right).

However, a receiver far enough removed so that sky wave was the dominant mode would see a much different picture. If both ground wave and sky wave were present, as could occur on 80 meters at a distance of 20 miles, the received dit might appear as shown on the right in **Fig 9.1**. Several things, however, distort the pulse. One is multipath, meaning that the signal arrives by more than one route. One route might be the ground wave, another the single-hop sky wave, and others, very likely considerably weaker, multiple-hop sky waves.

In addition, if the ionosphere is moving up or down, Doppler shifts will change the received frequency slightly. If operation is near the MUF, some energy may be greatly delayed by reflections from varying heights, further smearing the pulse. Noise of various types will be added to the pulse. Finally, fading effects may be noticed even on such a short time frame as a single Morse letter.

The successful decoding of a Morse signal consists of simply deciding whether a pulse was sent or not, and its length. Humans have been given a brain and sense organs that, when properly trained, can cope with all these problems to a high degree. Our brain's ability to filter out extraneous noise and signals is remarkable. The dynamic range and sensitivity of both vision and hearing are close to the theoretical limits. Human AGC (automatic gain control) operates to provide a continually variable "decision threshold." If all else fails during on-air contacts, repeats can be requested.

A satisfactory non-human Morse decoder would have to replicate all of the above human functions. As it turns out, Morse, using OOK, is one of the hardest digital systems to copy error free, although by using fixed speed, and with reasonably good propagation, it is possible. Those who have worked with software or hardware schemes for decoding Morse will point out that narrow filtering will clean up the above pulse considerably. This is really a form of averaging or integration, which smoothes out the abrupt changes in amplitude. A phase-locked loop detector provides averaging and threshold detection in one circuit. Even with these improvements, OOK signals are difficult to decode reliably.

The second oldest method of transmitting pulses is with frequency shift keying (FSK) where, instead of turning a signal on and off, it is made to jump between different frequencies to correspond to the "key-up" and "key-down" conditions. RTTY (radio teletype) does not *require* the use of FSK and some have used OOK for RTTY. However, because of long use, most of us associate FSK transmissions with RTTY, the mechanical or computer decoding of text (plus crude images) using the Baudot code. Digital text modes have used AM (of which OOK is an example, but not the only one), FSK or PSK (phase shift keying).

As with FM analog transmissions, FSK and PSK tend to discriminate against amplitude noise, which is common on HF. Thus, amplitude changes resulting from fading and/or static can be reduced. However, as discussed in this *Handbook's* **AC/RF Sources** chapter, AM noise can change into phase noise, so this is not a complete solution. Modes based on phase modulation, such as the PSK modes, are not noise free. You need only observe the little phase "compass" included in some PSK31 software to be aware that even when transmitting an idle tone, the received signal's phase will jump around.

Doppler shifts introduce noise into phase- and frequency-shift systems. If we are counting on seeing a certain frequency shift as a signal, you don't want the path adding any shifts that will show up as noise. Thus, Doppler shift places a lower limit on the amount of frequency or phase shift needed to distinguish between the various data bits. On HF, ionospheric Doppler, which can be up to 5 Hz, places a lower limit on the carrier spacing of multi carrier modes. Lengthening of pulses of up to several milliseconds is common on HF paths. The general solution for channel-produced lengthening of pulses is simply to use a slower rate of transmission. Thus, the pulse has time to settle before being interpreted.

The fastest CW operators slow down when unusual multipath propagation conditions smear the characters together. A machine can operate at speeds where even normal multipath can smear the characters, not just in extreme cases. This factor puts an upper limit on the *symbol rate*. If a bit rate is to be increased beyond this point, each pulse must carry more than one bit. This calls for the use of complex modulation schemes involving multiple states per pulse. These can be represented by a number of different frequencies in an FSK mode, many phases

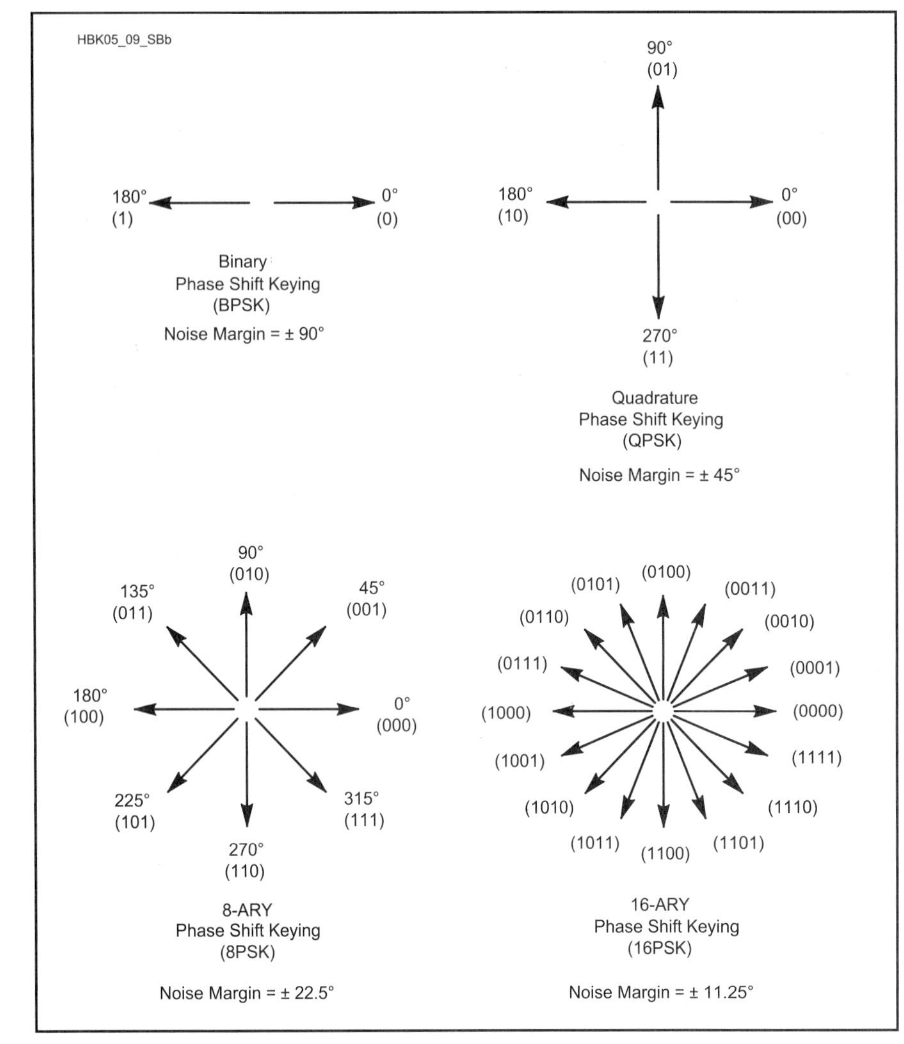

Fig 9.2—An example of multilevel phase modulation.

in a PSK mode, many levels in an AM mode or a combination of several or even all of the above. An example of multi-level phase modulation is illustrated in **Fig 9.2**. One type of a complex system would be QAM (quadrature amplitude modulation), which requires special equipment to observe.

With complex modulation forms, noise-induced changes in both amplitude and phase will cause some pulses to fall into oblivion, where you simply cannot say what was sent. Thus, a higher signal to noise ratio is needed for complex types of modulation such as 64QAM, whereas binary modes may work closer to the noise.

For high bit rates on HF, many carriers may be used, each one carrying multiple amplitude and phase levels. There are, however, regulatory and practical limits. Thus, while it is feasible to transmit digital speech in a 3-kHz bandwidth, full-motion, high-resolution full-motion TV signals would be very difficult to transmit on HF, even if an entire amateur band were used.

Schemes that use more than one frequency, such as FSK and multicarrier systems, must cope with *selective fading*, where not all frequencies fade at the same time. Users of RTTY FSK systems have long observed that either the mark or space frequency may momentarily fade away leaving only the other. A good decoder will work with only one of the two tones present.

All present and future digital modes must address these problems, and the development of new text modes has seen a steady progression from the original and hallowed CW and RTTY modes. As digital voice, image, text and control modes develop, they will all cope with the above channel limitations in various ways and with varying degrees of success. Some will be very resistant to QRM; others will be efficient in use of spectrum or in *throughput*—the amount of data that can be sent in a given time.

Modes that are optimum for a QRM-free VHF channel may work poorly on HF. Optimum HF modes will be too slow for some VHF applications. Error-correction schemes will always be useful for the more difficult channels, since on HF there is no such thing as a 100% reliable channel. On all frequencies, hams have a habit of "pushing the limits" so that marginal paths are often used. The design of digital communications systems will always be an exercise in trade-offs.

EFFECT OF THE HF PATH ON ANALOG SIGNALS

Analog signals undergo the same impairments as digital signals along an ionospheric transmission path. However, the signals are normally decoded by the combination of the human ear and brain, which overcome problems, often without much notice. Frequency-selective fading of double-sideband AM signals manifests itself as distortion or mushiness at the receiver audio that may be ignored by the human operator. Even if the signal becomes temporarily unreadable, often the operator can fill in the blanks because the information is familiar or expected. Single-sideband AM (SSB) usually suffers less mushiness in the receiver due to frequency selective fading because the signal occupies a narrower bandwidth, in which low and high-frequency audio tend to fade together. The human operator can usually cope with such temporary fadeouts or can request a retransmission for any part missed.

MULTIPLE ACCESS AND MULTIPLEXING TECHNOLOGIES

To appreciate some of the more complex communications systems, you need an understanding of the different methods of sharing a carrier or accessing the frequency spectrum. *Multiple access* refers to more than one originating source having use of the media. *Multiplexing* means combining of two or more information streams into one carrier or transmission path.

Frequency Division Multiple Access (FDMA)

FDMA is probably the oldest and most familiar method of accessing the frequency spectrum, since individual signals are on different frequency channels. It is also the least efficient, since each frequency occupies a slot that is reserved for one user at a time.

Frequency Division Multiplexing (FDM)

FDM uses more than one subcarrier, imposed on a carrier, to convey different information. It traditionally was used for multiplexed telephone systems but is rarely used in the Amateur Radio Service.

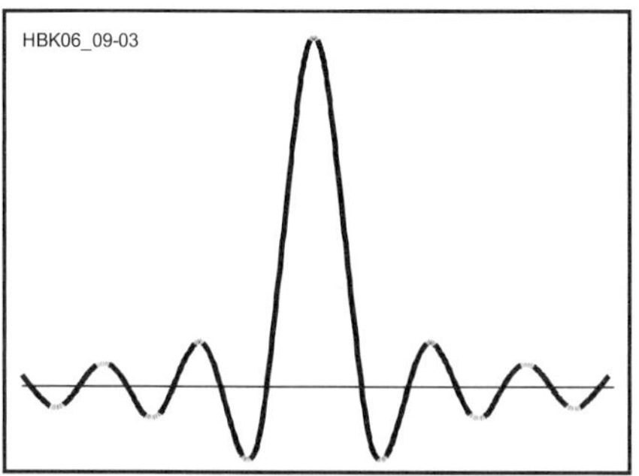

Fig 9.3—Spectrum of an individual OFDM carrier.

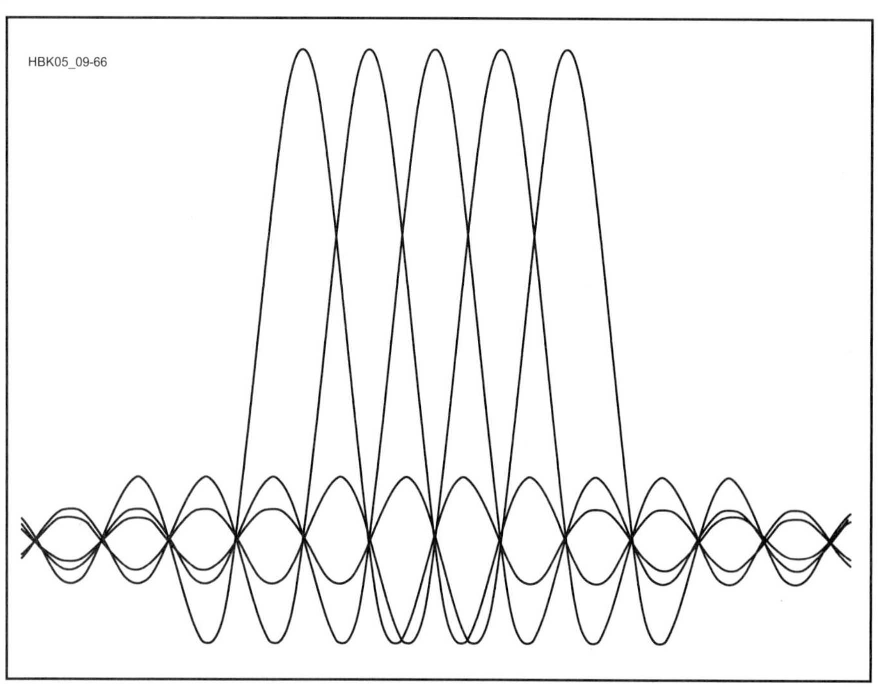

Fig 9.4—Overall OFDM spectrum.

Time Division Multiple Access (TDMA)

TDMA is simply time-sharing a frequency. In a general sense, this occurs naturally as stations in a QSO take turns transmitting. TDMA is also used in digital systems that reverse the direction of a circuit automatically to send information and acknowledgements.

Time Division Multiplex (TDM)

TDM is transmission of two or more signals over a common channel by interleaving so that the signals occur in different time slots. Some cellular telephone systems, such as Global System for Mobile Communications (GSM) use TDM. In the Amateur Radio Service it is used mostly for telemetry, such as from amateur satellites and remote repeaters.

Orthogonal Frequency Division Multiplexing (OFDM)

The term *orthogonal* is derived from the fact that multiple carriers are closely spaced in frequency, but positioned such that they do not interfere with one another. The center frequency of one carrier's signal falls within the nulls of the signals on either side of it. **Figs 9.3** and **9.4,** illustrate how the carriers are interleaved to prevent intercarrier interference.

Because each carrier is modulated at a relatively low rate, OFDM links suffer less intersymbol interference (ISI) on HF ionospheric paths than single-carrier modulation at a higher rate. See Smith, Doug, KF6DX, "Distortion and Noise in OFDM Systems," *QEX* Mar/Apr 2005.

Code Division Multiple Access (CDMA)

CDMA is a form of *spread spectrum* and is generated by modulating a carrier with a spreading code sequence known to both the sender and receiver. Unlike FDMA and TDMA, there is no fixed limit on the number of users but the number is not infinite.

Major Modulation Systems

The broadest category of modulation is how the *main carrier* is modulated. The major types are amplitude modulation, angle modulation and pulse modulation.

AMPLITUDE MODULATION

Amplitude modulation (AM) covers a class of modulation systems in which the amplitude of the *main* carrier is the characteristic that is varied. AM is sometimes simplistically described as varying the amplitude of the carrier from zero power to a peak power level. In fact, the carrier itself stays at the same amplitude when modulated by an analog (such as voice) baseband signal. The modulation itself produces sidebands, which are bands of frequencies on both sides of the carrier frequency. AM is basically a process of heterodyning or non-linear mixing. As in any mixer, when a carrier and baseband modulation are combined, there are three products in the frequency range of interest: (1) the carrier, (2) the lower sideband (LSB), and (3) the upper sideband (USB). Thus, if a carrier of 10 MHz were modulated by a 1-kHz sine wave, the outputs would be as shown in **Fig 9.5**.

The bandwidth of the modulated signal in this example would be 2 kHz, the difference between the lowest and highest frequencies. In AM, the difference between the carrier and farthest-away component of the sideband is determined by the highest frequency component contained in the baseband-modulating signal.

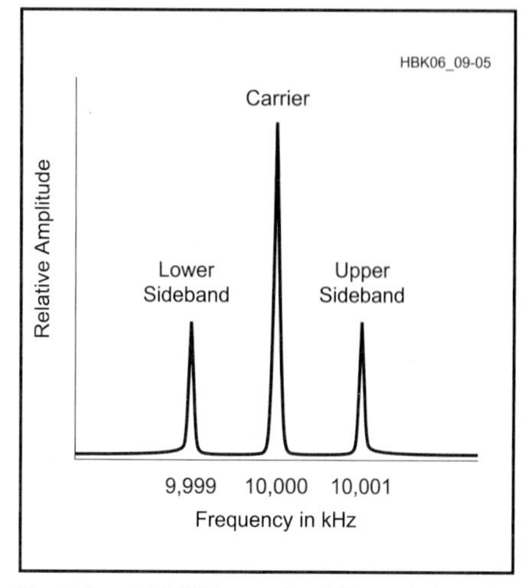

Fig. 9.5—A 10-MHz carrier AM-modulated by a 1-kHz sine wave.

ANGLE MODULATION

Two particular forms of angle modulation are *frequency modulation* (FM) and *phase modulation* (PM). Frequency and phase modulation are not independent, since the frequency cannot be varied without also varying the phase, and vice versa.

The communications effectiveness of FM and PM depends almost entirely on the receiving methods. If the receiver can respond to frequency and phase changes but is insensitive to amplitude changes, it will discriminate against most forms of noise, particularly impulse noise, such as that from ignition systems.

Frequency Modulation

Fig 9.6 is a representation of frequency modulation. When a modulating signal is applied, the carrier frequency is increased during one half cycle of the modulating signal and decreased during the half cycle of the opposite polarity. This is indicated in the drawing by the fact that the RF cycles occupy less time (higher frequency) when the modulating signal is negative.

The change in the carrier frequency (*frequency deviation*) is proportional to the instantaneous amplitude of the modulating signal. Thus, the deviation is small when the instantaneous amplitude of the modulating signal is small and is greatest when the modulating signal reaches its peak, either positive or negative. The drawing shows that the amplitude of the RF signal does not change during modulation. This is an oversimplification and is true only in the overall sense, as the amplitude of both the carrier and sidebands do vary with frequency modulation. FM is capable of conveying dc levels, as it can maintain a specific frequency.

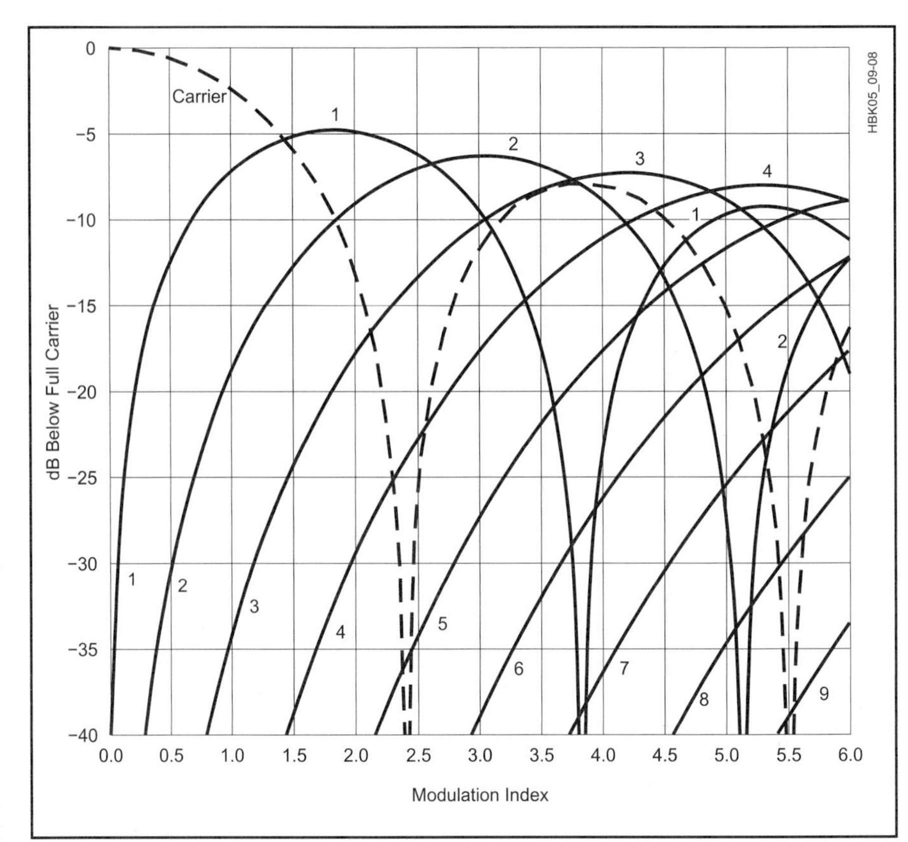

Fig 9.7—Amplitude of the FM carrier and sidebands with modulation index. This is a graphical representation of mathematical functions developed by F. W. Bessel. Note that the carrier completely disappears at modulation indexes of 2.405 and 5.52.

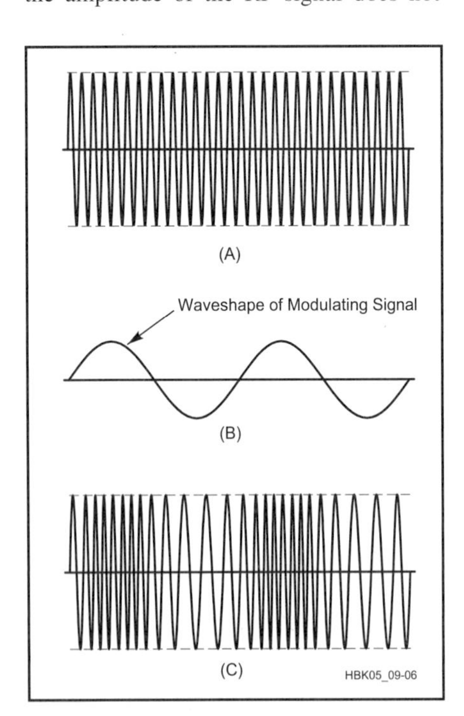

Fig 9.6—Graphical representation of frequency modulation. In the unmodulated carrier (A) each RF cycle occupies the same amount of time. When the modulating signal (B) is applied, the radio frequency is increased and decreased according to the amplitude and polarity of the modulating signal (C).

Phase Modulation

In phase modulation, the characteristic varied is the carrier phase from a reference value. In PM systems, the demodulator responds only to instantaneous changes in frequency. PM cannot convey dc levels unless special phase-reference techniques are used. The amount of frequency change, or deviation, is directly proportion to how rapidly the phase is changing and the total amount of the phase change.

Bessel Functions

Bessel functions are employed—using the carrier null method—to set deviation. Some version of the chart shown in **Fig 9.7** has appeared in the *ARRL Handbook* for 50 years. This chart is unlike previous ones in that the values are plotted here in dB, which is more familiar to anyone who uses an S meter or a spectrum analyzer to observe the various FM sidebands. This version also plots all values as positive because receivers, including spectrum analyzers, do not distinguish between positive and negative phase values. Thus, this plot will give values directly in dB below the unmodulated carrier of each component of a frequency-modulated wave, based on the modulation index.

Since the carrier and each sideband of a frequency-modulated signal change amplitude according to fixed rules as the deviation and modulating frequency change, we can use those rules to set deviation, provided we have a way to observe the FM spectrum. Based on a set of mathematical functions named after F.W. Bessel, who developed them, we know that a modulation index of 2.405 will produce what is called the "first carrier null." Thus, if we wish to set our deviation to 5 kHz, we can use an audio tone of 5000/2.405 or 2079 Hz. While observing the spectrum, we can then increase the deviation from zero until the carrier is in a null. This guarantees that the deviation is now 5000 Hz. If we use a frequency counter to set the audio tone accurately, the exactness of the deviation setting should be very high. Similarly, for setting the deviation to 3 kHz, we could use the audio frequency of 1247 kHz and adjust for the first carrier null. If a spectrum analyzer is not available, an all-mode receiver using a narrow CW filter could be used to detect the carrier null, using the S meter and carefully tuning to the carrier. Additional carrier nulls occur with modulation indices of 5.52 and 8.654. These would be useful for wideband FM or when using low audio frequencies for setting the deviation with narrow-band FM.

Other methods of setting deviation include observing the bandwidth on a spectrum analyzer using a very low frequency audio tone and using a deviation meter—an FM receiver whose audio output is metered on a scale calibrated directly in kHz of deviation. An FM service monitor may include both a deviation meter and a spectrum display. The carrier-null method is the most accurate of the three methods and can even be used to calibrate a deviation meter.

You can also use the plot in Fig 9.7 to predict the bandwidth of any given audio frequency and deviation combination. Consider this example. We wish to keep our bandwidth narrow enough to pass through a 15-kHz receive filter and we are transmitting a tone of 3 kHz. Since the third set of sidebands will be 18 kHz apart, we would do well to keep them, and all higher sidebands, below –40 dB. A quick look at the chart shows that this means the modulation index must be no more than about 0.7, meaning the deviation should be 0.7 × 3 = 2.1 kHz. If we are willing to allow the third set of sidebands to be only 34 dB down, we can use a modulation index of 1, meaning the deviation will be 3 kHz—close to the recommended value for 9600 bits/s digital signals on FM. The above calculations strictly apply only when the highest audio frequency (3 kHz) is present. If there is little chance of 100% modulation at the highest audio frequency as, for example, with a normal voice signal, higher deviation could be used. When new digital modems and modes are used on FM, the above procedures should be part of the design.

Operating Modes

This chapter examines various *operating modes* used in the Amateur Radio Service, including text modes, data, telemetry and telecommand, voice, image, spread spectrum and multimedia. While modes once fit into neat categories, there is now a blurring of the definitions. For example, data transmissions could include images.

TELEGRAPHY MODES

These are basically text modes; that is, transmission of letters, figures and punctuation, in a format suitable for printing at the receiving station. Morse telegraphy and radioteletype (Baudot and ASCII) are described, but you should be aware that the term "telegraphy" includes facsimile transmission as well.

Morse Telegraphy (CW)

Text messages sent by on-off keying (OOK, also known as amplitude-shift keying, or ASK) is the original mode for both amateur and commercial radio. It is alive and well today and is not expected to fade away. For many amateurs, it is the principal, or even the only, mode they use and many take great and justifiable pride in their proficiency with it. The complete international Morse (including the new @ character ·-−·-·) code itself is defined in ITU-R Recommendation M.1677, *International Morse code*.

CW continues in use, however, not just for reasons of nostalgia. When used by an experienced operator, it can rival most any mode for "getting the message through" under marginal conditions and is absolutely unrivaled in terms of the simplicity of the equipment needed. Methods of generating the code characters, and even of decoding them, have used the latest technology, but the straight key and "copy by ear" are still in use. Of all modes, CW is the most versatile in terms of signaling speed. It is used at speeds—measured in words per minute (WPM)—of less than one, and up to several hundred. Depending on ability of the operator, direct human copy works well between about 5 and 60 WPM, but for very slow or very fast speeds, the signals may be recorded and the speed adjusted to allow human decoding. Very slow speeds and extremely narrow filters make possible communication using signals below the noise, while very fast speeds are useful for meteor scatter communication where bandwidth is large, but the reflection path lasts only a second or less.

The bandwidth occupied by a CW signal depends on the keying rate (See the **Mixers, Modulators and Demodulators** chapter of this *Handbook*), with higher speeds requiring a wider filter to pass the sidebands. In addition, occupied bandwidth depends on the rise and fall time and the shape of the keyed RF envelope. That shape should be somewhat rounded (no abrupt transitions) in order to prevent "key clicks"—harmonics of the keying pulse. These can extend over several kHz and cause unnecessary interference. The ideal RF envelope of a code element would rise and fall in the shape of a sine wave. See **Figs 9.8** and **9.9**. ARRL has long recommended a 5-ms rise time for CW, up to 60 WPM, which keeps the signal within a 150-Hz bandwidth. Use of a nar-

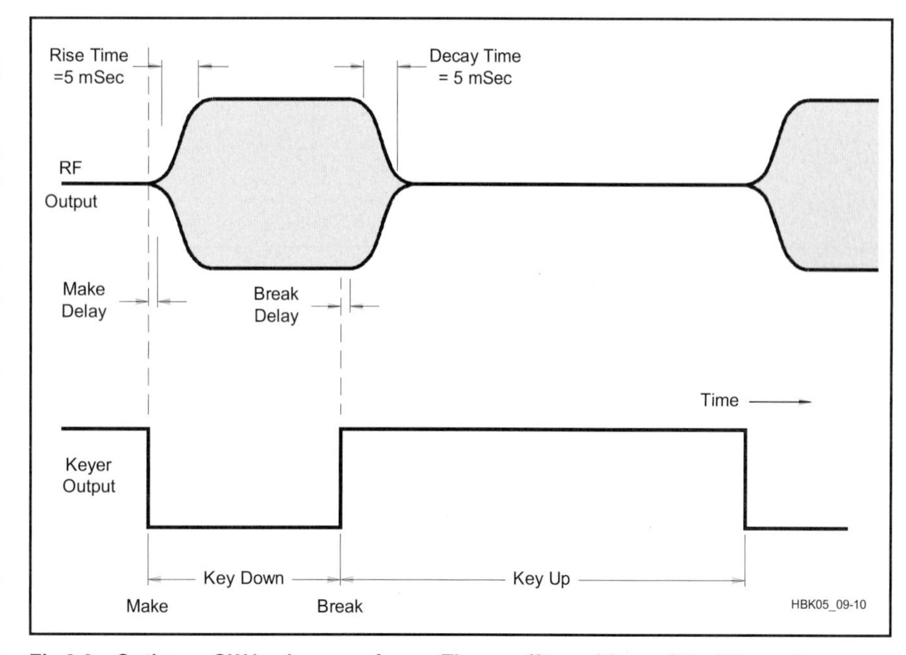

Fig 9.8—Optimum CW keying waveforms. The on-off transitions of the RF envelope should be smooth, ideally following a sine-wave curve. See text.

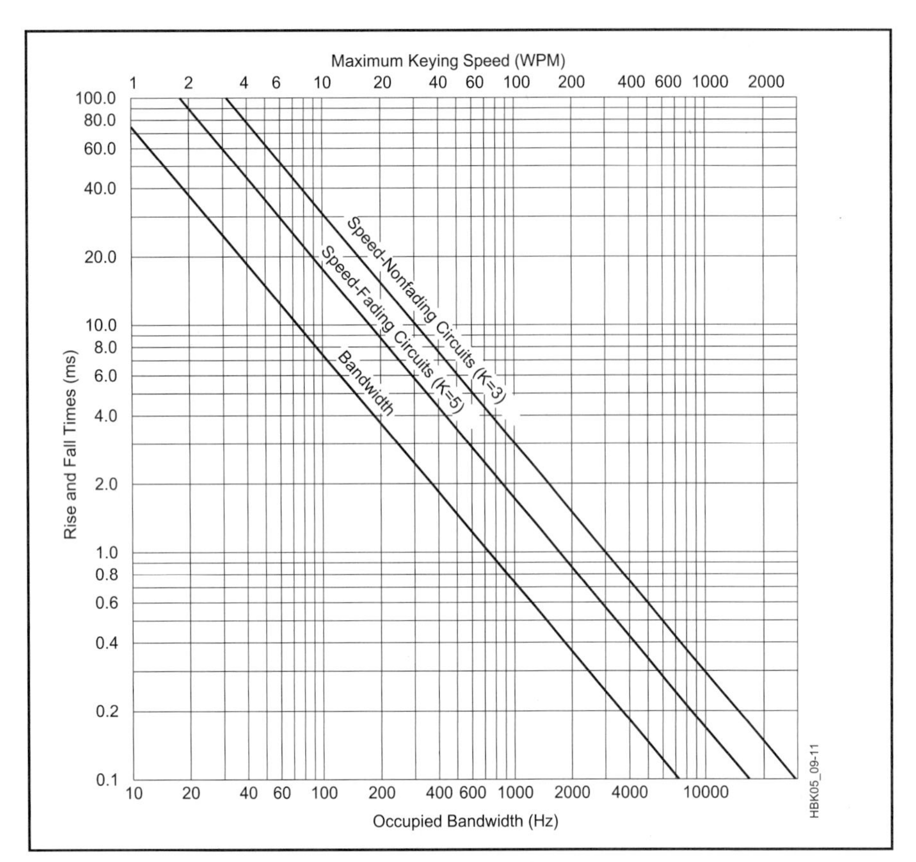

Fig 9.9—Keying speed vs rise and fall times vs bandwidth for fading and non-fading communications circuits. For example, to optimize transmitter timing for 25 WPM on a non-fading circuit, draw a vertical line from the WPM axis to the K = 3 line. From there draw a horizontal line to the rise/fall time axis (approximately 15 ms). Draw a vertical line from where the horizontal line crosses the bandwidth line and see that the bandwidth will be about 60 Hz.

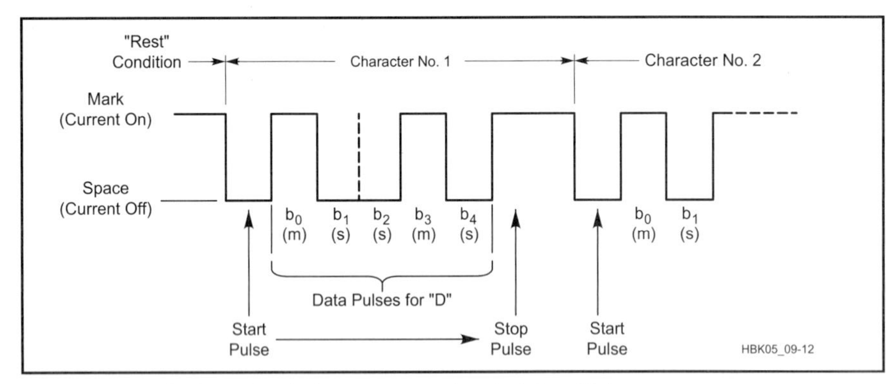

Fig 9.10—A typical Baudot timing sequence for the letter "D."

rower filter than this on receive end is uncommon for ear-copied CW; therefore, narrower bandwidth is unnecessary and would make the signal sound "mushy." Very fast pulses, such as would be used for High-Speed CW (HSCW) meteor scatter work, are computer generated and can occupy a normal SSB filter bandwidth.

Morse code is one of the most efficient modes in terms of information sent per baud. The commonly accepted ratio for bauds to WPM is WPM = 1.2 × *B*. Thus, a keying speed of 25 dots per second or 50 bauds is equal to 60 WPM. The efficiency of Morse text messages is based on the use of the shortest code combinations to represent the most commonly used letters and symbols. Efficiency is further achieved by extensive use of abbreviations and "Q signals." By making use of these multiple levels of universally recognized coding schemes, CW can get essential information across quickly. CW abbreviations are universal so that simple contacts can be made without the need of a shared language. In skilled hands, CW can achieve a QSO or traffic rate approaching that of phone operation while using a fraction of the bandwidth.

Baudot Radioteletype (RTTY)

One of the first data communications codes to receive widespread use had five bits (also called "levels") to present the alphabet, numerals, symbols and machine functions. In the US, we use International Telegraph Alphabet No. 2 (ITA2), commonly called *Baudot,* as specified in FCC §97.309(a)(1). The code is defined in the ITA2 Codes table on the CD-ROM included with this book. In the United Kingdom, the almost-identical code is called *Murray* code. There are many variations in five-bit coded character sets, principally to accommodate foreign-language alphabets.

Five-bit codes can directly encode only 2^5 = 32 different symbols. This is insufficient to encode 26 letters, 10 numerals and punctuation. This problem can be solved by using one or more of the codes to select from multiple code-translation tables. ITA2 uses a LTRS code to select a table of upper-case letters and a FIGS code to select a table of numbers, punctuation and special symbols. Certain symbols, such as carriage return, occur in both tables. Unassigned ITA2 FIGS codes may be used for the remote control of receiving printers and other functions.

FCC rules provide that ITA2 transmissions must be sent using start-stop pulses, as illustrated in **Fig 9.10**. The bits in the figure are arranged as they would appear on an oscilloscope.

Speeds and Signaling Rates

The signaling speeds for RTTY are those used by the old TTYs, primarily 60 WPM or 45.45 bauds. The *baud* (Bd) is a unit of signaling speed equal to one pulse (event) per second. The signaling rate, in bauds, is the reciprocal of the shortest pulse length.

Transmitter Keying

When TTYs and TUs (terminal units) roamed the airwaves, frequency-shift keying (FSK) was the order of the day. DC signals from the TU controlled some form of reactance (usually a capacitor or varactor) in a transmitter oscillator stage that shifted the transmitter frequency. Such direct FSK is still an option with some new radios.

AFSK

Multimode communications processors (MCPs), however, generally connect to the radio AF input and output, often through the speaker and microphone connectors, and sometimes through auxiliary connectors. They simply feed AF tones to the microphone input of an SSB transmitter or transceiver. This is called AFSK for "audio frequency-shift keying."

When using AFSK, make certain that audio distortion, carrier and unwanted sidebands do not cause interference. Particularly when using the low tones discussed later, the harmonic distortion of the tones should be kept to a few percent. Most modern AFSK generators are of the continuous-phase (CPFSK) type. Also remember that equipment is operating at a 100% duty cycle for the duration of a transmission. For safe operation, it is often necessary to reduce the transmitter power output (25 to 50% of normal) from the level that is safe for CW operation.

What are High and Low Tones?

US amateurs customarily use the same modems (2125 Hz mark, 2295 Hz space) for both VHF AFSK and HF via an SSB transmitter. Because of past problems (when 850-Hz shift was used), some amateurs use "low tones" (1275 Hz mark, 1445 Hz space). Both high and low tones can be used interchangeably on the HF bands because only the *amount* of shift is important. The frequency difference is unnoticed on the air because each operator tunes for best results. On VHF AFSK, however, the high and low tone pairs are not compatible.

Transmit Frequency

It is normal to use the lower sideband mode for RTTY on SSB radio equipment. In order to tune to an exact RTTY frequency, remember that most SSB radio equipment displays the frequency of its (suppressed) carrier, not the frequency of the mark signal. Review your MCP's manual to determine the tones used and calculate an appropriate display frequency. For example, to operate on 14,083 kHz with a 2125-Hz AFSK mark frequency, the SSB radio display (suppressed-carrier) frequency should be 14,083 kHz + 2.125 kHz = 14,085.125 kHz.

Receiving Baudot

TUs (Terminal Units) have been replaced by multi-mode communications processors (MCPs), which accept AF signals from a radio and translate them into common ASCII text or graphics file formats (see **Fig 9.11**). Because the basic interface is via ASCII, MCPs are compatible with virtually any PC running a simple terminal program. Many MCPs handle CW, RTTY, ASCII, packet, fax, SSTV and new digital modes as they come into amateur use. To an increasing extent, personal computer sound cards with appropriate software are a viable and low-cost alternative to MCPs. However, sound cards have their limitations and dedicated hardware can more efficiently perform some operations.

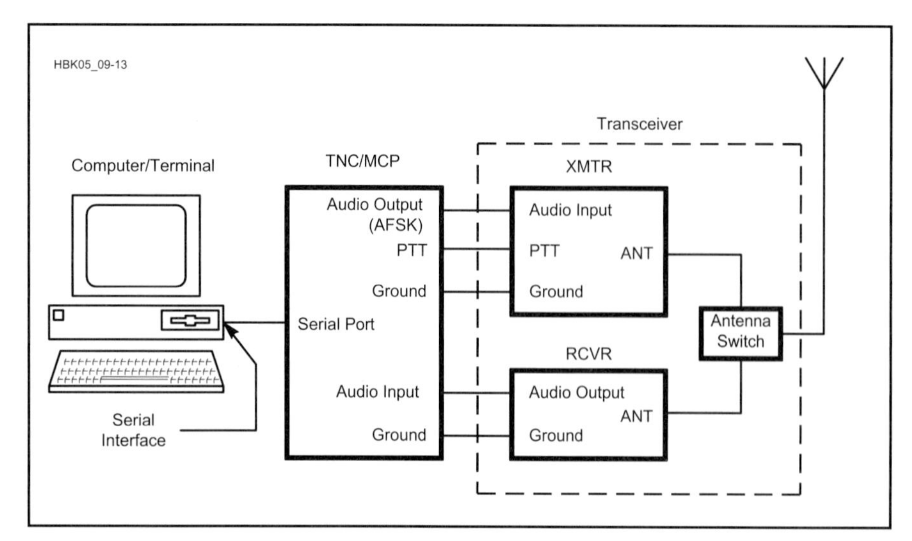

Fig 9.11—A typical multimode communications processor (MCP) station. MCPs can do numerous data modes as well as SSTV and fax.

AFSK Demodulators

An AFSK demodulator takes the shifting tones from the audio output of a receiver and produces TTY keying pulses. FM is a common AFSK demodulation method. The signal is first bandpass filtered to remove out-of-band interference and noise. It is then limited to remove amplitude variations. The signal is demodulated in a discriminator or a PLL. The detector output is low pass filtered to remove noise at frequencies above the keying rate. The result is fed to a circuit that determines whether it is a mark or a space.

AM (limiterless) detectors, when properly designed, permit continuous copy even when the mark or space frequency fades out completely. At 170-Hz shift, however, the mark and space frequencies tend to fade at the same time. For this reason, FM and AM demodulators are comparable at 170-Hz shift.

Diversity Reception

Although not restricted to RTTY, *diversity reception* can be achieved by using two antennas, two receivers and a dual demodulator. Some amateurs are using it with good results. One of the antennas would be the normal station antenna for that band. The second antenna could be either another antenna of the same polarization located at least 3/8-wavelength away, or an antenna of the opposite polarization located near the first antenna. A problem is to get both receivers on the same frequency without carefully tuning each one manually. Two demodulators are needed for this type of diversity. Also, some type of diversity combiner, selector or processor is needed. Many commercial or military RTTY demodulators are equipped for diversity reception.

The payoff for using diversity is a worthwhile improvement in copy. Depending on fading conditions, adding diversity may be equivalent to raising transmitter power severalfold.

BAUDOT RTTY BIBLIOGRAPHY

Ford, Steve, WB8IMY, ARRL's *HF Digital Handbook,* Third Ed., ARRL, 2004.

Henry, "Getting Started in Digital Communications," Part 3 (RTTY), *QST*, May 1992.

Hobbs, Yeomanson and Gee, *Teleprinter Handbook*, Radio Society of Great Britain.

Nagle, "Diversity Reception: an Answer to High Frequency Signal Fading," *Ham Radio*, Nov 1979, pp 48-55.

ASCII (IA5) RADIOTELETYPE

The American National Standard Code for Information Interchange (ASCII) is a coded character set used for information-processing systems, communications systems and related equipment. Current FCC regulations provide that amateur use of ASCII shall conform to ASCII as defined in ANSI Standard X3.4-1977. Its international counterparts are ISO 646-1983 and International Alphabet No. 5 (IA5) as specified in ITU-T Recommendation V.3.

ASCII uses 7 bits to represent letters, figures, symbols and control characters. Unlike ITA2 (Baudot), ASCII has both upper- and lower-case letters. A table of ASCII characters is presented as "ASCII Character Set" on the CD-ROM that accompanies this book.

Parity

While not strictly a part of the ASCII standard, an eighth bit may be added for parity (P) checking. FCC rules permit optional use of the parity bit. The applicable US and in-

Glossary of Digital Communications Terminology

ACK—Acknowledgment, the control signal sent to indicate the correct receipt of a transmission block.
Address—A character or group of characters that identifies a source or destination.
AFSK—Audio frequency-shift keying.
ALE—Automatic link establishment.
AMRAD—Amateur Radio Research and Development Corporation, a nonprofit organization dedicated to experimentation.
AMSAT—Radio Amateur Satellite Corporation.
AMTOR—Amateur teleprinting over radio, an amateur radioteletype transmission technique employing error correction as specified in several ITU-R Recommendations M.476-2 through M.476-4 and M.625.
ANSI—American National Standards Institute.
Answer—The station intended to receive a call. In modem usage, the called station or modem tones associated therewith.
APCO—Association of Public Safety Communications Officials.
ARQ—Automatic Repeat reQuest, an error-sending station, after transmitting a data block, awaits a reply (ACK or NAK) to determine whether to repeat the last block or proceed to the next.
ASCII—American National Standard Code for Information Interchange, a code consisting of seven information bits.
AX.25—Amateur packet-radio link-layer protocol. Copies of protocol specification are available from ARRL HQ.
Backwave—An unwanted signal emitted between the pulses of an on/off-keyed signal.
Balanced—A relationship in which two stations communicate with one another as equals; that is, neither is a primary (master) or secondary (slave).
Baud—A unit of signaling speed equal to the number of discrete conditions or events per second. (If the duration of a pulse is 20 ms, the signaling rate is 50 bauds or the reciprocal of 0.02, abbreviated Bd).
Baudot code—A coded character set in which five bits represent one character. Used in the US to refer to ITA2.
Bell 103—A 300-baud full-duplex modem using 200-Hz-shift FSK of tones centered at 1170 and 2125 Hz.
Bell 202—A 1200-baud modem standard with 1200-Hz mark, 2200-Hz space, used for VHF FM packet radio.
BER—Bit error rate.
BERT—Bit-error-rate test.
Bit stuffing—Insertion and deletion of 0s in a frame to preclude accidental occurrences of flags other than at the beginning and end of frames.
Bit—Binary digit, a single symbol, in binary terms either a one or zero.
Bit/s—Bits per second.
BLER—Block error rate.
BLERT—Block-error-rate test.
Break-in—The ability to hear between elements or words of a keyed signal.
Byte—A group of bits, usually eight.
Carrier detect (CD)—Formally, received line signal detector, a physical-level interface signal that indicates that the receiver section of the modem is receiving tones from the distant modem.
CDMA—Code division multiple access.
Chirp—Incidental frequency modulation of a carrier as a result of oscillator instability during keying.
CLOVER—Trade name of digital communications system developed by Hal Communications.
COFDM—Coded Orthogonal Frequency Division Multiplex, OFDM plus coding to provide error correction and noise immunity.
Collision—A condition that occurs when two or more transmissions occur at the same time and cause interference to the intended receivers.
Constellation—A set of points in the complex plane which represent the various combinations of phase and amplitude in a QAM or other complex modulation scheme.
Contention—A condition on a communications channel that occurs when two or more stations try to transmit at the same time.
Control field—An 8-bit pattern in an HDLC frame containing commands or responses, and sequence numbers.
CRC—Cyclic redundancy check, a mathematical operation. The result of the CRC is sent with a transmission block. The receiving station uses the received CRC to check transmitted data integrity.
CSMA—Carrier sense multiple access, a channel access arbitration scheme in which packet-radio stations listen on a channel for the presence of a carrier before transmitting a frame.
CTS—clear to send, a physical-level interface circuit generated by the DCE that, when on, indicates the DCE is ready to receive transmitted data (abbreviated CTS).
DARPA—Defense Advanced Research Projects Agency.
DBPSK—Differential binary phase-shift keying.
DQPSK—Differential quadrature phase-shift keying.
DCE—Data circuit-terminating equipment, the equipment (for example, a modem) that provides communication between the DTE and the line radio equipment.
Domino—A conversational HF digital mode similar in some respects to MFSK16.
DRM—Digital Radio Mondiale. A consortium of broadcasters, manufacturers, research and governmental organizations which developed a system for digital sound broadcasting in bands between 100 kHz and 30 MHz.
EIA—Electronic Industries Alliance.
EIA-232—An EIA standard physical-level interface between DTE (terminal) and DCE (modem), using 25-pin connectors. Formerly RS-232, a popular serial line standard, equivalent of ITU-T V.24 and V.28.
Envelope-delay distortion—In a complex waveform, unequal propagation delay for different frequency components.
Equalization—Correction for amplitude-frequency and/or phase-frequency distortion.
Eye pattern—An oscilloscope display in the shape of one or more eyes for observing the shape of a serial digital stream and any impairments.
Facsimile (fax)—A form of *telegraphy* for the transmission of fixed images, with or without half-tones, with a view to their reproduction in a permanent form.
FCS—Frame check sequence. (See CRC.)
FDM—Frequency division multiplexing
FDMA—Frequency division multiple access
FEC—Forward error correction, an error-control technique in which the transmitted data is sufficiently redundant to permit the receiving station to correct some errors.
FSK—Frequency-shift keying.
GNU—A project to develop a free UNIX style operating system.
G-TOR—A digital communications system developed by Kantronics.
HDLC—High-level data link control procedures as specified in ISO 3309.
Hellschreiber—A facsimile system for transmitting text.
Host—As used in packet radio, a computer with applications programs accessible by remote stations.
IA5—International Alphabet No. 5, a 7-bit coded character set, ITU-T version of ASCII.
IBOC—In Band On Channel. A method of using the same channel on the AM or FM broadcast bands to transmit simultaneous digital and analog modulation.
Information field—Any sequence of bits containing the intelligence to be conveyed.
ISI—Intersymbol interference; slurring of one symbol into the next as a result of multipath propagation.
ISO—International Organization for Standardization.
ITA2—International Telegraph Alphabet No. 2, a ITU-T 5-bit coded character set commonly called the Baudot or Murray code.

ITU—International Telecommunication Union, a specialized agency of the United Nations. (See **www.itu.int**.)
ITU-R—Radiocommunication Sector of the ITU, formerly CCIR.
ITU-T—Telecommunication Standardization Sector of the ITU, formerly CCITT.
Jitter—Unwanted variations in amplitude or phase in a digital signal.
Key clicks—Unwanted transients beyond the necessary bandwidth of a keyed radio signal.
LAP—Link access procedure, ITU-T Recommendation X.25 unbalanced-mode communications.
LAPB—Link access procedure, balanced, ITU-T Recommendation X.25 balanced-mode communications.
Layer—In communications protocols, one of the strata or levels in a reference model.
Level 1—Physical layer of the OSI reference model.
Level 2—Link layer of the OSI reference model.
Level 3—Network layer of the OSI reference model.
Level 4—Transport layer of the OSI reference model.
Level 5—Session layer of the OSI reference model.
Level 6—Presentation layer of the OSI reference model.
Level 7—Application layer of the OSI reference model.
Linux—A free Unix-type operating system originated by Linus Torvalds, et al. Developed under the GNU General Public License.
Loopback—A test performed by connecting the output of a modulator to the input of a demodulator.
LSB—Least-significant bit.
MFSK16—A multi-frequency shift communications system
Modem—Modulator-demodulator, a device that connects between a data terminal and communication line (or radio). Also called data set.
MSB—Most-significant bit.
MSK—Frequency-shift keying where the shift in Hz is equal to half the signaling rate in bits per second.
MT-63—A keyboard-to-keyboard mode similar to PSK31 and RTTY.
NAK—Negative acknowledge (opposite of ACK).
Node—A point within a network, usually where two or more links come together, performing switching, routine and concentrating functions.
NRZI—Nonreturn to zero. A binary baseband code in which output transitions result from data 0s but not from 1s. Formal designation is NRZ-S (nonreturn-to-zero-space).
Null modem—A device to interconnect two devices both wired as DCEs or DTEs; in EIA-232 interfacing, back-to-back DB25 connectors with pin-for-pin connections except that Received Data (pin 3) on one connector is wired to Transmitted Data (pin 3) on the other.
Octet—A group of eight bits.
OFDM—Orthogonal Frequency Division Multiplex. A method of using spaced subcarriers that are phased in such a way as to reduce the interference between them.
Originate—The station initiating a call. In modem usage, the calling station or modem tones associated therewith.
OSI-RM—Open Systems Interconnection Reference Model specified in ISO 7498 and ITU-T Recommendation X.200.
Packet radio—A digital communications technique involving radio transmission of short bursts (frames) of data containing addressing, control and error-checking information in each transmission.
PACTOR®—Trade name of digital communications protocols offered by Special Communications Systems GmbH & Co KG (SCS).
Parity check—Addition of non-information bits to data, making the number of ones in a group of bits always either even or odd.
PID—Protocol identifier. Used in AX.25 to specify the network-layer protocol used.
Primary—The master station in a master-slave relationship; the master maintains control and is able to perform actions that the slave cannot. (Compare secondary.)
Project 25—Digital voice system developed for APCO, also known as P25.
Protocol—A formal set of rules and procedures for the exchange of information within a network.
PSK—Phase-shift keying.
PSK31—A narrow-band digital communications system developed by Peter Martinez, G3PLX.
Q15X25—A DSP-intensive mode intended as an error-free mode more reliable on HF than packet.
QAM—Quadrature Amplitude Modulation. A method of simultaneous phase and amplitude modulation. The number that precedes it, eg, 64QAM, indicates the number of discrete stages in each pulse.
QPSK—Quadrature phase-shift keying.
RAM—Random access memory.
Router—A network packet switch. In packet radio, a network-level relay station capable of routing packets.
RTS—Request to send, physical-level signal used to control the direction of data transmission of the local DCE.
RTTY—Radioteletype.
RxD—Received data, physical-level signals generated by the DCE are sent to the DTE on this circuit.
SCAMP—Sound Card Automated Message Protocol, an inexpensive alternative to hardware for passing e-mail traffic on narrow-bandwidth channels.
Secondary—The slave in a master-slave relationship. Compare primary.
Source—In packet radio, the station transmitting the frame over a direct radio link or via a repeater.
SSID—Secondary station identifier. In AX.25 link-layer protocol, a multipurpose octet to identify several packet-radio stations operating under the same call sign.
TAPR—Tucson Amateur Packet Radio Corporation, a nonprofit organization involved in packet-radio development.
TDM—Time division multiplexing
TDMA—Time division multiple access
Telecommand—The use of telecommunication for the transmission of signals to initiate, modify or terminate functions of equipment at a distance.
Telemetry—The use of telecommunication for automatically indicating or recording measurements at a distance from the measuring instrument.
Telephony—A form of telecommunication primarily intended for the exchange of information in the form of speech.
Telegraphy—A form of telecommunication in which the transmitted information is intended to be recorded on arrival as a graphic document; the transmitted information may sometimes be presented in an alternative form or may be stored for subsequent use.
Teleport—A radio station that acts as a relay between terrestrial radio stations and a communications satellite.
Television—A form of telecommunication for the transmission of transient images of fixed or moving objects.
Throb—A multi-frequency shift mode like MFSK16.
TNC—Terminal node controller, a device that assembles and disassembles packets (frames); sometimes called a PAD.
TR switch—Transmit-receive switch to allow automatic selection between receive and transmitter for one antenna.
TTY—Teletypewriter.
Turnaround time—The time required to reverse the direction of a half-duplex circuit, required by propagation, modem reversal and transmit-receive switching time of transceiver.
TxD—Transmitted data, physical-level data signals transferred on a circuit from the DTE to the DCE.
UI—Unnumbered information frame.
V.24—An ITU-T Recommendation defining physical-level interface circuits between a DTE (terminal) and DCE (modem), equivalent to EIA-232.
V.28—An ITU-T Recommendation defining electrical characteristics for V.24 interface.
Virtual circuit—A mode of packet networking in which a logical connection that emulates a point-to-point circuit is established (compare Datagram).
Window—In packet radio at the link layer, the range of frame numbers within the control field used to set the maximum number of frames that the sender may transmit before it receives an acknowledgment from the receiver.
X.25—An ITU-T packet-switching protocol Recommendation.

ternational standards (ANSI X3.16-1976; ITU-T Recommendation V.4) recommend an even parity sense for asynchronous and odd parity sense for synchronous data communications. The standards, however, generally are not observed by hams. By sacrificing parity, the eighth bit can be used to extend the ASCII 128-character code to 256 characters.

ASCII Serial Transmission

Serial transmission standards for ASCII (ANSI X3.15 and X3.16; ITU-T Recommendation V.4 and X.4) specify that the bit sequence shall be least-significant bit (LSB) first to most-significant bit (MSB); that is, b0 through b6 (plus the parity bit, P, if used).

Serial transmission may be either *synchronous* or *asynchronous*. In synchronous transmissions, only the information bits (and optional parity bit) are sent, as shown in **Fig 9.12A**.

Asynchronous serial transmission adds a start pulse and a stop pulse to each character. The start pulse length equals that of an information pulse. The stop pulse may be one or two bits long. There is some variation, but one stop bit is the convention.

ASCII Data Rates

Data-communication signaling rates depend largely on the medium and the state of the art when the equipment was selected. The most-used rates tend to progress in 2:1 steps from 300 to 9600 bits/s and in 8 kbits/s increments from 16 kbits/s upward.

The "baud" (Bd) is a unit of signaling speed equal to one discrete condition or event per second. In single-channel transmission, such as the FCC prescribes for Baudot transmissions, the signaling rate in bauds equals the data rate in bits per second. However, the FCC does not limit ASCII to single-channel transmission. Some digital modulation systems have more than two (mark and space) states. In *dibit* (pronounced "die-bit") modulation, two ASCII bits are sampled at a time. The four possible states for a dibit are 00, 01, 10 and 11. In four-phase modulation, each state is assigned an individual phase of 0°, 90°, 180° and 270° respectively. For dibit phase modulation, the signaling speed in bauds is half the information-transfer rate in bits/s. Since the FCC specifies the digital sending speed in bauds, amateurs may transmit ASCII at higher information rates by using digital modulation systems that encode more bits per signaling element.

Amateur ASCII RTTY Operations

On April 17, 1980, the FCC first permitted ASCII in the Amateur Radio Service. Amateurs have been slow to abandon Baudot in favor of asynchronous serial ASCII. Rather than transmitting start-stop ASCII, this code has become embedded in more sophisticated data transmission modes, which are described later in this chapter.

ASCII BIBLIOGRAPHY

ANSI X3.4-1977, "Code for Information Interchange," American National Standards Institute.

ANSI X3.15-1976, "Bit sequencing of the American National Standard Code for Information Interchange in Serial-by-Bit Data Transmission."

ANSI X3.16-1976, "Character Structure and Character Parity Sense for Serial-by-Bit Data Communication Information Interchange."

ANSI X3.25-1976, "Character Structure and Character Parity Sense for Parallel-by-Bit Communication in American National Standard Code for Information Interchange."

Bemer, "Inside ASCII," *Interface Age*, May, June and July 1978.

ITU-T Recommendation V.3, "International Alphabet No. 5."

ITU-T Recommendation V.4, "General Structure of Signals of International Alphabet No. 5 Code for Data Transmission over the Public Telephone Network."

Mackenzie, *Coded Character Sets, History and Development*, Addison-Wesley Publishing Co, 1980.

AMTOR

AMTOR is derived from ITU-R Recommendation M.476, and is known "narrowband direct printing" (NBDP) and commercially as "SITOR." It has been largely overtaken by newer protocols.

AMTOR uses two forms of time diversity in either Mode A (ARQ, Automatic Repeat reQuest) or Mode B (FEC, Forward Error Correction). In Mode A, a repeat is sent only when requested by the receiving station. In Mode B, each character is sent twice. In Mode A or Mode B, the second type of time diversity is supplied by the redundancy of the code itself.

Mode B (FEC)

When transmitting to no particular station (for example calling CQ, in a net operation or during bulletin transmissions) there is no (one) receiving station to request repeats. Mode B uses a simple forward-error-control (FEC) technique: It sends each character twice. Burst errors are virtually eliminated by delaying the repetition for a period thought to exceed the duration of most noise bursts. In AMTOR, groups of five characters are sent (DX) and then repeated (RX). At 70 ms per character, there is 280 ms between the first and second transmissions of a character.

The Information Sending Station (ISS) transmitter must be capable of 100% duty-cycle operation for Mode B. Thus, it may be necessary to reduce power level to 25% to 50% of full rating.

Mode A (ARQ)

This synchronous system transmits blocks of three characters from the Information Sending Station (ISS) to the Information Receiving Station (IRS). After each block, the IRS either acknowledges correct receipt (based on the 4/3 mark/space ratio) or requests a repeat. The station that initiates the ARQ protocol is known as the Master Station (MS). The MS first sends the selective call of the called station in blocks of three characters, listening between blocks. Four-letter AMTOR calls are normally derived from the first character and the last three letters of the station call sign. For example, W1AW's AMTOR call would be WWAW. The Slave Station (SS) recognizes its selective call and answers that it is ready. The MS now becomes the ISS and will send traffic as soon as the IRS says it is ready.

On the air, AMTOR Mode A signals have a characteristic "chirp-chirp" sound. Because of the 210/240-ms on/off timing, Mode A can be used with some transmitters at full power levels.

AMTOR BIBLIOGRAPHY

Ford, Steve, WB8IMY, *Your RTTY/AMTOR Companion* (Newington, CT: ARRL, 1993.)

Henry, Bill, "Getting Started in Digital Communications-AMTOR," *QST*, Jun 1992.

ITU-R Recommendations M.476 and 625, "Direct-Printing Telegraph Equipment

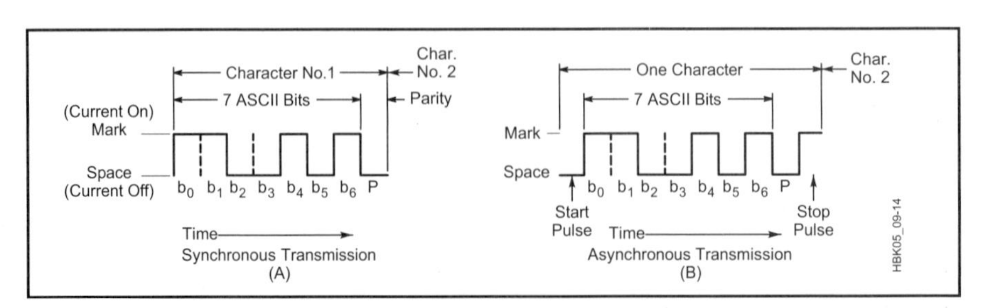

Fig 9.12—Typical serial synchronous and asynchronous timing for the ASCII character "S."

in the Maritime Mobile Service."

Martinez, Peter, "AMTOR, An Improved RTTY System Using a Microprocessor," *Radio Communication*, RSGB, Aug 1979.

Newland, Paul, "A User's Guide to AMTOR Operation," *QST*, Oct 1985.

PSK31

Peter Martinez, G3PLX, who was instrumental in bringing us AMTOR, also developed PSK31 for real time keyboard-to-keyboard QSOs. This section was adapted from an article in *RadCom*, Jan 1999. The name derives from the modulation type (phase-shift keying) and the data rate, which is actually 31.25 bauds. PSK31 is a robust mode for HF communications that features the 128 ASCII (Internet) characters and the full 256 ANSI character set. This mode works well for two-way QSOs and for nets. Time will tell if PSK31 will replace Baudot RTTY on the amateur HF bands.

Morse code uses a single carrier frequency keyed on and off as dits and dahs to form characters. RTTY code shifts between two frequencies, one for *mark* (1) the other for *space* (0). Sequences of marks and spaces comprise the various characters.

Martinez devised a new variable-length code for PSK31 that combines the best of Morse and RTTY. He calls it *Varicode* because a varying number of bits are used for each character (see **Fig 9.13**). Much like Morse code, the more commonly used letters in PSK31 have shorter codes.

As with RTTY, there is a need to signal the gaps between characters. The Varicode does this by using "00" to represent a gap. The Varicode is structured so that two zeros never appear together in any of the combinations of 1s and 0s that make up the characters. In on-the-air tests, Martinez has verified that the unique "00" sequence works significantly better than RTTY's stop code for keeping the receiver synchronized.

With Varicode, a typing speed of about 50 words per minute requires a 32 bit/s transmission rate. Martinez chose 31.25 bit/s because it can be easily derived from the 8-kHz sample rate used in many DSP systems.

The shifting carrier phase generates sidebands 31.25 Hz from the carrier. These are used to synchronize the receiver with the transmitter. The bandwidth of a PSK31 signal is shown in **Fig 9.14**.

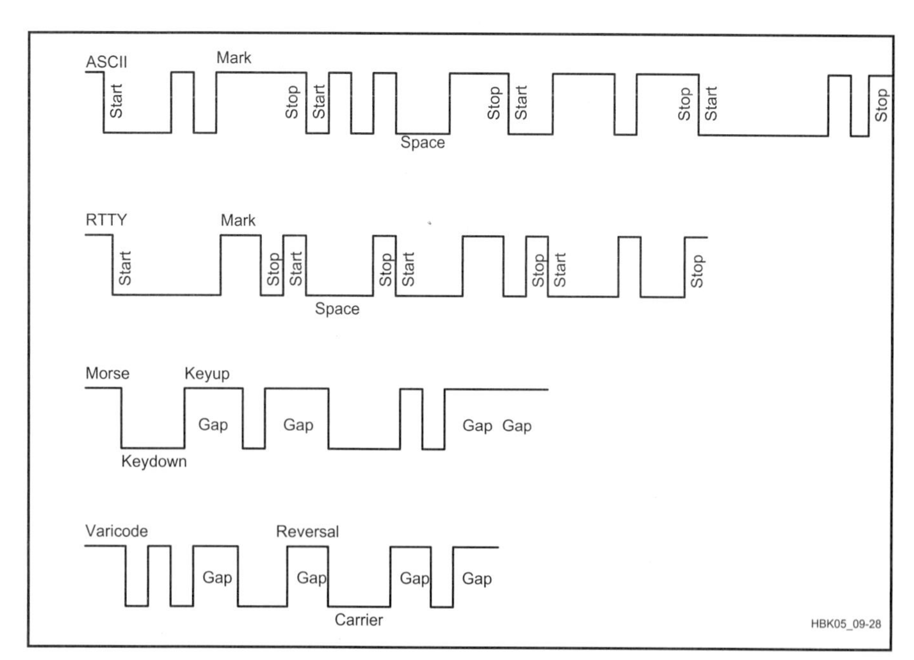

Fig 9.13—Codes for the word "ten" in ASCII, Baudot, Morse and Varicode.

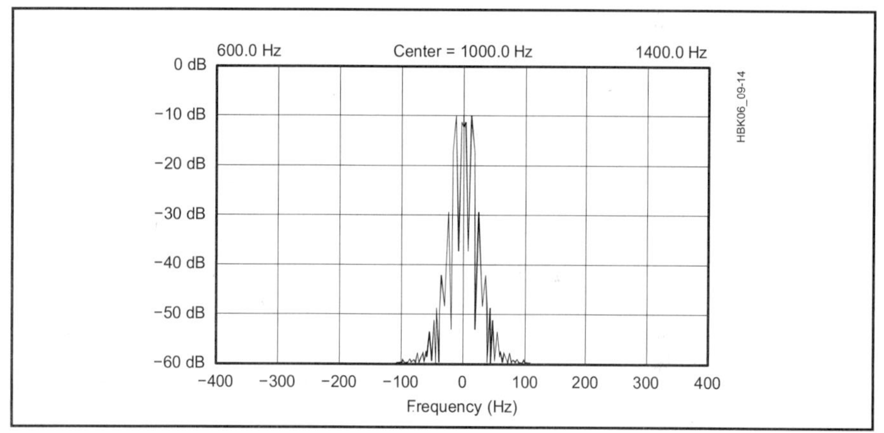

Fig 9.14—The spectrum of a PSK31 signal.

PSK31 Error Correction

Martinez added error correction to PSK31 by using QPSK (quaternary phase shift keying) and a *convolutional encoder* to generate one of four different phase shifts that correspond to patterns of five successive data bits. At the receiving end, a Viterbi decoder is used to correct errors. There are 32 possible sequences for five bits. The Viterbi decoder tracks these possibilities while discarding the least likely and retaining the most likely sequences. Retained sequences are given a score that is based on the running total. The most accurate sequence is reported, and thus errors are corrected.

Operating PSK31 in the QPSK mode should result in 100% copy under most conditions, but at a price. Tuning is twice as critical as it is with BPSK. An accuracy of less than 4 Hz is required for the Viterbi decoder to function properly.

Getting Started

In addition to a transceiver and antenna, you only need a computer with a Windows operating system and a 16-bit sound card to receive and transmit PSK31. Additional information and software is available for free download over the Web. Use a search engine to find PSK31 information and links to downloads.

An interesting wrinkle is to generate text, transmit it via PSK31 or some other RTTY or data mode, receive it and use a speech synthesizer to read the message. An example of this technique was described by W3NRG in the October 2004 issue of *CQ Magazine* (p 48). Synthesized speech takes some getting used to, as everybody sounds pretty much alike, and the personality of the speaker does not come through.

PSK31 BIBLIOGRAPHY

Ford, Steve, WB8IMY, ARRL's *HF Digital Handbook,* Third Ed., ARRL, 2004.

DATA MODES

The difference between text and data modes is not abrupt but a blur. *Data* could be used to mean text, numbers, telecommand, telemetry and in some cases images. The third letter of the emission symbol "D" is used in common for data, telecommand and telemetry.

Packet Radio

Data communications is telecommunications between computers. *Packet switching* is a form of data communications that transfers data by subdividing it into "packets," and *packet radio* is packet switching using the medium of radio. This description was written by Steve Ford, WB8IMY.

Packet radio has its roots in the Hawaiian Islands, where the University of Hawaii began using the mode in 1970 to transfer data to its remote sites dispersed throughout the islands. Amateur packet radio began in Canada after the Canadian Department of Communications permitted amateurs to use the mode in 1978. (The FCC permitted amateur packet radio in the US in 1980.)

In the first half of the 1980s, packet radio was the habitat of a small group of experimenters who did not mind communicating with a limited number of potential fellow packet communicators. In the second half of the decade, packet radio "took off" as the experimenters built a network that increased the potential number of packet stations that could intercommunicate and thus attracted tens of thousands of communicators who wanted to take advantage of this potential.

Packet radio provides error-free data transfer. The receiving station receives information exactly as the transmitting station sends it, so you do not waste time deciphering communication errors caused by interference or changes in propagation.

Packet uses time efficiently, since packet bulletin-board systems (PBBSs) permit packet operators to store information for later retrieval by other amateurs. And it uses the radio spectrum efficiently, since one radio channel may be used for multiple communications simultaneously, or one radio channel may be used to interconnect a number of packet stations to form a "cluster" that provides for the distribution of information to all of the clustered stations. The popular *DX PacketClusters* are typical examples (see **Fig 9.15**).

Each local channel may be connected to other local channels to form a network that affords interstate and international data communications. This network can be used by interlinked packet bulletin-board systems to transfer information, messages and third-party traffic via HF, VHF, UHF and satellite links. Primary node-to-node links are also active on the Internet.

It uses other stations efficiently, since any packet-radio station can use one or more other packet-radio stations to relay data to its intended destination. It uses current station transmitting and receiving equipment efficiently, since the same equipment used for voice communications may be used for packet communications. The outlay for the additional equipment necessary to make your voice station a packet-radio station may be as little as $100. It also allows you to use that same equipment as an alternative to costly landline data communications links for transferring data between computers.

The TNC

The terminal node controller—or *TNC*—is at the heart of every packet station. A TNC is actually a computer unto itself. It contains the AX.25 packet protocol firmware, along with other enhancements depending on the manufacturer. The TNC communicates with you through your computer or data terminal. It also allows you to communicate with other hams by feeding packet data to your transceiver.

The TNC accepts data from a computer or data terminal and assembles it into packets (see **Fig 9.16**). In addition, it translates the digital packet data into audio tones that can be fed to a transceiver. The TNC also functions as a receiving device, translating the audio tones into digital data a com-

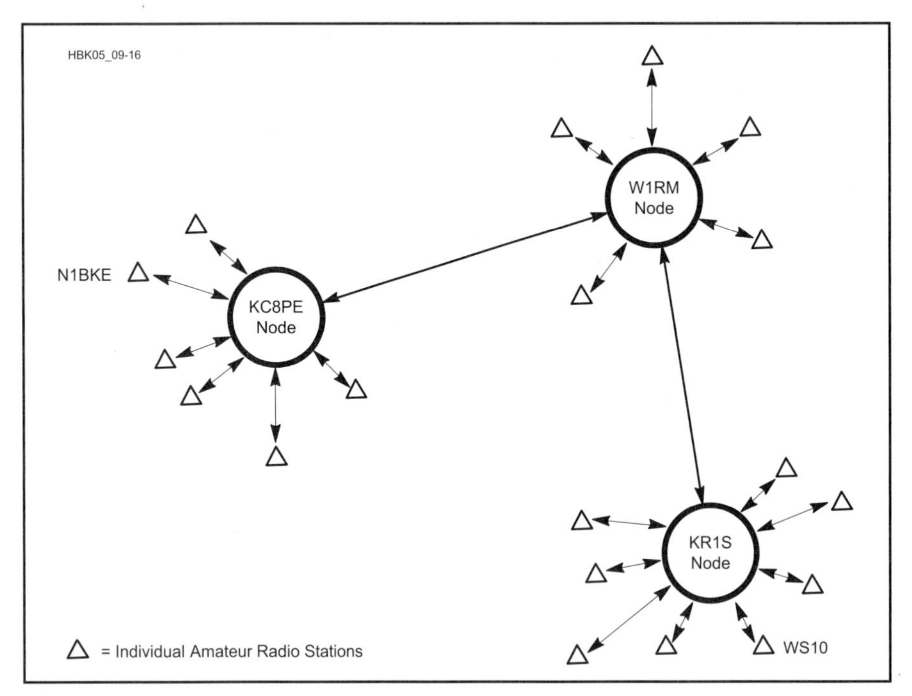

Fig 9.15—*DX PacketClusters* are networks comprised of individual nodes and stations with an interest in DXing and contesting. In this example, N1BKE is connected to the KC8PE node. If he finds a DX station on the air, he'll post a notice—otherwise known as a *spot*—which the KC8PE node distributes to all its local stations. In addition, KC8PE passes the information along to the W1RM node. W1RM distributes the information and then passes it to the KR1S node, which does the same. Eventually, WS1O—who is connected to the KR1S node—sees the spot on his screen. Depending on the size of the network, WS1O will receive the information within minutes after it was posted by N1BKE.

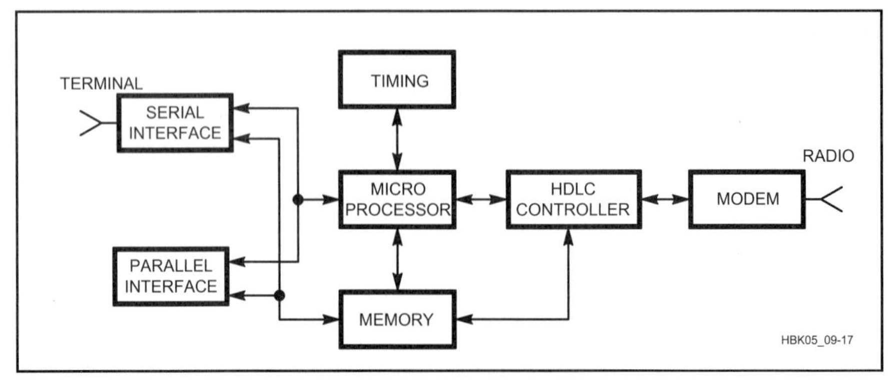

Fig 9.16—The functional block diagram of a typical TNC.

puter or terminal can understand. The part of the TNC that performs this tone-translating function is known as a *modem* (see **Fig 9.17**).

If you're saying to yourself, "These TNCs sound a lot like telephone modems," you're pretty close to the truth! The first TNCs were based on telephone modem designs. If you're familiar with so-called *smart* modems, you'd find that TNCs are very similar.

You have plenty of TNCs to choose from. The amount of money you'll spend depends directly on what you want to accomplish. Most TNCs are designed to operate at 300 and 1200 bit/s, or 1200 bit/s exclusively (see **Fig 9.18**). There are also TNCs dedicated to 1200 and 9600 bit/s operation, or 9600 bit/s exclusively. Many of these TNCs include convenient features such as personal packet mailboxes, where friends can leave messages when you're not at home. Some TNCs also include the ability to easily disconnect the existing modem and substitute another. This feature is very important if you wish to experiment at different data rates. For example, a 1200 bit/s TNC with a *modem disconnect header* can be converted to a 9600 bit/s TNC by disconnecting the 1200 bit/s modem and adding a 9600 bit/s modem.

If you're willing to spend more money, you can buy a complete *multimode communications processor*, or *MCP*. These devices not only offer packet, they also provide the capability to operate RTTY, CW, AMTOR, PACTOR, FAX and other modes. In other words, an MCP gives you just about every digital mode in one box.

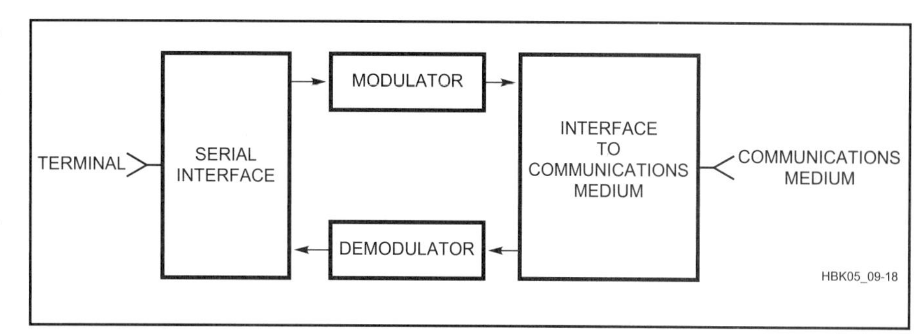

Fig 9.17—A block diagram of a typical modem.

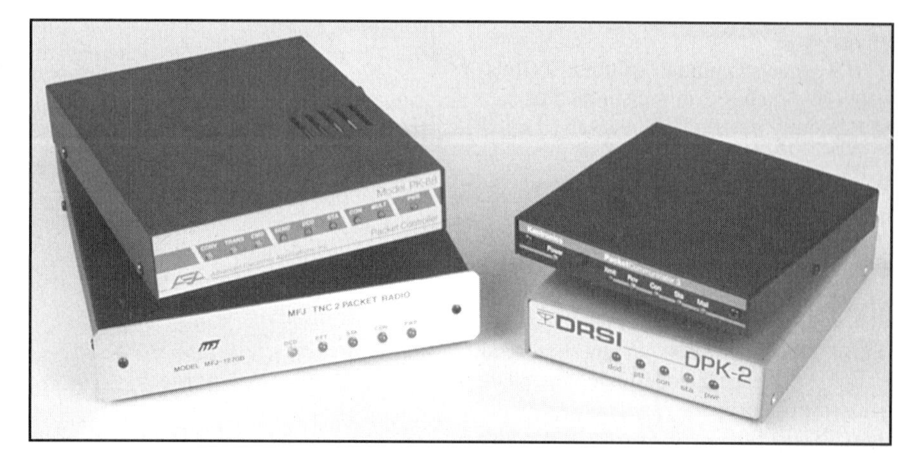

Fig 9.18—Four popular 1200 bit/s packet TNCs: (clockwise, from bottom left) the MFJ-1270C, AEA PK-88, Kantronics KPC-3 and the DRSI DPK-2.

TNC Emulation and Internal TNCs

TNC-emulation systems exist for IBM PCs and compatibles. One is known as *BayCom,* which uses the PC to emulate the functions of a TNC/terminal while a small external modem handles the interfacing. BayCom packages are available in kit form for roughly half the price of a basic TNC.

PC owners also have the option of buying full-featured TNCs that mount *inside* their computers. TNC *cards* are available on the market. They are complete TNCs that plug into card slots inside the computer cabinet. No TNC-to-computer cables are necessary. Connectors are provided for cables that attach to your transceiver. In many cases, specialized software is also provided for efficient operation.

Transceiver Requirements

Packet activity on the HF bands typically takes place at 300 bit/s using common SSB transceivers. The transmit audio is fed from the TNC to the microphone jack or auxiliary audio input. Receive audio is obtained from the radio's external speaker jack or auxiliary audio output. Tuning is critical for proper reception; a visual tuning indicator—available on some TNCs and all MCPs—is recommended.

These simple connections also work for 1200 bit/s packet, which is common on the VHF bands (2 m in particular). Almost any FM transceiver can be made to work with 1200 bit/s packet by connecting the transmit audio to the microphone jack and taking the receive audio from the external speaker (or earphone) jack.

At data rates beyond 1200 bit/s, transceiver requirements become more rigid. At 9600 bit/s (the most popular data rate above 1200 bit/s), the transmit audio must be injected at the modulator stage of the FM transceiver. Receive audio must be tapped at the discriminator. Most 9600 bit/s operators use modified Amateur Radio transceivers or commercial radios.

In the mid 1990s amateur transceiver manufacturers began incorporating data ports on some FM voice rigs. The new "data-ready" radios are not without problems, however. Their IF filter and discriminator characteristics leave little room for error. If you're off frequency by a small amount, you may not be able to pass data. In addition, the ceramic discriminator coils used in some transceivers have poor group delay, making it impossible to tune them for wider bandwidths. With this in mind, some amateurs prefer to make the leap to 9600 bit/s and beyond using dedicated amateur data radios.

Regardless of the transceiver used, setting the proper deviation level is extremely critical. At 9600 bit/s, for example, optimum performance occurs when the maximum deviation is maintained at 3 kHz. Deviation adjustments involve monitoring the transmitted signal with a deviation meter or service monitor. The output level of the TNC is adjusted until the proper deviation is achieved.

PACKET NETWORKING

Digipeaters

A *digipeater* is a packet-radio station capable of recognizing and selectively repeating packet frames. An equivalent term used in industry is *bridge*. Virtually any TNC can be used as a single-port digipeater, because the digipeater function is included in the AX.25 Level 2 protocol firmware. Although the use of digipeaters is waning today as network nodes take their place, the digipeater function is handy when you need a relay and no node is available, or for on-the-air testing.

TCP/IP

If you're an active packeteer, sooner or later someone will bring up the subject of TCP/IP—Transmission Control Protocol/ Internet Protocol. Despite its name, TCP/IP is more than two protocols; it's actually a set of several protocols. Together they provide a high level of flexible, "intelligent" packet networking. TCP/IP enthusiasts see a future when the entire nation, and perhaps the world, will be linked by high-speed TCP/IP systems using terrestrial microwave and satellites.

TCP/IP has a unique solution for busy networks. Rather than transmitting packets at randomly determined intervals, TCP/IP stations automatically *adapt* to network delays as they occur. As network throughput slows down, active TCP/IP stations sense the change and lengthen their transmission delays accordingly. As the network speeds up, the TCP/IP stations shorten their delays to match the pace. This kind of intelligent network sharing virtually guarantees that all packets will reach their destinations with the greatest efficiency the network can provide.

With TCP/IP's adaptive networking scheme, you can chat using the *telnet* protocol with a ham in a distant city and rest assured that you're not overburdening the system. Your packets simply join the constantly moving "freeway" of data. They might slow down in heavy traffic, but they *will* reach their destination eventually. (This adaptive system is used for all TCP/IP packets, no matter what they contain.)

TCP/IP excels when it comes to transferring files from one station to another. By using the TCP/IP *file transfer protocol* (ftp), you can connect to another station and transfer computer files—including software. As you can probably guess, transferring large files can take time. With TCP/IP, however, you can still send and receive mail (using the *SMTP* protocol) or talk to another ham while the transfer is taking place.

When you attempt to contact another station using TCP/IP, all network routing is performed automatically according to the TCP/IP address of the station you're trying to reach. In fact, TCP/IP networks are transparent to the average user.

To operate TCP/IP, all you need is a computer (it must be a computer, not a terminal), a 2-m FM transceiver and a TNC with *KISS* capability. As you might guess, the heart of your TCP/IP setup is software. The TCP/IP software set was written by Phil Karn, KA9Q, and is called *NOSNET* or just *NOS* for short.

There are dozens of *NOS* derivatives available today. All are based on the original *NOSNET*. The programs are available primarily for IBM-PC compatibles and Macintoshes. You can obtain *NOS* software from on-line sources such as the CompuServe *HAMNET* forum libraries, Internet ftp sites, Amateur Radio-oriented BBSs and elsewhere. *NOS* takes care of all TCP/IP functions, using your "KISSable" TNC to communicate with the outside world. The only other item you need is your own IP address. Individual IP Address Coordinators assign addresses to new TCP/IP users.

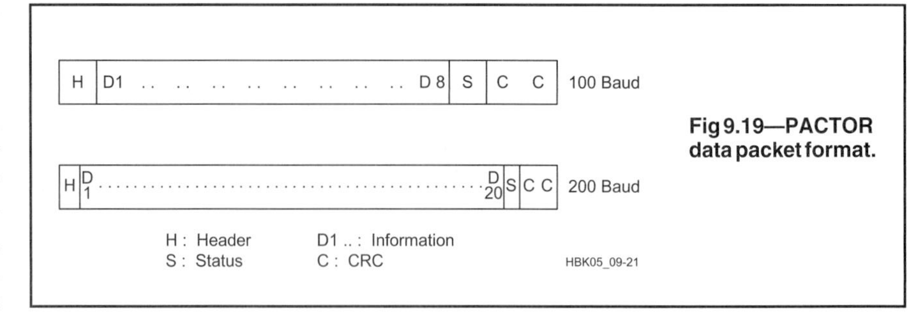

Fig 9.19—PACTOR data packet format.

PACKET BIBLIOGRAPHY

ARRL/TAPR, *proc.* Digital Communications Conferences, ARRL, annually 1983-present.

Ball, Bob, WB8WGA, "An Inexpensive Terminal Node Controller for Packet Radio," QEX, Mar/Apr 2005.

Fox, Terry, WB4JFI, *AX.25 Packet-Radio Link-Layer Protocol*, ARRL, 1984 (maintained by Tucson Amateur Packet Radio–TAPR)

Horzepa, Stan, WA1LOU, *Your Gateway to Packet Radio*, ARRL, 1989.

Roznoy, Rich, K1OF, Packet: Speed, More Speed and Applications, ARRL, 1997.

PACTOR

PACTOR (PT), now often referred to as PACTOR-I, is an HF radio transmission system developed by German amateurs Hans-Peter Helfert, DL6MAA, and Ulrich Strate, DF4KV. It was designed to overcome the shortcomings of AMTOR and packet radio. It performs well under both weak-signal and high-noise conditions. PACTOR-I has been overtaken by PACTOR-II and PACTOR-III but remains in use.

TRANSMISSION FORMATS

Information Blocks

All packets have the basic structure shown in **Fig 9.19**, and their timing is as shown in **Table 9.1**:

- *Header*: Contains a fixed bit pattern to simplify repeat requests, synchronization and monitoring. The header is also important for the Memory ARQ function. In each packet carrying new information, the bit pattern is inverted.
- *Data*: Any binary information. The format is specified in the status word. Current choices are 8-bit ASCII or 7-bit ASCII (with Huffman encoding). Characters are not broken across packets. ASCII RS (hex 1E) is used as an IDLE character in both formats.
- *Status word*: See **Table 9.2**
- *CRC*: The CRC is calculated according to the CCITT standard, for the data, status and CRC.

Table 9.1

PACTOR Timing

Object	*Length (seconds)*
Packet	0.96 (200 bd: 192 bits; 100 bd: 96 bits)
CS receive time	0.29
Control signals	0.12 (12 bits at 10 ms each)
Propagation delay	0.17
Cycle	1.25

Table 9.2

PACTOR Status Word

Bit	*Meaning*
0	Packet count (LSB)
1	Packet count (MSB)
2	Data format (LSB)
3	Data format (MSB)
4	Not defined
5	Not defined
6	Break-in request
7	QRT request

Data Format Bits

Format	*bit 3*	*bit 2*
ASCII 8 bit	0	0
Huffman code	0	1
Not defined	1	0
Not defined	1	1

Bits 0 and 1 are used as a packet count; successive packets with the same value are identified by the receiver as repeat packets. A modulus-4 count helps with unrecognized control signals, which are unlikely in practice.

Acknowledgment Signals

The PACTOR acknowledgment signals are shown in **Table 9.3**. Each of the signals is 12 bits long. The characters differ in pairs in 8 bits (Hamming offset) so that the chance of confusion is reduced. If the CS is not correctly received, the TX reacts by repeating the last packet. The request status can be uniquely recognized by the 2-bit packet number so that wasteful transmissions of pure RQ blocks are unnecessary.

Timing

The receiver pause between two blocks is 0.29 s. After deducting the CS lengths, 0.17 s remain for switching and propagation delays so that there is adequate reserve for DX operation.

CONTACT FLOW

Listening

In the listen mode, the receiver scans any received packets for a CRC match. This method uses a lot of computer processing resources, but it's flexible.

CQ

A station seeking contacts transmits CQ packets in an FEC mode, without pauses for acknowledgment between packets. The transmit time length, number of repetitions and speed are the transmit operator's choice. (This mode is also suitable for bulletins and other group traffic.) Once a listening station has copied the call, the listener assumes the TX station role and initiates a contact. Thus, the station sending CQ initially takes the RX station role. The contact begins as shown in **Table 9.4**

Speed Changes

With good conditions, PACTOR's normal signaling rate is 200 bauds, but the system automatically changes from 200 to 100 bauds and back, as conditions demand. In addition, Huffman coding can further increase the throughput by a factor of 1.7. There is no loss of synchronization speed changes; only one packet is repeated.

When the RX receives a bad 200-baud packet, it can acknowledge with CS4. TX immediately assembles the previous packet in 100-baud format and sends it. Thus, one packet is repeated in a change from 200 to 100 bauds.

The RX can acknowledge a good 100-baud packet with CS4. TX immediately switches to 200 bauds and sends the next packet. There is no packet repeat in an upward speed change.

Change of Direction

The RX station can become the TX station by sending a special change-over packet in response to a valid packet. RX sends CS3 as the first section of the changeover packet. This immediately changes the TX station to RX mode to read the data in that packet and responds with CS1 and CS3 (acknowledge) or CS2 (reject).

End of Contact

PACTOR provides a sure end-of-contact procedure. TX initiates the end of contact by sending a special packet with the QRT bit set in the status word and the call of the RX station in byte-reverse order at 100 bauds. The RX station responds with a final CS.

PACTOR-II

This is a significant improvement over PACTOR-I, yet it is fully compatible with the older mode. Also invented in Germany, PACTOR uses 16PSK to transfer up to 800 bit/s at a 100-baud rate. This keeps the bandwidth less than 500 Hz.

PACTOR-II uses digital signal processing (DSP) with Nyquist waveforms, Huffman *and* Markov compression, and powerful Viterbi decoding to increase transfer rate and sensitivity into the noise level. The effective transfer rate of text is over 1200 bit/s. Features of PACTOR II include:

- Frequency agility—It can automatically adjust or lock two signals together over a ±100-Hz window.
- Powerful data reconstruction based upon computer power—With over 2 MB of available memory.
- Cross correlation—Applies analog Memory ARQ to acknowledgment frames and headers.
- Soft decision making—Uses artificial intelligence (AI), as well as digital information received to determine frame validity.
- Extended data block length—When transferring large files under good conditions, the data length is doubled to increase the transfer rate.
- Automatic recognition of PACTOR-I, PACTOR-II and so on, with automatic mode switching.
- Intermodulation products are canceled by the coding system.
- Two long-path modes extend frame timing for long-path terrestrial and satellite propagation paths.

This is a fast, robust mode, possibly the most powerful in the ham bands. It has excellent coding gain as well. It can also communicate with all earlier PACTOR-I systems. PACTOR-II stations acknowledge each received transmission block. PACTOR-II employs computer logic as well as received data to reassemble defective data blocks into good frames. This reduces the number of transmissions and increases the throughput of the data.

Table 9.3
PACTOR Control Signals

Code	*Chars* (hex)	*Function*
CS1	4D5	Normal acknowledge
CS2	AB2	Normal acknowledge
CS3	34B	Break-in (forms header of first packet from RX to TX)
CS4	D2C	Speed change request

All control signals are sent only from RX to TX.

PACTOR-III

PACTOR-III is a software upgrade for existing PACTOR-II modems that provides a data transmission mode for improved speed and robustness. PACTOR-III is not a new modem or hardware device. Most current PACTOR-II modems are upgradeable to use PACTOR-III via a software update, since PACTOR-II firmware accommodates the new PACTOR-III software. Both the transmitting and receiving stations must support PACTOR-III for end-to-end communications using this mode.

PACTOR-III's maximum uncompressed speed is 2722 bit/s. Using online compression, up to 5.2 kbit/s is achievable. This requires an audio passband from 400 Hz to 2600 Hz (for PACTOR-III speed level 6).

On an average channel, PACTOR-III is more than three times faster than PACTOR-II. On good channels, the effective throughput ratio between PACTOR-III and PACTOR-II can exceed five. PACTOR-III is also slightly more robust than PACTOR-II at their lower SNR edges.

The ITU emission designator for PACTOR-III is 2K20J2D. Because PACTOR-III builds on PACTOR-II, most specifications like frame length and frame structure are adopted from PACTOR-II. The only significant difference is PACTOR III's multi-tone waveform that uses up to 18 carriers while PACTOR-II uses only two carriers. PACTOR-III's carriers are located in a 120-Hz grid and modulated with 100 symbols per second DBPSK or DQPSK. Channel coding is also adopted from PACTOR-II's Punctured Convolutional Coding.

PACTOR-III Link Establishment

The calling modem uses the PACTOR-I FSK connect frame for compatibility. When the called modem answers, the modems negotiate to the highest level of which both modems are capable. If one modem is only capable of PACTOR-II, then the 500 Hz PACTOR-II mode is used for the session. With the *MYLevel* (MYL) command a user may limit a modem's highest mode. For example, a user may set MYL to "1" and

Table 9.4

PACTOR Initial Contact

Master Initiating Contact

Size (bytes)	*1*	*8*	*6*
Content	/Header	/SLAVECAL	/SLAVECAL/
Speed (bauds)	100	100	200

Slave Response

The receiving station detects a call, determines mark/space polarity, decodes 100-bd and 200-bd call signs. It uses the two call signs to determine if it is being called and the quality of the communication path. The possible responses are:

First call sign does not match slave's (Master not calling this slave)	none
Only first call sign matches slave's (Master calling this slave, poor communications)	CS1
First and second call signs both match the slaves (good circuit, request speed change to 200 bd)	CS4

Table 9.5

CLOVER-II Modulation Modes

As presently implemented, CLOVER-II supports a total of 7 different modulation formats: 5 using PSM and 2 using a combination of PSM and ASM (Amplitude Shift Modulation).

Name	*Description*	*In-Block Data Rate*
16P4A	16 PSM, 4-ASM	750 bps
16PSM	16 PSM	500 bps
8P2A	8 PSM, 2-ASM	500 bps
8PSM	8 PSM	375 bps
QPSM	4 PSM	250 bps
BPSM	Binary PSM	125 bps
2DPSM	2-Channel Diversity BPSM	62.5 bps

only a PACTOR-I connection will be made, set to "2" and PACTOR-I and II connections are available, set to "3" and PACTOR-I through III connections are enabled. The default MYL is set to "2" with the current firmware and with PACTOR-III firmware it will be set to "3". If a user is only allowed to occupy a 500 Hz channel, MYL can be set to "2" and the modem will stay in its PACTOR-II mode. The PACTOR-III Protocol Specification is available on the Web at **www.scs-ptc.com/pactor.html**.

PACTOR Bibliography

ARRL Web, Technical Descriptions, **www.arrl.org/FandES/field/regulations/techchar/**.

Ford, Steve, WB8IMY, ARRL's *HF Digital Handbook,* Third Ed., ARRL. 2004.

G-TOR

This brief description has been adapted from "A Hybrid ARQ Protocol for Narrow Bandwidth HF Data Communication" by Glenn Prescott, WBØSKX, Phil Anderson, WØXI, Mike Huslig, KBØNYK, and Karl Medcalf, WK5M (May 1994 *QEX*).

G-TOR is short for *Golay-TOR*, an innovation of Kantronics, Inc. It was inspired by HF Automatic Link Establishment (ALE) concepts and is structured to be compatible with ALE.

The purpose of the G-TOR protocol is to provide an improved digital radio communication capability for the HF bands. The key features of G-TOR are:

Standard FSK tone pairs (mark and space)

- Link-quality-based signaling rate: 300, 200 or 100 bauds
- 2.4-second transmission cycle
- Low overhead within data frames
- Huffman data compression—two types, on demand
- Embedded run-length data compression
- Golay forward-error-correction coding
- Full-frame data interleaving
- CRC error detection with hybrid ARQ
- Error-tolerant "Fuzzy" acknowledgments.

The G-TOR Protocol

Since one of the objectives of this protocol is ease of implementation in existing TNCs, the modulation format consists of standard tone pairs (FSK), operating at 300, 200 or 100 bauds, depending upon channel conditions. (G-TOR initiates contacts and sends ACKs only at 100 bauds.) The G-TOR waveform consists of two phase-continuous tones (BFSK), spaced 200 Hz apart (mark = 1600 Hz, space = 1800 Hz); however, the system can still operate at the familiar 170-Hz shift (mark = 2125 Hz, space = 2295 Hz), or with any other convenient tone pairs. The optimum spacing for 300-baud transmission is 300 Hz, but you trade some performance for a narrower bandwidth.

Each transmission consists of a synchronous ARQ 1.92-s frame and a 0.48-s interval for propagation and ACK transmissions (2.4 s cycle). All advanced protocol features are implemented in the signal-processing software.

Data Compression

Data compression is used to remove redundancy from source data. Therefore, fewer bits are needed to convey any given message. This increases data throughput and decreases transmission time—valuable features for HF. G-TOR uses run-length coding and two types of Huffman coding during normal text transmissions. Run-length coding is used when more than two repetitions of an 8-bit character are sent. It provides an especially large savings in total transmission time when repeated characters are being transferred.

The Huffman code works best when the statistics of the data are known. G-TOR applies Huffman A coding with the upper- and lower-case character set, and Huffman B coding with upper-case-only text. Either type of Huffman code reduces the average number of bits sent per character. In some situations, however, there is no benefit from Huffman coding. The encoding process is then disabled. This decision is made on a frame-by-frame basis by the information-sending station.

Golay Coding

The real power of G-TOR resides in the properties of the (24,12) extended Golay error-correcting code, which permits correction of up to three random errors in three received bytes. The (24,12) extended Golay code is a half-rate error-correcting code: Each 12 data bits are translated into an additional 12 parity bits (24 bits total). Further, the code can be implemented to produce separate input-data and parity-bit frames.

The extended Golay code is used for G-TOR because the encoder and decoder are simple to implement in software. Also, Golay code has mathematical properties that make it an ideal choice for short-cycle synchronous communication.

G-TOR Bibliography

ARRL Web, Technical Descriptions, **www.arrl.org/FandES/field/regula-**

tions/techchar/.

Ford, Steve, WB8IMY, ARRL's *HF Digital Handbook,* Third Ed., ARRL. 2004.

CLOVER-II

The desire to send data via HF radio at high data rates and the problems encountered when using AX.25 packet radio on HF radio led Ray Petit, W7GHM, to develop a unique modulation waveform and data transfer protocol that is now called "CLOVER-II." Bill Henry, K9GWT, supplied this description of the Clover-II system. CLOVER modulation is characterized by the following key parameters:

- Very low base symbol rate: 31.25 symbols/second (all modes).
- Time-sequence of amplitude-shaped pulses in a very narrow frequency spectrum. Occupied bandwidth = 500 Hz at 50 dB below peak output level.
- Differential modulation between pulses.
- Multilevel modulation.

The low base symbol rate is very resistant to multipath distortion because the time between modulation transitions is much longer than even the worst-case time-smearing caused by summing of multipath signals. By using a time-sequence of tone pulses, Dolph-Chebychev "windowing" of the modulating signal and differential modulation, the total occupied bandwidth of a CLOVER-II signal is held to 500 Hz.

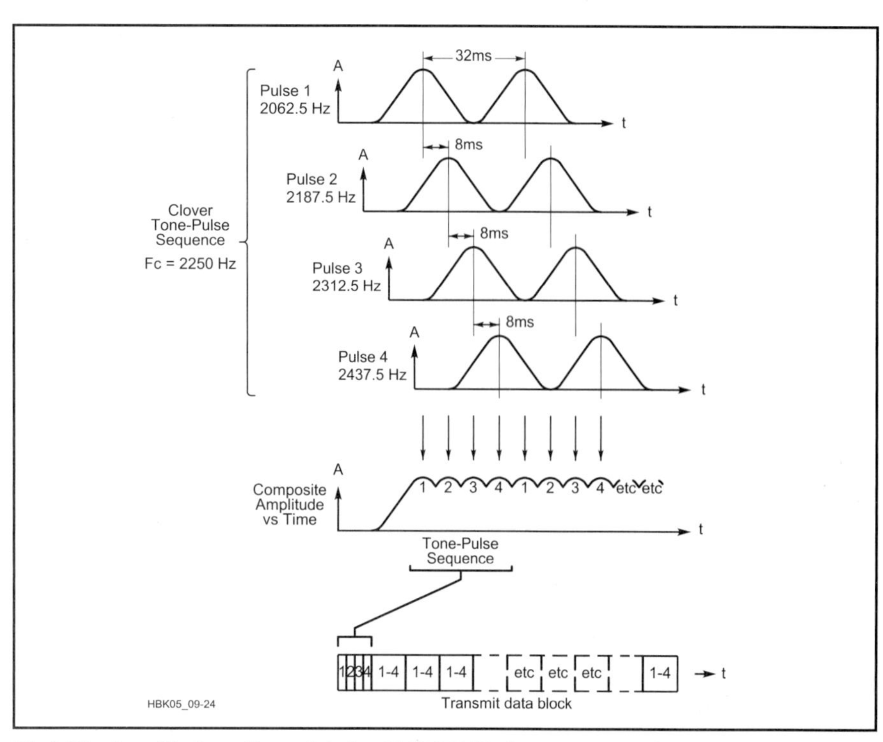

Fig 9.20—Amplitude vs time plots for CLOVER-II's four-tone waveform.

The CLOVER Waveform

Multilevel tone, phase and amplitude modulation give CLOVER a large selection of data modes that may be used (see **Table 9.5**). The adaptive ARQ mode of CLOVER senses current ionospheric conditions and automatically adjusts the modulation mode to produce maximum data throughput. When using the "Fast" bias setting, ARQ throughput automatically varies from 11.6 byte/s to 70 byte/s.

The CLOVER-II waveform uses four tone pulses that are spaced in frequency by 125 Hz. The time and frequency domain characteristics of CLOVER modulation are shown in **Figs 9.20**, **9.21** and **9.22**. The time-domain shape of each tone pulse is intentionally shaped to produce a very compact frequency spectrum. The four tone pulses are spaced in time and then combined to produce the composite output shown. Unlike other modulation schemes, the CLOVER modulation spectra is the same for all modulation modes.

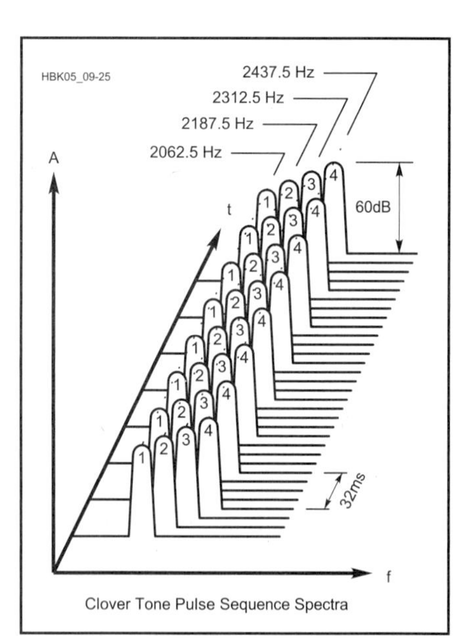

Fig 9.21—A frequency-domain plot of a CLOVER-II waveform.

Modulation

Data is modulated on a CLOVER-II signal by varying the phase and/or amplitude of the tone pulses. Further, all data modulation is differential on the same tone pulse—data is represented by the phase (or amplitude) difference from one pulse to the next. For example, when binary phase modulation is used, a data change from "0" to "1" may be represented by a change in the phase of tone pulse 1 by 180° between the first and second occurrence of that pulse. Further, the phase state is changed only while the pulse amplitude is zero. Therefore, the wide frequency spectra normally associated with PSK of a continuous carrier is avoided. This is true for all CLOVER-II modulation formats. The term "phase-shift modulation" (PSM) is used when describing CLOVER modes to emphasize this distinction.

Coder Efficiency Choices

CLOVER-II has four "coder efficiency" options: 60%, 75%, 90% and 100% ("efficiency" being the approximate ratio of real data bytes to total bytes sent). "60% efficiency" corrects the most errors but has the lowest net data throughput. "100% efficiency" turns the encoder off and has the highest throughput but fixes no errors. There is therefore a tradeoff between raw data throughput versus the number of errors that can be corrected without resorting to retransmission of the entire data block.

Note that while the "In Block Data Rate" numbers listed in the table go as high as 750 bit/s, overhead reduces the net throughput or overall efficiency of a CLOVER transmission. The FEC coder efficiency setting and protocol requirements of FEC and ARQ modes add overhead and reduce the net efficiency. **Table 9.6** and **Table 9.7** detail the relationships between block size, coder efficiency, data bytes per block and correctable byte errors per block.

CLOVER FEC

All modes of CLOVER-II use Reed-Solomon forward error correction (FEC) data encoding, which allows the receiving station to correct errors without requiring a

Table 9.6

Data Bytes Transmitted Per Block

Block Size	***Reed-Solomon Encoder Efficiency*** *60%*	*75%*	*90%*	*100%*
17	8	10	12	14
51	28	36	42	48
85	48	60	74	82
255	150	188	226	252

Table 9.7

Correctable Byte Errors Per Block

Block Size	***Reed-Solomon Encoder Efficiency*** *60%*	*75%*	*90%*	*100%*
17	1	1	0	0
51	9	5	2	0
85	16	10	3	0
255	50	31	12	0

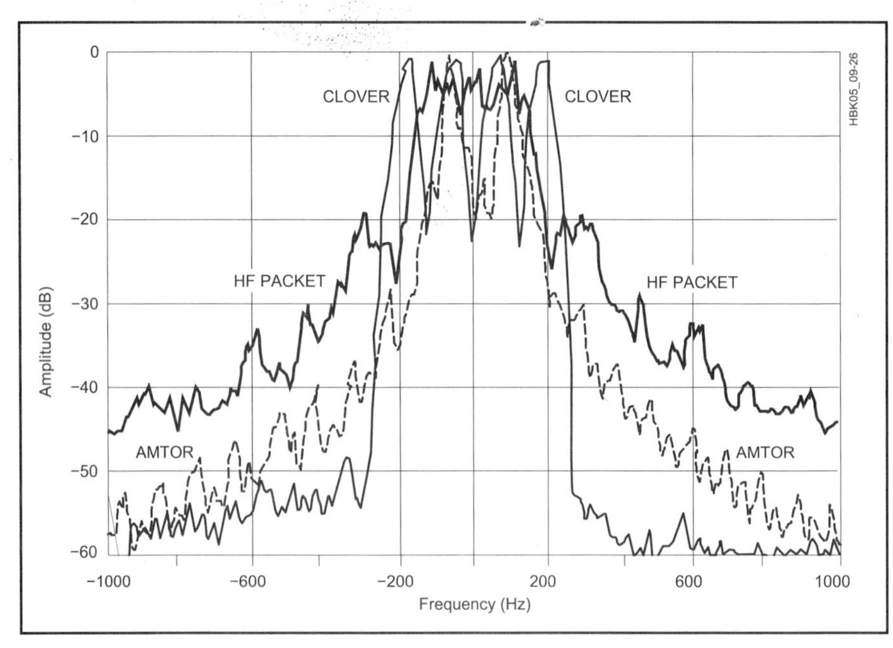

Fig 9.22—Spectra plots of AMTOR, HF packet-radio and CLOVER-II signals.

repeat transmission. This is a very powerful error-correction technique that is not available in some other common HF data modes.

CLOVER ARQ

Reed-Solomon data coding is the primary means by which errors are corrected in CLOVER "FEC" mode. In ARQ mode, CLOVER-II employs a three-step strategy to combat errors. First, channel parameters are measured and the modulation format is adjusted to minimize errors and maximize data throughput. This is called the "Adaptive ARQ Mode" of CLOVER-II. Second, Reed-Solomon encoding is used to correct a limited number of byte errors per transmitted block. Finally, only those data blocks in which errors exceed the capacity of the Reed-Solomon decoder are repeated (selective block repeat).

With seven different modulation formats, four data-block lengths (17, 51, 85 or 255 bytes) and four Reed-Solomon coder efficiencies (60%, 75%, 90% and 100%), there are 112 (7 × 4 × 4) different waveform modes that could be used to send data via CLOVER. Once all of the determining factors are considered, however, there are eight different waveform combinations that are actually used for FEC and/or ARQ modes.

CLOVER-2000

CLOVER-2000 is a faster version of CLOVER (about four times faster) that uses eight tone pulses, each of which is 250 Hz wide, spaced at 250-Hz centers, contained within the 2-kHz bandwidth between 500 and 2500 Hz. The eight tone pulses are sequential, with only one tone being present at any instant and each tone lasting 2 ms. Each frame consists of eight tone pulses lasting a total of 16 ms, so the base modulation rate of a CLOVER-2000 signal is always 62.5 symbols per second (regardless of the type of modulation being used). CLOVER-2000's maximum raw data rate is 3000 bit/s.

Allowing for overhead, CLOVER-2000 can deliver error-corrected data over a standard HF SSB radio channel at up to 1994 bit/s, or 249 characters (8-bit bytes) per second. These are the uncompressed data rates; the maximum throughput is typically doubled for plain text if compression is used. The effective data throughput rate of CLOVER-2000 can be even higher when binary file transfer mode is used with data compression.

The binary file transfer protocol used by HAL Communications operates with a terminal program explained in the HAL E2004 engineering document listed under references. Data compression algorithms tend to be context sensitive—compression that works well for one mode (eg, text), may not work well for other data forms (graphics, etc). The HAL terminal program uses the PK-WARE compression algorithm, which has proved to be a good general-purpose compressor for most computer files and programs. Other algorithms may be more efficient for some data formats, particularly for compression of graphic image files and digitized voice data. The HAL Communications CLOVER-2000 modems can be operated with other data compression algorithms in the users' computers.

CLOVER-2000 is similar to the previous version of CLOVER, including the transmission protocols and Reed-Solomon error detection and correction algorithm. The original descriptions of the CLOVER Control Block (CCB) and Error Correction Block (ECB) still apply for CLOVER-2000, except for the higher data rates inherent to CLOVER-2000. Just like CLOVER, all data sent via CLOVER-2000 is encoded as 8-bit data bytes and the error-correction coding and modulation formatting processes are transparent to the data stream—every bit of source data is delivered to the receiving terminal without modification.

Control characters and special "escape sequences" are not required or used by CLOVER-2000. Compressed or encrypted data may therefore be sent without the need to insert (and filter) additional control characters and without concern for data integrity. Five different types of modulation may be used in the ARQ mode—BPSM (Binary Phase Shift Modulation), QPSM (Quadrature PSM), 8PSM (8-level PSM), 8P2A (8PSM + 2-level Amplitude-Shift Modulation), and 16P4A (16 PSM plus 4 ASM).

The same five types of modulation used in ARQ mode are also available in Broadcast (FEC) mode, with the addition of 2-Channel Diversity BPSM (2DPSM). Each CCB is sent using 2DPSM modulation, 17-byte block size and 60% bias. The maximum ARQ data throughput varies from 336 bit/s for BPSM to 1992 bit/s for 16P4A modulation. BPSM is most useful for weak and badly distorted data signals, while the highest format (16P4A) needs extremely good channels, with high SNRs and almost no multipath.

Most ARQ protocols designed for use with HF radio systems can send data in only one direction at a time. CLOVER-2000 does not need an "OVER" command; data may flow in either direction at any time. The CLOVER ARQ time frame automatically adjusts to match the data volume sent in either or both directions. When first linked, both sides of the ARQ link exchange information using six bytes of the CCB. When one station has a large volume of data buffered and ready to send, ARQ mode automatically shifts to an expanded time frame during which one or more 255 byte data blocks are sent.

If the second station also has a large volume of data buffered and ready to send, its half of the ARQ frame is also expanded. Either or both stations will shift back to CCB level when all buffered data has been sent. This feature provides the benefit of full-duplex data transfer but requires use of only simplex frequencies and half-duplex radio equipment. This two-way feature of CLOVER can also provide a back-channel orderwire capability. Communications may be maintained in this "chat" mode at 55 words per minute, which is more than adequate for real-time keyboard-to-keyboard communications.

CLOVER Bibliography

ARRL Web, Technical Descriptions, **www.arrl.org/FandES/field/regulations/techchar/**.

Ford, Steve, WB8IMY, ARRL's *HF Digital Handbook,* Third Ed., ARRL. 2004.

SCAMP AND RDFT

SCAMP (Sound Card Amateur Message Protocol) is intended as a low-cost alternative to commercial modems (TNCs). A paper describing SCAMP was presented by Rick Muething, KN6KB, at the 2004 ARRL/TAPR Digital Communications Conference. It is a new digital sound card protocol suitable for both HF and VHF for transmission of text messages with binary attachments. It is compatible with Winlink 2000 and is designed for manually initiated message forwarding. SCAMP is not a keyboard (chat) mode.

SCAMP incorporates the work by Barry Sanderson, KB9VAK, on Redundant Digital File Transfer (RDFT) and adds an ARQ wrapper around RDFT to ensure error-free transmission. There are four redundancy levels: 10%, 20%, 40% and 70%, the latter being the most robust and requires the most transmission time. Audio data is sent at a standard rate of 11.025 kHz, with 16-bit samples using a PC sound card. SCAMP occupies a bandwidth of about 2 kHz and a net throughput of 2 to 4 kbytes/minute, depending on conditions. It employs an automated "channel-busy" detector for reduction of QRM and to protect against QRM from "hidden transmitters." For more details, please see the SCAMP Bibliography, below.

On-the-air peer-to-peer testing began in November 2004 and the first transcontinental transmission was made in December 2004 between N6KZB in Temecula, CA, and W3QA in West Chester, PA. Beta testing began of SCAMP with WinLink 2000 in March 2005.

SCAMP and RDFT Bibliography

ARRL Letter, The Vol 23, No 48, "SCAMP On-Air Testing Commences," Dec 10, 2005.

Muething, "SCAMP (Sound Card Amateur Message Protocol)," *proc., 2004 ARRL/TAPR Digital Communications Conference.*

Muething, "SCAMP Protocol Specification," **winlink.org/Presentations/SCAMPspec.pdf**.

AUTOMATIC LINK ESTABLISHMENT

The US military services have found it difficult to maintain a sufficient number of qualified radio operators to operate MF/HF radios. So the Defense Department contracted with MITRE Corporation for the development of a method of operating MF/HF radios without skilled operators. MITRE studied what skilled operators do and developed Automatic Link Establishment (ALE) to operate radios and make contact with another station without human intervention and under computer control. ALE automatically finds the best frequency among a prearranged list using techniques such as selective calling, handshaking, link quality analysis, polling, sounding, etc.

ALE is used by the Military Affiliate Radio Service (MARS). It has also been adopted by some radio amateurs.

ALE Waveform

The ALE waveform is designed to be compatible with the audio passband of a standard SSB radio. It has a robust waveform for reliability during poor path conditions. It consists of 8-ary frequency-shift keying (FSK) modulation with eight orthogonal tones, a single tone for a symbol. These tones represent 3 bits of data, with least significant bit to the right, as follows:

Frequency	*Data*
750 Hz	000
1000 Hz	001
1250 Hz	011
1500 Hz	010
1750 Hz	110
2000 Hz	111
2250 Hz	101
2500 Hz	100

The tones are transmitted at a rate of 125 tones per second, 8 ms per tone. The resultant transmitted bit rate is 375 bit/s. The basic ALE word consists of 24 bits of information. Details can be found in Federal Standard 1045, Detailed Requirements, **www.its.bldrdoc.gov/fs-1054a/45-detr.htm**.

ALE Bibliography

Adair, Robert, KAØCKS, et al, "A Federal Standard for HF Radio Automatic Link Establishment," *QEX* January 1990.

Adair, Robert, KAØCKS, et al, "The Growing Family of Federal Standards for HF Radio Automatic Link Establishment (ALE)—Part I," *QEX* July 1993; Part II, *QEX* August 1993; Part III, *QEX*, September 1993; Part IV, *QEX* October 1993; Part V, *QEX* November 1993; Part VI *QEX* December 1993.

Brain, Charles, G4GUO, *PC-ALE Project*, **www.chbrain.dircon.co.uk/peale.html**.

Menold, Ronald, AD4TB, "ALE—The Coming of Automatic Link Establishment," *QST* February 1995, p. 68 (Technical Correspondence).

National Communications System, "Telecommunications: HF Radio Automatic Link Establishment," Federal Standard 1045A, October 1993.

Internetworking

Although it has been a goal of some radio amateurs to develop a digital communications network independent of the Internet, interconnection with the Internet provides a good bridge between isolated amateur radio nets. Several methods of transferring data, e-mail or linking repeaters have been developed.

WINLINK 2000

WinLink 2000 is a *Windows* application that permits messages to be transferred automatically between the Internet and remote amateur stations, which may be on recreational vehicles or at sea. The Internet is used as a backbone to allow WinLink mailbox operation (MBO) stations to share their databases. Its original author was Victor Poor, W5SSM. See: **winlink.org/**.

IRLP

Created by David Camerpon, WE7LTD, the Internet Radio Linking Project (IRLP) uses Voice over Internet Protocol (VoIP) to form a voice communications network of servers and nodes between amateur repeaters and/or simplex stations. See: **www.irlp.net/**.

EchoLink

EchoLink was developed by Jonathan Taylor, K1RFD, to link a personal computer to communicate by VoIP with several thousand repeaters having *EchoLink* capabilities. Or, it can be used to permit amateur stations within range of your station to connect with the Internet. See: **www.echolink.org/**.

eQSO

eQSO, created by Paul Davies, MØZPD, was designed to operate like a worldwide amateur radio net. See: **www.eqso.net/**.

Internetworking Bibliography

Brone, Jeff, WB2JNA, "EchoLink for Beginners," *QST*, January 2005.

Ford, Steve, WB8IMY, *ARRL's HF Digital Handbook,* Third Ed., ARRL. 2004.

Ford, Steve, WB8IMY, "VoIP and Amateur Radio," *QST*, February 2003.

Horzepa, Stan, WA1LOU, "WinLink 2000: A Worldwide HF BBS," *QST*, March 2000.

Linden, Louis, KI5TO, "Winlink 2000 in the Jungle," *QST*, November 2004.

TELEMETRY, TRACKING AND TELECOMMAND

According to FCC Part 97 rules, *telemetry* is a one-way transmission of measurements at a distance from the measuring instrument, whereas *telecommand* is a one-way transmission to initiate, modify or terminate functions of a device at a distance. Actually, the two go hand in hand, since it is important to have telemetry first, then modify the remote device, then look once again in the telemetry to see if the desired action took place.

Telemetry, tracking and telecommand (often seen as TT&C) are attracting increasing attention because Amateur Radio rules permit higher power transmitters than allowed under Part 15 of the FCC rules. TT&C is distinct from traditional forms of Amateur Radio (telegraphy, voice and image intended to be heard or seen by human operators), since it receives information from an object and commands the object to take an action. Although pulse modulation systems are common, TT&C also uses familiar communications modes, such as television (in this case used as a form of telemetry), packet radio (such as ASCII used for telemetry coding, commands or uploading programs).

This section provides only a sampling of telemetry, tracking and telecommand systems involving Amateur Radio. *APRS* (Automatic Position Reporting System) is a marriage of an application of the Global Positioning System and Amateur Radio to relay position and tracking information. Telemetry and telecommand are also used to manage remote terrestrial stations, as well as amateur satellites. Radio Control (R/C) of remote objects has long been a part of Amateur Radio because of the versatility offered by Part 97 rules to licensed operators. R/C is not limited to model cars, boats and airplanes but is vital for the growing field of robotics.

APRS

Bob Bruninga, WB4APR, developed Automatic Position Reporting System (APRS) as a result of trying to use packet radio for real-time communications for public service events. Packet radio is not well suited for those real-time events, where information has a very short lifetime. APRS avoids the complexity and limitations of trying to maintain a connected network. It uses UI (unconnected) frames to permit any number of stations to participate and exchange data, just like voice users would on a voice net. Stations that have information to contribute simply transmit it, and all stations monitor and collect all data on frequency. APRS also recognizes that one of the greatest real-time needs at any special event or emergency is the knowledge of where all stations and other key assets are located. APRS accomplishes the real-time display of operational traffic via a split screen and map displays.

Since the object of APRS is the rapid dissemination of real-time information using packet UI frames, a fundamental precept is that old information is less important than new information. All beacons, position reports, messages and display graphics are redundantly transmitted, but at longer and longer repetition rates. Each new beacon is transmitted immediately, then again 20 seconds later. After every transmission, the period is doubled. After ten minutes only six packets have been transmitted. After an hour this results in only three more beacons; and only three more for the rest of the day! Using this redundant UI broadcast protocol, APRS is actually much more efficient than if a fully connected link had to be maintained between all stations.

The standard configuration for packet radio hardware (radio-to-TNC-to-computer) also applies to APRS until you add a GPS (Global Positioning System) receiver to the mix. You don't need a GPS receiver for a stationary APRS installation (nor do you need a computer for a mobile or tracker APRS installation). In these cases, an extra port or special cable is not necessary. It is necessary, however, when you desire both a computer and a GPS receiver in the same installation.

One way of accomplishing this is by using a TNC or computer that has an extra serial port for a GPS receiver connection. Alternatively, you can use a hardware single port switch (HSP) cable to connect a TNC and GPS receiver to the same serial port of your computer. The HSP cable is available from a number of sources including TNC manufacturers Kantronics, MFJ and PacComm.

Whichever GPS connection you use, make sure that you configure the APRS software so it is aware that a GPS receiver is part of the hardware configuration and how the GPS receiver connection is accomplished.

APRS also supports an optional weather station interface. The wind speed, direction, temperature and rainfall are inserted into the station's periodic position report. The station shows up on all APRS maps as a large blue dot, with a white line showing the wind speed and direction. Several automatic APRS weather reporting stations, supported with additional manual reporting stations, can form a real-time reporting network in support of SKYWARN activities. For additional information see the book, *APRS Tracks, Maps and Mobiles* by Stan Horzepa, WA1LOU, published by ARRL.

Radio Control (R/C)

Amateur Radio gave birth to the radio control (R/C) hobby as we know it today. FCC §97.215 rules specifically permit "remote control of model craft" as a licensed amateur station activity. Station identification is not required for R/C, and the transmitter power is limited to 1 W. FCC §97.215 states:

"Telemetry transmitted by an amateur station on or within 50 km of the Earth's surface is not considered to be codes or ciphers intended to obscure the meaning of communications."

This section was contributed by H. Warren Plohr, W8IAH. The simplest electronic control systems are currently used in low-cost toy R/C models. These toys often use simple on/off switching control that can be transmitted by on/off RF carrier or tone modulation. More expensive toys and R/C hobby models use more sophisticated con-

Fig 9.23—Photo of three R/C model electric cars.

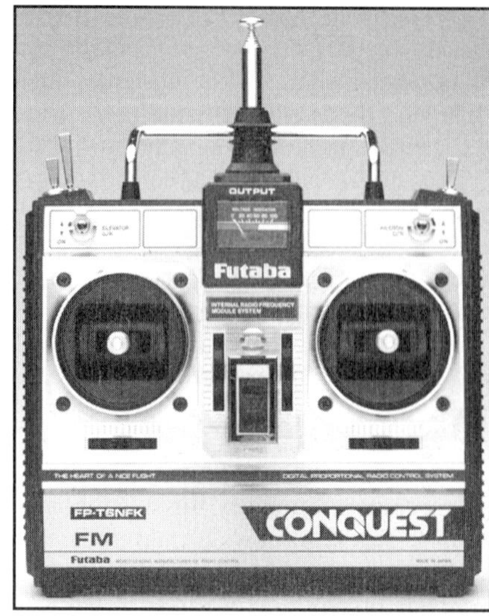

(A)

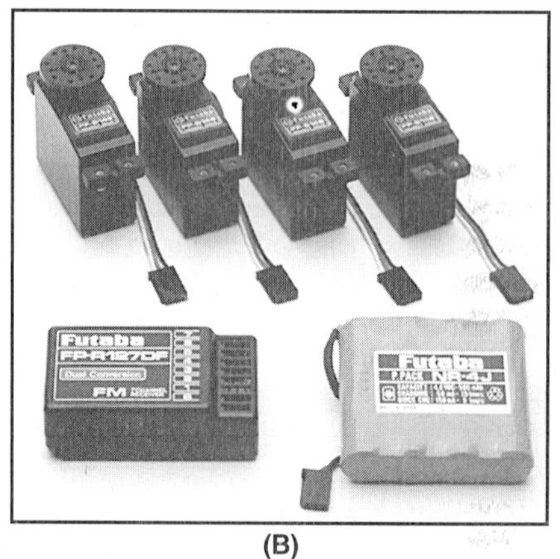

(B)

Fig 9.25—A, photo of Futaba's Conquest R/C aircraft transmitter. B shows the matching airborne system.

trol techniques. Several simultaneous proportional and switching controls are available, using either analog or digital coding on a single RF carrier.

R/C hobby sales records show that control of model cars is the most popular segment of the hobby. Battery powered cars like that shown in **Fig 9.23** are the most popular. Other popular types include models powered by small internal combustion gas engines.

R/C model aircraft are next in the line of popularity and include a wide range of styles and sizes. Fixed-wing models like those shown in **Fig 9.24** are the most popular. They can be unpowered (gliders) or powered by either electric or gas engines. The basic challenge for a new model pilot is to operate the model in flight without crashing. Once this is achieved, the challenge extends to operating detailed scaled models in realistic flight, performing precision aerobatics, racing other models or engaging in model-to-model combat.

The challenge for the R/C glider pilot is to keep the model aloft in rising air currents. The most popular rotary-wing aircraft models are helicopters. The sophistication of model helicopters and their control systems can only be appreciated when one sees a skilled pilot perform a schedule of precision flight maneuvers. The most exotic maneuver is sustained inverted flight, a maneuver not attainable by a full-scale helicopter.

R/C boats are another facet of the hobby. R/C water craft models can imitate full-scale ships and boats, from electric motor powered scale warships that engage in scale battles, to gas powered racing hydroplanes, model racing yachts and even submarines.

Most R/C operation is no longer on Amateur Radio frequencies. The FCC currently authorizes 91 R/C frequencies between 27 MHz and 76 MHz. Some frequencies are for all models, some are only for aircraft and others for surface (cars, boats) models only. Some frequencies are used primarily for toys and others for hobbyist models. Amateur Radio R/C

Fig 9.24—Photo of two R/C aircraft models.

operators use the 6-m band almost exclusively. Spot frequencies in the upper part of the band are used in geographical areas where R/C operation is compatible with 6-m repeater operation and TV Channel-2 signals that can interfere with control. Eight spot frequencies, 53.1 to 53.8 MHz, spaced 100 kHz apart, are used. There is also a newer 200 kHz R/C band from 50.8 to 51.0 MHz providing ten channels spaced 20 kHz apart. The close channel spacing in this band requires more selective receivers than do the 53-MHz channels. The Academy of Model Aeronautics (**www.modelaircraft.org**) *Membership Manual* provides a detailed list of all R/C frequencies in current use as well as other useful information. The *ARRL Repeater Directory* lists current Amateur Radio R/C frequencies.

Fig 9.25 shows a typical commercial R/C system, consisting of a hand-held aircraft transmitter (A), a multiple-control receiver, four control servos and a battery (B). This particular equipment is available for any of the ten R/C frequencies in the 50.8-51.0 MHz band. Other commercially available control devices include relays (solid-state and mechanical) and electric motor speed controllers.

Some transmitters are tailored to specific kinds of models. A helicopter, for example, requires simultaneous control of both collective pitch and engine throttle. A model helicopter pilot commands this response with a linear motion of a single transmitter control stick. The linear control stick signal

Fig 9.26—Photo of Airtronics Infinity 660 R/C aircraft transmitter.

is conditioned within the transmitter to provide the encoder with a desired combination of nonlinear signals. These signals then command the two servos that control the vertical motion of the helicopter.

Transmitter control-signal conditioning is provided by either analog or digital circuitry. The signal conditioning circuitry is often designed to suit a specific type of model, and it is user adjustable to meet an individual model's control need. (Low-cost transmitters use analog circuitry.) They are available for helicopters, sailplanes and pattern (aerobatic) aircraft.

More expensive transmitters use digital microprocessor circuitry for signal conditioning. **Fig 9.26** shows a transmitter that uses a programmable microprocessor. It is available on any 6-m Amateur Radio R/C frequency with switch-selectable PPM or PCM coding. It can be programmed to suit the needs of a helicopter, sailplane or pattern aircraft. Nonvolatile memory retains up to four user-programmed model configurations.

Many R/C operators use the Amateur Radio channels to avoid crowding on the non-ham channels. Others do so because they can operate home-built or modified R/C transmitters without obtaining FCC type acceptance. Still others use commercial R/C hardware for remote control purposes around the shack.

R/C RF MODULATION

The coded PPM or PCM information for R/C can modulate an RF carrier via either amplitude- or frequency-modulation techniques. Commercial R/C systems use both AM and FM modulation for PPM, but use FM exclusively for PCM.

The AM technique used by R/C is 100% "down modulation." This technique switches the RF carrier off for the duration of the PPM pulse, usually 250 to 350 µs. A typical transmitter design consists of a third-overtone transistor oscillator, a buffer amplifier and a power amplifier of about 1/2 W output. AM is achieved by keying the 9.6-V supply to the buffer and final amplifier.

The FM technique used by R/C is frequency shift keying (FSK). The modulation is applied to the crystal-oscillator stage, shifting the frequency about 2.5 or 3.0 kHz. The direction of frequency shift, up or down with a PPM pulse or PCM code, can be in either direction, as long as the receiver detector is matched to the transmitter. R/C manufacturers do not standardize, so FM receivers from different manufacturers may not be compatible.

SIGNALING TECHNIQUES

Radio control (R/C) of models has used many different control techniques in the past. Experimental techniques have included both frequency- and time-division multiplexing, using both electronic and mechanical devices. Most current systems use time-division multiplexing of pulse-width information. This signaling technique, used by hobbyist R/C systems, sends pulse-width information to a remotely located pulse-feedback servomechanism. Servos were initially developed for R/C in the 1950s and are still used today in all but low-cost R/C toys.

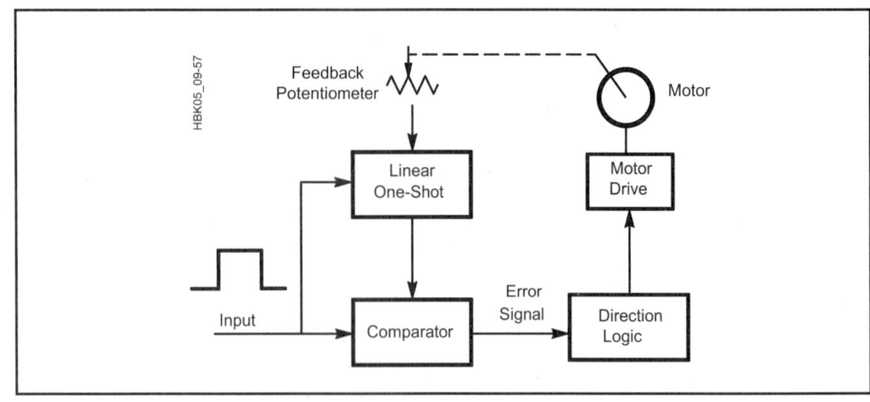

Fig 9.27—Diagram of a pulse-feedback servo.

Fig 9.27 is a block diagram of a pulse-feedback servo. The leading edge of the input pulse triggers a linear one-shot multivibrator. The width of the one-shot output pulse is compared to the input pulse. Any pulse width difference is an error signal that is amplified to drive the motor. The motor drives a feedback potentiometer that controls the one-shot timing. When this feedback loop reduces the error signal to a few microseconds, the drive motor stops. The servo position is a linear function of the input pulse width. The motor-drive electronics are usually timed for pulse repetition rates of 50 Hz or greater and a pulse width range of 1 to 2 ms. A significantly slower repetition rate reduces the servomechanism slew rate but not the position accuracy.

In addition to motor driven servos, the concept of pulse-width comparison can be used to operate solid-state or mechanical relay switches. The same concept is used in solid-state proportional electric motor speed controllers. These speed controllers are used to operate the motors powering model cars, boats and aircraft. Currently available model speed controllers can handle tens of amperes of direct current at voltages up to 40 V dc using MOSFET semiconductor switches.

Requirements

The signaling technique required by R/C is the transmission of 1- to 2-ms-wide pulses with an accuracy of ±1 µs at repetition rates of about 50 Hz. A single positive-going dc pulse of 3 to 5 V amplitude can be hard wired to operate a single control servomechanism. If such a pulse is used to modulate an RF carrier, however, distortion of the pulse width in the modulation/demodulation process is often unacceptable. Consequently, the pulse-width information is usually coded for RF transmission. In addition, most R/C systems require pulse-width information for more than one control. Time-division multiplexing of each control provides this multichannel capability. Two coding techniques are used to transfer the pulse-width information for multiple control channels: pulse-position modulation (PPM) and pulse-code modulation (PCM).

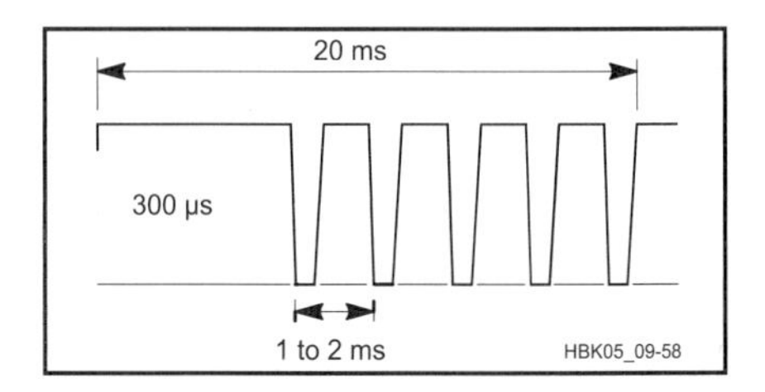

Fig 9.28—Diagram of a four-channel PPM RF envelope.

Pulse-Position Modulation

PPM is analog in nature. The timing between transmitted pulses is an analog of the encoded pulse width. A train of pulses encodes multiple channels of pulse-width information as the relative position or timing between pulses. Therefore the name, pulse-position modulation. The transmitted pulse is about 300 µs in width and uses slow rise and fall times to minimize the transmitter RF bandwidth. The shape of the received waveform is unimportant because the desired information is in the timing between pulses. **Fig 9.28** diagrams a frame of five pulses that transmits four control channels of pulse-width information. The frame of modulation pulses is clocked at 50 Hz for a frame duration of 20 ms. Four multiplexed pulse widths are encoded as the times between five 300-µs pulses. The long period between the first and the last pulse is used by the decoder for control-channel synchronization.

PPM is often incorrectly called *digital control* because it can use digital logic circuits to encode and decode the control pulses. A block diagram of a typical encoder is shown in **Fig 9.29**. The 50-Hz clock frame generator produces the first 300-µs modulation pulse and simultaneously triggers the first one-shot in a chain of multivibrators. The trailing edge of each one-shot generates a 300-µs modulation pulse while simultaneously triggering the succeeding multivibrator one-shot. In a four-channel system the fifth modulation pulse, which indicates control of the fourth channel, is followed by a modulation pause that is dependent on the frame rate. The train of 300-µs pulses are used to modulate the RF carrier.

Received pulse decoding can also use digital logic semiconductors. **Fig 9.30** shows a simple four-control-channel decoder circuit using a 74C95 CMOS logic IC. The IC is a 4-bit shift register operated in the right-shift mode. Five data pulses spaced 1 to 2 ms apart, followed by a synchronization pause, contain the encoded pulse-width information in one frame. During the sync pause, the RC circuit discharges and sends a logic-one signal to the 74C95 serial input terminal. Subsequent negative going data pulses remove the logic-one signal from the serial input and sequentially clock the logic one through the four D-flip-flops. The output of each flip-flop is a positive going pulse, with a width corresponding to the time between the clocking pulses. The output of each flip-flop is a demultiplexed signal that is used to control the corresponding servo.

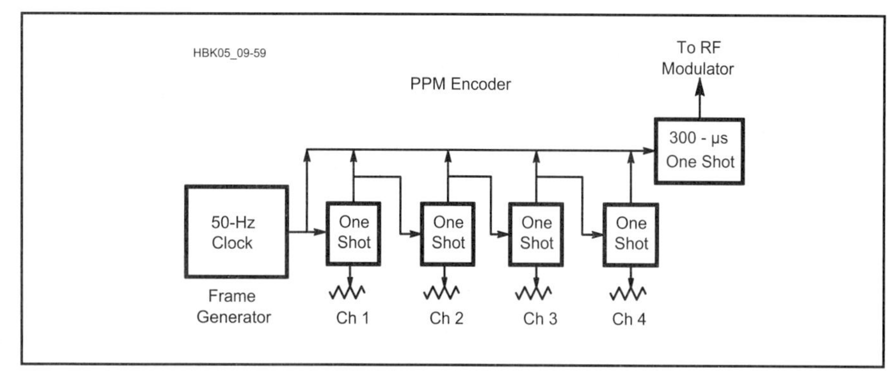

Fig 9.29—Diagram of a PPM encoder.

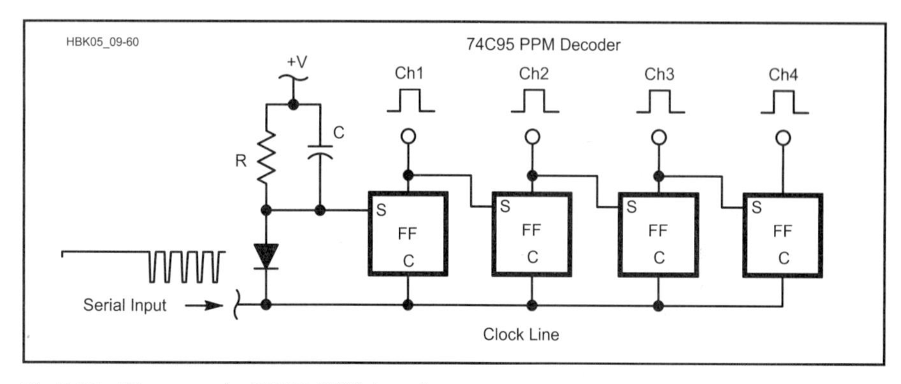

Fig 9.30—Diagram of a 74C95 PPM decoder.

Pulse Code Modulation

PCM uses true digital code to transfer R/C signals. The pulse width data of each control channel is converted to a binary word. The digital word information of each control channel is coded and multiplexed to permit transmission of multiple channels of control on a single RF carrier. On the receiving end, the process is reversed to yield the servo control signals.

There is no standard for how the digital word is coded for transmission. Therefore PCM R/C transmitters and receivers from different makers are not interchangeable. Some older PCM systems provide only 256 discrete positions for 90° of servo motion, thereby limiting servo resolution. Newer systems use more digital bits for each word and provide smooth servo motion with 512 and 1024 discrete positions. All PCM and PPM systems use the same servo input-signal and supply voltages. Therefore the servos of different manufacture are interchangeable once compatible wiring connectors have been installed.

AMATEUR SATELLITE TT&C

TT&C plays a vital part of the launching and management of amateur satellites. Satellites have onboard intelligence and are increasing able to make their own decisions but Article 25 of the international Radio Regulations requires the following:

> *"Administrations authorizing space stations in the amateur-satellite service shall ensure that sufficient earth command stations are established before launch to ensure that any harmful interference caused by emissions from a station in the amateur-satellite service can be terminated immediately."*

The U.S. implementation of the Radio Regulations in Part 15 of the FCC rules has these provisions:

> ***"§97.211 Space telecommand station.***
>
> *(a) Any amateur station designated by the licensee of a space station is eligible to transmit as a telecommand station for that space station, subject to the privileges of the class of operator license held by the control operator.*
>
> *(b) A telecommand station may transmit special codes intended to obscure the meaning of telecommand messages to the station in space operation.*
>
> .....
>
> *(d) A telecommand station may transmit one-way communications."*

Telemetry from amateur satellites, such as from "engineering beacons" is available to all amateurs. Computer programs are available from AMSAT for decoding the telemetry to monitor the health of the spacecraft and other measurements. However, telecommand of amateur satellites is closely held in order to maintain effective control.

Amateur Satellite TT&C References

AMSAT, **www.amsat.org**.

Davidoff, Martin, K2UBC, *The Radio Amateur's Satellite Handbook,* ARRL, Rev first ed, 2003.

Voice Modes

AMPLITUDE MODULATION (AM)

This material was written by John O. Stanley, K4ERO. The first AM broadcast of speech and music occurred nearly a century ago, when on Christmas eve of 1906, Fessenden, using a modulated high frequency alternator, surprised ship operators with a program of music, Bible readings and poetry. The development of a continuous wave transmitter, one that produced a constant sine wave output, rather than the rough spark signal, made AM practical. Thus, CW, as this pure wave was called, not only greatly enhanced Morse communications, but allowed voice transmissions as well. By changing the strength or amplitude of this smooth continuous wave, a voice could be superimposed on the radio frequency carrier.

The decade of the 1920s saw not only the rapid development of the broadcast industry, but also enabled many hams to try the new voice mode. Indeed, in those early years, there was sometimes little difference between a ham who used voice and a broadcaster. The situation was a mess, and QRM was king! By 1929 it was permissible to use AM voice in limited portions of our amateur spectrum and, on some bands, only the most qualified licensees had the privilege.

Users of AM had to learn that an RF wave could have only a certain amount of audio imposed upon it before overmodulation occurred. Trying to go above 100% modulation produced severe distortion and splatter. AM remained the dominant voice mode for ham operations well into the second half of the 20th century, when it was gradually eclipsed by SSB (SSB is actually a form of AM) and FM. We can still hear AM on the ham bands today, mostly coming from stations using vintage gear. AMers usually choose operating times when the bands are less crowded, and often take pride in a clean and clear signal.

The great advantage of AM, and one reason for its long history, is the ease with which a full carrier AM signal can be received. This was all important in broadcasting where, for every transmitter, there were thousands or even millions of receivers. With modern integrated circuits, complex detectors now cost very little. Therefore, the biggest reason for keeping AM broadcasting, at present, is to avoid obsolescing the billions of existing receivers. These will gradually have to be replaced when digital broadcasts begin in the AM and shortwave bands.

There are many ways to produce an AM signal, but all of them involve multiplying the amplitude of the information to be transmitted by the amplitude of the radio wave that will carry it. When multiplication of two signals takes place, as opposed to their simple addition, *mixing* is involved. The result is multiple signals, including the sum and difference of the AF and RF frequencies. These two "products" will appear as sidebands alongside what was the original RF frequency. Mixing, modulation, detection, demodulation, and heterodyning all refer to this multiplication process and can all be analyzed by the same mathematical treatment. See the **Mixers, Modulators and Demodulators** chapter of this *Handbook* for a more detailed discussion of this process.

If an RF signal is modulated by a single audio tone, and observed on an oscilloscope, it will appear as shown on the right in **Fig 9.31B**. Observing the same signal on a spectrum analyzer will show that the composite signal observed on the scope is composed of three discrete parts as shown on the left in Fig 9.31B. The center peak, which is identical with the original unmodulated wave shown in Fig 9.31A, is usually called the *carrier*, although this terminology is deceiving and imprecise. It is the composite RF signal, as seen on the oscilloscope, which actually carries the audio in the form of variations in its amplitude, so we might well have referred to the center frequency as a "reference" or some other such term.

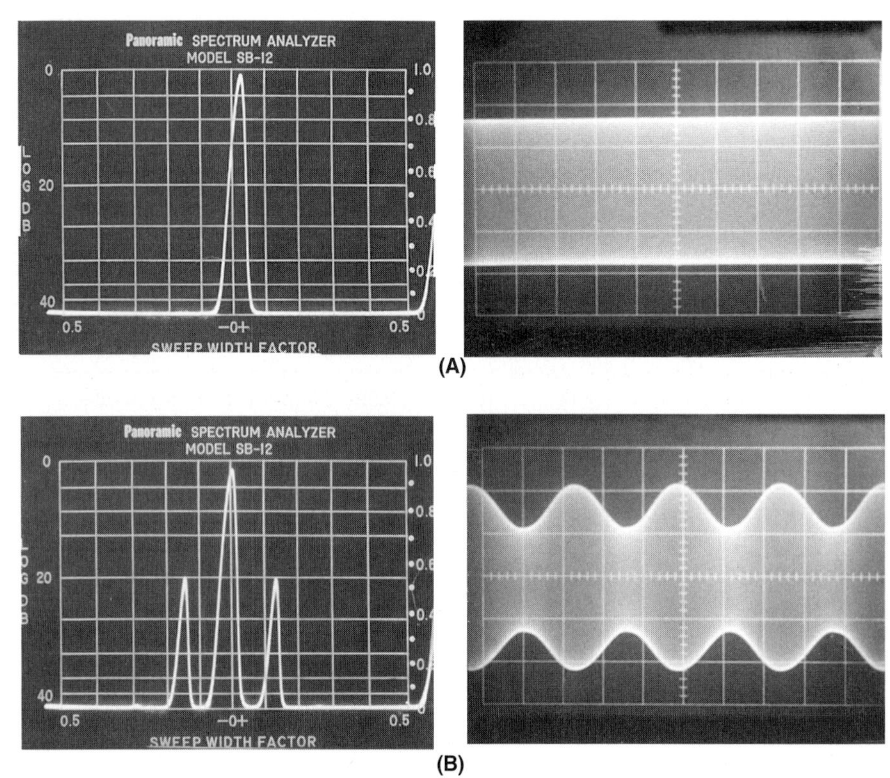

Fig 9.31—Electronic displays of AM signals in the frequency and time domains. A shows an unmodulated carrier or single-tone SSB signal. B shows a full-carrier AM signal modulated 20% with a sine wave.

As a reference signal, the carrier contains important, though not indispensable, information. For a signal with both sidebands present, it provides a very important frequency and phase reference that allows simple and undistorted detection, using nothing more than a diode. The carrier also provides an amplitude reference, which is used by AM receivers to set the gain of the receiver, using AGC or automatic gain control. The carrier also contains most of the power of the transmitted signal, while most of the important information is in the sidebands. See the **Mixers, Modulators and Demodulators** chapter in this *Handbook*, which gives details of power distribution in an AM signal.

SINGLE SIDEBAND (SSB)

Telephone engineers developed a system of using only one of the two sidebands, which, being mirror images of each other, contain the same information. SSB systems attracted the attention of hams soon after WWII and gradually became the voice mode of choice for the HF bands. SSB is considered a form of AM, in that it is identical to an AM signal with one sideband, and with all or part of the carrier removed. The complexity of generating a SSB signal, plus the difficulty of tuning the generally unstable receivers common in the 1950s, slowed the changeover to the new mode, but its adoption was inevitable. SSB became popular because of its greater power efficiency,

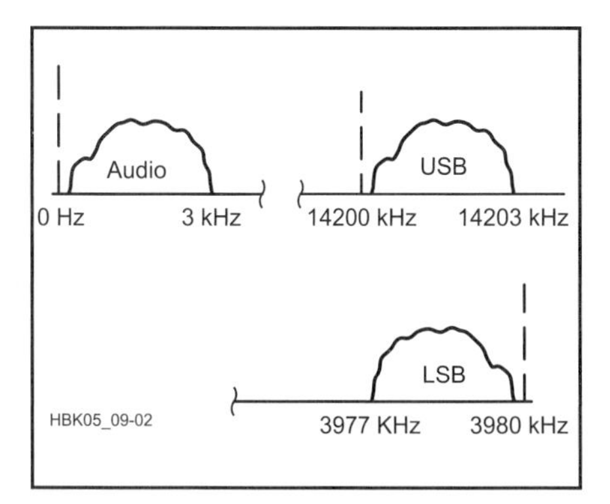

Fig 9.32—How an occupied radio frequency spectrum shifts with application of an audio (baseband) signal. The dotted line represents the RF carrier point, or in the case of the 3-kHz audio signal, the reference frequency, 0 kHz.

which allowed each watt of RF to go further. The fact that it occupied less bandwidth was a plus also and very welcome on the most crowded bands. See the sidebar **SSB on 20 and 75 Meters** in this chapter.

While systems used for telephone relays used pilot carriers so that the signal could be reproduced without distortion, hams chose to eliminate the carrier entirely. This required generating a reference frequency at the receiver, which, if accurate to within 20 Hz, allowed intelligible speech to be recovered. Since amateur regulations have long prohibited transmission of music, the distortion produced by loss of the exact phase and frequency reference was not serious. The loss of the amplitude reference was overcome with the development of the "hang" AGC, which works on the average value of the received sideband, which is constantly changing. While not as fast or accurate as the carrier-based AGC available in AM, this has proven satisfactory, if proper attention is given to its design (See the **Receivers and Transmitters** chapter of this *Handbook*.)

Thus, SSB, while giving up some fidelity and while increasing complexity, has proven superior to full-carrier AM for speech communication because of its power and bandwidth efficiency. And under certain circumstances, such as selective fading, it can actually have less distortion than DSB AM. On HF, it is possible for the carrier to fade in an DSB AM signal, leaving less than is needed for envelope detection. Medium wave AM broadcasts often have this problem at night. It can be overcome with "exalted carrier detection." Synchronous detection is a refinement of this method. (See the **Mixers, Modulators and Demodulators** chapter of this *Handbook*.) SSB, in effect, uses exalted carrier detection all the time.

An SSB signal is best visualized as an audio or *baseband* signal that has simply been shifted upwards into the radio frequency spectrum, as shown in **Fig 9.32**. The relative frequencies, phases and amplitudes of all the components will be the same as the original frequency components except for having had a fixed reference frequency added to them. Surprisingly, this process, called heterodyning, is not done by directly adding the signals together, but by multiplying them and subsequently filtering or phasing out the carrier and one of the sidebands. The **Mixers, Modulators and Demodulators** chapter of this *Handbook* explains this interesting process in detail.

The relative frequencies within the band of information being transmitted may appear inverted; that is, lower frequencies in the original audio signal are higher in the RF signal. When this happens, we call the signal *lower sideband* or LSB. LSB is produced when the final frequency is the result of subtraction rather than addition. If a tone of 1 kHz is heterodyned to 14201 kHz by mixing with a 14200 kHz carrier, the result will be upper sideband, since 14200 + 1 gives us that result. When the same tone appears at 3979 kHz by mixing it with a 3980 kHz carrier, we know that an LSB signal was produced since 3980 – 1 gives us the 3979 result. Whenever the audio tone needs a minus sign to find the result, we are on LSB.

SSB ON 20 AND 75 METERS —THE 9 TO 5 CONNECTION

SSB experiments began on 75 meters because it was the lowest frequency phone band in widespread use. Due to perpetual crowding and its DX potential, 20 meters also seemed to call for use of SSB. Some early rigs included only these two bands. The popular homebrew W2EWL rig was built on the chassis of a war surplus ARC-5 transmitter using its 5 MHz VFO, and generated the sideband signal on 9 MHz using the phasing method. Nine plus five is 14 MHz, and nine minus five is 4 MHz, yielding 75 or 20 meter coverage by choosing which of the two mix products we would filter out and amplify. Thus, two bands were covered with the same VFO/IF combination. Other rigs used a tunable IF from 5.0 to 5.5 MHz. This was subtracted from a 9-MHz crystal to obtain 4.0 to 3.5 MHz, and added to 9 MHz to cover 14.0 to 14.5 MHz. This process reversed the sidebands, and eventually led to the convention of using LSB on the lower bands and USB on the higher bands. This also explains why on some vintage rigs the 75-meter band dial reads backwards!—*K4ERO*

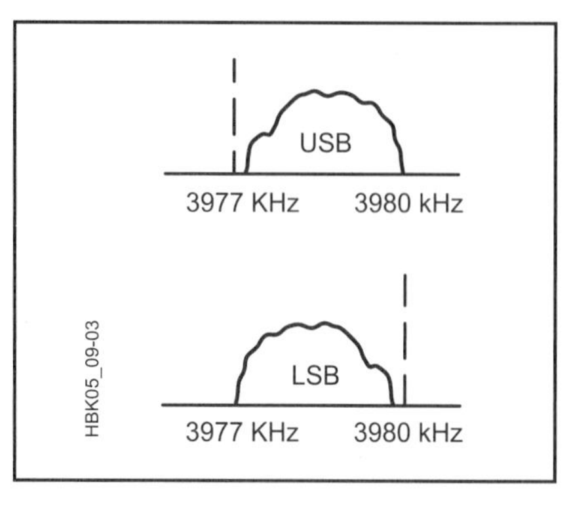

Fig 9.33—A method of changing sidebands with virtually no change in the frequency spectrum occupied. Note that the carrier point position *has* changed concurrent with the change from LSB to USB.

In most mixing schemes there will be three frequencies involved (carrier, VFO, and band select crystal) but the principle still holds.

The frequency of an SSB transmission is designated as that of the carrier, which is the frequency (or the sum of several frequencies) used to shift the baseband information into the RF spectrum. In a good SSB signal, little or no energy actually appears on the frequency we say we are using. It is strictly a reference. For this reason, some radio services have chosen to designate SSB channels by the center of the occupied bandwidth rather than the carrier frequency. Ham practice is to designate the carrier frequency and whether the upper or lower sideband is in use. An interesting exception is the new five-channel, 60-m amateur band (a secondary allocation) where the FCC specified a 2.8-kHz bandwidth on five center frequencies: 5332, 5348, 5368, 5373 and 5405 kHz. Only USB voice (2K8J3E emission) is permitted.

Most hams will find it more natural to remember USB at corresponding carrier frequencies of 5330.5, 5346.5, 5366.5, 5371.5 and 5403.5 kHz. Since the USB or the LSB is considered "normal" for each of our bands, it is assumed that the sideband in use is understood. We need to remember when switching sidebands that we will be occupying a different portion of the spectrum than before the switch, and we may inadvertently cause QRM, unless we check for a clear frequency. If you wish to change from LSB to USB without changing the spectrum occupied, you must retune your dial down about 3 kHz, as a careful study of **Fig 9.33** should make clear. This principle applies to digital as well as voice modes, but usually not to CW, where modern rigs make the above adjustment for us. This means that the frequency readout with a CW signal will be the actual frequency occupied, but with analog voice and digital modes this will probably not be the case.

Another need for understanding where sideband signals actually fall is in operating close to the edge of a band or subband. For example, on 20 meters where USB is used, you must not operate above approximately 14.347 MHz, since the transmission will be outside the band if you operates much higher. Operation with a suppressed carrier exactly on 4.0 MHz could be done on LSB if the signal is very clean, but is not recommended. Most modern rigs prevent out of allocated band transmissions but *do not* preclude the above cases of improper operation.

Today there are many new modes for text, speech and image transmission, and more will be developed in the future. Often these are transmitted using SSB. Knowing exactly where the signal will appear on the band depends on understanding how LSB and USB signals are produced. These modes use either a separate circuit or more recently a computer sound card to produce audio frequency tones that represent the information in coded form. This is then fed into the audio input of an SSB transmitter. They are then heterodyned to the desired amateur band for transmission. In a transceiver, the incoming signals are similarly heterodyned back to the audio range for processing in the computer sound card or other circuitry. Some computer-based digital modes allow reading the actual signal frequency off the screen, provided the transceiver dial is properly set.

Voice signals and some text and image modes require linear amplification. This means that the amplifiers in the transmitter must faithfully represent the amplitude as well as the frequency of the baseband signal. If they fail to do so, intermodulation distortion (IMD) products appear and the signal becomes much wider than it should be, producing interference (QRM) on nearby frequencies. CW and FM do not require a linear amplifier, but you can use one for these modes also, at a small price in efficiency. Some VHF "brick" amplifiers have a choice of either the more efficient class C amplification or the more linear class B amplification. The linear or SSB mode must be chosen if SSB voice and some digital modes are being used. Whenever linear amplification is needed, *flat-topping* must be prevented. This results from overdriving the amplifier so that it goes above the design power limit and becomes non-linear.

SSB transmitters and most linear amplifiers use automatic level control (ALC) to prevent overdrive and flat topping. However, there are limits to ALC and flat topping can still occur if the amplifier is grossly over driven. The surest way to create ill will on any band is to cause spatter by over driving

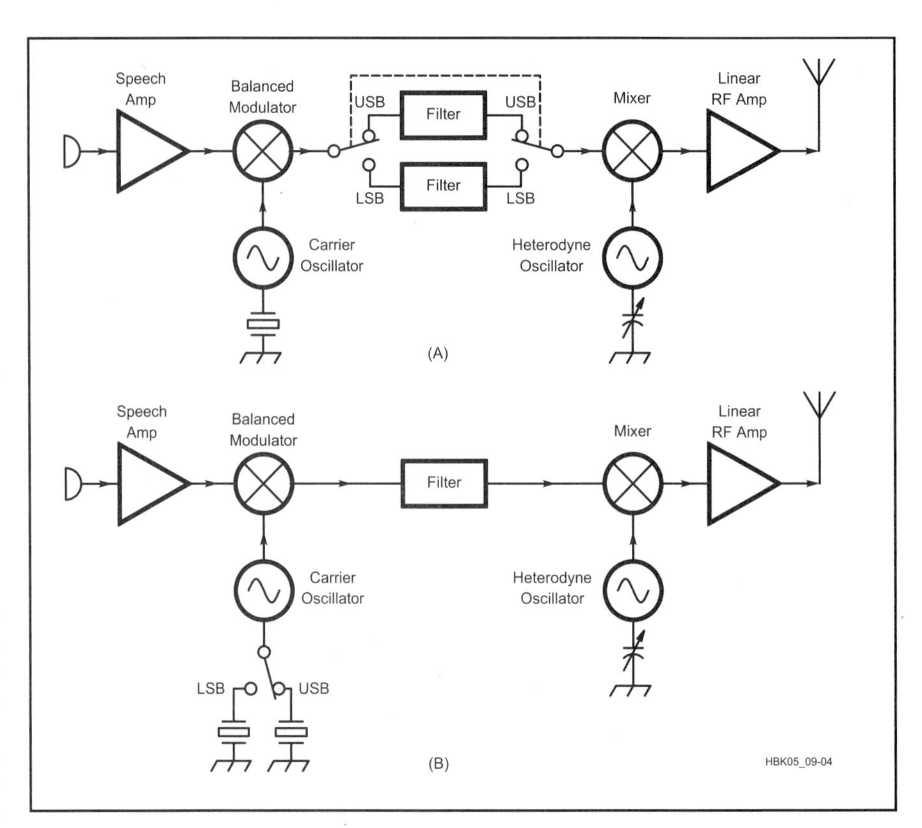

Fig 9.34—Block diagrams of filter-method SSB generators. They differ in the manner that the upper and lower sideband are selected.

your amplifier, regardless of the mode. Amplifiers suitable for both linear and non-linear signals are discussed in the **RF Power Amplifiers** chapter of this *Handbook*. The effects of non-linear amplification are also further treated in the **Mixers, Modulators and Demodulators** chapter.

How an SSB Signal is Produced

When the proper receiver bandwidth is used, an SSB signal will show an effective gain of up to 9 dB over an AM signal of the same peak power. Because the redundant information is eliminated, the required bandwidth is half that of a comparable AM (DSB) emission. Unlike DSB, the phase of the local carrier generated in the receiver is unimportant.

SSB Generation: The Filter Method

If the DSB signal from the balanced modulator is applied to a narrow bandpass filter, one of the sidebands can be greatly attenuated. Because a filter cannot have infinitely steep skirts, the response of the filter must begin to roll off within about 300 Hz of the phantom carrier to obtain adequate suppression of the unwanted sideband. This effect limits the ability to transmit bass frequencies, but those frequencies have little value in voice communications. The filter rolloff can be used to obtain an additional 20 dB of carrier suppression. The bandwidth of an SSB filter is selected for the specific application. For voice communications, typical values are 1.8 to 3.0 kHz.

Fig 9.34 illustrates two variations of the filter method of SSB generation. In A, the heterodyne oscillator is represented as a simple VFO, but may be a premixing system or synthesizer. The scheme at B is perhaps less expensive than that of A, but the heterodyne oscillator frequency must be shifted when changing sidebands in order to maintain dial calibration.

SSB Generation: The Phasing Method

Fig 9.35 shows another method to obtain an SSB signal. The audio and carrier signals are each split into equal components with a 90° phase difference (called *quadrature*) and applied to balanced modulators. When the DSB outputs of the modulators are combined, one sideband is reinforced and the other is canceled. The figure shows sideband selection by means of transposing the audio leads, but the same result can be achieved by switching the carrier leads. The phase shift and amplitude balance of the two channels must be very accurate if the unwanted sideband is to be adequately attenuated. **Table 9.8** shows the required phase accuracy of one channel (AF or RF) for various levels of opposite sideband suppression.

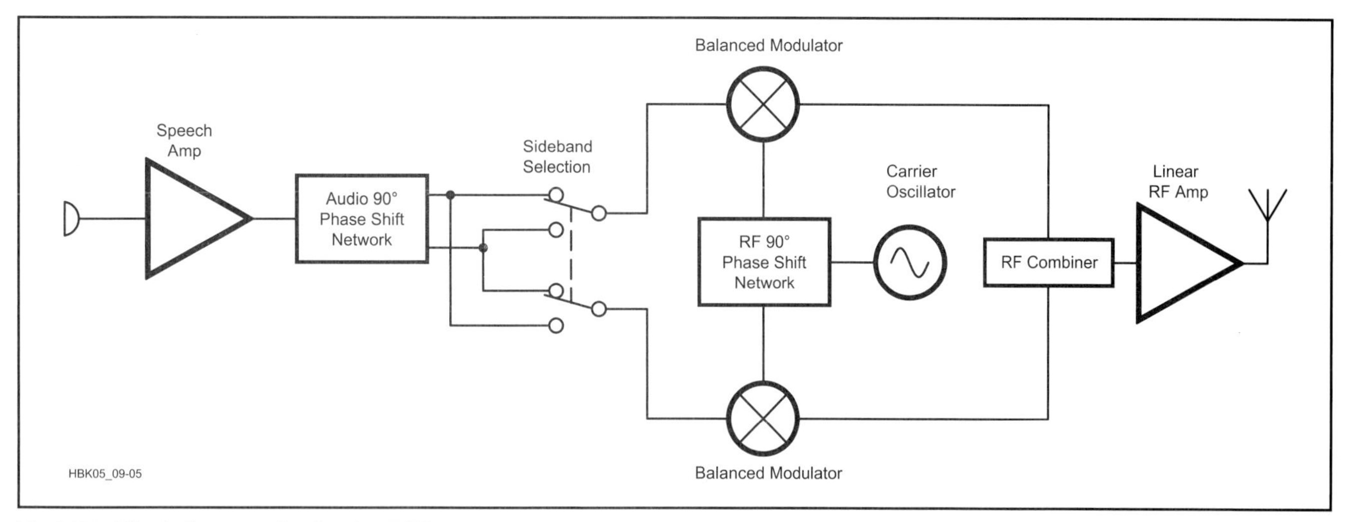

Fig 9.35—Block diagram of a phasing SSB generator.

Table 9.8
Unwanted Sideband Suppression as a Function of Phase Error

Phase Error (deg.)	*Suppression (dB)*
0.125	59.25
0.25	53.24
0.5	47.16
1.0	41.11
2.0	35.01
3.0	31.42
4.0	28.85
5.0	26.85
10.0	20.50
15.0	16.69
20.0	13.93
30.0	9.98
45.0	6.0

The numbers given assume perfect amplitude balance and phase accuracy in the other channel.

The shows that a phase accuracy of 1° is required to achieve unwanted sideband suppression of greater than 40 dB. It is difficult to achieve this level of accuracy over the entire speech band. The phase-accuracy tolerance can be loosened to 2° if the peak deviations can be made to occur within that spectral gap. The major advantage of the phasing system is that the SSB signal can be generated at the operating frequency without the need of heterodyning. Phasing can be used to good advantage even in fixed-frequency systems. A loose-tolerance (4°) phasing exciter followed by a simple two-pole crystal filter can generate a high-quality signal at low cost.

Audio Phasing Networks

Since the phasing method requires that all baseband signals be presented to the balanced modulators in both a normal (in phase) and quadrature (90° phase shifted) signal, we must provide, in the case of an audio signal, a network that can produce a constant 90° phase shift over a wide frequency range. Fortunately, the absolute phase shift is not as important as the relative phase between the two channels. Various circuits have been devised that will provide this relative shift. Robert Dome, W2WAM, pioneered a simple network using precision components that achieved this and his network was used in early SSB work. The polyphase network, which appeared in this *Handbook* for several editions, required more—but less precise—components. Methods using active filter techniques are also available.

With DSP (Digital Signal Processing), producing a 90° phase shift over a wide frequency range is easily accomplished using the Hilbert transformer. This will likely give new life to the phasing method of SSB generation since many new radios already have DSP capability present for other reasons. See the **Receivers and Transmitters** chapter of this *Handbook* for an example of an SSB receiver using DSP with the phasing method. See also the **Digital Signal Processing** chapter.

Producing 90° phase-shifted signals at RF frequencies has also used several approaches. For VHF and up, a quarter-wave section of coax is possible. Generating an RF signal at four times the desired frequency and dividing down with flip-flops generates quadrature signals accurate over a wide range of frequencies. Phase lock loops provide yet another approach.

The phasing method is useful not only for generating an SSB signal, but for any mixing or frequency-conversion task. In-Phase and Quadrature (I&Q) modulators, demodulators and mixers are in common use in modern communication technology. These allow elimination of image frequencies without filters, or greatly relax the specification of filters that are used. Digital modulation can be generated in an I&Q format that can be directly heterodyned into the RF spectrum using I&Q modulators. The **Digital Signal Processing** chapter of this *Handbook* discusses many of these concepts.

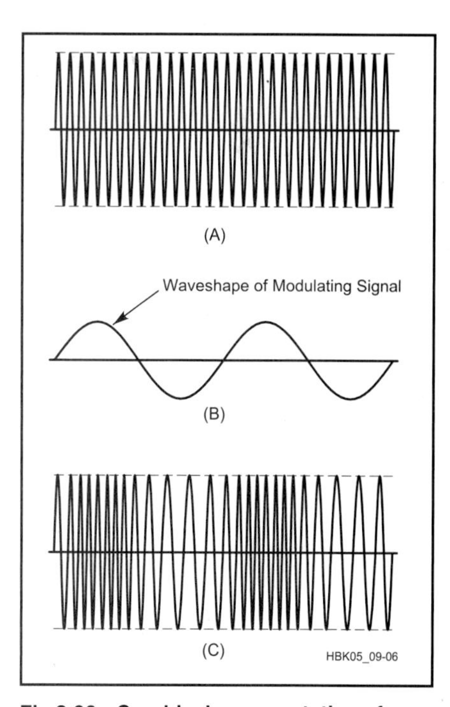

Fig 9.36 – Graphical representation of frequency modulation. In the unmodulated carrier (A) each RF cycle occupies the same amount of time. When the modulating signal (B) is applied, the radio frequency is increased and decreased according to the amplitude and polarity of the modulating signal (C).

FREQUENCY MODULATION (FM)

Unlike AM, which changes the amplitude of a radio wave in accordance with the strength of the modulation signal, FM changes the frequency of the wave so that the instantaneous value of frequency represents a voltage level in the modulating signal as is shown in **Fig 9.36**. This means that the demodulator must extract the information by generating an output whose amplitude is determined by the frequency of the received wave. Thus, FM transmission involves amplitude to frequency conversion and vice-versa. Producing these conversions was not as easy as it was in the case of AM, and thus FM was not employed as early as was AM.

As you can see in **Fig 9.37**, the circuits required for FM were especially difficult in the case of the receiver. See also the AM- and Angle-Demodulation subsections of the **Mixers, Modulators and Demodulators** chapter in this *Handbook*. In addition, mathematical analysis seemed to show that FM would require a very large bandwidth (theoretically infinite), and this discouraged early experimenters.

Edwin Armstrong was a ham before the days of call signs. While a young man, he invented the regenerative, super-regenerative and superheterodyne receivers. He went on to challenge the prevailing wisdom and developed a practical FM system. His "Yankee Network" provided high fidelity broadcasts throughout the northeastern United States in the late 1930s, using frequencies below our 6-meter band. After WWII, FM was moved to 88-108 MHz and became FM broadcasting as we now know it. Dependable day and night reception was a result of the frequency chosen, not the mode, but wideband FM, which had dictated the use of a VHF frequency where bandwidth was available, provided the wide audio response, high signal to noise ratio, and freedom from static that AM could never have provided, even at VHF. The advantages of FM were proven even when bandwidths were less than infinity. The math had not been wrong, but had just been taken a bit too literally.

Hams experimented with narrowband FM (NBFM) on the HF bands during the 1950s, but nothing much came of it. The explosion in the use of FM in the amateur bands came after surplus commercial FM equipment, using frequencies near 150 MHz, became available in the 1960s and 1970s. Two meters was the first to use this equipment and is still the workhorse of the VHF FM bands. Hams, like the commercial and public service users before them, discovered that FM has certain advantages—less noise, ease of operation, no fussy tuning and suitability for use through repeaters.

A mathematical analysis of FM is complex, and well beyond the scope of this chapter. Readers who are interested in more details can consult the **Mixers, Modulators and Demodulators** chapter of this *Handbook*. Unlike AM, where the occupied bandwidth is simple to calculate (twice the highest modulating frequency), FM bandwidth depends on both the modulating frequency and the *deviation*, which is equal to the peak frequency excursion above and below the central carrier frequency. As the math predicts, there are sidebands that extend to infinity but, fortunately, these drop off in amplitude rather quickly. As Armstrong surmised, ignoring sidebands that contain only a tiny portion of the total energy does not impair the quality of the received signal.

As a rule of thumb, adequate bandwidth for an FM voice system using narrowband modulation (5 kHz or so) is $Bn = 2\ (M+D)$ where Bn is the necessary bandwidth in hertz, M is the maximum modulation frequency in hertz, and D is the peak deviation in hertz. For narrowband FM with voice, the bandwidth equals $2 \times (3000+5000) =$ 16 kHz. This defines the filter through which the signal can be received without noticeable distortion.

Examples of FM spectra using various modulation indices are found in the **Mixers, Modulators and Demodulators** chapter of this *Handbook*. Note that as more and more sidebands appear, the amplitude of each is reduced. This is because all of the sidebands, plus the carrier, must add together (vectorially) to produce a total wave of constant amplitude. This is characteristic of an FM signal. This constant amplitude signal has the advantage of being easy to amplify without the need for a linear amplifier. Many VHF and UHF brick-type amplifiers have separate settings for FM and SSB. The FM setting is more efficient since, by giving up the requirement for linearity, we can bias the transistors for greater efficiency. Thus, an FM amplifier is easier to build than one suitable for AM or SSB.

However, this constant amplitude characteristic of FM comes at a price. The full power is being transmitted, even between words or when one is holding down the push to talk, but not actually speaking. For normal speech, the power advantage FM gains by amplifier efficiency is lost compared to SSB, where power is only transmitted when the voice requires it. One should not, however, conclude that the unmodulated FM signal serves no purpose. Its presence "quiets" the channel, opens the squelch of the receiver(s), and turns on any repeater(s) that might be in the circuit. There may also be various control tones (squelch, etc.) present, even though these may be inaudible because they are in a frequency range that the human ear does not easily perceive.

Using FM and PM with Digital Modes

Frequency-shift keying (FSK) is a means of producing frequency modulation that has discrete states; that is, the instantaneous frequency takes on definite values representing digital information. FSK is a form of FM and some of the same principles apply. FSK was covered earlier in the section on RTTY and other digital modes.

Phase modulation (PM) is very similar to FM in that it is not possible to change the frequency of a signal without impacting its phase, and vice versa. Instantaneous frequency can be considered to be the rate of change of phase of a signal. Some FM modulators have used this relationship to produce FM by phase modulation along with audio frequency shaping to convert the PM signal into the equivalent of an FM signal. This issue is discussed further in the **Mixers, Modulators and Demodulators** chapter of

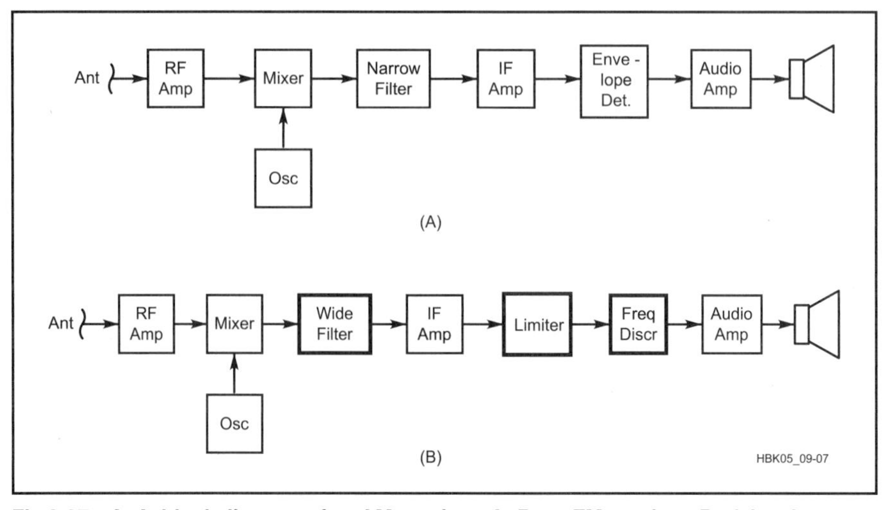

Fig 9.37—At A, block diagram of an AM receiver. At B, an FM receiver. Dark borders outline the sections that are different in the FM set.

this *Handbook*.

Phase shift keying (PSK) is a form of phase modulation suitable for digital transmissions. It is discussed further in the following pages of this chapter. Both FSK and PSK produce sidebands in accordance with the same principles discussed above. However, in order to control bandwidth, digital signals using PSK may depart from the requirement that an FM signal have a constant amplitude. Such signals are really a combination of FM and AM, and linear amplification must be used.

DIGITAL VOICE MODES

There is a risk in saying anything about an area that is developing rapidly both inside and outside Amateur Radio. Amateurs are watching digital voice developments in other radio services but not all are suitable models for Amateur Radio applications.

On MF and HF, transmission of digital voice is difficult owing to multipath propagation, QRM and noise. Several digital voice systems have been developed and more are expected. The most prominent contender is *Digital Radio Mondiale* (DRM). DRM is a non-profit consortium of broadcasters, manufacturers, educational and governmental organizations devoted to developing a single standard for digital sound broadcasting in long, medium and short wave bands. As of 2005, thousands of *software* radios were being used to hear regular DRM broadcasts from a growing number of countries. The software radio consists of a modified HF receiver, a sound card and computer software downloaded from DRM. See **www.drmx.org/** for details. As of mid 2005, there were few if any *hardware* DRM receivers available even at relatively high prices.

The DRM standard has several modes, some aimed at high fidelity music and others suitable for voice. The broadcaster can select the most appropriate mode, and the receiver will switch automatically to that mode. The various DRM modes occupy 4.5, 5, 9, 10 or 20 kHz according to the spectrum available and the quality desired. See **Fig 9.38**. DRM produces excellent quality but is more subject to the effects of interference and propagation than DSB AM.

Another digital sound system used in the broadcasting service is called *IBOC*—In-Band On-Channel. The basic idea is to send a digital signal underneath an existing AM or FM program without one interfering with the other. Although used in the United States, IBOC hasn't caught on for international broadcasting. An article on IBOC at New York station WOR was presented in the March 2003 issue of *QST* (p 28).

The International Telecommunication Union approved a standard known as ITU-R Recommendation BS.1514, *System for digital sound broadcasting in the broadcasting bands below 30 MHz*. It describes DRM and IBOC, and compares the systems.

Amateur Radio Digital Voice

For HF Amateur Radio, digital voice has the potential to provide better quality than SSB. It could have other yet-to-be-exploited possibilities, such as adapting to conditions from a "robotic" sounding speech under marginal propagation to "arm-chair copy" when conditions are good. It is possible to imbed some ancillary information in the digital stream so the receiver will be able to display call signs, graphics and other information of interest to the stations in QSO.

In 2000, the ARRL Board of Directors created a Digital Voice Working Group to investigate and promote digital voice in the Amateur Radio Service. The pioneering work done by Charles Brain, G4GUO, and Andy Talbot, G4JNT, was published in the May-June 2000 issue of *QEX*. Their system was based on use of the AMBE 2020 encoder-decoder. It uses Orthogonal Frequency Division Multiplexing (OFDM) with 36 carriers in a band of 300-2500 Hz. At least one commercial version of AMBE 2020/G4GUO system is available in the amateur market. See Hallas, Joel, W1ZR, "AOR ARD9800 Digital Voice Modem," *QST*, February 2004.

The January-February 2003 issue of *QEX* (p 49) described a special Amateur Radio adaptation of the DRM system to fit inside a 3-kHz bandwidth. This system has taken on the name *HamDream* and some information can be found on the Web at **www.qslnet.de/member/hb9tlk/**.

These OFDM standards use many carriers spaced about 50 Hz apart, each using 16QAM (Quadrature Amplitude Modulation with 16 discrete states in each symbol) or some similar modulation scheme. To mitigate the effects of multipath propagation, the symbol rate must be limited to a few hundred bauds. Thus, the high bit rate needed for voice requires both multiple carriers and complex modulation. There are tradeoffs between complexity, weak signal sensitivity, reliability under difficult conditions, speech quality and latency. The most obvious way to generate and demodulate such a signal is to use a computer and a sound card.

While the same digital voice encoder-decoders could be used at MF/HF as well as VHF and above, it may be desirable to optimize the system for best performance in each frequency range. At MF/HF, the emphasis is naturally on reliability in the presence of fading and interference, while at VHF and UHF, it is possible to design for quality of speech reproduction and possibly multimedia (voice/data/image). See **High Speed Multimedia Radio** later in this chapter.

There is much room for innovation and experimentation in this field. A great deal of work will go into developing whatever digital voice mode we will be using 10 years from now. Those interested in being a part of this exciting technology should begin by mastering the material in the **Electrical Signals and Components** and **DSP** chapters of this *Handbook*, and keeping up with *QST* and *QEX* material on digital speech. Also check the following Web sites: **www.arrl.org/tis/info/digivoice.html**; **www.dougsmith.net**; **www.temple.edu/k3tu/digital_voice.htm**; **www.tapr.org/tapr/dv/**; **www.rac.ca/opsinfo/infodig.htm#Digital%20Speech** and **www.DRM.org.**

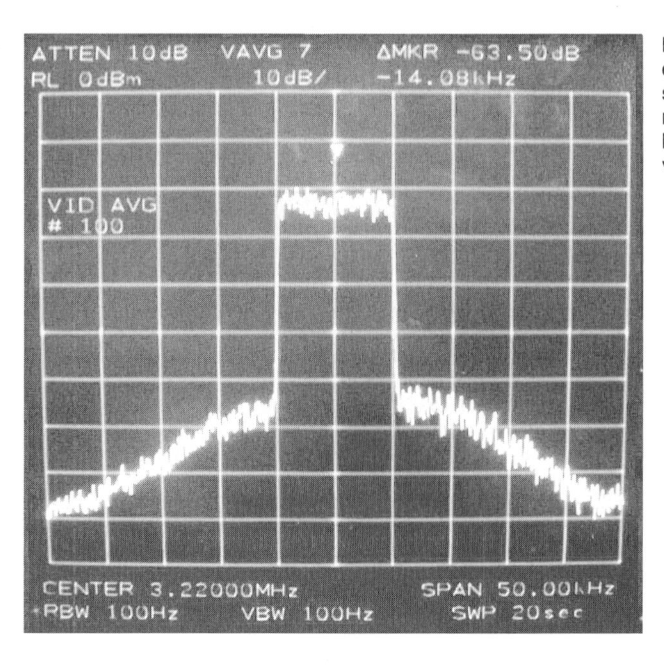

Fig 9.38—A DRM HF digital broadcast signal. Per-division resolution is 5 kHz horizontal and 10 dB vertical.

Image Modes

FACSIMILE

This section, by Dennis Bodson, W4PWF, Steven Karty, N5SK, and Ralph Taggart, WB8DQT, covers the several facsimile systems most commonly used in Amateur Radio today. For further information on the area of facsimile, its history and the development of related standards associated with this mode, refer to *FAX: Facsimile Technology and Systems*.[1] The subject of Weather fax, while of interest to many amateurs, is not a primary activity of the Amateur Radio Service. Information on this subject is contained in the *Weather Satellite Handbook*[2] and the *ARRL Image Communications Handbook*.[3]

Facsimile Overview

Facsimile (fax) is a method for transmitting very high resolution still pictures using voice-bandwidth radio circuits. The narrow bandwidth of the fax signal, equivalent to SSTV (Slow Scan TV), provides the potential for worldwide communications on the HF bands. Fax is the oldest of the image-transmitting technologies and has been for years the primary method of transmitting newspaper photos and weather charts. Fax is also used to transmit high-resolution cloud images from both polar-orbit and geostation-

Fig 9.39—Amateur Image Communications encompass a wide range of activities, a few of which are illustrated here. Narrowband Television (NBTV) experimenters explore the history and technology of the earliest days of television by restoring or recreating mechanical TV gear while exploring the possibilities of narrowband, full motion TV, primarily using computer technology. Amateur Television (ATV) operators use standard broadcast television, typically in color, to communicate on UHF and microwave frequencies. The scope of their operating activities ranges from point-to-point communication (simplex or via local ATV repeaters), roving or portable operation for a variety of reasons, including emergency and public service communications, and the application of ATV to remote sensing via aircraft, high-altitude balloons, and remote-control vehicles of all sorts. Slow-scan Television (SSTV) involves the transmission of medium and high-resolution images, usually in full-color, using standard Amateur voice equipment (typically SSB or FM). Most modern SSTV activity is computer-based, offering international DX on HF frequencies and local, regional, or space communications (satellite, MIR, and now the International Space Station) on VHF and UHF. Facsimile (Fax) encompasses the transmission and reception of very high-resolution still images (typically using computers) over a period of several to many minutes. One of the most popular areas of Amateur experimentation and operation has involved the reception of imagery from polar-orbit and geostationary weather satellite. While this *Handbook* will provide a brief introduction to some of these activities, all of them and more are covered in much greater detail in the *ARRL Image Communications Handbook*.

ary satellites. Many of these images are retransmitted using fax on the HF bands.

The resolution of typical fax images greatly exceeds what can be obtained using SSTV or even conventional television (typical images will be made up of 800 to 1600 scanning lines). This high resolution is achieved by slowing down the rate at which the lines are transmitted, resulting in image transmission times of 4 to 10 minutes.

Modern personal computers have virtually eliminated bulky mechanical fax recorders from most amateur installations. Now the incoming image can be stored in computer memory and viewed on a standard TV monitor or a high-resolution computer graphics display. The use of a color display system makes it entirely practical to transmit color fax images when band conditions permit.

The same computer-based system that handles fax images is often capable of SSTV operation as well, blurring what was once a clear distinction between the two modes. The advent of the personal computer has provided amateurs with a wide range of options within a single imaging installation. SSTV images of low or moderate resolution can be transmitted when crowded band conditions favor short-frame transmission times. When band conditions are stable and interference levels are low, the ability to transmit very high resolution fax images is just a few keystrokes away!

Hardware and Software

The computer allows reception and transmission of various fax modes, where parameters such as line-per-minute rates and indices of cooperation can be altered by simply pressing a key or by pointing and clicking a mouse. Many fax programs are available as either commercial software or shareware. Usually, the shareware packages (and often trial versions of the commercial packages) are available by downloading from the Internet.

A good starting point is the ARRL software repositories. To get to them, set your browser to the *ARRL Web* and go to the FTP (files) link in the site index. You can use any commercial search site to look for "fax" AND "software." Examples of several fax programs are as follows:

- *JVFAX* is a very popular fax program. It is DOS-based program with a large number of options for installation. It can receive and transmit several fax formats, black-and-white and color. Your computer's serial port, connected to a very simple interface, provides the connection to your transceiver.
- The *FAX 480* software program can also be used with fax as well as SSTV. For more information on this program and others including website addresses, see the July 1998 *QST* article "FAX 480 and SSTV Interfaces and Software," p 32. A copy for downloading of the free software program *vester_n.zip* for *FAX 480* can be found online at the Oakland University FTP site. This program also uses a simple interface almost identical to that for *JVFAX*.
- *Weatherman* is a DOS-based program, using a SoundBlaster (or compatible) card as the interface. The program is shareware and provides receive-only capability. A single, shielded wire from your receiver audio output to the computer audio input is the only connection needed.
- *WXSat* operates under *Windows 3.X*. While specifically set up to decode and store weather-satellite APT pictures, it can also be used for HF-fax reception.

Both *Weatherman* and *WXSat* are samples of what you can find during a search on the Internet. Often, programs are offered and then either withdrawn or improved over the versions previously distributed—To get the latest and greatest you have to periodically search and see what comes up. If you use an online service such as CompuServe or AOL, they are another source of fax software. Check their ham forums or sections for listings.

Many commercial multimode controllers either contain software to receive and transmit fax, or are compatible with PC-hosted software. Available controller suppliers include MFJ, Timewave, and Kantronics; additional software may be required for the Kam Plus. Check the advertising pages of *QST* for the latest units available.

One well-known fax page on the Internet, complete with downloadable software, is posted and maintained by Marius Rensen; it contains listings of commercial fax transmissions for you to test your software or just SWL for interest. Before using a program taken from any Internet source, check other sources for newer versions. It is not uncommon to have older versions posted on one place and newer versions in another. It is always a good idea to virus check software before and after unzipping.

Image transmission using voice bandwidth is a trade-off between resolution and time. In the section on slow-scan television, standards are described that permit 240-line black-and-white images to be transmitted in about 36 seconds, while color images of similar resolution require anywhere from 72 to 188 seconds, depending on the color format. In terms of resolution, 240-line SSTV images are roughly equivalent to what you would obtain with a standard broadcast TV signal recorded on a home VCR. This is more than adequate for routine video communication, but there are many situations that demand images with higher resolution.

HAL Communications Corporation has developed an interesting system that enables a standard fax machine (Group 3 or G3) to send commercial fax images over HF radio. HAL Communications accomplishes this with just two small ancillary devices, which connect between a standard fax machine and an ordinary HF radio transceiver. This method is frequently referred to as "G3 fax over radio." Any G3 fax machine can be connected to the HAL FAX-4100 controller with just a standard RJ-11 modular connector. The FAX-4100 controller connects directly to the HAL CLOVER-2000 (DSP-4100) radio data modem, which in turn connects to the HF transceiver. This entire setup is duplicated at the opposite end of the link.

A "call" is initiated from the fax machine keypad just as if the fax machine were connected to a phone line. The FAX-4100 controller includes a built-in 9600-baud G3 modem that emulates the telephone system. The controller at the initiating end answers the ring from the originating fax machine, establishes the HF radio link (based on the "phone number"), and handshakes with the controller at the other end to start the receiving fax machine. Fax image data then passes from the fax machine into the controller's memory at the originating end. The controller also establishes a data link between the CLOVER-2000 modems at both ends, then passes the fax data through them and the controller at the receiving end, and finally into the receiving G3 fax machine. HAL has automated the HF radio operating procedures. To the user, sending a fax over HF radio is a simple three-step process:

1. Lay the page(s) on the fax machine.
2. Enter the ID number of the other station.
3. Push GO on the fax machine.

Housekeeping control functions and indications are also automated, feeding messages back to the fax machine whenever possible (link failed, other station not available, etc). A full page can be sent in 2 to 6 minutes, depending upon ionospheric conditions and the image density of the page being transmitted. The entire link set up and maintenance procedure is transparent to the fax operator, who need not know nor care that an HF radio system is part of the fax link. It all works just like a standard fax telephone transmission. An additional piece of equipment is available from HAL to enable the same fax machine to be shared between HF radio and conventional telephone lines. The HAL LI-4100 Line Interface is a "smart switch" that can be connected between the fax machine, the FAX-4100 controller, and up to two telephone lines.

Courteous SSTV Operating

- Recommended frequencies: 3.845, 7.171, 14.230, 14.233, 21.340, 28.680, 145.5 MHz.
- 14.230 is the most active.
- Make contact by voice before sending SSTV.
- Not all systems recognize the VIS code, so it is good manners to announce the mode before transmitting.

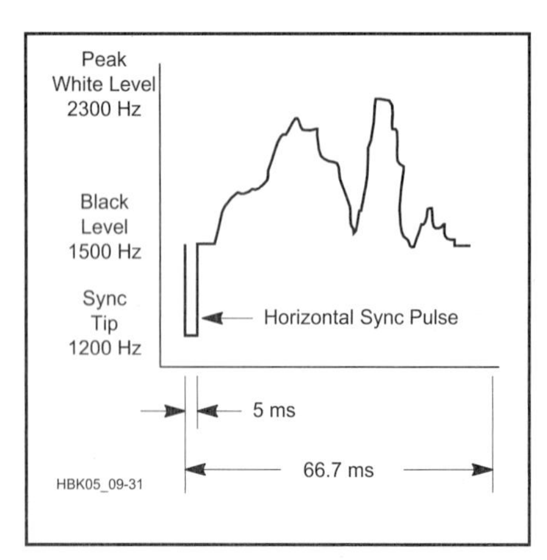

Fig 9.40—Early SSTV operators developed a basic 8-second black and white transmission format. The sync pulses are often called "blacker than black." A complete picture would have 120 lines (8 seconds at 15 ms per line). Horizontal sync pulses occur at the beginning of every line; a 30 ms vertical sync pulse precedes each frame.

Facsimile References

McConnell, Bodson, and Urban, *FAX: Facsimile Technology and Systems*, 3rd Ed., Artech House, 1999,

Taggart, Ralph, WB8DQT, *Weather Satellite Handbook*, 5th Ed. (Newington: ARRL, 1994).

Taggart, Ralph, WB8DQT, "A New Standard for Amateur Radio Facsimile," *QST*, Feb 1993.

Taggart, Ralph, WB8DQT, *ARRL Image Communications Handbook*, 1st Ed. (Newington: ARRL, 2002).

SLOW-SCAN TELEVISION (SSTV)

An ancient Chinese proverb states: *"A picture is worth a thousand words."* It's still true today. Sight is our highest bandwidth sense and the primary source of information about the world around us. What would you think about a TV news program without pictures about the stories? Would you enjoy reading the comics if there were no drawings with the text? Do you close your eyes when talking to someone in person? Many hams feel the same way about conversing with Amateur Radio: Sending images is a wonderful way to enhance communication. This material was written by John Langner, WB2OSZ.

For decades only a dedicated few kept SSTV alive. The small numbers of commercial equipment were very expensive and home-brewing was much too complicated for most people. Early attempts at computer-based systems were rather crude and frustrating to use.

The situation has changed dramatically in recent years. There is now a wide variety of commercial products and home-brew projects to fit every budget. SSTV activity is experiencing rapid growth. There is much software that uses computer sound cards for SSTV.

The early SSTV 8-second transmission standard is illustrated in **Fig 9.40**. Audio tones in the 1500 to 2300-Hz range represent black, white and shades of gray. A short 1200-Hz burst separates the scan lines and a longer 1200-Hz tone signals the beginning of a new picture.

Color SSTV History

The early experimenters weren't content with only black and white (B&W) images and soon devised a clever way to send color pictures with B&W equipment. The transmitting station sends the same image three times, one each with red, green and blue filters in front of the TV camera lens. The receiving operator took three long-exposure photographs of the screen, placing red, green and blue filters in front of the film camera's lens at the appropriate times. This was known as the "frame sequential" method.

In the 1970s, it became feasible to save these three images in solid-state memory and simultaneously display them on an ordinary color TV. The frame-sequential method had some drawbacks. As the first frame was received you'd see a red and black image. During the second frame, green and yellow would appear. Blue, white, and other colors wouldn't show up until the final frame. Any noise (QRM or QRN) could ruin the image registration (the overlay of the frames) and spoil the picture.

The next step forward was the "line sequential" method. Each line is electronically scanned three times before being transmitted: once each for the red, green, and blue picture components. Pictures could be seen in full color as they were received and registration problems were reduced. The Wraase SC-1 modes are examples of early line-sequential color transmission. They have a horizontal sync pulse for each of the color component scans. The major weakness here is that if the receiving end gets out of step, it won't know which scan represents which color.

Rather than sending color images with the usual RGB (red, green, blue) components, Robot Research used luminance and chrominance signals for their 1200C modes. The first half or two thirds of each scan line contains the luminance information, which is a weighted average of the R, G and B components. The remainder of each line contains the chrominance signals with the color information. Existing B&W equipment could display the B&W-compatible image on the first part of each scan line and the rest would go off the edge of the screen. This compatibility was very beneficial when most people still had only B&W equipment.

The luminance-chrominance encoding made more efficient use of the transmission time. A 120-line color image could be sent in 12 s, rather than the usual 24 s. Our eyes are more sensitive to details in changes of brightness than color, so the time could be used more efficiently by devoting more time to luminance than chrominance. The NTSC and PAL broadcast standards also take advantage of this vision characteristic and use less bandwidth for the color part of the signal.

The 1200C introduced another innovation: It encoded the transmission mode in the vertical sync signal. By using narrow FSK encoding around the sync frequency, compatibility was maintained. This new signal just looked like an extra-long vertical sync to older equipment. The luminance-chrominance encoding offers some benefits, but image quality suffers. It is acceptable for most natural images but looks bad for sharp, high-contrast edges, which are more and more common as images are altered via computer graphics. As a result, all newer modes have returned to RGB encoding.

The Martin and Scottie modes are essentially the same except for the timings. They have a single horizontal sync pulse for each set of RGB scans. Therefore, the receiving end can easily get back in step if synchronization is temporarily lost. Although they have horizontal sync, some implementations ignore them on receive. Instead, they rely on very accurate time bases at the transmitting and receiving stations to keep in step. The advantage of this "synchronous" strategy is that missing or corrupted sync pulses won't disturb the received image. The disadvantage is that even slight timing inaccuracies produce slanted pictures.

In the late 1980s, yet another incompatible mode was introduced. The AVT mode is different from all the rest in that it has *no horizontal sync*. It relies on very accurate oscillators at the sending and receiving stations to maintain synchronization. If the beginning-of-frame sync is missed, it's all over. There is no way to determine where a scan line begins. However, it's much harder to miss the 5-s header than the 300-ms VIS code. Redundant information is encoded 32 times and a more powerful error-detection scheme is used. It's only necessary to receive a small part of the AVT header in order to achieve synchronization. After this, noise can wipe out parts of the image, but image alignment and colors remain correct. **Table 9.9** lists characteristics of common modes.

Scan Converters

A scan converter is a device that converts signals from one TV standard to another. In this particular case we are interested in converting between SSTV, which can be sent through audio channels, and fast scan (broadcast or ATV), so we can use ordinary camcorders and color televisions to generate and display pictures. From about 1985 to 1992, the Robot 1200C was king.

Fig 9.41 shows a typical SSTV station built around a scan converter such as the Robot 1200C or a SUPERSCAN 2001. The scan converter has circuitry to accept a TV signal from a camera and store it in memory. It also generates a display signal for an ordinary television set. The interface to the radio is simply audio in, audio out and a push-to-talk (PTT) line. In the early days, pictures were stored on audio tape, but now computers store them in memory. Once a picture is in a computer, it can be enhanced with paint programs.

This is the easiest approach. Just plug in the cables, turn on the power and it works. Many people still prefer special dedicated hardware, but most of the recent growth of SSTV has been from these lower cost PC-based systems using sound cards and software.

SSTV with a Computer

There were many attempts to use early home computers for SSTV. Those efforts were hampered by very small computer memories, poor graphics capabilities and poor software development tools.

Surprisingly, little was available for the ubiquitous IBM PC until around 1992, when several systems appeared in quick succession. By this time, all new computers had a VGA display, which is required for this application. Most modern SSTV stations look like **Fig 9.42**. Some sort of interface is used to get audio in and out of the computer. These can be external interfaces connected to a serial or printer port, an internal computer card specifically designed for SSTV or even a peripheral audio card.

Table 9.9
SSTV Transmission Characteristics

Mode	*Designator*	*Color Type*	*Scan Time (sec)*	*Scan Lines*	*Notes*
AVT	24	RGB	24	120	D
	90	RGB	90	240	D
	94	RGB	94	200	D
	188	RGB	188	400	D
	125	BW	125	400	D
Martin	M1	RGB	114	240	B
	M2	RGB	58	240	B
	M3	RGB	57	120	C
	M4	RGB	29	120	C
HQ	HQ1	YC	90	240	G
	HQ2	YC	112	240	G
Pasokon TV	P3	RGB	203	16+480	
	P5	RGB	305	16+480	
	P7	RGB	406	16+480	
Robot	8	BW	8	120	A,E
	12	BW	12	120	E
	24	BW	24	240	E
	36	BW	36	240	E
	12	YC	12	120	
	24	YC	24	120	
	36	YC	36	240	
	72	YC	72	240	
Scottie	S1	RGB	110	240	B
	S2	RGB	71	240	B
	S3	RGB	55	120	C
	S4	RGB	36	120	C
	DX	RGB	269	240	B
Wraase SC-1	24	RGB	24	120	C
	48	RGB	48	240	B
	96	RGB	96	240	B
Wraase SC-2	30	RGB	30	128	
	60	RGB	60	256	
	120	RGB	120	256	
	180	RGB	180	256	
Pro-Skan	J120	RGB	120	240	
WinPixPro	GVA 125	BW	125	480	
	GVA 125	RGB	125	240	
	GVA 250	RGB	250	480	
JV Fax	JV Fax Color	RGB	variable	variable	F
FAX480	Fax 480	BW	138	480	
	Truscan	BW	128	480	H
	Colorfax 480	RGB	384	480	I

Notes

RGB—Red, green and blue components sent separately.
YC—Sent as Luminance (Y) and Chrominance (R-Y and B-Y).
BW—Black and white.
A—Similar to original 8-second black & white standard.
B—Top 16 lines are gray scale. 240 usable lines.
C—Top 8 lines are gray scale. 120 usable lines.
D—AVT modes have a 5-second digital header and no horizontal sync.
E—Robot 1200C doesn't really have B&W mode but it can send red, green or blue memory separately. Traditionally, just the green component is sent for a rough approximation of a b&w image.
F—JV Fax Color mode allows the user to set the number of lines sent, the maximum horizontal resolution is slightly less than 640 pixels. This produces a slow but very high resolution picture. SVGA graphics are required.
G—Available only on Martin 4.6 chipset in Robot 1200C.
H—Vester version of FAX480 (with VIS instead of start signal and phasing lines).
I—Trucolor version of Vester Truscan.

Perhaps the single most significant breakthrough in computer-based SSTV is the wide range of Windows- and DOS-based programs using the PC's soundcard as the main transmit/receive interface. Many operators nowadays use the popular freeware program *MMSSTV* by JE3HHT (see **mmhamsoft.ham-radio.ch/mmsstv/**), with a simple hardware interface to go into and to come out of the PC's soundcard. Information on current computer SSTV software is available at **www.tima.com/~djones**. The subject of computer SSTV software and interfacing is also discussed at length in the *Image Communications Handbook* published by ARRL.

A simple "clipper" hardware interface to the computer's soundcard can be built with less than $15 worth of RadioShack parts. **Fig 9.43** shows such an interface circuit used for receiving and transmitting. Connect the output of T2 to the phone patch input (sometimes labeled LINE INPUT) of your transceiver, if it has one. Otherwise, you'll have to use the microphone input. R3 is set to the proper level for the audio going to the transmitter. You must set the audio signal into the transceiver at a level it can handle without distortion.

There is no low-pass filtering in the audio line between the computer output and transmitter audio input. On-the-air checks with many stations reveal that no additional external filtering is required when using SSB transmitters equipped with mechanical or

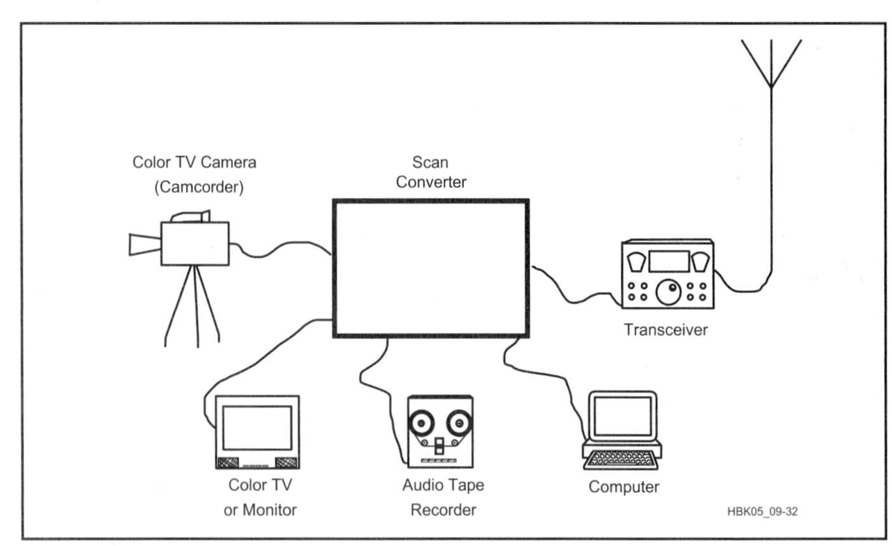

Fig 9.41—Diagram of an older SSTV station based on a scan converter.

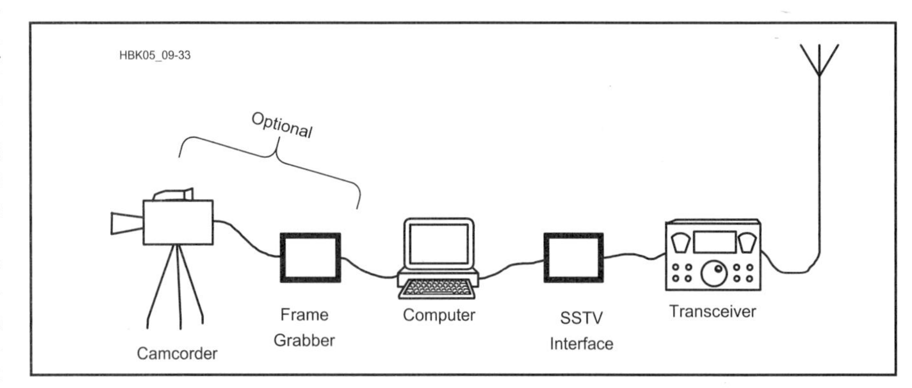

Fig 9.42—A modern SSTV station that utilizes the soundcard in a PC.

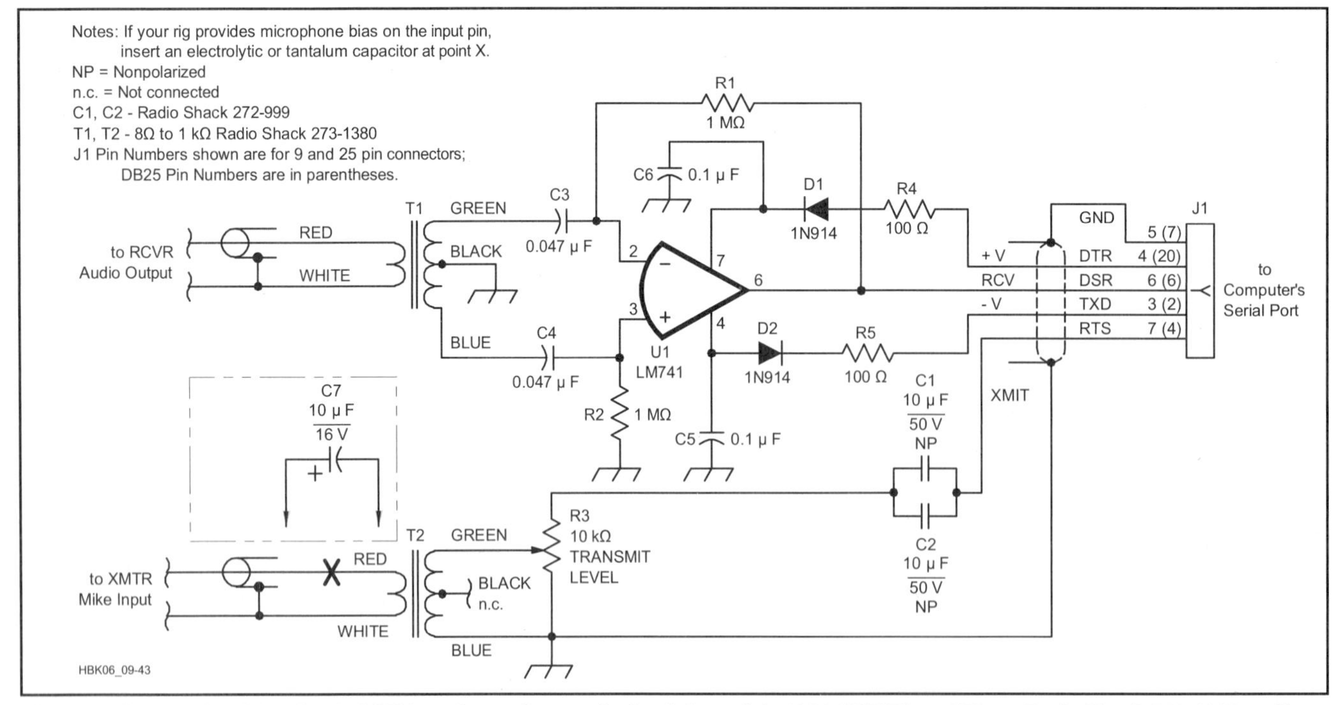

Fig 9.43—Schematic of the simple SSTV receive and transmit circuit from July 1998 *QST*. T1 and T2 are RadioShack 273-1380 audio-output transformers; the 20-µF, 50-V capacitor is a parallel combination of two RadioShack 272-999 10-µF, 50-V non-polarized capacitors; equivalent parts can be substituted. Unless otherwise specified, resistors are 1/4-W, 5%-tolerance carbon composition or film units. At J1, numbers in parentheses are for 25-pin serial port connectors; other numbers are for 9-pin connectors.

crystal filters. If you intend to use this circuit with an AM or phasing-type SSB rig (or with VHF/UHF FM transmitters), add audio filtering to provide the required spectral purity. An elliptical low-pass filter such as described by Campbell (see references) should be adequate for most cases.

Circuit component values aren't critical, nor is the circuit's physical construction. A PC board is available from FAR Circuits, but perf-board construction employing short leads works fine.

Digital Slow-Scan Television

DSSTV is a method of transmitting computer image files, such as JPEG or GIF over Amateur Radio, as described in an article by Ralph Taggart, WB8DQT, in the Feb 2004 issue of *QST*. The signal format phase modulates a total of eight subcarriers (ranging from 590 to 2200 Hz at intervals of 230 Hz. Each subcarrier has nine possible modulation states. This signal modulation format is known as *redundant digital file transfer* (RDFT) developed by Barry Sanderson, KB9VAK. RDFT is also used with SCAMP, described earlier in this chapter.

SSTV Summary

For decades there was a convenient excuse for not trying SSTV: it cost kilobucks to buy a specialized piece of equipment. But you can't use that excuse anymore. There are free programs that only require trivial hardware interfaces to receive and transmit slow-scan pictures. Once you get hooked, there are plenty of other home-brew projects and commercial products available at affordable prices. You need not be a computer wizard to install and use these systems.

SSTV Bibliography

Battles, B. and Ford, S., "Smile—You're on Ham Radio!" *QST*, Oct 1992.

Bodson, D., W4PWF, and Karty, S., N5SK, "FAX480 and SSTV Interfaces and Software," *QST*, Jul 1998.

Campbell, R, "High-Performance, Single-Signal Direct-Conversion Receivers," *QST*, Jan 1993. See also Feedback, *QST*, Apr 1993, p 75.

Langner, J. WB2OSZ, "Slow Scan Television—It isn't expensive anymore," *QST*, Jan 1993,.

SSTV Glossary

ATV—Amateur Television. Sending pictures by Amateur Radio. You'd expect this abbreviation to apply equally to fast-scan television (FSTV), slow-scan television (SSTV) and facsimile (fax), but it's generally applied only to FSTV.

AVT—Amiga Video Transceiver. 1) Interface and software for use with an Amiga computer, developed by Ben Blish-Williams, AA7AS, and manufactured by Advanced Electronic Applications (AEA); 2) a family of transmission modes first introduced with the AVT product.

Back porch—The blank part of a scan line immediately following the horizontal sync pulse.

Chrominance—The color component of a video signal. NTSC and PAL transmit color images as a black-and-white compatible luminance signal along with a color subcarrier. The subcarrier phase represents the hue and the subcarrier's amplitude is the saturation. Robot color modes transmit pixel values as luminance (Y) and chrominance (R-Y [red minus luminance] and B-Y [blue minus luminance]) rather than RGB (red, green, blue).

Demodulator—For SSTV, a device that extracts image and sync information from an audio signal.

Field—Collection of top to bottom scan lines. When interlaced, a field does not contain adjacent scan lines and there is more than one field per frame.

Frame—One complete scanned image. The Robot 36-second color mode has 240 lines per frame. NTSC has 525 lines per frame with about 483 usable after subtracting vertical sync and a few lines at the top containing various information.

Frame Sequential—A method of color SSTV transmission that sent complete, sequential frames of red, then green and blue. Now obsolete.

Front porch—he blank part of a scan line just before the horizontal sync.

FSTV—Fast-Scan TV. Same as common, full-color, motion commercial broadcast TV.

Interlace—Scan line ordering other than the usual sequential top to bottom. For example, NTSC sends a field with just the even lines in 1/60 second, then a field with just the odd lines in 1/60 second. This results in a complete frame 30 times a second. AVT "QRM" mode is the only SSTV mode that uses interlacing.

Line Sequential—A method of color SSTV transmission that sends red, green, and blue information for each sequential scan line. This approach allows full-color images to be viewed during reception.

Luminance—The brightness component of a video signal. Usually computed as Y (the luminance signal) = 0.59 G (green) + 0.30 R (red) + 0.11 B (blue).

Martin—A family of amateur SSTV transmission modes developed by Martin Emmerson, G3OQD, in England.

NTSC—National Television System Committee. Television standard used in North America and Japan.

PAL—Phase alteration line. Television standard used in Germany and many other parts of Europe.

Pixel—Picture element. The dots that make up images on a computer's monitor.

P7 monitor—SSTV display using a CRT having a very-long-persistence phosphor.

RGB—Red, Green, Blue. One of the models used to represent colors. Due to the characteristics of the human eye, most colors can be simulated by various blends of red, green, and blue light.

Robot—(1) Abbreviation for Robot 1200C scan converter; (2) a family of SSTV transmission modes introduced with the 1200C.

Scan converter—A device that converts one TV standard to another. For example, the Robot 1200C converts SSTV to and from FSTV.

Scottie—A family of amateur SSTV transmission modes developed by Eddie Murphy, GM3SBC, in Scotland.

SECAM—Sequential color and memory. Television standard used in France and the Commonwealth of Independent States.

SSTV—Slow Scan Television. Sending still images by means of audio tones on the MF/HF bands using transmission times of a few seconds to a few minutes.

Sync—That part of a TV signal that indicates the beginning of a frame (vertical sync) or the beginning of a scan line (horizontal sync).

VIS—Vertical Interval Signaling. Digital encoding of the transmission mode in the vertical sync portion of an SSTV image. This allows the receiver of a picture to automatically select the proper mode. This was introduced as part of the Robot modes and is now used by all SSTV software designers.

Wraase—A family of amateur SSTV transmission modes first introduced with the Wraase SC-1 scan converter developed by Volker Wraase, DL2RZ, of Wraase Electronik, Germany.

Montalbano, J., KA2PYJ, "The ViewPort VGA Color SSTV System," *73*, Aug 1992.

Taggart, R., WB8DQT, "Digital Slow-Scan Television," *QST*, Feb 2004, p 47-51.

Taggart, R., WB8DQT, *Image Communications Handbook*, Published by ARRL, Newington, CT, 2002. ARRL Order No. 8616.

Vester, B., K3BC "Vester SSTV/FAX80/Fax System Upgrades," Technical Correspondence, *QST*, Jun 1994.

Vester, B., K3BC, "SSTV: An Inexpensive System Continues to Grow," Dec 1994 *QST*.

Vester, B., K3BC, "K3BC's SSTV Becomes TRUSCAN," Technical Correspondence, *QST*, Jul 1996.

Table 9.10

Line-of-Sight Snow-Free 70-cm ATV Communication Distances

This table relates transmit and receive station antenna gains to communication distances in miles for 1/10/100 W PEP at 440 MHz. To find the possible snow-free distance under line-of-sight conditions, select the column that corresponds to transmit antenna gain and the row for the receive antenna gain. Read the distance where the row and column intersect. Multiply the result by 0.5 for 902 MHz and 0.33 for 1240 MHz.
The table assumes 2 dB of feed-line loss, a 3 dB system noise figure at both ends and snow-free is greater than 40 dB picture:noise ratio (most home cameras give 40 to 45 dB picture:noise; this is used as the limiting factor to define snow-free ATV pictures). The P unit picture rating system goes down about 6 dB per unit. For instance, P4 pictures would be possible at double the distances in the table.

RX Antenna / ***TX Antenna***	*0 dBd*	*4 dBd*	*9 dBd*	*15.8 dBd*
0 dBd	0.8/2.5/8	1/3.5/11	2/7/22	5/15/47
4 dBd	1/3.5/11	2/6/19	3.5/11/34	7.5/23/75
9 dBd	2/7/22	3.5/11/34	6/19/60	13/42/130
15.8 dBd	5/15/47	7.5/23/75	13/42/130	29/91/290

FAST-SCAN TELEVISION

Fast-scan amateur television (FSTV or just ATV) is a wide-band mode that uses standard broadcast, or NTSC, television scan rates. It is called "fast scan" only to differentiate it from slow-scan TV. In fact, no scan conversions or encoder/decoders are necessary with FSTV. Any standard TV set can display the amateur video and audio. Standard (1 V P-P into 75 Ω) composite video from home camcorders, cameras, VCRs or computers is fed directly into an AM ATV transmitter. The audio has a separate connector and goes through a 4.5 MHz FM subcarrier generator that is mixed with the video. This section was written by Tom O'Hara, W6ORG.

Fig 9.44—Students enjoy using ATV to communicate between science and computer classes.

Fig 9.45—The ATV view shows the aft end of the Space Shuttle cargo bay during mission STS-9.

Amateurs regularly show themselves in the shack, zoom in on projects, show home video tapes, computer programs and just about anything that can be shown live or by tape (see **Figs 9.44** and **9.45**). Whatever the camera "sees" and "hears" is faithfully transmitted, including color and sound information. Picture quality is about equivalent to that of a VCR, depending on video signal level and any interfering carriers. All of the sync and signal-composition information is present in the composite-video output of modern cameras and camcorders. Most camcorders have an accessory cable or jacks that provide separate video and audio outputs. Audio output may vary from one camera to the next, but usually it has been amplified from the built-in microphone to between 0.1 to 1 V P-P (into a 10-kΩ load).

ATV transmitters have been carried by helium balloons to above 100,000 ft, to the edge of space. The result is fantastic video transmissions, showing the curvature of the Earth, that have been received as far as 500 miles from the balloon. Small cameras have been put into the cockpits of R/C model airplanes to transmit a pilot's-eye view. Many ATV repeaters retransmit Space Shuttle video and audio from NASA during missions. This is especially exciting for schools involved with SAREX. ATV is used for public service events, such as parades, races, Civil Air Patrol searches and remote damage assessment.

Emergency service coordinators have found that live video from a site gives a better understanding of a situation than is possible from voice descriptions alone. Weather-radar video, WEFAX, or other computer generated video has also been carried by ATV transmitters for RACES groups during significant storms. This use enables better allocation of resources by presenting real-time information about the storm track. Computer graphics and video special effects are often transmitted to dazzle the viewers.

Table 9.11

Bit encoding for 5.5 Mbps and 11 Mbps CCK transmissions.

Data Rate Mbps	CCK encoded bit	DQPSK encoded bit
5.5	2	2
11	6	2

How Far Does ATV Go?

The theoretical snow-free line-of-sight distance for 10 W, given 15.8-dBd antennas and 2-dB feed-line loss at both ends, is 91 miles. (See **Table 9.11**.) However, except for temperature-inversion skip conditions, reflections, or through high hilltop repeaters, direct line-of-sight ATV contacts seldom exceed 25 miles. The RF horizon over flat terrain with a 50-ft tower is 10 miles. For best DX, use low-loss feed line and a broadband high-gain antenna, up as high as possible. The antenna system is the most important part of an ATV system because it affects both receive and transmit signal strength.

A snow-free, or "P5," picture rating (see **Fig 9.46**) requires at least 200 µV (–61 dBm) of signal at the input of the ATV receiver, depending on the system noise figure and bandwidth. The noise floor increases with bandwidth. Once the receiver system gain and noise figure reaches this

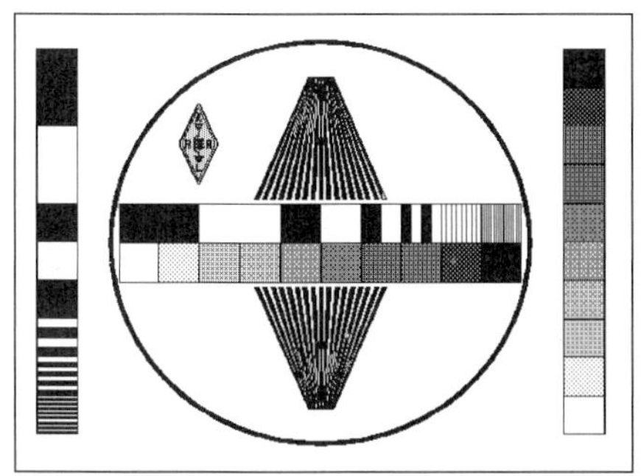

P5 —Excellent

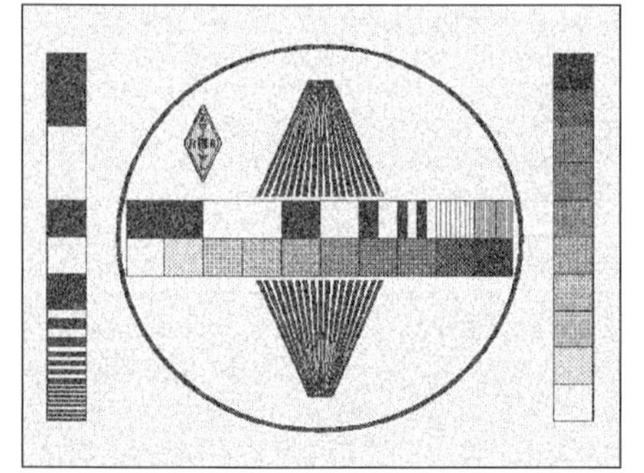

P4—Good

P3—Fair

P2—Poor

P1—Barely perceptible

Fig 9.46—An ATV quality reporting system.

floor, no additional gain will increase sensitivity. At 3-MHz bandwidth the noise floor is 0.8 µV (–109 dBm) at standard temperature. If you compare this to an FM voice receiver with 15 kHz bandwidth; there is a 23 dB difference in the noise floor. However the eye, much like the ear of experienced CW operators, can pick out sync bars in the noise below the noise floor. Sync lock and large well contrasted objects or lettering can be seen between 1 and 2 µV. Color and subcarrier sound come out of the noise between 2 and 8 µV depending on their injection level at the transmitter and TV-set differences.

Two-meter FM is used to coordinate ATV contacts. Operators must take turns transmitting on the few available channels and the 2-m link allows full-duplex audio from many receiving stations to the ATV transmitting station, who speaks on the sound subcarrier. This is great for interactive show and tell. It is also much easier to monitor a squelched 2-m channel using an omni antenna rather than searching out each station with a beam. Depending on the third-harmonic relationship to the video on 70 cm, 144.34 MHz and 146.43 MHz (simplex) are the most popular frequencies. They are often mixed with the subcarrier sound on ATV repeater outputs.

Getting the Picture

Since the 70-cm band corresponds to cable TV channels 57 through 61, seeing your first ATV picture may be as simple as connecting a good outside 70-cm antenna (aligned for the customary local polarization) to a cable-ready TV set's antenna input jack. Cable channel 57 is 421.25 MHz, and each channel is progressively 6 MHz higher. (Note that cable and broadcast UHF channel frequencies are different.) Check the *ARRL Repeater Directory* for a local ATV repeater output that falls on one of these cable channels. Cable-ready TVs may not be as sensitive as a low-noise downconverter designed just for ATV, but this technique is well worth a try.

Most stations use a variable tuned downconverter specifically designed to convert the whole amateur band down to a VHF TV channel. Generally the 400 and 900-MHz bands are converted to TV channel 3 or 4, whichever is not used in the area. For 1200 MHz converters, channels 7 through 10 are used to get more image rejection. The downconverter consists of a low-noise preamp, mixer and tunable or crystal-controlled local oscillator. Any RF at the input comes out at the lower frequencies. All signal processing is done in the TV set. A complete receiver with video and audio output would require all the TV sets circuitry, less the sweep and CRT components. There is no picture-quality gain by going direct from a receiver to a video monitor (as compared with a TV set) because IF and detector bandwidth are still the limiting factors.

A good low-noise amateur downconverter with 15 dB gain ahead of a TV set will give sensitivity close to the noise floor. A preamp located in the shack will not significantly increase sensitivity, but rather will reduce dynamic range and increase the probability of intermodulation interference. Sensitivity can be increased by increasing antenna-system gain:

- Reducing feed-line loss
- Increasing antenna gain
- Or adding an antenna mounted preamp (which will eliminate the coax loss, plus any loss through transmit linear amplifier TR relays).

Remember that each 6 dB increase in combination of transmitted power, reduced coax loss, antenna gain or receiver sensitivity can double the line-of-sight distance.

Foliage greatly attenuates the signal at UHF, so place antennas above the tree tops for the best results. Beams made for 432-MHz weak-signal work or 440-MHz FM may not have enough SWR bandwidth to cover all the ATV frequencies for transmitting, but they are okay for reception. A number of manufacturers now make ATV beam antennas to cover the whole band from 420 to 450 MHz. Use low-loss coax (such as Belden 9913: 2.5 dB/100 ft at 400 MHz) or Hardline for runs over 100 ft. All outside connectors must be weatherproofed with tape or coax sealer—Any water that gets inside the coax will greatly increase the attenuation. Almost all ATV antennas use N connectors, which are more resistant to moisture contamination than other types.

Antenna polarization varies from area to area. Technically, the polarization should be chosen to give additional isolation (up to 20 dB) from other users near the channel. It is more common to find that the polarity was

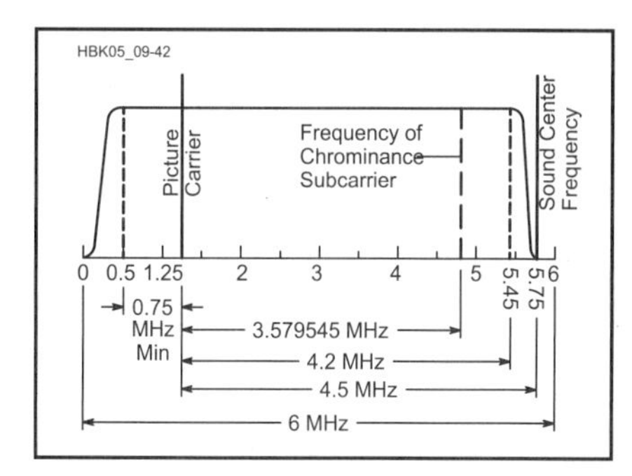

Fig 9.47—A 6-MHz video channel with the video carrier 1.25 MHz up from the lower edge. The color subcarrier is at 3.58 MHz and the sound subcarrier at 4.5 MHz above the video carrier.

determined by the first local ATV operators (which antennas they had in place for other modes). Generally, those on 432 MHz SSB and weak-signal DX have horizontally polarized antennas, and those into FM, public service or repeaters will have vertical antennas. Check with local ATV operators before permanently locking down the antenna-mast clamps. Circularly polarized antennas let you work all modes, including satellites, with only 3 dB sacrificed when working a fixed polarity.

ATV Frequencies

Standard broadcast TV channels are 6 MHz wide to accommodate the composite video, 3.58 MHz color and 4.5 MHz sound subcarriers. (See **Fig 9.47**.) Given the NTSC 525 horizontal line and 30 frames per second scan rates, the resulting horizontal resolution bandwidth is 80 lines per MHz. Therefore, with the typical TV set's 3-dB rolloff at 3 MHz (primarily in the IF filter), up to 240 vertical black lines can be seen. Color bandwidth in a TV set is less than this, resulting in up to 100 color lines. Lines of resolution are often confused with the number of horizontal scan lines per frame. The video quality should be every bit as good as on a home video recorder.

The lowest frequency amateur band wide enough to support a TV channel is 70 cm (420 - 450 MHz), and it is the most popular. With transmit power, antenna gains and coax losses equal, decreasing frequency increases communication range. The 33-cm band goes half the distance that 70 cm does, but this can be made up to some extent with high-gain antennas, which are physically smaller at the higher frequency. A Technician class or higher license is required to transmit ATV on this band, and Novices can transmit ATV only in the 1270 to 1295 MHz segment of the 23-cm band. Depending on local band plan options, there is room for no more than two simultaneous ATV channels in the 33- and 70-cm bands without interference. Generally, because only two channels are available in the 70-cm band, an ATV repeater input on 439.25 or 434.0 MHz is shared with simplex stations. 421.25 MHz is the most popular in-band repeater output frequency. At least 12 MHz of separation is necessary for in-band repeaters because of filter-slope attenuation characteristics and TV-set adjacent-channel rejection. Some repeaters have their output on the 33-cm or 23-cm bands (the 923.25 and 1253.25 MHz output frequencies are most popular). This frees up a channel on 70 cm for simplex. Such cross-band repeaters also make it easier for the transmitting operator to monitor the repeated video with only proper antenna separation needed to prevent receiver desensitization. 426.25 MHz is used for simplex, public service and R/C models in areas with cross-band repeaters, or as a alternative to the main ATV activities on 434.0 or 439.25 MHz. Before transmitting, check with local ATV operators, repeater owners and frequency coordinators listed in the *ARRL Repeater Directory* for the coordinated frequencies used in your area.

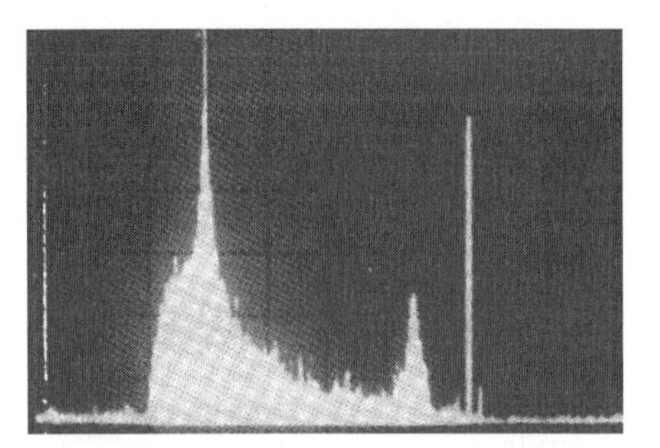

Fig 9.48—A spectrum-analyzer photo of a color ATV signal. Each vertical division represents 10 dB; horizontal divisions are 1 MHz. Spectrum power density varies with picture content, but typically 90% of the sideband power is within the first 1 MHz.

Fig 9.49—A photo of an ATV image of the Space Shuttle interior with K6KMN's repeater ID overlaid. Automatic video overlay in the picture easily solves the 10-minute ID requirement for Space Shuttle retransmissions and other long transmissions.

Since a TV set receives a 6-MHz bandwidth, ATV is more susceptible to interference from many other sources than are narrower modes. Interference 40 dB below the desired signal can be seen in video. Many of our UHF (and above) amateur bands are shared with radar and other government radio positioning services. These show up as horizontal bars in the picture. Interference from amateurs who are unaware of the presence of the ATV signal (or in the absence of a technically sound and publicized local band plan) can wipe out the sound or color or put diagonal lines in the picture.

DSB and VSB Transmission

While most ATV is double sideband (DSB) with the widest component being the sound subcarrier out ±4.5 MHz, over 90% of the spectrum power is in the first 1 MHz on both sides of the carrier for DSB or VSB (vestigial sideband). As can be seen in **Fig 9.48**, the video power density is down more than 30 dB at frequencies greater than 1 MHz from the carrier. DSB and VSB are both compatible with standard TV receivers, but the lower sound and color subcarriers are rejected in the TV IF filter as unnecessary. In the case of VSB, less than 5% of the lower sideband energy is attenuated. The other significant energy frequencies are the sound (set in the ATV transmitter at 15 dB below the peak sync) and the color at 3.58 MHz (greater than 22 dB down).

Narrowband modes operating greater than 1 MHz above or below the video carrier are rarely interfered with or know that the ATV transmitter is on unless the narrowband signal is on one of the subcarrier frequencies or the stations are too near one another. If the band is full and the lower sideband color and sound subcarrier frequencies need to be used by a dedicated link or repeater, a VSB filter in the antenna line can attenuate them another 20 to 30 dB, or the opposite antenna polarization can be used for more efficient packing of the spectrum. Since all amateur linear amplifiers reinsert the lower sideband to within 10 dB of DSB, a VSB filter in the antenna line is the only cost-effective way to reduce the unnecessary lower sideband subcarrier energy if more than 1 W is used. In the more populated areas, 2-m calling or coordination frequencies are often used to work out operating time shifts, and so on, between all users sharing or overlapping the same segment of the band.

ATV Identification

ATV identification can be on video or the sound subcarrier. A large high-contrast call-letter sign on the wall behind the operating table in view of the camera is the easiest way to fulfill the requirement. Transmitting stations fishing for DX during band openings often make up call-ID signs using fat black letters on a white background to show up best in the snow. Their city and 2-m monitoring frequency are included at the bottom of the sign to make beam alignment and contact confirmation easier.

Quite often the transmission time exceeds 10 minutes, especially when transmitting demonstrations, public-service events, space-shuttle video, balloon flights or a video tape. A company by the name of Intuitive Circuits makes a variety of boards that will overlay text on any video looped through them. Call letters and other information can be programmed into the board's non-volatile memory by on-board push buttons or an EIA-232 line from a computer (depending on the version and model of the OSD board). There is even a model that will accept NMEA-0183 GPRMC data from a GPS receiver and overlay latitude, longitude, altitude, direction and speed, as well as call letters, on the applied camera video. This is ideal for ATV rockets, balloons and R/C vehicles. The overlaid ID can be selected to be on, off or flashed on for a few seconds every 10 minutes to automatically satisfy the ID requirement of §97.119 (see **Fig 9.49**). The PC Electronics VOR-2 board has an automatic nine-minute timer, and it also has an end-of-transmission hang timer that switches to another video source for ID.

Driving Amplifiers with ATV

Wide-band AM video requires some special design considerations for linear amplifiers (as compared to those for FM and SSB amplifiers). Many high-power amateur amplifiers would oscillate (and possibly self destruct) from high gain at low frequencies if they were not protected by feedback networks and power RF chokes. These same stability techniques can affect some of the 5-MHz video bandwidth. Sync, color and sound can be very distorted unless the amplifier has been carefully designed for both stability and AM video modulation.

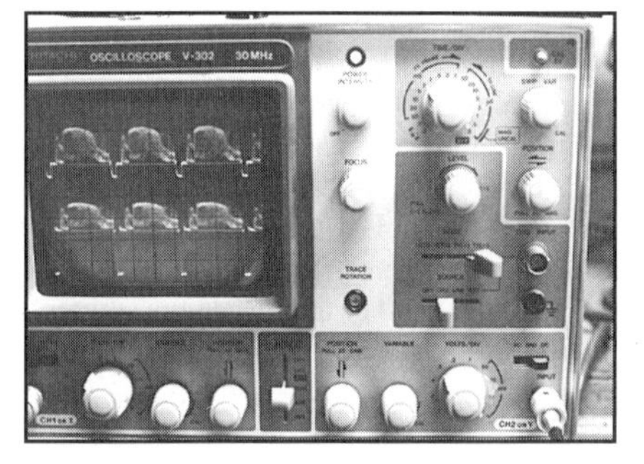

Fig 9.50—An oscilloscope used to observe a video waveform. The lower trace is the video signal as it comes out of the sync stretcher. The upper trace is the signal from the Mirage D1010-N amplifier.

Mirage, Teletec and Down East Microwave either make special ATV amplifiers or offer standard models that were designed for all modes, including ATV. Basically the collector and base bias supplies have a range of capacitors to keep the voltage constant under modulation, while at the same time using the minimum-value low-resistance series inductors or chokes to prevent self oscillation.

Almost all amateur linear power amplifiers have gain compression from half to their full rated peak envelope power. To compensate for this, the ATV exciter/modulator has a sync stretcher to maintain the proper transmitted video to sync ratio (see **Fig 9.50**). With both video and sound subcarrier disconnected, the pedestal control is set for maximum power output. Peak sync should first be set to 90% of the rated peak envelope power. (This is necessary to give some head room for the 4.5 MHz sound that is mixed and adds with the video waveform.) The TXA5-70 exciter/modulator has a RF power control to set this. Once this is done, the blanking pedestal control can be set to 60% of the peak sync value. For example, a 100-W amplifier would first be set for 90 W with the RF power control and then 54 W with the pedestal control. Then the sound subcarrier can be turned back on and the video plugged in and adjusted for best picture. If you could read it on a peak-reading power meter made for video, the power would be between 90 and 100 W PEP. On a dc oscilloscope connected to a RF diode detector in the antenna line, it can be seen that the sync and blanking pedestal power levels remain constant at their set levels regardless of video gain setting or average picture contrast. On an averaging meter like a Bird 43, however, it is normal to read something less than the pedestal set-up power.

ATV Repeaters

Basically there are two kinds of ATV repeaters: in band and cross band. 70-cm in-band repeaters are more difficult to build and use, yet they are more popular because equipment is more available and less expensive. Indeed, cable-ready TV sets tune the 70-cm band with no modifications.

Why are 70-cm repeaters more difficult to build? The wide bandwidth of ATV makes for special filter requirements. Response across the 6-MHz passband must be as flat as possible with minimum insertion loss, but also must sharply roll off to reject other users as little as 12 MHz away. Special multi-pole interdigital or combline VSB filters are used to meet the requirement. An ATV duplexer can be used to feed one broadband omnidirectional antenna, but an additional VSB filter is needed in the transmitter line for sufficient attenuation of noise and IMD products.

A cross-band repeater, because of the great frequency separation between the input and output, requires less sophisticated filtering to isolate the transmitter and receiver. In addition, a cross-band repeater makes it easier for users to see their own video (no duplexer is needed, only sufficient antenna spacing). Repeater linking is easier too, if the repeater outputs alternate between the 23- and 33-cm bands.

Fig 9.51 shows a block diagram for a simple 70-cm in-band repeater. No duplexer is shown because the antennas and VSB filters provide adequate isolation. The repeater

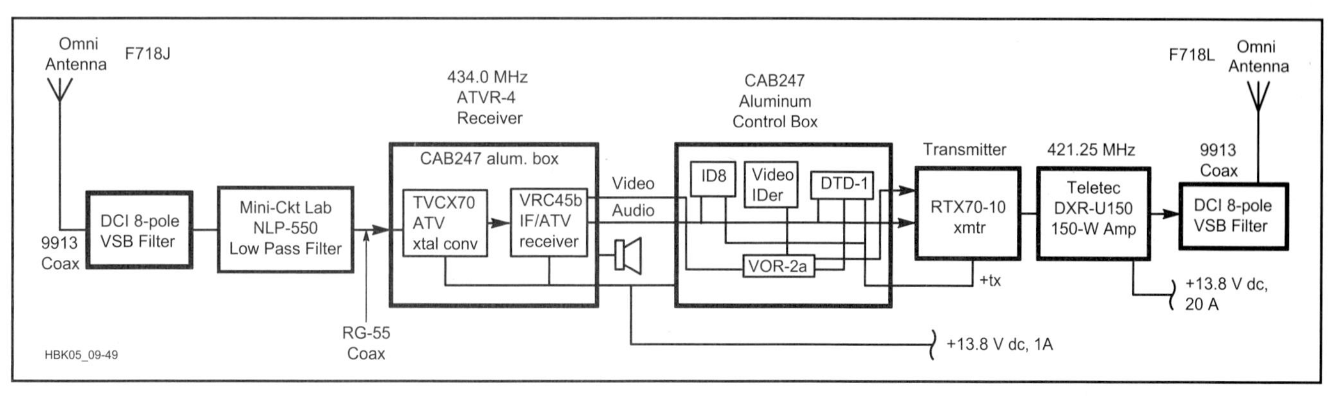

Fig 9.51—A block diagram of a 70-cm in-band ATV repeater. The antennas are Diamond omnidirectional verticals, which require 20 ft (minimum) of vertical separation to prevent receiver desensitization. The VSB filters are made by DCI; they have the proper band-pass characteristics and only 1 dB insertion loss. A low pass filter on the receiver is also necessary because cavity type filters repeat a pass-band at odd harmonics and the third-harmonic energy from the transmitter may not be attenuated enough. The receiver, 10-W transmitter and VOR are made by PC Electronics. The Communications Specialists DTD-1 DTMF decoder and ID8 Morse identifier (optional if a video ID is used) are used to remotely turn the repeater transmitter on or off and to create a CW ID, respectively. Alternatively, an Intuitive Circuits ATV4-4 ATV repeater controller board can do all the control box functions as well as remotely select from up to four video sources.

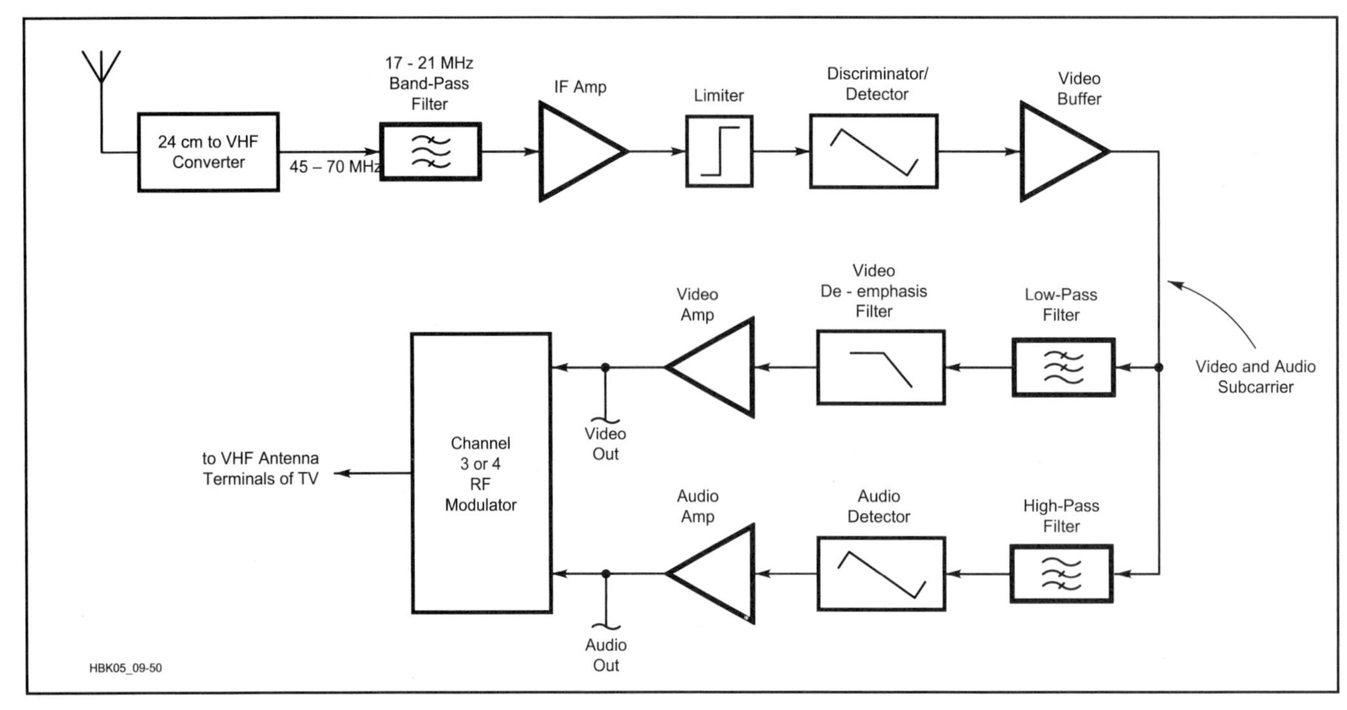

Fig 9.52—Block diagram of an FMATV receiver.

transmitter power supply should be separate from the receiver and exciter supply. ATV is amplitude modulated, therefore the current varies greatly from maximum at the sync tip to minimum during white portions of the picture. Power supplies are not generally made to hold tight regulation with such great current changes at rates up to several megahertz. Even the power supply leads become significant inductors at video frequencies; they will develop a voltage across them that can be transferred to other modules on the same power-supply line.

To prevent unwanted key up from other signal sources, ATV repeaters use a video operated relay (VOR). The VOR senses the horizontal sync at 15,734 Hz in much the same manner that FM repeaters use CTCSS tones. Just as in voice repeaters, an ID timer monitors VOR activity and starts the repeater video ID generator every nine minutes, or a few seconds after a user stops transmitting.

Frequency Modulated ATV (FMATV)

While AM is the most popular mode because of greater equipment availability, lower cost, less occupied bandwidth and use of a standard TV set, FMATV is gaining interest among experimenters and also repeater owners for links. FM on the 1200-MHz band is the standard in Europe because there is little room for video in their allocated portion of the 70-cm band. FMATV occupies 17 to 21 MHz depending on deviation and sound subcarrier frequency. The US 70-cm band is wide enough but has great interference potential in all but the less populated areas. Most available FMATV equipment is made for the 1.2, 2.4 and 10.25-GHz bands. **Fig 9.52** is a block diagram of an FMATV receiver.

The US standard for FMATV is 4 MHz deviation with the 5.8-MHz sound subcarrier set to 10 dB below the video level. 1252 or 1255 MHz are suggested frequencies in order to stay away from FM voice repeaters and other users higher in the band, while keeping sidebands above the 1240-MHz band edge. Using the US standard, with Carson's rule for FM occupied bandwidth, it comes out to just under 20 MHz. So 1250 MHz would be the lowest possible frequency. Almost all modern FMATV equip-

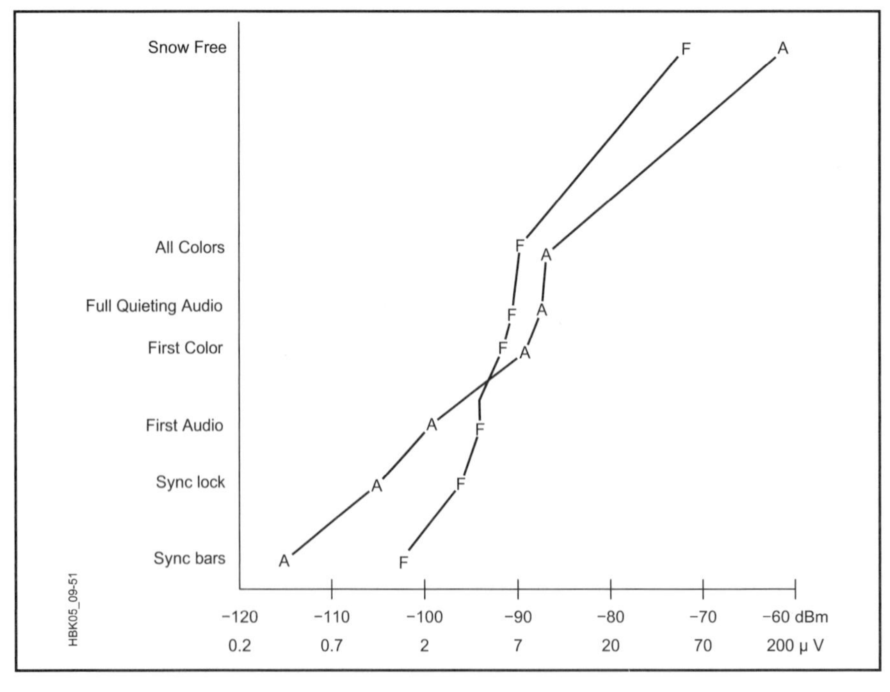

Fig 9.53—Two approaches to ATV receiving. This chart compares AM (A) and FM (F) ATV as seen on a TV receiver and monitor. Signal levels are into the same downconverter with sufficient gain to be at the noise floor. The FM receiver bandwidth is 17 MHz, using the US standard.

ment is synthesized, but if yours is not, use a frequency counter to monitor the frequency for warm up drift. Check with local frequency coordinators before transmitting because the band plan permits other modes in that segment.

Experimentally, using the US standard, FMATV gives increasingly better picture-to-noise ratios than AMATV at receiver input signals greater than 5 µV. Because of the wider noise bandwidth and FM threshold effect, AM video can be seen in the noise well before FM. For DX work, it has been shown that AM signals are recognizable signals in the snow at four times (12 dB) greater distance than FM signals, with all other factors equal. Above the FM threshold, however, FM rapidly overtakes AM. Snow-free pictures occur above 50 µV, or four times farther away than with AM signals. The crossover point is near the signal level where sound and color begin to appear for both systems. **Fig 9.53** compares AM and FMATV across a wide range of signal strengths.

There are a variety of methods to receive FMATV. Older satellite receivers have a 70 or 45-MHz input and require a down converter with 40 to 50 dB gain ahead of them. Also satellite receivers are made for wider deviation and need some video gain to give the standard 1 volt peak-to-peak video output when receiving a signal with standard 4-MHz deviation. Current satellite receivers directly tune anywhere from 900 to 2150 MHz and they only need a preamp added at the antenna for use on the 33 and 23-cm ham bands. The additional video gain can often be had by adjusting an internal pot or changing the gain with a resistor.

Some of the inexpensive Part 15 license-free wireless video receivers in the 33 cm band use 4-MHz deviation FM video, and most of the 2.4-GHz ones are FM, which can be used directly. However, they may or may not have the standard de-emphasis video network, which then may have to be added. On 2.4 GHz, some of the Part 15 frequencies are outside the band and care should be taken to use only those inside the 2390 to 2450 MHz ham band if modified. Wavecom Jr has been the most popular 2.4 GHz license-free video transmitter and receiver (available from ATV Research). These have been modified for higher power and other features, as well as having all four of the channels in the ham band using interface boards from PC Electronics.

Fig 9.54—N8QPJ mounted an ATV setup aboard this model Humvee.

Gunnplexers on 10.4 GHz make inexpensive point to point ATV links for public-service applications or between repeaters. A 10-mW Gunnplexer with 17-dB horn can cover over 2 miles line-of-sight when received on a G8OZP low noise 3-cm LNB and satellite receiver. An application note for construction of the 3-cm transmitter comes with the GVM-1 Gunnplexer video modulator board from PC Electronics.

For short distance ATV from R/C vehicles, low-power FM ATV modules with 50 to 100-mW output in the 33, 23 or 13-cm bands are often used. These offer less desense possibility to the R/C receiver. An example can be seen on the model Humvee in **Fig 9.54**.

DIGITAL AMATEUR TELEVISION (DATV)

German amateurs have lead the way in digital ATV. For the past few years, Uwe Kraus, DJ8DW, and others have had a stand at HamRadio—the large European Amateur Radio gathering in Friedrichshafen, Germany. The motivation for DATV is about the same as for commercial digital television, particularly high quality pictures even with weak signals and a distinctively smaller bandwidth than that occupied by analog TV. A breakthrough occurred in September 1998 when the DATV team transmitted digital pictures over a 62-mile path with a 2 MHz bandwidth at 434 MHz using MPEG-1 encoding.

See: **www.von-info-ch/hb9afo/histoire/news043.htm** and **www.von-info.ch/hb9afo/datv_e.htm**.

Further ATV Reading

Amateur Television Quarterly Magazine.

CQ-TV, British ATV Club, a quarterly publication available through *Amateur Television Quarterly Magazine.*

Kramer Klaus, DL4KCK, "AGAF e.V. DATV-Boards-Instructions for starting up," *Amateur Television Quarterly Magazine*, spring 2005.

Ruh, "ATV Secrets for the Aspiring ATVer," Vol 1, 1991 and Vol 2, 1992. Available through *Amateur Television Quarterly Magazine.*

Seiler, Thomas, HB9JNX/AE4WA, et al, "Digital Amateur TeleVision (D-ATV), *proc. ARRL/TAPR Digital Communications Conference*, **www.baycom.org/~tom/ham/dcc2001/datv.pdf**

Taggart, "An Introduction to Amateur Television," April, May and June 1993 *QST.*

Taggart, R., WB8DQT, *Image Communications Handbook*, Published by ARRL, Newington, CT, 2002. ARRL Order No. 8616. See also **www.arrl.org/catalog**.

Spread Spectrum

Contributors to this section were André Kesteloot, N4ICK, John Champa, K8OCL, and Kris Mraz, N5KM. *The ARRL Spread Spectrum Sourcebook* contains a more complete treatment of the subject. The following information takes the subject from early experiments by the Amateur Radio Research and Development Corporation (AMRAD) to contemporary Amateur Radio use of spread spectrum technology for high-speed multimedia (HSMM) applications.

Spread spectrum originated in the 1930s, shrouded in secrecy. In 1942, Hollywood movie actress Hedy Lamarr and composer George Antheil were granted a patent for spread spectrum. Despite the fact that John Costas, W2CRR, published a paper on non-military applications of spread spectrum communications in 1959, spread spectrum was used almost solely for military purposes until the late 1970s. In 1981, the FCC granted AMRAD a Special Temporary Authorization to conduct Amateur Radio spread spectrum experiments. In June 1986, the FCC authorized all US amateurs to use spread spectrum above 420 MHz. These FCC grants were intended to encourage the development of spread spectrum, which was an important element in commercial wireless systems that emerged in the 1990s.

WHY SPREAD SPECTRUM

Faced with increasing noise and interference levels on most RF bands, traditional wisdom still holds that the narrower the RF bandwidth, the better the chances that "the signal will get through." This is not so.

In 1948, Claude Shannon published his famous paper, "A Mathematical Theory of

Communication" in the *Bell System Technical Journal*, followed by "Communications in the Presence of Noise" in the *Proceedings of the IRE* for January 1949. A theorem that follows Shannon's, known as the Shannon-Hartley theorem, states that the channel capacity C of a band-limited Gaussian channel is:

$$C = W\log_2\left(1+\frac{S}{N}\right)\text{bits/s} \qquad \text{(Eq 1)}$$

where

W is the bandwidth,

S is the signal power and

N is the noise within the channel bandwidth.

This theorem states that should the channel be perfectly noiseless, the capacity of the channel is infinite. It should be noted, however, that making the bandwidth W of the channel infinitely large does *not* make the capacity infinite, because the channel noise increases proportional to the channel bandwidth.

Within reason, however, you can trade power for bandwidth. In addition, the power density at any point of the occupied bandwidth can be very small, to the point that it may be well *below* the noise floor of the receiver. The US Navy Global Positioning System (GPS) is an excellent example of the use of what is called direct-sequence spread spectrum. The average signal at the GPS receiver's antenna terminals is approximately –160 dBW (for the C/A code). Since most sources of interference are relatively narrowband, spread-spectrum users will also benefit, as narrowband interfering signals are rejected automatically during the despreading process, as will be explained later in this section.

These benefits are obtained at the cost of fairly intricate circuitry: The transmitter must spread its signal over a wide bandwidth in accordance with a certain prearranged code, while the receiver must somehow synchronize on this code and recombine the received energy to produce a usable signal. To generate the code, use is made of pseudonoise (PN) generators. The PN generators are selected for their correlation properties. This means that when two similar PN sequences are compared out of phase their correlation is nil (that is, the output is 0), but when they are exactly in phase their correlation produces a huge peak that can be used for synchronization purposes.

This synchronization process has been (and still is) the major complicating factor in any spread spectrum link, for how can one synchronize on a signal that can be well below the receiver's noise floor? Because of the cost associated with the complicated synchronization processes, spread spectrum applications were essentially military-related until the late 1970s. The development of ICs then allowed for the replacement of racks and racks of tube equipment by a few plug-in PC boards, although the complexity level itself did not improve. Amateur Radio operators could not afford such levels of complexity and had to find simpler solutions, at the cost of robustness in the presence of interference.

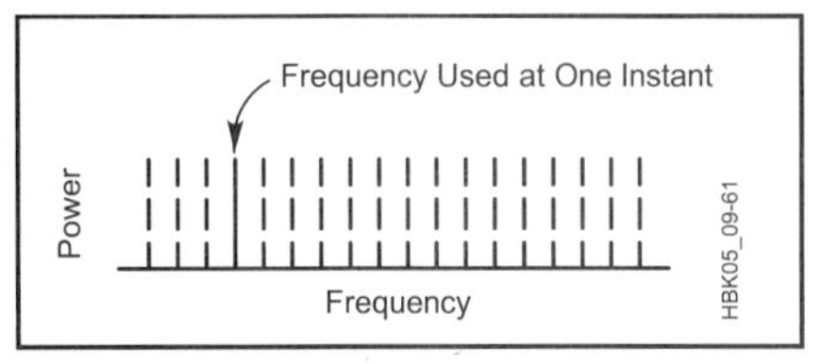

Fig 9.55—Power vs frequency for frequency-hopping spread spectrum signals. Emissions jump around to discrete frequencies in pseudo-random fashion.

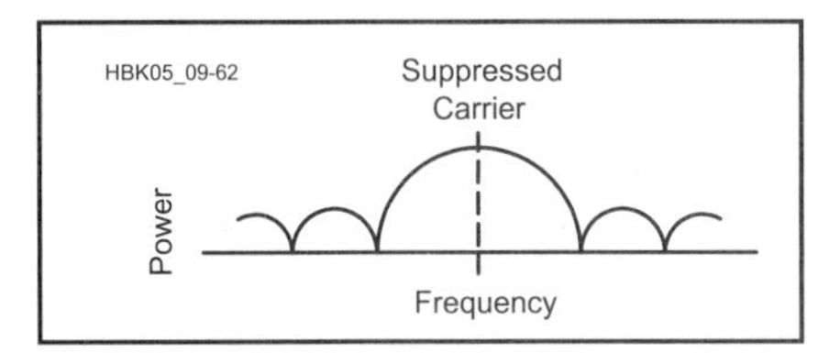

Fig 9.56—Power vs frequency for a direct-sequence-modulated spread spectrum signal. The envelope assumes the shape of a (sin x/x)^2 curve. With proper modulating techniques, the carrier is suppressed.

Spread-Spectrum Transmissions

A transmission can be called "spread spectrum" if the RF bandwidth used is (1) much larger than that needed for traditional modulation schemes and (2) independent of the modulation content. Although numerous spread spectrum schemes are in existence, amateurs can use any of them as long as the modulation scheme has been published, for example on the ARRL website. By far, frequency-hopping (FH) and direct-sequence spread spectrum (DSSS) are the most popular forms within the Amateur Radio community.

To understand FH, let us assume a transmitter is able to transmit on any one of 100 discrete frequencies F1 through F100. We now force this equipment to transmit for 1 second on each of the frequencies, but in an apparently random pattern (for example, F1, F62, F33, F47...) See **Fig 9.55**. Should some signal interfere with the receiver site on three of those discrete frequencies, the system will still have achieved reliable transmission 97% of the time. Because of the built-in redundancy in human speech, as well as the availability of error-correcting

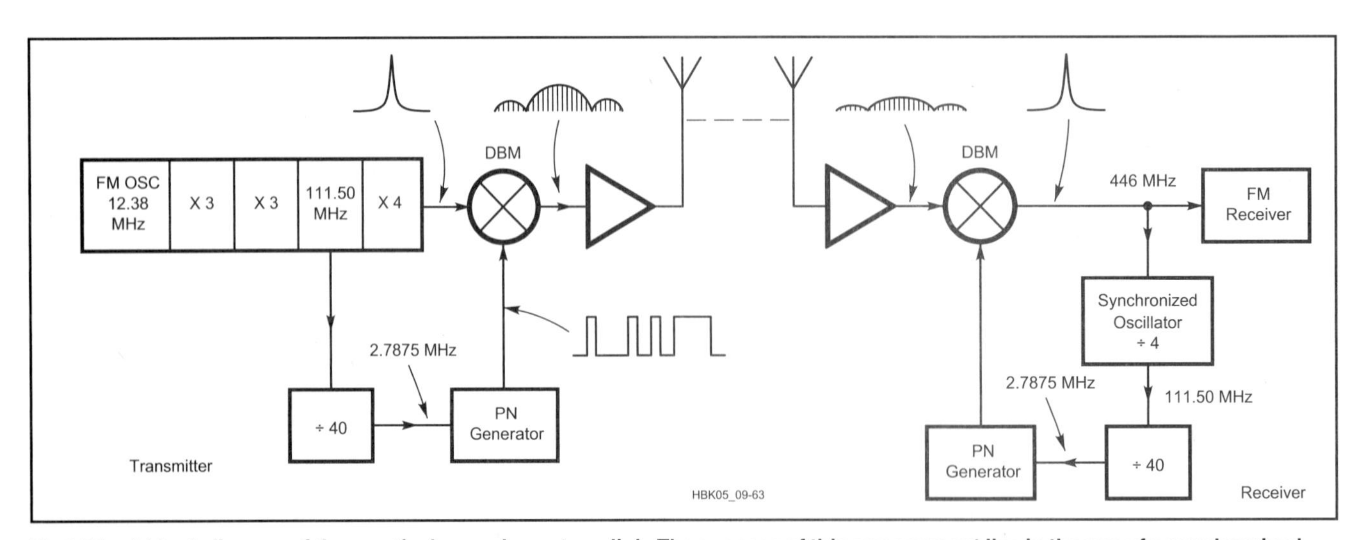

Fig 9.57—A block diagram of the practical spread spectrum link. The success of this arrangement lies in the use of a synchronized oscillator (right) to recover the transmitter clock signal at the receiving site.

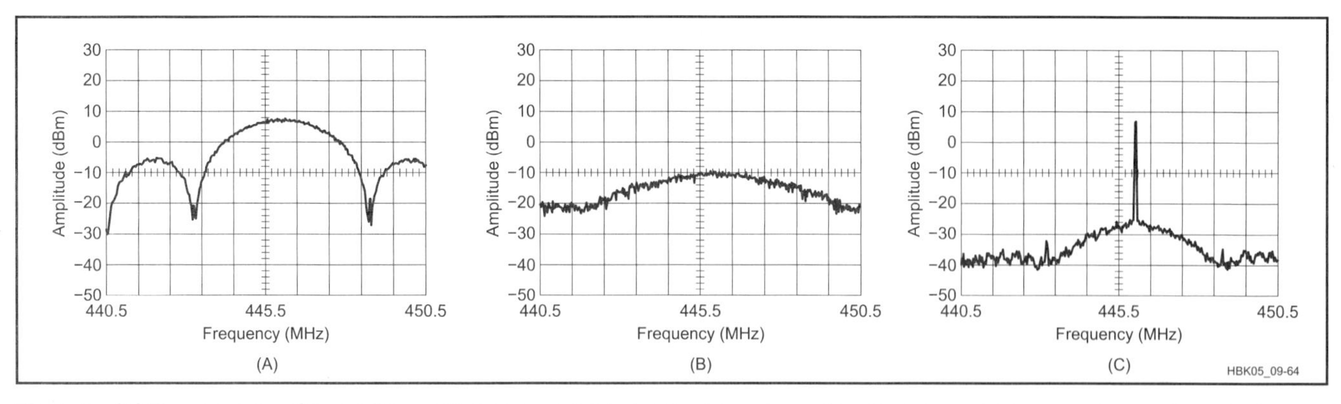

Fig 9.58—(A) The envelope of the unfiltered biphase-modulated spread spectrum signals as viewed on a spectrum analyzer. In this practical system, band-pass filtering is used to confine the spread spectrum signal to the amateur band. (B) At the receiver end of the line, the filtered spread spectrum signal is apparent only as a 10-dB hump in the noise floor. (C) The despread signal at the output of the receiver DBM. The original carrier—and any modulation components that accompany it—has been recovered. The peak carrier is about 45 dB above the noise floor—more than 30 dB above the hump shown at B. (These spectrograms were made at a sweep rate of 0.1 s/division and an analyzer bandwidth of 30 kHz; the horizontal scale is 1 MHz/division.)

codes in data transmissions, this approach is particularly attractive for systems that must operate in heavy interference.

In a DSSS transmitter, an RF carrier and a pseudo-random pulse train are mixed in a doubly balanced mixer (DBM). In the process, the RF carrier disappears and is replaced by a noise-like wideband transmission, as shown in **Fig 9.56**. At the receiver, a similar pseudo-random signal is reintroduced and the spread spectrum signal is correlated, or despread, while narrowband interference is spread simultaneously by the same process.

The technical complexity mentioned above is offset by several important advantages for military and space applications:

- *Interference rejection*. If the interference is not synchronized with the original spread spectrum signal, it will not appear after despreading at the receiver.
- *Security*. The length and sophistication of the pseudo-random codes used can be such as to make unauthorized recovery difficult, if not impossible.
- *Power density*. Low power density makes for easy hiding of the RF signal and a resulting lower probability of detection.

So far as the Amateur Radio community is concerned, particular benefit will be derived from the interference rejection just mentioned, since it offers both robustness and reliability of transmissions, as well as a low probability of interference to other users. Additionally, spread spectrum has the potential to allow better utilization of the RF spectrum allocated to amateurs. There is a limit as to how many conventional signals can be placed in a given band before serious transmission degradation takes place. Additional spread spectrum signals will not cause severe interference, but may instead only raise the background noise level. This becomes particularly important in bands shared with other users and in our VHF and UHF bands increasingly targeted by would-be commercial users. The utilization of a channel by many transmitters is essentially the concept behind CDMA (Code Division Multiple Access), a system in which several DSSS transmissions can share the same RF bandwidth, provided they utilize orthogonal pseudo-random sequences.

Amateur Radio Spread Spectrum

Experimentation sponsored by AMRAD began in 1981 led to the design and construction of a practical DSSS UHF link. This project was described in May 1989 *QST* and was reprinted in *The ARRL Spread Spectrum Sourcebook*. In it, André Kesteloot, N4ICK, offered a simple solution to the problem of synchronization. The block diagram is shown in **Fig 9.57**, and **Fig 9.58** shows the RF signals at the transmitter output, at the receiver antenna terminals and the recovered signal after correlation. James Vincent, G1PVZ, replaced the original FM scheme with a continuously variable delta modulation system, or CVSD. In 1989 in a paper titled *License-Free Spread Spectrum Packet Radio,* Al Broscius, N3FCT, suggested the use of Part 15 spread spectrum wireless local area network (WLAN) devices that were becoming available be put to use in amateur radio.

In 1997 TAPR started the development of a 1-W, 128-kbit/s, FHSS radio for the amateur radio 902 MHz band. In late 1999 the FCC considerably relaxed the Amateur Radio service rules regarding the use of spread spectrum. These changes allowed amateurs to use commercial off-the-shelf (COTS) Part 15 spread spectrum devices used under § 97.311 of the FCC rules.

Emergence of Commercial Part 15 Equipment

Just as military surplus radio equipment fueled Amateur Radio in the 1950s, and commercial FM radios and repeaters snowballed the popularity of VHF/UHF amateur repeaters in the 1960s and 1970s, the availability of commercial wireless LAN (WLAN) equipment is driving the direction and popularity of Amateur Radio use of spread spectrum in the 2000s. FCC Part 15 documents the technical rules for commercial spread-spectrum equipment. The Institute of Electrical and Electronics Engineers (IEEE) has provided the standards under which manufacturers have developed equipment for sale commercially. IEEE 802.11 standardized FHSS and DSSS for the 2.4 GHz band at data rates of 1 and 2 Mbit/s. Next came the release of 802.11b, which provided the additional data rates of 5.5 and 11 Mbit/s but only for DSSS. FHSS was not carried forward. This was followed by 802.11g, which does not use SS but uses OFDM for data rates of 6, 9, 12, 18, 24, 36, 48 and 54 Mbit/s as well as backward compatibility with 802.11b. As of this writing the most recent release of the standard is 802.11a. This release addresses the use of OFDM in certain parts of the 5 GHz band. It provides the same data rates as 802.11g. The currently unreleased 802.11n standard promises data rates in excess of 108 Mbit/s.

Frequency Hopping Spread Spectrum

FHSS radios, as specified in 802.11, hop among 75 of 79 possible non-overlapping frequencies in the 2.4 GHz band. Each hop occurs approximately every 400 ms with a hop time of 224 µs. Since these are Part 15 devices, the radios are limited to a maximum peak output power of 1 W and a maximum bandwidth of 1 MHz (–20 dB) at any given hop frequency. The rules allow using a smaller number of hop frequencies at wider bandwidths (and lower power: 125 mW) but most manufacturers have opted not to develop equipment using these options. Consequently, off-the-shelf equipment with this wider bandwidth capability is not readily available to the amateur.

The hopping sequences are well defined by 802.11. There are three sets of 26 such sequences (known as *channels*) consisting of 75 frequencies each. The ordering of the frequencies is designed as a pseudo-random sequence hopping at least 6 MHz higher or lower that the current carrier frequency such that no two channels are on the same frequency at the same time. Channel assignment can be coordinated among multiple collocated networks so that there is minimal interference among radios operating in the same band.

The FHSS radio can operate at data rates of 1 and 2 Mbit/s. The binary data stream modulates the carrier frequency using frequency shift keying. At 1 Mbit/s the carrier frequency is modulated using 2-Level Gaussian Frequency Shift Keying (2GFSK) with a shift of ±100 kHz. The data rate can be doubled to 2 Mbit/s by using 4GFSK modulation with shifts of ±75 kHz and ±225 kHz.

Direct Sequence Spread Spectrum

DSSS uses a fast digital sequence to accomplish signal spreading. That is, a well-known pseudo-random digital pattern of ones and zeros is used to modulate the data at a very high rate. In the simplest case of DSSS, defined in 802.11, an 11-bit pattern known as a Barker sequence (or Barker code) is used to modulate every bit in the input data stream. The Barker sequence is 10110111000. Specifically, a "zero" data bit is modulated with the Barker sequence resulting in an output sequence of 10110111000. Likewise, a "one" data bit becomes 01001000111 after modulation (the inverted Barker code). These output patterns are known as *chipping* streams; each bit of the stream is known as a *chip*. It can be seen that a 1 Mbit/s input data stream becomes an 11 Mbit/s output data stream.

The DSSS radio, like the FHSS radio, can operate at data rates of 1 and 2 Mbit/s. The chipping stream is used to phase modulate the carrier via phase shift keying. Differential Binary Phase Shift Keying (DBPSK) is used to achieve 1 Mbit/s, and Differential Quadrature Phase Shift Keying (DQPSK) is used to achieve 2 Mbit/s. Fig 9.58 shows a typical 1 or 2 Mbit/s DSSS signal having a major lobe bandwidth of ±11 MHz (–30 dB). The first minor sidelobe is down at least 30 dB and the second minor sidelobe is down 50 dB as required by Part 15 rules.

The higher data rates specified in 802.11b are achieved by using a different pseudo-random code known as a *Complimentary Sequence*. Recall the 11-bit Barker code can encode one data bit. The 8-bit Complimentary Sequence can encode 2 bits of data for the 5.5-Mbit/s data rate or 6 bits of data for the 11-Mbit/s data rate. This is known as *Complimentary Code Keying (CCK)*. Both of these higher data rates use DQPSK for carrier modulation. DQPSK can encode two data bits per transition. **Table 9.12** shows how four bits of the data stream are encoded to produce a 5.5- Mbit/s data rate and eight bits are encoded to produce an 11-Mbit/s data rate. There are 64 different combinations of the 8-bit Complimentary Sequence that have mathematical properties that allow easy demodulation and interference rejection. At 5.5 Mbit/s, only four of the combinations are used. At 11 Mbit/s, all 64 combinations are used.

As an example, for an input data rate of 5.5 Mbit/s, four bits of data are sampled at the rate of 1.375 million samples per second. Two input bits are used to select one of four eight-bit CCK sequences. These eight bits are clocked out at a rate of 11 Mbit/s. The two remaining input bits are used to select the phase at which the eight bits are transmitted. **Fig 9.59A** shows a conceptual block diagram of a 5.5-Mbit/s CCK transmitter modulator, while Fig 9.59B shows an 11-Mbit/s modulator.

Orthogonal Frequency Division Modulation

OFDM provides its spreading function by transmitting the data simultaneously on multiple carriers. 802.11g and 802.11a specify 20-MHz wide channels with 52 carriers spaced every 312.5 kHz. Of the 52 carriers, four are non-data pilot carriers that carry a known bit pattern to simplify demodulation. The remaining 48 carriers are modulated at 250 thousand transitions per

Table 9.12
Bit encoding for 5.5 Mbps and 11 Mbps CCK transmissions

Data Rate, Mbps	*CCK encoded bit*	*DQPSK encoded bits*
5.5	2	2
11	6	2

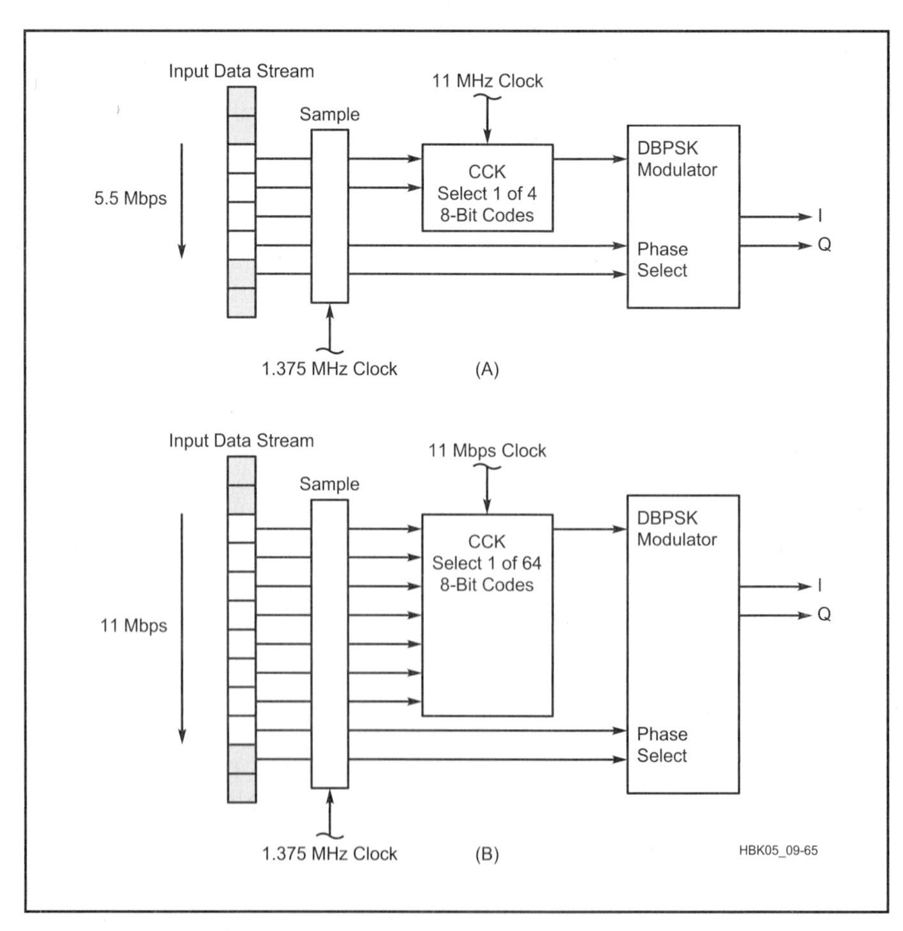

Fig 9.59—Conceptual block diagram of a modulator for a CCK Spread Spectrum transmitter. (A) 5.5 Mbit/s data rate. (B) 11 Mbit/s data rate. See text.

second. Taking all 48 transitions in parallel is known as a symbol. That is, at any given instant in time 48 bits of data are being transmitted.

The term *orthogonal* is derived from the fact that these carriers are positioned such that they do not interfere with one another. The center frequency of one carrier's signal falls within the nulls of the signals on either side of it.

OFDM radios can be used to transmit data rates of 6, 9, 12, 18, 24, 36, 48 and 54 Mbit/s as specified by both 802.11a and 802.11g. In order to transmit at faster and faster data rates in the same 20-MHz channel, different modulation techniques are employed: BPSK, QPSK, 16QAM and 64QAM. In addition, some of the bits transmitted are used for error correction, so the raw data rates could be reduced by up to half of what they would be without error correction. For instance, assuming BPSK (one bit per carrier) and assuming half the bits are used for error correction (known as the coding rate, R); the resulting data rate would be 6 Mbit/s.

48 carriers × 1 bit per carrier × 1/2 R = 24 bits (effective)

24 bits × 250 kilo transitions per second = 6 Mbit/s.

Table 9.13 shows a complete list of the modulation methods and coding rates employed by OFDM. The higher data rates will require better signal strength to maintain error free reception due to using few error correction bits and more complex modulation methods.

Table 9.13

Summary of the modulation techniques used by OFDM to achieve the different data rates.

Data Rate Mbps	*Modulation*	*Coding Rate, R*
6	BPSK	1/2
9	BPSK	3/4
12	QPSK	1/2
18	QPSK	3/4
24	16QAM	1/2
36	16QAM	3/4
48	64QAM	2/3
54	64QAM	3/4

Spread Spectrum References

Dixon, *Spread Spectrum Systems*, second edition, 1984, Wiley Interscience, New York.

Dixon, *Spread Spectrum Techniques*, 1976, IEEE Press, New York.

Kesteloot, Ed., *The ARRL Spread Spectrum Sourcebook* (Newington, CT: ARRL, 1990). Includes Hershey, *QST* and *QEX* material listed separately here.

"Poisson, Shannon and the Radio Amateur," *Proceedings of the IRE*, Dec 1959.

Multimedia Systems

In January 2001, the ARRL Board of Directors voted unanimously that the ARRL should proceed with the development of High Speed Digital Networks for the Amateur Service. The ARRL President appointed a group of individuals knowledgeable in the field from the international Amateur Radio community and industry. The group would report to the Technology Task Force (TTF). The TTF established the High Speed Multimedia (HSMM) Working Group, with John Champa, K8OCL, as its chairman. Champa identified two initial goals for the working group, so as to immediately begin the development of such high speed digital amateur radio networks:

1. Encourage the amateur adoption and modification of commercial off-the-shelf (COTS) IEEE 802.11 spread spectrum hardware and software for Part 97 uses.
2. Encourage or develop other high-speed digital radio networking techniques, hardware, and applications.

These efforts were rapidly dubbed *HSMM Radio*. Although initially dependent on adaptation of COTS 802.11 gear to Part 97, it is obvious from these goals that HSMM radio is not a specific operating mode, but more of a direction or driving force within amateur radio.

Furthermore, in HSMM radio, the emphasis has shifted away from primarily keyboard radio communication, as in conventional packet radio, to multimedia radio. This includes simultaneous voice, video, data and text over radio.

In HSMM radio these individual mediums have different names, much like their Internet counterparts. For example, voice modes, although technically digital voice, are most often called *streaming audio*. However, since it is two-way voice over an IP network similar to the direction being taken by contemporary commercial telephony technology, the same technology use to link many amateur radio repeaters over the Internet, the name *voice-over-IP* (VoIP) may be more appropriate.

Video modes, although sometimes called amateur digital video (ADV), are also known as *streaming video*. Again, perhaps the commercial term for such two-way video QSOs may be more appropriate: IPVC (*IP videoconferencing*).

Text exchanges via a keyboard are often used in HSMM radio, but they are similarly called by their Internet or Packet Radio name: *Chat mode*. File transfers using FTP can also be done, just as on the Internet. This combination of Internet terminology, coupled with this dramatic shift in emphasis within amateur radio from traditional analog point-to-point radio toward networked digital radios, has resulted in many amateurs nick naming HSMM radio *The Hinternet*. Although the name implies some under-dog status to some, the name seems to be sticking.

HSMM RADIO APPLICATIONS

HSMM radio has some unique ham radio networking applications and operational practices that differentiate the Hinternet from normal Wi-Fi hotspots at coffee houses and airports, which you may have read about in the popular press. HSMM radio techniques are used, for example, for system RC (remote control) of amateur radio stations.

In this day of environmentally sensitive neighborhoods, one of the greatest challenges, particularly in high density residential areas, is constructing ham radio antennas, particularly high, tower-mounted HF beam antennas. In addition, such amateur installations represent a significant investment in time and resources. This burden could be easily shared among a small group of friendly hams, a radio club or a repeater group.

Implementing a link to a remote HF station via HSMM radio is easy to do. Most computers now come with built-in multimedia support. Most amateur radio transceivers are capable of PC control. Adding the radio networking is relatively simple. Most HSMM radio links use small 2.4-GHz antennas mounted outdoors or pointed through a window. These UHF antennas are relatively small and inconspicuous when compared to a full-size 3-element HF Yagi on a tall steel tower.

For example, Darwin Thompson, K6USW, has performed remote control of a Kenwood TS-480SAT/HX transceiver, which can be controlled over a LAN and the Internet, or in this case the Hinternet. The Kenwood International website provides two programs for the TS-480SAT/HX at: **www.kenwood.net/indexKenwood.cfm?do=SupportFileCategory&FileCatID=3**.

The *ARHP-10* program is the radio host

program. It operates the computer attached to the transceiver. Just follow the instructions included with the software to make the cables to interface the radio to your computer. The *ARCP-480* program is the radio control software. *ARCP-480* operates the computer at the other end of the remote control link. By attaching a suitable headset to this remote PC, the operator now has full control of the transceiver via the HSMM radio link and can use voice-over-Internet-protocol (VoIP) to transmit and receive audio.

A ham does not have to have an antenna-unfriendly homeowners association (HOA) or a specific deed restriction problem to put RC via HSMM radio to good use. This system RC concept could be extended to other types of amateur radio stations. For example, it could be used to link a ham's home to a shared, high-performance amateur radio DX station, EME station or OSCAR satellite ground station for a special event, or on a regular basis.

SHARED HIGH-SPEED INTERNET ACCESS

Sharing high-speed Internet access (Cable, DSL, etc) with another ham is a popular application for HSMM radio. Half of the US population is restricted to slow dial-up Internet connections (usually around 20 to 40 kbit/s) over regular analog telephone lines. Getting a high-speed Internet connection, even a shared one, can dramatically change the surfing experience! Just remember that if you use an HSMM radio to share high speed access to the Internet, which Amateur Radio has content restrictions, for example no commercial for-profit business e-mails, etc. An example might be an amateur television station (ATV) transmitting an outdoor scene and inadvertently picking-up a billboard in the station camera. Such background sources are merely incidental to your transmission. They are not the primary purpose of your communications, plus they are not intended for rebroadcast to the public.

Just as on the Internet, it is possible to do such things as playing interactive games, complete with sound effects and full-motion animation with HSMM radio. This can be lots of fun for new and old hams alike, plus it can attract others in the "Internet Generation" to get interested in amateur radio and perhaps become new radio club members. In the commercial world these activities are called "WLAN Parties." Such e-games are also an excellent method for testing the true speed of your station's Hinternet link.

HSMM RADIO IN EMERGENCY COMMUNICATIONS

There are a number of significant reasons and exciting new examples why HSMM radio is the way of the future for many Emergency Communications (EmComm) situations. These may or may not be under ARES or RACES auspices.

1. The amount of digital radio traffic on

SS and HSMM Glossary

Ad Hoc Mode—An operating mode of a client RIC that allows it to associate directly with any other RIC without having to go through an Access Point. See Infrastructure mode.

AP—Access Point

APRS—Automatic Position Reporting System

Association—The service used to establish access point/station mapping and enable station use of the WLANs services in infrastructure mode.

Authentication—Process by which the wireless communications system verifies the identity of a user attempting to use a WLAN prior to the user associating with the AP.

Band-limited Gaussian Channel—A "brickwall" linear filter that is equal to a constant over some frequency band and equal to zero elsewhere, and by white Gaussian noise with a constant power spectrum over the channel bandwidth.

Barker Code—An 11-bit digital sequence used to modulate (spread) the input data stream. A one bit is represented by the sequence 10110111000 and a zero bit is represented by the sequence 01001000111.

CCK—Complimentary Code Keying. A spreading technique in which the input data stream is modulated with a digital sequence (the complimentary code) depending on the value of the data stream. In 802.11b, for example, the complimentary code consists of 64 eight-bit values. Six data bits from the input stream are used to select which of the complimentary codes is used to modulate the data. See Barker Code.

Correlation—A measure of how closely a signal matches a delayed version of itself shifted n units in time.

COTS—Commercial Off The Shelf equipment.

DBPSK—Differential Binary Phase Shift Keying. A method of modulating data onto a carrier by changing the phase of the carrier relative to its current phase. A binary "1" is represented by a +90 degree phase shift and a binary "0" is represented by a 0 degree phase shift.

DHCP—Dynamic Host Configuration Protocol. A protocol used by a client computer to obtain an IP address for use on a network.

DSSS—Direct Sequence Spread Spectrum. A spread spectrum system in which the carrier has been modulated by a high speed spreading code and an information data stream. The high speed code sequence dominates the "modulating function" and is the direct cause of the wide spreading of the transmitted signal. (Title 47, Chapter I, Part 2, subpart A, section 2.1 Terms and Definitions).

DQPSK—Differential Quadrature Phase Shift Keying. A method of modulating data onto a carrier by changing the phase of the carrier similar to DBPSK except that two bits can be represented by a single phase shift such as following this scheme:

2-Bit Value	*Phase Shift (degrees)*
00	0
01	+90
10	–90
11	180

FHSS—Frequency Hopping Spread Spectrum. A spread spectrum system in which the carrier is modulated with the coded information in a conventional manner causing a conventional spreading of the RF energy about the frequency carrier. The frequency of the carrier is not fixed but changes at fixed intervals under the direction of a coded sequence. The wide RF bandwidth needed by such a system is not required by spreading of the RF energy about the carrier but rather to accommodate the range of frequencies to which the carrier frequency can hop. The test of a frequency hopping system is that the near term distribution of hops appears random, the long term distribution appears evenly distributed over the hop set, and sequential hops are randomly distributed in both direction and magnitude of change in the hop set. (Title 47, Chapter I, Part 2, subpart A, section 2.1 Terms and Definitions).

GPS—Global Positioning System

IEEE—Institute of Electrical and Electronic Engineering

IEEE 802.11—An IEEE standard specifying FHSS and DSSS in the 2.4 GHz band at 1 Mbit/s and 2 Mbit/s data rates. 802.11 is also used as a general term for all spread spectrum devices operating under Part 15. For example "The 802.11 network" could be referring to a collection of RICs and APs using 802.11b and 802.11g based devices.

IEEE 802.11a—An IEEE standard specifying OFDM in the 5.8 GHz band at 6, 12, 16, 24, 36, 48, and 54 Mbit/s data rates.

IEEE 802.11b—An IEEE standard specifying DSSS in the 2.4 GHz band at 5.5 and 11 Mbit/s data rates in addition to being backward compatible with DSSS at 1 and 2 Mbit/s specified in 802.11.

IEEE 802.11g—An IEEE standard specifying OFDM in the 2.4 GHz band 6, 12, 16, 24, 36, 48, and 54 Mbit/s data

2.4 GHz is increasing and operating under low powered, unlicensed Part 15 limitations cannot overcome this noise.

2. EmComm organizations increasingly need high-speed radio networks that can simultaneously handle voice, video, data and text traffic.
3. The cost of a commercially installed high-speed data network can be more than emergency organizations and communities can collectively afford.
4. EmComm managers also know that they need to continuously exercise any emergency communications system and have trained operators for the system in order for it to be dependable.

Being able to send live digital video images of what is taking place at a disaster site to everybody on the HSMM radio network can be invaluable in estimating the severity of the situation, planning appropriate responding resources and other reactions. The Emergency Operations Center (EOC) can actually see what is happening while it is happening. Submitting a written report while simultaneously talking to the EOC using Voice over IP (VoIP) would provide additional details.

With HSMM radio, often all that is needed to accomplish such immediacy in the field is a laptop computer equipped with a wireless local area network card (PCMCIA) with an external antenna jack. In HSMM radio jargon such a card is simply called a RIC (radio interface card). Connect any digital camera with a video output port or any webcam, and a headset to the laptop's sound card. Then connect the RIC to a short Yagi antenna (typically 18 inches of antenna boom length) and point the antenna back to the EOC.

HSMM RADIO RELAY

There are a number of ways to extend the HSMM link. The most obvious means would be to run higher power and to place the antennas as high as possible, as is the case with VHF/UHF FM repeaters. In some densely populated urban areas of the country this approach with 802.11, at least in the 2.4 GHz band, may cause some interference with other users. Other means of getting greater distances using 802.11 on 2.4 GHz or other amateur bands should be considered. One approach is to use highly directive, high-gain antennas, or what is called the directive link approach.

Another approach used by some HSMM radio networks is what is called a low-profile radio network design. They depend on several low power sources and radio relays of various types. For example, two HSMM radio repeaters (known commercially as *access points*, or APs, about $100 devices) may be placed back-to-back in what is known as bridge mode. In this configuration they will simply act as an automatic radio relay for the high-speed data. Using a series of such radio

rates in addition to being backward compatible with DSSS at 1, 2, 5.5, and 11 Mbit/s specified in 802.11b.

IEEE 802.11n—An IEEE standard specifying data rates up to 250 Mbit/s and being backward compatible with 802.11a and 802.11g.

IEEE 802.16—An IEEE standard specifying wireless last-mile broadband access in the Metropolitan Area Network (MAN). Also known as WiMAX.

ISM—Industrial, Scientific, and Medical. Specific frequency bands authorized by Part 18 rules for non-communication equipment such as microwave ovens, RF lighting, etc. The ISM spectrum where spread spectrum is allowed is located at 2.4 – 2.5 GHz and 5.725 – 5.875 GHz band.

Infrastructure Mode—An operating mode of a client RIC that requires all communications to go through an Access Point.

NMEA 0183—National Marine Electronics Association interface standard which defines electrical signal requirements, data transmission protocol and time, and specific sentence formats for a 4800-baud serial data bus.

OFDM—Orthogonal Frequency Division Multiplexing. A modulation method in which the communication channel is divided into multiple subcarriers each being individually modulated. While not meeting the Part 2 definition of spread spectrum the FCC has given specific authorization for OFDM systems.

Orthogonal—A mathematical term derived from the Greek word orthos, which means straight, right, or true. In terms of RF, orthogonal applies to the frequencies of the subcarriers which are selected so that at each one of these subcarrier frequencies, all the other subcarriers do not contribute to the overall waveform. In other words, the subcarrier channel is independent of the other channels.

PCMIA—Personal Computer Manufacturer Interface Adaptor.

Pigtail—A short piece of coaxial cable with a appropriate connectors to match the RIC antenna port and an external antenna system.

QAM—Quadrature Amplitude Modulation. A method of modulating data onto a carrier by changing both the phase and amplitude of the carrier. In its simplest form, 2QAM, the modulation is identical to BPSK. 16QAM represents 4 bits by changing among 16 phase/amplitude states. 64QAM represents six bits by changing among 64 phase/amplitude states.

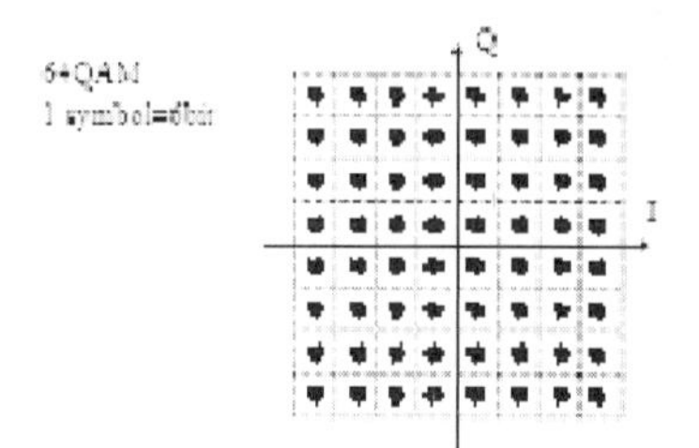

RIC—Radio Interface Card. The radio equivalent of a Network Interface Card (NIC).

RLAN—Radio Local Area Network. See also WLAN.

RMAN—Radio Metropolitan Area Network

Spread Spectrum—An information bearing communications system in which: (1) Information is conveyed by modulation of a carrier by some conventional means, (2) the bandwidth is deliberately widened by means of a spreading function over that which would be needed to transmit the information alone. (Title 47, Chapter I, Part 2, subpart A, section 2.1 Terms and Definitions).

SSID—Service Set Identifier. A unique alphanumeric string used to identify a WLAN, or in the case of HSMM, RLAN, by using the individual call sign and perhaps the name of the amateur radio club or repeater group.

UNII—Unlicensed National Information Infrastructure. The UNII spectrum is located at 5.15 - 5.35 GHz, 5.725 - 5.825 GHz, and the recently added 5.470-5.725 GHz band.

USB—Universal Serial Bus.

VPN—Virtual Private Network.

WEP—Wired Equivalent Privacy. An encryption algorithm used by the authentication process for authenticating users and for encrypting data payloads over a WLAN.

WEP Key—An alphanumeric character string used to identify an authenticating station and used as part of the data encryption algorithm.

Wi-Fi—Wireless Fidelity. Refers to products certified as compatible by the Wi-Fi Alliance. See **www.wi-fi.org**. This term is also applied in a generic sense to mean any 802.11 capability.

WiMAX—Familiar name for the IEEE 802.16 standard.

WISP—Wireless Internet Service Provider

WLAN—Wireless Local Area Network.

relays on a series of amateur towers between the end-points of the link, it is possible to cover greater distances with relatively low power and yet still move lots of multimedia data.

BASIC HSMM RADIO STATION

How do you set up an HSMM radio base station? It is really very easy. HSMM radio amateurs can go to any electronics outlet or office supply store and buy commercial off-the shelf (COTS) Wireless LAN gear, either IEEE 802.11b or IEEE 802.11g. They then connect external outdoor antennas. That is all there is to it.

There are some purchasing guidelines to follow. First, decide what interfaces you are going to need to connect to your computer. Equipment is available for all standard computer interfaces: Ethernet, USB and PCMCIA. If you use a laptop in your station, get the PCMCIA card. Make certain it is the type with an external antenna connection. If you have a PC, get the Wireless LAN adaptor type that plugs into either the USB port or the RJ45 Ethernet port. Make certain it is the type that has a removable rubber duck antenna or external antenna port! The included directions will explain how to install these devices.

The core of any HSMM radio station is a computer-operated HSMM 2.4-GHz radio transceiver, and it will probably cost about $60 to $80. Start off teaming up with a nearby ham radio operator. Do your initial testing in the same room together. Then as you increase distances going toward your separate station locations, you can coordinate using a suitable local FM simplex frequency. Frequently hams will use 146.52 MHz or 446.00 MHz, the National FM Simplex Calling Frequencies for the 2-m and 70-cm bands, for voice coordination. More recently, HSMM radio operators have tended to use 1.2-GHz FM transceivers and handheld (HT) radios. The 1.2-GHz amateur band more closely mimics the propagation characteristics of the 2.4-GHz amateur band. The rule of thumb is that if you cannot hear the other station on the 1.2-GHz FM radio, you probably will not be able to link up the HSMM radios either.

Hams frequently ask why 802.11 transmitter output and receiver sensitivity are stated typically in dBm. The simple answer is that this convention simplifies certain calculations. For transmitter output, convert dBm to power using the formula for dB. The reference power level is 1 mW. That means that +10 dBm = 10 mW and +20 dBm = 100 mW.

For receive, it's a bit more complicated if you want the more familiar units. First, calculate power level and then covert that to voltage across 50 Ω. A good RIC receiver is able to receive down to –96 dBm. That would be equal to 0.00000000025 mW, which is 3.54 µV across 50 Ω.

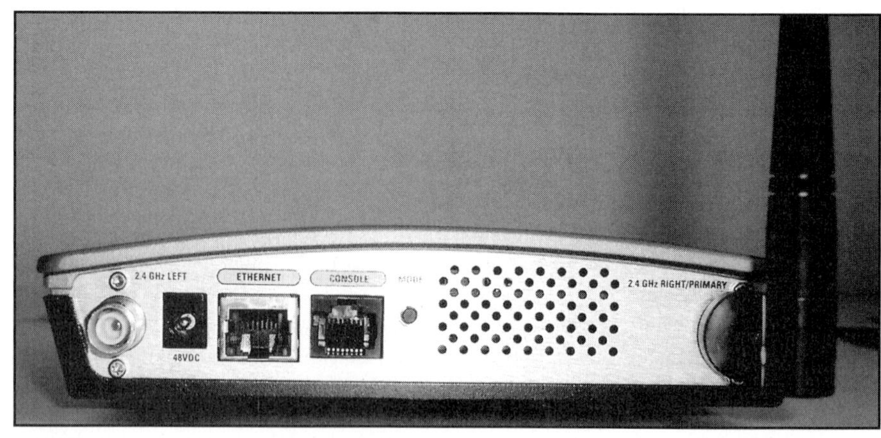

Fig 9.60—Back panel of a typical HSMM-style repeater. This device is known commercially as a wireless access point (AP). It is essentially a computer wireless network hub to enable multiple radio stations to share the various resources of the network. This particular model is a Cisco Model 1200. Note that the left, or secondary antenna's rubber duck has been removed from the TNC connector to show that the connector is of the reverse polarity (RP) type. This is designated as a female TNC/RP connector. Manufacturers of 802.11 gear typically install a RP-type of some type connector to prevent FCC Part 15 unlicensed users from employing their equipment in a non-certified manner. Of course, this is not an issue for licensed Part 97 users, however as in this case, a male RP-type plug will be required in order to connect the device to an outside antenna. The provision of a secondary antenna is to provide space diversity, which helps reduce the negative impact of multipath propagation of the radio signals. The secondary antenna may be ignored when connecting the primary antenna to a single outside antenna, especially if it is a highly directive antenna, which would help reduce multipath effects. *(Photo: John Champa, K8OCL).*

HSMM RADIO REPEATERS

Access Points

What hams would call a repeater, and computer buffs would call a hub, the WiFi industry refers to as a *wireless access point*, or simply *AP*. This is a device that allows several amateur radio stations to share the radio network and all the devices and circuits connected to it.

An 802.11b AP will sell for about $80 and an 802.11g AP for about $100. The AP acts as a central collection point for digital radio traffic, and can be connected to a single computer or to another radio or wired network.

The AP is provided with an *SSID*, which is the station identification it constantly broadcasts. For ham purposes, the SSID can be set as your call sign, thus providing automatic, and constant station identification. To use an AP in a radio network the wireless computer users have to exit ad-hoc mode and enter what is called the *infrastructure mode*, in their operating software. Infrastructure mode requires that you specify the radio network your computer station is intended to connect to, so set your computer station to recognize the SSID you assigned to the AP (yours or another ham's AP) to which you wish to connect.

Point-to-Point Links

The AP can also be used as one end of a point-to-point radio network. If you want to extend a radio network connection from one location to another, for example in order to remotely operate an HF station, you could use an AP at the network end and use it to communicate to a computer at the remote station location.

An AP allows for more network features and improved information security than is provided by ad-hoc mode. Most APs provide DHCP service, which is another way of saying they will automatically assign an Internet (IP) address to the wireless computers connected to the radio network. In addition, they can provide filtering, which allows only known users to access the network.

MOBILE HSMM OPERATING

When hams use the term mobile HSMM station what they are normally talking about is a wireless computer set-up in their vehicle to operate in a stationary portable fashion. Nobody is suggesting that you try to drive a vehicle and look at a computer screen at the same time! That would be very dangerous. So unless you have somebody else driving the vehicle, keep your eyes on the road and not on the computer screen.

What sort of equipment is needed to operate an HSMM mobile station?

• Some type of portable computer, such

as a laptop. Some hams use a PDA, notebook or other small computing device. The operating system can be *Microsoft Windows*, *Linux*, or *Mac OS*, although *Microsoft XP* offers some new and innovative WLAN functionality.

- Some type of radio software hams would call an *automatic monitor*, and computer buffs would call a *sniffer utility*. The most common type being used by hams is Marius Milner's *Network Stumbler for Windows* frequently just called, *NetStumbler*. All operating systems have monitoring programs that are available. *Linux* has *Kismet*; *MAC OS* has *MacStumbler*. Marius Milner has a version for the PocketPC, which he calls *MiniStumbler*.
- A RIC (*Radio Interface Card* = PCMCIA WiFi computer adapter card with external antenna port), which is supported by the monitoring utility you are using. The most widely supported RIC is the Orinoco line. The Orinoco line is inexpensive and fairly sensitive.
- An external antenna attached to your RIC. This is often a magnetically mounted omni-directional vertical antenna on the vehicle roof, but small directional antennas pointed out a window or mounted on a small ground tri-pod are also frequently used.
- A pigtail or short strain-relief cable will be needed to connect from the RIC antenna port to the N-series, RP/TNC or other type connector on the external antenna.
- A GPS receiver that provides NMEA 0183 formatted data and computer interface cable. This allows the monitoring utility to record where HSMM stations are located on a map, just as in APRS. GPS capability is optional, but just as with APRS capability, it makes the monitored information much more useful for locating HSMM stations.

Warning

While operating your HSMM mobile station, if you monitor an unlicensed Part 15 station (non-ham), some types of WiFi equipment will automatically associate or link to such stations, if they are not encrypted, and many are not (that is, WEP is not enabled). Although Part 15 stations share the 2.4-GHz band on a non-interfering basis with hams, they are operating in another service. In another part of this section we will provide various steps you can take to prevent Part 15 stations from automatically linking with HSMM stations. So in like manner, except in the case of a communications emergency, we recommend that you do not use a Part 15 station's Internet connection for any ham purpose.

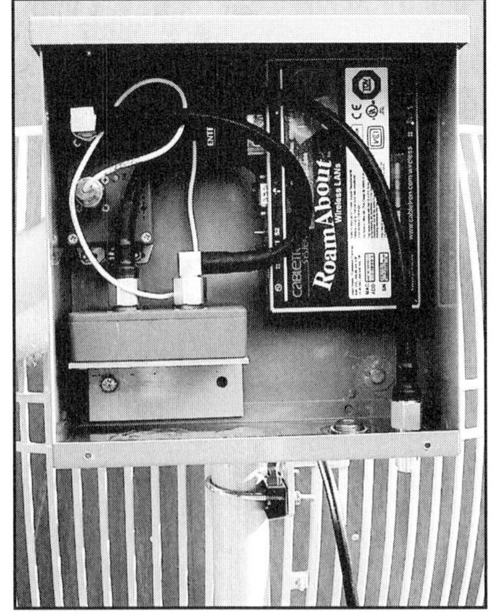

Fig 9.61—View of HSMM equipment (802.11b) inside an antenna-mounted NEMA-4 box. Mounting the equipment at the end of the dish antenna's pigtail significantly reduces feed line losses and greatly enhances the performance of the station. The box contains both a bridge, and a 500 mW bidirectional amplifier or BDA (lower left). Amplifier power is provided by the power insertion module seen in the upper left corner of the enclosure. *(Photo: John Champa, K8OCL).*

HSMM AREA SURVEYS

Both licensed amateurs and unlicensed (Part 15) stations use the 2.4-GHz band. To be a good neighbor, find out what others are doing in your area before designing your community HSMM radio network. This is easy to do using IEEE 802.11 modulation. Unless it has been disabled, an active repeater (AP) is constantly sending out an identification beacon known as the SSID. In HSMM practice this is simply the ham station call sign (and perhaps the local radio club name) entered into the software configuration supplied with the CD that comes with the repeater. So every HSMM repeater is also a continuous beacon.

A local area survey using appropriate monitoring software, for example the free *NetStumbler* software downloaded and running on your PC (**www.netstumbler.com/index.php**), is recommended prior to starting up any HSMM operations. Slew your station's directional antenna through a 360° arc, or drive your HSMM mobile station (described earlier) around your local area.

This HSMM area survey will identify and automatically log most other 802.11 station activity in your area. There are many different ways to avoid interference with other users of the band when planning your HSMM operating. For example, moving your operating frequency 2-3 channels away from the other stations is often sufficient. Why several channels and not just one? Because the channels have considerable overlap. Why this situation exists is beyond the scope of this section, but here is the situation: The channels are only 5 MHz wide, but the DSSS or OFDM modulation of 802.11 is 22 MHz wide. Commercial users often recommend moving 5 channels away from the nearest AP to completely avoid interference. There are six channels within the amateur 2.4 GHz band, but there are problems for hams with two of them. Channel 1 centered on 2412 MHz overlaps with OSCAR satellite downlink frequencies. Channel 6 centered on 2437 MHz is by far the most common out-of-the-box default channel for the majority of WLAN equipment sold in the US, so that often is not the best choice. Subsequently, most HSMM radio groups end up using either channel 3 or channel 4, depending on their local situation. Again, an area survey is recommended before putting anything on the air.

However, because of the wide sidebands used in these inexpensive broad banded 802.11 modulations, even moving 2-3 channels away from such activity may not be enough to totally avoid interference, especially if you are running what in HSMM is considered high power (typically 1800 mW RF output—more on that subject later). You may have to take other steps. For example, you may use a different polarization with your antenna system. Many HSMM stations use horizontal polarization because much 802.11 activity in their area is primarily vertically polarized.

HSMM ANTENNA SYSTEMS

There are a number of factors that determine the best antenna design for a specific HSMM radio application. Most commonly, HSMM stations use horizontal instead of vertical polarization.

Furthermore, most HSMM stations use highly directional antennas instead of omnidirectional antennas. Directional antennas provide significantly more gain and thus better signal-to-noise ratios, which in the case of 802.11 modulations means higher rate data throughput. Higher data throughput, in turn, translates into more multimedia radio capability.

Highly directional antennas also have many other advantages. Such antennas can allow two hams to shoot over, or shoot around, or even shoot between, other wireless stations on the band.

However, the nature of 802.11 modulations coupled with the various configurations of many COTS devices allows hams to economically experiment with many other fascinating antenna designs. Such unique antenna system designs can be used to simply help avoid interference, or to extend the range of HSMM links, or both.

Space Diversity

Some APs and some RICs have space-diversity capability built-into their design. However, it is not always operated in the same fashion, so check the literature or the website of your particular device's manufacturer to be certain how the dual antenna ports are used. For example, many APs come equipped with two rubber ducky antennas and two antenna ports. One antenna port may be the primary and the other port the secondary input to the transceiver. Which signal input is used may depend on which antenna is providing the best S/N ratio at that specific instant. Experimentation using two outside high-gain antennas spaced 10 or more wavelengths apart (that is only about one meter on the 2.4-GHz band) may be very worthwhile in improving data throughput on long links. Such extended radio paths tend to experience more multipath signal distortion. This multipath effect is caused by multiple signal reflections off various objects in the path of the linking signal. The use of space diversity techniques may help reduce this effect and thus improve the data rate throughput on the link. Again, the higher the date rates the more multimedia radio techniques that can be used on that network.

Circular Polarization

The use of circular polarization created using helical antennas, patch feed-points on dish antennas or other means, warrants further study by radio amateurs. Remember this is high-speed digital radio. To avoid symbol errors, circularly polarized antennas should be used at both ends of the link. Also, be certain that the antennas are of the same *handedness*, for example, right-hand circular polarization (RHCP). The ability of circular polarization to enhance propagation of long-path HSMM radio signals should not be overlooked.

Circularly Polarized Space Diversity

A combination or hybrid antenna design combining both circularly polarized antennas and space diversity could yield some extraordinary signal propagation results. For example, it has been suggested that perhaps using a RHCP for one antenna and LHCP for the other antenna, especially using spacing greater than 10 wavelengths, in such a system could provide a nearly "bullet-proof" design. Only actual field testing of such designs under different terrain features would reveal such potential.

Mixed Antenna Design Problem

In conventional wide-bandwidth analog radio antennas systems, so long as both antennas at both ends of a radio link have broad bandwidths and the same polarization, all is fine. While this may be true for wide bandwidth analog signals, such as amateur television VSB (vestigial sideband) signals or FM ATV signals, it may not be true for broad bandwidth high-speed digital signals.

Fig 9.62—FM voice repeater, amateur television (ATV), and HSMM antennas mounted on a hydraulically operated mast. This portable installation was used to provide shared high-speed Internet access and other special communications support to the many hams attending the 2003 Pacificon Hamvention in San Ramon, CA. The HSMM station is also used to provide streaming video or amateur digital video (ADV) to the Mount Diablo ARC's analog FM ATV repeater on the nearby mountain. *(Photo: John Champa, K8OCL)*

First, 802.11 modulations produce very broadband signals, typically 22 MHz. Secondly, the evidence to date indicates that the use of a same polarized antenna with one type of feed point at one end of the link and the use of a same polarized antenna with a different type of feed point at the other end of the link, may introduce a problem with high-speed digital signals. A common example of this potential mixed-antenna issue would be if one HSMM station uses a horizontally polarized linear Yagi, while the other HSMM station at the opposite end of the link uses a horizontally polarized loop Yagi.

Here is another typical situation. Let us say the ham at one end of the radio path uses a dish antenna with a horizontal dipole feed-point. The other ham at the opposite end of the path uses a horizontally polarized loop Yagi. Both antennas have gain, both antennas are broadband width designs, and both antennas are horizontally polarized. Nonetheless, the hams may experience higher BER (bit error rate) because of symbol errors caused by the different manner in which the two antennas manipulate the digital radio signal wave front. Further radio amateur experimentation with HSMM radio signals is warranted to determine the full impact on the radio link of using mixed antenna types.

RUNNING HIGHER POWER

Hams often ask why operate 802.11 modes under licensed Part 97 regulations when we may also operate such modes under unlicensed Part 15 regulations, and without the content restrictions imposed on the Amateur Radio service?

A major advantage of operating under Amateur Radio regulations is the feasibility of operating with more RF power output and larger, high-gain directive antennas. These added capabilities enable hams to increase the range of their operations. The enhanced signal-to-noise ratio provided by running high power will also allow better data packet throughput. This enhanced throughput, in turn, enables more multimedia experimentation and communication capability over such increased distances.

In addition, increasing the effective radiated power (ERP) of an HSMM radio link provides for more robust signal margins and consequently a more reliable link. These are important considerations in providing effective emergency communications services and accomplishing other important public service objectives in a band increasingly occupied by unlicensed stations and other noise sources.

It should be noted that the existing FCC amateur radio regulations covering spread spectrum (SS) at the time this is being written were implemented prior to 802.11 being available. The provision in the existing regulations calling for automatic power control (APC) for RF power outputs in excess of 1 W is not considered technologically feasible in the case of 802.11 modulations for various reasons. As a result the FCC has communicated to the ARRL that the APC provision of the existing SS regulations are therefore not applicable to 802.11 emissions under Part 97.

However, using higher than normal output power in HSMM radio, in the shared 2.4 GHz band, is also something that should be done with considerable care, and only after careful analysis of link path conditions and the existing 802.11 activity in your area. Using the minimum power necessary for the communications is the law and has always been a good operating practice for hams.

There are also other excellent and far less

expensive alternatives to running higher power when using 802.11 modes. For examples, amateurs are also allowed to use higher-gain directional antennas. Such antennas increase both the transmit and receive effectiveness of the transceiver. Also, by placing equipment as close to the station antenna as possible, a common amateur OSCAR satellite and VHF/UHF DXing technique, the feed-line loss is significantly reduced. This makes the HSMM station transceiver more sensitive to received signals, while also getting more of its transmitter power to the antenna.

Only after an HSMM radio link analysis (see the link calculations portion at **www.arrl.org/hsmm/** or go to **logidac.com/gfk/80211link/pathAnalysis.html**) clearly indicates that additional RF output power is required to achieve the desired path distance should more power output be considered.

At that point in the analysis showing that higher power is required, what is needed is called a bi-directional amplifier (BDA). This is a super fast switching pre-amplifier/amplifier combination that is usually mounted at the end of the antenna pig-tail near the top of the tower or mast. A reasonably priced 2.4-GHz 1800-mW watt output BDA is available from the FAB Corporation (**www.fab-corp.com**). It is specifically designed for amateur HSMM radio experimenters. Be certain to specify HSMM when placing your order. Also, to help prevent unauthorized use by unlicensed Part 15 stations, the FAB Corp may request a copy of your amateur license to accompany the order, and they will only ship the BDA to your licensee address as recorded in the FCC database.

This additional power output of 1800 mW should be sufficient for nearly all amateur operations. Even those supporting EmComm, which may require more robust signal margins than normally needed by amateurs, seldom will require more power output than this level. If still greater range is needed, there are other less expensive ways to achieve such ranges as described in the section HSMM Radio Relays.

When using a BDA and operating at higher than normal power levels on the channels 2 through 5 recommended for Amateur Radio use. These channels are arbitrary channels intended for Part 15 operation and are not required for Amateur Radio use, but they are hard-wired into the gear so we are stuck with them. You should also be aware of the sidebands produced by 802.11 modulation. These sidebands are in addition to the normal 22 MHz wide spread spectrum signal. Accordingly, if your HSMM radio station is next door to an OSCAR ground station or other licensed user of the band, you may need to take extra steps in order to avoid interfering with them.

The use of a tuned output filter may be necessary to avoid causing QRM. Even when operating on the recommended channels in the 2-5 range, whenever you use higher than normal power, some of your now amplified sidebands may go outside the amateur band, which stops at 2450 MHz. So from a practical point of view, whenever the use of a BDA is required to achieve a specific link objective, it is a good operating practice to install a tuned filter on the BDA output. Such filters are not expensive and they're readily available from several commercial sources. It should also be noted that most BDAs currently being marketed, while suitable for 802.11b modulation, they are often not suitable for the newer, higher speed 802.11g modulation.

There is another point to consider. Depending on what other 802.11 operating may be taking place in your area, it may be a good practice to only run higher power when using directional or sectional antennas. Such antennas allow hams to operate over and around other licensed stations, but also including unlicensed Part 15 activity in your area that you don't want to disrupt (a local school WLAN, WISP, etc). Again, before running high power, it is recommended that an area survey be conducted using a mobile HSMM rig as described earlier to determine what other 802.11 activity is in your area and what channels are already in use.

INFORMATION SECURITY

An HSMM radio station could be considered a form of software defined radio. Your computer running the appropriate software combined with the RIC makes a single unit, which is now your station HSMM transceiver. However, unlike other radios, your HSMM radio is now a networked radio device. It could be connected directly to other computers and to other radio networks and even to the Internet. So each HSMM radio (PC + RIC + software) needs to be protected. There are at least two basic steps that should be taken with regards to all HSMM radios:

The PC should be provided with an anti-virus program. This anti-virus software must be regularly updated to remain effective. Such programs may have come with the PC when it was purchased. If that is not the case, reasonably priced anti-virus programs are readily available from a number of sources.

Secondly, it is important to use a firewall software program on your HSMM radio. The firewall should be configured to allow all outgoing traffic, but to restrict all incoming traffic without specific authorization. Commercial personal computer firewall products are available from Symantec, ZoneLabs and McAfee Network Associates.

Check this URL for a list of freeware firewalls for your personal computer: **www.webattack.com/freeware/security/fwfirewall.shtml**.

Check this URL for a list of shareware firewalls for your personal computer: **www.webattack.com/Shareware/security/swfirewall.shtml**.

Once a group of HSMM stations has set-up and configured a repeater (AP) into a radio local area network (RLAN) then additional steps may need to be taken to restrict access to the repeater. Only Part 97 stations should be allowed to associate with the HSMM repeater. Remember, in the case of 802.11 modulations, the 2.4-GHz band is shared with Part 15 unlicensed 802.11 stations. How do you keep these unlicensed stations from automatically associating (auto-associate) with your licensed ham radio HSMM network?

Many times the steps taken to avoid interference with other stations also limits those other stations' capability to auto-associate with the HSMM repeater and to improve the overall security of the HSMM station. For example, you could use a different antenna polarization than the Part 15 station, or you could operate with a directional antenna oriented toward the desired coverage area rather than using an omnidirectional antenna.

The most effective method to keep unlicensed Part 15 stations off the HSMM repeater is to simply enable the *Wired Equivalent Protection* (WEP) already built into the 802.11 equipment. The WEP encrypts or scrambles the digital code on the HSMM repeater based on the instruction or "key" given to the software. Such encryption makes it impossible for unlicensed stations not using the specific code to accidentally auto-associate) with the HSMM repeater.

The primary purpose of this WEP implementation in the specific case of HSMM operating is to restrict access to the ham network by requiring all stations to authenticate themselves. Ham stations do this by using the WEP implementation with the appropriate ham key. Hams are permitted by FCC regulations to encrypt their transmission in specific instances; however, ironically at the time of this writing, this is not one of them. Accordingly, for hams to use WEP for authentication and not for encryption, the key used to implement the WEP must be published. The key must be published in a manner accessible by most of the amateur radio community. This fulfills the traditional ham radio role as a self-policing service. The current published ham radio WEP key is available at the home page of the ARRL Technology Task Force High Speed Multimedia Working Group: **www.arrl.org/hsmm/**.

Before implementing WEP on your HSMM repeater be certain that you have checked the website to ensure that you are using the current published WEP key. The key may need to be occasionally changed.

HSMM FREQUENCIES

Up to this point all the discussion has been regarding HSMM radio operations on the 2.4-GHz amateur band. However, 802.11 modulations can be used on any amateur band above 902 MHz.

On the 902 MHz band, using 802.11 modulations would occupy nearly the entire band. This may not be a problem in your area depending on the nature of the other existing users of the band in your area, either licensed or unlicensed. FM repeaters may not have a problem with sharing the frequency with 802.11 operations, since they would likely just hear an 802.11 modulated signal as weak background noise, and the 802.11 modulation, especially the OFDM channels used by 802.11g, would simply work around the FM interference with little negative impact. There is some older 802.11 gear (FHSS) available on the surplus market for amateur experimentation. Alternatively, some form of frequency transverter may be used to take 2.4 GHz to the 902-MHz band.

The 1.2-GHz band has some potential for 802.11 experimenting. Some areas have several FM voice repeaters and even ATV FM repeaters on the band. But again these relatively narrow bandwidth signals would likely hear any 802.11 modulations as simply background noise. Looking at the potential interference from the HSMM perspective, even in the case of the FM ATV, it is unlikely the signal would significantly disrupt the 802.11 modulation unless the two signals were on exactly the same center frequency or at least with complete overlap in bandwidth. Keep in mind that the FM ATV signal is only several megahertz wide, but the 802.11 modulation is 22 MHz wide. For the analog signal to wipe out the spread spectrum signal, it would need to overpower or completely swamp the 802.11 RIC receiver's front end.

The 3.5-GHz band offers some real possibilities for 802.11 developments. Frequency transverters are available to get to the band from 2.4 GHz and there is little other activity on the band at this time. Developments in Europe of 802.16 with 108 Mbit/s data throughput may make 3.5-GHz gear available for amateur experimentation in the US. Hams are investigating the feasibility of using such gear when it becomes available in the US for providing a RMAN or *radio metropolitan area networks*. The RMAN would be used to link the individual HSMM repeaters (AP) or RLANs together in order to provide county-wide or regional HSMM coverage, depending on the ham radio population density.

The 5-GHz band is also being investigated. The COTS 802.11a modulation gear has OFDM channels that operate in this Amateur Radio band. The 802.11a modulation could be used in a ham RLAN operating much as 802.11g is in the 2.4-GHz band. It is also being considered by some HSMM groups as a means of providing MAN links. This band is also being considered by AMSAT for what is known as a C-N-C transponder. This would be an HSMM transponder onboard probably a Phase-3 high-altitude or a Phase-4 geostationary OSCAR with uplink and downlink bandpass both within the 5-GHz amateur band. Some other form of modulation other than 802.11 would likely have to be used because of timing issues and other factors, but the concept is at least being seriously discussed.

RMAN link alternatives are also being tested by hams. One of these is the use of *virtual private networks* (VPN) similar to the method currently used to provide worldwide FM voice repeater links via the Internet. Mark Williams, AB8LN, of the HSMM Working Group is leading a team to test the use of various VPN technologies for linking HSMM repeaters.

HF is not being ignored either. It is possible that a modulation form that, while it is neither SS nor HSMM, might be able to produce data rates fast enough to efficiently handle e-mail type traffic on the HF bands, while still occupying an appropriate bandwidth. Such modulation would be helpful in an emergency with providing an outlet for RMAN e-mail traffic. Neil Sablatzky, K8IT, is leading a team of ham investigators on the HF and VHF bands.

Finally there are commercial products being developed such as the Icom D-STAR system that could readily be integrated into a RMAN infrastructure.

HSMM REFERENCES

Use of HSMM over Amateur Radio is a developing story. You can keep up with developments by visiting *ARRLWeb* at **arrl.org/hsmm/**.

For more details about using HSMM radio for remote control of stations, see the article "Remote-Control HF Operation over the Internet," by Brad Wyatt, K6WR, *QST*, November 2001 p 47-48.

For guidelines on using e-games on-the air in Amateur Radio, see the HSMM column titled "Is (sic) All Data Acceptable Data" by Neil Sablatzky, K8IT, in the Fall 2003 issue of *CQ VHF*.

For more information regarding HSMM on future OSCAR satellites, see the *Proceedings of the AMSAT-NA 21st Space Symposium*, November 2003, Toronto, Ontario, Canada, especially the paper by Clark, Tom, W3IWI, "C-C RIDER, A New Concept for Amateur Satellites," available from ARRL.

Burger, Michael W, AH7R, and John J. Champa, K8OCL, "HSMM in a Briefcase," *CQ VHF*, Fall 2003, p 32.

Champa, John, K8OCL, and Ron Olexa, KA3JIJ, "How To Get Into HSMM," *CQ VHF*, Fall 2003.

Champa, John, K8OCL, and Stephensen, John, KD6OZH, "28 kbps to 9 Mbps UHF Modems for Amateur Radio Stations," *QEX*, Mar/Apr 2005.

Cooper, G.R., and McGillem, C. D., *Modern Communications and Spread Spectrum*, New York, McGraw-Hill, 1986.

Duntemann, Jeff, K7JPD, *Jeff Duntemann's Wi-Fi Guide,* 2nd Ed, Paraglyph Press, 2004.

Flickenger, Rob, *Building Wireless Community Networks,* 2nd Ed, O'Reilly, 2003.

Flickenger, Rob, *Wireless Hacks*, O'Reilly, 2003.

Ford, Steve, WB8IMY, "VoIP and Amateur Radio," *QST*, February 2003, p 44-47.

Ford, Steve, WB8IMY, *ARRL's HF Digital Handbook*, American Radio Relay League, 2001.

Fordham, David, KD9LA, "802.11 Experiments in Virginia's Shenandoah Valley," *QST, July 2005.*

Gast, Matthew S., *802.11 Wireless Networks, The Definitive Guide*, O'Reilly, 2002.

Geier, Jim, *Wireless LANs, Implementing High Performance IEEE 802.11 Networks, 2nd Ed*, SAMS, 2002.

Husain, Kamran, and Parker, Timothy, PhD, et al, *Linux Unleashed*, SAMS, 1995.

McDermott, T., *Wireless Digital Communications: Design and Theory*, TAPR, 1996.

Mraz, Kris I, N5KM, "High Speed Multimedia Radio," *QST*, April 2003, pp 28-34.

Olexa, Ron, KA3JIJ, "Wi-Fi for Hams Part 1: Part 97 or Part 15," *CQ*, June 2003, pp 32-36.

Olexa, Ron, KA3JIJ, "Wi-Fi for Hams Part 2: Building a Wi-Fi Network," *CQ*, July 2003, p 34-38.

Patil, Basavaraj, et. Al,. *IP in Wireless Networks*, Prentice Hall, 2003.

Potter, Bruce and Fleck, Bob, *802.11 Security*, O'Reilly, 2003.

Reinhardt, Jeff, AA6JR, "Digital Hamming: A Need for Standards," *CQ*, January 2003, p 50-51.

Rinaldo, Paul L., W4RI, and Champa, John J., K8OCL, "On The Amateur Radio Use of IEEE 802.11b Radio Local Area Networks," *CQ VHF*, Spring 2003, p 40-42.

Rotolo, Don, N2IRZ, "A Cheap and Easy High-Speed Data Connection," *CQ*, February 2003, p 61-64.

Torrieri, D.J., *Principles of Secure Communication Systems*, Boston, Artech House, 1985.

Chapter 10

Oscillators and Synthesizers

Just say in public that oscillators are one of the most important, fundamental building blocks in radio technology and you will immediately be interrupted by someone pointing out that *tuned-RF* (*TRF*) receivers can be built without any form of oscillator at all. This is certainly true, but it shows how some things can be taken for granted. What use is any receiver without signals to receive? All intentionally transmitted signals trace back to some sort of signal generator—an oscillator or frequency synthesizer. In contrast with the TRF receivers just mentioned, a modern, all-mode, feature-laden MF/HF transceiver may contain in excess of a dozen RF oscillators and synthesizers, while a simple QRP CW transmitter may consist of nothing more than a single oscillator. (This chapter was written by David Stockton, GM4ZNX. Frederick J. Telewski, WA7TZY, also contributed to the Frequency Synthesizers section.)

In the 1980s, the main area of progress in the performance of radio equipment was the recognition of receiver intermodulation as a major limit to our ability to communicate, with the consequent development of receiver front ends with improved ability to handle large signals. So successful was this campaign that other areas of transceiver performance now require similar attention. One indication of this is any equipment review receiver dynamic range measurement qualified by a phrase like "limited by oscillator phase noise." A plot of a receiver's effective selectivity can provide another indication of work to be done: An IF filter's high-attenuation region may appear to be wider than the filter's published specifications would suggest—almost as if the filter characteristic has grown sidebands! In fact, in a way, it has: This is the result of local-oscillator (LO) or synthesizer *phase noise* spoiling the receiver's overall performance. Oscillator noise is the prime candidate for the next major assault on radio performance.

The sheer number of different oscillator circuits can be intimidating, but their great diversity is an illusion that evaporates once their underlying pattern is seen. Almost all RF oscillators share one fundamental principle of operation: an amplifier and a filter operate in a loop (**Fig 10.1**). There are plenty of filter types to choose from:

- LC
- Quartz crystal and other piezoelectric materials
- Transmission line (stripline, microstrip, troughline, open-wire, coax and so on)
- Microwave cavities, YIG spheres, dielectric resonators
- Surface-acoustic-wave (SAW) devices

Should any new forms of filter be invented, it's a safe guess that they will also be applicable to oscillators. There is an equally large range of amplifiers to choose from:

- Vacuum tubes of all types
- Bipolar junction transistors
- Field effect transistors (JFET, MOSFET, GaAsFET, in all their varieties)
- Gunn diodes, tunnel diodes and other negative-resistance generators

It seems superfluous to state that anything that can amplify can be used in an oscillator, because of the well-known propensity of all prototype amplifiers to oscillate! The choice of amplifier is widened further by the option of using single- or multiple-stage amplifiers and discrete devices versus integrated circuits. Multiply all of these options with those of filter choice and the resulting set of combinations is very large, but a long way from complete. Then there are choices of how to couple the amplifier into the filter and the filter into the amplifier. And then there are choices to make in the filter section: Should it be tuned by variable capacitor, variable inductor or some form of sliding cavity or line?

Despite the number of combinations that are possible, a manageably small number of types will cover all but very special requirements. Look at an oscillator circuit and "read" it: What form of filter—*resonator*—does it use? What form of amplifier? How have the amplifier's input and output been coupled into the filter? How is the filter tuned? These are simple, easily answered questions that put oscillator types into appropriate categories and make them understandable. The questions themselves may make more

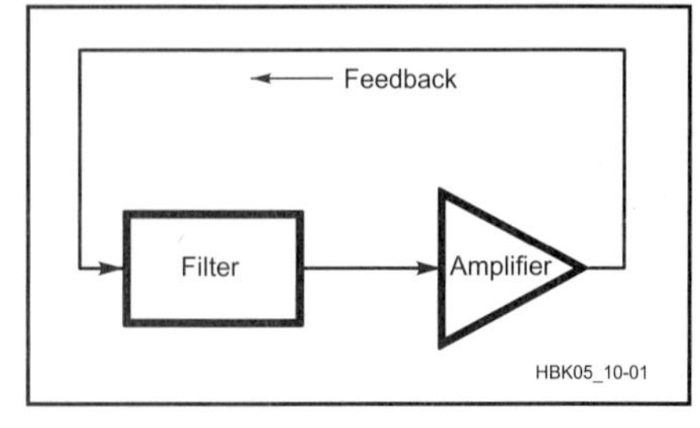

Fig 10.1—Reduced to essentials, an oscillator consists of a filter and an amplifier operating in a feedback loop.

sense if we understand the mechanics of oscillation, in which *resonance* plays a major role.

HOW OSCILLATORS WORK

Maintained Resonance

The pendulum, a good example of a resonator, has been known for millennia and understood for centuries. It is closely analogous to an electronic resonator, as shown in **Fig 10.2**. The weighted end of the pendulum can store energy in two different forms: The *kinetic energy* of its motion and the *potential energy* of it being raised above its rest position. As it reaches its highest point at the extreme of a swing, its velocity is zero for an instant as it reverses direction. This means that it has, at that instant, no kinetic energy, but because it is also raised above its rest position, it has some extra potential energy.

At the center of its swing, the pendulum is at its lowest point with respect to gravity and so has lost the extra potential energy. At the same time, however, it is moving at its highest speed and so has its greatest kinetic energy. Something interesting is happening: The pendulum's stored energy is continuously moving between potential and kinetic forms. Looking at the pendulum at intermediate positions shows that this movement of energy is smooth. Newton provided the keys to understanding this. It took his theory of gravity and laws of motion to explain the behavior of a simple weight swinging on the end of a length of string and calculus to perform a quantitative mathematical analysis. Experiments had shown the period of a pendulum to be very stable and predictable. Apart from side effects of air drag and friction, the length of the period should not be affected by the mass of the weight, nor by the amplitude of the swing.

A pendulum can be used for timing events, but its usefulness is spoiled by the action of drag or friction, which eventually stops it. This problem was overcome by the invention of the *escapement*, a part of a clock mechanism that senses the position of the pendulum and applies a small push in the right direction and at the right time to maintain the amplitude of its swing or oscillation. The result is a mechanical oscillator: The pendulum acts as the filter, the escapement acts as the amplifier and a weight system or wound-up spring powers the escapement.

Electrical oscillators are closely analogous to the pendulum, both in operation and in development. The voltage and current in the tuned circuit—often called *tank circuit* because of its energy-storage ability—both vary sinusoidally with time and are 90° out of phase. There are instants when the current is zero, so the energy stored in the inductor must be zero, but at the same time the voltage across the capacitor is at its peak, with all of the circuit's energy stored in the electric field between the capacitor's plates. There are also instants when the voltage is zero and the current is at a peak, with no energy in the capacitor. Then, all of the circuit's energy is stored in the inductor's magnetic field.

Just like the pendulum, the energy stored in the electrical system is swinging smoothly between two forms; electric field and magnetic field. Also like the pendulum, the tank circuit has losses. Its conductors have resistance, and the capacitor dielectric and inductor core are imperfect. Leakage of electric and magnetic fields also occurs, inducing currents in neighboring objects and just plain radiating energy off into space as radio waves. The amplitudes of the oscillating voltage and current decrease steadily as a result. Early intentional radio transmissions, such as those of Heinrich Hertz's experiments, involved abruptly dumping energy into a tuned circuit and letting it oscillate, or *ring*, as shown in **Fig 10.3**. This was done by applying a spark to the resonator. Hertz's resonator was a gapped ring, a good choice for radiating as much of the energy as possible. Although this looks very different from the LC tank of Fig 10.2, it has inductance *distributed* around its length and capacitance distributed across it and across its gap, as opposed to the *lumped* L and C values in Fig 10.2. The gapped ring therefore works just the same as the LC tank in terms of oscillating voltages and currents. Like the pendulum and the LC tank, its period, and

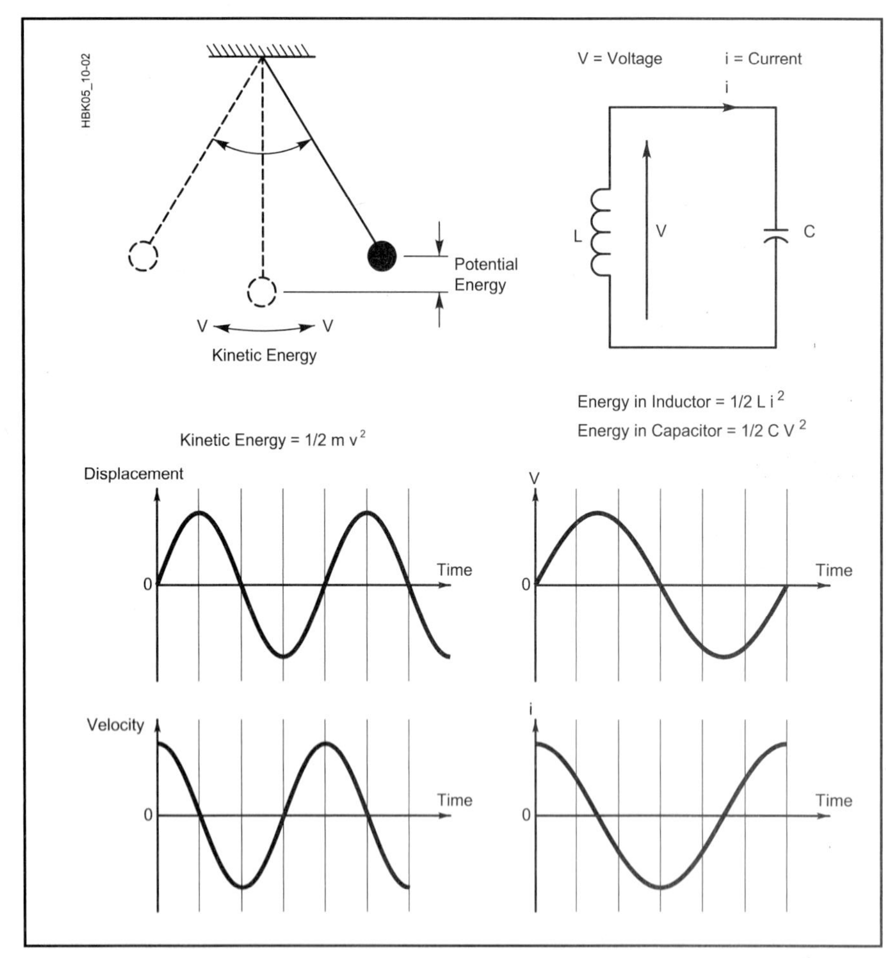

Fig 10.2—A resonator lies at the heart of every oscillatory mechanical and electrical system. A mechanical resonator (here, a pendulum) and an electrical resonator (here, a tuned circuit consisting of L and C in parallel) share the same mechanism: the regular movement of energy between two forms—potential and kinetic in the pendulum, electric and magnetic in the tuned circuit. Both of these resonators share another trait: Any oscillations induced in them eventually die out because of losses—in the pendulum, due to drag and friction; in the tuned circuit, due to the resistance, radiation and inductance. Note that the curves corresponding to the pendulum's displacement vs velocity and the tuned circuit's voltage vs current, differ by one quarter of a cycle, or 90°.

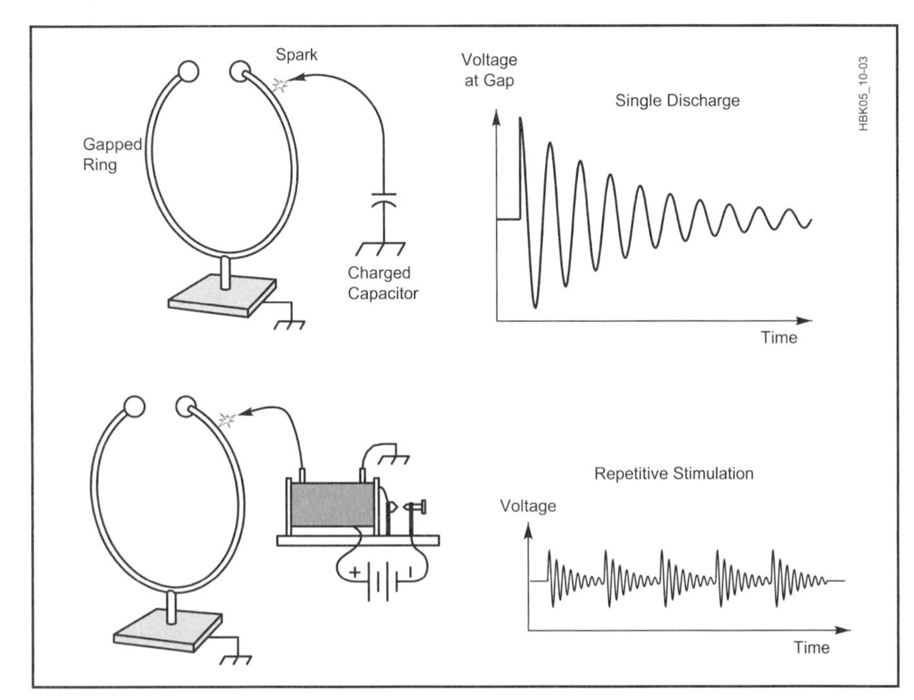

Fig 10.3—Stimulating a resonance, 1880s style. Shock-exciting a gapped ring with high voltage from a charged capacitor causes the ring to oscillate at its resonant frequency. The result is a damped wave, each successive alternation of which is weaker than its predecessor because of resonator losses. Repetitively stimulating the ring produces trains of damped waves, but oscillation is not continuous.

therefore the frequency at which it oscillates, is independent of the magnitude of its excitation.

Making a longer-lasting signal with the Fig 10.3 arrangement merely involves repeating the sparks. The problem is that a truly continuous signal cannot be made this way. The sparks cannot be applied often enough or always timed precisely enough to guarantee that another spark re-excites the circuit at precisely the right instant. This arrangement amounts to a crude spark transmitter, variations of which served as the primary means of transmission for the first generation of radio amateurs. The use of damped waves is now entirely forbidden by international treaty because of their great impurity. Damped waves look a lot like car-ignition waveforms and sound like car-ignition interference when received.

What we need is a *continuous wave* (*CW*) oscillation—a smooth, sinusoidal signal of constant amplitude, without phase jumps, a "pure tone." To get it, we must add to our resonator an equivalent of the clock's escapement—a means of synchronizing the application of energy and a fast enough system to apply just enough energy every cycle to keep each cycle at the same amplitude.

Amplification

A sample of the tank's oscillation can be extracted, amplified and reinserted. The gain can be set to exactly compensate the tank losses and perfectly maintain the oscillation. The amplifier usually need only give low gain, so active devices can be used in oscillators not far below their unity (unity = 1) gain frequency. The amplifier's output must be lightly coupled into the tank—the aim is just to replace lost energy, not forcibly drive the tank. Similarly, the amplifier's input should not heavily load the tank. It is a good idea to think of *coupling* networks rather than *matching* networks in this application, because a matched impedance extracts the maximum available energy from a source, and this would certainly spoil an oscillator.

Fig 10.4A shows the block diagram of an oscillator. Certain conditions must be met for oscillation. The criteria that separate oscillator loops from stable loops are often attributed to Barkhausen by those aiming to produce an oscillator and to Nyquist by those aiming for amplifier stability, although they boil down to the same boundary. Fig 10.4B shows the loop broken and a test signal inserted. (The loop can be broken anywhere; the amplifier input just happens to be the easiest place to do it.) The criterion for oscillation says that at a frequency at which the phase shift around the loop is exactly zero, the net gain around the loop must equal or exceed unity (that is, one). (Later, when we design phase-locked loops, we must revisit this concept in order to check that *those* loops *cannot* oscillate.) Fig 10.4C shows what happens to the amplitude of an oscillator if the loop gain is made a little higher or lower than one.

The loop gain has to be precisely one if we want a stable amplitude. Any inaccuracy will cause the amplitude to grow to clipping or shrink to zero, making the oscillator useless. Better accuracy will only slow, not stop this process. Perfect precision is clearly impossible, yet there are enough working oscillators in existence to prove that we are missing something important. In an amplifier, nonlinearity is a nuisance, leading to signal distortion and intermodulation, yet nonlinearity is what makes stable oscillation possible. All of the vacuum tubes and transistors used in oscillators tend to reduce their gain at higher signal levels. With such components, only a tiny change in gain can shift the loop's operation between amplitude growth and shrinkage. Oscillation stabilizes at that level at which the gain of the active device sets the loop gain at exactly one.

Another gain-stabilization technique involves biasing the device so that once some level is reached, the device starts to turn off over part of each cycle. At higher levels, it cuts off over more of each cycle. This effect reduces the effective gain quite strongly, stabilizing the amplitude. This badly distorts the signal (true in most common oscillator circuits) in the amplifying device, but provided the amplifier is lightly coupled to a high-Q resonant tank, the signal in the tank should not be badly distorted.

Many radio amateurs now have some form of circuit-analysis software, usually running on a PC. Attempts to analyze oscillators by this means often fail by predicting growing or shrinking amplitudes, and often no signal at all, in circuits that are known to work. Computer analysis of oscillator circuits can be done, but it requires a sophisticated program with accurate, nonlinear, RF-valid models of the devices used, to be able to predict operating amplitude. Often even these programs need some special tricks to get their modeled oscillators to start. Such software is likely to be priced higher than most private users can justify, and it still doesn't replace the need for the user to understand the circuit. With that understanding, some time, some parts and a little patience will do the job, unassisted.

Textbooks give plenty of coverage to the frequency-determining mechanisms of oscillators, but the amplitude-determining mechanism is rarely covered. It is often not even mentioned. There is a

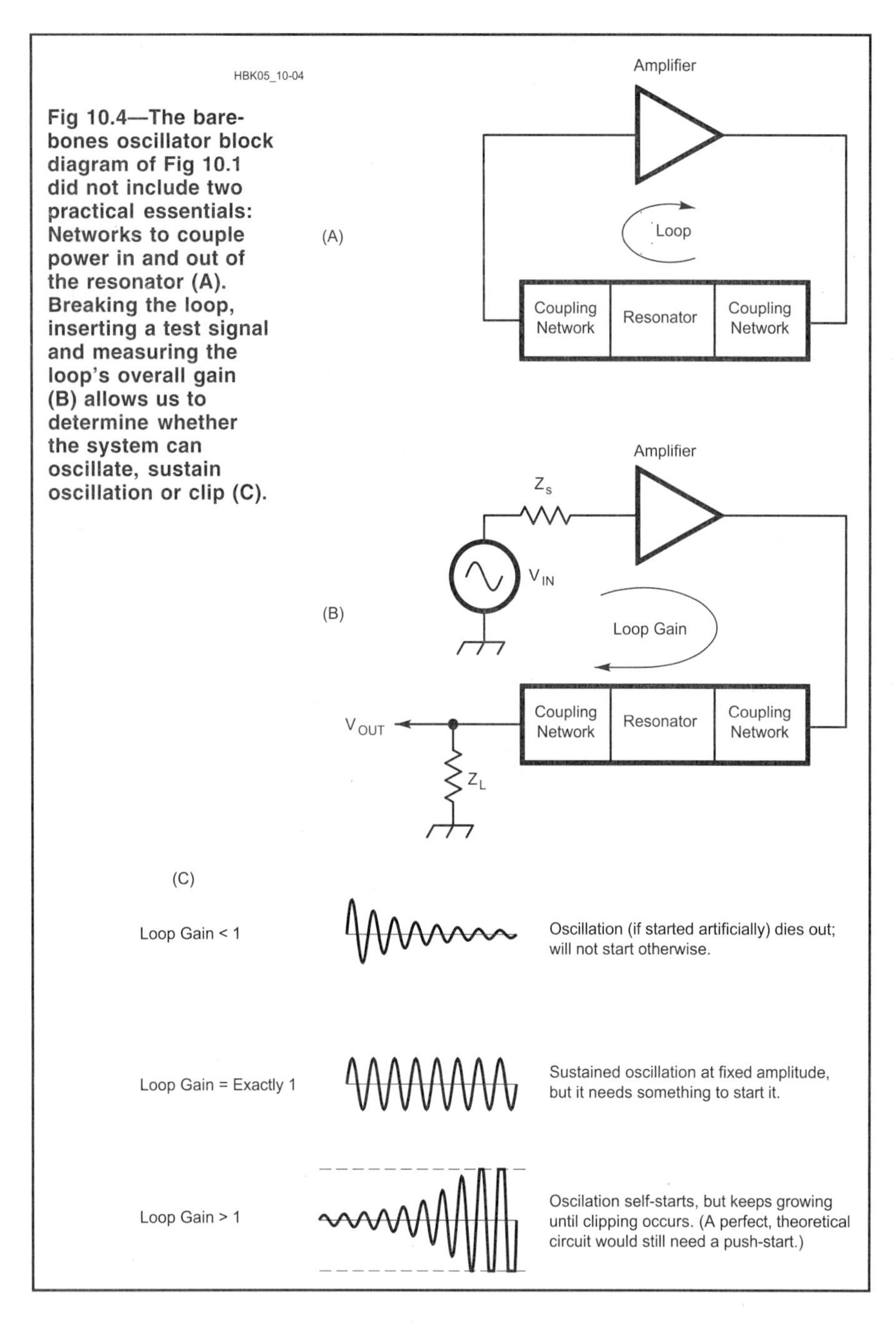

Fig 10.4—The bare-bones oscillator block diagram of Fig 10.1 did not include two practical essentials: Networks to couple power in and out of the resonator (A). Breaking the loop, inserting a test signal and measuring the loop's overall gain (B) allows us to determine whether the system can oscillate, sustain oscillation or clip (C).

good treatment in Clarke and Hess, *Communications Circuits: Analysis and Design*.

Start-Up

Perfect components don't exist, but if we could build an oscillator from them, we would naturally expect perfect performance. We would nonetheless be disappointed. We could assemble from our perfect components an oscillator that exactly met the criterion for oscillation, having slightly excessive gain that falls to the correct amount at the target operating level and so is capable of sustained, stable oscillation. But being capable of something is not the same as doing it, for there is another stable condition. If the amplifier in the loop shown in Fig 10.4 has no input signal, and is perfect, it will give no output! No signal returns to the amplifier's input via the resonator, and the result is a sustained and stable *lack* of oscillation. Something is needed to start the oscillator.

This fits the pendulum-clock analogy: A wound-up clock is stable with its pendulum at rest, yet after a push the system will sustain oscillation. The mechanism that drives the pendulum is similar to a Class C amplifier: It does not act unless it is driven by a signal that exceeds its threshold. An electrical oscillator based on a Class C amplifier can sometimes be kicked into action by the turn-on transient of its power supply. The risk is that this may not always happen, and also that should some external influence stop the oscillator, it will not restart until someone notices the problem and cycles the power. This can be very inconvenient!

A real-life oscillator whose amplifier does not lose gain at low signal levels can self-start due to noise. **Fig 10.5** shows an oscillator block diagram with the amplifier's noise shown, for our convenience, as a second input that adds with the true input. The amplifier amplifies the noise. The resonator filters the output noise, and this signal returns to the amplifier input. The importance of having slightly excessive gain until the oscillator reaches operating amplitude is now obvious. If the loop gain is slightly above one, the recirculated noise must, within the resonator's bandwidth, be larger than its original level at the input. More noise is continually summed in as a noise-like signal continuously passes around the loop, undergoing amplification and filtering as it does. The level increases,

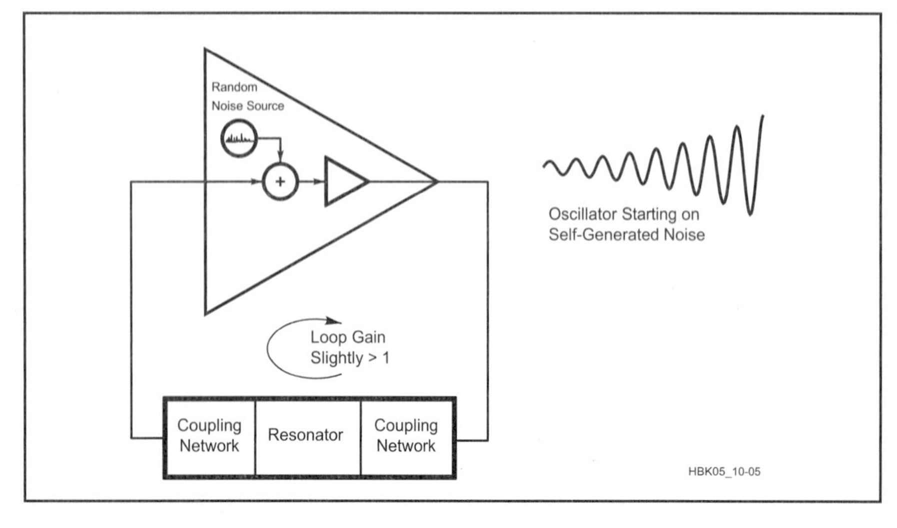

Fig 10.5—An oscillator with noise. Real-world amplifiers, no matter how quiet, generate some internal noise; this allows real-world oscillators to self-start.

causing the gain to reduce. Eventually, it stabilizes at whatever level is necessary to make the net loop gain equal to one.

So far, so good. The oscillator is running at its proper level, but something seems very wrong. It is not making a proper sine wave; it is recirculating and filtering a noise signal. It can also be thought of as a Q multiplier with a controlled (high) gain, filtering a noise input and amplifying it to a set level. Narrow-band filtered noise approaches a true sine wave as the filter is narrowed to zero width. What this means is that we cannot make a true sine-wave signal—all we can do is make narrow-band filtered noise as an approximation to one. A high-quality, low-noise oscillator is merely one that does tighter filtering. Even a kick-started Class-C-amplifier oscillator has noise continuously entering its circulating signal, and so behaves similarly.

A small-signal gain greater than one is absolutely critical for reliable starting, but having too much gain can make the final operating level unstable. Some oscillators are designed around limiting amplifiers to make their operation predictable. AGC systems have also been used, with an RF detector and dc amplifier used to servo-control the amplifier gain. It is notoriously difficult to design reliable crystal oscillators that can be published or mass-produced without having occasional individuals refuse to start without some form of shock.

Mathematicians have been intrigued by "chaotic systems" where tiny changes in initial conditions can yield large changes in outcome. The most obvious example is meteorology, but much of the necessary math was developed in the study of oscillator start-up, because it is a case of chaotic activity in a simple system. The equations that describe oscillator start-up are similar to those used to generate many of the popular, chaotic fractal images.

PHASE NOISE

Viewing an oscillator as a filtered-noise generator is relatively modern. The older approach was to think of an oscillator making a true sine wave with an added, unwanted noise signal. These are just different ways of visualizing the same thing: They are equally valid views, which are used interchangeably, depending which best makes some point clear. Thinking in terms of the signal-plus-noise version, the noise surrounds the carrier, looking like sidebands and so can also be considered to be equivalent to random-noise FM and AM on the ideal sine-wave signal. This gives us a third viewpoint. Strangely, these noise sidebands are called *phase noise*. If we consider the addition of a noise voltage to a sinusoidal voltage, we

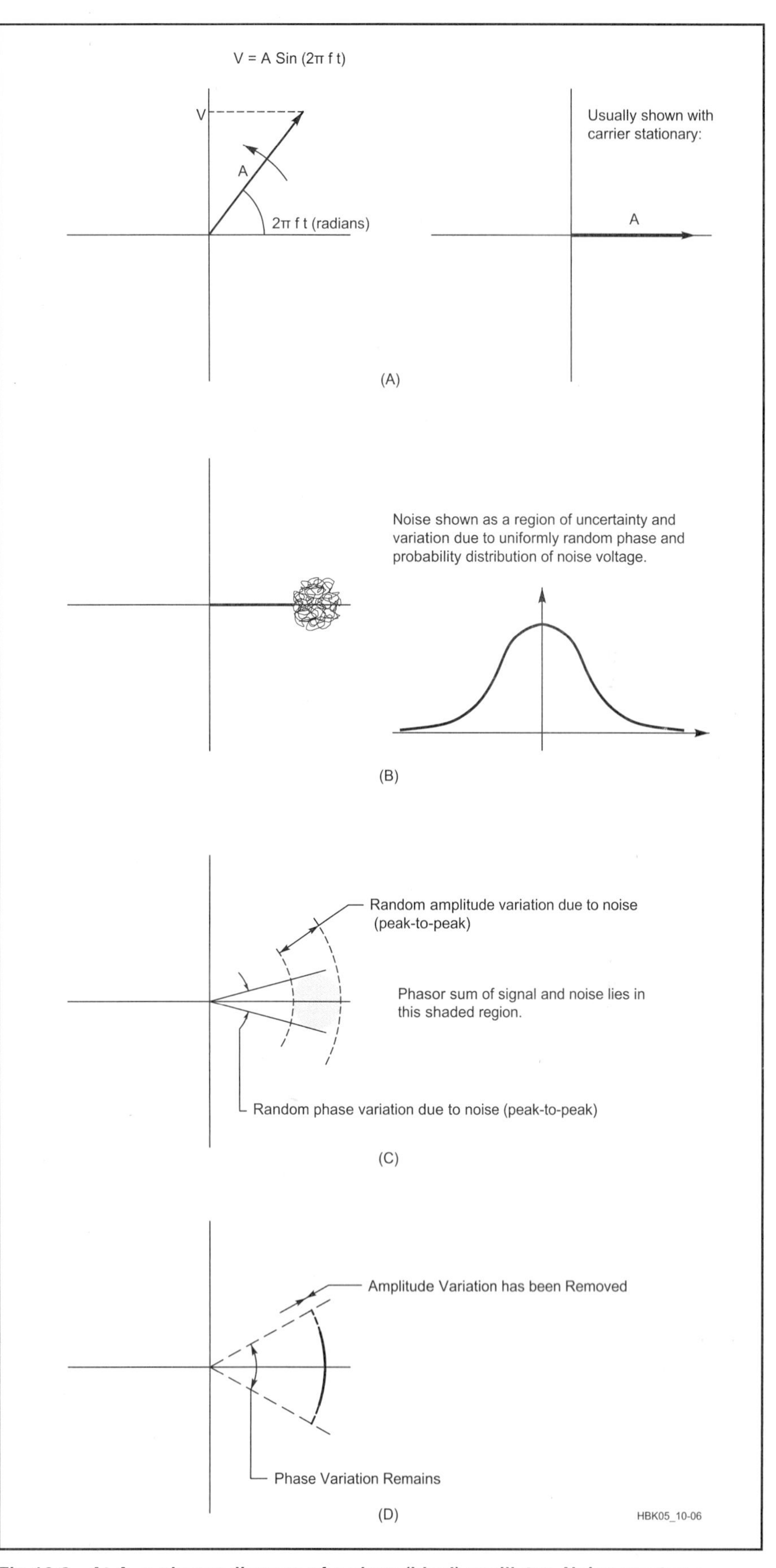

Fig 10.6—At A, a phasor diagram of a clean (ideal) oscillator. Noise creates a region of uncertainty in the vector's length and position (B). AM noise varies the vector's length; PM noise varies the vector's relative angular position (C) Limiting a signal that includes AM and PM noise strips off the AM and leaves the PM (D).

must take into account the phase relationship. A *phasor diagram* is the clearest way of illustrating this. **Fig 10.6A** represents a clean sine wave as a rotating vector whose length is equal to the peak amplitude and whose frequency is equal to the number of revolutions per second of its rotation. Moving things are difficult to depict on paper, so phasor diagrams are usually drawn to show the dominant signal as stationary, with other components drawn relative to this.

Noise contains components at many frequencies, so its phase with respect to the dominant, theoretically pure signal—the "carrier"—is random. Its amplitude is also random. Noise can only be described in statistical terms because its voltage is constantly and randomly changing, yet it does have an average amplitude that can be expressed in rms volts. Fig 10.6B shows noise added to the carrier phasor, with the noise represented as a fuzzy, uncertain region in which the sum phasor wanders randomly. The phase of the noise is uniformly random—no direction is more likely than any other—but the instantaneous magnitude of the noise obeys a probability distribution like that shown, higher values being progressively rarer. Fig 10.6C shows how the extremities of the noise region can be considered as extremes of phase and amplitude variation from the normal values of the carrier.

Phase modulation and frequency modulation are closely related. Phase is the integral of frequency, so phase modulation resembles frequency modulation in which deviation decreases with increasing modulating frequency. Thus, there is no need to talk of "frequency noise" because *phase noise* already covers it.

Fig 10.6C clearly shows AM noise as the random variation of the length of the sum phasor, yet "amplitude noise" is rarely discussed. The oscillator's amplitude control mechanism acts to reduce the AM noise by a small amount, but the main reason is that the output is often fed into some form of limiter that strips off AM components just as the limiting IF amplifier in an FM receiver removes any AM on incoming signals. The limiter can be obvious, like a circuit to convert the signal to logic levels, or it can be implicit in some other function. A diode ring mixer may be driven by a sine-wave LO of moderate power, yet this signal drives the diodes hard on and hard off, approximating square-wave switching. This is a form of limiter, and it removes the effect of any AM on the LO. Fig 10.6D shows the result of passing a signal with noise through a limiting amplifier. For these reasons, AM noise sidebands are rarely a problem in oscillators, and so are normally ignored. There is one subtle problem to beware of, however. If a sine wave drives some form of switching circuit or limiter, and the threshold is offset from the signal's mean voltage, any level changes will affect the exact switching times and cause *phase jitter*. In this way, AM sidebands are translated into PM sidebands and pass through the limiter. This is usually called *AM-to-PM conversion* and is a classic problem of limiters.

Effects of Phase Noise

You would be excused for thinking that phase noise is a recent discovery, but all oscillators have always produced it. Other changes have elevated an unnoticed characteristic up to the status of a serious impairment. Increased crowding and power levels on the ham bands, allied with greater expectations of receiver performance as a result of other improvements, have made phase noise more noticeable, but the biggest factor has been the replacement of VFOs in radios by frequency synthesizers. It is a major task to develop a synthesizer that tunes in steps fine enough for SSB and CW while competing with the phase-noise performance of a reasonable-quality LC

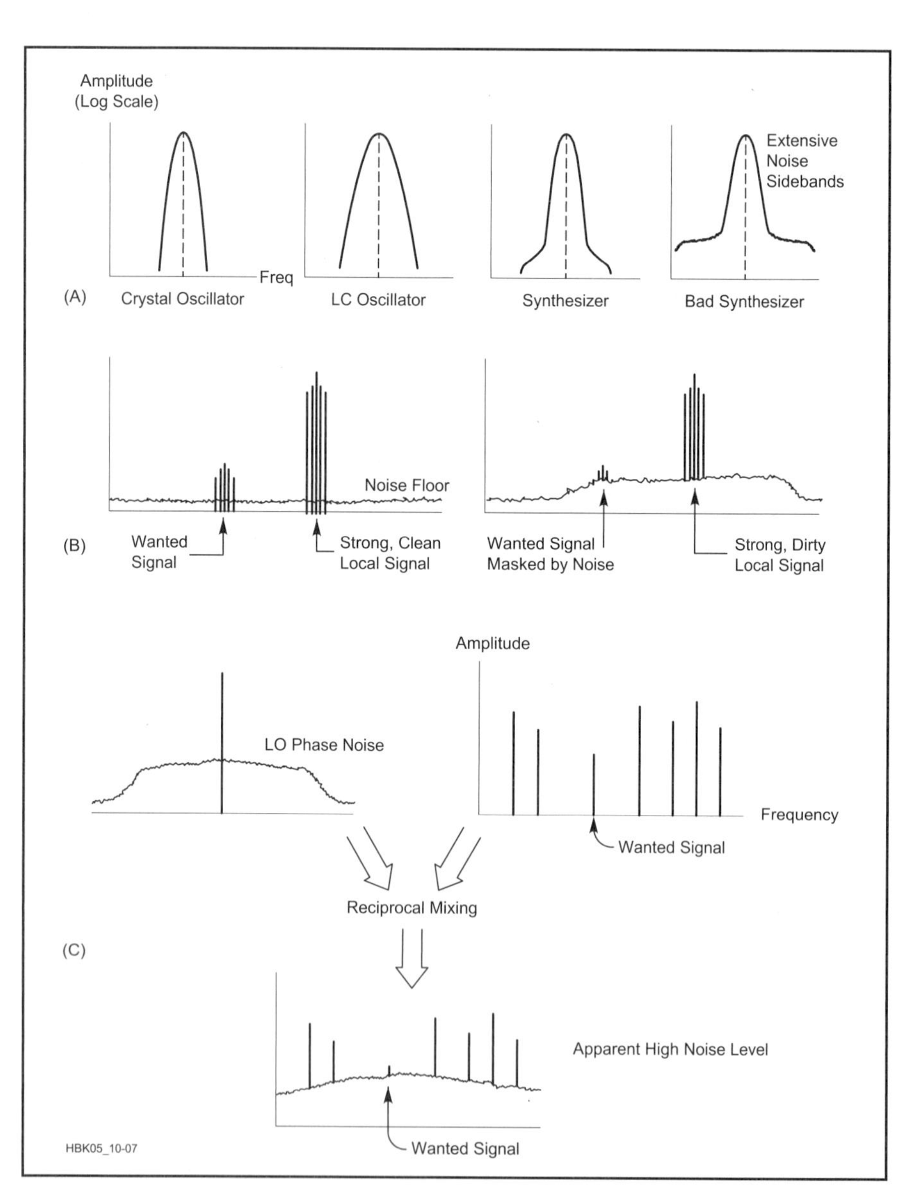

Fig 10.7—The effects of phase noise. At A, the relative phase-noise spectra of several different signal-generation approaches. At B, how transmitted phase noise degrades the weak-signal performance even of nearby receivers with phase-quiet oscillators, raising the effective noise floor. What is perhaps most insidious about phase noise in Amateur Radio communication is that its presence in a receiver LO can allow strong, clean transmitted signals to degrade the receiver's noise floor just as if the transmitted signals were dirty. This effect, reciprocal mixing, is shown at C.

VFO. Many synthesizers have fallen far short of that target. Phase noise is worse in higher-frequency oscillators and the trend towards general-coverage, upconverting structures has required that local oscillators operate at higher and higher frequencies. **Fig 10.7A** shows sketches of the relative phase-noise performance of some oscillators. The very high Q of the quartz crystal in a crystal oscillator gives it the potential for much lower phase noise than LC oscillators. A medium-quality synthesizer has close-in phase noise performance approaching that of a crystal oscillator, while further from the carrier frequency, it degrades to the performance of a modest-quality LC oscillator. There may be a small bump in the noise spectrum at the boundary of these two zones. A bad synthesizer can have extensive noise sidebands extending over many tens (hundreds in extreme cases) of kilohertz at high levels.

Phase noise on a transmitter's local oscillators passes through its stages, is amplified and fed to the antenna along with the intentional signal. The intentional signal is thereby surrounded by a band of noise. This radiated noise exists in the same proportion to the transmitter power as the phase noise was to the oscillator power if it passes through no narrow-band filtering capable of limiting its bandwidth. This radiated noise makes for a noisier band. In bad cases, nearby stations can be unable to receive over many tens of kilohertz. Fig 10.7B illustrates the difference between clean and dirty transmitters.

The effects of receiver-LO phase noise are more complicated, but at least it doesn't affect other stations' reception. The process is called *reciprocal mixing*.

Reciprocal Mixing

This is an effect that occurs in all mixers, yet despite its name, reciprocal mixing is an LO, not a mixer, problem. Imagine that the outputs of two supposedly unmodulated signal generators are mixed together and the mixer output is fed into an FM receiver. The receiver produces sounds, indicating that the resultant signal is modulated nonetheless. Which signal generator is responsible? This is a trick question, of course. A moment spent thinking about how a change in the frequency of either input signal affects the output signal will show that FM or PM on either input reaches the output. The best answer to the trick question is therefore "either or both."

The modulation on the mixer output is the combined modulations of the inputs. This means that modulating a receiver's local oscillator is indistinguishable from using a clean LO and having the same modulation present on *all* incoming signals. (This is also true for AM provided that a fully linear multiplier is used as the mixer, but mixers are commonly driven into switching, which strips any AM off the LO signal. This is the chief reason why the phase component of oscillator noise is more important than any AM component.)

The word *indistinguishable* is important in the preceding paragraph. It does not mean that the incoming signals are themselves modulated, but that the signals in the receiver IF and the noise in the IF, sound exactly as if they were. What really happens is that the noise components of the LO are extra LO signals that are offset from the carrier frequency. Each of them mixes other signals that are appropriately offset from the LO carrier into the receiver's IF. Noise is the sum of an infinite number of infinitesimal components spread over a range of frequencies, so the signals it mixes into the IF are spread into an infinite number of small replicas, all at different frequencies. This amounts to scrambling these other signals into noise. It is tedious to look at the effects of receiver LO phase noise this way. The concept of reciprocal mixing gives us an easier, alternative view that is much more digestible and produces identical results.

A poor oscillator can have significant noise sidebands extending out many tens of kilohertz on either side of its carrier. This is the same, as far as the signals in the receiver IF are concerned, as if the LO were clean and every signal entering the mixer had these noise sidebands. Not only will the wanted signal (and its noise sidebands) be received, but the noise sidebands added by the LO to signals near, but outside, the receiver's IF passband will overlap it. If the band is busy, each of many signals present will add its set of noise sidebands—and the effect is cumulative. This produces the appearance of a high background-noise level on the band. Many hams tend to accept this, blaming "conditions."

Hams now widely understand reception problems due to intermodulation, and almost everyone knows to apply RF attenuation until the signal gets no clearer. Intermodulation is a nonlinear effect, and the levels of the intermod products fall by greater amounts than the reduction in the intermodulating signals. The net result is less signal, but with the intermodulation products dropped still further. This improvement reaches a limit when more attenuation pushes the desired signal too close to the receiver's noise floor.

Reciprocal mixing is a linear process, and the mixer applies the same amount of noise "deviation" to incoming signals as that present on the LO. Therefore the ratio of noise-sideband power to signal power is the same for each signal, and the same as that on the LO. Switching in RF attenuation reduces the power of signals entering the mixer, but the reciprocal mixing process still adds the noise sidebands at the same *relative* power to each. Therefore, no reception improvement results. Other than building a quieter oscillator, the only way of improving things is to use narrow preselection to band-limit the receiver's input response and reduce the number of incoming signals subject to reciprocal mixing. This reduces the number of times the phase noise sidebands get added into the IF signal. Commercial *tracking preselectors*—selective front-end circuits that tune in step with a radio's band changes and tuning, are expensive, but one that is manually tuned would make a modest-sized home-brew project and could also help reduce intermodulation effects. When using a good receiver with a linear front end and a clean LO, amateurs accustomed to receivers with poor phase-noise performance report the impression is of a seemingly emptier band with gaps between signals—and then they begin to find readable signals in some of the gaps.

Fig 10.7C shows how a noisy oscillator affects transmission and reception. The effects on reception are worst in Europe, on 40 m, at night. Visitors from North America, and especially Asia, are usually shocked by the levels of background noise. In ITU Region 1, the Amateur Radio 40-m allocation is 7.0 to 7.1 MHz; above this, ultra-high-power broadcasters operate. The front-end filters in commercial ham gear are usually fixed band-pass designs that cover the wider 40-m allocations in the other regions. This allows huge signals to reach the mixer and mix large levels of LO phase noise into the IF. Operating collocated radios, on the same band, in a multioperator contest, requires linear front ends, preselection and state-of-the-art phase-noise performance. Outside of amateur circles, only warship operation is more demanding, with kilowatt transmitters and receivers sharing antennas on the same mast.

A Phase Noise Demonstration

Healthy curiosity demands some form of demonstration so the scale of a problem can be judged "by ear" before measurements are attempted. We need to be able to measure the noise of an oscillator alone (to aid in the development of quieter ones) and we also need to be able to measure the phase noise of the oscillators in a receiver (a transmitter can be treated as an oscillator). Conveniently, a receiver contains most of the functions needed to demonstrate its own phase noise.

No mixer has perfect port-to-port isolation, and some of its local-oscillator signal leaks through into the IF. If we tune a general-coverage receiver, with its antenna disconnected, to exactly 0 Hz, the local oscillator is exactly at the IF center frequency, and the receiver acts as if it is tuned to a very strong unmodulated carrier. A typical mixer might give only 40 dB of LO isolation and have an LO drive power of at least 10 mW. If we tune away from 0 Hz, the LO carrier tunes away from the IF center and out of the passband. The apparent signal level falls. Although this moves the LO carrier out of the IF passband, some of its noise sidebands will not be, and the receiver will respond to this energy as an incoming noise signal. To the receiver operator, this sounds like a rising noise floor as the receiver is tuned toward 0 Hz. To get good noise floor at very low frequencies, some professional/military receivers, like the Racal RA1772, use very carefully balanced mixers to get as much port-to-port isolation as possible, and they also may switch a crystal notch filter into the first mixer's LO feed.

This demonstration cannot be done if the receiver tunes amateur bands only. As it is, most general-coverage radios inhibit tuning in the LF or VLF region. It could be suggested by a cynic that how low manufacturers allow you to tune is an indication of how far they think their phase-noise sidebands could extend!

The majority of amateur transceivers with general-coverage receivers are programmed not to tune below 30 to 100 kHz, so means other than the "0 Hz" approach are needed to detect LO noise in these radios. Because reciprocal mixing adds the LO's sidebands to clean incoming signals, in the same proportion to the incoming carrier as they exist with respect to the LO carrier, all we need do is to apply a strong, clean signal wherever we want within the receiver's tuning range. This signal's generator must have lower phase noise than the radio being evaluated. A general-purpose signal generator is unlikely to be good enough; a crystal oscillator is needed.

It's appropriate to set the level into the receiver to about that of a strong broadcast carrier, say S9 + 40 dB. Set the receiver's mode to SSB or CW and tune around the test signal, looking for an increasing noise floor (higher hiss level) as you tune closer towards the signal, as shown in **Fig 10.8**. Switching in a narrow CW filter allows you to hear noise closer to the carrier than is possible with an SSB filter. This is also the technique used to measure a receiver's effective selectivity, and some equipment reviewers kindly publish their plots in this format. *QST* reviews, done by the ARRL Lab, often include the results of specific phase-noise measurements.

Measuring Receiver Phase Noise

There are several different ways of measuring phase noise, offering different tradeoffs between convenience, cost and effort. Some methods suit oscillators in isolation, others suit them in-situ (in their radios).

If you're unfamiliar with noise measurements, the units involved may seem strange. One reason for this is that a noise signal's power is spread over a frequency range, like an infinite number of infinitesimal sinusoidal components. This can be thought of as similar to painting a house. The area that a gallon of paint can cover depends on how thinly it's spread. If someone asks how much paint was used on some part of a wall, the answer would have to be in terms of paint volume per square foot. The wall can be considered to be an infinite number of points, each with an infinitesimal amount of paint applied to it. The question of what volume of paint has been applied at some specific point is unanswerable. With noise, we must work in terms of *power density*, of watts per hertz. We therefore express phase-noise level as a ratio of the carrier power to the noise's power density. Because of the large ratios involved, expression in decibels is convenient. It has been a convention to use *dBc* to mean "decibels with respect to the carrier."

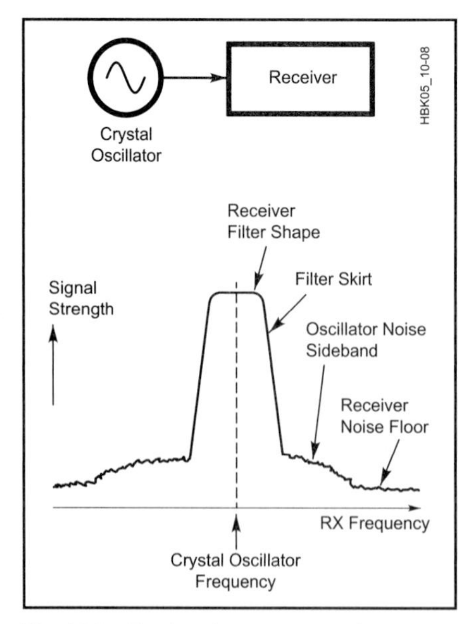

Fig 10.8—Tuning in a strong, clean crystal-oscillator signal can allow you to hear your receiver's relative phase-noise level. Listening with a narrow CW filter switched in allows you to get better results closer to the carrier.

For phase noise, we need to work in terms of a standard bandwidth, and 1 Hz is the obvious candidate. Even if the noise is measured in a different bandwidth, its equivalent power in 1 Hz can be easily calculated. A phase-noise level of –120 dBc in a 1-Hz bandwidth (often written as *–120 dBc/Hz*) translates into each hertz of the noise having a power of 10^{-12} of the carrier power. In a bandwidth of 3 kHz, this would be 3000 times larger.

The most convenient way to measure phase noise is to buy and use a commercial phase noise test system. Such a system usually contains a state-of-the-art, low-noise frequency synthesizer and a low-frequency spectrum analyzer, as well as some special hardware. Often, a second, DSP-based spectrum analyzer is included to speed up and extend measurements very close to the carrier by using the Fast Fourier Transform (FFT). The whole system is then controlled by a computer with proprietary software. With a good system like this costing about $100,000, this is not a practical method for amateurs, although a few fortunate individuals have access to them at work. These systems are also overkill for our needs, because we are not particularly interested in determining phase-noise levels very close to and very far from the carrier.

It's possible to make respectable receiver-oscillator phase-noise measurements with less than $100 of parts and a multimeter. Although it's time-consuming, the technique is much more in keeping with the amateur spirit than using a $100k system! An ordinary multimeter will produce acceptable results, a meter capable of indicating "true rms" ac voltages is preferable because it can give correct readings on sine waves *and* noise. **Fig 10.9** shows the setup. Measurements can only be made around the frequency of the crystal oscillator, so if more than one band is to be tested, crystals must be changed, or else a set of appropriate oscillators is needed. The oscillator should produce about +10 dBm (10 mW) and be *very* well shielded. (To this end, it's advisable to build the oscillator into a die-cast box and power it from internal batteries. A noticeable shielding improvement results even from avoiding the use of an external power switch; a reed-relay element inside the box can be positioned to connect the battery when a small permanent magnet is placed against a marked place outside the box.)

Likewise, great care must be taken with attenuator shielding. A total attenuation of around 140 dB is needed, and with so much attenuation in line, signal leakage can easily exceed the test signal that

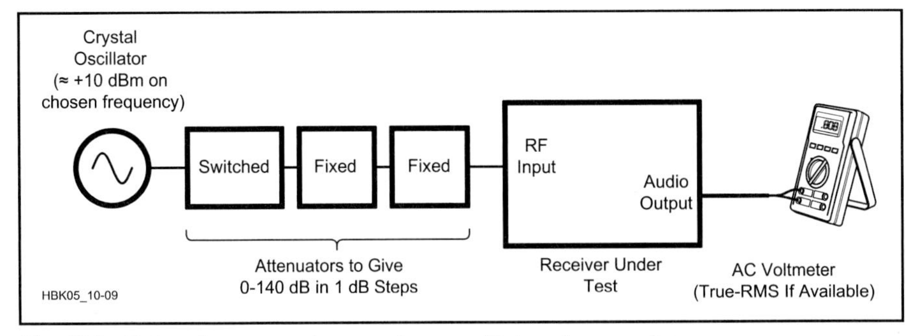

Fig 10.9—Setup for measuring receiver-oscillator phase noise.

reaches the receiver. It's not necessary to be able to switch-select all 140 dB of the attenuation, nor is this desirable, as switches can leak. (The 1995 and older editions of this *Handbook* contained a step attenuator that's satisfactory.) All of the attenuators' enclosure seams must be soldered. A pair of boxes with 30 dB of fixed attenuation each are needed to complete the set. With 140 dB of attenuation, coax cable leakage is also a problem. The only countermeasure against this is to minimize all cable lengths and to interconnect test-system modules with BNC plug-to-plug adapters (UG-491As) where possible.

Ideally, the receiver could simply be tuned across the signal from the oscillator and the response measured using its signal-strength (S) meter. Unfortunately, receiver S meters are notoriously imprecise, so an equivalent method is needed that does not rely on the receiver's AGC system.

The trick is not to measure the response to a fixed level signal, but to measure the changes in applied signal power needed to give a fixed response. Here is a step-by-step procedure based on that described by John Grebenkemper, KI6WX, in March and April 1988 *QST*:

1. Connect the equipment as shown in Fig 10.9, but with the crystal oscillator off. Set the step attenuator to maximum attenuation. Set the receiver for SSB or CW reception with its narrowest available IF filter selected. Switch out any internal preamplifiers or RF attenuators. Select AGC off, maximum AF and RF gain. It may be necessary to reduce the AF gain to ensure the audio amplifier is at least 10 dB below its clipping point. The ac voltmeter or an oscilloscope on the AF output can be used to monitor this.

2. To measure noise, it is important to know the bandwidth being measured. A true-RMS ac voltmeter measures the power in the noise reaching it. To calculate the noise density, we need to divide by the receiver's *noise bandwidth*. The receiver's –6-dB IF bandwidth can be used as an approximation, but purists will want to plot the top 20 dB of the receiver's bandwidth on linear scales and integrate the area under it to find the width of a rectangle of equal area and equal height. This accounts properly for the noise in the skirt regions of the overall selectivity. (The very rectangular shape of common receiver filters tends to minimize the error of just taking the approximation.)

Switch on the test oscillator and set the attenuators to give an AF output above the noise floor and below the clipping level with the receiver peaked on the signal. Tune the receiver off to each side to find the frequencies at which the AF voltage is half that at the peak. The difference between these is the receiver's –6-dB bandwidth. High accuracy is not needed: 25% error in the receiver bandwidth will only cause a 1-dB error in the final result. The receiver's published selectivity specifications will be close enough. The benefit of integration is greater if the receiver has a very rounded, low-ringing or low-order filter.

3. Retune the receiver to the peak. Switch the oscillator off and note the noise-floor voltage. Turn the oscillator back on and adjust the attenuator to give an AF output voltage 1.41 times (3 dB) larger than the noise floor voltage. This means that the noise power and the test signal power at the AF output are equal—a value that's often called the *MDS* (*minimum discernible signal*) of a receiver. Choosing a test-oscillator level at which to do this test involves compromise. Higher levels give more accurate results where the phase noise is high, but limit the lowest level of phase noise that can be measured because better receiver oscillators require a greater input signal to produce enough noise to get the chosen AF-output level. At some point, either we've taken all the attenuation out and our measurement range is limited by the test

Table 10.1
SSB Phase Noise of ICOM IC-745 Receiver Section

Oscillator output power = –3 dBm (0.5 mW)
Receiver bandwidth (Δf) = 1.8 kHz
Audio noise voltage = –0.070 V
Audio reference voltage (V_0) = 0.105 V
Reference attenuation (A_0) = 121 dB

Offset Frequency (kHz)	*Attenuation (A_1) (dB)*	*Audio V_1 (volts)*	*Audio V_2 (volts)*	*Ratio V_2/V_1*	*SSB Phase Noise (dBc/Hz)*
4	35	0.102	0.122	1.20	–119
5	32	0.104	0.120	1.15	–122*
6	30	0.104	0.118	1.13	–124*
8	27	0.100	0.116	1.16	–127*
10	25	0.106	0.122	1.15	–129*
15	21	0.100	0.116	1.16	–133*
20	17	0.102	0.120	1.18	–137
25	14	0.102	0.122	1.20	–140
30	13	0.102	0.122	1.20	–141
40	10	0.104	0.124	1.19	–144
50	8	0.102	0.122	1.20	–146
60	6	0.104	0.124	1.19	–148
80	4	0.102	0.126	1.24	–150
100	3	0.102	0.126	1.24	–151
150	3	0.102	0.124	1.22	–151
200	0	0.104			–154
250	0	0.100			–154
300	0	0.98			–154
400	0	0.96			–154
500	0	0.96			–154
600	0	0.97			–154
800	0	0.96			–154
1000	0	0.96			–154

*Asterisks indicate measurements possibly affected by receiver overload (see text).

oscillator's available power, or we overload the receiver's front end, spoiling the results.

Record the receiver frequency at the peak, (f_0), the attenuator setting (A_0) and the audio output voltage (V_0). These are the carrier measurements against which all the noise measurements will be compared.

4. Now you must choose the offset frequencies—the spacings from the carrier—at which you wish to make measurements. The receiver's skirt selectivity will limit how close to the carrier noise measurements can be made. (Any measurements made too close in are valid measurements of the receiver selectivity, but because the signal measured under these conditions is sinusoidal and not noise like, the corrections for noise density and noise bandwidth are not appropriate.) It is difficult to decide where the filter skirt ends and the noise begins, and what corrections to apply in the region of doubt and uncertainty. A good practical approach is to listen to the audio and tune away from the carrier until you can't distinguish a tone in the noise. The ear is superb at spotting sine tones buried in noise, so this criterion, although subjective, errs on the conservative side.

Tune the receiver to a frequency offset from f_0 by your first chosen offset and adjust the attenuators to get an audio output voltage as close as possible to V_0. Record the total attenuation, A_1 and the audio output voltage, V_1. The SSB phase noise (qualified as *SSB* because we're measuring the phase noise on only one side of the carrier, whereas some other methods cannot segregate between upper and lower noise sidebands and measure their sum, giving *DSB* phase noise) is now easy to calculate:

$$L(f) = A_1 - A_0 10\log(BW_{noise})$$

where

L(f) = SSB phase noise in dBc/Hz

BWnoise = receiver noise bandwidth, Hz.

5. It's important to check for overload. Decrease the attenuation by 3 dB, and record the new audio output voltage, V_2. If all is well, the output voltage should increase by 22% (1.8 dB); if the receiver is operating nonlinearly, the increase will be less. (An 18% increase is still acceptable for the overall accuracy we want.) Record V_2/V_1 as a check: a ratio of 1.22:1 is ideal, and anything less than 1.18:1 indicates a bad measurement.

If too many measurements are bad, you may be overdriving the receiver's AF amplifier, so try reducing the AF gain and starting again back at Step 3. If this doesn't help, reducing the RF gain and starting again at Step 3 should help if the compression is occurring late in the IF stages.

6. Repeat Steps 4 and 5 at all the other offsets you wish to measure. If measurements are made at increments of about half the receiver's bandwidth, any discrete (non-noise) spurs will be found. A noticeable tone in the audio can indicate the presence of one of these. If it is well clear of the noise, the measurement is valid, but the noise bandwidth correction should be ignored, giving a result in dBc.

Table 10.1 shows the results for an ICOM IC-745 as measured by KI6WX, and

Transmitter Phase-Noise Measurement in the ARRL Lab

Here is a brief description of the technique used in the ARRL Lab to measure transmitter phase noise. The system essentially consists of a direct-conversion receiver with very good phase-noise characteristics. As shown in Fig B, we use an attenuator after the transmitter, a Mini-Circuits ZAY-1 mixer, a Hewlett-Packard 8640B signal generator, a band-pass filter, an audio-frequency low-noise amplifier and a spectrum analyzer (HP 8563E) to make the measurements.

The transmitter signal is mixed with the output of the signal generator, and signals produced in the mixing process that are not required for the measurement process are filtered out. The spectrum analyzer then displays the transmitted phase-noise spectrum. The 100 mW output of the HP 8640B is barely enough to drive the mixer—the setup would work better with 200 mW of drive. To test the phase noise of an HP 8640B, we use a second '8640B as a reference source. It is quite important to be sure that the phase noise of the reference source is lower than that of the signal under test, because we are really measuring the combined phase-noise output of the signal generator and the transmitter. It would be quite embarrassing to publish phase-noise plots of the reference generator instead of the transmitter under test! The HP 8640B has much cleaner spectral output than most transmitters.

A sample phase-noise plot for an amateur transceiver is shown in Fig A. It was produced with the test setup shown in Fig B. Measurements from multiple passes are taken and averaged. A 3-Hz video bandwidth also helps average and smooth the plots. These plots do not necessarily reflect the phase-noise characteristics of all units of a particular model.

The log reference level (the top horizontal line on the scale in the plot) represents –60 dBc/Hz. It is common in industry to use a 0-dBc log reference, but such a reference level would not allow measurement of phase-noise levels below –80 dBc/Hz. The actual measurement bandwidth used on the spectrum analyzer is 100 Hz, but the reference is scaled for a 1-Hz bandwidth. This allows phase-noise levels to be read directly from the display in dBc/Hz. Because each vertical division represents 10 dB, the plot shows the noise level between –60 dBc/Hz (the top horizontal line) and –140 dBc/Hz (the bottom horizontal line). The horizontal scale is 2 kHz per division. The offsets shown in the plots are 2 through 20 kHz.

What Do the Phase-Noise Plots Mean?

Although they are useful for comparing different radios, plots can also be used to calculate the amount of interference you may receive from a nearby transmitter

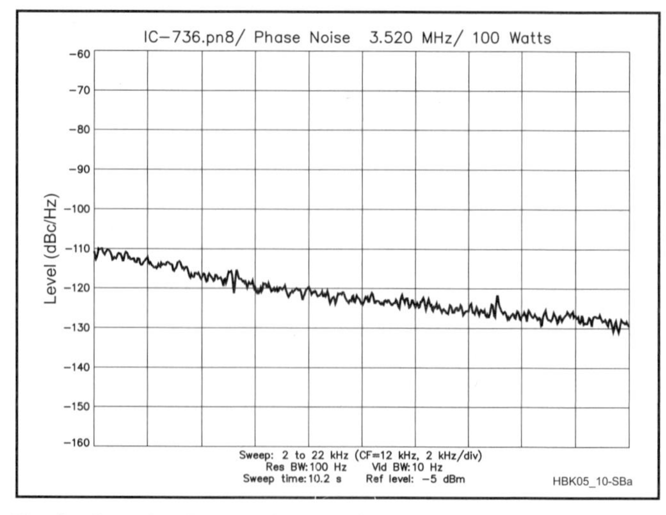

Fig A—Sample phase noise plot for an amateur transceiver.

Fig 10.10 shows this data in graphic form. His oscillator power was only –3 dBm, which limited measurements to offsets less than 200 kHz. More power might have allowed noise measurements to lower levels, although receiver overload places a limit on this. This is not important, because the real area of interest has been thoroughly covered. When attempting phase-noise measurements at large offsets, remember that any front-end selectivity, before the first mixer, will limit the maximum offset at which LO phase-noise measurement is possible.

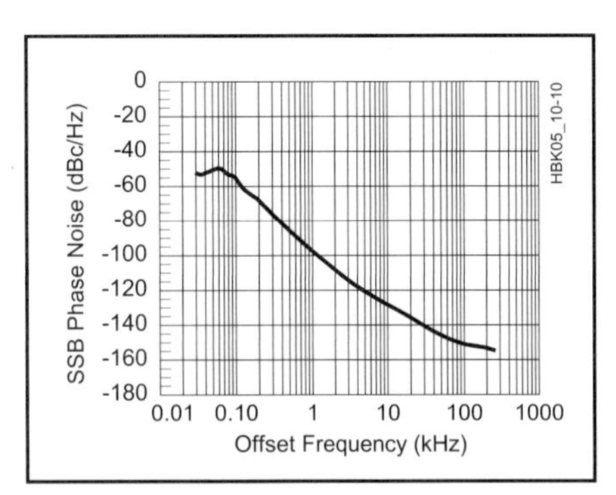

Fig 10.10—The SSB phase noise of an ICOM IC-745 transceiver (serial number 01528) as measured by KI6WX.

MEASURING OSCILLATOR AND TRANSMITTER PHASE NOISE

Measuring the composite phase noise of a receiver's LO requires a clean test oscillator. Measuring the phase noise of an incoming signal, whether from a single oscillator or an entire transmitter, requires the use of a clean receiver, with lower phase noise than the source under test. The sidebar, "Transmitter Phase-Noise Measurement in the ARRL Lab," details the method used to measure composite noise (phase noise and amplitude noise, the practical effects of which are indistinguishable on the air) for *QST* Product Reviews. Although targeted at measuring high power signals from entire transmitters, this approach can be used to measure lower-level signals simply by changing the amount of input attenuation used.

At first, this method—using a low-frequency spectrum analyzer and a low-phase-noise signal generator—looks unnecessarily elaborate. A growing number of radio amateurs have acquired good-quality spectrum analyzers for their shacks since older model Tektronix and Hewlett-Packard instruments have started to appear on the surplus market at affordable prices. The obvious question is, "Why not just use one of these to view the signal and read phase-noise levels directly off the screen?" Reciprocal mixing is the problem. Very few spectrum analyzers have clean enough local oscillators not to completely swamp the noise being measured. Phase-noise measurements involve the measurement of low-level components very close to a large carrier, and that carrier will mix the noise sidebands of *the analyzer's LO* into its IF. Some way of notching out the carrier is needed, so that the analyzer need only handle the noise sidebands. A crystal filter could be designed to do the job, but this would be expensive, and one would be needed for every different oscillator

---

with known phase-noise characteristics. An approximation is given by

$$A_{QRM} = NL + 10 \times \log(BW)$$

where

A_{QRM} = Interfering signal level, dBc
NL = noise level on the receive frequency, dBc
BW = receiver IF bandwidth, in Hz

For instance, if the noise level is –90 dBc/Hz and you are using a 2.5-kHz SSB filter, the approximate interfering signal will be –56 dBc. In other words, if the transmitted signal is 20 dB over S9, and each S unit is 6 dB, the interfering signal will be as strong as an S3 signal.

The measurements made in the ARRL Lab apply only to transmitted signals. It is reasonable to assume that the phase-noise characteristics of most transceivers are similar on transmit and receiver because the same oscillators are generally used in local-oscillator (LO) chain.

In some cases, the receiver may have better phase-noise characteristics than the transmitter. Why the possible difference? The most obvious reason is that circuits often perform less than optimally in strong RF fields, as anyone who has experienced RFI problems can tell you. A less obvious reason results from the way that many high-dynamic-range receivers work. To get good dynamic range, a sharp crystal filter is often placed immediately after the first mixer in the receive line. This filter removes all but a small slice of spectrum for further signal processing. If the desired filtered signal is a product of mixing an incoming signal with a noisy oscillator, signals far away from the desired one can end up in this slice. Once this slice of spectrum is obtained, however, unwanted signals cannot be reintroduced, no matter how noisy the oscillators used in further signal processing. As a result, some oscillators in receivers don't affect phase noise.

The difference between this situation and that in transmitters is that crystal filters are seldom used for reduction of phase noise in transmitting because of the high cost involved. Equipment designers have enough trouble getting smooth, click-free break-in operation in transceivers without having to worry about switching crystal filters in and out of circuits at 40-wpm keying speeds!—*Zack Lau, W1VT, ARRL Laboratory Engineer*

Fig B—ARRL transmitter phase-noise measurement setup.

frequency to be tested. The alternative is to build a direct-conversion receiver using a clean LO like the Hewlett-Packard HP8640B signal generator and spectrum-analyze its "audio" output with an audio analyzer. This scheme mixes the carrier to dc; the LF analyzer is then ac-coupled, and this removes the carrier. The analyzer can be made very sensitive without overload or reciprocal mixing being a problem. The remaining problem is then keeping the LO—the HP8640B in this example—at exactly the carrier frequency. 8640s are based on a shortened-UHF-cavity oscillator and can drift a little. The oscillator under test will also drift. The task is therefore to make the 8640B track the oscillator under test. For once we get something for free: The HP8640B's FM input is dc coupled, and we can use this as an electronic fine-tuning input. As a further bonus, the 8640B's FM deviation control acts as a sensitivity control for this input. We also get a phase detector for free, as the mixer output's dc component depends on the phase relationship between the 8640B and the signal under test (remember to use the dc coupled port of a diode ring mixer as the output). Taken together, the system includes everything needed to create a crude phase-locked loop that will automatically track the input signal over a small frequency range. **Fig 10.11** shows the arrangement.

The oscilloscope is not essential for operation, but it is needed to adjust the system. With the loop unlocked (8640B FM input disconnected), tune the 8640 off the signal frequency to give a beat at the mixer output. Adjust the mixer drive levels to get an undistorted sine wave on the scope. This ensures that the mixer is not being overdriven. While the loop is off-tuned, adjust the beat to a frequency within the range of the LF spectrum analyzer and use it to measure its level, "A_c" in dBm. This represents the carrier level and is used as the reference for the noise measurements. Connect the FM input of the signal generator, and switch on the generator's dc FM facility. Try a deviation range of 10 kHz to start with. When you tune the signal generator toward the input frequency, the scope will show the falling beat frequency until the loop jumps into lock. Then it will display a noisy dc level. Fine tune to get a mean level of 0 V. (This is a very-low-bandwidth, very-low-gain loop. Stability is not a problem; careful loop design is not needed. We actually want as slow a loop as possible; otherwise, the loop would track and cancel the slow components of the incoming signal's phase noise, preventing their measurement.)

When you first take phase-noise plots, it's a good idea to duplicate them at the generator's next lower FM-deviation range and check for any differences in the noise level in the areas of interest. Reduce the FM deviation range until you find no further improvement. Insufficient FM deviation range makes the loop's lock range narrow, reducing the amount of drift it can compensate. (It's sometimes necessary to keep gently trimming the generator's fine tune control.)

Set up the LF analyzer to show the noise. A sensitive range and 100-Hz resolution bandwidth are appropriate. Measure the noise level, "A_n" in dBm. We must now calculate the noise density that this represents. Spectrum-analyzer filters are normally *Gaussian*-shaped and bandwidth-specified at their –3-dB points. To avoid using integration to find their true-noise power bandwidth, we can reasonably assume a value of 1.2 × BW. A spectrum analyzer logarithmically compresses its IF signal ahead of the detectors and averaging filter. This affects the statistical distribution of noise voltage and causes the analyzer to read low by 2.5 dB. To produce the same scale as the ARRL Lab photographs, the analyzer reference level must be set to –60 dBc/Hz, which can be calculated as:

$$A_{ref} = A_c - 10\log(1.2 \times BW) + 62.5\,dBm \quad (2)$$

where

A_{ref} = analyzer reference level, dBm
A_c = carrier amplitude, dBm

This produces a scale of –60 dBc/Hz at the top of the screen, falling to –140 dBc/Hz at the bottom. The frequency scale is 0 to 20 kHz with a resolution bandwidth (BW in the above equation) of 100 Hz. This method combines the power of *both* sidebands and so measures DSB phase noise. To calculate the equivalent SSB phase noise, subtract 3 dB for noncoherent noise (the general "hash" of phase noise) and subtract 6 dB for coherent, discrete components (that is, single-frequency spurs). This can be done by setting the reference level 3 to 6 dB higher.

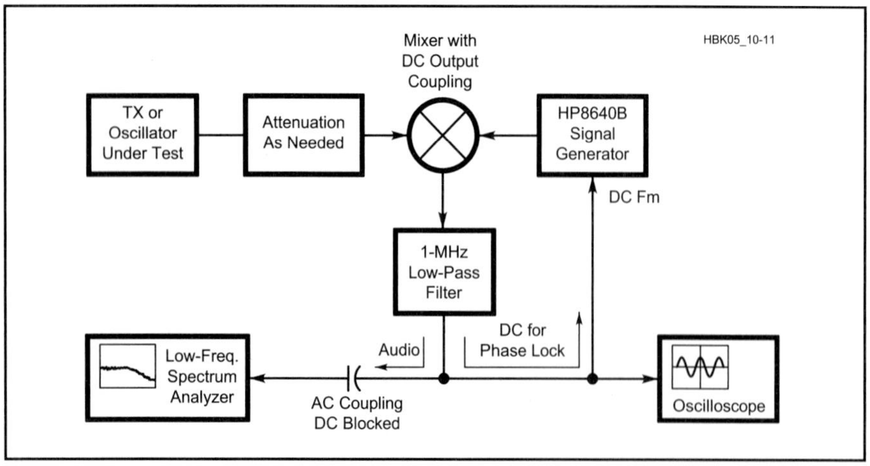

Low-Cost Phase Noise Testing

All that expensive equipment may seem far beyond the means of the average Amateur Radio experimenter. With careful shopping and a little more effort, alternative equipment can be put together for pocket money. (All of the things needed—parts for a VXO, a surplus spectrum analyzer and so on—have been seen on sale cheap enough to total less than $100.) The HP8640B is good and versatile, but for use at one oscillator frequency, you can build a VXO for a few dollars. It will only cover one oscillator frequency, but a VXO can provide even better phase-noise performance than the 8640B.

As referenced at the end of this chapter, *Pontius* has also demonstrated that signal-source phase-noise measurements can be accurately obtained without the aid of expensive equipment.

Fig 10.11—Arrangement for measuring phase noise by directly converting the signal under test to audio. The spectrum analyzer views the signal's noise sidebands as audio; the signal's carrier, converted to dc, provides a feedback signal to phase-lock the Hewlett-Packard HP8640B signal generator to the signal under test.

OSCILLATOR CIRCUITS AND CONSTRUCTION

LC Oscillator Circuits

The LC oscillators used in radio equipment are usually arranged as *variable frequency oscillators* (*VFOs*). Tuning is achieved by either varying part of the capacitance of the resonator or, less commonly, by using a movable magnetic core to vary the inductance. Since the early days of radio, there has been a huge quest for the ideal, low-drift VFO. Amateurs and professionals have spent immense effort on this pursuit. A brief search of the literature reveals a large number of designs, many accompanied by claims of high stability. The quest for stability has been solved by the development of low-cost frequency synthesizers, which can give crystal-controlled stability. Synthesizers have other problems though, and the VFO still has much to offer in terms of signal cleanliness, cost and power consumption, making it attractive for home construction. No one VFO circuit has any overwhelming advantage over any other—component quality, mechanical design and the care taken in assembly are much more important.

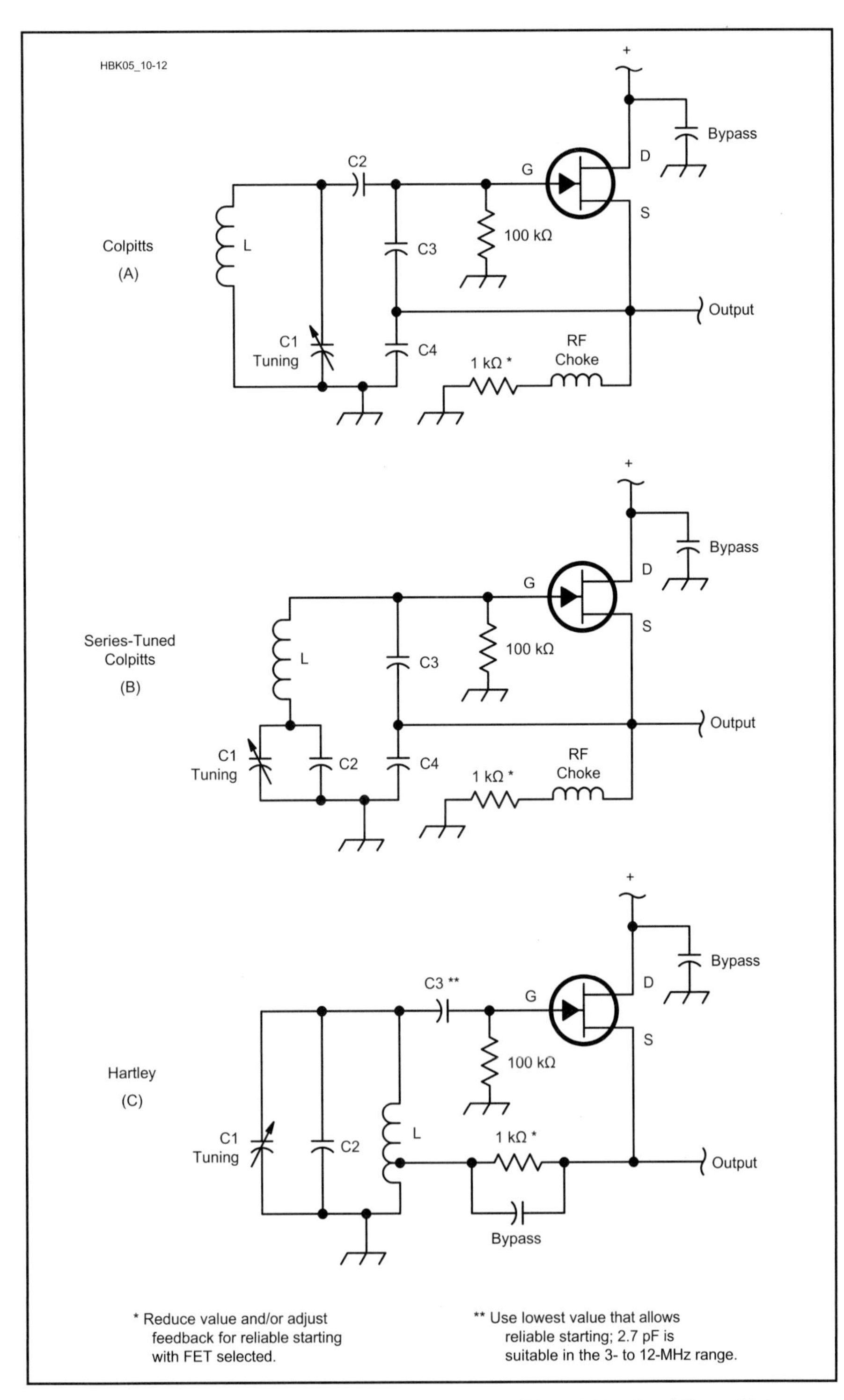

Fig 10.12—The Colpitts (A), series-tuned Colpitts (B) and Hartley (C) oscillator circuits. Rules of thumb: C3 and C4 at A and B should be equal and valued such that their X_c = 45 Ω at the operating frequency; for C2 at A, X_c = 100 Ω. For best stability, use C0G or NP0 units for all capacitors associated with the FETs' gates and sources. Depending on the FET chosen, the 1-kΩ source-bias-resistor value shown may require adjustment for reliable starting.

Fig 10.12 shows three popular oscillator circuits stripped of any unessential frills so they can be more easily compared. The original Colpitts circuit (Fig 10.12A) is now often referred to as the *parallel-tuned Colpitts* because its series-tuned derivative (Fig 10.12B) has become common. All three of these circuits use an amplifier with a voltage gain less than unity, but large current gain. The N-channel JFET source follower shown appears to be the most popular choice nowadays. In the parallel-tuned Colpitts, C3 and C4 are large values, perhaps 10 times larger than typical values for C1 and C2. This means that only a small fraction of the total tank voltage is applied to the FET, and the FET can be considered to be only lightly coupled into the tank. The FET is driven by the sum of the voltages across C3 and C4, while it drives the voltage across C4 alone. This means that the tank operates as a resonant, voltage-step-up transformer compensating for the less-than-unity-voltage-gain amplifier. The resonant circuit consists of L, C1, C2, C3 and C4. The resonant frequency can be calculated by using the standard formulas for capacitors in series and parallel to find the resultant capacitance effectively connected across the inductor, L, and then use the standard formula for LC resonance:

$$f = \frac{1}{2\pi\sqrt{LC}} \quad (3)$$

where

f = frequency in hertz
L = inductance in henries
C = capacitance in farads.

For a wide tuning range, C2 must be kept small to reduce the effect of C3 and C4 swamping the variable capacitor C1.

The series-tuned Colpitts circuit works in much the same way. The difference is that the variable capacitor, C1, is positioned so that it is well-protected from being swamped by the large values of C3 and C4. In fact, *small* values of C3, C4 would act to limit the tuning range. Fixed capacitance, C2, is often added across C1 to allow the tuning range to be reduced to that required, without interfering with C3 and C4, which set the amplifier coupling. The series-tuned Colpitts has a reputation for better stability than the parallel-tuned original. Note how C3 and C4 swamp the capacitances of the amplifier in both versions.

The Hartley is similar to the parallel-tuned Colpitts, but the amplifier source is tapped up on the tank inductance instead of the tank capacitance. A typical tap placement is 10 to 20% of the total turns up from the "cold" end of the inductor. (It's usual to refer to the lowest-signal-voltage end of an inductor as *cold* and the other, with the highest signal voltage as *hot*.) C2 limits the tuning range as required; C3 is reduced to the minimum value that allows reliable starting. This is necessary because the Hartley's lack of the Colpitts's capacitive divider would otherwise couple the FET's capacitances to the tank more strongly than in the Colpitts, potentially affecting the circuit's frequency stability.

In all three circuits, there is a 1 kΩ resistor in series with the source bias choke. This resistor does a number of desirable things. It spoils the Q of the inevitable low-frequency resonance of the choke with the tank tap circuit. It reduces tuning drift due to choke impedance and winding capacitance variations. It also protects against spurious oscillation due to internal choke resonances. Less obviously, it acts to stabilize the loop gain of the built-in AGC action of this oscillator. Stable operating conditions act to reduce frequency drift. Some variations of these circuits may be found with added resistors providing a dc bias to stabilize the system quiescent current. More elaborate still are variations characterized by a constant-current source providing bias. This can be driven from a separate AGC detector system to give very tight level control. The gate-to-ground clamping diode (1N914 or similar) long used by radio amateurs as a means of avoiding gate-source conduction has been shown by Ulrich Rohde, KA2WEU, to degrade oscillator phase-noise performance, and its use is virtually unknown in professional circles.

Fig 10.13 shows some more VFOs to illustrate the use of different devices. The triode Hartley shown includes *permeability tuning*, which has no sliding contact like that to a capacitor's rotor and can be made reasonably linear by artfully spacing the coil turns. The slow-motion drive can be done with a screw thread. The disadvantage is that special care is needed to avoid backlash and eccentric rotation of the core. If a nonrotating core is used, the slides have to be carefully designed to prevent motion in unwanted directions. The Collins Radio Company made extensive use of tube-based permeability tuners, and a semiconductor version can still be found in a number of Ten-Tec radios.

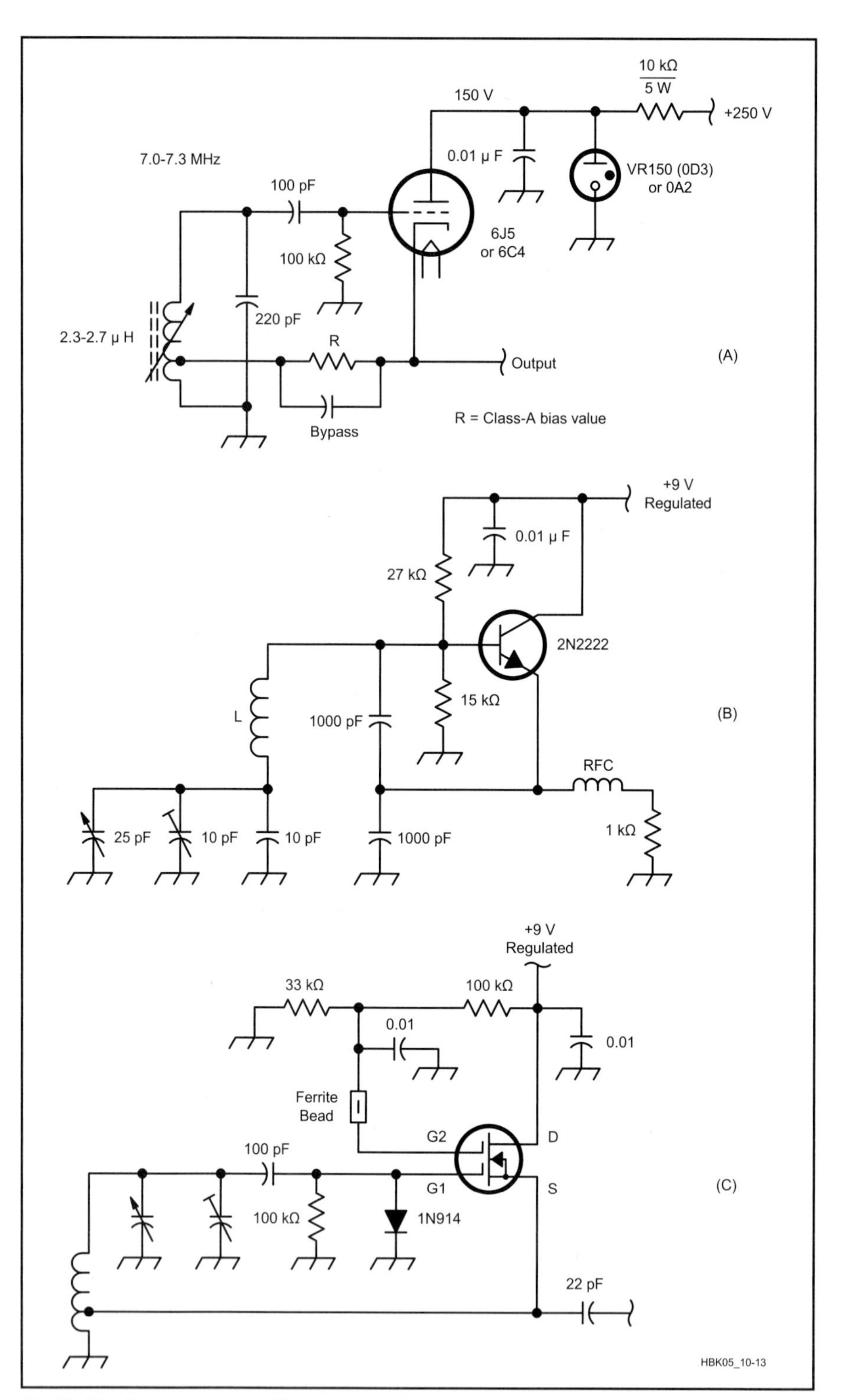

Fig 10.13—Three more oscillator examples: at A, a triode-tube Hartley; at B, a bipolar junction transistor in a series-tuned Colpitts; at C, a dual-gate MOSFET Hartley.

Vacuum tubes cannot run as cool as competitive semiconductor circuits, so care is needed to keep the tank circuit away from the tube heat. In many amateur and commercial vacuum-tube oscillators, oscillation drives the tube into grid current at the positive peaks, causing rectification and producing a negative grid bias. The oscillator thus runs in Class C, in which the conduction angle reduces as the signal amplitude increases until the amplitude stabilizes. As in the FET circuits of Fig 10.12, better stability and phase-noise performance can be achieved in a vacuum-tube oscillator by moving it out of true Class C—that is, by including a bypassed cathode-bias resistor (the resistance appropriate for Class A operation is a good starting value). A small number of people still build vacuum-tube radios partly to be different, partly for fun, but the semiconductor long ago achieved dominance in VFOs.

The voltage regulator (*VR*) tube shown in Fig 10.13A has a potential drawback: It is a gas-discharge device with a high striking voltage at which it starts to conduct and a lower extinguishing voltage, at which it stops conducting. Between these extremes lies a region in which decreasing voltage translates to increasing current, which implies negative resistance. When the regulator strikes, it discharges any capacitance connected across it to the extinguishing voltage. The capacitance then charges through the source resistor until the tube strikes again, and the process repeats. This *relaxation oscillation* demonstrates how negative resistance can cause oscillation. The oscillator translates the resultant sawtooth modulation of its power supply into frequency and amplitude variation. Because of VR tubes' ability to support relaxation oscillation, a traditional rule of thumb is to keep the capacitance directly connected across a VR tube to 0.1 µF or less. A value much lower than this can provide sufficient bypassing in Fig 10.13A because the dropping resistor acts as a decoupler, increasing the bypass's effectiveness.

There is a related effect called *squegging*, which can be loosely defined as oscillation on more than one frequency at a time, but which may also manifest itself as RF oscillation interrupted at an AF rate, as in a superregenerative detector. One form of squegging occurs when an oscillator is fed from a power supply with a high source impedance. The power supply charges up the decoupling capacitor until oscillation starts. The oscillator draws current and pulls down the capacitor voltage, starving itself of power until oscillation stops. The oscillator stops drawing current, and the decoupling capacitor then recharges until oscillation restarts. The process, the low-frequency cycling of which is a form of relaxation oscillation, repeats indefinitely. The oscillator output can clearly be seen to be pulse modulated if an oscilloscope is used to view it at a suitable time-base setting. This fault is a well-known consequence of poor design in battery-powered radios. As dry cells become exhausted, their internal resistance rises quickly and circuits they power can begin to misbehave. In audio stages, such misbehavior may manifest itself in the *putt-putt* sound of the slow relaxation oscillation called *motorboating*.

Compared to the frequently used JFET, bipolar transistors, Fig 10.13B, are relatively uncommon in oscillators because their low input and output impedances are more difficult to couple into a high-Q tank without excessively loading it. Bipolar devices do tend to give better sample-to-sample amplitude uniformity for a given oscillator circuit, however, as JFETs of a given type tend to vary more in their characteristics.

The dual-gate MOSFET, Fig 10.13C, is very rarely seen in VFO circuits. It imposes the cost of the components needed to bias and suppress VHF parasitic oscillation at the second gate, and its inability to generate its own AGC through gate-source conduction forces the addition of a blocking capacitor, resistor and diode at the first gate for amplitude control.

VFO Components and Construction

Tuning Capacitors and Reduction Drives

As most commercially made radios now use frequency synthesizers, it has become increasingly difficult to find certain key components needed to construct a good VFO. The slow-motion drives, dials, gearboxes, associated with names like Millen, National, Eddystone and Jackson are no longer available. Similarly, the most suitable silver-plated variable capacitors with ball races at both ends, although still made, are not generally marketed and are expensive. Three approaches remain: Scavenge suitable parts from old equipment; use tuning diodes instead of variable capacitors—an approach that, if uncorrected through phase locking, generally degrades stability and phase-noise performance; or use two tuning capacitors, one with a capacitance range 1/5 to 1/10 that of the other, in a bandset/bandspread approach.

Assembling a variable capacitor to a chassis and its reduction drive to a front panel can result in *backlash*—an annoying tuning effect in which rotating the capacitor shaft deforms the chassis and/or panel rather than tuning the capacitor. One way of minimizing this is to use the reduction drive to support the capacitor, and use the capacitor to support the oscillator circuit board.

Fixed Capacitors

Traditionally, silver-mica fixed capacitors have been used extensively in oscillators, but their temperature coefficient is not as low as can be achieved by other types, and some silver micas have been known to behave erratically. Polystyrene film has become a proven alternative. One warning is worth noting: Polystyrene capacitors exhibit a permanent change in value should they ever be exposed to temperatures much over 70°C; they do not return to their old value on cooling. Particularly suitable for oscillator construction are the low-temperature-coefficient ceramic capacitors, often described as *NP0* or *C0G* types. These names are actually temperature-coefficient codes. Some ceramic capacitors are available with deliberate, controlled temperature coefficients so that they can be used to compensate for other causes of frequency drift with temperature. For example, the code N750 denotes a part with a temperature coefficient of –750 parts per million per degree Celsius. These parts are now somewhat difficult to obtain, so other methods are needed.

In a Colpitts circuit, the two large-value capacitors that form the voltage divider for the active device still need careful selection. It would be tempting to use any available capacitor of the right value, because the effect of these components on the tank frequency is reduced by the proportions of the capacitance values in the circuit. This reduction is not as great as the difference between the temperature stability of an NP0 ceramic part and some of the low-cost, decoupling-quality X7R-dielectric ceramic capacitors. It's worth using low-temperature coefficient parts even in the seemingly less-critical parts of a VFO circuit—even as decouplers. Chasing the cause of temperature drift is more challenging than fun. Buy critical components like high-stability capacitors from trustworthy sources.

Inductors

Ceramic coil forms can give excellent results, as can self-supporting air-wound coils (Miniductor). If you use a magnetic core, make it powdered iron, never ferrite, and support it stably. Stable VFOs have been made using toroidal cores, but again, ferrite must be avoided. Micrometals mix number 6 has a low temperature coefficient and works well in conjunction with NP0 ceramic capacitors. Coil forms in other materials have to be assessed on an individual basis.

A material's temperature stability will not be apparent until you try it in an oscillator, but you can apply a quick test to identify those nonmetallic materials that are lossy enough to spoil a coil's Q. Put a sample of the coil-form material into a microwave oven along with a glass of water and cook it about 10 s on low power.

Do not include any metal fittings or ferromagnetic cores. Good materials will be completely unaffected; poor ones will heat and may even melt, smoke, or burst into flame. (This operation is a fire hazard if you try more than a tiny sample of an unknown material. Observe your experiment continuously and do not leave it unattended.)

Wes Hayward, W7ZOI, suggests annealing toroidal VFO coils after winding. Roy Lewallen, W7EL reports achieving success with this method by boiling his coils in water and letting them cool in air.

Voltage Regulators

VFO circuits are often run from locally regulated power supplies, usually from resistor/Zener diode combinations. Zener diodes have some idiosyncrasies that could spoil the oscillator. They are noisy, so decoupling is needed down to audio frequencies to filter this out. Zener diodes are often run at much less than their specified optimum bias current. Although this saves power, it results in a lower output voltage than intended, and the diode's impedance is much greater, increasing its sensitivity to variations in input voltage, output current and temperature. Some common Zener types may be designed to run at as much as 20 mA; check the data sheet for your diode family to find the optimum current.

True Zener diodes are low-voltage devices; above a couple of volts, so-called Zener diodes are actually avalanche types. The temperature coefficient of these diodes depends on their breakdown voltage and crosses through zero for diodes rated at about 5 V. If you intend to use nothing fancier than a common-variety Zener, designing the oscillator to run from 5 V and using a 5.1-V Zener, will give you a free advantage in voltage-versus-temperature stability. There are some diodes available with especially low temperature coefficients, usually referred to as *reference* or *temperature-compensated diodes*. These usually consist of a series pair of diodes designed to cancel each other's temperature drift. Running at 7.5 mA, the 1N829A gives 6.2 V ±5% and a temperature coefficient of just ±5 parts per million (ppm) maximum per degree Celsius. A change in bias current of 10% will shift the voltage less than 7.5 mV, but this increases rapidly for greater current variation. The 1N821A is a lower-grade version, at ±100 ppm/°C. The LM399 is a complex IC that behaves like a superb Zener at 6.95 V, ±0.3 ppm/°C. There are also precision, low-power, three-terminal regulators designed to be used as voltage references, some of which can provide enough current to run a VFO. There are comprehensive tables of all these devices between pages 334 and 337 of Horowitz and Hill, *The Art of Electronics*, 2nd ed.

Oscillator Devices

The 2N3819 FET, a classic from the 1960s, has proven to work well in VFOs, but, like the MPF102 also long-popular with ham builders, it's manufactured to wide tolerances. Considering an oscillator's importance in receiver stability, you should not hesitate to spend a bit more on a better device. The 2N5484, 2N5485 and 2N5486 are worth considering; together, their transconductance ranges span that of the MPF102, but each is a better-controlled subset of that range. The 2N5245 is a more recent device with better-than-average noise performance that runs at low currents like the 2N3819. The 2N4416/A, also available as the plastic-cased PN4416, is a low-noise device, designed for VHF/UHF amplifier use, which has been featured in a number of good oscillators up to the VHF region. Its low internal capacitance contributes to low frequency drift. The J310 (plastic; the metal-cased U310 is similar) is another popular JFET in oscillators.

The 2N5179 (plastic, PN5179 or MPS5179) is a bipolar transistor capable of good performance in oscillators up to the top of the VHF region. Care is needed because its absolute-maximum collector-emitter voltage is only 12 V, and its collector current must not exceed 50 mA. Although these characteristics may seem to convey fragility, the 2N5179 is sufficient for circuits powered by stabilized 6-V power supplies.

VHF-UHF devices are not really necessary in LC VFOs because such circuits are rarely used above 10 MHz. Absolute frequency stability is progressively harder to achieve with increasing frequency, so free-running oscillators are rarely used to generate VHF-UHF signals for radio communication. Instead, VHF-UHF radios usually use voltage-tuned, phase-locked oscillators in some form of synthesizer. Bipolar devices like the BFR90 and MRF901, with f_Ts in the 5-GHz region and mounted in stripline packages, are needed at UHF.

Integrated circuits have not been mentioned until now because few specific RF-oscillator ICs exist. Some consumer ICs—the NE602, for example—include the active device(s) necessary for RF oscillators, but often this is no more than a single transistor fabricated on the chip. This works just as a single discrete device would, although using such on-chip devices may result in poor isolation from the rest of the circuits on the chip. There is one specialized LC-oscillator IC: the Motorola MC1648. This device has been made since the early 1970s and is a surviving member of a long-obsolete family of emitter-coupled-logic (ECL) devices. Despite the MC1648's antiquity, it has no real competition and is still widely used in current military and commercial equipment. Market demand should force its continued production for some more years to come. Its circuitry is complex for an oscillator, with a multitransistor oscillator cell controlled by a detector and amplifier in an on-chip ALC system. The MC1648's first problem is that the ECL families use only about a 1-V swing between logic levels. Because the oscillator is made using the same ECL-optimized semiconductor manufacturing processes and circuit design techniques, this same limitation applies to the signal in the oscillator tuned circuit. It is possible to improve this situation by using a tapped or transformer-coupled tank circuit to give improved Q, but this risks the occurrence of the device's second problem.

Periodically, semiconductor manufacturers modernize their plants and scrap old assembly lines used to make old products. Any surviving devices then must undergo some redesign to allow their production by the new processes. One common result of this is that devices are shrunk, when possible, to fit more onto a wafer. All this increases the f_T of the transistors in the device, and such evolution has rendered today's MC1648s capable of operation at much higher frequencies than the specified 200-MHz limit. This allows higher-frequency use, but great care is needed in the layout of circuits using it to prevent spurious oscillation. A number of old designs using this part have needed reengineering because the newer parts generate spurious oscillations that the old ones didn't, using PC-board traces as parasitic tuned circuits.

The moral is that a UHF-capable device requires UHF-cognizant design and layout even if the device will be used at far lower frequencies. **Fig 10.14** shows the MC1648 in a simple circuit and with a tapped resonator. These more complex circuits have a greater risk of presenting a stray resonance within the device's operating range, risking oscillation at an unwanted frequency. This device is *not* a prime choice for an HF VFO because the physical size of the variable capacitor and the inevitable lead lengths, combined with the need to tap-couple to get sufficient Q for good noise performance, makes spurious oscillation difficult to avoid. The MC1648 is really intended for tuning-diode control in phase-locked loops operating at VHF. This difficulty is inherent in wideband devices, especially oscillator circuits connected to their tank by a single "hot" terminal, where there is simply no isolation between the amplifier's input and output paths. Any resonance in the associated circuitry can control the frequency of oscillation.

The popular NE602 mixer IC has a built-in oscillator and can be found in many published circuits. This device has separate input and output pins to the tank and has proved to be quite tame. It's still a relatively new part and may not have been "improved" yet (so far, it has progressed from the NE602 to the NE602A, the A version affording somewhat higher dynamic range than the original NE602). It might be a good idea for anyone laying out a board using one to take a little extra care to keep PCB traces short in the oscillator section to build in some safety margin so that the board can be used reliably in the future. Experienced (read: "bitten") professional designers know that their designs are going to be built for possibly more than 10 years and have learned to make allowances for the progressive improvement of semiconductor manufacture.

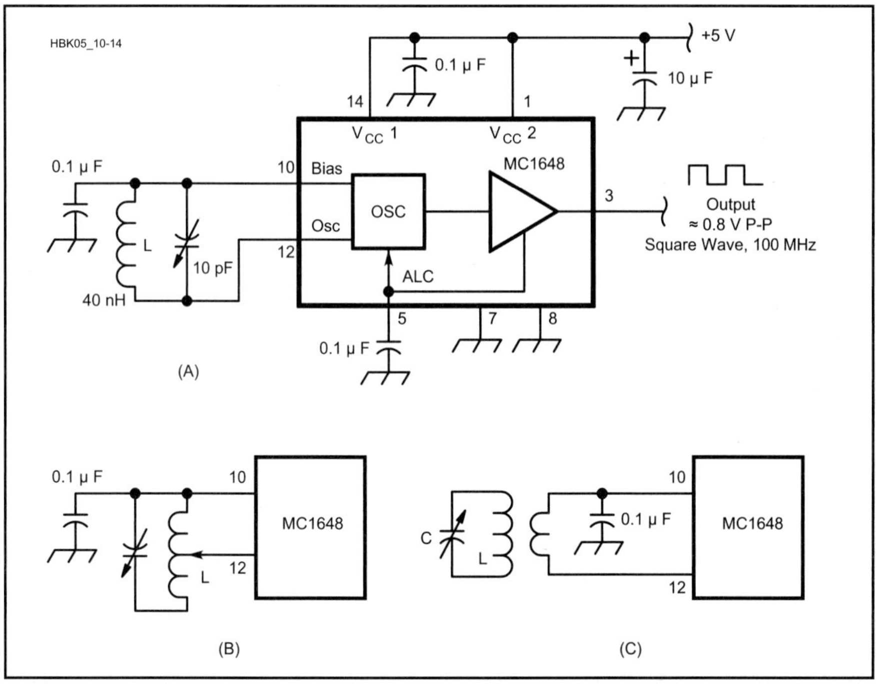

Fig 10.14—One of the few ICs ever designed solely for oscillator service, the ECL Motorola MC1648 (A) requires careful design to avoid VHF parasitics when operating at HF. Keeping its tank Q high is another challenge; B and C show means of coupling the IC's low-impedance oscillator terminals to the tank by tapping up on the tank coil (B) or with a link (C).

Three High-Performance HF VFOs

The G3PDM Vackar VFO

The Vackar VFO shown in **Fig 10.15** was developed over 20 years ago by Peter Martin, G3PDM, for the Mark II version of his high-dynamic-range receiver. This can be found in the Radio Society of Great Britain's *Radio Communication Handbook*, with some further comments on the oscillator in RSGB's *Amateur Radio Techniques*. This is a prime example of an oscillator that has been successfully optimized for maximum frequency stability. Not only does it work extremely well, but it still represents the highest stability that can be achieved. Its developer commented on a number of points, which also apply to the construction of different VFO circuits.

• Use a genuine Vackar circuit, with

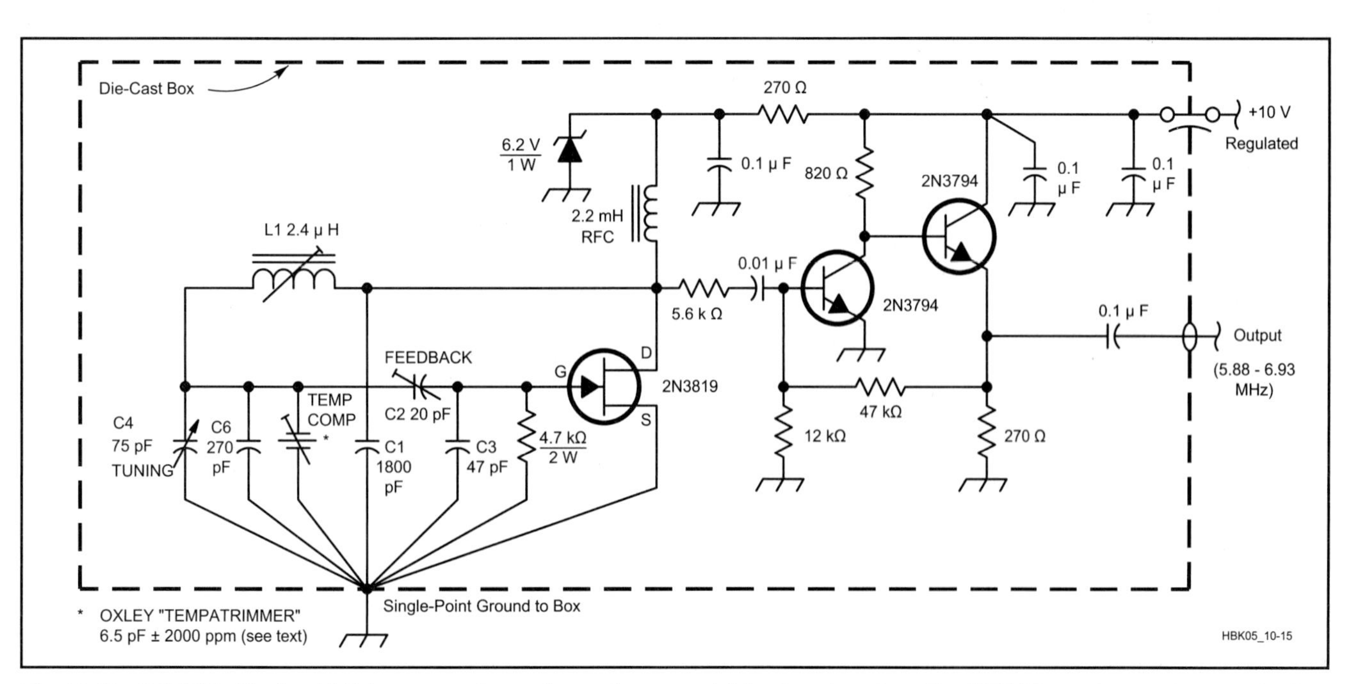

Fig 10.15—G3PDM's Vackar VFO has proved popular and successful for two decades. The MPF102 can be used as a substitute for the 2N3819. Generally, VFOs can be adapted to work at other frequencies (within the limits of the active device). To do so, compute an adjustment factor: f_{old} / f_{new}. Multiply the value of each frequency determining or feedback L or C by the factor. As frequency increases, it may help to increase feedback even more than indicated by the factor.

C1/(C4+C6) and C3/C2 = 6.

- Use a strong box, die-cast or milled from solid metal.
- Use a high-quality variable capacitor (double ball bearings, silver plated).
- Adjust feedback control C2, an air-dielectric trimmer, so the circuit just oscillates.
- Thoroughly clean all variable capacitors in an ultrasonic bath.
- Use an Oxley "Tempatrimmer," a fixed capacitor whose temperature coefficient is variable over a wide range, for adjustable temperature compensation. (The "Thermatrimmer" is a lower cost, smaller range alternative.)
- C1, C3 and C6 are silver-mica types glued to a solid support to reduce sensitivity to mechanical shock.
- The gate resistor is a 4.7-kΩ, 2-W carbon-composition type to minimize self-heating.
- The buffer amplifier is essential.
- Circuits using an added gate-ground diode seem to suffer increased drift.
- Its power supply should be well-regulated. Make liberal use of decoupling capacitors to prevent unintentional feedback via supply rails.
- Single-point grounding for the tank and FET is important. This usually means using one mounting screw of the tuning capacitor.
- The inductor used a ceramic form with a powdered-iron core mounted on a spring-steel screw.
- Short leads and thick solid wire (#16 to #18 gauge) are essential for mechanical stability.

Operating in the 6-MHz region, this oscillator drifted 500 Hz during the first minute as it warmed up and then drifted at 2 Hz in 30 minutes. It must be stressed that such performance does not indicate a wonderful circuit so much as care in construction, skillful choice of components, artful mechanical layout and diligence in adjustment. The 2-W gate resistor may seem strange, but 20 years ago many components were not as good as they are today. A modern, low-inductance component of low temperature coefficient and more modest power rating should be fine.

The buffer amplifier is an important part of a good oscillator system as it serves to prevent the oscillator seeing any impedance changes in the circuit being driven. This could otherwise affect the frequency.

The Oxley Tempatrimmer used by G3PDM was a rare and expensive component, used commercially only in very high-quality equipment, and it may no longer be made. An alternative temperature compensator can be constructed from currently available components, as will be described shortly. G3PDM also referred to his "standard mallet test," where a thump with a wooden hammer produced an average shift of 6 Hz, as an illustration of the benefits of solid mechanical design.

The K7HFD Low-Noise Oscillator

The other high performance oscillator example, shown in **Fig 10.16**, was designed for low-noise performance by Linley Gumm, K7HFD, and appears on page 126 of ARRL's *Solid State Design for the Radio Amateur*. Despite its publication in the homebrewer's bible, this circuit seems to have been overlooked by many builders. It uses no unusual components and looks simple, yet it is a subtle and sophisticated circuit. It represents the antithesis of G3PDM's VFO: In the pursuit of low noise sidebands, a number of design choices have been made that will degrade the stability of frequency over temperature.

The effects of oscillator noise have already been covered, and Fig 10.7 shows the effect of limiting on the signal from a noisy oscillator. Because AM noise sidebands can get translated into PM noise sidebands by imperfect limiting, there is an advantage to stripping off the AM as early as possible, in the oscillator itself. An ALC system in the oscillator will counteract and cancel only the AM components within its bandwidth, but an oscillator based on a limiter will do this over a broad bandwidth. K7HFD's oscillator uses a differential pair of bipolar transistors as a limiting amplifier. The dc bias voltage at the bases and the resistor in the common emitter path to ground establishes a controlled dc bias current. The ac voltage between the bases switches this current between the two collectors. This applies a rectangular pulse of current into link winding L2, which drives the tank. The output impedance of the collector is high in both the current on and current off states. Allied with the small number of turns of the link winding, this presents a very high impedance to the tank circuit, which minimizes the damping of the tank Q. The input impedance of the limiter is also quite high and is applied across only a one-turn tap of L1, which similarly minimizes the effect on the tank Q. The input transistor base is driven into conduction only on one peak of the tank waveform. The output transformer has the inverse of the current pulse applied to it, so the output is not a low distortion sine wave, although the output harmonics will not be as extensive as simple theory would suggest because the circuit's high output impedance allows stray capacitances to attenuate high-frequency components. The low-frequency transistors used also act to reduce the harmonic power.

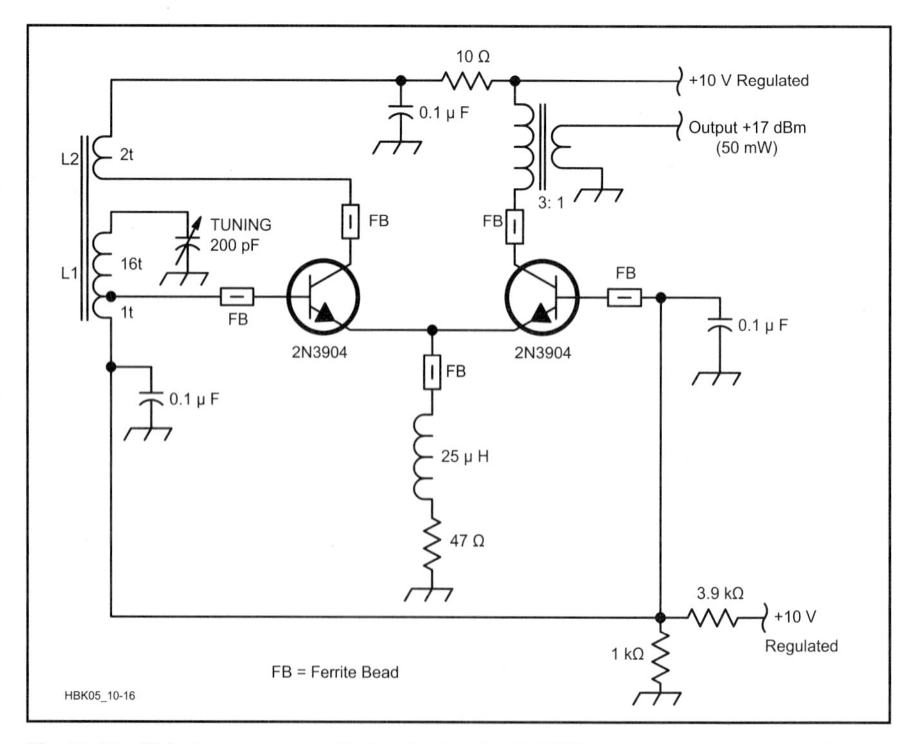

Fig 10.16—This low-noise oscillator design by K7HFD operates at an unusually high power level to achieve a high C/N (carrier-to-noise) ratio. Need other frequencies? See Fig 10.15 (caption) for a frequency-scaling technique.

With an output of +17 dBm, this is a power oscillator, running with a large dc input power, so appreciable heating results that can cause temperature-induced drift. The circuit's high-power operation is a deliberate ploy to create a high signal-to-noise ratio by having as high a signal power as possible. This also reduces the problem of the oscillator's broadband noise output. The limitation on the signal level in the tank is the transistors' base-emitter-junction breakdown voltage. The circuit runs with a few volts peak-to-peak across the one-turn tap, so the full tank is running at over 50 V P-P. The single easiest way to damage a bipolar transistor is to reverse bias the base-emitter junction until it avalanches. Most devices are only rated to withstand 5 V applied this way, the current needed to do damage is small, and very little power is needed. If the avalanche current is limited to less than that needed to perform immediate destruction of the transistor, it is likely that there will be some degradation of the device, a reduction in its bandwidth and gain along with an increase in its noise. These changes are irreversible and cumulative. Small, fast signal diodes have breakdown voltages of over 30 V and less capacitance than the transistor bases, so one possible experiment would be to try the effect of adding a diode in series with the base of each transistor and running the circuit at even higher levels.

The amplitude must be controlled by the drive current limit. The voltage on L2 must never allow the collector of the transistor driving it to go into saturation, if this happens, the transistor presents a very low impedance to L2 and badly loads the tank, wrecking the Q and the noise performance. The circuit can be checked to verify the margin from saturation by probing the hot end of L2 and the emitter with an oscilloscope. Another, less obvious, test is to vary the power-supply voltage and monitor the output power. While the circuit is under current control, there is very little change in output power, but if the supply is low enough to allow saturation, the output power will change significantly.

The use of the 2N3904 is interesting, as it is not normally associated with RF oscillators. It is a cheap, plain, general-purpose type more often used at dc or audio frequencies. There is evidence that suggests some transistors that have good noise performance *at RF* have worse noise performance at low frequencies, and that the low-frequency noise they create can modulate an oscillator, creating noise sidebands. Experiments with low-noise audio transistors may be worthwhile, but many such devices have very low f_T and high junction capacitances. In the description of this circuit in *Solid State Design for the Radio Amateur*, the results of a phase-noise test made using a spectrum analyzer with a crystal filter as a preselector are given. Ten kilohertz out from the carrier, in a 3-kHz measurement bandwidth, the noise was over 120 dB below the carrier level. This translates into better than –120 – 10 log (3000), which equals –154.8 dBc/Hz, SSB. At this offset, –140 dBc is usually considered to be excellent. This is state-of-the art performance by today's standards—in a 1977 publication.

A JFET Hartley VFO

Fig 10.17 shows an 11.1-MHz version of a VFO and buffer closely patterned after that used in 7-MHz transceiver designs published by Roger Hayward, KA7EXM, and Wes Hayward, W7ZOI (The Ugly Weekender) and Roy Lewallen, W7EL (the Optimized QRP Transceiver). In it, a 2N5486 JFET Hartley oscillator drives the two-2N3904 buffer attributed to Lewallen. This version diverges from the originals in that its JFET uses source bias (the bypassed 910-Ω resistor) instead of a gate-clamping diode and is powered from a low-current 7-V regulator IC instead of a Zener diode and dropping resistor. The 5-dB pad sets the buffer's output to a level appropriate for "Level 7" (+7-dBm-LO) diode ring mixers.

The circuit shown was originally built with a gate-clamping diode, no source bias and a 3-dB output pad. Rebiasing the oscillator as shown increased its output by 2 dB without degrading its frequency stability (200 to 300-Hz drift at power up, stability within ±20 Hz thereafter at a constant room temperature).

Temperature Compensation

The general principle for creating a high-stability VFO is to use components with minimal temperature coefficients in circuits that are as insensitive as possible to changes in components' secondary characteristics. Even after careful minimization of the causes of temperature sensitivity, further improvement can still be desirable. The traditional method was to split one of the capacitors in the tank so that it could be apportioned between low-temperature-coefficient parts and parts with deliberate temperature dependency. Only a limited number of different, controlled temperature coefficients are available, so the proportioning between low coefficient and controlled coefficient parts was varied to "dilute" the temperature sensitivity of a part more sensitive than needed. This was a tedious process, involving much trial and error, an undertaking made more complicated by the difficulty of arranging means of heating and cooling the unit being compensated. (Hayward described such a means in December 1993 *QST*.) As commercial and military equipment have been based on frequency synthesizers for some time, supplies of capacitors with controlled temperature sensitivity are drying up. An alternative approach is needed.

A temperature-compensated crystal oscillator (TCXO) is an improved-stability version of a crystal oscillator that is used widely in industry. Instead of using controlled-temperature coefficient capacitors, most TCXOs use a network of thermistors and normal resistors to control the bias of a tuning diode. See **Fig 10.18**. Manufacturers measure the temperature vs frequency characteristic of sample oscillators, and use a computer program to calculate the optimum normal resistor values for production. This can reliably achieve at least a tenfold improvement in stability. We are not interested in mass manufacture, but the idea of a thermistor tuning a varactor is worth stealing. The parts involved are likely to be available for a long time.

Browsing through component suppliers' catalogs shows ready availability of 4.5- to 5-kΩ-bead thermistors intended for temperature-compensation purposes, at less than a dollar each. **Fig 10.19** shows a circuit based on this form of temperature compensation. **Fig 10.20** illustrates a practical VCO using a tuning diode. Commonly available thermistors have negative temperature coefficients, so as temperature rises, the voltage at the counterclockwise (ccw) end of R8 increases, while that at the clockwise (cw) end drops. Somewhere near the center, there is no change. Increasing the voltage on the tuning diode decreases its capacitance, so settings toward R8's ccw end simulate a negative-temperature-coefficient capacitor; toward its clockwise end, a positive-temperature-coefficient part. Choose R1 to pass 8.5 mA from whatever supply voltage is available to the 6.2-V reference diode, D1. The 1N821A/1N829A-family diode used has a very low temperature coefficient and needs 7.5 mA bias for best performance; the bridge takes 1 mA. R7 and R8 should be good-quality multiturn trimmers. D2 and C1 are best determined by trial and error. Practical components aren't known well enough to rely on analytical models. Choose the least capacitance that provides enough compensation range. This reduces the noise added to the oscillator. (It is possible, though tedious, to solve for the differential varactor voltage with respect to R2 and R5, via differential calculus and circuit theory. The equations in Hayward's 1993 article can then be

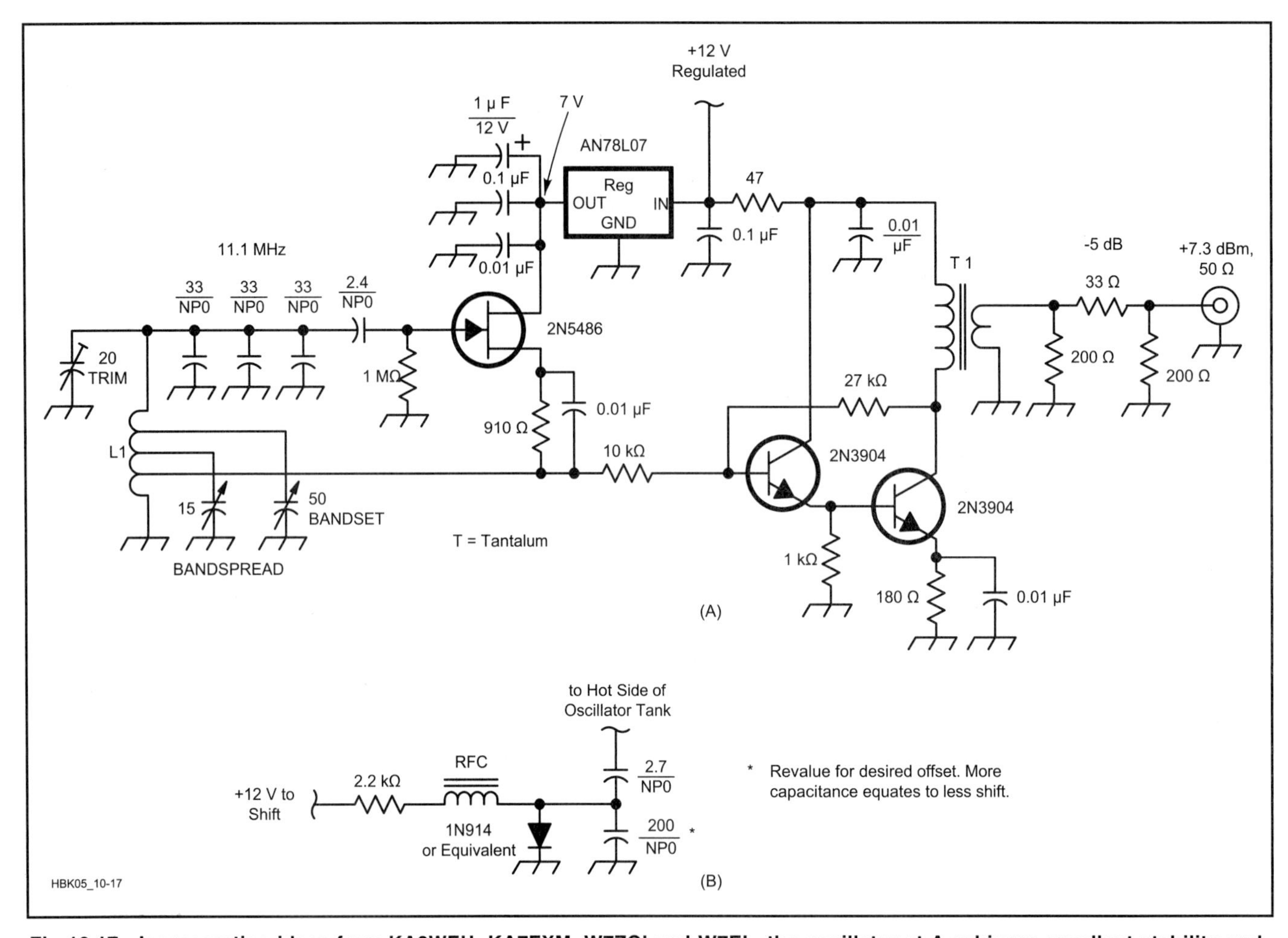

Fig 10.17—Incorporating ideas from KA2WEU, KA7EXM, W7ZOI and W7EL, the oscillator at A achieves excellent stability and output at 11.1 MHz without the use of a gate-clamping diode, as well as end-running the shrinking availability of reduction drives through the use of bandset and bandspread capacitors. L1 consists of 10 turns of B & W #3041 Miniductor (#22 tinned wire, 5/8 inch in diameter, 24 turns per inch). The source tap is 2 1/2 turns above ground; the tuning-capacitor taps are positioned as necessary for bandset and bandspread ranges required. T1's primary consists of 15 turns of #28 enameled wire on an FT-37-72 ferrite core; its secondary, 3 turns over the primary. B shows a system for adding fixed TR offset that can be applied to any LC oscillator. The RF choke consists of 20 turns of #26 enameled wire on an FT-37-43 core. Need other frequencies? See Fig 10.15 (caption) for a frequency-scaling technique.

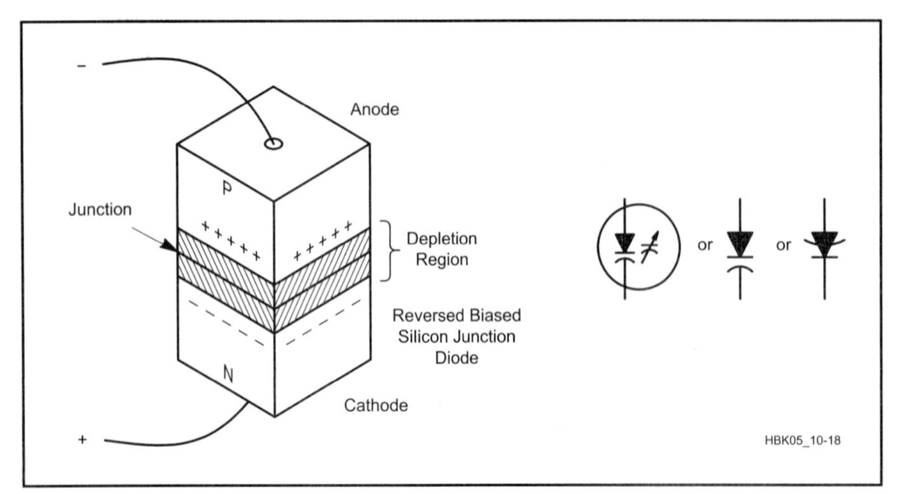

Fig 10.18—A modern voltage-controlled oscillator (VCO) uses the voltage-variable characteristic of a diode PN junction for tuning.

modified to accommodate the additional capacitors formed by D2 and C1.) Use a single ground point near D2 to reduce the influence of ground currents from other circuits. Use good-quality metal-film components for the circuit's fixed resistors.

The circuit requires two adjustments, one at each of two different temperatures, and achieving them requires a stable frequency counter that can be kept far enough from the radio so that the radio, not the counter, is subjected to the temperature extremes. (Using a receiver to listen to the oscillator under test can speed the adjustments.) After connecting the counter to the oscillator to be corrected, run the radio containing the oscillator and compensator in a room-temperature, draft-free environment until the oscillator's frequency

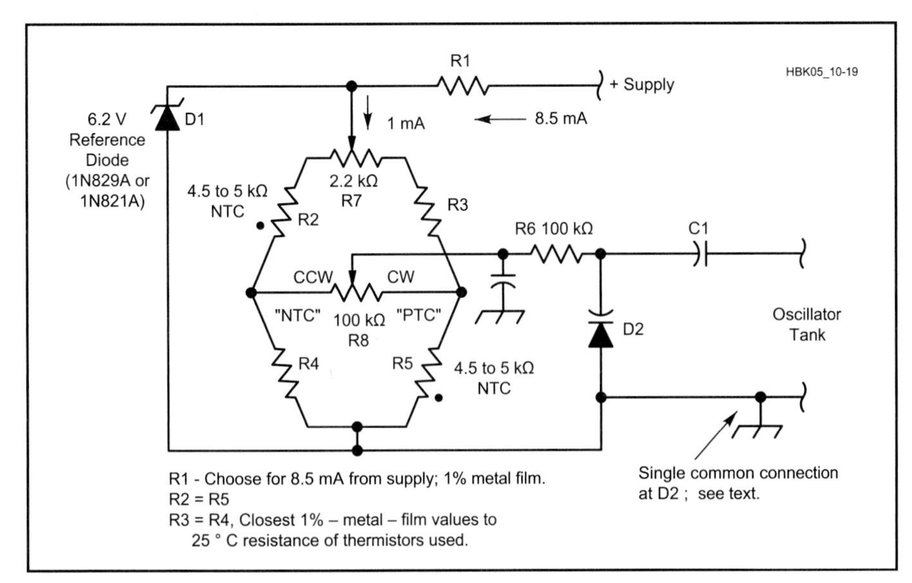

Fig 10.19—Oscillator temperature compensation has become more difficult because of the scarcity of negative-temperature-coefficient capacitors. This circuit, by GM4ZNX, uses a bridge containing two identical thermistors to steer a tuning diode for drift correction. The 6.2-V Zener diode used (a 1N821A or 1N829A) is a temperature-compensated part; just any 6.2-V Zener will not do.

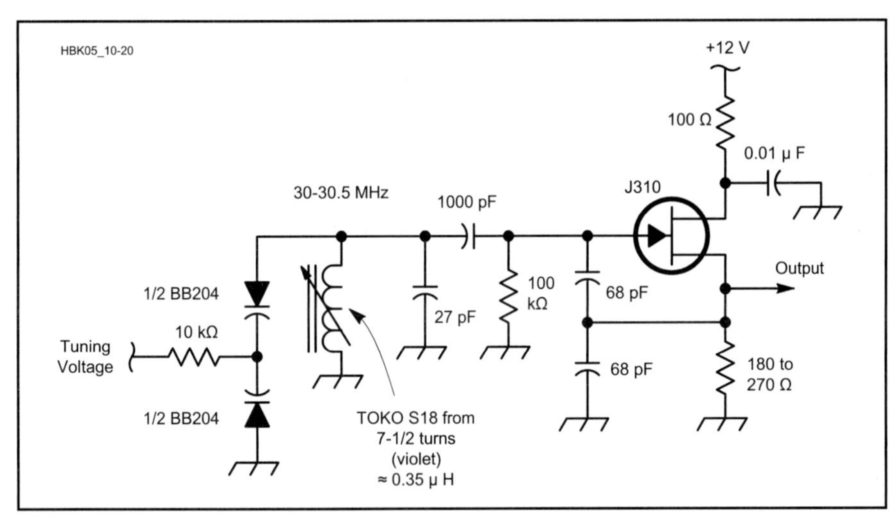

Fig 10.20—A practical VCO. The tuning diodes are halves of a BB204 dual, common-cathode tuning diode (capacitance per section at 3 V, 39 pF) or equivalent. The ECG617, NTE617 and MV104 are suitable dual-diode substitutes, or use pairs of 1N5451s (39 pF at 4 V) or MV2109s (33 pF at 4 V).

reaches its stable operating temperature rise over ambient. Lock its tuning, if possible. Adjust R7 to balance the bridge. This causes a drop of 0 V across R8, a condition you can reach by winding R8 back and forth across its range while slowly adjusting R7. When the bridge is balanced and 0 V appears across R8, adjusting R8 causes no frequency shift. When you've found this R7 setting, leave it there, set R8 is to the exact center of its range and record the oscillator frequency.

Run the radio in a hot environment and allow its frequency to stabilize. Adjust R8 to restore the frequency to the recorded value. The sensitivity of the oscillator to temperature should now be significantly reduced between the temperatures at which you performed the adjustments. You will also have somewhat improved the oscillator's stability outside this range.

For best results with any temperature-compensation scheme, it's important to group all the oscillator and compensator components in the same box, avoiding differences in airflow over components. A good oscillator should not dissipate much power, so it's feasible, even advisable, to mount all of the oscillator components in an unventilated box. In the real world, temperatures change, and if the components being compensated and the components doing the compensating have different thermal time constants, a change in temperature can cause a temporary change in frequency until the slower components have caught up. One cure for this is to build the oscillator in a thick-walled metal box that's slow to heat or cool, and so dominates and reduces the possible rate of change of temperature of the circuits inside. This is sometimes called a *cold oven.*

Shielding and Isolation

Oscillators contain inductors running at moderate power levels and so can radiate strong enough signals to cause interference with other parts of a radio, or with other radios. Oscillators are also sensitive to radiated signals. Effective shielding is therefore important. A VFO used to drive a power amplifier and antenna (to form a simple CW transmitter) can prove surprisingly difficult to shield well enough. Any leakage of the power amplifier's high-level signal back into the oscillator can affect its frequency, resulting in a poor transmitted note. If the radio gear is in the station antenna's near field, sufficient shielding may be even more difficult. The following rules of thumb continue to serve ham builders well:

- Use a complete metal box, with as few holes as possible drilled in it, with good contact around surface(s) where its lid(s) fit(s) on.
- Use feedthrough capacitors on power and control lines that pass in and out of the VFO enclosure, and on the transmitter or transceiver enclosure as well.
- Use *buffer amplifier* circuitry that amplifies the signal by the desired amount *and* provide sufficient attenuation of signal energy flowing in the reverse direction. This is known as *reverse isolation* and is a frequently overlooked loophole in shielding. Figs 10.15 and 10.17 include buffer circuitry of proven performance. As another (and higher-cost) option, consider using a high-speed buffer-amplifier IC (such as the LM6321N by National Semiconductor, a part that combines the high input impedance of an op amp with the ability to drive 50-Ω loads directly up into the VHF range).
- Use a mixing-based frequency-generation scheme instead of one that operates straight through or by means of multiplication. Such a system's oscillator stages can operate on frequencies with no direct frequency relationship to its output frequency.

- Use the time-tested technique of running your VFO at a *subharmonic* of the output signal desired—say, 3.5 MHz in a 7-MHz transmitter—and *multiply* its output frequency in a suitably nonlinear stage for further amplification at the desired frequency.

Quartz Crystals in Oscillators

Because crystals afford Q values and frequency stabilities that are orders of magnitude better than those achievable with LC circuits, fixed-frequency oscillators usually use quartz-crystal resonators. Master references for frequency counters and synthesizers are always based on crystal oscillators.

So glowing is the executive summary of the crystal's reputation for stability that newcomers to radio experimentation naturally believe that the presence of a crystal in an oscillator will force oscillation at the frequency stamped on the can. This impression is usually revised after the first few experiences to the contrary! There is no sure-fire crystal oscillator circuit (although some are better than others); reading and experience soon provide a learner with plenty of anecdotes to the effect that:

- Some circuits have a reputation of being temperamental, even to the point of not always starting.
- Crystals sometimes mysteriously oscillate on unexpected frequencies.

Even crystal manufacturers have these problems, so don't be discouraged from building crystal oscillators. The occasional uncooperative oscillator is a nuisance, not a disaster, and it just needs a little individual attention. Knowing how a crystal behaves is the key to a cure.

Quartz and the Piezoelectric Effect

Quartz is a crystalline material with a regular atomic structure that can be distorted by the simple application of force. Remove the force, and the distorted structure springs back to its original form with very little energy loss. This property allows *acoustic waves*—sound—to propagate rapidly through quartz with very little attenuation, because the velocity of an acoustic wave depends on the elasticity and density (mass/volume) of the medium through which the wave travels.

If you heat a material, it expands. Heating may cause other characteristics of a material to change—such as elasticity, which affects the speed of sound in the material. In quartz, however, expansion and change in the speed of sound are very small and tend to cancel, which means that the transit time for sound to pass through a piece of quartz is very stable.

The third property of this wonder material is that it is *piezoelectric*. Apply an electric field to a piece of quartz and the crystal lattice distorts just as if a force had been applied. The electric field applies a force to electrical charges locked in the lattice structure. These charges are captive and cannot move around in the lattice as they can in a semiconductor, for quartz is an insulator. A capacitor's dielectric stores energy by creating physical distortion on an atomic or molecular scale. In a piezoelectric crystal's lattice, the distortion affects the entire structure. In some piezoelectric materials, this effect is sufficiently pronounced that special shapes can be made that bend *visibly* when a field is applied.

Consider a rod made of quartz. Any sound wave propagating along it eventually hits an end, where there is a large and abrupt change in acoustic impedance. Just as when an RF wave hits the end of an unterminated transmission line, a strong reflection occurs. The rod's other end similarly reflects the wave. At some frequency, the phase shift of a round trip will be such that waves from successive round trips exactly coincide in phase and reinforce each other, dramatically increasing the wave's amplitude. This is *resonance*.

The passage of waves in opposite directions forms a standing wave with antinodes at the rod ends. Here we encounter a seeming ambiguity: not just one, but a *family* of different frequencies, causes standing waves—a family fitting the pattern of $^1/_2$, $^2/_3$, $^5/_2$, $^7/_2$ and so on, wavelengths into the length of the rod. And this *is* the case: A quartz rod *can* resonate at any and all of these frequencies.

The lowest of these frequencies, where the crystal is $^1/_2$ wavelength long, is called the *fundamental* mode. The others are named the third, fifth, seventh and so on, *overtones*. There is a small phase-shift error during reflection at the ends, which causes the frequencies of the overtone modes to differ slightly from odd integer multiplies of the fundamental. Thus, a crystal's third overtone is very close to, but not exactly, three times, its fundamental frequency. Many people are confused by overtones and harmonics. Harmonics are additional signals at *exact* integer multiples of the fundamental frequency. Overtones are not signals at all; they are additional resonances that can be exploited if a circuit is configured to excite them.

The crystals we use most often resonate in the 1- to 30-MHz region and are of the *AT cut, thickness shear* type, although these last two characteristics are rarely mentioned. A 15-MHz-fundamental crystal of this type is about 0.15 mm thick. Because of the widespread use of reprocessed war-surplus, pressure-mounted *FT-243* crystals, you may think of crystals as small rectangles on the order of a half inch in size. The crystals we commonly use today are discs, etched and/or doped to their final dimensions, with metal electrodes deposited directly on the quartz. A crystal's diameter does not directly affect its frequency; diameters of 8 to15 mm are typical.

AT cut is one of a number of possible standard designations for the orientation at which a crystal disc is sawed from the original quartz crystal. The crystal lattice atomic structure is asymmetric, and the orientation of this with respect to the faces of the disc influences the crystal's performance. *Thickness shear* is one of a number of possible orientations of the crystal's mechanical vibration with respect to the disc. In this case, the crystal vibrates perpendicularly to its thickness. This is not easy to visualize, and diagrams don't help much, but **Fig 10.21** is an attempt at illustrating this. Place a moist bathroom sponge between the palms of your hands, move one hand up and down, and you'll see thickness shear in action.

There is a limit to how thin a disc can be made, given requirements of accuracy and price. Traditionally, fundamental-mode crystals have been made up to 20 MHz, although 30 MHz is now common at a

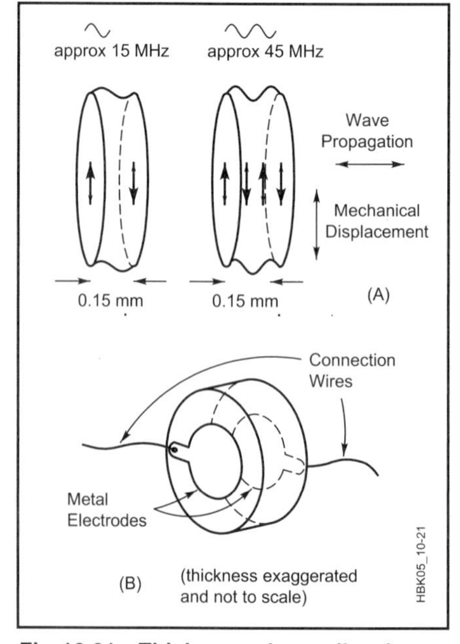

Fig 10.21—Thickness-shear vibration at a crystal's fundamental and third overtone (A); B shows how the modern crystals commonly used by radio amateurs consist of etched quartz discs with electrodes deposited directly on the crystal surface.

moderately raised price. Using techniques pioneered in the semiconductor industry, crystals have been made with a central region etched down to a thin membrane, surrounded by a thick ring for robustness. This approach can push fundamental resonances to over 100 MHz, but these are more lab curiosities than parts for everyday use. The easy solution for higher frequencies is to use a nice, manufacturably thick crystal on an overtone mode. All crystals have all modes, so if you order a 28.060-MHz, third-overtone unit for a little QRP transmitter, you'll get a crystal with a fundamental resonance somewhere near 9.353333 MHz, but its manufacturer will have adjusted the thickness to plant the third overtone exactly on the ordered frequency. An accomplished manufacturer can do tricks with the flatness of the disc faces to make the wanted overtone mode a little more active and the other modes a little less active. (As some builders discover, however, this does not *guarantee* that the wanted mode is the most active!)

Quartz's piezoelectric property provides a simple way of driving the crystal electrically. Early crystals were placed between a pair of electrodes in a case. This gave amateurs the opportunity to buy surplus crystals, open them and grind them a little to reduce their thickness, thus moving them to higher frequencies. The frequency could be reduced very slightly by loading the face with extra mass, such as by blackening it with a soft pencil. Modern crystals have metal electrodes deposited directly onto their surfaces (Fig 10.21B), and such tricks no longer work.

The piezoelectric effect works both ways. Deformation of the crystal produces voltage across its electrodes, so the mechanical energy in the resonating crystal can also be extracted electrically by the same electrodes. Seen electrically, at the electrodes, the mechanical resonances look like electrical resonances. Their Q is very high. A Q of 10,000 would characterize a *poor* crystal nowadays; 100,000 is often reached by high-quality parts. For comparison, a Q of over 200 for an LC tank is considered good.

Accuracy

A crystal's frequency accuracy is as outstanding as its Q. Several factors determine a crystal's frequency accuracy. First, the manufacturer makes parts with certain tolerances: ±200 ppm for a low-quality crystal for use as in a microprocessor clock oscillator, ±10 ppm for a good-quality part for professional radio use. Anything much better than this starts to get expensive! A crystal's resonant frequency is influenced by the impedance presented to its terminals, and manufacturers assume that once a crystal is brought within several parts per million of the nominal frequency, its user will perform fine adjustments electrically.

Second, a crystal ages after manufacture. Aging could give increasing or decreasing frequency; whichever, a given crystal usually keeps aging in the same direction. Aging is rapid at first and then slows down. Aging is influenced by the care in polishing the surface of the crystal (time and money) and by its holder style. The cheapest holder is a soldered-together, two-part metal can with glass bead insulation for the connection pins. Soldering debris lands on the crystal and affects its frequency. Alternatively, a two-part metal can be made with flanges that are pressed together until they fuse, a process called *cold-welding*. This is much cleaner and improves aging rates roughly fivefold compared to soldered cans. An all-glass case can be made in two parts and fused together by heating in a vacuum. The vacuum raises the Q, and the cleanliness results in aging that's roughly ten times slower than that achievable with a soldered can. The best crystal holders borrow from vacuum-tube assembly processes and have a *getter*, a highly reactive chemical substance that traps remaining gas molecules, but such crystals are used only for special purposes.

Third, temperature influences a crystal. A reasonable, professional quality part might be specified to shift not more than ±10 ppm over 0 to 70°C. An AT-cut crystal has an *S*-shaped frequency-versus-temperature characteristic, which can be varied by slightly changing the crystal cut's orientation. **Fig 10.22** shows the general shape and the effect of changing the cut angle by only a few seconds of arc. Notice how all the curves converge at 25°C. This is because this temperature is normally chosen as the reference for specifying a crystal. The temperature stability specification sets how accurate the manufacturer must make the cut. Better stability may be needed for a crystal used as a receiver frequency standard, frequency counter clock and so on. A crystal's temperature characteristic shows a little hysteresis. In other words, there's a bit of offset to the curve depending on whether temperature is increasing or decreasing. This is usually of no consequence except in the highest-precision circuits.

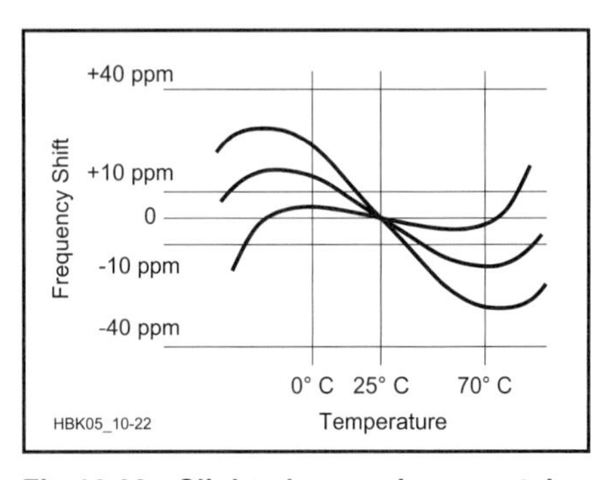

Fig 10.22—Slight changes in a crystal cut's orientation shift its frequency-versus-temperature curve.

It is the temperature of the quartz that is important, and as the usual holders for crystals all give effective thermal insulation, only a couple of milliwatts dissipation by the crystal itself can be tolerated before self-heating becomes troublesome. Because such heating occurs in the quartz itself and does not come from the surrounding environment, it defeats the effects of temperature compensators and ovens.

The techniques shown earlier for VFO for temperature compensation can also be applied to crystal oscillators. An after-compensation drift of 1 ppm is routine and 0.5 ppm is good. The result is a *temperature-compensated crystal oscillator* (*TCXO*). Recently, oscillators have appeared with built-in digital thermometers, microprocessors and ROM look-up tables customized on a unit-by-unit basis to control a tuning diode via a digital-to-analog converter (DAC) for temperature compensation. These *digitally temperature-compensated oscillators* (*DTCXO*s) can reach 0.1 ppm over the temperature range. With automated production and adjustment, they promise to become the cheapest way to achieve this level of stability.

Oscillators have long been placed in temperature-controlled *ovens*, which are typically held at 80°C. Stability of several parts per billion can be achieved over temperature, but this is a limited benefit as aging can easily dominate the accuracy. These are usually called *oven-controlled crystal oscillators* (*OCXOs*).

Fourth, the crystal is influenced by the impedance presented to it by the circuit in which it is used. This means that care is needed to make the rest of an oscillator circuit stable, in terms of impedance and phase shift.

Gravity can slightly affect crystal resonance. Turning an oscillator upside down usually produces a small frequency shift, usually much less than 1 ppm; turning the oscillator back over reverses this. This effect is quantified for the highest-quality reference oscillators.

The Equivalent Circuit of a Crystal

Because a crystal is a passive, two-terminal device, its electrical appearance is that of an impedance that varies with

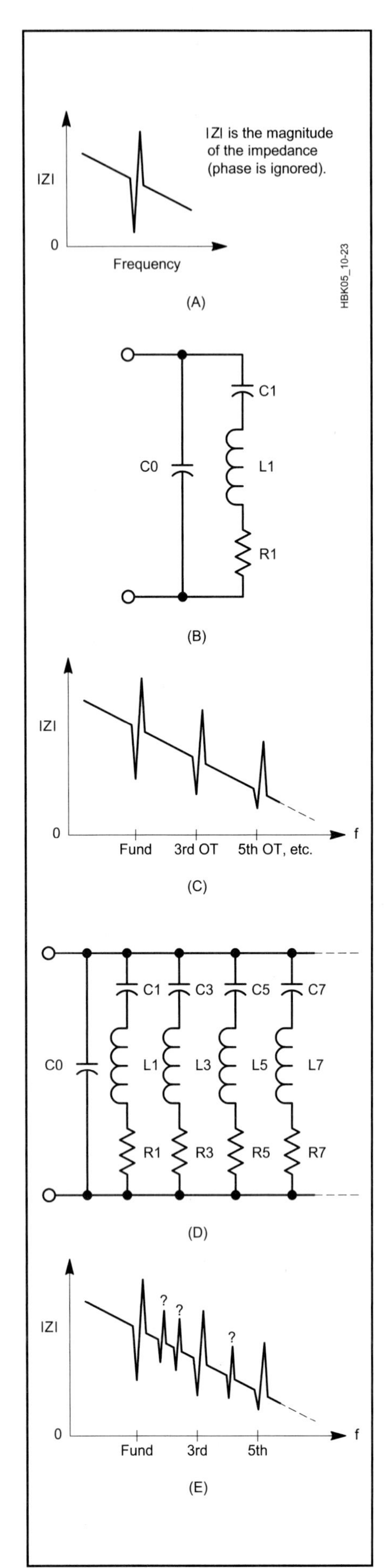

Fig 10.23—Exploring a crystal's impedance (A) and equivalent circuit (B) through simplified diagrams. C and D extend the investigation to include overtones; E, to spurious responses not easily predictable by theory or controllable through manufacture. A crystal may oscillate on any of its resonances under the right conditions.

frequency. **Fig 10.23A** shows a very simplified sketch of the magnitude (phase is ignored) of the impedance of a quartz crystal. The general trend of dropping impedance with increasing frequency implies capacitance across the crystal. The sharp fall to a low value resembles a series-tuned tank, and the sharp peak resembles a parallel-tuned tank. These are referred to as series and parallel resonances. Fig 10.23B shows a simple circuit that will produce this impedance characteristic. The impedance looks purely resistive at the exact centers of both resonances, and the region between them has impedance increasing with frequency, which looks inductive.

C1 (sometimes called *motional capacitance*, C_m, to distinguish it from the lumped capacitance it approximates) and L1 (*motional inductance,* L_m) create the series resonance, and as C0 and R1 are both fairly small, the impedance at the bottom of the dip is very close to R1. At parallel resonance, L1 is resonating with C1 and C0 in series, hence the higher frequency. The impedance of the parallel tank is immense, the terminals are connected to a capacitive tap, which causes them to see only a small fraction of this, which is still a very large impedance. The overtones should not be neglected, so Figs 10.23C and 10.23D include them. Each overtone has series and parallel resonances and so appears as a series tank in the equivalent circuit. C0 again provides the shifted parallel resonance.

This is still simplified, because real-life crystals have a number of spurious, unwanted modes that add yet more resonances, as shown in Fig 10.23E. These are not well controlled and may vary a lot even between crystals made to the same specification. Crystal manufacturers work hard to suppress these spurs and have evolved a number of recipes for shaping crystals to minimize them. Just where they switch from one design to another varies from manufacturer to manufacturer.

Always remember that the equivalent circuit is just a representation of crystal behavior and does not represent circuit components actually present. Its only use is as an aid in designing and analyzing circuits using crystals. **Table 10.2** lists typical equivalent-circuit values for a variety of crystals. It is impossible to build a circuit with 0.026 to 0.0006-pF capacitors; such values would simply be swamped by strays. Similarly, the inductor must have a Q orders of magnitude better than is practically achievable, and impossibly low stray C in its winding.

The values given in Table 10.2 are nothing more than rough guides. A crystal's frequency is tightly specified, but this still allows inductance to be traded for capacitance. A good manufacturer could hold these characteristics within a ±25% band or could vary them over a 5:1 range by special design. Similarly marked parts from different sources vary widely in motional inductance and capacitance.

Quartz is not the only material that behaves in this way, but it is the best. Resonators can be made out of lithium tantalate and a group of similar materials that have lower Q, allowing them to be *pulled* over a larger frequency range in VCXOs. Much more common, however, are ceramic resonators based on the technology of the well-known ceramic IF filters. These have much lower Q than quartz and much poorer frequency precision. They serve mainly as clock resonators for cheap microprocessor systems in which every last cent must be saved. A ceramic resonator could be used as the basis of a wide range, cheap VXO, but its frequency stability would not be as good as a good LC VFO.

Crystal Oscillator Circuits

Crystal oscillator circuits are usually categorized as series- or parallel-mode types, depending on whether the crystal's low- or high-impedance resonance comes

Table 10.2
Typical Equivalent Circuit Values for a Variety of Crystals

Crystal Type	*Series L*	*Series C (pF)*	*Series R (Ω)*	*Shunt C (pF)*
1-MHz fundamental	3.5 H	0.007	340	3.0
10-MHz fundamental	9.8 mH	0.026	7	6.3
30-MHz third overtone	14.9 mH	0.0018	27	6.2
100-MHz fifth overtone	4.28 mH	0.0006	45	7.0

into play at the operating frequency. The series mode is now the most common; parallel-mode operation was more often used with vacuum tubes. **Fig 10.24** shows a basic series-mode oscillator. Some people would say that it is an overtone circuit, used to run a crystal on one of its overtones, but this is not necessarily true. The tank (L-C1-C2) tunes the collector of the common-base amplifier. C1 is larger than C2, so the tank is tapped in a way that transforms to a lower impedance, decreasing signal voltage, but increasing current. The current is fed back into the emitter via the crystal. The common-base stage provides a current gain of less than unity, so the transformer in the form of the tapped tank is essential to give loop gain. There are *two* tuned circuits, the obvious collector tank and the series-mode one "in" the crystal. The tank kills the amplifier's gain away from its tuned frequency, and the crystal will only pass current at the series resonant frequencies of its many modes. The tank resonance is much broader than any of the crystal's modes, so it can be thought of as the crystal setting the frequency, but the tank selecting which of the crystal's modes is active. The tank could be tuned to the crystal's fundamental, or one of its overtones.

Fundamental oscillators can be built without a tank quite successfully, but there is always the occasional one that starts up on an overtone or spurious mode. Some simple oscillators have been known to change modes while running (an effect triggered by changes in temperature or loading) or to not always start in the same mode! A series-mode oscillator should present a low impedance to the crystal at the operating frequency. In Fig 10.24, the tapped collector tank presents a transformed fraction of the 1-kΩ collector load resistor to one end of the crystal, and the emitter presents a low impedance to the other. To build a practical oscillator from this circuit, choose an inductor with a reactance of about 300 Ω at the wanted frequency and calculate C1 in series with C2 to resonate with it. Choose C1 to be 3 to 4 times larger than C2. The amplifier's quiescent ("idling") current sets the gain and hence the operating level. This is not easily calculable, but can be found by experiment. Too little quiescent current and the oscillator will not start reliably; too much and the transistor can drive itself into saturation. If an oscilloscope is available, it can be used to check the collector waveform; otherwise, some form of RF voltmeter can be used to allow the collector voltage to be set to 2 to 3 V RMS. 3.3 kΩ would be a suitable starting point for the emitter bias resistor. The transistor type is not critical; 2N2222A or 2N3904 would be fine up to 30 MHz; a 2N5179 would allow operation as an overtone oscillator to over 100 MHz (because of the low collector voltage rating of the 2N5179, a supply voltage lower than 12 V is required). The ferrite bead on the base gives some protection against parasitic oscillation at UHF.

If the crystal is shorted, this circuit should still oscillate. This gives an easy way of adjusting the tank; it is even better to temporarily replace the crystal with a small-value (tens of ohms) resistor to simulate its *equivalent series resistance* (ESR), and adjust L until the circuit oscillates close to the wanted frequency. Then restore the crystal and set the quiescent current. If a lot of these oscillators were built, it would sometimes be necessary to adjust the current individually due to the different equivalent series resistance of individual crystals. One variant of this circuit has the emitter connected directly to the C1/C2 junction, while the crystal is a decoupler for the transistor base (the existing capacitor and ferrite bead not being used). This works, but with a greater risk of parasitic oscillation.

We commonly want to trim a crystal

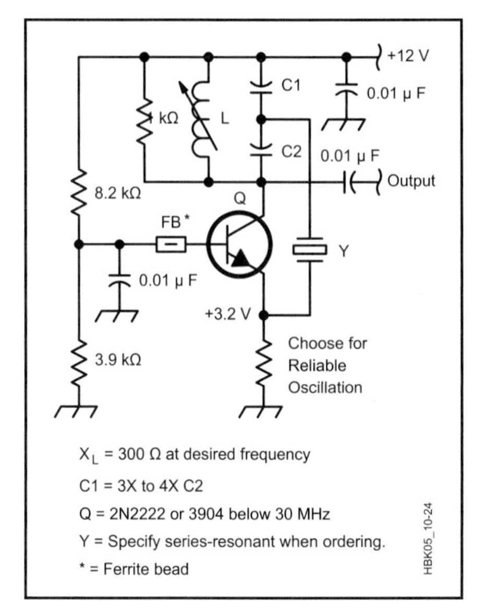

Fig 10.24—A basic series-mode crystal oscillator. A 2N5179 can be used in this circuit if a lower supply voltage is used; see text.

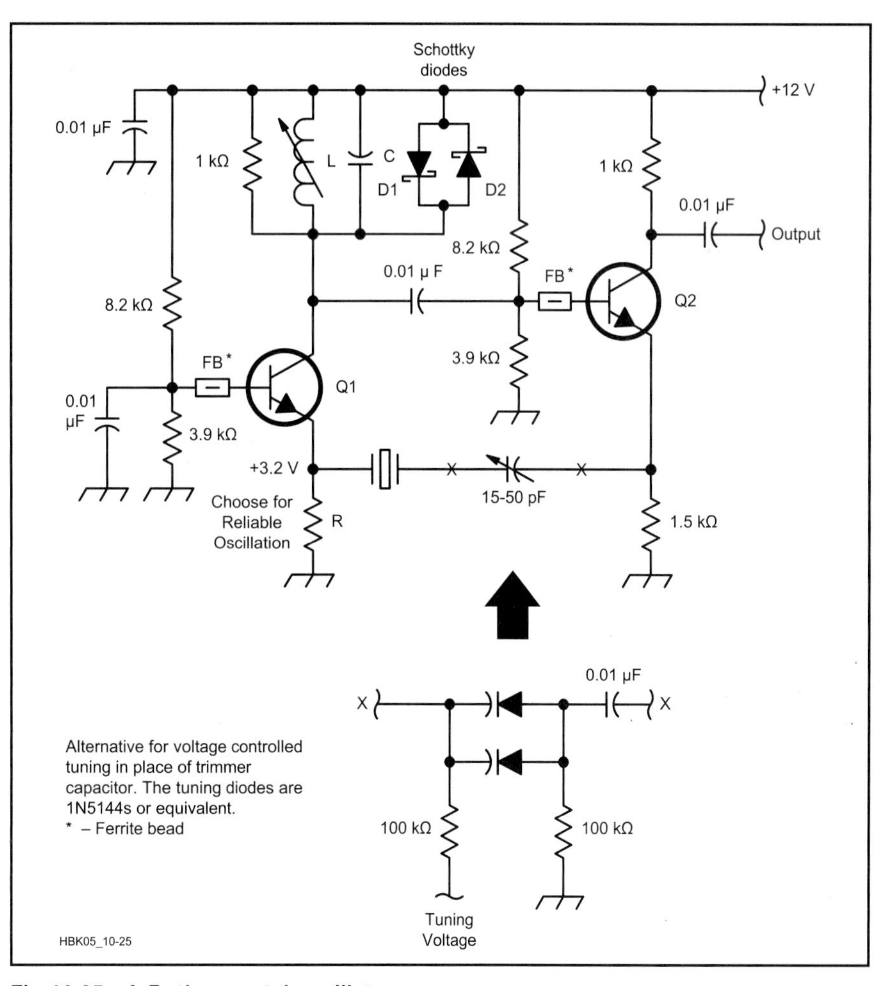

Fig 10.25—A Butler crystal oscillator.

oscillator's frequency. While off-tuning the tank a little will pull the frequency slightly, too much detuning spoils the mode control and can stop oscillation (or worse, make the circuit unreliable). The answer to this is to add a trimmer capacitor, which will act as part of the equivalent series tuned circuit, in series with the crystal. This will shift the frequency in one way only, so the crystal frequency must be respecified to allow the frequency to be varying around the required value. It is common to specify a crystal's frequency with a standard load (30 pF is commonly specified), so that the manufacturer grinds the crystal such that the series resonance of the specified mode is accurate when measured with a capacitor of this value in series. A 15- to 50-pF trimmer can be used in series with the crystal to give fine frequency adjustment. Too little capacitance can stop oscillation or prevent reliable starting. The Q of crystals is so high that marginal oscillators can take several seconds to start!

This circuit can be improved by reducing the crystal's driving impedance with an emitter follower as in **Fig 10.25**. This is the *Butler* oscillator. Again the tank controls the mode to either force the wanted overtone or protect the fundamental mode. The tank need not be tapped because Q2 provides current gain, although the circuit is sometimes seen with C split, driving Q2 from a tap. The position between the emitters offers a good, low-impedance environment to keep the crystal's in-circuit Q

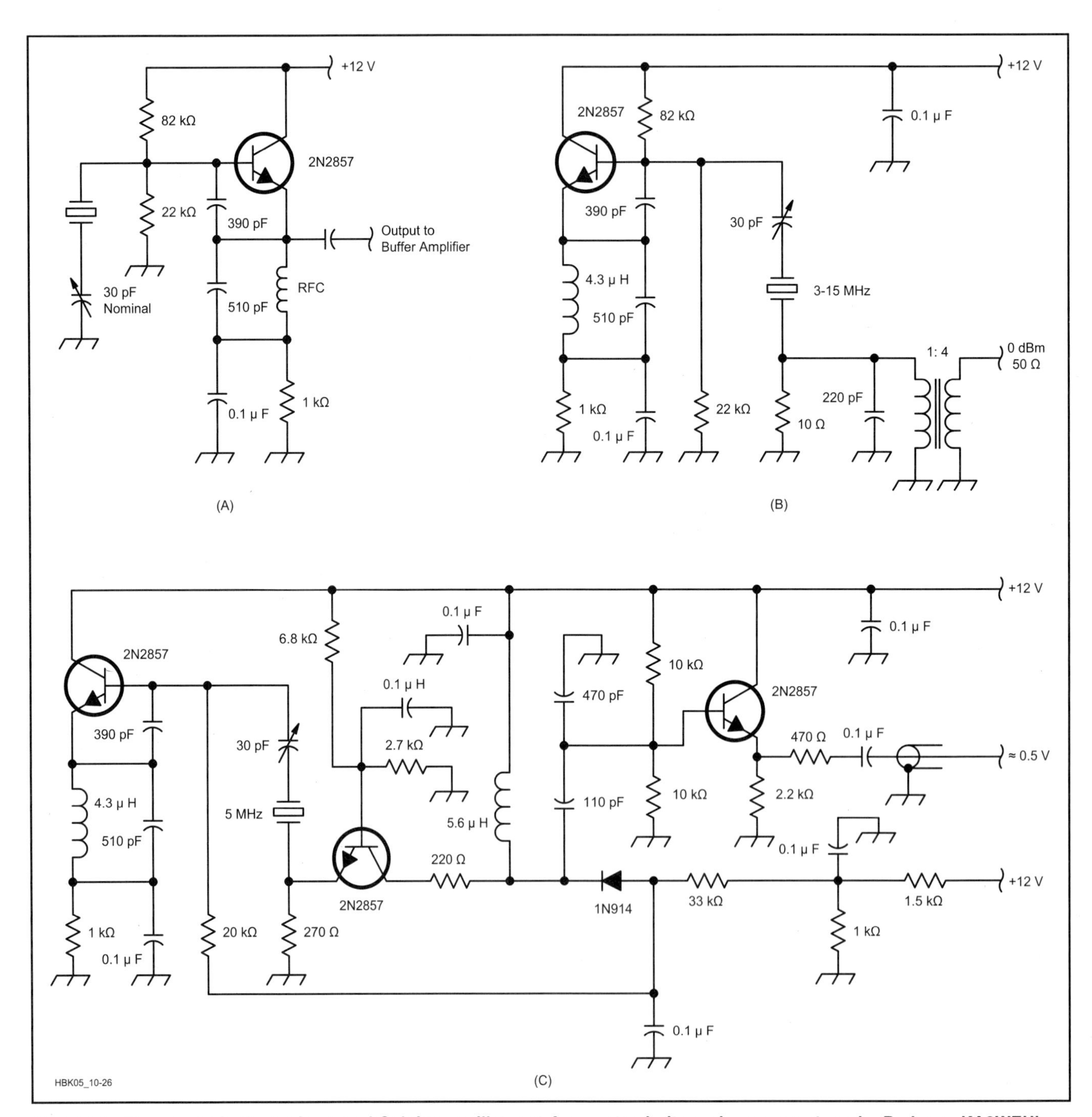

Fig 10.26—The crystal in the series-tuned Colpitts oscillator at A operates in its series-resonant mode. B shows KA2WEU's low-noise version, which uses the crystal as a filter and features high harmonic suppression. The circuit at C builds on the B version by adding a common-base output amplifier and ALC loop.

high. R, in the emitter of Q1, is again selected to give reliable oscillation. The circuit has been shown with a capacitive load for the crystal, to suit a unit specified for a 30-pF load. An alternative circuit to give electrical fine tuning is also shown. The diodes across the tank act as limiters to stabilize the operating amplitude and limit the power dissipated in the crystal by clipping the drive voltage to Q2. The tank should be adjusted to peak at the operating frequency, not used to trim the frequency. The capacitance in series with the crystal is the proper frequency trimmer.

The Butler circuit works well, and has been used in critical applications to 140 MHz (seventh-overtone crystal, 2N5179 transistor). Although the component count is high, the extra parts are cheap ones. Increasing the capacitance in series with the crystal reduces the oscillation frequency but has a progressively diminishing effect. Decreasing the capacitance pulls the frequency higher, to a point at which oscillation stops; before this point is reached, start-up will become unreliable. The possible amount of adjustment, called *pulling range*, depends on the crystal; it can range from less than ten to several hundred parts per million. Overtone crystals have much less pulling range than fundamental crystals on the same frequency; the reduction in pulling is roughly proportional to the square of the overtone number.

Low-Noise Crystal Oscillators

Fig 10.26A shows a crystal operating in its series mode in a series-tuned Colpitts circuit. Because it does not include an LC tank to prevent operation on unwanted modes, this circuit is intended for fundamental mode operation only and relies on that mode being the most active. If the crystal is ordered for 30-pF loading, the frequency trimming capacitor can be adjusted to compensate for the loading of the capacitive divider of the Colpitts circuit. An unloaded crystal without a trimmer would operate slightly off the exact series resonant frequency in order to create an inductive impedance to resonate with the divider capacitors. Ulrich Rohde, KA2WEU, in Fig 4-47 of his book *Digital PLL Frequency Synthesizers—Theory and Design*, published an elegant alternative method of extracting an output signal from this type of circuit, shown in Fig 10.26B. This taps off a signal from the current in the crystal itself. This can be thought of as using the crystal as a band-pass filter for the oscillator output. The RF choke in the emitter keeps the emitter bias resistor from loading the tank and degrading the Q. In this case (3-MHz operation), it has been chosen to resonate close to 3 MHz with the parallel capacitor (510 pF) as a means of forcing operation on the wanted mode. The 10-Ω resistor and the transformed load impedance will reduce the in-circuit Q of the crystal, so a further development substituted a common base amplifier for the resistor and transformer. This is shown in Fig 10.26C. The common-base amplifier is run at a large quiescent current to give a very low input impedance. Its collector is tuned to give an output with low harmonic content and an emitter follower is used to buffer this from the load. This oscillator sports a simple ALC system, in which the amplified and rectified signal is used to reduce the bias voltage on the oscillator transistor's base. This circuit is described as achieving a phase noise level of –168 dBc/Hz a few kilohertz out from the carrier. This may seem far beyond what may ever be needed, but frequency multiplication to high frequencies, whether by classic multipliers or by frequency synthesizers, multiplies the deviation of any FM/PM sidebands as well as the carrier frequency. This means that phase noise worsens by 20 dB for each tenfold multiplication of frequency. A clean crystal oscillator and a multiplier chain is still the best way of generating clean microwave signals for use with narrow-band modulation schemes.

It has already been mentioned that overtone crystals are much harder to pull than fundamental ones. This is another way of saying that overtone crystals are less influenced by their surrounding circuit, which is helpful in a frequency-standard oscillator like this one. Even though 5 MHz is in the main range of fundamental-mode crystals and this circuit will work well with them, an overtone crystal has been used. To further help stability, the power dissipated in the crystal is kept to about 50 µW. The common-base stage is effectively driven from a higher impedance than its own input impedance, under which conditions it gives a very low noise figure.

VXOs

Some crystal oscillators have frequency trimmers. If the trimmer is replaced by a variable capacitor as a front-panel control, we have a *variable crystal oscillator* (*VXO*): a crystal-based VFO with a narrow tuning range, but good stability and noise performance. VXOs are often used in small, simple QRP transmitters to tune a few kilohertz around common calling frequencies. Artful constructors, using optimized circuits and components, have achieved 1000-ppm tuning ranges. Poor-quality "soft" crystals are more pullable than high-Q ones. Overtone crystals are not suited to VXOs. For frequencies beyond the usual limit for fundamental mode crystals, use a fundamental unit and frequency multipliers.

ICOM and Mizuho made some 2-m SSB transceivers based on multiplied VXO local oscillators. This system is simple and can yield better performance than many expensive synthesized radios. SSB filters are available at 9 or 10.7 MHz, to yield sufficient image rejection with a single conversion. Choice of VXO frequency depends on whether the LO is to be above or below signal frequency and how much multiplication can be tolerated. Below 8 MHz multiplier filtering is difficult. Above 15 MHz, the tuning range per crystal narrows. A 50-200 kHz range per crystal should work with a modern front-end design feeding a good 9-MHz IF, for a

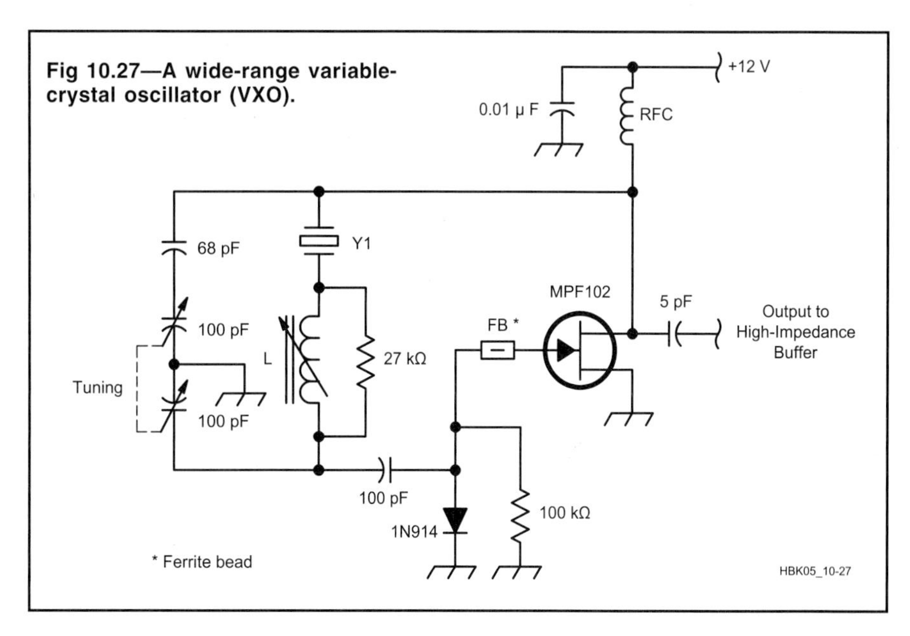

Fig 10.27—A wide-range variable-crystal oscillator (VXO).

contest quality 2-m SSB Receiver.

The circuit in **Fig 10.27** is a JFET VXO from Wes Hayward, W7ZOI, and Doug DeMaw, W1FB, optimized for wide-range pulling. Published in *Solid State Design for the Radio Amateur*, many have been built and its ability to pull crystals as far as possible has been proven. Ulrich Rohde, KA2WEU, has shown that the diode arrangement as used here to make signal-dependent negative bias for the gate confers a phase-noise disadvantage, but oscillators like this that pull crystals as far as possible need any available means to stabilize their amplitude and aid start-up. In this case, the noise penalty is worth paying. This circuit can achieve a 2000-ppm tuning range with amenable crystals. If you have some overtone crystals in your junk box whose fundamental frequency is close to the wanted value, they are worth trying.

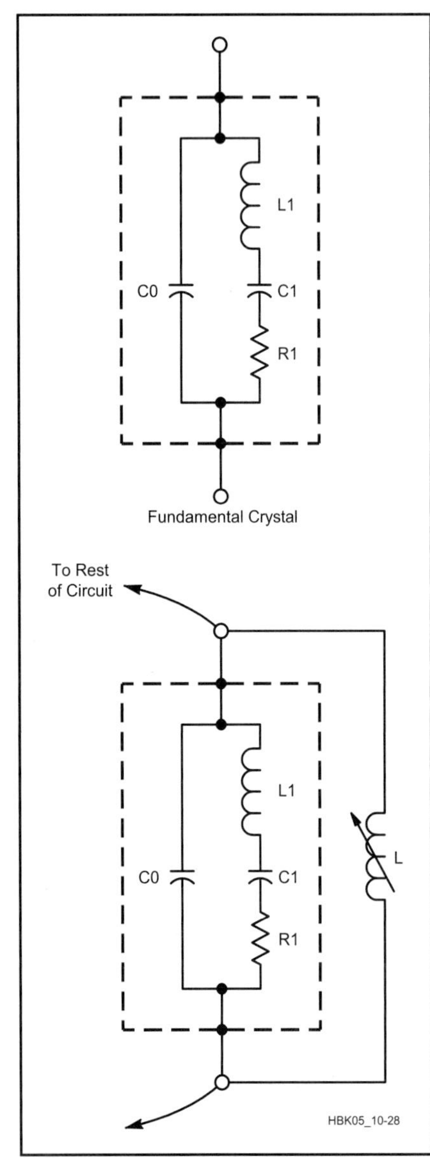

Fig 10.28—Using an inductor to "tune out" C0 can increase a crystal oscillator's pulling range.

This sort of circuit doesn't necessarily stop pulling at the extremes of the possible tuning range, sometimes the range is set by the onset of undesirable behavior such as jumping mode or simply stopping oscillating. L was a 16-µH slug-tuned inductor for 10-MHz operation. It is important to minimize the stray and interwinding capacitance of L since this dilutes the range of impedance presented to the crystal.

One trick that can be used to aid the pulling range of oscillators is to tune out the C0 of the equivalent circuit with an added inductor. **Fig 10.28** shows how. L is chosen to resonate with C0 for the individual crystal, turning it into a high-impedance parallel-tuned circuit. The Q of this circuit is orders of magnitude less than the Q of the true series resonance of the crystal, so its tuning is much broader. The value of C0 is usually just a few picofarads, so L has to be a fairly large value considering the frequency it is resonated at. This means that L has to have low stray capacitance or else it will self-resonate at a lower frequency. The tolerance on C0 and the variations of the stray C of the inductor means that individual adjustment is needed. This technique can also work wonders in crystal ladder filters.

Logic-Gate Crystal Oscillators

A 180° phase-shift network and an inverting amplifier can be used to make an oscillator. A single stage RC low-pass network cannot introduce more than 90° of phase shift and only approaches that as the signal becomes infinitely attenuated. Two stages can only approach 180° and then have immense attenuation. It takes three stages to give 180° and not destroy the loop gain. **Fig 10.29A** shows the basic form of the *phase-shift* oscillator; Fig 10.29B is an example the phase-shift oscillator commonly used in commercial Amateur Radio transceivers as an audio sidetone oscillator. The frequency-determining network of an RC oscillator has a Q of less than one, which is a massive disadvantage compared to an LC or crystal resonator. The Pierce crystal oscillator is a converted phase-shift oscillator, with the crystal taking the place of one series resistor, Fig 10.29C. At the exact series resonance, the crystal looks resistive, so by suitable choice of the capacitor values, the circuit can be made to oscillate. The crystal has a far steeper phase/frequency relationship than the rest of the network, so the crystal is the dominant controller of the frequency. The Pierce circuit is rarely seen in this full form. Instead, a cut-down version has become the most common circuit for crystal-clock oscillators for digital systems. Fig 10.29D shows this minimalist Pierce, using a logic inverter as the amplifier. R_{bias} provides dc negative feedback to bias the gate into its linear region. At first sight it appears that this arrangement should not oscillate, but the crystal is a resonator (not a simple resistor) and oscillation occurs offset from the series resonance, where the crystal appears inductive, which makes up the missing phase shift. This is one circuit that *cannot* oscillate exactly on the crystal's series resonance. This circuit is included in many microprocessors and other digital ICs that need a clock. It is also the usual circuit inside the miniature clock oscillator cans. It is not a very reliable circuit, as operation is dependent on the crystal's equivalent series resistance and the output impedance of the logic gate, occasionally the logic device or the crystal needs to be changed to start oscillation, sometimes playing with the capacitor values is necessary. It is doubtful whether these circuits are ever designed—values (the two capacitors) seem to be arrived at by experiment. Once going, the circuit is reliable, but its drift and noise are moderate, not good—acceptable for a clock oscillator. The commercial packaged oscillators are the same, but the manufacturers have handled the production foibles on a batch-by-batch basis.

RC Oscillators

Plenty of RC oscillators are capable of operating to several megahertz. Some of these are really constant-current-into-capacitor circuits, which are easier to make in silicon. Like the phase-shift oscillator above, the timing circuit Q is less than one, giving very poor noise performance that's unsuitable even to the least demanding radio application. One example is the oscillator section of the CD4046 phase-locked-loop IC. This oscillator has poor stability over temperature, large batch-to-batch variation and a wide variation in its voltage-to-frequency relationship. It is not recommended that this sort of oscillator is used at RF in radio systems. (The '4046 phase detector section is very useful, however, as we'll see later.) These oscillators are best suited to audio applications.

VHF AND UHF OSCILLATORS

A traditional way to make signals at higher frequencies is to make a signal at a lower frequency (where oscillators are easier) and multiply it up to the wanted range. Multipliers are still one of the easiest ways of making a clean UHF/microwave signal. The design of a multiplier depends on whether the multiplication factor is an odd or even number. For odd

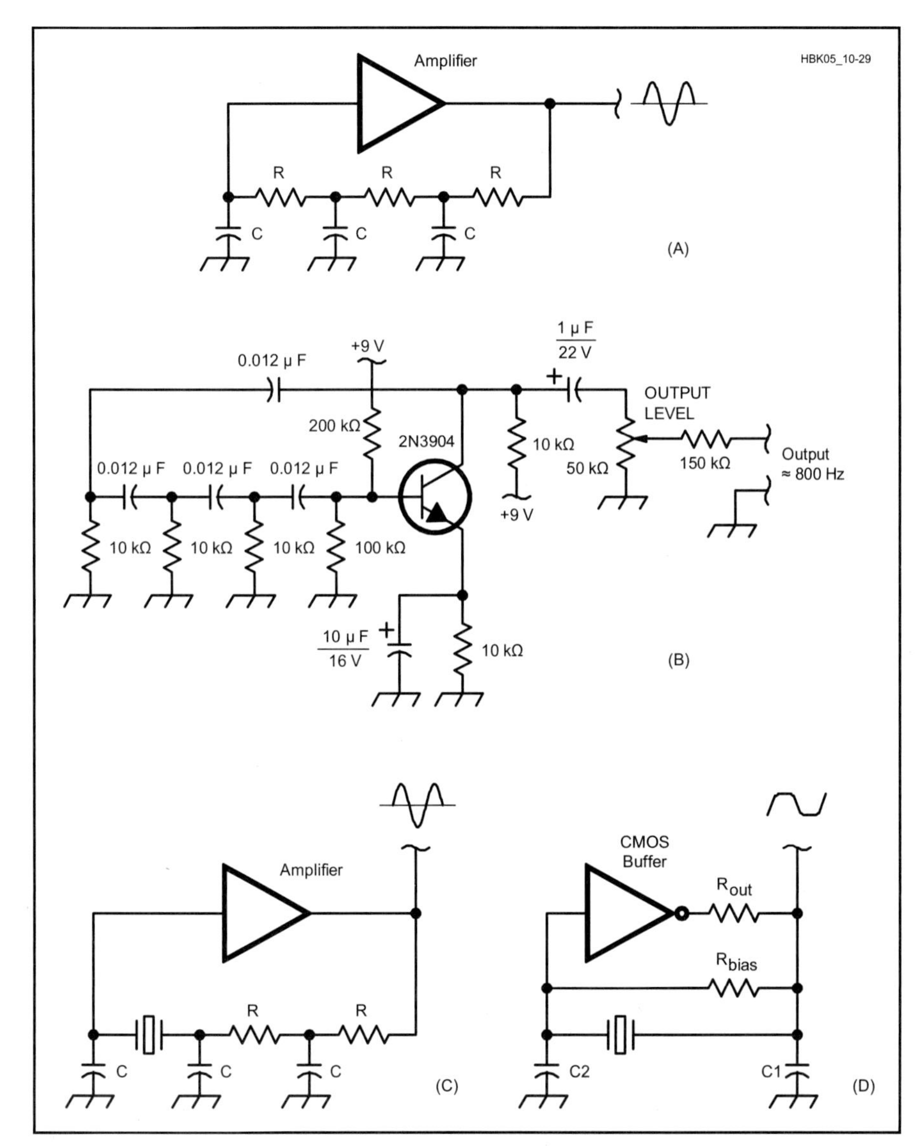

Fig 10.29—In a phase-shift oscillator (A) based on logic gates, a chain of RC networks—three or more—provide the feedback and phase shift necessary for oscillation, at the cost of low Q and considerable loop loss. Many commercial Amateur Radio transceivers have used a phase-lead oscillator similar to that shown at B as a sidetone generator. Replacing one of the resistors in A with a crystal produces a Pierce oscillator (C), a cut-down version of which (D) has become the most common clock oscillator configuration in digital systems.

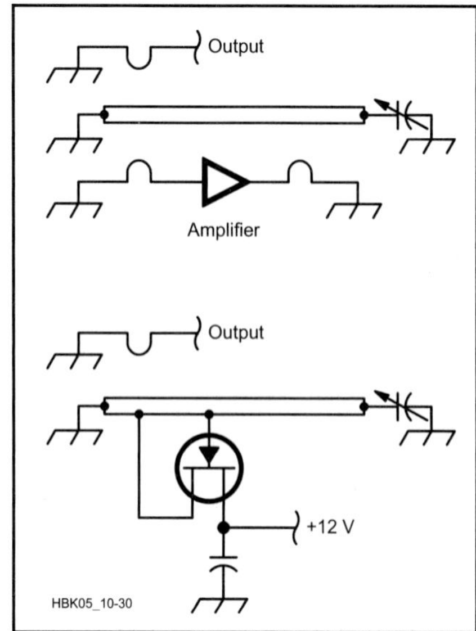

Fig 10.30—Oscillators that use transmission-line segments as resonators. Such oscillators are more common than many of us may think, as Fig 10.31 reveals.

multiplication, a Class-C biased amplifier can be used to create a series of harmonics; a filter selects the one wanted. For even multiplication factors, a full-wave-rectifier arrangement of distorting devices can be used to create a series of harmonics with strong even-order components, with a filter selecting the wanted component. At higher frequencies, diode-based passive circuits are commonly used. Oscillators using some of the LC circuits already described can be used in the VHF range. At UHF different approaches become necessary.

Fig 10.30 shows a pair of oscillators based on a resonant length of line. The first one is a return to basics: a resonator, an amplifier and a pair of coupling loops. The amplifier can be a single bipolar or FET device or one of the monolithic microwave integrated circuit (MMIC) amplifiers. The second circuit is really a Hartley, and one was made as a test oscillator for the 70-cm band from a 10-cm length of wire suspended 10 mm over an unetched PC board as a ground plane, bent down and soldered at one end, with a trimmer at the other end. The FET was a BF981 dual-gate device used as a source follower.

No free-running oscillator will be stable enough on these bands except for use with wideband FM or video modulation and AFC at the receiver. Oscillators in this range are almost invariably tuned with tuning diodes controlled by phase-locked-loop synthesizers, which are themselves controlled by a crystal oscillator.

There is one extremely common UHF oscillator that is rarely applied intentionally. The answer to this riddle is a configuration that is sometimes deliberately built as a useful wide-tuning oscillator covering say, 500 MHz to 1 GHz—and is also the modus operandi of a very common form of *spurious* VHF/UHF oscillation in circuitry intended to process lower-frequency signals! This oscillator has no generally accepted name. It relies on the creation of a small negative resistance in series with a series resonant LC tank. **Fig 10.31A** shows the circuit in its simplest form. This circuit is well-suited to construction with printed-circuit inductors. Common FR4 glass-epoxy board is lossy at these frequencies; better performance can be achieved by using the much more expensive glass-Teflon board. If you can get surplus offcuts of this type of material, it has many uses at UHF and microwave, but it is difficult to use, as the adhesion between the copper and the substrate can be poor. A high-UHF transistor with a 5-GHz f_T like the BFR90 is suitable; the base inductor can be 30 mm of 1-mm trace folded into a hairpin shape (inductance, less than 10 nH). Analyzing this circuit using a comprehensive model of the UHF transistor reveals that the emitter presents an impedance that is small, resistive and negative to the outside world. If this is large enough to more than cancel the effective series resistance of the emitter tank, oscillation will occur. Fig 10.31B shows a very basic emitter-follower circuit with some capacitance to ground on both the input and output. If the capacitor shunting the input is a distance away from the transistor, the trace can look like an inductor—and small capacitors at audio and low RF can look like very good decouplers at hundreds of MHz. The length to the capacitor shunting the output will behave as a series resonator at a frequency where it is a 1/4 wavelength long. This circuit is the same as in Fig 10.31A; it, too, can oscillate. The semiconductor manufacturers have steadily improved their small-signal transistors to give better gain and bandwidth so that any transistor circuit where all three electrodes find themselves decoupled to ground at UHF may oscillate at several hundred megahertz. The upshot of this is that there is no longer any branch of electronics where RF design and layout techniques can be safely ignored. A circuit must not just be designed to do what it *should* do, it must also be designed so that it cannot do what it *should not* do.

There are three ways of taming such a circuit; adding a small resistor, perhaps 50 to 100 Ω in the collector lead, close to the transistor, or adding a similar resistor in the base lead, or by fitting a ferrite bead over the base lead under the transistor. The resistors can disturb dc conditions, depending on the circuit and its operating currents. Ferrite beads have the advantage that they can be easily added to existing equipment and have no effect at dc and low frequencies. Beware of some electrically conductive ferrite materials that can short transistor leads. If an HF oscillator uses beads to prevent any risk of spurs (Fig 10.16), the beads should be anchored with a spot of adhesive to prevent movement that can cause small frequency shifts. Ferrite beads of Fair-Rite no. 43 material are especially suitable for this purpose; they are specified in terms of impedance, not inductance. Ferrites at frequencies above their normal usable range become very lossy and can make a lead look not inductive, but like a few tens of ohms, resistive.

Microwave Oscillators

Low-noise microwave signals are still best made by multiplying a very-low-noise HF crystal oscillator, but there are a number of oscillators that work directly at microwave frequencies. Such oscillators can be based on resonant lengths of stripline or microstrip, and are simply scaled-down versions of UHF oscillators, using microwave transistors and printed striplines on a low-loss substrate, like alumina. Techniques for printing metal traces

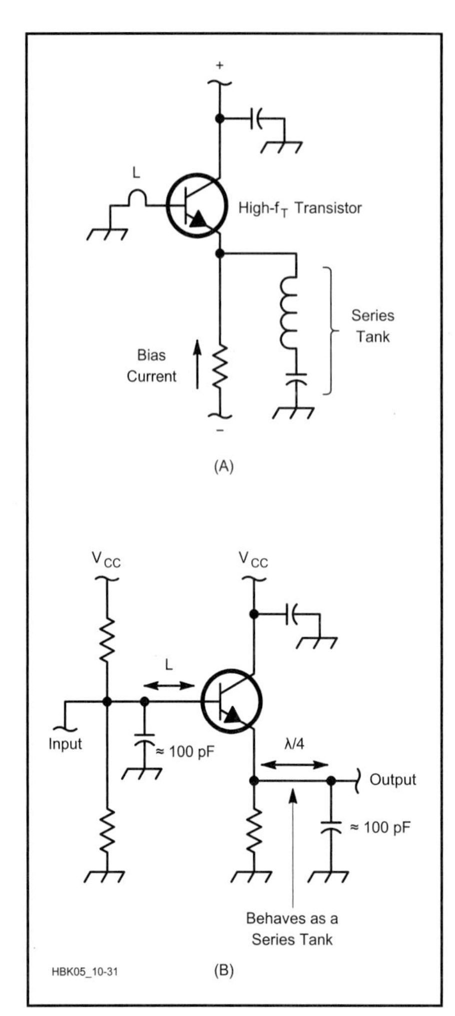

Fig 10.31—High device gain at UHF and resonances in circuit board traces can result in spurious oscillations even in non-RF equipment.

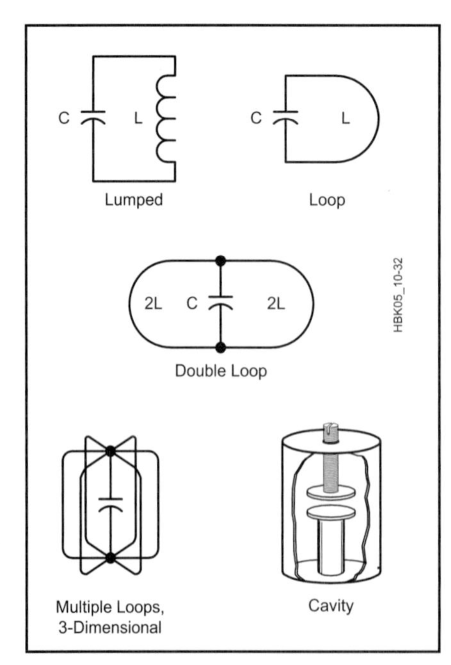

Fig 10.32—Evolution of the cavity resonator.

on substrates to form filters, couplers, matching networks and so on, have been intensively developed over the past two decades. Much of the professional microwave community has moved away from waveguide and now uses low-loss coaxial cable with a solid-copper shield—*semi-rigid cable*—to connect circuits made flat on ceramic or Teflon based substrates. Semiconductors are often bonded on as unpackaged chips, with their bond-wire connections made directly to the traces on the substrate. At lower microwave frequencies, they may be used in standard surface-mount packages. From an Amateur Radio viewpoint, many of the processes involved are not feasible without access to specialized furnaces and materials. Using ordinary PC-board techniques with surface-mount components allows the construction of circuits up to 4 GHz or so. Above this, structures get smaller and accuracy becomes critical; also PC-board materials quickly become very lossy.

Older than stripline techniques and far more amenable to home construction, cavity-based oscillators can give the highest possible performance. The dielectric constant of the substrate causes stripline structures to be much smaller than they would be in free air, and the lowest-loss substrates tend to have very high dielectric constants. Air is a very low-loss dielectric with a dielectric constant of 1, so it gives high Q and does not force excessive miniaturization. **Fig 10.32** shows a series of structures used by G. R. Jessup, G6JP, to illustrate the evolution of a cavity from a tank made of lumped components. All cavities have a number of different modes of resonance, the orientation of the currents and fields are shown in **Fig 10.33**. The cavity can take different shapes, but that shown has proven to suppress unwanted modes well. The gap need not be central and is often right at the top. A screw can be fitted through the top, protruding into the gap, to adjust the frequency.

To make an oscillator out of a cavity, an amplifier is needed. Gunn and tunnel diodes have regions in their characteristics where their current *falls* with increasing bias voltage. This is negative resistance. If such a device is mounted in a loop in a cavity and bias applied, the negative resistance can more than cancel the effective loss resistance of the cavity, causing oscillation. These diodes are capable of operating at extremely high frequencies and were discovered long before transistors were developed that had any gain at microwave frequencies.

A *Gunn-diode* cavity oscillator is the basis of many of the Doppler-radar modules used to detect traffic or intruders. **Fig 10.34** shows a common configuration. The coupling loop and coax output connector could be replaced with a simple aperture to couple into waveguide or a mixer cavity. **Fig 10.35** shows a transistor cavity oscillator version using a modern microwave transistor. FET or bipolar devices can be used. The two coupling loops are completed by the capacitance of the feedthroughs.

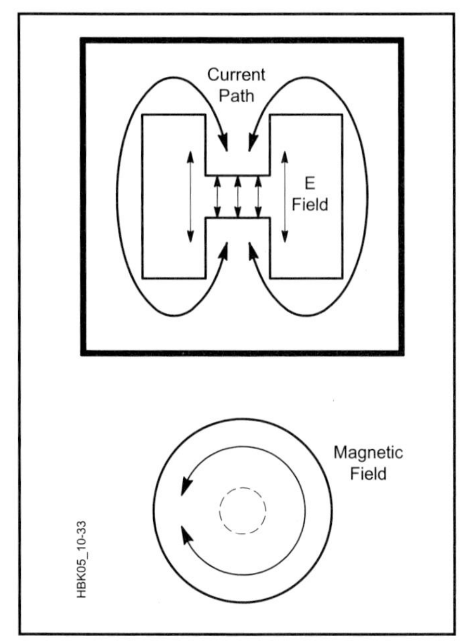

Fig 10.33—Currents and fields in a cavity.

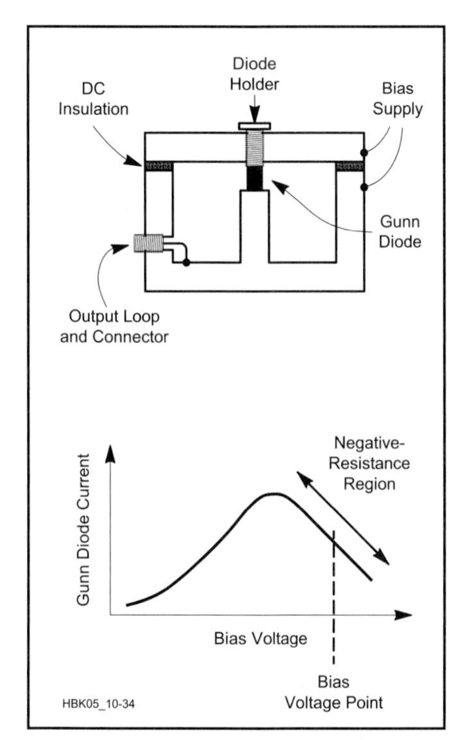

Fig 10.34—A Gunn diode oscillator uses negative resistance and a cavity resonator to produce radio energy.

The *dielectric-resonator oscillator* (*DRO*) may soon become the most common microwave oscillator of all, as it is used in the downconverter of satellite TV receivers. The dielectric resonator itself is a ceramic cylinder, like a miniature hockey puck, several millimeters in diameter. The ceramic has a very high dielectric constant, so the surface (where ceramic meets air in an abrupt mismatch) reflects electromagnetic waves and makes the ceramic body act as a resonant cavity. It is mounted on a substrate and coupled to the active device of the oscillator by a stripline that runs past it. At 10 GHz, a FET made of gallium arsenide (GaAsFET), rather than silicon, is normally used. The dielectric resonator elements are made at frequencies appropriate to mass applications like satellite TV. The set-up charge to manufacture small quantities at special frequencies is likely to be prohibitive for the foreseeable future. The challenge with these devices is to devise new ways of using oscillators on industry standard frequencies. Their chief attraction is their low cost in large quantities and compatibility with microwave stripline (microstrip) techniques. Frequency stability and Q are competitive with good cavities, but are inferior to that achievable with a crystal oscillator and chain of frequency multipliers. Satellite TV downconverters need free running oscillators with less than 1 MHz of drift at 10 GHz.

There are a number of thermionic (vacuum-tube) microwave sources, klystrons, magnetrons and *backwards wave oscillators* (*BWO*s). Available devices are either very old or designed for very high power.

The *yttrium-iron garnet* (*YIG*) oscillator was developed for a wide tuning range as a solid-state replacement for the BWO, and many of them can be tuned over more than an octave. They are complete, packaged units that appear to be a heavy block of metal with low-frequency connections for power supplies and tuning, and an SMA connector for the RF output. The manufacturer's label usually states the tuning range and often the power supply voltages. This is very helpful because, with new units priced in the kilodollar range, it is important to be able to identify surplus units. The majority of YIGs are in the 2- to 18-GHz region, although units down to 500 MHz and up to 40 GHz are occasionally found. In this part of the spectrum there is no octave-tunable device that can equal their cleanliness and stability. A 3-GHz unit drifting less than 1 kHz per second gives an idea of typical stability. This seems very poor—until we realize that this is 0.33 ppm per second. Nevertheless, any YIG application involving narrow-band modulation will usually require some form of frequency stabilizer.

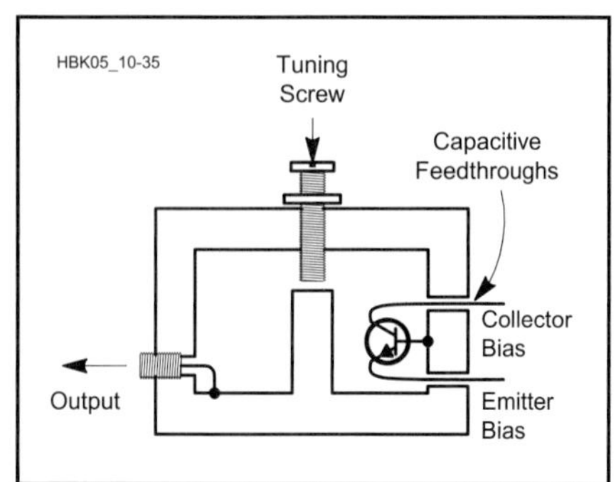

Fig 10.35—A transistor can also directly excite a cavity resonator.

Good quality, but elderly, RF spectrum analyzers have found their way into the workshops of a number of dedicated constructors. A 0- to 1500-MHz analyzer usually uses a 2- to 3.5-GHz YIG as its first local oscillator, and its tuning circuits are designed around it. A reasonable understanding of YIG oscillators will help in troubleshooting and repair.

YIG spheres are resonant at a frequency controlled not only by their physical dimensions, but also by any applied magnetic field. A YIG sphere is carefully oriented within a coupling loop connected to a negative-resistance device and the whole assembly is placed between the poles of an electromagnet. Negative-resistance diodes have been used, but transistor circuits are now common. The support for the YIG sphere often contains a thermostatically controlled heater to reduce temperature sensitivity.

The first problem with a magnetically tuned oscillator is that magnetic fields, especially at low frequencies, are extremely difficult to shield; the tuning will be influenced by any local fields. Varying fields will cause frequency modulation. The magnetic core must be carefully designed to be all-enclosing in an attempt at self-shielding and then one or more nested mu-metal cans are fitted around everything. It is still important to site the unit away from obvious sources of magnetic noise, like power transformers. Cooling fans are less obvious sources of fluctuating magnetic fields, since some are 20 dB worse than a well designed 200-W 50/60-Hz transformer.

The second problem is that the oscillator's internal tuning coils need significant current from the power supply to create strong fields. This can be eased by adding a permanent magnet as a fixed "bias" field, but the bias will shift as the magnet ages. The only solution is to have a coil with many turns, hence high inductance, which will require a high supply voltage to permit rapid tuning, thus increasing the power consumption. The usual compromise is to have dual coils: One with many turns allows slow tuning over a wide range; a second with much fewer turns allows fast tuning or FM over a limited range. The main coil can have a sensitivity in the 20 MHz/mA range; and the "FM" coil, perhaps 500 kHz/mA.

The frequency/current relationship can have excellent linearity. **Fig 10.36** shows the construction of a YIG oscillator.

For some insight into present and future trends in oscillator applications, see the sidebar beginning on page 52.

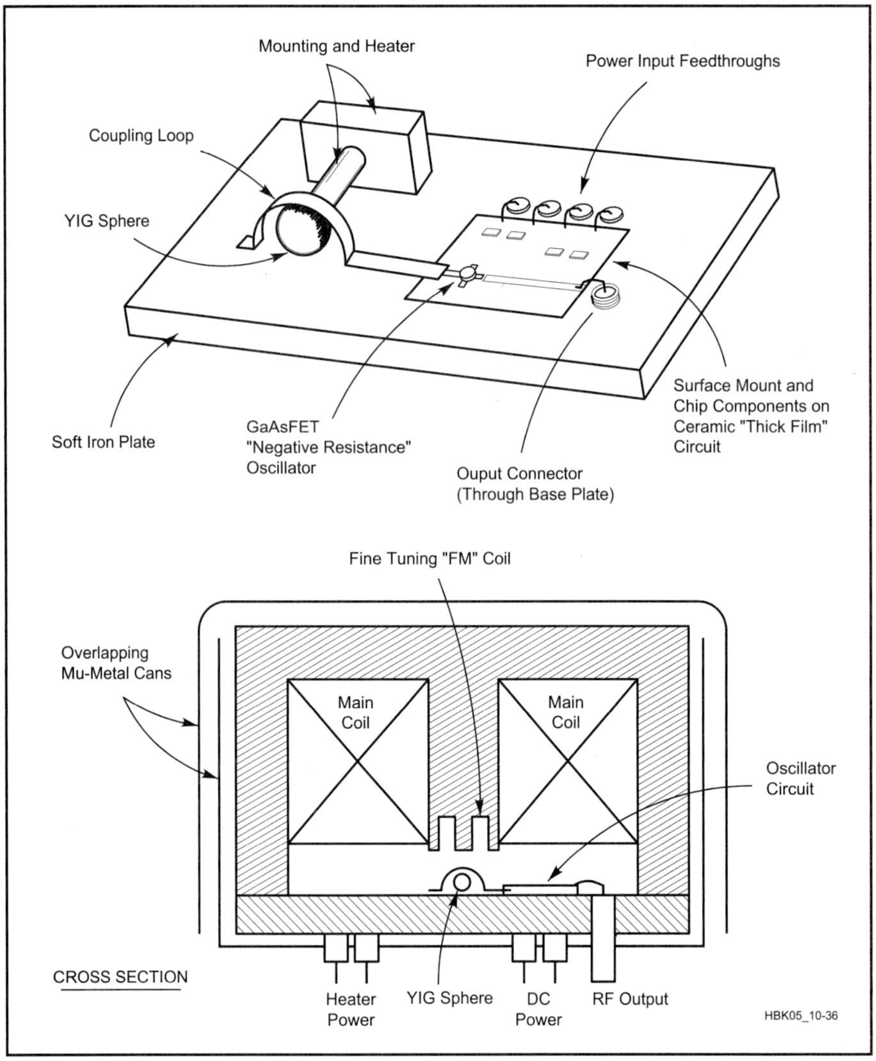

Fig 10.36—A yttrium-iron-garnet (YIG) sphere serves as the resonator in the sweep oscillators used in many spectrum analyzers.

Frequency Synthesizers

Like many of our modern technologies, the origins of Frequency Synthesis can be traced back to WW II. The driving force was the desire for stable, rapidly switchable and accurate frequency control technology to meet the demands of narrowband, frequency-agile HF communications systems without resorting to large banks of switched crystals. Early synthesizers were cumbersome and expensive, and therefore their use was limited to the most sophisticated communications systems. With the help of the same technologies that have taken computers from "rooms" to the palms of our hands, the role of frequency synthesis has gone far beyond its original purpose and has become one of the most enabling technologies in modern communications equipment.

Just about every communications device manufactured today, be it a handheld transceiver, cell phone, pager, AM/FM entertainment radio, scanner, television, HF communications equipment, or test equipment contains a synthesizer. Synthesis is the technology that allows an easy interface with both computers and microprocessors. It provides amateurs with many desirable features, such as the feel of an analog knob with 10-Hz frequency increments, accuracy and stability determined by a single precision crystal oscillator, frequency memories, and continuously variable splits. Now reduced in size to only a few small integrated circuits, frequency synthesizers have also replaced the cumbersome chains of frequency multipliers and filters in VHF, UHF and microwave equipment, giving rise to many of the highly portable communications devices we use today. Frequency synthesis has also had a major impact in lowering the cost of modern equipment, as well as reducing manufacturing complexity.

Frequency synthesizers have been categorized in two general types, *direct synthesizers* (not to be confused with direct digital synthesis) and *indirect synthesizers*. The architecture of the direct types consists mainly of multipliers, dividers, mixers, filters and copious amounts of shielding. They are also cumbersome and expensive, and have all but disappeared from the market. The indirect varieties make use of programmable dividers and phase-lock loops, thereby considerably reducing their complexity. There are a number of variations on the indirect theme. They include programmable division, variable or dual modulus division, and fractional N. Other techniques that have been developed since the direct/indirect classification are direct digital synthesis (DDS) and rate multiplier synthesis. Synthesizers used in equipment today are usually a hybrid of the indirect technique and some of the latter developments. There are entire textbooks devoted to treating these various methods in depth, and as such it is not practical to discuss each one in detail in this handbook. We will focus on the indirect approach, based on phase-locked loops. This approach came to dominate frequency synthesis long ago, and is still dominant today.

Phase-Locked Loops

To understand the indirect synthesizer, we need to understand the phase-lock loop. The principle of the *phase-locked loop* (*PLL*) synthesizer is very simple. An oscillator can be built to cover the required frequency range, so what is needed is a system to keep its tuning correct. This is done by continuously comparing the phase of the oscillator to a stable reference, such as a crystal oscillator and using the result to steer the tuning. If the oscillator frequency is too high, the phase of its output will start to lead that of the reference by an increasing amount. This is detected and is used to decrease the frequency of the oscillator. Too low a frequency causes an increasing phase lag, which is detected and used to increase the frequency of the oscillator. In this way, the oscillator is locked into a fixed-phase relationship with the reference, which means that their frequencies are held exactly equal.

The oscillator is now under control, but is locked to a fixed frequency. **Fig 10.37** shows the next step. The phase detector does not simply compare the oscillator frequency with the reference oscillator; both signals have now been passed through frequency dividers. Advances in digital integrated circuits have made frequency dividers up to microwave frequencies (over 10 GHz) commonplace. The divider on the reference path has a fixed division factor, but that in the VCO path is programmable, the factor being entered digitally as a (usually) binary word. The phase detector is now operating at a lower frequency, which is a submultiple of both the reference and output frequencies.

The phase detector, via the loop amplifier, steers the oscillator to keep both its inputs equal in frequency. The reference frequency, divided by M, is equal to the output frequency divided by N. The output frequency equals the reference frequency × N/M. N is programmable (and an integer), so this synthesizer is capable of tuning in steps of F_{ref}/M.

As an example, to make a 2-m radio covering 144 to 148 MHz with a 10.7-MHz IF, we need a local oscillator covering 154.7-158.7 MHz. If the channel spacing is 20 kHz, then F_{ref}/M is 20 kHz, so N is 7735 to 7935. There is a free choice of F_{ref} or M, but division by a round binary number is easiest. Crystals are readily available for 10.24 MHz, and one of these (divided by 512) will give 20 kHz at low cost. ICs are readily available containing most of the circuitry necessary for the reference oscillator (less the crystal), with the programmable divider and the phase detector.

The phase detector in this example compares the relative timing of 5-V pulses (CMOS logic level) at 20 kHz. Inevitably, the phase detector output will contain strong components at 20 kHz and its harmonics in addition to the wanted "steering" signal. The loop filter must block these unwanted signals; otherwise the loop amplifier will amplify them and apply them to the VCO, generating unwanted FM. No filtering can be perfect and the VCO is very sensitive, so most synthesizers have measurable sidebands spaced at the phase detector operating frequency (and harmonics) away from the carrier. These are called *reference-frequency sidebands*. (Exactly which is the reference frequency is ambiguous: Do we mean the frequency of the reference oscillator, or the frequency applied to the reference input of the phase detector? You must look carefully at context whenever *reference frequency* is mentioned.)

The loop filter is not usually built as a single block of circuitry, it is often made up of three areas: Some components directly after the phase detector, a shaped frequency response in the loop amplifier and some more components between the loop amplifier and the VCO.

The PLL is like a feedback amplifier, although the "signal" around its loop is represented by frequency in some places, by phase in others and by voltage in others. Like any feedback amplifier, there is the risk of instability and oscillations traveling around the loop, which can be seen as massive FM on the output and a strong ac signal on the tuning line to the VCO. The loop's filtering and gain has to be designed to prevent this.

The following example illustrates the loop action: Imagine that we shift the reference oscillator by 1 Hz. The reference applied to the phase detector would shift 1/M Hz, and the loop would respond by shifting the output frequency to produce a matching shift at the other phase detector port, so the output is shifted by N/M Hz. Imagine now that we apply a very small amount of FM to the reference oscillator. The amount of deviation will be amplified by N/M—but this is only true for low modulating frequencies. If the modulating frequency is increased, eventually the loop filter starts to reduce the gain around the loop, the loop ceases to track the modulation and the deviation at the output falls. This is referred to as the *closed-loop frequency response* or *closed-loop bandwidth*. Poorly designed loops can have poor closed-loop responses. For example, the gain can have large peaks above N/M at some reference modulation frequencies, indicating marginal stability and excessively amplifying any noise at those off-

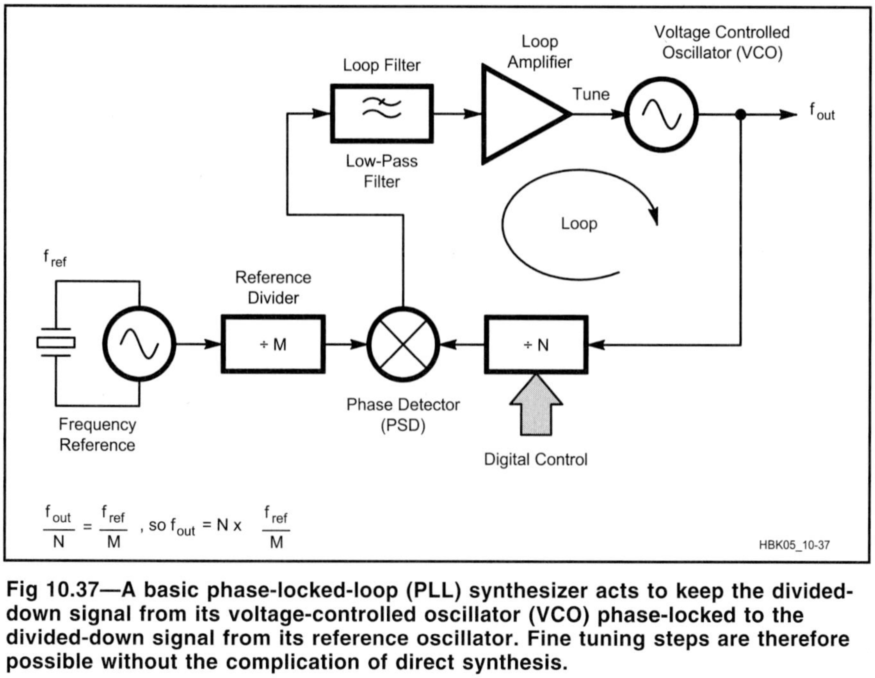

$$\frac{f_{out}}{N} = \frac{f_{ref}}{M}, \text{ so } f_{out} = N \times \frac{f_{ref}}{M}$$

HBK05_10-37

Fig 10.37—A basic phase-locked-loop (PLL) synthesizer acts to keep the divided-down signal from its voltage-controlled oscillator (VCO) phase-locked to the divided-down signal from its reference oscillator. Fine tuning steps are therefore possible without the complication of direct synthesis.

sets from the carrier. The design of the loop filtering and the gain around the loop sets the closed-loop performance.

In a single-loop synthesizer, there is a trade-off between how fine the step size can be versus the performance of the loop. A loop with a very fine step size runs the phase detector at a very low frequency, so the loop bandwidth has to be kept very, very low to keep the reference-frequency sidebands low. Also, the reference oscillator usually has much better phase-noise performance than the VCO, and (within the loop bandwidth) the loop acts to oppose the low-frequency components of the VCO phase-noise sidebands. This very useful cleanup activity is lost when loops have to be narrowed in bandwidth to allow narrow steps. Low-bandwidth loops are slow to respond—they exhibit overly long *settling times*—to the changes in N necessary to change channels. Absolutely everything seems to become impossible in any attempt to get fine resolution from a phase-locked loop.

Single-loop synthesizers are okay for the 2-m FM example for a number of reasons: Channel spacings in this band are not too small, FM is not as critical of phase noise as other modes, 2-m FM is rarely used for weak-signal DXing and the channelization involved does away with the desirability of simulating fast, smooth, continuous tuning. None of these excuses apply to MF/HF radios, however, and ways to circumvent the problems are needed.

A clue was given earlier, by carefully referring to *single*-loop synthesizers. It's possible to use *several* PLLs, or for that matter, some of the other mentioned techniques such as DDS, fractional N, variable- or dual-modulus etc, as components in a larger structure, dividing the frequency of one loop and adding it to the frequency of another that has not been divided. This represents a form of hybrid between the old direct synthesizer and the PLL. A form of PLL containing a mixer in place of the programmable divider can be used to perform the frequency addition.

Let us continue our discussion of phase-lock loop synthesizers by examining the role of each of the component pieces of the system. They are, the VCO, the dividers, prescalers, the phase detector and the loop-compensation amplifier.

Voltage-Controlled Oscillators

Voltage-controlled oscillators are commonly referred to as *VCOs*. (Voltage-tuned oscillator, VTO, more accurately describes the circuitry most commonly used in VCOs, but tradition is tradition! An exception to this is the YIG oscillator, described earlier as sometimes used in UHF and microwave PLLs, which is *current*-tuned.) In all the oscillators described so far, except for permeability tuning, the frequency is controlled or trimmed by a variable capacitor. These are modified for voltage tuning by using a tuning diode. The oscillator section of this chapter shows the schematic symbol and construction detail of a voltage-controlled or *varactor* (tuning) diode as is typically used in a VCO (see Fig 10.18).

When this oscillator is used in a PLL, the PLL adjusts the tuning voltage to completely cancel any drift in the oscillator (provided it does not drift further than it can be tuned) regardless of its cause, so no special care is needed to compensate a PLL VCO for temperature effects. Adjusting the inductor core will not change the frequency; the PLL will adjust the tuning voltage to hold the oscillator on frequency. This adjustment is important, though. It's set so that the tuning voltage neither gets too close to the maximum available, nor too low for acceptable Q, as the PLL is tuned across its full range.

In a high-performance, low noise synthesizer, the VCO may be replaced by a bank of switched VCOs, each covering a section of the total range needed, each VCO having better Q and lower noise because the lossy diodes constitute only a smaller part of the total tank capacity than they would in a single, full-range oscillator. Another method uses tuning diodes for only part of the tuning range and switches in other capacitors (or inductor taps) to extend the range. The performance of wide-range VCOs can be improved by using a large number of varactors in parallel in a high-C, low-L tank.

Programmable Dividers

Designing your own programmable divider is now a thing of the past, as complete systems have been made as single ICs for over 10 years. The only remaining reason to do so is when an ultra-low-noise synthesizer is being designed.

It may seem strange to think of a digital counter as a source of random noise, but digital circuits have propagation delays caused by analog effects, such as the charging of stray capacitances. Digital circuits are composed of the same sorts of components as analog circuits; there are no such things as digital and analog electrons! Signals are made of voltages and currents. The prime difference between analog and digital electronics lies in the different ways meaning is assigned to the magnitude of a signal. The differences in circuit design are consequences of this, not causes.

Thermal noise in the components of a digital circuit, added to the signal voltage will slightly change the times at which thresholds are crossed. As a result, the output of a digital circuit will have picked up some timing *jitter*. Jitter in one or both of the signals applied to a phase detector can be viewed as low-level random phase modulation. Within the loop bandwidth, the phase detector steers the VCO so as to oppose and cancel it, so phase jitter or noise is applied to the VCO in order to cancel out the jitter added by the divider. This makes the VCO noisier. The noise performance of the high-speed CMOS logic normally used in programmable dividers is good. For ultra-low-noise dividers, ECL devices are sometimes used.

As shown in **Fig 10.38A**, a programmable divider consists of a loadable counter and some control circuitry. The counter is designed to count downward. The programmed division factor is loaded into the counter, and the incoming frequency clocks the downward counting. When the count reaches zero, the counter is quickly reloaded with the division factor before the next clock edge. The maximum frequency of operation is limited by the minimum time needed to reliably perform the loading operation. One improvement is to have a circuit to recognize a state a few clock cycles before the zero count is reached, so the reload can be performed at a more leisurely pace during those cycles. This imposes a minimum division ratio, but that is rarely a problem. An output pulse is given during the loading cycle. In this way a frequency is divided by the number loaded.

The maximum input frequency that can be handled is set by the speed of available logic. The synchronous reset cycle forces the entire divider to be made of equally fast logic. Just adding a fast *prescaler*—a fixed-ratio divider ahead of the programmable one—will increase the maximum frequency that can be used, but it also scales up the loop step size. Equal division has to be added to the reference input of the phase detector to restore the step size, reducing the phase detector's operating frequency. The loop bandwidth has to be reduced to restore the filtering needed to suppress reference frequency sidebands. This makes the loop much slower to change frequency and degrades the noise performance.

Variable or Dual-Modulus Prescalers

A plain programmable divider for VHF use would need to be built from ECL devices. It would be expensive, hot and power-hungry. The idea of a fast front end

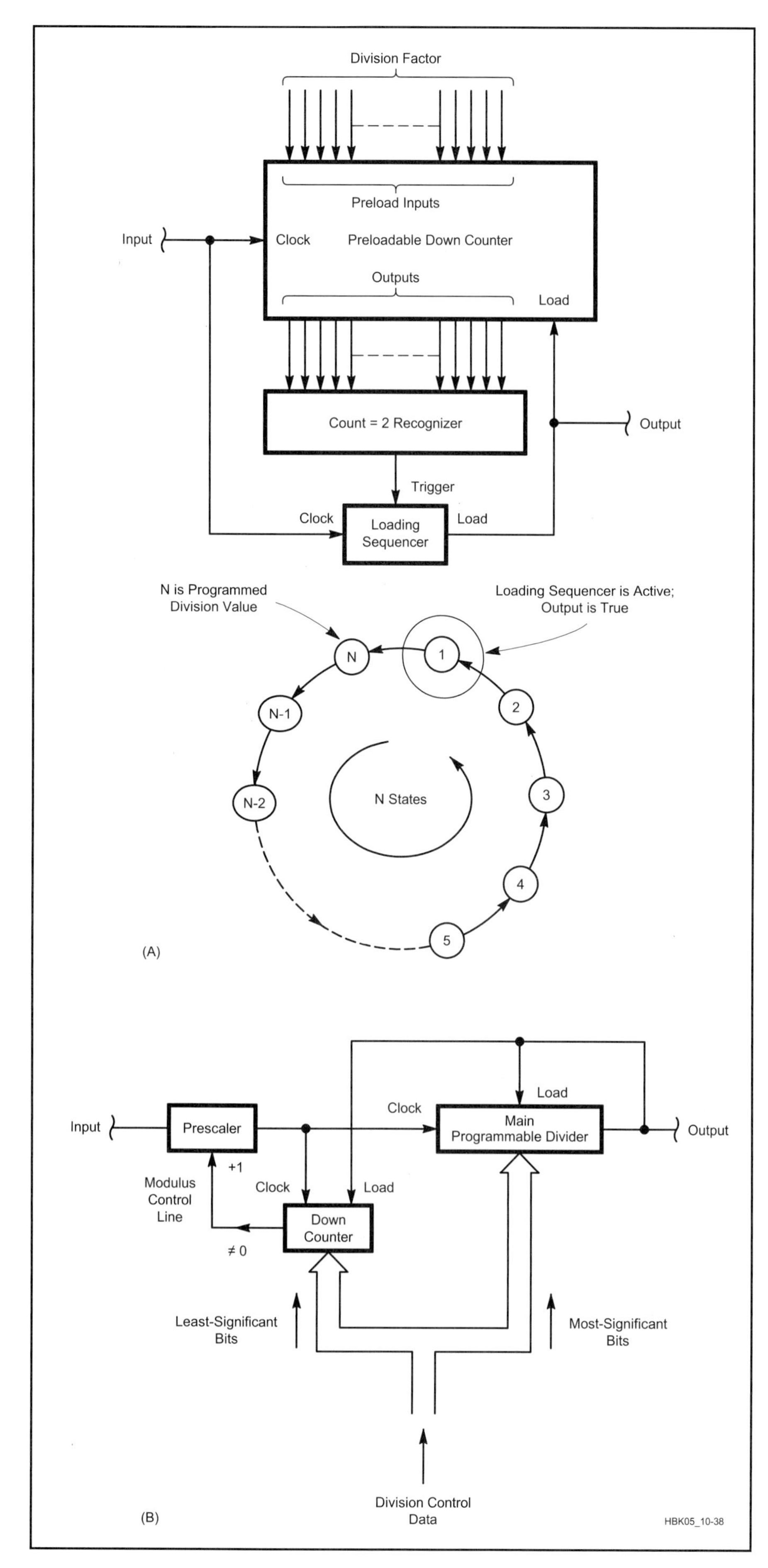

Fig 10.38—(A) Shows the mechanism of a programmable frequency divider. (B) shows the function of a dual-modulus prescaler. The counter is reloaded with N when the count reaches 0 or 1, depending on the sequencer action.

ahead of a CMOS programmable divider would be perfect if these problems could be circumvented. Consider a fast divide-by-ten prescaler ahead of a programmable divider where division by 947 is required. If the main divider is set to divide by 94, the overall division ratio is 940. The prescaler goes through its cycle 94 times and the main divider goes through its cycle once, for every output pulse. If the prescaler is changed to divide by 11 for 7 of its cycles for every cycle of the entire divider system, the overall division ratio is now $[(7 \times 11) + (87 \times 10)] = 947$. At the cost of a more elaborate prescaler and the addition of a slower programmable counter to control it, this prescaler does not multiply the step size and avoids all the problems of fixed prescaling.

Fig 10.38B shows the general block diagram of a dual-modulus prescaled divider. The down counter controls the modulus of the prescaler. The numerical example, just given, used decimal arithmetic, although binary is now usual. Each cycle of the system begins with the last output pulse having loaded the frequency control word, into both the main divider and the prescaler controller. If the part of the word loaded into the prescaler controller is not zero, the prescaler is set to divide by 1 greater than its normal ratio. Each cycle of the prescaler clocks the down counter. Eventually, it reaches zero, and two things happen: The counter is designed to freeze at zero (and it will remain frozen until it is next reloaded) and the prescaler is switched back to its normal ratio, at which it will remain until the next reload. One way of visualizing this is to think of the prescaler as just being a divider of its normal ratio, but with the ability to "steal" a number of input pulses controlled by the data loaded into its companion down counter. Note that a dual-modulus prescaler system has a *minimum* division ratio, needed to ensure there are enough cycles of the prescaler to allow enough input pulses to be stolen.

Dual-modulus prescaler ICs are widely used and widely available. Devices for use to a few hundred megahertz are cheap, and devices in the 2.5-GHz region are commonly available. Common prescaler IC division ratio pairs are: 8-9, 10-11, 16-17, 32-33, 64-65 and so on. Many ICs containing programmable dividers are available in versions with and without built-in prescaler controllers.

Phase Detectors

A phase detector (PSD) produces an output voltage that depends on the phase relationship between its two input signals. If two signals, *in phase on exactly the same frequency*, are mixed together in a conventional diode-ring mixer with a dc-coupled output port, one of the products is direct current (0 Hertz). If the phase relationship between the signals changes, the mixer's dc output voltage changes. With both signals in phase, the output is at its most positive; with the signals 180° out of phase, the output is at its most negative. When the phase difference is 90° (the signals are said to be *in quadrature*), the output is 0 V.

Applying sinusoidal signals to a phase detector causes the detector's output voltage to vary sinusoidally with phase angle, as in **Fig 10.39A**. This nonlinearity is not a problem, as the loop is usually arranged to run with phase differences close to 90°. What might seem to be a more serious complication is that the detector's phase-voltage characteristic repeats every 180°, not 360°. Two possible input phase differences can therefore produce a given output voltage. This turns out not to be a problem, because the two identical voltage points lie on opposing slopes on the detector's output-voltage curve. One direction of slope gives positive feedback (making the loop unstable and driving the VCO away from what otherwise would be the lock angle) over to the other slope, to the true and stable lock condition.

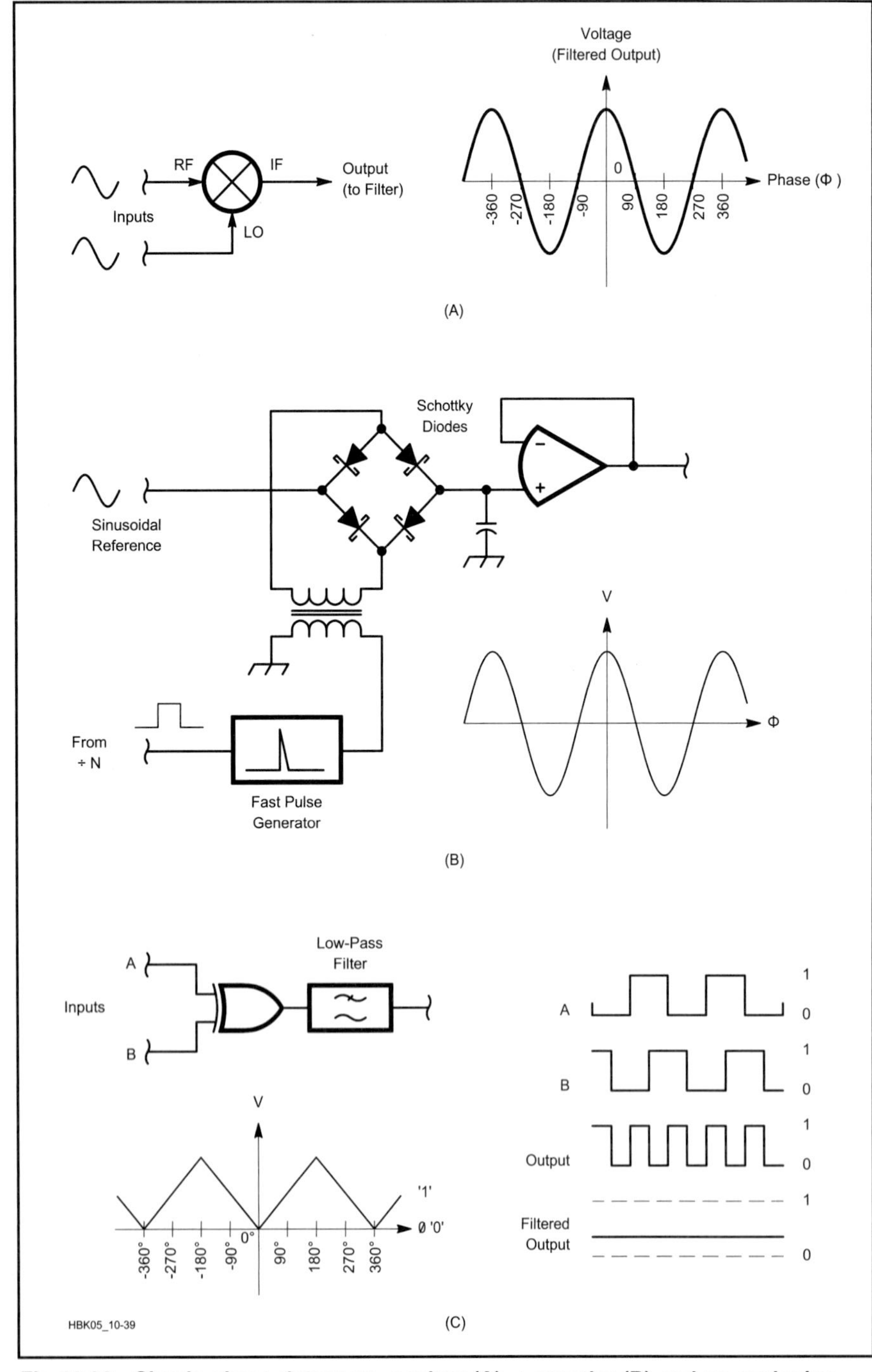

Fig 10.39—Simple phase detectors: a mixer (A), a sampler (B) and an exclusive-OR gate (C).

MD108, SBL-1 and other diode mixers can be used as phase detectors, as can active mixers like the MC1496. Mixer manufacturers make some parts (Mini-Circuits RPD-1 and so on) that are specially optimized for phase-detector service. All these devices can make excellent, low-noise phase detectors but are not commonly used in ham equipment. A high-speed sample-and-hold circuit, based on a Schottky-diode bridge, can form a very low-noise phase detector and is sometimes used in specialized instruments. This is just a variant on the basic mixer; it produces a similar result, as shown in Fig 10.39B.

The most commonly used simple phase detector is just a single exclusive-OR (XOR) logic gate. This circuit gives a logic 1 output only when *one* of its inputs is at logic 1; if both inputs are the same, the output is a logic 0. If inputs A and B in Fig 10.39C are almost in phase, the output will be low most of the time, and its average filtered value will be close to the logic 0 level. If A and B are almost in opposite phase, the output will be high most of the time, and its average voltage will be close to logic 1. This circuit is very similar to the mixer. In fact, the internal circuit of ECL XOR gates is the same transistor tree as found in the MC1496 and similar mixers, with some added level shifting. Like the other simple phase detectors, it produces a cyclic output, but because of the square-wave input signals, produces a triangular output signal. To achieve this circuit's full output-voltage range, it's important that the reference and VCO signals applied operate at a 50% duty cycle.

Phase-Frequency Detectors

All the simple phase detectors described so far are really specialized mixers. If the loop is out of lock and the VCO is far off frequency, such phase detectors give a high-frequency output midway between zero and maximum. This provides no information to steer the VCO towards lock, so the loop remains unlocked. Various solutions to this problem, such as crude relaxation oscillators that start up to sweep the VCO tuning until lock is acquired, or laboriously adjusted "pretune" systems in which a DAC, driven from the divider control data, is used to coarse tune the VCO to within locking range, have been used in the past. Many of these solutions have been superseded, although pretune systems are still used in synthesizers that must change frequency very rapidly.

The *phase-frequency detector* is the usual solution to lock-acquisition problems. It behaves like a simple phase detector over an extended phase range, but its characteristic is not repetitive. Because its output voltage stays high or low, depending on which input is higher in frequency, this PSD can steer a loop towards lock from anywhere in its tuning range. **Fig 10.40** shows the internal logic of the phase-frequency detector in the CD4046 PLL chip. (The CD4046 also contains an XOR PSD). When the phase of one input leads that of the other, one of the output MOSFETs is pulsed on with a duty cycle proportional to the phase difference. Which MOSFET receives drive depends on which input is leading. If both inputs are in phase, the output will include small pulses due to noise, but their effects on the average output voltage will cancel. If one signal is at a higher frequency than the other, its phase will lead by an increasing amount, and the detector's output will be held close to either V_{DD} or ground, depending on which input signal is higher in frequency. To get a usable voltage output, the MOSFET outputs can be terminated in a high-value resistor to $V_{DD}/2$, but the CD4046's output stage was really designed to drive current pulses into a capacitive load, with pulses of one polarity charging the capacitor and pulses of the other discharging it. The capacitor integrates the pulses. Simple phase detectors are normally used to lock their inputs in quadrature, but phase-frequency detectors are used to lock their input signals in phase.

Traditional textbooks and logic-design courses give extensive coverage to avoiding *race hazards* caused by near-simultaneous signals racing through parallel paths in a structure to control a single output. Avoiding such situations is important in many circuits because the outcome is strongly dependent on slight differences in gate speed. The phase-frequency detector is the one circuit whose entire function depends on its built-in ability to *make* both its inputs race. Consequently, it's not easy to home-brew a phase-frequency detector from ordinary logic parts. Fortunately, they are available in IC form, usually combined with other functions. The MC4044 is an old stand-alone TTL phase-frequency detector; the MC12040 is an ECL derivative, and the much faster and compatible MCH 12140 can be used to over 600 MHz. The Hittite HMC403S8G can be used to 1.3 GHz. CMOS versions can be found in the CD4046 PLL chip and in almost all current divider-PLL synthesizer chips.

The race-hazard tendency does cause one problem in phase-frequency detectors: A device's delays and noise, rather than its input signals, control its phase-voltage characteristic in the zero-phase-difference region. This degrades the loop's noise performance and makes its phase-to-voltage coefficient uncertain and variable in a small region, a "dead zone"—unfortunately, in the detector's normal operating range! It's therefore normal to bias operation slightly away from *exact* phase equality to avoid this problem. Fortunately, the newer, faster phase detectors like the MCH 12140 and HMC403S8G tend to minimize this "dead zone" problem.

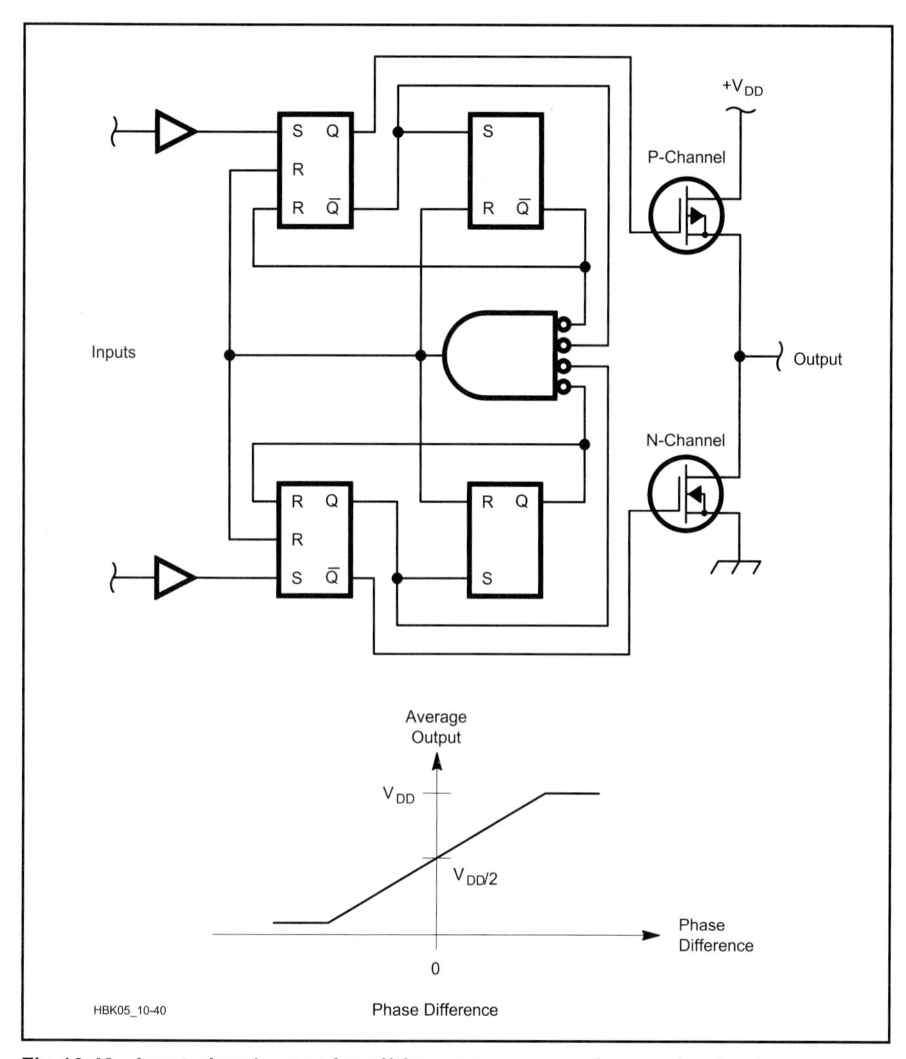

Fig 10.40—Input signals very far off frequency can confuse a simple phase-detector; a phase-frequency detector solves this problem.

The Loop Compensation Amplifier

Phase-locked loops have acquired a troublesome reputation for a variety of reasons. In the past, a number of commercially made radios have included poorly designed synthesizers that produced excessive noise sidebands, or wouldn't lock reliably. Some un-producible designs have been published for home construction that could not be made to work. Many experimenters have toiled over an unstable loop, desperately trying anything to get it to lock stably. So the PLL has earned a shady reputation. Because of all of this aversion therapy, very few amateurs are now prepared to attempt to build a PLL.

Luckily, things are not as bad as they used to be. The proliferation of synthesis and PLLs in contemporary equipment have lead to a number of excellent integrated circuits, as well as fine application notes to support them.

The **first** task is to take action to ensure that all previously discussed components of the loop are functioning properly and stably before attempting to close the loop. This means that the oscillators, dividers and amplifiers must all be unconditionally stable. Any instability or squegging in these components must be dealt with prior to attempting to close the loop.

The **second** task is to produce a closed-loop characteristic that best suits the application. Herein is where the greatest difficulty frequently lies. The selection of the closed-loop bandwidth and phase margin or peaking is usually one of great compromise and thought. The process frequently begins with an educated approximation that is then evaluated and modified. There are many factors to consider, including the degree of reference frequency suppression, switching speed, noise, microphonics and modulation, if any. Once the optimum loop characteristic is established, the next problem is to maintain it with respect to variations of the division resulting from the synthesis, and also gain variations associated with the nonlinear tuning curve of the VCO. While not always required in amateur applications, loops in sophisticated synthesizers frequently employ programmable multiplying DACs to maintain constant loop bandwidths and phase margin. In an example later in this chapter, we will actually describe the process for establishing the loop bandwidth for a simple synthesizer.

The **third** task is to design a loop compensation amplifier (filter) that will ensure stable operation when the loop is closed. The design of this compensation network or filter requires knowledge of the VCO gain and linearity, the phase-detector gain, and the effective division ratio and its variation in the loop. The foregoing requirements imply the necessity for measurement and calculation to have any reasonable hope of producing a successful outcome.

Note carefully that we are designing a *loop*. Trial and error, intuitive component choice, or "reusing" a loop amplifier design from a different synthesizer may lead to a low probability of success. All oscillators are loops, and any loop will oscillate when certain conditions are met. Look again at the RC phase-shift oscillator from which the "Pierce" crystal oscillator is derived (Fig 10.29A). Our PLL is a loop, containing an amplifier and a number of RC sections. It has all the parts needed to make an oscillator. If a loop is unstable and oscillates, there will be an oscillatory voltage superimposed on the VCO tuning voltage. This will produce massive unwanted FM on the output of the PLL, which is absolutely undesirable.

Before you turn to a less demanding chapter, consider this: It's one of nature's jokes that easy procedures often hide behind a terror-inducing facade. The math you need to design a good PLL may look weird when you see it for the first time and seem to involve some obviously impossible concepts, but it ends in a very simple procedure that allows you to calculate the response and stability of a loop. Professional designers must handle these things daily, and because they like differential equations no more than anyone else does, they use a graphical method based on the work of the mathematician, Laplace, to generate PLL designs. (We'll leave proofs in the textbooks, where they belong! Incidentally, if you can get the math required for PLL design under control, as a free gift you will also be ready to do modern filter synthesis and design beam antennas, since the math they require is essentially the same.)

The easy solutions to PLL problems can only be performed during the design phase. We must deliberately design loops with sufficient stability safety margins so that all foreseeable variations in component tolerances, from part-to-part, over temperature and across the tuning range, cannot take the loop anywhere near the threshold of oscillation. More than this, it is important to have adequate stability margin because loops with lesser margins will exhibit amplified phase-noise sidebands, and that is another PLL problem that can be designed out.

At the beginning of this chapter (following the text cite of Fig 10.4A), the criterion for stability/oscillation of a loop was mentioned: Barkhausen's for oscillation or Nyquist's for stability. This time Harry Nyquist is our hero.

A PLL is a negative feedback system, with the feedback opposing the input, this opposition or inversion around the loop amounts to a 180° phase shift. This is the frequency-*independent* phase shift round the loop, but there are also frequency-*dependent* shifts that will add in. These will inevitably give an increasing, lagging phase with increasing frequency. Eventually an extra 180° phase shift will have occurred, giving a total of 360° around the loop at some frequency. We are in trouble if the gain around the loop has not dropped below unity (0 dB) by this frequency.

Note that we are not concerned only with the loop operating frequency. Consider a 30-MHz low-pass filter passing a 21-MHz signal. Just because there is no signal at 30 MHz, our filter is no less a 30-MHz circuit. Here we use the concept of frequency to describe "what would happen *if* a signal was applied at that frequency."

The next sections show us how to use the graphical concepts of poles and zeros to calculate the gain and phase response around a loop. The graphical concepts are quite useful in that they provide insight into the effect of various component choices on the loop performance.

Poles and Zeros

Let's define some terms first. We all think of frequency as stretching from zero to infinity, but let's imagine that additional numbers could be used. This is pure imagination—we can't make signals at *complex* values (with *real* and *imaginary* components) of hertz—but if we try using complex numbers for frequency, some "impossible" things happen. For example, take a simple RC low-pass network shown in **Fig 10.41**. The frequency response of such a network is well known, but its phase response is *not* so well known. The equation shown for the *transfer function*, the "gain" of this simple circuit, is pretty standard, although the j is included to make it complete and accurately describe the output in phase as well as magnitude (j is an impossible number, the square root of –1, usually called an *imaginary* number and used to represent a 90° phase shift).

Although the equation shown in Fig 10.41 was constructed to show the behavior of the circuit at real-world frequencies, there is nothing to stop us from using it to explore how the circuit would behave at impossible frequencies, just out of curiosity. At one such frequency, the circuit goes crazy. At $f = -j\ 2\pi RC$, the denominator of the equation is zero, and the gain is infinite! Infinite gain is a pretty amazing thing to achieve with a passive circuit—but because this can only happen at an impossible frequency, it cannot happen in the real world. This frequency is equal to the network's –3 dB frequency multiplied by j; *it* is called a *pole*. Other circuits, which produce real amplitude responses that start to *rise* with frequency, act similarly, but their crazy effect is called a *zero*—an imaginary frequency at which gain goes to zero. The frequency of a circuit's zero is again j times its 3-dB frequency (+3 dB in this case).

These things are clearly impossible, but we can map all the poles and zeros of a complex circuit and look at the patterns to determine if the circuit possesses unwanted gain and/or phase bumps across a frequency range of interest. Poles are associated with 3-dB roll-off frequencies and 45° phase lags, zeros are associated with 3-dB "roll-up" frequencies and 45° phase leads. We can plot the poles and zeros on a two-dimensional chart of real and imaginary frequency. (The names of this chart and its

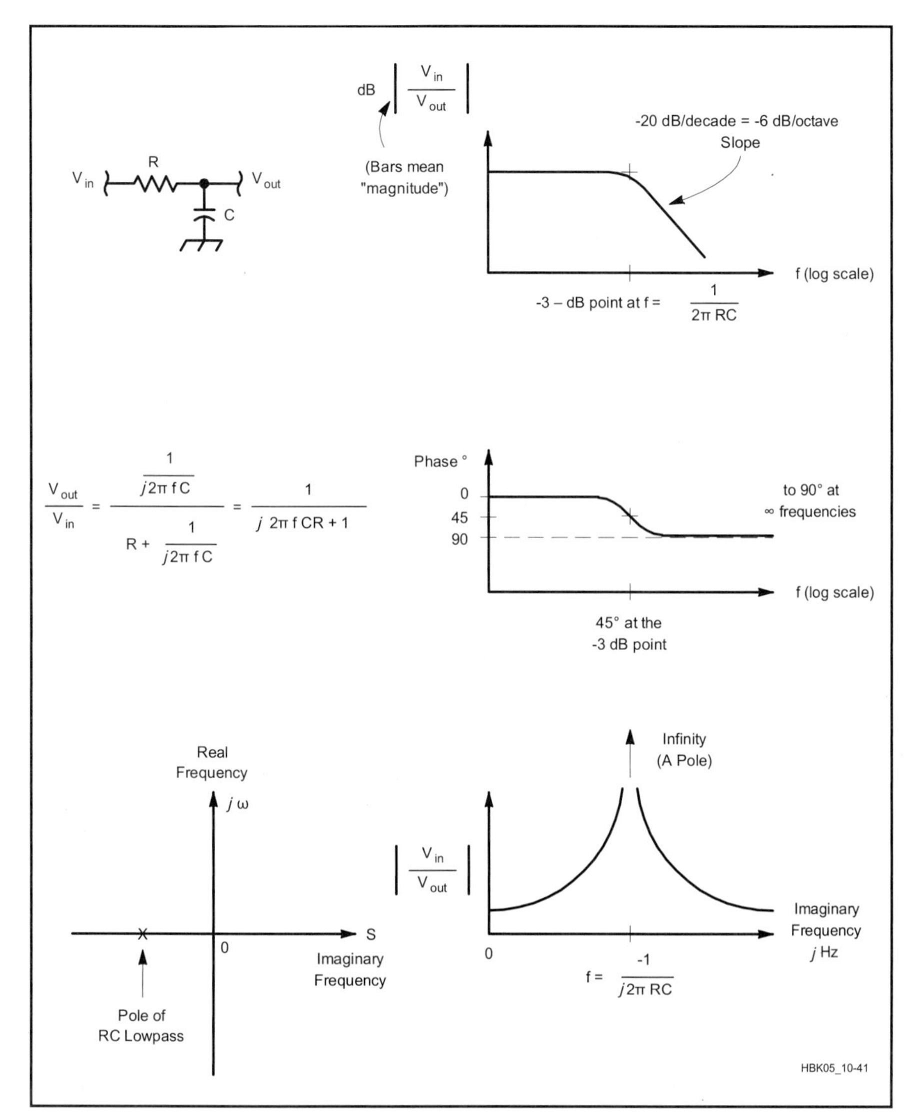

Fig 10.41—A simple RC filter is a "pole."

axes are results of its origins in *Laplace transforms*, which we don't need to touch in this discussion.) The traditional names for the axes are used, but we can add labels with clearer meanings.

What is a Pole?—A pole is associated with a bend in a frequency response plot where attenuation with increasing frequency increases by 6 dB per octave (20 dB per decade; an octave is a 2:1 frequency ratio, a decade is a 10:1 frequency ratio).

There are four ways to identify the existence and frequency of a pole:

1. A downward bend in a gain vs frequency plot, the pole is at the –3 dB point for a single pole. If the bend is more than 6 dB/octave, there must be multiple poles at this frequency.

2. A 90° change in a phase vs frequency plot, where lag increases with frequency. The pole is at the point of 45° added lag. Multiple poles will add their lags, as above.

3. On a circuit diagram, a single pole looks like a simple RC low-pass filter. The pole is at the –3 dB frequency ($1/(2\pi RC)$ Hz). Any other circuit that gives the same response will produce a pole at the same frequency.

4. In an equation for the transfer function of a circuit, a pole is a theoretical value of frequency, which would make the equation predict infinite gain. This is clearly impossible, but as the value of frequency will either be absolute zero, or will have an imaginary component, it is impossible to make a signal at a pole frequency.

What is a Zero?—A zero is the complement of a pole. Each zero is associated with an upward bend of 6 dB per octave in a gain vs frequency plot. The zero is at the +3 dB point. Each zero is associated with a transition on a phase-versus-frequency plot that *reduces* the lag by 90°. The zero is at the 45° point of this S-shaped transition.

In math, it is a frequency at which the transfer-function equation of a circuit predicts zero gain. This is not impossible in real life (unlike the pole), so zeros can be found at real-number frequencies as well as complex-number frequencies.

On a circuit diagram, a pure zero would need gain that increases with frequency forevermore above the zero frequency. This implies active circuitry that would inevitably run out of gain at some frequency, which implies one or more poles up there. In real-world circuits, zeros are usually not found making gain go up, but rather in conjunction with a pole, giving a gain slope between two frequencies and flat gain beyond them. Real-world zeros are only found chaperoned by a greater or equal number of poles.

Consider a classic RC high-pass filter. The gain increases at 6 dB per octave from 0 Hz (so there must be a zero at 0 Hz) and then levels off at $1/(2\pi RC)$ Hz. This leveling off is really a pole, it adds a 6 dB per octave roll off to cancel the roll-up of the zero.

Poles and Zeros in the Loop Amplifier—The loop-amplifier circuit used in the example loop has a blocking capacitor in its feedback path. This means there is no dc feedback. At higher frequencies the reactance of this capacitor falls, increasing the feedback and so reducing the gain. This is an *integrator*. The gain is immense at 0 Hz (dc) and falls at 6 dB per octave. This points to a pole at 0 Hz. This is true whatever the value of the series resistance feeding the signal into the amplifier inverting input and whatever the value of the feedback capacitor. The values of these components scale the gain rather than shape it. The shape of the integrator gain is fixed at –6 dB per octave, but these components allow us to move it to achieve some wanted gain at some wanted frequency.

At high frequencies, the feedback capacitor in an ordinary integrator will have very low reactance, giving the circuit very low gain. In our loop amplifier, there is a resistor in series with the capacitor. This limits how low the gain can go, in other words the integrator's downward (with increasing frequency) slope is leveled off. This resistor and capacitor make a zero, and its +6 dB per octave slope cancels the –6 dB per octave slope created by the pole at 0 Hz.

Recipe for A PLL Pole-Zero Diagram

We can choose a well-tried loop-amplifier circuit and calculate values for

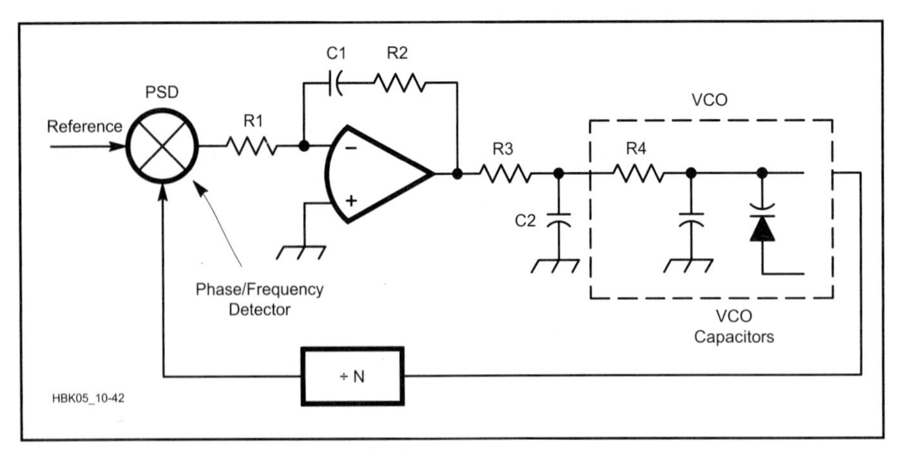

Fig 10.42—A common loop-amplifier/filter arrangement.

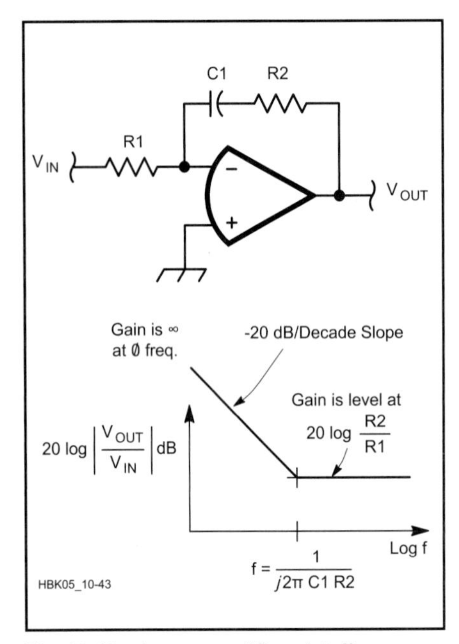

Fig 10.43—Loop-amplifier detail.

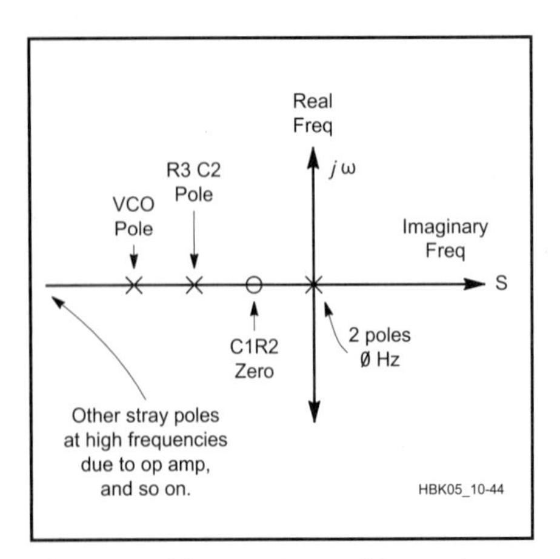

Fig 10.44—The loop's resulting pole-zero "constellation."

its components to make it suit our loop. **Fig 10.42** shows a common synthesizer loop arrangement. The op amp operates as an integrator, with R2 added to level off its falling gain. An integrator converts a dc voltage at its input to a ramping voltage at its output. Greater input voltages yield faster ramps. Reversing the input polarity reverses the direction of the ramp.

The system's phase-frequency detector (connected to work in the right sense) steers the integrator to ramp in the direction that tunes the VCO toward lock. As the VCO approaches lock, the phase detector reduces the drive to the integrator, and the ramping output slows and settles on the right voltage to give the exact, locked output frequency. Once lock is achieved, the phase detector outputs short pulses that "nudge" the integrator to keep the divided VCO frequency exactly locked in phase with the reference. Now we'll take a look around the entire loop and find the poles and zeros of the circuits.

The integrator produces a –20 dB per decade roll off from 0 Hz, so it has a pole at 0 Hz. It also includes R2 to cancel this slope, which is the same as adding a rising slope that exactly offsets the falling one. This implies a zero at $f = 1/(2\pi C1R2)$ Hz, as **Fig 10.43** shows. R3 and C2 make another pole at $f = 1/(2\pi C2R3)$ Hz. A VCO usually includes a series resistor that conveys the control voltage to the tuning diode, which also is loaded by various capacitors. This creates another pole.

The VCO generates frequency, while the phase detector responds to phase. Phase is the integral of frequency, so together they act as another integrator and add another pole at 0 Hz.

What About the Frequency Divider?—The frequency divider is like a simple attenuator in its effect on loop response. Imagine a signal generator that is switched between 10 and 11 MHz on alternate Tuesdays. Imagine that we divide its output by 10. The output of the divider will change between 1 and 1.1 MHz, still on alternate Tuesdays. The divider divides the deviation of the frequency (or phase) modulated signal, but it cannot affect the modulating frequency.

A possible test signal passing around the example loop is in the form of frequency modulation as it passes through the divider, so it is simply attenuated. A divide by N circuit will give $20 \log_{10}(N)$ dB of attenuation (reduction of loop gain). This completes the loop. We can plot all this information on one *s-plane*, **Fig 10.44**, a plot sometimes referred to as a PLL system's *pole-zero constellation*.

Open-Loop Gain and Phase

Just putting together the characteristics of the blocks around the loop, without allowing for the loop itself, gives us the system's *open-loop* characteristic, which is all we really need.

What Is Stability?—Here "open-loop response" means the gain around the loop that would be experienced by a signal at some frequency *if* such a signal were inserted. We do not insert such signals in the actual circuit, but the concept of frequency-dependent gain is still valid. We need to ensure that there is insufficient gain, at *all* frequencies, to ensure that the loop cannot create a signal and begin oscillating. More than this, we want a good safety margin to allow for component variations and because loops that are *close* to instability perform poorly.

The loop gain and phase are doing interesting things on our plots at frequencies in the AF part of the spectrum. This does not mean that there is visible operation at those frequencies. Imagine a bad, unstable loop. It will have a little too much loop gain at some frequency, and unavoidable noise will build up until a strong signal is created. The amplitude will increase until nonlinearities limit it. A 'scope view of the tuning voltage input to the VCO will show a big signal, often in the hertz to tens of kilohertz region, often large enough to drive the loop amplifier to the limits of its output swing, close to its supply voltages. As this big signal is applied to the tuning voltage input to the VCO, it will modulate the VCO frequency across a wide range. The output will look like that of a sweep generator on a spectrum analyzer. What is wrong? How can we fix this loop? Well, the problem may be excessive loop gain, or improper loop time constants. Rather than work directly with the loop time constants, it is far easier to work with the pole and zero frequencies of the loop response. It's

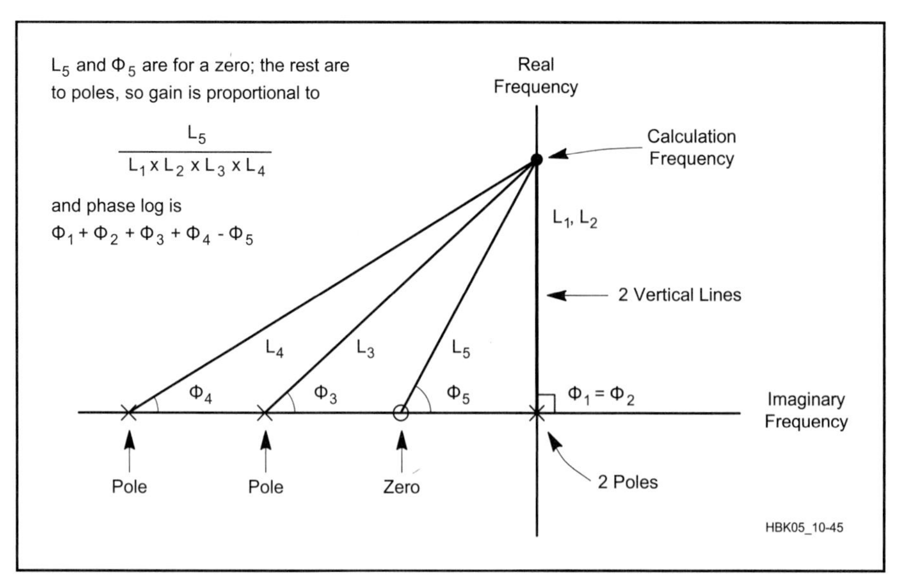

Fig 10.45—Calculating loop gain and phase characteristics from a pole-zero diagram.

just a different view of the same things.

Now imagine a good, stable loop, with an adequate stability safety margin. There will be some activity around the loop: the PSD (phase detector) will demodulate the phase noise of the VCO and feed the demodulated noise through the loop amplifier in such a way as to cancel the phase noise. This is how a good PLL should give *less* phase noise than the VCO alone would suggest. In a good loop, this noise will be such a small voltage that a 'scope will not show it. In fact, connecting test equipment in an attempt to measure it can usually *add* more noise than is normally there.

Finally, imagine a poor loop that is only just stable. With an inadequate safety margin, the action will be like that of a Q multiplier or a regenerative receiver close to the point of oscillation: There will be an amplified noise peak at some frequency. This spoils the effect of the phase detector trying to combat the VCO phase noise and gives the opposite effect. The output spectrum will show prominent bumps of exaggerated phase noise, as the excess noise frequency modulates the VCO.

Now, let's use the pole-zero diagram as a graphic tool to find the system open-loop gain and phase. As we do so, we need to keep in mind that the frequency we have been discussing in designing a loop response is the frequency of a theoretical test signal passing around the loop. In a real PLL, the loop signal exists *in two forms*: As sinusoidal voltage between the output of the phase detector and the input of the VCO, and as a sinusoidal modulation of the VCO frequency in the remainder of the loop.

With our loop's pole-zero diagram in hand, we can pick a frequency at which we want to know the system's open-loop gain and phase. We plot this value on the graph's vertical (Real Frequency) axis and draw lines between it and each of the poles and zeros, as shown in **Fig 10.45**. Next, we measure the lengths of the lines and the "angles of elevation" of the lines. The loop gain is proportional to the product of the lengths of all the lines to zeros *divided by* the product of all the lengths of the lines to the poles. The phase shift around the loop (lagging phase equates to a positive phase shift) is equal to the sum of the pole angles minus the sum of the zero angles. We can repeat this calculation for a number of different frequencies and draw graphs of the loop gain and phase versus frequency.

All the lines to poles and zeros are hypotenuses of right triangles, so we can use Pythagoras's rule and the tangents of angles to eliminate the need for scale drawings. Much tedious calculation is involved because we need to repeat the whole business for each point on our open-loop response plots. This much tedious calculation is an ideal application for a computer.

The procedure we've followed so far gives only *proportional* changes in loop gain, so we need to calculate the loop gain's *absolute* value at some (chosen to be easy) frequency and then relate everything to this. Let's choose 1 Hz as our reference. (Note that it's usual to express angles in radians, not degrees, in these calculations and that this normally renders frequency in peculiar units of *radians per second*. We can keep frequencies in hertz if we remember to include factors of 2π in the right places. A frequency of 1 Hz = 2π radians/second, because 2π radians = 360°.)

We must then calculate the loop's proportional gain at 1 Hz from the pole-zero diagram, so that the constant of proportion can be found. For starters, we need a reasonable estimate of the VCO's *voltage-to-frequency gain*—how much it changes frequency per unit change of tuning voltage. As we are primarily interested in stability, we can just take this number as the slope, in Hertz per volt, at the steepest part of voltage-versus-frequency tuning characteristic, which is usually at the low-frequency end of the VCO tuning range. (You can characterize a VCO's voltage-versus-frequency gain by varying the bias on its tuning diode with an adjustable power supply and measuring its tuning characteristic with a voltmeter and frequency counter.)

The loop divide-by-N stage divides our theoretical modulation—the tuning corrections provided through the phase detector,

$$\text{Open loop gain (dB)} = 20\log\left[\frac{\sqrt{P_1^2+1}\sqrt{P_2^2+1}\sqrt{P_3^2+1}\sqrt{P_4^2+1}\sqrt{Z^2+f^2}}{\sqrt{P_1^2+f^2}\sqrt{P_2^2+f^2}\sqrt{P_3^2+f^2}\sqrt{P_4^2+f^2}\sqrt{Z^2+1}}\times 10^{\frac{\text{unity freq gain}}{20}}\right]$$

$$\text{Phase (lead)} = \tan^{-1}\left(\frac{f}{Z}\right) - \tan^{-1}\left(\frac{f}{P_2}\right) - \tan^{-1}\left(\frac{f}{P_3}\right) - \tan^{-1}\left(\frac{f}{P_4}\right)$$

f is the frequency of the point to be characterized. P_1, P_2, P_3, P_4, Z are the frequencies of the poles and zero (all in the same units!)

Fig 10.46—Pole-zero frequency-response equations capable of handling up to four poles and one zero.

loop amplifier and the VCO tuning diode—by its programmed ratio. The worst case for stability occurs at the divider's lowest N value, where the divider's "attenuation" is least. The divider's voltage-versus-frequency gain is therefore 1/N, which, in decibels, equates to –20 log (N).

The change from frequency to phase has a voltage gain of one at the frequency of 1 radian/second, $1/(2\pi)$, which equates to –16 dB, at 1 Hz. The phase detector will have a specified "gain" in volts per radian. To finish off, we then calculate the gain of the loop amplifier, including its feedback network, at 1 Hz.

There is no need even to draw the pole-zero diagram. **Fig 10.46** gives the equations needed to compute the gain and phase of a system with up to four poles and one zero. They can be extended to more singularities and put into a simple computer program. Alternatively, you can type them into your favorite spreadsheet and get printed plots. The computational power needed is trivial, and a listing to run these equations on a Hewlett-Packard HP11C (or similar pocket calculator, using RPN) is available from ARRL HQ for an SASE. (Contact the Technical Department Secretary and ask for the 1995 *Handbook* PLL design program.)

Fig 10.47 shows the sort of gain- and phase-versus-frequency plots obtained from a "recipe" loop design. We want to know where the loop's phase shift equals, or becomes more negative than, –180°. Added to the –180° shift inherent in the loop's feedback (the polarity of which must be negative to ensure that the phase detector drives the VCO toward lock instead of away from it), this will be the point at which the loop itself oscillates—*if* any loop gain remains at this point. The game is to position the poles and zero, and set the unit frequency gain, so that the loop gain falls to below 0 dB before the –180° line is crossed.

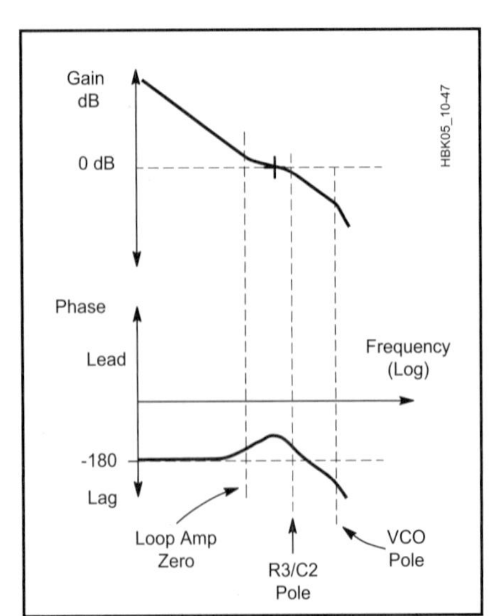

Fig 10.47—An open-loop gain/phase plot.

At the low-frequency end of the Fig 10.47 plots, the two poles have had their effect, so the gain falls at 40 dB/decade, and the phase remains just infinitesimally on the safe side of –180°. The next influence is the zero, which throttles the gain's fall back to 20 dB/decade and peels the phase line away from the –180° line. The zero would eventually bring the phase back to 90° of lag, but the next pole bends it back before that and returns the gain slope back to a fall of 20 dB/decade. The last pole, that attributable to the VCO, pushes the phase over the –180° line.

It's essential that there not be a pole between the 0-Hz pair and the zero, or else the phase will cross the line early. In this example, the R3-C2 pole and the zero do all the work in setting this critical portion of the loop's response. Their frequency spacing should create a phase bump of 30 to 45°, and their particular frequency positions are constrained as a compromise between sufficient loop bandwidth and sufficient suppression of reference-frequency sidebands. We know the frequency of the loop amplifier zero (that contributed by R1 and C1), so all we need to do now is design the loop amplifier to exhibit a frequency response such that the open-loop gain passes through 0 dB at a frequency close to that of the phase-plot bump.

This loop-design description may have been hard to follow, but a loop amplifier can be designed in under 30 minutes with a pocket calculator and a little practice. The high output impedance of commonly used CMOS phase detectors favors loop amplifiers based on FET-input op amps and allows the use of high-impedance RC networks (large capacitors can therefore be avoided in their design). LF356 and TL071 op amps have proven successful in loop-amplifier service and have noise characteristics suited to this environment.

To sum up, the recipe gives a proven pole zero pattern (listed in order of increasing frequency):

1. Two poles at 0 Hz (one is the integrator, the other is implicit in all PLs)
2. A zero controlled by the RC series in the loop amplifier feedback path
3. A simple RC low-pass pole
4. A second RC low-pass pole formed by the resistor driving the capacitive load of the tuning voltage inputs of the VCO

To design a loop: Design the VCO, divider, PSD and find their coefficients to get the loop gain at unit frequency (1 Hz), then, keeping the poles and zeros in the above order, move them about until you get the same 45° phase bump at a frequency close to your desired bandwidth. Work out how much loop-amplifier unit-frequency gain is needed to shift the gain-frequency-frequency plot so that 0-dB loop gain occurs at the center of the phase bump. Then calculate R and C values to position the poles and zero and the feedback-RC values to get the required loop amplifier gain at 1 Hz.

To Cheat—You can scale the example loop to other frequencies: Just take the reactance values at the zero and pole frequencies and use them to scale the components, you get the nice phase plot bump at a scaled frequency. Even so, you still must do the loop-gain design.

Don't forget to do this for your lowest division factor, as this is usually the least stable condition because there is less attenuation. Also, VCOs are usually most sensitive at the low end of their tuning range.

Noise in Phase-Locked Loops

Differences in Q usually make the phase-noise sidebands of a loop's reference oscillator much smaller than those of the VCO. Within its loop bandwidth, a PLL acts to correct the phase-noise components of its VCO and impose those of the reference. Dividing the reference oscillator to produce the reference signal applied to the phase detector also divides the deviation of the reference oscillator's phase-noise sidebands, translating to a 20-dB reduction in phase noise per decade of division, a factor of 20 log (M) dB, where M is the reference divisor. Offsetting this, within its loop bandwidth the PLL acts as a frequency *multiplier*, and this multiplies the deviation, again by 20 dB per decade, a factor of 20 log (N) dB, where N is the loop divider's divisor. Overall, the reference sidebands are increased by 20 log (N/M) dB. Noise in the dividers is, in effect, present at the phase detector input, and so is increased by 20 log (N) dB. Similarly, op-amp noise can be calculated into an equivalent phase value at the input to the phase detector, and this can be increased by 20 log (N) to arrive at the effect it has on the output.

Phase noise can be introduced into a PLL by other means. Any amplifier stages between the VCO and the circuits that follow it (such as the loop divider) will contribute some noise, as will microphonic effects in loop- and reference-filter components (such as those due to the piezoelectric properties of ceramic capacitors

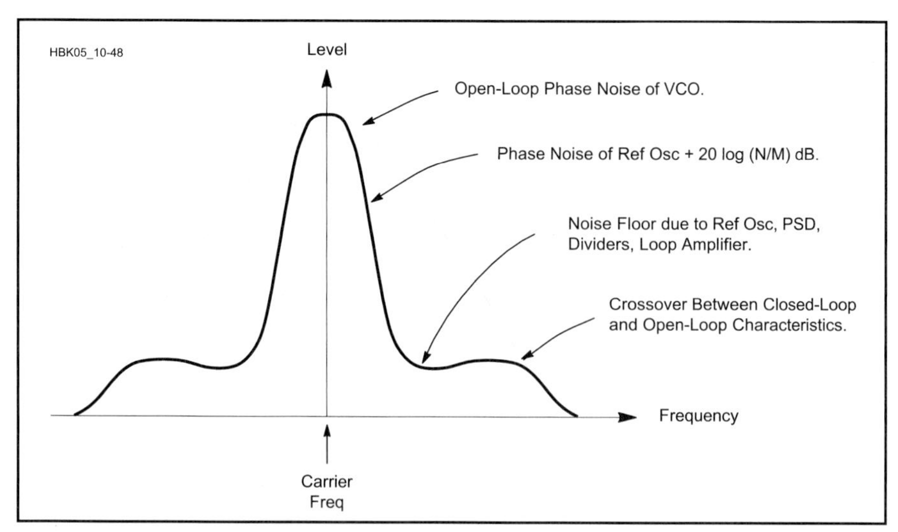

Fig 10.48—A PLL's open- and closed-loop phase-noise characteristics. The noise bumps at the crossover between open- and closed-loop characteristics are typical of PLLs; the severity of the bumps reflects the quality of the system's design.

and the crystal filters sometimes used for reference-oscillator filtering). Noise on the power supply to the system's active components can modulate the loop. The fundamental and harmonics of the system's ac line supply can be coupled into the VCO directly or by means of ground loops.

Fig 10.48 shows the general shape of the PLL's phase noise output. The dashed curve shows the VCO's noise performance when unlocked; the solid curve shows how much locking the VCO to a cleaner reference improves its noise performance. The two noise bumps are a classic characteristic of a phase-locked loop. If the loop is poorly designed and has a low stability margin, the bumps may be exaggerated—a sign of noise amplification due to an overly peaky loop response.

Exaggerated noise bumps can also occur if the loop bandwidth is less than optimum. Increasing the loop bandwidth in such a case would widen the band over which the PLL acts, allowing it to do a better job of purifying the VCO—but this might cause other problems in a loop that's deliberately bandwidth-limited for better suppression of reference-frequency sidebands. Immense loop bandwidth is not desirable, either: Farther away from the carrier, the VCO may be so quiet on its own that widening the loop would make it *worse*.

A BASIC DESIGN EXAMPLE

In this section, we will explore the application of the design principles previously covered, as well as some of the tradeoffs required in a practical design. We will also cover measurement techniques designed to give the builder confidence that the loop design goals have been met. Finally, we will cover some common troubleshooting problems.

As our design example, a synthesized local oscillator chain for a 10-GHz Transverter will be considered. **Fig 10.49** is a simplified block diagram of a 10-GHz converter. This example was chosen as it represents a departure from the traditional multi-stage multiply and filter approach. It permits realization of the oscillator system with two very simple loops and minimal RF hardware. It is also representative of what is achievable with current hardware, and can fit in a space of 2 to 3 square inches. This example is intended as a vehicle to explore the loop design aspects and is not offered as a "construction project." The additional detail required would be beyond the scope of this chapter.

In this example, two synthesized frequencies, 10 GHz and 340 MHz, are required. Since 10.368 GHz is one of the popular traffic frequencies, we will mix this with the 10-GHz LO to produce an IF of 368 MHz. The 368-MHz IF signal will be subsequently mixed with a 340-MHz LO to produce a 28-MHz final IF, which can be fed into the 10-meter input of any amateur transceiver. We will focus our attention on the design of the 10-GHz synthesizer only. Once this is done, the same principles may be applied to the 340-MHz section. Our goal is to attempt to design a low-noise LO system (this implies minimum division) with a loop reference oscillator that is an integer multiple of 10 MHz. Using this technique will allow the entire system to be locked at a later time to a 10-MHz standard for precise frequency control.

Ideally, we would like to start with a low-noise 100-MHz crystal reference and a 10-GHz oscillator divided by 100. It is here that we are already confronted with

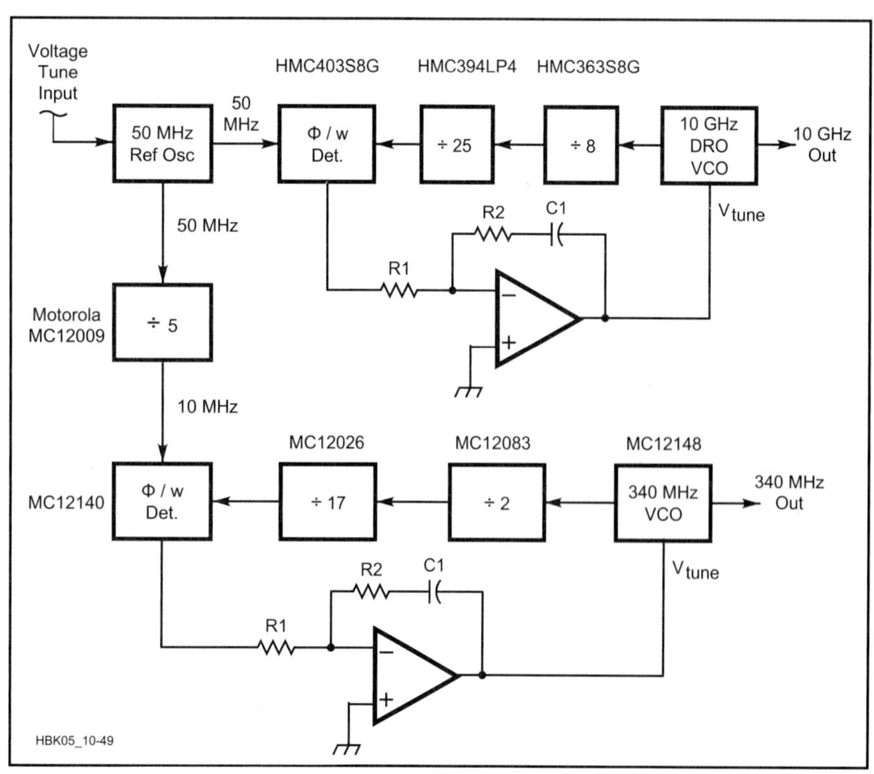

Fig 10.49—A simplified block diagram of a local oscillator, for a 10-GHz converter.

our first practical design tradeoff. We are considering using a line of microwave integrated circuits made by Hittite Microwave. These include a selection of prescalers operating to 12 GHz, a 5-bit counter that will run to 2.2 GHz and a phase/frequency detector that will run up to 1.3 GHz. The 5-bit counter presents the first problem. Our ideal scheme would be to use a divide-by-4 prescaler and enter the 5-bit counter at 2.5 GHz, subsequently dividing by 25 to 100 MHz. The problem is that the 5-bit counter is only rated to 2.2 GHz and not 2.5 GHz. This forces us to use a divide-by-8 prescaler and enter the 5-bit divider at 1.25 GHz. This is well in the range of the divider, but we can no longer use an integer division (ie, 1.25 GHz/12.5 = 100 MHz) to get to 100 MHz. The next easiest option is to reduce the reference frequency to 50 MHz and to let the 5-bit divider run as a divide-by-25, giving us a total division ratio of 200.

Having already faced our first design tradeoff, a number of additional aspects must be considered to minimize the conflicts in the design. **First**, this is a "static" synthesizer—That is, it will not be required to change frequency during operation. As a result, switching time considerations are irrelevant, as are the implications that the switching time would have had on the loop bandwidth. The effects of variable division ratios are also eliminated, as well as any problems associated with nonlinear tuning of the VCOs. **Second**, the reference frequencies (50 MHz and 10 MHz) are very large with respect to any practically desirable loop bandwidths. This makes the requirement of the loop filter to eliminate reference sidebands quite easy to achieve. In fact, reference sideband suppression will more likely be a function of board layout and shielding effectiveness rather than suppression in the loop filter. This also allows placement of the reference suppression poles far out in the pole zero constellation and well outside the loop bandwidth (about 10 × BW3 of the loop). This placement of the reference suppression poles will give us the option of using a "type 2, 2nd order" loop approximation, which will greatly simplify the calculation process. **Third**, based on all of the foregoing tradeoffs, the loop bandwidth can be chosen almost exclusively on the basis of noise performance.

Before we can proceed with the loop calculations we need some additional information: specifically VCO gain, VCO noise performance, divider noise performance, phase detector gain, phase detector noise performance and finally reference noise performance. The phase detector and divider information is available from specification sheets. Now we need to select the 50-MHz reference and the 10-GHz VCO. An excellent choice for a low noise reference is the one described by John Stephensen on page 13 in Nov/Dec 1999 *QEX*. The noise performance of this VCXO is in the order of –160 dBc/Hz at 10-kHz offset at the fundamental frequency. For the 10-GHz VCO, there are many possibilities, including cavity oscillators, YIG oscillators and dielectric resonator oscillators. Always looking for parts that are easily available, useable and economical, salvaging a dielectric resonator from a Ku band LNB (see Jan-Feb 2002 *QEX*, p 3) appears promising. These high-Q oscillators can be fitted with a varactor and tuned over a limited range with good results. The tuning sensitivity of our dielectric resonator VCO is about 10 MHz per volt and the phase noise at 10-kHz offset is –87dBc/Hz. We are now in a position to plot the phase noise of the reference, the dividers, the division effect (46 dB) and the VCO. The phase noise plots are shown in **Fig 10.50**. The noise of the reference is uniformly about 25 dB below the divider and phase detector noise floor. This means that the noise of the phase detector and dividers will be the dominant contribution within the loop bandwidth. The cross over point is at approximately 9 kHz. At frequencies lower than 9 kHz, the loop will reduce the noise of the VCO (eg, at 1 kHz the reduction will be about 20 dB). At frequencies above 9 kHz, the natural noise roll off of the VCO will dominate. At this point, we can also see the effect of having chosen a reference frequency of 50 MHz. Had we been able to use a 100-MHz reference instead of 50 MHz and reduce the division factor from 200 to 100, we could have put the loop bandwidth at about 30 kHz and picked up an additional 6-dB improvement in close-in noise performance. Nevertheless, even with the 6-dB penalty, this will be quite a respectable 10-GHz LO. It should be apparent that a fair amount of thought and effort was required to be able to determine an applicable bandwidth for this loop.

Now that we have some estimate of the noise performance and the required loop bandwidth, it is time to design the loop. The second consideration cited above enabling the application of the "type-2, 2nd order" approximation is now going to pay off. The math for this loop is really quite simple. There are two main variables, the "natural frequency" of the loop, ω_n, (omega-n in radians/second) and the "damping factor" D (delta is dimensionless). The closed-loop bandwidth of the loop is a function of ω_n and D. **Fig 10.51**

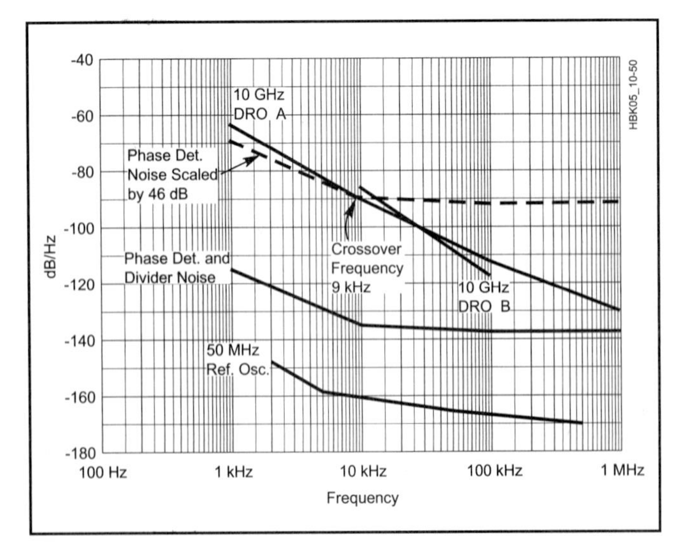

Fig 10.50—Phase noise plots of the 10-GHz transverter design example.

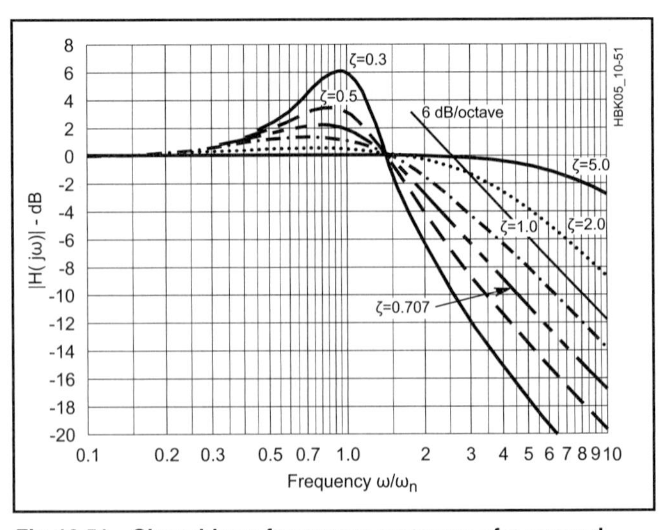

Fig 10.51—Closed-loop frequency response of a second-order loop as a function of the loop's natural frequency and damping factor (ω_n).

shows loop response as a function of ω_n and D. Values of D less than 0.5 are not desirable, as they tend towards instability and poor phase margin. A value for D of 0.707 is referred to as "critically damped" and is favored in applications where settling time is critical (see Gardner). For our application, we will choose a D of 1.0. This will give acceptable phase margin and further simplify the math. For a D of 1, $\omega_n = (6.28 \times \text{BW3}) / 2.48$. Substituting the 9-kHz loop bandwidth for BW3 yields 22790 radians per second for ω_n. Our loop filter will take the basic form shown in Fig 10.43. We can compute values for R1, R2 and C1 as follows.

$$R1 = (Kpd \times Kvco)/(N \times \omega_n^2 \times C1)$$

where Kpd is the phase detector gain, 1/6.28 volts/radian in this case, Kvco is the VCO gain in radians/Hz, 6.28×10^7 radians/volt in this case, N is the division factor, 200 in this case. C1 is the feedback capacitor value in Farads. To proceed, we need to start with an estimate for C1. Practically speaking, since odd values of capacitors are more difficult to obtain, let us try a value for C1 of 0.01 µF.

$$R2 = 2 \times D/(\omega_n \times C1)$$

R1 computes as 9626 Ω and R2 is 8775 Ω. Since we will be using a phase/frequency detector with differential outputs, we will have to modify the circuit of Fig 10.43 to become a differential amplifier. We will also add a passive input filter to keep the inputs of the op amp from being stressed by the very fast and short pulses emanating from the phase/frequency detector. We will also add a passive "hash filter" at the output of the op amp to limit the amount of out of band noise delivered to the VCO tuning port. Both of these filters will be designed for a cutoff frequency of 90 kHz (ie,10 times the 9-kHz closed-loop bandwidth, as cited above in the second consideration. Both of these filters will also aid in the rejection of reference frequency sidebands. For the input filter, we simply divide the input resistor in half and add a capacitor to ground. The value of this capacitor is determined as follows:

$$C2 = 4/(62.8 \times BW3 \times R1)$$

$$C2 = 735\ \text{pF}$$

Using 680 or 750 pF will be adequate. The output filter is also quite simple, with one minor stipulation. Op Amps can exhibit stability problems when asked to drive a capacitive load through too low a value of resistor. One very safe way around this is not to use a value for R3 lower than the recommended minimum impedance for full output of the amplifier. For many amplifiers, this value is around 600 Ω. For the output filter, we compute C3 as follows, for R3 = 600 Ω:

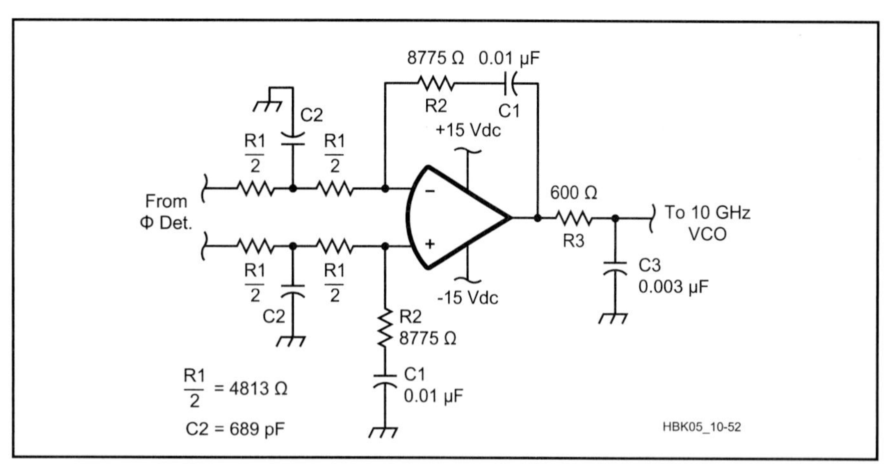

Fig 10.52—A complete loop-compensation amplifier for the design example 10-GHz transverter.

$$C3 = 1/(R3 \times 62.8 \times BW3)$$

$$C3 = 2950\ \text{pF}$$

Using 3000 pF will be adequate. This completes the computations for the loop compensation amplifier. The complete amplifier is shown in **Fig 10.52**.

MEASUREMENTS

One of the first things we need to measure when designing a PLL is the VCO gain. The tools we will need include a voltmeter, some kind of frequency measuring device like a receiver or frequency counter and a source of clean variable DC voltage. The setup in **Fig 10.53**, containing one or more 9-V batteries and a 10-turn, 10-kΩ pot will due nicely. One simply varies the voltage some amount and then records the associated frequency change of the VCO. The gain of the VCO is then delta F over delta V. The phase detector gain constant can usually be found on the specification sheets of the components selected.

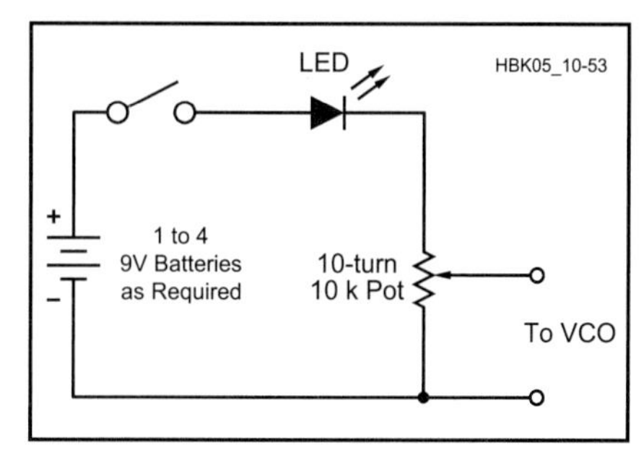

Fig 10.53—Clean, variable d-c voltage source used to measure VCO gain in a PLL design.

Measuring closed loop bandwidth is slightly more complicated. The way it is commonly done in the laboratory is to replace the reference oscillator with a DCFM-able (ie, a signal generator whose FM port is DC coupled) signal generator and then feed the tracking generator output of a low-frequency spectrum analyzer into the DCFM-able generator while observing the spectrum of the tuning voltage on the spectrum analyzer (see **Fig 10.54**). While this approach is quite straightforward, few amateurs have access to or can afford the test equipment to do this. Today however, thanks to the PC sound card-based spectrum analyzer and tracking generator programs (software by Interflex for example), amateurs can measure the closed-loop response of loops that are less than 20 kHz in bandwidth for significantly less than a king's ransom in test equipment! Here's how.

One approach is to build the reference oscillator with some "built-in test equipment" (BITE) already in the design. This BITE takes the form of some means of DCFM-ing the reference oscillator. This is one of the things that make John Stephenson's oscillator (previously mentioned) attractive. The oscillator includes a varactor input for the purpose of phase locking it to a high-stability low-frequency standard oscillator. This same varactor tuning input takes the place of the aforementioned DCFM-able signal generator. Now all we need to complete our test setup is some batteries. First, we need to make sure the input of our sound card is AC coupled (when connected to the VCO tuning port, there will be a DC component present and the sound card may not like this) and that the AC coupling is flat down to at least 10%

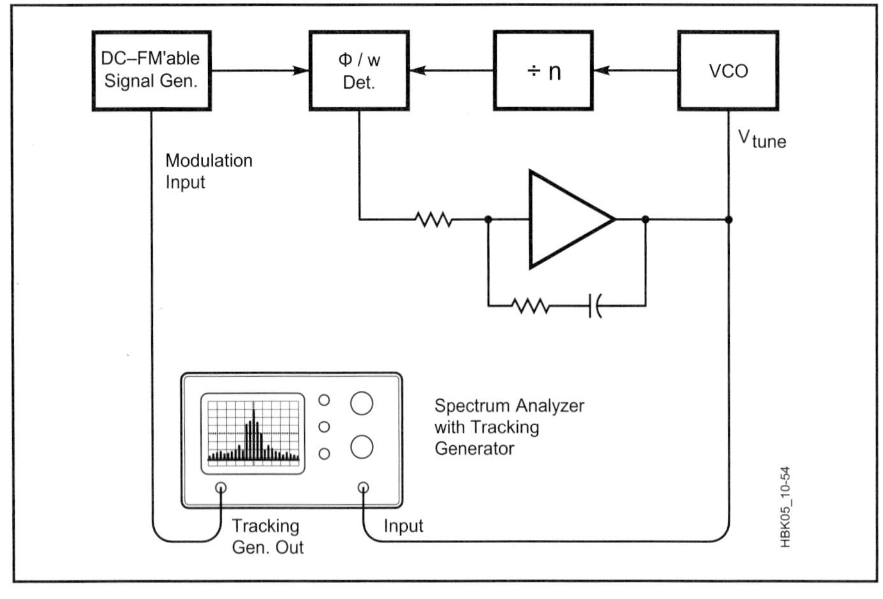

Fig 10.54—Common laboratory setup for measuring the closed-loop bandwidth of loops.

of the natural frequency of the loop. If the sound card is not AC coupled, a capacitor will be needed at the input to block the DC on the tuning voltage.

The next step is to determine the tuning sensitivity of the reference oscillator. This is done in the same manner as was done for the VCO described above. Once this has been established, we will set the audio oscillator for an output voltage that will produce between 100 Hz and 1 kHz of deviation. We will also need to establish that the output of the sound card is DC coupled so that we can place a small battery in series with the audio generator and the tuning port of the reference oscillator. We can provide a DC return on the output by putting a resistive 10-dB attenuator of the appropriate impedance on the output of the sound card and passing the varactor bias through the attenuator to ground. This is required to bias the varactors and also assure that the audio voltage will not drive the varactors into conduction. We can now connect the AC coupled input of the sound card to the VCO tuning voltage and turn on the loop.

Do not be surprised to see signal components at multiples of the power line frequency, as well as the signal from the audio oscillator. The presence of line frequency components is an indication that the loop is doing its job and removing these components from the spectrum of the VCO. We can now "sweep" the audio oscillator and, by plotting the amplitude of the audio oscillators response, determine the closed loop bandwidth of the PLL. In the case of the loop described above, (ω_n = 22790, D = 1.0) we should expect to see slightly over 1 dB of peaking at $0.7\omega_n$ and the 3-dB down point should fall at $2.5\omega_n$. In any event, if the peaking exceeds 3 dB, the loop phase margin is growing dangerously small and steps should be taken to improve it.

COMMON PROBLEMS

Here are some frequently encountered problems in PLL designs:

- The outputs of the phase detector are inverted. This results in the loop going to one or the other rail. The loop cannot possibly lock in this condition. Solution: Swap the phase detector outputs.
- The loop cannot comply with the tuning voltage requirements of the VCO. If the loop runs out of tuning voltage before the required voltage for a lock is reached, the locked condition is not possible. Solution: Re-center the VCO at a lower tuning voltage or increase the rail voltages on the op amp.
- The loop is very noisy and the tuning voltage is very low. The tuning voltage on the varactor diodes should not drop below the RF voltage swing in the oscillator tank circuit. Solution: Adjust the VCO so that the loop locks with a higher tuning voltage.

PLL SYNTHESIZER ICS

Now that we have learned how to deal with the loop design aspects of an indirect synthesizer, it is time to look at just a few of the many PLL synthesizers available today. Simple PLL ICs have been available since the early 1970s, but most of these contain a crude, low-Q VCO and a phase detector. They were intended for general use as tone detectors and demodulators, not as major elements in communication-quality frequency synthesizers.

One device well worth noting in this group, however, is the CD4046, which contains a VCO and a pair of phase detectors. The CD4046's VCO is useless for our purposes, but its phase-frequency detector is quite good. Better yet, the CD4046 is a low-cost part and one of the cheapest ways of getting a good phase-frequency detector. Its VCO-disable pin is a definite design plus. Later CD4046 derivatives, the 74HC4046 (CMOS input levels) and 74HCT4046 (TTL compatible input levels) are usable to much higher frequencies and seem to be more robust, but note that these versions are for +5 V supplies only, whereas the original CD4046 can be used up to 15 V.

Since then, many more complex ICs specifically intended for frequency synthesizers have been introduced by companies like Motorola, National (National LMX series) and others as a result of the growth in popularity of wireless devices. They normally contain a programmable divider, a phase-frequency detector and a reference divider that usually allows a small choice of division ratios. Usually, the buffer amplifier on these parts' reference input is arranged so that it can be used as part of a simple crystal oscillator. This is adequate for modest frequency accuracy, but an independent TCXO or OCXO is better.

Among Motorola's more popular parts is the MC145151. The MC145151 brings all of its division-control bits out to individual pins and needs no sequencing for control. It is the best choice if only a few output frequencies are needed, because they can be programmed via a diode matrix. The MC145151's divide-by-*N* range is 3 to 16383, controlled by a 14-bit word in binary format. The reference can be divided by 8, 128, 256, 512, 1024, 2048, 2410 and 8192. It's possible to operate the MC145151 to 30 MHz, and its phase-frequency detector is similar to that in the CD4046, lacking the MOSFET "Tri-State" output, but with the added benefit of a lock-detector output. The MC145152 is a variant that includes control circuitry for an external dual-modulus prescaler. The choice of external prescaler sets the maximum operating frequency, as well as maximum and minimum division ratios. For some unknown reason, the reference division choices differ from those of the MC145151: 8, 64, 128, 256, 512, 1024, 1160 and 2048. There is also a Tri-State

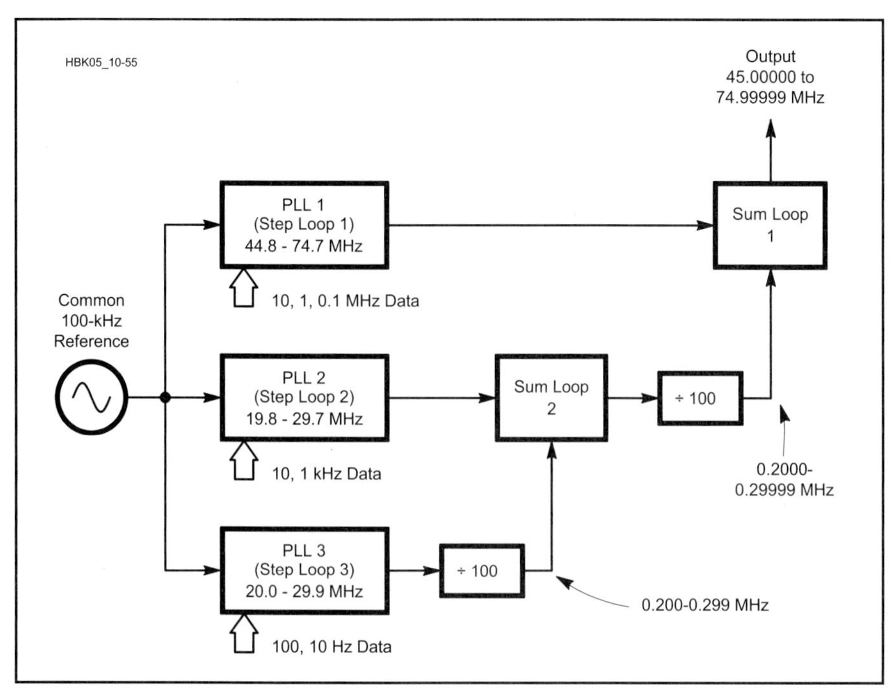

Fig 10.55—A five-loop synthesizer.

phase-frequency detector output.

The MC145146 is simply an MC145152 with its control data formatted as a 4-bit bus writing to eight addresses. This reduces the number of pins required and allows more data to be input, so the reference divider is fully programmable from 3 to 4095.

Multiloop Synthesizers

The trade-off between small step size (resolution) and all other PLL performance parameters has already been mentioned. The *multiloop synthesizer*, a direct synthesizer constructed from two or more PLL synthesizers, is one way to break away from this trade-off. **Fig 10.55**, the block diagram of a five-loop synthesizer, reflects the complexity found in some professional receivers. All of its synthesis loops run with a 100-kHz step size, which allows about a 10-kHz loop bandwidth. Such a system's noise performance, settling time after a frequency change and reference-sideband suppression can all be very good.

Cost concerns generally render such elaboration beyond reach for consumer-grade equipment, so various cut-down versions, all of which involve trade-offs in performance, have been used. For example, a three-loop machine can be made by replacing the lower three loops with a single loop that operates with a 1-kHz step size, accepting the slower frequency stepping determined by the narrow loop bandwidth that this involves.

DIRECT DIGITAL SYNTHESIS

Direct digital synthesis (DDS) is covered in this *Handbook's* **DSP and Software Radio Design** chapter. DDS technology is on the verge of covering all the local oscillator requirements of an HF transceiver. Very economical chipsets that cover partial requirements are available and can be exploited with the help of PLLs. For an example, see **Fig 10.56** and **Table 10.3**. A single loop with a programmable divider could be used in place of the crystal-oscillator bank. The result would be very close to the approach used in the latest generation of commercial Amateur Radio gear. That manufacturers advertise them as using direct digital synthesis has given some people the impression that their synthesizers consist entirely of DDS. In fact, the DDS in modern ham transceivers replaces the lower-significance "interpolation" loops in what otherwise would be regular multiloop synthesizers.

This is not to say that the DDS in our current ham gear is not a great improvement over the size, complexity and cost of what it replaces. Its random noise sidebands are usually excellent, and it can execute fast, clean frequency changes. The latest devices do 32-bit phase arithmetic and so offer over a billion frequency steps, which translates into a frequency resolution of a few *millihertz* with the usual clock rates for our applications—so the old battle for better resolution and low cost has already been won. Direct digital synthesis is not without a few problems, but fortunately, hybrid structures using PLL and DDS can allow one technique to compensate for the weaknesses of the other.

The prime weakness of the direct digital synthesizer is its quantization noise. A DDS cannot construct a perfect sine wave because each sample it outputs must be the nearest available voltage level from the set its DAC can make. So a DDS's output waveform is really a series of steps that only approximate a true sine wave.

We can view a DDS's output as an ideal sine wave plus an irregular "error" waveform. The spectrum of the error waveform is the set of unwanted frequency components found on the output of the DDS. Quantization noise is not the only source of unwanted DAC outputs. DACs can also give large output spikes, called glitches, as they transit from one level to another. In some DACs, glitches cause larger unwanted components than quantization. These components are scattered over the full frequency range passed by the DDS's low-pass filter, and their frequencies shift as the DDS tunes to different frequencies. At some frequencies, a number of components may coincide and form a single, larger component.

A summing loop PLL acts as a tracking filter, with a bandwidth measured in kilohertz. Acting on a DDS's output, a summing loop passes only quantization components close to the carrier. A system designer seeking to minimize the DDS's noise contribution must choose between using a loop bandwidth narrow enough to filter the DDS (thereby reducing the loop's ability to purify its VCO) or a more expensive low "glitch energy" DAC with more bits of resolution (allowing greater loop bandwidth and better VCO-noise control).

Complex ICs containing most of a DDS have been on the market for over 15 years, steadily getting cheaper, adding functions and occasionally taking onboard the functions of external parts (first the sine ROM, now the DAC as well). The Analog Devices AD7008 is an entire DDS (just add a low-pass filter...) on one CMOS chip. It includes a 10-bit DAC (fast enough for the whole system to clock at 50 MHz) and some digital-modulation hardware.

For the home-brewing amateur, DDS's first drawback is that these devices must receive their frequency data in binary form, loaded via a serial port, and this really forces the use of a microprocessor system in the radio. DDS's second drawback for experimenters is that surface-mount packages are becoming the norm for DDS ICs. Offsetting this, the resulting

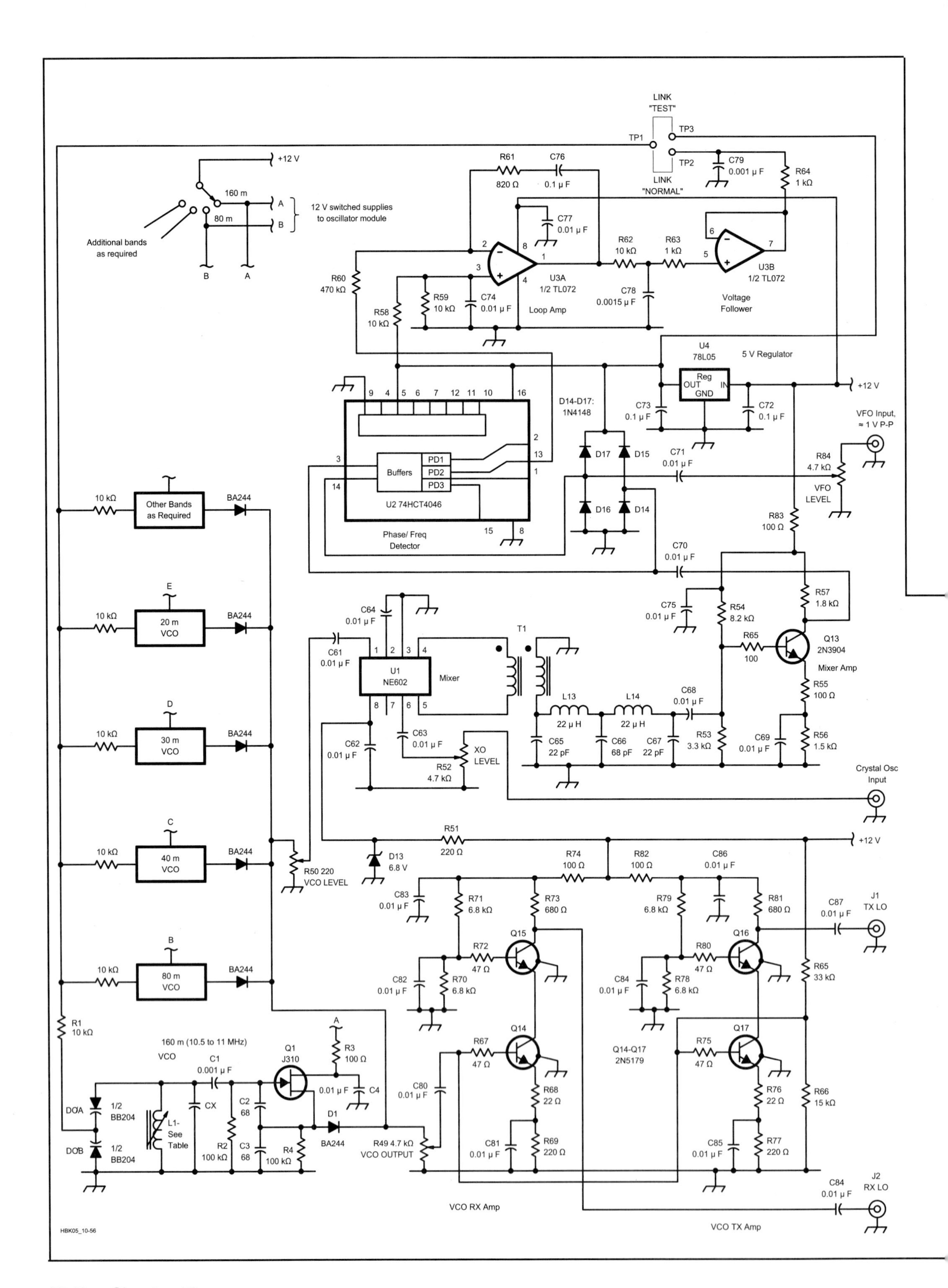
LINK
"TEST"
TP1
TP3
TP2
LINK
"NORMAL"
+12 V
160 m
80 m
12 V switched supplies
to oscillator module
Additional bands
as required
R61
820 Ω
C76
0.1 μ F
C79
0.001 μ F
R64
1 kΩ
C77
0.01 μ F
R62
10 kΩ
R63
1 kΩ
U3A
1/2 TL072
Loop Amp
U3B
1/2 TL072
Voltage
Follower
C78
0.0015 μ F
R60
470 kΩ
R59
10 kΩ
C74
0.01 μ F
R58
10 kΩ
U4
78L05
5 V Regulator
Reg
OUT
IN
GND
C73
0.1 μ F
C72
0.1 μ F
D14-D17:
1N4148
VFO Input,
≈ 1 V P-P
Buffers
PD1
PD2
PD3
U2 74HCT4046
Phase/ Freq
Detector
C71
0.01 μ F
R84
4.7 kΩ
VFO
LEVEL
R83
100 Ω
C70
0.01 μ F
Other Bands
as Required
BA244
10 kΩ
20 m
VCO
30 m
VCO
40 m
VCO
80 m
VCO
C64
0.01 μ F
C61
0.01 μ F
U1
NE602
Mixer
T1
C75
0.01 μ F
R54
8.2 kΩ
R57
1.8 kΩ
R65
100
Q13
2N3904
Mixer Amp
L13
22 μ H
L14
22 μ H
C68
0.01 μ F
R55
100 Ω
C62
0.01 μ F
C63
0.01 μ F
XO
LEVEL
R52
4.7 kΩ
C65
22 pF
C66
68 pF
C67
22 pF
R53
3.3 kΩ
C69
0.01 μ F
R56
1.5 kΩ
Crystal Osc
Input
R51
220 Ω
D13
6.8 V
R50 220
VCO LEVEL
R74
100 Ω
R82
100 Ω
C86
0.01 μ F
C83
0.01 μ F
R71
6.8 kΩ
R73
680 Ω
R79
6.8 kΩ
R81
680 Ω
C87
0.01 μ F
J1
TX LO
Q15
Q16
R72
47 Ω
R80
47 Ω
C82
0.01 μ F
R70
6.8 kΩ
C84
0.01 μ F
R78
6.8 kΩ
R65
33 kΩ
R1
10 kΩ
160 m (10.5 to 11 MHz)
VCO
C1
0.001 μ F
Q1
J310
R3
100 Ω
0.01 μ F
C4
Q14
R67
47 Ω
Q14-Q17
2N5179
Q17
R75
47 Ω
C80
0.01 μ F
R68
22 Ω
R76
22 Ω
R66
15 kΩ
D𝐶A
1/2
BB204
L1-
See
Table
CX
C2
68
D1
BA244
R2
100 kΩ
C3
68
R4
100 kΩ
R49 4.7 kΩ
VCO OUTPUT
C81
0.01 μ F
R69
220 Ω
C85
0.01 μ F
R77
220 Ω
C84
0.01 μ F
J2
RX LO
D𝐶B
VCO RX Amp
VCO TX Amp
HBK05_10-56

Table 10.3
Band-Specific Component Data for the Summing Loop in Fig 10.56

Band (MHz)	*Output Range (MHz)*	*Crystal Frequency (MHz)*	C_x *(pF)*	*VCO Coil*
1.5-2.0	10.5-11.0	16.0	0	20 turns on Toko 10K-series inductor form (≈4.29 µH)
3.5-4.0	12.5-13.0	18.0	0	16 turns on Toko 10K-series inductor form (≈3.03 µH)
7.0-7.5	16.0-16.5	21.5	0	12 turns on Toko 10K-series inductor form (≈1.85 µH)
10.0-10.5	19.0-19.5	24.5	27	8 turns on Toko 10K-series inductor form (≈0.87 µH)
14.0-14.5	23.0-23.5	28.5	56	8¹/2 turns white Toko S-18-series (≈0.435 µH)
18.0-18.5	27.0-27.5	32.5	39	7¹/2 turns violet Toko S-18-series (≈0.375 µH)
21.0-21.5	30.0-30.5	35.5	27	7¹/2 turns violet Toko S-18-series (≈0.350 µH)
24.5-25.0	33.5-34.0	39.0	22	6¹/2 turns blue Toko S-18-series (≈0.300 µH)
28.0-28.5	37.0-37.5	42.5	22	5¹/2 turns green Toko S-18-series (≈0.245 µH)
28.5-29.0	37.5-38.0	43.0	22	5¹/2 turns green Toko S-18-series (≈0.239 µH)
29.0-29.5	38.0-38.5	43.5	22	5¹/2 turns green Toko S-18-series (≈0.232 µH)
29.5-30.0	38.5-39.0	44.0	22	5¹/2 turns green Toko S-18 series (≈0.227 µH)

The Toko 10K-series forms have four-section bobbins. The VCO-coil windings for 160 through 30 m are therefore split into four equal sections (for example, 5 + 5 + 5 + 5 turns for the 160-m coil).

Fig 10.56—The G3ROO/GM4ZNX summing-loop PLL phase-locks its VCO to the frequency difference between a crystal oscillator and 5.0- to 5.5-MHz VFO (these circuits are not shown; see text). Table 10.3 lists the conversion-crystal frequency, VCO tuned-circuit padding capacitance (C_x) and VCO inductor data required for each band. Any 5.0- to 5.5-MHz VFO capable at least 2.5 mW (4 dBm) output with a 50-Ω load—1 V P-P—can drive the circuit's VFO input.

C_x, C2, C3—NP0 or C0G ceramic, 10% or tighter tolerance.
C65, C66, C67—NP0, C0G or general-purpose ceramic, 10% tolerance or tighter.
D0A, D0B—BB204 dual, common-cathode tuning diode (capacitance per section at 3 V, 39 pF) or equivalent. The ECG617, NTE617 and MV104 are suitable dual-diode substitutes, or use pairs of 1N5451s (39 pF at 4 V) or MV2109s (33 pF at 4V).
D1—BA244 switching diode. The 1N4152 is a suitable substitute.
L1—Variable inductor; see Table 10.3 for value.
L13, L14—22-µH choke, 20% tolerance or better (Miller 70F225AI, 78F270J, 8230-52, 9250-223; Mouser 43LQ225; Toko 144LY-220J [Digi-Key TK4232] suitable).
T1—6 bifilar turns of #28 enameled wire on an FT-37-72 ferrite toroid (≈30 µH per winding).
U2—74HC4046 or 74HCT4046 PLL IC (CD4046 unsuitable; see text).

simplification of an entire synthesizer to a few ICs should cause more people to experiment with them.

Fractional-N Synthesis

A single loop would be capable of any step size if its programmable divider weren't tied to integer numbers. Some designers long ago tried to use such *fractional-N* values by switching the division ratio between two integers, with a duty cycle that set the fractional part. This whole process was synchronized with the divider's operation. Averaged over many cycles, the frequency really did come out as wanted, allowing interpolation between the steps mandated by integer-N division. This approach was largely abandoned because it added huge sidebands (at the fractional frequency and its harmonics) to the loop's output. One such synthesizer design—which has been applied in a large amount of equipment and remains in use—uses a hybrid digital/analog system to compute a sawtooth voltage waveform of just the right amplitude, frequency and phase that can be added to the VCO tuning voltage to cancel the fractional-frequency FM sidebands. This system is complex, however, and like all cancellation processes, it can never provide complete cancellation. It is appreciably sensitive to changes due to tolerances, aging and alignment. Even when applied in a highly developed form using many tight-tolerance components, such a system cannot reduce its fractional frequency sidebands to a level much below –70 to –80 dBc.

A new approach, which has been described in a few articles in professional and trade journals, does not try to cancel its sidebands at all. Its basic principle is delightfully simple. A digital system switches the programmable divider of a normal single loop around a set of division values. This set of values has two properties: First, its average value is controllable in very small steps, allowing fine interpolation of the integer steps of the loop; second, the FM it applies to the loop is huge, but the resultant sideband energy is strongly concentrated in very-high-order sidebands. The loop cannot track such fast FM and so filters off this modulation!

The key to this elegant approach is that it deliberately shapes the spectrum of its loop's unwanted components such that they're small at frequencies at which the loop will pass them and large at frequencies filterable by the loop. The result is a reasonably clean output spectrum.

EXPLORING THE SYNTHESIZER IN A COMMERCIAL MF/HF TRANSCEIVER

Few people would contemplate building an entire synthesized transceiver, but far more will need to understand enough of a commercially built one to be able to fix it or modify it. Choosing a radio to use as an example was not too difficult. The ICOM IC-765 received high marks for its clean

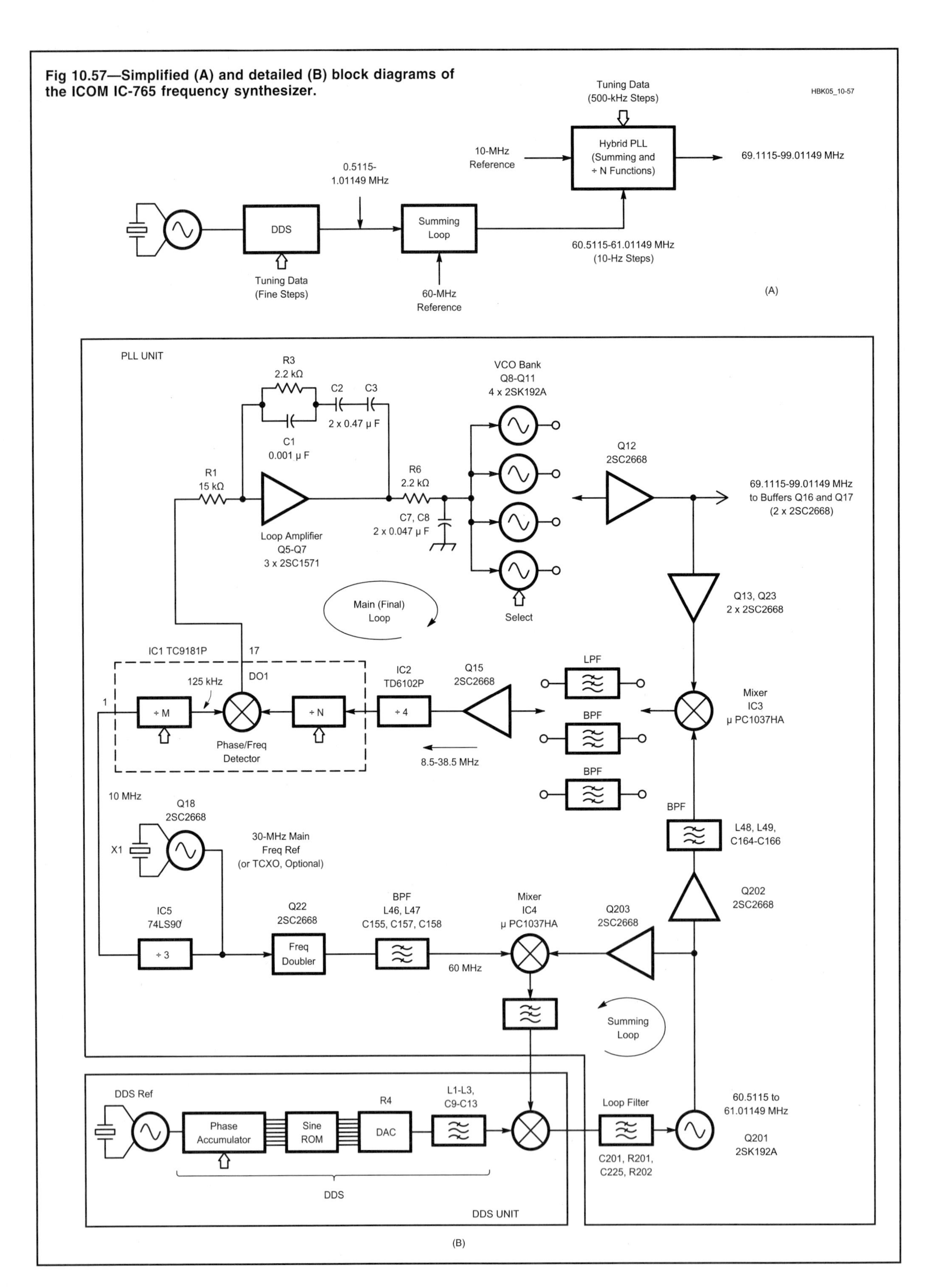

Fig 10.57—Simplified (A) and detailed (B) block diagrams of the ICOM IC-765 frequency synthesizer.

synthesizer in its *QST* Product Review, so it certainly has a synthesizer worth examining. We are grateful to ICOM America for their permission to reprint the IC-765's schematic in the discussion to follow.

Fig 10.57A shows a simplified block diagram of the IC-765's synthesizer. It contains one DDS and two PLLs. Notice that the DDS has its own frequency-reference oscillator. DDSs usually use binary arithmetic in their phase accumulators, so their step size is equal to their clock (reference) frequency divided by a large, round, binary number. The latest 32-bit machines give a step size of $1/(2^{32})$ of their clock frequency, that is, a ratio of 1/(4,294,967,296). This means that if we want, say, a 10-Hz step size, we must have a peculiar reference frequency so that the increment of the DDS is a submultiple of 10 Hz. This is what ICOM designers chose. One alternative would be to use a convenient reference frequency (say 10 MHz) and accept a strange synthesizer step size. A 32-bit machine clocked at 10 MHz will give a step size of 0.002328 Hz. It would be simple to have the radio's microprocessor select the nearest of these very fine steps to the frequency set by the user, giving the user the appearance of a 10-Hz step size. An error of ±1.2 millihertz is trivial compared to the accuracy of all usual reference oscillators.

Installing the IC-765's optional high-stability 30-MHz TCXO does nothing to improve the accuracy of the radio's DDS section, since it uses its own reference. The effect of the stability and accuracy of the DDS reference on the overall tuned frequency, however, is smaller than that of the main reference.

The DDS runs at a comfortably low frequency (0.5115 to 1.01149 MHz) to ease the demands placed on DAC settling time. The final loop needs a signal near 60 MHz as a summing input. Just mixing the DDS with 60 MHz would require a complex filter to reject the image. The IC-765 avoids this by using a summing loop. The summing loop VCO must only tune from 60.5115 to 61.01149 MHz, plus some margin for aging and temperature, but care is needed in this circuit because, as in the home-brew summing loop described earlier, the loop may latch up if this VCO goes below 60 MHz.

The final loop is a normal PLL with a programmable divider, but instead the VCO is not fed directly into the divider—it is mixed down by the 60.5115- to 61.01149-MHz signal first. The final loop

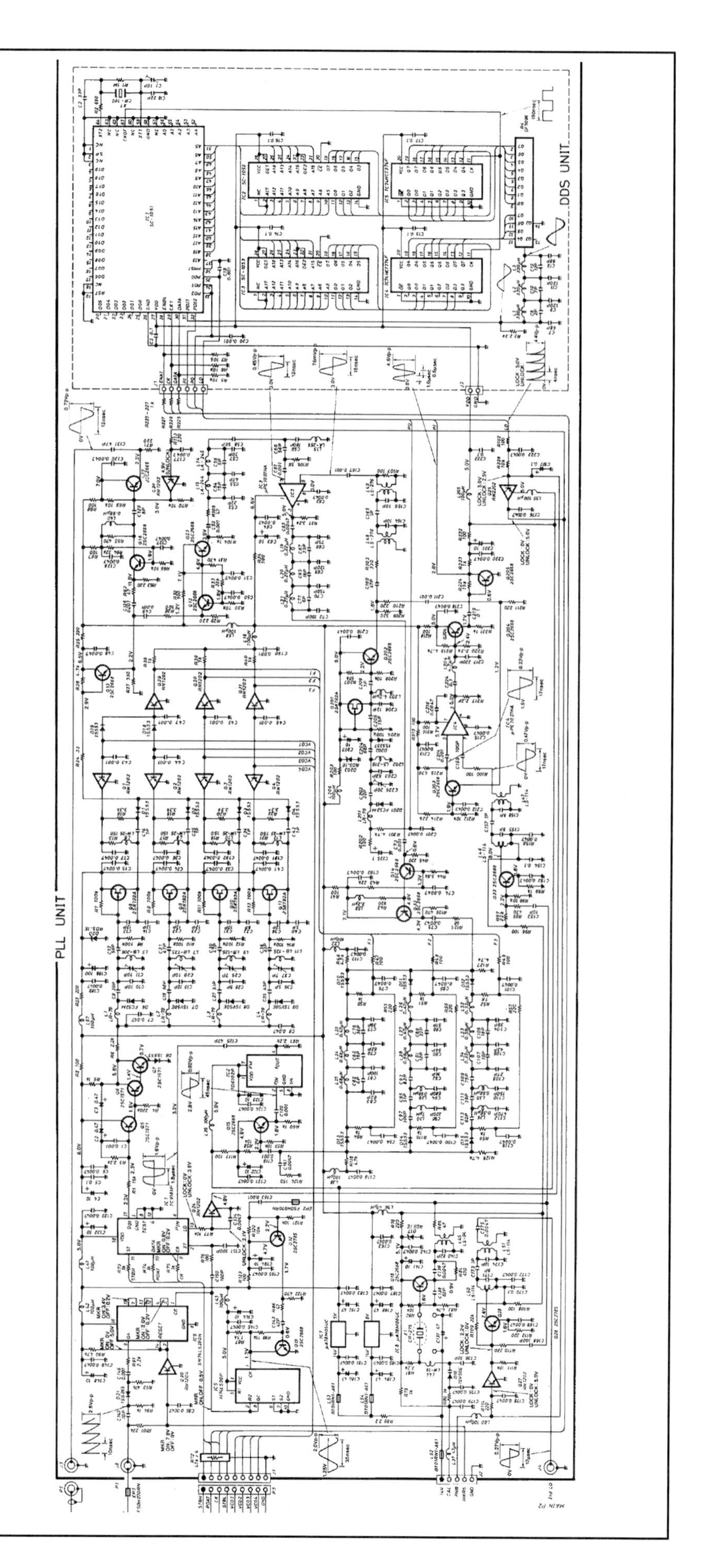

Fig 10.58—The IC-765 frequency synthesizer down to the component level. The key ICs in the radio's DDS (IC1, IC2 and IC3, right) appear to be custom components made especially for this application.

uses a bank of four switched VCOs, each one covering a one fourth of the system's full output range of 69.1115 to 99.01149 MHz. This mix-down feeds an 8.5- to 38.5-MHz signal into the programmable divi-der. To remove unwanted mixer outputs over this range, three different, switched filters are needed before the signal is amplified to drive the divider.

Fig 10.57B shows the synthesizer in greater detail. The programmable divider/PSD chip used (IC1, a TC9181P) will not operate up to 38 MHz, so a prescaler (IC2, a TD6102P) has been added before its input. This is a fixed divide-by-four prescaler, which forces the phase detector to be run at one-fourth of the step size, at 125 kHz. (A dual-modulus prescaler would have been very desirable here because it would allow the PSD to run at 500 kHz, easing the trade-off between reference-frequency suppression and loop bandwidth. Unfortunately, the lowest division factor necessary is too low for the simple application of any of the common ICs with prescaler controllers.)

Fig 10.58 shows the full circuit diagram of the IC-765's synthesizer. The DDS is on a small sub board, DDS unit (far right). IC1, an SC-1051, appears to be a custom IC, incorporating the summing loop's phase accumulator and phase detector. The DDS's sine ROM is split into two parts: IC2, an SC-1052; and IC3, an SC-1053. IC4 and IC5, both 74HCT374s, are high-speed-CMOS flip-flops used as latches to minimize glitches by closely synchronizing all 12 bits of data coming from the ROMs. The DAC is simply a binary-weighted resistor ladder network, R4, which relies on the CMOS latch outputs all switching exactly between the +5 V supply and ground.

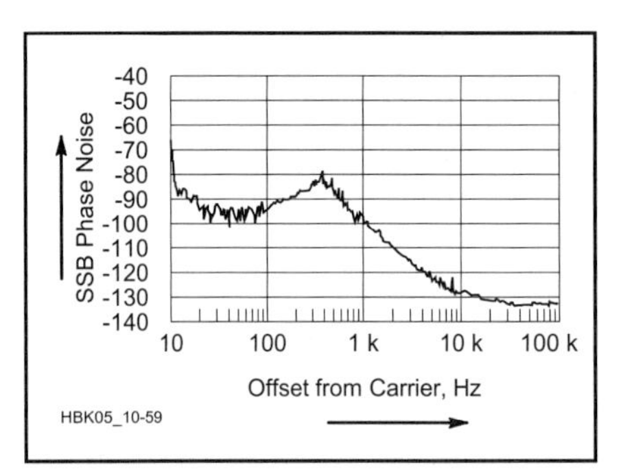

Fig 10.59—The phase noise of an ICOM IC-765 (serial no. 03077) as measured by a Hewlett-Packard phase-noise measurement system.

Look at the output of the main loop phase detector (pin 17 of IC1 on the PLL unit), find Q5, Q6 and Q7 and look at the RC network around them. It's the recipe loop design again! R3 creates the zero, but notice C1 across it. This is a neat and economical way of creating one of the other poles. R6 and C7 create the final pole.

Fig 10.59 shows the phase noise of an IC-765 measured on a professional phase-noise-measurement system. The areas shown in Fig 10.48 can be seen clearly. The slope above 500 Hz is the phase noise of the final VCO and exactly tracks the *Good* curve in KI6WX's *QST* article on the effects of phase noise. Below the noise peak, the phase noise falls to levels 20 dB better than KI6WX's *Excellent* curve. This low-noise area complements the narrow CW filter skirts and the narrow notch filter.

Many technically inclined amateurs may be wondering what could be done to improve the IC-765's synthesizer, or to design an improved one. The peak in the IC-765's phase noise at 500 Hz is quite prominent. It does not appear to be due to a marginally unstable loop design, but rather to a mismatch between the choice of loop bandwidth and the noise performance of the oscillators involved. The DDS's resistor-based DAC is unlikely to perform as well as a purpose-designed

Present and Future Trends in Oscillator Application

In this chapter a wide variety of oscillator types and frequency-synthesis schemes are discussed. Which techniques are the most important today and which ones will likely become important in the future?

As we follow radio technology from the invention of the vacuum tube, we can see a continuous evolution in the types of oscillators deemed most useful at any given time. Broadly speaking, communications systems started out with LC oscillators, and when the need for greater frequency stability arose designers moved to crystal oscillators. As the need to vary frequency became apparent, low-drift VFOs were developed to replace the crystals. As higher frequencies began to be exploited, stability again became the dominant problem. Multi-conversion systems that used low-frequency stable VFOs with crystal oscillators were developed to establish the desired high frequency stability. Today with stability, variability, and programmability all being requirements, frequency-synthesis techniques have been adopted as the norm.

When we look inside today's synthesized transceivers, we usually see only two types of oscillators. The first is a temperature-compensated crystal oscillator (TCXO), whose main purpose is to set the frequency calibration of the transceiver. The second type is the low phase-noise voltage-controlled oscillator (VCO) used in synthesizer phase lock loops.

Today, formerly popular mechanically tuned VFOs have fallen by the wayside, with the exception perhaps being their use in QRP projects and nostalgia radios. Even this could change in the near future in QRP projects, as very low power consumption synthesizers are already employed in cell phones and handhelds.

As we look to the future, we can see several emerging trends. Audio DSP has been around for a number of years, and some of the more contemporary radios are now employing IF DSP. As A-D/D-A technology approaches 16 bits at sample rates exceeding 60 MHz, it becomes feasible to produce an almost entirely DSP 1.8 to 30-MHz transceiver. The only remaining analog RF sections will be input filtering and gain compensation, and the output power amplifier and filtering. All of the traditional local oscillator synthesis hardware, IF filters, IF amplifiers, mixers and demodulation will be done in a DSP engine that resides behind an A-D/D-A converter pair. For the frequency ranges covered by the A-D/D-A pair, the synthesis process will be substantially "hidden." It will likely look like the software for direct digital synthesizer (DDS) minus the DA converter. What kinds of oscillators will be required for this architecture and what parameters will be most important?

Again, two types of oscillators will be necessary. At first glance, it would be reasonable to have a TCXO for frequency accuracy, just as in today's synthesized radios. While this would be perfectly acceptable, another option now exists. Current cell phones are being equipped with low-cost Global Positioning Satellite (GPS) chip sets to meet the federally mandated 911 emergency location requirements. An appropriate GPS chip set could give amateur transceivers frequency accuracy on a par with the cesium frequency standards employed by the GPS system. An additional benefit

12-bit DAC, and it also lets power-supply noise modulate the output. The system's designers appear to have deliberately reduced the loop bandwidth to better filter DDS spurs. If we were free to increase the cost and complexity of the unit, we could buy a high-performance DAC designed for low DDS spurs and trade this improvement off against noise by increasing the loop bandwidth—redesigning the loop poles, zero and gain. To help with the loop bandwidth, and get a 12-dB improvement in any noise from the main loop's phase detector, divider and loop amplifier, we could remove the fixed prescaler (IC2, the TD6102P) and operate the PSD at 500 kHz. This would require the design of a faster programmable divider that can handle 38 MHz.

Finally, the IC-765's VCOs could be improved. Their tuned circuits could be changed to low-L, high-C, multiple-tuning-diode types. If we added a higher-voltage power supply to allow higher diode tuning voltages than the 5.6-V level at which the design now operates (only little current is needed), higher-Q tuning diodes can be used—and we can avoid using them at lower tuning voltages, where tuning-diode Q degrades. This could reduce the phase noise above the noise peak by several decibels. Changing the loop bandwidth would reduce the noise bump's height and move it farther from the carrier.

It's important to keep all of these what-ifs in perspective: The IC-765's synthesizer was one of the best available in amateur MF/HF transceivers when this chapter was written, and its complexity is about half that of old multiloop, non-DDS examples. Like any product of mass production, its design involved trade-offs between cost, performance and component availability. As better components become less expensive, we can expect excellent synthesizer designs like the IC-765's to come down in price and even better synthesizer designs to become affordable in Amateur Radio gear.

BIBLIOGRAPHY AND REFERENCES

Clarke and Hess, *Communications Circuits; Analysis and Design* (Addison-Wesley, 1971; ISBN 0-201-01040-2). Wide coverage of transistor circuit design, including techniques suited to the design of integrated circuits. Its age shows, but it is especially valuable for its good mathematical treatment of oscillator circuits, covering both frequency- and amplitude-determining mechanisms. Look for a copy at a university library or initiate an interlibrary loan.

J. Grebenkemper, "Phase Noise and Its Effects on Amateur Communications," *Part 1*, *QST*, Mar 1988, pp 14-20; *Part 2*, Apr 1988, pp 22-25. Also see Feedback, *QST*, May 1988, p 44. This material also appears in volumes 1 (1991) and 2 (1993) of *The ARRL Radio Buyer's Sourcebook*. Covers the effects of phase noise and details the measurement techniques used in the ARRL Lab. Those techniques are also described in a sidebar earlier in this chapter.

W. Hayward and D. DeMaw, *Solid State Design for the Radio Amateur* (Newington, CT: ARRL, 1986). A good sourcebook of RF design ideas, with good explanation of the reasoning behind design decisions.

"Spectrum Analysis...Random Noise Measurements," Hewlett-Packard Application Note 150-4. Source of correction factors for noise measurements using spectrum analyzers. HP notes should be available through local HP dealers.

U. Rohde, *Digital PLL Frequency Synthesizers* (Englewood Cliffs, NJ: Prentice-

would be that the transceiver would now know its location anywhere in the world. This would be of value to DXpeditioners and contesters!

The second oscillator would most likely take the form of a very low phase noise voltage-controlled crystal oscillator (VCXO). Some variant of the Butler oscillator would be a likely candidate. This oscillator would be disciplined by the output of the GPS chipset or the onboard frequency standard TCXO. The requirement for low phase noise will remain very important, since any clock jitter in the DSP engine or the A-D/D-A clocks will degrade the phase-noise performance of the transceiver.

The possibility of an all-DSP HF radio raises the question of whether or not frequency synthesizers, as we now know them, will go the way of the VFO? For the most demanding applications, such as in signal generators and test equipment, the answer is likely no. The reason is that the most current synthesis techniques offer better performance than is possible with the basic DDS approach, although at additional cost. With respect to HF amateur radio equipment—where cost performance tradeoffs are important—the answer is likely yes.

Now that we have discussed the possibility of a DSP based HF radio, what about VHF, UHF and microwave amateur equipment? It is reasonable to assume that the direct DSP approach could be extended up in frequency as the A-D/D-A technologies improve in speed, however, there are some limiting factors regarding phase noise performance. The present method of using a tunable IF, such as the HF transceiver and a transverter, is likely to remain the practical option for some time to come. What we are likely to see is a different approach to the generation of the LO frequency in transverters. Many of today's transverters employ a crystal oscillator and multiplier chain to produce the LO injection frequency. This architecture has some disadvantages in that the crystal oscillator often has an offset and drift that must be taken up in the HF transceiver. There is also the issue of spurious responses due to the multiplying process and adequacy of the attendant filters. An alternative would be to use a simplified static synthesizer whose frequency reference is the same as that used in the HF transceiver. This approach, implemented with a low-noise UHF or microwave VCO and phase-lock loop, could yield reasonable phase noise, greater frequency accuracy, fewer spurious responses and smaller size as opposed to the crystal/multiplier chain technique. Again, we can see that the low-noise VCO will be an important element in these designs. These UHF and microwave VCOs would likely employ ceramic or dielectric resonators like those now used extensively in cell phones and TVRO front ends.

It is extremely likely that voltage-controlled oscillators, whether they are crystal, LC, ceramic or dielectric resonator types, will dominate the communications electronics landscape. Very low phase noise will be the cardinal performance specification for these VCOs. High-stability frequency-standard oscillators, while still viable, may give way to GPS as a preferred means of accurate frequency deployment. Mechanically tuned VFOs have all but disappeared from modern equipment and are likely to continue to diminish in application.

Hall, Inc). Now available solely from Compact Software, 201 McLean Blvd, Paterson, NJ 07504, this book contains the textbook-standard mathematical analyses of frequency synthesizers combined with unusually good insight into what makes a better synthesizer and a *lot* of practical circuits to entertain serious constructors. A good place to look for low-noise circuits and techniques.

U. Rohde, "Key Components of Modern Receiver Design," *Part 1, QST*, May 1994, pp 29-32; *Part 2, QST*, Jun 1994, pp 27-31; *Part 3, QST*, Jul 1994, pp 42-45. Includes discussion on phase-noise reduction techniques in synthesizers and oscillators.

Pappenfus and others, *Single Sideband Circuits and Systems* (McGraw-Hill). A book on HF SSB transmitters, receivers and accoutrements by Rockwell-Collins staff. Contains chapters on synthesizers and frequency standards. The frequency synthesizer chapter predates the rise of the DDS and other recent techniques, but good information about the effects of synthesizer performance on communications is spread throughout the book.

Motorola Applications Note AN-551, "Tuning Diode Design Techniques," included in Motorola RF devices databooks. A concise explanation of varactor diodes, their characteristics and use.

I. Keyser, "An Easy to Set Up Amateur Band Synthesizer," *RADCOM* (RSGB), Dec 1993, pp 33-36.

B.E. Pontius, "Measurement of Signal Source Phase Noise with Low-Cost Equipment," *QEX*, May-June 1998, pp 38-49.

F.M. Gardner, *Phase Lock Techniques* (John Wiley and Sons, 1966, 1975; ISBN 0-471-29156-0).

Chapter 11

Mixers, Modulators and Demodulators

At base, radio communication involves translating information into radio form, letting it travel for a time, and translating it back again. Translating information into radio form entails the process we call *modulation*, and *demodulation* is its reverse. One way or another, every transmitter used for radio communication, from the simplest to the most complex, includes a means of modulation; one way or another, every receiver used for radio communication, from the simplest to the most complex, includes a means of demodulation.

Modulation involves varying one or both of a radio signal's basic characteristics — amplitude and frequency (or phase) — to convey information. A circuit, stage or piece of hardware that modulates is called a *modulator*.

Demodulation involves reconstructing the transmitted information from the changing characteristic(s) of a modulated radio wave. A circuit, stage or piece of hardware that demodulates is called a *demodulator*.

Many radio transmitters, receivers and transceivers also contain *mixers* — circuits, stages or pieces of hardware that combine two or more signals to produce additional signals at sums of and differences between the original frequencies. Amateur Radio textbooks have traditionally handled mixers separately from modulators and demodulators, and modulators separately from demodulators.

This chapter, by David Newkirk, ex-W9VES, and Rick Karlquist, N6RK, examines mixers, modulators and demodulators together because the job they do is essentially the same. Modulators and demodulators translate information into radio form and back again; mixers translate one frequency to others and back again. All of these translation processes can be thought of as forms of frequency translation or frequency shifting — the function traditionally ascribed to mixers. We'll therefore begin our investigation by examining what a mixer is (and isn't), and what a mixer does.

THE MECHANISM OF MIXERS AND MIXING

What is a Mixer?

Mixer is a traditional radio term for a circuit that shifts one signal's frequency up or down by combining it with another signal. The word *mixer* is also used to refer to a device used to blend multiple audio inputs together for recording, broadcast or sound reinforcement. These two mixer types differ in one very important way: A radio mixer makes new frequencies out of the frequencies put into it, and an audio mixer does not.

Mixing Versus Adding

Radio mixers might be more accurately called "frequency mixers" to distinguish them from devices such as "microphone mixers," which are really just signal *combiners*, *summers* or *adders*. In their most basic, ideal forms, both devices have two inputs and one output. The combiner simply *adds* the instantaneous voltages of the two signals together to produce the output at each point in time (**Fig 11.1**). The mixer, on the other hand, *multiplies* the instantaneous voltages of the two signals together to produce its output signal from instant to instant (**Fig 11.2**). Comparing the output spectra of the combiner and mixer, we see that the combiner's output contains only the frequencies of the two inputs, and nothing else, while the mixer's output contains *new* frequencies. Because it combines one energy with another, this process is sometimes called *heterodyning*, from the Greek words for *other* and *power*.

Mixing as Multiplication

Since a mixer works by means of multiplication, a bit of math can show us how they work. To begin with, we need to represent the two signals we'll mix, A and B, mathematically. Signal A's instantaneous amplitude equals

$$A_a \sin 2\pi f_a t \quad (1)$$

in which A is peak amplitude, f is frequency, and t is time. Likewise, B's instantaneous amplitude equals

$$A_b \sin 2\pi f_b t \quad (2)$$

Since our goal is to show that multiplying two signals generates sum and difference frequencies, we can simplify these signal definitions by assuming that the peak amplitude of each is 1. The equation for Signal A then becomes

$$a(t) = A \sin\left(2\pi f_a t\right) \quad (3)$$

and the equation for Signal B becomes

$$b(t) = B \sin\left(2\pi f_b t\right) \quad (4)$$

Each of these equations represents a sine wave and includes a subscript letter to help us keep track of where the signals go.

Merely combining Signal A and Signal

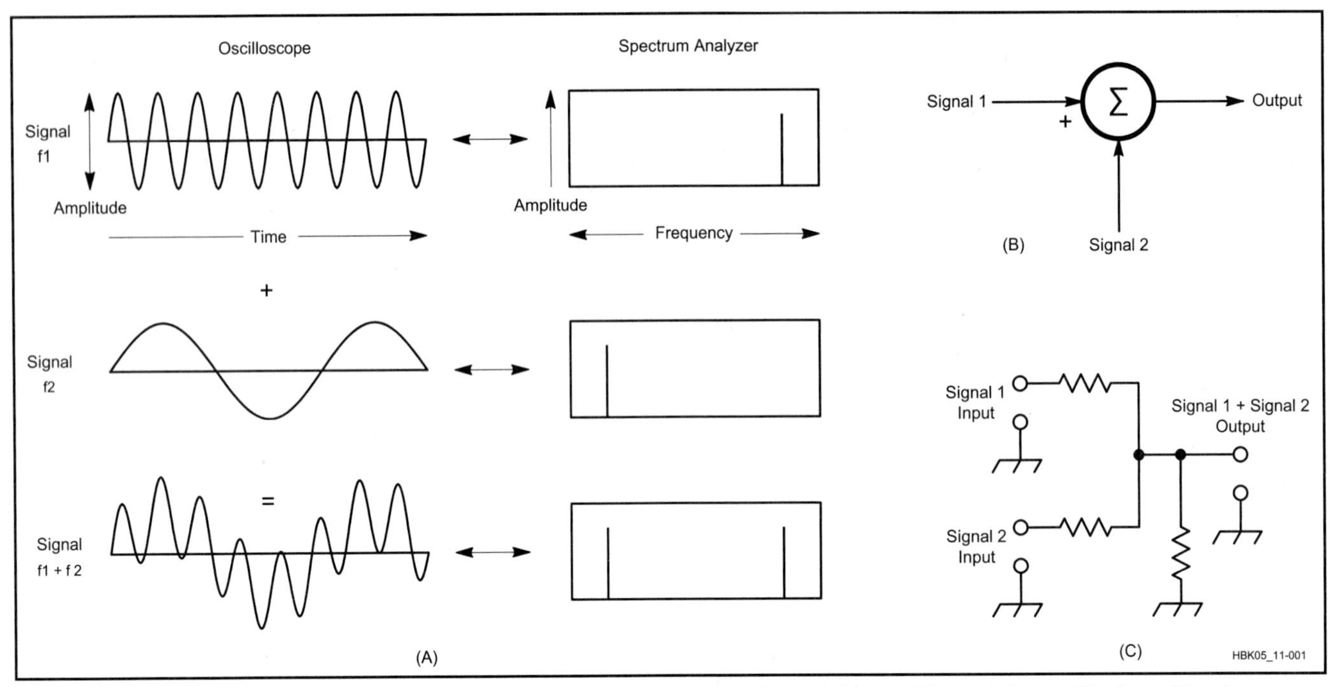

Fig 11.1 — *Adding or summing* two sine waves of different frequencies (f1 and f2) combines their amplitudes without affecting their frequencies. Viewed with an *oscilloscope* (a real-time graph of amplitude versus time), adding two signals appears as a simple superimposition of one signal on the other. Viewed with a *spectrum analyzer* (a real-time graph of signal amplitude versus frequency), adding two signals just sums their spectra. The signals merely coexist on a single cable or wire. All frequencies that go into the adder come out of the adder, and no new signals are generated. Drawing B, a block diagram of a summing circuit, emphasizes the stage's mathematical operation rather than showing circuit components. Drawing C shows a simple summing circuit, such as might be used to combine signals from two microphones. In audio work, a circuit like this is often called a mixer — but it does not perform the same function as an RF mixer.

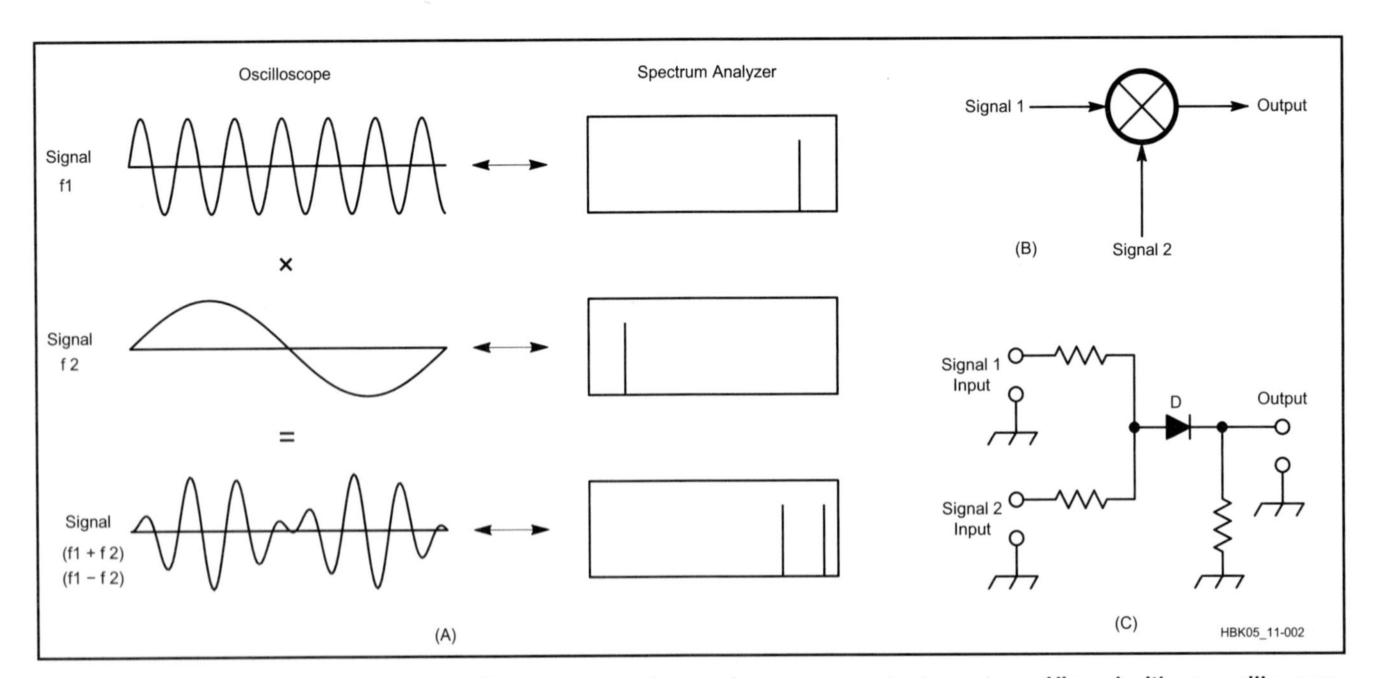

Fig 11.2 — *Multiplying* two sine waves of different frequencies produces a new output spectrum. Viewed with an oscilloscope, the result of multiplying two signals is a composite wave that seems to have little in common with its components. A spectrum-analyzer view of the same wave reveals why: The original signals disappear entirely and are replaced by two new signals — at the *sum* and *difference* of the original signals' frequencies. Drawing B diagrams a multiplier, known in radio work as a mixer. The *X* emphasizes the stage's mathematical operation. (The circled *X* is only one of several symbols you may see used to represent mixers in block diagrams, as Fig 11.3 explains.) Drawing C shows a very simple multiplier circuit. The diode, D, does the mixing. Because this circuit does other mathematical functions and adds them to the sum and difference products, its output is more complex than f1 + f2 and f1 – f2, but these can be extracted from the output by filtering.

B by letting them travel on the same wire develops nothing new:

$$a(t) + b(t) = A\sin(2\pi f_a t) + B\sin(2\pi f_b t) \quad (5)$$

As needlessly reflexive as equation 5 may seem, we include it to highlight the fact that multiplying two signals is a quite different story. From trigonometry, we know that multiplying the sines of two variables can be expanded according to the relationship

$$\sin x \sin y = \frac{1}{2}\left[\cos(x - y) - \cos(x + y)\right] \quad (6)$$

Conveniently, Signals A and B are both sinusoidal, so we can use equation 6 to determine what happens when we multiply Signal A by Signal B. In our case, $x = 2\pi f_a t$ and $y = 2\pi f_b t$, so plugging them into equation 6 gives us

$$a(t) \bullet b(t) = \frac{AB}{2}\cos(2\pi[f_a - f_b]t) - \frac{AB}{2}\cos(2\pi[f_a + f_b]t) \quad (7)$$

Now we see two momentous results: a sine wave at the frequency *difference* between Signal A and Signal B $2\pi(f_a - f_b)t$, and a sine wave at the frequency *sum* of Signal A and Signal B $2\pi(f_a + f_b)t$. (The products are cosine waves, but since equivalent sine and cosine waves differ only by a phase shift of 90°, both are called *sine waves* by convention.)

This is the basic process by which we translate information into radio form and translate it back again. If we want to transmit a 1-kHz audio tone by radio, we can feed it into one of our mixer's inputs and feed an RF signal — say, 5995 kHz — into the mixer's other input. The result is two radio signals: one at 5994 kHz (5995 – 1) and another at 5996 kHz (5995 + 1). We have achieved modulation.

Converting these two radio signals back to audio is just as straightforward. All we do is feed them into one input of another mixer, and feed a 5995-kHz signal into the mixer's other input. Result: a 1-kHz tone. We have achieved demodulation; we have communicated by radio.

The key principle of a radio mixer is that in mixing multiple signal voltages together, it adds and subtracts their frequencies to produce new frequencies. (In the field of signal processing, this process, *multiplication in the time domain,* is recognized as equivalent to the process of *convolution in the frequency domain.* Those interested in this alternative approach to describing the generation of new frequencies through mixing can find more information about it in the many textbooks available on this fascinating subject.) The difference between the mixer we've been describing and any mixer, modulator or demodulator that you'll ever use is that it's ideal. We put in two signals and got just two signals out. *Real* mixers, modulators and demodulators, on the other hand, also produce *distortion* products that make their output spectra "dirtier" or "less clean," as well as putting out some energy at input-signal frequencies and their harmonics. Much of the art and science of making good use of multiplication in mixing, modulation and demodulation goes into minimizing these unwanted multiplication products (or their effects) and making multipliers do their frequency translations as efficiently as possible.

Putting Multiplication to Work

Piecing together a coherent picture of how multiplication works in radio communication isn't made any easier by the fact that traditional terms applied to a given multiplication approach and its products may vary with their application. If, for instance, you're familiar with standard textbook approaches to mixers, modulators and demodulators, you may be wondering why we didn't begin by working out the math involved by examining *amplitude modulation*, also known as *AM*. "Why not tell them about the *carrier* and how to get rid of it in a *balanced modulator*?" A transmitter enthusiast may ask "Why didn't you mention *sidebands* and how we conserve spectrum space and power by getting rid of one and putting all of our power into the other?" A student of radio receivers, on the other hand, expects any discussion of the same underlying multiplication issues to touch on the topics of *LO feedthrough*, *mixer balance* (*single* or *double*?), *image rejection* and so on.

You likely expect this book to spend some time talking to you about these things, so it will. But *this* radio-amateur-oriented discussion of mixers, modulators and demodulators will take a look at their common underlying mechanism *before* turning you loose on practical mixer, modulator and demodulator circuits. Then you'll be able to tell the forest from the trees. **Fig 11.3** shows the block symbol for a traditional mixer along with several IEC symbols for other functions mixers may perform.

It turns out that the mechanism underlying multiplication, mixing, modulation and demodulation is a pretty straightforward thing: Any circuit structure that *nonlinearly distorts* ac waveforms acts as a multiplier to some degree.

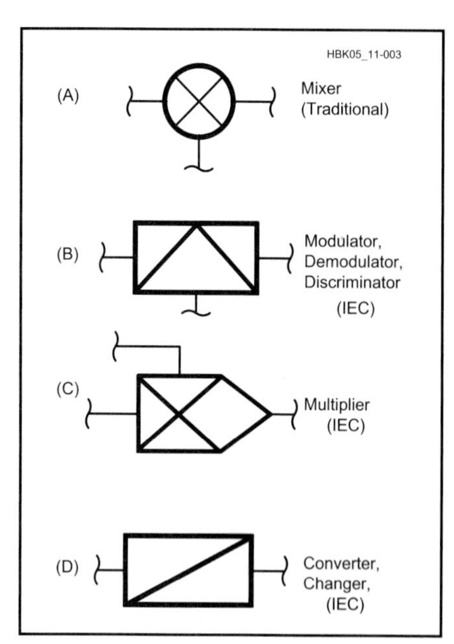

Fig 11.3 — We commonly symbolize mixers with a circled *X* (A) out of tradition, but other standards sometimes prevail (B, C and D). Although the converter/changer symbol (D) can conceivably be used to indicate frequency changing through mixing, the three-terminal symbols are arguably better for this job because they convey the idea of two signal sources resulting in a new frequency. (*IEC* stands for *International Electrotechnical Commission*.)

Nonlinear Distortion?

The phrase *nonlinear distortion* sounds redundant, but isn't. Distortion, an externally imposed change in a waveform, can be linear; that is, it can occur independently of signal amplitude. Consider a radio receiver front-end filter that passes only signals between 6 and 8 MHz. It does this by *linearly distorting* the single complex waveform corresponding to the wide RF spectrum present at the radio's antenna terminals, reducing the amplitudes of frequency components below 6 MHz and above 8 MHz relative to those between 6 and 8 MHz. (Considering multiple signals on a wire as one complex waveform is just as valid, and sometimes handier, than considering them as separate signals. In this case, it's a bit easier to think of distortion as something that happens to a waveform rather than something that happens to separate signals relative to each other. It would be just as valid — and certainly more in keeping with the consensus view — to say merely that the filter attenuates signals at frequencies below 6 MHz and above 8 MHz.) The filter's output waveform certainly differs from its input waveform; the waveform has been distorted.

But because this distortion occurs independently of signal level or polarity, the distortion is linear. No new frequency components are created; only the amplitude relationships among the wave's existing frequency components are altered. This is *amplitude* or *frequency* distortion, and all filters do it or they wouldn't be filters.

Phase or *delay distortion*, also linear, causes a complex signal's various component frequencies to be delayed by different amounts of time, depending on their frequency but independently of their amplitude. No new frequency components occur, and amplitude relationships among existing frequency components are not altered. Phase distortion occurs to some degree in all real filters.

The waveform of a non-sinusoidal signal can be changed by passing it through a circuit that has only linear distortion, but only *nonlinear distortion* can change the waveform of a simple sine wave. It can also produce an output signal whose output waveform changes as a function of the input amplitude, something not possible with linear distortion. Nonlinear circuits often distort excessively with overly strong signals, but the distortion can be a complex function of the input level.

Nonlinear distortion may take the form of *harmonic distortion*, in which integer multiples of input frequencies occur, or *intermodulation distortion (IMD)*, in which different components multiply to make new ones.

Any departure from absolute linearity results in some form of nonlinear distortion, and this distortion can work for us or against us. Any so-called linear amplifier distorts nonlinearly to some degree; any device or circuit that distorts nonlinearly can work as a mixer, modulator, demodulator or frequency multiplier. An amplifier optimized for linear operation will nonetheless mix, but inefficiently; an amplifier biased for nonlinear amplification may be practically linear over a given tiny portion of its input-signal range. The trick is to use careful design and component selection to maximize nonlinear distortion when we want it, and minimize it when we don't. Once we've decided to maximize nonlinear distortion, the trick is to minimize the distortion products we don't want, and maximize the products we desire.

Keeping Unwanted Distortion Products Down

Ideally, a mixer multiplies the signal at one of its inputs by the signal at its other input, but does not multiply a signal at the same input by itself, or multiple signals at the same input by themselves or by each other. (Multiplying a signal by itself — squaring it — generates harmonic distortion [specifically, *second-harmonic* distortion] by adding the signal's frequency to itself per equation 7. Simultaneously squaring two or more signals generates simultaneous harmonic and intermodulation distortion, as we'll see later when we explore how a diode demodulates AM.)

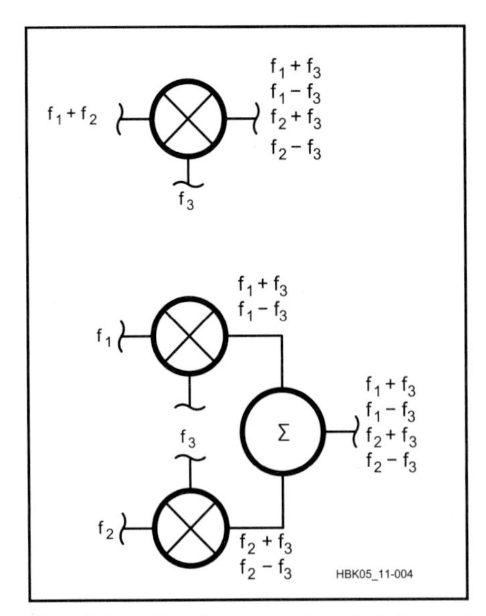

Fig 11.4 — Feeding two signals into one input of a mixer results in the same output as if f_1 and f_2 are each first mixed with f_3 in two separate mixers, and the outputs of these mixers are combined.

Consider what happens when a mixer must handle signals at two different frequencies (we'll call them f_1 and f_2) applied to its first input, and a signal at a third frequency (f_3) applied to its other input. Ideally, a mixer multiplies f_1 by f_3 and f_2 by f_3, but does not multiply f_1 and f_2 by each other. This produces output at the sum and difference of f_1 and f_3, and the sum and difference of f_2 and f_3, but *not* the sum and difference of f_1 and f_2. **Fig 11.4** shows that feeding two signals into one input of a mixer results in the same output as if f_1 and f_2 are each first mixed with f_3 in two separate mixers, and the outputs of these mixers are combined. This shows that a mixer, even though constructed with nonlinearly distorting components, actually behaves as a *linear frequency shifter*. Traditionally, we refer to this process as mixing and to its outputs as *mixing products*, but we may also call it frequency *conversion*, referring to a device or circuit that does it as a *converter*, and to its outputs as *conversion products*.

Real mixers, however, at best act only as *reasonably* linear frequency shifters, generating some unwanted IMD products — spurious signals, or *spurs* — as they go. Receivers are especially sensitive to unwanted mixer IMD because the signal-level spread over which they must operate without generating unwanted IMD is often 90 dB or more, and includes infinitesimally weak signals in its span. In a receiver, IMD products so tiny that you'd never notice them in a transmitted signal can easily obliterate weak signals. This is why receiver designers apply so much effort to achieving "high dynamic range."

The degree to which a given mixer, modulator or demodulator circuit produces unwanted IMD is often *the* reason why we use it, or don't use it, instead of another circuit that does its wanted-IMD job as well or even better.

Other Mixer Outputs

In addition to desired sum-and-difference products and unwanted IMD products, real mixers also put out some energy at their input frequencies. Some mixer implementations may *suppress* these outputs — that is, reduce one or both of their input signals by a factor of 100 to 1,000,000, or 20 to 60 dB. This is good because it helps keep input signals at the desired mixer-output sum or difference frequency from showing up at the IF terminal — an effect reflected in a receiver's *IF rejection* specification. Some mixer types, especially those used in the vacuum-tube era, suppress their input-signal outputs very little or not at all.

Input-signal suppression is part of an overall picture called *port-to-port isolation*. Mixer input and output connections are traditionally called *ports*. By tradition, the port to which we apply the shifting signal is the *local-oscillator (LO)* port. The convention for naming the other two ports (one of which must be an output, and the other of which must be an input) is usually that the higher-frequency port is called the *RF (radio frequency)* port and the lower-frequency port is called the *IF (intermediate frequency)* port. If a mixer's output frequency is lower than its input frequency, then the RF port is an input and the IF port is an output. If the output frequency is higher than the input frequency, the IF port may be the input and the RF port may be the output. (We hedge with *may be* because usage varies. When in doubt, check a diagram carefully to determine which port is the "gozinta" and which port is the "gozouta.")

It's generally a good idea to keep a mixer's input signals from appearing at its output port because they represent energy that we'd rather not pass on to subsequent circuitry. It therefore follows that it's usually a good idea to keep a mixer's LO-port energy from appearing at its RF port, or its RF-port energy from making it through to the IF port. But there are some notable exceptions.

Mixers and Amplitude Modulation

Now that we've just discussed what a fine thing it is to have a mixer that doesn't let its input signals through to its output port, we can explore a mixing approach that outputs one of its input signals so strongly that the fed-through signal's amplitude at least equals the combined amplitudes of the system's sum and difference products! This system, *amplitude modulation*, is the oldest means of translating information into radio form and back again. It's a frequency-shifting system in which the original unmodulated signal, traditionally called the *carrier*, emerges from the mixer along with the sum and difference products, traditionally called *sidebands*.

We can easily make the carrier pop out of our mixer along with the sidebands merely by building enough *dc level shift* into the information we want to mix so that its waveform never goes negative. Back at equations 1 and 2, we decided to keep our mixer math relatively simple by setting the peak voltage of our mixer's input signals directly equal to their sine values. Each input signal's peak voltage therefore varies between +1 and –1, so all we need to do to keep our modulating-signal term (provided with a subscript m to reflect its role as the modulating or information waveform) from going negative is add 1 to it. Identifying the carrier term with a subscript c, we can write

$$\text{AM signal} = \left(1 + m \sin 2\pi f_m t\right) \sin 2\pi f_c t \quad (8)$$

Notice that the modulation ($2\pi f_m t$) term has company in the form of a coefficient, m. This variable expresses the modulating signal's varying amplitude — variations that ultimately result in amplitude modulation. Expanding equation 8 according to equation 6 gives us

$$\begin{aligned}\text{AM signal} &= \sin 2\pi f_c t \\ &+ \frac{1}{2} m \cos\left(2\pi f_c - 2\pi f_m\right)t \\ &- \frac{1}{2} m \cos\left(2\pi f_c + 2\pi f_m\right)t\end{aligned} \quad (9)$$

The modulator's output now includes the carrier ($\sin 2\pi f_c t$) in addition to sum and difference products that vary in strength according to m. According to the conventions of talking about modulation, we call the sum product, which comes out at a frequency higher than that of the carrier, the *upper sideband (USB)*, and the difference product, which comes out a frequency lower than that of the carrier, the *lower sideband (LSB)*. We have achieved amplitude modulation.

Why We Call It Amplitude Modulation

We call the modulation process described in equation 8 *amplitude modulation* because the complex waveform consisting of the sum of the sidebands and carrier varies with the information signal's magnitude (m). Concepts long used to illustrate AM's mechanism may mislead us into thinking that the *carrier* varies in strength with modulation, but careful study of equation 9 shows that this doesn't happen. The carrier, $\sin 2\pi f_c t$, goes into the modulator — we're in the modulation business now, so it's fitting to use the term *modulator* instead of *mixer* — as a sinusoid with an unvarying maximum value of $|1|$. The modulator multiplies the carrier by the dc level (+1) that we added to the information signal ($m \sin 2\pi f_m t$). Multiplying $\sin 2\pi f_c t$ by 1 merely returns $\sin 2\pi f_c t$. We have proven that the carrier's amplitude does not vary as a result of amplitude modulation. The carrier is, however, used by many circuits as a reference signal.

Overmodulation

Since the audio we are transmitting in AM shows up entirely as energy in its sidebands, it follows that the more energetic we make the sidebands, the more information energy will be available for an AM receiver to "recover" when it demodulates the signal. Even in an ideal modulator, there's a practical limit to how strong we can make an AM signal's sidebands relative to its carrier, however. Beyond that limit, we severely distort the waveform we want to translate into radio form.

We reach AM's distortion-free modulation limit when the sum of the sidebands and carrier at the modulator output *just reaches zero* at the modulating waveform's most negative peak (**Fig 11.5**). We call this condition *100% modulation*, and it occurs when m equals 1. (We enumerate *modulation percentage* in values from 0 to 100%. The lower the number, the less information energy in the sidebands. You may also see modulation enumerated in terms of a *modulation factor* from 0 to 1, which directly equals m; a modulation factor of 1 is the same as 100% modulation.) Equation 9 shows that each sideband's voltage is half that of the carrier. Power varies as the square of voltage, so the power in each sideband of a 100%-modulated signal is therefore $(1/2)^2$ times, or 1/4, that of the carrier. A transmitter

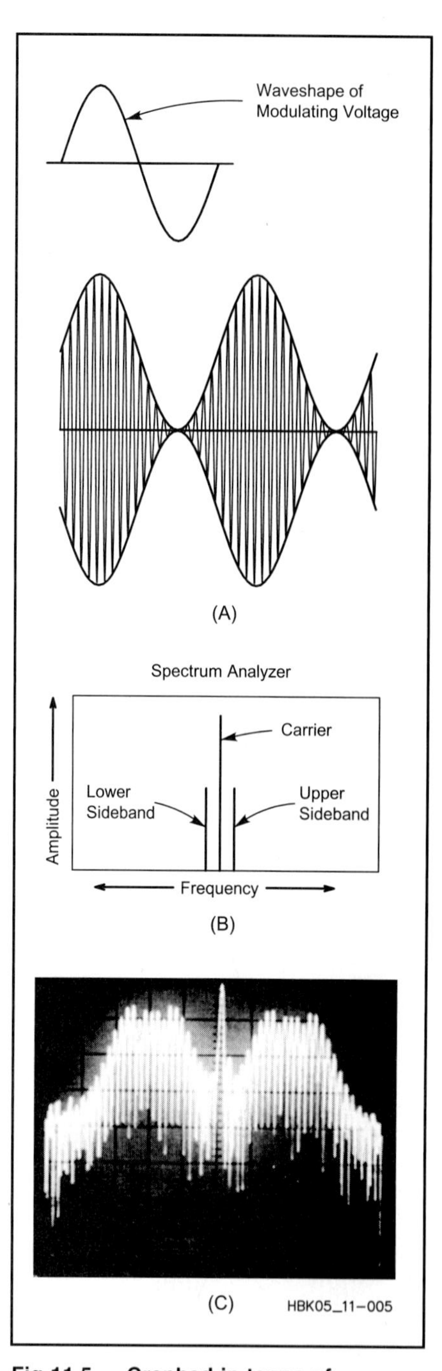

Fig 11.5 — Graphed in terms of amplitude versus time (A), the *envelope* of a properly modulated AM signal exactly mirrors the shape of its modulating waveform, which is a sine wave in this example. This AM signal is modulated as fully as it can be — 100% — because its envelope *just* hits zero on the modulating wave's negative peaks. Graphing the same AM signal in terms of amplitude versus frequency (B) reveals its three spectral components: Carrier, upper sideband and lower sideband. B shows sidebands as single-frequency components because the modulating waveform is a sine wave. With a complex modulating waveform, the modulator's sum and difference products really do show up as *bands* on either side of the carrier (C).

capable of 100% modulation when operating at a carrier power of 100 W therefore puts out a 150-W signal at 100% modulation, 50 W of which is attributable to the sidebands. (The *peak* envelope power [PEP] output of a double-sideband, full-carrier AM transmitter at 100% modulation is four times its carrier PEP. This is why our solid-state, "100-W" MF/HF transceivers are usually rated for no more than about 25 W carrier output at 100% amplitude modulation.)

One-hundred-percent modulation is a brick-wall limit because an amplitude modulator can't reduce its output to less than zero. Trying to increase modulation beyond the 100% point results in *overmodulation* (**Fig 11.6**), in which the modulation envelope no longer mirrors the shape of the modulating wave (Fig 11.6A). An overmodulated wave contains more energy than it did at 100% modulation, but some of the added energy now exists as *harmonics of the modulating waveform* (Fig 11.6B). This distortion makes the modulated signal take up more spectrum space than it needs. In voice operation, overmodulation commonly happens only on syllabic peaks, making the distortion products sound like crashy, transient noise we refer to as *splatter*.

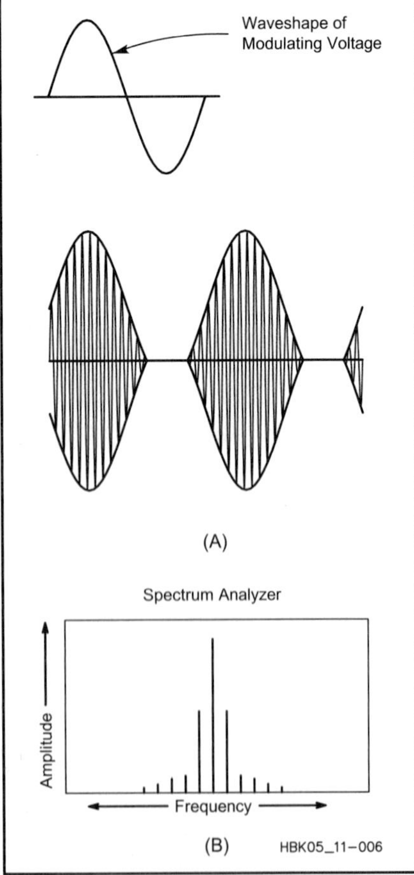

Fig 11.6 — Overmodulating an AM transmitter results in a modulation envelope (A) that doesn't faithfully mirror the modulating waveform. This distortion creates additional sideband components that broaden the transmitted signal (B).

Modulation Linearity

If we increase an amplitude modulator's modulating-signal input by a given percentage, we expect a proportional modulation increase in the modulated signal. We expect good *modulation linearity*. Suboptimal amplitude modulator design may not allow this, however. Above some modulation percentage, a modulator may fail to increase modulation in proportion to an increase its input signal (**Fig 11.7**). Distortion, and thus an unnecessarily wide signal, results.

Using AM to Send Morse Code

Fig 11.8A closely resembles what we see when a properly adjusted CW transmitter sends a string of dots. Keying a carrier on and off produces a wave that varies in amplitude and has double (upper and lower) sidebands that vary in spectral composition according to the duration and envelope shape of the on-off transitions. The emission mode we call *CW* is therefore a form of AM. The concepts of modulation percentage and overmodulation are usually not applied to generating an on-off-keyed Morse signal, however. This is related to how we copy CW by ear, and the fact that, in CW radio communication, we usually don't translate the received signal all the way back into its original pre-modulator (*baseband*) form, as a closer look at the process reveals.

In CW transmission, we usually open and close a keying line to make dc transitions that turn the transmitted carrier on and off. See Fig 11.8B. CW reception usually does not entirely reverse this process, however. Instead of demodulating a CW signal all the way back to its baseband self — a shifting dc level — we want the presences and absences of its carrier to create long and short audio tones. Because the carrier is RF and not AF, we must mix it

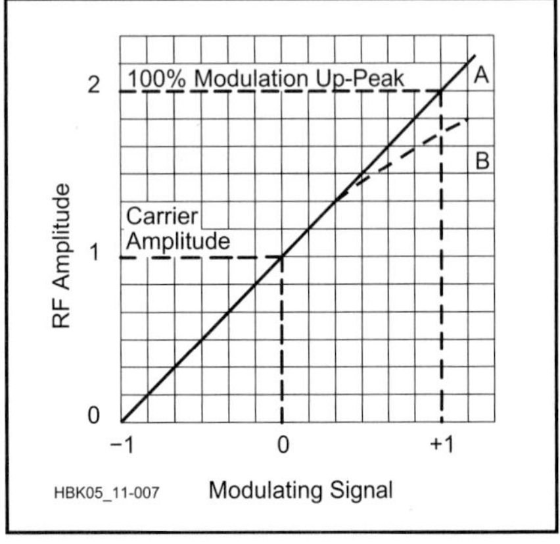

Fig 11.7 — An ideal AM transmitter exhibits a straight-line relationship (A) between its instantaneous envelope amplitude and the instantaneous amplitude of its modulating signal. Distortion, and thus an unnecessarily wide signal, results if the transmitter cannot respond linearly across the modulating signal's full amplitude range.

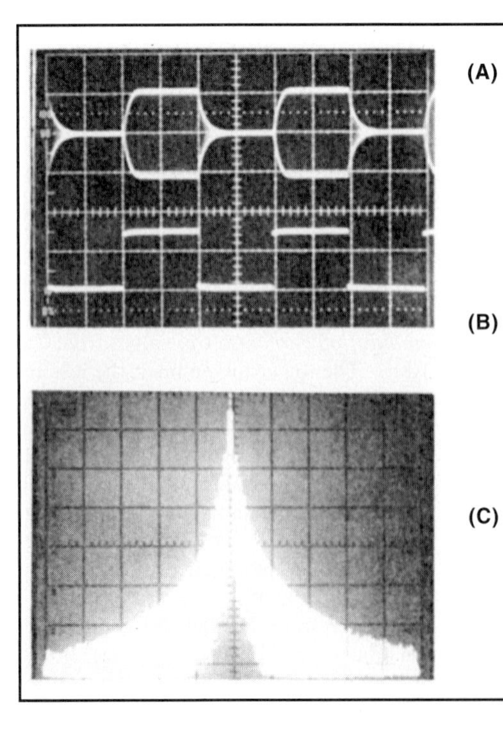

(A)

(B)

(C)

Fig 11.8 — Telegraphy by on-off-keying a carrier is also a form of AM, called *CW* (short for *continuous wave*) for reasons of tradition. *Waveshaping* in a CW transmitter often causes a CW signal's RF envelope (the amplitude-versus-time display at A) to contain less harmonic energy than the faster transitions of its keying waveform (B) suggest should be the case. C, an amplitude-versus-frequency display, shows that even a properly shaped CW signal has many sideband components.

with a locally generated RF signal — from a *beat-frequency oscillator (BFO)* — that's close enough in frequency to produce a difference signal at AF. What goes into our transmitter as shifting dc comes out of our receiver as thump-delimited tone bursts of dot and dash duration. We have achieved CW communication.

The dots and dashes of a CW signal must start and stop abruptly enough so we can clearly distinguish the carrier's presences and absences from noise, especially when fading prevails. The keying sidebands, which sound like little more than thumps when listened to on their own, help our brains be sure when the carrier tone starts and stops.

It so happens that we always need to hear one or more harmonics of the fundamental keying waveform for the code to sound sufficiently crisp. If the transmitted signal will be subject to propagational fading — a safe assumption for any long-distance radio communication — we *harden* our keying by making the transmitter's output rise and fall more quickly. This puts more energy into more keying sidebands and makes the signal more copiable in the presence of fading — in particular, *selective fading*, which linearly distorts a modulated signal's complex waveform and randomly changes the sidebands' strength and phase relative to the carrier and each other. The appropriate keying hardness also depends on the keying speed. The faster the keying in WPM, the faster the on-off times — the harder the keying — must be for the signal to remain ear-readable through noise and fading.

Instead of thinking of this process in terms of modulation percentage, we just ensure that a CW transmitter produces sufficient keying-sideband energy for solid reception. Practical CW transmitters rarely do their keying with a modulator stage as such. Instead, one or more stages are turned on and off to modulate the carrier with Morse, with rise and fall times set by R and C values associated with the stages' keying and/or power supply lines. A transmitter's CW *waveshaping* is therefore usually hardwired to values appropriate for reasonably high-speed sending (35 to 55 WPM or so) in the presence of fading. As a result, we generally cannot vary keying hardness at will as we might vary a voice transmitter's modulation with a front-panel control. Rise and fall times of 1 to 5 ms (5 ms rise and fall times equate to a keying speed of 36 WPM in the presence of fading and 60 WPM if fading is absent) are common.

The faster a CW transmitter's output changes between zero and maximum, the more bandwidth its carrier and sidebands occupy. See Fig 11.8C. Making a CW signal's keying too hard is therefore spectrum-wasteful and unneighborly because it makes the signal wider than it needs to be. Keying sidebands that are stronger and wider than necessary are traditionally called *clicks* because of what they sound like on the air. There is a more detailed discussion of keying waveforms in the **Receivers and Transmitters** chapter of this *Handbook*.

The Many Faces of Amplitude Modulation

We've so far examined mixers, multipliers and modulators that produce complex output signals of two types. One, the action of which equation 7 expresses, produces only the frequency sum of and frequency difference between its input signals. The other, the amplitude modulator characterized by equations 8 and 9, produces carrier output in addition to the frequency sum of and frequency difference between its input signals. Exploring the AM process led us to a discussion on-off-keyed CW, which is also a form of AM.

Amplitude modulation is nothing more and nothing less than varying an output signal's amplitude according to a varying voltage or current. All of the output signal types mentioned above are forms of amplitude modulation, and there are others. Their names and applications depend on whether the resulting signal contains a carrier or not, and both sidebands or not. Here's a brief overview of AM-signal types, what they're called, and some of the jobs you may find them doing:

- *Double-sideband (DSB), full-carrier AM* is often called just *AM*, and often what's meant when radio folk talk about just *AM*. (When the subject is broadcasting, *AM* can also refer to broadcasters operating in the 525- to 1705-kHz region, generically called *the AM band* or *the broadcast band* or *medium wave*. These broadcasters used only double-sideband, full-carrier AM for many years, but many now use combinations of amplitude modulation and *angle modulation*, which we'll explore shortly, to transmit stereophonic sound.) Equations 8 and 9 express what goes on in generating this signal type. What we call *CW* — Morse code done by turning a carrier on and off — is a form of DSB, full-carrier AM.
- *Double-sideband, suppressed-carrier AM* is what comes out of a circuit that does what equation 7 expresses — a sum (upper sideband), a difference (lower sideband) and no carrier. We didn't call its sum and difference outputs upper and lower sidebands earlier in equation 7's neighborhood, but we'd do so in a transmitting application. In a transmitter, we call a circuit that suppresses the carrier while generating upper and lower sidebands a *balanced modulator*, and we quantify its *carrier suppression*, which is always less than infinite. In a receiver, we call such a circuit a *balanced mixer*, which may be *single-balanced* (if it lets either its RF signal or its LO [carrier] signal through to its output) or *double-balanced* (if it suppresses both its input signal and LO/carrier in its output), and we quantify its *LO suppression* and *port-to-port isolation*, which are always less than infinite. (Mixers [and amplifiers] that afford no balance whatsoever are sometimes said to be *single-ended*.) Sometimes, DSB suppressed-carrier AM is called just *DSB*.
- *Vestigial sideband (VSB), full-carrier AM* is like the DSB variety with one sideband partially filtered away for bandwidth reduction. Commercial television systems that transmit AM video use VSB AM.
- *Single-sideband, suppressed-carrier AM* is what you get when you generate a DSB, suppressed carrier AM signal and throw away one sideband with filtering or phasing. We usually call this signal type just *single sideband (SSB)* or, as appropriate, *upper sideband (USB)* or *lower sideband (LSB)*. In a modulator or demodulator system, the *unwanted sideband* — that is, the sum or difference signal we don't want — may be called just that, or it may be called the *opposite sideband*, and we refer to a system's *sideband rejection* as a measure of how well the opposite sideband is suppressed. In receiver mixers not used for demodulation and transmitter mixers not used for modulation, the unwanted sum or difference signal, or the input signal that produces the unwanted sum or difference, is the *image*, and we refer to a system's *image rejection*. A pair of mixers specially configured to suppress either the sum or the difference output is an *image-reject mixer (IRM)*. In receiver demodulators, the unwanted sum or difference signal may just be called the opposite sideband, or it may be called the *audio image*. A receiver capable of rejecting the opposite sideband or audio image is said to be capable of *single-signal* reception.
- *Single-sideband, full-carrier AM* is akin to full-carrier DSB with one sideband missing. Commercial and military communicators may call it *AM equivalent*

(AME) or *compatible AM (CAM)* — *compatible* because it can be usefully demodulated in AM and SSB receivers and because it occupies about the same amount of spectrum space as SSB.)

- *Independent sideband (ISB) AM* consists an upper sideband and a lower sideband containing different information (a carrier of some level may also be present). Radio amateurs sometimes use ISB to transmit simultaneous slow-scan-television and voice information; international broadcasters sometimes use it for point-to-point audio feeds as a backup to satellite links.

Mixers and AM Demodulation

Translating information from radio form back into its original form — demodulation — is also traditionally called *detection*. If the information signal we want to detect consists merely of a baseband signal frequency-shifted into the radio realm, almost any low-distortion frequency-shifter that works according to equation 7 can do the job acceptably well.

Sometimes we recover a radio signal's information by shifting the signal right back to its original form with no intermediate frequency shifts. This process is called *direct conversion*. More commonly, we first convert a received signal to an *intermediate frequency* so we can amplify, filter and level-control it prior to detection. This is *superheterodyne* reception, and most modern radio receivers work in this way. Whatever the receiver type, however, the received signal ultimately makes its way to one last mixer or demodulator that completes the final translation of information back into audio, or into a signal form suitable for device control or computer processing. In this last translation, the incoming signal is converted back to recovered-information form by mixing it with one last RF signal. In heterodyne or *product* detection, that final frequency-shifting signal comes from a BFO. The incoming-signal energy goes into one mixer input port, BFO energy goes into the other, and audio (or whatever form the desired information takes) results.

If the incoming signal is full-carrier AM and we don't need to hear the carrier as a tone, we can modify this process somewhat, if we want. We can use the carrier itself to provide the heterodyning energy in a process called *envelope detection*.

Envelope Detection and Full-Carrier AM

Fig 11.5 graphically represents how a full-carrier AM signal's *modulation envelope* corresponds to the shape of the modulating wave. If we can derive from the

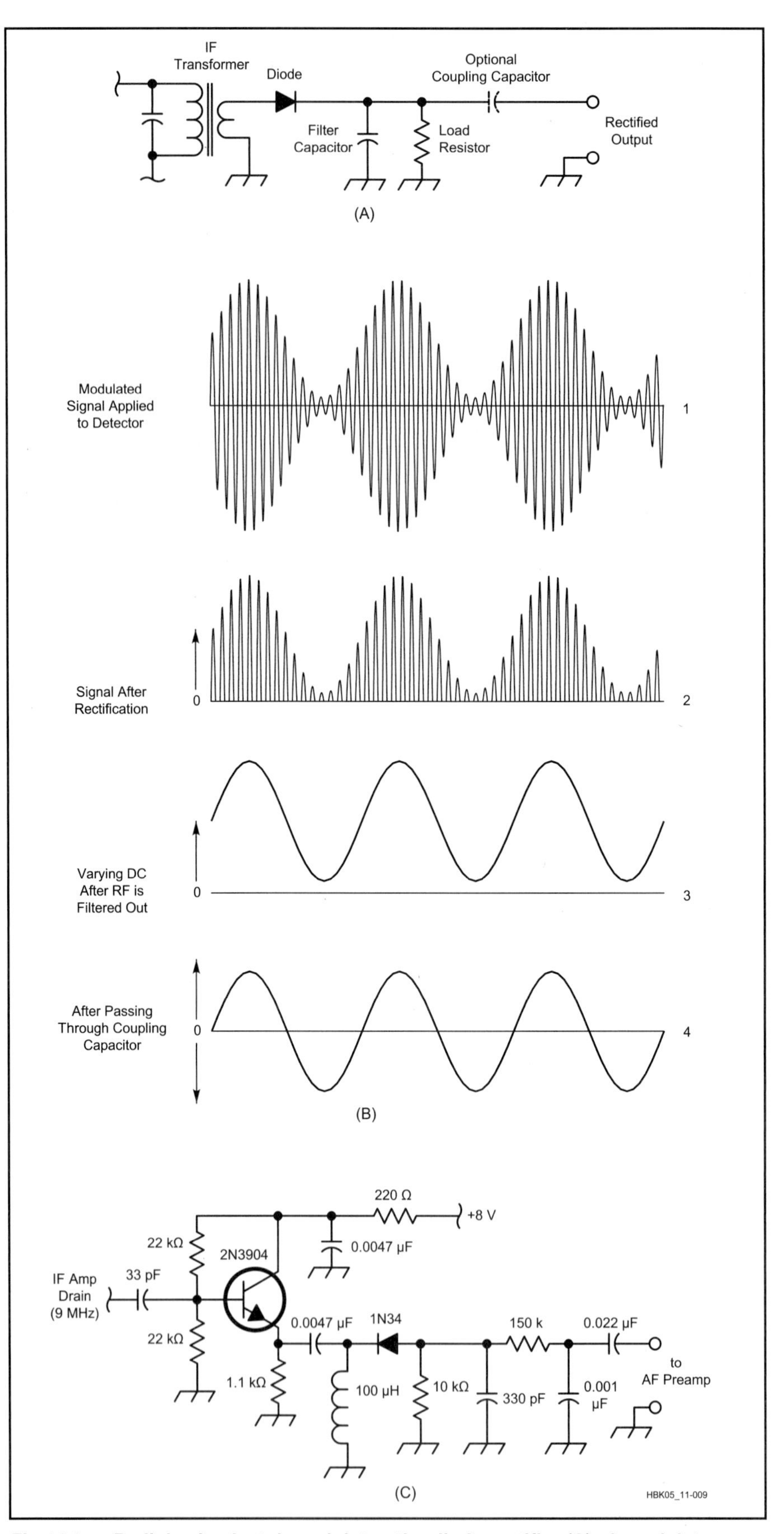

Fig 11.9 — Radio's simplest demodulator, the diode rectifier (A), demodulates an AM signal by multiplying its carrier and sidebands to produce frequency sums and differences, two of which sum into a replica of the original modulation (B). Modern receivers often use an emitter follower to provide low-impedance drive for their diode detectors (C).

modulated signal a voltage that varies according to the modulation envelope, we will have successfully recovered the information present in the sidebands. This process is called envelope detection, and we can achieve it by doing nothing more complicated than half-wave-rectifying the modulated signal with a diode (**Fig 11.9**).

That a diode demodulates an AM signal by allowing its carrier to multiply with its sidebands may jar those long accustomed to seeing diode detection ascribed merely to "rectification." But a diode is certainly nonlinear. It passes current in only one direction, and its output voltage is (within limits) proportional to the square of its input voltage. These nonlinearities allow it to multiply.

Exploring this mathematically is tedious with full-carrier AM because the process squares three summed components (carrier, lower sideband and upper sideband). Rather than fill the better part of a page with algebra, we'll instead characterize the outcome verbally: In "just rectifying" a DSB, full-carrier AM signal, a diode detector produces

- Direct current (the result of rectifying the carrier);
- A second harmonic of the carrier;
- A second harmonic of the lower sideband;
- A second harmonic of the upper sideband;
- Two difference-frequency outputs (upper sideband minus carrier, carrier minus lower sideband), each of which is equivalent to the modulating waveform's frequency, and both of which sum to produce the recovered information signal; and
- A second harmonic of the modulating waveform (the frequency difference between the two sidebands).

Three of these products are RF. Lowpass filtering, sometimes little more than a simple RC network, can remove the RF products from the detector output. A capacitor in series with the detector output line can block the carrier-derived dc component. That done, only two signals remain: the recovered modulation and, at a lower level, its second harmonic — in other words, second-harmonic distortion of the desired information signal.

Mixers and Angle Modulation

Amplitude modulation served as our first means of translating information into radio form because it could be implemented as simply as turning an electric noise generator on and off. (A spark transmitter consisted of little more than this.) By the 1930s, we had begun experimenting with translating information into radio form and back again by modulating a radio wave's angular velocity (frequency or phase) instead of its overall amplitude. The result of this process is *frequency modulation (FM)* or *phase modulation (PM)*, both of which are often grouped under the name *angle modulation* because of their underlying principle.

A change in a carrier's frequency or phase for the purpose of modulation is called *deviation*. An FM signal deviates according to the amplitude of its modulating waveform, independently of the modulating waveform's frequency; the higher the modulating wave's amplitude, the greater the deviation. A PM signal deviates according to the amplitude *and frequency* of its modulating waveform; the higher the modulating wave's amplitude *and/or frequency*, the greater the deviation.

An angle-modulated signal can be mathematically represented as

$$\begin{aligned} f_c(t) &= \cos\left(2\pi f_c t + m \sin\left(2\pi f_m t\right)\right) \\ &= \cos\left(2\pi f_c t\right)\cos\left(m \sin\left(2\pi f_m t\right)\right) \\ &\quad - \sin\left(2\pi f_c t\right)\sin\left(m \sin\left(2\pi f_m t\right)\right) \end{aligned} \tag{10}$$

In it, we see the carrier frequency ($2\pi f_c t$) and modulating signal ($\sin 2\pi f_m t$) as in the equation for AM (equation 8). We again see the modulating signal associated with a coefficient, m, which relates to degree of modulation. (In the AM equation, *m* is the modulation factor; in the angle-modulation equation, m is the *modulation index* and, for FM, equals the deviation divided by the modulating frequency.) We see that angle-modulation occurs as the cosine of the sum of the carrier frequency ($2\pi f_c t$) and the modulating signal ($\sin 2\pi f_m t$) times the modulation index (m). In its expanded form, we see the appearance of sidebands above and below the carrier frequency.

Angle modulation is a multiplicative process, so, like AM, it creates sidebands on both sides of the carrier. Unlike AM, however, angle modulation creates an *infinite* number of sidebands on either side of the carrier! This occurs as a direct result of modulating the carrier's angular velocity, to which its frequency and phase directly relate. If we continuously vary a wave's angular velocity according to another periodic wave's cyclical amplitude variations, the rate at which the modulated wave repeats *its* cycle — its frequency — passes through an infinite number of values. (How many individual amplitude points are there in one cycle of the modulating wave? An infinite number. How many corresponding discrete frequency or phase values does the corresponding angle-modulated wave pass through as the modulating signal completes a cycle? An infinite number!) In AM, the carrier frequency stays at one value, so AM produces two sidebands — the sum of its carrier's unchanging frequency value and the modulating frequency, and the difference between the carrier's unchanging frequency value and the modulating frequency. In angle modulation, the modulating wave shifts the frequency or phase of the carrier through an infinite number of different frequency or phase values, resulting in an infinite number of sum and difference products.

Wouldn't the appearance on the air of just a few such signals result in a bedlam of mutual interference? No, because most of angle modulation's uncountable sum and difference products are vanishingly weak in practical systems, and because they don't show up just anywhere in the spectrum. Rather, they emerge from the modulator spaced from the average ("resting," unmodulated) carrier frequency by integer multiples of the modulating frequency (**Fig 11.10**). The strength of the sidebands relative to the carrier, and the strength and

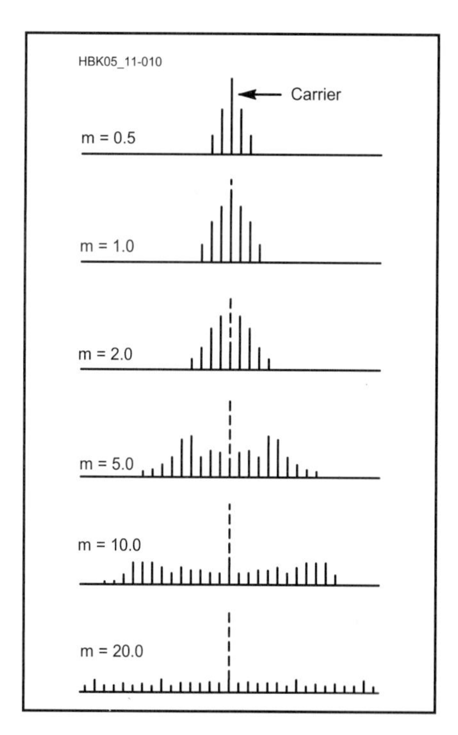

Fig 11.10 — Angle-modulation produces a carrier and an infinite number of upper and lower sidebands spaced from the average ("resting," unmodulated) carrier frequency by integer multiples of the modulating frequency. (This drawing is a simplification because it only shows relatively strong, close-in sideband pairs; space constraints prevent us from extending it to infinity.) The relative amplitudes of the sideband pairs and carrier vary with modulation index, *m*.

phase of the carrier itself, vary with the degree of modulation — the modulation index. (The *overall* amplitude of an angle-modulated signal does not change with modulation, however; when energy goes out of the carrier, it shows up in the sidebands, and vice versa.) In practice, we operate angle-modulated transmitters at modulation indexes that make all but a few of their infinite sidebands small in amplitude. (A mathematical tool called *Bessel functions* help determine the relative strength of the carrier and sidebands according to modulation index. The **Modes and Modulation Sources** chapter includes a graph to illustrate this relationship.) Selectivity in transmitter and receiver circuitry further modify this relationship, especially for sidebands far away from the carrier.

Angle Modulators

Vary a reactance in or associated with an oscillator's frequency-determining element(s), and you vary the oscillator's frequency. Vary the tuning of a tuned circuit through which a signal passes, and you vary the signal's phase. A circuit that does this is called a *reactance modulator*, and can be little more than a tuning diode or two connected to a tuned circuit in an oscillator or amplifier (**Fig 11.11**). Varying a reactance through which the signal passes (**Fig 11.12**) is another way of doing the same thing.

The difference between FM and PM depends solely on how, and not how much, deviation occurs. A modulator that causes deviation in proportion to the modulating wave's amplitude and frequency is a phase modulator. A modulator that causes deviation only in proportion to the modulating signal's amplitude is a frequency modulator.

Increasing Deviation by Frequency Multiplication

Maintaining modulation linearity is just as important in angle modulation as it is in AM, because unwanted distortion is always our enemy. A given angle-modulator circuit can frequency- or phase-shift a carrier only so much before the shift stops occurring in strict proportion to the amplitude (or, in PM, the amplitude and frequency) of the modulating signal.

If we want more deviation than an angle modulator can linearly achieve, we can operate the modulator at a suitable subharmonic — submultiple — of the desired frequency, and process the modulated signal through a series of *frequency multipliers* to bring it up to the desired frequency. The deviation also increases by the overall multiplication factor, relieving the modulator of having to do it all directly. A given FM or PM radio design may achieve its final output frequency through a combination of mixing (frequency shift, no deviation change) and frequency multiplication (frequency shift *and* deviation change).

The Truth About "True FM"

Something we covered a bit earlier bears closer study:

[An FM signal deviates according to the amplitude of its modulating waveform, independently of the modulating waveform's frequency; the higher the modulating wave's amplitude, the greater the deviation. A PM signal deviates according to the amplitude *and frequency* of its modulating waveform; the higher the modulating wave's amplitude *and/or frequency*, the greater the deviation.]

The practical upshot of this excerpt is that we can use a phase modulator to generate FM. All we need to do is run a PM transmitter's modulating signal through a low-pass filter that (ideally) halves the signal's amplitude for each doubling of frequency (a reduction of "6 dB per octave," as we sometimes see such responses characterized) to compensate for its phase modulator's "more deviation with higher frequencies" characteristic. The result is an FM, not PM, signal. FM achieved with a phase modulator is sometimes called *indirect FM* as opposed to the *direct FM* we get from a frequency modulator.

We sometimes see radio gear manufacturers claim that one piece of gear is better than another solely because it generates "true FM" as opposed to indirect FM. We can immunize ourselves against such claims by keeping in mind that direct and indirect FM *sound exactly alike in a receiver* when done correctly.

Depending on the nature of the modulation source, there *is* a practical difference between a frequency modulator and a phase modulator. Answering two questions can tell us whether this difference matters: Does our modulating signal contain a dc level or not? If so, do we need to accurately preserve that dc level through our radio communication link for successful communication? If both answers are *yes*, we must choose our hardware and/or information-encoding approach carefully,

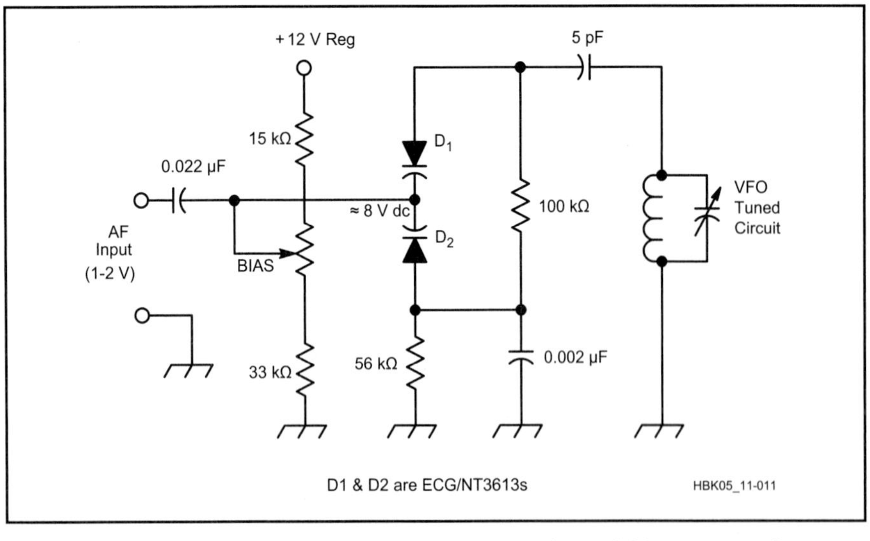

Fig 11.11 — One or more tuning diodes can serve as the variable reactance in a reactance modulator. This HF reactance modulator circuit uses two diodes in series to ensure that the tuned circuit's RF-voltage swing cannot bias the diodes into conduction. D1 and D2 are "30-volt" tuning diodes that exhibit a capacitance of 22 pF at a bias voltage of 4. The BIAS control sets the point on the diode's voltage-versus-capacitance characteristic around which the modulating waveform swings.

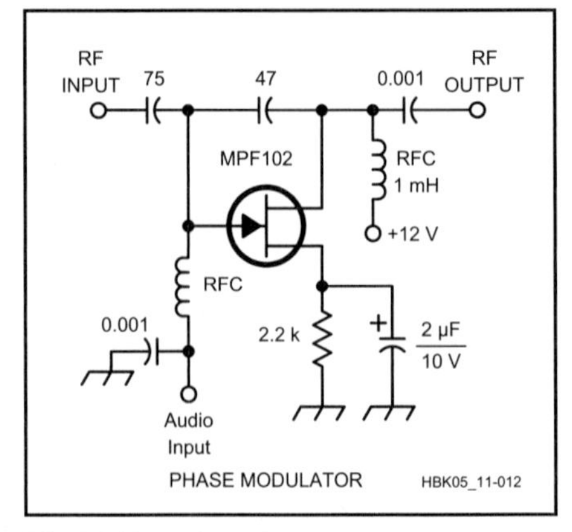

Fig 11.12 — A series reactance modulator acts as a variable shunt around a reactance — in this case, a 47-pF capacitor — through which the carrier passes.

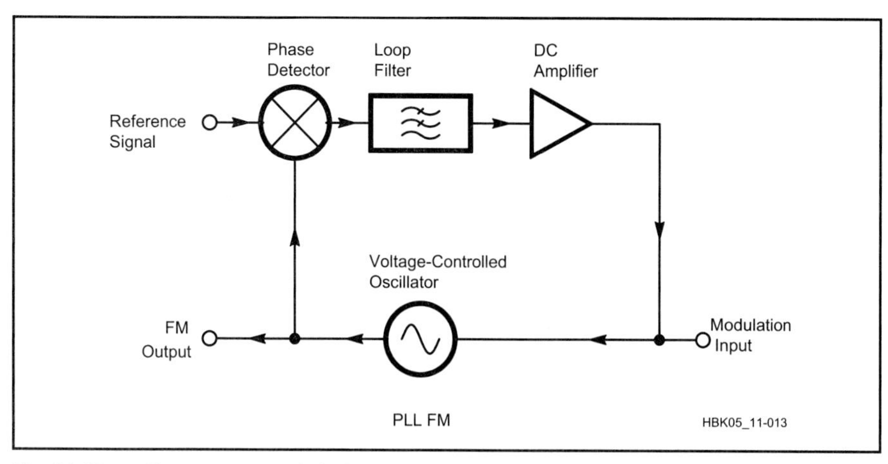

Fig 11.13 — Frequency modulation, PLL-style.

because a frequency modulator can convey shifts in its modulating wave's dc level, while a phase modulator, which responds only to instantaneous changes in frequency and phase, cannot.

Consider what happens when we want to frequency-modulate a phase-locked-loop-synthesized transmitted signal. **Fig 11.13** block-diagrams a PLL frequency modulator. Normally, we modulate a PLL's VCO because it's the easy thing to do. As long as our modulating frequency results infrequency excursions too fast for the PLL to follow and correct — that is, as long as our modulating frequency is outside the PLL's *loop bandwidth* — we achieve the FM we seek. Trying to modulate a dc level by pushing the VCO to a particular frequency and holding it there fails, however, because a PLL's loop response includes dc. The loop sees the modulation's dc component as a correctable error and dutifully "fixes" it. FMing a PLL's VCO therefore can't buy us the dc response "true FM" is supposed to allow.

We *can* dc-modulate a PLL modulator, but we must do so by modulating the loop's *reference*. The PLL then adjusts the VCO to adapt to the changed reference, and our dc level gets through. In this case, the modulating frequency must be *within* the loop bandwidth — which dc certainly is — or the VCO won't be corrected to track the shift.

Mixers and Angle Demodulation

With the awesome prospect of generating an infinite number of sidebands still fresh in our minds, we may be a bit disappointed to learn that we commonly demodulate angle modulation by doing little more than turning it into AM and then envelope- or product-detecting it! But this is what happens in many of our FM receivers and transceivers, and we can get a handle on this process by realizing that a form of angle-modulation-to-AM conversion begins quite early in an angle-modulated signal's life because of distortion of the modulation by amplitude-linear circuitry — something that happens to angle-modulated signals, it turns out, in any linear circuit that doesn't have an amplitude-versus-frequency response that's utterly flat out to infinity.

Think of what happens, for example, when we sweep a constant-amplitude signal up in frequency — say, from 1 kHz to 8 kHz — and pass it through a 6-dB-per-octave filter (**Fig 11.14A**). The filter's rolloff causes the output signal's amplitude to decrease as frequency increases. Now imagine that we linearly sweep our constant-amplitude signal *back and forth* between 1 kHz and 8 kHz at a constant rate of 3 kHz per second (Fig 11.14B). The filter's output *amplitude* now varies cyclically over time as the input signal's *frequency* varies cyclically over time. Right before our eyes, a frequency change turns into an amplitude change. The process of converting angle modulation to amplitude modulation has begun.

This is what happens whenever an angle-modulated signal passes through circuitry with an amplitude-versus-frequency response that isn't flat out to infinity. As the signal deviates across the frequency-response curves of whatever circuitry passes it, its angle modulation is, to some degree, converted to AM — a form of crosstalk between the two modulation types, if we wish to look at it that way. (Variations in system phase linearity also cause distortion and FM-to-AM conversion, because the sidebands do not have the proper phase relationship with respect to

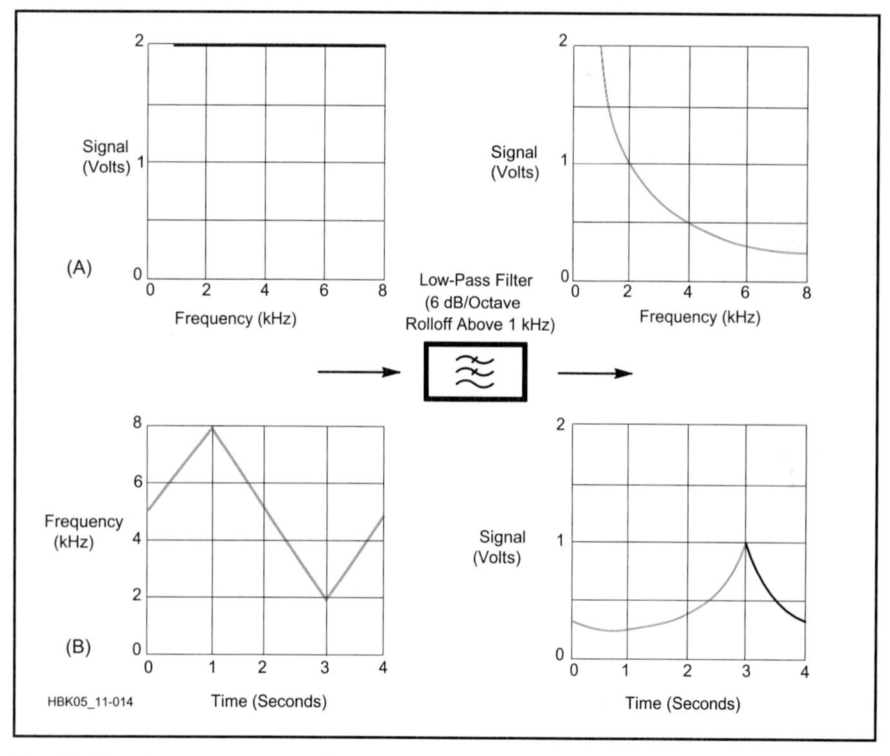

Fig 11.14 — Frequency-sweeping a constant-amplitude signal and passing it through a low-pass filter results in an output signal that varies in amplitude with frequency (A). Sweeping the input signal back and forth between two frequency limits causes the output signal's amplitude to vary between two limits (B). This is the principle behind the angle-demodulation process called *frequency discrimination*.

each other and with respect to the carrier.)

All we need to do to put this effect to practical use is develop a circuit that does this frequency-to-amplitude conversion linearly (and, since more output is better, steeply) across the frequency span of the modulated signal's deviation. Then we envelope-demodulate the resulting AM, and we're in.

Fig 11.15 shows such a circuit — a *discriminator* — and the sort of amplitude-versus-frequency response we expect from it. (It's possible to use an AM receiver to recover understandable audio from a narrow angle-modulated signal by "off-tuning" the signal so its deviation rides up and down on one side of the receiver's IF selectivity curve. This *slope detection* process served as an early, suboptimal form of frequency discrimination.)

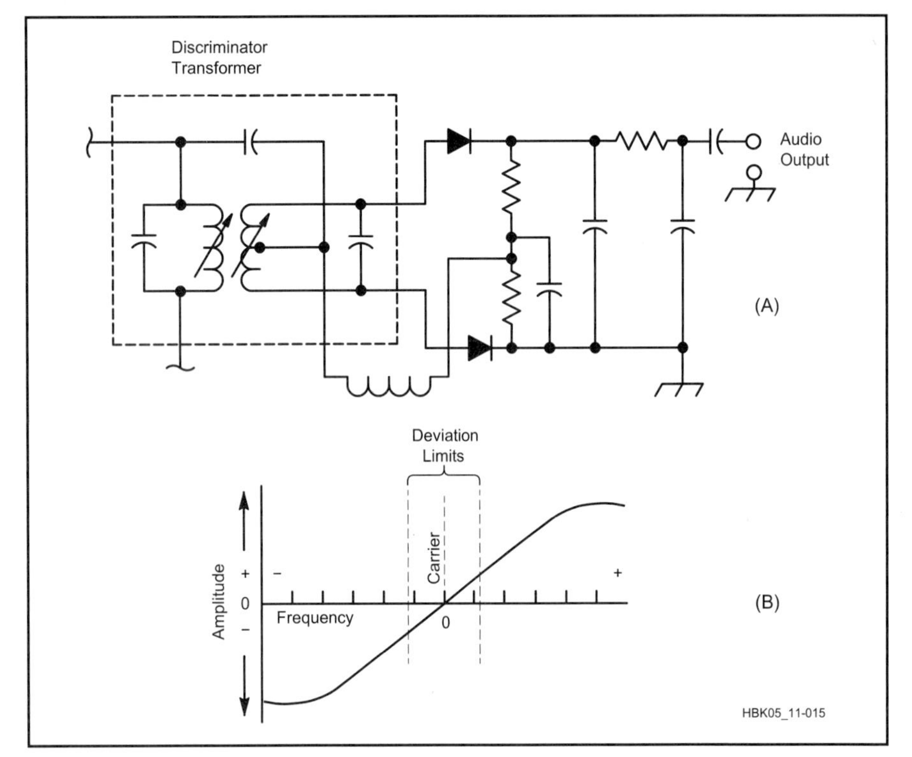

Fig 11.15 — A *discriminator* (A) converts an angle-modulated signal's deviation into an amplitude variation (B) and envelope-detects the resulting AM signal. For undistorted demodulation, the discriminator's amplitude-versus-frequency characteristic must be linear across the input signal's deviation. A *crystal discriminator* uses a crystal as part of its frequency-selective circuitry.

Quadrature Detection

It's also possible to demodulate an angle-modulated signal merely by multiplying it with a time-delayed copy of itself in a double-balanced mixer (**Fig 11.16**). For simplicity's sake, we'll represent the mixer's RF input signal as just a sine wave with an amplitude, *A*

$$A\sin(2\pi ft) \tag{11}$$

and its time-delayed twin, fed to the mixer's LO input, as a sine wave with an amplitude, *A*, and a time delay of d:

$$A\sin[2\pi f(t+d)] \tag{12}$$

Setting this special mixing arrangement into motion, we see

$$A\sin(2\pi ft)\bullet A\sin(2\pi ft+d)$$
$$=\frac{A^2}{2}\cos(2\pi fd)-\frac{A^2}{2}\cos(2\pi fd)\cos(2\bullet 2\pi ft)$$
$$+\frac{A^2}{2}\sin(2\pi fd)\sin(2\bullet 2\pi ft) \tag{13}$$

Two of the three outputs — the second and third terms — emerge at twice the input frequency; in practice, we're not interested in these, and filter them out. The remaining term — the one we're after — varies in amplitude and sign according to how far and in what direction the carrier shifts away from its resting or center frequency (at which the time delay, d, causes the mixer's RF and LO inputs to be exactly 90° out of phase — in *quadrature* — with each other). We can examine this effect by replacing f in equations 11 and 12 with the sum term $f_c + f_s$, where f_c is the center frequency and f_s is the frequency shift. A 90° time delay is the same as a quarter cycle of f_c, so we can restate d as

$$d=\frac{1}{4f_c} \tag{14}$$

The first term of the detector's output then becomes

$$\frac{A^2}{2}\cos\left(2\pi(f_c+f_s)d\right)$$
$$=\frac{A^2}{2}\cos\left(2\pi(f_c+f_s)\frac{1}{4f_c}\right)$$
$$=\frac{A^2}{2}\cos\left(\frac{\pi}{2}+\frac{\pi f_s}{2f_c}\right) \tag{15}$$

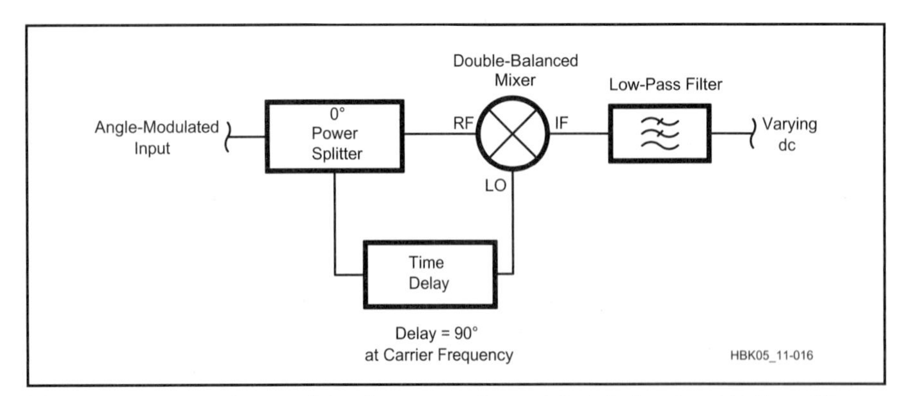

Fig 11.16 — In *quadrature detection*, an angle-modulated signal multiplies with a time-delayed copy of itself to produce a dc voltage that varies with the amplitude and polarity of its phase or frequency excursions away from the carrier frequency. A practical quadrature detector can be as simple as a 0° power splitter (that is, a power splitter with in-phase outputs), a diode double-balanced mixer, a length of coaxial cable ¼-λ (electrical) long at the carrier frequency, and a bit of low-pass filtering to remove the detector output's RF components. IC quadrature detectors achieve their time delay with one or more resistor-loaded tuned circuits (Fig 11.17).

When f_s is zero (that is, when the carrier is at its center frequency), this reduces to

$$\frac{A^2}{2}\cos\left(\frac{\pi}{2}\right)=0 \qquad (16)$$

As the input signal shifts higher in frequency than f_c, the detector puts out a positive dc voltage that increases with the shift. When the input signal shifts lower in frequency than f_c, the detector puts out a negative dc voltage that increases with the shift. The detector therefore recovers the input signal's frequency or phase modulation as an amplitude-varying dc voltage that shifts in sign as f_s varies around f_c — in other words, as ac. We have demodulated FM by means of quadrature detection.

An ideal quadrature detector puts out 0 V dc when no modulation is present (with the carrier at f_c). The output of a real quadrature detector may include a small *dc offset* that requires compensation. If we need the detector's response all the way down to dc, we've got it; if not, we can put a suitable blocking capacitor in the output line for ac-only coupling.

Quadrature detection is more common than frequency discrimination nowadays because it doesn't require a special discriminator transformer, and because the necessary balanced-detector circuitry can easily be implemented in IC structures along with limiters and other receiver circuitry. The synchronous AM detector project later in this chapter uses such a chip, the Philips Components-Signetics NE604A, to do limiting and phase detection as part of a phase-locked loop (PLL); **Fig 11.17** shows another example.

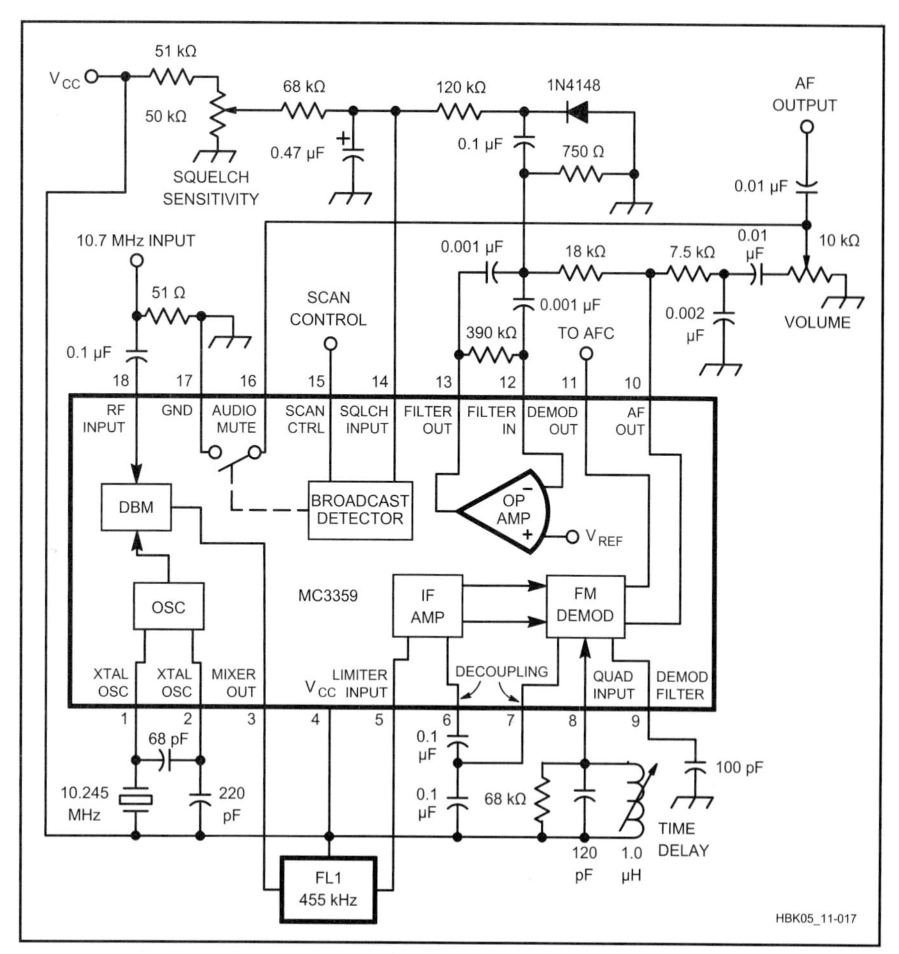

Fig 11.17 — The Motorola MC3359 is one of many FM subsystem ICs that include limiter and quadrature-detection circuitry. The TIME DELAY coil is adjusted for minimum recovered-audio distortion.

PLL Angle Demodulation

Back at Fig 11.14, we saw how a PLL can be used as an angle modulator. A PLL also makes a fine angle *de*modulator. Applying an angle-modulated signal to a PLL keeps its phase detector and VCO hustling to maintain loop lock through the input signal's angle variations. The loop's error voltage therefore tracks the input signal's modulation, and its variations mirror the modulation signal. Turning the loop's varying dc error voltage into audio is just a blocking capacitor away.

Although we can't convey a dc level by directly modulating the VCO in a PLL angle modulator, a PLL demodulator can respond down to dc quite nicely. A constant frequency offset from f_c (a dc component) simply causes a PLL demodulator to swing its VCO over to the new input frequency, resulting in a proportional dc offset on the VCO control-voltage line. Another way of looking at the difference between a PLL angle modulator and a PLL angle demodulator is that a PLL demodulator works with a varying reference signal (the input signal), while a PLL angle modulator generally doesn't.

Amplitude Limiting Required

By now, it's almost household knowledge that FM radio communication systems are superior to AM in their ability to suppress and ignore static, manmade electrical noise and (through an angle-modulation-receiver characteristic called *capture effect*) co-channel signals sufficiently weaker than the desired signal. AM-noise immunity is not intrinsic to angle modulation, however; it must be designed into the angle-modulation receiver.

If we note the progress of A from the left to the right side of the equal sign in equation 13, we realize that the amplitude of a quadrature detector's input signal affects the amplitude of a quadrature detector's three output signals. A quadrature detector therefore responds to AM, and so does a frequency discriminator. To achieve FM's storied noise immunity, then, these angle demodulators must be preceded by *limiting* circuitry that removes all amplitude variations from the incoming signal.

PRACTICAL BUILDING BLOCKS FOR MIXING, MODULATION AND DEMODULATION

So far, we've tended to look at mixing as a process that frequency-shifts one sinusoidal wave by mixing it with another. We need to expand our thinking to other cases, however, since it turns out that many practical mixers work best with *square-wave* signals applied to their LO inputs.

Sine-Wave Mixing, Square-Wave Mixing

Thinking of mixers as multiplying sine waves implies that mixers act like tiny analog computers performing millions of multiplications per second. It's certainly possible to build mixers this way by using an IC mixer circuit (**Fig 11.18**) conceived in 1967 by Barrie Gilbert and widely

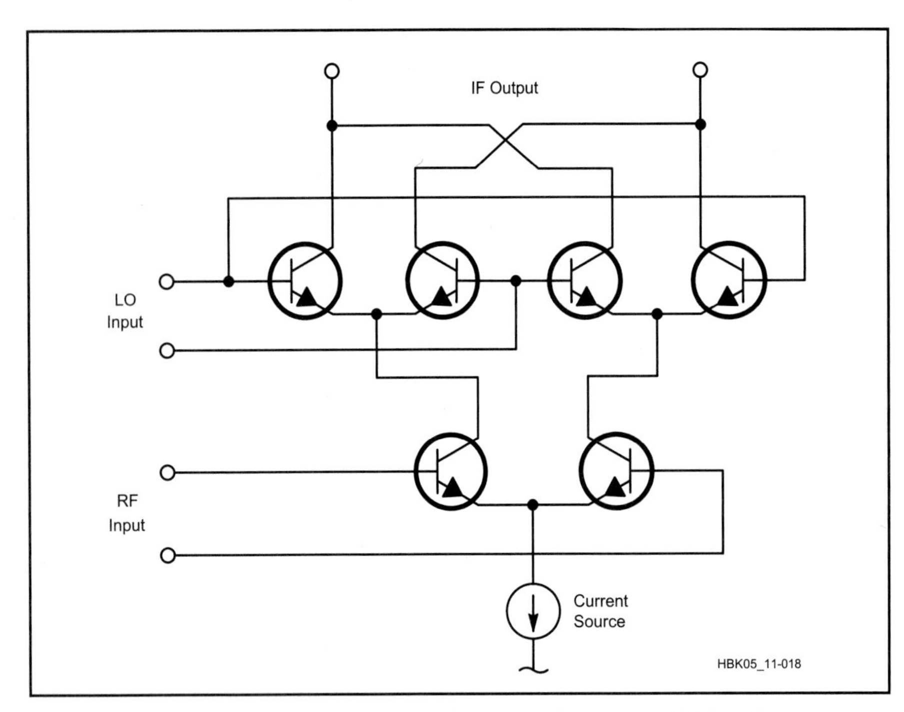

Fig 11.18 — The Gilbert cell mixer. The Motorola MC1496 and Philips Components-Signetics NE602A are based on this circuit.

known as the Gilbert cell. (Gilbert himself was not responsible for this eponym; indeed, he has noted that a prior art search at the time found that essentially the same idea — used as a "synchronous detector" and not as a true mixer — had already been patented by H. E. Jones.) A Gilbert cell consists of two differential transistor pairs whose bias current is controlled by one of the input signals. The other signal drives the differential pairs' bases, but only after being "predistorted" in a diode circuit. (This circuit distorts the signal equally and oppositely to the inherent distortion of the differential pair.) The resulting output signal is an accurate multiplication of the input voltages.

Early Gilbert-cell ICs, such as the Motorola MC1495 multiplier, had a number of disadvantages, including critical external adjustments, narrow bandwidth, and limited dynamic range. Modern Gilbert cells, such as the Burr-Brown MPY600 and Analog Devices AD834, overcome most of these disadvantages, and have led to an increase in usage of analog multipliers as mixers. Most practical radio mixers do not work exactly as analog multipliers, however. In practice, they act more like fast analog *switches*.

In using a mixer as a fast switching device, we feed its LO input with a single-frequency square wave rather than a sine wave, and feed sine waves, audio, or other complex signals to the mixer's RF input. The RF port serves as the mixer's "linear" input, and therefore must preferably exhibit low intermodulation and harmonic distortion. Feeding a ±1-V square wave into the LO input alternately multiplies the linear input by +1 or –1. Multiplying the RF-port signal by +1 just transfers it to the output with no change. Multiplying the RF-port signal by –1 does the same thing, except that the signal inverts (flips 180° in phase). The LO port need not exhibit low intermodulation and harmonic distortion; all it has to do is preserve the switching signal's fast rise and fall times.

Using square-wave LO drive allows us to simplify the Gilbert multiplier by dispensing with its predistortion circuitry. The Motorola MC1496, still in wide use despite its age, is an example of this. The Philips Components-Signetics NE602A and its relatives, popular with Amateur Radio experimenters, are modern MC1496 descendants. The vast majority of Gilbert-cell mixers in current use are square-wave LO-drive types. In practice, though, we don't have to square the LO signals we apply to them to make them work well. All we need to do is drive their LO inputs with a sine wave of sufficient amplitude to overdrive the associated transistors' bases. This clips the LO waveform, effectively resulting in square-wave drive.

Reversing-Switch Mixers

We can multiply a signal by a square wave without using an analog multiplier at all. All we need is a pair of balun transformers and four diodes (**Fig 11.19A**).

With no LO energy applied to the circuit, none of its diodes conduct. RF-port energy (1) can't make it to the LO port because there's no direct connection between the secondaries of T1 and T2, and (2) doesn't produce IF output because T2's secondary balance results in energy cancellation at its center tap, and because no complete IF-energy circuit exists through T2's secondary with both of its ends disconnected from ground.

Applying a square wave to the LO port biases the diodes so that, 50% of the time, D1 and D2 are on and D3 and D4 are reverse-biased off. This unbalances T2's secondary by leaving its upper wire floating and connecting its lower wire to ground through T1's secondary and center tap. With T2's secondary unbalanced, RF-port energy emerges from the IF port.

The other 50% of the time, D3 and D4 are on and D1 and D2 are reverse-biased off. This unbalances T2's secondary by leaving its lower wire floating, and connects its upper wire to ground through T1's secondary and center tap. With T2's secondary unbalanced, RF-port energy again emerges from the IF port — shifted 180° relative to the first case because T2's active secondary wires are now, in effect, transposed relative to its primary.

A reversing switch mixer's output spectrum is the same as the output spectrum of a multiplier fed with a square wave. This can be analyzed by thinking of the square wave in terms of its Fourier series equivalent, which consists of the sum of sine waves at the square wave frequency and all of its odd harmonics. The amplitude of the equivalent series' fundamental sine wave is $4/\pi$ times (2.1 dB greater than) the amplitude of the square wave. The amplitude of each harmonic is inversely proportional to its harmonic number, so the third harmonic is only $1/3$ as strong as the fundamental (9.5 dB below the fundamental), the 5th harmonic is only $1/5$ as strong (14 dB below the fundamental) and so on. The input signal mixes with each harmonic separately from the others, as if each harmonic were driving its own separate mixer, just as we illustrated with two sine waves in Fig 11.4. Normally, the harmonic outputs are so widely removed from the desired output frequency that they are easily filtered out, so a reversing-switch mixer is just as good as a sine-wave-driven analog multiplier for most practical purposes, and usually better — for radio purposes — in terms of dynamic range and noise.

An additional difference between multiplier and switching mixers is that a switching mixer's signal flow is reversible. It really only has one dedicated input (the LO input). The other terminals can be thought of as I/O (input/output) ports, since either one can be the input as long as the other is the output.

Conversion Loss

Fig 11.19B shows a perfect *multiplier* mixer. That is, the output is the product of the input signal and the LO. The LO is a perfect square wave. Its peak amplitude is ±1.0 V and its frequency is 8 MHz. Fig 11.19C shows the output waveform (the product of two inputs) for an input signal whose value is 0 dBm and whose frequency is 2 MHz. Notice that for each transition of the square-wave LO, the sine-wave output-waveform polarity reverses. There are 16 transitions during the interval shown, at each zero-crossing point of the output waveform. Fig 11.19D shows the mixer output spectrum. The principle components are at 6 MHz and 10 MHz, which are the sum and difference of the signal and LO frequencies. Each of these is –3.9 dBm. Numerous other pairs of output frequencies occur that are also spaced 4 MHz apart and centered at 24 MHz, 40 MHz and 56 MHz and higher odd harmonics of 8 MHz. The ones shown are at –13.5 dBm, –17.9 dBm and –20.9 dBm. Because the mixer is lossless, the sum of all of the outputs must be exactly equal to the value of the input signal. As explained previously, this output spectrum can also be understood in terms of each of the odd-harmonic components of the square-wave LO operating independently.

If the mixer were a lossless switching mixer, such as Fig 11.19A, with diodes that are perfect switches, the results would be mathematically identical to the above example. The diodes would commutate the input signal exactly as shown in Fig 11.19C.

Now consider the perfect multiplier mixer of Fig 11.19B with an LO that is a perfect sine wave with a peak amplitude of ±1.0 V. In this case the dashed lines of Fig 11.19D show that only two output frequencies are present, at 6 MHz and 10 MHz (see also Fig 11.2). Each component now has a –6 dBm level. The product of the 0 dBm sine-wave input at one frequency and the ±1.0V sine-wave LO at another frequency (see Eq 6 in this chapter) is the –3 dBm total output.

These examples illustrate the difference between the square-wave LO and the sine-wave LO, for a perfect multiplier. For the same peak value of both LO waves, the square-wave LO delivers 2.1 dB more output at 6 MHz and 10 MHz than the sine-wave LO. An actual diode mixer such as Fig 11.19A behaves more like a switching mixer. Its sine-wave LO waveform is considerably flattened by interaction between

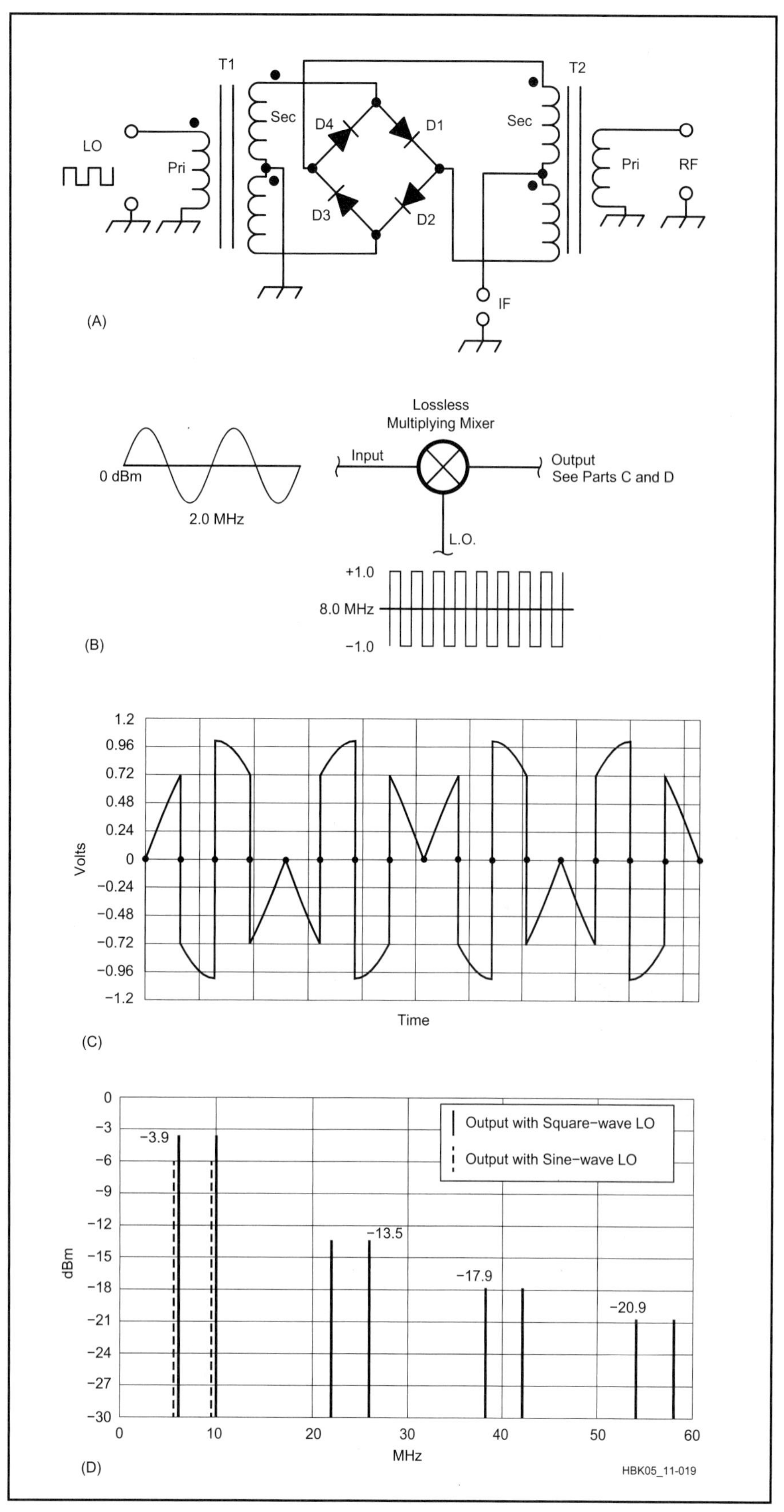

Fig 11.19 — Part A shows a general-purpose diode *reversing-switch* mixer. This mixer uses a square-wave LO and a sine-wave input signal. The action of this mixer is described in the text. Part B is an ideal *multiplier* mixer. The square-wave LO and a sine-wave input signal produce the output waveform shown in part C. The solid lines of part D show the output spectrum with the square-wave LO. The dashed lines show the output spectrum with a sine-wave LO.

the diodes and the LO generator, so that it looks somewhat like a square wave. The diodes have nonlinearities, junction voltages, capacitances, resistances and imperfect parameter matching. (See the **Real-World Component Characteristics** chapter.) Also, "re-mixing" of a diode mixer's output with the LO and the input is a complicated possibility. The practical end result is that diode double-balanced mixers have a conversion loss, from input to each of the two major output frequencies, in the neighborhood of 5 to 6 dB.

The Diode Double-Balanced Mixer: A Basic Building Block

The most common implementation of a reversing switch mixer is the diode *double-balanced mixer* (DBM). DBMs can serve as mixers (including image-reject types), modulators (including single- and double-sideband, phase, biphase, and quadrature-phase types) and demodulators, limiters, attenuators, switches, phase detectors, and frequency doublers. In some of these applications, they work in conjunction with power dividers, combiners and hybrids.

The Basic DBM Circuit

We have already seen the basic diode DBM circuit (Fig 11.19). In its simplest form, a DBM contains two or more unbalanced-to-balanced transformers and a Schottky diode ring consisting of 4 × n diodes, where n is the number of diodes in each leg of the ring. Each leg commonly consists of up to four diodes.

As we've seen, the degree to which a mixer is *balanced* depends on whether either, neither or both of its input signals (RF and LO) emerge from the IF port along with mixing products. An unbalanced mixer suppresses neither its RF nor its LO; both are present at its IF port. A single-balanced mixer suppresses its RF or LO, but not both. A double-balanced mixer suppresses its RF *and* LO inputs. Diode and transformer uniformity in the Fig 11.19 circuit results in equal LO potentials at the center taps of T1 and T2. The LO potential at T1's secondary center tap is zero (ground); therefore, the LO potential at the IF port is zero.

Balance in T2's secondary likewise results in an RF null at the IF port. The RF potential between the IF port and ground is therefore zero — except when the DBM's switching diodes operate, of course!

The Fig 11.19 circuit normally also affords high RF-IF isolation because its balanced diode switching precludes direct connections between T1 and T2. A diode DBM can be used as a current-controlled switch or attenuator by applying dc to its IF port, albeit with some distortion. This causes opposing diodes (D2 and D4, for instance) to conduct to a degree that depends on the current magnitude, connecting T1 to T2.

One extension of the single-diode-ring DBM is a *double* double-balanced mixer (DDBM) with high dynamic range and larger signal handling capability than a single-ring design. **Fig 11.20** diagrams such a DDBM, which uses transmission-line transformers and two diode rings. This type of mixer has higher 1-dB compression point (usually 3 to 4 dB lower than the LO drive) than a DBM. Low distortion is a typical characteristic of DDBMs. Depending on the ferrite core material used (ferrites with a magnetic permeability — μ — of 100 to 15,000), frequencies as low as a few hundred hertz and as high as a few gigahertz can be covered.

Diode DBM Components

Commercially manufactured diode DBMs generally consist of: (A) a supporting base; (B) a diode ring; (C) two or more ferrite-core transformers commonly wound with two or three twisted-pair wires; (D) encapsulating material; and (E) an enclosure.

Diodes

Hot-carrier (Schottky) diodes are the devices of choice for diode-DBM rings because of their low ON resistance, although ham-built DBMs for non-critical MF/HF use commonly use switching diodes like the 1N914 or 1N4148. The forward voltage drop, V_f, across each diode in the ring determines the mixer's optimum local-oscillator drive level. Depending on the forward voltage drop of each of its diodes and the number of diodes in each ring leg, a diode DBM may be categorized as a Level 0, 3, 7, 10, 13, 17, 23 or 27 device. The numbers indicate the mixer's optimal LO drive level in dBm. As a rule of thumb, the LO signal must be 20 dB larger than the RF and IF signals for proper operation. This ensures that the LO signal, rather than the RF or IF signals, switches the mixer's diodes on and off — a critical factor in minimizing IMD and maximizing dynamic range.

Schottky diodes are characterized by loss and contact resistance (R_s), junction capacitance (C_j), and forward voltage drop (V_f) at a known current, typically 1 mA or 10 mA. The lower the diode-to-diode V_f difference in millivolts, the better the diode match at dc. (Some early diode DBM designs used diodes in series with a parallel resistor/capacitor combination for automatic bias-

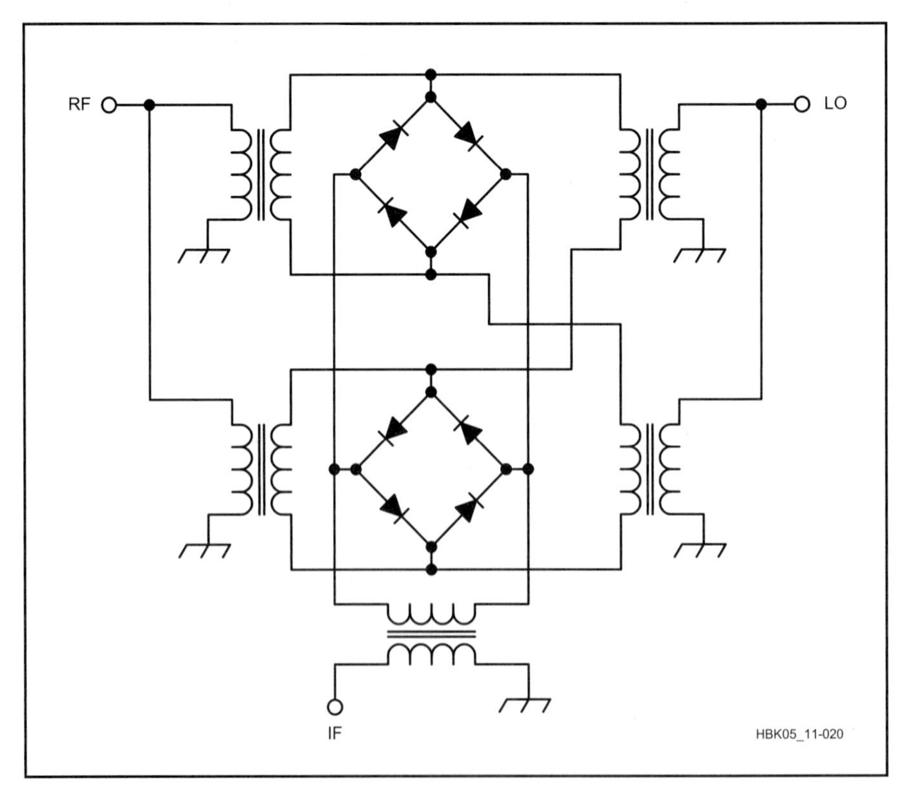

Fig 11.20 — Five transmission-line transformers and two Schottky-quad rings form this double double-balanced mixer (DDBM). Such designs can provide lower distortion, better signal-handling capability and higher interport isolation than single-ring designs.

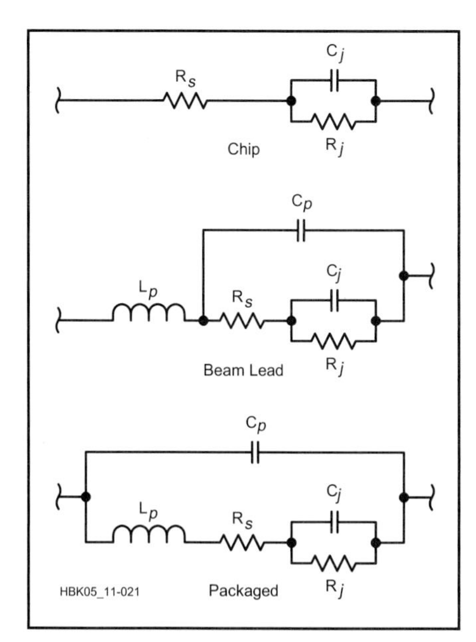

Fig 11.21 — Its semiconductive properties aside, a Schottky diode can be represented as a network consisting of resistance, capacitance and/or inductance. Of these, the junction capacitance (C_j) plays an especially critical role in a double-balanced mixer's high-end response. (R_j = junction resistance, R_s = contact resistance, L_p = parasitic inductance and C_p = parasitic capacitance.)

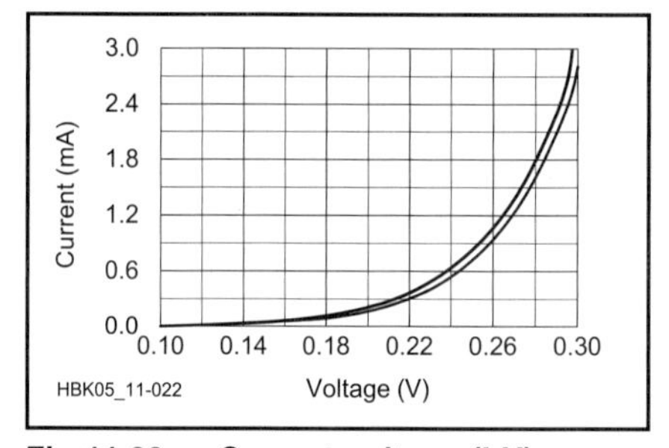

Fig 11.22 — Current-voltage (I-V) characteristic for Schottky diode quad, showing worst-case voltage imbalance (the spread between the two curves) among the four diodes.

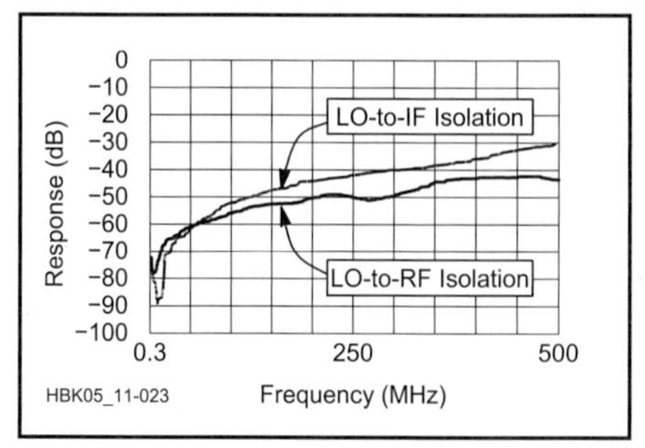

Fig 11.23 — A diode DBM's port-to-port isolation depends on how well its diodes match and how well its transformers are balanced. This graph shows LO-IF and LO-RF isolation versus frequency for a Synergy Microwave CLP-4A3 mixer.

ing.) Better diode matching (in V_f and C_j) results in higher isolation among the ports. Diodes capable of operating at higher frequencies have lower junction capacitance and lower parasitic inductance. **Fig 11.21** shows the equivalent circuit for Schottky diodes of three package types.

Manufacturers of diodes suitable for DBMs characterize their diodes as low-barrier, medium-barrier, high-barrier and very-high-barrier (usually two or more diodes in each leg), with typical V_f values of 220 mV, 350 mV, 600 mV and 1 V or more, respectively. **Fig 11.22** shows a typical current-voltage (I-V) characteristic for a low-barrier Schottky quad capable of operating up to 4 GHz. Note that as current through the diodes increases, the V_f difference among the ring's diodes also increases, affecting the balance.

At higher frequencies, diode packaging becomes critical and expensive. As the frequency of operation increases, the effect of junction capacitance and package capacitance cannot be ignored. Part or all of the capacitance can be compensated at the mixer's highest operating frequency by properly designing the unbalanced-to-balanced transformers. The transformer inductance and diode junction capacitance form a low-pass network with its cutoff frequency higher than the frequency of operation. Compensated in this way, diodes with a junction capacitance of 0.2 pF can be used up to 8 to 10 GHz.

Transformers

From the DBM schematic shown in Fig 11.19, it's clear that the LO and RF transformers are unbalanced on the input side and balanced on the diode side. The diode ends of the balanced ports are 180° out of phase throughout the frequency range of interest. This property causes signal cancellations that result in higher port-to-port isolation. **Fig 11.23** plots LO-RF and LO-IF isolation versus frequency for Synergy Microwave's CLP-4A3 DBM. Isolations on the order of 70 dB occur at the lower end of the band as a direct result of the balance among the four diode-ring legs and the RF phasing of the balanced ports.

As we learned in our discussion of generic switching mixers, transformer efficiency plays an important role in determining a mixer's conversion loss and drive-level requirement. Core loss, copper loss and impedance mismatch all contribute to transformer losses.

Ferrite in toroidal, bead, balun (multihole), or rod form can serve as DBM transformer cores. Radio amateurs commonly use Fair-rite Mix 43 ferrite (μ = 950), but if the mixer will be used over a wide temperature range, the core material must be evaluated in terms of temperature coefficient and curie temperature (the temperature at which a ferromagnetic material loses its magnetic properties). In some materials, μ may change drastically across the desired temperature range, causing a frequency-response shift with temperature. Once a suitable core material and form have been selected, frequency requirements determine the necessary core size. For a given core shape and size, the number of turns, wire size, and the number of twists determine transformer performance. Wire placement also plays an important role.

RF transformers combine lumped and distributed capacitance and inductance. The interwinding capacitance and characteristic impedance of a transformer's twisted wires sets the transformer's high-frequency response. The core's μ and size, and the number of winding turns, determine the transformer's lower frequency limit. Covering a specific frequency range requires a compromise in the number of turns used with a given core. Increasing a transformer's core size and number of turns improves its low-frequency response. Cores may be stacked to meet low-frequency performance specs.

Inexpensive mixers operating up to 2 GHz most commonly use twisted trifilar (three-wire) windings made of a wire size between #36 and #32. The number of twists per unit length of wire determines a winding's characteristic impedance. Twisted wires are analogous to transmission lines, and can be analyzed in terms of distributed interwinding capacitance. Decreasing the number of twists lowers the interwinding capacitance and increases the frequency of operation. On the other hand, using fewer twists per inch than four makes winding difficult because the wires tend to separate instead of behaving as a single cable.

The transmission-line effect predominates at the higher end of a transformer's frequency range. If two impedances, Z_1 and Z_2, need to be matched through a transmission line of characteristic impedance, Z_0, then

$$Z_0 = \sqrt{Z_1 \times Z_2} \quad (17)$$

Fig 11.24 shows two types of transformers using twisted wires: (A) a three-wire type in which the primary winding is isolated from the secondary winding with a center tap, and (B) a two-wire (transmission-line) type in which two sets of transmission lines are interconnected to form a center tap at the secondary with a direct connection between primary and second-

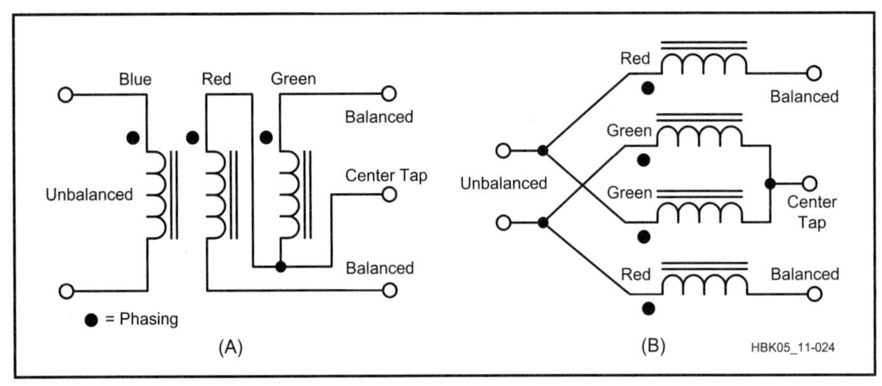

Fig 11.24 — Transformers for DBMs: three-wire (A), and transmission-line (B).

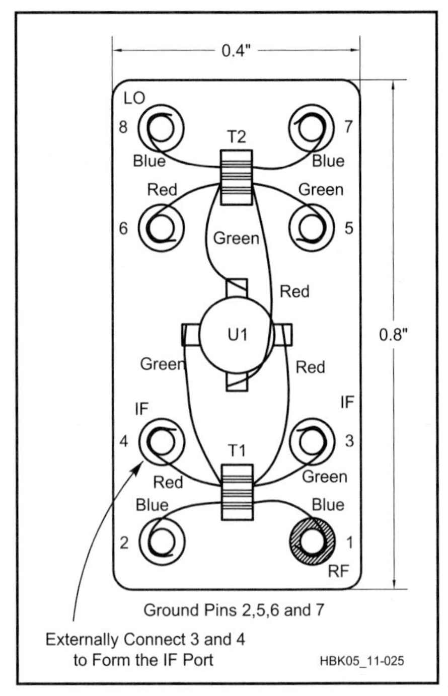

Fig 11.25 — How a typical commercial DBM is wired. The use of different wire colors for the transformers' various windings speeds assembly and minimizes error. U1, a Schottky-diode quad, contains D1-D4 of Fig 11.19.

ary. The primary-to-secondary turns ratio determines the impedance match as shown in equation 17. The properties of these two transformer types can be summarized as follows:

1. By virtue of its construction, the three-wire transformer is less symmetrical at higher frequencies than the transmission-line type.

2. The transformers' lower cutoff frequency (f_L) is determined by the equation

$$\omega L > 4R \tag{18}$$

where

L = inductance of the winding

R = system impedance; for example, 50 Ω, 75 Ω and so on; and

$\omega = 2\pi f_L$.

3. The transmission-line transformer's upper cutoff frequency (f_H) is determined by the highest frequency at which its wires' twists (that is, the coupling between them) allow it to function as a transmission line of the proper characteristic impedance.

4. Transformers convert one impedance, Z_1 (primary) to another, Z_2 (secondary) according to the relationship

$$Z_2 = Z_1 (N)^2 \tag{19}$$

where

N = secondary to primary turns ratio. Within certain limits, if Z_1 is varied, Z_2 also varies to a new value multiplied by N^2. Thus, a mixer designed for a 50-Ω system may work in a 75-Ω system with minor modifications.

Diode DBMs in Practice

Fig 11.25 shows the wiring of a typical commercial DBM made with toroidal cores. The wires are wrapped around the package pins and diode leads, and then soldered. In this unit, the LO transformer's primary winding connects across pins 7 and 8; the RF-transformer primary, across pins 1 and 2. The pin pairs 3-4 and 5-6 are connected externally to form the transformers' secondary center taps, one of which (5-6, that of the LO transformer) connects to a common ground point while the other (3-4, that of the RF transformer) serves as the IF port.

The DBM shown in Fig 11.25 has a dc-coupled IF port. If necessary, this DBM can be operated at a particular polarity (positive or negative) by appropriately connecting the LO, RF, IF and common ground points.

DBM Specifications

Most of the parameters important in building or selecting a diode DBM also apply to other mixer types. They include: conversion loss and amplitude flatness across the required IF bandwidth; variation of conversion loss with input frequency; variation of conversion loss with LO drive, 1-dB compression point; LO-RF, LO-IF and RF-IF isolation; intermodulation products; noise figure (usually within 1 dB of conversion loss); port SWR; and dc offset, which is directly related to isolation among the RF, LO and IF ports.

Conversion Loss

Fig 11.26 shows a plot of conversion loss versus intermediate frequency in a typical DBM. The curves show conversion loss for two fixed RF-port signals, one at 100 kHz and the another at 500 MHz, while varying the LO frequency from 100 kHz to 500 MHz.

Fig 11.27 plots a diode DBM's simulated output spectrum. Note that the RF input is –20 dBm and the IF output (the frequency difference between the RF and LO signals) is –25 dBm, implying a conversion loss of 5 dB. This figure also applies to the sum of both signals (RF + LO).

We minimize a diode DBM's conversion loss, noise figure and intermodulation by keeping its LO drive high enough to switch its diodes on fully and rapidly. **Fig 11.28** plots noise figure for the DDBM shown in Fig 11.20. Its 4-dB noise figure assumes ideal transformers and somewhat idealized diodes; typical mixers have a noise figure of 5 to 6 dB. **Fig 11.29** plots conversion gain (loss) for the same mixer circuit.

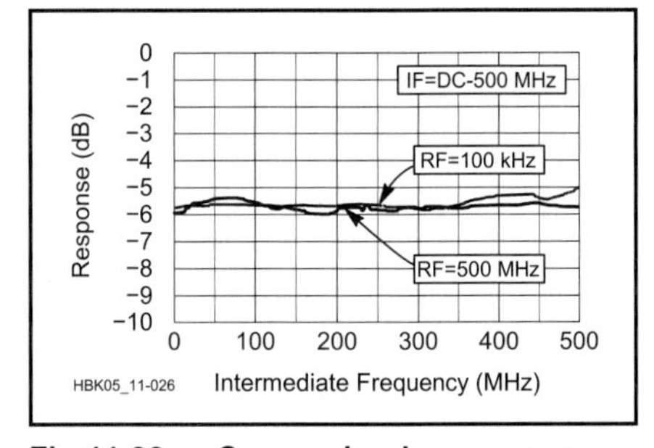

Fig 11.26 — Conversion loss versus frequency for a typical diode DBM. The LO drive level is +7 dBm.

Insufficient LO drive results in increased noise figure and conversion loss. IMD also increases because RF-port signals have a greater chance to control the mixer diodes when the LO level is too low.

Dynamic Range: Compression, Intermodulation and More

The output of a linear stage — including a mixer, which we want to act as a

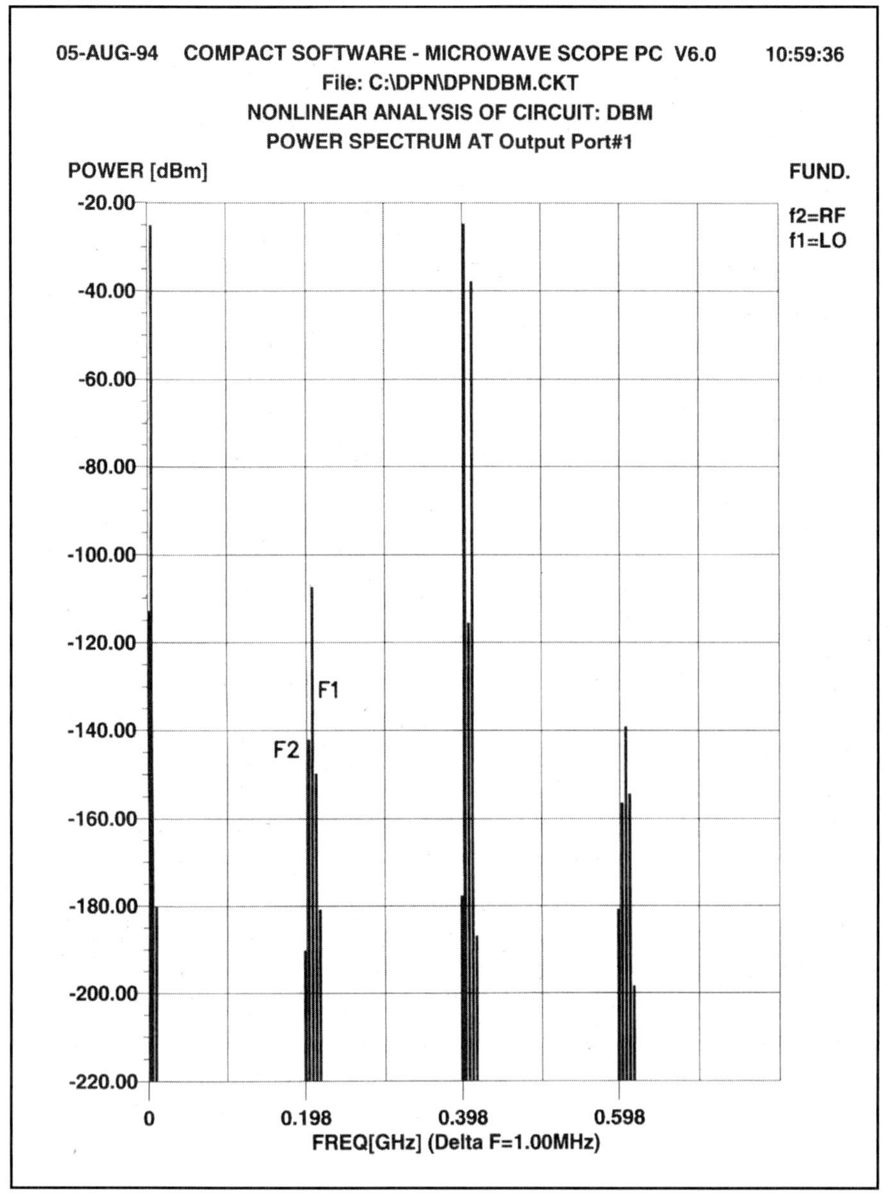

Fig 11.27 — Simulated diode-DBM output spectrum. Note that the desired output products (the highest two products, RF – LO and RF + LO) emerge at a level 5 dB below the mixer's RF input (–20 dBm). This indicates a mixer conversion loss of 5 dB. (*Microwave SCOPE* simulation)

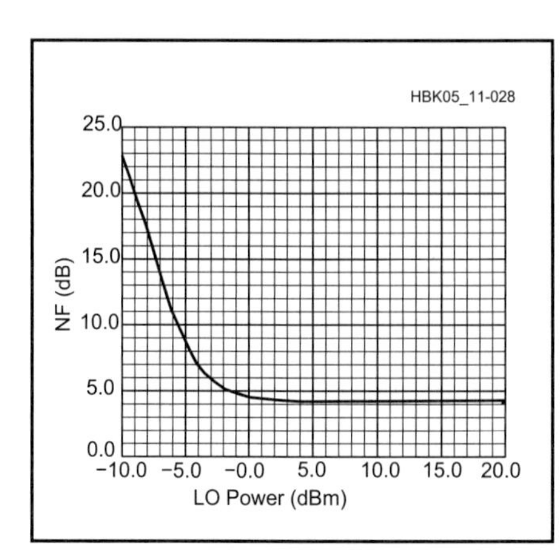

Fig 11.28 — Noise figure versus LO drive for a DDBM built along the lines of Fig 11.20.

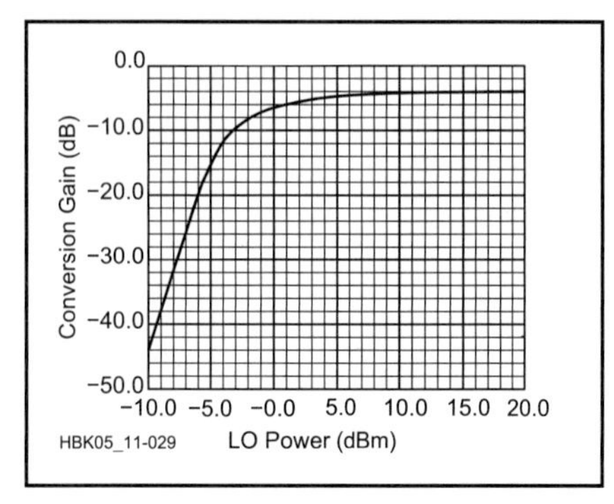

Fig 11.29 — Conversion gain (loss) for a DDBM. Increasing a mixer's LO level beyond that sufficient to turn its switching devices all the way on merely makes them dissipate more LO power and does not improve performance.

linear frequency shifter — tracks its input signal decibel by decibel, every 1-dB change in its input signal(s) corresponding to an identical 1-dB output change. This is the stage's *first-order* response.

Because no device is perfectly linear, however, two or more signals applied to it generate sum and difference frequencies. These IMD products occur at frequencies and amplitudes that depend on the order of the IMD response as follows:

- *Second-order* IMD products change 2 dB for every decibel of input-signal change (this figure assumes that the IMD comes from equal-level input signals), and appear at frequencies that result from the simple addition and subtraction of input-signal frequencies. For example, assuming that its input bandwidth is sufficient to pass them, an amplifier subjected to signals at 6 and 8 MHz will produce second-order IMD products at 2 MHz (8–6) and 14 MHz (8+6).
- *Third-order* IMD products change 3 dB for every decibel of input-signal change (this also assumes equal-level input signals), and appear at frequencies corresponding to the sums and differences of twice one signal's frequency plus or minus the frequency of another. Assuming that its input bandwidth is sufficient to pass them, an amplifier subjected to signals at 14.02 MHz (f_1) and 14.04 MHz (f_2) produces third-order IMD products at 14.00 ($2f_1 - f_2$), 14.06 ($2f_2 - f_1$), 42.08 ($2f_1 + f_2$) and 42.10 ($2f_2 + f_1$) MHz. The subtractive products (the 14.02 and 14.04-MHz products in this example) are close to the desired signal and can cause significant interference. Thus, third-order-IMD performance is of great importance in receiver mixers and RF amplifiers.

It can be seen that the IMD order determines how rapidly IMD products change level per unit change of input level. *N*th-order IMD products therefore change by *n* dB for every decibel of input-level change.

IMD products at orders higher than three can and do occur in communication systems, but the second- and third-order products are most important in receiver front ends. In transmitters, third- and higher-odd-order products are important because they widen the transmitted signal.

Intercept Point

The second type of dynamic range concerns the receiver's *intercept point*, sometimes simply referred to as *intercept*. Intercept point is typically measured by applying two or three signals to the antenna input, tuning the receiver to count the number of resulting spurious

Testing and Calculating Intermodulation Distortion in Receivers

Second and third-order IMD can be measured using the setup of **Fig A**. The outputs of two signal generators are combined in a 3-dB hybrid coupler. Such couplers are available from various companies, and can be homemade. The 3-dB coupler should have low loss and should itself produce negligible IMD. The signal generators are adjusted to provide a known signal level at the output of the 3-dB coupler, say, –20 dBm for each of the two signals. This combined signal is then fed through a calibrated variable attenuator to the device under test. The shielding of the cables used in this system is important: At least 90 dB of isolation should exist between the high-level signal at the input of the attenuator and the low-level signal delivered to the receiver.

The measurement procedure is simple: adjust the variable attenuator to produce a signal of known level at the frequency of the expected IMD product ($f_1 \pm f_2$ for second-order, $2f_1 - f_2$ or $2f_2 - f_1$ for third-order IMD).

To do this, of course, you have to figure out what equivalent input signal level at the receiver's operating frequency corresponds to the level of the IMD product you are seeing. There are several ways of doing this. One way — the way used by the ARRL Lab in their receiver tests — uses the minimum discernible signal. This is defined as the signal level that produces a 3-dB increase in the receiver audio output power. That is, you measure the receiver output level with no input signal, then insert a signal at the operating frequency and adjust the level of this input signal until the output power is 3 dB greater than the no-signal power. Then, when doing the IMD measurement, you adjust the attenuator of Fig A to cause a 3-dB increase in receiver output. The level of the IMD product is then the same as the MDS level you measured.

There are several things I dislike about doing the measurement this way. The problem is that you have to measure noise power. This can be difficult. First, you need an RMS voltmeter or audio power meter to do it at all. Second, the measurement varies with time (it's noise!), making it difficult to nail down a number. And third, there is the question of the audio response of the receiver; its noise output may not be flat across the output spectrum. So I prefer to measure, instead of MDS, a higher reference level. I use the receiver's S meter as a reference. I first determine the input signal level it takes to get an S1 reading. Then, in the IMD measurement, I adjust the attenuator to again give an S1 reading. The level of the IMD product signal is now equal to the level I measured at S1. Note that this technique gives a different IMD level value than the MDS technique. That's OK, though. What we are trying to determine is the *difference* between the level of the signals applied to the receiver input and the level of the IMD product. Our calculations will give the same result whether we measure the IMD product at the MDS level, the S1 level or some other level.

An easy way to make the reference measurement is with the setup of Fig A. You'll have to switch in a lot of attenuation (make sure you have an attenuator with enough range), but doing it this way keeps all of the possible variations in the measurement fairly constant. And this way, the difference between the reference level and the input level needed to produce the desired IMD product signal level is simply the difference in attenuator settings between the reference and IMD measurements.

Calculating Intercept Points

Once we know the levels of the signals applied to the receiver input and the level of the IMD product, we can easily calculate the intercept point using the following equation:

$$IP_n = \frac{n \bullet P_A - P_{IM_n}}{n-1} \qquad \text{(A)}$$

Here, n is the order, P_A is the receiver input power (of one of the input signals), P_{IMn} is the power of the IMD product signal, and IP_n is the nth-order intercept point. All powers should be in dBm. For second and third-order IMD, equation A results in the equations:

$$IP_2 = \frac{2 \bullet P_A - P_{IM_2}}{2-1} \qquad \text{(B)}$$

$$IP_3 = \frac{3 \bullet P_A - P_{IM_3}}{3-1} \qquad \text{(C)}$$

You can measure higher-order intercept points, too.

Example Measurements

To get a feel for this process, it's useful to consider some actual measured values.

The first example is a Rohde & Schwarz model EK085 receiver with digital preselection. For measuring second-order IMD, signals at 6.00 and 8.01 MHz, at –20 dBm each, were applied at the input of the attenuator. The difference in attenuator settings between the reference measurement and the level needed to produce the desired IMD product signal level

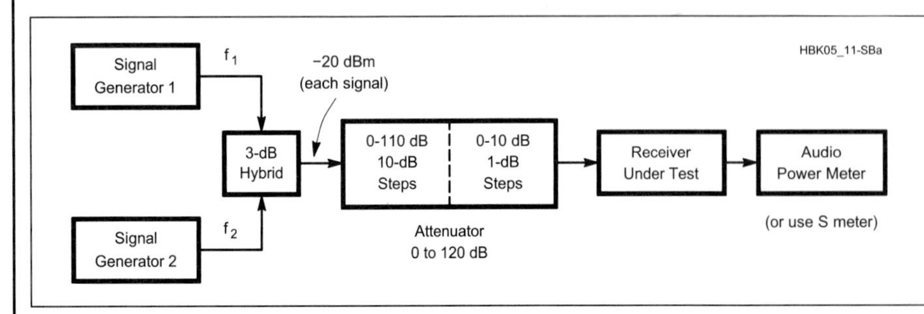

Fig A — Test setup for measurement of IMD performance. Both signal generators should be types such as HP 608, HP 8640, or Rhode & Schwarz SMDU, with phase-noise performance of –140 dBc/Hz or better at 20 kHz from the signal frequency.

was found to be 125 dB. The calculation of the second-order IP is then:

$$IP_2 = \frac{2(-20\text{ dBm}) - (-20\text{ dBm} - 125\text{ dB})}{2-1}$$

$$= -40\text{ dBm} + 20\text{ dBm} + 125\text{ dB} = +105\text{ dBm}$$

For IP_3, we set the signal generators for 0 dBm at the attenuator input, using frequencies of 14.00 and 14.01 MHz. The difference in attenuator settings between the reference and IMD measurements was 80 dB, so:

$$IP_3 = \frac{3(0\text{ dBm}) - (0\text{ dBm} - 80\text{ dB})}{3-1}$$

$$= \frac{0\text{ dBm} + 80\text{ dB}}{2} = +40\text{ dBm}$$

We also measured the IP_3 of a Yaesu FT-1000D at the same frequencies, using attenuator-input levels of –10 dBm. A difference in attenuator readings of 80 dB resulted in the calculation:

$$IP_3 = \frac{3(-10\text{ dBm}) - (-10\text{ dBm} - 80\text{ dB})}{3-1}$$

$$= \frac{-30\text{ dBm} + 10\text{ dBm} + 80\text{ dB}}{2}$$

$$= \frac{-20\text{ dBm} + 80\text{ dB}}{2}$$

$$= +30\text{ dBm}$$

Synthesizer Requirements

To be able to make use of high third-order intercept points at these close-in spacings requires a low-noise LO synthesizer. You can estimate the required noise performance of the synthesizer for a given IP_3 value. First, calculate the value of receiver input power that would cause the IMD product to just come out of the noise floor, by solving equation A for P_A, then take the difference between the calculated value of P_A and the noise floor to find the dynamic range. Doing so gives the equation:

$$ID_3 = \frac{2}{3}(IP_3 + P_{min}) \qquad \text{(D)}$$

Where ID_3 is the third-order IMD dynamic range in dB and P_{min} is the noise floor in dBm. Knowing the receiver bandwidth, BW (2400 Hz in this case) and noise figure, NF (8 dB) allows us to calculate the noise floor, P_{min}:

$$P_{min} = -174\text{ dBm} + 10\log(BW) + NF$$

$$= -174\text{ dBm} + 10\log(2400) + 8$$

$$= -132\text{ dBm}$$

The synthesizer noise should not exceed the noise floor when an input signal is present that just causes an IMD product signal at the noise floor level. This will be accomplished if the synthesizer noise is less than:

$$ID + 10\log(BW) = 114.7\text{ dB} + 10\log(2400)$$

$$= 148.5\text{ dBc/Hz}$$

in the passband of the receiver. Such synthesizers hardly exist.—*Dr. Ulrich L. Rohde, KA2WEU*

responses, and measuring their level relative to the input signal.

Because a device's IMD products increase more rapidly than its desired output as the input level rises, it might seem that steadily increasing the level of multiple signals applied to an amplifier would eventually result in equal desired-signal and IMD levels at the amplifier output. Real devices are incapable of doing this, however. At some point, every device *overloads*, and changes in its output level no longer equally track changes at its input. The device is then said to be operating in *compression*; the point at which its first-order response deviates from linearity by 1 dB is its *1-dB compression point*. Pushing the process to its limit ultimately leads to *saturation*, at which point input-signal increases no longer increase the output level.

The power level at which a device's second-order IMD products equal its first-order output (a point that must be extrapolated because the device is in compression by this point) is its *second-order intercept point*. Likewise, its *third-order intercept point* is the power level at which third-order responses equal the desired signal. **Fig 11.30** graphs these relationships.

Input filtering can improve second-order intercept point; device non-linearities

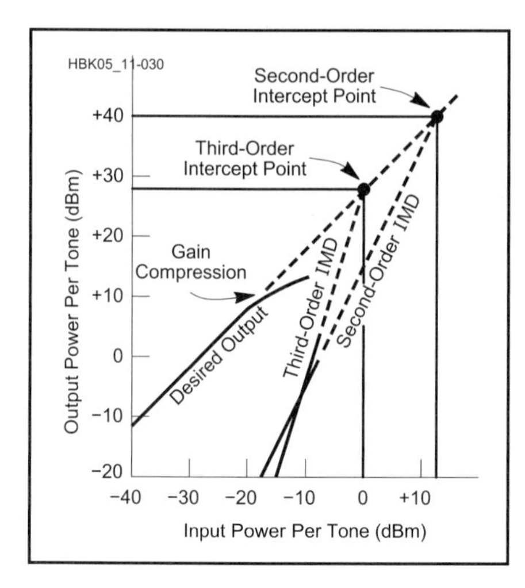

Fig 11.30 — A linear stage's output tracks its input decibel by decibel on a 1:1 slope — its *first-order* response. *Second-order intermodulation distortion (IMD) products* produced by two equal-level input signals ("tones") rise on a 2:1 slope — 2 dB for every 1 dB of input increase. *Third-order IMD products* likewise increase 3 dB for every 1 dB of increase in two equal tones. For each IMD order *n*, there is a corresponding *intercept point* IP_n at which the stage's first-order and *n*th order products are equal in amplitude. The first-order output of real amplifiers and mixers falls off (the device overloads and goes into *compression*) before IMD products can intercept it, but intercept point is nonetheless a useful, valid concept for comparing radio system performance. The higher an amplifier or mixer's intercept point, the stronger the input signals it can handle without overloading. The input and output powers shown are for purposes of example; every device exhibits its own particular IMD profile. (After W. Hayward, *Introduction to Radio Frequency Design*, Fig 6.17)

Fig 11.31 — A diplexer resistively terminates energy at unwanted frequencies while passing energy at desired frequencies. This band-pass diplexer (A) uses a series-tuned circuit as a selective pass element, while a high-C parallel-tuned circuit keeps the network's terminating resistor R1 from dissipating desired-frequency energy. Computer simulation of the diplexer's response with *ARRL Radio Designer 1.0* characterizes the diplexer's insertion loss and good input match from 8.8 to 9.2 MHz (B) and from 1 to 100 MHz (C); and the real and imaginary components of the diplexer's input impedance from 8.8 to 9.2 MHz with a 50-Ω load at the diplexer's output terminal (D). The high-C, low-L nature of the L2-C2 circuit requires that C2 be minimally inductive; a 10,000-pF chip capacitor is recommended. This diplexer was described by Ulrich L. Rohde and T. T. N. Bucher in *Communications Receivers: Principles and Design.*

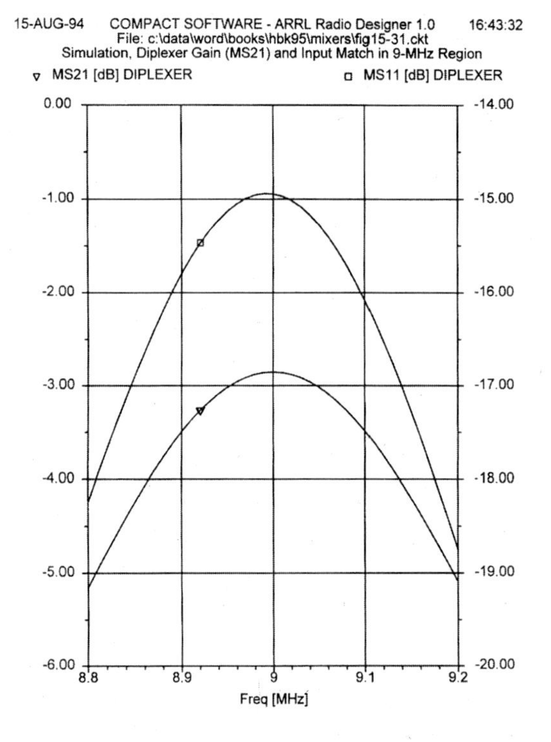

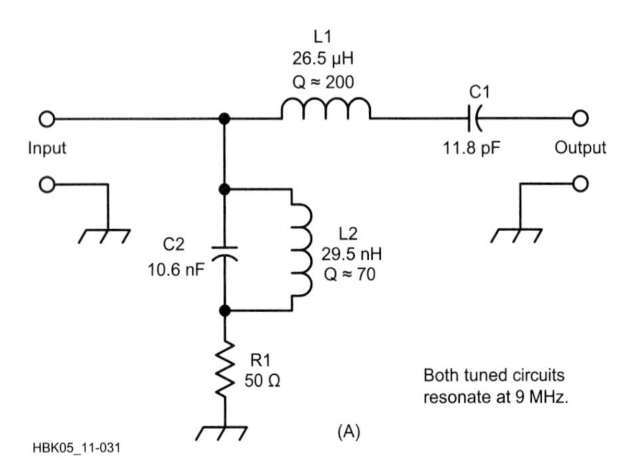

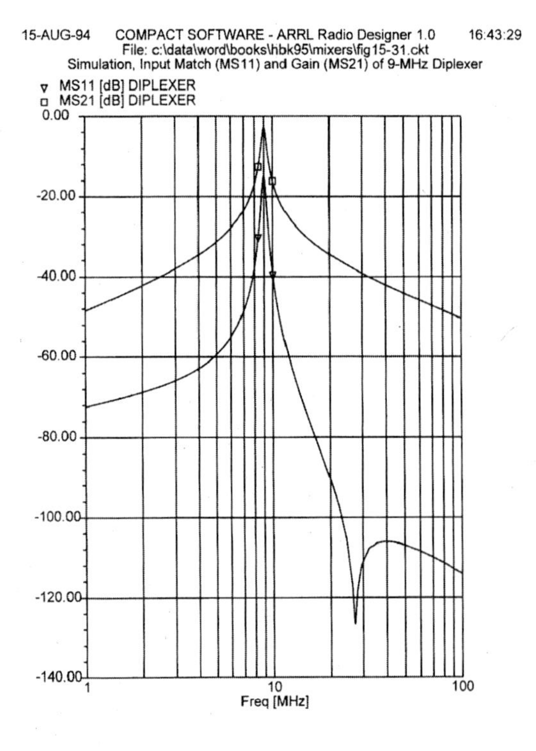

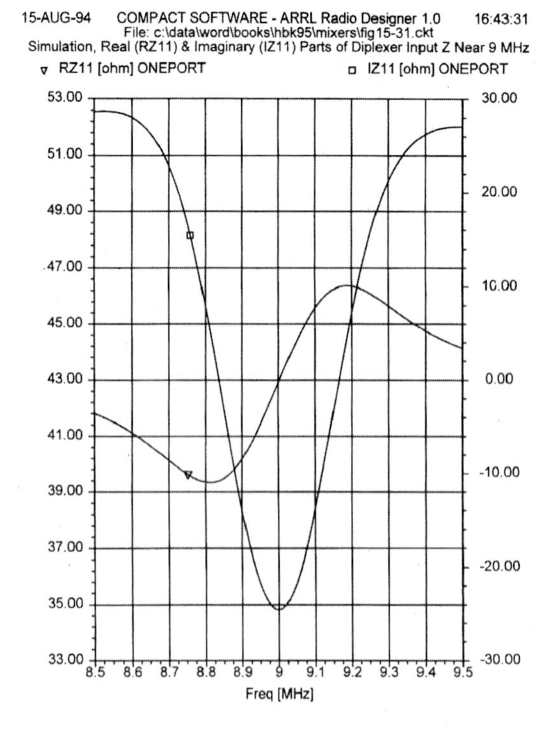

determine the third, fifth and higher-odd-number intercept points. In preamplifiers, third-order intercept point is directly related to dc input power; in mixers, to the local-oscillator power applied.

Applying Diode DBMs

At first glance, applying a diode DBM is easy: We feed the signal(s) we want to frequency-shift (at or below the maximum level called for in the mixer's specifications, such as –10 dBm for the Mini-Circuits SBL-1 and Synergy Microwave S-1, popular Level 7 parts) to the DBM's RF port, feed the frequency-shifting signal (at the proper level) to the LO port, and extract the sum and difference products from the mixer's IF port.

There's more to it than that, however, because diode DBMs (along with most other modern mixer types) are *termination-sensitive*. That is, their ports — particularly their IF (output) ports — must be resistively terminated with the proper impedance (commonly 50 Ω, resistive). A wideband, resistive output termination is particularly critical if a mixer is to achieve its maximum dynamic range in receiving applications. Such a load can be achieved by

- terminating the mixer in a 50-Ω resistor or attenuator pad (a technique usually avoided in receiving applications because it directly degrades system noise figure);
- terminating the mixer with a low-noise, high-dynamic-range *post-mixer amplifier* designed to exhibit a wideband resistive input impedance; or
- terminating the mixer in a *diplexer*, a frequency-sensitive signal splitter that appears as a two-terminal resistive load at its input while resistively dissipating unwanted outputs and passing desired outputs through to subsequent circuitry.

Termination-insensitive mixers are available, but this label can be misleading. Some termination-insensitive mixers are nothing more than a termination-sensitive mixer packaged with an integral post-mixer amplifier. True termination-insensitive mixers are less common and considerably more elaborate. Amateur builders will more likely use one of the

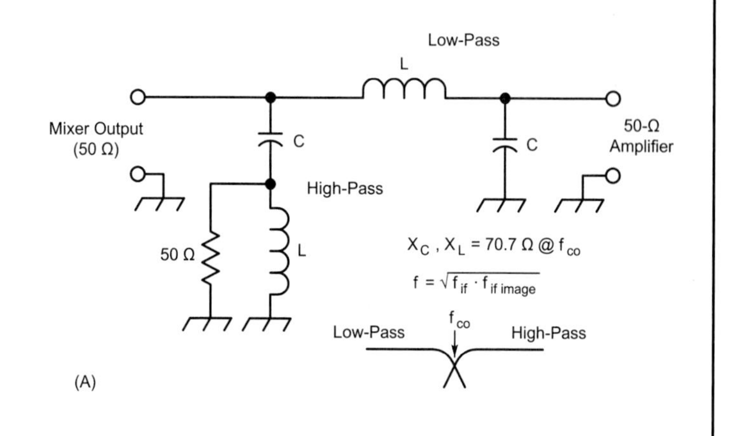

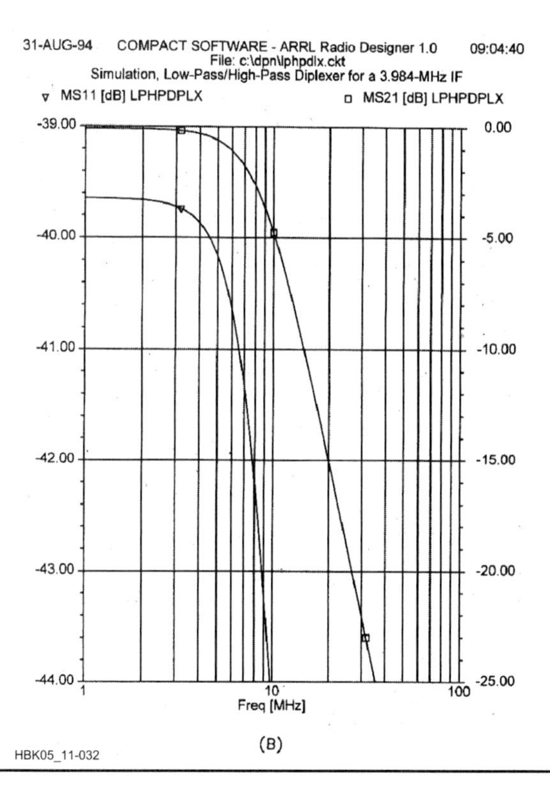

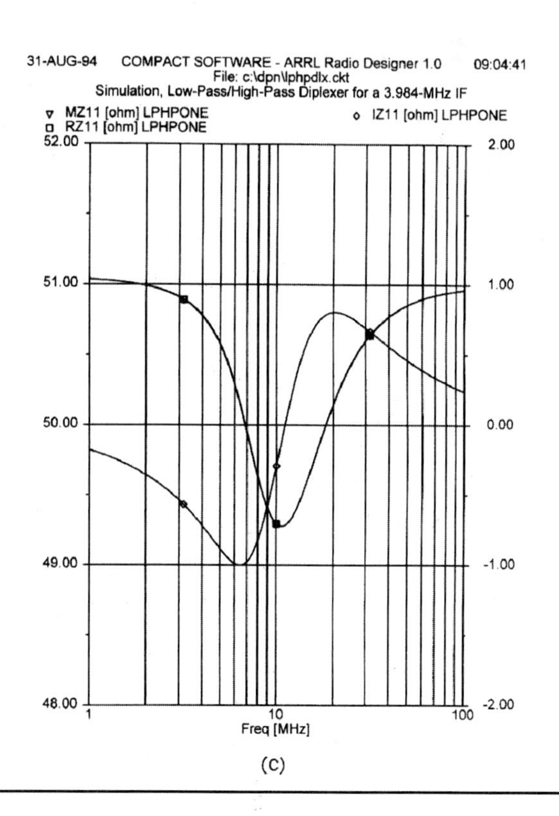

Fig 11.32 — All of the inductors and capacitors in this high-pass/low-pass diplexer (A) exhibit a reactance of 70.7 Ω at its tuned circuits' 3-dB cutoff frequency (the geometric mean of the IF and IF image). B and C show *ARRL Radio Designer* simulations of this circuit configured for use in a receiver that converts 7 MHz to 3.984 MHz using a 10.984-MHz LO. The IF image is at 17.984 MHz, giving a 3-dB cutoff frequency of 8.465 MHz. The inductor values used in the simulation were therefore 1.33 µH (Q = 200 at 25.2 MHz); the capacitors, 265 pF (Q = 1000). This drawing shows idler load and "50-Ω Amplifier" connections suitable for a receiver in which the IF image falls at a frequency *above* the desired IF. For applications in which the IF image falls below the desired IF, interchange the 50-Ω idler load resistor and the diplexer's "50-Ω Amplifier" connection so the idler load terminates the diplexer low-pass filter and the 50-Ω amplifier terminates the high-pass filter.

many excellent termination-sensitive mixers available in connection with a diplexer, post-mixer amplifier or both.

Fig 11.31 shows one diplexer implementation. In this approach, L1 and C1 form a series-tuned circuit, resonant at the desired IF, that presents low impedance between the diplexer's input and output terminals at the IF. The high-impedance parallel-tuned circuit formed by L2 and C2 also resonates the desired IF, keeping desired energy out of the diplexer's 50-Ω load resistor, R1.

The preceding example is called a *band-pass diplexer.* **Fig 11.32** shows another type: a *high-pass/low-pass diplexer* in which each inductor and capacitor has a reactance of 70.7 Ω at the 3-dB cutoff frequency. It can be used after a "difference" mixer (a mixer in which the IF is the difference between the signal frequency and LO) if the desired IF and its image frequency are far enough apart so that the image power is "dumped" into the network's 51-Ω resistor. (For a "summing" mixer — a mixer in which the IF is the sum of the desired signal and LO — interchange the 50-Ω idler load resistor and the diplexer's "50-Ω Amplifier" connection.) Richard Weinreich, KØUVU, and R. W. Carroll described this circuit in November 1968 *QST* as one of several absorptive TVI filters.

Fig 11.33 shows a BJT post-mixer amplifier design made popular by Wes Hayward, W7ZOI, and John Lawson, K5IRK. RF feedback (via the 1-kΩ resistor) and emitter degeneration (the ac-coupled 5.6-Ω emitter resistor) work together to keep the stage's input impedance low and uniformly resistive across a wide bandwidth.

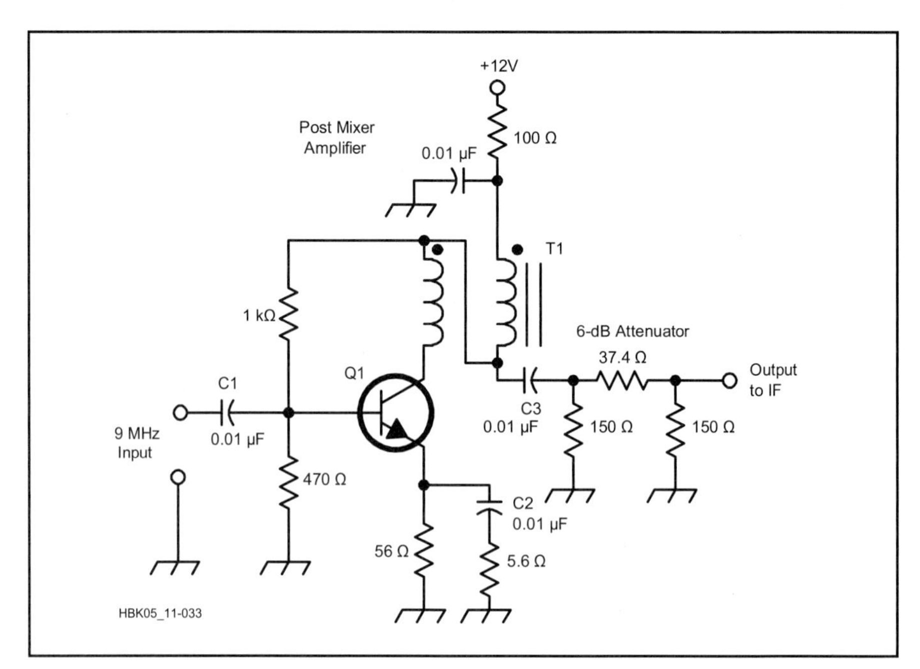

Fig 11.33 — The post-mixer amplifier from Hayward and Lawson's Progressive Communications Receiver (November 1981 *QST*). This amplifier's gain, including the 6-dB loss of the attenuator pad, is about 16 dB; its noise figure, 4 to 5 dB; its output intercept, 30 dBm. The 6-dB attenuator is essential if a crystal filter follows the amplifier; the pad isolates the amplifier from the filter's highly reactive input impedance. This circuit's input match to 50 Ω below 4 MHz can be improved by replacing 0.01-µF capacitors C1, C2 and C3 with low-inductance 0.1-µF units (chip capacitors are preferable).

Q1 — TO-39 CATV-type bipolar transistor, f_T = 1 GHz or greater. 2N3866, 2N5109, 2SC1252, 2SC1365 or MRF586 suitable. Use a small heat sink on this transistor.

T1 — Broadband ferrite transformer, ≈42 µH per winding: 10 bifilar turns of #28 enameled wire on an FT 37-43 core.

Amplitude Modulation with a DBM

We can generate DSB, suppressed-carrier AM with a DBM by feeding the carrier to its RF port and the modulating signal to the IF port. This is a classical *balanced modulator*, and the result — sidebands at radio frequencies corresponding to the carrier signal plus audio and the RF signal minus audio — emerges from the DBM's LO port. If we also want to transmit some carrier along with the sidebands, we can dc-bias the IF port (with a current of 10 to 20 mA) to upset the mixer's balance and keep its diodes from turning all the way off. (This technique is sometimes used for generating CW with a balanced modulator otherwise intended to generate DSB as part of an SSB-generation process.) **Fig 11.34** shows a more elegant approach to generating full-carrier AM with a DBM.

As we saw earlier when considering the many faces of AM, two DBMs, used in

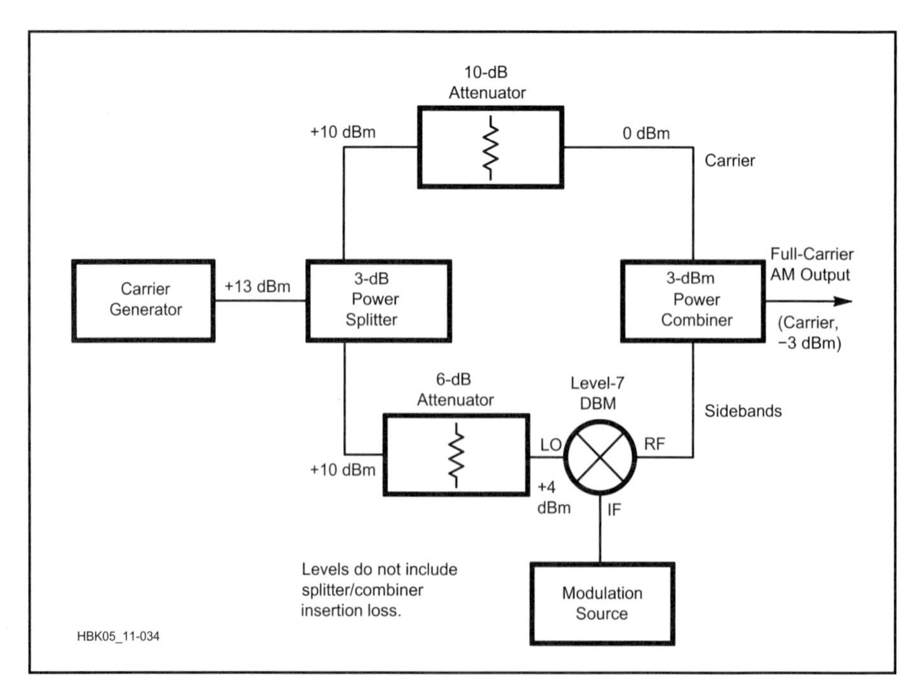

Fig 11.34 — Generating full-carrier AM with a diode DBM.

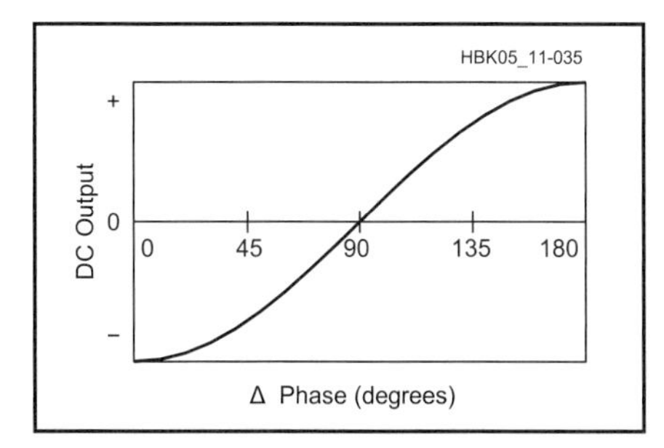

Fig 11.35 — A phase detector's dc output is the cosine of the phase difference between its input and reference signals.

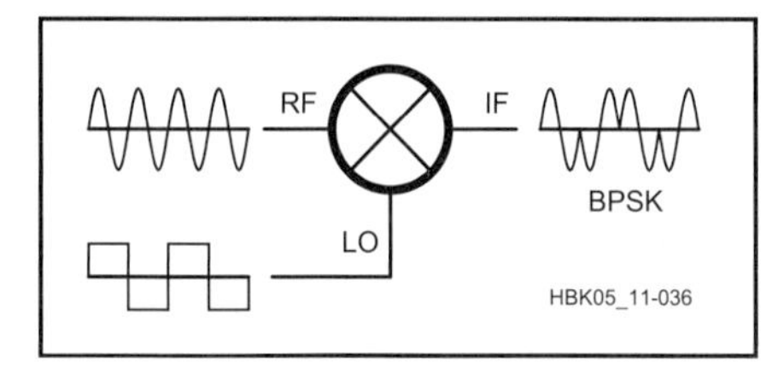

Fig 11.36 — Mixing a carrier with a square wave generates biphase-shift keying (BPSK), in which the carrier phase is shifted 180° for data transmission. In practice, as in this drawing, the carrier and data signals are phase-coherent so the mixer switches only at carrier zero crossings.

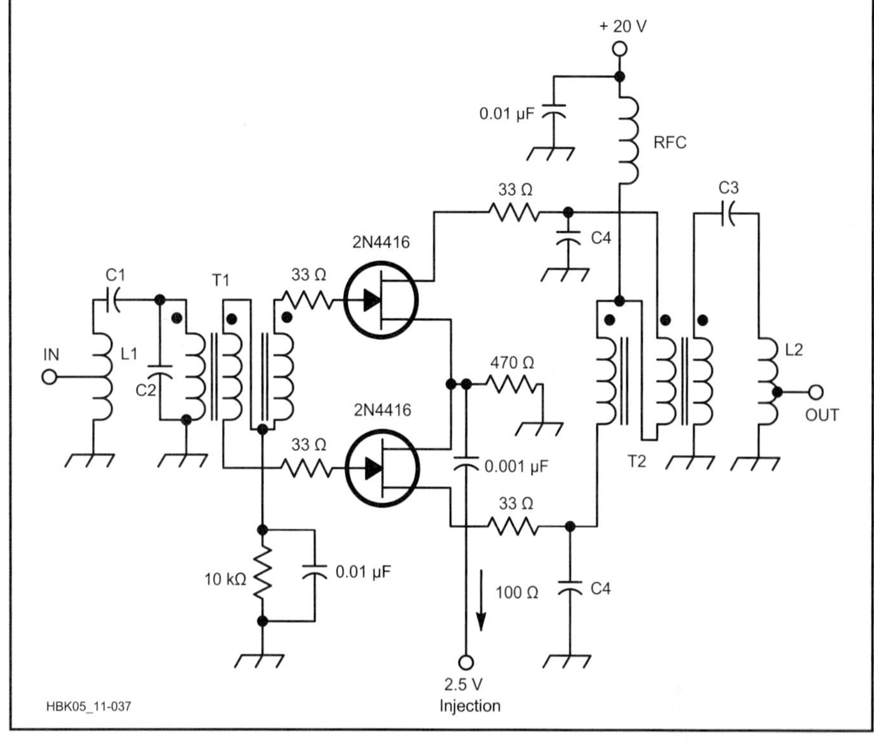

Fig 11.37 — Two 2N4416 JFETs provide high dynamic range in this mixer circuit from Sabin, *QST*, July 1970. L1, C1 and C2 form the input tuned circuit; L2, C3 and C4 tune the mixer output to the IF. The trifilar input and output transformers are broadband transmission-line types.

conjunction with carrier and audio phasing, can be used to generate SSB, suppressed-carrier AM. Likewise, two DBMs can be used with RF and LO phasing as an image-reject mixer.

Phase Detection with a DBM

As we saw in our exploration of quadrature detection, applying two signals of equal frequency to a DBM's LO and RF ports produces an IF-port dc output proportional to the cosine of the signals' phase difference (**Fig 11.35**). (This assumes that the DBM has a dc-coupled IF port, of course. If it doesn't — and some DBMs don't — phase-detector operation is out.) Any dc output offset introduces error into this process, so critical phase-detection applications use low-offset DBMs optimized for this service.

Biphase-Shift Keying Modulation with a DBM

Back in our discussion of square-wave mixing, we saw how multiplying a switching mixer's linear input with a square wave causes a 180° phase shift during the negative part of the square wave's cycle. As **Fig 11.36** shows, we can use this effect to produce *biphase-shift keying (BPSK)*, a digital system that conveys data by means of carrier phase reversals. A related system, *quadrature phase-shift keying (QPSK)* uses two DBMs and phasing to convey data by phase-shifting a carrier in 90° increments.

Transistors as Mixer Switching Elements

We've covered diode DBMs in depth because their home-buildability, high performance and suitability for direct connection into 50-Ω systems makes them attractive to Amateur Radio builders. The abundant availability of high-quality manufactured diode mixers at reasonable prices makes them excellent candidates for home construction projects. Although diode DBMs are common in telecommunications as a whole, their conversion loss and relatively high LO power requirement have usually driven the manufacturers of high-performance MF/HF Amateur Radio receivers and transceivers to other solutions. Those solutions have generally involved single- or double-balanced FET mixers — MOSFETs in the late 1970s and early 1980s, JFETs from the early 1980s to date. Many of the JFET designs are variations of a single-balanced mixer circuit introduced to *QST* readers in 1970!

Fig 11.37 shows the circuit as it was presented by William Sabin in "The Solid-State Receiver," *QST*, July 1970. Two 2N4416 JFETs operate in a common-source configuration, with push-pull RF input and parallel LO drive. **Fig 11.38** shows a similar circuit as implemented in the ICOM IC-765 transceiver. In this version, the JFETs (2SK125s) operate in common-gate, with the LO applied across a 220-Ω resistor between the gates and ground.

Bipolar junction transistors can also work well in switching mixers. In June 1994 *QST*, Ulrich Rohde, KA2WEU, published a medium-frequency mixer well suited to shortwave applications and home-built projects. Shown in **Fig 11.39**, it consists of two transistors in a push-pull, single-balanced configuration. Because of the degenerative feedback introduced by the 20-Ω emitter resistors, the two transistors need not be matched. This mixer's advantage lies in its achievement of a 33-dBm-output intercept with only 17 dBm of local oscillator drive. (Typically, a diode DBM with the same IMD performance requires 25 to 27 dBm of LO drive.) Tests indicate that the upper frequency limit of this mixer lies in the

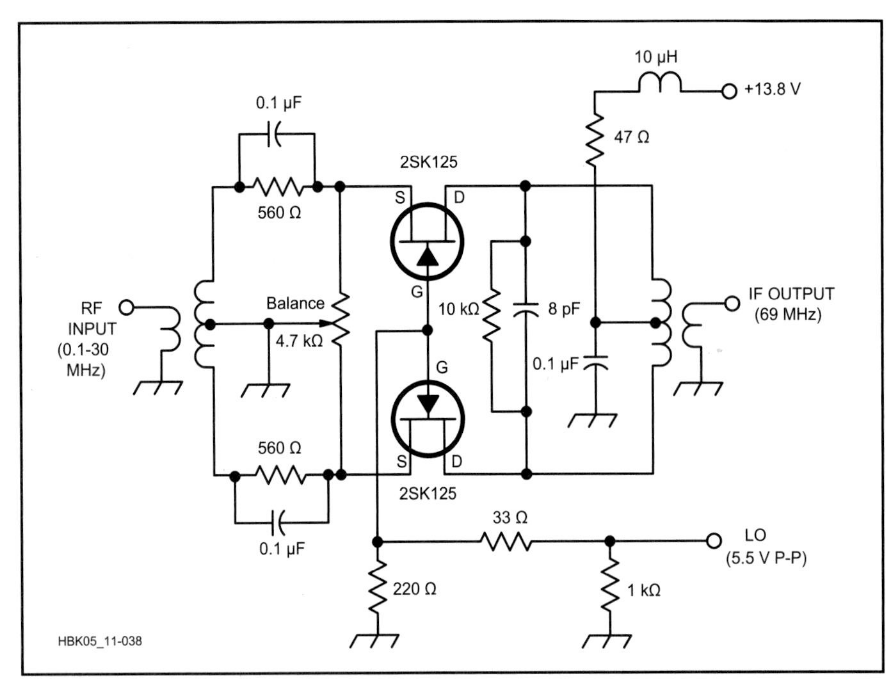

Fig 11.38 — The ICOM IC-765's single-balanced 2SK125 mixer achieves a high dynamic range (per *QST* Product Review, an IP_3 of 10.5 dBm at 14 MHz with preamp off) with arrangement very much like Sabin's. The first receive mixer in many commercial Amateur Radio transceiver designs of the 1980s and 1990s used a 2SK125 pair in much this way.

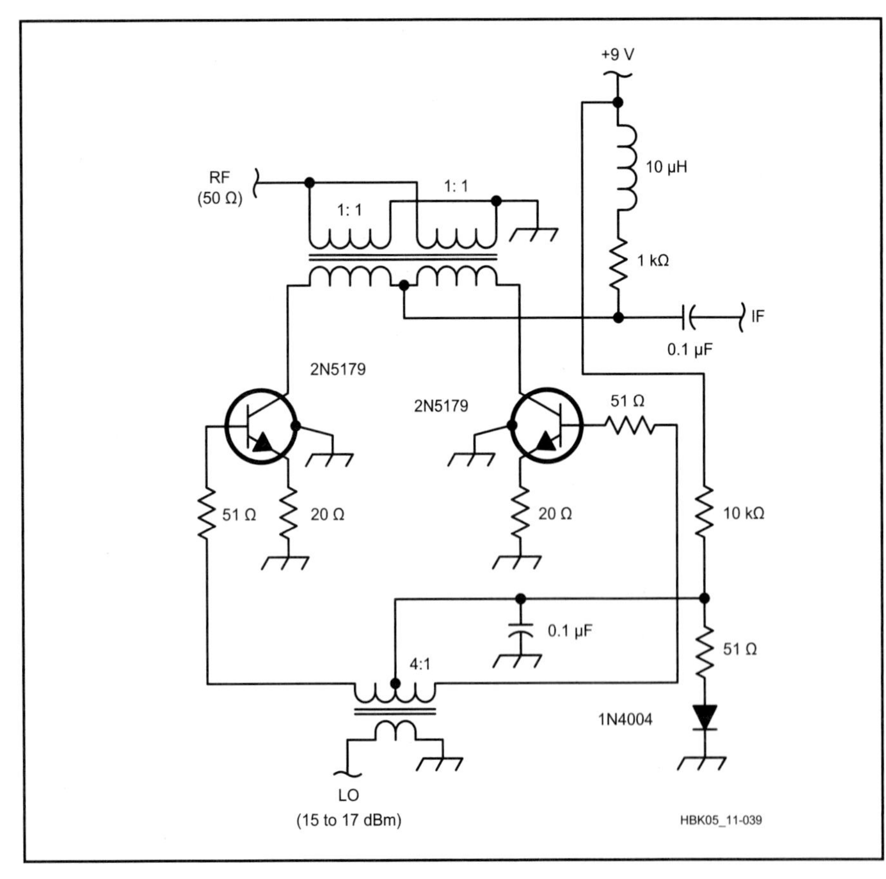

Fig 11.39 — This single-balanced, push-pull BJT mixer achieves a high dynamic range (IP_3 ≈33 dBm) with 15 to 17 dBm of LO drive. Its insertion loss is approximately 6 dB. A diode-ring mixer would require 25 to 27 dBm of LO drive to achieve the same IP_3 .

500-MHz region. The circuit's lower frequency limit depends on the transformer inductances and the ferrites used for the transformer cores.

Examining the state of the art we find that the best receive-mixer dynamic ranges are achieved with quads of RF MOSFETs operating as passive switches, with no drain voltage applied. The best of these techniques involves following a receiver's first mixer with a diplexer and low-loss roofing crystal filter, rather than terminating the mixer in a strong wideband amplifier.

High-Performance Mixer Experiments

Colin Horrabin's (G3SBI) experimentations with variations of an original high-performance mixer circuit by Jacob Mahkinson, N6NWP, led to the development of a new mixer configuration, called an H-mode mixer. This name comes from the signal path through the circuit. (See **Fig 11.40A**.) Horrabin is a professional scientist/engineer at the Science and Engineering Research Council's Darebury Laboratory, which has supported his investigative work on the H-mode switched-FET mixer, and consequently holds intellectual title to the new mixer. This does not prevent readers from taking the development further or using the information presented here.

Inputs A and B are complementary square-wave inputs derived from the sine-wave local oscillator at twice the required square-wave frequency. If A is ON, then FETs F1 and F3 are ON and F2 and F4 are OFF. The direction of the RF signal across T1 is given by the E arrows. When B is ON, FETs F2 and F4 are ON and F1 and F3 are OFF. The direction of the RF signal across T1 reverses, as shown by the F arrows.

This is still the action of a switching mixer, but now the source terminal of each FET switch is grounded, so that the RF signal switched by the FET cannot modulate the gate voltage. In this configuration the transformers are important: T1 is a Mini-Circuits type T4-1 and T2 is a pair of these same transformers with their primaries connected in parallel.

The Ubiquitous NE602: A Popular Gilbert Cell Mixer

Introduced as the NE602 in the mid-1980s, the Philips Components-Signetics NE602A mixer-oscillator IC has become greatly popular with amateur experimenters for transmit mixers, receive mixers and balanced modulators. **Fig 11.41** shows its equivalent circuit. The NE602A's typical current drain is 2.4 mA; its supply voltage

Fig 11.40 — A shows the operation of the H mode switched mixer developed by Colin Horrabin, G3SBI. B shows the actual mixer circuit implemented with the Siliconix SD5000 DMOS FET quad switch IC and a 74AC74 flip-flop.

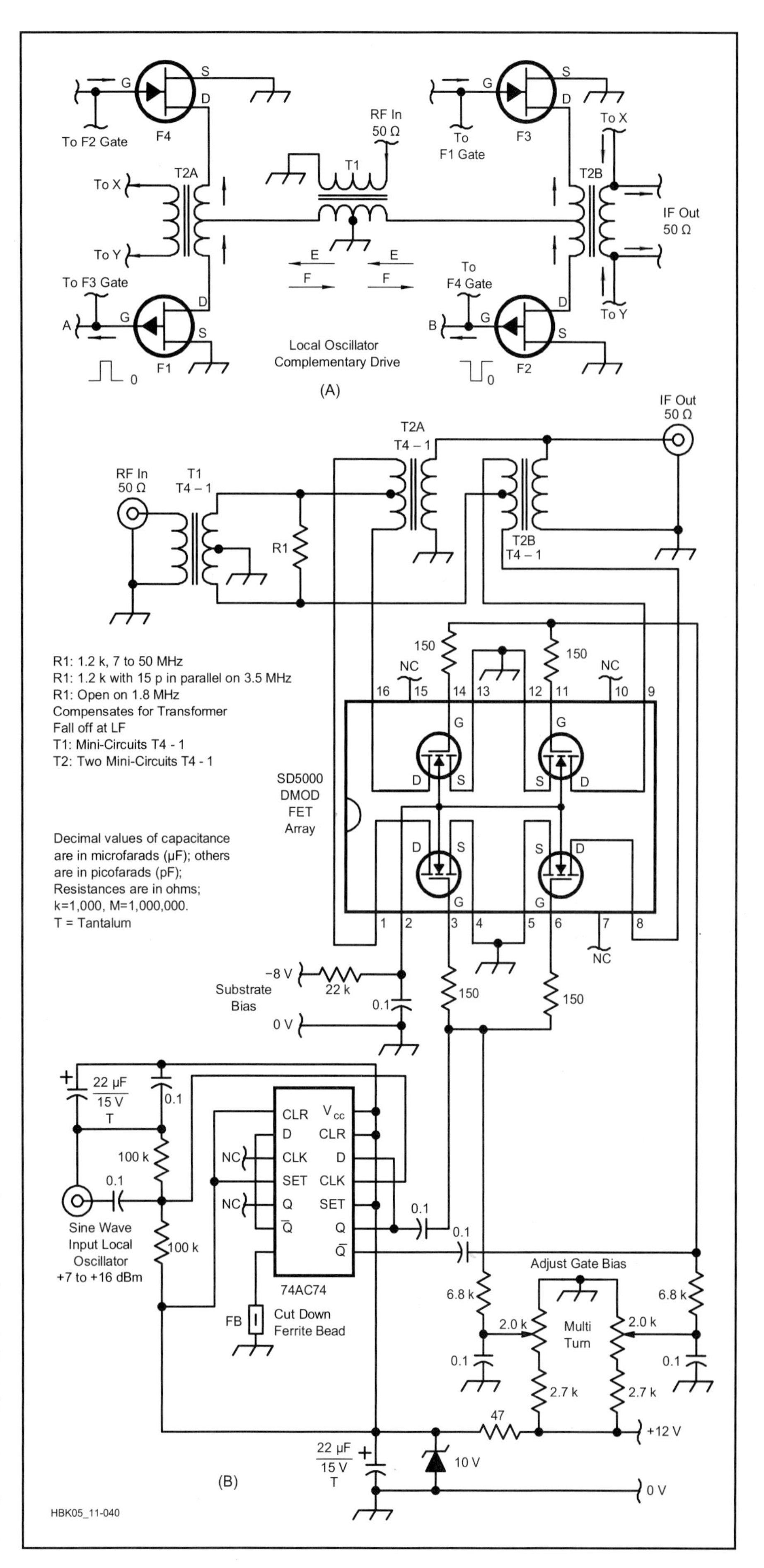

range is 4.5 to 8.0.

As we learned in exploring sine-wave versus square-wave mixing, the NE602's mixer is a Gilbert cell multiplier. Its inputs (RF) and outputs (IF) can be single- or double-ended (balanced) according to design requirements (**Fig 11.42**). Each input's equivalent ac impedance is approximately 1.5 kΩ in parallel with 3 pF; each output's resistance is 1.5 kΩ. The mixer can typically handle signals up to 500 MHz. At 45 MHz, its noise figure is typically 5.0 dB; its typical conversion gain, 18 dB. Considering the NE602A's low current drain, its input IP_3 (measured at 45 MHz with 60-kHz spacing) is usefully good at –15 dBm. Factoring in the mixer's conversion gain results in an equivalent output IP_3 of about 5 dBm.

The NE602A's on-board oscillator can operate up to 200 MHz in LC and crystal-controlled configurations (**Fig 11.43** shows three possibilities). Alternatively, energy from an external LO can be applied to the chip's pin 6 via a dc blocking capacitor. At least 200 mV P-P of external LO drive is required for proper mixer operation.

NE602A Usage Notes

The '602 was intended to be used as the second mixer in double-conversion FM cellular radios, in which the first IF is typically 45 MHz, and the second IF is typically 455 kHz. Such a receiver's second mixer can be relatively weak in terms of dynamic range because of the adjacent-signal protection afforded by the high selectivity of the first-IF filter preceding it. When used as a first mixer, the '602 can provide a two-tone third-order dynamic range between 80 and 90 dB, but this figure is greatly diminished if a pre-amplifier is used ahead of the '602 to improve the system's noise figure.

When the '602 is used as a second mixer, the sum of the gains preceding it should not exceed about 10 dB. NE602 product detection therefore should not follow a high-gain IF section unless appropriate attenuation is inserted between the '602 and the IF strip.

The '602 is generally *not* a good choice for VHF and higher-frequency mixers because of its input noise and diminishing

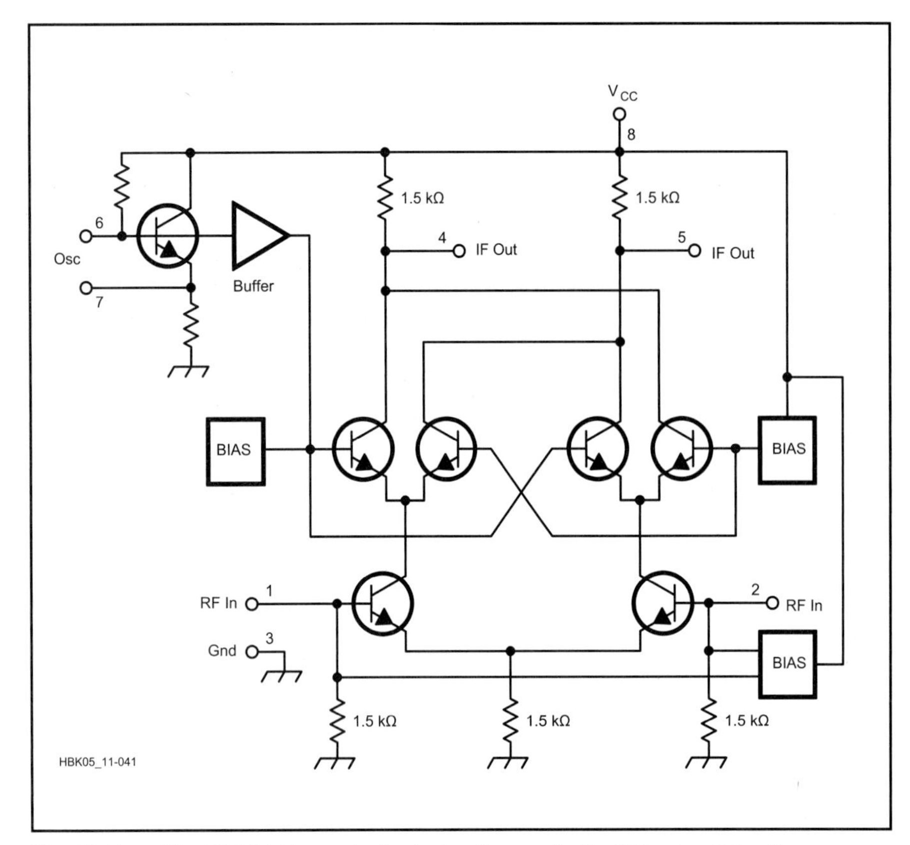

Fig 11.41 — The NE602A's equivalent circuit reveals its Gilbert-cell heritage.

IMD performance at high frequencies. There are applications, however, where 6-dB noise figures and 60- to 70-dB dynamic range performance is adequate. If your target specifications exceed these numbers, you should consider other mixers at VHF and up.

NE602A Relatives

The NE602A (SA602A for operation over a wider temperature range) began life as the NE602/SA602, a part with a slightly lower IP_3 than the A version. The pinout-identical NE612A/SA612A costs less as a result of wider tolerances. All of these parts should nonetheless work satisfactorily in most "NE602" experimenter projects. The same mixer/oscillator topology, modified for slightly higher dynamic range at the expense of somewhat less mixer gain, is also available in the Philips Components-Signetics mixer/oscillator/FM IF chips NE/SA605 (input IP_3, typically –10 dBm) and NE/SA615 (input IP_3, typically –13 dBm).

REFERENCES

J. Dillon, "The Neophyte Receiver," *QST*, Feb 1988, pp 14-18. Describes a VFO-tuned, NE602-based direct-conversion

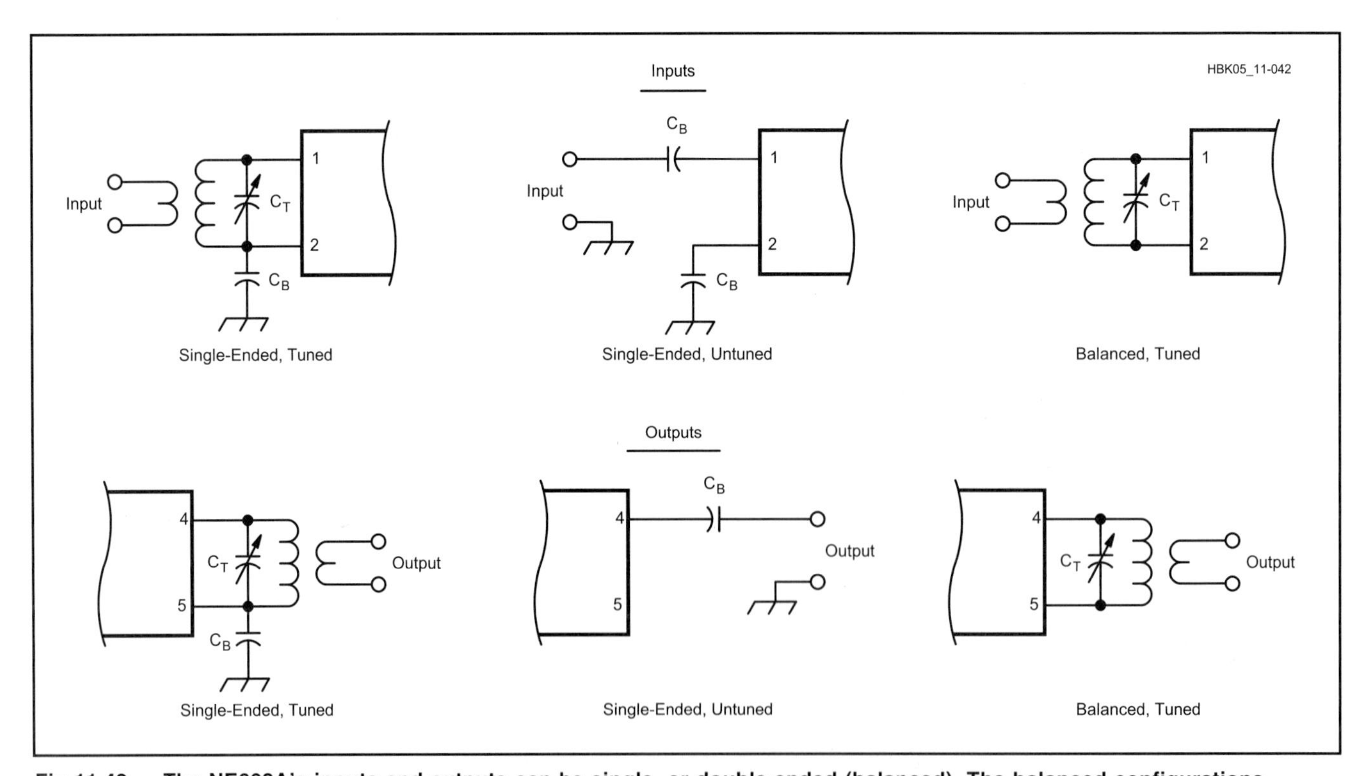

Fig 11.42 — The NE602A's inputs and outputs can be single- or double-ended (balanced). The balanced configurations minimize second-order IMD and harmonic distortion, and unwanted envelope detection in direct-conversion service. C_T tunes its inductor to resonance; C_B is a bypass or dc-blocking capacitor. The arrangements pictured don't show all the possible input/output configurations; for instance, you can also use a center-tapped broadband transformer to achieve a balanced, untuned input or output.

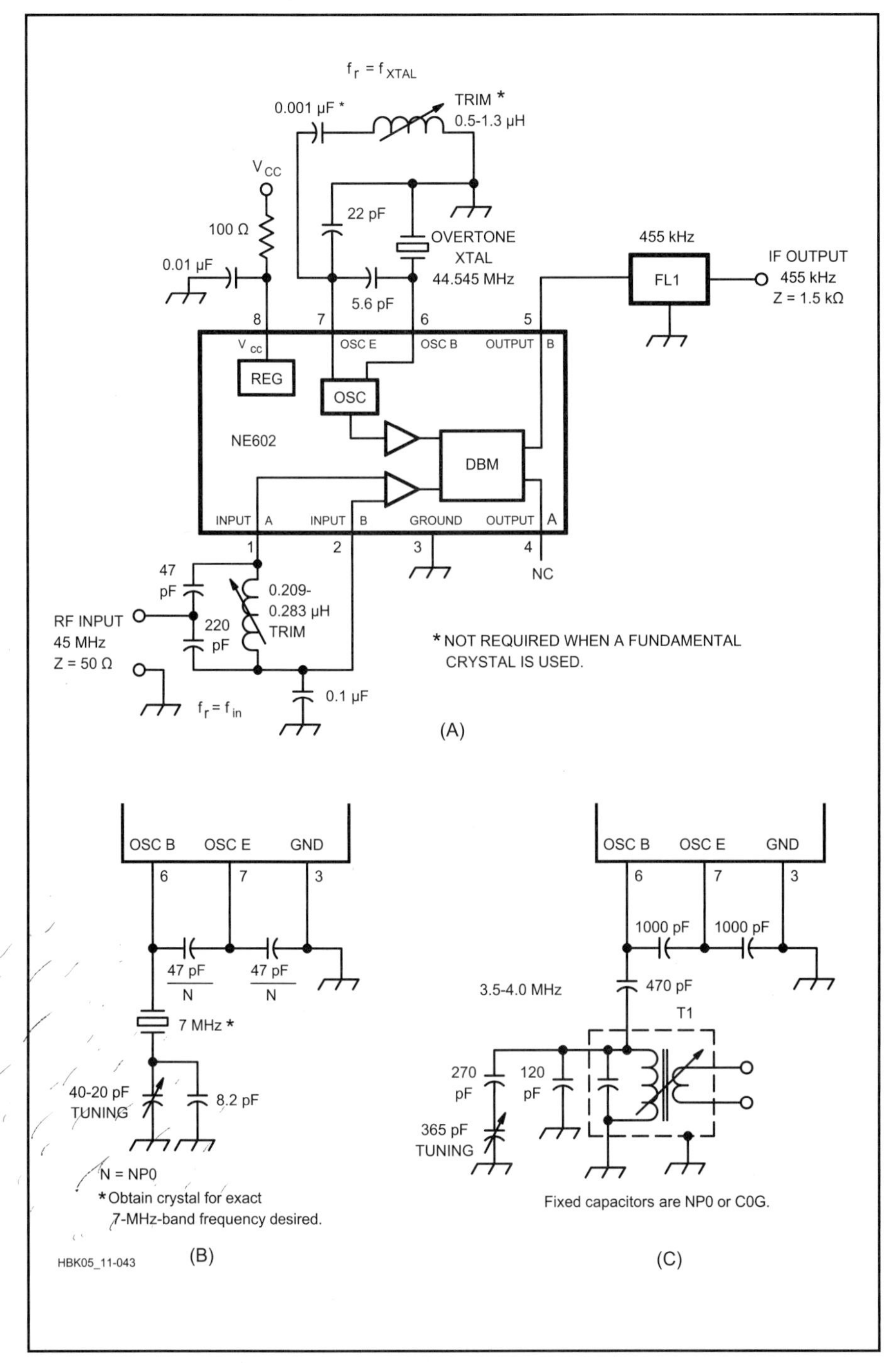

Fig 11.43 — Three NE602A oscillator configurations: crystal overtone (A); crystal fundamental (B); and LC-controlled (C). T1 in C is a Mouser 10.7-MHz IF transformer, green core, 7:1 turns ratio, part no. 42IF123.

receiver for 80 and 40 m.

B. Gilbert, "Demystifying the Mixer," self-published monograph, 1994.

P. Hawker, ed, "Super-Linear HF Receiver Front Ends," Technical Topics, *Radio Communication,* Sep 1993, pp 54-56.

W. Hayward, *Introduction to Radio Frequency Design* (Newington, CT: ARRL, 1994).

P. Horowitz and W. Hill, *The Art of Electronics*, 2nd ed (New York: Cambridge University Press, 1989).

S. Joshi, "Taking the Mystery Out of Double-Balanced Mixers," *QST*, Dec 1993, pp 32-36.

D. Kazdan "What's a Mixer?" *QST*, Aug 1992, pp 39-42.

J. Makhinson, "High Dynamic Range MF/HF Receiver Front End," *QST*, Feb 1993, pp 23-28. Also see Feedback, *QST*, Jun 1993, p 73.

RF/IF Designer's Handbook (Brooklyn, NY: Scientific Components, 1992).

RF Communications Handbook (Philips Components-Signetics, 1992).

L. Richey, "W1AW at the Flick of a Switch," New Ham Horizons, *QST*, Feb 1993, pp 56-57. Describes a crystal-controlled NE602 receiver buildable for 80, 40 and 20 m.

U. Rohde and T. Bucher, *Communications Receivers: Principles and Design* (New York: McGraw-Hill Book Co, 1988).

U. Rohde, "Key Components of Modern Receiver Design," Part 1, *QST*, May 1994, pp 29-32; Part 2, *QST*, Jun 1994, pp 27-31; Part 3, *QST*, Jul 1994, 42-45.

U. Rohde, "Testing and Calculating Intermodulation Distortion in Receivers," *QEX*, Jul 1994, pp 3-4.

W. Sabin, "The Solid-State Receiver," *QST*, Jul 1970, pp 35-43.

Synergy Microwave Corporation product handbook (Paterson, NJ: Synergy Microwave Corp, 1992).

R. Weinreich and R. W. Carroll, "Absorptive Filter for TV Harmonics," *QST,* Nov 1968, pp 20-25.

R. Zavrel, "Using the NE602," Technical Correspondence, *QST*, May 1990, pp 38-39.

Chapter 12

RF and AF Filters

This chapter contains basic design information and examples of the most common filters used by radio amateurs. It was prepared by Reed Fisher, W2CQH, and includes a number of design approaches, tables and filters by Ed Wetherhold, W3NQN, and others. The chapter is divided into two major sections. The first section contains a discussion of filter theory with some design examples. It includes the tools needed to predict the performance of a candidate filter before a design is started or a commercial unit purchased. Extensive references are given for further reading and design information. The second section contains a number of selected practical filter designs for immediate construction.

Basic Concepts

A filter is a network that passes signals of certain frequencies and rejects or attenuates those of other frequencies. The radio art owes its success to effective filtering. Filters allow the radio receiver to provide the listener with only the desired signal and reject all others. Conversely, filters allow the radio transmitter to generate only one signal and attenuate others that might interfere with other spectrum users.

The simplified SSB receiver shown in **Fig 12.1** illustrates the use of several common filters. Three of them are located between the antenna and the speaker. They provide the essential receiver filter functions. A preselector filter is placed between the antenna and the first mixer. It passes all frequencies between 3.8 and 4.0 MHz with low loss. Other frequencies, such as out-of-band signals, are rejected to prevent them from overloading the first mixer (a common problem with shortwave broadcast stations). The preselector filter is almost always built with LC filter technology.

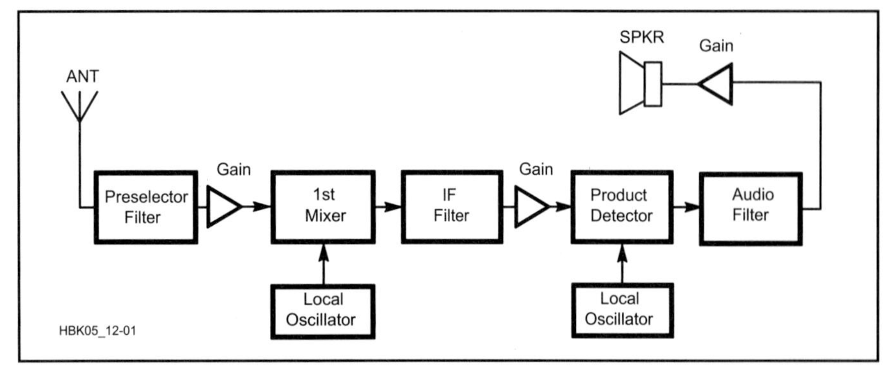

Fig 12.1 — One-band SSB receiver. At least three filters are used between the antenna and speaker.

An intermediate frequency (IF) filter is placed between the first and second mixers. It is a band-pass filter that passes the desired SSB signal but rejects all others. The age of the receiver probably determines which of several filter technologies is used. As an example, 50-kHz or 455-kHz LC filters and 455-kHz mechanical filters were used through the 1960s. Later model receivers usually use quartz crystal filters with center frequencies between 3 and 9 MHz. In all cases, the filter bandwidth must be less than 3 kHz to effectively reject adjacent SSB stations.

Finally, a 300-Hz to 3-kHz audio band-pass filter is placed somewhere between the detector and the speaker. It rejects unwanted products of detection, power supply hum and noise. Today this audio filter is usually implemented with active filter technology.

The complementary SSB transmitter block diagram is shown in **Fig 12.2**. The same array of filters appear in reverse order.

First is a 300-Hz to 3-kHz audio filter, which rejects out-of-band audio signals such as 60-Hz power supply hum. It is placed between the microphone and the balanced mixer.

The IF filter is next. Since the balanced mixer generates both lower and upper sidebands, it is placed at the mixer output

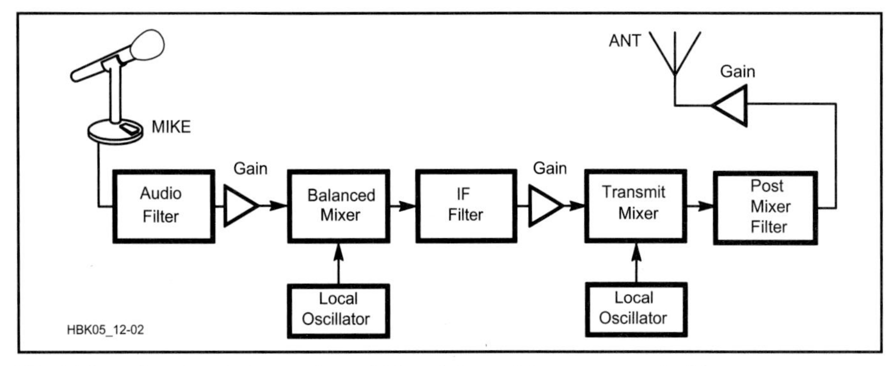

Fig 12.2 — One-band transmitter. At least three filters are needed to ensure a clean transmitted signal.

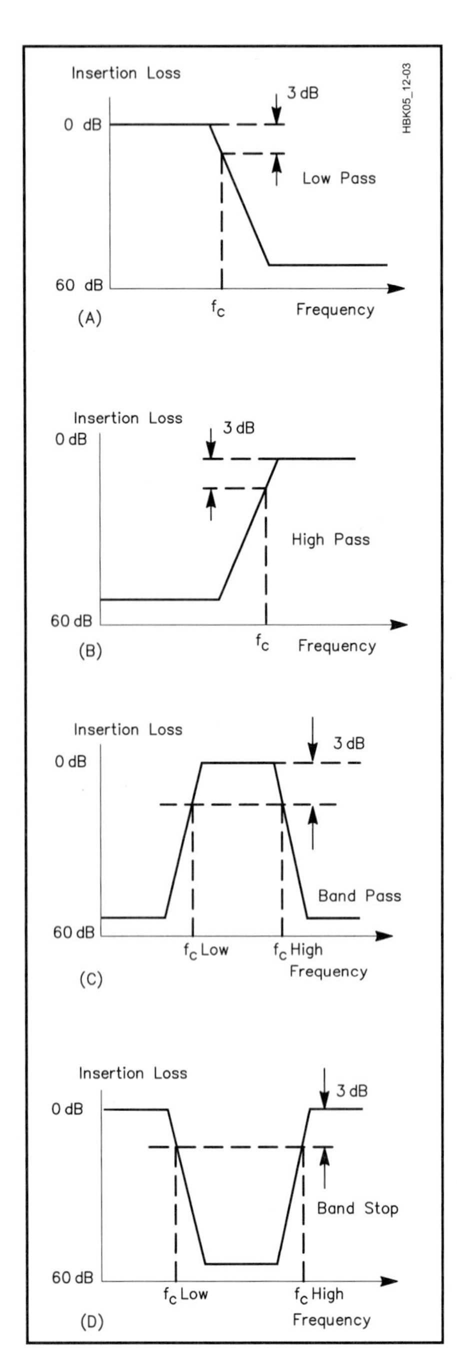

Fig 12.3 — Idealized filter responses. Note the definition of f_c is 3 dB down from the break points of the curves.

to pass only the desired lower (or upper) sideband. In commercial SSB transceivers this filter is usually the same as the IF filter used in the receive mode.

Finally, a 3.8 to 4.0-MHz band-pass filter is placed between transmit mixer and antenna to reject unwanted frequencies generated by the mixer and prevent them from being amplified and transmitted.

This chapter will discuss the four most common types of filters: low-pass, high-pass, band-pass and band-stop. The idealized characteristics of these filters are shown in their most basic form in **Fig 12.3**.

A low-pass filter permits all frequencies below a specified cutoff frequency to be transmitted with small loss, but will attenuate all frequencies above the cutoff frequency. The "cutoff frequency" is usually specified to be that frequency where the filter loss is 3 dB.

A high-pass filter has a cutoff frequency above which there is small transmission loss, but below which there is considerable attenuation. Its behavior is opposite to that of the low-pass filter.

A band-pass filter passes a selected band of frequencies with low loss, but attenuates frequencies higher and lower than the desired passband. The passband of a filter is the frequency spectrum that is conveyed with small loss. The transfer characteristic is not necessarily perfectly uniform in the passband, but the variations usually are small.

A band-stop filter rejects a selected band of frequencies, but transmits with low loss frequencies higher and lower than the desired stop band. Its behavior is opposite to that of the band-pass filter. The stop band is the frequency spectrum in which attenuation is desired. The attenuation varies in the stop band rising to high values at frequencies far removed from the cutoff frequency.

FILTER FREQUENCY RESPONSE

The purpose of a filter is to pass a desired frequency (or frequency band) and reject all other undesired frequencies. A simple single-stage low-pass filter is shown in **Fig 12.4**. The filter consists of an inductor, L. It is placed between the voltage source e_g and load resistance R_L. Most generators have an associated "internal" resistance, which is labeled R_g.

When the generator is switched on, power will flow from the generator to the load resistance R_L. The purpose of this low-pass filter is to allow maximum power flow at low frequencies (below the cutoff frequency) and minimum power flow at high frequencies. Intuitively, frequency filtering is accomplished because the inductor has reactance that vanishes at dc but becomes large at high frequency. Thus, the current, I, flowing through the load resistance, R_L, will be maximum at dc and less at higher frequencies.

The mathematical analysis of Fig 12.4 is as follows: For simplicity, let $R_g = R_L = R$.

$$i = \frac{e_g}{2R + jX_L} \qquad (1)$$

where

$X_L = 2\pi f L$

f = generator frequency.

Power in the load, P_L, is:

$$P_L = \frac{{e_g}^2 R_L}{4R^2 + {X_L}^2} \qquad (2)$$

Available (maximum) power will be delivered from the generator when:

$X_L = 0$ and $R_g = R_L$

$$P_O = \frac{{E_g}^2}{4R_g} \qquad (3)$$

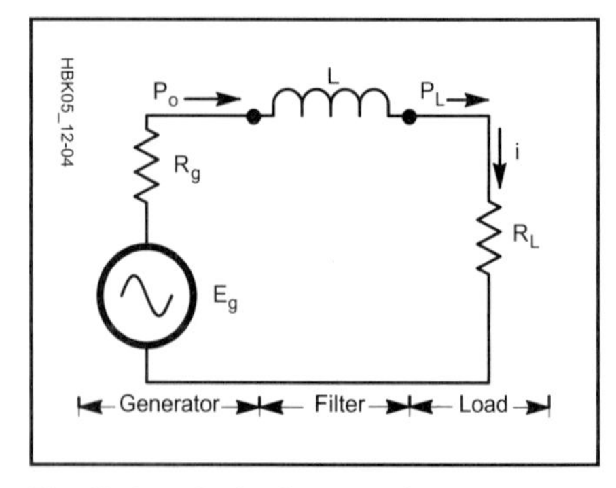

Fig 12.4 — A single-stage low-pass filter consists of a series inductor. DC is passed to the load resistor unattenuated. Attenuation increases (and current in the load decreases) as the frequency increases.

The filter response is:

$$\frac{P_L}{P_O} = \frac{\text{power in the load}}{\text{available generator power}} \qquad (4)$$

The filter cutoff frequency, called f_c, is the generator frequency where

$$2R = X_L \text{ or } f_c = \frac{R}{\pi L} \qquad (5)$$

As an example, suppose $R_g = R_L = 50\ \Omega$ and the desired cutoff frequency is 4 MHz. Equation 4 states that the cutoff frequency is where the inductive reactance $X_L = 100\ \Omega$. At 4 MHz, using the relationship $X_L = 2\pi f L$, L = 4 µH. If this filter is constructed, its response should follow the curve in **Fig 12.5**. Note that the gentle rolloff in response indicates a poor filter. To obtain steeper rolloff a more sophisticated filter, containing more reactances, is necessary. Filters are designed for specific value of purely resistive load impedance called the *terminating resistance*. When such a resistance is connected to the output terminals of a filter, the impedance looking into the input terminals will equal the load resistance throughout most of the passband. The degree of mismatch across the passband is shown by the SWR scale at the left-hand side of Fig 12.5. If maximum power is to be extracted from the generator driving the filter, the generator resistance must equal the load resistance. This condition is called a "doubly terminated" filter. Most passive filters, including the LC filters described in this chapter, are designed for double termination. If a filter is not properly terminated, its passband response changes.

Certain classes of filters, called "transformer filters" or "matching networks" are specifically designed to work between unequal generator and load resistances. Band-pass filters, described later, are easily designed to work between unequal terminations.

All passive filters exhibit an undesired nonzero loss in the passband due to unavoidable resistances associated with the reactances in the ladder network. All filters exhibit undesired transmission in the stop band due to leakage around the filter network. This phenomenon is called the "ultimate rejection" of the filter. A typical high-quality filter may exhibit an ultimate rejection of 60 dB.

Band-pass filters perform most of the important filtering in a radio receiver and transmitter. There are several measures of their effectiveness or *selectivity*. Selectivity is a qualitative term that arose in the 1930s. It expresses the ability of a filter (or the entire receiver) to reject unwanted adjacent signals. There is no mathematical measure of selectivity.

The term *Q* is quantitative. A band-pass filter's *quality factor* or Q is expressed as Q = (filter center frequency)/(3-dB bandwidth). *Shape factor* is another way some filter vendors specify band-pass filters. The shape factor is a ratio of two filter bandwidths. Generally, it is the ratio (60-dB bandwidth) / (6-dB bandwidth), but some manufacturers use other bandwidths. An ideal or *brick-wall* filter would have a shape factor of 1, but this would require an infinite number of filter elements. The IF filter in a high-quality receiver may have a shape factor of 2.

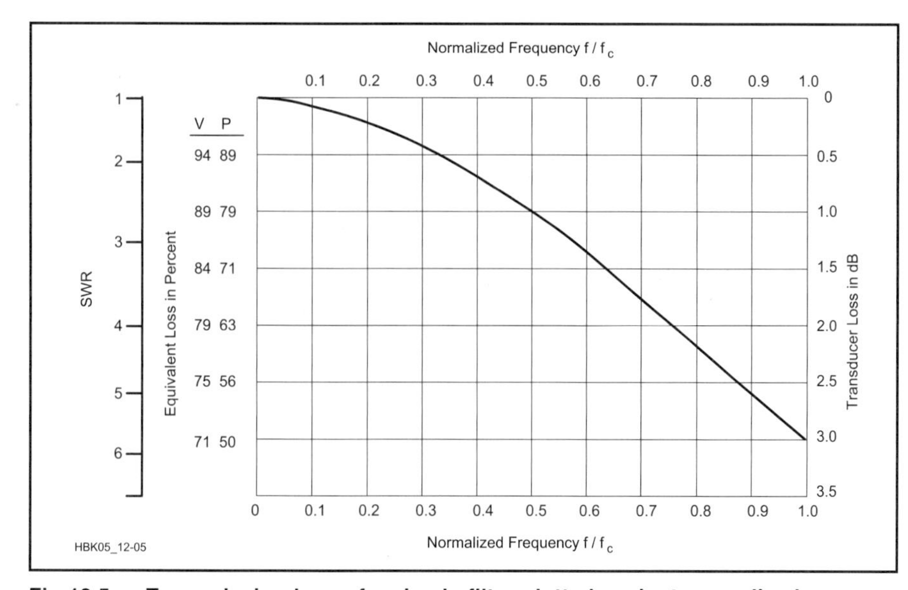

Fig 12.5 — Transmission loss of a simple filter plotted against normalized frequency. Note the relationship between loss and SWR.

POLES AND ZEROS

In equation 1 there is a frequency called the "pole" frequency that is given by $f_p = 0$.

In equation 1 there also exists a frequency where the current i becomes zero. This frequency is called the *zero frequency* and is given by: f_0 = infinity. Poles and zeros are intrinsic properties of all networks. The poles and zeros of a network are related to the values of inductances and capacitances in the network.

Poles and zero locations are of interest to the filter theorist because they allow him to predict the frequency response of a proposed filter. For low-pass and high-pass filters the number of poles equals the number of reactances in the filter network. For band-pass and band-stop filters the number of poles specified by the filter vendors is usually taken to be half the number of reactances.

LC FILTERS

Perhaps the most common filter found in the Amateur Radio station is the inductor-capacitor (LC) filter. Historically, the LC filter was the first to be used and the first to be analyzed. Many filter synthesis techniques use the LC filter as the mathematical model.

LC filters are usable from dc to approximately 1 GHz. Parasitic capacitance associated with the inductors and parasitic inductance associated with the capacitors make applications at higher frequencies impractical because the filter performance will change with the physical construction and therefore is not totally predictable from the design equations. Below 50 or 60 Hz, inductance and capacitance values of LC filters become impractically large.

Mathematically, an LC filter is a linear, lumped-element, passive, reciprocal network. Linear means that the ratio of output to input is the same for a 1-V input as for a 10-V input. Thus, the filter can accept an input of many simultaneous sine waves without intermodulation (mixing) between them.

Lumped-element means that the inductors and capacitors are physically much smaller than an operating wavelength. In this case, conductor lengths do not contribute significant inductance or capacitance, and the time that it takes for signals to pass through the filter is insignificant. (Although the different times that it takes for different frequencies to pass through the filter — known as group delay — is still significant for some applications.)

The term *passive* means that the filter

does not need any internal power sources. There may be amplifiers before and/or after the filter, but no power is necessary for the filter's equations to hold. The filter alone always exhibits a finite (nonzero) insertion loss due to the unavoidable resistances associated with inductors and (to a lesser extent) capacitors. Active filters, as the name implies, contain internal power sources.

Reciprocal means that the filter can pass power in either direction. Either end of the filter can be used for input or output.

TIME DOMAIN VS FREQUENCY DOMAIN

Humans think in the time domain. Life experiences are measured and recorded in the stream of time. In contrast, Amateur Radio systems and their associated filters are often better understood when viewed in the frequency domain, where frequency is the relevant system parameter. *Frequency* may refer to a sine-wave voltage, current or electromagnetic field. The sine-wave voltage, shown in **Fig 12.6**, is a waveform plotted against time with equation $V = A \sin(2\pi f t)$. The sine wave has a peak amplitude A (measured in volts) and frequency, f (measured in cycles/second or Hertz). A graph showing frequency on the horizontal axis is called a spectrum. A filter response curve is plotted on a spectrum graph.

Historically, radio systems were best analyzed in the frequency domain. The radio transmitters of Hertz (1865) *and* Marconi (1895) consisted of LC resonant circuits excited by high-voltage spark gaps. The transmitters emitted packets of damped sine waves. The low-frequency (200-kHz) antennas used by Marconi were found to possess very narrow bandwidths, and it seemed natural to analyze antenna performance using sine-wave excitation. In addition, the growing use of 50 and 60-Hz alternating current (ac) electric power systems in the 1890s demanded the use of sine-wave mathematics to analyze these systems. Thus engineers trained in ac power theory were available to design and build the early radio systems.

In the frequency domain, the radio world is imagined to be composed of many sine waves of different frequencies flowing endlessly in time. It can be shown by the Fourier transform (Ref 7) that all periodic waveforms can be represented by summing sine waves of different frequencies. For example, the square-wave voltage shown in **Fig 12.7** can be represented by a "fundamental" sine wave of frequency f = 1/t and all its odd harmonics: 3f, 5f, 7f and so on. Thus, in the frequency domain a sine wave is a *narrowband* signal (zero bandwidth) and a square wave is a "wideband" signal.

If the square-wave voltage of Fig 12.7 is passed through a low-pass filter, which removes some of its high-frequency components, the waveform of **Fig 12.8** results. The filtered square wave now has a rise time, which is the time required to rise from 10% to 90% of its peak value (A). The rise time is approximately:

$$\tau_R = \frac{0.35}{f_c} \tag{6}$$

where f_c is the cutoff frequency of the low-pass filter.

Thus a filter distorts a time-domain signal by removing some of its high-frequency components. Note that a filter cannot distort a sine wave. A filter can only change the amplitude and phase of sine waves. A linear filter will pass multiple sine waves without producing any intermodulation or "beats" between frequencies — this is the definition of *linear.*

The purpose of a radio system is to convey a time-domain signal originating at a source to some distant point with minimum distortion. Filters within the radio system transmitter and receiver may intentionally or unintentionally distort the source signal. A knowledge of the source signal's frequency-domain bandwidth is required so that an appropriate radio system may be designed.

Table 12.1 shows the minimum neces-

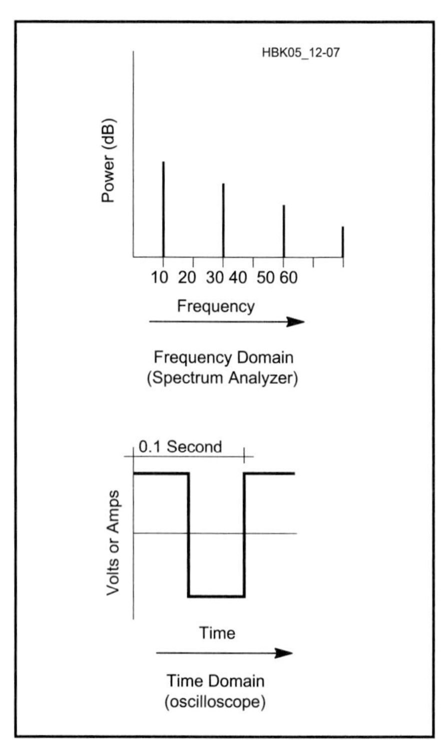

Fig 12.7 — Square-wave voltage. Many frequencies are present, including f = 1/t and odd harmonics 3f, 5f, 7f with decreasing amplitudes.

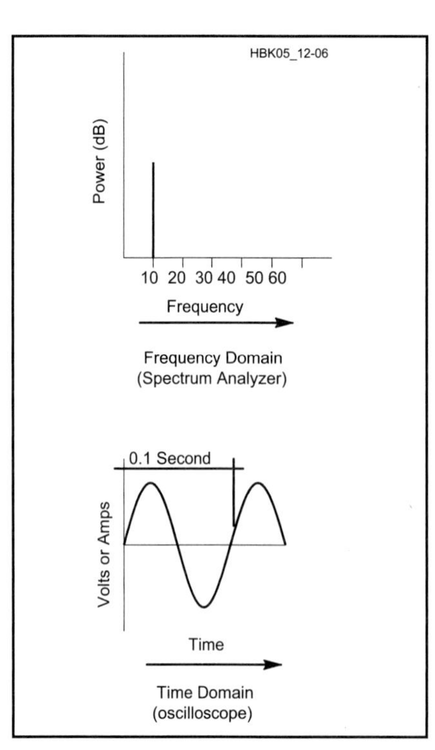

Fig 12.6 — Ideal sine-wave voltage. Only one frequency is present.

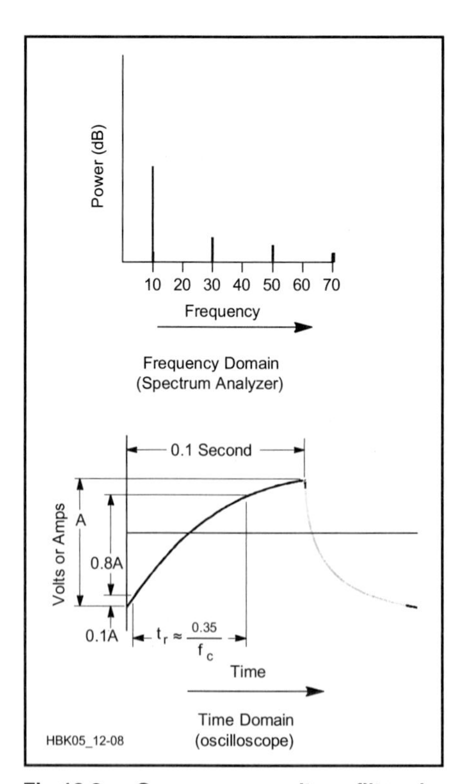

Fig 12.8 — Square-wave voltage filtered by a low-pass filter. By passing the square wave through a filter, the higher frequencies are attenuated. The rectangular shape (fast rise and fall items) are *rounded* because the amplitude of the higher harmonics is decreased.

Table 12.1

Typical Filter Bandwidths for Typical Signals.

Source	*Required Bandwidth*
High-fidelity speech and music	20 Hz to 15 kHz
Telephone-quality speech	200 Hz to 3 kHz
Radiotelegraphy (Morse code, CW)	200 Hz
HF RTTY	1000 Hz (varies with frequency shift)
NTSC television	60 Hz to 4.5 MHz
SSTV	200 Hz to 3 kHz
1200 bit/s packet	200 Hz to 3 kHz

sary bandwidth of several common source signals. Note that high-fidelity speech and music requires a bandwidth of 20 Hz to 15 kHz, which is that transmitted by high-quality FM broadcast stations. However, telephone-quality speech requires a bandwidth of only 200 Hz to 3 kHz. Thus, to minimize transmit spectrum, as required by the FCC, filters within amateur transmitters are required to reduce the speech source bandwidth to 200 Hz to 3 kHz at the expense of some speech distortion. After modulation the transmitted RF bandwidth will exceed the filtered source bandwidth if inefficient (AM or FM) modulation methods are employed. Thus the post-modulation *emission bandwidth* may be several times the original filtered source bandwidth. At the receiving end of the radio link, band-pass filters are required to accept only the desired signal and sharply reject noise and adjacent channel interference.

As human beings we are accustomed to operation in the time domain. Just about all of our analog radio connected design occurs in the frequency domain. This is particularly true when it comes to filters. Although the two domains are convertible, one to the other, most filter design is performed in the frequency domain.

Filter Synthesis

The image-parameter method of filter design was initiated by O. Zobel (Ref 1) of Bell Labs in 1923. Image-parameter filters are easy to design and design techniques are found in earlier editions of the ARRL *Handbook*. Unfortunately, image parameter theory demands that the filter terminating impedances vary with frequency in an unusual manner. The later addition of "m-derived matching half sections" at each end of the filter made it possible to use these filters in many applications. In the intervening decades, however, many new methods of filter design have brought both better performance and practical component values for construction.

MODERN FILTER THEORY

The start of modern filter theory is usually credited to S. Butterworth and S. Darlington (Refs 3 and 4). It is based on this approach: Given a desired frequency response, find a circuit that will yield this response.

Filter theorists were aware that certain known mathematical polynomials had "filter like" properties when plotted on a frequency graph. The challenge was to match the filter components (L, C and R) to the known polynomial poles and zeros. This pole/zero matching was a difficult task before the availability of the digital computer. Weinberg (Ref 5) was the first to publish computer-generated tables of normalized low-pass filter component values. ("Normalized" means 1-Ω resistor terminations and cutoff frequency $\omega_c = 2\pi f_c = 1$ radian/s.)

An ideal low-pass filter response shows no loss from zero frequency to the cutoff frequency, but infinite loss above the cutoff frequency. Practical filters may approximate this ideal response in several different ways.

Fig 12.9 shows the Butterworth or "maximally-flat" type of approximation. The Butterworth response formula is:

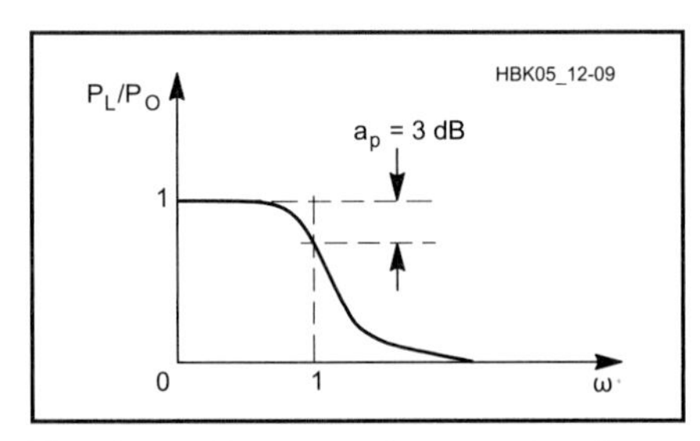

Fig 12.9 — Butterworth approximation of an ideal low-pass filter response. The 3-dB attenuation frequency (f_c) is normalized to 1 radian/s.

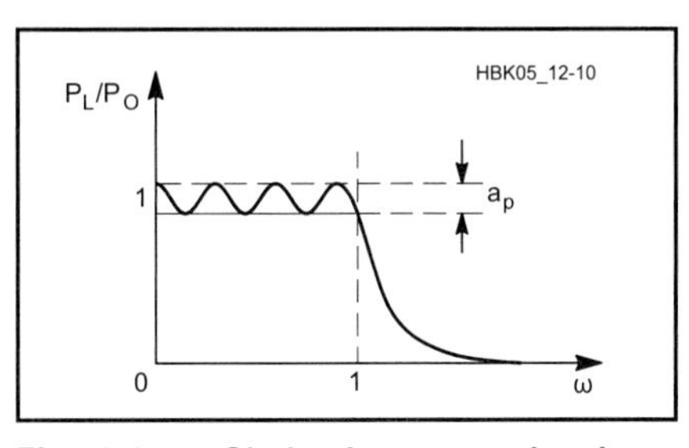

Fig 12.10 — Chebyshev approximation of an ideal low-pass filter. Notice the ripple in the passband.

$$\frac{P_L}{P_O} = \frac{1}{1 + \left(\frac{\omega}{\omega_c}\right)^{2n}} \qquad (7)$$

where

ω = frequency of interest
ω_c = cutoff frequency
n = number of poles (reactances)
P_L = power in the load resistor
P_O = available generator power

The passband is exceedingly flat near zero frequency *and* very high attenuation is experienced at high frequencies, but the approximation for both pass and stop bands is relatively poor in the vicinity of cutoff.

Fig 12.10 shows the Chebyshev approximation. Details of the Chebyshev response formula can be found in (Ref 24). Use of this reference as well as similar references for Chebyshev filters requires detailed familiarity with Chebyshev polynomials.

IMPEDANCE AND FREQUENCY SCALING

Fig 12.11A shows normalized component values for Butterworth filters up to ten poles. Fig 12.11B shows the schematic diagrams of the Butterworth low-pass filter. Note that the first reactance in Fig 12.11B is a shunt capacitor C1, whereas in Fig 12.11C the first reactance is a series inductor L1. Either configuration can be used, but a design using fewer inductors is usually chosen.

In filter design, the use of *normalized*

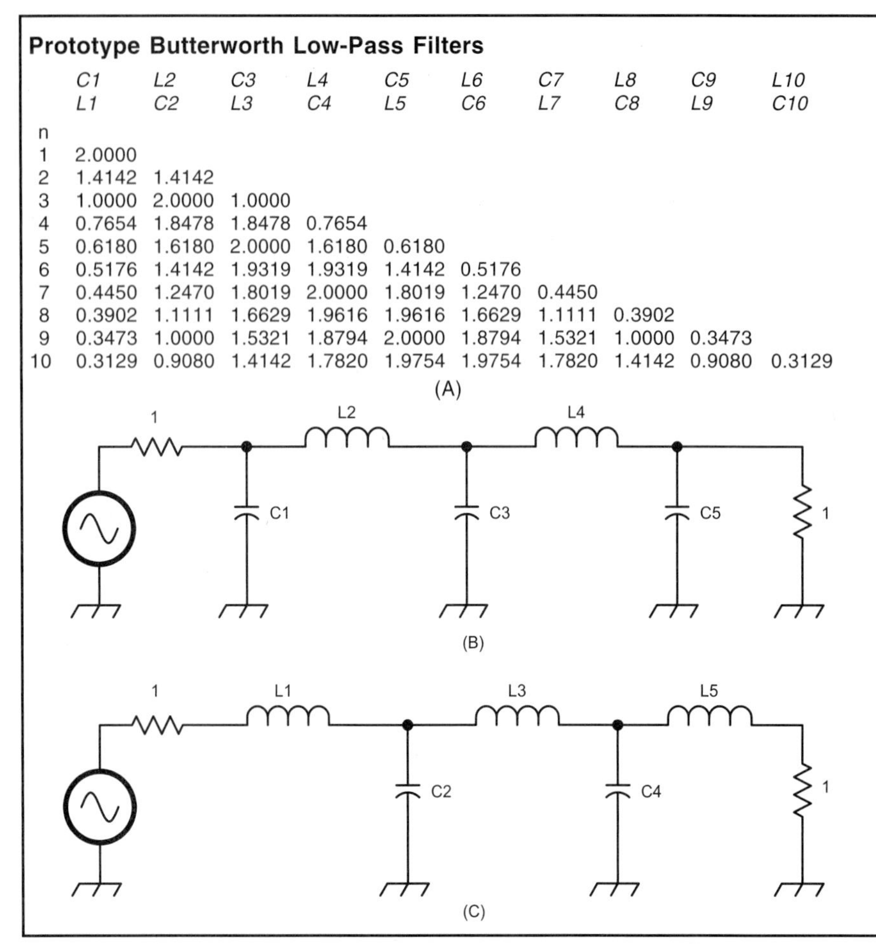

Prototype Butterworth Low-Pass Filters

n	C1 L1	L2 C2	C3 L3	L4 C4	C5 L5	L6 C6	C7 L7	L8 C8	C9 L9	L10 C10
1	2.0000									
2	1.4142	1.4142								
3	1.0000	2.0000	1.0000							
4	0.7654	1.8478	1.8478	0.7654						
5	0.6180	1.6180	2.0000	1.6180	0.6180					
6	0.5176	1.4142	1.9319	1.9319	1.4142	0.5176				
7	0.4450	1.2470	1.8019	2.0000	1.8019	1.2470	0.4450			
8	0.3902	1.1111	1.6629	1.9616	1.9616	1.6629	1.1111	0.3902		
9	0.3473	1.0000	1.5321	1.8794	2.0000	1.8794	1.5321	1.0000	0.3473	
10	0.3129	0.9080	1.4142	1.7820	1.9754	1.9754	1.7820	1.4142	0.9080	0.3129

Fig 12.11 — Component values for Butterworth low-pass filters. Greater values of n require more stages.

values is common. Normalized generally means a design based on 1-Ω terminations and a cutoff frequency (passband edge) of 1 radian/second. A filter is *denormalized* by applying the following two equations:

$$L' = \left(\frac{R'}{R}\right)\left(\frac{\omega}{\omega'}\right)L \quad (8)$$

$$C' = \left(\frac{R}{R'}\right)\left(\frac{\omega}{\omega'}\right)C \quad (9)$$

where

L', C', ω' and R' are the new (desired) values

L and C are the values found in the filter tables

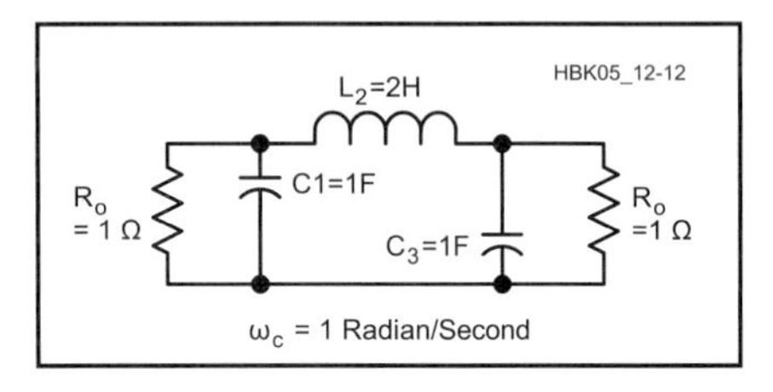

Fig 12.12 — A 3-pole Butterworth filter designed for a normalized frequency of 1 radian/s.

R = 1 Ω

ω = 1 radian/s.

For example, consider the design of a 3-pole Butterworth low-pass filter for a transmitter speech amplifier. Let the desired cutoff frequency be 3000 Hz and the desired termination resistances be 1000 Ω. The normalized prototype, taken from Fig 12.11B is shown in **Fig 12.12**. The new (desired) inductor value is:

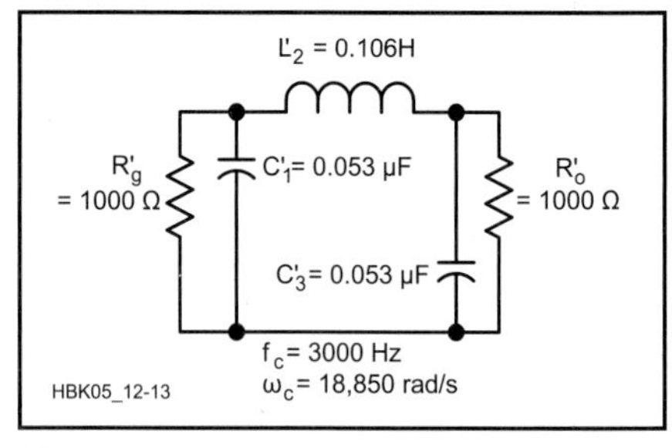

Fig 12.13 — A 3-pole Butterworth filter scaled to 3000 Hz.

$$L' = \left(\frac{1000\,\Omega}{1\Omega}\right)\left(\frac{1\text{ radian/second}}{2\pi(3000)\text{Hz}}\right)2\text{ H}$$

or L' = 0.106 H.

The new (desired) capacitor value is:

$$C' = \left(\frac{1\Omega}{1000\,\Omega}\right)\left(\frac{1\text{ radian/second}}{2\pi(3000)\text{Hz}}\right)1\text{ F}$$

or C' = 0.053 µF.

The final denormalized filter is shown in **Fig 12.13**. The filter response, in the passband, should obey curve n = 3 in **Fig 12.14**. To use the normalized frequency response curves in Fig 12.14 calculate the frequency ratio f/f_c where f is the desired frequency and f_c is the cutoff frequency. For the filter just designed, the loss at 2000 Hz can be found as follows: When f is 2000 Hz, the frequency ratio is: f/f_c = 2000/3000 = 0.67. Therefore the predicted loss (from the n = 3

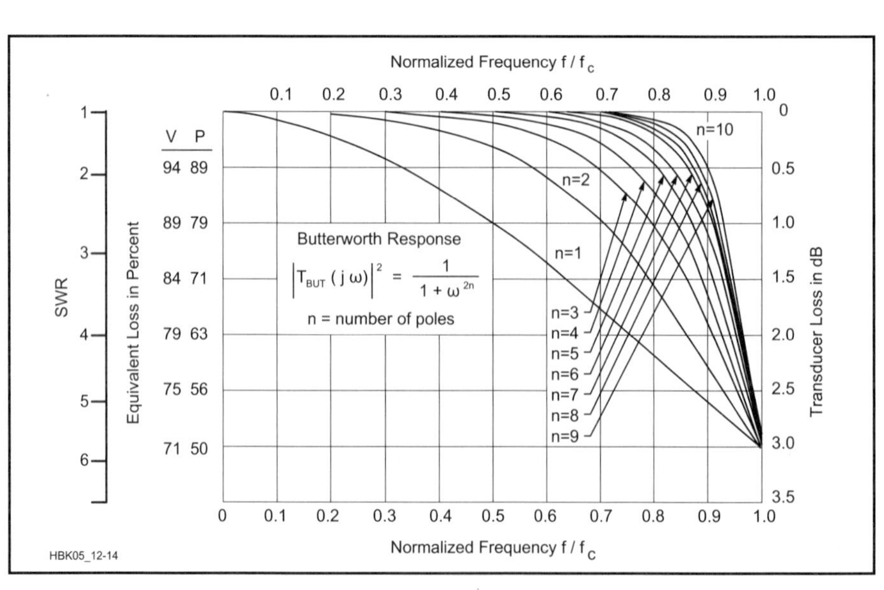

Fig 12.14 — Passband loss of Butterworth low-pass filters. The horizontal axis is normalized frequency (see text).

curve) is about 0.37 dB.

When f is 4000 Hz, the filter is operating in the stop-band (**Fig 12.17**). The resulting frequency ratio is: $f/f_c = 4000/3000 = 1.3$. Therefore the expected loss is about 8 dB. Note that as the number of reactances (poles) increases the filter response approaches the low-pass response of Fig 12.3A.

BAND-PASS FILTERS—SIMPLIFIED DESIGN

The design of band-pass filters may be directly obtained from the low-pass prototype by a frequency translation. The low-pass filter has a "center frequency" (in the parlance of band-pass filters) of 0 Hz. *The frequency translation from 0 Hz to the band-pass filter center frequency, f, is obtained by replacing in the low-pass prototype all shunt capacitors with parallel tuned circuits and all series inductors with series tuned circuits.*

As an example, suppose a band-pass filter is required at the front end of a home-brew 40-m QRP receiver to suppress powerful adjacent broadcast stations. The proposed filter has these characteristics:

- Center frequency, f_c = 7.15 MHz
- 3-dB bandwidth = 360 kHz
- terminating resistors = 50 Ω
- 3-pole Butterworth characteristic.

Start the design for the normalized 3-pole Butterworth low-pass filter (shown in Fig 12.11). First determine the center frequency from the band-pass limits. This frequency, f_O, is found by determining the geometric mean of the band limits. In this case the band limits are 7.15 + 0.360/2 = 7.33 MHz and 7.15 –0.360/2 = 6.97 MHz; then

$$f_O = \sqrt{f_{lo} \times f_{hi}} = \sqrt{6.97 \times 7.33} = 7.14 \text{ MHz} \quad (10)$$

where

f_{lo} = low frequency end of the band-pass (or band-stop)

f_{hi} = high frequency end of the band-pass (or band-stop)

[Note that in this case there is little difference between 7.15 (bandwidth center) and 7.147 (band-edge geometric mean) because the bandwidth is small. For wide-band filters, however, there can be a significant difference.]

Next, denormalize to a new interim low-pass filter having R' = 50 Ω and f' = 0.36 MHz.

$$L' = \left(\frac{50}{1}\right)\left(\frac{1}{2 \times \pi \times 0.36 \times 10^6}\right) 2\,\text{H} = 44.2\,\mu\text{H}$$

$$C' = \left(\frac{1}{50}\right)\left(\frac{1}{2 \times \pi \times 0.36 \times 10^6}\right) 1\,\text{F} = 8842\,\text{pF}$$

This interim low-pass filter, shown in **Fig 12.15**, has a cutoff frequency f_c = 0.36 MHz and is terminated with 50-Ω resistors. The desired 7.147-MHz band-pass filter is achieved by parallel resonating the shunt capacitors with inductors and series resonating the series inductor with a series capacitor. All resonators must be tuned to the center frequency. Therefore, variable capacitors or inductors are required for the resonant circuits. Based on the L' and C' just calculated the parallel-resonating inductor values are:

$$L1 = L3 = \frac{1}{C'\left(2 \times \pi \times f_O\right)^2} = 0.056\,\mu\text{H}$$

The series-resonating capacitor value is:

$$C2 = \frac{1}{L'\left(2 \times \pi \times f_O\right)^2} = 11.2\,\text{pF}$$

The final band-pass filter is shown in **Fig 12.16**. The filter should have a 3-dB

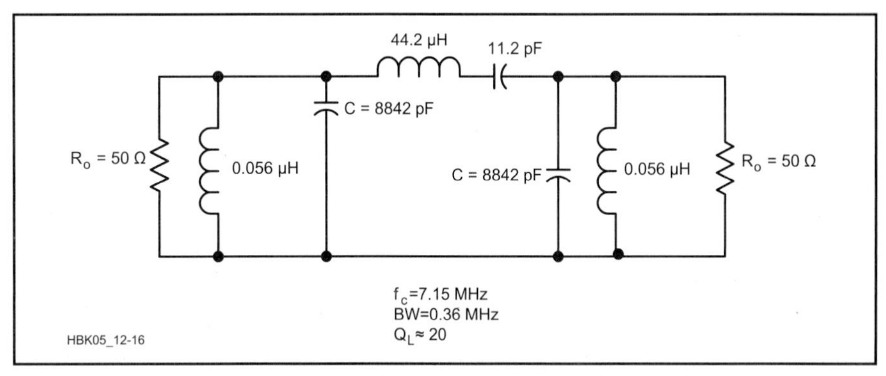

Fig 12.16 — Final filter design consists of the low-pass filter scaled to a center frequency of 7.15 MHz.

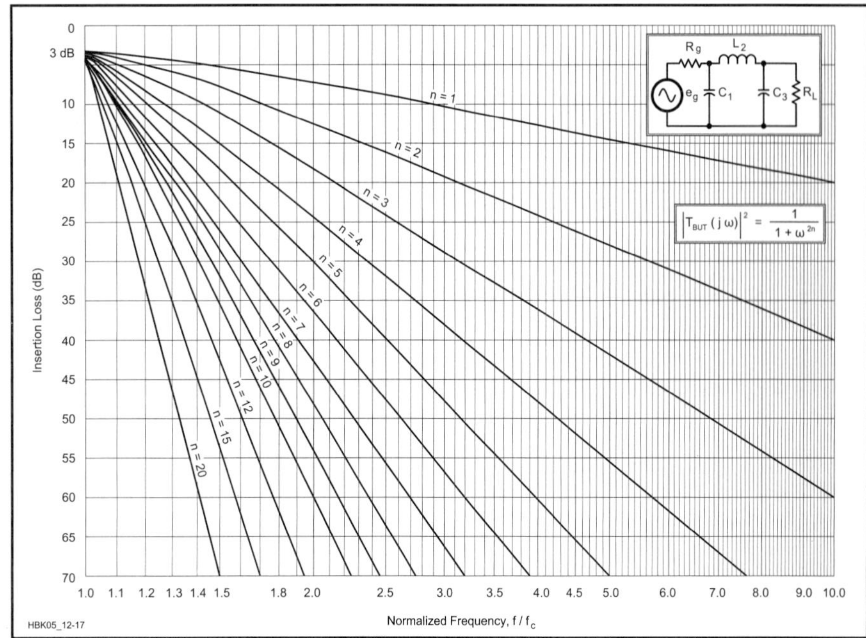

Fig 12.17 — Stop-band loss of Butterworth low-pass filters. The almost vertical angle of the lines representing filters with high values of n (10, 12, 15, 20) show the slope of the filter will be very high (sharp cutoff).

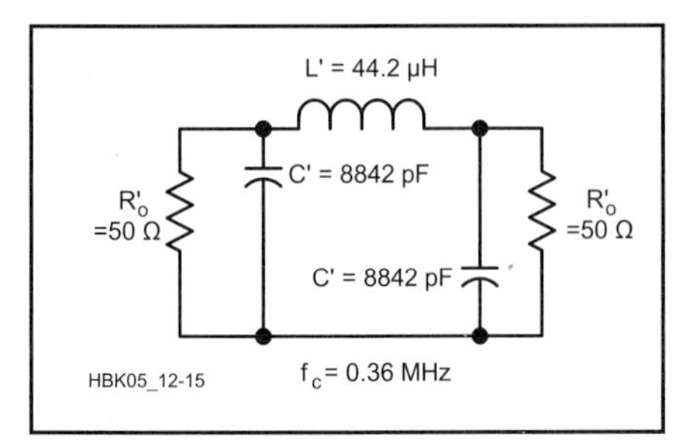

Fig 12.15 — Interim 3-pole Butterworth low-pass filter designed for cutoff at 0.36 MHz.

bandwidth of 0.36 MHz. That is, the 3-dB loss frequencies are 6.97 MHz and 7.33 MHz. The filter's loaded Q is: Q = 7.147/0.36 or approximately 20.

The filter response, in the passband, falls on the "n = 3" curve in **Fig 12.17**. To use the normalized frequency response curves, calculate the frequency ratio f/f_c. For this band-pass case, f is the difference between the desired attenuation frequency and the center frequency, while f_c is the upper 3-dB frequency minus the center frequency. As an example the filter loss at 7.5 MHz is found by using the normalized frequency ratio given by:

$$\frac{f}{f_c} = \frac{7.5 - 7.147}{7.33 - 7.147} = 1.928$$

Therefore, from Fig 12.17 the expected loss is about 17 dB.

At 6 MHz the loss may be found by:

$$\frac{f}{f_c} = \frac{7.147 - 6}{7.33 - 7.147} = 6.26$$

The expected loss is approximately 47 dB. Unfortunately, awkward component values occur in this type of band-pass filter. The series resonant circuit has a very large LC ratio and the parallel resonant circuits have very small LC ratios. The situation worsens as the filter loaded Q_L ($Q_L = f_0/BW$) increases. Thus, this type of band-pass filter is generally used with a loaded Q less than 10.

Good examples of low-Q band-pass filters of this type are demonstrated by W3NQN's High Performance CW Filter and Passive Audio Filter for SSB in the 1995 and earlier editions of this *Handbook* (Ref 25).

[Note: This analysis used the geometric f_c with the assumption that the filter response is symmetrical about f_c, which it is not. A more rigorous analysis yields 16.9 dB at 7.5 MHz and 50.7 dB at 6 MHz. — Ed.]

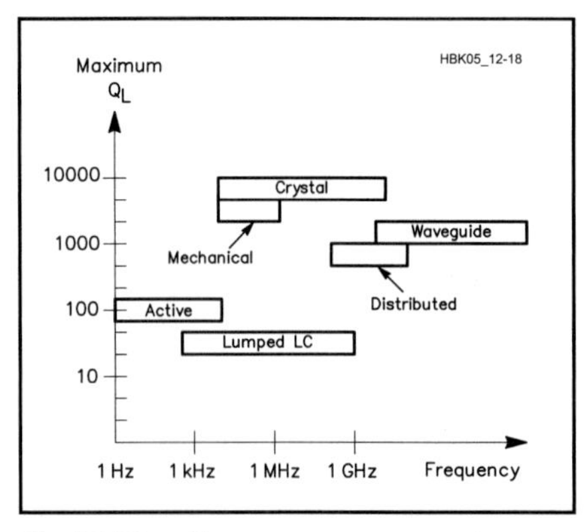

Fig 12.18 — Frequency range and maximum loaded Q of band-pass filters. Crystal filters are shown with the highest Q_L and LC filters the lowest.

Q Restrictions—Band-pass Filters

Most filter component value tables assume lossless reactances. In practice, there are always resistance losses associated with capacitors and inductors (especially inductors). Lossy reactances in low-pass filters modify the response curve. There is finite loss at zero frequency and the cutoff "knee" at f_c will not be as sharp as predicted by theoretical response curves.

The situation worsens with band-pass filters. As loaded Q is increased, the midband insertion loss may become intolerable. Therefore, before a band-pass filter design is started, estimate the expected loss.

An approximate estimate of band-pass filter midband response is given by:

$$\frac{P_L}{P_O} = \left(1 - \frac{Q_L}{Q_U}\right)^{2N} \quad (11A)$$

where:

P_L = power delivered to load resistor R_L
P_O = power available from generator:

$$P_O = \frac{{e_g}^2}{4R_L} \quad (11B)$$

Q_U = unloaded Q of inductor:

$$Q_U = \frac{2\pi \times f_0 \times L}{R} \quad (11C)$$

R = inductor series resistance
L = inductance
Q_L = filter loaded Q

$$Q_L = \frac{f_0}{BW_3} \quad (11D)$$

BW_3 = 3-dB bandwidth
N = number of filter stages.

This equation assumes that all losses are in the inductors. For example, the expected loss of the 7.15-MHz filter shown in Fig 12.16 is found by assuming Q_U = 150. Q_L is found by equation 11D to be = 7.147/0.36 = 19.8 or approximately 20. Since N = 3 then:

$$\frac{P_L}{P_O} = \left(1 - \frac{20}{150}\right)^6$$

from equation (11A), which equals 0.423. Expressed as dB this is equal to 10 log (0.423) = – 3.73 dB.

Therefore this filter may not be suitable for some applications. If the insertion loss is to be kept small there are severe restrictions on Q_L/Q_U. With typical lumped inductors Q_U seldom exceeds 200. Therefore, LC band-pass filters are usually designed with Q_L not exceeding 20 as shown in **Fig 12.18**.

This loss vs bandwidth trade-off is usually why the final intermediate frequency (IF) in older radio receivers was very low. These units used the equivalent of LC filters in their IF coupling. Generally, for SSB reception the desired receiver bandwidth is about 2.5 kHz. Then 50 kHz was often chosen as the final IF since this implies a loaded Q_L of 20. AM broadcast receivers require a 10-kHz bandwidth and use a 455-kHz IF, which results in Q_L = 45. FM broadcast receivers require a 200-kHz bandwidth and use a 10.7-MHz IF and Q_L = 22.

Filter Design Using Standard Capacitor Values/Software

Practical filters must be designed using commercially available components. Modern computer programs are available to aid in filter design. Originally, however, tables based upon *standard value capacitors* (SVC) were used to facilitate this design process. These SVC tables are now located on the CD-ROM included with this *Handbook*. It is instructive to understand how filters are designed using tables so that you will more easily understand how to use modern computer-based design techniques. To illustrate the process of filter design using filter design tables, the procedure presented here uses computer-calculated tables of performance parameters and component values for 5-element Chebyshev 50-Ω filters. The tables permit the quick and easy selection of an equally terminated passive LC filter for applications where the attenuation response is of primary interest. All of the capacitors in the Chebyshev designs have standard, off-the-shelf values to simplify construction. Although the tables cover only the 1 to 10-MHz frequency range, a simple scaling procedure gives standard-value capacitor (SVC) designs for any impedance level and virtually any cutoff frequency.

Extracts from filter design tables are

> ## Try ELSIE for "LC"!
>
> This *Handbook* includes an *ELSIE.EXE* file as companion software designed and provided courtesy of Jim Tonne, WB6BLD (see the *Handbook* CD-ROM contents page). The *ELSIE.EXE* software (freeware) is a student version of the larger commercial version (to the 21st Order and up to 42 Stages!) which allows the user to design a variety of filter configurations and response characteristics up to the 7th Order, 7th Stage level. *ELSIE* is also a *Windows* program. *ELSIE* software and some other interesting programs for hams can also be found at: **http://tonnesoftware.com/**

reprinted in this section to illustrate the design procedure.

The following text by Ed Wetherhold, W3NQN, is adapted from his paper entitled *Simplified Passive LC Filter Design for the EMC Engineer*. It was presented at an IEEE International Symposium on Electromagnetic Compatibility in 1985.

The approach is based upon the fact that for most nonstringent filtering applications, it is not necessary that the actual cutoff frequency exactly match the desired cutoff frequency. A deviation of 5% or so between the actual and desired cutoff frequencies is acceptable. This permits the use of design tables based on standard capacitor values instead of passband ripple attenuation or reflection coefficient.

STANDARD VALUES IN FILTER DESIGN CALCULATIONS

Capacitors are commercially available in special series of preferred values having designations of E12 (10% tolerance) and E24 (5% tolerance; Ref 22) The reciprocal of the E-number is the power to which 10 is raised to give the step multiplier for that particular series.

First the normalized Chebyshev and elliptic component values are calculated based on many ratios of standard capacitor values. Next, using a 50-Ω impedance level, the parameters of the designs are calculated and tabulated to span the 1-10 MHz decade. Because of the large number of standard-value capacitor (SVC) designs in this decade, the increment in cutoff frequency from one design to the next is sufficiently small so that virtually any cutoff frequency requirement can be satisfied. Using such a table, the selection of an appropriate design consists of merely scanning the cutoff frequency column to find a design having a cutoff frequency that most closely matches the desired cutoff frequency.

CHEBYSHEV FILTERS[1]

Low-pass and high-pass 5-element Chebyshev designs were selected for tabulation because they are easy to construct and will satisfy the majority of nonstringent filtering requirements where the amplitude response is of primary interest. The precalculated 50-Ω designs are presented in extracts from tables of low-pass and high-pass designs with cutoff frequencies covering the 1-10 MHz decade. In addition to the component values, attenuation vs frequency data and SWR are also included in the table. The passband attenuation ripples are so low in amplitude that they are swamped by the filter losses and are not measurable.

LOW-PASS TABLES

Fig 12.19 is an extract from filter design tables for the low-pass 5-element Chebyshev capacitor input/output configuration. This filter configuration is generally preferred to the alternate inductor input/output configuration because it requires fewer inductors. Generally, decreasing input impedance with increasing frequency in the stop band presents no problems. **Fig 12.20** shows the corresponding information for low-pass appli-

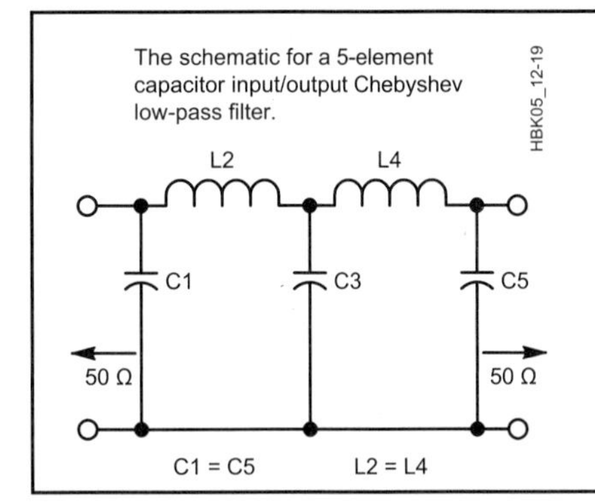

The schematic for a 5-element capacitor input/output Chebyshev low-pass filter.

Fig 12.19 — A portion of a 5-element Chebyshev low-pass filter design table for 50-Ω impedance, C-in/out and standard E24 capacitor values.

	—FREQUENCY (MHz)—				MAX	C1,5	L2,4	C3
No.	F_{CO}	3 dB	20 dB	40 dB	SWR	(pF)	(µH)	(pF)
1	1.01	1.15	1.53	2.25	1.355	3600	10.8	6200
2	1.02	1.21	1.65	2.45	1.212	3000	10.7	5600
3	1.15	1.29	1.71	2.51	1.391	3300	9.49	5600
4	1.10	1.32	1.81	2.69	1.196	2700	9.88	5100
5	1.25	1.41	1.88	2.75	1.386	3000	8.67	5100
6	1.04	1.37	1.94	2.94	1.085	2200	9.82	4700
7	1.15	1.41	1.95	2.92	1.155	2400	9.37	4700

Schematic for a 5-element inductor-input/output Chebyshev low-pass filter. See the corresponding table in References chapter for the attenuation response curve.

Fig 12.20 — A portion of a 5-element Chebyshev low-pass filter design table for 50-Ω impedance, L-in/out and standard-value L and C.

	—FREQUENCY (MHz)—				MAX	L1,5	C2,4	L3
No.	F_{CO}	3 dB	20 dB	40 dB	SWR	(µH)	(pf)	(µH)
1	0.744	1.15	1.69	2.60	1.027	5.60	4700	13.7
2	0.901	1.26	1.81	2.76	1.055	5.60	4300	12.7
3	1.06	1.38	1.94	2.93	1.096	5.60	3900	11.8
4	1.19	1.47	2.05	3.07	1.138	5.60	3600	11.2
5	1.32	1.58	2.17	3.23	1.192	5.60	3300	10.6
6	0.911	1.39	2.03	3.12	1.030	4.70	3900	11.4
7	1.08	1.50	2.16	3.29	1.056	4.70	3600	10.6
8	1.25	1.63	2.30	3.48	1.092	4.70	3300	9.92
9	1.42	1.77	2.46	3.68	1.142	4.70	3000	9.32
10	1.61	1.92	2.63	3.90	1.209	4.70	2700	8.79
11	1.05	1.64	2.41	3.72	1.025	3.90	3300	9.63

cations, but with an inductor input/output configuration. This configuration is useful when the filter input impedance in the stop band must rise with increasing frequency. For example, some RF transistor amplifiers may become unstable when terminated in a low-pass filter having a stop-band response with a decreasing input impedance. In this case, the inductor-input configuration may eliminate the instability. (Ref 23) Because only one capacitor value is required in the designs of Fig 12.20, it was feasible to have the inductor value of L1 and L5 also be a standard value.

HIGH-PASS TABLES

A high-pass 5-element Chebyshev capacitor input/output configuration is shown in the table extract of **Fig 12.21**. Because the inductor input/output configuration is seldom used, it was not included.

SCALING TO OTHER FREQUENCIES AND IMPEDANCES

The tables shown are for the 1-10 MHz decade and for a 50-Ω equally terminated impedance. The designs are easily scaled to other frequency decades and to other equally terminated impedance levels, however, making the tables a universal design aid for these specific filter types.

Frequency Scaling

To scale the frequency and the component values to the 10-100 or 100-1000 MHz decades, multiply all tabulated frequencies by 10 or 100, respectively. Then divide all C and L values by the same number. The A_s and SWR data remain unchanged. To scale the filter tables to the 0.1-1 kHz, 1-10 kHz or the 10-100 kHz decades, divide the tabulated frequencies by 1000, 100 or 10, respectively. Next multiply the component values by the same number. By changing the "MHz" frequency headings to "kHz" and the "pF" and "µH" headings to "nF" and "mH," the tables are easily changed from the 1-10 MHz decade to the 1-10 kHz decade and the table values read directly. Because the impedance level is still at 50 Ω, the component values may be awkward, but this can be corrected by increasing the impedance level by ten times using the impedance scaling procedure described below.

Impedance Scaling

All the tabulated designs are easily scaled to impedance levels other than 50 Ω, while keeping the convenience of standard-value capacitors and the "scan mode" of design selection. If the desired new impedance level differs from 50 Ω by a factor of 0.1, 10 or 100, the 50-Ω designs are scaled by shifting the decimal points of the component values. The other data remain unchanged. For example, if the impedance level is increased by ten or one hundred times (to 500 or 5000 Ω), the decimal point of the capacitor is shifted to the left one or two places and the decimal point of the inductor is shifted to the right one or two places. With increasing impedance the capacitor values become smaller and the inductor values become larger. The opposite is true if the impedance decreases.

When the desired impedance level differs from the standard 50-Ω value by a factor such as 1.2, 1.5 or 1.86, the following scaling procedure is used:

1. Calculate the impedance scaling ratio:

$$R = \frac{Z_X}{50} \qquad (12)$$

where Z_X is the desired new impedance level, in ohms.

2. Calculate the cutoff frequency (f_{50co}) of a "trial" 50-Ω filter,

$$f_{50co} = R \times f_{xco} \qquad (13)$$

where R is the impedance scaling ratio and f_{xco} is the desired cutoff frequency of the filter at the new impedance level.

3. From the appropriate SVC table select a design having its cutoff frequency closest to the calculated f_{50co} value. The tabulated capacitor values of this design are taken directly, but the frequency and inductor values must be scaled to the new impedance level.

4. Calculate the exact f_{xco} values, where

$$f_{xco} = \frac{f'_{50co}}{R} \qquad (14)$$

and f'_{50co} is the tabulated cutoff frequency of the selected design. Calculate the other frequencies of the design in the same way.

5. Calculate the inductor values for the new filter by multiplying the tabulated inductor values of the selected design by the square of the scaling ratio, R.

Notes

[1]The Chebyshev filter is named after Pafnuty Lvovitch Chebyshev (1821-1894), a famous Russian mathematician and academician. While touring Europe in 1852 to inspect various types of machinery, Chebyshev became interested in the mechanical linkage used in Watt's steam engine. This linkage converted the reciprocating motion of the piston rod into rotational motion of a flywheel needed to run factory machinery. Chebyshev noted that Watt's piston had zero lateral discrepancy at three points in its cycle. He concluded that a somewhat different linkage would lead to a discrepancy of half of Watt's and would be zero at five points in the piston cycle. Chebyshev then wrote a paper — now considered a mathematical classic — that laid the foundation for the topic of **best approximation of functions by means of polynomials**. It is these same polynomials that were originally developed to improve the reciprocating-to-rotational linkage in a steam engine that now find application in the design of the Chebyshev passive LC filters. From Philip J. Davis, *The Thread, A Mathematical Yarn*, 2nd edition, Harcourt Brace Jovanovitch, Publishers, New York, 1989, 1983; 124-page paperback.

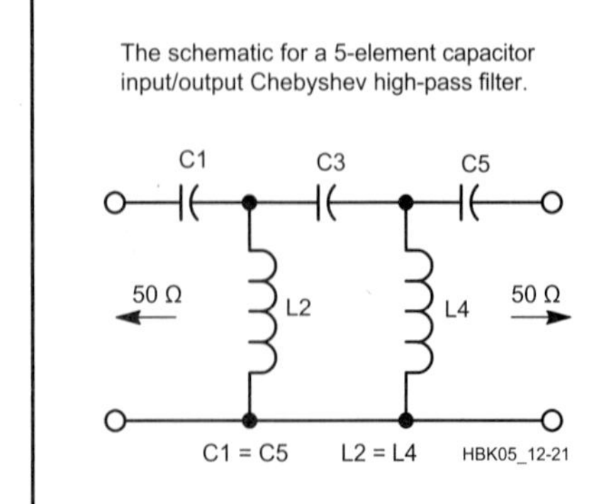

The schematic for a 5-element capacitor input/output Chebyshev high-pass filter.

Fig 12.21 — A portion of a 5-element Chebyshev high-pass filter design table for 50-Ω impedance, C-in/out and standard E24 capacitor values.

—FREQUENCY (MHz)—				*Max*	*C1,5*	*L2,4*	*C3*	
No.	*F_{co}*	*3 dB*	*20 dB*	*40 dB*	*SWR*	*(pF)*	*(H)*	*(pF)*
1	1.04	0.726	0.501	0.328	1.044	5100	6.45	2200
2	1.04	0.788	0.554	0.366	1.081	4300	5.97	2000
3	1.17	0.800	0.550	0.359	1.039	4700	5.85	2000
4	1.07	0.857	0.615	0.410	1.135	3600	5.56	1800
5	1.17	0.877	0.616	0.406	1.076	3900	5.36	1800
6	1.33	0.890	0.609	0.397	1.034	4300	5.26	1800
7	1.12	0.938	0.686	0.461	1.206	3000	5.20	1600
8	1.25	0.974	0.693	0.461	1.109	3300	4.86	1600
9	1.38	0.994	0.691	0.454	1.057	3600	4.71	1600
10	1.54	1.00	0.683	0.444	1.028	3900	4.67	1600

Chebyshev Filter Design (Normalized Tables)

The figures and tables in this section provide the tools needed to design Chebyshev filters including those filters for which the previously published *standard value capacitor* (SVC) designs might not be suitable. **Table 12.2** lists normalized low-pass designs that, in addition to low-pass filters, can also be used to calculate high-pass, band-pass and band-stop filters in either the inductor or capacitor input/output configurations for equal impedance terminations. **Table 12.3** provides the attenuation for the resultant filter.

This material was prepared by Ed Wetherhold, W3NQN, who has been the author of a number of articles and papers on the design of LC filters. It is a complete revision of his previously published filter design material and provides both insight to the design and actual designs in just a few minutes.

For a given number of elements (N), increasing the filter reflection coefficient (RC or ρ) causes the attenuation slope to increase with a corresponding increase in both the passband ripple amplitude (a_p) and SWR and with a decrease in the filter return loss. All of these parameters are mathematically related to each other. If one is known, the others may be calculated. Filter designs having a low RC are preferred because they are less sensitive to component and termination impedance variations than are designs having a higher RC. The RC percentage is used as the independent variable in Table 12.2 because it is used as the defining parameter in the more frequently used tables, such as those by Zverev and Saal (see Refs 17 and 18).

The return loss is tabulated instead of passband ripple amplitude (a_p) because it is easy to measure using a return loss bridge. In comparison, ripple amplitudes less than 0.1 dB are difficult to measure accurately. The resulting values of attenuation are contained in Table 12.3 and corresponding values of a_p and SWR may be found by referring to the Equivalent Values of Reflection Coefficient, Attenuation, SWR and Return Loss table in the **Component Data and References** chapter. The filter used (low pass, high-pass, band-pass and so on) will depend on the

Table 12.2

Element values of Chebyshev low-pass filters normalized for a ripple cutoff frequency (Fa_p) of one radian/sec ($^1/_{2\pi}$ Hz) and 1-Ω terminations.

Use the top column headings for the low-pass C-in/out configuration and the bottom column headings for the low-pass L-in/out configuration. Fig 12.22 shows the filter schematics.

N	RC (%)	Ret Loss (dB)	$F3/F_{ap}$ Ratio	C1 (F)	L2 (H)	C3 (F)	L4 (H)	C5 (F)	L6 (H)	C7 (F)	L8 (H)	C9 (F)
3	1.000	40.00	3.0094	0.3524	0.6447	0.3524						
3	1.517	36.38	2.6429	0.4088	0.7265	0.4088						
3	4.796	26.38	1.8772	0.6292	0.9703	0.6292						
3	10.000	20.00	1.5385	0.8535	1.104	0.8535						
3	15.087	16.43	1.3890	1.032	1.147	1.032						
5	0.044	67.11	2.7859	0.2377	0.5920	0.7131	0.5920	0.2377				
5	0.498	46.06	1.8093	0.4099	0.9315	1.093	0.9315	0.4099				
5	1.000	40.00	1.6160	0.4869	1.050	1.226	1.050	0.4869				
5	1.517	36.38	1.5156	0.5427	1.122	1.310	1.122	0.5427				
5	2.768	31.16	1.3892	0.6408	1.223	1.442	1.223	0.6408				
5	4.796	26.38	1.2912	0.7563	1.305	1.577	1.305	0.7563				
5	6.302	24.01	1.2483	0.8266	1.337	1.653	1.337	0.8266				
5	10.000	20.00	1.1840	0.9732	1.372	1.803	1.372	0.9732				
5	15.087	16.43	1.1347	1.147	1.371	1.975	1.371	1.147				
7	1.000	40.00	1.3004	0.5355	1.179	1.464	1.500	1.464	1.179	0.5355		
7	1.427	36.91	1.2598	0.5808	1.232	1.522	1.540	1.522	1.232	0.5808		
7	1.517	36.38	1.2532	0.5893	1.241	1.532	1.547	1.532	1.241	0.5893		
7	3.122	30.11	1.1818	0.7066	1.343	1.660	1.611	1.660	1.343	0.7066		
7	4.712	26.54	1.1467	0.7928	1.391	1.744	1.633	1.744	1.391	0.7928		
7	4.796	26.38	1.1453	0.7970	1.392	1.748	1.633	1.748	1.392	0.7970		
7	8.101	21.83	1.1064	0.9390	1.431	1.878	1.633	1.878	1.431	0.9390		
7	10.000	20.00	1.0925	1.010	1.437	1.941	1.622	1.941	1.437	1.010		
7	10.650	19.45	1.0885	1.033	1.437	1.962	1.617	1.962	1.437	1.033		
7	15.087	16.43	1.0680	1.181	1.423	2.097	1.573	2.097	1.423	1.181		
9	1.000	40.00	1.1783	0.5573	1.233	1.550	1.632	1.696	1.632	1.550	1.233	0.5573
9	1.517	36.38	1.1507	0.6100	1.291	1.610	1.665	1.745	1.665	1.610	1.291	0.6100
9	2.241	32.99	1.1271	0.6679	1.342	1.670	1.690	1.793	1.690	1.670	1.342	0.6679
9	2.512	32.00	1.1206	0.6867	1.357	1.688	1.696	1.808	1.696	1.688	1.357	0.6867
9	4.378	27.17	1.0915	0.7939	1.419	1.786	1.712	1.890	1.712	1.786	1.419	0.7939
9	4.796	26.38	1.0871	0.8145	1.427	1.804	1.713	1.906	1.713	1.804	1.427	0.8145
9	4.994	26.03	1.0852	0.8239	1.431	1.813	1.712	1.913	1.712	1.813	1.431	0.8239
9	8.445	21.47	1.0623	0.9682	1.460	1.936	1.692	2.022	1.692	1.936	1.460	0.9682
9	10.000	20.00	1.0556	1.025	1.462	1.985	1.677	2.066	1.677	1.985	1.462	1.025
9	15.087	16.43	1.0410	1.196	1.443	2.135	1.617	2.205	1.617	2.135	1.443	1.196
N	RC (%)	Ret Loss (dB)	$F3/F_{ap}$ Ratio	L1 (H)	C2 (F)	L3 (H)	C4 (F)	L5 (H)	C6 (F)	L7 (H)	C8 (F)	L9 (H)

Table 12.3

Normalized Frequencies at Listed Attenuation Levels for Chebyshev Low-Pass Filters with N = 3, 5, 7 and 9.

		Attenuation Levels (dB)										
N	*RC(%)*	*1.0*	*3.01*	*6.0*	*10*	*20*	*30*	*40*	*50*	*60*	*70*	*80*
3	1.000	2.44	3.01	3.58	4.28	6.33	9.27	13.59	19.93	29.25	42.92	63.00
3	1.517	2.15	2.64	3.13	3.74	5.52	8.08	11.83	17.35	25.46	37.36	54.83
3	4.796	1.56	1.88	2.20	2.60	3.79	5.53	8.08	11.83	17.35	25.45	37.35
3	10.000	1.31	1.54	1.78	2.08	3.00	4.34	6.33	9.26	13.57	19.90	29.20
3	15.087	1.20	1.39	1.59	1.85	2.63	3.79	5.52	8.06	11.81	17.32	25.41
5	1.000	1.46	1.62	1.76	1.94	2.39	2.97	3.69	4.62	5.79	7.27	9.13
5	1.517	1.38	1.52	1.65	1.80	2.22	2.74	3.41	4.26	5.33	6.69	8.40
5	4.796	1.19	1.29	1.39	1.50	1.82	2.22	2.74	3.41	4.26	5.33	6.69
5	6.302	1.16	1.25	1.34	1.44	1.74	2.12	2.61	3.24	4.04	5.05	6.34
5	10.000	1.11	1.18	1.26	1.35	1.61	1.95	2.39	2.96	3.69	4.61	5.78
5	15.087	1.07	1.13	1.20	1.28	1.51	1.82	2.22	2.74	3.41	4.25	5.33
7	1.000	1.23	1.30	1.37	1.45	1.65	1.89	2.18	2.53	2.95	3.44	4.04
7	1.517	1.19	1.25	1.32	1.39	1.57	1.80	2.07	2.39	2.79	3.25	3.81
7	4.796	1.10	1.15	1.19	1.25	1.39	1.57	1.80	2.07	2.39	2.79	3.25
7	8.101	1.07	1.11	1.15	1.19	1.32	1.49	1.69	1.94	2.24	2.60	3.03
7	10.000	1.05	1.09	1.13	1.18	1.30	1.45	1.65	1.89	2.18	2.53	2.94
7	15.087	1.04	1.07	1.10	1.14	1.25	1.39	1.57	1.80	2.07	2.39	2.78
9	1.000	1.13	1.18	1.22	1.26	1.38	1.51	1.67	1.85	2.07	2.32	2.61
9	1.517	1.11	1.15	1.19	1.23	1.34	1.46	1.61	1.78	1.99	2.22	2.50
9	4.796	1.06	1.09	1.11	1.15	1.23	1.34	1.46	1.61	1.78	1.99	2.22
9	8.445	1.04	1.06	1.09	1.11	1.19	1.28	1.40	1.53	1.69	1.88	2.10
9	10.000	1.03	1.06	1.08	1.10	1.18	1.27	1.38	1.51	1.67	1.85	2.07
9	15.087	1.02	1.04	1.06	1.08	1.15	1.23	1.34	1.46	1.61	1.78	1.99

application and the stop-band attenuation needed.

The filter schematic diagrams shown in **Fig 12.22** are for low-pass and high-pass versions of the Chebyshev designs listed in Table 12.2. Both low-pass and high-pass equally terminated configurations and component values of the C-in/out or L-in/out filters can be derived from this single table. By using a simple procedure, the low-pass and high-pass designs can be transformed into corresponding band-pass and band-stop filters. The normalized element values of the low-pass C-in/out and L-in/out designs, Fig 12.22A and B, are read directly from the table using the values associated with either the top or bottom column headings, respectively.

The first four columns of Table 12.2 list *N* (the number of filter elements), *RC* (reflection coefficient percentage), *return loss* and the ratio of the 3-dB-to-F_{ap} frequencies. The passband maximum ripple amplitude (a_p) is not listed because it is difficult to measure. If necessary it can be calculated from the reflection coefficient. The $F3/F_{ap}$ ratio varies with N and RC; if both of these parameters are known, the $F3/F_{ap}$ ratio may be calculated. The remaining columns list the normalized Chebyshev element values for equally terminated filters for Ns from 3 to 9 in increments of 2.

The Chebyshev passband ends when the passband attenuation first exceeds the maximum ripple amplitude, a_p. This frequency is called the "ripple cutoff frequency, F_{ap}" and it has a normalized value of unity. All Chebyshev designs in Table 12.2 are based on the ripple cutoff frequency instead of the more familiar 3-dB frequency of the Butterworth response. However, the 3-dB frequency of a Chebyshev design may be obtained by multiplying the ripple cutoff frequency by the $F3/F_{ap}$ ratio listed in the fourth column.

The element values are normalized to a ripple cutoff frequency of 0.15915 Hz (one radian/sec) and 1-Ω terminations, so that the low-pass values can be transformed directly into high-pass values. This is done by replacing all Cs and Ls in the low-pass configuration with Ls and Cs and by replacing all the low-pass element values with their reciprocals. The normalized values are then multiplied by the appropriate C and L scaling factors to obtain the final values based on the desired ripple cutoff frequency and impedance level. The listed C and L element values are in farads and henries and become more reasonable after the values are scaled to the desired cutoff frequency and impedance level.

The normalized designs presented are a mixture: Some have integral values of *reflection coefficient (RC)* (1% and 10%) while others have "integral" values of *passband ripple amplitude* (0.001, 0.01 and 0.1 dB). These ripple amplitudes correspond to reflection coefficients of 1.517, 4.796 and 15.087%, respectively. By having tabulated designs based on integral values of both reflection coefficient and passband ripple amplitude, the correctness of the normalized component values may be checked against those same values published in filter handbooks whichever parameter, RC or a_p, is used.

In addition to the customary normalized design listings based on integral values of reflection coefficient or ripple amplitude, Table 12.2 also includes unique designs having special element ratios that make them more useful than previously published tables. For example, for N = 5 and RC = 6.302, the ratio of C3/C1 is 2.000. This ratio allows 5-element low-pass filters to be realized with only one capacitor value because C3 may be obtained by using parallel-connected capacitors each having the same value as C1 and C5.

In a similar way, for N = 7 and RC = 8.101, C3/C1 and C5/C1 are also 2.000. Another useful N = 7 design is that for RC = 1.427%. Here the L4/L2 ratio is

1.25, which is identical to 110/88. This means a seventh-order C-in/out low-pass audio filter can be realized with four surplus 88-mH inductors. Both L2 and L6 can be 88 mH while L4 is made up of a series connection of 22 mH and 88 mH. The 22-mH value is obtained by connecting the two windings of one of the four surplus inductors in parallel. Other useful ratios also appear in the N = 9 listing for both C3/C1 and L4/L2.

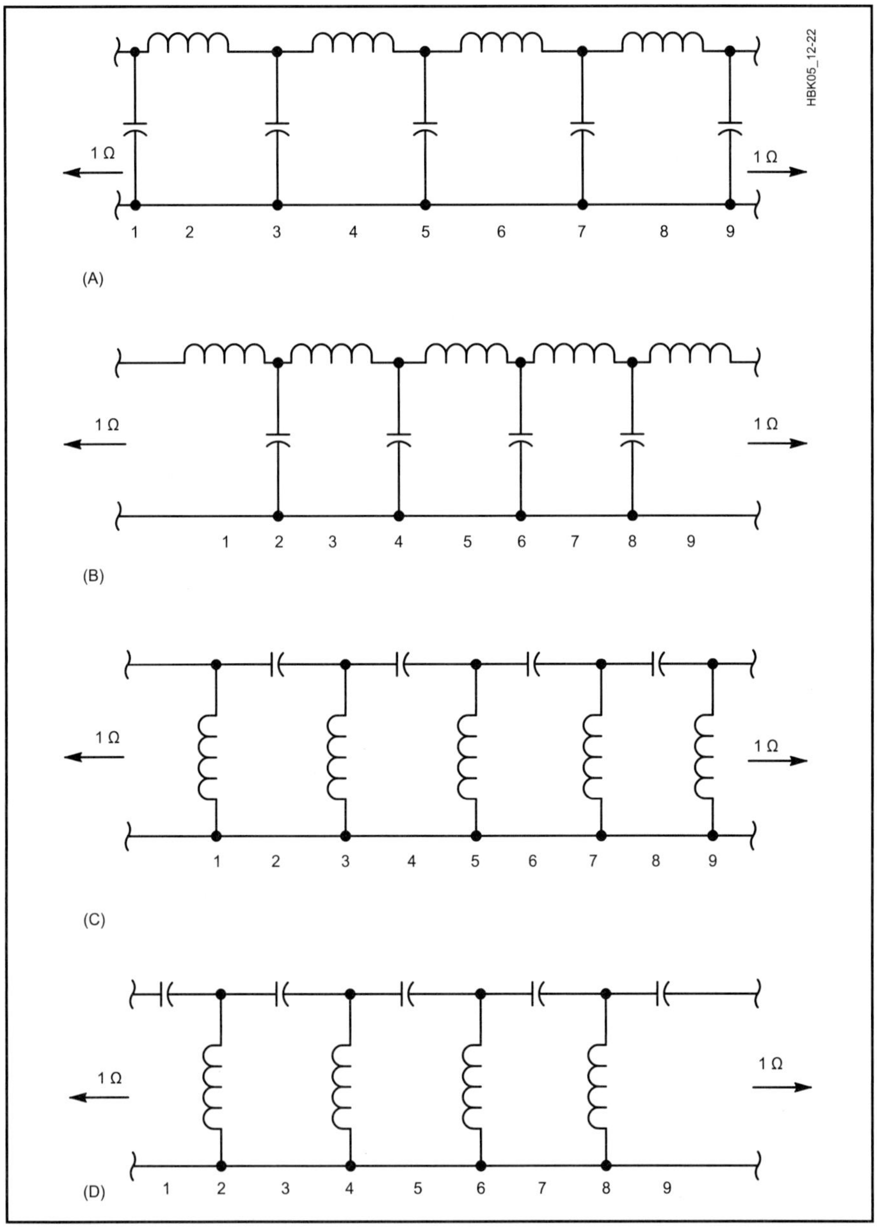

Fig 12.22 — The schematic diagrams shown are low-pass and high-pass Chebyshev filters with the C-in/out and L-in/out configurations. For all normalized values see Table 12.2.

A: C-in/out low-pass configuration. Use the C and L values associated with the top column headings of the Table.

B: L-in/out low-pass configuration. For normalized values, use the L and C values associated with the bottom column headings of the Table.

C: L-in/out high-pass configuration is derived by transforming the C-in/out low-pass filter in A into an L-in/out high-pass by replacing all Cs with Ls and all Ls with Cs. The reciprocals of the lowpass component values become the highpass component values. For example, when n = 3, RC = 1.00% and C1 = 0.3524 F, L1 and L3 in C become 2.838 H.

D: The C-in/out high-pass configuration is derived by transforming the L-in/out low-pass in B into a C-in/out high-pass by replacing all Ls with Cs and all Cs with Ls. The reciprocals of the low-pass component values become the high-pass component values. For example, when n = 3, RC = 1.00% and L1 = 0.3524 H, C1 and C3 in D become 2.838 F.

Except for the first two N = 5 designs, all designs were calculated for a reflection coefficient range from 1% to about 15%. The first two N = 5 designs were included because of their useful L3/L1 ratios. Designs with an RC of less than 1% are not normally used because of their poor selectivity. Designs with RC greater than 15% yield increasingly high SWR values with correspondingly increased objectionable reflective losses and sensitivity to termination impedance and component value variations.

Low-pass and high-pass filters may be realized in either a C-input/output or an L-input/output configuration. The C-input/output configuration is usually preferred because fewer inductors are required, compared to the L-input/output configuration. Inductors are usually more lossy, bulky and expensive than capacitors. The selection of the filter order or number of filter elements, N, is determined by the desired stop-band attenuation rate of increase and the tolerable reflection coefficient or SWR. A steeper attenuation slope requires either a design having a higher reflection coefficient or more circuit elements. Consequently, to select an optimum design, the builder must determine the amount of attenuation required in the stop band and the permissible maximum amount of reflection coefficient or SWR.

Table 12.3 shows the theoretical normalized frequencies (relative to the ripple cutoff frequency) for the listed attenuation levels and reflection coefficient percentages for Chebyshev low-pass filters of 3, 5, 7 and 9 elements. For example, for N = 5 and RC = 15.087%, an attenuation of 40 dB is reached at 2.22 times the ripple cutoff frequency (slightly more than one octave). The tabulated data are also applicable to high-pass filters by simply taking the reciprocal of the listed frequency. For example, for the same previous N and RC values, a high-pass filter attenuation will reach 40 dB at 1/2.22 = 0.450 times the ripple cutoff frequency.

The attenuation levels are theoretical and assume perfect components, no coupling between filter sections and no signal leakage around the filter. A working model should follow these values to the 60 or 70-dB level. Beyond this point, the actual response will likely degrade somewhat from the theoretical.

Fig 12.23 shows four plotted attenuation vs normalized frequency curves for N = 5 corresponding to the normalized frequencies in Table 12.3 At two octaves above the ripple cutoff frequency, f_c, the attenuation slope gradually becomes 6 dB per octave per filter element.

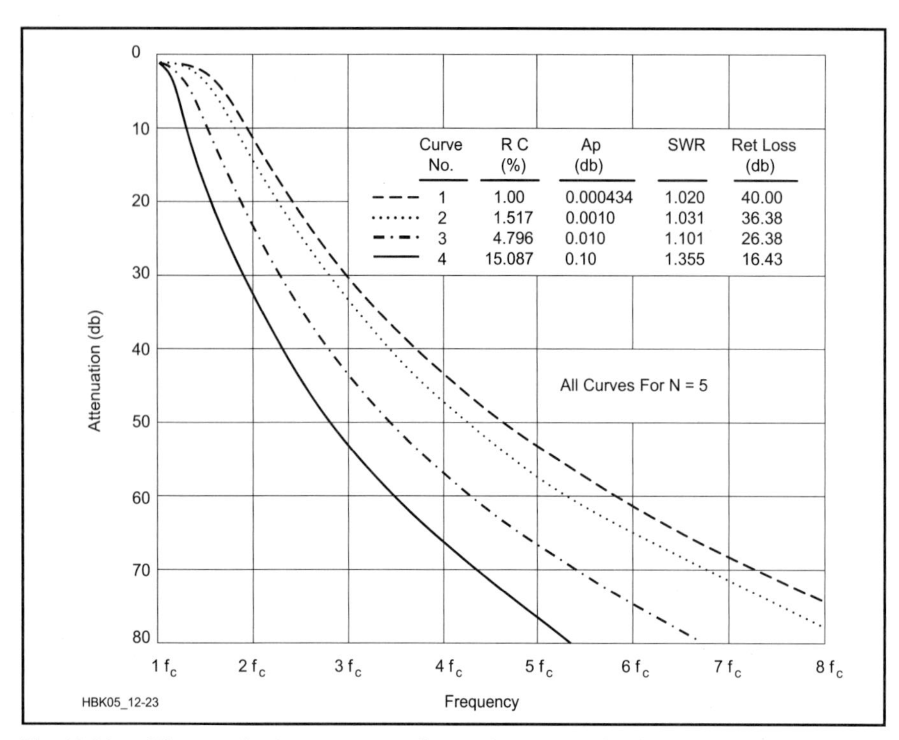

Fig 12.23 — The graph shows attenuation vs frequency for four 5-element low-pass filters designed with the information obtained from Table 12.2. This graph demonstrates how reflection coefficient percentage (RC), maximum passband ripple amplitude (a_p), SWR, return loss and attenuation rolloff are all related. The exact frequency at a specified attenuation level can be obtained from Table 12.3.

LOW-PASS AND HIGH-PASS FILTERS

Low-Pass Filter

Let's look at the procedure used to calculate the capacitor and inductor values of low-pass and high-pass filters by using two examples. Assume a 50-Ω low-pass filter is needed to give more than 40 dB of attenuation at $2f_c$ or one octave above the ripple-cutoff frequency of 4.0 MHz. Referring to Table 12.3, we see from the 40-dB column that a filter with 7 elements (N = 7) and a RC of 4.796% will reach 40 dB at 1.80 times the cutoff frequency or 1.8 × 4 = 7.2 MHz. Since this design has a reasonably low reflection coefficient and will satisfy the attenuation requirement, it is a good choice. Note that no 5-element filters are suitable for this application because 40 dB of attenuation is not achieved one octave above the cutoff frequency.

From Table 12.2, the normalized component values corresponding to N = 7 and RC = 4.796% for the C-in/out configuration are: C1, C7 = 0.7970 F, L2, L6 = 1.392 H, C3, C5 = 1.748 F and L4 = 1.633 H. See Fig 12.23A for the corresponding configuration. The C and L normalized values will be scaled from a ripple cutoff frequency of one radian/sec and an impedance level of 1 Ω to a cutoff frequency of 4.0 MHz and an impedance level of 50 Ω. The C_s and L_s scaling factors are calculated:

$$C_s = \frac{1}{2\pi R f} \qquad (15)$$

$$L_s = \frac{R}{2\pi f} \qquad (16)$$

where:
R = impedance level
f = cutoff frequency.

In this example:

$$C_s = \frac{1}{2\pi R f} = \frac{1}{2\pi \times 50 \times 4 \times 10^6} = 795.8 \times 10^{-12}$$

$$L_s = \frac{R}{2\pi f} = \frac{50}{2\pi \times 4 \times 10^6} = 1.989 \times 10^{-6}$$

Using these scaling factors, the capacitor and inductor normalized values are scaled to the desired cutoff frequency and impedance level:

C1, C7 = 0.797 × 795.8 pF = 634 pF
C3, C5 = 1.748 × 795.8 pF = 1391 pF
L2, L6 = 1.392 × 1.989 µH = 2.77 µH
L4 = 1.633 × 1.989 µH = 3.25 µH

High-Pass Filter

The procedure for calculating a high-pass filter is similar to that for a low-pass filter, except a low-pass-to-high-pass transformation must first be performed. Assume a 50-Ω high-pass filter is needed to give more than 40 dB of attenuation one octave below ($f_c/2$) a ripple cutoff frequency of 4.0 MHz. Referring to Table 12.3, we see from the 40-dB column that a 7-element low-pass filter with RC of 4.796% will give 40 dB of attenuation at $1.8f_c$. If this filter is transformed into a high-pass filter, the 40-dB level is reached at $f_c/1.80$ or at $0.556f_c$ = 2.22 MHz. Since the 40-dB level is reached before one octave from the 4-MHz cutoff frequency, this design will be satisfactory.

From Fig 12.22, we choose the low-pass L-in/out configuration in B and transform it into a high-pass filter by replacing all inductors with capacitors and all capacitors with inductors. Fig 12.22D is the filter configuration after the transformation. The reciprocals of the low-pass values become the high-pass values to complete the transformation. The high-pass values of the filter shown in Fig 12.22D are:

$$C1, C7 = \frac{1}{0.7970} = 1.255\ F$$

$$L2, L6 = \frac{1}{1.392} = 0.7184\ H$$

$$C3, C5 = \frac{1}{1.748} = 0.5721\ F$$

and

$$L4 = \frac{1}{1.633} = 0.6124\ H$$

Using the previously calculated C and L scaling factors, the high-pass component values are calculated the same way as before:

C1, C7 = 1.255 × 795.8 pF = 999 pF
C3, C5 = 0.5721× 795.8 pF = 455 pF
L2, L6 = 0.7184 × 1.989 µH = 1.43 µH
L4 = 0.6124 × 1.989 µH = 1.22 µH

BAND-PASS FILTERS

Band-pass filters may be classified as either narrowband or broadband. If the ratio of the upper ripple cutoff frequency to the lower cutoff frequency is greater than two, we have a wideband filter. For wideband filters, the band-pass filter (BPF) requirement may be realized by simply cascading separate high-pass and low-pass filters having the same design impedance. (The assumption is that the filters maintain their individual responses even though they are cascaded.) *For this to be true, it is important that both filters have a relatively low reflection coefficient percentage (less than 5%) so the SWR variations in the passband will be small.*

For narrowband BPFs, where the separation between the upper and lower cutoff frequencies is less than two, it is necessary to transform an appropriate low-pass filter into a BPF. That is, we use the low-pass normalized tables to design narrowband BPFs.

We do this by first calculating a low-pass filter (LPF) with a cutoff frequency equal to the desired bandwidth of the BPF. The LPF is then transformed into the desired BPF by resonating the low-pass components at the geometric center frequency of the BPF.

For example, assume we want a 50-Ω BPF to pass the 75/80-m band and attenuate all signals outside the band. Based on the passband ripple cutoff frequencies of 3.5 and 4.0 MHz, the geometric center frequency $= (3.5 \times 4.0)^{0.5} = (14)^{0.5} = 3.741657$ or 3.7417 MHz. Let's slightly extend the lower and upper ripple cutoff frequencies to 3.45 and 4.058 MHz to account for possible component tolerance variations and to maintain the same center frequency. We'll evaluate a low-pass 3-element prototype with a cutoff frequency equal to the BPF passband of (4.058–3.45)MHz = 0.608 MHz as a possible choice for transformation.

Further, assume it is desired to attenuate the second harmonic of 3.5 MHz by at least 40 dB. The following calculations show how to design an N = 3 filter to provide the desired 40-dB attenuation at 7 MHz and above.

The bandwidth (BW) between 7 MHz on the upper attenuation slope (call it "f+") of the BPF and the corresponding frequency at the same attenuation level on the lower slope (call it "f–") can be calculated based on $(f+)(f-) = (f_c)^2$ or

$$f- = \frac{14}{7} = 2\text{ MHz}$$

Therefore, the bandwidth at this unknown attenuation level for 2 and 7 MHz is 5 MHz. This 5-MHz BW is normalized to the ripple cutoff BW by dividing 5.0 MHz by 0.608 MHz:

$$\frac{5.0}{0.608} = 8.22$$

We now can go to Table 12.3 and search for the corresponding normalized frequency that is closest to the desired normalized BW of 8.22. The low-pass design of N = 3 and RC = 4.796% gives 40 dB for a normalized BW of 8.08 and 50 dB for 11.83. Therefore, a design of N = 3 and RC = 4.796% with a normalized BW of 8.22 is at an attenuation level somewhere between 40 and 50 dB. Consequently, a low-pass design based on 3 elements and a 4.796% RC will give slightly more than the desired 40 dB attenuation above 7 MHz. The next step is to calculate the C and L values of the low-pass filter using the normalized component values in Table 12.2.

From this table and for N = 3 and RC = 4.796%, C1, C3 = 0.6292 F and L2 = 0.9703 H. We calculate the scaling factors as before and use 0.608 MHz as the ripple cutoff frequency:

$$C_s = \frac{1}{2\pi R f}$$

$$= \frac{1}{2\pi \times 50 \times 0.608 \times 10^6} = 5235 \times 10^{-12}$$

$$L_s = \frac{R}{2\pi f} = \frac{50}{2\pi \times 0.608 \times 10^6} = 13.09 \times 10^{-6}$$

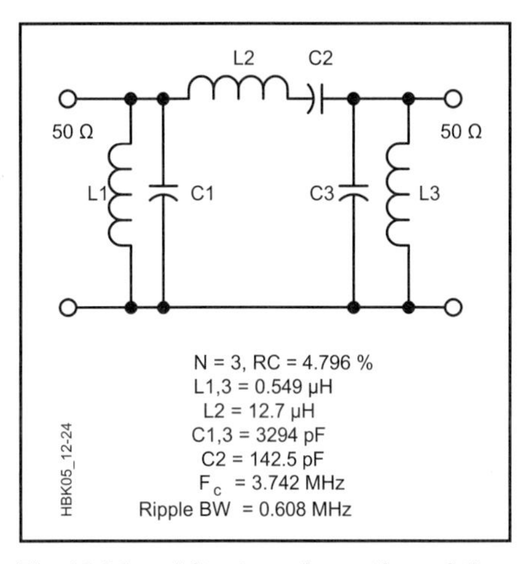

Fig 12.24 — After transformation of the band-pass filter, all parallel elements become parallel LCs and all series elements become series LCs.

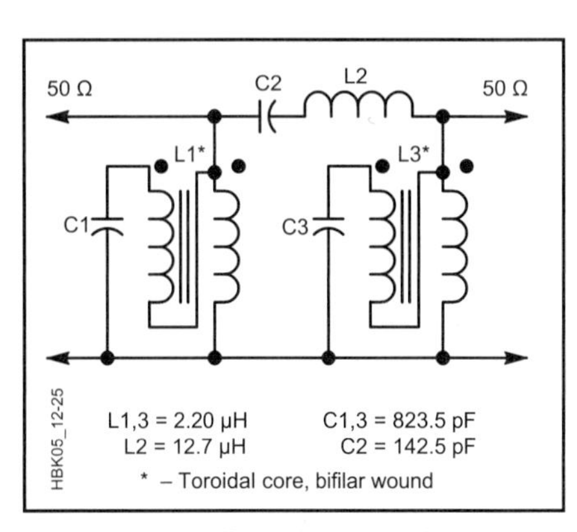

Fig 12.25 — A filter designed for 200-Ω source and load provides better values. By tapping the inductors, we can use a 200-Ω filter design in a 50-Ω system.

C1, C3 = 0.6292 × 5235 pF = 3294 pF and L2 = 0.9703 × 13.09 µH = 12.70 µH.

The LPF (in a pi configuration) is transformed into a BPF with 3.7417-MHz center frequency by resonating the low-pass elements at the center frequency. The resonating components will take the same identification numbers as the components they are resonating.

$$L1, L3 = \frac{25330}{\left(F_c^2 \times C1\right)} = \frac{25330}{(14 \times 3294)} = 0.5493\ \mu\text{H}$$

$$C2 = \frac{25330}{\left(F_c^2 \times L2\right)} = \frac{25330}{14 \times 12.7} = 142.5\text{ pF}$$

where L, C and f are in µH, pF and MHz respectively.

The BPF circuit after transformation (for N = 3, RC = 4.796%) is shown in **Fig 12.24**.

The component-value spread is

$$\frac{12.7}{0.549} = 23$$

and the reactance of L1 is about 13 Ω at the center frequency. For better BPF performance, the component spread should be reduced and the reactance of L1 and L3 should be raised to make it easier to achieve the maximum possible Q for these two inductors. This can be easily done by designing the BPF for an impedance level of 200 Ω and then using the center taps on L1 and L3 to obtain the desired 50-Ω terminations. The result of this approach is shown in **Fig 12.25**

The component spread is now a more reasonable

$$\frac{12.7}{2.20} = 5.77$$

and the L1, L3 reactance is 51.6 Ω. This higher reactance gives a better chance to achieve a satisfactory Q for L1 and L3 with a corresponding improvement in the BPF performance.

As a general rule, keep reactance values between 5 Ω and 500 Ω in a 50-Ω circuit. When the value falls below 5 Ω, either the equivalent series resistance of the inductor or the series inductance of the capacitor degrades the circuit Q. When the inductive reactance is greater than 500 Ω, the inductor is approaching self-resonance and circuit Q is again degraded. In practice, both L1 and L3 should be bifilar wound on a powdered-iron toroidal core to assure that optimum coupling is obtained between turns over the entire winding. The junction of the bifilar winding serves as a center tap.

Side-Slope Attenuation Calculations

The following equations allow the calculation of the frequencies on the upper and lower sides of a BPF response curve at any given attenuation level if the bandwidth at that attenuation level and the geometric center frequency of the BPF are known:

$$f_{lo} = -X + \sqrt{f_c^{\,2} + X^2} \qquad (17)$$

$$f_{hi} = f_{lo} + BW \qquad (18)$$

where

BW = bandwidth at the given attenuation level,

f_c = geometric center frequency

$$X = \frac{BW}{2}$$

For example, if f = 3.74166 MHz and BW = 5 MHz, then

$$\frac{BW}{2} = X = 2.5$$

and:

$$f_{lo} = -2.5 + \sqrt{3.74166^2 + 2.5^2} = 2.00 \text{ MHz}$$

$$f_{hi} = f_{lo} + BW = 2 + 5 = 7 \text{ MHz}$$

BAND-STOP FILTERS

Band-stop filters may be classified as either narrowband or broadband. If the ratio of the upper ripple cutoff frequency to the lower cutoff frequency is greater than two, the filter is considered wideband. A wideband band-stop filter (BSF) requirement may be realized by simply paralleling the inputs and outputs of separate low-pass and high-pass filters having the same design impedance and with the low-pass filter having its cutoff frequency one octave or more below the high-pass cutoff frequency.

In order to parallel the low-pass and high-pass filter inputs and outputs without one affecting the other, it is essential that each filter have a high impedance in that portion of its stop band that lies in the passband of the other. *This means that each of the two filters must begin and end in series branches.* In the low-pass filter, the input/output series branches must consist of inductors and in the high-pass filter, the input/output series branches must consist of capacitors.

When the ratio of the upper to lower cutoff frequencies is less than two, the BSF is considered to be narrowband, and a calculation procedure similar to that of the narrowband BPF design procedure is used. However, in the case of the BSF, the design process starts with the design of a high-pass filter having the desired impedance level of the BSF and a ripple cutoff frequency the same as that of the desired ripple bandwidth of the BSF. After the HPF design is completed, every high-pass element is resonated to the center frequency of the BSF in the same manner as if it were a BPF, except that all shunt branches of the BSF will consist of series-tuned circuits, and all series branches will consist of parallel-tuned circuits — just the opposite of the resonant circuits in the BPF. The reason for this becomes obvious when the impedance characteristics of the series and parallel circuits at resonance are considered relative to the intended purpose of the filter, that is, whether it is for a band-pass or a band-stop application.

The design, construction and test of band-stop filters for attenuating high-level broadcast-band signals is described in reference 30 at the end of this chapter.

Quartz Crystal Filters

Practical inductor Q values effectively set the minimum achievable bandwidth limits for LC band-pass filters. Higher-Q circuit elements must be employed to extend these limits. These high-Q resonators include PZT ceramic, mechanical and coaxial devices. However, the quartz crystal provides the highest Q and best stability with temperature and time of all available resonators. Quartz crystals suitable for filter use are fabricated over a frequency range from audio to VHF.

The quartz resonator has the equivalent circuit shown in **Fig 12.26**. L_s, C_s and R_s represent the *motional* reactances and loss resistance. C_p is the parallel plate capacitance formed by the two metal electrodes separated by the quartz dielectric. Quartz has a dielectric constant of 3.78. **Table 12.4** shows parameter values for typical moderate-cost quartz resonators. Q_U is the resonator unloaded Q.

$$Q_U = 2\pi f_s r_s \qquad (19)$$

Q_U is very high, usually exceeding 25,000. Thus the quartz resonator is an ideal component for the synthesis of a high-Q band-pass filter.

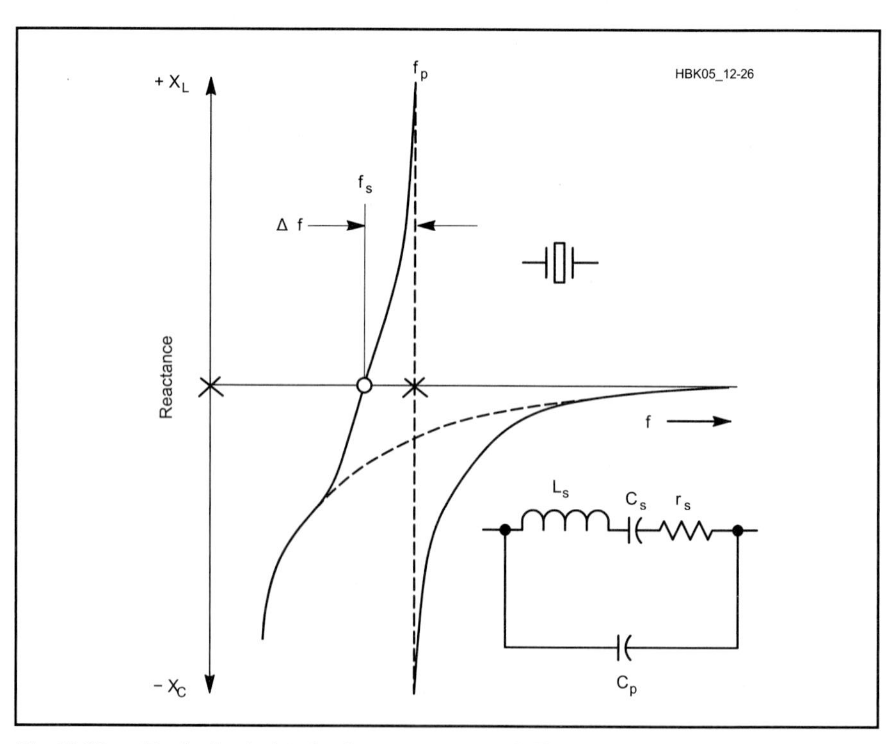

Fig 12.26 — Equivalent circuit of a quartz crystal. The curve plots the crystal reactance against frequency. At f_p, the resonance frequency, the reactance curve goes to infinity.

Table 12.4
Typical Parameters for AT-Cut Quartz Resonators

Freq (MHz)	*Mode n*	*rs (Ω)*	*Cp (pF)*	*Cs (pF)*	*L (mH)*	Q_U
1.0	1	260	3.4	0.0085	2900	72,000
5.0	1	40	3.8	0.011	100	72,000
10.0	1	8	3.5	0.018	14	109,000
20	1	15	4.5	0.020	3.1	26,000
30	3	30	4.0	0.002	14	87,000
75	3	25	4.0	0.002	2.3	43,000
110	5	60	2.7	0.0004	5.0	57,000
150	5	65	3.5	0.0006	1.9	27,000
200	7	100	3.5	0.0004	2.1	26,000

Courtesy of Piezo Crystal Co, Carlisle, Pennsylvania

A quartz resonator connected between generator and load, as shown in **Fig 12.27A**, produces the frequency response of Fig 12.27B. There is a relatively low loss at the series resonant frequency f_s and high loss at the parallel resonant frequency f_p. The test circuit of Fig 12.27A is useful for determining the parameters of a quartz resonator, but yields a poor filter.

A crystal filter developed in the 1930s is shown in **Fig 12.28A**. The disturbing effect of C_p (which produces f_p) is canceled by the *phasing capacitor,* C1. The voltage-reversing transformer T1 usually consists of a bifilar winding on a ferrite core. Voltages V_a and V_b have equal magnitude but 180° phase difference. When C1 = C_p, the effect of C_p will disappear and a well-behaved single resonance will occur as shown in Fig 12.28B. The band-pass filter will exhibit a loaded Q given by:

$$Q_L = \frac{2 \pi f_S L_S}{R_L} \qquad (20)$$

This single-stage "crystal filter," operating at 455 kHz, was present in almost all high-quality amateur communications receivers up through the 1960s. When the filter was switched into the receiver IF amplifier the bandwidth was reduced to a few hundred Hz for Morse code reception.

The half-lattice filter shown in **Fig 12.29** is an improvement in crystal filter design. The quartz resonator parallel-plate capacitors, C_p, cancel each other. Remaining series resonant circuits, if properly offset in frequency, will produce an approximate 2-pole Butterworth or Chebyshev response. Crystals A and B are usually chosen so that the parallel resonant frequency (f_p) of one is the same as the series resonant frequency (f_s) of the other.

Half-lattice filter sections can be cascaded to produce a composite filter with many poles. Until recently, most vendor- supplied commercial filters were lattice types. Ref 11 discusses the computer design of half-lattice filters.

Fig 12.27 — A: Series test circuit for a crystal. In the test circuit the output of a variable frequency generator, e_g, is used as the test signal. The frequency response in B shows the highest attenuation at resonance (f_p). See text.

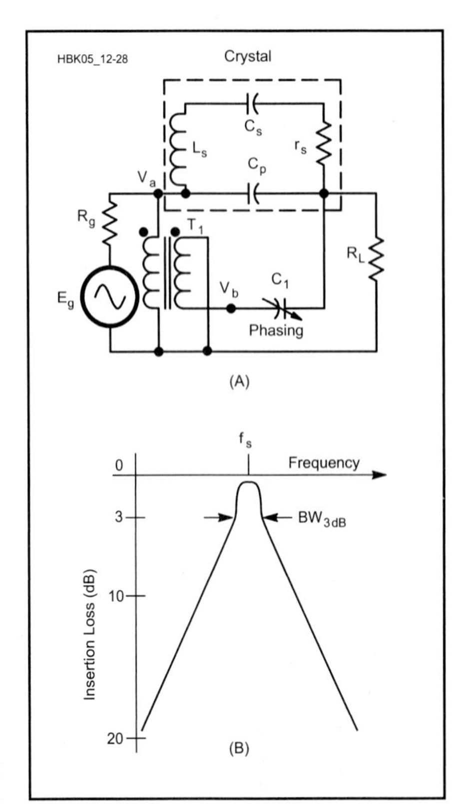

Fig 12.28 — The practical one-stage crystal filter in A has the response shown in B. The phasing capacitor is adjusted for best response (see text).

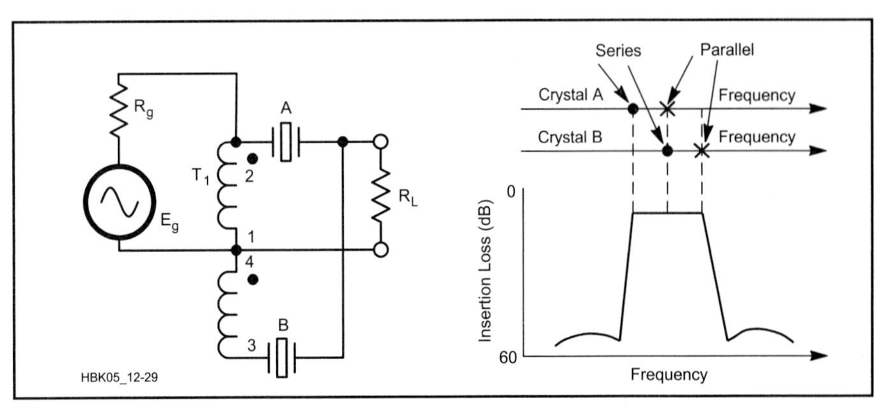

Fig 12.29 — A half-lattice crystal filter. No phasing capacitor is needed in this circuit.

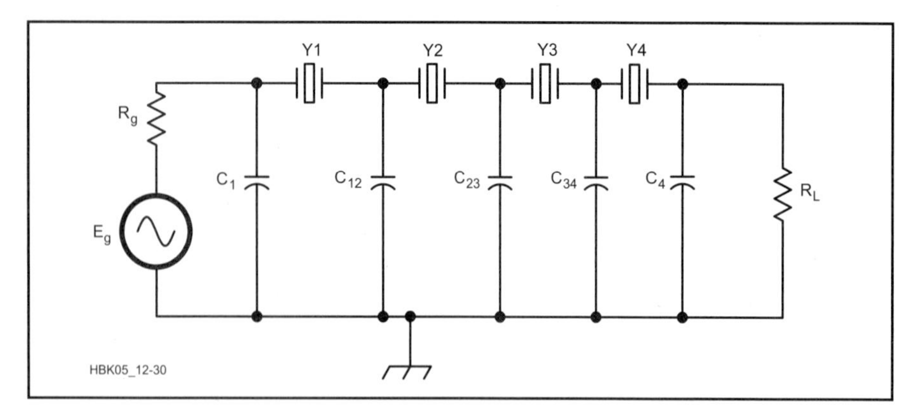

Many quartz crystal filters produced today use the ladder network design shown in **Fig 12.30**. In this configuration, all resonators have the same series resonant frequency f_s. Inter-resonator coupling is provided by shunt capacitors such as C12 and C23. Refs 12 and 13 provide good ladder filter design information. A test set for evaluating crystal filters is presented elsewhere in this chapter.

Fig 12.30 — A four-stage crystal ladder filter. The crystals must be chosen properly for best response.

Monolithic Crystal Filters

A monolithic (Greek: one-stone) crystal filter has two sets of electrodes deposited on the same quartz plate, as shown in **Fig 12.31**. This forms two resonators with acoustic (mechanical) coupling between them. If the acoustic coupling is correct, a 2-pole Butterworth or Chebyshev response will be achieved. More than two resonators can be fabricated on the same plate yielding a multipole response. Monolithic crystal filter technology is popular because it produces a low parts count, single-unit filter at lower cost than a lumped-element equivalent. Monolithic crystal filters are typically manufactured in the range from 5 to 30 MHz for the fundamental mode and up to 90 MHz for the third-overtone mode. Q_L ranges from 200 to 10,000.

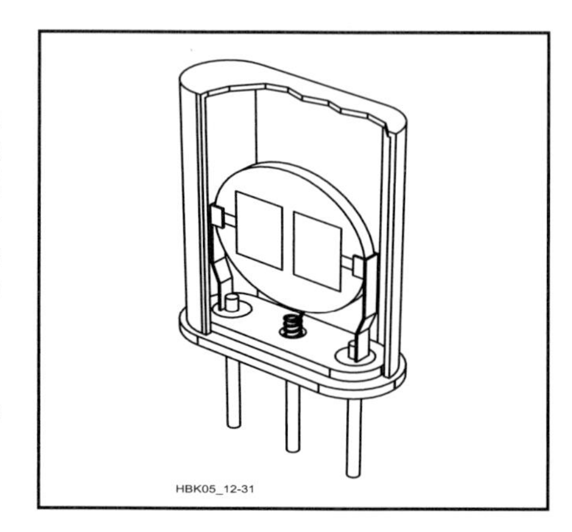

Fig 12.31 — Typical two-pole monolithic crystal filter. This single small (1/2 to 3/4-inch) unit can replace 6 to 12, or more, discrete components.

SAW Filters

The resonators in a monolithic crystal filter are coupled together by bulk acoustic waves. These acoustic waves are generated and propagated in the interior of a quartz plate. It is also possible to launch, by an appropriate transducer, acoustic waves that propagate only along the surface of the quartz plate. These are called "surface-acoustic-waves" because they do not appreciably penetrate the interior of the plate.

A surface-acoustic-wave (SAW) filter consists of thin aluminum electrodes, or fingers, deposited on the surface of a piezoelectric substrate as shown in **Fig 12.32**. Lithium Niobate ($LiNbO_3$) is usually favored over quartz because it yields less insertion loss. The electrodes make

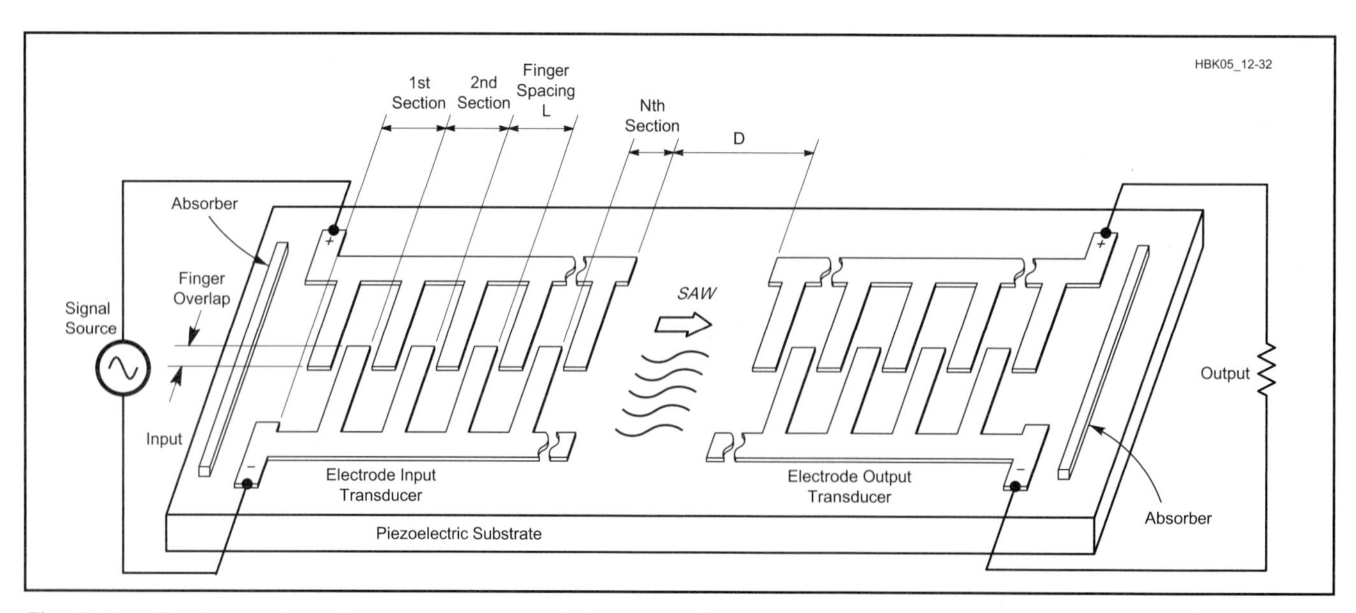

Fig 12.32 — The *interdigitated* transducer, on the left, launches SAW energy to a similar transducer on the right (see text).

up the filter's transducers. RF voltage is applied to the input transducer and generates electric fields between the fingers. The piezoelectric material vibrates launching an acoustic wave along the surface. When the wave reaches the output transducer it produces an electric field between the fingers. This field generates a voltage across the load resistor.

Since both input and output transducers are not entirely unidirectional, some acoustic power is lost in the acoustic absorbers located behind each transducer. This lost acoustic power produces a midband electrical insertion loss typically greater than 10 dB. The SAW filter frequency response is determined by the choice of substrate material and finger pattern. The finger spacing, (usually one-quarter wavelength) determines the filter center frequency. Center frequencies are available from 20 to 1000 MHz. The number and length of fingers determines the filter loaded Q and shape factor.

Loaded Qs are available from 2 to 100, with a shape factor of 1.5 (equivalent to a dozen poles). Thus the SAW filter can be made broadband much like the LC filters that it replaces. The advantage is substantially reduced volume and possibly lower cost. SAW filter research was driven by military needs for exotic amplitude-response and time-delay requirements. Low-cost SAW filters are presently found in television IF amplifiers where high midband loss can be tolerated.

Transmission-Line Filters

LC filter calculations are based on the assumption that the reactances are *lumped*—the physical dimensions of the components are considerably less than the operating wavelength. Therefore the unavoidable interturn capacitance associated with inductors and the unavoidable series inductance associated with capacitors are neglected as secondary effects. If careful attention is paid to circuit layout and miniature components are used, lumped LC filter technology can be used up to perhaps 1 GHz.

Transmission-line filters predominate from 500 MHz to 10 GHz. In addition they are often used down to 50 MHz when narrowband (Q_L > 10) band-pass filtering is required. In this application they exhibit considerably lower loss than their LC counterparts.

Replacing lumped reactances with selected short sections of TEM transmission lines results in transmission-line filters. In TEM, or *Transverse Electromagnetic Mode,* the electric and magnetic fields associated with a transmission line are at right angles (transverse) to the direction of wave propagation. Coaxial cable, stripline and microstrip are examples of TEM components. Waveguides and waveguide resonators are not TEM components.

TRANSMISSION LINES FOR FILTERS

Fig 12.33 shows three popular transmission lines used in transmission-line filters. The circular coaxial transmission line (coax) shown in Fig 12.33A consists of two concentric metal cylinders separated by dielectric (insulating) material.

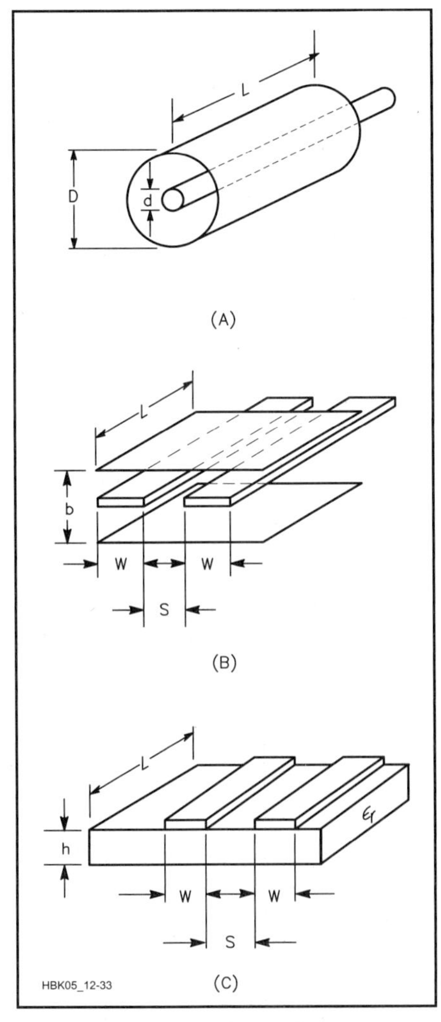

Fig 12.33 — Transmission lines. A: Coaxial line. B: Coupled stripline, which has two ground planes. C: Microstripline, which has only one ground plane.

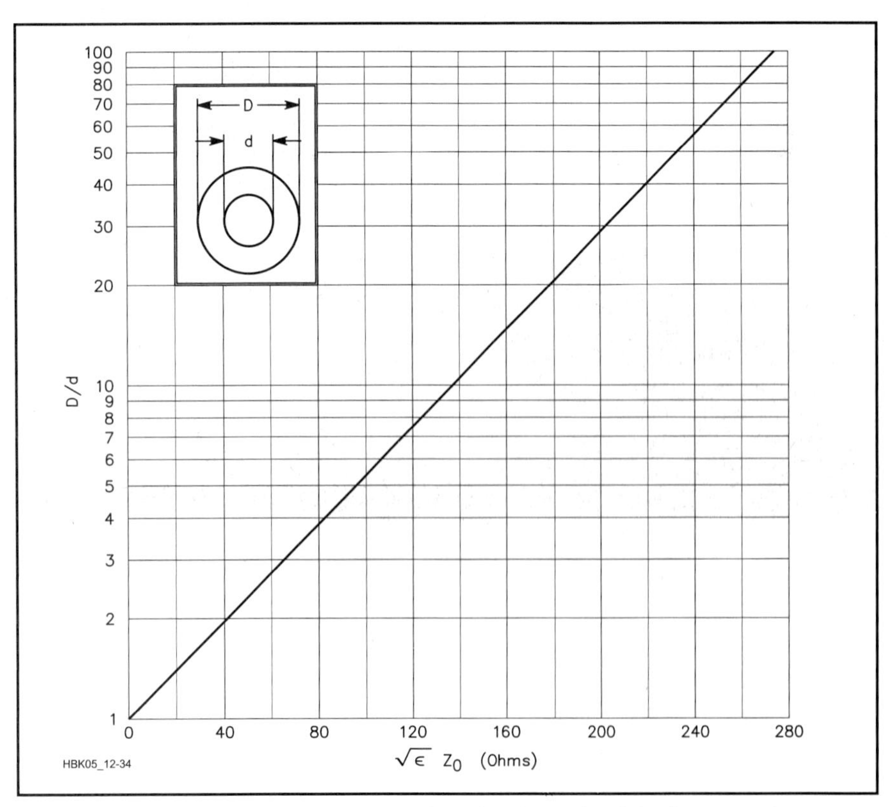

Fig 12.34 — Coaxial-line impedance varies with the ratio of the inner- and outer-conductor diameters. The dielectric constant, ε, is 1.0 for air and 2.32 for polyethylene.

Coaxial transmission line possesses a characteristic impedance given by:

$$Z_0 = \frac{138}{\sqrt{\varepsilon}} \log\left(\frac{D}{d}\right) \qquad (21)$$

A plot of Z_0 vs D/d is shown in **Fig 12.34**. At RF, Z_0 is an almost pure resistance. If the distant end of a section of coax is terminated in Z_0, then the impedance seen looking into the input end is also Z_0 at all frequencies. A terminated section of coax is shown in **Fig 12.35A**. If the distant end is not terminated in Z_0, the input impedance will be some other value. In Fig 12.35B the distant end is short-circuited and the length is less than $^1/_4$ λ. The input impedance is an inductive reactance as seen by the notation $+j$ in the equation in part B of the figure.

The input impedance for the case of the open-circuit distant end, is shown in Fig 12.35C. This case results in a capacitive reactance ($-j$). Thus, short sections of coaxial line (stubs) can replace the inductors and capacitors in an LC filter. Coax line inductive stubs usually have lower loss than their lumped counterparts.

X_L vs ℓ for shorted and open stubs is shown in **Fig 12.36**. There is an optimum value of Z_0 that yields lowest loss, given by

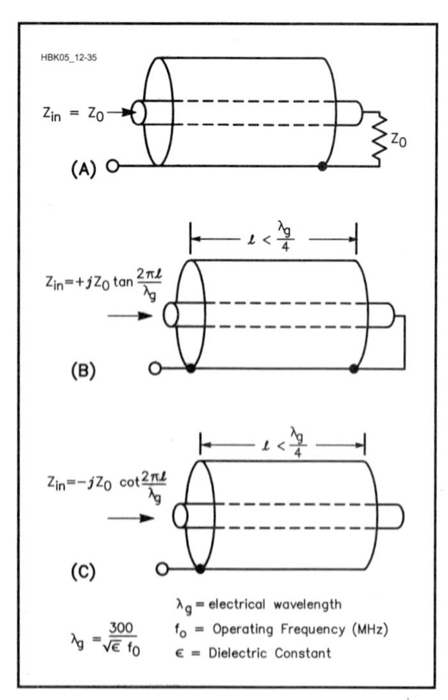

Fig 12.35 — Transmission line stubs. A: A line terminated in its characteristic impedance. B: A shorted line less than $^1/_4$-λ long is an *inductive* stub. C: An open line less than $^1/_4$-λ long is a *capacitive* stub.

HBK05_12-36

ℓ/λ_g	Shorted Stub X_L Ω	Open Stub X_C Ω
0	0	∞
0.05	16.2	154
0.10	36.3	68.8
0.125	50	50
0.15	68.8	36.3
0.20	154	16.2
0.25	∞	0

Fig 12.36 — Stub reactance for various lengths of transmission line. Values are for Z_0 = 50 Ω. For Z_0 = 100 Ω, double the tabulated values.

$$Z_0 = \frac{75}{\sqrt{\varepsilon}} \qquad (22)$$

If the dielectric is air, $Z_0 = 75$ Ω. If the dielectric is polyethylene ($\varepsilon = 2.32$) $Z_0 = 50$ Ω. This is the reason why polyethylene dielectric flexible coaxial cable is usually manufactured with a 50-Ω characteristic impedance.

The first transmission-line filters were built from sections of coaxial line. Their mechanical fabrication is expensive and it is difficult to provide electrical coupling between line sections. Fabrication difficulties are reduced by the use of shielded strip transmission line (stripline) shown in Fig 12.33B. The outer conductor of stripline consists of two flat parallel metal plates (ground planes) and the inner conductor is a thin metal strip. Sometimes the inner conductor is a round metal rod. The dielectric between ground planes and strip can be air or a low-loss plastic such as polyethylene. The outer conductors (ground planes or shields) are separated from each other by distance *b*.

Striplines can be easily coupled together by locating the strips near each other as shown in Fig 12.33B. Stripline Z_0 vs width (w) is plotted in **Fig 12.37**. Air-dielectric stripline technology is best for low bandwidth ($Q_L > 20$) band-pass filters.

The most popular transmission line is microstrip (unshielded stipline), shown in Fig 12.33C. It can be fabricated with standard printed-circuit processes and is the least expensive configuration. Unfortunately, microstrip is the lossiest of the three lines; therefore it is not suitable for narrow band-pass filters. In microstrip the outer conductor is a single flat metal ground-plane. The inner conductor is a thin metal strip separated from the ground-plane by a solid dielectric substrate. Typical substrates are 0.062-inch G-10 fiberglass ($\varepsilon = 4.5$) for the 50- MHz to 1-GHz frequency range and 0.031-inch Teflon ($\varepsilon = 2.3$) for frequencies above 1 GHz.

Conductor separation must be minimized or free-space radiation and unwanted coupling to adjacent circuits may become problems. Microstrip characteristic impedance and the effective dielectric constant (ε) are shown in **Fig 12.38**. Unlike coax and stripline, the effective dielectric constant is less than that of the substrate since a portion of the electromagnetic wave propagating along the microstrip "sees" the air above the substrate.

The least-loss characteristic impedance

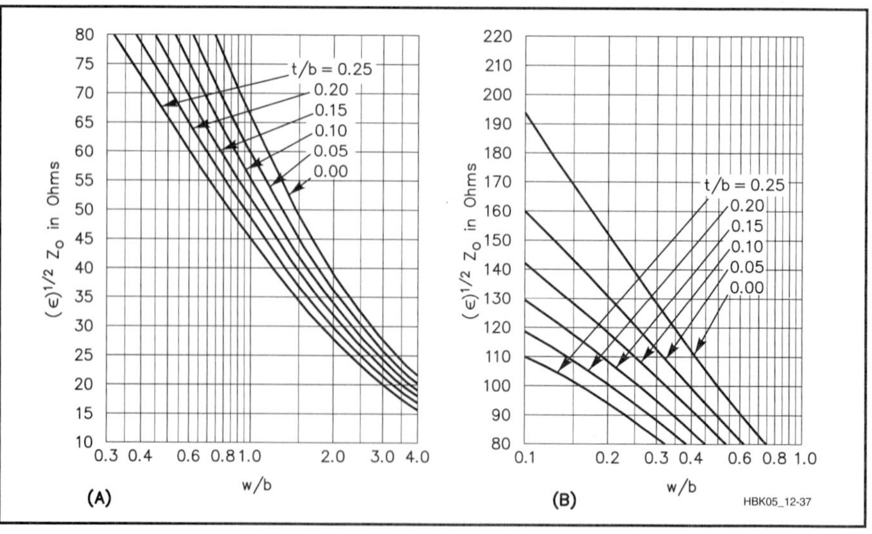

Fig 12.37 — The Z_0 of stripline varies with *w*, *b* and *t* (conductor thickness). See Fig 12.33B. The conductor thickness is *t* and the plots are normalized in terms of *t/b*.

$Z_0 \Omega$	ε=1 (AIR) W/h	ε=2.3 (RT/Duroid) W/h	$\sqrt{\varepsilon_e}$	ε=4.5 (G-10) W/h	$\sqrt{\varepsilon_e}$
25	12.5	7.6	1.4	4.9	2.0
50	5.0	3.1	1.36	1.8	1.85
75	2.7	1.6	1.35	0.78	1.8
100	1.7	0.84	1.35	0.39	1.75
	$\sqrt{\varepsilon}=1$				

HBK05_12-38

Fig 12.38 — Microstrip parameters (after H. Wheeler, *IEEE Transactions on MTT*, March 1965, p 132). ε_e is the effective ε.

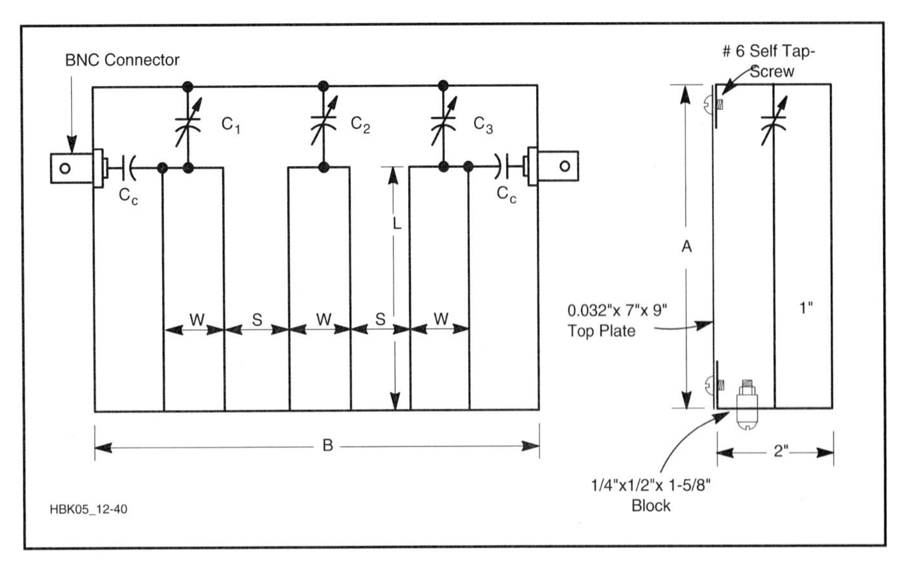

Fig 12.40 — This Butterworth filter is constructed in combline. It was originally discussed by R. Fisher in December 1968 *QST*.

Dimension	*52 MHz*	*146 MHz*	*222 MHz*
A	9"	7"	7"
B	7"	9"	9"
L	7 3/8"	6"	6"
S	1"	1 1/16"	1 3/8"
W	1"	1 5/8"	1 5/8"

Capacitance (pF)	*52 MHz*	*146 MHz*	*222 MHz*
C1	110	22	12
C2	135	30	15
C3	110	22	12
C_c	35	6.5	2.8
Q_L	10	29	36
Performance			
BW3 (MHz)	5.0	5.0	6.0
Loss (dB)	0.6	0.7	—

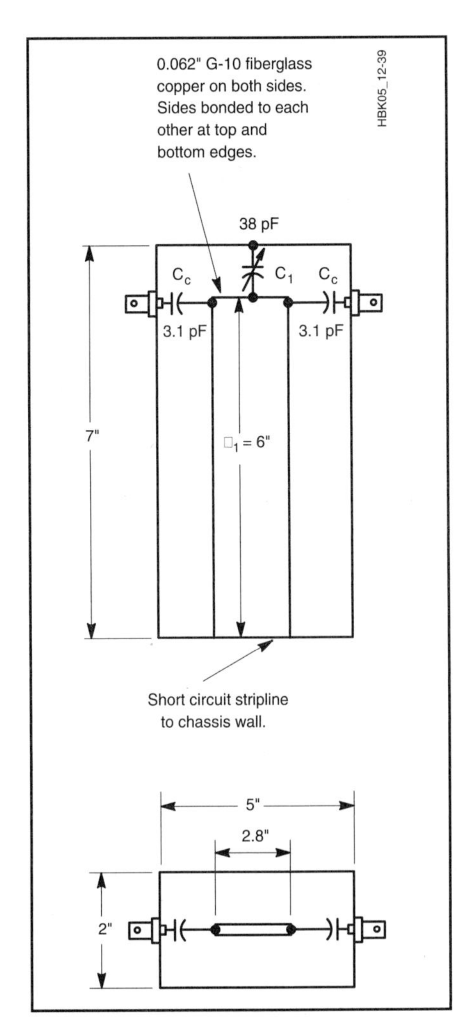

Fig 12.39 — This 146-MHz stripline band-pass filter has been measured to have a Q_L of 63 and a loss of approximately 1 dB.

for stripline and microstrip-lines is not 75 Ω as it is for coax. Loss decreases as line width increases, which leads to clumsy, large structures. Therefore, to conserve space, filter sections are often constructed from 50-Ω stripline or microstrip stubs.

Transmission-Line Band-Pass Filters

Band-pass filters can also be constructed from transmission-line stubs. At VHF the stubs can be considerably shorter than a quarter wavelength yielding a compact filter structure with less midband loss than its LC counterpart. The single-stage 146-MHz stripline band-pass filter shown in **Fig 12.39** is an example. This filter consists of a single inductive 50-Ω strip-line stub mounted into a 2 × 5 × 7-inch aluminum box. The stub is resonated at 146 MHz with the "APC" variable capacitor, C1. Coupling to the 50-Ω generator and load is provided by the coupling capacitors C_c. The measured performance of this filter is: f_o = 146 MHz, BW = 2.3 MHz (Q_L = 63) and midband loss = 1 dB.

Single-stage stripline filters can be coupled together to yield multistage filters. One method uses the capacitor coupled band-pass filter synthesis technique to design a 3-pole filter. Another method allows closely spaced inductive stubs to magnetically couple to each other. When the coupled stubs are grounded on the same side of the filter housing, the structure is called a "combline filter." Three examples of combline band-pass filters are shown in **Fig 12.40**. These filters are constructed in 2 × 7 × 9-inch chassis boxes.

Quarter-Wave Transmission-Line Filters

Fig 12.41 shows that when $\ell = 0.25\ \lambda_g$, the shorted-stub reactance becomes infinite. Thus, a 1/4-λ shorted stub behaves like a parallel-resonant LC circuit. Proper input and output coupling to a 1/4-λ resonator yields a practical band-pass filter. Closely spaced 1/4-λ resonators will couple together to form a multistage band-pass filter. When the resonators are grounded on opposite walls of the filter housing, the structure is called an "interdigital filter" because the resonators look like interlaced fingers. Two examples of 3-pole UHF interdigital filters are shown in **Fig 12.41**. Design graphs for round-rod interdigital filters are given in Ref 16. The 1/4-λ resonators may be tuned by physically changing their lengths or by tuning the screw opposite each rod.

If the short-circuited ends of two 1/4-λ resonators are connected to each other, the

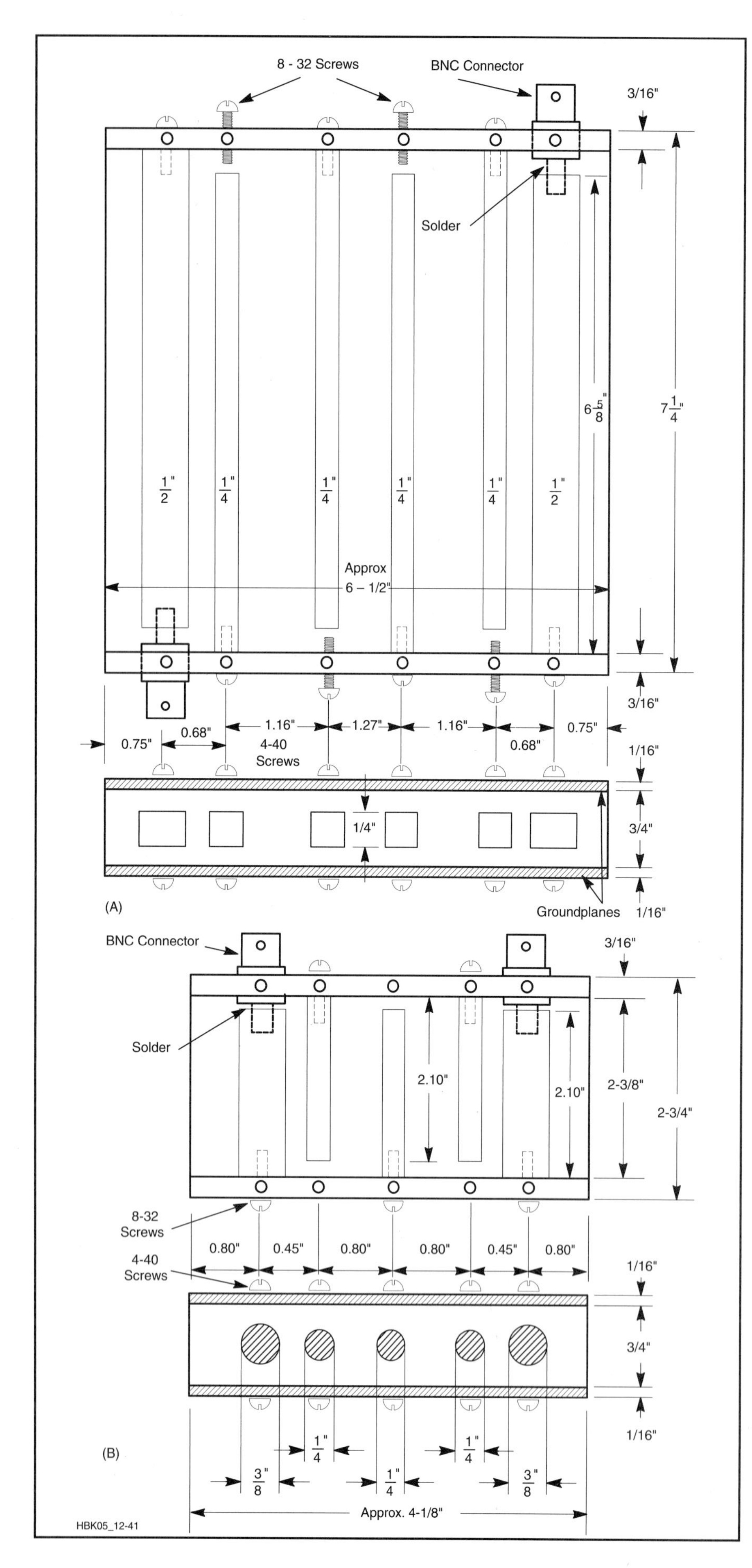

Fig 12.41 — These 3-pole Butterworth filters (upper: 432 MHz, 8.6 MHz bandwidth, 1.4 dB pass-band loss; lower: 1296 MHz, 110 MHz bandwidth, 0.4 dB pass-band loss) are constructed as interdigitated filters. The material is from R. E. Fisher, March 1968 *QST*.

resulting $^1/_2$-λ stub will remain in resonance, even when the connection to ground-plane is removed. Such a floating $^1/_2$-λ microstrip line, when bent into a U-shape, is called a "hairpin" resonator. Closely coupled hairpin resonators can be arranged to form multistage band-pass filters. Microstrip hairpin band-pass filters are popular above 1 GHz because they can be easily fabricated using photo-etching techniques. No connection to the ground-plane is required.

Transmission-Line Filters Emulating LC Filters

Low-pass and high-pass transmission-line filters are usually built from short sections of transmission lines (stubs) that emulate lumped LC reactances. Sometimes low-loss lumped capacitors are mixed with transmission-line inductors to form a hybrid component filter. For example, consider the 720-MHz, 3-pole microstrip low-pass filter shown in **Fig 12.42A** that emulates the LC filter shown in Fig 12.42B. C1 and C3 are replaced with 50-Ω open-circuit shunt stubs ℓ_C long. L2 is replaced with a short section of 100-Ω line ℓ_L long. The LC filter, Fig 12.42B, was designed for f_c = 720 MHz. Such a filter could be connected between a 432-MHz transmitter and antenna to reduce harmonic and spurious emissions. A reactance chart shows that X_C is 50 Ω, and the inductor reactance is 100 Ω at f_c. The microstrip version is constructed on G-10 fiberglass 0.062-inch thick, with ε = 4.5. Then, from Fig 12.38, *w* is 0.11 inch and $\ell_C = 0.125\ \lambda_g$ for the 50-Ω capacitive stubs. Also, from Fig 12.38, *w* is 0.024 inch and ℓ_L is 0.125 ℓ_g for the 100-Ω inductive line. The inductive line length is approximate because the far end is not a short circuit. ℓ_g is 300/(720)(1.75) = 0.238 m, or 9.37 inches. Thus ℓ_C is 1.1 inch and ℓ_L is 1.1 inch.

This microstrip filter exhibits about 20 dB of attenuation at 1296 MHz. Its response rises again, however, around 3 GHz. This is because the fixed-length transmission-line stubs change in terms of wavelength as the frequency rises. This particular filter was designed to eliminate third-harmonic energy near 1296 MHz from a 432-MHz transmitter and does a better job in this application than the Butterworth filter in Fig 12.41, which has spurious responses in the 1296-MHz band.

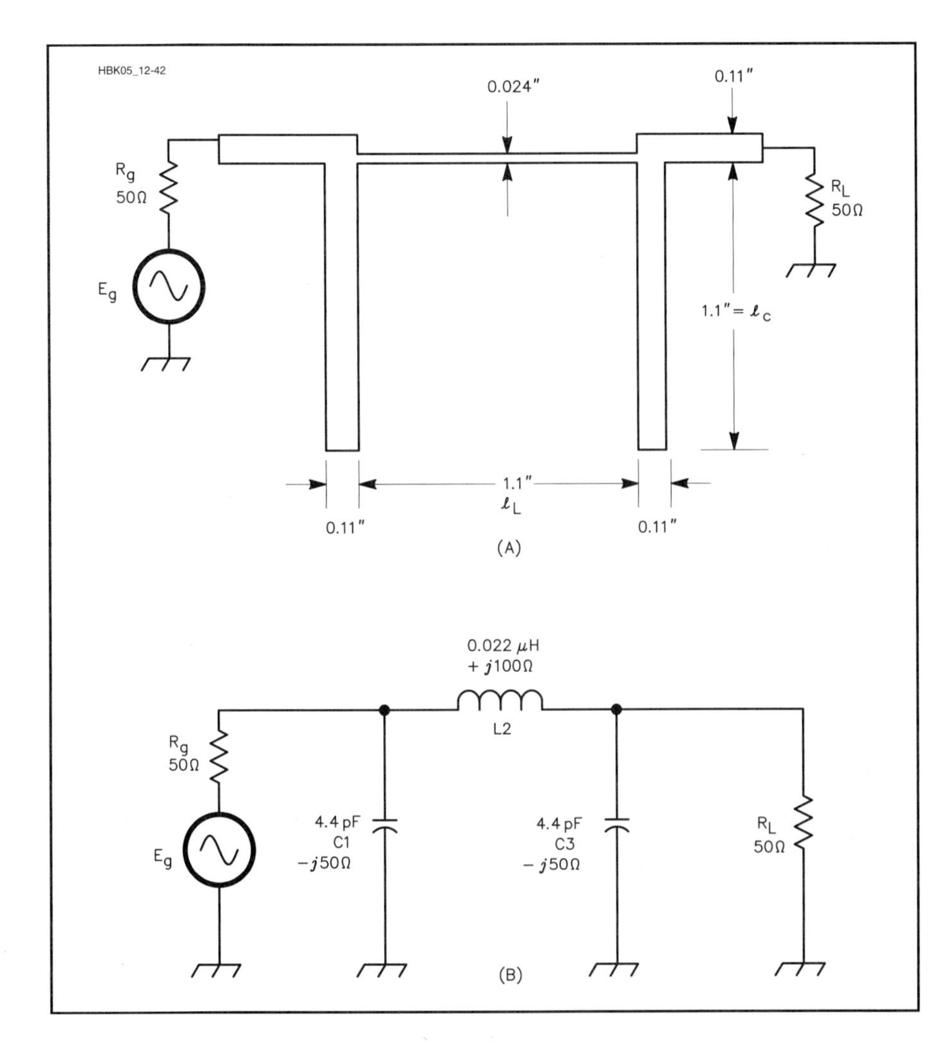

Fig 12.42 — A microstrip 3-pole emulated-Butterworth low-pass filter with a cutoff frequency of 720 MHz. A: Microstrip version built with G-10 fiberglass board (ε = 4.5, h = 0.062 inches). B: Lumped LC version of the same filter. To construct this filter with lumped elements very small values of L and C must be used and stray capacitance and inductance must be reduced to a tiny fraction of the component values.

Helical Resonators

Ever-increasing occupancy of the radio spectrum brings with it a parade of receiver overload and spurious responses. Overload problems can be minimized by using high-dynamic-range receiving techniques, but spurious responses (such as the image frequency) must be filtered out before mixing occurs. Conventional tuned circuits cannot provide the selectivity necessary to eliminate the plethora of signals found in most urban and many suburban neighborhoods. Other filtering techniques must be used.

Helical resonators are usually a better choice than ¼-λ cavities on 50, 144 and 222 MHz to eliminate these unwanted inputs. They are smaller and easier to build. In the frequency range from 30 to 100 MHz it is difficult to build high-Q inductors, and coaxial cavities are very large. In this frequency range the helical resonator is an excellent choice. At 50 MHz for example, a capacitively tuned, ¼-λ coaxial cavity with an unloaded Q of 3000 would be about 4 inches in diameter and nearly 5 ft long. On the other hand, a helical resonator with the same unloaded Q is about 8.5 inches in diameter and 11.3 inches long. Even at 432 MHz, where coaxial cavities are common, the use of helical resonators results in substantial size reductions.

The helical resonator was described by W1HR in a *QST* article as a coil surrounded by a shield, but it is actually a shielded, resonant section of helically wound transmission line with relatively high characteristic impedance and low axial propagation velocity. The electrical length is about 94% of an axial ¼-λ or 84.6°. One lead of the helical winding is connected directly to the shield and the other end is open circuited as shown in **Fig 12.43**. Although the shield may be any shape, only round and square shields will be considered here.

DESIGN

The unloaded Q of a helical resonator is determined primarily by the size of the shield. For a round resonator with a copper coil on a low-loss form, mounted in a copper shield, the unloaded Q is given by

$$Q_U = 50D\sqrt{f_0} \quad (23)$$

where

D = inside diameter of the shield, in inches

f_0 = frequency, in MHz.

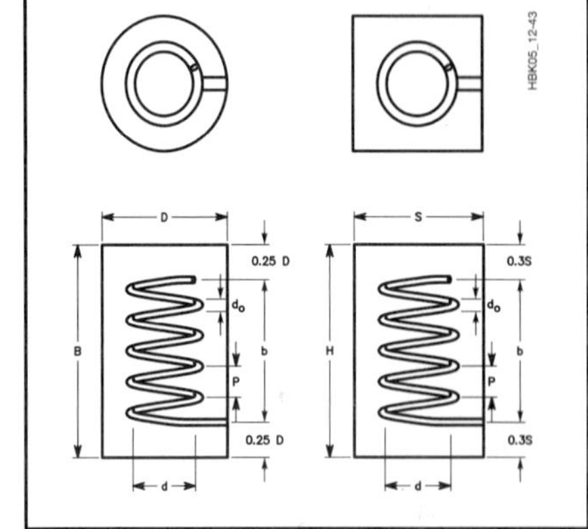

Fig 12.43 — Dimensions of round and square helical resonators. The diameter, D (or side, S) is determined by the desired unloaded Q. Other dimensions are expressed in terms of D or S (see text).

D is assumed to be 1.2 times the width of one side for square shield cans. This formula includes the effects of losses and imperfections in practical materials. It yields values of unloaded Q that are easily attained in practice. Silver plating the shield and coil increases the unloaded Q

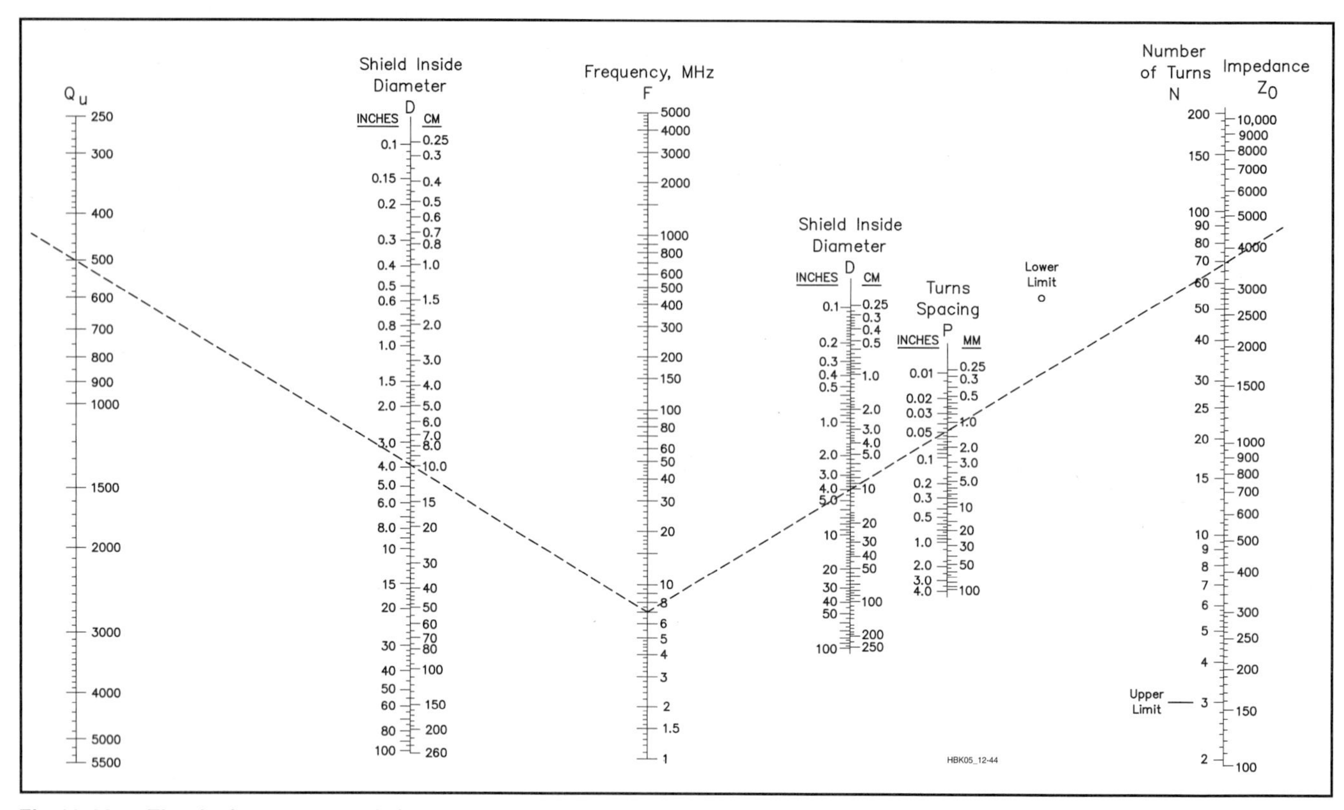

Fig 12.44 — The design nomograph for round helical resonators starts by selecting Q_U and the required shield diameter. A line is drawn connecting these two values and extended to the frequency scale (example here is for a shield of about 3.8 inches and Q_U of 500 at 7 MHz). Finally the number of turns, N, winding pitch, P, and characteristic impedance, Z_0, are determined by drawing a line from the frequency scale through selected shield diameter (but this time to the scale on the right-hand side. For the example shown, the dashed line shows P ≈ 0.047 inch, N = 70 turns, and Z_n = 3600 Ω).

by about 3% over that predicted by the equation. At VHF and UHF, however, it is more practical to increase the shield size slightly (that is, increase the selected Q_U by about 3% before making the calculation). The fringing capacitance at the open-circuit end of the helix is about 0.15 D pF (that is, approximately 0.3 pF for a shield 2 inches in diameter). Once the required shield size has been determined, the total number of turns, N, winding pitch, P and characteristic impedance, Z_0, for round and square helical resonators with air dielectric between the helix and shield, are given by:

$$N = \frac{1908}{f_0 D} \quad (24A)$$

$$P = \frac{f_0 D^2}{2312} \quad (24B)$$

$$Z_0 = \frac{99{,}000}{f_0 D} \quad (24C)$$

$$N = \frac{1590}{f_0 S} \quad (24D)$$

$$P = \frac{f_0 S^2}{1606} \quad (24E)$$

$$Z_0 = \frac{82{,}500}{f_0 S} \quad (24F)$$

In these equations, dimensions D and S are in inches and f_0 is in megahertz. The design nomograph for round helical resonators in **Fig 12.44** is based on these formulas.

Although there are many variables to consider when designing helical resonators, certain ratios of shield size to length and coil diameter to length, provide optimum results. For helix diameter, d = 0.55 D or d = 0.66 S. For helix length, b = 0.825D or b = 0.99S. For shield length, B = 1.325 D and H = 1.60 S.

Fig 12.45 simplifies calculation of these dimensions. Note that these ratios result in a helix with a length 1.5 times its diameter, the condition for maximum Q. The shield is about 60% longer than the helix — although it can be made longer — to completely contain the electric field at the top of the helix and the magnetic field at the bottom.

The winding pitch, P, is used primarily to determine the required conductor size. Adjust the length of the coil to that given by the equations during construction. Conductor size ranges from 0.4 P to 0.6 P for both round and square resonators and are plotted graphically in **Fig 12.46**.

Obviously, an area exists (in terms of frequency and unloaded Q) where the designer must make a choice between a conventional cavity (or lumped LC circuit) and a helical resonator. The choice is affected by physical shape at higher frequencies. Cavities are long and relatively small in diameter, while the length of a helical resonator is not much greater than its diameter. A second consideration is that point where the winding pitch, P, is less than the radius of the helix (otherwise the structure tends to be nonhelical). This condition occurs when the helix has fewer than three turns (the "upper limit" on the design nomograph of Fig 12.44).

CONSTRUCTION

The shield should not have any seams parallel to the helix axis to obtain as high an unloaded Q as possible. This is usually not a problem with round resonators because large-diameter copper tubing is used

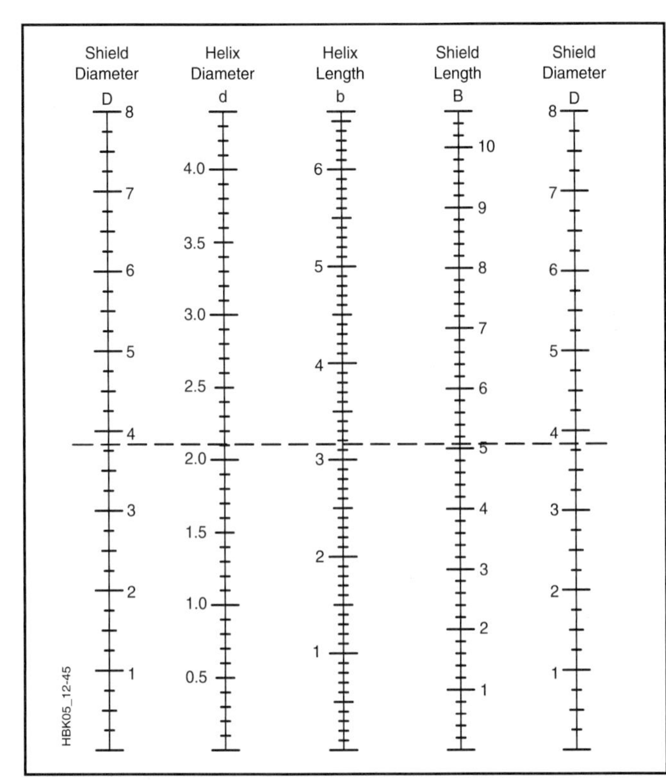

Fig 12.45 — The helical resonator is scaled from this design nomograph. Starting with the shield diameter, the helix diameter, d, helix length, b, and shield length, B, can be determined with this graph. The example shown has a shield diameter of 3.8 inches. This requires a helix mean diameter of 2.1 inches, helix length of 3.1 inches, and shield length of 5 inches.

Fig 12.46 — This chart provides the design information of helix conductor size vs winding pitch, P. For example, a winding pitch of 0.047 inch results in a conductor diameter between 0.019 and 0.028 inch (#22 or #24 AWG).

for the shield, but square resonators require at least one seam and usually more. The effect on unloaded Q is minimum if the seam is silver soldered carefully from one end to the other.

Results are best when little or no dielectric is used inside the shield. This is usually no problem at VHF and UHF because the conductors are large enough that a supporting coil form is not required. The lower end of the helix should be soldered to the nearest point on the inside of the shield.

Although the external field is minimized by the use of top and bottom shield covers, the top and bottom of the shield may be left open with negligible effect on frequency or unloaded Q. Covers, if provided, should make electrical contact with the shield. In those resonators where the helix is connected to the bottom cover, that cover must be soldered solidly to the shield to minimize losses.

TUNING

A carefully built helical resonator designed from the nomograph of Fig 12.44 will resonate very close to the design frequency. Slightly compress or expand the helix to adjust resonance over a small range. If the helix is made slightly longer than that called for in Fig 12.45, the resonator can be tuned by pruning the open end of the coil. However, neither of these methods is recommended for wide frequency excursions because any major deviation in helix length will degrade the unloaded Q of the resonator.

Most helical resonators are tuned by means of a brass tuning screw or high-quality air-variable capacitor across the open end of the helix. Piston capacitors also work well, but the Q of the tuning capacitor should ideally be several times the unloaded Q of the resonator. Varactor diodes have sometimes been used where remote tuning is required, but varactors can generate unwanted harmonics and other spurious signals if they are excited by strong, nearby signals.

When a helical resonator is to be tuned by a variable capacitor, the shield size is based on the chosen unloaded Q at the operating frequency. Then the number of turns, N *and* the winding pitch, P, are based on resonance at 1.5 f_0. Tune the resonator to the desired operating frequency, f_0.

INSERTION LOSS

The insertion loss (dissipation loss), I_L, in decibels, of all single-resonator circuits is given by

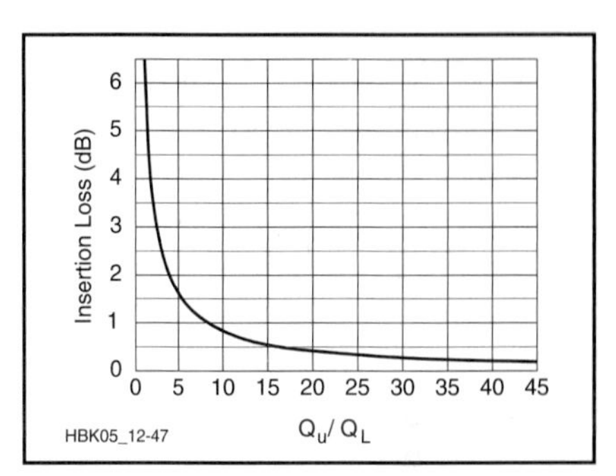

Fig 12.47 — The ratio of loaded (Q_L) to unloaded (Q_U) Q determines the insertion loss of a tuned resonant circuit.

$$I_L = 20 \log_{10} \left(\frac{1}{1 - \frac{Q_L}{Q_U}} \right) \qquad (25)$$

where

Q_L = loaded Q

Q_U = unloaded Q

This is plotted in **Fig 12.47**. For the most practical cases ($Q_L > 5$), this can be closely approximated by $I_L \approx 9.0 (Q_L/Q_U)$ dB. The selection of Q_L for a tuned circuit is dictated primarily by the required selectivity of the circuit. However, to keep dissipation loss to 0.5 dB or less (as is the case for low-noise VHF receivers), the unloaded Q must be at least 18 times the Q_L.

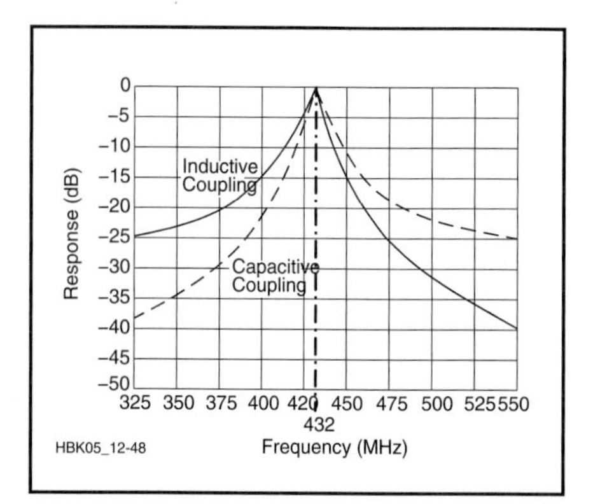

Fig 12.48 — This response curve for a single-resonator 432-MHz filter shows the effects of capacitive and inductive input/output coupling. The response curve can be made symmetrical on each side of resonance by combining the two methods (inductive input and capacitive output, or vice versa).

COUPLING

Signals are coupled into and out of helical resonators with inductive loops at the bottom of the helix, direct taps on the coil or a combination of both. Although the correct tap point can be calculated easily, coupling by loops and probes must be determined experimentally.

The input and output coupling is often provided by probes when only one resonator is used. The probes are positioned on opposite sides of the resonator for maximum isolation. When coupling loops are

Unless otherwise specified, values of R are in ohms, C is in farads, F in hertz and ω in radians per second. Calculations shown here were performed on a scientific calculator.

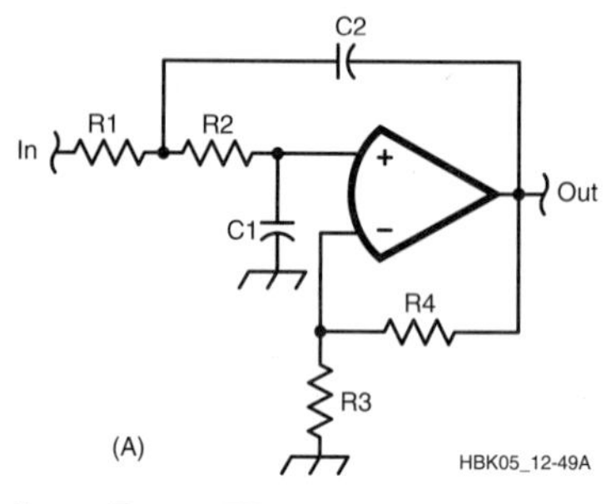

Low-Pass Filter

$$C_1 \le \frac{\left[a^2 + 4(K-1)\right]C_2}{4}$$

$$R_1 = \frac{2}{\left[aC_2 + \sqrt{\left[a^2 + 4(K-1)\right]C_2^{\,2} - 4C_1C_2}\right]\omega_C}$$

$$R_2 = \frac{1}{C_1C_2R_1\omega_C^{\,2}}$$

$$R_3 = \frac{K(R_1 + R_2)}{K-1} \quad (K > 1)$$

$$R_4 = K(R_1 + R_2)$$

where
K = gain
f_c = –3 dB cutoff frequency
$\omega_c = 2\pi f_c$
C_2 = a standard value near $10/f_c$ (in µF)

Note: For unity gain, short R4 and omit R3.

Example:
a = 1.414 (see table, one stage)
K = 2
f = 2700 Hz
ω_c = 16,964.6 rads/sec
C_2 = 0.0033 µF
C1 ≤ 0.00495 µF (use 0.0050 µF)
R1 ≤ 25,265.2 Ω (use 24 kΩ)
R2 = 8,420.1 Ω (use 8.2 kΩ)
R3 = 67,370.6 Ω (use 68 kΩ)
R4 = 67,370.6 Ω (use 68 kΩ)

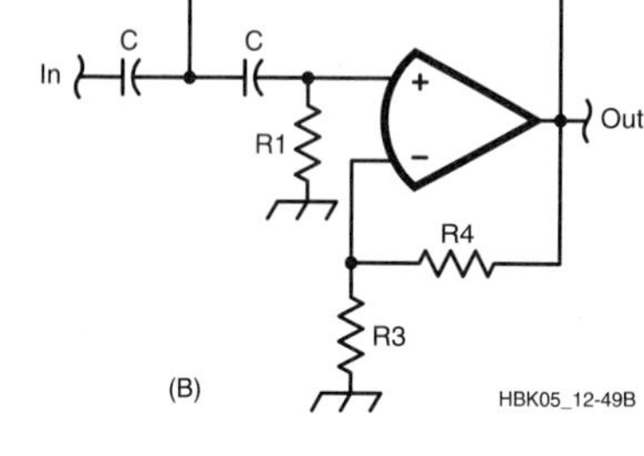

High-Pass Filter

$$R_1 = \frac{4}{\left[a + \sqrt{a^2 + 8(K-1)}\right]\omega_C\, C}$$

$$R_2 = \frac{1}{\omega_C^{\,2}C^2R_1}$$

$$R_3 = \frac{KR_1}{K-1} \quad (K > 1)$$

$$R_4 = KR_1$$

where
K = gain
f_c = –3 dB cutoff frequency
$\omega_c = 2\pi f_c$
C = a standard value near $10/f_c$ (in µF)

Note: For unity gain, short R4 and omit R3.

Example:
a = 0.765 (see table, first of two stages)
K = 4
f = 250 Hz
ω_c = 1570.8
C = 0.04 µF (use 0.039 µF)
R1 = 11,123.2 Ω (use 11 kΩ)
R2 = 22,722 Ω (use 22 kΩ)
R3 = 14,830.9 Ω (use 15 kΩ)
R4 = 44,492.8 Ω (use 47 kΩ)

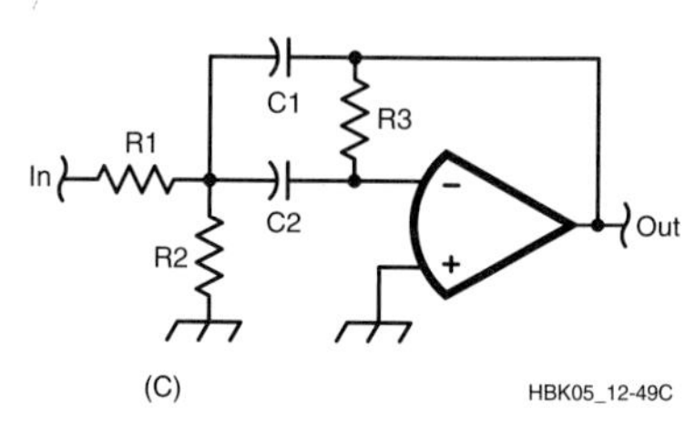

Band-Pass Filter

Pick K, Q, $\omega_o = 2\pi f_c$
where f_c = center freq.
Choose C
Then

$$R1 = \frac{Q}{K_0\omega_0 C}$$

$$R2 = \frac{Q}{\left(2Q^2 - K_0\right)\omega_0 C}$$

$$R3 = \frac{2Q}{\omega_0 C}$$

Example:
K = 2, f_o = 800 Hz, Q = 5 and C = 0.022 µF
R1 = 22.6 kΩ (use 22 kΩ)
R2 = 942 Ω (use 910 Ω)
R3 = 90.4 kΩ (use 91 kΩ)

Fig 12.49 — Equations for designing a low-pass RC active audio filter are given at A. B, C and D show design information for high-pass, band-pass and band-reject filters, respectively. All of these filters will exhibit a Butterworth response. Values of K and Q should be less than 10.

used, the plane of the loop should be perpendicular to the axis of the helix and separated a small distance from the bottom of the coil. For resonators with only a few turns, the plane of the loop can be tilted slightly so it is parallel with the slope of the adjacent conductor.

Helical resonators with inductive coupling (loops) exhibit more attenuation to signals above the resonant frequency (as compared to attenuation below resonance), whereas resonators with capacitive coupling (probes) exhibit more attenuation below the passband, as shown for a typical 432-MHz resonator in **Fig 12.48**. Consider this characteristic when choosing a coupling method. The passband can be made more symmetrical by using a combination of coupling methods (inductive input and capacitive output, for example).

If more than one helical resonator is required to obtain a desired band-pass characteristic, adjacent resonators may be coupled through apertures in the shield wall between the two resonators. Unfortunately, the size and location of the aperture must be found empirically, so this method of coupling is not very practical unless you're building a large number of identical units.

Since the loaded Q of a resonator is determined by the external loading, this must be considered when selecting a tap (or position of a loop or probe). The ratio of this external loading, R_b, to the characteristic impedance, Z_0, for a $^1/_4$-λ resonator is calculated from:

$$K = \frac{R_b}{Z_0} = 0.785\left(\frac{1}{Q_L} - \frac{1}{Q_U}\right) \qquad (26)$$

Even when filters are designed and built properly, they may be rendered totally ineffective if not installed properly. Leakage around a filter can be quite high at VHF and UHF, where wavelengths are short. Proper attention to shielding and good grounding is mandatory for minimum leakage. Poor coaxial cable shield connection into and out of the filter is one of the greatest offenders with regard to filter leakage. Proper dc-lead bypassing throughout the receiving system is good practice, especially at VHF and above. Ferrite beads placed over the dc leads may help to reduce leakage. Proper filter termination is required to minimize loss.

Most VHF RF amplifiers optimized for noise figure do not have a 50-Ω input impedance. As a result, any filter attached to the input of an RF amplifier optimized for noise figure will not be properly terminated and filter loss may rise substantially. As this loss is directly added to the RF amplifier noise figure, carefully choose and place filters in the receiver.

ACTIVE FILTERS

Passive HF filters are made from combinations of inductors and capacitors. These may be used at low frequencies, but the inductors often become a limiting factor because of their size, weight, cost and losses. The active filter is a compact, low-cost alternative made with op amps, resistors and capacitors. They often occupy a fraction of the space required by an LC filter. While active filters have been traditionally used at low and audio frequencies, modern op amps with small-signal bandwidths that exceed 1 GHz have extended their range into MF and HF.

Active filters can perform any common filter function: low pass, high pass, bandpass, band reject and all pass (used for phase or time delay). Responses such as Butterworth, Chebyshev, Bessel and elliptic can be realized. Active filters can be designed for gain, and they offer excellent stage-to-stage isolation.

Despite the advantages, there are also some limitations. They require power, and performance may be limited by the op amp's finite input and output levels, gain and bandwidth. While LC filters can be designed for high-power applications, active filters usually are not.

The design equations for various filters are shown in **Fig 12.49**. **Fig 12.50** shows a typical application of a two-stage, bandpass filter. A two-stage filter is considered the minimum acceptable for CW, while three or four stages will prove more effective under some conditions of noise and interference.

CRYSTAL-FILTER EVALUATION

Crystal filters, such as those described earlier in this chapter, are often constructed of surplus crystals or crystals

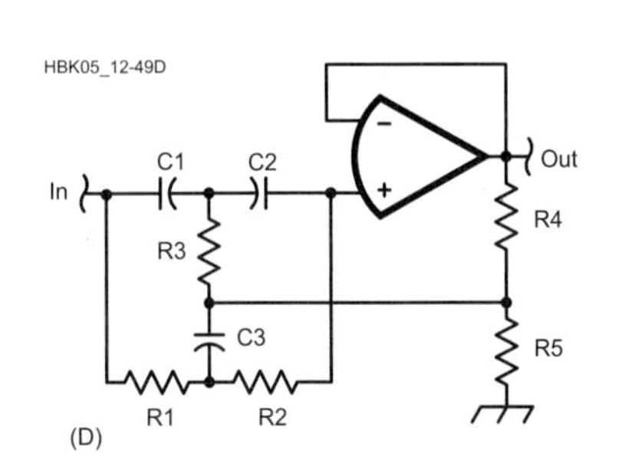

Band-Reject Filter

$$F_0 = \frac{1}{2\pi R1C1}$$

$$K = 1 - \frac{1}{4Q}$$

$$R >> (1-K)R1$$

where

$$C1 = C2 = \frac{C3}{2} = \frac{10\,\mu F}{f_0}$$

R1 = R2 = 2R3
R4 = (1 – K)R
R5 = K × R

Example:
f_0 = 500 Hz, Q=10
K = 0.975
C1 = C2 = 0.02 μF (or use 0.022 μF)
C3 = 0.04 μF (or use 0.044 μF)
R1 = R2 = 15.92 kΩ (use 15 kΩ)
R3 = 7.96 kΩ (use 8.2 kΩ)
R >> 398 Ω (390 Ω)
R4 = 9.95 Ω (use 10 Ω)
R5 = 388.1 Ω (use 390 Ω)

Factor "a" for Low- and High-Pass Filters

No. of Stages	*Stage 1*	*Stage 2*	*Stage 3*	*Stage 4*
1	1.414	—	—	—
2	0.765	1.848	—	—
3	0.518	1.414	1.932	—
4	0.390	1.111	1.663	1.962

These values are truncated from those of Appendix C of Ref 21, for even-order Butterworth filters.

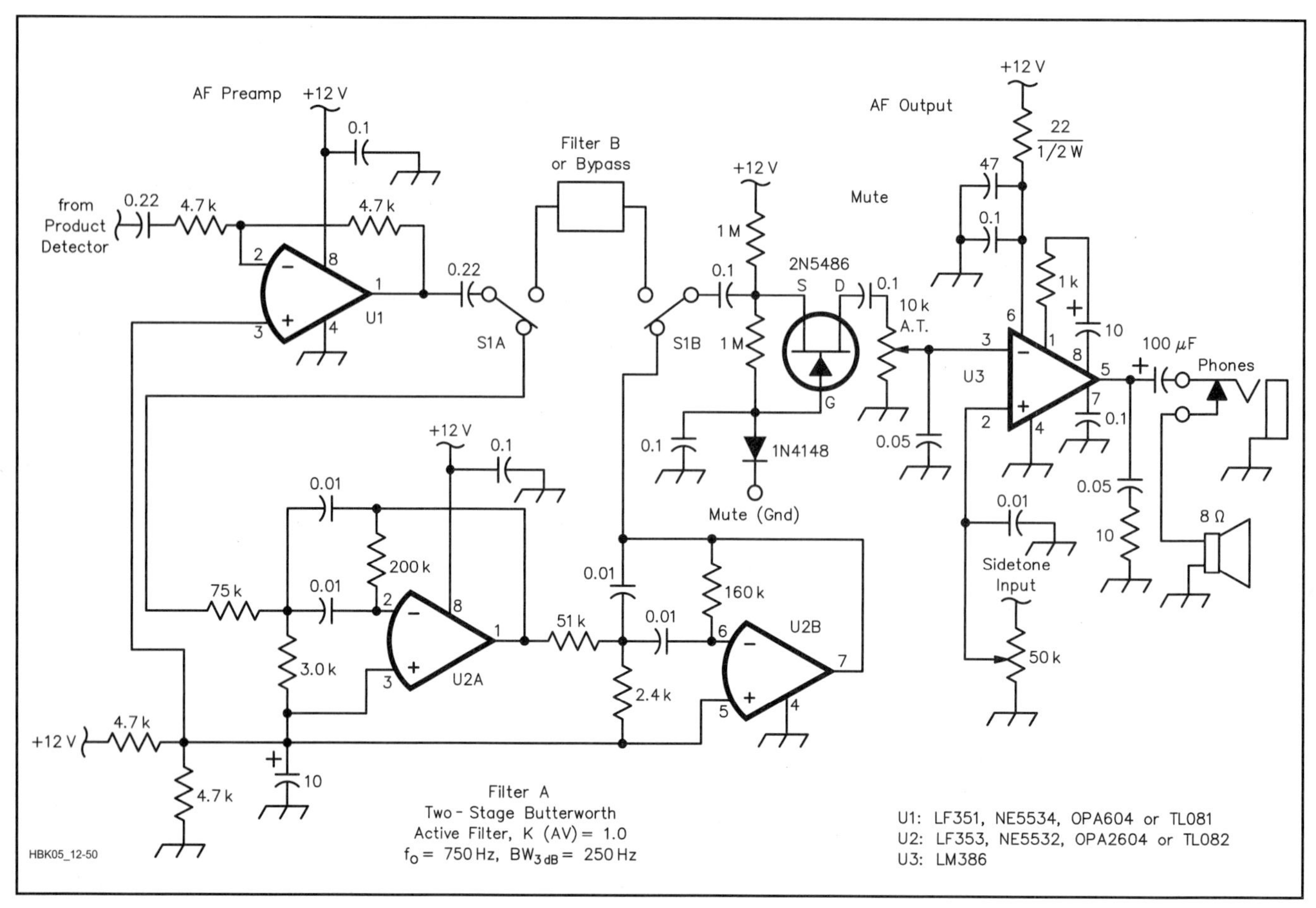

Fig 12.50 — Typical application of a two-stage active filter in the audio chain of a QRP CW tranceiver. The filter can be bypassed, or another filter can be switched in by S1.

whose characteristics are not exactly known. Randy Henderson, WI5W, developed a swept frequency generator for testing these filters. It was first described in March 1994 *QEX*. This test instrument adds to the ease and success in quickly building filters from inexpensive microprocessor crystals.

A template, containing additional information, is available on the CD included with this book.

An Overview

The basic setup is shown in **Fig 12.51A**. The VCO is primarily a conventional LC-tuned Hartley oscillator with its frequency tuned over a small range by a varactor diode (MV2104 in part B of the figure). Other varactors may be used as long as the capacitance specifications aren't too different. Change the 5-pF coupling capacitor to expand the sweep width if desired.

The VCO signal goes through a buffer amplifier to the filter under test. The filter is followed by a wide-bandwidth amplifier and then a detector. The output of the detector is a rectified and filtered signal. This varying dc voltage drives the vertical input of an oscilloscope. At any particular time, the deflection and sweep circuitry commands the VCO to "run at this frequency." The same deflection voltage causes the oscilloscope beam to deflect left or right to a position corresponding to the frequency.

Any or all of these circuits may be eliminated by the use of appropriate commercial test equipment. For example, a commercial sweep generator would eliminate the need for everything but the wide-band amplifier and detector. Motorola, Mini-Circuits Labs and many others sell devices suitable for the wide-band amplifiers and detector.

The generator/detector system covers approximately 6 to 74 MHz in three ranges. Each tuning range uses a separate RF oscillator module selected by switch S1. The VCO output and power-supply input are multiplexed on the "A" lead to each oscillator. The tuning capacitance for each VCO is switched into the appropriate circuit by a second set of contacts on S1. C_T is the coarse tuning adjustment for each oscillator module.

Two oscillator coils are wound on PVC plastic pipe. The third, for the highest frequency range, is self-supporting #14 copper wire. Although PVC forms with Super Glue dope may not be "state of the art" technology, frequency stability is completely adequate for this instrument.

The oscillator and buffer stage operate at low power levels to minimize frequency drift caused by component heating. Crystal filters cause large load changes as the frequency is swept in and out of the passband. These large changes in impedance tend to "pull" the oscillator frequency and cause inaccuracies in the passband shape depicted by the oscilloscope. Therefore a buffer amplifier is a necessity. The wideband amplifier in **Fig 12.52** is derived from one in ARRL's *Solid State Design for the Radio Amateur*.

S2 selects a 50-Ω, 10-dB attenuator in the input line. When the attenuator is in the line, it provides a better output match for the filter under test. The detector uses some forward bias for D2. A simple unbiased diode detector would offer about 50 dB of dynamic range. Some dc bias increases the dynamic range to almost 70 dB. D3, across the detector output (the

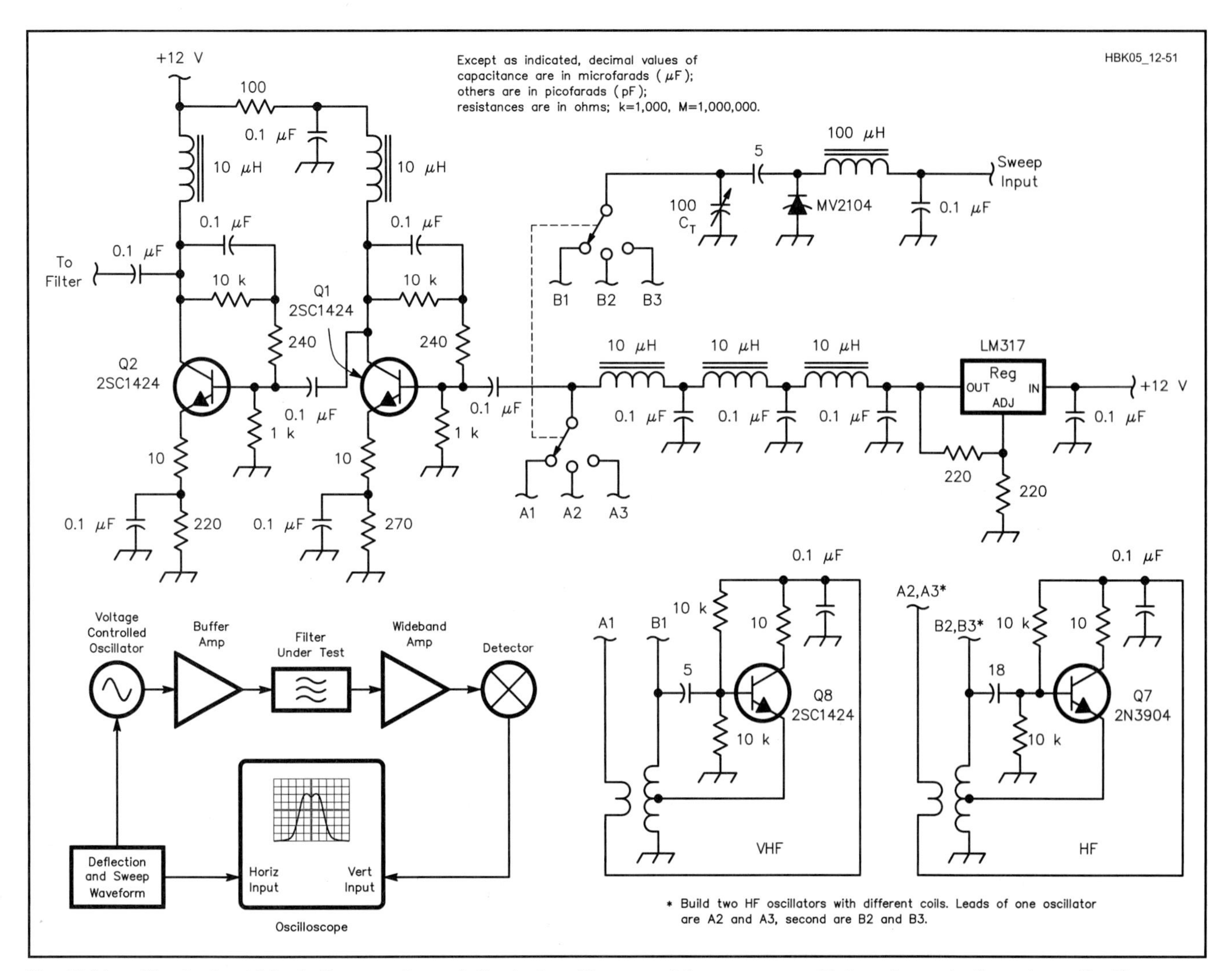

Fig 12.51 — The test set block diagram, lower left, starts with a swept frequency oscillator, shown in the schematic. If a commercial swept-frequency oscillator is available, it can be substituted for the circuit shown.

scope input), increases the vertical-amplifier sensitivity while compressing or limiting the response to high-level signals. With this arrangement, high levels of attenuation (low-level signals) are easier to observe and low attenuation levels are still visible on the CRT. The diode only kicks in to provide limiting at higher signal levels.

The horizontal-deflection sweep circuit uses a dual op-amp IC (see **Fig 12.53**). One section is an oscillator; the other is an integrator. The integrator output changes linearly with time, giving a uniform brightness level as the trace is moved from side to side. Increasing C1 decreases the sweep rate. Increasing C2 decreases the slope of the output waveform ramp.

Operation

The CRT is swept in both directions, left to right and right to left. The displayed curve is a result of changes in frequency, not time. Therefore it is unnecessary to incorporate the usual right-to-left, snap-back and retrace blanking used in oscilloscopes.

S3 in Fig 12.53 disables the automatic sweep function when opened. This permits manual operation. Use a frequency counter to measure the VCO output, from which bandwidth can be calculated. Turn

Table 12.5
VCO Coils

Coil	*Inside Diameter (inches)*	*Length (inches)*	*Turns, Wire*	*Inductance (μH)*
large	0.85	1.1	18 t, #28	5.32
medium	0.85	0.55	7 t, #22	1.35
small	0.5	0.75	5 t, #14	0.27

The two larger coils are wound on 3/4-inch PVC pipe and the smaller one on a 1/2-inch drill bit. Tuning coverage for each oscillator is obtained by squeezing or spreading the turns before gluing them in place. The output windings connected to A1, A2 and A3 are each single turns of #14 wire spaced off the end of the tapped coils. The taps are approximate and 25 to 30% of the full winding turns — up from the cold (ground) end of the coils.

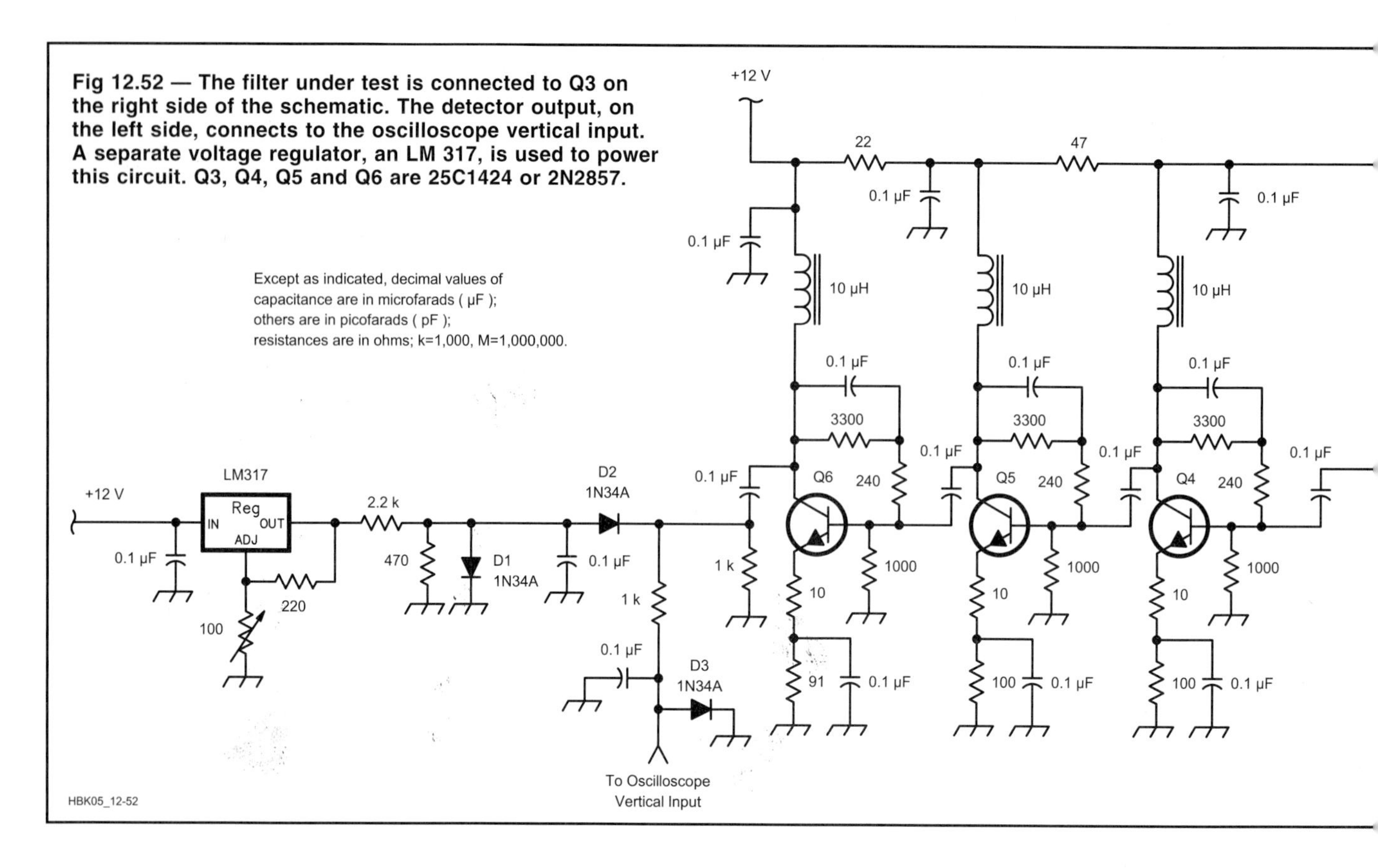

Fig 12.52 — The filter under test is connected to Q3 on the right side of the schematic. The detector output, on the left side, connects to the oscilloscope vertical input. A separate voltage regulator, an LM 317, is used to power this circuit. Q3, Q4, Q5 and Q6 are 25C1424 or 2N2857.

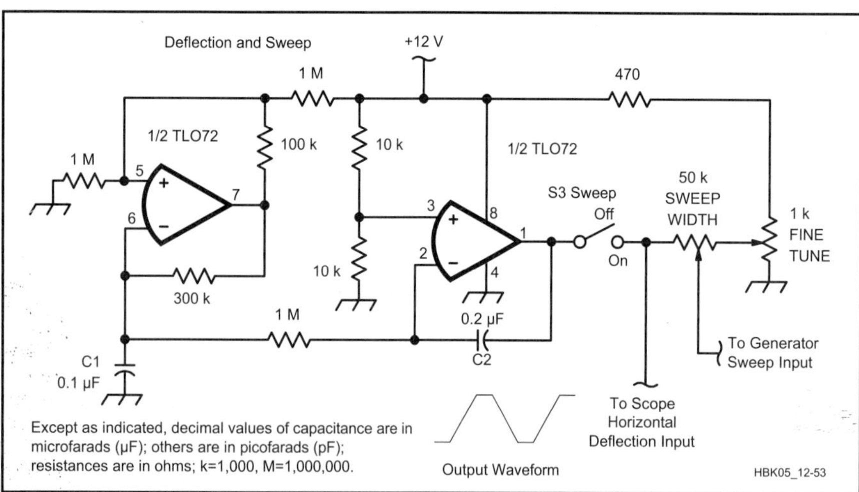

Fig 12.53 — The sweep generator provides both an up and down sweep voltage (see text) for the swept frequency generator and the scope horizontal channel.

the fine-tune control to position the CRT beam at selected points of the pass-band curve. The difference in frequency readings is the bandwidth at that particular point or level of attenuation.

Substitution of a calibrated attenuator for the filter under test can provide reference readings. These reference readings may be used to calibrate an otherwise uncalibrated scope vertical display in dB.

The buffer amplifier shown here is set up to drive a 50-Ω load, and the wide-bandwidth amplifier input impedance is about 50 Ω. If the filter is not a 50-Ω unit, however, various methods can be used to accommodate the difference. For example, a transformer may be used for widely differing impedance levels, whereas a minimum-loss resistive pad may be preferable where impedance levels differ by a factor of approximately 1.5 or less — presuming some loss is acceptable.

References

A. Ward, "Monolithic Microwave Integrated Circuits," Feb 1987 *QST*, pp 23-29.

Z. Lau, "A Logarithmic RF Detector for Filter Tuning," Oct 1988 *QEX*, pp 10-11.

BAND-PASS FILTERS FOR 144 OR 222 MHZ

Spectral purity is necessary during transmitting. Tight filtering in a receiving system ensures the rejection of out-of-band signals. Unwanted signals that lead to receiver overload and increased intermodulation-distortion (IMD) products result in annoying in-band "birdies." One solution is the double-tuned band-pass filters shown in **Fig 12.54.** They were designed by Paul Drexler, WB3JYO. Each includes a resonant trap coupled between the resonators to provide increased rejection of undesired frequencies.

Many popular VHF conversion schemes use a 28-MHz intermediate frequency (IF), yet proper filtering of the image frequency is often overlooked in amateur designs. The low-side injection

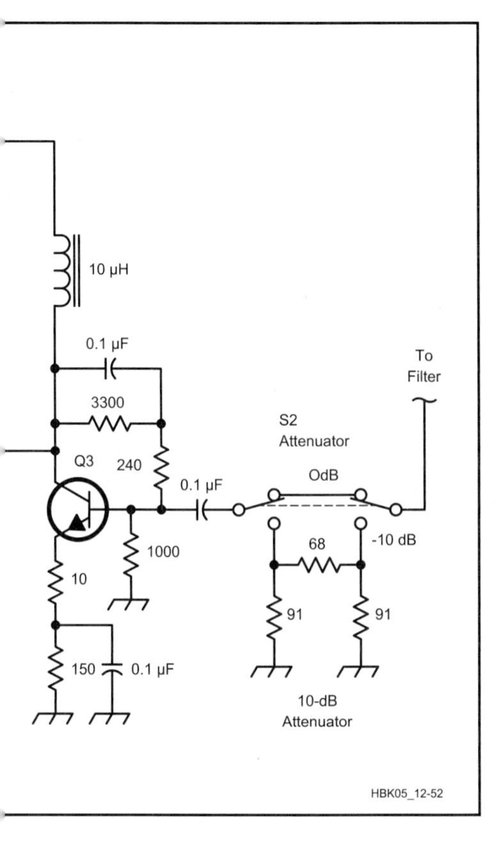

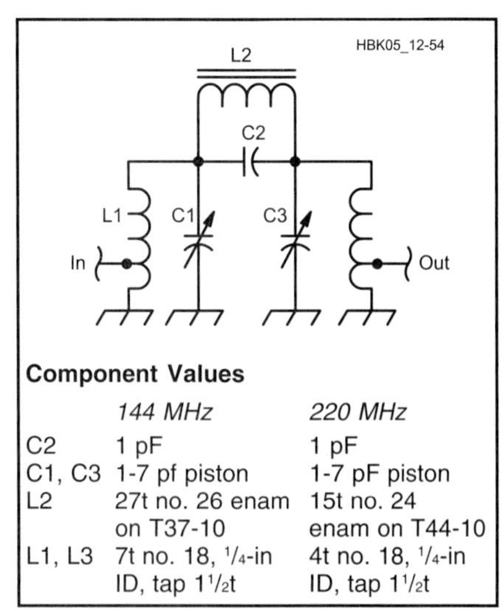

Component Values

	144 MHz	220 MHz
C2	1 pF	1 pF
C1, C3	1-7 pf piston	1-7 pF piston
L2	27t no. 26 enam on T37-10	15t no. 24 enam on T44-10
L1, L3	7t no. 18, 1/4-in ID, tap 1 1/2t	4t no. 18, 1/4-in ID, tap 1 1/2t

Fig 12.54 — Schematic of the band-pass filter. Components must be chosen to work with the power level of the transmitter.

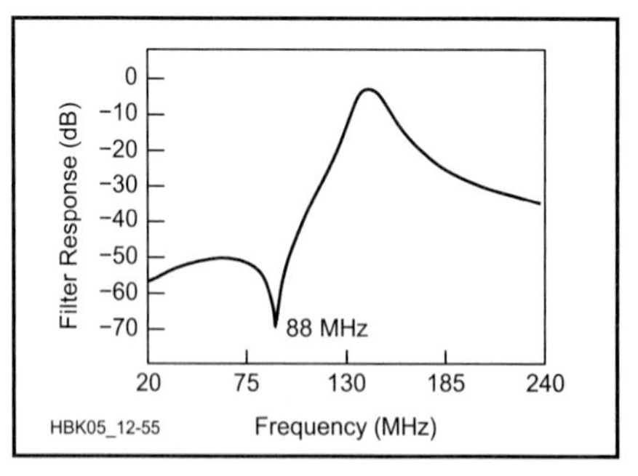

Fig 12.55 — Filter response plot of the 144-MHz band-pass filter, with an image-reject notch for a 28 MHz IF.

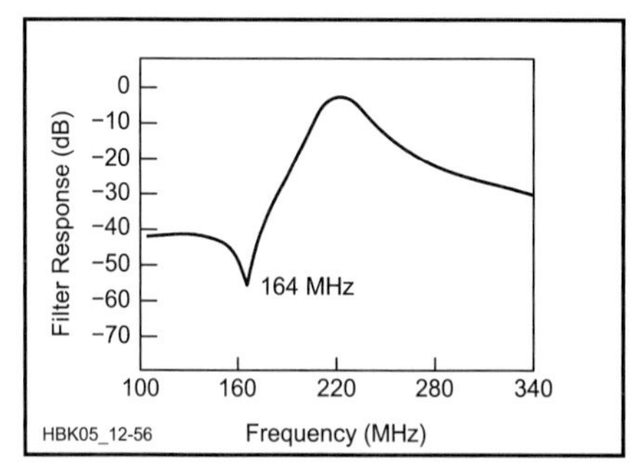

Fig 12.56 — Filter response plot of the 222-MHz band-pass filter, with an image-reject notch for a 28 MHz IF.

frequency used in 144-MHz mixing schemes is 116 MHz and the image frequency, 88 MHz, falls in TV channel 6. Inadequate rejection of a broadcast carrier at this frequency results in a strong, wideband signal at the low end of the 2-m band. A similar problem on the transmit side can cause TVI. These band-pass filters have effectively suppressed undesired mixing products. See **Fig 12.55** and **12.56**.

The circuit is constructed on a double-sided copper-clad circuit board. Minimize component lead lengths to eliminate resistive losses and unwanted stray coupling. Mount the piston trimmers through the board with the coils soldered to the opposite end, parallel to the board. The shield between L1 and L3 decreases mutual coupling and improves the frequency response. Peak C1 and C3 for optimum response.

L1, C1, L3 and C3 form the tank circuits that resonate at the desired frequency. C2 and L2 reject the undesired energy while allowing the desired signal to pass. The tap points on L1 and L3 provide 50-Ω matching; they may be adjusted for optimum energy transfer. Several filters have been constructed using a miniature variable capacitor in place of C2 so that the notch frequency could be varied.

Switched Capacitor Filters

The *switched capacitor filter,* or SCF, uses an IC to synthesize a high-pass, low-pass, band-pass or notch filter. The performance of multiple-pole filters is available, with Q and bandwidth set by external resistors. An external clock frequency sets the filter center frequency, so this frequency may be easily changed or digitally controlled. Dynamic range of 80 dB, Q of 50, 5-pole equivalent design and maximum usable frequency of 250 kHz are available for such uses as audio CW and RTTY filters. In addition, all kinds of digital tone signaling such as DTMF and modem encoding and decoding are being designed with these circuits.

AN EASY-TO-BUILD, HIGH-PERFORMANCE PASSIVE CW FILTER

Modern commercial receivers for amateur radio applications have featured CW filters with digital signal processing (DSP) circuits. These DSP filters provide exceptional audio selectivity with the added advantages of letting the user change the filter's center frequency and bandwidth. Yet in spite of these improvements, many hams are dissatisfied with DSP filters due to increased distortion of the CW signal and the presence of a constant low-level, wide-band noise at the audio output. One way to avoid this distortion and noise is to switch to a selective passive filter that generates no noise! Although the center frequency and bandwidth of the passive filter is fixed and cannot be changed, this is not a serious problem once a center frequency preferred by the user is chosen. The bandwidth can be made narrow enough for good selectivity with no ringing that frequently occurs when the bandwidth is too narrow. This passive CW filter project was designed, built and refined over many years by ARRL Technical Advisor Edward E. Wetherhold, W3NQN.

The effectiveness of an easy-to-build, high-performance passive CW filter in providing distortion-free and noise-free

CW reception — when compared with several commercial amateur receivers using DSP filtering — was experienced by Steve Root, KØSR. He reported that when he replaced his DSP filter with the passive CW filter that he assembled, he had the impression that the signals in the filter passband were amplified. In reality, the noise floor appeared to drop one or two dB. When attempting to hear low-level DX CW signals, Steve now prefers the passive CW filter over DSP filters.[1] The CW filter assembled and used by KØSR is the passive five-resonator CW filter that has been widely published in many Handbooks and magazines since 1980, and most recently in Rich Arland's, K7SZ, QRP column in the May 2002 issue of *QST* (see references 2-11 at the end of this text).

If you want to build the high-performance passive five-resonator CW filter and experience no-distortion and no-noise CW reception, this article will show you how.

This inductor-capacitor CW filter uses one stack of 85-mH inductors and two modified separate inductors in a five-resonator circuit that is easy to assemble, gives high performance and is low cost. Although these inductors have been referred to as "88 mH" over the past 25 years, their actual value is closer to 85 mH, and for that reason the designs presented in this article are based on an inductor value of 85 mH.

Five bandpass filter designs for center frequencies between 546 Hz and 800 Hz are listed in **Table 12.6**. Select the center frequency that matches your transceiver sidetone frequency. If you are using a direct conversion receiver or an old receiver with a BFO, you may select any of the designs having a center frequency that you find easy on your ears. The author can provide a kit of parts with detailed instructions for assembling this filter at a nominal cost. For contact information, see the end of this text.

The actual 3-dB bandwidth of the filters is between 250 and 270 Hz depending on the center frequency. This bandwidth is narrow enough to give good selectivity, and yet broad enough for easy tuning with no ringing. Five high-Q resonators provide good skirt selectivity that is adequate for interference-free CW reception. Simple construction, low cost and good performance make this filter an ideal first project for anyone interested in putting together a useful station accessory, provided you operate CW mode of course!

DESIGNS AND INTERFACING

Fig 12.57 shows the filter schematic diagram. Component values are given in Table 12.6 for five center-frequency designs. All designs are to be terminated in an impedance between 200 and 230 Ω and standard commercial 8-Ω to 200-Ω audio transformers are used to match the filter input and output to the 8-Ω audio output jack on your receiver — and to an 8-Ω headset. Details are discussed a bit later in this text to interface using headphones with other than 8-Ω impedances that are now quite common.

Table 12.6

CW Filter Using One 85-mH Inductor Stack and Two Modified 85-mH Inductors

Center Freq. (Hz)	*546*	*600*	*700*	*750*	*800*
C1, C5 (nF)	1000	828	608	530	466
C2, C4 (μF)	1.0	1.0	1.0	1.0	1.0
C3 (nF)	333	276	202.7	176.5	155
L2, L4 (mH)	85	70.36	51.69	45.0	39.6
Remove Turns*	NONE	66	160	200	232

*The total number of turns removed, split equally from each of the two windings of L2. Do the same also for L4. (E.g., for a 700-Hz center frequency, remove 80 turns from each of the two windings of L2, for a total of 160 turns removed from L2. Repeat exactly for L4.)

For all designs: L1, L5 = 85 mH; L3 = 3 (85 mH) = 255 mH; T1,T2 = 8/200 Ω CT; R1 = 6.8 to 50 Ω. Although the surplus inductors are commonly considered to be 88 mH, the actual value is closer to 85 mH. For this reason, all designs are based on the 85-mH value. L2 and L4 have white cores, Magnetic Part No. 55347, OD Max = 24.3 mm, ID Min = 13.77 mm, HT = 9.70 mm; μ = 200, AL = 169 mH/1000T ±8%. The calculated 3-dB BW is 285 Hz and is the same for all designs; however, the actual bandwidth is 5 to 10-percent narrower depending on the inductor Q at the edges of the filter passband.

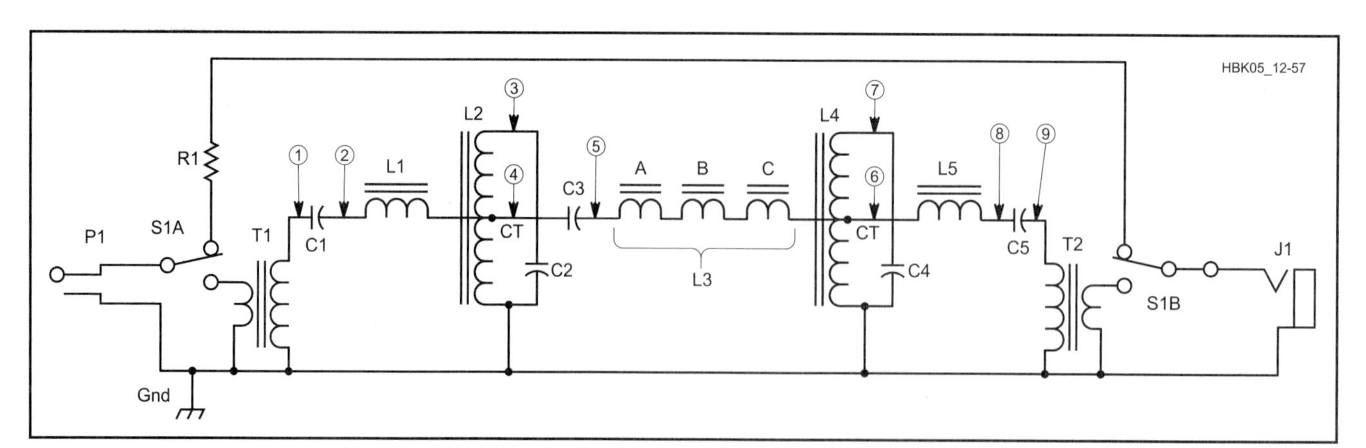

Fig 12.57 — Schematic diagram of the five-resonator CW filter. See Table 12.6 for capacitor and inductor values to build a filter with a center frequency of 546, 600, 700, 750 or 800 Hz.

P1 — Phone plug to match your receiver audio output jack.
J1 — Phone jack to match your headphone.
R1 — 6.8 to 50 ohms, 1/4-W, 10% resistor (see text).
S1 — DPDT switch.
T1, T2 — 200 to 8-Ω impedance-matching transformers, 0.4-W, Miniature Core Type EI-24, Mouser No. 42TU200.

Note: The circled numbers identify the circuit nodes corresponding to the same nodes labeled in the pictorial diagram in Fig 12.58.

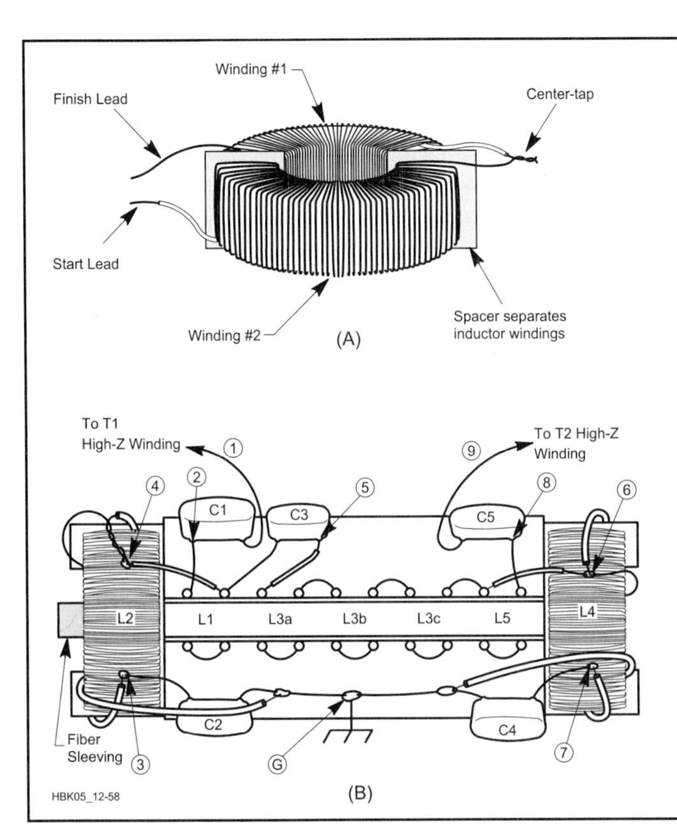

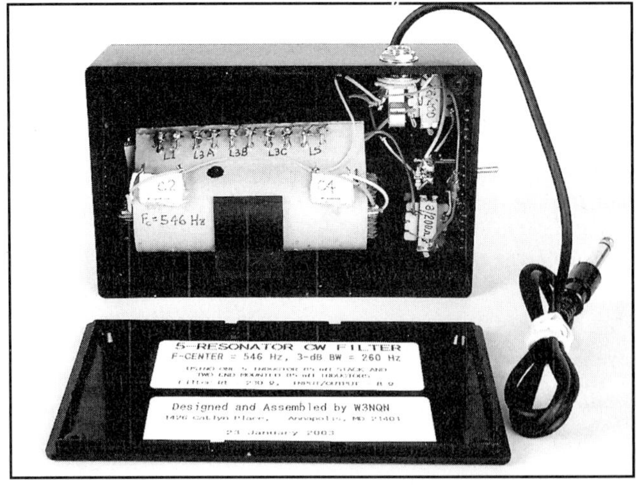

(C)

Fig 12.58 — Part A shows a pictorial diagram of the lead-connection details for L2 and L4. Part B shows the filter wiring diagram, including the inductor stack wiring of L1, L3 and L5. Part C is a photo of the assembled filter installed in a Jameco H2581 plastic box. The bypass switch (S1) and input/output transformers (T1, T2) are on the right side of the box.

CONSTRUCTION

The encircled numbers in Fig 12.57 indicate the filter circuit nodes for reference. **Fig 12.58A** shows the L2 and L4 inductor lead connections for the 546-Hz design where no turns need to be removed; the two inductors are used in their original condition. For all other designs, turns need to be removed from each of the windings. The number of turns requiring removal from the L2 and L4 windings is listed in Table 12.6.

Fig 12.58B shows a pictorial of the filter assembly and the connections between the capacitors and the 85-mH stack terminals. Inductors L1, L3 and L5 are contained within the inductor stack and are interconnected using the terminal lugs on the stack as shown in the pictorial diagram. The encircled numbers show the circuit nodes corresponding to those in Fig 12.57.

After the correct number of turns are removed from L2 and L4, the leads are gently scraped until you see copper and then the start lead (with sleeving) of one winding is connected to the finish lead of the other winding to make the center tap. The center tap lead and the other start and finish leads of L2 and L4 are connected as indicated in Fig 12.58B. L2 and L4 are fastened to opposite ends of the stack with clear silicone sealant that is available in a small tube at low cost from your local hardware store. Use the silicone sealant to fasten C2 and C4 to the side of the stack. The capacitor leads of C1, C3 and C5 are adequate to support the capacitors when their leads are soldered to the stack terminals. Fig 1C is a photo of the assembled filter installed in a Jameco plastic box. Transformers T1 and T2 are secured to the bottom of the plastic box with more silicone sealant and are placed on opposite sides of the DPDT switch. See the photograph for the placement of the phone jack and plug.

After the stack and capacitor wiring is completed, the correctness of the wiring is checked before installing the stack in the box. To do this, check the measured node-to-node resistances of the filter with the values listed in **Table 12.7**.

Table 12.7
Node-to-Node Resistances for the 546-Hz CW Filter

Nodes *From*	*To*	*Component Designation*	*Resistance (ohms ±20%)*
1	GND	T1 hi-Z winding	12
2	GND	L1 + 1/2(L2)	12
3	GND	L2	8
4	GND	1/2(L2)	4
5	GND	L3 + 1/2(L4)	28
6	GND	1/2(L4)	4
7	GND	L4	8
8	GND	L5 + 1/2(L4)	12
9	GND	T2 hi-Z winding	12
2	4	L1	8
5	6	L3	24
6	8	L5	8
2	3	L1 + 1/2(L2)	12
8	7	L5 + 1/2(L4)	12

Notes

1. See Figs 12.57 and 12.58 for the filter node locations.
2. Check your wiring using the resistance values in this table. If there is a significant difference between your measured values and the table values, you have a wiring error that must be corrected!
3. The resistances of L2 and L4 in the four other filters will be somewhat less than the 546-Hz values.

For accurate measurements, use a high-quality digital ohmmeter.

INTERFACING TO SOURCE AND LOAD

The T1 and T2 transformers match the filter to the receiver low-impedance audio output and to an 8-ohm headset or speaker. If your headset impedance is greater than 200 Ω, omit T2 and connect a 1/2-watt resistor from node 9 (C5 output lead) to ground. Choose the resistor so the parallel combination of the headset impedance and the resistor gives the correct filter termination impedance (within about 10% of 230 Ω).

PERFORMANCE

The measured 30-dB and 3-dB bandwidths of the 750-Hz filter are about 567 and 271 Hz, respectively. The 30/3-dB shape factor is 2.09. Use this factor to compare the selectivity performance of this filter with others. **Fig 12.59** shows the measured relative attenuation responses of the 546-Hz and 750-Hz filters. These responses were measured in a 200-Ω system without the transformers. All attenuation levels were measured relative to a zero-dB attenuation level at the filter center frequency.

The measured insertion loss of these passive filters with transformers is slightly less than 3 dB and this is typical of filters of this type. This small loss is compensated by slightly increasing the receiver audio gain.

R1 is selected to maintain a relatively constant audio level when the filter is switched in or out of the circuit. The correct value of R1 for your audio system should be determined by experiment and probably will be between 6.8 and 50 Ω. Start with a short circuit across the S1A and B terminals and gradually increase the resistance until the audio level appears to be the same with the filter in or out of the circuit.

Thousands of hams have constructed this five-resonator filter, and many have commented on its ease of assembly, excellent performance and lack of hiss and ringing!

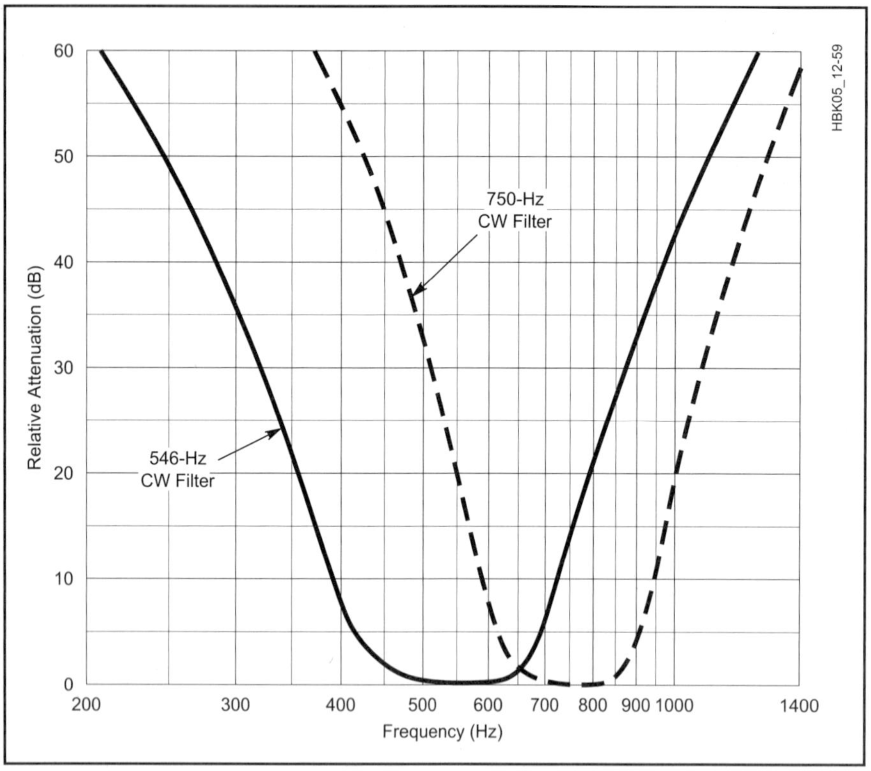

Fig 12.59 — Measured attenuation responses of the 546- and 750-Hz filters. The responses are plotted relative to the zero dB attenuation levels at the center frequencies of the filters. The other filter response curves are similar, but centered at their design frequency.

ORDERING PARTS/CONTACTING THE AUTHOR

The author can provide a kit of parts with detailed instructions for assembling this filter at a nominal cost. The kit includes an inductor stack and two inductors, a pre-punched plastic box with a plastic mounting clip for the inductor stack, five matched capacitors, two transformers, a phone plug and jack and a miniature DPDT switch. Write to Ed Wetherhold, W3NQN, 1426 Catlyn Place, Annapolis, MD 21401-4208 for details about parts and prices. Be sure to include a self-addressed, stamped 9 1/2 × 4-inch envelope with your request.

Notes

[1] Private correspondence from Steve Root, KØSR, in his letter to the author dated 5 September 2002. (Permission was received to publish his comments.)

[2] 1994 *ARRL Handbook*, 71st edition, Robert Schetgen, KU7G, Editor, pp 28-1,-2, Simple High-Performance CW Filter.

[3] *Radio Handbook*, 23rd edition, W. Orr, W6SAI, Editor, Howard W. Sams & Co., 1987 (1-Stack CW Filter), p 13-4,-5,-6.

[4] Wetherhold, "Modern Design of a CW Filter using 88- and 44-mH Surplus Inductors," *QST*, Dec. 1980, pp 14-19 and Feedback, *QST*, Jan 1981, p 43.

[5] Wetherhold, "High-Performance CW Filter," *Ham Radio*, Apr 1981, pp 18-25.

[6] Wetherhold, "CW and SSB Audio Filters Using 88-mH Inductors," *QEX*, Dec 1988, pp 3-10.

[7] Wetherhold, "A CW Filter for the Radio Amateur Newcomer," *RADIO COMMUNICATION* (Radio Society of Great Britain), Jan 1985, pp 26-31.

[8] Wetherhold, "Easy-to-Build One-Stack CW Filter Has High Performance and Low Cost," *SPRAT* (Journal of the G-QRP Club), Issue No. 54, Spring 1988, p 20.

[9] Piero DeGregoris, I3DGF, "Un Facile Filtro CW ad alte prestazioni e basso costo," *Radio Rivista* 12-93, pp 44, 45.

[10] QRP Power, Rich Arland, K7SZ, contributing editor, May 2002 *QST*, p 96, Passive CW Filters.

[11] Ken Kaplan, WB2ART, "Building the W3NQN Passive Audio Filter," *The Keynote*, Issue 7, 2002, pp 16-17, Newsletter of FISTS CW Club.

[12] MPP Cores for Filter and Inductor Applications, MAGNETICS 1991 Catalog, Butler, PA, p 64.

A BC-BAND ENERGY-REJECTION FILTER

Inadequate front-end selectivity or poorly performing RF amplifier and mixer stages often result in unwanted cross-talk and overloading from adjacent commercial or amateur stations. The filter shown is inserted between the antenna and receiver. It attenuates the out-of-band signals from broadcast stations but passes signals of interest (1.8 to 30 MHz) with little or no attenuation.

The high signal strength of local broadcast stations requires that the stop-band attenuation of the high-pass filter also be high. This filter provides about 60 dB of stop-band attenuation with less than 1 dB of attenuation above 1.8 MHz. The filter input and output ports match 50 Ω with a maximum SWR of 1.353:1 (reflection coefficient = 0.15). A 10-element filter yields adequate stop-band attenuation and a reasonable rate of attenuation rise. The design uses only standard-value capacitors.

BUILDING THE FILTER

The filter parts layout, schematic diagram, response curve and component values are shown in **Fig 12.60**. The standard capacitor values listed are within 2.8% of the design values. If the attenuation peaks (f2, f4 and f6) do not fall at 0.677, 1.293 and 1.111 MHz, tune the series-resonant circuits by slightly squeezing or separating the inductor windings.

Construction of the filter is shown in **Fig 12.61**. Use Panasonic NP0 ceramic disk capacitors (ECC series, class 1) or equivalent for values between 10 and 270 pF. For values between 330 pF and 0.033 µF, use Panasonic P-series polypropylene (type ECQ-P) capacitors. These capacitors are available through Digi-Key and other suppliers. The powdered-iron T-50-2 toroidal cores are available through Amidon, Palomar Engineers and others.

For a 3.4-MHz cutoff frequency, divide the L and C values by 2. (This effectively doubles the frequency-label values in Fig 12.60.) For the 80-m version, L2 through L6 should be 20 to 25 turns each, wound on T-50-6 cores. The actual turns required may vary one or two from the calculated values. Parallel-connect capacitors as needed to achieve the nonstandard capacitor values required for this filter.

FILTER PERFORMANCE

The measured filter performance is shown in Fig 12.60. The stop-band attenuation is more than 58 dB. The measured cutoff frequency (less than 1 dB attenuation) is under 1.8 MHz. The measured passband loss is less than 0.8 dB from 1.8 to 10 MHz. Between 10 and 100 MHz, the insertion loss of the filter gradually increases to 2 dB. Input impedance was measured between 1.7 and 4.2 MHz. Over the range tested, the input impedance of the filter remained within the 37 to 67.7-Ω input-impedance window (equivalent to a maximum SWR of 1.353:1).

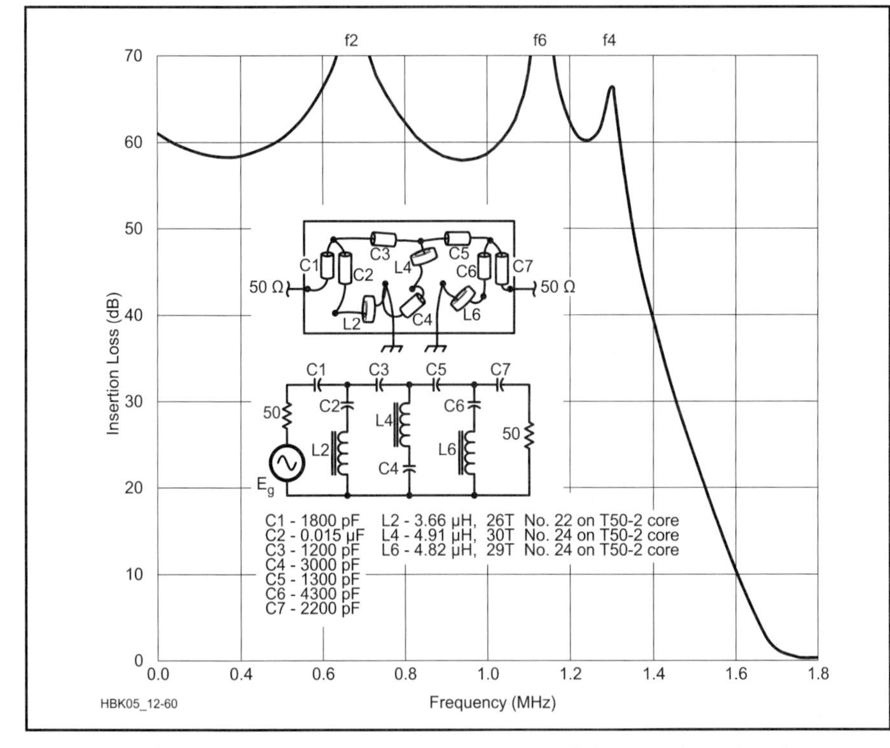

Fig 12.60 — Schematic, layout and response curve of the broadcast band rejection filter.

Fig 12.61 — The filter fits easily in a 2 × 2 × 5-inch enclosure. The version in the photo was built on a piece of perfboard.

A WAVE TRAP FOR BROADCAST STATIONS

Nearby medium-wave broadcast stations can sometimes cause interference to HF receivers over a broad range of frequencies. This being the case, set a trap to catch the unwanted frequencies.

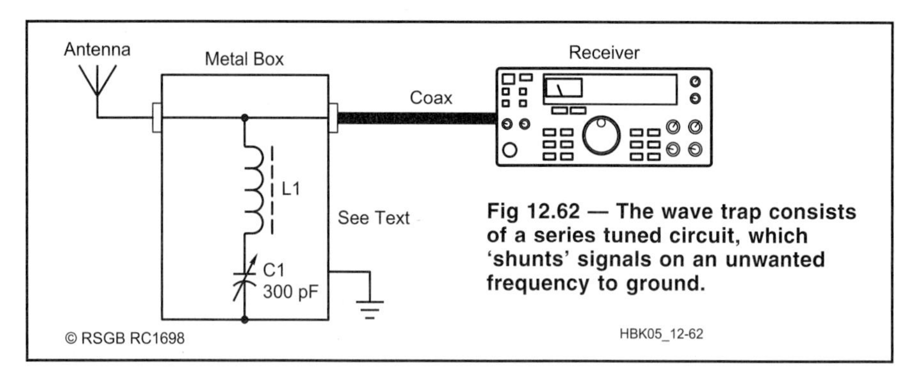

Fig 12.62 — The wave trap consists of a series tuned circuit, which 'shunts' signals on an unwanted frequency to ground.

OPERATION

The way the circuit works is quite simple. Referring to Fig **12.62**, you can see that it consists essentially of only two components, a coil L1 and a variable capacitor C1. This series-tuned circuit is connected in parallel with the antenna circuit of the receiver. The characteristic of a series-tuned circuit is that the coil and capacitor have a very low impedance (resistance) to frequencies very close to the frequency to which the circuit is tuned. All other frequencies are almost unaffected. If the circuit is tuned to 1530 kHz, for example, the signals from a broadcast station on that frequency will flow through the filter to ground, rather than go on into the receiver. All other frequencies will pass straight into the receiver. In this way, any interference caused in the receiver by the station on 1530 kHz is significantly reduced.

CONSTRUCTION

This is a series-tuned circuit that is adjustable from about 540 kHz to 1600 kHz. It is built into a metal box, **Fig 12.63**, to shield it from other unwanted signals and is connected as shown in Fig 12.62. To make the inductor, first make a *former* by winding two layers of paper on the ferrite rod. Fix this in place with black electrical tape. Next, lay one end of the wire for the coil on top of the former, leaving about an inch of wire protruding beyond the end of the ferrite rod. Use several turns of electrical tape to secure the wire to the former. Now, wind the coil along the former, making sure the turns are in a single layer and close together. Leave an inch or so of wire free at the end of the coil. Once again, use a couple of turns of electrical tape to secure the wire to the former. Finally, remove half an inch of enamel from each end of the wire.

Alternatively, if you have an old AM transistor radio, a suitable coil can usually be recovered already wound on a ferrite rod. Ignore any small coupling coils. Drill the box to take the components, then fit them in and solder together as shown in Fig **12.64**. Make sure the lid of the box is fixed securely in place, or the wave trap's performance will be adversely affected by pick-up on the components.

CONNECTION AND ADJUSTMENT

Connect the wave trap between the antenna and the receiver, then tune C1 until the interference from the offending broadcast station is a minimum. You may not be able to eliminate interference completely, but this handy little device should reduce it enough to listen to the amateur bands. Lets say you live near an AM transmitter on 1530 kHz, and the signals break through on your 1.8-MHz receiver. By tuning the trap to 1530 kHz, the problem is greatly reduced. If you have problems from more than one broadcast station, the problem needs a more complex solution.

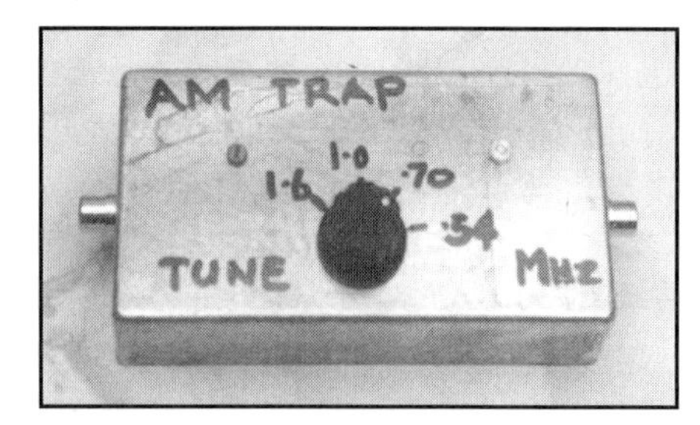

Fig 12.63 — The wave trap can be roughly calibrated to indicate the frequency to which it is tuned.

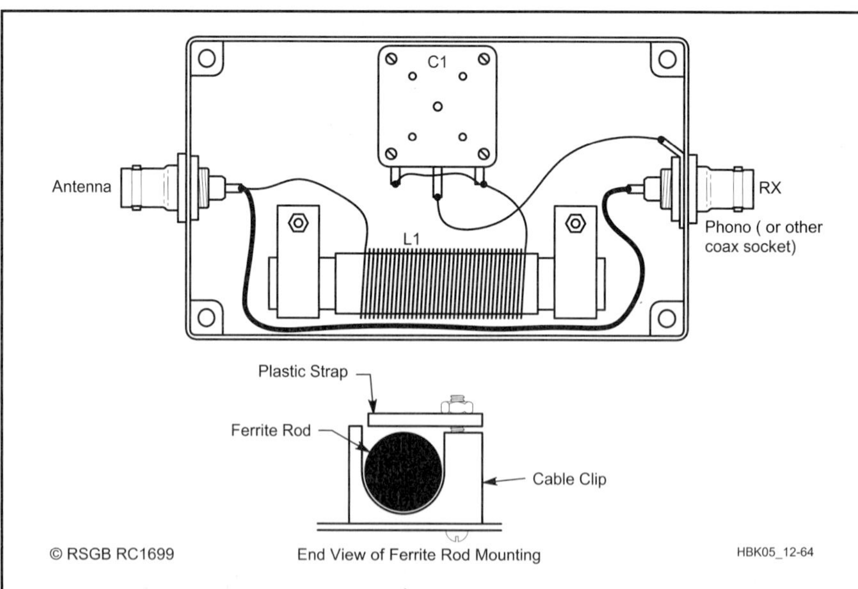

Fig 12.64 — Wiring of the wave trap. The ferrite rod is held in place with cable clips.

Components List

Inductor	Capacitor	Associated items
L1 80 turns of 30 SWG enamelled wire, wound on a ferrite rod	C1 300 pF polyvaricon variable	Case (die-cast box) Knob to suit Sockets to suit Nuts and bolts Plastic cable clips

SECOND-HARMONIC-OPTIMIZED (CWAZ) LOW-PASS FILTERS[1]

The FCC requires transmitter spurious outputs below 30 MHz to be attenuated by 40 dB or more for power levels between 5 and 500 W. For power levels greater than 5 W, the typical second-harmonic attenuation (40-dB) of a seven-element Chebyshev low-pass filter (LPF) is marginal. An additional 10 dB of attenuation is needed to ensure compliance with the FCC requirement.

Jim Tonne, WB6BLD, solved the problem of significantly increasing the second-harmonic attenuation of the seven-element Chebyshev LPF while maintaining an acceptable return loss (> 20 dB) over the amateur passband. Jim's idea was presented in February 1999 *QST* by Ed Wetherhold, W3NQN. These filters are most useful with single-band, single-device transmitters. Common medium-power multiband transceivers use push-pull power amplifiers because such amplifiers inherently suppress the second harmonic.

Tonne modified a seven-element Chebyshev standard-value capacitor (SVC) LPF to obtain an additional 10 dB of stop-band loss at the second-harmonic frequency. He did this by adding a capacitor across the center inductor to form a resonant circuit. Unfortunately, return loss (RL) decreased to an unacceptable level, less than 12.5 dB. He needed a way to add the resonant circuit, while maintaining an acceptable RL level over the passband.

The typical LPF, and the Chebyshev SVC designs listed in this chapter all have acceptable RL levels that extend from the filter ripple-cutoff frequency down to dc. For many Amateur Radio applications, we need an acceptable RL only over the amateur band for which the LPF is designed. We can trade RL levels below the amateur band for improved RL in the passband, and simultaneously increase the stop-band loss at the second-harmonic frequency.

THE CWAZ LOW-PASS FILTER

This new eight-element LPF has a topology similar to that of the seven-element Chebyshev LPF, with two exceptions: The center inductor is resonated at the second harmonic in the filter stop band, and the component values are adjusted to maintain a more than acceptable RL across the amateur passband. To distinguish this new LPF from the SVC Chebyshev LPF, Wetherhold named it the "Chebyshev with Added Zero" or "CWAZ" LPF design.

You should understand that CWAZ LPFs are *output filters for single-band transmitters*. They provide optimum second and higher harmonic attenuation while maintaining a suitable level of

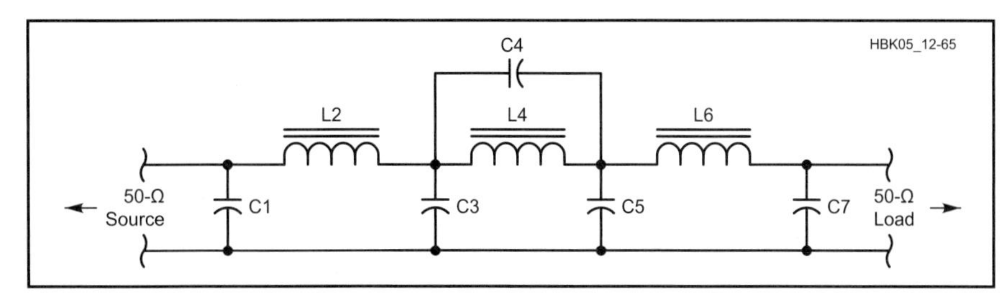

Fig 12.65 — Schematic diagram of a CWAZ low-pass filter designed for maximum second-harmonic attenuation. See Table 12.8 for component values of CWAZ 50-Ω designs. L4 and C4 are tuned to resonate at the F4 frequency given in Table 12.8. For an output power of 10 W into a 50-Ω load, the RMS output voltage is $\sqrt{10 \times 50} = 22.4$ V. Consequently, a 100 V dc capacitor derated to 60 V (for RF filtering) is adequate for use in these LPFs if the load SWR is less than 2.5:1.

Table 12.8

CWAZ 50-Ω Low-Pass Filters

Designed for second-harmonic attenuation in amateur bands below 30 MHz.

Band (m)	*Start Frequency (MHz)*	*C1,7 (pF)*	*C3,5 (pF)*	*C4 (pF)*	*L2,6 (μH)*	*L4 (μH)*	*F4 (MHz)*
—	1.00	2986	4556	680.1	9.377	8.516	2.091
		1659	2531	378			3.76
160	1.80	1450 + 220	2100 + 470		5.21	4.73	
		1500 + 150	2200 + 330	330 + 47			3.78
		853	1302	194			7.32
80	3.50		1150 + 150		2.68	2.43	
		470 + 390	1200 + 100	150 + 47			7.27
		427	651	97.2			14.6
40	7.00				1.34	1.22	
		330 + 100	330 + 330	100			14.4
		296	451	67.3			21.1
30	10.1				0.928	0.843	
		150 + 150	470	68			21.0
		213	325	48.6			29.3
20	14.0				0.670	0.608	
		220	330	47			29.8
		165	252	37.6			37.8
17	18.068				0.519	0.471	
		82 + 82	100 + 150	39			37.1
		142	217	32.4			43.9
15	21.0				0.447	0.406	
		150	220	33			43.5
		120	183	27.3			52.0
12	24.89				0.377	0.342	
		120	180	27			52.4
		107	163	24.3			58.5
10	28.0				0.335	0.304	
		100	82 + 82	27			55.6

NOTE: The CWAZ low-pass filters are designed for a single amateur band to provide more than 50-dB attenuation to the second harmonic of the fundamental frequency and to the higher harmonics. All component values for any particular band are calculated by dividing the 1-MHz values in the first row (included for reference only) by the start frequency of the selected band. The upper capacitor values in each row show the calculated design values obtained by dividing the 1-MHz capacitor values by the amateur-band start frequency in megahertz. The lower standard-capacitor values are suggested as a convenient way to realize the design values. The middle capacitor values in the 160- and 80-meter-band designs are suggested values when the high-value capacitors (greater than 1000 pF) are on the low side of their tolerance range. The design F4 frequency (see upper value in the F4 column) is calculated by multiplying the 1-MHz F4 value by the start frequency of the band. The lower number in the F4 column is the F4 frequency based on the suggested lower capacitor value and the listed L4 value.

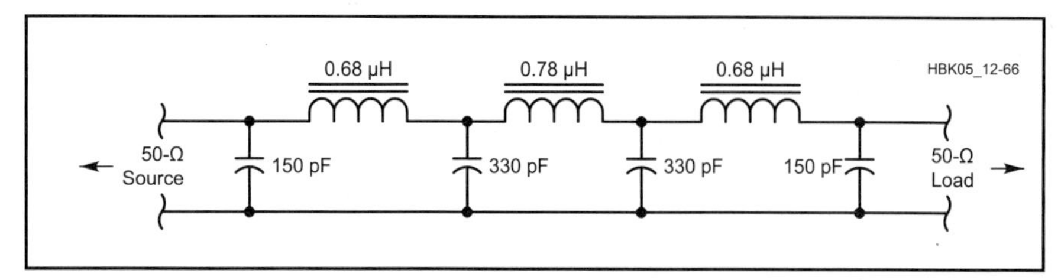

Fig 12.66 — Schematic diagram of a 20-meter SVC Chebyshev LPF.

return loss over the amateur band for which they're designed.

Fig 12.65 shows a schematic diagram of a CWAZ LPF design. **Table 12.8** lists suggested capacitor and inductor values for all amateur bands from 160 through 10 meters. If you want to calculate CWAZ values for different bands, simply divide the first-row C and L values (for 1 MHz) by the start frequency of the desired band. For example, C1, 7 for the 160-meter design is equal to 2986/1.80 = 1659 pF. The other component values for the 160-meter LPF are calculated in a similar manner.

CWAZ VERSUS SEVENTH-ORDER SVC[1]

The easiest way to demonstrate the superiority of a CWAZ LPF over the Chebyshev LPF is to compare the RL and insertion-loss responses of these two designs. **Fig 12.66** shows a 20-meter SVC Chebyshev LPF design based on the SVC tables on the *Handbook* CD-ROM. **Fig 12.67** shows the computer-calculated return- and insertion-loss responses of the LPF shown in Fig 12.66. The plotted responses were made using Jim Tonne's *ELSIE* filter design and analysis software. The Windows-based program is available from this web site — **http://tonnesoftware.com.**[1] **Fig 12.68** shows the computer-calculated return- and insertion-loss responses of a CWAZ LPF intended to replace the seven-element 20-meter Chebyshev SVC LPF. The stop-band attenuation of the CWAZ LPF in the second-harmonic band is more than 60 dB and is substantially greater than that of the Chebyshev LPF. Also, the pass-band RL of the CWAZ LPF is quite satisfactory, at more than 25 dB. The disadvantages of the CWAZ design are that an extra capacitor is needed across L4, and several of the designs listed in Table 12.8 require paralleled capacitors to realize the design values. Nevertheless, these disadvantages are minor in comparison to the increased second-harmonic stop-band attenuation that is possible with a CWAZ design.

Notes

[1]Those seriously interested in passive LC filter design can experience the capabilities of *ELSIE* software. The *student version* of this software permits filter design configurations up to the 7th Order, 7th Stage level. The software can be downloaded from the web site of Jim Tonne (WB6BLD) at this URL: **http://tonnesoftware.com/**.

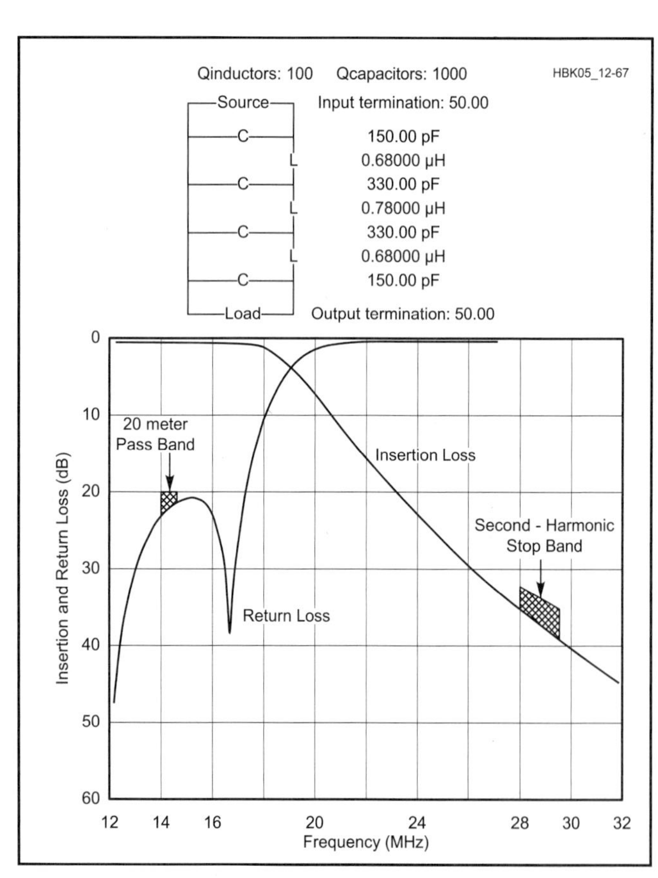

Fig 12.67 — The plots show the *ELSIE* computer-calculated return- and insertion-loss responses of the seventh-order Chebyshev SVC low-pass filter shown in Fig 12.66. The 20-meter passband RL is about 21 dB, and the insertion loss over the second-harmonic frequency band ranges from 35 to 39 dB. A listing of the component values is included.

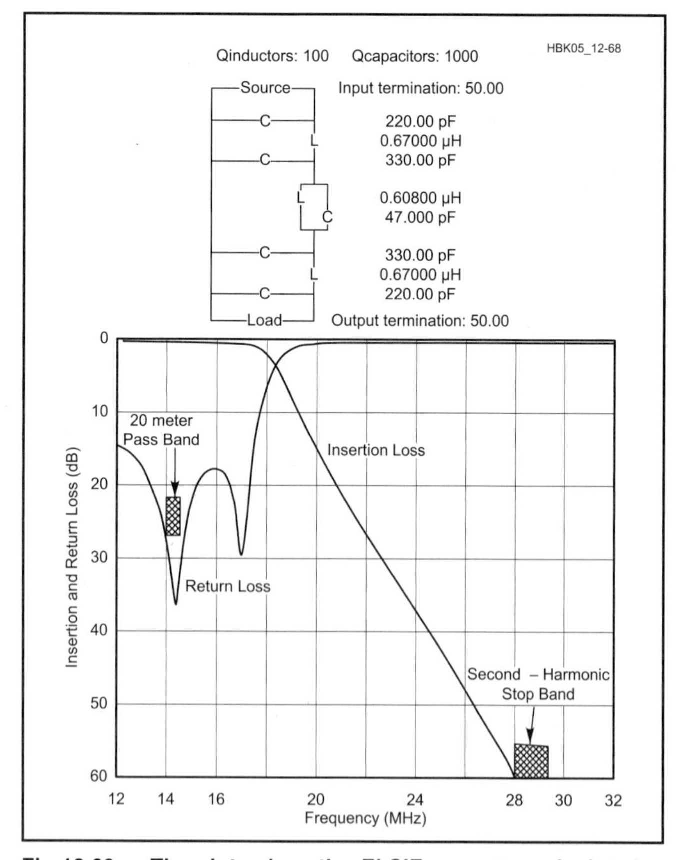

Fig 12.68 — The plots show the *ELSIE* computer-calculated return- and insertion-loss responses of the eight-element low-pass filter using the CWAZ capacitor and inductor values listed in Table 12.8 for the 20-meter low-pass filter. Notice that the calculated attenuation to second-harmonic signals is greater than 60 dB, while RL over the 20-meter passband is greater than 25 dB.

THE DIPLEXER FILTER

This section, covering diplexer filters, was written by William E. Sabin, WØIYH. The diplexer is helpful in certain applications, and Chapter 11 shows them used as frequency mixer terminations.

Diplexers have a constant filter-input resistance that extends to the stop band as well as the passband. Ordinary filters that become highly reactive or have an open or short-circuit input impedance outside the passband may degrade performance.

Fig 12.69 shows a *normalized* prototype 5-element, 0.1-dB Chebyshev low-pass/high-pass (LP/HP) filter. This idealized filter is driven by a voltage generator with zero internal resistance, has load resistors of 1.0 Ω and a cutoff frequency of 1.0 radian per second (0.1592 Hz). The LP prototype values are taken from standard filter tables.[1] The first element is a series inductor. The HP prototype is found by:

a) replacing the series L (LP) with a series C (HP) whose value is 1/L, and

b) replacing the shunt C (LP) with a shunt L (HP) whose value is 1/C.

For the Chebyshev filter, the return loss is improved several dB by multiplying the prototype LP values by an experimentally derived number, K, and dividing the HP values by the same K. You can calculate the LP values in henrys and farads for a 50-Ω RF application with the following formulas:

$$L_{LP} = \frac{K L_{P(LP)} R}{2\pi f_{CO}};\; C_{LP} = \frac{K C_{P(LP)}}{2\pi f_{CO} R}$$

where

$L_{P(LP)}$ and $C_{P(LP)}$ are LP prototype values

K = 1.005 (in this specific example)

R = 50 Ω

f_{CO} = the cutoff (–3-dB response) frequency in Hz.

For the HP segment:

$$L_{HP} = \frac{L_{P(HP)} R}{2\pi f_{CO} K};\; C_{HP} = \frac{C_{P(HP)}}{2\pi f_{CO} KR}$$

where $L_{P(HP)}$ and $C_{P(HP)}$ are HP prototype values.

Fig 12.70 shows the LP and HP responses of a diplexer filter for the 80-meter band. The following items are to be noted:

- The 3 dB responses of the LP and HP meet at 5.45 MHz.
- The input impedance is close to 50 Ω at all frequencies, as indicated by the high value of return loss (SWR <1.07:1).
- At and near 5.45 MHz, the LP input reactance and the HP input reactance are conjugates; therefore, they cancel and produce an almost perfect 50-Ω input resistance in that region.

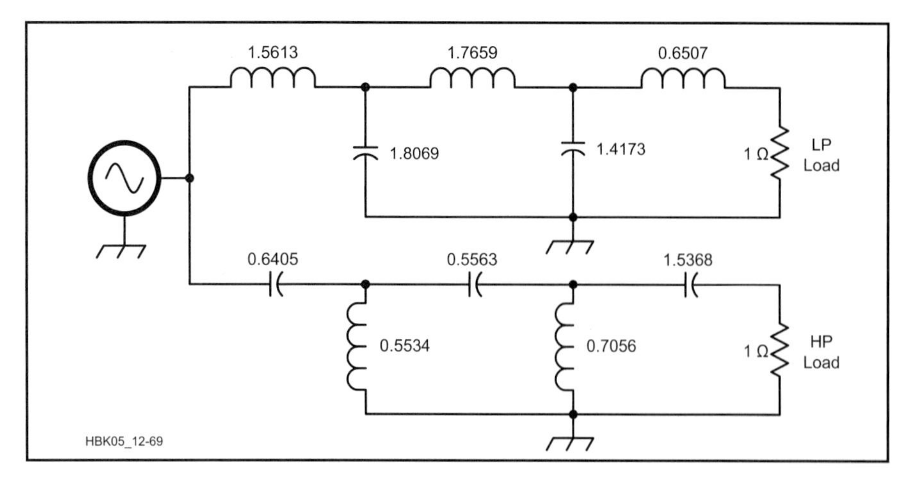

Fig 12.69 — Low-pass and high-pass prototype diplexer filter design. The low-pass portion is at the top, and the high-pass at the bottom of the drawing. See text.

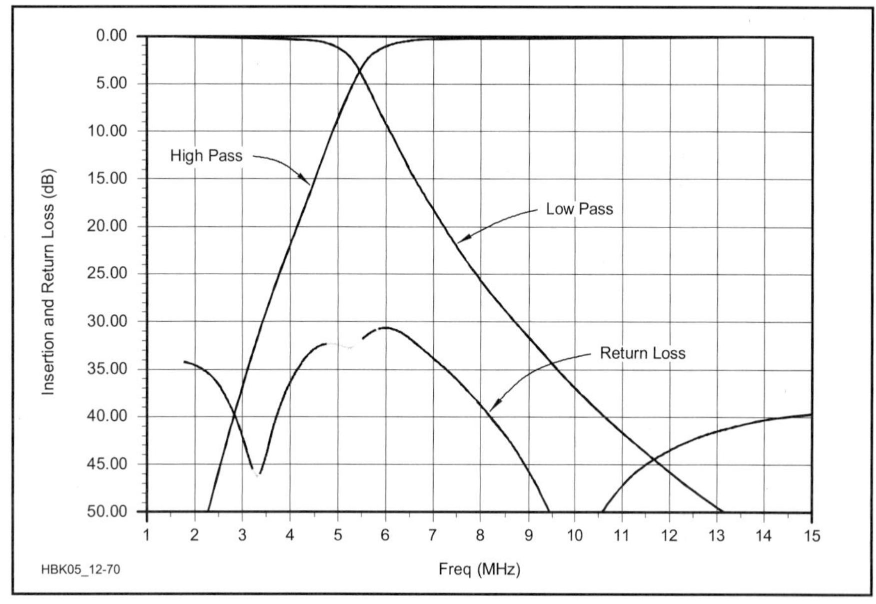

Fig 12.70 — Response for the low-pass and high-pass portions of the 80-meter diplexer filter. Also shown is the return loss of the filter.

- Because of the way the diplexer filter is derived from synthesis procedures, the transfer characteristic of the filter is mostly independent of the actual value of the amplifier dynamic output impedance.[2] This is a useful feature, since the RF power amplifier output impedance is usually not known or specified.
- The 80-meter band is well within the LP response.
- The HP response is down more than 20 dB at 4 MHz.
- The second harmonic of 3.5 MHz is down only 18 dB at 7.0 MHz. Because the second harmonic attenuation of the LP is not great, it is necessary that the amplifier itself be a well-balanced push-pull design that greatly rejects the second harmonic. In practice this is not a difficult task.
- The third harmonic of 3.5 MHz is down almost 40 dB at 10.5 MHz.

Fig 12.71A shows the unfiltered of a solid-state push-pull power amplifier for the 80-meter band. In the figure you can see that:

- The second harmonic has been suppressed by a proper push-pull design.
- The third harmonic is typically only 15 dB or less below the fundamental.

The amplifier output goes through our diplexer filter. The desired output comes from the LP side, and is shown in Fig 12.71B. In it we see that:

- The fundamental is attenuated only about 0.2 dB.
- The LP has some harmonic content; however, the attenuation exceeds FCC requirements for a 100-W amplifier.

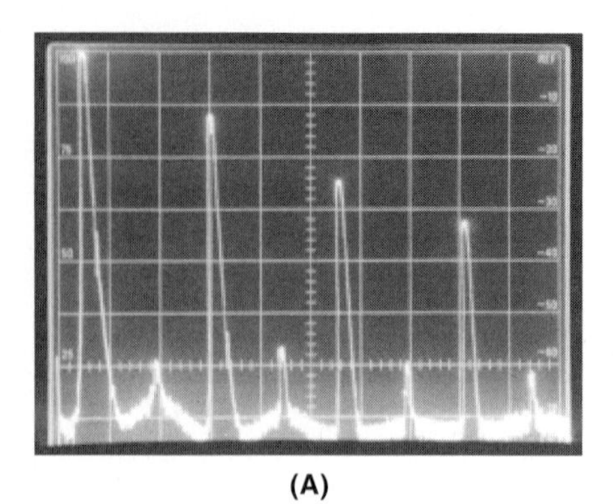

(A)

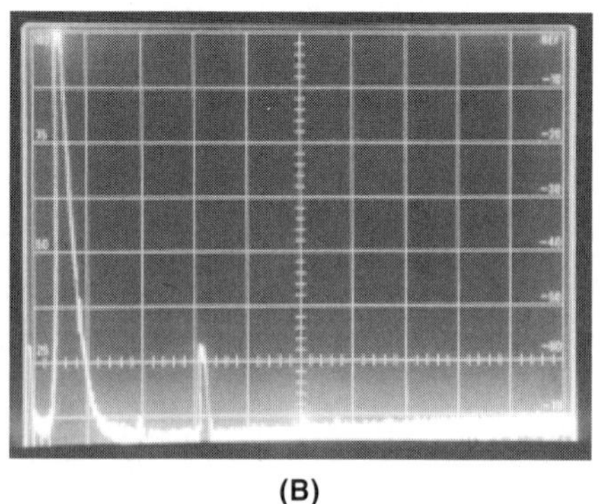

(B)

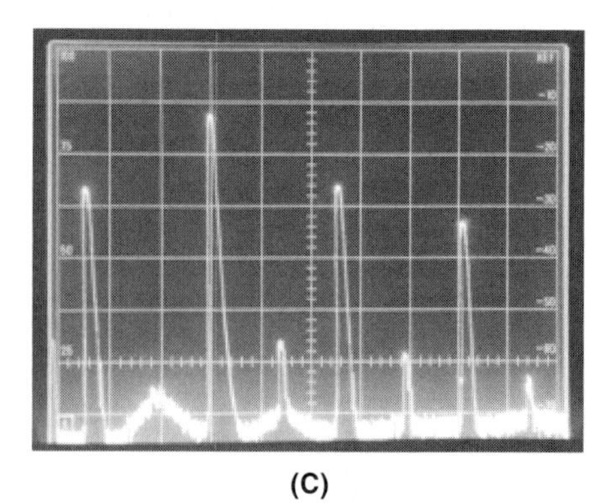

(C)

Fig 12.71 — At A, the output spectrum of a push-pull 80-meter amplifier. At B, the spectrum after passing through the low-pass filter. At C, the spectrum after passing through the high-pass filter.

Fig 12.71C shows the HP output of the diplexer that terminates in the HP load or *dump* resistor. A small amount of the fundamental frequency (about 1%) is also lost in this resistor. Within the 3.5 to 4.0 MHz band, the filter input resistance is almost exactly the correct 50-Ω load resistance value. This is because power that would otherwise be *reflected* back to the amplifier is absorbed in the dump resistor.

Solid state power amplifiers tend to have stability problems that can be difficult to debug.[3] These problems may be evidenced by level changes in: load impedance, drive, gate or base bias, B+, etc. Problems may arise from:

- The reactance of the low-pass filter outside the desired passband. This is especially true for transistors that are designed for high-frequency operation.
- Self resonance of a series inductor at some high frequency.
- A stopband impedance that causes voltage, current and impedance reflections back to the amplifier, creating instabilities within the output transistors.

Intermodulation performance can also be degraded by these reflections. The strong third harmonic is especially bothersome for these problems.

The diplexer filter is an approach that can greatly simplify the design process, especially for the amateur with limited PA-design experience and with limited home-lab facilities. For these reasons, the amateur homebrew enthusiast may want to consider this solution, despite its slightly greater parts count and expense.

The diplexer is a good technique for narrowband applications such as the HF amateur bands.[4] From Fig 12.70, we see that if the signal frequency is moved beyond 4.0 MHz the amount of desired signal lost in the dump resistor becomes large. For signal frequencies below 3.5 MHz the harmonic reduction may be inadequate. A single filter will not suffice for all the HF amateur bands.

This treatment provides you with the information to calculate your own filters. A *QEX* article has detailed instructions for building and testing a set of six filters for a 120-W amplifier. These filters cover all nine of the MF/HF amateur bands.[5] Check *ARRLWeb* at: **www.arrl.org/qex/**.

You can use this technique for other filters such as Bessel, Butterworth, linear phase, Chebyshev 0.5, 1.0, etc.[6] However, the diplexer idea does *not* apply to the elliptic function types.

The diplexer approach is a resource that can be used in any application where a constant value of filter input resistance over a wide range of passband and stopband frequencies is desirable for some reason. The *ARRL Radio Designer* program is an ideal way to finalize the design before the actual construction.[7] The coil dimensions and the dump resistor wattage need to be determined from a consideration of the power levels involved, as illustrated in Fig 12.71.

Another significant application of the diplexer is for elimination of EMI, RFI and TVI energy. Instead of being reflected and very possibly escaping by some other route, the unwanted energy is dissipated in the dump resistor.[7]

Notes

[1]Williams, A. and Taylor, F., *Electronic Filter Design Handbook*, any edition, McGraw-Hill.

[2]Storer, J.E., *Passive Network Synthesis*, McGraw-Hill 1957, pp 168-170. This book shows that the input resistance is ideally constant in the passband and the stopband and that the filter transfer characteristic is ideally independent of the generator impedance.

[3]Sabin, W. and Schoenike, E., *HF Radio Systems and Circuits*, Chapter 12, Noble Publishing, 1998. This publication is available from ARRL as Order no. 7253. It can be ordered at: **www.arrl.org/catalog/**. Also the previous edition of this book, *Single-Sideband Systems and Circuits*, McGraw-Hill, 1987 or 1995.

[4]Dye, N. and Granberg, H., *Radio Frequency Transistors, Principles and Applications*, Butterworth-Heinemann, 1993, p 151.

[5]Sabin, W.E. WØIYH, "Diplexer Filters for the HF MOSFET Power Amplifier," *QEX*, Jul/Aug, 1999. Also check ARRLWeb at: **www.arrl.org/qex/**.

[6]See note 1. *Electronic Filter Design Handbook* has LP prototype values for various filter types, and for complexities from 2 to 10 components.

[7]Weinrich, R. and Carroll, R.W., "Absorptive Filters for TV Harmonics," *QST*, Nov 1968, pp 10-25.

OTHER FILTER PROJECTS

Filters for specific applications may be found in other chapters of this *Handbook*. Receiver input filters, transmitter filters, interstage filters and others can be separated from the various projects and built for other applications. Since filters are a first line of defense against *electromagnetic interference* (EMI) problems, the following filter projects appear in the EMI/Direction Finding chapter:

- Differential-mode high-pass filter for 75-Ω coax (for TV reception)
- *Brute-force* ac-line filter
- Loudspeaker common-mode choke
- LC filter for speaker leads
- Audio equipment input filter

REFERENCES

1. O. Zobel, "Theory and Design of Electric Wave Filters," *Bell System Technical Journal*, Jan 1923.
2. ARRL *Handbook*, 1968, p 50.
3. S. Butterworth, "On the Theory of Filter Amplifiers," *Experimental Wireless and Wireless Engineer*, Oct 1930, pp 536-541.
4. S. Darlington, "Synthesis of Reactance 4-Poles Which Produce Prescribed Insertion Loss Characteristics," *Journal of Mathematics and Physics*, Sep 1939, pp 257-353.

5. L. Weinberg, "Network Design by use of Modern Synthesis Techniques and Tables," *Proceedings of the National Electronics Conference*, vol 12, 1956.
6. Laplace Transforms: P. Chirlian, *Basic Network Theory,* McGraw Hill, 1969.
7. Fourier Transforms: *Reference Data for Engineers*, Chapter 7, 7th edition, Howard Sams, 1985.
8. Cauer Elliptic Filters: *The Design of Filters Using the Catalog of Normalized Low-Pass Filters,* Telefunken, 1966. Also Ref 7, pp 9-5 to 9-11.
9. M. Dishal, "Top Coupled Band-pass Filters," *IT&T Handbook*, 4th edition, American Book, Inc, 1956, p 216.
10. W. E. Sabin, WØIYH, "Designing Narrow Band-Pass Filters with a BASIC Program," May 1983, *QST*, pp 23-29.
11. U. R. Rohde, DJ2LR, "Crystal Filter Design with Small Computers" May 1981, *QST*, p 18.
12. J. A. Hardcastle, G3JIR, "Ladder Crystal Filter Design," Nov 1980, *QST*, p 20.
13. W. Hayward, W7ZOI, "A Unified Approach to the Design of Ladder Crystal Filters," May 1982, *QST*, p 21.
13a. J. Makhinson, N6NWP, "Designing and Building High-Performance Crystal Ladder Filters," Jan 1995, *QEX*, pp 3-17.
13b. W. Hayward, W7ZOI, "Refinements in Crystal Ladder Filter Design," June 1995, *QEX*, pp 16-21.
14. R. Fisher, W2CQH, "Combline VHF Band-pass Filters," Dec 1968, *QST*, p 44.
15. R. Fisher, W2CQH, "Interdigital Band-pass Filters for Amateur VHF/UHF Applications," Mar 1968, *QST*, p 32.
16. W. S. Metcalf, "Graphs Speed Interdigitated Filter Design," *Microwaves*, Feb 1967.
17. A. Zverev, *Handbook of Filter Synthesis*, John Wiley and Sons.
18. R. Saal, *The Design of Filters Using the Catalog of Normalized Low-Pass Filters*, Telefunken.
19. P. Geffe, *Simplified Modern Filter Design* (New York: John F. Rider, a division of Hayden Publishing Co, 1963).
20. *A Handbook on Electrical Filters* (Rockville, Maryland: White Electromagnetics, 1963).
21. A. B. Williams, *Electronic Filter Design Handbook* (New York: McGraw-Hill, 1981).
22. *Reference Data for Radio Engineers*, 6th edition, Table 2, p 5-3 (Indianapolis, IN: Howard W. Sams & Co, 1981).
23. R. Frost, "Large-Scale S Parameters Help Analyze Stability," *Electronic Design*, May 24, 1980.
24. Edward E. Wetherhold, W3NQN, "Modern Design of a CW Filter Using 88 and 44-mH Surplus Inductors," Dec 1980, *QST*, pp14-19. See also Feedback in Jan 1981, *QST*, p 43.
25. E. Wetherhold, W3NQN, " CW and SSB Audio Filters Using 88-mH Inductors," *QEX*-82, pp 3-10, Dec 1988; *Radio Handbook*, 23rd edition, W. Orr, editor, p 13-4, Howard W. Sams and Co., 1987; and, "A CW Filter for the Radio Amateur Newcomer," *Radio Communication*, pp 26-31, Jan 1985, RSGB.
26. M. Dishal, "Modern Network Theory Design of Single Sideband Crystal Ladder Filters," *Proc IEEE*, Vol 53, No 9, Sep 1965.
27. J.A. Hardcastle, G3JIR, "Computer-Aided Ladder Crystal Filter Design," *Radio Communication*, May 1983.
28. John Pivnichny, N2DCH, "Ladder Crystal Filters," MFJ Publishing Co, Inc.
29. D. Jansson, WD4FAB and E. Wetherhold, W3NQN, "High-Pass Filters to Combat TVI," Feb 1987, *QEX*, pp 7-8 and 13.
30. E. Wetherhold, W3NQN, "Band-Stop Filters for Attenuating High-Level Broadcast-Band Signals," Nov 1995, *QEX*, pp 3-12.
31. T. Moliere, DL7AV, "Band-Reject Filters for Multi-Multi Contest Operations," Feb 1996, *CQ Contest*, pp 14-22.
32. T. Cefalo Jr, WA1SPI, "Diplexers, Some Practical Applications," Fall 1997, *Communications Quarterly*, pp 19-24.
33. R. Lumachi, WB2CQM. "How to Silverplate RF Tank Circuits," Dec 1997, *73 Amateur Radio Today*, pp 18-23.
34. W. Hayward, W7ZOI, "Extending the Double-Tuned Circuit to Three Resonators," Mar/Apr 1998, *QEX*, pp 41-46.
35. P. R. Cope, W2GOM/7, "The Twin-T Filter," July/Aug 1998, *QEX*, pp 45-49.
36. J. Tonne, WB6BLD, "Harmonic Filters, Improved," Sept/Oct 1998, *QEX*, pp 50-53.
37. E. Wetherhold, W3NQN, "Second-Harmonic-Optimized Low-Pass Filters," Feb 1999, *QST*, pp 44-46.
38. F. Heemstra, KT3J, "A Double-Tuned Active Filter with Interactive Coupling," Mar/Apr 1999, *QEX*, pp 25-29.
39. E. Wetherhold, W3NQN, "Clean Up Your Signals with Band-Pass Filters," Parts 1 and 2, May and June 1998, *QST*, pp 44-48 and 39-42.
40. B. Bartlett, VK4UW and J. Loftus, VK4EMM, "Band-Pass Filters for Contesting," Jan 2000, *National Contest Journal*, pp 11-15.
41. W. Sabin, WØIYH, "Diplexer Filters for the HF MOSFET Power Amplifier," July/Aug 1999, *QEX*, pp 20-26.
42. W. Sabin, WØIYH, "Narrow Band-Pass Filters for HF," Sept/Oct 2000, *QEX*, pp 13-17.
43. Z. Lau, W1VT, "A Narrow 80-Meter Band-Pass Filter," Sept/Oct 1998, *QEX*, p 57.
44. E. Wetherhold, W3NQN, "Receiver Band-Pass Filters Having Maximum Attenuation in Adjacent Bands," July/Aug 1999, *QEX*, pp 27-33.

Chapter 13

EMI/Direction Finding

THE SCOPE OF THE PROBLEM

As our lives become filled with technology, the likelihood of electronic interference increases. Every lamp dimmer, garage-door opener or other new technical "toy" contributes to the electrical noise around us. Many of these devices also "listen" to that growing noise and may react unpredictably to their electronic neighbors.

Sooner or later, nearly every Amateur Radio operator will have a problem with interference. Most cases of interference can be cured! The proper use of "diplomacy" skills and standard cures will usually solve the problem.

This section of Chapter 13, by Ed Hare, W1RFI, is only an overview. *The ARRL RFI Book* contains detailed information on the causes of and cures for nearly every type of interference problem.[1]

Pieces of the Problem

Every interference problem has two components — the equipment that is involved and the people who use it. A solution requires that we deal with both the equipment and the people effectively.

First, define the term "interference" without emotion. The ARRL recommends that the hams and their neighbors cooperate to find solutions. The FCC shares this view.

Important Terms

Bypass capacitor — a capacitor used to provide a low-impedance radio-frequency path around a circuit element.

Common-mode signals — signals that are in phase on both (or several) conductors in a system.

Conducted signals — signals that travel by electron flow in a wire or other conductor.

Decibel (dB) — a logarithmic unit of relative power measurement that expresses the ratio of two power levels.

Differential-mode signals — Signals that arrive on two or more conductors such that there is a 180° phase difference between the signals on some of the conductors.

Electromagnetic compatibility (EMC) — the ability of electronic equipment to be operated in its intended electromagnetic environment without either causing interference to other equipment or systems, or suffering interference from other equipment or systems.

Electromagnetic interference (EMI) — any electrical disturbance that interferes with the normal operation of electronic equipment.

Emission — electromagnetic energy propagated from a source by radiation.

Filter — a network of resistors, inductors and/or capacitors that offer little resistance to certain frequencies while blocking or attenuating other frequencies.

Fundamental overload — interference resulting from the fundamental signal of a radio transmitter.

Ground — a low-impedance electrical connection to the earth. Also, a common reference point in electronic circuits.

Harmonics — signals at exact integral multiples of the operating (or *fundamental*) frequency.

High-pass filter — a filter designed to pass all frequencies above a cutoff frequency, while rejecting frequencies below the cutoff frequency.

Immunity — the ability of electronic equipment to reject interference from external sources of electromagnetic energy. This is the conjugate of the term "susceptibility" and is the term typically used in the commercial world.

Induction — the transfer of electrical signals via magnetic coupling.

Interference — the unwanted interaction between electronic systems.

Intermodulation — the undesired mixing of two or more frequencies in a nonlinear device, which produces additional frequencies.

Low-pass filter — a filter designed to pass all frequencies below a cutoff frequency, while rejecting frequencies above the cutoff frequency.

Noise — any signal that interferes with the desired signal in electronic communications or systems.

Nonlinear — having an output that is not in linear proportion to the input.

Notch filter — a filter that rejects or suppresses a narrow band of frequencies within a wider band of frequencies.

Passband — the band of frequencies that a filter conducts with essentially no attenuation.

Radiated emission — radio-frequency energy that is coupled between two systems by electromagnetic fields.

Radio-frequency interference (RFI) — interference caused by a source of radio-frequency signals. This is a subclass of EMI.

Spurious emission — An emission, on frequencies outside the necessary bandwidth of a transmission, the level of which may be reduced without affecting the information being transmitted.

Susceptibility — the characteristic of electronic equipment that permits undesired responses when subjected to electromagnetic energy.

TVI — interference to television systems.

Responsibility

When an interference problem occurs, we may ask "Who is to blame?" The ham and the neighbor often have different opinions. It is almost natural (but unproductive) to fix blame instead of the problem.

No amount of wishful thinking (or demands for the "other guy" to solve the problem) will result in a cure for interference. Each individual has a unique perspective on the situation, and a different degree of understanding of the personal and technical issues involved. On the other hand, each person has certain responsibilities to the other and should be prepared to address those responsibilities fairly.

FCC Regulations

A radio operator is responsible for the proper operation of the radio station. This responsibility is spelled out clearly in Part 97 of the FCC regulations. If interference is caused by a spurious emission from your station, you *must* correct the problem there.

Fortunately, most cases of interference are *not* the fault of the transmitting station. Most interference problems involve some kind of electrical noise or fundamental overload.

Personal Diplomacy

What happens when you first talk to your neighbor sets the tone for all that follows. Any technical solutions cannot help if you are not allowed in your neighbor's house to explain them! If the interference is not caused by spurious emissions from your station, however, you should be a locator of solutions, not a provider of solutions.

Your neighbor will probably *not* understand all of the technical issues — at least not at first. Understand that, regardless of fault, an interference problem is annoying to your neighbor. Let your neighbor know that you want to help find a solution and that you want to begin by talking things over.

Talk about some of the more important technical issues, in non-technical terms. Interference can be caused by unwanted signals from your transmitter. Assure your neighbor that you will check your station thoroughly and correct any problems. You should also discuss the possible susceptibility of consumer equipment. You may want to print a copy of the RFI information found on *ARRLWeb* at: **www.arrl.org/news/rfi/neighbors.html**.

Here is a good analogy: If you tune your TV to channel 3, and see channel 8 instead, would you blame channel 8? No. You might check another set to see if it has the same problem, or call channel 8 to see if the station has a problem. If channel 8 was operating properly, you would likely decide that your TV set is broken. Now, if you tune your TV to channel 3, and see your local shortwave radio station (quite possibly Amateur Radio), don't blame the shortwave station without some investigation. In fact, many televisions respond to strong signals outside the television bands. They may be working as designed, but require added filters and/or shields to work properly near a strong, local RF signal.

Your neighbor will probably feel much better if you explain that you will help *find* a solution, even if the interference is *not* your fault. This offer can change your image from neighborhood villain to hero, especially if the interference is not caused by your station. (This is often the case.)

PREPARE YOURSELF

Learn About EMI

In order to troubleshoot and cure EMI, you need to learn more than just the basics. This is especially important when dealing with your neighbor. If you visit your neighbor's house and try a few dozen things that don't work (or make things worse), your neighbor may lose confidence in your ability to help cure the problem. If that happens, you may be asked to leave.

Local Help

If you are not an expert (and even experts can use moral support), you should find some local help. Fortunately, such help is often available from your Section Technical Coordinator (TC). The TC knows of any local RFI committees and may have valuable contacts in the local utility companies. Even an expert can benefit from a TC's help.

The easiest way to find your TC is through your ARRL Section Manager (SM). There is a list of SMs on page 16 of any recent *QST*. He or she can quickly put you in contact with the best source of local help.

Even if you can't secure the help of a local expert, a second ham can be a valuable asset. Often a second party can help defuse any hostility. It is also helpful to have someone to operate your station while you and your neighbor run through troubleshooting steps and try various cures.

Prepare Your Home

The first step toward curing an interference problem is to make sure your own signal is clean. You must eliminate all interference in your own house to be sure you are not causing the interference! This is also a valuable troubleshooting tool: If you know your station is clean, you have cut the size of the problem in half! If the FCC ever gets involved, you can demonstrate that you are not interfering with your own equipment.

Apply EMI cures to your own consumer electronics equipment. When your neighbor sees your equipment working well, it demonstrates that filters work and cause no harm.

To clean up your station, clean up the mess! A rat's nest of wires, unsoldered connections and so on in your station can contribute to EMI. To help build a better relationship, you may want to show your station to your neighbor. A clean station looks professional; it inspires confidence in your ability to solve the EMI problem.

Install a transmit filter (low-pass or band-pass) and a reasonable station ground. (If the FCC becomes involved, they will ask you about both items.) Show your neighbor that you have installed the necessary filter on your transmitter and explain that if there is still interference, it is necessary to try filters on the neighbor's equipment, too.

Operating practices and station-design considerations can affect EMI. Don't overdrive a transmitter or amplifier; that can increase its harmonic output. You can take steps to reduce the strength of your signal at the victim equipment. This might include reducing transmit power. Locate the antenna as far as possible from susceptible equipment or its wiring (ac line, telephone, cable TV). Antenna orientation may be important. For example, if your HF dipole at 30 ft is coupling into the neighbor's overhead cable-TV drop, that coupling could be reduced 20 dB by changing to a vertical antenna — even more by orienting the antenna so that the drop is off its end. Try different modes; CW or FM usually do not generate nearly as much telephone interference as AM or SSB, for example.

Call Your Neighbor

Now that you have learned more about EMI, located some local help (we'll assume it's the TC) and done all of your homework, make contact with your neighbor. First, arrange an appointment convenient for you, the TC and your neighbor. After you introduce the TC, allow him or her to explain the issues to your neighbor. Your TC will be able to answer most questions, but be prepared to assist with support and additional information as required.

Invite the neighbor to visit your station. Show your neighbor some of the things you do with your radio equipment. Point out any test equipment you use to keep your station in good working order. Of course, you want to show the filters you have installed on your transmitter.

Next, have the TC operate your station on several different bands. Show your neighbor that your home electronics equipment is working properly while your station is in operation. Point out the filters you have installed to correct any susceptibility problems.

At this point, tell your neighbor that the next step is to try some of these cures on his or her equipment. This is a good time to emphasize that the problem is probably not your fault, but that you and the TC will try to help find a solution anyway.

Table 13.1 is a list of the things needed to troubleshoot and solve most EMI problems. Decide ahead of time which of these items are needed and take them with you.

Table 13.1
EMI Survival Kit

Filters:

(2) 300-Ω high-pass filter (different brands recommended)
(2) 75-Ω high-pass filter (different brands recommended)
(2) Commercially available common-mode chokes (optional)
(12) Assorted ferrite cores: 43, 63 and 75 material, FT-140 and FT-240 size
(3) Telephone RFI filters (different brands recommended)
(2) Brute-force ac line filters
(6) 0.01-µF ceramic capacitors
(6) 0.001-µF ceramic capacitors
(2) Speaker-lead filters (optional)

Miscellaneous:

- Hand tools, assorted screwdrivers, wire cutters, pliers
- Hookup wire
- Electrical tape
- Soldering iron and solder (use with caution!)
- Assorted lengths 75-Ω coaxial cable with connectors
- Spare F connectors, male
- F-connector female-female "barrel"
- Alligator clips
- Notebook and pencil
- Portable multimeter

At Your Neighbor's Home

You and the TC should now visit the neighbor's home. Inspect the equipment installation and ask when the interference occurs, what equipment is involved and what frequencies or channels are affected. The answers are valuable clues. Next, either you or the TC should operate your station while the other observes the effects. Try all bands and modes that you use. Ask the neighbor to demonstrate the problem.

The tests may show that your station isn't involved at all. You may immediately recognize electrical noise or some kind of equipment malfunction. If so, explain your findings to the neighbor and suggest that he or she contact appropriate service personnel.

EMC Fundamentals

Knowledge is one of the most valuable tools for solving EMI problems. A successful EMI cure usually requires familiarity with the relevant technology and troubleshooting procedures.

SOURCE-PATH-VICTIM

All cases of EMI involve a *source* of electromagnetic energy, a device that responds to this electromagnetic energy (*victim*) and a transmission path that allows energy to flow from the source to the victim. Sources include radio transmitters, receiver local oscillators, computing devices, electrical noise, lightning and other natural sources.

There are three ways that EMI can travel from the source to the victim: radiation, conduction and induction. Radiated EMI propagates by electromagnetic radiation from the source, through space to the victim. A conducted signal travels over wires connected to the source and the victim. Induction occurs when two circuits are magnetically (and in some cases, electrically) coupled. Most EMI occurs via conduction, or some combination of radiation and conduction. For example, a signal is radiated by the source and picked up by a conductor attached to the victim (or directly by the victim's circuitry) and is then conducted into the victim. EMI from induction is rare.

DIFFERENTIAL VS COMMON-MODE

It is important to understand the differences between differential-mode and common-mode conducted signals (see **Fig 13.1**). Each of these conduction modes requires different EMI cures. Differential-mode cures, (the typical high-pass filter, for example) do not attenuate common-mode signals. On the other hand, a typical common-mode choke does not affect interference resulting from a differential-mode signal.

Differential-mode currents usually have two easily identified conductors. In a two-wire transmission line, for example, the signal leaves the generator on one line and returns on the other. When the two conductors are in close proximity, they form a transmission line and there is a 180° phase difference between their respective signals. It's relatively simple to build a filter that passes desired signals and shunts unwanted signals to the return line. Most *desired* signals, such as the TV signal inside a coaxial cable are differential-mode signals.

In a common-mode circuit, many wires of a multiwire system act as if they were a single wire. The result can be a good antenna, either as a radiator or as a receptor of unwanted energy. The return path is usually earth ground. Since the source and return conductors are usually well separated, there is no reliable phase difference between the conductors and no convenient place to shunt unwanted signals. Toroid chokes are the answer to common-mode interference. (The following explanation applies to rod cores as well as toroids, but since rod cores may couple into nearby circuits, use them only as a last resort.)

Toroids work differently, but equally well, with coaxial cable and paired conductors. A common-mode signal on a coaxial cable is usually a signal that is present on the *outside* of the cable *shield*. When we wrap the cable around a ferrite-toroid core, the choke appears as a reactance in series with the outside of the shield, but it has no

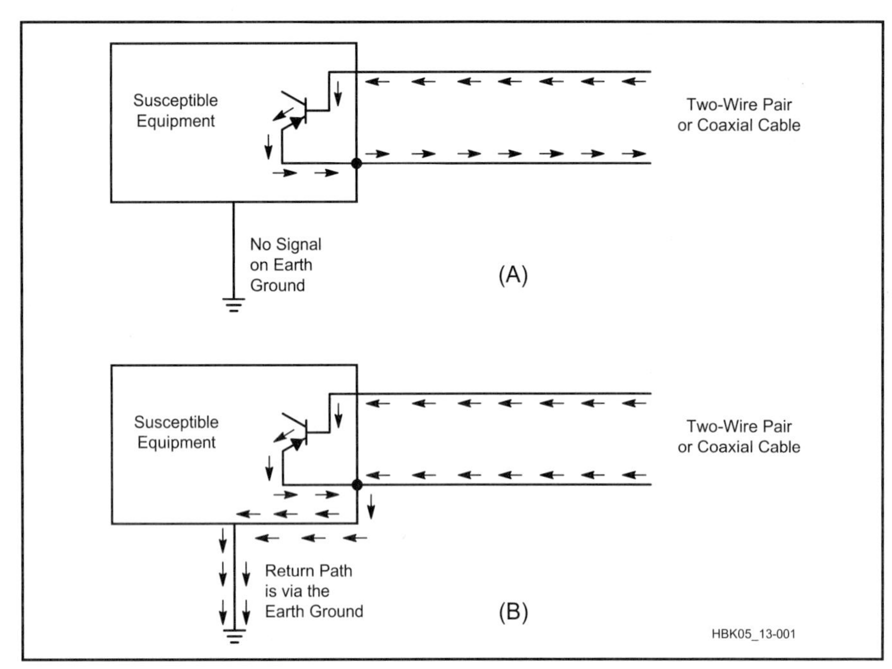

Fig 13.1 — A shows a differential-mode, while B shows a common-mode signal. The two kinds of signals are described in the text.

effect on signals inside the cable because their field is (ideally) confined inside the shield. With paired conductors such as zip-cord, signals with opposite phase set up magnetic fluxes of opposite phase in the core. These "differential" fluxes cancel each other, and there is no net reactance for the differential signal. To common-mode signals, however, the choke appears as a reactance in series with the line.

Toroid chokes work less well with single-conductor leads. Because there is no return current to set up a canceling flux, the choke appears as a reactance in series with *both* the desired and undesired signals.

SOURCES OF EMI

The basic causes of EMI can be grouped into several categories:

- Fundamental overload effects
- External noise
- Spurious emissions from a transmitter
- Intermodulation distortion or other external spurious signals

As an EMI troubleshooter, you must determine which of these are involved in your interference problem. Once you do, it is easy to select the necessary cure.

Fundamental Overload

Most cases of interference are caused by fundamental overload. The world is filled with RF signals. Properly designed equipment should be able to select the desired signal, while rejecting all others. Unfortunately, because of design deficiencies such as inadequate shields or filters, some equipment is unable to reject strong out-of-band signals.

A strong fundamental signal can enter equipment in several different ways. Most commonly, it is conducted into the equipment by wires connected to it. Possible conductors include antennas and feed lines, interconnecting cables, power lines and ground wires. TV antennas and feed lines, telephone or speaker wiring and ac power leads are the most common points of entry.

The effect of an interfering signal is directly related to its strength. The strength of a radiated signal diminishes with the square of the distance from the source. When the distance from the source doubles, the strength of the electromagnetic field decreases to one-fourth of its strength at the original distance from the source. This characteristic can often be used to help solve EMI cases. You can often make a significant improvement by moving the victim equipment and the antenna farther away from each other.

External Noise

Most cases of interference reported to the FCC involve some sort of external noise source. The most common of these noise sources are electrical. External "noise" can also come from transmitters or from unlicensed RF sources such as computers, video games, electronic mice repellers and the like. Typically, such devices are legal under Part 15 of the FCC's rules.

Electrical noise is fairly easy to identify by looking at the picture of a susceptible TV or listening on an HF receiver. A photo of electrical noise on a TV screen is shown in the TVI section of this chapter. On a receiver, it usually sounds like a buzz, sometimes changing in intensity as the arc or spark sputters a bit. If you determine the problem to be caused by external noise, it must be cured at the source. Refer to the Electrical Noise section of this chapter and *The ARRL RFI Book*.

Spurious Emissions

All transmitters generate some (hopefully few) RF signals that are outside their allocated frequency bands. These out-of-band signals are called spurious emissions, or *spurs*. Spurious emissions can be

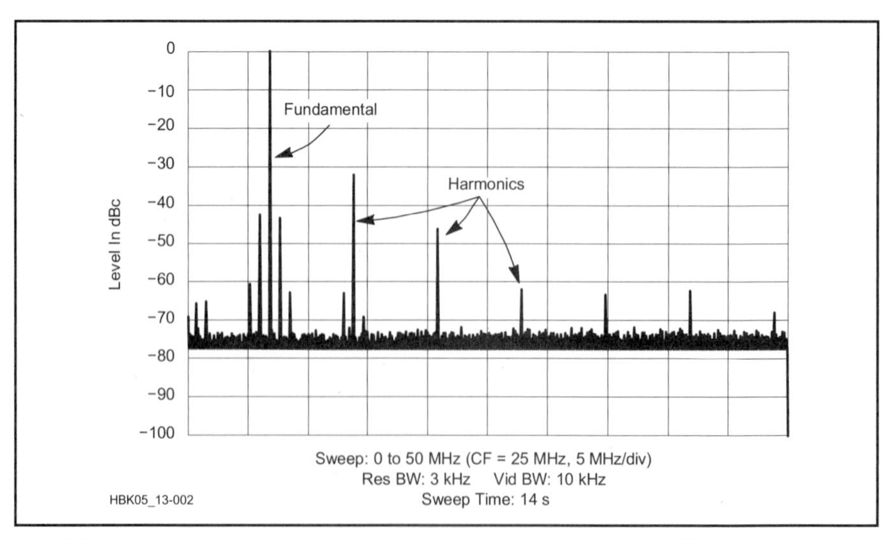

Fig 13.2 — The spectral output of a typical amateur transmitter. The fundamental is at 7 MHz. There are visible harmonics at 14, 21 and 28 MHz. Unlabeled lines are non-harmonic spurious emissions. This transmitter complies with the stringent FCC spectral-purity regulations regarding amateur transmitters with less than 5 W of RF output.

discrete signals or wideband noise. Harmonics, the most common spurious emissions, are signals at exact multiples of the operating (or *fundamental*) frequency. Other discrete spurious signals are usually caused by the superheterodyne mixing process used in most modern transmitters. **Fig 13.2** shows the spectral output of a transmitter, including harmonics and mixing products.

Transmitters may also produce broadband noise and/or "parasitic" oscillations. (Parasitic oscillations are discussed in the **RF Power Amplifiers** chapter.) If these unwanted signals cause interference to another radio service, FCC regulations require the owner to correct the problem.

TROUBLESHOOTING EMI

Most EMI cases are complex. They involve a source, a path and a victim. Each of these main components has a number of variables: Is the problem caused by harmonics, fundamental overload, conducted emissions, radiated emissions or a combination of all of these factors? Should it be fixed with a low-pass filter, high-pass filter, common-mode chokes or ac-line filter? How about shielding, isolation transformers, a different ground or antenna configuration?

By the time you finish with these questions, the possibilities could number in the millions. You probably will not see your exact problem and cure listed in this book or any other. You must diagnose the problem!

Troubleshooting an EMI problem is a three-step process, and all three steps are equally important:

- Identify the problem
- Diagnose the problem
- Cure the problem.

Identify the Problem

Is It Really EMI? — Before trying to solve a suspected case of EMI, verify that the symptoms actually result from external causes. A variety of equipment malfunctions or external noise can look like interference. "Your" EMI problem might be caused by another ham or a radio transmitter of another radio service, such as a local CB or police transmitter.

Is It Your Station? — If it appears that your station is involved, operate your station on each band, mode and power level that you use. Note all conditions that produce interference. If no transmissions produce the problem, your station *may* not be the cause. (Although some contributing factor may have been missing in the test.) Have your neighbor keep notes of when and how the interference appears: what time of day, what station, what other appliances were in use, what was the weather? You should do the same whenever you operate. If you can readily reproduce the problem with your station, you can start to troubleshoot the problem.

Diagnose the Problem

Look Around — Aside from the brain, eyes are a troubleshooter's best tool. Look around. Installation defects contribute to many EMI problems. Look for loose connections, shield breaks in a cable-TV installation or corroded contacts in a telephone installation. Fix these first.

Problems that occur only on harmonics of the fundamental signal usually indicate the transmitter. Harmonics can also be generated in nearby semiconductors, such as an unpowered VHF receiver left connected to an antenna, or a corroded connection in a tower guy wire. Harmonics can also be generated in the front-end components of the TV or radio experiencing interference.

Is the wiring connected to the victim equipment resonant on one or more amateur bands? If so, a common-mode choke placed at the middle of the wiring may be an easy cure.

These are only a few of the questions you might need to ask. Any information you gain about the systems involved will help find the EMI cause and cure.

Cures

At Your Station — Make sure that your own station and consumer equipment are clean. This cuts the size of the problem in half! Once this is done, you won't need to diagnose or troubleshoot your station later. Also, any cures successful at your house may work at your neighbor's as well. If you do have problems in your own house, refer first to the Transmitter section of this chapter, or continue through the troubleshooting steps and specific cures and take care of your own problem first.

Simplify the Problem — Don't tackle a complex system — such as a telephone system in which there are two lines running to 14 rooms — all at once. You could spend the rest of your life running in circles and never find the true cause of the problem.

There's a better way. In our hypothetical telephone system, first locate the telephone jack closest to the telephone service entrance. Disconnect the lines to more remote jacks and connect one EMI-resistant telephone at the remaining jack. If the interference remains, try cures until the problem is solved, then start adding lines and equipment back one at a time, fixing the problems as you go along. If you are lucky, you will solve all of the problems in one pass. If not, at least you can point to one piece of equipment as the source of the problem.

Multiple Causes — Many EMI problems have multiple causes. These are usually the ones that give new EMI troubleshooters the most trouble. If, for example, a TVI problem is caused by harmonics from the transmitter, an arc in the transmitting antenna, an overloaded TV preamp, differential-mode fundamental overload generating harmonics in the TV tuner, induced and conducted RF in the ac-power system and a common-mode signal picked up on the shield of the TV's coaxial feed line, you would never find a cure by trying only one at a time!

In this case, the solution requires that you apply all of the cures at the same time. When troubleshooting, if you try a cure, leave it in place. When you finally try a cure that really works, start removing the "temporary" attempts one at a time. If the interference returns, you know that there were multiple causes.

OVERVIEW OF TECHNIQUES

Shields

Shields are used to set boundaries for radiated energy. Thin conductive films, copper braid and sheet metal are the most common shield materials. Maximum shield effectiveness usually requires solid sheet metal that completely encloses the source or susceptible circuitry or equipment. Small discontinuities, such as holes or seams, decrease shield effectiveness. In addition, mating surfaces between different parts of a shield must be conductive. To ensure conductivity, file or sand off paint or other nonconductive coatings on mating surfaces.

Filters

A major means of separating signals relies on their frequency differences. Filters offer little opposition to certain frequencies while blocking others. Filters vary in attenuation characteristics, frequency characteristics and power-handling capabilities. The names given to various filters are based on their uses.

Low-pass filters pass frequencies below some cutoff frequency, while attenuating frequencies above that cutoff frequency. A typical low-pass filter curve is shown in **Fig 13.3**. A schematic is shown in **Fig 13.4**. These filters are difficult to construct properly so you should buy one. Many retail Amateur Radio stores that advertise in *QST* stock low-pass filters.

High-pass filters pass frequencies above some cutoff frequency while attenuating frequencies below that cutoff frequency. A typical high-pass filter curve is shown in **Fig 13.5**. **Fig 13.6** shows a schematic of a typical high-pass filter. Again, it is best to buy one of the commercially available filters.

Bypass capacitors can be used to cure EMI problems. A bypass capacitor is usually placed between a signal or power lead and circuit ground. It provides a low-impedance path to ground for RF signals. Bypass capacitors for HF signals are usually 0.01 µF, while VHF bypass capacitors are usually 0.001 µF.

AC-line filters, sometimes called "brute-force" filters, are used to filter RF energy from power lines. A schematic is shown in **Fig 13.7**. Only *ac-rated* capacitors as specified in Fig 13.7 should be used for the filter. These RFI capacitors must be used in applications where a hazard could be present to a person who touches the associated equipment — if such a capacitor were to fail. Y-Class capacitors are designed for connection between power lines and from line to ground. There are several sub-classes of Y-class capacitors, Y2 being the most common. Y2 capaci-

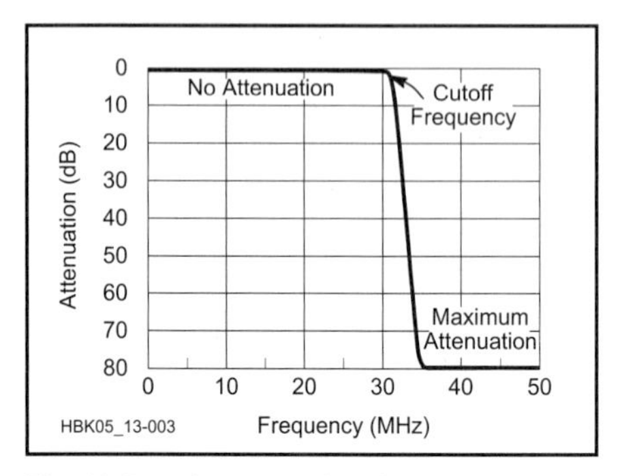

Fig 13.3 — An example of a low-pass filter-response curve.

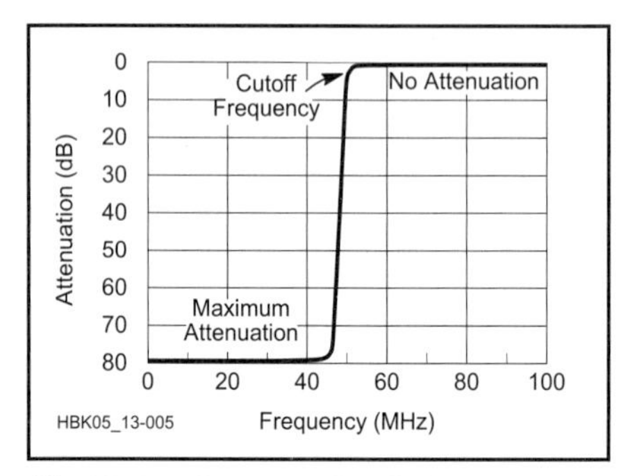

Fig 13.5 — An example of a high-pass filter response curve.

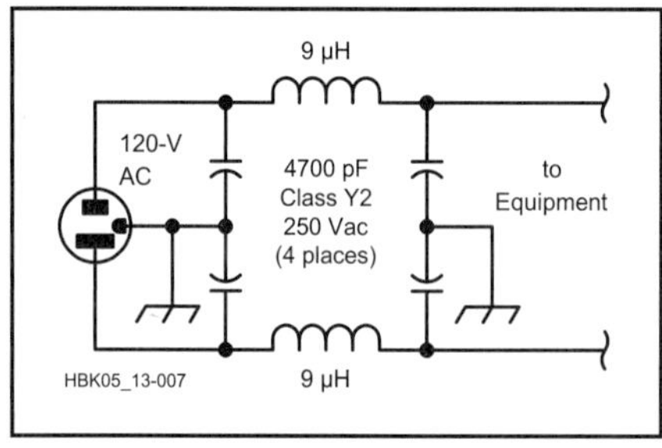

Fig 13.7 — A "brute-force" ac-line filter.

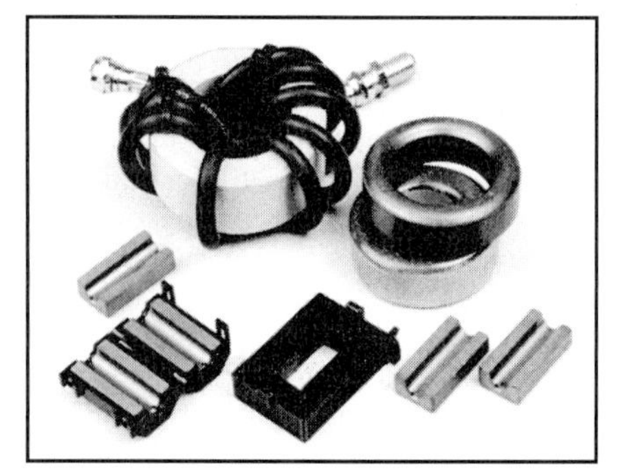

Fig 13.8 — Several styles of common-mode chokes.

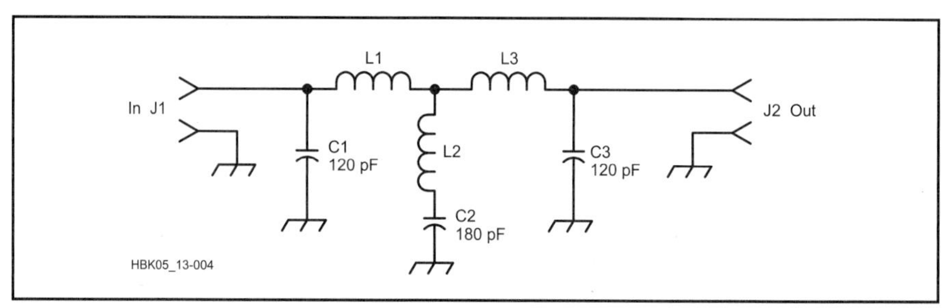

Fig 13.4 — A low-pass filter for amateur transmitting use. Complete construction information appears in the Transmitters chapter of *The ARRL RFI Book*.

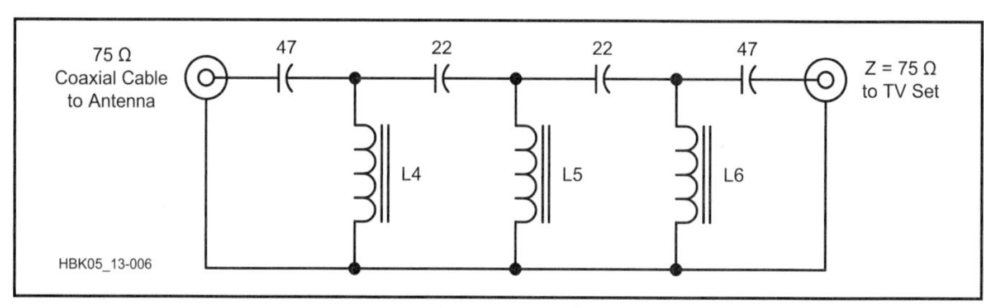

Fig 13.6 — A differential-mode high-pass filter for 75-Ω coax. It rejects HF signals picked up by a TV antenna or that leak into a cable-TV system. It is ineffective against common-mode signals. All capacitors are high-stability, low-loss, NP0 ceramic discs. Values are in pF. The inductors are all #24 enameled wire on T-44-0 toroid cores. L4 and L6 are each 12 turns (0.157 µH). L5 is 11 turns (0.135µH).

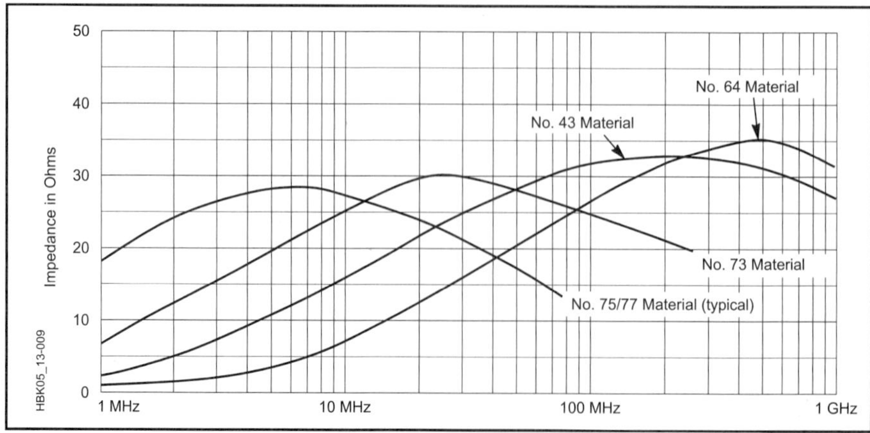

Fig 13.9 — Impedance vs. frequency plots for "101" size ferrite beads.

tors are rated for nominal working voltages less than or equal to 250 Vac. Their peak impulse voltage rating is considerably higher however, up to 5 kV.

Common-Mode Chokes

Common-mode chokes may be the best-kept secret in Amateur Radio. The differential-mode filters described earlier are *not* effective against common-mode signals. To eliminate common-mode signals properly, you need common-mode chokes. They may help nearly any interference problem, from cable TV to telephones to audio interference caused by RF picked up on speaker leads.

Common-mode chokes usually have ferrite core materials. These materials are well suited to attenuate common-mode currents. Several kinds of common-mode chokes are shown in **Fig 13.8**.

The optimum size and ferrite material are determined by the application and frequency. For example, an ac cord with a plug attached cannot be easily wrapped on a small ferrite core. The characteristics of ferrite materials vary with frequency, as shown by the graph in **Fig 13.9**.

Grounds

An electrical ground is not a huge sink that somehow swallows noise and unwanted signals. Ground is a *circuit* concept, whether the circuit is small, like a radio receiver, or large, like the propagation path between a transmitter and cable-TV installation. Ground forms a universal reference point between circuits.

This chapter deals with the EMC aspects of grounding. While grounding is not a cure-all for EMI problems, ground is an important safety component of any electronics installation. It is part of the lightning protection system in your station and a critical safety component of your house wiring. Any changes made to a grounding system must not compromise these important safety considerations. Refer to the **Safety** chapter for important information about grounding.

Many amateur stations have several grounds: a safety ground that is part of the ac-wiring system, another at the antenna for lightning protection and perhaps another at the station for EMI control. These grounds can interact with each other in ways that are difficult to predict.

Ground Loops

All of these station grounds can form a large ground loop. This loop can act as a large loop antenna, with increased susceptibility to lightning or EMI problems. **Fig 13.10** shows a ground loop and a proper single-point ground system.

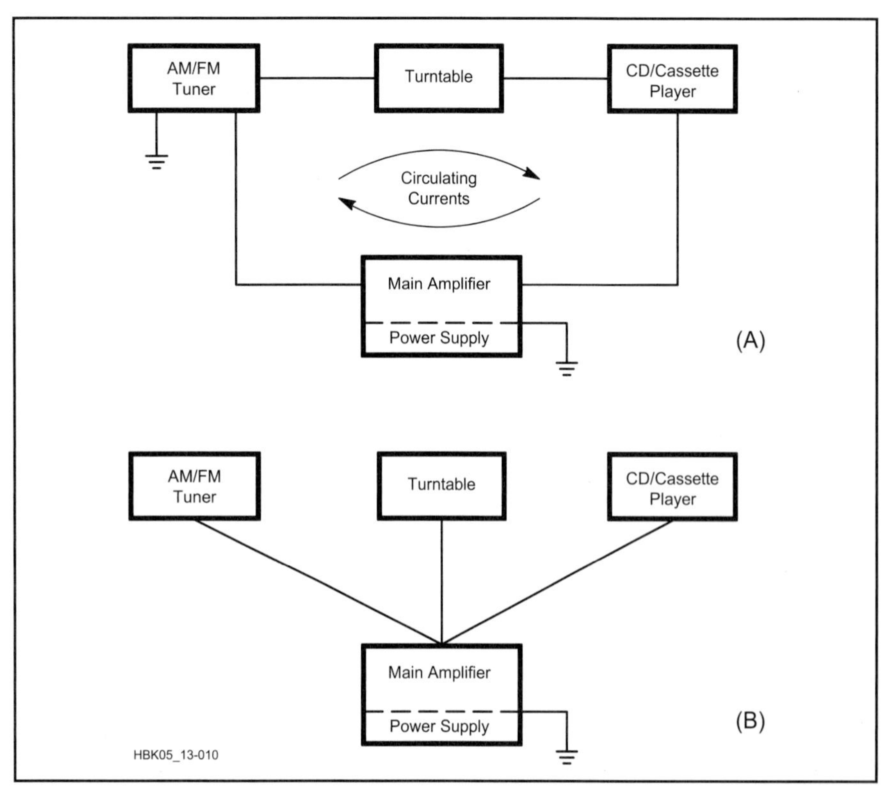

Fig 13.10 — A shows a stereo system grounded as an undesirable "ground loop." B is the proper way to ground a multiple-component system.

When is Ground not a Ground?

In many stations, it is impossible to get a good RF connection to earth ground. Most practical installations require several feet of wire between the station ground connection and an outside ground rod. Many troublesome harmonics are in the VHF range. At VHF, a ground wire length can be several wavelengths long — a very effective long-wire antenna! Any VHF signals that are put on a long ground wire will be radiated. This is usually not the intended result of grounding.

Take a look at the station shown in **Fig 13.11**. In this case, the ground wire could very easily contribute to an interference problem in the downstairs TV set.

While a station ground may cure some transmitter EMI problems — either by putting the transmitter chassis at a low-impedance reference point or by rearranging the problem so the "hot spots" are farther away from susceptible equipment — it is not the cure-all that some literature has suggested. A ground is easy to install, and it may reduce stray fundamental or harmonic currents on your antenna lead; it is worth a try.

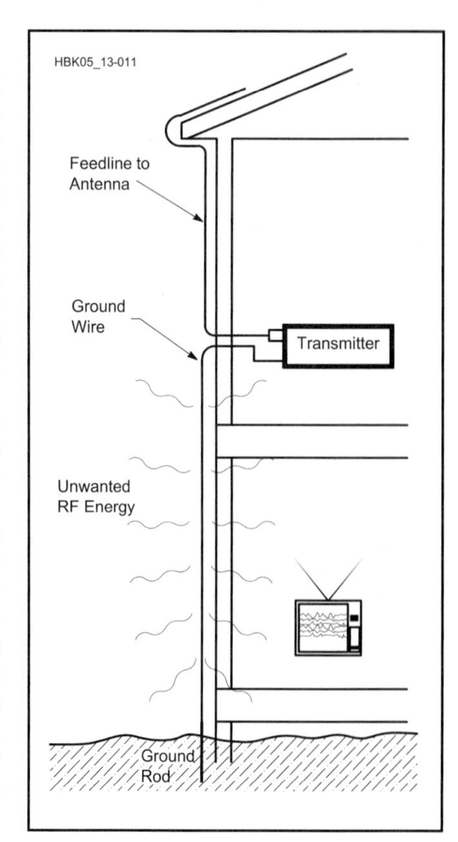

Fig 13.11 — When a transmitter is located on an upper floor, the ground lead may act as an antenna for VHF/UHF energy. It may be better to not use a normal ground.

SPECIFIC CURES

Now that you have learned some EMI fundamentals, you can work on technical solutions. A systematic approach will identify the problem and suggest a cure. Armed with your EMI knowledge, a kit of filters and tools, your local TC and a determination to solve the problem, it is time to diagnose the problem.

Most EMI problems can be solved by the application of standard cures. If you try these cures and they work, you may not need to troubleshoot the problem at all. Perhaps if you can install a low-pass filter on your transmitter or a common-mode choke on a TV, the problem will be solved.

Here are some specific cures for different interference problems. You should also get a copy of *The ARRL RFI Book*. It's comprehensive and picks up where this chapter leaves off. Here are several standard cures.

Transmitters

We start with transmitters not because most interference comes from transmitters, but because your station transmitter is under your direct control. Many of the troubleshooting steps in other parts of this chapter assume that your transmitter is "clean" (free of unwanted RF output).

Controlling Spurious Emissions — Start by looking for patterns in the interference. If the interference is only on frequencies that are multiples of your operating frequency, you clearly have interference from harmonics. (Although these harmonics may *not* come from your station!)

If HF-transmitter spurs are interfering with a VHF service, a low-pass filter on the transmitter will usually cure the problem. Install it after the amplifier (if used) and *before* the antenna tuner. (A second filter between the transmitter and amplifier may occasionally help as well.) Install a low-pass filter as your first step in any interference problem that involves another radio service.

Interference from non-harmonic spurious emissions is extremely rare in commercially built radios. Any such problem indicates a malfunction that should be repaired.

Television Interference (TVI)

For a TV signal to look good, it must have about a 45 to 50 dB signal-to-noise ratio. This requires a good signal at the TV antenna-input connector. This brings up an important point: to have a good signal, you must be in a good signal area. The FCC does not protect fringe-area reception.

TVI, or interference to any radio service, can be caused by one of several things:

- Spurious signals within the TV channel coming from your transmitter or station.
- The TV set may be overloaded by your transmitter's fundamental signal.
- Signals within the TV channel from some source other than your station, such as electrical noise, an overloaded mast-mounted TV preamplifier or a transmitter in another service.
- The TV set might be defective or misadjusted, making it look like there is an interference problem.

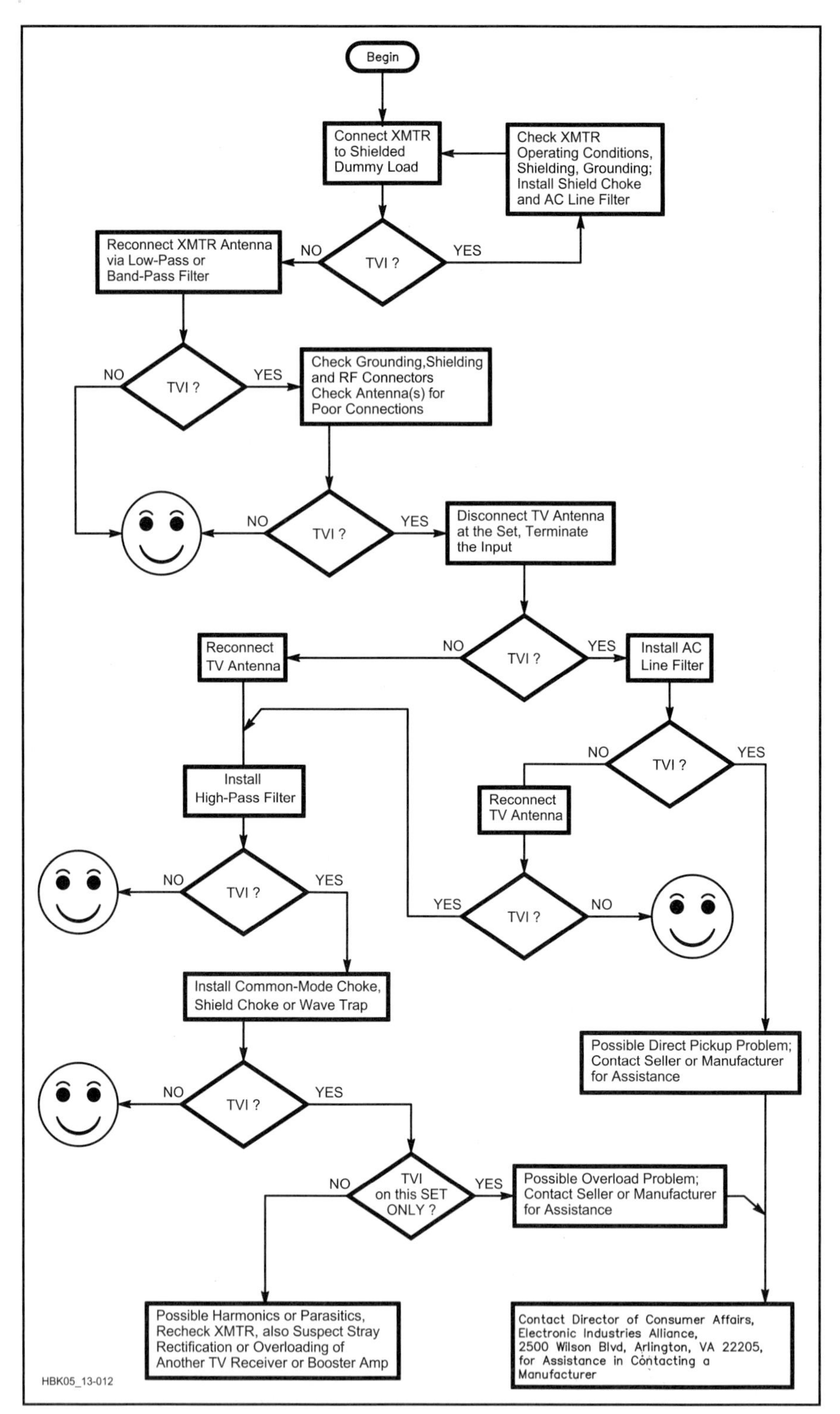

Fig 13.12 — TVI Troubleshooting Flowchart

All of these potential problems are made more severe because the TV set is hooked up to *two* antenna systems: (1) the incoming antenna and its feed line and (2) the ac power lines. These two "long-wire" antennas can couple *a lot* of fundamental or harmonic energy into the TV set! The TVI Troubleshooting Flowchart in **Fig 13.12** is a good starting point.

> **Warning: Performing Repairs**
>
> You are the best judge of a local situation, but the ARRL strongly recommends that you do not work on your neighbor's equipment. The minute you take the back off a TV or open up a telephone, you may become liable for problems. Internal modifications to your neighbor's equipment may cure the interference problem, but months later, when that 25-year-old clunker gives up the ghost, you may be held to blame. In some states, it is *illegal* for you to do *any* work on electronic equipment other than your own. — *Ed Hare, W1RFI, ARRL Laboratory Supervisor*

Fundamental Overload

A television set can be overloaded by a strong, local RF signal. This happens because the manufacturer did not install the necessary filters and shields to protect the TV set from other signals present on the air. These design deficiencies can sometimes be corrected externally.

Start by determining if the interference is affecting the video, the sound or both. If it is present only on the sound, it is probably a case of audio rectification. (See the Stereos section of this chapter.) If it is present on the video, or both, it could be getting into the video circuitry or affecting either the tuner or IF circuitry.

The first line of defense for an antenna-connected TV is a high-pass filter. Install a high-pass filter directly on the back of the TV set. You may also have a problem with common-mode interference. The second line of defense is a common-mode choke on the antenna feed line — try this first in a cable-television installation. These two filters can probably cure most cases of TVI!

Fig 13.13 shows a "bulletproof" installation. If this doesn't cure the problem, the TV circuitry is picking up your signal directly. In that case, don't try to fix it yourself — it is a problem for the TV manufacturer.

VHF Transmitters — A VHF transmitter can interfere with over-the-air TV reception. Most TV tuners are not very selective and a strong VHF signal can overload the tuner easily. In this case, a VHF notch or stop-band filter at the TV can help by reducing the VHF fundamental signal that gets to the TV tuner. Star Circuits is one company that sells tunable notch filters.[2]

The Electronic Industries Alliance (EIA) can help you contact equipment manufacturers. Contact them directly for assistance in locating help at **www.eia.org**.

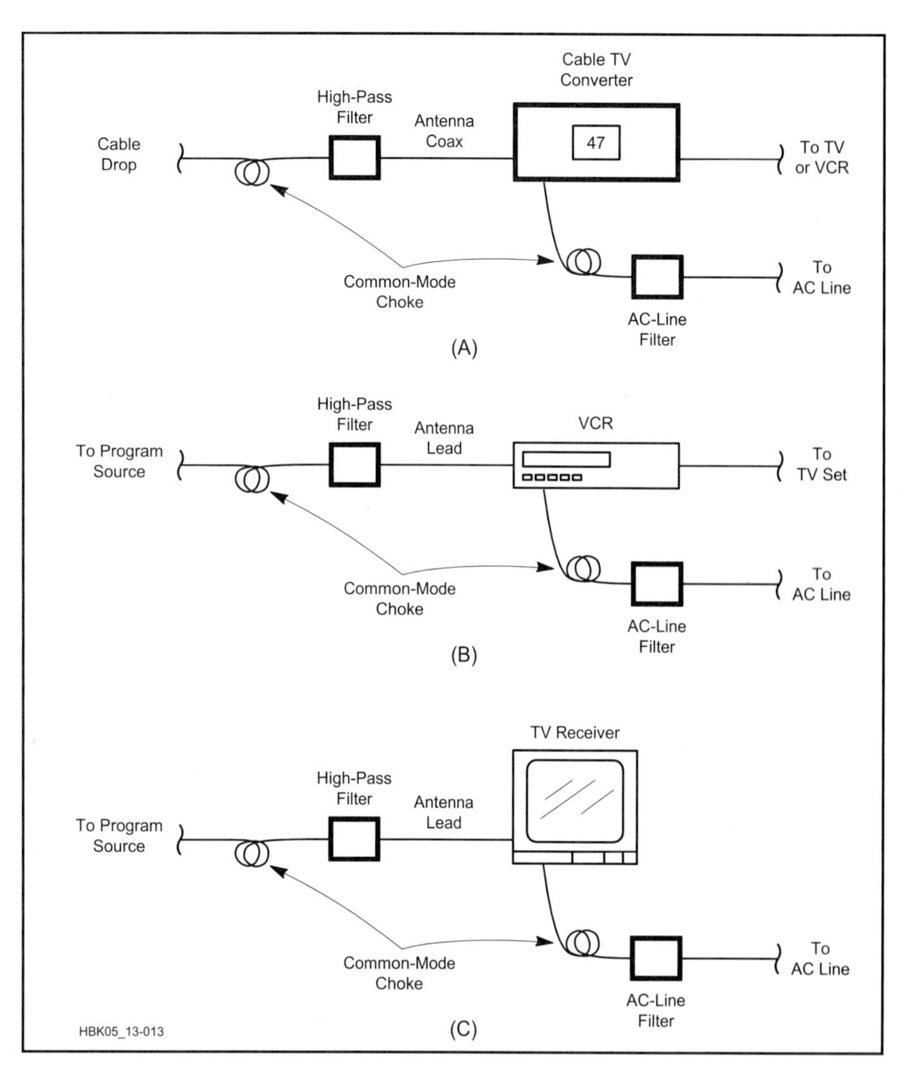

Fig 13.13 — This sort of installation should cure any kind of conducted TVI. It will not cure direct-pickup or spurious-emission problems.

Spurious Emissions

Start by analyzing which TV channels are affected. The TV Channel Chart, **Fig 13.14**, shows the relationship of the ham allocations and their harmonics to over-the-air and cable channels. Each channel is 6-MHz wide. If the interference is only on channels that are multiples of your operating frequency, you clearly have interference from harmonics. (It is not certain that these harmonics are coming from your station, however.)

You are responsible for spurious signals produced by your station. If your station is generating any interfering spurious sig-

> **Warning: Surplus Toroidal Cores**
>
> Don't use an unknown core or an old TV yoke core to make a common-mode choke. Such cores may not be suitable for the frequency you want to remove. If you try one of these "unknowns" and it doesn't work, you may incorrectly conclude that a common-mode choke won't help. Perhaps the correct material would have done the job.
>
> Ferrite beads are also used for EMI control, both as common-mode chokes and low-pass filters. It takes quite a few beads to be effective at the lower end of the HF range, though. It is usually better to form a common-mode choke by wrapping about 10 to 20 turns of wire or coaxial cable around an FT-140 (1.4-inch OD) or FT-240 (2.4-inch OD) core of the correct material. Mix 43 is a good material for most of the HF and VHF ranges. — *Ed Hare, W1RFI, ARRL Laboratory Supervisor*

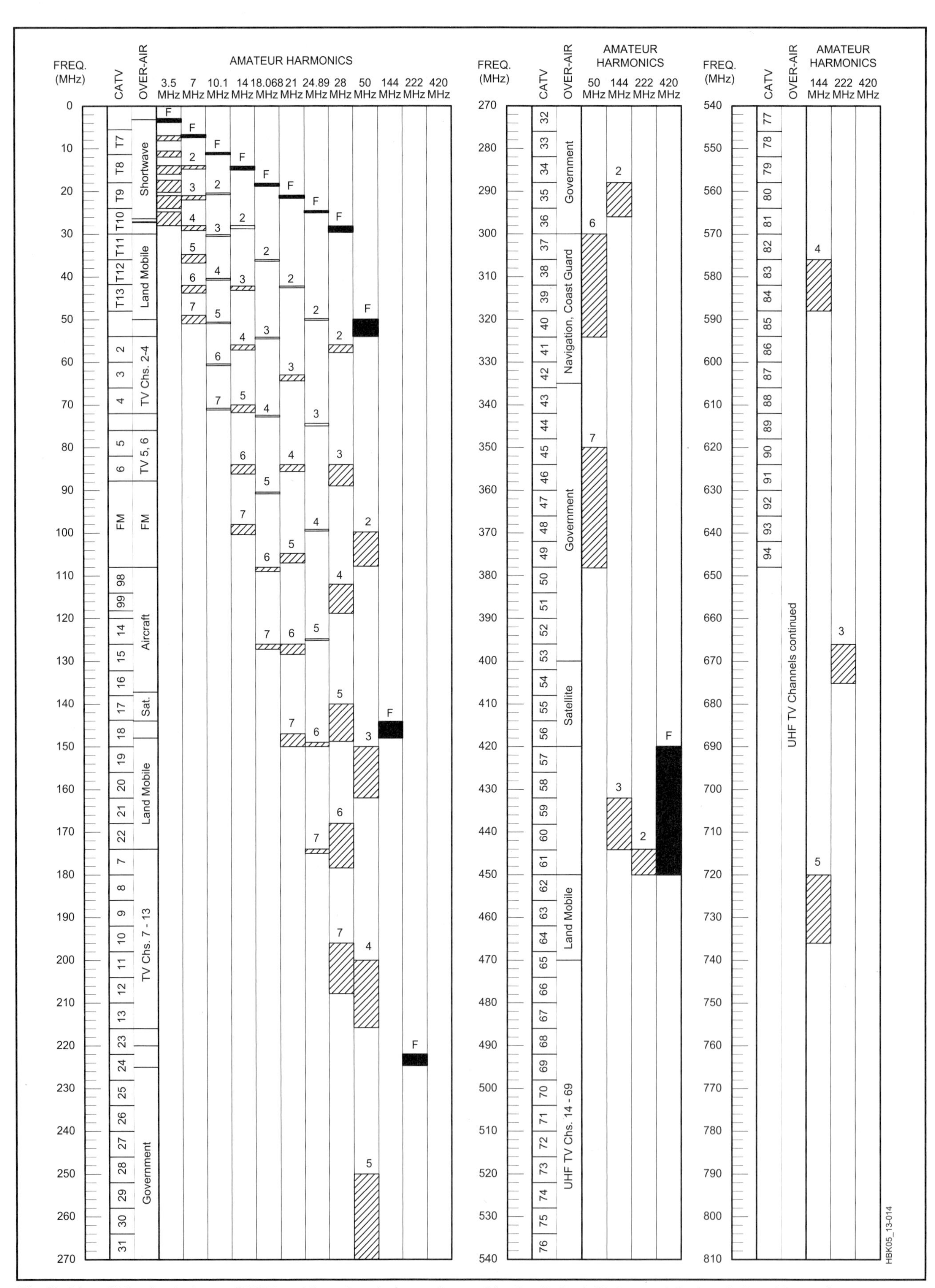

Fig 13.14 — This chart shows CATV and broadcast channels used in the United States and their relationship to the harmonics of MF, HF, VHF and UHF amateur bands.

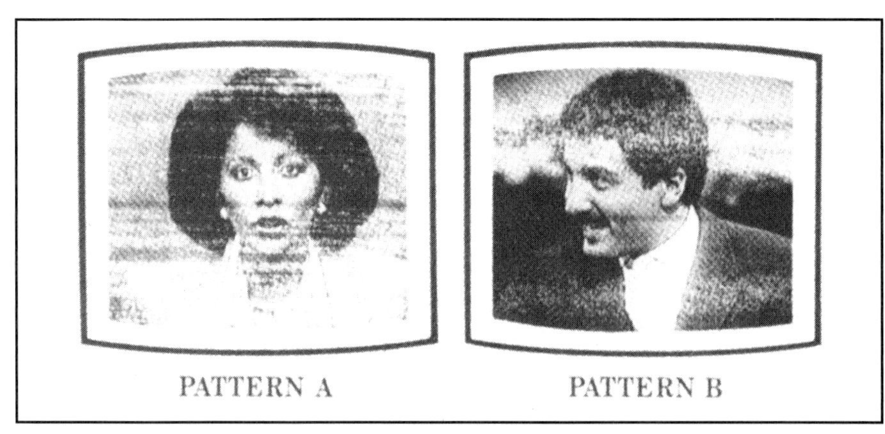

Fig 13.15 — Two examples of TVs experiencing electrical noise.

Fig 13.16 — Several turns of coax on a ferrite core eliminate HF and VHF signals from the outside of a coaxial cable.

nals, the problem must be cured there. So, if the problem occurs only when you transmit, go back and check your station. Refer to the section on Transmitters. You must first find out if the transmitter has any spurs.

If your transmitter and station check "clean," then you must look elsewhere. The most likely cause is TV susceptibility — fundamental overload. This is usually indicated by interference to all channels, or at least all VHF channels. If the problem is fundamental overload, see that section earlier in this chapter. If not, read on.

Electrical Noise

Electrical noise is fairly easy to identify by looking at the picture or listening on an HF receiver. Electrical noise on a TV screen is shown in **Fig 13.15**. On a receiver, it usually sounds like a buzz, sometimes changing in intensity as the arc or spark sputters a bit. If you have a problem with electrical noise, go to the Electrical Noise section.

Cable TV

Cable TV has been a blessing and a curse for Amateur Radio TVI problems. On the plus side, the cable delivers a strong, consistent signal to the TV receiver. It is also (in theory) a shielded system, so an external signal can't get in and cause trouble. On the minus side, the cable forms a large, long-wire antenna that can pick up lots of external signals on its shield (in the common mode). Many TVs and VCRs and even some cable set-top converters are easily overloaded by such common-mode signals.

Leakage into a cable-TV system is called ingress. Leakage out is called egress. If the cable isn't leaking, there should be no external signals getting inside the cable. So, an in-line filter such as a high-pass filter is not usually necessary. For a cable-connected TV, the first line of defense is a common-mode choke. Only in rare cases is a high-pass filter necessary. It is important to remember this, because if your neighbor has several TVs connected to cable and you suggest the wrong filter (at $15 each), you may have a personal diplomacy problem of a whole new dimension. **Fig 13.16** shows a common-mode choke.

Fig 13.13 shows a bulletproof installation for cable TV. (The high-pass filter is usually not needed.) If all of the cures shown have been tried, the interference probably results from direct pickup inside the TV. In this case, contact the TV manufacturer through the EIA.

Interference to cable-TV installations from VHF transmitters is a special case. Cable TV uses frequencies allocated to over-the-air services, such as Amateur Radio. When the cable shielding is less than perfect, interference can result.

The TV Channel Chart in Fig 13.14 shows which cable channels coincide with

> **PART II**
>
> **INTERFERENCE TO OTHER EQUIPMENT**
>
> **CHAPTER 6**
>
> **TELEPHONES, ELECTRONIC ORGANS, AM/FM RADIOS, STEREO AND HI-FI EQUIPMENT**
>
> Telephones, stereos, computers, electronic organs and home intercom devices can receive interference from nearby radio transmitters. When this happens, the device improperly functions as a radio receiver. Proper shielding or filtering can eliminate such interference. The device receiving interference should be modified in your home while it is being affected by interference. This will enable the service technician to determine where the interfering signal is entering your device.
>
> The device's response will vary according to the interference source. If, for example, your equipment is picking up the signal of a nearby two-way radio transmitter, you likely will hear the radio operator's voice. Electrical interference can cause sizzling, popping or humming sounds.

Fig 13.17 — Part of page 18 from FCC *Interference Handbook* (1990 edition) explains the facts and places responsibility for interference to non-radio equipment.

ham bands. If, for example, you have interference to cable channel 18 from amateur 2-m operation, suspect cable ingress. Contact the cable company; it may be their responsibility to locate and correct the problem. The cable company is not responsible, however, for leakage occurring in customer-owned, cable-ready equipment that is tuned to the same frequency as the over-the-air signal. If there is interference to a cable-TV installation, the cable company should be able to demonstrate interference-free reception when using a cable-company supplied set-top converter.

TV Preamplifiers

Some television owners use a preamplifier — sometimes when it's not needed. Preamplifiers are only needed in weak-signal areas, and they often cause more trouble than they prevent. They are subject to the same overload problems as TVs, and their location on the antenna mast usually makes it difficult to install the appropriate cures. You may need to install a high-pass or notch filter at the *input* of the preamplifier, as well as a common-mode choke on the input, output and power-supply wiring (if separate) to affect a complete cure.

VCRs

A VCR usually contains a television tuner, or has a TV channel output, so it is subject to all of the interference problems of a TV receiver. It is also hooked up to an antenna or cable system and the ac-line wiring. The video baseband signal extends from 30 Hz to 3.5 MHz, with color information centered around 3.5 MHz and the FM sound subcarrier at 4.5 MHz. The entire video baseband is frequency modulated onto the tape at frequencies up to 10 MHz. It is no wonder that some VCRs are quite susceptible to EMI.

Many cases of VCR EMI can be cured. Start by proving that the VCR is the susceptible device. Temporarily disconnect the VCR from the television. If there is no interference to the TV, then the VCR is the most likely culprit.

You need to find out how the interfering signal is getting into the VCR. Temporarily disconnect the antenna or cable feed line from the VCR. If the interference goes away, then the antenna line is involved. In this case, you can probably fix the problem with a common-mode choke or high-pass filter.

Fig 13.13 shows a bulletproof VCR installation. If you have tried all of the cures shown and still have a problem, the VCR is probably subject to direct pickup. In this case, contact the manufacturer through the EIA.

Non-radio Devices

Interference to non-radio devices is not the fault of the transmitter. (A portion of the *FCC Interference Handbook,* 1990 Edition, is shown in **Fig 13.17**.[3]) In essence, the FCC views non-radio devices that pick up nearby radio signals as improperly functioning; contact the manufacturer and return the equipment. The FCC does not require that non-radio devices include EMI protection and they don't offer legal protection to users of these devices that are susceptible to interference.

Telephones

Telephones have probably become the number one interference problem of Amateur Radio. However, most cases of telephone interference can be cured by correcting any installation defects and installing telephone EMI filters where needed.

Telephones can improperly function as radio receivers. There are devices inside many telephones that act like diodes. When such a telephone is connected to the telephone wiring (a large antenna), an AM radio receiver can be formed. When a nearby transmitter goes on the air, these telephones can be affected.

Troubleshooting techniques were discussed earlier in the chapter. The suggestion to simplify the problem applies especially to telephone interference. Disconnect all telephones except one, right at the service entrance if possible, and start troubleshooting the problem there.

If any one device, or bad connection in the phone system, detects RF and puts the detected signal back onto the phone line as audio, that audio cannot be removed with filters. Once the RF has been detected and turned into audio, it cannot be filtered out because the interference is at the same frequency as the desired audio signal. To affect a cure, you must locate the detection point and correct the problem there.

The telephone company lightning arrestor may be defective. Defective arrestors can act like diodes, rectifying any nearby RF energy. Telephone-line amplifiers or other electronic equipment may also be at fault. Leave the telephone company equipment to the experts, however. There are important safety issues that are the sole responsibility of the telephone company.

Inspect the installation. Years of exposure in damp basements, walls or crawl spaces may have caused deterioration. Be suspicious of anything that is corroded or discolored. In many cases, homeowners have installed their own telephone wiring, often using substandard wiring. If you find sections of telephone wiring made from nonstandard cable, replace it with standard twisted-pair wire. Radio Shack, among others, sells several kinds of telephone wire.

Next, evaluate each of the telephone instruments. If you find a susceptible telephone, install a telephone EMI filter on that telephone. Several *QST* advertisers sell small, attractive telephone EMI filters.

If you determine that you have interference only when you operate on one particular ham band, the telephone wiring is probably resonant on that band. If possible, install a few strategically placed in-line telephone EMI filters to break up the resonance.

Telephone Accessories — Answering machines, fax machines and some alarm systems are also prone to interference problems. All of the troubleshooting techniques and cures that apply to telephones also apply to these telephone devices. In addition, many of these devices connect to the ac mains. Try a common-mode choke and/or ac-line filter on the power cord (which may be an ac cord set, a small transformer or power supply).

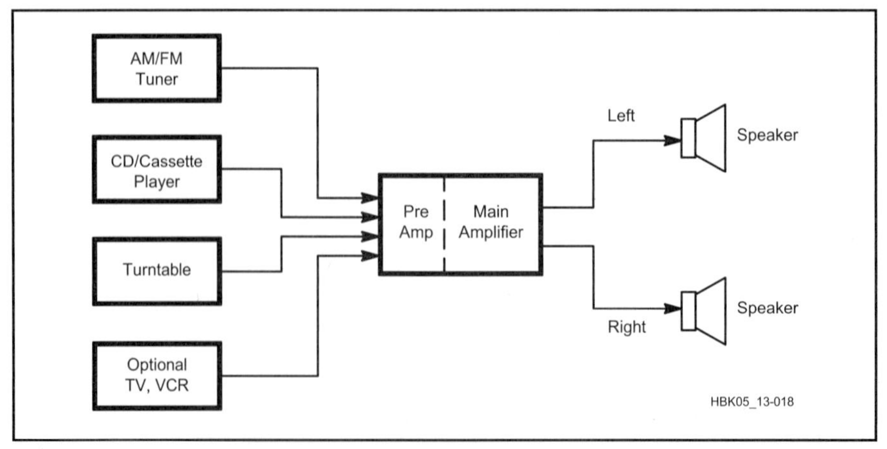

Fig 13.18 — A typical modern stereo system.

Cordless Telephones — A cordless telephone is an unlicensed *radio* device that is manufactured and used under Part 15 of the FCC regulations. The FCC does not intend Part 15 devices to be protected from interference. These devices usually have receivers with very wide front-end filters, which make them very susceptible to interference. A label on the telephone or a paragraph in the owner's manual should explain that the telephone must not cause interference to other services and must tolerate any interference caused to it.

It's worthwhile to try a telephone filter on the base unit and properly filter its ac line cord. (You might get lucky!) The best source of help is the manufacturer, but they may point out that the Part 15 device is not protected from interference. These kinds of problems are difficult to fix after the fact. The necessary engineering should be done when the device is designed.

Other Audio Devices

Other audio devices, such as stereos, intercoms and public-address systems can also pick up and detect strong nearby transmitters. The FCC considers these non-radio devices and does not protect them from licensed radio transmitters that may interfere with their operation. See Fig 13.17 for the FCC's point of view.

Use the standard troubleshooting techniques discussed earlier in this chapter to isolate problems. In a multi-component stereo system (as in **Fig 13.18**), for example, you must determine what combination of components is involved with the problem. First, disconnect all auxiliary components to determine if there is a problem with the main receiver/amplifier. (Long speaker/interconnect cables are prime suspects.)

Stereos — If the problem remains with the main amplifier isolated, determine if the interference level is affected by the volume control. If so, the interference is getting into the circuit *before* the volume control, usually through accessory wiring. If the volume control has no effect on the level of the interfering sound, the interference is getting in *after* the control, usually through speaker wires.

Speaker wires are often resonant on the HF bands. In addition, they are often connected directly to the output transistors, where RF can be detected. Most amplifier designs use a negative feedback loop to improve fidelity. This loop can conduct the detected RF signal back to the high-gain stages of the amplifier. The combination of all of these factors makes the speaker leads the usual indirect cause of interference to audio amplifiers.

There is a simple test that will help determine if the interfering signal is being coupled into the amplifier by the speaker leads. Temporarily disconnect the speaker leads from the amplifier, and plug in a test set of headphones with short leads. If there is no interference with the headphones, filtering the speaker leads will cure the problem.

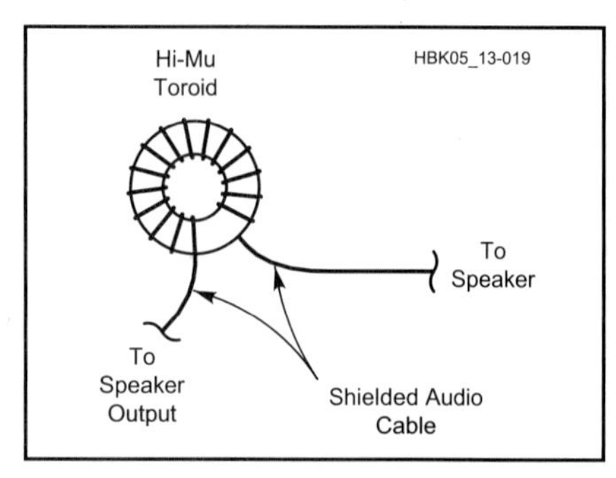

Fig 13.19 — This is how to make a speaker-lead common-mode choke. Be sure to use the correct ferrite material.

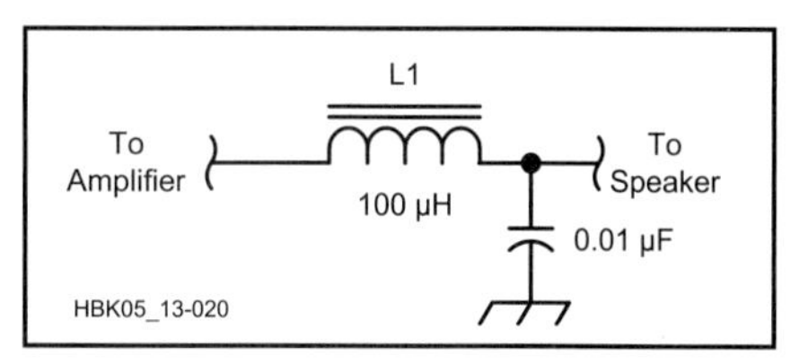

Fig 13.20— An LC filter for speaker leads.

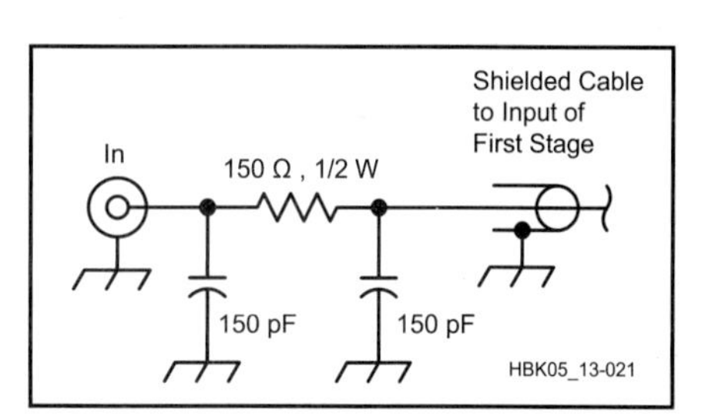

Fig 13.21 — A filter for use at the input of audio equipment. The components should be installed inside of the chassis at the connector by a qualified technician.

> **Warning: Bypassing Speaker Leads**
>
> Older amateur literature might tell you to put a 0.01-µF capacitor across the speaker terminals to cure speaker-lead interference. *Don't do this!* Some modern solid-state amplifiers can break into a destructive, full-power, sometimes ultrasonic oscillation if they are connected to a highly capacitive load. If you do this to your neighbor's amplifier, you will have a whole new kind of personal diplomacy problem! *— Ed Hare, W1RFI, ARRL Laboratory Supervisor*

The best way to eliminate RF signals from speaker leads is with common-mode chokes. **Fig 13.19** shows how to wrap speaker wires around an FT-140-43 ferrite core to cure speaker-lead EMI. Use the correct core material for the job. See the information about common-mode chokes earlier in this chapter.

Another way to cure speaker-lead interference is with an LC filter as shown in **Fig 13.20**.

Interconnect cables can couple interfering signals into an amplifier or accessories. The easiest cure here is also a common-mode choke. However, it may also be necessary to add a differential-mode filter to the input of the amplifier or accessory. **Fig 13.21** shows a home-brew version of such a filter.

Intercoms and Public-Address Systems — All of these problems also apply to intercoms, public-address (PA) systems and similar devices. These systems usually have long speaker leads or interconnect cables that can pick up a lot of RF energy from a nearby transmitter. The cures discussed above do apply to these systems, but you may also need to contact the manufacturer to see if they have any additional, specific information.

Computers and Other Unlicensed RF Sources

Computers and microprocessors can be sources, or victims, of interference. These devices contain oscillators that can, and do, radiate RF energy. In addition, the internal functions of a computer generate different frequencies, based on the various data rates as software is executed. All of these signals are digital — with fast rise and fall times that are rich in harmonics.

Don't just think "computer" when thinking of computer systems. Many household appliances contain microprocessors: digital clocks, video games, calculators and more.

Computing devices are covered under Part 15 of the FCC regulations as unintentional emitters. The FCC has set up absolute radiation limits for these devices. FCC regulations state that the operator or owner of Part 15 devices must take whatever steps are necessary to reduce or eliminate any interference they cause to a licensed radio service. This means that if your neighbor's video game interferes with your radio, the neighbor is responsible for correcting the problem. (Of course, your neighbor may appreciate your help in locating a solution!)

The FCC has set up two levels of type

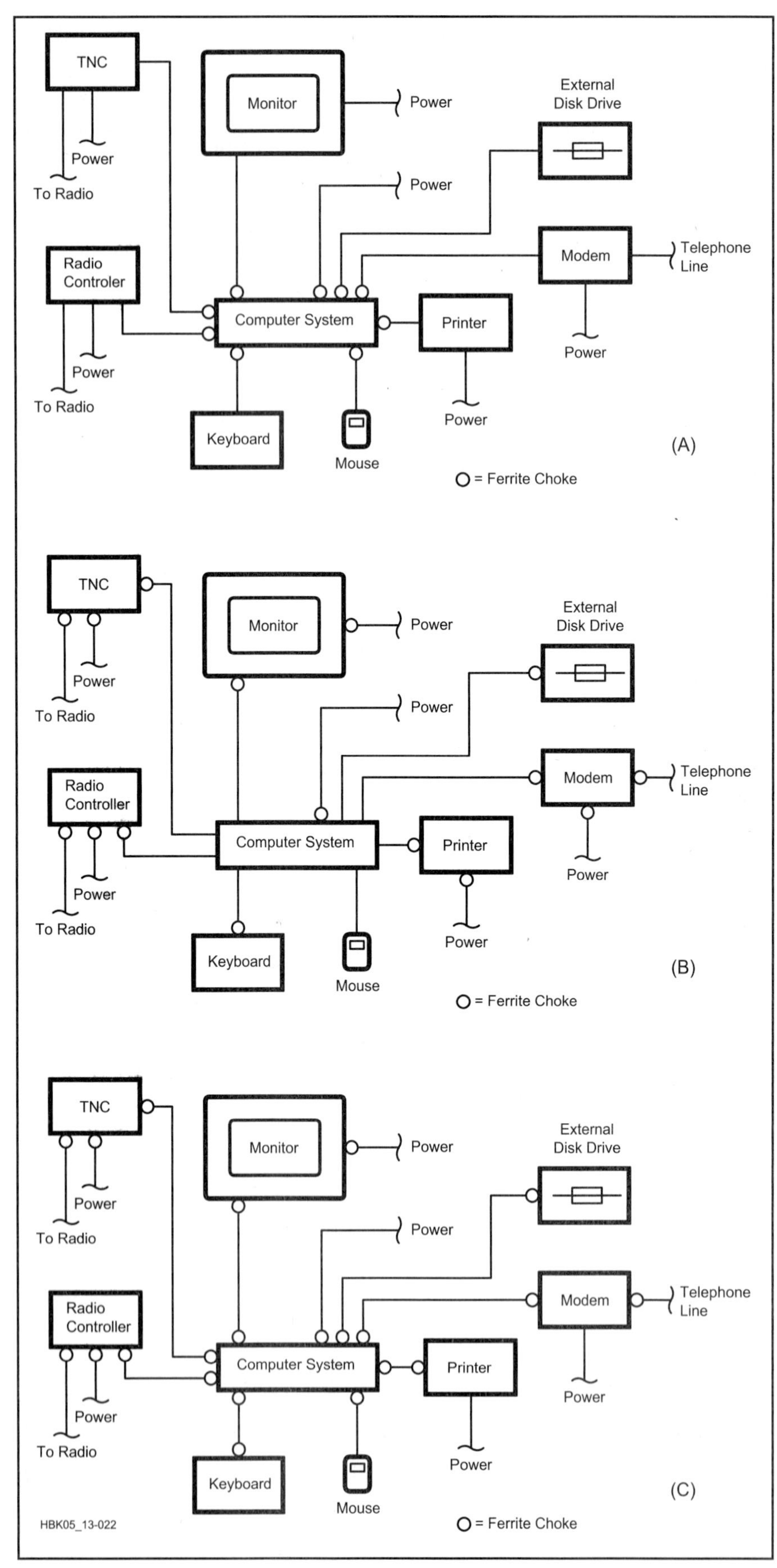

Fig 13.22 — Where to locate ferrites in a computer system. At A, the computer is noisy, but the peripherals are quiet. At B, the computer is quiet, but external devices are noisy. At C, both the computer and externals are noisy.

acceptance for computing devices. Class A is for computers used in a commercial environment. FCC Class B requirements are more stringent — for computers used in residential environments. If you buy a computer or peripheral, be sure that it is Class B certified or it will probably generate interference to your amateur station or home-electronics equipment.

If you find that your computer system is interfering with your radio (not uncommon in this digital-radio age), start by simplifying the problem. Temporarily switch off as many peripherals as possible and disconnect their cables from the back of the computer. If possible, use just the computer, keyboard and monitor. This test may indicate one or more peripherals as the source of the interference.

When seeking cures, first ensure that all interconnection cables are shielded. Replace any unshielded cables with well-shielded ones; this often significantly reduces RF noise from computer systems. The shield must also be terminated properly at the connectors. Unfortunately, quite often the only way to find out is to take it apart. The second line of defense is the common-mode choke, made from a ferrite toroid. The toroids should be installed as close to the computer and/or peripheral device as practical. **Fig 13.22** shows the location of common-mode chokes in a complete computer system where both the computer and peripherals are noisy.

In some cases, a switching power supply may be a source of interference. A common-mode choke and/or ac-line filter may cure this problem. In extreme cases of computer interference you may need to improve the shielding of the computer. Refer to *The ARRL RFI Book* for more information about how to do this. Don't forget that some peripherals (such as modems) are connected to the phone line, so you may need to treat them like telephones.

Automobiles

As automobiles have become more technologically sophisticated, questions about the compatibility of automobiles and amateur transmitters have increased in number and scope. The use of microprocessors in autos makes them computer systems on wheels, subject to all of the same problems as any other computer. Installation of ham equipment can cause problems, ranging from nuisances like a dome light coming on every time you transmit to serious ones such as damage to the vehicle electronic control module (ECM).

Only qualified service personnel should work on automotive EMC problems.

Many critical safety systems on modern cars should not be handled by amateurs. Even professionals can meet with mixed results. The ARRL (TIS) contacted each of the automobile manufacturers and asked about their EMC policies, service bulletins and best contacts to resolve EMI problems. About 20% of the companies never answered, and answers from the rest ranged from good to poor. One company even said that the answers to those questions were "proprietary."

Some of the companies *do* have reasonable EMC policies, but these policies often fall apart at the dealer level. The ARRL has reports of problems with nearly every auto manufacturer. Check with your dealer before you install a transceiver in a car. The dealer can direct you to any service bulletins or information that is applicable to your model. If you are not satisfied with the dealer's response, contact the regional or factory customer service representatives.

For additional information about automotive EMC, refer to the Automobiles chapter in *The ARRL RFI Book*, or see the automotive pages of the RFI section on *ARRLWeb*: **www.arrl.org/tis/info/rfigen.html**.

Electrical Noise

Many electrical appliances and power lines can generate electrical noise. On a receiver, electrical noise usually sounds like a rough buzz, heard across a wide frequency range. The buzz will either have a strong 60- or 120-Hz component, or its pitch will vary with the speed of a motor that generates the noise. The appearance of electrical noise on a television set is shown in the TVI section of this chapter. This kind of noise can come from power lines, electrical motors or switches, to name just a few. Here is one quick diagnostic trick — if electrical noise seems to come and go with the weather, the source is probably outside, usually on the power lines. If electrical noise varies with the time of day, it is usually related to what people are doing, so look to your own, or your neighbors', house and lifestyle. *The ARRL RFI Book* describes techniques for locating RFI sources.

Filters usually cure electrical noise. At its source, the noise can usually be filtered with a differential-mode filter. A differential-mode filter can be as simple as a 0.01-µF ac-rated capacitor, such as Panasonic part ECQ-U2A103MN, or it can be a pi-section filter like that shown in Fig 13.7.

For removing signals that arrive via power lines, a common-mode choke is usually the best defense. Wrap about 10 turns of the ac-power cord around an FT-240-43 ferrite core; do this as close as possible to the device you are trying to protect.

Electrical noise can also indicate a dangerous electrical condition that needs to be corrected. The ARRL has recorded several cases where defective or arcing doorbell transformers caused widespread neighborhood electrical interference. This subject is well covered in the *The ARRL RFI Book*.[4]

Power Lines — Electrical noise frequently comes from lines and equipment owned by the power company. DO NOT hammer on poles, shake guy wires, or otherwise disturb any other utility-owned equipment in an effort to locate the source of the problem. Doing so is a potentially FATAL action! Leave any investigation of these sources to the power company. If you have a problem and are not getting help from the power company, contact us at ARRL and we can help you work it out with them. Many power companies have qualified, knowledgeable people on staff to correct any EMI problems that occur with their equipment. However, if they seem confused and unsure as how to proceed with your concern, ARRL can provide them with information on noise-locating techniques. All they need to do is ask. You can be encouraged from the fact that such noise sources are usually simple to locate and correct — it will typically cost the power company a lot less time and money than they might expect.

In Conclusion

Remember that EMI problems can be cured. With the proper technical knowledge and interpersonal skills, you can deal effectively with the people and hardware that make up any EMI problem.

Notes

[1]ARRL Order no. 6834 is available from ARRL Publication Sales or your local Amateur Radio equipment dealer.

[2]Star Circuits Model 23H tunes 6 m, Model 1822 tunes 2 m and Model 46FM tunes the FM broadcast band. Their address is Star Circuits, PO Box 94917, Las Vegas, NV 89193.

[3]The FCC Interference to Home Electronic Entertainment Equipment Handbook is available from the US Government Printing Office. See **http://bookstore.gpo.gov**.

[4]*The ARRL RFI Book*, ARRL Order no. 6834, is available from ARRL Publication Sales or your local Amateur Radio equipment dealer.

FINDING NOISE SOURCES IN THE SHACK

The radio amateur of yesteryear is to be envied to some degree because of the relatively small amount of electrical noise that caused problems. How different it is today, with every house full of electrical equipment capable of emitting electromagnetic radiation to interfere with the poor radio amateur who is trying to listen to signals on the bands.

This project detects the radiation that causes problems to the amateur, and the noise can be heard. When we say to other members of the household "Please don't turn on that computer, vacuum cleaner or TV" they cannot understand why we are complaining, but this little device will allow you to show them and let them hear the 'noise' with which we have to contend.

Table 13.2
Components List

Resistors	*Value*
R1	1 kΩ
R2, R6	100 Ω
R3, R4	47 kΩ
R5	100 kΩ
R7	10 Ω
R8	10 kΩ, with switch
Capacitors	*Value*
C1, C6	4.7 µF, 16 V electrolytic
C2, C5	0.01 µF
C3, C4	22 µF, 16 V electrolytic
C7	0.047 µF
C8	10 µF, 16 V electrolytic
Capacitors	*Value (continued)*
C9, C11	330 µF, 16 V electrolytic
C10	0.1 µF
Semiconductors	
U1	LM741
U2	LM386
Additional Items	
LS1	Small 8-Ω loudspeaker
Perforated board, 7 × 7 cm	
PP3 battery and clip	
3.5 mm mono-jack socket	
Case	
Telephone pick-up coil	

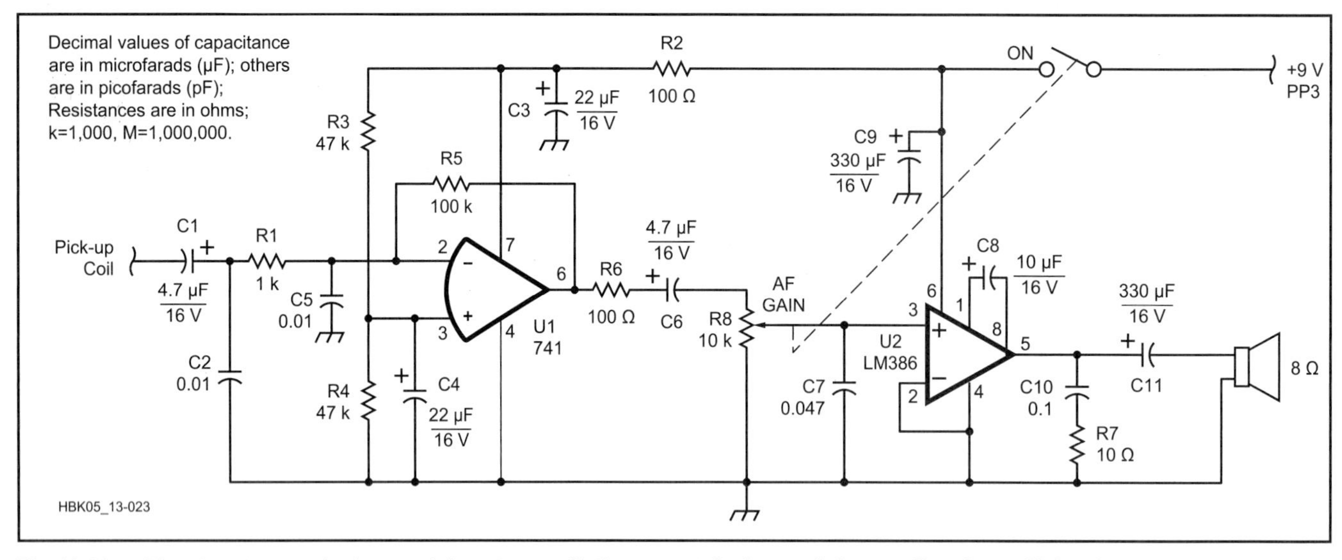

Fig 13.23 — The detector works by receiving stray radiation on a telephone pick-up coil and amplifying it to loudspeaker level.

CONSTRUCTION

The circuit (**Fig 13.23**) uses a telephone pick-up coil as a detector, the output of which is fed into a LM741 IC preamplifier, followed by a LM386 IC power amplifier. See **Table 13.2** for the complete component list.

The project is built on a perforated board (**Fig 13.24**), with the component leads pushed through the holes and joined with hook-up wire underneath. There is a wire running around the perimeter of the board to form an earth bus.

Build from the loudspeaker backwards to R8, apply power and touch the wiper of R8. If everything is OK you should hear a loud buzz from the speaker. Too much gain may cause a feedback howl, in which case you will need to adjust R8 to reduce the gain. Complete the rest of the wiring and test with a finger on the input, which should produce a click and a buzz. The pick-up coil comes with a lead and 3.5 mm jack, so you will need a suitable socket.

RELATIVE NOISES

Place a high-impedance meter set to a low-ac-voltage range across the speaker leads to give a comparative readout between different items of equipment in the home. Sample readings are shown in **Table 13.3**.

Table 13.3
Readings (pick-up coil near household items)

Item	Reading
29-MHz oscilloscope	0.56 V
Old computer monitor	0.86 V
Old computer with plastic case	1.53 V
New computer monitor	0.45 V
New tower PC with metal case	0.15 V
Old TV	1.2 V
New TV	0.4 V
Plastic-cased hairdryer	4.6 V
Vacuum cleaner	3.6 V
Drill	4.9 V

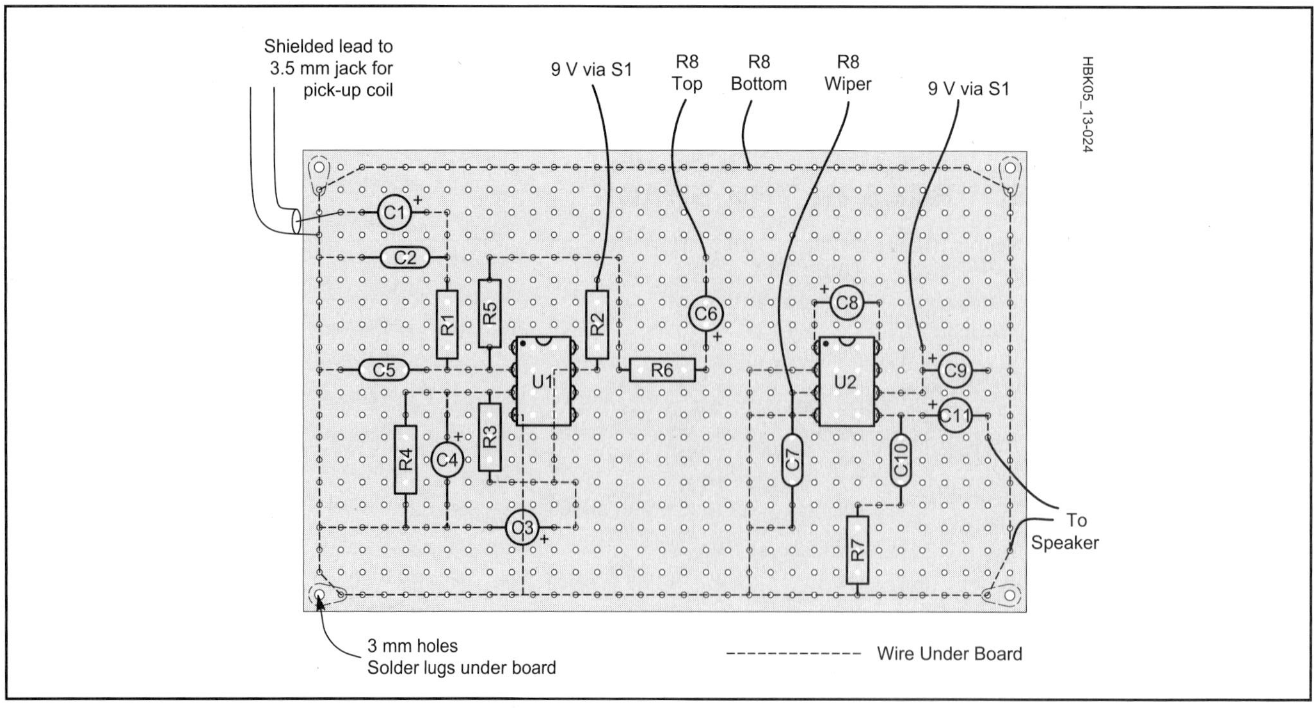

Fig 13.24 — The project is built on perforated board with point-to-point wiring underneath.

Radio Direction Finding

Far more than simply finding the direction of an incoming radio signal, radio direction finding (RDF) encompasses a variety of techniques for determining the exact location of a signal source. The process involves both art and science. RDF adds fun to ham radio, but has serious purposes, too.

This section was written by Joe Moell, KØOV.

RDF is almost as old as radio communication. It gained prominence when the British Navy used it to track the movement of enemy ships in World War I. Since then, governments and the military have developed sophisticated and complex RDF systems. Fortunately, simple equipment, purchased or built at home, is quite effective in Amateur Radio RDF.

In European and Asian countries, direction-finding contests are foot races. The object is to be first to find four or five transmitters in a large wooded park. Young athletes have the best chance of capturing the prizes. This sport is known as *foxhunting* (after the British hill-and-dale horseback events) or *ARDF* (Amateur Radio direction finding).

In North America and England, most RDF contests involve mobiles — cars, trucks, and vans, even motorcycles. It may be possible to drive all the way to the transmitter, or there may be a short hike at the end, called a *sniff*. These competitions are also called foxhunting by some, while others use *bunny hunting, T-hunting* or the classic term *hidden transmitter hunting*.

In the 1950s, 3.5 and 28 MHz were the most popular bands for hidden transmitter hunts. Today, most competitive hunts worldwide are for 144-MHz FM signals, though other VHF bands are also used. Some international foxhunts include 3.5-MHz events.

Even without participating in RDF contests, you will find knowledge of the techniques useful. They simplify the search for a neighborhood source of power-line interference or TV cable leakage. RDF must be used to track down emergency radio beacons, which signal the location of pilots and boaters in distress. Amateur Radio enthusiasts skilled in transmitter hunting are in demand by agencies such as the Civil Air Patrol and the US Coast Guard Auxiliary for search and rescue support.

The FCC's Field Operations Bureau has created an Amateur Auxiliary, administered by the ARRL Section Managers, to deal with interference matters. In many areas of the country, there are standing agreements between Local Interference Committees and district FCC offices, permitting volunteers to provide evidence leading to prosecution in serious cases of malicious amateur-to-amateur interference. RDF is an important part of the evidence-gathering process.

The most basic RDF system consists of a directional antenna and a method of detecting and measuring the level of the radio signal, such as a receiver with signal strength indicator. RDF antennas range from a simple tuned loop of wire to an acre of antenna elements with an electronic beam-forming network. Other sophisticated techniques for RDF use the Doppler effect, or measure the time of arrival difference of the signal at multiple antennas.

All of these methods have been used from 2 to 500 MHz and above. However, RDF practices vary greatly between the HF and VHF/UHF portions of the spectrum. For practical reasons, high gain beams, Dopplers and switched dual antennas find favor on VHF/UHF, while loops and phased arrays are the most popular choices on 6 m and below. Signal propagation differences between HF and VHF also affect RDF practices. But many basic transmitter-hunting techniques, discussed later in this chapter, apply to all bands and all types of portable RDF equipment.

RDF ANTENNAS FOR HF BANDS

Below 50 MHz, gain antennas such as Yagis and quads are of limited value for RDF. The typical installation of a tribander on a 70-ft tower yields only a general direction of the incoming signal, due to ground effects and the antenna's broad forward lobe. Long monoband beams at greater heights work better, but still cannot achieve the bearing accuracy and repeatability of simpler antennas designed specifically for RDF.

RDF Loops

An effective directional HF antenna can be as uncomplicated as a small loop of wire or tubing, tuned to resonance with a capacitor. When immersed in an electromagnetic field, the loop acts much the same as the secondary winding of a transformer. The voltage at the output is proportional to the amount of flux passing through it and the number of turns. If the loop is oriented such that the greatest amount of area is presented to the magnetic field, the induced voltage will be the highest. If it is rotated so that little or no area is cut by the field lines, the voltage induced in the loop is zero and a null occurs.

To achieve this transformer effect, the loop must be small compared with the signal wavelength. In a single-turn loop, the conductor should be less than 0.08-λ long. For example, a 28-MHz loop should be less than 34 inches in circumference, giving a diameter of approximately 10 inches. The loop may be smaller, but that will reduce its voltage output. Maximum output from a small loop antenna is in directions corresponding to the plane of the loop; these lobes are very broad. Sharp nulls, obtained at right angles to that plane, are more useful for RDF.

For a perfect bidirectional pattern, the loop must be balanced electrostatically with respect to ground. Otherwise, it will exhibit two modes of operation, the mode of a perfect loop and that of a non-directional vertical antenna of small dimensions. This dual-mode condition results in mild to severe inaccuracy, depending on the degree of imbalance, because the outputs of the two modes are not in phase.

The theoretical true loop pattern is illustrated in **Fig 13.25A**. When properly balanced, there are two nulls exactly 180° apart. When the unwanted antenna effect is appreciable and the loop is tuned to resonance, the loop may exhibit little directivity, as shown in Fig 13.25B. By detuning the loop to shift the phasing, you may obtain a useful pattern similar to Fig 13.25C. While not symmetrical, and not necessarily at right angles to the plane of the loop, this pattern does exhibit a pair of nulls.

By careful detuning and amplitude balancing, you can approach the unidirectional pattern of Fig 13.25D. Even though there may not be a complete null in the pattern, it resolves the 180° ambiguity of Fig 13.25A. Korean War era military loop antennas, sometimes available on today's

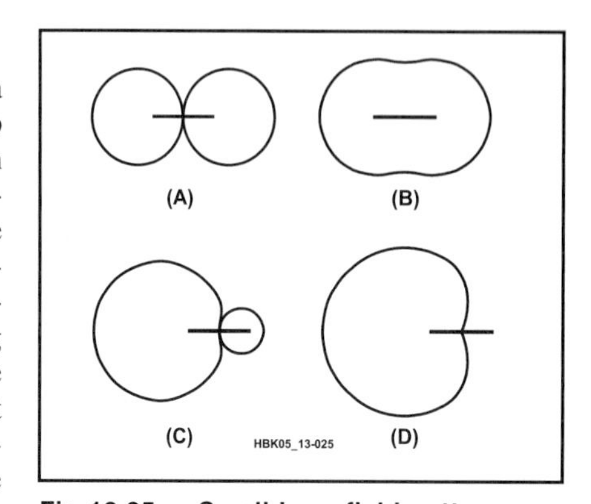

Fig 13.25 — Small loop field patterns with varying amounts of antenna effect — the undesired response of a loop acting merely as a mass of metal connected to the receiver antenna terminals. The horizontal lines show the plane of the loop turns.

surplus market, use this controlled-antenna-effect principle.

An easy way to achieve good electrostatic balance is to shield the loop, as shown in **Fig 13.26**. The shield, represented by the dashed lines in the drawing, eliminates the antenna effect. The response of a well-constructed shielded loop is quite close to the ideal pattern of Fig 13.25A.

For 160 through 30 m, single-turn loops that are small enough for portability are usually unsatisfactory for RDF work. Multi-turn loops are generally used instead. They are easier to resonate with practical capacitor values and give higher output voltages. This type of loop may also be shielded. If the total conductor length remains below 0.08 λ, the directional pattern is that of Fig 13.25A.

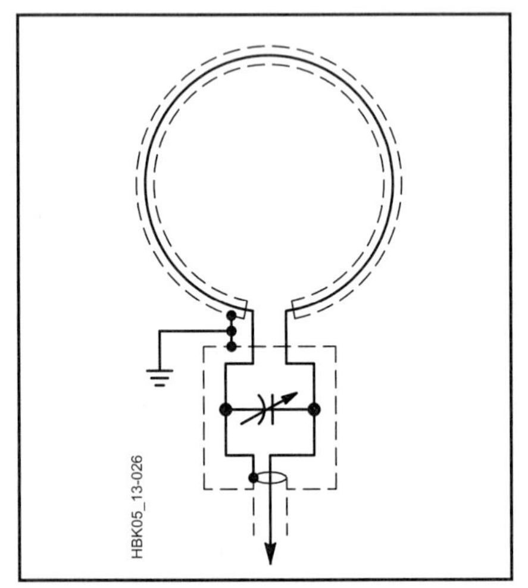

Fig 13.26 — Electrostatically-shielded loop for RDF. To prevent shielding of the loop from magnetic fields, leave the shield unconnected at one end.

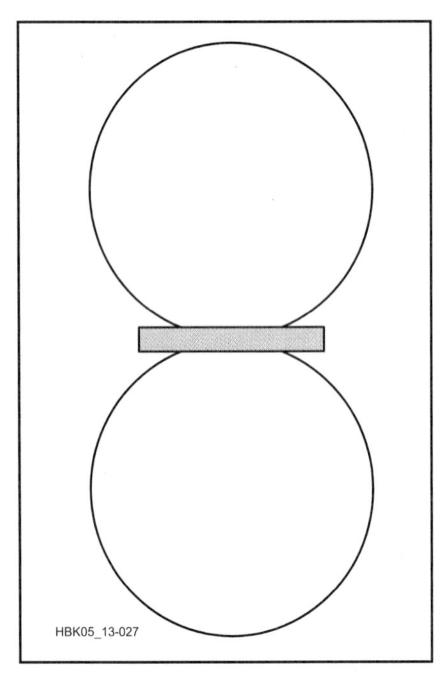

Fig 13.27 — Field pattern for a ferrite-rod antenna. The dark bar represents the rod on which the loop turns are wound.

Ferrite Rod Antennas

Another way to get higher loop output is to increase the permeability of the medium in the vicinity of the loop. By winding a coil of wire around a form made of high-permeability material, such as ferrite rod, much greater flux is obtained in the coil without increasing the cross-sectional area.

Modern magnetic core materials make compact directional receiving antennas practical. Most portable AM broadcast receivers use this type of antenna, commonly called a *loopstick*. The loopstick is the most popular RDF antenna for portable/mobile work on 160 and 80 m.

As does the shielded loop discussed earlier, the loopstick responds to the magnetic field of the incoming radio wave, and not to the electrical field. For a given size of loop, the output voltage increases with increasing flux density, which is obtained by choosing a ferrite core of high permeability and low loss at the frequency of interest. For increased output, the turns may be wound over two rods taped together. A practical loopstick antenna is described later in this chapter.

A loop on a ferrite core has maximum signal response in the plane of the turns, just as an air core loop. This means that maximum response of a loopstick is broadside to the axis of the rod, as shown in **Fig 13.27**. The loopstick may be shielded to eliminate the antenna effect; a U-shaped or C-shaped channel of aluminum or other form of "trough" is best. The shield must not be closed, and its length should equal or slightly exceed the length of the rod.

Sense Antennas

Because there are two nulls 180° apart in the directional pattern of a small loop or loopstick, there is ambiguity as to which null indicates the true direction of the target station. For example, if the line of bearing runs east and west from your position, you have no way of knowing from this single bearing whether the transmitter is east of you or west of you.

If bearings can be taken from two or more positions at suitable direction and distance from the transmitter, the ambiguity can be resolved and distance can be estimated by triangulation, as discussed later in this chapter. However, it is almost always desirable to be able to resolve the ambiguity immediately by having a unidirectional antenna pattern available.

You can modify a loop or loopstick antenna pattern to have a single null by adding a second antenna element. This element is called a sense antenna, because it senses the phase of the signal wavefront for comparison with the phase of the loop output signal. The sense element must be omnidirectional, such as a short vertical. When signals from the loop and the sense antenna are combined with 90° phase shift between the two, a heart-shaped (cardioid) pattern results, as shown in **Fig 13.28A**.

Fig 13.28B shows a circuit for adding a sense antenna to a loop or loopstick. For the best null in the composite pattern, signals from the loop and sense antennas must be of equal amplitude. R1 adjusts the level of the signal from the sense antenna.

In a practical system, the cardioid pattern null is not as sharp as the bidirectional null of the loop alone. The usual procedure when transmitter hunting is to use the loop alone to obtain a precise line of bearing, then switch in the sense antenna and take another reading to resolve the ambiguity.

Phased Arrays and Adcocks

Two-element phased arrays are popular for amateur HF RDF base station installations. Many directional patterns are possible, depending on the spacing and phasing of the elements. A useful example is two 1/2-λ elements spaced 1/4-λ apart and fed 90° out of phase. The resultant pattern is a cardioid, with a null off one end of the axis of the two antennas and a broad peak in the opposite direction. The directional frequency range of this antenna is limited to one band, because of the critical length of the phasing lines.

The best-known phased array for RDF is the Adcock, named after the man who invented it in 1919. It consists of two vertical elements fed 180° apart, mounted so the array may be rotated. Element spacing is not critical, and may be in the range from 0.1 to 0.75 λ. The two elements must be of identical lengths, but need not be self-resonant; shorter elements are commonly used. Because neither the element spacing nor length is critical in terms of wavelengths, an Adcock array may operate over more than one amateur band.

Fig 13.29 is a schematic of a typical Adcock configuration, called the H-Adcock because of its shape. Response to a vertically polarized wave is very similar to a conventional loop. The passing wave induces currents I1 and I2 into the vertical members. The output current in the trans-

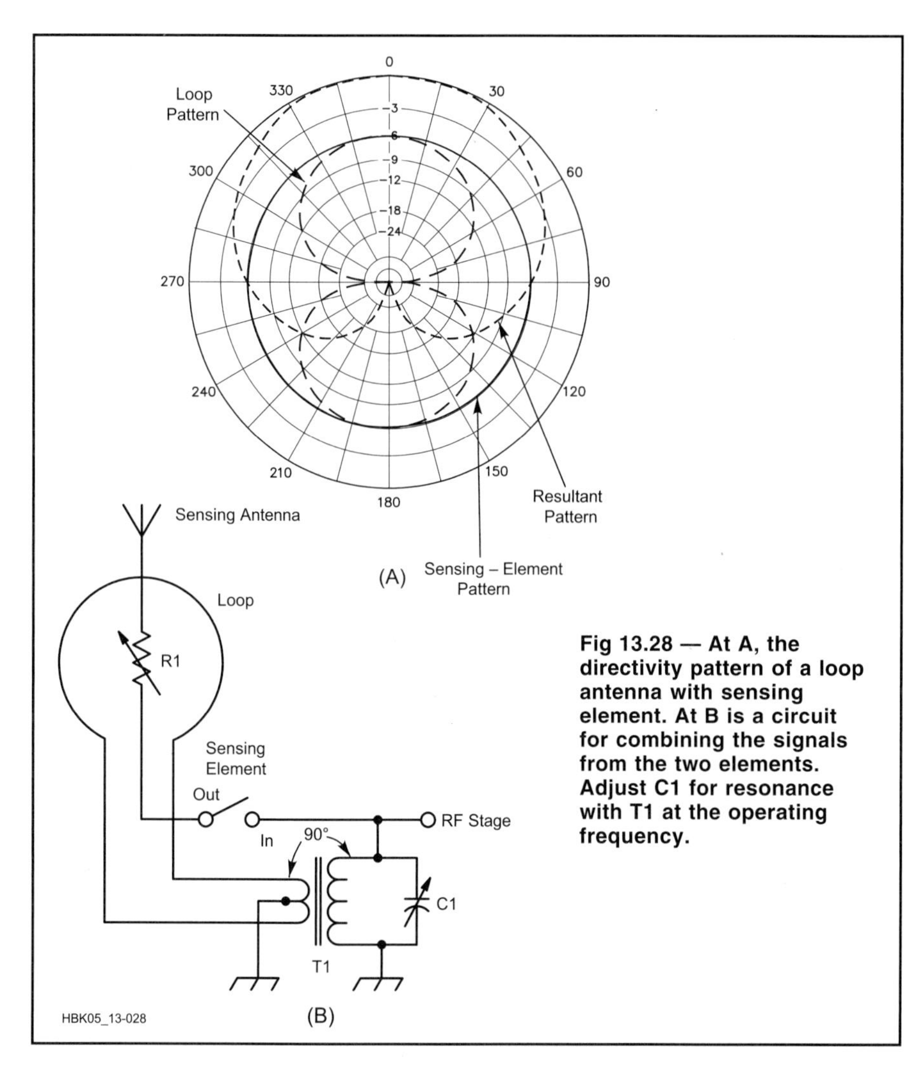

Fig 13.28 — At A, the directivity pattern of a loop antenna with sensing element. At B is a circuit for combining the signals from the two elements. Adjust C1 for resonance with T1 at the operating frequency.

mission line is equal to their difference. Consequently, the directional pattern has two broad peaks and two sharp nulls, like the loop. The magnitude of the difference current is proportional to the spacing (d) and length (l) of the elements. You will get somewhat higher gain with larger dimensions. The Adcock of **Fig 13.30**, designed for 40 m, has element lengths of 12 ft and spacing of 21 ft (approximately 0.15 λ).

Fig 13.31 shows the radiation pattern of the Adcock. The nulls are broadside to the axis of the array, becoming sharper with increased element spacing. When element spacing exceeds $^3/_4$ λ, however, the antenna begins to take on additional unwanted nulls off the ends of the array axis.

The Adcock is a vertically polarized antenna. The vertical elements do not respond to horizontally polarized waves, and the currents induced in the horizontal members by a horizontally polarized wave (dotted arrows in Fig 13.29) tend to balance out regardless of the orientation of the antenna.

Since the Adcock uses a balanced feed system, a coupler is required to match the unbalanced input of the receiver. T1 is an air-wound coil with a two-turn link wrapped around the middle. The combination is resonated with C1 to the operating frequency. C2 and C3 are null-clearing capacitors. Adjust them by placing a low-power signal source some distance from the antenna and exactly broadside to it. Adjust C2 and C3 until the deepest null is obtained.

While you can use a metal support for the mast and boom, wood is preferable because of its non-conducting properties. Similarly, a mast of thick-wall PVC pipe gives less distortion of the antenna pattern than a metallic mast. Place the coupler on

Fig 13.29 — A simple Adcock antenna and its coupler.

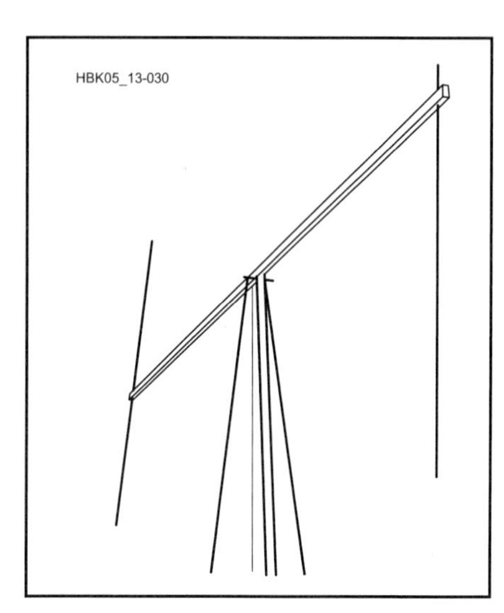

Fig 13.30 — An experimental Adcock antenna on a wooden frame.

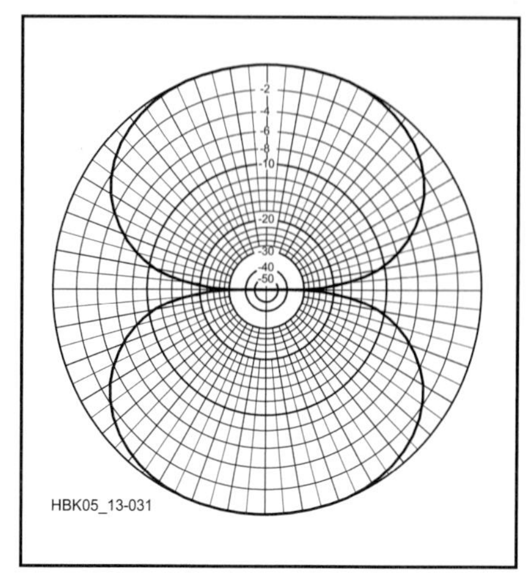

Fig 13.31 — The pattern of an Adcock array with element spacing of ½ wavelength. The elements are aligned with the vertical axis.

the ground below the wiring harness junction on the boom and connect it with a short length of 300-Ω twin-lead-feed line.

Loops vs. Phased Arrays

Loops are much smaller than phased arrays for the same frequency, and are thus the obvious choice for portable/mobile HF RDF. For base stations in a triangulation network, where the 180° ambiguity is not a problem, Adcocks are preferred. In general, they give sharper nulls than loops, but this is in part a function of the care used in constructing and feeding the individual antennas, as well as of the spacing of the elements. The primary construction considerations are the shielding and balancing of the feed line against unwanted signal pickup and the balancing of the antenna for a symmetrical pattern. Users report that Adcocks are somewhat less sensitive to proximity effects, probably because their larger aperture offers some space diversity.

Skywave Considerations

Until now we have considered the directional characteristics of the RDF loop only in the two-dimensional azimuthal plane. In three-dimensional space, the response of a vertically oriented small loop is doughnut-shaped. The bidirectional null (analogous to a line through the doughnut hole) is in the line of bearing in the azimuthal plane and toward the horizon in the vertical plane. Therefore, maximum null depth is achieved only on signals arriving at 0° elevation angle.

Skywave signals usually arrive at nonzero wave angles. As the elevation angle increases, the null in a vertically oriented loop pattern becomes shallower. It is possible to tilt the loop to seek the null in elevation as well as azimuth. Some amateur RDF enthusiasts report success at estimating distance to the target by measurement of the elevation angle with a tilted loop and computations based on estimated height of the propagating ionospheric layer. This method seldom provides high accuracy with simple loops, however.

Most users prefer Adcocks to loops for skywave work, because the Adcock null is present at all elevation angles. Note, however, that an Adcock has a null in all directions from signals arriving from overhead. Thus for very high angles, such as under-250-mile skip on 80 and 40 m, neither loops nor Adcocks will perform well.

Electronic Antenna Rotation

State-of-the-art fixed RDF stations for government and military work use antenna arrays of stationary elements, rather than mechanically rotatable arrays. The best-known type is the Wullenweber antenna. It has a large number of elements arranged in a circle, usually outside of a circular reflecting screen. Depending on the installation, the circle may be anywhere from a few hundred feet to more than a quarter of a mile in diameter. Although the Wullenweber is not practical for most amateurs, some of the techniques it uses may be applied to amateur RDF.

The device, which permits rotating the antenna beam without moving the elements, has the classic name *radio goniometer*, or simply *goniometer*. Early goniometers were RF transformers with fixed coils connected to the array elements and a moving pickup coil connected to the receiver input. Both amplitude and phase of the signal coupled into the pickup winding are altered with coil rotation in a way that corresponded to actually rotating the array itself. With sufficient elements and a goniometer, accurate RDF measurements can be taken in all compass directions.

Beam Forming Networks

By properly sampling and combining signals from individual elements in a large array, an antenna beam is electronically rotated or steered. With an appropriate number and arrangement of elements in the system, it is possible to form almost any desired antenna pattern by summing the sampled signals in appropriate amplitude and phase relationships. Delay networks and/or attenuation are added in line with selected elements before summation to create these relationships.

To understand electronic beam forming, first consider just two elements, shown as A and B in **Fig 13.32**. Also shown is the wavefront of a radio signal arriving from a distant transmitter. The wavefront strikes element A first, then travels somewhat farther before it strikes element B. Thus, there is an interval between the times that the wavefront reaches elements A and B.

We can measure the differences in arrival times by delaying the signal received at element A before summing it with that from element B. If two signals are combined directly, the amplitude of the sum will be maximum when the delay for element A exactly equals the propagation delay, giving an in-phase condition at the summation point. On the other hand, if one of the signals is inverted and the two are added, the signals will combine in a 180° out-of-phase relationship when the element A delay equals the propagation delay, creating a null. Either way, once the time delay is determined by the amount of delay required for a peak or null, we can convert it to distance. Then trigonometry calculations provide the direction from which the wave is arriving.

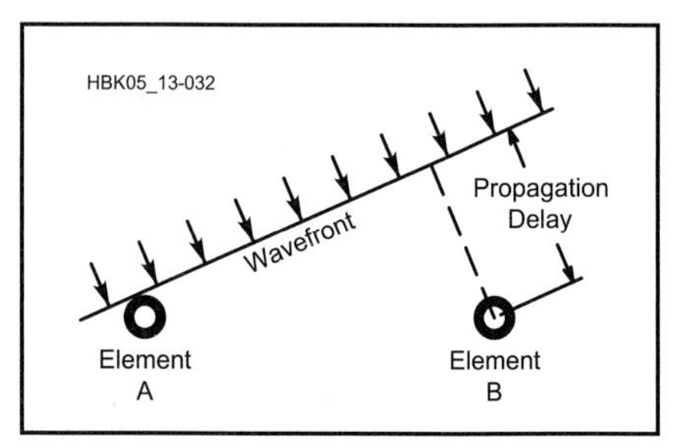

Fig 13.32 — One technique used in electronic beam forming. By delaying the signal from element A by an amount equal to the propagation delay, two signals are summed precisely in phase, even though the signal is not in the broadside direction.

Altering the delay in small increments steers the peak (or null) of the antenna. The system is not frequency sensitive, other than the frequency range limitations of the array elements. Lumped-constant networks are suitable for delay elements if the system is used only for receiving. Delay lines at installations used for transmitting and receiving employ rolls of coaxial cable of various lengths, chosen for the time delay they provide at all frequencies, rather than as simple phasing lines designed for a single frequency.

Combining signals from additional elements narrows the broad beamwidth of the pattern from the two elements and suppress unwanted sidelobes. Electronically switching the delays and attenuations to the various elements causes the formed beam to rotate around the compass. The package of electronics that does this, including delay lines and electronically switched attenuators, is the beam-forming network.

METHODS FOR VHF/UHF RDF

Three distinct methods of mobile RDF are commonly in use by amateurs on VHF/UHF bands: directional antennas, switched dual antennas and Dopplers. Each has advantages over the others in certain situations. Many RDF enthusiasts employ more than one method when transmitter hunting.

Directional Antennas

Ordinary mobile transceivers and handhelds work well for foxhunting on the popular VHF bands. If you have a lightweight beam and your receiver has an easy-to-read S-meter, you are nearly ready to start. All you need is an RF attenuator and some way to mount the setup in your vehicle.

Amateurs seldom use fractional wavelength loops for RDF above 60 MHz because they have bidirectional characteristics and low sensitivity, compared to other practical VHF antennas. Sense circuits for loops are difficult to implement at VHF, and signal reflections tend to fill in the nulls. Typically VHF loops are used only for close-in sniffing where their compactness and sharp nulls are assets, and low gain is of no consequence.

Phased Arrays

The small size and simplicity of 2-element driven arrays make them a common choice of newcomers at VHF RDF. Antennas such as phased ground planes and ZL Specials have modest gain in one direction and a null in the opposite direction. The gain is helpful when the signal is weak, but the broad response peak makes it difficult to take a precise bearing.

As the signal gets stronger, it becomes possible to use the null for a sharper S-meter indication. However, combinations of direct and reflected signals (called *multipath*) will distort the null or perhaps obscure it completely. For best results with this type of antenna, always find clear locations from which to take bearings.

Parasitic Arrays

Parasitic arrays are the most common RDF antennas used by transmitter hunters in high competition areas such as Southern California. Antennas with significant gain are a necessity due to the weak signals often encountered on weekend-long T-hunts, where the transmitter may be over 200 miles distant. Typical 144-MHz installations feature Yagis or quads of three to six elements, sometimes more. Quads are typically home-built, using data from *The ARRL Antenna Book* and *Transmitter Hunting* (see Bibliography).

Two types of mechanical construction are popular for mobile VHF quads. The model of **Fig 13.33** uses thin gauge wire (solid or stranded), suspended on wood dowel or fiberglass rod spreaders. It is lightweight and easy to turn rapidly by hand while the vehicle moves. Many hunters prefer to use larger gauge solid wire (such as AWG 10) on a PVC plastic pipe frame (**Fig 13.34**). This quad is more rugged and has somewhat wider frequency range, at the expense of increased weight and wind resistance. It can get mashed going under a willow, but it is easily reshaped and returned to service.

Yagis are a close second to quads in popularity. Commercial models work fine for VHF RDF, provided that the mast is attached at a good balance point. Lightweight and small-diameter elements are desirable for ease of turning at high speeds.

Fig 13.33 — The mobile RDF installation of WB6ADC features a thin wire quad for 144 MHz and a mechanical linkage that permits either the driver or front passenger to rotate the mast by hand.

Fig 13.34 — KØOV uses this mobile setup for RDF on several bands, with separate antennas for each band that mate with a common lower mast section, pointer and 360° indicator. Antenna shown is a heavy gauge wire quad for 2 m.

A well-designed mobile Yagi or quad installation includes a method of selecting wave polarization. Although vertical polarization is the norm for VHF-FM communications, horizontal polarization is allowed on many T-hunts. Results will be poor if a VHF RDF antenna is cross-polarized to the transmitting antenna, because multipath and scattered signals (which have indeterminate polarization) are enhanced, relative to the cross-polarized direct signal. The installation of Fig 13.33 features a slip joint at the boom-to-mast junction, with an actuating cord to rotate the boom, changing the polarization. Mechanical stops limit the boom rotation to 90°.

Parasitic Array Performance for RDF

The directional gain of a mobile beam (typically 8 dB or more) makes it unexcelled for both weak signal competitive hunts and for locating interference such as TV cable leakage. With an appropriate receiver, you can get bearings on any signal mode, including FM, SSB, CW, TV, pulses and noise. Because only the response peak is used, the null-fill problems and proximity effects of loops and phased arrays do not exist.

You can observe multiple directions of arrival while rotating the antenna, allowing you to make educated guesses as to which signal peaks are direct and which are from non-direct paths or scattering. Skilled operators can estimate distance to the transmitter from the rate of signal strength increase with distance traveled. The RDF beam is useful for transmitting, if necessary, but use care not to damage an attenuator in the coax line by transmitting through it.

The 3-dB beamwidth of typical mobile-mount VHF beams is on the order of 80°. This is a great improvement over 2-element driven arrays, but it is still not possible to get pinpoint bearing accuracy. You can achieve errors of less than 10° by carefully reading the S-meter. In practice, this is not a major hindrance to successful mobile RDF. Mobile users are not as concerned with precise bearings as fixed station operators, because mobile readings are used primarily to give the general direction of travel to "home in" on the signal. Mobile bearings are continuously updated from new, closer locations.

Amplitude-based RDF may be very difficult when signal level varies rapidly. The transmitter hider may be changing power, or the target antenna may be moving or near a well-traveled road or airport. The resultant rapid S-meter movement makes it hard to take accurate bearings with a quad. The process is slow because the antenna must be carefully rotated by hand to "eyeball average" the meter readings.

Switched Antenna RDF Units

Three popular types of RDF systems are relatively insensitive to variations in signal level. Two of them use a pair of vertical dipole antennas, spaced 1/2 λ or less apart, and alternately switched at a rapid rate to the input of the receiver. In use, the indications of the two systems are similar, but the principles are different.

Switched Pattern Systems

The switched pattern RDF set (**Fig 13.35**) alternately creates two cardioid antenna patterns with lobes to the left and the right. The patterns are generated in much the same way as in the phased arrays described above. PIN RF diodes select the alternating patterns. The combined antenna outputs go to a receiver with AM detection. Processing after the detector output determines the phase or amplitude difference between the patterns' responses to the signal.

Switched pattern RDF sets typically have a zero center meter as an indicator. The meter swings negative when the signal is coming from the user's left, and positive when the signal source is on the right. When the plane of the antenna is exactly perpendicular to the direction of the signal source, the meter reads zero.

The sharpness of the zero crossing indication makes possible more precise bearings than those obtainable with a quad or Yagi. Under ideal conditions with a well-built unit, null direction accuracy is within 1°. Meter deflection tells the user which way to turn to zero the meter. For example, a negative (left) reading requires turning the antenna left. This solves the 180° ambiguity caused by the two zero crossings in each complete rotation of the antenna system.

Because it requires AM detection of the switched pattern signal, this RDF system finds its greatest use in the 120-MHz aircraft band, where AM is the standard mode. Commercial manufacturers make portable RDF sets with switched pattern antennas and built-in receivers for field portable use. These sets can usually be adapted to the amateur 144-MHz band. Other designs are adaptable to any VHF receiver that covers the frequency of interest and has an AM detector built in or added.

Switched pattern units work well for RDF from small aircraft, for which the two vertical antennas are mounted in fixed positions on the outside of the fuselage or simply taped inside the windshield. The left-right indication tells the pilot which way to turn the aircraft to home in. Since street vehicles generally travel only on roads, fixed mounting of the antennas on them is undesirable. Mounting vehicular switched-pattern arrays on a rotatable mast is best.

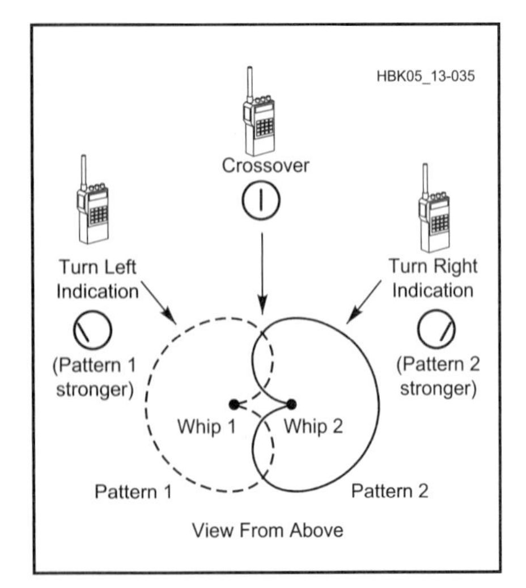

Fig 13.35 — In a switched pattern RDF set, the responses of two cardioid antenna patterns are summed to drive a zero center indicator.

Time-of-Arrival Systems

Another kind of switched antenna RDF set uses the difference in arrival times of the signal wavefront at the two antennas. This narrow-aperture Time-Difference-of-Arrival (TDOA) technology is used for many sophisticated military RDF systems. The rudimentary TDOA implementation of **Fig 13.36** is quite effective for amateur use. The signal from transmitter 1 reaches antenna A before antenna B. Conversely, the signal from transmitter 3 reaches antenna B before antenna A. When the plane of the antenna is perpendicular to the signal source (as transmitter 2 is in the figure), the signal arrives at both antennas simultaneously.

If the outputs of the antennas are alternately switched at an audio rate to the receiver input, the differences in the arrival times of a continuous signal produce phase changes that are detected by an FM discriminator. The resulting short pulses sound like a tone in the receiver output. The tone disappears when the antennas are equidistant from the signal source, giving an audible null.

The polarity of the pulses at the discriminator output is a function of which antenna is closer to the source. Therefore, the pulses can be processed and used to drive a left-right zero-center meter in a manner similar to the switched pattern units described above. Left-right LED indicators may replace the meter for economy and visibility at night.

RDF operations with a TDOA dual antenna RDF are done in the same manner as with a switched antenna RDF set. The main difference is the requirement for an FM receiver in the TDOA system and an AM receiver in the switched pattern case. No RF attenuator is needed for close-in work in the TDOA case.

Popular designs for practical do-it-yourself TDOA RDF sets include the Simple Seeker (described elsewhere in this chapter) and the W9DUU design (see article by Bohrer in the Bibliography). Articles with plans for the Handy Tracker, a simple TDOA set with a delay line to resolve the dual-null ambiguity instead of LEDs or a meter, are listed in the Bibliography.

Performance Comparison

Both types of dual antenna RDFs make good on-foot "sniffing" devices and are excellent performers when there are rapid amplitude variations in the incoming signal. They are the units of choice for airborne work. Compared to Yagis and quads, they give good directional performance over a much wider frequency range. Their indications are more precise than those of beams with broad forward lobes.

Dual-antenna RDF sets frequently give inaccurate bearings in multipath situations, because they cannot resolve signals of nearly equal levels from more than one direction. Because multipath signals are a combined pattern of peaks and nulls, they appear to change in amplitude and bearing as you move the RDF antenna along the bearing path or perpendicular to it, whereas a non-multipath signal will have constant strength and bearing.

The best way to overcome this problem is to take large numbers of bearings while moving toward the transmitter. Taking bearings while in motion averages out the effects of multipath, making the direct signal more readily discernible. Some TDOA RDF sets have a slow-response mode that

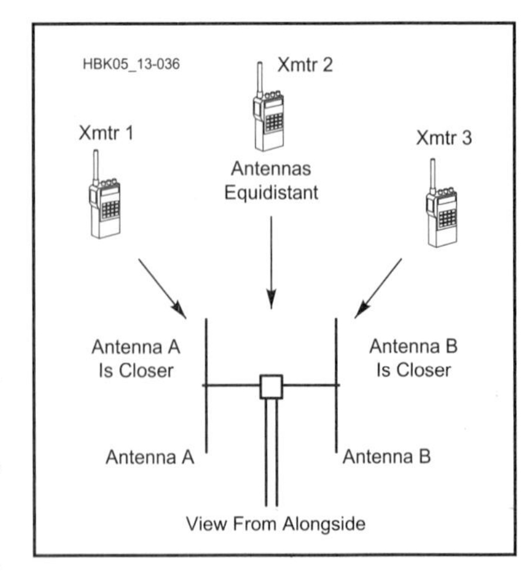

Fig 13.36 — A dual-antenna TDOA RDF system has a similar indicator to a switched pattern unit, but it obtains bearings by determining which of its antennas is closer to the transmitter.

aids the averaging process.

Switched antenna systems generally do not perform well when the incoming signal is horizontally polarized. In such cases, the bearings may be inaccurate or unreadable. TDOA units require a carrier type signal such as FM or CW; they usually cannot yield bearings on noise or pulse signals.

Unless an additional method is employed to measure signal strength, it is easy to "overshoot" the hidden transmitter location with a TDOA set. It is not uncommon to see a TDOA foxhunter walk over the top of a concealed transmitter and walk away, following the opposite 180° null, because there is no display of signal amplitude.

Doppler RDF Sets

RDF sets using the Doppler principle are popular in many areas because of their ease of use. They have an indicator that instantaneously displays direction of the signal source relative to the vehicle heading, either on a circular ring of LEDs or a digital readout in degrees. A ring of four, eight or more antennas picks up the signal. Quarter-wavelength monopoles on a ground plane are popular for vehicle use, but half-wavelength vertical dipoles, where practical, perform better.

Radio signals received on a rapidly moving antenna experience a frequency shift due to the Doppler effect, a phenomenon well known to anyone who has observed a moving car with its horn sounding. The horn's pitch appears higher than normal as the car approaches, and lower as the car recedes. Similarly, the received radio frequency increases as the antenna moves toward the transmitter and vice versa. An FM receiver will detect this frequency change.

Fig 13.37 shows a $^1/_4$-λ vertical antenna being moved on a circular track around point P, with constant angular velocity. As the antenna approaches the transmitter on its track, the received frequency is shifted higher. The highest instantaneous frequency occurs when the antenna is at point A, because tangential velocity toward the transmitter is maximum at that point. Conversely, the lowest frequency occurs when the antenna reaches point C, where velocity is maximum away from the transmitter.

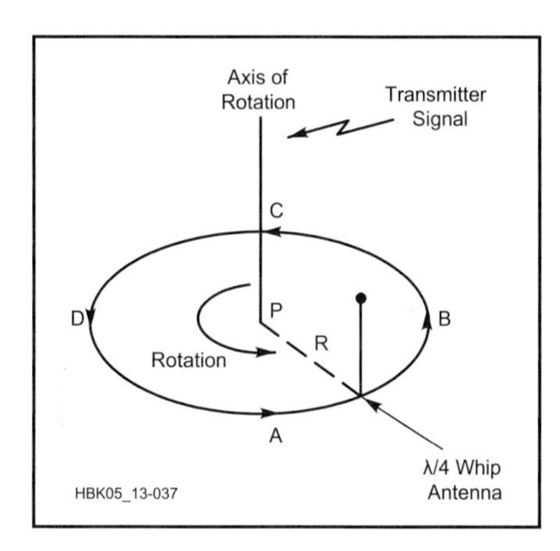

Fig 13.37 — A theoretical Doppler antenna circles around point P, continuously moving toward and away from the source at an audio rate.

Fig 13.38 shows a plot of the component of the tangential velocity that is in the direction of the transmitter as the antenna moves around the circle. Comparing Figs 13.37 and 13.38, notice that at B in Fig 13.38, the tangential velocity is crossing zero from the positive to the negative and the antenna is closest to the transmitter. The Doppler shift and resulting audio output from the receiver discriminator follow the same plot, so that a negative-slope zero-crossing detector, synchronized with the antenna rotation, senses the incoming direction of the signal.

The amount of frequency shift due to the Doppler effect is proportional to the RF frequency and the tangential antenna velocity. The velocity is a function of the radius of rotation and the angular velocity (rotation rate). The radius of rotation must be less than ¼ λ to avoid errors. To get a usable amount of FM deviation (comparable to typical voice modulation) with this radius, the antenna must rotate at approximately 30,000 RPM (500 Hz). This puts the Doppler tone in the audio range for easy processing.

Mechanically rotating a whip antenna at this rate is impractical, but a ring of whips, switched to the receiver in succession with RF PIN diodes, can simulate a rapidly rotating antenna. Doppler RDF sets must be used with receivers having FM detectors. The DoppleScAnt and Roanoke Doppler (see Bibliography) are mobile Doppler RDF sets designed for inexpensive home construction.

Doppler Advantages and Disadvantages

Ring-antenna Doppler sets are the ultimate in simplicity of operation for mobile RDF. There are no moving parts and no manual antenna pointing. Rapid direction indications are displayed on very short signal bursts.

Many units lock in the displayed direction after the signal leaves the air. Power variations in the source signal cause no difficulties, as long as the signal remains above the RDF detection threshold. A Doppler antenna goes on top of any car quickly, with no holes to drill. Many Local Interference Committee members choose Dopplers for tracking malicious interference, because they are inconspicuous (compared to beams) and effective at tracking the strong vertically polarized signals that repeater jammers usually emit.

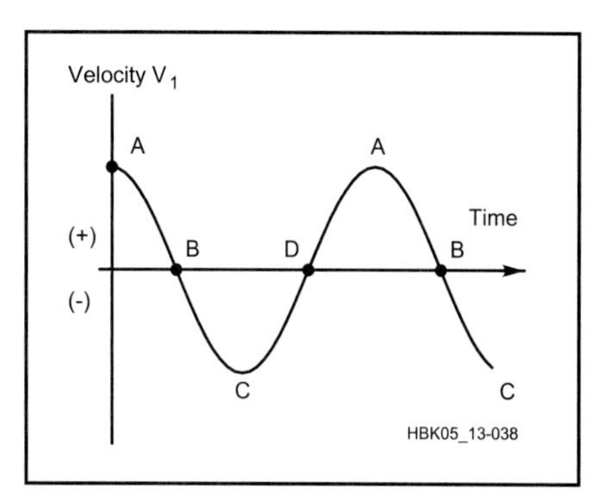

Fig 13.38 — Frequency shift versus time produced by the rotating antenna movement toward and away from the signal source.

A Doppler does not provide superior performance in all VHF RDF situations. If the signal is too weak for detection by the Doppler unit, the hunt advantage goes to teams with beams. Doppler installations are not suitable for on-foot sniffing. The limitations of other switched antenna RDFs also apply: (1) poor results with horizontally polarized signals, (2) no indication of distance, (3) carrier type signals only and (4) inadvisability of transmitting through the antenna.

Readout to the nearest degree is provided on some commercial Doppler units. This does not guarantee that level of accuracy, however. A well-designed four-monopole set is typically capable of ±5° accuracy on 2 m, if the target signal is vertically polarized and there are no multipath effects.

The rapid antenna switching can introduce cross modulation products when the user is near strong off-channel RF sources. This self-generated interference can temporarily render the system unusable. While not a common problem with mobile Dopplers, it makes the Doppler a poor choice for use in remote RDF installations at fixed sites with high power VHF transmitters nearby.

Mobile RDF System Installation

Of these mobile VHF RDF systems, the Doppler type is clearly the simplest from a mechanical installation standpoint. A four-whip Doppler RDF array is easy to implement with magnetic mount antennas. Alternately, you can mount all the whips on a frame that attaches to the vehicle roof with suction cups. In either case, setup is rapid and requires no holes in the vehicle.

Fig 13.39 — A set of TDOA RDF antennas is light weight and mounts readily through a sedan window without excessive overhang.

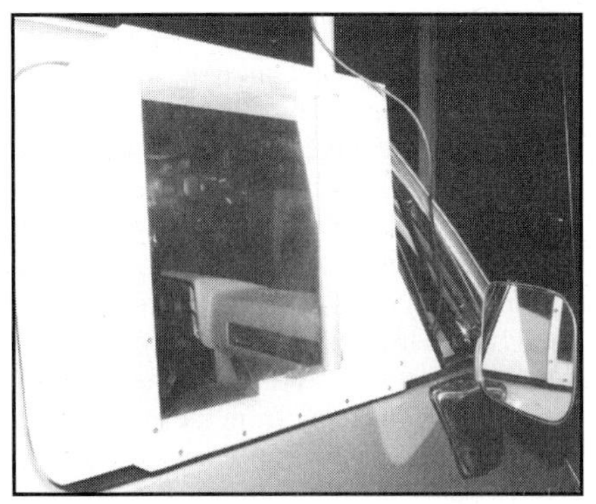

Fig 13.40 — A window box allows the navigator to turn a mast mounted antenna with ease while remaining dry and warm. No holes in the vehicle are needed with a properly designed window box.

You can turn small VHF beams and dual-antenna arrays readily by extending the mast through a window. Installation on each model vehicle is different, but usually the mast can be held in place with some sort of cup in the arm rest and a plastic tie at the top of the window, as in **Fig 13.39**. This technique works best on cars with frames around the windows, which allow the door to be opened with the antenna in place. Check local vehicle codes, which limit how far your antenna may protrude beyond the line of the fenders. Larger antennas may have to be put on the passenger side of the vehicle, where greater overhang is generally permissible.

The window box (**Fig 13.40**) is an improvement over through-the-window mounts. It provides a solid, easy-turning mount for the mast. The plastic panel keeps out bad weather. You will need to custom-design the box for your vehicle model. Vehicle codes may limit the use of a window box to the passenger side.

For the ultimate in convenience and versatility, cast your fears aside, drill a hole through the center of the roof and install a waterproof bushing. A roof-hole mount permits the use of large antennas without overhang violations. The driver, front passenger and even a rear passenger can turn the mast when required. The installation in Fig 13.34 uses a roof-hole bushing made from mating threaded PVC pipe adapters and reducers. When it is not in use for RDF, a PVC pipe cap provides a watertight cover. There is a pointer and 360° indicator at the bottom of the mast for precise bearings.

DIRECTION-FINDING TECHNIQUES AND PROJECTS

The ability to locate a transmitter quickly with RDF techniques is a skill you will acquire only with practice. It is very important to become familiar with your equipment and its limitations. You must also understand how radio signals behave in different types of terrain at the frequency of the hunt. Experience is the best teacher, but reading and hearing the stories of others who are active in RDF will help you get started.

Verify proper performance of your portable RDF system before you attempt to track signals in unknown locations. Of primary concern is the accuracy and symmetry of the antenna pattern. For instance, a lopsided figure-8 pattern with a loop, Adcock, or TDOA set leads to large bearing errors. Nulls should be exactly 180° apart and exactly at right angles to the loop plane or the array boom. Similarly, if feedline pickup causes an off-axis main lobe in your VHF RDF beam, your route to the target will be a spiral instead of a straight line.

Perform initial checkout with a low-powered test transmitter at a distance of a few hundred feet. Compare the RDF bearing indication with the visual path to the transmitter. Try to "find" the transmitter with the RDF equipment as if its position were not known. Be sure to check all nulls on antennas that have more than one.

If imbalance or off-axis response is found in the antennas, there are two options available. One is to correct it, insofar as possible. A second option is to accept it and use some kind of indicator or correction procedure to show the true directions of signals. Sometimes the end result of the calibration procedure is a compromise between these two options, as a perfect pattern may be difficult or impossible to attain.

The same calibration suggestions apply for fixed RDF installations, such as a base station HF Adcock or VHF beam. Of course it does no good to move it to an open field. Instead, calibrate the array in its intended operating position, using a portable or mobile transmitter. Because of nearby obstructions or reflecting objects, your antenna may not indicate the precise direction of the transmitter. Check for imbalance and systemic error by taking readings with the test emitter at locations in several different directions.

The test signal should be at a distance of 2 or 3 miles for these measurements, and should be in as clear an area as possible during transmissions. Avoid locations where power lines and other overhead wiring can conduct signal from the transmitter to the RDF site. Once antenna adjustments are optimized, make a table of bearing errors noted in all compass directions. Apply these error values as corrections when actual measurements are made.

Preparing to Hunt

Successfully tracking down a hidden transmitter involves detective work — examining all the clues, weighing the evidence and using good judgment. Before setting out to locate the source of a signal, note its general characteristics. Is the frequency constant, or does it drift? Is the signal continuous, and if not, how long are transmissions? Do transmissions occur at regular intervals, or are they sporadic? Irregular, intermittent signals are the most difficult to locate, requiring patience and quick action to get bearings when the transmitter comes on.

Refraction, Reflections and the Night Effect

You will get best accuracy in tracking ground wave signals when the propagation path is over homogeneous terrain. If there is a land/water boundary in the path, the different conductivities of the two media can cause bending (refraction) of the wave front, as in **Fig 13.41A**. Even the most sophisticated RDF equipment will not indicate the correct bearing in this situation, as the equipment can only show the

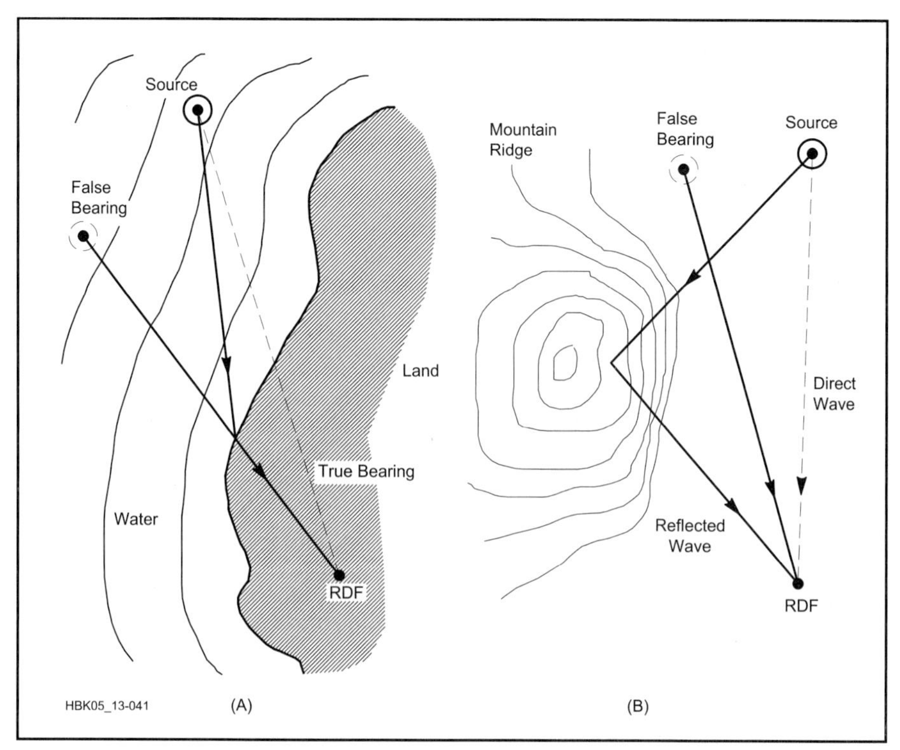

Fig 13.41 — RDF errors caused by refraction (A) and reflection (B). The reading at A is false because the signal actually arrives from a direction that is different from that to the source. At B, a direct signal from the source combines with a reflected signal from the mountain ridge. The RDF set may average the signals as shown, or indicate two lines of bearing.

direction from which the signal is arriving. RDFers have observed this phenomenon on both HF and VHF bands.

Signal reflections also cause misleading bearings. This effect becomes more pronounced as frequency increases. T-hunt hiders regularly achieve strong signal bounces from distant mountain ranges on the 144-MHz band.

Tall buildings also reflect VHF/UHF signals, making mid-city RDF difficult. Hunting on the 440-MHz and higher amateur bands is even more arduous because of the plethora of reflecting objects.

In areas of signal reflection and multipath, some RDF gear may indicate that the signal is coming from an intermediate point, as in Fig 13.41B. High gain VHF/UHF RDF beams will show direct and reflected signals as separate S-meter peaks, leaving it to the operator to determine which is which. Null-based RDF antennas, such as phased arrays and loops, have the most difficulty with multi-path, because the multiple signals tend to make the nulls very shallow or fill them in entirely, resulting in no bearing indication at all.

If the direct path to the transmitter is masked by intervening terrain, a signal reflection from a higher mountain, building, water tower, or the like may be much stronger than the direct signal. In extreme cases, triangulation from several locations will appear to "confirm" that the transmitter is at the location of the reflecting object. The direct signal may not be detectable until you arrive at the reflecting point or another high location.

Objects near the observer such as concrete/steel buildings, power lines and chain-link fences will distort the incoming wavefront and give bearing errors. Even a dense grove of trees can sometimes have an adverse effect. It is always best to take readings in locations that are as open and clear as possible, and to take bearings from numerous positions for confirmation. Testing of RDF gear should also be done in clear locations.

Locating local signal sources on frequencies below 10 MHz is much easier during daylight hours, particularly with loop antennas. In the daytime, D-layer absorption minimizes skywave propagation on these frequencies. When the D layer disappears after sundown, you may hear the signal by a combination of ground wave and high-angle skywave, making it difficult or impossible to obtain a bearing. RDFers call this phenomenon the *night effect*.

While some mobile T-hunters prefer to go it alone, most have more success by teaming up and assigning tasks. The driver concentrates on handling the vehicle, while the assistant (called the "navigator" by some teams) turns the beam, reads the meters and calls out bearings. The assistant is also responsible for maps and plotting, unless there is a third team member for that task.

Maps and Bearing-Measurements

Possessing accurate maps and knowing how to use them is very important for successful RDF. Even in difficult situations where precise bearings cannot be obtained, a town or city map will help in plotting points where signal levels are high and low. For example, power line noise tends to propagate along the power line and radiates as it does so. Instead of a single source, the noise appears to come from a multitude of sources. This renders many ordinary RDF techniques ineffective. Mapping locations where signal amplitudes are highest will help pinpoint the source.

Several types of area-wide maps are suitable for navigation and triangulation. Street and highway maps work well for mobile work. Large detailed maps are preferable to thick map books. Contour maps are ideal for open country. Aeronautical charts are also suitable. Good sources of maps include auto clubs, stores catering to camping/hunting enthusiasts and city/county engineering departments.

A *heading* is a reading in degrees relative to some external reference, such as your house or vehicle; a *bearing* is the target signal's direction relative to your position. Plotting a bearing on a hidden transmitter from your vehicle requires that you know the vehicle location, transmitter heading with respect to the vehicle and vehicle heading with respect to true north.

First, determine your location, using landmarks or a navigation device such as a GPS receiver. Next, using your RDF equipment, determine the bearing to the hidden transmitter (0 to 359.9°) with respect to the vehicle. Zero degrees heading corresponds to signals coming from directly in front of the vehicle, signals from the right indicate 90°, and so on.

Finally, determine your vehicle's true heading, that is, its heading relative to true north. Compass needles point to magnetic north and yield magnetic headings. Translating a magnetic heading into a true heading requires adding a correction factor, called *magnetic declination*[1], which is a positive or negative factor that depends on your location.

Declination for your area is given on US Geological Survey (USGS) maps, though it undergoes long-term changes.

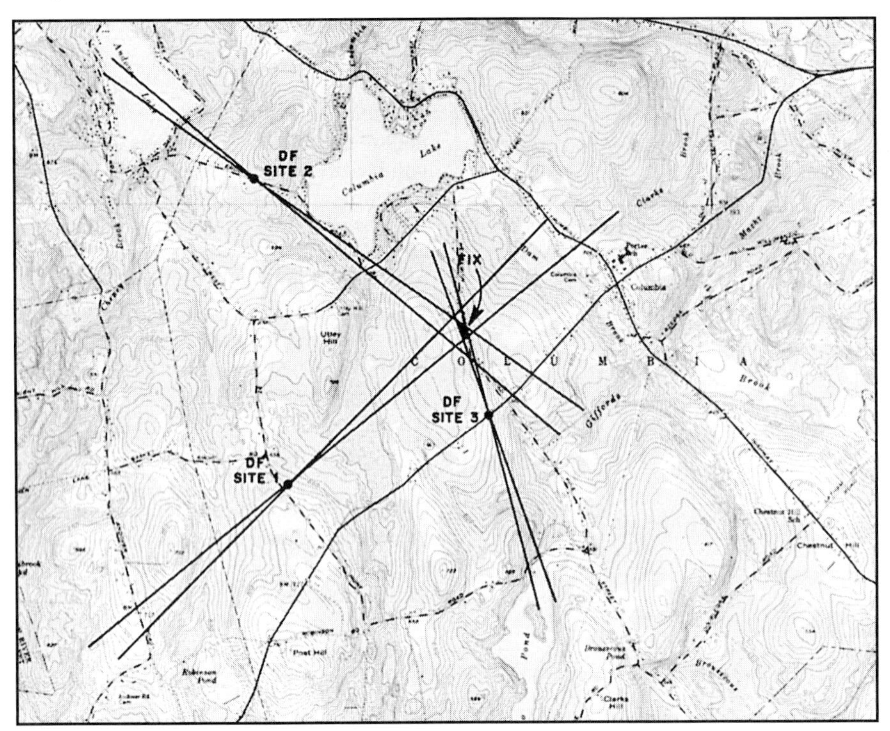

Fig 13.42 — Bearing sectors from three RDF positions drawn on a map for triangulation. In this case, bearings are from loop antennas, which have 180° ambiguity.

Add the declination to your magnetic heading to get a true heading.

As an example, assume that the transmitted signal arrives at 30° with respect to the vehicle heading, that the compass indicates that the vehicle's heading is 15°, and the magnetic declination is +15°. Add these values to get a true transmitter bearing (that is, a bearing with respect to true north) of 60°.

Because of the large mass of surrounding metal, it is very difficult to calibrate an in-car compass for high accuracy at all vehicle headings. It is better to use a remotely mounted flux-gate compass sensor, properly corrected, to get vehicle headings, or to stop and use a hand compass to measure the vehicle heading from the outside. If you T-hunt with a mobile VHF beam or quad, you can use your manual compass to sight along the antenna boom for a magnetic bearing, then add the declination for true bearing to the fox.

Triangulation Techniques

If you can obtain accurate bearings from two locations separated by a suitable distance, the technique of *triangulation* will give the expected location of the transmitter. The intersection of the lines of bearing from each location provides a *fix*. Triangulation accuracy is greatest when stations are located such that their bearings intersect at right angles. Accuracy is poor when the angle between bearings approaches 0° or 180°.

There is always uncertainty in the fixes obtained by triangulation due to equipment limitations, propagation effects and measurement errors. Obtaining bearings from three or more locations reduces the uncertainty. A good way to show the probable area of the transmitter on the triangulation map is to draw bearings as a narrow sector instead of as a single line. Sector width represents the amount of bearing uncertainty. **Fig 13.42** shows a portion of a map marked in this manner. Note how the bearing from Site 3 has narrowed down the probable area of the transmitter position.

Computerized Transmitter Hunting

A portable computer is an excellent tool for streamlining the RDF process. Some T-hunters use one to optimize VHF beam bearings, generating a two-dimensional plot of signal strength versus azimuth. Others have automated the bearing-taking process by using a computer to capture signal headings from a Doppler RDF set, vehicle heading from a flux-gate compass, and vehicle location from a GPS receiver (**Fig 13.43**). The computer program can compute averaged headings from a Doppler set to reduce multipath effects.

Provided with perfect position and bearing information, computer triangulation could determine the transmitter location within the limits of its computational accuracy. Two bearings would exactly locate a fox. Of course, there are always uncertainties and inaccuracies in bearing and position data. If these uncertainties can be determined, the program can compute the uncertainty of the triangulated bearings. A "smart" computer program can evaluate bearings, triangulate the bearings of multiple hunters, discard those that appear erroneous, determine which locations have particularly great or small multipath problems and even "grade" the performance of RDF stations.

By adding packet radio connections to a group of computerized base and mobile RDF stations, the processed bearing data from each can be shared. Each station in the network can display the triangulated bearings of all. This requires a common map coordinate set among all stations. The USGS Universal Transverse Mercator (UTM) grid, consisting of 1×1-km grid squares, is a good choice.

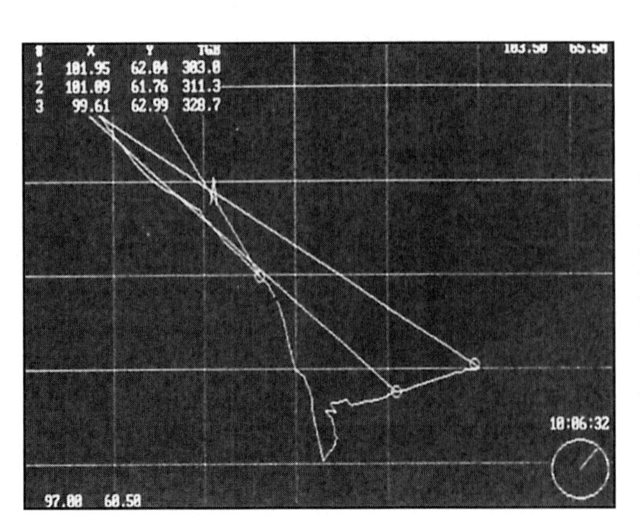

Fig 13.43 — Screen plot from a computerized RDF system showing three T-hunt bearings (straight lines radiating from small circles) and the vehicle path (jagged trace). The grid squares correspond to areas of standard topographic maps.

The computer is an excellent RDF tool, but it is no substitute for a skilled "navigator." You will probably discover that using a computer on a high-speed T-hunt requires a full-time operator in the vehicle to make full use of its capabilities.

Skywave Bearings and Triangulation

Many factors make it difficult to obtain accuracy in skywave RDF work. Because of Faraday rotation during propagation, skywave signals are received with random polarization. Sometimes the vertical component is stronger, and at other times the horizontal. During periods when the vertical component is weak, the signal may appear to fade on an Adcock RDF system. At these times, determining an accurate signal null direction becomes very difficult.

For a variety of reasons, HF bearing accuracy to within 1 or 2° is the exception rather than the rule. Errors of 3 to 5° are common. An error of 3° at a thousand miles represents a distance of 52 miles. Even with every precaution taken in measurement, do not expect cross-country HF triangulation to pinpoint a signal beyond a county, a corner of a state or a large metropolitan area. The best you can expect is to be able to determine where a mobile RDF group should begin making a local search.

Triangulation mapping with skywave signals is more complex than with ground or direct waves because the expected paths are great-circle routes. Commonly available world maps are not suitable, because the triangulation lines on them must be curved, rather than straight. In general, for flat maps, the larger the area encompassed, and the greater the error that straight line triangulation procedures will give.

A highway map is suitable for regional triangulation work if it uses some form of conical projection, such as the Lambert conformal conic system. This maintains the accuracy of angular representation, but the distance scale is not constant over the entire map.

One alternative for worldwide areas is the azimuthal-equidistant projection, better known as a great-circle map. True bearings for great-circle paths are shown as straight lines from the center to all points on the Earth. Maps centered on three or more different RDF sites may be compared to gain an idea of the general geographic area for an unknown source.

For worldwide triangulation, the best projection is the *gnomonic*, on which all great circle paths are represented by straight lines and angular measurements with respect to meridians are true. Gnomonic charts are custom maps prepared especially for government and military agencies.

Skywave signals do not always follow the great-circle path in traveling from a transmitter to a receiver. For example, if the signal is refracted in a tilted layer of the ionosphere, it could arrive from a direction that is several degrees away from the true great-circle bearing.

Another cause of signals arriving off the great-circle path is termed *sidescatter*. It is possible that, at a given time, the ionosphere does not support great-circle propagation of the signal from the transmitter to the receiver because the frequency is above the MUF for that path. However, at the same time, propagation may be supported from both ends of the path to some mutually accessible point off the great-circle path. The signal from the source may propagate to that point on the Earth's surface and hop in a sideways direction to continue to the receiver.

For example, signals from Central Europe have propagated to New England by hopping from an area in the Atlantic Ocean off the northwest coast of Africa, whereas the great-circle path puts the reflection point off the southern coast of Greenland. Readings in error by as much as 50° or more may result from sidescatter. The effect of propagation disturbances may be that the bearing seems to wander somewhat over a few minutes of time, or it may be weak and fluttery. At other times, however, there may be no telltale signs to indicate that the readings are erroneous.

Closing In

On a mobile foxhunt, the objective is usually to proceed to the hidden T with minimum time and mileage. Therefore, do not go far out of your way to get off-course bearings just to triangulate. It is usually better to take the shortest route along your initial line of bearing and "home in" on the signal. With a little experience, you will be able to gauge your distance from the fox by noting the amount of attenuation needed to keep the S-meter on scale.

As you approach the transmitter, the signal will become very strong. To keep the S-meter on scale, you will need to add an RF attenuator in the transmission line from the antenna to the receiver. Simple resistive attenuators are discussed in another chapter.

In the final phases of the hunt, you will probably have to leave your mobile and continue the hunt on foot. Even with an attenuator in the line, in the presence of a strong RF field, some energy will be coupled directly into the receiver circuitry. When this happens, the S-meter reading changes only slightly or perhaps not at all as the RDF antenna rotates, no matter how much attenuation you add. The cure is to shield the receiving equipment. Something as simple as wrapping the receiver in foil or placing it in a bread pan or cake pan, covered with a piece of copper or aluminum screening securely fastened at several points, may reduce direct pickup enough for you to get bearings.

Alternatively, you can replace the receiver with a field-strength meter as you close in, or use a heterodyne-type active attenuator. Plans for these devices are at the end of this chapter.

The Body Fade

A crude way to find the direction of a VHF signal with just a hand-held transceiver is the body fade technique, so named because the blockage of your body causes the signal to fade. Hold your HT close to your chest and turn all the way around slowly. Your body is providing a shield that gives the hand-held a cardioid sensitivity pattern, with a sharp decrease in sensitivity to the rear. This null indicates that the source is behind you (**Fig 13.44**).

If the signal is so strong that you can't find the null, try tuning 5 or 10 kHz off frequency to put the signal into the skirts of the IF passband. If your hand-held is dual-band (144/440 MHz) and you are hunting on 144 MHz, try tuning to the much weaker third harmonic of the signal in the 440-MHz band.

The body fade null, which is rather shallow to begin with, can be obscured by reflections, multipath, nearby objects, etc. Step well away from your vehicle before trying to get a bearing. Avoid large buildings, chain-link fences, metal signs and the like. If you do not get a good null, move to a clearer location and try again.

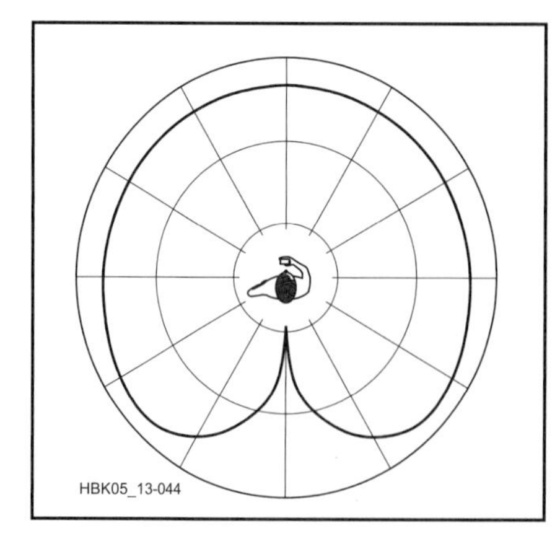

Fig 13.44 — When performing the body fade maneuver, a hand-held transceiver exhibits this directional pattern.

Air Attenuators

In microwave parlance, a signal that is too low in frequency to be propagated in a waveguide (that is, below the *cutoff frequency*) is attenuated at a predictable logarithmic rate. In other words, the farther inside the waveguide, the weaker the signal gets. Devices that use this principle to reduce signal strength are commonly known as *air attenuators*. Plans for a practical model for insertion in a coax line are in *Transmitter Hunting* (see Bibliography).

With this principle, you can reduce the level of strong signals into your hand-held transceiver, making it possible to use the body fade technique at very close range. Glen Rickerd, KC6TNF, documented this technique for *QST*. Start with a pasteboard mailing tube that has sufficient inside diameter to accommodate your hand-held. Cover the outside of the tube completely with aluminum foil. You can seal the bottom end with foil, too, but it probably will not matter if the tube is long enough. For durability and to prevent accidental shorts, wrap the foil in packing tape. You will also need a short, stout cord attached to the hand-held. The wrist strap may work for this, if long enough.

Fig 13.45 — The air attenuator for a VHF hand-held in use. Suspend the radio by the wrist strap or a string inside the tube.

To use this air attenuation scheme for body fade bearings, hold the tube vertically against your chest and lower the hand-held into it until the signal begins to weaken (**Fig 13.45**). Holding the receiver in place, turn around slowly and listen for a sudden decrease in signal strength. If the null is poor, vary the depth of the receiver in the tube and try again. You do not need to watch the S-meter, which will likely be out of sight in the tube. Instead, use noise level to estimate signal strength.

For extremely strong signals, remove the "rubber duck" antenna or extend the wrist strap with a shoelace to get greater depth of suspension in the tube. The depth that works for one person may not work for another. Experiment with known signals to determine what works best for you.

Note

[1] *Declination* is the term as denoted on land USGS topographic maps. *Deviation* and *Variation* are terms used on nautical and aviation charts, respectively.

THE SIMPLE SEEKER

The Simple Seeker for 144 MHz is the latest in a series of dual-antenna TDOA projects by Dave Geiser, W5IXM. Fig 13.36 and accompanying text shows its principle of operation. It is simple to perform rapid antenna switching with diodes, driven by a free-running multivibrator. For best RDF performance, the switching pulses should be square waves, so antennas are alternately connected for equal times. The Simple Seeker uses a CMOS version of the popular 555 timer, which demands very little supply current. A 9-V alkaline battery will give long life. See **Fig 13.46** for the schematic diagram.

PIN diodes are best for this application because they have low capacitance and handle a moderate amount of transmit power. Philips ECG553, NTE-555, Motorola MPN3401 and similar types are suitable. Ordinary 1N4148 switching diodes are acceptable for receive-only use.

Off the null, the polarity of the switching pulses in the receiver output changes (with respect to the switching waveform), depending on which antenna is nearer the source. Thus, comparing the receiver output phase to that of the switching waveform determines which end of the null line points toward the transmitter. The common name for a circuit to make this comparison is a *phase detector*, achieved in this unit with a simple bridge circuit. A phase detector balance control is included, although it may not be needed. Serious imbalance indicates incorrect receiver tuning, an off-frequency target signal, or misalignment in the receiver IF stages.

Almost any audio transformer with approximately 10:1 voltage step-up to a center-tapped secondary meets the requirements of this phase detector. The output is a positive or negative indication, applied to meter M1 to indicate left or right.

ANTENNA CHOICES

Dipole antennas are best for long-distance RDF. They ensure maximum signal pickup and provide the best load for transmitting. **Fig 13.47** shows plans for a pair of dipoles mounted on an H frame of 1/2-inch PVC tubing. Connect the 39-inch elements to the switcher with coaxial cables of *exactly* equal length. Spacing between dipoles is about 20 inches for 2 m, but is not critical. To prevent external currents flowing on the coax shield from disrupting RDF operation, wrap three turns (about 2 inch diameter) of the incoming coax to form a choke balun.

For receive-only work, dipoles are effective over much more than their useful transmit bandwidth. A pair of appropriately spaced 144-MHz dipoles works from 130 to 165 MHz. You will get greater tone amplitude with greater dipole spacing, making it easier to detect the null in the presence of modulation on the signal. But do not make the spacing greater than one-half free-space wavelength on any frequency to be used.

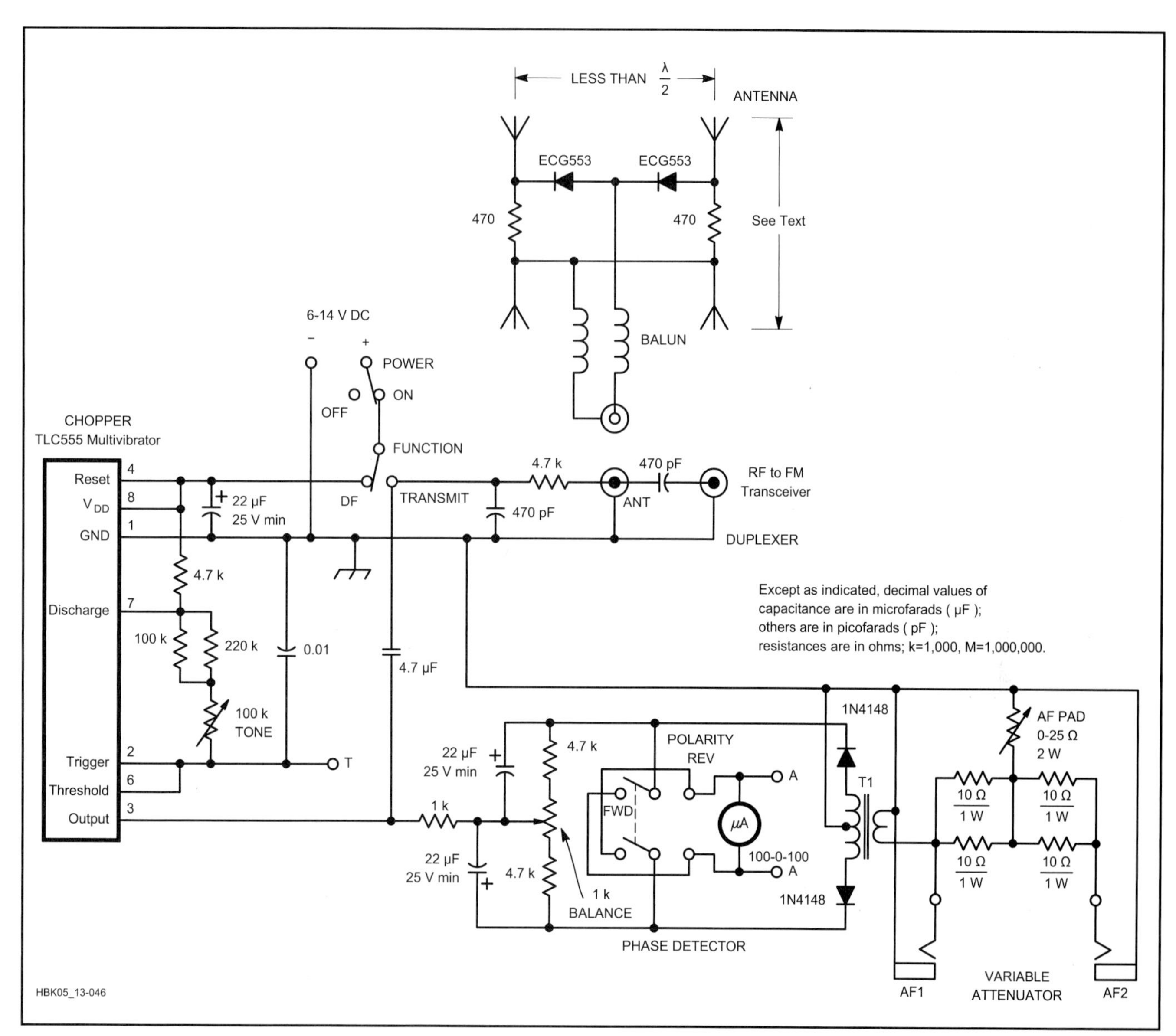

Fig 13.46 — Schematic of the Simple Seeker. A capacitor from point T to ground will lower the tone frequency, if desired. A single SPDT center-off toggle switch can replace separate power and function switches.

Best bearing accuracy demands that signals reach the receiver only from the switched antenna system. They should not arrive on the receiver wiring directly (through an unshielded case) or enter on wiring other than the antenna coax. The phase detecting system is less amplitude sensitive than systems such as quads and Yagis, but if you use small-aperture antennas such as "rubber duckies," a small signal leak may have a big effect. A wrap of aluminum foil around the receiver case helps block unwanted signal pickup, but tighter shielding may be needed.

Fig 13.48 shows a "sniffer" version of the unit with helix antennas. The added RDF circuits fit in a shielded box, with the switching pulses fed through a low-pass filter (the series 4.7-kΩ resistor and shunt 470-pF capacitor) to the receiver. The electronic switch is on a 20-pin DIP pad, with the phase detector on another pad (see **Fig 13.49**).

Because the phase detector may behave differently on weak and strong signals, the Simple Seeker incorporates an audio attenuator to allow either a full-strength audio or a lesser, adjustable received signal to feed the phase detector. You can plug headphones into jack AF2 and connect receiver audio to jack AF1 for no attenuation into the phase detector, or reverse the external connections, using the pad to control level to both the phones and the phase detector.

Convention is that the meter or other indicator deflects left when the signal is to the left. Others prefer that a left meter indication indicates that the antenna is rotated too far to the left. Whichever your choice, you can select it with the DPDT polarity switch. Polarity of audio output varies between receivers, so test the unit and receiver on a known signal source and mark the proper switch position on the unit before going into the field.

PIN diodes, when forward biased, exhibit low RF resistance and can pass up to approximately 1 W of VHF power without damage. The transmit position on the function switch applies steady dc bias to one of the PIN diodes, allowing communications from a hand-held RDF transceiver.

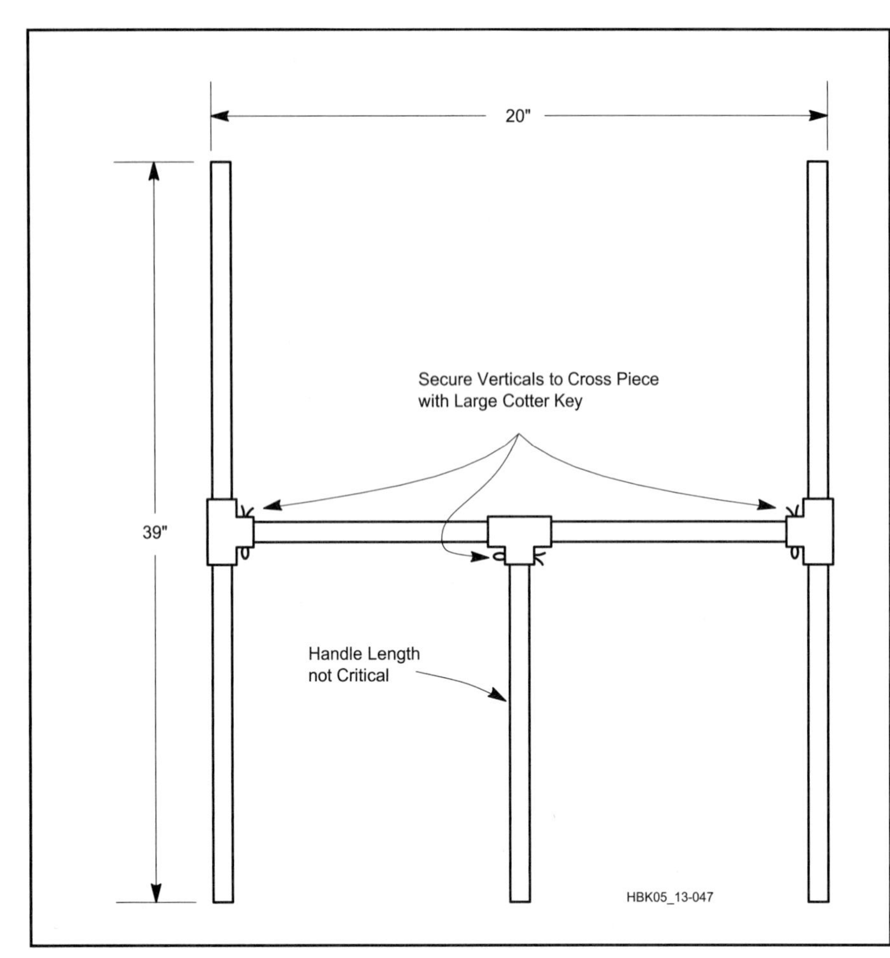

Fig 13.47 — "H" frame for the dual dipole Simple Seeker antenna set, made from 1/2-in. PVC tubing and tees. Glue the vertical dipole supports to the tees. Connect vertical tees and handle to the cross piece by drilling both parts and inserting large cotter pins. Tape the dipole elements to the tubes.

Fig 13.48 — Field version of the Simple Seeker with helix antennas.

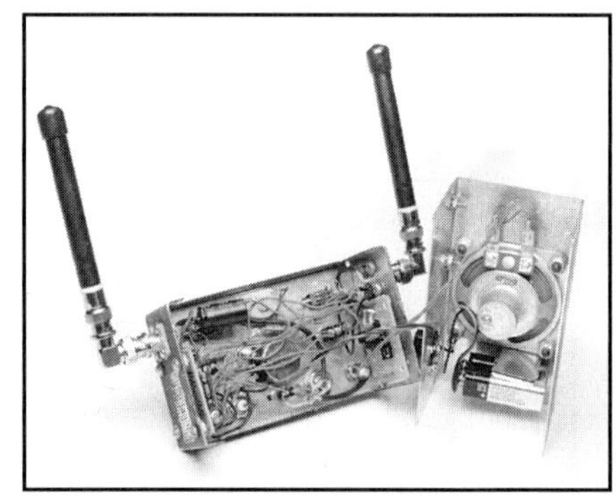

Fig 13.49 — Interior view of the Simple Seeker. The multivibrator and phase detector circuits are mounted at the box ends. This version has a convenient built-in speaker.

AN ACTIVE ATTENUATOR FOR VHF-FM

During a VHF transmitter hunt, the strength of the received signal can vary from roughly a microvolt at the starting point to nearly a volt when you are within an inch of the transmitter, a 120-dB range. If you use a beam or other directional array, your receiver must provide accurate signal-strength readings throughout the hunt. Zero to full scale range of S-meters on most hand-held transceivers is only 20 to 30 dB, which is fine for normal operating, but totally inadequate for transmitter hunting. Inserting a passive attenuator between the antenna and the receiver reduces the receiver input signal. However, the usefulness of an external attenuator is limited by how well the receiver can be shielded.

Anjo Eenhoorn, PAØZR, has designed a simple add-on unit that achieves continuously variable attenuation by mixing the received signal with a signal from a 500-kHz oscillator. This process creates mixing products above and below the input frequency. The spacing of the closest products from the input frequency is equal to the local oscillator (LO) frequency. For example, if the input signal is at 146.52 MHz, the closest mixing products will appear at 147.02 and 146.02 MHz.

The strength of the mixing products varies with increasing or decreasing LO signal level. By DFing on the mixing product frequencies, you can obtain accurate headings even in the presence of a very strong received signal. As a result, any hand-held transceiver, regardless of how poor it's shielding may be, is usable for transmitter hunting, up to the point where complete blocking of the receiver front end occurs. At the mixing product frequencies, the attenuator's range is greater than 100 dB.

Varying the level of the oscillator signal provides the extra advantage of controlling the strength of the input signal as it passes through the mixer. So as you close in on the target, you have the choice of monitoring and controlling the level of the input signal or the product signals, whichever provides the best results.

The LO circuit (**Fig 13.50**) uses the easy-to-find 2N2222A transistor. Trimmer capacitor C1 adjusts the oscillator's frequency. Frequency stability is only a minor concern; a few kilohertz of drift is tolerable. Q1's output feeds an emitter-follower buffer using a 2N3904 transistor, Q2. A linear-taper potentiometer (R6) controls the oscillator signal level present at the cathode of the mixing diode, D1. The diode and coupling capacitor C7 are in series with the signal path from antenna

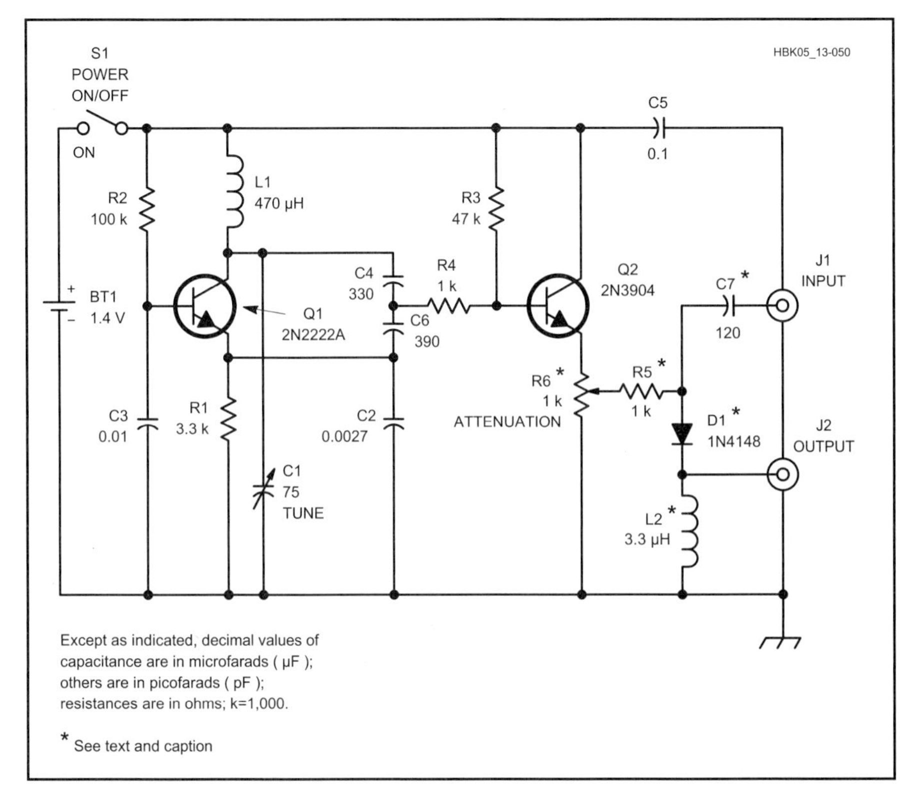

Fig 13.50 — Schematic of the active attenuator. Resistors are 1/4-W, 5%-tolerance carbon composition or film.

BT1 — Alkaline hearing-aid battery, Duracell SP675 or equivalent.
C1 — 75-pF miniature foil trimmer.
J1, J2 — BNC female connectors.
L1 — 470-µH RF choke.
L2 — 3.3-µH RF choke.
R6 — 1-kΩ, 1-W linear taper (slide or rotary).
S1 — SPST toggle.

input to attenuator output.

This frequency converter design is unorthodox; it does not use the conventional configuration of a doubly balanced mixer, matching pads, filters and so on. Such sophistication is unnecessary here. This approach gives an easy to build circuit that consumes very little power. PAØZR uses a tiny 1.4-V hearing-aid battery with a homemade battery clip. If your enclosure permits, you can substitute a standard AAA-size battery and holder.

CONSTRUCTION AND TUNING

For a template for this project, including the PC board layout and parts overlay, see the **Templates** section of the *Handbook* CD-ROM. A circuit board is available from FAR Circuits. The prototype (**Fig 13.51**) uses a plated enclosure with female BNC connectors for RF input and output. C7, D1, L2 and R5 are installed with point-to-point wiring between the BNC connectors and the potentiometer. S1 mounts on the rear wall of the enclosure.

Most hams will find the 500-kHz frequency offset convenient, but the oscillator can be tuned to other frequencies. If VHF/UHF activity is high in your area, choose an oscillator frequency that creates mixing products in clear portions of the band. The attenuator was designed for 144-MHz RDF, but will work elsewhere in the VHF/UHF range.

You can tune the oscillator with a frequency counter or with a strong signal of known frequency. It helps to enlist the aid of a friend with a hand-held transceiver a short distance away for initial tests. Connect a short piece of wire to J1, and cable your hand-held transceiver to J2. Select a simplex receive frequency and have your assistant key the test transmitter at its lowest power setting. (Better yet, attach the transmitter to a dummy antenna.)

With attenuator power on, adjust R6 for mid-scale S-meter reading. Now retune the hand-held to receive one of the mixing products. Carefully tune C1 and R6 until you hear the mixing product. Watch the S-meter and tune C1 for maximum reading.

If your receiver features memory channels, enter the hidden transmitter frequency along with both mixing product frequencies before the hunt starts. This allows you to jump from one to the other at the press of a button.

When the hunt begins, listen to the fox's frequency with the attenuator switched on. Adjust R6 until you get a peak reading. If the signal is too weak, connect your quad or other RDF antenna directly to your transceiver and hunt without the attenuator until the signal becomes stronger.

As you get closer to the fox, the attenuator will not be able to reduce the on-frequency signal enough to get good bearings. At this point, switch to one of the mixing product frequencies, set R6 for on-scale reading and continue. As you make your final approach, stop frequently to adjust R6 and take new bearings. At very close range, remove the RDF antenna altogether and replace it with a short piece of wire. It's a good idea to make up a short length of wire attached to a BNC fitting in advance, so you do not damage J1 by sticking random pieces of wire into the center contact.

While it is most convenient to use this system with receivers having S-meters,

Fig 13.51 — Interior view of the active attenuator. Note that C7, D1 and L2 are mounted between the BNC connectors. R5 (not visible in this photograph) is connected to the wiper of slide pot R6.

the meter is not indispensable. The active attenuator will reduce signal level to a point where receiver noise becomes audible. You can then obtain accurate fixes with null-seeking antennas or the "body fade" technique by simply listening for maximum noise at the null.

RDF BIBLIOGRAPHY

Bohrer, "Foxhunt Radio Direction Finder," *73 Amateur Radio*, Jul 1990, p 9.

Bonaguide, "HF DF — A Technique for Volunteer Monitoring," *QST*, Mar 1984, p 34.

DeMaw, "Maverick Trackdown," *QST*, Jul 1980, p 22.

Dorbuck, "Radio Direction Finding Techniques," *QST*, Aug 1975, p 30.

Eenhoorn, "An Active Attenuator for Transmitter Hunting," *QST*, Nov 1992, p 28.

Flanagan and Calabrese, "An Automated Mobile Radio Direction Finding System," *QST*, Dec 1993, p 51.

Geiser, "A Simple Seeker Direction Finder," *ARRL Antenna Compendium, Volume 3*, p 126.

Gilette, "A Fox-Hunting DF Twin'Tenna," *QST*, Oct 1998, pp 41-44.

Johnson and Jasik, *Antenna Engineering Handbook*, Second Edition, New York: McGraw-Hill.

Kossor, "A Doppler Radio-Direction Finder," *QST*, Part 1: May 1999, pp 35-40; Part 2: June 1999, pp 37-40.

McCoy, "A Linear Field-Strength Meter," *QST*, Jan 1973, p 18.

Moell and Curlee, *Transmitter Hunting: Radio Direction Finding Simplified*, Blue Ridge Summit, PA: TAB/McGraw-Hill. (This book, available from ARRL, includes plans for the Roanoke Doppler RDF unit and in-line air attenuator, plus VHF quads and other RDF antennas.)

Moell, "Transmitter Hunting — Tracking Down the Fun," *QST*, Apr 1993, p 48 and May 1993, p 58.

Moell, "Build the Handy Tracker," *73 Magazine*, Sep 1989, p 58 and Nov 1989, p 52.

O'Dell, "Simple Antenna and S-Meter Modification for 2-Meter FM Direction Finding," *QST*, Mar 1981, p 43.

O'Dell, "Knock-It-Down and Lock-It-Out Boxes for DF," *QST*, Apr 1981, p 41.

Ostapchuk, "Fox Hunting is Practical and Fun!" *QST*, Oct 1998, pp 68-69.

Rickerd, "A Cheap Way to Hunt Transmitters," *QST*, Jan 1994, p 65.

The "Searcher" (SDF-1) Direction Finder, Rainbow Kits.

Chapter 14

Receivers and Transmitters

In this chapter, William E. Sabin, WØIYH, discusses the "system design" of Amateur Radio receivers and transmitters. "A Single-Stage Building Block" reviews briefly a few of the basic properties of the various individual building block circuits, described in detail in other chapters, and the methods that are used to combine and interconnect them in order to meet the requirements of the completed equipment. "The Amateur Radio Communication Channel" describes the relationships between the equipment system design and the electromagnetic medium that conveys radio signals from transmitter to receiver. This understanding helps to put the radio equipment mission and design requirements into perspective. Then we discuss receiver, transmitter, transceiver and transverter design techniques in general terms. At the end of the theory discussion is a list of references for further study on the various topics. The projects section contains several hardware descriptions that are suitable for amateur construction and use on the ham bands. They have been selected to illustrate system-design methods. The emphasis in this chapter is on analog design. Those functions that can be implemented using digital signal processing (DSP) can be explored in other chapters, but an initial basic appreciation of analog methods and general system design is very valuable.

A SINGLE-STAGE BUILDING BLOCK

We start at the very beginning with **Fig 14.1**, a generic single-stage module that would typically be part of a system of many stages. A signal source having an "open-circuit" voltage V_{gen} causes a current Igen to flow through Z_{gen}, the impedance of the generator, and Z_{in}, the input impedance of the stage. This input current is responsible for an open-circuit output voltage V_d (measured with a high-impedance voltmeter) that is proportional to I_{gen}. V_d produces a current Iout and a voltage drop across Z_{out}, the output impedance of the stage and Z_{load}, the load impedance of the stage. Observe that the various Zs may contain reactance and resistance in various combinations. Let's first look at the different types of gain and power relationships that can be used to describe this stage.

Actual Power Gain

Current I_{gen} produces a power dissipation P_{in} in the resistive component of Z_{in} that is equal to $I_{gen}^2 R_{in}$. The current I_{out} produces a power dissipation P_{load} in the resistive component of Z_{load} that is equal to $I_{out}^2 R_{load}$. The actual power gain in dB is 10 log (P_{load} / P_{in}). This is the conventional usage of dB, to describe a power ratio.

Voltage Gain

The current I_{gen} produces a voltage drop across Z_{in}. V_d produces a current I_{out} and a voltage drop V_{out} across Z_{load}. The voltage gain is the ratio

$$V_{out} / V_{in} \quad (1)$$

In decibels (dB) it is

$$20 \log (V_{out} / V_{in}) \quad (2)$$

This alternate usage of dB, to describe a voltage ratio, is common practice. It is *different* from the power gain mentioned in the previous section because it does *not* take into account the power ratio or the resistance values involved. It is a voltage ratio only. It is used in troubleshooting and other instances where a rough indication of operation is needed, but precise measurement is unimportant. Voltage gain is often used in high-impedance circuits such as pentode vacuum tubes and is also sometimes convenient in solid-state circuits. Its improper usage often creates errors in radio circuit design because many calculations, for correct answers, require power ratios rather than voltage ratios. We will see several examples of this throughout this chapter.

Available Power

The maximum power, in watts, that can be obtained from the generator is V_{gen}^2 / (4 R_{gen}). To see this, suppose temporarily that X_{gen} and X_{in} are both zero. Then let R_{in} increase from zero to some large value. The maximum power in R_{in} occurs when $R_{in} = R_{gen}$ and the power in R_{in} then has the value mentioned above (plot a graph of power in R_{in} vs R_{in} to verify this, letting V_{gen} = 1 V and R_{gen} = 50 Ω). This is called the "available power" (we're assuming sine-wave signals). If X_{gen} is an inductive (or capacitive) reactance and if X_{in} is an equal value of capacitive (or inductive) reactance, the net series reactance is nullified and the above discussion holds true. If the net reactance is not zero the current I_{gen} is reduced and the power in R_{in} is less than maximum. The process of tuning out the reactance and then transforming the resistance of R_{in} is called "conjugate matching." A common method for doing this conjugate matching is to put an impedance transforming circuit of some kind, such as a transformer or a tuned circuit, between the generator and the stage input that "transforms" R_{in} to the value R_{gen} (as seen by the generator) and at the same time nullifies the reactance.

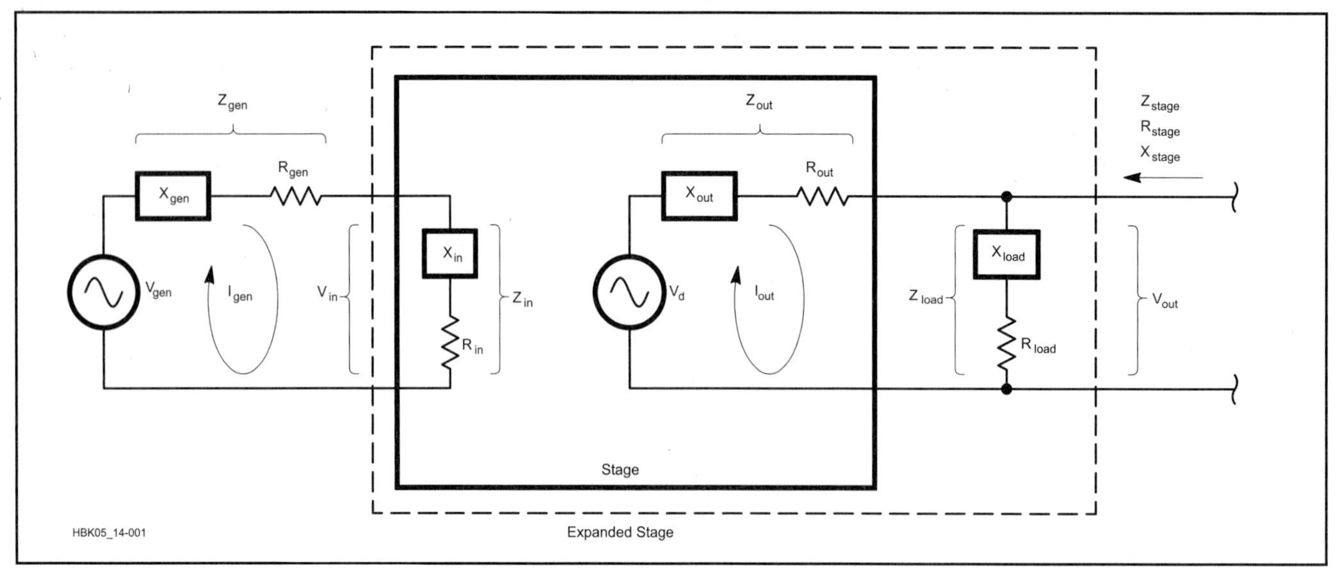

Fig 14.1—A single-stage building-block signal processor. The properties of this stage are discussed in the text.

Fig 14.2 illustrates this idea and later discussion gives more details about these interstage networks. A small amount of power is lost within any lossy elements of the matching network. This same technique can be used between the output of the stage in Fig 14.1 and the load impedance Z_{load}. In this case, the stage delivers the maximum amount of power to the load resistance. If both input and output are processed in this way, the stage utilizes the generator signal to the maximum extent possible. It is very important to note, however, that in many situations we do not want this maximum utilization. We deliberately "mismatch" in order to achieve certain goals that will be discussed later (Ref 1).

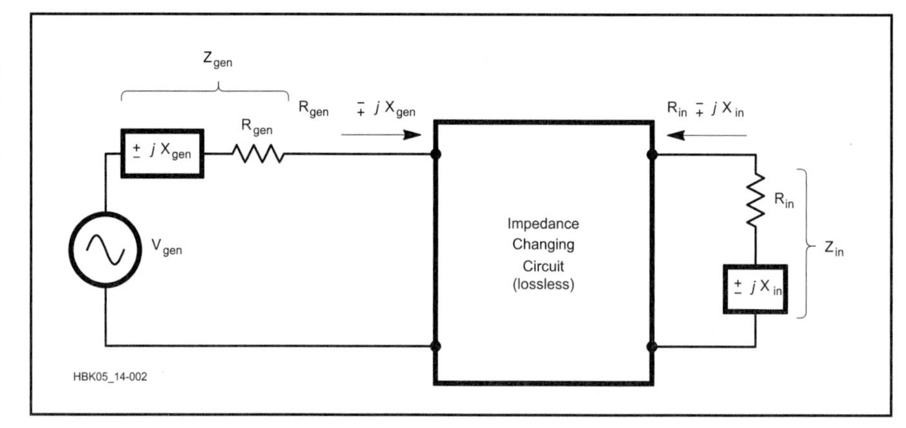

Fig 14.2—The conjugate impedance match of a generator to a stage input. The network input impedance is $R_{gen} \pm jX_{gen}$ and its output impedance is $R_{in} \pm jX_{in}$ (where either R term may represent a dynamic impedance). Therefore the generator and the stage input are both impedance matched for maximum power transfer.

The dBm Unit of Power

In low-level radio circuitry, the watt (W) is inconveniently large. Instead, the milliwatt (mW) is commonly used as a reference level of power. The dB with respect to 1 mW is defined as

$$dBm = 10 \log (P_W / 0.001) \quad (3)$$

where

dBm = Power level in dB with respect to 1 mW

P_W = Power level, watts.

For example, 1 W is equivalent to 30 dBm. Also

$$P_W = 0.001 \times 10^{dBm/10} \quad (4)$$

Maximum Available Power Gain

The ratio of the power that is available from the stage, $V_d^2 / (4\ R_{out})$, to the power that is available from the generator, $V_{gen}^2 / (4\ R_{gen})$, is called the maximum available power gain. In some cases the circuit is adjusted to achieve this value, using the conjugate-match method described above. In many cases, as mentioned before, less than maximum gain is acceptable, perhaps more desirable.

Available Power Gain

Consider that in Fig 14.1 the stage and its output load Z_{load} constitute an "expanded" stage as defined by the dashed box. The power available from this new stage is determined by V_{out} and by R_{stage}, the resistive part of Z_{stage}. The available power gain is then $V_{out}^2 / (4\ R_{stage})$ divided by $V_{gen}^2 / (4\ R_{gen})$. This value of gain is used in a number of design procedures. Note that Z_{load} can be a physical network of some kind, or it may be partly or entirely the input impedance of the stage following the one shown in Fig 14.1.

In the latter case it is sometimes convenient to "detach" this input impedance from the next stage and make it part of the expanded first stage, as shown in Fig 14.1, but we note that Z_{out} is still the generator (source) impedance that the input of the next stage "sees."

Transducer Power Gain

The transducer gain is defined as the ratio of the power actually delivered to R_{load} in Fig 14.1 to the power that is available from the generator V_{gen} and R_{gen}. In other words, how much more power does the stage deliver to the load than the generator could deliver if the generator were impedance matched to the load? We will discuss how to use this kind of gain later.

Feedback (Undesired)

One of the most important properties of the single-stage building block in Fig 14.1 is that changes in the load impedance Z_{load} cause changes in the input impedance Z_{in}. Changes in Z_{gen} also affect Z_{out}. These effects are due to reverse coupling, within the stage, from output to input. For many kinds of circuits (such as networks, filters, attenuators, transformers and so on) these effects cause no unexpected problems.

But, as the chapter on **RF Power Amplifiers** explains in detail, in active circuits such as amplifiers this reverse coupling within one stage can have a major impact not only on that stage but also on other stages that follow and precede. It is the effect on system performance that we discuss here. In particular, if a stage is expected to have certain gain, noise factor and distortion specifications, all of these can be changed either by reverse coupling (undesired feedback) within the stage or adjacent stages. For example, internal feedback can cause the input impedance of a certain stage "A" (Fig 14.1) to become very large. If this impedance is the load impedance for the preceding stage, the gain of the preceding stage can become excessive, creating problems in both stages. This same feedback can cause the gain of stage "A" to become greater, thereby causing the next stage to be driven into heavy distortion. A very common event is that stage "A" goes into oscillation. All of these occurrences are common in poorly designed radio equipment. Changes in temperature and variations in component tolerances are major contributors to these problems.

One particular example is shown in **Fig 14.3**, a transistor amplifier, shown in skeleton form, with sharply tuned resonators at input and output.

Because of reverse coupling, the two tuned circuits interact, making adjustments difficult or even impossible. The likelihood of oscillation is very high. There are two solutions: drastically reduce the gain of the amplifier, or use an amplifier circuit that has very little reverse coupling. Usually, both methods are used simultaneously (in the right amount) in order to get predictable performance. The object lesson for the system designer is that a combination of reduced gain and low reverse coupling is the safe way to go when designing a radio system. More stages may be required, but the price is well worthwhile. The cascode amplifier, grounded-gate amplifier, dual-gate FET and many types of IC amplifiers are examples of circuits that have little reverse coupling and good stability. "Neutralization" methods are used to cancel reverse coupling that causes instability. All such circuits are said to be "unilateral," which means "in one direction" and both input and output can be independently tuned as in Fig 14.3 if the gain is not too high.

Feedback (Desired)

The **RF Power Amplifiers** chapter explains how negative feedback (good feedback) can be used to stabilize a circuit and make it much more predictable over a range of temperature and component tolerances. Here we wish to point out some system implications of negative feedback. One is that the gain, noise-figure and distortion performances within a stage are made much more constant and predictable. Therefore a system designer can put building blocks together with more confidence and less guesswork.

There are some problems, though. In some circuits the amount of feedback depends on both the output impedance of the driving circuit and the input impedance of the next stage. A classic example is the cascadable amplifier shown in **Fig 14.4**.

In this circuit, if the output load impedance becomes very low the amplifier input impedance becomes high, and vice versa (a "teeter-totter" effect). Other amplifier properties also can change. With amplifiers of this type it is important to maintain the correct impedances at the input and output interfaces. Any building block should be examined for effects of this kind. Data sheets frequently specify the reverse transfer values as well as those for forward transfer. Often, lab measurements are needed. Apply a signal to the output and measure the reverse coupling to the input. Where varying load and source impedances are involved, look for a circuit that is less vulnerable (that is, has less reverse coupling).

Another problem is that feedback networks often add thermal noise sources to a circuit and so degrade its noise figure. In systems where this is a consideration, use so-called "lossless feedback" circuits. These circuits use very efficient transformers instead of resistors or lossy networks that introduce thermal noise into a system.

Noise Factor and Noise Figure

The output resistance of the signal generator that drives a typical signal processing block such as shown in Fig 14.4 is a source of thermal noise power, which is a natural phenomenon occurring in the resistive component of any impedance. It is caused by random motion of electrons within a conducting (or semiconducting) material. Note that the reactive part of an impedance is not a source of thermal noise power because the voltage across a pure reactance and the current through the reactance are in phase quadrature (90°) at any one frequency. The average value of the product of these two (the power) is zero. If this is true at any frequency, then it is true at all frequencies. Also, a purely

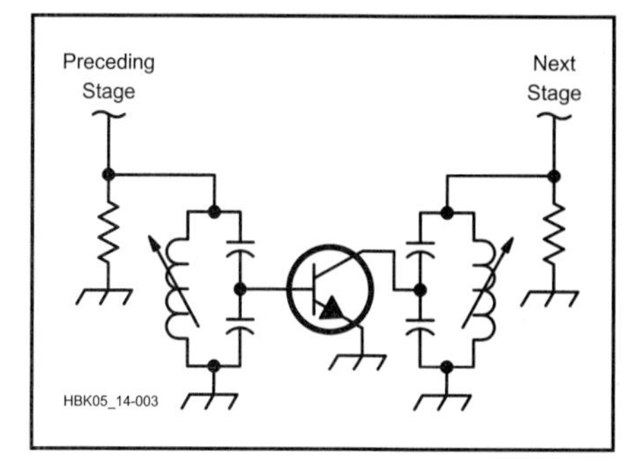

Fig 14.3—A double tuned transistor amplifier circuit that may oscillate due to excessive amplification and reverse coupling.

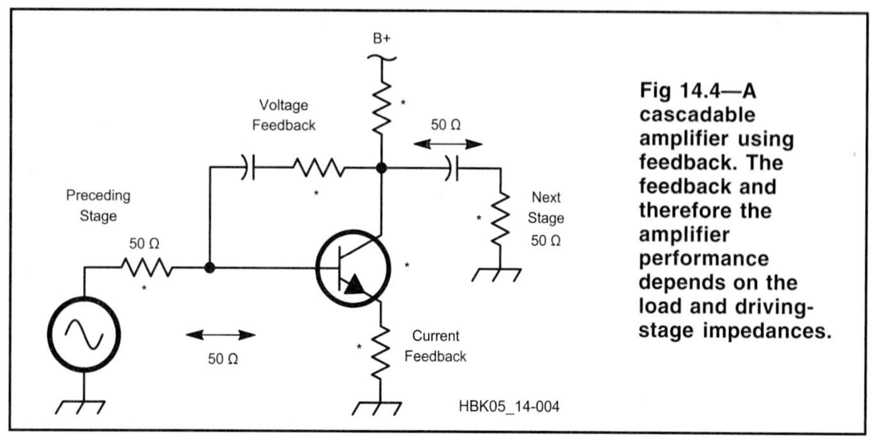

Fig 14.4—A cascadable amplifier using feedback. The feedback and therefore the amplifier performance depends on the load and driving-stage impedances.

Negative Feedback in RF Circuit Design

This sidebar shows how negative feedback can enhance the performance of the RF amplifier circuits that are used in homebrew receivers, transmitters or transceivers. The sidebar lists the advantages of negative feedback, and presents examples for the advantages. We will not consider automatic gain control (AGC) or automatic level control (ALC) circuits, often referred to as "envelope feedback" (see later sections of this chapter).

1. All amplifiers have nonlinearities that produce output frequencies (harmonics, intermodulation distortion and adjacent channel interference) of a certain amount that are not present in the input signal. Feedback can reduce the *percentage*.

2. Feedback can be used to increase or decrease the input impedance or output impedance of an amplifier stage. Actual L, C and R values are modified by feedback to "effective" values. The concept of a dynamic or lossless resistance that does not dissipate power or create thermal noise is introduced.

3. Feedback can improve the stability (freedom from a tendency to oscillate) of an amplifier. This includes the methods of feedback network compensation and also "neutralization" or "unilateralization" which means the reduction of internal feedback within the amplifier.

4. Feedback can make the frequency response, the input impedance and the output impedance of an amplifier more constant over a wide frequency band.

5. Feedback can reduce gain and frequency response changes due to temperature, supply voltage, value tolerances of inductors, capacitors and resistors and transistor parameter spreads. The performance variations over a large number of identical circuits are greatly reduced.

6. By controlling the performance of each stage in a multistage system, using feedback, the overall performance can be more accurately predicted and maintained with little or no "tweaking" of the individual stages.

Feedback in an amplifier stage nearly always reduces the gain (ratio of output to input voltage, current or power) to a lower value. For a certain overall gain requirement, more stages are required. This is in nearly all cases a mild penalty. Each extra stage is an additional source of the imperfections that were enumerated above. This means that the entire chain must be designed as a system in order to meet the system goals. Feedback can also be applied over two, or sometimes three cascaded stages, but is more difficult. In this brief overview we will focus mainly on single-stage feedback.

The Negative Feedback Concept

Fig A shows an amplifier with negative feedback. Prior to feedback, the amplifier has gain G and a phase reversal that we assume for simplicity is 180°. A portion of the output *voltage*, or perhaps a portion of the output *current*, goes through a feedback network to the input where it combines, possibly in *series* with or in *parallel* with the generator signal to produce a *modified* input signal that is smaller than it was with no feedback. This new input consists partly of generator signal, assumed to be perfect, and partly of an output signal, assumed to be imperfect. Note that the generator signal and the fed-back signal are in opposite phase. When this modified input signal is amplified (even though it is amplified imperfectly itself) the original imperfections that were

(continued on page 14.6)

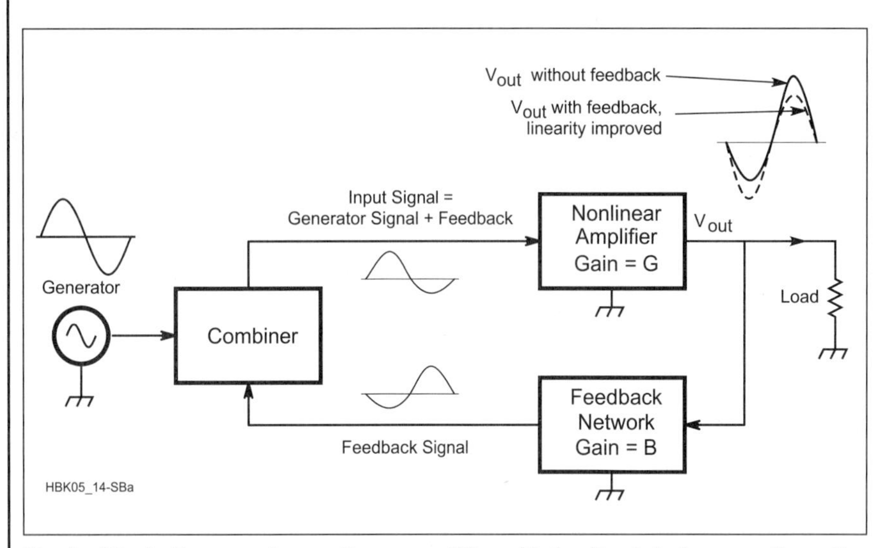

Fig A—Block diagram of a nonlinear amplifier with feedback to improve linearity.

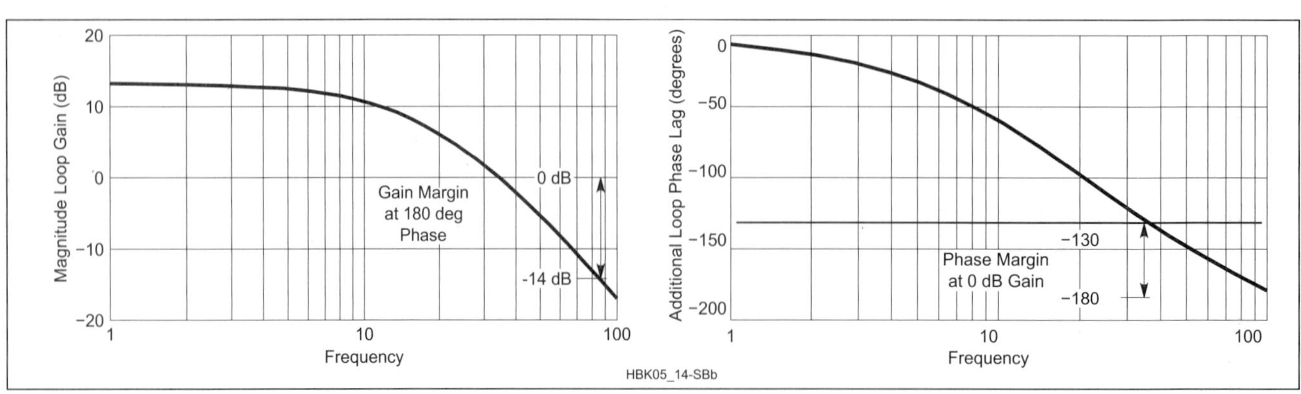

Fig B—Graph 1 shows the gain margin of the feedback loop when the loop phase shift is 180°. Graph 2 shows the phase margin of the feedback look when the loop gain is 0 dB.

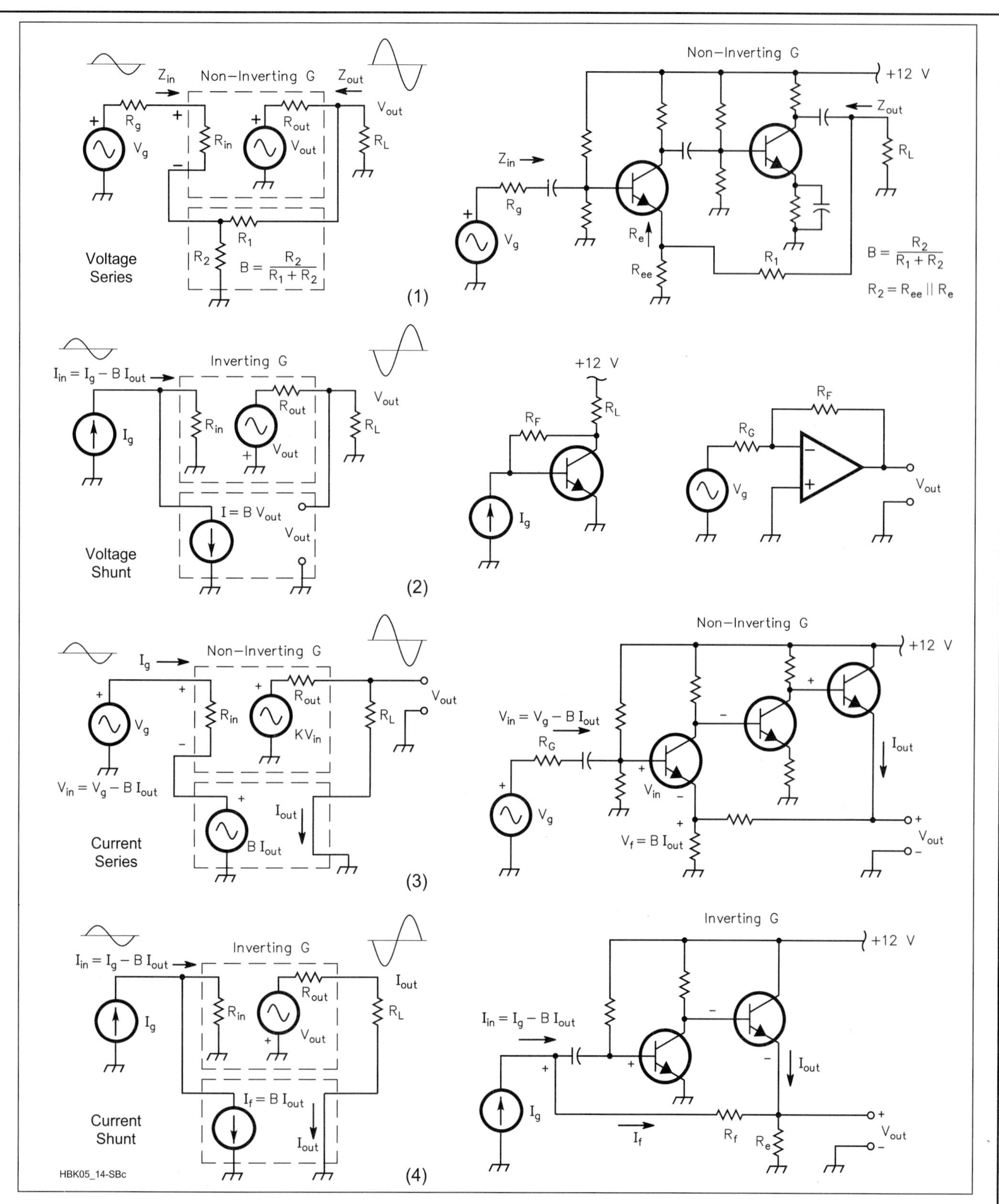

Fig C—Part 1 is an example of voltage-series feedback. A fraction, B, of the noninverted output voltage is in series with the input-signal voltage. Z_{out} is reduced and Z_{in} is increased. R_2 consists of R_{ee} in parallel with R_e (= 1/gm).
Part 2 is an example of voltage-shunt feedback. The signal generator is a constant-current source, I_g. Current B V_{out} is in parallel with I_g and in phase-opposition to I_g. The two sample circuits are a one-stage trans-resistance amplifier and an op-amp as an inverting voltage amplifier. Input impedance and output impedance are both reduced.
Part 3 is an example of current-series feedback, or a transconductance amplifier. The noninverted output current produces a feedback voltage of V_f = B I_{out}, which is in series opposition with the signal V_g. Z_{out} is increased and Z_{in} is also increased.
Part 4 is an example of current-shunt feedback. The inverted output current produces a feedback current If = B I_{out}, which is in shunt opposition with the signal I_g. Z_{out} is increased and Z_{in} is decreased.

created in the amplifier are reduced. That is, the fed-back imperfections are amplified and phase reversed, and at the output are in the correct phase that they *counteract* the amplifier's inherent imperfections. Also, we can now increase the generator level so that the output returns to the same value that it had with no feedback. The feedback amplifier is then less imperfect, but some residual imperfection always remains.

Gain and Stability

In Fig A as the product G B (known as the "loop gain") becomes large, the gain AFB with feedback is nearly 1/B where B is the feedback gain, usually of a linear, passive feedback network. In other words, the total gain AFB of the feedback amplifier is:

$$A_{FB} = \frac{G}{1-GB} \approx \frac{1}{B} \text{ if } -GB >> +1.0 \qquad \text{(Eq A1)}$$

In an inverting amplifier (Fig A) G and AFB are negative numbers, due to the phase inversion in G. In a noninverting amplifier, G and AFB are positive numbers and the feedback network B must provide the phase inversion (the negative number). If G B in Eq A1 reaches the value +1.0, due to *additional* phase shift, the amplifier becomes unstable (the denominator becomes zero). The design of G and B must prevent this. We must make |G B|—the magnitude of G B—less than 1.0 (0 dB) at the frequency at which an additional phase shift of 180° occurs. This is called "gain margin." We must also make this additional phase shift of G B less than 180° at the frequency at which |G B| equals 1.0 (0 dB). This is called "phase margin." For small values of these margins, you can see erratic amplifier behavior and possibly oscillation on an oscilloscope and on a spectrum analyzer. In practice, a phase margin of 45° is usually a safe value that also provides a good gain margin. An important task is to perform a graphical plot, over a wide frequency range, of the magnitude and phase of the product G B, examine the margins and then modify the design as needed. **Fig B** shows typical loop gain and phase plots. The most desirable response is one that approaches unity (0 dB) loop gain at –6 dB per octave because this provides good gain and phase margins. As a follow-up, plot the frequency response of AFB (Eq A1) to verify the final result and see some of the benefits of the negative feedback. Equations very similar to Eq A1 apply to the other benefits of negative feedback that were previously mentioned. The "cost" of the feedback is that a larger generator signal is needed, or may require a pre-amplifier stage. This additional stage operates at a lower signal level and may not need as much feedback.

Four Negative Feedback Topologies

The advantages of negative feedback can be achieved in four basic ways, according to how the feedback affects the input impedance of the amplifier, which can be increased or decreased, and the output impedance of the amplifier, which can be increased or decreased. At the output, the feedback can be *derived* from the voltage across the load or the current through the load. At the input, the feedback can be *applied* in series with the generator or in parallel with the generator. These options provide the following four variations. **Fig C** shows the block diagrams for these options, and also a simple example of each type.

1. Voltage-Series. Reduces output impedance, increases input impedance.
2. Voltage-Shunt. Reduces output impedance, reduces input impedance.
3. Current-Series. Increases output impedance, increases input impedance.
4. Current-Shunt. Increases output impedance, reduces input impedance.

Feedback can be used to match impedances. For example, the 1000-Ω output resistance of an amplifier can be reduced by feedback to 50 Ω. It then correctly terminates a 50-Ω filter, transmission line or the next amplifier stage.—*William E. Sabin, WØIYH*

"dynamic resistance" such as R_e, the dynamic resistance (ΔV/ΔI) of a perfect forward conducting PN junction, is also not a source of thermal noise. However, the junction is a source of "shot noise" power that, by the way, is only 50% as great as the thermal noise that R_e would have if it were an actual resistor (Ref 2).

Each "*" in Fig 14.4 indicates a noise source. Passive elements generate thermal noise. Active components such as transistors generate thermal noise and other types, such as shot noise and flicker (1/f) noise, internally. These "excess" noises are all superimposed on the signal from the generator. Therefore the noise factor of a single stage is a measure of how much the signal to noise ratio is degraded as a signal passes through that stage.

Refer now to the diagram and equations in **Fig 14.5**. F is noise factor and S_i / N_i is the input signal to noise ratio from the signal generator. S_o / N_o is signal to noise ratio at the output and kTB is the thermal noise power that is available from any value of resistance (kT = –174 dBm in a 1-Hz bandwidth at room temperature). G is S_o / S_i, the available power gain of the stage and B is the noise bandwidth at the *output* of the stage, assumed to be not wider than the noise bandwidth at the input. The case where the output noise bandwidth is wider will be considered in a later section.

Noise bandwidth is defined in Fig 14.5. An ideal rectangular frequency response has a maximum value that is defined at the reference frequency. The area under the rectangle is the same as the area under the actual filter response, therefore the noise within the rectangle and within the actual filter response are equal. The width of the rectangle is called the noise bandwidth. Various kinds of filters have certain ratios of signal bandwidth to noise bandwidth that can be measured or calculated.

Part of the output noise is amplified thermal noise from the signal generator. To find the noise that is generated within the stage, we must subtract the amplified signal generator noise from the total output noise. Fig 14.5 shows the equation that performs this operation and the quantity (F–1)kTBG is the excess noise that the stage contributes.

In general, the excess noise of a stage is

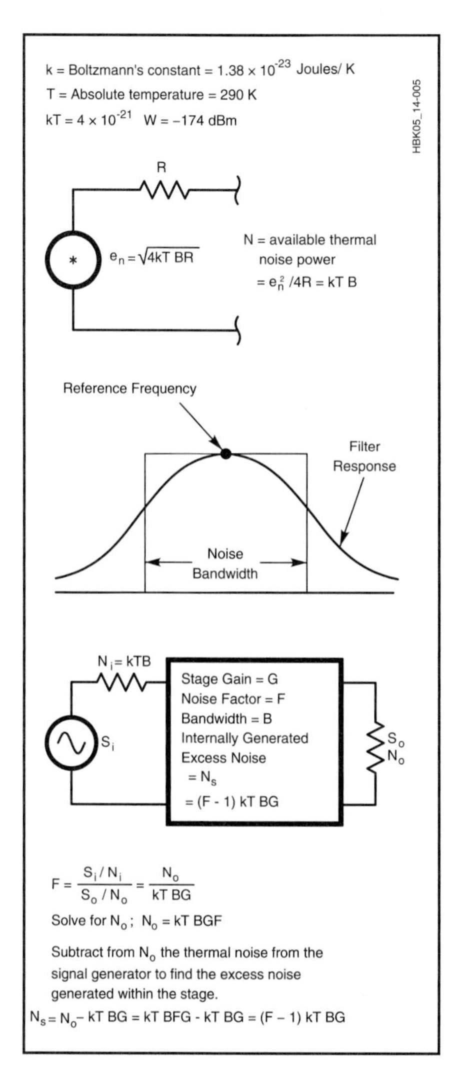

Fig 14.5—Diagram and equations that explain F, the noise factor of a single stage. The excess noise generated within the stage is also indicated. The definition of noise bandwidth is included.

the output noise minus the amplified noise from the previous stage. A thorough understanding of this concept is very important for any one who designs radio systems that employ low-level signals. Finally, noise figure, NF, is 10 times the logarithm of F, the noise factor (Ref 3).

Noise Factor of a Passive Device

Often the stage in Fig 14.5 is a filter, an attenuator or some other passive device (no amplification) that contains only thermal noise sources. In a device of this kind, the output noise is thermal noise of the same value as the thermal noise of the generator alone. That is all the thermal noise sources inside the device and also the generator resistance can be combined into a single resistor whose available noise power is kTB, the same as that of the generator alone. Therefore no additional noise is added by the device. But the available signal power is reduced by the attenuation (loss of signal) of the device. Therefore, using the equation for noise factor in Fig 14.5, the noise factor F of the device is numerically equal to its attenuation.

For example, a 3-dB attenuator has a 3-dB noise figure, or a noise factor of 2. This important fact is very useful in radio design. It applies only when there is no amplification and no shot noise or 1/f noise sources within the device. This discussion assumes that all components and the generator are at the same temperature. If not, a slightly more complicated procedure involving Equivalent Noise Temperatures (T_E), to be discussed later in this chapter, can be used.

Sensitivity

Closely related to the concept of noise figure (or noise factor) is the idea of sensitivity. Suppose a circuit, a component or a complete system has a noise figure NF (dB) and therefore a noise output N_o (dBm). Then the value in dBm of a signal generator input that increases the total output (signal + noise) by 10 dB is defined as the "sensitivity." That is, 10 × log [(signal + noise) / (noise)] = 10 dB. The ratio (signal) / (noise) is then equal to 9.54 dB (= 10 × log(9)). N_o is equal to kTBFG as shown in Fig 14.5. B is noise bandwidth. Using this information, the sensitivity is

$$S(\text{dBm}) = -174\ (\text{dBm}) + 9.54\ (\text{dB}) + \text{NF}\ (\text{dB}) + 10\log(B) \quad (5)$$

In terms of the "open circuit voltage" from a 50-Ω signal generator (twice the reading of the generator's voltmeter) the sensitivity is

$$E\ (\text{volts open circuit}) = 0.4467 \times 10^{S/20} \quad (6)$$

N_o is 9.54 dB below the sensitivity value. This is sometimes referred to in specifications as the "noise floor." The signal level that is equal to the noise floor is sometimes referred to as the minimum detectable signal (MDS). Also associated with No is the concept of "noise temperature" which we discuss later under Microwave Receivers.

Distortion in a Single Stage

Suppose the input to a stage is called X. If the stage is perfectly linear the output is Y, and Y = AX, where A is a constant of proportionality. That is, Y is a perfect replica of X, possibly changed in size. But if the stage contains something nonlinear such as a diode, transistor, magnetic material or other such device, then $Y = AX + BX^2 + CX^3 + \ldots$ The additional terms are "distortion" terms that deliver to the output artifacts that were not present in the generator. Without getting too mathematical at this point, if the input is a pure sine wave at frequency f, the output will contain "harmonic distortion" at frequencies 2f, 3f and so on. If the input contains two signals at f1 and f2, the output contains *intermodulation distortion* (IMD) products at f1+f2, f1–f2, 2f1+f2, 2f2–f1, just to name a few. All semiconductors, vacuum tubes and magnetic materials create distortion and the radio designer's job is to limit the distortion products to acceptable levels. We wish to look at distortion from a system-design standpoint.

There are several ways to reduce distortion. One is to use a high-power device operated well below its maximum ratings. This leads to devices that dissipate more power in the form of heat. Unfortunately, these devices also tend to be noisier; so high power levels and low noise tend to be incompatible goals in most cases. (Some modern devices, such as certain GaAsFETs, achieve improved values of dynamic range.) Also, a large reduction in distortion is not always assured with this method, especially in transmitters.

Second, reduce the signal level into the device. This allows a lower power device to be used that will tend to be less noisy. To get the same output level, though, we must increase the gain of the stage. Then we run into another problem: if the signal at the output of this lower power stage becomes too large, distortion is generated at the output. Also, as mentioned before, high-gain stages tend to be unstable at RF.

Third, reduce the stage gain. But then we must add another stage in order to get the required output level. This additional stage turns out to be a high-power stage. The addition of another stage adds more noise and distortion contamination to the signal.

Fourth, use negative feedback. This is a powerful technique that is discussed in detail in the **RF Power Amplifiers** chapter. In general, if we increase the stage gain and perhaps make it more powerful, we can use feedback to reduce distortion and stabilize performance with respect to component variations. The feedback stage may be noisier, although the use of loss-less feedback can improve this situation. Negative feedback is the preferred method for reducing distortion in radio design, but the gain reduction due to feedback means that more stages are needed. This tends to reintroduce some noise and distortion.

A fifth way reduces distortion by increasing selectivity. For example, harmonics of an RF amplifier can be eliminated by a tuned circuit. Products such as f1 + f2 and f1 – f2 can often also be eliminated. Third-order products such as 2f1 – f2 and so on (and higher odd-order products) frequently are sufficiently close to f1 and f2 that selectivity does not help much, but if they are somewhat removed in frequency these so-called "adjacent channel" products can be greatly reduced.

A sixth way is to use push-pull circuits (see the **RF Power Amplifiers** chapter) that tend to greatly reduce "even-order" products such as 2f, 4f, f1 + f2, f1 – f2, and so on. But "odd-orders" such as 3f, 5f, 2f1 + f2, 2f1–f2 are not reduced by this method except as noted later in the Modules in Combination section.

A seventh way uses diplexers to absorb undesired harmonics or other spurious products. So there are compromises to be made. The designer must look for the compromise that gets the job done in an acceptable manner and is optimal in some sense. For example, devices are available that are optimized for linearity

IMD Ratio

If a pair of equal-amplitude signals creates IMD products, the IMD ratios (IMR) are the differences, in dB, between each of the two tones and each of the IMD products (see **Fig 14.6**).

Intercept Point

The intercept point is a figure of merit that is commonly used to describe the IMD performance of an individual stage or a complete system. For example, third-order products increase at the rate of 3:1. That is, a 1-dB increase in the level of each of the two-tone input signals produces (ideally, but not always exactly true) a 3-dB increase in third-order IMD products. As the input levels increase, the distortion products seen at the output on a spectrum analyzer could catch up to, and equal, the level of the two desired signals, if the circuit did not go into a limiting process (see next topic). The input level at which this occurs is the input intercept point. Fig 14.6 shows the concept graphically, and also derives from the geometry an equation that relates signal level, distortion and intercept point. A similar process is used to get a second-order intercept point for second-order IMD. These formulas are very useful in designing radio systems and circuits. If the input intercept point (dBm) and the gain of the stage (dB) are added the result is an output intercept point (dBm). Receivers are specified by input intercept point, referring distortion back to the receive antenna input. Transmitter specifications use output intercept, referring distortion to the transmit antenna output.

Gain Compression

The gain of a circuit that is linear and has little distortion products deteriorates rapidly when the instantaneous input or output level reaches a critical point where the peak or trough of the waveform begins to "clip" or "saturate." The 1-dB compression point occurs when the output is 1 dB less than it would be if the stage were still linear. Some circuits do not need to be linear (and should not be linear), and we will look at several examples. In many applications linearity is necessary, especially in SSB receivers and transmitters. The situation for a linear circuit is optimum when the input and output become nonlinear simultaneously. This means that the gain, bias and load impedance are all properly coordinated. We will study this more closely in later sections.

Dynamic Range

There is a relationship between noise factor, IMD, gain compression and bandwidth in a building block stage. In general, an active circuit that has a low noise factor tends to have a poor intercept point and vice versa. A well-designed transistor or circuit tries to achieve the best of both worlds. Dynamic range is a measure of this capability. Suppose that a circuit has a third-order input intercept of +10 dBm, a noise factor of 6 dB and a noise bandwidth of 1000 Hz. We want to determine its dynamic range. At a certain level per tone of a two-tone input signal the third-order IMD products are equal to the noise level in the 1000-Hz band. The ratio, in dB, of each of the two tones to the noise level is called the "spurious free dynamic range" (SFDR). **Fig 14.7** illustrates the problem and derives the proper formula. Note that the bandwidth is an important player. For the example above, the dynamic range is DR = 0.67 (10 – (–174 + 10 log(1000) + 6)) = 99 dB. Often the dynamic range is calculated using a 1.0-Hz bandwidth. This is called "normalization." Another kind of dynamic range compares the 1-dB compression level with the noise level. This is the CFDR (compression-free dynamic range). Fig 14.7 illustrates this also.

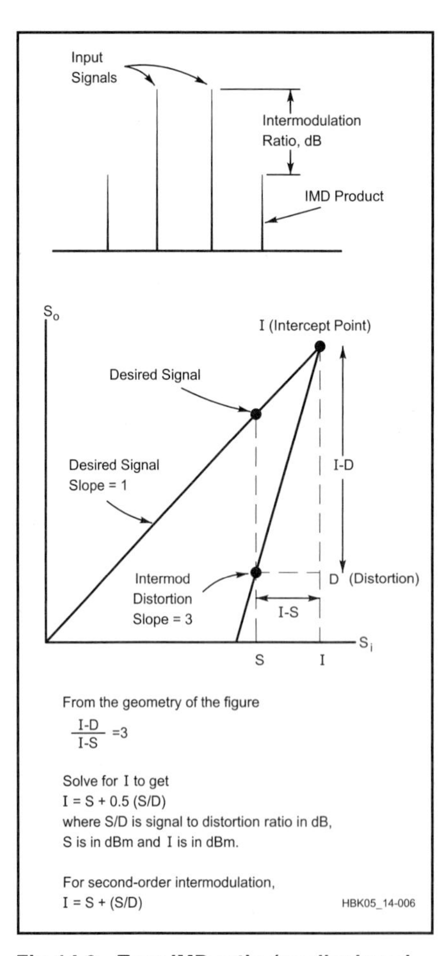

Fig 14.6—Top: IMD ratio (as displayed on a spectrum analyzer). Bottom: intercept point.

Modules in Combination

Quite often the performance of a single stage can be greatly improved by combining two identical modules. Because the input power is split evenly between the two modules the drive source power can be twice as great and the output power will also be twice as great. In transmitters, especially, this often works better than a single transistor with twice the power rating. Or, for the same drive and output power, each module need supply only one-half as much power, which usually means better distortion performance. Often, the total number of stages can be reduced in this manner, with resulting cost savings. If the combining is performed properly, using hybrid transformers, the modules interact with each other much less, which can avoid certain problems. These are the system-design implications of module combining.

Three methods are commonly used to combine modules: parallel (0°), push-pull (180°) and quadrature (90°). In RF circuit design, the combining is often done with special types of "hybrid" transformers called splitters and combiners. These are both the same type of transformer that can perform either function. The splitter is at the input, the combiner at the output. We will only touch very briefly on these topics in this chapter and suggest that the reader consult the **RF Power Amplifiers** chapter and the very considerable literature for a deeper understanding and for techniques

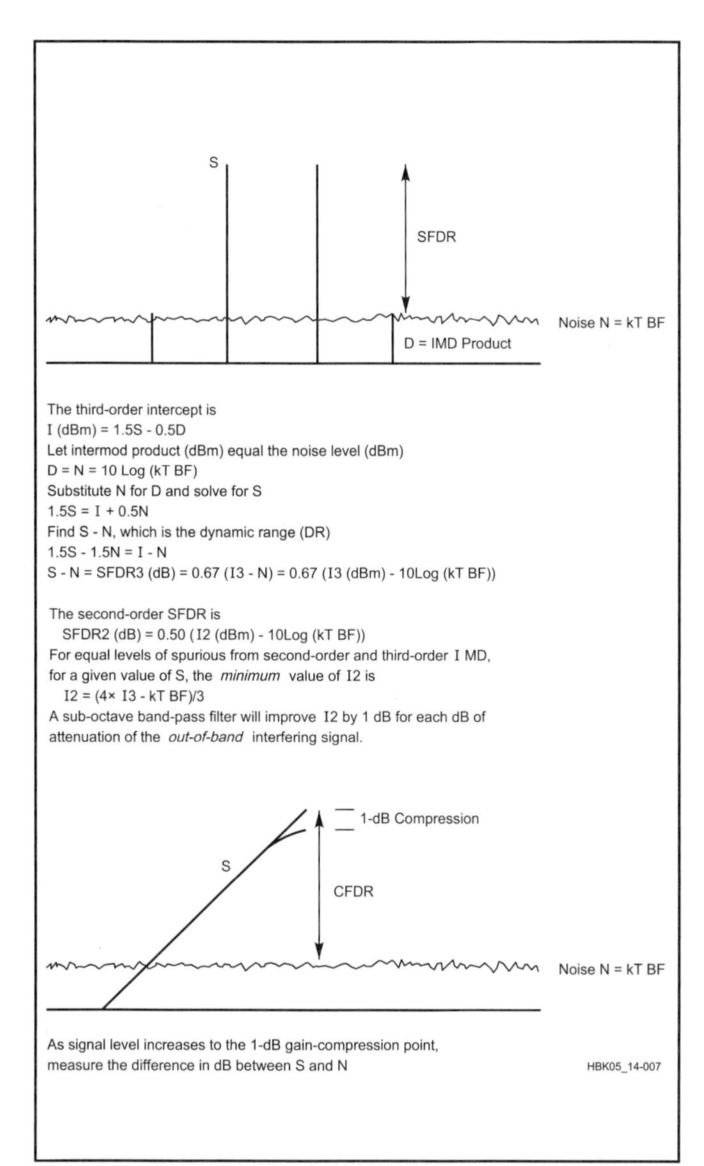

Fig 14.7—The definitions of spurious-free dynamic range (SFDR) and compression-free dynamic range (CFDR). The derivation yields a very useful equation for SFDR.

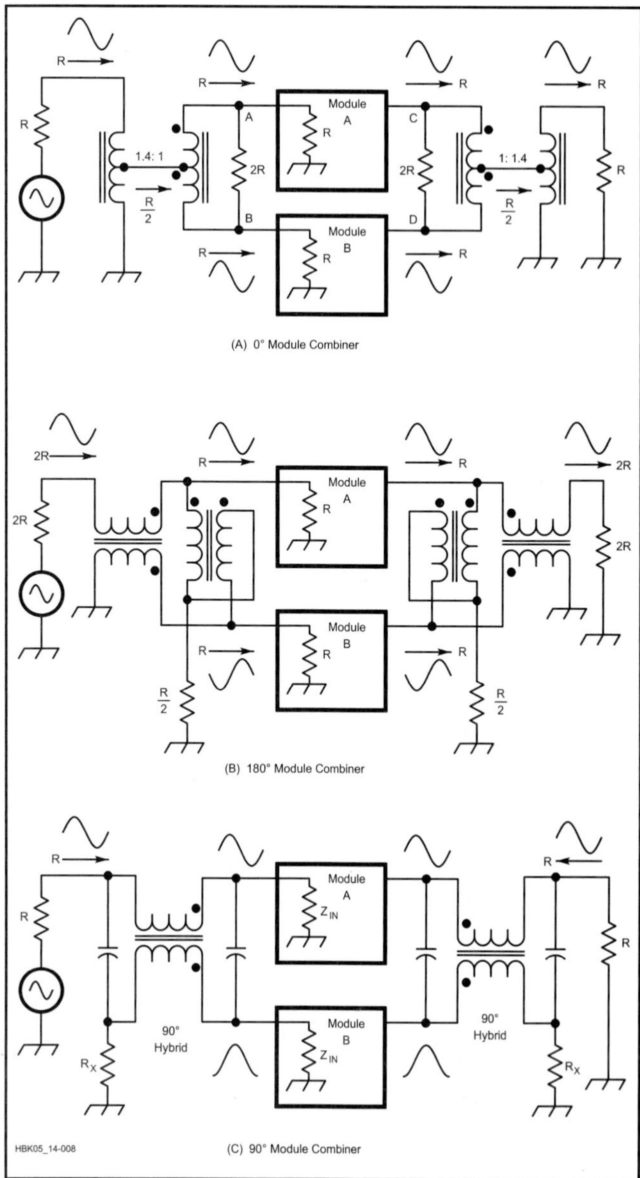

Fig 14.8—The three basic techniques for combining modules.

used at different frequency ranges.

Fig 14.8 illustrates one example of each of the three basic types. In a 0° hybrid splitter at the input the tight coupling between the two windings forces the voltages at A and B to be equal in amplitude and also equal in phase if the two modules are identical. The 2R resistor between points A and B greatly reduces the transfer of power between A and B via the transformer, but only if the generator resistance is closely equal to R. The output combiner separates the two outputs C and D from each other in the same manner, if the output load is equal to R, as shown. No power is lost in the 2R resistor if the module output levels are identical.

The 180° hybrid produces push-pull operation. The advantages of push pull were previously discussed. The horizontal transformers, 1:1 balun transformers, allow one side of the input and output to be grounded. The R/2 resistors improve isolation between the two modules if the 2R resistors are accurate, and dissipate power if the two modules are not identical.

In a 90° hybrid splitter, if the two modules are identical but their identical input impedance values may not be equal to R, the hybrid input impedance is nevertheless R Ω, a fact that is sometimes very useful in system design. The power that is "reflected" from the mismatched module input impedance is absorbed in RX, the "dump" resistor, thus creating a virtual input impedance equal to R. The two module inputs are 90° apart. At the output, the two identical signals, 90° apart, are combined as shown and the output resistance is also R. This basic hybrid is a narrowband device, but methods for greatly extending the frequency range are in the literature (Ref 3). One advantage of the 90° hybrid is that catastrophic failure in one module causes a loss of only one half of the power output.

MULTISTAGE SYSTEMS

As the next step in studying system design we will build on what we've learned about single stages, and look at the methods for organizing several building block circuits and their interconnecting networks so that they combine and interact in a desirable and predictable manner. These methods are applicable to a wide variety of situations. Further study of this chapter will reveal how these methods can be adapted to various situations. We will consider typical receiving circuits and typical transmitting circuits.

Properties of Cascaded Stages

Fig 14.9 shows a simple receiver "front end" circuit consisting of a preselector filter, an RF amplifier, a second filter and a double balanced diode mixer. We want to know the gain, bandwidth, noise factor, second and third-order intercept points, SFDR and CFDR for this combination, when the circuitry following these stages has the values shown. Let's consider one item at a time.

Gain of Cascaded Stages

The antenna tuned circuit L1, C1, C2 has some resistive loss; therefore the power that is available from it is less than the power that is available from the generator. Let's say this loss is 2.0 dB. Next, find the available power gain of the RF amplifier. First, note that the generator voltage V_s is transformed up to a larger voltage V_g by the input tuned circuit, according to the behavior of this kind of circuit. This step-up increases the gain of the RF amplifier because the FET now has a larger gate voltage to work with. (A bit of explanation: The FET has a high input impedance therefore, since the generator resistance R_s is only 50 Ω, a voltage step-up will utilize the FET's capabilities much better. But an excessive step-up opens up the possibility that the FET and other "downstream" circuits can be overdriven by a moderately large signal. So this step-up process should not be carried too far). The gain also depends on the drain load resistance, which is the mixer input impedance, stepped up by the circuit L2, C3, C4. Again, there is some loss within this tuned circuit, say 2.0 dB. If the drain load is too large the FET drain voltage swing can become excessive, creating distortion. The RF amplifier can become unstable due to excessive gain. Note also that the unbypassed source resistor provides negative feedback, to help make the RF amplifier more predictable. Dual-gate FETs have relatively little reverse coupling.

We come now to the mixer, whose available gain is about –6 dB. This is the difference between its available IF output power and its available RF input power. This is a fairly low-level mixer, so it can be easily overdriven if the RF gain is too high. Harmonic IMD and two-tone IMD can become excessive (see later discussion in this chapter). On the other hand, as we will discuss later, too little RF gain will yield a poor receiver noise figure.

The concepts of available gain and transducer gain were introduced earlier. If we multiply the available gains of the input filter, RF amplifier, interstage filter and mixer, we have the available gain of the entire combination. The transducer gain is the ratio of the power actually delivered to R_L to the power that is available from the generator. To get the transducer gain of the combination, multiply the available gain of the first three circuits by the transducer gain of the last circuit (the mixer). This concept may require some thought on your part, but it is one that is frequently used and it adds understanding to how circuits are cascaded. One example, the transducer gain of a receiver, compares the signal power available from the antenna with the power into the loudspeaker (a perfectly linear receiver is assumed).

Fig 14.9 also shows an example of a commonly used graphical method for the available gain of a cascade. The loss or increase of available power at each step is shown. As the input increases the other values follow. But at some point, measurements of linearity or IMD will show that some circuit is being driven excessively, as the example indicates. To improve performance at that point, we may want to make gain changes or take some other action. If the overload is premature, a more powerful amplifier or a higher-level mixer may be needed. It may be possible to reduce the gain of the RF amplifier by reducing the step-up in the input LC circuit or the drain load circuit, but this may degrade noise figure too much. This is where the "optimization" process begins.

A method that is often used in the lab is to plot the voltage levels at various points in the system. These voltages are easily measured with an RF voltmeter or spectrum analyzer, using a high-impedance probe. This is a convenient way to make comparative measurements, with the understanding that voltage values are not the same thing as power-gain values, although they may be mathematically related. Many times, these voltage measurements quickly locate excessive or deficient drive conditions during the design or troubleshooting process. Comparisons of measured values with previous measurements of the same kind on properly functioning equipment are used to locate problems.

Selectivity of Cascaded Stages

The simplified receiver example of

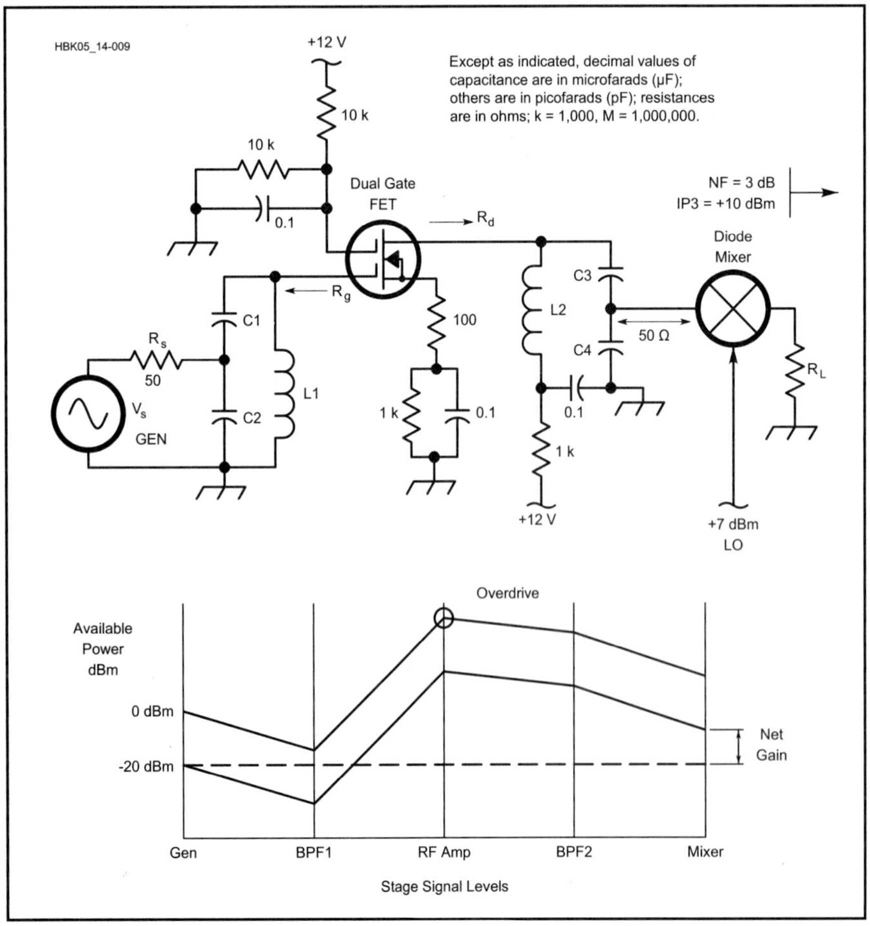

Fig 14.9—An example of cascaded stage design, a simple receiver front end.

Fig 14.9 shows two resonant circuits (filters) tuned to the signal frequency. They attenuate strong signals on adjacent frequencies so that these signals will not disturb the reception of a desired weak signal at center frequency. **Fig 14.10** shows the response of the first filter and also the composite response of both filters at the mixer input.

Consider first the situation at the output of the first filter. If a strong signal is present, somewhat removed from the center frequency, the selectivity of the first filter may just barely prevent excessive signal level in the RF amplifier. When this signal is amplified and filtered again by the second filter, its level at the input of the mixer may be excessive. Our system design problem is to coordinate the amplifier gain and second filter selectivity so that the mixer level is not too great. (A computer simulation tool, such as *ARRL Radio Designer,* can be instructive and helpful.) Then we can say that for that level of undesired signal at that frequency offset the cascade is properly designed.

The decisions regarding the "expected" maximum level and minimum frequency offset of the undesired signal are based on the operating environment for the equipment, with the realistic understanding that occasionally both of these values may be violated. If improvement is needed, it may then be necessary to (a) improve the selectivity, (b) use a more robust amplifier and mixer or (c) reduce amplifier gain. Very often, increases in cost, complexity and system noise factor are the byproducts of these measures.

Cascaded signal filters are often used to obtain a selectivity shape that has a flat top response and rapid or deep attenuation beyond the band edges. This method is often preferred over a single, more complex "brick wall" filter that has a very steep rate of attenuation outside the passband.

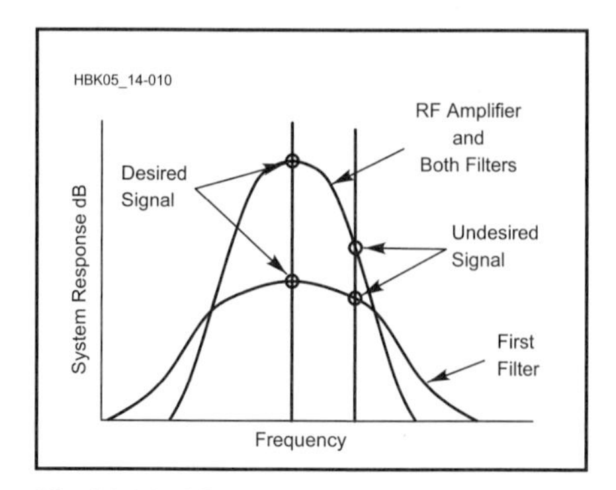

Fig 14.10—The gain and cumulative selectivity, between the generator and the mixer, of the example circuit in Fig 14.9.

Noise Factor of a Cascade

In the example of Fig 14.9, the overall noise factor of the two-stage circuit is defined in the same way as for a single stage. It is the degradation in signal-to-noise ratio (S/N) from the signal generator to the output. This total noise factor can be found by direct measurement or by a stage-by-stage analysis. If we wish to optimize the total noise factor or look for trade-offs between it and other things such as gain and distortion, a stage-by-stage analysis is needed.

The definition of noise factor for a single stage applies as well to each stage in the chain. For each stage there is a signal and thermal noise generator, internal sources of excess noise and a noise bandwidth. In a cascade, the signal and thermal noise sources for a particular stage are found in the previous stage, as shown in **Fig 14.11A**. But this thermal noise source has already been accounted for as part of the excess noise for the previous stage. Therefore, this thermal noise must not be counted twice in the calculation. On this basis, Fig 14.11A derives the formula for the noise factor of a two-stage system. This formula can then be used to find the noise factor of the multi-segment system in Fig 14.11B by applying it repetitively, first to stage N + 1 and N,

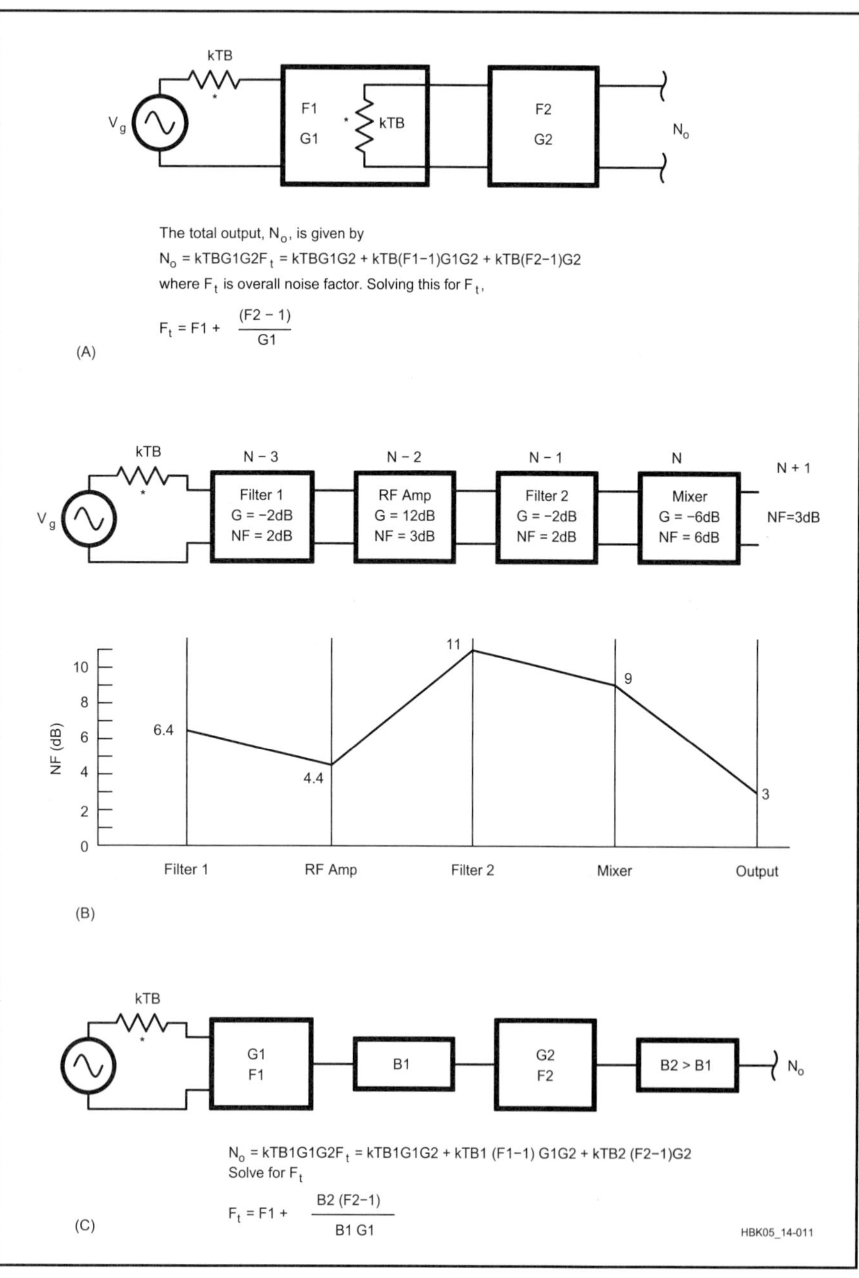

Fig 14.11—A: the noise factor of a two-stage network. B: cumulative noise factor for the example in Fig 14.10. C: noise factor when the bandwidth increases toward the output.

then to N and N – 1, then to N – 1 and N – 2 and so on, where N is the total number of stages and N + 1 is the rest of the system after the last stage. Fig 14.11B shows the cumulative noise figure (dB) at each point in the example of Fig 14.9. The graphical method aids the analysis visually.

Note that the diode mixer's noise figure approximately equals its gain loss. In applying the formula, if G1 is a lossy device, not an amplifier, then F1 equals its attenuation factor and G1 = 1/F1. Also, observe the critical role that values of RF amplifier gain and noise figure play in establishing the overall noise factor (or noise figure), despite the high noise figure that follows it. In Fig 14.11A and B, we assumed that the noise bandwidth does not increase toward the output. If the noise bandwidth does increase toward the output a complication occurs. Fig 14.11C provides a modified formula that is more accurate under these conditions. This situation is often encountered in practice, as we will see, especially in the discussion of receiver design (Ref 4).

Distortion in Cascaded Circuits

The IMD created in one stage combines with the distortion generated in following stages to produce a cumulative effect at the output of the cascade. The phase relationships between the distortion products of one stage and those of another stage can vary from 0° (full addition) to 180° (full subtraction). It is customary to assume that they add in-phase as a worst case. Under these conditions, **Fig 14.12** shows how to determine distortion at the input of a stage. Formulas are given for finding the third-order and second-order input intercept points in dBm. These formulas can be applied repetitively, in a manner similar to the noise-factor formula, to get the cumulative intercept point at each stage of the cascade. The output intercept point, in dBm, of a stage is equal to its input intercept point, in dBm, plus the gain, in dB, of the stage. When a purely passive, linear stage is part of the analysis, use a large value of intercept such as 100 dBm (10^7 W, Ref 5).

THE AMATEUR RADIO COMMUNICATION CHANNEL

In order to design radio equipment it is first necessary to know what specifications the equipment must have in order to establish and maintain communication. This is a very large and complex subject that we cannot fully explore here, however, it is possible to point out certain properties of the communication channel, especially as it pertains to Amateur Radio, and to discuss equipment requirements for successful communication. The "channel" is:

- the frequency band that is being transmitted and to which the distant receiver is tuned, and
- the electromagnetic medium that conveys the signal.

The Amateur Radio bands are, in fact, a very difficult arena for communications and a severe test of radio-equipment design. The very wide range of received signal levels, the high density of signals whose channels often overlap or are closely adjacent, the relatively low power levels and the randomness (the lack of formal operating protocols) are the main challenges for Amateur Radio equipment designers. An additional challenge is to design the equipment for moderate cost, which often implies technical specifications that are somewhat below commercial and military standards. These relaxed standards sometimes add to the amateur's problems.

Received Noise Levels

There are three major sources of noise arriving at the receive antenna:

- atmospheric noise generated by disturbances in the Earth's environment,
- galactic noise from outer space and
- noise from transmitters other than the desired signal.

Let's briefly discuss each of these kinds of noise.

Atmospheric noise (including man-made noise) is maximum at frequencies below 10 MHz, where it has *average* values about 40 dB above the thermal noise at 290 K (K = kelvins, absolute temperature). Above 10 MHz, its strength decreases at 20 dB per *octave.* At VHF and above, it is of little importance (Ref 6).

However, various studies have found that at certain times and locations and in certain directions this noise approaches the level of thermal noise at 290 K, even at the lower frequencies. Therefore the conventional wisdom that a low receiver noise figure is not important at low HF is not completely true. Amateurs, in particular, exploit these occurrences, and most amateur HF receivers have noise figures in the 8 to 12-dB range for this reason, among others. A very efficient antenna at a low frequency can modify this conclusion, though, because of its greater signal and noise gathering power (for example, a half-wave dipole gathers about 12 dB more power at 1.8 MHz than a half-wave dipole at 30 MHz (Ref 7)). When the noise level is high, an attenuator in the antenna lead can reduce receiver vulnerability to strong interfering signals without reducing the S/N ratio of weaker signals. In other words, the system (that is, receiver plus noisy antenna) dynamic range is improved (the receiver intercept point increases and the system noise is reduced). The antenna noise, after attenuation, should be several dB above the receiver internal noise. This is a typical example of a communication-link design consideration that may not be necessary if the receiver is of high quality.

Receive Antenna Directivity

If the receive antenna has gain, and can be aimed in a certain direction, it often happens that atmospheric noise is less in that direction. A lower receiver noise figure may then help. Or, if the noise arrives uniformly from all directions but the desired signal is increased by the antenna gain, then the S/N ratio is increased. That is, the noise is constant but the signal is greater. (Explanation: if the noise is the same from all directions the high-gain antenna receives more noise from the desired direction but rejects noise from other directions; therefore the total received noise tends to remain constant.) This is one of the advantages of the HF rotary beam antenna. The same gain can

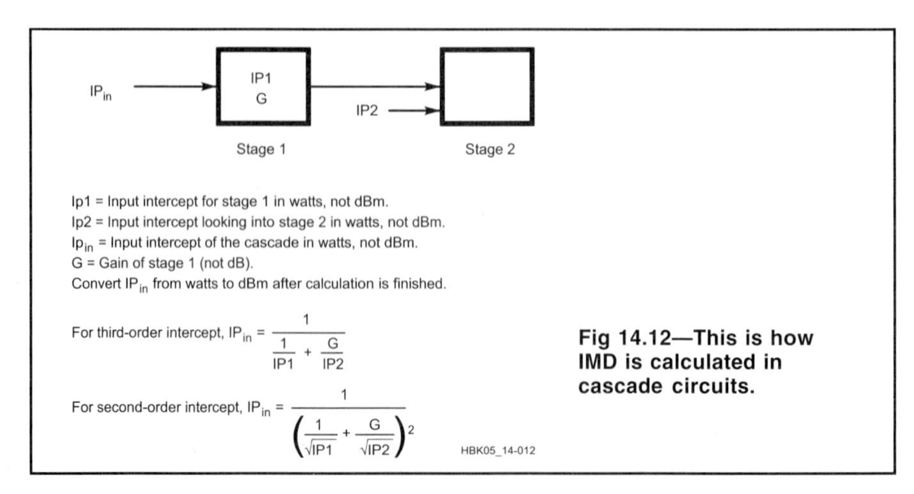

Fig 14.12—This is how IMD is calculated in cascade circuits.

also cause strong undesired signals to challenge the receiver's dynamic range (or null out an undesired signal).

Galactic Noise

The *average* noise level from outer space is about 20 dB above that of thermal noise at 20 MHz and decreases at about 20 dB per *decade* of frequency (Ref 6). But at microwave frequencies, high-gain antennas with very-low-noise-figure receivers are able to locate sources of relatively intense (and very low) galactic noise. Amateurs working at microwave frequencies up to 10 GHz go to great lengths to get their antenna gains and receiver sensitivities good enough to take advantage of the high and low noise levels.

Transmitter Noise

Fig 14.13A shows the spectral output of a typical amateur transmitter. The desired modulation lies within a certain well defined bandwidth, which is determined by the type of modulation. Because of unavoidable imperfections in transmitter design, there are some out-of-band modulation artifacts such as high-order IMD products. The signal filter (SSB, CW and so on) response also has some slope outside the passband. There is also a region of phase noise generated in the various mixers and local oscillators (LOs). These phase noise sidebands are "coherent." That is, the upper frequency sidebands have a definite phase relationship to the lower frequency sidebands. At higher values of frequency offset, a noncoherent "additive" noise shelf may become greater than phase noise and it can extend over a considerable frequency band. Other outputs such as harmonics and other transmitter-generated spurious emissions are problems.

The general design goals for the transmitter are:

1. Make the unavoidable out-of-band distortion products as small as technology and equipment cost and complexity will reasonably allow,
2. Design the synthesizers and other local oscillators and mixers so that phase noise, as measured in a bandwidth equal to that of the desired modulation, is less than the out-of-band distortion products in goal 1 and,
3. Make the wideband noise sufficiently small that the noise will be less than any unavoidable receiver noise at nearby receivers with the same bandwidth.

If the additive noise is very small, LO phase noise may come back into the picture. In narrowband systems such as Morse code (CW) it can be very difficult or impractical to make transmitted phase noise less than the normal Morse code sidebands (see later discussion of this topic). The general method to reduce wideband noise from the transmitter is to place the narrowband modulation band-pass filter at as high a signal level as possible and to follow that with a high-level mixer and then a low-noise first-stage RF amplifier.

Phase-noise amplitude varies with modulation. That is, the LO phase noise is modulated onto the outgoing signal by the "reciprocal mixing" process (the signal becomes the "LO" and the LO phase noise becomes the "signal"). If the actual LO to phase noise ratio is X dB, the ratio of the transmit signal to its phase noise is also X dB. In SSB the magnitude of the phase-noise sidebands is maximum only on modulation peaks. In CW it exists only when the transmitter is "key down." The additive noise, on the other hand, may be much more constant. If the power amplifiers are Class A or Class AB, additive noise does not require any actual signal and tends to remain more nearly constant with modulation.

In a communication link design, the receiver's culpability must also be considered. The receiver's LOs also generate phase noise that is modulated onto an adjacent-channel signal (reciprocal mixing) to produce an in-band noise interference, as shown in Fig 14.13B. In view of this, the transmitter and receiver share equal responsibility regarding phase noise, and there is little point in making either one a great deal better unless the other is improved also. Nevertheless, high-quality receivers with low phase noise exist, and they are vulnerable to transmitter phase noise. The converse situation also exists; receiver phase noise can contaminate a clean incoming signal (Ref 8).

Receiver Gain and Transmitter Power Requirements

The minimum level of a received signal is a function of the antenna noise level and the bandwidth. As just one example, for an HF SSB system with a 2.0-kHz bandwidth and a noise level 10 dB above thermal (–131 dBm in a 2.0-kHz band) the minimum readable signal, say 3 dB above the noise level, would be –128 dBm. Assume the receiver-generated noise is negligible. If the audio output to a loudspeaker is, say +20 dBm (0.1 W), then the total required receiver gain is 148 dB, which is an enormous amount of amplification. For a CW receiver with 200-Hz bandwidth, the minimum signal would be –138 dBm and the gain would be 158 dB. Receivers at other frequencies with lower noise levels can re-

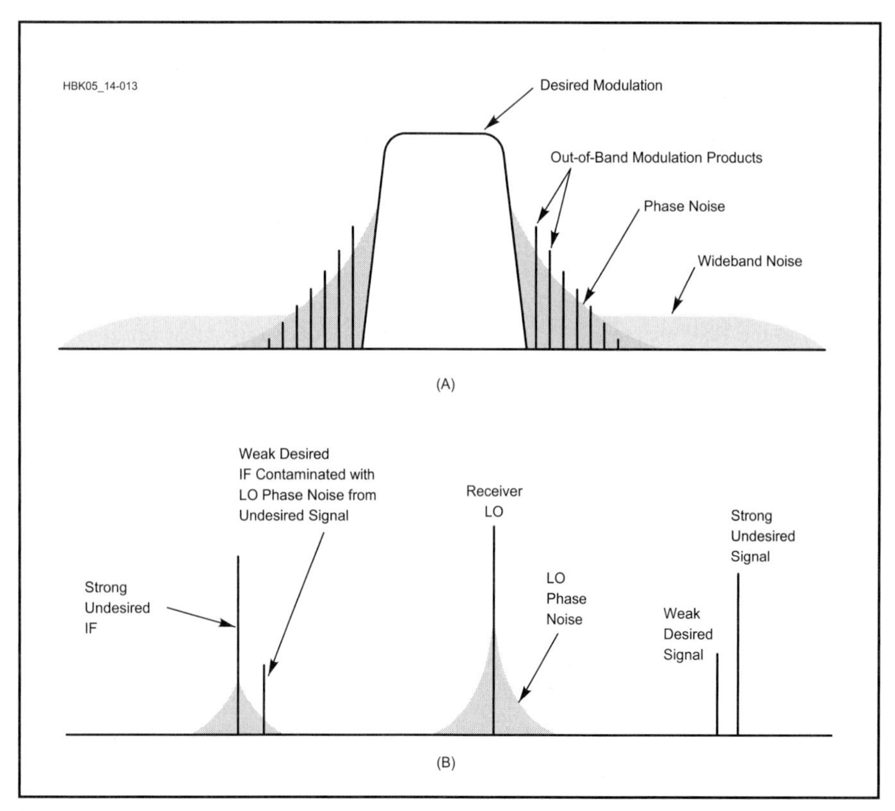

Fig 14.13—A: transmitter spectrum with discrete out-of-band products, phase noise and white noise. B: reciprocal mixing of LO phase noise onto an incoming signal.

quire even higher gains to get the desired audio output level. If the transmission path attenuation can be predicted or calculated, the required transmitter power can be estimated. These kinds of calculations are often done in UHF and microwave amateur work, but less often at HF (see the microwave receiver section for an example).

We do, however, get some "feel" for the receiver gain requirements, how the receiver interacts with the "channel," and that the minimum signal power is an almost incredibly small 1.6×10^{-16} W. On the other hand, amateur receiver S-meters are calibrated up to an input signal level of –13 dBm (60 dB above 100 µV from a 50-Ω source). Therefore the receiver must deal with a desired signal range of at least 115 dB (128 – 13) for the SSB example or at least 125 dB for the CW example (assuming that AGC limits the signal levels within the receiver).

Fading

Radio signals very often experience changes in strength due either to reflections from nearby objects (multipath) or, in the case of HF, to multiple reflections in the ionosphere. At a particular frequency and at a certain time, a signal arriving by multipath may decrease severely. The effect is noticed over a narrow band of frequencies called the "fading bandwidth." At HF, the center frequency of this fade band drifts slowly across the spectrum. Communication links are degraded by these effects, so equipment design and various communication modes are used to minimize them. For example, SSB is less vulnerable than conventional AM. In AM, loss of carrier or phase shift of the carrier, relative to the sidebands, causes distortion and reduces audio level.

The UHF/Microwave Channel

At frequencies above about 300 MHz, we need to account for the interaction of the Earth environment with the transmitted and received signals. Here are some of the things to consider:

1. Line-of-sight communications distance, as a function of receiver and transmitter antenna height.
2. Losses from atmospheric gasses and water vapor (above several gigahertz).
3. Temperature effects on paths: reflections, refractions, diffractions and transmission "ducts."
4. Atmospheric density inversions due to atmospheric pressure variations and weather fronts.
5. Tropospheric reflections and scattering.
6. Meteor scattering (mostly at VHF but occasionally at UHF).
7. Receive-antenna sky temperature.

Competitive amateur operators who are active at these frequencies become proficient at recognizing and dealing with these communication channel effects and learn how they affect equipment design. They become proficient at estimating channel performance, including path loss, receive system noise figure (or noise temperature), antenna radiation patterns and gain.

Receiver Design Techniques

We will now look at the various kinds of receivers that are used by amateurs and at specific circuit designs that are commonly used in these receivers. The emphasis is entirely on analog approaches. Methods that use digital signal processing (DSP) for various signal processing functions are covered in the **DSP** chapter.

EARLY RECEIVER DESIGN METHODS

Fig 14.14 shows some early types of receivers. We will look briefly at each. Each discussion contains information that has wider applicability in modern circuit design, and is therefore not merely of historical interest. A lot of good old ideas are still around, with new faces.

The Crystal Set

In Fig 14.14A the antenna circuit (capacitive at low frequencies) is series resonated by the primary coil to maximize the current through both, which also maximizes the voltage across the secondary. The semiconductor crystal rectifier then demodulates the AM signal. This demodulation process utilizes the carrier of the AM signal as a "LO" and frequency translates (mixes) the RF signal down to baseband (audio). The rectifier and its output load impedance (the headphones) constitute a loading effect on the tuned circuit. For maximum audio output, a certain amount of coupling to the primary coil provides an optimum impedance match between the rectifier circuit and the tuning circuit. The selectivity is then somewhat less than the maximum obtainable. To improve the selectivity, reduce the secondary L/C ratio and/or decrease the coupling to the primary. Some decrease in audio output will usually result.

This basic mechanism for demodulating an AM signal by using a rectifier is identical to that used in nearly all modern AM receivers. One important feature of this rectifier is a signal level "threshold effect" below which rectification quickly ceases. Therefore the crystal set, without RF amplification, is not very good for very weak signals. Early crystal receivers used large antennas to partially solve this problem, but they were vulnerable to strong signals (their dynamic range was not very good). However, a large antenna does make greater tuner selectivity possible (if needed) because looser coupling can be used in the tuner. That is, the loading of the secondary resonator by the antenna and rectifier can both be reduced somewhat.

There is one other interesting property of this detector. The two AM sidebands add in phase (coherently) at the audio output, but noise above and below the carrier frequency add in random phase (non-coherently). Therefore the detector provides the same signal-to-noise ratio as a single sideband (SSB) signal (Ref 9).

The Tuned Radio Frequency (TRF) Receiver

As its name implies, the TRF receiver uses one or more tuned RF stages followed by a detector stage and audio amplifier. The variable capacitors (or sometimes variable coils) of each tuned stage track each other as the receiver is tuned. Each LC tuned circuit provides an additional band pass filter and so increases the overall selectivity of the receiver. One important benefit of the TRF approach is very high quality audio when receiving AM signals. The selectivity varies with tuning however and is reduced at the high frequency end of the tuning range, where the capacitance is low and the L/C ratio is high. At the low end of the tuning range, selectivity can become too high and roll-off the modulation sidebands. This variation in selectivity (and gain) are the TRF's main drawbacks. A related type of receiver, the "Neutrodyne" was an early triode design that used neutralization capacitors to prevent its RF stages from oscillating. See Fig 14.14B. Multigrid tubes or dual-gate FETs usually do not need neutralization.

The Regenerative Receiver

Edwin Howard Armstrong invented the regenerative circuit around 1914. Fig 14.14C shows a modern version. Q2 is basically a modified JFET oscillator that uses positive feedback (termed "regeneration"), to greatly increase both gain and selectivity.

A portion of the detector's amplified RF output is fed back to its input, in phase, by tickler winding L3. The signal is then amplified over and over, providing a gain of about 20,000 (86 dB). To minimize drift, this circuit uses a regulated detector supply voltage and a "throttle" capacitor regeneration control. Regeneration also introduces negative resistance into the circuit, greatly increasing its selectivity. This avoids the necessity of using several tuned stages. As regeneration is increased above the oscillation point, both gain and selectivity go down.

The regenerative receiver is easy for beginners to "home-brew" and different coils (or capacitors) can simply be switched-in to provide a very wide tuning range. Like the TRF architecture, the regen also tends to provide hi-fi, low distortion audio. The tradeoff, however, is that regeneration must be user adjusted and this requires both practice and patience. When receiving AM signals, feedback is set to a point just below self-oscillation. For CW

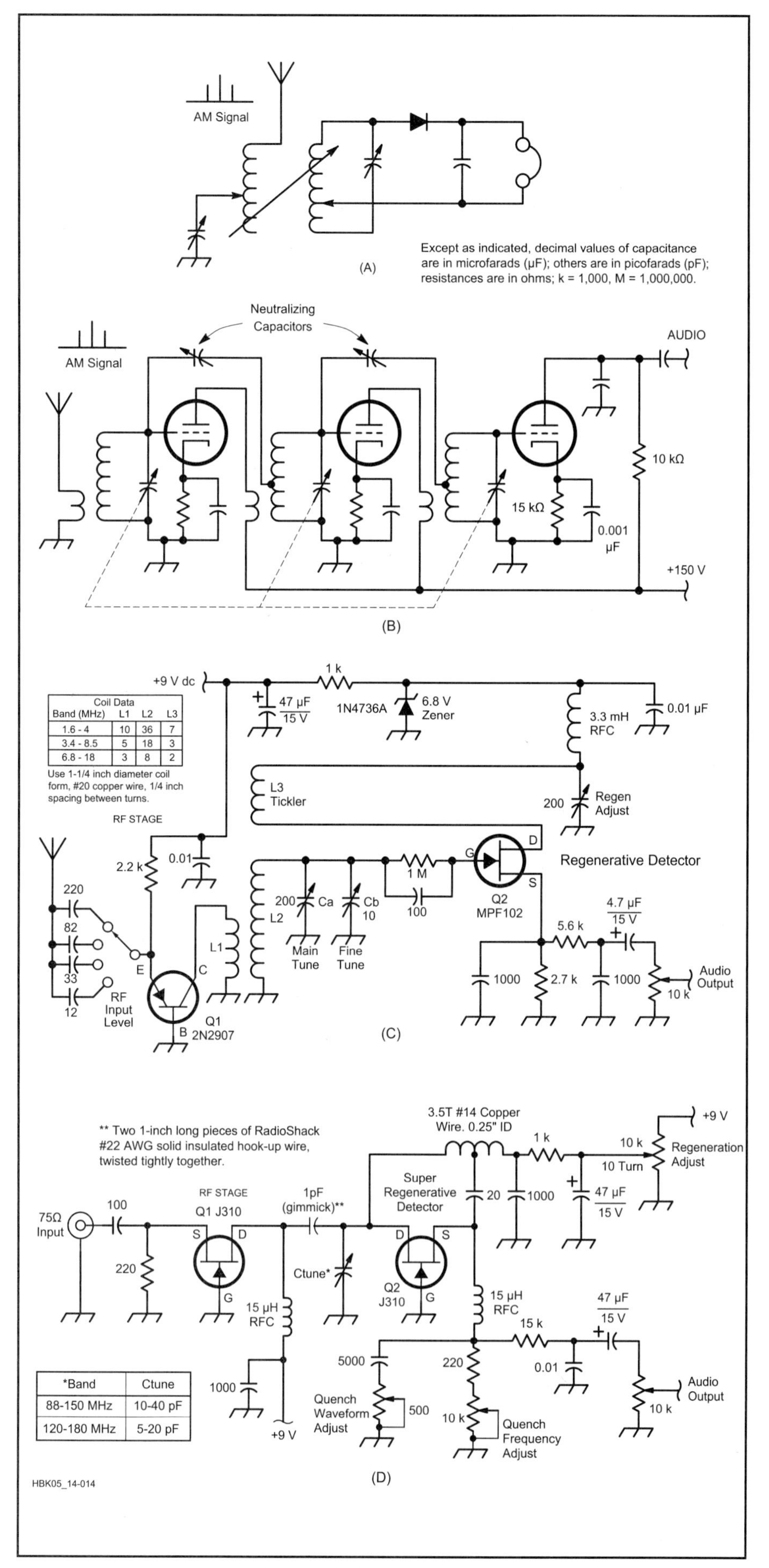

Fig 14.14—A shows a simple crystal set receiver that is as much fun to build and use today as it was in the early days of radio. B is an example of a "Neutrodyne" receiver, which was a variation of the tuned radio-frequency (TRF) receiver. C is an example of a modern regenerative receiver circuit, using transistors as the active elements. D is a modern self-quenched superregenerative receiver circuit.

and SSB reception, the detector is adjusted so that it is operating above the oscillation threshold, thus providing a beat note.

An RF stage (Q1) is used to isolate the 1 mW detector from the antenna and for additional gain. An input attenuator allows the operator to reduce the RF input level, increasing selectivity and preventing the detector from "blocking" on strong CW or SSB transmissions. For more details see: Kitchin, "High Performance Regenerative Receiver Design," November/December 1998 *QEX*.

The Superregenerative Receiver

Invented by Edwin Howard Armstrong around 1922, the superregenerative circuit is an oscillating regenerative detector that is periodically shut down or "quenched" by a second oscillation, usually between 20 and 30 kHz. It is essentially an amplitude-modulated oscillator whose quenching oscillations allow the input signal to build-up to the oscillation threshold repeatedly, providing typical detector gains of one million (120 dB). Superregenerative detectors can employ a separate quench oscillator (separately quenched) or produce their own secondary relaxation oscillations (self quenched).

The "superregen" receiver can be used from the lower VHF range all the way up into the microwave region and provides a simple receiver of great sensitivity. Fig 14.14D shows a modern circuit design for Amateur Radio experimentation.

Although much easier for a home builder to construct than a VHF superhet, the superregenerative receiver's extremely high gain is often difficult to control, requiring careful design and construction. The superregen also has a high nonsignal background noise; however, a simple squelch circuit will cure this. If high selectivity is needed, several controls must be carefully adjusted while the receiver is tuned. But for wide-band AM and FM reception, the controls can be preset. A superregen is sometimes operated in the "linear mode" for low audio distortion, although standard operation is sufficient for communications quality audio.

The classic Amateur Radio article about superregenerative receivers is Ross Hull's July 1931 *QST* article, "Five Meter Receiver Progress." This article describes a successful superregnerative receiver for 56 to 60 MHz.

Nat Bradley, ZL3VN, has discovered a new use for the superregen. If a separate local oscillation is mixed with the received signal, so that the frequency difference between the two is equal to the quench frequency, the detector will provide direct narrow-band FM demodulation. For more details, see: Kitchin, "New Super Regenerative Circuits for Amateur VHF and UHF Experimentation," September/October 2000 *QEX*.

MODERN RECEIVER DESIGN METHODS

The superheterodyne and the direct-conversion receiver are the most popular modern receivers and the chief topic of this discussion. Both were conceived in the 1915-1922 time frame. Direct conversion was used by Bell Labs in 1915 SSB experiments; it was then called the "homodyne" detector. E. H. Armstrong devised the superhet in about 1922, but for about 12 years it was considered too expensive for the (at that time) financially strapped amateur operators. The advent of "single signal reception," pioneered by J. Lamb at ARRL, and the gradual end of the Great Depression era brought about the demise of the regenerative receiver. We begin with a discussion of the direct conversion receiver, which has been rediscovered by amateur equipment builders and experimenters in recent years.

Direct Conversion (D-C) Receivers

The direct conversion (D-C) receiver, in its simplest form, has some similarities to the regenerative receiver:

- The signal frequency is converted to audio in a single step.
- An oscillator very near the signal frequency produces an audible beat note.
- Signal bandwidth filtering is performed at baseband (audio).
- Signals and noise (both receiver noise and antenna noise) on both sides of the oscillation frequency appear equally in the audio output. The image (on the other side of zero beat) noise is an "excess" noise that degrades the noise factor and dynamic range.

There are three major differences favoring the D-C receiver:

- There is no delicate state of regeneration involved. A low-gain or passive mixer of high stability is used instead.
- The oscillator is a separate and very stable circuit that is buffered and coupled to the mixer.
- The D-C (that uses modern circuit design) has much better dynamic range.

The regen has enormous RF gain (and Q multiplication) and therefore little audio gain is needed. The D-C delivers a very low-level audio that must be greatly amplified and filtered. RF amplification, band-pass filtering and automatic gain control (AGC) can be easily placed ahead of the mixer with beneficial results.

An enhancement of the D-C concept can perform a fairly large reduction of the signal and noise image responses mentioned above. It is a major technical problem, however, to get a degree of reduction over a wide frequency range, say 1.6-30 MHz, that compares with that easily obtainable using superheterodyne methods.

D-C Receiver Design Example

Fig 14.15 is a schematic diagram of a simple D-C receiver that utilizes all of the principles mentioned above except image rejection. The emphasis is on simplicity for both SSB and CW reception on the 14-MHz band. The LO is standard *Handbook* circuitry and is not shown. **Fig 14.16** shows a simple example of an active CW band-pass filter centered at 450 Hz.

The input RF filter shields the receiver from large out-of-ham-band signals and has a noise figure of 2 dB. The grounded-gate RF amplifier has a gain of 8 dB, a noise figure of 3 dB and an input third-order-intercept point of 18 dBm. Its purpose is to improve a 14-dB noise figure (at the antenna input) without the amplifier to about 8.5 dB. It also eliminates any significant LO conduction to the antenna and provides opportunities for RF AGC. The total gain ahead of the mixer is about 6 dB, which degrades the IMD performance of the mixer and subsequent circuitry somewhat. However, the receiver still has a third-order intercept (IP3) of about 6 dBm for two tones within the range of the audio filter. The IP3 is 11 dBm (quite respectable) when one of the two tones is outside the range of the low-pass audio filter that precedes the first audio amplifier. The low-pass filter protects the audio amplifier from wideband signals and noise. The intercept point could be improved by eliminating the RF amplifier, but the antenna input noise figure would then be much worse. This is a common trade-off decision that receiver designers must make.

The above analysis would be correct for a conventional receiver, but in this case there is a small complication that we will mention only briefly. The noise sources ahead of the mixer, both thermal and excess, that are on the image side (the side of the carrier opposite a weak desired signal) are translated to the baseband and appear as an increased noise level at the input of the first audio amplifier. If the SFDR (previously defined) in a 1000-Hz bandwidth were ordinarily 95 dB, using the above numbers, the actual SFDR would be perhaps 2.5 dB less. An image-reject mixer would correct this problem. Observe that the low noise figure of the RF amplifier minimizes the gain needed to get the desired overall noise figure. Also, its good intercept point minimizes strong-signal degradation contributed by the amplifier.

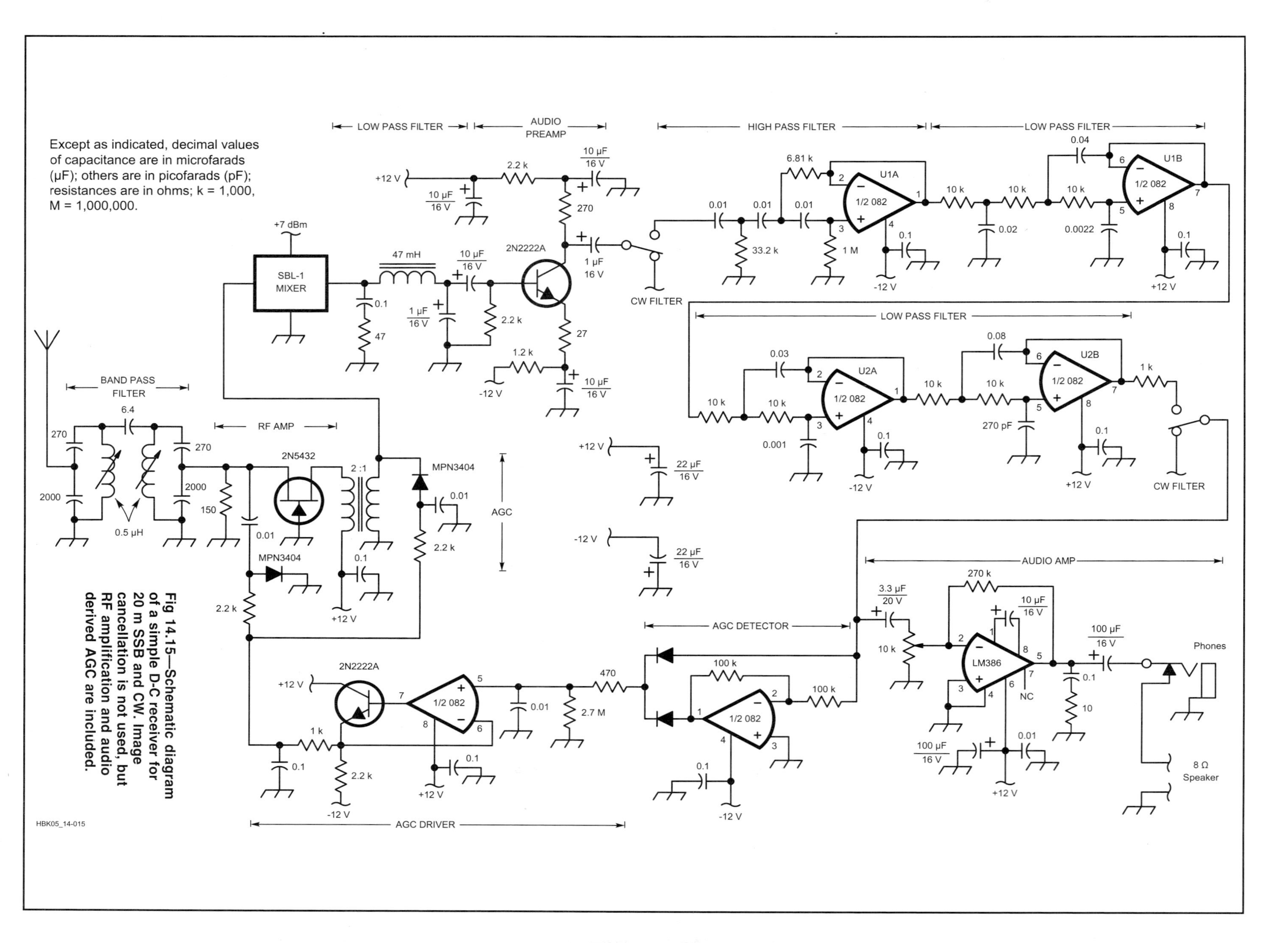

Fig 14.15—Schematic diagram of a simple D-C receiver for 20 m SSB and CW. Image cancellation is not used, but RF amplification and audio derived AGC are included.

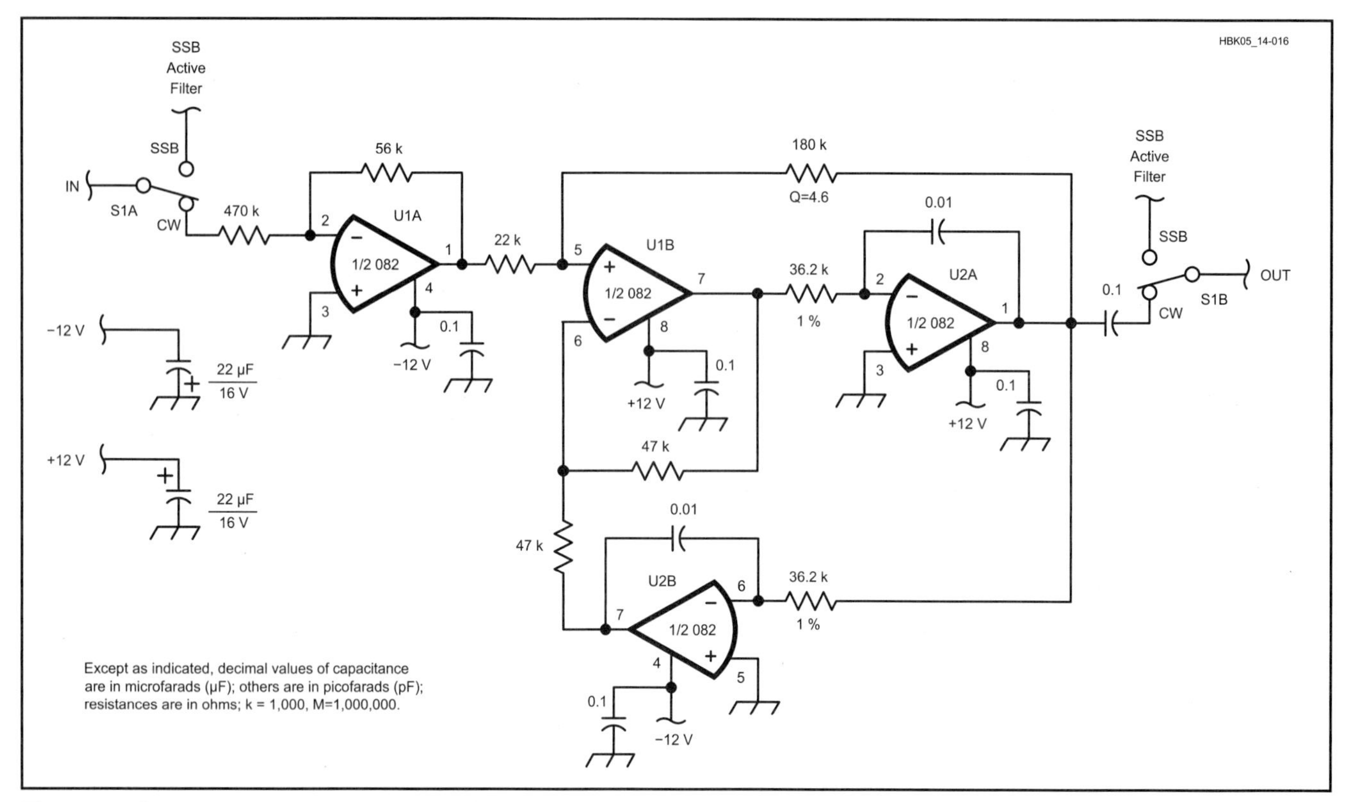

Fig 14.16—Schematic diagram of an active CW band-pass filter for the D-C receiver in Fig 14.15.

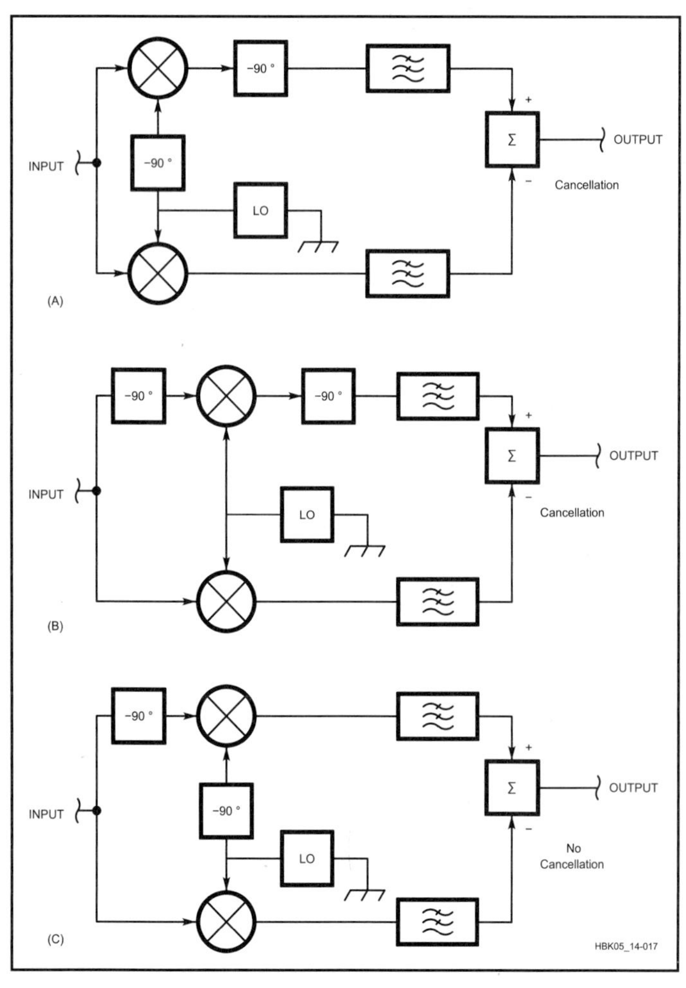

Fig 14.17—Both A and B are workable image cancelling mixer stages. The scheme at C will not cancel image signals.

The input BPF eliminates problems from second-order IMD. Flicker-effect (1/f) noise in the mixer audio output may also be a problem, which the RF amplification reduces and the mixer design should minimize.

The preceding analysis illustrates the kind of thinking that goes into receiver design. If we can quantify performance in this manner we have a good idea of how well we have designed the receiver.

For the circuit in Fig 14.15 and the numbers given above, the gain ahead of the first audio amplifier is 0 dB. As stated before, this amplifier is protected from wideband interference by the 2-element low-pass audio filter ahead of it, which attenuates at a rate of 12 dB per octave. This filter could have more elements if desired. By minimizing front-end gain, the tendency for the audio stages to overload before the earlier stages do so is minimized—if the audio circuitry is sufficiently "robust." This should be checked out using two-tone and gain-compression tests on the audio circuits. Audio-derived AGC helps prevent signal-path overload by strong desired (in-band) signals. Additional AGC can be applied in the audio section by using a variable-gain audio op amp (MC3340P).

The audio SSB and CW band-pass filters are simplified active op-amp filters that could be improved, if desired, by using methods mentioned in some of the

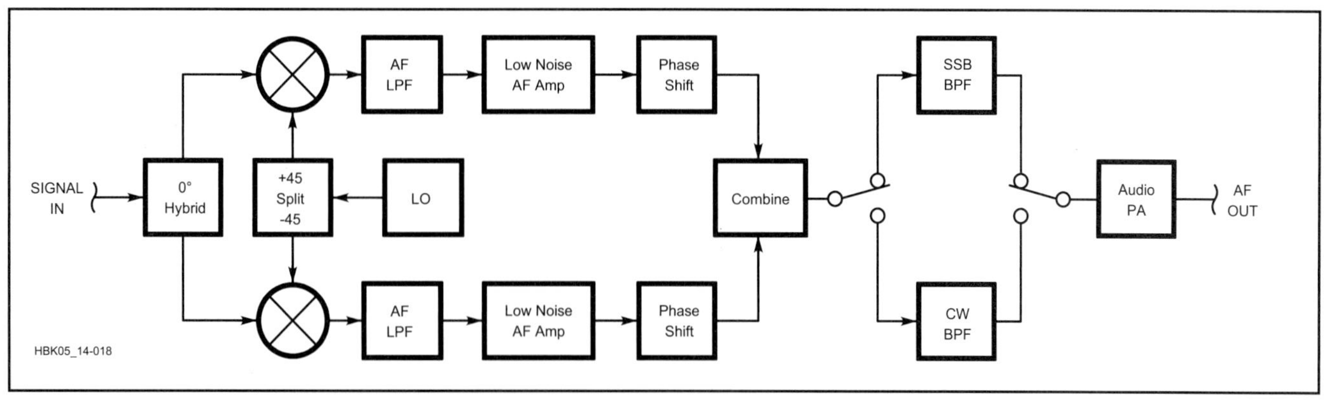

Fig 14.18—Typical block diagram of an image cancelling D-C receiver.

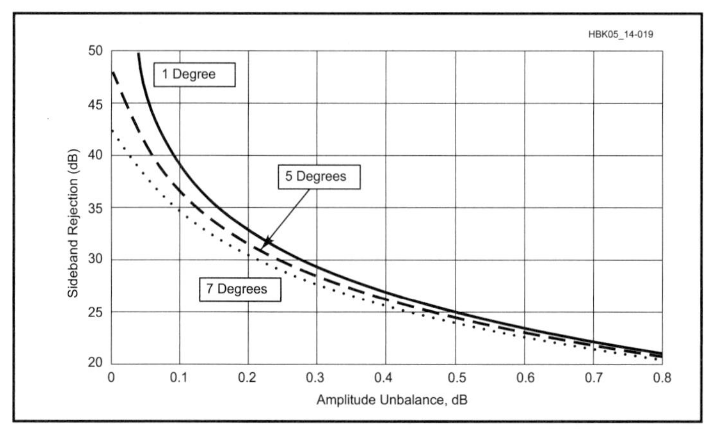

Fig 14.19—A plot of sideband rejection versus phase error and amplitude unbalance.

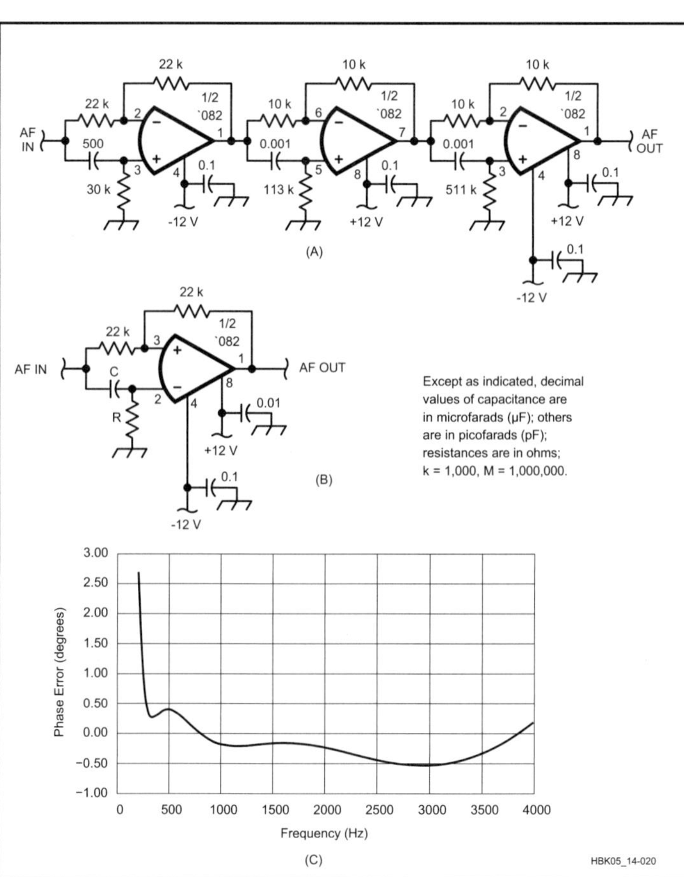

Fig 14.20—A: an example of an audio phase-shift circuit. B: a single all-pass stage. C: phase error vs audio frequency for a pair of circuits like that in A with appro-priate values of R and C.

references. In an advanced design, digital signal processing (DSP) could be used. Good shape factor is the main requirement for good adjacent-channel rejection. Good transient response (maximally flat group delay) would be a "runner-up" consideration. Digital FIR filters and analog elliptic filters are good choices.

Image Rejection in the D-C Receiver

Rejection of noise and signals on one side of the LO is a major enhancement and also a major complication of the D-C receiver (Refs 11, 12). **Fig 14.17** shows two correct ways to build an image canceling mixer and one incorrect way. The third way does not perform the required phase cancellations for image reduction. In practice, two ±45° phase shifters are used, rather than one 90° stage. As mentioned before, it is very difficult to get close phase tracking over a wide band of signal and LO frequencies. In amateur equipment, front panel "tweaker" controls would be practical.

The block diagram in **Fig 14.18** is a typical approach to an image-canceling D-C receiver. The two channels, including RF, mixers and audio must be very closely matched in amplitude and phase. The audio phase-shift networks must have equal gain and very close to a 90° phase difference. **Fig 14.19** relates phase error in degrees and amplitude error in dB to the rejection in dB of the opposite (image) sideband. For 30 or 40 dB of rejection, the need for close matching is apparent.

AUDIO PHASE SHIFTERS

Fig 14.20A shows an example of an audio phase-shift network. The stage in Fig 14.20B is one section, an active "all-pass" network that has these properties:

- The gain is exactly 1.0 at all frequencies and

• The phase shift changes from 180° at very low frequency to 0° at very high frequency.

The shift of this single stage is +90° at $f = 1/(2\pi RC)$. By cascading several of these with carefully selected values of RC the set of stages has a smooth phase shift across the audio band. A second set of stages is chosen such that the phase difference between the two sets is very close to 90°. The choices of R and C values have been worked out using computer methods; you can also find them in other handbooks (Ref 13). Fig 14.20C shows the phase error for two circuits like the one shown in Fig 14.20A. Note the rapid increase in error at very low audio frequencies (an improvement would be desirable for CW work). These frequencies should be greatly attenuated by the audio band-pass filters that follow.

D-C Receiver Problem Areas

Because of the high audio gain, microphonic reactions due to vibration of low-level audio stages are common. Good, solid construction is necessary. Another problem involves leakage of the LO into the RF signal path by conduction and/or radiation. The random fluctuations in phase of the leakage signal interact with the LO to produce some unpleasant modulation and microphonic effects. Hum in the audio can be caused by interactions between the LO and the power supply; good bypassing and lead filtering of the power supply are needed. A small amount of RF amplification is beneficial for all of these problems.

The Superheterodyne Receiver

GENERAL DISCUSSION

The superheterodyne ("superhet") method is by far the most widely used approach to receiver and transmitter design. **Fig 14.21A** shows the basic elements as applied to an SSB/CW receiver, which we will consider first. We will consider a superhet transmitter later in this chapter.

RF from the antenna is filtered (preselected) by a band-pass filter of some kind to reduce certain kinds of spurious responses and then (possibly) RF amplified. A mixer, or frequency converter circuit, *multiplies* (in the time domain) its two inputs, the signal and the LO. The result of this multiplication process is a pair of output intermediate frequencies (IFs) that are the sum and difference of the signal and LO frequencies. If the mixer is a perfect multiplier, as the equation in Fig 14.21 suggests, it is a linear mixer, these are the only output frequencies present and it has all the properties of any other linear circuit except for the change of frequency. If the mixer is a commutating (switching) mode mixer it is still a perfect mixer but additional frequencies of lesser amplitude are present. See the **Mixers, Modulators and Demodulators** chapter for a detailed discussion.

One of these outputs is selected to be the "desired" IF by the designer. It is then band-pass filtered and amplified. The bandwidths and shape factors of these filters are optimized for the kind of signal being received (AM, SSB, CW, FM, digital data). Two of the main attributes of the superhet are that this signal filtering band shape and also the IF amplification are constant for any value of the receive signal frequency. An excessively narrow preselector filter could, however, have

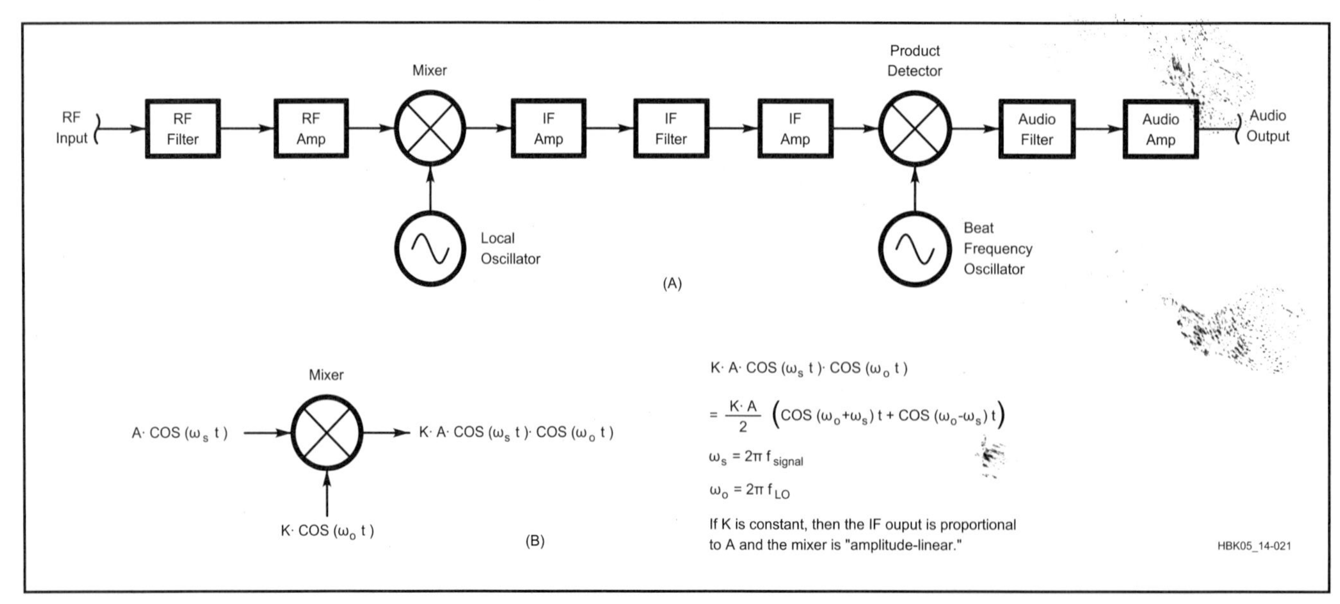

Fig 14.21—(A) Basic block diagram of a superhet receiver. (B) Showing how the input signal and a constant LO input produce a linear mixing action.

some effect on the desired signal, as we saw in the case of the TRF receiver.

A second mixer, or "detector" as it is usually called, translates the IF signal to baseband (audio) where it is further amplified, possibly filtered, and applied to an output transducer (headphones, loudspeaker, some other signal processor or display).

A superhet receiver may also contain multiple frequency conversions (IFs). Later discussion will focus on strategies used to select these IFs. Let's begin with a detailed discussion of the classic down-conversion superhet. Almost all of the topics apply as well to the various other kinds of receiver designs in subsequent sections.

Superhet Characteristics: A Down-Conversion Example

The desirability of the superhet approach is offset somewhat by certain penalties and problem areas. As a vehicle for mentioning these difficulties, seeing how to deal with them and discussing analysis and design methods; we use the tutorial example in **Fig 14.22**. That is a "down converting" single-conversion 14-MHz superhet with a 1.5-MHz IF. This receiver is simple and capable of fairly good performance in the 1.8 to 30-MHz frequency range. Fig 14.22 is intentionally incomplete and meant for instructional purposes only; do not attempt to duplicate it as a project.

Block Diagram

The block labels of Fig 14.22 show that a preselector and RF amplifier are followed by downward frequency conversion to 1.5 MHz. This is followed by IF amplification and crystal filtering, a product detector, audio band-pass filters and an audio power output stage. Equal emphasis is given to SSB and CW. AGC circuitry is included. The audio and AGC circuits are the same as those in Fig 14.15 and Fig 14.16. As a first step, let's look at spurious responses of the mixer.

Mixer Spurious Responses

Mixers and their spurious responses are covered in detail in the **Mixers** chapter, but we will present a brief overview of the subject for our present purposes. We then will see how this information is used in the design process.

The mixer is vulnerable to RF signals other than the desired signal. Various harmonics of any undesired RF signal and harmonics of the LO combine to produce spurious IF outputs (called harmonic IMD). If these spurious outputs are within the IF passband they appear at the receiver output. The strength of these outputs depends on: harmonic number and strength of the RF signal as it appears at the mixer input, the harmonic number of the LO, the LO power rating (7 dBm, 17 dBm, and so on) and the design of the mixer.

Commercially available double-balanced diode mixers are so convenient, easy-to-use and of such low cost and high quality that they are used in many Amateur Radio receiver and transmitter projects. These mixers also do a good job of rejecting certain kinds of spurious responses. Our numerical examples will be based on typical published data for one of these mixers (Ref 14).

Fig 14.23A shows an example for the mixer tuned to a desired signal at 14.00 MHz with the LO at 15.50 MHz. The locations of undesired signals that cause a spurious response are shown in Fig 14.22B; they are at 14.75, 15.00, 16.00 and 17.00 MHz (there are many others of lesser importance). Each of these undesired signals produces a 1.50-MHz output from the mixer. The figure indicates the harmonics of the undesired signal and the harmonics of the LO that are involved in each instance. The "order" of the spurious product is the sum of these harmonic numbers, for example the one at 16.00 MHz is a sixth-order product. The spurious at 17.00 MHz is called the "image" because it is also 1.50 MHz away from the LO, just as the 14.00 desired signal is 1.5 MHz away from the LO. It is a second-order response, as is the response at 14.00 MHz.

Fig 14.23C is a chart that shows the relative responses for various orders of harmonic IMD products for a signal level (desired or undesired) of 0 dBm and an LO level of +7 dBm. The values are typical for a great many +7-dBm mixers having various brand names and they improve greatly for higher level mixers (at the same RF levels). The second-order (desired and image) both have a reference value of 0 dB and the others are in dB below those two.

We can now consider the receiver design that suppresses these spurious responses so that they do not interfere with a weak desired signal at 14.00 MHz. If an interfering signal is reduced in amplitude at the mixer RF input by 1.0 dB, the suppression of that spur is improved by 1.0 × Signal Harmonic Number dB. This is true in principle, but in reality the reduction may be somewhat less. For example, the spur produced by 15.00 MHz is reduced 3 dB for each dB that we reduce its level. We accomplish this task by choosing the right mixer, limiting the amount of RF amplification and designing adequate selectivity into the preselector circuitry. With respect to selectivity, though, note that in many other mixing schemes the interfering signal is so close to the desired frequency that selectivity does little good. Then we must use a mixer with a higher LO level and/or reduce RF gain.

The design method is illustrated by the following numerical example. Suppose that a signal at 14.75 MHz (the IF/2 spur) is at –20 dBm (very strong) at the antenna and –10 dBm at the mixer RF port. From the chart, this spur will be reduced by 71 + 2 (0 – (–10)) = 91 dB to a level of –10 – 91 = –101 dBm. If this is not enough then a preselector will help. If the preselector attenuates 14.75 MHz by 5 dB, the total spur reduction will be 71 + 2(0 – (–15)) = 101 dB to a level of –15 – 101 = –116 dBm, a 15-dB improvement. Notice that spurs involving high harmonics of the signal frequency attenuate more quickly as the input RF level is reduced.

On the other hand, if we consider the image signal at 17 MHz, all of the reduction of this spur must come from the preselector. In other words, selectivity is the only way to reduce the image response unless an image reducing mixer circuit is used. In this example, additional spur reduction is obtained by using a preselector circuit topology that has improved attenuation *above* the passband.

In designing *any* receiver we must be reasonable about spur and image reduction. Receiver cost and complexity can increase dramatically if we are not willing to accept an occasional spurious response due to some very strong and seldom occurring signal. In the case of a certain persistent interference some specific cure for that source can usually be devised. A sharply tuned "trap" circuit, a special preselector or a temporary antenna attenuator are a few examples. In practice, for down-conversion superhets, 90 dB of image reduction is excellent and 80 dB is usually plenty good enough for amateur work.

In classical down-conversion superhets, the preselection circuits are tuned and bandswitched in unison with the LO. They must all "track" each other across the dial. The cost and complexity of this arrangement have made this approach prohibitive in modern commercial multiband designs (Ref 15). For amateur work the approach in Fig 14.22 is more practical, using switched or even plug-in band-pass preselectors and oscillator coils. A frequency counter, offset by the 1.5 MHz IF and connected to the LO, eliminates the need for a calibrated dial.

Two-Tone Intermodulation Distortion

Another important mixer spurious response is two-tone IMD. This distortion has been covered previously in this chapter,

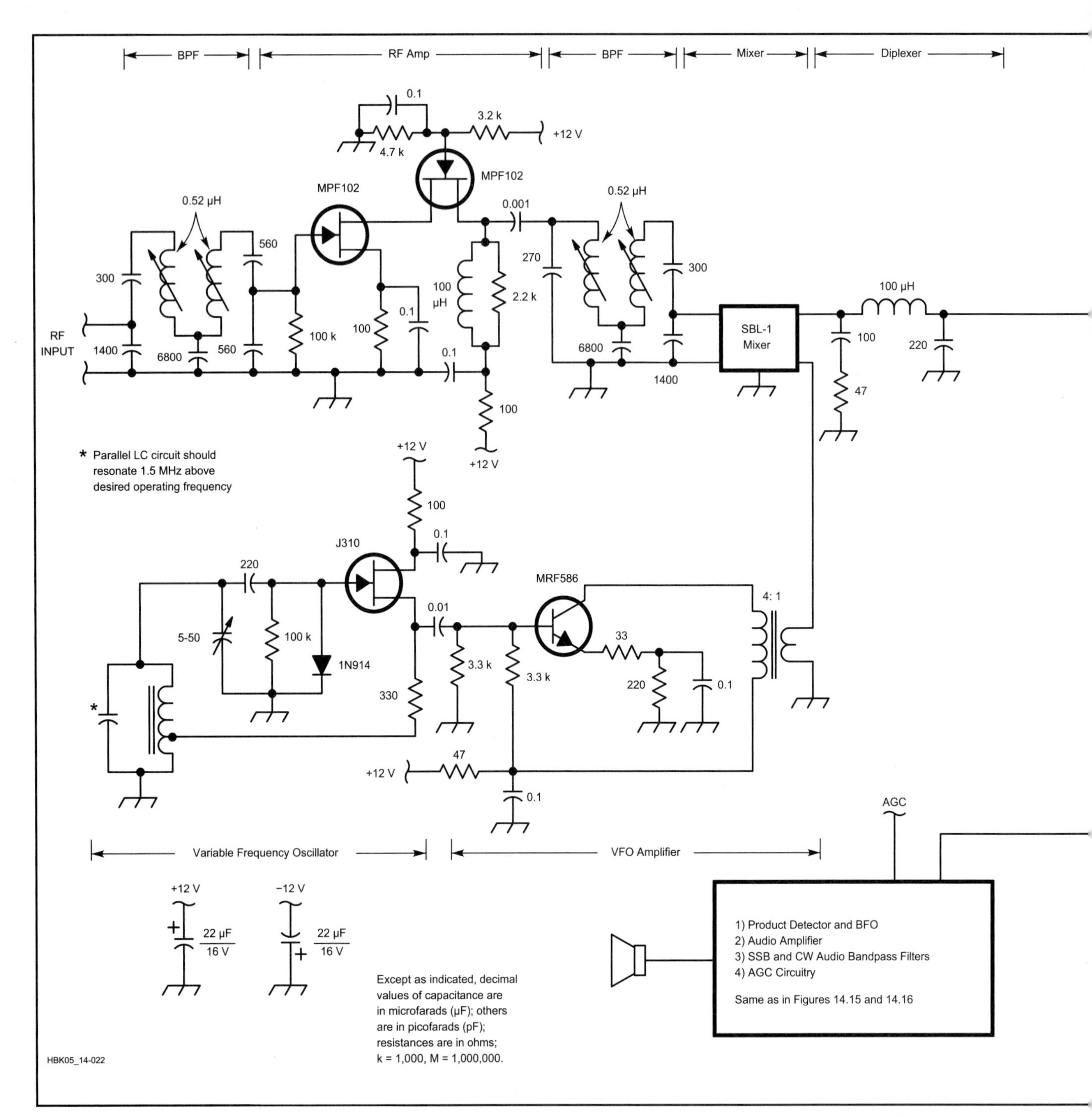

Fig 14.22—Specific example of a down-conversion superhet that is used to explain and analyze superhet behavior and design.

and the **Mixers** chapter gives more detail. From a system design standpoint, the trade-offs between receiver noise figure and IMD have been covered in this chapter, and the choices of mixer, RF gain (if any) and selectivity are decided in a study exercise of performance, cost and complexity.

A receiver that has a 10 to 20-dBm third-order intercept point for two signals 20 kHz and 40 kHz removed is an excellent receiver in many applications. Some advanced experimenters have built receivers with 25 to 40-dBm values of IP3. Values of 40 dBm are near the state of the art (Ref 16).

A matter of considerable interest concerns the way that IMD varies as the separation between the two tones increases. In Fig 14.22, for example, if one tone is 1.0 kHz (or 100 kHz) above 14.00 the other is 2.0 kHz (or 200 kHz) above. We see that for very close tone separations the IF filter may not prevent the tones from reaching the circuits following the IF filter. As the separation increases, first one, then both, tones fall outside the IF filter passband and the IMD becomes much less. However the mixer and the amplifier after the mixer are still vulnerable. At greater separations the preselector starts to protect these two stages, but the RF amplifier is not well protected by the first RF filter until the tone separation becomes greater, perhaps 200 kHz. It is a common procedure to plot a graph of receiver third-order input intercept point vs tone separation and then look for ways to improve the

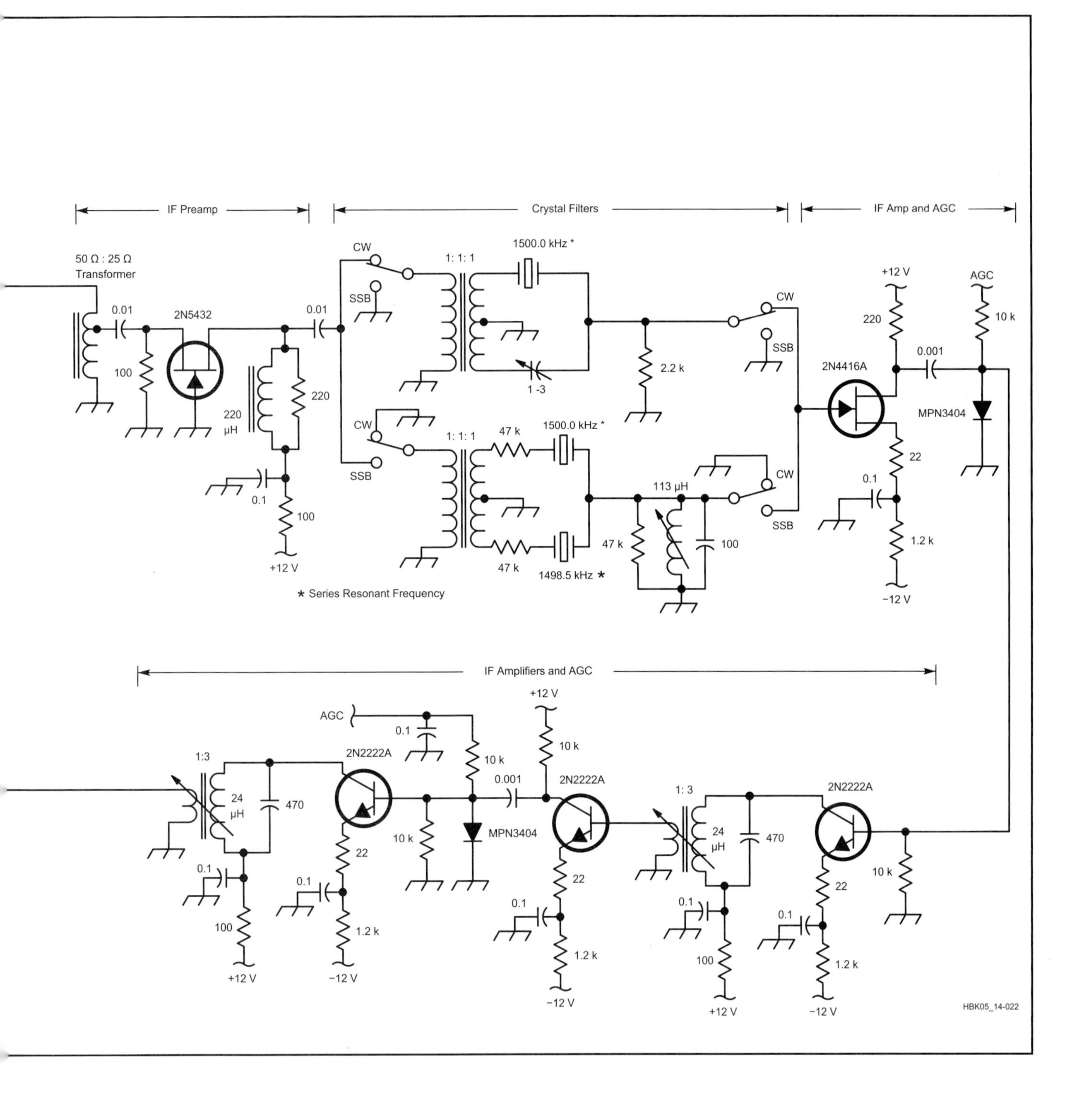

overall performance.

The stages after the IF filter are protected by AGC so that, hopefully, tones in the IF passband do not overdrive the circuits after the IF filter. But in the example of Fig 14.22 there is also a narrowband audio filter and the AGC is derived from the output of this filter. This means that circuits *after* the IF filter but *ahead of* the audio filter may not always be as well protected as we would like. Strong tones that get through the IF filter may be stopped by the audio filter and not affect the AGC. This particular example illustrates a very common problem in all kinds of receivers that have *distributed* selectivity. It is also found universally in multiple conversion receivers, as we will discuss later.

GAIN AND NOISE FIGURE DISTRIBUTION

Based on the information given so far, the approach to designing a superhet receiver, whether a downconverter or any other kind, can now be summarized by the following guidelines:

1. Try to keep the gain ahead of the mixer and the narrow band-pass filters (SSB, CW and so on) as low as possible. For a fixed components cost (such as mixers and amplifiers), this minimizes the IMD, both two tone and harmonic.
2. Reducing the gain implies that the noise figure may be a little higher. It is always best to avoid making the noise

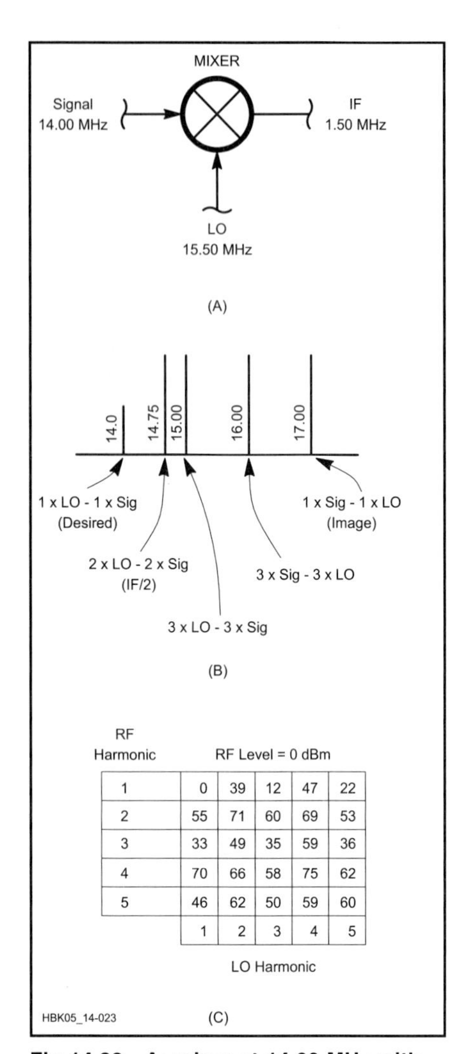

Fig 14.23—A: mixer at 14.00 MHz with LO at 15.50 and IF at 1.50. B: locations of strong signals that interfere with desired signal at 14.00 MHz due to harmonic IMD. C: typical chart of harmonic IMD products for a double balanced diode mixer with 0 dBm signal and 7 dBm LO.

figure any lower than necessary. Noise figure is usually more important at microwave frequencies than at HF, and strong signal interference is usually less important. Where interference is a problem an increase in noise figure is almost always mandatory, except possibly when a higher-level mixer is used. A narrowband preselector, for example, will increase the noise figure (and also the intercept point) because of its passband attenuation.

3. Amplifier circuits and modules always involve a trade-off of some kind between intercept point and noise figure. Designers look for devices and circuits that optimize the SFDR for the particular kind of receiver under design.
4. If the receiver has distributed selectivity, make the first IF filter good enough that the AGC/IF-overload problem mentioned above is minimized.
5. To *minimize* the gain ahead of the mixer, *follow* the mixer with a low-noise, high-dynamic-range amplifier with no more gain than necessary, say 10 dB or so (see Fig 14.22).
6. Terminate the mixer in such a way that its IMD is minimized. Fig 14.22 shows a simple IF diplexer that absorbs the output image at 29.5 MHz (14.0 + 15.5).
7. The RF terminal of the mixer should be short circuited at the image frequency so that noise at the image frequency (from the preceding circuitry) is minimized.
8. Because a large amount of overall gain is needed, reducing front-end gain implies that the gain after the first IF filter must be very large. The problem of IF and audio noise then arises. It is very desirable to use a low-noise amplifier right after the first IF filter (see Fig 14.22) and to restrict the bandwidth of the IF/AF amplifiers. A second IF/AF filter downstream, and also possibly an image-reducing product detector, are excellent ways to accomplish this. This step also minimizes the degradation of receiver noise figure that can be caused by this wideband noise.
9. The LO must have very low phase noise to reduce reciprocal mixing. Also, the mixer must have good balance (meaning isolation or rejection) from LO port to RF and IF ports so that broadband additive noise from the LO amplifiers does not degrade the mixer noise figure. This is especially important when the RF amplifier gain has been minimized. If the mixer is not balanced in this sense at the LO port, a band-pass filter between LO and mixer is very desirable.

AUTOMATIC GAIN CONTROL (AGC)

The amplitude of the desired signal at each point in the receiver is controlled by AGC. Each stage has a distortion vs signal-level characteristic that must be known, and the stage input level must not become excessive. The signal being received has a certain signal-to-distortion ratio that must not be degraded too much by the receiver. For example, if an SSB signal has –30 dB distortion products the receiver should have –40 dB quality. A correct AGC design ensures that each stage gets the right input level. It is often necessary to redesign some stages in order to accomplish this (Ref 17).

The AGC Loop

Fig 14.24A shows a typical AGC loop that is often used in amateur receivers. The AGC is applied to the stages through RF decoupling circuits that prevent the stages from interacting with each other. The AGC amplifier helps to provide enough AGC loop gain so that the gain-control characteristic of Fig 14.24B is achieved. The AGC action does not begin until a certain level, called the AGC threshold, is reached. The Threshold Volts input in Fig 14.24A serves this purpose. After that level is exceeded, the audio level slowly increases. The audio rise beyond the threshold value is usually in the 5 to 10-dB range. Too much or too little audio rise are both undesirable for most operators.

As an option, the AGC to the RF amplifier is held off, or "delayed," by the 0.6-V forward drop of the diode so that the RF gain does not start to decrease until larger signals appear. This prevents a premature increase of the receiver noise figure. Also, a time constant of one or two seconds after this diode helps keep the RF gain steady for the short term.

Fig 14.25 is a typical plot of the signal levels at the various stages of a certain ham band receiver. Each stage has the proper level and a 115-dB change in input level produces a 10-dB change in audio level. A manual gain control would produce the same effect.

AGC Time Constants

In Fig 14.24, following the precision rectifier, R1 and C1 set an "attack" time, to prevent excessively fast application of AGC. One or two milliseconds is a good value for the R1 × C1 product. If the antenna signal suddenly disappears, the AGC loop is opened because the precision rectifier stops conducting. C1 then discharges through R2 and the C1 × R2 product can be in the range of 100 to 200 ms. At some point the rectifier again becomes active, and the loop is closed again.

An optional modification of this behavior is the "hang AGC" circuit (Ref 18). If we make R2 × C1 much longer, say 3 seconds or more, the AGC voltage remains almost constant until the R5, C2 circuit decays with a switch selectable time constant of 100 to 1000 ms. At that time R3 quickly discharges C1 and full receiver gain is quickly restored. This type of control is appreciated by many operators because of the lack of AGC "pumping" due to modulation, rapid fading and other sudden signal level changes.

AGC Loop Problems

If the various stages have the property that each 1-V change in AGC voltage changes the gain by a constant amount (in dB), the AGC loop is said to be "log lin-

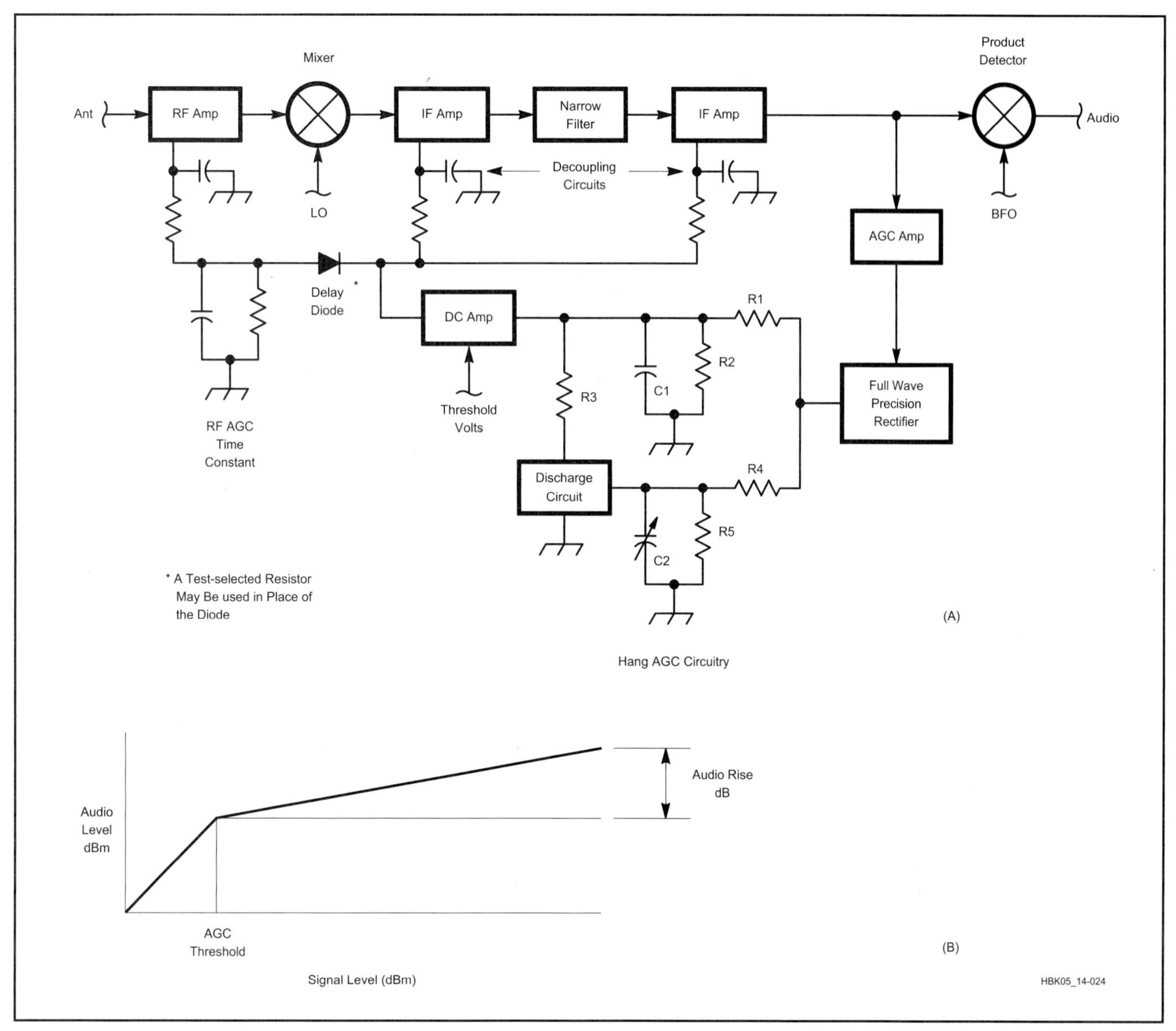

Fig 14.24—AGC principles. A: typical superhet receiver with AGC applied to multiple stages of RF and IF. B: audio output as a function of antenna signal level.

ear" and regular feedback principles can be used to analyze and design the loop. But there are some difficulties that complicate this textbook model. One has already been mentioned, that when the signal is rapidly decreasing the loop becomes "open loop" and the various capacitors discharge in an open-loop manner. When the signal is increasing beyond the threshold, or if it is decreasing slowly enough, the feedback theory applies more accurately. In SSB and CW receivers rapid changes are the rule and not the exception.

Another problem involves the narrow band-pass IF filter. The group delay of this filter constitutes a time lag in the loop that can make loop stabilization difficult. Moreover, these filters nearly always have much greater group delay at the edges of the passband, so that loop problems are aggravated at these frequencies. Overshoots and undershoots, called "gulping," are very common. Compensation networks that advance the phase of the feedback help to offset these group delays. The design problem arises because some of the AGC is applied before the filter and some after the filter. It is a good idea to put as much fast AGC as possible after the filter and use a slower decaying AGC ahead of the filter. The delay diode and RC in Fig 14.24A are helpful in that respect. Complex AGC designs using two or more compensated loops are also in the literature. If a second cascaded narrow filter is used in the IF it is usually a lot easier to leave the second or "downstream" filter out of the AGC loop.

Another problem is that the control characteristic is often not log-linear. For example, dual-gate MOSFETs tend to have much larger dB/V at large values of gain reduction. Many IC amplifiers have the same problem. The result is that large signals cause instability because of excessive loop gain. There are variable gain op amps and other ICs available that are intended for gain control loops.

Audio frequency components on the AGC bus can cause problems because the amplifier gains are modulated by the audio and distort the desired signal. A hang AGC circuit can reduce or eliminate this problem.

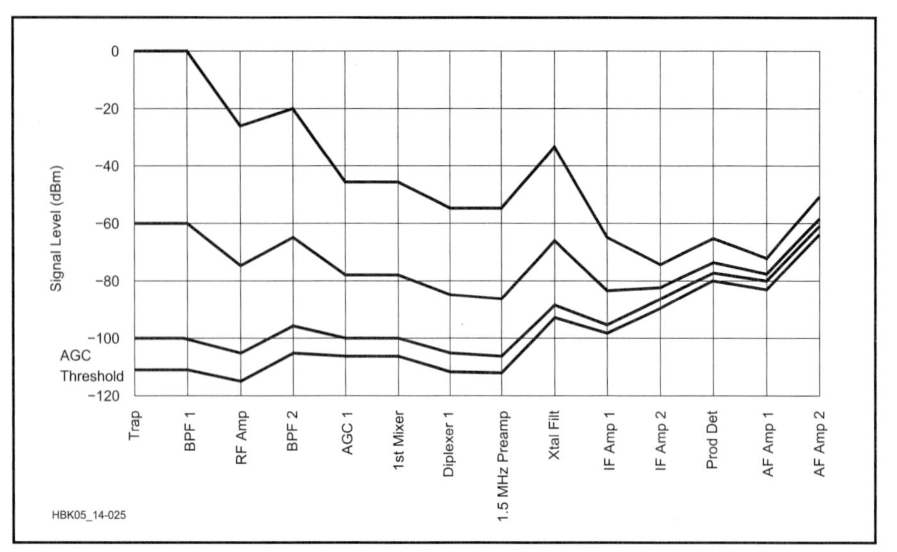

Fig 14.25—Gain control of a ham-band receiver using AGC. A manual gain control could produce the same result.

Finally, if we try to reduce the audio rise to a very low value, the required loop gain becomes very large, and stability problems become very difficult. It is much better to accept a 5 to 10 dB variation of audio output.

Because many parameters are involved and many of them are not strictly log-linear, it is best to achieve good AGC performance through an initial design effort and finalize the design experimentally. Use a signal generator, attenuator and a signal pulser (2-ms rise and fall times, adjustable pulse rate and duration) at the antenna and a synchronized oscilloscope to look at the IF envelope. Tweak the time constants and AGC distribution by means of resistor and capacitor decade boxes. Be sure to test throughout the passband of each filter. The final result should be a smooth and pleasant sounding SSB/CW response, even with maximum RF gain and strong signals. Patience and experience are helpful.

Audio-Derived AGC

The example in Fig 14.15 shows audio-derived AGC. There is a problem with this approach also. At low audio frequencies the AGC can be slow to develop. That is, low-frequency audio sine waves take a long time to reach their peaks. During this time the RF/IF/AF stages can be overdriven. If the RF and IF gains are kept at a low level this problem can be reduced. Also, attenuating low audio frequencies prior to the first audio amplifier should help. With audio AGC, it is important to avoid so-called "charge pump" rectifiers or other slow-responding circuits that require multiple cycles to pump up the AGC voltage. Instead, use a peak-detecting circuit that responds accurately on the first positive or negative transition.

AGC Circuits

Fig 14.26 shows some gain controllable circuits. Fig 14.26A shows a two-stage 455-kHz IF amplifier with PIN-diode gain control. This circuit is a simplified adaptation from a production receiver, the Collins 651S. The IF amplifier section shown is preceded and followed by selectivity circuits and additional gain stages with AGC. The 1.0-µF capacitors aid in loop compensation. The favorable thing about this approach is that the transistors remain biased at their optimum operating point. Right at the point where the diodes start to conduct, a small increase in IMD may be noticed, but that goes away as diode current increases slightly. Two or more diodes can be used in series, if this is a problem (it very seldom is).

Fig 14.26B is an audio-derived AGC circuit using a full-wave rectifier that responds to positive or negative excursions of the audio signal. The RC circuit follows the audio closely.

Fig 14.26C shows a typical circuit for the MC1350P RF/IF amplifier. The graph of gain control vs AGC volts shows the change in dB/V. If the control is limited to the first 20 dB of gain reduction this chip should be favorable for good AGC transient response and good IMD performance. Use multiple low-gain stages rather than a single high-gain stage for these reasons. The gain control within the MC1350P is accomplished by diverting signal current from the first amplifier stage into a "current sink." This is also known as the "Gilbert multiplier" architecture. Another chip of this type is the NE/SA5209. This type of approach is simpler to implement than discrete-circuit approaches, such as dual-gate MOSFETs that are now being replaced by IC designs.

Fig 14.26D shows the high-end performance National Semiconductor CLC520AJP (14-pin DIP plastic package) voltage controlled amplifier. It is specially designed for accurate log-linear AGC from 0 to 40 dB with respect to a preset maximum voltage gain from 6 to 40 dB. Its frequency range is dc to 150 MHz. It costs about $11.50 in small quantities and is an excellent IF amplifier for high-performance receiver or transmitter projects.

IF FILTERS

There are some aspects of IF-filter design that influence the system design of receivers and transmitters. The influence of group delay, especially at the band-pass edges, on AGC-loop performance has been mentioned. Shape factor is also significant (the ratio of two bandwidths, usually 60-dB:6-dB widths). To get good adjacent-channel rejection, the transition-band response should fall very quickly. Unfortunately, this goal aggravates group-delay problems at the passband edges. It also causes poor transient response, especially in CW filters. Another filter phenomenon can cause problems: at sharp passband edges signals and noise produce a raspy sound that is annoying and interferes with weak signals.

A desirable filter response would be slightly rounded at the edges of the passband, say to –6 dB, with a steep rolloff after that. This is known as a "transitional filter" (Ref 19). Cascaded selectivity with two filters, each having fewer "poles" (than a single filter would) is also a good approach. Both methods have a smoother group delay across the passband and reduce the problems mentioned above.

Ultimate Attenuation

In a high-gain receiver with as much as 110 dB of AGC the ultimate attenuation of the filter is important. Low-level leakage through or around the filter produces high-pitch interference that is especially noticeable on CW. Give special attention to parts layout, wiring and shielding. (Filter selector switches are often leakage culprits.) Cascaded IF filters also help very considerably.

Audio Filter Supplement

An audio band-pass filter can be used to supplement IF filtering. This can help to improve signal-to-noise ratio and reduce

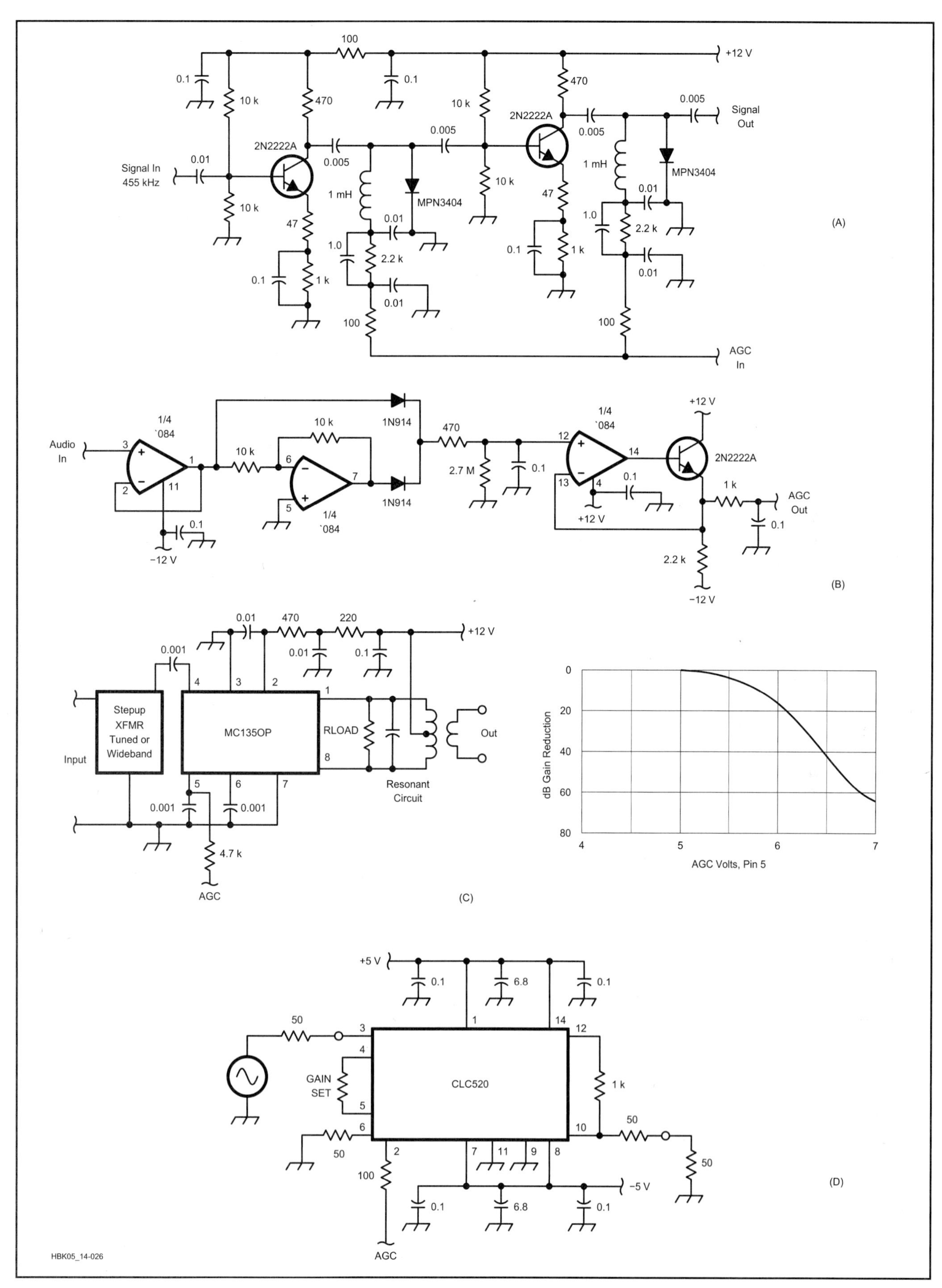

Fig 14.26—Some gain controllable amplifiers and a rectifier suitable for audio derived AGC.

adjacent-channel interference. Supplementary audio filtering also helps reduce the high-frequency leakage problem mentioned above. Another significant problem: If AGC is made in the IF section, strong signals inside the IF passband but outside the audio passband can "pump" or modulate the AGC, rendering weak desired signals hard to copy. This is especially noticeable during periods of high band activity, such as in a contest. These filters can use analog (see Fig 14.15 and Fig 14.16) or digital (DSP) technology.

Some Simple Crystal Filters

Fig 14.27 and **Fig 14.28** present two crystal filters to consider for a simple down-conversion receiver with a 1.5-MHz IF (see Fig 14.23). The crystals are a set of three available from JAN Crystals. The filters are both driven from a low-impedance source (200 Ω, for example).

CW Filter

Fig 14.27A is a "semi-lattice" filter using a single crystal for CW work (Ref 19). Capacitor C_c balances the bridge circuit at the crystal's parallel-resonant frequency because it is equal to the holder capacitance C_0 of the crystal. The response is then symmetrical around the series resonant frequency of the crystal. The selectivity is determined by the value of R_{out}. As the value decreases the selectivity sharpens as shown in Fig 14.27B. If this filter is combined with an audio band-pass filter as in Fig 14.16, pretty good CW selectivity is possible. In Fig 14.27C, the capacitor is increased to 8.3 pF and a notch appears at –1.7 kHz. This is the "single signal" adjustment. Also, note that the response on the high side is degraded quite a bit. The notch can be located above or below center frequency by adjusting the capacitor value; the degradation is on the opposite side of center.

SSB Filter

Fig 14.28 is a "half-lattice" filter (Ref 19). The schematic diagram shows the LCR values and the series resonant frequencies of the two crystals. One of these (1.4998 MHz) is the same type as the one used in the CW filter. The trimmer capacitor equalizes the two values of C_0, the crystal shunt capacitance (very important) in case they are not already closely matched. Place the trimmer across the crystal that has the lowest value of C_0. The response curve shows good symmetry and modest adjacent-channel rejection. The output tuned circuit absorbs load capacitance to get a pure R_{load} (also important). The follow-up audio speech band-pass filter in Fig 14.15 will improve the overall response considerably.

Mechanical Filters

Mechanical filters use transducers and the magnetostriction principle of certain materials to obtain a multiresonator narrow band-pass filter in the 100 to 500-kHz range. They are very frequency stable, accurate and reliable. An interesting example, for radio amateurs, is the Rockwell-Collins "Low Cost Series" of miniaturized torsional-mode filters for 455.0 kHz. They come in four styles with 3 dB/60 dB bandwidths of 0.3/0.5/1.5/2.0, 2.5/5.2 and 5.5/11 kHz. Used filters are sometimes available from various sources (Ref 15).

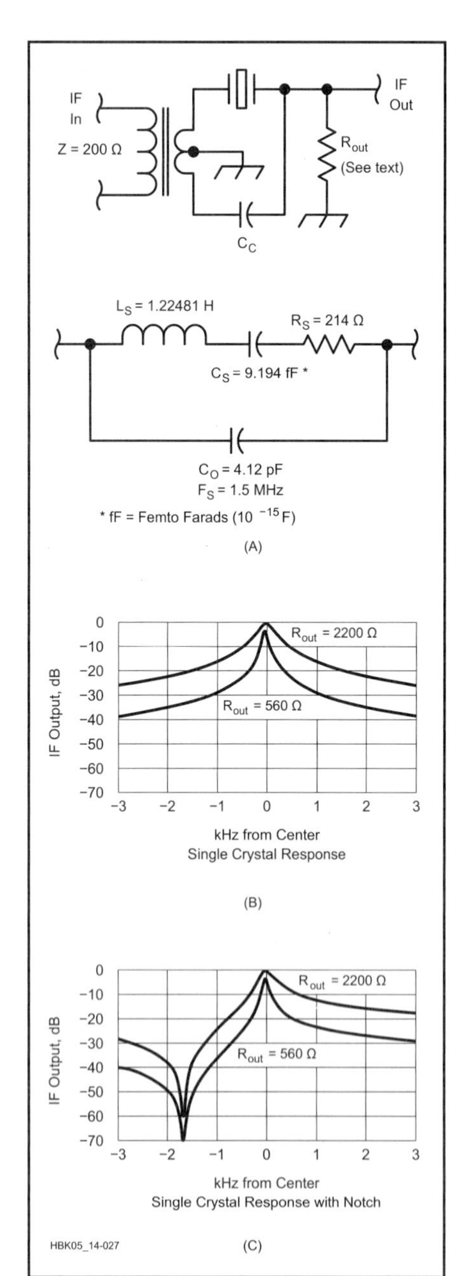

Fig 14.27—A single-crystal filter circuit for a simple CW receiver design. See also Fig 14.22.

Multielement Crystal Filters

A discussion of more complex crystal filters appears in the **RF and AF Filters** chapter of this *Handbook*. In this chapter we have considered only two very simple examples that might appeal especially to student designers and builders of a receiver that downconverts to an IF less than 2 MHz or so.

From the system design standpoint, note that for voice reception amateurs often use optional IF filters with less than the conventional bandwidth for SSB (for example 1.8 kHz), even though they reduce higher frequency speech components. This helps to improve adjacent-channel interference, which is a severe problem on some amateur bands.

It is common practice to use multipole crystal filters in the range from 5 to 10 MHz, because they can be economically designed for that frequency range. It is also common to cascade these filters with other types, such as mechanical or LC filters, at lower IFs (more about this later).

Filter Switching

Filter switching for different modes (AM, SSB, CW, RTTY and so on) requires some careful design to prevent impedance mismatching, leakage (discussed before) and spurious coupling to other circuitry. There are three general methods for switching: mechanical, relays, solid-state (diodes or transistors). **Fig 14.29** shows examples of relay and diode switching that work quite well. The relays can be inexpensive miniature RadioShack 275-241 SPDT units, one at each input and one at each output. The diodes can be inexpensive Motorola MPN3404 PIN diodes. These circuits assume that all filters are terminated with the same impedance values (Ref 15).

In PIN diode applications, IMD *can* be a problem with inadequate bias or excessive signal levels. The application (PIN diode and circuit) should be tested at the highest expected signal level.

One major problem involves high-level IF-output-signal leakage or BFO leakage into the input of the filter, which can produce high passband ripple and other unpleasant problems such as AGC malfunctions.

THE VLF IF RECEIVER

An approach to IF selectivity that has been used frequently over the years in both home-built and factory-made amateur receivers uses a second down conversion from an IF at, say 4 or 5 MHz or even

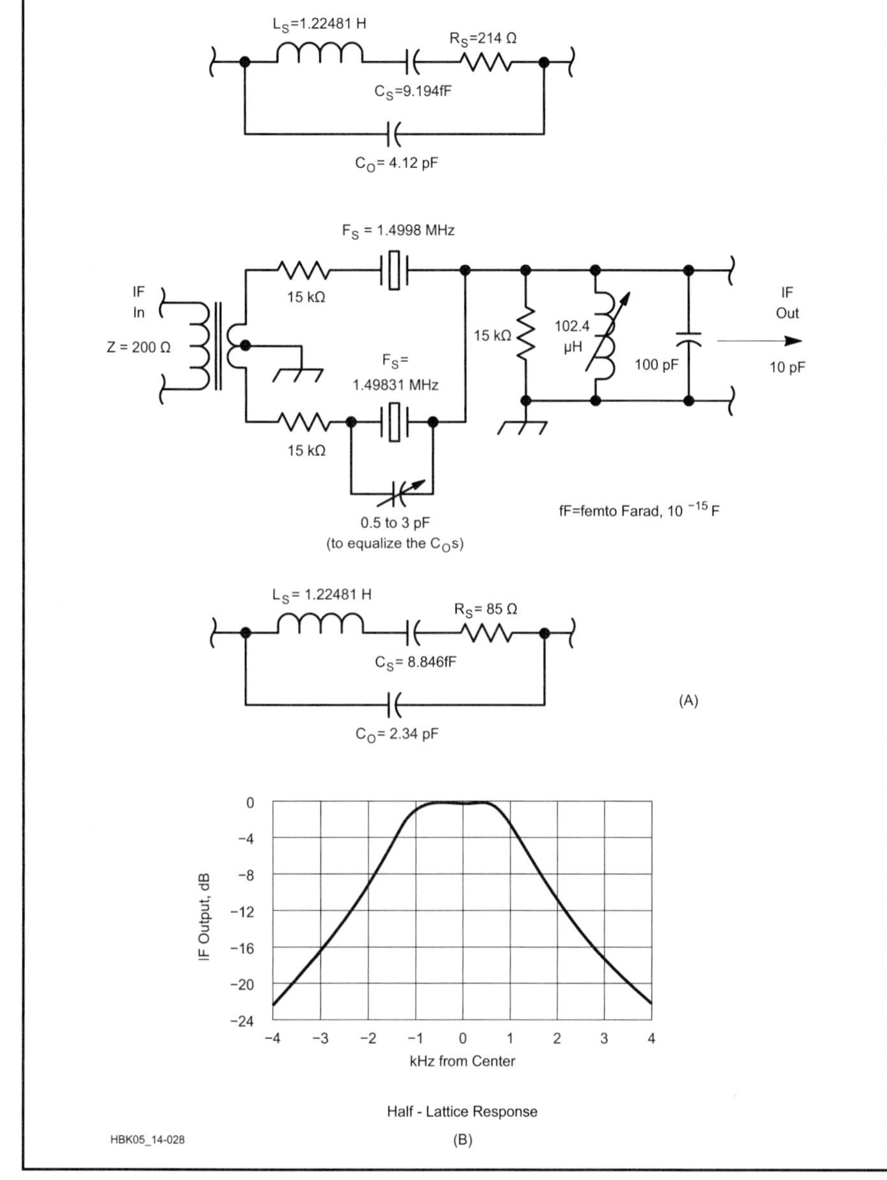

Fig 14.28—A two-crystal half-lattice filter for a simple SSB receiver. See also Fig 14.22.

455 kHz, to a very low frequency, usually 50 to 85 kHz. At these frequencies several double tuned LC filters, separated by amplifier stages, make possible excellent improvements in SSB/CW frequency response and ultimate attenuation along with a relatively flat group delay. These amplifier stages can also have AGC. Low-cost (four-pole) crystal filters (SSB and CW) at the higher IF followed by two lower-IF channels (SSB and CW) make a very desirable combination. This is also an effective way to assure a narrow noise bandwidth for the overall receiver. One requirement is that the circuitry ahead of the VLF downconverter must provide good rejection of an image frequency that is only 100 to 170 kHz away (Ref 20).

AM DEMODULATION

There is some interest among amateurs in double-sideband AM reception on the HF broadcast bands. Coherent AM detection is a way to reduce audio distortion that is caused by a temporary reduction of the carrier. This "selective fading" is due to phase cancellations caused by multipath propagation. By inserting a large, locally created carrier onto the signal this effect is reduced. The term "exalted carrier reception" is sometimes used. In reception of a double-sideband AM signal, the phase of the inserted carrier must be identical to that of the incoming carrier. If not, reduced audio and also audio distortion result. Therefore the common method is to use a phase-locked loop (PLL) to coordinate the phases of the incoming carrier and the locally generated carrier. This requires a Type II PLL, which drives, or integrates, the phase difference to zero degrees (Ref 21).

MULTIPLE CONVERSION SUPERHETS

There are a couple of drawbacks to the downconverting receiver just described. First, the LO must be bandswitched. Also, its tuning must track with the preselector tuning even though the preselector is offset from the signal frequency by the amount of the IF. A tuning dial scale is required for each band, and the receiver must be fitted to it at the factory. This adds a lot of cost and complexity.

A solution to these problems is shown in **Fig 14.30**. A crystal controlled first mixer is preceded by a gang-tuned preselector and is followed by a wideband first IF that is 200-kHz wide. The second mixer has a VFO that tunes a 200-kHz range. To change bands, the crystal is switched and the preselector is band-switched. An additional tuned circuit removes the wideband additive noise from the crystal LO, so that it does not degrade the noise figure of the *unbalanced* mixer circuit.

One of the main design problems is to select the first IF, its bandwidth and the second mixer design so that harmonic IMD products (involving the signal, crystal frequency, first IF, second IF and VFO frequency) do not cause appreciable interference. In the example of Fig 14.30, a first IF at 2.9275 MHz (the signal frequency would be 14.2275 MHz) and a VFO at 2.7 MHz produce a fourth-order spurious response at 455 kHz, therefore the first IF filter must attenuate 2.9275 MHz sufficiently and the second mixer must reject the fourth-order response sufficiently. We have discussed the fourth-order (IF/2) response previously.

One of the main bonuses of this approach is that the tunable second LO can be very stable and accurately calibrated. This calibration is the same for any signal band. Another advantage is that the first crystal LO is very stable and has little phase noise. A third bonus is that the high value of the first IF simplifies the preselector design for good image rejection in the first mixer (Ref 22).

The second mixer is vulnerable to two-tone IMD caused by strong interfering signals that lie within, or near, the 200-kHz-wide first-IF bandwidth, and that have been amplified by the circuitry preceding it. They do not make AGC because they are outside the narrow signal filters.

This cascaded-selectivity problem, which we have discussed previously,

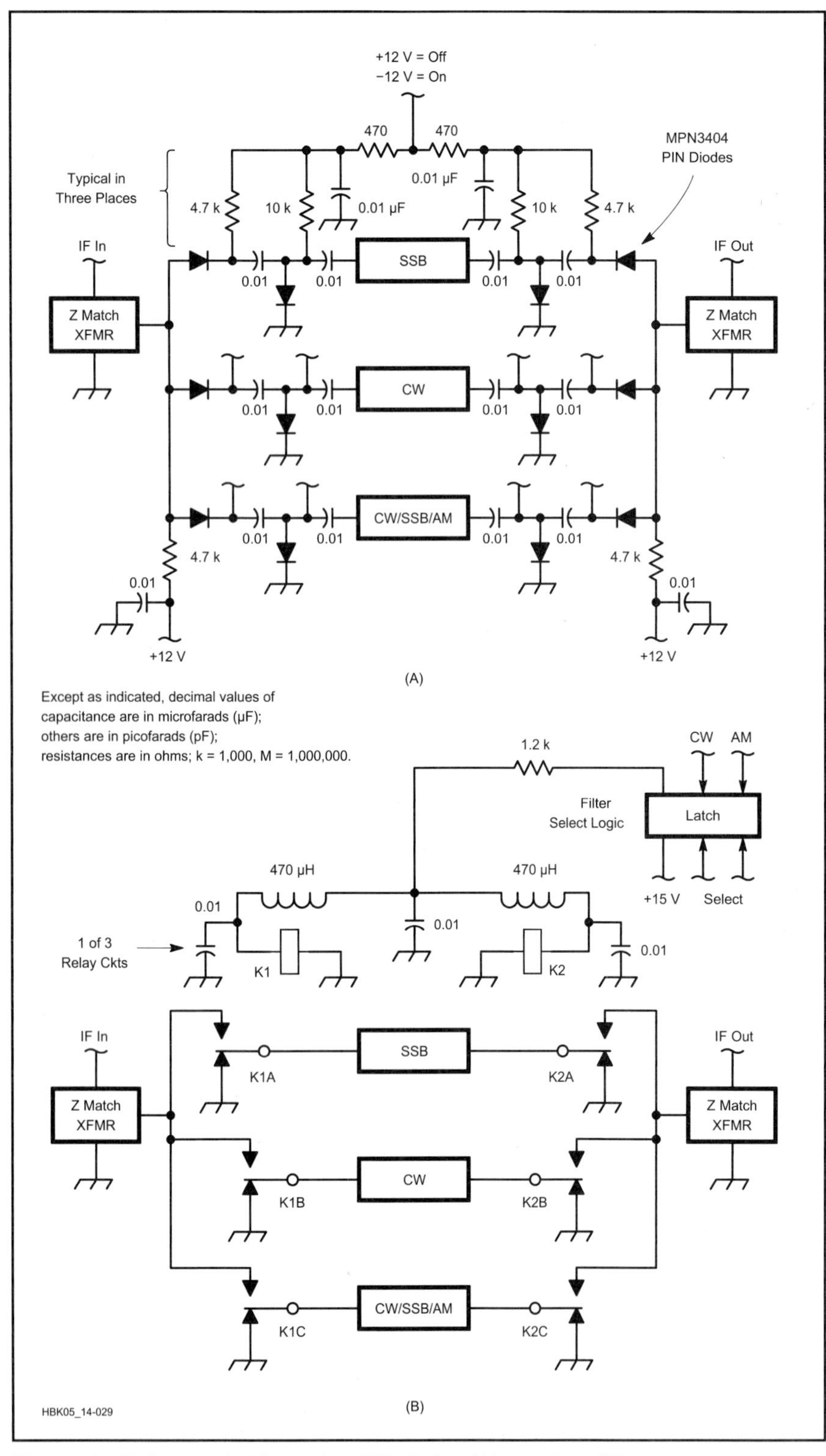

Fig 14.29—IF filter switching using PIN diodes (A) or relays (B).

makes it necessary to very carefully control the gain and noise-figure distribution ahead of the second mixer. Also, put the narrow signal filter right after the second mixer and follow that with a low-noise IF amplifier, so that "front end" gain can be minimized. In more expensive receivers of this kind, the first IF is sharply gang-tuned along with the second LO in order to reduce this problem (Ref 23).

This general approach has been extended in order to make a general-coverage receiver that has acceptable spurious responses. The first IF can be switched between two different frequency ranges and various combinations of up conversion and down conversion are used. This subject is interesting, but more complex than we can cover here. This approach is also not frequently used at this time.

THE UP CONVERSION SUPERHET

The most common approach to superhet design today is the "up converter." This designation is reserved for receivers in which the first IF is greater than the highest receive frequency. First IF values can be as low as 35 MHz for low-cost HF receivers or as high as 3 GHz for wideband receivers (and spectrum analyzers) that cover the 1 MHz to 2.5 GHz range. Let's begin by discussing the general properties of all up conversion receivers.

An Up Converter Example

The block diagram in **Fig 14.31** is one example for HF amateur SSB/CW use. The input circuit responds uniformly to a wide frequency band, 1.8 to 30 MHz. A 1.8 to 30-MHz band-pass filter is at the input. The absence of any narrow pre-selection is typical, but in difficult environments an electronically tuned or electromechanically tuned preselector is often used. Another option is a set of "half octave" (2 to 3 MHz, 3 to 4.5 MHz and so on) filters switched by PIN diodes or relays. This type of filter eliminates second-order IMD. For example, if we are listening to a weak signal at 2.00 MHz, two strong signals at 2.01 and 4.01 MHz would not create a spur at 2.00 MHz because the one at 4.01 MHz would be greatly attenuated.

Wideband Interference

The wideband circuitry in the front end is vulnerable to strong signals over the entire frequency range if no preselection is used. Therefore the strong-signal performance is a major consideration. Total receiver noise figure is usually allowed to increase somewhat in order to achieve this goal. Double balanced passive (or often active) mixers with high intercept points (second and third-order) and high LO levels are common. A typical high-quality up conversion HF receiver has a third-order intercept (IP3) of 20 to 30 dBm and a noise figure of 10 to 14 dB. High-end performers will have an IP3 of 32 to 40 dBm and a noise figure of 8 to 12 dB.

As a practical matter, in all but the most severe situations with collocated transmitters, there is very little need in Amateur Radio for the most advanced receiver specifications. One reason for this involves statistics. To get two-tone IMD interference on a *high-quality* receiver at some particular frequency there must be two strong sig-

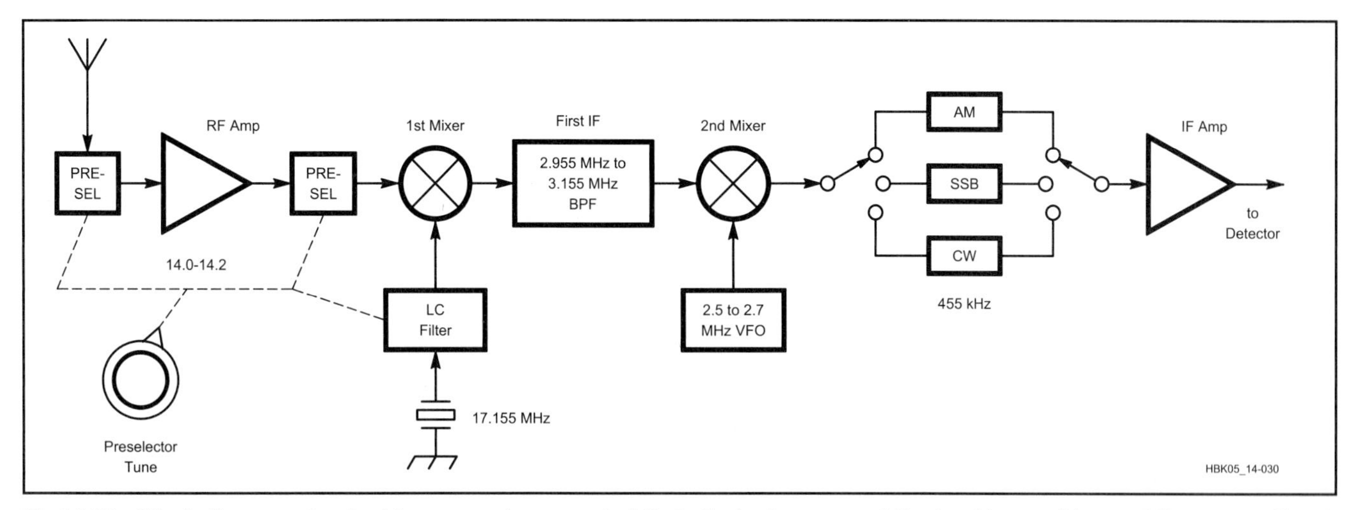

Fig 14.30—Block diagram of a double-conversion superhet that eliminates some of the tracking problems of the conventional superhet.

nals, or perhaps one very strong and a second weaker signal, on just the right pair of frequencies and at the same time. In nearly all cases, the "chances" of this are small. In Amateur Radio contest situations, these kinds of interactions are more probable. For persistent cases, other remedies are usually available.

After the Mixer

We would really like to go from the mixer directly into an SSB or CW filter, but at the high frequency of the first IF this is not realistic. Therefore we run into a major compromise: It is necessary to have at least one additional wide-band frequency conversion before getting to the narrow filters. The first IF filter can be as narrow as cost and technology will permit. In the 35 to 110-MHz range crystal filters with bandwidths of 10 to 20 kHz are available, but they are somewhat expensive in small quantities. Fig 14.31 shows an option with far less cost. The LC filter in the first IF is about 1.0-MHz wide but it has enough attenuation at 50 MHz to yield excellent image rejection in the second mixer. If we use a high-input-intercept, low-gain, low-noise amplifier followed by a strong second mixer (minimize the gain ahead of the second mixer and let the receiver noise figure go up a couple of dB) the overall receiver performance will be excellent, especially with the kinds of efficient antennas that amateurs use.

Terminating the Mixers

In the upconversion receiver, getting a pure wideband resistive termination for the mixer IF port is a problem. The output of the first mixer in Fig 14.31 contains undesired frequencies. For example, a 10-MHz signal produces 70-MHz (desired IF), 90-MHz (image) and 80-MHz (LO leakage). For a 2-MHz signal there would be 70, 72 and 74-MHz outputs. A filter that passes 70-MHz, rejects the others and at the same time terminates the mixer resistively over a wide band is a complicated band-pass diplexer.

Usually the termination is an amplifier input impedance plus a much simpler band-pass diplexer. The amplifier input should be a pure resistance, and it may then be required to deal with the vector sum of all three products. Diode upcon-verter mixers have typically 30 to 40 dB LO-to-IF isolation. If the LO level is 23 dBm, the amplifier may be looking at –7 to –17 dBm of LO feed-through, which is fairly strong. The output of the amplifier and also the next stage must deal with these amplified values. The second mixer is much easier to terminate with a diplexer, as Fig 14.31 shows.

At lower signal and LO levels (7 dBm or less), MMIC amplifiers like the MAV-11 may provide a good termination across a wide bandwidth. However, susceptibility to IMD must be checked carefully.

At the RF terminal of the mixer, any noise at the image frequency from previous stages (such as RF amplifier, antenna or even thermal noise) must not be allowed to enter the mixer because it degrades the mixer noise figure. The RF terminal should be short circuited at the image frequency if possible.

Choosing the First IF

The choice of the first IF is a compromise between cost and performance. First, consider harmonic IMD. Published data for several high-level diode-mixer models show that if the IF is greater than three times the highest signal frequency (greater than 90 MHz for a 0 to 30-MHz receiver) the rejection of harmonics of the signal frequency increases considerably. For example 3 times an interferer at 33 MHz produces a 99 MHz IF. The input 30-MHz low-pass filter would attenuate the 33-MHz signal and so would help considerably. On the other hand, 24.75 times 4 is also 99 MHz, but the mixer does a better job of rejecting this fourth harmonic. Other spurious responses tend to improve also.

However, other factors are involved, most important of which are the LO designs for the first and second mixers. In up-converters, the LOs are invariably synthesizers whose output frequencies are phase locked to a low-frequency reference crystal oscillator. As the LO frequencies increase, two other things increase: cost and quantity of high-frequency synthesizer components, and synthesizer phase noise. Also, the exact choice is interwoven with the details of the synthesizer design. Special IFs, such as 109.350 MHz, are chosen after complex trade-off studies. The cost of the first-IF signal-path components, especially filters, tends to increase also.

For all of these reasons, the IF is quite often chosen at a lower frequency. In Fig 14.31, a 70-MHz IF is shown. Crystal filters at this frequency are widely available at reasonable cost. LC filters with a 1.0 MHz (or less) bandwidth are easy to construct and get working. A 45-MHz IF is also popular. Helical resonator filters are excellent candidates at higher IF frequencies, although they can be a bit large.

The problems associated with lower IFs can be greatly improved by using a higher performance mixer. The costs are a better mixer and a more powerful LO amplifier.

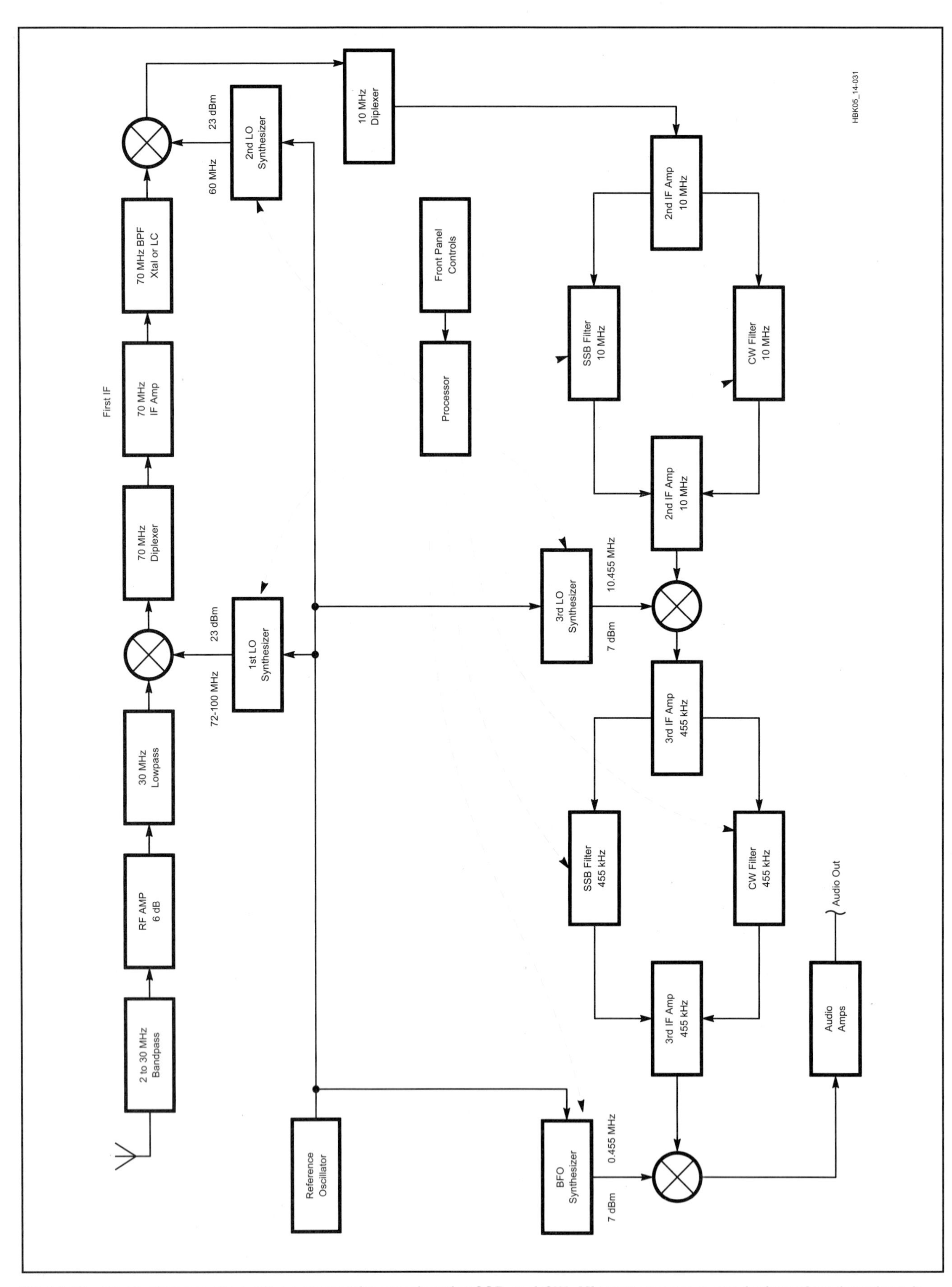

Fig 14.31—Block diagram of an HF up-conversion receiver for SSB and CW. Microprocessor control of receiver functions is included. LOs are from synthesizers.

The "Gray Area"

In the upconverter front end we encounter the "cascaded selectivity" problem that was mentioned previously in this chapter. Strong signals that are *within* the first IF passband but *outside* the SSB/CW passband (the gray area) do not make AGC and therefore are not controlled by AGC. These signals intermodulate in the mixers and in the amplifiers that precede the SSB/CW filters. It is important to realize that the receiver gray-area IMD performance is the composite of all these stages, and not just the first mixer alone. It is not until the first IF filter takes effect that things get a lot better. This degradation for strong adjacent-channel signals is an artifact of the upconverting superhet receiver. In practice, it often happens that reciprocal mixing with the adjacent channel signals (caused by receiver synthesizer phase noise) also gets into the act. Phase noise on the interfering signals is also not uncommon. The design problem is to make IMD in the gray area as small as reasonably possible. The example of Fig 14.31 is an acceptable compromise in this respect for a fairly high quality receiver.

Triple Conversion

The block diagram of Fig 14.31 shows a 10-MHz second IF. This IF is selected for two reasons:

- There are narrowband crystal filters available (including homemade ladder filters) in that vicinity.
- The image frequency for the second mixer is at 50 MHz, which can be highly attenuated by the first IF filter.

It is entirely possible to let the rest of the IF remain at this frequency, and many receivers do just that. Nonetheless, there are some advantages to having a third conversion to a lower IF. For example, it is desirable to get the large amount of IF gain needed at two different frequencies, so that stability problems and leakage from the output to input are reduced. Also, a wide variety of excellent filters are readily available at 455 kHz and other low frequencies.

Note that there is a cascade of the SSB and CW filters at the two frequencies. The desirability of this was discussed earlier and it is a powerful concept for narrowband receiver design. Placing cascaded filters at two different frequencies has another advantage that we will look into presently. For the third mixer the image frequency is at 10.910 MHz, therefore the 10-MHz filter must have good rejection at that frequency. If the first IF filter has good rejection at 70.910 MHz it will help reduce this image also. We are looking for 90 to 100 dB rejection of this potentially serious image problem. If the first IF uses a 1.0 MHz wide LC filter, as we have suggested, some additional 10 MHz LC filtering will probably be needed.

The 50 kHz IF

The third IF could be the cascaded LC filter circuits at VLF, one for SSB and one for CW, which we discussed earlier. This is an excellent approach that has been used often. The image frequency for the third mixer is only 100 kHz away, so that problem needs careful attention. Image canceling mixers have been used to help get the required 90 to 100 dB (Ref 24).

Local Oscillator (LO) Leakage

It is easy to see that with four LOs running, some at levels of 0.2 W or more, interactions between the mixers can occur. In a multiple-conversion receiver mechanical packaging, shielding, circuit placement and lead filtering are very critical areas. As one example of a problem, if the 60-MHz second LO leaks into the first mixer, a vulnerability to strong 10-MHz input signals results. It is called "IF feedthrough" because the second IF is 10 MHz. Other audible "tweets" are very common; they occur at various frequencies that involve harmonics of LOs beating together in various mixers. It is a major exercise to devise a "frequency scheme" that minimizes tweets, or at least puts them where they do not cause too much trouble (for amateurs, outside the ham bands). After that the "dog work" of reducing the remaining tweets below the noise level begins. It is a very educational experience. Synthesizers produce numerous artifacts that can also be very troublesome. It is a very common dilemma to build a receiver using "cheap" construction and poorly conceived packaging, and then try to bully the thing into good behavior.

Frequency Tuning

The synthesized first and second LOs present several different ways to tune the receiver frequency. This chapter cannot get into the details of synthesizer design, so these are only a few brief remarks. Let's discuss two options from a system-design standpoint:

- Do all of the tuning in the first LO. If steps of 10 Hz (or 1 Hz) are needed, a single-loop synthesizer that tunes in, say 500-kHz or 1.0-MHz steps, can be used. Then, a direct digital synthesizer (DDS) that tunes in 10-Hz (or 1-Hz) steps is included in the main loop in what is termed a "translation loop." The DDS frequency is added into the loop in such a manner that its imperfections are not increased by frequency multiplication. Because the reference frequency for the loop is high (500 kHz or 1.0 MHz) the phase noise of the main loop is quite small, if the loop and the circuitry are correctly designed and if the LO frequency is not extremely high. The digital frequency readout is obtained from the bits that program the synthesizer. A simpler approach might use a free-running VFO plus a low-cost frequency counter instead of the DDS. The counter can be designed to display the receiver signal frequency.
- If the first IF filter is sufficiently wideband it is possible to tune the first LO in steps of 500 kHz or 1.0 MHz and tune the second LO in 10-Hz steps. This may be a simpler method because the second LO need only be tuned over a small range.

With this second approach the first LO could be a crystal oscillator with switched overtone crystals, one for each ham band. The second LO could be a combination of crystal and VFO. One disadvantage (not extremely serious for amateur work) is that the LOs are not locked to a very accurate reference. Another is that a separate crystal (easy to get) is needed for each band. A frequency counter on the second LO could be used to get a close approximation to the signal frequency. This approach might be of interest to the home builder who is not yet ready to get involved with synthesizers. A 500-kHz crystal calibrator in the receiver would mark the band edges accurately.

Passband Tuning

While listening to a desired signal in the presence of another partially overlapping and interfering signal, whether in SSB or CW mode, it is often possible to "move" the interference at least somewhat out of the receiver passband without affecting the tune frequency (pitch) of the desired signal. In Fig 14.31, if the processor has independent fine tuning control of the second and third LOs and also the BFO, it is a matter of software design to accomplish this. It is done by controlling the overlap or intersection of the passbands of the 10-MHz filters and the 455-kHz filters. There are three things that can be done: the bandwidth can be decreased, the center frequency of the passband can be moved and both can be done simultaneously.

This scheme works best when both SSB or both CW filters are of high quality and have the same bandwidth (for example, 2.5 kHz and 500 Hz), fairly flat response and the same shape factor. As the passband is made narrower by decreasing the overlap, however, the composite shape factor is degraded somewhat (it gets larger). For CW especially, this is not detrimental. A very steep-sided response at narrow bandwidths is not desirable from a transient-response standpoint. The effect is not serious for SSB either.

Later discussion in the **Transceivers** chapter will present another method of passband tuning using a variable frequency mixer scheme rather than software control. This method is commonly found in manufactured equipment.

Noise Blanking

The desire to eliminate impulse noise from the receiver audio output has led to the development of special IF circuits that detect the presence of a noise impulse and open the signal path just long enough to prevent the impulse from getting through. Most often, a diode switch is used to open the signal path. An important design requirement is that the desired IF signal must be delayed slightly, ahead of the switch, so that the switch is opened precisely when the noise arrives at the switch. The circuitry that detects the impulse and operates the switch has a certain time delay, so the signal in the mainline IF path must be delayed also. The **Transceivers** chapter describes how a noise blanker is typically implemented. (See also Ref 17.)

VHF and UHF Receivers

The basic ideas presented in previous sections all operate equally well in receivers that are intended for the VHF and UHF regions. One difference, however, is that narrow-band frequency modulation (NBFM) is commonly used. Yes, hams do use SSB for longer distance communications because of its better weak-signal performance. However, most voice communication in this range is done using FM. This section will focus on the differences between VHF/UHF and HF receivers.

NARROWBAND FM (NBFM) RECEIVERS

Fig 14.32A is a block diagram of an NBFM receiver for the VHF/UHF amateur bands.

Front End

A low-noise front end is desirable because of the decreasing atmospheric noise level at these frequencies and also because portable gear uses short rod antennas at ground level. Nonetheless, the possibilities for gain compression and harmonic IMD, multitone IMD and cross modulation are also substantial. So dynamic range is an important design consideration, especially if large, high-gain antennas are used. FM limiting should not occur until after the NBFM signal filter. Because of the high occupancy of the VHF/UHF spectrum by powerful broadcast transmitters and nearby two-way radio services, front-end preselection is desirable, so that low noise figure can be achieved economically within the amateur band. Fig 14.32B is an example of a simple front end for the 144- to 148-MHz band.

Downconversion

Downconversion to the final IF can occur in one or two stages. Favorite IFs are in the 5 to 10-MHz region, but at the higher frequencies rejection of the image 10 to 20-MHz away can be difficult, requiring considerable preselection. So at the higher frequencies an intermediate IF in the 30 to 50-MHz region is a better choice. Fig 14.32A shows dual down-conversion.

IF Filters

The customary frequency deviation in amateur NBFM is about 5 kHz RMS (7-kHz peak) and the audio speech band extends to 5 kHz. This defines a minimum modulation index (defined as the deviation ratio) of 7/5 = 1.4. An inspection of the Bessel function plots shows that this condition confines most of the 300 to 5000-Hz speech information sidebands within a 12 to 15-kHz-wide bandwidth filter. With this bandwidth, channel separations of 20 or 25 kHz are achievable.

Many amateur NBFM transceivers are channelized in steps that can vary from 5 to 25 kHz. For low distortion of the audio output (after FM detection), this filter should have good phase linearity across the bandwidth. This would seem to preclude filters with very steep descent outside the passband, that tend to have very nonlinear phase near the band edges. But since the amount of energy in the higher speech frequencies is naturally less, the actual distortion due to this effect may be acceptable for speech purposes. A possible qualifier to this may be when pre-emphasis of the higher speech frequencies occurs at the transmitter and de-emphasis compensates at the receiver (a commonly found feature).

Limiting

After the filter, hard limiting of the IF is needed to remove any amplitude components. In a high-quality receiver, special attention is given to any nonlinear phase shift that might result from the limiter circuit design. This is especially important in phase-coherent data receivers. In amateur receivers for speech it may be less important. Also, the "ratio detector" (see the **Mixers, Modulators and Demodulators** chapter) largely eliminates the need for a limiter stage, although the limiter approach is probably still preferred.

FM Detection

The discussion of this subject is deferred to the **Mixers, Modulators and Demodulators** chapter. Quadrature detection is used on some popular NBFM multistage ICs. An example receiver chip will be presented later. Also see the **Transceivers** chapter.

NBFM Receiver Weak-Signal Performance

The noise bandwidth of the IF filter is not much greater than twice the audio bandwidth of the speech modulation, as it would be in wideband FM. Therefore such things as capture effect, the threshold effect and the noise quieting effect so familiar to wideband FM are still operational, but somewhat less so, in NBFM. For NBFM receivers, sensitivity is specified in terms of a SINAD (see the **Test Procedures** chapter) ratio of 12 dB. Typical values are –110 to –125 dBm, depending on the low-noise RF preamplification that often can be selected or deselected (in strong signal environments).

LO Phase Noise

In an FM receiver, LO phase noise superimposes phase modulation, and therefore frequency modulation, onto the

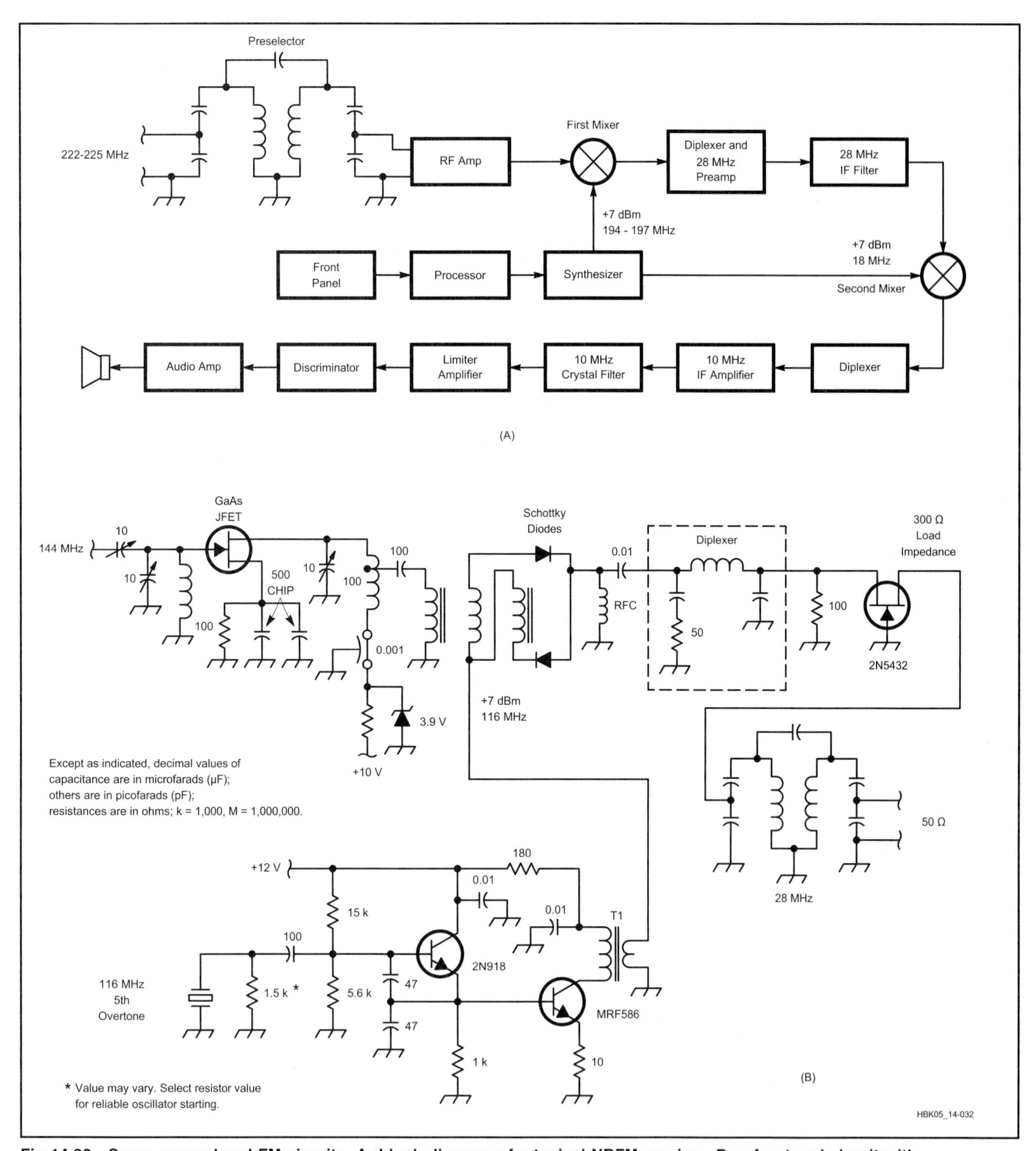

Fig 14.32—Some narrowband FM circuits. A: block diagram of a typical NBFM receiver. B: a front-end circuit with preselection and down conversion.

desired signal. This reduces the ultimate signal-to-noise ratio within the passband. This effect is called "incidental FM (IFM)." The power density of IFM (W/Hz) is equal to the phase noise power density (W/Hz) multiplied by the square of the frequency offset from the carrier (the familiar parabolic effect in FM). If the receiver uses high-frequency deemphasis at the audio output (–6 dB per octave from 300 to 3000 Hz, a common practice), the IFM level at higher audio frequencies can be reduced. Levels of total (integrated) IFM from 10 to 50 Hz are high quality for amateur voice work. Ordinarily, as the signal increases the noise would be "quieted" (that is, "captured") in an FM receiver, but in this case the signal and the phase noise riding "piggy back" on the signal increase in the same proportion. As the signal becomes large the signal-to-noise ratio therefore approaches some final value (Ref 8). A similar ultimate SNR effect occurs in SSB receivers. On the other hand, a perfect AM receiver tends to suppress LO phase noise (Ref 9).

NBFM ICs

A wide variety of special ICs for NBFM receivers are available. Many of these were designed for "cordless" or cellular telephone applications and are widely used. **Fig 14.33** shows some popular versions for a 50-MHz NBFM receiver. One is an RF amplifier chip (NE/SA5204A) for 50-Ω input to 50-Ω output with 20 dB of gain. This gain should be reduced to perhaps 8 dB. The second chip (NE/SA602A) is a front-end device (Ref 25) with an RF amplifier, mixer and LO. The third is an IF amplifier, limiter and quadrature NBFM detector (NE/SA604A) that also has a very useful RSSI (logarithmic received signal strength indicator) output and also a "mute" function. The fourth is the LM386, a widely used audio-amplifier chip. Another NBFM receiver chip, complete in one package, is the MC3371P.

The NE/SA5204A plus the two tuned circuits help to improve image rejection. An alternative would be a single double tuned filter with some loss of noise figure. The Mini-Circuits MAR/ERA series of MMIC amplifiers are excellent devices also. The crystal filters restrict the noise bandwidth as well as the signal bandwidth. A cascade of two low-cost filters is suggested by the vendors. Half-lattice filters at 10 MHz are shown, but a wide variety of alternatives, such as ladder networks, are possible.

Another recent IC is the MC13135, which features double conversion and two IF amplifier frequencies. This allows more gain on a single chip with less of the cross-coupling that can degrade stability. This desirable feature of multiple down-conversion was mentioned previously in this chapter.

The diagram in Fig 14.33 is (intentionally) only a general outline that shows how chips can be combined to build complete equipment. The design details and specific parts values can be learned from a careful study of the data sheets and application notes provided by the IC vendors. Amateur designers should learn how to use these data sheets and other information. The best places to learn about data sheets are data books and application notes.

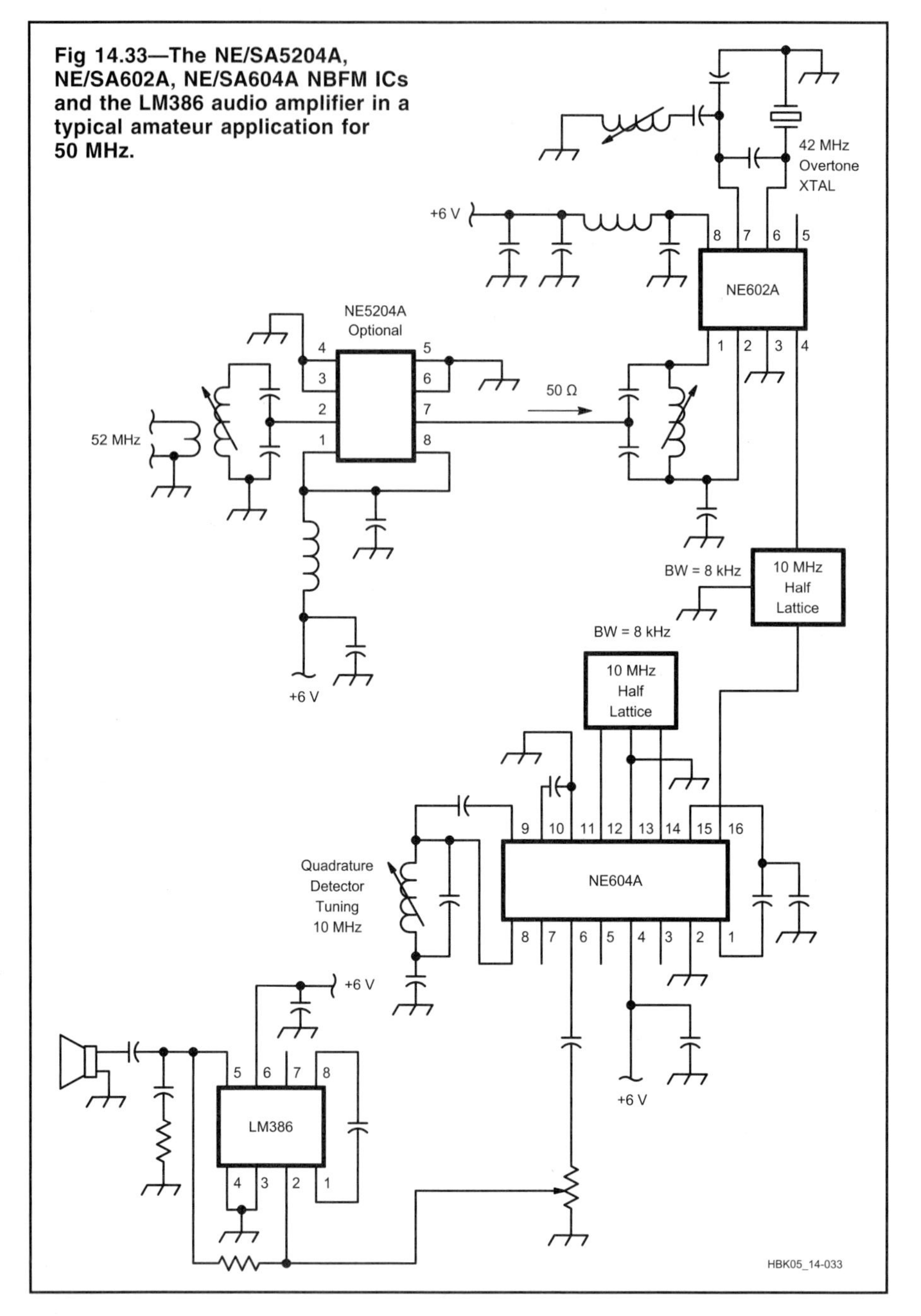

Fig 14.33—The NE/SA5204A, NE/SA602A, NE/SA604A NBFM ICs and the LM386 audio amplifier in a typical amateur application for 50 MHz.

UHF TECHNIQUES

The Ultra High Frequency spectrum comprises the range from 300 MHz to 3 GHz. All of the basic principles of radio system design and circuit design that have been discussed so far apply as well in this range, but the higher frequencies require some special thinking about the methods of circuit design and the devices that are used.

GAAS FET PREAMP FOR 430 MHZ

Fig 14.34 shows the schematic diagram and the physical construction of a typical RF circuit at 430 MHz. It is a GaAsFET preamplifier intended for low noise Earth-Moon-Earth or satellite reception. The construction uses chip capacitors, small helical inductors and a stripline surface-mount GaAsFET, all mounted on a G10 (two layers of copper) glass-epoxy PC board. The very short length of interconnection leads is typical. The bottom of the PC board is a ground plane. At this frequency, lumped components are still feasible, while microstrip circuitry tends to be rather large.

At higher frequencies, microstrip methods become more desirable in most cases because of their smaller dimensions. However, the advent of tiny chip capacitors and chip resistors have extended the frequency range of discrete components. For example, the literature shows methods of building LC filters at as high as 2 GHz, using chip capacitors and tiny helical inductors. Commercially available amplifier and mixer circuits operate at 2 GHz,

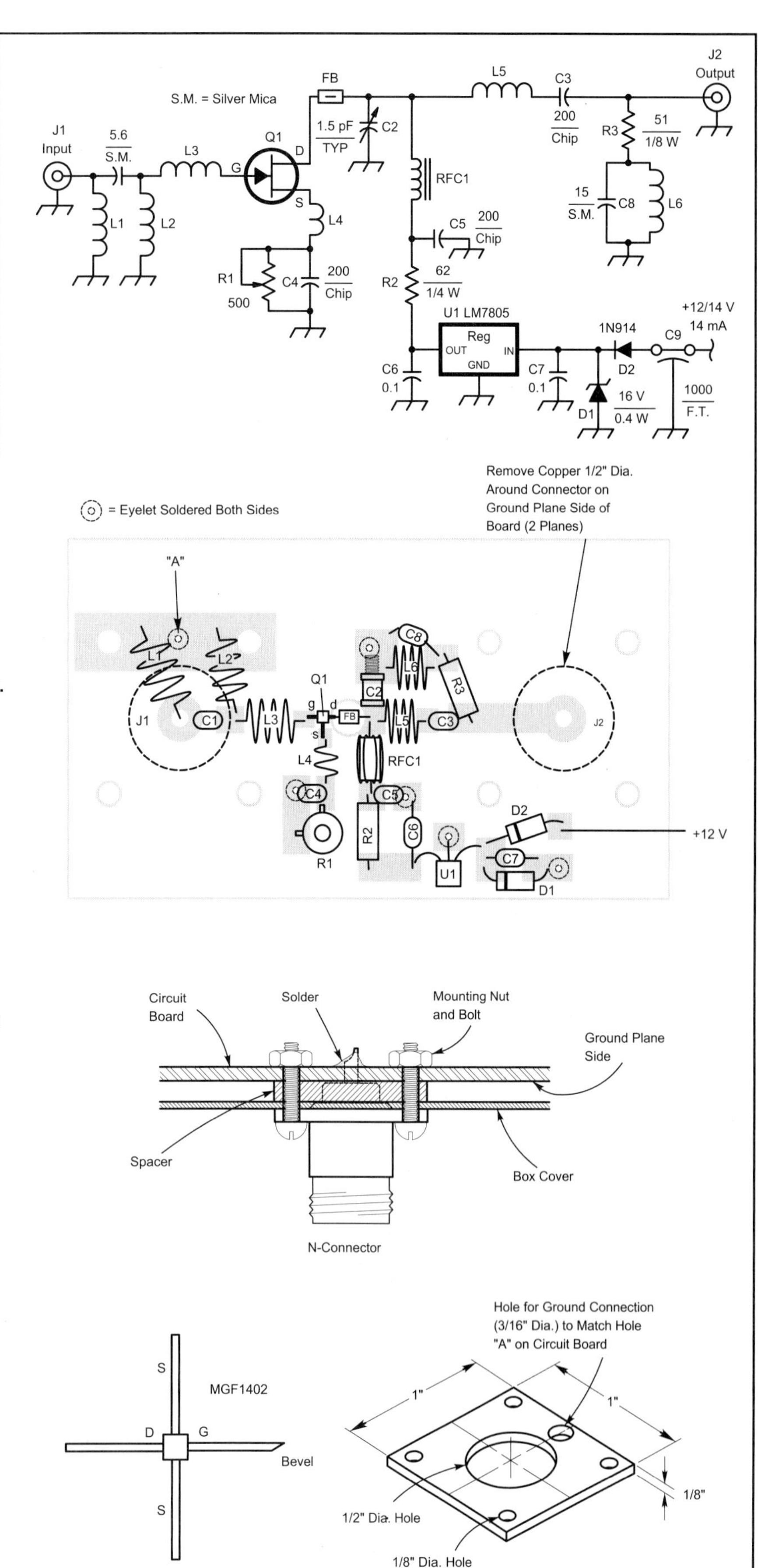

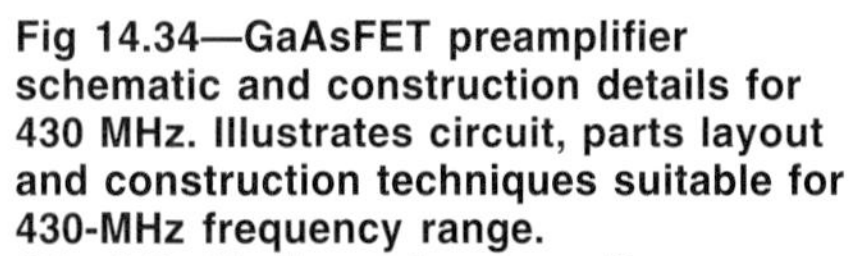
Fig 14.34—GaAsFET preamplifier schematic and construction details for 430 MHz. Illustrates circuit, parts layout and construction techniques suitable for 430-MHz frequency range.

C1—5.6-pF silver-mica capacitor or same as C2.
C2—0.6- to 6-pF ceramic piston trimmer capacitor (Johanson 5700 series or equiv).
C3, C4, C5—200-pF ceramic chip capacitor.
C6, C7—0.1-µF disc ceramic capacitor, 50 V or greater.
C8—15-pF silver-mica capacitor.
C9—500- to 1000-pF feedthrough capacitor.
D1—16- to 30-V, 500-mW Zener diode (1N966B or equiv).
D2—1N914, 1N4148 or any diode with ratings of at least 25 PIV at 50 mA or greater.
J1, J2—Female chassis-mount Type-N connectors, PTFE dielectric (UG-58 or equiv).
L1, L2—3t, #24 tinned wire, 0.110-inch ID spaced 1 wire diam.
L3—5t, #24 tinned wire, 3/16-inch ID, spaced 1 wire diam. or closer. Slightly larger diameter (0.010 inch) may be required with some FETs.
L4, L6—1t #24 tinned wire, 1/8-inch ID.
L5—4t #24 tinned wire, 1/8-inch ID, spaced 1 wire diam.
Q1—Mitsubishi MGF1402.
R1—200- or 500-Ω cermet potentiometer (initially set to midrange).
R2—62-Ω, 1/4-W resistor.
R3—51-Ω, 1/8-W carbon composition resistor, 5% tolerance.
RFC1—5t #26 enameled wire on a ferrite bead.
U1—5-V, 100-mA 3-terminal regulator (LM78L05 or equiv. TO-92 package).

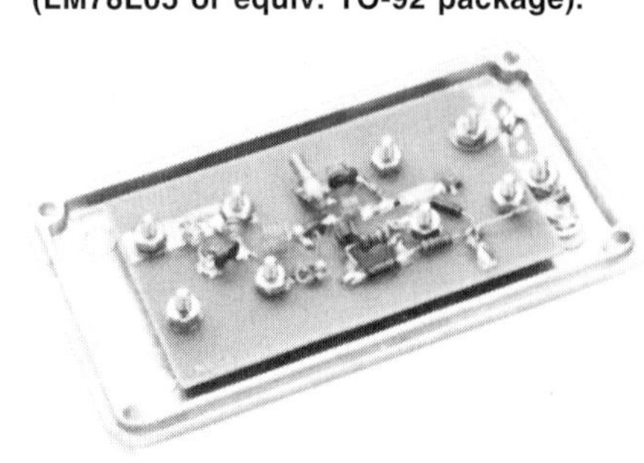

using these types of components on ceramic substrates.

ICs for UHF

In recent years a wide variety of highly miniaturized monolithic microwave ICs (MMIC) have become available at reasonable cost. Among these are the Avantek MODAMP and the Mini Circuits MAR and MAV/ERA lines. Designer kits containing a wide assortment of these are available at reasonable cost. They come in a wide variety of gains, intercepts and noise figures for frequency ranges from dc to 2 GHz. A more expensive option, for more sophisticated receiver applications, are the hybrid "cascadable" amplifiers, built on ceramic substrates and mounted in TO5 or TO8 metal cans. Most of these circuits are intended for a 50-Ω to 50-Ω interface.

A wide variety of hybrid amplifiers, designed for the Cable TV industry, are available for the frequency range from 1 to 1200 MHz (for example, the Motorola CA series, in type 714x packages). These have gains from 15 to 35 dB, output 1-dB compression points from 22 to 30 dBm and noise figures from 4.5 to 8.5 dB. Such units are excellent alternatives to discrete home-brew circuits for many applications where very low noise figures are not needed. In small quantities they may be a bit expensive sometimes, but the total cost of a home-built circuit, including labor (even the amateur experimenter's time is not really "free") is often at least as great. Home-built circuits do, however, have very important educational value.

UHF Design Aids

Circuit design and evaluation at the higher frequencies usually require some kind of minimal lab facilities, such as a signal generator, a calibrated noise generator and, hopefully, some kind of simple (or surplus) spectrum analyzer. This is true because circuit behavior and stability depend on a number of factors that are difficult to "guess at," and intuition is often unreliable. The ideal instrument is a vector network analyzer with all of the attachments (such as an S-parameter measuring setup), but very few amateurs can afford this. Another very desirable thing would be a circuit design and analysis program for the personal computer. Software packages created especially for UHF and microwave circuit design are available. They tend to be somewhat expensive, but worthwhile for a serious designer. Inexpensive SPICE programs are a good compromise. *ARRL Radio Designer* is an excellent, low cost choice.

A 902 to 928-MHz (33-cm) Receiver

Fig 14.35A is a block diagram of a 902-MHz down-converting receiver. A cavity resonator at the antenna input provides high selectivity with low loss. The first RF amplifier is a GaAsFET. Two additional 902-MHz band-pass microstrip filters and a BFR96 transistor provide more gain and image rejection (at RF—56 MHz) for the Mini Circuits SRA12 mixer. The output is at 28.0 MHz.

Cumulative Noise Figure

Fig 14.35B shows the cumulative noise figure (NF) of the signal path, including the 28-MHz receiver. The 1.5-dB cumulative NF of the input cavity and first-RF-amplifier combination, considered by itself, is degraded to 1.9 dB by the rest of the system following the first RF amplifier. The NF values of the various components for this example are reasonable, but may vary somewhat for actual hardware. Also, losses prior to the input, such as transmission-line losses (very important), are not included. They would be part of the complete receive-system analysis, however. It is common practice to place a low-noise preamp outdoors, right at the antenna, to overcome coax loss (and to permit use of inexpensive coax).

Local Oscillator (LO)Design

The +7 dBm LO at 874 to 900 MHz is derived from a set of crystal oscillators and frequency multipliers, separated by band-pass filters. These filters prevent a

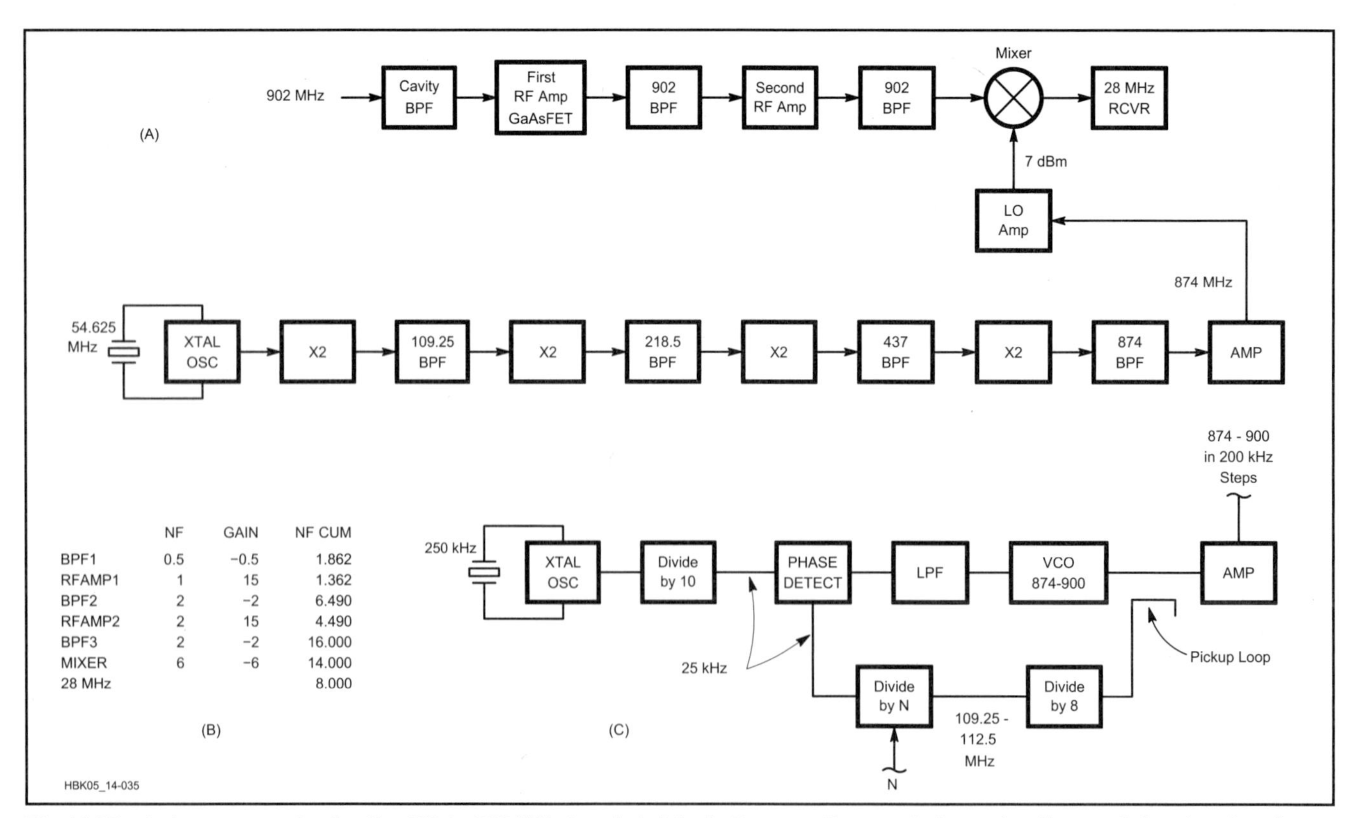

Fig 14.35—A down converter for the 902 to 928 MHz band. A: block diagram; B: cumulative noise figure of the signal path; C: alternative LO multiplier using a phase locked loop.

wide assortment of spurious frequencies from appearing at the mixer LO port. They also enhance the ability of the doubler stage to generate the second harmonic. That is, they have a very low impedance at the input frequency, thereby causing a large current to flow at the fundamental frequency. This increases the nonlinearity of the circuit, which increases the second-harmonic component. The higher filter impedance at the second harmonic produces a large harmonic output.

For very narrow-bandwidth use, such as EME, the crystal oscillators are often oven controlled or otherwise temperature compensated. The entire LO chain must be of low-noise design and the mixer should have good isolation from LO port to RF port (to minimize noise transfer from LO to RF).

A phase-locked loop using GHz-range prescalers (as shown in Fig 14.35C) is an alternative to the multiplier chain. The divide-by-N block is a simplification; in practice, an auxiliary dual-modulus divider in a "swallow count" loop would be involved in this segment. The cascaded 902-MHz band-pass filters in the signal path should attenuate any image frequency noise (at RF–56 MHz) that might degrade the mixer noise figure.

Summary

This example is fairly typical of receiver design methods for the 500 to 3000 MHz range, where down-conversion to an existing HF receiver is the most-convenient and cost-effective approach for amateurs. At higher frequencies a double down conversion with a first IF of 200 MHz or so, to improve image rejection, might be necessary. Usually, though, the presence of strong signals at image frequencies is less likely. Image-reducing mixers plus down conversion to 28 MHz is also coming into use, when strong interfering signals are not likely at the image frequency.

"No-Tune" Techniques

In recent years, a series of articles have appeared that emphasize simplicity of construction and adjustment. The use of printed-circuit microstrip filters that require little or no adjustment, along with IC

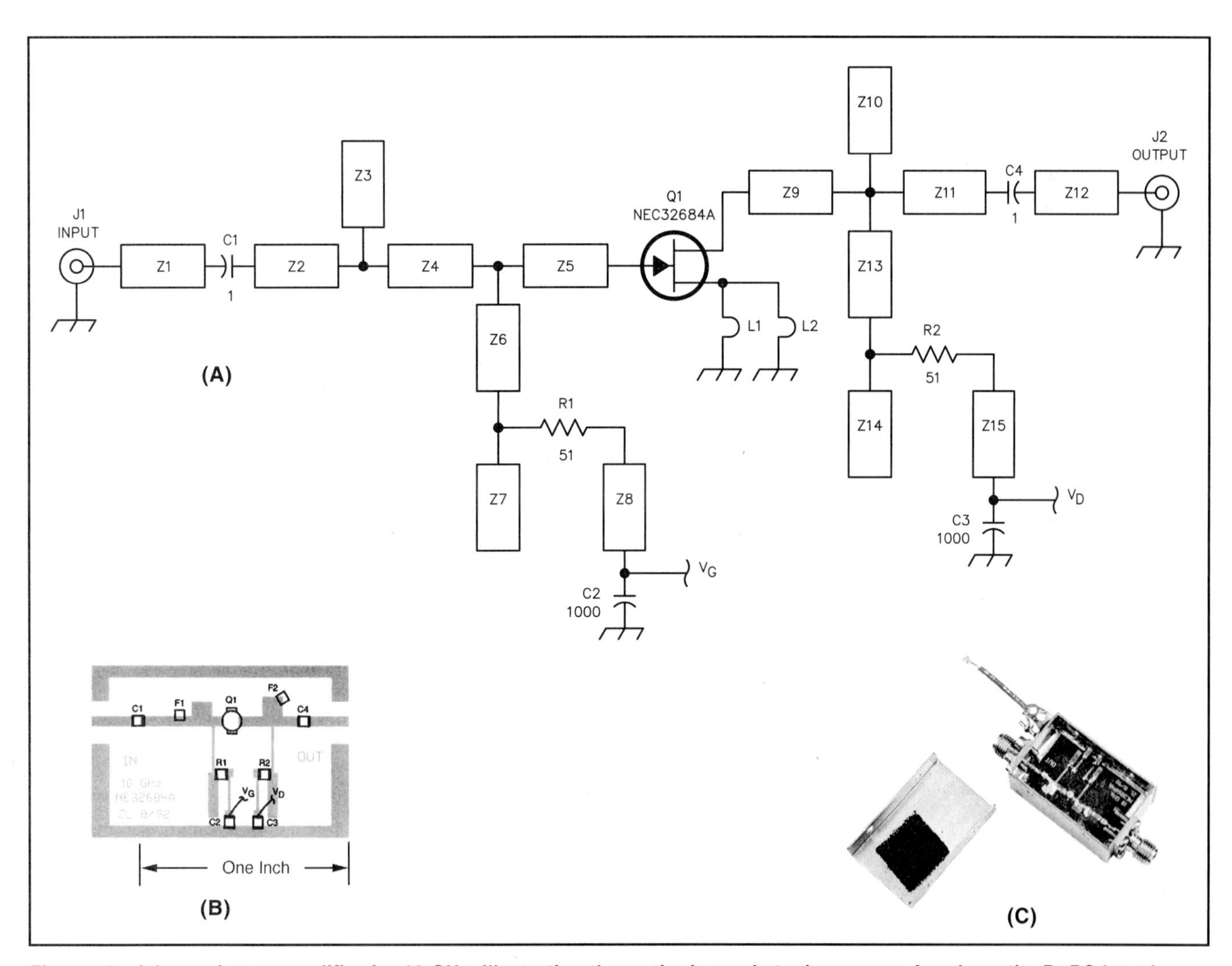

Fig 14.36—A low-noise preamplifier for 10 GHz, illustrating the methods used at microwaves. A: schematic. B: PC board layout. Use 15-mil 5880 Duroid, dielectric constant of 2.2 and a dissipation factor of 0.0011. For a negative of the board write, phone or e-mail the Technical Department Secretary at ARRL HQ and request the template from the December 1992 *QEX* "RF" column. C: A photograph of the completed preamp.

C1, C4—1-pF ATC 100-A chip capacitors. C1 must be very low loss.
C2, C3—1000-pF chip capacitors. (Not critical.) The ones from Mini Circuits work fine.
F1, F2—Pieces of copper foil used to tune the preamp.
J1, J2—SMA jacks. Ideally these should be microstrip launchers. The pin should be flush against the board.
L1, L2—The 15-mil lead length going through the board to the ground plane.
R1, R2—51-Ω chip resistors.
Z1-Z15—Microstriplines etched on the PC board.

or MMIC devices, or discrete transistors, in precise PC-board layouts that have been carefully worked out, make it much easier to "get going" on the higher frequencies. Several of the **References** at the end of this discussion show how these UHF and microwave units are designed and built (Refs 26, 27). See also the projects at the end of this chapter.

Microwave Receivers

The world above 3 GHz is a vast territory with a special and complex technology and an "art form" that are well beyond the scope of this chapter. We will scratch the surface by describing a specific receiver for the 10-GHz frequency range and point out some of the important special features that are unique to this frequency range.

A 10-GHz Preamplifier

Fig 14.36A is a schematic and parts list, B is a PC-board parts layout and C is a photograph for a 10-GHz preamp, designed by Zack Lau at ARRL HQ. With very careful design and packaging techniques a noise figure approaching the 1 to 1.5-dB range was achieved. This depends on an accurate 50-Ω generator impedance and noise matching the input using a microwave circuit-design program such as *Touchstone* or *Harmonica*. Note that microstrip capacitors, inductors and transmission-line segments are used almost exclusively. The circuit is built on a 15-mil Duroid PC board. In general, this kind of performance requires some elegant measurement equipment that few amateurs have. A detailed discussion appears in Ref 28; it is recommended reading. On the other hand, preamp noise figures in the 2 to 4-dB range are much easier to get (with simple test equipment) and are often satisfactory for terrestrial communication.

Articles written by those with expertise and the necessary lab facilities almost always include PC board patterns, parts lists and detailed instructions that are easily duplicated by readers. So it is possible to "get going" on microwaves using the material supplied in the articles. Microwave ham clubs and their publications are a good way to get started in microwave amateur technology.

Because of the frequencies involved, dimensions of microstrip circuitry must be very accurate. Dimensional stability and dielectric constant reliability of the boards must be very good.

System Performance

At microwaves, an estimation of system performance can often be performed using known data about the signal path terrain, atmosphere, transmitter and receivers systems. **Fig 14.37** shows a simplified example of how this works. This example is adapted from Dec 1980 *QST* (also see ARRL *UHF/Microwave Experimenter's Manual,* p 7-55). In the present context of receiver design we wish to establish an approximate goal for the receiver system, including the antenna and transmission line.

A more detailed analysis includes terrain variations, refraction effects, the Earth's curvature, diffraction effects and interactions with the atmosphere's chemical constituents and temperature gradients. *The ARRL UHF/Microwave Experimenter's Manual* is a good text for these matters.

In microwave work, where very low

Analysis of a 10.368 GHz communication link with SSB modulation:

Free space path loss (FSPL) over a 50 mile line-of-sight path (S) at F = 10.368 GHz:
FSPL = 36.6 (dB) + 20 log F (MHz) + 20 log S (Mi) = 36.6 + 80.3 + 34 = 150.9 dB.

Effective isotropic radiated power (EIRP) from transmitter:
EIRP (dBm) = P_{XMIT} (dBm) + Antenna Gain (dBi)

The antenna is a 2 ft diameter (D) dish whose gain GA (dBi) is:
GA = 7.0 + 20 log D (ft) + 20 log F (GHz) = 7.0 + 6.0 + 20.32 = 33.3 dBi

Assume a transmission-line loss L_T, of 3 dB
The transmitter power P_T = 0.5 (mW PEP) = –3 (dBm PEP)
$P_{XMIT} = P_T$ (dBm PEP) – L_T (dB) = –3 – 3 = –6 (dBm PEP)
EIRP = $P_{XMIT} + G_A$ = –6 + 33.3 = 27.3 (dBm PEP)
Using these numbers the received signal level is:
P_{RCVD} = EIRP (dBm) – Path loss (dB) = 27.3 (dBm PEP) – 150.9 (dB) = –123.6 (dBm PEP)
Add to this a receive antenna gain of 17 dB. The received signal is then P_{RCVD} = –123.6 +17 = –106.6 dBm

Now find the receiver's ability to receive the signal:
The antenna noise temperature T_A is 200 K. The receiver noise figure NF_R is 6 dB (FR=3.98, noise temperature T_R = 864.5 K) and its noise bandwidth (B) is 2400 Hz. The feedline loss L_L is 3 dB (F = 2.00, noise temperature T_L = 288.6 K). The system noise temperature is:
$T_S = T_A + T_L + (L_L)(T_R)$
T_S = 200 + 288.6 + (2.0) (864.5) = 2217.6 K
$N_S = kT_SB = 1.38 \times 10^{-23} \times 2217.6 \times 2400 = 7.34 \times 10^{-17}$ W = –131.3 dBm
This indicates that the PEP signal is –106.6 –(–131.3) = 24.7 dB above the noise level. However, because the average power of speech, using a speech processor, is about 8 dB less than PEP, the average signal power is about 16.7 dB above the noise level.

To find the system noise factor F_S we note that the system noise is proportional to the system temperature T_S and the "generator" (antenna) noise is proportional to the antenna temperature T_A. Using the idea of a "system noise factor":

$F_S = T_S / T_A$ = 2217.6 / 200 = 11.09 = 10.45 dB.

If the antenna temperature were 290 K the system noise figure would be 9.0 dB, which is precisely the sum of receiver and receiver coax noise figures (6.0 + 3.0).

Fig 14.37—Example of a 10-GHz system performance calculation. Noise temperature and noise factor of the receiver are considered in detail.

noise levels and low noise figures are encountered, experimenters like to use the "effective noise temperature" concept, rather than noise factor. The relationship between the two is given by

$$T_E = 290\ (F - 1) \quad (7)$$

T_E is a measure, in terms of temperature, of the "excess noise" of a component (such as an amplifier). A resistor at 290 + T_E would have the same available noise power as the device (referred to the device's input) specified by T_E. For a lossy device (such as a lossy transmission line) T_E is given by $T_E = 290\ (L - 1)$, where L is the loss factor (same as its noise factor). The cascade of noise temperatures is similar to the formula for cascaded noise factors.

$$TS = TG + TE1 + TE2/G1 + TE3/(G1G2) + TE4/(G1G2G3) + \ldots \quad (8)$$

where TS is the system noise temperature (including the generator, which may be an antenna) and TG is the temperature of the antenna.

The 290 number in the formulas for T_E is the standard ambient temperature (kelvins) at which the noise factor of a two-port transducer is defined and measured, according to an IEEE recommendation. So those formulas relate a noise factor F, measured at 290 K, to the temperature T_E. In general, though, it is perfectly correct to say that the ratio $(S_I/N_I)/(S_O/N_O)$ can be thought of as the ratio of total system output noise to that system output noise attributed to the "generator" alone, regardless of the temperature of the equipment or the nature of the generator, which may be an antenna at some arbitrary temperature, for example. This ratio is, in fact, a special "system noise factor (or figure), F_S" that need not be tied to any particular temperature such as 290 K. The use of the F_S notation avoids any confusion. As the example of Fig 14.37 shows, the value of this system noise factor F_S is just the ratio of the total system temperature to the antenna temperature.

Having calculated a system noise temperature, the receive system noise floor (that is, the antenna input level of a signal that would exactly equal system noise, both observed at the receiver output) associated with that temperature is:

$$N = k\ T_S\ B_N \quad (9)$$

where
$k = 1.38 \times 10^{-23}$ and
B_N = noise bandwidth

The system noise figure FS is indicated in the example also. It is higher than the sum of the receiver and coax noise figures.

The example includes a loss of 3 dB in the receiver transmission line. The formula for T_S in the example shows that this loss has a double effect on the system noise temperature, once in the second term (288.6) and again in the third term (2.0). If the receiver (or high-gain preamp with a 6 dB NF) were mounted at the antenna, the receive-system noise temperature would be reduced to 1064.5 K and a system noise figure, FS, of 7.26 dB, a very substantial improvement. Thus, it is the common practice to mount a preamp at the antenna.

MICROWAVE RECEIVER FOR 10 GHZ

Ref 29 provides a good example of modern amateur experimenter techniques for the 10-GHz band. The intended use for the radio is narrowband CW and SSB work, which requires extremely good frequency stability in the LO. Here, we will discuss the receiver circuit.

Block Diagram

Fig 14.38 is a block diagram of the receiver. Here are some important facets of the design.

1. The antenna should have sufficient gain. At 10 GHz, gains of 30 dBi are not difficult to get, as the example of Fig 14.37 demonstrates. A 4-ft dish might be difficult to aim, however.
2. For best results a very low-noise preamp at the antenna reduces loss of system sensitivity when antenna temperature is low. For example, if the antenna temperature at a quiet direction of the sky is 50 K and the receiver noise figure is 4 dB (due in part to transmission-line loss), the system temperature is 488 K for a system noise figure of 9.9 dB. If the receiver noise figure is reduced to 1.5 dB by adding a preamp at the antenna the system temperature is reduced to 170 K for a system noise figure of 3.4 dB, which is a very big improvement.
3. After two stages of RF amplification using GaAsFETs, a probe-coupled cavity resonator attenuates noise at the mixer's image frequency, which is 10.368 – 0.288 = 10.080 GHz. An image reduction of 15 to 20 dB is enough to prevent image frequency noise generated by the RF amplifiers from affecting the mixer's noise figure.
4. The singly balanced diode mixer uses a "rat-race" 180° hybrid. Each terminal of the ring is 1/4 wavelength (90°) from its closest neighbors. So the anodes of the two diodes are 180° (1/2 wavelength) apart with respect to the LO port, but in-phase with respect to the RF port. The inductors (L1, L2) connected to ground present a low impedance at the IF frequency. The mixer microstrip circuit is carefully "tweaked" to improve system performance. Use the better mixer in the transmitter.
5. The crystal oscillator is a fifth-overtone Butler citcuit that is capable of high stability. The crystal frequency error and drift are multiplied 96 times (10.224/0.1065), so for narrowband SSB or CW work it may be difficult to get on (and stay on) the "calling frequency" at 10.368 GHz. One acceptable (not perfect) solution might be to count the 106.5 MHz with a frequency counter whose internal clock is constantly compared with WWV. Adjust to 106.5 MHz as required. At times there may be a small Doppler shift on the WWV signal. It may be necessary to switch to a different WWV frequency, or WWV's signals may not be strong enough. Surplus frequency standards of high quality are sometimes available. Many operators just "tune" over the expected range of uncertainty.
6. The frequency multiplier chain has numerous band-pass filters to "purify" the harmonics by reducing various frequency components that might affect the signal path and cause spurious responses. The final filter is a tuned cavity resonator that reduces spurs from previous stages. Oscillator phase noise amplitude is multiplied by 96.0 also, so the oscillator must have very good short term stability to prevent contamination of the desired signal.
7. A second hybrid splitter provides an LO output for the transmitter section of the radio. The 50-Ω resistor improves isolation between the two output ports. The two-part *QST* article (Ref 29) is recommended reading for this very interesting project, which provides a fairly straightforward (but not extremely simple) way to get started on 10 GHz.

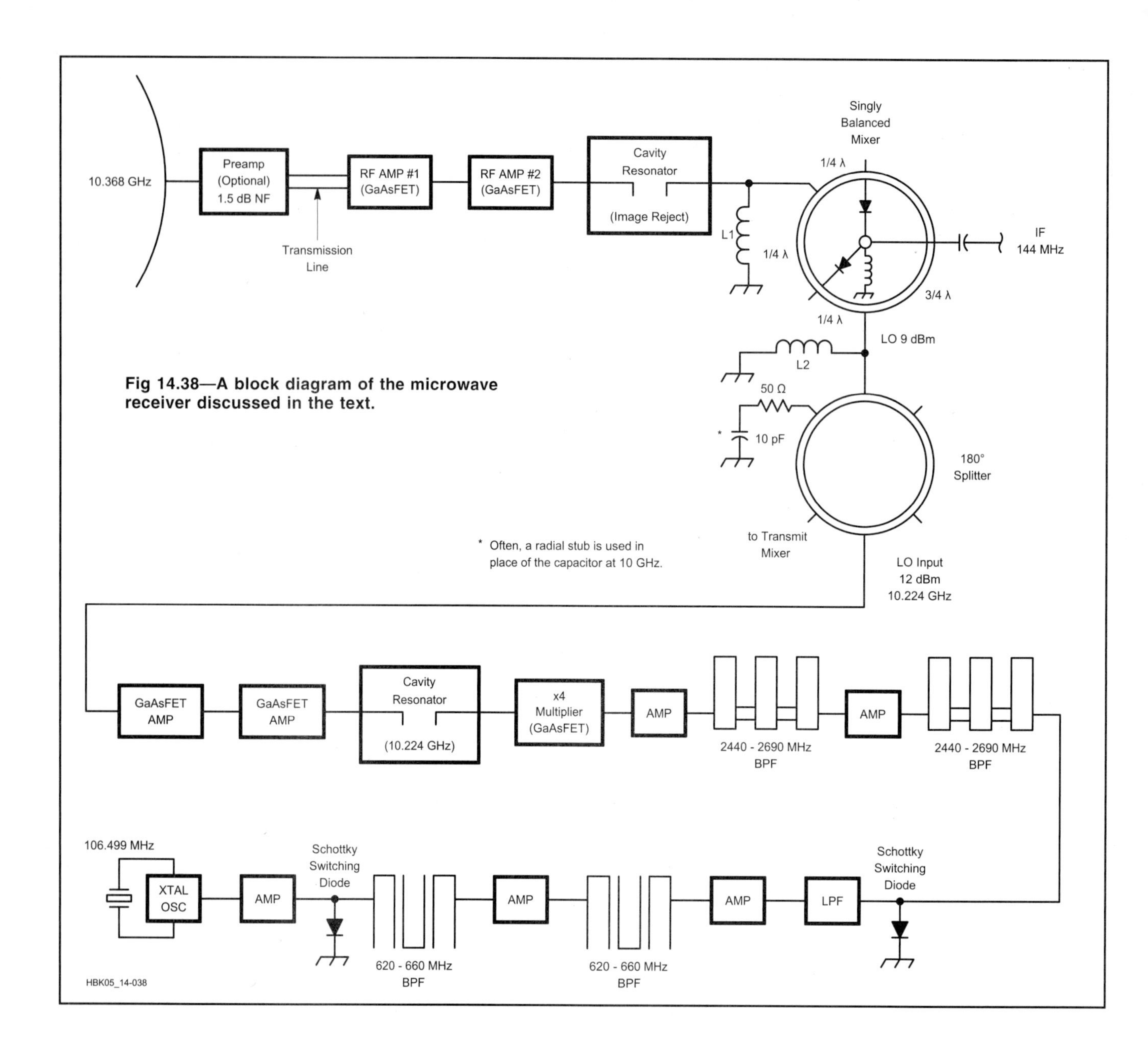

Fig 14.38—A block diagram of the microwave receiver discussed in the text.

Transmitter Design

TRANSMITTER DESIGN VS RECEIVER DESIGN

Many of the building blocks used in transmitter design are either identical to or very similar to those used in receiver design. Such things as mixers, oscillators, low-level RF/IF/AF amplifiers are the same. There is one major difference in the usage of these items, though. In a transmitter, the ratio of maximum to minimum signal levels for each of these is much less than in a receiver, where a very large ratio exists routinely. In a transmitter the signal, as it is developed to its final frequency and power level, is carefully controlled at each stage so that the stage is driven close to some optimum upper limit. The noise figures and dynamic ranges of the various stages are somewhat important, but not as important as in a receiver.

The transmitter design is concerned with the development of the desired high level of output power as cleanly, efficiently and economically as possible. Spurious outputs that create interference are a major concern. Protection circuitry that prevents self-destruction in the event of parts failures or mishandling by the operator help the reliability in ways that are unimportant in receivers.

THE SUPERHET SSB/CW TRANSMITTER

The same mixing schemes, IF frequencies and IF filters that are used for superhet receivers can be, and very often are, used for a transmitter. **Fig 14.39** is a block diagram of one approach. Let's discuss the various elements in detail,

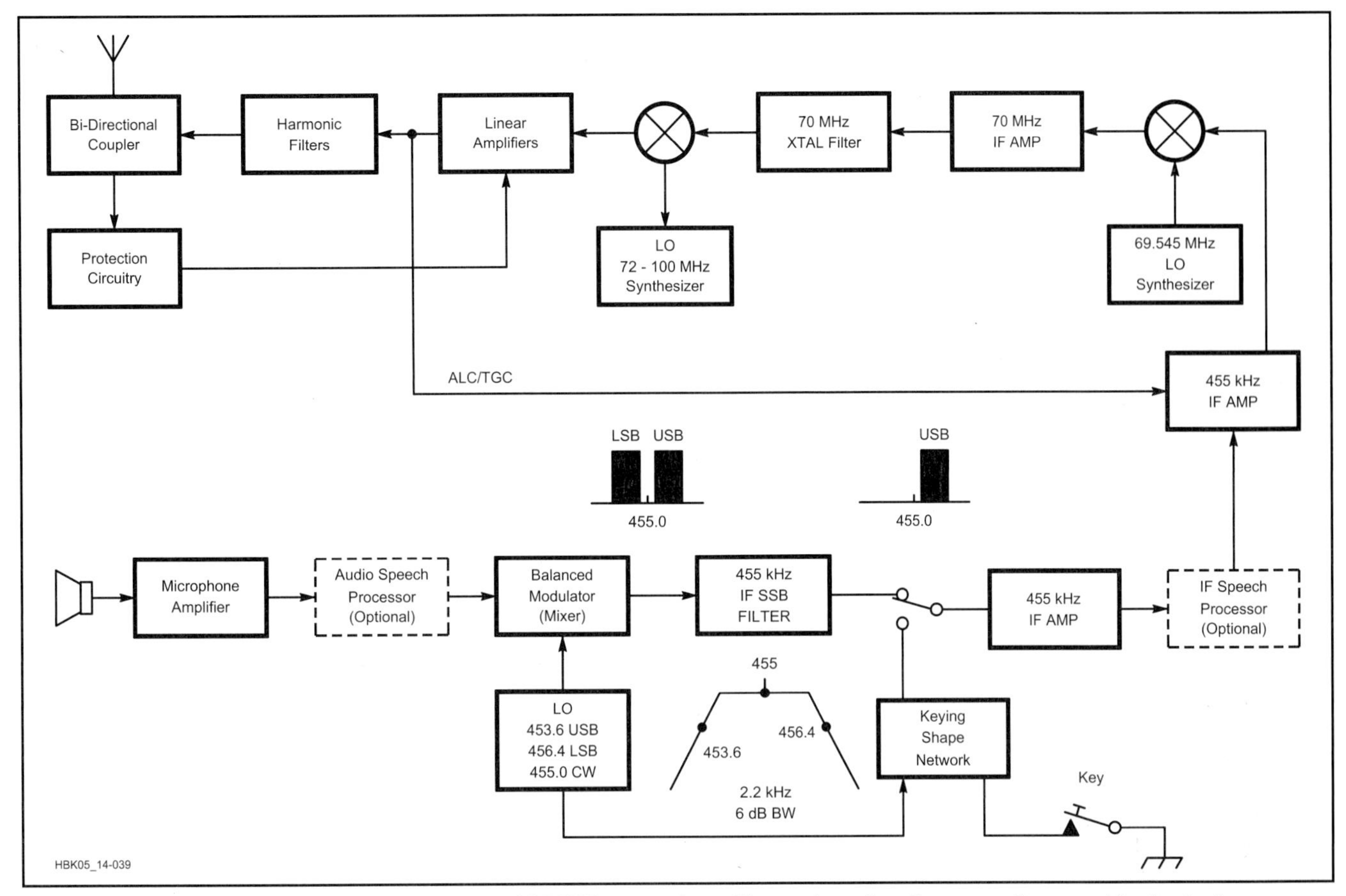

Fig 14.39—Block diagram of an up conversion SSB/CW transmitter. The various functions are discussed in the text.

starting at the microphone.

Microphones (Mics)

A microphone is a transducer that converts sound waves into electrical signals. For speech, its frequency response should be as flat as possible from below 200 to above 3500 Hz. Response peaks in the microphone can increase the peak to average ratio of speech, which then degrades (increases) the peak to average ratio of the transmitted signal. If a transmitter uses speech processing, most microphones pick up a lot of background ambient noises because the speech amplification, whether it be at audio or IF/RF, may be as much as 20 dB greater than without speech processing. A "noise canceling" microphone is recommended to reduce this background pickup. Microphone output levels vary, depending on the microphone type. Typical amateur mics produce about 10 to 100 mV.

Ceramic

Ceramic mics have high output impedances but low level outputs. They require a high-resistance load (usually about 50 kΩ) for flat frequency response and lose low-frequency response as this resistance is reduced (electrically, the mic "looks like" a small capacitor). These mics vary widely in quality, so a "cheap" mic is not a good bargain because of its effect on the transmitted power level and speech quality.

Dynamic

A dynamic mic resembles a small loudspeaker, with an impedance of about 680 Ω and an output of about 12 mV on voice peaks. In many cases a transformer (possibly built-in) transforms the impedance to 100 kΩ or more and delivers about 100 mV on voice peaks. Dynamic mics are widely used by amateurs.

Electret

"Electret" mics use a piece of special insulator material that contains a "trapped" polarization charge (Q) at its surfaces and a capacitance (C). Sound waves modulate the capacitance (dC) of the material and cause a voltage change (dV) according to the law $dV/V = -dC/C$. For small changes (dC) the change (dV) is almost linear. A polarizing voltage of about 4 V is required to maintain the charge. The mic output level is fairly low, and a preamp is sometimes required. These mics have been greatly improved in recent years.

Microphone Amplifiers

The balanced modulator and (or) the audio speech processor need a certain optimum level, which can be in the range of 0.3 to 0.6 V ac into perhaps 1 kΩ to 10 kΩ Excess noise generated within the microphone amplifier should be minimized, especially if speech processing is used. The circuit in **Fig 14.40** uses a low-noise BiFET op amp. The 620-Ω resistor is selected for a low impedance microphone, and switched out of the circuit for high-impedance mics. The amplifier gain is set by the 100-kΩ potentiometer.

It is also a good idea to experiment with the low-and high-frequency responses of the mic amplifier to compensate for the frequency response of the mic and the voice of the operator.

Audio Speech Clipping

If the audio signal from the microphone amplifier is further amplified, say by as much as 12 dB, and then if the peaks are clipped (sometimes called "slicing") by 12 dB by a speech clipper, the output peak value is the same as before the clipper, but the average value is increased considerably. The resulting signal contains harmonics and IMD but the speech intelligibility, especially in a white-noise background, is improved by 5 or 6 dB.

The clipped waveform frequently tends to have a square-wave appearance, especially on voice peaks. It is then band-pass filtered to remove frequencies below 300 and above 3000 Hz. The filtering of this signal can create a "repeaking" effect. That is, the peak value tends to increase noticeably above its clipped value.

An SSB generator responds poorly to a square-wave audio signal. The Hilbert Transform effect, well known in mathematics, creates significant peaks in the RF envelope. These peaks cause out-of-band splatter in the linear amplifiers unless Automatic Level Control (ALC, to be discussed later) cuts back on the RF gain. The peaks increase the peak-to-average ratio and the ALC reduces the average SSB power output, thereby reducing some of the benefit of the speech processing. The square-wave effect is also reduced by band-pass filtering (300 to 3000 Hz) the input to the clipper as well as the output.

Fig 14.41 is a circuit for a simple audio speech clipper. A CLIP LEVEL potentiometer before the clipper controls the amount of clipping and an OUTPUT LEVEL potentiometer controls the drive level to the balanced modulator. The correct adjustment of these potentiometers is done with a two-tone audio input or by talking into the microphone, rather than a single tone, because single tones don't exhibit the repeaking effect.

Audio Speech Compression

Although it is desirable to keep the voice level as high as possible, it is difficult to maintain constant voice intensity when speaking into the microphone. To overcome this variable output level, it is possible to use an automatic gain control that follows the average variations in speech amplitude. This can be done by rectifying and filtering some of the audio output and applying the resultant dc to a control terminal in an early stage of the amplifier.

The circuit of **Fig 14.42A** works on this AGC principle. One section of a Signetics 571N IC is used. The other section can be connected as an expander to restore the dynamic range of received signals that have been compressed in transmission. Operational transconductance amplifiers such as the CA3080 are also well suited for speech compression.

When an audio AGC circuit derives control voltage from the output signal the system is a closed loop. If short attack time is necessary, the rectifier-filter bandwidth must be opened up to allow syllabic modulation of the control voltage. This allows some of the voice frequency signal to enter the control terminal, causing distortion and instability. Because the syllabic frequency and speech-tone frequencies have relatively small separation, the simpler feedback AGC systems compromise fidelity for fast response.

Problems with loop dynamics in audio AGC can be side-stepped by eliminating the loop and using a forward-acting system. The control voltage is derived from the input of the amplifier, rather than from the output. Eliminating the feedback loop allows unconditional stability, but the trade-off between response time and fidelity remains. Care must be taken to avoid excessive gain between the signal input and the control voltage output. Otherwise the transfer characteristic can reverse; that is, an increase in input level can cause a decrease in output. A simple forward-acting compressor is shown in Fig 14.42B.

Balanced Modulators

A balanced modulator is a mixer. A more complete discussion of balanced modulator design is provided in the **Mixers** chapter. Briefly, the IF frequency LO (455 kHz in the example of Fig 14.39) translates the audio frequencies up to a pair of IF frequencies, the LO plus the audio frequency and the LO minus the audio frequency. The balance from the LO port to the IF output causes the LO frequency to

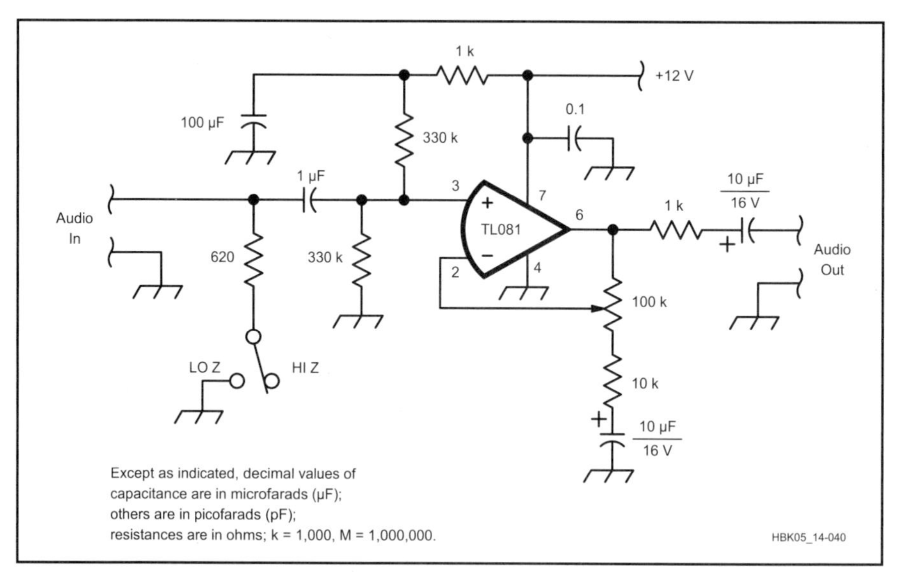

Fig 14.40—Schematic diagram of a simple op-amp microphone amplifier for low- and high-impedance microphones.

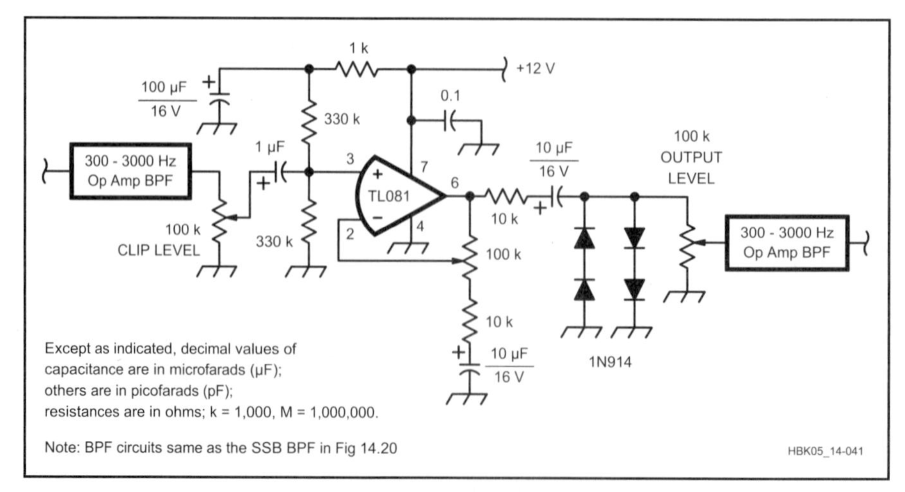

Fig 14.41—A simple audio speech clipper. The input signal is band pass filtered, amplified by 20 dB, clipped and band pass filtered again.

be suppressed by 30 to 40 dB. Adjustments are provided to improve the LO null.

The filter method of SSB generation uses an IF band-pass filter to pass one of the sidebands and block the other. In Fig 14.39 the filter is centered at 455.0 kHz. The LO is offset to 453.6 kHz or 456.4 kHz so that the upper sideband or the lower sideband (respectively) can pass through the filter. This creates a problem for the other LOs in the radio, because they must now be properly offset so that the final transmit output's carrier (suppressed) frequency coincides with the frequency readout on the front panel of the radio. Various schemes have been used to do this. One method uses two crystals for the 69.545-MHz LO that can be selected. In synthesized radios the programming of the microprocessor control moves the various LOs. Some synthesized radios use two IF filters at two different frequencies, one for USB and one for LSB, and a 455.0-kHz LO. These radios can be designed to transmit two independent sidebands (ISB, Ref 17).

In times past, balanced modulators using diodes, balancing potentiometers and numerous components spread out on a PC board were universally used. These days it doesn't make sense to use this approach. ICs and packaged diode mixers do a much better job and are less expensive. The most famous modulator IC, the MC1496, has been around for more than 20 years and is still one of the best and least expensive. **Fig 14.43** is a typical balanced modulator circuit using the MC1496.

The data sheets for balanced modulators and mixers specify the maximum level of audio for a given LO level. Higher audio levels create excessive IMD. The IF filter after the modulator removes higher-order IMD products that are outside its passband but the in-band IMD products should be at least 40 dB below each of two equal test tones. Speech clipping (AF or IF) can degrade this to 10 dB or so, but in the absence of speech processing the signal should be clean, in-band.

IF Filters

The desired IF filter response is shown in **Fig 14.44A**. The reduction of the carrier frequency is augmented by the filter response. It is common to specify that the filter response be down 20 dB at the carrier frequency. Rejection of the opposite sideband should (hopefully) be 60 dB, starting at 300 Hz below the carrier frequency, which is the 300-Hz point on the opposite sideband. The ultimate attenuation should be at least 70 dB. This would represent a very good specification for a high quality transmitter. The filter passband should be as flat as possible (ripple

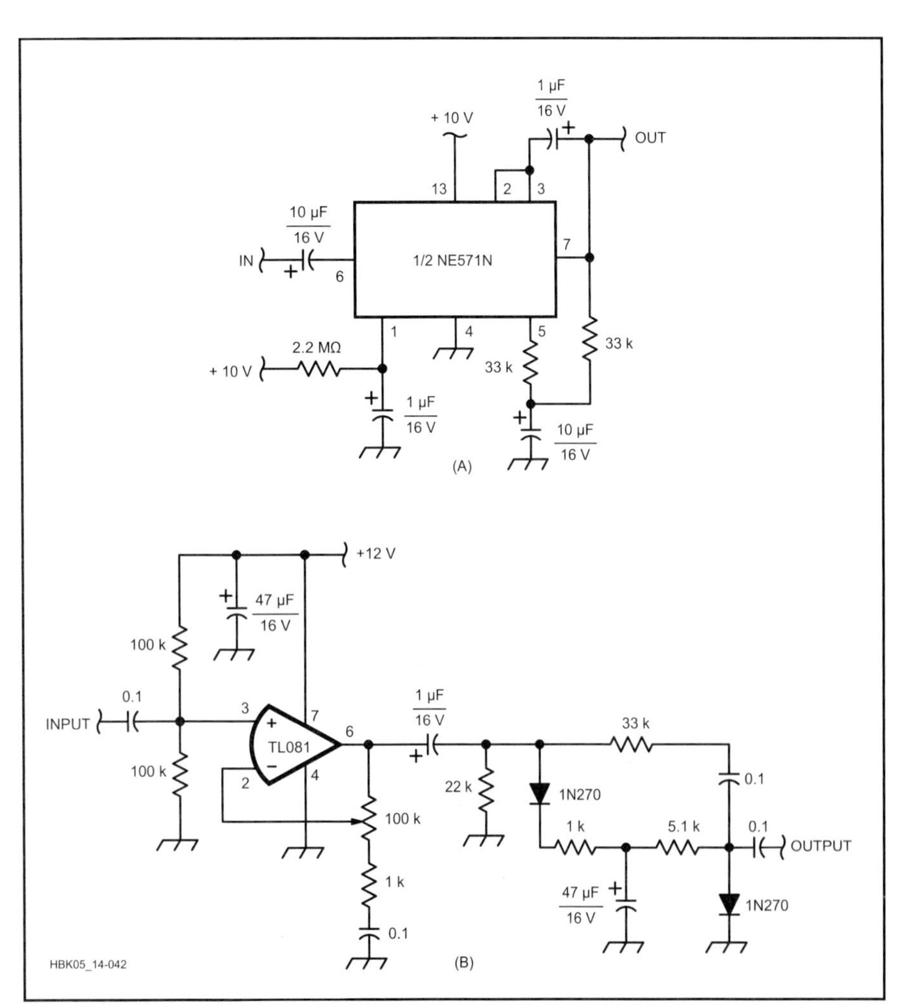

Fig 14.42—Typical solid-state compressor circuits. The circuit at A works on the AGC principle, while that at B is a forward-acting compressor.

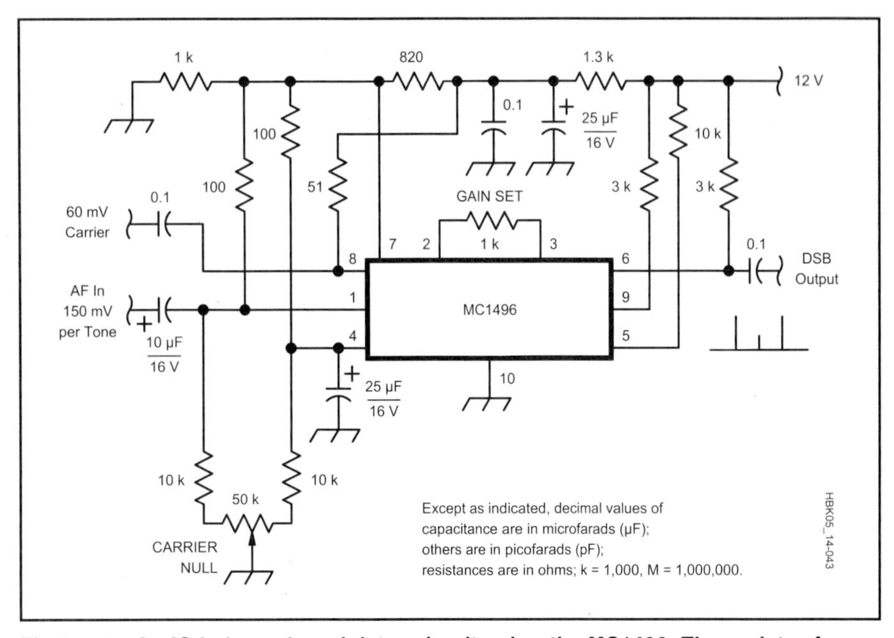

Fig 14.43—An IC balanced modulator circuit using the MC1496. The resistor from pin 2 to pin 3 sets the conversion gain.

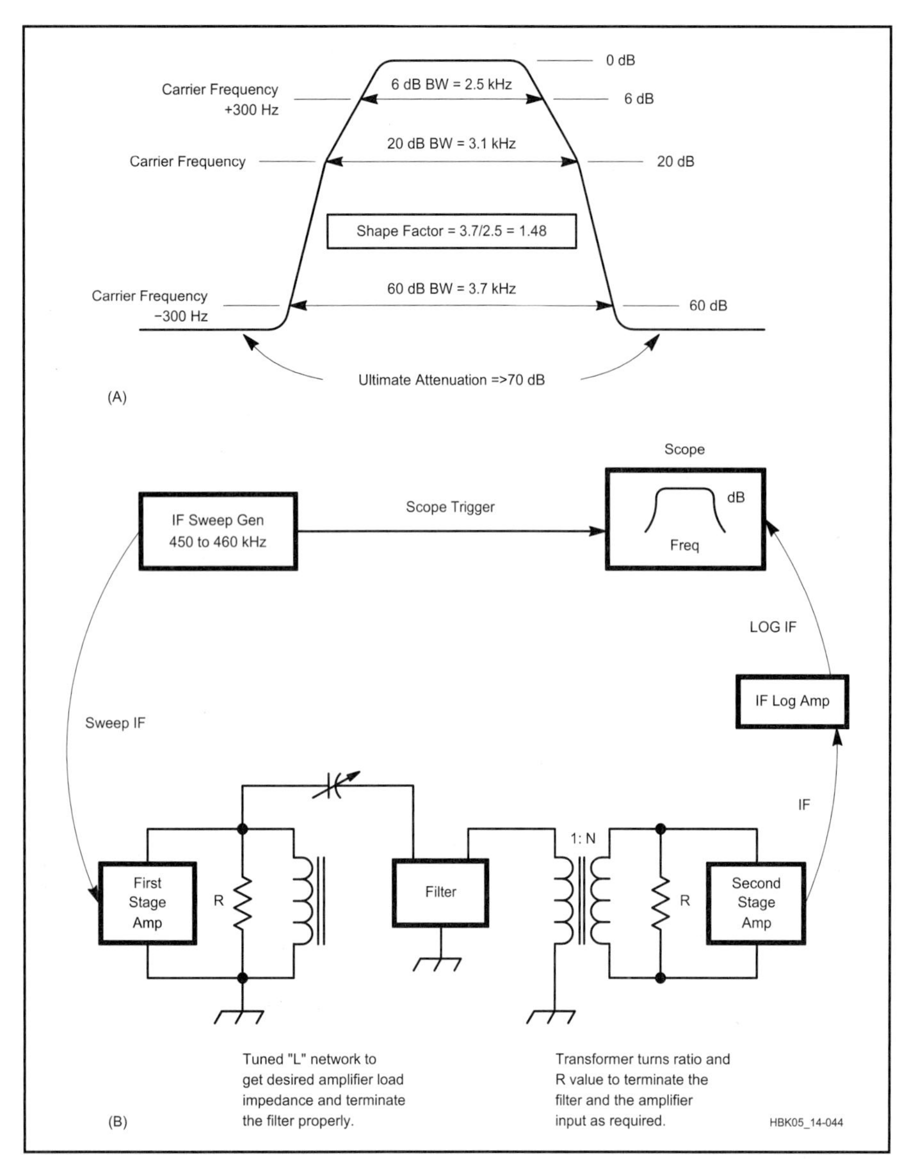

Fig 14.44—A: desired response of a SSB IF filter. B: one method of terminating a mechanical filter that allows easy and accurate tuning adjustment and also a possible test setup for performing the adjustments.

less than 1 dB or so).

Special filters, designated as USB or LSB, are designed with a steeper rolloff on the carrier frequency side, in order to improve rejection of the carrier and opposite sideband. Mechanical filters are available that do this. Crystal-ladder filters (see the **Filters** chapter) are frequently called "single-sideband" filters because they also have this property. The steep skirt can be on the low side or the high side, depending on whether the crystals are across the signal path or in series with the signal path, respectively.

Filters require special attention to their terminations. The networks that interface the filter with surrounding circuits should be accurate and stable over temperature. They should be easy to adjust. One very good way to adjust them is to build a narrowband sweep generator and look at the output IF envelope with a logarithmic amplifier, as indicated in Fig 14.44B. There are three goals: The driver stage must see the desired load impedance: the stage after the filter must see the desired source (generator) impedance and the filter must be properly terminated at both ends. Fig 14.44B shows two typical approaches. This kind of setup is a very good way to make sure the filters and other circuitry are working properly.

Finally, overdriven filters (such as crystal or mechanical filters) can become nonlinear and generate distortion. So it is necessary to heed the manufacturer's instructions. Magnetic core materials used in the tuning networks must be sufficiently linear at the signal levels encountered. They should be tested for IMD separately.

IF Speech Clipper

Audio-clipper speech processors generate a considerable amount of in-band harmonics and IMD (involving different simultaneously occurring speech frequencies). The total distortion detracts somewhat from speech intelligibility. Other problems were mentioned in the section on audio processing. IF clippers overcome most of these problems, especially the Hilbert Transform problem (Ref 17).

Fig 14.45A is a diagram of a 455-kHz IF clipper using high-frequency op amps. 20 dB of gain precedes the diode clippers. A second amplifier establishes the desired output level. The clipping produces a wide band of IMD products close to the IF freuqency. Harmonics of the IF frequency are generated that are easily rejected by subsequent selectivity. "Close-in" IMD distortion products are band limited by the 2.5-kHz-wide IF filter so that out-of-band splatter is eliminated. The in-band IMD products are at least 10 dB below the speech tones.

Fig 14.46 shows oscilloscope pictures of an IF clipped two-tone signal at various levels of clipping. The level of clipping in a radio can be estimated by comparing with these photos. Listening tests verify that the IMD does not sound nearly as bad as harmonic distortion. In fact, processed speech sounds relatively clean and crisp. Tests also verify that speech intelligibility in a noise background is improved by 8 dB.

The repeaking effect from band-pass filtering the clipped IF signal occurs, and must be accounted for when adjusting the output level. A two-tone audio test signal or a speech signal should be used. The ALC circuitry (discussed later) will cut back the IF gain to prevent splattering in the power amplifiers. If the IF filter is of high quality and if subsequent amplifiers are "clean," the transmitted signal is of very high quality, very effective in noisy situations and often also in "pile-ups."

The extra IF gain implies that the IF signal entering the clipper must be free of noise, hum and spurious products. The cleanup filter also helps reduce the carrier frequency, which is outside the passband.

An electrically identical approach to the IF clipper can be achieved at audio frequencies. If the audio signal is translated to, say 455 kHz, processed as described, and translated back to audio, all the desirable effects of IF clipping are retained. This output then plugs into the transmitter's microphone jack. Fig 14.45B shows the basic method. The mic amplifier and the MC1496 circuits have been previously shown and the clipper circuit is the same as in Fig 14.45A.

Another method, performed at audio, synthesizes mathematically the function of the IF clipper. This method is mentioned in Ref 17, and was an accssory for the Collins KWM380 transceiver.

The interesting operating principle in all of these examples is that the characteristics of the IF-clipped (or equivalent) speech signal do not change during frequency translation, even when translated down to audio and then back up to IF in a balanced modulator.

IF Linearity and Noise

Fig 14.39 indicates that after the last SSB filter, whether it is just after the SSB modulator or after the IF clipper, subsequent BPFs are considerably wider. For example, the 70-MHz crystal filter may be 15 to 30 kHz wide. This means that there is a "gray region" in the transmitter just like the one that we saw in the up conversion receiver, where out-of-band IMD that is generated in the IF amplifiers and mixers can cause adjacent-channel interference.

A possible exception, not shown in Fig 14.39, is that there may be an intermediate IF in the 10-MHz region that also contains a narrow filter, as we saw in the triple-conversion receiver in Fig 14.31.

The implication is that special attention must be paid to the linearity of these circuits. It's the designer's job to make sure that distortion in this gray area is much less than distortion generated by the PA and also less than the phase noise generated by the final mixer. Recall also that the total IMD generated in the exciter stages is the resultant of several amplifier and mixer stages in cascade; therefore, each element in the chain must have at least 40 to 50-dB IMD quality. The various drive levels should be chosen to guarantee this. This requirement for multistage linearity is one of the main technical and cost burdens of the SSB mode.

Of interest also in the gray region are white, additive thermal and excess noises originating in the first IF amplifier after the SSB filter and highly magnified on their way to the output. This noise can be comparable to the phase noise level if the phase noise is low, as it would be in a high-quality radio. Recall also that phase noise is at its worst on modulation peaks, but additive noise may be (and often is) present even when there is no modulation. This is a frequent problem in colocated transmitting and receiving environments. Many transmitter designs do not have the benefit of the narrow filter at 70 MHz, so the amplified noise can extend over a much wider frequency range.

CW Mode

Fig 14.39 shows that in the CW mode a carrier is generated at the center of the SSB filter passband. There are two ways to make this carrier available. One way is to unbalance the balanced modulator so that the LO can pass through. Each kind of balanced modulator circuit has its own

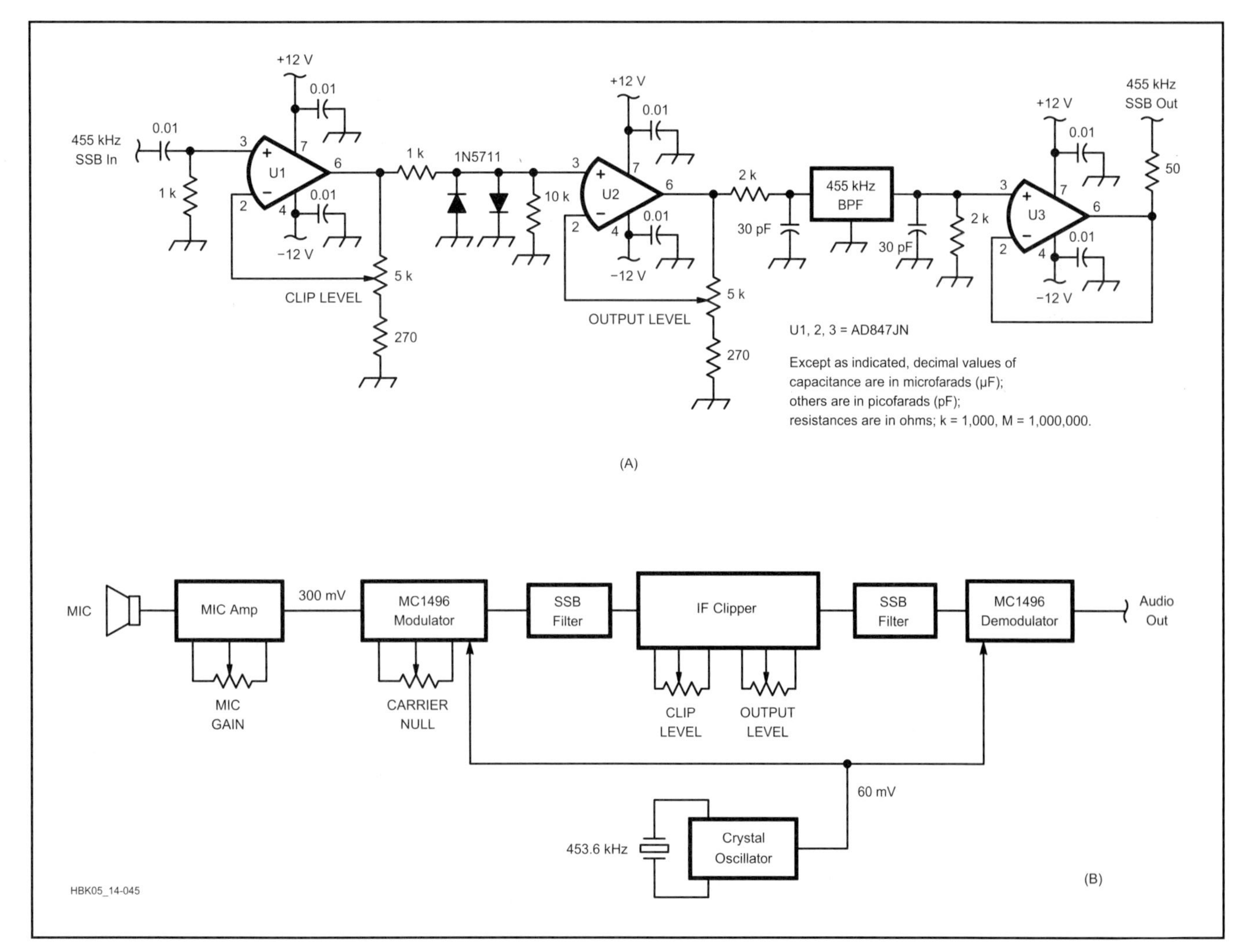

Fig 14.45—IF speech clipping. A: an IF clipper circuit approach. B: the audio signal is translated to 455 kHz, processed, and translated back to audio.

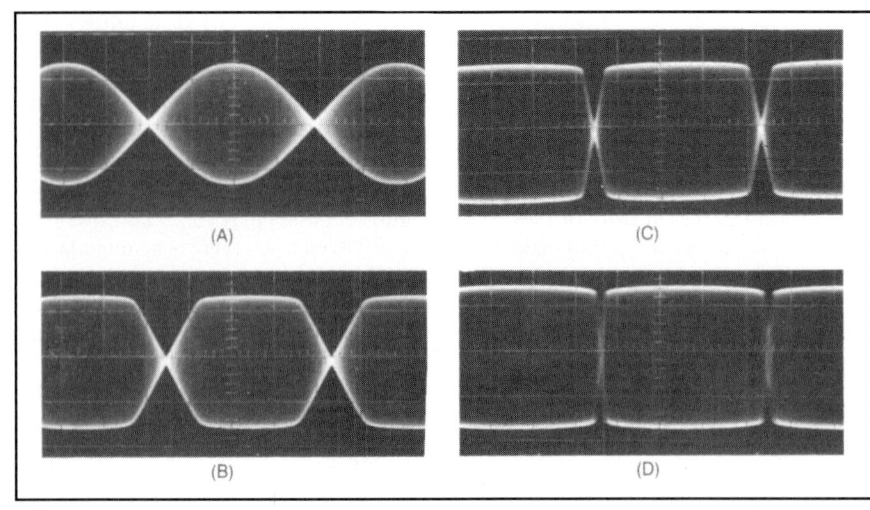

Fig 14.46—Two-tone envelope patterns with various degrees of RF clipping. All envelope patterns are formed using tones of 600 and 1000 Hz. At A, clipping threshold; B, 5 dB of clipping; C, 10 dB of clipping; D, 15 dB of clipping (from "RF Clippers for SSB," by W. Sabin, July 1967 *QST*, pp 13-18).

method of doing this. The approach chosen in Fig 14.39 is to go around the modulator and the SSB filter.

A shaping network controls the envelope of the IF signal to accomplish two things: control the shape of the Morse code character in a way that limits wideband spectrum emissions that can cause interference, and makes the Morse code signal easy and pleasant to copy.

RF Envelope Shaping

On-off keying (CW) is a special kind of low-level amplitude modulation (a low-signal-level stage is turned on and off). It is special because the sideband power is subtracted from the carrier power, and not provided by a separate "modulator" circuit, as in high-level AM. It creates a spectrum around the carrier frequency whose amplitude and bandwidth are influenced by the rates of signal amplitude rise and fall and by the curvature of the keyed waveform. Refer to the graph of keying speed, rise and fall times, and bandwidth in the **Modes and Modulation Sources** chapter for some information about this spectrum. (That figure is repeated here as **Fig 14.47**.) The vertical axis is labeled Rise and Fall Times (ms). For a rise/fall time of 6 ms (between the 10% and 90% values) go horizontally to the line marked Bandwidth. A –20 dB bandwidth of roughly 120 Hz is indicated on the lower horizontal axis. Continuing to the K = 5 and K = 3 lines, the upper horizontal axis suggests code speeds of 30 wpm and 50 wpm respectively, as described in the text that accompanies the figure in the **Modes and Modulation Sources** chapter. These code speeds can be accommodated by the rise and fall times displayed on the vertical axis. For code speeds greater than these the Morse code characters become "soft" sounding and difficult to copy, especially under less-than-ideal propagation conditions.

For a narrow spectrum and freedom from adjacent channel interference, a further requirement is that the spectrum must fall off very rapidly beyond the –20 dB bandwidth indicated in Fig 14.47. A sensitive narrowband CW receiver that is tuned to an adjacent channel that is only 1 or 2 kHz away can detect keying sidebands that are 80 to 100 dB below the key-down level of a strong CW signal. An additional consideration is that during key-up a residual signal, called "backwave" should not be noticeable in a nearby receiver. A backwave level at least 90 dB below the key-down carrier is a desirable goal.

Fig 14.48 is the schematic of one waveshaping circuit that has been used successfully. A Sallen-Key third-order op amp low-pass filter (0.1 dB Chebyshev response) shapes the keying waveform, produces the rate of rise and fall and also softens the leading and trailing corners just the right amount. The key closure activates the CMOS switch, U1, which turns on the 455-kHz IF signal. At the key-up time, the input to the waveshaping filter is turned off, but the IF signal switch remains closed for an additional 12 ms.

The keying waveform is applied to the gain control pin of a CLC5523 amplifier IC (similar to the CLC520, shown in Fig 14.26D). This device, like nearly all gain-control amplifiers, has a *logarithmic* control of gain, therefore some experimental "tweaking" of the capacitor values was used to get the result shown in **Fig 14.49A**. The top trace shows the on/off operation of the IF switch, U1. The signal is turned on shortly before the rise of the keying pulse begins and remains on for about 12 ms after the keying pulse is turned off, so that the waveform falls smoothly to a very low value. The result is an excellent spectrum and an almost complete absence of backwave. The bottom trace shows the resulting keyed RF-output waveshape. It has an excellent spectrum, as verified by critical listening tests. The thumps and clicks that are found in some CW transmitters are virtually absent. The rise and fall intervals have a waveshape that is approximately a cosine. Spread-spectrum frequency-hop waveforms have used this approach to minimize wideband interference.

Fig 14.49B is an accurate SPICE simulation of the wave shaping circuit output before the signal is processed by the CLC5523 amplifier. To assist in adjusting the circuit, create a steady stream of 40 ms dots that can be seen on an RF oscilloscope that is looking at the final PA output envelope. It is important to make sure that the excellent waveshape is not degraded on its way to the transmitter output. Single-Sideband linear power amplifiers are well suited for a CW transmitter, but they must stay within their linear range, and the backwave problem must be resolved.

When evaluating the spectrum of an incoming CW signal during on-the-air operations, a poor receiver design can contribute problems caused by its vulnerability to a strong but clean adjacent channel signal. Clicks, thumps, front end overload, reciprocal mixing, etc can be created in the receiver. It is important to put the blame where it really belongs.

For additional information see "A 455-kHz IF Signal Processor for SSB/CW," William Sabin, WØIYH, *QEX*, March/April 2002, pp 11-16.

Wideband Noise

In the block diagram of Fig 14.39 the last mixer and the amplifiers after it are wideband circuits that are limited only by the harmonic filters and by any selectivity that may be in the antenna system. Wideband phase noise transferred onto the transmitted modulation by the last LO can extend over a wide frequency range, therefore LO (almost always a synthesizer of some kind) cleanliness is always a matter of great concern (Ref 8).

The amplifiers after this mixer are also sources of wide-band "white" or additive noise. This noise can be transmitted even

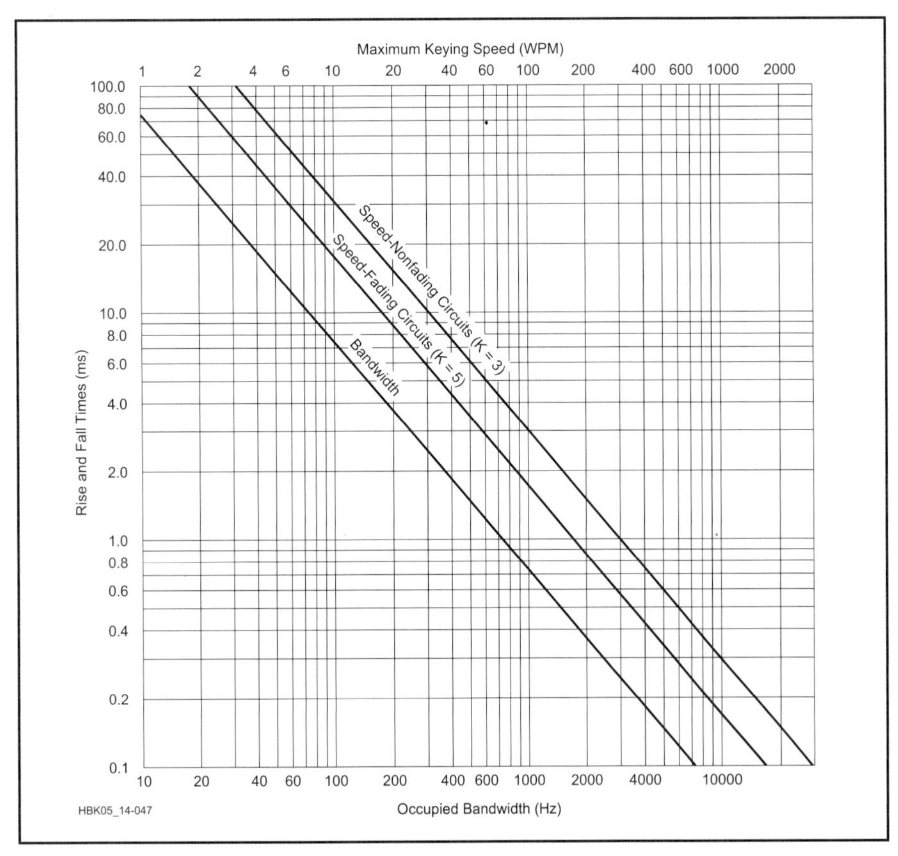

Fig 14.47—Keying speed vs rise and fall times vs bandwidth for fading and nonfading communications circuits. For example, for transmitter output waveform rise and fall times of approximately 6 ms, draw a horizontal line from 6.0 ms on the Rise and Fall Times scale to the Bandwidth line. Then draw a vertical line to the occupied bandwidth scale at the bottom of the graph. In this case the bandwidth is about 130 Hz. Also extend the 6.0 ms horizontal line to the K = 3 line for a nonfading circuit. Finally draw a vertical line from the K = 3 line to the wpm axis. The 6 ms rise and fall time should be suitable for keying speeds up to about 50 wpm in this example.

during times when there is no modulation, and it can be a source of local interference. To reduce this noise: use a high-level mixer with as much signal output as possible, and make the noise figure of the first amplifier stage after the mixer as low as possible.

Transmitters that are used in close proximity to receivers, such as on shipboard, are always designed to control wideband emissions of both additive noise and phase noise, referred to as "composite" noise.

Transmit Mixer Spurious Signals

The last IF and the last mixer LO in Fig 14.39 are selected so that, as much as possible, harmonic IMD products are far enough away from the operating frequency that they fall outside the passband of the low-pass filters and are highly attenuated. This is difficult to accomplish over the transmitter's entire frequency range. It helps to use a high-level mixer and a low enough signal level to minimize those products that are unavoidable. Low-order crossovers that cannot be sufficiently reduced are unacceptable, however; the designer must "go back to the drawing board."

Automatic Level Control (ALC)

The purpose of ALC is to prevent the various stages in the transmitter from being overdriven. Over-drive can generate too much out-of-band distortion or cause excessive power dissipation, either in the amplifiers or in the power supply. ALC does this by sampling the peak amplitude of the modulation (the envelope variations) of the output signal and then developing a dc gain-control voltage that is applied to an early amplifier stage, as suggested in Fig 14.39.

ALC is usually derived from the last stage in a transmitter. This ensures that this last stage will be protected from overload. However, other stages prior to the last stage may not be as well protected; they may generate excessive distortion. It is possible to derive a composite ALC from more than one stage in a way that would prevent this problem. But designers usually prefer to design earlier stages conservatively enough so that, given a temperature range and component tolerances, the last stage can be the one source of ALC. The gain control is applied to an early stage so that all stages are aided by the gain reduction.

Speech Processing with ALC

A fast response to the leading edge of the modulation is needed to prevent a transient overload. After a strong peak, the control voltage is "remembered" for some time as the voltage in a capacitor. This voltage then decays partially through a resistor between peaks. An effective practice provides two capacitors and two time constants. One capacitor decays quickly with a time constant of, say 100 ms, the other with a time constant of several seconds. With this arrangement a small amount of speech processing, about 1 or 2 dB, can be obtained. (Explanation: The dB of improvement mentioned has to do with the improvement in speech intelligibility in a random noise background. This improvement is equivalent to what could be achieved if the transmit power were increased that same number of dB).

The gain rises a little between peaks so that weaker speech components are enhanced. But immediately after a peak it takes a while for the enhancement to take place, so weak components right after a strong peak are not enhanced very much. **Fig 14.50A** shows a complete ALC circuit that performs speech processing.

ALC in Solid-State Power Amplifiers

Fig 14.50B shows how a dual directional coupler can be used to provide ALC for a solid-state power amplifier (PA). The basic idea is to protect the PA transistors from excessive SWR and dissipation by monitoring both the forward power and the reflected power.

Transmit Gain Control (TGC)

This is a widely used feature in commercial and military equipment. A calibrated "tune-up" test carrier of a certain known level is applied to the transmitter. The output carrier level is sampled, using a diode detector. The resulting dc voltage is used to set the gain of a low-level stage. This control voltage is digitized and stored in memory so that it is semipermanent. A new voltage may be generated and stored after each frequency change, or the stored

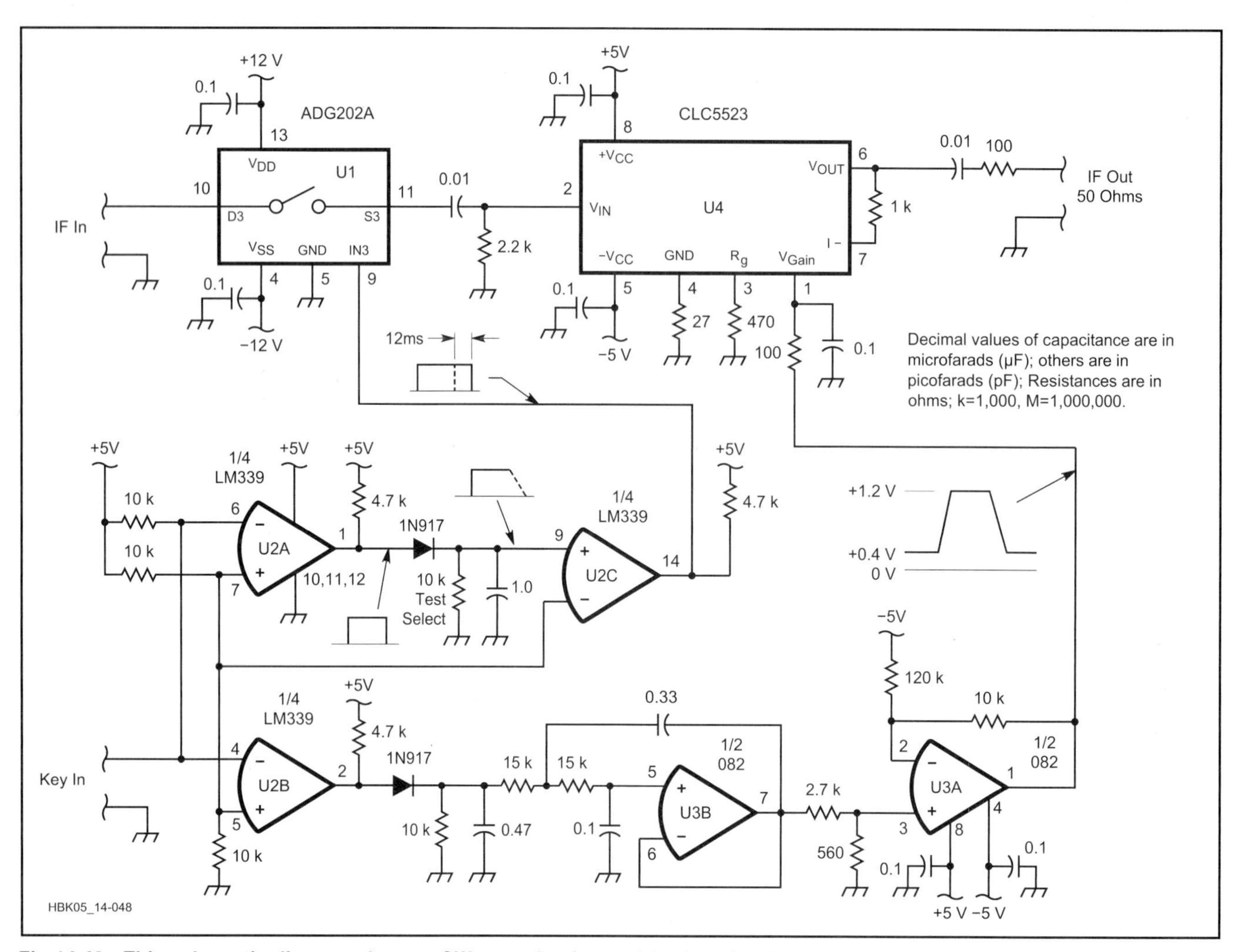

Fig 14.48—This schematic diagram shows a CW waveshaping and keying circuit suitable for use with an SSB/CW transmitter such as is shown in Fig 14.39.

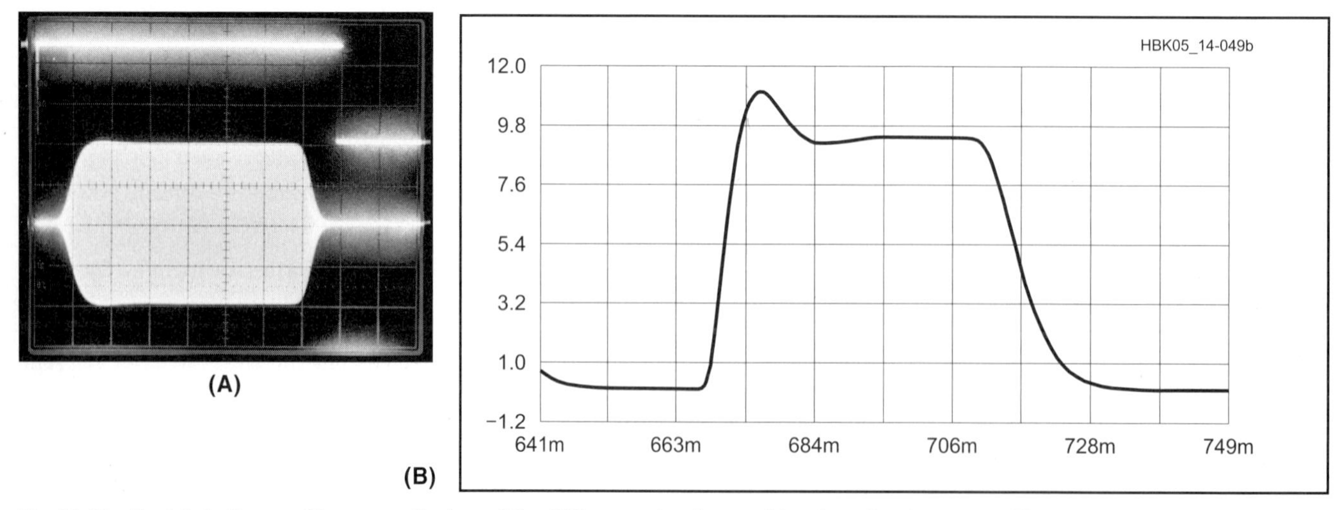

Fig 14.49—Part A is the oscilloscope display of the CW waveshaping and keying circuit output. The top trace is the IF keying signal applied to S1 of Fig 14.48. The bottom trace is the transmitter output RF spectrum. Part B is a SPICE simulation of the waveshaping network. When this signal is applied to the logarithmic control characteristic of the CLC5523 amplifier, the RF envelope is modified slightly to the form shown in A.

value may be "fetched." A test signal is also used to do automatic antenna tuning. A dummy load is used to set the level and a low-level signal (a few mW) is used for the antenna tune-up.

Transmitter Output Load Impedance

The following logical processes are used to tune and load the final PA of a transmitter:

1. The RF input power requirement to the input terminal of the PA has been determined.
2. The desired load impedance of the plate/collector/drain of the PA has been determined, either graphically or by calculation, from the power to be delivered to the load, the dc power supply voltage and the ac voltage on the plate/collector/drain (see **RF Power Amplifiers** chapter).
3. The input impedance, looking toward the antenna, of the transmission line that is connected to the transmitter is adjusted by a network of some kind to its Z_0 value (if it is not already equal to that value).
4. A network of some kind is designed, which transforms the transmission-line Z_0 to the impedance required in step 2. This may be a sharply tuned resonator with impedance transforming capability, or it may be a wideband transformer of some kind.

Under these conditions the PA is performing as intended. Note that a knowledge of the output impedance of the PA is not needed to get these results. That is, we are interested mostly in the actual power gain of the PA, which does not require a knowledge of the amplifier's output impedance.

The output impedance, looking backward from the plate/collector/drain terminal of the network, in step 4, will have some influence on the selectivity of a resonant tuned circuit or the frequency response of a low-pass filter. This must (or should) be considered during the design process, but it is not needed during the "tune and load" process.

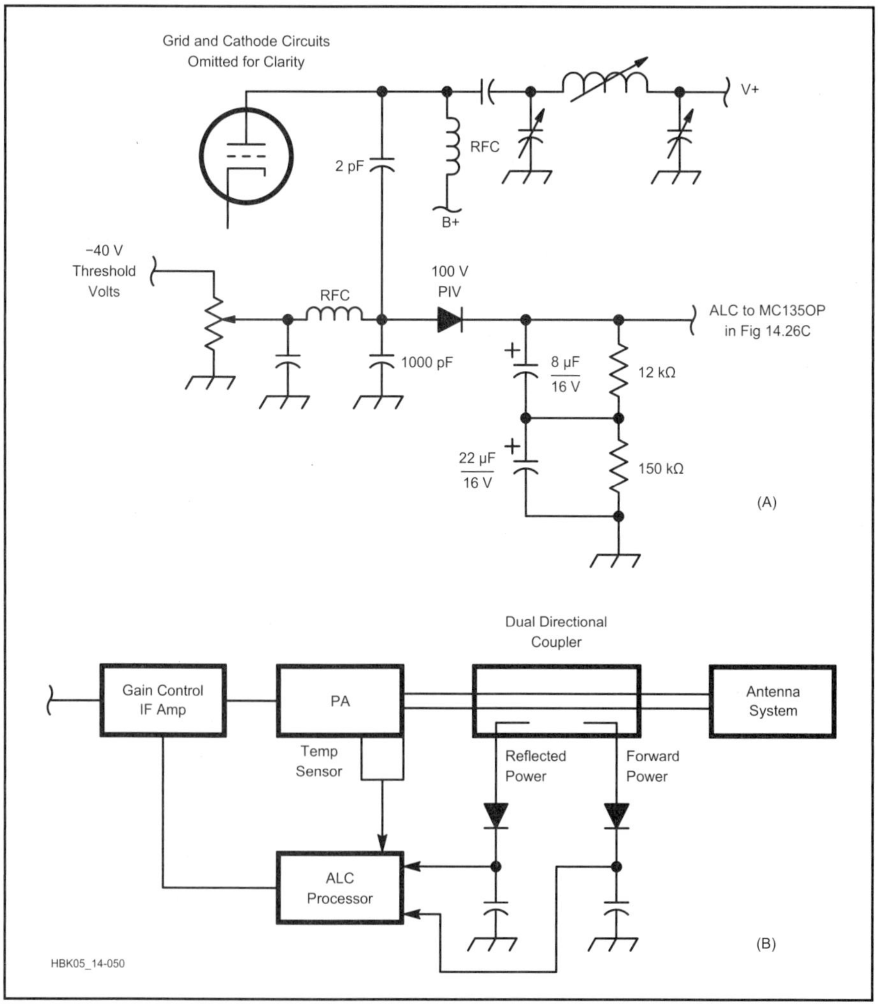

Frequency Multipliers

A passive multiplier using diodes is shown in **Fig 14.51A**. The full-wave rectifier circuit can be recognized, except that the dc component is shorted to ground. If the fundamental frequency ac input is 1.0 V RMS the second harmonic is 0.42 V RMS or 8 dB below the input, including some small diode losses. This value is found by calculating the Fourier Series coefficients for the full-wave-rectified sine wave, as shown in many textbooks.

Transistor and vacuum-tube frequency multipliers operate on the following principle: if a sine wave input causes the plate/collector/drain current to be distorted (not a sine wave) then harmonics of the input are generated. If an output resonant circuit is tuned to a harmonic the output at the harmonic is emphasized and other frequencies are attenuated. For a particular harmonic the current pulse should be distorted in a way that maximizes that harmonic. For example, for a doubler the current pulse should look like a half sine wave (180° of conduction). A transistor with Class B bias would be a good choice. For a tripler use 120° of conduction (Class C).

An FET, biased at a certain point, is very nearly a "square law" device. That is, the drain-current change is proportional to the square of the gate-voltage change. It is then an efficient frequency doubler that also deemphasizes the fundamental.

A push-push doubler is shown in Fig 14.51B. The FETs are biased in the square-law region and the BALANCE potentiometer minimizes the fundamental frequency. Note that the gates are in push-pull and the drains are in parallel. This causes second harmonics to add in-phase at the output and fundamental components to cancel.

Fig 14.51C shows an example of a bipolar-transistor doubler. The efficiency of a doubler of this type is typically 50%, a tripler 33% and a quadrupler 25%. Harmonics other than the one to which the output tank is tuned will appear in the output unless effective band-pass filtering is applied. The collector tap on L1 is placed at the point that offers the best compromise between power output and spectral purity.

A push-pull tripler is shown in Fig 14.51D. The input and output are both push-pull. The balance potentiometer minimizes even harmonics. Note that the transistors have no bias voltage in the base

Fig 14.50—A: an ALC circuit with speech processing capability. B: protection method for a solid-state transmitter.

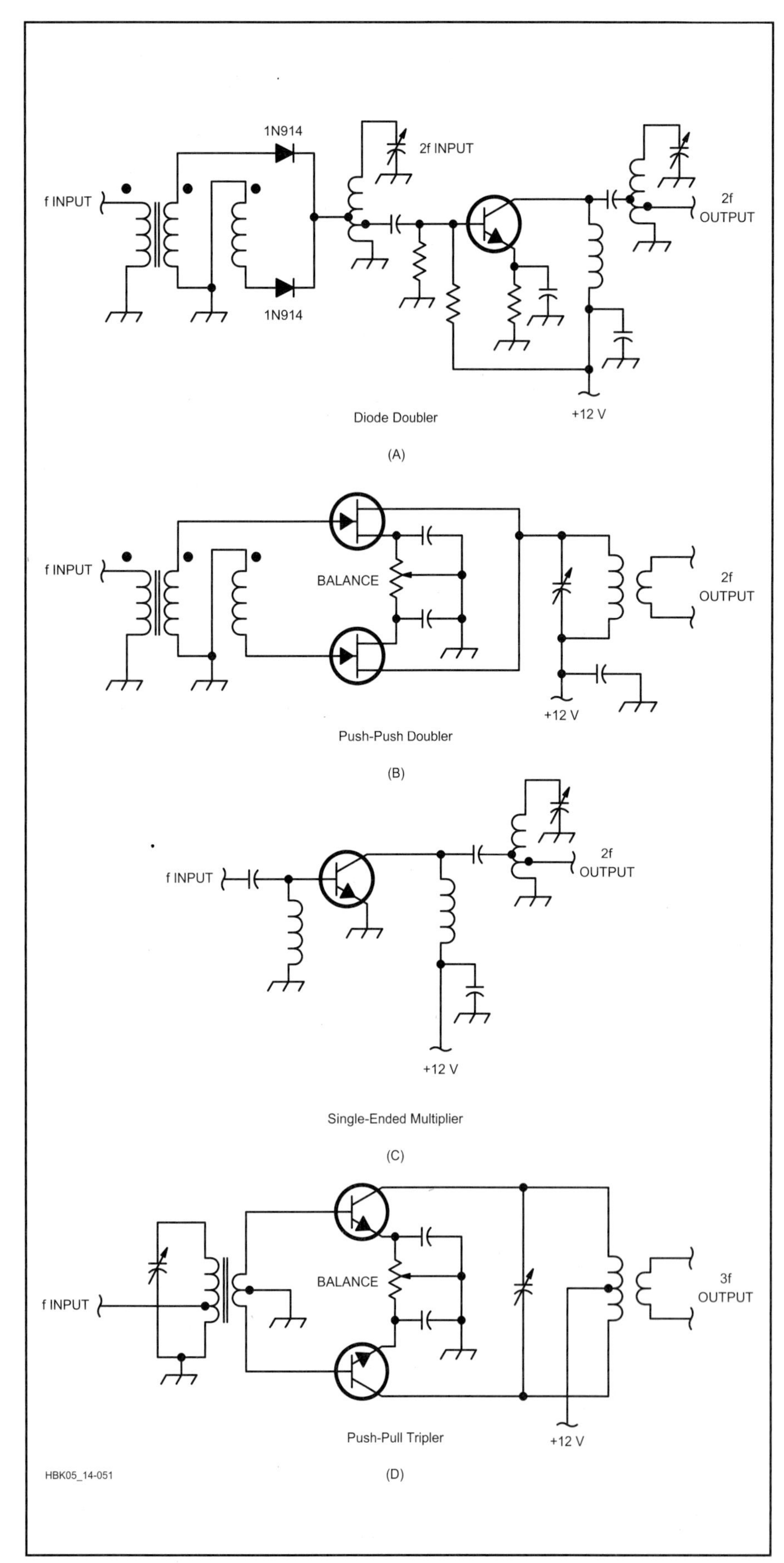

Fig 14.51—A: diode doubler. B: push-push doubler using JFETS. C: single-ended multiplier using a BJT. D: push-pull tripler using BJTs.

circuit; this places the transistors in Class C for efficient third-harmonic production. Choose an input drive level that maximizes harmonic output.

The step recovery diode (SRD) is an excellent device for harmonic generation, especially at microwave frequencies. The basic idea of the SRD is as follows: When the diode is forward conducting, a charge is stored in the diode's diffusion capacitance; and if the diode is quickly reverse-biased, the stored charge is very suddenly released into an LC harmonic-tuned circuit. The circuit is also called a "comb generator" because of the large number of harmonics that are generated. (The spectral display looks like a comb.) Phase-locked loops (PLLs) can be made to lock onto these harmonics. A typical low-cost SRD is the HP 5082-0180, found in the HP Microwave & RF Designer's Catalog. **Fig 14.52A** is a typical schematic. For more information regarding design details there are two References: Hewlett-Packard application note AN-920 and Ref 30.

The varactor diode can also be used as a multiplier. Fig 14.52B shows an example. This circuit depends on the fact that the capacitance of a varactor changes with the instantaneous value of the RF excitation voltage. This is a nonlinear process that generates harmonic currents through the diode. Power levels up to 25 W can be generated in this manner.

IMPEDANCE TRANSFORMATION BETWEEN CASCADED CIRCUITS

One of the most common tasks the electronics designer encounters is to correctly interface between the output of one circuit and the input of an adjacent circuit, so that both are operating in the desired manner. We introduced this in Figs 14.1 and 14.2. In this segment we will present a unified overview of the general topic that should be helpful to the designer of RF circuits. This is a very large topic, so we must stick to basic ideas and give References. The networks we will consider are in two categories, broadband and narrowband. The modern trend in Amateur Radio is to employ the personal computer, using software such as *ARRL Radio Designer*, *SPICE*, *Mathcad*, the Smith Chart and associated design programs.

Impedance "Matching" and "Transformation"

Fig 14.53A shows a network connected between a generator with internal resistance R_{IN} and a load R_{OUT}. In this case the load and source are "matched" because each sees itself, looking into the network. As Figs 14.1 and 14.2 explained, this idea extends to the idea of "conjugate match." There are

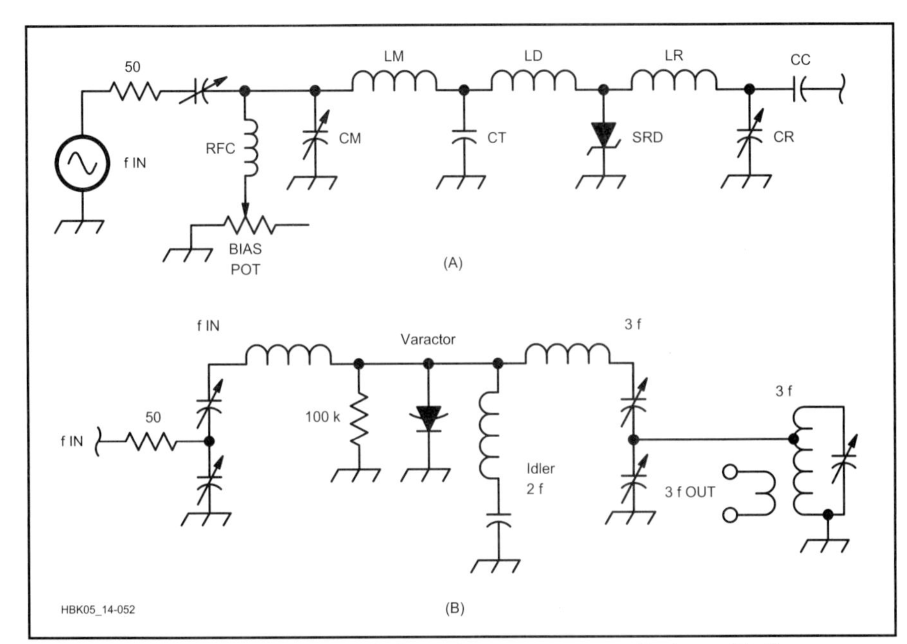

Fig 14.52—Diode frequency multipliers. A: step-recovery diode multiplier. B: varactor diode multiplier.

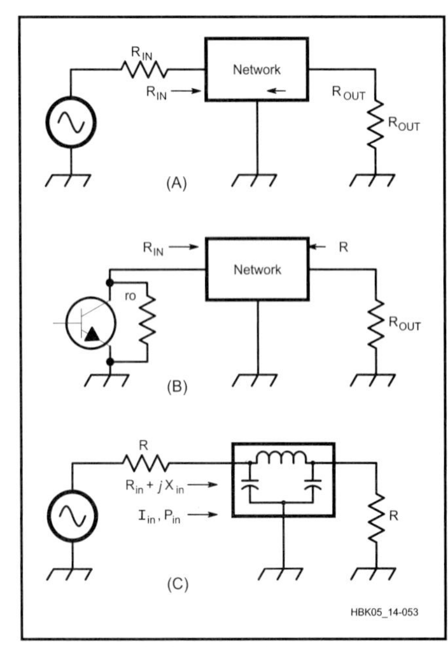

Fig 14.53—At A, matching network driven by generator. At B, matching network driven by transistor. At C, low-pass filter network.

many circuits that require this kind of impedance matching for textbook operation. In Fig 14.53B, showing a transistor amplifier, a different situation may exist. The transistor sees the R_{IN} that it needs for correct operation of the transistor.

That is, for a certain value of collector DC voltage and current and a certain allowed maximum value of collector RF voltage and current, a certain value of R_{IN} is required. However, the load R_{OUT}, looking back, may not see itself, but something much different, R, because the output resistance r_o of the transistor is not guaranteed to be the same as R_{IN} in Fig 14.53A. In this situation we could say that the load R_{OUT} is "transformed" to R_{IN}. This difference is important to understand in many applications. In some small-signal transistor circuits the dynamic output resistance r_o of the transistor and R_{OUT} are actually "matched" by a network as in Fig 14.53A. [Note: "dynamic" means that r_o is not a physical resistor but an internal, lossless *negative-feedback* property of the transistor]. In many other situations correct circuit performance requires that R_{OUT} see some specific value other than itself, looking into the output of the network.

A frequent practice for networks, including the *resonant* coupling networks (see Fig 14.59) is to combine a physical resistor with r_o so that the network is properly terminated at the input end. If the resistor is used, the transistor then sees the network and resistor combination; this combination should be the load that we want. For example, suppose r_o is 10 kΩ and an R_{IN} of 1 kΩ (including r_o) is required. A 2 kΩ network and a 2.5 kΩ resistor in parallel could be used and the network is then correctly double-terminated as in Fig 14.53A.

Of course, some signal power will be lost in the 2.5 kΩ resistor, but in low-level circuits we often accept that. An elegant alternative is to use additional negative feedback in the transistor circuit to reduce r_o to 1 kΩ and use a 1 kΩ network. In some circuits, especially RF power amplifiers, the dissipation in an extra resistor cannot be tolerated, and in this case the network internal impedance R, looking into the output terminals, is usually ignored and the main goal is to get the right R_{IN}.

The **RF Power Amplifiers** chapter shows how Pi and Pi-L networks for power amplifiers are calculated. Use simulation to verify that the circuit is behaving as you want and to get network component losses. Simulations also show that the loading by a resistor or tube/transistor dynamic r_o at the R_{IN} end changes the selectivity, and this can be monitored by actual measurement. The Q and LC values can be modified as needed.

Mismatch in Lossless Networks

Fig 14.53C shows a generator with resistance R and a lowpass filter (LPF) connected to an R load resistor. At very low frequencies the filter is "transparent" and the maximum power is delivered to the load. The inductances and capacitances are assumed to be lossless and the generator sees the R load. But at high frequencies the reactances cause the output power to be attenuated. The input impedance is $R_{IN} \pm j X_{IN}$. The power (real power) delivered to the filter input is $P_{IN} = I_{IN}^2 \times R_{IN}$.

Because L and C are lossless, this power P_{IN} must be identical to the power that is actually delivered to the R load. I_{IN}, X_{IN} and R_{IN} in the low-pass filter (LPF) are modified by the reactances in such a way that P_{IN}, therefore P_{LOAD}, is reduced. These changes are equivalent to an impedance mismatch between the generator and the filter. Or, we can say the generator no longer sees the correct load resistance R. This is the basic mechanism by which lossless networks control the frequency response.

BROADBAND TRANSFORMERS

Conventional Transformers

Fig 14.54A shows a push-pull amplifier that we will use to point out the main properties and the problems of conventional transformers. The medium of signal transfer from primary to secondary is magnetic flux in the core. If the core material is ferromagnetic then this is basically a nonlinear process that becomes increasingly nonlinear if the flux becomes too large or if there is a dc current through the winding that biases the core into a nonlinear region. Nonlinearity causes harmonics and IMD.

Push-pull operation eliminates the dc biasing effect if the stage is symmetrical.

The magnetic circuit can be made more linear by adding more turns to the windings. This reduces the ac volts per turn, increases the reactance of the windings and therefore reduces the flux. For a given physical size, however, the wire resistance, distributed capacitance and leakage reactance all tend to increase as turns are added. This reduces efficiency and bandwidth. Higher permeability core materials and special winding techniques can improve things up to a point, but eventually linearity becomes more difficult to maintain.

Fig 14.54B is an approximate equivalent circuit of a typical transformer. It shows the leakage reactance and winding capacitance that affect the high-frequency response and the coil inductance that affects the low-frequency response. Fig 14.54C shows how these elements determine the frequency response, including a resonant peak at some high frequency.

The transformers in a system are correctly designed and properly coordinated when the total distortion caused by them is at least 10 dB less than the total distortion due to all other nonlinearities in the system. Do not over-design them in relation to the rest of the equipment. During the design process, distortion measurements are made on the transformers to verify this.

The main advantage of the conventional transformer, aside from its ability to transform between widely different impedances over a fairly wide frequency band, is the very high resistance between the windings. This isolation is important in many applications and it also eliminates coupling capacitors, which can sometimes be large and expensive.

In radio-circuit design, conventional transformers with magnetic cores are often used in high-impedance RF/IF amplifiers, in high-power solid-state amplifiers and in tuning networks such as antenna couplers. They are seldom used any more in audio circuits. Hybrid transformers, such as those in Fig 14.8, are often "conventional."

Fig 14.54D considers a typical application of a conventional transformer in a linear Class-A RF power amplifier. The load is 50 Ω and the maximum allowable transistor collector voltage and current excursions for linear operation are shown. The value of DC current and the sinewave limits are determined by studying the collector voltage-current curves (or constant-current curves) in the data manual to find the most linear region. To deliver this power to the 50 Ω load, the turns ratio is calculated from the equations. In an RF amplifier a ferrite or powdered-iron core would be used. The efficiency in this ex-

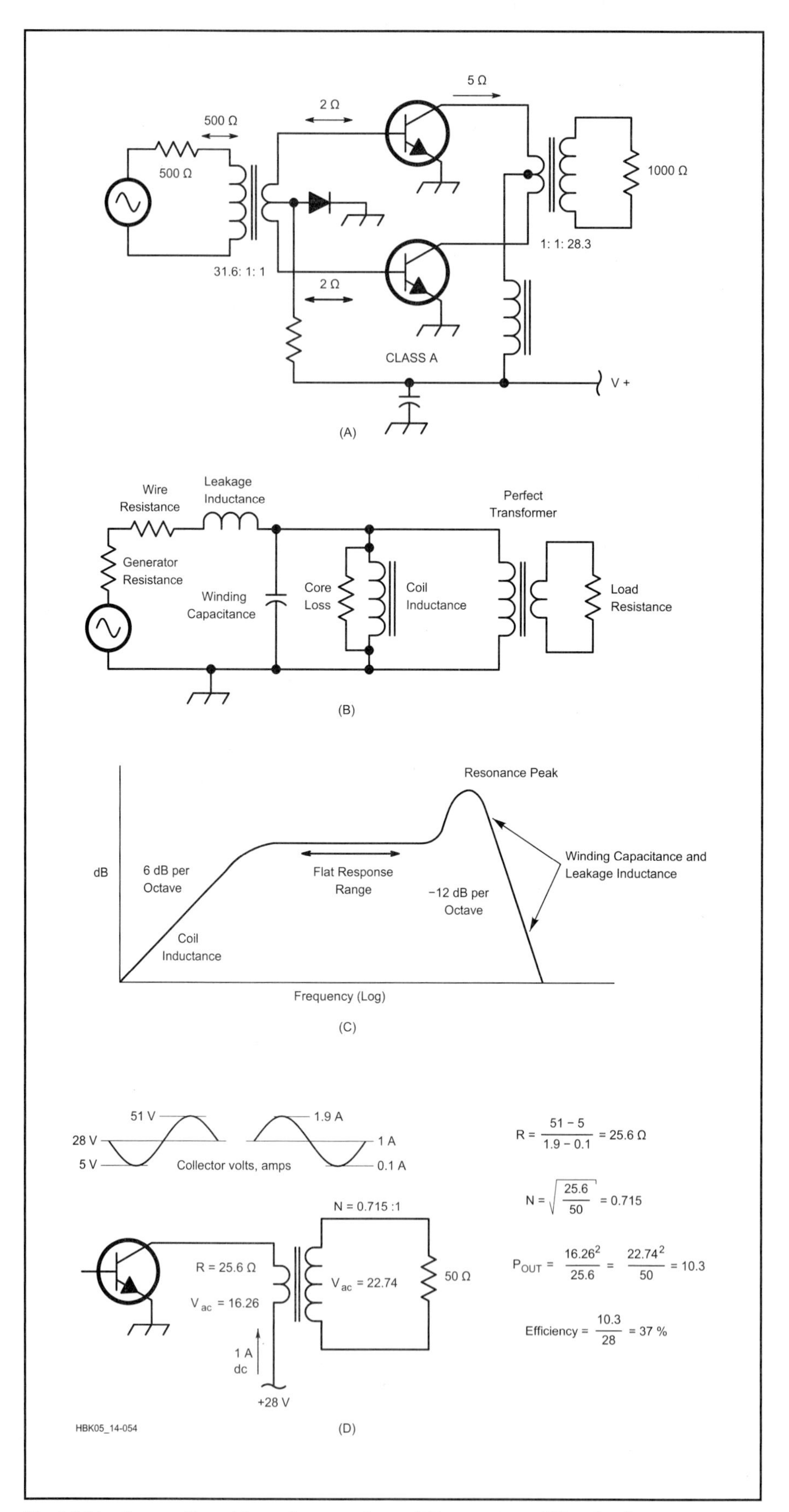

Fig 14.54—Conventional transformers in an RF power amplifier. Leakage reactances, stray capacitances and core magnetizations limit the bandwidth and linearity, and also create resonant peaks. D shows design example of transformer-coupled RF power amplifier.

ample is 37%.

Transmission Line Transformers

The basic transmission line transformer, from which other transformers are derived, is the 1:1 choke (or current) balun, shown in **Fig 14.55A**. We consider the following basic properties:

- A pair of close-spaced wires or a length of coax (ie, a transmission line) wraps around a ferrite rod or toroid or through a number of beads. For the 3.5 to 29.7 MHz band, type 43 ferrite ($\mu = 850$), or equivalent, is usually preferred. Other types such as 77 (at 1.8 MHz, $\mu = 2000$) or 61 (at VHF bands, $\mu = 120$) are used. The Z_0 of the line should equal R.
- Because of the ferrite, a large impedance exists between points A and C and a virtually identical impedance between B and D. This is true for parallel wires and it is also true for coax. The ferrite affects the A to C impedance of the coax inner conductor and the B to D impedance of the outer braid equally.
- The conductors (two wires or coax braid and center-wire) are tightly coupled by electromagnetic fields and therefore constitute a good conventional transformer with a turns ratio of 1:1. The voltage from A to C is equal to and in-phase with that from B to D. These are called the *common-mode voltages* (CM).
- A common-mode (CM) current is one that has the same value and direction in both wires (or braid and center wire). Because of the ferrite, the CM current encounters a high impedance that acts to reduce (choke) the current. The normal differential-mode (DM) signal does not encounter this CM impedance because the electromagnetic fields due to equal and opposite currents in the two conductors cancel each other at the ferrite, so the magnetic flux in the ferrite is virtually zero.
- The main idea of the transmission line transformer is that although the CM impedance may be very large, the DM signal is virtually unopposed, especially if the line length is a small fraction of a wavelength.
- A common experience is a CM current that flows on the outside of a coax braid due to some external field, such as a nearby antenna or noise source. The balun reduces (chokes) the CM current due to these sources. But it is very important to keep in mind that the common-mode voltage across the ferrite winding that is due to this current is efficiently coupled to the center wire by conventional transformer action, as mentioned before and easily verified. This equality of CM voltages, and also CM impedances, reduces the *conversion* of a CM signal to an *undesired* DM signal that can interfere with the *desired* DM signal in both transmitters and receivers.
- The CM current, multiplied by the CM impedance due to the ferrite, produces a CM voltage. The CM impedance has L and C reactance and also R. So L, C and R cause a broad parallel self-resonance at some frequency. The R component also produces some dissipation (heat) in the ferrite. This dissipation is an excellent way to dispose of a small amount of unwanted CM power.
- The main feature of the ferrite is that the choke is effective over a bandwidth of one, possibly two decades of frequency. In addition to the ferrite choke balun, straight or coiled lengths of coax (no core and almost no CM dissipation)

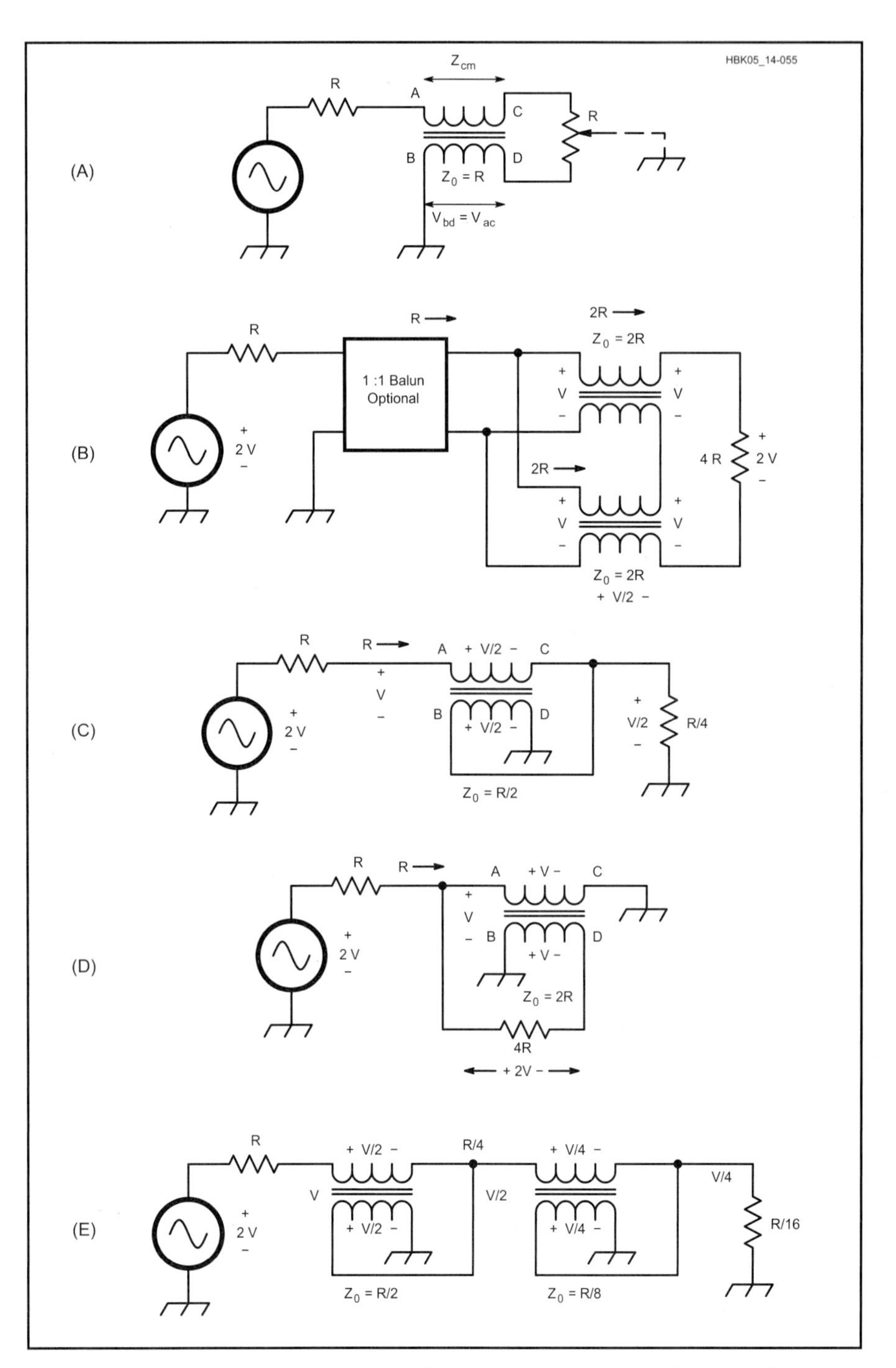

Fig 14.55—At A, basic balun. At B, 1:4 Guanella transformer. At C, Ruthroff transformer, 4:1 unbalanced. At D, Ruthroff 1:4 balanced transformer. At E, Ruthroff 16:1 unbalanced transformer.

are used within narrow frequency bands. A one-quarter-wave length of transmission line is a good choke balun at a single frequency or within a narrow band.
- The two output wires of the balun in Fig 14.55A have a high impedance with respect to, and are therefore "isolated" from, the generator. This feature is very useful because now any point of R at the output can be grounded. In a well-designed balun circuit *almost* all of the current in one conductor returns to the generator through the other conductor, despite this ground connection. Note also that the ground connection introduces some CM voltage across the balun cores and this has to be taken into account. This CM voltage is maximum if point C is grounded. If point D is grounded and if all "ground" connections are at the same potential, which they often are not, the CM voltage is zero and the balun may no longer be needed. In a coax balun the return current flows on the inside surface of the braid.

We now look briefly at a transmission line transformer that is based on the choke balun. Fig 14.55B shows two identical choke baluns whose inputs are in parallel and whose outputs are in series. The output voltage amplitude of each balun is identical to the common input, so the two outputs add in-phase (equal time delay) to produce twice the input voltage. It is the high CM impedance that makes this voltage addition possible. If the power remains constant the load current must be one-half the generator current, and the load resistor is 2V/0.5I = 4V/I = 4R.

The CM voltage in each balun is V/2, so there is some flux in the cores. The right side floats. This is named the *Guanella* transformer. If Z_0 of the lines equals 2R and if the load is pure resistance 4R then the input resistance R is independent of line length. If the lines are exactly one-quarter wavelength, then $Z_{IN} = (2R)^2 / Z_L$, an impedance inverter, where Z_{IN} and Z_L are complex. The quality of balance can often be improved by inserting a 1:1 balun (Fig 14.55A) at the left end so that both ends of the 1:4 transformer are floating and a ground is at the far left side as shown. The Guanella can also be operated from a grounded right end to a floating left end. The 1:1 balun at the left then allows a grounded far left end.

Fig 14.55C is a different kind, the *Ruthroff* transformer. The input voltage V is divided into two equal in-phase voltages AC and BD (they are tightly coupled), so the output is V/2. And because power is constant, $I_{OUT} = 2I_{IN}$ and the load is R/4. There is a CM voltage V/2 between A and C and between B and D, so in normal operation the core is not free of magnetic flux. The input and output both return to ground so it can also be operated from right to left for a 1:4 impedance stepup. The Ruthroff is often used as an amplifier interstage transformer, for example between 200 Ω and 50 Ω. To maintain low attenuation the line length should be much less than one-fourth wavelength at the highest frequency of operation, and its Z_0 should be R/2. A balanced version is shown in Fig 14.55D, where the CM voltage is V, not V/2, and transmission is from left-to-right only. Because of the greater flux in the cores, no different than a conventional transformer, this is not a preferred approach, although it could be used with air wound coils (for example in antenna tuner circuits) to couple 75 Ω unbalanced to 300 Ω balanced. The tuner circuit could then transform 75 Ω to 50 Ω.

Fig 14.56 illustrates, in skeleton form, how transmission-line transformers can be used in a push-pull solid state power amplifier. The idea is to maintain highly balanced stages so that each transistor shares equally in the amplification in each stage. The balance also minimizes even-order harmonics so that low-pass filtering of the output is made much easier. In the diagram, T1 and T5 are current (choke) baluns that convert a grounded connection at one end to a balanced (floating) connection at the other end, with a high impedance to ground at both wires. T2 transforms the 50 Ω generator to the 12.5 Ω (4:1 impedance) input impedance of the first stage. T3 performs a similar step-down transformation from the collectors of the first stage to the gates of the second stage. The MOSFETs require a low impedance from gate to ground. The drains of the output stage require an impedance step up from 12.5 Ω to 50 Ω, performed by T4. Note how the choke baluns and the transformers collaborate to maintain a high degree of balance throughout the amplifier. Note also the various feedback and loading networks that help keep the amplifier frequency response flat.

Tips on Toroids and Coils

Some notes about toroid coils: Toroids do have a small amount of leakage flux, despite rumors to the contrary. Toroid coils are wound in the form of a helix (screw thread) around the circular length of the core. This means that there is a small component of the flux from each turn that is perpendicular to the circle of the toroid (parallel to the axis through the hole) and is therefore not adequately linked to all the other turns. This effect is responsible for a small leakage flux and the effect is called the "one-turn" effect, since the result is equivalent to one turn that is wound around the outer edge of the core and not through the hole. Also, the inductance of a toroid can be adjusted, also despite rumors to the contrary. If the turns can be pressed closer together or separated a little, inductance variations of a few percent are possible.

A grounded aluminum shield between adjacent toroidal coils can eliminate any significant capacitive or inductive (at high frequencies) coupling. These effects are most easily noticed if a network analyzer is available during the checkout procedure, but how many of us are that lucky? Spot checks with an attenuator ahead of a receiver that is tunable to the harmonics are also very helpful.

There are many transformer schemes that use the basic ideas of Fig 14.55. Several of them, with their toroid winding instructions, are shown in **Fig 14.57**. Because of space limitations, for a comprehensive treatment we suggest Jerry Sevick's books *Transmission Line Transformers* and *Building and Using Baluns and Ununs*, both available from ARRL. For applications in solid-state RF power amplifiers, see Sabin and Schoenike, *HF Radio Systems and Circuits*, Chapter 12, also available from ARRL.

Tuned (Resonant) Networks

There is a large class of LC networks that utilize resonance at a single frequency to transform impedances over a narrow band. In many applications the circuitry that the network connects to has internal reactances, inductive or capacitive, combined with resistance. We want to absorb these reactances, if possible, to become an integral part of the network design. By looking at the various available network possibilities we can identify those that will do this at one or both ends of the network. Some networks must operate between two different values of resistance, others can also operate between equal resistances. As mentioned before, nearly all networks also allow a choice of selectivity, or Q, where Q is (approximately) the resonant frequency divided by the 3-dB bandwidth.

As a simple example that illustrates the method, consider the generator and load of **Fig 14.58A**. We want to absorb the 20 pF and the 0.1 µH into the network. We use the formulas to calculate L and C for a 500 Ω to 50 Ω L-network, then subtract 20 pF from C and 0.1 µH from L. As a second iteration we can improve the design by considering the resistance of the L that we just found. Suppose it is 2 Ω. We can recalculate new values L′ and C′ for a network from 500 Ω to 52 Ω, as shown in

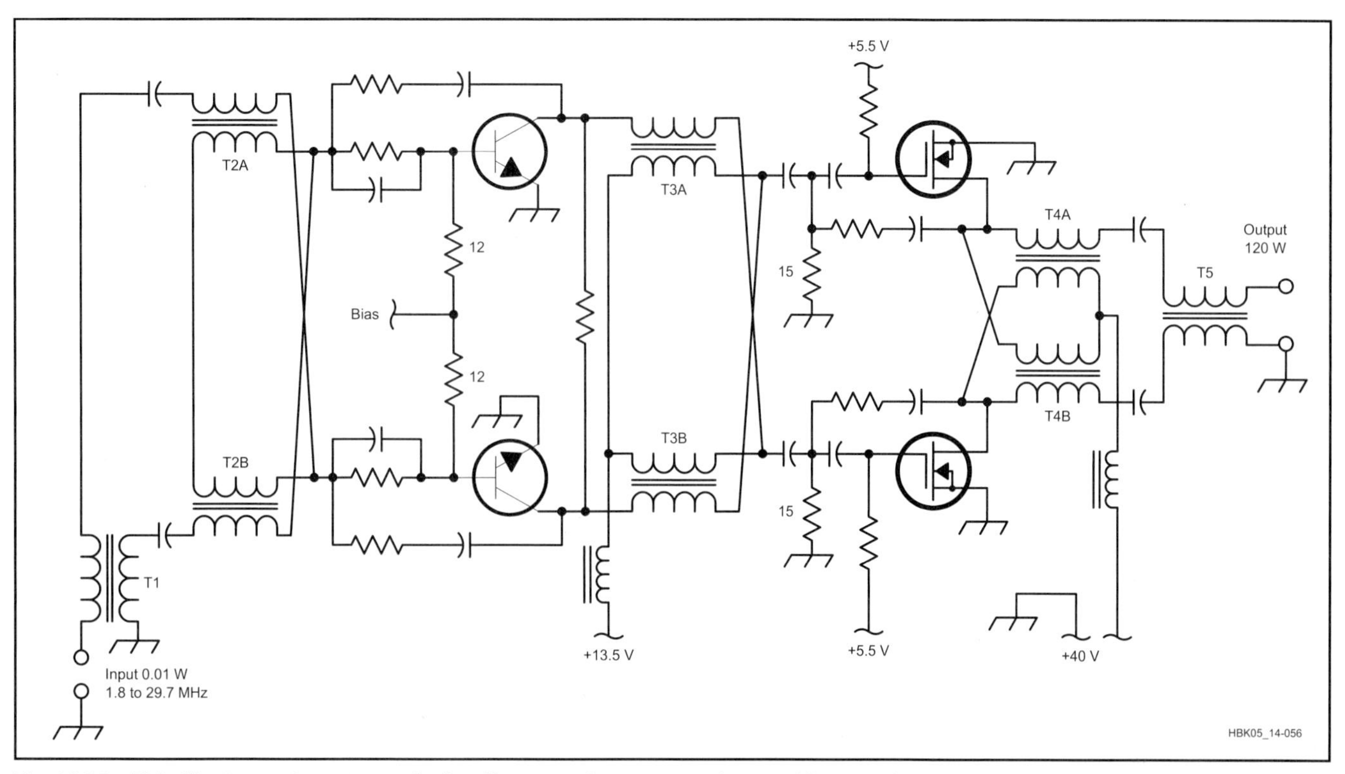

Fig 14.56—This illustrates how transmission-line transformers can be used in a push-pull power amplifier.

Fig 14.58B.

Further iterations are possible but usually trivial. More complicated networks and more difficult problems can use a computer to expedite absorbing process. Always try to absorb an inductance into a network L and a capacitance into a network C in order to minimize spurious LC resonances and undesired frequency responses. Inductors and capacitors can be combined in series or in parallel as shown in the example. Fig 14.58C shows useful formulas to convert series to parallel and vice versa to help with the designs.

A set of 14 simple resonant networks, and their equations, is presented in **Fig 14.59**. Note that in these diagrams RS is the low impedance side and RL is the high impedance side and that the X values are calculated in the top-down order given. The program *MATCH.EXE* can perform the calculations.

ARRL Radio Designer can also help a lot with special circuit-design problems and some approaches to resonant network design. It can graph the frequency response, compute insertion loss and also tune the capacitances and inductances across a frequency band. You may select the selectivity (Q) in such programs based on frequency-response requirements. The program can also be trimmed to help realize realistic or standard component values. A math program such as *Mathcad* can also make this a quick and easy process. You can find additional information for Pi and Pi-L networks in the **RF Power Amplifiers** chapter.

NBFM Transmitter Block Diagram

Fig 14.60 shows the phase-modulation method, also known as indirect FM. It is the most widely used approach to NBFM. Phase modulation is performed at low IF, say 455 kHz. Prior to the phase modulator, speech filtering and processing are performed to achieve four goals:

1. Convert phase modulation to frequency modulation (see below),
2. Preemphasize higher speech frequencies for improved signal-to-noise ratio at the receive end,
3. Perform speech processing to emphasize the weaker speech components and
4. Adjust for the microphone's frequency response and possibly also the operator's voice characteristics.

Multiplier stages move the signal to some desired higher IF and also multiply the frequency deviation to the desired final value. If the FM deviation generated in the 455-kHz modulator is 250 Hz, the deviation at 9.1 MHz is 20 × 250, or 5 kHz. A second conversion to the final output frequency is then performed. Prior to this final translation, IF band-pass filtering is performed in order to minimize adjacent-channel interference that might be caused by excessive frequency deviation. This filter needs good phase linearity to assure that the FM sidebands maintain the correct phase relationships. If this is not done, an AM component is introduced to the signal, which can cause nonlinear distortion problems in the PA stages. The final frequency translation retains a constant value of FM deviation for any value of the output signal frequency.

The IF/RF amplifiers are Class C amplifiers because the signal in each amplifier contains, at any one instant, only a single value of instantaneous frequency and not multiple simultaneous frequencies as in SSB. These amplifiers are not sources of IMD, so they need not be "linear." The sidebands that appear in the output are a result only of the FM process (the Bessel functions).

In phase modulation, the frequency deviation is directly proportional to the frequency of the audio signal. To make the deviation independent of the audio frequency, an audio-frequency response that rolls off at 6 dB per octave is needed. An op-amp integrator circuit in the audio amplifier accomplishes this. This process

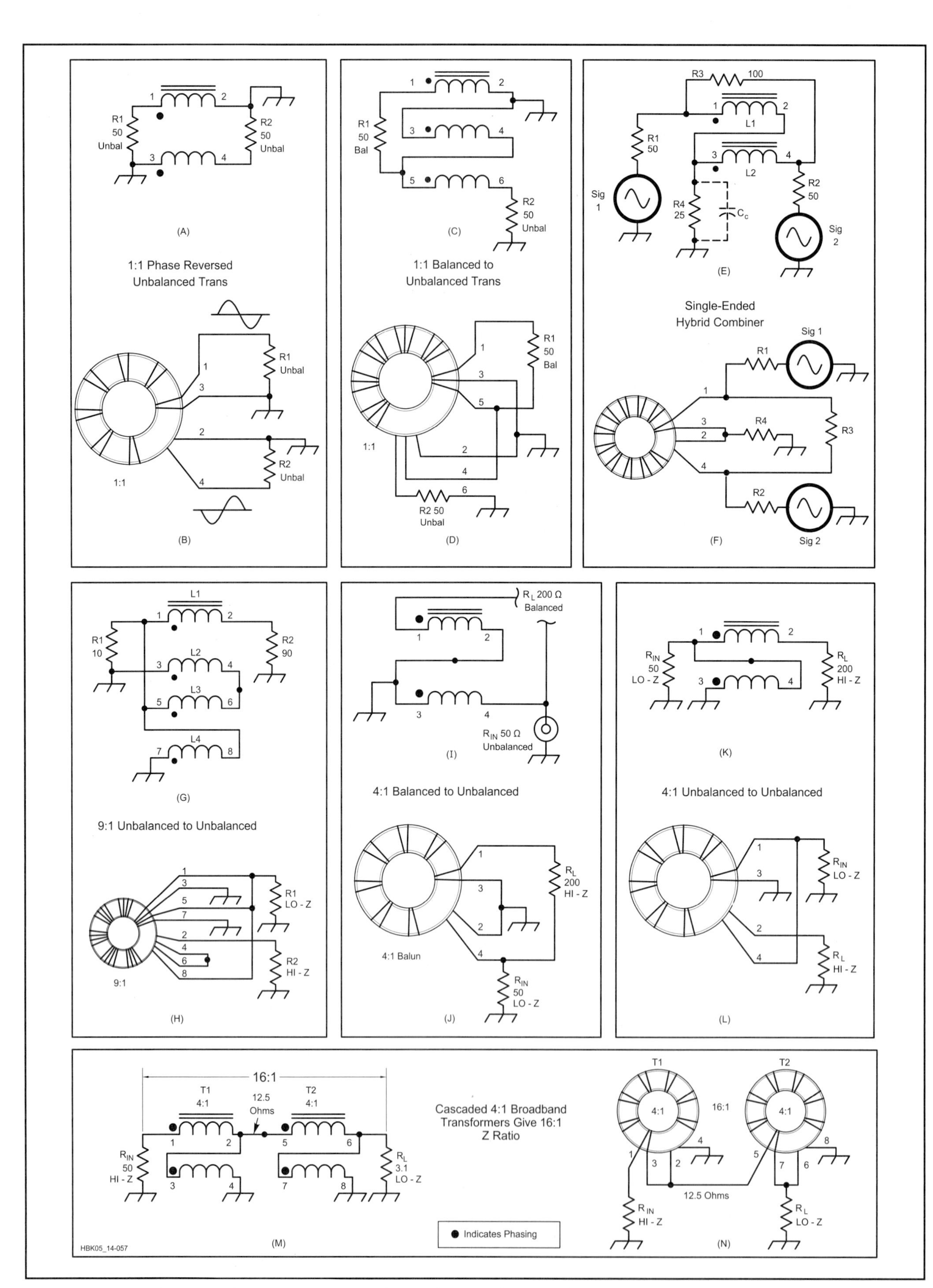

Fig 14.57—Assembly instructions for some transmission-line transformers. See text for typical magnetic materials used.

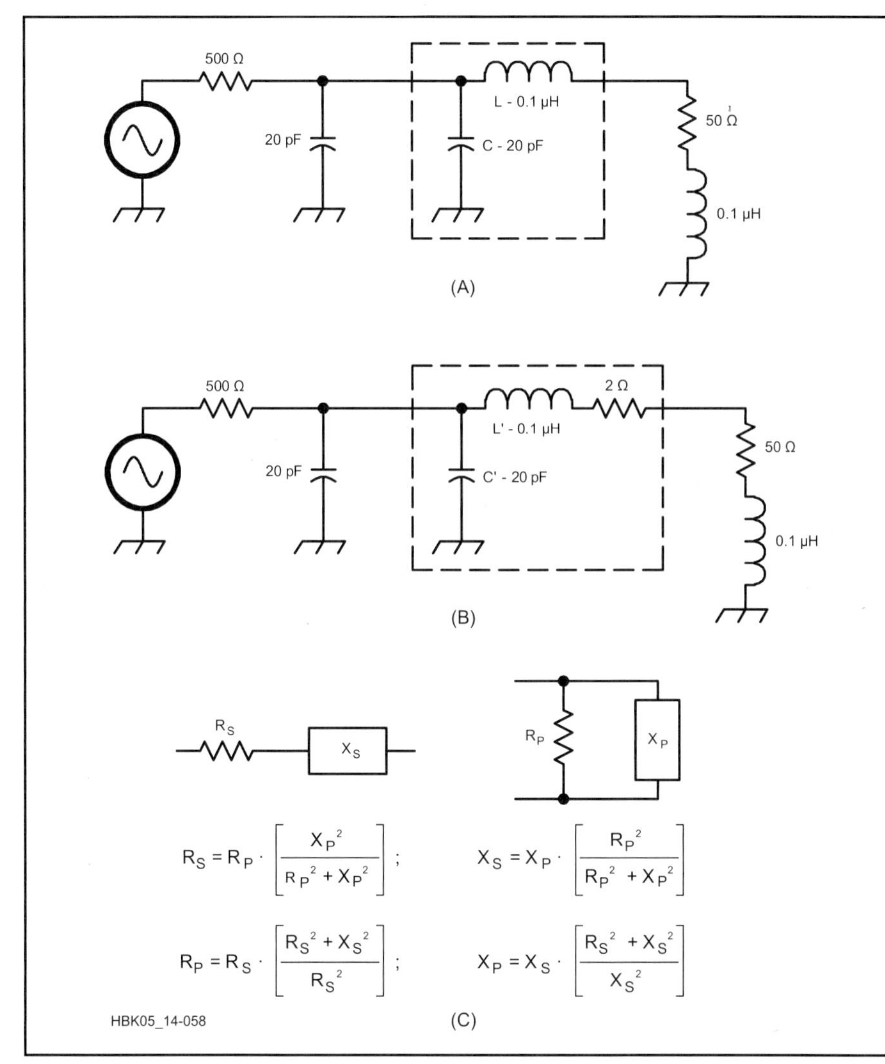

Fig 14.58—At A, impedance transformation, first iteration. At B, second iteration compensates L and C values for coil resistance. At C, series-parallel conversions.

converts phase modulation to frequency modulation. In addition, audio speech processing helps to maintain a constant value of speech amplitude, therefore constant IF deviation, with respect to audio speech levels. Also, preemphasis of the speech frequencies (6 dB per octave from 300 to 3000 Hz) is commonly used to improve the signal-to-noise ratio at the receive end. Analysis shows that this is especially effective in FM systems when the corresponding deemphasis is used at the receiver (Ref 31).

An IF limiter stage may be used to ensure that any amplitude changes created during the modulation process are removed. The indirect-FM method allows complete frequency synthesis to be used in all the transmitter LOs, so that the channelization of the output frequency is very accurate. The IF and RF amplifier stages are operated in a highly efficient Class-C mode, which is helpful in portable equipment operating on small internal batteries.

NBFM is more tolerant of frequency misalignments, between the transmitter and receiver, than is SSB. In commercial SSB communication systems, this problem is solved by transmitting a pilot carrier that is 10 or 12 dB below PEP. The receiver phase locks to this pilot carrier. The pilot carrier is also used for squelch and AGC purposes. A short-duration "memory" feature in the receiver bridges across brief pilot-carrier dropouts, caused by multipath nulls.

"Direct FM" frequency modulates a high-frequency (say, 9 MHz or so) crystal oscillator by varying the voltage on a varactor. The audio is preemphasized and processed ahead of the frequency modulator. The **Transceivers** chapter describes such a system.

References

[1]G. Gonzalez, *Microwave Transistor Amplifiers*, Englewood Cliffs, NJ, 1984, Prentice-Hall.

[2]Motchenbacher and Fitchen, *Low-Noise Electronic Design*, pp 22-23, New York NY, 1973, John Wiley & Sons.

[3]W. Hennigan, W3CZ, "Broadband Hybrid Splitters and Summers," Oct 1979 *QST*.

[4]W. Sabin, "Measuring SSB/CW Receiver Sensitivity," Oct 1992, *QST*. See also Technical Correspondence, Apr 1993 *QST*.

[5]W. Sabin, WØIYH, "A BASIC Approach to Calculating Cascaded Intercept Points and Noise Figure," Oct 1981 *QST*.

[6]Howard Sams & Co, Inc, *Reference Data for Radio Engineers*, Indianapolis, IN, p 29-2.

[7]J. Kraus and K. Carver, *Electromagnetics*, second edition, 1973, section 14-5, McGraw-Hill, NY.

[8]J. Grebenkemper, KI6WX, "Phase Noise and its Effects on Amateur Communications," Mar and Apr 1988 *QST*.

[9]W. Sabin, "Envelope Detection and AM Noise-Figure Measurement," Nov 1988 *RF Design*, p 29.

[10]H. Hyder, "A 1935 Ham Receiver," Sep 1986 *QST*, p 27.

[11]G. Breed, "A New Breed of Receiver," Jan 1988 *QST*, p 16.

[12]R. Campbell, "High-Performance, Single-Signal Direct-Conversion Receivers," Jan 1993 *QST*, p 32.

[13]A. Williams and F. Taylor, *Electronic Filter Design Handbook*, second edition, Chapter 7, McGraw-Hill, NY 1988.

[14]*RF Designer's Handbook*, Mini-Circuits Co, Brooklyn, NY.

[15]W. Sabin, "The Mechanical Filter in HF Receiver Design," Mar 1996 *QEX*.

[16]J. Makhinson, "A High-Dynamic-Range MF/HF Receiver Front End," Feb 1993 *QST*.

[17]W. Sabin and E. Schoenike, Eds., *Single-Sideband Systems and Circuits*, McGraw-Hill, 1987.

[18]B. Goodman, "Better AGC For SSB and Code Reception," Jan 1957 *QST*.

[19]A. Zverev, *Handbook of Filter Synthesis*, Wiley & Sons, 1967.

[20]R. Pittman and G. Summers, "The Ultimate CW Receiver," Sep 1952 *QST*.

[21]J. Vermasvuori, "A Synchronous Detector for AM Reception," Jul 1993 *QST*.

[22]"The Collins 75S-3 Receiver," Product Review, Feb 1962 *QST*.

[23]"The 75A-4 Receiver," Product Review, Apr 1955 *QST*.

[24]S. Prather, "The Drake R-8 Receiver,"

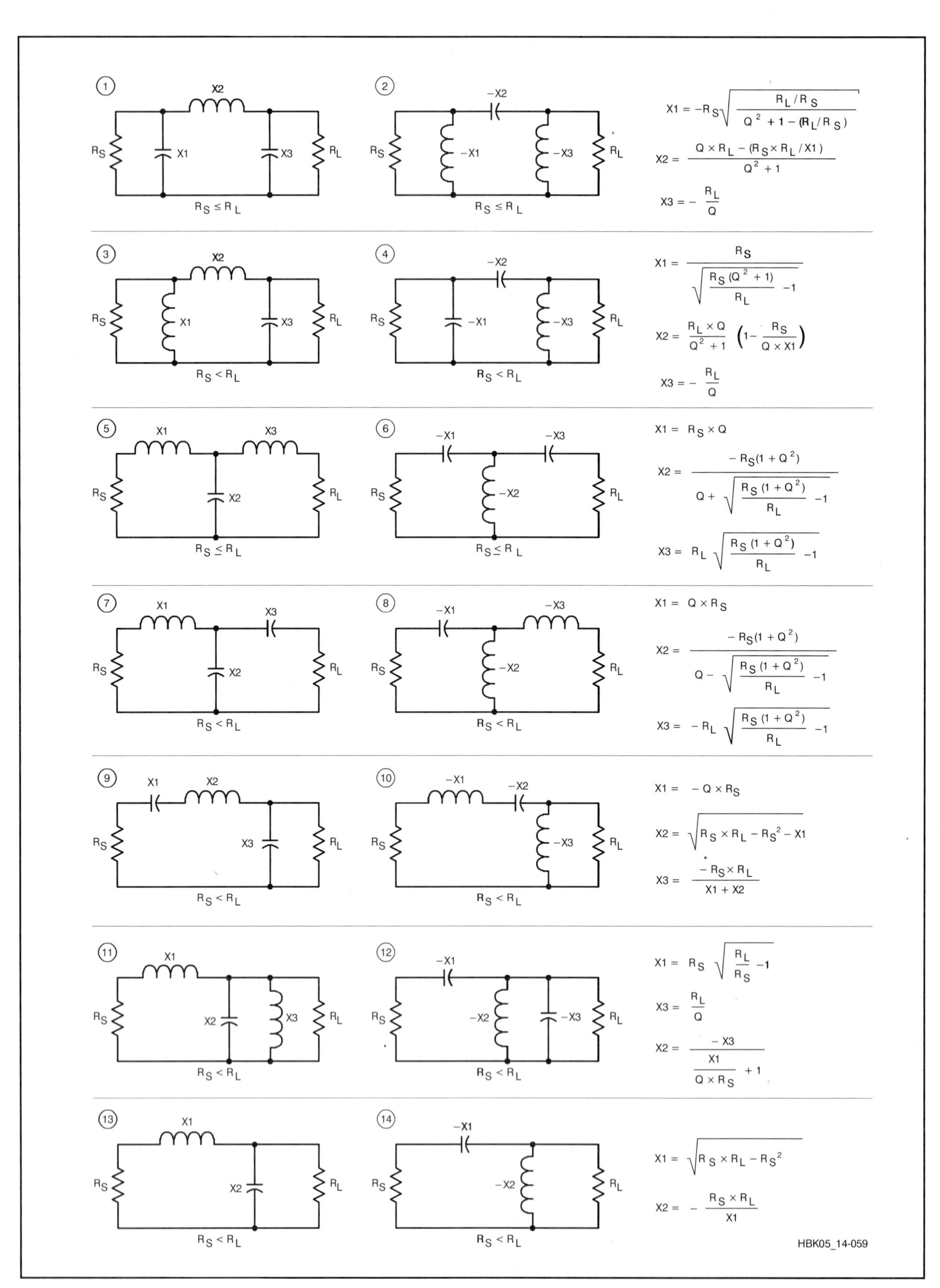

Fig 14.59—Fourteen impedance transforming networks with their design equations (for lossless components).

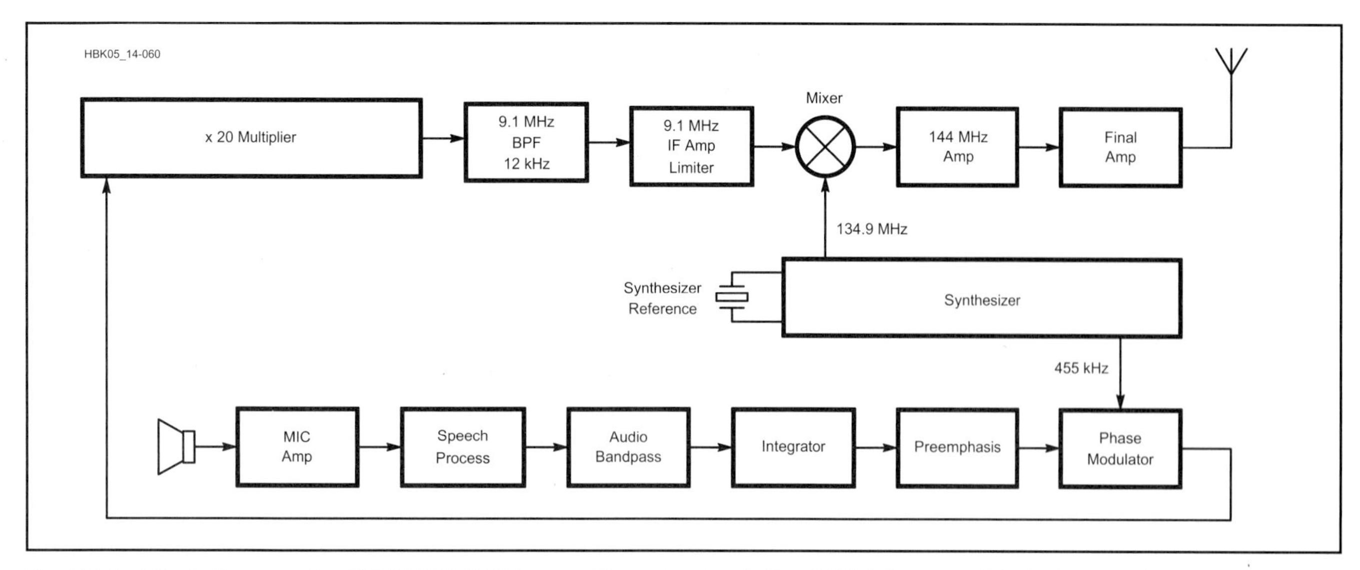

Fig 14.60—Block diagram of a VHF/UHF NBFM transmitter using the indirect FM (phase modulation) method.

Fall 1992 *Communications Quarterly*. Also see J. Kearman, "The Drake R-8 Shortwave Receiver," Mar 1992 *QST*, p 72.

[25]R. Zavrel, "Using the NE602," Technical Correspondence, May 1990 *QST*.

[26]R. Campbell, "A Single-Board, No-Tune 902 MHz Transverter," Jul 1991 *QST*.

[27]Z. Lau, "A No-Tune 222 MHz Transverter," Jul 1993 *QEX*.

[28]Z. Lau, "The Quest for 1 dB NF on 10 GHz," Dec 1992 *QEX*.

[29]Z. Lau, "Home-Brewing a 10-GHz SSB/CW Transverter," May and Jun 1993 *QST*.

[30]ARRL, *UHF/Microwave Experimenter's Manual*, 1990, p 6-50.

[31]M. Şchwartz, *Information Transmission, Modulation and Noise,* third edition, McGraw-Hill, 1980.

[32]R. Healy, "The Omni VI Transceiver," Product Review, Jan 1993 *QST*, p 65.

A Rock-Bending Receiver for 7 MHz

This simple receiver by Randy Henderson, WI5W, originally published in Aug 1995 *QST*, is a direct-conversion type that converts RF directly to audio. Building a stable oscillator is often the most challenging part of a simple receiver. This one uses a tunable crystal-controlled oscillator that is both stable and easy to reproduce. All of its parts are readily available from multiple sources and the fixed-value capacitors and resistors are common components available from many electronics parts suppliers.

THE CIRCUIT

This receiver works by mixing two radio-frequency signals together. One of them is the signal you want to hear, and the other is generated by an oscillator circuit (Q1 and associated components) in the receiver. In **Fig 14.61**, mixer U1 puts out sums and differences of these signals and their harmonics. We don't use the sum of the original frequencies, which comes out of the mixer in the vicinity of 14 MHz. Instead, we use the frequency *difference* between the incoming signal and the receiver's oscillator—a signal in the audio range if the incoming signal and oscillator frequencies are close enough to each other. This signal is filtered in U2, and amplified in U2 and U3. An audio transducer (a speaker or headphones) converts U3's electrical output to audio.

How the Rock Bender Bends Rocks

The oscillator is a tunable crystal oscillator—a variable crystal oscillator, or *VXO*. Moving the oscillation frequency of a crystal like this is often called *pulling*. Because crystals consist of precisely sized pieces of quartz, crystals have long been called *rocks* in ham slang—and receivers, transmitters and transceivers that can't be tuned around due to crystal frequency control have been said to be *rockbound*. Widening this rockbound receiver's tuning range with crystal pulling made *rock bending* seem just as appropriate!

L2's value determines the degree of pulling available. Using FT-243-style crystals and larger L2 values, the oscillator reliably tunes from the frequency marked on the holder to about 50 kHz below that point with larger L2 values. (In the author's receiver a 25-kHz tuning range was achieved.) The oscillator's frequency stability is very good.

Inductor L2 and the crystal, Y1, have more effect on the oscillator than any other components. Breaking up L2 into two or three series-connected components often works better than using one RF choke. (The author used three molded RF chokes in series—two 10-µH chokes and one 2.7-µH unit.) Making L2's value too large makes the oscillator stop.

The author tested several crystals at Y1. Those in FT-243 and HC-6-style holders seemed more than happy to react to adjustment of C7 (TUNING). Crystals in the smaller HC-18 metal holders need more inductance at L2 to obtain the same tuning range. One tiny HC-45 unit from International Crystals needed 59 µH to eke out a mere 15 kHz of tuning range.

Input Filter and Mixer

C1, L1, and C2 form the receiver's input filter. They act as a peaked *low-pass* network to keep the mixer, U1, from responding to signals higher in frequency than the 40-meter band. (This is a good idea because it keeps us from hearing video buzz from local television transmitters, and signals that might mix with harmonics of the receiver's VXO.) U1, a Mini-Circuits SBL-1, is a passive diode-ring mixer. Diode-ring mixers usually perform better if the output is terminated properly. R11 and C8 provide a resistive termination at RF without disturbing U2A's gain or noise figure.

Audio Amplifier and Filter

U2A amplifies the audio signal from U1. U2B serves as an active low-pass filter. The values of C12, C13 and C14 are

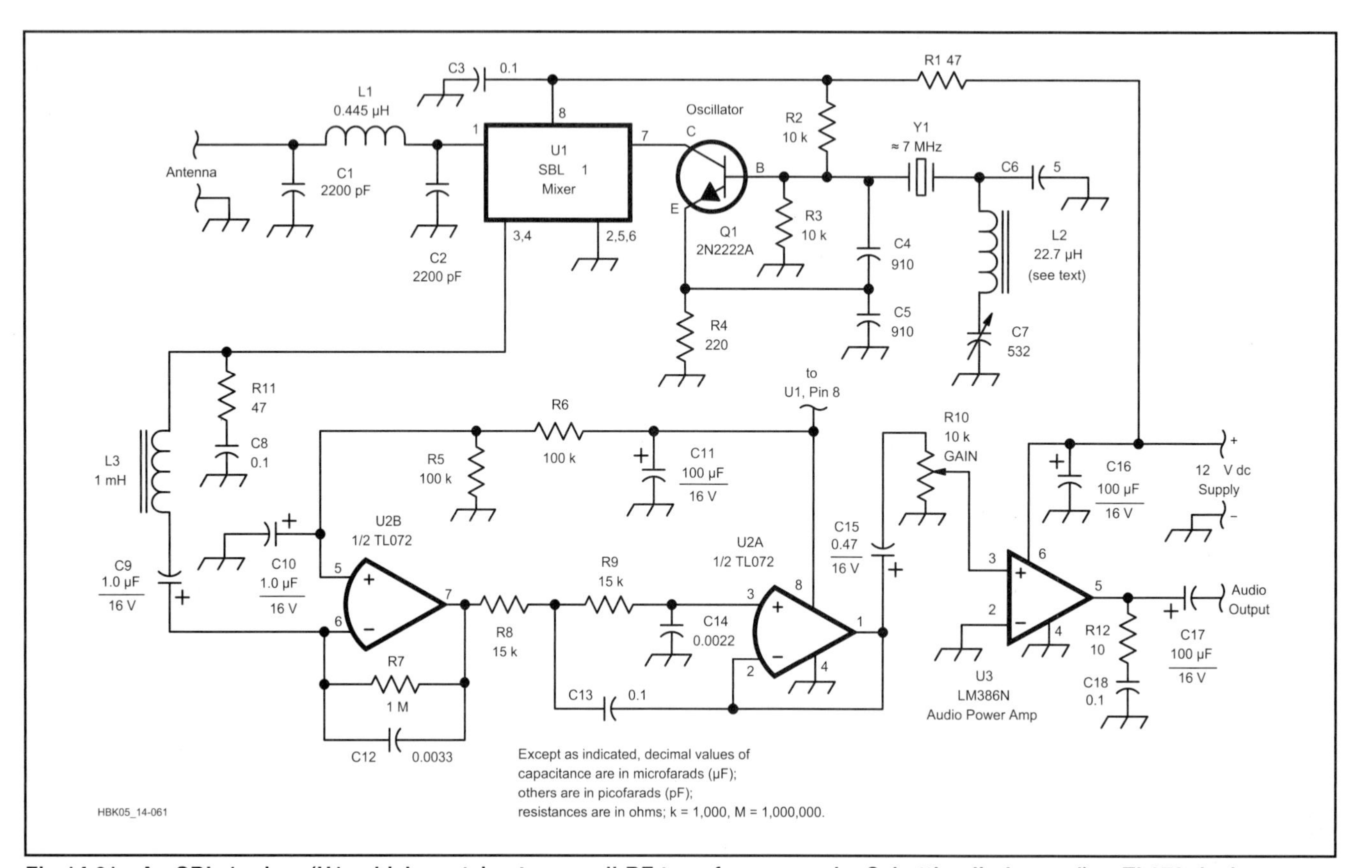

Fig 14.61—An SBL-1 mixer (U1, which contains two small RF transformers and a Schottky-diode quad), a TL072 dual op-amp IC (U2) and an LM386 low-voltage audio power amplifier IC (U3) do much of the Rock-Bending Receiver's magic. Q1, a variable crystal oscillator (VXO), generates a low-power radio signal that shifts incoming signals down to the audio range for amplification in U2 and U3. All of the circuit's resistors are 1/4-W, 5%-tolerance types; the circuit's polarized capacitors are 16-V electrolytics, except C10, which can be rated as low as 10 V. The 0.1-µF capacitors are monolithic or disc ceramics rated at 16 V or higher.

C1, C2—Ceramic or mica, 10% tolerance.
C4, C5, and C6—Polystyrene, dipped silver mica, or C0G (formerly NP0) ceramic, 10% tolerance.
C7—Dual-gang polyethylene-film variable (266 pF per section) available as #24TR218 from Mouser Electronics (800-346-6873, 817-483-4422). Screws for mounting C7 are Mouser #48SS003. A rubber equipment foot serves as a knob. (Any variable capacitor with a maximum capacitance of 350 to 600 pF can be substituted; the wider the capacitance range, the better.)
C12, C13, C14—10% tolerance. For SSB, change C12, C13 and C14 to 0.001 µF.
U2—TL072CN or TL082CN dual JFET op amp.
L1—4 turns of AWG #18 wire on 3/4-inch PVC pipe form. Actual pipe OD is 0.85 inch. The coil's length is about 0.65 inch; adjust turns spacing for maximum signal strength. Tack the turns in place with cyanoacrylic adhesive, coil dope or Duco cement. (As a substitute, wind 8 turns of #18 wire around 75% of the circumference of a T-50-2 powdered-iron core. Once you've soldered the coil in place and have the receiver working, expand and compress the coil's turns to peak incoming signals, and then cement the winding in place.)
L2—Approximately 22.7 µH; consists of one or more encapsulated RF chokes in series (two 10-µH chokes [Mouser #43HH105 suitable] and one 2.7-µH choke [Mouser #43HH276 suitable] used by author). See text
L3—1-mH RF choke. As a substitute, wind 34 turns of #30 enameled wire around an FT-37-72 ferrite core.
Q1—2N2222, PN2222 or similar small-signal, silicon NPN transistor.
R10—5 or 10-kΩ audio-taper control (RadioShack No. 271-215 or 271-1721 suitable).
U1—Mini-Circuits SBL-1 mixer.
Y1—7-MHz fundamental-mode quartz crystal. Ocean State Electronics carries 7030, 7035, 7040, 7045, 7110 and 7125-kHz units.
PC boards for this project are available from FAR Circuits.

appropriate for listening to CW signals. If you want SSB stations to sound better, make the changes shown in the caption for Fig 14.61.

U3, an LM386 audio power amplifier IC, serves as the receiver's audio output stage. The audio signal at U3's output is more than a billion times more powerful than a weak signal at the receiver's input, so don't run the speaker/earphone leads near the circuit board. Doing so may cause a squealy audio oscillation at high volume settings.

CONSTRUCTION

If you're already an accomplished builder, you know that this project can be built using a number of construction techniques, so have at it! If you're new to building, you should consider building the Rock-Bending Receiver on a printed circuit (PC) board. (The parts list tells where you can buy one ready-made.) See **Fig 14.62** for details on the physical layout of several important components used in the receiver. **Fig 14.63** shows photos of two different receivers using two different approaches to construction—one using a PC board and the other using "ugly" techniques.

If you use a homemade double-sided circuit board based on the PC pattern on the accompanying CD, you'll notice that it has more holes than it needs to. The extra holes (indicated in the part-placement diagram with square pads) allow you to connect its ground plane to the ground traces on its foil side. (Doing so reduces the inductance of some of the board's ground paths.) Pass a short length of bare wire (a clipped-off component lead is fine) into each of these holes and solder on both sides. Some of the circuit's components (C1, C2 and others) have grounded leads accessible on both sides of the board. Solder these leads on both sides of the board.

Another important thing to do if you use a homemade double-sided PC board is to countersink the ground plane to clear all ungrounded holes. (Countersinking clears copper away from the holes so components won't short-circuit to the ground plane.) A $^1/_4$-inch-diameter drill bit works well for this. Attach a control knob to the bit's shank and you can safely use the bit as a manual countersinking tool. If you countersink your board in a drill press, set it to about 300 rpm or less, and use very light pressure on the feed handle.

Mounting the receiver in a metal box or cabinet is a good idea. Plastic enclosures can't shield the TUNING capacitor from the presence of your hand, which may slightly affect the receiver tuning. You don't have to completely enclose the receiver—a flat aluminum panel screwed to a wooden base is an acceptable alternative. The panel supports the tuning capacitor, GAIN control and your choice of audio connector. The base can support the circuit board and antenna connector.

CHECKOUT

Before connecting the receiver to a power source, thoroughly inspect your work to spot obvious problems like solder bridges, incorrectly inserted components or incorrectly wired connections. Using the schematic (and PC-board layout if you built your receiver on a PC board), recheck every component and connection one at a time. If you have a digital voltmeter (DVM), use it to measure the resistance between ground and everything that should be grounded. This includes things like pin 4 of U2 and U3, pins 2, 5, 6 of U1, and the rotor of C7.

If the grounded connections seem all right, check some supply-side connections with the meter. The connection between pin 6 of U3 and the positive power-supply lead should show less than 1 Ω of resistance. The resistance between the supply lead and pin 8 of U1 should be about 47 Ω because of R1.

If everything seems okay, you can apply power to the receiver. The receiver will work with supply voltages as low as 6 V and as high as 13.5 V, but it's best to stay within the 9 to 12-V range. When first testing your receiver, use a current-limited power supply (set its limiting between 150 and 200 mA) or put a 150-mA fuse in the connection between the receiver and its power source. Once you're sure that everything is working as it should, you can remove the fuse or turn off the current limiting.

If you don't hear any signals with the antenna connected, you may have to do some troubleshooting. Don't worry; you can do it with very little equipment.

TROUBLE?

The first clue to look for is noise. With the GAIN control set to maximum, you should hear a faint rushing sound in the speaker or headphones. If not, you can use a small metallic tool and your body as a

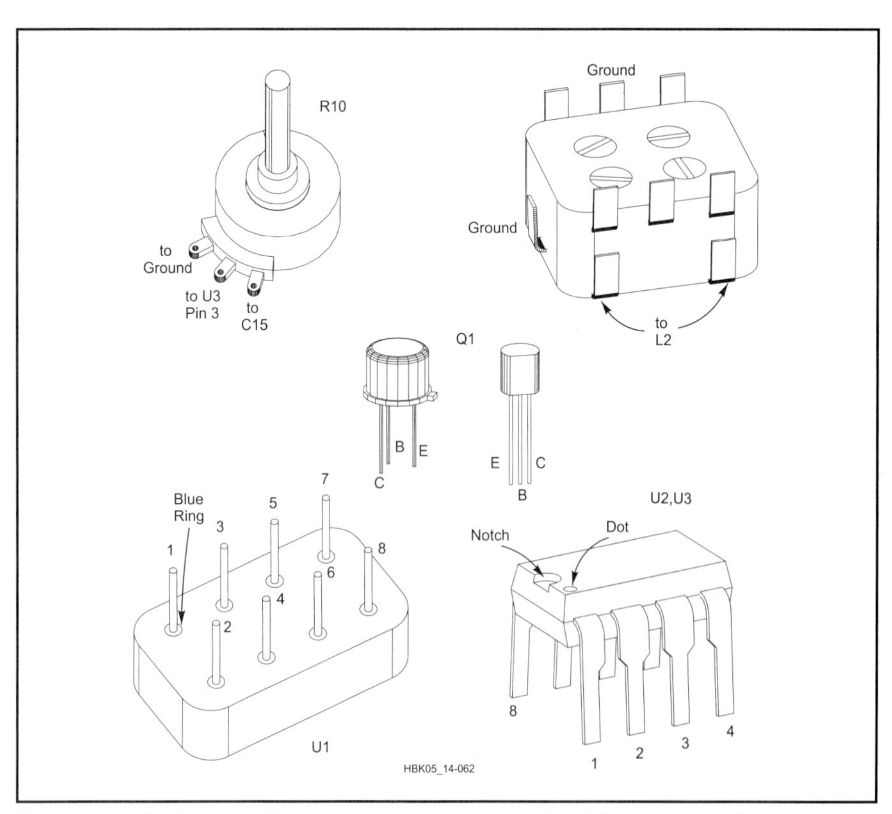

Fig 14.62—The Mouser Electronics part suggested for C7 has terminal connections as shown here. (You can use any variable capacitor with a maximum capacitance of 350 to 600 pF for C7, but its terminal configuration may differ from that shown here.) Two Q1-case styles are shown because plastic or metal transistors will work equally well for Q1. If you build your Rock-Bending Receiver using a prefab PC board, you should mount the ICs in 8-pin mini-DIP sockets rather than just soldering the ICs to the board.

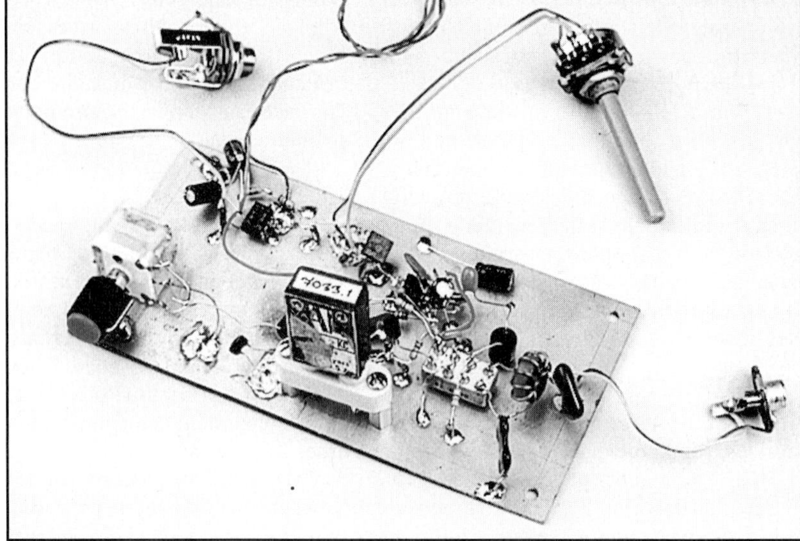

Fig 14.63—Ground-plane construction, PC-board construction—both of these approaches can produce the same good Rock Bending Receiver performance. (WI5W built the one that looks nice, and ex-W9VES—who wrote this caption—built the one that doesn't.)

sort of test-signal generator. (If you have any doubt about the safety of your power supply, power the Rock-Bending Receiver from a battery during this test.) Turn the GAIN control to maximum. Grasp the metallic part of a screwdriver, needle or whatever in your fingers, and use the tool to touch pin 3 of U3. If you hear a loud scratchy popping sound, that stage is working. If not, then something directly related to U3 is the problem.

You can use this technique at U2 (pin 3, then pin 5) and all the way to the antenna. If you hear loud pops when touching either end of L3 but not the antenna connector, the oscillator is probably not working. You can check for oscillator activity by putting the receiver near a friend's transceiver (both must be in the same room) and listening for the VXO. Be sure to adjust the TUNING control through its range when checking the oscillator.

The dc voltage at Q1's base (measured without the RF probe) should be about half the supply voltage. If Q1's collector voltage is about equal to the supply voltage, and Q1's base voltage is about half that value, Q1 is probably okay. Reducing the value of L2 may be necessary to make some crystals oscillate.

OPERATION

Although the Rock-Bending Receiver uses only a handful of parts and its features are limited, it performs surprisingly well. Based on tests done with a Hewlett-Packard HP 606A signal generator, the receiver's minimum discernible signal (by ear) appears to be 0.3 μV. The author could easily copy 1-μV signals with his version of the Rock-Bending Receiver.

Although most HF-active hams use transceivers, there are advantages in using separate receivers and transmitters. This is especially true if you are trying to assemble a simple home-built station.

A Wideband MMIC Preamp

This project illustrates construction techniques used in the microwave region (at and beyond 1 GHz). It also results in a neat "dc to daylight" preamplifier with many uses around your shack, not the least of which is monitoring the downlinks from Amateur Radio satellites. The original article was written by William Parmley, KR8L, in Nov 1997 *QST*.

The preamplifier uses the MAR-6 monolithic microwave integrated circuit (MMIC) manufactured by Mini-Circuits Labs. The MAR-6 is a four terminal, surface mount device (SMD) with an operating frequency range from dc to 2 GHz, a noise figure of 3 dB, a gain of up to 20 dB, and input and output impedances of 50 Ω. The basic concept for the preamplifier and the construction techniques used to build it came from *The ARRL UHF/Microwave Experimenter's Manual*. The parts and circuit board material in this project are readily available from sources such as Ocean State Electronics.

CIRCUIT DESCRIPTION

Fig 14.64 is the schematic for the preamplifier. C1 and C2 are dc blocking capacitors. The device receives at the output lead, through RF choke L1 and limiting resistor R1. The only other components used are the bypass capacitors on the lead. C1 and C2 should present a low impedance at the lowest signal frequency of interest. The author designed his preamplifier for 435 MHz, a downlink frequency for many amateur satellites. Two 220 pF disc ceramic capacitors were used for C1 and C2. To use the preamplifier at 29 MHz for downlink signals from Russian RS-series satellites, C1 and C2 become 0.001 μF disc ceramic capacitors.

The power-supply voltage determines R1's value. The MAR-6 draws about 16 mA, and needs a V_{cc} of about 3.5 V. Use Ohm's Law to calculate the necessary voltage drop from your power supply voltage down to 3.5 V. The author's power

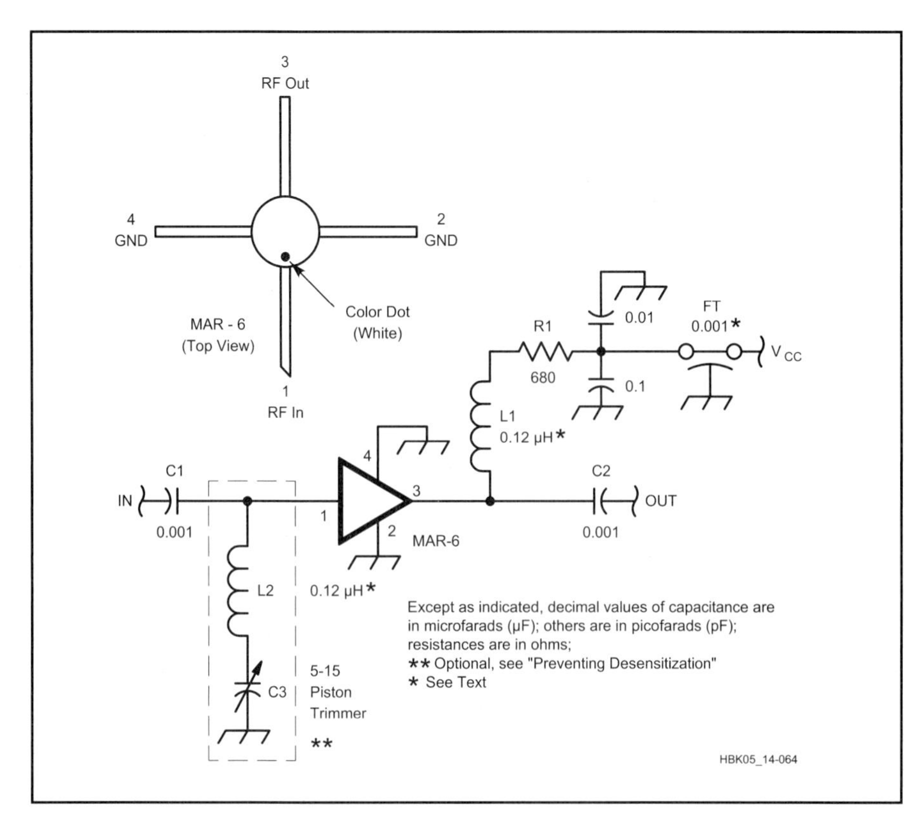

Fig 14.64—A schematic of the preamp circuit. Equivalent parts may be substituted.

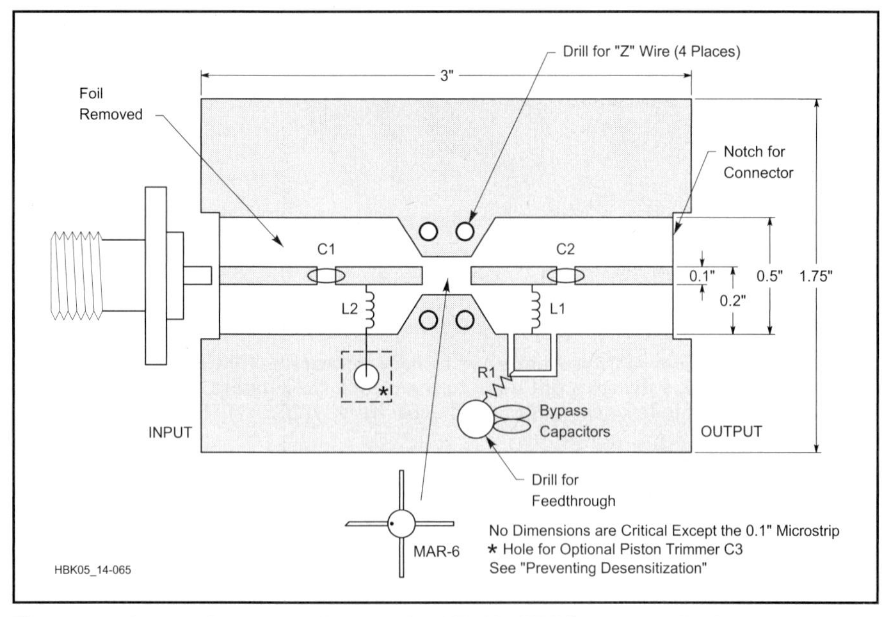

Fig 14.65—A part-placement diagram for KR8L's MMIC preamp. Dark areas are copper on the component (near) side. The reverse is a copper ground plane.

supply provided about 14.6 V, so a 680 Ω, 1/2-W resistor was used for R1. RF choke L1 helps isolate the power supply from the MMIC output. L1 is a homemade 0.12 µH choke, consisting of 8 turns of #30 enameled wire around the shank of a 3/16-inch drill bit, spaced for a total length of 0.3 inches. (Remove the drill bit; it's only a winding mandrel!)

This value of L1 was left in place when the preamplifier was used at lower frequencies. The remaining three essential parts are bypass capacitors. Because capacitors have self-resonant frequencies (resulting from unavoidable inductances in the devices and their leads), it is a common practice to use capacitors of several different values in parallel. This design uses a 0.001 µF feedthrough capacitor passing through the circuit board ground plane. The parallel 0.01 µF and 0.1 µF capacitors are disc ceramics. L2 and C3 are optional components for 432 MHz, used to provide some selectivity against desensitization when transmitting on 144 MHz for satellite Mode J, at the expense of wideband coverage, of course.

CIRCUIT CONSTRUCTION

Fig 14.65 shows the circuit-board layout. The material used is double-sided, glass-epoxy board with a thickness of 0.0625 inches, known as FR-4 or G-10. This is the least expensive board material suited for microwave use. (The board I used is a product of GC/Thorsen in Rockford, Illinois.) Notice that most of the top of the board, and all the bottom side of the board, serves as circuit ground.

The signal-conducting part of the circuit is a "microstrip." (That is a strip-type transmission line: a conductor above or between extended conducting surfaces.—*Ed.*) The line width, board thickness and board dielectric constant determine the microstrip's characteristic impedance. A 0.1-inch-wide line and the ground plane on 0.0625-inch-thick G-10 form a 50-Ω transmission line, which matches the MMIC's input and output impedance.

The author fabricated his board by laying out the traces with a machinist's rule. Then he cut through the copper foil with a knife and lifted off the unwanted copper areas while heating them with a 100-W soldering gun. You could etch the board if you prefer, or use any other method you like. The single mounting pad was "etched" by grinding away the copper with a hand-held grinder.

The MMIC is tiny. Connect it to the traces with the shortest possible distances between the traces and the body. (The author managed to achieve about 0.03 inch.) Also, the device leads are very delicate—if possible, do not bend them at all. To fit the MMIC leads flat on the PC-board traces without bending, a small depression was ground in the board dielectric for the MMIC body. Remember that, viewed from the top, the colored dot (white on the MAR-6) on the body marks pin 1, the input lead. The other leads are numbered counterclockwise; pin 3 is the output lead.

Mount the blocking capacitors as close to the board as possible. To do this, the capacitor leads were cut to about 1/16-inch long. Both the capacitor leads and circuit traces were tinned and then the capacitors were soldered in place. This method of mounting minimizes lead inductance.

The author installed N connectors for his unit. To achieve a "zero lead length," he notched the ends of the board to fit the profile of the connectors and installed the connectors directly to the board. The center pin was laid on top of the microstrip and soldered. Then the connector body was soldered to the ground foil in four places: two on the top of the board and two on the bottom. Another very good technique is to drill a hole in the microstrip and insert the center pin from the bottom of the board. The center pin is then soldered to the microstrip, and the body is soldered to the ground foil or mounted with machine screws. (If you do this, be sure to remove a bit of foil from around the hole on the bottom side so the center pin doesn't short to ground.) The latter approach is much better if you want to mount the preamplifier into a box. You can mount the board on the inside of the lid with the connectors projecting through.

It is important that all portions of the ground foil be at equal potential, particularly near the MMIC and the board edges. To achieve this, wrap the long edges of the board with pieces of 0.003-inch-thick brass shim stock and solder them on both top and bottom. Thinner or thicker material is suitable (up to about 0.005 inch), as is copper flashing. Two small holes were drilled on either side of each MMIC ground lead, and a small Z-shaped wire was soldered to each side of the board. (A Z wire is a short, small-gauge, solid copper wire bent 90°, inserted through the hole, bent 90° again and soldered on both sides of the board.)

The inductor is also mounted using minimal-length leads. One lead connects to the microstrip and the other to the square pad. The resistor connects from the pad to the feedthrough capacitor, and the other two bypass capacitors connect from the feed-through to the ground foil.

HOOKUP AND OPERATION

For the basic preamplifier design there is nothing to align or adjust. Simply connect the preamplifier between your antenna and receiver and apply power. If you connect the preamp to a transceiver, take precautions to prevent transmitting through the preamp! This preamplifier is very handy for many uses: adding gain to an older 10-meter receiver or scanner, boosting signal-generator output or for casual monitoring of the Amateur Radio satellites on 29, 145 and 435 MHz. A commercial metal box, home-made PC board or thin sheet-metal boxes make suitable cabinets for this project.

A Binaural I-Q Receiver

This little receiver was designed and built by Rick Campbell, KK7B. It was first described in the March 1999 issue of *QST*. It replaces the narrow filters and interference-fighting hardware and software of a conventional radio with a wide-open *binaural I-Q detector*. If you liken a conventional receiver to a high-powered telescope, this receiver is a pair of bright, wide-field binoculars. The receiver's classic junk-box-available-parts construction approach achieves better RF integrity than that of much commercial ham gear. A PC board and parts kit is available for those who prefer to duplicate a proven design.[1] The total construction time was only 17 hours. There are a number of toroids to wind, and performance was not compromised to simplify construction or reduce parts count. **Fig 14.66** is a photo of the front panel built by KK7B.

Fig 14.66—A receiver with presence . . . to fully appreciate this receiver, you've got to hear it! "Once my ears got used to the effect, they had to drag me away from this radio. This is one I gotta have!"—*Ed Hare, W1RFI, ARRL Lab Supervisor*

BINAURAL I-Q RECEPTION

Modern receivers use a combination of band-pass filters and digital signal processing (DSP) to select a single signal that is then amplified and sent to the speaker or headphones. When DSP is used, the detector often takes the form shown in **Fig 14.67**. The incoming signal is split into two paths, then mixed with a pair of local oscillators (LOs) with a relative 90° phase shift. This results in two baseband signals: an in-phase, or *I* signal, and a quadrature, or *Q* signal. Each of the two baseband signals contains all of the information in the upper and lower sidebands. The baseband pair also contains all of the information needed to determine whether a signal is on the upper or lower sideband before multi-

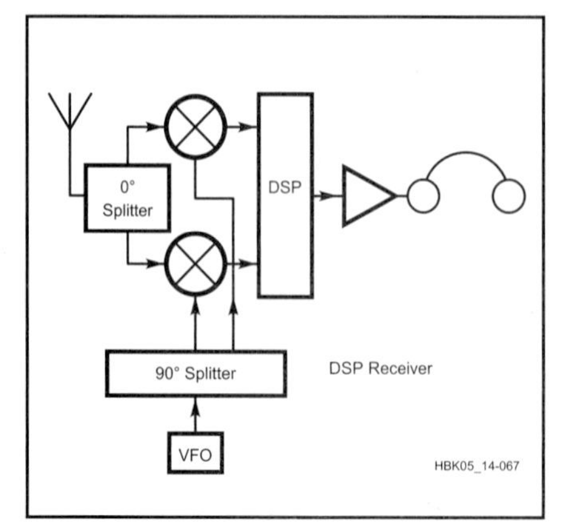

Fig 14.67—The simplified block diagram of a receiver using a DSP detector; see text.

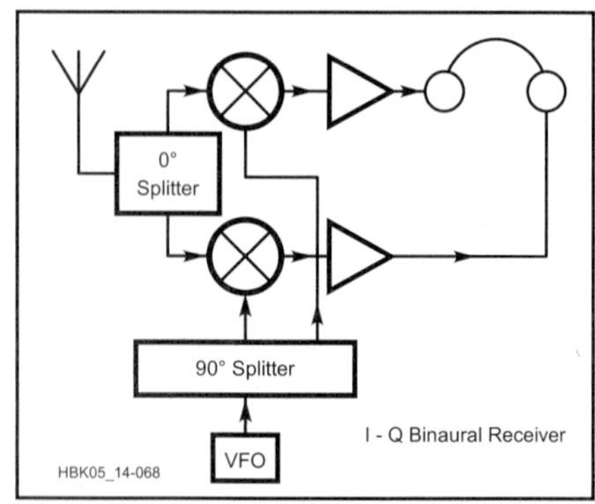

Fig 14.68—The block diagram of a binaural I-Q receiver that allows the ear/brain combination to process the detector output, resulting in stereo-like reception.

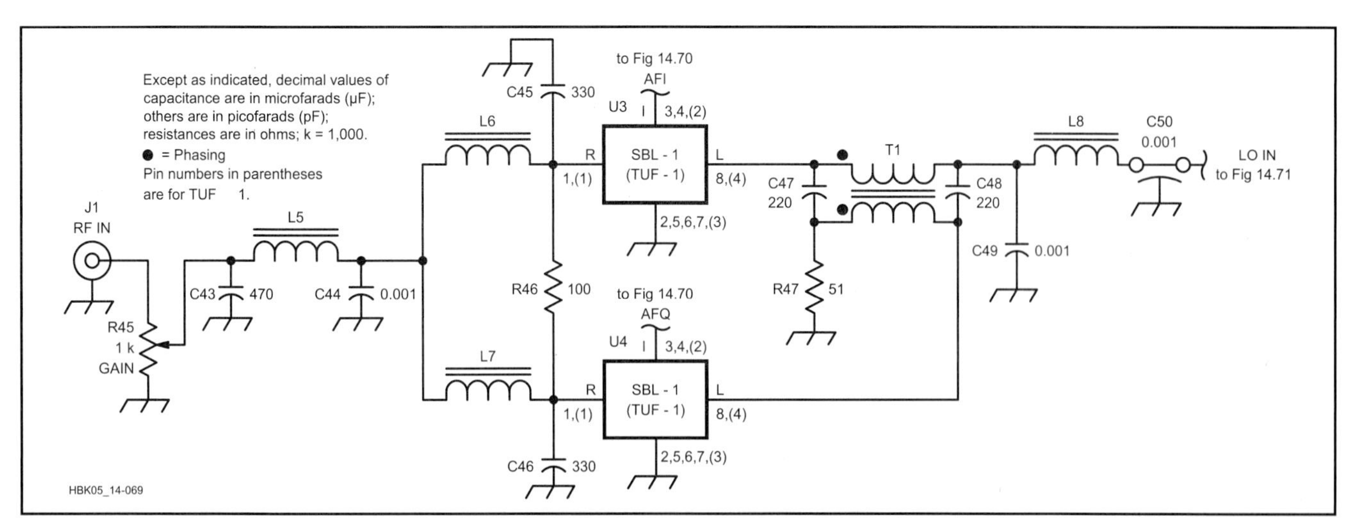

Fig 14.69—This diagram shows the front end and *I* and *Q* demodulators of the Binaural Weekender receiver. Unless otherwise specified, resistors are 1/4 W, 5% tolerance carbon-composition or film units. Equivalent parts can be substituted. Pin connections for the SBL-1 and TUF-1 mixers at U3 and U4 are shown; the TUF-1 pin numbers are in parentheses. A kit is available (see Note 1). Parts are available from several distributors including Digi-Key Corp, Mouser Electronics, and Newark Electronics.

C43—470 pF disc ceramic.
C44, C49—0.001 µF metal polyester.
C45, C46—330 pF disc ceramic.
C47, C48—220 pF disc ceramic.
C50—0.001 µF feed-through capacitor.
J1—Chassis-mount female BNC connector.
L5—1.6 µH, 24 turns #28 enameled wire on T-30-6 powdered-iron core.
L6, L7—1.3 µH, 21 turns #28 enameled wire on T-30-6 powdered-iron core.
L8—350 nH, 11 turns #28 enameled wire on T-30-6 powdered-iron core.
R45—1 kΩ panel-mount pot.
T1—17 bifilar turns #28 enameled wire on T-30-6 powdered-iron core.
U3, U4—Mini-Circuits SBL-1 or TUF-1 mixer.

plication. An analog signal processor consisting of a pair of audio phase-shift networks and a summer could be used to reject one sideband. In a DSP receiver, the *I* and *Q* baseband signals are digitized and the resulting sets of numbers are phase-shifted and added.

The human brain is a good processor for information presented in pairs. We have two eyes and two ears. Generally speaking, we prefer to observe with both eyes open, and listen with both ears. This gives us depth of field and three-dimensional hearing that allows us to sort out the environment around us. The ear/brain combination can be used to process the output of the I-Q detectors as shown in **Fig 14.68**.

The sound of CW signals on a binaural I-Q receiver is like listening to a stereo recording made with two identical microphones spaced about six inches apart. The same information is present on each channel, but the *relative phase* provides a stereo effect that is perceived as three-dimensional space. Signals on different sidebands—and at different frequencies—appear to originate at different points in space. Because SSB signals are composed of many audio frequencies, they sound a little spread in the perceived three-dimensional sound space. This spreading also occurs with most sounds encountered in nature, and is pleasant to hear.

To keep the receiver as simple as possible, a single-band direct-conversion (D-C) approach is used. A crystal-controlled converter can be added for operation on other bands, changing the receiver to a single-conversion superhet. Alternatively, the binaural I-Q detector can be used in a conventional superhet, with a tunable first converter and fixed-frequency BFO. If proper receiver design rules are followed, there is no advantage to either design over the other.

THE RECEIVER

Figs 14.69, **14.70** and **14.71** show the complete receiver schematic. In Fig 14.69, signals from the antenna are connected directly to a 1-kΩ GAIN pot on the front panel. J1 is a BNC antenna connector, popular with QRP builders. Adjusting the gain before splitting the signal path avoids the need for a two-gang volume control, and eliminates having to use separate RF and AF-gain adjustments. This volume-control arrangement leaves the "stereo background noise" constant and varies the signal-to-noise ratio. The overall gain is selected so that the volume is all the way up when the band is quiet. Resistor values R9 and R31 may be changed to modify the overall gain if required. After the volume control, the signal is split with a Wilkinson divider and connected to two SBL-1 diode-ring mixers. (The TUF-1 is a better mixer choice, but I had more SBL-1s in my junk box.) The VFO signal is fed to the two mixers through a quadrature hybrid, described by Reed Fisher.[2] All of the circuitry under the chassis is broadband, and there are *no* tuning adjustments.

The audio-amplifier design of Fig 14.70 is derived from that used in the R1 High-Performance Direct-Conversion Receiver,[3] with appropriate simplifications. The R1 high-power audio output is not needed to drive headphones, the low-pass filter is eliminated, and the diplexer has fewer components. Distortion performance is not compromised—well over 60 dB of in-band two-tone dynamic range is available. The original article, and the additional notes in Technical Correspondence for February 1996,[4] describe the audio-amplifier chain in detail.

THE VFO

Fig 14.71 is the schematic of the receiver VFO, a JFET Hartley oscillator with a JFET buffer amplifier. Components for the VFO tuned circuit are chosen for linear tuning from 7.0 to 7.3 MHz with the available junk-box variable capacitor. Setting up the VFO is best done with a frequency counter, receiver and oscilloscope. The frequency counter makes it easy to select the parallel NPO capacitors and

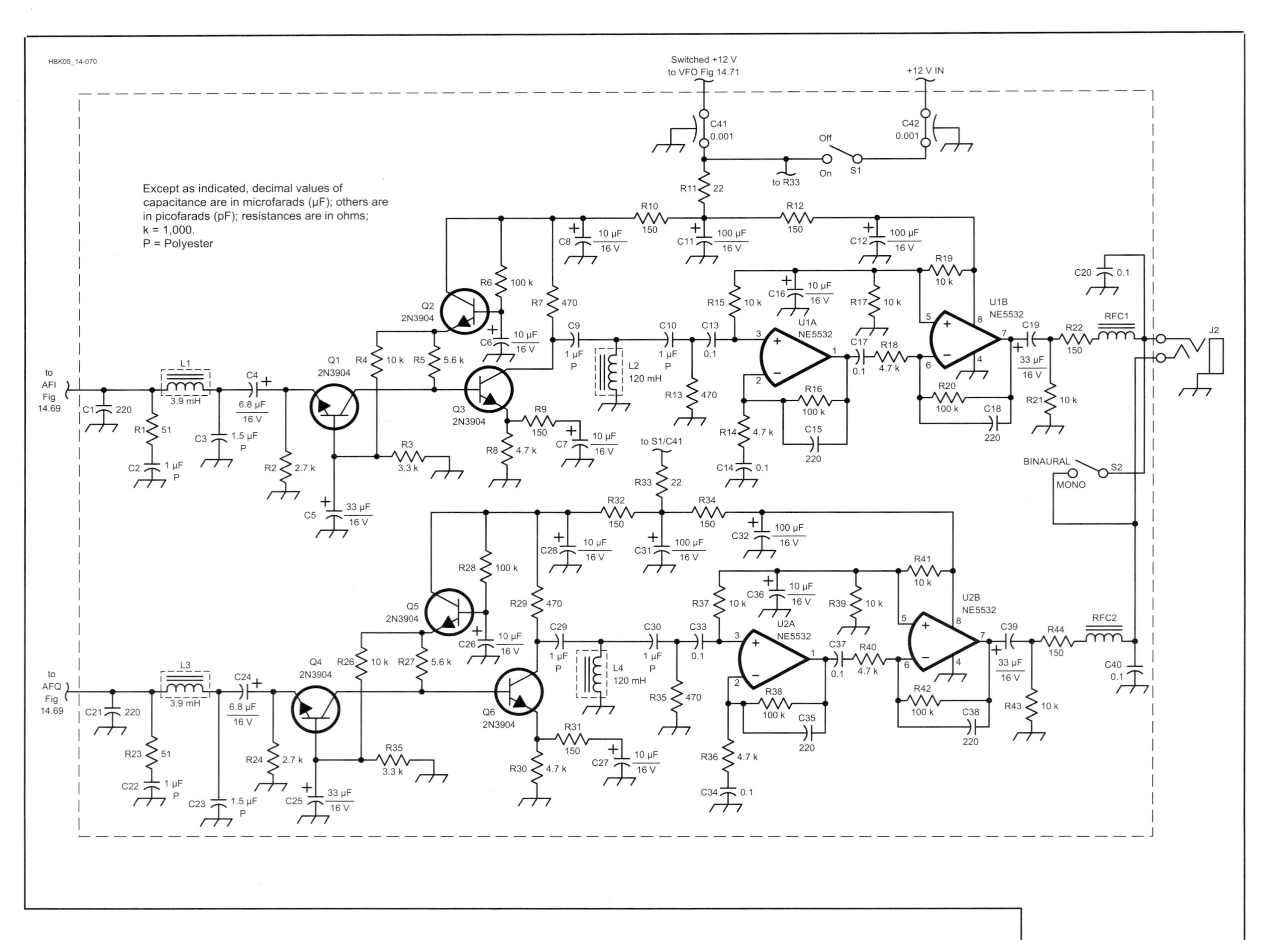
HBK05_14-070
Except as indicated, decimal values of capacitance are in microfarads (µF); others are in picofarads (pF); resistances are in ohms;
k = 1,000.
P = Polyester
Switched +12 V to VFO Fig 14.71
+12 V IN
to R33
to S1/C41
Off
On
S1
S2
BINAURAL
MONO
J2
to AFI Fig 14.69
to AFQ Fig 14.69
U1A NE5532
U1B NE5532
U2A NE5532
U2B NE5532
Q1 2N3904
Q2 2N3904
Q3 2N3904
Q4 2N3904
Q5 2N3904
Q6 2N3904
L1 3.9 mH
L2 120 mH
L3 3.9 mH
L4 120 mH
RFC1
RFC2

Fig 14.70—This diagram shows the receiver audio-amplifier design.
C1, C15, C18, C21, C35, C38—220 pF disc ceramic.
C2, C9, C10, C22, C29, C30—1 µF metal polyester (Panasonic ECQ-E(F) series).
C3, C23—1.5 µF metal polyester (Panasonic ECQ-E(F) series).
C4, C24—6.8 µF, 16 V electrolytic (Panasonic KA series).
C5, C19, C25, C39—33 µF, 16 V electrolytic (Panasonic KA series).
C6, C7, C8, C16, C26, C27, C28, C36—10 µF, 16 V electrolytic (Panasonic KA series).
C11, C12, C31, C32—100 µF, 16 V electrolytic (Panasonic KA series).
C13, C14, C17, C20, C33, C34, C37, C40—0.1 µF metal polyester (Panasonic V series).
C41, C42, C50—0.001 µF feed-through capacitor.
J2—1/8-inch stereo phone jack.
L1, L3—3.9 mH Toko 10RB shielded inductor.
L2, L4—120 mH Toko 10RB shielded inductor.
Q1 through Q6—2N3904.
RFC1, RFC2—10 turns #28 enameled wire on Amidon ferrite bead FB 43-2401 (six-hole bead).
S1, S2—SPST toggle switch.
U1, U2—NE5532 dual low-noise high-output op amp.

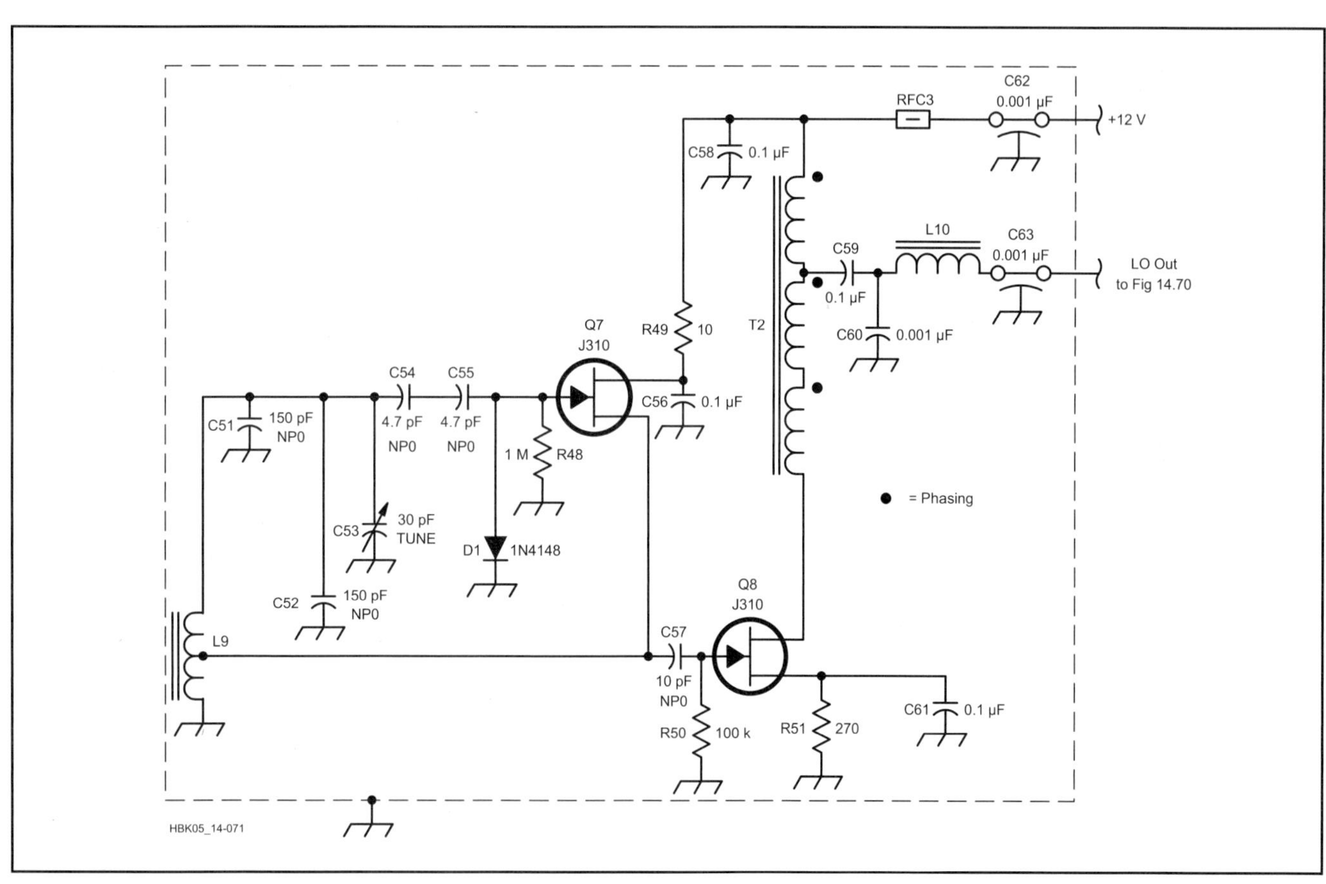

Fig 14.71—The diagram shows the prototype binaural receiver's VFO. The LO output is +10 dBm. This simple VFO works exceptionally well, but must be completely shielded for good D-C receiver performance. A receiver with an open PC-board VFO will work better if the variable oscillator is not running on the received frequency. As noted elsewhere, the kit version of the receiver uses a different VFO.

C51, C52—150 pF, NPO disc ceramic.
C53—30 pF air-dielectric variable.
C54, C55—4.7 pF NPO disc ceramic.
C56, C57, C59, C61—0.1 µF metal polyester (Panasonic V series).
C57—10 pF NPO disc ceramic.
C60—0.001 pF metal polyester.
C62, C63—0.001 µF feedthrough capacitor.
D1—1N4148.
L9—1.5 µH, 22 turns #22 enameled wire on T-37-6 powdered-iron core; tap 5 turns from ground end.
L10—350 nH, 11 turns #28 on T-30-6 powdered-iron core.
Q7, Q8—J310 (U310 used in prototype).
RFC3—10 turns #28 enameled wire on Amidon ferrite bead FB 43-2401 (six-hole bead used in prototype).
T2—10 trifilar turns #28 enameled wire on Amidon ferrite bead FB 43-2401 (six-hole bead used in prototype).

squeezing and spreading the wire turns on L1 achieves the desired tuning range. After the tuning range is set, listen to the VFO signal with a receiver to make sure the VFO tunes smoothly and has a good note. Interrupt the power to hear its start-up chirp. The signal may sound ratty with the frequency counter on, so turn it off. The VFO is one area where craftsmanship pays off. Solid construction, a self-aligning variable-capacitor mounting, complete RF and air shielding and good capacitor bearings all contribute to a receiver that is a joy to tune.

Both connections to the VFO compartment are made with feed-through capacitors. The power supply connection is self-explanatory, but passing RF through a feed-through capacitor (at LO Out) may seem a bit unusual. Electrically, the capacitor is one element of a low-pass pi network. Using feed-through capacitors keeps local VHF signals (high-powered FM broadcast and TV signals near my location) out of the VFO compartment. A second pi network feeds the VFO signal to the detector circuit below the chassis. The use of VHF construction techniques in a 40-meter receiver may seem like overkill, but the present KK7B location is line-of-sight to broadcast towers serving the Portland, Oregon area. Using commercial HF gear with conventional bypassing under these circumstances provided disappointing results.

Fig 14.66 shows the prototype receiver front panel. Receiver controls are simple and intuitive. The ear/brain adjusts so naturally to binaural listening that I added a BINAURAL/MONO switch to provide a quick reminder of how signals sound on a conventional receiver. The switch acts

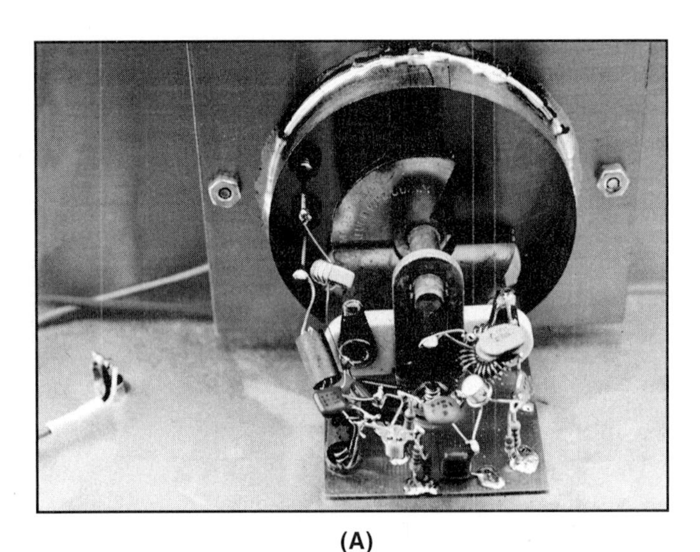

(A)

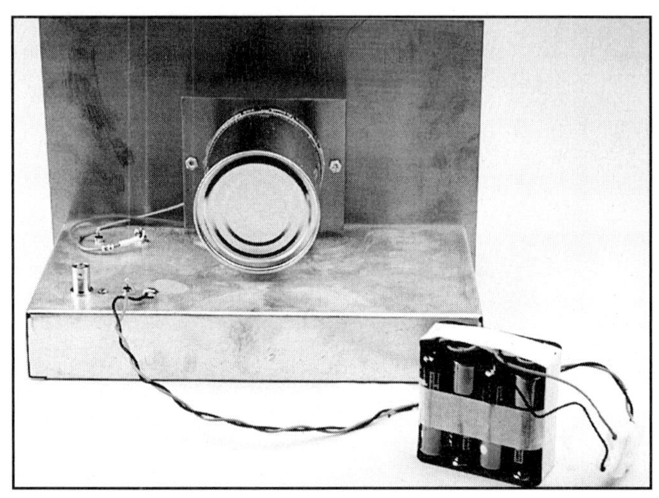

(B)

Fig 14.72—A shows a close-up of the VFO. The simple VFO used in the prototype works exceptionally well, but must be completely shielded for good D-C receiver performance. B shows how an empty mushroom can can live again as a VFO shield in the prototype receiver.

What Do You Hear?

Even the earliest solid-state direct-conversion (D-C) receivers had a *presence* or *clarity* that is rarely duplicated in more elaborate receivers. Many of us remember the first time we heard this crispness in a "homebrewed" D-C receiver. As we try to "enhance" our rigs through the addition of IF filters and other "features," we still hope that the result will be as clean as that first D-C receiver.

This binaural D-C receiver is such an experience—but even better. The binaural processing supplies the ears with additional information without compromising what was already there, enhancing the presence.

As you tune through a CW signal on a quiet band (best done with your eyes closed while sitting in a solid chair), a centered signal enters, but moves to the left background, undergoes circular motions at the back of your head as you tune through zero beat, repeats the previous gyrations on the right side, fades to the right background, and finally drops away in the center. Multiple signals within the receiver passband are distributed throughout this perceived space. With training, concentration on one signal allows it to be copied among the many. An SSB signal seems to occupy parts of the space, left and right, with clarity when properly tuned, leaving others vacant. Static crashes and white noise appear distributed throughout the entire space without well defined position. Receiver noise, although present, has no perceived position.

It's vital that this receiver include a front-panel switch to shift between binaural and monaural output. Although useful during the learning process, it becomes indispensable for the demonstrations that you will want to do. I used the switch to set up my son, Roger, KA7EXM, for the experience. We entered the shack and I handed him the headphones. He put one phone to just one ear, but I told him that he had to use both, that it would not work with just one. He put the phones on his head, casually tuned the receiver through the 40-meter CW band, removed the phones and commented, "Well, it sounds just like a direct-conversion receiver: A good one, but still just a direct-conversion receiver." I smiled and asked him to put the headphones on again. As I flipped the switch to the binaural position his hand reached out, seeking the support of the workbench. His facial expression became more serious. He eased into the chair and began tuning the receiver, very slowly at first. After a minute he took the headphones off, but remained speechless for a while—an unusual condition for Roger. Finally, he commented, "Wow! The appliance guys have never heard that!"

A builder of the Binaural Weekender should prepare for some truly unusual experiences.—*Wes Hayward, W7ZOI*

much like the STEREO-MONO switch on an FM broadcast receiver—given the choice, it always ends up in the STEREO position!

The author uses a pair of Koss SG-65 headphones with his receiver. They are not necessary, but have some useful features. First, at about $32, they are relatively inexpensive. Second, they have relatively high-impedance drivers, (90 Ω) so they can be driven at reasonable volume directly from an op amp. Finally, they make an attempt at low distortion. Other headphones in the same price bracket are acceptable, but some have much lower impedance and won't provide a very loud audio signal using the component values given in the schematic. Those $2.95 bubble-packed, throw-away headphones are not a good choice! Audiophile headphones are fine, but don't really belong on an experimenter's bench. A stray clip-lead brushing across the wrong wire in the circuit can instantly burn out a driver and seriously ruin your day.

BUILDING A BINAURAL WEEKENDER

A few construction details are generally important, while others were determined by the components that happened to be in my junk box. The big reduction drive is delightful to use, but doesn't contribute to electrical performance. I purchased it at a radio flea market. The steel chassis provides a significant reduction in magnetic hum pickup, something that can be a problem if the receiver is operated near a power transformer. (Steel chassis are available from parts houses that cater to audio experimenters.) The VFO mounting and mushroom-can shield shown in **Fig 14.72** are a simple way to eliminate mechanical backlash, keep radiated VFO energy off the antenna, prevent hand capacitance from shifting the tuning, and reduce VFO drift caused by air currents.

Experienced builders can duplicate this receiver simply using the schematic and construction techniques described here. Unlike a phasing receiver, there is no need to precisely duplicate the exact amplitudes and phases between the two channels. The ear/brain combination is the ultimate adaptive processor, and it quickly learns to focus on a desired signal and ignore interference. Small errors in phase and amplitude balance are heard as slight shifts in a signal's position. Standard-tolerance components may be used throughout.

One note about the kit version: A very good VFO can be built on an open PC board if the variable oscillator is not running on the desired output frequency. The Kanga kit VFO runs at one-half the desired frequency, and is followed by a balanced frequency doubler and driver amplifier.

OTHER EXPERIMENTS

My earliest experiments with binaural detectors feeding stereo audio amplifiers were done in 1979, using two antennas. The technique works very well, but requires two antennas either physically spaced some distance apart, or of different polarization. Listening to OSCAR 13 on a binaural receiver with cross-polarized Yagis was an unsettling experience. The need for two antennas is a liability—these days most of us struggle to put up one. A number of experiments have also been done with binaural independent sideband (ISB) reception. These are profoundly interesting for AM broadcast reception, and could be used for amateur AM or DSB reception using a Costas Loop for carrier recovery. Binaural ISB detection of shortwave AM broadcasting can be analyzed as a form of spread spectrum with the ear/brain combination serving the despreading function, or as a form of frequency diversity, with the ear/brain as an optimal combiner.

The binaural techniques described here are analogous to binocular vision: They present the same information to each ear, but from a slightly different angle. This provides a very natural sound environment that the brain interprets as three-dimensional space. There are other "binaural" techniques that involve the use of different filter responses for the right and left ears. My experiments with different filter responses for the left and right ears have not been particularly interesting, and I have not pursued them.

SUMMARY

This little receiver is a joy to tune around the band. It is a serious *listening* receiver, and allows digging for weak signals in a whole new way. Digging for weak signals in a three-dimensional sound field is sometimes referred to as the "cocktail party effect." It is difficult to quantify the performance of a binaural receiver, because the final signal processing occurs in the brain of the listener—you. The experimental literature of psycho-acoustics suggests that the ear/brain combination provides a signal-to-noise advantage of approximately 3 dB when listening to speech or a single tone in the presence of uncorrelated binaural noise. The amount of additional noise in the opposite sideband is also 3 dB, so it appears that the binaural I-Q detector breaks even. In some applications, such as UHF weak-signal work, the binaural I-Q detector may have an advantage, as it permits listening to a larger slice of the band without a noise penalty. In other situations, such as CW sweepstakes, the "cocktail party" may get entirely out of hand. Binoculars and telescopes both have their place.

Notes

[1]The complete kit version, available from Kanga US, uses a different VFO circuit than the one shown here. The kit VFO runs at one-half the desired output frequency, and is followed by a balanced frequency doubler and driver amplifier.

Steel chassis such as the Hammond 1441-12 (2 × 7 × 5 inches [HWD]) with 1431-12 bottom plate and the Hammond 1441-14 (2 × 9 × 5 inches [HWD]) with 143-14 bottom plate are suitable enclosures. These chassis and bottom plates are not available in single quantities directly from Hammond, but are available from Allied Electronics and Newark Electronics.

[2]Reed Fisher, W2CQH, "Twisted-Wire Quadrature Hybrid Directional Couplers," *QST*, Jan 1978, pp 21-23. See also IEEE Transactions MTT, Vol MTT-21, No. 5, May 1973, pp 355-357.

[3]Rick Campbell, KK7B, "High-Performance Direct-Conversion Receivers," *QST*, Aug 1992, pp 19-28.

[4]Rick Campbell, KK7B, "High-Performance, Single-Signal Direct-Conversion Receivers," *QST*, Jan 1993, pp 32-40. See also Feedback, *QST*, Apr 1993, p 75.

References

Campbell, Rick, KK7B, "Direct Conversion Receiver Noise Figure," Technical Correspondence, *QST*, Feb 1996, pp 82-85.

Campbell, Richard L., "Adaptive Array with Binaural Processor," *Proceedings of the IEEE Antennas and Propagation Society International Symposium*, Philadelphia, PA, June 1986, pp 953-956.

Campbell, Rick, KK7B, "Binaural Presentation of SSB and CW Signals Received on a Pair of Antennas," *Proceedings of the 18th Annual Conference of the Central States VHF Society*, Cedar Rapids, IA, July 1984, pp 27-33.

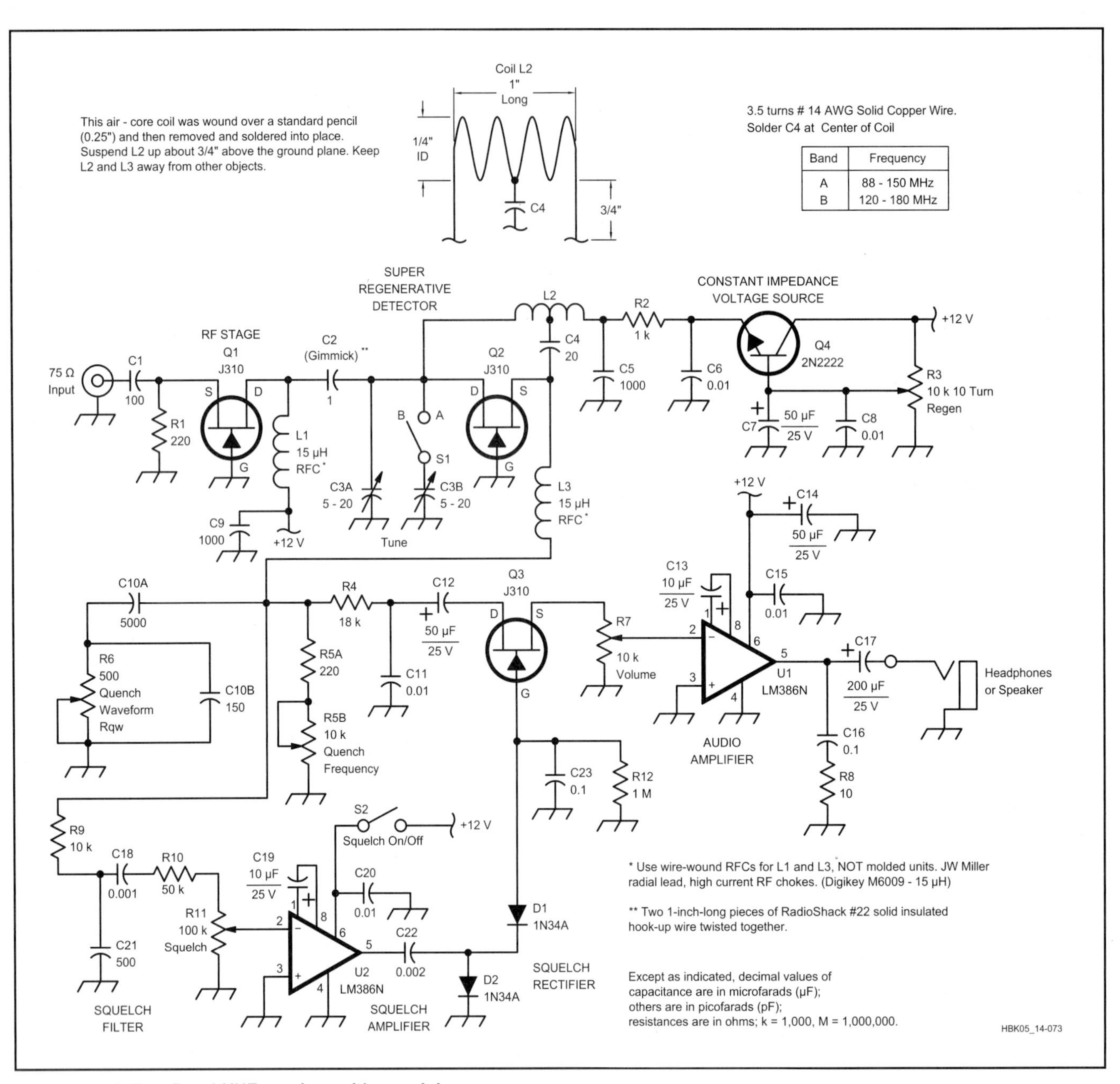

Fig 14.73—A Two-Band VHF receiver with squelch.

A SUPERREGENERATIVE VHF RECEIVER WITH SQUELCH

Introduction

The complexity of many published receiver circuits has "scared off" some would-be home builders. Yet, making a receiver from scratch is an extremely rewarding experience. Charles Kitchin, N1TEV, put together this VHF receiver that can easily be built by the average person, and does not require any special components or test equipment. It covers roughly 88 to 180 MHz in two bands. This covers the FM broadcast band, the aircraft band and the 2-meter ham bands. Above 2 meters, it also receives fire, police, marine and weather frequencies. The circuit draws about 20 mA.

Don't expect to squeeze out the ultimate in selectivity or stability with this receiver, although it does provide a sensitivity of around 0.5 µV. It uses the principle of superregeneration for this high sensitivity with a low parts count.

For narrow bandwidth FM reception, this receiver does require careful adjustment of several controls as it is tuned. A low-cost PC board is available from FAR Circuits, which greatly simplifies construction and helps prevent any layout or wiring errors.

Circuit Description

Fig 14.73 shows the circuit. RF signals from the antenna enter via a 75Ω coaxial cable and are ac coupled to the source of JFET Q1. This untuned RF amplifier provides excellent stability, good output-to-input isolation, and modest voltage gain over a wide frequency range. It also has a low input impedance (to match the coax line) and a high output impedance, which minimizes loading on the detector. R1 provides protective dc bias for the JFET. L1 is an RF choke, which extracts the amplified RF signal from the JFET drain.

Q2 operates as a superregenerative detector in a modified Hartley oscillator configuration. Capacitor C3 and coil L2 set the tuning (and detector oscillation) frequency. L2 should be stretched so that it is about one inch in length. After the circuit has been built and the detector is operating correctly, the turns on L2 can be compressed or expanded to raise or lower the tuning range. A gimmick capacitor, C2, couples the signal from the RF stage to the detector. Its value is somewhere around 1 pF. The gimmick lightly couples signals into the detector. Any over-coupling would reduce selectivity and might create "dead spots" in the tuning range of the receiver.

Band switching is accomplished simply and easily, by using S1 to switch-in either one or two gangs of tuning capacitor C3. S1 was wired directly between the two "hot" terminals of C3 using two *very* short lengths of #14 AWG copper wire. These support the switch quite well. With this arrangement, it is necessary to build the receiver with an open top, so you can reach in to change bands.

The value of C3 is not critical. A two or three-gang capacitor salvaged from an old FM radio will work nicely. Other small variable capacitors may be substituted, as long as their maximum capacity is not too great. Small mica capacitors may be wired in series or parallel with C3, or turns may be added or subtracted from L2, to change the tuning range. Capacitors C4 and C5 should be mica (or NP0 ceramic), as they need to be low-drift, high-Q devices.

The RC components on the Q2 source set up a secondary relaxation oscillation that self-quenches the receiver at a super audio rate. Resistor R6 slows the build-up and decay of this oscillation and modifies its waveform into a sine wave for high selectivity. Q4 reduces any variations in quench waveform shape as the regeneration control is varied. The audio output from Q2 is low pass filtered by R4 and C11. This prevents the quenching oscillations from reaching the audio stage and also improves the audio quality. The filtered audio signal is ac coupled to Q3 and then on to the volume control and LM386 audio amplifier.

A second filter circuit consisting of R9, C21 and C18, R10 passes mainly the upper audio frequencies to amplifier U2. This drives diodes D1 and D2, and the negative voltage output then squelches the receiver, by turning off Q3 during no-signal periods.

Construction Guidelines

Be sure to solder all ground leads of the FAR Circuits' board to both the top and bottom of the board, because these are not plated-through holes. When using a hand-wired board, it is essential that the drains of Q1 and Q2 be about one-quarter inch apart, with the gimmick capacitor connected between them. All wiring should use the shortest leads possible. The use of a rigid metal enclosure greatly improves the receiver's frequency stability. A vernier dial and a 10-turn potentiometer are recommended for easy TUNING and REGEN control. The metal ground plane of the PC board is connected to the metal enclosure using a single short wire. Use shielded wiring for all controls. Mount C3 directly on the PC board ground plane, then pass its shaft through an oversized hole in the metal front panel, without touching the panel. Use plastic knobs for the controls.

Testing and Operation

Test the audio stage first; then check the detector and RF stage. Check the audio by placing your finger on the wiper of R7. To test the detector, leave out C2. With the SQUELCH turned-off, set R6 to mid-position, turn the REGEN level up high and adjust R5B until the receiver oscillates. You should hear a loud "rushing" noise. Check that oscillation occurs over the entire tuning range adjusting R5B and R6 as needed. If there are any dead spots, move L2 and L3 farther away from all other objects. Then insert C2 and twist the turns together as much as possible without

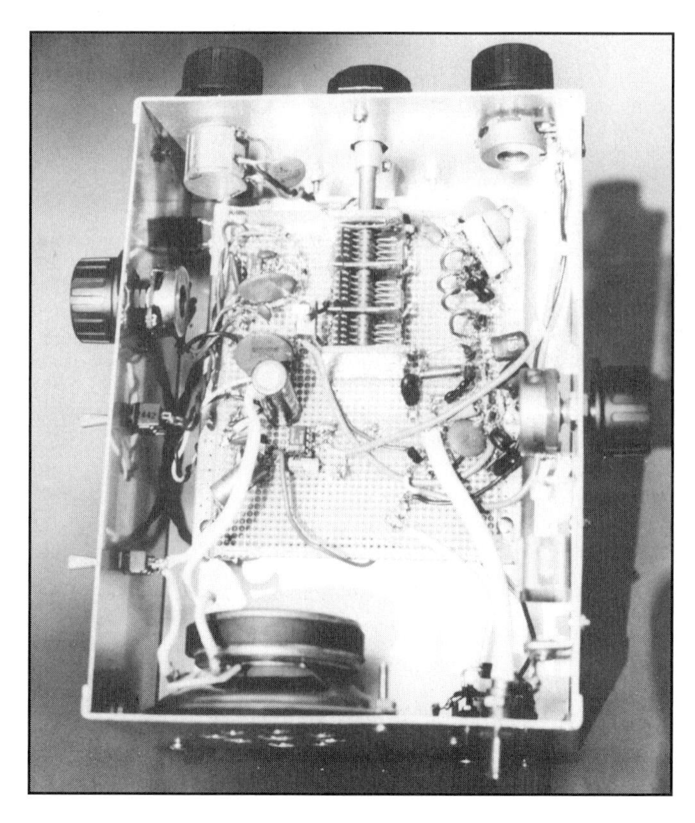

Fig 14.74—Photos of the completed Two-Band VHF receiver.

affecting detector operation.

Turn the REGEN up high on the 88-108 MHz FM broadcast band (wide-band FM mode). With QUENCH WAVEFORM (R6) set to 0 Ω; adjust the REGEN control for minimum distortion. Use the same settings for the 118-136 MHz aircraft band (AM mode) and set the SQUELCH so that the no-signal background noise is just muted. For operation on the 2-meter band (NBFM mode), increase R6 to about midrange and set REGEN fairly high. Tune-in a station, reduce the REGEN level until the audio level increases, and then carefully adjust TUNING and REGEN for maximum volume. Higher R6 resistance will be needed when receiving very strong signals. When receiving NBFM, headphone level is more than adequate but speaker volume will be low. Additional audio gain can be added between R7 and U1, if desired.

This receiver can be modified to cover the 6-meter band by changing a few components. Use a 50-pF air variable capacitor for C3, with a 40-pF mica capacitor in parallel. Increase L1 and L3 to 33 µH, L2 to 7 turns, C10A to 7000 pF, C4 to 50 pF; C5 to 2000 pF, omit C10B, twist C2 more tightly to approximately 2 pF.

For further reading, see Kitchin "New Super Regenerative Circuits for Amateur VHF and UHF Experimentation," September/October 2000 *QEX*. **Fig 14.74** shows photos of a finished receiver.

A BROADBAND HF AMPLIFIER USING LOW-COST POWER MOSFETS

Many articles have been written encouraging experimenters to use power MOSFETs to build HF RF amplifiers.[1-8] That's because power MOSFETs—popular in the design of switching power supplies—cost as little as $1 each, whereas RF MOSFET prices start at about $35 each!

Mike Kossor, WA2EBY, designed and built this amplifier after hundreds of hours experimenting with power MOSFETs. The construction projects described in Notes 1 to 8, provide useful information about MOSFETs and general guidelines for working with them, including biasing, parasitic-oscillation suppression, broadband impedance-matching techniques and typical amplifier performance data.

With the design described here, 1 W of input power produces over 40 W of output (after harmonic filtering) from 160 through 10 meters. In addition to the basic amplifier, there is an RF-sensed TR relay and a set of low-pass filters designed to suppress harmonic output and comply with FCC requirements. The amplifier is built on double-sided PC board and requires *no tuning*. Another PC board contains the low-pass filters. Power-supply requirements are 28 V dc at 5 A, although the amplifier performs well at 13.8 V dc.

There are no indications of instability, no CW key clicks and no distortion on SSB has been reported by stations contacted while using the amplifier. To make it easy for you to duplicate this project, PC boards and parts kits are available, all at a cost of about $100![9] Etching patterns and parts-placement diagrams are included on the accompanying CD.

AN OVERVIEW OF MOSFETS

MOSFETs operate very differently from bipolar transistors. MOSFETs are voltage-controlled devices and exhibit a very high input impedance at dc, whereas bipolar transistors are current-controlled devices and have a relatively low input impedance. Biasing a MOSFET for linear operation only requires applying a fixed voltage to its gate via a resistor. With MOSFETs, no special bias or feedback circuitry is required to maintain the bias point over temperature as is required with bipolar transistors to prevent thermal runaway.[10] With MOSFETs, the gate-threshold voltage increases with increased drain current. This works to turn off the device, especially at elevated temperatures as transconductance decreases and R_{DSon} (static drain-to-source *on* resistance) increases.

These built-in self-regulating actions prevent MOSFETs from being affected by thermal runaway. MOSFETs do not require negative feedback to suppress low-frequency gain as is often required with bipolar RF transistors. Bipolar transistor gain increases as frequency decreases. Very high gain at dc and low frequencies can cause unwanted, low-frequency oscil-

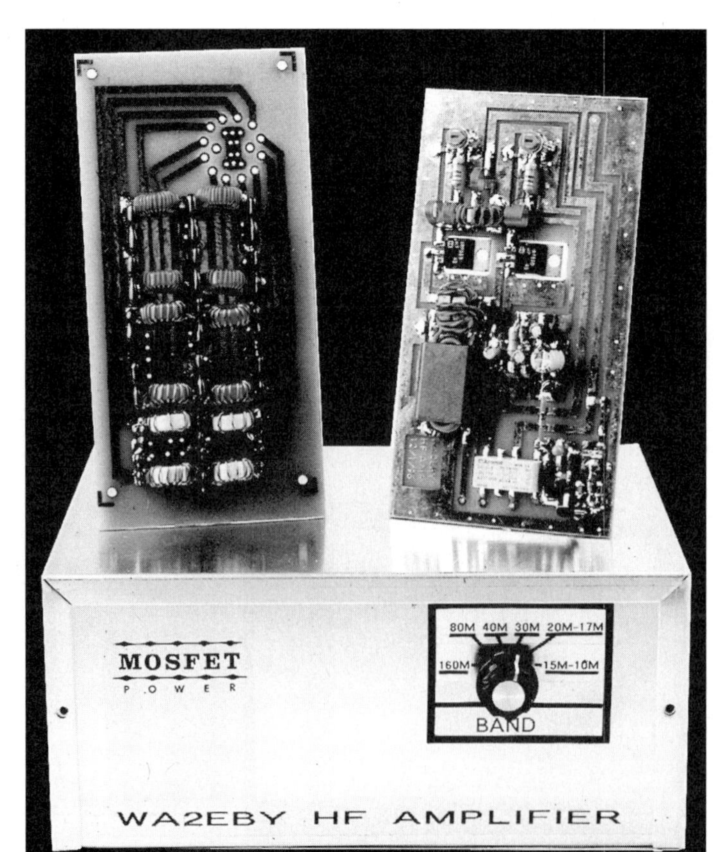

Fig 14.75—The WA2EBY MOSFET HF Amplifier produces over 40 W output with 1 W of input power.

lation to occur in bipolar transistor RF amplifiers unless negative feedback is employed to prevent it. Low-frequency oscillation can damage bipolar transistors by causing excess power dissipation, leading to thermal runaway.

MOSFET LIMITATIONS

Of course, MOSFETs do have their limitations. The high gate impedance and the device structure make them susceptible to electrostatic discharge (ESD) damage. Some easily applied precautions prevent this: Use a soldering iron with grounded tip; use a wrist strap connected to ground through a 1 MΩ resistor to bleed off excess body charge while handling MOSFETs and do all work on an antistatic mat connected to ground via a 1 MΩ resistor.

The sensitivity of a MOSFET's gate to static and high-voltage spikes also makes it vulnerable to damage resulting from parasitic oscillation. This undesired self-oscillation could result in excessive gate-to-source voltage that permanently damages the MOSFET's gate insulation. Another MOSFET limitation is gate capacitance. This parameter limits the frequency at which a MOSFET can operate effectively as an RF amplifier. The author recommends reviewing the referents of Notes 1 to 3 if you are interested in more detailed information about MOSFETs.

POWER MOSFET RF AMPLIFIERS

The author built several power MOSFET amplifiers to check their performance. His experiments underscore the need to observe *exact* construction techniques and physical layout if similar performance is to be expected. Although he used PC board construction, his results differed significantly in several of the experiments because the circuit layout was not the same as the original layout. A photo of the WA2EBY amplifier is shown in **Fig 14.75**.

Considerable experimentation with several designs resulted in the circuit shown in **Fig 14.76**. This amplifier consists of two power MOSFETs operating in push-pull, and employs an RF-sensed TR relay.

During receive, TR relay K1 is de-energized. Signals from the antenna are connected to J2 and routed through K1 to a transceiver connected to J1. (This path loss is less than 0.3 dB from 1.8 MHz through 30 MHz.) In transmit, RF voltage from the transceiver is sampled by C17 and divided by R6 and R7. D2 and D3 rectify the RF voltage and charge C16. Q3 begins to conduct when the detected RF voltage across C16 reaches approximately 0.7 V. This energizes K1, which then routes the transmitted RF signal from J1 to the amplifier input and switches the amplifier output to the low-pass filter block and then to the antenna at J2. RF-sensed relay response is very fast. No noticeable clipping of the first CW character has been reported.

An RF attenuator (consisting of R8, R9 and R10) allows you to adjust the amplifier input power to 1 W. (The parts list contains resistor values to reduce the output of 2 or 5 W drivers to 1 W.) The 1 W signal is then applied to the primary of T1 via an input impedance-matching network consisting of L3.

T1 is a 1:1 balun that splits the RF signal into two outputs 180° out of phase. One of these signals is applied by C1 to the gate of Q1. The other signal is routed via C2 to the gate of Q2. The drains of Q1 and Q2 are connected to the primary of output transformer T3, where the two signals are recombined in phase to produce a single output. T3 also provides impedance transformation from the low output impedance of the MOSFETs to the 50-Ω antenna port. DC power is provided to the drains of Q1 and Q2 by a phase-reversal choke, T2. This is a very effective method to provide power to Q1 and Q2 while presenting a high impedance to the RF signal over a broad range of frequencies. The drain chokes for Q1 and Q2 are wound on the same core, and the phase of one of the chokes (see the phasing-dot markings on T2) is reversed. C9 increases the bandwidth of the impedance transformation provided by T3, especially at 21 MHz.

The 5 V bias supply voltage is derived from 28 V by Zener diode D1 and current-limiting resistor R11. Bypass capacitors C3, C4, C5, C6 and C13 remove RF voltages from the bias supply voltage. Gate bias for Q1 and Q2 is controlled independently. R1 adjusts the gate-bias voltage to Q1 via R3 and L1. R2 works similarly for Q2 via R4 and L2.

At low frequencies, the amplifier input impedance is essentially equal to the series value of R3 and R4. L1 and L2 improve the input-impedance match at higher frequencies. The low value of series resistance provided by R3 and R4 also reduces the Q of impedance-matching inductors L1 and L2, which improves stability. DC blocking capacitors C1 and C2 prevent loading the gate bias-supply voltage.

C14 keeps transistor Q3 conducting and K1 energized between SSB voice syllables or CW elements. Without C14, K1 would chatter in response to the SSB modulation envelope and fast keying. Increasing the value of C14 increases the time K1 remains energized during transmit. The reverse voltage generated by K1 when the relay is deenergized is clamped to a safe level by D4. D5 drops the 28 V supply to 13 V to power 12 V relay K1. D5 can be replaced with a jumper if K1 has a 28 V dc coil or if you intend to operate the amplifier with a 13.8 V dc supply.

HARMONIC FILTERING

Although biased for class AB linear operation, this amplifier (like others of its type) exhibits some degree of non-linearity, resulting in the generation of harmonics. This push-pull amplifier design cancels even-order harmonics (2f, 4f, 6f, etc) in the output transformer, T3. Odd-order harmonics are not canceled. Second-order harmonics generated by the amplifier are typically less than –30 dBc (30 dB below the carrier) whereas third-order harmonics are typically only –10 dBc. FCC regulations require all HF RF-amplifier harmonic output power to be at least –40 dBc at power levels between 50 to 500 W. To meet this requirement, it is common practice for HF amplifiers to use low-pass filters. Separate low-pass filters are needed for the 160, 80, 40 and 30 meter bands. The 20 and 17 meter bands can share the same low-pass filter. So, too, the 15, 12 and 10 meter bands can share a common low-pass filter; see Fig 14.76.

Switching between the six filters can be a messy wiring problem, especially on the higher-frequency bands where lead lengths should be kept short for optimum performance. This problem is solved by mounting all six low-pass filters on a PC board. A two-pole, six-position rotary switch (S1) mounted directly on the same PC board manages all filter interconnections. One pole of S1 connects the amplifier output to one of the six filter inputs, while the other pole of S1 simultaneously connects the corresponding filter output to the TR relay, K1. Only two coaxial-cable connections are required between the RF amplifier and the low-pass filter board.

AMPLIFIER CONSTRUCTION

The amplifier is constructed on a double-sided PC board with plated through holes to provide top-side ground connections. Chip resistors and capacitors were used to simplify construction, but leaded capacitors may work if lead lengths are kept short. First, assemble all chip capacitors and resistors on the PC board. Tweezers help to handle chip components. Work with only one component value at a time. (Chip caps and resistors can be very difficult to identify!) Chip capacitor and resistor mounting is simplified by tinning one side of the PC board trace with solder before positioning the capacitor or resistor. Touch the soldering iron tip to the capacitor or resistor to tack it in place. Finish

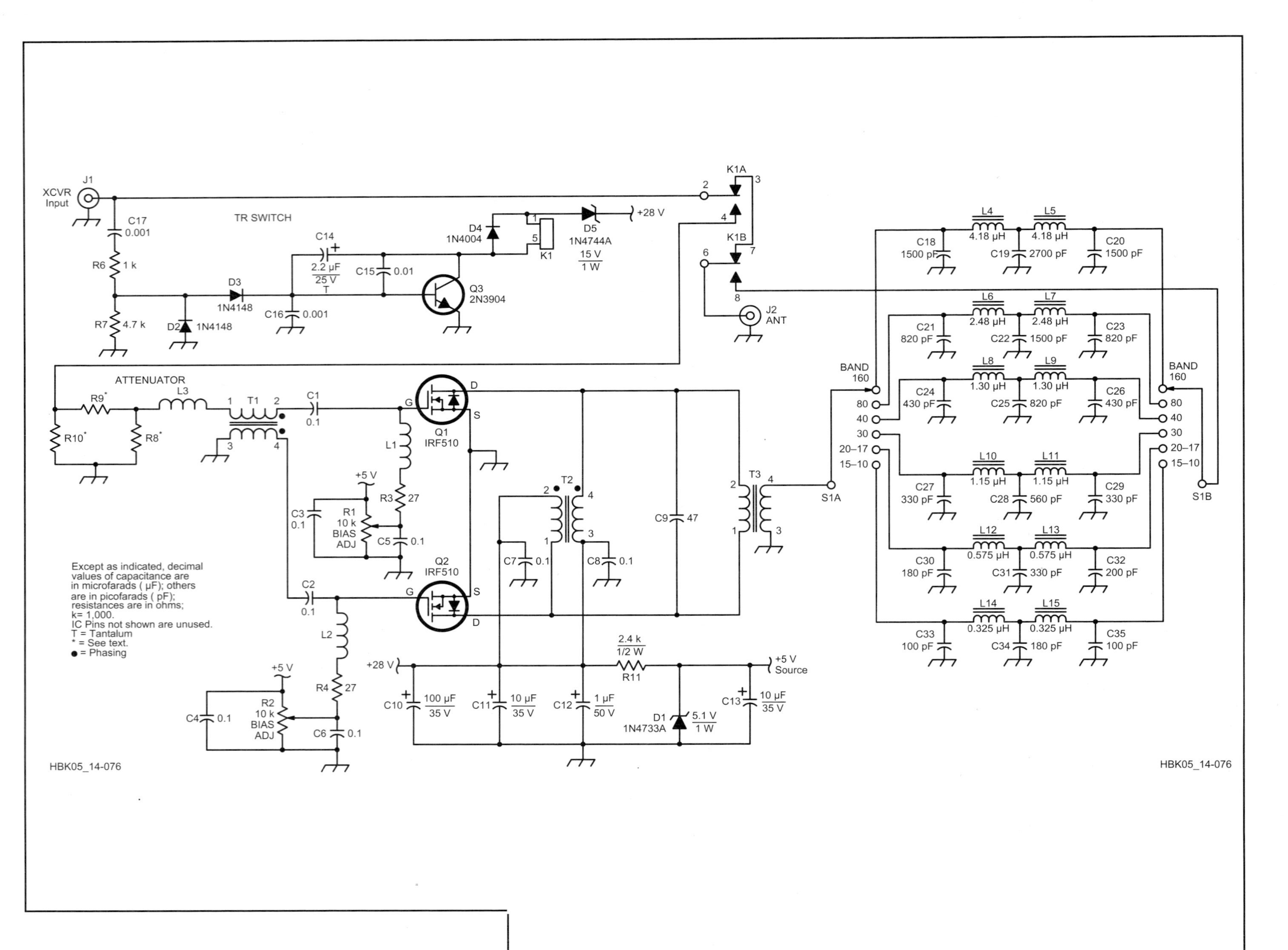
XCVR Input
J1
TR SWITCH
C17 0.001
R6 1 k
R7 4.7 k
D2 1N4148
D3 1N4148
C16 0.001
C14 2.2 µF 25 V T
C15 0.01
Q3 2N3904
D4 1N4004
K1
D5 1N4744A 15 V 1 W
+28 V
K1A
K1B
J2 ANT
ATTENUATOR
R9*
R10*
R8*
L3
T1
C1 0.1
L1
Q1 IRF510
+5 V
R3 27
C3 0.1
R1 10 k BIAS ADJ
C5 0.1
Q2 IRF510
C2 0.1
L2
R4 27
R2 10 k BIAS ADJ
C4 0.1
C6 0.1
T2
C7 0.1
C8 0.1
C9 47
T3
S1A
S1B
BAND 160 80 40 30 20–17 15–10
+28 V
C10 100 µF 35 V
C11 10 µF 35 V
C12 1 µF 50 V
R11 2.4 k 1/2 W
D1 1N4733A 5.1 V 1 W
C13 10 µF 35 V
+5 V Source
L4 4.18 µH
L5 4.18 µH
C18 1500 pF
C19 2700 pF
C20 1500 pF
L6 2.48 µH
L7 2.48 µH
C21 820 pF
C22 1500 pF
C23 820 pF
L8 1.30 µH
L9 1.30 µH
C24 430 pF
C25 820 pF
C26 430 pF
L10 1.15 µH
L11 1.15 µH
C27 330 pF
C28 560 pF
C29 330 pF
L12 0.575 µH
L13 0.575 µH
C30 180 pF
C31 330 pF
C32 200 pF
L14 0.325 µH
L15 0.325 µH
C33 100 pF
C34 180 pF
C35 100 pF
Except as indicated, decimal values of capacitance are in microfarads (µF); others are in picofarads (pF); resistances are in ohms; k= 1,000.
IC Pins not shown are unused.
T = Tantalum
* = See text.
● = Phasing
HBK05_14-076

Fig 14.76—Schematic of the MOSFET all-band HF amplifier. Unless otherwise specified, resistors are 1/4 W, 5% tolerance carbon-composition or film units. The low-pass filter section shows some filter component values that differ from the calculated values of a standard 50 Ω-input filter. Such differences improve the impedance matching between the amplifier and the load. Capacitors in the filter section are all dipped mica units. Equivalent parts can be substituted. Part numbers in parentheses are Mouser; see Note 9 and the References chapter for contact information.

C1-C8—0.1 µF chip (140-CC502Z104M).
C9—47 pF chip (140-CC502N470J).
C10—100 µF, 35 V (140-HTRL35V100).
C11, C13—15 µF, 35 V (140MLR35V10).
C12—1 µF, 50 V (140-MLRL50V1.0).
C14—2.2 µF, 35 V tantalum (581-2.2M35V).
C15—0.01 µF chip (140-CC502B103K)
C16, C17—0.001 µF chip (140-CC502B102K).
C18, C20, C22—1500 pF (5982-19-500V1500).
C19—2700 pF (5982-19-500V2700).
C21, C23, C25—820 pF (5982-19-500V820).
C24, C26—430 pF (5982-15-500V430)
C27, C29, C31—330 pF (5982-19-500V330).
C28—560 pF (5982-19-500V560).
C30, C34—180 pF (5982-15-500V180).
C32—200 pF (5982-15-500V200).
C33, C35—100 pF (5982-10-500V100).
D1—1N4733A, 5.1 V, 1 W Zener diode (583-1N4733A).
D4—1N4004A (583-1N4004A).
D2, D3—1N4148 (583-1N4148).
D5—1N4744A, 15 V, 1 W Zener diode (583-1N4744A).
J1, J2—SO-239 UHF connector (523-81-120) or BNC connector (523-31-10).
K1—12 V DPDT, 960 Ω coil, 12.5 mA (431-OVR-SH-212L).
L1, L2—9 1/2 turns #24 enameled wire, closely wound 0.25-in. ID.
L3—3 1/2 turns #24 enameled wire, closely wound 0.190-in. ID
Q1, Q2—IRF510 power MOSFET (570-IRF510).
Q3—2N3904 (610-2N3904).
R1, R2—10 kΩ trim pot (323-5000-10K).
R3, R4—27 Ω, 1/2 W (293-27).
R6—1 kΩ chip (263-1K).
R7—4.7 kΩ chip (263-4.7K).
R8—130 Ω, 1 W (281-130); for 7 dB pad (5 W in, 1 W out).
R9—43 Ω, 2 W (282-43); for 7 dB pad (5 W in, 1 W out).
R10—130 Ω, 3 W (283-130); for 7 dB pad (5 W in, 1 W out).
R8, R10—300 Ω, 1/2 W (273-300); for 3 dB pad (2 W in, 1 W out).
R9—18 Ω, 1 W (281-18); for 3 dB pad (2 W in, 1 W out).
R11—2.4 kΩ, 1/2 W (293-2.4K).
S1—2 pole, 6 position rotary (10YX026).
T1—10 bifilar turns #24 enameled wire on an FT-50-43 core.
T2—10 bifilar turns #22 enameled wire on two stacked FT-50-43 cores.
T3—Pri 2 turns, sec 3 turns #20 Teflon-covered wire on BN-43-3312 balun core.
Misc: Aluminum enclosure 3.5×8×6 inches (HWD) (537-TF-783), two TO-220 mounting kits (534-4724), heat-sink compound (577-1977), amplifier and low-pass filter PC boards (see Note 9), heat sink (AAVID [Mouser 532-244609B02]; see text), about two feet of RG-58 coax, #24 enameled wire and #20 Teflon-insulated wire.

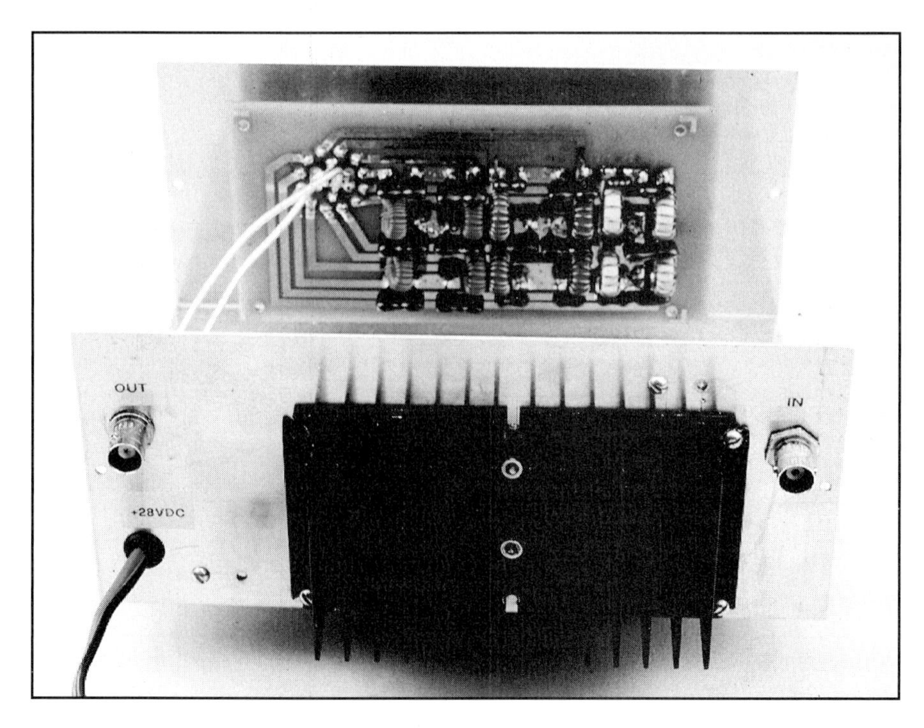

Fig 14.77—This rear-panel view of the amplifier shows the heat sink. The filter board mounts on the back side of the front panel.

mounting by soldering the opposite side of the component. *Don't apply too much heat to chip capacitors.* The metalized contacts on the capacitor can be damaged or completely removed if too much heat is applied. Use a 15 to 20 W soldering iron and limit soldering time to five seconds.

Mount axial-leaded resistors, diodes and remaining capacitors next. To avoid damaging them, mount inductors and transformers last. L1 and L2 are wound on a 1/4-inch drill-bit shaft. By wrapping the wire around the shaft 10 times, you'll get 9 1/2 turns. The last turn arcs only a half-turn before entering the PC board. L3 is wound on a 0.190-inch diameter drill bit with 3 1/2 turns wound the same way as L1 and L2. Mounting K1 is simplified by first bending all its leads 90° outward so it lies flat on the PC board. Be sure to follow the anti-static procedures mentioned at the beginning of this project while handling MOSFETs. The gate input can be damaged by electrostatic discharge!

When winding T3, wind the primary first and add the secondary winding over the primary. Be sure to use Teflon-insulated wire for these windings; the high operating temperatures encountered will likely melt standard hook-up wire insulation.

Heat Sinking

Together, Q1 and Q2 dissipate up to 59 W. A suitable heat sink is required to prevent the transistors from overheating and damage. I used an AAVID 244609B02 heat sink originally designed for dc-to-dc power converters. The amplifier PC board and heat sink are attached to an aluminum enclosure by two #4-40 screws drilled through the PC board, enclosure and heat sink at diagonally opposite corners. See **Fig 14.77**. A rectangular cutout in the enclosure allows Q1 and Q2 direct access to the heat sink. This is essential because of the large thermal impedance associated with the TO-220 package (more on this topic later). Mark the locations of the transistor-tab mounting-hole location in the center of the heat sink, between the cooling fins. Disassemble the heat sink to drill 0.115 inch holes for #4-40 mounting screws, or tap #4-40 mounting holes in the center of the heat-sink fins.

Use mica insulators and grommets when mounting Q1 and Q2 to prevent the #4-40 mounting screws from shorting the TO-220 package drain connections (tabs) to ground. Coat both sides of the mica insulator with a *thin* layer of thermal compound to improve the thermal conduction between the transistor tab and the heat sink. Be sure to install the mica insulator on the heat sink *before* assembling the amplifier PC board to the enclosure and heat sink. The mica insulators are larger than the cut outs in the PC board, making it impossible to install them after the PC board is mounted.

LOW-PASS FILTER CONSTRUCTION

Inductor winding information for the low-pass filters is provided in **Table 14.1**. A PC-board trace is available on the amplifier PC board next to amplifier output (J3) to allow the installation of a single-band low-pass filter between the terminals of J3 and the J4 input to K1. This is handy if you intend to use the amplifier on one band only. The input inductor of the low-pass filter connects from J3 to the single PC trace adjacent to J3. The output inductor connects in series between the single PC trace to J4. The three filter capacitors connect from J3, J4 and the PC-board trace near J3 to ground. *This single trace is not used when multiple filters are required.* Remember to remove the single trace adjacent to J3 on the amplifier PC board before attaching the amplifier board between the RF connectors on the rear panel of the enclosure.

Table 14.1
Low-Pass Filter Inductor Winding Information
(Refer to Fig 14.76)

Inductor Number	*No. of Turns*	*Core*
L4, L5	30	T-50-2
L6, L7	22	T-50-2
L8, L9	16	T-50-2
L10, L11	14	T-50-2
L12, L13	11	T-50-6
L14, L15	8	T-50-6

Note: All inductors are wound with #22 enameled wire except for L4 to L7, which are wound with #24 enameled wire.

Multiple-Band Filters

Using the amplifier on more than one band requires a different approach. A set of six low-pass filters is built on a double-sided PC board with plated through holes to provide top-side ground connections. A PC-board mount, two-pole, six-position rotary switch does all low-pass filter selection. Silver-mica, leaded capacitors are used in all the filters. On 160 through 30 meters, T-50-2 toroids are used in the inductors. T-50-6 toroids are used for inductors on 20 through 10 meters. The number of turns wound on a toroid core are counted on the toroid's OD as the wire passes

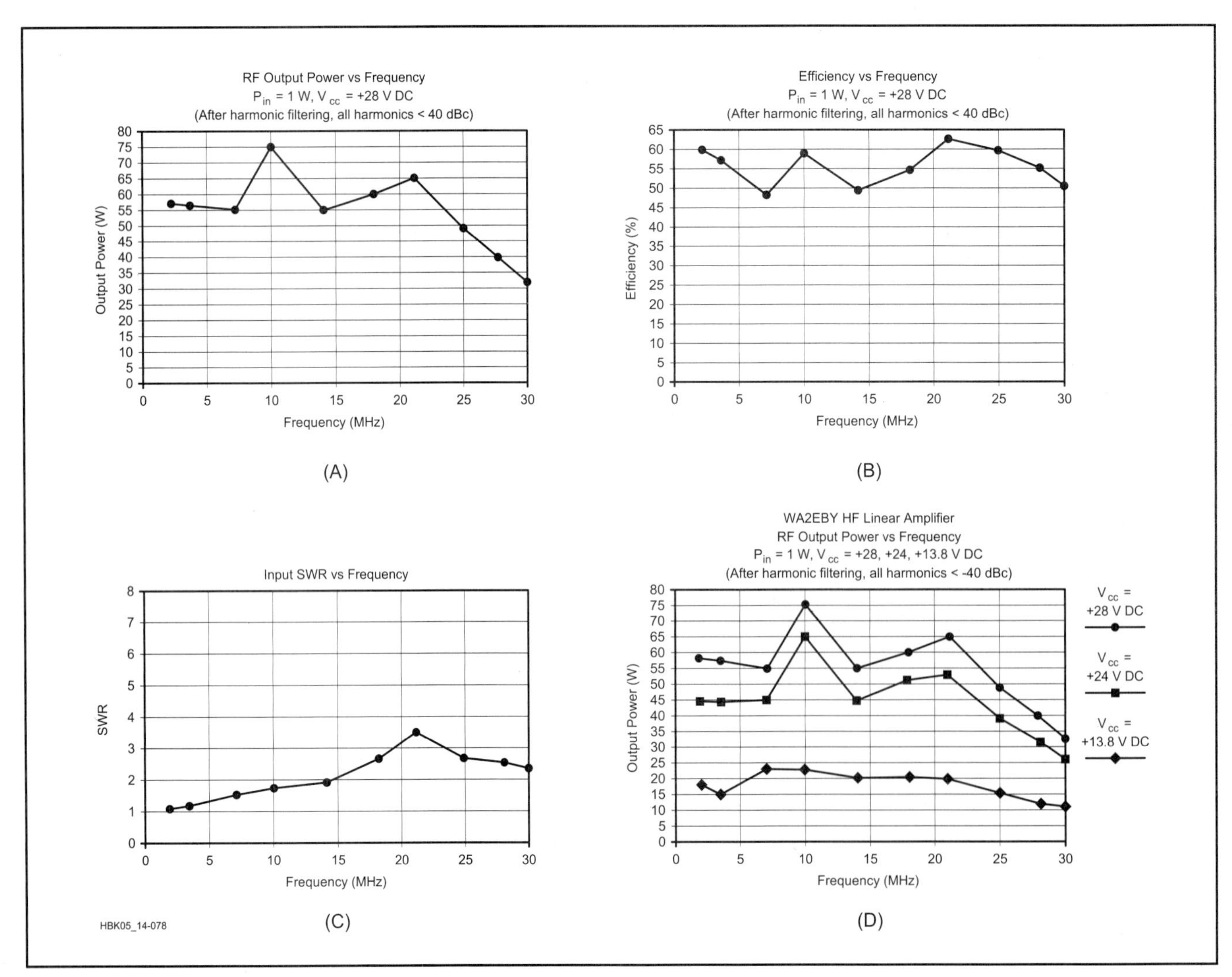

Fig 14.78—A shows the amplifier RF output power. B shows the amplifier efficiency. C shows the amplifier input SWR. D shows the amplifier RF output power versus supply voltage.

through the core center. (The **Circuit Construction** chapter provides complete details for winding toroids.) Assemble one filter section at a time starting with the 160, 80, 40 and 30-meter filters. With the switch mounting position at your upper left, the 160-m filter input (C18) is near the top edge of the board and the filter output (C20) is near the bottom edge. *The last two filters are out of sequence;* the 15-10 meter filter comes *before* the 20-17 meter filter) and the inputs/outputs are reversed to simplify the PC-board layout. The input capacitors, C30 and C33, are mounted on the board *bottom edge,* and output capacitors, C32 and C35, are on the *top edge.*

Use care when assembling the rotary switch. All 14 terminals must fit through the PC board without damaging or bending the pins. Make sure there are no bent pins before you attempt assembly. Insert the rotary switch into the PC board. Do *not* press the rotary switch all the way into the PC-board holes flush with the ground plane! If you do, the top flange of the signal pins may short to the ground plane.

BIAS ADJUSTMENT

The biasing procedure is straightforward and requires only a multimeter to complete. First, set R1 and R2 fully counterclockwise, (0 V on the gates of Q1 and Q2). Terminate the RF input and output with 50-Ω loads. Next, connect the 28 V supply to the amplifier in series with a multimeter set to the 0-200 mA current range. Measure and record the idling current drawn by the 5 V bias supply. (The value should be approximately 9.5 mA (28 – 5.1 V) / 2.4 kΩ = 9.5 mA). Set the Q1 drain current to 10 mA by adjusting R1 until the 28 V supply current increases by 10 mA above the idling current (9.5 + 10 = 19.5 mA). Next, adjust R2 for a Q2 drain current of 10 mA. This is accomplished by adjusting R2 until the 28 V supply current increases by an additional 10 mA (to 29.5 mA).

AMPLIFIER PERFORMANCE

With a 28 V power supply and 1 W of drive, the RF output power of this amplifier exceeds 40 W from 1.8 MHz through 28 MHz. Peak performance occurs at 10 MHz, providing about 75 W after filtering! A performance graph for this amplifier is shown in **Fig 14.78A**.

As shown in Fig 14.78B, this amplifier achieves an efficiency of better than 50% over its frequency range, except at 7 MHz where the efficiency drops to 48%.

Fig 14.78C shows the input SWR of the amplifier. It exceeds 2:1 above 14 MHz. The input SWR can be improved to better than 2:1 on all bands by adding a 3 dB pad (R8-R10 of Fig 14.76) at the input and supplying 2 W to the pad input. This keeps the amplifier drive at 1 W.

Fig 14.78D graphs the amplifier RF output power as a function of drain supply voltage. During this test, the amplifier RF drive level was kept constant at 1 W. As you can see, even when using a 13.8 V dc supply, the amplifier provides over 10 W output (a gain of more than 10 dB) from 1.8 to 30 MHz.

OPERATION

The amplifier requires no tuning while operating on any HF amateur band. You must, however, *be sure to select the proper low-pass filter prior to transmitting.* If the wrong low-pass filter is selected, damage to the MOSFETs may result. Damage will likely result if you attempt to operate the amplifier on a band with the low-pass filter selected for a lower frequency. For example, driving the amplifier with a 21 MHz signal while the 1.8 MHz low-pass filter is selected will likely destroy Q1 and/or Q2.

The amplifier can also be damaged by overheating. This limitation is imposed by the TO-220 packages in which Q1 and Q2 are housed. The thermal resistance from junction to case is a whopping 3.5°C/W. This huge value makes it virtually impossible to keep the junction temperature from exceeding the +150°C target for good reliability. Consider the following conditions: key down, 1 W input, 53 W output on 7 MHz (worst-case band for efficiency). The amplifier consumes 28 V × 4 A = 112 W, of which 53 W are sent to the antenna, so 59 W (112 W – 53 W = 59 W) are dissipated in Q1 and Q2. Assuming equal current sharing between Q1 and Q2, each transistor dissipates 29.5 W. To keep the transistor junction temperature below +150°C requires preventing the transistor case temperature from exceeding 46.8°C (150 –[3.5 × 29.5]) while dissipating 29.5 W. Also, there is a temperature rise across the mica insulator between the transistor case and heat sink of 0.5°C/W. That makes the maximum allowable heat-sink temperature limited to 46.8 (0.5 × 29.5) = 32°C. In other words, the heat sink must dissipate 59 W (29.5 from each transistor)

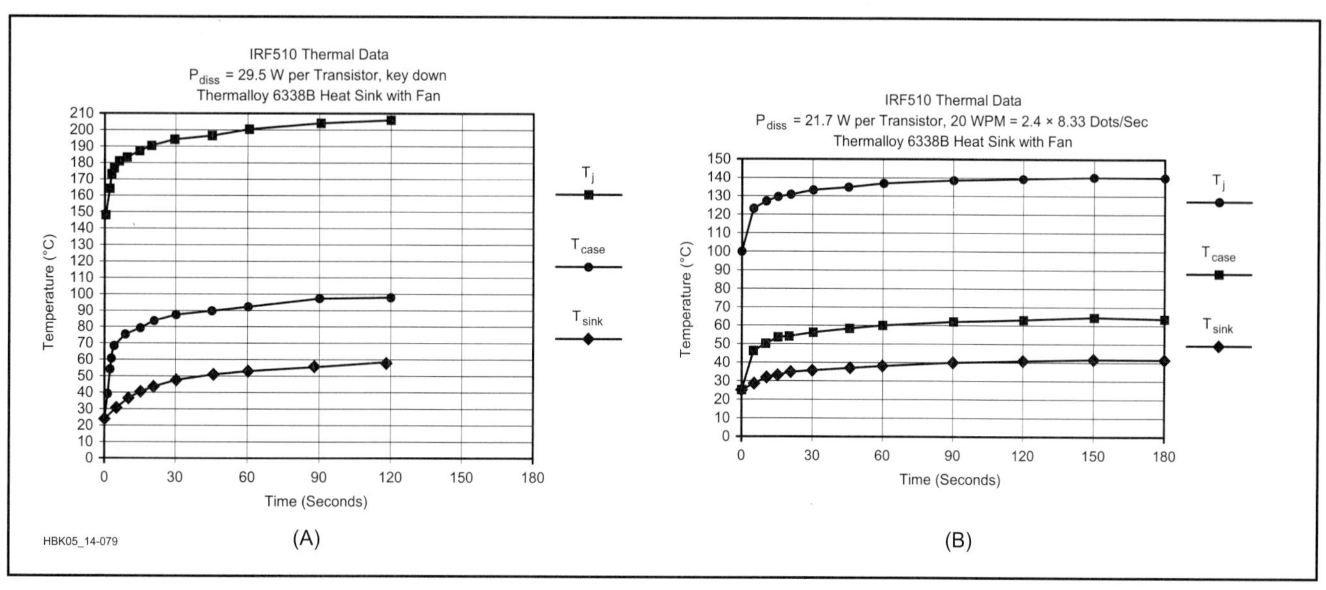

Fig 14.79—A is a graph of the amplifier thermal performance during key-down conditions. B is a graph of the amplifier thermal performance during simulated CW conditions.

with only a 7°C rise above room temperature (25°C). Even if the junction temperatures were allowed to reach the absolute maximum of 175°C, the heat sink temperature must not exceed 57°C. Accomplishing this requires a heat sink with a thermal resistance of (57 – 25) / 59 = 0.54°C/W. This is far less than the 1.9°C/W rating of the AAVID 244609B02 heat sink I used. The situation may seem bleak, but all is not lost. These calculations make it clear that the amplifier should not be used for AM, FM or any other continuous-carrier operation. The amplifier should be used only for CW and SSB operation where the duty cycle is significantly reduced.

Thermal performance of the amplifier is illustrated in **Fig 14.79A**. Data was taken under dc operating conditions with power-dissipation levels set equal to conditions under RF operation. A Radio Shack brushless 12 V dc fan (RS 273-243A) blows air across the heat sink. Key down, the maximum rated junction temperature is reached in as little as five seconds. Prolonged key-down transmissions should be avoided for this reason.

Under intermittent CW conditions, the situation is very different. Transistor-case temperatures reached 66°C after operating four minutes under simulated CW conditions at 20 WPM (60 ms on, 60 ms off). The corresponding junction temperature is +141°C (based on an equivalent RMS power dissipation of 21.7 W per transistor). This keeps the junction temperature under the 150°C target (see Fig 14.79B). One simple way to reduce power dissipation is to reduce the power-supply voltage to 24 V. RF output power will decrease about 10 W from the maximum levels achieved with a 28 V supply.

From a thermal standpoint, the IRF510 power MOSFET is a poor choice for this RF amplifier application. Although I must say I am impressed with the robustness of these devices considering the times I spent testing them key down, five minutes at a time, without failure. Q1 and/or Q2 may need to be replaced after a year or so of operation because of the compromise in reliability. Considering their low cost, that is not a bad trade-off.

STABILITY

High gain, broad bandwidth and close input/output signal routing (within the TR relay) all work against stability. With a good load (< 2:1 SWR) the amplifier is stable from 1.8 MHz through 39 MHz. Oscillation was observed when the transmitter frequency was increased to 40 MHz. The output load match also affects stability. I spent a great deal of time trying to make this design unconditionally stable even with loads exceeding 3:1 SWR without sacrificing output power (gain) at 28 MHz without success. I did identify some reasonable compromises.

One of the easiest ways to improve stability and the input SWR seen by the RF source is to add an RF attenuator (pad) at the amplifier input. An attenuator is absolutely required if the transmitter (driver) provides more than 1 W to the amplifier. R8, R9 and R10 form an RF attenuator that attenuates the transmitter drive level, but does not attenuate received signals because it is only in the circuit when K1 is energized. To drive this amplifier with a 2W-output transmitter requires use of a 3-dB pad. The pad improves the amplifier input SWR and the isolation between the amplifier's input and output. The drawback is that 1 W is wasted in the pad. Likewise, a 5-W driver requires use of a 7-dB pad, and 4 W are wasted in the pad. (Values for R8, R9 and R10 to make a 3-dB pad and a 7-dB pad are given in the parts list of the caption for Fig 14.76.) Installing a pad requires cutting the PC-board trace *under R9,* otherwise R9 would be shorted out by the trace. Make a small cut (0.1 inch wide) in the trace under R9 before soldering R9 in position. R8 and R10 have the same values, but may have different power ratings. Connect R10 between the RF input side of R9 and ground. Install R8 between the amplifier side of R9 and ground.

An impedance mismatch between the output of a 1-W driver and the amplifier input can be a source of instability. (Obviously, if the driving transmitter output power is only 1 W, you can't use a pad as described earlier.) If you encounter stability problems, try these remedies: Place a resistor in parallel with L1 and L2 to decrease the Q of the amplifier matching network (try values between 50 and 220 Ω). Try reducing the value of L3 or eliminating L3 entirely. Both of these modifications improve stability, but reduce the amplifier output power above 21 MHz.

SUMMARY

This project demonstrates how inexpensive power MOSFETs can be used to build an all-band linear HF power amplifier. Frequency of operation is extended beyond the limits of previous designs using the IRF510 and improved input-impedance matching. Long-term reliability is recognized as a compromise because of the poor thermal performance of the low-cost TO-220 package.

If you have been thinking about adding an amplifier to your QRP station, this project is a good way to experiment with amplifier design and is an excellent way to become familiar with surface-mount "chip" components. Amidon, Inc provides parts kits for this project (see Note 9).

ACKNOWLEDGMENTS

The author thanks the following individuals associated with this project: Harry Randel, WD2AID, for his untiring support in capturing the schematic diagram and parts layout of this project; Al Roehm, W2OBJ, for his continued support and encouragement in developing, testing, editing and publishing this project; Larry Guttadore, WB2SPF, for building, testing and photographing the project; Dick Jansson, WD4FAB, for thermal-design suggestions; Adam O'Donnell, N3RCS, for his assistance building prototypes; and his wife, Laura, N2TDL, for her encouragement and support throughout the project.

Notes

[1]Doug DeMaw, W1FB, "Power-FET Switches as RF Amplifiers," *QST,* Apr 1989, pp 30-33. See also Feedback, *QST,* May 1989, p 51.

[2]Wes Hayward, W7ZOI, and Jeff Damm, WA7MLH, "Stable HEXFET RF Power Amplifiers," Technical Correspondence, *QST,* Nov 1989, pp 38-40; also see Feedback, *QST,* Mar 1990, p 41.

[3]Jim Wyckoff, AA3X, "1 Watt In, 30 Watts Out with Power MOSFETs at 80 Meters," Hints and Kinks, *QST,* Jan 1993, pp 50-51.

[4]Doug DeMaw, W1FB, "Go Class B or C with Power MOSFETs," *QST,* March 1983, pp 25-29.

[5]Doug DeMaw, W1FB, "An Experi-mental VMOS Transmitter", *QST,* May 1979, pp 18-22.

[6]Wes Hayward, W7ZOI, "A VMOS FET Transmitter for 10-Meter CW," *QST,* May 1979, pp 27-30.

[7]Ed Oxner, KB6QJ (ex-W9PRZ), "Build a Broadband Ultralinear VMOS Amplifier," *QST,* May 1979, pp 23-26.

[8]Gary Breed, K9AY, "An Easy-to-Build 25-Watt MF/HF Amplifier," *QST,* Feb 1994, pp 31-34.

[9]The following two kits are available from Amidon Inc: Amplifier ferrite kit (Amidon P/N HFAFC) containing the ferrite cores, balun core and magnet and Teflon wire to wind the transformers for the HF amplifier. Price: $3.50 plus shipping. Low-pass filter cores kit (Amidon P/N HFFLT) containing all iron cores and wire for the low-pass filters. Price: $4.50 plus shipping.

[10]Motorola Application Reports Q1/95, HB215, *Application Report AR346.* Thermal runaway is a condition that occurs with bipolar transistors because bipolar transistors conduct more as temperature increases, the increased conduction causes an increase in temperature, which further increases conduction, etc. The cycle repeats until the bipolar transistor overheats and is permanently damaged.

A DRIFT-FREE VFO

By following several design guidelines, Jacob Makhinson, N6NWP, built a low-cost, easy-to-construct LC VFO with a very low level of phase noise. The article originally appeared in December 1996 *QST*.

The method shown makes the oscillator essentially drift-free, with very little phase noise. VFOs built with these techniques are viable in applications where low overall noise level and wide dynamic range is of great importance. The technique can also spare VFO designers the drudgery of more conventional drift-compensating techniques.

Many VFO designs have appeared in the Amateur Radio literature, and the quest for a low-drift VFO hasn't ceased. If the frequency-stability requirements are stringent, the thermal-drift compensation can be very tedious. Wes Hayward's *QST* article[1] devoted to VFO drift compensation is an excellent example of this difficult pursuit.

DESIGN CRITERIA

To avoid degradation of the receiver's front end, several requirements should be imposed on the phase noise level of the VFO. An excessively high level of close-in phase noise (within the bandwidth of the SSB signal) may reduce the receiver's ability to separate closely spaced signals. As an example, a 14-pole crystal filter described in Note 2 provides adjacent-signal rejection of 103 dB at a 2-kHz offset. This requires the use of a VFO with –139 dBc/Hz phase noise at a 2-kHz offset.

$$Pn = P - 10\log(BW) = -103 - 10\log(4000) = -139 \text{ dBc/Hz}$$

where

Pn = VFO phase-noise spectral density, in decibels relative to the carrier output power, in a 1-Hz bandwidth (dBc/Hz)

P = VFO power level (dBc) in a given bandwidth (BW)

BW = test bandwidth, in Hertz

In addition, excessive close-in phase noise may lead to reciprocal mixing, where the noise sidebands of a VFO mix with strong off-channel signals to produce unwanted IF signals. Excessive far-out phase noise may degrade the receiver dynamic range. In a properly designed receiver, the phase-noise-governed dynamic range (PNDR) should be equal to or better than the spurious-free dynamic range (SFDR). We can calculate the PNDR:[3]

$$PNDR = -Pn - 10\log(BW)$$

Assuming the PNDR equals the SFDR at 112 dB in a 2.5-kHz IF noise bandwidth, the required far-out phase noise level is –146 dBc/Hz:

$$Pn = -SFDR - 10\log(BW) = -112 - 34 = -146 \text{ dBc/Hz}$$

Another form of VFO instability—frequency drift—has always been a nuisance and a great concern to the amateur community. The objective of this project was to keep the long-term frequency drift (seconds, minutes, hours) under 20 Hz. This includes thermal drift from both internal heating and environmental changes.

BLOCK DIAGRAM

The block diagram of **Fig 14.80** shows the LC VFO and the frequency stabilizer. The stabilizer monitors the VFO frequency and forms an error signal that is applied to the VFO to compensate for frequency drift. This technique, which is capable of stabilizing a VFO to within a few hertz, was devised by Klaas Spaargaren, PA0KSB, and first described in *RadCom* magazine in 1973.[4] This project builds upon Spaargaren's idea and presents a few refinements.

The stabilizer converts a free-running VFO into an oscillator that can be tuned in the usual fashion, but then locks to the nearest of a series of small frequency steps. Unlike traditional PLL frequency synthesizers, the stabilizer has no effect on the phase-noise performance of the VFO; it only compensates for thermal drift.

The timing signal (2.6 Hz) is derived from a crystal oscillator via a frequency divider. The timing signal drives a NAND gate to provide a crystal-controlled time window, during which the binary counter counts the VFO output. When the gate closes, the final digit of the count remains in the counter. For counts 0 to 3, the Q3 output of the counter is a logic 0; for counts of 4 to 7, a logic 1.

The result is stored in a D flip-flop memory cell: When the 2.6-Hz timing signal goes low, the first of three one-shots triggers. The second follows and clocks the binary counter Q3 output into the memory cell. The negative-going pulse from the third resets the counter for the next counting sequence.

The output of the memory cell is applied to an RC integrating circuit with a time constant of several minutes. This slowly changing dc voltage controls the VFO frequency via a couple of varicaps connected to a tap on the VFO coil.

If the counter output is 0, the memory-cell output is 1, which charges C and increases the VFO frequency. A counter output of 1 discharges C and decreases the VFO frequency. The stabilizer constantly searches for equilibrium, so the VFO frequency slowly swings a few hertz around the lock frequency. The circuit limits the frequency swing to a maximum of ±2 Hz, typically ±1 Hz.

A difficulty arises when the operator changes frequency because the control voltage is disturbed. If the memory-cell output connects directly to the RC integrator, the frequency correction that occurs immediately after tuning results in a frequency hop. To overcome this problem, an analog switch disconnects the integrator from the memory during tuning. The tuning detector—an infrared interrupter switch and a one-shot—controls the analog switch.

VFO CIRCUIT DESCRIPTION

The VFO is a tapped-coil Hartley oscillator that is optimized for low phase noise (see Fig 14.80B). It follows the design rules compiled by Ulrich Rohde, intended to minimize the phase noise in oscillators.[5]

The tank coil, L1, has an iron-powder toroidal core; coil Q exceeds 300. C1, C4, C5 and C7 are NP0 (C0G) ceramic capacitors (5% or 10% tolerance). C2 is the main tuning capacitor, and C3 is a small ceramic trimmer capacitor.

The VFO frequency range is set from 6.0 MHz to 6.4 MHz (to accommodate a 20-meter receiver with an 8-MHz IF). The loaded Q of the resonator is kept high by using a tapped coil and loose coupling to the gate of the FET through C7 (more than 8 kΩ at 6 MHz). The RF voltage swing across the resonator exceeds 50 V, P-P. Varicaps D1 and D2, which compensate for thermal drift, are connected across the coil's lower tap (less than 14% of the total turns) and have a negligible effect on overall phase-noise. J310 is the TO-92 version of U310—a very low-noise FET in HF applications.

An ALC loop limits the voltage swing. The signal is sampled at the primary of T1, rectified by the D5-C21 network and fed to the inverting input of an integrator, U1A, where it is compared against the reference voltage at the junction of R18 and R19. The dc voltage at the integrator output sets Q1's drain current so that the signal swing at T1's primary is always 2.5 V, P-P. The ALC loop also makes VFO performance independent of Q1's pinch-off voltage. The signal at Q1's source is a 6.5-V, P-P, sinusoid with almost no distortion.

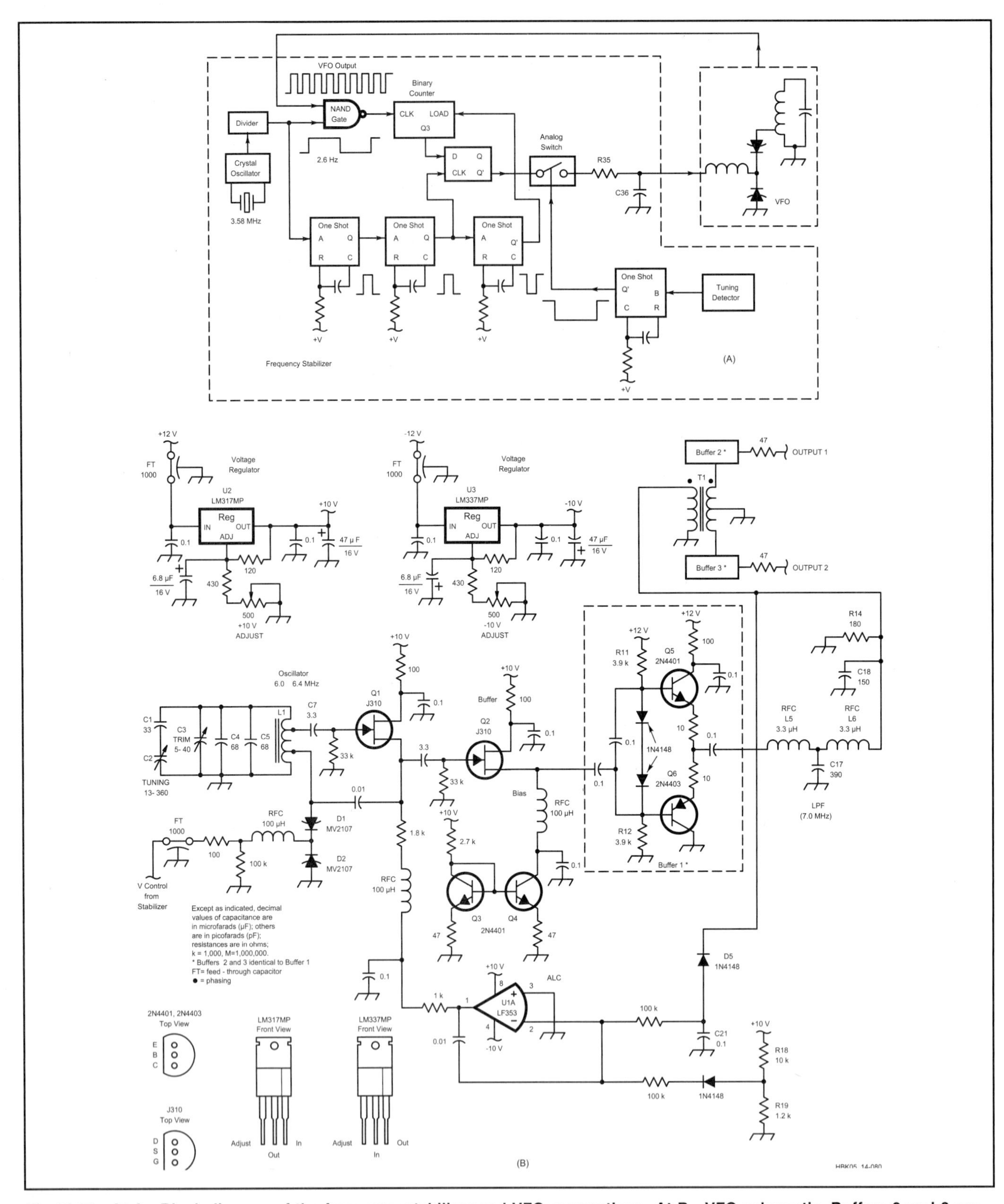

Fig 14.80—At A—Block diagram of the frequency stabilizer and VFO connections. At B—VFO schematic. Buffers 2 and 3 are identical to Buffer 1. Most of the parts are available from Mouser Electronics, Digi-Key Corporation or Allied Electronics. The cores for L1 and T1 are from Amidon Associates. Use 1/4-W, 5%-tolerance carbon-composition or film resistors and ceramic, 20%-tolerance capacitors unless otherwise indicated. RF chokes or encapsulated inductors may be used for those labeled "RFC."

Q1, Q2—J310, N-channel JFET (Allied).
D1, D2—MV2107 or ECG/NTE613 tuning diode (Varicap, Allied).
L1—29 turns of #18 AWG enameled copper wire on a T-80-6 iron-powder toroidal core tapped at 4 turns and 20 turns from the cold end (Amidon).
T1—#32 AWG enameled copper wire on a BN-43-2402 two-hole ferrite balun core (Amidon) primary: 5 turns; secondary: 16 turns, center tapped
Vector part #8007 circuit board (Digi-Key).
Vector part #T44 terminals (Digi-Key).

Q2 is a high-impedance buffer that is loosely coupled to Q1. Q2's drain current is set to 3.4 mA (by the constant-current source, Q3-Q4) regardless of Q2's pinch-off voltage. Buffer 1 is a push-pull stage biased into slight conduction by resistors R11 and R12. It has excellent linearity and a very low output impedance, which is required to drive an LC filter. The filter (L5, L6, C17, C18 and R14) is a four-pole, 0.1-dB Chebyshev low-pass filter with a ripple frequency of 7 MHz. All harmonics at the VFO output are at least 45 dB below the fundamental.

T1 provides the two complementary outputs required for a commutation mixer and raises the voltage swing at the VFO output. Buffers 2 and 3 are electrically identical to buffer 1. They further decouple the VFO from its load and serve as low-distortion 50-Ω. drivers. The signal level at each output is 4 V, P-P, when driving a high-impedance load (eg, a CMOS gate), +10 dBm when driving a 50-Ω load.

FREQUENCY STABILIZER

NAND gates U4A and B (see **Fig 14.81**) comprise a Pierce crystal oscillator. The timing signal (2.6 Hz) appears at the output of the frequency divider (U5, U6, U7 and U8A). The exact frequency of the crystal and the timing signal is unimportant, but the stabilizer has been optimized for 2.3 to 2.7 Hz.

There are two requirements for the crystal oscillator: No harmonics should fall in the IF passband, and the crystal should have a low temperature coefficient. Crystal-oscillator thermal drift should not exceed 10 Hz within the temperature operating range. Crystals in HC-33 cases with frequencies between 2.0 and 3.58 MHz worked best for me. The frequency divider is sufficiently flexible to provide the desired timing-signal frequency.

U4C, biased into a linear range, converts the sinusoidal signal from one of the two VFO outputs into a square wave. U4D gates the VFO signal bursts into the clock input of the binary counter, U9. At the end of every burst, the final digit is held by the counter.

The falling edge of the timing signal triggers U10A, the first of three cascaded one-shots. The pulse at the output of U10B clocks the data from the counter into U8B. The pulse at the output of U11A resets U9.

If the number of pulses in each successive burst is equal (no VFO drift), U9 constantly counts the same number, and the output of U8B never changes. In practice, however, U8B constantly toggles between two states. The integrating circuit, R35-C36 (time constant = 6.5 minutes), converts the toggling into a slowly changing voltage. Varicaps D1 and D2 transform a few millivolts of change into ±1 or 2 Hz change of VFO frequency.

U13A, a high-input-impedance buffer, prevents the discharge of C36. U13B, a noninverting amplifier with a gain of 1.5, ensures compliance between the control-voltage range and the capacitance-per-volt ratio of the Varicaps (1 to 6 V for best performance). Network R36, R37, C37 and D7 establishes the initial dc voltage applied to the varicaps; the value is set by the C37-C36 voltage divider.

An infrared interrupter switch, U14, serves as sensor in the tuning-detector circuit. The slotted interrupter detects the movements of a serrated disc (see **Fig 14.82**) on the VFO reduction-drive shaft. U15A and B, a two-level limit comparator, converts the signal at its input into pulses. U16A produces trigger pulses for the one-shot, U11B, by detecting both leading and falling edges of the signal at its input. U11B is retriggerable—its Q output stays low during manual tuning and for 3.6 seconds after tuning stops. Analog switch U12A disconnects C36 from the flip-flop during tuning, thus preserving the capacitor charge. This system does not provide for an RIT control.

CONSTRUCTION

The VFO and the stabilizer are in separate boxes. Mount components within the enclosures on the perf board's foil side. Make ground connections to the foil plane. Use Vector pins as terminal posts for the input and output signals.

The VFO box is a die-cast aluminum enclosure ($4\,^{11}/_{16} \times 3\,^{11}/_{16} \times 2\,^{1}/_{16}$ inches) to ensure mechanical rigidity. The two RF outputs exit the box via BNC connectors and coax. DC enters via feedthrough capacitors. Rigidly attach C2 to the enclosure wall. Cover L1 with a low-loss polystyrene Q dope and place it as far as possible from the ground plane and enclosure walls. The layout is not critical, but observe standard RF building methods: use short leads, dress them for minimum coupling and solder bypass capacitors directly to the ground plane close to the terminal they bypass.

The stabilizer is in a $5\,^{1}/_{2} \times 3 \times 1\,^{1}/_{4}$-inch LMB aluminum enclosure. Component placement and layout is not critical, but keep component leads short around the crystal oscillator. Use a BNC connector for the signal from the VFO module. Solder the ground pins of all ICs directly to the ground plane, and decouple each power-supply pin of ICs U4 through U9 to ground via a 0.1 μF capacitor. Route all dc voltages to the module via feedthrough capacitors to avoid RF leakage. Mount U14 so that the serrated disc is in the middle of the slot.

Mount C37 in a socket in case you need to adjust its value: For unknown or varying VFO drift direction, use a 22 μF capacitor to place the initial varicap control voltage at midrange ($V_c \approx 2.9$ V). Use 10-μF if VFO drift is predominantly negative ($V_c \approx 1.5$ V), and 33 μF if it's predominantly positive ($V_c \approx 4.0$ V).

MEASUREMENTS

The VFO thermal drift without the stabilizer was under 800 Hz at room temperature (after 90 minutes) and under 1500 Hz when the ambient temperature was raised 20°C. There was no attempt to compensate for thermal drift.

With the frequency stabilizer connected, the thermal drift did not exceed 10 Hz at room temperature, 20 Hz when raised 20°C. In one of the experiments, power was on for several days, and drift was under 10 Hz at room temperature. Frequency lock is attained in less than 10 seconds after the power is switched on.

With the components shown in the schematic, the stabilizer can compensate for a maximum 1800-Hz drift with a 25°C temperature rise. To compensate for a greater frequency drift, select varicaps with higher diode capacitances; the frequency swing will increase from ±1 or 2 Hz to a higher value.

Notes

[1]Wes Hayward, W7ZOI, "Measuring and Compensating Oscillator Frequency Drift," *QST,* Dec 1993, pp 37-41.

[2]Jacob Makhinson, N6NWP, "Designing and Building High-Performance Crystal Ladder Filters," *QEX,* Jan 1995, pp 3-17.

[3]Peter Chadwick, G3RZP, "Phase Noise Intermodulation and Dynamic Range," *Frequency Dividers and Synthesizers IC Handbook,* Plessey Semiconductors, 1988, p 151.

[4]Klaas Spaargaren, PA0KSB, "Technical Topics: Crystal-Stabilized VFO" *RadCom,* Jul 1973, pp 472-473. Comments followed in later "Technical Topics" columns. Also see, "Frequency Stabilization of L-C Oscillators," *QEX,* February 1996, pp 19-23.

[5]Ulrich Rohde, KA2WEU/DJ2LR, *Digital PLL Frequency Synthesizers* (Englewood Cliffs: Prentice-Hall, 1983) p 78.

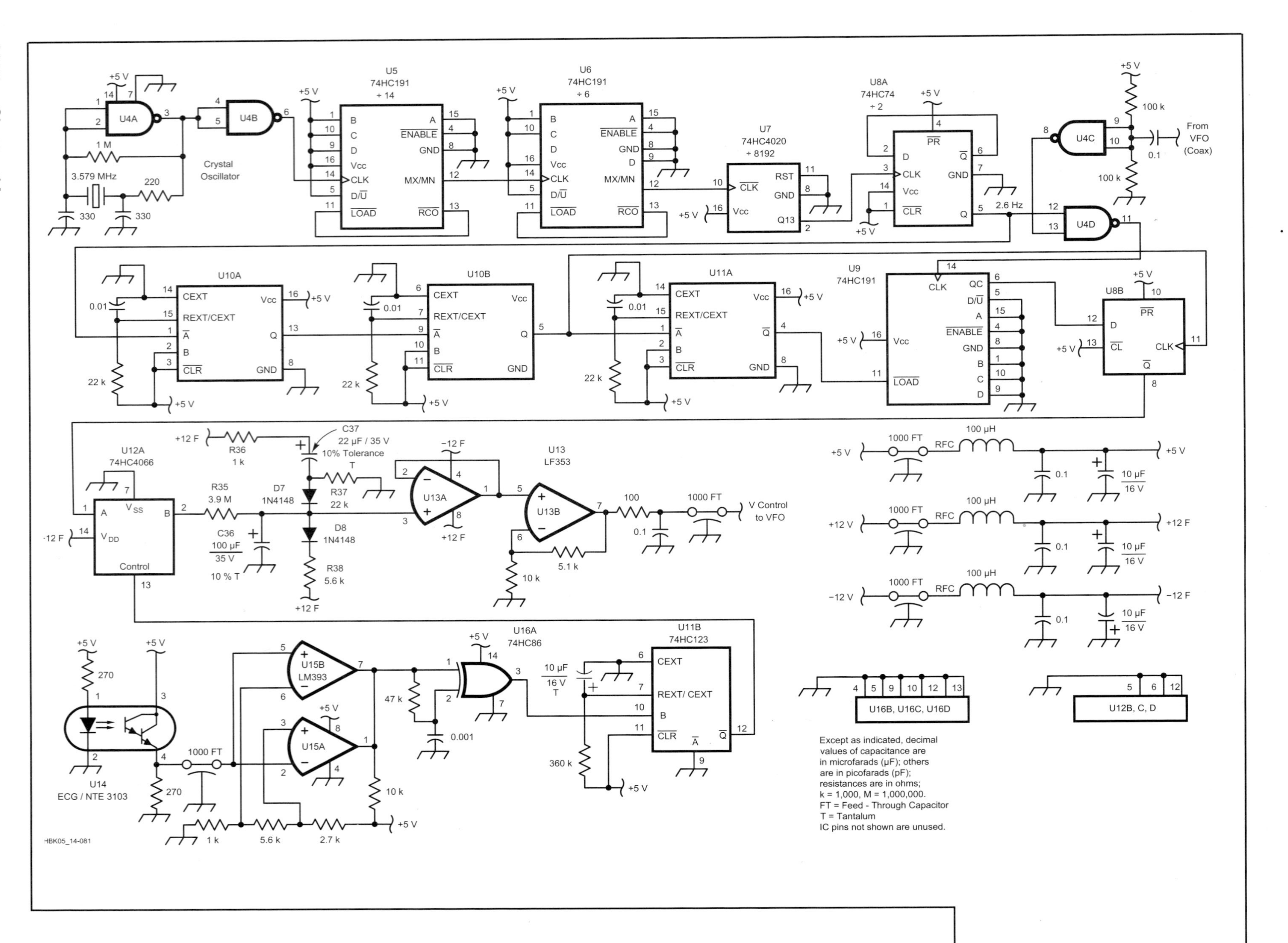
U4A
U4B
Crystal Oscillator
3.579 MHz
U5 74HC191 ÷ 14
U6 74HC191 ÷ 6
U7 74HC4020 ÷ 8192
U8A 74HC74 ÷ 2
U4C
U4D
From VFO (Coax)
2.6 Hz
U10A
U10B
U11A
U9 74HC191
U8B
U12A 74HC4066
C37 22 µF / 35 V 10% Tolerance T
R36 1 k
R35 3.9 M
D7 1N4148
R37 22 k
D8 1N4148
C36 100 µF 35 V 10 % T
R38 5.6 k
U13A
U13 LF353
U13B
V Control to VFO
1000 FT
RFC
100 µH
U14 ECG / NTE 3103
U15B LM393
U15A
U16A 74HC86
U11B 74HC123
U16B, U16C, U16D
U12B, C, D
Except as indicated, decimal values of capacitance are in microfarads (µF); others are in picofarads (pF); resistances are in ohms; k = 1,000, M = 1,000,000.
FT = Feed - Through Capacitor
T = Tantalum
IC pins not shown are unused.
HBK05_14-081

Figure 14.81—Stabilizer schematic. Use 1/4-W, 5%-tolerance carbon-composition or film resistors and ceramic, 20%-tolerance capacitors unless otherwise indicated.

U4—74HC00 Quad NAND gate.
U5, U6, U9—74HC191 presettable 4-bit binary counter.
U7—74HC4020 14-bit binary ripple counter.
U8—74HC74 dual D flip-flop.
U10, U11—Dual 74HC123 one-shot.
U12—4066 quad analog switch.
U14—ECG/NTE3103 optical interrupter (Darlington output).

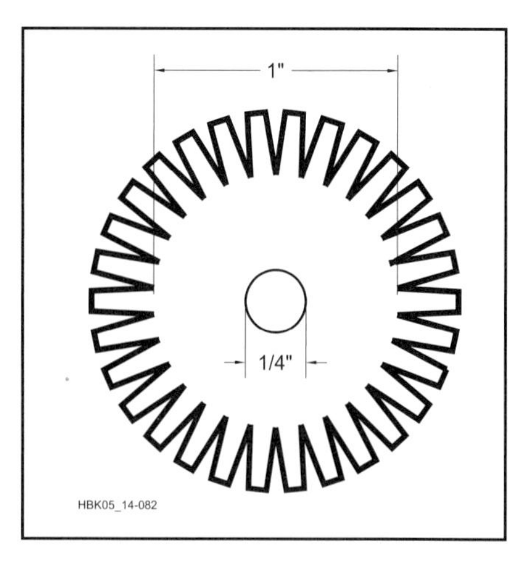

Figure 14.82—Mechanical details of interrupter wheel. Use a good reduction drive and make one tooth for every 20 to 40 Hz of frequency change. Use any rigid, opaque material.

A SIMPLE REGENERATIVE HF RECEIVER FOR BEGINNERS

This project was designed by Charles Kitchin, N1TEV, and originally published in the September 2000 issue of *QST*. It features a low cost, portable shortwave receiver that's ideal for a Scout radio merit badge or for learning basic radio theory. This project is fun to build and easy to get working. This little radio is a fun way to discover ham-band QSOs, news, music and all the other things the shortwave bands have to offer. With it, you can receive dozens of international shortwave broadcast stations at night. Although this little receiver is quite sensitive, it naturally won't match the performance of a commercial HF rig. If you have not used a regenerative receiver before, you'll have to practice adjusting the controls, but that's part of the adventure. Many of today's experienced "homebrewers" got their start by building simple circuits just like this one. You'll gain experience in winding a coil and following a schematic. As your interest in radio communication develops, you can build a more complex receiver later.

This little set requires only a single hand-wound coil and consumes just 5 mA from a 9-V battery. At that rate, an alkaline battery can provide approximately 40 hours of operation. The sound quality of this receiver is excellent when using Walkman headphones. The radio can also drive a small speaker. To simplify construction, a low-cost PC board is available from FAR Circuits.[1] You can house the receiver in a readily available RadioShack plastic project box.

CIRCUIT DESCRIPTION

Fig 14.83 shows the schematic. L1 and C1 tune the input signal from the antenna. Regenerative RF amplifier Q1 operates as grounded-base Hartley oscillator. Its positive feedback provides a signal amplification of around 100,000. The very low operating power of this stage, only 30 µW, makes this receiver very portable and prevents interference to other sets in the area. R2 controls the amount of positive feedback (regeneration). D1 and C4 comprise a floating detector that provides high sensitivity with little loading on Q1. The relatively low back-resistance of the 1N34 germanium diode (don't use a silicon diode here!) provides the necessary dc return path for the detector.

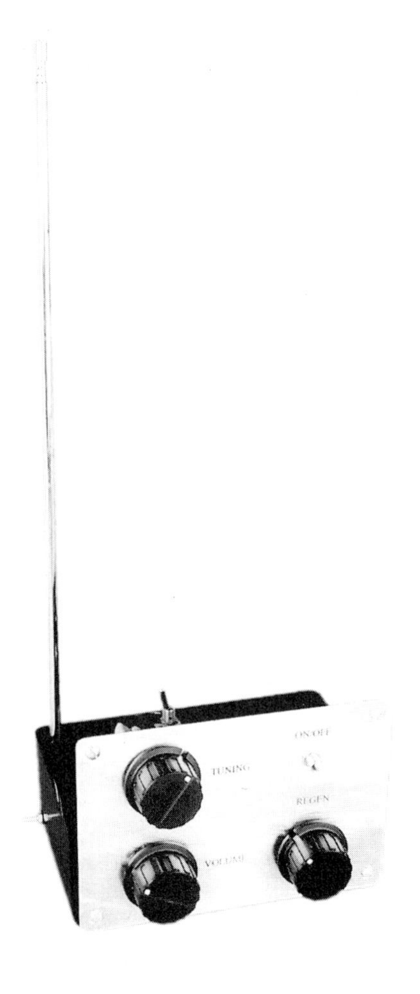

VOLUME control R5 sets the level of detected audio driving U1, an LM386 audio amplifier. C5 provides low-pass filtering that keeps RF out of the audio amplifier. R4 isolates the low-pass filter from the detector circuit when the volume control is at the top of its range. The bottom of the VOLUME control, R5, and pin 3 of the LM386 float above ground, so that both inputs of the IC are coupled. This allows the use of a 100-kΩ VOLUME control; this high resistance value prevents excessive loading of the detector. D5 protects the receiver from an incorrectly connected battery. L1 is wound on a coil form using a standard 35-mm plastic film can or 1-inch-diameter pill bottle. C1 can be any air-dielectric variable capacitor with a maximum capacitance of 100 to 365 pF. A two ganged capacitor from an old AM radio will work well. Total frequency coverage varies with the capacitance value used, but any capacitor in that range should cover the 40-meter ham band and several international broadcast bands. If you use a capacitor with a max. capacity larger than 150 pF, the receiver will cover a very wide frequency range but it will be more difficult to tune-in an individual station. In that event, the optional fine-tuning control (see the inset of Fig 14.83) is recommended. **Figs 14.84** and **14.85** show the construction details.

D6 functions as a poor man's Varactor (voltage-variable capacitor). As the volt-

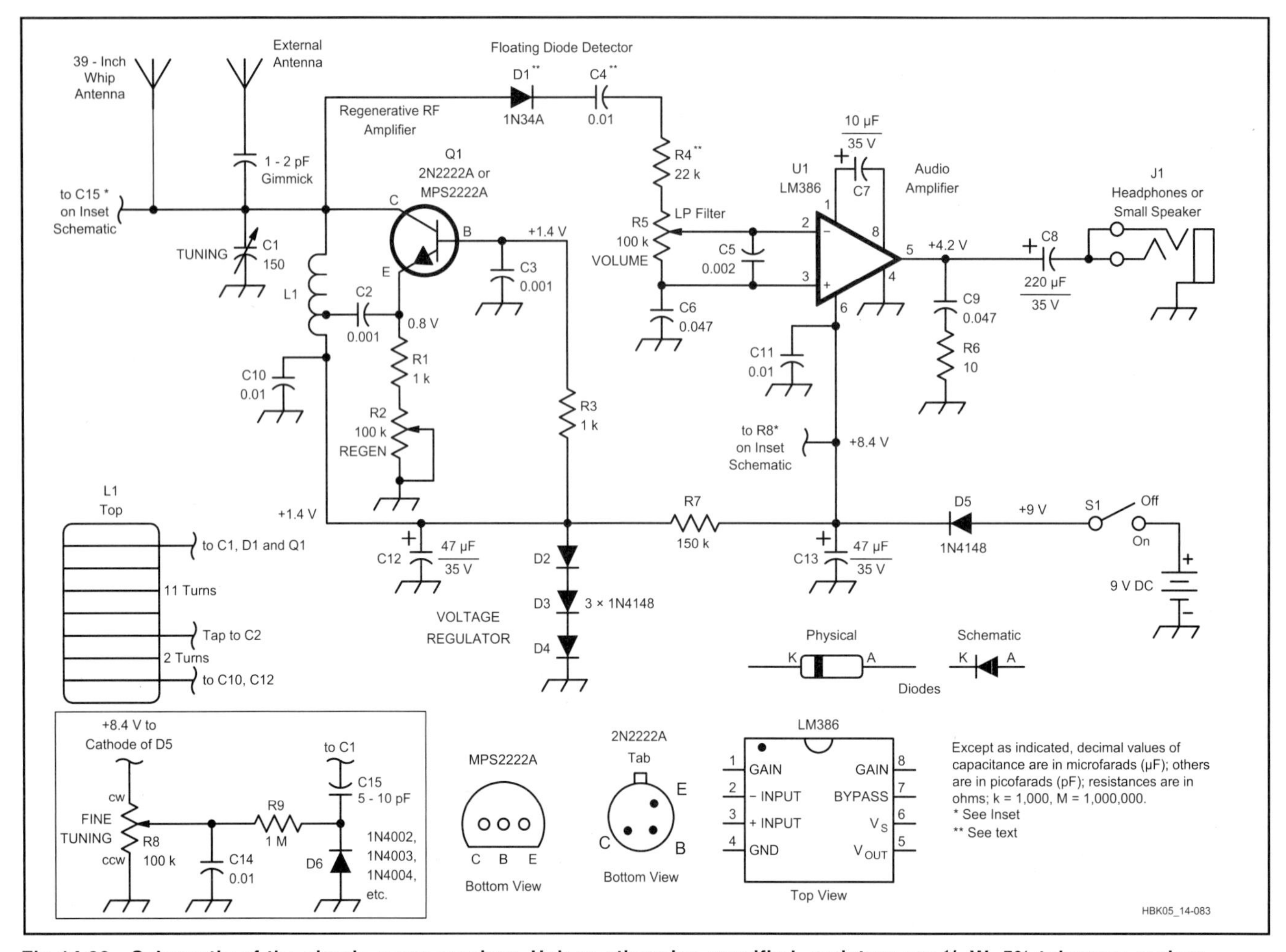

Fig 14.83—Schematic of the simple regen receiver. Unless otherwise specified, resistors are 1/4-W, 5%-tolerance carbon-composition or metal-film units. Part numbers in parentheses are RadioShack. Equivalent parts can be substituted; n.c. indicates no connection.

C1—150 pF (maximum value) air-dielectric variable capacitor; see text.
C2, C3—0.001 µF, 50 V (or more) disc ceramic (RS 272-126).
C4, C10, C11, C14—0.01 µF, 50 V (or more) disc ceramic (RS 272-131).
C5—0.002 µF, 50 V (or More) disc ceramic (use two RS 272-126 connected in parallel).
C6, C9—0.047 µF, 50 V disc ceramic (RS 272-134).
C7—10 µF, 35 V electrolytic (RS 272-1025).
C8—220 µF, 35 V electrolytic (RS 272-1017).
C12, C13—47 µF, 35 V electrolytic (RS 272-1027).
C15—5 to 10 pF, 50 V (or more) mica (RS 272-120).
D1—1N34A *germanium* diode (RS 276-1123); don't use a silicon diode here.
D2-D5—1N4148 or any similar diode (RS 276-1122).
D6—1N4003 silicon diode (RS 276-1102).
J1—1/8-inch, three-circuit jack (RS 274-246).
L1—See text.
Q1—2N2222A NPN transistor (RSU11328507) or MPS2222A (RS 276-2009).
R1, R3—1 kΩ (RS 271-1321).
R2, R5—100 kΩ potentiometer, linear taper (RS 272-092).
R4—22 kΩ (RS 271-1339).
R6—10 Ω (RS 271-1301).
R7—150 kΩ (RSU11345287) or use series-connected 100 kΩ (RS 271-1347) and 47 kΩ (RS 271-1342) resistors.
R8—100 kΩ audio-taper pot (RS 271-1722); connect so that clockwise rotation increases the voltage at the junction of the pot arm, R9 and C14.
R9—1 MΩ (RS 271-1356).
S1—SPST miniature toggle (RS 275-612).
U1—LM386N-1 audio amplifier (RS 276-1731).
Misc—PC board (see Note 1); 8-pin DIP socket for U1 (RS 276-1995A); 9-V battery clip (RS 270-325); three knobs (RS 274-402A); project box (RS 270-1806); #6-32 screws and nuts, rubber feet; 9-V battery, RadioShack 22-gauge solid, insulated hook-up wire.

age from FINE-TUNING control R8 is increased, the diode is reverse biased and its capacitance decreases. For two-band operation, use a 150-pF capacitor for C1 and install a miniature toggle switch, right on the capacitor itself, to add an additional 250-pF mica capacitor in parallel with C1. With the capacitor switched-in, the receiver will now tune the 80-meter band.

COIL WINDING

For the coil winding, use 22-gauge solid-conductor insulated hook-up wire. Drill a mounting hole in the bottom of the coil form. Then, drill two small holes in the side of the coil form, near the top. Wind the coil starting from the top of the form going to the bottom, keeping the turns well above the PC board. Feed one end of the wire through the first hole to the inside of the form, then out through the second. Tie

Fig 14.84—In this view of the receiver, the TUNING capacitor is at the left. Immediately behind it is the coil, L1. Just to the left of the TUNING capacitor, you can see D1, C4 and R4, as discussed in the text.

Fig 14.85—This close-up shows the interconnection of series-connected D1, C4 and R4 between the TUNING capacitor and VOLUME control.

a knot in the wire where it enters the form—this keeps the wire from loosening-up later on. Be sure to leave a two to three inch length of wire at each end of the coil so you can make connections to the PC board. You can wind the coil in either direction, clockwise or counterclockwise. Tightly wind the the wire onto the form, counting the turns as you go. Keep the turns close together and don't let the wire loosen as you wind. To make the coil tap, wind 11 turns on the coil form. While holding the wire with your thumb and index finger, mark the tap point and remove the insulation at that point. Solder a two to three-inch piece of wire to the tap. Continue winding turns until the coil is finished (13 turns total). Keep the free end of the wire in place using a piece of tape and drill two more holes in the coil form where the winding ends. Feed the wire end in and out of the coil as before and tie a knot at the end to hold the winding in place. When the coil is finished, remove the tape then carefully solder the three wires from the coil (bottom, tap and top) to their points on the PC board keeping the wire lengths as short as possible. Mount the coil away from any metal objects.

CONSTRUCTION DETAILS

The prototype receiver used a Radio Shack plastic project box, with a metal front panel. For best performance, the floating detector must be wired using short, direct connections. Therefore, these components are not mounted on the PC board. Mount the VOLUME control, R5, close to the TUNING capacitor, C1. Connect D1, C4 and R4 in series between the "hot" side of C1 (the stator) and the top of the VOLUME control. See Fig 14.85. Mount the TUNING capacitor and the regeneration (REGEN) control on opposite sides of the front panel. The VOLUME and REGEN controls are best mounted near the bottom of the front panel to keep their connecting wires to the PC board as short as possible. Use very short twisted wires or shielded wires for the VOLUME and REGEN control connections. Be sure to connect a wire between the metal front panel and the PC board ground. If you use the RadioShack jack specified for J1 (RS 274-276), connect pins 2 and 5 together and attach that common lead to C8. Ground pin 1 of the jack. If you intend to use a small speaker, connect it between pins 1 and 3. Then, when headphones are plugged in, the speaker will be disconnected automatically.

This receiver will pick up lots of stations using a 39-inch whip antenna connected directly to the "hot" side of C1 (the junction of C1, L1 and the collector of Q1). The set will be easily detuned by hand and body movements, however. For best performance, use an external 20 foot, or longer, length of insulated wire run outside to a tree. But be sure not to connect this wire directly to the set, as this will load down Q1 and ruin selectivity. Instead, use a small "gimmick" capacitor, approximately 1 to 2 pF, between the external antenna and C1. This is easy to do: simply twist together two pieces of solid insulated hook-up wire, each about two inches long. Solder one wire to C1, the other to a banana jack on the back of the set. Twist the wire's other (insulated) ends together approximately three times, depending on antenna length. Twist the wires enough to bring in lots of stations but not enough to load down Q1.

TESTING AND OPERATING THE RECEIVER

Set the VOLUME and REGEN controls to midrange, plug in the headphones, attach the battery and turn on the receiver. Check to see that the audio stage is working by placing a finger on the wiper of the VOLUME control and listening for a buzz. Then use just three feet of hook-up wire connected directly to C1, as a temporary antenna. Adjust the REGEN control until the set produces a "live" sound, indicating that Q1 is oscillating. If not, then carefully recheck the wiring and measure the voltages labeled on the schematic, using a high-impedance DVM or multimeter. Common problems are Q1 being wired backwards (emitter and collector connections reversed) and the wires from coil L1 connected to the wrong places on the PC board.

Use two hands when operating the receiver: one for tuning, the other for controlling regeneration. For international broadcast stations or AM phone operation, carefully, adjust the REGEN control so that Q1 operates just below oscillation. For CW and SSB, increase the REGEN level so that the set *just* oscillates, providing the required local oscillation for these modes. If you operate this receiver close to another radio, the regen's 30-µW oscillator might cause interference. For details on building a higher-performance regen receiver for serious CW and SSB reception see: Kitchin, "High Performance Regenerative Receiver Design," Nov/Dec 1998 *QEX*.

[1]A PC board for this radio is available from FAR Circuits. Price: $5 each plus $1.50 shipping for up to three boards.

Chapter 15

Transceivers, Transverters and Repeaters

TRANSCEIVERS

In recent years the transceiver has become the most popular type of purchased equipment among amateurs. The reasons for this popularity are:

1. It is economical to use LOs (especially synthesizers), IF amplifiers and filters, power supplies, DSP modules and microprocessor controls for both transmit and receive.
2. It is simpler to perform transmit-receive (T/R) switching functions smoothly and with the correct timing within the same piece of equipment.
3. It is convenient to set a receive frequency and the identical (or properly offset) transmit frequency simultaneously.

In addition, transceivers have acquired very impressive arrays of operator aids that help the operator to communicate more easily and effectively. The complex design, numerous features and the very compact packaging have made the transceiver essentially a "store bought" item that is extremely difficult for the individual amateur to duplicate at home. The complexity of the work done by teams of design specialists at the factories is incompatible with the technical backgrounds of nearly all individual amateur operators.

The result of this modern trend is that amateur home-built equipment tends to be simpler; with less power output and more specialized in design (one-band, QRP, CW only, direct conversion, no-tune, receive only, transmit only and so on). Or, the amateur designs and builds add-on devices such as antenna couplers, active adaptive filters, computer interfaces and such.

TRANSCEIVER EXAMPLE

As a way of providing a detailed, in-depth description of modern high-quality transceiver design, let's discuss one example, the Ten-Tec Omni VI Plus, an HF ham-band-only solid-state 100-W (output) transceiver, shown in **Fig 15.1**. Let's consider first the signal-path block diagram in **Fig 15.2**, one section at a time.

Receiver Front End

The receive antenna can be either the same as the transmitting antenna or an auxiliary receive antenna. A 20-dB attenuator can be switched in as needed. A 1.6-MHz high-pass filter attenuates the broadcast band. A 9.0-MHz trap attenuates very strong signals at 9.0 MHz that might create interference in the form of

Fig 15.1 — Photograph of Ten-Tec Omni VI Plus HF transceiver.

The Many Flavors of Transceivers

Today, commercial transceivers can be found in the vast majority of Amateur Radio stations in the United States — and increasingly throughout the world. This is not to suggest, however, that hams cannot design and build their own transceivers. Many here and around the globe have done just that with great success! Such homebrew designs cover the range of circuit complexity and performance.

Whether homemade or commercial, and regardless of individual design complexity, the idea and practice of packaging a receiver and transmitter (and often, even a common power supply for both functions) into one unit has eclipsed the once-standard separate *receiver* and *transmitter* combination. Those venerable pairs were characteristic of 20th century ham shacks up to and even well after World War II. During the 1960s and 1970s, more transceiver designs were offered, and correspondingly more hams accepted them due to their efficiency, convenience and other beneficial features.

For the class of ham transceivers available today — in terms of sophistication and technology — what was once thought to be nearly impossible to design and build is now not only quite common, but expected in most medium and high-end commercial models! Here are some of the more common and recent commercial radios in the ham transceiver category.

Elecraft K2

The basic K2 transceiver package, shown in **Fig A**, is a QRP kit that covers CW from 80-10 Meters. With inclusion of an SSB adapter option, transmitter power output ranges from 100 mW (QRP_P level) to 12 W. Other options include a 160-m module, DSP module, audio filter, two-stage noise blanker, auto antenna tuner and the KPA100 — an RF power module that delivers 100-W output on SSB or CW. *Manufacturer*: Elecraft, PO Box 69, Aptos, CA 95001-0069; **www.elecraft.com**.

Fig A — Elecraft K2.

ICOM IC-706MKIIG

Fig B shows a versatile radio that can serve in either a mobile or base installation. This rig covers 160-10, 6 and 2 Meters, 70 cm and also sports a general coverage receiver! DSP equipped. Specified power output (SSB, CW and FM) is 100 W from 160-6 Meters and 40 W on AM. Power output capabilities are lower on the operational transmit frequencies above 50 MHz. *Manufacturer*: ICOM America, 2380 116th Avenue NE, Bellevue, WA 98004; **www.icomamerica.com**.

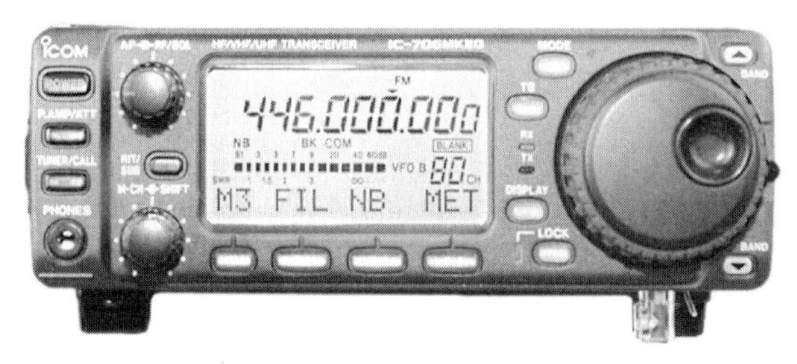

Fig B — ICOM IC-706MKIIG.

blocking or harmonic IMD, especially when tuned to the 10.1-MHz (30-m) or 7-MHz (40-m) bands.

A set of band-pass filters, one for each HF amateur band, eliminates image responses and other spurs in the first mixer. These filters are also used in the low-level transmit stages. A low-noise, high-dynamic-range, grounded-gate JFET RF amplifier with about 9 dB of gain precedes the double balanced diode mixer, which uses 17 dBm of LO in a high-side mixer.

First IF

The first IF is 9.0 MHz. Because the LO is on the high side, there is a sideband inversion (USB becomes LSB and so on) after the first mixer. A grounded-gate, low-noise JFET amplifier terminates the first mixer in a resistive load and provides 6 dB of gain. This preamp helps to establish the receiver sensitivity (0.15 µV) with minimum gain preceding the mixer. The preamp is followed by a 15-kHz-wide two-pole filter, which is used for NBFM reception. It is also a roofing filter for the IF amplifier and the noise-blanker circuit that follow it.

The noise blanker gathers impulse energy from the 15-kHz filter, amplifies and rectifies it, and opens a balanced diode noise gate. The IF signal ahead of the gate is delayed slightly by a two-pole filter so that the IF noise pulse and the blanking pulse arrive at the gate at the same time.

The standard IF filter for SSB/CW has 8 poles, is centered at 9.0015 MHz and is 2.4 kHz wide at the –6 dB points. Following this filter, two optional 9.0-MHz filters with the following bandwidths can be installed: 1.8 kHz, 500 Hz, 250 Hz or a 500 Hz RTTY filter. The optional filters are in cascade with the standard filter, for improved ultimate attenuation.

Passband Tuning Section

A mixer converts 9.0 MHz to 6.3 MHz and drives a standard 2.4 kHz wide filter. One of three optional filters, 1.8 kHz,

Kenwood TS-870S

When this DSP-at-IF radio (see **Fig C**) was first introduced, it was considered revolutionary due to the ability of the rig to achieve good selectivity without the traditional standard and optional crystal IF filters. It covers 160-10 Meters, and also has a general coverage receiver from 30 kHz to 30 MHz. RF output is continually adjustable up to 100 W (SSB, CW, FSK and FM), and up to 25 W on AM. The TS-870S requires a separate power supply. *Manufacturer*: Kenwood USA Corp, 3975 Johns Creek Ct, Suwanee, GA 30024-1265; **www.kenwood.net**.

Fig C — Kenwood TS-870S.

Ten-Tec Orion (Model 565)

This advanced transceiver, shown in **Fig D**, is the current top-of-the-line offering from Ten-Tec. A variety of automatically- or manually-selectable front-end crystal roofing filters, and dual 32-bit floating-point processors characterize the Orion's formidable high-tech receive weaponry. Excellent Software-Defined Radio (SDR) techniques are employed which include user-installable software upgrades via the Internet. RF output (SSB, CW, FM) is 100 W, and AM mode yields a 19.1-W carrier. An external power supply is required. *Manufacturer*: Ten-Tec Inc., 1185 Dolly Parton Parkway, Sevierville, TN 37862; **www.tentec.com**.

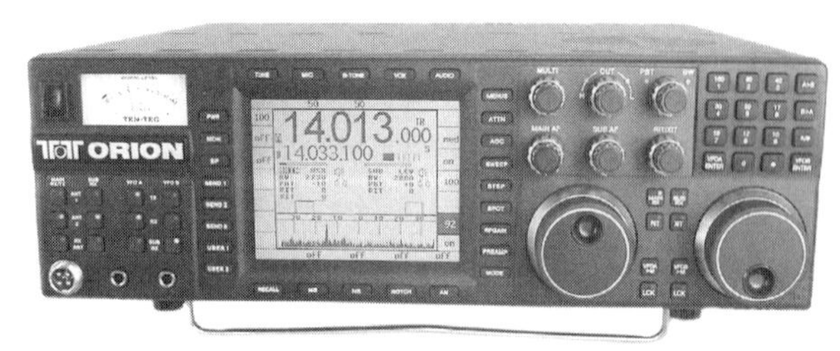

Fig D — Ten-Tec Orion (Model 565).

Yaesu FT-1000MP Mark V Field

Fig E shows the Mark V *Field* — the 100-W, internal-power-supply version of the larger 200-W Mark V. This technologically-advanced transceiver continues a model-number heritage with roots back to the analog FT-1000 and 1000D base radios. The *Field* features enhanced DSP (EDSP) and offers final amplifier operation at Class A on SSB (at reduced output) as does the regular Mark V. A variable RF front-end filter (preselector) provides for additional rejection of strong nearby signals. *Manufacturer*: Vertex Standard, 10900 Walker Street, Cypress, CA 90630; **www.vxstdusa.com**.

Fig E — Yaesu FT-1000MP Mark V Field.

500 Hz (CW) or 250 Hz can be selected instead. A second mixer translates back to the 9.0-MHz frequency. A voltage tuned crystal oscillator at 15.300 MHz (tunable ±1.5 kHz) is the LO for both mixers. This choice of LO and the 6.3-MHz IF results in very low levels of harmonic IMD products that might cross over the signal frequency and cause spurious outputs. The passband can be tuned ±1.5 kHz.

The composite passband is the intersection of the fixed 9.0-MHz passband and the tunable 6.3-MHz passband. If the first filter is wide and the second much narrower the passband width remains constant over most of the adjustment range. If both have the same bandwidth the resultant bandwidth narrows considerably as the second filter is adjusted. This can be especially helpful in CW mode. **Fig 15.3** shows how passband tuning works.

IF Amplifiers after Passband Tuning

A low-noise grounded-gate JFET amplifier, with PIN-diode AGC, establishes a low noise figure and a low level of IF noise after the last IF filter. Two IC IF amplifiers (MC1350P) provide most of the receive IF gain. These three stages provide all of the AGC for the receiver. The AGC loop does not include the narrow-band IF filters. Two AGC recovery times (Fast and Slow) are available. AGC can be switched off for manual RF gain control as well. The AGC drives the S-meter, which is calibrated at 50 µV for S9 and 0.8 µV for S3.

Product Detector

The IC product detector (CA3053E) uses LO frequencies of 9.000 MHz for LSB and CW (in receive only), 9.003 MHz for USB. When switching between USB and LSB, for a constant value of signal carrier frequency (such as 14.20000 MHz), the LO of the first mixer is moved 3.00 kHz in order to keep the signal within the passband of the IF SSB filters. More about this later.

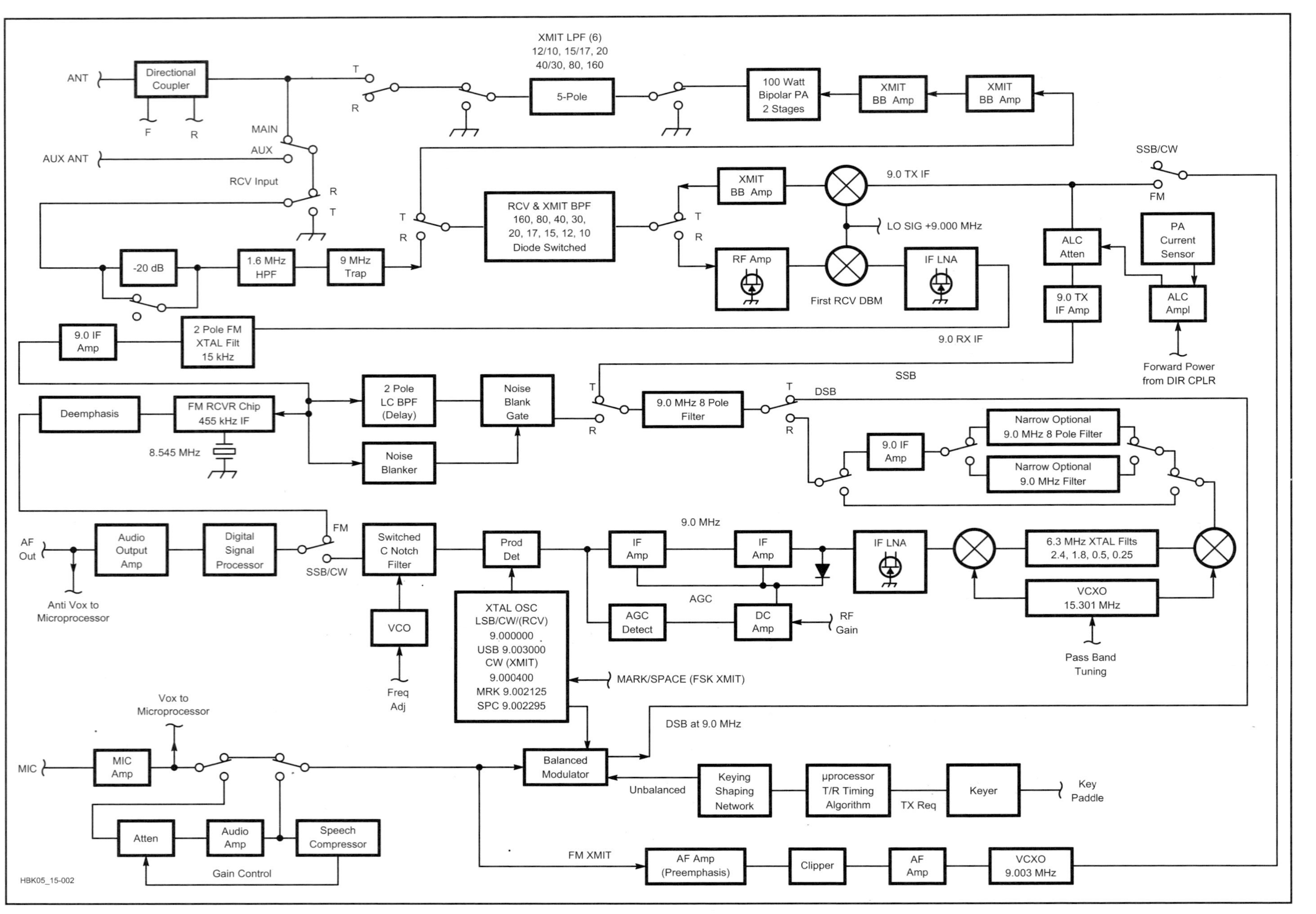

Fig 15.2 — Signal path block diagram, receive and transmit, for the Omni VI Plus transceiver.

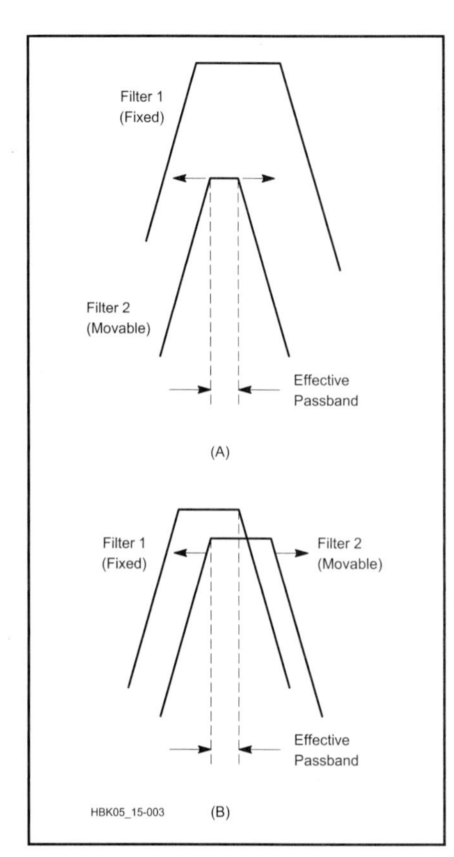

Fig 15.3 — Explanation of passband tuning. A: wide first filter and narrow second filter. B: two narrow filters.

Audio Notch Filter

In the CW and digital modes, a switched-capacitor notch filter (MF5CN) places a narrow notch in the audio band. The location of the notch is determined by the clock rate applied to the chip. This is determined by a VCO (CD4046BE) whose frequency is controlled by the front panel NOTCH control.

NBFM Reception

After the 15 kHz wide IF filter at 9.0 MHz and before the noise blanker, the IF goes to the NBFM receiver chip (MC3371P). A mixer (8.545-MHz LO) converts it to 455 kHz. The signal goes through an off-chip ceramic band-pass filter, and then goes back on-chip to the limiter stages and a quadrature detector. A received signal strength indicator (RSSI) output provides a dc voltage that is proportional to the dB level of the signal. This voltage goes to the front panel meter when in the NBFM mode. A squelch function (NBFM only) is controlled from a potentiometer on the front panel.

Audio Digital Signal Processing (DSP)

The DSP is based on the Analog Devices ADSP 2105 processor. The DSP program is stored in an EPROM and loaded into the 2105's RAM on power-up. DSP can be used in both SSB and CW. In USB or LSB (not CW or data) the DSP automatically locates and notches out one or several interfering carriers. In SSB or CW the manual audio notch filter described previously is also available, either as a notch filter or to reduce high frequency response (hiss filter). In the CW mode the DSP can be instructed to low-pass filter the audio with several corner-frequency values. A DSP noise reduction function tracks desired signals and attenuates broadband noise by as much as 15 dB, depending on conditions.

Audio Output

The 1.5-W audio output uses a TDA2611 chip. Either FM audio or SSB/CW audio or, in transmit, a CW sidetone, can be fed to the speaker or headphones. The sidetone level (a software adjustment) is separate from the volume control. The audio output, after A/D conversion, is also fed to the Anti-VOX algorithm in the microprocessor.

Transmit Block Diagram

Now, let's look at the path from microphone or key to the antenna, one stage at a time.

Microphone Amplifier

The suggested microphone is 200 Ω to 50 kΩ at 5 mV (–62 dB). A polarizing voltage for electret mics is provided. The Mic Amp drives the balanced modulator, either directly or through the speech compressor. It also supplies VOX information to the microprocessor, via an A/D converter. The microprocessor software sets Vox hang time and sensitivity, as well as the Anti-VOX, via the keypad. Timing and delays for T/R switching are also in the software.

Speech Processor

The audio speech processor is a compressor, as discussed previously. A dc voltage that is proportional to the amount of compression is sent to the front panel meter so that compression can be set to the proper level. Clipper diodes limit any fast transients that might overdrive the signal path momentarily.

Balanced Modulator

The balanced modulator generates a double-sideband, suppressed carrier IF at 9.0 MHz. The LO is that used for the receive product detector. There is a carrier nulling adjustment. In CW and FSK modes, the modulator is unbalanced to let the LO pass through. A built-in iambic keyer (Curtis style A or B) is adjustable from 10 to 50 WPM. An external key or keyer can also be plugged in.

IF Filter

The standard 9.0015 MHz, 2.4-kHz-wide 8-pole filter (also used in receive) removes either the lower or upper sideband. The output is amplified at 9.0 MHz.

ALC

The forward-power measurement from the PA output directional coupler is used for ALC, which is applied at the output of the first 9.0-MHz IF amplifier by a PIN diode attenuator. A front panel LED lights when ALC is operating. An additional circuit monitors dc current in the PA and cuts back on RF drive if the PA current exceeds 22 A.

Output Mixer

This mixer translates to the output signal frequency. The same LO frequency is used for each transmit mode (USB/LSB) and the same 3.00000-kHz frequency shift is used to assure that the frequency readout is always correct.

Band-pass Filters

The mixer output contains the image, at 9.0 plus LO and low values of harmonics and harmonic IMD products. A band-pass filter for each ham band, the same ones used in receive, eliminates all out of band products from the transmitter output.

PAs

Four stages of amplification, culminating in push-pull bipolar MRF454s in Class AB, supply 100 W output, CW or PEP from 160 to 10 m. Temperature sensing of the transistors in the last two stages helps to prevent thermal damage. Full output can be maintained, key down, for 20 minutes. Forced air heat-sink cooling allows unlimited on-time.

LO Frequency Management

The LO goals are to achieve low levels of phase noise, high levels of frequency stability and at the same time keep equipment cost within the reach of as many amateurs as possible. As part one of the LO analysis we look at the method used to adjust the LOs in order to keep the speech spectrum of a USB or LSB signal within the 9.0-MHz IF filter passband. This method is somewhat typical for many equipment designs. Refer to **Fig 15.4**.

First Mixer and Product Detector

An SSB signal whose carrier frequency is at 14.20000 MHz, which may be LSB or USB, is translated so that the modula-

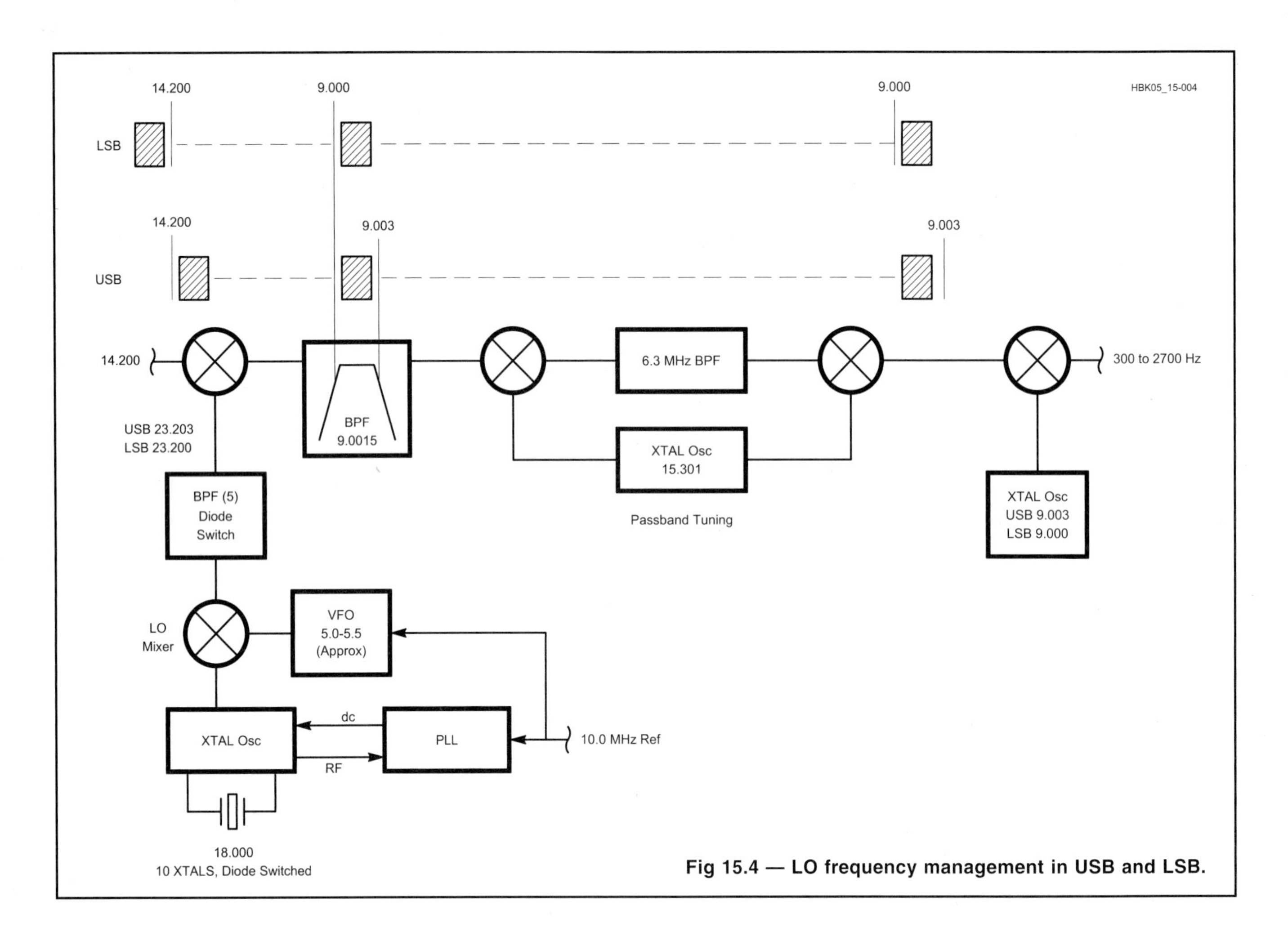

Fig 15.4 — LO frequency management in USB and LSB.

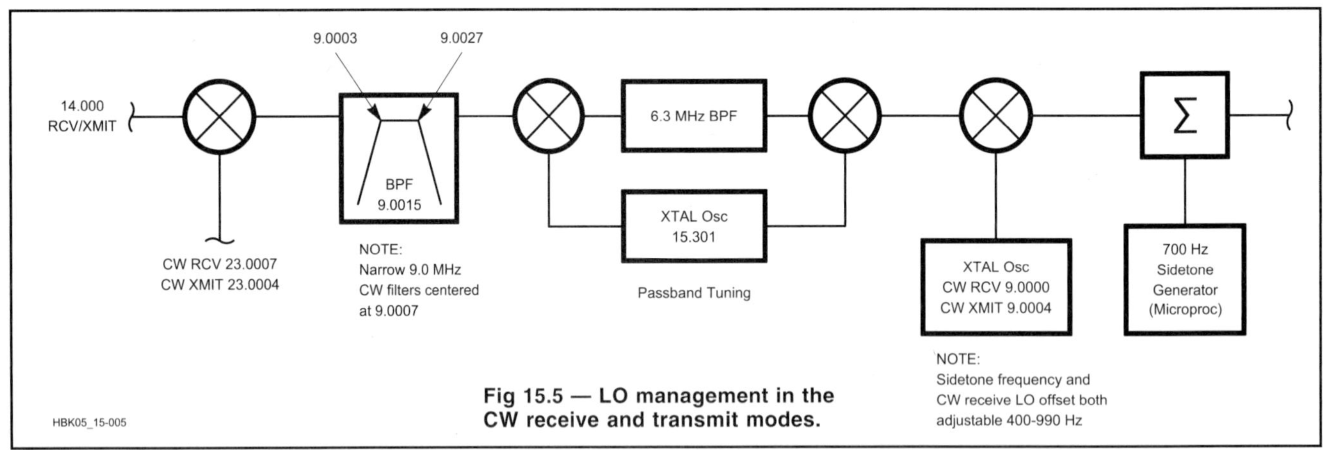

Fig 15.5 — LO management in the CW receive and transmit modes.

tion (300 to 2700 Hz) in either case falls within the passband of the 9.0015-MHz 8-pole crystal filter. For a USB signal this is accomplished by increasing the first LO 3 kHz, as indicated. Since the LO is on the high side of the signal frequency there is a sideband "inversion" at the first mixer. The passband tuning module does not change this relationship. At the product detector, the LO is increased 3 kHz in USB so that it is the same as the carrier (suppressed) frequency of the IF signal. Note that the designators "USB" or "LSB" at the product detector LO refer to the antenna signal, not the IF signal. The jog of the first mixer LO is accomplished partly within the crystal oscillator and partly within the VFO. The microprocessor sends the frequency instructions to both of these oscillators. Despite the frequency offsets, the digital readout displays the correct carrier frequency, in this example 14.20000 MHz. And, of course, the same procedures apply to the transmit mode.

Another interesting idea involves the first LO mixer. The 18-MHz crystal and the VFO are added to get 23.0 to 23.5 MHz, for the 20-m band. But the output of the LO mixer also contains the difference, 12.5 to 13.0 MHz, which is just right for the 80-m band. For the 20-m band one BPF selects 23.0 to 23.5 MHz. For the 80-m band the

12.5 to 13.0-MHz BPF is selected. A problem occurs, though, because now the direction of frequency tuning is reversed, from high to low. The microprocessor corrects this by reversing the direction of the tuning knob (an optical encoder). Other "book keeping" is performed so that the operation is transparent to the operator. A similar trick is used on the 17-m band and the 28.5 to 29.0-MHz segment of the 10-m band.

CW-Mode LO Frequencies

In CW mode the "transceiver problem" shows up. See **Fig 15.5** for a discussion of this problem. If the received carrier is on exactly 14.00000 MHz and if we want to transmit our carrier on that same exact frequency then the transmitter and the receiver are both "zero beat" at 14.00000 MHz. In receive we would have to retune the receiver, say up 700 Hz, to get an audible 700-Hz beat note. But then when we transmit we are no longer on 14.0000 MHz but are at 14.00070 MHz. We would then have to reset to 14.00000 when we transmit.

The transceiver's microprocessor performs all of these operations automatically. Fig 15.5 shows that in receive the first-LO frequency is increased 700 Hz. This puts the first IF at 9.0007 MHz, which is inside the passband of the 9.0 MHz IF filter. The BFO is at 9.0000 MHz and an audio beat note at 700 Hz is produced. This 700-Hz pitch is compared to a 700-Hz audio oscillator (from the microprocessor). When the two pitches coincide the signal frequency display of our transceiver coincides almost exactly with the frequency of the received signal. The digital frequency display reads "14.00000" at all times. The value of the 700-Hz reference beat can be adjusted between 400 and 950 Hz by the user. The receive LO shift matches that value.

When the optional 500/250 Hz CW 9.0 MHz IF filters are used, these are centered at 9.0007 MHz. These filters are used in receive but not in transmit.

When we transmit, the transmit frequency is that which the frequency display indicates, 14.00000 MHz. However, there is a slight problem. The 9.0-MHz transmit IF must be increased slightly to get the speech signal within the passband of the 9.0 MHz IF filter. The transmit BFO is therefore at 9.0004 MHz. The mixer LO is also moved up 400 Hz so that the transmit output frequency will be exactly 14.0000 MHz.

In addition to the above actions, the RIT (receive incremental tuning) and the XIT (transmit incremental tuning) knobs permit up to ±9.9 kHz of independent control of the receive and transmit frequencies, relative to the main frequency readout.

Local Oscillators

Fig 15.4 indicates that the crystal oscillator for the first mixer is phase locked. Each of the 10 crystals is locked to a 100-kHz reference inside the PLL chip. This reference is derived from a 10-MHz system reference.

Let's go into some detail regarding the very interesting 5.0 to 5.5 MHz VFO circuitry. **Fig 15.6** is a block diagram. The VCO output, 200 to 220 MHz, is divided by 40 to get 5.0 to 5.5 MHz. The reference frequency for the PLL loop is 10 kHz, so each increment of the final output is 10 kHz / 40, which is 250 Hz. Phase noise in the PLL is also divided by 40, which is equivalent to 32 dB (20 × log (40)).

To get 10-Hz steps at the output, the voltage-tuned crystal oscillator at 39.94 MHz is tuned in 200-Hz steps (the division ratio from oscillator to final output is 20 instead of 40 because of the frequency doubler). To get 200-Hz steps in this oscillator, serial data from the microprocessor is fed into a latch. This data is sent from a RAM lookup table that has the correct values to get the 200-Hz increments very accurately. The outputs of the latch are fed into an R/2R ladder (D/A conversion) and the dc voltage tunes the VCXO. Adjustment potentiometers calibrate the tuning range of the oscillator over 5000 Hz in 200-Hz increments. At the final output, this tuning range fills in the 10-Hz steps between the 250-Hz increments of the PLL. Although this circuit is not phase locked to a reference (it's an open-loop), the resulting frequency steps are very accurate, especially after the division by 20. This economical approach reduces the complexity and cost of the VFO considerably, but performs extremely well (very low levels of phase noise and frequency drift).

CW Break-In (Fast QSK) Keying

The radio is capable of break-in keying at rates up to 25 WPM when it is in the FAST QSK mode. This mode is also used

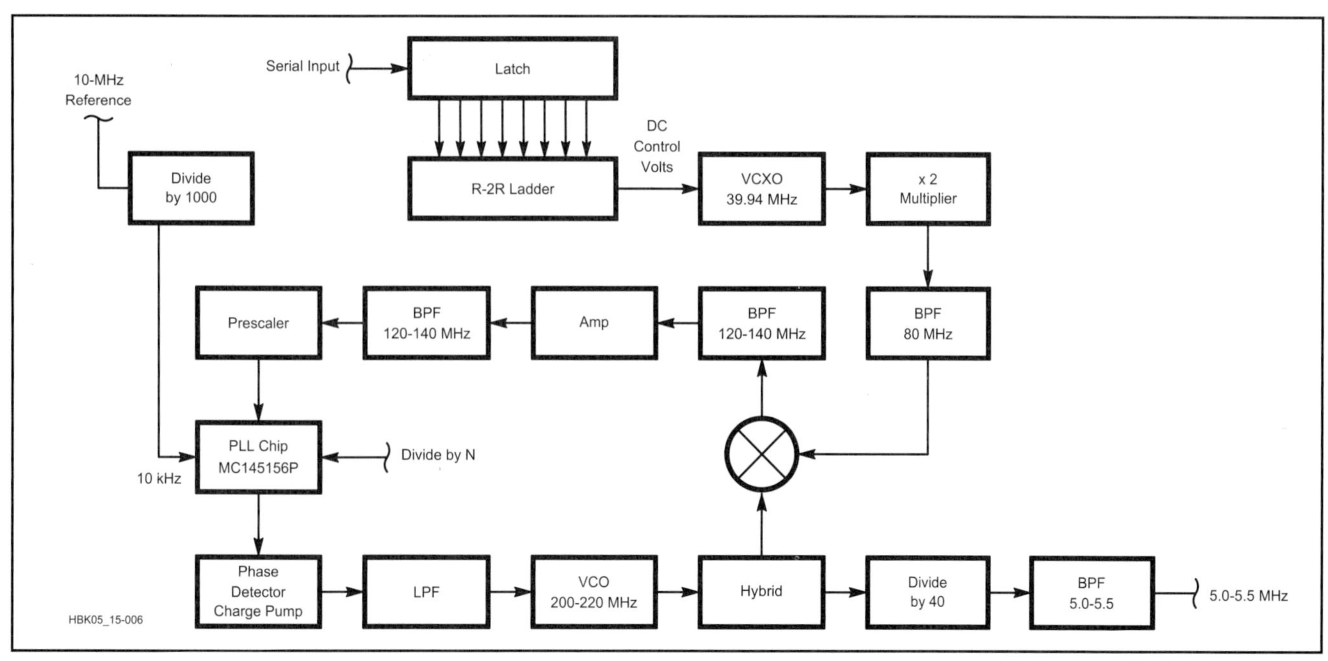

Fig 15.6 — Block diagram of the VFO.

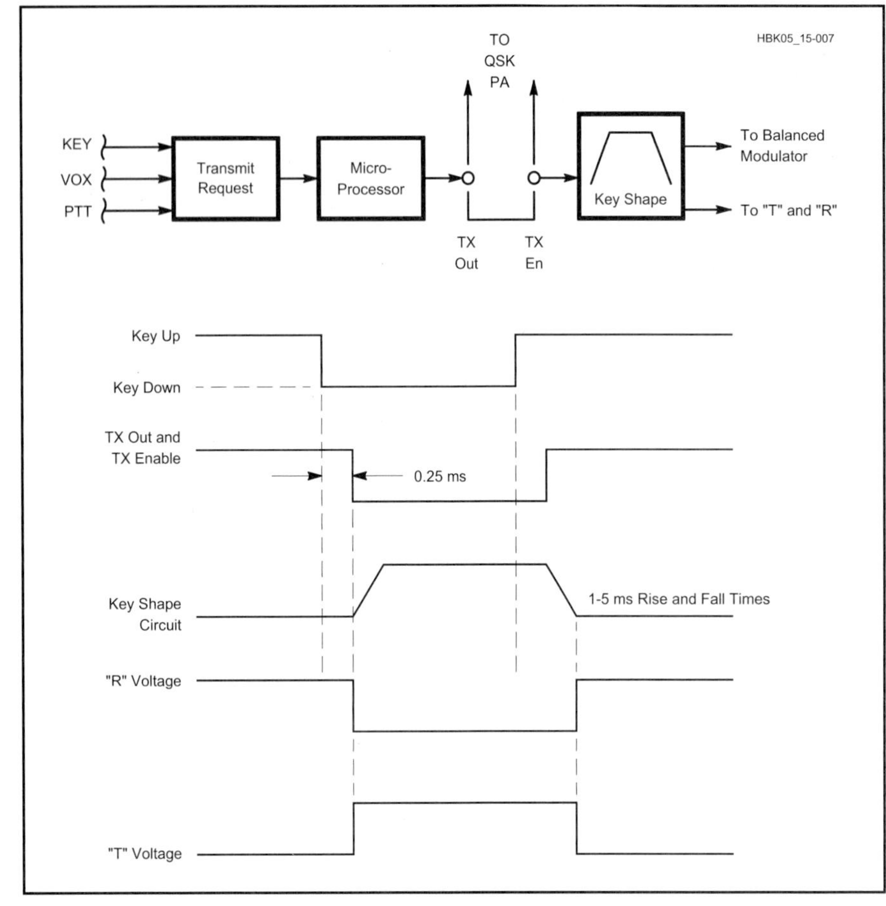

Fig 15.7 — Fast QSK operation of the Omni VI transceiver.

for AMTOR FSK. **Fig 15.7** explains the action and the timing involved in this T/R switching. The sequence of events is as follows:

1. The key is pressed.

2. A Transmit Request is sent to the microprocessor.

3. The microprocessor changes the LO and BFO to their transmit frequencies.

4. After a 0.25-ms delay a transmit out logic level is sent to a jack on the rear panel.

5. The Transmit Out signal is jumpered to the TRANSMIT ENABLE jack. If an external Fast QSK PA is being used, an additional short delay is introduced while it is being switched to the transmit mode.

6. The Transmit Enable signal starts the keying-waveform circuit, which ramps up in 3 ms.

7. Near the bottom of this ramp, the "T" (transmit B+) voltage goes high and the "R" (receive B+) voltage goes low. The T/R reed relay (very fast) at the Omni VI PA output switches to transmit.

8. The shaped keying waveform goes to the balanced modulator and the RF envelope builds up. There is a very brief delay from balanced modulator to Omni VI output. The T/R relay is switched before the RF arrives.

9. The key is opened.

10. The Transmit Out and Transmit Enable lines go high.

11. The keying waveform ramps down. RF ramps down to zero.

12. After 5 ms (a fixed delay set by the microprocessor) "T" goes low and "R" goes high and the microprocessor returns the LO and BFO to their receive frequencies.

Slow QSK

In the Slow QSK mode the action is as described above. The radio reverts quickly to the receive mode. However, the receive audio is muted until the end of an extended (adjustable) delay time.

There is also a relay in the Omni VI that can be used to T/R switch a conventional (not Fast QSK) external PA. This option is selected from the operator's menu and is available only in the Slow QSK mode. The relay is held closed for the duration of the delay time in the slow QSK mode. When using this option the operator must ensure that the external PA switches fast enough on the first "key-down" so that it is not "hot switched" or that the first dot is missed. Many older PAs do not respond well at high keying rates. If the PA is slow, we can still use the Fast QSK mode with the Omni VI ("barefoot") and the PA will be bypassed because the optional relay is not energized in the Fast QSK mode.

VOX

In SSB mode, VOX and PTT perform the same functions and in the same manner as the CW Slow QSK described above. The VOX hang-time adjustment is separate from the CW hang-time adjustment. A MUTE jack allows manual switching (foot switch, and so on) to enable the transmit mode without applying RF. The key or the VOX then subsequently applies RF to the system. This arrangement helps if the external PA has slow T/R switching.

Operating Features

Modern transceivers have, over the years, acquired a large ensemble of operator's aids that have become very popular. Here are some descriptions of them:

Key Pad

The front panel key pad is the means for configuring a wide assortment of operating preferences and for selecting bands and modes.

Frequency Change

1. Use the tuning knob. The tuning rate can be programmed to 5.12, 2.56, 1.71, 1.28, 1.02 or 0.85 kHz per revolution. The knob has adjustable drag.

2. UP and DOWN arrows give 100 kHz per step.

3. Band selection buttons.

4. Keyboard entry of an exact frequency.

Mode Selection

1. ***Tune***: Places the rig in CW mode, key down, for various "tune-up" operations.

2. ***CW***: An optional DSP low-pass filter can be selected. Cutoff frequencies of 600, 800, 1000, 1200 or 1400 Hz can be designated. A SPOT function generates a 700-Hz audio sidetone that can be used for precise frequency setting (the received signal pitch matches the 700-Hz tone). The pitch of the sidetone can be adjusted from 400 to 950 Hz. Audio level is adjustable also. FAST QSK and SLOW QSK are available as previously described. Cascaded CW filters are available, with passband tuning of one of the filters. CW filter options: 500/250 Hz at 6.3 MHz IF and 500 Hz at 9.0-MHz IF.

3. ***USB or LSB***: Standard SSB IF bandwidth is 2.4 kHz (two 8-pole filters in cascade). The second filter can be passband-

tuned ±1.5 kHz. Additional IF filters with 1.8-kHz BW are available.

4. ***FSK and AFSK***: Special FSK filter for receive. AMTOR operation with FAST QSK capability. AFSK generator can be plugged into microphone jack.

5. ***FM***: 15-kHz IF filters at 9.0 MHz and 455 kHz. Quadrature detection, RSSI output and squelch. Adjustable transmit deviation. FM transmit uses the direct method.

6. ***VOX***: VOX sensitivity and hang time adjustable via the key pad. Anti-VOX level adjustable.

Time of Day Clock

There is a digital readout on front panel.

Built-In Iambic Keyer

Curtis type A or B, front panel speed knob. Adjustable dot-dash ratio. Also external key or keyer.

Dimmer

Adjusts brightness of front panel display.

Dual VFOs

Select A or B. Independent frequency, mode, RIT and filter choices stored for each VFO. Used for split-frequency operation.

Receiver Incremental Tuning (RIT)

Each VFO has its own stored RIT value.

Frequency Offset Display

RIT value adjusted with knob. RIT can be toggled on and off, or cleared to zero.

Transmitter Incremental Tuning (XIT)

Same comments as RIT. Simultaneous RIT and XIT.

Cross-Band and Cross-Mode Operation

For cross-band operation, use PTT for SSB and manual switch for CW.

Scratch-Pad Memory

Stores a displayed frequency. Restores that frequency to the VFO on command.

Band Register

Allows toggling between two frequencies on each band.

Memory Store

Store 100 values of frequency, band, mode, filter, RIT, XIT. Memory channels can be recalled by channel number (key pad), "scrolling" the memory channels or "memory tune" using the main tuning knob.

Lock

Locks the main tuning dial.

User Option Menu

Enables configuration of the radio via the keypad.

Meter

Select between receive signal strength (on SSB/CW or NBFM), speech processor level, forward power, SWR and PA dc current.

AGC

Fast, slow, off and manual RF gain control.

FM Squelch Adjust

Passband Tuning Knob

Notch

Automatically notch out several heterodynes on SSB/FM or manually notch on CW/digital modes. Adjustable low-pass filter in CW mode.

Antenna Switch

Auxiliary antenna may be selected in receive mode.

Interface Port

25-pin D connector for interface to personal computer.

OTHER TRANSCEIVERS

Other transceivers vary in cost, complexity and features. The one just described is certainly one of the best, at a reasonable price. For reviews of other transceivers (to see the differences in cost, features and performance specs) refer to the Product Reviews in *QST*.

THE NORCAL SIERRA: AN 80-15 M CW TRANSCEIVER

Most home-built QRP transceivers cover a single band, for good reason: complexity of the circuit and physical layout can increase dramatically when two or more bands are covered. This holds for most approaches to multiband design, including the use of multipole switches, transverters and various forms of electronic switching.[1]

If the designer is willing to give up instant band switching, then plug-in band modules can be used. Band modules are especially appropriate for a transceiver that will be used for extended portable operation, for example: back-packing. The reduced circuit complexity improves reliability, and the extra time it takes to change bands usually isn't a problem. Also, the operator need take only the modules needed for a particular outing.

The Sierra transceiver shown in **Fig 15.8** uses this technique, providing coverage of all bands from 80 through 15 m with good performance and relative simplicity.[2] The name Sierra was inspired by the mountain range of the same name — a common hiking

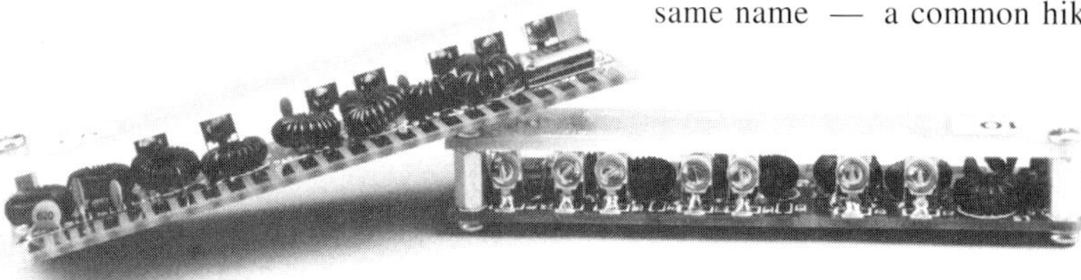

Fig 15.8 — The Sierra transceiver. One band module is plugged into the center of the main PC board; the remaining boards are shown below the rig. Quick-release latches on the top cover of the enclosure make it easy to change bands.

destination for West Coast QRPers. The transceiver was designed and built by Wayne Burdick, N6KR, and field tested by members of NorCal, the Northern California QRP Club.[3]

FEATURES

One of the most important features of the Sierra for the portable QRP operator is its low current drain. Because it has no relays, switching diodes or other active band-switching circuitry, the Sierra draws only 30 mA on receive.[4] Another asset for field operation is the Sierra's low-frequency VFO and premixing scheme, which provides 150 kHz of coverage and good frequency stability on all bands.

The receiver is a single-conversion superhet with audio-derived AGC and RIT. It has excellent sensitivity and selectivity, and will comfortably drive a speaker. Transmit features include full break-in keying, shaped keying and power output averaging 2 W, with direct monitoring of the transmitted signal in lieu of sidetone. Optional circuitry allows monitoring of relative power output and received signal strength.

Table 15.1
Crystal Oscillator and Premix (PMO) Frequencies in MHz

The premixer (U7) subtracts the VFO (2.935 to 3.085 MHz) from the crystal oscillator to obtain the PMO range shown. The receive mixer (U2) subtracts the RF input from the PMO signal, yielding 4.915 MHz. The transmit mixer (U8) subtracts 4.915 MHz from the PMO signal to produce an output in the RF range.

RF Range	*Crystal Oscillator*	*PMO Range*
3.500-3.650	11.500	8.415-8.565
7.000-7.150	15.000	11.915-12.065
10.000-10.150	18.000	14.915-15.065
14.000-14.150	22.000	18.915-19.065
18.000-18.150	26.000	22.915-23.065
21.000-21.150	29.000	25.915-26.065

Physically, the Sierra is quite compact — the enclosure is 2.7 × 6.2 × 5.3 inches (HWD) — yet there is a large amount of unused space both inside and on the front and rear panels. This results from the use of PC board-mounted controls and connectors. The top cover is secured by quick-release plastic latches, which provide easy access to the inside of the enclosure. Band changes take only a few seconds.

CIRCUIT DESCRIPTION

Fig 15.9 is a block diagram of the Sierra. The diagram shows specific signal frequencies for operation on 40 m. **Table 15.1** provides a summary of crystal oscillator and premix frequencies for all bands.

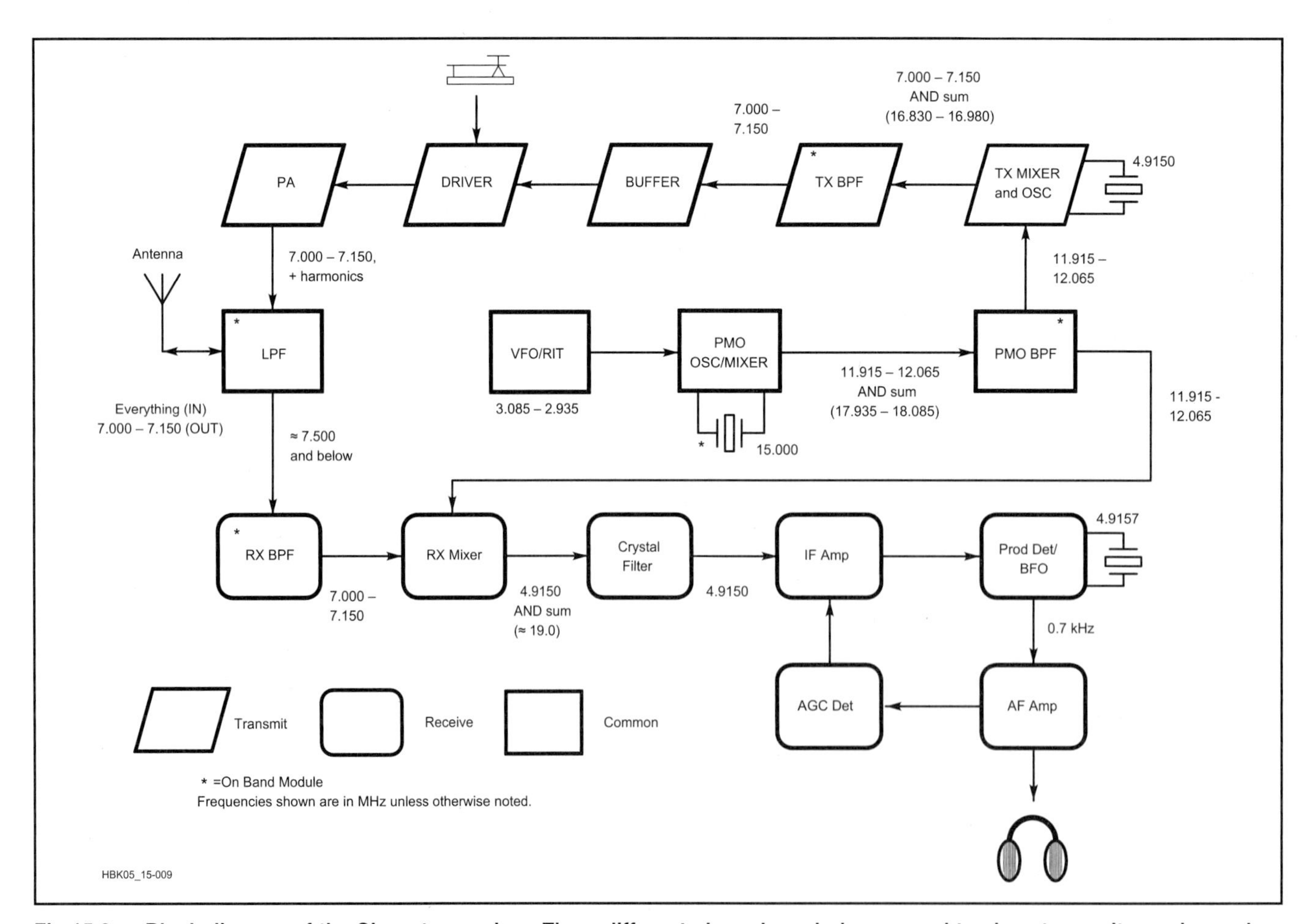

Fig 15.9 — Block diagram of the Sierra transceiver. Three different-shaped symbols are used to show transmit, receive and common blocks. Those blocks with an asterisk (*) are part of the band module. Signal frequencies shown are for 40 m; see Table 15.1 for a list of crystal oscillator and premix frequencies for all bands.

Table 15.2

Band Module Components

All crystals are fundamental, 15-pF load capacitance, 0.005% frequency tolerance, in HC-49 holders. Fixed capacitors over 5 pF are 5% tolerance. All coils are wound with enameled wire.

	Band					
Part	*80 m*	*40 m*	*30 m*	*20 m*	*17 m*	*15 m*
C32, C35	33 pF, 5%	47 pF, 5%	not used	not used	not used	not used
C34	5 pF, 5%	5 pF, 5%	2 pF, 5%	2 pF, 5%	2 pF, 5%	2 pF, 5%
C47, C49	820 pF, 5%	330 pF, 5%	330 pF, 5%	220 pF, 5%	150 pF, 5%	150 pF, 5%
C48	1800 pF, 5%	820 pF, 5%	560 pF, 5%	470 pF, 5%	330 pF, 5%	330 pF, 5%
C65	5 pF, 5%	5 pF, 5%	2 pF, 5%	1 pF, 5%	1 pF, 5%	1 pF, 5%
L1	50 µH, 30 t #28 on FT-37-61	14 µH, 16 t #26 on FT-37-61	5.2 µH, 36 t #28 on T-37-2	2.9 µH, 27 t #28 on T-37-2	1.7 µH, 24 t #28 on T-37-6	1.9 µH, 25 t #28 on T-37-6
L3, L4	32 µH, 24 t #26 on FT-37-61	5.2 µH, 36 t #28 on T-37-2	4.4 µH, 33 t #28 on T-37-2	2.9 µH, 27 t #28 on T-37-2	1.7 µH, 24 t #28 on T-37-6	1.9 µH, 25 t #28 on T-37-6
L5, L6	2.1 µH, 23 t #26 on T-37-2	1.3 µH, 18 t #26 on T-37-2	1.0 µH, 16 t #26 on T-37-2	0.58 µH, 12 t #26 on T-37-2	0.43 µH, 12 t #26 on T-37-6	0.36 µH, 11 t #26 on T-37-6
L8, L9	8.0 µH, 12 t #26 on FT-37-61	2.5 µH, 25 t #28 on T-37-2	1.6 µH, 20 t #28 on T-37-2	1.3 µH, 18 t #26 on T-37-2	0.97 µH, 18 t #26 on T-37-6	0.87 µH, 17 t #28 on T-37-6
T1 (Sec same as L1)	Pri: 2 t #26 on FT-37-61	Pri: 1 t #26 on FT-37-61	Pri: 3 t #26 on T-37-2	Pri: 2 t #26 on T-37-2	Pri: 2 t #26 on T-37-6	Pri: 2 t #26 on T-37-6
X8	11.500 MHz (ICM 434162)	15.000 MHz (ICM 434162)	18.000 MHz (ICM 434162)	22.000 MHz (ICM 435162)	26.000 MHz (ICM 436162)	29.000 MHz (ICM 436162)

The schematic is shown in **Fig 15.10**. See **Table 15.2** for band-module component values.

On all bands, the VFO range is 2.935 MHz to 3.085 MHz. The VFO tunes "backwards": At the low end of each band, the VFO frequency is 3.085 MHz. U7 is the premixer and crystal oscillator, while Q8 buffers the premix signal prior to injection into the receive mixer (U2) and transmit mixer (U8).

A low-pass filter, three band-pass filters and a premix crystal make up each band module. To make the schematic easier to follow, this circuitry is integrated into Fig 15.10, rather than drawn separately. J5 is the band module connector (see the note on the schematic).

The receive mixer is an NE602, which draws only 2.5 mA and requires only about 0.6 V (P-P) of oscillator injection at pin 6. An L network is used to match the receive mixer to the first crystal filter (X1-X4). This filter has a bandwidth of less than 400 Hz. The single-crystal second filter (X5) removes some of the noise generated by the IF amplifier (U7), a technique W7ZOI described.[5] This second filter also introduces enough loss to prevent the IF amplifier from overdriving the product detector (U4).

The output of the AF amplifier (U3) is dc-coupled to the AGC detector. U3's output floats at $V_{cc}/2$, about 4 V, which happens to be the appropriate no-signal AGC voltage for the IF amplifier when it is operated at 8 V. C26, R5, R6, C76 and R7 provide AGC loop filtering. Like all audio-derived AGC schemes, this circuit suffers from pops or clicks at times.

Transmit signal monitoring is achieved by means of a separate 4.915 MHz oscillator for the transmitter; the difference between this oscillator and the BFO determines the AF pitch. Keying is exponentially shaped, with the rise time set by the turn-on delay of transmit mixer U8 and the fall time determined by C51, in the emitter of driver Q6.

CONSTRUCTION

The Sierra's physical layout and packaging make it relatively easy to build and align, although this isn't a project for the first-time builder. The boards and custom enclosure described here are included as part of an available kit.[6] Alternative construction methods are discussed below.

With the exception of the components on the band module, all of the circuitry for the Sierra is mounted on a single 5 × 6 inch PC board. This board contains not only the components, but all of the controls and connectors as well. The board is double-sided with plated-through holes, which permits flexible arrangement of the circuitry while eliminating nearly all hand-wiring. The only two jumpers on the board, W1 and W2, are short coaxial cables between the RF GAIN control and the receiver input filters.

A dual-row edge connector (J5) provides the interface between the main board and the band module. The 50 pins of J5 are used in pairs, so there are actually only 25 circuits (over half of which are ground connections).

The band module boards are 1.25 × 4 inches (HW). They, too, are double-sided, maximizing the amount of ground plane. Because the band modules might be inserted and removed hundreds of times over the life of the rig, the etched fingers that mate with J5 are gold-plated. Each etched finger on the front is connected to the corresponding finger on the back by a plated through hole, which greatly improves reliability over that of a single finger contact.

Each band module requires eight toroids: two for the low-pass filter, and two each for the receive, transmit and premix band-pass filters. The builder can secure the toroids to the band module with silicone adhesive or Q-dope. Right-angle-mount trimmer capacitors allow alignment from above the module. Each band module has a top cover made of PC board material. The cover protects the components during insertion, removal and storage.

The VFO capacitor is a 5-40 pF unit with a built-in 8:1 vernier drive. The operating frequency is read from a custom dial fabricated from 0.060-inch Lexan. The dial mounts on a hub that comes with the capacitor.

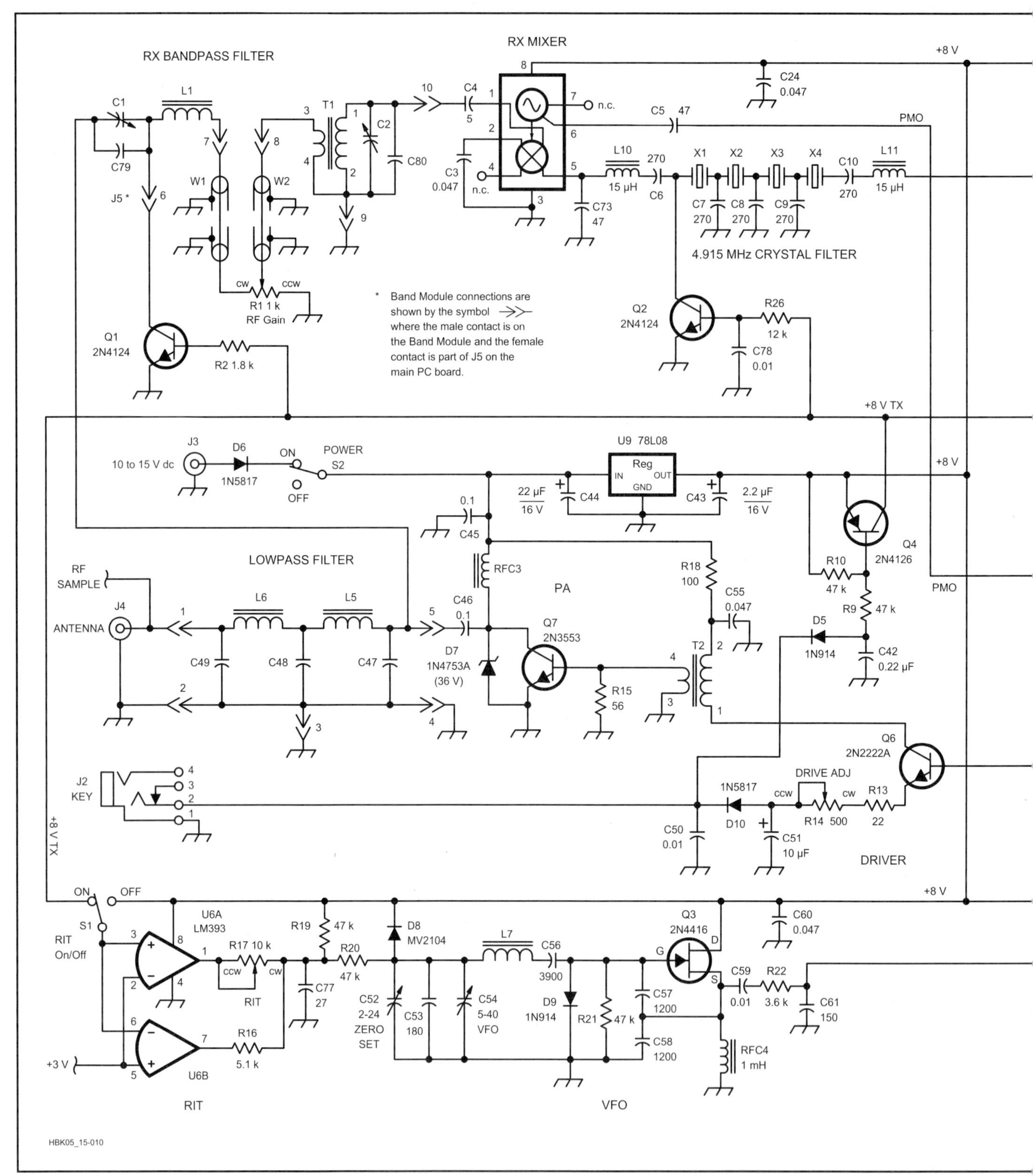

Fig 15.10 — Schematic of the Sierra transceiver. Parts that change for each band are shown in Table 15.2.

C1, C2, C33, C36, C64, C66, C70 — 9-50 pF right-angle-mount ceramic trimmer (same for all band modules, Mouser 24AA084)
C16, C38 — Ceramic trimmer, 8-50 pF (Mouser 24AA024)
C52 — Air variable, 2-24 pF (Mouser 530-189-0509-5)
C53 — Disc, 180 pF, 5%, NP0
C54 — 5-40 pF air variable with 8:1 vernier drive
C56 — Polystyrene, 3900 pF, 5%
C57, C58 — Polystyrene, 1200 pF, 5%
D6, D10 — 1N5817, 1N5819 or similar
D7 — 36 V, 1 W Zener diode (Mouser 333-1N4753A)
D8 — MV2104 varactor diode, or equivalent
J1, J2 — PC-mount 3.5-mm stereo jack with switch (Mouser 161-3500)
J3 — 2.1-mm dc power jack (Mouser 16PJ031)
J4 — PC-mount BNC jack (Mouser 177-3138)
J5 — 50 pin, dual-row edgeboard connector with 0.156-inch spacing (Digi-Key S5253-ND)

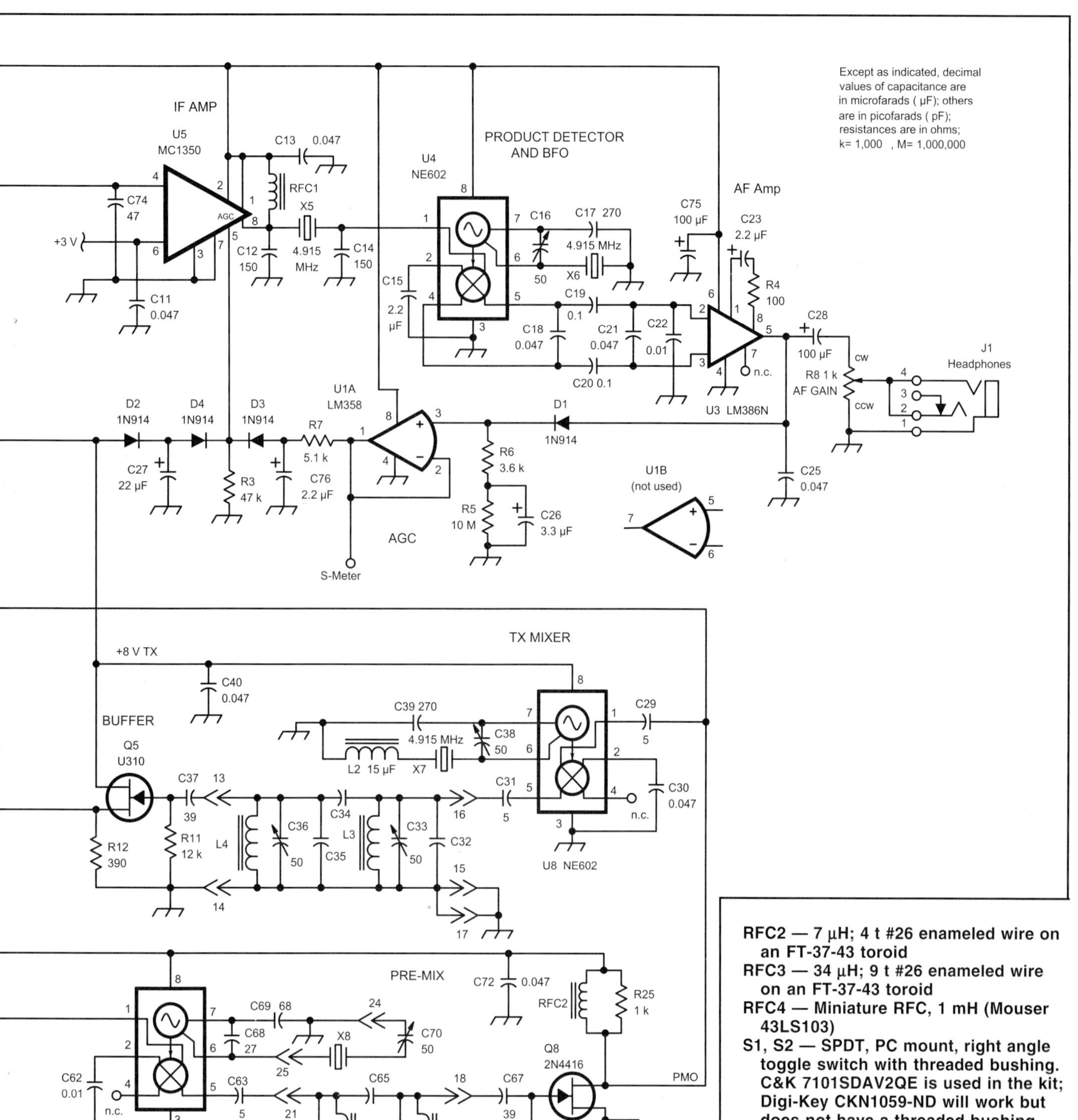

L10, L11 — 18 µH; 18 t #28 enameled wire on an FT-37-61 toroid
L2 — Miniature RFC, 15 µH (Mouser, 43LS185)
L7 — 19 µH; 58 t #28 enameled wire on a T-68-7 toroid
Q5 — U310, J310, 2N4416 or other high-transconductance device
R1, R8 — PC-mount 1-kΩ pot (Mouser 31CW301)
R14, R101 — 500 Ω trimmer (Mouser 323-4295P-500)
R17 — PC-mount 10-kΩ pot (Mouser 31CW401)
RFC1 — 3.5 µH; 8 t #26 enameled wire on an FT-37-61 toroid
RFC2 — 7 µH; 4 t #26 enameled wire on an FT-37-43 toroid
RFC3 — 34 µH; 9 t #26 enameled wire on an FT-37-43 toroid
RFC4 — Miniature RFC, 1 mH (Mouser 43LS103)
S1, S2 — SPDT, PC mount, right angle toggle switch with threaded bushing. C&K 7101SDAV2QE is used in the kit; Digi-Key CKN1059-ND will work but does not have a threaded bushing
T2 — Primary: 12 t #26 enameled wire; secondary: 3 t on an FT-37-43 toroid
U1 — LM358N dual op-amp IC
U2, U4, U7, U8 — NE602AN mixer-oscillator IC
U3 — LM386N-1 audio amplifier IC
U5 — MC1350P IF amplifier IC
U6 — LM393N dual comparator IC
U9 — 8 V regulator, TO-92 package (Digi-Key AN78L08-ND)
W1, W2 — RG-174 coaxial jumper, about 3 inches long (see text)
X1-X7 — 4.915 MHz, HC-49 (Digi-Key CTX050). X1 through X5 should be matched (their series-resonant frequencies within 50 Hz)

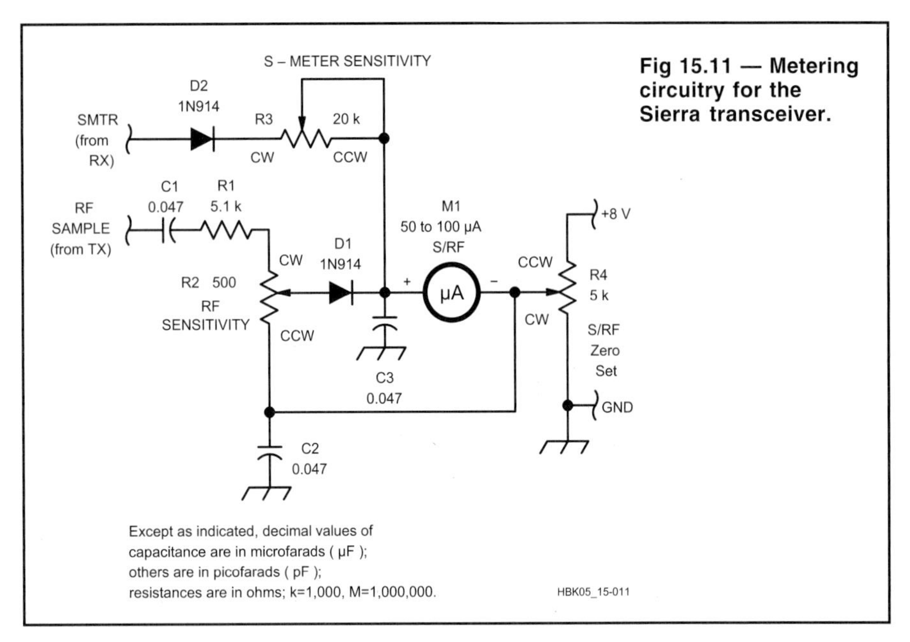

Fig 15.11 — Metering circuitry for the Sierra transceiver.

The Sierra's custom 0.060-inch aluminum enclosure offers several benefits in both construction and operation. Its top and bottom covers are identical U-shaped pieces. The bottom is secured to the main board by two 0.375-inch standoffs, while the top is secured to the bottom by two long-life, quick-release plastic latches. As a result, the builder can easily remove both covers to make "live" adjustments or signal measurements without removing any controls, connectors or wires. The front and rear panels attach directly to the controls and connectors on the main board. This keeps the panels rigid and properly oriented.

As can be seen in the photograph, the interior of the rig is uncluttered. NorCal QRP Club members have taken advantage of this, building in keyers, frequency counters and other accessories — and even storing up to four band modules in the top cover. One popular addition is an S/RF meter, the circuit shown in **Fig 15.11**.

The construction techniques described above represent only one way to build the Sierra; other physical layouts may better suit your needs. For example: If no built-ins are needed, the rig could be built in a smaller enclosure. You could replace the VFO capacitor with a small 10-turn pot and a varactor diode. If necessary, eliminate RIT and metering.

If a different physical layout is required, determine the orientation and mounts for the band module connector first, and then arrange the various circuit blocks around it. Use short leads and good ground-plane techniques to avoid instability, especially on the band modules. Point-to-point or "dead-bug" construction are possible, but in some cases shields and additional decoupling may be required. Use a reliable connector if band modules will be repeatedly inserted and removed.

ALIGNMENT

The minimum recommended equipment for aligning the rig is a DMM with homemade RF probe and a ham-band transceiver. Better still is a general-coverage receiver or frequency counter.[7] Start with a 40- or 20-m module; these are usually the easiest to align.

First, set the VFO to the desired band edge by adjusting C52. If exactly 150 kHz of range is desired, squeeze or spread the windings of L7 and readjust C52 iteratively until this range is obtained. RIT operation can also be checked at this time. Reduce the value of R19 if more RIT range is desired.

Prepare each band module for alignment by setting all of its trim caps to midrange. (The final settings will be close to midpoint in most cases.)

Receiver alignment is straightforward. Set BFO trimmer C16 to midrange, RF GAIN (R1) to maximum and AF GAIN (R8) so that noise can be heard on the phones or speaker. On the band module, peak the premix trimmers (C64 and C66) for maximum signal level measured at Q8's drain. Set the fine frequency adjustment (C70) by lightly coupling a frequency counter to U7, pin 7. Next, connect an antenna to J4 and adjust the receiver filter trimmers (C1 and C2) for maximum signal. The AGC circuitry normally requires no adjustment, but the no-signal gain of the IF amplifier can be increased by decreasing the value of R3.

Before beginning transmitter alignment, set the drive-level control, R14, to minimum. Key the rig while monitoring the transmitted signal on a separate receiver and peak the transmit band-pass filter using C33 and C36. Then, with a dummy load or well-matched antenna connected to J4, set R14 to about 90% of maximum and check the output power level. It may be necessary to stagger-tune C33 and C36 on the lower bands in order to obtain constant output power across the desired tuning range. On 80 m the –3 dB transmit bandwidth will probably be less than 150 kHz.

Typically, output on 80, 40 and 20 m is 2.0-2.5 W, and on the higher bands 1.0-2.0 W. Some builders have obtained higher outputs on all bands by modifying the band-pass filters. However, filter modification may compromise spectral purity of the output, so the results should be checked with a spectrum analyzer. Also, note that the Sierra was designed to be a 2-W rig: additional RF shielding and decoupling may be required if the rig is operated at higher power levels.

PERFORMANCE

The Sierra design uses a carefully selected set of compromises to keep complexity low and battery life long. An example is the use of NE602 mixers, which affects both receive and transmit performance. On receive, the RF gain will occasionally need reduction when strong signals overload the receive mixer. On transmit, ARRL Lab tests show that the rig complies with FCC regulations for its power and frequency ranges.

Aside from the weak receive mixer, receiver performance is very good. There are no spurious signals (birdies) audible on any band. ARRL Lab tests show that the Sierra's receiver has a typical MDS of about –139 dBm, blocking dynamic range of up to 112 dB and two-tone dynamic range of up to 90 dB. AGC range is about 70 dB.

The Sierra's transmitter offers smooth break-in keying, along with direct transmit signal monitoring. There are two benefits to direct monitoring:

- the clean sinusoidal tone is easier on the ears than most sidetone oscillators and
- the pitch of the monitor tone is the correct receive-signal pitch to listen for when calling other stations.

The TR mute delay capacitor, C27, can be reduced to as low as 4.7 µF to provide faster break-in keying if needed.

The prototype Sierra survived its christening at Field Day, 1994, where members of the Zuni Loop Expeditionary Force used it on 80, 40, 20 and 15 m. There, Sierra compared favorably to the Heath HW-9 and several older Ten-Tec rigs, having as good or better sensitivity and selectivity — and in most cases better-sounding sidetone and break-in keying. While the other rigs had higher output power, they couldn't touch the Sierra's small size, light weight and low power consumption. The Sierra has consistently received high marks from stations worked too, with reports of excellent keying and stability.

CONCLUSION

At the time this article was written, over 100 Sierras had been built. Many have been used extensively in the field, where the rig's unique features are an asset. For some builders, the Sierra has become the primary home station rig.

The success of the Sierra is due, in large part, to the energy and enthusiasm of the members of NorCal, who helped test and refine early prototypes, procured parts for the field-test units and suggested future modifications.[8] This project should serve as a model for other clubs who see a need for an entirely new kind of equipment, perhaps something that is not available commercially.

Notes

[1]One of N6KR's previous designs, the Safari-4, is a good example of how complex a band-switched rig can get. See "The Safari-4...." Oct through Dec 1990 *QEX*.

[2]Band modules for 160, 12 and 10 m have also been built. Construction details for these bands are provided in the Sierra information packet available from the ARRL. Go to the ARRL Web page: **www.arrl.org/notes/** to download a template package or write to the ARRL Technical Department Secretary and request the '96 *Handbook* Sierra template package.

[3]For information about NorCal, write to Jim Cates, WA6GER. Please include an SASE.

[4]Most multiband rigs draw from 150 to 500 mA on receive, necessitating the use of a larger battery. A discussion of battery life considerations can be found in "A Solar-Powered Field Day," May 1995 *QST*.

[5]Solid-State Design, p 87.

[6]Full and partial kits are available. The full kit comes with all components, controls, connectors, and a detailed assembly manual. Complete band modules kits are available for 80, 40, 30, 20, 17 and 15 m. For information, contact Wilderness Radio in Los Altos, CA.

[7]The alignment procedure given here is necessarily brief. More complete instructions are provided with the ARRL template package and the kit. See note 2 to obtain a template.

[8]The author would like to acknowledge the contributions of several NorCal members: Doug Hendricks, KI6DS; Jim Cates, WA6GER; Bob Dyer, KD6VIO; Dave Meacham, W6EMD; Eric Swartz, WA6HHQ, Bob Warmke, W6CYX; Stan Cooper, K4DRD; Vic Black, AB6SO; and Bob Korte, KD6KYT.

A 10-WATT SSB TRANSCEIVER FOR THE 60-METER BAND

Photo by Dan Wolfgang

It is not often we get a new amateur allocation in the HF spectrum! This 60-m transceiver design provides the builder with an approach that emphasizes a minimum number of adjustments during construction. This should be considered an advanced homebrew project. It makes use of a select number of surface-mount (SMT) components not readily duplicated using traditional construction techniques.[1] This project was designed by Dave Benson, K1SWL (ex-NN1G).

The new 60-Meter band departs from the traditional amateur allocations in that operation is channelized on five discrete frequencies. Operation is restricted to Upper sideband (USB), voice only. These channels are specified as being within a 2.8-kHz window centered on 5332, 5348, 5368, 5373 and 5405 kHz.

The multiple fixed-channel allocation suggests that the traditional analog Local Oscillator (LO) scheme would be unattractive from repeatability and stability standpoints. While a crystal-controlled LO approach is perfectly feasible, cost and complexity increase rapidly for more than several channels. This design makes use of the Direct-Digital-Synthesis (DDS) approach. It uses a MicroChip 16F628A, a 2K × 14 CMOS microcontroller that performs DDS frequency management, channel selection and fine-tuning (RIT) control and channel indication. Microcontroller source code and hex files for this project are both available.[2] **Fig 15.12** shows the schematic diagram and parts list for this 60-m project. The DDS IC (U14) contains two sets of frequency registers, loaded at power-up with two copies of the LO frequency. One is used during *transmit* and is fixed in frequency. The second frequency register set is used during *receive* and is fine-tuned by use of the RIT function. RIT is mechanized using an inexpensive electromechanical shaft encoder. Each detented step of the encoder yields a 10 Hz frequency step, with limit stops in firmware at ±500 Hz, sufficient to fine-tune stations slightly off-frequency within the channel window. The microcontroller steers the DDS IC between the two register sets via the T/R* signal (U14, pin 10).

Each time the frequency is initialized or changed, the DDS IC receives a sync signal and a burst of 40 clock pulses and corresponding serial data. This data is in the form of an 8-bit control word and a 32-bit frequency control word. The IC generates a sinusoidal waveform with 10 bits of D/AC resolution. This is an accuracy of 1 part in 1024 (2^{10}) — maximum error is one part in 2048, which corresponds to roughly 66 dB below the desired output. The LO frequency varies by channel and is at approximately 2.3 MHz. The output of the DDS is applied to a low-pass filter to removed unwanted high-frequency energy. This consists of both the 25-MHz clock energy and an alias signal. The alias signal is at the difference of the system clock and desired outputs, i.e., ~22.7 MHz. The resulting spectrum at the output of this filter is quite clean. Other components (spurs) of this waveform are present at low levels and their presence establishes the Spurious-Free Dynamic Range (SFDR). Those close-in spurs, i.e., nearest the desired LO frequency are of most interest. Spurs farther from the desired frequency are less critical, as their effect on receiver performance is mitigated by receiver front-end selectivity. This SFDR is reasonable for the application, but would normally be augmented with a cleanup loop (PLL) for higher-performance applications.

The microcontroller and DDS use their own local voltage regulator (U12). The primary contributor to digital noise in the

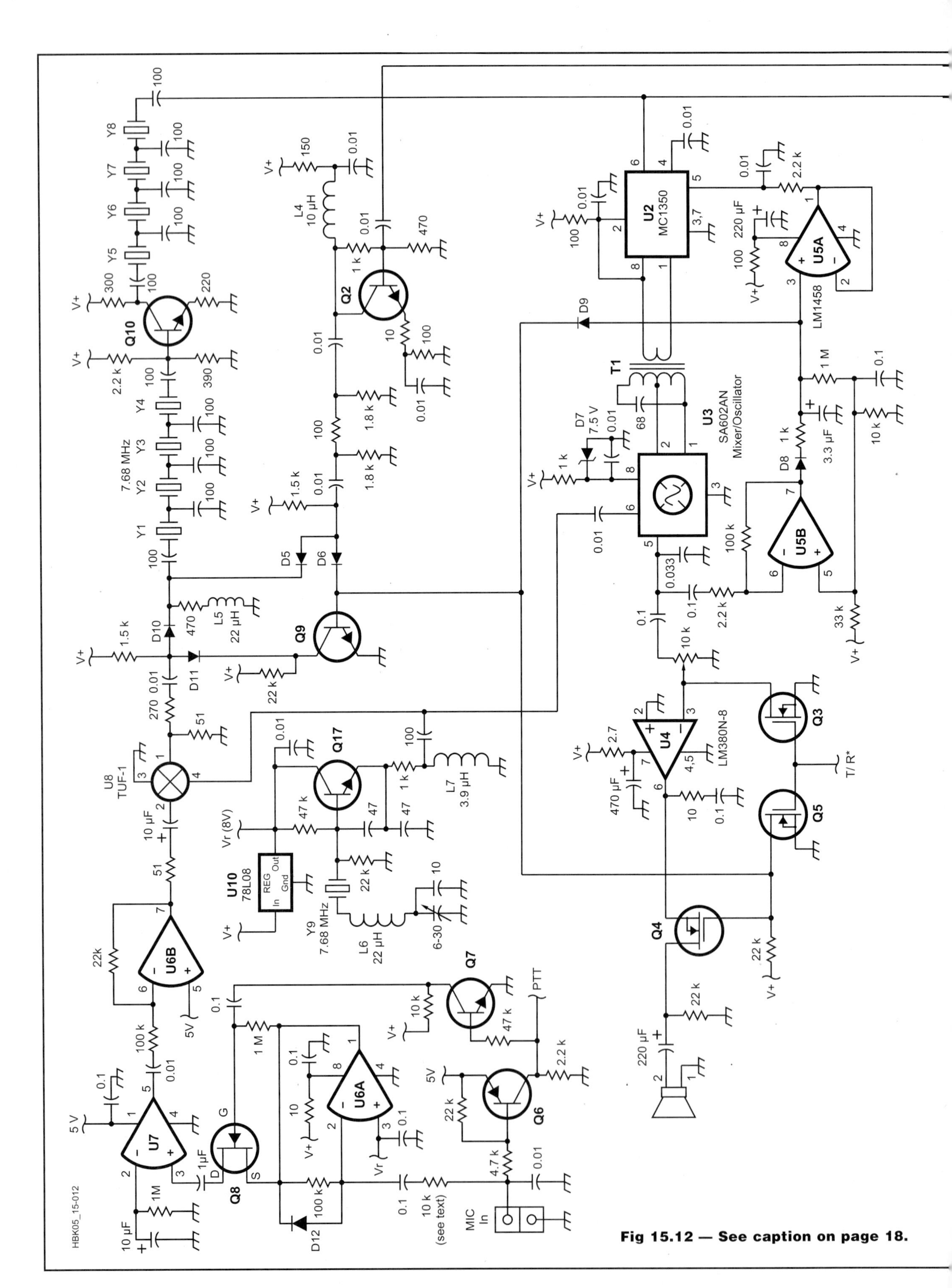

Fig 15.12 — See caption on page 18.

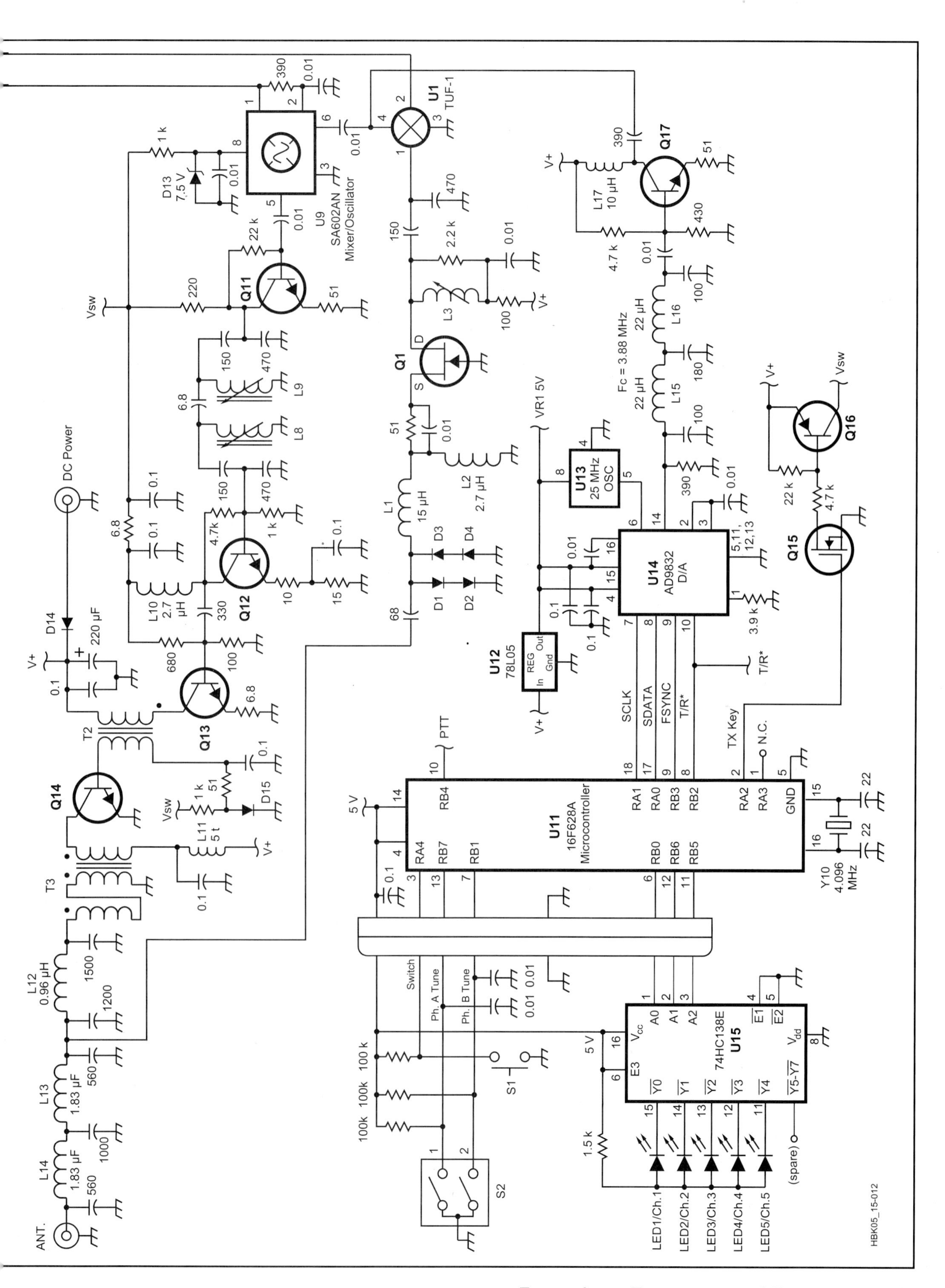

ANT.
DC Power
D14
220 µF
V+
0.1
T2
T3
Q14
Q13
Q12
Q11
Q1
Q15
Q16
Q17
L10
2.7 µH
L11
5 t
L12
0.96 µH
L13
1.83 µF
L14
1.83 µF
1500
1200
560
1000
Vsw
D15
L8
L9
D13
7.5 V
U9
SA602AN
Mixer/Oscillator
U1
TUF-1
L1
15 µH
L2
2.7 µH
L3
D1
D2
D3
D4
VR1 5V
U12
78L05
REG
In
Out
Gnd
U13
25 MHz
OSC
U14
AD9832
D/A
L15
22 µH
L16
22 µH
Fc = 3.88 MHz
L17
10 µH
SCLK
SDATA
FSYNC
T/R*
TX Key
N.C.
PTT
U11
16F628A
Microcontroller
RA4
RB7
RB1
RB4
RA1
RA0
RB3
RB2
RA2
RA3
GND
RB0
RB6
RB5
Y10
4.096
MHz
Switch
Ph. A Tune
Ph. B Tune
S1
S2
U15
74HC138E
5 V
LED1/Ch.1
LED2/Ch.2
LED3/Ch.3
LED4/Ch.4
LED5/Ch.5
(spare)
HBK05_15-012

Fig 15.12 — Schematic of the 10-W, USB Transceiver for the new 60-m band. All resistors are ¼-W, 5% carbon film. Small value (<.01 uF) fixed capacitors are NPO/C0G characteristic; polarized capacitors are aluminum electrolytic. One possible source of RF power transistors is RF Parts Company, San Marcos, CA.

D1-D6, D8-D12 — 1N4148 small-signal diode
D7, D13 — 7.5V Zener diode, Mouser 625-1N5236B
D14 — 1N5822, 3A Schottky rectifier diode
D15 — 1N4001, 1A rectifier diode
L1 — 15 uH submin. RF choke, Mouser 434-22-150
L2 — 2.7 uH submin. RF choke, Mouser 434-22-2R7
L3, L8, L9 — 10.7 MHz IF transformer, secondary unused, Mouser 42IF123
L4, L17 — 10 uH submin. RF choke, Mouser 434-22-100
L5, L6 — 22 uH submin. RF choke, Mouser 434-22-220
L7 — 3.9 uH submin. RF choke, Mouser 434-22-3R9
L10 — 2.7 uH submin. RF choke, Mouser 434-22-2R7
L11 — 5 turns #22 AWG on FT37-43 core
L12 — 0.96 uH, 14 turns #22 AWG on T50-2 core
L13, L14 — 1.83 uH, 19 turns #22 AWG on T50-2 core
L15, L16 — 22 uH submin. RF choke, Mouser 434-22-220
Q1 — J309 or J310 FET
Q2 — 2N2222A transistor
Q3, Q4, Q5, Q15 — 2N7000 MOSFET
Q6, Q16 — 2N3906 PNP transistor
Q7, Q9, Q10,Q11, Q17 — 2N4401 transistor
Q8 — 2N5485 FET
Q12 — 2N2219A transistor
Q13 — 2SC2078 or 2SC2166 transistor
Q14 — 2SC2312C transistor
S1 — pushbutton switch, single pole-normally open
S2 — rotary shaft encoder, Mouser 318 ENC16024P
T1 — 10.7 MHz IF transformer, Mouser 42IF123
T2 — 4 turns #22 AWG bifilar on FT50-43 core
T3 — 4 turns #22 AWG trifilar on FT50-43 core, observe polarity indications on schematic
U1, U8 — TUF-1, +7 dBm mixer, Minicircuits Labs
U2 — MC1350P IF Amplifier, Jameco Electronics 24942CF
U3, U9 — SA602AN Mixer/Oscillator IC, Future Electronics
U4 — LM380N-8, AF Amplifier, 8-pin. DigiKey LM380N-8
U5, U6 — LM1458, MC1458, or equivalent, dual operational amplifier
U7 — Speech Amplifier IC, AN6123. Digikey AN6123MSTXLCT
U10 — 8V, 100 mA 3-terminal voltage reg., Mouser 511-L78L08ACZ
U11 — 16F628A Microcontroller, Digikey PIC16F628A-04/P. Note: this is a blank (unprogrammed) device requires programming. See notes
U12 — 5V, 100 mA 3-terminal voltage reg, Mouser 511-L78L05ACZ
U13 — CMOS Clock oscillator, 25.00 MHz, Mouser 73-X052B2500
U14 — 10-bit Direct-digital synthesis (DDS) IC, Analog Devices AD9832BRU, Avnet Electronics Marketing
U15 — 74HC138E IC, 3:8 Demultiplexer
Y1-Y9 — 7.68 MHz microprocessor crystal, 20-pF load calibration group within 100 Hz. Mouser. 520-HCA768-20

receiver proved to be those ICs — this separation was necessary to keep the receive quiet.

The crystal (IF) filter is central to the transceiver. The IF is 7.68 MHz, a frequency chosen over several others based on spurious emissions consideration based on a birdy analysis. The filter (Y1-Y8 and associated capacitors) is mechanized as a pair of 4-pole filters for a total of 8 poles of filtering. Although the filter skirt slope is steeper on the high-frequency side, the slope is sufficient for good residual-carrier and alternate sideband suppression with oscillator injection on the low-frequency side. This has the important result of permitting upper-sideband generation at a much lower (2.3 MHz) LO frequency. This considerably eases the frequency-accuracy issue.

The IF passband width is 1.9 kHz. Although somewhat narrow for ragchewing, it is representative of the filter widths used for contesting. This narrow width proved necessary to provide sufficient suppression of out-of-channel energy, and also eased the frequency-tolerance issue by guard-banding both edges of the frequency window. Loss through the filter is minimal; passband ripple is a maximum of 2.4 dB. The source and termination impedances are 330 Ω. The crystals are readily-available microprocessor types (see the parts list in Fig 15.12) and should be selected by frequency within a grouping of 100 Hz in frequency.

The IF filter is shared between transmit and receive functions to avoid a significant duplication of hardware. Diodes D10/D11 (transmit side) and D5/D6 (receive) serve to isolate Receive and Transmit circuit functions as needed. For the network that is turned off, the series diode is back-biased (non-conducting) and the shunt diode is conducting (low-impedance) for good isolation performance. Rather than replicate this hardware at the downstream end of the filter, the filter drives both receiver product detector and transmit second mixer in parallel. This is a tradeoff based on the cost of the associated components (the two mixers) and printed circuit-board layout complexity. While the SA602A IC used for the transmit second mixer is inferior to the diode-ring mixers for 3rd-order intercept and balance (suppression between ports), it is well adequate in this narrow-band application.

The receiver design uses a +7dBm diode-ring mixer for its first mixer. This yields higher odd-order intercept performance than the popular SA602/SA12 mixer found in many minimum-parts-count designs. Grounded-gate JFET Q1 provides modest gain ahead of the mixer. While not necessary for signal-to-noise ratio purposes, it provides additional isolation between antenna and any LO leakage from the receive first mixer. Note that the presence of this preamp stage degrades the receiver intercept to reduce the superiority of the diode-ring mixer over the SA602/SA612 — a mixed blessing.

This transceiver design uses the readily-available HT speaker-mics simply to afford a standardized approach. Any of a number of ICOM/Yaesu-compatible products is usable — you'll want the twin-plug (2.5mm and 3.5mm) version with two contacts on each prong. Aftermarket offerings include the Pryme SPM-100 and the RadioShack 291-314, among others. This choice of microphone has several ramifications to the design, discussed below.

Both microphone element and PTT switch are combined in series (single-signal and ground) on the 2.5-mm connector plug. Although it reduces the number of connections to a speaker/mic, its practical effect is to introduce a sizable audio transient into the speech-amplifier circuitry on key-down. Op-amp U6, section A, provides some gain but also serves to clamp this audio transient to within one diode drop of reference voltage V_r via D12. Q7 is configured as a one-shot and provides a brief gating pulse to FET Q8 to open the audio path briefly during key-down to avoid upsetting subsequent speech-amplifier stages.

U7 is a speech-amplifier IC recently available, and designed with the cell-phone industry in mind. It features approximately 45 dB of AGC action, reducing that variation in input signal range into a peak output range of about 2 dB with modest distortion figures.

The remaining section of U6 serves to buffer this ICs output and provide the necessary low-source impedance to the transmit first mixer. Speaker/mics vary noticeably in terms of microphone output level. The value shown at the microphone input path (see schematic) assumes a fairly low microphone output level. This resistance value should be revised upwards, reducing gain, if you receive reports of excessive audio background noise while transmitting.

The speaker/mic approach also requires high isolation between receive and transmit audio signals. During receive, the microphone element is physically disconnected and the speech amplifier circuitry is further isolated by the filter-switching network (D10/D11). During transmit, the receiver output must be positively disconnected from the speaker to prevent audio feedback through the speaker/mic. This is typically manifested as a hollow sound or by squealing at higher feedback levels. Several MOSFET switches around the audio amplifier IC provide the function. Transistor Q3 provides a low-resistance shunt to ground during transmit. Additionally, Q4 is employed as a series switch to disconnect the path between AF amplifier and speaker. A variation of configurations was tested in the quest for high isolation, this one proving the most successful. Upstream, i.e., IF gain control was tried unsuccessfully — the supply feed to the audio amp IC itself proved to be a source of unwanted feedback via power-supply sag under high-current conditions.

For a channelized frequency allocation, the task of frequency annunciation is considerably simplified. Five LED indicators corresponding to the 5 allocated channels accomplish this. The three LED logic signals are binary encoded and are expanded into one-of-eight to illuminate individual LEDs. Three spare lines are available in the event that additional channels become available. For the advanced homebrewer, firmware for several of these three signals could be recoded as serial data to an LCD or LED display. Jameco Electronics offers a serial-to-parallel-port expansion IC compatible with many parallel-LCD display modules. While the LCD has largely overtaken the LED display in popularity for reasons of power consumption, the recent availability of super-bright, 7-segment LED displays renders that approach worthy of reconsideration. (The incentive to upgrade in this way would of course be strengthened by any assignment of additional frequencies within the 60-meter band.)

Alignment is straightforward due to the low number of adjustments in this design. The receiver is peaked via L3 and T1 for maximum signal level. The firmware contains a built-in calibration mode for fine-tuning the LO frequency. Powering up the transceiver with the channel-select switch depressed yields a loud audio tone. Its frequency is adjusted to 800 Hz using a frequency counter applied to the transceiver speaker output. If a frequency counter is not available, this critical step must still be performed and can be easily accomplished using currently-available computer soundcard applications.[3] The transceiver stays in this mode until power is removed. Upon key-down into a dummy load, inductors L8 and L9 are adjusted for maximum indicated output power.

Going *on the air* is as simple as one could hope for. The transceiver powers up at the lowest frequency channel and illuminates one of five LEDs. Depressing pushbutton switch S1 advances the tuning to the next channel. If the switch is held down, the channel selection scans through the bank of five frequencies until the switch is released. With each change in channel selection, the RIT function offset is reset to the neutral position.

Spectral purity measurements taken at a 10-W carrier show spur content at –50dBc or better. Harmonic content at the second harmonic is –46 dBc and –50 dBc or better for all higher harmonics.

Notes

[1]A double-sided printed-circuit board is available from the author with SMT components preinstalled and functional-tested. $35 US ($40 DX) includes shipping. See **www.smallwonderlabs.com/projects.htm** for ordering information.

[2]Documented source code for the 16CF628 microcontroller, as well as hex (programming) files are available for free download. See **www.smallwonderlabs.com/projects.htm**.

[3]A variety of Amateur-Radio applications with soundcard/spectrographic displays are available. See **www.psk31.com** software links. DigiPan, among others, is freeware. See also **www.visualizationsoftware.com** for the popular Spectrogram application. This has a registration fee but provides a free trial period.

Transverters

At VHF, UHF and microwave frequencies, transverters that interact with factory-made transceivers in the HF or VHF range are common and are often homebuilt. These units convert the transceiver transmit signal up to a higher frequency and convert the receive frequency down to the transceiver receive frequency. The resulting performance and signal quality at the higher frequencies are enhanced by the frequency stability and the signal processing capabilities of the transceiver. For example, SSB and narrowband CW from 1.2 to 10 GHz are feasible, and becoming more popular. Some HF and VHF transceivers have special provisions such as connectors, signal-path switching and T/R switching that facilitate use with a transverter.

VHF TRANSVERTERS

The methods of individual circuit design for a transverter are not much different than methods that have already been described. The most informative approach would be to study carefully an actual project description.

The interface between the transceiver and transverter requires some careful planning. For example, the transceiver power output must be compatible with the transverter's input requirements. This may require an attenuator or some modifications to a particular transverter or transceiver.

When receiving, the gain of the transverter must not be so large that the transceiver front-end is overdriven (system IMD is seriously degraded). On the other hand, the transverter gain must be high enough and its noise figure low enough so that the overall system noise figure is within a dB or so of the transverter's own noise figure. The formulas in the **Receivers and Transmitters** chapter for cascaded noise figure and cascaded third-order intercept points should be used during the design process to assure good system performance. The transceiver's performance should be either known or measured to assist in this effort.

MICROWAVE TRANSVERTERS

The microwave receiver section of the **Receivers and Transmitters** chapter in this *Handbook* discusses a 10-GHz transverter project and provides references to the *QST* articles that give a detailed description. The reader is encouraged to refer to these articles and to review the previous material in that chapter.

Other Information

The ARRL UHF/Microwave Experimenter's Manual and ARRL's *Microwave Projects* contain additional interesting and valuable descriptions of transverter and transponder requirements.

Repeaters

This section was written by Paul M. Danzer, N1II.

In the late 1960s two events occurred that changed the way radio amateurs communicated. The first was the explosive advance in solid-state components — transistors and integrated circuits. A number of new "designed for communications" integrated circuits became available, as well as improved high-power transistors for RF power amplifiers. Vacuum tube-based equipment, expensive to maintain and subject to vibration damage, was becoming obsolete.

At about the same time, in one of its periodic reviews of spectrum usage, the Federal Communications Commission (FCC) mandated that commercial users of the VHF spectrum reduce the deviation of truck, taxi, police, fire and all other commercial services from 15 kHz to 5 kHz. This meant that thousands of new narrowband FM radios were put into service and an equal number of wideband radios were no longer needed.

As the new radios arrived at the front door of the commercial users, the old radios that weren't modified went out the back door, and hams lined up to take advantage of the newly available "commercial surplus." Not since the end of World War II had so many radios been made available to the ham community at very low or at least acceptable prices. With a little tweaking, the transmitters and receivers were modified for ham use, and the great repeater boom was on.

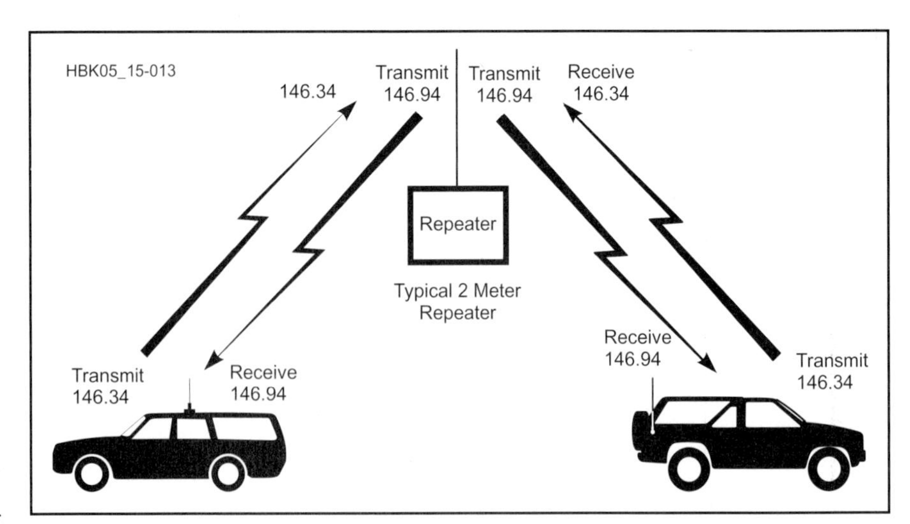

Fig 15.13 — Typical 2-m repeater, showing mobile-to-mobile communication through a repeater station. Usually located on a hill or tall building, the repeater amplifies and retransmits the received signal on a different frequency.

Table 15.3
Types of Repeaters

ATV — Amateur TV — Same coverage advantages as voice repeaters to hams using wideband TV in the VHF and UHF bands. Often consist of pairs of repeaters — one for the ATV and the other for the voice coordination.

AM and SSB — There is no reason to limit repeaters to FM. There are a number of other modulation-type repeaters, some experimental and some long-established.

Digipeaters — Digital repeaters used primarily for packet communications. Can use a single channel (single port) or several channels (multi-port) on one or more VHF and UHF bands.

Multi-channel (wideband) — Amateur satellites are best-known examples. Wide bandwidth (perhaps 50 to 200 kHz) is selected to be received and transmitted so all signals in bandwidth are heard by the satellite (repeater) and retransmitted, usually on a different VHF or UHF band. Satellites are discussed elsewhere in this chapter.

Although not permitted or practical for terrestrial use in the VHF or UHF spectrum, there is no reason wideband repeaters cannot be established in the microwave region where wide bandwidths are allowed. This would be known as frequency multiplexing.

WHAT IS A REPEATER?

Trucking companies and police departments learned long ago that they could get much better use from their mobile radios by using an automated relay station called a *repeater*. Not all radio dispatchers are located near the highest point in town or have access to a 300-ft tower. But a repeater, whose basic idea is shown in **Fig 15.13**, can be more readily located where the antenna system is as high as possible and can therefore cover a much greater area.

Types of Repeaters

The most popular and well-known type of amateur repeater is an FM voice system on the 29, 144, 222 or 440-MHz bands. Tens of thousands of hams use small 12-V powered radios in their vehicles for both casual ragchewing and staying in touch with what is going on during heavy traffic or commuting times. Others have low-power battery-operated handheld units for 144, 222 or 440 MHz. Some mobile and handheld transceivers operate on two bands. But there are several other types of ham radio repeaters. **Table 15.3** describes them.

FM is the mode of choice, as it was in commercial service, since it provides a high degree of immunity to mobile noise pulses. Operations are *channelized* — all stations operate on the same transmit frequency and receive on the same receive frequency. In addition, since the repeater receives signals from mobile or fixed stations and retransmits these signals simultaneously, the transmit and receive frequencies are different, or *split*. Direct contact between two or more stations that listen and transmit on the same frequency is called operating *simplex*.

Individuals, clubs, amateur civil defense support groups and other organizations all sponsor repeaters. Anyone with a valid amateur license for the band can establish a repeater in conformance with the FCC rules. No one owns specific repeater frequencies, but nearly all repeaters are *coordinated* to minimize repeater-to-repeater interference. Frequency coordination and interference are discussed later in this chapter.

Block Diagrams

Repeaters normally contain at least the sections shown in **Fig 15.14**. After this, the sky is the limit on imagination. As an example, a remote receiver site can be used to try to eliminate interference (**Fig 15.15**).

The two sites can be linked either by telephone ("hard wire") or a VHF or UHF link. Once you have one remote receiver site it is natural to consider a second site to better hear those "weak mobiles" on the other side of town (**Fig 15.16**). Some of the stations using the repeater are on 2 m while others are on 440. Just link the two repeaters! (**Fig 15.17**).

Want to help the local Civil Air Patrol (CAP)? Add a receiver for aircraft emergency transmitters (ELT). Tornados? It is now legal to add a weather channel receiver (**Fig 15.18**).

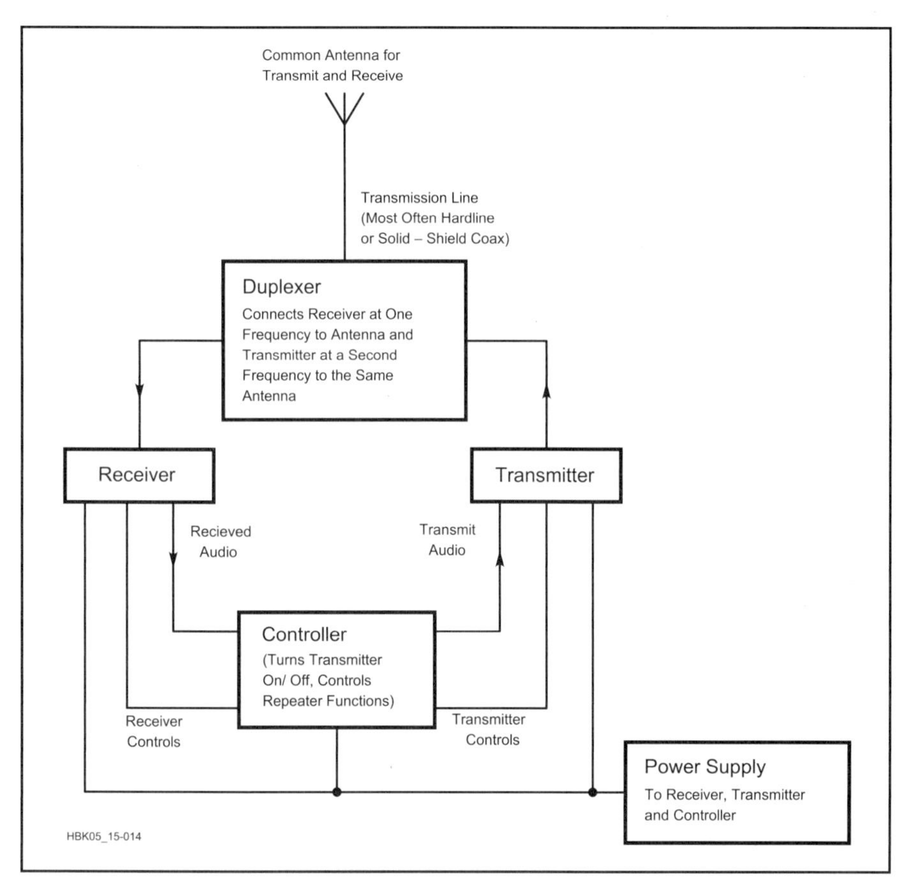

Fig 15.14 — The basic components of a repeater station. In the early days of repeaters, many were home-built. Today, most are commercial, and are far more complex than this diagram suggests.

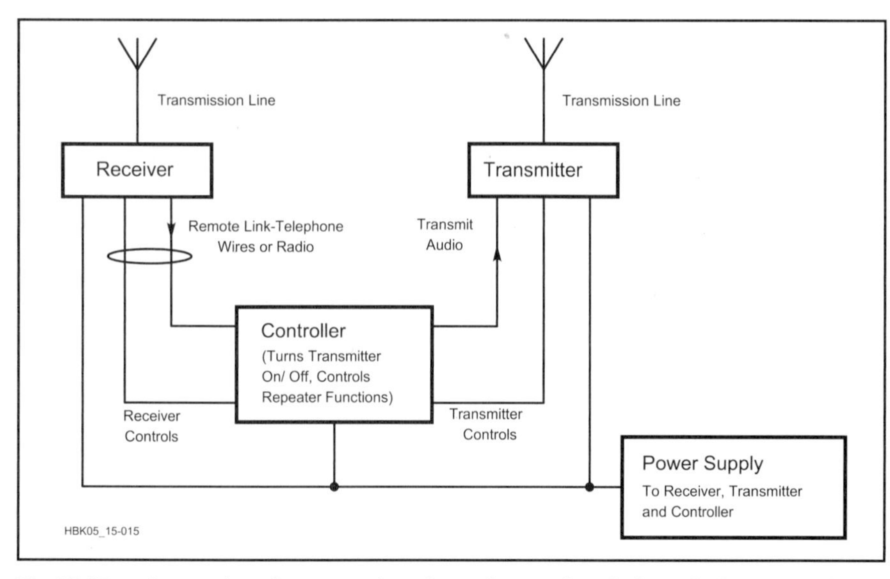

Fig 15.15 — Separating the transmitter from the receiver helps eliminate certain types of interference. The remote receiver can be located on a different building or hill, or consist of a second antenna at a different height on the tower.

The list goes on and on. Perhaps that is why so many hams have put up repeaters.

Repeater Terminology

Here are some definitions of terms used in the world of Amateur Radio FM and repeaters:

access code — one or more numbers and/or symbols that are keyed into the repeater with a DTMF tone pad to activate a repeater function, such as an autopatch.

autopatch — a device that interfaces a repeater to the telephone system to permit repeater users to make telephone calls. Often just called a "patch."

break — the word used to interrupt a conversation on a repeater *only* to indicate that there is an emergency.

carrier-operated relay (COR) — a device that causes the repeater to transmit in response to a received signal.

channel — the pair of frequencies (input and output) used by a repeater.

closed repeater — a repeater whose access is limited to a select group (see ***open repeater***).

control operator — the Amateur Radio operator who is designated to "control" the operation of the repeater, as required by FCC regulations.

courtesy beep — an audible indication that a repeater user may go ahead and transmit.

coverage — the geographic area within which the repeater provides communications.

CTCSS — abbreviation for continuous tone-controlled squelch system, a series of subaudible tones that some repeaters use to restrict access. (see **closed repeater**).

digipeater — a packet radio (digital) repeater.

DTMF — abbreviation for dual-tone multifrequency, the series of tones generated from a keypad on a ham radio transceiver (or a regular telephone).

duplex or **full duplex** — a mode of communication in which a user transmits on one frequency and receives on another frequency simultaneously (see ***half duplex***).

duplexer — a device that allows the repeater transmitter and receiver to use the same antenna simultaneously.

frequency coordinator — an individual or group responsible for assigning frequencies to new repeaters without causing interference to existing repeaters.

full quieting — a received signal that contains no noise.

half duplex — a mode of communication in which a user transmits at one time and receives at another time.

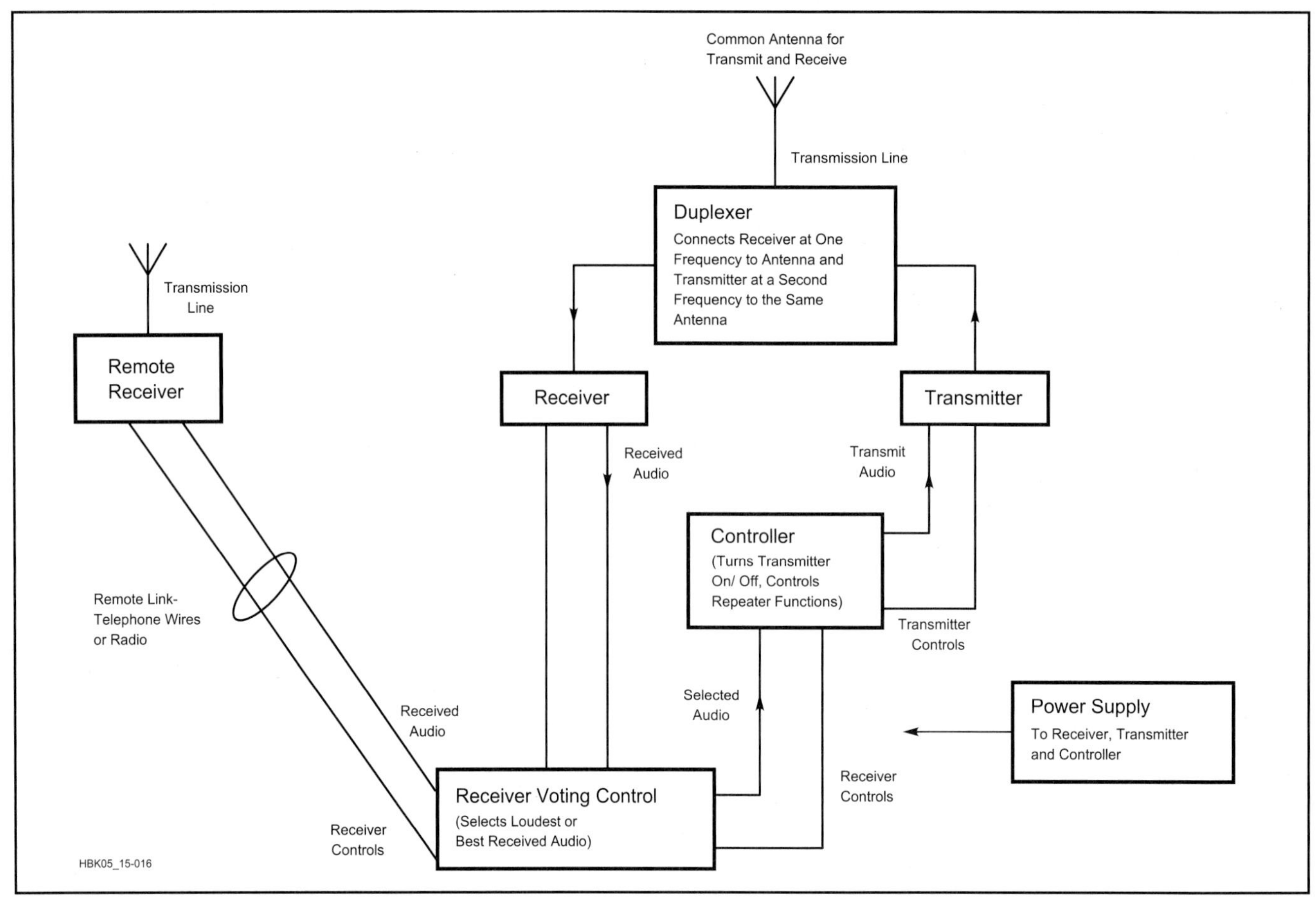

Fig 15.16 — A second remote receiver site can provide solid coverage on the other side of town.

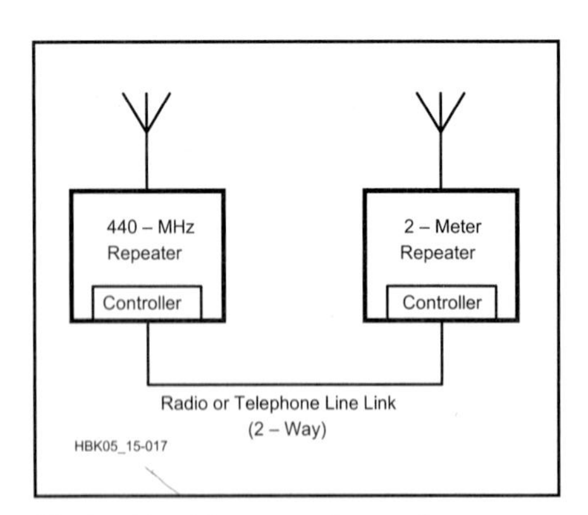

Fig 15.17 — Two repeaters using different bands can be linked for added convenience.

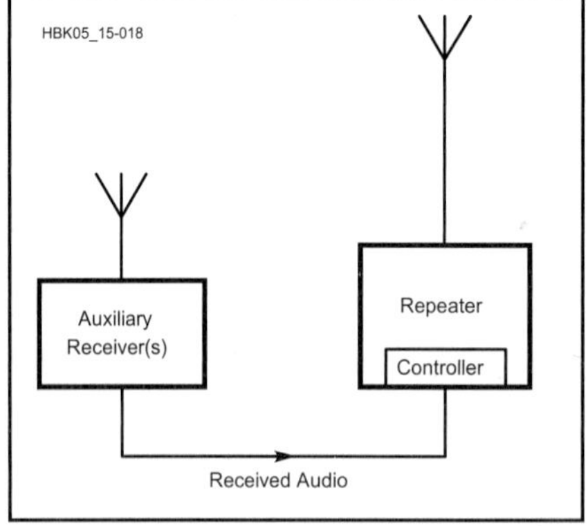

Fig 15.18 — For even greater flexibility, you can add an auxiliary receiver.

handheld — a small, lightweight portable transceiver small enough to be carried easily.

hang time — the short period following a transmission that allows others who want to access the repeater a chance to do so; a *courtesy beep* sounds when the repeater is ready to accept another transmission.

input frequency — the frequency of the repeater's receiver (and your transceiver's transmitter).

intermodulation distortion (IMD) — the unwanted mixing of two strong RF signals that causes a signal to be transmitted on an unintended frequency.

key up — to turn on a repeater by transmitting on its input frequency.

machine — a repeater system.

magnetic mount or **mag-mount** — an antenna with a magnetic base that permits quick installation and removal from a motor vehicle or other metal surface.

NiCd — a nickel-cadmium battery that may be recharged many times; often used to power portable transceivers. Pronounced "NYE-cad."

open repeater — a repeater whose access is not limited.

output frequency — the frequency of the repeater's transmitter (and your transceiver's receiver).

over — a word used to indicate the end of a voice transmission.

Repeater Directory — an annual ARRL publication that lists repeaters in the US, Canada and other areas.

separation or **split** — the difference (in kHz) between a repeater's transmitter and receiver frequencies. Repeaters that use unusual separations, such as 1 MHz on 2 m, are sometimes said to have "oddball splits."

simplex — a mode of communication in which users transmit and receive on the same frequency.

time-out — to cause the repeater or a repeater function to turn off because you have transmitted for too long.

timer — a device that measures the length of each transmission and causes the repeater or a repeater function to turn off after a transmission has exceeded a certain length.

tone pad — an array of 12 or 16 numbered keys that generate the standard telephone dual-tone multifrequency (***DTMF***) dialing signals. Resembles a standard telephone keypad. (see ***autopatch***).

Advantages of Using a Repeater

When we use the term *repeater* we are almost always talking about transmitters and receivers on VHF or higher bands, where radio-wave propagation is normally line of sight. Sometimes a hill or building in the path will allow refraction or other types of edge effects, reflections and bending. But for high quality, consistently solid communications, line of sight is the primary mode.

We know that the effective range of VHF and UHF signals is related to the height of each antenna. Since repeaters can usually be located at high points, one great advantage of repeaters is the extension of coverage area from low-powered mobile and portable transceivers.

Fig 15.19 illustrates the effect of using a repeater in areas with hills or mountains. The same effect is found in metropolitan areas, where buildings provide the primary blocking structures.

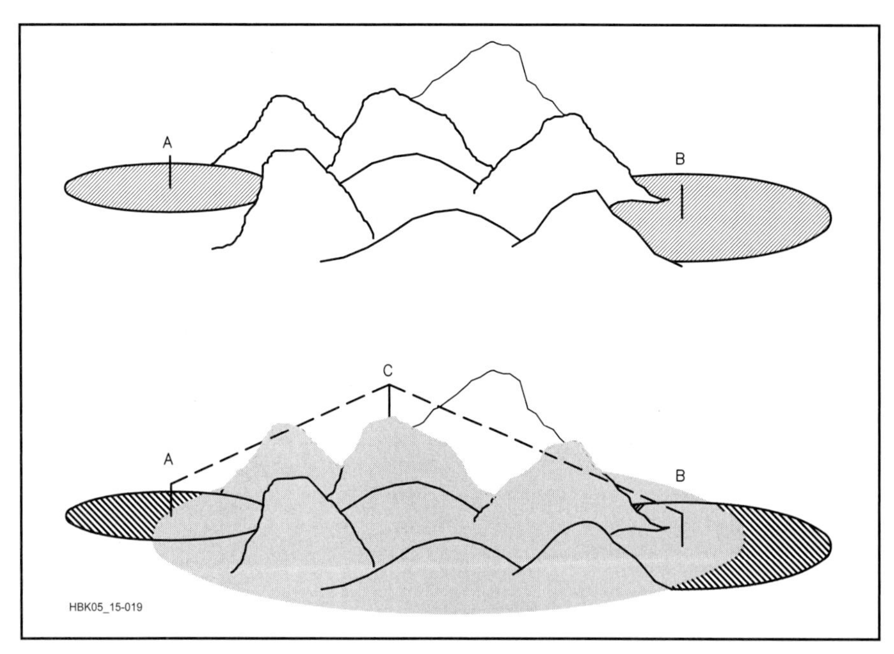

Fig 15.19 — In the upper diagram, stations A and B cannot communicate because their mutual coverage is limited by the mountains between them. In the lower diagram, stations A and B can communicate because the coverage of each station falls within the coverage of repeater C, which is on a mountaintop.

Fig 15.20 — In the Rocky Mountain west, handheld transceivers can often cover great distances, thanks to repeaters located atop high mountains. *(photo courtesy WBØKRX and NØIET)*

Siting repeaters at high points can also have disadvantages. When two nearby repeaters use the same frequencies, your transceiver might be able to receive both. But since it operates FM, the *capture effect* usually ensures that the stronger signal will capture your receiver and the weaker signal will not be heard — at least as long as the stronger repeater is in use.

It is also simpler to provide a very sensitive receiver, a good antenna system, and a slightly higher power transmitter at just one location — the repeater — than at each mobile, portable or home location. A superior repeater system compensates for the low power (5 W or less), and small, inefficient antennas that many hams use to operate through them. The repeater maintains the range or coverage we want, despite our equipment deficiencies. If both the handheld transceiver and the repeater are at high elevations, for example, communication is possible over great distances, despite the low output power and inefficient antenna of the transceiver (see **Fig 15.20**).

Repeaters also provide a convenient meeting place for hams with a common interest. It might be geographic — your town — or it might be a particular interest such as DX or passing traffic. Operation is channelized, and usually in any area you can find out which channel — or repeater — to pick to ragchew, get highway information, or whatever your need or interest is. The fact that operation is channelized also provides an increased measure of driving safety — you don't have to tune around and call CQ to make a contact, as on the HF bands. Simply call on a repeater frequency — if someone is there and they want to talk, they will answer you.

EMERGENCY OPERATIONS

When there is a weather-related emergency or a disaster (or one is threatening), most repeaters in the affected area immediately spring to life. Emergency operation and traffic always take priority over other ham activities, and many repeaters are equipped with emergency power sources just for these occasions.

Almost all Amateur Radio emergency organizations use repeaters to take advantage of their extended range, uniformly good coverage and visibility. Most repeat-

Fig 15.21 — During disasters like the Mississippi River floods of 1993, repeaters over a wide area are used solely for emergency-related communication until the danger to life and property is past. ***(photo courtesy WA9TZL)***

ers are well known — everyone active in an area with suitable equipment knows the local repeater frequencies. For those who don't, many transceivers provide the ability to scan for a busy frequency. See **Fig 15.21**.

REPEATERS AND THE FCC

The law in the United States changes over time to adapt to new technology and changing times. Since the early 1980s, the trend has been toward deregulation, or more accurately in the case of radio amateurs, self-regulation. Hams have established band plans, calling frequencies, digital protocols and rules that promote efficient communication and interchange of information.

Originally, repeaters were licensed separately with detailed applications and control rules. Repeater users were forbidden to use their equipment in any way that could be interpreted as commercial. In some cases, even calling a friend at an office where the receptionist answered with the company name was interpreted as a problem.

The rules have changed, and now most nonprofit groups and public service events can be supported and businesses can be called — as long as the participating radio amateurs are not earning a living from this specific activity.

We can expect this trend to continue. For the latest rules and how to interpret them, see *QST* and *The FCC Rule Book*, published by the ARRL.

FM REPEATER OPERATION AND EQUIPMENT

Operating Techniques

There are almost as many operating procedures in use on repeaters as there are repeaters. Only by listening can you determine the customary procedures on a particular machine. A number of common operating techniques are found on many repeaters, however.

One such common technique is the transmission of *courtesy tones*. Suppose several stations are talking in rotation — one following another. The repeater detects the end of a transmission of one user, waits a few seconds, and then transmits a short tone or beep. The next station in the rotation waits until the beep before transmitting, thus giving any other station wanting to join in a brief period to transmit their call sign. Thus the term *courtesy tone* — you are politely pausing to allow other stations to join in the conversation.

Another common repeater feature that encourages polite operation is the *repeater timer*. Since repeater operation is channelized — allowing many stations to use the same frequency — it is polite to keep your transmissions short. If you forget this little politeness many repeaters simply cut off your transmission after 2 or 3 minutes of continuous talking. After the repeater "times out," the timer is reset and the repeater is ready for the next transmission. The timer length is often set to 3 minutes or so during most times of the day and 1 or 1½ minutes during commuter rush hours when many mobile stations want to use the repeater.

A general rule, in fact law — both internationally and in areas regulated by the FCC — is that emergency transmissions always have priority. These are defined as relating to life, safety and property damage. Many repeaters are voluntarily set up to give mobile stations priority, at least in checking onto the repeater. If there is going to be a problem requiring help, the request will usually come from a mobile station. This is particularly true during rush hours; some repeater owners request that fixed stations refrain from using the repeater during these hours. Since fixed stations usually have the advantages of fixed antennas and higher power, they can operate simplex more easily. This frees the repeater for mobile stations that need it.

A chart of suggested operating priorities is given in **Fig 15.22**. Many but not all repeaters conform to this concept, so it can be used as a general guideline.

The figure includes a suggested priority control for *closed repeaters*. These are repeaters whose owners wish, for any number of reasons, not to have them listed as available for general use. Often they require transmission of a *subaudible* or *CTCSS* tone (discussed later). Not all repeaters requiring a CTCSS tone are closed. Other closed repeaters require the transmission of a coded telephone push-button *(DTMF)* tone sequence to turn on. It is desirable that all repeaters, including generally closed repeaters, be made available at least long enough for the presence of emergency information to be made known.

Repeaters have many uses. In some areas they are commonly used for formal traffic nets, replacing or supplementing the nets usually found on 75-m SSB. In other areas they are used with tone alerting for severe-weather nets. Even when a particular repeater is generally used for ragchewing it can be linked for a special purpose. As an example, an ARRL volunteer official may hold a periodic section meeting across her state, with linked repeaters allowing both announcements and questions directed back to her.

One of the most common and important uses of a repeater is to aid visiting hams. Since repeaters are listed in the *ARRL Repeater Directory* and other directories, hams traveling across the country with

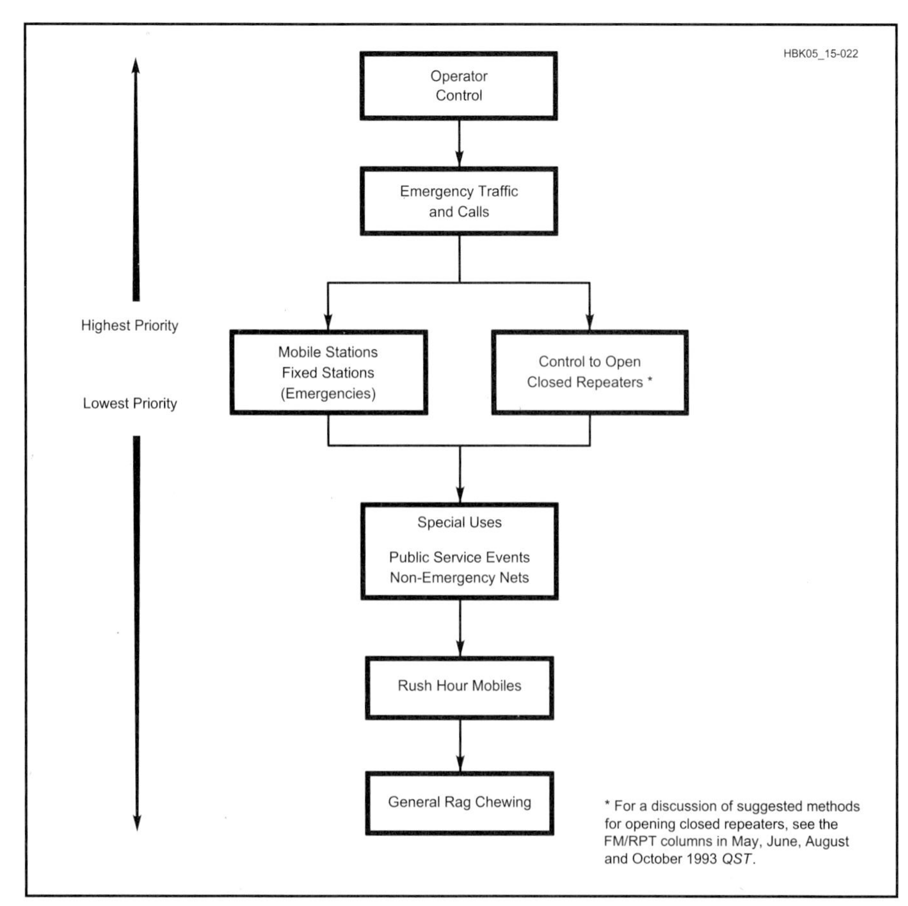

Fig 15.22 — The chart shows recommended repeater operating priorities. Note that, in general, priority goes to mobile stations.

mobile or handheld radios often check into local repeaters asking for travel route, restaurant or lodging information. Others just come on the repeater to say hello to the local group. In most areas courtesy prevails — the visitor is given priority to say hello or get the needed help.

Detailed information on repeater operating techniques is included in a full chapter of the *ARRL Operating Manual*.

Home and Mobile Equipment

There are many options available in equipment used on repeaters — both home-built and commercial. It is common to use the same radio for both home station and mobile, or mobile and handheld use. A number of these options are shown in **Fig 15.23**.

Handheld Transceivers

A basic handheld radio with 100 mW to 5 W output can be mounted in an automobile with or without a booster amplifier or "brick."

Several types of antennas can be used in the handheld mode. The smallest and most convenient is a rubber flex antenna, known as a "rubber duckie," a helically wound antenna encased in a flexible tube. Unfortunately, to obtain the small size the use of a wire helix or coil often produces a very low efficiency.

A quarter-wave whip, which is about 19 inches long for the 2-m band, is a good choice for enhanced performance. The rig and your hand act as a ground plane and a reasonably efficient result is obtained. A longer antenna, consisting of several electrical quarter-wave sections in series, is also commercially available. Although this antenna usually produces extended coverage, the mechanical strain of 30 or more inches of antenna mounted on the radio's antenna connector can cause problems. After several months, the strain may require replacement of the connector.

Selection of batteries will change the output power from the lowest generally available — 0.1 or 0.5 W — to the 5-W level. Charging is accomplished either with a "quick" charger in an hour or less or with a trickle charger overnight.

Power levels higher than 7 W may cause a safety problem on handheld units, since the antenna is usually close to the operator's head and eyes. See the **Safety** chapter for more information.

For mobile operation, a 12-V power cord plugs into the auto cigarette lighter. In addition, commercially available brick amplifiers — available either assembled or as kits — can be used to raise the output power level of the handheld radio to 10 to 70 W. These amplifiers often come with transmit-receive sensing and optional preamplifiers. One such unit is shown in block form in **Fig 15.24**.

Mobile Equipment

Mobile antennas range from quick and easy "clip-it-on" mounting to "drill through the car roof" assemblies. The four general classes of mobile antennas shown in the center section of Fig 15.23 are the most popular choices. Before experimenting with antennas for your vehicle, there are some precautions to be taken.

Through-the-glass antennas: Rather than trying to get the information from your dealer or car manufacturer, test any such antenna first using masking tape or some other temporary technique to hold the antenna in place. Some windshields are metallicized for defrosting, tinting and car radio reception. Having this metal in the way of your through-the-glass antenna will seriously decrease its efficiency.

Magnet-mount antennas are convenient, but only if your car has a metal roof. The metal roof serves as the ground plane.

Through-the-roof antenna mounting: Drilling a hole in your car roof may not be the best option unless you intend to keep the car for the foreseeable future. This mounting method provides the best efficiency, however, since the (metal) roof serves as a ground plane. Before you drill, carefully plan and measure how you intend to get the antenna cable down under the interior car headliner to the radio.

Trunk lid and clip-on antennas: These antennas are good compromises. They are usually easy to mount and they perform acceptably. Cable routing must be planned. If you are going to run more than a few watts, do not mount the antenna close to one of the car windows — a significant portion of the radiated power may enter the car interior.

Mobile rigs used at home can be powered either from rechargeable 12-V batteries or fixed power supplied from the 120-V ac line. Use of 12-V batteries has the advantage of providing back-up communications ability in the event of a power interruption. When a storm knocks down power lines and telephone service, it is common to hear hams using their mobile or 12-V powered rigs making autopatch calls to the power and telephone company to advise them of loss of service.

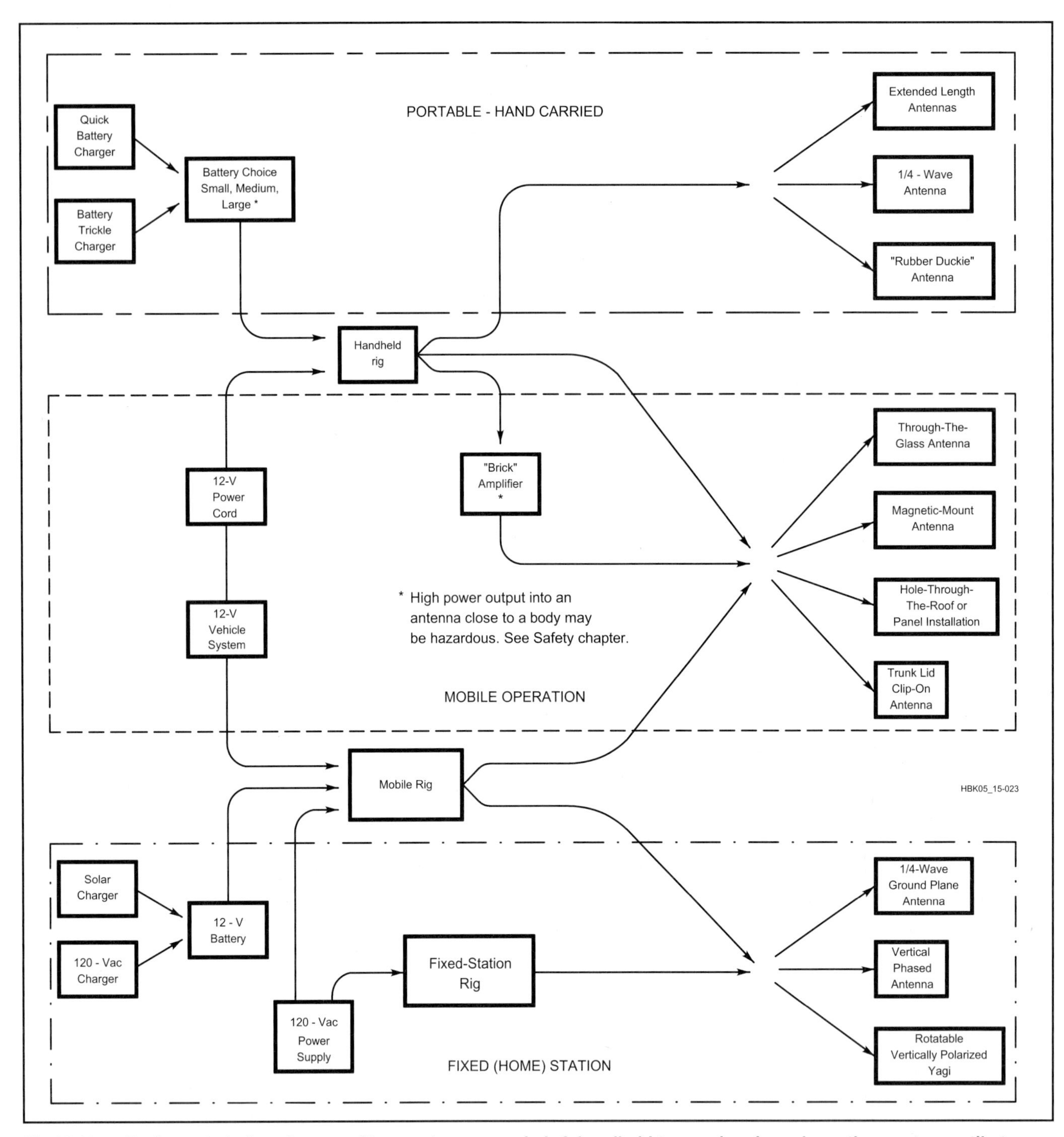

Fig 15.23 — Equipment choices for use with repeaters are varied. A handheld transceiver is perhaps the most versatile type of radio, as it can be operated from home, from a vehicle and from a mountaintop.

Home Station Equipment

The general choice of fixed-location antennas is also shown in Fig 15.23. A rotatable Yagi is normally not only unnecessary but undesirable for repeater use, since it has the potential of extending your transmit range into adjacent area repeaters on the same frequency pair. All antennas used to communicate through repeaters should be vertically polarized for best performance.

Both commercial and homemade $^1/_4$-λ and larger antennas are popular for home use. A number of these are shown in the **Antennas** chapter. Generally speaking, $^1/_4$-λ sections may be stacked up to provide more gain on any band. As you do so, however, more and more power is concentrated toward the horizon. This may be desirable if you live in a flat area. See **Fig 15.25**.

While most hams do not to try to build transceivers for use on repeaters, accessories provide a fertile area for construction and experimentation. What mobile operator has not wished that there was no need to hold the microphone continuously?

A single-pole, single-throw switch can be mounted in a small box and Velcro used to attach it temporarily to your seat.

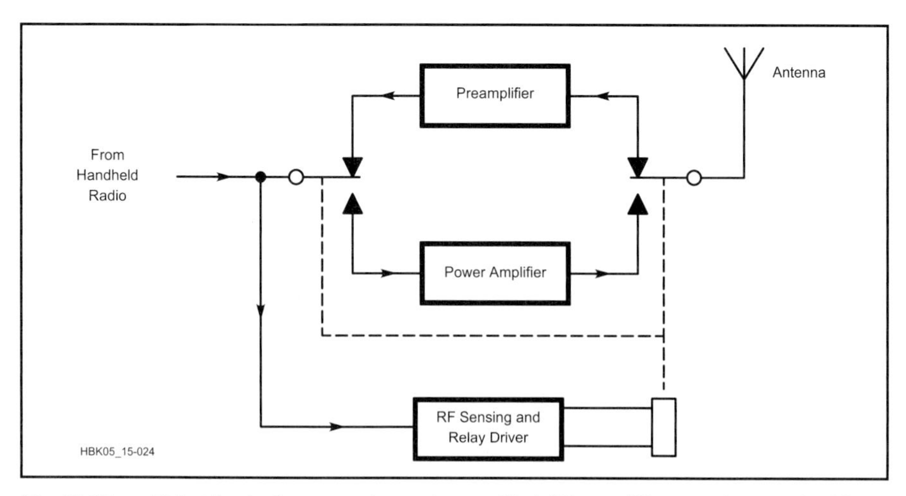

Fig 15.24 — This block diagram shows how a "brick" amplifier can be used with a receiver preamplifier. RF energy from the transceiver is detected, turns on the relay, and puts the RF power amplifier in line with the antenna. When no RF is sensed from the transceiver, the receiving preamp is in line.

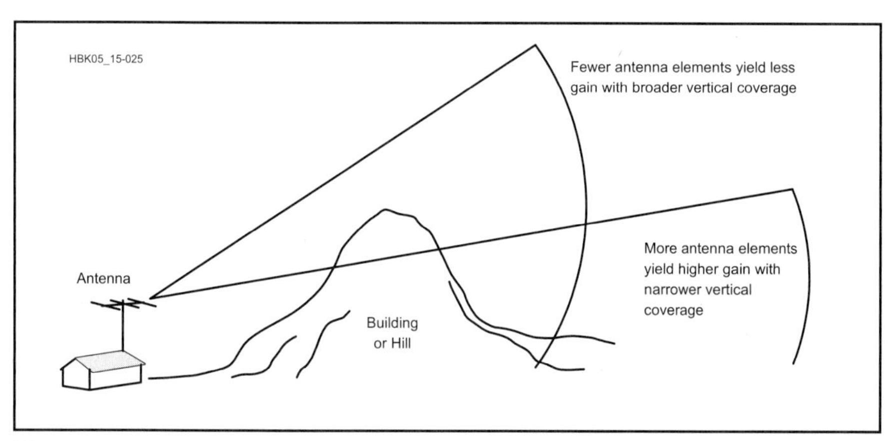

Fig 15.25 — As with all line-of-sight communications, terrain plays an important role in how your signal gets out.

Flip the lever on to transmit and flip it off to listen — in the meantime your hands are free.

Autopatches and Tones

One of the most attractive features of repeaters is the availability of autopatch services. This allows the mobile or portable station to use a standard telephone key pad to connect the repeater to the local telephone line and make outgoing calls.

Table 15.4 shows the tones used for these services. Some keyboards provide the standard 12 sets of tones corresponding to the digits 0 through 9 and the special signs # and *. Others include the full set of 16 pairs, providing special keys A through D. The tones are arranged in two groups, usually called the low tones and high tones. Two tones, one from each group, are required to define a key or digit. For example, pressing 5 will generate a 770-Hz tone and a 1336-Hz tone simultaneously.

The standards used by the telephone company require the amplitudes of these two tones to have a certain relationship. Fortunately, most tone generators used for this purpose have the amplitude relationship as part of their construction. Initially, many hams used surplus telephone company keypads. These units were easily installed — usually just two or three wires were connected. Unfortunately they were constructed with wire contacts and their reliability was not great when used in a moving vehicle.

Many repeaters require pressing a code number sequence or the special figures * or # to turn the autopatch on and off. Out-of-area calls are usually locked out, as are services requiring the dialing of the prefix 0 or 1. "Speed dial" is often available, although occasionally this can conflict with the use of * or # for repeater control, since these special symbols are used by the telephone company for its own purposes.

Table 15.4
Standard Telephone (DTMF) Tones

	Low Tone Group		*High Tone Group*	
	1209 Hz	*1336 Hz*	*1477 Hz*	*1633 Hz*
697 Hz	1	2	3	A
770 Hz	4	5	6	B
852 Hz	7	8	9	C
941 Hz	*	0	#	D

Some repeaters require the use of *subaudible* or CTCSS tones to utilize the autopatch, while others require these tones just to access the repeater in normal use. Taken from the commercial services, subaudible tones are not generally used to keep others from using a repeater but rather are a method of minimizing interference from users of the same repeater frequency.

For example, in **Fig 15.26** a mobile station on hill A is nominally within the normal coverage area of the Jonestown repeater (146.16/76). The Smithtown repeater, also on the same frequency pair, usually cannot hear stations 150 miles away but since the mobile is on a hill he is in the coverage area of both Jonestown and Smithtown. Whenever the mobile transmits both repeaters hear him.

The common solution to this problem, assuming it happens often enough, is to equip the Smithtown repeater with a CTCSS decoder and require all users of the repeater to transmit a CTCSS tone to access the repeater. Thus, the mobile station on the hill does not come through the Smithtown repeater, since he is not transmitting the required CTCSS tone.

Table 15.5 shows the available CTCSS tones. They are usually transmitted by adding them to the transmitter audio but at an amplitude such that they are not readily heard by the receiving station. It is common to hear the tones described by their code designators — a carryover from their use by Motorola in their commercial communications equipment.

Listings in the *ARRL Repeater Directory* include the CTCSS tone required, if any.

Frequency Coordination and Band Plans

Since repeater operation is channelized, with many stations sharing the same frequency pairs, the amateur community has

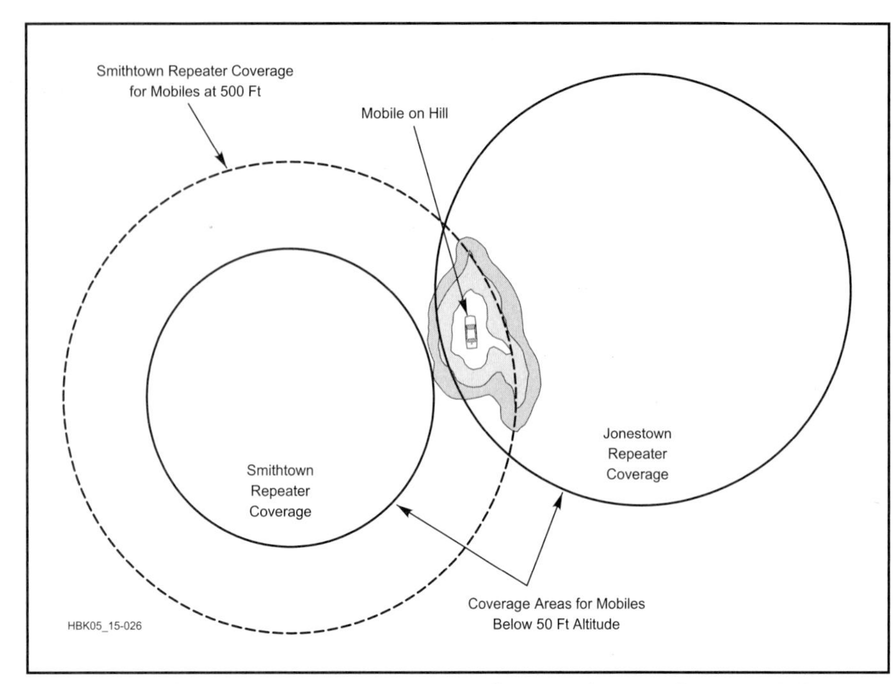

Fig 15.26 — When two repeaters operate on the same frequencies, a well-situated operator can key up both repeaters simultaneously. Frequency coordination prevents this occurrence.

Table 15.5
CTCSS (PL) Tone Frequencies

The purpose of CTCSS (PL) is to reduce cochannel interference during band openings. CTCSS (PL) equipped repeaters would respond only to signals having the CTCSS tone required for that repeater. These repeaters would not respond to weak distant signals on their inputs and correspondingly not transmit and repeat to add to the congestion.

The standard ANSI/EIA frequency codes, in hertz, with their Motorola alphanumeric designators, are as follows:

67.0 — XZ	91.5 — ZZ	118.8 — 2B	156.7 — 5A	179.9 — 6B	210.7 — M2
69.3 — WZ	94.8 — ZA	123.0 — 3Z	159.8	183.5	218.1 — M3
71.9 — XA	97.4 — ZB	127.3 — 3A	162.2 — 5B	186.2 — 7Z	225.7 — M4
74.4 — WA	100.0 — 1Z	131.8 — 3B	165.5	189.9	229.1 — 9Z
77.0 — XB	103.5 — 1A	136.5 — 4Z	167.9 — 6Z	192.8 — 7A	233.6 — M5
79.7 — WB	107.2 — 1B	141.3 — 4A	171.3	199.5	241.8 — M6
82.5 — YZ	110.9 — 2Z	146.2 — 4B	173.8 — 6A	203.5 — M1	250.3 — M7
85.4 — YA	114.8 — 2A	151.4 — 5Z	177.3	206.5 — 8Z	254.1 — 0Z
88.5 — YB					

formed coordinating groups to help minimize conflicts between repeaters and among repeaters and other modes. Over the years, the VHF bands have been divided into repeater and nonrepeater subbands. These frequency-coordination groups maintain lists of available frequency pairs in their areas. A complete list of frequency coordinators, band plans and repeater pairs is included in the *ARRL Repeater Directory*.

Each VHF and UHF repeater band has been subdivided into repeater and nonrepeater channels. In addition, each band has a specific *offset* — the difference between the transmit frequency and the receive frequency for the repeater. While most repeaters use these standard offsets, others use "oddball splits." These nonstandard repeaters are generally also coordinated through the local frequency coordinator. **Table 15.6** shows the standard frequency offsets for each repeater band.

The 10-m repeater band offers an additional challenge for repeater users. It is the only repeater band where ionospheric propagation is a regular factor. Coupled with the limited number of repeater frequency assignments available, the standard in this band is to use CTCSS tones on a regional basis. **Table 15.7** lists the coordinated tone assignments. As can be seen, 10-m repeaters in the 4th call area will use either the 146.2 or 100.0 (4B or 1Z) CTCSS tone.

Table 15.6
Standard Frequency Offsets for Repeaters

Band	*Offset*
29 MHz	100 kHz
52 MHz	1 MHz
144 MHz	600 kHz
222 MHz	1.6 MHz
440 MHz	5 MHz
902 MHz	12 MHz
1240 MHz	12 MHz

Table 15.7
10-M CTCSS Frequencies

In 1980 the ARRL Board of Directors adopted the 10-m CTCSS (PL) tone-controlled squelch frequencies listed below for voluntary incorporation into 10-m repeater systems to provide a uniform national system.

Call Area	*Tone 1*		*Tone 2*	
W1	131.8 Hz	-3B	91.5 Hz	-ZZ
W2	136.5	-4Z	94.8	-ZA
W3	141.3	-4A	97.4	-ZB
W4	146.2	-4B	100.0	-1Z
W5	151.4	-5Z	103.5	-1A
W6	156.7	-5A	107.2	-1B
W7	162.2	-5B	110.9	-2Z
W8	167.9	-6Z	114.8	-2A
W9	173.8	-6A	118.8	-2B
WØ	179.9	-6B	123.0	-3Z
VE	127.3	-3A	88.5	-YB

Chapter 16

DSP and Software Radio Design

Digital signal processing (DSP) is one of the great technological innovations of the last hundred years. It has found a permanent place not only in radio, but also in the exploration for oil and other fossil fuels, high-definition television (HDTV), compact-disc (CD) recording and many other facets of our lives. Its popularity stems from certain advantages: DSP filters do not need tuning and may be exactly duplicated from unit to unit; temperature variations are virtually non-existent; and DSP represents the ultimate in flexibility, since general-purpose DSP hardware can be programmed to perform many different functions, often eliminating other hardware. This chapter was written by Doug Smith, KF6DX.

DSP FUNDAMENTALS

In this chapter, you will see that DSP is about rapidly measuring analog signals, recording the measurements as a series of numbers, processing those numbers, then converting the new sequence back to analog signals. How we process the numbers depends on which of many possible functions we are performing. We will take a look at some of those functions and explore how real DSP systems are implemented in software and hardware.

Sampling

The process of generating a sequence of numbers that represent periodic measurements of a continuous analog waveform is called *sampling*. Each number in the sequence is a single measurement of the instantaneous amplitude of the waveform at a sampling time. When we make the measurements continually at regular intervals, the result is a sequence of numbers representing the amplitude of the signal at evenly spaced times.

This process is illustrated in **Fig 16.1**. Note that the frequency of the sine wave being sampled is much less than the *sampling frequency*, f_s. In other words, we are taking many samples during each cycle of

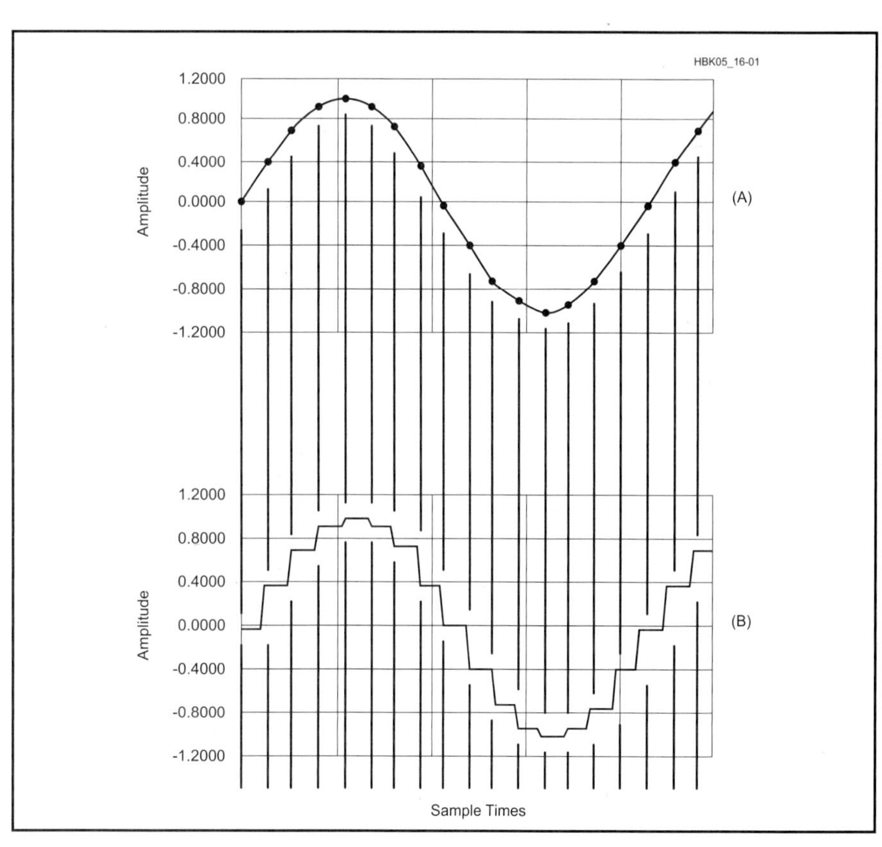

Fig 16.1—Sine wave of frequency much less than the sampling frequency (A). The sampled sine wave (B).

the sine wave. The sampled waveform does not contain information about what the analog signal did between samples, but it still roughly resembles the sine wave. Were we to feed the analog sine wave into a spectrum analyzer, we would see a single spike at the sine wave's frequency. Pretty obviously, the spectrum of the sampled waveform is not the same, since it is a step-wise representation.

The sampled signal's spectrum can be predicted and interpreted in the following way. The analog sine wave's spectrum is shown in **Fig 16.2A**, above the spectrum of the sampling function in Fig 16.2B. The sampled signal is just the *product* of the two signals; its spectrum is the *convolution* of the two input spectra, as shown in Fig 16.2C. The sampling process is equivalent to a mixing process: They each perform a multiplication of the two input signals.

Note that the sampled spectrum repeats at intervals of f_s. These repetitions are called *aliases* and are as real as the fundamental in the sampled signal. Each contains all the information necessary to fully describe the original signal. In general, we are only interested in the fundamental, but let's see what happens when the sampling frequency is *less than* that of the analog input.

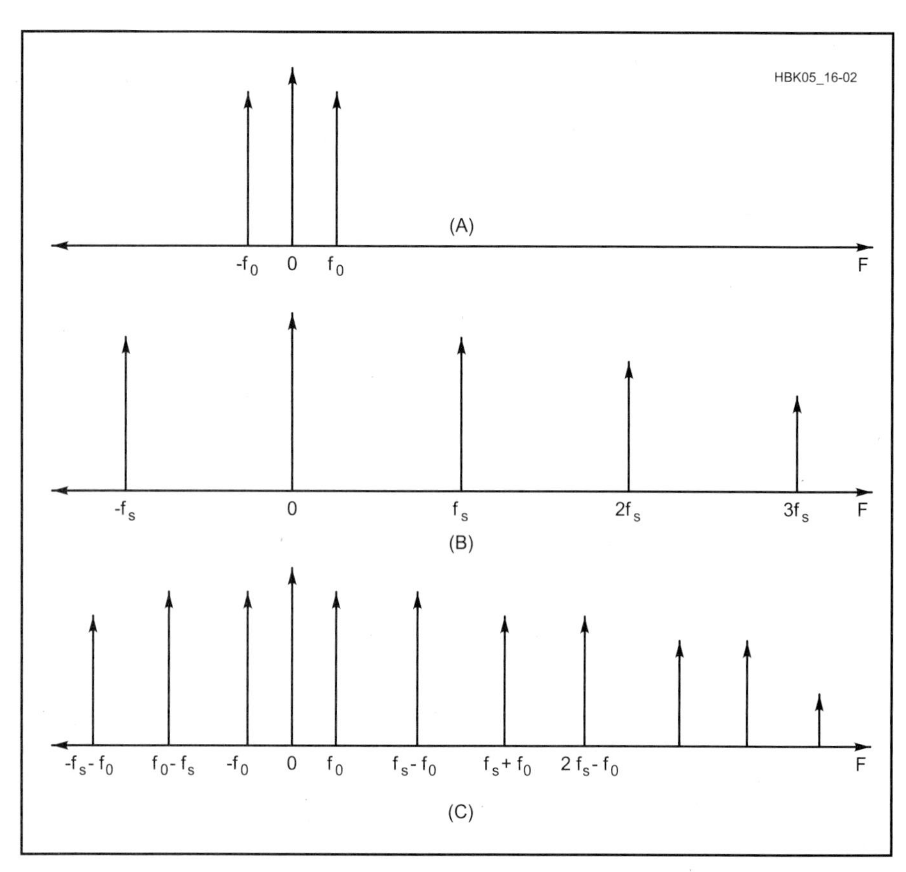

Fig 16.2—Spectrum of an analog sine wave (A). The spectrum of a sampling function (B). The spectrum of the sampled sine wave (C).

Sine Wave, Alias Sine Wave: Harmonic Sampling

Take the case wherein the sampling frequency is less than that of the analog sine wave. See **Fig 16.3**. The sampled output no longer matches the input waveform. Notice that the sampled signal retains the shape of a sine wave at a frequency lower than that of the input. Ordinarily, this would not be a happy situation.

A downward frequency translation is useful, though, in the design of IF-DSP receivers. In addition, lower sampling frequencies are good because they allow more time between samples for signal processing algorithms to do their work; that is, lower sampling rates ease the processing burden. Caution is required, though: An input signal near twice the sampling frequency would produce the same output as that of Fig 16.3. To use this technique, then, we must first limit the bandwidth (BW) of the input: A band-pass filter (BPF) is called for. This is known as *harmonic sampling*. The BPF is referred to as an *anti-aliasing* filter.

Input signals must fall between the fundamental (or some harmonic) of the sampling frequency and the point half way to the next higher harmonic. A frequency translation will take place, but no information about the shape of the input signal will be lost. A spectral representation of harmonic sampling is

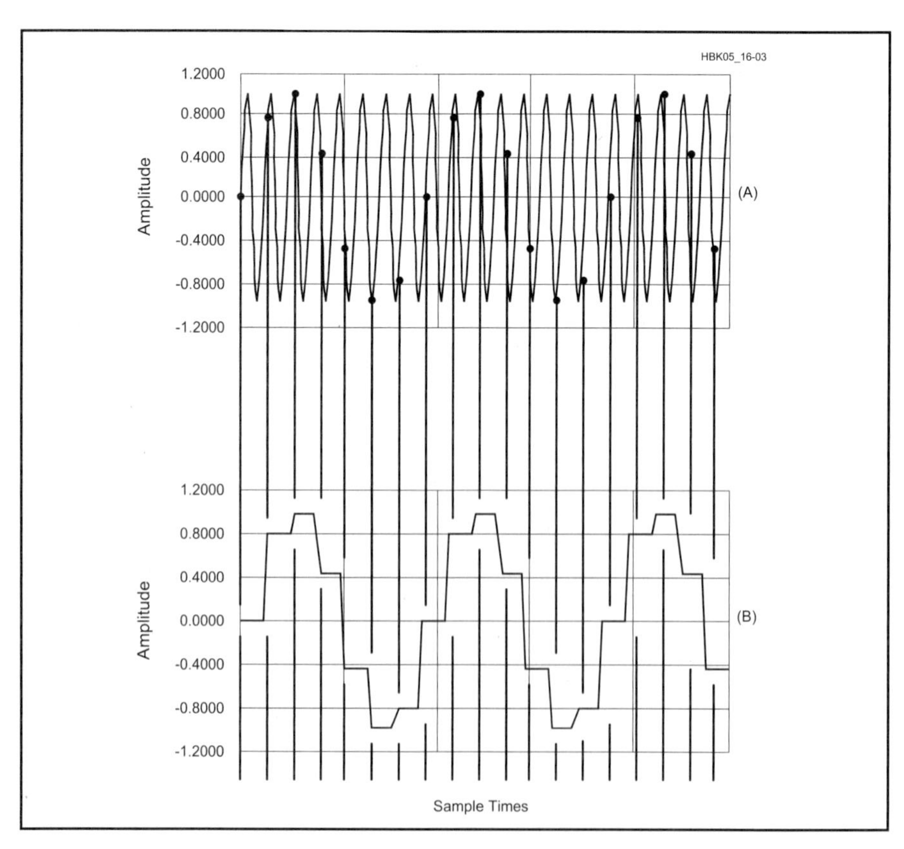

Fig 16.3—Sine wave of frequency greater than the sampling frequency (A). Harmonically sampled sine wave (B).

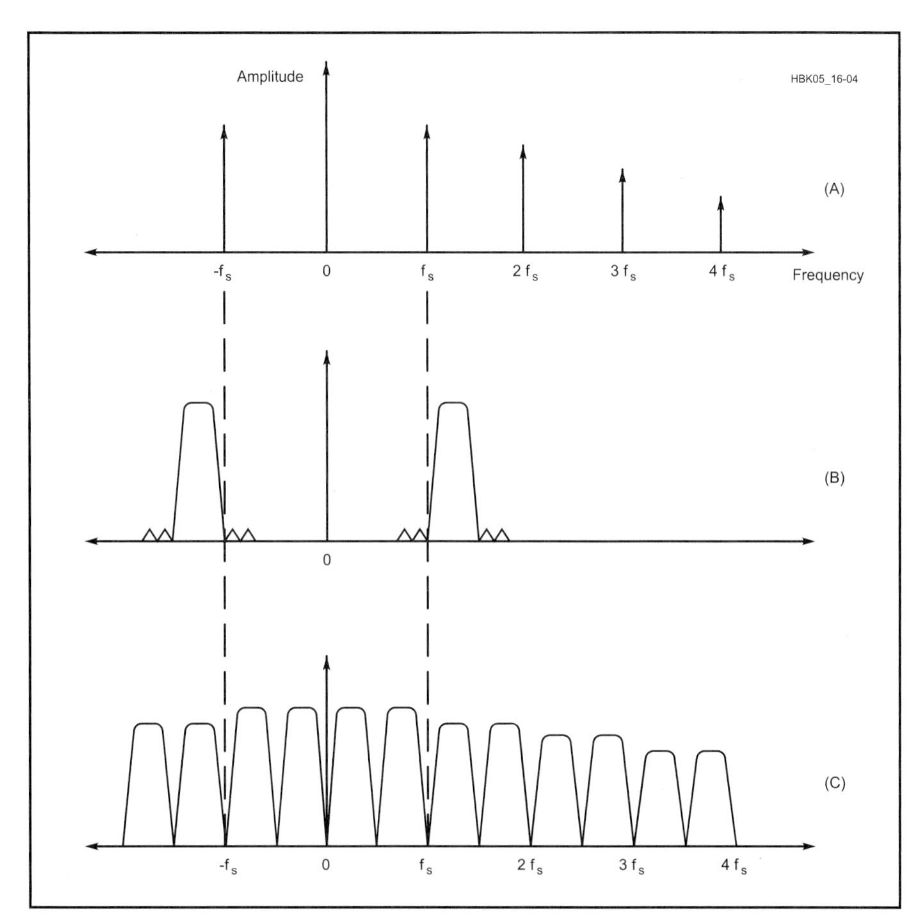

Fig 16.4—Spectrum of a sampling function (A). Spectrum of a band of real signals (B). Spectrum of a harmonically sampled band of real signals (C).

shown in **Fig 16.4**. It reveals the basis for the often-misquoted *Nyquist sampling theorem*: The sampling frequency must be at least twice the input BW to avoid aliasing. Such aliasing would destroy information; once incurred, nothing can remedy it.

Data Converters and Quantization Noise

The device used to perform sampling is called an *analog-to-digital converter* (ADC). For each sample, an ADC produces a binary number that is directly proportional to the input voltage. The number of bits in its binary output limits the number of discrete voltage levels that can be represented. An 8-bit ADC, for example, can only give one of 256 values. This means the amplitude reported is not the exact amplitude of the input, but only the closest value of those available. The difference is called the *quantization error.*

The amplitude reported by the ADC can, therefore, be thought of as the sum of two signals: the desired input and the quantization error. In a perfect ADC, the error cannot exceed ±½ of the value of the least-significant bit of the converter—this is the error signal's peak-to-peak amplitude. Assuming the desired input is changing and covers a large range of quantization levels, the error is just as likely to be negative as positive, and just as likely to be small as large. Hence, the error signal is pseudo-random and appears as *quantization noise*.

This noise is spread uniformly over the entire input BW of $f_s/2$. Taking this and the maximum signal the ADC can handle into account, the maximum signal-to-quantization-noise ratio produced by the ADC is:

$$SNR_{max} \approx 6.02b + 1.76\,dB \quad (1)$$

where b is the number of bits used by the converter.

For a simple 16-bit ADC, the SNR cannot exceed about 98 dB. The reason we wrote that the quantization noise was pseudo-random and not truly random is the following: If there were a harmonic relationship between the input signal and the sampling frequency, the noise might tend to concentrate itself at discrete frequencies.

Aperture Jitter

In addition to quantization noise, noise is introduced in ADCs by slight variations in the exact times of sampling. Phase noise in the ADC's clock source, as well as other inaccuracies in the sampling mechanisms, produce undesired phase modulation of the sampled signal. Again, assuming it is uncorrelated with the input signal, this *aperture jitter noise* will be distributed across the entire input BW. Its amplitude is proportional to the squares of both the desired signal's frequency and the RMS time jitter in the sampling rate, and inversely proportional to the sampling rate itself. With contemporary crystal-derived clock sources, aperture jitter is usually not a significant factor until the sampling frequencies reach VHF; even at those frequencies, the effect may be small compared with quantization noise.

Over-Sampling and Sigma-Delta ADCs

The nature of the above-mentioned noise sources is such that if we could increase the sampling frequency by some factor N, then digitally filter the output back down to a lower rate, we could improve the SNR by almost the factor N. This is because the noise would be spread over a larger BW; much of the high-frequency noise would be eliminated by the digital filter. This technique is called *over-sampling.*

So-called *sigma-delta* converters use this method to achieve the best possible dynamic range. They employ one-bit quantizers at very high speed and digital decimation filters (described later) to reduce the sampling frequency, thus improving SNR. They represent the state of the art in ADC technology. Other factors, such as the noise figure of analog stages inside an ADC, tend to limit the SNR of real converters to within a few dB of that calculated by Eq 1.

Non-Linearity in ADCs

The quantization steps of a real converter are not perfectly spaced; conversion results are contaminated by the inaccuracy. In general, two types of non-linearity are characterized by manufacturers: *differential non-linearity* (DNL) and *integral non-linearity* (INL).

DNL is the measure of the output non-uniformity from one input step to the next. It is expressed as the maximum error in the output between adjacent input steps as measured over the entire input range of the device. The worst errors usually occur near the middle of the scale. Since we are talking about the accuracy of the smallest steps the converter can resolve, noisy low-order distortion products caused by this effect limit dynamic range. Current technology uses correction systems to compensate for temperature variations that would otherwise further degrade performance.

An ADC is considered *monotonic* if a steady increase in the input signal always results in an increase in the output. Device manufacturers hold DNL to ±0.5 bits or better so that monotonicity is maintained.

INL is a measure of an ADC's large-signal handling capability. To measure it, we first inject a signal of amplitude A and measure the output; when we inject a signal of amplitude 100A, we expect the output to grow in exact proportion. INL represents the maximum error in the output between *any two* input levels. Another way to think about this is to plot the input against the output and see how straight the line is. INL produces harmonic distortion and IMD; values for typical converters are ±1 or 2 bits over the entire range.

Spurious-Free Dynamic Range and Dithering

Spurious-free dynamic range (SFDR) is defined as the ratio of the largest signal the converter can accurately handle to the largest source of noise and distortion caused by effects mentioned above. Quite often, undesired components may appear in unexpected parts of the input spectrum; spurious responses may be found without apparent explanation. It turns out there are explanations, of course, but we will defer that discussion. Suffice it to write here that manufacturers test for SFDR and usually specify it on their data sheets, especially for high-speed devices.

Sometimes noise and distortion effects conspire to add at discrete frequencies. It is found that the addition of random noise at the clock input helps dissipate these spurious responses. This technique is known as *dithering*. It may seem strange, but artificial noise—usually several bits in amplitude and high enough in frequency to be eliminated by the decimation filter—actually reduces quantization noise and improves performance rather than degrading it.

Digital-to-Analog Converters: Additional Distortion Sources

Digital-to-analog converters (DACs) perform the conversion of binary numbers back into analog voltages—the reverse operation of ADCs. They suffer from all the inadequacies described earlier, as well as a few of their own. The first unique distortion of DACs is one of frequency response: *zero-order sample-and-hold distortion.*

Typical converters are sample-and-hold devices: They continue to output the last sampled value throughout the sample period. This effect acts as a low-pass filter having a frequency response:

$$H_r = \frac{\sin\left(\frac{\pi f}{f_s}\right)}{\left(\frac{\pi f}{f_s}\right)} \qquad (2)$$

Note the classical (sin x)/x form. The high-frequency roll-off is quite undesirable in many circumstances. For example, if the output frequency is one quarter the sampling frequency, an attenuation of about 1 dB will occur. Correction can be made for this, but an increase in sampling frequency reduces the attenuation. Interpolation of the sampled output signal (described later) is called for in many cases.

Settling Time and Glitch Energy

When the output of a DAC changes from one voltage to another, it obviously cannot do so instantaneously; a finite time is required for the voltage to reach its new value. This is known as the *settling time.* It is usually defined as the time required to settle to within some number of voltage-equivalent bits of the final value.

Glitch energy or *glitch area* is defined as the product of the voltage error during the settling time and the settling time itself. While volt-seconds are not units of energy, it is assumed the DAC is driving some kind of load; thus, these units can be translated into units of energy (watt-seconds), performing work on that load. The settling mechanism is an important factor in the production of spurious outputs in DACs. Manufacturers usually specify the glitch energy for their high-speed devices. It is an especially important number for direct-digital-synthesis (DDS) applications.

Note also that DACs produce aliases, again repeating at intervals of f_s. These must usually be removed using an analog LPF. Occasionally, a BPF may be used, and one of the aliases taken as the desired output. This can be a clever way of getting an upward frequency translation under certain conditions.

Reducing the Sampling Frequency: Decimation

As we have seen, sampling at high rates is beneficial because it eases the design of the analog filters we must use to avoid aliasing. It also reduces quantization noise and aperture jitter. We have also noted that lower sampling rates help reduce the computational burden in DSP systems. In addition, we will discover that when it is time to digitally filter some signals, making the filter's BW a large fraction of the sampling frequency makes it easier to build sharp-skirted filters—exactly what DSP is famous for.

Reduction of the sampling frequency is usually called *decimation*. Decimation is normally done by integer factors (although it does not have to be) and is equivalent to resampling an already-sampled signal at a lower rate. The resampled signal has a family of aliases, repeating at intervals of the lower sampling frequency; we have to reduce the BW to less than half this lower sampling frequency to avoid the aliasing that would destroy information.

The process of decimation is simple: Just throw away the unwanted samples. To decimate by two, for example, only every other sample is retained. A *decimation filter*, operating, at the higher sampling rate, f_s, reduces signal BW to less than $f_s/4$ prior to discarding the samples to avoid aliasing. But why spend time computing filter outputs that we are only going to discard? We may compute only those we intend to keep. This is exactly the same as running the decimation filter at the lower rate. This method is typical of those used by DSP designers to save time and effort. See the chapter Appendix for a software project (Project A) that demonstrates decimation using Alkin's *PC-DSP* program. This program is included with the book listed in the **Bibliography**.

Increasing the Sampling Frequency: Interpolation

We learned that when it is time to convert back to analog, an artificial increase in sampling rate may be advantageous. It will push aliases higher in frequency where they are easier to remove by analog filtering, and it will relieve some of the sample-and-hold distortion. So, even having decimated the data at some earlier stage in our designs, we may later employ the process of *interpolation.*

Decimation was performed by deleting samples. Interpolation is performed by inserting them. The inserted samples have a value of zero and are placed between the existing samples. While this increases the sampling frequency, the information in the original samples is not destroyed; however, new information is added in the form of aliases, and an *interpolation filter* is usually required. This filter, most often a low-pass, operates at the higher sampling frequency, f_s, and eliminates components in the interpolated data above half the original sampling frequency.

The way numbers are represented in DSPs is a major consideration. Let's take a look at this before moving on to filtering algorithms.

Representation of Numbers: Floating-Point vs Fixed-Point

One of the things that makes general-purpose computers so useful is their ability to perform *floating-point* calculations. In this form of numeric representation, numbers are stored in two pieces: a fractional part, or *mantissa*, and an exponent. The mantissa is assumed to be a binary number representing an absolute value less than unity, and the exponent, a binary integer. This approach allows the computer to handle a large range of numbers, from very small to very large. Some DSP chips support floating-point calculations, but it is not as great an advantage in signal processing as it is in general-purpose computing because the range of values we are dealing with in DSP is limited anyway. For this reason, *fixed-point* processors are common in DSP.

A fixed-point processor treats numbers as just the mantissa and does away with the exponent. The radix point—the separation between the integer and fractional parts of a number—is usually assumed to reside to the left of the most-significant bit. This is convenient, since the product of two fractions less than unity is always another fraction less than unity. The *sum* of two fractions, though, may be greater than unity: *overflow* would be the result. Overflow is a constant concern for fixed-point DSP programmers and leads to considerations for *scaling* of data, as discussed further below, which may limit system dynamic range to less than the data converters' capabilities.

DSP ALGORITHMS FOR RADIO

Digital Filters

The ability to construct high-performance filters is probably the most important rationale for using DSP in radio transceivers. An expensive crystal or mechanical filter having a single BW can be replaced by a set of superior digital filters, offering as many BWs as the associated on-board memory can support.

As shape-factor requirements get more stringent, filters get more complex. As a filter gets more complex—with additional inductors and capacitors in the analog case, or additional delay elements in the digital case—the sensitivity of the filter's response to errors in the element values becomes more severe. Thus, for analog filters, precise values of resistance, inductance and capacitance must be maintained if the filter is to operate as designed. Establishing those values is difficult; holding them within tolerances over temperature variations and aging is more so. DSP filters, on the other hand, are unchanging. The "component" values are numbers stored in a computer that are not susceptible to temperature changes or aging. Filters that would be impractical or impossible in the analog realm are easily implemented by DSP algorithms.

We can build digital filters having linear phase responses, which is very difficult in the analog world. This is an advantage mainly for digital communication modes such as FSK and PSK. Also, filters may be combined numerically to yield composite responses without the need for adding hardware. This is useful for passband tuning or graphic-equalizer applications.

DSP filters are usually characterized by their *impulse responses*. The impulse response of a digital filter is the output of the filter when the input is a one-sample, unity-amplitude impulse. Impulse response is directly related to frequency response by a *Fourier transform*, about which we will learn more later. Suffice it to write for now that digital filters may be broadly divided into two classes: finite impulse response (FIR) and infinite impulse response (IIR). The presence or absence of feedback separates the two.

FIR Filters

Take a look at the block diagram of the FIR filter shown in **Fig 16.5**. The string of boxes labeled z^{-1} is simply a delay line, with each box representing a one-sample delay. Programmers will note that with one input sample in each position, this is just a buffer of length five. Each buffer location may be referred to as a *tap* in the delay line. The datum at each tap, x(n), is multiplied by one of the filter *coefficients*, h(n). All the products are summed at each sample time to produce the filter output. At the next sample time, samples are shifted down the delay line by one position and the *multiply-and-accumulate* (MAC) operation is performed again. Coefficients remain in place and do not shift. The mathematical expression describing this repetitive MAC operation is also called a *convolution sum*:

$$y(k) = \sum_{n=0}^{L-1} h(n)\, x(k-n) \qquad (3)$$

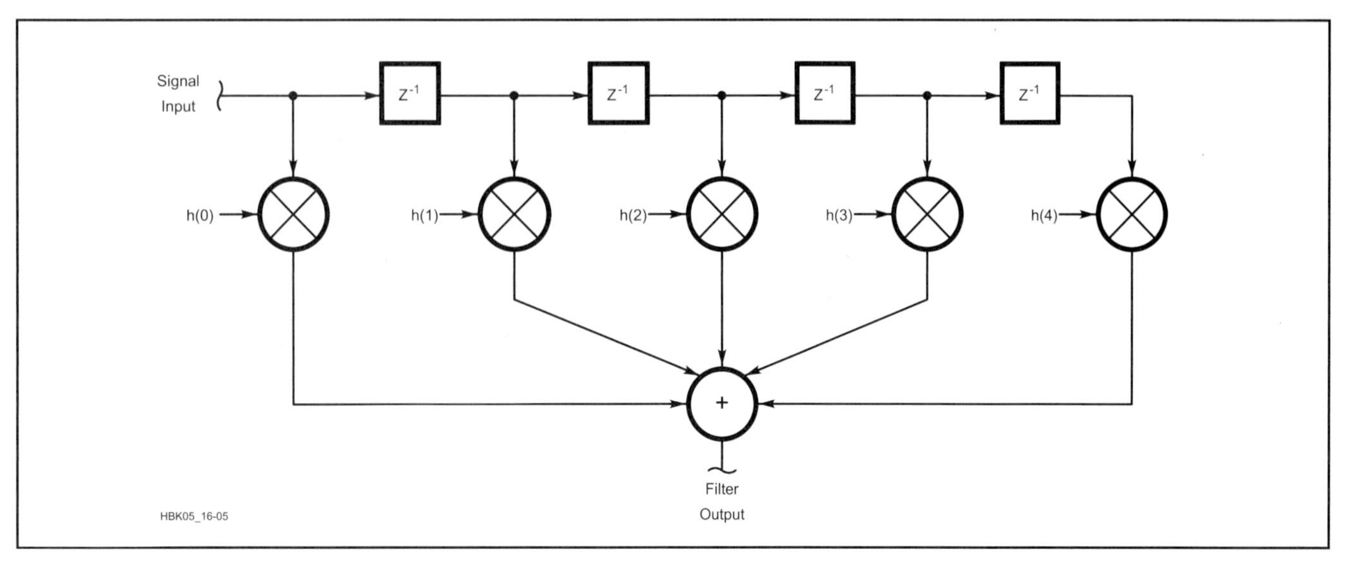

Fig 16.5—Block diagram of an FIR filter for L = 5.

where x(k-n) represents the input data in the buffer.

Since the output depends only on past input values, the filter is said to be a *causal process*. Since no feedback is employed, it is unconditionally stable.

In an FIR filter, the set of coefficients, h(n), is identical to the impulse response of the filter. The trick, then, is to find the impulse response that gives us the frequency response we want. Almost any frequency response can be generated if we use enough taps. In general, low shape factors (steeper roll-offs) require more taps. Most filter-design methods begin with an estimate of the number of taps needed. Rabiner and Gold indicate the estimate may be taken as:

$$L = 1 - \frac{10 \log(\delta_1 \delta_2) - 15}{14\left(\frac{f_T}{f_s}\right)} \quad (4)$$

where δ_1 is the passband ripple, δ_2 is the stopband attenuation, f_T is the transition BW (the bandwidth between the edge of the passband and the edge of the stopband (ie, the filter skirt), f_s is the sampling frequency, and L, the number of taps, is called the *length* of the filter. This equation assumes that enough bits of resolution are used to achieve the required accuracy. In practice, filters of over 100 taps are used to realize shape factors of less than 1.15:1.

Normally, an FIR filter's impulse response has a symmetry about center; that is, h(0) = h(L–1), h(1) = h(L–2), and so forth. It turns out this is sufficient to ensure a linear phase response and flat group-delay characteristics. The total delay through an FIR filter of length L is:

$$t = \frac{L}{2f_s} \quad (5)$$

As noted, this delay is *independent* of frequency. Remember that longer filters demand more processing than shorter filters.

When personal computers are used to design FIR filters, coefficients are usually represented in floating-point format to the full accuracy of the computer—often with 12 or more decimal digits in the mantissa. Embedded, fixed-point DSP implementations ordinarily achieve only 16-bit accuracy. The *truncation* of coefficients and data to this accuracy affects the frequency response and ultimate attenuation of filters, and may be the factor that determines dynamic range. Also notice that when we multiply a 16-bit coefficient by a 16-bit datum, the product is a 32-bit number. We are then adding several 32-bit numbers in the final accumulator of an FIR filter. The result may grow by several more bits to 35 or so by the time we are done. At some stage, the result may overflow the accumulator, especially in FIR filters with small transition BWs (sharp skirts). The worst-case output can grow as large as the sum of the absolute value of all the coefficients:

$$y_{max} = \pm \sum_{n=0}^{L-1} \left| h(n) \right| \quad (6)$$

We might have to scale the data, the coefficients, or both by the reciprocal of this number to avoid overflow.

The filter output at each sample time is usually rounded back down to the bit-resolution of the DAC; say, to 16 bits. The rounding operation introduces a small error in the result. This rounding error is directly analogous to quantization noise; it is computed in almost exactly the same way. A trade-off exists between the possibility of overflow, which is catastrophic, and loss of accuracy because of rounding. It is interesting to note that truncation of filter coefficients affects the frequency response of the filter but not the amount of noise in the output. On the other hand, truncation and rounding of data do not affect the frequency response but add quantization noise to the output.

One FIR filter-design approach takes advantage of the fact that a filter's frequency response is the Fourier transform of its impulse response. Thus, we may start with a sampled version of the frequency response and apply an *inverse* Fourier transform to obtain the impulse response. All filter-design software is capable of using this method. Better designs may be obtained in many cases by using an algorithm developed by Parks and McClellan. This approach produces an *equi-ripple* design in which all of the passband ripples are the same amplitude, as are all the stopband ripples. Another popular algorithm is the *least-squares* method. Its claim to fame is that it minimizes the error in the desired frequency response.

Since finding coefficient sets for a given filter design is so computationally intensive, it is a good job for a computer program. DSP filter-design programs are readily available at low cost. Refer to the **DSP System Software** section toward the end of this chapter for further discussion of filter design and the **Bibliography** for a list of software design tools. The article by Kossor has a practical circuit example of a commutating BPF that employs principles of DSP. Also see Project B in the chapter **Appendix** for examples of FIR filter designs.

IIR Filters

While FIR filters have a lot going for them, they tend to require a large number of taps and a proportional amount of processing power. As opposed to that, an IIR (infinite impulse response) filter can provide sharp transition BWs with relatively few calculations. What it will not provide, in general, is a linear phase response. In circumstances where the computational burden is of more concern than the phase response, IIR filters may be desirable.

Unlike FIR filters, IIR filters employ feedback: That is what makes their impulse responses infinite. For this reason, IIR filters are usually designed by converting traditional analog filter designs, such as Chebyshev and elliptical types. See the **RF and AF Filters** chapter of this book for a description of those designs. The transfer function of an analog Chebyshev low-pass filter can be written as the ratio of a constant to an nth-order polynomial:

$$H_S = \frac{K}{a_0 s^n + a_1 s^{n-1} + a_2 s^{n-2} + \cdots + a_n s^0} \quad (7)$$

Tables in the literature, such as in Zverev, list the values of the coefficients, a_n, related to the cutoff frequency; these are used to derive actual component values for the filter. The low-pass design can be transformed to band-pass or band-stop responses. Two popular methods exist for deriving the digital transfer function from the analog: These are known as the *impulse-invariant* and *bilinear transform* methods.

The impulse-invariant method assures that the digital filter will have an impulse response equivalent to its analog counterpart, and thus the same phase response. Problems arise, though, if the bands of interest are near half the sampling frequency; the digital filter's response can develop serious errors in this case. Because of this problem, the impulse-invariant method is not as good as the bilinear transform method. As indicated by Sabin and Schoenike, the bilinear transform method makes a convenient substitution for s in Eq 7 above. The filter output comes out as:

$$y(k) = \sum_{n=0}^{L-1} \alpha(n)\, x(k-n) - \sum_{n=1}^{L-1} \beta(n)\, y(k-n) \quad (8)$$

This filter has L zeros and L–1 poles. The block diagram of such a filter for L = 5 is shown in **Fig 16.6**. Feedback is evident in the diagram: The paths involving coefficients β loop back and are added to

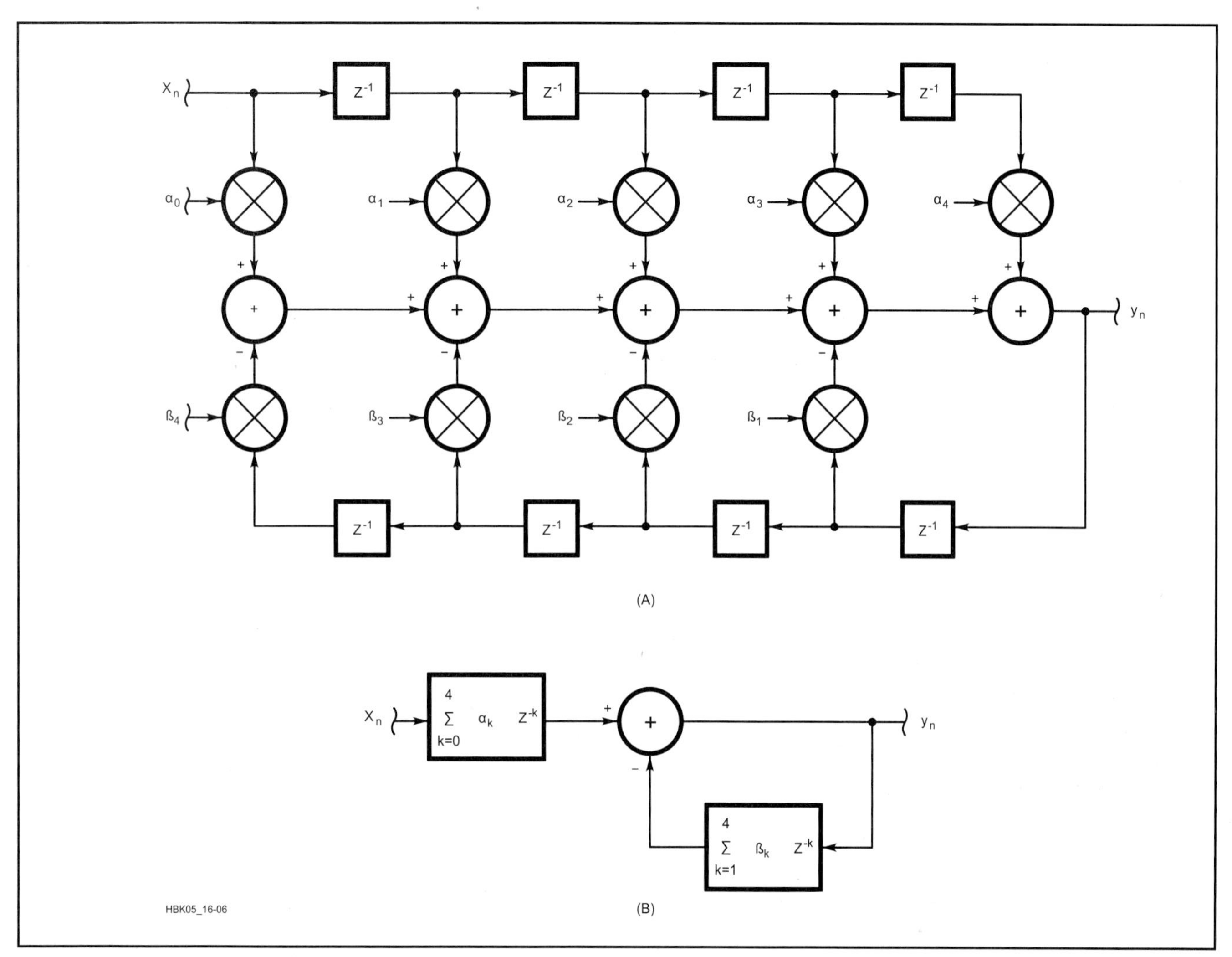

Fig 16.6—Block diagram of an IIR filter for L = 5.

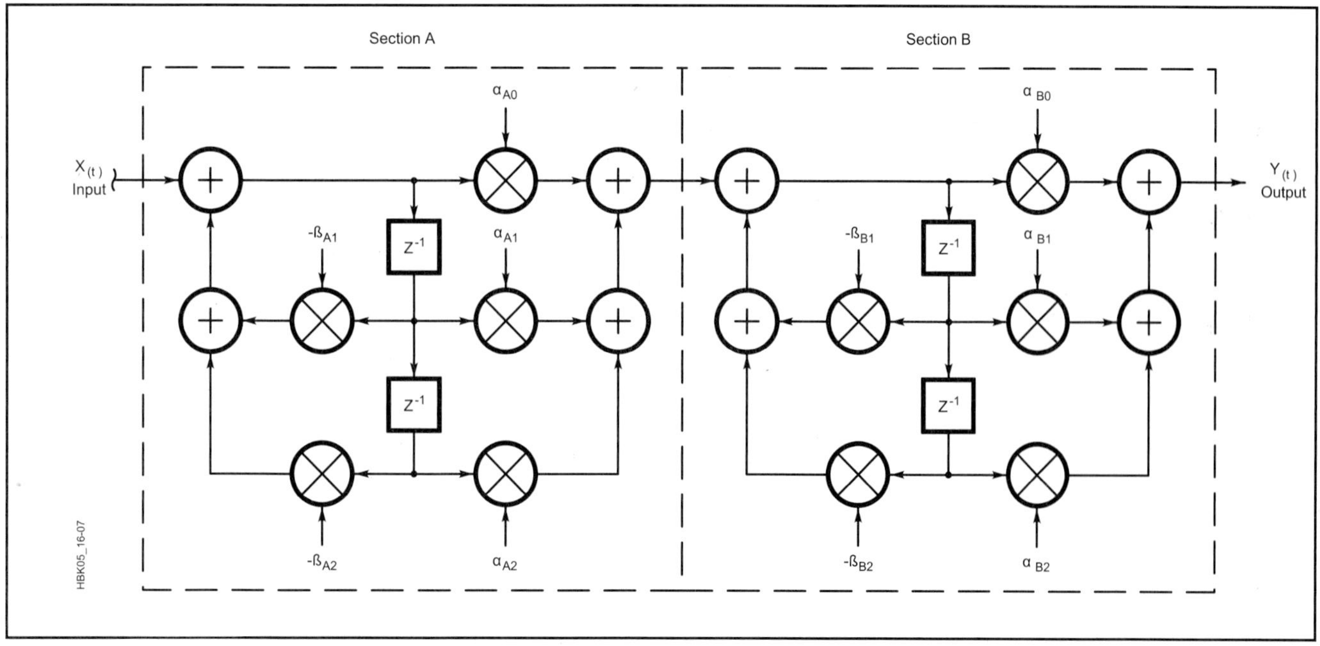

Fig 16.7—Block diagram of a cascade-form IIR filter.

the signal path.

The *direct form* of Eq 8 may be factored into 2-pole sections and implemented in cascaded form. The output of each section serves as the input to the next. See **Fig 16.7**. This configuration requires a few more multiplications than the direct form, but suffers less from instability problems that may plague IIR filters. Since feedback is being used, IIR filters are not necessarily unconditionally stable. They also tend to be prone to *limit cycles*, low-level oscillations that arise near the lower end of the dynamic range. For these and other reasons, data and coefficient storage should be cleared or set to zero before processing begins.

A Simple Digital Notch Filter

Along with common LPFs, HPFs and BPFs, radio designers are interested in one other type of filter, the *notch*. While most filter-design software can generate notch filters using FIR methods discussed above, Widrow and Stearns have described an unusual type in which the number of taps is minimized. In fact, they were able to prove that only two taps are needed for each frequency to be notched. This is great, since it reduces computation to almost nil. We will take a look at it here and touch briefly on some of the theory of *adaptive signal processing*, treated in depth later.

The situation is this: We want to copy a broadband signal, such as an SSB phone signal, and suddenly a dreadful carrier appears in the passband. Our notch filter will remove it and we will have complete control over the notch width, as well as a notch depth limited only by the bit resolution of our system. Dr Widrow found that one can build a filtering system that minimizes repetitive signal energy by altering the filter coefficients "on the fly" using a certain algorithm. Known as the *least-mean-squares* (LMS) method, it describes a way to adjust filter coefficients over time to remove undesired, steady tones in the input. A complex reference signal is used at the exact frequency of the offending tone. The algorithm then forms a BPF centered at the tone frequency whose output is subtracted from the input to create the notch. The block diagram of a two-tap system is shown in **Fig 16.8**.

The broadband input is called x(t). The reference input consists of two signals, cos $(\omega_0 t)$ and sin $(\omega_0 t)$. These signals feed multipliers having coefficients h(1) and h(2), which in turn feed an accumulator just as in a normal FIR filter. This is the BPF output; it is subtracted from the input to form the notch output, e(t). Note that the BPF output is also available at no additional overhead. While the initial values of the coefficients are unimportant to the steady state, the procedure for updating them with the LMS algorithm is:

$$h_{t+1}(1) = h_t(1) + 2\mu e(t) x_t(1)$$
$$h_{t+1}(2) = h_t(2) + 2\mu e(t) x_t(2) \qquad (9)$$

where $0 < \mu < 1$. Analysis shows that as the reference inputs are sinusoidal, the system is linear and time-invariant for output e(t), although the coefficient values do not necessarily approach any fixed value. The 3-dB BW of the notch is:

$$BW = \frac{2\mu A^2}{t_s} \text{ rad/s} \qquad (10)$$

The Q of the filter may be readily computed. Thus, we have control over the BW by varying the factor μ and the amplitude of the reference signal. The depth of the null is, in general, superior to that of a fixed filter because the algorithm tracks the correct phase relationship for ideal cancellation, even if the reference frequency is changing slowly with the offending tone. Each additional tone to be notched demands two additional taps in the filter. Noise in the input may cause us to have to add more taps to maintain sufficient accuracy. Additional detail of adaptive signal processing will be found below and in material shown in the **Bibliography**.

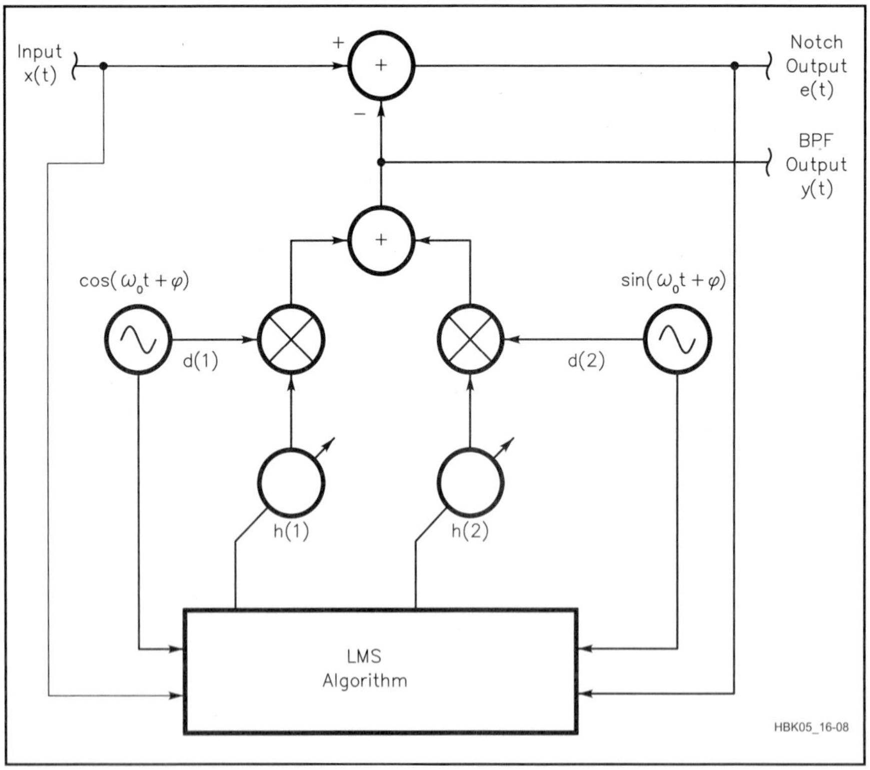

Fig 16.8—Block diagram of a two-tap, adaptive notch filter.

Lattice and Other Structures

While many filter-design software packages do not have the capability to work with them, *lattice structures* and other types of digital filters have seen use, especially in adaptive signal processing. Crystal and mechanical lattice filters are common elements of many transceivers. A digital lattice or ladder filter is a lot like its analog brother. The design of digital lattice filters is similar as well. Digital lattice filters may be either FIR or IIR. Also note that from the IIR cascade form above, we can derive a *parallel form* that may be computationally beneficial in some cases. The design of this kind of filter is a very complicated session in partial fraction expansion. Widrow and Stearns provide more information on these and other exotic concepts.

ANALYTIC SIGNALS AND MODULATION

DSP implementations of transceiver functions, such as modulation and demodulation, compel designers to examine the mathematics behind them. Computers are good at crunching numbers, but they do exactly what they are told! If we expect a DSP system to generate an SSB signal, for example, we had better know which calculations to perform and which to avoid.

Mathematics of Complex Signals

Because DSP makes it easy to build frequency-independent phase shifters—a fantasy in the analog world—the *phasing* or "I/Q" method has dominated other modulation techniques. Complex signals are not generally well understood and quite often form a stumbling block to those wishing to grasp DSP concepts. The idea of negative frequency is especially troublesome. The key to understanding these concepts lies in the theory of complex numbers. A real signal, such as a cosine wave, is normally thought of as a positive frequency. It can be transmitted and detected normally; however, we shall see that such a signal actually consists of positive *and* negative frequencies when examined in the complex domain.

A real cosine wave embodies the relation:

$$x(t) = \cos(wt) \tag{11}$$

where $\omega = 2\pi f$ and t is time. In the complex domain, the cosine wave is really the sum of two complex signals:

$$x(t) = \frac{1}{2}\left\{\begin{matrix}[\cos(\omega t) + j\sin(\omega t)] + \\ [\cos(\omega t) - j\sin(\omega t)]\end{matrix}\right\} \tag{12}$$

This signal has both positive and negative frequency components. The real parts add and the imaginary parts cancel to make the equation true. In the complex plane, where the real part is one axis and the imaginary part the other, this signal can be represented as two vectors rotating in opposite directions. See **Fig 16.9**.

While this depiction is beautiful and elegant to the mathematician, what does it really mean to you and me? Well, it means that signals represented in complex form can have a one-sided spectrum—that is, only a positive or a negative frequency component. This is useful as we mix our signals upward to their final frequency positions in a modulator.

As our first example, let's select the task of taking a real input signal, such as the audio from a microphone, and converting it to an SSB signal that can be transmitted. We obviously have to translate the audio signal upward in frequency and preserve its spectral content within the band we want the transmitted signal to occupy. If we wish to produce an upper-sideband (USB) signal, we want the carrier and lower sideband to be suppressed as much as possible. Were we able to translate the spectrum of our cosine wave—with its symmetrical positive- and negative-frequency components—upward in frequency far enough, we would have two positive frequencies separated by twice the original signal's frequency. For a real signal, this is exactly what happens when it is applied to an analog mixer: Both sum and difference frequencies are generated. See the **Receivers, Transmitters and Transverters** chapter for more detail of the operation of mixers as multipliers.

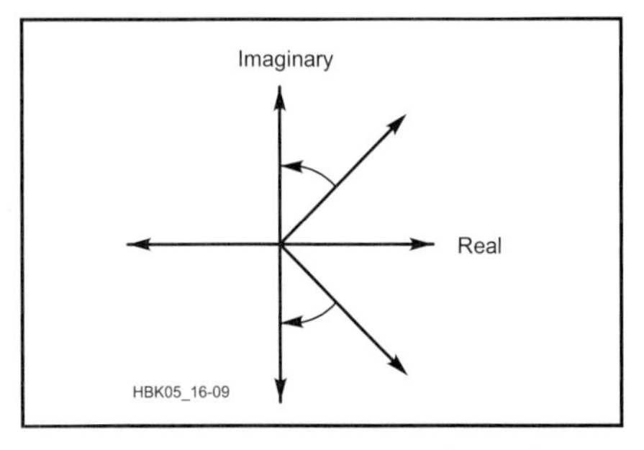

Fig 16.9—Vector representation of a real cosine wave.

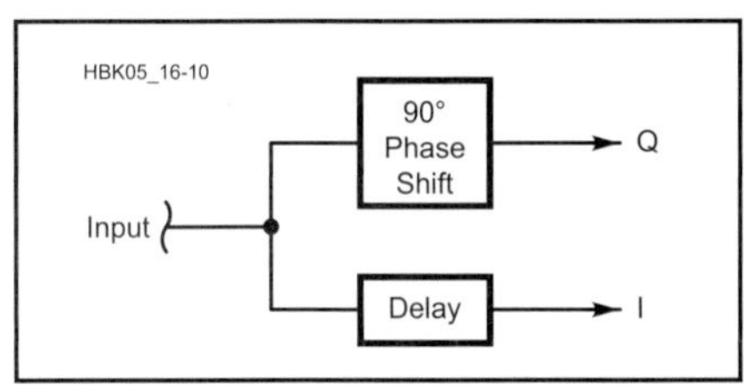

Fig 16.10—Hilbert transformer producing an analytic signal.

To move our sampled audio signal upward in frequency, we must multiply it by (mix it with) a local oscillator. The local-oscillator function can be implemented in DSP software using direct digital synthesis (DDS) techniques. In this case, though, the local oscillator must be complex; that is, it must have two outputs with a 90° phase relationship between them. This is the same as saying there must be both a sine and a cosine output from it. This will enable us to mix signals having a one-sided spectrum.

When we implement a complex mixer in DSP, we are multiplying complex numbers by complex numbers. Note that the calculations for the real and imaginary parts are carried out separately; each part is treated as if it were a single, real multiplication. Two complex numbers a + *j* b and c + *j* d, when multiplied, produce:

$$(a + jb)(c + jd) = (ac - bd) + j(ad + bc) \tag{13}$$

Four real multiplications and two real additions are required.

Hilbert Transformers and an SSB Modulator

If we want to create a signal having a one-sided spectrum from a real input signal, such as from the microphone, we need to shift all the frequency components in

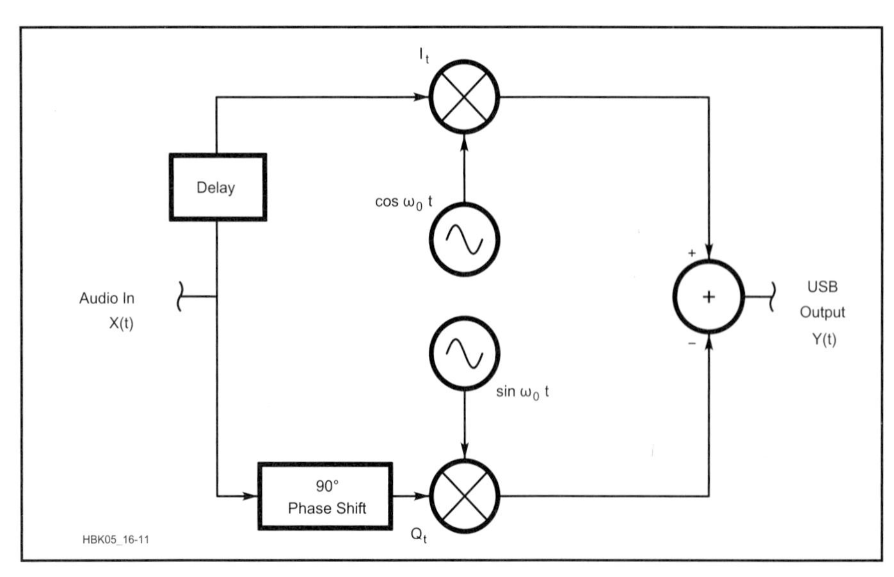

Fig 16.11—Block diagram of a half-complex mixer.

the sampled signal by 90°. Fortunately, in DSP, we have a way to do that: the *Hilbert transformer*. Recall that an FIR filter with a symmetrical impulse response exhibits a constant, frequency-independent delay. It turns out a filter with an *anti-symmetrical* impulse response—that is, with h(0) = –h(L –1), h(1) = –h(L –2), and so forth—produces a linear phase response, too, but with a phase response exactly 90° different from the symmetrical-impulse-response filter. This is exactly the type of filter we need to generate the components of an *analytic signal*.

Fig 16.10 shows a system using a Hilbert transformer to create an analytic signal from the microphone audio. Since the Hilbert transformer includes not only a 90° phase shift, but also a fixed delay of L/2 sample periods, we need an L/2 delay in the leg that does not contain a phase shift. The delay through the two paths is then equal and the only difference between the two signals produced is the 90° phase shift. The non-phase-shifted signal is called I, the phase-shifted signal is called Q. Together, these signals form our analytic signal I + *j* Q. Now let's see what it looks like when we multiply this signal by a complex local oscillator. In this case, we are performing the multiplication:

$$\begin{aligned}&[\cos(\omega t) + j\sin(\omega t)][I(t) + jQ(t)] = \\ &\quad [I(t)\cos(\omega t) - Q(t)\sin(\omega t)] + \\ &\quad j[I(t)\sin(\omega t) + Q(t)\cos(\omega t)]\end{aligned} \qquad (14)$$

This is the equation for a USB signal. We are only interested in the real part of the result, since we only have one real channel on which to transmit. For this reason, the system is really a half-complex mixer.

A block diagram of such a mixer is shown in **Fig 16.11**. This is, in fact, the phasing

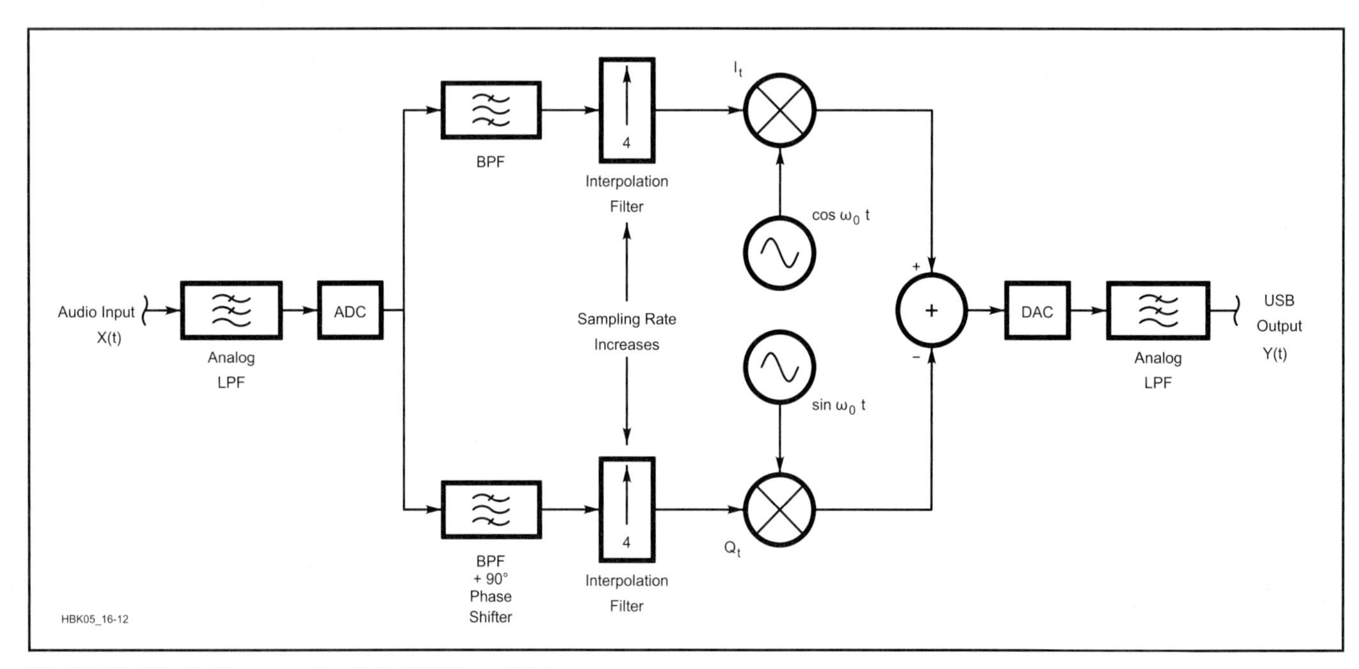

Fig 16.12—Block diagram of a digital SSB modulator.

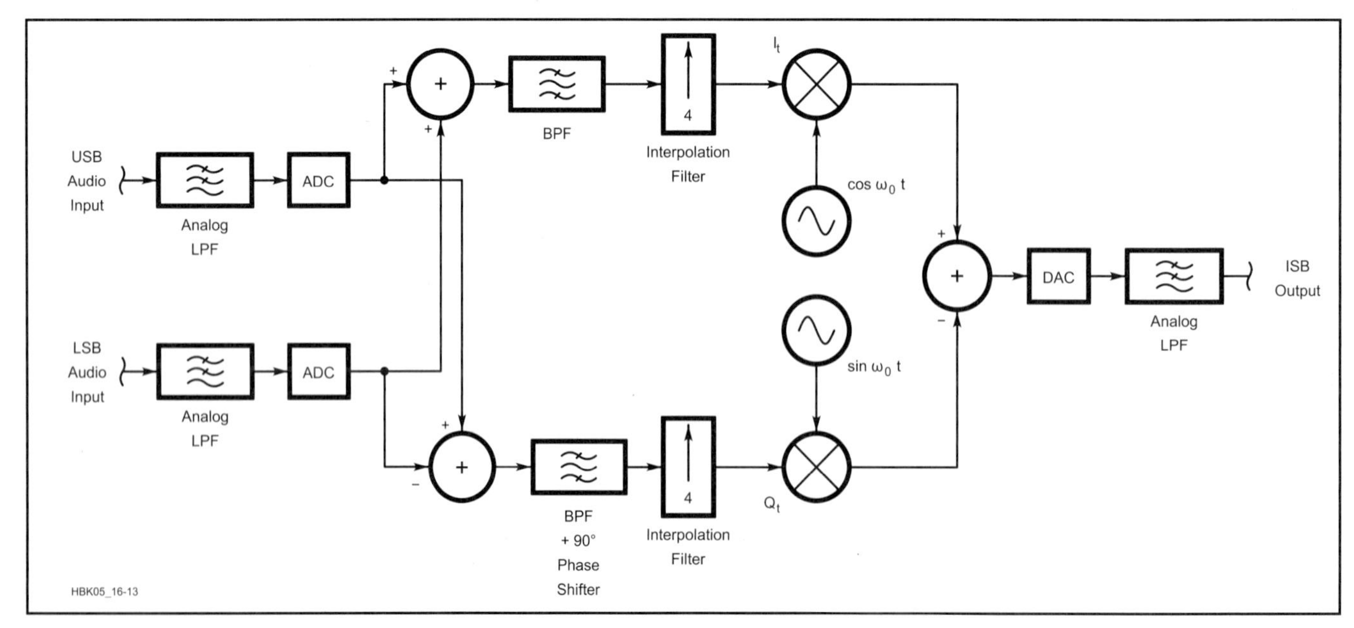

Fig 16.13—Block diagram of a digital ISB modulator.

method. Output signals are translated upward by the frequency of the local oscillator, ω_0 radians per second, or $\omega_0/2\pi$ hertz. Most transmitter designs will translate signals to an IF significantly higher in frequency than audio, so it is wise to include an increase in the sampling rate prior to mixing. An interpolation filter is naturally needed. It is particularly convenient to choose an interpolation factor of 4, because the cosine LO produces values of 1, 0, –1 and 0 during a full cycle; the sine LO produces values of 0, 1, 0 and –1. No actual multiplications need take place, saving time and accuracy. The Hilbert transformer can operate at the lower, original sampling rate, but we would like to include band-pass filtering to limit the spectrum to about 3 kHz BW. In fact, we can build a pair of DSP filters that provide the BPF response and the 90° phase relationship, as described below. Our SSB modulator then matches that shown in **Fig 16.12**.

Before discussing how to generate analytic filter pairs, it is worth noting a few properties of SSB signals created in this way. First, were we to add the I and Q signals instead of subtract them in the summation block of Fig 16.11, we would have an LSB signal instead of USB. It is not too hard to see that we could easily both add and subtract to produce a DSB, suppressed-carrier signal. We can even pre-add and subtract *two* audio signals to produce an independent-sideband (ISB) signal, as shown in **Fig 16.13**. More than two channels can be combined in this way. Second, since the amplitude of the carrier, $\cos(\omega_0 t) \pm j \sin(\omega_0 t)$, is constant, the amplitude of an SSB signal can be specified as some function of the modulating signal. If we think of the analytic audio signal as a vector in the complex plane, its length is equal to the signal's instantaneous amplitude:

$$A(t) = \left[I^2(t) + Q^2 t(1)\right]^{1/2} \qquad (15)$$

Finally, the phase of the signal is the instantaneous angle of this rotating vector:

$$\phi(t) = \arctan\left[\frac{Q(t)}{I(t)}\right] \qquad (16)$$

Now we can rewrite the real part of Eq 14 as:

$$y(t) = A(t)\cos\left\{\left[\omega + \frac{d\phi(t)}{dt}\right]t\right\} \qquad (17)$$

$\frac{d\phi(t)}{dt}$ is the rate of change of phase (the frequency) of the baseband signal (the audio). Eq 17 shows that a USB signal is just an upward frequency translation of the baseband signal by some RF of angular frequency ω. We may also write:

$$\left[I(t) + j\,Q(t)\right] = A(t)\left\{\cos\left[\phi(t)\right] + j\sin\left[\phi(t)\right]\right\} \qquad (18)$$

which shows that while the envelope, A(t), of an SSB signal is identical to that of the baseband signal producing it, A(t) is not the same as the baseband signal's waveform, represented by x(t) in Figs 16.11 and 16.12. An SSB signal preserves the amplitude and phase information of the baseband signal and occupies identical bandwidth.

Analytic Filter-Pair Synthesis

We have seen how complex mixing translates signals in frequency with a one-sided spectrum. We will use this fact to our advantage in creating an analytic filter pair. Each filter will have the same frequency response as the other. They will differ only in their phase responses.

We begin by designing a low-pass filter having the desired transition-band characteristic, $H(\omega)$; we obtain its impulse response, h(t). Multiplying the impulse response by a complex sinusoid of angular frequency ω_0 results in two sets of coefficients—one for the real part, and one for the imaginary part:

$$h_I(t) = h(t)\cos(\omega_0 t)$$
$$h_Q(t) = h(t)\sin(\omega_0 t) \qquad (19)$$

The frequency response of either one of these filters is given by:

$$H_\omega = \frac{H_{(\omega-\omega_0)} + H_{(\omega+\omega_0)}}{2} \qquad (20)$$

which is a BPF centered at ω_0. The I filter has a phase response differing 90° at every frequency from the Q filter. The frequency translation theorem works just as well on the responses of filters as it does on real signals. To perform this transformation of the L co-

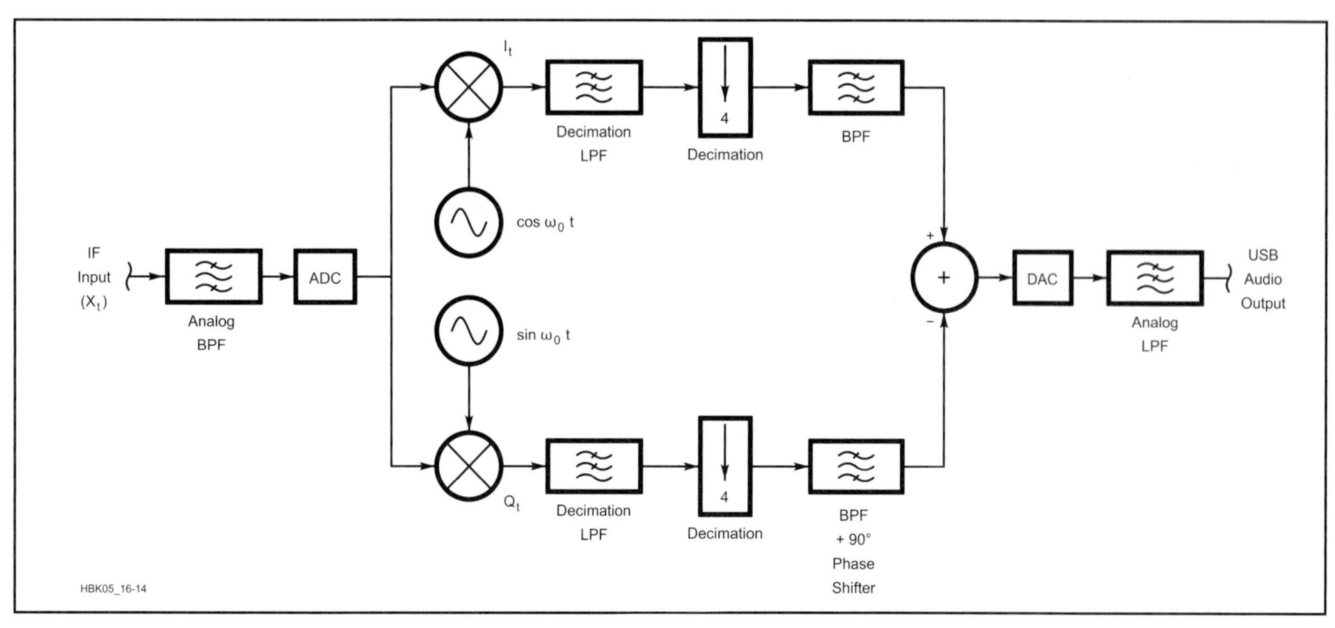

Fig 16.14—Block diagram of a digital SSB demodulator.

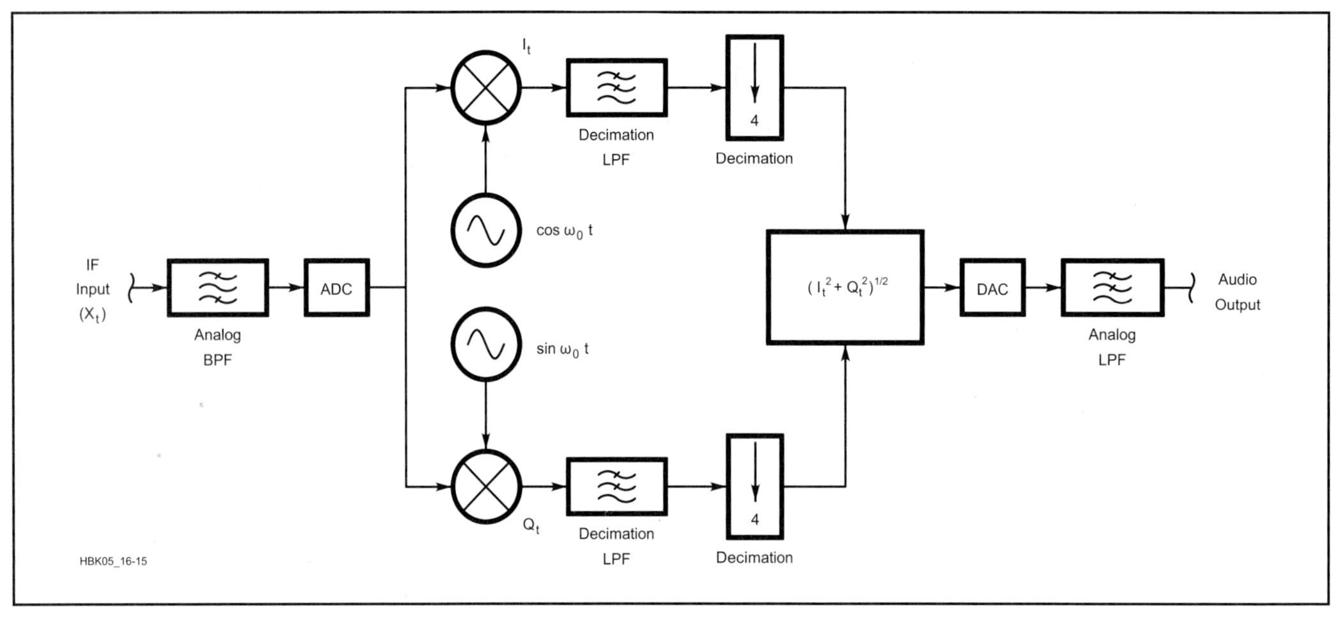

Fig 16.15—Block diagram of a digital AM demodulator.

efficients of the prototype LPF, we calculate new coefficients according to:

For $0 \le k \le L-1$,

$$h_I(k) = h(k)\cos\left[\omega_0\left(k - \frac{L}{2} + \frac{1}{2}\right)t_s\right]$$
$$h_Q(k) = h(k)\sin\left[\omega_0\left(k - \frac{L}{2} + \frac{1}{2}\right)t_s\right] \quad (21)$$

where t_s is the sampling period. When the low-frequency transition band is placed near zero frequency, as we would like for SSB, the BW of each BPF is approximately twice that of the prototype LPF. A very interesting thing sometimes happens when the number of taps is odd: The odd-numbered coefficients are zero. This allows reduction in computation by a factor of two. Refer to Project C in the **Appendix** for a practical example of how analytic filter pairs are generated.

We can alter the exciter's frequency response by convolving the impulse response of our analytic filter pair with that of a filter having the desired characteristic. New coefficients are calculated using the same convolution sum as in Eq 3. Graphic or parametric equalizers may be implemented in this way.

Demodulation: SSB

As in digital exciters, phasing methods prevail in receivers; the process is almost exactly the reverse of the modulator's. **Fig 16.14** presents the block diagram of a digital SSB receiver. After the IF signal is digitized, we wish to reduce the sampling rate and the filtered BW as soon as possible. This is because we need as much time as possible between input samples for the intense filtering and other computations we must perform. As noted above, reduced sampling rates also ease the design of the digital filters that provide the final selectivity. We therefore include a decimation filter and decimate by a factor of 4. Again, the LO signals take on only values of 1, 0, –1 and 0, eliminating multiplications. Digitized signals are translated to baseband using the complex mixing algorithms outlined above. Since the input signal, x(t), is real, only two multiplications are necessary:

$$I(t) = x(t)\cos(\omega t)$$
$$Q(t) = x(t)\sin(\omega t) \quad (22)$$

Now we have an analytic signal as before; the frequency of the BFO, ω_0 rad/s, is chosen to beat the carrier frequency to zero hertz. An analytic filter pair precedes the summation in which we select the sideband we want. The equations work precisely in reverse: That is why they are Hilbert *transforms*.

AM Demodulation

One's first inclination is to demodulate an AM signal by rectifying it. A better way is to use the I and Q signals we have already developed using Eq 15. Now we are stuck with computing square roots. Lucky for us, a fellow named Isaac Newton figured out a slick way almost 400 years ago. In the 17th century, these calculations were

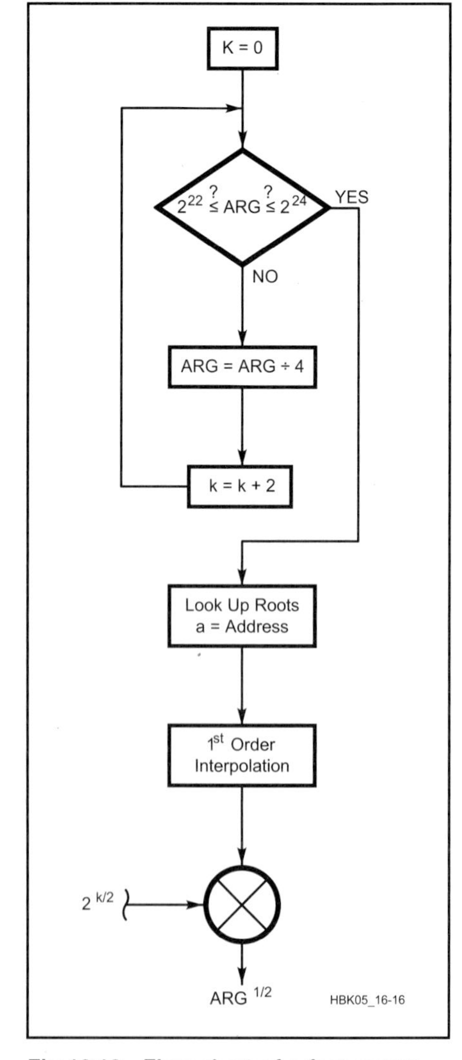

Fig 16.16—Flow chart of a fast square-root algorithm.

quite a burden—everything had to be done by hand. Because this is such a common problem in computing, a lot of additional effort has gone into finding faster algorithms since that time. A very fast look-up-table method is also presented here that may be more attractive where enough memory is available.

An I/Q AM demodulator dodges problems associated with rectification methods. It also can use the decimation filters for final selectivity, obviating much of the computations found in the SSB demodulator. **Fig 16.15** shows the circuit. Newton's method for square roots goes like this: Take a crude guess at the square root of the number in question. Divide the number by the crude guess. Add the crude guess to this ratio and divide it all by 2. Use this result as the new crude guess and repeat the process until the desired accuracy is obtained:

$$\text{let GUESS}_{new} = \left(\frac{\frac{\text{Number}}{\text{GUESS}_{old}} + \text{GUESS}_{old}}{2} \right)$$

$$\text{let GUESS}_{old} = \text{GUESS}_{new} \qquad (23)$$

REPEAT

In practice, the accuracy of the result reaches the limit of 16-bit representations in five or six iterations when the first guess is good. It is about half an order of magnitude slower than the following look-up table method, but is still among the best where memory is at a premium. Project D in the **Appendix** describes a *QuickBasic 4.5* example of Newton's method.

A very fast look-up-table method for computing integer square roots has been discovered. It employs a short (256-entry) table and first-order interpolation between table entries. First-order interpolation is described in detail in the DDS section below. To preserve accuracy, the algorithm also uses the process of argument normalization. The algorithm serves as our fifth software project in DSP in the **Appendix**.

The argument of this function—the number of which we must find the square root—is a 32-bit integer. The result is a 16-bit integer. Refer to **Fig 16.16**, a flow chart of the process. In the first step, the argument is normalized to within the range 2^{22}-2^{24}. Arguments greater than 2^{24} are divided by an even integral power of two, 2^k, where:

$$k = \Im\left[\log_2(\text{arg}) - 23\right] \qquad (24)$$

The script I indicates the integer part, and k—which takes on values of 0, 2, 4 or 6—is saved for later processing. Now the normalized argument is split into integer and fractional parts, with the radix point residing to the left of bit 15:

$$a = \Im\left(\frac{\text{arg}}{2^{k+8}}\right)$$

$$b = \mathcal{F}\left(\frac{\text{arg}}{2^{k+8}}\right) \qquad (25)$$

where a is the integer part and b is the fractional part. In other words, a comprises bits 16-23 of the normalized argument, and b is bits 0-15, as shown in the flow chart. Next, we use a as the address into the look-up table, fetching a 16-bit value, x_a. This value is the nearest table entry lower than the actual root. Fractional part b is used to interpolate between this value and the next higher table entry, x_{a+1}:

$$\text{root} = b\left(x_{a+1} - x_a\right) + x_a \qquad (26)$$

This is the square root of the normalized argument.

Finally, this result must be multiplied by the square root of 2^k, which is of course $2^{k/2}$. The result is then "de-normalized" and ready for use. Restricting k to an even integer (as we did) makes this a simple bit-shifting operation, as in the normalization process above. The 16-bit result produced by this algorithm is accurate to within several least-significant bits over the entire range of 32-bit arguments. It is quite a bit faster than the 5 or 6 iterations of Newton's method required for the same accuracy; this is because it avoids the divisions that Newton's method employs. Most DSPs take 3 or 4 times the processing time for a fractional division as they take for multiplication or look-up table indexing. Project E in the **Appendix** describes an assembly-language implementation of this square-root algorithm.

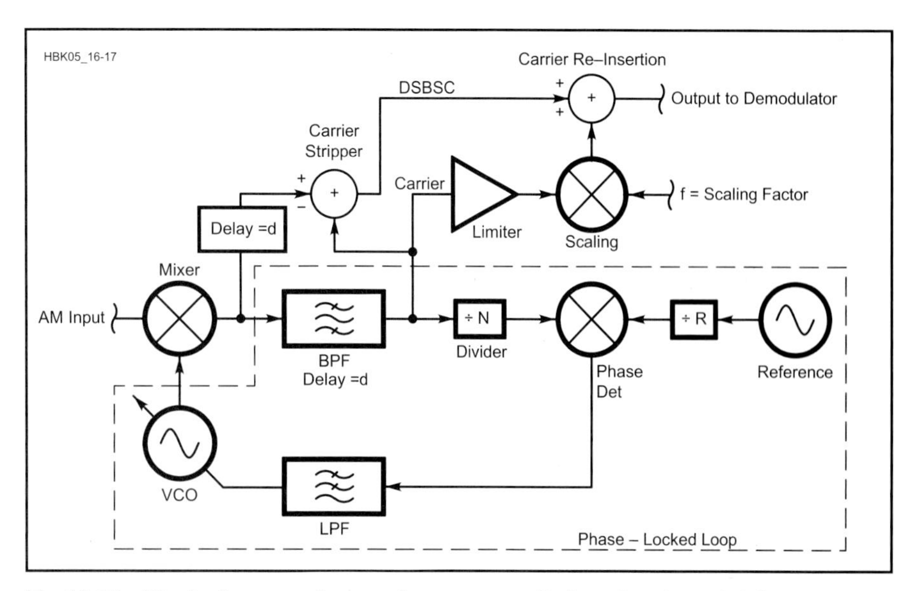

Fig 16.17—Block diagram of a synchronous, exalted-carrier demodulator.

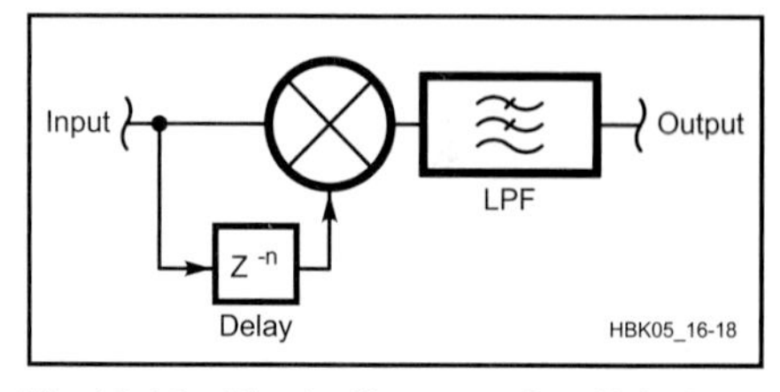

Fig 16.18—Block diagram of a digital quadrature detector.

Additional threshold extension and distortion-avoidance procedures may be employed in an AM demodulator. Of particular interest is the *synchronous, exalted-carrier* demodulator. Synchronous, in this case, means that the demodulator's frequency standard is phase-locked to the received carrier. This forces the phases of modulation components into their correct relationships and therefore minimizes phase distortion. A small advantage in SNR performance of up to 3 dB is also gained. DSP makes it relatively easy to build a narrow BPF, centered on the carrier, that strips the modulation prior to application to the PLL used to achieve lock. The exalted-carrier technique is a way of avoiding distortion caused by selective fading of the carrier. Ordinarily, when the received carrier's amplitude drops, the signal becomes over-modulated, even though it was not transmitted that way. Distortion can be severe. Exalted carrier strips the carrier from the signal using the narrow BPF and it is used to drive the PLL. A copy of the limited carrier is then added back to the carrier-stripped signal, in its original phase prior to demodulation at an amplitude that avoids over-modulation. See Fig 16.17 for a block diagram of this type of demodulator. Refer to the chapter on **Modes and Modulation Sources** for more discussion of AM waveforms, and to the chapter on **Mixers, Modulators and Demodulators** for an implementation of this scheme.

FM and PM Demodulation

Traditional FM and PM demodulators, such as discriminators (filters) and PLLs may be implemented in DSP. But again, the I/Q method carries distinct advantages as it exploits mathematical relationships. We already defined the phase of an analytic signal in Eq 16 and so we can build a PM demodulator directly by finding arctangents. Possibilities include look-up tables and Taylor series. For an FM demodulator, we would then differentiate the string of phase samples using the technique of *first differencing*. We simply take the difference between adjacent samples by subtracting them:

$$f(t) = \phi(t) - \phi(t-1) \quad (27)$$

and this is the FM demodulator's output.

One common analog technique that stands out among DSP implementations is the *quadrature detector*. It is certainly simple and convenient to generate delays and multipliers, such as are required. The input signal is multiplied by a time-delayed copy of itself to produce a voltage proportional to its phase excursions away from the center frequency. This voltage is also proportional to the amount of delay inserted. See **Fig 16.18.** When the delay is an odd integral multiple of one quarter the input period, the output is zero. Longer delays produce greater output-voltage sensitivities; that is, $dV/d\phi$ increases.

Digital BFO Generation: Direct Digital Synthesis

Synthesizers have come a long way since first becoming popular in HF transceivers of the 1970s. Availability of components then lagged well behind the development of theory. Now, hardware capabilities have nearly caught up—which is the case for DSP in general—and are driving the very rapid advancement of equipment we are now experiencing. Paralleling breakthroughs in the microprocessor and data-acquisition fields, progress in *direct digital synthesis* (DDS) has enabled performance levels only dreamed of a decade ago. Virtually all new designs may profit from this technology. Below, we will cover quite a few issues having impact on transceiver performance: phase noise, spectral purity, frequency stability, lock times and tuning resolution. A DDS circuit using dedicated hardware is described; discussion of BFO and LO generation in software follows.

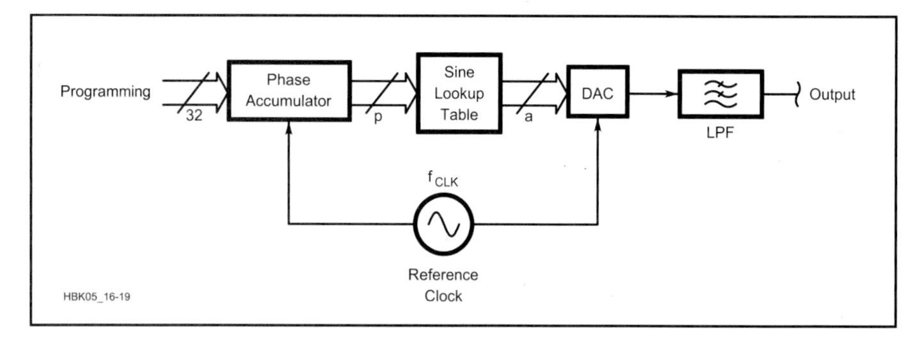

Fig 16.19—DDS block diagram.

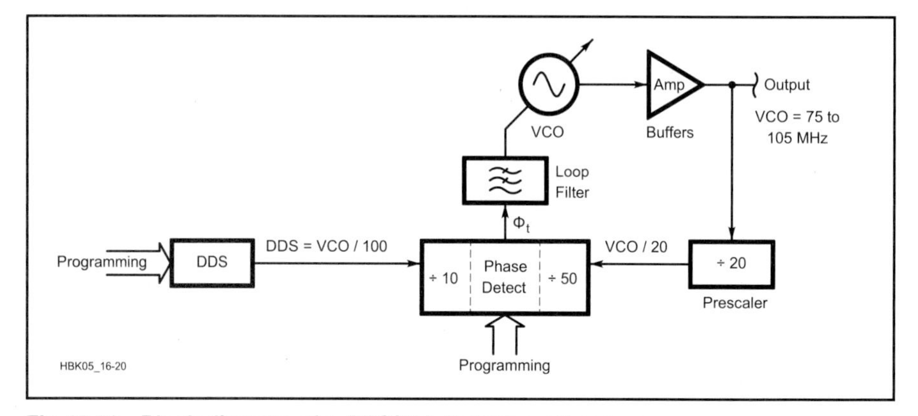

Fig 16.20—Block diagram of a DDS/PLL hybrid synthesizer.

Synthesizer performance affects receiver dynamic range. Phase-noise and spectral-purity issues are in play. Phase noise is the unwanted phase modulation of transceiver frequency-control elements by circuit noise. It appears at and near the transmitter's output frequency and may cause interference to stations on adjacent frequencies. In addition, it may cause interference in one's own receiver—even if the signals received are phase-noise free—through the process of *reciprocal mixing*. See the **Oscillators and Synthesizers** chapter for a discussion of this effect. The spectral purity of a synthesizer may also affect receiver dynamic range by introducing spurious responses where spurs exist on the synthesizer's output. This may be true especially for the first LO in a receiver across the entire range of frequencies present. It is extremely important that this LO be clean.

Radio amateurs are free to operate anywhere within large frequency bands, so it might seem that frequency accuracy is not very critical. Prevalent narrow-band communication modes require it, though, and operators have come to expect excellent stability from their rigs. It is reasonable to expect ±20-Hz stability over a range of –10 to +50° C. Digital compensation techniques currently achieve this. We wish to attain a tuning speed that does not impose limitations on typical use. "Cross-band" or split-frequency operation ought to be considered. For a frequency step of ±600 kHz, an upper limit of 25 ms on the lock time of a synthesizer is a reasonable goal. Lock time is defined as the time required to settle within the stability limits we already set. The smallest frequency steps should be such that they do not impede performance. 10 Hz used to be good enough, but now certain digital modes benefit from finer tuning. In addition, the digital notch filter described before is so sharp that it occasionally needs to be within 1 Hz!

A DDS system generates digital samples of a sine wave and converts them to an analog signal using a DAC. See **Fig 16.19**. In a DDS chip, a phase accumulator is incremented at each clock time; the phase information is used to look up a sine-wave amplitude from a table. This value is passed to the DAC, which outputs a step-wise sine wave. As we saw before, the spectrum of this sine wave is seasoned with aliases and contains other minor pollutants. Since the phase is represented by a binary number with a fixed number of bits, p, errors develop because the number is truncated to that number of bits. Truncation generates PM spurs in the DDS output. This occurs prior to the DAC. Further errors are related to the output resolution of the look-up table. Table values representing the amplitudes are truncated to some number of bits, a. This mechanism produces AM spurs in the output. According to Cercas *et al*, the largest PM spurs have amplitude:

$$P_{spur} = -(6.02p - 5.17)\ \text{dBc} \quad (28)$$

and maximum AM spurs can rise to:

$$P_{spur} = -(6.02a + 1.75)\ \text{dBc} \quad (29)$$

Phase noise at the output is that of the DDS clock source times the ratio of the output frequency to the clock frequency, as limited by divider noise. Spurious levels also tend to grow as the DDS output frequency approaches the Nyquist limit. Strange spurs at the output are usually related to IMD and harmonics of the desired signal and their aliases. Remember that frequencies exceeding half the sampling frequency "fold back" into the signal spectrum at a position determined by their frequency, modulo $f_s/2$. High-order harmonics are liable to find their way into one's band of interest. Traps at the DAC output have been known to suppress these responses. See Project F in the **Appendix** for the schematic of a DDS project.

In the analog signal we generate, the DAC introduces more AM spurs, harmonics and IMD because of its inherent nonlinearity, as discussed above. Spurs are also likely at the clock frequency, its harmonics and sub-harmonics. A higher-order LPF will take care of these, but we must see what we can do about the others. It turns out we may eliminate *all* the AM spurs by squaring the DDS output. We can do nothing about the remaining PM spurs. Cranking through Eq 28 will show that they can be made very low: –113 dBc for a 20-bit-address sine look-up table and 32-bit phase accumulator. This parameter is critical in case we want to use the DDS as the reference to a high-frequency PLL circuit. The PLL will multiply the phase noise and PM spurs by the ratio of the PLL output frequency to the PLL reference frequency within the PLL loop BW. Outside the loop BW, the VCO itself is responsible for establishing spectral purity. So while dividing the DDS to the PLL reference frequency lowers phase noise and PM spurs, the PLL multiplies them back upward. A trade-off exists between spur levels and reference frequency, hence lock time.

A PLL reference frequency near 100 kHz has been found to be sufficient for the desired lock times, with an output-to-reference ratio of 1000. Such a loop should achieve very fast lock times, as it can be expected to lock within 500 cycles of the reference input. The DDS tuning time is at least three orders of magnitude faster than this. In the example, the VCO output is near 100 MHz. DDS energy is injected at the reference input to the PLL chip, squaring it and dividing it by 10; the DDS runs near 1000 kHz. The block diagram of a PLL using a DDS as its reference is shown in **Fig 16.20**. Spurs and phase noise inside a loop BW of, say, 1 kHz are amplified by the PLL by the factor:

$$N = 20\log\left(\frac{f_{VCO}}{f_{REF}}\right) = 40\ \text{dB} \quad (30)$$

Of course, we tune the hybrid synthesizer by programming the DDS; the PLL programming is fixed. Let's say we want 1-Hz tuning resolution at the VCO output. As the DDS frequency is 1/100 of the output, we must tune the DDS in 10 *millihertz* steps! Tuning resolution in a DDS circuit is determined by the phase accumulator's bit resolution, p, and the DDS clock's frequency, f_{clk}:

$$df_{DDS} = \frac{f_{clk}}{2^p} \quad (31)$$

A clock frequency around 10 MHz and p = 32 easily satisfy our conditions, producing a step size of 2.3 millihertz. As noted above, making the DDS output frequency a small fraction of the clock frequency makes it easier to get a clean output. A range of about half an octave eases the design of the LPF or BPF used at the DDS output to limit spurs, aliases and clock feed-through.

The phase-accumulator/look-up-table approach is equally useful in generating numeric BFOs in software. One of the first things to emerge when considering this scheme is the potentially large size of the look-up table. To maintain the full dynamic range of a DSP system requires BFO phase and amplitude performance, as limited by Eqs 28 and 29, at least as good as the rest of the system. In 16-bit systems, we are shooting for about 90-100 dB of dynamic range. A table with 2^{16} = 65,536 entries is not much of a problem for DDS chip manufacturers to include on-board, but it may tax available memory space in embedded systems.

Fortunately, a couple of ways around the problem have been uncovered. The first involves the process of *interpolation*, very much like the artificial increase of sampling frequencies we examined above. In this method, we restrict the number of table entries to some arbitrary number, $M << 2^{16}$, while keeping the bit-resolution of the entries themselves, a, high enough to satisfy the limits of Eq 29 for the spur levels we can tolerate. Take the case where $M = 2^8 = 256$ and a = 16. The phase accumulator, incremented at each sample time by an amount df that is directly proportional to the output frequency, forms the address into the look-up table. Let this address have bit-resolution p = 16. According to Eq 28, PM spurs will not exceed –91 dBc. Since there are only 256 table entries, we may use the most-significant byte (MSB) of the address to find the table entries that straddle the correct output value. We then use the least-significant byte (LSB) as an unsigned fraction to find out how far between the two table entries we must go to reach the correct output value. If, in order of increasing address in the table, our two adjacent table entries are d_1 and d_2, we may perform a first-order interpolation between the entries using:

$$d_{int} = d_1\left(\frac{256 - LSB}{256}\right) + d_2\left(\frac{LSB}{256}\right) \quad (32)$$

This results in a linear, piece-wise representation of the data, as shown in **Fig 16.21**. The worst-case amplitude

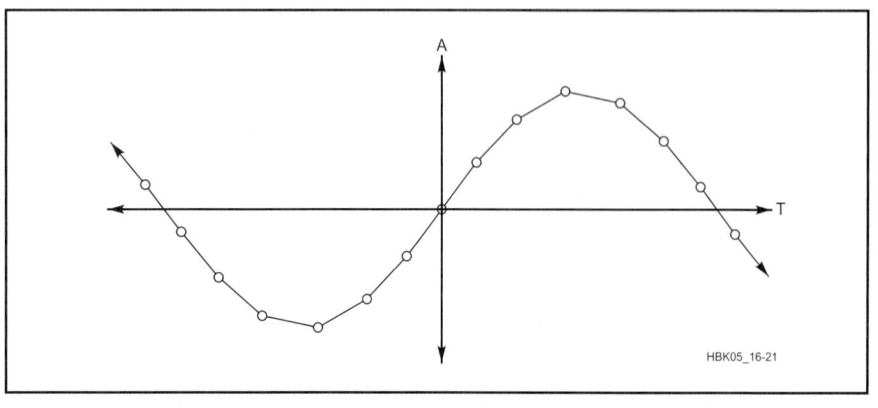

Fig 16.21—Linear piece-wise representation of data resulting from first-order interpolation.

errors caused by this straight-line approximation place total harmonic distortion (THD) at the output at around 0.03% or –70 dBc. Much of this harmonic distortion is concentrated near half the sampling frequency, though, and may not be of much concern in actual systems. Doubling M would reduce THD to around 0.01%. Second and higher-order interpolation algorithms are available that outperform the first-order approximations by a long way.

In systems where an even smaller look-up table must be used, computation of sines and cosines using Taylor series might be attractive. THD is less than 0.008% when using four or five terms from the polynomials:

$$\sin(x) = x - \frac{1}{3!}x^3 + \frac{1}{5!}x^5 - \frac{1}{7!}x^7 \ldots \quad (33)$$

and

$$\cos(x) = 1 - \frac{1}{2!}x^2 + \frac{1}{4!}x^4 - \frac{1}{6!}x^6 + \frac{1}{8!}x^8 \ldots \quad (34)$$

DIGITAL SPEECH PROCESSING

Virtually all modern transmitters employ fast-attack, slow-decay RF compression: It is called automatic level control (ALC). Because transmitters are usually peak-power limited, some form of gain control is necessary to prevent overdrive of the final RF power amplifier.

RF Compression

A typical ALC system detects the transmitter's envelope with a rectifier and filter, applying this control signal to some gain-controlled stage or stages in the exciter. An increasing level from the envelope-detector results in decreasing gain such that the peak envelope power (PEP) is regulated. ALC is a servo loop employing negative feedback, usually developed only on voice peaks. As the decay time of the detector is decreased, some amplification of parts of speech falling between peaks is achieved. Enhancement cannot exceed the total gain reduction occurring at the voice peaks and usually falls in the range of 3-6 dB. The increase in the transmitter's average output power (talk power) may be quite a bit less than this depending on the characteristics of the voice, especially the *peak-to-average ratio*. In a digital exciter, we may eliminate the need for an analog gain-controlled stage by employing a numeric gain control factor in software and simply regulating the modulator's output level.

Human voices have peak-to-average ratios as high as 15 dB. This does not utilize a peak-limited transmitter very well in SSB mode: At the 100-W PEP level, the average output power might be as little as 3 W! RF compression raises the average output power and tends to further improve intelligibility by bringing out subtle parts of speech. In a digital I/Q modulator, we have a distinct advantage in designing an RF compressor: The RF envelope can be calculated before the modulation is performed. Once the microphone audio has been sampled and converted to an analytic signal, Eq 15 may be used to compute the envelope. To avoid the time-consuming square-root calculation, we may use an approximation:

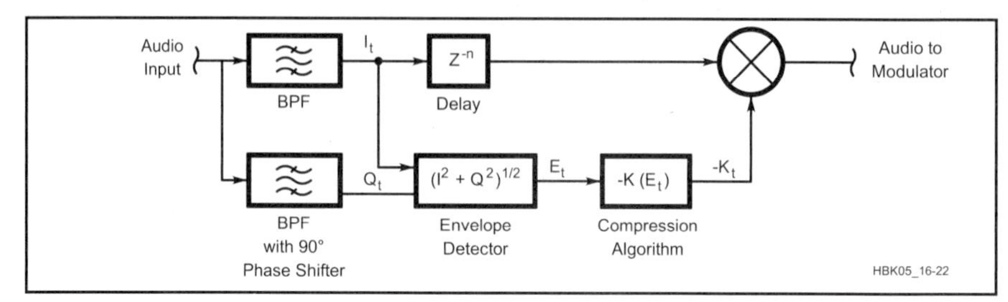

Fig 16.22—Digital RF compressor block diagram.

$$\begin{Bmatrix} \text{For}: |I| > |Q|, \left(I^2 + Q^2\right)^{1/2} \approx |I| + 0.4|Q| \\ \text{For}: |Q| > |I|, \left(I^2 + Q^2\right)^{1/2} \approx |Q| + 0.4|I| \end{Bmatrix} \quad (35)$$

The envelope signal is used to compress the range of baseband levels prior to modulation so that the peak-to-average ratio is reduced. A block diagram of this system is shown in **Fig 16.22**. The net effect of the system can be shown to be identical to that of a direct RF compressor. This naturally involves distortion, since the transmitter is no longer linear; however, the distortion produced enhances the syllabic and formant energy in speech without introducing the "mushy" sound caused by heavy audio compression or clipping. As the attack and decay times of an RF compressor are made faster, it approaches the performance of an RF clipper, known to be the most effective form of processing. Because the baseband audio is processed prior to filtering and modulation, occupied BW does not increase much; low-order IMD products will be created, though, that fall within the desired transmit BW. These products ultimately limit the effectiveness of the compressor. This technique may also be applied to receivers.

Audio Compression: Building an AM Transmitter

It has long been a problem to hold the carrier and modulation levels constant in AM transmitters covering several octaves of frequency, such as at HF. Because a baseband signal may not have symmetrical positive and negative amplitudes about its average value, a suitable analog ALC system would be incredibly complex.

In DSP, we may prevent *carrier shift* using adaptive techniques; we prevent over-modulation using an audio compressor. (Refer to **Fig 16.23**.) First, the ratio of drive level to output level, d(t)/y(t), is easily computed by a DSP when the transmit-

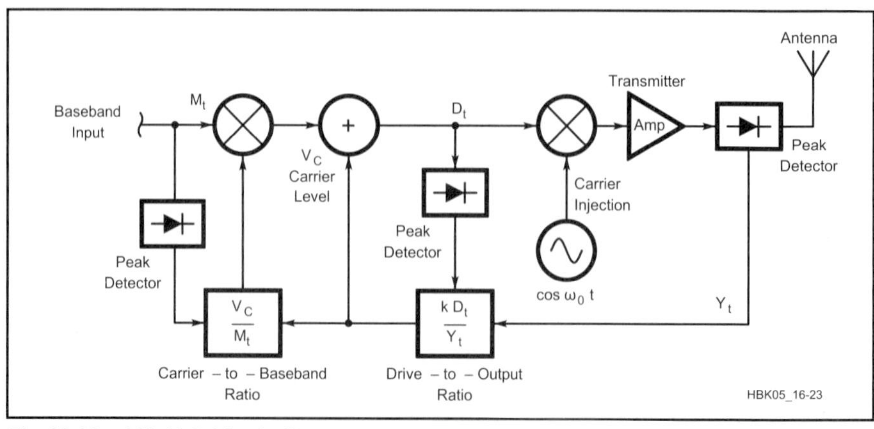

Fig 16.23—AM ALC block diagram.

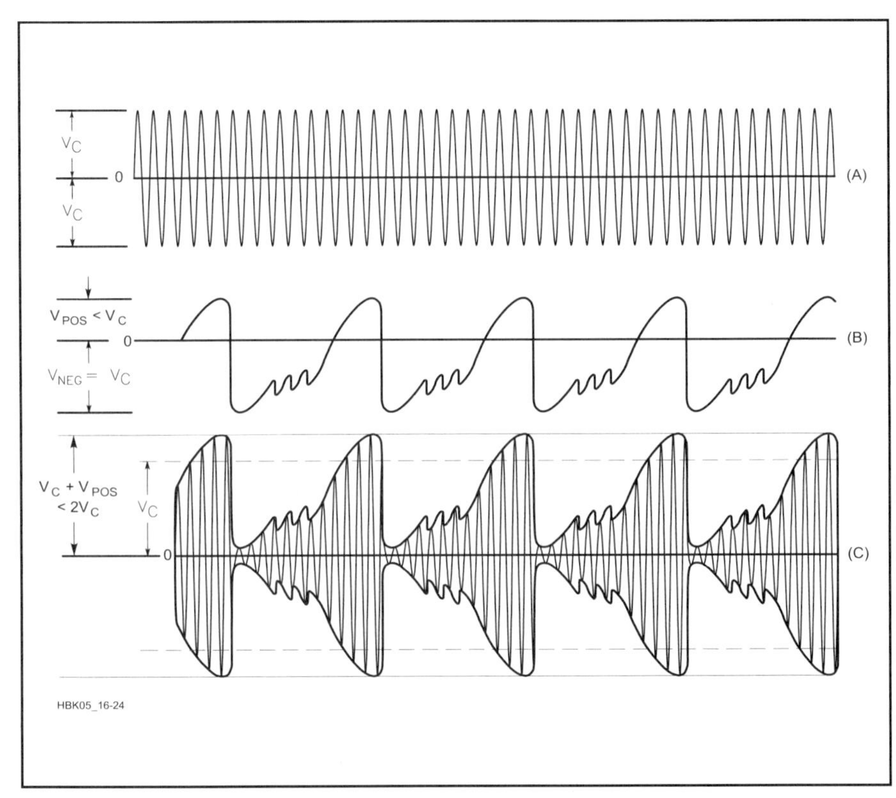

Fig 16.24—AM carrier (A). Baseband input with asymmetrical amplitudes (B). AM modulator output (C).

ter is on. From this, we can calculate what drive level is required to reach exactly 25% of the peak-power setting. We want the carrier to have this amplitude, regardless of modulation (or lack of it). Second, the baseband signal applied to the modulator must have a maximum peak level equal to the carrier's drive level established above. When the carrier and compressed baseband levels are added, the result is a 100%-modulated AM signal.

Fig 16.24 shows this situation, using a baseband signal whose negative excursions are greater than its positive excursions about the average value. Now two servomechanisms are operating in our AM ALC: One continually computes the drive-to-output ratio and sets the carrier level; the other compresses the peak baseband signal to that same peak level. Since the baseband peak detector has to find either the highest negative or highest positive peak, asymmetrical audio inputs may produce an unexpected result: Either the upward or downward modulation may reach 100% before the other can do so. If the downward modulation limits baseband amplitude first, the upward modulation would not cause the transmitter to reach its set PEP level without introducing a carrier shift.

INTERFERENCE-REDUCTION TECHNIQUES

We touched on the idea of a manually tuned adaptive notch filter using the LMS (least-mean-squares) algorithm. These principles are explored in more detail here, especially as they apply to interference- and noise-reduction systems. The nature of information-bearing signals is that they are in some way coherent; that is, they have some features that distinguish them from noise. For example, voice signals have attributes related to the pitch, syllabic content and impulse response of a person's voice.

Adaptive Filtering

We will find it possible to build an adaptive filter that accentuates those repetitive components and suppresses the non-repetitive (noise). Much research has been done about detection of a sinusoidal signal buried in noise. Adaptive filtering methods are based on the exploitation of the statistical properties of the sampled input signal, specifically, *autocorrelation*. Simply put, autocorrelation refers to how recent samples of a waveform resemble past input samples. We will discuss an *adaptive predictor*, which actually makes a reasonable guess at what the next sample will be based on past samples. This leads directly to an adaptive noise-reduction system.

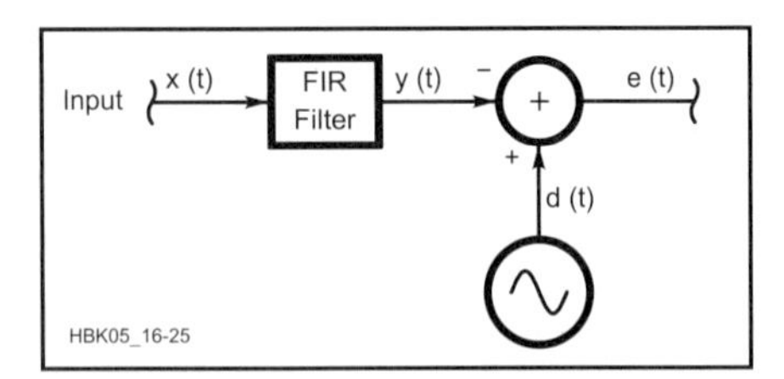

Fig 16.25—An adaptive modeling system.

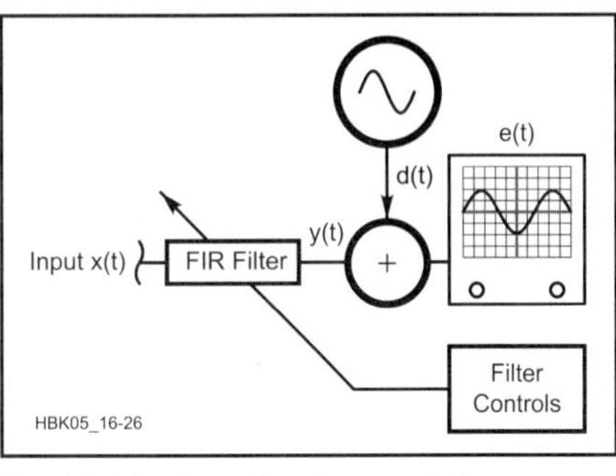

Fig 16.26—An adaptive modeling system, which requires a person at the filter controls.

An Adaptive Interference Canceler

Imagine we have some sampled input signal, x(t), that we want to adaptively filter to enhance its repetitive content. In the case of a CW signal, all that is required is a BPF centered on the desired frequency. We know that this signal takes the form of a sine wave and that its amplitude will change markedly. Its frequency may not be absolutely constant, either, but we will assume it is fixed for now. We set up an FIR filter structure and an error-measurement system to compare a reference sine wave, d(t), with the output of the filter, y(t). See **Fig 16.25**. Sine wave d(t) is the same frequency we expect the CW tone to be. The difference output, e(t), is known as the *error signal*.

Now imagine some person is watching the error signal and has their hands on the controls that change the filter coefficients. (See **Fig 16.26**.) Minimizing the error signal by tweaking the coefficients forces the filter to converge to a BPF centered at the frequency of d(t). The speed and accuracy of that convergence is going to depend on how well the person analyzes and reacts to the error data. If it is difficult to tell that a sine wave is present, then adjusting the filter will also be difficult. Further, if the sampling rate is high enough, a person will not be able to keep up; they can check the error only so often or can generate long-term averages of the error.

Using the typical processes of the human mind, the person will soon discover that if they turn the controls the wrong way, the error increases. This information is used to reverse the direction of adjustment. The person then turns the controls the other way. It soon emerges that the person is on a *performance surface*, with an "uphill" and a "downhill," and they know the goal is to go only downhill. So they thrash about with the controls, sometimes making mistakes, but ultimately making headway overall down the hill. At some point, the error gets rather small: They know they are near the "bottom of the bowl." Once at the bottom, it is uphill no matter which way they go. The goal of minimizing the error e(t) has been achieved. They continue gently flailing about with the controls, but always staying near the bottom. This situation is analogous to aligning an analog BPF with an adjustment tool.

After doing this whole thing several times, the person finds that certain rules help speed up the process. First, there is a relationship between the magnitude of the error and the amount they must tweak the controls. If the total error is large, a lot of tweaking must be done; if small, then it is better to make small adjustments to stay near the bottom of the performance surface. Second, there is a correlation between the error, e(t); the input samples, x(t); and the coefficient set, h(t) they need to adjust. Derivation of algorithms providing for steepest descent down the hill is a long and tedious exercise in linear algebra. Let's just say the person goes to school, becomes an expert in matrix mathematics and discovers that one of the fastest and most accurate ways down the hill is to adjust coefficients at sample time t according to:

$$h_{t+1}(k) = h_t(k) + 2\mu e(t)x(t) \quad (36)$$

This is the LMS algorithm. It was developed by Widrow and Hoff in the late 1950s.

Replacing the person with the LMS algorithm, as shown in **Fig 16.27**, we have our

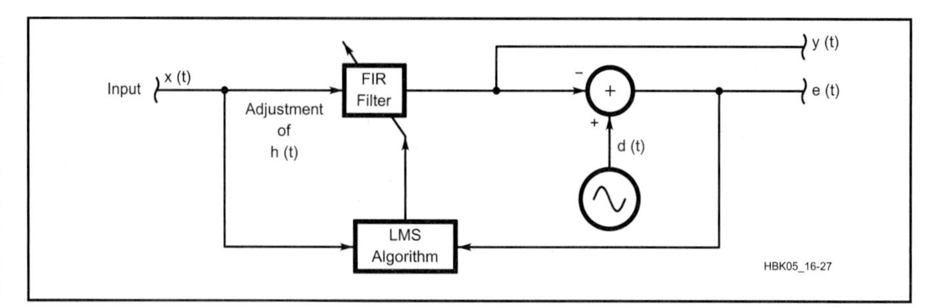

Fig 16.27— An adaptive interference canceler.

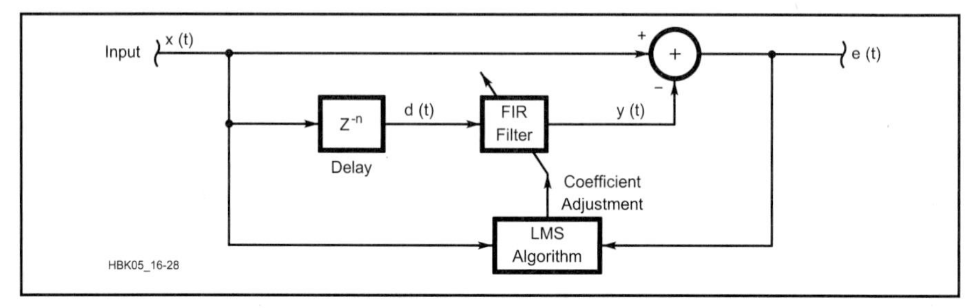

Fig 16.28—An adaptive predictor.

manually tuned adaptive interference canceler. Note that both the desired output, y(t), and the undesired, e(t), are available. This is nice in case we want to take only the broadband component and reject the tone. An obvious application of such tone rejection would be as a *notch filter*, and conversely, reception of a desired tone (signal) while rejecting the broadband (noise) is also possible; that is, *noise reduction*! Performance issues of interest include the adjustment error near the bottom of the performance surface and the speed of adaptation. One of the first things we notice about the LMS algorithm is that each of these factors is directly proportional to μ. We select its value, which ranges from 0 to 1, to set the desired properties. A trade-off exists between speed and misadjustment. Large values of μ result in fast convergence, but large misadjustment in the steady state. Total misadjustment is also proportional to the number of filter taps, L, and this may place a limitation on the complexity of the filter that may be used. The total delay through the filter also grows with its length; it may become unacceptably large under certain conditions. As in Eq 10, the BW of the adaptive BPF is:

$$BW = \frac{2\mu A^2}{t_s} \text{ rad/s} \quad (37)$$

Small values of μ result in narrower filters that take longer to adapt. Attempts may be made to adjust μ on the fly by using a value that changes in proportion to the error, e(t). A large value is selected initially for rapid convergence, then it is decreased to minimize the long-term misadjustment. This works fine so long as the characteristics of the input signal are not rapidly changing.

An Adaptive Interference Canceler Without An External Reference: An Adaptive Predictor

In the above example, we knew pretty much what to expect at the output: a sine wave of known frequency. What happens when we do not know much about the nature of the input signal, except that it contains coherent components? Quite a few circumstances like this arise in practice. It might seem at first that adaptive processing could not be applied; however, if a delay, z^n is inserted in the *primary input*, x(k), to create the *reference input*, d(k), periodic signals may be detected and thereby enhanced (or eliminated). See **Fig 16.28**. This delay forms an *autocorrelation offset*, representing the time difference used to compare past input samples with present samples. The amount of delay must be chosen so that the desired components in the input signal correlate with themselves, and the undesired components do not. This is an *adaptive predictor*: Predictable components are enhanced, while the unpredictable parts are removed. Experiments show that for any given value of m, the filter converges quickest when the delay, z^{-n}, is set between one half and one times the filter's total delay.

We may predict this circuit's noise-reduction performance using the ratio of the prefiltered BW to that of the converged filter:

$$\Delta SNR = 10\log\left(\frac{BW_{input}}{BW_{filter}}\right) = 10\log\left(\frac{BW_{input}}{2\mu A^2 f_s}\right) \quad (38)$$

As an example, for μ = 0.005, A = 1, BW_{input} = 3 kHz and f_s = 15 kHz, the SNR improvement is about 13 dB. When adaptive filters with many taps are used, multiple tones may be either enhanced or notched. Under most conditions, the undesired components are large compared to the desired; enhancement of signals is needed most when the input SNR is low. This situation may not give us enough thrashing about to find our way down the performance surface to convergence. Adding artificial noise to satisfy this condition is tempting, but it turns out we can alter the algorithm slightly to improve our lot without actually adding such noise. These additional terms in the algorithm are known as *leakage terms*.

The unique feature of *leaky LMS algorithms* is a continual "nudging" of the filter coefficients toward zero. The effect of a leakage term can be striking, especially when applied to noise-reduction of voice signals. The SNR increases because the filter coefficients tend toward a lower throughput gain in the absence of coherent input signals. More significantly, leakage helps the filter adapt under low-SNR conditions—exactly when we need noise-reduction the most. One way to implement leakage is to add a small constant of the appropriate sign to each coefficient at every sample time:

$$h_{t+1}(k) = h_t(k) + 2\mu e(t)x(t) - \lambda\{\text{sign}[h_t(k)]\} \quad (39)$$

The value of λ may be altered to vary the amount of leakage. Large values prevent the filter from converging on *any* input components, and things get very quiet indeed. Small values are useful in extending the noise floor of the system. In the absence of coherent input signals, the coefficients move linearly toward zero; during convergent conditions, the total misadjustment is increased to at least λ, but this is not usually serious enough to affect signal quality.

An alternate way to implement leakage is to scale the coefficients at each sample time by some factor, γ, thus also nudging them toward zero:

$$h_{t+1}(k) = \gamma\, h_t(k) + 2\mu e(t)x(t) \quad (40)$$

For values of γ just less than unity, leakage is small; values near zero represent large leakage and again prevent the filter from converging. It can be shown that the leaky LMS algorithm is equivalent to adding normalized noise power to the input x(t) equal to:

$$\sigma^2 = \frac{1-\gamma}{2\mu} \quad (41)$$

The leaky LMS algorithm must adapt to survive, much as a hummingbird must flap its wings. Were the factor μ suddenly set to zero, the coefficients would all die away, never to recover. Therefore, it is perhaps unwise to use these algorithms with adaptive values of m. Although values for γ and μ of greater than unity have been tried, the inventors refer to these procedures as "the dangerous LMS algorithm." Enough said.

FOURIER TRANSFORMS

While Fourier transforms are not used exclusively for interference reduction, we present them under that heading here because they are generally superior to adaptive-filtering algorithms in that application. The penalty for this greater effectiveness is an increased computational burden. The relationship Joseph Fourier (pronounced **foor**-ee-ay, 1768-1830) formulated between the application of heat to a solid body and its propagation has direct analogy to the behavior of electrical signals as they pass though filters and other networks. The laws he wrote define the connection between time- and frequency-domain descriptions of signals. They form the basis for DSP spectral analysis, which makes them extremely valuable tools for many functions, including digging signals out of the noise, as we will see.

A Fourier transform is a mathematical technique for determining the frequency content of a signal. Applied to a signal over some finite period of time, it produces an output that describes frequency content by assuming that the section of the signal being analyzed repeats itself indefinitely. The idea of applying Fourier transforms to noise reduction is that if we can analyze an input signal at many frequencies and exclude those results not meeting certain criteria, we can eliminate undesired signals. Noise reduction may be accomplished by applying the transform results at frequencies for which a preset amplitude threshold is not met. What remains are the frequencies where the energy is greatest, and that means signal-to-noise ratio is improved.

Originally, the Fourier transform was developed for continuous signals. In DSP, we use a variant of it called the *discrete Fourier transform* (DFT). It is the discrete version because it operates on sampled signals. It is a *block transform* because it converts a block of N input samples into a block of N output *bins*. The input block may be any N contiguous samples. A DFT makes use of complex sinusoids and produces a complex result. When the input data are real, meaning they lack an imaginary part, half the output block consists of the *complex conjugates* of the other half, and so is redundant. When a complex input is used, none of the output bins is redundant.

We learned before that a complex sinusoid is just a pair of waves: a cosine wave and sine wave of the same frequency. Since we will be dragging around a lot of these in the equations below, we introduce a little mathematical shorthand for them called the *Euler identity*:

$$e^{j\omega t} = \cos(\omega t) + j\sin(\omega t) \quad (42)$$

where e is base of natural logarithms. We will shorten this even more later. For each output bin k, where $0 \le k \le N-1$, the DFT is computed as:

$$X(t) = \sum_{n=0}^{N-1} x(n) e^{\frac{-j2\pi nk}{N}} \quad (43)$$

Expanding Eq 43 using the Euler identity yields:

$$X(k) = \sum_{n=0}^{N-1} x(n) \cos\left(\frac{2\pi nk}{N}\right) - j \sum_{n=0}^{N-1} x(n) \sin\left(\frac{2\pi nk}{N}\right) \quad (44)$$

So each bin has a real part and an imaginary part. Note that each part is calculated using the same convolution sum we saw in Eq 3. Eq 44 is in normal complex-number form: $a + j\,b$. These coefficients a and b yield the amplitude and phase of the signal x(t) at frequency $\omega = (k\, f_s)/N$:

$$A_k = \left(a_k^2 + b_k^2\right)^{1/2} \quad (45)$$

$$\phi_k = \arctan\left(\frac{b}{a}\right) \quad (46)$$

k is directly proportional to the frequency of its bin according to:

$$f_k = \frac{kf_s}{N}, \text{ for } k < \frac{N}{2} \quad (47)$$

The bins are evenly spaced in frequency by the amount $f_1 = f_s/N$, but there are actually only N/2 real frequencies represented. As mentioned above, half the DFT bins produced from a real input are redundant. Complex inputs may analyze positive and negative frequencies separately.

Working in reverse, we may reconstruct time-domain signal x(t) by summing X(k) for all values of k:

$$x(t) = \frac{1}{N} \sum_{k=0}^{N-1} X(k) e^{\frac{j2\pi kn}{N}} \quad (48)$$

This is the *inverse discrete Fourier transform* (IDFT or DFT^{-1}). It is important to note the duality of the DFT/IDFT relationship. The transforms are not really altering the signal in any way, they are only different ways of representing it mathematically. The strength of the DFT in noise-reduction systems is that it evaluates the amplitude and phase of each frequency component to the exclusion of others.

As far as we can reduce the *resolution BW*, f_s/N, we can eliminate additional noise by artificially zeroing frequency bins not meeting a pre-defined amplitude threshold. Finer resolution BW is obtained by increasing the number of bins, N, decreasing the sampling frequency, or both. Increasing the number of bins, N, involves taking a larger block of N input samples; the larger block represents a longer time span. Obviously we have to wait for N samples to be taken before we can Fourier transform a complete block: A delay of N samples is the result.

Since the DFT assumes the input block repeats indefinitely, we have discontinuities at the beginning and end of the block where the data have been chopped out of the continuous string of input samples. These abrupt discontinuities cause unexpected spectral components to appear, just as fast on-off keying of a CW transmitter does. This phenomenon is known as *spectral leakage*. Discrete signal components in the input "leak" some of their energy into adjacent frequency bins, smearing the spectrum slightly. Increasing the number of bins, N, helps alleviate this problem. Increasing N moves the bins closer together; a signal that falls between two bins will still cause leakage into adjacent bins, but since the bins are closer together, the spread in frequency will be less. Even so, input components are still spreading their energy over several bins and this overlap makes it difficult to determine their exact amplitudes and phases.

To minimize that problem, we use a technique known as *windowing* on the input data prior to transformation. The data block is multiplied by a *window function*, then used as input to the DFT normally. Window functions are chosen to shape the block of data by removing the sharp transitions in its envelope. Examples of window functions and their DFTs are shown in **Fig 16.29**.

The rectangular window is equivalent to not using a window at all, as all the samples are multiplied by a constant. The other window functions achieve various amounts of side-lobe reduction. These window functions are also used to design filters using the Fourier transform method. In fact, these sequences can be used as the impulse responses of prototype LPFs, as should be evident from their frequency responses. Notice that they each involve a trade-off between transition BW and ultimate attenuation. Also note that in the figure values of ultimate attenuation are plotted without regard to dynamic-range limitations that may be imposed by the bit-resolution of actual systems.

Fast Fourier Transforms

In the years before computers, reduction of computational burden was extremely desirable. Many excellent mathematicians, including Runge, studied the problem of calculating DFTs more rapidly than the direct form of Eq 43. They recognized that the direct form requires N complex multiplications and additions per bin and that N bins are to be calculated, for a total computational burden proportional to N^2. The first breakthrough was achieved when they realized that the complex sinusoid $e^{-j2\pi kn/N}$ is periodic with period N, so a reduction in computations is possible through the symmetry property:

$$e^{\frac{-j2\pi k(N-n)}{N}} = e^{\frac{j2\pi kn}{N}} \quad (49)$$

This led to the construction of algorithms that effectively break any N DFT computations of length N, into N computations of length $\log_2 N$. Thus, the computational burden is reduced to be proportional to N $\log_2 N$. Because even this much calculation was not practical by hand, the usefulness of the faster algorithms was overlooked until Cooley and Tukey revived it in the 1960s.

To exploit the symmetry referred to, we have to break the DFT computations of length N into successively smaller calculations. This is done by *decomposing* either the input or output sequence. Algorithms wherein the input sequence, x(t), is decomposed into smaller sub-sequences are called *decimation-in-time* FFT algorithms; output decompositions result in *decimation-in-frequency* FFTs. The decomposition is based on the fact that for some convenient number of samples, N, many of the sine and cosine values are the same and products can be combined prior to computing the convolution sums. In addition, other products have factors that are other sine and cosine values. It turns out that electing to decompose by successive factors of two produces a very compact and efficient algorithm: a *radix-2* FFT algorithm.

Now for that additional bit of complex-sinusoidal shorthand mentioned earlier. Lots of complex sinusoids will appear in the diagrams to follow, so it sure would be nice to reduce the clutter a bit more. Let's follow the popular DSP text of Oppenheim and Schafer and select the notation:

$$e^{\frac{-j2\pi kn}{N}} = W_N^{kn} \quad (50)$$

This is used in **Fig 16.30** in a flow chart for a complete FFT calculation, for N = 8. Multiplication symbols represent complex multiplications, addition symbols represent complex additions. Note that each complex multiplication requires four real multiplications and two real additions. Complex additions need two real additions.

We have eight input points and eight output points. Observe that the diagram could not be drawn without crossing many signal paths—there is a lot of calculation

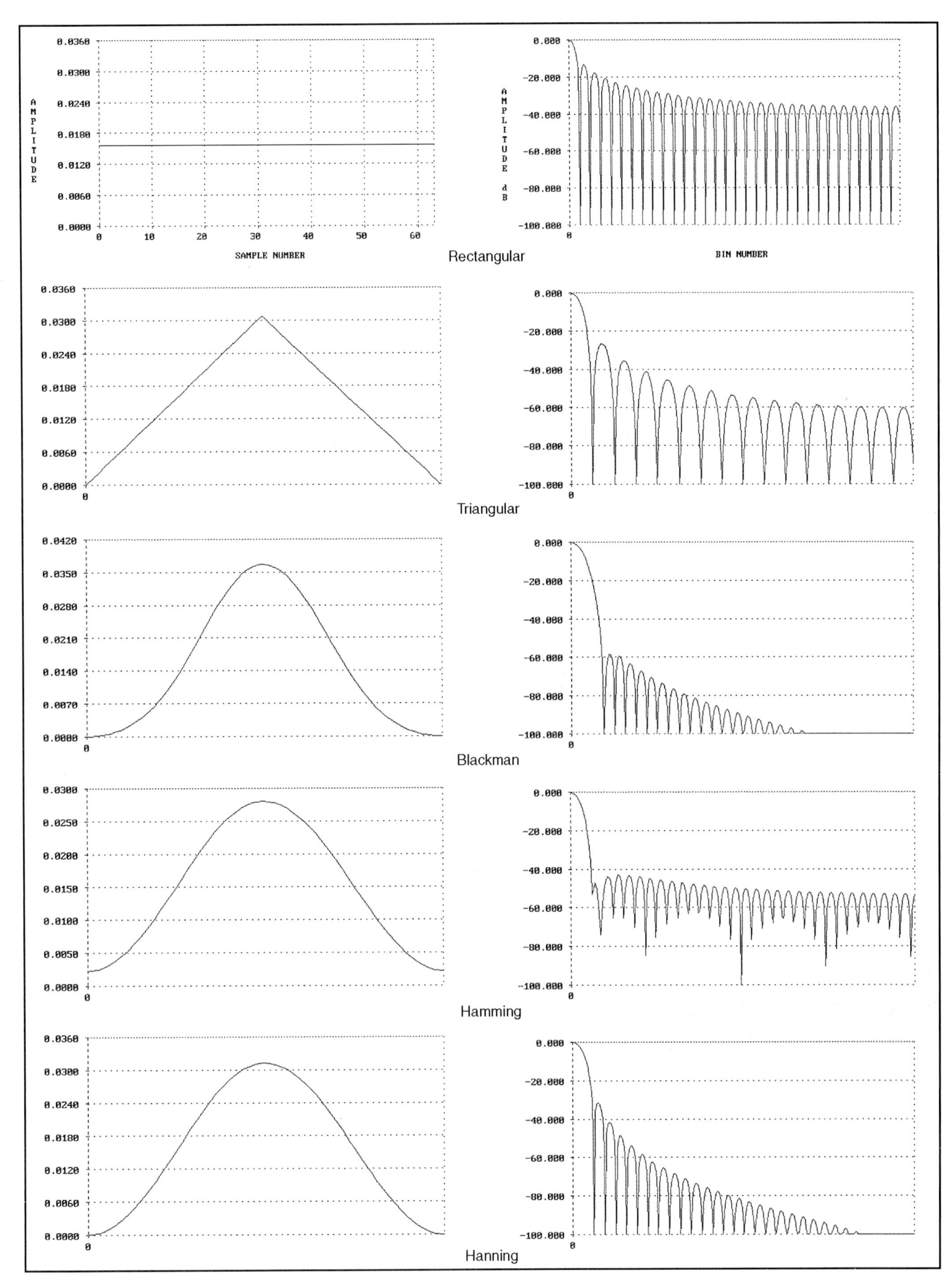

Fig 16.29—Various window functions and their Fourier transforms.

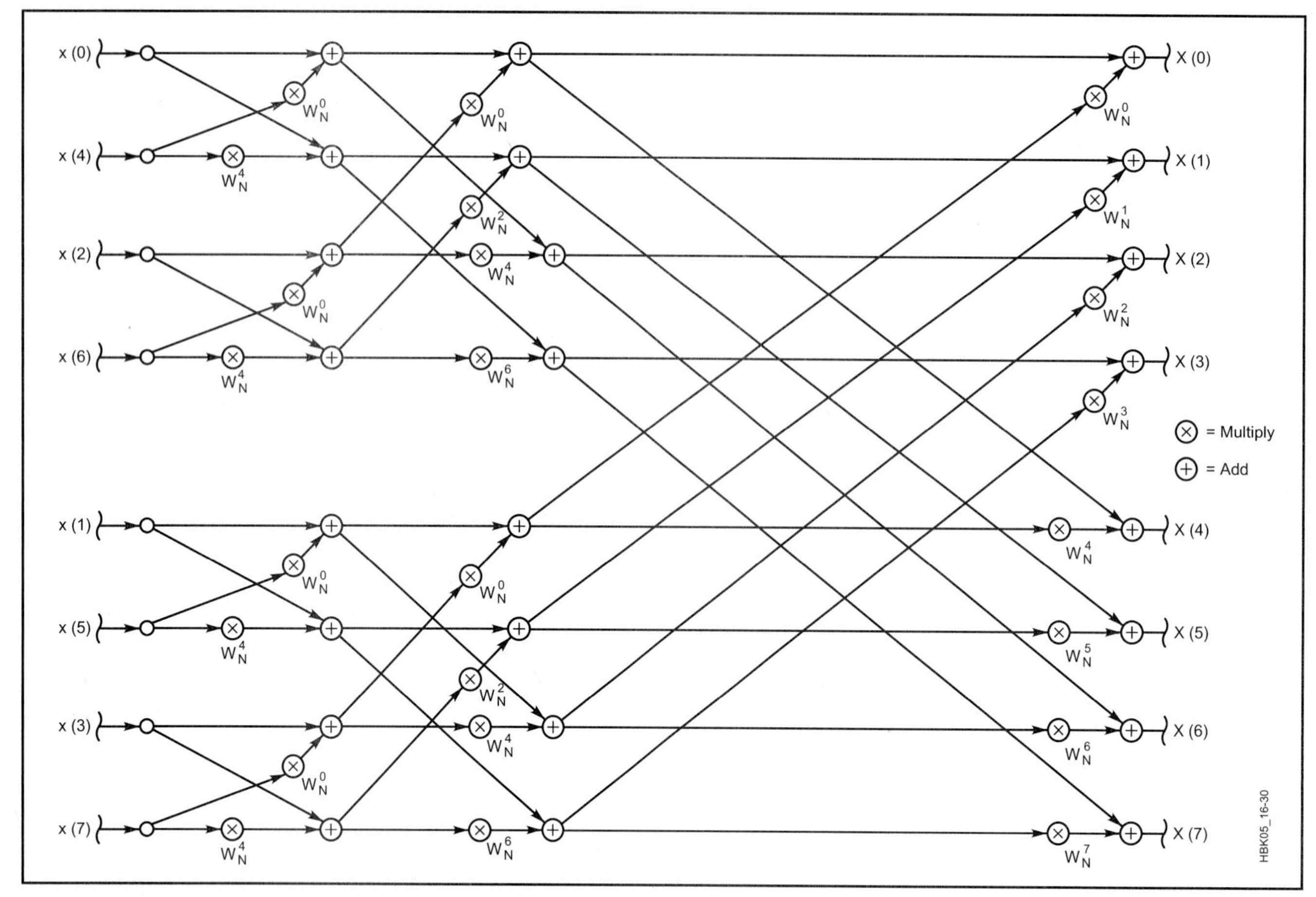

Fig 16.30—Flow chart of an 8-sample FFT.

going on! Computations progress from left to right in $\log_2 N = 3$ stages; each stage requires N complex multiplications and additions, so the total burden is proportional to N $\log_2 N$. Further, each stage transforms N complex numbers into another set of N complex numbers. This suggests we should use a complex array of size N to store the inputs and outputs of each stage as we go along.

An examination of the branching of terms in the diagram reveals that pairs of intermediate results are linked by pairs of calculations like that shown in **Fig 16.31**. Because of the appearance of this diagram, it is known as a *butterfly computation*.

Making use of another symmetry of complex sinusoids, we can reduce the total multiplications of the butterfly by another factor of two. A modified butterfly flow diagram is shown in **Fig 16.32**. This calculation can be performed *in place* because of the one-to-one correspondence between the inputs and outputs of each butterfly. The nodes are connected horizontally on the diagram. The data from locations a and b are required to compute the new data to be stored in those same locations, hence only one array is needed during calculation. A complete

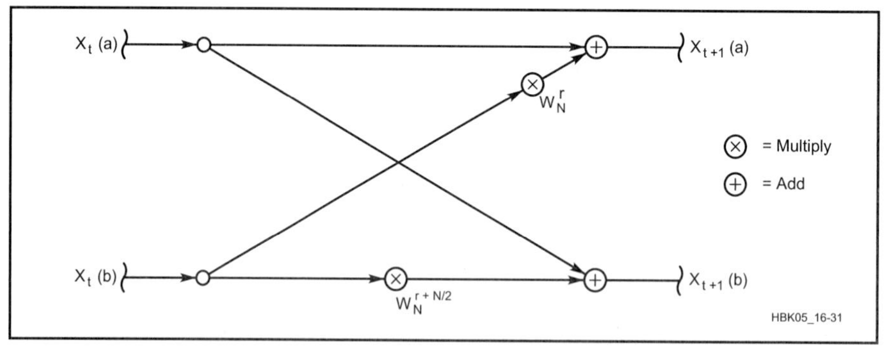

Fig 16.31—Butterfly calculation in a decimation-in-time FFT.

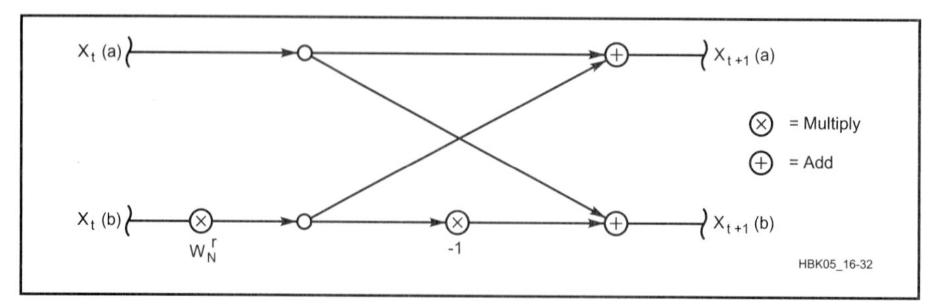

Fig 16.32—Modified butterfly calculation.

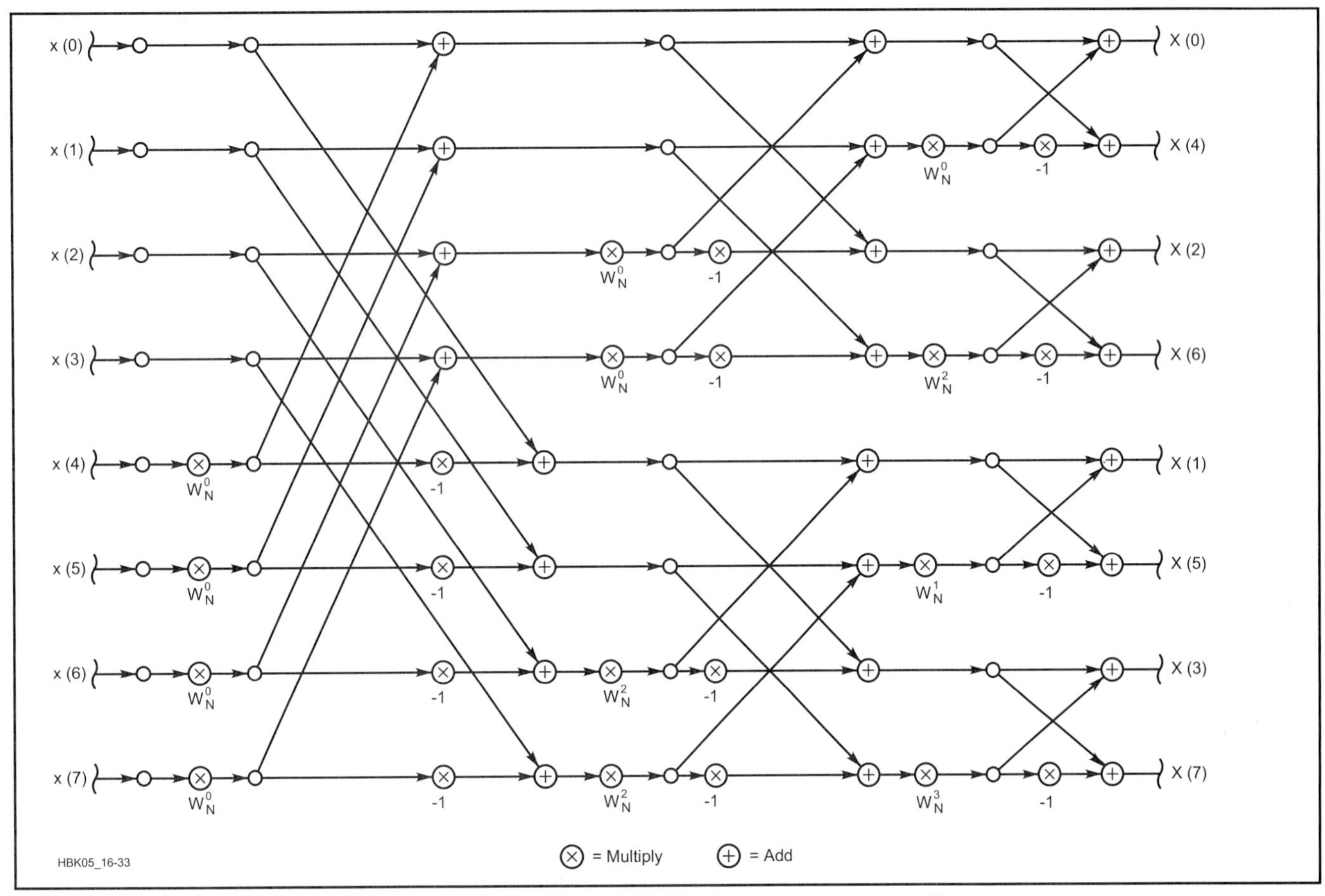

Fig 16.33—Decimation-in-time FFT with different input/output order and modified butterflies.

8-point FFT with the modified butterflies is shown in **Fig 16.33**.

An interesting result of our decomposition of the input sequence is that in the diagram, the input samples are no longer in ascending order; in fact, they are in *bit-reversed* order. It turns out this is a necessity for doing the calculation in place. To see why this is so, let's review briefly what happens in the decomposition process. We first separate the input samples into even- and odd-numbered samples. Naturally, all the even-numbered samples appear in the top half of the diagram, the odds in the bottom. Next, we separated each of these sets into their even- and odd-numbered parts. This process was repeated until we had N subsequences of length one. It resulted in the sorting of the input data in a bit-reversed way. This is not very convenient for us in setting up the calculation, but at least the output arrives in the correct order.

General FFT Computational Considerations

While we are on the subject, this business of bit-reversed indexing is the first thing that ties one's brain in knots during coding of these algorithms, so let's have at it. Several approaches are feasible to translate a normally ordered index to a bit-reversed one: a look-up table, the bit-polling method, reverse bit-shifting and the reverse counter approach.

The look-up table is perhaps the most straightforward approach. The table may be calculated ahead of time and the index used as an address into the table. Most systems do not require very large values of N, so the space taken by the table is not objectionable.

For more space-sensitive applications, the bit-polling method may be attractive. Since the bit-reversed indices were generated through successive divisions by two and determination of odd or even, a tree

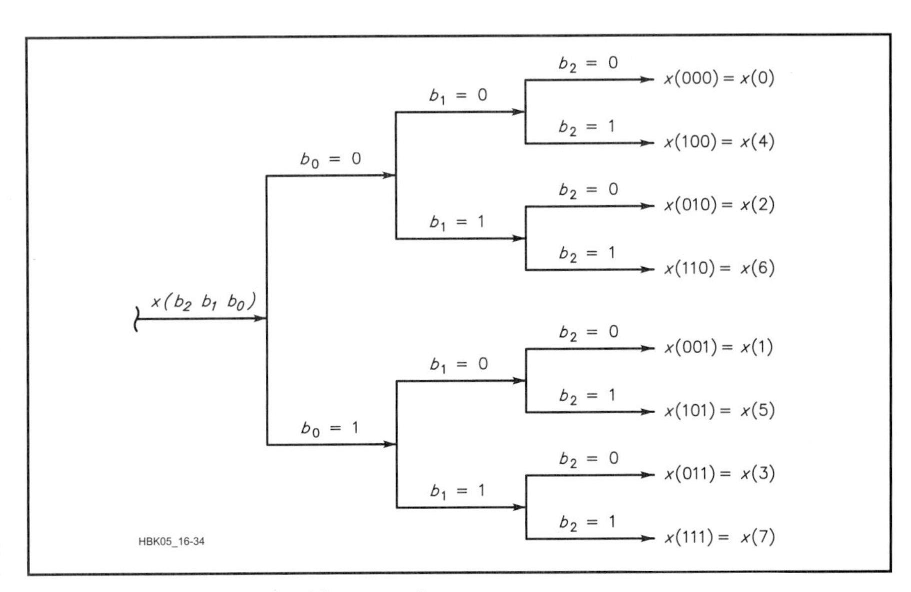

Fig 16.34—Polling tree for bit reversal.

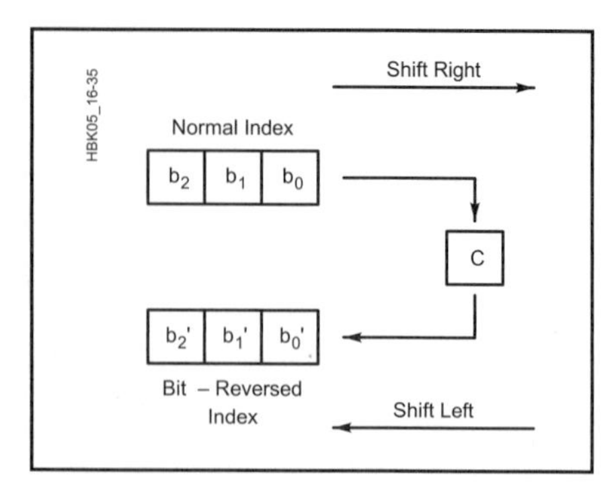

Fig 16.35—Register arrangement for bit-reversal shifting.

structure can be devised that leads us to the correct translation, based on bit-polling. See **Fig 16.34**. The algorithm examines the least-significant bit, then branches either upward or downward based on the state of the bit. Then the second least-significant bit is examined and another branch taken, and so forth, until all bits have been examined.

The bit-shifting method requires about the same computation time as bit-polling. Two registers are used: one for the input index shifting right through the carry bit, the other shifting left through carry. After all the bits have been shifted, the left-shifting register contains the result. See **Fig 16.35**.

Finally, Gold and Rader have described a flow diagram for a bit-reversal counter than can be "decremented" each time the index is to change. If data are actually to be moved during sorting, the exchange is made between data at input index n and bit-reversed index m, but only once. That is, only N/2 exchanges need be performed.

During the actual calculations, indexing of data and coefficients requires attention to many details. In particular, several symmetries about offsets of the index may be exploited. At the first stage of Fig 16.33, all the multipliers are equal to $W_N^0 = 1$, so no actual multiplications need take place; all the butterfly inputs are adjacent elements of the input array x(t). At the second stage, all the multipliers are either W_N^0 r integral powers of $W_N^{N/4}$ and the butterfly inputs are two samples apart, and so forth.

The coefficients are indexed in ascending order. These are normally calculated ahead of time and stored in a table. Another way is to use a *recursion formula* to generate them on the fly, but this is discouraged because of numerical-accuracy effects that destroy the efficiency of the technique.

All those multiplications and additions take their toll on the numerical accuracy of our final result. Quantization noise is multiplied and added as well, and at the output of a DFT, the noise power grows by N times.

In an FFT calculation, the situation is roughly the same; however, the requirement to avoid overflow at intermediate stages may force us to scale the data, the coefficients, or both. This further reduces the dynamic range of any FFT. Results have been offered indicating noise increases in the vicinity of 12N. In addition, the quantization-noise contribution of the coefficients increases in inverse proportion to p, the number of bits used to represent them. This, in turn, means that the noise increase with respect to N is slow.

In FFT-based noise-reduction systems, we perform some modification of the frequency-domain data, such as zeroing bins not meeting a pre-defined amplitude

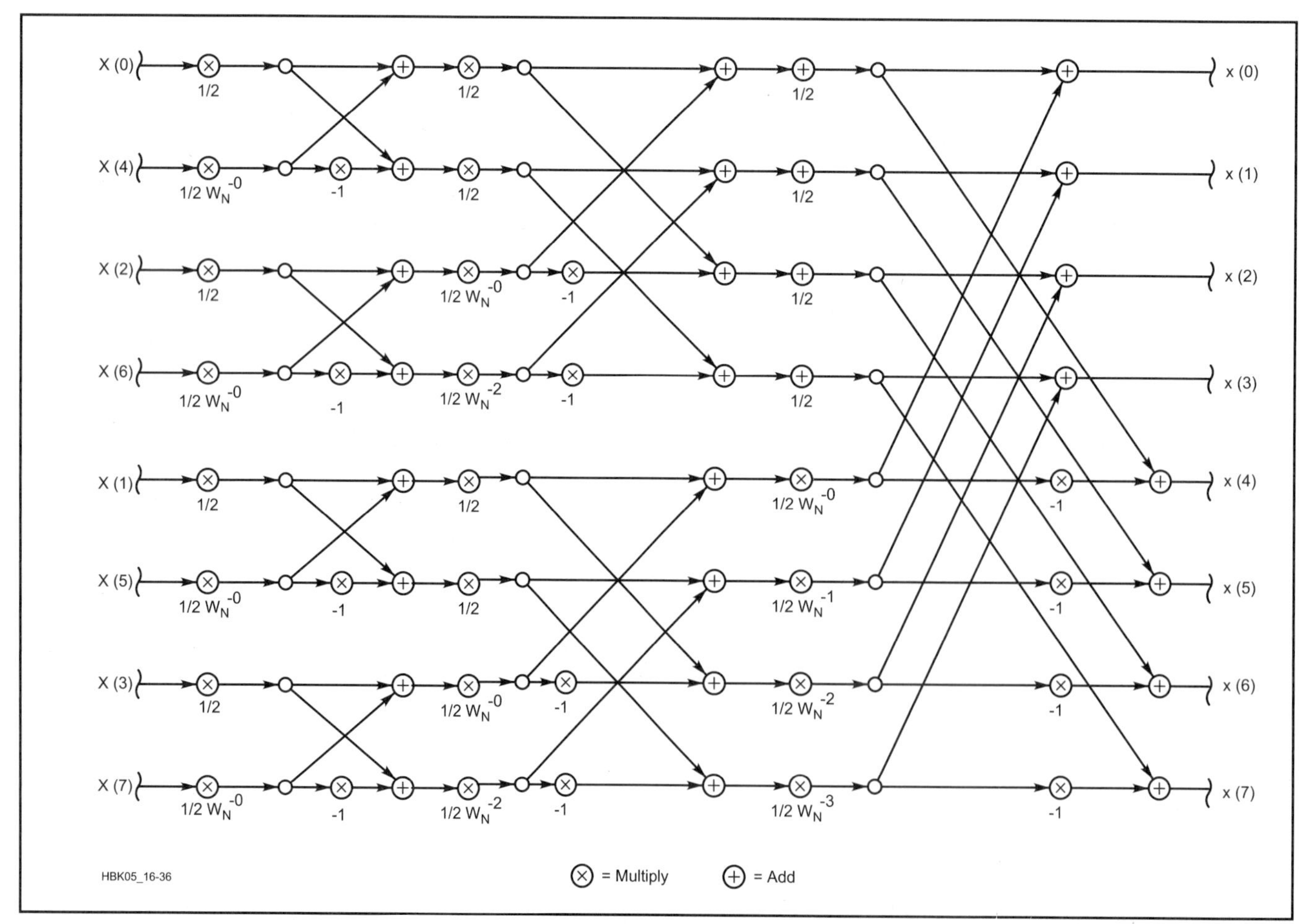

Fig 16.36—FFT^{-1} implemented by interchange of inputs, outputs, and coefficients.

threshold. Then we transform the modified data back to the time domain. The duality of the Fourier transform and its inverse can be shown in the flow diagram of a FFT^{-1} as in **Fig 16.36**. This diagram was produced from Fig 16.33 by simply substituting ½ W_N^{-kn} for W_N^{kn} at each stage and, of course, using X(k) as the input to obtain x(t) as the output.

Alternatively, we may compute the FFT^{-1} by using the FFT flow diagram and swapping the inputs and outputs and reversing the direction of signal flow. It is important to note that this is a consequence of that fact that we can rearrange the nodes of the flow diagrams however we want, so long as we do not alter the result. The transforms work just as well in reverse as they do in the forward direction.

Damn-Fast Fourier Transforms

When it is necessary to compute Fourier transforms on a sample-by-sample basis, or where frequency resolution must be non-uniform across the sampling BW, even traditional FFTs may be too computationally intensive for the processing horsepower available. A class of algorithms that computes the next transform output very rapidly—based solely on current transform output and the next input sample—has been discovered. A method is included here for controlling its inherent divergence problem by brute force.

The derivation begins by looking at how the Fourier transform results change for each bin at each sample time. Say we start with some discrete Fourier transform output bins $X_r(k)$ at sample time r. Then we compute the DFT for the next sample time r + 1 and examine the sequences to see what has changed. For r = 0, each DFT sequence expands to:

$$\begin{aligned} x_0(k) &= W_N^{0k}\,x(0) + W_N^{1k}\,x(1) + \\ &\quad W_N^{2k}\,x(2) + \cdots + W_N^{(N-1)k}\,x(N-1) \\ x_1(k) &= W_N^{0k}\,x(1) + W_N^{1k}\,x(2) + \\ &\quad W_N^{2k}\,x(3) + \cdots + W_N^{(N-1)k}\,x(N) \end{aligned} \qquad (51)$$

What is evident is that each input sample x(n) that was multiplied by W_N^{nk} in the summation for $X_0(k)$ is now multiplied by $W_N^{(n-1)k}$ in the summation for $X_1(k)$. The *ratio* of the two sequences is nearly:

$$\frac{X_1(k)}{X_0(k)} \approx \frac{W_N^{(n-1)k}}{W_N^{nk}} = W_N^{-k} \qquad (52)$$

We still have two terms "hanging out" of the relationship, namely the first and the last:

$$W_N^{0k}\,x(0) \text{ and } W_N^{(N-1)}\,x(N) \qquad (53)$$

that have not been accounted for in the ratio. If we first subtract x(0) from $X_0(k)$ before taking the ratio, then add the new term $W_N^{(N-1)k}\,x(N)$ after, we have the correct result:

$$x_1(k) = W_N^{-k}\left[x_0(k) - x(0)\right] + W_N^{(N-1)}\,x(N) \qquad (54)$$

Now this may be simplified a little, since:

$$\begin{aligned} W_N^{(N-1)k} &= e^{\frac{-j2\pi(N-1)k}{N}} \\ &= e^{\frac{-j2\pi N}{N}} \bullet e^{\frac{j2\pi k}{N}} = W_N^{-k} \end{aligned} \qquad (55)$$

and substituting:

$$X_1(k) = W_N^{-k}\left[X_0(k) - x(0) + x(N)\right] \qquad (56)$$

This is the *damn-fast* Fourier transform (DFFT). It means: For N values of k, we can compute the new DFT from the old with N complex multiplications and 2N complex additions, or a computational burden proportional to N. If we begin with $X_0(k) = 0$ and take the first N value of x(n) = 0, we can start the thing rolling. It saves computation over the FFT by a factor of:

$$\frac{N\log_2 N}{2N} = \frac{\log_2 N}{2} \qquad (57)$$

which for large values of N is very significant indeed. For example, if N = 1024, the improvement is by a factor of five. Over the direct-form DFT, it is a factor of N^2/N faster. But there is a catch: An error term will grow in the output because the truncation and rounding noise discussed previously is cumulative. The error will continue to grow unless we do something about it.

The simplest way to handle the situation is to compute two DFFTs for all the output bins k, resetting every other block of N input samples to zero. In other words, one DFFT begins at some time with an input buffer that has been zeroed, the other continues to operate on the continuous stream of real input samples. As sample-taking continues, DFFT output is taken from the second calculation. As the buffer of the first DFFT gradually fills with real samples, the block of zeroes it originally held disappears. At this point, each DFFT produces the same result except for the greater error in the second DFFT because of truncation and rounding effects. Output is then taken from the first DFFT and the buffer of the second is zeroed; the calculations continue for another N iterations, at which time the exchange and reset are again done, and so forth, continually. This places an upper bound on the cumulative error to that associated with 2N iterations and increases the computational burden by a factor of two. Now the savings over the FFT is only:

$$\frac{\log_2 N}{4} \qquad (58)$$

which for N > 16 still represents an improvement. DFFT output quantization noise is at least twice that of the DFT.

Frequency resolution of DFFTs is controlled by the block length, N, used in the calculations, just as in DFTs or FFTs. Resolution may be set differently, though, for each bin; further, not all bins need be computed to compute any particular bin, unlike the Cooley-Tukey FFT. Is there an inverse DFFT? Well, because inverse Fourier transforms map into the time domain, it is simple enough to just compute the next output sample rather than the next N output samples. The easiest output term to compute is x(0), since all coefficients are $W_N^0 = 1$. The output is then just:

$$x(0) = \frac{1}{N}\sum_{k=0}^{N-1} X(k) \qquad (59)$$

and only one multiplication is involved.

In radio transceivers, DSP may be applied at baseband or audio, at an IF stage, or directly at RF. This section examines general approaches for each of those topologies. As we move the analog-to-digital interface closer to the antenna, we eliminate an increasing amount of analog hardware.

DSP at Baseband

It is reasonably straightforward to apply DSP at baseband or audio with any kind of traditional analog transceiver design. Such an arrangement is shown in **Fig 16.37** as the combination of a regular analog receiver and an outboard DSP unit. Many features typically associated with DSP may be obtained that way, such as noise reduction, automatic notch and speech processing. The one feature that is a bit difficult to obtain with that configuration is additional bandwidth reduction in the receiver through DSP filtering.

Let's say the receiver bandwidth is 3.0 kHz and we wish to implement an RTTY filter having a bandwidth of 500 Hz. It follows that some of the signals we digitize will be outside the final bandwidth. When the desired signal is strong relative to the undesired, everything is fine; but when a strong undesired signal appears within the receiver's bandwidth and outside the DSP filter, it may actuate the receiver's analog AGC. That would reduce the level of our desired signal as the receiver keeps the level of the undesired signal constant. Our desired signal's amplitude would go up and down with the level of the interference.

Without some form of gain compensation, such a system would be unusable. Of course, we could turn off the analog AGC in the receiver; but then, the total dynamic range would be severely compromised and distortion would become likely for strong signals. Instead, we may elect to implement a *digital AGC system* in DSP that compensates for the gain variations and provides its own timing.

A block diagram of part of a digital AGC system is shown in **Fig 16.38**. It consists of a gain-control block (multiplier) and a ratio detector. In the diagram, the peak undesired signal amplitude is called m; the peak desired signal amplitude is called n. The signal that is digitized is naturally the sum of the desired and undesired signals, or μ + n. The ratio detector computes the ratio of that sum to n:

$$k = \frac{m+n}{n} \qquad (60)$$

where k is the factor by which the filtered output must be digitally boosted to remain at constant peak amplitude. Note that is true only when μ + n is constant and the receiver's analog AGC is working. Below the receiver's analog AGC threshold, no digital gain boost would be required because no gain reduction occurs. Thus, a separate digital AGC subsystem is required to hold n constant. That part is shown at the right of Fig 16.38. Holding n constant means that this digital AGC has no threshold or "knee." All signals down to the noise floor of the receiver are amplified to the same peak amplitude.

Decay times of the two parts of the algorithm must be identical. To make the thing work properly, they must also be equal to or less than that of the receiver's analog AGC. A delay is inserted in the detector path μ + n to compensate the delay through the DSP filter. Scaling might be necessary to prevent overflow in the algorithm. Special attention must be paid to what happens during the attack time. Some receivers exhibit *AGC overshoot*, which may cause spikes on incoming signals, resulting in rapid gain excursions. A good approach is to allow gain adjustment in proportion to the attack time of μ + n, but only if it persists at a higher level for several milliseconds. That avoids reaction to noise pulses.

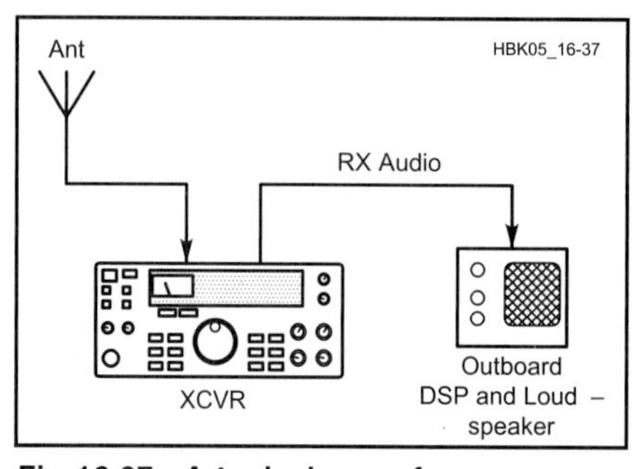

Fig 16.37—A typical use of an outboard, baseband DSP processor.

Factor k is always greater than one, hence the multiplication is not the simple fractional type described above. We may now have a need to extend fixed-point math to values greater than unity. It is tedious but not too difficult. We just handle the integer and fractional parts separately. We have to multiply k by a fractional decay factor δ at each sample time and also multiply k by another fraction—the filtered signal. Separating the integer and fractional parts by a radix point, we adopt the notation k = (a.b), where:

$$a \in \Im . b \in \mathcal{F} \qquad (61)$$

meaning that we treat a as an integer and b as a fraction.

A number like decay factor δ has a zero integer part: δ=(0.d). The result of the multiplication k δ=(a.b)(0.d) is:

$$(a.b)(0.d) = ([\Im ad + \Im bd] \times [\mathcal{F} ad + \mathcal{F} bd]) \qquad (62)$$

where the carry is from the addition of the fractional products, which must occur first.

In practice, baseband DSP filtering may

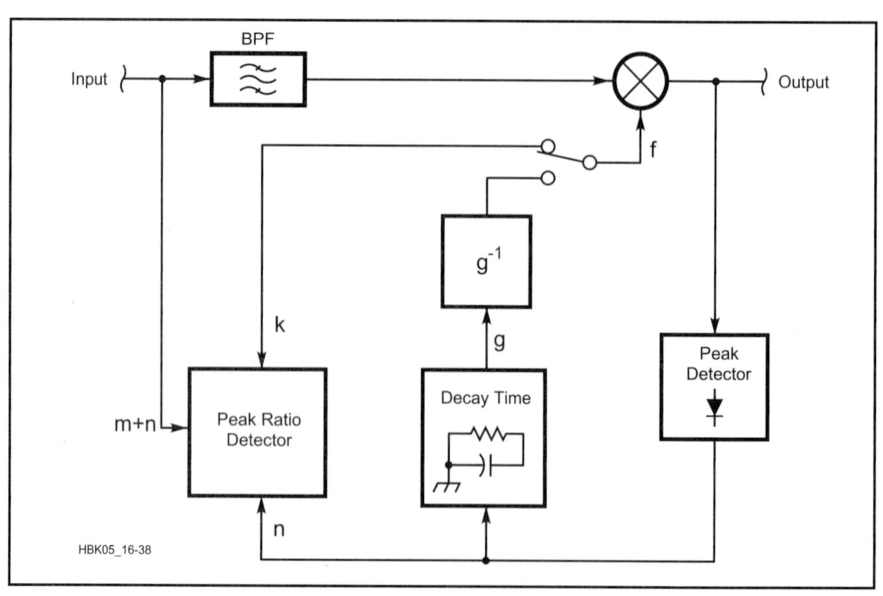

Fig 16.38—Digital AGC system block diagram.

be limited by *in-band IMD* and synthesizer *phase-noise* effects that plague the analog transceiver itself. Those may cause unwanted signals to appear in the passband, masking the desired signal. With an otherwise perfect receiver, performance is limited by the available SNR and SFDR of the ADC and by the phase noise of its clock. As we move the digital-to-analog interface closer to the antenna, converter noise and phase-noise issues become more critical. Other factors actually help us as we go to IF-DSP.

IF-DSP at a Low IF

The primary reason for wanting to digitize signals closer to the antenna is to eliminate expensive filters and other hardware whose functions may be performed in DSP. By going to a low IF, we can get rid of analog balanced modulators, product detectors, squelch and even multiple crystal or mechanical filters. Many things judged quite difficult or impossible in the analog world may be included, as well. On the other hand, many designers are returning to crystal or mechanical filters ahead of the digitization point in DSP receivers. That still ensures maximum protection against interfering signals for those who want top performance.

To do IF-DSP in a receiver, we apply harmonic sampling and a fast sigma-delta ADC at an IF in or near the audio range. 16- and even 24-bit ADCs are common at the time of this writing because they are widely used in digital audio applications such as CD players.

Recall that in harmonic sampling, the sampling frequency may be as low as the IF minus half its bandwidth; but the sampling frequency cannot be less than twice the bandwidth, lest aliasing occur. An IF bandwidth of 20 kHz, for example, requires a sampling rate of at least 40 kHz. We ought to consider, however, what *image rejection* we are going to get based on such a low IF. Roofing filters in the IF strips and other frequency-selective circuits will determine the image rejection by their attenuation at a frequency offset equal to twice the IF. If we intend to use the same IF in transmit mode, the second LO will appear at a frequency offset equal to the IF. Quite a few poles of analog filtering are required around this arrangement. See **Fig 16.39**.

From the antenna, signals are band-pass filtered to attenuate first-mixer image responses, to reduce LO leakage and to improve second-order IMD response. Then they are mixed to a VHF first IF to dodge as many spurious responses as possible. A VHF first IF may be chosen above twice or even three times the highest RF to get away from second- and third-order products. Six to eight poles of crystal filtering may be used in the strip, with several gain-controlled stages interspersed. In any receiver design, it is best to distribute gain and loss evenly to avoid degradation of SNR under reduced-gain conditions. We would like the received SNR to continue increasing as the input signal strength increases. Therefore, gain reduction is usually made to occur in the stages farthest away from the antenna first, followed by earlier stages nearest the antenna.

First-IF signals may be converted directly to the low IF or an intermediate IF may be used. The second-IF strip amplifies them and possibly filters them further. Enough gain is included to raise minimum signal levels to about 10 dB above the noise floor of the DSP section. That assures that ADC itself will not affect the sensitivity of the receiver. A traditional analog AGC is employed.

The analog AGC prevents very large signals from exceeding the maximum allowable ADC input level. It thereby extends the dynamic range of the receiver. Only a few years ago, state-of-the-art ADCs did not exhibit sufficient dynamic range for high-performance HF rigs and range extension was absolutely necessary. The analog AGC sets the IF output to 6-10 dB below the ADC maximum input level. That margin allows the *headroom* necessary to accommodate AGC overshoot and noise spikes. ADC overload is catastrophic and must not be allowed. Finally, an analog AGC makes it easier to keep analog stages linear over the range of signals encountered. These days, receivers may be called on to handle signals are large as one watt! Recent designs employing 18- to 24-bit ADCs exhibit 100 dB or more of SFDR so that analog AGC need not come into play until signals reach S9+30 dB or so.

IF-DSP receivers typically must have digital AGCs as well as analog. Embedded DSP systems have information about what the analog AGC is doing, so it is possible to make the two AGCs work together to achieve the desired characteristics. Those desired characteristics include an AGC threshold that resides well above the noise floor of the receiver. Traditionally, receivers have been designed with AGC thresholds around 3 mV. Signals below that level are not gain-controlled

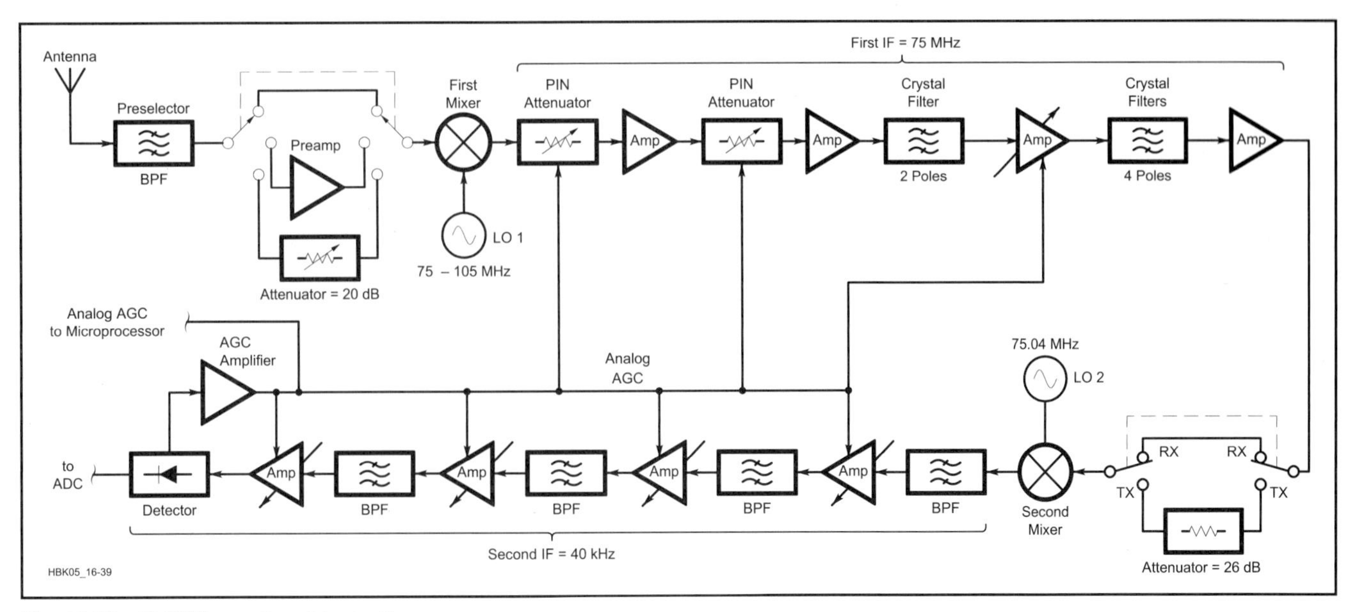

Fig 16.39—IF-DSP receiver block diagram.

and the receiver gets quiet.

In digital AGC systems, it is relatively easy to provide a variable threshold. The net effect of a variable threshold is very much like that of an IF gain control. It is also fairly easy to implement a peak-hold or "hang" function that retains the most recent peak for an adjustable period of time. The decay time of the AGC may be readily set in software to yield settings from slow to fast. A very fast decay time essentially turns the digital AGC off, allowing large signals to be clipped at the set output level. The attack time is generally fixed.

For traditional analog AGC systems not under the control of the DSP, analog gain-reduction information may be obtained by digitizing the AGC voltage, as shown in Fig 16.39. The voltage value is used to look up a gain-reduction factor from a table stored in non-volatile memory. Such a table may be built using measurements of the actual hardware. Minor unit-to-unit variations are readily handled by placing the digital gain-compensation point inside the main digital AGC loop, as described below.

An alternative approach involves generating the analog AGC voltage in the DSP itself. See **Fig 16.40**. A digital-to-analog converter develops a voltage for application to analog gain-controlled stages. The chief drawback to the scheme is a significant delay between peak detection and gain change, since signals must propagate all the way through the DSP section before being detected. That can be compensated with a delay in the analog IF strip; but typically, the required delays of several ms are impractical.

In any case, call the analog AGC gain-reduction factor g, where $0 < g < 1$. For example, were $g = {}^1/_2$, analog gain reduction would be $-20 \log ({}^1/_2)$ or about 6 dB. Now it remains for the DSP to compute how much of that gain reduction was caused by in-band signals and how much by interference. If all of it were caused by in-band signals, no gain compensation would be necessary and we would use digital gain-boost factor $f = 1$. If all of it were caused by interference, in-band signals would have to be boosted by a factor $f = g^{-1} = 2$. For cases in between those two extremes, the procedure is a little tricky because f cannot be described by a single equation.

As in the baseband case of Fig 16.38, the DSP calculates the ratio $k = (m + n)/n$. To restore a variable threshold to the digital AGC, the next step is to determine whether n by itself was large enough to actuate analog AGC. The DSP does that by comparing k with g^{-1}. The algorithm accounts for three cases in the comparison.

- Case 1: If $k < g^{-1}$, then n by itself is large enough to actuate analog AGC and the gain-boost factor used is $f = k$. The ratio of signals solely determines the boost factor.
- Case 2: If $k > g^{-1}$, then n by itself is not large enough to actuate analog AGC and the gain-boost factor is $f = g^{-1}$. Analog gain reduction solely determines the boost factor.
- Case 3: When $k = g^{-1}$, it obviously does not matter which is used as the gain-boost factor since they are equal.

Remember that when analog AGC is inactive, no gain boost need be applied. Note that g depends only on the characteristics of the analog gain-controlled stage or stages; k depends on the ratio of in-band and interfering signals, irrespective of the analog section. The two possible gain-boost variables therefore produce different functions and curves. The curves are guaranteed to meet where $k = g^{-1}$.

Placing the digital gain boost inside the AGC loop assures that a constant peak output level will be maintained even in the face of minor variations in analog gain control. Inside the loop, we apply digital gain boost to signals before they are peak-detected. Therefore, the main digital AGC loop prevents them from exceeding the set output level when interference—and k or g^{-1}—rapidly increase. In addition, IF gain may be manually reduced by artificially increasing the analog AGC voltage without deleterious effects.

Finally, gain-boost factor f may be directly used to compensate a signal-strength meter by the appropriate amount. Below the onset of analog AGC, the DSP makes a measurement of the peak IF level to find signal strength, along with factor f; above the analog AGC threshold, the look-up table mentioned above must be used to add to the S-meter reading since the IF peak level remains constant. So just as the receiver output level remains constant in the presence of interference, so does the S meter. When IF gain is manually reduced, the S meter goes down— not up, as in so many rigs.

Conversion schemes used in IF-DSP receivers may also be used in the transmitter by simply swapping the LOs, inputs and outputs. One switching arrangement for that is shown in **Fig 16.41**. Isolation between the ports of the LO DPDT switch

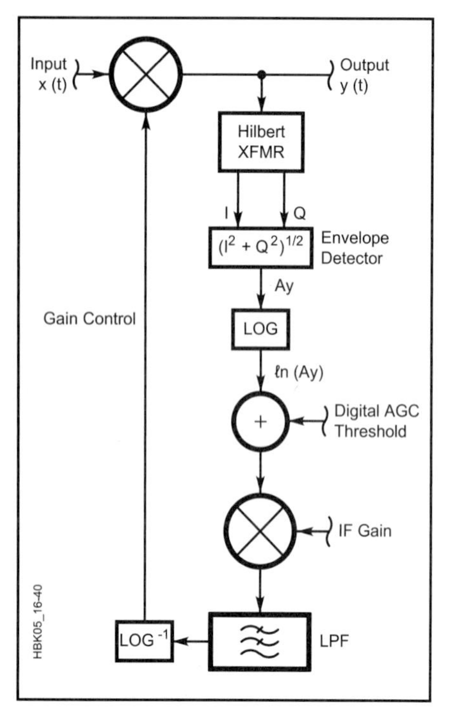

Fig 16.40—IF-DSP receiver with digitally derived analog AGC (after Frerking).

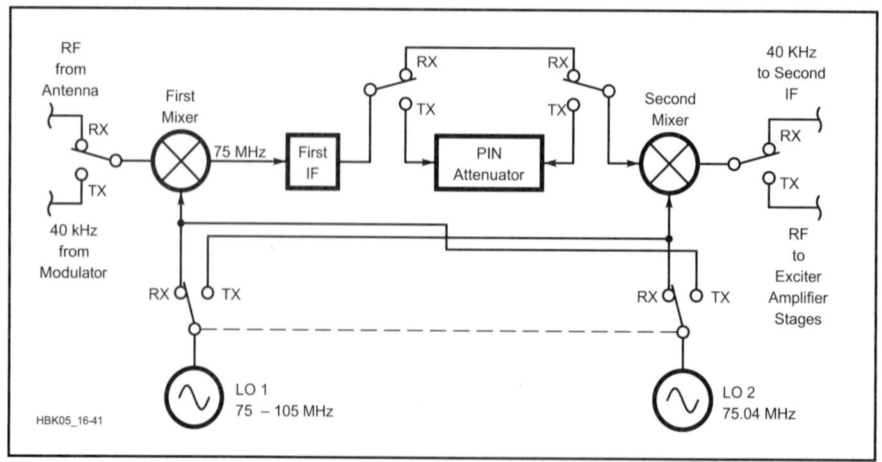

Fig 16.41—Block diagram of IF-DSP conversion scheme with T/R switching added.

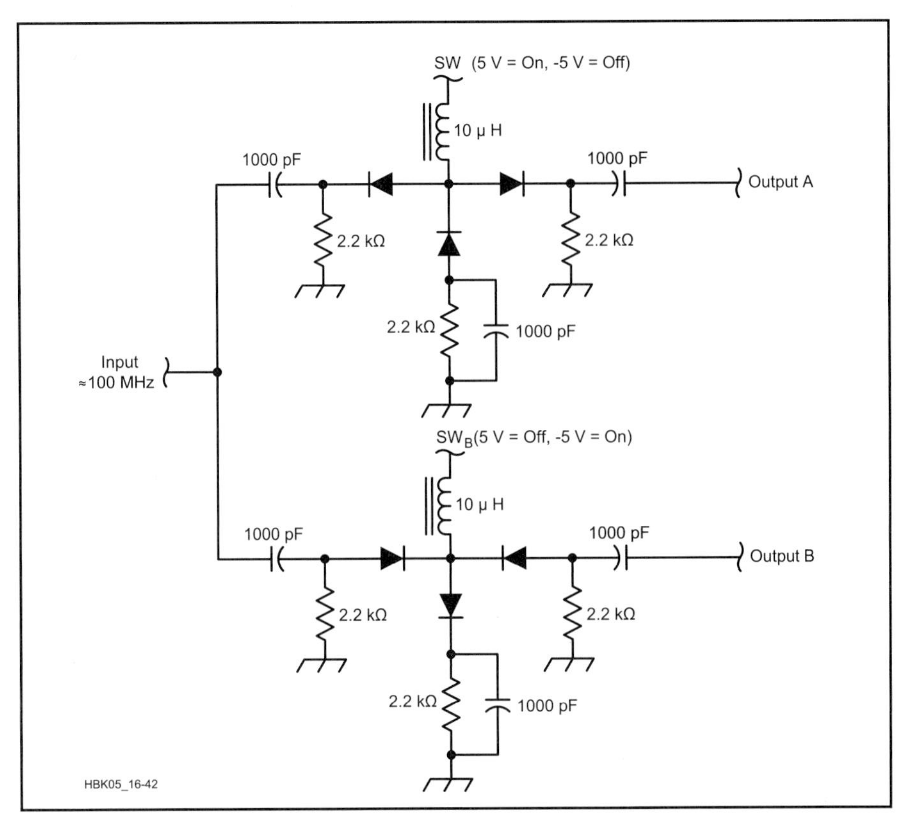

Fig 16.42—SPDT LO switch using PIN diodes. Diodes are Philips BA682 or equiv.

must be better than or equal to the desired level of transmitter spectral purity. An example of such a DPDT switch is shown in **Fig 16.42** using PIN diodes at VHF. Switch control voltages swing between +5 Vdc and –5 Vdc. When the series diodes are on, the shunt diodes are off, and vice versa. This particular circuit was designed for a 75-105-MHz first LO and a 75.04-MHz second LO. It achieves around 80 dB of isolation in the worst case with careful PCB layout.

Switching of the first mixer's input may be achieved using a relay or PIN diodes. The second mixer's output may be switched using commercially available ICs, such as the Signetics SE630. Isolation in these switches is important because it determines spurious responses in both receive and transmit modes.

Gain-controlled stages or step attenuators may have to employed to provide a change in IF-strip gain between transmit and receive. To see why that might be necessary, let's examine the difference in gain between a transmitter and receiver in typical service.

A receiver takes as little as –132 dBm from an antenna and amplifies it to around 1 W, or +30 dBm. The gain is therefore:

$$A_{RX} = 30 - (-132) \text{ dBm} = 162 \text{ dB} \quad (63)$$

In a transmitter, a typical dynamic microphone might produce 5 mV RMS into 600 Ω, or –44 dBm. To get to 100 Ω or +50 dBm, the gain is:

$$A_{TX} = 50 - (-44) \text{ dBm} = 94 \text{ dB} \quad (64)$$

The receiver has the far more difficult task, but the transmitter is still doing yeoman's duty. Considering a maximum path loss of:

$$LOSS_{PATH} = 50 - (-132) \text{ dBm} = 182 \text{ dB} \quad (65)$$

it is a wondrously large amount of enhancement we get from our electronics, since the total power gain from the microphone on one end to the loudspeaker on the other end must be:

$$A_{TOTAL} = 162 + 94 = 256 \text{ dB} \quad (66)$$

or a factor of 4×10^{25}!

Some recent commercial receiver designs have gone to a front-end AGC system that reduces the RF gain instead of or along with the IF gain under large-signal conditions. That is fine so long as the subsequent increase in noise figure can be tolerated.

Transmitters are likely to have gains that vary quite a bit with frequency, temperature and supply voltage. Like receivers, they may be called on to handle a large range of input levels without exceeding a set output level. ALC serves that purpose.

It is plausible to arrange for ALC in an IF-DSP transmitter by digitizing an indication of forward power, such as from a bridge, and adjusting the drive signal applied to the exciter. In that case, no analog gain-controlled stages are needed; but it does reduce the available dynamic range of the transmitter somewhat, but it is not usually enough to worry about.

The other possibility is to employ a traditional analog ALC with gain-controlled stages. Still, some adjustment of drive from the DSP is called for to maintain optimum performance over wide ranges of frequency and output power.

Transmit gain control (TGC) is a neat concept that was evidently first practiced at Collins Radio. It is a secondary ALC system that slowly changes the maximum drive applied to a transmitter so that the main ALC does not have to work so hard. The benefits include a minimum of overshoot on SSB and CW and prevention of ALC pumping.

We must apply sufficient drive to achieve desired output power; but we do not want to apply more drive than absolutely necessary. When a DSP can get information about the required level, it can optimize drive. One reason to do so is to maintain optimal RF rise and fall times and RF envelope shapes that minimize interference to others.

When an ALC-controlled transmitter is driven hard, it rises rapidly to its set power level. After it gets there, the ALC loop attempts to reduce gain. If all that happens too fast, it becomes very difficult to avoid spikes and other artifacts in the output. Digital TGC forces a DSP to examine ALC voltage to determine the amount of gain reduction occurring in analog. As in the receiver case, it does that by digitizing the voltage and using it as an address into a look-up table. When analog gain reduction is excessive, the DSP is programmed to reduce drive. In the absence of ALC, it is programmed to increase drive to a preset maximum. TGC usually changes quite slowly, although it is often set to reduce drive more quickly than to increase it.

TGC is set to achieve a drive level slightly higher by a fixed margin than what is necessary to attain rated power. A 1-dB margin is common. Note that no matter what the set power level, TGC will alter drive to match. That is handy in transmitters that use ALC over a wide range of power levels.

Direct-Conversion Transceivers

In a direct-conversion receiver, signals are converted directly to baseband without intervening IFs. An increasingly popular method these days is to use some kind of image-canceling quadrature mixer at the front end, coupled with a DDS-controlled, low-noise LO and baseband filtering. The quadrature mixer converts RF signals to an analytic pair which, in turn, is digitized by a sound card on a PC. The PC then does the demodulation, spectral analysis and so forth in DSP.

One implementation of a quadrature mixer having outstanding performance is the so-called *Tayloe detector*, popularized by Dan Tayloe, N7VE. It is a commutating, sampling mixer and detector that uses four LO phases. It reportedly achieves good large-signal performance, low conversion loss (< 1 dB) and with proper clock generation, good noise figure.

Gerald Youngblood, AC5OG, has chosen the Tayloe detector for his SDR-1000 project. A block diagram of the hardware portion of his receiver is shown in **Fig 16.43**. Refer to Gerald's *QEX* articles or visit the ARRL TIS pages on software radios for details (**www.arrl.org/tis**).

Leif Åsbrink, SM5BSZ, uses a similar but more traditional quadrature-mixer approach in his Linrad system, which runs under *Linux*. His *QEX* articles, listed in the **Bibliography** at the end of this chapter, contain the specifics. They also may be downloaded from ARRL's TIS site.

In the direct-conversion technique the LO is, in effect, placed very close to the desired signal and through sampling and decimation, translates it to baseband. The closeness of the LO to the desired signal accentuates phase-noise effects such as reciprocal mixing and makes short-term clock stability an issue. Fortunately, low-noise, crystal-derived synthesizer and clock designs are becoming available. RMS clock jitter is usually specified in units of time (picoseconds), but a clock's phase-noise-versus-frequency characteristic tells the whole tale.

The Nyquist criterion compactly determines the sampling rate required for any given signal or group of signals. If the digitized bandwidth is 50 kHz, the minimum sampling frequency must be at least 100 kHz, even if the signal frequencies lie in the VHF range or beyond. Ancillary sample-and-hold devices may be employed in a direct-conversion receiver to ease the requirements of an ADC. Digitized bandwidth must remain within half the final sampling frequency to avoid aliasing; thus, interest in narrow pre-selector filters has been renewed.

In the example of a 50-kHz received bandwidth, any increase in sampling rate above 100 kHz is called *over-sampling*. Over-sampling may be important because it provides an SNR gain by spreading quantization noise over a larger bandwidth, then filtering out some of the noise, as discussed above. When we use harmonic sampling, however, we may also be *under-sampling* our signals. We can be both over-sampling and under-sampling simultaneously because one is defined with respect to sampled bandwidth and the other by the frequencies of interest.

Frequency planning is of special concern in direct-conversion architectures. Quite commonly, spurious responses appear in high-speed data converters that we must plan to avoid. Problems may also be created in supposedly linear analog stages that generate significant harmonic content. Sometimes, those harmonics show up as aliases in the digitized spectrum that may appear in one's passband or mix with other signals present. Careful selection of sampling frequency and IF may place those responses where they are harmless: outside the band of interest. Over-sampling generally moves us toward the goal of high SFDR by providing more spectrum into which spurs may harmlessly fall.

The technique known as *dithering* further improves SFDR in general by spreading the energy in discrete spurs over greater bandwidths. Dithering artificially adds noise to the data-converter clock, to the input, or both to achieve spreading. Spurious reduction on the order of 20 dB has been attained with modern high-speed (> 40 MHz) data converters.

Digital Direct Conversion

In the ultimate digital receiver, signals are sampled directly at RF without any analog mixers or conversion stages. In practice, some gain is required ahead of the ADC because of the current limitations of technology. So far as gain stages can be designed with high dynamic ranges and good large-signal handling capacities, direct digital conversion (DDC) comes within reach.

In a DDC receiver, RF signals are translated to baseband using a numerically controlled digital oscillator or DDS. Such a DDS produces only digital samples of the LO, since it mixes samples of the RF digitally. Harmonic sampling may be employed to capture only a small portion of the spectrum available, or high-speed sampling may be used to capture large chunks of the band of interest. In a narrow-band situation such as SSB or CW on HF, several stages of decimation are implemented in hardware to reduce the sampling rate as bandwidth is decreased. Analog pre-selectors may still be a wise addition to the design to preserve second-order dynamic range. See **Fig 16.44**.

Note that a DDC receiver may separately down-convert and demodulate more than one channel at a time. Digital down-converter ICs are now appearing that assist the designer toward that goal. Cellular-telephone and other commercial systems have been exploiting that DSP advantage for several years now and it is expected to appear in Amateur Radio.

Digital direct conversion is easier to

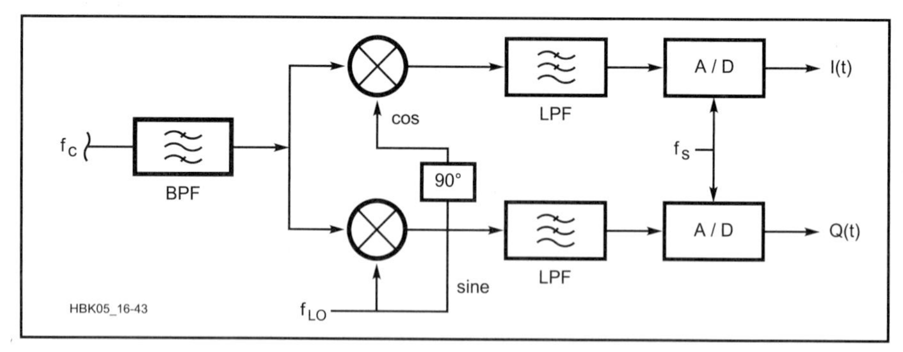

Fig 16.43—Block diagram of AC5OG's hardware.

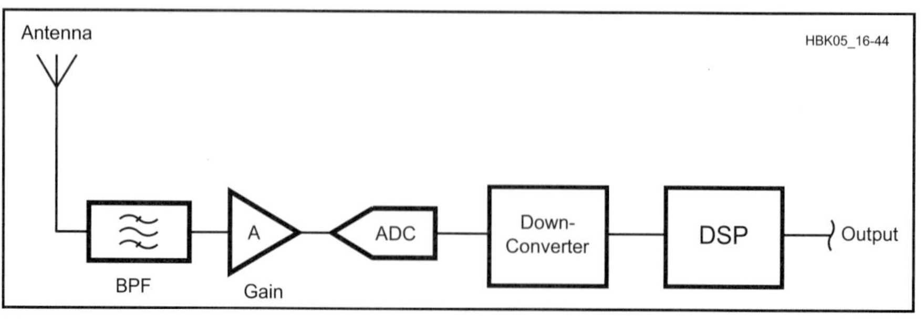

Fig 16.44—Block diagram of digital direct-conversion hardware.

implement in transmitters than in receivers. Advances in DAC technology have come faster than for ADCs. ICs such as the Analog Device AD9854 (300 MHz) and AD9858 (1 GHz) make it possible to directly generate virtually any kind of waveform directly at RF. Filtering and amplification yield a simple yet very accurate DSP system for exciters.

All this flexibility has led to a universally accepted concept that opens vast possibilities for amateurs: the *software radio*. Let's look at what's possible now with software radios and what they hold in store for the future.

SOFTWARE RADIOS

What is a software radio? Well, to be as comprehensive as possible, we can state that a software radio is a radio:

1. Whose hardware is so ubiquitous as to be able to handle almost any modulation format, signal bandwidth and frequency desired.
2. Whose functionality may be altered at will by downloading new software.
3. That replaces traditional analog subsystems with DSP implementations.
4. That may be commanded to perform adaptive signal processing and other operations with the goals of finding clear channels on which to communicate and avoiding interference with other users.
5. That may be instructed how to independently recognize various communications signal formats and conform to them.

The first three definitions may be considered primary and the last two, secondary; but they all illustrate certain possibilities. One virtue of software radios is their flexibility. Only software stands between the status quo and a new set of functions or a new level of performance. Writing software is not for everyone; but once it is written, it is readily ported among compatible platforms, such as PCs.

A software radio that uses a PC for its DSP functions and a standard hardware interface is attractive for that and other reasons. The newest designs incorporate a high-speed digital interface between the head-end hardware and PC, using USB 2.0, IEEE 1394 or 100BaseT, for example, providing access to digitized signals at an early stage in the signal-processing chain. It is possible to write a software program that not only allows the user to perform the usual radio functions, but that also allows configuration of DSP algorithms at the various levels. Processing elements may be customized and rearranged, thereby facilitating experimentation.

A PC-controlled software radio may be commanded and controlled through what is called an *applications programming interface* or API. The job of the API is to translate a standard set of commands or protocols to a radio-specific set. Using the API technique, programmers may use a standard software interface to program and communicate with the radio and be assured that an API, usually written by the radio designer, will interpret commands accordingly. Hams have caught on to that idea and developed APIs for their units. See the article by Larry Dobranski, VA3LGD, in the **Bibliography** for more information.

One area in which Amateur Radio rigs have made little progress in the last 40 years is that of transmitter IMD reduction. Software radio technology and DSP present the possibility of pre-distortion of the drive signals applied to final power amplifiers for that purpose. Drive signals may be purposefully distorted using the inverse of the response of the power amplifier. The net result is some compensation for the amplitude and phase non-linearities of the amplifier and therefore reduction of IMD and interference. Designers, however, are finding that the bandwidth required by the pre-distorted signal is at least five times that of a regular, uncompensated signal. That and the difficulties of measuring power amplifier distortion from unit to unit in production have rendered the method largely impractical to date, but that is expected to change as research continues.

Adaptive beamforming, or the creation of antenna systems with automatically varying radiation patterns, extends the concept of adaptive signal processing to the spatial domain. The goal of such "smart" antenna systems is to condition the radiation pattern of an array to maximize reception of desired signals and to minimize interference and noise on an adaptive basis.

A simple form of adaptive array consists of two or more omnidirectional antennas, like verticals, connected to a multi-channel adaptive receiver and signal processor. At least one of the antennas in the array feeds the signal processor through an adaptive filter. The DSP controls the impulse response of the filter to either enhance or cancel received signals based on their direction of arrival or certain other criteria. For example, the DSP may be programmed to accept only sinusoidal signals and reject broadband signals such as noise. See the article "Introduction to Adaptive Beamforming," listed in the **Bibliography**, for more information.

HARDWARE FOR EMBEDDED DSP SYSTEMS

What is it about a microprocessor that makes it a DSP? Well, DSPs are special because they include facilities uniquely designed for the type of calculations common in signal-processing algorithms. They are almost all 16-bit machines, or better, and so are very powerful even without their special facilities. DSPs may be classified primarily by their representation of numbers (fixed-point vs floating-point), also by their data-path width (16-bit, 32-bit), by their programmability (general-purpose vs dedicated co-processor) and their speed.

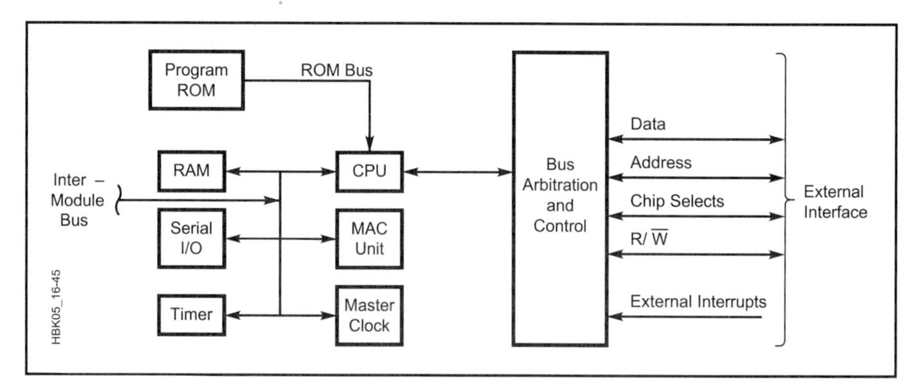

Fig 16.45—Fixed-point DSP block diagram.

Fixed-Point DSPs

Fixed-point DSPs are generally simpler than floating-point units, so they are typically less expensive. Fixed-point processors are common in embedded systems, especially for radio. Special software instructions and separate high-speed computational units are included to accelerate the processing of those common DSP calculations already mentioned. Perhaps the most-used operation is the convolution sum, performed as a series of MAC instructions (see the section on FIR Filters). Designers are interested in how many MACs per second a DSP can execute, because for anything beyond simple audio processing, only a small amount of time is available between samples for filtering and other functions.

A typical 16-bit, fixed-point DSP is shown in the block diagram of **Fig 16.45**. It employs what is called the *Harvard architecture*: It has separate program and data memory paths and also includes a *pipeline* for holding instructions waiting to be executed. This arrangement speeds things along because the CPU can fetch future instructions even when it is executing the current instruction or fetching data from another path.

Consider how this affects an FIR filter algorithm, for example. For each tap in the filter, the processor must multiply a constant (a filter coefficient) by a data value (a stored sample). When the processor can fetch both values simultaneously, an entire cycle time is saved. The subsequent addition of the product to the accumulator and the incrementing of indices for the next MAC instruction may also be executed in a single cycle. When large filters are being implemented, time savings quickly mount. Contrast this with the many cycles needed to perform the same operations in a general-purpose computer and you will see why specialized processors are so much more capable of handling sampled signals.

This business of execution speed is a large factor in the selection of a DSP for any particular use. System planning must begin by reckoning how many instructions can be executed between sample times. In a system with a 30-kHz sampling rate, only 33 µs are available, so a fixed-point DSP that can execute two million MACs per second (2 M-MACs/s) can only get 66 of these in the space between samples. For all but the simplest of systems, this is generally insufficient power for good filtering and other requirements and a separate filter *co-processor* must be employed. This is discussed further below. DSPs are now available having over 200 M-MACs/s performance.

Many fixed-point DSPs are available that also have undedicated parallel and serial input/output (I/O) on board. These may be very useful for embedded applications by obviating the need for other hardware. Processors embedded in radios have traditionally been shut off during times when no user input is present, stopping their clocks. This is done to eliminate the digital-circuit noise that otherwise would be difficult to remove. With a DSP in critical signal paths, this luxury is not possible. Careful attention to shielding, grounding and bypassing must therefore be paid. A DSP and associated support components humming along at 25 MHz—or more—tend to generate lots of noise and discrete spectral elements. They also tend to draw significant current, although dissipations in the one-watt range are typical; for base-station equipment, this is not usually a big concern.

Table 16.1
Fixed and Floating-Point DSPs

Part Number	*Manufacturer*	*# of bits*	*Fixed/Floating*
TMS320Cxx	Texas Instr.	16	Fixed
DSP320Cxx	Microchip	16	"
DSP16	ATT	16	"
ADSP21xx	Analog Dev.	16	"
MC68HC16	Motorola	16	"
MC5600x	Motorola	24	"
MB862xx	Fujitsu	24	Floating
MC9600x	Motorola	32	"
DSP32x	ATT	32	"
TMS320Cxx	Texas Instr.	32	"
ADSP21xxx	Analog Dev.	32	"

Fixed-point math brings with it a limitation on the range of numbers that can be represented, notwithstanding the extended integer/fractional representation demonstrated above. This limitation may form an obstacle to achieving the highest possible dynamic range. For this reason, floating-point DSPs are also widely available for use where greater boundaries must be set on the range of numbers handled. **Table 16.1** shows a listing of popular fixed-point DSPs, along with their floating-point cousins. Manufacturers supply evaluation boards, some of which include data converters and other support circuitry. Control software running on a desk-top computer is available for downloading *object code*—the DSP instructions that make up the program—as well as for debugging by use of tools such as break-points and register dumps.

Floating-Point DSPs

Representation of numbers is a critical decision to be made early in the system design process. A decision to use a floating-point DSP, at generally higher cost than fixed-point, is usually made either to remove dynamic-range barriers or to grant greater flexibility to algorithms that require scaling of data and coefficients, such as the FFT algorithms discussed above. We saw that each floating-point number requires two storage locations: one for the mantissa and one for the exponent. One would expect the processing of these numbers to be slowed by having to handle twice the data, but floating-point architectures are devised in such a way as to minimize or even eliminate this apparent handicap.

Multiplying two floating-point numbers involves multiplying the mantissas, then adding the exponents and any carry (or borrow) from the multiplication. Since multiplications generally require more time than additions, summing the exponents does not really slow the machine very much. Adding two floating-point numbers, though, requires the addition of the mantissas and a possible adjustment to the exponent, and this is always a bit slower than can be done on fixed-point numbers. With an optimized MAC unit, even this restriction can be removed for the bulk of calculations in typical DSP applications. Other than for these points, the block diagram of a floating-point DSP does not look very different from that of the fixed-point unit in Fig 16.45.

Selecting Data Converters

Complete DSP systems almost always include data converters in the form of one or more ADCs and DACs. Selection of these devices for any particular application is made with regard to cost, bit-resolution, speed, SFDR and digital interface. Manufacturers characterize devices on these bases and obviously, we must choose them so they will handle the highest sampling rate at our analog interface. In general, bit-resolution and speed determine SFDR. Dual 16-bit ADCs and DACs are now very common because they are used in compact-disc (CD) recorders and playback units at a sampling rate of 44.1 kHz. Note that 44.1 kilosamples/s of two channels in a stereo system is equal to $(2) \times (44{,}100) \times (16) \approx 1.41$ megabits per second. Devices with 20 and even 24-bit capability are catching on. This is a lot of data and the bit-resolution of data converters is most often chosen to match that of the DSP, although there may be advantages in having slightly more bit-resolution in the DSP to mitigate round-off errors, as noted in the FIR Filters section above.

We noted before that over-sampling of input signals brings significant advantages for the DSP designer. For this reason, sigma-delta ADCs are the "top of the crop" for use in IF-DSP and DDC receivers. As sampling frequencies increase, over-sampling becomes more difficult to achieve. Engineers working in cellular radio and similar technologies deal with much wider BWs than most of those found in Amateur Radio, and so must grapple with reduced dynamic range—fortunately, they also require less. ADCs that handle 12 to 16 bits at speeds exceeding 65 MHz are available. Viable DDC designs are finding their ways into many commercial services worldwide.

Converters must interface with DSPs through a high-speed digital connection of some kind. Parallel transfer—all 16 bits at once, for example—is more common among DACs than ADCs. High-speed, three-line serial interfaces are popular among converter manufacturers and several standards have evolved. Some of these are compatible with one another. Bearing in mind the amount of data being transferred, realize that these serial links may run at clock speeds in excess of 100 MHz. ADC/DAC evaluation boards may be connected to DSP evaluation boards to form a prototype DSP system. Some data converters are listed in **Table 16.2**.

Table 16.2
Data Converters

Part Number	*Manufacturer*	*# bits*	*Speed*	*ADC/DAC*
HI1276	Harris/Intersil	8	500 Ms/s	ADC
AD6645	Analog Dev.	14	80/105 Ms/s	"
AD7722	Analog Dev.	16	200 ks/s	"
AD9854	Analog Dev.	12	300 Ms/s	NCO + DAC
AD9858	Analog Dev.	12	300 Ms/s	NCO + DAC
ADC76	Burr Brown	16	50 ks/s	ADC
PCM1750	"	18	44 ks/s	dual ADC
CS5322	Crystal	24	2 ks/s	ADC
CS5360	Crystal	24	50 ks/s	"
AD1871	Analog Dev.	24	96 ks/s	"
PCM1802	TI	24	96 ks/s	"
UDA1361TS	Philips	24	96 ks/s	"
AK5394A	AKM	24	192 ks/s	"
BT254	Brooktree	24	30 Ms/s	"
Note: Also see Maxim, National, Sipex, Analogic				
CA3338A	Harris/Int.	8	50 Ms/s	DAC
HI1171	"	8	40 Ms/s	"
HI5780	"	10	40 Ms/s	"
HI20201	"	10	160 Ms/s	"
AD9777	Analog Dev.	16	160 Ms/s	"
PCM56	Burr Brown	16	93 ks/s	"
PCM66	"	16	44 ks/s	dual DAC
Note: See also National, Maxim, etc.				

Table 16.3
Co-processors and DDC Chips

Part Number	*Manufacturer*	*# bits*	*Speed*	*Function*
AD6620/34	Analog Dev.	14	100Ms/s	DSP down-conv.
HSP50016	Harris/Int.	16	52 Ms/s	DSP down-conv.
HSP50110	"	10	60 Ms/s	Quadr. tuner
HSP50210	"	10	52 Ms/s	DSP Costas loop
HSP50306	"	6	2 Mbit/s	QPSK demod
HSP43xxx	"	10-24	Var.	DSP filters
510	Harris et al	16	10 Ms/s	Mult/Acc
LMA2010	Logic Dev, IDT	16	40 Ms/s	Mult/Acc
HSP4510x	Harris/Int.	20-32	33 Ms/s	DDS
Various	Xylinx, Altera, Atmel, etc.	8-32	>100 Ms/s	FPGAs

Extra Processing Power: DSP Co-Processors

Quite often, a single, general-purpose DSP by itself is not sufficient to handle the computational load in a project. This may be determined early in the system design by evaluating the number of MACs required by filters and other algorithms. Several solutions present themselves: adding one or more general-purpose DSPs, adding specialized co-processor chips, or designing a custom co-processor using programmable-logic chips.

More than one general-purpose DSP may be used to augment net data capacity. The trend these days, though, is to use dedicated co-processor chips that are op-

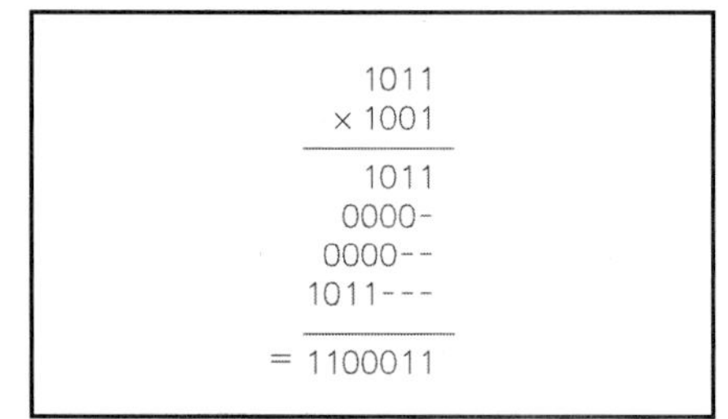

Fig 16.46—Long multiplication of two 4-bit binary numbers.

timized for the function they are to perform. This is especially true of FFT and other operations that do not lend themselves well to the MAC procedures for which general-purpose chips are optimized. Whatever the algorithm, it seems that multiplication of numbers takes the most time, so a co-processor that incorporates a astmultiplication algorithm is desirable. A lot of effort has gone into fast multipliers since the 1980s and for the IF-DSP or DDC designer, a knowledge of how it is done may bring plentiful results.

The multiplication of two binary numbers may be decomposed into an addition of several binary numbers. We know that fast binary addition is readily achieved by relatively simple logic. Let's take a look at this, since it forms the basis for most fast multipliers. Shown in **Fig 16.46** is the long multiplication of two 4-bit binary numbers. It is performed in base two the same way as it is in base ten: First, take the least-significant digit of the lower multiplicand and multiply it by the other multiplier. Since in binary, this digit is either one or zero, the digits we write under the line is either a copy of the top multiplicand, or all zeros. Then, the next-significant digit of the lower multiplicand is used, with the result written below the first and shifted one digit to the left. This process continues until all bits of the lower multiplicand have been used. Finally, all the interim results are added. This last result is the product of the two numbers. Note that the result may contain a number of bits as high as the sum of the number of bits in both the multiplicands. Project G in the **Appendix** shows how simple logic is used to implement a fast multiplier. Pipe-lining and latency issues are discussed there.

Refinements of this technique that use look-up tables and combinatorial methods yield speed increases. Field-programmable gate array (FPGA) manufacturers have worked out the details of these algorithms and routinely provide them to users. FPGAs are available now in very-high-speed versions ($f_{clk} \geq 200$ MHz) that may be used for DSP co-processing. FPGA designs may also employ the Harvard architecture using external, *dual-port memory* to provide a register-based interface to host DSPs. Normally, one sample is passed to the co-processor and one retrieved at each sample time. Filters exceeding 100 taps may be implemented this way, saving processing time in the host DSP for other housekeeping tasks.

Entire down-conversion and I/Q modulation sub-systems have been incorporated on a single chip. These chip sets may be advantageous where FPGA-based designs either do not meet requirements or are too expensive. A sampling of ready-to-use co-processors and DDC chips is given in **Table 16.3**. Also read some of the reference material listed at the end of this chapter for more information on dedicated DSP co-processors.

DSP SYSTEM SOFTWARE

Assembly Language and Timing Requirements

Embedded-DSP application software is most often written in *assembly language*, the native language of the DSP in use. Instructions to be executed are arranged in order, according to the *von Neumann model*, and entered as lines in a text file, using the mnemonics provided by the DSP manufacturer. When this *source code* is ready, an *assembler* program is invoked that translates the source code into object code—the numbers that the DSP understands as instructions. The object code is then transferred to the program memory of the target system for execution.

The reason assembly language is so prevalent in embedded applications is the critical timing involved. Programs compiled in high-level languages do not always handle interrupt-driven events well (the input or output samples) and may bog down. To minimize the required hardware speed, processing of some second-line tasks such as squelch and ALC must have reduced sampling rates to fit into the whole picture. Only a part of their processing burden may be performed at each sample time. This is a form of *time-distributed processing* and is just one in the DSP designer's bag of tricks.

Someone will always think of something more for a transceiver to do and it is better to err on the side of higher speed and more memory at the start than to run out later. Even so, DSP designers must carefully evaluate all the functions included at the outset. Other shortcuts—like the assumption of only integer values by a BFO at one-fourth the sampling frequency—may present themselves, but one cannot always count on it; one must plan diligently to avoid roadblocks. In addition, *unexpected things can occur* if due thought is not given to quantization and scaling effects, especially where adaptive processing is applied, no matter the representation of numbers used. DSP-chip manufacturers provide assemblers and instruction details free of charge. Their applications engineers are ordinarily ready to assist. A plethora of information is available on the Web.

Filter-Design Software

Several software packages for DSP filter design are listed at the end of this chapter. Many more are available. You can expect to find reasonably priced software that will design FIR and IIR filters, as well as let you perform convolution, multiplication, addition, logarithms and other calculations on numeric sequences.

FIR filters usually may be designed with a choice of method (Fourier, Parks-McClellan, least-squares), length, frequency response, and ripple magnitude; they may use various window functions to achieve different shape factors and passband/stopband attenuations. Some are able to take coefficient and data quantization into account and some are not. Large filters may deviate significantly from their theoretical responses because of these effects, so if you are contemplating reasonably long filters, check into this capability.

IIR filter design usually includes a choice of various analog-filter prototypes. Software packages may vary in their ability to display, print, or plot responses and write coefficient files to disk. Filter coefficients are generally part of system firmware and must be transferred from the host DSP to a filter co-processor on demand. It must be possible to translate the filter-design software's output to a format the compiler software understands. A translation program may have to be written to accomplish this.

Longer and more-complex FIR filters may be implemented by convolving the impulse responses of several different filters. This allows the alteration of the frequency response of standard filters to include graphic or parametric equalization and IF shift. Such filtering systems are already being employed in Amateur Radio and commercial transceivers.

Other DSP Design Tools

FPGA design software is generally available from chip manufacturers. In addition, many schematic-capture and PCB-layout software vendors provide interfaces to popular FPGAs and other programmable devices. Hardware Design Language (HDL) and Verilog Hardware Design Language (VHDL) have become popular for translating user requirements into programming code for FPGAs. Most FPGA programmers understand HDL or VHDL.

A rich variety of flow-chart software exists in both the public and private domains. It may be especially useful for time-sensitive applications in DSP.

BIBLIOGRAPHY

(Key: **D** = disk included, **A** = disk available, **F** = filter design software)

DSP Software Tools

Alkin, O., *PC-DSP*, Prentice Hall, Englewood Cliffs, NJ, 1990 (**DF**).

Kamas, A. and Lee, E., *Digital Signal Processing Experiments*, Prentice Hall, Englewood Cliffs, NJ, 1989 (**DF**).

Momentum Data Systems, Inc., *QEDesign*, Costa Mesa, CA, 1990 (**DF**).

Stearns, S. D. and David, R. A., *Signal Processing Algorithms in FORTRAN and C*, Prentice Hall, Englewood Cliffs, NJ, 1993 (**DF**).

Textbooks

Frerking, M. E., *Digital Signal Processing in Communication Systems*, Van Nostrand Reinhold, New York, NY, 1994.

Ifeachor, E. and Jervis, B., *Digital Signal Processing: A Practical Approach*, Addison-Wesley, 1993 (**AF**).

Madisetti, V. K. and Williams, D. B., Editors, *The Digital Signal Processing Handbook*, CRC Press, Boca Raton, FL, 1998 (**D**).

Oppenheim, A. V. and Schafer, R. W., *Digital Signal Processing*, Prentice Hall, Englewood Cliffs, NJ, 1975.

Parks, T. W. and Burrus, C.S., *Digital Filter Design*, John Wiley and Sons, New York, NY, 1987.

Proakis, J. G. and Manolakis, D., *Digital Signal Processing*, Macmillan, New York, NY, 1988.

Proakis, J. G., Rader, C. M., et. al., *Advanced Digital Signal Processing*, Macmillan, New York, NY, 1992.

Rabiner, L. R. and Schafer, R. W., *Digital Processing of Speech Signals*, Prentice Hall, Englewood Cliffs, NJ, 1978.

Rabiner, L. R. and Gold, B., *Theory and Application of Digital Signal Processing*, Prentice Hall, Englewood Cliffs, NJ, 1975.

Sabin, W. E. and Schoenike, E. O., Eds., *HF Radio Systems and Circuits*, rev. 2nd ed., Noble Publishing Corp, Norcross, GA, 1998.

Smith, D., *Digital Signal Processing Technology: Essentials of the Communications Revolution*, ARRL, 2001.

Widrow, B. and Stearns, S. D., *Adaptive Signal Processing*, Prentice Hall, Englewood Cliffs, NJ, 1985.

Zverev, A. I., *Handbook of Filter Synthesis*, John Wiley and Sons, New York, NY, 1967.

Articles

Albert, J. and Torgrim, W., "Developing Software for DSP," *QEX*, March, 1994, pp 3-6.

Anderson, P. T., "A Simple SSB Receiver Using a Digital Down-Converter," *QEX*, March, 1994, pp 17-23.

Anderson, P. T., "A Faster and Better ADC for the DDC-Based Receiver," *QEX*, Sep/Oct 1998, pp 30-32.

Applebaum, S. P., "Adaptive arrays," *IEEE Transactions Antennas and Propagation*, Vol. PGAP-24, pp 585-598, September, 1976

Åsbrink, L., "Linrad: New Possibilities for the Communications Experimenter," *QEX*, Part 1, Nov/Dec 2002; Part 2, Jan/Feb 2003; Part 3 May/Jun 2003.

Ash, J. et al., "DSP Voice Frequency Compandor for Use in RF Communi-cations," *QEX*, July, 1994, pp 5-10.

Beals, K., "A 10-GHz Remote-Control System for HF Transceivers," *QEX*, Mar/Apr, 1999, pp 9-15.

Bloom, J., "Measuring SINAD Using DSP," *QEX*, June, 1993, pp 9-18.

Bloom, J., "Negative Frequencies and Complex Signals," *QEX*, September, 1994.

Brannon, B., "Basics of Digital Receiver Design," *QEX*, Sep/Oct, 1999, pp 36-44.

Cahn, H., "Direct Digital Synthesis—An Intuitive Introduction," *QST*, August, 1994, pp 30-32.

Cercas, F. A. B., Tomlinson, M. and Albuquerque, A. A., "Designing With Digital Frequency Synthesizers," *Proceedings of RF Expo East*, 1990.

de Carle, B., "A Receiver Spectral Display Using DSP," *QST*, January, 1992, pp 23-29.

Dick, R., "Tune SSB Automatically," *QEX*, Jan/Feb, 1999, pp 9-18.

Dobranski, L., "The Need for Applications Programming Interfaces (APIs) in Amateur Radio," *QEX*, Jan/Feb 1999, pp 19-21.

Emerson, D., "Digital Processing of Weak Signals Buried in Noise," *QEX*, January, 1994, pp 17-25.

Forrer, J., "Programming a DSP Sound Card for Amateur Radio," *QEX*, August, 1994.

Green, R., "The Bedford Receiver: A New Approach," *QEX*, Sep/Oct, 1999, pp 9-23.

Hale, B., "An Introduction to Digital Signal Processing," *QST*, September, 1992, pp 43-51.

Kossor, M., "A Digital Commutating Filter," *QEX*, May/Jun, 1999, pp 3-8.

Morrison, F., "The Magic of Digital Filters," *QEX*, February, 1993, pp 3-8.

Olsen, R., "Digital Signal Processing for the Experimenter," *QST*, November, 1994, pp 22-27.

Reyer, S. and Herschberger, D., "Using the LMS Algorithm for QRM and QRN Reduction," *QEX*, September, 1992, pp 3-8.

Rohde, D., "A Low-Distortion Receiver Front End for Direct-Conversion and DSP Receivers," *QEX*, Mar/Apr, 1999, pp 30-33.

Runge, C., *Z. Math. Physik*, Vol 48, 1903; also Vol 53, 1905.

Scarlett, J., "A High-Performance Digital Transceiver Design," *QEX*, Part 1, Jul/Aug 2002; Part 2, Mar/Apr 2003.

Smith, D., "Introduction to Adaptive Beamforming," *QEX*, Nov/Dec 2000.

Smith, D., "Signals, Samples and Stuff: A DSP Tutorial, Parts 1-4," *QEX*, Mar/Apr-Sep/Oct, 1998.

Ulbing, Sam, "Surface-Mount Technology—You Can Work With It! Parts 1-4," *QST*, April-July, 1999

Ward, R., "Basic Digital Filters," *QEX*, August, 1993, pp 7-8.

Youngblood, G., "A Software-Defined Radio for the Masses," *QEX*; Part 1, Jul/Aug 2002; Part 2, Sep/Oct 2002; Part 3, Nov/Dec 2002; Part 4, Mar/Apr 2003.

Appendix: DSP Projects

Project A: Decimation
Project B: FIR Filter Design Variations
Project C: Analytic Filter-Pair Generation
Project D: Newton's Method for Square Roots in QuickBasic 4.5
Project E: A Fast Square-Root Algorithm Using a Small Look-Up Table in Assembly Language
Project F: A High-Performance DDS
Project G: A Fast Binary Multiplier in High-Speed CMOS Logic

PROJECT A: DECIMATION

This project illustrates the concept of decimation using Alkin's *PC-DSP* program, included with the book of that name listed in the **Bibliography**. First, generate 40 samples of the sinusoid y(n) = sin(n/4), where $0 \leq n \leq 39$. This sequence may be generated using the "Sine" function of the "Generate" sub-menu under the "Data" menu, with parameters Var1 = SIN, A = 1, B = 0.25, C = 0 and #Samples = 40. Press F2 to display the data, which should match **Fig 16.A1**.

Next, decimate the sequence by a factor of 2 using the "Decimate" function found in the "Process" sub-menu under the "Data" menu. Use parameters Var1 = SIN2, Var2 = SIN, Factor = 2. Display the new sequence by pressing F2. It should match **Fig 16.A2**.

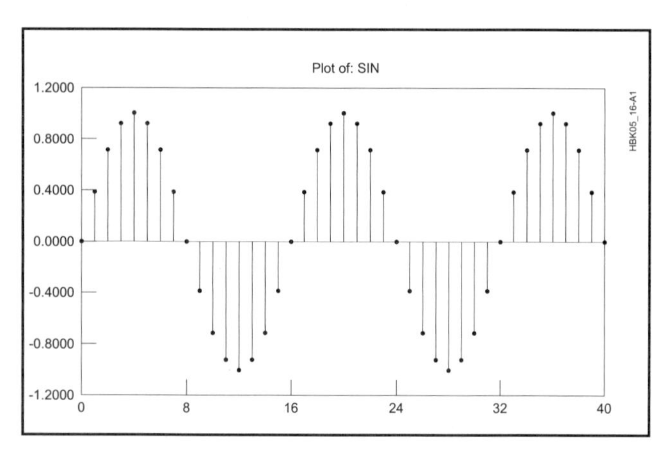

Fig 16.A1—A 40-sample sine wave.

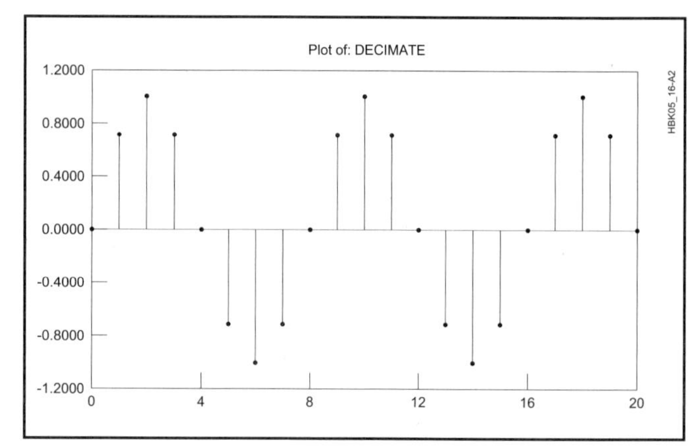

Fig 16.A2—Decimated, 20-sample sine wave.

PROJECT B: FIR FILTER DESIGN VARIATIONS

An FIR filter's ultimate attenuation and its transition BW are largely determined by the filter's length: the number of taps used in its design. Fourier and other design methods do not always readily optimize the trade-off among transition BW, ultimate attenuation and ripple. One way to achieve better ultimate attenuation at the expense of passband ripple is to convolve the impulse responses of two short filters to obtain a longer filter. The two impulse-response sequences are processed by precisely the same convolution sum that is used to compute FIR filter outputs (Eq 3 in the main text).

A filter obtained by convolving two filters of length L has length 2L –1. In one example, two LPFs of length 31 may be convolved to produce a filter of length 61. The resulting frequency response, plotted against that of a LPF designed with Fourier methods for an identical length of 61 taps, would show that the ultimate attenuation of the convolved filter is 20 dB or 10 times greater than that of the plain, Fourier-designed filter. Also, the convolved filter would have a greater passband ripple and a narrower transition region. Quite often, filters that were designed using different window functions may be convolved to get some of the benefits of each in the final filter.

A look back at Fig 16.29 reveals that different window functions achieve different transition BWs and values of ultimate attenuation. The rectangular window attains a narrow transition BW, but a poor ultimate attenuation; the Blackman window, on the other hand, has nearly optimal ultimate attenuation and a moderate transition BW. Let's see what happens when we convolve the impulse responses of filters designed using each method. We will constrain ourselves to filters with odd numbers of taps so that the convolved impulse response will also have an odd number of taps.

Using your favorite filter-design software, first design a LPF by the Fourier method with a length of 31, using a rectangular window, and a cut-off frequency (–6 dB point) of $0.25f_s$. Its frequency response is shown in **Fig 16.B1A**. We produce a second filter having the same cut-off frequency of $0.25f_s$ using a Blackman window, whose response is shown in **Fig 16.B1B**. The response of the filter formed by the convolution of the two filters is shown in **Fig 16.B1C**, along with that of a standard Fourier-designed LPF. The final filter has length 61 taps. Notice that the filter obtains the benefits of the rectangular window's sharp transition region and those of the Blackman window's good ultimate attenuation.

A second advantage may be garnered by convolving two different filters in that their responses may be governed separately, while producing desired changes in frequency (or phase) response. A good example of this arises when it is desired to alter the audio response of an SSB transmitter (or receiver), but keep the ultimate attenuation characteristics the same. A long BPF with excellent transition properties may be convolved with a much shorter filter that is manipulated to provide the desired passband response.

FIR filters used in Amateur Radio transceivers must usually have at least 60 dB ultimate attenuation. This generally requires at least 63 taps. As our second FIR filter variation, let's consider a case wherein we want to customize an IF-DSP transmitter's frequency response without impacting opposite-sideband rejection. We will use a 99-tap BPF in each leg of a Hilbert transformer (as part of an SSB modulator) whose response is convolved with that of a 31-tap filter describing the variation in frequency response we want. The 99-tap fixed filter has the frequency response shown in **Fig 16.B2A**. The 31-tap filter has been designed using Fourier methods to have a 6 dB/octave rise in its frequency response, as shown in **Fig 16.B2B**.

The frequency response of the convolution of the two filters' impulse responses is shown in **Fig 16.B2C**. It is important to note that the net response is that of the *product* of the two filters' frequency responses; that is, if $H_1(\omega)$ and $H_2(\omega)$ are the two frequency response functions, the final response is simply:

$$H_{composite}(\omega) = H_1(\omega)H_2(\omega) \qquad (B1)$$

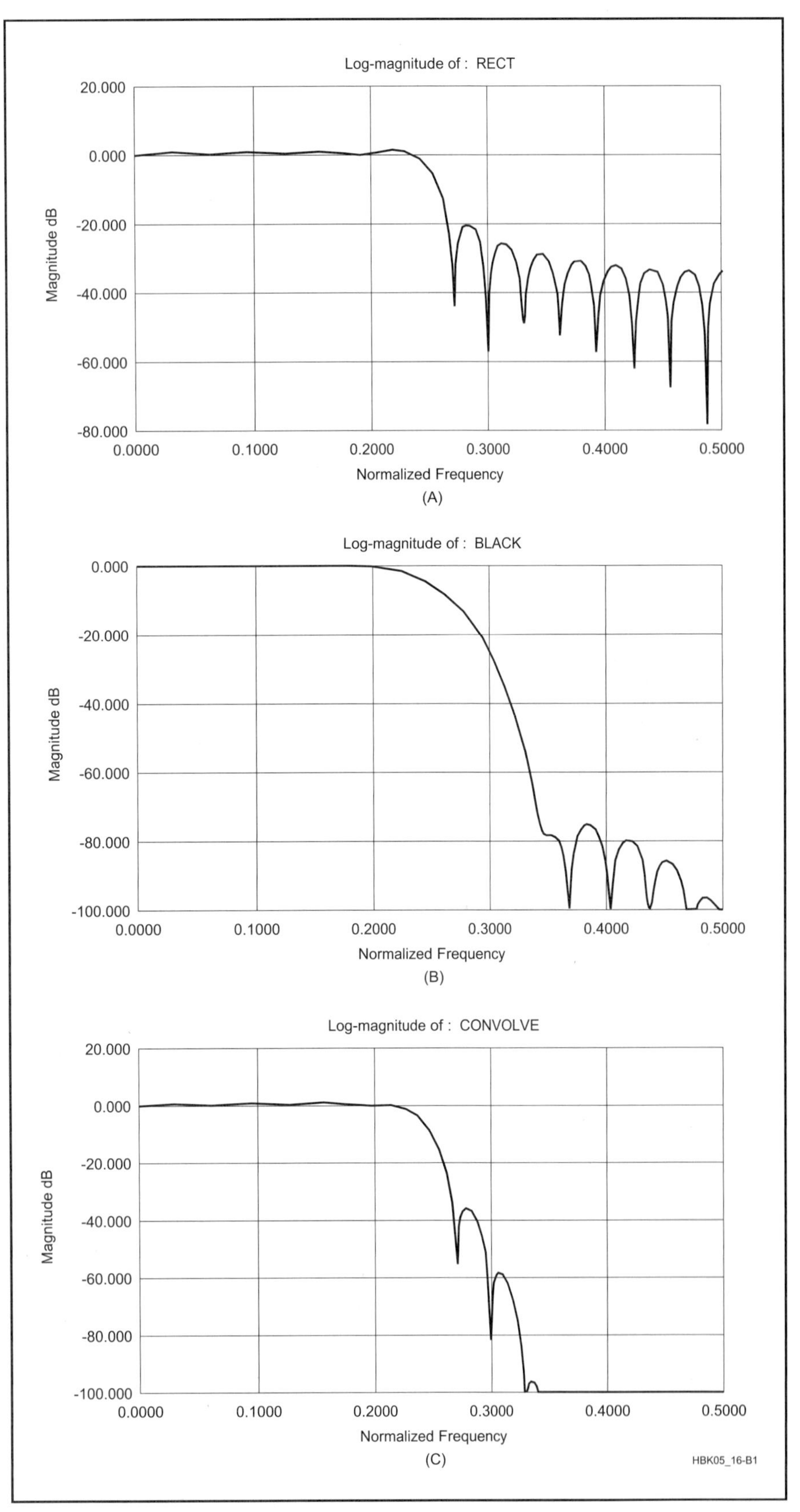

Fig 16.B1—LPF frequency response, rectangular window (A). LPF frequency response, Blackman window (B). LPF frequency response, convolution of filters shown in A and B (C).

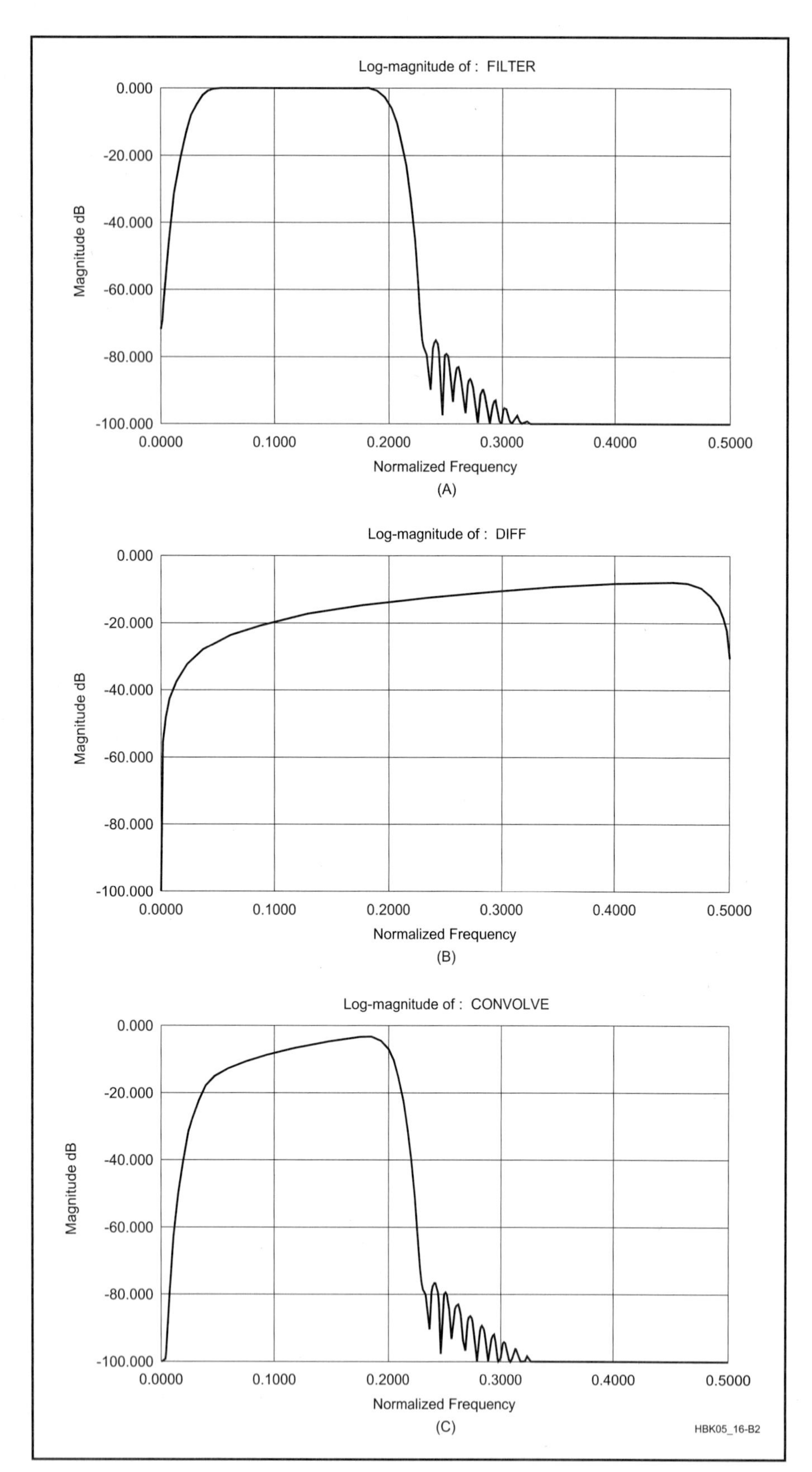

Fig 16.B2—BPF for SSB use, L = 99 (A). LPF having rising frequency response, L = 31 (B). Frequency response of convolution of filters shown in A and B (C).

PROJECT C: ANALYTIC FILTER PAIR GENERATION

Frequency-translation properties of complex multiplication work just as well on the responses of filters as they do on real signals. In this project, we will explore just how these properties are applied to the generation of analytic filter pairs. Analytic filter pairs are used to produce complex signals from real signals for the purposes of modulation, demodulation, and other processing algorithms.

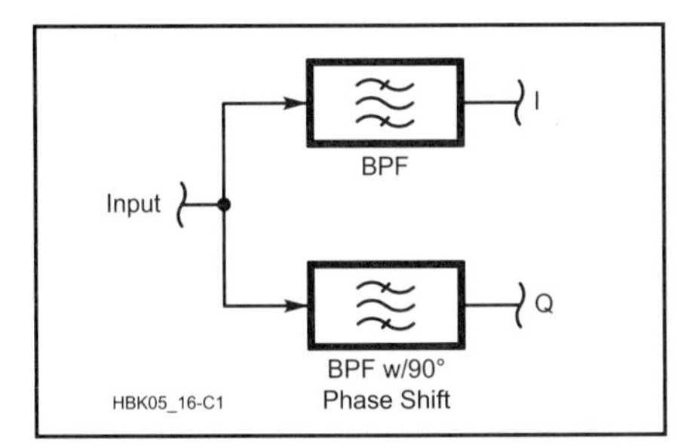

Fig 16.C1—Hilbert transformer using an analytic filter pair.

An analytic filter pair consists of two filters (usually BPFs) whose frequency responses are identical, but whose phase responses differ at every frequency by 90°. These filters are used in legs of a Hilbert transformer, as shown in **Fig 16.C1**. The creation of these filters begins with the design of a LPF prototype having the desired passband, transition-band, and stopband characteristics. Such a prototype filter, as might suffice for an SSB receiver, would have a frequency response such as that shown in **Fig 16.C2A**.

The filter's impulse response (L = 63) is then multiplied by a sine-wave sequence (also L = 63) whose frequency represents the amount of upward translation applied to the LPF's frequency response. If the sine wave is high enough in frequency, the resulting impulse response is a BPF filter centered on ω_0, the sine wave's frequency. See **Fig 16.C2B**. Likewise, the prototype LPF's impulse response is multiplied by a cosine-wave sequence to produce a filter having the same frequency response as that of the sine-wave filter, but with a phase response differing by 90°. Sample-by-sample multiplication occurs according to Eq 21 in the main text.

When an analytic filter pair is used in a demodulator, IF shift may be included by varying the frequency of w_0. In combination with various filter BWs, IF shift is useful in avoiding interference by modifying a receiver's frequency response. Further modification may be obtained by convolving each filter in the analytic pair with a filter having the desired characteristic. The phase relation between the filters in the pair will not be altered by the convolution.

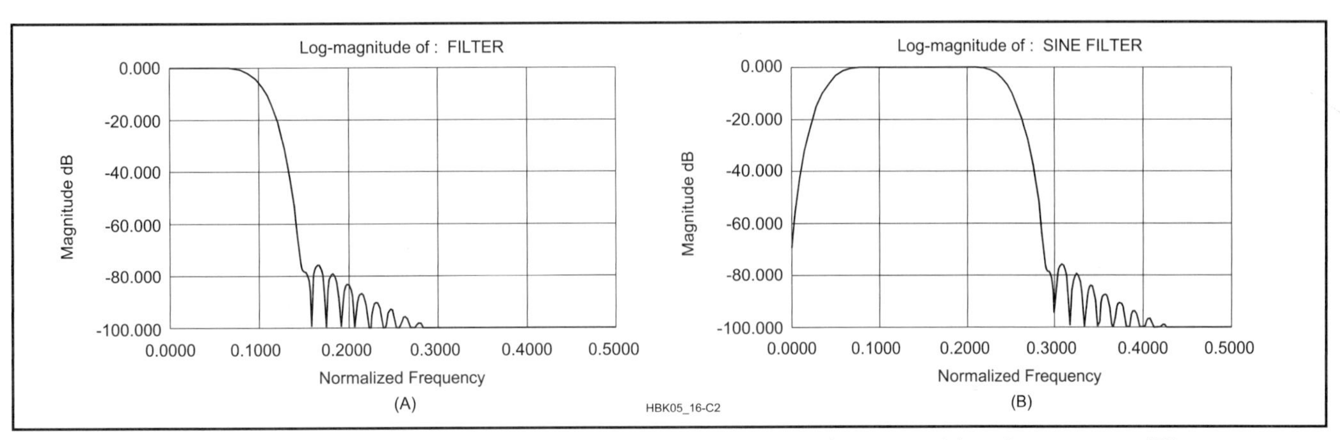

Fig 16.C2—LPF prototype frequency response (A). BPF frequency response of processed impulse response (B).

PROJECT D: NEWTON'S METHOD FOR SQUARE ROOTS IN QUICKBASIC 4.5

In this example of Newton's method, a generic *BASIC* program is given that computes the root of a 32-bit integer to within an error margin, DERROR. The root of a 32-bit integer is naturally a 16-bit integer. Emphasis is placed in what follows on speed of execution and accuracy as influenced by truncation and rounding. 32-bit integer variables are defined DEFLONG, 16-bit integers are DEFINT. Integer math in *QuickBasic* is much faster than floating-point math.

As described in the **AM Demodulation** section in the main text, Newton's method iteratively converges on a result. Experience has shown that three to six iterations are necessary to obtain best accuracy for a 16-bit result, but here we execute as many iterations as necessary to obtain accuracy DERROR, initially defined to be one least-significant bit or $1/(2^{15}) \approx 30 \times 10^{-6}$. Note that if DERROR is small or zero, convergence may never be reached because of quantization noise. A loop counter, K, is established to count iterations. The program displays on the computer screen the argument, its root and the iteration count. Users may readily modify the program to use random numbers as arguments to time the number of roots per second it calculates.

The program is included in the *2002 ARRL Handbook* companion software. The software is available for free download from *ARRLWeb* at: **www.arrl.org/notes**.

PROJECT E: A FAST SQUARE-ROOT ALGORITHM USING A SMALL LOOP-UP TABLE

This project is a machine-language example of a fast square-root algorithm. The target processor in this case is the Motorola MC68HC16Z1, a 16-bit, fixed-point DSP. The method is depicted in **Fig 16.16** in the main text. Like the previous project, this is included in the *2002 ARRL Handbook* companion software. The software is available for free download from *ARRLWeb* at **www.arrl.org/notes**.

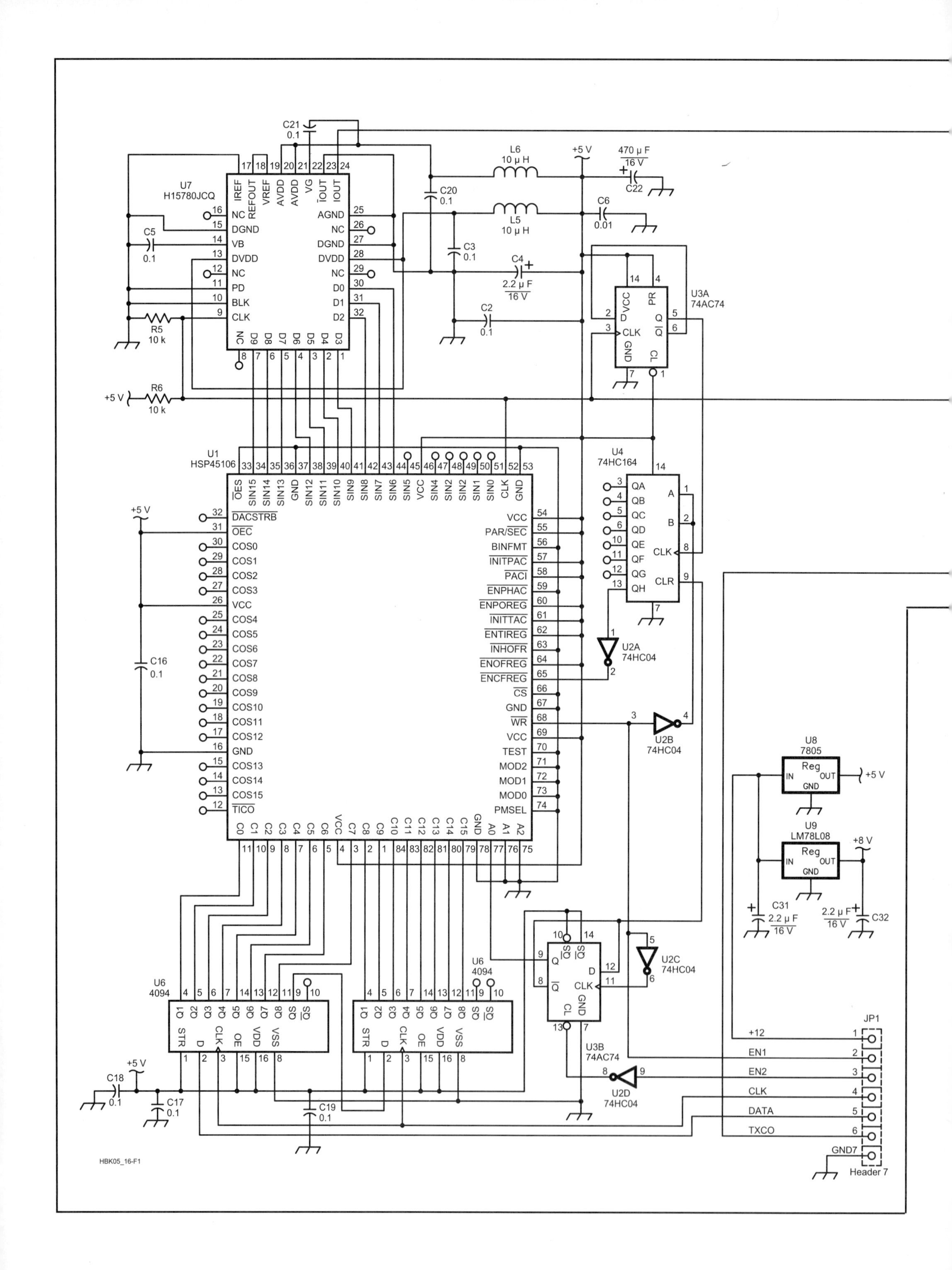

C21
0.1
U7
H15780JCQ
IREF
REFOUT
VREF
AVDD
AVDD
VG
IOUT
IOUT
NC
DGND
VB
DVDD
NC
PD
BLK
CLK
AGND
NC
DGND
DVDD
NC
D0
D1
D2
NC
D9
D8
D7
D6
D5
D4
D3
C5
0.1
R5
10 k
+5 V
R6
10 k
C20
0.1
L6
10 μ H
L5
10 μ H
+5 V
470 μ F
16 V
C22
C6
0.01
C3
0.1
C4
2.2 μ F
16 V
C2
0.1
U3A
74AC74
VCC
PR
D
Q
CLK
Q
GND
CL
U1
HSP45106
OES
SIN15
SIN14
SIN13
GND
SIN12
SIN11
SIN10
SIN9
SIN8
SIN7
SIN6
SIN5
VCC
SIN4
SIN3
SIN2
SIN1
SIN0
CLK
GND
DACSTRB
OEC
COS0
COS1
COS2
COS3
VCC
COS4
COS5
COS6
COS7
COS8
COS9
COS10
COS11
COS12
GND
COS13
COS14
COS15
TICO
VCC
PAR/SEC
BINFMT
INITPAC
PACI
ENPHAC
ENPOREG
INITTAC
ENTIREG
INHOFR
ENOFREG
ENCFREG
CS
GND
WR
VCC
TEST
MOD2
MOD1
MOD0
PMSEL
C0
C1
C2
C3
C4
C5
C6
VCC
C7
C8
C9
C10
C11
C12
C13
C14
C15
GND
A0
A1
A2
C16
0.1
U4
74HC164
QA
QB
QC
QD
QE
QF
QG
QH
A
B
CLK
CLR
U2A
74HC04
U2B
74HC04
U8
7805
Reg
IN
OUT
GND
+5 V
U9
LM78L08
Reg
IN
OUT
GND
+8 V
C31
2.2 μ F
16 V
2.2 μ F
16 V
C32
U6
4094
Q1
Q2
Q3
Q4
Q5
Q6
Q7
Q8
QS
QS
STR
D
CLK
OE
VDD
VSS
U6
4094
Q1
Q2
Q3
Q4
Q5
Q6
Q7
Q8
QS
QS
STR
D
CLK
OE
VDD
VSS
SQ
SQ
Q
D
Q
CLK
CL
GND
U2C
74HC04
U3B
74AC74
U2D
74HC04
C18
0.1
+5 V
C17
0.1
C19
0.1
JP1
+12
EN1
EN2
CLK
DATA
TXCO
GND7
Header 7
HBK05_16-F1

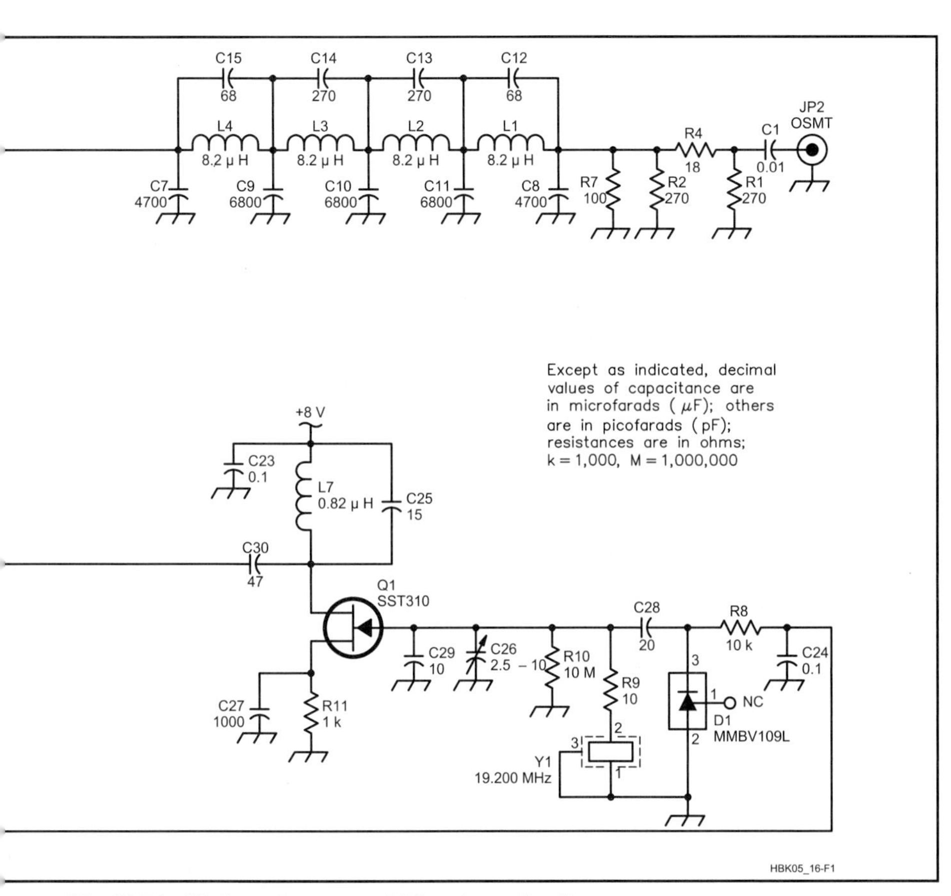

Fig 16.F1—High-performance DDS schematic diagram.

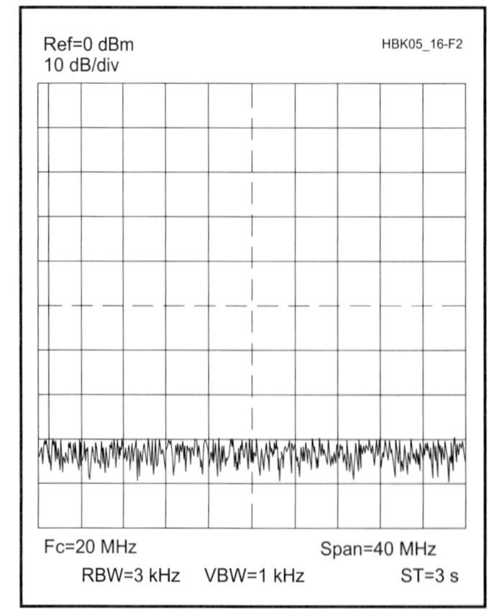

Fig 16.F2—Typical output spectrum of DDS.

PROJECT F: A HIGH-PERFORMANCE DDS

A DDS is described below that is used as a reference for a PLL. See **Fig 16.F1**. This DDS is designed to cover a small range of frequencies near 1 MHz. A crystal-oscillator clock at 19.2 MHz is applied to both the DDS, a Harris/Intersil HSP45106, and the DAC, a Harris/Intersil HI5780. Making the DDS output frequency a small fraction of the clock frequency makes it relatively easy to obtain excellent spurious performance. PM spurs are limited to –90 dBc and AM spurs to about –60 dBc. If the output is not squared at the input to a PLL chip, an external Schmitt-trigger squaring stage may be added, eliminating virtually all the AM spurs prior to the LPF.

The LPF at the output of the circuit is a 4-section elliptical type. Design impedance is 100 Ω. This filter cuts out many high-frequency spurs and stops clock feed-through. The DAC's 10 input lines are fed from the 10 most-significant bits of one of the DDS's outputs. The HSP45106 has two 16-bit outputs (sine and cosine) to accommodate the needs of complex-mixer designs, but only one is being used here.

The DDS chip itself is programmed using a 16-bit parallel interface. This is transformed into a serial interface by shift registers U5 and U6, divider U3 and counter U4. Each time the frequency is changed, an internal 32-bit phase-increment accumulator must be updated. The phase increment is just f_{out}/f_{clk}, expressed as a 32-bit, unsigned fraction. This value is written into the chip in two 16-bit segments, most-significant bit of the most-significant word first.

During serial programming, a data bit is placed on the DATA line by the host microprocessor; the clock line is toggled high, then low to shift the bit into the shift registers. After the first 16 bits have been shifted, they are written into the DDS by toggling the ENABLE line. Counter U4 supplies the necessary write pulse with appropriate timing. The remaining 16 bits are then shifted and written to the chip, completing the operation.

An example of the output spectrum of this circuit is shown in **Fig 16.F2**. Components are surface-mount types and care must be exercised during construction. See Ulbing's article in the **Bibliography** for information on surface-mount soldering techniques.

PROJECT G: A FAST BINARY MULTIPLIER IN HIGH-SPEED CMOS LOGIC

In this project, a fast 4-bit binary multiplier is described that may be constructed from 'HC-series logic gates or programmed into an FPGA. Two variations are explored: one without pipelining, and one with pipelining. Pipelining is employed where the propagation delays of gates limit throughput.

As seen in Fig 16.45 in the main text, a 4-bit multiplication may be broken into several 4-bit additions. In our circuit, 4-bit adders are used to add rows of bits in the summation, each one producing a single output bit. The diagram of a fast, 4-bit adder with look-ahead carry is shown in **Fig 16.G1**.

In this multiplier, 4-bit adders are used to add adjacent rows of bits in the traditional way. A multiplier connected this way is shown in **Fig 16.G2**. Not all bits in each addend have mates in the other, so 4-bit adders suffice. In the case where execution speed exceeds the reciprocal of the total propagation delay, pipelining must be employed to avoid error.

To use pipelining, we place storage registers between the stages of addition and one interim result is held by each stage at each clock time. See **Fig 16.G3**. The result is the same, but appears only after a latency of three clock times. When maximum gate delays are well known, this approach also yields more predictable performance because the latency is independent of the input data.

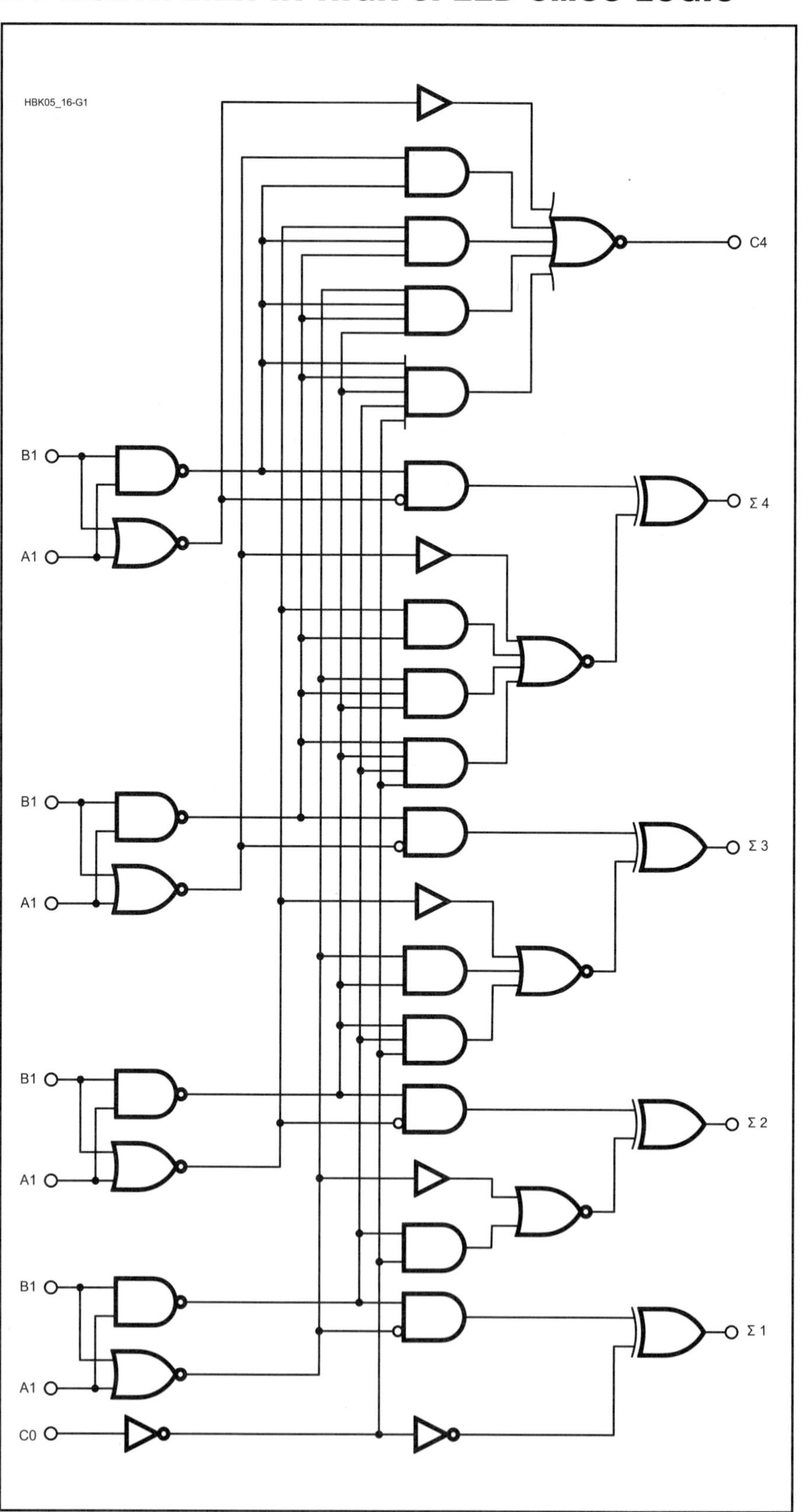

Fig 16.G1—A 4-bit adder schematic diagram.

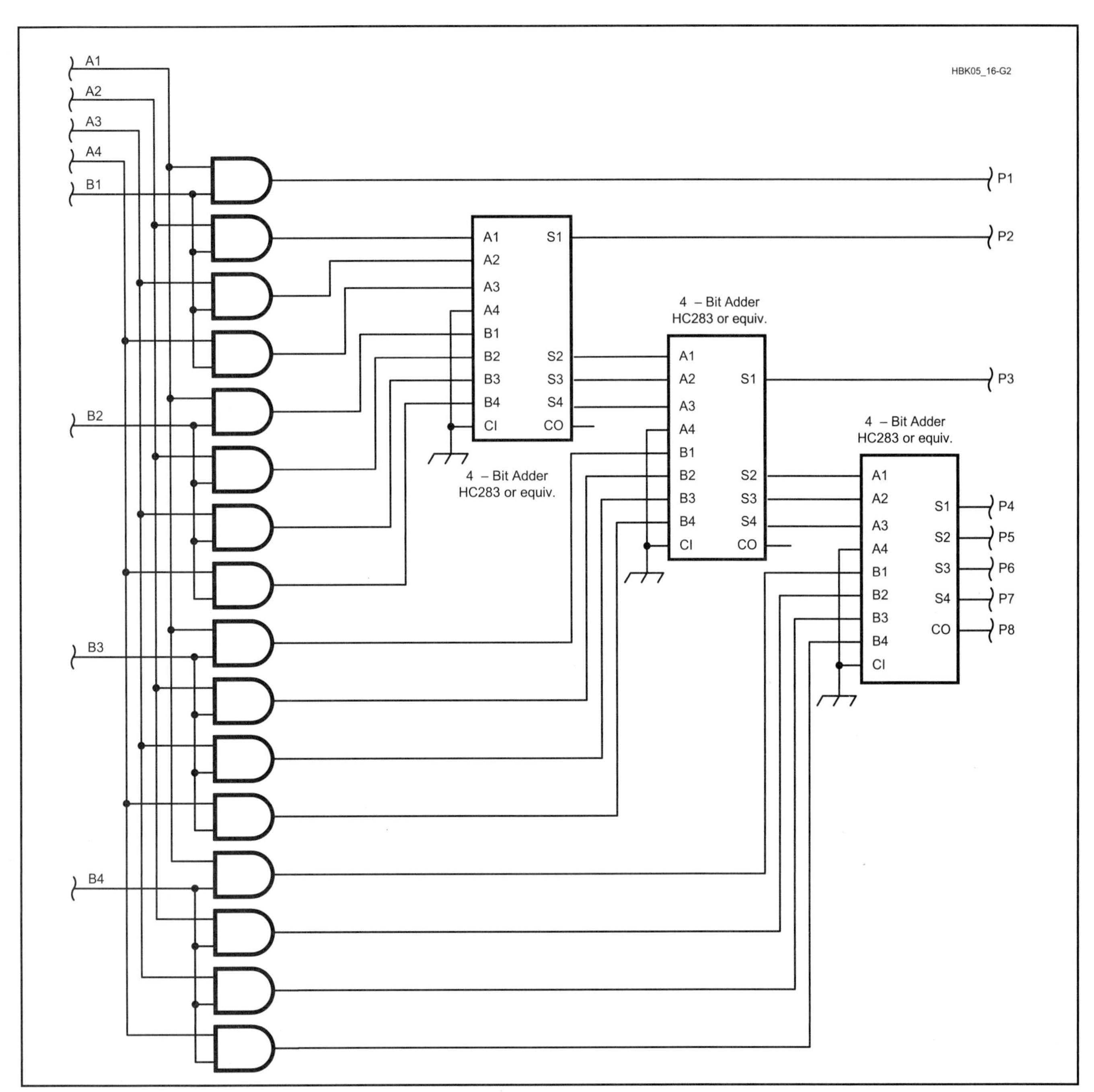

Fig 16.G2—Complete 4-bit multiplier, no pipelining.

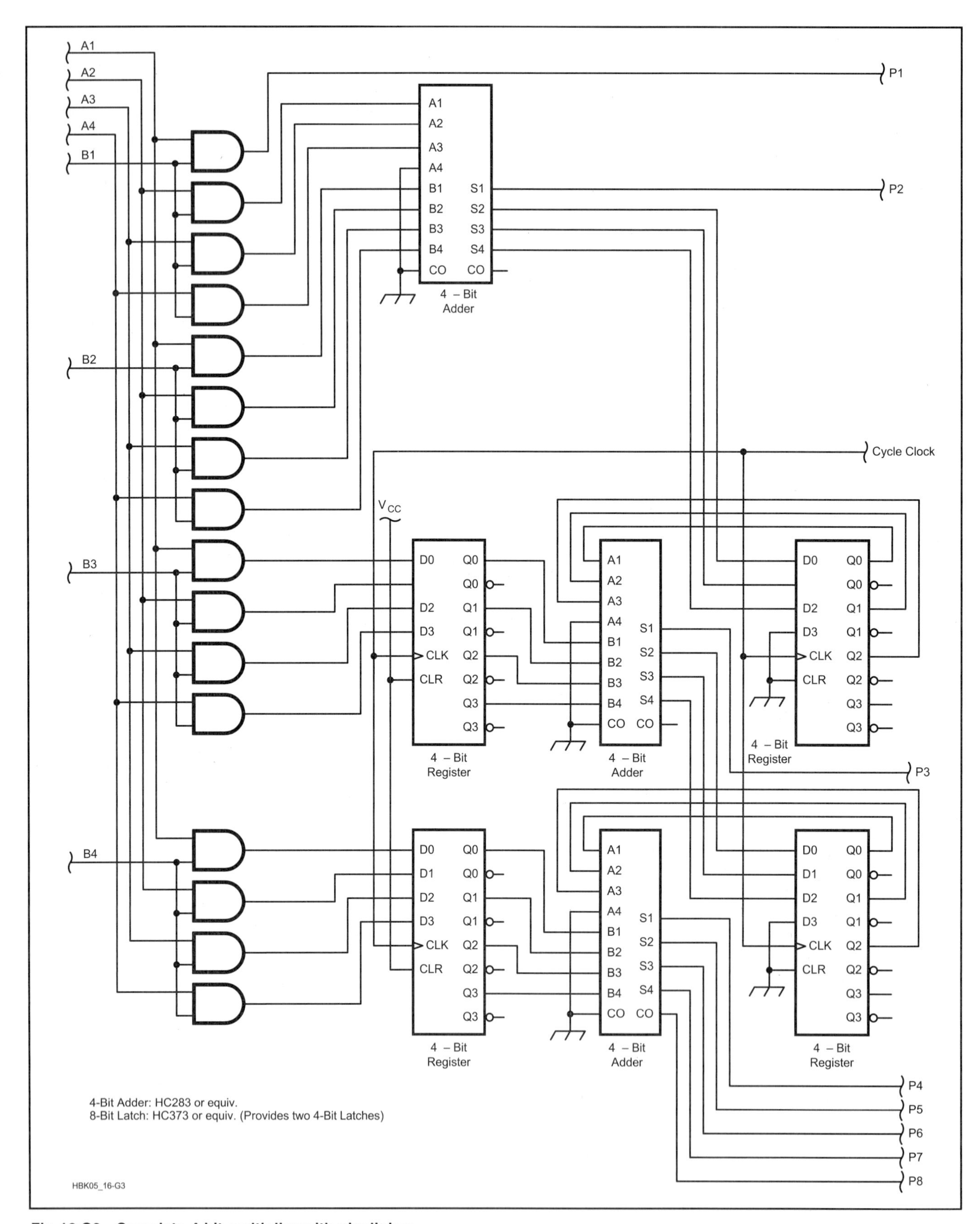

Fig 16.G3—Complete 4-bit multiplier with pipelining.

Chapter 17

Power Supplies

Glossary

Bipolar Transistor — A term used to denote the common two junction transistor types (NPN, PNP) as opposed to the field effect families of devices (JFET, MOSFET and so on).

Bleeder — A resistive load across the output or filter of a power supply, intended to quickly discharge stored energy once the supply is turned off.

C-Rate — The charging rate for a battery, expressed as a ratio of the battery's ampere-hour rating.

Circular Mils — A convenient way of expressing the cross-sectional area of a round conductor. The area of the conductor in circular mils is found by squaring its diameter in mils (thousandths of an inch), rather than squaring its radius and multiplying by pi. For example, the diameter of 10-gauge wire is 101.9 mils (0.1019 inch). Its cross-sectional area is 10380 CM, or 0.008155 square inches.

Core Saturation (Magnetic) — That condition whereby the magnetic flux in a transformer or inductor core is more than the core can handle. If the flux is forced beyond this point, the permeability of the core will decrease, and it will approach the permeability of air.

Crowbar — A last-ditch protection circuit included in many power supplies to protect the load equipment against failure of the regulator in the supply. The crowbar senses an overvoltage condition on the supply's output and fires a shorting device (usually an SCR) to directly short-circuit the supply's output and protect the load. This causes very high currents in the power supply, which blow the supply's input-line fuse.

Darlington Transistor — A package of two transistors in one case, with the collectors tied together, and the emitter of one transistor connected to the base of the other. The effective current gain of the pair is approximately the product of the individual gains of the two devices.

DC-DC Converter — A circuit for changing the voltage of a dc source to ac, transforming it to another level, and then rectifying the output to produce direct current.

Fast Recovery Rectifier — A specially doped rectifier diode designed to minimize the time necessary to halt conduction when the diode is switched from a forward-biased state to a reverse-biased state.

Foldback Current Limiting — A special type of current limiting used in linear power supplies, which reduces the current through the supply's regulator to a low value under short circuited load conditions in order to protect the series pass transistor from excessive power dissipation and possible destruction.

Ground Fault (Circuit) Interrupter (GFI or GFCI) — A safety device installed between the household power mains and equipment where there is a danger of personnel touching an earth ground while operating the equipment. The GFI senses any current flowing directly to ground and immediately switches off all power to the equipment to minimize electrical shock. GFIs are now standard equipment in bathroom and outdoor receptacles.

Input-Output Differential — The voltage drop appearing across the series pass transistor in a linear voltage regulator. This term is usually stated as a minimum value, which is that voltage necessary to allow the regulator to function and conduct current. A typical figure for this drop in most three-terminal regulator ICs is about 2.5 V. In other words, a regulator that is to provide 12.5 V dc will need a source voltage of at least 15.0 V at all times to maintain regulation.

Inverter — A circuit for producing ac power from a dc source.

Peak Inverse Voltage — The maximum reverse-biased voltage that a semiconductor is rated to handle safely. Exceeding the peak inverse rating can result in junction breakdown and device destruction.

Power Conditioner — Another term for a power supply.

Regulator — A device (such as a Zener diode) or circuitry in a power supply for maintaining a constant output voltage over a range of load currents and input voltages.

Resonant Converter — A form of dc-dc converter characterized by the series pass switch turning on into an effective series-resonant load. This allows a zero current condition at turn-on and turn-off. The resonant converter normally operates at frequencies between 100 kHz and 500 kHz and is very compact in size for its power handling ability.

Ripple — The residual ac left after rectification, filtration and regulation of the input power.

RMS — *R*oot of the *M*ean of the *S*quares. Refers to the effective value of an alternating voltage or current, corresponding to the dc voltage or current that would cause the same heating effect.

Secondary Breakdown — A runaway failure condition in a transistor, occurring at higher collector-emitter voltages, where hot spots occur due to (and promoting) localization of the collector current at that region of the chip.

Series Pass Transistor, or Pass Transistor — The transistor(s) that controls the passage of power between the unregulated dc source and the load in a regulator. In a linear regulator, the series pass transistor acts as a controlled resistor to drop the voltage to that needed by the load. In a switch-mode regulator, the series pass transistor switches between its ON and OFF states.

SOAR (*S*afe *O*perating *AR*ea) — The range of permissible collector current and collector-emitter voltage combinations where a transistor may be safely operated without danger of device failure.

Spike — An extremely short perturbation on a power line, usually lasting less than a few microseconds.

Surge — A moderate-duration perturbation on a power line, usually lasting for hundreds of milliseconds to several seconds.

Transient — A short perturbation on a power line, usually lasting for microseconds to tens of milliseconds.

Varistor — A surge suppression device used to absorb transients and spikes occurring on the power lines, thereby protecting electronic equipment plugged into that line. Frequently, the term MOV (*M*etal *O*xide *V*aristor) is used instead.

Volt-Amperes — The product obtained by multiplying the current times the voltage in an ac circuit without regard for the phase angle between the two. This is also known as the *apparent power* delivered to the load as opposed to the actual or *real power* absorbed by the load, expressed in watts.

Voltage Multiplier — A type of rectifier circuit that is arranged so as to charge a capacitor or capacitors on one half-cycle of the ac input voltage waveform, and then to connect these capacitors in series with the rectified line or other charged capacitors on the alternate half-cycle. The voltage doubler and tripler are commonly used forms of the voltage multiplier.

Alternating-Current Power

Ken Stuart, W3VVN, wrote the text for the theory portion of this chapter.

In most residences, three wires are brought in from the outside electrical-service mains to the house distribution panel. In this three-wire system, one wire is neutral and should be at earth ground. The voltage between the other two wires is 60-Hz alternating current with a potential difference of approximately 240 V RMS. Half of this voltage appears between each of these wires and the neutral, as indicated in **Fig 17.1A**. In systems of this type, the 120-V household loads are divided at the breaker panel as evenly as possible between the two sides of the power mains. Heavy appliances such as electric stoves, water heaters, central air conditioners and so forth, are designed for 240-V operation and are connected across the two ungrounded wires.

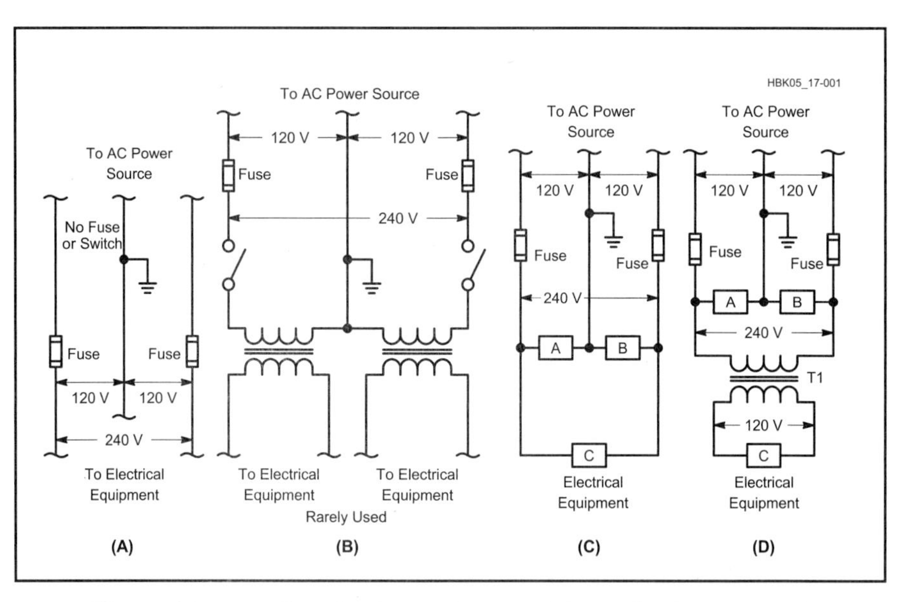

17.1 — Three-wire power-line circuits. At A, normal three-wire-line termination. No fuse should be used in the grounded (neutral) line. The ground symbol is the power company's ground, not yours! Do *not* connect *anything*, including the equipment chassis, to the power neutral wire. At B, the "hot" lines each have a switch, but a switch in the neutral line would not remove voltage from either side of the line and should never be used. At C, connections for both 120 and 240-V transformers. At D, operating a 120-V plate transformer from the 240-V line to avoid light blinking. T1 is a 2:1 step-down transformer.

Both ungrounded wires should be fused. A fuse or switch should never be used in the neutral wire, however. Opening the neutral wire does not disconnect the equipment from an active or "hot" line, creating a potential shock hazard between that line and earth ground.

Another word of caution should be given at this point. Since one side of the ac line is grounded to earth, all communications equipment should be reliably connected to the ac-line ground through a heavy ground braid or bus wire of #14 or heavier-gauge wire. This wire must be a separate conductor. You must not use the power-wiring neutral conductor for this safety ground. (A properly wired 120-V outlet with a ground terminal uses one wire for the ac *hot* connection, one wire for the ac *neutral* connection and a third wire for the safety *ground* connection.) This not only places the chassis of the equipment at earth ground for minimal RF energy on the chassis, but also provides a measure of safety for the operator in the event of accidental short or leakage of one side of the ac line to the chassis.

Remember, the antenna system is almost always bypassed to the chassis via an RF choke or tuning circuit, which could make the antenna electrically "live" with respect to the earth ground and create a potentially lethal shock hazard. A Ground Fault Circuit Interrupter (GFCI or GFI) is also desirable for safety reasons, and should be a part of the shack's electrical power wiring.

FUSES AND CIRCUIT BREAKERS

All transformer primary circuits should be fused properly, and multiple secondary outputs should also be individually fused. To determine the approximate current rating of the fuse or circuit breaker to be used, multiply each current being drawn by the load or appliance, in amperes, by the voltage at which the current is being drawn. In the case of linear regulated power supplies, this voltage has to be the voltage appearing at the output of the rectifiers before being applied to the regulator stage. Include the current taken by bleeder resistors and voltage dividers. Also include filament power if the transformer is supplying filaments. The National Electrical Code also specifies maximum fuse ratings based on the wire sizes used in the transformer and connections.

After multiplying the various voltages and currents, add the individual products. This will be the total power drawn from the line by the supply. Then divide this power by the line voltage and add 10 or 20%. Use a fuse or circuit breaker with the nearest larger current rating. Remember that the charging of filter capacitors can take large surges of current when the supply is turned on. If turn on is a problem, use slow-blow fuses, which allow for high initial surge currents.

For low-power semiconductor circuits, use fast-blow fuses. As the name implies, such fuses open very quickly once the current exceeds the fuse rating by more than 10%.

ELECTRICAL POWER CONDITIONING

We often use the term "power supply" to denote a piece of equipment that will process the electrical power from a source, such as the ac power mains, by manipulating it so the device output will be acceptable to other equipment that we want to power. The common form of the power supply is the familiar direct-current power supply, which will power a transmitter, receiver or other of a wide variety of electronic devices.

In the strictest terms, however, the power supply is not actually a source, or "supply" of power, but is actually a processor of already existing energy. Therefore, the old term of "power supply" is becoming obsolete, and a new term has arisen to refer to the technology of the processing of electrical power: "power conditioning." By contrast, the term "power supply" is now used to refer to devices for chemical to electrical energy conversion (batteries) or mechanical to electrical conversion (generators). Other varieties include thermoelectric generators (TEGs) and radioactive thermoelectric generators (RTGs).

In this chapter, we shall examine the traditional forms of power conditioning, which consists of the following component parts in various combinations: transformer, rectifier, filter and regulator. We will also look briefly at true power supplies as we examine battery technology and emergency power generation.

POWER TRANSFORMERS

Numerous factors are considered in order to match a transformer to its intended use. Some of these parameters are listed below:

1. Output voltage and current (volt-ampere rating)
2. Power source voltage and frequency
3. Ambient temperature
4. Duty cycle and temperature rise of the transformer at rated load
5. Mechanical shape and mounting

Volt-Ampere Rating

In alternating-current equipment, the term "volt-ampere" is often used rather than the term "watt." This is because ac components must handle reactive power as well as real power. If this is confusing, consider a capacitor connected directly across the secondary of a transformer. The capacitor appears as a reactance that permits current to flow, just as if the load were a resistor. The current is at a 90° phase angle, however. If we assume a perfect capacitor, there will be no heating of the capacitor, so no real power (watts) will be delivered by the transformer. The transformer must still be capable of supplying the voltage, and be able to handle the current required by the reactive load. The current in the transformer windings will heat the windings as a result of the I^2R losses in the winding resistances. The product of the voltage and current is referred to as "volt-amperes", since "watts" is reserved for the real, or dissipated, power in the load. The volt-ampere rating will always be equal to, or greater than, the power actually being drawn by the load.

The number of volt-amperes (VA) delivered by a transformer depends not only upon the dc load requirements, but also upon the type of dc output filter used (capacitor or choke input), and the type of rectifier used (full-wave center tap or full-wave bridge). With a capacitive-input filter, the heating effect in the secondary is higher because of the high peak-to-average current ratio. The volt-amperes handled by the transformer may be several times the power delivered to the load. The primary winding volt-amperes will be somewhat higher because of transformer losses.

Source Voltage and Frequency

A transformer operates by producing a magnetic field in its core and windings. The intensity of this field varies directly with the instantaneous voltage applied to the transformer primary winding. These variations, coupled to the secondary windings, produce the desired output voltage. Since the transformer appears to the source as an inductance in parallel with the (equivalent) load, the primary will appear as a short circuit if dc is applied to it. The unloaded inductance of the primary must be high enough so as not to draw an excess amount of input current at the design line frequency (normally 60 Hz). This is achieved by providing sufficient turns on the primary and enough magnetic core material so that the core does not saturate during each half-cycle.

The magnetic field strength produced in the core is usually referred to as the *flux density*. It is set to some percentage of the maximum flux density that the core can stand without saturating, since at saturation the core becomes ineffective and causes the inductance of the primary to plummet to a very low level and input current to rise rapidly. This causes high primary currents and extreme heating in the primary windings. For this reason, transformers and other electromagnetic equipment designed for 60-Hz systems must not be used on 50-Hz power systems unless specifically designed to handle the lower frequency.

How to Evaluate an Unmarked Power Transformer

Many hams that regularly visit hamfests eventually end up with a junk box filled with used and unmarked transformers. After years of use, transformer labels or markings on the coil wrappings may come off or be obscured. There is a good possibility that the transformer is still useable. The problem is to determine what voltages and currents the transformer can supply. First consider the possibility that you may have an audio transformer or other impedance-matching device rather than a power transformer. If you aren't sure, don't connect it to ac power!

If the transformer has color-coded leads, you are in luck. There is a standard for transformer lead color-coding, as is given in the **Component Data and References** chapter. Where two colors are listed, the first one is the main color of the insulation; the second is the color of the stripe.

Check the transformer windings with an ohmmeter to determine that there are no shorted (or open) windings. The primary winding usually has a resistance higher than a filament winding and lower than a high-voltage winding.

A convenient way to test the transformer is to rig a pair of test leads to an electrical plug with a 25-W household light bulb in series to limit current to safe (for the transformer) levels. See **Fig 17.2**. Use an isolation transformer, and be sure to insulate all connections before you plug into the ac mains. Switch off the power while making or changing any connections. Connect the test leads to each winding separately. BE CAREFUL! YOU ARE DEALING WITH HAZARDOUS VOLTAGES! The filament/heater windings will cause the bulb to light to full brilliance. The high-voltage winding will cause the bulb to be extremely dim or to show no light at all, and the primary winding will probably cause a small glow.

When you are connected to what you think is the primary winding, measure the voltages at the low-voltage windings with an ac voltmeter. If you find voltages close to 6-V ac and 5-V ac, you know that you have found the primary. Label the primary and low voltage windings.

Even with the light bulb, a transformer can be damaged by connecting ac mains power to a low-voltage or filament winding. In such a case the insulation could break down in a high-voltage winding.

Connect the voltmeter to the high-voltage windings. Remember that the old TV transformers will typically put out as much as 800 V or so across the winding, so make sure that your meter can withstand these potentials without damage. Divide 6.3 by the voltage you measured across the 6.3-V winding in this test setup. This gives a multiplier that you can use to determine the actual no-load voltage rating of the high-voltage secondary. Simply multiply the ac voltage measured across the winding by the multiplier.

The current rating of the windings can be determined by loading each winding with the primary connected directly (no bulb) to the ac line. Using power resistors, increase loading on each winding until its voltage drops by about 10% from the no-load figure. The current drawn by the resistors is the approximate winding load-current rating.

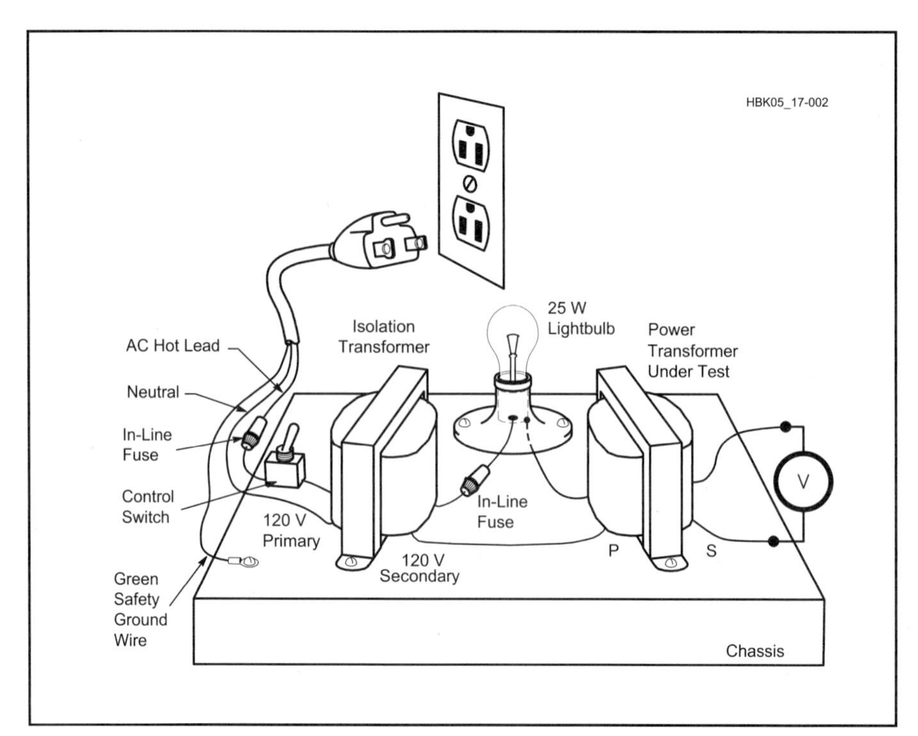

Fig 17.2 — Use a test fixture like this to test unknown transformers. Don't omit the isolation transformer, and be sure to insulate all connections before you plug into the ac mains.

Rectifier Types

VACUUM TUBE

Once the mainstay of the rectifier field, the vacuum-tube rectifier has largely been supplanted by the silicon diode, but it may be found in vintage receivers still in use. Vacuum-tube rectifiers were characterized by high forward voltage drops and inherently poor regulation, but they were immune to ac line transients that can destroy other rectifier types.

MERCURY VAPOR

The mercury-vapor rectifier was an improvement over the vacuum tube rectifier in that the electron stream from cathode to plate would ionize the vaporized mercury in the tube and greatly reduce the forward voltage drop. Since ionized mercury is a much better conductor of current than a vacuum, these tubes can carry relatively high currents. As a result, they were popular in transmitters and RF power amplifiers.

Mercury rectifiers had to be treated with special care, however. When power was initially applied, the tube filament had to be turned on first to vaporize condensed mercury before the high-voltage ac could be applied to the plate. This could take from one to two minutes. Also, if the tube was handled or the equipment transported, filament power would have to be applied for about a half hour to vaporize any mercury droplets that might have been shaken onto tube insulating surfaces. Mercury vapor rectifiers have mostly been replaced by silicon diodes.

SELENIUM

The selenium rectifier was the first of the solid-state rectifiers to find its way into commercial electronic equipment. Offering a relatively low forward voltage drop, selenium rectifiers found their way into the plate supplies of test equipment and accessories, which needed only a few tens of milliamperes of current at about a hundred volts, such as grid-dip meters, VTVMs and so forth.

Selenium rectifiers had a relatively low reverse resistance and were therefore inefficient. Voltage breakdown per rectifying junction was only about 20 V.

GERMANIUM

Germanium diodes were the first of the solid-state semiconductor rectifiers. They have an extremely low forward voltage drop. Germanium diodes are relatively temperature sensitive, however. They can be easily destroyed by overheating during soldering, for instance. Also, they have some degree of back resistance, which varies with temperature.

Germanium diodes are used for special applications where the very low forward drop is needed, such as signal diodes used for detectors and ring modulators.

SILICON

Silicon diodes are the main choice today for virtually all rectifier applications. They are characterized by extremely high reverse resistance, forward drops of usually a volt or less and operation at high temperatures.

FAST RECOVERY

DC-DC converters regularly operate at 25 kHz and higher frequencies. Switch-mode regulators also operate in these same frequency ranges. When the switching transistors in these devices switch, voltage transitions take place within time periods usually much less than one microsecond, and the new FET switching transistors cause transitions that are often less than 100 ns.

When the transitions in these circuits occur, the previously conducting diodes see a reversal of current direction. This change tends to reverse bias those diodes, and thereby put them into an open-circuit condition. Unfortunately, solid-state rectifiers cannot be made to cease conduction instantaneously. As a result, when the opposing diodes in a bridge rectifier or full-wave rectifier become conductive at the time the converter switches states, the diodes being turned off will actually conduct in the reverse direction for a brief time, and effectively short circuit the converter for several microseconds. This puts excessive strain on the switching transistors and creates high current spikes, leading to electromagnetic interference. As the switching frequency of the converter or regulator increases, more of these transitions happen each second, and more power is lost due to this diode cross-conduction.

Semiconductor manufacturers have recognized this as a problem for some time. Many companies have product lines of specially doped diodes designed to minimize this storage time. These diodes are called fast-recovery rectifiers and are commonly used in high-frequency dc-dc converters and regulators. Diodes are available that can recover in about 50 ns and less, as compared to standard-recovery diodes that can take several microseconds to cease conduction in the reverse direction.

Amateurs building their own switching power supplies and dc-dc converters will find greatly improved performance with the use of these diodes in their output rectifiers. Fast-recovery rectifiers are not needed for 60-Hz rectification because the source voltage is a sine wave (no fast transitions) and the input frequency is too slow for transitions to be of significance.

Rectifier Circuits

HALF-WAVE RECTIFIER

Fig 17.3 shows a simple half-wave rectifier circuit. A rectifier (in this case a semiconductor diode) conducts current in one direction but not the other. During one half of the ac cycle, the rectifier conducts and there is current through the rectifier to the load (indicated by the solid line in Fig 17.3B). During the other half cycle, the rectifier is reverse biased and there is no current (indicated by the broken line in Fig 17.3B) to the load. As shown, the output is in the form of pulsed dc, and current always flows in the same direction. A filter can be used to smooth out these variations and provide a higher average dc voltage from the circuit. This idea will be covered in the section on filtration further on in this chapter.

The average output voltage — the voltage read by a dc voltmeter — with this circuit (no filter connected) is $0.45 \times E_{RMS}$ of the ac voltage delivered by the transformer secondary. Because the frequency of the pulses is low (one pulse per cycle), considerable filtering is required to pro-

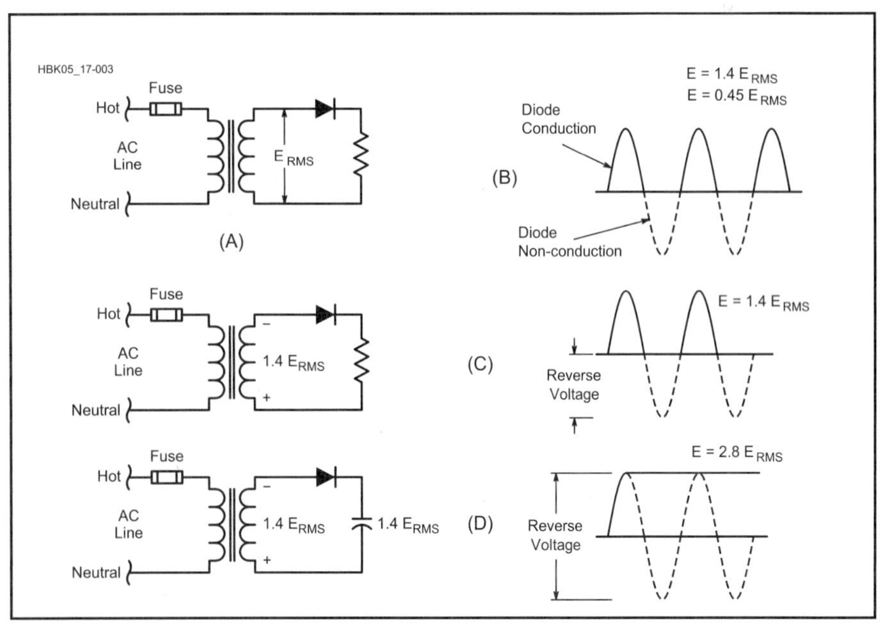

Fig 17.3 — Half-wave rectifier circuit. A illustrates the basic circuit, and B displays the diode conduction and nonconduction periods. The peak-inverse voltage impressed across the diode is shown at C and D, with a simple resistor load at C and a capacitor load at D. E_{PIV} is 1.4 E_{RMS} for the resistor load and 2.8 E_{RMS} for the capacitor load.

vide adequately smooth dc output. For this reason the circuit is usually limited to applications where the required current is small, as in a transmitter bias supply.

The peak inverse voltage (PIV), the voltage that the rectifier must withstand when it isn't conducting, varies with the load. With a resistive load, it is the peak ac voltage ($1.4 \times E_{RMS}$); with a capacitor filter and a load drawing little or no current, it can rise to $2.8 \times E_{RMS}$. The reason for this is shown in parts C and D of Fig 17.3. With a resistive load as shown at C, the voltage applied to the diode is that voltage on the lower side of the zero-axis line, or $1.4 \times E_{RMS}$. A capacitor connected to the circuit (shown at D) will store the peak positive voltage when the diode conducts on the positive pulse. If the circuit is not supplying any current, the voltage across the capacitor will remain at that same level. The peak inverse voltage impressed across the diode is now the sum of the voltage stored in the capacitor plus the peak negative swing of voltage from the transformer secondary. In this case the PIV is $2.8 \times E_{RMS}$.

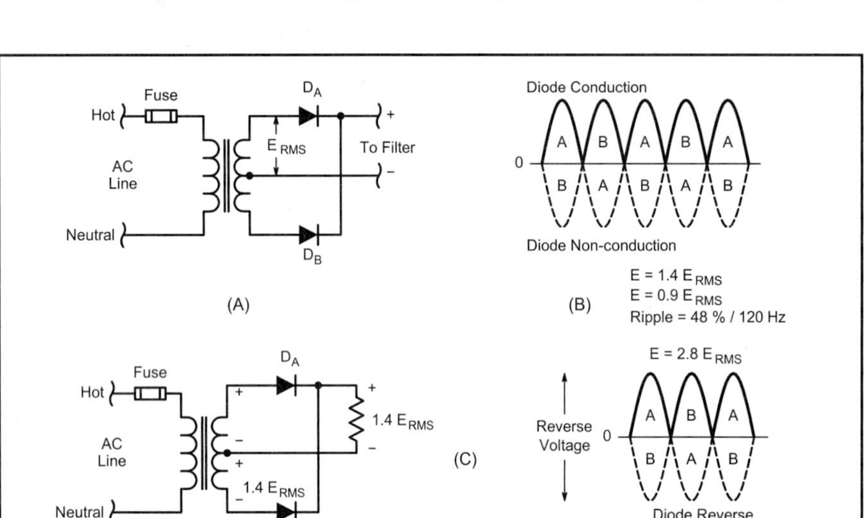

Fig 17.4 — Full-wave center-tap rectifier circuit. A illustrates the basic circuit. Diode conduction is shown at B with diodes A and B alternately conducting. The peak-inverse voltage for each diode is 2.8 E_{RMS} as depicted at C.

FULL-WAVE CENTER-TAP RECTIFIER

A commonly used rectifier circuit is shown in **Fig 17.4**. Essentially an arrangement in which the outputs of two half-wave rectifiers are combined, it makes use of both halves of the ac cycle. A transformer with a center-tapped secondary is required with the circuit.

The average output voltage is $0.9 \times E_{RMS}$ of half the transformer secondary; this is the maximum that can be obtained with a suitable choke-input filter. The peak output voltage is $1.4 \times E_{RMS}$ of half the transformer secondary; this is the maximum voltage that can be obtained from a capacitor-input filter.

As can be seen in Fig 17.4C, the PIV impressed on each diode is independent of the type of load at the output. This is because the peak inverse voltage condition occurs when diode A conducts and diode B does not conduct. The positive and negative voltage peaks occur at precisely the same time, a condition different from that in the half-wave circuit. As the cathodes of diodes A and B reach a positive peak ($1.4\ E_{RMS}$), the anode of diode B is at a negative peak, also $1.4\ E_{RMS}$, but in the opposite direction. The total peak inverse voltage is therefore $2.8\ E_{RMS}$.

Fig 17.4B shows that the frequency of the output pulses is twice that of the half-wave rectifier. Comparatively less filtering is required. Since the rectifiers work alternately, each handles half of the load current. The current rating of each rectifier need be only half the total current drawn from the supply.

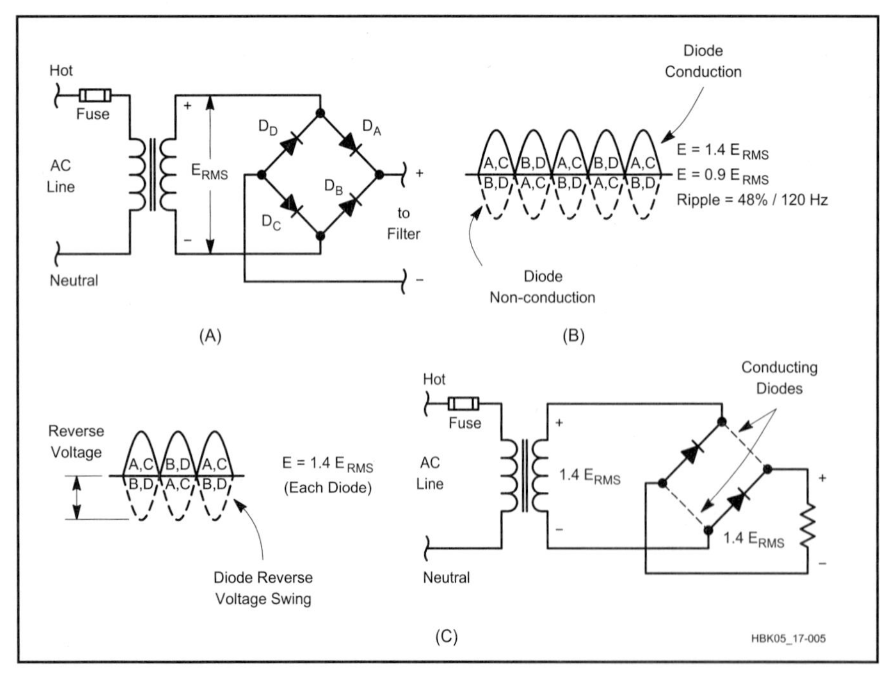

Fig 17.5 — Full-wave bridge rectifier circuit. The basic circuit is illustrated at A. Diode conduction and nonconduction times are shown at B. Diodes A and C conduct on one half of the input cycle, while diodes B and D conduct on the other. C displays the peak inverse voltage for one half cycle. Since this circuit reverse-biases two diodes essentially in parallel, 1.4 E_{RMS} is applied across each diode.

FULL-WAVE BRIDGE RECTIFIER

Another commonly used rectifier circuit is illustrated in **Fig 17.5**. In this arrangement, two rectifiers operate in series on each half of the cycle, one rectifier being in the lead to the load, the other being the return lead. As shown in Fig 17.5A and B, when the top lead of the transformer secondary is positive with respect to the bottom lead, diodes A and C will conduct while diodes B and D are reverse biased. On the next half cycle, when the top lead of the transformer is negative with respect to the bottom, diodes B and D will conduct while diodes A and C are reverse biased.

The output wave shape is the same as that from the simple full-wave center-tap rectifier circuit. The average dc output voltage into a resistive load or choke-input filter is 0.9 times the RMS voltage delivered by the transformer secondary;

with a capacitor filter and a light load, the maximum output voltage is 1.4 times the secondary RMS voltage.

Fig 17.5C shows the inverse voltage to be 1.4 E_{RMS} for each diode. When an alternate pair of diodes (such as D_A and D_C) is conducting, the other diodes are essentially connected in parallel in a reverse-biased direction. The reverse stress is then 1.4 E_{RMS}. Each pair of diodes conducts on alternate half cycles, with the full load current through each diode during its conducting half cycle. Since each diode is not conducting during the other half cycle the average current is one half the total load current drawn from the supply.

PROS AND CONS OF THE RECTIFIER CIRCUITS

Comparing the full-wave center-tap rectifier circuit and the full-wave bridge-rectifier circuit, we can see that both circuits have almost the same rectifier requirement, since the center tap has half the number of rectifiers as the bridge. These rectifiers have twice the inverse voltage rating requirement of the bridge diodes, however. The diode current ratings are identical for the two circuits. The bridge makes better use of the transformer's secondary than the center-tap rectifier, since the transformer's full winding supplies power during both half cycles, while each half of the center-tap circuit's secondary provides power only during its positive half-cycle. This is usually referred to as the *transformer utilization factor*, which is unity for the bridge configuration and 0.5 for the full-wave, center-tapped circuit.

The bridge rectifier often takes second place to the full-wave center tap rectifier in high-current low-voltage applications. This is because the two forward-conducting series-diode voltage drops in the bridge introduce a volt or more of additional loss, and thus more heat to be dissipated, than does the single diode drop of the full-wave rectifier.

The half-wave configuration is rarely used in 60-Hz rectification for other than bias supplies. It does see considerable use, however, in high-frequency switching power supplies in what are called *forward converter* and *flyback converter* topologies.

VOLTAGE MULTIPLIERS

Other rectification circuits of interest are the so-called *voltage multipliers*. These circuits function by the process of charging one or more capacitors on one half cycle of the ac waveform, and then connecting that capacitor or capacitors in series with the opposite polarity of the ac waveform on the alternate half cycle. With full-wave multipliers, this charging occurs during both half-cycles.

Voltage multipliers, particularly doublers, find considerable use in high-voltage supplies. When a doubler is employed, the secondary winding of the power transformer need only be half the voltage that would be required for a bridge rectifier. This reduces voltage stress in the windings and decreases the transformer insulation requirements. It also reduces the chance of corona in the windings, prolonging the life of the transformer. This is not without cost, however, because the transformer-secondary current rating has to be correspondingly doubled.

Half-Wave Doubler

Fig 17.6 shows the circuit of a half-wave voltage doubler. Parts B, C and D illustrate the circuit operation. For clarity, assume the transformer voltage polarity at the moment the circuit is activated is that shown at B. During the first negative half cycle, D_A conducts (D_B is in a nonconductive state), charging C1 to the peak rectified voltage (1.4 E_{RMS}). C1 is charged with the polarity shown at B. During the positive half cycle of the secondary voltage, D_A is cut off and D_B conducts, charging capacitor C2. The amount of voltage delivered to C2 is the sum of the transformer peak secondary voltage plus the voltage stored in C1 (1.4 E_{RMS}). On the next negative half cycle, D_B is non-conducting and C2 will discharge into the load. If no load is connected across C2, the capacitors will remain charged — C1 to 1.4 E_{RMS} and C2 to 2.8 E_{RMS}. When a load is connected to the circuit output, the voltage across C2 drops during the negative half cycle and is recharged up to 2.8 E_{RMS} during the positive half cycle.

The output waveform across C2 resembles that of a half-wave rectifier circuit because C2 is pulsed once every cycle. Fig 17.6D illustrates the levels to which the two capacitors are charged throughout the cycle. In actual operation, the capacitors will not discharge all the way to zero as shown.

Full-Wave Doubler

Fig 17.7 shows the circuit of a full-wave voltage doubler. The circuit operation can best be understood by following Parts B, C and D. During the positive half cycle of the transformer secondary voltage, as shown at B, D_A conducts charging capacitor C1 to 1.4 E_{RMS}. D_B is not conducting at this time.

During the negative half cycle, as shown at C, D_B conducts, charging capacitor C2 to 1.4 E_{RMS}, while D_A is non-conducting. The output voltage is the sum of the two capacitor voltages, which will be 2.8 E_{RMS} under no-load conditions. Fig 17.7D illustrates that each capacitor alternately receives a charge once per cycle. The effective filter capacitance is that of C1 and C2 in series, which is less than the capacitance of either C1 or C2 alone.

Resistors R1 and R2 in Fig 17.7A are used to limit the surge current through the rectifiers. Their values are based on the transformer voltage and the rectifier surge-current rating, since at the instant

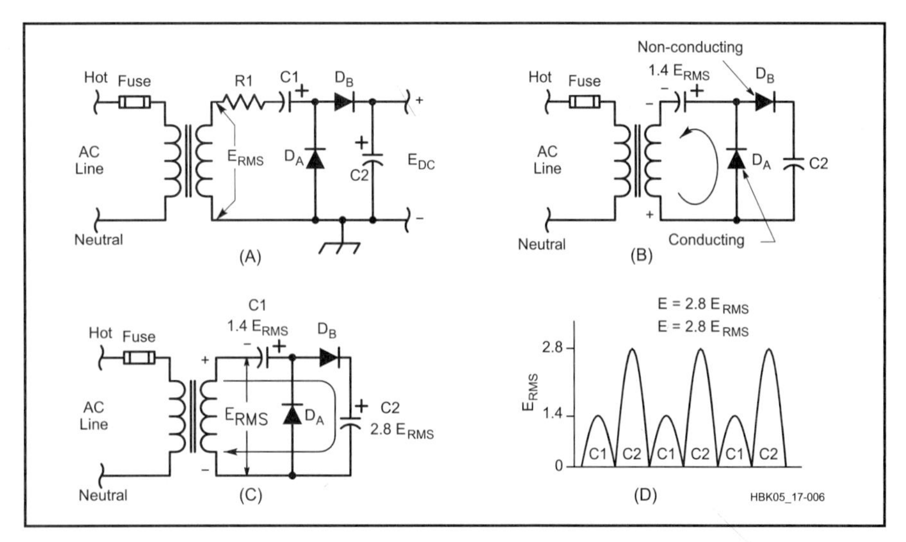

Fig 17.6 — Part A shows a half-wave voltage-doubler circuit. B displays how the first half cycle of input voltage charges C1. During the next half cycle (shown at C), capacitor C2 charges with the transformer secondary voltage plus that voltage stored in C1 from the previous half cycle. The arrows in parts B and C indicate the conventional current. D illustrates the levels to which each capacitor charges over several cycles.

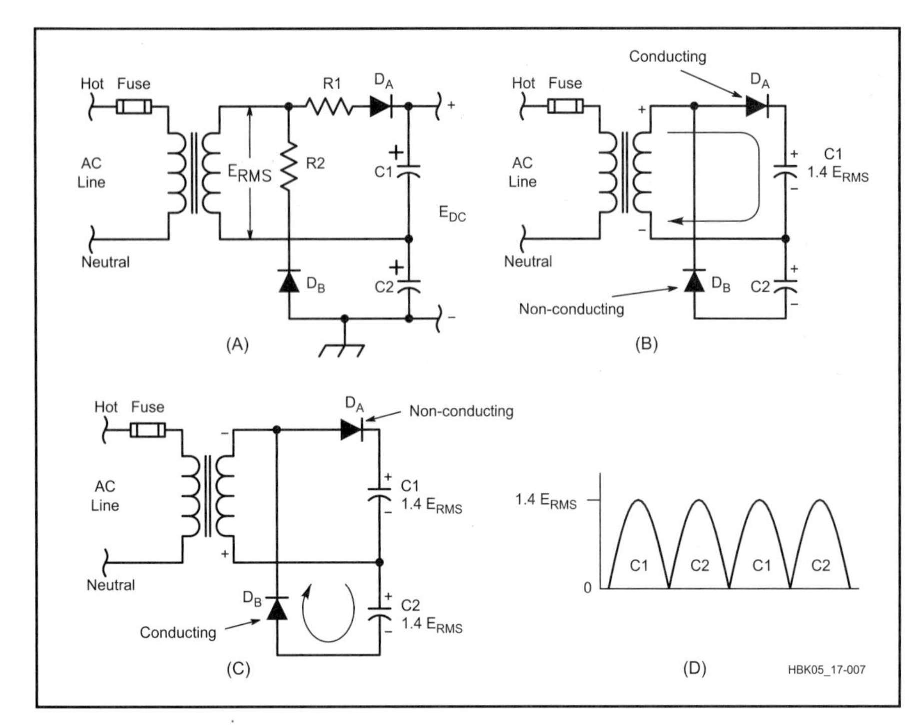

Fig 17.7 — Part A shows a full-wave voltage-doubler circuit. One-half cycle is shown at B and the next half cycle is shown at C. Each capacitor receives a charge during every input-voltage cycle. D illustrates how each capacitor is charged alternately.

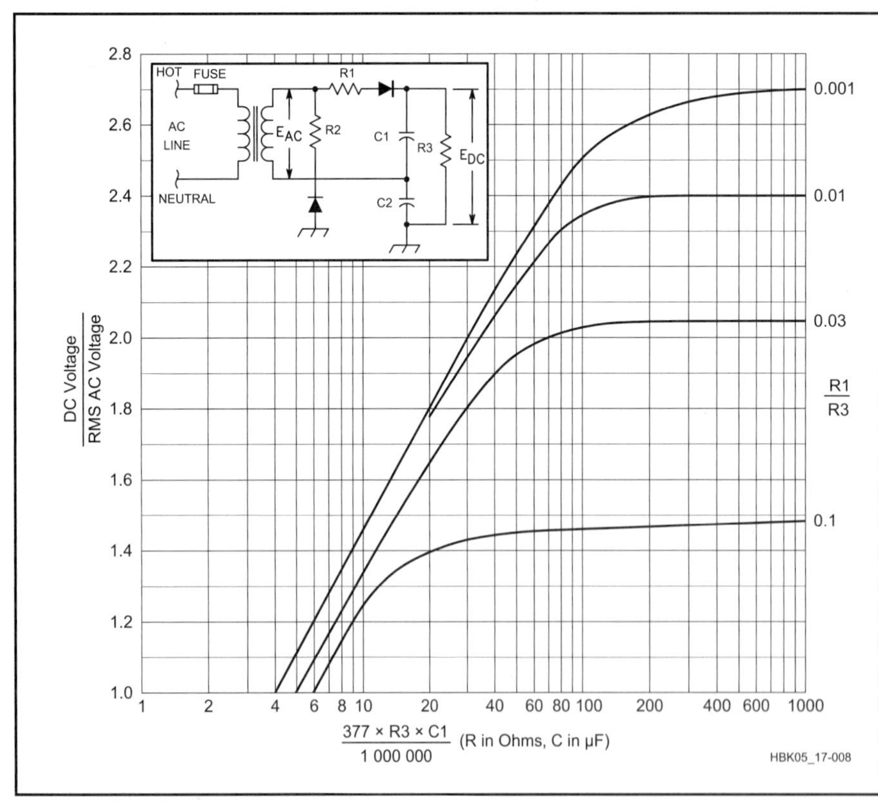

Fig 17.8 — DC output voltages from a full-wave voltage-doubler circuit as a function of the filter capacitances and load resistance. For the ratio R1 / R3 and for the R3 × C1 product, resistance is in ohms and capacitance is in microfarads. Equal resistance values for R1 and R2, and equal capacitance values for C1 and C2 are assumed. These curves are adapted from those published by Otto H. Schade in "Analysis of Rectifier Operation," ***Proceedings of the I. R. E.*****, July 1943.**

the power supply is turned on, the filter capacitors look like a short-circuited load. Provided the limiting resistors can withstand the surge current, their current-handling capacity is based on the maximum load current from the supply. Output voltages approaching twice the peak voltage of the transformer can be obtained with the voltage doubling circuit shown in Fig 17.7. **Fig 17.8** shows how the voltage depends upon the ratio of the series resistance to the load resistance, and the load resistance times the filter capacitance. The peak inverse voltage across each diode is 2.8 E_{RMS}.

Tripler and Quadrupler

Fig 17.9A shows a voltage-tripling circuit. On one half of the ac cycle, C1 and C3 are charged to the source voltage through D1, D2 and D3. On the opposite half of the cycle, D2 conducts and C2 is charged to twice the source voltage, because it sees the transformer plus the charge in C1 as its source (D1 is cut off during this half cycle). At the same time, D3 conducts, and with the transformer and the charge in C2 as the source, C3 is charged to three times the transformer voltage.

The voltage-quadrupling circuit of Fig 17.9B works in similar fashion. In either of the circuits of Fig 17.9, the output voltage will approach an exact multiple of the peak ac voltage when the output

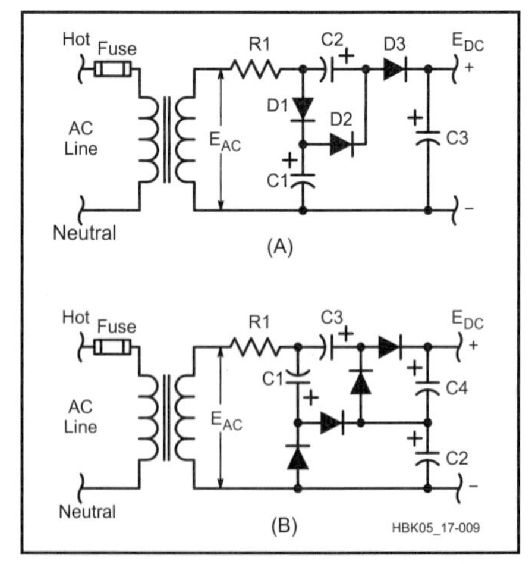

Fig 17.9 — Voltage-multiplying circuits with one side of the transformer secondary used as a common connection. A shows a voltage tripler and B shows a voltage quadrupler. Capacitances are typically 20 to 50 µF, depending on the output current demand. Capacitor dc ratings are related to E_{PEAK} (1.4 E_{RMS}):
C1 — Greater than E_{PEAK}
C2 — Greater than 2 E_{PEAK}
C3 — Greater than 3 E_{PEAK}
C4 — Greater than 2 E_{PEAK}

current drain is low and the capacitance values are high.

RECTIFIER RATINGS VERSUS OPERATING STRESS

Power supplies designed for amateur equipment use silicon rectifiers almost exclusively. These rectifiers are available in a wide range of voltage and current ratings. In peak inverse voltage (PIV) ratings of 600 or less, silicon rectifiers carry current ratings as high as 400 A. At 1000 PIV, the current ratings may be several amperes. It is possible to stack several units in series for higher voltages. Stacks are available commercially that will handle peak inverse voltages up to 10 kV at a load current of 1 A or more.

RECTIFIER STRINGS OR STACKS

Diodes in Series

When the PIV rating of a single diode is not sufficient for the application, similar diodes may be used in series. (Two 500 PIV diodes in series will withstand 1000 PIV and so on.) There used to be a general recommendation to place a resistor across each diode in the string to equalize the PIV drops. With modern diodes, this practice is no longer necessary.

Modern silicon rectifier diodes are constructed to have an avalanche characteristic. Simply put, this means that the diffusion process is controlled so the diode will exhibit a Zener characteristic in the reverse biased direction before destructive breakdown of the junction can occur. This provides a measure of safety for diodes in series. A diode will go into Zener conduction before it self destructs. If other diodes in the chain have not reached their avalanche voltages, the current through the avalanched diode will be limited to the leakage current in the other diodes. This should normally be very low. For this reason, shunting resistors are generally not needed across diodes in series rectifier strings. In fact, shunt resistors can actually create problems because they can produce a low-impedance source of damaging current to any diode that may have reached avalanche potential.

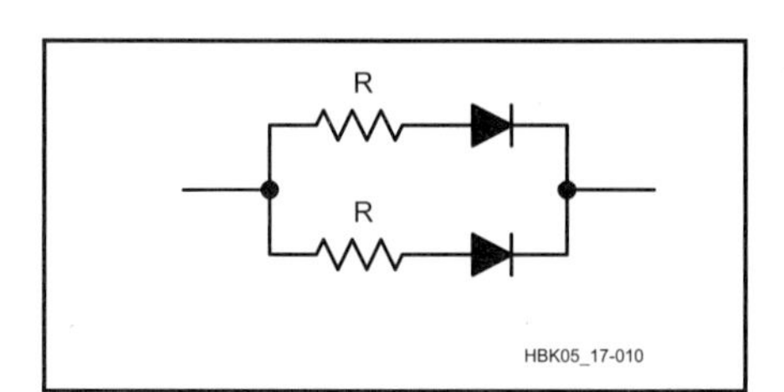

Fig 17.10 — Diodes can be connected in parallel to increase the current-handling capability of the circuit. Each diode should have a series current-equalizing resistor, with a value selected to provide a few tenths of a volt drop at the expected current.

Diodes in Parallel

Diodes can be placed in parallel to increase current-handling capability. Equalizing resistors should be added as shown in **Fig 17.10**. Without the resistors, one diode may take most of the current. The resistors should be selected to have several tenths of a volt drop at the expected peak current.

RECTIFIER PROTECTION

The important specifications of a silicon diode are:

1. PIV, the peak inverse voltage.
2. I_0, the average dc current rating.
3. I_{REP} — the peak repetitive forward current.
4. I_{SURGE}, a non-repetitive peak half-sine wave of 8.3 ms duration (one-half cycle of 60-Hz line frequency).
5. Switching speed.
6. Power dissipation and thermal resistance.

The first two specifications appear in most catalogs. I_{REP} and I_{SURGE} often are not specified in catalogs, but they are very important. Because the rectifier never allows current to flow more than half the time, when it does conduct it has to pass at least twice the average direct current. With a capacitor-input filter, the rectifier conducts much less than half the time, so that when it does conduct, it may pass as much as 10 to 20 times the average dc current, under certain conditions. This is shown in **Fig 17.11**. Part A shows a simple half-wave rectifier with a resistive load. The waveform to the right of the drawing shows the output voltage along with the diode current. Parts B and C show conditions for circuits with "low" capacitance and "high" capacitance to filter the output.

After the capacitor is charged to the peak-rectified voltage, a period of diode non-conduction elapses while the output voltage discharges through the load. As the voltage begins to rise on the next positive pulse, a point is reached where the rectified voltage equals the stored voltage in the capacitor. As the voltage rises beyond that point, the diode begins to supply current. The diode will continue to conduct until the waveform reaches the crest, as shown. Since the diode has only that short time in which to charge the capacitor with enough energy to provide power to the load for the non-conducting balance of the cycle, the current will be high. The larger the capacitor for a given load, the shorter the diode conduction time and the higher the peak repetitive current (I_{REP}).

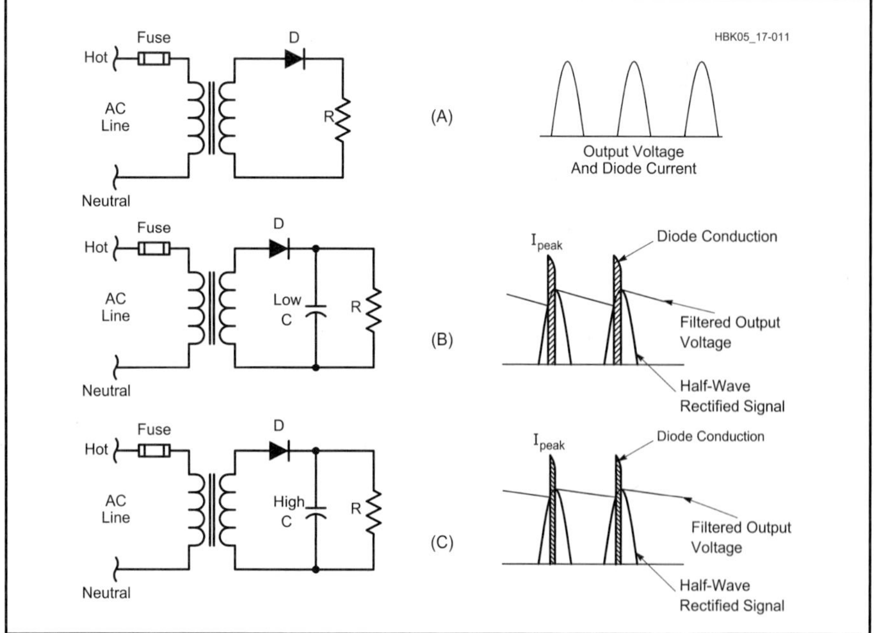

Fig 17.11 — The circuit shown at A is a simple half-wave rectifier with a resistive load. The waveform shown to the right is the output voltage and current. B illustrates how the diode current is modified by the addition of a capacitor filter. The diode conducts only when the rectified voltage is greater than the voltage stored in the capacitor. Since this time is usually only a short portion of a cycle, the peak current will be quite high. C shows an even higher peak current. This is caused by the larger capacitor, which effectively shortens the diode conduction period.

Current Inrush

When the supply is first turned on, the discharged input capacitor looks like a dead short, and the rectifier passes a very heavy current. This current transient is called I_{SURGE}. The maximum surge current rating for a diode is usually specified for a duration of one-half cycle (at 60 Hz), or about 8.3 ms. Some form of surge protection is usually necessary to protect the diodes until the input capacitor becomes nearly charged, unless the diodes used have a very high surge-current rating (several hundred amperes). If a manufacturer's data sheet is not available, an educated guess about a diode's capability can be made by using these rules of thumb for silicon diodes commonly used in Amateur Radio power supplies:

Rule 1. The maximum I_{REP} rating can be assumed to be approximately four times the maximum I_0 rating.

Rule 2. The maximum I_{SURGE} rating can be assumed to be approximately 12 times the maximum I_0 rating. (This figure should provide a reasonable safety factor. Silicon rectifiers with 750-mA dc ratings, for example, seldom have 1-cycle surge ratings of less than 15 A; some are rated up to 35 A or more.) From this you can see that the rectifier should be selected on the basis of I_{SURGE} and not on I_0 ratings.

Although you can sometimes rely on the dc resistance of the transformer secondary to provide ample surge-current limiting, this is seldom true in high-voltage power supplies. Series resistors are often installed between the secondary and the rectifier strings or in the transformer's primary circuit, but these can be a deterrent to good voltage regulation.

Voltage Spikes

Vacuum-tube rectifiers had little problem with voltage spikes on the incoming power lines — the possibility of an internal arc was of little consequence, since the heat produced was of very short duration and had little effect on the massive plate and cathode structures. Unfortunately, such is not the case with silicon diodes.

Silicon diodes, because of their forward voltage drop of about one volt, create very little heat with high forward currents and therefore have tiny junction areas. Conduction in the reverse direction, however, can cause junction temperatures to rise extremely rapidly with the resultant melting of the silicon and migration of the dopants into the rectifying junction. Destruction of the semiconductor junction is the end result.

To protect semiconductor rectifiers, special surge-absorption devices are available for connection across the incoming ac bus or transformer secondary. These devices operate in a fashion similar to a Zener diode, by conducting heavily when a specific voltage level is reached. Unlike Zener diodes, however, they have the ability to absorb very high transient energy levels without damage. With the clamping level set well above the normal operating voltage range for the rectifiers, these devices normally appear as open circuits and have no effect on the power-supply circuits. When a voltage transient occurs, however, these protection devices clamp the spike and thereby prevent destruction of the rectifiers.

Transient protectors are available in three basic varieties:

1. Silicon Zener diodes — large junction Zeners specifically made for this purpose and available as single junction for dc (unipolar) and back-to-back junctions for ac (bipolar). These silicon protectors are available under the trade name of TransZorb from General Semiconductor Corporation and are also made by other manufacturers. They have the best transient suppressing characteristics of the three varieties mentioned here, but are expensive and have the least energy absorbing capability per dollar of the group.

2. Varistors — made of a composition metal-oxide material that breaks down at a certain voltage. Metal-oxide varistors, also known as MOVs, are cheap and easily obtained, but have a higher internal resistance, which allows a greater increase of clamped voltage than the Zener variety. Varistors can also degrade with successive transients within their rated power handling limits (this is not usually a problem in the ham shack where transients are few and replacement of the varistor is easily accomplished).

Varistors usually become short circuited when they fail. Large energy dissipation can result in device explosion. Therefore, it is a good idea to include a fuse that limits the short-circuit current through the varistor, and to protect people and circuitry from debris.

3. Gas tube — similar in construction to the familiar neon bulb, but designed to limit conducting voltage rise under high transient currents. Gas tubes can usually withstand the highest transient energy levels of the group. Gas tubes suffer from an ionization time problem, however. A high voltage across the tube will not immediately cause conduction. The time required for the gas to ionize and clamp the spike is inversely proportional to the level of applied voltage in excess of the device ionization voltage. As a result, the gas tube will let a little of the transient through to the equipment before it activates.

In installations where reliable equipment operation is critical, the local power is poor and transients are a major problem, the usual practice is to use a combination of protectors. Such systems consist of a varistor or Zener protector, combined with a gas-tube device. Operationally, the solid-state device clamps the surge immediately, with the beefy gas tube firing shortly thereafter to take most of the surge from the solid-state device.

Heat

The junction of a diode is quite small; hence, it must operate at a high current density. The heat-handling capability is, therefore, quite small. Normally, this is not a prime consideration in high-voltage, low-current supplies. Use of high-current rectifiers at or near their maximum ratings (usually 2-A or larger stud-mount rectifiers) requires some form of heat sinking. Frequently, mounting the rectifier on the main chassis — directly, or with thin mica insulating washers — will suffice. If insulated from the chassis, a thin layer of silicone grease should be used between the diode and the insulator, and between the insulator and the chassis, to assure good heat conduction. Large, high-current rectifiers often require special heat sinks to maintain a safe operating temperature. Forced-air cooling is sometimes used as a further aid. Safe case temperatures are usually given in the manufacturer's data sheets and should be observed if the maximum capabilities of the diode are to be realized. See the thermal design section in the **Real- World Component Characteristics** chapter for more information.

Filtration

The pulsating dc waves from the rectifiers are not sufficiently constant in amplitude to prevent hum corresponding to the pulsations. Filters are required between the rectifier and the load to smooth out the pulsations into an essentially constant dc voltage. The design of the filter depends to a large extent on the dc voltage output, the voltage regulation of the power supply and the maximum load current rating of the rectifier. Power-supply filters are low-pass devices using series inductors and shunt capacitors.

LOAD RESISTANCE

In discussing the performance of power-supply filters, it is sometimes convenient to express the load connected to the output terminals of the supply in terms of resistance. The load resistance is equal to the output voltage divided by the total current drawn, including the current drawn by the bleeder resistor.

VOLTAGE REGULATION

The output voltage of a power supply always decreases as more current is drawn, not only because of increased voltage drops in the transformer and filter chokes, but also because the output voltage at light loads tends to soar to the peak value of the transformer voltage as a result of charging the first capacitor. Proper filter design can eliminate the soaring effect. The change in output voltage with load is called voltage regulation, and is expressed as a percentage.

$$\text{Percent Regulation} = \frac{(E1 - E2)}{E2} \times 100\% \qquad (1)$$

where:
E1 = the no-load voltage
E2 = the full-load voltage.

A steady load, such as that represented by a receiver, speech amplifier or unkeyed stages of a transmitter, does not require good (low) regulation as long as the proper voltage is obtained under load conditions. The filter capacitors must have a voltage rating safe for the highest value to which the voltage will soar when the external load is removed.

A power supply will show more (higher) regulation with long-term changes in load resistance than with short temporary changes. The regulation with long-term changes is often called the static regulation, to distinguish it from the dynamic regulation (short temporary load changes). A load that varies at a syllabic or keyed rate, as represented by some audio and RF amplifiers, usually requires good dynamic regulation (15% or less) if distortion products are to be held to a low level. The dynamic regulation of a power supply can be improved by increasing the value of the output capacitor.

When essentially constant voltage regardless of current variation is required (for stabilizing an oscillator, for example), special voltage regulating circuits described later in this chapter are used.

BLEEDER RESISTOR

A bleeder resistor is a resistance connected across the output terminals of the power supply. Its functions are to discharge the filter capacitors as a safety measure when the power is turned off and to improve voltage regulation by providing a minimum load resistance. When voltage regulation is not of importance, the resistance may be as high as 100-W per volt of power supply output voltage. The resistance value to be used for voltage-regulating purposes is discussed in later sections. From the consideration of safety, the power rating of the resistor should be as conservative as possible, since a burned-out bleeder resistor is dangerous!

RIPPLE FREQUENCY AND VOLTAGE

Pulsations at the output of the rectifier can be considered to be the result of an alternating current superimposed on a steady direct current. From this viewpoint, the filter may be considered to consist of shunt capacitors that short circuit the ac component while not interfering with the flow of the dc component. Series chokes will readily pass dc but will impede the flow of the ac component.

The alternating component is called ripple. The effectiveness of the filter can be expressed in terms of percent ripple, which is the ratio of the RMS value of the ripple to the dc value in terms of percentage.

$$\text{Percent Ripple (RMS)} = \frac{E1}{E2} \times 100\% \qquad (2)$$

where:
E1 = the RMS value of ripple voltage
E2 = the steady dc voltage.

Any frequency multiplier or amplifier supply in a CW transmitter should have less than 5% ripple. A linear amplifier can tolerate about 3% ripple on the plate voltage. Bias supplies for linear amplifiers should have less than 1% ripple. VFOs, speech amplifiers and receivers may require a ripple no greater than to 0.01%.

Ripple frequency refers to the frequency of the pulsations in the rectifier output waveform — the number of pulsations per second. The ripple frequency of half-wave rectifiers is the same as the line-supply frequency — 60 Hz with a 60-Hz supply. Since the output pulses are doubled with a full-wave rectifier, the ripple frequency is doubled — to 120 Hz with a 60-Hz supply.

The amount of filtering (values of inductance and capacitance) required to give adequate smoothing depends on the ripple frequency. More filtering is required as the ripple frequency is reduced.

CAPACITOR-INPUT FILTERS

Capacitor-input filter systems are shown in **Fig 17.12**. Disregarding voltage drops in the chokes, all have the same characteristics except with respect to ripple. Better ripple reduction will be obtained when LC sections are added as shown in Fig 17.12B and C.

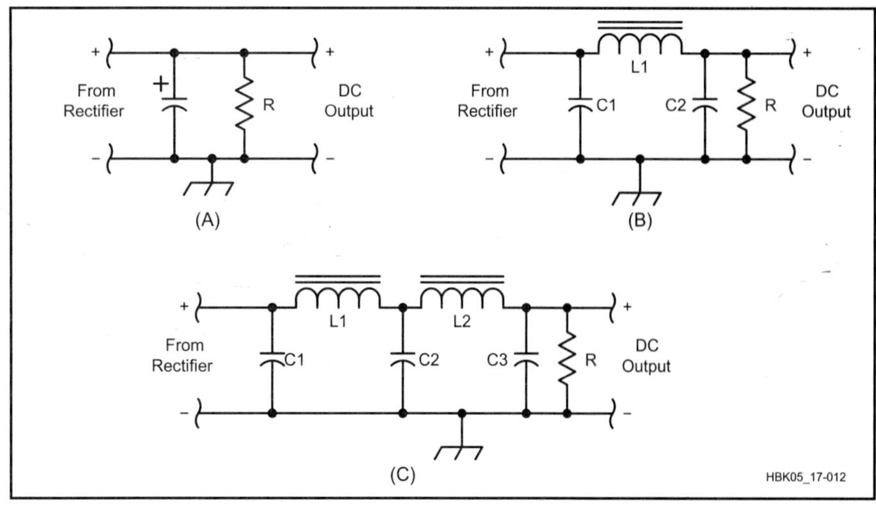

Fig 17.12 — Capacitor-input filter circuits. At A is a simple capacitor filter. B and C are single- and double-section filters, respectively.

Input Versus Output Voltage

The average output voltage of a capacitor-input filter is generally poorly regulated with load-current variations. This is because the rectifier diodes conduct for only a small portion of the ac cycle to charge the filter capacitor to the peak value of the ac waveform. When the instantaneous voltage of the ac passes its peak, the diode ceases to conduct. This forces the capacitor to support the load current until the ac voltage on the opposing diode in the bridge or full wave rectifier is high enough to pick up the load and recharge the capacitor. For this reason, the diode currents are usually quite high.

Since the cyclic peak voltage of the capacitor-filter output is determined by the peak of the input ac waveform, the minimum voltage and, therefore, the ripple amplitude, is determined by the amount of voltage discharge, or "droop," occurring in the capacitor while it is discharging and supporting the load. Obviously, the higher the load current, the proportionately greater the discharge, and therefore the lower the average output.

Although not exactly accurate, an easy way to determine the peak-to-peak ripple for a certain capacitor and load is to assume a constant load current. We can calculate the droop in the capacitor by using the relationship:

$$C \times E = I \times t \quad (3)$$

where:

C = the capacitance in microfarads,
E = the voltage droop, or peak-to-peak ripple voltage,
I = the load current in milliamperes and
t = the length of time per cycle when the rectifiers are not conducting, and the filter capacitor must support the load current. For 60-Hz, full-wave rectifiers, t is about 7.5 ms.

As an example, let's assume that we need to determine the peak to peak ripple voltage at the dc output of a full-wave rectifier/filter combination that produces 13.8 V dc and supplies a transceiver drawing 2.0 A. The filter capacitor in the power supply is 5000 µF. Using the above relationship:

$$C \times E = I \times t$$

$$5000\ \mu F \times E = 2000\ mA \times 7.5\ ms$$

$$E = \frac{2000\ mA \times 7.5\ ms}{5000\ \mu F} = 3\ V\ P\text{-}P$$

Obviously, this is too much ripple. A capacitor value of about 20000 mF would be better suited for this application. If a linear regulator is used after this rectifier/filter combination, however, and the source voltage raised to produce a dc voltage of about 20 V, the 5000-µF capacitor with its 3-V peak-to-peak ripple would work well, since the regulator would remove the ripple content before the output power was applied to the transceiver.

CHOKE-INPUT FILTERS

Choke-input filters have become less popular than they once were, because of the high surge current capability of silicon rectifiers. Choke-input filters provide the benefits of greatly improved output voltage stability over varying loads and low peak-current surges in the rectifiers. On the negative side, however, the choke is bulky and heavy, and the output voltage is lower than that of a capacitor-input filter.

As long as the inductance of the choke is large enough to maintain a continuous current over the complete cycle of the input ac waveform, the filter output voltage will be the average value of the rectified output. The average dc value of a full-wave rectified sine wave is 0.637 times its peak voltage. Since the RMS value is 0.707 times the peak, the output of the choke input filter will be (0.637 / 0.707), or 0.90 times the RMS ac voltage. For light loads, however, there may not be enough energy stored in the choke during the input waveform crest to allow continuous current over the full cycle. When this happens, the filter output voltage will rise as the filter assumes more and more of the characteristics of a capacitor-input filter.

Choke-input filters see extensive use in the energy-storage networks of switch-mode regulators.

Regulation

The output of a rectifier/filter system may be usable for some electronic equipment, but for today's transceivers and accessories, further measures may be necessary to provide power sufficiently clean and stable for their needs. Voltage regulators are often used to provide this additional level of conditioning.

Rectifier/filter circuits by themselves are unable to protect the equipment from the problems associated with input-power-line fluctuations, load-current variations and residual ripple voltages. Regulators can eliminate these problems, but not without costs in circuit complexity and power-conversion efficiency.

ZENER DIODES

A Zener diode (named after American physicist Dr. Clarence Zener) can be used to maintain the voltage applied to a circuit at a practically constant value, regardless of the voltage regulation of the power supply or variations in load current. The typical circuit is shown in **Fig 17.13**. Note that the cathode side of the diode is connected to the positive side of the supply. The electrical characteristics of a Zener diode under conditions of forward and reverse voltage are given in the **Real-World Component Characteristics** chapter.

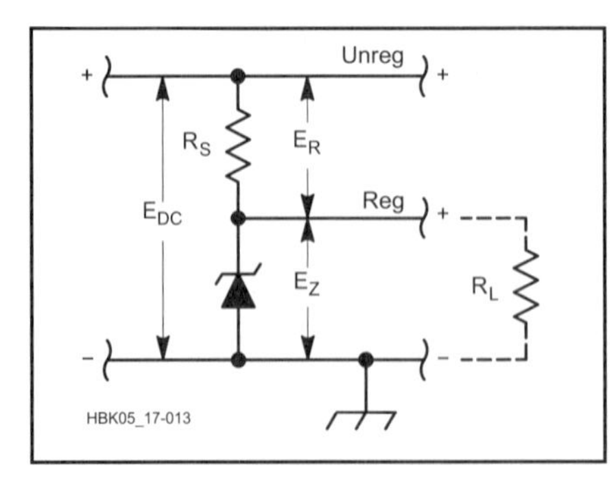

Fig 17.13 — Zener-diode voltage regulation. The voltage from a negative supply may be regulated by reversing the power-supply connections and the diode polarity.

Zener diodes are available in a wide variety of voltages and power ratings. The voltages range from less than two to a few hundred, while the power ratings (power the diode can dissipate) run from less than 0.25 W to 50 W. The ability of the Zener diode to stabilize a voltage depends on the diode's conducting impedance. This can be as low as 1 Ω or less in a low-voltage, high-power diode or as high as 1000 Ω in a high-voltage, low-power diode.

LINEAR REGULATORS

Linear regulators come in two varieties: *series* and *shunt*. The shunt regulator is simply an electronic (also called "active") version of the Zener diode. For the most part, the active shunt regulator is rarely used since the series regulator is a superior choice for most applications. **Fig 17.14B** shows a shunt regulator.

The series regulator consists of a stable voltage reference, which is usually estab-

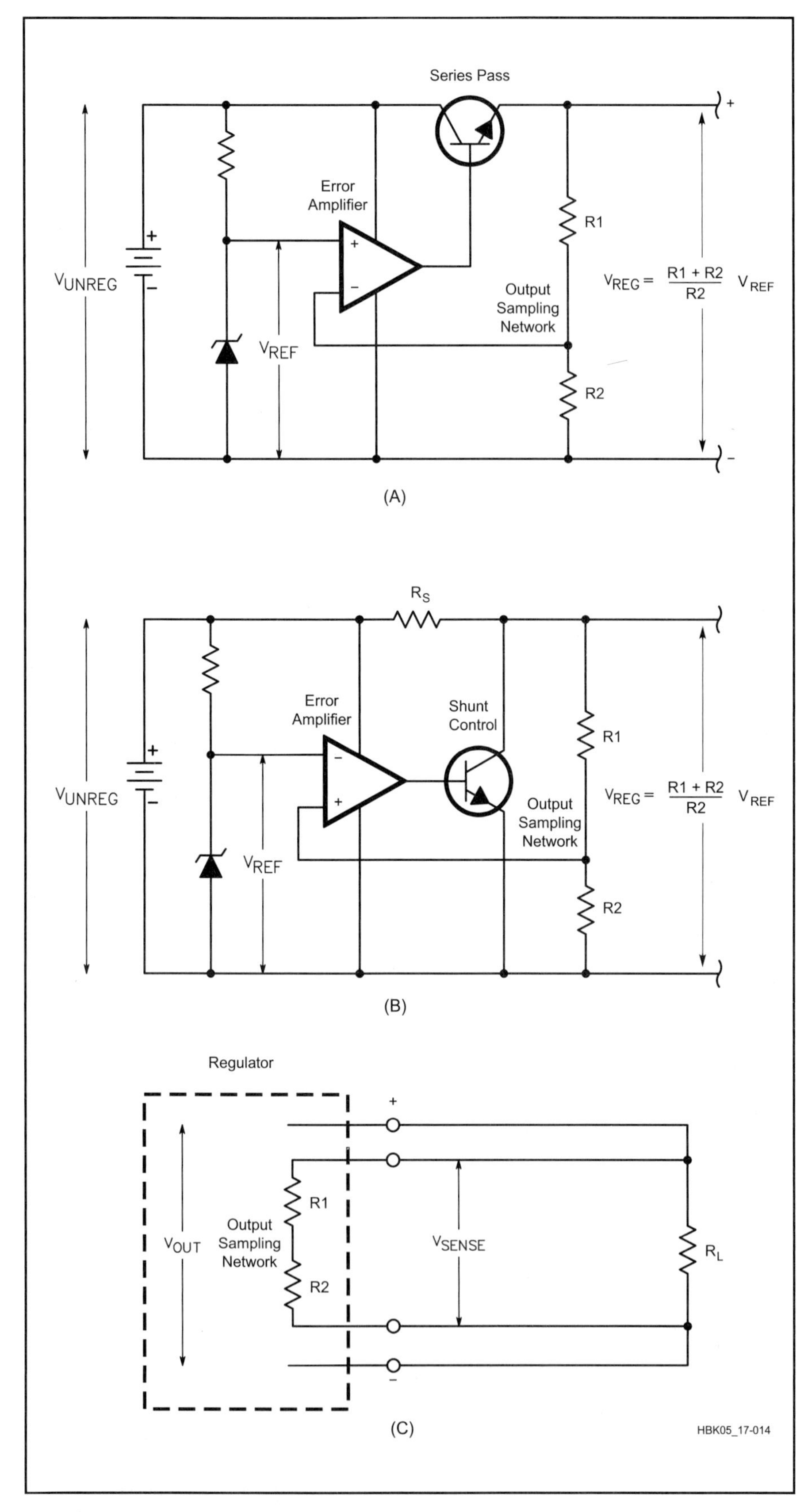

Fig 17.14 — Linear electronic voltage-regulator circuits. In these diagrams, batteries represent the unregulated input-voltage source. A transformer, rectifier and filter would serve this function in most applications. Part A shows a series regulator and Part B shows a shunt regulator. Part C shows how remote sensing overcomes poor load regulation caused by the I R drop in the connecting wires by bringing them inside the feedback loop.

lished by a Zener diode, a transistor in series with the power source and the load (called a *series pass transistor*), and an error amplifier. In critical applications a temperature-compensated reference diode would be used instead of the Zener diode. (See Fig 17.14A.)

The output voltage is sampled by the error amplifier, which compares the output (usually scaled down by a voltage divider) to the reference. If the scaled-down output voltage becomes higher than the reference voltage, the error amplifier reduces the drive current to the pass transistor, thereby allowing the output voltage to drop slightly. Conversely, if the load pulls the output voltage below the desired value, the amplifier drives the pass transistor into increased conduction.

The "stiffness" or tightness of regulation of a linear regulator depends on the gain of the error amplifier and the ratio of the output scaling resistors. In any regulator, the output is cleanest and regulation stiffest at the point where the sampling network or error amplifier is connected. If heavy load current is drawn through long leads, the voltage drop can degrade the regulation at the load. To combat this effect, the feedback connection to the error amplifier can be made directly to the load. This technique, called *remote sensing*, moves the point of best regulation to the load by bringing the connecting loads inside the feedback loop. This is shown in Fig 17.14C.

Input Versus Output Voltage

In a series regulator, the pass-transistor power dissipation is directly proportional to the load current and input/output voltage differential. The series pass element can be located in either leg of the supply. Either NPN or PNP devices can be used, depending on the ground polarity of the unregulated input.

The differential between the input and output voltages is a design tradeoff. If the input voltage from the rectifiers and filter is only slightly higher than the required output voltage, there will be minimal voltage drop across the series pass transistor resulting in minimal thermal dissipation and high power-supply efficiency. The supply will have less capability to provide regulated power in the event of power line brownout and other reduced line voltage conditions, however. Conversely, a higher input voltage will provide operation over a wider range of input voltage, but at the expense of increased heat dissipation.

Pass Transistors

Darlington Pairs

A simple Zener-diode reference or IC op-amp error amplifier may not be able to source enough current to a pass transistor that must conduct heavy load current. The Darlington configuration of **Fig 17.15A** multiplies the pass-transistor beta, thereby extending the control range of the error amplifier. If the Darlington arrangement is implemented with discrete transistors, resistors across the base-emitter junctions may be necessary to prevent collector-to-base leakage currents in Q1 from being amplified and turning on the transistor pair. These resistors are contained in the envelope of a monolithic Darlington device.

When a single pass transistor is not available to handle the current required from a regulator, the current-handling capability may be increased by connecting two or more pass transistors in parallel. The circuit of Fig 17.15B shows the method of connecting these pass transistors. The resistances in the emitter leads of each transistor are necessary to equalize the currents.

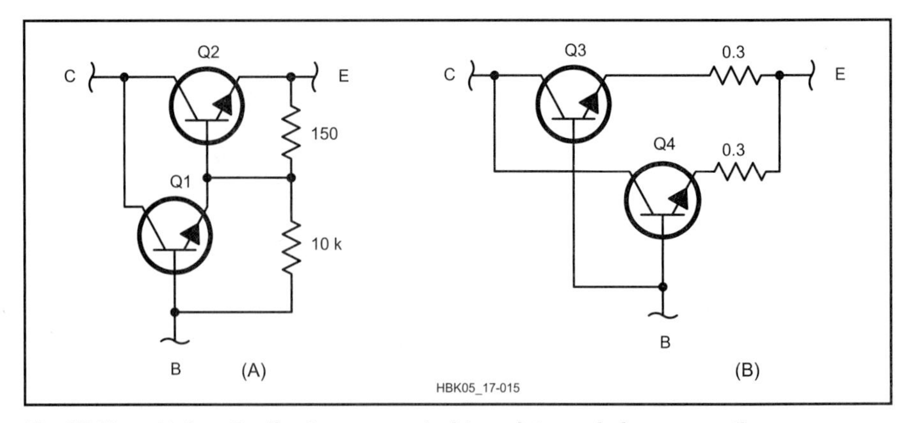

Fig 17.15 — At A, a Darlington-connected transistor pair for use as the pass element in a series-regulating circuit. At B, the method of connecting two or more transistors in parallel for high-current output. Resistances are in ohms. The circuit at A may be used for load currents from 100 mA to 5 A, and the one at B may be used for currents from 6 A to 10 A.

Q1 — Motorola MJE 340 or equivalent

Q2 - Q4 — Power transistor such as 2N3055 or 2N3772

Transistor Ratings

When bipolar (NPN, PNP) power transistors are used in applications in which they are called upon to handle power on a continuous basis, rather than switching, there are four parameters that must be examined to see if any maximum limits are being exceeded. Operation of the transistor outside these limits can easily result in device failure. Unfortunately, not many hams (nor, sometimes, equipment manufacturers) are aware of all these parameters. Yet, for a transistor to provide reliable operation, the circuit designer must be sure not to allow his power supply or amplifier to cause overstress.

The four limits are maximum collector current (I_C), maximum collector-emitter voltage (V_{CEO}), maximum power and second breakdown (I_{SB}). All four of these parameters are graphically shown on the transistor's data sheet on what is known as a Safe Operating ARea (SOAR) graph. (See **Fig 17.16**.) The first three of these limits are usually also listed prominently with the other device information, but it is often the fourth parameter that is responsible for the "sudden death" of the power transistor after an extended operating period.

The maximum current limit of the transistor (I_C MAX) is usually the current limit for fusing of the bond wire connected to the emitter, rather than anything pertaining to the transistor chip itself. When this limit is exceeded, the bond wire can melt and open circuit the emitter. On the operating curve, this limit is shown as a horizontal line extending out from the Y-axis (zero volts between collector and emitter) and ending at the voltage point where the constant power limit begins.

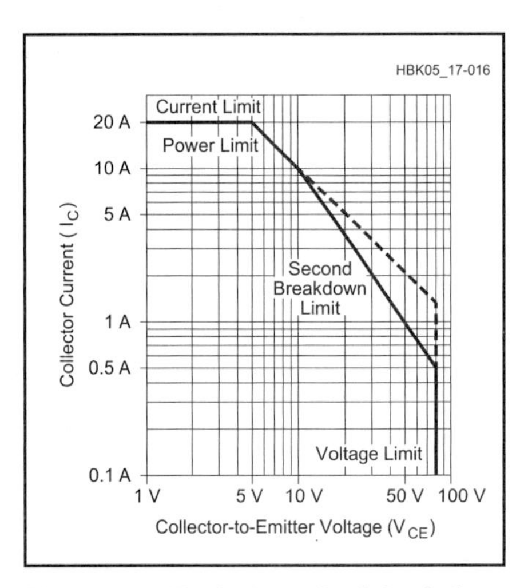

Fig 17.16 — Typical graph of the Safe Operating ARea (SOAR) of a transistor. See text for details. Safe operating conditions for specific devices may be quite different from those shown here.

The maximum collector-emitter voltage limit of the transistor (V_{CE} MAX) is the point at which the transistor can no longer stand off the voltage between collector and emitter.

With increasing collector-emitter voltage drop at maximum collector current, a point is reached where the power in the transistor will cause the junction temperature to rise to a level where the device leakage current rapidly increases and begins to dominate. In this region, the product of the voltage drop and the current would be constant and represent the maximum power (P_t) rating for the transistor; that is, as the voltage drop continues to increase, the collector current must decrease to maintain the power dissipation at a constant value.

With most of the higher voltage rated transistors, a point is reached on the constant power portion of the curve whereby, with further increased voltage drop, the maximum power rating is *not* constant, but decreases as the collector to emitter voltage increases. This decrease in power handling capability continues until the maximum voltage limit is reached.

This special region is known as the forward bias second breakdown (FBSB) area. Reduction in the transistor's power handling capability is caused by localized heating in certain small areas of the transistor junction ("hot spots"), rather than a uniform distribution of power dissipation over the entire surface of the device.

The region of operating conditions contained within these curves is called the Safe Operating ARea, or SOAR. If the transistor is always operated within these limits, it should provide reliable and continuous service for a long time.

MOSFET Transistors

The bipolar junction transistor (BJT) is rapidly being replaced by the MOSFET transistor in new power supply designs due to the latter's ease of drive. Just as the BJT comes in both NPN and PNP varieties, the MOSFET is available in N-channel and P-channel types, with the N-channel being the more popular of the two. The N-channel MOSFET is equivalent to the NPN bipolar, and the P-channel is equivalent to the PNP.

There are some considerations that should be observed when using a MOSFET as a linear regulator series pass transistor. Several volts of gate drive are needed in order to start conduction of the device, as opposed to less than one volt for the BJT. MOSFETs are inherently very-high-frequency devices, and will readily oscillate with stray-circuit capacitances. In order to prevent oscillation in the transistor and surrounding circuits, it is common practice to insert a small resistor of about 100 ohms directly in series with the gate of the series-pass transistor to reduce the gate circuit Q.

Overcurrent Protection

Damage to a pass transistor can occur when the load current exceeds the safe amount. **Fig 17.17A** illustrates a simple current-limiter circuit that will protect Q1. All of the load current is routed through R1. A voltage difference will exist across R1; the value will depend on the exact load current at a given time. When the load current exceeds a predetermined safe value, the voltage drop across R1 will forward-bias Q2 and cause it to conduct. Because Q2 is a silicon transistor, the voltage drop across R1 must exceed 0.6 V to turn Q2 on. This being the case, R1 is chosen for a value that provides a drop of 0.6 V when the maximum safe-load current is drawn. In this instance, the drop will be 0.6 V when I_L reaches 0.5 A. R2 protects the base-emitter junction of Q2 from current spikes, or from destruction in the event Q1 fails under short-circuit conditions.

When Q2 turns on, some of the current through R_S flows through Q2, thereby depriving Q1 of some of its base current. This action, depending upon the amount of Q1 base current at a precise moment, cuts off Q1 conduction to some degree, thus limiting the current through it.

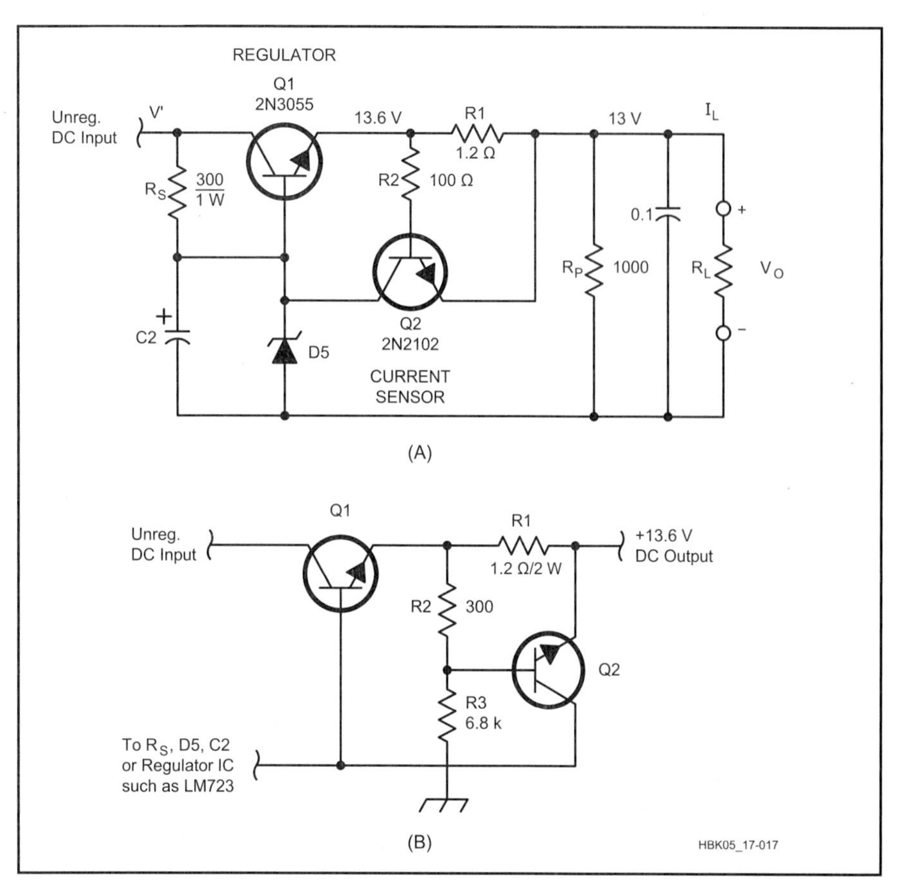

Fig 17.17 — Overload protection for a regulated supply can be implemented by addition of a current-overload-protective circuit, as shown at A. At B, the circuit has been modified to employ current-foldback limiting.

Foldback Current Limiting

Under short-circuit conditions, a constant-current-type current limiter must still withstand the full source voltage and limited short-circuit current simultaneously, which can impose a very high-power dissipation or second-breakdown stress on the series pass transistor. For example, a 12-V regulator with current limiting set for 10 A and having a source of 16 V will have a dissipation of 40 W [(16 V – 12 V) × 10 A] at the point of current limiting (knee). But its dissipation will rise to 160 W under short-circuit conditions (16 V × 10 amps).

A modification of the limiter circuit can cause the regulated-output current to decrease with decreasing load resistance after the over-current knee. With the output shorted, the output current is only a fraction of the knee-current value, which protects the series-pass transistor from excessive dissipation and possible failure. Using the previous example of the 12-V, 10-A regulator, if the short-circuit current is designed to be 3 A (the knee is still 10 A), the transistor dissipation with a short circuit will be only 16 V × 3 A = 48 W.

Fig 17.17B shows how the current-limiter example given in the previous section would be modified to incorporate foldback limiting. The divider string formed by R2 and R3 provides a negative bias to the base of Q2, which prevents Q2 from turning on until this bias is overcome by the drop in R1 caused by load current. Since this hold-off bias decreases as the output voltage drops, Q2 becomes more sensitive to current through R1 with decreasing output voltage. See **Fig 17.18**.

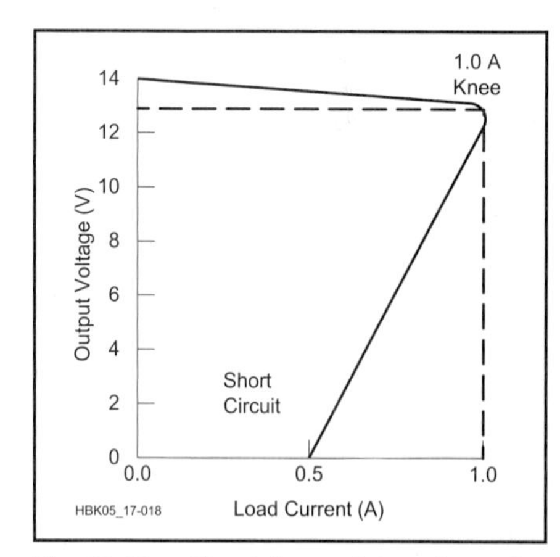

Fig 17.18 — The 1-A regulator shown in Fig 17.17B will fold back to 0.5 A under short-circuit conditions. See text.

The circuit is designed by first calculating the value of R1 for short-circuit current. For example, if 0.5 A is chosen, the value for R1 is simply 0.6 V / 0.5 A = 1.2 Ω (with the output shorted, the amount of hold-off bias supplied by R2 and R3 is very small and can be neglected). The knee current is then chosen. For this example, the selected value will be 1.0 A. The divider string is then proportioned to provide a base voltage at the knee that is just sufficient to turn on Q2 (a value of 13.6 V for 13.0 V output). With 1.0 A flowing through R1, the voltage across the divider will be 14.2 V. The voltage dropped by R2 must then be 14.2 V –13.6 V, or 0.6 V. Choosing a divider current of 2 mA, the value of R2 is then 0.6 V / 0.002 A = 300 Ω. R3 is calculated to be 13.6 V / 0.002 A = 6800 Ω.

"Crowbar" Circuits

Electronic components *do* fail from time to time. In a regulated power supply, the only component standing between an elevated dc source voltage and your rig is one transistor, or a group of transistors wired in parallel. If the transistor, or one of the transistors in the group, happens to short internally, your rig could suffer lots of damage.

To safeguard the rig or other load equipment against possible overvoltage, some power-supply manufacturers include a circuit known as a crowbar. This circuit usually consists of a silicon-controlled rectifier (SCR) connected directly across the output of the power supply, with a voltage-sensing trigger circuit tied to its gate. In the event the output voltage exceeds the trigger set point, the SCR will fire, and the output is short circuited. The resulting high current in the power supply (shorted output in series with a series pass transistor failed short) will blow the power supply's line fuses. This is both a protection for the supply as well as an indicator that something has malfunctioned internally. For these reasons, never replace blown fuses with ones that have a higher current rating.

IC VOLTAGE REGULATORS

The modern trend in regulators is toward the use of three-terminal devices commonly referred to as *three-terminal regulators*. Inside each regulator is a voltage reference, a high-gain error amplifier, temperature-compensated voltage sensing resistors and a pass element. Many currently available units have thermal shutdown, overvoltage protection and current foldback, making them virtually destruction-proof.

Three-terminal regulators (a connection for unregulated dc input, regulated dc output and ground) are available in a wide range of voltage and current ratings. It is easy to see why regulators of this sort are so popular when you consider the low price and the number of individual components they can replace. The regulators are available in several different package styles, depending on current ratings. Low-current (100 mA) devices frequently use the plastic TO-92 and DIP-style cases. TO-220 packages are popular in the 1.5-A range, and TO-3 cases house the larger 3-A and 5-A devices.

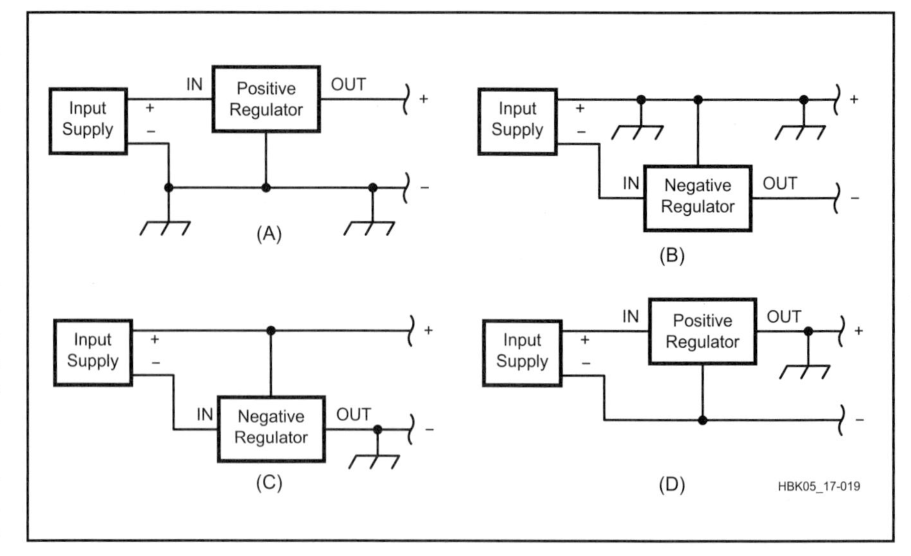

Fig 17.19 — Parts A and B illustrate the conventional manner in which three-terminal regulators are used. Parts C and D show how one polarity regulator can be used to regulate the opposite-polarity voltage.

Three-terminal regulators are available as positive or negative types. In most cases, a positive regulator is used to regulate a positive voltage and a negative regulator a negative voltage. Depending on the system ground requirements, however, each regulator type may be used to regulate the "opposite" voltage.

Fig 17.19A and B illustrate how the regulators are used in the conventional mode. Several regulators can be used with a common-input supply to deliver several voltages with a common ground. Negative regulators may be used in the same manner. If no other common supplies operate from the input supply to the regulator, the circuits of Fig 17.19C and D may be used to regulate positive voltages with a negative regulator and vice versa. In these configurations the input supply is floated; neither side of the input is tied to the system ground.

Manufacturers have adopted a system of family numbers to classify three-terminal regulators in terms of supply polarity, output current and regulated voltage. For example, National uses the number LM7805C to describe a positive 5-V, 1.5-A regulator; the comparable unit from Texas Instruments is a UA7805KC. LM7812C describes a 12-V regulator of similar characteristics. LM7905C denotes a negative 5-V, 1.5-A device. There are many such families with widely varied ratings available from manufacturers. Fixed-voltage regulators are available with output ratings in most common values between 5 and 28 V. Other families include devices that can be adjusted from 1.25 to 50 V.

Regulator Specifications

When choosing a three-terminal regulator for a given application, the most important specifications to consider are device output voltage, output current, input-to-output differential voltage, line regulation, load regulation and power dissipation. Output voltage and current requirements are determined by the load with which the supply will ultimately be used.

Input-to-output differential voltage is one of the most important three-terminal regulator specifications to consider when designing a supply. The differential value (the difference between the voltage applied to the input terminal and the voltage on the output terminal) must be within a specified range. The minimum differential value, usually about 2.5 V, is called the dropout voltage. If the differential value is less than the dropout voltage, no regulation will take place. At the other end of the scale, maximum input-output differential voltage is generally about 40 V. If this differential value is exceeded, device failure may occur.

Increases in either output current or differential voltage produce proportional increases in device power consumption. By employing a safety feature called current foldback, some manufacturers ensure that maximum dissipation will never be exceeded in normal operation. **Fig 17.20** shows the relationship between output current, input-output differential and cur-

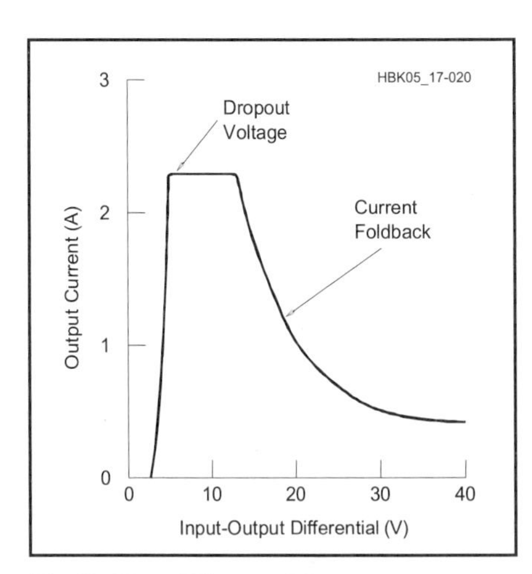

Fig 17.20 — Effects of input-output differential voltage on three-terminal regulator current.

rent limiting for a three-terminal regulator nominally rated for 1.5-A output current. Maximum output current is available with differential voltages ranging from about 2.5 V (dropout voltage) to 12 V. Above 12 V, the output current decreases, limiting the device dissipation to a safe value. If the output terminals are accidentally short circuited, the input-output differential will rise, causing current foldback, and thus preventing the power-supply components from being over stressed. This protective feature makes three-terminal regulators particularly attractive in simple power supplies.

When designing a power supply around a particular three-terminal regulator, input-output voltage characteristics of the regulator should play a major role in selecting the transformer-secondary and filter-capacitor component values. The unregulated voltage applied to the input of the three-terminal device should be higher than the dropout voltage, yet low enough that the regulator does not go into current limiting caused by an excessive differential voltage. If, for example, the regulated output voltage of the device shown in Fig 17.20 was 12, then unregulated input voltages of between 14.5 and 24 would be acceptable if maximum output current is desired.

In use, all but the lowest current regulators generally require an adequate external heat sink because they may be called on to dissipate a fair amount of power. Also, because the regulator chip contains a high-gain error amplifier, bypassing of the input and output leads is essential for stable operation.

Most manufacturers recommend bypassing the input and output directly at the leads where they protrude through the heat sink. Solid tantalum capacitors are usually recommended because of their good high-frequency capabilities.

External capacitors used with IC regulators may discharge through the IC junctions under certain circuit conditions, and high-current discharges can harm ICs. Look at the regulator data sheet to see whether protection diodes are needed, what diodes to use and how to place them in any particular application.

In addition to fixed-output-voltage ICs, high-current, adjustable voltage regulators are available. These ICs require little more than an external potentiometer for an adjustable output range from 5 to 24 V at up to 5 A. The unit price on these items is only a few dollars, making them ideal for test-bench power supplies. A very popular low current, adjustable output voltage three terminal regulator, the LM317, is shown in **Fig 17.21**. It develops a steady 1.25-V reference, V_{REF}, between the output and adjustment terminals. By installing R1 between these terminals, a constant current, I1, is developed, governed by the equation:

$$I1 = \frac{V_{REF}}{R1} \quad (4)$$

Both I1 and a 100-µA error current, I2, flow through R2, resulting in output voltage V_O. V_O can be calculated using the equation:

$$V_O = V_{REF}\left(1 + \frac{R2}{R1}\right) + I2 \times R2 \quad (5)$$

Any voltage between 1.2 and 37 V may be obtained with a 40-V input by changing the ratio of R2 to R1. At lower output voltages, however, the available current will be limited by the power dissipation of the regulator.

Fig 17.22 shows one of many flexible applications for the LM317. By adding only one resistor with the regulator, the voltage regulator can be changed into a constant-current source capable of charging NiCd batteries, for example. Design equations are given in the figure. The same precautions should be taken with adjustable regulators as with the fixed-voltage units. Proper heat sinking and lead bypassing are essential for proper circuit operation.

Increasing Regulator Output Current

When the maximum output current from an IC voltage regulator is insufficient to operate the load, discrete power transistors may be connected to increase the current capability. **Fig 17.23** shows two methods for boosting the output current of a positive regulator, although the same techniques can be applied to negative regulators.

In A, an NPN transistor is connected as an emitter follower, multiplying the output current capacity by the transistor beta. The shortcoming of this approach is that the base-emitter junction is not inside the feed-back loop. The result is that the output voltage is reduced by the base-emitter

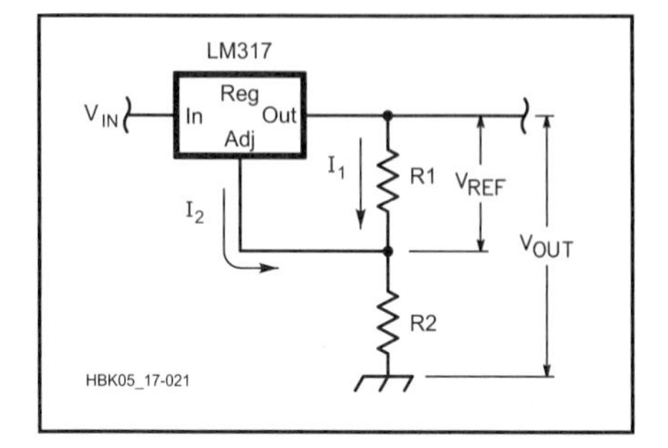

Fig 17.21 — By varying the ratio of R2 to R1 in this simple LM317 schematic diagram, a wide range of output voltages is possible. See text for details.

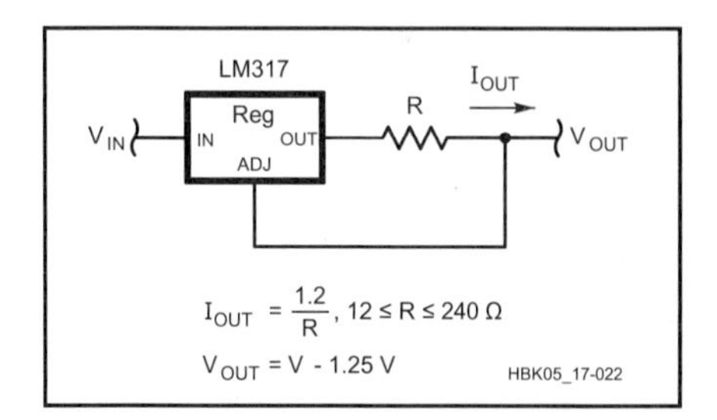

Fig 17.22 — The basic LM317 voltage regulator is converted into a constant-current source by adding only one resistor.

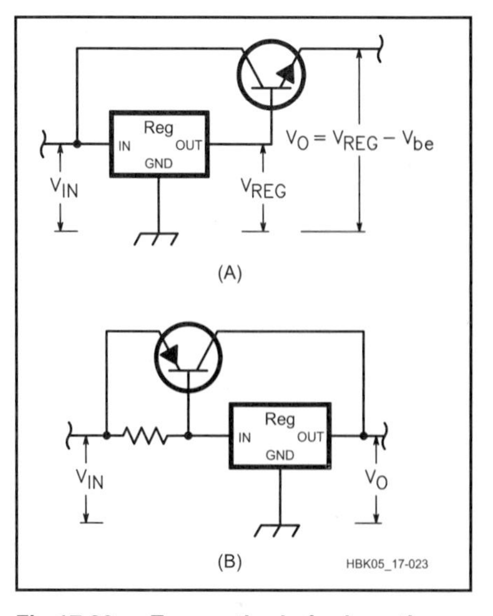

Fig 17.23 — Two methods for boosting the output-current capacity of an IC voltage regulator. Part A shows an NPN emitter follower and B shows a PNP "wrap-around" configuration. Operation of these circuits is explained in the text.

drop, and the load regulation is degraded by variations in this drop.

The circuit at B has a PNP transistor "wrapped around" the regulator. The regulator draws current through the base-emitter junction, causing the transistor to conduct. The IC output voltage is unchanged by the transistor because the collector is connected directly to the IC output (sense point). Any increase in output voltage is detected by the IC regulator, which shuts off its internal-pass transistor, and this stops the boost-transistor base current.

HIGH-PRECISION SERIES REGULATOR LOOP DESIGN

This regulator-loop-design material was contributed by William E. Sabin, WØIYH. (A similar discussion appears in May 1991 *QEX*, pp 3-9.) The discussion here concerns some of the important factors in the design and testing of a series-regulator feedback loop. These techniques were used to design "A Series Regulated 4.5- to 25-V 2.5-A Power Supply," which appears later in this chapter. The values and measurements discussed here are from that circuit. **Fig 17.24** is a simplified version of that supply, which is adequate for this discussion.

The series regulator is a good example of a feedback-control system. Open-loop gain and bandwidth, the phase and gain margins and the transient response are important factors. The goal of the design is to maximize the regulator closed-loop performance. One approach is to use a high value of open-loop gain and establish the open-loop frequency response in two ways: (1) an RC low-pass filter consisting of C6, R2, R3, ½ R_e and the output resistances of Q1 and Q2; and (2) a single small capacitor (C4) at the regulator IC. Note that the voltage drop across R2 and R3 is applied to pins 2 and 3 of the LM723 regulator IC, which are used for current limiting. R2 sets the current limit value, and R2's value affects the RC filter.

Test Circuits

Fig 17.24 shows three test circuits. One is an adjustable Load-Test Circuit that can be modulated linearly (almost) by: (1) a sine or triangle wave from a function generator with a dc-offset adjustment (so that the waveform always has positive polarity), or (2) by a bidirectional square wave. This circuit is used to test the loop response to various load fluctuations. It has proved to be very informative, as discussed later.

The second test circuit (Loop Gain Tester) is inserted into the regulator loop, so that a test signal can be injected into the loop in order to measure the open-loop gain and frequency response. Notice that the loop is closed through the feedback path of R4, R_b and R_a at dc and very low frequencies (less than 0.5 Hz), and the dc-output voltage is reasonably well regulated (which is essential to loop testing). By observing the magnitude (and rate of change) of the frequency response, it is possible to deduce information about phase shift. With this information available, the gain and phase margins (and therefore the regulation, stability, transient response and output impedance) of the closed-loop regulator can be estimated.

The third test circuit is a two-stage op amp preamplifier and oscilloscope. It is used to measure very small signals in the 0.1-Hz to 400-kHz range.

Open-Loop Tests

The test signal applied to points A and

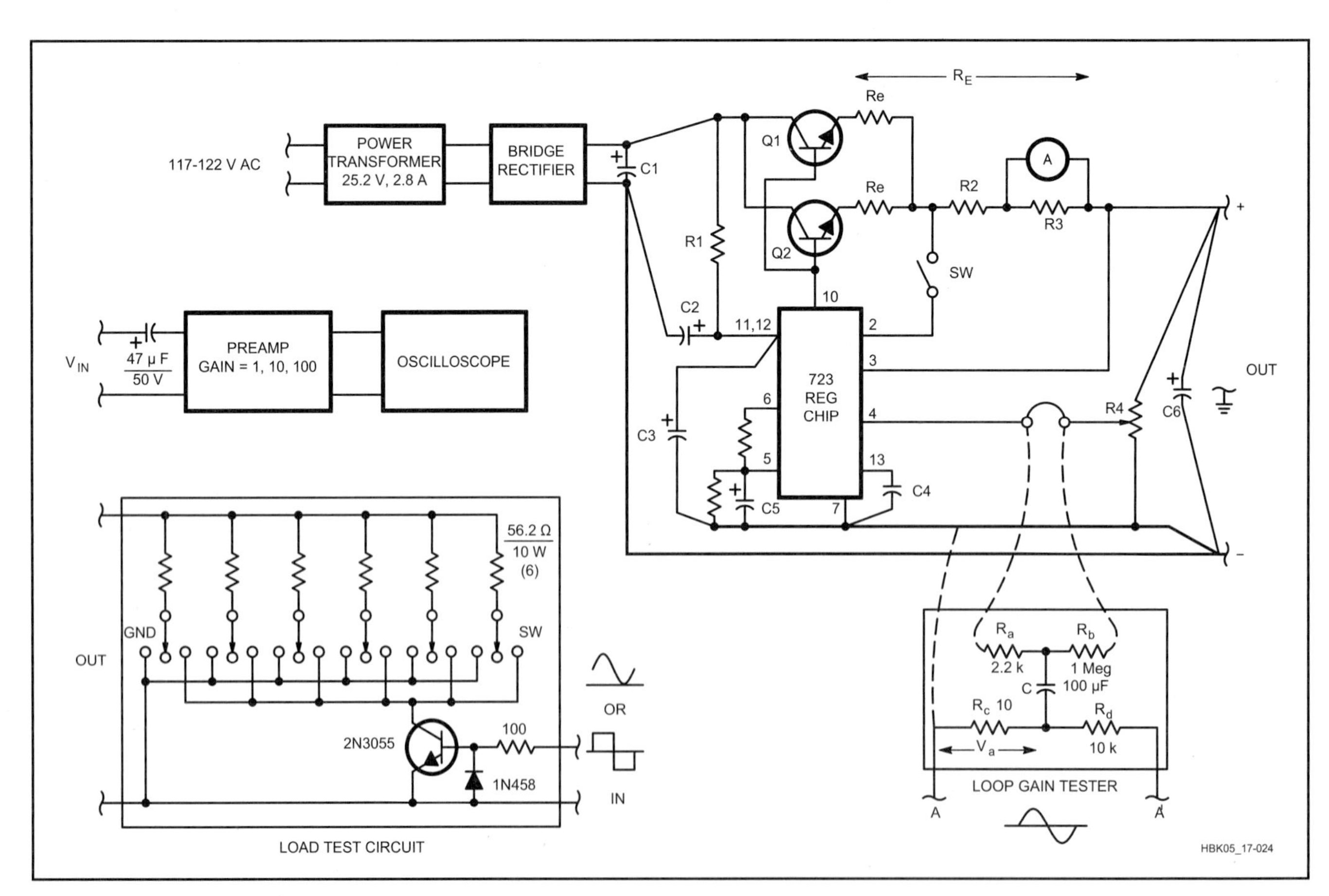

Fig 17.24 — A simplified voltage-regulator feedback loop. The Loop-Gain Tester, Load Tester, Load-Test Circuit and oscilloscope circuits are used to test open-loop response and gain. The heavy lines indicate critical low-impedance circuit paths.

A' is reduced 60 dB by a voltage divider: R_d and R_c. (The voltage division converts convenient input voltages, which we can measure with ordinary equipment, to microvolt levels in the loop.) Capacitor C couples V_a, the voltage across R_c, to the '723 through R_a. R_a is roughly the resistance that the '723 sees in normal operation. The test signal is amplified by at most 74 dB on its way (clockwise around the loop and through the regulator) to the right-hand end of R_b. It is then attenuated 100 dB by R_b and R_c. This means that the "leak-through" back to the '723 input is much smaller than the V_a that we started with, if the frequency is 2 Hz or greater. At dc (and very low frequencies) the regulator functions somewhat normally. Above 2 Hz, then, the magnitude of the open-loop gain at the test frequency is very nearly the ratio $|V_{out}| / |V_a|$.

The first benefit of this test circuit was that it isolated an instability in the '723. The oscillation, at several hundred kHz, was cured by adding C4 (33 pF) and C3 (100 µF / 50 V with very short leads). Normally, one would suspect an oscillation to involve the overall loop, but this was not the case. This kind of instability is common in feedback control systems: Everything appears to function (usually not to full specification), but an embedded element is not stable.

Open-Loop Frequency Response

Referring to Fig 17.24, the open-loop gain is the product of three factors. The test signal is:

1. Voltage amplified by the regulator IC (about 74 dB for the '723) on its way (clockwise) to the emitters of Q1 and Q2.
2. Low-pass filtered by C6 and R_E (the combination of ½ R_e + R2 + R3).
3. Divided by potentiometer R4.

For the 2.5-A supply, the greatest open-loop gain values are 59 dB at 25 V output and 74 dB at 4.5 V.

Fig 17.25 shows the open-loop frequency response to the top of R4 when R_E is set for a 2.5-A current limit. At very low frequencies, the drop-off is due to the gradual closure of the feedback loop, as mentioned above. At higher frequencies, the roll off results from the combined effects of C4 and C6; it occurs at a 6-dB-per-octave rate (within the errors of instrumentation). The cutoff frequency is about 280 Hz, which is:

$$f_{co} = \frac{1}{2 \times \pi \times R_E \times C6} \qquad (6)$$

where:
f_{co} = cutoff frequency in Hz
R_E = 0.57 Ω
C6 = 1000 µF

For comparison, a reference curve (6 dB per octave at the high and low ends) is superimposed. At about 1.2 kHz or so, the reactance of C6 is roughly equal to its equivalent series resistance (ESR), which is about 0.13 Ω for a small 1000-µF aluminum-electrolytic capacitor. Beyond this frequency the impedance of C6 does not diminish, and C4 takes over (thereby maintaining the 6-dB-per-octave roll-off rate). Careful measurements and computer simulations of the regulator loop verified that C4 and C6 do, in fact, collaborate quite well in this manner.

This characteristic is the desired effect. At 120 Hz (the major ripple frequency) the loop gain is maximum, so the regulator loop works hard to suppress output ripple. At higher frequencies, the roll-off rate implies a loop phase shift in the neighborhood of 90°, which assures closed-loop stability and good transient response. Closed-loop transient response tests (using a square-wave signal injected into the Load Test Circuit) verify the absence of ringing or large overshoot.

When R_E is increased to limit at smaller currents, the cutoff frequency decreases. In the example supply, the 0.5-A and 0.1-A range cutoff frequencies are 58 Hz and 12 Hz, respectively, and the roll-off rate remains 6 dB per octave as before. Hence, the loop gain at 120 Hz is reduced. The ripple voltage across C1, however, is also greatly reduced at lighter load currents. In the end, output ripple remains very low.

Closed-Loop Response

The closed-loop gain of the regulator is:

$$G_{CL} = 20 \log \left(\frac{V_{out}}{V_{ref}} \right) \qquad (7)$$

where:
G_{CL} = closed-loop gain, in dB
V_{out} = output voltage
V_{ref} = reference voltage.

Here, V_{out} is 4.5 V, minimum (25.0 V, maximum) and V_{ref} is 4.5 V. Fig 17.25 shows the locations of the minimum and maximum gain values and also the corresponding closed-loop bandwidths. By locating the 280-Hz cutoff frequency fairly close to the 120-Hz ripple frequency, the closed-loop bandwidth is minimized, which is desirable in a voltage regulator.

Another important regulator parameter is closed-loop output impedance. **Fig 17.26** shows a computer simulation of this parameter. Mathematical analysis and actual measurements using the Load Test Circuit with a sine-wave test signal corroborate the simulations quite well. Two results are shown. For curve A, the C6 component (of Fig 17.24) is removed from the circuit, and C4 is increased so that the 280-Hz cutoff frequency is maintained, as

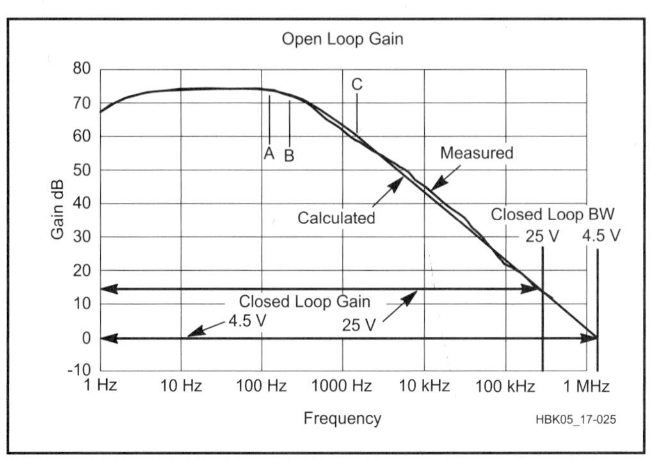

Fig 17.25 — The open-loop frequency-response curves of the voltage-regulator feedback loop. Point A is 120 Hz; point B is 280 Hz; point C is 1.2 kHz (above 1.2 kHz, C6 is no longer effective).

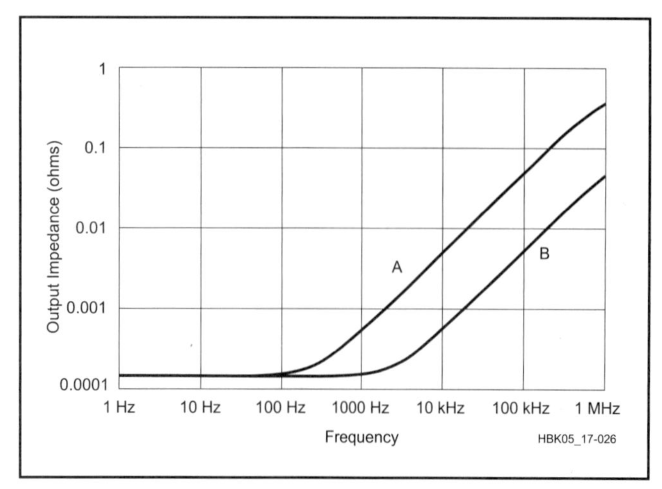

Fig 17.26 — Output impedance magnitude for the voltage-regulator feedback loop.

we discussed before. At low frequencies, the output impedance should be R_E (0.57 Ω) divided by the open-loop voltage gain (5000 maximum), about 0.11 mΩ. Above 280 Hz, though, the output impedance increases rapidly because the open-loop gain is decreasing. It will eventually reach the value of R_E.

For curve B, the original values of C6 and C4 are used, and the output impedance remains low up to about 1.2 kHz. It then increases, but remains much lower than curve A. This happens because C4 is smaller than in curve A and C4 mainly determines the frequency characteristic. In other words, the impedance of C6 (its reactance plus its ESR) is in parallel with the relatively small output impedance of a high-gain feedback amplifier. Therefore, C4 is much less influential in determining the power-supply output impedance. This situation gradually changes as frequency increases.

This discussion shows that the output-impedance characteristic of the power supply is reduced at frequencies that are significant in certain applications. Furthermore, it can be reduced by feedback to levels that are impossible with practical capacitors. To take advantage of this lower impedance, the regulator must be located extremely close to the load (remote sensing is another possibility).

When extremely tight regulation and low-output impedance are important, the leads to the load must be very short, heavy straps. Multiple loads should be connected in parallel directly at the binding posts: A "daisy-chain" connection scheme does not assure equal regulation for each load; it counteracts precision regulation.

High-Voltage Techniques

The construction of high-voltage supplies poses special considerations in addition to the normal design and construction practices used for lower-voltage supplies. In general, the constructor needs to remember that physical spacing between leads, connections, parts and the chassis must be sufficient to prevent arcing. Also, the series connection of components such as capacitor and resistor strings needs to be done with consideration for voltage stresses in the components.

CAPACITORS

Capacitors will usually need to be connected in series strings to form an equivalent capacitor with the capability to withstand the applied voltage. When this is done, equal-value bypassing resistors need to be connected across each capacitor in the string in order to distribute the voltage equally across the capacitors. The *equalizing resistors* should have a value low enough to equalize differences in capacitor leakage resistance between the capacitors, while high enough not to dissipate excessive power. Also, capacitor bodies need to be insulated from the chassis and from each other by mounting them on insulating panels, thereby preventing arcing to the chassis or other capacitors in the string.

For high voltages, oil-filled paper-dielectric capacitors are superior to electrolytics because they have lower internal impedance at high frequencies, lower leakage resistance and are available with higher working voltages. These capacitors are available in values of several microfarads and have working voltage ratings of thousands of volts. Avoid older oil-filled capacitors. They may contain polychlorinated biphenyls (PCBs), a known cancer-causing agent. Newer capacitors have eliminated PCBs and have a notice on the case to that effect.

BLEEDER RESISTORS

Bleeder resistors should be given careful consideration. These resistors provide protection against shock when the power supply is turned off and dangerous wiring is exposed. A general rule is that the bleeder should be designed to reduce the output voltage to 30 V or less within 2 seconds of turning off the power supply. Take care to ensure that the maximum voltage rating of the resistor is not exceeded. The bleeder will consist of several resistors in series. One additional recommendation is that two separate bleeder strings be used, to provide safety in the event one of the strings fails.

METERING TECHNIQUES

Special considerations should be observed for metering of high-voltage supplies, such as the plate supplies for linear amplifiers. This is to provide safety to both personnel and also to the meters themselves.

To monitor the current, it is customary to place the ammeter in the supply return (ground) line. This ensures that both meter terminals are close to ground potential, as compared to the hazard created by placing the meter in the positive output line — in which case the voltage on each meter terminal would be near the full high-voltage potential. Also, there is the strong possibility that an arc could occur between the wiring and coils inside the meter and the chassis of the amplifier or power supply itself. This hazardous potential cannot exist with the meter in the negative leg. Another good safety practice is to place a low-voltage Zener diode across the terminals of the ammeter. This will bypass the meter in the event of an internal open circuit in the meter.

For metering of high voltage, the builder should remember that resistors to be used in multiplier strings have voltage-breakdown ratings. Usually, several resistors need to be used in series to reduce voltage stress across each resistor. A basic rule of thumb is that resistors should be limited to a maximum of 200 V, unless rated otherwise. Therefore, for a 2000-V power supply, the voltmeter would have a string of 10 resistors connected in series to distribute the voltage equally.

Batteries and Charging

The availability of solid-state equipment makes it practical to use battery power under portable or emergency conditions. Hand-held transceivers and instruments are obvious applications, but even fairly powerful transceivers (100 W or so output) may be practical users of battery power (for example, emergency power for the home station for ARES operation).

Lower-power equipment can be powered from two types of batteries. The "primary" battery is intended for one-time use and is then discarded; the "storage" (or "secondary") battery may be recharged many times.

A battery is a group of chemical cells, usually series-connected to give some desired multiple of the cell voltage. Each assortment of chemicals used in the cell gives a particular nominal voltage. This must be taken into account to make up a particular battery voltage. For example, four 1.5-V carbon-zinc cells make a 6-V battery and six 2-V lead-acid cells make a 12-V battery. The **Electrical Fundamentals** chapter has more information about energy storage in batteries. In addition, the **Real-World Component Characteristics** chapter has information about battery capacity and charge/discharge rates.

PRIMARY BATTERIES

One of the most common primary-cell types is the alkaline cell, in which chemical oxidation occurs during discharge. When there is no current, the oxidation essentially stops until current is required. A slight amount of chemical action does continue, however, so stored batteries eventually will degrade to the point where the battery will no longer supply the desired current. The time taken for degradation without battery use is called *shelf life.*

The alkaline battery has a nominal voltage of 1.5 V. Larger cells are capable of producing more milliampere hours and less voltage drop than smaller cells. Heavy-duty and industrial batteries usually have a longer shelf life.

Lithium primary batteries have a nominal voltage of about 3 V per cell and by far the best capacity, discharge, shelf-life and temperature characteristics. Their disadvantages are high cost and the fact that they cannot be readily replaced by other types in an emergency.

The lithium-thionyl-chloride battery is a primary cell, and should not be recharged under any circumstances. The charging process vents hydrogen, and a catastrophic hydrogen explosion can result. Even accidental charging caused by wiring errors or a short circuit should be avoided.

Silver oxide (1.5 V) and mercury (1.4 V) batteries are very good where nearly constant voltage is desired at low currents for long time periods. Their main use (in subminiature versions) is in hearing aids, though they may be found in other mass-produced devices such as household smoke alarms.

SECONDARY OR RECHARGEABLE BATTERIES

Many of the chemical reactions in primary batteries are theoretically reversible if current is passed through the battery in the reverse direction.

Primary batteries should not be recharged for two reasons: It may be dangerous because of heat generated within sealed cells, and even in cases where there may be some success, both the charge and life are limited. One type of alkaline battery is rechargeable, and is so marked.

Nickel Cadmium

The most common type of small rechargeable battery is the nickel-cadmium (NiCd), with a nominal voltage of 1.2 V per cell. Carefully used, these are capable of 500 or more charge and discharge cycles. For best life, the NiCd battery must not be fully discharged. Where there is more than one cell in the battery, the most-discharged cell may suffer polarity reversal, resulting in a short circuit, or seal rupture. All storage batteries have discharge limits, and NiCd types should not be discharged to less than 1.0 V per cell. There is a popular belief that it is necessary to completely discharge NiCd cells in order to recharge them to full capacity. Called the "memory effect," professional engineers have proved this to be a myth.

Nickel-cadmium cells are not limited to "D" cells and smaller sizes. They also are available in larger varieties ranging to mammoth 1000 Ah units having carrying handles on the sides and caps on the top for adding water, similar to lead-acid types. These large cells are sold to the aircraft industry for jet-engine starting, and to the railroads for starting locomotive diesel engines. They also are used extensively for uninterruptible power supplies. Although expensive, they have very long life. Surplus cells are often available through surplus electronics dealers, and these cells often have close to their full rated capacity.

Advantages for the ham in these vented-cell batteries lie in the availability of high discharge current to the point of full discharge. Also, cell reversal is not the problem that it is in the sealed cell, since water lost through gas evolution can easily be replaced. Simply remove the cap and add distilled water. By the way, tap water should never be added to either nickel-cadmium or lead-acid cells, since dissolved minerals in the water can hasten self discharge and interfere with the electrochemical process.

Lead Acid

The most widely used high-capacity rechargeable battery is the lead-acid type. In automotive service, the battery is usually expected to discharge partially at a very high rate, and then to be recharged promptly while the alternator is also carrying the electrical load. If the conventional auto battery is allowed to discharge fully from its nominal 2 V per cell to 1.75 V per cell, fewer than 50 charge and discharge cycles may be expected, with reduced storage capacity.

The most attractive battery for extended high-power electronic applications is the so-called "deep-cycle" battery, which is intended for such uses as powering electric fishing motors and the accessories in recreational vehicles. Size 24 and 27 batteries furnish a nominal 12 V and are about the size of small and medium automotive batteries. These batteries may furnish between 1000 and 1200 W-hr per charge at room temperature. When properly cared for, they may be expected to last more than 200 cycles. They often have lifting handles and screw terminals, as well as the conventional truncated-cone automotive terminals. They may also be fitted with accessories, such as plastic carrying cases, with or without built-in chargers.

Lead-acid batteries are also available with gelled electrolyte. Commonly called *gel cells,* these may be mounted in any position if sealed, but some vented types are position sensitive.

Lead-acid batteries with liquid electrolyte usually fall into one of three classes — conventional, with filling holes and vents to permit the addition of distilled water lost from evaporation or during high-rate charge or discharge; maintenance-free, from which gas may escape but water cannot be added; and sealed. Generally, the deep-cycle batteries have filling holes and vents.

Nickel Metal Hydride

This battery type is quite similar to the NiCd, but the Cadmium electrode is replaced by one made from a porous metal alloy that traps hydrogen; therefore the name of metal hydride. Many of the basic characteristics of these cells are similar to NiCds. For example, the voltage is very nearly the same, they can be slow-charged from a constant current source, and they can safely be deep cycled. There are also some important differences: The most attractive feature is a much higher capacity for the same cell size — often nearly twice as much as the NiCd types! The typical size AA NiMH cell has a capacity between 1000 and 1300 mAh, compared to the 600 to 830 mAh for the same size NiCd. Another advantage of these cells is a complete freedom from memory effect. We can also find comfort in the fact that NiMH cells do not contain any dangerous substances, while both NiCd and lead-acid cells do contain quantities of toxic heavy metals.

The internal resistance of NiMH cells is somewhat higher than that of NiCd cells, resulting in reduced performance at very high discharge current. This can cause slightly reduced power output from an HT powered by a NiMH pack, but the effect is barely noticeable, and the higher capacity and resulting longer run time far outweigh this. At least one manufacturer warns that the self-discharge of NiMH cells is higher than for NiCd, but again, in practice this can hardly be noticed. The fast-charge process is different for NiMH batteries. A fast charger designed for NiCd will not correctly charge NiMH batteries. But many commercial fast chargers are designed for both types of batteries. NiMH batteries outperform NiCd batteries whenever high capacity is desired, while NiCd batteries still have advantages when delivering very high peak currents.

At the time of this writing, many cell phones and portable computers use NiMH batteries, and several manufacturers offer NiMH packs for Amateur Radio applications. Standard-sized NiMH cells are widely available from the major electronic parts suppliers.

Lithium-Ion Cells

The lithium-ion cell is another possible alternative to NiCd cells. It features, for the same energy storage, about one third the weight and one half the volume of a NiCd. It also has a lower self-discharge rate. Typically, at room temperature, a NiCd cell will lose from 0.5 to 2% of its charge per day. The Lithium-ion cell will lose less than 0.5% per day and even this loss rate decreases after about 10% of the charge has been lost. At higher temperatures the difference is even greater. The result is that Lithium-ion cells are a much better choice for standby operation where frequent recharge is not available.

One major difference between NiCd and Li-ion cells is the cell voltage. The nominal voltage for a NiCd cell is about 1.2 V. For the Li-ion cell it is 3.6 V with a maximum cell charging voltage of 4 V. You cannot substitute Li-ion cells directly for NiCd cells. You will need one Li-ion cell for three NiCd cells. Chargers intended for NiCd batteries must not be used with Li-ion batteries, and vice versa.

PROS AND CONS

Chemical Hazards of Each Battery Type

In addition to the precautions given above, the following precautions are recommended. (Always follow the manufacturer's advice.)

Gas escaping from storage batteries may be explosive. Keep flames or lighted tobacco products away.

Dry-charged storage batteries should be given electrolyte and allowed to soak for at least half an hour. They should then be charged at about a 15 A rate for 15 minutes or so. The capacity of the battery will build up slightly for the first few cycles of charge and discharge, and then have fairly constant capacity for many cycles. Slow capacity decrease may then be noticed.

No battery should be subjected to unnecessary heat, vibration or physical shock. The battery should be kept clean. Frequent inspection for leaks is a good idea. Electrolyte that has leaked or sprayed from the battery should be cleaned from all surfaces. The electrolyte is chemically active and electrically conductive, and may ruin electrical equipment. Acid may be neutralized with sodium bicarbonate (baking soda), and alkalis may be neutralized with a weak acid such as vinegar. Both neutralizers will dissolve in water, and should be quickly washed off. Do not let any of the neutralizer enter the battery.

Keep a record of the battery use, and include the last output voltage and (for lead-acid storage batteries) the hydrometer reading. This allows prediction of useful charge remaining, and the recharging or procuring of extra batteries, thus minimizing failure of battery power during an excursion or emergency.

Internal Resistance

Cell internal resistance is very important to handheld-transceiver users. This is because the internal resistance is in series with the battery's output and therefore reduces the available battery voltage at the high discharge currents demanded by the transmitter. The result is reduced transmitter output power and power wasted in the cell itself by internal heating. Because of cell-construction techniques and battery chemistry, certain types of cells typically have lower internal resistance than others.

The NiCd cell is the undisputed king of cell types for high discharge current capability. Also, the NiCd maintains this low internal resistance throughout its discharge curve, because the specific gravity of its potassium-hydroxide electrolyte does not change.

Next in line is probably the alkaline primary cell. When these cells are used with handheld transceivers, it is not uncommon to have lower output power, and often to have the low battery indicator come on, even with fresh cells.

The lead-acid cell, which is finding popularity in belt-hung battery packs, is pretty close to the alkaline cell for internal resistance, but this is only at full charge. Unlike the NiCd, the electrolyte in the lead-acid cell enters into the chemical reaction. During discharge, the specific gravity of the electrolyte gradually drops as it approaches water, and the conductivity decreases. Therefore, as the lead-acid cell approaches a discharged state, the internal resistance increases. For the belt pack, larger cells are used (approximately 2 Ah) and the internal resistance is consequently reduced.

The worst cell of all is the common carbon-zinc flashlight cell. With the transmit current demand levels of handheld radios, these cells are pretty much useless.

BATTERY CAPACITY

The common rating of battery capacity is ampere hours (Ah), the product of current drain and time. The symbol "C" is commonly used; C/10, for example, would be the current available for 10 hours continuously. The value of C changes with the discharge rate and might be 110 at 2 A but only 80 at 20 A. **Fig 17.27** gives capacity-to-discharge rates for two standard-size lead-acid batteries. Capacity may vary from 35 mAh for some of the small hearing-aid batteries to more than 100 Ah for a size 28 deep-cycle storage battery.

Sealed primary cells usually benefit from intermittent (rather than continuous) use. The resting period allows completion of chemical reactions needed to dispose of by-products of the discharge.

The output voltage of all batteries drops as they discharge. "Discharged" condition for a 12-V lead-acid battery, for instance, should not be less than 10.5 volts. It is also good to keep a running record of hydrometer readings, but the conventional readings of 1.265 charged and 1.100 discharged apply only to a long, low-rate discharge. Heavy loads may discharge the battery with little reduction in the hydrometer reading.

Batteries that become cold have less of their charge available, and some attempt to keep a battery warm before use is worthwhile. A battery may lose 70% or more of its capacity at cold extremes, but it will recover with warmth. All batteries have

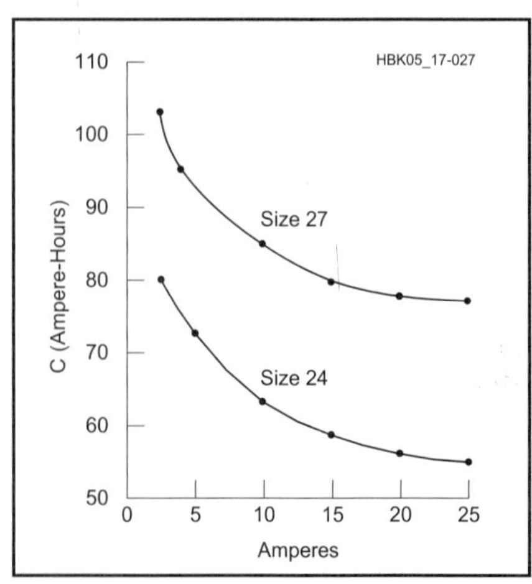

Fig 17.27 — Output capacity as a function of discharge rate for two sizes of lead-acid batteries.

some tendency to freeze, but those with full charges are less susceptible. A fully charged lead-acid battery is safe to –30°F (–34°C) or colder. Storage batteries may be warmed somewhat by charging. Blowtorches or other flame should never be used to heat any type of battery.

A practical discharge limit occurs when the load will no longer operate satisfactorily on the lower output voltage near the "discharged" point. Much gear intended for "mobile" use may be designed for an average of 13.6 V and a peak of perhaps 15 V, but will not operate well below 12 V. For full use of battery charge, the gear should operate well (if not at full power) on as little as 10.5 V with a nominal 12 to 13.6-V rating.

Somewhat the same condition may be seen in the replacement of carbon-zinc cells by NiCd storage cells. Eight carbon-zinc cells will give 12 V, while 10 of the same size NiCd cells are required for the same voltage. If a 10-cell battery holder is used, the equipment should be designed for 15 V in case the carbon-zinc units are plugged in.

Discharge Planning

Transceivers usually drain a battery at two or three rates: one for receiving, one for transmit standby and one for key-down or average voice transmit. Considering just the first and last of these (assuming the transmit standby is equal to receive),average two-way communication would require the low rate ¾ of the time and the high rate ¼ of the time. The ratio may vary somewhat with voice. The user may calculate the percentage of battery charge used in an hour by the combination (sum) of rates. If, for example, 20% of the battery capacity is used in an hour, the battery will provide five hours of communications per charge. In most actual traffic and DX-chasing situations the time spent listening should be much greater than that spent transmitting.

Charging/Discharging Requirements

The rated full charge of a battery, C, is expressed in ampere-hours. No battery is perfect, so more charge than this must be offered to the battery for a full-charge. If, for instance, the charge rate is 0.1 C (the 10-hour rate), 12 or more hours may be needed for the charge.

Basically NiCd batteries differ from the lead-acid types in the methods of charging. It is important to note these differences, since improper charging can drastically shorten the life of a battery. NiCd cells have a flat voltage-versus-charge characteristic until full charge is reached; at this point the charge voltage rises abruptly. With further charging, the electrolyte begins to break down and oxygen gas is generated at the positive (nickel) electrode and hydrogen at the negative (cadmium) electrode.

Since the cell should be made capable of accepting an overcharge, battery manufacturers typically prevent the generation of hydrogen by increasing the capacity of the cadmium electrode. This allows the oxygen formed at the positive electrode to reach the metallic cadmium of the negative electrode and reoxidize it. During overcharge, therefore, the cell is in equilibrium. The positive electrode is fully charged and the negative electrode less than fully charged, so oxygen evolution and recombination "wastes" the charging power being supplied.

In order to ensure that all cells in a NiCd battery reach a fully-charged condition, NiCd batteries should be charged by a constant current at about a 0.1-C current level. This level is about 50 mA for the AA-size cells used in most hand-held radios. This is the optimum rate for most NiCds since 0.1 C is high enough to provide a full charge, yet it is low enough to prevent over-charge damage and provide good charge efficiency.

Although fast-charge-rate (3 to 5 hours typically) chargers are available for hand-held transceivers, they should be used with care. The current delivered by these units is capable of causing the generation of large quantities of oxygen in a fully charged cell. If the generation rate is greater than the oxygen recombination rate, pressure will build in the cell, forcing the vent to open and the oxygen to escape. This can eventually cause drying of the electrolyte, and then cell failure. The cell temperature can also rise, which can shorten cell life. To prevent overcharge from occurring, fast-rate chargers should have automatic charge-limiting circuitry that will switch or taper the charging current to a safe rate as the battery reaches a fully-charged state.

Gelled-electrolyte lead-acid batteries provide 2.4 V/cell when fully charged. Damage results, however, if they are overcharged. (Avoid constant-current or trickle charging unless battery voltage is monitored and charging is terminated when a full charge is reached.) Voltage-limited charging is best for these batteries. A proper charger maintains a safe charge-current level until 2.3 V/cell is reached (13.8 V for a 12-V battery). Then, the charge current is tapered off until 2.4 V/cell is reached. Once charged, the battery may be safely maintained at the "float" level, 2.3 V/cell. Thus, a 12-V gel-cell battery can be "floated" across a regulated 13.8-V system as a battery backup in the event of power failure.

Deep-cycle lead-acid cells are best charged at a slow rate, while automotive and some NiCd types may safely be given quick charges. This depends on the amount of heat generated within each cell, and cell venting to prevent pressure build-up. Some batteries have built-in temperature sensing, used to stop or reduce charging before the heat rise becomes a danger. Quick and fast charges do not usually allow gas recombination, so some of the battery water will escape in the form of gas. If the water level falls below a certain point, acid hydrometer readings are no longer reliable. If the water level falls to plate level, permanent battery damage may result.

Overcharging NiCds in moderation causes little loss of battery life. Continuous overcharge, however, may generate a voltage depression when the cells are later discharged. For best results, charging of NiCd cells should be terminated after 15 hours at the slow rate. Better yet, circuitry may be included in the charger to stop charging, or reduce the current to about 0.02 C when the 1.43-V-per-cell terminal voltage is reached. For lead-acid batteries, a timer may be used to run the charger to make up for the recorded discharge, plus perhaps 20%. Some chargers will switch over automatically to an acceptable standby charge.

Solar Charging Systems

Price and availability make solar panels an attractive way to maintain the charge on your batteries. Relatively small, low-power solar arrays provide a convenient way to charge a NiCd or sealed-lead-acid battery for emergency and portable operation. This is especially popular with QRP operators. You should always connect some type of charge controller between the battery and the solar array. This will prevent overcharging the battery, and the possible resulting battery damage. The Micro M+ described in the projects section of this chapter is a suitable charge controller.

Emergency Operations

CARE AND FEEDING OF GENERATORS

For long-term-emergency operation, a generator is a must, as anyone who has operated field day can attest to. The generator will provide power as long as the fuel supply holds out. Proper care is necessary to keep the generator operating reliably, however.

When the generator runs out of fuel, the operator may be tempted to rush over with the gasoline can and begin refueling. This is very hazardous, since the engine's manifold and muffler are at temperatures that can ignite spilled gasoline. The operator should wait a few minutes to allow hot surfaces to cool sufficiently to ensure safety. For these periods when the generator is shut down, plan on having battery power available to support station operation until the generator can be brought back on line.

Check the level of the engine's lubricating oil from time to time. If the oil sump becomes empty, the engine can seize, putting the station out of operation and necessitating costly engine repairs.

Remember that the engine will produce carbon monoxide gas while it is running. The generator should never be run indoors, and should be placed away from open windows and doors to keep exhaust fumes from coming inside.

INVERTERS

For battery-powered operation of alternating-current loads, inverters are available. An inverter is a dc-to-ac converter, switching at 60 Hz to provide 120-V ac to the loads.

Inverters come in varying degrees of sophistication. The simplest, as mentioned above, produces a square-wave output. This is no problem for lighting and other loads that don't care about the input waveform. Lots of equipment using motors will work poorly or not at all when supplied with square wave power. Therefore, many higher-power inverters use waveform shaping to approximate a sine-wave output. The simplest of these methods is a resonant inductor and capacitor filter. Higher-power units employ pulse-width modulation of the converter switches to create a sinusoidal output waveform.

Power-Supply Projects

Construction of a power supply can be one of the most rewarding projects undertaken by a radio amateur. Whether it's a charger for the NiCds in a VHF handheld transceiver, a low-voltage, high-current monster for a new 100-W solid-state transceiver, or a high-voltage supply for a new linear amplifier, a power supply is basic to all of the radio equipment we operate and enjoy. Final testing and adjustment of most power-supply projects requires only a voltmeter, and perhaps an oscilloscope — tools commonly available to most amateurs.

General construction techniques that may be helpful in building the projects in this chapter are outlined in the **Circuit Construction** chapter. Other chapters in this *Handbook* contain basic information about the components that make up power supplies.

Safety must always be carefully considered during design and construction of any power supply. Power supplies contain potentially lethal voltages, and care must be taken to guard against accidental exposure. For example, electrical tape, insulated tubing (spaghetti) or heat-shrink tubing is recommended for covering exposed wires, components leads, component solder terminals and tie-down points. Whenever possible, connectors used to mate the power supply to the outside world should be of an insulated type designed to prevent accidental contact.

Connectors and wire should be checked for voltage and current ratings. Always use wire with an insulation rating higher than the working voltages in the power supply. Special high-voltage wire is available for use in B+ supplies. The **Component Data and References** chapter contains a table showing the current-carrying capability of various wire sizes. Scrimping on wire and connectors to save money could result in flashover, meltdown or fire.

All fuses and switches should be placed in the hot leg(s) only. The neutral leg should not be interrupted. Use of a three-wire (grounded) power connection will greatly reduce the chance of accidental shock. The proper wiring color code for 120-V circuits is: black — hot; white — neutral; and green — ground. For 240-V circuits, the second hot lead generally uses a red wire.

POWER-SUPPLY PRIMARY CIRCUIT CONNECTOR STANDARD

The International Commission on Rules for the Approval of Electrical Equipment (CEE) standard for power-supply primary-circuit connectors for use with detachable cable assemblies is the CEE-22 (see **Fig 17.28**). The CEE-22 has been recognized by the ARRL and standards agencies of many countries. Rated for up to 250 V, 6 A at 65°C, the CEE-22 is the most commonly used three-wire (grounded), chassis-mount primary circuit connector for electronic equipment in North America and Europe. It is often used in Japan and Australia as well.

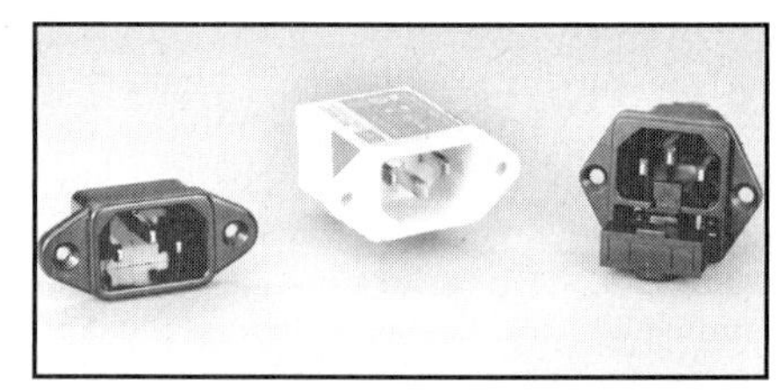

Fig 17.28 — CEE-22 connectors are available with built-in line filters and fuse holders.

When building a power supply requiring 6 A or less for the primary supply, a builder would do well to consider using a CEE-22 connector and an appropriate cable assembly, rather than a permanently installed line cord. Use of a detachable line cord makes replacement easy in case of damage. CEE-22 compatible cable assemblies are available with a wide variety of power plugs including most types used overseas.

Some manufacturers even supply the CEE-22 connector with a built-in line filter. These connector/filter combinations are especially useful in supplies that are operated in RF fields. They are also useful in digital equipment to minimize conducted interference to the power lines.

CEE-22 connectors are available in many styles for chassis or PC-board mounting. Some have screw terminals; others have solder terminals. Some styles even contain built-in fuse holders.

A SERIES-REGULATED 4.5- TO 25-V, 2.5-A POWER SUPPLY

For home-laboratory requirements, a series-regulator supply is simpler and less expensive than a switching power supply. Series-regulated supplies are also free of electrical switching noise, which is a problem with some switching supplies. During tests of sensitive low-level circuitry, the power supply output should be pure dc; it should not contribute to problems in the circuit under test.

The power supply in **Fig 17.29** was designed and built by William E. Sabin, WØIYH. See "High-Precision Series Regulator Loop Design," earlier in this chapter, for a discussion of the design, analysis and tests of the feedback control loop used in this power supply.

FEATURES

This supply was designed to meet the following objectives:

- Continuously variable output voltage from 4.5 to 25.0 V.
- Tight load voltage regulation, better than 0.03%, for load currents from 0 to 2.0 A, 0.1% to 2.5 A; 0.01% into a 2.0-A load for ac-line voltages from 117 to 122 V.
- Excellent response to load fluctuations and transients (low output impedance).
- Very low ac ripple, less than 2 µV RMS with a 2.0-A load.
- Very low random noise, less than 2 µV RMS from 0.1 Hz to 500 kHz.
- Use an off-the-shelf transformer and other easily obtainable parts.
- Load currents up to 2.5 A, continuous duty.
- Switch selectable current limiting, at 0.1, 0.5 or 2.5 A, to protect delicate circuits under test.

THE CIRCUIT

Fig 17.30 is a schematic of the power-supply circuit. An LM723 regulator was selected because it's simple, and its reference voltage is available (pin 6) for filtering. Reference noise (typical of zener diodes) is reduced to a very low level via C5, as suggested in the data sheet for the LM723. The current-limiting circuitry is also accessible (at pins 2 and 3). It senses the voltage drop across R2 and R3.

The combination of R2 and R3 sets the current limit at 0.62 / (R2 + R3) A. R3 (R3a and R3b together) forms an 0.11-Ω, 4-W resistor, which acts as a shunt for the digital meter that measures load current. R8 provides meter adjustment without affecting R3 significantly. S3 selects from three values of R2 to set current limits at approximately 0.1, 0.5 or 2.5 A.

C1 has a large value to reduce output ripple voltage. Smaller values would increase ripple and require a higher trans-former voltage to prevent regulator drop out. Other voltage drops between the Q1/Q2 emitters and the output terminal are mini-mized to assure that a standard 25.2-V trans-former can do the job. The R1-C2 combination reduces ac ripple at the LM723 by a factor of 25; this eliminates the need for an extremely large value of C1.

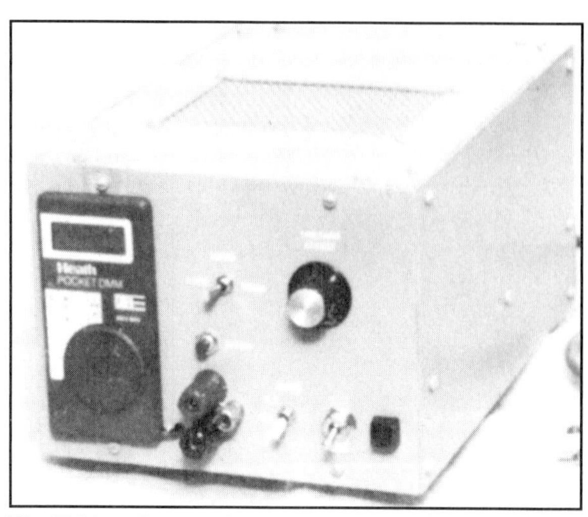

Fig 17.29 — An exterior view of the 2.5-A power supply. The DMM is mounted on the control panel as the output meter.

The circuitry of Q3, R10, C7 and D1 prevents the voltage on pins 11 and 12 of U2 from exceeding the 40-V maximum rating of the LM723 (especially with light loads and high line voltage). C7 eliminates a very small ac ripple at the dc output. As load current increases, the voltage at C1 decreases and Q3 is saturated.

When R4 is adjusted to reduce output voltage, U2 and Q1/Q2 are switched off until C6 discharges to the lower voltage. This causes the emitter-base junctions of Q1 and Q2 to break down (at about 2.0 V) so that C6 could discharge through R5. D2 provides an alternate path and prevents this breakdown. R5 provides a minimum load for U2.

The foldback circuit is interesting: Q4, D4 and R13 provide a constant current through, and therefore a constant voltage drop across, R16. This voltage makes the current limiting (pins 2 and 3 of U2) work properly over the entire 4.5- to 25.0-V range. As the current limiting action pulls the voltage at the top of R16 below about 4.0 V, D3 quickly stops conducting, the voltage across R16 approaches zero and the load current is limited at about 1.9 A. This limits Q1/Q2 and T1 dissipation and provides short-circuit protection. This is a regenerative positive feedback process. R14 sets the current at which foldback begins, 2.6 A. There is further discussion of foldback circuitry earlier in this chapter.

An inexpensive Heath SM-2300-A auto ranging DMM (mounted on the front panel) displays the supply output levels. The dedicated DMM can display very small output changes. S2 selects either voltage or current (divided by ten) for display. (The DMM is left in its voltage range for both measurements.) The voltage across R3 is 0.11 times the load current. R8 provides the required ammeter adjustment without significantly affecting R3. The meter shows 0.2 for 2 A. Ordinary wire-wound resistors are adequate for R3 because they do not heat significantly at 2.5 A.

R9 sets the reference voltage on pin 5 of U2 at 4.5 V to establish the minimum output voltage. R11 sets the 25.0-V upper limit. A three-wire line cord assures that the supply chassis is always tied to the ac-line ground, for safety reasons. The dc output is not referenced to chassis ground, and performance is independent of the ac-ground connection. If the load current is very small, it may take a long time for C6 to discharge when the power is switched off; a press of S4 quickly discharges the capacitors (through R15) after turnoff. The mechanical construction emphasizes heat removal, so a cooling fan is not needed.

CONSTRUCTION

Several areas are critical to good performance. Pay particular attention to these:

- Wire C1 with short, heavy leads to present a minimum impedance to ac ripple. Also, connect C1 directly to the negative binding post with a heavy lead. This provides a low inductance path for load-current fluctuations and keeps them off the PC-board ground plane.
- Connect C2 directly to the negative lead of C1.
- Connect C6 directly across the binding posts.
- Connect R4 to R11 with a low-impedance lead and connect R11 directly to the PC board ground plane.

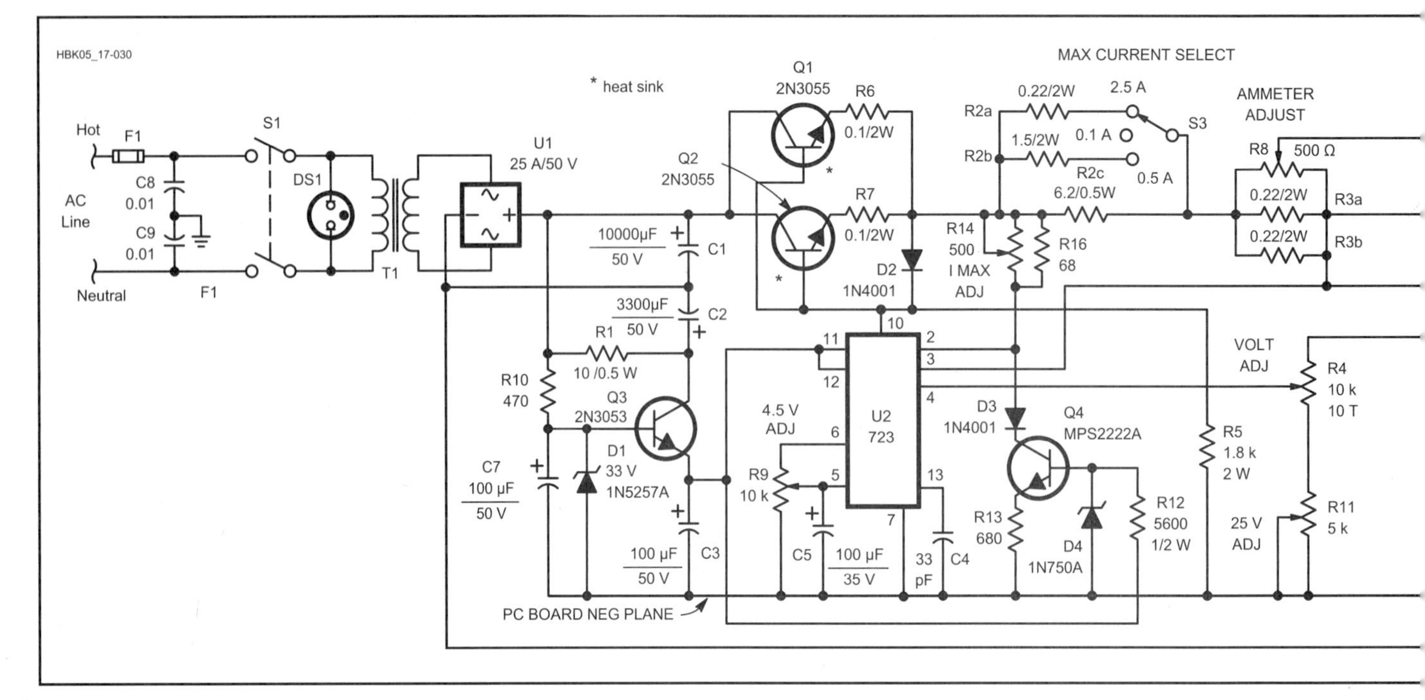

Fig 17.30 — Schematic of the 4.5 to 25-V, 2.5-A regulated dc power supply. Heavy lines indicate critical low-impedance conductors; use short conductors with large surface areas. "RS" signifies a RadioShack part number; "CDE" is a Cornell Dubilier Electric part.

C1 — 10000 μF, 50 V (CDE 10000-50-AC, or equiv)
C2 — 3300 μF, 50 V (CDE 3300-50-M, or equiv)
C3, C7 — 100 μF, 50 V (RS 272-1044, or equiv)
C4 — 33 pF, 50 V
C5 — 100 μF, 35 V (RS 272-1028, or equiv)
C6 — 1000 μF, 35 V (RS 272-1032, or equiv)
C8, C9 — 0.01 μF, ac-rated capacitor.
D1 — 1N5257A 33-V Zener diode.
D2, D3 — 1N4001 (RS 276-1101, or equiv)
D4 — 1N750A
DS1 — Neon lamp, 120-V ac (RS 272-704, or equiv)
F1 — 1.5 A slow-blow (RS 270-1284, or equiv)
M1 — DMM (Heath SM-2300-A, or equiv)
P1 — Three-wire power cord and plug
Q1, Q2 — 2N3055 (RS 276-2041, or equiv)
Q3 — 2N3053 (RS 276-2030, or equiv).
Q4 — MPS2222A (RS 276-2009, or equiv)
R1 — 10 Ω, 0.5 W
R2a — 0.22 Ω, 2 W
R2b — 1.5 Ω, 2 W
R2c — 6.2 Ω, 5%, 0.5 W
R3 — 0.15 Ω, 2 W
R4 — 10 kΩ, 10-turn potentiometer (Bourns 3540S, or equiv)
R5 — 1.8 kΩ, 2 W
R6, R7 — 0.1 Ω, 2 W
R8, R14 — 500 Ω potentiometer (RS 271-226, or equiv)
R9 — 10 kΩ potentiometer.
R10 — 470 Ω, 0.25 W
R11 — 5 kΩ potentiometer
R12 — 5.6 kΩ, 0.5 W
R13 — 680 Ω, 0.25 W
R16 — 68 Ω 0.25 W
S1, S2 — DPDT (RS 275-652, or equiv).
S3 — SPDT (center off) (RS 275-654, or equiv)
S4 — Normally open, momentary-contact, push-button switch (RS 275-1547, or equiv)
T1 — 120-V primary, 25.2-V, 2.8-A secondary (Stancor P-8388, or equiv)
U1 — Rectifier bridge, 25 A, 50 V (RS 276-1185, or equiv)
U2 — LM723 regulator (RS 27-1740, or equiv)
Heat sink — Wakefield 403A, or equiv (two required).

(A)

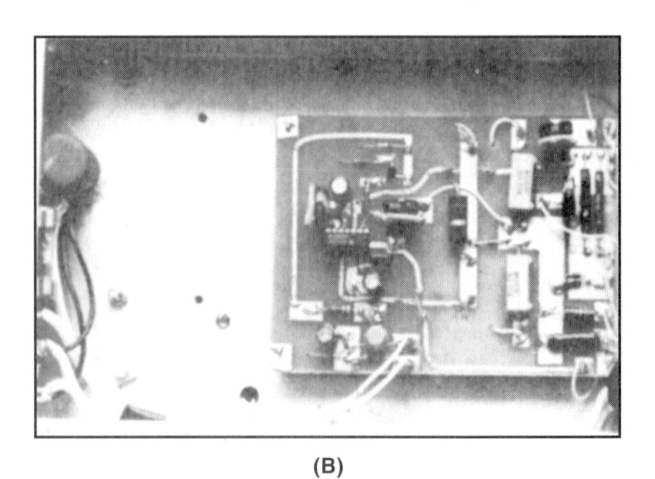

(B)

Fig 17.31 — A shows a top view of the open power-supply chassis. The 1/8-inch plate holding C1 is visible at the left edge of the heat sinks. B shows a bottom view. Notice that the components are mounted on the etched side of the board. The circuit pattern is simple enough for freehand layout. The front panel is at right in both views.

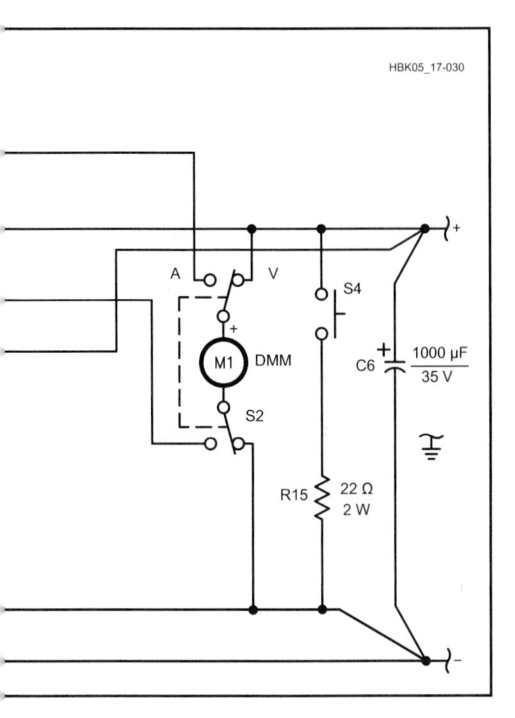

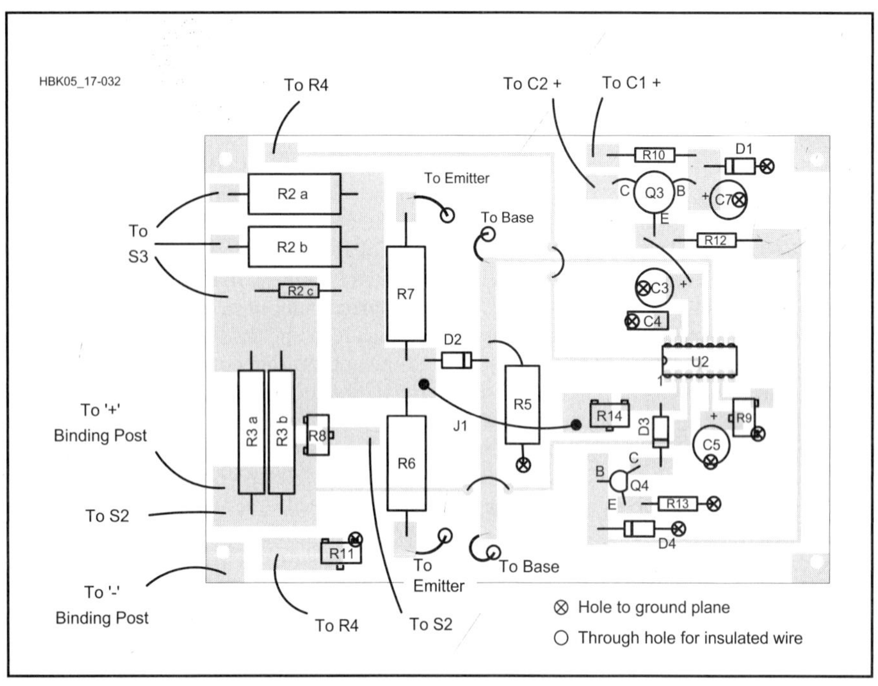

Fig 17.32 — A parts-placement diagram for the 2.5-A power supply. The component side is shown, and the etching is on the component side. Cross hairs indicate holes through the board. Component leads that terminate at cross hairs are grounded on the far side of the board. The emitter and base leads of Q1 and Q2 do not connect to the ground plane. Use insulated wire to pass through the PC board and holes in the chassis to the transistors on the heat sinks. Wire Jumper J1 is soldered to the pad surfaces without drilling holes. If you do drill holes for the jumper be sure to remove the ground plane around the holes on the non-component side. A full-size etching pattern is in the **Templates** section of the *Handbook* CD-ROM.

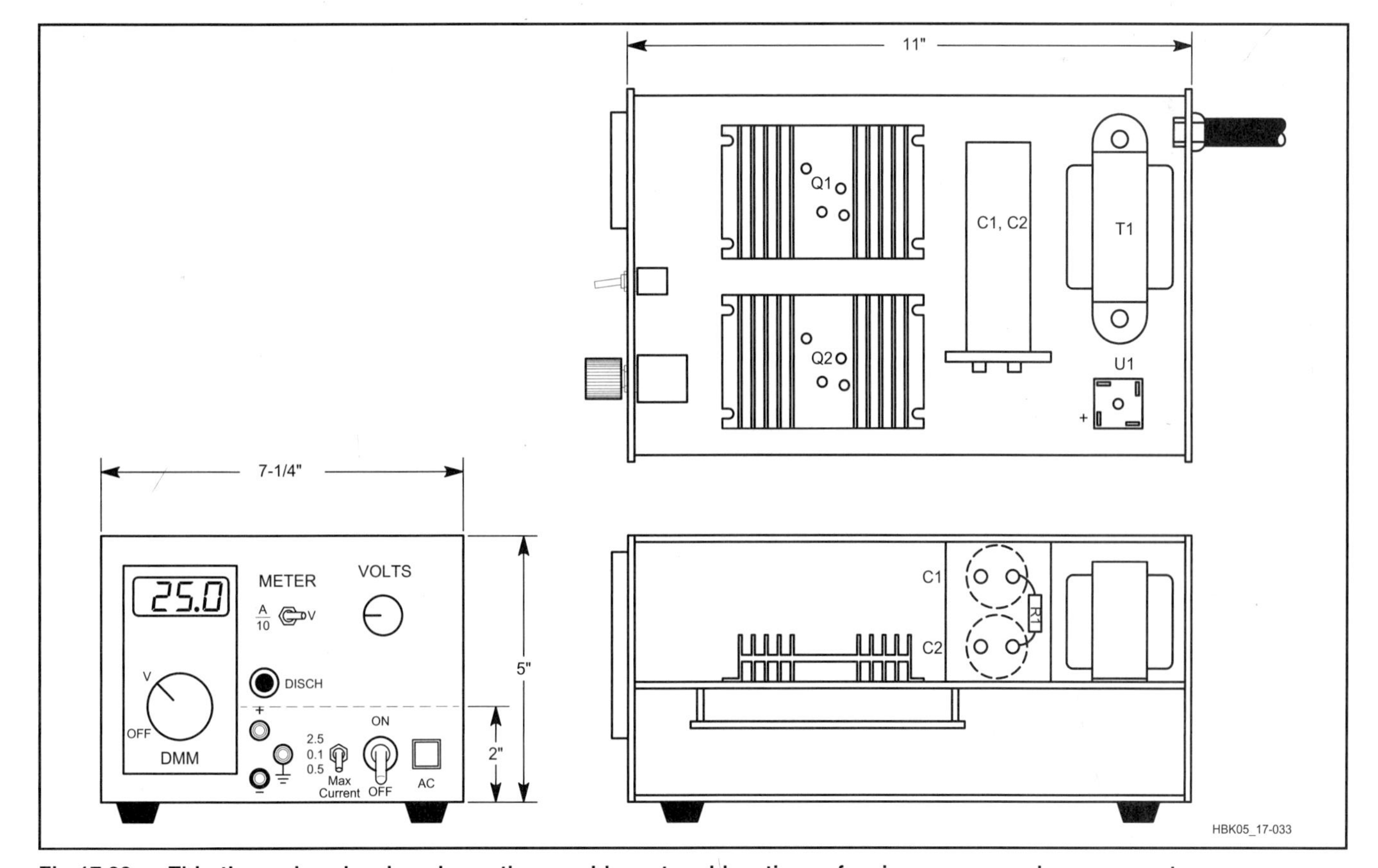

Fig 17.33 — This three-view drawing shows the panel layout and locations of major power-supply components.

- Connect the regulator PC board ground plane to the negative binding post at a single point.
- Use heavy-duty binding posts to reduce a small but significant voltage drop (from the rear to the front) at the front panel.

Fig 17.31 shows the general construction and **Fig 17.32** shows the front-panel layout. The cabinet and chassis surface are 1/16-inch aluminum plates connected by aluminum angle stock, drilled and tapped for #6-32 screws. Ventilation screens at the top and rear provide an excellent chimney effect. The chassis plate is tightly joined to the side plates for good heat transfer. Construction of the cabinet shown is somewhat labor intensive; it involves a lot of metal work. Any enclosure of similar size with good ventilation is suitable. For example, a 7 × 11 ×2 -inch chassis with a bottom cover and rubber feet, a 7½ × 6-inch front panel and a U-shaped perforated metal cover would serve well.

Q1 and Q2 each have a Wakefield 403A heat sink. The sinks were selected according to the thermal design discussion in the **Real-World Component Characteristics** chapter. The worst-case 2N3055 junction temperature was calculated to be 145°C, based on a measured case temperature of 95°C and dissipation of 32 W (each) for Q1 and Q2. This temperature is a little higher than recommended, but it's acceptable for intermittent lab use.

The DMM is epoxied to a narrow aluminum strip, which is then screwed to the front panel. This allows easy removal to replace the DMM battery.

Fig 17.31B shows the supply underside. The PC board is mounted on standoff insulators and positioned so that the positive and negative outputs are close to the binding posts.

Fig 17.33 is a parts-placement diagram for the regulator board. A full-size etching pattern is included in the **Templates** section of the CD-ROM included with this *Handbook*. All components and wiring are on the etched side of the board (with the help of a few jumper wires). The other side is entirely ground plane. The components are surface mounted by bending a section of each lead and soldering it to the surface of the appropriate copper pad. The socket for U2 is mounted by bending the pins out 90° and soldering pins 11 and 12 to the pad for C3. Some leads pass through holes (marked with cross hairs in Fig 17.33) in the PC board and solder to the ground plane. The emitter and base leads of Q1 and Q2 are isolated from ground; use insulated-wire to pass from the PC-board pads through holes in the PC board and chassis to the transistors.

Fig 17.32 shows the placement of major parts in front, top and side views. Silastic and pieces of Kraft paper cover any exposed 120-V wiring.

C1, C2 and R1 are mounted on a ⅛-in thick plastic board as shown in Fig 17.32. In the supply shown, it's a piece of ⅛-in PC board with wide traces. A piece of angle stock mounts the board to the chassis.

The main voltage control, R4, is a 10-turn potentiometer, for ease of adjustment. C8 and C9 are ceramic capacitors rated for ac-line use. Once the supply is complete, double check all wiring and connections.

ADJUSTMENTS

Several adjustments are needed: the dedicated AMMETER ADJUST (R8), the foldback limit (R 14, I MAX ADJ), the lower (R9, 4.5 VOLT ADJ) and upper (R11, 25 V ADJ) voltage limits. First, obtain an accurate external ammeter and a test load that can draw at least 3 A at 12 V or so (automobile sealed beam headlights work well). Series connect the load and external meter to the supply binding posts. Set S3 to the 2.5-A range; set R4 for minimum output; set S2 to A and switch on the power. Advance R4 until the ammeter reads 2.0 A, and adjust R8 until the panel meter also reads 0.200. Next, increase R4 until the current slightly exceeds 2.6 A, and adjust R14 so that current begins to decrease (foldback). Switch off the supply and remove the external meter and load.

Adjust the voltage limits with no load connected and S2 set to V. Switch on the supply, set R4 for minimum voltage and adjust R9 until the meter reads 4.5. Increase R4 to maximum voltage and adjust R11 so that the meter reads 25. The supply is ready for use.

PERFORMANCE AND USE

The example supply met the design goals: When the line voltage was varied from 117 to 122 V (with a 25-V dc 2.0-A load), the output voltage varied less than 0.01%. When the dc load was varied from 0 to 2.0 A, the output voltage changed less than 0.03%.

When extremely tight regulation and low output impedance are important, the leads to the load must be very short, heavy straps. Multiple loads should be parallel-connected directly at the binding posts. A "Daisy chain" connection scheme does not assure equal regulation for each load; it counteracts the precision regulation of this supply.

A 13.8-V, 40-A SWITCHING POWER SUPPLY

Switching power supplies ("switchers," as they are often called) offer very attractive features — small size, low weight, high efficiency and low heat dissipation. Although some early switchers produced objectionable amounts of RF noise, nowadays you can build very quiet switchers using proper design techniques and careful EMI filtering. This power supply produces 13.8 V, regulated to better than 1%, at a continuous load current up to 40 A and with an efficiency of 88%. No minimum load is required and the ripple on the output is about 20 mV.

The supply produces no detectable RF noise at any frequency higher than the main switching frequency of 50 kHz. The author, Manfred Mornhinweg, XQ2FOD, checked this with a wire looped around his supply, tuning his TS-450 from 30 kHz to 40 MHz. The completed supply weighs only 2.8 kg (6.2 pounds)! **Fig 17.34** shows the unit built by HQ staffer Larry Wolfgang, WR1B, for the ARRL Lab.

Fig 17.34 — Photo of the completed 13.8-V, 40-A switching supply built by WR1B.

LINEAR VERSUS SWITCHING SUPPLIES

A typical linear regulated power supply is simple and uses few parts — but several of these parts are big, heavy and expensive. The efficiency is usually only around 50%, producing *lots* of heat that must be removed by a big heat sink and often fans.

In this switching power supply, the line voltage is directly rectified and filtered at

300 V dc, which feeds a power oscillator operating at 25 kHz. This relatively high frequency allows the use of a small, lightweight and low-cost transformer. The output is then rectified and filtered. The control circuit steers the power oscillator so that it delivers just the right amount of energy needed; little energy is wasted.

While MOSFETs can switch faster, bipolar switching devices have lower conduction losses. Since very fast switching was undesirable because of RF noise, the author used bipolar transistors. These tend to be *too* slow, however, if the driving current is heavier than necessary. If the transistors must switch at varying current levels, the drive to them must also be varied. This is called *proportional driving* and is used in this project.

The switching topology used is called a *half-bridge forward* converter design (also known as a *single-ended push pull* converter — *Ed.*). The converter is controlled using pulse-width modulation, using the generic 3524 IC.

CIRCUIT DESCRIPTION

Refer to the schematic diagram in **Fig 17.35**. Line voltage enters through P1, a connector that includes EMI filtering. It then goes through fuse F1, a 2-pole power switch and an additional common-mode noise filter (C1, L1, C2). Two NTC (negative temperature coefficient) resistors limit the inrush current. Each exhibits a resistance of about 2.5 Ω when cold and then loses most of its resistance as it heats up. A rectifier delivers the power to C3A and C3B, big electrolytic capacitors working at the 300-V dc level. The power oscillator is formed by Q1, Q2, the components near them, and the feedback and control transformer T3. T2 and associated components act as a primary-current sensor.

T1 is the power transformer, delivering a 20-V square wave to the Schottky rectifiers (D6 through D9). A toroidal inductor L2 and six low equivalent-series-resistance (ESR) electrolytic capacitors form the main filter, while L3, C23 and C24 are there for additional ripple reduction. The 13.8 V is delivered to the output through a string of ferrite beads with RF decoupling capacitors mounted directly on the output terminals.

The control circuit IC U1 is powered from an auxiliary rectifier D17. U1 senses the output voltage and the current level and controls the power oscillator through Q3 and Q4. C37, C35 and R23 are used to implement a full PID (*proportional-integral-derivative*) response in the control loop.

A quad operational amplifier U2 controls the cooling fan according to the average current level and also drives the voltage indicating tricolor LED, which glows green if the voltage is okay, orange if the voltage is too low and red if it is too high.

MORE DESIGN DETAILS

When the unit is powered up, the operating voltage builds up on C3A and C3B, and R2 and R6 bias the two power transistors Q1 and Q2 into their active zones. They start conducting a few mA, but for only a short time, because the positive feedback introduced by T3 quickly throws the system out of balance. One of the two transistors receives an increased base current from T3, while the other one sees its base drive reduced. It takes just a fraction of a microsecond for one of the transistors to become saturated and the other cut off. Which transistor will start first is unpredictable, but for this analysis let's suppose it is Q1. Because the control circuit is not yet powered, Q3 and Q4 are off at startup.

T1 sees about 150 V ac across its primary, producing about 20 V ac on the secondary. Schottky rectifiers D6 through D9 rectify this, so L2 sees 20 V across it. The current in L2 will start rising and this is reflected back to the primary side of T1. The primary current passes through the one-turn winding of T3, forcing one-eighth as much current to flow into the base of Q1, the transistor assumed to be conducting at this moment. After some time, the ferrite core of T3 will saturate, causing the base drive of Q1 to decrease sharply. Q1 will stop and Q2 will start conducting. Now the flux in T3's core decreases, crosses zero and increases in the other direction until it saturates the core again, shutting Q2 off and turning Q1 back on. Meanwhile, the current in L2 continues to build up and the filter capacitors C17 through C22 are charged.

For safe startup, it is essential that T3 saturates completely before T1 starts to do so. If this were not the case, the transistors would have to switch under a very high and potentially destructive current. The power supply will oscillate freely for only a few cycles, because D17 is already charging C32 and C33, powering up the control circuit so that it takes over the control of the power oscillator. Note that the self-oscillation frequency must be lower than the operating frequency for the feedback loop to be able to control things properly.

Q3 and Q4, together with D13 and D14, can place a short on T3's control winding. This holds the voltage across that transformer close to zero, regardless of any current that may be flowing in the windings. When U1 wishes to switch Q1 on, it simply switches pin 12 to ground, switching off Q4 and ending the short circuit on T3. Through R14 and D12, about 15 mA flow into the control winding center tap, returning to ground via Q3. This puts about 50 mA into the base of Q1, which quickly switches on. Now the heavy collector current (up to 8 A at full load) adds up to the total current flowing in T3 and puts enough drive into Q1 to keep it saturated at that heavy current. Note that by this method the strong drive current for the power transistors comes from the collector current through T3 so the control circuit does not have to provide any substantial driving power.

If U1 now determines that Q1 has been conducting long enough, it simply switches off pin 12. Q4 starts conducting again, shorting out T3. The current in T3 is dumped into Q4, which may have to take up to 300 mA. The voltage on T3 falls and Q1 switches off. Some time later, U1 grounds pin 13, starting the conduction cycle for Q2.

U1 uses two input signals to decide what to do with its outputs. One is a sample of the output voltage, taken through R25 and nearby components, while the other is a current sample taken through the primary of T2. This current transformer produces 200 times less current from its secondary than what goes through its one-turn primary. At full load, about 40 mA goes into R12, producing a maximum voltage drop of about 7 V. This is rectified and half of it is taken at the center tap, divided down by R13 and VR1 and smoothed by C31. When VR1 is properly adjusted, there will be 200 mV at pin 4 of U1 with the power supply running at full load.

A second amplifier inside U1 is used for current limiting. Its inputs are at pins 4 and 5. This amplifier is ground-referenced and has an internal offset of 200 mV. The amplifier will pull down the main error amplifier's output if the difference between pin 4 and pin 5 reaches 200 mV.

U1 also contains an internal oscillator, whose frequency is set by R24 and C36 to approximately 50 kHz. The sawtooth output of this oscillator is connected to an internal comparator, which has its other input internally connected to the output of the error amplifier. The output of the comparator is a square wave whose duty cycle depends on the dc voltage at the output of the error amplifier.

During operation at medium to high loads, the duty cycle is about 70%. At the cathodes of the Schottky rectifiers you will see a square wave that stays at about 20 V for some 14 µs, and then goes slightly below ground level for 6 µs. L2, which has its output end at a constant 13.8 V, will therefore see about 6 V for 14 µs, followed

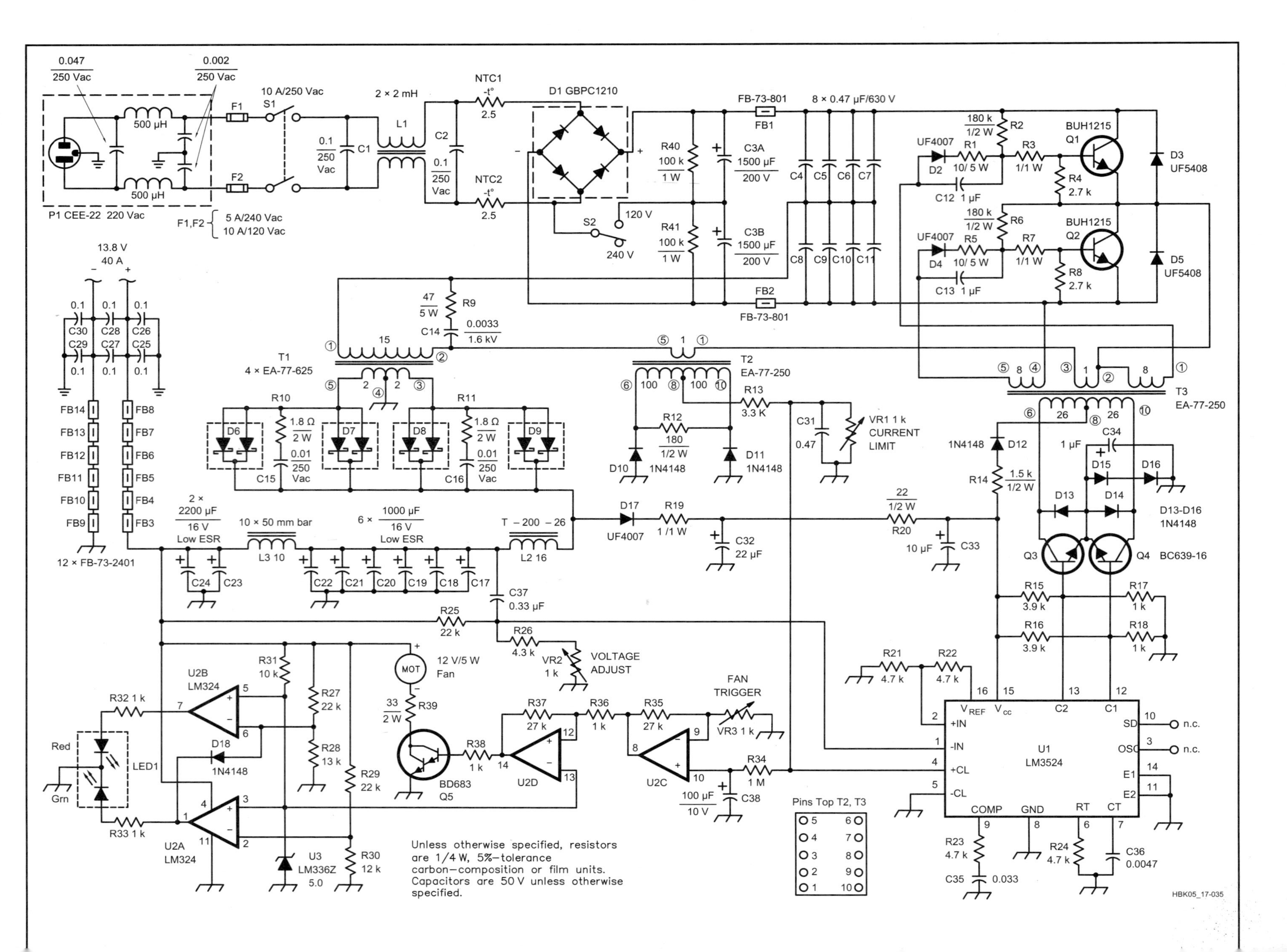

0.047
250 Vac
0.002
250 Vac
500 µH
500 µH
P1 CEE-22 220 Vac
F1
F2
F1,F2 5 A/240 Vac 10 A/120 Vac
S1
10 A/250 Vac
C1 0.1 250 Vac
L1 2 × 2 mH
C2 0.1 250 Vac
NTC1 2.5
NTC2 2.5
D1 GBPC1210
S2
120 V
240 V
R40 100 k 1 W
R41 100 k 1 W
C3A 1500 µF 200 V
C3B 1500 µF 200 V
FB1
FB2
FB-73-801
8 × 0.47 µF/630 V
C4
C5
C6
C7
C8
C9
C10
C11
Q1 BUH1215
Q2 BUH1215
D3 UF5408
D5 UF5408
R1 10/5 W
R2 180 k 1/2 W
R3 1/1 W
R4 2.7 k
R5 10/5 W
R6 180 k 1/2 W
R7 1/1 W
R8 2.7 k
D2 UF4007
D4 UF4007
C12 1 µF
C13 1 µF
T3 EA-77-250
R9 47 5 W
C14 0.0033 1.6 kV
T1 4 × EA-77-625
D6
D7
D8
D9
R10 1.8 Ω 2 W
R11 1.8 Ω 2 W
C15 0.01 250 Vac
C16 0.01 250 Vac
T2 EA-77-250
R12 180 1/2 W
R13 3.3 K
D10 1N4148
D11 1N4148
C31 0.47
VR1 1 k CURRENT LIMIT
D17 UF4007
R19 1/1 W
C32 22 µF
R20 22 1/2 W
C33 10 µF
R14 1.5 k 1/2 W
D12 1N4148
C34 1 µF
D13-D16 1N4148
Q3
Q4 BC639-16
R15 3.9 k
R16 3.9 k
R17 1 k
R18 1 k
U1 LM3524
VREF
Vcc
+IN
-IN
+CL
-CL
COMP
GND
RT
CT
C1
C2
E1
E2
SD
OSC
n.c.
R21 4.7 k
R22 4.7 k
R23 4.7 k
C35 0.033
R24 4.7 k
C36 0.0047
Pins Top T2, T3
L2 16
T – 200 – 26
C37 0.33 µF
R26 4.3 k
VR2 1 k
VOLTAGE ADJUST
R25 22 k
6 × 1000 µF 16 V Low ESR
C17
C18
C19
C20
C21
C22
L3 10
10 × 50 mm bar
2 × 2200 µF 16 V Low ESR
C23
C24
13.8 V 40 A
C25
C26
C27
C28
C29
C30
0.1
FB3
FB4
FB5
FB6
FB7
FB8
FB9
FB10
FB11
FB12
FB13
FB14
12 × FB-73-2401
FAN TRIGGER
VR3 1 k
R34 1 M
C38 100 µF 10 V
U2C
R35 27 k
R36 1 k
R37 27 k
U2D
R38 1 k
Q5 BD683
MOT
12 V/5 W Fan
R39 33 2 W
R29 22 k
R30 12 k
R27 22 k
R28 13 k
R31 10 k
U3 LM336Z 5.0
D18 1N4148
U2B LM324
U2A LM324
R32 1 k
R33 1 k
LED1
Red
Grn
Unless otherwise specified, resistors are 1/4 W, 5%-tolerance carbon-composition or film units. Capacitors are 50 V unless otherwise specified.
HBK05_17-035

Fig 17.35 — Schematic diagram and parts list for the 13.8-V, 40-A switching power supply.

C1, 2 — 0.1 µF, 250 V ac polypropylene, Digi-Key P4610ND
C3A, C3B — 1500 µF, 200 V electrolytic
C4 to C11 — 0.47 µF, 400 V polypropylene, Digi-Key P3496ND
C12, C13 — 1 µF, 50 V ceramic multilayer
C14 — 0.0033 µF, 1.6 kV polypropylene
C15, C16 — 0.01 µF, 250 V ac polypropylene
C17 to C22 — 1000 µF, 25 V low-impedance (low-ESR) electrolytic
C23, C24 — 2200 µF, 16 V low-impedance (low-ESR) electrolytic
C25 to C30 — 0.1 µF, 50 V ceramic
C31 — 0.47 µF, 50 V ceramic multilayer
C35 — 0.033 µF, 50 V polyester
C36 — 0.0047 µF, 50 V polyester
C37 — 0.33 µF, 50 V polyester or ceramic multilayer
D1 — Rectifier bridge, 1 kV, 12 A, GBPC1210 or similar
D2, D4, D17 — Ultrafast diode, 1 kV, 1 A. UF4007 or similar. Lower voltage (down to 100 V) is acceptable. The ARRL Lab used UF1007 diodes, Digi-Key UF1007DICT-ND
D3, D5 — Ultrafast diode 1 kV, 3 A. UF5408 or similar, from Techsonic
D6 to D9 — Dual Schottky diode, 100 V, 30 A total. PBYR30100CT or similar. Single diode would also be suitable. ARRL Lab used International Rectifier 30CPQ100, Digi-Key 30CPQ100-ND
D10 to D16, D18 — 1N4148 switching diode
F1, F2 — Fuse, 10 A for 120-V ac operation; 5 A for 240-V ac operation
FB1, FB2 — Amidon FB-73-801 ferrite bead, slipped over wire. Available from Bytemark
FB3 to FB14 — Amidon FB-73-2401 ferrite beads, slipped six each over the two 13.8 V dc output cables. Available from Bytemark
L1 — Common mode choke, approximately 2 mH each winding, 6 A. Author used junk box specimen. We used a Magnatek CMT908-V1 choke (Digi-Key part 10543-ND) in the supply built in the ARRL Lab
L2 — 20 µH, 60 A choke. 16 turns on Amidon T-200-26 toroid, wound with ten #16 enameled wires in parallel
L3 — 5 µH (uncritical), 60 A choke. 10 turns on ferrite solenoid, 10 mm diameter, 50 mm long. Wound with two #12 wires in parallel. Amidon #33-050-200 used in ARRL Lab
LED1—Dual LED, green-red, common cathode, Digi-Key LU204615-ND (pin 1 is red; pin 3 is green)
M1 — 12 V, 5 W brushless dc fan, approximately 120 × 120 × 25 mm, Digi-Key P9753-ND is 120 × 120 × 38 mm and 5.5 W
NTC1, NTC2 — Inrush current limiter, 2.5 Ω cold resistance, Digi-Key KC003L-ND
P1 — Male ac connector with integrated EMI filter, 250 V ac, 10 A, Newark 07H8844
Q1, Q2 — High voltage switching transistor, BUH1215 or similar. Motorola MJW16010 was used in ARRL Lab; Newark 08TMJW16012
Q3, Q4 — BC639-16 transistor, available from Newark. Must resist 100 V and 0.5 A
Q5 — BD683 Darlington transistor, from Techsonic. The back of the transistor should be facing the outside of the board
R1, R5 — 10 Ω, 5 W, low inductance preferred. For the supply built in the ARRL Lab, we used three 30 Ω, 2 W film resistors wired in parallel
R9 — 47 Ω, 5 W, low inductance preferred. For the supply built in the ARRL Lab, we used three 150 Ω, 2 W film resistors wired in parallel
R10, R11 — 1.8 Ω, 2 W, low inductance preferred
S1 — 2-pole power switch, 250 V ac, 10 A
S2 — 120/240 V ac power selector slide switch, 250 V ac, 10 A. A locking tab made of aluminum locks the switch in either 240 or 120-V position
T1 — Primary 15 turns, secondary 2+2 turns. Wound with copper foil and mylar sheet. Uses four Amidon EA-77-625 ferrite E-cores (8 halves). Equivalents include Thompson GER42x21x15A, Phillips 768E608, TDK EE42/42/15
T2 — Secondary is 100+100 turns #36 enamel wire. Primary is one turn #14 plastic insulated cable, wound on secondary. Wound on Amidon EE24-25-B bobbin. Uses an Amidon EA-77-250 core. Equivalents are Thompson GER25x10x6, Phillips 812E25Q, TDK EE25/19
T3 — Control winding is 26+26 turns #28 enamel wire. Base windings are 8 turns #20 each. Collector winding is one turn #14 plastic insulated wire. Bobbin and core same as T2
U1 — Pulse-width modulator IC, LM3524, SG3524, UC3524 or similar
U2 — Quad single-supply operational amplifier, LM324 or similar
U3 — 5 V voltage reference, LM336Z-5.0 or similar
VR1 to VR3 — 1 kΩ PCB-mounted trimpot, Digi-Key #3309P-102-ND
Cabinet — Hammond Manufacturing, PN 1426Y-B, 12 × 6 × 5.5 inches, and internal case mounting rails, Hammond Manufacturing 1448R12, used in ARRL Lab

by –14 V for the rest of the time. Given its inductance of about 20 µH, the current in L2 will increase by about 4 A during each conduction cycle and decrease by that same amount during rest time. As long as the current drawn from the power supply is more than 2 A, the current in L2 will never cease completely. For example, if the current is 20 A on average, the current in L2 will vary between about 18 and 22 A. As the ripple current stays basically constant while operating at up to the maximum current of the power supply, filter capacitors C17 to C22 are never exposed to more than about 1.5 A RMS total ripple current, assuring that they have a long lifetime. This is an advantage over some other types of switching power supplies, where the ripple current is much higher, forcing the designer to use more expensive capacitors or to accept reduced lifetime in these components.

If the load is less than about 2 A, the current flow in L2 is no longer continuous. The duty cycle of the power transistors starts to drop, until at zero load the duty cycle almost becomes zero too.

C37 serves several purposes. For higher frequencies it couples the first filter stage (L2 and C17 through C22) to the error amplifier, while for lower frequencies (and at dc) the output of the supply is sampled. This is necessary because each filter stage introduces 180° of phase shift at the higher frequencies. After two stages the phase shift goes through a full 360°, making it impossible to stabilize the control loop without additional circuitry. But for dc, sampling the output is desirable to compensate for the voltage drop in L3. C37 gives the error amplifier a nice PID response, together with R23 and C35. This affords the best possible transient behavior with unconditional stability. In addition, C37 provides some measure of soft starting, so the voltage does not overshoot too much when first switching on the power supply.

R34 and C38 average out the current level over a period of about 2 minutes. U2C amplifies the resulting voltage by an amount that can be adjusted. U2D acts as a Schmidt trigger to switch the fan cleanly on and off when the current average crosses the trigger level set by VR3. R39 limits the speed of the fan to a rather low value that is more than enough to keep the power supply cool. At this low speed the fan produces almost no noise and it will probably last longer than its owner.

Snubbers and EMI Filters

No transformer is perfect. Each winding has some inductance that is not magnetically coupled to the others. There is also the magnetizing current, which can be a considerable part of the total current in small transformers. At the end of a conduction cycle, a strong current flows in T1. After switching the power transistors off, some means must be provided to discharge the energy stored in the magnetic field of the core and in the leakage inductances. D3 and D5 are included for this purpose. They recover most of this energy and dump it back into C3. Another portion flows through the Schottky diodes into L2, but this cannot be more than the current flowing in L2 at the moment of switchoff.

A problem arises if the magnetizing current is bigger than the actual load current, a situation that can occur during startup. Also it must be taken into account that diodes, even fast ones, take some time to switch, and the transformer cannot wait to start dumping its energy. So some absorbing RC networks have to be included. These are commonly called *snubbers*. R9 and C14 form the primary snubber, absorbing energy during the switching of D3, D5, Q1 and Q2. On the secondary side of T1, R10, C15, R11 and C16 protect the Schottky rectifiers from inductive spikes.

Some RF noise is generated and it must be cleaned up. Between C3 and the power oscillator, two Type-73 ferrite beads FB1 and FB2 perform a critical noise-absorbing task. On the output side, L2 already absorbs most of the noise. It is wound on a high-permeability iron-powder toroid that is very lossy in the HF range. The main filter capacitors have low equivalent-series-resistances for good filtering.

L3 is another noise absorber. To minimize capacitive coupling, a ferrite solenoid was used instead of a toroid so that the input windings are well separated from the output ones. The ferrite used starts absorbing at HF, so this coil not only blocks but also absorbs RF energy. Finally, the output leads are passed through a dozen 73-material ferrite beads. The filtering is completed by bypass capacitors on the output leads to the cabinet. Note that the ground on the printed circuit board is floating to reduce stray HF currents on the enclosure.

Running on 240/120 V ac

The author lives in a country where the mains supply is 220 V at 50 Hz. This supply will accept input voltages between about 95 to 250 V ac, using S2 to switch from 240 to 120 V ac operation. For 120-V ac operation the fuse F1 should be rated at 10 A, 5 A for 240 V ac operation.

THE PCB

The exact size of the pc board is 120 × 272 mm (4.72 × 10.71 inches). It must be made from good quality, single-sided glass epoxy board — don't try to use a cheaper grade of board. The heavy components would stress it too much and the copper adhesion is not good enough for the heavy soldering required. A circuit board is available from FAR Circuits.

BUILDING THE MAGNETIC COMPONENTS

The biggest challenge for most home builders will be the magnetic components. To keep things simple, Amidon cores were used. The only exceptions are L1 and L3, which were made from materials found in the author's junk box. Both of these inductors are not critical, and suitable Amidon part numbers are included in the parts list.

T1, the main power transformer, is the heart of this circuit. T1 was built using a tape-winding technique, stacking four

Parts Substitution

Don't be afraid to substitute parts when you can't find the exact one specified. Here is some information for hard-to-find parts:

D1: Any rectifier bridge that can handle 8 A at 240 V ac (or 12 A at 120 V ac), with enough headroom for spikes, will do the job. Try to find one that fits the PCB or modify the board accordingly. You may also use single diodes, but mount them close to the board to get suitable heatsinking through their terminals.

D2, D4, D17: Any ultrafast diode rated for at least 100 V and 1 A is suitable. The author used the UF4007, which is an ultrafast equivalent to the 1N4007 (1 kV, 1 A). *Do not* use 1N4007 diodes! They are not fast enough for this job. You need a switching speed in the 50-ns class.

D3, D5: You can use any ultrafast diode rated for 600 V, 3 A or higher. The UF5408 is rated at 1 kV, somewhat of an overkill here. Again, *do not* use the low speed 1N5408.

D6, D7, D8, D9: PBYR30100CT dual Schottky diodes were used. A good replacement is any single or dual Schottky rectifier rated at least at 100 V and 30 A total current, that comes in a TO-218 or similar package. If you use single diodes, you may have to bend the pins to fit the board properly. These 100-V Schottky diodes have been widely available only for a few years, although they are becoming more common.

Q1, Q2: BUH1215 transistors were used, which can work at a higher voltage than actually necessary in this circuit. If you need to replace them, look for any NPN power switching transistors that have a V_{CEO} of at least 400 V, I_C of at least 15 A, an h_{FE} of at least 12 at 8 A, and come in a TO-218 or similar package. The power transistors *must* maintain their beta up to at least 8 A; otherwise they will cut short the conduction cycles when the load increases. Motorola MJW16010 transistors are a suitable alternative.

Some power switching transistors have a reverse protection diode and a base-to-emitter resistor built in. Beware of these! The resistor would not allow this power supply to start. If in doubt, take a multimeter and measure the resistance between base and emitter. If you get the same low resistance (typically 50 Ω) in *both* senses, the transistor is unsuitable for this project. If you get a diode behavior, the transistor is okay.

Q3, Q4: Instead of the BC639-16 you can use any small TO-92 cased NPN transistor that has a V_{CEO} rating of 100 V and an I_C of 1 A. Be careful with the pinout, because not all TO-92 transistors use the same pinout. You may have to bend the leads to fit the printed circuit board.

Q5: Instead of the BD683 you can use any small NPN Darlington transistor that has an I_C of at least 1 A.

U3: If you have trouble finding the LM336Z-5.0 voltage reference, you have several options. You may use a reference at another voltage (2.5 V is typical) and modify the values of R27 to R30 accordingly. Or you may replace U3 with a 3-terminal regulator like the 7805, modifying the circuit as necessary. Finally, you could completely eliminate U3 and R31 and use the 5V reference provided by U1 at pin 16. In this case you would lose the voltage indicator's independence from U1. Note that using a simple zener diode instead of U3 is not suitable because zeners are not stable enough for this application.

If you cannot find low-ESR electrolytic capacitors, simply use normal capacitors. The circuit is designed to place a low ripple current on these capacitors, so standard components can be used. The noise at the output will be slightly higher, however.

pairs of ferrite E cores to obtain the necessary magnetic capabilities. Comments from WR1B as he constructed the transformers and inductors are included in the construction details below.

Making T1

Because four cores are stacked there is no factory-made bobbin available for this transformer, so the author made a paper bobbin. He wound the transformer using 0.1-mm thick copper strips interleaved with Mylar sheets, because a thick wire needed for the heavy current would be impossible to bend around the sharp corners of the bobbin. Instead of using a lot of thin wires in parallel, it is better to use copper strips. The whole assembly is sealed in epoxy resin, with the magnetic cores also glued in place with epoxy.

Cut a piece of hardwood to serve as a form when making the bobbin. As the center legs of the four stacked cores measure 62 × 12 mm (2.44 × 0.47 inches), the wood block must be 63 mm (2.48 inches) wide and 12.5 mm (0.49 inches) thick, to allow for some play. The length of the block should be around 100 mm (4 inches). The height of the bobbin will be 28 mm (1.10 inches), so make your block long enough to hold it with the bobbin in place with room for holding onto it. The author used a belt sander to trim his wood block to the exact dimensions. Try to be precise — if the bobbin is too big you will waste valuable winding space, running the risk of not being able to fit the windings. If the bobbin comes out too small your finished winding assembly may not fit the ferrite cores, making it unusable.

Now wrap the wood block with one layer of plastic film, such as that used in the kitchen to preserve food. This material allows you to remove the bobbin from the wood block easily. Cut a strip of strong packing paper, 28 mm (1.10 inches) wide and about 1 m (39.4 inches) long. A brown-paper grocery bag is a good source of suitable paper. Mix some 5-minute epoxy glue (the author used the type sold in airplane modeling shops, which comes in good sized bottles) and apply a layer of epoxy to the paper strip. Now wind 6 layers of the paper strip very tightly around the plastic-wrapped wood block. Wrap another sheet of plastic film around your work and press it between two wooden blocks held together with strong rubber bands or wood clamps so the long sides of the bobbin are flat and smooth against the wood. Now place the bobbin assembly in an oven for about 15 minutes at 50ºC (122ºF). The epoxy sets much more quickly and becomes somewhat stronger at that temperature.

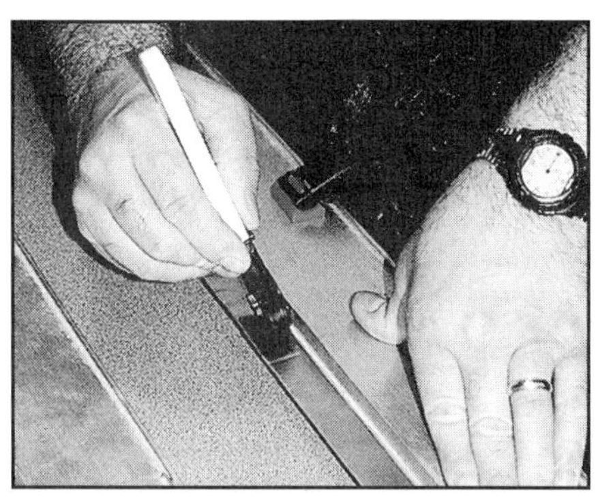

Fig 17.36 — Larry Wolfgang, WR1B, using a 4-foot straightedge designed as a guide for hand-held circular saws to clamp the copper-foil tape to a board on a tabletop. After carefully measuring to ensure a uniform 22-mm width, he cut the foil tape using a Fiskars rotary cutter. Be careful to keep the cutter wheel against the straightedge for the entire length. Move the tape in 4-foot intervals to cut the entire length. *(Photo by Dan Wolfgang.)*

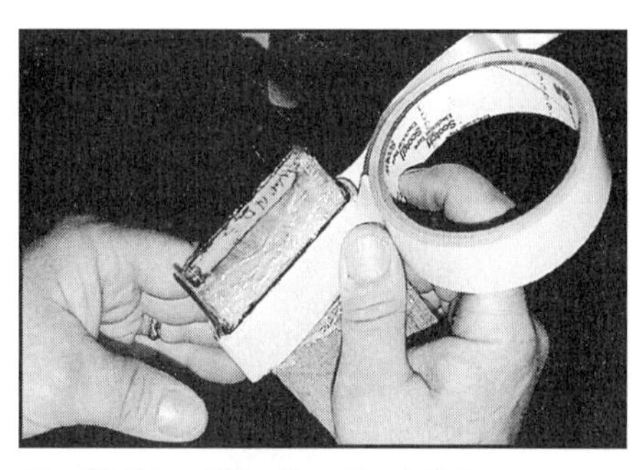

Fig 17.37 — Winding the foil tape tightly on the epoxy-coated-paper bobbin on the wooden block. The Mylar tape is unrolled and positioned over the foil layer as you wind. *(Photo by Dan Wolfgang.)*

[Comments from Larry Wolfgang, WR1B: The paper I used for my T1 bobbin was cut from a 36-inch-wide length of Kraft paper. This had been used to wrap some paper my wife had purchased at an art-supply store. It was about as heavy as the paper used for grocery bags. I used 30-minute epoxy for this step, providing a bit more "working time" than 5-minute epoxy allows. It takes *lots* of epoxy, because so much soaks into the paper. My epoxy was the kind with the double plunger, and equal amounts come out of both tubes as you push in the plunger. Wear rubber or plastic gloves to protect your hands. I squeezed out an amount that made a puddle of resin and a puddle of hardener each about 1½ inches across and ⅛ inch or so deep. This was not enough, and I had to mix more. I used a spring clamp to hold the paper to my workbench and then held the paper in one hand while spreading epoxy with a heavy toothpick. I coated the entire length and then wrapped my plastic-covered wooden block. My electronic-controlled gas oven only allows me to set the temperature as low as 170ºF, so I had to watch the temperature and shut the oven off as the temp rose to about 150ºF, then let it cool down. I ran it twice this way to "cure" the 30-minute epoxy I used for the bobbin. — *WR1B*]

Now you will need some 0.1 mm (0.004 inches = 4 mils) thick copper tape, and some Mylar sheet of a similar thickness. Cut the copper in strips 22 mm (0.87 inches) wide, and the Mylar in strips 28 mm (1.10 inches) wide. (The Wireman has suitable copper foil available.) If you can make long strips, say 2 m (6.56 feet), this is an advantage. Otherwise, you will have to solder individual copper strips together. In total, you will need about 7 m (23 feet) of copper tape and slightly less Mylar tape. [I made 7 meters of "double-thickness" tape, using two 3-mil thick, sticky-backed copper tapes that we had in the ARRL Lab. After making the 15-turn winding, I cut the leftovers in four equal lengths to make the "four-layer tape" used in the secondary. There was less than a foot of left-over tape after the transformer was completed. The Mylar tape I used was made by 3M and was 2-mil thick and 1-inch wide with adhesive backing. This thickness is sufficient for the voltages involved, provided that care is taken so that the Mylar isn't punctured by accident. If, like the author, you cut strips from a sheet of copper, you should file down the edges to remove burrs. See **Fig 17.36**. — *WR1B*]

Once the epoxy has had ample time to harden and has cooled, remove the rubber bands, the outer wood blocks, and the outer plastic wrapping (don't worry if it doesn't come off completely). Do not remove the plastic wrapping that separates the bobbin from the wood. The wrapped wooden core and epoxy-paper bobbin subassembly is now complete.

With a 60 mm (2.36 inch) length of #12 bare copper wire, wrap the end of one of your copper strips around the wire, so that the wire protrudes out from one side of the copper loop. Use a big soldering iron to flow some solder into the junction. Try to avoid getting solder on the outside, because this could later puncture the Mylar insulation. [I scraped the adhesive from the back of the sticky-backed tape where I soldered the wire. Otherwise, the solder won't stick to the back of the copper, and the layers may not have good conductivity between them. — *WR1B*]

Now place the copper wire on one of the narrow sides of the bobbin, so that the copper strip is centered on the width of the bobbin, leaving 3 mm (0.12 inches) room on each side. Seal the start of the copper strip to the bobbin with some thin adhesive tape. See **Fig 17.37**.

Position the start of a Mylar strip so that it covers all the copper and is centered on the bobbin, and then tape it in place. Wind 15 turns of this copper-Mylar sandwich as tightly as possible, keeping the Mylar aligned with the bobbin sides and the copper nicely centered. Don't lose your grip, or the whole thing will spring apart! If the copper strip is not long enough, fix everything with strong rubber bands or a clamp, and solder another copper strip to the end of the first one, allowing 2 mm of overlap. Before doing this, cut the first copper sheet so that the joint will be on one of the narrow sides of the bobbin, because here you have space, while the wide sides will have to fit inside the ferrite core's window. If the Mylar strip runs out, just use adhesive tape to add another strip. Make the overlap 5 mm to avoid risk of creepage between the sheets and also try to locate the joint on one of the narrow sides of the bobbin. See **Fig 17.38**.

When the 15 turns are complete, cut the copper strip so that the second terminal will be on the same narrow side of the bobbin as the first terminal. Solder the second terminal (another 60 mm piece of bare copper wire) to the strip, position it and wind three or four layers of Mylar to make the insulation safe between the primary and secondary. [I started my primary winding with the bulge of the wire on the corner, so that I was immediately winding along the wide side. When I finished the 15 turns, I positioned the end wire so it is on the narrow side, just beyond the corner of the long side. This way, the two bulges meet at the middle, but don't cross each other. — *WR1B*]

If you think this is a messy business, you are right. But it's fun too! The secondary is just a little bit messier: It is wound with a five-layer sandwich — four layers of copper and the Mylar topping layer. But it's only four turns total, so take a deep breath and do it. Solder the four copper strips together around a piece of #12 copper wire. Don't be overly worried if the outcome is not very clean; the author's was quite a mess too, yet it worked well on the first try. Just be sure you don't create sharp edges or pointed solder mounds, because these may damage the insulation. See Fig 17.38 for details.

Now position the start of your secondary conductor so the terminal wire will come out on the same side as those of the primary, but on the other narrow side of the coil assembly. The goal is to end up with a transformer with its primary leads on one extreme and the secondary on the other, and that will also fit the printed circuit board nicely. Wind two turns, solder the center tap wire between the four copper strips, wind the other two turns, solder the last terminal wire, and then wind a finishing layer of Mylar and fix it in place with adhesive tape. This finishes the worst part of making T1.

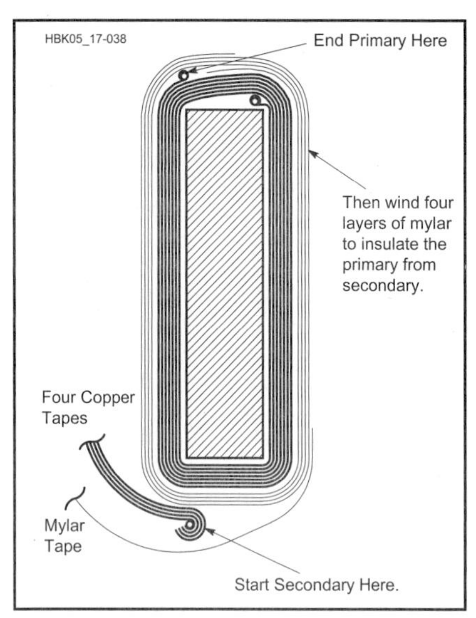

Fig 17.38 — Primary 15 turns on bobbin, with start of 4-turn, center-tapped secondary winding.

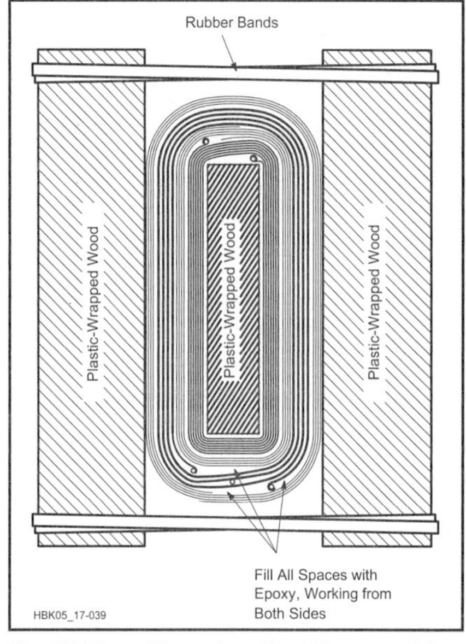

Fig 17.39 — Clamping the T1 assembly and filling with epoxy.

Fig 17.40 — Photo showing how the core halves must fit into the completed transformer after it is removed from the wooden block. You will have to file off the rough edges of epoxy to allow the cores to meet properly. The top E-cores have not been inserted into the bobbin yet. *(Photo by Larry Wolfgang.)*

What you have now is a springy, messy coil assembly that will fall apart if you let it go. You have to seal it, but this is easy to do. Temporarily hold things together with some stout rubber bands. Wrap the two wooden blocks, the same blocks you used to press together the bobbin, in plastic film. Place them against the sides of the coil assembly, and apply hard pressure, using a clamp or a lot of rubber bands, so that the long sides of the coil straighten out completely and any slack is displaced to the narrow sides. Now mix a fair quantity of epoxy glue, place the coil assembly so that the pins face up, and let the epoxy run into the coil. Continue supplying epoxy until it starts to set. If it drips out from the other side, no problem. (Just don't do this work over your best carpet!) When the epoxy doesn't flow any longer, turn the coil assembly over, mix a new batch of epoxy and fill the other side completely, forming a smooth surface. As the lower side is now sealed, the epoxy will not flow out there. When this epoxy has set, turn the assembly over again, mix some more epoxy and apply it to form a smooth surface there. The idea is to replace all the air between the copper and Mylar sheets with epoxy, and especially to fill the room left by the copper strip, which is narrower than the Mylar. This filling is necessary both for mechanical and for electric safety reasons. See **Fig 17.39**. [My wooden "screw clamps" worked well for applying strong even pressure to the sides. I don't think rubber bands would apply enough pressure to minimize the air space inside the transformer. — *WR1B*]

Now place the assembly in the oven again. Let the epoxy harden completely, then remove the coil from the oven, remove the clamp, rubber bands, wooden blocks, wooden core and all remains of plastic film. You will be surprised how your messy and springy assembly changed into a very robust, hard and strong coil. Now test-fit the ferrite cores. See **Fig 17.40**. Determine if they can be installed easily, so that each pair of facing E-cores comes together in intimate contact, without pressing on the winding. If everything is right, the winding should have some play room in the assembled core. But it is easy to get too much epoxy on the coil. If this happens, work the epoxy down with a file so that it doesn't disturb the ferrite. The ferrite core *must* close properly, otherwise you risk power transistor failure.

When the sides fit, prepare some more epoxy, apply a very thin layer to all contact faces of the ferrite cores and mount them onto the coil assembly. You can hold them in place with adhesive tape until the epoxy sets. Again, use the oven to speed up the hardening. The last thing you have to do is bend the copper wires into the proper shape to fit the printed circuit board holes. Be sure that on the secondary winding the center tap is actually in the center position. The polarity of the other pins doesn't matter. This completes the manufacture of T1. All the other transformers and coils are just child's play after making T1!

Making T2

The current sense transformer T2 has a lot of turns but they needn't be wound nicely side-by-side. You can use a winding machine with a turns counter, or you can just wind T2 by hand. Get some #36 or other thin enameled wire, solder the end to one of the outer pins of the EE24-25-B bobbin, and wind 100 turns. Don't worry if your winding is criss-crossed and ugly, and don't feel guilty if you lose count and wind a few turns more or less. As long as you don't overdo it, it will just affect the position of VR1 when you adjust the completed power supply later. Solder the wire to the center pin on the same side, then wind another 100 turns in the same sense. Solder to the other outer pin on the same bobbin side, and apply one or two layers of Mylar to protect the thin wire.

With #14 plastic insulated wire, wind one single turn over the Mylar, and solder the two ends to the two outer pins of the other side of the bobbin. It doesn't matter which end goes to which side. Install the EA77-250 core with a small amount of epoxy cement, and T2 is finished. [I used #14 AWG house wire here. The insulation made it a bit tight for the core, but it fit. — *WR1B*]

Making T3

T3 is made using the same kind of bobbin and core as T2. Wind 26 turns of #28 enameled wire. The 26 turns should fit nicely in a single layer. Study the schematic diagram to see how the windings connect to the bobbin pins. Bring the wire back to the starting side over the last half turn, for connection to the center-tap pin. Wind one layer of Mylar sheet, then put on the next 26 turns. Again, bring the wire back to the starting side over the last half turn for connection to the bobbin pin.

Wind 3 layers of Mylar tape, to insulate the primary and secondary properly. Wind 8 turns of #20 wire, and solder the ends to the bobbin pins. Look at the printed circuit board drawing to determine which wire is soldered to which pin. Wind a single layer of Mylar, then wind the other 8-turn winding over the first one. This will leave a space at one side of the bobbin big enough to take the single turn of #14 plastic insulated wire. This completes the assembly. See **Fig 17.41** for a cross-sectional view of the windings. Now glue the core in place with epoxy cement and T3 is finished.

Making L2

L2 is wound on an Amidon T-200-26 iron powder toroid core. As it is too difficult to bend thick wire through a toroid, and tape winding it is not practical either, the author chose to make this coil with 10 pieces of #16 enameled wire in parallel.

Cut the wires to about 1.5 m (59 inches) in length and lightly twist them together. Then insert the bundle into the core, and starting from the middle of the wire bundle, wind 8 turns, using half of the core's circumference. Now wind another 7 turns, starting from the middle toward the other end of the wire bundle. The 16th turn is the one you made when you inserted the wire bundle into the core to start.

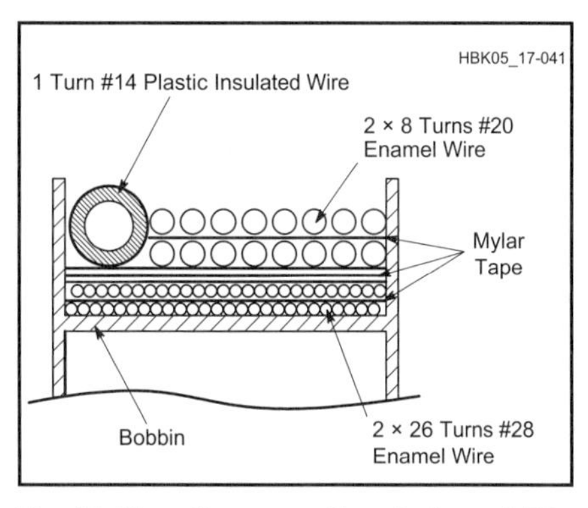

Fig 17.41 — Cross-sectional view of T3 (not to scale), showing distribution of windings.

Making L3

To make L3 you must first find a suitable rod. I used a part of an old ferrite antenna rod about 10 mm in diameter (0.39 inches) and 50 mm long (1.97 inches). (An Amidon number 33-050-200 rod is just the right size.) Wind 10 bifilar turns of #12 enameled wire. This wire is quite stiff, but it is still no problem to handle. You should wind the coil on a 12 mm (15/32 inches) drill bit, allow it to spring open and place it on the ferrite core. Otherwise you could crack the ferrite trying to wind directly on it. A tapered "drift punch" helps to open the turns just enough to fit the core. Fix the core to the winding with some epoxy. Bend the wires so that all four of them point down with the core pointing straight up. That's the position in which L3 is mounted on the PCB.

PUTTING IT TOGETHER

Install and solder all parts except for Q1, Q2, and D6 to D9. Before installing D1, fashion a simple heatsink from a 30 × 80 mm (1.18 × 3.15 inches) piece of 1 mm (0.039 inches) thick aluminum sheet, bent into U shape. Drill a hole and screw the rectifier bridge onto the heatsink together with a lock washer. Then solder D1 to the board.

The author made his own enclosure, using two 3-mm (0.12 inches) aluminum

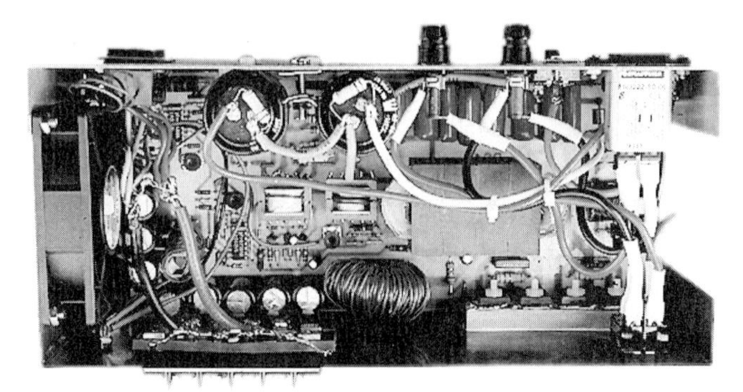

Fig 17.42 — Photo of top of PCB mounted in cabinet.

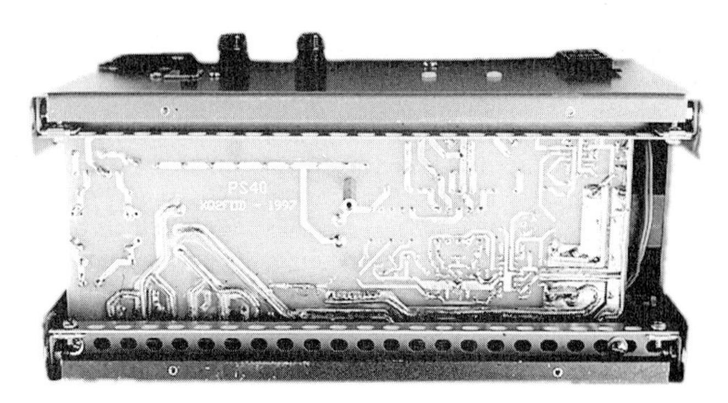

Fig 17.43 — Photo of bottom of PCB mounted in cabinet.

plates, measuring 300 × 120 mm (11.81 × 4.72 inches) for the front and rear walls. They are screwed to the fan, the PCB and to a 120-mm (4.72-inch) long spreader tube of 6-mm (0.24-inch) diameter, so that these parts become integral to the structure. The connections between the PCB, aluminum plates and fan were made with small pieces of 10 × 10 mm (0.39 × 0.39 inches) aluminum angle stock. The assembly is surprisingly rigid.

The top and bottom covers were made from 1 mm (0.04 inch) aluminum sheet and measure 126 × 300 mm (4.96 × 11.81 inches). The bottom cover has a hole for the PCB's center mount. The side covers were cut from wire mesh to allow unrestricted airflow, and measure 122 × 126 mm (4.80 × 4.96 inches). The panels are held together with 10 × 10 mm (0.39 × 0.39 inches) aluminum angle stock, running along all edges and held with small sheet-metal screws. These covers are not installed until the power supply is complete, tested and adjusted.

All the panels were painted flat black on the outside, which looks nice together with the anodized aluminum angle stock. The edges and insides were kept free of paint, in order to get proper electrical contact between the panels for good shielding.

The version made by WR1B (see **Fig 17.42** and **Fig 17.43**) used a Hammond Manufacturing ventilated, low-profile instrument case, catalog number 1426Y-B. This is a rugged case that also looks very nice. Larry mounted the circuit board inside the case using a pair of steel mounting rails, also from Hammond, catalog number 1448R12.

The components external to the PCB (P1, SW1, C3, the LED and the output screw terminal block) are mounted to the front and rear panels. Q1 and Q2 are mounted to the rear panel, using M3 Nylon screws and 3 mm (0.12-inch) thick ceramic insulators. These thick insulators were used not only for safety reasons but also because they reduce the capacitive coupling of the transistors to the enclosure. Do not use metal screws with plastic washers, because this approach does not give enough safety margin to operate at the input line voltage. [The author's junk-box ceramic insulators proved difficult to duplicate for the supply we built in the ARRL Lab. Equivalent new parts would have nearly doubled the cost of the supply! Instead, for good heat-transfer properties, we used thin rubber insulators manufactured by Wakefield Engineering as PN 175-6-250-P, available from Newark Electronics as PN 46F7884. Individual aluminum spacers milled from aluminum blocks were used between Q1, Q2 and the Schottky diodes and the metal chassis. Care must be taken to make sure the surfaces of the spacers are parallel and free of burrs to ensure low thermal resistance.]

The Schottky diodes are mounted using the same kind of insulators and screws, but there is a heat spreader made from 6-mm (0.24-inch) aluminum plate between those insulators and the case. All surfaces requiring thermal contact are covered with heat-transfer compound before assembly. When installing the diodes and transistors, first do all the mechanical assembly, and then solder the pins. Otherwise you could stress them too much while fastening the screws.

All wire connections are made next, and the output filter is assembled by sliding the ferrite beads over the output cables and soldering the bypass capacitors C25 through C30. Be sure to use thick wire for the output. A 40-A, continuous-duty current is no joke.

The tracks on the PCB cannot be trusted to carry 40 A without some help. Use a big soldering iron (100 to 150 W) to solder lengths of #12 bare copper wire cut and bent to fit the shape of all the high-current paths. To prevent any failures due to vibration from the fan, place some drops of hot-melt glue anywhere a wire is connected to the board. Hot-melt glue is also excellent for fixing anything that would otherwise rattle, like ferrite beads.

Testing and Adjusting

Make sure you do a thorough visual check. Set the three potentiometers to mid position. Check that there is no continuity between the ac input and ground, between the ac input and the dc output, or between the dc output and chassis ground.

Connect a variable voltage supply (you need 12 to 15 V for the tests) to the output leads, without plugging the switcher into the ac line. You should see the LED light up. Change the voltage fed into your project to see how the LED changes color. If you have a dual-channel oscilloscope, connect its two channels to the base-emitter junctions of the power transistors. [Since you are not connected to the ac power line, you will not be grounding it through the oscilloscope's ground leads connected to the emitter leads. — *Ed.*] With the external voltage at about 12 V, you should see small pulses. As you increase the voltage, the pulses will suddenly disappear. You can pre-adjust VR2 by setting your lab power supply to exactly 13.8 V and then setting VR2 to where the pulses just disappear.

Now it's time to start up the switcher. Remove your lab supply and the oscilloscope leads, and connect the supply to the ac line in series with a 60-W light bulb. This will avoid most or all damage if something is really wrong. Connect a voltmeter to the output and switch on your supply. If everything is right, the bulb will light up, then slowly dim while the power supply starts up and delivers about 13.8 V.

Now, connect a load of about 2 A to the output — a car brake-light bulb makes a good load. At 2-A output, the bulb in the ac line will probably glow, with 13.8-V dc at the output. If everything is okay so far, now comes the big moment. Remove the series bulb from the ac circuit. Startup of the supply should be fast and you can now connect a heavier load to it. With a load of 2 to 10 A connected (the value is uncritical, given the good regulation of this supply), adjust VR2 so that you have exactly 13.8 V at the output.

Next adjust the current shutdown point. For this you need a load that can handle 40 A. You could make one by connecting a lot of car headlamps in parallel or you could use some resistance wire to build a big power resistor. The author made a 13.8-V, 550-W heater for his supply. Connect the load and adjust VR1 so that the output voltage is just at the limit of shutting down.

The last adjustment is for the fan trigger point. Connect a 65-W car headlamp or similar load that consumes about 5 A. Let the supply run for several minutes, then move VR3 to the point where the fan switches on. Now check out the trigger function by changing the load several times between about 2 and 10 A. The fan should switch off and on between 30 to 60 seconds after each load change. You may have to readjust VR3 until you get the fan to switch on at no more than 7-A continuous load and switch off at about 4 A.

And If It Doesn't Work?

If you used substitute parts for the magnetic cores and made a bad choice, the results could be dramatic. If either T1 or L2 saturates, the power transistors could burn out before the fuse has a chance to open. The protective light bulb in the ac line will avoid damage in this case, so by all means use that bulb for initial testing!

Another possible error is reversing the phase of a winding in T3. If you get one of the 8-turn windings reversed, the results will be explosive unless you have the light bulb in series. If you reverse the 1-turn winding, the power supply will simply not start.

28-V, HIGH-CURRENT POWER SUPPLY

Many modern high-power transistors used in RF power amplifiers require 28-V dc collector supplies, rather than the traditional 12-V supply. By going to 28 V (or even 50 V), designers significantly reduce the current required for an amplifier in the 100-W or higher output class. The power supply shown in **Fig 17.44** through **Fig 17.48** is conservatively rated for 28 V at 10 A (enough for a 150-W output amplifier) — continuous duty! It was designed with simplicity and readily-available components in mind. Mark Wilson, K1RO, built this project in the ARRL lab.

CIRCUIT DETAILS

The schematic diagram of the 28-V supply is shown in **Fig 17.45**. T1 was designed by Avatar Magnetics specifically for this project. The primary requires 120-V ac, but a dual-primary (120/240 V) version is available. The secondary is rated for 32 V at 15 A, continuous duty. The primary is bypassed by two 0.01-µF capacitors and protected from line transients by an MOV.

U1 is a 25-A bridge module available

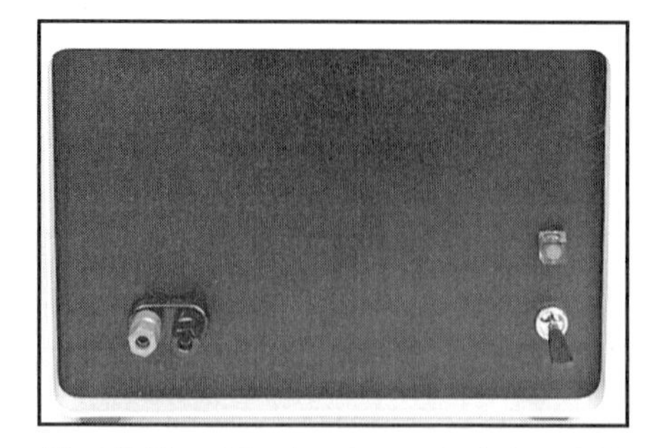

Fig 17.44 — The front panel of the 28-V power supply sports only a power switch, pilot lamp and binding posts for the voltage output. There is room for a voltmeter, should another builder desire one.

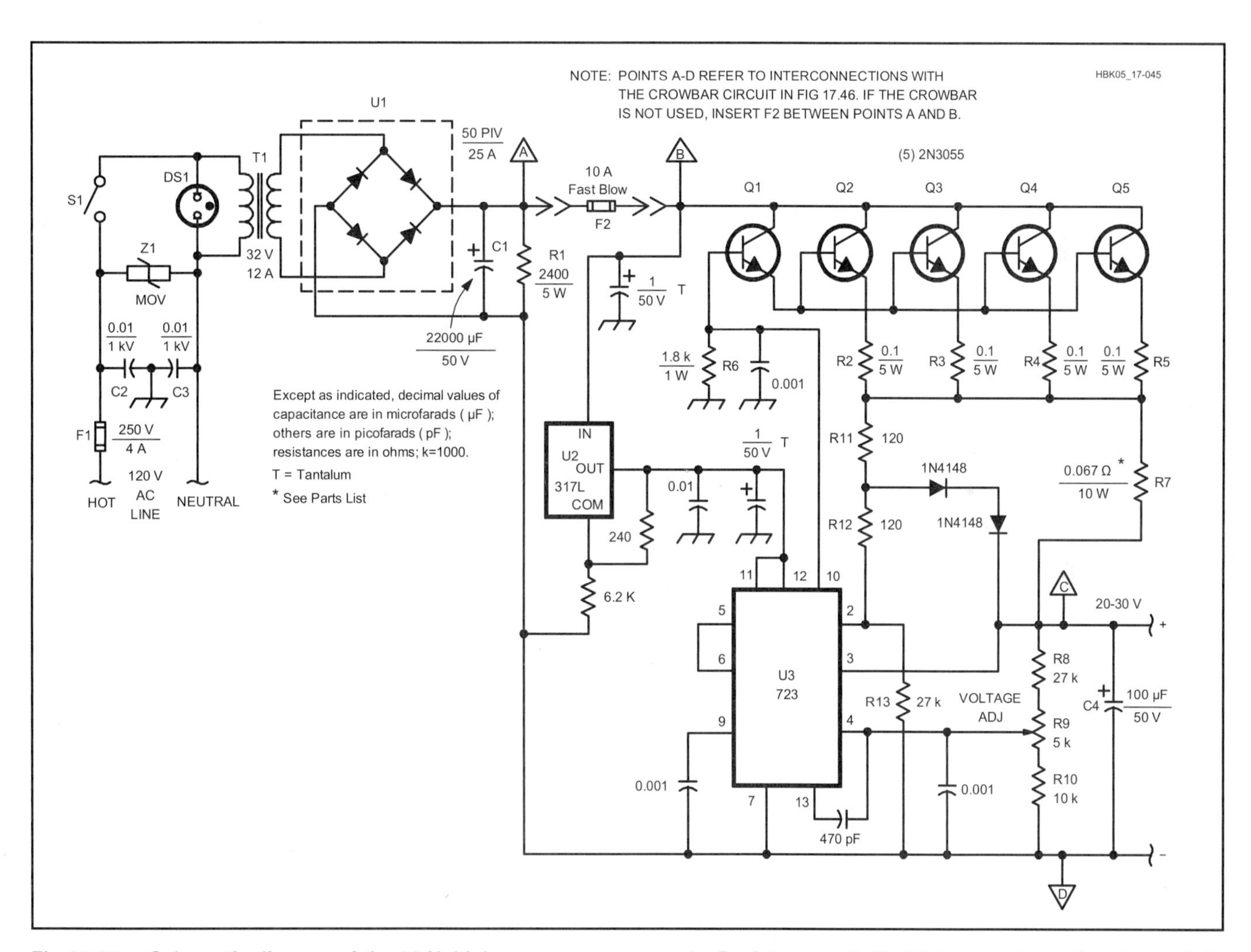

Fig 17.45 — Schematic diagram of the 28-V, high-current power supply. Resistors are ¼-W, 5% types unless otherwise noted. Capacitors are disc ceramic unless noted; capacitors marked with polarity are electrolytic. Parts numbers given in parentheses that are preceded by the letters RS are RadioShack catalog numbers.

C1 — Electrolytic capacitor, 22000 µF, 50 V (Mallory CGS223U050X4C or equiv., available from Mouser Electronics)
C2, C3 — AC-rated bypass capacitors.
C4 — Electrolytic capacitor, 100 µF, 50 V
DS1 — Pilot lamp, 120-V ac (RS 272-705)
Q1-Q5 — NPN power transistor, 2N3055 or equiv. (RS 276-2041)
R2-R5 — Power resistor, 0.1 Ω, 5 W (or greater), 5% tolerance
R7 — Power resistor, 0.067 Ω, 10 W (or greater), made from three 0.2-Ω, 5-W resistors in parallel
T1 — Power transformer. Primary, 120-V ac; secondary, 32 V, 15 A. (Avatar Magnetics AV-430 or equiv. Dual primary version is part #AV-431. Available from Avatar Magnetics.)
U1 — Bridge rectifier, 50 PIV, 25 A (RS 276-1185)
U2 — Three-terminal adjustable voltage regulator, 100 mA (LM-317L or equiv.). See text
U3 — 723-type adjustable voltage regulator IC, 14-pin DIP package (LM-723, MC1723, etc. RS 276-1740)
Z1 — 130-V MOV (RS 276-570)

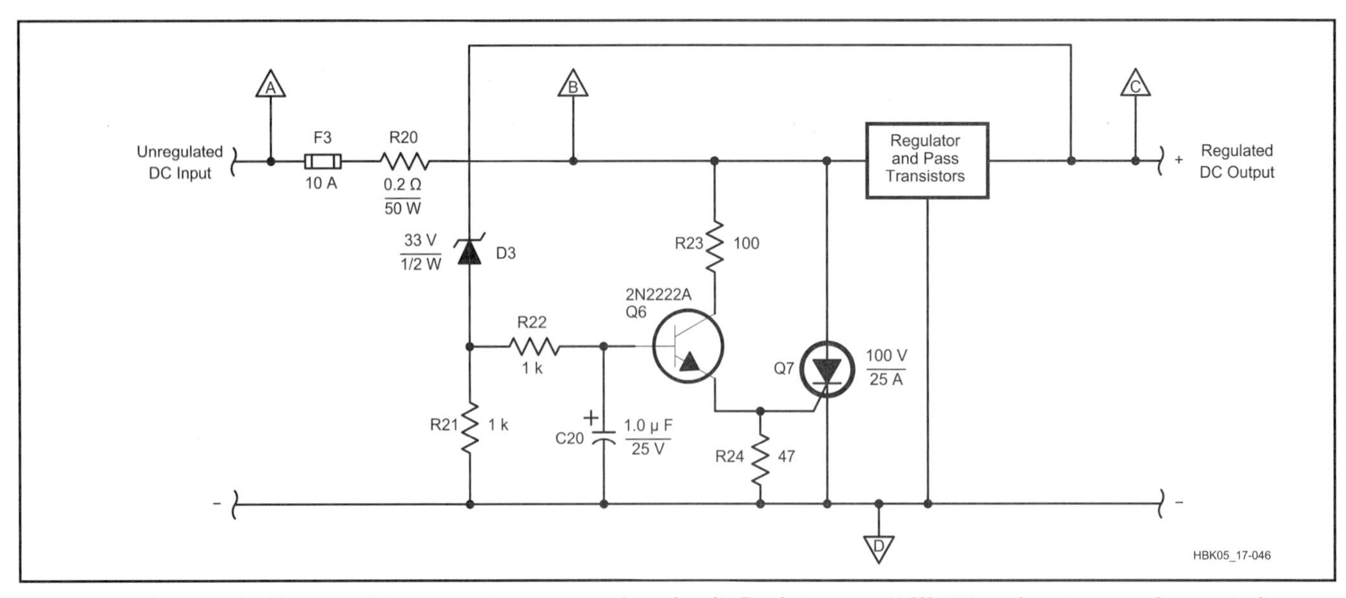

Fig 17.46 — Schematic diagram of the overvoltage protection circuit. Resistors are ¼-W, 5% carbon types unless noted.
D3 — 33 V, ½-W Zener (NTE 5036A or equiv.)
Q6 — NPN Transistor (2N2222A or equiv.)
Q7 — 100 V, 25A SCR (NTE 5522 or equiv.)

from RadioShack or a number of other suppliers. It requires a heat sink in this application. Filter capacitor C1 is a computer-grade 22,000-μF electrolytic. Bleeder resistor R1 is included for safety because of the high value of C1; bleeder current is about 12 mA.

There is a tradeoff between the transformer secondary voltage and the filter-capacitor value. To maintain regulation, the minimum supply voltage to the regulator circuitry must remain above approximately 31 V. Ripple voltage must be taken into account. If the voltage on the bus drops below 31 V in ripple valleys, regulation may be lost.

In this supply, the transformer secondary voltage was chosen to allow use of a commonly available filter value. The builder found that 50-V electrolytic capacitors of up to about 25,000 μF were common and the prices reasonable; few dealers stocked capacitors above that value, and the prices increased dramatically. If you have a larger filter capacitor, you can use a transformer with a lower secondary voltage; similarly, if you have a transformer in the 28- to 35-V range, you can calculate the size of the filter capacitor required. Equation 3, earlier in this chapter in the Filtration section, shows how to calculate ripple for different filter-capacitor and load-current values.

The regulator circuitry takes advantage of commonly available parts. The heart of the circuit is U3, a 723 voltage regulator IC. The values of R8, R9 and R10 were chosen to allow the output voltage to be varied from 20 to 30 V. The 723 has a maximum input voltage rating of 40 V,

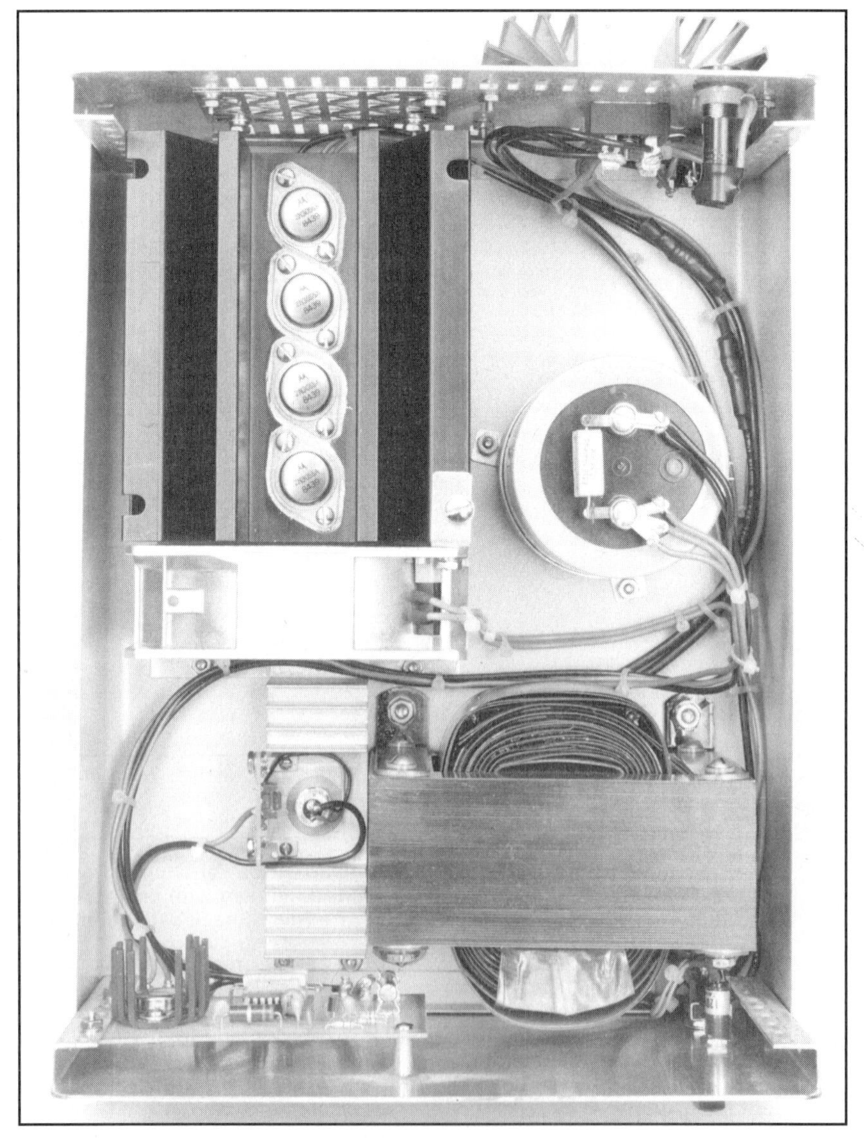

Fig 17.47 — Interior of the 28-V, high-current power supply. The cooling fan is necessary only if the pass transistors and heat sink are mounted inside the cabinet. See text.

somewhat lower than the filtered bus voltage. U2 is an adjustable 3-terminal regulator; it is set to provide approximately 35 V to power U3. U3 drives the base of Q1, which in turn drives pass transistors Q2-Q5. This arrangement was selected to take advantage of common components. At first glance, the number of pass transistors seems high for a 10-A supply. Input voltage is high enough that the pass transistors must dissipate about 120 W (worst case), so thermal considerations dictate the use of four transistors. See the **Real-World Component Characteristics** chapter for a complete discussion of thermal design. If you use a transformer with a significantly different secondary potential, refer to the thermal-design tutorial to verify the size heat sink required for safe operation.

R9 is used to adjust supply output voltage. Since this supply was designed primarily for 28-V applications, R9 is a "set and forget" control mounted internally. A 25-turn potentiometer is used here to allow precise voltage adjustment. Another builder may wish to mount this control, and perhaps a voltmeter, on the front panel to easily vary the output voltage.

The 723 features current foldback if the load draws excessive current. Foldback current, set by R7, is approximately 14 A, so F2 should blow if a problem occurs. The output terminals, however, may be shorted indefinitely without damage to any power-supply components.

If the regulator circuitry should fail, or if a pass transistor should short, the unregulated supply voltage will appear at the output terminals. Most 28-V RF transistors would fail with 40-plus volts on the collector, so a prospective builder might wish to incorporate the overvoltage protection circuit shown in **Fig 17.46** in the power supply. This circuit is optional. It connects across the output terminals and may be added or deleted with no effect on the rest of the supply. If you choose to use the "crowbar," make the interconnections as shown. Note that R20 and F3 of Fig 17.46 are added between points A and B of Fig 17.45. If the crowbar is not used, connect F2 between points A and B of Fig 17.45.

The crowbar circuit functions as follows: The Zener-hold off diode (D3) blocks the positive regulated voltage from appearing at the base of Q6 until its avalanche voltage is exceeded. In the case of the device selected, this voltage level is 33 V, which provides for small overshoots that might occur with sudden removal of the output load (switching off a load, for instance).

In the event the output voltage exceeds 33 V, D3 will conduct, and forward bias Q6 through R22 and C20, which eliminates short duration transients and noise. When Q6 is biased on, trigger current flows through R23 and Q6 into the gate of SCR Q7, turning it on and shorting the raw dc source, forcing F3 to blow. Since some SCRs have a tendency to turn themselves on at high temperature, resistor R24 shunts any internal leakage current to ground.

CONSTRUCTION

Fig 17.47 shows the interior of the 28-V supply. It is built in a Hammond 1401K enclosure. All parts mount inside the box. The regulator components are mounted on a small PC board attached to the rear of the front panel. See **Fig 17.48**. Most of the parts were purchased at local electronics stores or from major national suppliers. Many parts, such as the heat sink, pass transistors, 0.1-Ω power resistors and filter capacitor can be obtained from scrap computer power supplies found at flea markets.

Q2-Q5 are mounted on a Wakefield model 441K heat sink. The transistors are mounted to the heat sink with insulating washers and thermal heat-sink compound to aid heat transfer. RadioShack TO-3 sockets make electrical connections easier. The heat-sink surface under the transistors must be absolutely smooth. Carefully deburr all holes after drilling and lightly sand the edges with fine emery cloth.

A five-inch fan circulates air past the heat sink inside the cabinet. Forced-air cooling is necessary only because the heat sink is mounted inside the cabinet. If the heat sink was mounted on the rear panel with the fins vertical, natural convection would provide adequate cooling and no fan would be required.

U1 is mounted to the inside of the rear panel with heat-sink compound. Its heat sink is bolted to the outside of the rear panel to take advantage of convection cooling.

U2 may prove difficult to find. The 317L is a 100-mA version of the popular 317-series 1.5-A adjustable regulator. The 317L is packaged in a TO-92 case, while the normal 317 is usually packaged in a larger TO-220 case. Many electronics suppliers sell them, and RCA SK7644 or Sylvania ECG1900 direct replacements are available from many local electronics shops. If you can't find a 317L, you can use a regular 317 (available from RadioShack, among others).

R7 is made from two 0.1-Ω, 5-W resistors connected in parallel. These resistors get warm under sustained operation, so they are mounted approximately 1/16 inch above the circuit board to allow air to circulate and to prevent the PC board from becoming discolored. Similarly, R6 gets warm to the touch, so it is mounted away from the board to allow air to circulate. Q1 becomes slightly warm during sustained operation, so it is mounted to a small TO-3 PC board heat sink.

Not obvious from the photograph is the use of a single-point ground to avoid ground-loop problems. The PC-board ground connection and the minus lead of the supply are tied directly to the minus terminal of C1, rather than to a chassis ground.

The crowbar circuit is mounted on a small heat sink near the output terminals. Q7 is a

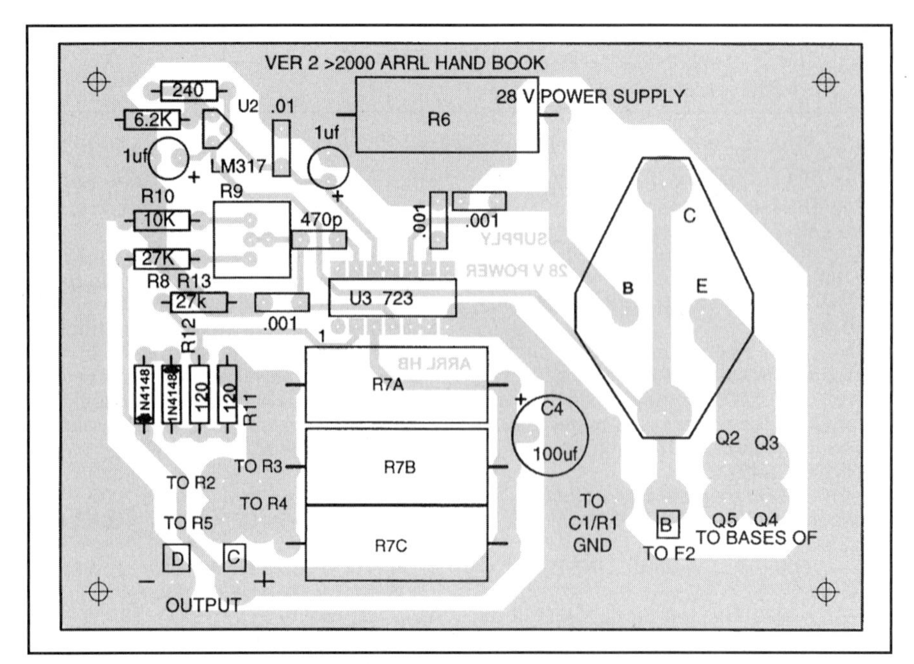

Fig 17.48 — Parts placement diagram for the 28-V power supply. A full-size etching pattern is in the **Templates** section of the *Handbook* CD-ROM.

stud-mount SCR and is insulated from the heat sink. The other components are mounted on a small circuit board attached to the heat sink with angle brackets.

Although the output current is not extremely high, #14 or #12 wire should be used for all high-current runs, including the wiring between C1 and the collectors of Q2-Q5; between R2-R5 and R7; between F2 and the positive output terminal; and between C1 and the negative output terminal. Similar wire should be used between the output terminals and the load.

TESTING

First, connect T1, U1 and C1 and verify that the no-load voltage is approximately 44 V dc. Then, connect unregulated voltage to the PC board and pass transistors. Leave the gate lead of Q6 disconnected from pin 8 of U4 at this time. You should be able to adjust the output voltage between approximately 20 and 30 V. Set the output to 28 V.

Next, short the output terminals to verify that the current foldback is working. Voltage should return to 28 when the shorting wire is disconnected. This completes testing and setup.

The supply shown in the photographs dropped approximately 0.1 V between no load and a 12-A resistive load. During testing in the ARRL lab, this supply was run for four hours continuously with a 12-A resistive load on several occasions, without any difficulty.

A COMMERCIAL-QUALITY, HIGH-VOLTAGE POWER SUPPLY

This two-level, high-voltage power supply was designed and built by ARRL *Handbook* Editor Dana G. Reed, W1LC. It was designed primarily for use with an RF power amplifier using a triode in class AB_2 grounded-grid operation. The supply is rated at a continuous output current of 1.5 A, and will easily handle intermittent peak currents of 2 A. The 12-V control circuitry, and the low-tap setting of the plate transformer secondary, can both be used with the N7ART 2-meter amplifier. See the **RF Power Amplifiers** chapter for details.

The step-start circuit is straightforward and ensures that the rectifier diodes are current-limited when the power supply is first turned on. A 6-kV meter is used to monitor high-voltage output.

Fig 17.49 is a schematic diagram of the bi-level supply. An ideal power supply for

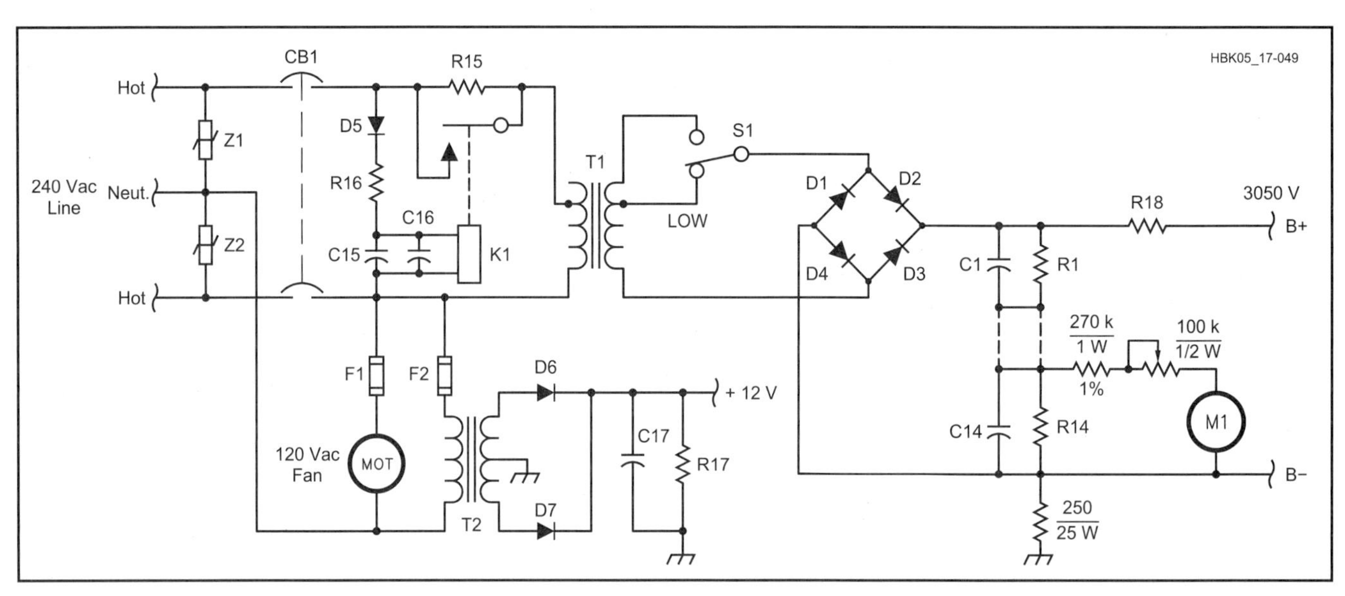

Fig 17.49 — Schematic diagram of the 3050-V/5400-V high-voltage power supply.

C1 to C14 — Electrolytic capacitor, 800 µF, 450 V (Mallory CGS801T450V4L or equiv.)
C15-C16 — Electrolytic capacitor, 4700 µF, 50 V
C17 — Electrolytic capacitor, 1000 µF, 50 V
CB1 — 20-A hydraulic/magnetic circuit breaker (Potter and Brumfield W68X2Q12-20 or equiv). 40-A version required for commercial applications/service (Potter and Brumfield W92X112-40)
D1-D4 — Commercial diode block assembly: K2AW HV14-1, from K2AW's Silicon Alley
D5 — 1000-PIV, 3-A diode, 1N5408 or equiv
D6-D7 — 200-PIV, 3-A diode, 1N5402 or equiv
F1-F2 — Fuse, 0.5 A, 250 V (Littelfuse® 313 Series, 3AG Glass Body or equiv)
K1 — DPDT power relay, 24-V dc coil; both poles of 240-V ac/25-A contacts in parallel (Potter & Brumfield PRD-11DY0-24 or equiv)
M1 — High-voltage meter, 6-kV dc full scale. (Important: Use a 1-mA or smaller meter movement to minimize parallel-resistive loading at R14. Also, select series meter-resistor and adjustment-potentiometer values to calibrate your specific meter. Values shown are for a 1-mA meter movement.)
MOT1 — Cooling fan, 119mm, 110-120-V ac, 30-60 CFM, (EBM 4800Z or equiv)
R1 to R14 — Bleeder resistor, 100 kΩ, 3 W, Metal Oxide Film (MOF)
R15 — Power resistor, 50 Ω, 100 W
R16 — Power resistor, 1.2 kΩ, 25 W
R17 — Power resistor, 30 Ω, 25 W
R18 — Power resistor, 20 Ω, 50 W
S1 — Ceramic rotary, 2-pos. tap-select switch (optional). Voltage rating between tap positions should be at least 2.5 kV. Mount switch on insulated or ungrounded material such as a metal plate on standoff insulators, or an insulating plate, and use only a *non-conductive* or otherwise *electrically-isolated* shaft through the front panel for safety
T1 — High-voltage plate transformer, 220/230-V primary, 2000/3500-V, 1.5-A CCS JK secondary. (Peter W. Dahl Company, Inc., Hipersil C-Core.) Primary 220-V tap fed with nominal 240-V ac line voltage to obtain modest increase in specified secondary voltage levels
T2 — Power transformer, 120-V Pri., 18-VCT, 2-A Sec. (Hi-Q Magnetics; Mouser 41FJ020)
Z1-Z2 — 130-V MOV (AVX VE24M00131K or equiv)

a high-power linear amplifier should operate from a 240-V circuit, for best line regulation. A special, hydraulic/magnetic circuit breaker also serves as the disconnect for the plate transformer primary. Don't substitute a standard circuit breaker, switch or fuses for this breaker; fuses won't operate quickly enough to protect the amplifier or power supply in case of an operating abnormality. The bleeder resistors are each 100 kΩ, 3 W and of stable MOF design. These resistors are wired across each of the 14 capacitors to equalize voltage drops in the series-connected bank. This choice of bleeder resistor value provides a lighter load (less than 25 watts total under high-tap output) and benefits mainly the capacitor-bank filter by yielding much less heat as a result. A reasonable, but longer bleed-down time to fully discharge the capacitors results — about nine minutes after power is removed. A small fan is included to remove any excess heat from the power supply cabinet during operation.

POWER SUPPLY CONSTRUCTION

The power supply can be built into a 23 ½ × 10 ¾ × 16-inch cabinet. The plate transformer is quite heavy at 67 lbs, so use ⅛-inch aluminum for the cabinet bottom and reinforce it with aluminum angle for extra strength and stability. The capacitor bank will be sized for the specific capacitors used. This project employed ⅜-inch thick polycarbonate for reasonable mechanical stability and excellent high-voltage isolation. The full-wave bridge consists of four commercial diode block assemblies.

POWER SUPPLY OPERATION

When the front-panel breaker is turned on, a single 50-Ω, 100-W power resistor limits primary inrush current to a conservative value as the capacitor bank charges. After approximately two seconds, step-start relay K1 actuates, shorting the 50-Ω resistor and allowing full line voltage to be applied to the plate transformer. No-load output voltages under low- and high-tap settings as configured and shown in Fig 17.49 are 3050 V and 5400 V, respectively. Full-load levels are somewhat lower, approximately 2800 V and 4900 V. If a tap-select switch is used as described in the schematic parts list, it should only be switched when the supply is off.

THE MICRO M+

The Micro M+ is an ideal photovoltaic (PV) controller for use at home or in the field. It's an easy-to-build, one-evening project even a beginner can master. This project was designed by Mike Bryce, WB8VGE. An earlier charge controller called the "Micro M" proved to be a very popular project.[1]

Hams really do like to operate their rigs from solar power. Many have found solar power to be very addictive. I had dozens of requests for information on how to increase the current capacity of the original "Micro M" controller. The Micro M would handle up to 2 A of current. I wanted to improve the performance of the Micro M while I was at it. Because the Micro M switched the negative lead of the solar panel on and off, that lead had to be insolated from the system ground. While that's not a problem with portable use, it may cause trouble with a home station where all the grounds should be connected. Here's what I wanted to do:

- Reduce the standby current at night
- Increase current handling capacity to 4 A
- Change the charging scheme to high (positive) side switching
- Improve the charging algorithm
- Keep the size as small as possible, but large enough to easily construct.

I called the end result the Micro M+. You can assemble one in about an hour. Everything mounts on one double-sided PC board. It's small enough to mount inside your rig yet large enough so you won't misplace it! You can stuff four of them in your shirt pocket! And, you need not worry about RFI being generated by the Micro M+. It's completely silent and makes absolutely no RFI!

Fig 17.50 — This photo shows the Micro M+ charge controller circuit board. Leads solder to the board and connect to a solar panel and to the battery being charged.

The Micro M+ will handle up to 4 A of current from a solar panel. That's equal to a 75-W solar panel.[2] I've reduced the standby current to less than one milliamp. I've also introduced a new charging algorithm to the Micro M+. All the current switching is done on the positive side. Now, you can connect the photovoltaic array, battery and load grounds together.

A complete kit of parts is available as well as just the PC board.[3] The Micro M+ is easy to build, making it a perfect first time project.

HOW IT WORKS

Fig 17.50 shows the complete Micro M+. **Fig 17.51** shows the schematic diagram. Let's begin with the current handling part of the Micro M+. Current from the solar panel is controlled by a power MOSFET. Instead of using a common N-channel MOSFET, however, the Micro M+ uses an International Rectifier IRF4905 P-channel MOSFET. This P-channel FET has a current rating of 64 A with an RDS_{on} of 0.02 Ω. It comes in a TO-220 case. Current from the solar panel is routed directly to the MOSFET source lead.

N-channel power MOSFETs have very low RDS_{on} and even lower prices. To switch current on and off in a high-side application, though, the gate of an N-channel MOSFET must be at least 10 V higher than the rail it is switching. In a typical 12-V system, the gate voltage must be at least 22 V to ensure the MOSFET is turned completely on. If the gate voltage is less than that required to fully enhance the MOSFET, it will be almost on and somewhat off (the MOSFET is operating in its linear region). Hence, the device will likely be destroyed at high current levels.

Normally, to produce this higher gate voltage, some sort of oscillator is used to charge a capacitor via a voltage doubler. This charge pump generates harmonics that may ride on the dc flowing into the battery under charge. Normally, this would not cause any problem, and in most cases, a filter or two on the dc bus will eliminate most of the harmonics generated. Even the best filter won't get rid of all the harmonics, however. To compound the problem, long wire runs to and from the solar panels and batteries act like antennas.

The P-channel MOSFET eliminates the need for a charge pump altogether. To turn on a P-channel MOSFET, all we have to do is pull the gate lead to ground! Since the Micro M+ does not have a charge pump, it generates NO RFI!

Now, you may be wondering if the P-channel MOSFET is so great, why have you not seen them in applications like this before? The answer is twofold. First, the RDS_{on} of a P-channel MOSFET has always been much higher than its N-channel cousin. Several years ago, a P-channel

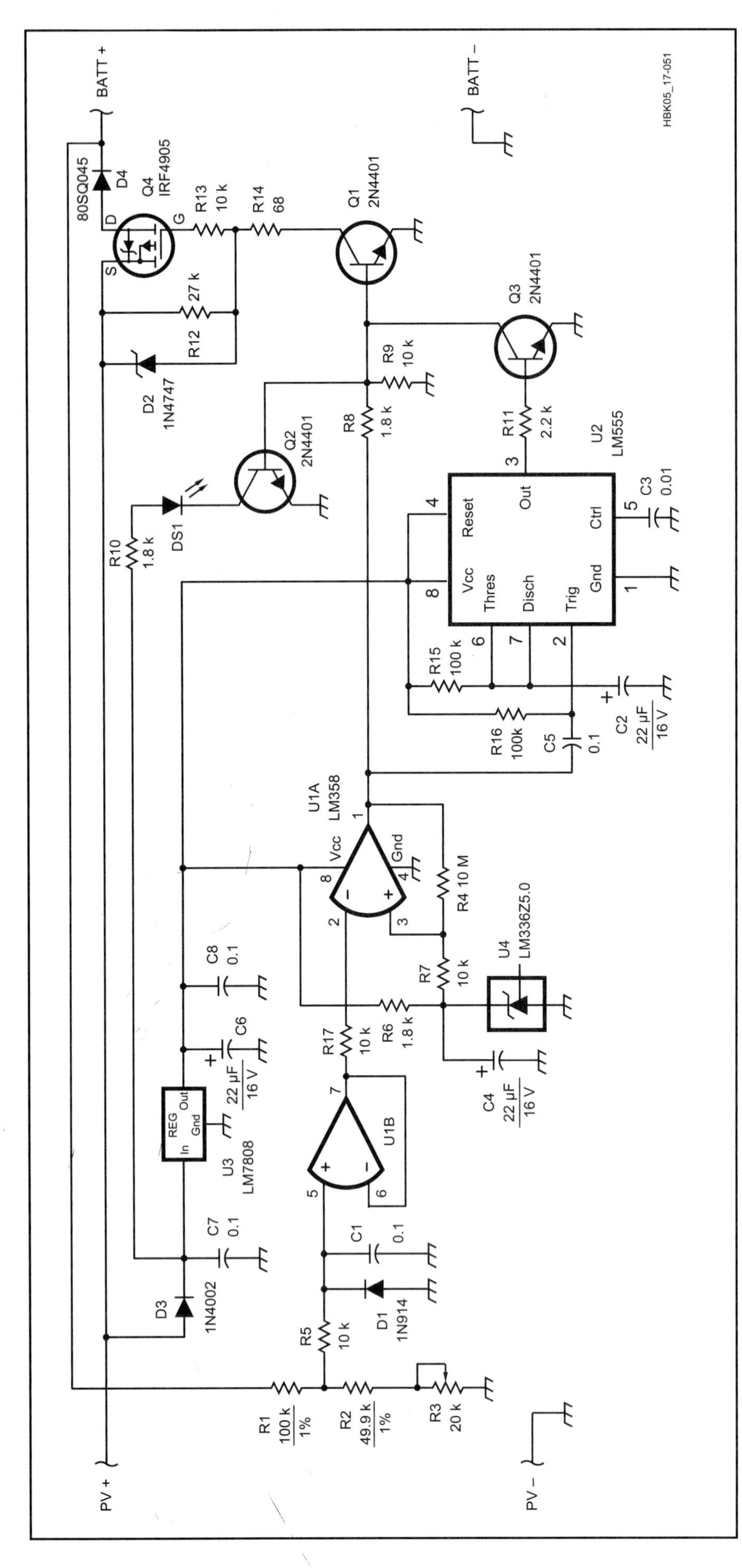

Fig 17.51 — The schematic diagram of the Micro M+ charge controller.
C1, C5, C7, C8 — 0.1 µF
C2, C4, C6 — 22 µF, 16 V electrolytic
C3 — 0.01 µF
D1 — 1N914, small signal silicon switching diode
D2 — 1N4747, 20-V, 1-W Zener
D3 — 1N4002, silicon rectifier diode
D4 — 80SQ045, 45-V, 8A Schottky diode
DS1 — LED, junkbox variety
Q1, Q2, Q3 — 2N4401 NPN small-signal transistor (2N2222 or 2N3904 will also work.)
Q4 — IRF4905 P channel MOSFET in TO-220 case. You will also need a small clip-on heat sink for this case.
R1 — 100 kΩ, 1%
R2 — 49.9 kΩ, 1%
R3 — 20 kΩ trimmer
U1 — LM358AN, Dual op-amp
U2 — LM555AN timer
U3 — LM78L08, 8-V regulator
U4 — LM336Z-5.0, 5.0-V Zener diode in TO-92 case. The adjust terminal allows control of the temperature coefficient and voltage over a range. The adjust terminal is not used for the Micro M+.

MOSFET with an RDS_{on} of 0.12 Ω was considered very low. At that time an N channel MOSFET had an RDS_{on} of 0.009 Ω. Suppose you want to control 10 A of current from your solar panel. Using the N-channel MOSFET above we find the MOSFET will dissipate less than a watt of power. On the other hand, the P-channel MOSFET will dissipate 12 W of power! Current generated by our solar panels is way too precious (and expensive) to have 12 W go up as heat from the charge controller.

The second factor was price. The P-channel MOSFET I described above would have easily sold for $19 each. The N-channel device would have been a few dollars.

More recently, the RDS_{on} of a typical P-channel MOSFET has fallen to 0.028 Ω. The price, while still a bit expensive, has dropped to about $8 each.

With the P channel MOSFET controlling the current, diode D4 — a 80SQ045 Schottky — prevents battery current from flowing into the solar panel at night. This diode also provides reverse polarity protection to the battery in the event you connect the solar panel backwards. This protects the expensive P channel MOSFET.

Zener diode D2, a 1N4747, protects the gate from damage due to spikes on the solar panel line. Resistor R12 pulls the gate up, ensuring the power MOSFET is off when it is supposed to be.

THE MICRO M+ LIKES TO SLEEP

The Micro M+ never draws current from the battery. The solar panel provides all the power the Micro M+ needs, which means the Micro M+ goes to sleep at night. When the sun rises, the Micro M+ will start up again. As soon as the solar panel is producing enough current and voltage to start charging the battery, the Micro M+ will pass current into the battery.

To reduce the amount of stand-by current, diode D3 passes current from the solar panel to U3, the voltage regulator. U3, an LM78L08 regulator, provides a steady +8 V to the Micro M+ controller. Bypass capacitors, C6, C7 and C8 are used to keep everything happy. As long as there is power being produced by the solar panel, the Micro M+ will be awake. At sun down, the Micro M+ will go to sleep. Sleep current is on the order of less than 1 mA!

BATTERY SENSING

The battery terminal voltage is divided down to a more usable level by resistors, R1, R2 and R3. Resistor R3, a 20 kΩ trimmer, sets the state-of-charge for the Micro M+. A filter consisting of R5 and C1 helps keep the input clean from noise picked up by the wires to and from the solar panel. Diode D1 protects the op-amp input in case the battery sense line was connected backwards.

An LM358 dual op-amp is used in the Micro M+. One section (U1B) buffers the divided battery voltage before passing it along to the voltage comparator, U1A. Here the battery sense voltage is compared to the reference voltage supplied by U4. U4 is an LM336Z-5.0 precision diode. To prevent U1A from oscillating, a 10-MΩ resistor is used to eliminate any hysteresis.

As long as the voltage of the battery under charge is below the reference point, the output of U1A will be high. This saturates transistors Q1 and Q2. Q2 conducts and lights LED DS1, the CHARGING LED. Q1, also fully saturated, pulls the gate of the P channel MOSFET to ground. This effectively turns on the FET, and current flows from the solar panel into the battery via D4.

As the battery begins to take up the charge, its terminal voltage will increase. When the battery reaches the state-of-charge set point, the output of U1A goes low. With Q1 and Q2 now off, the P channel MOSFET is turned off, stopping all current into the battery. With Q2 off, the CHARGING LED goes dark.

Since we have eliminated any hysteresis in U1A, as soon as the current stops, the output of U1A pops back up high again. Why? Because the battery terminal voltage will fall back down as the charging current is removed. If left like this, the Micro M+ would sit and oscillate at the state-of-charge set point.

To prevent that from happening, the output of U1A is monitored by U2, an LM555 timer chip. As soon as the output of U1A goes low, this low trips U2. The output of U2 goes high, fully saturating transistor Q3. With Q3 turned on, it pulls the base of Q1 and Q2 low. Since both Q1 and Q2 are now deprived of base current, they remain off.

With the values shown for R15 and C2, charging current is stopped for about four seconds after the state-of-charge has been reached.

After the four second delay, Q1 and Q2 are allowed to have base drive from U1A. This lights up the charging LED and allows Q4 to pass current once more to the battery.

As soon as the battery hits the state-of-charge once more, the process is repeated. As the battery becomes fully charged, the "on" time will shorten up while the "off" time will always remain the same four seconds. In effect, a pulse of current will be sent to the battery that will shorten over time. I call this charging algorithm "Pulse Time Modulation."

As a side benefit of the pulse time modulation, the Micro M+ won't go nuts if you put a large solar panel onto a small battery. The charging algorithm will always keep the off time at four seconds allowing the battery time to rest before being hit by higher current than normal for its capacity.

BUILDING YOUR OWN MICRO M+

There's nothing special about the circuit. The use of a PC board makes the assembly of the Micro M+ quick and easy. It also makes it much easier if you need to troubleshoot the circuit. The entire circuit can be built on a piece of perf board.

The power MOSFET must be protected against static discharges. A dash of common sense and standard MOSFET handling procedures will work best. Don't handle the MOSFET until you need to install it in the circuit. A wrist strap is a good idea to prevent static damage. Once installed in the PC board, the device is quite robust.

A small clip-on heat sink is used for the power MOSFET. If you desire, the MOSFET could be mounted to a metal chassis. If you do this, make sure you electrically insulate the MOSFET tab from the chassis.

If you plan to use the Micro M+ outside, then consider soldering the IC directly onto the board. I've found that cheap solder-plated IC sockets corrode. If you want to use an IC socket, use one with gold plated contacts.

Feel free to substitute part values. There's nothing really critical. I do suggest you stick with 1% resistors for both R1 and R2. This isn't so important for their closer tolerance but for the 50-PPM temperature compensation they have. You can use standard off-the-shelf parts for either or both R1 and R2, but the entire circuit should then be located in an environment with a stable temperature.

ADJUSTMENTS

You'll need a good digital voltmeter and a variable power supply. Set the power supply to 14.3 V. Connect the Micro M+ battery negative lead to the power supply negative lead. Connect the Micro M+ PV positive and battery positive leads to the power supply positive lead. The charging LED should be on. If not, adjust trimmer R3 until it comes on. Check for +8 V at the VCC pins of the LM358 and the LM555. You should also see +5 V from the LM336Z5.0 diode.

Quickly move the trimmer from one end of its travel to the other. At one point the

LED will go dark. This is the switch point. To verify that the "off pulse" is working, as soon as the LED goes dark quickly reverse the direction of the trimmer. The LED should remain off for several seconds and then come back on. If everything seems to be working, it's time to set the state-of-charge trimmer.

Now, slowly adjust the trimmer until the LED goes dark. You might want to try this adjustment more than once as the closer you get the comparator to switch at exactly 14.3 V, the more accurate the Micro M+ will be. Here's a hint I've learned after adjusting hundreds of Micro M+ controllers. Set the power supply to slightly above the cut-off voltage that you want. If you want 14.3 V, then set the supply to 14.5 V. I've found that in the time it takes to react to the LED going dark, you overshoot the cut-off point. Setting the supply higher takes this into account and usually you can get the trimmer set to exactly what you need in one try. That's all you need to do. Disconnect the supply from the Micro M+ and you're ready for the solar panel.

ODDS AND ENDS

The 14.3-V terminal voltage will be correct for just about all sealed and flooded-cell lead-acid batteries. You can change the state-of-charge set point if you want to recharge NiCds or captive sealed lead-acid batteries.

Keep the current from the solar panel within reason for the size of the battery you're going to be using. If you have a 7-amp hour battery, then don't use a 75-W solar panel. You'll get much better results and smoother operation with a smaller panel.

The tab of the power MOSFET is electrically hot. If you plan on using the Micro M+ without a protective case, make sure you insulate the tab from the heatsink. A misplaced wire touching the heatsink could cause real damage to both the Micro M+ and your equipment. A small plastic box from RadioShack works great.

MORE CURRENT?

Well yes, you can get the Micro M+ to handle more current. You must increase the capacity of the blocking diode and mount the power MOSFET on a larger heat sink. I've used an MBR2025 diode and a large heatsink for the MOSFET and can easily control 12 A of current.

BATTERY CHARGING WITHOUT A SOLAR PANEL?

Yes, it's possible. The trick is to use a power supply for which you can limit the output current. A discharged lead-acid battery will draw all the current it can from the charging source. In a solar panel setup, if the panel produces 3 A, that's all it will do. With an ac-powered supply, the current can be excessive. To use the Micro M+ with an ac-powered supply, set the voltage to 15.5 V. Then limit the current to 2 or 3 A.

No matter if you're camping in the outback, or storing photons just in case of an emergency, the Micro M+ will provide your battery with the fullest charge. The Micro M+ is simple to use and completely silent. Just like the sun!

Notes

[1]The Micro M, September 1996 *QST*, p 41.

[2]A 75-W module produces 4.4 A at 17 V. The Micro M+ can easily handle the extra 400 ma.

[3]A complete kit of parts is available from Sunlight Energy Systems, also known as The Heathkit Shop. Visa, MC accepted. **www.theheathkitshop.com**.

THE UPS — A UNIVERSAL POWER SUPPLY

If you have spent much time around personal computers, you have probably heard of the uninterruptible power supply (UPS). This supply is designed to provide a continuous source of power for a computer in the event of an ac power-line failure. The UPS plugs into the ac house current and the computer plugs into the UPS. Under normal operating conditions, the UPS passes the 120-V house current to the computer. Surge protection and ac power conditioning circuitry is included in the UPS, protecting your computer from voltage spikes and other electrical conditions that could cause damage.

A UPS contains a battery and dc-to-ac inverter circuitry. The battery-charging circuitry in the UPS will maintain the battery at a full charge during normal operation. If the ac power goes off for any reason, the inverter automatically turns on to maintain the 120-V ac supply. You can continue working on the computer, either until the UPS battery discharges or until the ac power comes back on. At the very least, this will give you time to save files and shut down your computer normally.

This project is adapted from the January 1999 *QST* Technical Correspondence column, page 64, by Robert Whitaker, KI5PG. It describes how you can modify a UPS to supply 120 V ac and 12 V dc for a wide variety of ham-shack applications.

WHAT TO LOOK FOR, WHERE TO FIND IT

Computer salvage dealers, computer shows and hamfests are good places to shop for a used UPS. Try calling the service department of some computer dealers and computer-repair services to find out what they have on hand.[1] You may pick up one or more older supplies that were taken in trade or with failed batteries, for much less than you would pay for a new

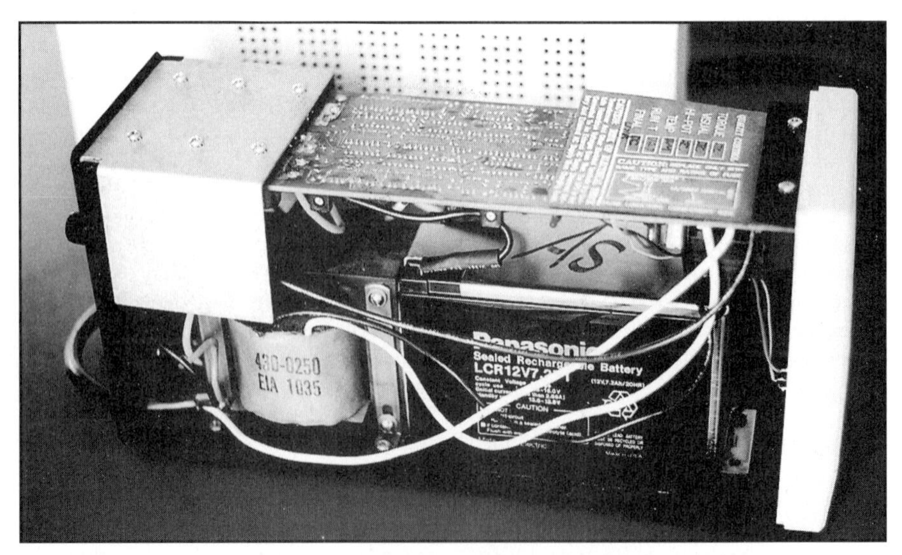

Fig 17.52 — UPS with cover removed to show internal layout. All UPS's have ac-line conditioning components, a battery, and a dc-to-ac inverter.

UPS. Computer users often throw these supplies away when the battery fails, even though many of the supplies have user-replaceable batteries!

Look for a UPS that can be forced into the inverting mode without the need to disconnect it from an active ac line. Most medium-sized units have an on/off switch and a test/alarm disable switch. Some American Power Conversion (APC)[2] UPS's have secondary DIP switches labeled TEST and ALARM DISABLE. On these models, the unit switches to the inverter mode while connected to ac power when you switch the unit on and press the ALARM DISABLE button.

Higher grade UPS's, such as the APC Back-UPS Pro series have a single on/off power pushbutton. These models can usually be forced into the inverter mode by pressing and holding the pushbutton for a few seconds.

Most medium or small UPS's use a single 12-V gel cell. Larger supplies may use two 6-V batteries in series. Some use two 12-V batteries to form a 24-V system. You will want a UPS with a 12-V system for maximum utility. A 12-V system can be easily configured to work with an external deep-discharge or marine battery. **Fig 17.52** shows a small UPS with the cover removed.

MODIFYING A UPS

If you have a UPS with a battery that will no longer hold a charge, you can either replace the battery or simply use your Universal Power Supply with an external battery. Remember that used batteries are considered hazardous materials, and must be taken to a recycling center or otherwise disposed of properly. Used or new batteries for these supplies are usually not too expensive. Check with battery suppliers like E. H. Yost & Company, W & W Manufacturing and B. G. Micro.

With the battery removed, test the charging circuit by plugging the UPS into a 120-V ac outlet and checking the voltage at the battery leads. With a 12-V system, the charging voltage should be around 13.85 V.

To attach an external battery to the UPS, simply bring the battery-lead connections outside the case. **Fig 17.53** shows how you can add a set of terminal posts on the back

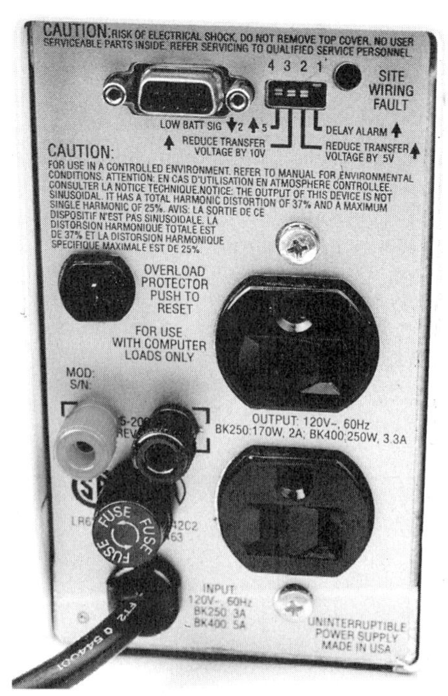

Fig 17.53 — This rear-panel view of a converted UPS shows a pair of binding posts for the 12 V dc connections and a fuse holder for the 12 V output. *Carefully* position and drill the mounting holes for the binding posts and fuse holder, so that you don't damage existing parts. Ensure that the placement of these parts will not interfere with any existing components. A Unibit or KWIK STEPPER is ideal for cutting through the 1/8-inch thick steel rear panel. (See "Tool Tips," Technical Correspondence, *QST*, Dec 1998, p 63.)

panel. You should also install a chassis-mounted fuse holder in series with the positive terminal post. A 15- or 20-A fuse should be sufficient to provide adequate current with a margin of safety against short circuiting the output.

Many UPS's have a DIP switch, one section of which disables the power-failure alarm. If your supply doesn't have such a switch, you may want to permanently disable the alarm by unsoldering the alarm lead or cutting a PC board trace.

Keep extra fuses handy by taping spares to the inside or outside of the case. Be sure to include extra fuses for the 120-V, ac-input and 12-V, dc-output lines. Don't let a careless mistake and a blown fuse deprive you of power when you need it most!

USING YOUR UPS

An uninterruptible power supply converted to a universal power supply has many uses:

- Portable 120-V, ac-power source using the internal battery.
- Portable 12-V, dc-power source using the internal battery.
- 120-V, ac-power source with dc for the inverter taken from an external 12-V, dc-automotive or deep-cycle battery.
- Base station 12-V, dc-power supply and 120-V, ac-backup supply.
- Battery charger (12 V) using 120-V, ac-line input.

The UPS is intended for medium power output for short-term use. Don't expect the internal battery-driven inverter or the battery alone to power your 100-W HF rig for a week. You can expect to power a VHF/UHF-mobile radio at medium power for a day or more during an emergency. If you are using an HT, you can probably operate it on high power for a week or more.

Adding a heavy gel-cell or deep-cycle marine battery in parallel with — or independently from — the internal back-up battery will prolong the power-delivery cycle. Be sure that you don't draw more power than your UPS's rated output. If you must draw power at or near the rated power output, use a fan to force air inside the case to help get rid of the heat.

HOW LONG WILL IT LAST?

A test may help you estimate how long you can expect to use the UPS's battery-driven inverter with a given load. Connect a lamp with a 60-W light bulb, which will draw 0.5 A at 120 V. Two lamps in parallel will draw 1 A. (Just plug in two lamps to the ac output of your supply.) Use a voltmeter to record the battery voltage at the beginning of the test and every 5 to 15 minutes. Plot your data on a graph, with time along the horizontal (X) axis and the battery voltage along the vertical (Y) axis.

Notes

[1]ATCI Consultants is one possible source for used uninterruptible power supplies. **www.dallas.net/~atci**

[2]American Power Conversion manufactures a variety of supplies and also offers repair and battery-replacement services. **www.apcc.com**

A PORTABLE POWER SUPPLY

This project was developed and the text written by Tony Jarvis, G6TTL. It originally appeared in the May, 1997 issue of *RadCom*, the monthly publication of the Radio Society of Great Britain (RSGB).

When time permits, I enjoy participating in the Backpacker series of RSGB contests,[1] operating in the 3-W category. The power supply I use for these outings is described here and shown in **Fig 17.54**. It is ideal for the purpose and can also be used as a simple, uninterruptible power supply (UPS) for the shack.

DESIGN CRITERIA

- A weight maximum of 10 lb (4.5 kg)
- Able to be carried in a small backpack
- Able to be plugged into the nearest power outlet to be recharged
- When at home, to run in 'float-charge mode' to operate low-current equipment
- To be of reasonable cost
- To use readily available components

At a Rainham Rally, I found several sealed, lead-acid cells rated at a nominal 12 V and 7-Ah capacity. They weighed in at a little over 2.2 kg (5 lb) — just what I needed! To prevent gassing, care must be exercised to not overcharge sealed cells, which requires a charge voltage limit of 13.8 V at the terminals. Thus, a stabilized supply is essential.

THE CIRCUIT

The float charger is hardly original; it was adapted from an excellent series of articles by John Case.[2] One major consideration was the need to over-engineer it, as I would leave it plugged in and switched on almost continuously.

Fig 17.55 shows the circuit diagram. Many of the components were salvaged from redundant equipment or the junk box; however, the critical components — transformer, pass transistor, reservoir capaci-

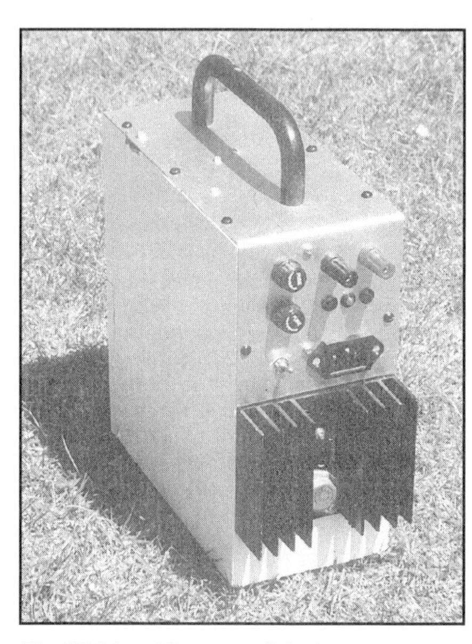

Fig 17.54 — The completed power supply.

Table 17.1
Components List

Resistors	
R1	1.2 kΩ
R2	1.5 kΩ
R3	2.7 kΩ
R4	0.5 Ω, 2W
R5	8.2 kΩ
R6	7.5 kΩ
R7	820 Ω
R8	500 Ω linear preset potentiometer
R9	1 kΩ linear preset potentiometer
VDR1	V275LA40A
All resistors ½-W, metal film, 5% tolerance, unless otherwise specified.	
Capacitors	
C1	10,000 µF, 40 V electrolytic
C2	4.7 µF, 40 V electrolytic
C3	500 pf ceramic
Semiconductors	
D1	Red LED
D2	Yellow LED
D3	Green LED
D4	MR752 (or similar)
D5	50 PIV, 25A
U1	LM723
Q1	2N3055
Additional items	
F1	1A fuse and holder
F2	3A fuse and holder
S1	DPST toggle
T1	Power transformer with 2 ea, 15V @ 0.75A secondaries
	7Ah sealed lead-acid battery
	IEC socket
	Matrix board
	Screw terminals, insulated
	Case to suit

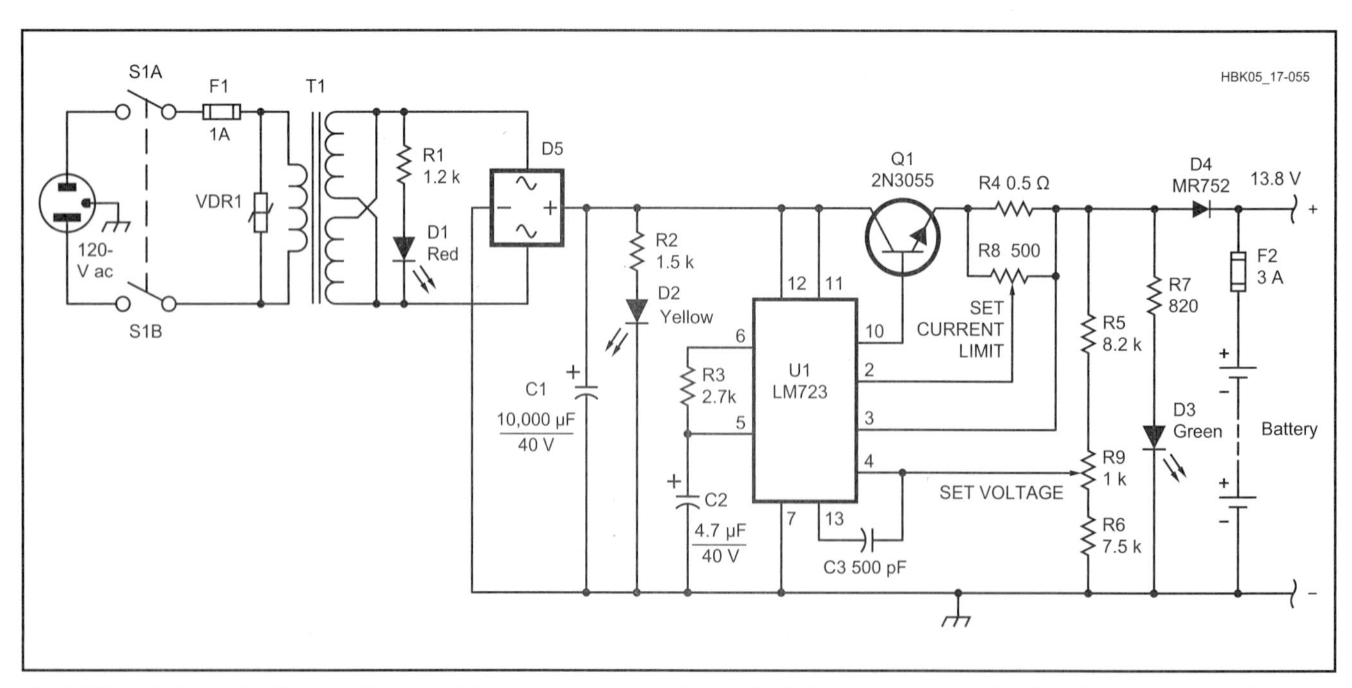

Fig 17.55 — Schematic diagram. The portable power supply works by float-charging a sealed lead-acid battery to provide an uninterruptible supply.

tor and regulator chip — were all purchased new.

There are many options available for layout; mine were dictated by the size and shape of the heatsink. In my case, the box that holds it all is made from half-inch plywood, with the major components mounted on an L-shaped aluminum plate that forms the front and part of one side. A voltage-dependent resistor (varistor) is located across the primary of the transformer, and an over-current control is provided to limit the current to 1.5A.

A heavy-duty diode, D4, is incorporated in the feed to the battery, and LEDs are placed at strategic points as a confidence feature and for ease of fault-finding. The voltage controller, an LM723, is mounted, along with its components, on a small piece of matrix board, the remaining components being wired point-to-point using substantial cable for the heavy-current paths. Setup is quite straightforward — connect everything together to a fully charged battery, connect a voltmeter to the battery's terminals, turn-on and adjust the output to 13.8V as measured at the battery terminals. Set the current limit to 1.5A, which in practice is rarely exceeded.

RESULTS

Does it meet the criteria set forth previously? That is for you to decide:

- Weight — 11 lb.
- Size — approximately 10 × 10 × 4 inches, including heatsink and handle.
- Cost — about $40.
- Capacity — enough for a full backpacking session.

Even when left on continuously, the temperature of the portable power pack hardly rises above room temperature.

Notes

[1]Backpacking — Summertime Delights, G6TTL, *RadCom*, May, 1997.

[2]Power Supplies on a Shoestring, GW4HWR, *RadCom*, July and August 1986.

Chapter 18

RF Power Amplifiers

This chapter describes the design and construction of power RF amplifiers for use in an Amateur Radio station. Dick Ehrhorn, WØID (ex-W4ETO), contributed materially to this section.

An amplifier may be required to develop as much as 1500 W of RF output power, the legal maximum in the United States.

YOU CAN BE KILLED by coming in contact with the high voltages inside a commercial or homebrew RF amplifier. Please don't take foolish chances. Remember that you CANNOT GO WRONG by treating each amplifier as potentially lethal! For a more thorough treatment of this all-important subject, please review the applicable sections of the *Safety* chapter in this *Handbook*.

Every component in an RF power amplifier must be carefully selected to endure high electrical stress levels without failing. Large amounts of heat are produced in the amplifier and must be dissipated safely. Generation of spurious signals must be minimized, not only for legal reasons, but also to preserve good neighborhood relationships. Every one of these challenges must be overcome to produce a loud, clean signal from a safe and reliable amplifier.

Types of Power Amplifiers

Power amplifiers are categorized by their power level, intended frequencies of operation, device type, class of operation and circuit configuration. Within each of these categories there are almost always two or more options available. Choosing the most appropriate set of options from all those available is the fundamental concept of design.

SOLID STATE VERSUS VACUUM TUBES

With the exception of high-power amplifiers, nearly all items of amateur equipment manufactured commercially today use solid-state (semiconductor) devices exclusively. Semiconductor diodes, transistors and integrated circuits (ICs) offer several advantages in designing and fabricating equipment:

- Compact design—Even with their heat sinks, solid-state devices are smaller than functionally equivalent tubes, allowing smaller packages.
- No-tune-up operation—By their nature, transistors and ICs lend themselves to low impedance, broadband operation. Fixed-tuned filters made with readily available components can be used to suppress harmonics and other spurious signals. Bandswitching of such filters is easily accomplished when necessary; it often is done using solid-state switches. Tube amplifiers, on the other hand, usually must be retuned on each band, and even for significant frequency movement within a band.
- Long life—Transistors and other semiconductor devices have extremely long lives if properly used and cooled. When employed in properly designed equipment, they should last for the entire useful life of the equipment—commonly 100,000 hours or more. Vacuum tubes wear out as their filaments (and sometimes other parts) deteriorate with time in normal operation; the useful life of a typical vacuum tube may be on the order of 10,000 to 20,000 hours.
- Manufacturing ease—Most solid-state devices are ideally suited for printed-circuit-board fabrication. The low voltages and low impedances that typify transistor and IC circuitry work very well on printed circuits (some circuits use the circuit board traces themselves as circuit elements); the high impedances found with vacuum tubes do not. The IC or transistor's physical size and shape also lends itself well to printed circuits and the devices usually can be soldered right to the board.

These advantages in fabrication mean reduced manufacturing costs. Based on all these facts, it might seem that there would be no place for vacuum tubes in a solid-state world. Transistors and ICs do have significant limitations, however, especially in a practical sense. Individual RF power transistors available today cannot develop more than approximately 150 W output; this figure has not changed much in the past two decades.

Individual present-day transistors cannot generally handle the combination of current and voltage needed, nor can they safely dispose of the amount of heat dissipated, for RF amplification to higher power levels. So pairs of transistors, or even pairs of pairs, are usually employed in practical power amplifier designs, even at the 100-W level. Beyond the 300-W output level, somewhat exotic (at least for most radio amateurs) techniques of power combination from multiple amplifiers ordinarily must be used. Although this is commonly done in commercial equipment, it is an expensive proposition.

It also is far easier to ensure safe cooling of vacuum tubes, which operate satisfactorily at surface temperatures as high as 150-200°C and may be cooled by simply blowing sufficient ambient air past or through their relative large cooling surfaces. The very small cooling surfaces of power transistors should be held to 75-100°C to avoid drastically shortening their life expectancy. Thus, assuming worst-case 50°C ambient air temperature, the large cooling surface of a vacuum tube can be allowed to rise 100-150°C above ambient, while the small surface of a transistor must not be allowed to rise more than about 50°C. Moreover, power tubes are considerably more likely than transistors to survive, without significant damage, the rare instance of severe overheating.

Furthermore, RF power transistors are much less tolerant of electrical abuse than are most vacuum tubes. An overvoltage spike lasting only microseconds can—and is likely to—destroy transistors costing $75 to $150 each. A comparable spike is unlikely to have any effect on a tube. So the important message is this: designing with expensive RF power transistors demands using extreme caution to ensure that adequate thermal and electrical protection is provided. It is an area best left to knowledgeable designers.

Even if one ignores the challenge of the RF portions of a high-power transistor amplifier, there is the dc power supply to consider. A solid-state amplifier capable of delivering 1 kW of RF output might require regulated (and transient-free) 50 V at more than 40 A. Developing that much current is a challenging and expensive task. These limitations considered, solid-state amplifiers have significant practical advantages up to a couple of hundred watts output. Beyond that point, and certainly at the kilowatt level, the vacuum tube still reigns for amateur constructors because of its cost-effectiveness and ease of equipment design.

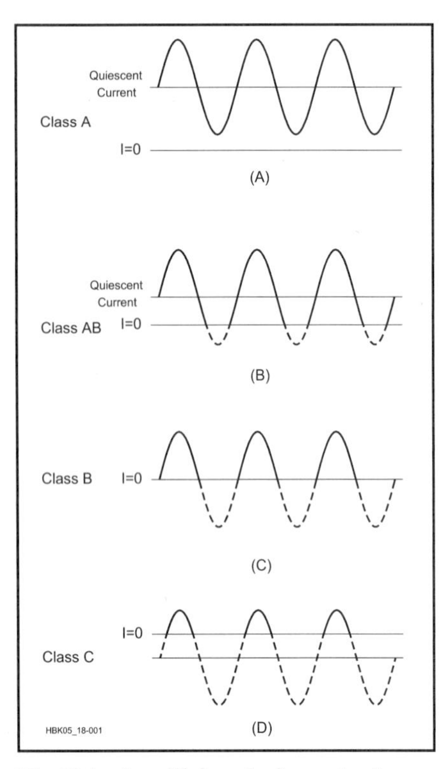

Fig 18.1—Amplifying device output current for various classes of operation. All assume a sinusoidal drive signal.

CLASSES OF OPERATION

The class of operation of an amplifier stage is defined by its conduction angle, the angular portion of each RF drive cycle, in degrees, during which plate current (or collector or drain current in the case of transistors) flows. This, in turn, determines the amplifier's gain, efficiency, linearity and input and output impedances.

- Class A: The conduction angle is 360°. DC bias and RF drive level are set so that the device is not driven to output current cutoff at any point in the driving-voltage cycle, so some device output current flows throughout the complete 360° of the cycle (see **Fig 18.1A**). Output voltage is generated by the variation of output current flowing through the load resistance. Maximum linearity and gain are achieved in a Class A amplifier, but the efficiency of the stage is low. Maximum theoretical efficiency is 50%, but 25 to 30% is more common in practice.
- Class AB: The conduction angle is greater than 180° but less than 360° (see Fig 18.1B). In other words, dc bias and drive level are adjusted so device output current flows during appreciably more than half the drive cycle, but less than the whole drive cycle. Efficiency is much better than Class A, typically reaching 50-60% at peak output power. Class AB linearity and gain are not as good as that achieved in Class A, but are very acceptable for even the most rigorous high-power SSB applications in Amateur Radio.

  Class AB vacuum tube amplifiers are further defined as class AB1 or AB2. In class AB1, the grid is not driven positive so no grid current flows. Virtually no drive power is required, and gain is quite high, typically 15-20 dB. The load on the driving stage is relatively constant throughout the RF cycle. Efficiency typically exceeds 50% at maximum output.

  In Class AB2, the grid is driven positive on peaks and some grid current flows. Efficiency commonly reaches 60%, at the expense of greater demands placed on the driving stage and slightly reduced linearity. Gain commonly reaches 15 dB.
- Class B: Conduction angle = 180°. Bias and RF drive are set so that the device is just cut off with no signal applied (see Fig 18.1C), and device output current flows during one half of the drive cycle. Efficiency commonly reaches as high as 65%, with fully acceptable linearity.
- Class C: The conduction angle is much less than 180°—typically 90°. DC bias is adjusted so that the device is cut off when no drive signal is applied. Output current flows only during positive crests in the drive cycle (see Fig 18.1D), so it consists of pulses at the drive frequency. Efficiency is relatively high—up to 80%—but linearity is extremely poor. Thus Class C amplifiers are not suitable for amplification of amplitude-modulated signals such as SSB or AM, but are quite satisfactory for use in on-off keyed stages or with frequency or phase modulation. Gain is lower than for the previous classes of operation, typically 10-13 dB.
- Classes D through H use various switched mode techniques and are not commonly found in amateur service. Their prime virtue is high efficiency, and they are used in a wide range of specialized audio and RF applications to reduce power-supply requirements and dissipated heat. These classes of RF amplifiers require fairly sophisticated design and adjustment techniques, particularly at high-power levels. The additional complexity and cost could rarely if ever be justified for amateur service.

Class of operation is independent of device type and circuit configuration (see **Electrical Signals and Components** chapter). The active amplifying device and the circuit itself must be uniquely applied for each operating class, but amplifier linearity and efficiency are determined by the class of operation. Clever amplifier design cannot improve on these fundamental limits. Poor design and implementation, though, can certainly prevent an amplifier from approaching its potential in efficiency and linearity.

MODELING THE ACTIVE DEVICE

It is very useful to have a model for the active devices used in a real-world RF power amplifier. Although the actual active device used in an amplifier might be a vacuum tube, a transistor or an FET, each model has certain common characteristics.

See **Fig 18.2A**, where a vacuum tube is modeled as a current generator in parallel with a dynamic plate resistance Rp and a load resistance RL. In this simplified model, any residual reactances (such as the inductance of connecting leads and the output capacity of the tube) are not specifically shown. The control-grid voltage in a vacuum tube controls the stream of electrons moving between the cathode and the plate. An important measure for a tube is its transconductance, which is the change in plate current caused by a change in grid-cathode voltage. The plate current is:

$$i_p = g_m \times e_g \quad (1)$$

where

i_p = plate current

g_m = transconductance (also called mutual conductance) of tube $\Delta i_p/\Delta e_g$

e_g = grid RF voltage.

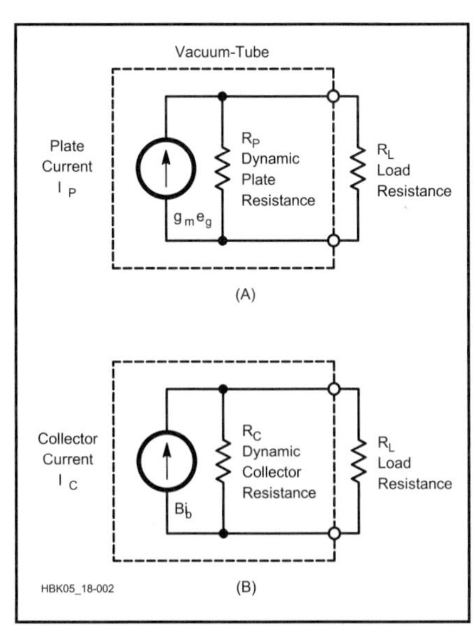

Fig 18.2—At A, the current-generator model for a vacuum-tube amplifier operating linearly. Typical values for R_p and R_L for small-signal vacuum tubes are 100 kΩ and 5 kΩ respectively. The plate current I_p is equal to the product of the tube tranconductance g_m times the grid voltage. At B, the current-generator model for a transistor. Typical values for R_C and R_L are on the same order as those for a small-signal vacuum tube.

The concept of dynamic plate resistance is sometimes misunderstood. It is a measure of how the plate current changes with a change in plate voltage, given a constant grid voltage. The control-grid voltage is by far the major determinant of the plate current in a triode. In a tetrode or pentode vacuum tube, the screen grid "screens" the plate current even further from the effect of changes in the plate voltage. For small-signal operation (where the plate voltage does not swing below the screen voltage) the plate current in a pentode or tetrode changes remarkably little when the plate voltage is changed. Thus the dynamic plate resistance is very high in a tetrode or pentode that is operating linearly, and only somewhat less for a triode. The plate current delivered into the load resistance R_L creates RF power.

An FET operates much like the vacuum-tube model. Obviously, there is no vacuum inside the case of an FET, and the FET electrodes are called gate, drain and source instead of grid, plate and cathode, but the current-generator model is just as viable for an FET as for a vacuum tube. In a transistor, the base current controls the flows of electrons (or holes) in the collector circuit. See Fig 18.2B. A transistor operating in a linear fashion resembles the operation of a tetrode or pentode vacuum tube since the equivalent collector dynamic output resistance is also high. This is so because the collector current is not affected greatly by the collector voltage—it is mainly determined by the base current. The collector current in the current-generator model for a transistor is:

$$i_c = \beta \times i_b \quad (2)$$

where

i_c = collector current

β = current gain of transistor

i_b = base current.

IMPEDANCE TRANSFORMATION—"MATCHING NETWORKS"

Over the years, some confusion in the amateur ranks has resulted from imprecise use of the terms *matching* and *matching network*. The term "matching" was first used in the technical literature in connection with transmission lines. When a matching network such as an antenna tuner is tuned properly, it "matches" (that is, makes equal) a particular load impedance to the fixed characteristic impedance of the transmission line used at the tuner input.

In this chapter, we are concerned with using active devices to generate useful RF power. For a given active device, RF power is generated most efficiently, and with the least distortion for a linear amplifier, when it delivers RF current into an *optimum value of load resistance*. For an amplifier, the output network transforms the load impedance (such as an antenna) into an optimum value of load resistance for the active device. In part to differentiate active power amplifiers from passive transmission lines, we prefer to call such a transforming network an *output network*, rather than a matching network.

Output Networks and Class AB, B and C Amplifiers

In Class AB, B and C amplifiers, we select a load resistance that will keep the tube or transistor from dissipating too much power or, in the case of Class AB or B amplifiers, to achieve the desired linearity. In these classes of amplifiers, the device output current is zero for large parts of the RF cycle. Because of this, the effective source resistance is no longer the simple dynamic plate resistance of a Class A amplifier. In fact, the value of R_p varies with the drive level. This means that, since the load resistance (of an antenna, for example) is constant, the efficiency of the amplifier also varies with the drive level.

It may at first appear contradictory that Class AB and B amplifiers use nonlinear devices but achieve "linear" operation nevertheless. The explanation is that the peak amplitude of device output current faithfully follows that of the drive voltage, even though its waveform does not. In tuned amplifiers, the flywheel effect of the resonant output network restores the missing part of each RF input cycle, as well as its sinusoidal waveform. In broadband transistor amplifiers, balanced push-pull circuitry commonly is used to restore the missing RF cycles, and low-pass filters on the output remove harmonics and thereby restore the sinusoidal RF waveform. The result in both cases is linear amplification of the input signal—by the clever application of nonlinear devices.

The usual practice in RF power amplifier design is to select an optimum load resistance that will provide the highest power output consistent with required linearity, while staying within the amplifying device's ratings. The optimum load resistance is determined by the amplifying device's current transfer characteristics and the amplifier's class of operation. For a transistor amplifier, the optimum load resistance is approximately:

$$R_L = \frac{V_{CC}^2}{2P_O} \quad (3)$$

where

R_L = the load resistance

V_{CC} = the collector dc voltage

P_O = the amplifier power output in watts.

Vacuum tubes have complex current transfer characteristics, and each class of operation produces different RMS values of RF current through the load impedance. The optimum load resistance for vacuum-tube amplifiers can be approximated by the ratio of the dc plate voltage to the dc plate current at maximum signal, divided by a constant appropriate to each class of operation. The load resistance, in turn, determines the maximum power output and efficiency the amplifier can provide. The optimum tube load resistance is

$$R_L = \frac{V_P}{K \times I_P} \quad (4)$$

where

R_L = the appropriate load resistance, in ohms

V_P = the dc plate potential, in V

I_P = the dc plate current, in A

K = a constant that approximates the RMS current to dc current ratio appropriate for each class. For the different classes of operation:

Class A, $K \approx 1.3$

Class AB, $K \approx 1.5 - 1.7$

Class B, $K \approx 1.57 - 1.8$

Class C, $K \approx 2$.

Graphical or computer-based analytical methods may be used to calculate more precisely the optimum plate load resistance for specific tubes and operating conditions, but the above "rules of thumb" generally provide satisfactory results for design.

The ultimate load for an RF power amplifier usually is a transmission line connected to an antenna or the input of another amplifier. It usually isn't practical, or even possible, to modify either of these load impedances to the optimum value needed for high-efficiency operation. An output network is thus used to transform the real load impedance to the optimum load resistance for the amplifying device. Two basic types of output networks are found in RF power amplifiers: tank circuits and transformers.

TANK CIRCUITS

Parallel-resonant circuits and their equivalents have the ability to store energy. Capacitors store electrical energy in the electric field between their plates; inductors store energy in the magnetic field induced by the coil winding. These circuits are referred to as tank circuits, since they act as storage "tanks" for RF energy.

The energy stored in the individual tank circuit components varies with time. Consider for example the tank circuit shown in **Fig 18.3**. Assuming that R is zero, the tank circuit dissipates no power. Therefore, no power need be supplied by the source; hence no line current I_{LINE} flows. Only circulating current I_{CIRC} flows, and it is exactly the same through both L and C at any instant. Similarly, the voltage across L and C is always exactly the same. At some point the capacitor is fully charged, and the current through both the capacitor and inductor is zero. So the inductor has no magnetic field and therefore no energy stored in its field. All the energy in the tank is stored in the capacitor's electric field.

At this instant, the capacitor starts to discharge through the inductor. The current flowing in the inductor creates a magnetic field, and energy transferred from the capacitor is stored in the inductor's magnetic field. Still assuming there is no loss in the tank circuit, the increase in energy stored in the inductor's magnetic field is exactly equal to the decrease in energy stored in the capacitor's electric field. The total energy stored in the tank circuit stays constant; some is stored in the inductor, some in the capacitor. Current flow into the inductor is a function of both time and of the voltage applied by the capacitor, which decreases with time as it discharges into the inductor. Eventually, the capacitor's charge is totally depleted and all the tank circuit's energy is stored in the magnetic field of the inductor. At this instant, current flow through L and C is maximum and the voltage across the terminals of both L and C is zero.

Since energy no longer is being transferred to the inductor, its magnetic field begins to collapse and becomes a source of current, still flowing in the same direction as when the inductor was being driven by the capacitor. When the inductor becomes a current source, the voltage across its terminals reverses and it begins to recharge the capacitor, with opposite polarity from its previous condition. Eventually, all energy stored in the inductor's magnetic field is depleted as current decreases to zero. The capacitor is fully charged, and all the energy is then stored in the capacitor's electric field. The exchange of energy from capacitor to inductor and back to capacitor is then repeated, but with opposite voltage polarities and direction of current flow from the previous exchange. It can be shown mathematically that the "alternating" current and voltage produced by this process are sinusoidal in waveform, with a frequency of

$$f = \frac{1}{2\pi\sqrt{LC}} \quad (5)$$

which of course is the resonant frequency of the tank circuit. In the absence of a load or any losses to dissipate tank energy, the tank circuit current would oscillate forever.

In a typical tank circuit such as shown in Fig 18.3, the values for L and C are chosen so that the reactance (X_L) of L is equal to the reactance (X_C) of C at the frequency of the signal generated by the ac voltage source. If R is zero (since X_L is equal to X_C), the line current I_{LINE} measured by M1 is close to zero. However, the circulating current in the loop made up of L, R and C is definitely not zero. Examine what would happen if the circuit were suddenly broken at points A and B. The circuit is now made up of L, C and R, all in series. X_L is equal to X_C, so the circuit is resonant. If some voltage is applied between points A and B, the magnitude of circulating current is limited only by resistance R. If R were equal to zero, the circulating current would be infinite!

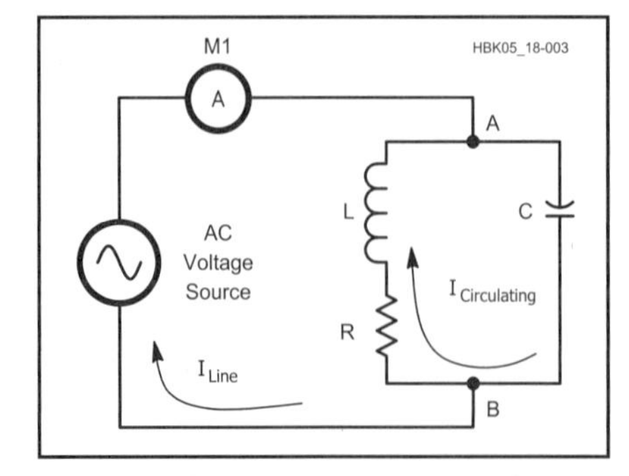

Fig 18.3—There are two currents in a tank circuit: the line current (I_{LINE}) and the circulating current (I_{CIRC}). The circulating current is dependent on tank Q.

THE FLYWHEEL EFFECT

A tank circuit can be likened to a flywheel—a mechanical device for storing energy. The energy in a flywheel is stored in the angular momentum of the wheel. As soon as a load of some sort is attached, the wheel starts to slow or even stop. Some of the energy stored in the spinning flywheel is now transferred to the load. In order to keep the flywheel turning at a constant speed, the energy drained by the load must be replenished. Energy has to be added to the flywheel from some external source. If sufficient energy is added to the flywheel, it maintains its constant rotational speed.

In the real world, of course, flywheels

and tank circuits suffer from the same fate; system losses dissipate some of the stored energy without performing any useful work. Air resistance and bearing friction slow the flywheel. In a tank circuit, resistive losses drain energy.

TANK CIRCUIT Q

In order to quantify the ability of a tank circuit to store energy, a quality factor, Q, is defined. Q is the ratio of energy stored in a system during one complete RF cycle to energy lost.

$$Q = 2\pi \frac{W_S}{W_L} \qquad (6)$$

where

W_S = is the energy stored

W_L = the energy lost to heat and the load.

By algebraic substitution and appropriate integration, the Q for a tank circuit can be expressed as

$$Q = \frac{X}{R} \qquad (7)$$

where

X = the reactance of either the inductor or the capacitor

R = the series resistance.

Since both circulating current and Q are proportional to 1/R, circulating current is therefore proportional to Q. The tank circulating current is equal to the line current multiplied by Q. If the line current is 100 mA and the tank Q is 10, then the circulating current through the tank is 1 A. (This implies, according to Ohm's Law, that the voltage potentials across the components in a tank circuit also are proportional to Q.)

When there is no load connected to the tank, the only resistances contributing to R are the losses in the tank circuit. The unloaded Q (Q_U) in that case is:

$$Q_U = \frac{X}{R_{Loss}} \qquad (8)$$

where

X = the reactance of either the inductor or capacitor

R_{Loss} = the effective series loss resistance in the circuit.

A load connected to a tank circuit has exactly the same effect on tank operation as circuit losses. Both consume energy. It just happens that energy consumed by circuit losses becomes heat rather than useful output. When energy is coupled out of the tank circuit into a load, the loaded Q (Q_L) is:

$$Q_L = \frac{X}{R_{Loss} + R_{Load}} \qquad (9)$$

where R_{Load} is the load resistance. Energy dissipated in R_{Loss} is wasted as heat. Ideally, all the tank circuit energy should be delivered to R_{Load}. This implies that R_{Loss} should be as small as practical, to yield the highest reasonable value of unloaded Q.

TANK CIRCUIT EFFICIENCY

The efficiency of a tank circuit is the ratio of power delivered to the load resistance (R_{Load}) to the total power dissipated by losses (R_{Load} and R_{Loss}) in the tank circuit. Within the tank circuit, R_{Load} and R_{Loss} are effectively in series, and the circulating current flows through both. The power dissipated by each is therefore proportional to its resistance. The loaded tank efficiency can therefore be defined as

$$\text{Tank Efficiency} = \frac{R_{Load}}{R_{Load} + R_{Loss}} \times 100 \qquad (10)$$

where efficiency is stated as a percentage. By algebraic substitution, the loaded tank efficiency can also be expressed as

$$\text{Tank Efficiency} = \left(1 - \frac{Q_L}{Q_U}\right) \times 100 \qquad (11)$$

where

Q_L = the tank circuit loaded Q

Q_U = the unloaded Q of the tank circuit.

It follows then that tank efficiency can be maximized by keeping Q_L low, which keeps the circulating current low and the I^2R losses down. Q_U should be maximized for best efficiency; this means keeping the circuit losses low.

The selectivity provided by a tank circuit helps suppress harmonic currents generated by the amplifier. The amount of harmonic suppression is dependent upon circuit loaded Q_L, so a dilemma exists for the amplifier designer. A low Q_L is desirable for best tank efficiency, but yields poorer harmonic suppression. High Q_L keeps amplifier harmonic levels lower at the expense of some tank efficiency. At HF, a compromise value of Q_L can usually be chosen such that tank efficiency remains high and harmonic suppression is also reasonable. At higher frequencies, tank Q_L is not always readily controllable, due to unavoidable stray reactances in the circuit. However, unloaded Q_U can always be maximized, regardless of frequency, by keeping circuit losses low.

TANK OUTPUT CIRCUITS

Tank circuit output networks need not take the form of a capacitor connected in parallel with an inductor. A number of equivalent circuits can be used to match the impedances normally encountered in a power amplifier. Most are operationally more flexible than a parallel-resonant tank. Each has its advantages and disadvantages for specific applications, but the final choice usually is based on practical construction considerations and the component values needed to implement a particular network. Some networks may require unreasonably high or low inductance or capacitance values. In that case, use another network, or a different value of Q_L. Several different networks may be investigated before an acceptable final design is reached.

The impedances of RF components and amplifying devices frequently are given in terms of a parallel combination of a resistance and a reactance, although it is often easier to use a series R-X combination to design networks. Fortunately, there is a series impedance equivalent to every parallel impedance and vice versa. The equivalent circuits, and equations for conversion from one to the other, are given in **Fig 18.4**. In order to use most readily available design equations for computing matching networks, the parallel impedance must first be converted to its equivalent series form.

The Q_L of a parallel impedance can be derived from the series form as well. Substitution of the usual formula for calculating Q_L into the equations from Fig 18.4 gives

$$Q_L = \frac{R_P}{X_P} \qquad (12)$$

where

R_P = the parallel equivalent resistance

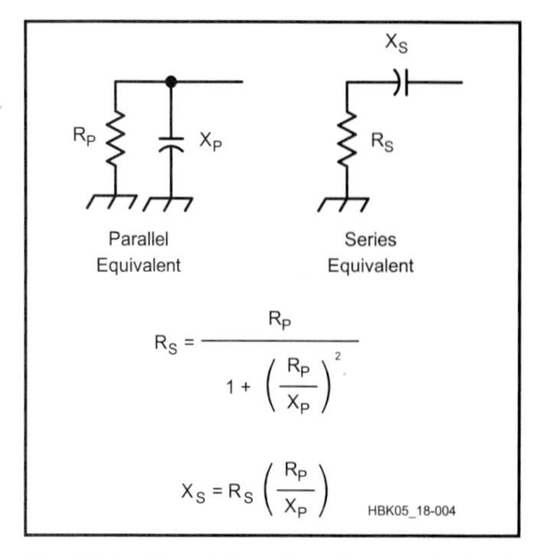

Fig 18.4—Parallel and series equivalent circuits and the formulas used for conversion.

X_P = the parallel equivalent reactance.

Several impedance-matching networks are shown in the **Receivers and Transmitters** chapter. A low-pass T network and two low-pass L networks are possible matching networks. Both types of matching networks provide good harmonic suppression. The pi network is also commonly used for amplifier matching. Harmonic suppression of a pi network is a function of the impedance transformation ratio and the Q_L of the circuit. Second-harmonic attenuation is approximately 35 dB for a load impedance of 2000 Ω in a pi network with a Q_L of 10. The third harmonic is typically 10 dB lower and the fourth approximately 7 dB below that. A typical pi network as used in the output circuit of a tube amplifier is shown in **Fig 18.5**.

You can calculate Pi-network matching-circuit values using the following equations. These equations are from Elmer (W5FD) Wingfield's August 1983 *QST* article, "New and Improved Formulas for the Design of Pi and Pi-L Networks," and Feedback in January 1984 *QST*. (See the Bibliography at the end of this chapter.) **Table 18.1** shows some data from a computer program Wingfield wrote to calculate these values. This program (PI-CMIN.EXE) and a similar program to calculate Pi-L network values (PI-LCMIN.EXE) are available from ARRLWeb (see page viii), along with several other useful Wingfield programs. The programs are for IBM PC and compatible computers. A more complete set of tables is also available from ARRL as a template package. See the **Component Data and References** chapter for ordering information.

The computer programs take into account the minimum practical capacitance (C_{min}) you can expect to achieve with your circuit, based on your knowledge of the tube output capacitance, stray circuit capacitance, the minimum capacitance of the variable tuning capacitors and a reasonable amount of capacitance for tuning. (Start with a minimum capacitance of about 35 pF for vacuum variable capacitors and about 45 to 50 pF for air variable capacitors.)

If the following equations lead to a capacitor value less than the minimum capacitance you expect to achieve, use the minimum value to recalculate the other quantities as shown in the Q_1 Based Pi-Network Equations. This will result in a final circuit operating Q value that is larger than the selected value. (Wingfield uses Q_0 to represent this output Q, which is the same as Q_L referred to earlier in this chapter. We will use Q_0 in the equations.)

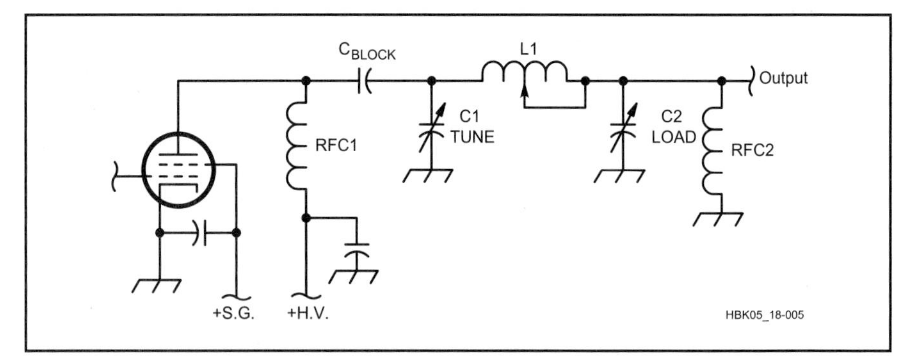

Fig 18.5—A pi matching network used at the output of a tetrode power amplifier. RFC2 is used for protective purposes in the event C_{BLOCK} fails.

The program output includes this new calculated Q_0 value.

Use the following equations to calculate specific component values for a Pi-network matching circuit. Select the desired circuit operating Q, Q_0, to satisfy these relationships, depending on whether the load resistance is higher or lower than the transformed resistance presented to the plate:

$$Q_0^2 > \frac{R_1}{R_2} - 1 \text{ and } Q_0^2 > \frac{R_2}{R_1} - 1 \qquad (13)$$

where:

R_1 is the input resistance to be matched, in ohms

R_2 is the load (output) resistance to be matched, in ohms.

Calculate the value of the input Q, Q1:

$$Q_1 = \frac{R_1 Q_0 - \sqrt{R_1 R_2 Q_0^2 - (R_1 - R_2)^2}}{R_1 - R_2} \qquad (14)$$

We will work through an example as the equations are presented. Let's select Q_0 = 12, R_1 = 1500 Ω and R2 = 50 Ω.

$$Q_1 = \frac{1500 \times 12 - \sqrt{1500 \times 50 \times 12^2 - (1500 - 50)^2}}{1500 - 50}$$

$$Q_1 = \frac{1.80 \times 10^4 - \sqrt{8.6975 \times 10^6}}{1450} = 10.38$$

Next calculate the value of the output Q, Q_2:

$$Q_2 = Q_0 - Q_1 \qquad (15)$$

Q2 = 12 – 10.38 = 1.62

Now calculate the reactance of the input capacitor, output capacitor and inductor.

$$X_{C1} = \frac{R_1}{Q_1} \qquad (16)$$

$$X_{C1} = \frac{1500}{10.38} = 144.5\,\Omega$$

$$X_{C2} = \frac{R_2}{Q_2} \qquad (17)$$

$$X_{C2} = \frac{50}{1.62} = 30.86\,\Omega$$

$$X_L = \frac{R_1 Q_0}{Q_1^2 + 1} \qquad (18)$$

$$X_L = \frac{1500 \times 12}{10.38^2 + 1} = \frac{1.80 \times 10^4}{108.74} = 165.5\,\Omega$$

Finally, calculate the component values:

$$C1 = \frac{1}{2\pi X_{C1}} \qquad (19)$$

where f is in Hz and X_{C1} is in ohms.

For our example, let's find the component values at 3.75 MHz.

$$C1 = \frac{1}{2\pi\, 3.75 \times 10^6 \times 144.5} = 294 \text{ pF}$$

$$C2 = \frac{1}{2\pi f X_{C2}} \qquad (20)$$

$$C2 = \frac{1}{2\pi\, 3.75 \times 10^6 \times 30.86} = 1375 \text{ pF}$$

$$L = \frac{X_L}{2\pi f} \qquad (21)$$

$$L = \frac{165.5}{2\pi\, 3.75 \times 10^6} = 7.02\,\mu\text{H}$$

As an alternate method, after selecting the values for Q_0, R_1 and R_2, you can use

Table 18.1
Pi-Network Values for Various Plate Impedances
(Sample Output from PI-CMIN.EXE by W5FD)
C in pF and L in µH
Pi-Net Values
R2=50 Ω, Q_0 = 12, C(min) = 35 pF

R1=1500 ohms

Band	*C1*	*C2*	*L*	
160	580	2718	13.9	
80	294	1378	7.0	
40	154	721	3.7	
30	109	511	2.7	
20	78	364	1.86	Q_0=12.0
17	61	285	1.45	Q_0=12.0
15	52	243	1.24	Q_0=12.0
12	44	207	1.06	Q_0=12.0
10	38	179	0.91	Q_0=12.0

R1=1600 ohms

Band	*C1*	*C2*	*L*	
160	547	2619	14.6	
80	278	1328	7.4	
40	145	695	3.9	
30	103	492	2.8	
20	73	351	1.96	Q_0=12.0
17	57	274	1.53	Q_0=12.0
15	49	234	1.31	Q_0=12.0
12	42	199	1.11	Q_0=12.0
10	36	172	0.96	Q_0=12.0

R1=1700 ohms

Band	*C1*	*C2*	*L*	
160	518	2527	15.4	
80	263	1281	7.8	
40	137	671	4.1	
30	97	475	2.9	
20	69	338	2.06	Q_0=12.0
17	54	265	1.61	Q_0=12.0
15	46	226	1.38	Q_0=12.0
12	39	192	1.17	Q_0=12.0
10	35	173	0.99	Q_0=12.3

R1=1800 ohms

Band	*C1*	*C2*	*L*	
160	491	2441	16.1	
80	249	1238	8.2	
40	130	648	4.3	
30	92	459	3.0	
20	66	327	2.16	Q_0=12.0
17	51	256	1.69	Q_0=12.0
15	44	218	1.44	Q_0=12.0
12	37	186	1.23	Q_0=12.0
10	35	180	0.99	Q_0=13.0

R1=1900 ohms

Band	*C1*	*C2*	*L*	
160	468	2360	16.9	
80	237	1197	8.6	
40	124	626	4.5	
30	88	443	3.2	
20	63	316	2.26	Q_0=12.0
17	49	247	1.77	Q_0=12.0
15	42	211	1.51	Q_0=12.0
12	36	180	1.29	Q_0=12.0
10	35	186	0.99	Q_0=13.7

R1=2000 ohms

Band	*C1*	*C2*	*L*	
160	446	2284	17.6	
80	226	1158	8.9	
40	118	606	4.7	
30	84	429	3.3	
20	60	306	2.36	Q_0=12.0
17	47	239	1.85	Q_0=12.0
15	40	204	1.58	Q_0=12.0
12	35	184	1.29	Q_0=12.5
10	35	193	0.98	Q_0=14.4

R1=2100 ohms

Band	*C1*	*C2*	*L*	
160	427	2213	18.4	
80	216	1122	9.3	
40	113	587	4.9	
30	80	416	3.5	
20	57	296	2.46	Q_0=12.0
17	45	232	1.92	Q_0=12.0
15	38	198	1.64	Q_0=12.0
12	35	189	1.30	Q_0=13.0
10	35	199	0.98	Q_0=15.1

R1=2200 ohms

Band	*C1*	*C2*	*L*	
160	409	2145	19.1	
80	207	1088	9.7	
40	109	569	5.1	
30	77	403	3.6	
20	55	287	2.56	Q_0=12.0
17	45	232	2.00	Q_0=12.0
15	37	192	1.71	Q_0=12.0
12	35	197	1.29	Q_0=13.7
10	35	205	0.98	Q_0=15.8

R1=2300 ohms

Band	*C1*	*C2*	*L*	
160	392	2081	19.8	
80	199	1055	10.1	
40	104	552	5.3	
30	74	391	3.7	
20	53	279	2.65	Q_0=12.0
17	41	218	2.08	Q_0=12.0
15	35	186	1.77	Q_0=12.0
12	35	210	1.30	Q_0=12.0
10	35	211	0.98	Q_0=16.5

R1=2400 ohms

Band	*C1*	*C2*	*L*	
160	377	2020	20.5	
80	191	1024	10.4	
40	100	536	5.5	
30	71	379	3.9	
20	51	270	2.75	Q_0=12.0
17	40	212	2.15	Q_0=12.0
15	35	192	1.78	Q_0=12.5
12	35	207	1.30	Q_0=14.8
10	35	216	0.98	Q_0=17.2

R1=2500 ohms

Band	*C1*	*C2*	*L*	
160	363	1961	21.3	
80	184	994	10.8	
40	96	520	5.6	
30	68	368	4.0	
20	49	262	2.85	Q_0=12.0
17	38	205	2.23	Q_0=12.0
15	35	198	1.78	Q_0=13.0
12	35	215	1.29	Q_0=15.5
10	35	222	0.98	Q_0=17.9

the following equations:

$$X_L = \frac{Q_0(R_1+R_2)+2\sqrt{R_1R_2\left(Q_0^2+4\right)-\left(R_1+R_2\right)^2}}{Q_0^2+4} \quad (22)$$

$$X_L = \frac{1.86\times10^4+2\sqrt{1.11\times10^7-2.4025\times10^6}}{148} = 165.5\,\Omega$$

$$Q_1 = \frac{Q_0 R_1}{X_L}-1 \quad (23)$$

$$Q_1 = \sqrt{\frac{12\times1500}{165.5}-1} = 10.38$$

$$Q_2 = Q_0 - Q_1 \quad (24)$$

or

$$Q_2 = \sqrt{\frac{Q_0 R_2}{X_L}-1} \quad (25)$$

$$Q_2 = \sqrt{\frac{12\times50}{165.5}-1} = 1.62$$

Use equations 16 and 17 to calculate the reactances of capacitors C1 and C2. Equations 19, 20 and 21 give the capacitance and inductance values for the pi network.

The *pi-L network* is a combination of a pi network followed by an L network. The pi network transforms the load resistance to an intermediate impedance level called the *image impedance*. Typically, the image impedance is chosen to be between 300 and 700 Ω. The L section then transforms from the image impedance down to 50 Ω. The output capacitor of the pi network is combined with the input capacitor for the L network, as shown in **Fig 18.6**. The pi-L configuration attenuates harmonics better than a pi network. Second harmonic level for a pi-L network with a Q_L of 10 is approximately 52 dB below the fundamental. The third harmonic is attenuated 65 dB and the fourth harmonic approximately 75 dB.

The following equations help you calculate pi-L matching-network values. Select an image resistance value (R_m) that the L network will supply as a load for the pi network. This value must be between the desired pi-L network input resistance (R_1) and the output load resistance (R_2). For example, you can use the value given:

$$R_m = \sqrt{R_1 \times R_2} \quad (26)$$

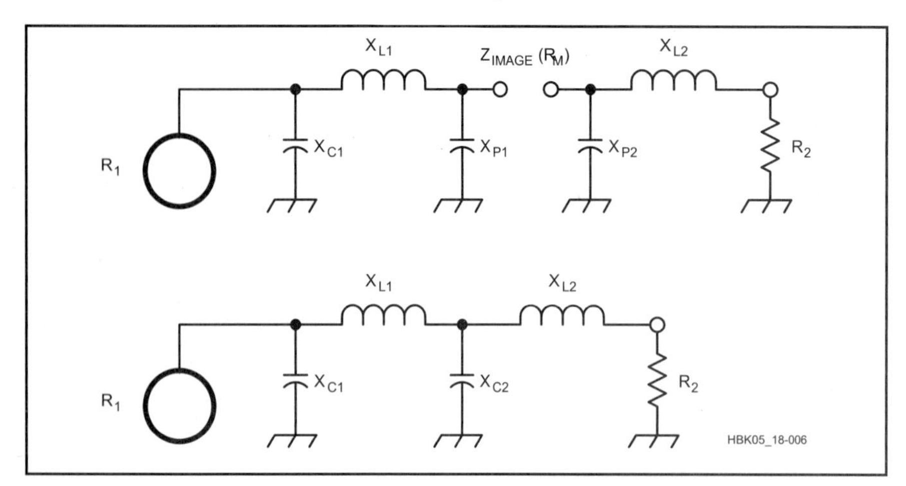

Fig 18.6—The pi-L network uses a pi network to transform the input impedances (R1) to the image impedance (Z_{IMAGE}). An L network transforms Z_{IMAGE} to R2.

The computer program, PI-LCMIN.EXE, uses 300 Ω for Rm in its calculations. Changing the image resistance results in a different network solution. Use this equation to compute the L network Q value, Q_L:

$$Q_L = \sqrt{\frac{R_m}{R_2} - 1} \tag{27}$$

We will work through an example, using R_1 = 1500 Ω, R_2 = 50 Ω and the desired pi-L network output Q, Q_0 = 12.

$$Q_L = \sqrt{\frac{300}{50} - 1} = 2.24$$

Use equations 28 and 29 to calculate the L-network reactances.

$$X_{L2} = Q_L R_2 \tag{28}$$

$$X_{L2} = 2.24 \times 50 = 112\ \Omega \tag{29}$$

$$X_{P2} = \frac{R_m}{Q_L} \tag{30}$$

$$X_{P2} = \frac{300}{2.24} = 134\ \Omega$$

Next calculate the desired Q of the pi-network section ($Q_{0\pi}$).

$$Q_{0p} = Q_0 - Q_L \tag{31}$$

$$Q_{0\pi} = 12 - 2.24 = 9.76$$

Use equations 14 through 18 or 22 through 26 to calculate the pi-network reactances, X_{C1}, X_{L1} and X_{P1} as shown in Fig 18.6. Be sure to use the value specified for Rm as R2 in these calculations. Also use the value just calculated for $Q_{0\pi}$ as Q_0. Notice that X_{P1} is X_{C2} in Eq 20.

$$Q_1 \frac{R_1 Q_{0\pi} - \sqrt{R_1 R_m Q_{0\pi} - (R_1 - R_m)^2}}{R_1 - R_m} \tag{32}$$

$$Q_1 = \frac{1500 \times 9.76 - \sqrt{1500 \times 300 \times 9.76^2 - (1500 - 300)^2}}{1500 - 300}$$

$$Q_1 = \frac{1.464 \times 10^4 - \sqrt{4.287 \times 10^7 - 1.44 \times 10^6}}{1200} = 6.84$$

$$Q_2 = Q_{0\pi} - Q_1 \tag{33}$$

$$Q_2 = 9.76 - 6.84 = 2.92$$

$$X_{C1} = \frac{R_1}{Q_1} \tag{34}$$

$$X_{C1} = \frac{1500}{6.84} = 219.3\ \Omega$$

$$X_{P1} = \frac{R_m}{Q_2} \tag{35}$$

$$X_{P1} = \frac{300}{2.92} = 102.7\ \Omega$$

$$X_{L1} = \frac{R_1 Q_{0\pi}}{Q_1^2 + 1} \tag{36}$$

$$X_{L1} = \frac{1500 \times 9.76}{6.84^2 + 1} = 306.3\ \Omega$$

Combine the two parallel capacitors, X_{P1} and X_{P2} to find the Pi-L network X_{C2} value.

$$X_{C2} = \frac{X_{P1} X_{P2}}{X_{P1} + X_{P2}} \tag{37}$$

$$X_{C2} = \frac{102.7 \times 134}{102.7 + 134} = 58.3\ \Omega$$

Finally, calculate the capacitance and inductance values using equations 22 through 24. **Table 18.2** shows some data from Wingfield's program, PI-LCMIN. For the sample calculation shown here, we choose a frequency of 3.75 MHz.

$$C_1 = \frac{1}{2\pi f X_{C1}} = \frac{1}{2\pi 3.75 \times 10^6 \times 219.3} = 193.5\ \text{pF}$$

$$C_2 = \frac{1}{2\pi f X_{C2}} = \frac{1}{2\pi 3.75 \times 10^6 \times 58.3} = 730\ \text{pF}$$

$$L_1 = \frac{X_{L1}}{2\pi f} = \frac{306.3}{2\pi 3.75 \times 10^6} = 13.0\ \mu\text{H}$$

$$L_2 = \frac{X_{L2}}{2\pi f} = \frac{112}{2\pi 3.75 \times 10^6} = 4.75\ \mu\text{H}$$

The values for L and C in Tables 18.1 and 18.2 are based on purely resistive load impedances and assume ideal capacitors and inductors. Any other circuit reactances will modify these values.

Stray circuit reactances, including tube capacitances and capacitor stray inductances, should be included as part of the matching network. It is not uncommon for such reactances to render the use of certain matching circuits impractical, because they require either unacceptable loaded Q values or unrealistic component values. If all matching network alternatives are investigated and found unworkable, some compromise solution must be found.

Above 30 MHz, transistor and tube reactances tend to dominate circuit impedances. At the lower impedances found in transistor circuits, the standard networks can be applied so long as suitable components are used. Above 50 MHz, capacitors often exhibit values far different from their marked values because of stray internal reactances and lead inductance, and this requires compensation. Tuned circuits are frequently fabricated in the form of strip lines or other transmission lines in order to circumvent the problem of building "pure" inductances and capacitances. The choice of components is often more significant than the type of network used.

The high impedances encountered in VHF tube-amplifier plate circuits are not easily matched with typical networks. Tube output capacitance is usually so large that most matching networks are unsuit-

Table 18.2
Pi-L Network Values for Various Plate Impedances
(Sample Output from PI-LCMIN.EXE by W5FD)
C in pF and L in µH
Pi-L Network Values
Rm = 300 Ω, Q_0 = 12, R2 = 50 Ω
C(Min) = 35 pF

R1=1500 ohms

Band	C1	C2	L1	L2	
160	382	1443	25.7	9.38	
80	194	732	13.0	4.76	
40	102	383	6.83	2.49	
30	72	270	4.82	1.76	
20	51	193	3.44	1.26	Q_0=12.0
17	40	151	2.69	0.98	Q_0=12.0
15	35	131	2.25	0.84	Q_0=12.2
12	35	123	1.64	0.71	Q_0=14.0
10	35	118	1.24	0.62	Q_0=15.9

R1=1600 ohms

Band	C1	C2	L1	L2	
160	362	1423	26.9	9.38	
80	184	722	13.6	4.76	
40	96	378	7.13	2.49	
30	68	267	5.04	1.76	
20	48	190	3.60	1.26	Q_0=12.0
17	38	149	2.81	0.98	Q_0=12.0
15	35	134	2.23	0.84	Q_0=12.8
12	35	126	1.63	0.71	Q_0=14.7
10	35	120	1.23	0.62	Q_0=16.7

R1=1700 ohms

Band	C1	C2	L1	L2	
160	344	1404	28.0	9.38	
80	175	712	14.2	4.76	
40	92	373	7.44	2.49	
30	65	263	5.25	1.76	
20	46	188	3.75	1.26	Q_0=12.0
17	36	147	2.94	0.98	Q_0=12.0
15	35	136	2.22	0.84	Q_0=13.4
12	35	129	1.62	0.71	Q_0=15.4
10	35	123	1.22	0.62	Q_0=17.5

R1=1800 ohms

Band	C1	C2	L1	L2	
160	328	1387	29.2	9.38	
80	166	703	14.8	4.76	
40	87	368	7.74	2.49	
30	61	260	5.47	1.76	
20	44	186	3.90	1.26	Q_0=12.0
17	35	147	3.01	0.98	Q_0=12.2
15	35	139	2.21	0.84	Q_0=13.9
12	35	131	1.61	0.71	Q_0=16.0
10	35	125	1.21	0.62	Q_0=18.2

R1=1900 ohms

Band	C1	C2	L1	L2	
160	313	1371	30.3	9.38	
80	159	695	15.4	4.76	
40	83	364	8.04	2.49	
30	59	257	5.68	1.76	
20	42	184	4.06	1.26	Q_0=12.0
17	35	149	2.99	0.98	Q_0=12.7
15	35	141	2.20	0.84	Q_0=14.5
12	35	133	1.60	0.71	Q_0=16.7
10	35	128	1.20	0.62	Q_0=19.0

R1=2000 ohms

Band	C1	C2	L1	L2	
160	300	1356	31.4	9.38	
80	152	687	15.9	4.76	
40	80	360	8.34	2.49	
30	56	254	5.89	1.76	
20	40	181	4.20	1.26	Q_0=12.0
17	35	152	2.97	0.98	Q_0=13.2
15	35	143	2.18	0.84	Q_0=15.1
12	35	136	1.59	0.71	Q_0=17.4
10	35	130	1.19	0.62	Q_0=19.7

R1=2100 ohms

Band	C1	C2	L1	L2	
160	288	1341	32.5	9.38	
80	146	680	16.5	4.76	
40	76	356	8.63	2.49	
30	54	251	6.09	1.76	
20	39	180	4.35	1.26	Q_0=12.0
17	35	154	2.97	0.98	Q_0=13.6
15	35	146	2.17	0.84	Q_0=15.6
12	35	138	1.58	0.71	Q_0=18.0
10	35	132	1.19	0.62	Q_0=20.5

R1=2200 ohms

Band	C1	C2	L1	L2	
160	277	1327	33.6	9.38	
80	140	673	17.0	4.76	
40	73	352	8.92	2.49	
30	52	249	6.30	1.76	
20	37	178	4.50	1.26	Q_0=12.0
17	35	156	2.95	0.98	Q_0=14.1
15	35	148	2.16	0.84	Q_0=16.2
12	35	140	1.57	0.71	Q_0=18.7
10	35	134	1.18	0.62	Q_0=21.3

R1=2300 ohms

Band	C1	C2	L1	L2	
160	266	1315	34.7	9.38	
80	135	667	17.6	4.76	
40	71	349	9.21	2.49	
30	50	246	6.50	1.76	
20	36	176	4.65	1.26	Q_0=12.0
17	35	158	2.93	0.98	Q_0=14.6
15	35	150	2.15	0.84	Q_0=16.7
12	35	142	1.57	0.71	Q_0=19.3
10	35	137	1.17	0.62	Q_0=22.0

R1=2400 ohms

Band	C1	C2	L1	L2	
160	257	1302	/35.8	9.38	
80	130	660	18.2	4.76	
40	68	346	9.50	2.49	
30	48	244	6.71	1.76	
20	35	176	4.71	1.26	Q_0=12.2
17	35	161	2.92	0.98	Q_0=15.0
15	35	152	2.13	0.84	Q_0=17.3
12	35	145	1.56	0.71	Q_0=20.0
10	35	139	1.17	0.62	Q_0=22.8

R1=2500 ohms

Band	C1	C2	L1	L2	
160	248	1291	36.9	9.38	
80	126	654	18.7	4.76	
40	66	343	9.79	2.49	
30	46	242	6.91	1.76	
20	35	178	4.70	1.26	Q_0=12.6
17	35	163	2.91	0.98	Q_0=15.5
15	35	154	2.13	0.84	Q_0=17.8
12	35	147	1.55	0.71	Q_0=20.6
10	35	141	1.16	0.62	Q_0=23.5

able. The usual practice is to resonate the tube output capacitance with a low-loss inductance connected in series or parallel. The result can be a very high-Q tank circuit. Component losses must be kept to an absolute minimum in order to achieve reasonable tank efficiency. Output impedance transformation is usually performed by a link inductively coupled to the tank circuit or by a parallel transformation of the output resistance using a series capacitor.

TRANSFORMERS

Broadband transformers are often used in matching to the input impedance or optimum load impedance in a power amplifier. Multioctave power amplifier performance can be achieved by appropriate application of these transformers. The input and output transformers are two of the most critical components in a broadband amplifier. Amplifier efficiency, gain flatness, input SWR, and even linearity all are affected by transformer design and application. There are two basic RF transformer types, as described elsewhere in this *Handbook*: the conventional transformer and the transmission-line transformer.

The conventional transformer is wound much the same way as a power transformer. Primary and secondary windings are wound around a high-permeability core, usually made from a ferrite or powdered-iron material. Coupling between the secondary and primary is made as tight as possible to minimize leakage inductance. At low frequencies, the coupling between windings is predominantly magnetic. As the frequency rises, core permeability decreases and leakage inductance increases; transformer losses increase as well.

Typical examples of conventional transformers are shown in **Fig 18.7**. In Fig 18.7A, the primary windings consist of brass or copper tubes inserted into ferrite sleeves. The tubes are shorted together at one end by a piece of copper-clad circuit board material. The secondary winding is threaded through the tubes. Since the low-impedance winding is only a single turn, the transformation ratio is limited to the squares of integers; for example, 1, 4, 9, 16, and so on. The lowest effective transformer frequency is determined by the inductance of the one-turn winding. It should have a reactance, at the lowest frequency of intended operation, at least four times greater than the impedance it is connected to.

The coupling coefficient between the two windings is a function of the primary tube diameter and its length, and the diameters and insulation thickness of the wire used in the high-impedance winding. High impedance ratios, greater than 36:1, should use large-diameter secondary windings. Miniature coaxial cable (using only the braid as the conductor) works well. Another use for coaxial cable braid is illustrated in Fig 18.7B. Instead of using tubing for the primary winding, the secondary winding is threaded through copper braid. Performance of the two units is almost identical.

The cores used must be large enough so the core material will not saturate at the power level applied to the transformer. Core saturation can cause permanent changes to the core permeability, as well as overheating. Transformer nonlinearity also develops at core saturation. Harmonics and other distortion products are produced, clearly an undesirable situation. Multiple cores can be used to increase the power capabilities of the transformer.

Transmission-line transformers are similar to conventional transformers, but can be used over wider frequency ranges. In a conventional transformer, high-frequency performance deterioration is caused primarily by leakage inductance, which rises with frequency. In a transmission-line transformer, the windings are arranged so there is tight capacitive coupling between the two. A high coupling coefficient is maintained up to considerably higher frequencies than with conventional transformers.

(A)

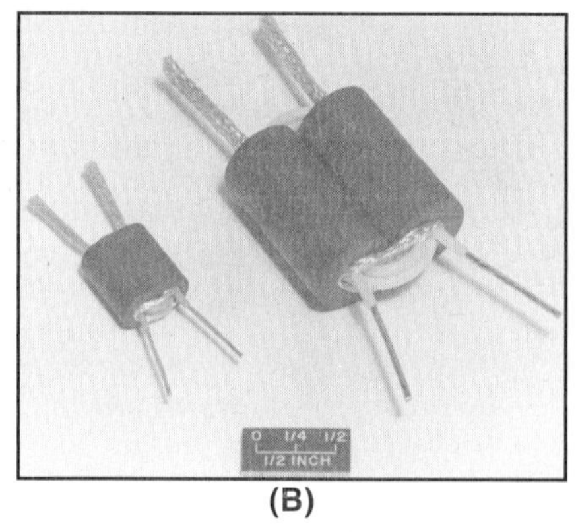

(B)

Fig 18.7—The two methods of constructing the transformers outlined in the text. At A, the one-turn loop is made from brass tubing; at B, a piece of coaxial cable braid is used for the loop.

OUTPUT FILTERING

Amplifier output filtering is sometimes necessary to meet spurious signal requirements. Broadband amplifiers, by definition, provide little if any inherent suppression of harmonic energy. Even amplifiers using output tank circuits often require further attenuation of undesired harmonics. High-level signals from one transmitter, particularly at multiple transmitter sites, can be intercepted by an antenna connected to another transmitter, conducted down the feed line and mixed in a power amplifier, causing spurious outputs. For example, an HF transceiver signal radiated from a triband beam may be picked up by a VHF FM antenna on the same mast. The signal saturates the low-power FM transceiver output stage, even with power off, and is reradiated by the VHF antenna. Proper use of filters can reduce such spurious energy considerably.

The filter used will depend on the application and the level of attenuation needed. Band-pass filters attenuate spurious signals above and below the passband for which they are designed. Low-pass filters attenuate only signals above the cutoff frequency, while high-pass filters reduce energy below the design cutoff frequency.

The **RF and AF Filters** chapter includes detailed information about designing suitable filters. Tables of component values in the **Component Data and References** chapter allow you to select a particular design and scale the values for different frequencies and impedance ranges as needed.

TRANSMITTING DEVICE RATINGS

Plate Dissipation

The ultimate factor limiting the power-handling capability of a tube is often (but not always) its maximum plate dissipation rating. This is the measure of how many watts of heat the tube can safely dissipate, if it is cooled properly, without exceeding critical temperatures. Excessive temperature can damage or destroy internal tube components or vacuum seals, resulting in tube failure. The same tube may have different voltage, current and power ratings depending on the conditions under which it is operated, but its safe temperature ratings must not be exceeded in any case!

Important cooling considerations are discussed in more detail in the Amplifier Cooling section of this chapter.

The efficiency of a power amplifier may range from approximately 25% to 75%, depending on its operating class, adjustment, and circuit losses. The efficiency indicates how much of the dc power supplied to the stage is converted to useful RF output power; the rest is dissipated as heat, mostly by the plate. By knowing the plate-dissipation limit of the tube and the efficiency expected from the class of operation selected, the maximum power input and output levels can be determined. The maximum safe power output is

$$P_{OUT} = \frac{P_D \, N_P}{100 - N_P} \quad (38)$$

where

P_{OUT} = the power output in W

P_D = the plate dissipation in W

N_P = the efficiency (10% = 10).

The dc input power would simply be

$$P_{IN} = \frac{100 \, P_D}{100 - N_P} \quad (39)$$

Almost all vacuum-tube power amplifiers in amateur service today operate as linear amplifiers (Class AB or B) with efficiencies of approximately 50% to 65%. That means that a useful power output of approximately 1 to 2.0 times the plate dissipation generally can be achieved. This requires, or course, that the tube is cooled enough to realize its maximum plate dissipation rating and that no other tube rating, such as maximum plate current or grid dissipation, is exceeded.

Type of modulation and duty cycle also influence how much output power can be achieved for a given tube dissipation. Some types of operation are less efficient than others, meaning that the tube must dissipate more heat. Some forms of modulation, such as CW or SSB, are intermittent in nature, causing less average heating than modulation formats such as RTTY in which there is continuous transmission. Power-tube manufacturers use two different rating systems to allow for the variations in service. CCS (Continuous Commercial Service) is the more conservative rating and is used for specifying tubes that are in constant use at full power. The second rating system is based on intermittent, low-duty-cycle operation, and is known as ICAS (Intermittent Commercial and Amateur Service). ICAS ratings are normally used by commercial manufacturers and individual amateurs who wish to obtain maximum power output consistent with reasonable tube life in CW and SSB service. CCS ratings should be used for FM, RTTY and SSTV applications. (Plate power transformers for amateur service are also rated in CCS and ICAS terms.)

Maximum Ratings

Tube manufacturers publish sets of maximum values for the tubes they produce. No maximum rated value should ever be exceeded. As an example, a tube might have a maximum plate-voltage rating of 2500 V, a maximum plate-current rating of 500 mA, and a maximum plate dissipation rating of 350 W. Although the plate voltage and current ratings might seem to imply a safe power input of 2500 V × 500 mA = 1250 W, this is true only if the dissipation rating will not be exceeded. If the tube is used in class AB2 with an expected efficiency of 60%, the maximum safe dc power input is

$$P_{IN} = \frac{100 P_D}{100 - N_D} = \frac{100 \times 350}{100 - 60} = 875 \text{ W}$$

In this case, any combination of plate voltage and current whose product does not exceed 875 W (and which allows the tube to achieve the expected 60% efficiency) is acceptable. A good compromise might be 2000 V and 437 mA: 2000 × 0.437 = 874 W input. If the maximum plate voltage of 2500 is used, then the plate current should be limited to 350 mA (not 500 mA) to stay within the maximum plate dissipation rating of 350 W.

TRANSISTOR POWER DISSIPATION

RF power-amplifier transistors are limited in power-handling capability by the amount of heat the device can safely dissipate. Power dissipation for a transistor is abbreviated P_D. The maximum rating is based on maintaining a case temperature of 25°C (77°F), which is seldom possible if a conventional air-cooled heat sink is used in an ambient air temperature of 70°F or higher. For higher temperatures, the device must be derated (in terms of milliwatts or watts per degree C) as specified by the manufacturer for that particular device. The efficiency considerations described earlier in reference to plate dissipation apply here also. A rule of thumb for selecting a transistor suitable for a given RF power output level is to choose one that has a maximum dissipation (with the heat sink actually to be used) of twice the desired output power.

Maximum Transistor Ratings

Transistor data sheets specify the maximum operating voltage for several conditions. Of particular interest is the V_{CEO} specification (collector to emitter voltage, with the base open). In RF amplifier service the collector to emitter voltage can rise to twice the dc supply potential. Thus, if a 12-V supply is used, the transistor should have a V_{CEO} of 24 V or greater to preclude damage.

The maximum collector current is also specified by the manufacturer. This specification is actually limited by the current-carrying capabilities of the internal bonding wires. Of course, the collector current must stay below the level that generates heat higher than the allowable device power dissipation. Many transistors are also rated for the load mismatch they can safely withstand. A typical specification might be for a transistor to tolerate a 30:1 SWR at all phase angles.

Transistor manufacturers publish data sheets that describe all the appropriate device ratings. Typical operating results are also given in these data sheets. In addition, many manufacturers publish application notes illustrating the use of their devices in practical circuits. Construction details are usually given. Perhaps owing to the popularity of Amateur Radio among electrical engineers, many of the notes describe applications especially suited to the Amateur Service. Specifications for some of the more popular RF power transistors are found in the **Component Data and References** chapter.

PASSIVE COMPONENT RATINGS

Output Tank Capacitor Ratings

The tank capacitor in a high-power amplifier should be chosen with sufficient spacing between plates to preclude high-voltage breakdown. The peak RF voltage present across a properly loaded tank circuit, without modulation, may be taken conservatively as being equal to the dc plate or collector voltage. If the dc supply voltage also appears across the tank capacitor, this must be added to the peak RF voltage, making the total peak voltage twice the dc supply voltage. At the higher voltages, it is usually desirable to design the tank circuit so that the dc supply voltages do not appear across the tank capacitor, thereby allowing the use of a smaller capacitor with less plate spacing. Capacitor manufacturers usually rate their products in terms of the peak voltage between plates. Typical plate spacings are given in **Table 18.3**.

Output tank capacitors should be mounted as close to the tube as temperature considerations will permit, to make possible the shortest path with the lowest possible inductive reactance from plate to cathode. Especially at the higher frequen-

Table 18.3

Typical Tank-Capacitor Plate Spacings

Spacing Inches	*Peak Voltage*	*Spacing Inches*	*Peak Voltage*	*Spacing Inches*	*Peak Voltage*
0.015	1000	0.07	3000	0.175	7000
0.02	1200	0.08	3500	0.25	9000
0.03	1500	0.125	4500	0.35	11000
0.05	2000	0.15	6000	0.5	13000

Table 18.4

Copper Conductor Sizes for Transmitting Coils for Tube Transmitters

Power Output (Watts)	*Band (MHz)*	*Minimum Conductor Size*
1500	1.8-3.5	10
	7-14	8 or 1/8"
	18-28	6 or 3/16"
500	1.8-3.5	12
	7-14	10
	18-28	8 or 1/8"
150	1.8-3.5	16
	7-14	12
	18-28	10

*Whole numbers are AWG; fractions of inches are tubing ODs.

cies, where minimum circuit capacitance becomes important, the capacitor should be mounted with its stator plates well spaced from the chassis or other shielding. In circuits in which the rotor must be insulated from ground, the capacitor should be mounted on ceramic insulators of a size commensurate with the plate voltage involved and—most important of all, from the viewpoint of safety to the operator—a well-insulated coupling should be used between the capacitor shaft and the knob. The section of the shaft attached to the control knob should be well grounded. This can be done conveniently by means of a metal shaft bushing at the panel.

Tank Coils

Tank coils should be mounted at least half their diameter away from shielding or other large metal surfaces, such as blower housings, to prevent a marked loss in Q. Except perhaps at 24 and 28 MHz, it is not essential that the coil be mounted extremely close to the tank capacitor. Leads up to 6 or 8 inches are permissible. It is more important to keep the tank capacitor, as well as other components, out of the immediate field of the coil.

The principal practical considerations in designing a tank coil usually are to select a conductor size and coil shape that will fit into available space and handle the required power without excessive heating. Excessive power loss as such is not necessarily the worst hazard in using too-small a conductor: it is not uncommon for the heat generated to actually unsolder joints in the tank circuit and lead to physical damage or failure. For this reason it's extremely important, especially at power levels above a few hundred watts, to ensure that all electrical joints in the tank circuit are secured mechanically as well as soldered. **Table 18.4** shows recommended conductor sizes for amplifier tank coils, assuming loaded tank circuit Qs of 15 or less on the 24 and 30 MHz bands and 8 to 12 on the lower frequency bands. In the case of input circuits for screen-grid tubes where driving power is quite small, loss is relatively unimportant and almost any physically convenient wire size and coil shape is adequate.

The conductor sizes in Table 18.4 are based on experience in continuous-duty amateur CW, SSB, and RTTY service and assume that the coils are located in a reasonably well ventilated enclosure. If the tank area is not well ventilated and/or if significant tube heat is transferred to the coils, it is good practice to increase AWG wire sizes by two (for example, change from #12 to #10) and tubing sizes by 1/16 inch.

Larger conductors than required for current handling are often used to maximize unloaded Q, particularly at higher frequencies. Where skin depth effects increase losses, the greater surface area of large diameter conductors can be beneficial. Small-diameter copper tubing, up to 3/8-inch outer diameter, can be used successfully for tank coils up through the lower VHF range. Copper tubing in sizes suitable for constructing high-power coils is generally available in 50-ft rolls from plumbing and refrigeration equipment suppliers. Silver-plating the tubing further reduces losses. This is especially true as the tubing ages and oxidizes. Silver oxide is a much better conductor than copper oxides, so silver-plated tank coils maintain their low-loss characteristics even after years of use.

At VHF and above, tank circuit inductances do not necessarily resemble the familiar coil. The inductances required to resonate tank circuits of reasonable Q at these higher frequencies are small enough that only strip lines or sections of transmission line are practical. Since these are constructed from sheet metal or large-diameter tubing, current-handling capabilities normally are not a relevant factor.

RF Chokes

The characteristics of any RF choke vary with frequency. At low frequencies the choke presents a nearly pure inductance. At some higher frequency it takes on high impedance characteristics resembling those of a parallel-resonant circuit. At a still higher frequency it goes through a series-resonant condition, where the impedance is lowest—generally much too low to perform satisfactorily as a shunt-feed plate choke. As frequency increases further, the pattern of alternating parallel and series resonances repeats. Between resonances, the choke will show widely varying amounts of inductive or capacitive reactance.

In series-feed circuits, these characteristics are of relatively small importance because the RF voltage across the choke is negligible. In a shunt-feed circuit such as is used in most high-power amplifiers, however, the choke is directly in parallel with the tank circuit, and is subject to the full tank RF voltage. If the choke does not present a sufficiently high impedance, enough power will be absorbed by the choke to burn it out. To avoid this, the choke must have a sufficiently high reactance to be effective at the lowest frequency (at least equal to the plate load resistance), and yet have no series resonances near any of the higher frequency bands. A resonant-choke failure in a high-power amplifier can be very dramatic and damaging!

Thus any choke intended for shunt-feed use should be carefully investigated with a dip meter. The choke must be shorted end-to-end with a direct, heavy braid or strap. Because nearby metallic objects affect the resonances, it should be mounted in its intended position, but disconnected from the rest of the circuit. A dip meter coupled an inch or two away from one end of the choke nearly always will show a deep, sharp dip at the lowest series-resonant frequency and shallower dips at higher series resonances.

Any choke to be used in an amplifier for the 1.8 to 28 MHz bands requires careful (or at least lucky!) design to perform well on all amateur bands within that range. Most simply put, the challenge is to achieve sufficient inductance that the choke doesn't "cancel" a large part of tuning capacitance on 1.8 MHz. At the same time, try to position all its series resonances where they can do no harm. In general, close wind enough #20 to #24 magnet wire to provide about 135 µH inductance on a 3/4 to 1-inch diameter cylindrical form of ceramic, Teflon, or fiberglass. This gives a reactance of 1500 Ω at 1.8 MHz and yet yields a first series resonance in the vicinity of 25 MHz. Before the advent of the 24.9-MHz band this worked fine. But trying to "squeeze" the resonance into the narrow gaps between the 21, 24, and/or 28-MHz bands is quite risky unless sophisticated instrumentation is available. If the number of turns on the choke is selected to place its first series resonance at 23.2 MHz, midway between 21.45 and 24.89 MHz, the choke impedance will typically be high enough for satisfactory operation on the 21, 24 and 28 MHz bands. The choke's first series resonance should be measured very carefully as described above using a dip meter and calibrated receiver or RF impedance bridge, with the choke mounted in place on the chassis.

Investigations with a vector impedance meter have shown that "trick" designs, such as using several shorter windings spaced along the form, show little if any improvement in choke resonance characteristics. Some commercial amplifiers circumvent the problem by bandswitching the RF choke. Using a larger diameter (1 to 1.5 inch) form does move the first series resonance somewhat higher for a given value of basic inductance. Beyond that, it is probably easiest for an all-band amplifier to add or subtract enough turns to move the first resonance to about 35 MHz and settle for a little less than optimum reactance on 1.8 MHz.

Blocking Capacitors

A series capacitor is usually used at the input of the amplifier output circuit. Its purpose is to block dc from appearing on matching circuit components or the antenna. As mentioned in the section on tank capacitors, output-circuit voltage requirements are considerably reduced when only RF voltage is present.

To provide a margin of safety, the voltage rating for a blocking capacitor should be at least 25% to 50% greater than the dc voltage applied. A large safety margin is desirable, since blocking capacitor failure can bring catastrophic results.

To avoid affecting the amplifier's tuning and matching characteristics, the blocking capacitor should have a low impedance at all operating frequencies. Its reactance at the lowest operating frequency should be not more than about 5% of the plate load resistance.

The capacitor also must be capable of handling, without overheating or significantly changing value, the substantial RF current that flows through it. This current usually is greatest at the highest frequency of operation, where tube output capacitance constitutes a significant part of the total tank capacitance. A significant portion of circulating tank current therefore flows through the blocking capacitor. As a conservative and very rough rule of thumb, the maximum RF current in the blocking capacitor (at 28 MHz) is

$$I_{CBlock} \approx I_p + 0.15 \times C_{OUT} \times V_{dc} \quad (40)$$

where

I_{CBlock} = maximum RMS current through blocking capacitor, in A

C_{OUT} = output capacitance of the output tubes, in pF

V_{dc} = dc plate voltage, in kV

I_p = dc plate current at full output, in A.

Transmitting capacitors are rated by their manufacturers in terms of their RF current-carrying capacity at various frequencies. Below a couple hundred watts at the high frequencies, ordinary disc ceramic capacitors of suitable voltage rating work well in high-impedance tube amplifier output circuits. Some larger disk capacitors rated at 5 to 8 kV also work well for higher power levels at HF; for example, two inexpensive Centralab type DD-602 discs (0.002 µF, 6 kV) in parallel have proved to be a reliable blocking capacitor for 1.5-kW amplifiers operating at plate voltages to about 2.5 kV. At very high power and voltage levels and at VHF, ceramic "doorknob" transmitting capacitors are needed for their low losses and high current handling capabilities. So-called "TV doorknobs" may break down at high RF current levels and should be avoided.

The very high values of Q_L found in many VHF and UHF tube-type amplifier tank circuits often require custom fabrication of the blocking capacitor. This can usually be accommodated through the use of a Teflon "sandwich" capacitor. Here, the blocking capacitor is formed from two parallel plates separated by a thin layer of Teflon. This capacitor often is part of the tank circuit itself, forming a very low-loss blocking capacitor. Teflon is rated for a minimum breakdown voltage of 2000 V per mil of thickness, so voltage breakdown should not be a factor in any practically realized circuit. The capacitance formed from such a Teflon sandwich can be calculated from the information presented elsewhere in this *Handbook* (use a dielectric constant of 2.1 for Teflon). In order to prevent any potential irregularities caused by dielectric thickness variations (including air gaps), Dow-Corning DC-4 silicone grease should be evenly applied to both sides of the Teflon dielectric. This grease has properties similar to Teflon, and will fill in any surface irregularities that might cause problems.

The very low impedances found in transistorized amplifiers present special problems. In order to achieve the desired low blocking-capacitor impedance, large-value capacitors are required. Special ceramic chips and mica capacitors are available that meet the requirements for high capacitance, large current carrying capability and low associated inductance. These capacitors are more costly than standard disk-ceramic or silver-mica units, but their level of performance easily justifies their price. Most of these special-purpose capacitors are either leadless or come with wide straps instead of normal wire leads. Disc-ceramic and other wire-lead capacitors are generally not suitable for transistor power-amplifier service.

SOURCES OF OPERATING VOLTAGES

Tube Filament or Heater Voltage

The heater voltage for the indirectly heated cathode tubes found in low-power classifications may vary 10% above or below rating without seriously reducing the life of the tube. A power vacuum tube can use either a directly heated filament or an indirectly heated cathode. The filament voltage for either type should be held within 5% of rated voltage. Because of internal tube heating at UHF and higher, the manufacturers' filament voltage rating often is reduced at these higher frequencies. The derated filament voltages should be followed carefully to maximize tube life. Series dropping resistors may be required in the filament circuit to attain the correct voltage. The voltage should be measured at the filament pins of the tube socket while the amplifier is running. The filament choke and interconnecting wiring all have voltage drops associated with them. The high current drawn by a power-tube heater circuit causes substantial voltage drops to occur across even small resistances. Also, make sure that the plate power drawn from the power line does not cause the filament voltage to drop below the proper value when plate power is applied.

Thoriated filaments lose emission when the tube is overloaded appreciably. If the overload has not been too prolonged, emission sometimes may be restored by operating the filament at rated voltage, with all other voltages removed, for a period of 30 to 60 minutes. Alternatively, you might try operating the tube at 20% above rated filament voltage for five to ten minutes.

VACUUM-TUBE PLATE VOLTAGE

DC plate voltage for the operation of RF amplifiers is most often obtained from a transformer-rectifier-filter system (see the **Power Supplies** chapter) designed to deliver the required plate voltage at the required current. It is not unusual for a power tube to arc over internally (generally from the plate to the screen or control grid) once or twice, especially soon after it is first placed into service. The flashover by itself is not normally dangerous to the tube, provided that instantaneous maximum plate current to the tube is held to a safe value and the high-voltage plate supply is shut off very quickly.

A good protective measure against this is the inclusion of a high-wattage power resistor in series with the plate high-voltage circuit. The value of the resistor, in ohms, should be approximately 10 to 15 times the no-load plate voltage in kV. This will limit peak fault current to 67 to 100 A. The series resistor should be rated for 25 or 50-W power dissipation; vitreous enamel coated wire-wound resistors of the common Ohmite or Clarostat types have been found to be capable of handling repeated momentary fault-current surges without damage. Aluminum-cased resistors such as some made by Dale are not recommended for this application. Each resistor also must be large enough to safely handle the maximum value of normal plate current; the wattage rating required may be calculated from $P = I^2R$. If the total filter capacitance exceeds 25 µF, it is a good idea to use 50-W resistors in any case. Even at high plate-current levels, the addition of the resistors does little to affect the dynamic regulation of the plate supply.

Since tube (or other high-voltage circuit) arcs are not necessarily self-extinguishing, a fast-acting plate overcurrent relay or primary circuit breaker also is recommended to quickly shut off ac power to the HV supply when an arc begins. Using this protective system, a mild HV flashover may go undetected, while a more severe one will remove ac power from the HV supply. (The cooling blower should remain energized however, since the tube may be hot when the HV is removed due to an arc.) If effective protection is not provided, however, a "normal" flashover, even in a new tube, is likely to damage or destroy the tube, and also frequently destroys the rectifiers in the power supply as well as the plate RF choke. A power tube that flashes over more than about 3 to 5 times in a period of several months likely is defective and will have to be replaced before long.

Grid Bias

The grid bias for a linear amplifier should be highly filtered and well regulated. Any ripple or other voltage change in the bias circuit modulates the amplifier. This causes hum and/or distortion to appear on the signal. Since most linear amplifiers draw only small amounts of grid current, these bias-supply requirements are not difficult to achieve.

Fixed bias for class AB1 tetrode and pentode amplifiers is usually obtained from a variable-voltage regulated supply. Voltage adjustment allows setting bias level to give the desired resting plate current. **Fig 18.8A** shows a simple Zener-diode-regulated bias supply. The dropping resistor is chosen to allow approximately 10 mA of Zener current. Bias is then reasonably well regulated for all drive conditions up to 2 or 3 mA of grid current. The potentiometer allows bias to be adjusted between Zener and approximately 10 V higher. This range is usually adequate to allow for variations in the characteristics of different tubes. Under standby conditions, when it is desirable to cut off the tube entirely, the Zener ground return is interrupted so the full bias supply voltage is applied to the grid.

In Fig 18.8B and C, bias is obtained from the voltage drop across a Zener diode in the cathode (or filament center-tap) lead. Operating bias is obtained by the voltage drop across D1 as a result of plate (and screen) current flow. The diode voltage drop effectively raises the cathode potential relative to the grid. The grid is therefore negative with respect to the cathode by the Zener voltage of the diode. The Zener-diode wattage rating should be twice the product of the maximum cathode current times the rated zener voltage. Therefore, a tube requiring 15 V of bias with a maximum cathode current of 100 mA would dissipate 1.5 W in the Zener diode. To allow a suitable safety factor, the diode rating should be 3 W or more. The circuit of Fig 18.8C illustrates how D1 would be used with a cathode driven (grounded grid) amplifier as opposed to the grid driven example at B.

In all cases, the Zener diode should be bypassed by a 0.01-µF capacitor of suitable voltage. Current flow through any type of diode generates shot noise. If not bypassed, this noise would modulate the amplified signal, causing distortion in the amplifier output.

Screen Voltage For Tubes

Power tetrode screen current varies widely with both excitation and loading. The current may be either positive or negative, depending on tube characteristics and amplifier operating conditions. In a linear amplifier, the screen voltage should be well regulated for all values of screen current. The power output from a tetrode is very sensitive to screen voltage, and any dynamic change in the screen potential can cause distorted output. Zener diodes are commonly used for screen regulation.

Fig 18.9 shows a typical example of a regulated screen supply for a power tetrode amplifier. The voltage from a fixed dc supply is dropped to the Zener stack voltage by the current-limiting resistor. A screen bleeder resistor is connected in parallel with the zener stack to allow for the negative screen current developed under certain tube operating conditions. Bleeder current is chosen to be roughly 10 to 20 mA greater than the expected maximum negative screen current, so that screen voltage is regulated for all values of current between maximum negative screen current and maximum positive screen current. For external-anode tubes in the 4CX250 family, a typical screen bleeder current value would be 20 mA. For the 4CX1000 family, a screen-bleeder current of 70 mA is required.

Screen voltage should never be applied to a tetrode unless plate voltage and load also are applied; otherwise the screen will act like an anode and will draw excessive current. Supplying the screen through a series-dropping resistor from the plate supply affords a measure of protection, since the screen voltage only appears when there is plate voltage. Alternatively, a fuse can be placed between the regulator and the bleeder resistor. The fuse should not be installed between the bleeder resistor and the tube, because the tube should never be operated without a load on the screen. Without a load, the screen potential tends to rise to the anode voltage. Any screen bypass capacitors or other associated circuits are likely be damaged by this high voltage.

In Fig 18.9, a varistor is connected from screen to ground. If, because of some circuit failure, the screen voltage should rise substantially above its nominal level, the

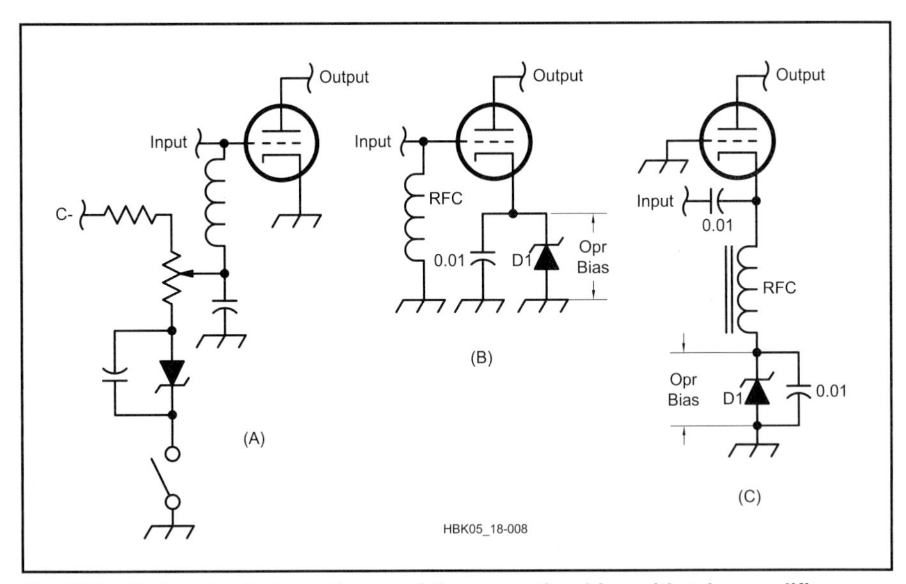

Fig 18.8—Various techniques for providing operating bias with tube amplifiers.

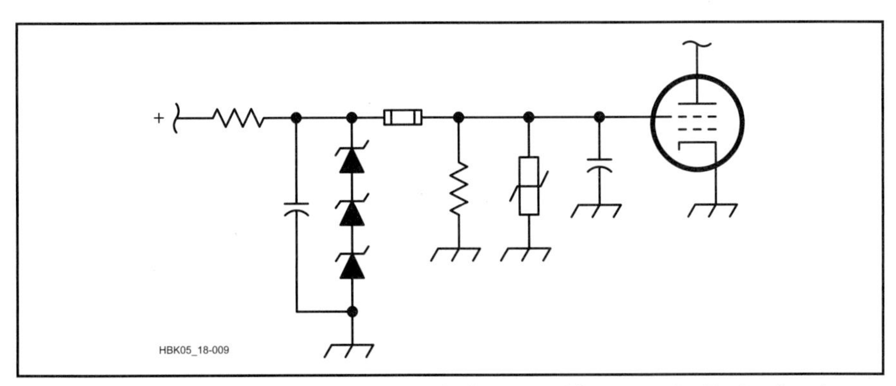

Fig 18.9—A Zener-regulated screen supply for use with a tetrode. Protection is provided by a fuse and a varistor.

varistor will conduct and clamp the screen voltage to a low level. If necessary to protect the varistor or screen dropping resistors, a fuse or overcurrent relay may be used to shut off the screen supply so that power is interrupted before any damage occurs. The varistor voltage should be approximately 30% to 50% higher than normal screen voltage.

Transistor Biasing

Solid-state power amplifiers generally operate in Class C or AB. When some bias is desired during Class C operation (**Fig 18.10A**), a resistance of the appropriate value can be placed in the emitter return as shown. Most transistors will operate in Class C without adding bias externally, but in some instances the amplifier efficiency can be improved by means of emitter bias. Reverse bias supplied to the base of the Class-C transistor should be avoided because it will lead to internal breakdown of the device during peak drive periods. The damage is frequently a cumulative phenomenon, leading to gradual destruction of the transistor junction.

A simple method for Class AB biasing is shown in Fig 18.10B. D1 is a silicon diode that acts as a bias clamp at approximately 0.7 V. This forward bias establishes linear-amplification conditions. That value of bias is not always optimum for a specified transistor in terms of IMD. Variable bias of the type illustrated in Fig 18.10C permits the designer sufficient flexibility to position the operating point for best linearity. The diode clamp or the reference sensor for another type of regulator is usually thermally bonded to the power transistor or its heat sink. The bias level then tracks the thermal characteristics of the output transistor. Since a transistor's current transfer characteristics are a function of temperature, thermal tracking of the bias is necessary to maintain device linearity and, in the case of bipolar devices, to prevent thermal runaway and the subsequent destruction of the transistor.

AMPLIFIER COOLING

Tube Cooling

Vacuum tubes must be operated within the temperature range specified by the manufacturer if long tube life is to be achieved. Tubes having glass envelopes and rated at up to 25-W plate dissipation may be used without forced-air cooling if the design allows a reasonable amount of convection cooling. If a perforated metal enclosure is used, and a ring of 1/4 to 3/8-inch-diameter holes is placed around the tube socket, normal convective airflow can be relied on to remove excess heat at room temperatures.

For tubes with greater plate dissipation ratings, and even for very small tubes operated close to maximum rated dissipation, forced-air cooling with a fan or blower is needed. Most manufacturers rate tube-cooling requirements for continuous-duty operation. Their literature will indicate the required volume of airflow, in cubic feet per minute (CFM), at some particular backpressure. Often this data is given for several different values of plate dissipation, ambient air temperature and even altitude above sea level.

One extremely important consideration is often overlooked by power-amplifier designers and users alike: a tube's plate dissipation rating is only its maximum potential capability. The power that it can actually dissipate safely depends directly on the cooling provided. The actual power capability of virtually all tubes used in high-power amplifiers for amateur service depends on the volume of air forced through the tube's cooling structure.

This requirement usually is given in terms of cubic feet of air per minute, (CFM), delivered into a "back pressure" representing the resistance of the tube cooler to air flow, stated in inches of water. Both the CFM of airflow required and the pressure needed to force it through the cooling system are determined by ambient air temperature and altitude (air density), as well as by the amount of heat to be dissipated. The cooling fan or blower must be capable of delivering the specified airflow into the corresponding backpressure. As a result of basic air flow and heat transfer principles, the volume of airflow required through the tube cooler increases considerably faster than the plate dissipation, and backpressure increases even faster than airflow. In addition, blower air output decreases with increasing backpressure until, at the blower's so-called

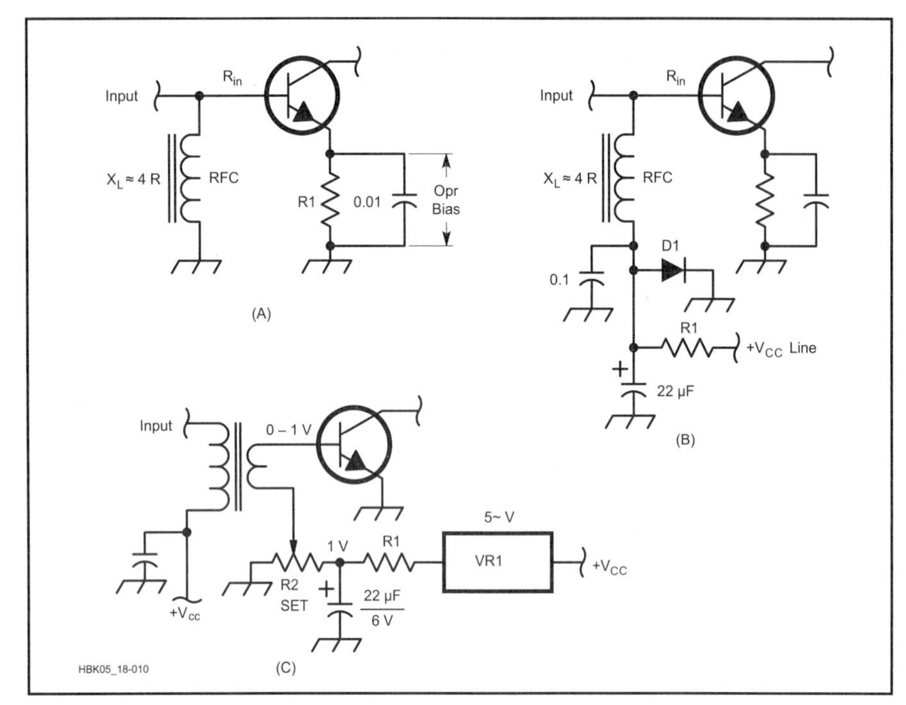

Fig 18.10—Biasing methods for use with transistor amplifiers.

Table 18.5

Specifications of Some Popular Tubes, Sockets and Chimneys

Tube	*CFM*	*Back Pressure (inches)*	*Socket*	*Chimney*
3-500Z	13	0.13	SK-400, SK-410	SK-416
3CX800A7	19	0.50	SK-1900	SK-1906
3CX1200A7	31	0.45	SK-410	SK-436
3CX1200Z7	42	0.30	SK-410	—
3CX1500/8877	35	0.41	SK-2200, SK-2210	SK-2216
4-400A/8438	14	0.25	SK-400, SK-410	SK-406
4-1000A/8166	20	0.60	SK-500, SK-510	SK-506
4CX250R/7850	6.4	0.59	SK602A, SK-610, SK-610A SK-611, SK-612, SK-620, SK-620A, SK-621, SK-630	
4CX400/8874	8.6	0.37	SK1900	SK606
4CX400A	8	0.20	SK2A	—
4CX800A	20	0.50	SK1A	—
4CX1000A/8168	25	0.20	SK-800B, SK-810B, SK-890B	SK-806
4CX1500B/8660	34	0.60	SK-800B, SK-1900	SK-806
4CX1600B	36	0.40	SK3A	CH-1600B

These values are for sea-level elevation. For locations well above sea level (5000 ft/1500 m, for example), add an additional 20% to the figure listed.

"cutoff pressure," actual air delivery is zero. Larger and/or faster-rotating blowers are required to deliver larger volumes of air at higher backpressure.

Values of CFM and backpressure required to realize maximum rated plate dissipation for some of the more popular tubes, sockets and chimneys (with 25°C ambient air and at sea level) are given in **Table 18.5**. Backpressure is specified in inches of water and can be measured easily in an operational air system as indicated in **Figs 18.11** and **18.12**. The pressure differential between the air passage and atmospheric pressure is measured with a device called a *manometer*. A manometer is nothing more than a piece of clear tubing, open at both ends and fashioned in the shape of a "U." The manometer is temporarily connected to the chassis and is removed after the measurements are completed. As shown in the diagrams, a small amount of water is placed in the tube. At Fig 18.12A the blower is "off" and the water seeks its own level, because the air pressure (ordinary atmospheric pressure) is the same at both ends of the manometer tube. At B, the blower is "on" (socket, tube and chimney in place) and the pressure difference, in terms of inches of water, is measured. For most applications a standard ruler used for measurement will yield sufficiently accurate results.

Table 18.6 gives the performance specifications for a few of the many Dayton blowers, which are available through Grainger catalog outlets in all 50 states. Other blowers having wheel diameters, widths and rotational speeds similar to any in Table 18.6 likely will have similar flow and backpressure characteristics. If in doubt about specifications, consult the manufacturer. Tube temperature under actual operating conditions is the ultimate criterion for cooling adequacy and may be determined using special crayons or lacquers that melt and change appearance at specific temperatures. The setup of Fig 18.12, however, nearly always gives sufficiently accurate information.

As an example, consider the cooling design of a linear amplifier to use one 3CX800A7 tube, to operate near sea level with the air temperature not above 25°C. The tube, running 1150-W dc input, easily delivers 750-W continuous output, resulting in 400-W plate dissipation ($P_{DIS} = P_{IN} - P_{OUT}$). According to the manufacturer's data, adequate tube cooling at 400 W P_D requires at least 6 CFM of air at 0.09 inches of water back pressure. In Table 18.6, a Dayton no. 2C782 will do the job with a good margin of safety.

If the same single tube were to be operated at 2.3 kW dc input to deliver 1.5 kW output (substantially exceeding its maximum electrical ratings!), P_{IN} would be about 2300 W and $P_D \approx 800$ W. The minimum cooling air required would be about 19 CFM at 0.5 inches of water pressure—doubling P_{DIS}, more than tripling the CFM of air flow required and increasing back pressure requirements on the blower by a factor of 5.5!

However, two 3CX800A7 tubes are needed to deliver 1.5 kW of continuous maximum legal output power in any case. Each tube will operate under the same conditions as in the single-tube example above, dissipating 400 W. The total cooling air requirement for the two tubes is therefore 12 CFM at about 0.09 inches of water, only two-thirds as much air volume and one-fifth the back pressure required by a single tube. While this may seem surprising, the reason lies in the previously mentioned fact that both the airflow required by a tube and the resultant back pressure increase much more rapidly than P_D of the tube. Blower air delivery capability, conversely, decreases as back-pres-

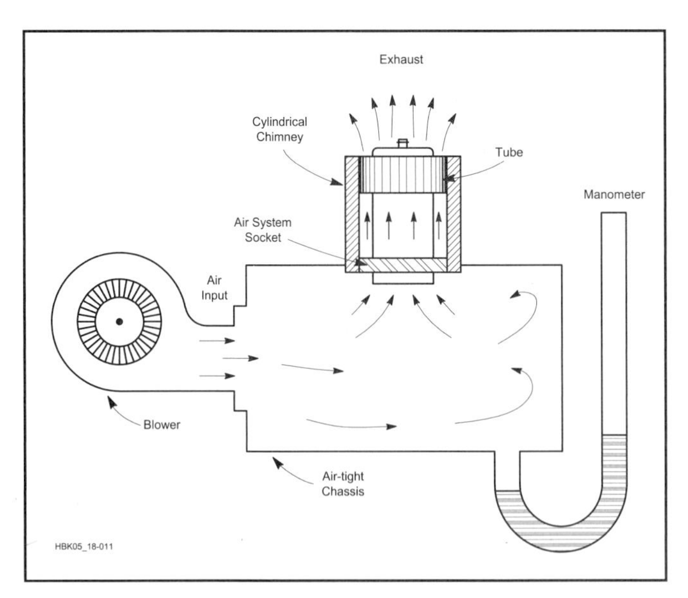

Fig 18.11—Air is forced into the chassis by the blower and exits through the tube socket. The manometer is used to measure system back pressure, which is an important factor in determining the proper size blower.

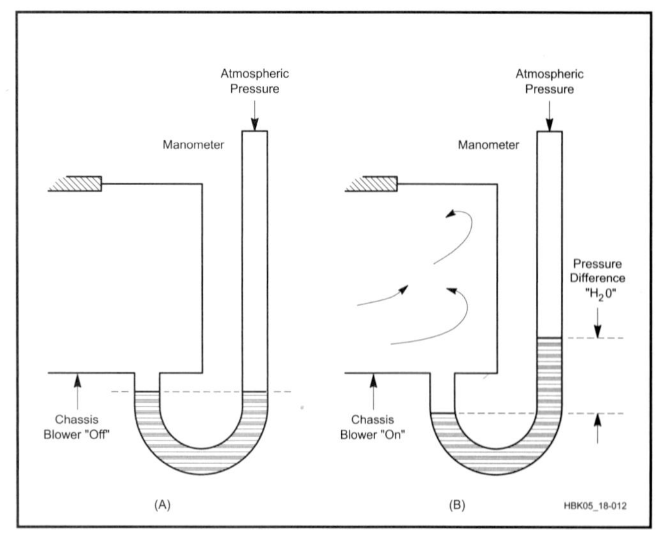

Fig 18.12—At A the blower is "off" and the water will seek its own level in the manometer. At B the blower is "on" and the amount of back pressure in terms of inches of water can be measured as indicated.

Table 18.6

Blower Performance Specifications

Wheel Dia	*Wheel Width*	*RPM*	*Free Air CFM*	*CFM for Back Pressure (inches) 0.1*	*0.2*	*0.3*	*0.4*	*0.5*	*Cutoff*	*Stock No.*
2"	1"	3160	15	13	4	—	—	—	0.22	2C782
3"	1 15/32"	3340	54	48	43	36	25	17	0.67	4C012
3"	1 7/8"	3030	60	57	54	49	39	23	0.60	4C440
3"	1 7/8"	2880	76	70	63	56	45	8	0.55	4C004
3 13/16"	1 7/8"	2870	100	98	95	90	85	80	0.80	4C443
3 13/16"	2 1/2"	3160	148	141	135	129	121	114	1.04	4C005

sure is increased. Thus a Dayton 2C782 blower can cool two 3CX800A7 tubes dissipating 800 W total, but a much larger (and probably noisier) no. 4C440 would be required to handle the same power with a single tube.

In summary, three very important considerations to remember are these:

- A tube's actual safe plate dissipation capability is totally dependent on the amount of cooling air forced through its cooling system. Any air-cooled power tube's maximum plate dissipation rating is meaningless unless the specified amount of cooling air is supplied.
- Two tubes will always safely dissipate a given power with a significantly smaller (and quieter) blower than is required to dissipate the same power with a single tube of the same type. A corollary is that a given blower can virtually always dissipate more power when cooling two tubes than when cooling a single tube of the same type.
- Blowers vary greatly in their ability to deliver air against backpressure so blower selection should not be taken lightly.

A common method for directing the flow of air around a tube involves the use of a pressurized chassis. This system is shown in Fig 18.11. A blower attached to the chassis forces air around the tube base, often through holes in its socket. A chimney is used to guide air leaving the base area around the tube envelope or anode cooler, preventing it from dispersing and concentrating the flow for maximum cooling.

A less conventional approach that offers a significant advantage in certain situations is shown in **Fig 18.13**. Here the anode compartment is pressurized by the blower. A special chimney is installed between the anode heat exchanger and an exhaust hole in the compartment cover. When the blower pressurizes the anode compartment, there are two parallel paths for airflow: through the anode and its chimney, and through the air system socket. Dissipation, and hence cooling air required, generally is much greater for the anode than for the tube base. Because high-volume anode airflow need not be forced through restrictive air channels in the base area, backpressure may be very significantly reduced with certain tubes and sockets. Only airflow actually needed is bled through the base area. Blower backpressure requirements may sometimes be reduced by nearly half through this approach.

Table 18.5 also contains the part numbers for air-system sockets and chimneys available for use with the tubes that are listed. The builder should investigate which of the sockets listed for the 4CX250R, 4CX300A, 4CX1000A and 4CX1600A best fit the circuit needs. Some of the sockets have certain tube elements grounded internally through the socket. Others have elements bypassed to ground through capacitors that are integral parts of the sockets.

Depending on your design philosophy and tube sources, some compromises in the cooling system may be appropriate. For example, if glass tubes are available inexpensively as broadcast pulls, a shorter life span may be acceptable. In such a case, an increase of convenience and a reduction in cost, noise, and complexity can be had by using a pair of "muffin" fans. One fan may be used for the filament seals and one for the anode seal, dispensing with a blower and air-system socket and chimney. The airflow with this scheme is not as uniform as with the use of a chimney. The tube envelope mounted in a cross flow has flow stagnation points and low heat transfer in certain regions of the envelope. These points become hotter than the rest of the envelope. The use of multiple fans to dis-

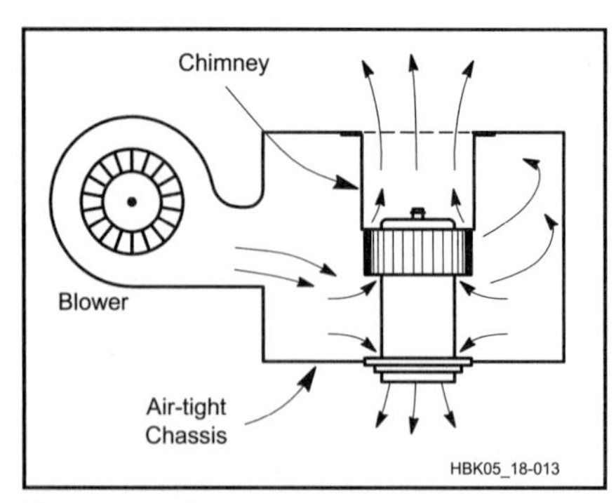

Fig 18.13—Anode compartment pressurization may be more efficient than grid compartment pressurization. Hot air exits upwards through the tube anode and through the chimney. Cool air also goes down through the tube socket to cool tube's pins and the socket itself.

turb the cross airflow can significantly reduce this problem. Many amateurs have used this cooling method successfully in low-duty-cycle CW and SSB operation but it is not recommended for AM, SSTV or RTTY service. The true test of the effectiveness of a forced air-cooling system is the amount of heat carried away from the tube by the air stream. The power dissipated can be calculated from the airflow temperatures. The dissipated power is

$$P_D = Q_A \left[\frac{T_2}{T_1} - 1 \right] \qquad (41)$$

where

P_D = the dissipated power, in W

Q_A = the air flow, in CFM (cubic feet per minute)

T_1 = the inlet air temperature, kelvin (normally quite close to room temperature)

T_2 = the amplifier exhaust temperature, kelvin.

The exhaust temperature can be measured with a cooking thermometer at the air outlet. The thermometer should not be placed inside the anode compartment because of the high voltage present.

Transistor Cooling

Transistors used in power amplifiers dissipate significant amounts of power, and the heat so generated must be effectively removed to maintain acceptable device temperatures. Some bipolar power transistors have the collector connected directly to the case of the device, as the collector creates most of the heat generated when the transistor is in operation. Others have the emitter connected to the case. However, if operated close to maximum rated dissipation, even the larger case designs cannot normally conduct heat away fast enough to keep the operating temperature of the device within the safe area—the maximum temperature that a device can stand without damage. Safe area is usually specified in a device data sheet, often in graphical form. Germanium power transistors theoretically may be operated at internal temperatures up to 100°C, while silicon devices may be run at up to 200°C. However, to assure long device lifetimes much lower case temperatures—not greater than 50° to 75°C for germanium and 75° to 100°C for silicon—are highly desirable. Leakage currents in germanium devices can be very high at elevated temperatures; thus, silicon transistors are preferred for most power applications.

A properly chosen heat sink often is essential to help keep the transistor junction temperature in the safe area. For low-power applications a simple clip-on heat sink will suffice, while for 100 W or higher input power a massive cast-aluminum finned radiator usually is necessary. The appropriate size heat sink can be calculated based on the thermal resistance between the transistor case and ambient air temperature. The first step is to calculate the total power dissipated by the transistor:

$$P_D = P_{DC} + P_{RFin} - P_{RFout} \qquad (42)$$

where

P_D = the total power dissipated by the transistor in W

P_{DC} = the dc power into the transistor, in W

P_{RFin} = the RF (drive) power into the transistor in W

P_{RFout} = the RF output power from the transistor in W.

The value of P_D is then used to obtain the θ_{CA} value from

$$\theta_{CA} = (T_C - T_A)/P_D \qquad (43)$$

where

θ_{CA} = the thermal resistance of the device case to ambient

T_C = the device case temperature in °C

T_A = the ambient temperature (room temperature) in °C.

A suitable heat sink, capable of transferring all of the heat generated by the transistor to the ambient air, can then be chosen from the manufacturer's specifications for θ_{CA}. A well-designed heat-sink system minimizes thermal path lengths and maximizes their cross-sectional areas. The contact area between the transistor and heat sink should have very low thermal resistance. The heat sink's mounting surface must be flat and the transistor firmly attached to the heat sink so intimate contact—without gaps or air voids—is made between the two. The use of silicone-based heat sink compounds can provide considerable improvement in thermal transfer. The thermal resistance of such grease is considerably lower than that of air, but not nearly as good as that of copper or aluminum. The quantity of grease should be kept to an absolute minimum. Only enough should be used to fill in any small air gaps between the transistor and heat sink mating surfaces. The maximum temperature rise in the transistor junction may easily be calculated by using the equation

$$T_J = (\theta_{JC} + \theta_{CA})\, P_D + TA \qquad (44)$$

where

T_J = the transistor junction temperature in °C

θ_{JC} = the manufacturer's published thermal resistance of the transistor

θ_{CA} = the thermal resistance of the device case to ambient

P_D = the power dissipated by the transistor

T_A = the ambient temperature in °C.

The value of T_J should be kept well below the manufacturer's recommended maximum to prevent premature transistor failure. Measured values of the ambient temperature and the device case temperature can be used in the preceding formulas to calculate junction temperature. The **Real-World Component Characteristics** chapter contains a more detailed discussion of transistor cooling.

Most of the problems facing an amplifier designer are not theoretical, but have to do with real-world component limitations. The **Real World Component Characteristics** chapter discusses the differences between ideal and real components.

A simplified equivalent schematic of an amplifying device is shown in **Fig 18.14A**. The input is represented by a series (parasitic) inductance feeding a resistance in parallel with a capacitance. The output consists of a current generator in parallel with a resistance and capacitance, followed by a series inductance. This is a reasonably accurate description of both transistors and vacuum tubes, regardless of circuit configuration (as demonstrated in Figs 18.17B and C). Both input and output impedances have a resistive component in parallel with a reactive component. Each also has a series inductive reactance, which represents connecting leads within the device. These inductances, unlike the other components of input and output impedance, often are not characterized in manufacturers' device specifications.

The amplifier input and output-matching networks must transform the complex impedances of the amplifying device to the source and load impedances (often 50-Ω transmission lines). Impedances associated with other parts of the amplifier circuit, such as a dc-supply choke, must also be considered in designing the matching networks. The matching networks and other circuit components are influenced by each other's presence, and these mutual effects must be given due consideration.

Perhaps the best way to clarify the considerations that enter into designing various types of RF power amplifiers is through example. The following examples illustrate common problems associated with power-amplifier design. They are not intended as detailed construction plans, but only demonstrate typical approaches useful in designing similar projects.

DESIGN EXAMPLE 1: A HIGH-POWER VACUUM-TUBE HF AMPLIFIER

Most popular HF transceivers produce approximately 100-W output. The EIMAC 8877 can deliver 1500-W output for approximately 60 W of drive when used in a grounded grid circuit. Grounded-grid operation is usually the easiest tube amplifier circuit to implement. Its input impedance is relatively low, often close to 50 Ω. Input/output shielding provided by the grid and negative feedback inherent in the grounded-grid circuit configuration reduce the likelihood of amplifier instability and provide excellent linearity without critical adjustments. Fewer supply voltages are needed in this configuration compared to others: Often just high-voltage dc for the plate and low-voltage ac for the filament.

The first step in the amplifier design process is to verify that the tube is actually capable of producing the desired results while remaining within manufacturer's ratings. The plate dissipation expected during normal operation of the amplifier is computed first. Since the amplifier will be used for SSB, a class of operation producing linear amplification must be used. Class AB2 provides a very good compromise between linearity and good efficiency, with effective efficiency typically exceeding 60%. Given that efficiency, an input power of 2500 W is needed to produce the desired 1500-W output. Operated under these conditions, the tube will dissipate about 1000 W—well within the manufacturer's specifications, provided

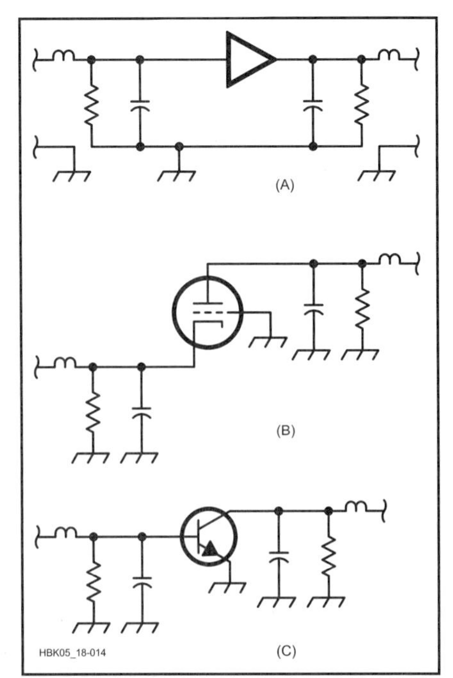

Fig 18.14—The electrical equivalents for power amplifiers. At A, the input is represented by a series stray inductance, then a resistor in parallel with a capacitor. The output is a current source in parallel with a resistor and capacitor, followed by a series stray inductance. These effects are applied to tubes and transistors in B and C.

adequate cooling airflow is supplied.

The grid in modern high-mu triodes is a relatively delicate structure, closely spaced to the cathode and carefully aligned to achieve high gain and excellent linearity. To avoid shortening tube life or even destruction of the tube, the specified maximum grid dissipation must not be exceeded for more than a few milliseconds under any conditions. For a given power output, the use of higher plate voltages tends to result in lower grid dissipation. It is important to use a plate voltage that is high enough to result in safe grid current levels at maximum output. In addition to maximum ratings, manufacturers' data sheets often provide one or more sets of "typical operation" parameters. This makes it even easier for the builder to achieve optimum results.

The 8877, operating at 3500 V, can produce 2075 W of RF output with excellent linearity and 64 W of drive. Operating at 2700 V it can deliver 1085 W with 40 W of drive. To some extent, the ease and cost of constructing a high-power amplifier, as well as its ultimate reliability, are enhanced by using the lowest plate voltage that will yield completely satisfactory performance. Interpolating between the two sets of typical operating conditions suggests that the 8877 can comfortably deliver 1.5 kW output with a 3100-V plate supply and 50 to 55 W of drive. Achieving 2500-W input power at this plate voltage requires 800 mA of plate current—well within the 8877's maximum rating of 1.0 A.

The next step in the design process is to calculate the optimum plate load resistance at this plate voltage and current for Class AB2 operation and design an appropriate output-matching network. From the earlier equations, R_L is calculated to be 2200 Ω.

Several different output networks might be used to transform the nominal 50-Ω resistance of the actual load to the 2200-Ω load resistance required by the 8877, but experience shows that pi and pi-L networks are most practical. Each can provide reasonable harmonic attenuation, is relatively easy to build mechanically and uses readily available components. The pi-L gives significantly greater harmonic attenuation than the pi and usually is the better choice—at least in areas where there is any potential for TVI or crossband interference. In a multiband amplifier, the extra cost of using a pi-L network is the "L" inductor and its associated bandswitch section.

To simplify and avoid confusion with terminology previously used in the pi and pi-L network design tables, in the remainder of this chapter Q_{IN} is the loaded Q of the amplifier's input matching tank, Q_{OUT} is the loaded Q of the output pi-L tank, Q_{PI} is the loaded Q of the output pi section only, and Q_L is the loaded Q of the output L section only.

The input impedance of a grounded-grid 8877 is typically on the order of 50 to 55 Ω, shunted by input capacitance of about 38 pF. While this average impedance is close enough to 50 Ω to provide negligible input SWR, the instantaneous value varies greatly over the drive cycle—that is, it is nonlinear. This nonlinear impedance is reflected back as a nonlinear load impedance at the exciter output, resulting in increased intermodulation distortion, reduced output power, and often meaningless exciter SWR meter indications. In addition, the tube's parallel input capacitance, as well as parasitic circuit reactances, often are significant enough at 28 MHz to create significant SWR. A tank circuit at the amplifier input can solve both of these problems by tuning out the stray reactances and stabilizing (linearizing) the tube input impedance through its flywheel effect. The input tank should have a loaded Q (called Q_{Lin} in this discussion) of at least two for good results. Increasing Q_{Lin} to as much as five results in a further small improvement in linearity and distortion, but at the cost of a narrower operating bandwidth. Even a Q_{Lin} of 1.0 to 1.5 yields significant improvement over an untuned input. A pi network commonly is used for input matching at HF.

Fig 18.15 illustrates these input and output networks applied in the amplifier circuit. The schematic shows the major components in the amplifier RF section, but with band-switching and cathode dc-return circuits omitted for clarity. C1 and C2 and L1 form the input pi network. C3 is a blocking capacitor to isolate the exciter from the cathode dc potential. Note that when the tube's average input resistance is close to 50 Ω, as in the case of the 8877, a simple parallel-resonant tank often can successfully perform the tuning and flywheel functions, since no impedance transformation is necessary. In this case it is important to minimize stray lead inductance between the tank and tube to avoid undesired impedance transformation.

The filament or "heater" in indirectly heated tubes such as the 8877 must be very close to the cathode to heat the cathode efficiently. A capacitance of several picofarads exists between the two. Particularly at very high frequencies, where these few picofarads represent a relatively low reactance, RF drive intended for the cathode can be capacitively coupled to the lossy filament and dissipated as heat. To avoid this, above about 50 MHz, the filament must be kept at a high RF impedance above ground. The high impedance (represented by choke RFC1 in Fig 18.15) minimizes RF current flow in the filament circuit so that RF dissipated in the filament becomes negligible. The choke's low-frequency resistance should be kept to a minimum to lessen voltage drops in the high-current filament circuit.

The choke most commonly used in this application is a pair of heavy-gauge insulated wires, bifilar-wound over a ferrite rod. The ferrite core raises the inductive reactance throughout the HF region so that a minimum of wire is needed, keeping filament-circuit voltage drops low. The bifilar winding technique assures that both filament terminals are at the same RF potential.

Below 30 MHz, the use of such a choke seldom is necessary or beneficial, but actually can introduce another potential problem. Common values of cathode-to-heater capacitance and heater-choke inductance often are series resonant in the 1.8 to 29.7 MHz HF range. A capacitance of 5 pF and an inductance of 50 µH, for example, resonate at 10.0 MHz; the actual components are just as likely to resonate near 7 or 14 MHz. At resonance, the circuit constitutes a relatively low impedance shunt from cathode to ground, which affects input impedance and sucks out drive signal. An unintended resonance like this near any operating frequency usually increases input SWR and decreases gain on that one particular band. While aggravating, the problem rarely completely disables or damages the amplifier, and so is seldom pursued or identified.

Fortunately, the entire problem is easily avoided—below 30 MHz the heater choke can be deleted. At VHF-UHF, or wherever a heater isolation choke is used for any reason, the resonance can be moved below the lowest operating frequency by connecting a sufficiently large capacitance (about 1000 pF) between the tube cathode and one side of the heater. It is good practice also to connect a similar capacitor between the heater terminals. It also would be good practice in designing other VHF/UHF amplifiers, such as those using 3CX800A7 tubes, unless the builder can ensure that the actual series resonance is well outside of the operating frequency range.

Plate voltage is supplied to the tube through RFC2. C5 is the plate blocking capacitor. The output pi-L network consists of tuning capacitor C6, loading capacitor C7, pi coil L2, and output L coil L3. RFC3 is a high-inductance RF choke placed at the output for safety purposes. Its value, usually 100 µH to 2 mH, is high enough so that it appears as an open circuit across the output connector for RF. However, should the plate blocking capacitor fail and allow high voltage onto the output matching network, RFC3 would short the dc to ground and blow the power-supply fuse or breaker. This prevents dangerous high voltage from appearing on the feed line or antenna. It also prevents electrostatic charge—from the antenna or from blocking capacitor leakage—from building up on the tank capacitors and causing periodic dc discharge arcs to ground. If such a dc discharge occurs while the amplifier is transmitting, it can trigger a potentially damaging RF arc.

Our next step is designing the input matching network. As stated earlier, tube input impedance varies moderately with plate voltage and load resistance as well as bias, but is approximately 50 to 55 Ω paralleled by C_{IN} of 38 pF, including stray capacitance. A simple parallel-resonant tank of Q_{IN} = 2 to 3 can provide an input SWR not exceeding 1.5:1, provided all

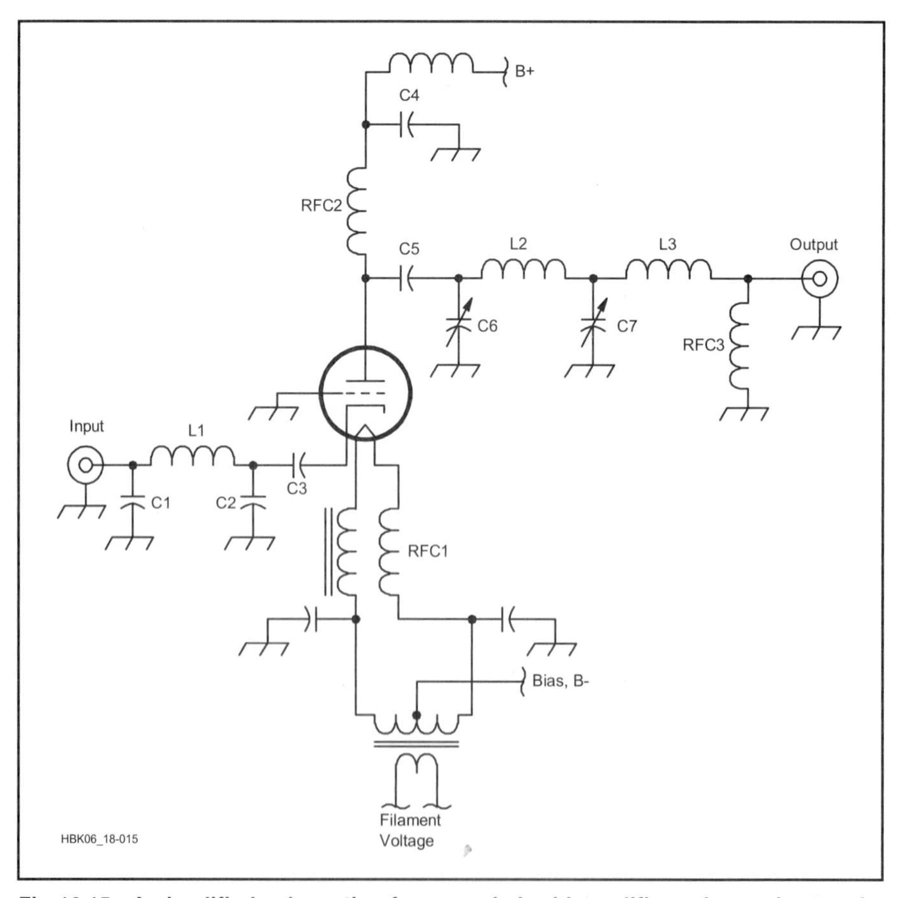

Fig 18.15—A simplified schematic of a grounded-grid amplifier using a pi network input and pi-L network output.

wiring from RF input connector to tank to cathode is heavy and short. On each band a Q_{IN} between 2 to 3 requires an $X_{Ctot} = X_{Lin}$ between 25 and 17 Ω.

A more nearly perfect match, with greater tolerance for layout and wiring variations, may be achieved by using the pi input tank as shown in Fig 18.15. Design of this input-matching circuit is straightforward. Component values are computed using a Q_{IN} between 2 or 3. Higher Q_{IN} values reduce the network's bandwidth, perhaps even requiring a front-panel tuning control for the wider amateur bands. The purpose of this input network is to present the desired input impedance to the exciter, not to add selectivity. As with a parallel tank, the value of the capacitor at the tube end of the pi network should be reduced by 38 pF; stray capacity plus tube C_{IN} is effectively in parallel with the input pi network's output.

The output pi-L network must transform the nominal 50-Ω amplifier load to a pure resistance of 2200 Ω. We previously calculated that the 8877 tube's plate must see 2200 Ω for optimum performance. In practice, real antenna loads are seldom purely resistive or exactly 50 Ω; they often exhibit SWRs of 2:1 or greater on some frequencies. It's desirable that the amplifier output network be able to transform any complex load impedance corresponding to an SWR up to about 2:1 into a resistance of 2200 Ω. The network also must compensate for tube C_{OUT} and other stray plate-circuit reactances, such as those of interconnecting leads and the plate RF choke. These reactances, shown in **Fig 18.16**, must be taken into account when designing the matching networks. Because the values of most stray reactances are not accurately known, the most satisfactory approach is to estimate them, and then allow sufficient flexibility in the matching network to accommodate modest errors.

Fig 18.16 shows the principal reactances in the amplifier circuit. C_{OUT} is the actual tube output capacitance of 10 pF plus the stray capacitance between its anode and the enclosure metalwork. This stray C varies with layout; we will approximate it as 5 pF, so C_{OUT} is roughly 15 pF. L_{OUT} is the stray inductance of leads from the tube plate to the tuning capacitor (internal to the tube as well as external circuit wiring.) External-anode tubes like the 8877 have essentially no internal plate leads, so L_{OUT} is almost entirely external. It seldom exceeds about 0.3 µH and is not very significant below 30 MHz. L_{CHOKE} is the reactance presented by the plate choke, which usually is significant only below 7 MHz. C_{STRAY} represents the combined stray capacitances to ground of the tuning capacitor stator and of interconnecting RF plate circuit leads. In a well-constructed, carefully thought out power amplifier, C_{STRAY} can be estimated to be approximately 10 pF. Remaining components C_{TUNE}, C_{LOAD}, and the two tuning inductors, form the pi-L network proper.

The tables presented earlier in this chapter greatly simplify the task of selecting output circuit values. Both the pi and pi-L design tables are calculated for a Q_{OUT} value of 12. A pi network loaded Q much lower than 10 does not provide adequate harmonic suppression; a value much higher than 15 increases matching network losses caused by high circulating currents. For pi networks, a Q_{OUT} of 12 is a good compromise between harmonic suppression and circuit losses. In practice, it often is most realistic and practical with both pi and pi-L output networks to accept somewhat higher Q_{OUT} values on the highest HF frequencies—perhaps as large as 18 or even 20 at 28 MHz. When using a pi-L on the 1.8 and 3.5 MHz bands, it often is desirable to choose a moderately lower Q_{OUT}, perhaps 8 to 10, to permit using a more reasonably-sized plate tuning capacitor.

Nominal pi-L network component values for 2200-Ω plate impedance can be taken directly from Table 18.2. These values can then be adjusted to allow for circuit reactances outside the pi-L proper. First, low-frequency component values should be examined. At 3.5 MHz, total tuning capacitance C1 value from Table 18.2 is 140 pF. From Fig 18.16 we know that three other stray reactances are directly in parallel with C_{TUNE} (assuming that L_{OUT} is negligible at the operating frequency, as it should be). The tube's internal and external plate capacitance to ground, C_{OUT}, is about 15 pF. Strays in the RF circuit, C_{STRAY}, are roughly 10 pF.

The impedance of the plate choke, X_{CHOKE}, is also in parallel with C_{TUNE}. Plate chokes with self-resonance characteristics suitable for use in amateur HF amplifiers typically have inductances of about 90 µH. At 3.5 MHz this is an inductive reactance of +1979 Ω. This appears in parallel with the tuning capacitance, effectively canceling an equal value of capacitive reactance. At 3.5 MHz, an X_C of 1979 Ω corresponds to 23 pF of capacitance—the amount by which tuning capacitor C_{TUNE} must be increased at 3.5 MHz to compensate for the effect of the plate choke.

The pi-L network requires an effective capacitance of 140 pF at its input at 3.5 MHz. Subtracting the 25 pF provided by C_{OUT} and C_{STRAY} and adding the 23 pF canceled by X_{CHOKE}, the actual value of C_{TUNE} must be 140 – 25 + 23 = 138 pF. It is good practice to provide at least 10% extra capacitance range to allow matching loads having SWRs up to 2:1. So, if 3.5 MHz is the lower frequency limit of the amplifier, a variable tuning capacitor with a maximum value of at least 150 to 160 pF should be used.

Component values for the high end of the amplifier frequency range also must be examined, for this is where the most losses will occur. At 29.7 MHz, the values in Table 18.2 are chosen to accommodate a minimum pi-L input capacitance of 35 pF, yielding Q_{OUT} = 21.3. Since C_{OUT} and C_{STRAY} contribute 25 pF, C_{TUNE} must have a minimum value no greater than 10 pF. A problem exists, because this value is not readily achievable with a 150 to 160-pF air variable capacitor suitable for operation with a 3100 V plate supply. Such a capacitor typically has a minimum capacitance of 25 to 30 pF. Usually, little or nothing can be done to reduce the tube's C_{OUT} or the circuit C_{STRAY}, and in fact the estimates of these may even be a little low. If 1.8 MHz capability is desired, the maximum tuning capacitance will be at least 200 to 250 pF, making the minimum-

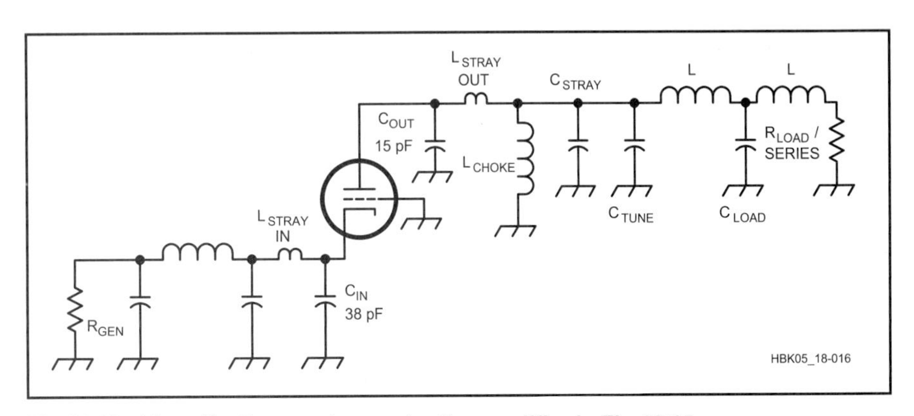

Fig 18.16—The effective reactances for the amplifier in Fig 18.15.

capacitance problem at 29.7 MHz even more severe.

There are three potential solutions to this dilemma. We could accept the actual minimum value of pi-L input capacitance, around 50 to 55 pF, realizing that this will raise the pi-L network's loaded Q to about 32. This results in very large values of circulating tank current. To avoid damage to tank components—particularly the bandswitch and pi inductor—from heat due to I^2R losses, it will be necessary to either use oversize components or reduce power on the highest-frequency bands. Neither option is appealing.

A second potential solution is to reduce the minimum capacitance provided by C_{TUNE}. We could use a vacuum variable capacitor with a 300-pF maximum and a 5-pF minimum capacity. These are rated at 5 to 15 kV, and are readily available. This reduces the minimum effective circuit capacitance to 30 pF, allowing use of the pi-L table values for Q_{OUT} = 12 on all bands from 1.8 through 29.7 MHz. While brand-new vacuum variables are quite expensive, suitable models are widely available in the surplus and used markets for prices not much higher than the cost of a new air variable. A most important caveat in purchasing a vacuum capacitor is to ensure that its vacuum seal is intact and that it is not damaged in any way. The best way to accomplish this is to "hi-pot" test the capacitor throughout its range, using a dc or peak ac test voltage of 1.5 to 2 times the amplifier plate supply voltage. For all-band amplifiers using plate voltages in excess of about 2500 V, the initial expense and effort of securing and using a vacuum-variable input tuning capacitor often is well repaid in efficient and reliable operation of the amplifier.

A third possibility is the use of an additional inductance connected in series between the tube and the tuning capacitor. In conjunction with C_{OUT} of the tube, the added inductor acts as an L network to transform the impedance at the input of the pi-L network up to the 2200-Ω load resistance needed by the tube. This is shown in **Fig 18.17A**. Since the impedance at the input of the main pi-L matching network is reduced, the loaded Q_{OUT} for the total capacitance actually in the circuit is lower. With lower Q_{OUT}, the circulating RF currents are lower, and thus tank losses are lower.

C_{OUT} in Fig 18.17 is the output capacitance of the tube, including stray C from the anode to metal enclosure. X_L is the additional series inductance to be added. As determined previously, the impedance seen by the tube anode must be a 2200 Ω resistance for best linearity and efficiency, and we have estimated C+ of the tube as 15 pF. If the network consisting of C_{OUT} and X_L is terminated at A by 2200 Ω, we can calculate the equivalent impedance at point B, the input to the pi-L network, for various values of series X_L. The pi-L network must then transform the nominal 50-Ω load at the transmitter output to this equivalent impedance.

We work backwards from the plate of the tube towards the C_{TUNE} capacitor. First, calculate the series-equivalent impedance of the parallel combination of the desired 2200-Ω plate load and the tube X_{OUT} (15 pF at 29.7 MHz = $-j$ 357 Ω). The series-equivalent impedance of this parallel combination is 56.5 – j 348 Ω, as shown in Fig 18.17B. Now suppose we use a 0.5 µH inductor, having an impedance of + j 93 Ω at 29.7 MHz, as the series inductance X_L. The resulting series-equivalent impedance is 56.5 – j 348 + j 93, or 56.5 – j 255 Ω. Converting back to the parallel equivalent gives the network of Fig 18.17C: 1205 Ω resistance in parallel with – j 267 Ω, or 20 pF at 29.7 MHz. The pi-L tuning network must now transform the 50-Ω load to a resistive load of 1205 Ω at B, and absorb the shunt capacity of 20 pF.

Using the pi-L network formulas in this chapter for R1 = 1205 Ω and Q_{OUT} = 15 at 29.7 MHz yields a required total capacitive reactance of 1205/15 = 80.3 Ω, which is 66.7 pF at 29.7 MHz. Note that for the same loaded Q_{OUT} for a 2200-Ω load line without the series inductor, the capacitive reactance is 2200/15 = 146.7 Ω, which is 36.5 pF. When the 20 pF of transformed input capacity is subtracted from the 66.7 pF total needed, the amount of capacity is 46.7 pF. If the minimum capacity in C_{TUNE} is 25 pF and the stray capacity is 10 pF, then there is a margin of 46.7 – 35 = 10.7 pF beyond the minimum capacity for handling SWRs greater than 1:1 at the load.

The series inductor should be a high-Q coil wound from copper tubing to keep losses low. This inductor has a decreasing, yet significant effect, on progressively lower frequencies. A similar calculation to the above should be made on each band to determine the transformed equivalent plate impedance, before referring to Table 18.2. The impedance-transformation effect of the additional inductor decreases rapidly with decreasing frequency. Below 21 MHz, it usually may be ignored and pi-L network values taken directly from the pi-L tables for R1 = 2200 Ω.

The nominal 90-µH plate choke remains in parallel with C_{TUNE}. It is rarely possible to calculate the impedance of a real HF plate choke at frequencies higher than about 5 MHz because of self-resonances. However, as mentioned previously, the choke's reactance should be sufficiently high that the tables are useful if the choke's first series-resonance is at 23.2 MHz.

This amplifier is made operational on multiple bands by changing the values of inductance at L2 and L3 for different bands. The usual practice is to use inductors for the lowest operating frequency, and short out part of each inductor with a switch, as necessary, to provide the inductance needed for each individual band. Wiring to the switch and the switch itself add stray inductance and capacitance to the circuit. To minimize these effects at the higher frequencies, the unswitched 10-m L2 should be placed closest to the high-impedance end of the network at C6.

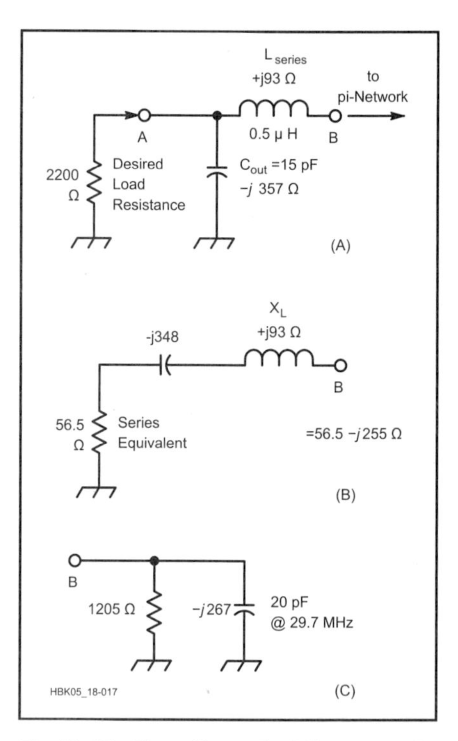

Fig 18.17—The effect of adding a small series inductance in vacuum tube output circuit. At A, a 0.5-µH coil L_{SERIES} is connected between anode and the output pi network, and this represents a reactance of + j 93 Ω at 29.7 MHz. The 15-pF output capacity (C_{OUT}) of the tube has a reactance of – j 357 Ω at 29.7 MHz. At B, the equivalent series network for the parallel 2200-Ω desired load resistance and the – j 357 Ω C_{OUT} is 56.5 Ω in series with – j 348 Ω. At C, this series-equivalent is combined with the series + j 93 Ω X_{SERIES} and converted back to the parallel equivalent, netting an equivalent parallel network of 1205 Ω shunted by a 20-pF capacitor. The pi tuning network must transform the load impedance (usually 50 Ω) into the equivalent parallel combination and absorb the 20-pF parallel component. The series inductor has less effect as the operating frequency is lowered from 29.7 MHz.

Stray capacitance associated with the switch then is effectively in parallel with C7, where the impedance level is around 300 Ω. The effects of stray capacitance are relatively insignificant at this low impedance level. This configuration also minimizes the peak RF voltage that the switch insulation must withstand.

Pi and L coil tap positions that yield desired values of inductance may be determined with fairly good accuracy by using a dip meter and a small mica capacitor of 5% tolerance. The pi and L coils and bandswitch should be mounted in the amplifier and their common point connected only to the bandswitch rotors. Starting at the highest-frequency switch position, lightly tack solder a short length of copper braid or strap to the pi or L switch stator terminal for that band. Using the shortest leads possible, tack a 50 to 100 pF, 5% dipped mica capacitor between the braid and a trial tap position on the appropriate coil. Lightly couple the dip meter and find the resonant frequency. The inductance then may be calculated from the equation

$$L = \frac{1{,}000{,}000}{(2\pi f)^2 C} \quad (45)$$

where

L = inductance in µH

C = capacitor value in pF

f = resonant frequency in MHz.

As each tap is located, it should be securely wired with strap or braid and the process repeated for successively lower bands.

The impedance match in both the input and output networks can be checked without applying dc voltage, once the amplifier is built. In operation, the tube input and output resistances are the result of current flow through the tube. Without filament power applied, these resistances are effectively infinite but C_{IN} and C_{OUT} are still present because they are passive physical properties of the tube. The tube input resistance can be simulated by an ordinary 5% ¼-W to 2-W composition or film resistor (don't use wirewound, though; they are more inductive than resistive at RF). A resistor value within 10% of the tube input resistance, connected in parallel with the tube input, presents approximately the same termination resistance to the matching network as the tube does in operation.

With the input termination resistor temporarily soldered in place using very short leads, input matching network performance can be determined by means of a noise bridge or an SWR meter that does not put out more RF power than the temporary termination resistor is capable of dissipating. Any good self- or dipper-powered bridge or analyzer should be satisfactory. Connect the bridge to the amplifier input and adjust the matching network, as necessary, for lowest SWR. Be sure to remove the terminating resistor before powering up the amplifier!

The output matching network can be evaluated in exactly the same fashion, even though the plate load resistance is not an actual resistance in the tube like the input resistance. According to the reciprocity principle, if the impedance presented at the output of the plate matching network is 50-Ω resistive when the network input is terminated with R_L, then the tube plate will "see" a resistive load equal to R_L at the input when the output is terminated in a 50-Ω resistance (and vice versa). In this case, a suitable 2200-Ω resistor should be connected as directly as possible from the tube plate to chassis. If the distance is more than a couple of inches, braid should be used to minimize stray inductance. The bridge is connected to the amplifier output. If coil taps have been already been established as described previously, it is a simple matter to evaluate the output network by adjusting the tune and load capacitors, band by band, to show a perfect 50-Ω match on the SWR bridge.

When these tests are complete, the amplifier is ready to be tested for parasitic oscillations in preparation for full-power operation. Refer to Amplifier Stabilization, later in this chapter.

DESIGN EXAMPLE 2: A MEDIUM-POWER 144-MHZ AMPLIFIER

For decades the 4CX250 family of power tetrodes has been used successfully up through 500 MHz. They are relatively inexpensive, produce high gain and lend themselves to relatively simple amplifier designs. In amateur service at VHF, the 4CX250 is an attractive choice for an amplifier. Most VHF exciters used now by amateurs are solid state and often develop 10 W or less output. The drive requirement for the 4CX250 in grounded cathode, Class AB operation ranges between 2 and 8 W for full power output, depending on frequency. At 144 MHz, manufacturer's specifications suggest an available output power of over 300 W. This is clearly a substantial improvement over 10 W, so a 4CX250B will be used in this amplifier.

The first design step is the same as in the previous example: Verify that the proposed tube will perform as desired while staying within the manufacturer's ratings. Again assuming a basic amplifier efficiency of 60% for Class AB operation, 300 W of output requires a plate input power of 500 W. Tube dissipation is rated at 250 W, so plate dissipation is not a problem, as the tube will only be dissipating 200 W in this amplifier. If the recommended maximum plate potential of 2000 V is used, the plate current for 500-W input will be 250 mA, which is within the manufacturer's ratings. The plate load resistance can now be calculated. Using the same formula as before, the value is determined to be 5333 Ω.

The next step is to investigate the output circuit. The manufacturer's specification for C_{OUT} is 4.7 pF. The inevitable circuit strays along with the tuning capacitor add to the circuit capacitance. A carefully built amplifier might only have 7 pF of stray capacitance, and a specially made tuning capacitor can be fabricated to have a midrange value of 3 pF. The total circuit capacitance adds up to about 15 pF. At 144 MHz this represents a capacitive reactance of only 74 Ω. The Q_L of a tank circuit with this reactance with a plate load resistance of 5333 Ω is 5333/74 = 72. A pi output matching network would be totally impractical, because the L required would be extremely small and circuit losses would be prohibitive. The simplest solution is to connect an inductor in parallel with the circuit capacitance to form a parallel-resonant tank circuit.

To keep tank circuit losses low with such a high Q_L, an inductor with very high unloaded Q must be used. The lowest-loss inductors are formed from transmission-line sections. These can take the form of either coaxial lines or strip lines. Both have their advantages and disadvantages, but the strip line is so much easier to fabricate that it is almost exclusively used in VHF tank circuits today.

The reactance of a terminated transmission line section is a function of both its characteristic impedance and its length (see the **Transmission Lines** chapter). The reactance of a line terminated in a short circuit is

$$X_{IN} = Z_0 \tan \ell \quad (46)$$

where

X_{IN} = is the circuit reactance

Z_0 = the line's characteristic impedance

ℓ = the transmission line length in degrees.

For lines shorter than a quarter wavelength (90°) the circuit reactance is inductive. In order to resonate with the tank-circuit capacitive reactance, the transmission-line reactance must be the same value, but inductive. Examination of the formula for transmission-line circuit reactance suggests that a wide range of lengths can yield the same inductive reactance, so long as the line Z_0 is appropriately scaled. Based on circuit Q considerations, the best bandwidth for a tank circuit results when the ratio of Z_0 to X_{IN} is between one and two. This implies that transmission line lengths between 26.5° and 45° give the best bandwidth. Between these two limits, and with some adjustment of Z_0, practical transmission lines can be designed. A transmission-line length of 35° is 8 inches long at 144 MHz, a workable dimension mechanically. Substitution of this value into the transmission-line equation gives a Z_0 of 105 Ω.

The width of the strip line and its placement relative to the ground planes determine the line impedance. Other stray capacitances such as mounting standoffs also affect the impedance. Accurate calculation of the line impedance for most physical configurations requires extensive application of Maxwell's equations and is beyond the scope of this book. The specialized case in which the strip line is parallel to and located halfway between two ground planes has been documented in *Reference Data for Radio Engineers* (see Bibliography). According to charts presented in that book, a 105-Ω strip line impedance is obtained by placing a line with a width of approximately 0.4 times the ground plane separation halfway between the ground planes. Assuming the use of a standard 3-inch-deep chassis for the plate compartment, this yields a stripline width of 1.2 inches. A strip line 1.2 inches wide located 1.5 inches above the chassis floor and grounded at one end has an inductive reactance of 74 Ω at 144 MHz.

The resulting amplifier schematic diagram is shown in **Fig 18.18**. L2 is the stripline inductance just described. C3 is the tuning capacitor, made from two parallel brass plates whose spacing is adjustable. One plate is connected directly to the strip line while the other is connected to ground through a wide, low-inductance strap. C2 is the plate blocking capacitor. This can be either a ceramic doorknob capacitor such as the Centralab 850 series or a homemade "Teflon sandwich." Both are equally effective at 144 MHz.

Impedance matching from the plate resistance down to 50 Ω can be either through an inductive link or through capacitive reactance matching. Mechanically, the capacitive approach is simpler to implement. **Fig 18.19** shows the development of reactance matching through a series capacitor (C4 in Fig 18.19). By using the parallel equivalent of the capacitor in series with the 50-Ω load, the load resistance can be transformed to the 5333-Ω plate resistance. Substitution of the known values into the parallel-to-series equivalence formulas reveals that a 2.15 pF capacitor at C4 matches the 50-Ω load to the plate resistance. The resulting parallel equivalent for the load is 5333 Ω in parallel with 2.13 pF. The 2.13-pF capacitor is effectively in parallel with the tank circuit.

A new plate line length must now be calculated to allow for the additional capacitance. The equivalent circuit diagram containing all the various reactances is shown in **Fig 18.20**. The total circuit capacitance is now just over 17 pF, which is a reactance of 64 Ω. Keeping the stripline width and thus its impedance constant at 105 Ω dictates a new resonant line length of 31°. This calculates to be 7.14 inches for 144 MHz.

The alternative coupling scheme is through the use of an inductive link. The link can be either tuned or untuned. The length of the link can be estimated based on the amplifier output impedance, in this case, 50 Ω. For an untuned link, the inductive reactance of the link itself should be approximately equal to the output impedance, 50 Ω. For a tuned link, the length depends on the link loaded Q, Q_L. The link Q_L should generally be greater than two, but usually less than five. For a Q_L of three this implies a capacitive reactance of 150 Ω, which at 144 MHz is just over 7 pF. The self-inductance of the link should of course be such that its impedance at 144 MHz is 150 Ω (0.166 mH). Adjustment of the link placement determines the transformation ratio of the circuit line. Some fine adjustment of this parameter can be made through adjustment of the link series tuning capacitor. Placement of the link relative to the plate inductor is an empirical process.

The input circuit is shown in Figs 18.21 and 18.23. C_{IN} is specified to be 18.5 pF for the 4CX250. This is only $-j$ 60 Ω at 2 m, so the pi network again is unsuitable. Since a surplus of drive is available with a 10-W exciter, circuit losses at the amplifier input are not as important as at the output. An old-fashioned "split-stator" tuned input can be used. L1 in Fig 18.18 is series tuned by C_{IN} and C1. The two capacitors are effectively in series (through the ground return). A 20-pF variable at C1 set to 18.5 pF gives an effective circuit capacitance of 9.25 pF. This will resonate at 144 MHz with an inductance of 0.13 µH at L1. L1 can be wound on a toroid core for mechanical convenience. The 50-Ω input impedance is then matched by link coupling to the toroid. The grid impedance is primarily determined by the value for R1, the grid bias feed resistor.

DESIGN EXAMPLE 3: A BROADBAND HF SOLID-STATE AMPLIFIER

Linear power-amplifier design using transistors at HF is a fundamentally simple process, although a good understanding of application techniques is important to insure that the devices are effectively protected against damage or destruction due to parasitic self-oscillations, power transients, load mismatch and/or overdrive.

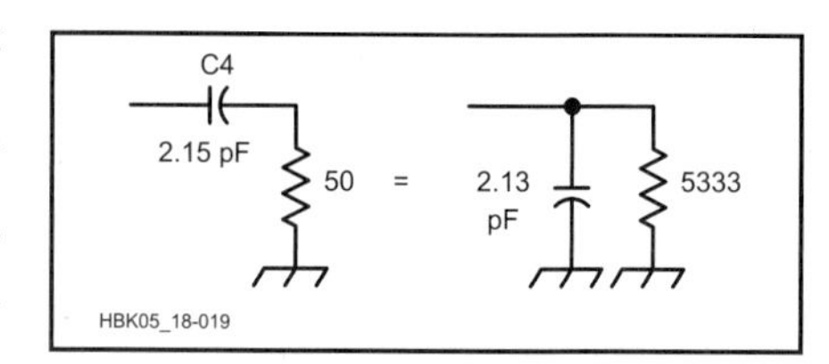

Fig 18.19—Series reactance matching as applied to the amplifier in Fig 18.18.

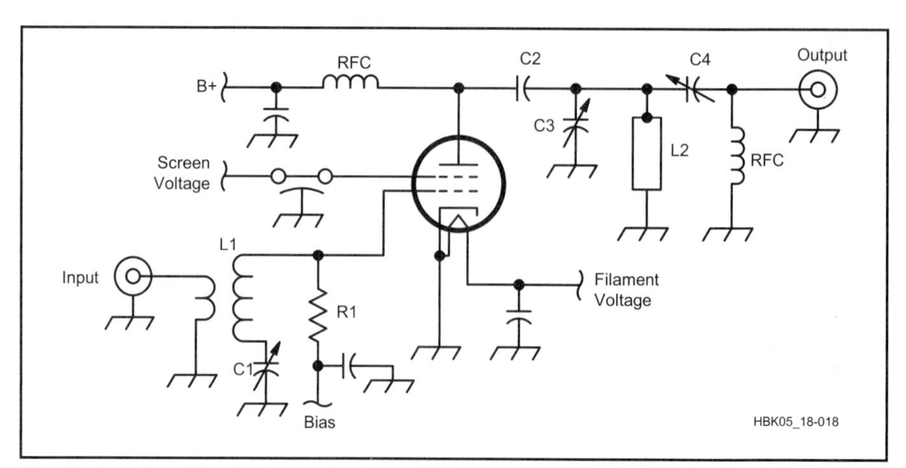

Fig 18.18—Simplified schematic for a VHF power amplifier using a power tetrode. The output circuit is a parallel-tuned tank circuit with series capacitive-reactance output matching.

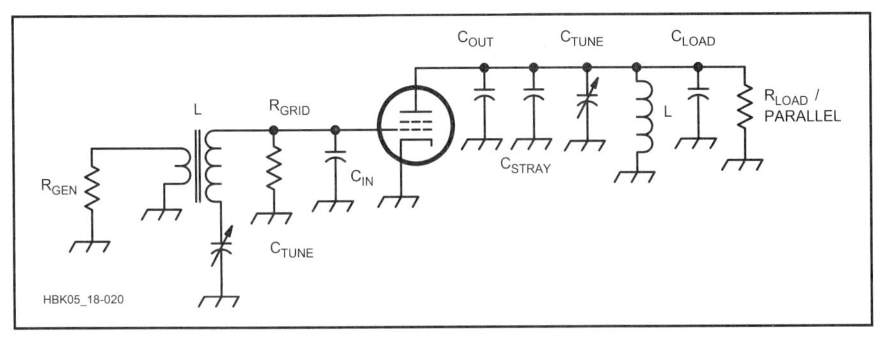

Fig 18.20—The reactances and resistances for the amplifier in Fig 18.18.

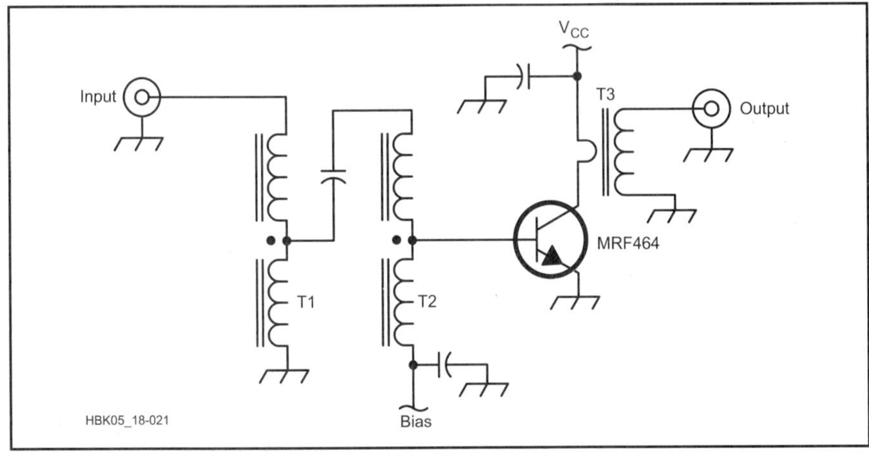

Fig 18.21—A simplified schematic of a broadband HF transistorized power amplifier. T1 and T2 are 4:1 broadband transformers to match the low input impedance of the transistor.

An appropriate transistor meeting the desired performance specifications is selected on the basis of dissipation and power output. Transistor manufacturers greatly simplify the design by specifying each type of power transistor according to its frequency range and power output. The amplifier designer need only provide suitable impedance matching to the device input and output, along with appropriate dc bias currents to the transistor.

The Motorola MRF464 is an RF power transistor capable of 80 W PEP output with low distortion. Its usable frequency range extends through 30 MHz. At a collector potential of 28 V, a collector efficiency of 40% is possible. **Fig 18.21** shows the schematic diagram of a 2 to 30-MHz broadband linear amplifier using the MRF464. The input impedance of the transistor is specified by the manufacturer to be 1.4 $- j\,0.30\ \Omega$ at 30 MHz and increases to 9.0 $- j\,5.40\ \Omega$ at 2 MHz. Transformers T1 and T2 match the 50-Ω amplifier input impedance to the median value of the transistor input impedance. They are both 4:1 step-down ratio transmission-line transformers. A single 16:1 transformer could be used in place of T1 and T2, but 16:1 transformers are more difficult to fabricate for broadband service.

The specified transistor load impedance is approximately 6 Ω (in parallel with a corresponding output capacitance) across the frequency range. T3 is a ferrite-loaded conventional transformer with a step-up ratio of approximately 8:1. This matches the transistor output to 50 Ω.

The amplifier has a falling gain characteristic with rising frequency. To flatten out gain across the frequency range, negative feedback could be applied. However, most power transistors have highly reactive input impedances and large phase errors would occur in the feedback loop. Instability could potentially occur.

A better solution is to use an input correction network. This network is used as a frequency-selective attenuator for amplifier drive. At 30 MHz, where transistor gain is least, the input power loss is designed to be minimal (less than 2 dB). The loss increases at lower frequencies to compensate for the increased transistor gain. The MRF464 has approximately 12 dB more gain at 1.8 MHz than at 30 MHz; the compensation network is designed to have 12 dB loss at 1.8 MHz. A properly designed compensation network will result in an overall gain flatness of approximately 1 dB.

AMPLIFIER STABILIZATION

Stable Operating Conditions

Purity of emissions and the useful life (or even survival) of the active devices in a tube or transistor circuit depend heavily on stability during operation. Oscillations can occur at the operating frequency or far from it, because of undesired positive feedback in the amplifier. Unchecked, these oscillations pollute the RF spectrum and can lead to tube or transistor over-dissipation and subsequent failure. Each type of oscillation has its own cause and its own cure.

In a linear amplifier, the input and output circuits operate on the same frequency. Unless the coupling between these two circuits is kept to a small enough value, sufficient energy from the output may be coupled in phase back to the input to cause the amplifier to oscillate. Care should be used in arranging components and wiring of the two circuits so that there will be negligible opportunity for coupling external to the tube or transistor itself. A high degree of shielding between input and output circuits usually is required. All RF leads should be kept as short as possible and particular attention should be paid to the RF return paths from input and output tank circuits to emitter or cathode.

In general, the best arrangement using a tube is one in which the input and output circuits are on opposite sides of the chassis. Individual shielded compartments for the input and output circuitry add to the isolation. Transistor circuits are somewhat more forgiving, since all the impedances are relatively low. However, the high currents found on most amplifier circuit boards can easily couple into unintended circuits. Proper layout, the use of double-sided circuit board (with one side used as a ground plane and low-inductance ground return), and heavy doses of bypassing on the dc supply lines often are sufficient to prevent many solid-state amplifiers from oscillating.

VHF and UHF Parasitic Oscillations

RF power amplifier circuits contain parasitic reactances that have the potential to cause so-called parasitic oscillations at frequencies far above the normal operating frequency. Nearly all vacuum-tube amplifiers designed for operation in the 1.8 to 29.7-MHz frequency range exhibit tendencies to oscillate somewhere in the VHF-UHF range—generally between about 75 and 250 MHz depending on the type and size of tube. A typical parasitic resonant circuit is highlighted by bold lines in **Fig 18.22**. Stray inductance between the tube plate and the output tuning capacitor forms a high-Q resonant circuit with the tube's C_{OUT}. C_{OUT} normally is much smaller (higher X_C) than any of the other circuit capacitances shown. The tube's C_{IN} and the tuning capacitor C_{TUNE} essentially act as bypass capacitors, while the various chokes and tank inductances shown have high reactances at VHF. Thus the values of these components have little influence on the parasitic resonant frequency.

Oscillation is possible because the VHF resonant circuit is an inherently high-Q parallel-resonant tank that is not coupled to the external load. The load resistance at the plate is very high, and thus the voltage gain at the parasitic frequency can be quite high, leading to oscillation. The parasitic frequency, f_r, is approximately:

$$f_r = \frac{1000}{2\pi\sqrt{L_P\, C_{OUT}}} \tag{47}$$

where

f_r = parasitic resonant frequency in MHz

L_P = total stray inductance between tube plate and ground via the plate tuning capacitor (including tube internal plate lead) in µH

C_{OUT} = tube output capacitance in pF.

In a well-designed HF amplifier, L_P might be in the area of 0.2 µH and C_{OUT} for an 8877 is about 10 pF. Using these figures, the equation above yields a potential parasitic resonant frequency of

$$f_r = \frac{1000}{2\pi\sqrt{0.2 \times 10}} = 112.5 \text{ MHz}$$

For a smaller tube, such as the 3CX800A7 with C_{OUT} of 6 pF, f_r = 145 MHz. Circuit details affect f_r somewhat, but these results do in fact correspond closely to actual parasitic oscillations experienced with these tube types. VHF-UHF parasitic oscillations can be prevented (*not* just minimized!) by reducing the loaded Q of the parasitic resonant circuit so that gain at its resonant frequency is insufficient to support oscillation. This is possible with any common tube, and it is especially easy with modern external-anode tubes like the 8877, 3CX800A7, and 4CX800A.

Z1 of Fig 18.22B is a parasitic suppressor. Its purpose is to add loss to the parasitic circuit and reduce its Q enough to prevent oscillation. This must be accomplished without significantly affecting normal operation. L_Z should be just large enough to constitute a significant part of the total parasitic tank inductance (originally represented by L_P), and located right at the tube plate terminal(s). If L_Z is made quite lossy, it will reduce the Q of the parasitic circuit as desired.

The inductance and construction of L_Z depend substantially on the type of tube used. Popular glass tubes like the 3-500Z and 4-1000A have internal plate leads made of wire. This significantly increases L_P when compared to external-anode tubes. Consequently, L_Z for these large glass tubes usually must be larger in order to constitute an adequate portion of the total value of L_P. Typically a coil of 3 to 5 turns of #10 wire, 0.25 to 0.5 inches in diameter and about 0.5 to 1 inches long is sufficient. For the 8877 and similar tubes it usually is convenient to form a "horseshoe" in the strap used to make the plate connection. A "U" about 1 inch wide and 0.75 to 1 inch deep usually is sufficient. In either case, L_Z carries the full operating-frequency plate current; at the higher frequencies this often includes a substantial amount of circulating tank current, and L_Z must be husky enough to handle it without overheating even at 29 MHz.

Regardless of the form of L_Z, loss may be introduced as required by shunting L_Z with one or more suitable noninductive resistors. In high-power amplifiers, two composition or metal film resistors, each 100 Ω, 2 W, connected in parallel across L_Z usually are adequate. For amplifiers up to perhaps 500 W a single 47-Ω, 2-W resistor may suffice. The resistance and power capability required to prevent VHF/UHF parasitic oscillations, while not overheating as a result of normal plate circuit current flow, depend on circuit parameters. Operating-frequency voltage drop across L_Z is greatest at higher frequencies, so it is important to use the minimum necessary value of L_Z in order to minimize power dissipation in R_Z.

The parasitic suppressors described above very often will work without modification, but in some cases it will be necessary to experiment with both L_Z and R_Z to find a suitable combination. Some designers use nichrome or other resistance wire for L_Z.

In exceptionally difficult cases, particularly when using glass tetrodes or pentodes, additional parasitic suppression may be attained by connecting a low value

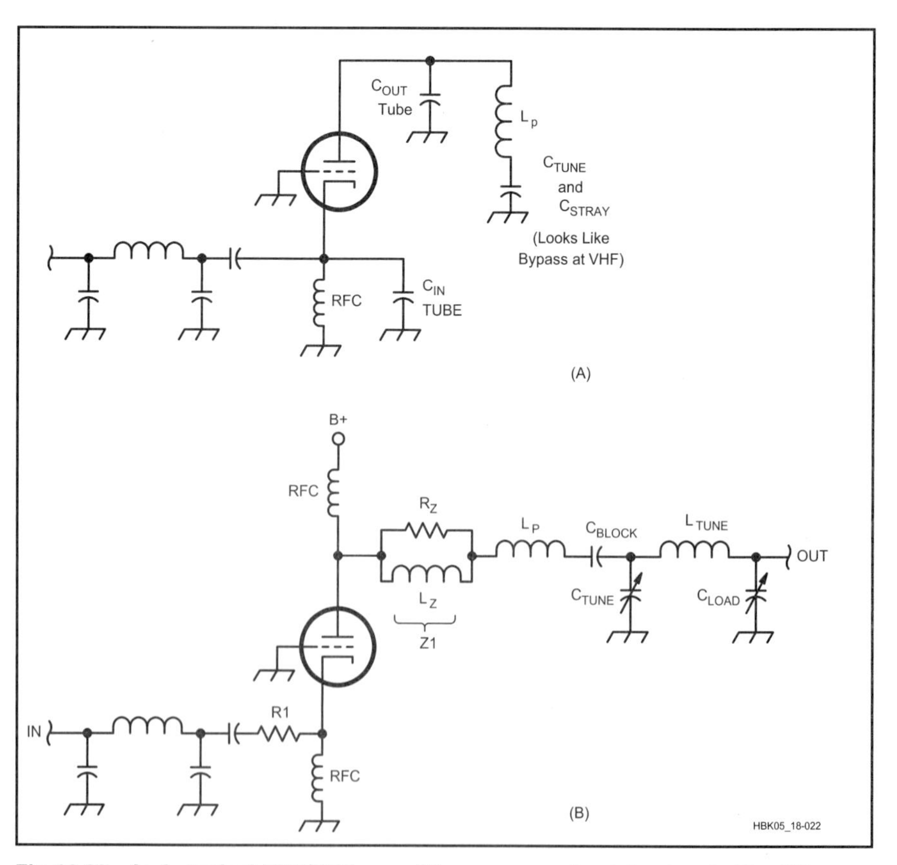

Fig 18.22—At A, typical VHF/UHF parasitic resonance in plate circuit. The HF tuning inductor in the pi network looks like an RF choke at VHF/UHF. The tube's output capacity and series stray inductance combine with the pi-network tuning capacity and stray circuit capacity to create a VHF/UHF pi network, presenting a very high impedance to the plate, increasing its gain at VHF/UHF. At B, Z1 lowers the Q and therefore gain at parasitic frequency.

resistor (about 10 to 15 Ω) in series with the tube input, near the tube socket. This is illustrated by R1 of Fig 18.22B. If the tube has a relatively low input impedance, as is typical of grounded-grid amplifiers and some grounded-cathode tubes with large C_{IN}, R1 may dissipate a significant portion of the total drive power.

Testing Tube Amplifiers for VHF-UHF Parasitic Oscillations

Every high-power amplifier should be tested before being placed in service, to insure that it is free of parasitic oscillations. For this test, nothing is connected to either the RF input or output terminals, and the bandswitch is first set to the lowest-frequency range. If the input is tuned and can be bandswitched separately, it should be set to the highest-frequency band. The amplifier control system should provide monitoring for both grid current and plate current, as well as a relay, circuit breaker or fast-acting fuse to quickly shut off high voltage in the event of excessive plate current. To further protect the tube grid, it is a good idea to temporarily insert in series with the grid current return line a resistor of approximately 1000 Ω to prevent grid current from soaring in the event a vigorous parasitic oscillation breaks out during initial testing.

Apply filament and bias voltages to the amplifier, leaving plate voltage off and/or cutoff bias applied until any specified tube warm-up time has elapsed. Then apply the lowest available plate voltage and switch the amplifier to transmit. Some idling plate current should flow. If it does not, it may be necessary to increase plate voltage to normal or to reduce bias so that at least 100 mA or so does flow. Grid current should be zero. Vary the plate tuning capacitor slowly from maximum capacitance to minimum, watching closely for any grid current or change in plate current, either of which would indicate a parasitic oscillation. If a tunable input network is used, its capacitor (the one closest to the tube if a pi circuit) should be varied from one extreme to the other in small increments, tuning the output plate capacitor at each step to search for signs of oscillation. If at any time either the grid or plate current increases to a large value, shut off plate voltage immediately to avoid damage! If moderate grid current or changes in plate current are observed, the frequency of oscillation can be determined by loosely coupling an RF absorption meter or a spectrum analyzer to the plate area. It will then be necessary to experiment with parasitic suppression measures until no signs of oscillation can be detected under any conditions. This process should be repeated using each bandswitch position.

When no sign of oscillation can be found, increase the plate voltage to its normal operating value and calculate plate dissipation (idling plate current times plate voltage). If dissipation is at least half of, but not more than, its maximum safe value, repeat the previous tests. If plate dissipation is much less than half of maximum safe value, it is desirable (but not absolutely essential) to reduce bias until it is. If no sign of oscillation is detected, the temporary grid resistor should be removed and the amplifier is ready for normal operation.

Parasitic Oscillations in Solid-State Amplifiers

In low-power solid-state amplifiers, parasitic oscillations can be prevented by using a small amount of resistance in series with the base or collector lead, as shown in **Fig 18.23A**. The value of R1 or R2 typically should be between 10 and 22 Ω. The use of both resistors is seldom necessary, but an empirical determination must be made. R1 or R2 should be located as close to the transistor as practical.

At power levels in excess of approximately 0.5 W, the technique of parasitic suppression shown in Fig 18.23B is effective. The voltage drop across a resistor would be prohibitive at the higher power levels, so one or more ferrite beads placed over connecting leads can be substituted (Z1 and Z2). A bead permeability of 125 presents a high impedance at VHF and above without affecting HF performance. The beads need not be used at both circuit locations. Generally, the terminal carrying the least current is the best place for these suppression devices. This suggests that the resistor or ferrite beads should be connected in the base lead of the transistor.

C3 of **Fig 18.24** can be added to some power amplifiers to dampen VHF/UHF parasitic oscillations. The capacitor should be low in reactance at VHF and UHF, but must present a high reactance at the operating frequency. The exact value selected will depend upon the collector impedance. A reasonable estimate is to use an X_C of 10 times the collector impedance at the operating frequency. Silver-mica or ceramic chip capacitors are suggested for this application. An additional advantage is the resultant bypassing action for VHF and UHF harmonic energy in the collector circuit. C3 should be placed as close to the collector terminal as possible, using short leads.

The effects of C3 in a broadband amplifier are relatively insignificant at the operating frequency. However, when a narrow-band collector network is used, the added capacitance of C3 must be absorbed into the network design in the same manner as the C_{OUT} of the transistor.

Low-Frequency Parasitic Oscillations

Bipolar transistors exhibit a rising gain characteristic as the operating frequency is lowered. To preclude low-frequency instabilities because of the high gain, shunt and degenerative feedback are often used. In the regions where low-frequency self-oscillations are most likely to occur, the feedback increases by nature of the feedback network, reducing the amplifier gain. In the circuit of Fig 18.24, C1 and R3 provide negative feedback, which increases progressively as the frequency is lowered. The network has a small effect at the desired operating frequency but has a pronounced effect at the lower frequencies. The values for C1 and R3 are usually chosen experimentally. C1 will usually be between 220 pF and 0.0015 µF for HF-band amplifiers while R3 may be a value from 51 to 5600 Ω.

R2 of Fig 18.24 develops emitter degeneration at low frequencies. The bypass capacitor, C2, is chosen for adequate RF bypassing at the intended operating frequency. The impedance of C2 rises progressively as the frequency is lowered, thereby increasing the degenerative feedback caused by R2. This lowers the amplifier gain. R2 in a power stage is seldom greater than 10 Ω, and may be as low as 1 Ω. It is important to consider that under some operating and layout conditions R2 can cause instability. This form of feedback should be used only in those circuits in which unconditional stability can be achieved.

R1 of Fig 18.24 is useful in swamping the input of an amplifier. This reduces the chance for low-frequency self oscillations, but has an effect on amplifier performance in the desired operating range. Values from 3 to 27 Ω are typical. When connected in shunt with the normally low base impedance of a power amplifier, the resistors lower the effective device input impedance slightly. R1 should be located as close to the transistor base terminal as possible, and the connecting leads must be kept short to minimize stray reactances. The use of two resistors in parallel reduces the amount of inductive reactance introduced compared to a single resistor.

Although the same concepts can be applied to tube-type amplifiers, the possibility of self-oscillations at frequencies lower than VHF is significantly lower than in solid-state amplifiers. Tube amplifiers will usually operate stably as long as the input-to-output isolation is greater than the stage gain. Proper shielding and dc-power-lead bypassing essentially

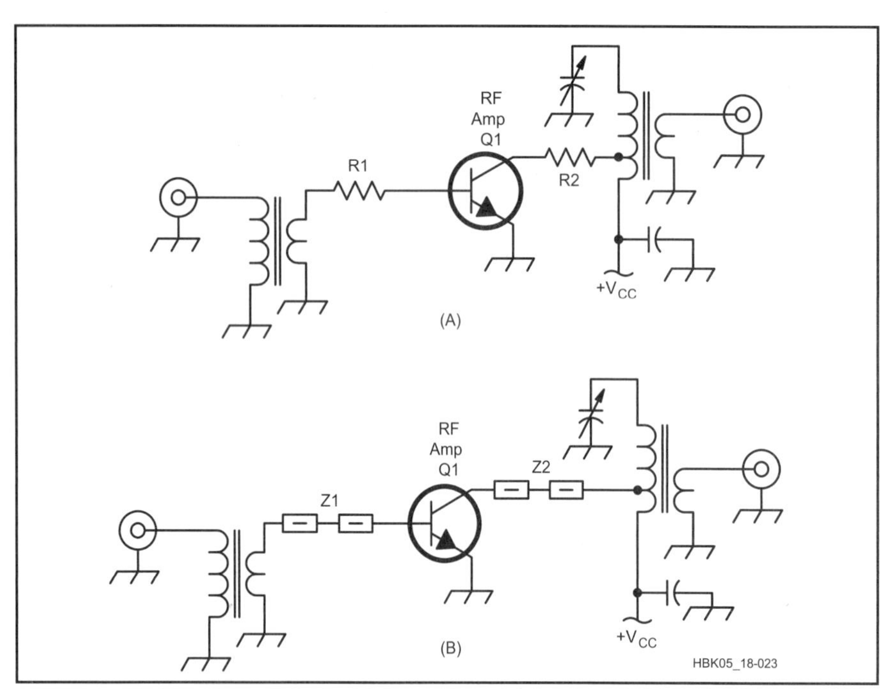

Fig 18.23—Suppression methods for VHF and UHF parasitics in solid-state amplifiers.

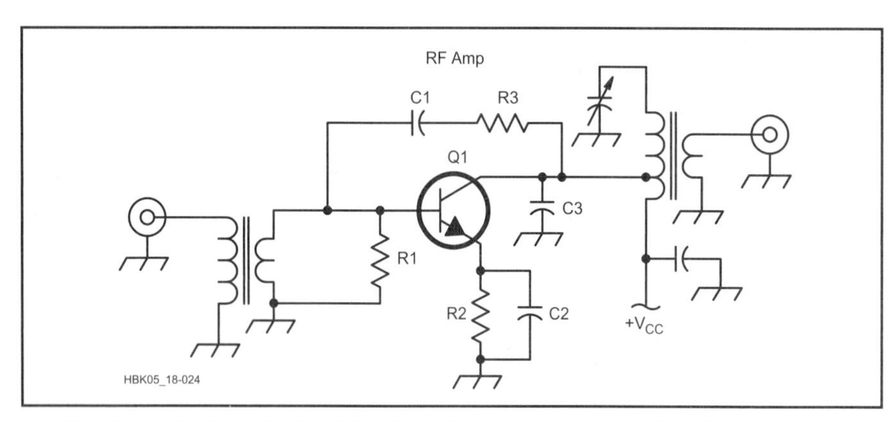

Fig 18.24—Illustration of shunt feedback in a transistor amplifier. C1 and R3 make up the feedback network.

eliminate feedback paths, except for those through the tube itself.

On rare occasions tube-type amplifiers will oscillate at frequencies in the range of about 50 to 500 kHz. This is most likely with high-gain tetrodes using shunt feed of dc voltages to both grid and plate through RF chokes. If the resonant frequency of the grid RF choke and its associated coupling capacitor occurs close to that of the plate choke and its blocking capacitor, conditions may support a tuned-plate tuned-grid oscillation. For example, using typical values of 1 mH and 1000 pF, the expected parasitic frequency would be around 160 kHz.

Make sure that there is no low-impedance, low-frequency return path to ground through inductors in the input matching networks in series with the low impedances reflected by a transceiver output transformer. Usually, oscillation can be prevented by changing choke or capacitor values to insure that the input resonant frequency is much lower than that of the output.

Amplifier Neutralization

Depending on stage gain and inter-electrode capacitances, sufficient positive feedback may occur to cause oscillation at the operating frequency. This should not occur in well-designed grounded-grid amplifiers, nor with tetrode or pentodes operating at gains up to about 15 dB as is current practice at HF where 50 to 100 W of drive is almost always available. If triodes are grid-driven, however, and under certain other circumstances, neutralization may be necessary because of output energy capacitively coupled back to the input as shown in **Fig 18.25**. Neutralization involves coupling a small amount of output energy back to the amplifier input out-of-phase, to cancel the unwanted in-phase (positive) feedback. A typical circuit is given in **Fig 18.26**. L2 provides a 180° phase reversal because it is center tapped. C1 is connected between the plate and the lower half of the grid tank. C1 is then adjusted so that the energy coupled from the tube output through the neutralization circuit is equal in amplitude and exactly 180° out-of-phase with the energy coupled from the output back through the tube. The two signals then cancel and oscillation is impossible.

The easiest way to adjust a neutralization circuit is to connect a low-level RF source to the amplifier output tuned to the amplifier operating frequency. A sensitive RF detector like a receiver is then connected to the amplifier input. The amplifier must be turned off for this test. The amplifier tuning and loading controls, as well as any input network adjustments are then peaked for maximum indication on the RF detector connected at the input. C1 is then adjusted for minimum response on the detector. This null indicates that the neutralization circuit is canceling energy coupled from the amplifier output to its input through tube, transistor or circuit capacitances.

Screen-Grid Tube Stabilization

The plate-to-grid capacitance in a screen-grid tube is reduced to a fraction of a picofarad by the interposed grounded screen. Nevertheless, the power gain of these tubes may be so great in some circuits that only a very small amount of feedback is necessary to start oscillation. To assure a stable tetrode amplifier, it is usually necessary to load the grid circuit, or to use a neutralizing circuit.

Grid Loading

The need for a neutralizing circuit may often be avoided by loading the grid circuit to reduce stage gain, provided that the driving stage has some power capacity to spare. Loading by tapping the grid down on the grid tank coil, or by placing a "swamping" resistor from grid to cathode, is effective to stabilize an amplifier. Either measure reduces the gain of the amplifier, lessening the possibility of oscillation. If a swamping resistor is connected between grid and cathode with very short leads, it may help reduce any tendency toward VHF-UHF parasitic oscillations as well. In a class AB1 amplifier, which draws no grid current, a swamping resistor can be used to replace the bias supply choke if parallel feed is used.

Often, reducing stage gain to the value required by available drive power is suffi-

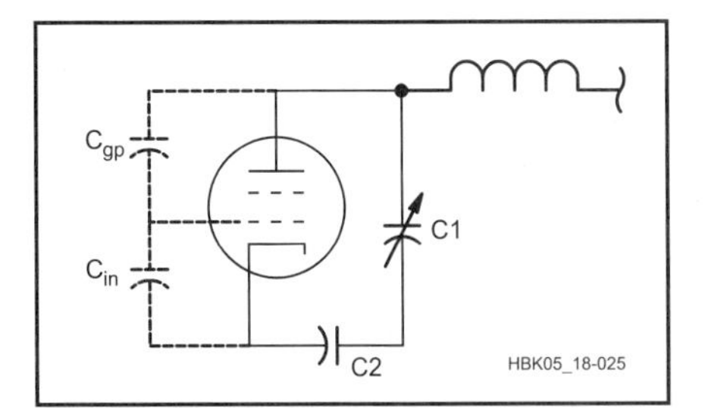

Fig 18.25—The equivalent feedback path due to the internal capacitance of the tube grid-plate structure in a power amplifier. Also see Fig 18.27.

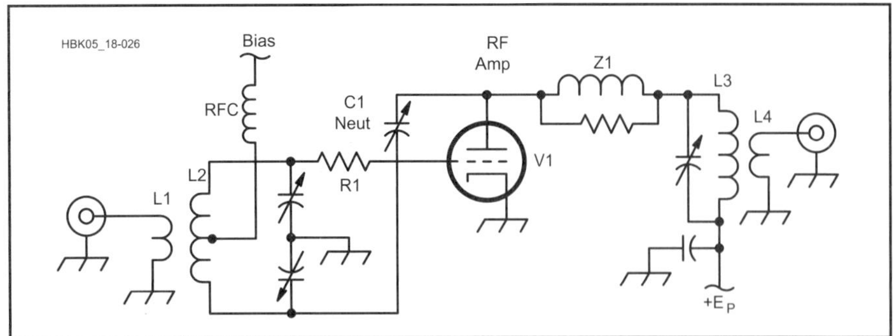

Fig 18.26—Example of neutralization of a single-ended RF amplifier.

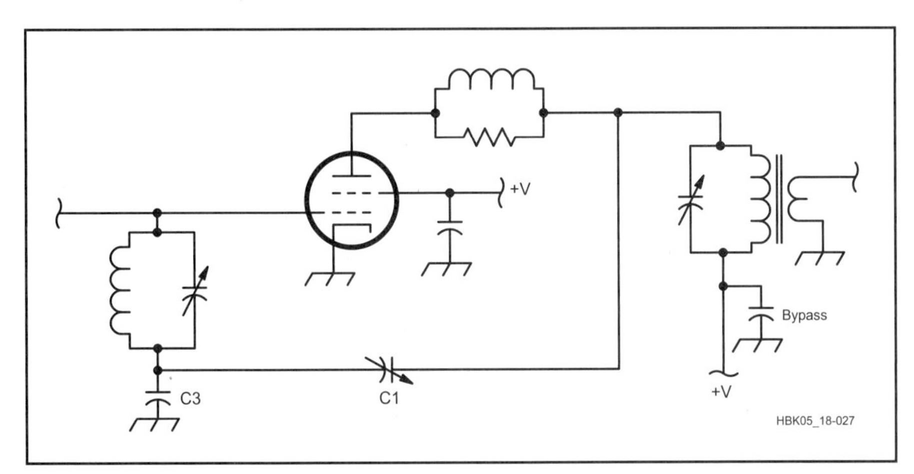

Fig 18.27—A neutralization circuit uses C1 to cancel the effect of the tube internal capacitance.

cient to assure stability. If this is not practical or effective, the bridge neutralizing system for screen-grid tubes shown in **Fig 18.27** may be used. C1 is the neutralizing capacitor. The value of C1 should be chosen so that at some adjustment of C1,

$$\frac{C1}{C3} = \frac{C_{gp}}{C_{IN}} \tag{48}$$

where

C_{gp} = tube grid-plate capacitance

C_{IN} = tube input capacitance.

The grid-to-cathode capacitance must include all strays directly across the tube capacitance, including the capacitance of the tuning capacitor.

3CX1500D7 RF Linear Amplifier

The following describes a 10-to-160-meter RF linear amplifier that uses the new compact Eimac 3CX1500D7 metal ceramic triode. It was designed and constructed by Jerry Pittenger, K8RA.

The amplifier features instant-on operation and provides a solid 1500 W RF output with less than 100 W drive. Specifications for this rugged tube include 1500-W anode dissipation, 50-W grid dissipation and plate voltages up to 6000 V. A matching 4000-V power supply is included. The amplifier can be easily duplicated and provides full output in key-down service with no time constraints in any mode. **Fig 18.28** shows the RF deck and power supply cabinets.

DESIGN OVERVIEW

The Eimac 3CX1500D7 was designed as a compact, but heavy-duty, alternative to the popular lineup of a pair of 3-500Z tubes. It

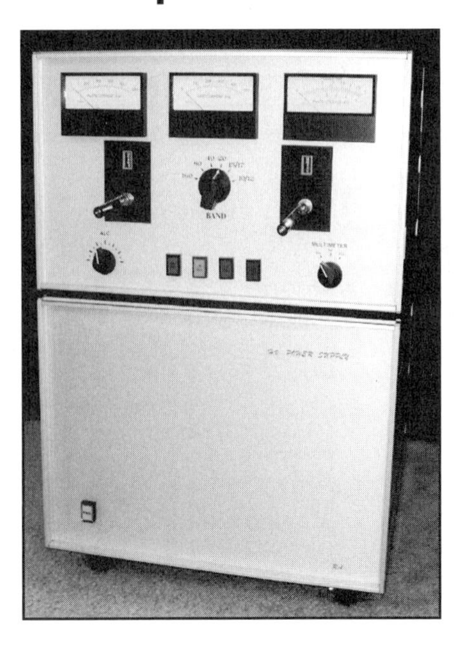

Fig 18.28—At A, front panel view of RF Deck and Power Supply for 3CX1500D7 amplifier. At B, rear view of RF Deck and Power Supply.

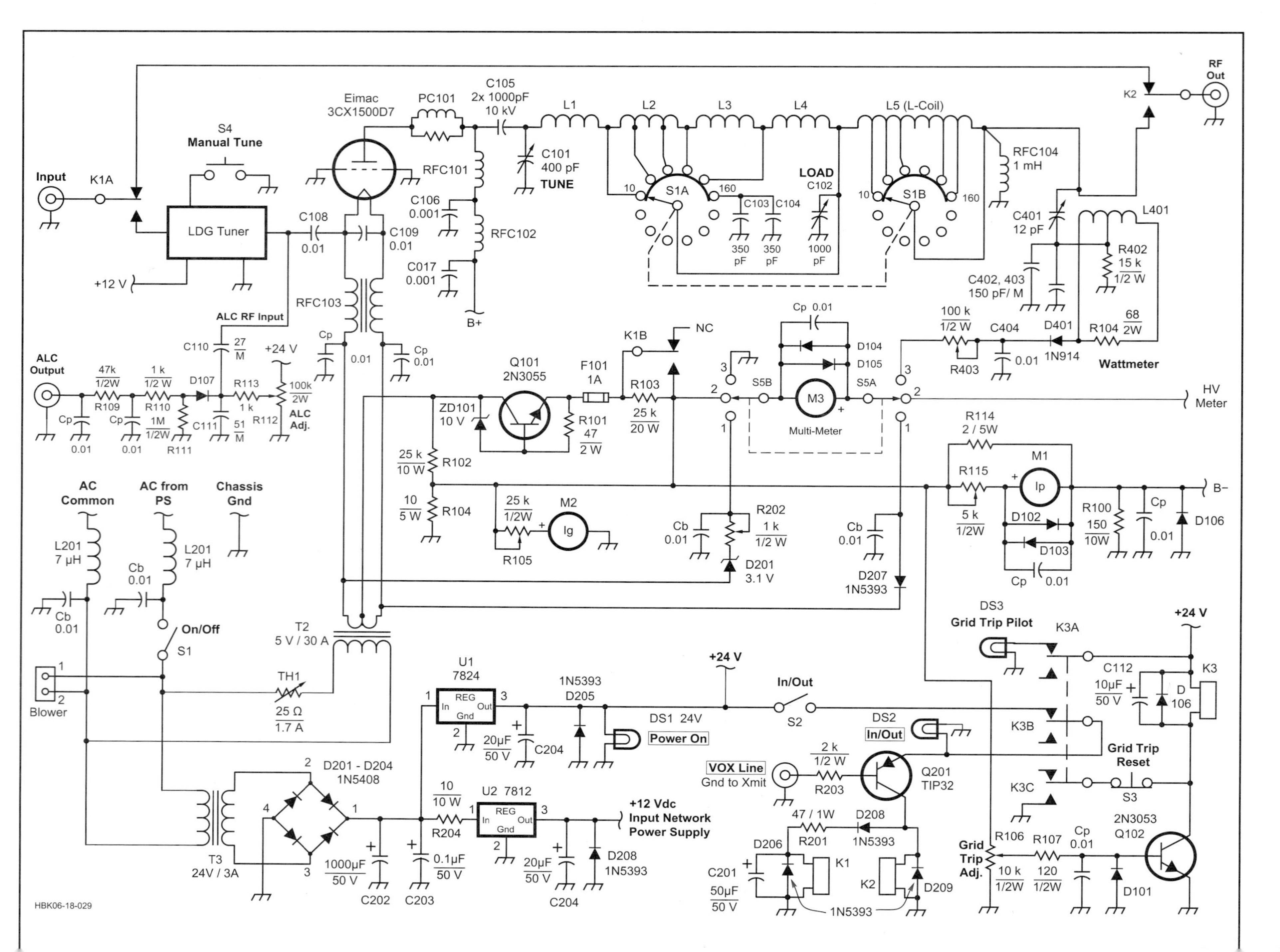
Input
K1A
S4
Manual Tune
LDG Tuner
+12 V
Eimac
3CX1500D7
PC101
C105
2x 1000pF
10 kV
L1
L2
L3
L4
L5 (L-Coil)
RF
Out
K2
RFC101
C101
400 pF
TUNE
S1A
LOAD
C102
S1B
RFC104
1 mH
C106
0.001
C108
0.01
C109
0.01
RFC102
C103
C104
350 pF
350 pF
1000 pF
C401
12 pF
L401
R402
15 k
1/2 W
C017
0.001
C402, 403
150 pF/ M
RFC103
B+
ALC RF Input
C110
27 M
+24 V
Cp
0.01
Cp 0.01
100 k
1/2 W
C404
D401
R104
68
2W
ALC
Output
47k
1/2W
R109
1 k
1/2 W
R110
D107
R113
1 k
100k
2W
ALC
Adj.
R112
C111
51 M
1M
1/2W
R111
K1B
NC
Q101
2N3055
F101
1A
R103
25 k
20 W
S5B
M3
S5A
Multi-Meter
D104
D105
R403
1N914
Wattmeter
HV
Meter
ZD101
10 V
R101
47
2 W
R114
2 / 5W
M1
R115
Ip
25 k
10 W
R102
10
5 W
R104
25 k
1/2W
M2
Ig
R105
Cb
0.01
R202
1 k
1/2 W
D201
3.1 V
D207
1N5393
5 k
1/2W
D102
D103
R100
150
10W
D106
B–
AC
Common
AC from
PS
Chassis
Gnd
L201
7 µH
Cb
0.01
On/Off
S1
T2
5 V / 30 A
TH1
25 Ω
1.7 A
Blower
U1
7824
REG
In
Out
Gnd
+24 V
1N5393
D205
DS1 24V
Power On
In/Out
S2
DS3
Grid Trip Pilot
K3A
K3
C112
10µF
50 V
D
106
K3B
DS2
In/Out
Grid Trip
Reset
VOX Line
Gnd to Xmit
2 k
1/2 W
R203
Q201
TIP32
K3C
S3
D201 - D204
1N5408
T3
24V / 3A
1000µF
50 V
C202
C203
10
10 W
R204
0.1µF
50 V
U2 7812
20µF
50 V
C204
D208
1N5393
+12 Vdc
Input Network
Power Supply
47 / 1W
R201
D206
C201
50µF
50 V
K1
1N5393
K2
D209
Grid
Trip
Adj.
R106
10 k
1/2W
R107
120
1/2W
2N3053
Q102
D101
HBK06-18-029

Fig 18.29—At A, schematic of RF Deck. At B, schematic of control unit.
B1—Dayton 4C763 squirrel-cage blower.
Cabinet—Buckeye Shapeform DSC-1054-16 (10×17×16-inch H×W×D), **www.buckeyeshapeform.com**.
Chimney (Teflon)—A. Howell, KB8JCY, PO Box 5842, Youngstown, OH 44504.
Cp—0.01 µF, 1 kV bypass disc ceramic.
C101—400 pF, 10 kV Jennings vacuum variable, UCSL-400.
C102—1000 pF, 5 kV Jennings vacuum variable, UCSL-1000.
C103, C104—350 pF, 5 kV ceramic doorknob.
C105—two parallel 0.001 µF, 7.5 kV disc ceramic (Ukrainian mfg).
C106, C107—0.001 µF, 7.5 kV disc ceramic.
C108, C109—0.01 µF, 3 kV transmitting mica (1 kV disc ceramics can be used).
C401—12 pF piston trimmer.
C403, C403—150 pF silver mica.
D101, D107, D205-D209—1N5393 (1 kV, 1.5 A).
D102-D106, D201-D204—1N5408 (1 kV, 3 A).
K1—4PDT, 24 Vdc KHP style (gold contacts).
K2—SPST vacuum relay, Kilovac H8/S4.
K3—4PDT, 24 V dc KHP style (gold contacts).
L1-L5—See Table 1.
L201, L202—Line chokes, 7 µH.
L401—24 t #22 enamel wire, center tapped on T50-6 core.
LDG Tuner—Modified AT-100Pro Autotuner, **www.ldgelectronics.com**.
M1-M3—Simpson Designer Series, Model 523, 1 mA movement.
PC101—2 t 3/4-inch diameter × 2-inch long, 1/2-inch brass strap with two 150 Ω, 2 W non-inductive carbon resistors in parallel.
Q101—2N3055 TO-220 case on heat sink.
Q102—2N3053 TO-18 case.
R103—25 kΩ, 25 W wire-wound.
R104—10 Ω, 5 W.
R108—150 Ω, 10 W wire-wound.
R112—100 kΩ, 2 W.
R403—100 kΩ, 0.5 W trim pot.
RFC101—90 µH, 3 A Plate Choke, Peter W. Dahl p/n CKRF000100, **pwdahl.com/cgi-bin/store/commerce.cgi**.
RFC102—14 t #18 enamel wire wound on 100 Ω, 2 W resistor.
RFC103—Bifilar 30 A filament choke, Peter W. Dahl p/n CKRF000080, **pwdahl.com/cgi-bin/store/commerce.cgi**.
RFC104—1 mH, 300 mA RF choke.
S1-S4—Alco 164TL2 momentary DPDT, **www.alliedelec.com/**.
S5—2P3P rotary switch.
SW1—RadioSwitch model 86, double-pole 12-position (30° indexing) with 6-finger wiper on each deck, p/n R86R1130001, **www.multi-tech-industries.com**.
T2—5 V, 30 A transformer, Peter W. Dahl EI-150 × 1.5 core, primary 115/230 V ac, **www.pwdahl.com**.
TH1—Thermistor, Thermometrics CL-200 (Mouser 527-CL200).
Tube socket—Eimac SK-410.
ZD101—10 V, 1 W zener 1N4740A.

Table 18.7
Pi-L Component Values

Frequency (MHz)	*C1 (pF)*	*C2 pF*	*L1 µH*	*L2 µH*	*Q*
1.850	211	1262	44.3	9.6	12
3.700	105	631	22.2	4.8	12
7.150	65	364	9.7	2.5	14
14.150	33	184	4.9	1.26	14
18.100	45	208	2.23	0.98	23
21.200	33	159	2.21	0.84	20
24.900	36	161	1.48	0.71	25
28.250	29	133	1.43	0.63	23

Tank Circuit Coils

Coil	*Band*	*Inductance*	*Construction*
L1	10/12-15/17 m	2.3 µH	7 1/2 t, 1/4-in. copper tube, 2-in. ID silver-plated 10/12-m tap @ 3 1/2 t 15/17-m tap @ 7 1/2 t
L2	20-40 m	7.4 µH	19 t, 3/16-in. copper tube, 2-in. ID silver plated 20 tap@ 8 t 40 tap @ 19 t
L3	80 m	12.4 µH	17 t on 3xT225-2 cores, #10 Teflon silver wire
L4	160 m	22.0 µH	23 t on 3xT300-2 cores, #10 Teflon silver wire
L5	L-Coil	9.6 µH	19 t on 2xT225-2 cores, #12 tinned wire w/Teflon sleeve 10/12-m tap @ 2 t 15/17-m tap @ 4 t 20-m tap @ 5 t 40-m tap @ 7 t 80-m tap @ 12 t 160-m tap @ 19 t

has a 5-V/30-A filament and a maximum plate dissipation of 1500W, compared to the 1000-W dissipation for a pair of 3-500Zs. The 3CX1500D7 uses the popular Eimac SK510 socket and requires forced air through the anode for cooling. The amplifier uses a conventional grounded-grid design with an adjustable grid-trip protection circuit. See the RF Deck schematic in **Fig 18.29**.

Output impedance matching is accomplished using a pi-L tank circuit for good harmonic suppression. The 10 to 40-meter coils are hand wound from copper tubing, and they are silver plated for efficiency. Toroids are used for the 80- and 160-meter coils for compactness. The amplifier incorporates a heavy-duty shorting-type bandswitch. Vacuum variable capacitors are used for pi-L tuning and loading.

A unique feature of this amplifier is the use of a commercial computer-controlled input network module from LDG Electronics (**www.ldgelectronics.com**). This greatly simplifies the amplifier design by eliminating the need for complex ganged switches and sometimes frustrating setup adjustments. The computer-controlled input network is reasonably priced and basically plug-and-play.

An adjustable ALC circuit is also included to control excess drive power. The amplifier metering circuits allow simultaneous monitoring of plate current, grid current, and a choice of RF output, plate voltage or filament voltage.

The blower was sized to allow full 1500-watt plate dissipation (65 cfm at 0.45 inches H_2O hydrostatic backpressure). The design provides for blower mounting on the rear of the RF deck or optionally in a remote location to reduce ambient blower noise in the shack.

The power supply is built in a separate cabinet with casters and is connected to the RF deck using a 6-conductor control cable, with a separate high voltage (HV) cable. The power transformer has multiple primary taps (220/230/240 V ac) and multiple secondary taps (2300/2700/3100 V ac). No-load HV ranges can be selected from 3200 to 4600 V dc using different primary-secondary combinations. The amplifier is designed to run at 4000 V dc under load to maintain a reasonable plate resistance and component size. A step-start circuit is included to protect against current surge at turn on that can damage the diode bank. The power supply schematic is shown in **Fig 18.30** and a photo of the inside of the power supply is shown in **Fig 18.31**.

Both +12-V and +24-V regulated power supplies are included in the power supply. The +12 V is required for the computer-con-

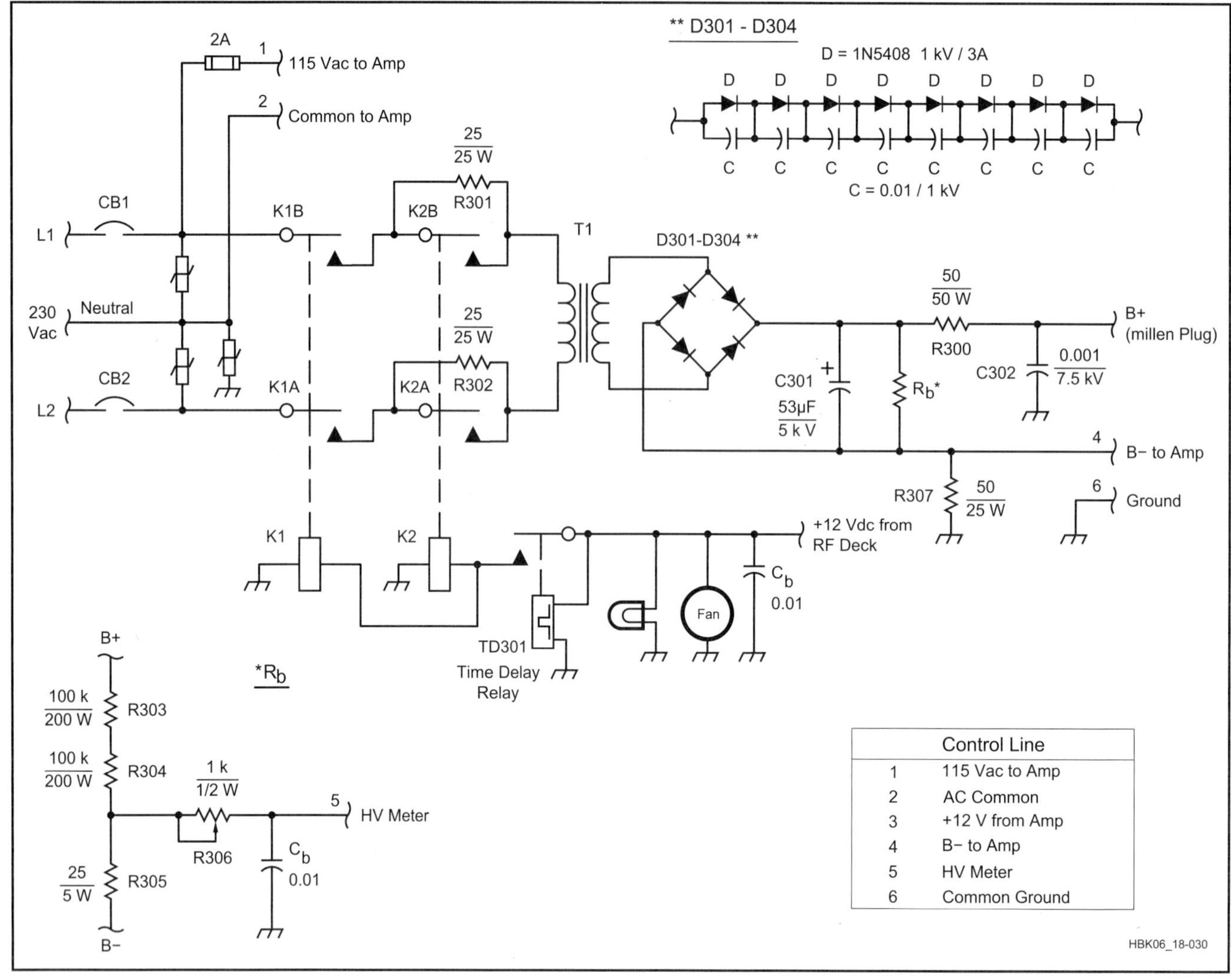

Fig 18.30—Schematic for Power Supply for 3CX1500D7 amplifier.

B1—Pilot lamp Alco 164-TZ, 12 V.
Cabinet—Buckeye Shapeform DSC-1204-16 (12×18×16-inch H×W×D), www.buckeyeshapeform.com.
C301—53 µF, 5 kV oil-filled, Peter W. Dahl p/n CDCF007100, pwdahl.com/cgi-bin/store/commerce.cgi.
Fan—12 Vdc brushless, 2¼ inch (Mouser 432-31432).
R303, R304—100 kΩ, 200 W wirewound, Peter W. Dahl p/n RP002000, pwdahl.com/cgi-bin/store/commerce.cgi.
SRY1-SRY4—Potter & Brumfield solid-state relay, SSR-240D25R.
T1—Peter W. Dahl Co, 220/230/240 pri : 2300/2700/3100 sec, 1.5 A, CCS, pwdahl.com.

trolled input network and +24 V is needed for the output vacuum relay. The input and output relays are time sequenced to avoid amplifier drive without a 50-Ω load. Relay actuation from the exciter uses a low-voltage/low-current circuit to accommodate the amplifier switching constraints imposed by many new solid-state radios.

Much thought was put into the physical appearance of the amplifier. The goal was to obtain a unit that looks commercial and that would look good sitting on the operating table. To accomplish the desired look, commercial cabinets were used. Not only does this help obtain a professional look but it eliminates a large amount of the metal work required in construction. Careful attention was taken making custom meter scales and cabinet labeling. The results are evident in the pictures provided.

GENERAL CONSTRUCTION NOTES

The amplifier was constructed using basic shop tools and does not require access to a sophisticated metal shop or electronics test bench. Basic tools included a band saw, a jig saw capable of cutting thin aluminum sheet, a drill press and common hand tools. Some skill in using tools is needed to obtain good results and insure safety, but most people can accomplish this project with careful planning and diligence.

Metal work can be a laborious activity. Building cabinets is an art within itself. This part of the project can be greatly simplified by using commercial cabinets. However, commercial cabinets are expensive (~$250 each) and could be a place where some dollars could be saved.

The amplifier is built in modules. This breaks the project into logical steps and facilitates testing the circuits along the way. For example, modules include the HV power supply, LV power supply, input network, control circuits, tank circuit and wattmeter.

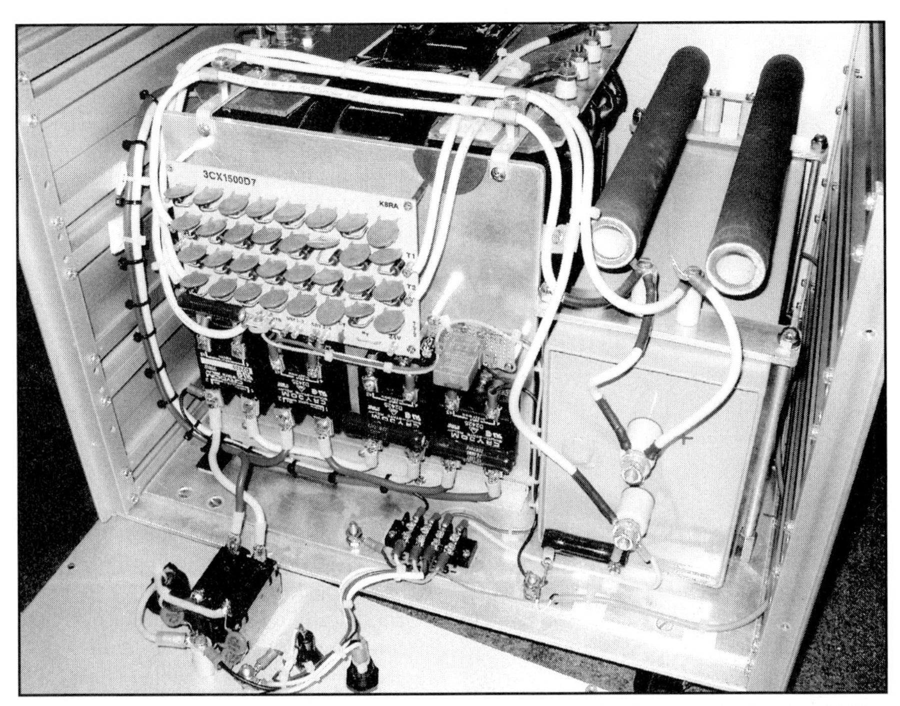

Fig 18.31—Inside view of Power Supply, showing rectifier stack, control relays and HV filter capacitor with bleeder resistors. The heavy-duty Peter W. Dahl transformer is at the upper left in this photo.

Each module can be tested prior to being integrated into the amplifier.

The project also made extensive use of computer tools in the design stage. The basic layout of all major components was done using the *Visio* diagramming software package. The printed-circuit boards were designed using a free layout program called *ExpressPCB* (**www.expresspcb.com**). Masks were developed and the iron-on transfer technique was used to transfer the traces to copper-clad board. The boards were then etched with excellent results. The layout underneath the RF Deck is shown in **Fig 18.32A** and the top side of the RF Deck is shown in Fig 18.32B.

Meter scales were made using an excellent piece of software called *Meter*, available by download from James Tonne, at **tonnesoftware.com**. Also, K8RA wrote an *Excel* spreadsheet to calculate the pi-L tank parameters. (A copy of the spreadsheet is on the CD-ROM that accompanies this book, as are the pc board templates.)

Although using computer tools simplifies the design step, all design work can be done without the use of a computer. Be creative and use the tools and resources at hand! There are many different ways to construct this design. The key secret is diligence and not compromising until it is done right. Note that the tank coils in this amplifier were wound at least three times, the inside side panels were cut twice and many printed circuit boards ended in the trash before acceptable boards were fabricated.

CABINET METAL WORK

By purchasing commercial cabinets, metal work required was minimized but not eliminated. The power supply components are very heavy. The transformer weighs about 70 pounds by itself. Therefore the base plate of the power supply cabinet needed to be reinforced. The original base plate for the cabinet was not used. One-eighth-inch plate was purchased from a local aluminum scrap company. Two pieces were sandwiched to provide a 1/4-inch plate. Of course 1/4-inch material could have been used but it was not available at the time of purchase.

The plate can be cut on a metal band saw using a guide or on a radial arm saw. Metal blades are readily available from Sears for both saws. If using a radial arm saw, multiple passes are required, lowering the blade slightly with each pass. Be sure to wear eye protection because the metal chips fly. The edges were then cleaned and straightened using a 4-inch belt grinder. If a belt sander is not available, a large file will work.

The two metal plates were held together with the mounting bolts on the four casters. The power supply base plate exactly matches the original base plate and fastened to the cabinet using the original tapped screw holes. All the heavy components are mounted on the base plate. The power supply must always be handled by lifting the base plate, since the cabinet does not have the structural integrity to bear the weight by itself.

The RF deck needed both a chassis plate and a front sub-panel. See Fig 18.32B. The sub-panel is used to mount the load and tune capacitors, the bandswitch and also provides RF shielding for the meters. Side plates were needed because of the cabinet configuration. The side plates, chassis plate and sub-panel all use 1/16-inch aluminum plate. After the side plates are cut and mounted to the cabinet sides, the chassis plate and front sub-panel are mounted using 1/2-inch aluminum angle to join the edges.

Cutting holes can often be a challenge. If a drill bit is the correct size, drilling a hole is easy, of course. But large-size round holes and square holes can be a challenge. This was especially true in this project since the front and rear panels are 1/8-inch aluminum plate.

The large meter holes can be cut using a hole saw on a drill press. For odd sizes, a "fly cutter" can be used. Fly cutters are available from Sears but a special warning is in order. These devices work well but are extremely dangerous. Make sure the cutting bit and the placement into the drill chuck are secure.

Large square holes are required for the turn counters. Mark the square hole to be cut. Drill a hole in each corner. The hole must be at least the size of the saw blade if a jigsaw is used to finish the hole. Note that the jigsaw must have a removable straight blade. If a metal-cutting jigsaw is not available, a series of small holes can be drilled in a straight line on all four sides and the edges smoothed with a file. Almost any hole can be custom cut by making a hole the approximate size and finishing it to the exact dimension with a file. It is slow and laborious but it works. When using a file on panels, be very careful that the file does not slip out of the hole and put an undesired scratch in the panel!

Once panel holes are cut, carefully label the panels before mounting the components. Dry transfers are used on both the power supply and the RF deck. Dry transfers of all sizes and fonts are available at graphics art stores and hobby shops. The author has found that hobby shops carry an excellent selection of dry transfers in the model railroad section.

RF DECK CONTROL CIRCUITS

The control circuits in the amplifier are

(A)

(B)

Fig 18.32—At A, under the chassis of the RF Deck. The autotuner used as the input network for this amplifier is at the upper right. At B, view of the Pi-L output network in the RF Deck.

not complex due to the simplicity of the grounded-grid design and the instant-on capability of the 3CX1500D7 tube. 120 V ac is routed from the HV power supply to the RF Deck in the 6-conductor control cable. When the on/off switch (S1) is pressed, 120 V ac is sent to the primary of the low voltage transformer (T2) and the filament transformer (T3). The surge current to the filament of the tube is suppressed by the thermistor (TH1) in one leg of the filament transformer primary. These are excellent current limiting devices that have a resistance of approximately 25 Ω cold but decrease to less then 1 Ω as they heat. Keep the thermistor in open air away from other components since they are designed to run hot.

The low voltage supply provides regulated +12 V dc and +28 V dc. The voltages are regulated using simple three-terminal regulators. Pilot lights are included in each push button switch, S1-S4, and a power indicator on the HV power supply. When the low-voltage power supply first comes on, +12 V dc is directed through the control cable back to the HV supply. High voltage is applied immediately to the instant-on tube. Therefore the amplifier is turned on and ready to go instantly—You don't have to listen to your friends working that rare one for three or four tense minutes while you wait for your amplifier to time in!

The amplifier is switched in and out of the circuit using a 4PDT KHP style relay (K1) for the input and a SPDT vacuum relay (K2) for the output. It is important to select the timing constants for the input relay (C201 and R201) so the input relay closes a few milliseconds after the vacuum relay. This avoids hot switching the output, which could fuse the vacuum-relay contacts. This is a balancing act since the brief time the input relay is open will present an open circuit to the exciter. Many modern radios now have exciter-timing circuits that close the amplifier relay circuit a few milliseconds before RF is transmitted.

It is recommended that timing components for the input relay be located in a place where they can be easily changed. Another approach is to build a breadboard circuit that feeds the relay coils in parallel but places the contacts in series. Feed a low voltage through the contacts of the two relays and monitor the timing with a dual-trace oscilloscope. This technique allows precise timing of contact closure as the two relays work together. Note that different relays will need different timing-circuit component values. A set of contacts on input relay, K1, is used to short across bias resistor, R103. The resistor biases the tube to cutoff in standby.

Approximately +10-V bias is provided to the center tap of the filament transformer to limit the idle current of the tube to approximately 125 mA. The bias is developed using the three components D101, R101 and Q101. These components could be replaced with a single 10-V/50-W zener diode. However, 50-W zeners are expensive and they are difficult to obtain. Using the circuit shown, the bias is provided by a common NPN transistor (Q101) and a one-watt zener (D101) you can obtain from RadioShack.

TUBE PROTECTION CIRCUIT

The main protection for the tube is a plate-current surge resistor and a grid-trip circuit. The current surge resistor (R308, 50-Ω/50-W) is in series with the B+ line and acts as a fuse should excessive current be drawn from the HV power supply. Ohm's law says that up to 1-A plate current can be drawn through the resistor and still stay within the 50-W rating of R308. However, let's assume a problem occurs and 5 A flows through the resistor. Resistor R3 must now dissipate 1250 W. The resistor will quickly fail and will shut down the HV to the 3CX1500D7 tube.

Q102 is a grid-trip circuit that snaps the amplifier offline if the grid current exceeds 400 mA. The grid current is drawn through the 10-Ω resistor (R104) connected between the B– line and chassis ground. The current creates a voltage across R104 that is fed to the grid-trip adjustment potentiometer, R106. Q102 is turned on when the base voltage reaches 0.6 V and actuates the grid-trip relay K3. K3 contacts break the +28 V dc input and output relay lines (K3B), locks the relay closed (K3C) and extinguishes the pilot bulb (K3A) of the GRID-TRIP RESET normally closed push-button switch (S3) located on the front panel. Pushing the GRID TRIP RESET switch (S3) breaks the current path for the grid-trip relay K3 and resets the relay. The reason the grid trip was actuated should be determined prior to attempting to use the amplifier again. Usually, this is caused by improper setting of the load capacitor or transmitting into the wrong antenna.

INPUT NETWORK

As mentioned before, this amplifier uses a unique concept for the input-matching network, getting rid of a switched network mechanically ganged to the main bandswitch. Not only can such a switching arrangement be awkward mechanically, but obtaining a reasonable network Q and a low SWR over an entire band can be difficult.

Thus the author decided to use a commercial automatic tuner integrated into the RF deck (see Fig 18.32A). The tuner is made by LDG and is based on their popular AT-100 Pro Autotuner, but was supplied without the front panel or rear panel connections and switches. This application is simple but elegant. The unit automatically initiates a retune if the input SWR exceeds approximately 1.5:1. The tuning cycle takes three to five seconds to execute. But retuning does not happen often because the tuner has over 4000 memories and remembers the settings for different frequency ranges. As the amplifier is used on each band, the tuner *learns* and stores settings into the memory. When switching bands, it only takes milliseconds to retrieve the data from memory and actuate the correct tuner relays.

Integration of the tuning network requires connections for RF input and output, +12 V and ground. RF input goes to the center of T1 and ground goes to J2 (clearly marked on the board). RF output goes to J3 and ground goes to J6. The + 12-V dc connection is the larger of the three holes at J10 (next to L10). The other two holes are grounds for dc connections.

A momentary contact switch (S4) is mounted on the front panel to provide manual control of the tuner. A normally open contact on S4 is connected to the input pin J9 (next to L12) and ground. (The pin is marked as the ring for the connector that is not installed.) The correct hole is on the C56 side. If the switch is pushed for less than 1/2 second, the tuner alternates between bypass and in-line modes. If S4 is pressed between 0.5 to 2.5 seconds, it does a memory tune from the stored data tables. If S4 is pressed for more than 2.5 seconds with RF applied, it skips the memory access, retunes and stores the new settings into the memory table. The manual retune function is seldom, if ever, used.

The tuner works perfectly and it really simplified the input-network design and construction. The SWR never exceeds 1.5:1 (typically it is 1.2:1). LDG provides the unit as a commercial product to amplifier builders.

PI-L NETWORK

A pi-L network is used to insure good harmonic attenuation. The pi-L circuit is actually a pi-network, followed by an L-network that provides additional harmonic attenuation. The L-section transforms the load of 50 Ω up to an intermediate resistance of 300 Ω. The pi section then transforms 300 Ω up to the desired plate load resistance of 3100 Ω. The plate load is calculated using the following formula:

$$R_L = \frac{E_P}{1.7 \times I_P} = \frac{4000}{1.7 \times 0.750} = 3137\,\Omega$$

A nominal Q of 12 is used for the network. But as with most RF amplifiers, the capacitance needed for the higher-frequency bands is less than is physically possible using variable capacitors. For Q = 12, the tune capacitance (C101) for 10 meters is 14 pF. Using a vacuum variable capacitor for C1 helps because the minimum achievable capacitance (12 pF) is substantially less than with an air variable. But the tune capacitance is the sum of variable C101, 7.1 pF for the output capacitance of the 3CX1500D7 tube and any stray capacitance resulting from the physical layout of the amplifier.

The minimum obtainable capacitance is thus on the order of 30 pF, which yields a higher value of loaded Q than optimum. The solution is two fold. First, connect the plate-tune capacitor (C101) one turn into the 10-meter coil. This actually forms an L-pi-L circuit. Second, accept a higher value of loaded Q so that the variable capacitor can still be tuned. **Table 18.7** shows the loaded Q finally used for each band setting. The disadvantages of higher loaded Qs are high circulating currents in the tank circuit and the need to retune during excursions across the higher-frequency bands. This amplifier works fine on all bands, delivering a solid 1500 W output even on 10 meters.

Another pi-L tank circuit design constraint in this amplifier is the bandswitch. Many amplifiers use a single-pole, 12-position, non-shorting switch. Although this type of switch is easier to find, it can be problematic because high voltages are generated that could result in arcing in the bandswitch—usually from the wiper to the high frequency taps. You should use a switch with a multiple-finger wiper (see Fig 18.29) that shorts out lower-frequency coil taps not being used. For example, when the amplifier is used on 20 meters, the 40-, 80- and 160-meter taps are shorted to the wiper.

However, shorting switches only allow for six connections with 30° indexing. The common shorting wiper consumes 180° of switch deck on 160 meters. This results in having to design the 10/12-meter and 15/17-meter bands to use single taps for each frequency pair. Again, this is accomplished by adjusting the loaded Q for each band so that shared bands so they require nearly the same inductance. From Table 18.7, the same band switch position is shared on the 10/12-meter bands (1.4 μH) and the 15/18-meter bands (2.2 μH).

In actual construction of the tank circuit, it is very useful to have access to both a capacitance meter and an inductance meter. The author used an Elenco LCM-1950 meter that measures both capacitance and inductance and is available for under $100 (**www.elenco.com**). With the tune and load capacitors mounted and connected to calibrated knobs or turns counters, make a table of capacitance verses knob settings. This is useful to estimate the initial setting for each band during setup and test. Also, measure the inductance of each coil turn to determine initial coil taps for each band. On this amplifier, only the 10-meter tap had to be adjusted from the predetermined settings.

As mentioned above, the pi-L tank circuit was designed for 3100-Ω plate-load resistance. Such a high plate resistance demands higher inductance values to obtain reasonable tank circuit Qs. Table 18.7 shows that 160 meters requires 42 μH. If air-wound coils were used exclusively, the coils would require many turns and would take up a lot of cabinet space. To maintain a reasonable physical coil size, therefore, toroidal coils were used for 80 (L3) and 160 (L4) meters in addition to the output coil (L5) (see Fig 18.32B). You should use substantial core material for high-power operation to avoid core heating. Core sizes were increased by using multiple cores taped together. Each ferrite core is wrapped with three layers of high temperature fiberglass tape, available from RF Parts (**www.rfparts.com**). Teflon-insulated #10 wire was wound to obtain the desired inductance in L3 and L4. Both coils are mounted on ceramic standoffs and held in place with Teflon blocks.

The output coil is wound on a pair of T225-2 cores using #12 tinned wire covered with a Teflon sleeve. Taps onto the coil are made by carefully trimming a small 1/8-inch space from the Teflon sleeve on the inner edge of the core facing the bandswitch. Taps are then made from the back section of the 2-pole bandswitch using #12 tinned wire. The proper placement of each tap is determined by first winding #12 insulated wire around the core. A small slit is carved into each turn and the inductance was measured. The copper wire is removed and the final Teflon-covered #12 tinned wire is wound onto the core. Using the output L-coil (L102) design values in Table 18.7, permanent taps were made.

Note that the taps for the output coil are not extremely critical. Select the closest turn to the value needed. The output coil is mounted on the back of the bandswitch on one of the switch wafer screws using a threaded 1-inch diameter Teflon rod. The Teflon rod holds the position of the coil. The weight is carried by the wire taps from the coil to the bandswitch contacts. Table 18.7 also gives the inductance and construction instructions for each coil.

L1 and L2 (10-40 meters) are silver plated. They were wound using a 2-inch aluminum pipe as a form. Clean the copper tubing with #0000 steel wool prior to winding. Wind the copper tubing close spaced on the pipe. Leave plenty of pigtail on each end of the coil. The ends can be trimmed to fit the mounting positions precisely. After winding the desired number of turns, plug the ends by closing the tube ends with a hammer, spread the coil windings and rinse the coil in acetone to remove any oil. Allow a few minutes to dry. The coil is now ready to plate.

Go to any photo shop and beg/buy a gallon of used photographic fixer solution. Note that used fixer solution has silver remnant. The more the solution has been used, the more the silver content. The coils can be silver plated by dipping the clean coil into the solution. Do not leave the coil in the solution too long or it will turn black. A thin but bright silver coat will be deposited on the copper tube. This is called *flash plating*. After dipping the coil into the solution, immediately rinse in a bath of clean water and blow dry under pressure with an air compressor, heat gun or hair dryer. If a thicker silver coat is desired, electroplating is necessary, a subject beyond the scope of this article.

A #10 lug is crimped and soldered onto the end of each coil and used to mount the coil. The L2 coil is mounted using a Teflon block that is held in place to the front subpanel with small screws. The block is carefully drilled with 3/16 inch holes the desired spacing of the coil about 3/4-inch from one edge. The block is sawed down through the holes creating two matching blocks. At each end of the block a hole is drilled and tapped (6-32 tap). The silver plated coil is sandwiched between the two blocks for secure support. The tapped screws serve as the connecting points for the ends of the coil.

METERING

The amplifier uses three separate meters to simultaneously monitor plate current, grid current and a choice of plate voltage, power output or filament voltage. Each meter is identical with a 1-mA full-scale movement. As mentioned previously, the custom scales for each movement were designed using the *Meter* software from Tonne Software. This allows up to three scales on each meter. Scales can be designed as either linear or log and the number of major and minor tick marks can be specified. Each scale can be labeled using different font size and color. The author printed the scales using a color inkjet printer onto glossy photo paper. The scales were carefully cut to match the meter faceplate and glued into place using a thin coat of adhesive.

Plate current is measured by M1 in series in the B− line using a current divider (R114 and R115) as shown in Fig 18.30. Adjust R114 to obtain full scale with 1-A of plate current. The meter was calibrated prior to installation using a low-voltage power supply with adjustable current limiting in series with an accurate digital meter.

M2 monitors grid current by measuring the voltage drop created by grid current flow through the 10-Ω resistor, R104. Connecting a voltage source (ie, small variable power supply) across R104 and measuring the actual current flow with an external meter provides a way to set the calibration pot, R105.

M3 is a multimeter that reads HV, RF power or filament voltage. The metering circuit is selected using a 2P3T rotary switch (S5). The HV metering circuit is in the HV power supply and fed to the RF deck through the control cable. The filament-voltage detect circuit is shown on the control circuit diagram (Fig 18.29: D207, D201 and R202). Adjust R202 for the proper reading on Meter M3. The 3.1-V zener (D201) expands the meter scale for more precise reading.

The RF wattmeter circuit is also shown in Fig 18.29. Only forward power is measured and potentiometer R403 is used for calibration. The wattmeter is not a precise instrument but gives a relative output reading. It is adequate for peaking power output when tuning. The meter provides good accuracy through 40 meters and then begins to read lower on the higher-frequency bands. This is due to the simplicity of the circuit and the toroid used. Quite honestly, don't expect much accuracy from this wattmeter.

HV POWER SUPPLY

The matching HV Power Supply (Fig 18.30) provides approximately 4000 V under load. It uses a full-wave bridge rectifier and is filtered using a single 53 μF/5000-V oil filled capacitor (C301). Whenever the HV supply is plugged into the 240-V line, live 120 V ac is routed to the RF deck through the control cable. The 120-V ac line is obtained from L1 and neutral of the 240-V ac line. The neutral line is isolated from ground for safety.

Actuating the on/off switch S1 on the front panel of the RF deck provides ac power to the low-voltage power supply. In turn, +12 V is returned to the power supply through the control cable and routed to a pair of solid state power relays (K1,

K2). Also, +12 V is routed to a timer relay that provides a two-second delay in applying +12 V to the second pair of solid-state relays (K3, K4). During the two-second delay, each leg of the 240-V ac primary voltage is routed through a 25-Ω resistor (R301, R302) to reduce the current surge when charging the filter capacitor, C301.

HV is metered at the bottom of the two series 100,000-Ω bleeder resistors (R303, R304). A current divider is created using a small potentiometer (R306) in parallel with a 25-Ω/5-W resistor (R305). The current divider is in series with the bleeder resistors and tied to the B– line. R306 is set to allow 1 mA of current to flow to the HV meter located on the front panel of the RF deck with 5000 V HV dc. The potentiometer R306 and the paralleled fixed resistor R305 need handle only a small amount of power, since the voltage and current flow is quite small at this point in the circuit.

The HV cable between the RF Deck and power supply is made from a length of automotive-ignition cable that has a #20 wire and 60,000-V insulation. Be sure to get a solid-wire center conductor and not the resistive carbon material. Also use high-quality HV connectors that are intended for such an application. Millen HV connectors (50001) were used in this amplifier. Coax boots intended for coaxial cable are used on each connector for added insulation and physical strength. The mounting holes for the Millen connectors are oversized and plastic screws were used for safety.

TUNING AND OPERATION

The amplifier is very easy to tune after the initial settings of the tune (C101) and load (C102) controls are determined. The correct settings are determined with a plate current of 700 mA with a corresponding grid current of 200 mA. The turn counters provide excellent resetability once the proper settings have been found initially. Required drive power is about 75 W for 1500 W output.

Thanks to CPI Eimac Division, LDG Electronics, Peter W Dahl Co, and MTI Inc (Radio Switch) for their support in this project.

A 6-Meter Kilowatt Amplifier Using the Svetlana 4CX1600B

The Svetlana 4CX1600B tube has attracted a lot of attention because of its potent capabilities and relatively low cost. Because of its high gain and its large anode dissipation capabilities, the tube has relatively large input and output capacitances—85 pF at the input and 12 pF at the output. Stray capacitance of about 10 pF must be added in as well. On bands lower than 50 MHz, these capacitances can be dealt with satisfactorily with a broadband 50-Ω input resistor and conventional out-

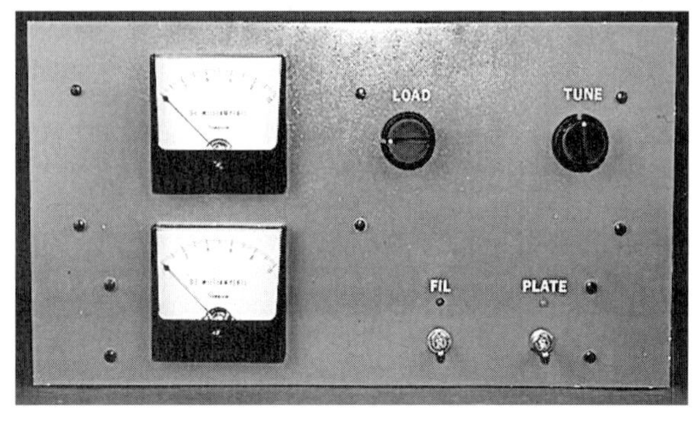

Fig 18.33—Photo of the front panel of W1QWJ's 6-meter 4CX1600B amplifier.

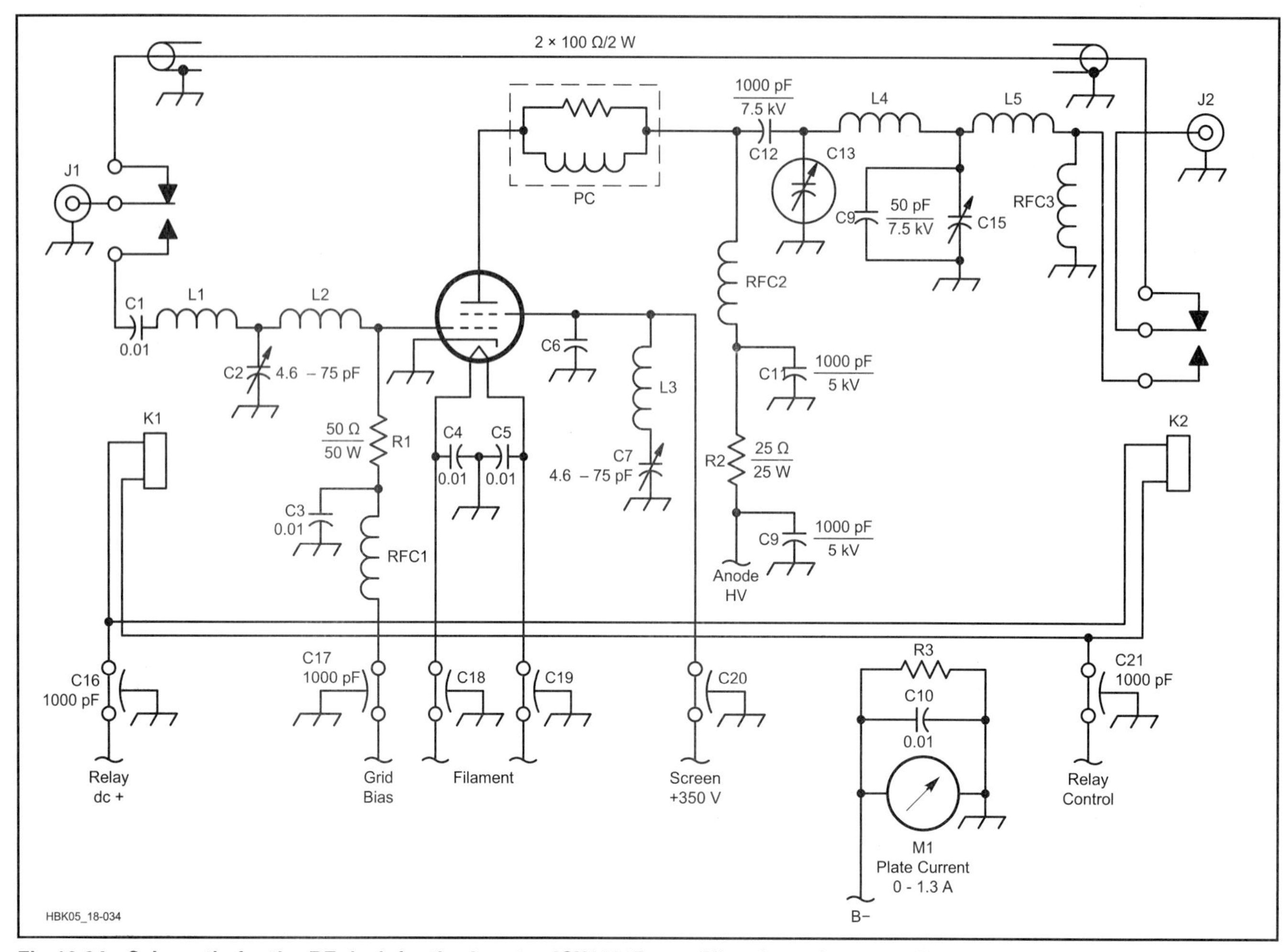

Fig 18.34—Schematic for the RF deck for the 6-meter 4CX1600B amplifier. Capacitors are disc ceramic unless noted. Addresses for parts suppliers can be found using *TISFind* and other search engines.

C2, C7—4.6-75 pF, 500-V air-variable trimmer capacitor, APC style.
C6—Screen bypass capacitor, built into SK-3A socket.
C13—1-45 pF, 5 kV, Jennings CHV1-45-5S vacuum-variable capacitor.
C14—50 pF, 7.5 kV, NPO ceramic doorknob capacitor.
C15—4-102 pF, 1100V, HFA-100A type air-variable capacitor.
C16, 17, 18, 19, 20, 21—1000 pF, 1 kV feedthrough capacitors.
L1—11 turns, #16, 3/8-inch diameter, 1-inch long.
L2—9 turns #16, 3/8-inch diameter, close-wound.
L3—8 turns #16, 3/8-inch diameter, 7/8-inch long.
L4—1/4-inch copper tubing, 4 1/2 turns, 1 1/4 inches diameter, 4 1/2 inches long.
L5—5 turns #14, 1/2 inch diameter, 1 3/8 inches long.
M1—0-1.3 A meter, with homemade shunt resistor, R3, across 0-10 mA movement meter.
PC—Parasitic suppressor, 2 turns #14, 1/2 inch diameter, shunted by two 100-Ω, 2-W carbon composition resistors in parallel.
RFC1—10 µH, grid-bias choke.
RFC2—Plate choke, 40 turns #20, 1/2-inch diameter, close-wound.
RFC3—Safety choke, 20 turns #20, 3/8 inch diameter.

put tuning circuitry.

See the article by George Daughters, K6GT, "The Sunnyvale/Saint Petersburg Kilowatt-Plus" in 2005 for details on suitable control and power-supply circuitry. This 6-meter amplifier uses the same basic design as K6GT's, except for modified input and output circuits in the RF deck. See **Fig 18.33**, a photograph of the front panel of the 6-meter amplifier.

On the 50-MHz band the tube's high input capacitance must be tuned out. Author Dick Stevens, W1QWJ, used a T network so that the input impedance looks like a nonreactive 50 Ω to the transceiver. To keep the output tuning network's loaded Q low enough for efficient power generation, he used a 1.5 to 46 pF Jennings CHV1-45-5S vacuum-variable capacitor, in a Pi-L configuration to keep harmonics low. You should use a quarter-wave shorted coaxial stub in parallel with the output RF connector to make absolutely sure that the second harmonic is reduced well below the FCC specification limits.

To guarantee stability, the author had to make sure the screen grid was kept as close as possible to RF ground. This allows the screen to do its job "screening"—this minimizes the capacitance between the control grid and the anode. He used the Svetlana SK-3A socket, which includes a built-in screen bypass capacitor, and augmented that with a 50-MHz series-tuned circuit to ground. In addition, to prevent VHF parasitics, he used a parasitic suppressor in the anode circuit.

Unlike the K6GT HF amplifier, this 6-meter amplifier uses no cathode degeneration. W1QWJ wanted maximum stable power gain, with less drive power needed on 6 m. He left the SK-3A socket in stock form, with the cathode directly grounded. This amplifier requires about 25 W of drive power to produce full output.

Fig 18.34 is a schematic of the RF deck built by W1QWJ. The control and power supply circuitry are basically the same as

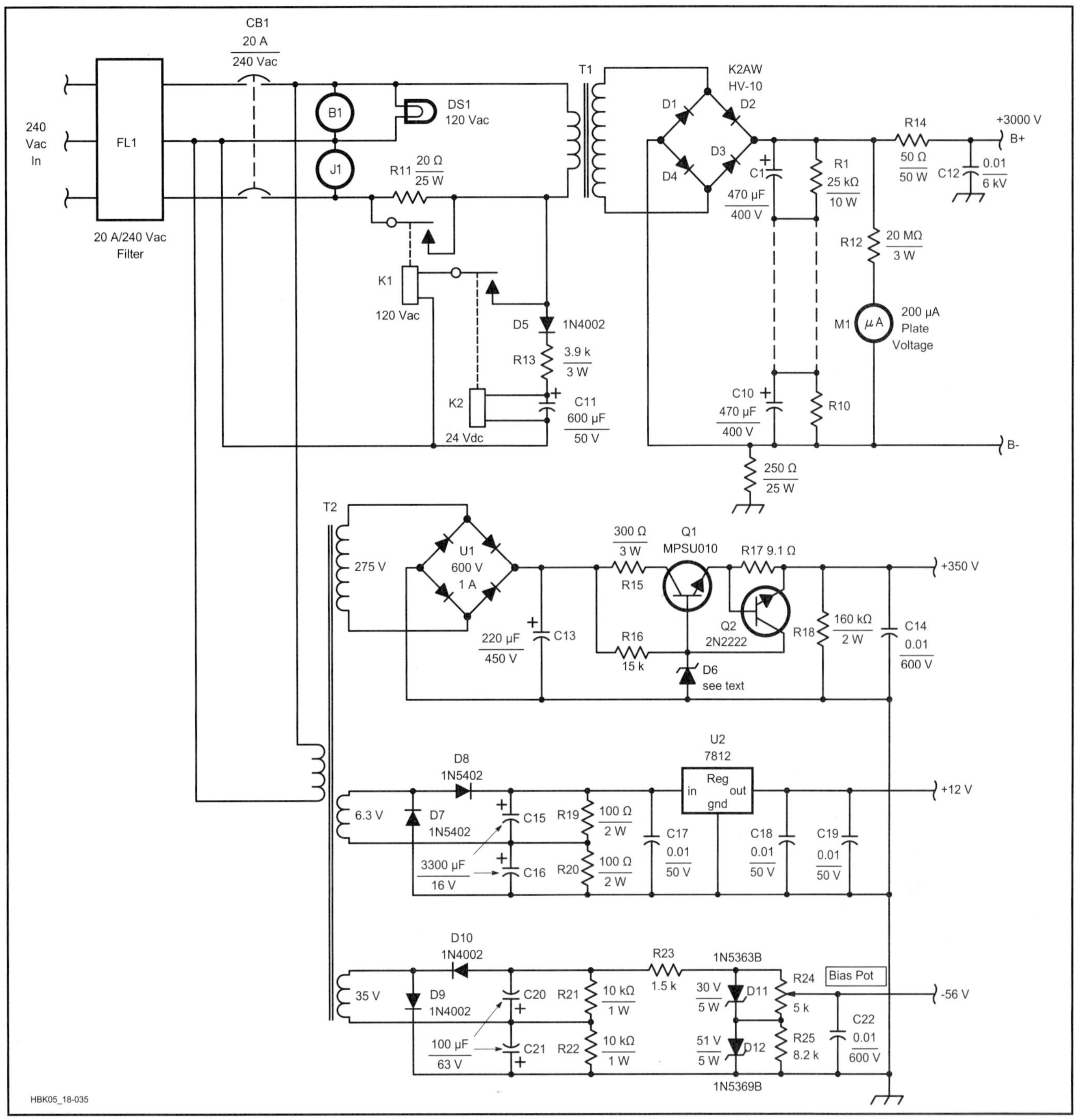

Fig 18.35—Partial schematic of K6GT HV power supply, showing modification with 250-Ω, 25-W power resistor to ground on B– line, allowing for metering of the plate current in the amplifier.

that used in Fig 18.29 and Fig 18.30 in the K6GT HF amplifier, except that plate current is monitored with a meter in series with the B– lead, since the cathode in this amplifier is grounded directly. The K6GT power supply is modified by inserting a 250-Ω, 25-W power resistor to ground in place of the direct ground connection. See **Fig 18.35**. In Fig 18.34, C1 blocks grid-bias dc voltage from appearing at the transceiver, while L1, L2 and C2 make up the T-network that tunes out the input capacitance of V1. R1 is a non-reactive 50-Ω 50-W resistor.

C6 is the built-in screen bypass capacitor in the SK-3A socket, while L3 and C7 make up the series-tuned screen bypass circuit. RFC3 is a safety choke, in case blocking capacitor C12 should break down and short, which would otherwise place high voltage at the output connector.

CONSTRUCTION

Like the K6GT amplifier, this W1QWJ amplifier is constructed in two parts: an RF deck and a power supply. Two aluminum chassis boxes bolted together and mounted to a front panel are used to make the RF deck. **Fig 18.36** shows the 4CX1600B tube and the 6-meter output tank circuit.

Fig 18.37 shows the underside of the RF deck, with the input circuitry shown in more detail in **Fig 18.38**. The 50-Ω, 50-W noninductive power resistor is shown at the bottom of Fig 18.38. Note

Fig 18.36 (left)—Close-up photo of the anode tank circuit for 6-meter kW amplifier. The air-cooling chimney has been removed in this photo.

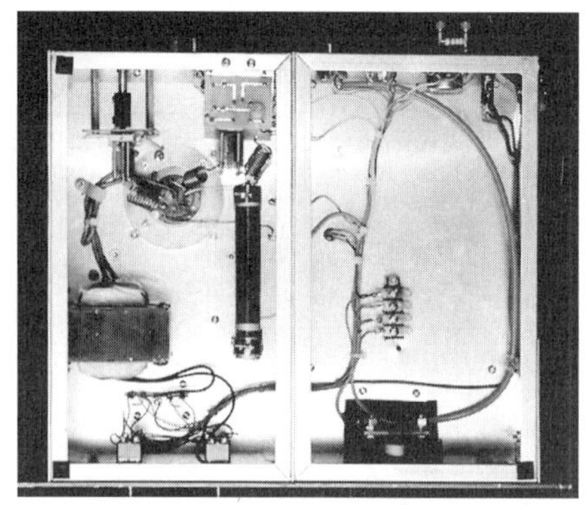

Fig 18.37—Underneath the 6-meter kW amplifier RF deck, showing on the left the tube socket and input circuitry.

Fig 18.38—Close-up photo of the input circuitry for the 6-meter kW amplifier. Input tuning capacitor C2 is adjusted from the rear panel during operation, if necessary. The series-tuning capacitor C7 used to thoroughly ground the screen for RF is shown at the lower right. It is adjusted through a normally plugged hole in the rear panel during initial adjustment only.

that the tuning adjustment for the input circuit is accessed from the rear of the RF deck.

AMPLIFIER ADJUSTMENT

The tune-up adjustments can be done without power applied to the amplifier and with the top and bottom covers removed. You can use readily available test instruments: an MFJ-259 SWR Analyzer and a VTVM with RF probe.

1. Activate the antenna changeover relay, either mechanically or by applying control voltage to it. Connect a 2700-Ω, 1/2-W carbon composition resistor from anode to ground using short leads. Connect the SWR analyzer, tuned to 50 MHz, to the output connector. Adjust plate tuning and loading controls for a 1:1 SWR. You are using the Pi-L network in reverse this way.
2. Now, connect the MFJ-259 to the input connector and adjust the input T-network for a 1:1 SWR. Some spreading of the turns of the inductor may be required.
3. Disconnect the Pi-L output network from the tube's anode, leaving the 2700-Ω carbon composition resistor from the anode still connected. Connect the RF probe of the VTVM to the anode and run your exciter at low power into the amplifier's input connector. Tune the screen series-tuned bypass circuit for a distinct dip on the VTVM. The dip will be sharp and the VTVM reading should go to zero.
4. Now, disconnect the 2700-Ω carbon resistor from the anode and replace the covers. Connect the power supply and control circuitry. When you apply power to the amplifier, you should find that only a slight tweaking of the output controls will be needed for final adjustment.

A 144-MHz Amplifier Using the 3CX1200Z7

This 2-meter, 1-kW amplifier uses the Eimac 3CX1200Z7 triode. The original article by Russ Miller, N7ART, appeared in December 1994 *QST*. The tube requires a warm up of about 10 seconds after applying filament voltage—no more waiting for three agonizingly long minutes until an amplifier can go on-line!

The 3CX1200Z7 is different from the earlier 3CX1200A7 by virtue of its external grid ring, redesigned anode assembly and a 6.3-V ac filament. One advantage to the 3CX1200Z7 is the wide range of plate voltages that can be used, from 2000 to 5500 V. This amplifier looks much like the easily duplicated W6PO design. The RF deck is a compact unit, designed for table-top use (See **Fig 18.39** and schematic in **Fig 18.40**.).

Fig 18.39—This table-top 2-meter power amplifier uses a quick-warm-up tube, a real plus when the band suddenly opens for DX and you want to join in.

Table 18.8 gives some data on the 3CX1200Z7 and **Table 18.9** lists CW operating performance for this amplifier.

Input Circuit

The author didn't use a tube socket. Instead, he bolted the tube directly to the top plate of the subchassis, using the four holes (drilled to clear a #6 screw) in the grid flange. Connections to the heater pins are via drilled and slotted brass rods. The

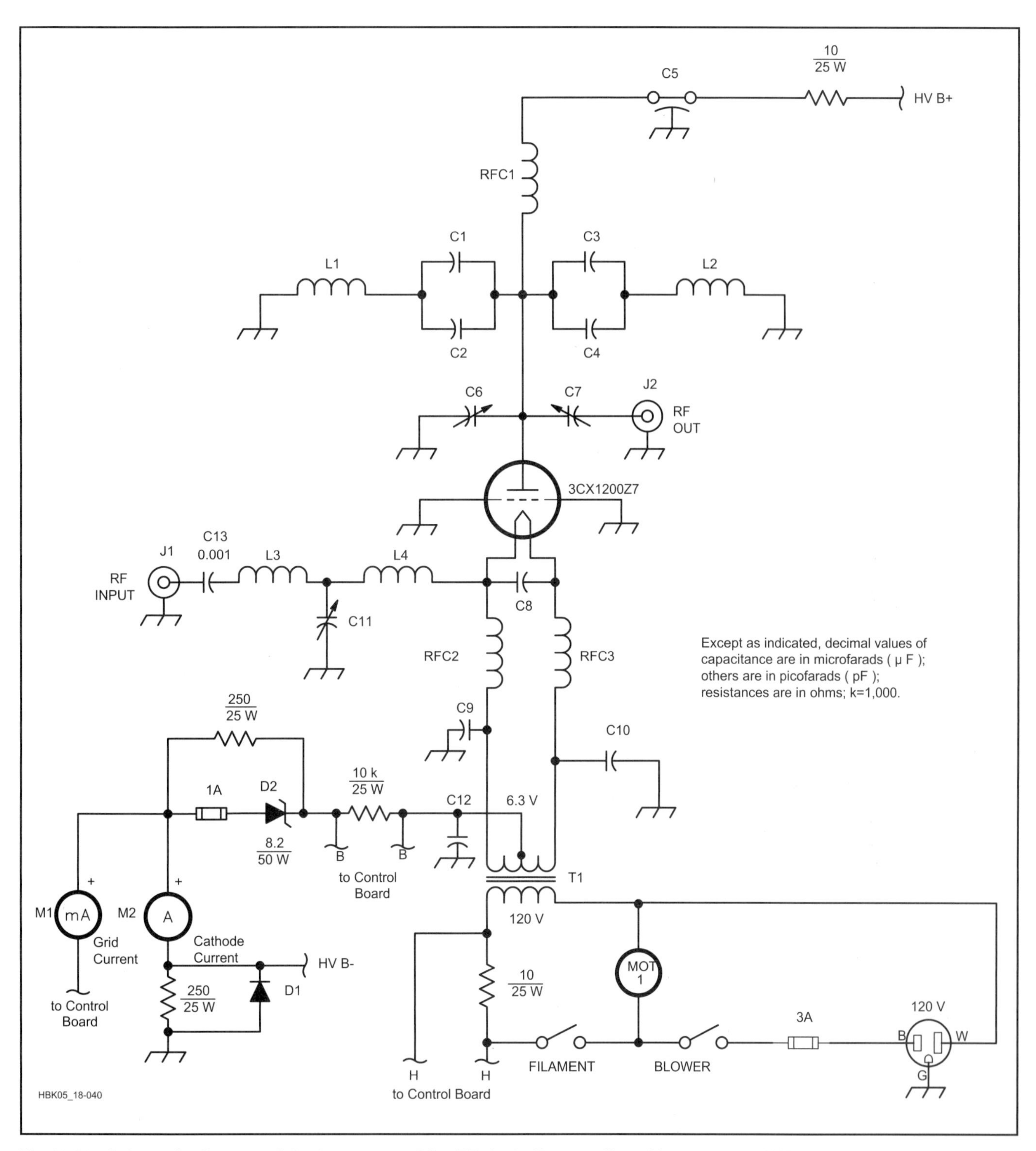

Fig 18.40—Schematic diagram of the 2-meter amplifier RF deck. For supplier addresses, use *TISFind* or other search engines.

C1-C4—100 pF, 5 kV, Centralab 850.
C5—1000 pF, 5 kV.
C6—Anode-tuning capacitor; see text and Fig 18.46 for details.
C7—Output-loading capacitor; see text and Fig 18.47 for details.
C8-C10, C13—1000-pF silver mica, 500 V.
C11—30-pF air variable.
C12—0.01 µF, 1 kV.
D1—1000 PIV, 3-A diode, 1N5408 or equiv.
D2—8.2-V, 50-W Zener diode, ECG 5249A.
J1—Chassis-mount BNC connector.
J2—Type-N connector fitted to output coupling assembly (see Fig 18.47).
L1, L2—Plate lines; see text and Fig 18.45 for details.
L3—5 t #14 enameled wire, 1/2-inch diameter, close wound.
L4—3 t #14, 5/8-inch diameter, 1/4-inch spacing.
RFC1—7 t #14, 5/8-inch diameter, 1 3/8 inch long.
RFC2, RFC3—10 t #12, 5/8-inch diameter, 2 inches long.
T1—Filament transformer. Primary: 120 V; secondary: 6.3 V, 25 A, center tapped. Available from Avatar Magnetics; part number AV-539.
M1—Grid milliammeter, 200 mA dc full scale.
M2—Cathode ammeter, 2 A dc full scale.
MOT1—140 free-air cfm, 120-V ac blower, Dayton 4C442 or equivalent.
Sources for some of the hard to get parts include Fair Radio Sales and Surplus Sales of Nebraska.

Table 18.8
3CX1200Z7 Specifications

Maximum Ratings
Plate voltage: 5500 V
Plate current: 800 mA
Plate dissipation: 1200 W
Grid dissipation: 50 W

Table 18.9
CW Operating Data

Plate voltage: 3200 V
Plate current (operating): 750 mA
Plate current (idling): 150 mA
Grid current: 165 mA
DC Power input: 2400 W
RF Power output: 1200 W
Plate dissipation: 1200 W
Efficiency: 50%
Drive power: 85 W
Input reflected power: 1 W

Fig 18.41—This view of the cathode-circuit compartment shows the input tuned circuit and filament chokes.

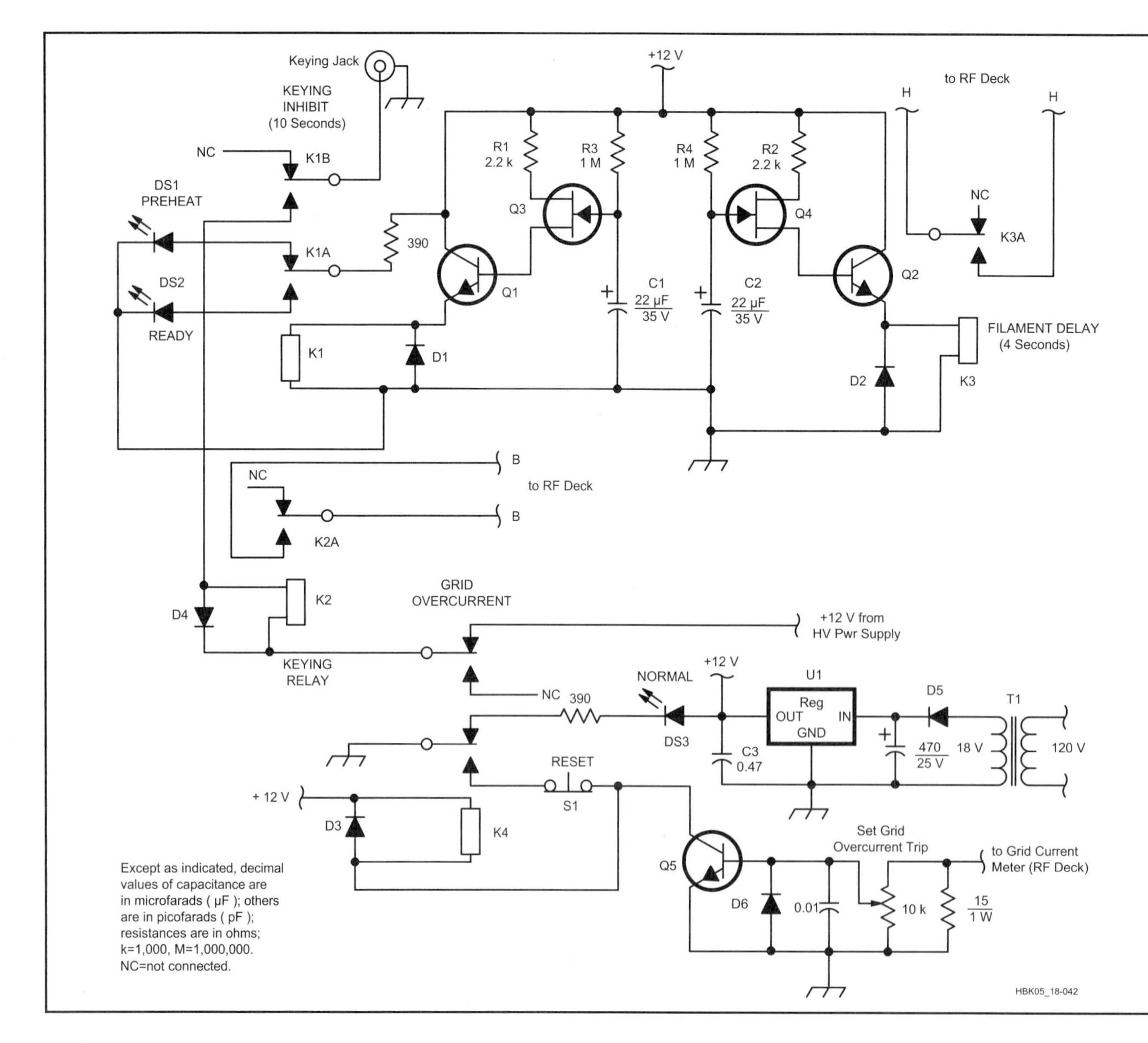

input circuit is contained within a $3^1/_2 \times 6 \times 7^1/_4$-inch (HWD) subchassis (Fig 18.40).

Control Circuit

The control circuit (**Fig 18.42**) is a necessity. It provides grid overcurrent protection, keying control and filament surge control. To protect the tube filament from stressful surge current, a timer circuit places a resistor in series with the primary of the filament transformer. After four seconds, the timer shorts the resistor, allowing full filament voltage to be applied. C2 and R4 establish the time delay.

Another timer inhibits keying for a total of 10 seconds, to give the internal tube temperatures a chance to stabilize. C1 and R3 determine the time constant of this timer. After 10 seconds, the amplifier can be keyed by grounding the keying line. When the amplifier is not keyed, it draws

Fig 18.43—This top view of the plate compartment shows the plate-line arrangement, C1-C4 and the output coupling assembly.

Fig 18.42—(Schematic diagram of the amplifier-control circuits.
C3—0.47-µF, 25-V tantalum capacitor.
D1-D5—1N4001 or equiv.
D6—1N4007 or equiv.
DS1—Yellow LED.
DS2—Green LED.
DS3—Red LED.
K1—Keying-inhibit relay, DPDT, 12-V dc coil, 1-A contact rating (RadioShack 275-249 or equiv).
K2—Amplifier keying relay, SPDT, 12-V dc coil, 2-A contact rating (RadioShack 275-248 or equiv).
K3—Filament delay relay, SPST, 12-V dc coil, 2-A contact rating (RadioShack 275-248 or equiv).
K4—Grid-overcurrent relay, DPDT, 12-V dc coil, 1-A contact rating (RadioShack 275-249 or equiv).
Q1, Q2, Q5—2N2222A or equiv.
Q3—MPF102 or equiv.
Q4—2N3819 or equiv.
S1—Normally closed, momentary pushbutton switch (RadioShack 275-1549 or equiv).
T1—Power transformer, 120-V primary, 18-V, 1-A secondary.
U1— +12 V regulator, 7812 or equiv.

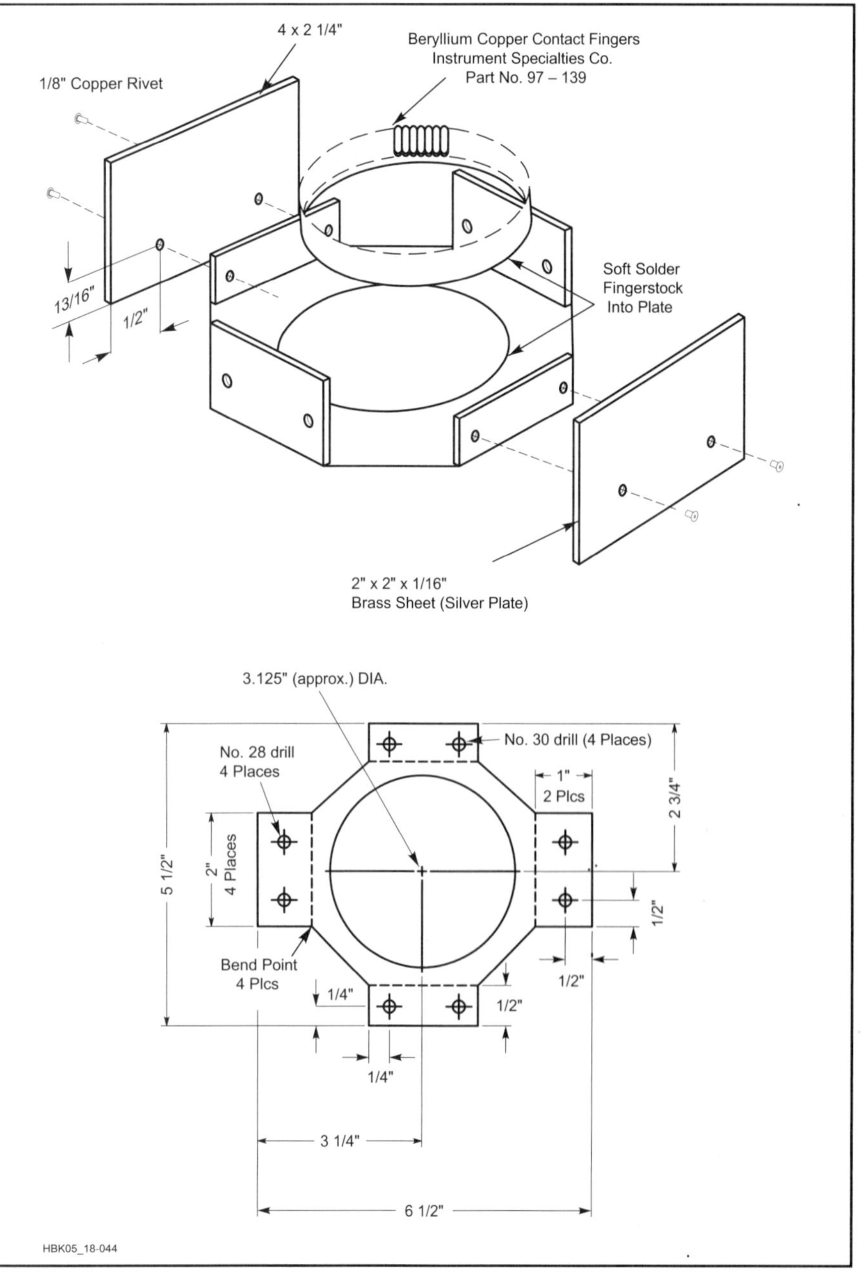

Fig 18.44—Anode collet details.

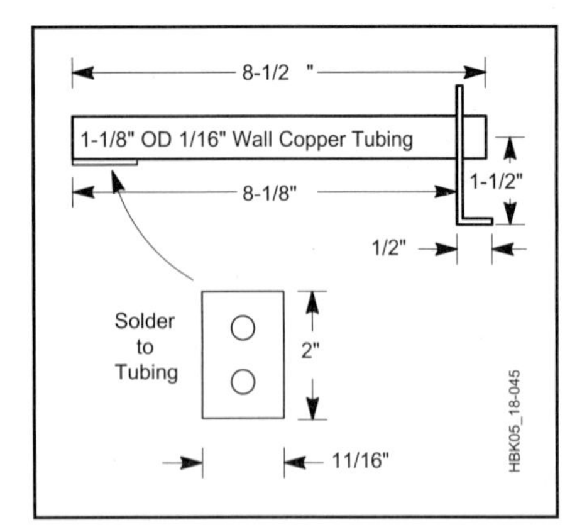

Fig 18.45—Plate line details.

no plate current. When keyed, idle current is approximately 150 mA, and the amplifier only requires RF drive to produce output. A safety factor is built in: The keying circuit requires +12 V from the high-voltage supply. This feature ensures that high voltage is present before the amplifier is driven.

The grid overcurrent circuit should be set to trip if grid current reaches 200 mA. When it trips, the relay latches and the NORMAL LED extinguishes. Restoration requires the operator to press the RESET switch.

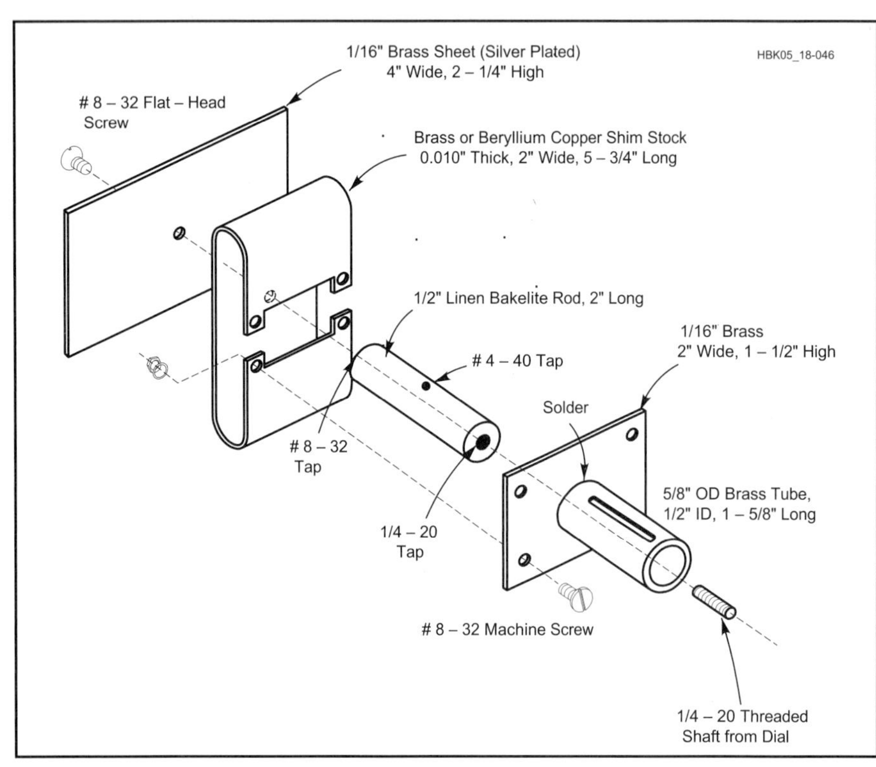

Fig 18.46—Plate tuning capacitor details.

Plate Circuit

Fig 18.43 shows an interior view of the plate compartment. A 4 × 2¼-inch tuning capacitor plate and a 2 × 2-inch output coupling plate are centered on the anode collet. See **Fig 18.44**. Sufficient clearance in the collet hole for the 3CX1200Z7 anode must be left for the fingerstock. The hole diameter will be approximately 3⅝ inches. **Fig 18.45** is a drawing of the plate line, **Fig 18.46** is a drawing of the plate tuning capacitor assembly, and **Fig 18.47** shows the output coupling assembly.

Cooling

The amplifier requires an air exhaust through the top cover, as the plate compartment is pressurized. Fashion a chimney from a 3½-inch waste-water coupling (black PVC) and a piece of 1/32-inch-thick Teflon sheet. The PVC should extend down from the underside of the amplifier cover plate by 1⅛ inches, with the Teflon sheet extending down ¾ inch from the bottom of the PVC.

The base of the 3CX1200Z7 is cooled using bleed air from the plate compartment. This is directed at the tube base, through a ⅞-inch tube set into the subchassis wall at a 45° angle. The recommended blower will supply more than enough air for any temperature zone. A smaller blower is not recommended, as it is doubtful that the base area will be cooled adequately. The 3CX1200Z7 filament draws 25 A at 6.3 V! It alone generates a great deal of heat around the tube base seals and pins, so good air flow is critical.

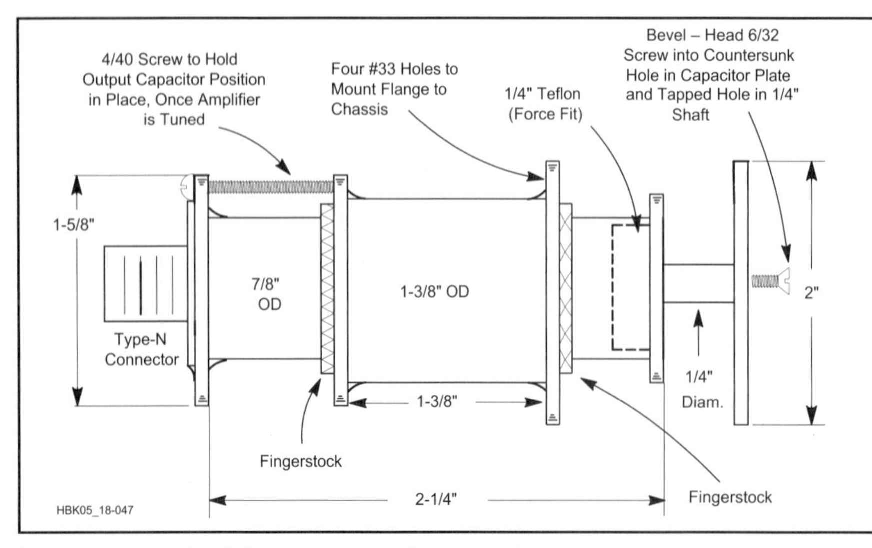

Fig 18.47—Details of the output coupling assembly.

Construction

The amplifier is built into a 12 × 12 × 10-inch enclosure. A 12 × 10-inch partition is installed 7¼ inches from the rear panel. The area between the partition and the front panel contains the filament transformer, control board, meters, switches, Zener diode and miscellaneous small parts. Wiring between the front-panel area and the rear panel is through a ½-inch brass tube, located near the shorted end of the right-hand plate line.

High voltage is routed from an MHV jack on the rear panel, through a piece of solid-dielectric RG-59 (not foam dielectric!), just under the shorted end of the left-hand plate line. The cable then passes through the partition to a high-voltage standoff insulator

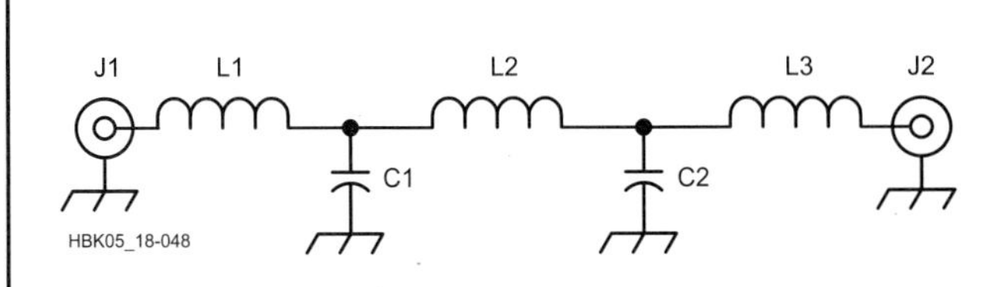

Fig 18.48—Schematic diagram of output harmonic filter.
C1, C2—27-pF Centralab 850 series ceramic transmitting capacitor.
J1, J2—Female chassis-mount N connector (UG-58 or equiv).
L1, L3—2 t #14 wire, 0.3125 inch ID, 0.375 inch long.
L2—3 t #14 wire, 0.3125 inch ID, 0.4375 inch long.

made from nylon. This insulator is fastened to the partition near the high-voltage feedthrough capacitor. A 10-Ω, 25-W resistor is connected between the insulator and the feedthrough capacitor.

The plate lines are connected to the dc-blocking capacitors on the plate collet with 1³/₄ × 2-inch phosphor-bronze strips. The bottom of the plate lines are attached to the sides of the subchassis, with the edge of the L-shaped mounting bracket flush with the bottom of the subchassis.

When preparing the subchassis top plate for the 3CX1200Z7, cut a 2¹¹/₁₆-inch hole in the center of the plate. This hole size allows clearance between the tube envelope and the top plate, without putting stress on the envelope in the vicinity of the grid flange seal.

Exercise care in placing the movable tuning plate and the movable output coupling disc, to ensure they cannot touch their fixed counterparts on the plate collet.

Operation

When the amplifier is first turned on, it cannot be keyed until:

- 10 seconds has elapsed.
- High voltage is available, as confirmed by presence of +12 V to the keying circuit.

Table 18.10
Power Supply Specifications

High voltage: 3200 V
Continuous current: 1.2 A
Intermittent current: 2 A
Step/Start delay: 2 secs

Connect the amplifier to a dummy load through an accurate power meter capable of indicating 1500 W full scale. Key the amplifier and check the idling plate current. With 3200-V plate voltage, it should be in the vicinity of 150 mA. Now, apply a small amount of drive and adjust the input tuning for maximum grid current. Adjust the output tuning until you see an indication of RF output. Increase drive and adjust the output coupling and tuning for the desired output. Do not overcouple the output; once desired output is reached, do not increase loading. Insert the hold-down screw to secure the output coupling capacitor from moving. One setting is adequate for tuning across the 2-meter band if the SWR on the transmission line is reasonably low.

When you shut down the amplifier, leave the blower running for at least three minutes after you turn off the filament voltage. The 3CX1200Z7 is an excellent tube. The author tried it with excessive drive, plate-current saturation, excessive plate dissipation—all the abuse it's likely to encounter in amateur applications. There were no problems, but that doesn't mean you should repeat these torture tests!

A Companion Power Supply

A well-designed and constructed high-voltage power supply is necessary to ensure linearity in SSB operation. Specifications of the power supply for this amplifier are given in **Table 18.10**. A schematic and parts list for a rugged power supply—usable with this project—are in the **Power Supplies** chapter. Although bi-level, it is otherwise similar to the author's design described in the December 1994 issue of *QST*.

Conclusion

This amplifier is a reliable and cost-effective way to generate a big 2-meter signal—almost as quickly as a solid-state amplifier. To ensure that the output of the amplifier meets current spectral purity requirements, a high-power output filter, as shown in **Fig 18.48**, should be used. The author reports that he can run full output while his wife watches TV in a nearby room.

Chapter 19

Station Layout and Accessories

Although many hams never try to build a major project, such as a transmitter, receiver or amplifier, they do have to assemble the various components into a working station. There are many benefits to be derived from assembling a safe, comfortable, easy-to-operate collection of radio gear, whether the shack is at home, in the car or in a field. This chapter, written by Wally Blackburn, AA8DX, covers the many aspects of setting up an efficient station.

This chapter will detail some of the "how tos" of setting up a station for fixed, mobile and portable operation. Such topics as station location, finding adequate power sources, station layout and cable routing are covered, along with some of the practical aspects of antenna erection and maintenance.

Regardless of the type of installation you are attempting, good planning greatly increases your chances of success. Take the time to think the project all the way through, consider alternatives, and make rough measurements and sketches during your planning and along the way. You will save headaches and time by avoiding "shortcuts." What might seem to save time now may come back to haunt you with extra work when you could be enjoying your shack.

One of the first considerations should be to determine what type of operating you intend to do. While you do not want to strictly limit your options later, you need to consider what you want to do, how much you have to spend and what room you have to work with. There is a big difference between a casual operating position and a "big gun" contest station, for example.

Fixed Stations

SELECTING A LOCATION

Selecting the right location for your station is the first and perhaps the most important step in assembling a safe, comfortable, convenient station. The exact location will depend on the type of home you have and how much space can be devoted to your station. Fortunate amateurs will have a spare room to devote to housing the station; some may even have a separate building for their exclusive use. Most must make do with a spot in the cellar or attic, or a corner of the living room is pressed into service.

Examine the possibilities from several angles. A station should be comfortable; odds are good that you'll be spending a lot of time there over the years. Some unfinished basements are damp and drafty—not an ideal environment for several hours of leisurely hamming. Attics have their drawbacks, too; they can be stifling during warmer months. If possible, locate your station away from the heavy traffic areas of your home. Operation of your station should not interfere with family life. A night of chasing DX on 80 m may be exciting to you, but the other members of your household may not share your enthusiasm.

Keep in mind that you must connect your station to the outside world. The location

Fig 19.1—Danny, KD4HQV, appreciates the simplicity that his operating position affords. (*Photo courtesy Conard Murray, WS4S*)

Fig 19.2—VE6AFO's QSL card reveals an impressive array of gear. Although many hams would appreciate having this much space to devote to a station, most of us must make do with less.

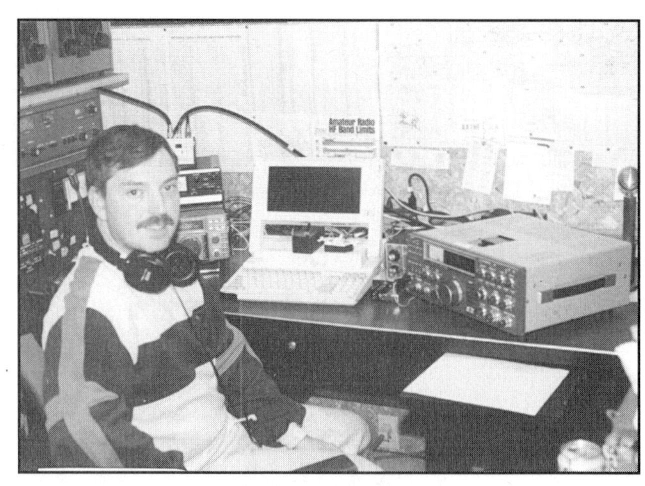
Fig 19.3—Scott, KA9FOX, operated this well laid out station, W9UP, during a recent contest. (*Photo courtesy NØBSH*)

you choose should be convenient to a good power source and an adequate ground. If you use a computer and modem, you may need access to a telephone jack. There should be a fairly direct route to the outside for running antenna feed lines, rotator control cables and the like.

Although most homes will not have an "ideal" space meeting all requirements, the right location for you will be obvious after you scout around. The amateurs whose stations are depicted in **Figs 19.1** through **19.3** all found the right spot for them. Weigh the trade-offs and decide which features you can do without and which are necessary for your style of operation. If possible pick an area large enough for future expansion.

THE STATION GROUND

Grounding is an important factor in overall station safety, as detailed in the **Safety** chapter. An effective ground system is necessary for every amateur station. The mission of the ground system is twofold. First, it reduces the possibility of electrical shock if something in a piece of equipment should fail and the chassis or cabinet becomes "hot." If connected to a properly grounded outlet, a three-wire electrical system grounds the chassis. Much amateur equipment still uses the ungrounded two-wire system, however. A ground system to prevent shock hazards is generally referred to as *dc ground*.

The second job the ground system must perform is to provide a low-impedance path to ground for any stray RF current inside the station. Stray RF can cause equipment to malfunction and contributes to RFI problems. This low-impedance path is usually called *RF ground*. In most stations, dc ground and RF ground are provided by the same system.

Ground Noise

Noise in ground systems can affect our sensitive radio equipment. It is usually related to one of three problems:

1) Insufficient ground conductor size
2) Loose ground connections
3) Ground loops

These matters are treated in precise scientific research equipment and certain industrial instruments by attention to certain rules. The ground conductor should be at least as large as the largest conductor in the primary power circuit. Ground conductors should provide a solid connection to both ground and to the equipment being grounded. Liberal use of lock washers and star washers is highly recommended. A loose ground connection is a tremendous source of noise, particularly in a sensitive receiving system.

Ground loops should be avoided at all costs. A short discussion of what a ground loop is and how to avoid them may lead you down the proper path. A ground loop is formed when more than one ground current is flowing in a single conductor. This commonly occurs when grounds are "daisy-chained" (series linked). The correct way to ground equipment is to bring all ground conductors out radially from a common point to either a good driven earth ground or a cold-water system. If one or more earth grounds are used, they should be bonded back to the service entrance panel. Details appear in the **Safety** chapter.

Ground noise can affect transmitted and received signals. With the low audio levels required to drive amateur transmitters, and the ever-increasing sensitivity of our receivers, correct grounding is critical.

STATION POWER

Amateur Radio stations generally require a 120-V ac power source. The 120-V ac is then converted to the proper ac or dc levels required for the station equipment. Power supply theory is covered in the **Power Supplies** chapter, and safety issues are covered in the **Safety** chapter. If your station is located in a room with electrical outlets, you're in luck. If your station is located in the basement, an attic or another area without a convenient 120-V source, you will have to run a line to your operating position.

Surge Protection

Typically, the ac power lines provide an adequate, well-regulated source of electrical power for most uses. At the same time, these lines are fraught with frequent power surges that, while harmless to most household equipment, may cause damage to more sensitive devices such as computers or test equipment. A common method of protecting these devices is through the use of surge protectors. More information on these and lightning protection is in the **Safety** chapter.

STATION LAYOUT

Station layout is largely a matter of personal taste and needs. It will depend mostly on the amount of space available, the equipment involved and the types of operating to be done. With these factors in mind, some basic design considerations apply to all stations.

The Operating Table

The operating table may be an office or computer desk, a kitchen table or a custom-

Fig 19.4—The basement makes a good location if it is dry. A ready-to-assemble computer desk makes an ideal operating table at a reasonable price. This setup belongs to WK8H. (*Photo courtesy AA8DX*)

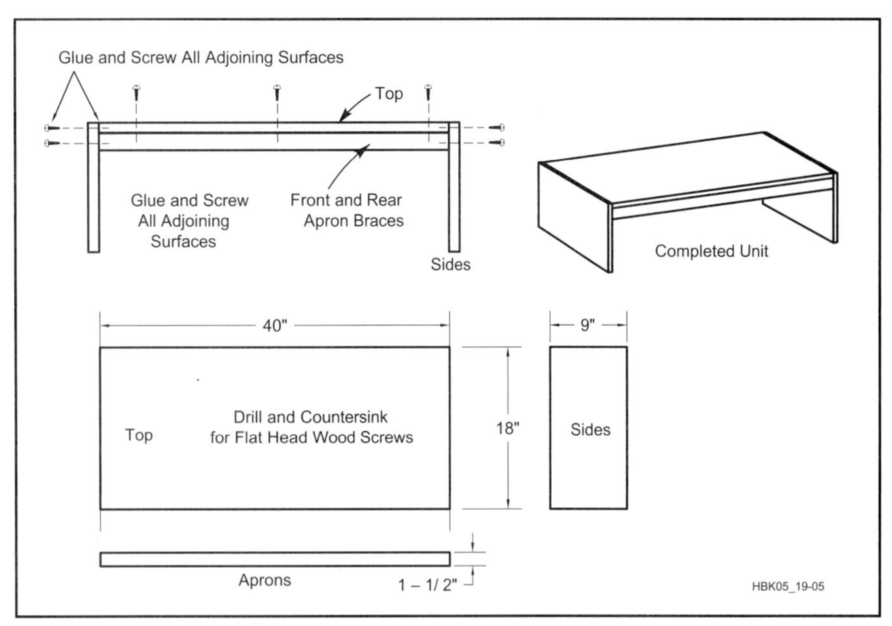

Fig 19.5—A simple but strong equipment shelf can be built from readily available materials. Use ¾-inch plywood along with glue and screws for the joints for adequate strength.

made bench. What you use will depend on space, materials at hand and cost. The two most important considerations are height and size of the top. Most commercial desks are about 29 inches above the floor. This is a comfortable height for most adults. Heights much lower or higher than this may cause an awkward operating position.

The dimensions of the top are an important consideration. A deep (36 inches or more) top will allow plenty of room for equipment interconnections along the back, equipment about midway and room for writing toward the front. The length of the top will depend on the amount of equipment being used. An office or computer desk makes a good operating table. These are often about 36 inches deep and 60 inches wide. Drawers can be used for storage of logbooks, headphones, writing materials, and so on. Desks specifically designed for computer use often have built-in shelves that can be used for equipment stacking. Desks of this type are available ready-to-assemble at most discount and home improvement stores. The low price and adaptable design of these desks make them an attractive option for an operating position. An example is shown in **Fig 19.4.**

Stacking Equipment

No matter how large your operating table, some vertical stacking of equipment may be necessary to allow you to reach everything from your chair. Stacking pieces of equipment directly on top of one another is not a good idea because most amateur equipment needs airflow around it for cooling. A shelf like that shown in **Fig 19.5** can improve equipment layout in many situations. Dimensions of the shelf can be adjusted to fit the size of your operating table.

Arranging the Equipment

When you have acquired the operating table and shelving for your station, the next task is arranging the equipment in a convenient, orderly manner. The first step is to provide power outlets and a good ground as described in a previous section. Be conservative in estimating the number of power outlets for your installation; radio equipment has a habit of multiplying with time, so plan for the future at the outset.

Fig 19.6 illustrates a sample station layout. The rear of the operating table is spaced about 1½ ft from the wall to allow easy access to the rear of the equipment. This installation incorporates two separate operating positions, one for HF and one for VHF. When the operator is seated at the HF operating position, the keyer and transceiver controls are within easy reach. The keyer, keyer paddle and transceiver are the most-often adjusted pieces of equipment in the station. The speaker is positioned right in front of the operator for the best possible reception. Accessory equipment not often adjusted, including the amplifier, antenna switch and rotator control box, is located on the shelf above the transceiver. The SWR/power meter and clock, often consulted but rarely touched, are located where the operator can view them without head movement. All HF-related equipment can be reached without moving the chair.

This layout assumes that the operator is right-handed. The keyer paddle is operated with the right hand, and the keyer speed and transceiver controls are operated with the left hand. This setup allows the operator to write or send with the right hand without having to cross hands to adjust the controls. If the operator is left-handed, some repositioning of equipment is necessary, but the idea is the same. For best results during CW operation, the paddle should be weighted to keep it from "walking" across the table. It should be oriented such that the operator's entire arm from wrist to elbow rests on the tabletop to prevent fatigue.

Some operators prefer to place the station transceiver on the shelf to leave the table top clear for writing. This arrangement leads to fatigue from having an unsupported arm in the air most of the time. If you rest your elbows on the tabletop, they will quickly become sore. If you rarely operate for prolonged periods, however, you may not be inconvenienced by having the transceiver on the shelf. The real secret to having a clear table top for logging, and so on, is to make the operating table deep enough that your entire arm from elbow to wrist rests on the table with the front panels of the equipment at your fingertips. This leaves plenty of room for paperwork, even with a microphone and keyer paddle on the table.

The VHF operating position in this sta-

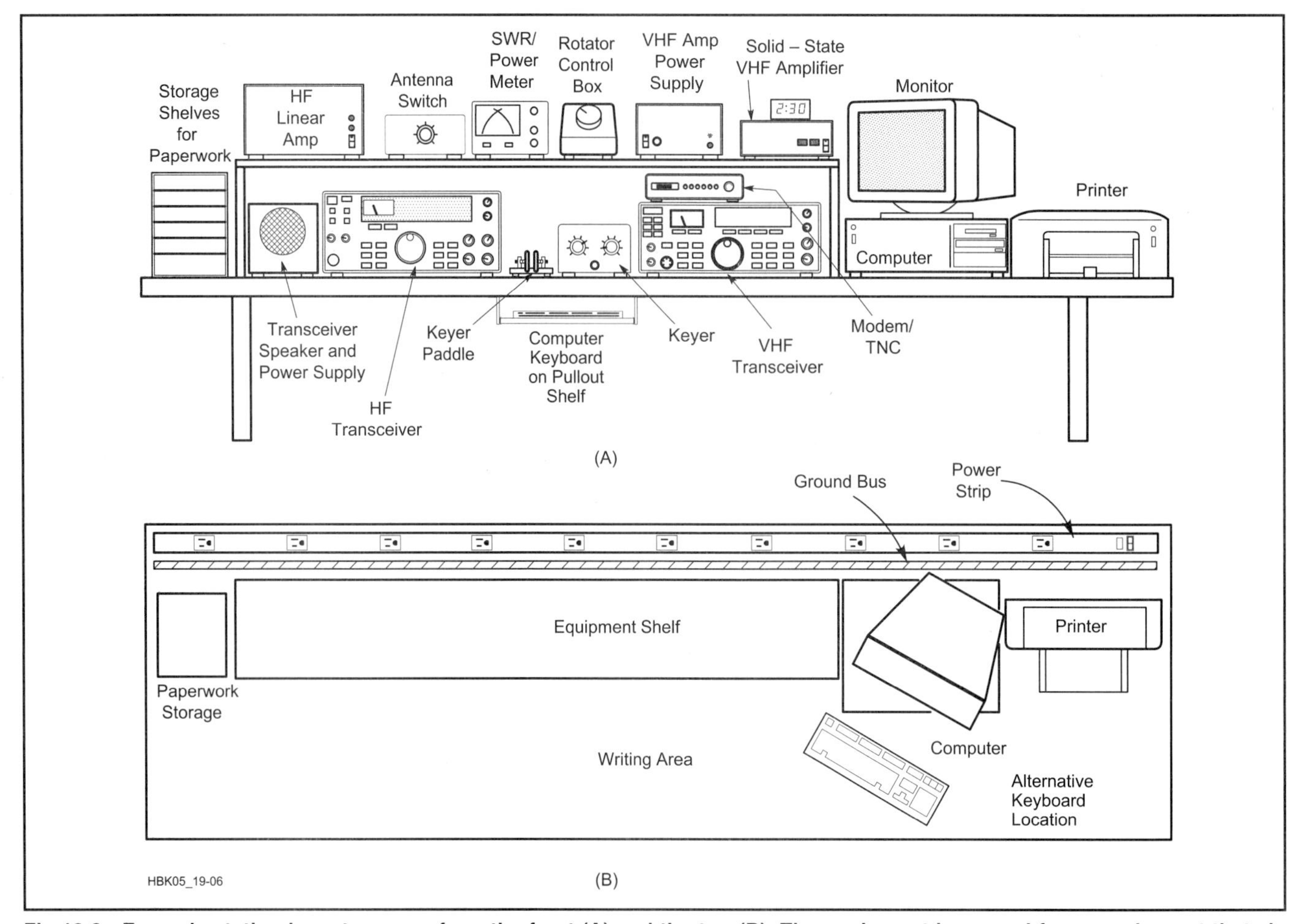

Fig 19.6—Example station layout as seen from the front (A) and the top (B). The equipment is spaced far enough apart that air circulates on all sides of each cabinet.

tion is similar to the HF position. The amplifier and power supply are located on the shelf. The station triband beam and VHF beam are on the same tower, so the rotator control box is located where it can be seen and reached from both operating positions. This operator is active on packet radio on a local VHF repeater, so the computer, printer, terminal node controller and modem are all clustered within easy reach of the VHF transceiver.

This sample layout is intended to give you ideas for designing your own station. Study the photos of station layouts presented here, in other chapters of this *Handbook* and in *QST*. Visit the shacks of amateur friends to view their ideas. Station layout is always changing as you acquire new gear, dispose of old gear, change operating habits and interests or become active on different bands. Configure the station to suit your interests, and keep thinking of ways to refine the layout. **Figs 19.7** and **19.8** show station arrangements tailored for specific purposes.

Equipment that is adjusted frequently sits on the tabletop, while equipment requiring infrequent adjustment is perched on a shelf. All equipment is positioned so the operator does not have to move the chair to reach anything at the operating position.

Fig 19.7—It was back to basics for Elias, K4IX, during a recent Field Day.

Fig 19.8—Richard, WB5DGR, uses a homebrew 1.5-kW amplifier to seek EME contacts from this nicely laid out station.

Aids for Hams with Disabilities

A station used by an amateur with physical disabilities or sensory impairments may require adapted equipment or particular layout considerations. The station may be highly customized to meet the operator's needs or just require a bit of "tweaking."

The myriad of individual needs makes describing all of the possible adaptive methods impractical. Each situation must be approached individually, with consideration to the operator's particular needs. However, many types of situations have already been encountered and worked through by others, eliminating the need to start from scratch in every case.

An excellent resource is the Courage Handi-Ham System. The Courage Handi-Ham System, a part of the Courage Center, provides a number of services to hams (and aspiring hams) with disabilities. These include study materials, equipment loans, adapted equipment, a newsletter and much more. Information needed to reach the Courage Handi-Hams is in the **Component Data and References** chapter.

INTERCONNECTING YOUR EQUIPMENT

Once you have your equipment and get it arranged, you will have to interconnect it all. No matter how simple the station, you will at least have antenna, power and microphone or key connections. Equipment such as amplifiers, computers, TNCs and so on add complexity. By keeping your equipment interconnections well organized and of high quality, you will avoid problems later on.

Often, ready-made cables will be available. But in many cases you will have to make your own cables. A big advantage of making your own cables is that you can customize the length. This allows more flexibility in arranging your equipment and avoids unsightly extra cable all over the place. Many manufacturers supply connectors with their equipment along with pinout information in the manual. This allows you to make the necessary cables in the lengths you need for your particular installation.

Always use high quality wire, cables and connectors in your shack. Take your time and make good mechanical and electrical connections on your cable ends. Sloppy cables are often a source of trouble. Often the problems they cause are intermittent and difficult to track down. You can bet that they will crop up right in the middle of a contest or during a rare DX QSO! Even worse, a poor quality connection could cause RFI or even create a fire hazard. A cable with a poor mechanical connection could come loose and short a power supply to ground or apply a voltage where it should not be. Wire and cables should have good quality insulation that is rated high enough to prevent shock hazards.

Interconnections should be neatly bundled and labeled. Wire ties, masking tape or paper labels with string work well. See **Fig 19.9**. Whatever method you use, proper labeling makes disconnecting and reconnecting equipment much easier. **Fig 19.10** illustrates the number of potential interconnections in a modern, full-featured transceiver.

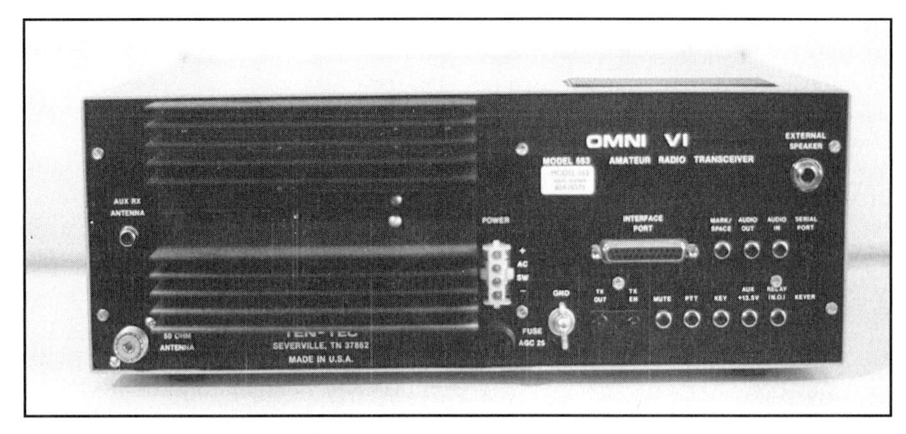

Fig 19.10—The back of this Ten-Tec Omni VI HF transceiver shows some of the many types of connectors encountered in the amateur station. Note that this variety is found on a single piece of equipment. (*Photo courtesy AA8DX*)

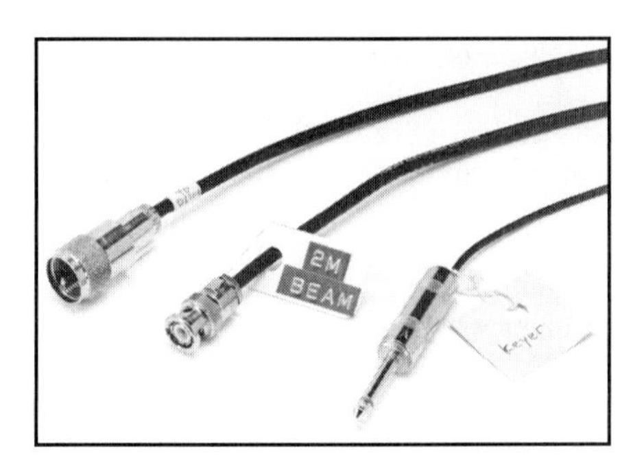

Fig 19.9—Labels on the cables make it much easier to rearrange things in the station. Labeling ideas include masking tape, card board labels attached with string and labels attached to fasteners found on plastic bags (such as bread bags).

Wire and Cable

The type of wire or cable to use depends on the job at hand. The wire must be of sufficient size to carry the necessary current. Use the tables in the **Component Data and References** chapter to find this information. Never use underrated wire; it will be a fire hazard. Be sure to check the insulation too. For high-voltage applications, the insulation must be rated at least a bit higher than the intended voltage. A good rule of thumb is to use a rating at least twice what is needed.

Use good quality coaxial cable of sufficient size for connecting transmitters, transceivers, antenna switches, antenna tuners and so on. RG-58 might be fine for a short patch between your transceiver and SWR bridge, but is too small to use between your legal-limit amplifier and Transmatch.

Hookup wire may be stranded or solid. Generally, stranded is a better choice since it is less prone to break under repeated flexing. Many applications require shielded wire to reduce the chances of RF getting into the equipment. RG-174 is a good choice for control, audio and some low-power applications. Shielded microphone or computer cable can be used where more conductors are necessary. For more information, see the **Transmission Lines** chapter.

Connectors

While the number of different types of connectors is mind-boggling, many manufacturers of amateur equipment use a few standard types. If you are involved in any group activities such as public service or emergency-preparedness work, check to see what kinds of connectors others in the group use and standardize connectors wherever possible. Assume connectors are not waterproof, unless you specifically buy one clearly marked for outdoor use (and assemble it correctly).

Audio, Power and Control Connectors

The simplest form of connector is found on terminal blocks. Although it is possible to strip the insulation from wire and wrap it around the screw, this method is not ideal. The wire tends to "squirm" out from under the screw when tightening, allowing strands to hang free, possibly shorting to other screws.

Terminal lugs, such as those in **Fig 19.11**, solve the problem. These lugs may be crimped (with the proper tool), soldered or both. Terminal lugs are available in different sizes. Use the appropriate size for your wire to get the best results.

Some common multipin connectors are

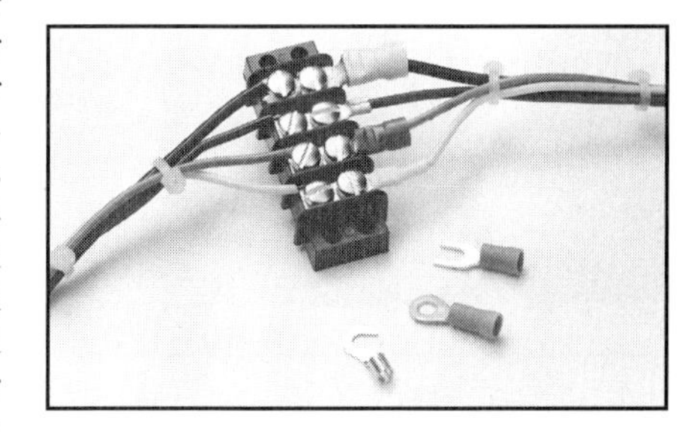

Fig 19.11—The wires on one side of this terminal block have connectors; the others do not. The connectors make it possible to secure different wire sizes to the strip and also make it much easier to change things around.

shown in **Fig 19.12**. The connector in Fig 19.12A is often referred to as a "Cinch-Jones connector." It is frequently used for connections to power supplies from various types of equipment. Supplying from two to eight conductors, these connectors are keyed so that they go together only one way. They offer good mechanical and electrical connections, and the pins are large enough to handle high current. If your cable is too small for the strain relief fitting, build up the outer jacket with a few layers of electrical tape until the strain relief clamps securely. The strain relief will keep your wires from breaking away under flexing or from a sudden tug on the cable.

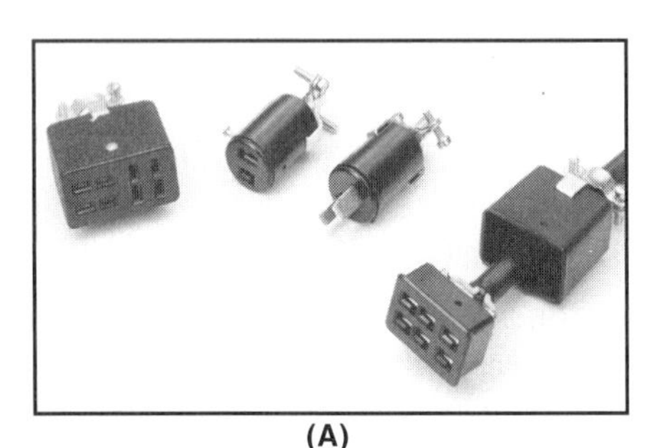

(A)

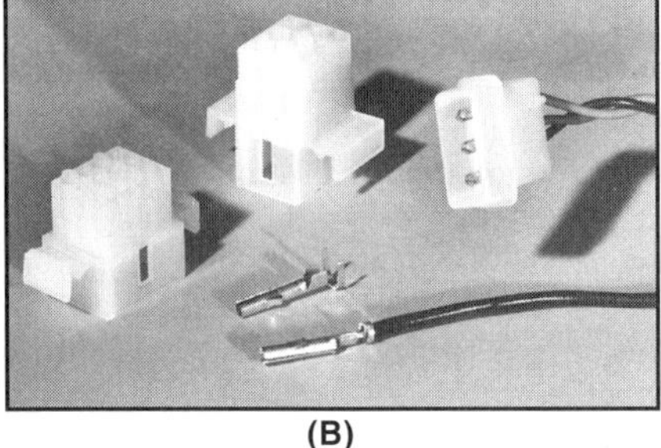

(B)

(C)

Fig 19.12—The plugs shown at A are often used to connect equipment to remote power supplies. The multipin connectors at B are used for control, signal and power lines. The DIN plug at C offers shielding and is often used for connecting accessories to transceivers.

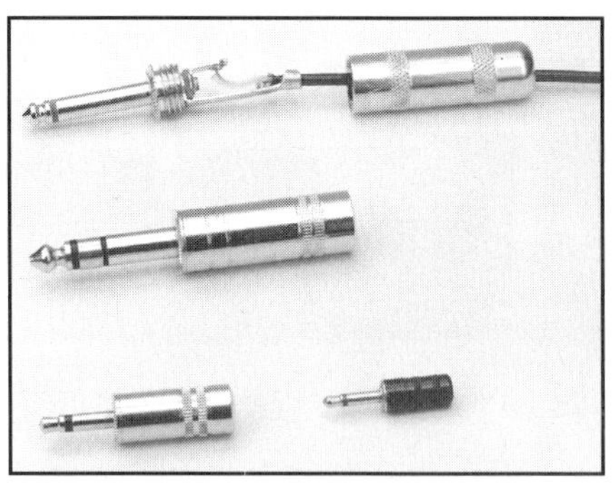

Fig 19.13—The phone-plug family. The 1/4-inch type is often used for headphone and key connections on amateur equipment. The three-circuit version is used with stereo headphones. The mini phone plug is commonly used for connecting external speakers to receivers and transceivers. A subminiphone plug is shown in the foreground for comparison. The shielded style with metal barrel is more durable than the plastic style.

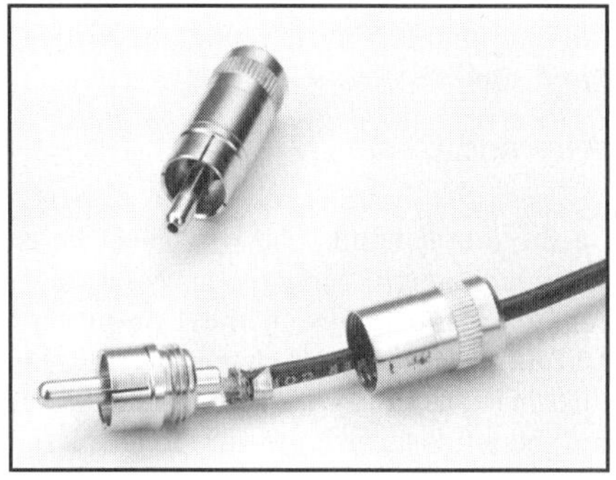

Fig 19.14—Phono plugs have countless uses around the shack. They are small and shielded; the type with the metal body is easy to grip. Be careful not to use too much heat when soldering the ground (outer) conductor—you may melt the insulation.

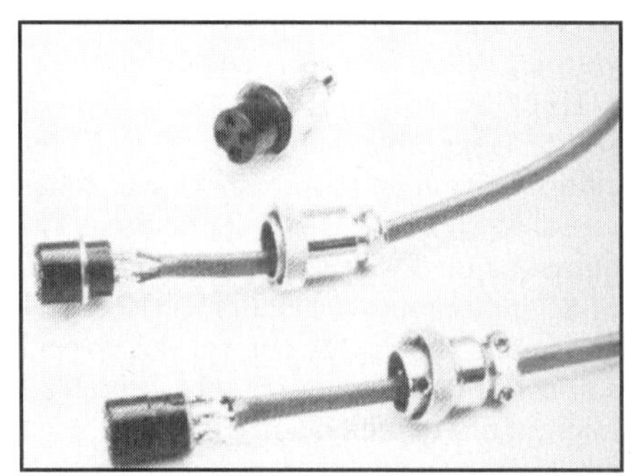

Fig 19.15—The four-pin mike connector is common on modern transmitters and receivers. More elaborate rigs use the eight-pin type. The extra conductors may be used for switches to remotely control the frequency or to power a preamplifier built into the mike case.

The plug in Fig 19.12B is usually called a "molex" connector. This plug consists of an insulated outer shell that houses the individual male or female "fingers." Each finger is individually soldered or crimped onto a conductor of the cable and inserted in the shell, locking into place. These connectors are used on many brands of amateur gear for power and accessory connections.

Fig 19.12C shows a DIN connector. Commonly having five to eight pins, these connectors are a European standard that have found favor with amateur equipment manufacturers around the world. They are generally used for accessory connections. A smaller version, the Miniature DIN, is becoming popular. It is most often used in portable gear but can be found on some full-size equipment as well.

Various types of phone plugs are shown in **Fig 19.13**. The 1/4-inch (largest) is usually used on amateur equipment for headphone and Morse key connections. They are available with plastic and metal bodies. The metal is usually a better choice because it provides shielding and is more durable.

Fig 19.13 also shows the 1/8-inch phone plug. These plugs, sometimes called miniature phone plugs, are used for earphone, external speaker, key and control lines. There is also a subminiature (3/32-inch) phone plug that is not common on amateur gear.

The phono, or RCA, plug shown in **Fig 19.14** is popular among amateurs. It is used for everything from amplifier relay-control lines, to low-voltage power lines, to low-level RF lines, to antenna lines. Several styles are available, but the best choice is the shielded type with the screw-on metal body. As with the phone plugs, the metal bodies provide shielding and are very durable.

Nowhere is there more variation than among microphone connectors. Manufacturers seem to go out of their way to use incompatible connectors! The most popular types of physical connectors are the four- and eight-pin microphone connectors shown in **Fig 19.15**. The simplest connectors provide three connections: audio, ground and push-to-talk (PTT). More complex connectors allow for such things as control lines from the microphone for frequency changes or power to the microphone for a preamplifier. When connecting a microphone to your rig, especially an after-market one, consult the manual. Follow the manufacturer's recommendations for best results.

If the same microphone will be used for multiple rigs with incompatible connectors, one or more adapters will be neces-

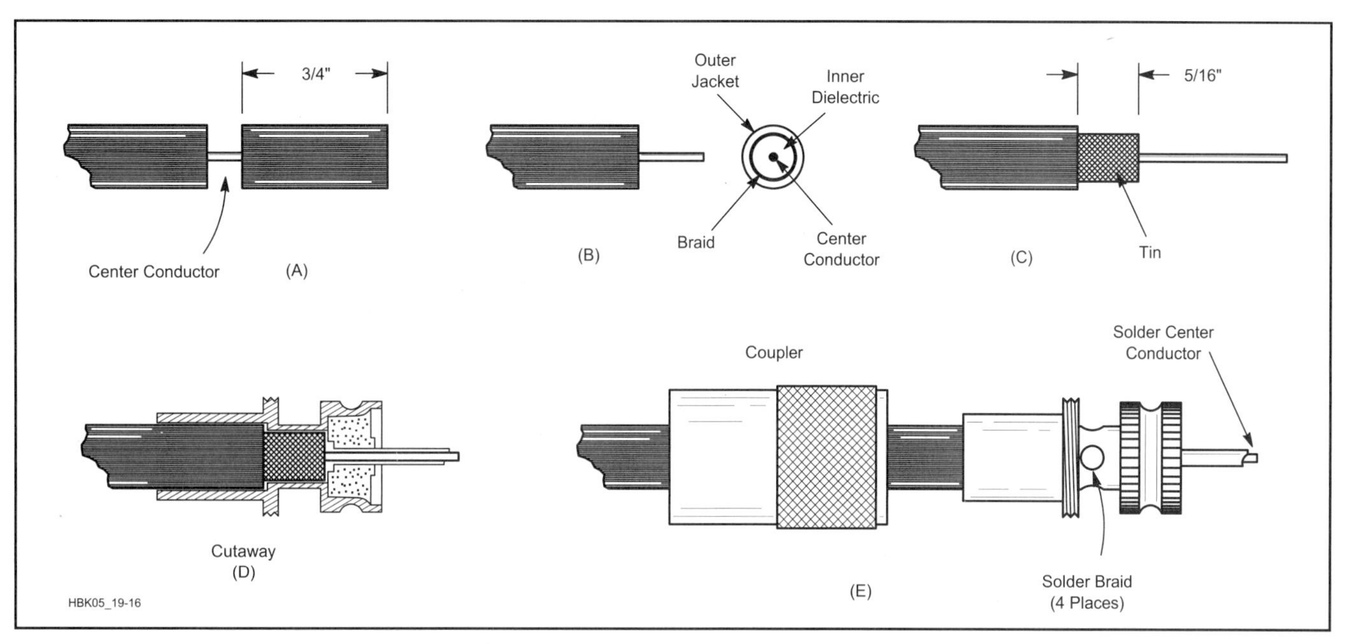

Fig 19.16—The PL-259, or UHF, connector is almost universal for amateur HF work and is popular for equipment operating in the VHF range. Steps A through E are described in detail in the text.

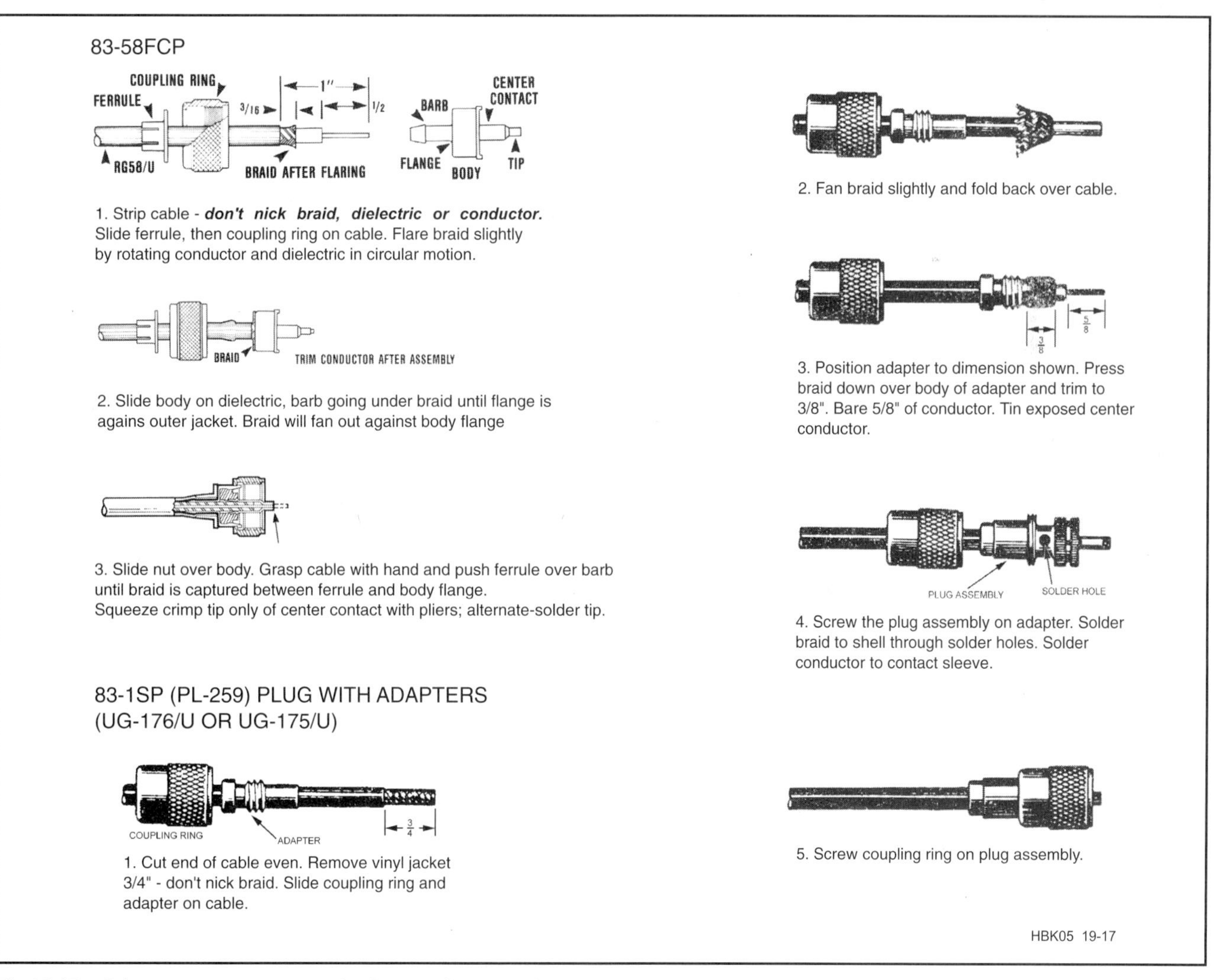

Fig 19.17—Crimp-on connectors and adapters for use with standard PL-259 connectors are popular for connecting to RG-58 and RG-59 type cable. (*Courtesy Amphenol Electronic Components, RF Division, Bunker Ramo Corp*)

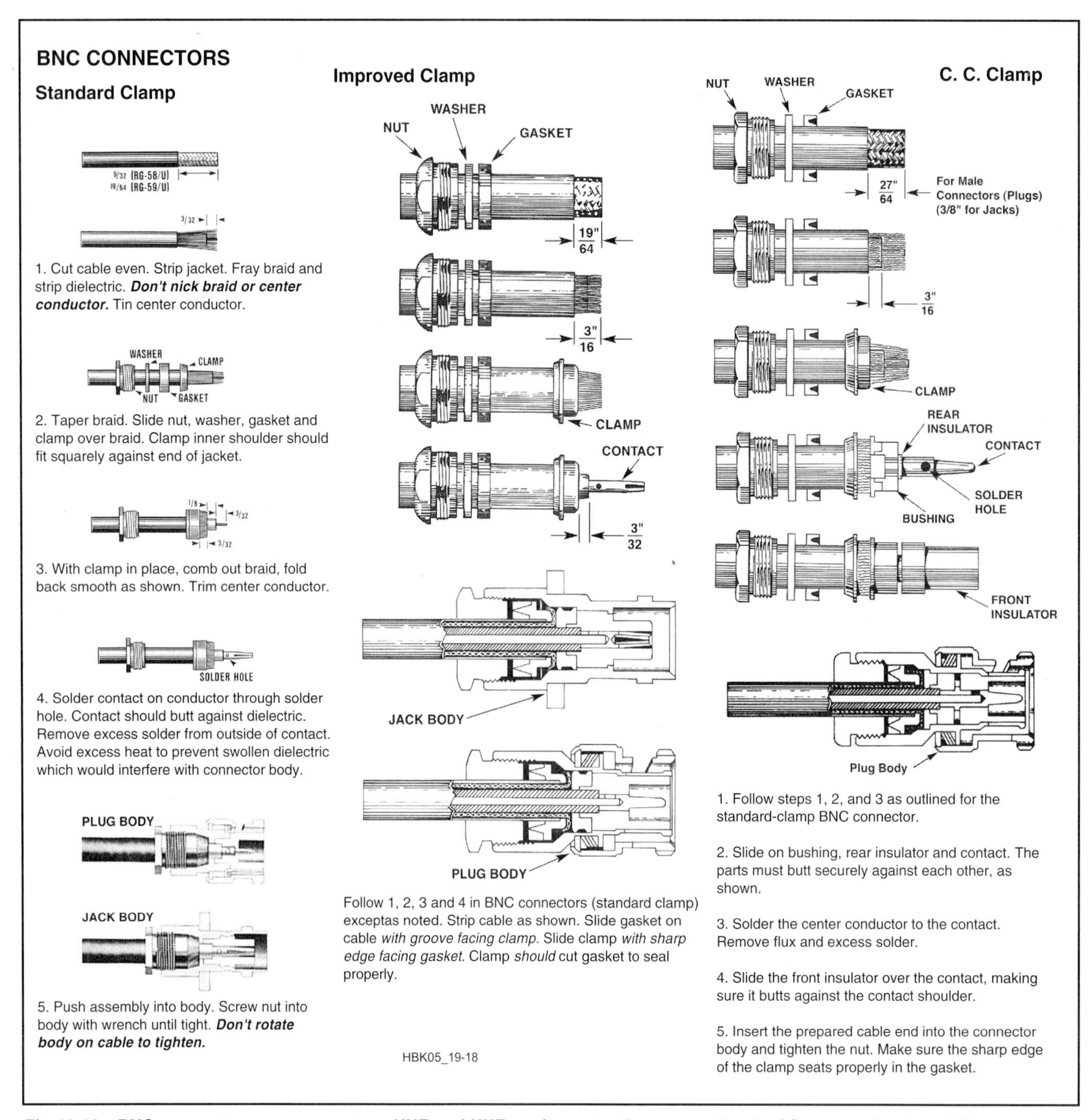

Fig 19.18—BNC connectors are common on VHF and UHF equipment at low power levels. (*Courtesy Amphenol Electronic Components, RF Division, Bunker Ramo Corp*)

sary. Adapters can be made with short pieces of cable and the necessary connectors at each end.

RF Connectors

There are many different types of RF connectors for coaxial cable, but the three most common for amateur use are the *UHF*, *Type N* and *BNC* families. The type of connector used for a specific job depends on the size of the cable, the frequency of operation and the power levels involved.

The so-called UHF connector is found on most HF and some VHF equipment. It is the only connector many hams will ever see on coaxial cable. PL-259 is another name for the UHF male, and the female is also known as the SO-239. These connectors are rated for full legal amateur power at HF. They are poor for UHF work because they do not present a constant impedance, so the UHF label is a misnomer. PL-259 connectors are designed to fit RG-8 and RG-11 size cable (0.405-inch OD). Adapters are available for use with smaller RG-58, RG-59 and RG-8X size cable. UHF connectors are not weatherproof.

Fig 19.16 shows how to install the solder type of PL-259 on RG-8 cable. Proper preparation of the cable end is the key to success. Follow these simple steps. Measure back about 3/4-inch from the cable end and slightly score the outer jacket around its circumference. With a sharp knife, cut through the outer jacket, through the braid, and through the dielectric, right down to the center conductor. Be careful not to score the center conductor. Cutting

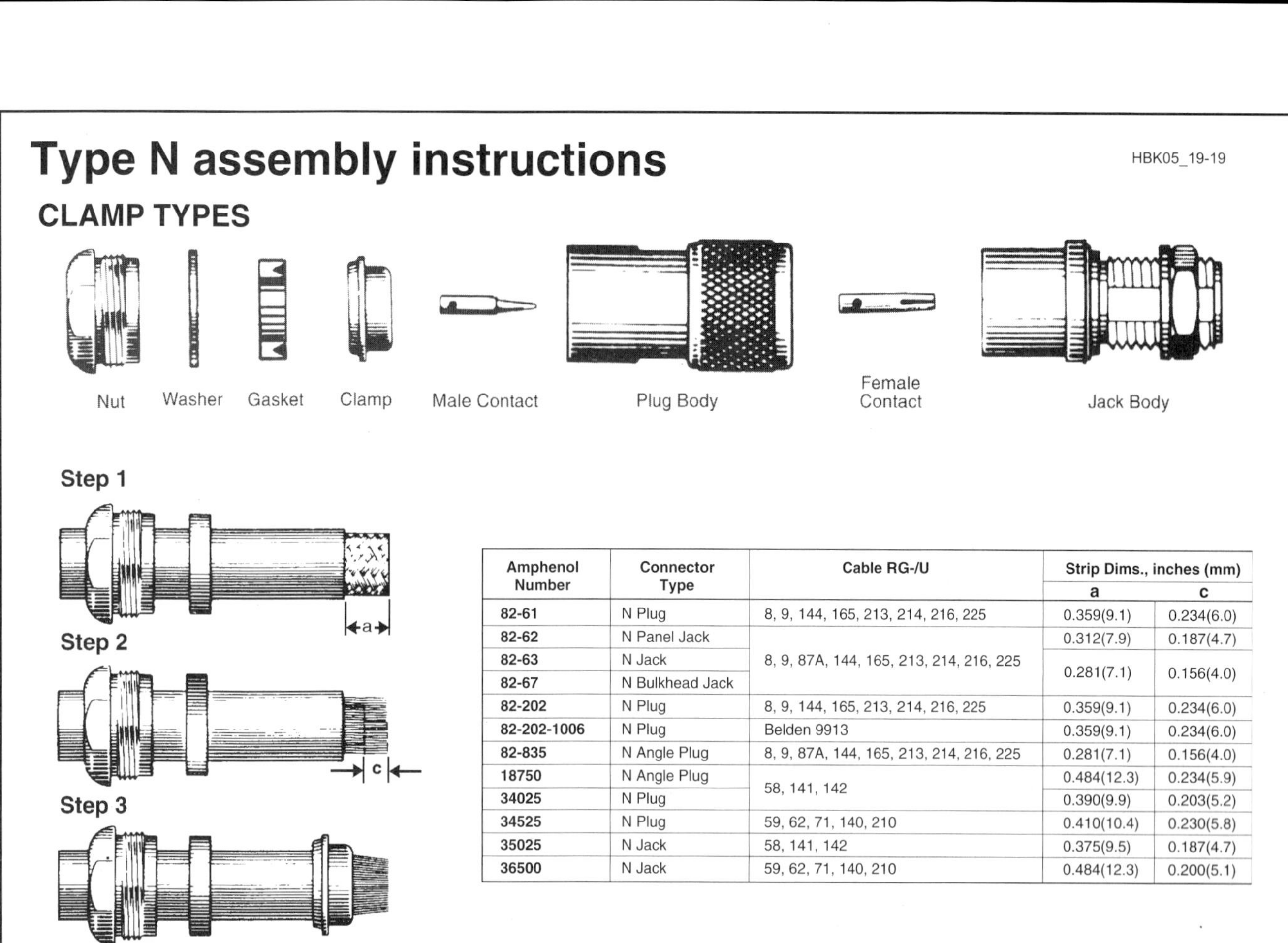

Amphenol Number	Connector Type	Cable RG-/U	Strip Dims., inches (mm)	
			a	c
82-61	N Plug	8, 9, 144, 165, 213, 214, 216, 225	0.359(9.1)	0.234(6.0)
82-62	N Panel Jack	8, 9, 87A, 144, 165, 213, 214, 216, 225	0.312(7.9)	0.187(4.7)
82-63	N Jack		0.281(7.1)	0.156(4.0)
82-67	N Bulkhead Jack			
82-202	N Plug	8, 9, 144, 165, 213, 214, 216, 225	0.359(9.1)	0.234(6.0)
82-202-1006	N Plug	Belden 9913	0.359(9.1)	0.234(6.0)
82-835	N Angle Plug	8, 9, 87A, 144, 165, 213, 214, 216, 225	0.281(7.1)	0.156(4.0)
18750	N Angle Plug	58, 141, 142	0.484(12.3)	0.234(5.9)
34025	N Plug		0.390(9.9)	0.203(5.2)
34525	N Plug	59, 62, 71, 140, 210	0.410(10.4)	0.230(5.8)
35025	N Jack	58, 141, 142	0.375(9.5)	0.187(4.7)
36500	N Jack	59, 62, 71, 140, 210	0.484(12.3)	0.200(5.1)

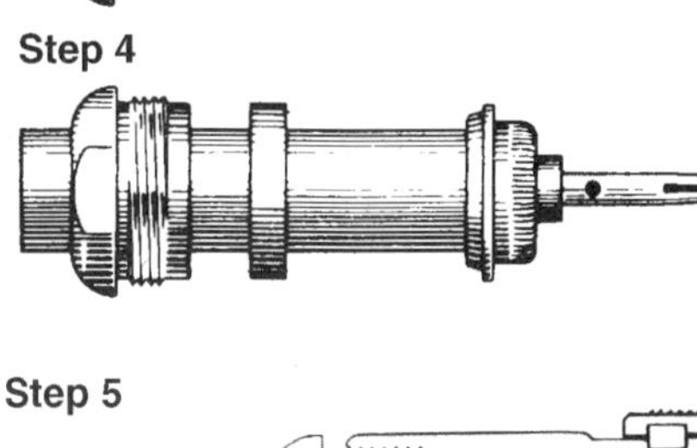

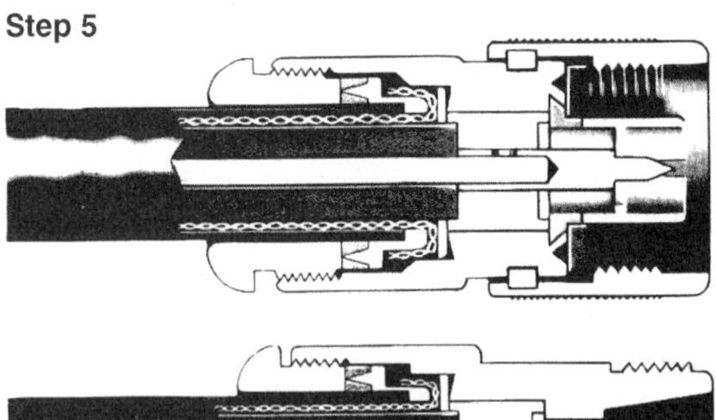

Step 1 Place nut and gasket, with "V" groove toward clamp, over cable and cut off jacket to dim. a.

Step 2 Comb out braid and fold out. Cut off cable dielectric to dim. c as shown.

Step 3 Pull braid wires forward and taper toward center conductor. Place clamp over braid and push back against cable jacket.

Step 4 Fold back braid wires as shown, trim braid to proper length and form over clamp as shown. Solder contact to center conductor.

Step 5 Insert cable and parts into connector body. Make sure sharp edge of clamp seats properly in gasket. Tighten nut.

Fig 19.19—Type N connectors are a must for high-power VHF and UHF operation. (*Courtesy Amphenol Electronic Components, RF Microwave Operations*)

through all outer layers at once keeps the braid from separating. Pull the severed outer jacket, braid and dielectric off the end of the cable as one piece. Inspect the area around the cut, looking for any strands of braid hanging loose and snip them off. There won't be any if your knife was sharp enough. Next, score the outer jacket about 5/16-inch back from the first cut. Cut through the jacket lightly; do not score the braid. This step takes practice. If you score the braid, start again. Remove the outer jacket.

Tin the exposed braid and center conductor, but apply the solder sparingly and avoid melting the dielectric. Slide the coupling ring onto the cable. Screw the connector body onto the cable. If you prepared the cable to the right dimensions, the center conductor will protrude through the center pin, the braid will show through the solder holes, and the body will actually thread onto the outer cable jacket.

Solder the braid through the solder

holes. Solder through all four holes; poor connection to the braid is the most common form of PL-259 failure. A good connection between connector and braid is just as important as that between the center conductor and connector. Use a large soldering iron for this job. With practice, you'll learn how much heat to use. If you use too little heat, the solder will bead up, not really flowing onto the connector body. If you use too much heat, the dielectric will melt, letting the braid and center conductor touch. Most PL-259s are nickel plated, but silver-plated connectors are much easier to solder and only slightly more expensive.

Solder the center conductor to the center pin. The solder should flow on the inside, not the outside, of the center pin. If you wait until the connector body cools off from soldering the braid, you'll have less trouble with the dielectric melting. Trim the center conductor to be even with the end of the center pin. Use a small file to round the end, removing any solder that built up on the outer surface of the center pin. Use a sharp knife, very fine sandpaper or steel wool to remove any solder flux from the outer surface of the center pin. Screw the coupling ring onto the body, and you're finished.

Fig 19.17 shows two options available if you want to use RG-58 or RG-59 size cable with PL-259 connectors. The crimp-on connectors manufactured specially for the smaller cable work very well if installed correctly. The alternative method involves using adapters for the smaller cable with standard RG-8 size PL-259s. Prepare the cable as shown. Once the braid is prepared, screw the adapter into the PL-259 shell and finish the job as you would a PL-259 on RG-8 cable.

The BNC connectors illustrated in **Fig 19.18** are popular for low power levels at VHF and UHF. They accept RG-58 and RG-59 cable, and are available for cable mounting in both male and female versions. Several different styles are available, so be sure to use the dimensions for the type you have. Follow the installation instructions carefully. If you prepare the cable to the wrong dimensions, the center pin will not seat properly with connectors of the opposite gender. Sharp scissors are a big help for trimming the braid evenly.

The Type N connector, illustrated in **Fig 19.19**, is a must for high-power VHF and UHF operation. N connectors are available in male and female versions for cable mounting and are designed for RG-8 size cable. Unlike UHF connectors, they are designed to maintain a constant impedance at cable joints. Like BNC connectors, it is important to prepare the cable to the right dimensions. The center pin must be positioned correctly to mate with the center pin of connectors of the opposite gender. Use the right dimensions for the connector style you have.

Computer Connectors

As if the array of connectors related to amateur gear were not enough, the prevalence of the computer in the shack has brought with it another set of connectors to consider. Most connections between computers and their peripherals are made with some form of multiconductor cable. Examples include shielded, unshielded and ribbon cable. Common connectors used are the 9- and 25-pin D-Subminiature connector, the DIN and Miniature DIN and the 36-pin Amphenol connector. Various edge-card connectors are used internally (and sometimes externally) on many computers. **Fig 19.20** shows a variety of computer connectors. See the **Component Data and References** chapter for other computer-connector pinout diagrams.

EIA-232 Serial Connections

The serial port on a computer is arguably the most used, and often most troublesome, connector encountered by the amateur. The serial port is used to connect modems, TNCs, computer mice and some printers to the computer. As the name implies, the data is transmitted serially.

The EIA-232-D (commonly referred to as RS-232) standard defines a system used to send data over relatively long distances. It is commonly used to send data anywhere from a few feet to 50 feet or more. The standard specifies the physical connection and signal lines. The serial ports on most computers comply with the EIA-232-D standard only to the degree necessary to operate with common peripherals. **Fig 19.21** shows the two most common connectors used for computer serial ports. A 9-pin connector can be adapted to a 25-pin by connecting like signals. Earth ground is not provided in the 9-pin version.

Equipment connected via EIA-232-D is usually classified in one of two ways: DTE (data terminal equipment) or DCE (data communication equipment). Terminals and computers are examples of DTE, while modems and TNCs are DCE.

The binary data is represented by specific voltage levels on the signal line. The EIA-232-D standard specifies that a binary one is represented by a voltage ranging from –3 to –25 V. A binary zero ranges from 3 to 25 V. ±12 V is a common level in many types of equipment, but anything within the specified ranges is just as valid.

The RTS (request to send), CTS (clear to send), DTR (data terminal ready) and DSR (data set ready) lines are used for handshaking signals. These signals are used to coordinate the communication between the DTE and DCE. The RTS and DTR line are used by the DTE to indicate to the DCE that it is ready to receive data from the DCE. The DCE uses the CTS/DSR lines to signal the DTE as to whether or not it is

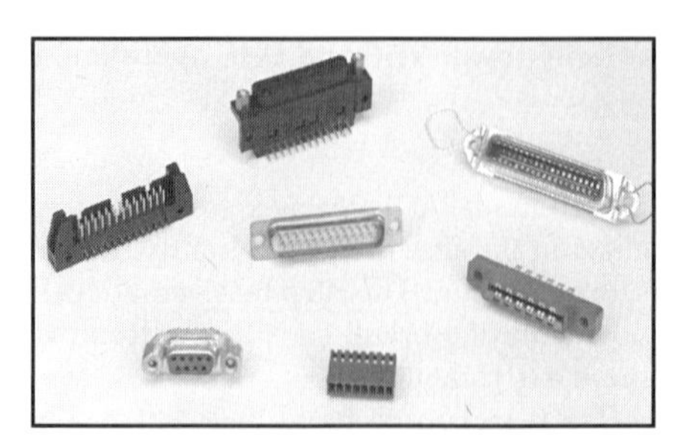

Fig 19.20—Various computer connectors.

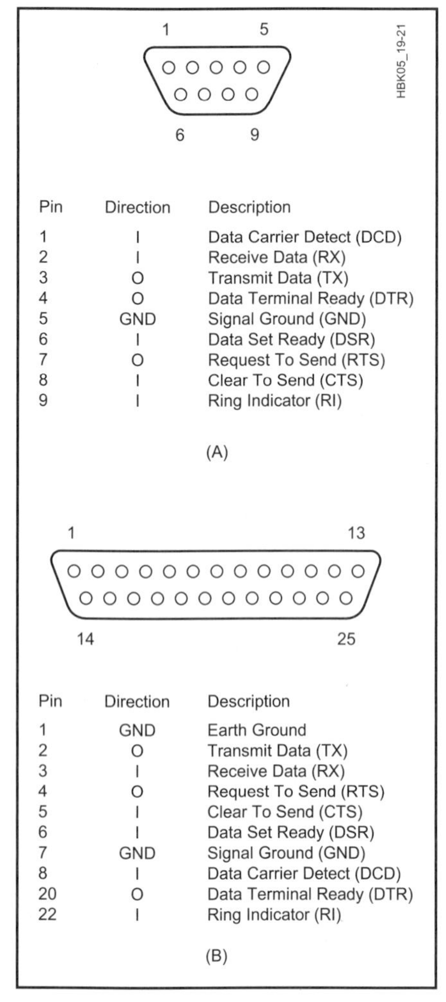

Fig 19.21—The two most common implementations of EIA-232-D serial connections on personal computers use 9- and 25-pin connectors.

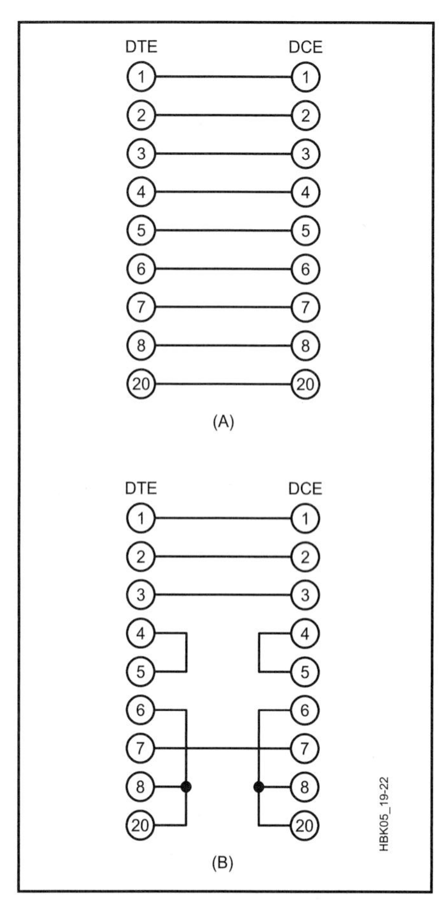

Fig 19.22—EIA-232-D serial connections for normal DTE/DCE (A) and those that ignore handshaking (B).

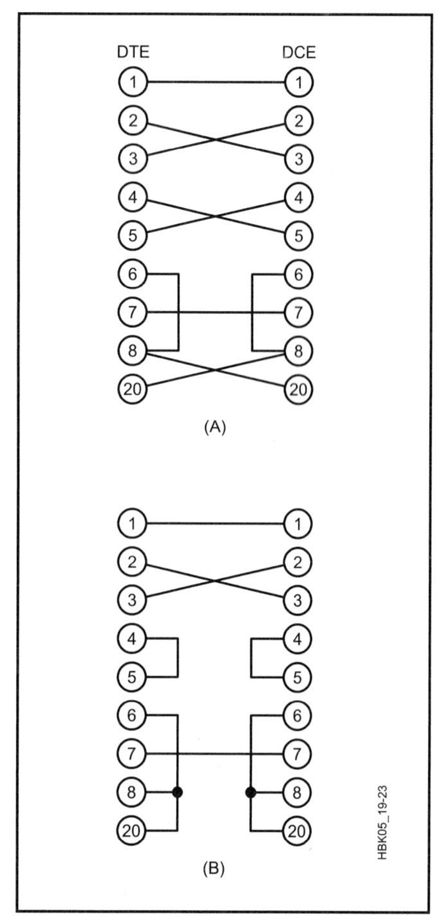

Fig 19.23—EIA-232-D null modem serial connection.

ready to accept data. DCD (data carrier detect) is also sometimes used by the DCE to signal the DTE that an active carrier is present on the communication line. A +12-V signal represents an active handshaking signal. The equipment "drops" the line to –12 V when it is unable to receive data.

You may notice that the name "ready to send" is sort of a misnomer for the DTE since it actually uses it to signal that it is ready to receive. This is a leftover from when communication was mostly one-way—DTE to DCE. Note also that the signal names really only make sense from the DTE point of view. For example, pin 2 is called TD on both sides, even though the DCE is receiving data on that pin. This is another example of this one-way terminology.

It would be much too simple if all serial devices implemented all of the EIA-232-D specifications. Some equipment ignores some or all of the handshaking signals. Other equipment expects handshaking signals to be used as specified. Connecting these two types of equipment together will result in a frustrating situation. One side will blindly send data while the other side blindly ignores all data sent to it!

Fig 19.22 shows the different possible ways to connect equipment. Fig 19.22A shows how to connect a "normal" DTE/DCE combination. This assumes both sides correctly implement all of the handshaking signals. If one or both sides ignore handshaking signals, the connections shown in Fig 19.22B will be necessary. In this scheme, each side is sending the handshaking signals to itself. This little bit of deceit will almost always work, but handshaking signals that are present will be ineffective.

Null Modem Connections

Some equipment does not fall completely in the DTE or DCE category. Some serial printers, for example, act as DCE while others act as DTE. Whenever a DTE/DTE or DCE/DCE connection is needed a special connection, known as a *null modem* connection, must be made. An example might be connecting two computers together so they can transmit data back and forth. A null modem connection simply crosses the signal and handshaking lines. **Fig 19.23** shows a normal null modem connection (A) and one for equipment that ignores handshaking (B).

Parallel Connections

Another common computer port is the parallel port. The most popular use for the parallel port by far is for printer connections. As the name implies, data is sent in a parallel fashion. There are eight data lines accompanied by a number of control and handshaking lines. A parallel printer connection typically uses a 25-pin D-Subminiature connector at the computer end and a 36-pin Amphenol connector (often called an Epson connector) on the printer.

Connecting Computers to Amateur Equipment

Most modern transceivers provide a serial connection that allows external control of the rig, typically with a computer. Commands sent over this serial control line can cause the rig to change frequency, mode and other parameters. Logging and contest software running on the computer often takes advantage of this capability.

The serial port of most radios operates with the TTL signal levels of 0 V for a binary 0 and 5 V for a binary 1. This is incompatible with the ±12 V of the serial port on the computer. For this reason, level shifting is required to connect the radio to the computer.

A couple of level shifter projects appear in the projects section at the end of this chapter. One of these two examples will work in most rig control situations, although some minor modifications may be necessary. Use the manual and technical documentation to find out what signals your radio requires and choose the circuit that fits the bill. The important factors to note are whether handshaking is implemented and what polarity the radio expects for the signals. In some cases, a 5-V level represents a logic 1 (active high) and in others a logic 0 (active low).

CSMA/CD Bus

Some equipment, notably ICOM rigs with the CI-V interface and recent Ten-Tec gear, use a CSMA/CD (carrier-sense multiple access/collision detect) bus that can interconnect a number of radios and computers simultaneously. This bus basically consists of a single wire, on which the devices transmit and receive data, and a ground wire. **Fig 19.24** illustrates the CSMA/CD scheme.

Each device connected to the bus has its own unique digital address. A radio comes from the manufacturer with a default address that can be changed if desired, usually by setting dip switches inside the

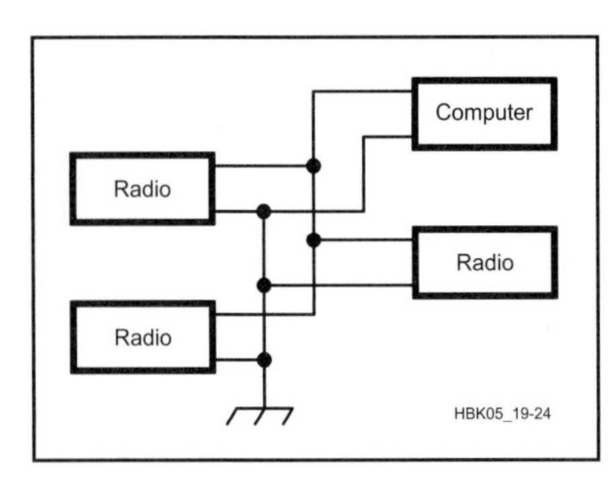

Fig 19.24—The basic two-wire bus system that ICOM and some Ten-Tec radios share among several radios and computers. In its simplest form, the network would include only one radio and one computer.

radio. Information is sent on the bus in the form of packets that include the control data and the address of the device (radio or computer) for which they are intended. A device receives every packet but only acts on the data when its address is embedded in the packet.

A device listens to the bus before transmitting to make sure it is idle. A problem occurs when both devices transmit on the bus at the same time: They both listen, hear nothing and start to send. When this happens, the packets garble each other. This is known as a collision. That is where the CSMA/CD bus collision-detection feature comes in: The devices detect the collision and each sender waits a random amount of time before resending. The sender waiting the shorter random time will get to send first.

Computer/TNC Connections

TNCs (terminal node controllers) also connect to the computer (or terminal) via the serial port. A TNC typically implements handshaking signals. Therefore, a connector like the one in Fig 19.22A will be necessary. Connectors at the TNC end vary with manufacturer. The documentation included with the TNC will provide details for hooking the TNC to the computer.

DOCUMENTING YOUR STATION

An often neglected but very important part of putting together your station is properly documenting your work. Ideally, you should diagram your entire station from the ac power lines to the antenna on paper and keep the information in a special notebook with sections for the various facets of your installation. Having the station well documented is an invaluable aid when tracking down a problem or planning a modification. Rather than having to search your memory for information on what you did a long time ago, you'll have the facts on hand.

Besides recording the interconnections and hardware around your station, you should also keep track of the performance of your equipment. Each time you install a new antenna, measure the SWR at different points in the band and make a table or plot a curve. Later, if you suspect a problem, you'll be able to look in your records and compare your SWR with the original performance.

In your shack, you can measure the power output from your transmitter(s) and amplifier(s) on each band. These measurements will be helpful if you later suspect you have a problem. If you have access to a signal generator, you can measure receiver performance for future reference.

INTERFACING HIGH-VOLTAGE EQUIPMENT TO SOLID-STATE ACCESSORIES

Many amateurs use a variety of equipment manufactured or home brewed over a considerable time period. For example, a ham might be keying a '60s-era tube rig with a recently built microcontroller-based electronic keyer. Many hams have modern solid-state radios connected to high-power vacuum-tube amplifiers.

Often, there is more involved in connecting HV (high-voltage) vacuum-tube gear to solid-state accessories than a cable and the appropriate connectors. The solid-state switching devices used in some equipment will be destroyed if used to switch the HV load of vacuum tube gear. The polarity involved is important too. Even if the voltage is low enough, a key-line might bias a solid-state device in such a way as to cause it to fail. What is needed is another form of level converter.

MOSFET Level Converters

While relays can often be rigged to interface the equipment, their noise, slow speed and external power requirement make them an unattractive solution in some cases. An alternative is to use power MOSFETs. Capable of handling substantial voltages and currents, power MOSFETs have become common design items. This has made them inexpensive and readily available.

Nearly all control signals use a common ground as one side of the control line. This leads to one of four basic level-conversion scenarios when equipment is interconnected:

1) A positive line must be actuated by a negative-only control switch.
2) A negative line must be actuated by a positive-only control switch.

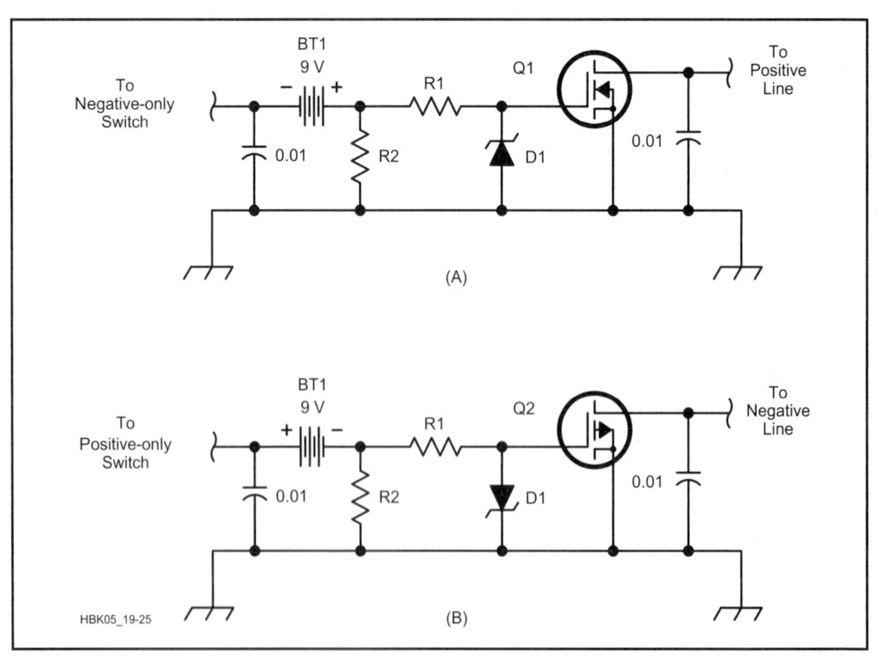

Fig 19.25—Level-shifter circuits for opposite input and output polarities. At A, from a negative-only switch to a positive line; B, from a positive-only switch to a negative line.

BT1—9-V transistor-radio battery.
D1—15-V, 1-W Zener diode (1N4744 or equiv).
Q1—IRF620.
Q2—IRF220 (see text).
R1—100 Ω, 10%, 1/4 W.
R2—10 kΩ, 10%, 1/4 W.

3) A positive line must be actuated by a positive-only control switch.
4) A negative line must be actuated by a negative-only control switch.

In cases 3 and 4 the polarity is not the problem. These situations become important when the control-switching device is incapable of handling the required open-circuit voltage or closed-circuit current.

Case 1 can be handled by the circuit in **Fig 19.25A**. This circuit is ideal for interfacing keyers designed for grid-block keying to positive CW key lines. A circuit suitable for case 2 is shown in Fig 19.25B. This circuit is simply the mirror image of that in Fig 19.25A with respect to circuit polarity. Here, a P-channel device is used to actuate the negative line from a positive-only control switch.

Cases 3 and 4 require the addition of an inverter, as shown in **Fig 19.26**. The inverter provides the logic reversal needed to drive the gate of the MOSFET high, activating the control line, when the control switch shorts the input to ground.

Almost any power MOSFET can be used in the level converters, provided the voltage and current ratings are sufficient to handle the signal levels to be switched. A wide variety of suitable devices is available from most large mail-order supply houses.

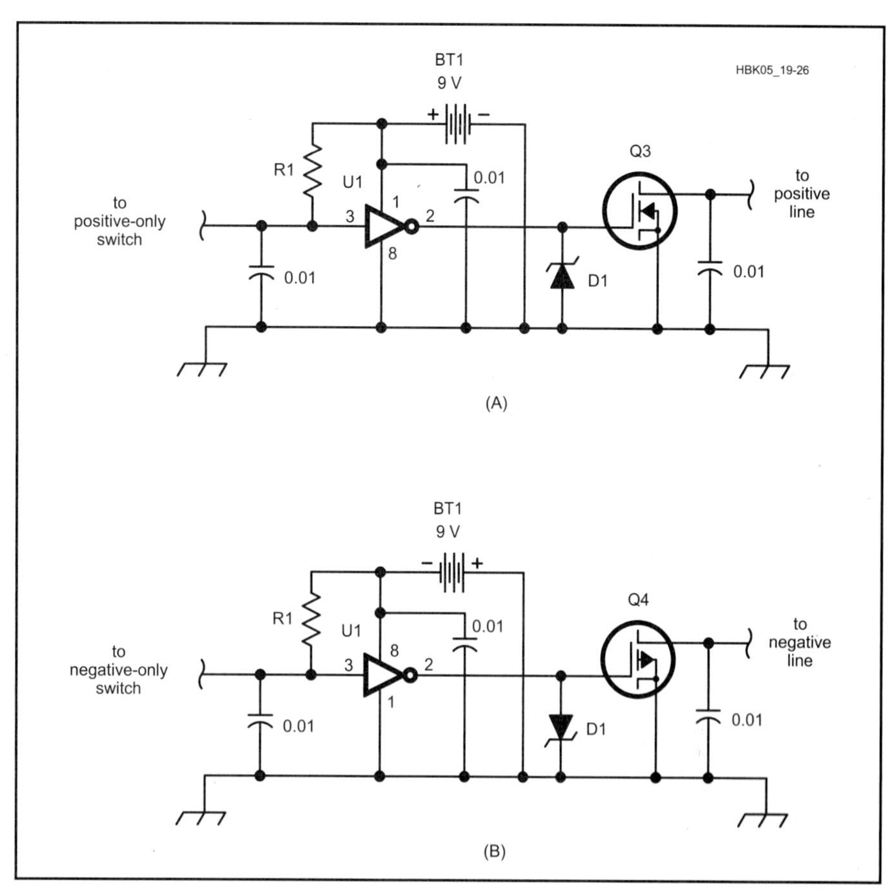

Fig 19.26—Circuits for same-polarity level shifters. At A, for positive-only switches and lines; B, for negative-only switches and lines.

BT1—9-V transistor-radio battery.
D1—15-V, 1-W Zener diode (1N4744 or equiv).
Q3—IRF620.
Q4—IRF220 (see text).
R1—10 kΩ, 10%, ¼ W.
U1—CD4049 CMOS inverting hex buffer, one section used (unused sections not shown; pins 5, 7, 9, 11 and 14 tied to ground).

Mobile and Portable Installations

Time and again, radio amateurs have been pressed into service in times of need. New developments outside of Amateur Radio (cellular phones, for example) often bring with them predictions that amateurs will no longer be needed to provide emergency communications. Just as often, a disaster proves beyond doubt the falseness of that exclamation. When the call for emergency communication is voiced by government and disaster relief organizations, mobile and portable equipment is pressed into service where needed. In addition to the occasional emergency or disaster type of communications, mobile and portable operation under normal conditions can challenge and reward the amateur operator.

Most mobile operation today is carried out by means of narrow-band repeaters. Major repeater frequencies reside in the 146 and 440-MHz bands. As these bands become increasingly congested, the 222 and 1240-MHz bands are being used for this reliable service mode as well. Many amateurs also enjoy mobile and portable HF operation because of the challenge and possibilities of worldwide communication.

MOBILE STATIONS

Installation and setup of mobile equipment can be considerably more challenging than for a fixed station. Tight quarters, limited placement options and harsher environments require innovation and attention to detail for a successful installation. The equipment should be placed so that operation will not interfere with driving. Driving safely is always the primary consideration; operating radio equipment is secondary. See **Fig 19.27** for one neat solution. If your vehicle has an airbag, be sure it can deploy unimpeded.

Mobile operation is not confined to lower power levels than in fixed stations. Many modern VHF FM transceivers are capable of 25 to 50 W of output. Compact HF rigs usually have outputs in the 100-W range and run directly from the 13.6-V supply.

If a piece of equipment will draw more than a few amps, it is best to run a heavy cable directly to the battery. Few circuits in an automobile electrical system can safely carry the more than 20 A required for a 100-W HF transceiver. Check the table in the **Component Data and References** chapter to verify the current-handling capabilities of various gauges of wire and cable. Adequate and well-placed fuses are necessary to prevent fire hazards. For maximum safety, fuse both the hot and ground lines near the battery. Automobile fires are costly and dangerous.

The limited space available makes antennas for mobile operation quite different than those for fixed stations. This is especially true for HF antennas. The **Antennas** chapter contains information for building and using mobile antennas.

Fig 19.27—N8KDY removed the ashtray for his mobile installation. The old faceplate for the ashtray is used as a cover for the rig when not in use. This may help reduce the temptation for would-be thieves. (*Photo courtesy AA8DX*)

Interference

In the past, interference in mobile installations almost always concerned interference to the radio equipment. Examples include ignition noise and charging-system noise. Modern automobiles—packed with arrays of sensors and one or more on-board computers—have made interference a two-way street. The original type of interference (to the radio) has also increased with the proliferation of these devices. An entire chapter in the *ARRL RFI Book* is devoted to ways to prevent and cure this problem.

PORTABLE STATIONS

Many amateurs experience the joys of portable operation once each year in the annual emergency exercise known as Field Day. Setting up an effective portable station requires organization, planning and some experience. For example, some knowledge of propagation is essential to picking the right band or bands for the intended communications link(s). Portable operation is difficult enough without dragging along excess equipment and antennas that will never be used.

Some problems encountered in portable operation that are not normally experienced in fixed-station operation include finding an appropriate power source and erecting an effective antenna. The equipment used should be as compact and lightweight as possible. A good portable setup is simple. Although you may bring gobs of gear to Field Day and set it up the day before, during a real emergency speed is of the essence. The less equipment to set up, the faster it will be operational.

Portable AC Power Sources

There are two popular sources of ac power for use in the field. One is referred to as a dc-to-ac converter, or more commonly, an *inverter*. The ac output of an inverter is a square wave. Therefore, some types of equipment cannot be operated from the inverter. Certain types of motors are among those devices that require a sine-wave output. **Fig 19.28** shows a typical commercial inverter. This model delivers 120 V of ac at 175 W continuous power rating. It requires 6 or 12 V dc input.

Besides having a square-wave output, inverters have some other traits that make them less than desirable for field use. Commonly available models do not provide a great deal of power. The 175-W model shown in Fig 19.28 could barely power a few light bulbs, let alone a number of transceivers. Higher-power models are available but are quite expensive. Another problem is that the batteries supplying the inverter with primary power are discharged as power is drawn from the inverter.

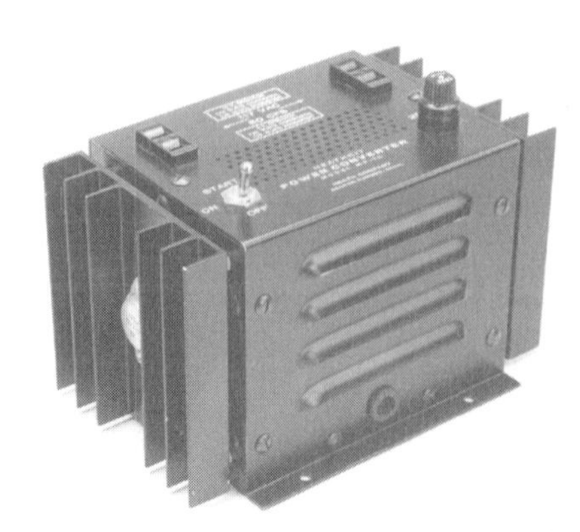

Fig 19.28—Photograph of a commercial dc-to-ac inverter that operates from 6 to 12 V dc and delivers 120 V ac (square wave) at 175 W.

Popularity and a number of competing manufacturers have caused gasoline generators to come down considerably in price. For a reliable, adequate source of ac (with sine-wave output), the gasoline-engine-driven generator is the best choice. (While still referred to as generators, practically all modern units actually use alternators to generate ac power.) Generators have become smaller and lighter as manufacturers have used aluminum and other lightweight materials in their construction. **Fig 19.29** shows the type of generator often used during Field Day.

Generators in the 3 to 5-kW range are easily handled by two people and can provide power for a relatively large multioperator field site. Most generators provide 12 V dc output in addition to 120/240 V ac.

Generator Maintenance

Proper maintenance is necessary to obtain rated output and a decent service life from a gasoline generator. A number of simple measures will prolong the life of the equipment and help maintain reliability.

It is a good idea to log the dates the unit is used and the operating time in hours. Many generators have hour-meters to make this simple. Include dates of maintenance and the type of service performed. The manufacturer's manual should be the

Fig 19.29—Modern gasoline engine-powered generators offer considerable ac power output in a relatively compact and lightweight package.

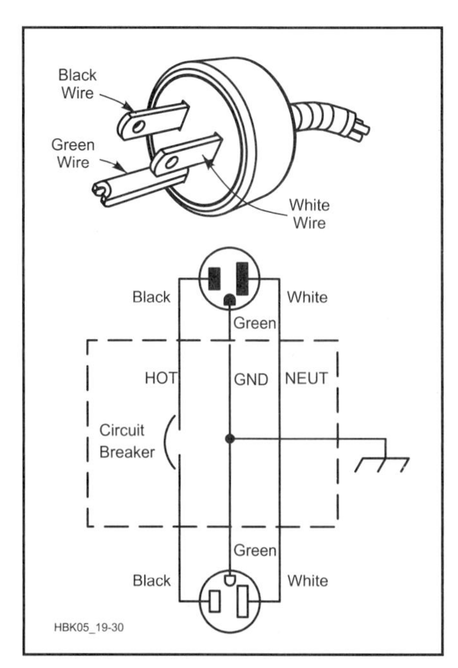

Fig 19.30—A simple accessory that provides overload protection for generators that do not have such provisions built in.

primary source of maintenance information and the final word on operating procedures and safety. The manual should be thoroughly covered by all persons who will operate and maintain the unit.

Particular attention should be paid to fuel quality and lubricating oil. A typical gasoline generator is often used at or near its rated capacity. The engine driving the alternator is under a heavy load that varies with the operation of connected electrical equipment. For these reasons, the demands on the lubricating oil are usually greater than for most gasoline-engine powered equipment such as lawn mowers, tractors and even automobiles. Only the grades and types of oil specified in the manual should be used. The oil should be changed at the specified intervals usually given as a number of operating hours.

Fuel should be clean, fresh and of good quality. Many problems with gasoline generators are caused by fuel problems. Examples include dirt or water in the fuel and old, stale fuel. Gasoline stored for any length of time changes as the more volatile components evaporate. This leaves excess amounts of varnish-like substances that will clog carburetor passages. If the generator will be stored for a long period, it is a good idea to run it until all of the fuel is burned. Another option is the use of fuel stabilizers added to the gasoline before storage.

Spark plugs should be changed as specified. Faulty spark plugs are a common cause of ignition problems. A couple of spare spark plugs should always be kept with the unit, along with tools needed to change them. Always use the type of spark plug recommended by the manufacturer.

Generator Ground

A proper ground for the generator is absolutely necessary for both safety reasons and to ensure proper operation of equipment powered from the unit. Most generators are supplied with a three-wire outlet, and the ground should connect to the plug as shown in **Fig 19.30**. Some generators require that the frame be grounded also. An adequate pipe or rod should be driven into the ground near the generator and connected to the provided clamp or lug. If no connection is provided, a clamp can be used to connect the ground lead to the frame of the generator. As always, follow the manufacturer's recommendations.

Portable Antennas

An effective antenna system is essential to all types of operation. Effective portable antennas, however, are more difficult to devise than their fixed-station counterparts. A portable antenna must be light, compact and easy to assemble. It is also important to remember that the portable antenna may be erected at a variety of sites, not all of which will offer ready-

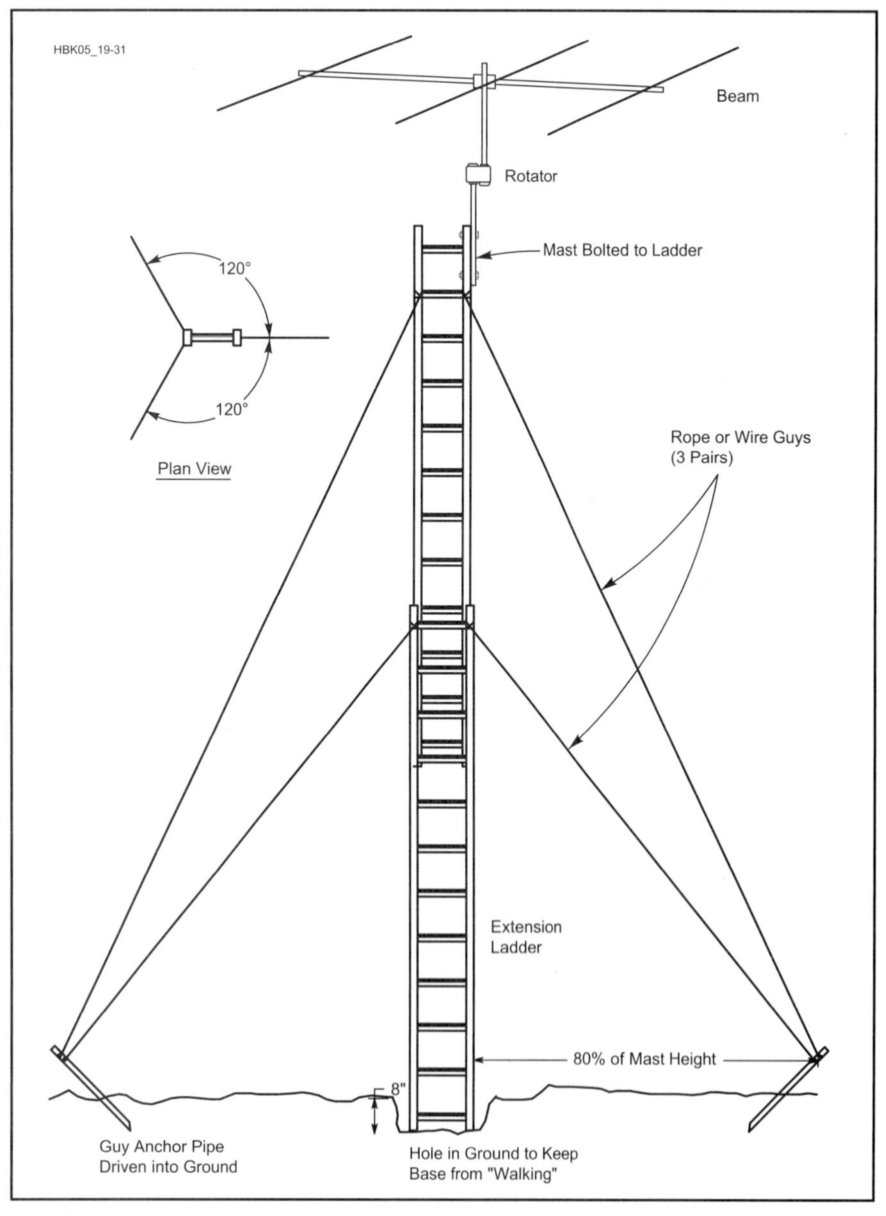

Fig 19.31—An aluminum extension ladder makes a simple but sturdy portable antenna support. Attach the antenna and feed lines to the top ladder section while it is nested and laying on the ground. Push the ladder vertical, attach the bottom guys and extend the ladder. Attach the top guys. Do not attempt to climb this type of antenna support.

made supports. Strive for the best antenna system possible because operations in the field are often restricted to low power by power supply and equipment considerations. Some antennas suitable for portable operation are described in the **Antennas** chapter.

(A)

(B)

Fig 19.32—The portable tower mounting system by WA7LYI. At A, a truck is "parked" on the homemade base plate to weigh it down. At B, the antennas, mast and rotator are mounted before the tower is pushed up. Do not attempt to climb a temporary tower installation.

Antenna Supports

While some amateurs have access to a truck or trailer with a portable tower, most are limited to what nature supplies, along with simple push-up masts. Select a portable site that is as high and clear as possible. Elevation is especially important if your operation involves VHF. Trees, buildings, flagpoles, telephone poles and the like can be pressed into service to support wire antennas. Drooping dipoles are often chosen over horizontal dipoles because they require only one support.

An aluminum extension ladder makes an effective antenna support, as shown in **Fig 19.31**. In this installation, a mast, rotator and beam are attached to the top of the second ladder section with the ladder near the ground. The ladder is then pushed vertical and the lower set of guy wires attached to the guy anchors. When the first set of guy wires is secured, the ladder may be extended and the top guy wires attached to the anchors. Do not attempt to climb a guyed ladder.

Figs 19.32 and **19.33** illustrate two methods for mounting portable antennas described by Terry Wilkinson, WA7LYI. Although the antennas shown are used for VHF work, the same principles can be applied to small HF beams as well.

In Fig 19.32A, a 3-ft section of Rohn 25 tower is welded to a pair of large hinges, which in turn are welded to a steel plate measuring approximately 18×30 inches. One of the rear wheels of a pickup truck is "parked" on the plate, ensuring that it will not move. In Fig 19.32B, quad array antennas for 144 and 222 MHz are mounted on a Rohn 25 top section, complete with rotator and feed lines. The tower is then pushed up into place using the hinges, and guy ropes anchored to heavy-duty stakes driven into the ground complete the installation. This method of portable tower installation offers an exceptionally easy-to-erect, yet sturdy, antenna support. Towers installed in this manner may be 30 or 40 ft high; the limiting factor is the number of "pushers" and "rope pullers" needed to get into the air. A portable station located in the bed of the pickup truck completes the installation.

The second method of mounting portable beams described by WA7LYI is shown in Fig 19.33. This support is intended for use with small or medium-sized VHF and UHF arrays. The tripod is available from any dealer selling television antennas; tripods of this type are usually mounted on the roof of a house. Open the tripod to its full size and drive a pipe into the ground at each leg. Use a hose clamp or small U-bolt to anchor each leg to its pipe.

The rotator mount is made from a 6-inch-long section of 1½-inch-diameter pipe welded to the center of an "X" made from two 2-ft-long pieces of concrete reinforcing rod (rebar). The rotator clamps onto the pipe, and the whole assembly is placed in the center of the tripod. Large rocks placed on the rebar hold the rotator in place, and the antennas are mounted on

(A)

(B)

(C)

Fig 19.33—The portable mast and tripod by WA7LYI. At A, the tripod is clamped to stakes driven into the ground. The rotator is attached to a homemade pipe mount. At B, rocks piled on the rotator must keep the rotator from twisting and add weight to stabilize the mast. At C, a 10-ft mast is inserted into the tripod/rotator base assembly. Four 432-MHz Quagis are mounted at the top.

a 10 or 15-ft mast section. This system is easy to make and set up.

Tips for Portable Antennas

Any of the antennas described in the **Antennas** chapter or available from commercial manufacturers may be used for portable operation. Generally, though, big or heavy antennas should be passed over in favor of smaller arrays. The couple of decibels of gain a 5-element, 20-m beam may have over a 3-element version is insignificant compared to the mechanical considerations. Stick with arrays of reasonable size that are easily assembled.

Wire antennas should be cut to size and tuned prior to their use in the field. Be careful when coiling these antennas for transport, or you may end up with a tangled mess when you need an antenna in a hurry. The coaxial cable should be attached to the center insulator with a connector for speed in assembly. Use RG-58 for the low bands and RG-8X for higher-band antennas. Although these cables exhibit higher loss than standard RG-8, they are far more compact and weigh much less for a given length.

Beam antennas should be assembled and tested before taking them afield. Break the beam into as few pieces as necessary for transportation and mark each joint for speed in reassembly. Hex nuts can be replaced with wing nuts to reduce the number of tools necessary.

Ground Rod Installation

A large sledgehammer, a small stepladder and a lot of elbow grease. That's the usual formula for driving in an 8-foot ground rod. Michael Goins, WB5YKX, reports success using fluid hydraulics to ease the task in very dense clay soil.

He suggests digging a small hole, about a foot deep, just enough to hold a few gallons of water. When the hole is complete, pour water into the hole, and then push the ground rod in as far as it will go. Next, pull it out completely. Some of the water will run into the smaller hole made by the ground rod.

Repeat the process, allowing water to run into the small hole each time you remove the rod. Continue pushing and removing until the rod is sunk as far as you want. About 6 inches of rod above the ground is usually enough to allow convenient connection of bonding clamps.

THE TiCK-2—A TINY CMOS KEYER 2

TiCK-2 stands for "Tiny CMOS Keyer 2." It is based on an 8-pin DIP microcontroller from Microchip Corporation, the PIC 12C509. This IC is a perfect candidate for all sorts of Amateur Radio applications because of its small size and high performance capabilities. This project was described fully in Oct 1997 *QST* by Gary M. Diana, Sr, N2JGU, and Bradley S. Mitchell, WB8YGG. The keyer has the following features:

- One memory message—a single memory message, capable of at least playing back "CQ CQ DE *callsign callsign* K"
- Mode A and B iambic keying
- Low current requirement—to support portable use.
- Low parts count—consistent with a goal for small physical size.
- Simple interfaces—this includes the rig and user interfaces. The user interface must be simple; ie, the operator shouldn't need a manual. The rig interface should be simple as well: paddles in, key line out.
- Sidetone—supply an audible sidetone for user-feedback functions and to support transceivers that do not have a built-in sidetone.
- Paddle select—allow the operator to swap the dot and dash paddles without having to rewire the keyer (or flip the paddles upside down!).
- Manual keying—permit interfacing a straight key (or external keyer) to the TiCK.

Design

Fig 19.34 shows the schematic for the keyer. The PIC 12C509 has two pages of 512 bytes of program read-only memory (ROM) and 41 bytes of random-access memory (RAM). This means that all the keyer *functions* have to fit within the ROM. The keyer *settings* such as speed, paddle selection, iambic mode and sidetone enable are stored in RAM. The RAM in this microcontroller is *volatile*, that is, the values stored in memory are lost if the power to the chip is cycled off, then on—but there's not much of a need to do that because of the low power requirements for the chip.

A 12C509 has eight pins. Two pins are needed for the dc input and ground con-

Table 19.1
TiCK-2 User Interface Description

Action	*TiCK-1 and 2 Response*	*Function*
Press pushbutton	S (dit-dit-dit)	Speed adjust: Press dit to decrease, dah to increase speed.
Hold pushbutton down	M (dah-dah)	Memory playback: Plays the message from memory, using the key line and sidetone (if enabled).
Hold pushbutton down	T (dah)	Tune: To unkey rig, press either paddle or pushbutton.
Hold pushbutton down	A (dit-dah)	ADMIN mode: Allows access to various TiCK-2 IC setup parameters.
Hold pushbutton down	I (dit-dit)	Input mode: Allows message entry. Press pushbutton when input is complete.
Hold pushbutton down	P (dit-dah-dah-dit)	Paddle select: Press paddle desired to designate as dit paddle.
Hold pushbutton down	A (dit-dah)	Audio select: Press dit to enable sidetone, dah to disable. Default: enabled.
Hold pushbutton down	SK (dit-dit-dit dah-dit-dah)	Straight key select: Pressing either paddle toggles the TiCK to/from straight key/keyer mode. Default: keyer mode.
Hold pushbutton down	M (dah-dah)	Mode select: Pressing the dit paddle puts the TiCK into iambic mode A; dah selects iambic mode B (the default).
Hold pushbutton down	K (dah-dit-dah)	Keyer mode: If pushbutton is released, the keyer returns to normal operation.
Hold pushbutton down	S (dit-dit-dit)	Cycle repeats with Speed adjust.

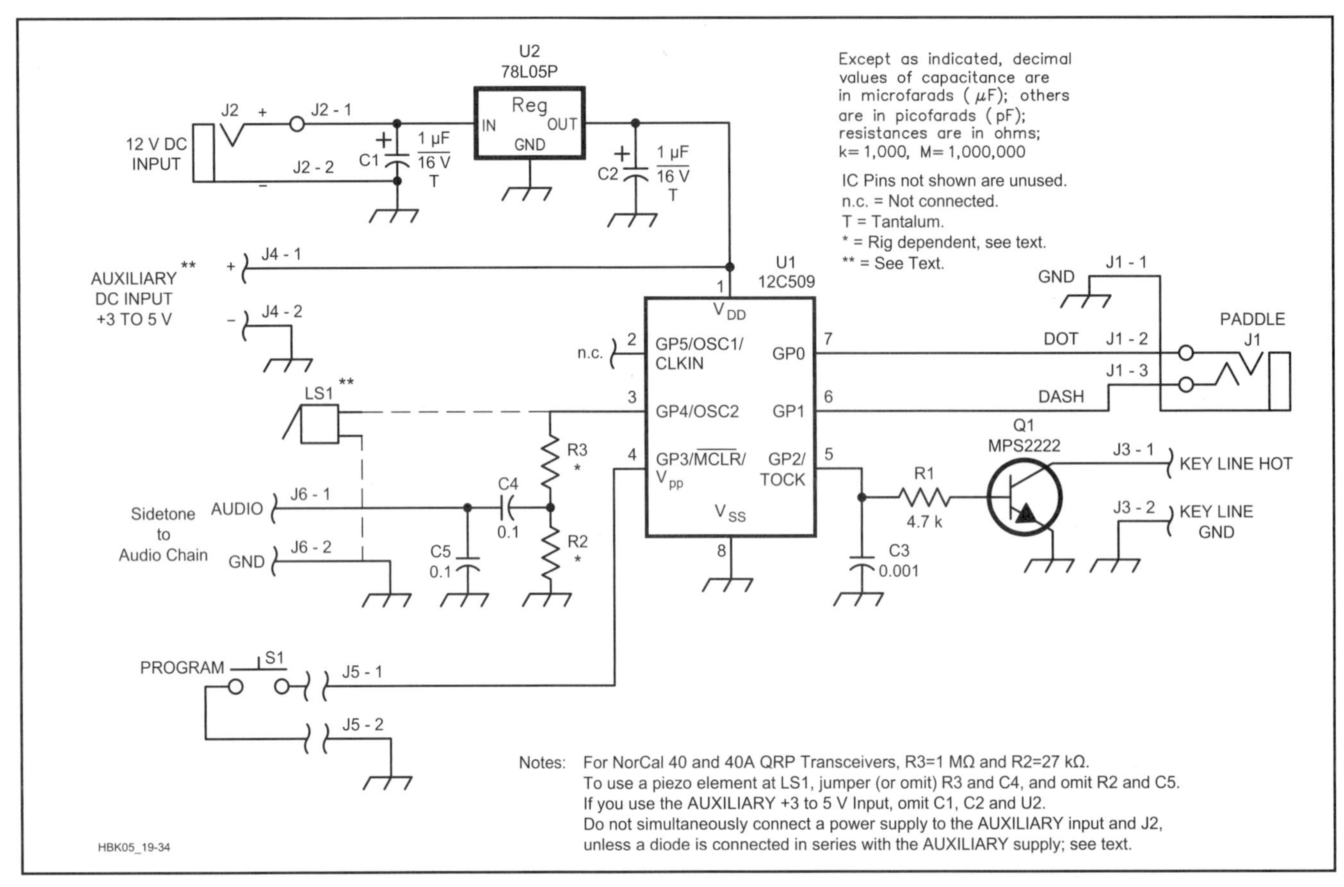

Fig 19.34—Schematic of the TiCK-2 keyer. Equivalent parts can be substituted. Unless otherwise specified, resistors are ¼ W, 5%-tolerance carbon-composition or film units. The PIC12C509 IC must be programmed before use; see Note in the parts list for U1. RS part numbers in parentheses are RadioShack; M = Mouser (Mouser Electronics).

C1, C2—1 µF, 16 V tantalum (RS 272-1434; M 581-1.0K35V).
J1—3-circuit jack (RS 274-249; M 161-3402).
J2—2-circuit jack (RS 274-251) or coaxial (RS 274-1563 or 274-1576).
J3—2-circuit jack (RS 274-251; M16PJ135).
LS1—Optional piezo element (RS 273-064).
Q1—MPS2222A, 2N2222, PN2222, NPN (RS 276-2709; M 333-PN2222).
S1—Normally open pushbutton (RS 275-1571; M 10PA011).
U1—Programmed PIC 12C509, available from Embedded Research. (Use the *TIS Find* program for address information.) The TiCK-2 chip/data sheet, $10; TiCK-2 programmed IC, PC board and manual, $15; TiCK-2 programmed IC, PC board, parts and manual, $21. All prices are postpaid within the continental US. Canadian residents please add $5; all others add $6 for shipping. New York state residents please add 8% sales tax. The DIP and SOIC (surface-mount) chips and kits are the same price. Please specify which one you prefer when ordering. Note: Components not included with the SMD version kit are the voltage regulator and voltage divider components for the audio output. Source code is not available.
U2-5 V, 100 mA regulator (RS 276-1770; M 333-ML78L05A).
Misc: PC board, 8-pin DIP socket, hardware, wire (use stranded #22 to #28, Teflon insulated for heat/solder resistance).

nections. The IC requires a clock signal. Several clock-source options are available; you can use: a crystal, RC (resistor and capacitor) circuit, resonator, or the IC's internal oscillator. The authors chose the internal 4-MHz oscillator to reduce the external parts count. Two I/O lines are used for the paddle input. One output feeds the key line, another output is required for the audio feedback (sidetone) and a third I/O line is assigned to a pushbutton.

User Interface

Using the two paddles and a pushbutton, you can access all of the TiCK's functions. Certain user-interface functions need to be more easily accessible than others; a prioritized list of functions (from most to least accessible) is presented in **Table 19.1**.

The TiCK employs a *single button interface* (SBI). This simplifies the TiCK PC board, minimizes the part count and makes for ease of use. Most other electronic keyers have multibutton user interfaces, which, if used infrequently, make it difficult to remember the commands. Here, a single button push takes you through the functions, one at a time, at a comfortable pace (based on the current speed of the keyer). Once the code for the desired function is heard, you simply let up on the button. The TiCK then executes the appropriate function, and/or waits for the appropriate input, either from the paddles or the pushbutton itself, depending on the function in question. Once the function is complete, the TiCK goes back into keyer mode, ready to send code through the key line.

The TiCK-2 IC generates a sidetone signal that can be connected to a piezoelectric element or fed to the audio chain of a transceiver. The latter option is rig-specific, but can be handled by more experienced builders.

The TiCK Likes to Sleep

To meet the low-current requirement, the authors took advantage of the 12C509's ability to *sleep*. In sleep mode, the processor shuts down and waits for input from either of the two paddles. While sleeping, the TiCK-2 consumes just a few microamperes. The TiCK-2 doesn't wait long to go to sleep either: As soon as there is no input from the paddles, it's snoozing! This feature should be especially attractive to amateurs who want to use the TiCK-2 in a portable station.

Assembling the TiCK

The TiCK-2's PC board size (1×1.2 inches) supports its use as an embedded and stand-alone keyer. The PC board has two dc input ports, one at J2 for 7 to 25 V and another (J4, AUXILIARY) for 2.5 to 5.5 V. The input at J2 is routed to an on-board 5-V regulator (U2), while the AUXILIARY input feeds the TiCK-2 directly. When making the dc connections, observe proper polarity: There is no built-in reverse-voltage protection at either dc input port.

The voltage regulator's bias current is quite high and will drain a 9 V battery quickly, even though the TiCK itself draws very little current. For this reason, the "most QRP way" to go may be to power the chip via the AUXILIARY power input and omit U2, C1 and C2.

When using the AUXILIARY dc input port, or if *both* dc inputs are used, connect a diode between the power source and the AUXILIARY power input pin, attaching the diode anode to the power source and the cathode to the AUXILIARY power input pin. This provides IC and battery protection and can also be used to deliver battery backup for your keyer settings.

Sidetone

A piezo audio transducer can be wired directly to the TiCK-2's audio output: pads and board space are available for voltage-divider components. This eliminates the need to interface the TiCK-2 with a transceiver's audio chain. Use a piezo *element*, not a piezo *buzzer*. A piezo *buzzer* contains an internal oscillator and requires only a dc voltage to generate the sound, whereas a piezo *element* requires an external oscillator signal (available at pin 3 of U1).

If you choose to embed the keyer in a rig and want to hear the keyer's sidetone instead of the rig's sidetone, you may choose to add R2, R3, C4 and C5. Typically R3 should be 1 MΩ to limit current. R2's value is dictated by the amount of drive required. A value of 27 kΩ is a good start. C4 and C5 values of 0.1 µF work quite well. C4 and C5 soften the square wave and capacitively couple J6-1 to the square-wave output of pin 3. Decreasing the value of R2 decreases the amount of drive voltage, especially below 5 kΩ. Use a 20 kΩ to 30 kΩ trimmer potentiometer at R2 when experimenting.

In Use

To avoid RF pickup, keep all leads to and from the TiCK as short as possible. The authors tested the TiCK in a variety of RF environments and found it to be relatively immune to RF. Make sure your radio gear is well grounded and avoid situations that cause an RF-hot shack.

The TiCK keys low-voltage positive lines, common in today's solid-state rigs. Don't try to directly key a tube rig because you will likely—at a minimum—ruin output transistor Q1.

To use the TiCK as a code-practice oscillator, connect a piezo element to the audio output at pin 3. If more volume is needed, use an audio amplifier, such as Radio Shack's 277-1008.

The higher the power-supply voltage (within the specified limits), the greater the piezo element's volume. Use 5 V (as opposed to 3 V) if more volume is desired. Also, try experimenting with the location of the piezo element to determine the proper mounting for maximum volume.

Just When You Thought It Was Small Enough...

In addition to the DIP version of the TiCK-2, there is also a surface-mount version of the keyer. This uses a 12C509 IC, which resides in a medium-size SOIC package measuring roughly 5×5 mm! This is approximately *two-thirds* the size of the DIP version of the IC. (For simplicity, the surface-mount keyer does not include the regulator circuitry.) See **Fig 19.35**.

The surface-mount version of the TiCK PC board has no provisions for standoffs, but it can easily be mounted in an enclosure (or on the back of a battery!) using double-stick foam tape. The authors used this method to put the surface-mount TiCK into some really tiny enclosures. To assist the builder, the pads on the TiCK board are larger than necessary.

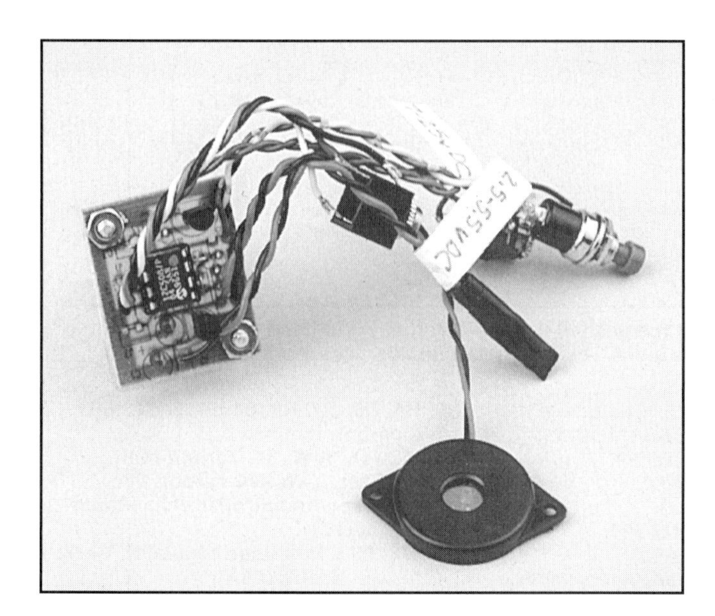

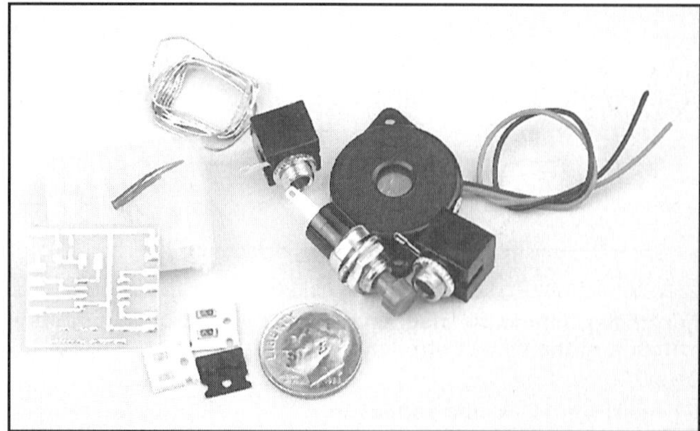

Fig 19.35—At left, photo of DIP version, and above, SMD version of the TiCK-2.

VINTAGE RADIO T/R ADAPTER

This T/R Adapter provides automatic transmit/receive switching for many vintage transmitters and receivers. It provides time-sequenced antenna transfer, receiver muting and transmitter keying in semi and full break-in CW modes or with push-to-talk systems in AM or SSB.

General Description

The vintage radio adapter consists of two assemblies, the control unit and a remotely located antenna relay. The control unit accepts key, keyer or push-to-talk input and produces time-sequenced outputs to transfer the antenna between receiver and transmitter, mute the receiver and key either cathode keyed or grid block keyed transmitters. The antenna relay is

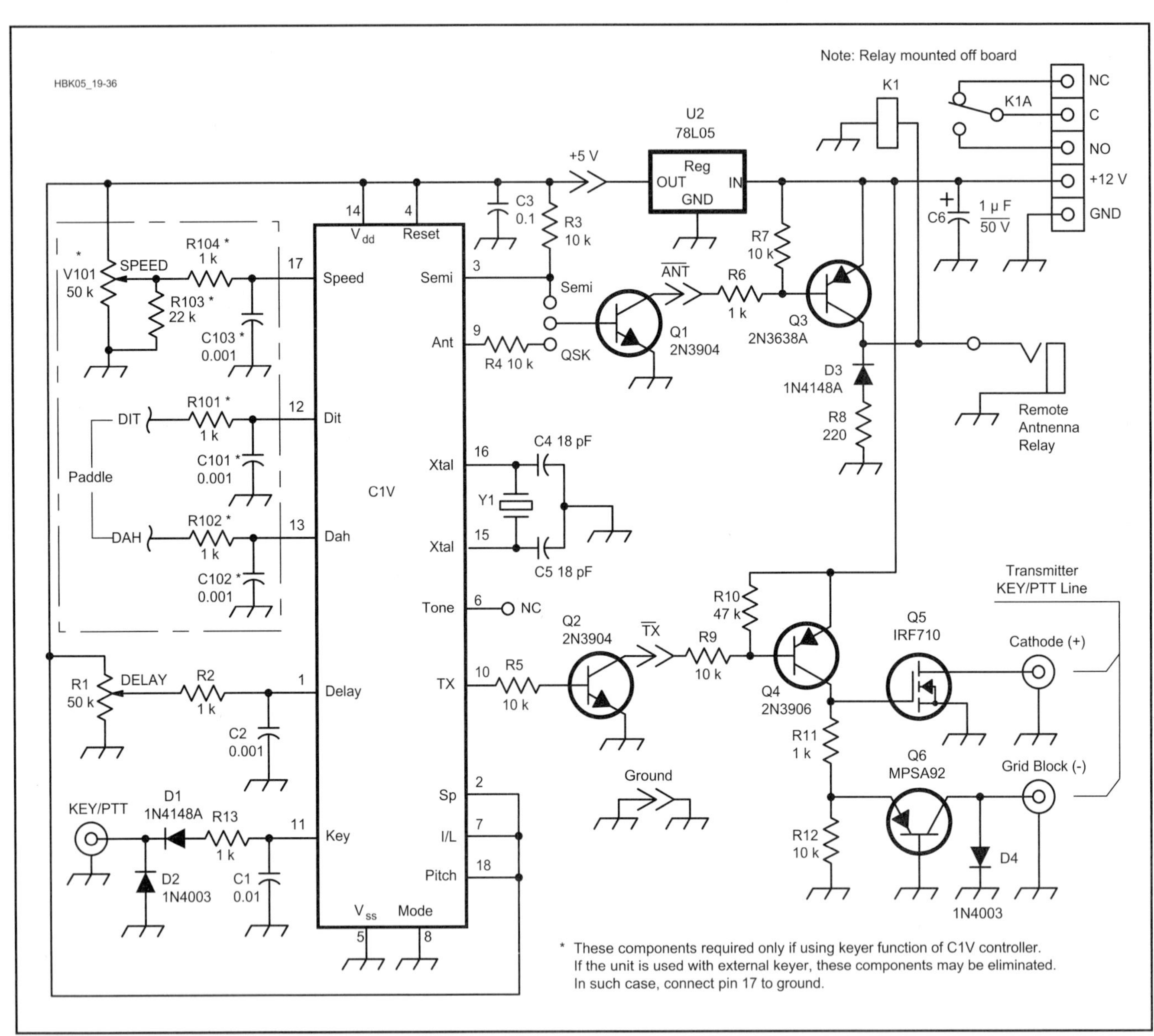

Fig 19.36—Schematic diagram of the T/R switch. The circuit within the dashed lines is optional—it is used with the keyer function of the C1V controller. If an external keyer is used, this circuit can be eliminated; in this case, connect pin 17 to ground.

C1—0.01-μF, 50-V disc capacitor (Mouser CD50Z6-103M).
C2—0.001-μF, 50-V disc capacitor (Mouser CD50P6-102M).
C3—0.1-μF disc capacitor (Mouser CD100U5-104M).
C4, C5—18-pF, 100-V, 10% NPO disc capacitor (Mouser 100N2-018J).
C6—1-μF, 50-V, vertical electrolytic (Mouser XRL50V1.0).
D1, D3—1N4148A switching diode.
D2, D4—1N4003 power diode.
K1—SPDT relay, 2 A, 12-V dc coil (Radio Adventures BAS111DC12).
Q1, Q2—2N3904 NPN, TO-92.
Q3—2N3638A PNP, TO-92.
Q4—2N3906 PNP, TO-92.
Q5—IRF710 HEXFET, TO-220.
Q6—PNP, TO-92 (MPSA92).
R1—50-kΩ potentiometer, linear taper (Mouser 31CN405).
R2, R6, R11, R13—1 kΩ, 1/4-W, 5% carbon film.
R3, R4, R5, R7, R9, R12—10-kΩ, 1/4-W, 5% carbon film.
R8—220-Ω, 1/4-W, 5% carbon film.
R10—47 kΩ, 1/4-W, 5% carbon film.
U1—C1V keyer/controller chip (Radio Adventures).
U2—78L05 5-V voltage regulator TO-92 (Mouser NJM78L05A).
Y1—2.0-MHz resonator (supplied with C1V).
Misc—Vintage Radio Adapter PC board (Radio Adventures 090-0112).

remotely mounted to aid in running coaxial cable and to minimize relay noise. The antenna relay is rated to handle over 100 W. If high power is contemplated, an antenna relay with a higher power rating is needed. A complete circuit board kit is available, making the unit very easy to duplicate. If you choose to build the unit from scratch, a template package that includes the PC board layout and part placement diagram is available from ARRL HQ.[1]

How it Works

The control circuit of the adapter is based on the C1V keyer/controller chip from Radio Adventures Corp. See **Fig 19.36**. In this circuit, the controller features of the chip are used to provide the sequencing outputs to various circuits. By adding a few components, the keyer functions of the C1V can be utilized if desired.

When the KEY input line is pulled low by a key, bug, keyer or push-to-talk switch, the C1V controller chip immediately raises the ANT signal pin, pin 9, and SEMI signal, pin 3, switching the antenna from receiver to transmitter and muting the receiver. About 5.5 ms later the C1V raises the TX signal pin, pin 10, keying the transmitter. The delay provides time for the relays to transfer and for the receiver to quiet.

When the KEY line is released, the C1V controller delays 5.5 ms and then lowers the TX signal, pin 10, unkeying the transmitter. This 5.5-ms delay compensates for the 5.5-ms transmit delay at the start of the keying sequence, hence preserving the keying waveform. 5.5 ms after lowering the TX pin, the C1V controller lowers the ANT pin, pin 9, switching the antenna back to the receiver and unmuting the receiver. This delay allows time for the transmitter power to decay before switching the antenna, preventing "hot" switching the antenna relay. This process provides full break-in keying up to about 40 WPM. See **Fig 19.37**.

Although ANT signal, pin 9, follows the keying, SEMI, pin 3, delays on release of the key. The amount of delay is determined by the setting of the DELAY potentiometer, R1. The delay can be varied from

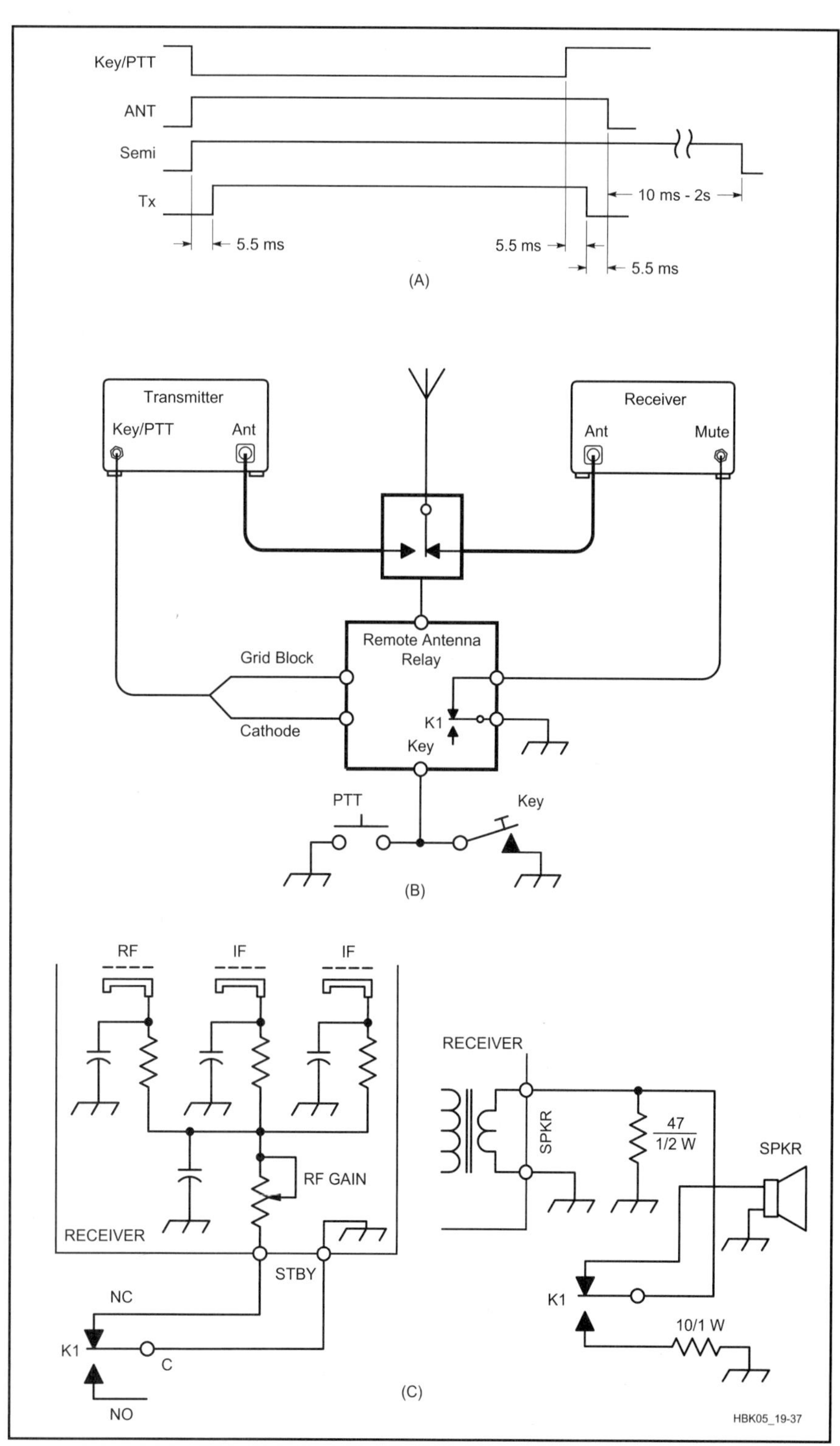

Fig 19.38—(A) Timing diagram of the C1V keyer/controller chip. The text explains how the control circuit works. (B) Block diagram showing how the antenna switch can connect to a vintage station. (C) Sample receiver mute circuits. Left: The receiver can be muted in the cathode or RF gain circuit. Right: If you prefer, the mute relay can also be connected to the speaker circuit.

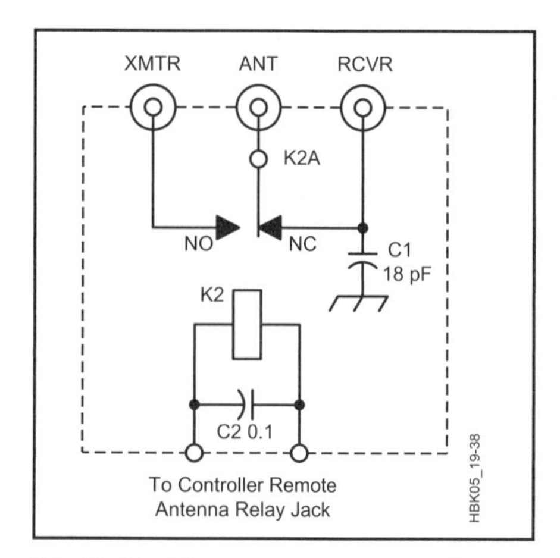

Fig 19.37—The remote antenna relay schematic diagram.
C1—18-pF disc ceramic, #100N2-018J
C2—0.1-µF disc ceramic, #CD100U5-104M.
K2—SPDT relay, 2 A, 12 V dc coil.
Misc:
Bud Econobox, 1$^1/_{16}$×3$^5/_8$×1$^1/_2$, #CU-123.
SO-239 coaxial connectors.

about 10 ms to about 1.5 seconds. If KEY line is again lowered before the delay is completed, the delay timer is reset and the full delay time is available after KEY is released. The output of SEMI, pin 3, provides semi break-in timing for use in situations where QSK is not desired.

Level shifter transistor Q1 can be connected to either ANT or SEMI as desired. Q1 drives relay amplifier Q3, which in turn energizes relay K1 and the remote antenna relay K2 (see **Fig 19.38**).

Level shifter transistor Q2 drives Q4, which provides drive for the HEXFET transistor used for cathode (positive) keying and the high voltage PNP transistor used for grid block (negative) keying. The HEXFET power transistor selected for the cathode keying output will key currents beyond 500 mA and open circuit voltages of beyond 200 V. These ratings should be adequate for most cathode-keyed rigs.

The PNP transistor is rated at over 200 V and 10 mA. If your application requires higher voltage or current, a transistor with adequate ratings must be substituted and emitter resistor R11 must be reduced in value to provide sufficient current. The value of the resistor can be found with the following formula: R11 = 10/I. For example, to switch 20 mA, R11 = 10/0.02 = 500 Ω, a 470-Ω resistor would be used.

Construction

Construction is straightforward if the circuit board kit is used. If you choose to hand wire your unit on perf board or lay out your own circuit board, layout is not critical. The finished boards can be mounted in an enclosure of your choice. It is convenient to mount the antenna relay in its own enclosure to reduce the possibility of RF getting into the logic circuits and to make routing of coaxial cables more direct. Note that the relays are mounted on small carrier PC boards and are suspended by their leads only. This method of mounting greatly reduces relay noise that occurs when the relays are mounted on the main PC board.

Installation

Power requirements for the adapter are 12 V dc at approximately 120 mA. Antenna connections are straightforward and need no further discussion. Select either the cathode or grid block output for connection to your transmitter as appropriate. See Figs 19.36 and 19.37.

There are a couple of options for muting the receiver. Fig 19.37C shows a typical circuit for muting the receiver in the cathode or RF gain circuit. Many receivers bring the mute terminals to the rear apron. In some cases it may be necessary to go inside the receiver to bring out the cathode circuit.

If your receiver does not provide mute terminals and/or you do not want to mute in the cathode circuit, you can connect the mute relay K1 into the speaker circuit as shown in Fig 19.37C (right). The resistors ensure that the speaker terminals of the receiver are always terminated by a load close to 8 Ω, preventing possible damage to the receiver if the resistors were not used.

Note: Some receivers, especially those built in the '30s and '40s, put the receiver into a standby condition by opening the center tap of the power transformer high-voltage winding. *Do not* attempt to mute your receiver by this method. Use cathode or speaker muting instead.

For AM or SSB systems, connect the receiver as discussed, choose the QSK connection for level shifter transistor Q1 and connect the microphone PTT switch to the key/PTT input.

Notes

[1]See the CD-ROM for template.

QUICK AND EASY CW WITH YOUR PC

A couple of chips and a few hours work will yield this CW only terminal for a PC. Designed by Ralph Taggart, WB8DQT, the software transforms a computer into a Morse machine that's a full-function CW keyboard *and* a receive display terminal.

The circuit works with IBM-compatible PCs and uses the printer port to communicate with the computer. Parts cost is generally less than $50, and a printed-circuit board is available to make construction easier.

Circuit Description

Each stage of the circuit in **Fig 19.39A** is labeled with its function. Ferrite beads are used to keep RF from entering the unit. K1 provides isolation, so any transmitter may be keyed without worrying about polarity. Fig 19.39B shows the power and computer interconnections for the circuit board.

Power Supply Options

Three voltage sources are required (+12 V, +5 V and –9 V) at relatively low current. The simplest approach is to use a wall-mount power transformer/supply (200 mA minimum) to provide the +12 V.

Fig 19.39—Schematic of the CW interface. All fixed value resistors are ¼-W, 5%-tolerance carbon film. Capacitance values are in microfarads (µF). RS indicates RadioShack part numbers. IC sections not shown are not used.

C1-C3, C5, C7-C13—0.1 µF monolithic or disc ceramic, 50 V.
C4—0.047 µF Polypropylene (dipped Mylar), 50 V.
C6—0.22 µF Polypropylene (dipped Mylar), 50 V.
C14—1 µF Tantalum or electrolytic, 50 V.
C15—0.47 µF Tantalum or electrolytic, 50 V.
C16—10 µF Tantalum or electrolytic, 50 V.
D1, D3—1N4004.
D2—1N270 germanium.
DS1—Green panel-mount LED.
DS2—Red panel-mount LED.
FB—Ferrite beads (11 total).
K1—12 V dc SPST reed relay (RS-275-233).
J1, J2—RCA phono jacks.
P1—4-pin microphone jack (RS-274-002).
Q1, Q3—2N4401.
Q2—MPF102.
R1—1 kΩ.
R2, R3—10 kΩ.
U1—NE567CN PLL tone decoder (8 pin).
U2—74LS14N hex Schmitt trigger (14 pin).
U3, U4—LM741CN op amp (8 pin).
U5—74LS00N quad NAND gate (14 pin).
4-pin microphone plug (RS-274-001).
4-pin microphone socket (chassis mount).
DB-25M Connector (RS-276-1547).
DB-25 Shell (RS-276-1549).
Coaxial power connector (RS-274-1563).
8-pin DIP IC sockets.
14-pin DIP IC sockets.
DPDT miniature toggle switch.
J3 is a panel-mounting coaxial power jack to match your wall-mount/ transformer power supply.
BT1 is a 9-V alkaline battery. See text. C17 and C18 are 0.1 µF, 50 V monolithic or disc ceramic bypass capacitors. The +5 V regulator chip should be mounted to the grounded wall of the cabinet. Off-board components are duplicated in section B of this drawing (J1, J2, P1, the CW LED indicator, and K1). The CW and POWER indicators are panel-mounting LED indicators (red for POWER and green for CW). FB indicates optional ferrite beads used to prevent RF interference with the interface circuits.

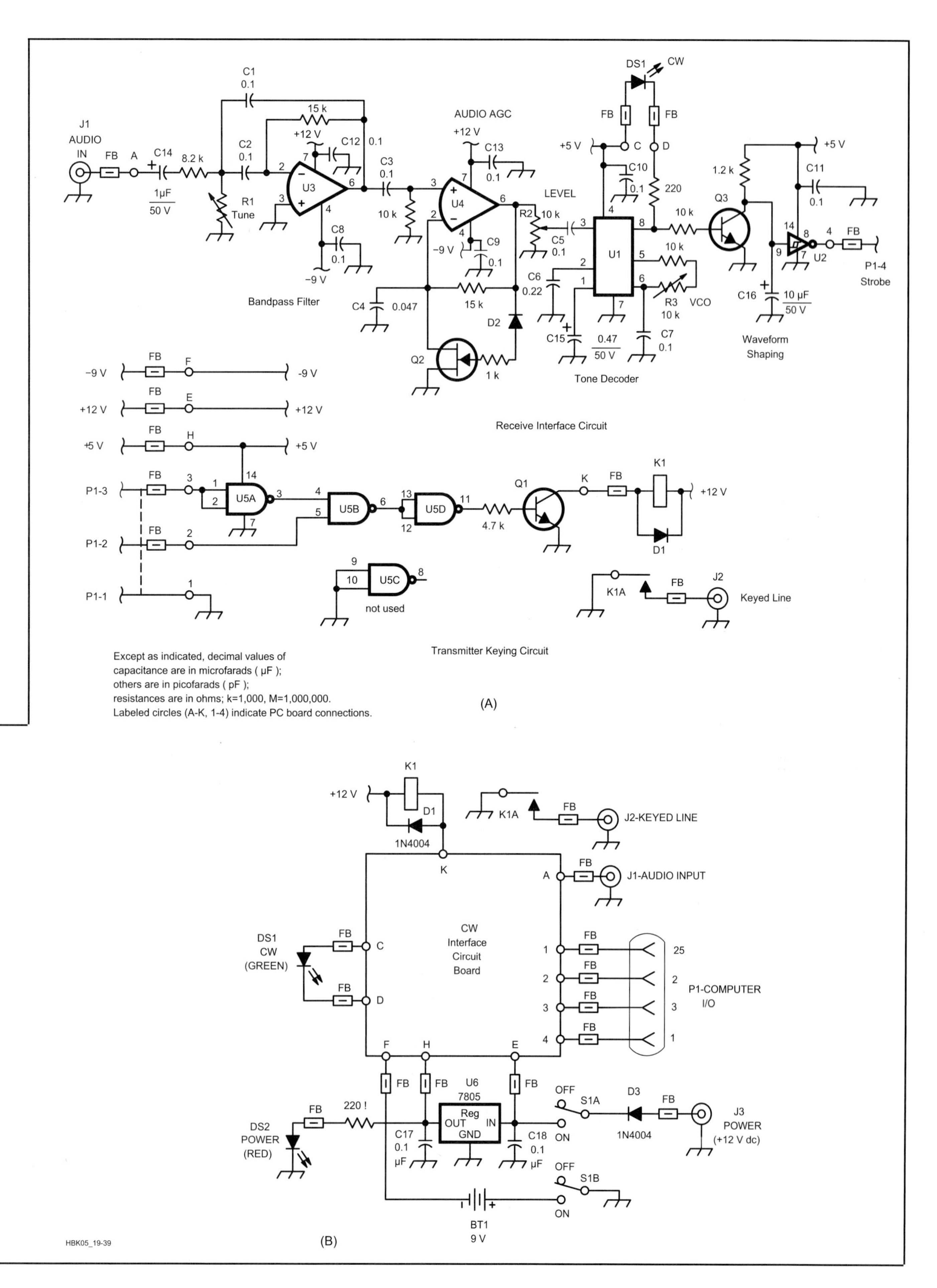
J1
AUDIO
IN
FB
A
C14
8.2 k
1µF
50 V
C1
0.1
15 k
+12 V
C12
0.1
C2
0.1
R1
Tune
U3
C8
0.1
-9 V
Bandpass Filter
AUDIO AGC
+12 V
C13
0.1
C3
0.1
10 k
U4
-9 V
C9
0.1
C4
0.047
15 k
D2
Q2
1 k
LEVEL
R2
10 k
C5
0.1
C6
0.22
C15
0.47
50 V
DS1
CW
FB
FB
+5 V
C
D
C10
0.1
220
U1
10 k
10 k
R3
10 k
VCO
C7
0.1
Tone Decoder
Q3
1.2 k
+5 V
C11
0.1
U2
FB
P1-4
Strobe
C16
10 µF
50 V
Waveform
Shaping
Receive Interface Circuit
-9 V
+12 V
+5 V
P1-3
P1-2
P1-1
U5A
U5B
U5D
U5C
not used
4.7 k
Q1
K
K1
+12 V
D1
K1A
J2
Keyed Line
Transmitter Keying Circuit
Except as indicated, decimal values of
capacitance are in microfarads (µF);
others are in picofarads (pF);
resistances are in ohms; k=1,000, M=1,000,000.
Labeled circles (A-K, 1-4) indicate PC board connections.
(A)
K1
+12 V
D1
1N4004
K1A
J2-KEYED LINE
J1-AUDIO INPUT
CW
Interface
Circuit
Board
DS1
CW
(GREEN)
25
P1-COMPUTER
I/O
U6
7805
Reg
OUT
IN
GND
DS2
POWER
(RED)
220 !
C17
0.1
µF
C18
0.1
µF
OFF
S1A
ON
D3
1N4004
J3
POWER
(+12 V dc)
OFF
S1B
ON
BT1
9 V
(B)
HBK05_19-39

A 7805 voltage regulator chip (U6) produces +5 V from the +12 V bus for the 74LS TTL ICs. Since the –9 V current requirements are very low, a 9 V alkaline transistor battery was used in WB8DQT's unit. This battery is switched in and out using one set of contacts on the POWER switch and will last a long time—unless you forget to turn the unit off between operating sessions!

The Computer Connection

The circuit connects to the PC parallel printer port, which is usually a DB-25F (female) connector on the rear of the computer. A standard cable with a mating DB-25M (male) connector on one end and a DB-25F on the other end could be used. This cable is available at any computer store, but it would require a DB-25M to be mounted on your project box.

Unfortunately, DB-25 connectors need an odd-shaped mounting hole, which is difficult to make with standard shop tools. Since only four conductors are needed (ground, printer data bits 0 and 1, and the strobe data bit), it's easier to make a cable. Use a 4-pin microphone connector at one end and a DB-25M at the other end. Wire the cable as follows:

Function	*Microphone Plug*	*DB-25M Connector*
Ground	1	25
Printer data 0	2	2
Printer data 1	3	3
Printer strobe	4	1

Drill a 5/8-inch round hole on the rear apron of the project enclosure for the mating chassis-mount 4-pin microphone socket.

Keying Options

For equipment with a positive, low-voltage keying line, point K on the board can be connected directly to the keying jack. In this case, omit K1 and its 1N4004 diode. For a wider range of transmitting equipment, use the keying relay. Mount it anywhere in the cabinet using a dab of silicone adhesive or a piece of double-sided foam mounting tape.

Construction

The simplest way to construct this circuit is on a single PC board. The PC-board pattern and parts overlay are on the CD-ROM. Make a PC board, or use the overlay to wire the circuit using perf-board. An etched and drilled PC board, with a silk-screened parts layout, is available from FAR Circuits.[1]

Any cabinet or enclosure that can accommodate the circuit board can be used. The POWER switch and POWER and CW LED indicators are the only front-panel items. J2 (KEYED LINE), J1 (AUDIO IN), J3 (+12 V DC POWER) and P1 (COMPUTER) are on the rear apron of the enclosure.

Alignment

There are three alignment adjustments, all of which are for the receive mode. Start by loading (and running) the software and turning the unit on. Switch the receiver to a dummy antenna to eliminate any interfering signals, and tune the receiver to a strong signal from a frequency calibrator or any other stable signal source. Carefully adjust the receiver for peak audio output. You may need a Y connector so the receiver can feed the interface and a speaker.

Connect a pair of headphones to the junction of the 0.1-µF capacitor and 10-kΩ resistor at pin 3 of U4. Adjust the Tune (R1) control on the PC board for the loudest signal. The filter is sharp, so make the adjustment carefully.

Set the PC board Level pot (R2) to midrange and adjust the VCO pot (R3) until the CW LED (DS1) comes on. Decrease the Level setting slightly (adjust the control in a counterclockwise direction) and readjust the VCO pot, if required, to cause the CW LED to light. Continue to reduce the Level setting in small steps, each time readjusting the VCO setting, until you reach the point where operation of the CW indicator becomes erratic.

Now turn the Level control back (clockwise) to just past the point where the LED comes on with no sign of erratic operation. The Level threshold setting is critical for best operation of the receive demodulator. If the control is advanced too far, the LED will trigger on background noise and copy will be difficult. If you reduce the setting too far, the interface will trigger erratically, even with a clean beat note. If you have a reasonably good CW receiver (CW bandwidth crystal filters and/or good audio filtering), you can back down the Level control until the LED stops flickering on all but the strongest noise pulses, but where it will still key reliably on a properly tuned CW signal.

Software Installation

The software for this project can be accessed from the CD-ROM. The distribution files include *MORSE2.EXE*, a sample set-up file (CW.DAT), a sample logging file (LOG.DAT), the HELP text file (CWHELP.DAT) and the program Quick-BASIC source code (MORSE2.BAS). To run the program log into the directory holding these files and type *MORSE <CR>*. The symbol <CR> stands for *Return* or *Enter*, depending on your keyboard.

The program menu permits you to enter or change the following items:

- SPEED—Select a transmitting speed from 5 to 60 WPM. The program auto-calibrates to your computer clock speed, and transmitting speeds are accurate to within 1%. On receive, the system automatically tracks the speed of the station you are copying up to 50 or 60 WPM.
- YOUR CALL—You can enter your call sign so you never have to type it in routine exchanges. The call can be changed at any time if you want to use the program for contests, special events, or any other situation where you will be using another call.
- OTHER CALL—If you enter the call of the station you are working (or would like to work), you can send all standard call exchanges at the beginning and end of a transmission with a single keystroke.
- CQ OPTIONS—Select one of two CQ formats. The "standard" format is a 3×3 call using your call sign. The program also lets you store a custom CQ format, which is useful for contests.
- MESSAGE BUFFERS—There are two message buffers. Either can be used for transmitting.
- SIDETONE—Select on or off and a frequency of 400 to 1200 Hz.
- WEIGHTING—Variable from 0.50 through 1.50.
- DEFAULT SETUP—All the information discussed up to this point can be saved into a default disk file (CW.DAT). These choices will then be selected whenever you boot the program. Any setup can be saved at anytime.
- LOGGING—The program supports a range of logging functions. It even includes the ability to check the log and let you know if you have worked that station before. If you have fully implemented the logging options, it will tell you the operator's name and QTH.
- HELP FILES—If you forget how to use a function or are using the program for the first time, you can call up on-screen HELP files that explain every function.

Notes

[1] See the CD-ROM for template. Use *TIS Find* program for address information.

AN EXPANDABLE HEADPHONE MIXER

Fig 19.40—The 3-channel headphone mixer is built on a small PC board. Lead length was kept to a minimum to aid stability.

From time to time, active amateurs find themselves wanting to listen to two or more rigs simultaneously with one set of headphones. For example, a DXer might want to comb the bands looking for new ones while keeping an ear on the local 2-m DX repeater. Or, a contester might want to work 20 m in the morning while keeping another receiver tuned to 15 m waiting for that band to open. There are a number of possible uses for a headphone mixer in the ham shack.

The mixer shown in **Figs 19.40** and **19.41** will allow simultaneous monitoring of up to three rigs. Level controls for each channel allow the audio in one channel to be prominent, while the others are kept in the background. Although this project was built for operation with three different rigs, the builder may vary the number of input sections to suit particular station requirements. This mixer was built in the ARRL Lab by Mark Wilson, K1RO.

CIRCUIT DETAILS

The heart of the mixer is an LM386 low-power audio amplifier IC. This 8-pin device is capable of up to 400-mW output at 8 Ω—more than enough for headphone listening. The LM386 will operate from 4- to 12-V dc, so almost any station power supply, or even a battery, will power it.

As shown in Fig 19.41, the input circuitry for each channel consists of an 8.2-Ω resistor (R1-R3) to provide proper termination for the audio stage of each transceiver, a 5000-Ω level control (R4-R6) and a 5600-Ω resistor (R7-R9) for isolation between channels. C1 sets the gain of the LM386 to 46 dB. With pins 1 and 8 open, the gain would be 26 dB. Feedback resistor R10 was chosen experimentally for minimum amplifier total harmonic distortion (THD). C2 and R11 form a "snubber" to prevent high-frequency oscillation, adding to amplifier stability. None of the parts values are particularly critical, except R1-R3, which should be as close to 8 Ω as possible.

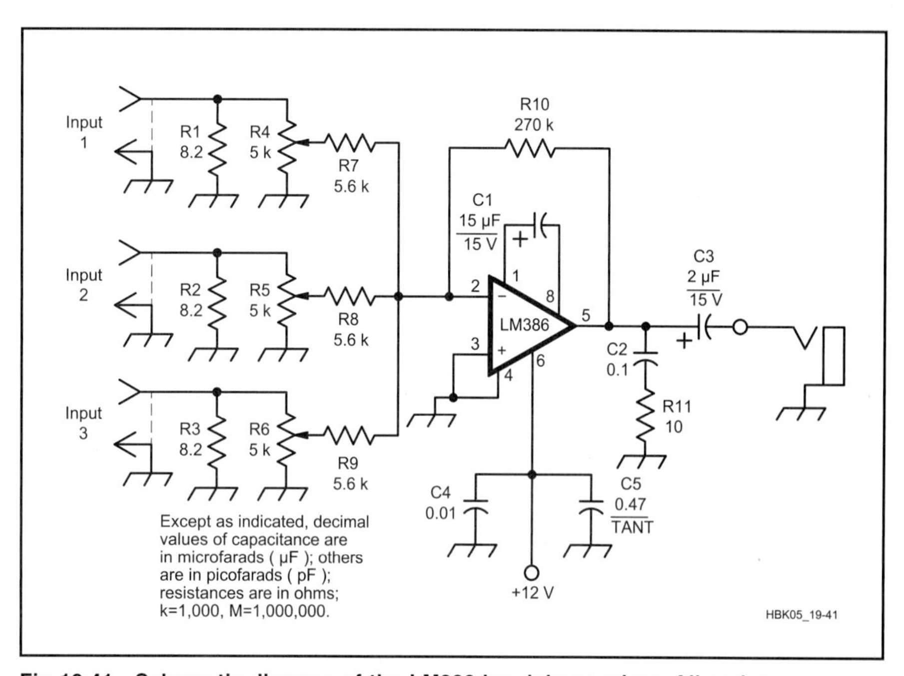

Fig 19.41—Schematic diagram of the LM386 headphone mixer. All resistors are 1/4 W. Capacitors are disc ceramic unless noted.

CONSTRUCTION

Most of the components are arranged on a small PC board.[1] Perfboard will work fine also, but some attention to detail is necessary because of the high gain of the LM386. Liberal use of ground connections, short lead lengths and a bypass capacitor on the power-supply line all add to amplifier stability.

The mixer was built in a small diecast box. Tantalum capacitors and 1/4-W resistors were used to keep size to a minimum. The '386 IC is available from RadioShack (cat. no. 276-1731). A 0.01-µF capacitor and a ferrite bead on the power lead help keep RF out of the circuit. In addition, shielded cable is highly recommended for all connections to the mixer. The output jack is wired to accept stereo headphones.

Output power is about 250 mW at 5% THD into an 8-Ω load. The output waveform faithfully reproduces the input waveform, and no signs of oscillation or instability are apparent.

Notes

[1]See CD-ROM for template.

A SIMPLE 10-MINUTE ID TIMER

This project was originally described in "Hints and Kinks" in the November 1993 issue of *QST* by John Conklin, WDØO. It is an update to an earlier WDØO design for which parts are no longer available.

This simple and effective timer can be built in an evening and uses inexpensive and easily obtained parts. Its timing cycle is independent of supply voltage and resets automatically upon power-up, as well as at the end of each cycle.

Construction and Adjustment

Assembly is straightforward and parts layout is not critical. The circuit of **Fig 19.42** can be built on a small piece of perfboard and housed in an inexpensive enclosure. To calibrate the time, set R1, TIME ADJ, at midpoint initially, and then adjust it by trial and error to achieve a 10-minute timing cycle. The buzzer will sound for about 1 second at the end of each cycle.

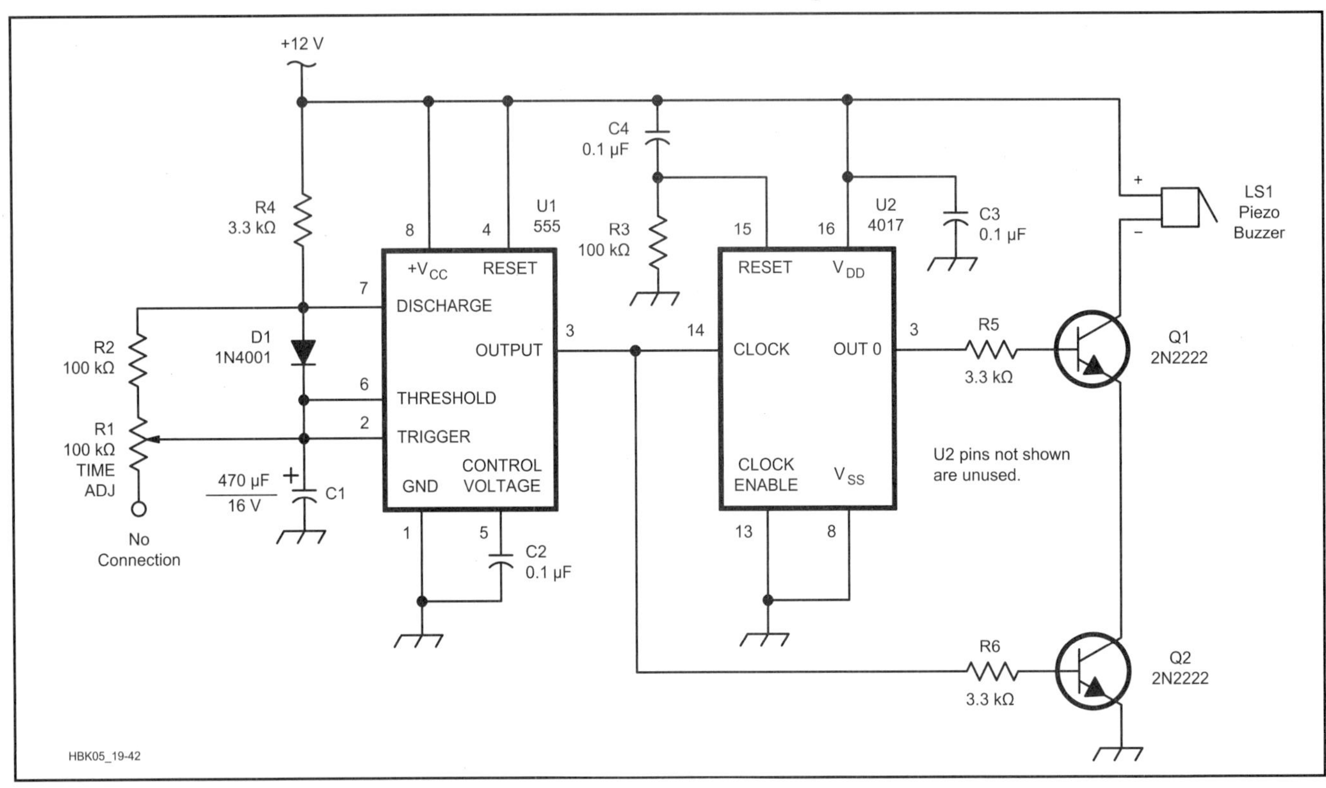

Fig 19.42—The new and improved 10-minute reminder uses easy-to-get components. Part numbers in parentheses are RadioShack; equivalent parts can be substituted. Resistors are 1/4-W, 5% or 10% tolerance units.

C1—470-µF electrolytic, 16 V or more (272-957).
C2, C3, C4—0.1 µF, 16 V or more (272-109).
D1—1N4001, 50 PIV, 1 A (276-1101).
LS1—12-V piezo buzzer (273-074).
Q1, Q2—N2222 or MPS2222A (276-2009).
R1—100-k trimmer potentiometer (271-284).
R2, R3—100 k (271-1347).
R4, R5, R6—3.3 k (271-1328).
U1—555 timer IC (276-1723).
U2—4017 decade counter IC (276-2417).

AUDIO BREAK-OUT BOX

Two integrated circuits and a small PC board are all you need to solve the problem of feeding one receiver into several add-ons, such as a TNC, a PC interface or a speaker. Ben Spencer, G4YNM, described this project in March 1995 *QST*. It takes the audio output from a receiver and applies it to the inputs of four identical, independent, low-level AF amplifiers and one high-level (1-W output) AF amplifier.

See **Fig 19.43**. Each low-level output channel can provide up to 20 dB of gain that's independently adjustable. You can apply audio to each of your accessories at a selected level without changing the level to the other accessories. In addition you can set the level to the speaker independently. Turn the speaker volume up to tune in the signal, and then turn the speaker volume down once tuning is finished and the mode is operating.

Circuit Description

Four identical low-level channels, each feeding an amplifier (U1A, B, C and D), are shown in Fig 19.43. Using the top channel as an example, C1 connects the input jack J1 to the noninverting input of U1A. R3 and R4 set U1A's voltage gain. R4 is the gain control, and when set fully clockwise (maximum resistance), the amplifier's gain is 10 (20 dB). At a counterclockwise (minimum resistance) setting, the amplifier's gain is 1 (0 dB).

The lower cut-off frequency (set by C2

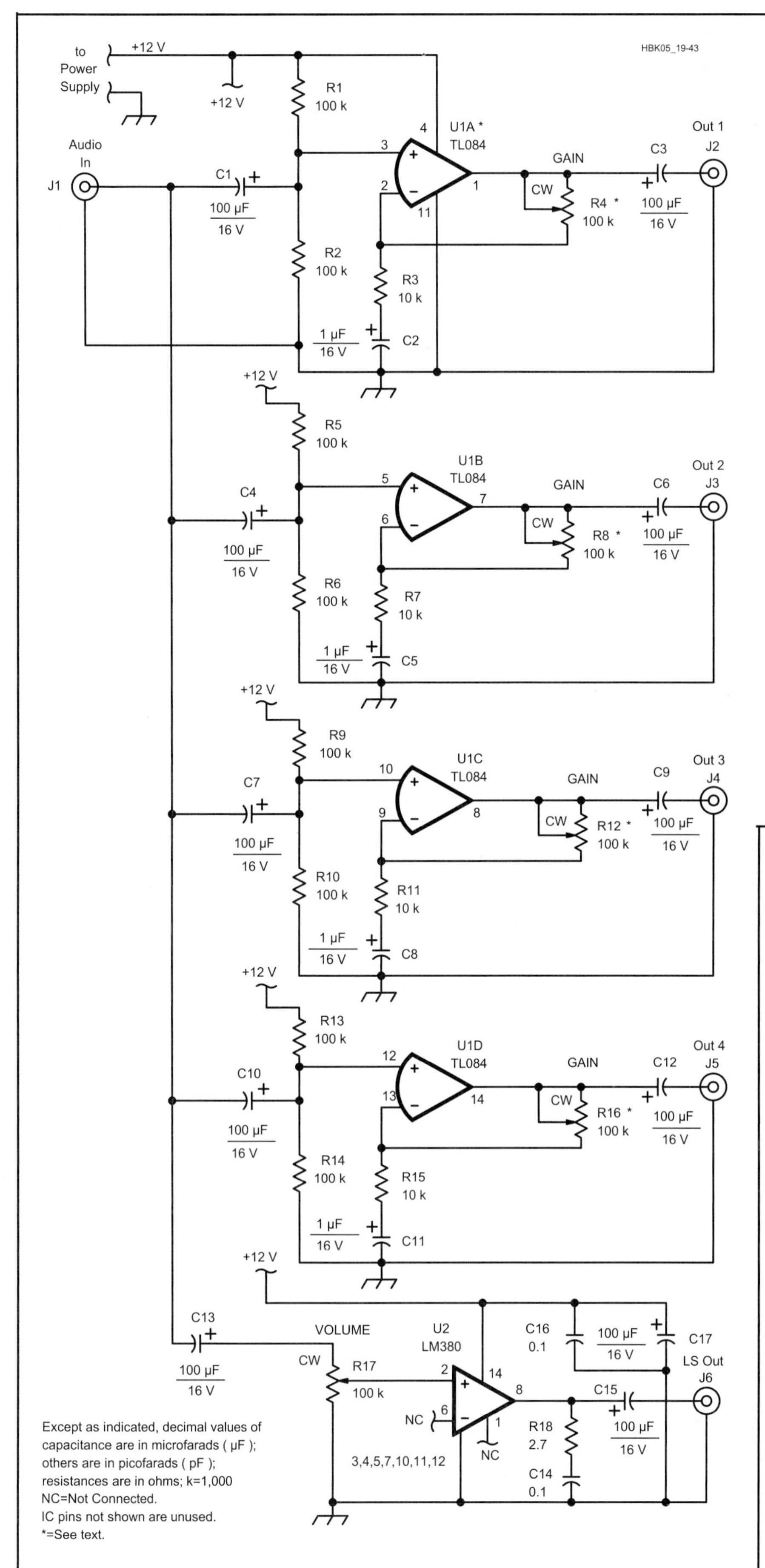

Fig 19.43—This audio break-out box requires less than 500 mA from a 12-VDC supply. All resistors are 1/4-W, 5%-tolerance carbon composition or film units unless otherwise specified. RS numbers in parentheses are RadioShack stock numbers.

C1, C3, C4, C6, C7, C9, C10, C12, C13, C15, C17—100 µF, 16-V radial electrolytic or tantalum (RS 272-1028).

C2, C5, C8, C11—1 µF, 16-V radial electrolytic or tantalum (RS 272-1434).

C14, C16—0.1 µF, 50 V disc ceramic (RS 272-135).

R1, R2, R5, R6, R9, R10, R13, R14—100 k.

R3, R7, R11, R15—10 kΩ.

R4, R8, R12, R16, R17—100-kΩ log or audio taper, panel-mount potentiometer (RS 271-1722) or PC-board vertical-mount trimmer potentiometer; see text.

R18—2.7 Ω, 1/2 W.

U1—TL084, TL074, or LM324 quad op amp (RS 276-1711).

U2—LM380N 2-W audio power amplifier. The LM380 is available in several packages. Be sure to use the 14-pin DIP if you are going to build this project on the PC board from FAR or from the ARRL template.

Misc: Single-sided PC board (see Note 1), enclosure, knobs, IC sockets, input and outpu t connectors of choice, hook-up wire.

and R3) is 16 Hz. The upper cut-off frequency of each channel is well beyond the audio frequency range. Each channel's output is dc isolated from its load; for example, U1A's output is dc isolated by C3.

R17 is the volume control for AF power amplifier U2. This stage will drive a low-impedance load such as a loudspeaker (4 to 16 Ω) at a level up to 1 W.

Construction

A single-sided PC board is available,[1] but the unit will work equally well built on perf-board. A template is on the CD-ROM and includes a PC board layout and a parts layout. This parts layout also can be used as a guide for construction on perf-board. The PC board directly accepts vertical and horizontal-mount single-turn potentiometers, but you can run wires from the mounting holes to front-panel-mount potentiometers. Since the project uses high-gain audio circuits, enclose it in a metal box. Place the input and output jacks on the rear panel to keep the interconnecting leads out of the way.

Checkout

After rechecking your wiring and soldering, connect the circuit to a 12-V power supply. The current drawn should be less than 50 mA when no audio is applied. Connect J1 to the AF output of your receiver and a speaker to J6. Adjust R17, VOLUME, for a comfortable listening level. Next check the operation of the low-power outputs by connecting J2, J3, J4 and J5 to a small earplug. Vary the four gain controls to check their operation. Each gain control can now be set to provide the audio level needed for each add-on.

Notes

[1]PC boards are available for $6, plus $1.50 shipping, from FAR Circuits. Use *TIS Find* program for latest address information.

AN SWR DETECTOR AUDIO ADAPTER

This SWR detector audio adapter is designed specifically for blind or vision-impaired amateurs, but anyone can use it. The basic circuit can be adapted to any application where you want to use an audio tone rather than a meter to give an indication of the value of a dc voltage.

Usually a meter (or meters) is used to display SWR by measuring the feed line forward and reflected voltages. This adapter generates two tones with frequencies that are proportional to these voltages. The tones are fed to a pair of stereo headphones (the miniature types are ideal) so one ear hears the forward-voltage tone and the other ear hears the reflected-voltage tone. Ben Spencer, G4YNM, described this system in the July 1994 *QST*. He connected the forward voltage tone to his left earphone and reverse voltage tone to his right earphone. Thus, tuning up a transmitter is simply a matter of tuning for the highest pitched tone in the left ear, and the lowest pitched tone in the right ear.

The PC board can be installed in existing SWR detectors, and the forward and reflected voltages obtained by tapping into the lines that currently connect the voltage sensors to the existing meters or meter selector switch.

Circuit Description

The audio-adapter circuit is shown in **Fig 19.44**. Each half of the adapter circuit operates identically. Most SWR detectors consist of two RF voltage sensors, one for forward voltage and one for reverse. These voltages are diode-rectified. The resulting dc voltages are fed to meters that indicate relative forward and reflected power. With

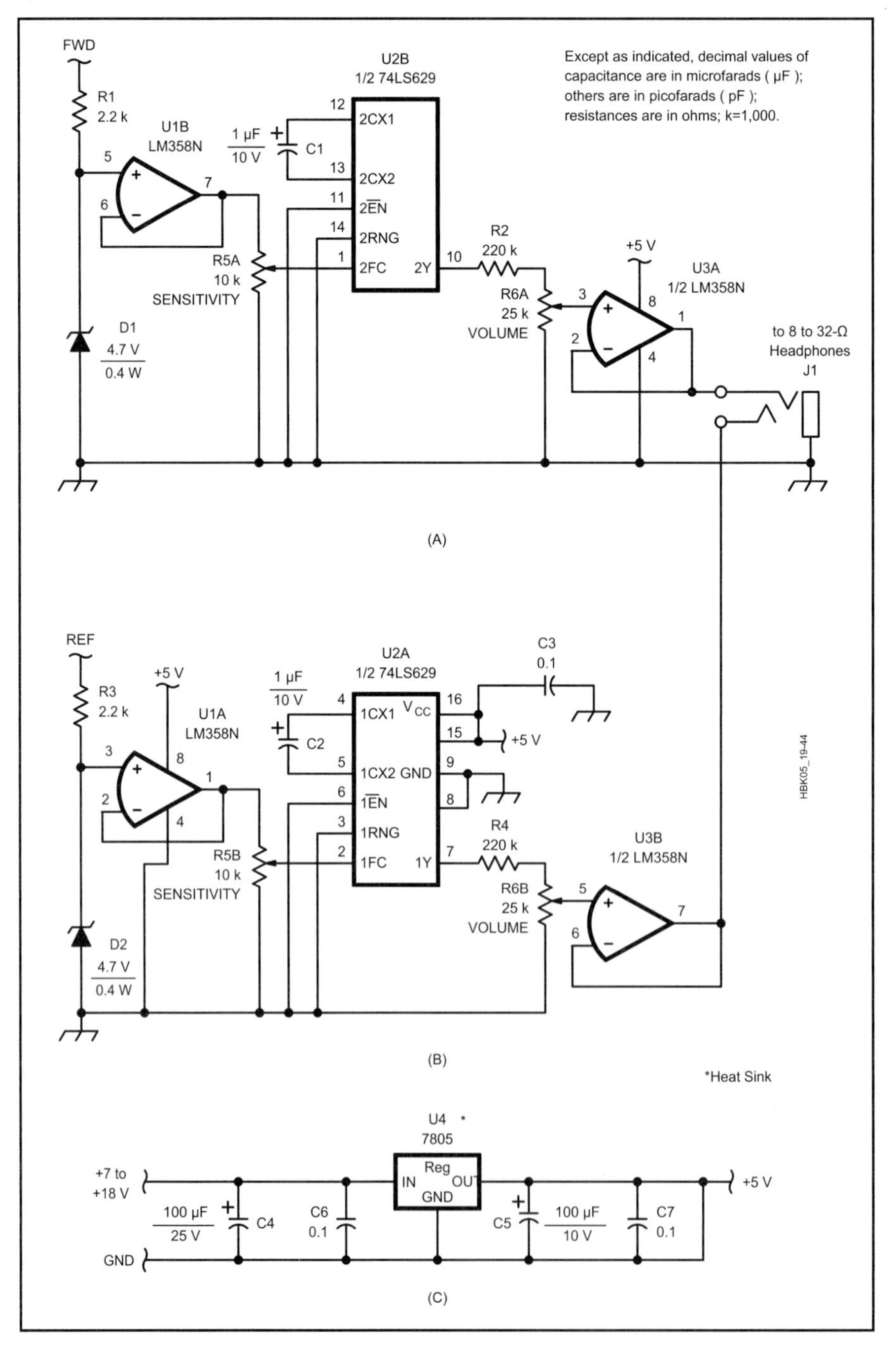

Fig 19.44—Schematic of the SWR detector audio-adapter circuit. Unless otherwise specified, resistors are 1/4-W, 5%-tolerance carbon-composition or film units. All capacitors are disc ceramic unless otherwise stated. The circuits of A and B are identical, each driving one earphone of an 8- to 32-Ω stereo headset. At A, the forward-voltage circuit; at B, the reflected-voltage circuit. A voltage regulator that provides 5-V dc is shown at C.

R5A, R5B—10-kΩ horizontal-mount trimmer potentiometer; optionally, a 25-kΩ dual-gang, panel-mount potentiometer can be used.

R6—25-kΩ dual-gang, panel-mount potentiometer.

U1, U3—LM358N dual op amp (available from Jameco. Use *TIS Find* program for address information or substitute an NTE928M (available from Hosfelt Electronics).

U2—74LS629 dual VCO (available from Jameco) or a NTE74LS629 (available from Hosfelt Electronics).

Misc: PC board, stereo headphone jack, 8- to 32-Ω stereo headphones, mounting hardware.

this circuit, the forward and reflected voltages are applied to the audio adapter board and drive voltage-controlled oscillators (VCOs).

The forward dc voltage from the SWR detector is routed to R1, buffered by U1B, fed to SENSITIVITY control R5A and applied to VCO U2B. As the voltage on pin 1 of U2B increases, so does the frequency of the tone output at U2B pin 10.

Adjusting the Sensitivity control sets the range of audio tones produced by the audio adapter. This signal is fed via VOLUME control R6A to the audio amplifier (U3A) to drive the left headphone. Zener diode Dl limits the maximum input voltage, partly to protect U1B, but also to limit the upper VCO frequency to about 3 kHz.

Without any dc input, each VCO runs at a low frequency (approximately 380 Hz) to tell you the unit is operating. Increased voltage on the transmission line—even from a low-power transmitter—is sufficient to cause the tone frequency to increase noticeably. As the voltage decreases, so does the frequency of the tone.

Construction

A single-sided PC board and template package are available.[1,2] The PC board is small enough to fit inside most existing SWR detectors and the circuit can be battery operated if required. Mount a stereo headphone jack on the SWR detector's front panel to accept the headphone plug.

R6A and R6B are parts of a dual-section, panel-mount potentiometer. R5A and R5B are PC-board mounted trimmer potentiometers. For those who want a panel-mounted SENSITIVITY control, a dual-gang potentiometer can be substituted for R5A and R5B.

Testing and Calibration

Once the unit is installed in an SWR detector, connect a 5-V power supply to the audio adapter board. When power is applied you should hear two identical low frequency tones in the headphones. Adjust the VOLUME controls to provide a comfortable listening level.

Next, connect your transmitter to a dummy load via the SWR detector. When you key your transmitter, the tone in the left earpiece should increase in frequency quite dramatically, representing increasing forward power. Theoretically, with a matched line and load, there should be no reflected voltage and therefore, the right-headphone tone shouldn't change. In all probability, however, the tone frequency will increase, but only slightly.

If you use an antenna tuner for matching your antenna system, you'll hear the two tones change frequency according to the degree of mismatch. The best match is indicated by the forward headphone tone reaching its maximum frequency while the reflected (right) headphone tone frequency decreases to its minimum.

Notes

[1]PC boards are available for $2.50, plus $1.50 shipping, from FAR Circuits. Use *TIS Find* program for latest address information.

[2]See CD-ROM for template.

PC VOLTMETER AND SWR BRIDGE

Personal computers are very good at doing arithmetic. To use this capability around the shack, the first thing to do is convert whatever you want to measure (voltage, power, SWR) to numbers. Next you have to find a way to put these numbers into your computer. Paul Danzer, N1II, took a single chip A/D (analog to digital converter) and built this unit to connect to a computer printer port. Construction and testing is just a few evenings' work. The software to run the chip is accessible from the *Handbook* CD-ROM.

Circuit Description

The circuit consists of a single-chip A/D converter, U2, and a DB-25 male plug (**Fig 19.45**). Pins 2 and 3 are identical volt-

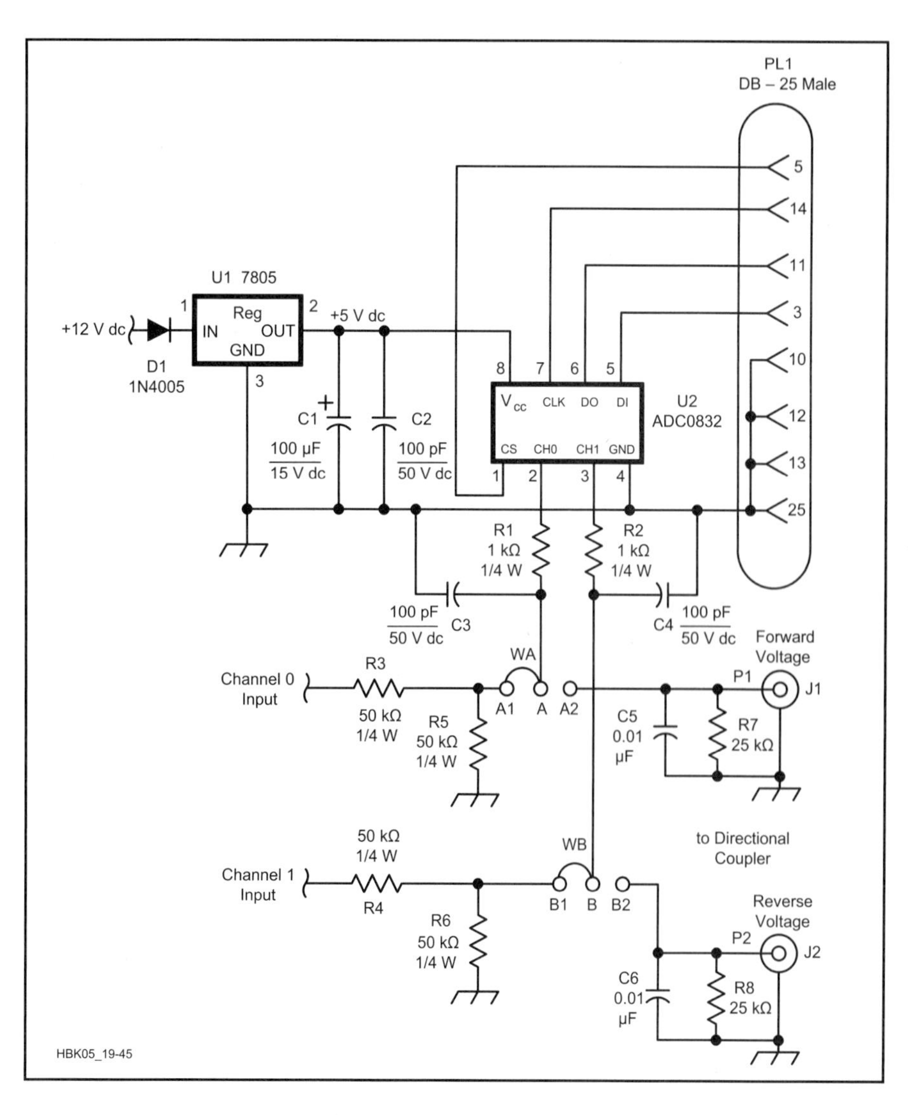

Fig 19.45—Only two chips are used to provide a dual-channel voltmeter. PL1 is connected through a standard 25-pin cable to a computer printer port. U2 requires an 8-pin IC socket. All resistors are 1/4 W. You can use the A/D as an SWR display by connecting it to a sensor such as the one used in the Tandem Match described in this chapter (see text). A few more resistors are all that are needed to change the voltmeter scale. The 50 kΩ resistors form 2:1 voltage dividers, extending the voltmeter scale (on both channels) to almost 10-V dc.

Fig 19.46—Construction of this model took only one evening. No special tools are required. All parts except U2 are available from most suppliers as well as RadioShack. The A/D converter chip can be purchased from any National Semiconductor dealer such as Digi-Key. Use *TIS Find* program for address information.

age inputs, with a range from 0 to slightly less than the supply voltage V_{CC} (+5 V). R1, R2, C3 and C4 provide some input isolation and RF bypass. There are four signal leads on U2—DO is the converted data from the A/D out to the computer, DI and CS are control signals from the computer and CLK is a computer generated clock signal sent to pin 7 of U2.

The +5 V supply is obtained from a +12 V source and regulator U1. One favorite accident, common in many ham shacks, is to connect power supply leads backwards. Diode D1 prevents any damage from this action. Current drain is usually less than 20 mA, so any 5-V regulator may be used for U1. The power supply ground, circuit ground and computer ground are all tied together.

In this form the circuit gives two identical dc voltmeters. To extend the range, a 2:1 divider, using 50-kΩ resistors, is shown. Resistor accuracy is not important, since the circuit is calibrated in the accompanying software.

The breadboard circuit, built on a universal PC board (RadioShack 276-150), is shown in **Fig 19.46** The voltage regulator is on the top left and the converter chip, U2, in an 8-pin socket. Power is brought in through a MOLEX plug. Signal input and ground are on the wire stubs. Two strips of soft aluminum, bent into L-shapes, hold the male DB-25 connector (RadioShack 276-1547) to the PC board.

Use It As An SWR Bridge

Most analog SWR measuring devices use a meter, which has a nonlinear scale calibration. An SWR of 3:1 is usually close to center scale, and values above this are rarely printed. To use the PC voltmeter as an SWR bridge indicator, move jumpers WA and WB from the A1 and B1 positions to the A2 and B2 positions. Disconnect the cathodes (banded end) of the diode detectors in your SWR bridge and connect them to J1 and J2.

The current that flows out of these diodes, and into J1 and J2, goes through the 25-kΩ resistors R7 and R8, to provide voltages of less than 5 V. These voltages are proportional to the forward and reverse voltages developed in the directional coupler. The software in the PC takes the sum and difference of these forward and reverse voltages, and calculates the SWR.

Software

The software, including a voltmeter function and an SWR function, is written in GW-BASIC and saved as an ASCII file. Therefore you can read it on any word processor, but if you modify it make sure you resave it as an ASCII file. It can be imported into QBasic and most other BASIC dialects.

It was written to be understandable rather than to be most efficient. Each line of basic code has a comment or explanation. It can be modified for most computers. The printer port used is LPT1, which is at a hex address of 378, 379 and 37A. If you wish to use LPT2 (printer port 2) try changing these addresses to 278, 279 and 27A.

Gary Sutcliffe, W9XT, wrote a small BASIC program to help you find the addresses of your printer ports. Run FINDLPT.BAS, which is included with the *Handbook* software accessible from the CD-ROM.

The CONV.BAS program was written to run on computers as slow as 4.7-MHz PC/XTs. If you get erratic results with a much faster computer, set line 1020 (CD=1) to a higher value to increase the width of the computer generated clock pulses.

The CONV.BAS program operates by first reading the value of voltage at point A into the computer, followed by the voltage at point B. It then prints on the screen these two values, and computes their sum and difference to derive the SWR. If you use the project as a voltmeter, simply ignore the SWR reading on the screen or suppress it by deleting lines 2150, 2160 and 2170. If the two voltages are very close to each other (within 1 mV) the program declares a bad reading for SWR.

Calibration

Lines 120 and 130 in the program independently set the calibration for the two voltage inputs. To calibrate a channel, apply a known voltage to the input point A. Read the value on the PC screen. Now multiply the constant in line 120 by the correct value and divide the result by the value you previously saw on the screen. Repeat the procedure for input point B and line 130.

THE TANDEM MATCH—AN ACCURATE DIRECTIONAL WATTMETER

Most SWR meters are not very accurate at low power levels because the detector diodes do not respond to low voltage in a linear fashion. This design uses a compensating circuit to cancel diode nonlinearity. It also provides peak detection for SSB operation and direct SWR readout that does not vary with power level. **Fig 19.47** is a photo of the completed project. The following information is condensed from an article by John Grebenkemper, KI6WX, in January 1987 *QST*. Some modifications by KI6WX were detailed in the "Technical Correspondence" column of July 1993 *QST*. A PC Board is available from FAR Circuits.[1]

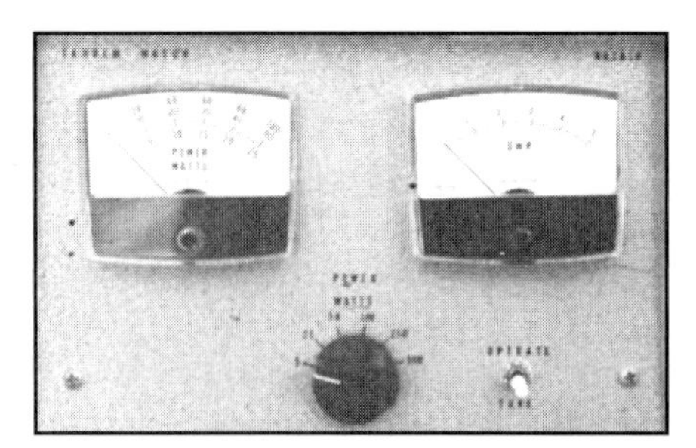

Fig 19.47—The Tandem Match uses a pair of meters to display net forward power and true SWR simultaneously.

Circuit Description

A directional coupler consists of an input port, an output port and a coupled port. Ideally, a portion of the power flowing from the input to the output appears at the coupled port, but *none* of the power flowing from the output to the input appears at the coupled port.

The coupler used in the Tandem Match consists of a pair of toroidal transformers connected in tandem. The configuration was patented by Carl G. Sontheimer and Raymond E. Fredrick (US Patent no. 3,426,298, issued February 4, 1969). It has been described by Perras, Spaudling and others. With coupling factors of 20 dB greater, this coupler is suitable to sample both forward and reflected power.

The configuration used in the Tandem Match works well over the frequency range of 1.8 to 54 MHz, with a nominal coupling factor of 30 dB. Over this range, insertion loss is less than 0.1 dB. The coupling factor is flat to within $\pm$0.1 dB from 1.8 to 30 MHz, and increases to only $\pm$0.3 dB at 50 MHz. Directivity exceeds 35 dB from 1.8 to 30 MHz and exceeds 26 dB at 50 MHz.

The low-frequency limit of this directional coupler is determined by the inductance of the transformer secondary windings. The inductive reactance should be greater than 150 Ω (three times the line characteristic impedance) to reduce insertion loss. The high-frequency limit of this directional coupler is determined by the length of the transformer windings. When the winding length approaches a significant fraction of a wavelength, coupler performance deteriorates.

The coupler described here may overheat at 1500 W on 160 m (because of the high circulating current in the secondary of T2). The problem could be corrected by using a larger core or one with greater permeability. A larger core would require longer windings; that option would decrease the high-frequency limit.

Most amateur directional wattmeters use a germanium-diode detector to minimize the forward voltage drop. Detector voltage drop is still significant, however, and an uncompensated diode detector does not respond to small signals in a linear fashion. Many directional wattmeters compensate for diode nonlinearity by adjusting the meter scale.

The effect of underestimating detected power worsens at low power levels. Under these conditions, the ratio of the forward power to the reflected power is overestimated because the reflected power is always less than the forward power. This results in an instrument that underestimates SWR, particularly as power is educed. A directional wattmeter can be checked for this effect by measuring SWR at several power levels. The SWR should be independent of power level.

The Tandem Match uses a feedback circuit to compensate for diode nonlinearity. Transmission-line SWR is displayed on a linear scale. Since the displayed SWR is not affected by changes in transmitter power, a matching network can be simply adjusted to minimize SWR. Transmatch adjustment requires only a few watts.

Construction

The schematic diagram for the Tandem Match is shown in **Fig 19.48**. The circuit is designed to operate from batteries and draws very little power. Much of the circuitry is of high impedance, so take care to isolate it from RF fields. House the circuit in a metal case. Most problems in the prototype were caused by stray RF in the op-amp circuitry.

The schematic shows two construction options. Connect jumpers W1, W2 and W3 to use the circuit as it was originally designed (with two 9-V batteries and TLC27L4 or TLC27M4 op amps). By omitting these jumpers, any quad FET-input op amps can be used instead of the TLC27x4s. Possible substitutes include the TL064, TL074, TL084, LF347 and LF444. In that case you should also omit the 9-V batteries and the automatic turn-on circuitry of Q1, Q2 and Q3 (everything to the left of the jumpers on the top row of the diagram). Now you will have to connect an external + 15 V supply between the + V line and chassis ground and a –15 V supply to the –V line.

The FAR Circuits Tandem Match circuit board is double sided, but does not have plated-through holes. The component side is mainly the chassis and circuit ground planes, although there are a few signal traces. You will have to install "jumper posts" in a few locations, and solder them to both sides of the board to connect these traces. Carefully follow the schematic diagram and parts-placement diagram supplied with the board to identify these "posts." Check the board carefully to ensure that none of the ground traces pass too close to a circuit lead. You may have to scrape a bit of foil away from a few places around the component holes. This is easy with an X-ACTO knife.

The trimmer pots must be square multiturn units with top adjustment screws for use with the FAR Circuits board. Mount the ferrite beads so they don't touch any board trace; the beads have sufficient leakage to cause problems in the high impedance parts of the circuit. Before mounting the SO-239 connectors to the circuit board, enlarge the center location holes to $\frac{5}{8}$-inch diameter to accept the connector body. The components connected to the SO-239 are soldered directly between the center pin and the board traces.

Directional Coupler

The directional coupler is constructed in its own small ($2\frac{3}{4} \times 2\frac{3}{4} \times 2\frac{1}{4}$-inch) aluminum box (see **Fig 19.49**). Two pairs of SO-239 connectors are mounted on opposite sides of the box. A piece of PC board is run diagonally across the box to improve coupler directivity. The pieces of RG-8X coaxial cable pass through holes in the PC board. (Note: Some brands of "mini 8" cable have extremely low breakdown voltage ratings and are unsuitable to carry even 100 W when the SWR exceeds 1:1. See "High-Power Operation" for details of a coupler made with RG-8 cable.)

Begin by constructing T1 and T2, which are identical except for their end connections. (Refer to Fig 19.49.) The primary for each transformer is the center conduc-

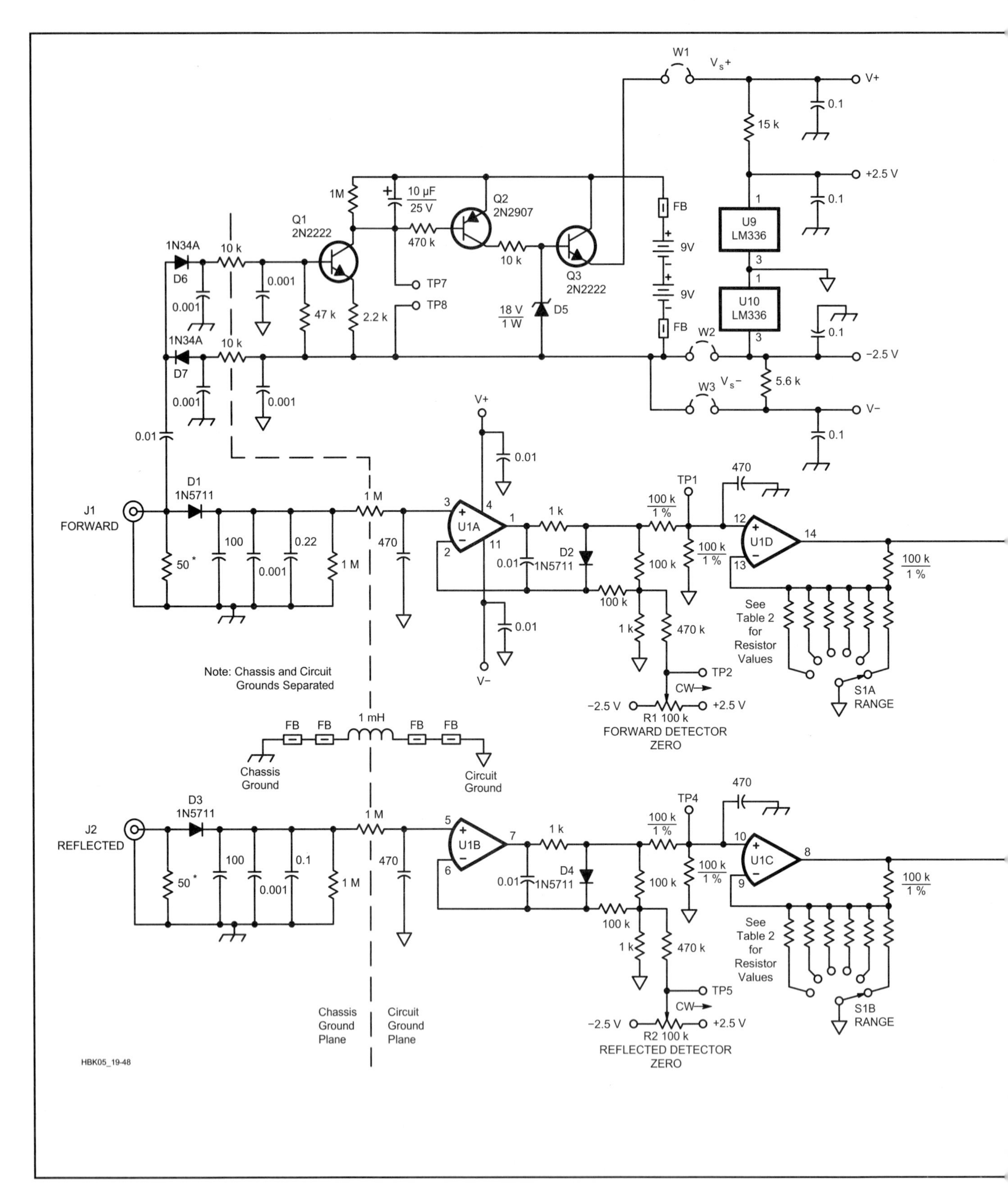

Fig 19.48—Schematic diagram of the Tandem Match directional wattmeter. Parts identified as RS are from RadioShack. Contact information for parts suppliers can be found using the *TIS Find* program.

D1-D4—1N5711.
D6, D7—1N34A or 1N271.
D8-D14-1N914.
FB-Ferrite bead, Amidon FB-73-101 or equiv.
J1, J2—SO-239 connector.
J3, J4—Open-circuit jack.
M1, M2—1 mA panel meter.
Q1, Q3, Q4—2N2222 metal case only.
Q2—2N2907 metal case or equiv.
R1, R2, R5—100-kΩ, 10-turn cermet Trimpot.
R3, R4—10-kΩ, 10-turn, cermet Trimpot.
U1-U3—TLC27M4 op amp.
U4—TLC27L2 or TLC27M2.
U5-U7—CA3146.
U9, U10—LM336.

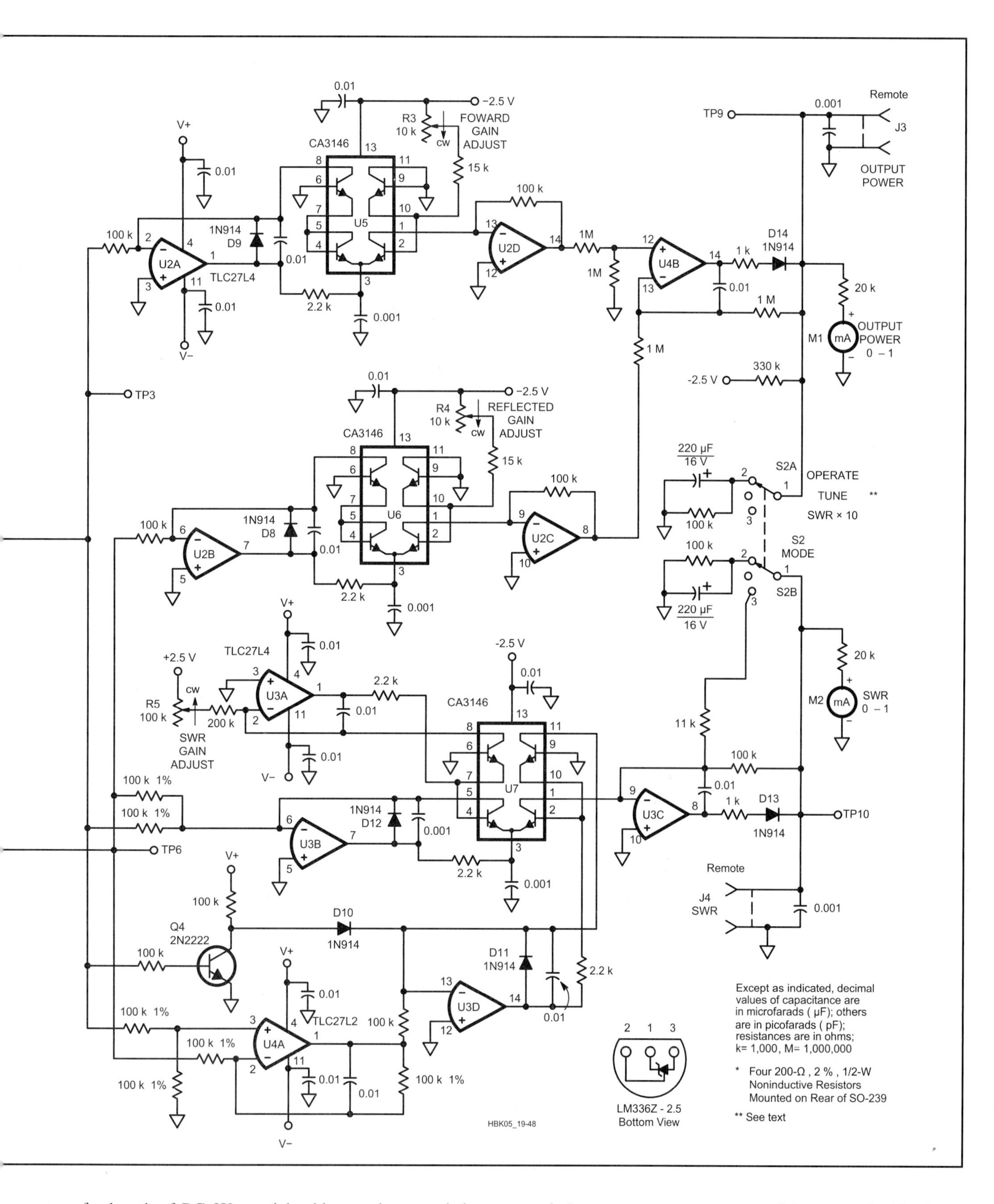

tor of a length of RG-8X coaxial cable. Cut two cable lengths sufficient for mounting as shown in the figure. Strip the cable jacket, braid and dielectric as shown. The cable braid is used as a Faraday shield between the transformer windings, so it is only grounded at one end. *Important—connect the braid only at one end or the directional-coupler circuit will not work properly!* Wind two transformer secondaries, each 31 turns of #24 enameled wire on a T-50-3 iron-powder core. Slip each core over one of the prepared cable pieces (including both the shield and the outer insulation). Mount and connect the transformers as shown in Fig 19.48, with the wires running through separate holes in the copper-clad PC board.

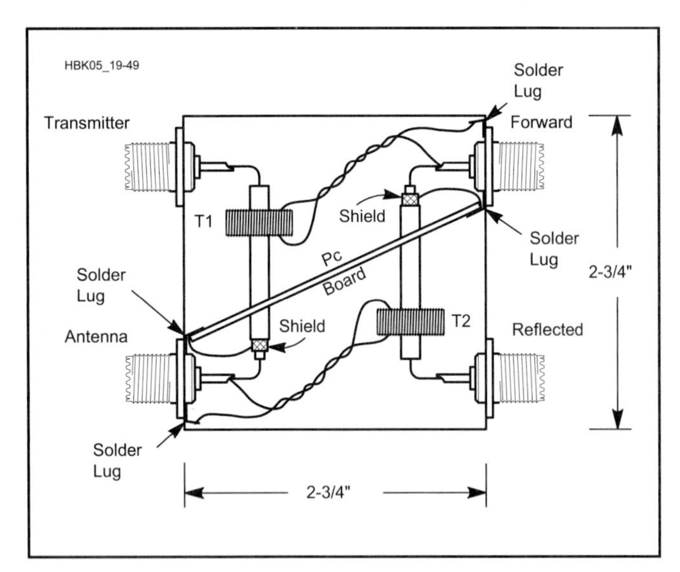

Fig 19.49—Construction details for the directional coupler. A metal case is required.

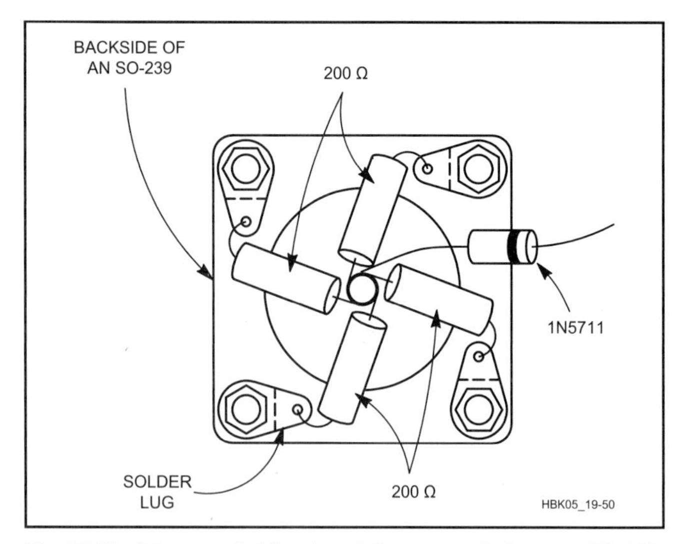

Fig 19.50—The parallel load resistors mounted on an SO-239 connector. Four 200-Ω resistors are mounted in parallel to provide a 50-Ω detector load.

The directional coupler can be mounted separately from the rest of the circuitry if desired. If so, use two coaxial cables to carry the forward- and reflected-power signals from the directional coupler to the detector inputs. Be aware, however, that any losses in the cables will affect power readings.

This directional coupler has not been used at power levels in excess of 100 W. For more information about using Tandem Match at high power levels, see "High-Power Operation."

Detector and Signal-processing Circuits

The detector and signal-processing circuits were constructed on a perforated, copper-clad circuit board. These circuits use two separate grounds—*it is extremely important to isolate the grounds as shown in the circuit diagram*. Failure to do so may result in faulty circuit operation. Separate grounds prevent RF currents on the cable shield from affecting the op-amp circuitry.

The directional coupler requires good 50-Ω loads. They are constructed on the back of the female UHF chassis connectors where the cables from the directional coupler enter the wattmeter housing. Each load consists of four 200-Ω resistors connected from the center conductor of the UHF connector to the four holes on the mounting flange, as shown in **Fig 19.50**. The detector diode is then mounted from the center conductor of the connector to the 100-pF and 1000-pF bypass capacitors, which are located next to the connector. The response of this load and detector combination measures flat to beyond 500 MHz.

Schottky-barrier diodes (type 1N5711) were used in this design because they were readily available. Any RF-detector diode with a low forward voltage drop (less than 300 mV) and reverse breakdown voltage greater than 30 V could be used. (Germanium diodes could be used in this circuit, but performance will suffer. If germanium diodes are used, reduce the values of the detector-diode and feedback-diode load resistors by a factor of 10.)

The rest of the circuit layout is not critical, but keep the lead lengths of 0.001- and 0.01-µF bypass capacitors short. The capacitors provide additional bypass paths for the op-amp circuitry.

D6 and D7 form a voltage doubler to detect the presence of a carrier. When the forward power exceeds 1.5 W, Q3 switches on and stays on until about 10 seconds after the carrier drops. (A connection from TP7 to TP9 forces the unit on, even with no carrier present.) The regulated references of +2.5 V and –2.5 V generated by the LM334 and LM336 are critical. Zener-diode substitutes would significantly degrade performance.

The four op amps in U1 compensate for nonlinearity of the detector diodes. D1-D2 and D3-D4 are the matched diode pairs discussed above. A RANGE switch selects the meter range. (A six-position switch was used here because it was handy.) The resistor values for the RANGE switch are shown in **Table 19.2**. Full-scale input power gives an output at U1C or U1D of 7.07 V. The forward- and reflected-power detectors are zeroed with R1 and R2.

The forward- and reflected-detector voltages are squared by U2, U5 and U6 so that the output voltages are proportional to forward and reflected power. The gain constants are adjusted using R3 and R4 so that an input of 7.07 V to the squaring circuit gives an output of 5 V. The difference between these two voltages is used by U4B to yield an output that is proportional to the power delivered to the transmission line. This voltage is peak detected (by an RC circuit connected to the OPERATE position of the MODE switch) to indicate and hold the maximum power measurement during CW or SSB transmissions.

SWR is computed from the forward and reflected voltages by U3, U4 and U7. When no carrier is present, Q4 forces the

Table 19.2
Performance Specifications for the Tandem Match

Power range:	1.5 to 1500 W
Frequency range:	1.8 to 54 MHz
Power accuracy:	Better than ±10% (±0.4 dB)
SWR accuracy:	Better than ±5%
Minimum SWR:	Less than 1.05:1
Power display:	Linear, suitable for use with either analog or digital meters
SWR display:	Linear, suitable for use with either analog or digital meters
Calibration:	Requires only an accurate voltmeter

SWR reading to be zero (that is, when the forward power is less than 2% of the full-scale setting of the RANGE switch). The SWR computation circuit gain is adjusted by R5. The output is peak detected in the OPERATE mode to steady the SWR reading during CW or SSB transmissions.

Transistor arrays (U5, U6 and U7) are used for the log and antilog circuits to guarantee that the transistors will be well matched. Discrete transistors may be used, but accuracy may suffer.

A three-position toggle switch selects the three operating modes. In the OPERATE mode, the power and SWR outputs are peak detected and held for a few seconds to allow meter reading during actual transmissions. In the TUNE mode, the meters display instantaneous output power and SWR.

A digital voltmeter is used to obtain more precise readings than are possible with analog meters. The output power range is 0 to 5 V (0 V = 0 W and 5 V = full scale). SWR output varies from 1 V (SWR = 1:1) to 5 V (SWR = 5:1). Voltages above 5 V are unreliable because of voltage limiting in some of the op amp circuits.

Calibration

The directional wattmeter can be calibrated with an accurate voltmeter. All calibration is done with dc voltages. The directional-coupler and detector circuits are inherently accurate if correctly built. To calibrate the wattmeter, use the following procedure:

1) Set the MODE switch to TUNE and the RANGE switch to 100 W or less.
2) Jumper TP7 to TP8. This turns the unit on.
3) Jumper TP1 to TP2. Adjust R1 for 0 V at TP3.
4) Jumper TP4 to TP5. Adjust R2 for 0 V at TP6.
5) Adjust R1 for 7.07 V at TP3.
6) Adjust R3 for 5.00 V at TP9, or a full-scale reading on M1.
7) Adjust R2 for 7.07 V at TP6.
8) Adjust R4 for 0 V at TP9, or a zero reading on M1.
9) Adjust R2 for 4.71 V at TP6.
10) Adjust R5 for 5.00 V at TP10, or a full-scale reading on M2.
11) Set the RANGE switch to its most sensitive scale.
12) Remove jumpers from TP1 to TP2 and TP4 to TP5.
13) Adjust R1 for 0 V at TP3.
14) Adjust R2 for 0 V at TP6.
15) Remove jumper from TP7 to TP8.

This completes the calibration procedure. This procedure has been found to equal calibration with expensive laboratory equipment. The directional wattmeter should now be ready for use.

Table 19.3
Range-Switch Resistor Values

Full-Scale Power Level (W)	*Range Resistor (1% Precision) (kΩ)*
1	2.32
2	3.24
3	4.02
5	5.23
10	7.68
15	9.53
20	11.0
25	12.7
30	15.0
50	18.7
100	28.7
150	37.4
200	46.4
250	54.9
300	63.4
500	100.0
1000	237.0
1500	649.0
2000	open

Accuracy

Performance of the Tandem Match has been compared to other well-known directional couplers and laboratory test equipment, and it equals any amateur directional wattmeter tested. Power measurement accuracy of the Tandem Match compares well to a Hewlett-Packard HP-436A power meter. The HP meter has a specified measurement error of less than ±0.05 dB. The Tandem Match tracked the 436A within ±0.5 dB from 10 mW to 100 W and within ±0.1 dB from 1 W to 100 W. The unit was not tested above 1200 W because a transmitter with a higher power rating was not available.

SWR performance was equally good when compared to the SWR calculated from measurements made with 436A and a calibrated directional coupler. The Tandem Match tracked the calculated SWR within ±5% for SWR values from 1:1 to 5:1. SWR measurements were made at 8 W and 100 W.

Operation

Connect the Tandem Match in the 50-Ω line between the transmitter and the antenna-matching network (or antenna if no matching network is used). Set the RANGE switch to a range greater than the transmitter output rating and the MODE switch to TUNE. When the transmitter is keyed, the Tandem Match automatically switches on and indicates both power delivered to the antenna and SWR on the transmission line. When no carrier is present, the output power and SWR meters indicate zero.

The OPERATE mode includes RC circuitry to momentarily hold the peak-power and SWR readings during CW or SSB transmissions. The peak detectors are not ideal, so there could be about 10% variation from the actual power peaks and the SWR reading. The SWR×10 mode increases the maximum readable SWR to 50:1. This range should be sufficient to cover any SWR value that occurs in amateur use. (A 50-ft open stub of RG-8 yields a measured SWR of only 43:1, or less, at 2.4 MHz because of cable loss. Higher frequencies and longer cables exhibit a smaller maximum SWR.)

It is easy to use the Tandem Match to adjust an antenna-matching network. Adjust the transmitter for minimum output power (at least 1.5 W). With the carrier on and the MODE switch set to TUNE or SWR×10, adjust the matching network for minimum SWR. Once minimum SWR is obtained, set the transmitter to the proper operating mode and output power. Place the Tandem Match in the OPERATE mode.

Parts

Few parts suppliers carry all the components needed for these couplers. Each may stock different parts. Good sources include Digi-Key, Surplus Sales of Nebraska, Newark Electronics and Anchor Electronics. Use the *TIS Find* program for latest address information.

High-power Operation

This material was condensed from a letter by Frank Van Zant, KL7IBA, that appears in July 1989 *QST* (pp 42-43). In April 1988, Zack Lau, W1VT, described a directional-coupler circuit (based on the same principle as Grebenkemper's circuit) for a QRP transceiver. The main advantage of Lau's circuit is very low parts count.

Grebenkemper uses complex log-antilog amplifiers to provide good measurement accuracy. This application gets away from complex circuitry, but retains reasonable measurement accuracy over the 1 to 1500-W range. It also forfeits the SWR-computation feature.

Lau's coupler uses ferrite toroids. It works great at low power levels, but the ferrite toroids heat excessively with high power, causing erratic meter readings and the potential for burned parts.

The Revised Design

Powdered-iron toroids are used for the transformers in this version of Lau's basic circuit. The number of turns on the secondaries was increased to compensate for the lower permeability of powdered iron.

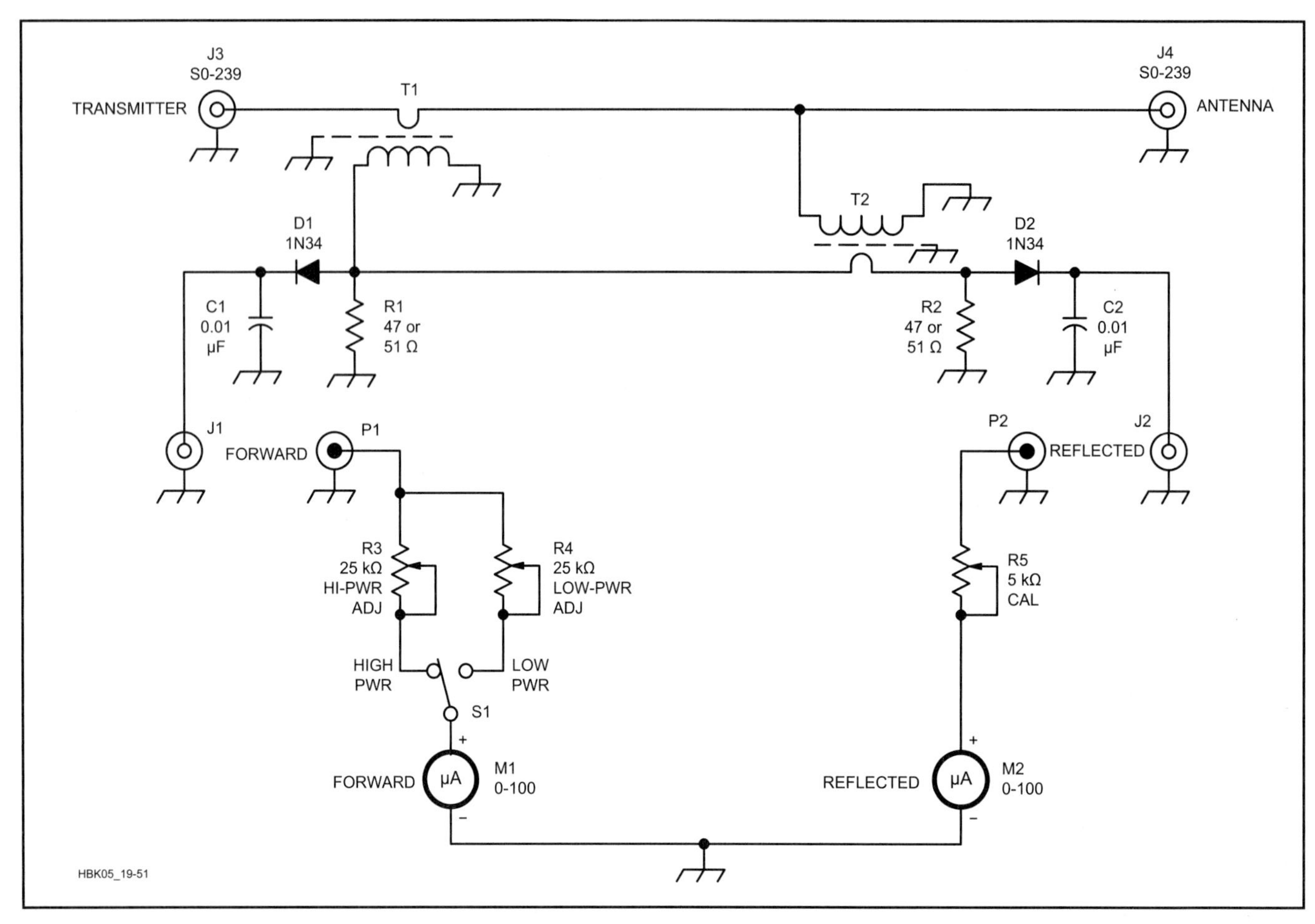

Fig 19.51—Schematic diagram of the high-power directional coupler. D1 and D2 are germanium diodes (1N34 or equiv). R1 and R2 are 47- or 51-Ω ½-W resistors. C1 and C2 have 500-V ratings. The secondary windings of T1 and T2 each consist of 40 turns of #26 to 30 enameled wire on T-68-2 powdered-iron toroid cores. If the coupler is built into an existing antenna tuner, the primary of T1 can be part of the tuner coaxial output line. The remotely located meters (M1 and M2) are connected to the coupler box at J1 and J2 via P1 and P2.

Two meters display reflected and forward power (see **Fig 19.51**). The germanium detector diodes (D1 and D2—1N34) provide fairly accurate meter readings particularly if the meter is calibrated (using R3, R4 and R5) to place the normal transmitter output at midscale. If the winding sense of the transformers is reversed, the meters are transposed (the forward-power meter becomes the reflected-power meter, and vice versa).

Construction

Fig 19.52 shows the physical layout of this coupler. The pickup unit is mounted in a 3½×3½×4-inch box. The meters, PC-mount potentiometers and HIGH/LOW power switch are mounted in a separate box or a compartment in an antenna tuner.

The primary windings of T1 and T2 are constructed much as Grebenkemper described, but use RG-8 with its jacket removed so that the core and secondary winding may fit over the cable. The braid is

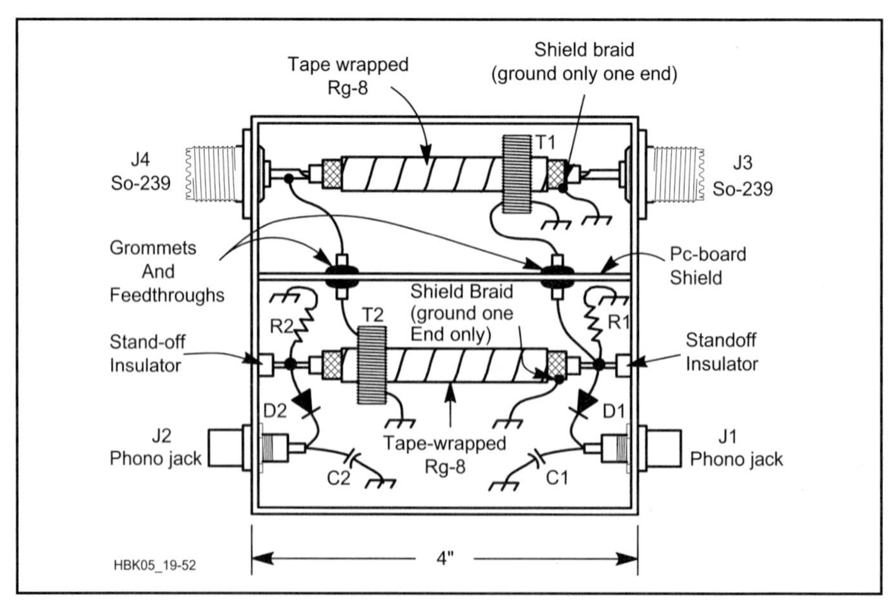

Fig 19.52—Directional-coupler construction details. Grommets or standoff insulators can be used to route the secondary windings of T1 and T2 through the PC-board shield. A 3½×3½×4-inch metal box serves as the enclosure.

wrapped with fiberglass tape to insulate it from the secondary winding. An excellent alternative to fiberglass tape—with even higher RF voltage-breakdown characteristics—is ordinary plumber's Teflon pipe tape, available at most hardware stores.

The transformer secondaries are wound on T-68-2 powdered-iron toroid cores. They are 40 turns of #26 to 30 enameled wire spread evenly around each core. By using #26 to 30 wire on the cores, the cores slip over the tape-wrapped RG-8 lines. With #26 wire on the toroids, a single layer of tape (slightly more with Teflon tape) over the braid provides an extremely snug fit for the core. Use care when fitting the cores onto the RG-8 assemblies. After the toroids are mounted on the RG-8 sections, coat the assembly with General Cement Corp Polystyrene Q Dope, or use a spot or two of RTV sealant to hold the windings in place and fix the transformers on the RG-8 primary windings.

Mount a PC-board shield in the center of the box, between T1 and T2, to minimize coupling between transformers. Suspend T1 between SO-239 connectors and T2 between two standoff insulators. The detector circuits (C1, C2, D1, D2, R1 and R2) are mounted inside the coupler box as shown.

Calibration, Tune up and Operation

The coupler has excellent directivity. Calibrate the meters for various power levels with an RF ammeter and a 50-Ω dummy load. Calculate I^2R for each power level, and mark the meter faces accordingly. Use R3, R4 and R5 to adjust the meter readings within the ranges. Diode nonlinearities are thus taken into account, and Grebenkemper's signal-processing circuits are not needed for relatively accurate power readings.

Start the tune-up process using about 10 W, adjust the antenna tuner for minimum reflected power, and increase power while adjusting the tuner to minimize reflected power.

This circuit has been built into several antenna tuners with good success. The bridge worked well at 1.5-kW output on 1.8 MHz. It also worked fine from 3.5 to 30 MHz with 1.2- and 1.5-kW output. The antenna is easily tuned for a 1:1 SWR using the null indication provided.

Amplifier settings for a matched antenna, as indicated with the wattmeter, closely agreed with those for a 50-Ω dummy load. Checks with a Palomar noise bridge and a Heath Antenna Scope also verified these findings. This circuit should handle more than 1.5 kW, *as long as the SWR on the feed line through the wattmeter is kept at or near 1:1*. (On one occasion high power was applied while the antenna tuner was not coupled to a load. Naturally the SWR was extremely high, and the output transformer secondary winding opened like a fuse. This resulted from the excessively high voltage across the secondary. The damage was easily and quickly repaired.)

Notes

[1]Use *TIS Find* program for latest address information. See CD-ROM for template.

AN EXTERNAL AUTOMATIC ANTENNA SWITCH FOR USE WITH YAESU OR ICOM RADIOS

This antenna-switching-control project involves a combination of ideas from several earlier published articles.[1,2,3] This system was designed to mount the antenna relay box outside the shack, such as on a tower. With this arrangement, only a single antenna feed line needs to be brought into the shack. The lead photo shows the control unit and relay box, designed and built by Joe Carcia, NJ1Q. As the W1AW chief operator, Joe has plans for the switch at W1AW. Either an ICOM or Yaesu HF radio will automatically select the proper antenna. In addition, a manual switch can override the ICOM automatic selection. That feature also provides a way to use the antenna with other radios. The antenna switch is not a two-radio switch, though. It will only work with one radio at a time.

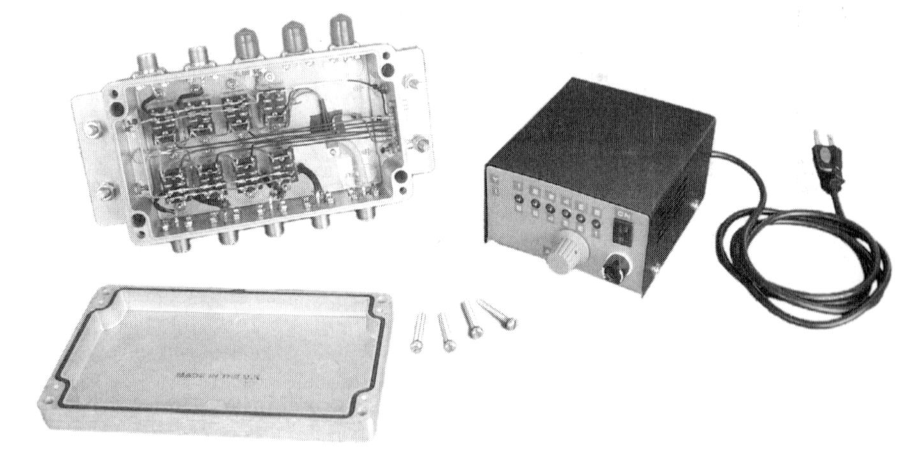

Many builders may want to use only the ICOM or only the Yaesu portion of the interface circuitry, depending on the brand of radio they own. The project is a "hacker's dream." It can be built in a variety of forms, with the only limitations being the builder's imagination.

Circuit Description

Fig 19.53 is a block diagram of the complete system. An ICOM or Yaesu HF radio connects to the appropriate decoder via the

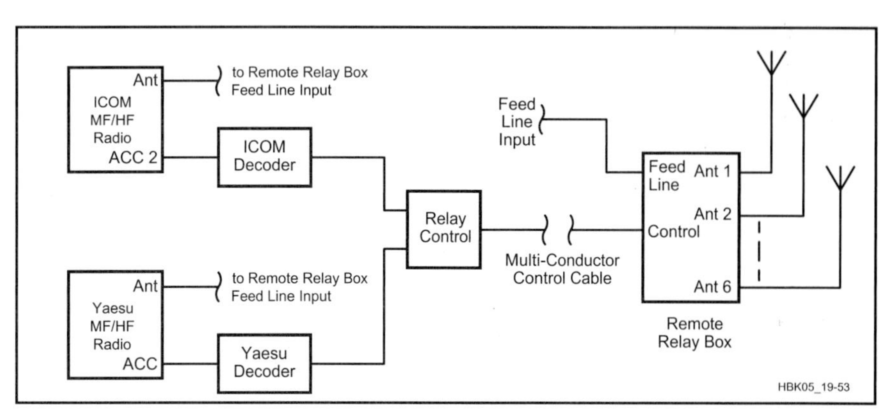

Fig 19.53—Block diagram of the remotely controlled automatic antenna switch.

Table 19.4

ICOM Accessory Connector Output Voltages By Band

Band (MHz)	*Output Voltage*
1.8	7 – 8.0
3.5	6 – 6.5
7	5 – 5.5
14	4 – 4.5
18, 21	3 – 3.5
24, 28	2 – 2.5
10	0 – 1.2

Note: The voltage step between bands is not constant, but close to 1.0 V, and the 10-MHz band is not in sequence with the others.

Table 19.5

Yaesu Band Data Voltage Output (BCD)

Band	*A (1)*	*B (2)*	*C (4)*	*D (8)*	*(BCD Equiv.)*
1.8	5V	0V	0V	0V	1
3.5	0V	5V	0V	0V	2
7.0	5V	5V	0V	0V	3
10.1	0V	0V	5V	0V	4
14	5V	0V	5V	0V	5
18.068	0V	5V	5V	0V	6
21	5V	5V	5V	0V	7
24.89	0V	0V	0V	5V	8
28	5V	0V	0V	5V	9

accessory connector on the back of the radio. Some other modern rigs have an accessory connector used for automatic bandswitching of amplifiers, tuners and other equipment. For example, Ten-Tec radios apply a 10 to 14-V dc signal to pins on the DB-25 interface connector for the various bands. Other radios use particular voltages on one of the accessory-connector pins to indicate the selected band. Check the owner's manual of your radio for specific information, or contact the manufacturer's service department for more details. You may be able to adapt the ideas presented in this project for use with other radios.

A single length of coax and a multiconductor control cable run from the rig and decoder/control box to the remotely located switch unit. The remote relay box is equipped with SO-239 connectors for the input as well as the output to each antenna. You can use any type of connectors, though.

ICOM radios use an 8-V reference and a voltage divider system to provide a stepped band-data output voltage. **Table 19.4** shows the output voltage at the accessory socket when the radio is switched to the various bands. Notice that seven voltage steps can be used to select different antennas. The ICOM accessory connector pin assignments needed for this project are:

Pin 1 +8 V reference
Pin 2 Ground
Pin 4 Band signal voltage
Pin 7 +12 (13.8) V supply

Yaesu radios provide the band information as binary coded decimal (BCD) data on four lines. Nine different BCD values allow you to select a different antenna for each of the MF/HF bands. **Table 19.5** shows the BCD data from Yaesu radios for the various bands. The Yaesu 8-pin DIN accessory connector pin assignments needed for this project are:

Pin 1 +12 (13.8) V supply
Pin 3 Ground
Pin 4 Band Data A
Pin 5 Band Data B
Pin 6 Band Data C
Pin 7 Band Data D

Fig 19.54 is the schematic diagram for the control box. We will discuss each part of the control circuit later in this description. First, let's turn our attention to the external antenna box.

EXTERNAL ANTENNA BOX

Only the number of control lines going out to the relay box limits the number of antennas this relay box will switch. The unit shown in the lead photo has ten SO-239 connectors, to switch the common feed line to any of nine antennas. Many hams will use an eight-conductor rotator cable (such as *Belden 9405*) to the relay box. Using eight wires, we can control seven relays (six for antennas and one to ground the feed line for lightning protection) with the relay coil B+ supply, as well as a ground lead. The lead photo in Fig 19.53 shows eight relays, and I plan to add two additional relays so I can select between all nine antennas when I install the unit at W1AW. The box contains a 12 V dc voltage regulator (LM7812 or equivalent), which I bolted to the aluminum box. I used an insulating spacer (TO-220 mounting hardware) between the back of the regulator and the box, and applied a layer of heat-sink compound on both sides of the insulating wafer. There is also a connector for power and control lines. I used a DB-15 connector because I plan to add more relays and control lines later, so I can switch between 9 antennas. A DB-9 connector would be suitable for use with the eight-conductor control cable, or you may wish to use a weatherproof connector. **Fig 19.55** shows the relay box schematic diagram.

Since the box will be located outside, I used a weatherproof metal box—a *Hammond Manufacturing*, type 1590Z150, watertight aluminum box. It's about 8½×4¼×3⅛ inches. This is a rather hefty box, meant to be exposed to years of various weather conditions. You can, however, use almost anything.

The coax connectors are mounted so each particular antenna connector is close to the relay, without too much crowding. I used flange-mount SO-239s (although the single hole type will also work). For added weather protection (and conductivity), I applied Penetrox® to the connector flange mount, including the threads of the mounting screws. On the power/control line connector, I used Coax Seal®.

I attached aluminum angle stock on either side of the box to mount it on a tower leg. The U-bolts should be of the proper size to fit the tower leg. They should also be galvanized or made of stainless steel.

Antenna Relays

One of the more difficult parts of this project was the modification of the relays (DPDT Omron LY2F-DC12). To improve isolation, the moveable contacts (armature) are wired in parallel and the connecting wire is routed through a hole in the relay case.

Remove the relay from its plastic case. Unsolder and remove the small wires from the armatures. Carefully solder a jumper across the armature lugs. I used #20 solid copper. Then solder a piece of very flexible wire (such as braid from RG-58 cable) to either armature lug. Obviously, the location of the wire depends on which side you wish to connect the SO-239. You will also need to make a hole in the plastic case that is large enough to accommodate the armature wire without placing any strain on the free movement of the armature. I slipped a length of insulating tubing over this wire to prevent it from shorting to the aluminum box.

The normally open and normally closed contacts are also wired in parallel. This can be done on the lugs themselves. For this, I used #12 solid copper wire.

I mounted the relays in the aluminum box, oriented so they could be wired together without difficulty. (See **Fig 19.56.**) With the exception of the wire used for the relay coils (#22 solid wire), I used #12 solid copper wire for the rest of the connections.

To eliminate the possibility of spikes or "back emf," a 1N4007 diode is soldered across the coil contacts of each relay. In addition, 0.01-µF capacitors across the diodes will reduce the possibility of stray RF causing problems with the relay operation.

Since the cable run from the shack to

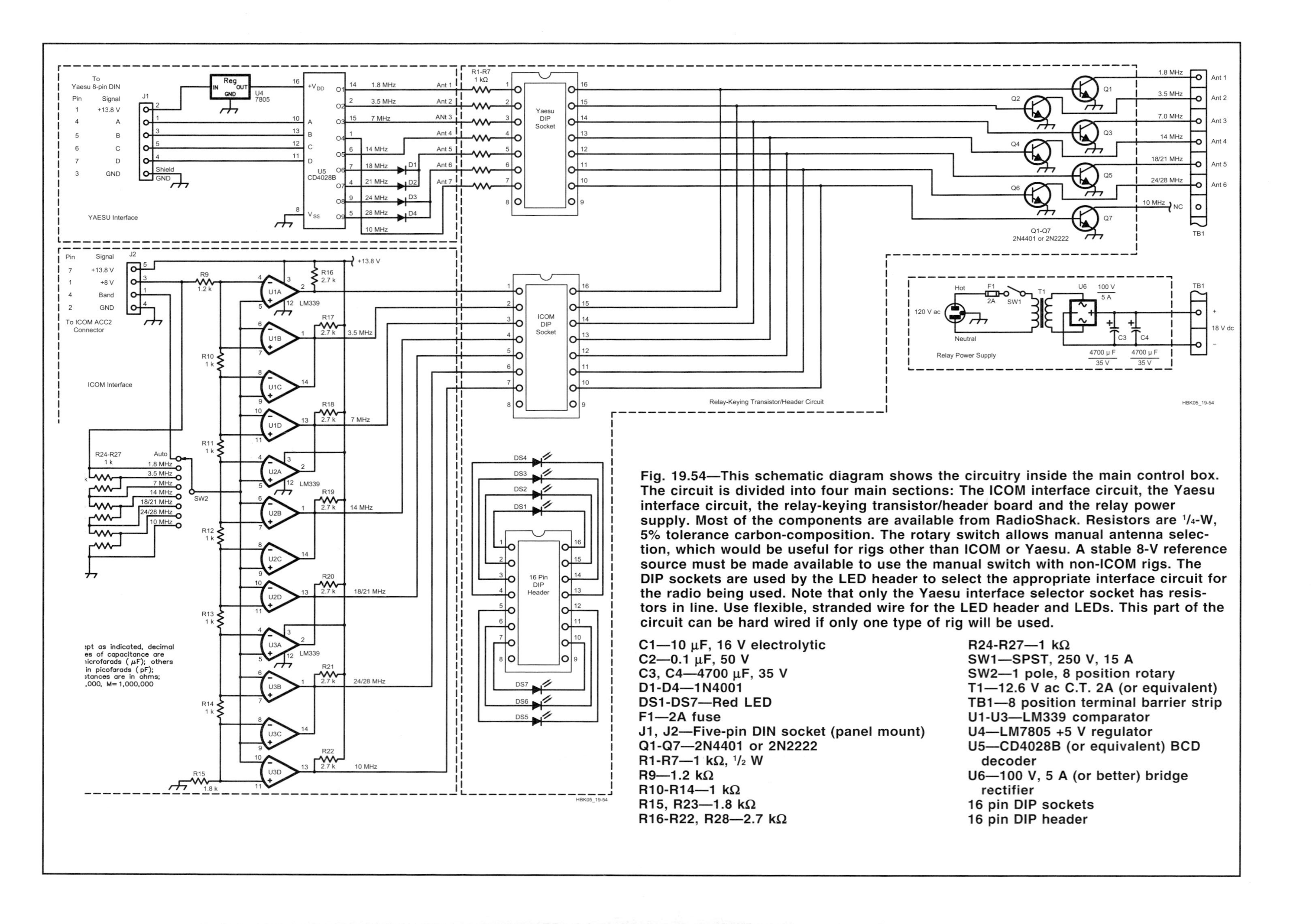

Fig. 19.54—This schematic diagram shows the circuitry inside the main control box. The circuit is divided into four main sections: The ICOM interface circuit, the Yaesu interface circuit, the relay-keying transistor/header board and the relay power supply. Most of the components are available from RadioShack. Resistors are 1/4-W, 5% tolerance carbon-composition. The rotary switch allows manual antenna selection, which would be useful for rigs other than ICOM or Yaesu. A stable 8-V reference source must be made available to use the manual switch with non-ICOM rigs. The DIP sockets are used by the LED header to select the appropriate interface circuit for the radio being used. Note that only the Yaesu interface selector socket has resistors in line. Use flexible, stranded wire for the LED header and LEDs. This part of the circuit can be hard wired if only one type of rig will be used.

C1—10 μF, 16 V electrolytic
C2—0.1 μF, 50 V
C3, C4—4700 μF, 35 V
D1-D4—1N4001
DS1-DS7—Red LED
F1—2A fuse
J1, J2—Five-pin DIN socket (panel mount)
Q1-Q7—2N4401 or 2N2222
R1-R7—1 kΩ, 1/2 W
R9—1.2 kΩ
R10-R14—1 kΩ
R15, R23—1.8 kΩ
R16-R22, R28—2.7 kΩ
R24-R27—1 kΩ
SW1—SPST, 250 V, 15 A
SW2—1 pole, 8 position rotary
T1—12.6 V ac C.T. 2A (or equivalent)
TB1—8 position terminal barrier strip
U1-U3—LM339 comparator
U4—LM7805 +5 V regulator
U5—CD4028B (or equivalent) BCD decoder
U6—100 V, 5 A (or better) bridge rectifier
16 pin DIP sockets
16 pin DIP header

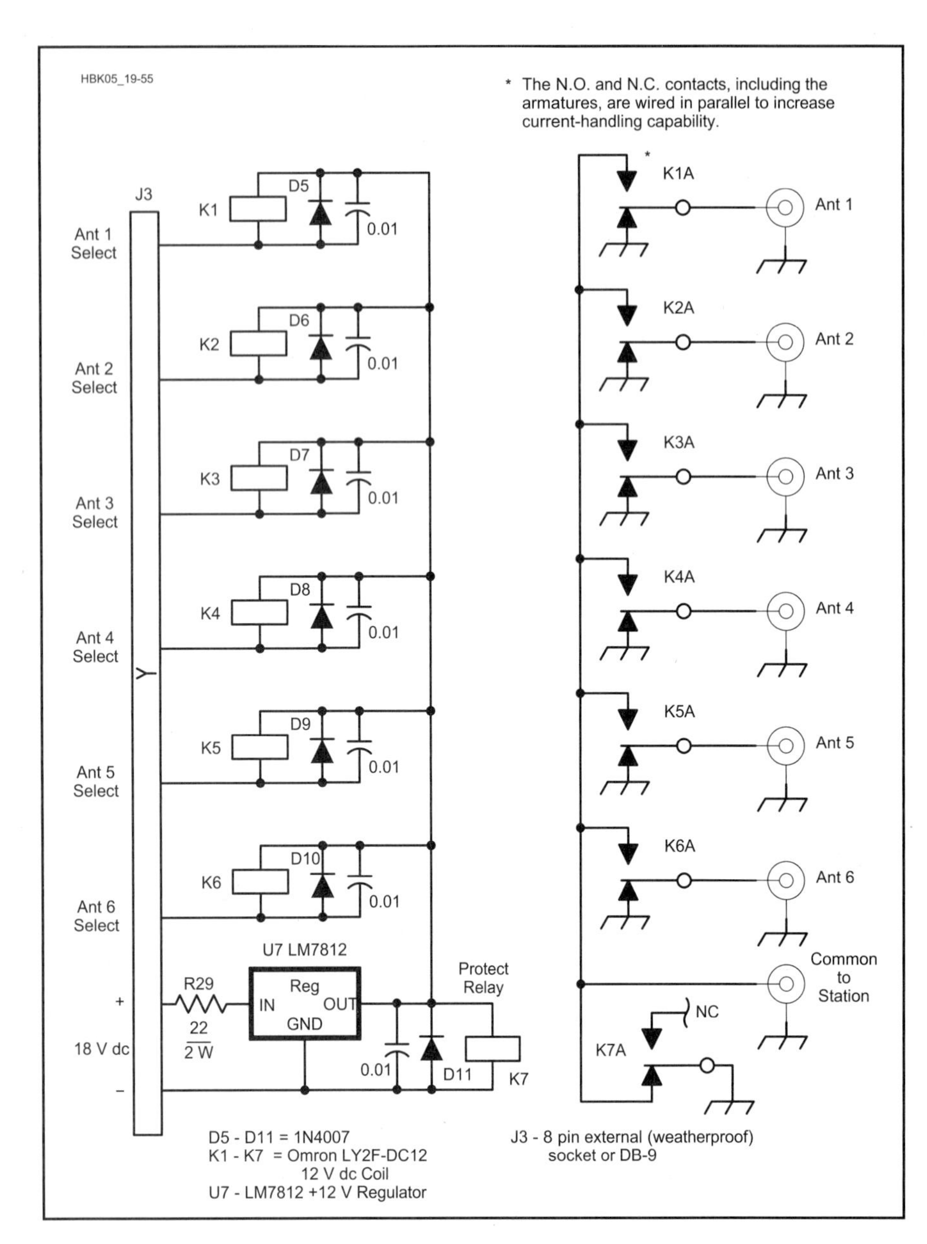

Fig 19.55—The schematic diagram for the external antenna relay box. All relays are DPDT, 250 V ac, 15 A contacts. R29 is used to limit the regulator current. Mount the regulator using TO-220 mounting hardware, with heatsink compound. With the exception of the normally closed and normally open contacts, all wiring is #22 solid copper wire.
D5-D11—1N4007
J3—8 pin external weatherproof connector (or DB-9 with appropriate weather sealant)
K1-K7—Omron LY2F-DC12 with 12 V coil (Allied Electronics Stock number 821-2019)
U7—LM7812 +12 V regulator

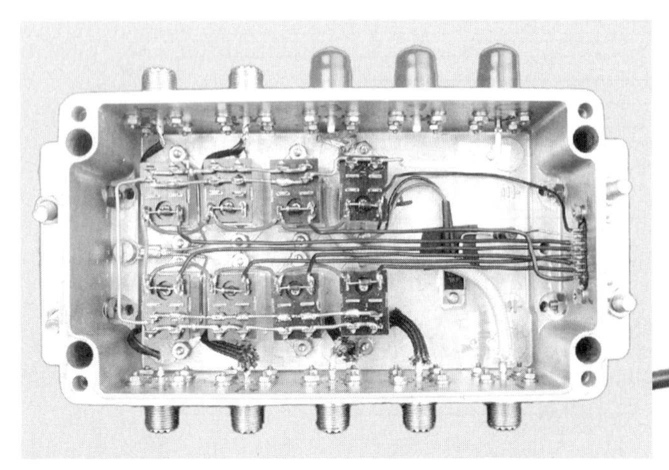

Fig 19.56—This photo shows the external relay box. The LM7812 regulator is mounted to the bottom of the box. The relay normally open and normally closed contacts are wired in parallel using #12 solid copper. Joe Carcia plans to install two additional relays to use this project at W1AW, so he included the extra flange-mount SO-239 connectors when he was drilling holes in the box. Unused SO-239s can be capped off.

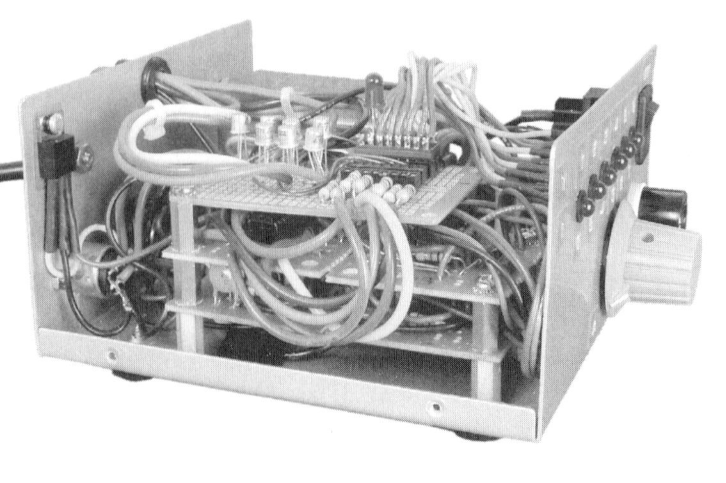

Fig 19.57 (below)—This photo shows the inside of the control box. A RadioShack aluminum enclosure holds all of the components. The ICOM interface is built on a two-section RadioShack Universal Project Board (276-159B) and another single section of the same board. The bottom board in the enclosure as well as the right half of the middle section hold the ICOM circuit. The Yaesu interface is built on another section of the Universal Project Board, and is on the left side of the middle section. The top circuit board is a RadioShack Universal Project Board (276-150), which holds the DIP sockets and relay-selection transistors. The circuit boards use point-to-point wiring. All high voltage leads are insulated. The LEDs are mounted in holders on the front panel. The 7805 5-V regulator is mounted on the back panel using TO-220 mounting hardware, with heatsink compound.

the tower can be quite long, consideration has to be given to the voltage drop that may occur. The relays require 12 V dc. As such, I installed a 12 V dc regulator in the box, and fed it with 18 V dc (at 2 amps) from the control box. If the cable run is not that long, however, you could just use a 12-V supply.

One of the relays is used for lightning protection. When not in use, the relay grounds the line coming in from the shack. When the control box is activated, it applies power to this relay, thus removing the ground on the station feed line. All the antenna lines are grounded through the normally closed relay contacts. They remain grounded until the relay receives power from the control box.

CONTROL BOX

This is the heart of the system. The 18 V dc power supply for the relays is located in this box, in addition to the Yaesu and ICOM decoder circuits and the relay-control circuitry. All connections to the relay box are made via an 8-position terminal barrier strip mounted on the back of the control box.

The front of the box has LEDs that indicate the selected antenna. A rotary switch can be used for manual antenna selection. The power switch and fuse are also located on the front panel.

The wiring schemes on the Yaesu and ICOM ACC sockets are so different, I opted to have a 5-pin DIN connector for each rig on the control box. Since there is only one set of LEDs, I used an 8-pin DIP header to select the appropriate control circuit for each radio. See **Fig 19.57**.

ICOM CIRCUITRY

This circuit originally appeared in April 1993 *QST*.[2] I've modified the circuit slightly to fit my application. The original circuit allowed for switching between seven antennas (from 160 to 10 meters). The Band Data signal from the ICOM radios goes to a string of LM339 comparators. Resistors R9 through R15 divide the 8-V reference signal from the rig to provide midpoint references between the band signal levels. The LM339 comparators decide which band the radio is on. A single comparator selects the 1.8 or 10 MHz band because those bands are at opposite ends of the range. The other bands each use two comparators. One determines if the band signal is above the band level and the other determines if it is below the band level. If the signal is between those two levels, the appropriate LED and relay-selection transistor switch is turned on.

I used point-to-point wiring on Radio Shack Universal Project Boards to build the various circuit sections. The ICOM interface uses both sections of a 276-159B project board shown at the bottom of the stack in Photo B for U1 and U2. Another section of project board holds U3, located on the right side of the middle section.

The ICOM circuit allows for manual antenna selection. The 8-V reference is normally taken directly from the ICOM ACC socket. If this circuit is to be used with other equipment, then a regulated 8-V source should be provided.

YAESU CIRCUITRY

The neat thing about Yaesu band data is that it's in a binary format. This means you can use a simple BCD decoder for band switching. The BCD output ranges from 1 to 9. In essence, you can switch between 9 antennas (or bands). Since the relay box switches just six antennas, I incorporated steering diodes (D1 through D4 in Fig 19.54) so I can use one antenna connection for multiple bands. In this regard, I opted to use one antenna connection for 17 and 15 meters, and another connection for 12 and 10 meters because the ICOM band data combines those bands. I did not include the control line or relay for a 30-meter antenna with this version of the project.

One section of the RadioShack 276-159B project board holds the Yaesu interface circuit. That board is shown on the left side of the middle layer of the stack shown in Fig 19.57.

DIP Sockets and Header

A RadioShack Universal Project Board, 276-150 holds the DIP sockets along with the relay keying transistors. This board is shown as the top layer in Photo B. The Yaesu socket has 1-kΩ resistors wired in series with each input pin. The other header connects directly to the ICOM circuitry.

The DIP header is used to switch the keying transistors between the ICOM and Yaesu circuitry. The LEDs are used to indicate antenna number. Use stranded wire (for its flexibility) when connecting to the LEDs.

Relay Keying Transistors

Both circuits use the same transistor-keying scheme, so I only needed one set of transistors. Each transistor collector connects to the terminal barrier strip. The emitters are grounded, and the bases are wired in parallel to the two 16-pin DIP sockets. The band data turns on one of the transistors, effectively grounding that relay-control lead. Current flows through the selected relay coil, switching that relay to the normally open position and connecting the station feed line to the proper antenna.

Power supply

The power supply is used strictly for the relays. Other power requirements are taken from the rig used. There is room here for variations on the power supply theme. In this case, I used a 12.6 V, center-tapped, 2 A power transformer. I feed the output to a bridge rectifier, and two 4700 µF, 35-V electrolytics. (I happened to have these parts on hand.)

Notes

[1]"An Antenna Switching System for Multi-Two and Single-Multi Contesting," by Tony Brock-Fisher, K1KP, January 1995 *NCJ*.
[2]"A Remotely Controlled Antenna Switch," by Nigel Thompson, April 1993 *QST*.
[3]"*NA* Logging Program© Section 11"

A TRIO OF TRANSCEIVER/COMPUTER INTERFACES

Virtually all modern Amateur Radio transceivers (and many general-coverage receivers) have provisions for external computer control. Most hams take advantage of this feature using software specifically developed for control, or primarily intended for some other purpose (such as contest logging), with rig control as a secondary function.

Unfortunately, the serial port on most radios cannot be directly connected to the serial port on most computers. The problem is that most radios use TTL signal levels while most computers use RS-232-D.

The interfaces described here simply convert the TTL levels used by the radio to the RS-232-D levels used by the computer, and vice versa. Interfaces of this type are often referred to as level shifters. Two basic designs, one having a couple of variations, cover the popular brands of radios. This article, by Wally Blackburn, AA8DX, first appeared in February 1993 *QST*.

TYPE ONE: ICOM CI-V

The simplest interface is the one used for the ICOM CI-V system. This interface

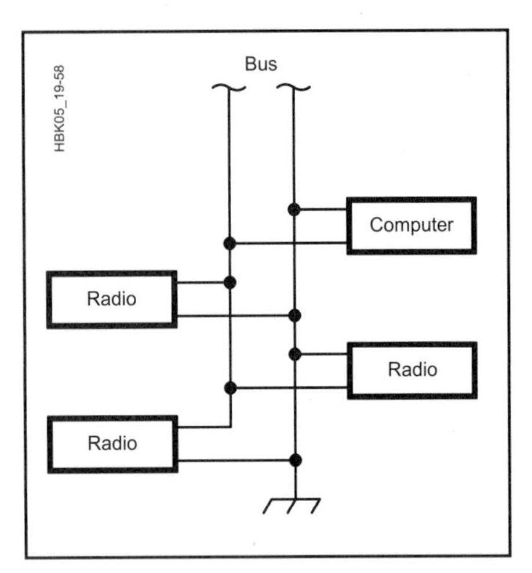

Fig 19.58—The basic two-wire bus system that ICOM and newer Ten-Tec radios share among several radios and computers. In its simplest form, the bus would include only one radio and one computer.

works with newer ICOM and Ten-Tec rigs. **Fig 19.58** shows the two-wire bus system used in these radios.

This arrangement uses a CSMA/CD (carrier-sense multiple access/collision detect) bus. This refers to a bus that a number of stations share to transmit and receive data. In effect, the bus is a single wire and common ground that interconnect a number of radios and computers.

The single wire is used for transmitting and receiving data. Each device has its own unique digital address. Information is transferred on the bus in the form of packets that include the data and the address of the intended receiving device.

The schematic for the ICOM/Ten-Tec interface is shown in **Fig 19.59**. It is also the Yaesu interface. The only difference is that the transmit data (TxD) and receive data (RxD) are jumpered together for the ICOM/Ten-Tec version.

The signal lines are active-high TTL. This means that a logical one is represented by a binary one (+5 V). To shift this to RS-232-D it must converted to –12 V while a binary zero (0 V) must be converted to +12 V. In the other direction, the opposites are needed: –12 V to +5 V and +12 V to 0 V.

U1 is used as a buffer to meet the interface specifications of the radio's circuitry and provide some isolation. U2 is a 5-V-powered RS-232-D transceiver chip that translates between TTL and RS-232-D levels. This chip uses charge pumps to obtain ±10 V from a single +5-V supply. This device is used in all three interfaces.

A DB25 female (DB25F) is typically used at the computer end. Refer to the discussion of RS-232-D earlier in the chapter for 9-pin connector information. The interface connects to the radio via a 1/8-inch phone plug. The sleeve is ground and the

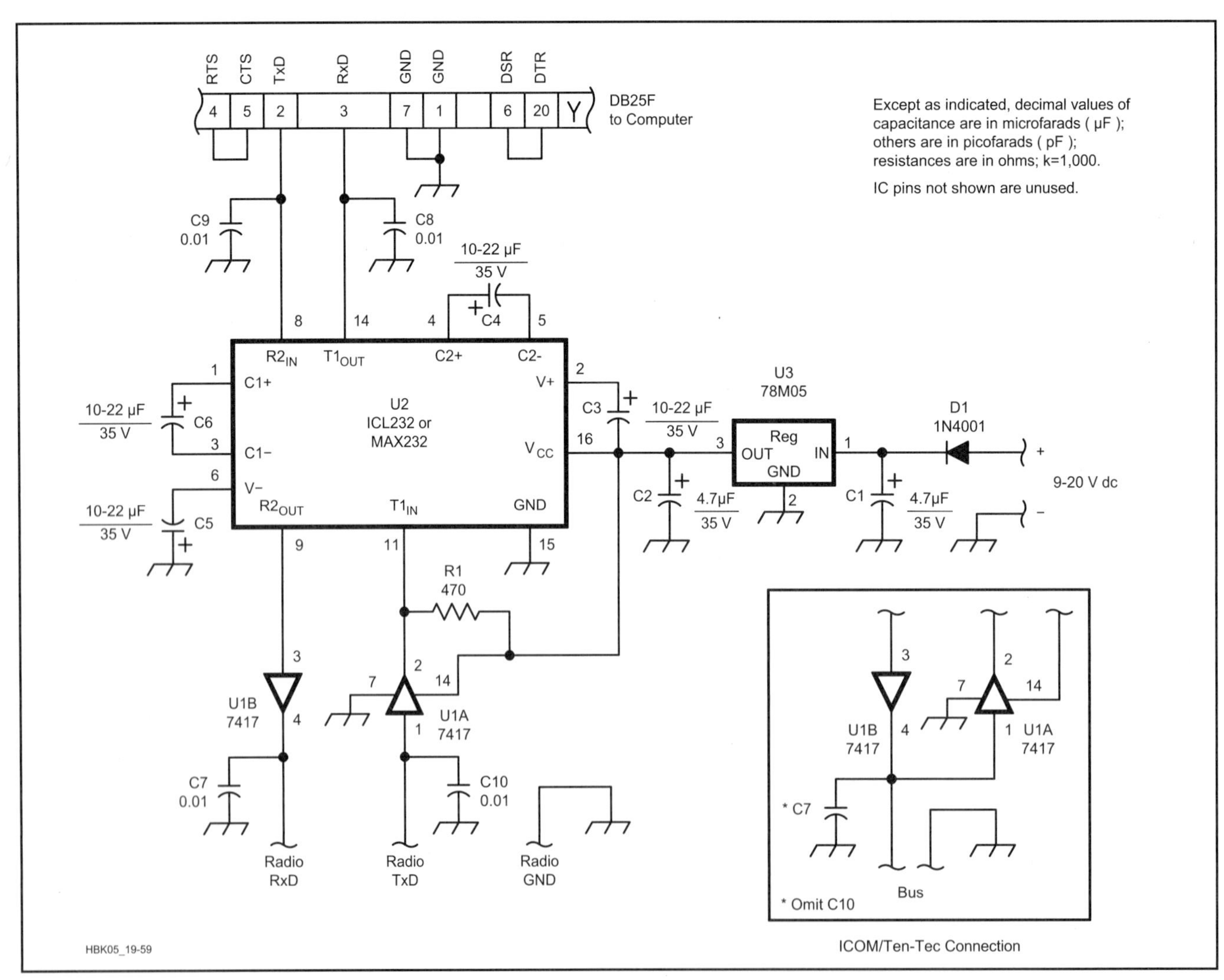

Fig 19.59—ICOM/Ten-Tec/Yaesu interface schematic. The insert shows the ICOM/Ten-Tec bus connection, which simply involves tying two pins together and eliminating a bypass capacitor.

C7-C10—0.01-µF ceramic disc.
U1—7417 hex buffer/driver.
U2—Harris ICL232 or Maxim MAX232.

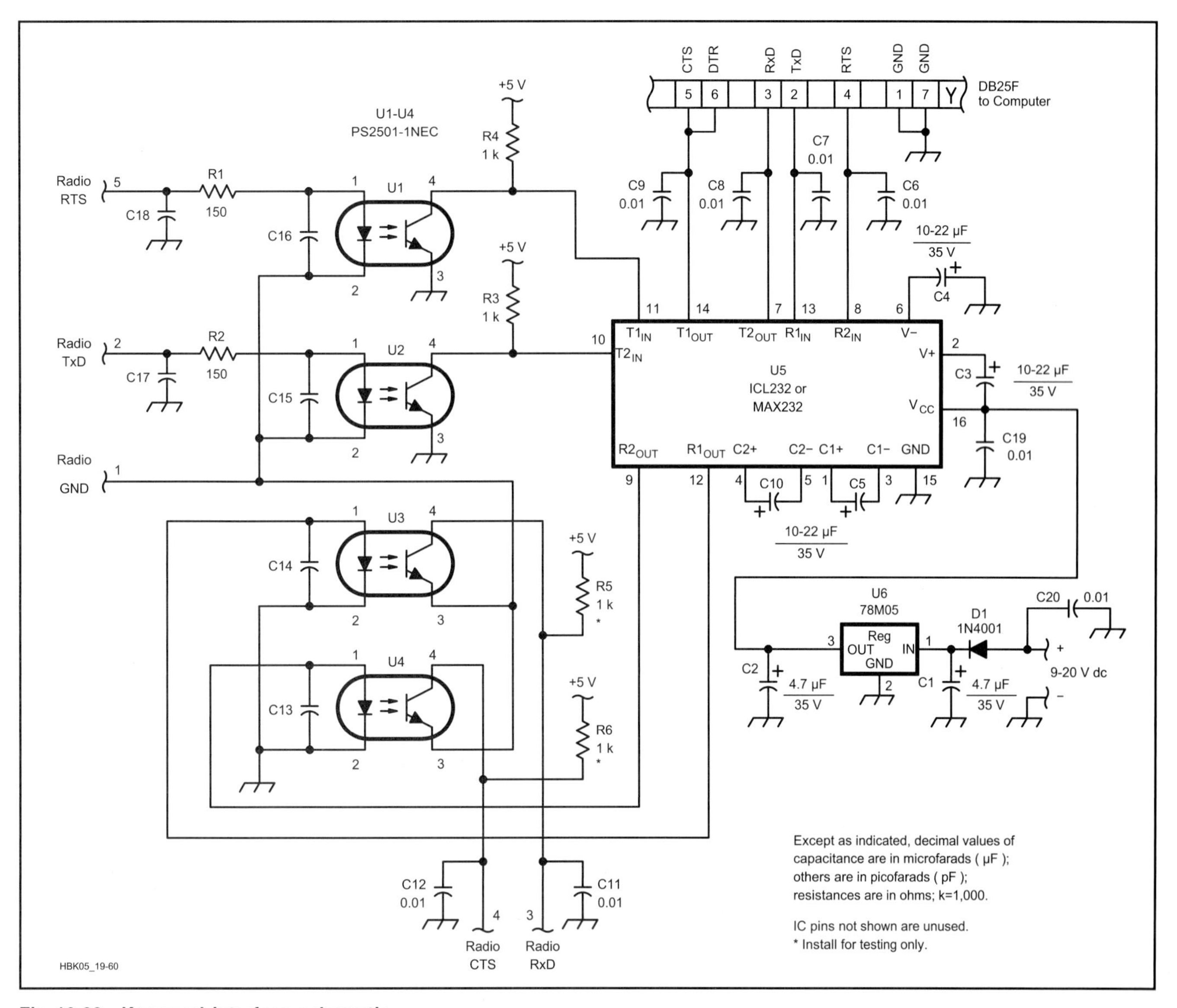

Fig 19.60—Kenwood interface schematic.
C6-C9, C11, C12, C17, C18—0.01-µF ceramic disc.
C13-C16, C19-C21—0.01 µF ceramic disc.
U1-U4—PS2501-1NEC (available from Digi-Key).
U5—Harris ICL232 or Maxim MAX232.

tip is the bus connection.

It is worth noting that the ICOM and Ten-Tec radios use identical basic command sets (although the Ten-Tec includes additional commands). Thus, driver software is compatible. The manufacturers are to be commended for working toward standardizing these interfaces somewhat. This allows Ten-Tec radios to be used with all popular software that supports the ICOM CI-V interface. When configuring the software, simply indicate that an ICOM radio (such as the IC-735) is connected.

TYPE TWO: YAESU INTERFACE

The interface used for Yaesu rigs is identical to the one described for the ICOM/Ten-Tec, except that RxD and TxD are not jumpered together. Refer to Fig 19.59. This arrangement uses only the RxD and TxD lines; no flow control is used.

The same computer connector is used, but the radio connector varies with model. Refer to the manual for your particular rig to determine the connector type and pin arrangement.

TYPE THREE: KENWOOD

The interface setup used with Kenwood radios is different in two ways from the previous two: Request-to-Send (RTS) and Clear-to-Send (CTS) handshaking is implemented and the polarity is reversed on the data lines. The signals used on the Kenwood system are active-low. This means that 0 V represents a logic one and +5 V represents a logic zero. This characteristic makes it easy to fully isolate the radio and the computer since a signal line only has to be grounded to assert it. Optoisolators can be used to simply switch the line to ground.

The schematic in **Fig 19.60** shows the Kenwood interface circuit. Note the different grounds for the computer and the radio. This, in conjunction with a separate power supply for the interface, provides excellent isolation.

The radio connector is a 6-pin DIN plug. The manual for the rig details this connector and the pin assignments.

Some of the earlier Kenwood radios

Table 19.6

Kenwood Interface Testing

Apply	Result
GND to Radio-5	–8 to –12 V at PC-5
+5 V to Radio-5	+8 to +12 V at PC-5
+9 V to PC-4	+5 V at Radio-4
–9 V to PC-4	0 V at Radio-4
GND to Radio-2	–8 to –12 at PC-3
+5 V to Radio-2	+8 to +12 V at PC-3
+9 V to PC-2	+5 V to Radio-3
–9 V to PC-2	0 V at Radio-3

Table 19.7

ICOM/Ten-Tec Interface Testing

Apply	Result
GND to Bus	+8 to +12 V at PC-3
+5 V to Bus	–8 to –12 V at PC-3
–9 V to PC-2	+5 V on Bus
+9 V to PC-2	0 V on Bus

Table 19.8

Yaesu Interface Testing

Apply	Result
GND to Radio TxD	+8 to +12 V at PC-3
+5 V to Radio TxD	–8 to –12 V at PC-3
+9 V to PC-2	0 V at Radio RxD
–9 V to PC-2	+5 V at Radio RxD

require additional parts before their serial connection can be used. The TS-440S and R-5000 require installation of a chipset and some others, such as the TS-940S require an internal circuit board.

Construction and Testing

The interfaces can be built using a PC board, breadboarding, or point-to-point wiring. PC boards and MAX232 ICs are available from FAR Circuits. Use the *TIS Find* program for the latest address information. The PC board template is available on the CD-ROM.

It is a good idea to enclose the interface in a metal case and ground it well. Use of a separate power supply is also a good idea. You may be tempted to take 13.8 V from your radio—and it works well in many cases: but you sacrifice some isolation and may have noise problems. Since these interfaces draw only 10 to 20 mA, a wall transformer is an easy option.

The interface can be tested using the data in **Tables 19.6, 7** and **8**. Remember, all you are doing is shifting voltage levels. You will need a 5-V supply, a 9-V battery and a voltmeter. Simply supply the voltages as described in the corresponding table for your interface and check for the correct voltage on the other side. When an input of –9 V is called for, simply connect the positive terminal of the battery to ground.

During normal operation, the input signals to the radio float to 5 V because of pullup resistors inside the radio. These include RxD on the Yaesu interface, the bus on the ICOM/Ten-Tec version, and RxD and CTS on the Kenwood interface. To simulate this during testing, these lines must be tied to a 5-V supply through 1-kΩ resistors. Connecting these to the supply without current-limiting resistors will damage the interface circuitry. R5 and R6 in the Kenwood schematic illustrate this. They are not shown (but are still needed) in the ICOM/Ten-Tec/Yaesu schematic. Also, be sure to note the separate grounds on the Kenwood interface during testing.

Another subject worth discussing is the radio's communication configuration. The serial ports of both the radio and the computer must be set to the same baud rate, parity, and number of start and stop bits. Check your radio's documentation and configure your software or use the PC-DOS/MS-DOS MODE command as described in the computer manual.

A COMPUTER-CONTROLLED TWO-RADIO SWITCHBOX

This versatile computer-controlled two-radio switchbox was designed by Dean Straw, N6BV, who made it primarily for contest operations using one of the popular computer logging programs, such as *CT*, *NA* or *TR*. The switchbox was built into two boxes, a main unit and a hard-wired remote head. **Fig 19.61** shows the back of the main unit, and **Fig 19.62** shows the small wired-remote head. The remote head is compact enough to place almost anywhere on a crowded operating desk. Besides toggle switches, it uses red and green LED annunciators to tell the operator exactly what is happening.

RadioShack components were used throughout the project as much as possible so that parts availability should not be a hurdle for potential builders. The overall cost using all-new parts was about $160.

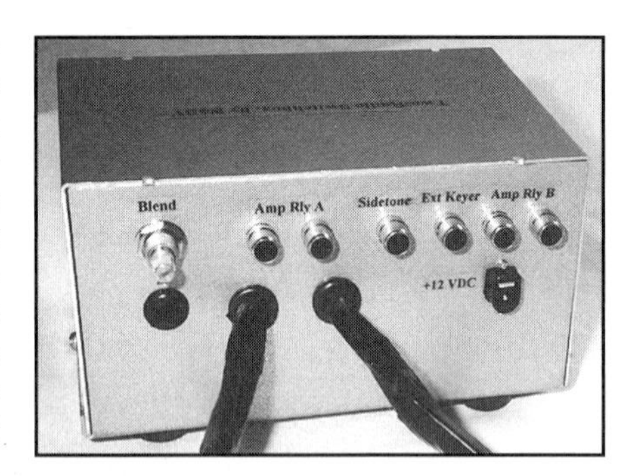

Fig 19.61—A view of the rear panel of the main box of the Two-Radio Switchbox.

Fig 19.62—The remote head for the Two-Radio Switchbox.

OVERVIEW OF FEATURES

The switchbox controls both transmitting and receiving functions for either phone or CW modes. (Data modes that connect through the transceiver's microphone input or that use direct FSK could also be controlled through the switchbox, using additional external switching.) This particular switchbox was built to work with two ICOM IC-765 HF transceivers, but you can easily wire the microphone, PTT, headphone and CW key-line connections to match your own radios.

Receiving Features

For this discussion, assume that Radio A is located to the left in front of the operator and Radio B is to the right. Assume also that the two radios are connected to separate antennas (and perhaps linear amplifiers), and that interaction and overload between the two radios has been minimized by good engineering. In other words, Radio B can receive effectively on one frequency band, even while Radio A is transmitting full power on another band—and vice versa. Here we'll assume that you are using stereo headphones. You can select:

1. Radio A in both ears (monaural)—for both transmit and receive, in the RX A switch position.
2. Radio B in both ears (monaural)—for both transmit and receive, in the RX B switch position.
3. Radio A in the left ear; Radio B in the right ear—for both transmit and re-

ceive, in the STEREO switch position.

4. Radio A in both ears in receive; Radio B in both ears while Radio A is transmitting, toggling automatically while in the AUTO TX switch position.

5. Radio B in both ears in receive; Radio A in both ears while Radio B is transmitting, toggling automatically while in the AUTO TX switch position.

6. Green LEDs on the remote head give instant indication of the source(s) of audio in the stereo headphones.

The AUTO TX facility in 4 and 5 above allows you to call CQ on one radio, while devoting full attention to listening to the second radio. You could, for example, look for new multipliers in a contest or to check whether another band is open or not. Late in a contest, you can easily become mesmerized listening to your own voice from a voice recorder calling CQ, or listening to the computer automatically calling CQ on CW. The AUTO TX facility forces you to pay attention to the second radio—but this function can be switched off, if you like, from the remote head. If you choose the STEREO receiving mode, the AUTO TX function is automatically disabled, since it would make it pretty confusing to have the right and left audio sources shift automatically.

Another useful feature in STEREO is a BLEND control on the main box. This allows you to shift the apparent position of the right-hand receive audio somewhere between full-right and near the middle of your spatial hearing range. Some operators claim that this helps cut down on fatigue during long operating sessions when using stereo reception.

There is a second stereo headphone jack on the main box, with a switch labeled FOLLOW A or B ONLY. A second operator can either monitor what the first operator is doing (perhaps for training or coordination during a contest), or else the second operator can pay full attention to the second receiver.

The switchbox also has a separate SIDETONE input jack for audio from an external keyer. This connects to the B channel so that you can still hear sidetone when you have selected AUTO TX (with Radio A as transmitter) and use the paddle instead of the computer to send CW, perhaps to send a fill for a missed report.

The wiring bundles going to each radio are set up to accommodate external DSP filters, something that can bring a high-quality older transceiver up to "modern" status, comparable to the newer radios with all their DSP bells and whistles.

Transmitting Features

1. The microphone "hot" line, the microphone "cold" line, the CW key line and the PTT line are all switched between Radio A and Radio B. Both microphone "hot" and "cold" leads are switched to reduce the possibility of ground-loop-induced 60-Hz hum on your transmitted signal.

2. In the TX A switch position on the remote head, Radio A is selected manually, or the computer program can control transmitter selection through one of its parallel ports. Placing the switch in position TX B overrides computer control and selects Radio B manually.

3. A manual T/R CONTROL switch, S5, on the main box can disable automatic computer control of the transmitter selection. While the *TR* program allows this function to be set by software control, *CT* doesn't have this ability, so S5 was added.

4. An external paddle may be connected to J10 to send CW, using the *TR* computer program as a keyer.

5. Two sets of paralleled, diode-isolated RCA phono jacks (J3 through J6) are mounted on the main box. You can connect the AMPLIFIER RELAY control lines from each transceiver to both the switchbox and two external amplifiers without needing Y-connectors. By the way, only one radio can transmit at a time with this switchbox. This keeps you completely legal in any single-operator or multi-single contest category.

6. An RCA phono connector J9 is available for an external foot-switch input to the *TR* program.

7. A separate EXT KEYER RCA phono connector J13 mounts on the main box. This parallels the LPT CW keyer output from the computer program.

8. Red LEDs on the remote head indicate which transmitter is active while it is transmitting.

THE SCHEMATIC

Fig 19.63A shows the schematic of the PCB used in the two-radio switchbox, and Fig 19.63B shows the interconnection diagram from the PCB to the other components in the main box and the remote head. Switching relays K1 and K2 are the heart of the receive-audio circuitry. K1A selects audio from either Receiver A or B and applies it to the left-ear terminal of headphone output jack J11. Relay K2B normally connects the right-ear terminal of J11 in parallel with the left-ear terminal, for monaural operation. In the STEREO mode, K2B puts the output from Receiver B into the right ear. Note that S3A on the remote head disables K1 when STEREO mode is selected, so that Receiver A output remains in the left ear, with Receiver B output in the right ear.

Relay contacts K1B are used to turn on the Green receive LEDs in the control head to indicate whether audio is coming from Receiver A alone, B alone, or both A and B together in stereo. Many contest operators have trained themselves to listen to two radios simultaneously in STEREO mode. But many of us are only endowed with a single brain, and we get easily distracted in stereo, especially on SSB! Thus the AUTO CQ feature was added to the switchbox.

In this mode, when S4 in the control head is set to AUTO TX and S3 is set to MONO, closure to ground on the AMPLIFIER RELAY line by either transceiver will toggle between the audio outputs from Receiver A and B automatically. When the radio stops transmitting, the receive audio will toggle back. Both AMPLIFIER RELAY control lines are isolated from each other by summing diodes.

Just so you won't be surprised, both the *CT* and *TR* software programs energize "Radio 2" as what we call "Radio A," and "Radio 1" as "Radio B." This is slightly non-intuitive but it seems to be a function of the default state of the computer parallel port. The operator, however, quickly becomes accustomed to this and toggles between the radios, while watching the computer screen to see which frequency band is active.

The author borrowed most of the transmitting-selection circuitry from an N6TR design. (N6TR is the creator of the *TR* program.)

CONSTRUCTION

Fig 19.64 shows a view of the inside of the completed main unit box. The author chose to use a standard Bud LMB 6×4×3½-inch aluminum box, but things got pretty crowded, as you can see. You will probably want to use a bigger box—and you will probably choose to use smaller cables in the bundles going to each transceiver. N6BV taped together three standard "zip cords" for the key-line, PTT and receiver audio cables, plus a shielded cable for the microphone line. This bundle was definitely overkill in terms of current-handling capacity and made the service loops inside the main box quite bulky. Smaller-gauge "speaker wire" would have been far easier to handle.

Most of the switchbox circuitry was built on a single-sided copper PCB using "wired-traces construction" (also called the "Lazy PC Board" technique), described in the **Circuit Construction** chapter. A large-diameter drill was used to ream out the copper around holes where the ground plane was supposed to be removed.

A 25-pin DB25F connector was mounted

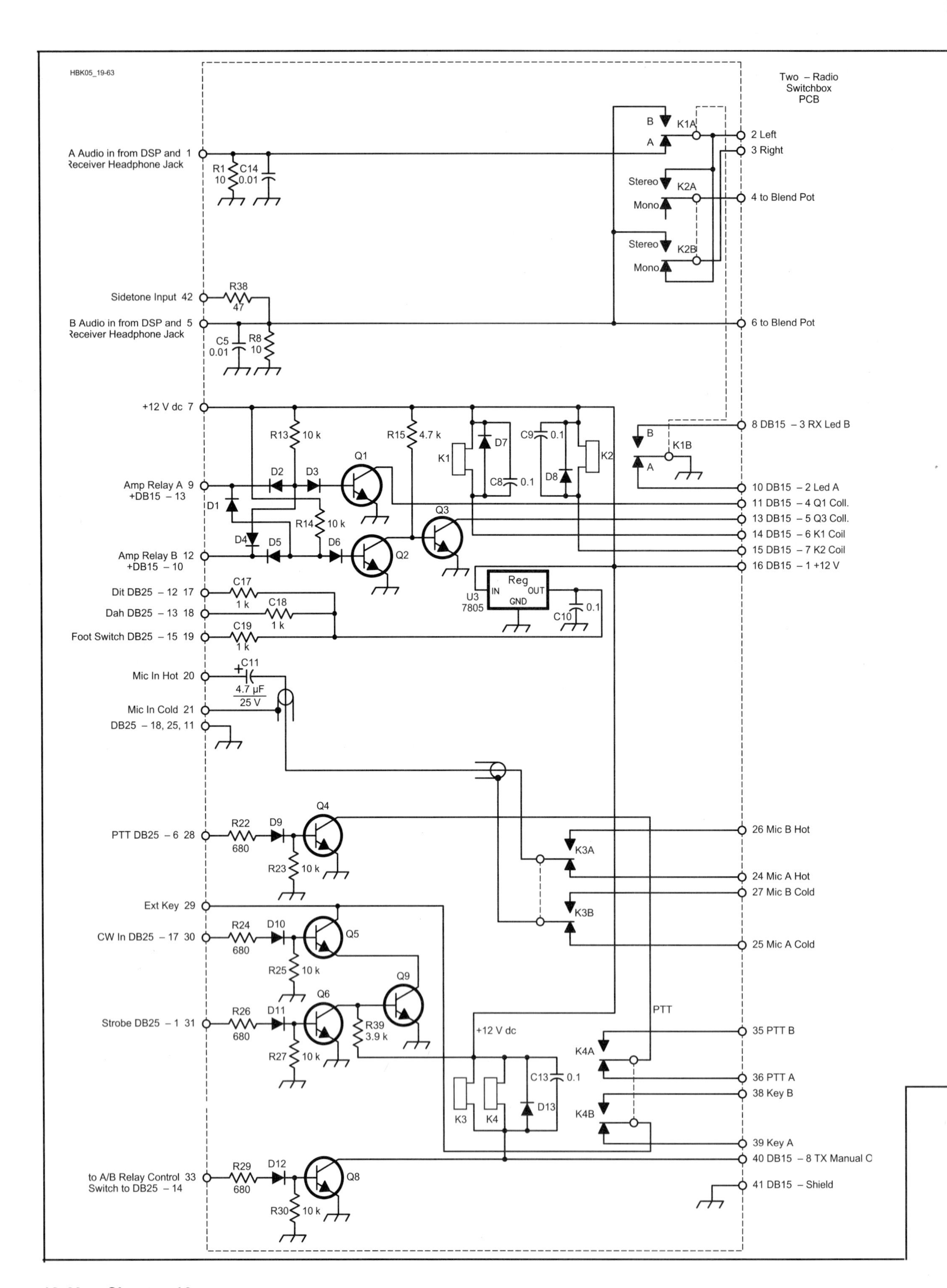
HBK05_19-63
Two – Radio
Switchbox
PCB
A Audio in from DSP and 1
Receiver Headphone Jack
R1
10
C14
0.01
B
A
K1A
2 Left
3 Right
Stereo
Mono
K2A
4 to Blend Pot
Stereo
Mono
K2B
Sidetone Input 42
R38
47
B Audio in from DSP and 5
Receiver Headphone Jack
C5
0.01
R8
10
6 to Blend Pot
+12 V dc 7
R13
10 k
R15
4.7 k
D7
C9
0.1
K1
K2
C8
0.1
D8
8 DB15 – 3 RX Led B
K1B
10 DB15 – 2 Led A
11 DB15 – 4 Q1 Coll.
13 DB15 – 5 Q3 Coll.
14 DB15 – 6 K1 Coil
15 DB15 – 7 K2 Coil
16 DB15 – 1 +12 V
Q1
D2
D3
Amp Relay A 9
+DB15 – 13
D1
R14
10 k
Q3
D4
D5
D6
Q2
Amp Relay B 12
+DB15 – 10
Reg
IN
OUT
GND
U3
7805
C10
0.1
Dit DB25 – 12 17
C17
1 k
Dah DB25 – 13 18
C18
1 k
Foot Switch DB25 – 15 19
C19
1 k
Mic In Hot 20
C11
4.7 µF
25 V
Mic In Cold 21
DB25 – 18, 25, 11
Q4
PTT DB25 – 6 28
R22
680
D9
R23
10 k
26 Mic B Hot
K3A
24 Mic A Hot
27 Mic B Cold
K3B
25 Mic A Cold
Ext Key 29
CW In DB25 – 17 30
R24
680
D10
Q5
R25
10 k
Q9
Q6
Strobe DB25 – 1 31
R26
680
D11
R39
3.9 k
R27
10 k
+12 V dc
PTT
35 PTT B
K4A
36 PTT A
38 Key B
C13
0.1
D13
K3
K4
K4B
39 Key A
40 DB15 – 8 TX Manual C
to A/B Relay Control 33
Switch to DB25 – 14
R29
680
D12
Q8
R30
10 k
41 DB15 – Shield

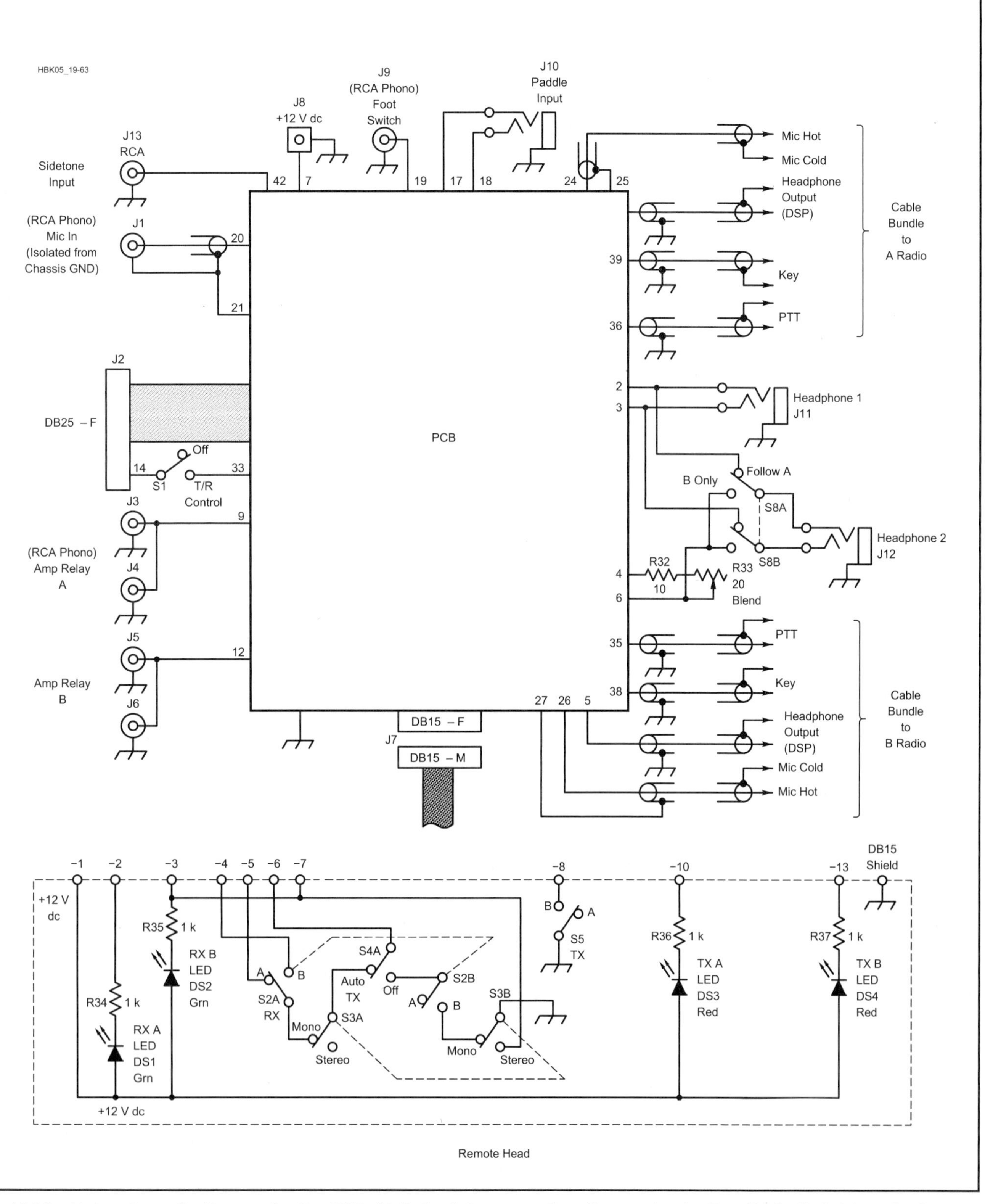

Fig 19.63—At A, schematic diagram for the PCB. At B, the interconnection diagram for the Two-Radio Switchbox. Resistors are 1/4 W. Capacitors are disk ceramic. Capacitors marked with polarity are electrolytic.

D1-D13—General-purpose silicon switching diodes, such as 1N4148 or 1N914 (RS 276-1620).
DS1, DS2—Green LED (RS 276-069A).
DS3, DS4—Red LED (RS 276-068A).
J1—Insulated RCA phono jack for microphone input.
J3-6, J6, J9, J13—Chassis-mounting, grounded RCA phono jack (RS 274-346).
J10-12—1/4-inch stereo phone jack (RS 274-312B).
J8—DC power jack (RS 274-1565A).
K1-K4—DPDT 12-V PC-mount relay (RS 275-49A).
Q1-Q9—General-purpose NPN switching transistor, 2N2222 or 2N3904 style (RS 276-1617).
S1-S8—Flat lever switch, DPDT (RS-275-636B).
S6, S7—Momentary contact SPST (RS 275-1556A).
U3—7805 5-V regulator (RS 271-281).

Fig 19.64—Inside the main box. With service loops for the cable bundles going to the two radios, the inside of the main box is cramped and a larger box would be a good idea!

on the main box for J2 and an inexpensive 6-foot long 25-conductor DB25M-to-DM25M cable (bought at a local computer store) went to the computer's parallel port. Much of the point-to-point wiring to the DB25 and DB15 connectors used ribbon wires from a scavenged surplus computer.

The remote-control head required a total of 10 wires from the main box, and a 15-pin DB15M connector J7 was used so that an inexpensive commercial VGA computer DB15F cable could be used. The connector at one end was cut off the 6-foot long cable, which was then hard-wired into the remote head. Tie wraps were used to provide mechanical strain reliefs for cables entering into the main box and into the remote head.

The author created labels using a word processing program in 12-point Times Roman typeface. These were laser printed onto a thin mylar sheet used for creating overhead transparency films. Clear nail polish ("Hard As Nails") was brushed lightly over the labels on the mylar sheet to protect them from wear. After the polish had dried, the labels were cut out using a paper cutter and they were then stuck onto the boxes using more clear nail polish as glue.

OPERATION

The switchbox and remote head can be operated entirely manually, with no connection to a computer, if you like. However, it is a lot more fun when you can control each radio from the computer keyboard, especially if you've interfaced the radio to read and write the frequency to the transceivers! Then you can "point-and-shoot" from packet DX spots or you can type in a desired frequency and press the [Enter] key to tune your radio.

You must set up your computer program to control the two-radio switchbox through a parallel (LPT) port. If you control your radios' frequencies, you will do that through serial ports. Follow the directions for your software carefully. In general, you've got to get everything *exactly right* in order for all functions to work properly, particularly the frequency and mode controls for your radio. Connect the switchbox to a source of +12 V dc that can source about 1 A. The ICOM radios have a jack on the back that will provide this.

With the control head switches set to TX A, MONO and TX AUTO OFF, toggle between Radio 2 and Radio 1. In the *CT* program press ALT-. (press the ALT key and the period key together) or in *TR* press ALT-R to toggle between transmitter A and B. You should hear relays changing in the switchbox and when you go into transmit the red Transmit LEDs on the remote head will light for the appropriate transmitter.

Now, switch between the RX A and the RX B positions. The green receiver RX A and RX B LEDs on the control head should alternately light. Try the STEREO switch to see both green LEDs light up. Switch back to MONO and then switch the TX AUTO switch on and key the transmitter. The RX A and RX B LEDs should toggle as you key and unkey the transmitter.

Adjust the receiver front-panel volume controls for normal audio levels. Now, check out STEREO headphone operation and adjust the BLEND pot for the spatial placement you prefer.

You should now be all set to enjoy computer-controlled two-radio operation!

TR TIME-DELAY GENERATOR FOR VHF/UHF SYSTEMS

If you've ever blown up your new GaAsFET preamp or hard-to-find coaxial relay, or are just plain worried about it, this transmit/receive (TR) time-delay generator is for you. This little circuit makes it simple to put some reliability into your present station or to get that new VHF or UHF transverter on the air fast, safe and simple. Its primary application is for VHF/UHF transverter, amplifier and antenna switching, but it can be used in any amplifier-antenna scheme. An enable signal to the TR generator will produce sequential output commands to receive relay, a TR relay, an amplifier and a transverter—automatically. All you do is sit back and work DX! This project was designed and built by Chip Angle, N6CA.

WHY SEQUENCE?

Several problems may arise in stations using transverters, extra power amplifiers and external antenna-mounted TR relays. The block diagram of a typical station is shown in **Fig 19.65**. When the HF exciter is switched into transmit by the PTT or VOX line, it immediately puts out a ground (or in some cases a positive voltage) command for relay control, and an RF signal.

If voltage is applied to the transverter, amplifier and antenna relays simultaneously, RF can be applied as the relay

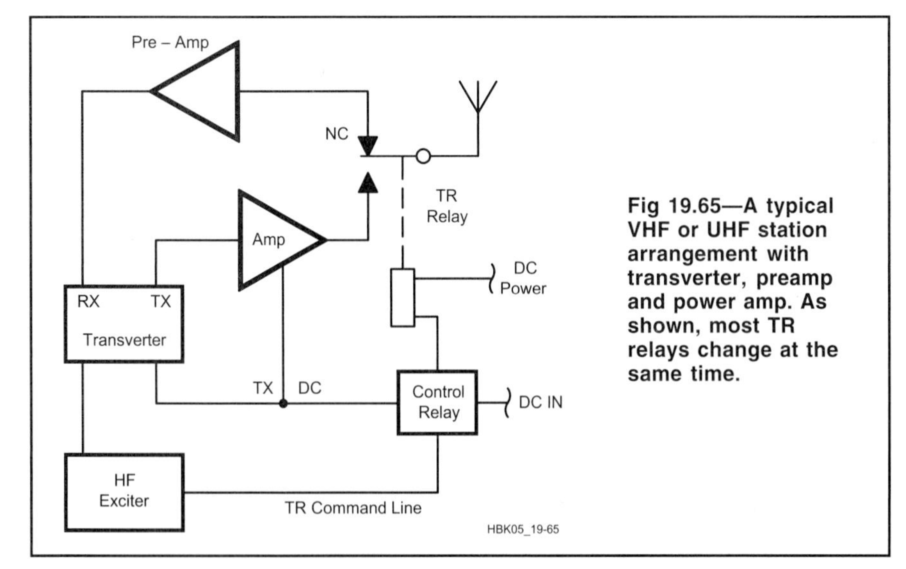

Fig 19.65—A typical VHF or UHF station arrangement with transverter, preamp and power amp. As shown, most TR relays change at the same time.

contacts bounce. In most cases, RF will be applied before a relay can make full closure. This can easily arc contacts on dc and RF relays and cause permanent damage. In addition, if the TR relay is not fully closed before RF from the power amplifier is applied, excessive RF may leak into the receive side of the relay. The likely result—preamplifier failure!

Fig 19.66 is a block diagram of a station with a remote-mounted preamp and antenna relays. The TR time-delay generator supplies commands, one after another, going into transmit and going back to receive from transmit, to turn on all station relays in the right order, eliminating the problems just described.

CIRCUIT DETAILS

Here's how it works. See the schematic diagram in **Fig 19.67**. Assume we're in receive and are going to transmit. A ground command to Q2 (or a positive voltage command to Q1) turns Q2 off. This allows C1 to charge through R1 plus 1.5 kΩ. This rising voltage is applied to all positive (+) inputs of U1, a quad comparator. The ladder network on all negative (–) inputs of U1 sets the threshold point of each comparator at a successively higher level. As C1 charges up, each comparator, starting with U1A, will sequentially change output states.

The comparator outputs are fed into U2, a quad exclusive-OR gate. This was included in the design to allow "state programming" of the various relays throughout the system. Because of the wide variety of available relays, primarily coaxial, you may be stuck with a relay that's exactly what you need—except its contacts are open when it's energized. To use this relay, you merely invert the output state of the delay generator by using a jumper between the appropriate OR-gate input and ground. Now, the relay will be "on" during receive and "off" during transmit. This might seem kind of strange; however, high-quality coaxial relays are hard to come by and if "backwards" relays are all you have, you'd better use them.

The outputs of U2 drive transistors Q3-Q6, which are "on" in the receive mode. Drive from the OR gates turns these transistors "off." This causes the collectors of Q3-Q6 to go high, allowing then base-to-emitter junctions of Q7-Q10 to be forward-biased through the LEDs to turn on the relays in sequential order. The LEDs serve as built-in indicators to check performance and sequencing of the generator. This is convenient if any state changes are made.

When the output transistors (Q7-Q10) are turned on, they pull the return side of the relay coils to ground. These output transistors were selected because of their high beta, a very low saturation voltage (V_{CE}) and low cost. They can switch (and have been tested at) 35 V at 600 mA for many days of continuous operation. If substitutions are planned, test one of the new transistors with the relays you plan to use to be sure that the transistor will be able to power the relay for long periods.

To go from transmit to receive, the sequencing order is reversed. This gives additional protection to the various system components. C1 discharges through

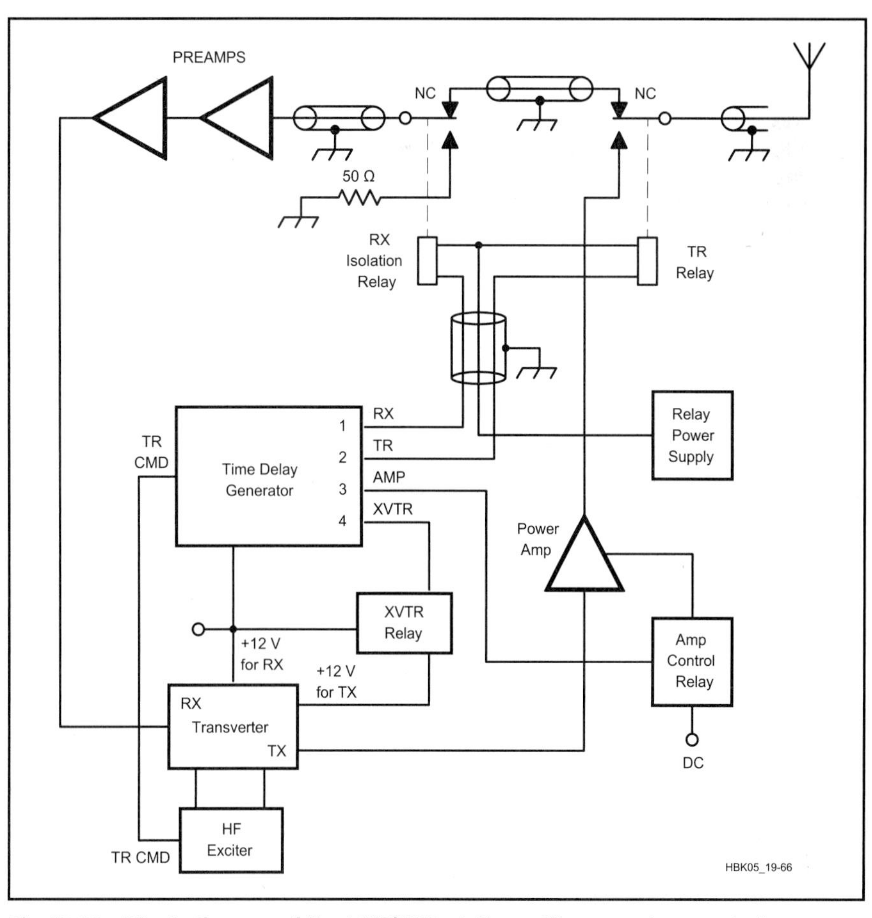

Fig 19.66—Block diagram of the VHF/UHF station with a remote-mounted preamp and antenna relays. The TR time-delay generator makes sure that everything switches in the right order.

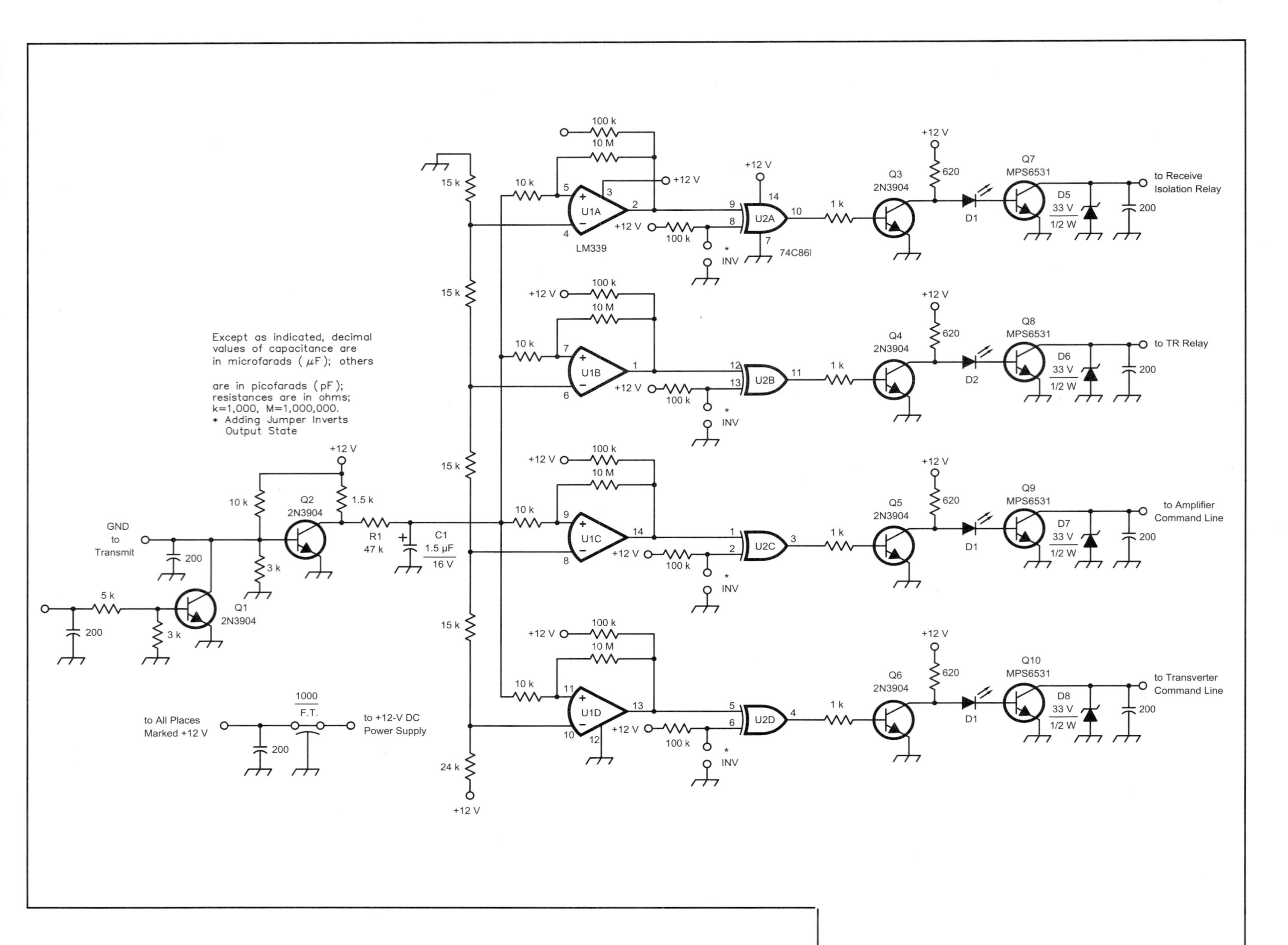

100 k
10 M
15 k
10 k
+12 V
U1A
U1B
U1C
U1D
LM339
U2A
U2B
U2C
U2D
74C86I
* INV
1 k
Q3 2N3904
Q4 2N3904
Q5 2N3904
Q6 2N3904
620
D1
D2
Q7 MPS6531
Q8 MPS6531
Q9 MPS6531
Q10 MPS6531
D5 33 V 1/2 W
D6 33 V 1/2 W
D7 33 V 1/2 W
D8 33 V 1/2 W
200
to Receive Isolation Relay
to TR Relay
to Amplifier Command Line
to Transverter Command Line
Except as indicated, decimal values of capacitance are in microfarads (µF); others are in picofarads (pF); resistances are in ohms; k=1,000, M=1,000,000.
* Adding Jumper Inverts Output State
GND to Transmit
Q2 2N3904
Q1 2N3904
1.5 k
3 k
5 k
R1 47 k
C1 1.5 µF 16 V
24 k
1000 F.T.
to All Places Marked +12 V
to +12-V DC Power Supply

Fig 19.67—Schematic diagram of the TR time-delay generator. Resistors are 1/4 W. Capacitors are disc ceramic. Capacitors marked with polarity are electrolytic.

D1-D4—Red LED (MV55, HP 5082-4482 or equiv.)
D5-D8—33-V, 500-mW Zener diode (1N973A or equiv.)
C1—1.5-µF, 16-V or greater, axial-lead electrolytic capacitor. See text.
Q1-Q6—General purpose NPN transistor (2N3904 or equiv.)
Q7-Q10—Low-power NPN amplifier transistor, MPS6531 or equiv. Must be able to switch up to 35 V at 600 mA continuously. See text.
R1—47-kΩ, 1/4-W resistor. This resistor sets the TR delay time constant and may have to be varied slightly to achieve the desired delay. See text.
U1—Quad comparator, LM339 or equiv.
U2—Quad, 2-input exclusive or gate (74C86N, CD4030A or equiv.)

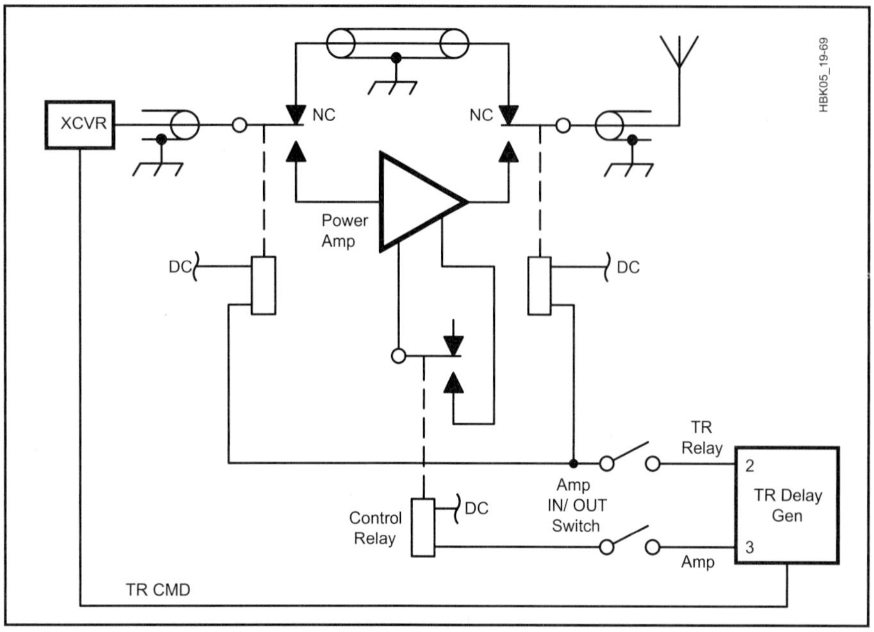

Fig 19.69—The TR time-delay generator can also be used to sequence the relays in an HF power amplifier.

R1 and Q2 to ground.

Fig 19.68 shows the relative states and duration of the four output commands when enabled. With the values specified for R1 and C1, there will be intervals of 30 to 50 milliseconds between the four output commands. Exact timing will vary because of component tolerances. Most likely everything will be okay with the values shown, but it's a good idea to check the timing with an oscilloscope just to be sure. Minor changes to the value of R1 may be necessary.

Most relays, especially coaxial, will require about 10 ms to change states and stop bouncing. The 30-ms delay will give adequate time for all closures to occur.

Construction and Hookup

One of the more popular antenna changeover schemes uses two coaxial relays: one for actual TR switching and one for receiver/preamplifier protection. See **Fig 19.69**.

Many RF relays have very poor isolation especially at VHF and UHF frequencies. Some of the more popular surplus relays have only 40-dB isolation at 144 MHz or higher. If you are running high power, say 1000 W (+60 dBm) at the relay, the receive side of the relay will see +20 dBm (100 mW) when the station is transmitting. This power level is enough to inflict fatal damage on your favorite preamplifier.

Adding a second relay, called the RX isolation relay here, terminates the preamp in a 50-Ω load during transmission and increases the isolation significantly. Also, in the event of TR relay failure, this extra relay will protect the receive preamplifier.

As shown in Fig 19.69, both relays can be controlled with three wires. This scheme provides maximum protection for the receiver. If high-quality relays are used and verified to be in working order, relay losses can be kept well below 0.1 dB, even at 1296 MHz. The three-conductor cable to the remote relays should be shielded to eliminate transients or other interference.

By reversing the RX-TX state of the TR relays (that is, connecting the transmitter Hardline and 50-Ω preamp termination to the normally open relay ports instead of the normally closed side), receiver protection can be provided. When the station is not in use and the system is turned off, the receive preamplifier will be terminated in 50 Ω instead of being connected to the antenna. The relays must be energized to receive. This might seem a little backward; however, if you are having static-charge-induced preamplifier failures, this may solve your problem.

Most coaxial relays aren't designed to be energized continuously. Therefore, adequate heat sinking of coaxial relays must be considered. A pair of Transco Y relays can be energized for several hours

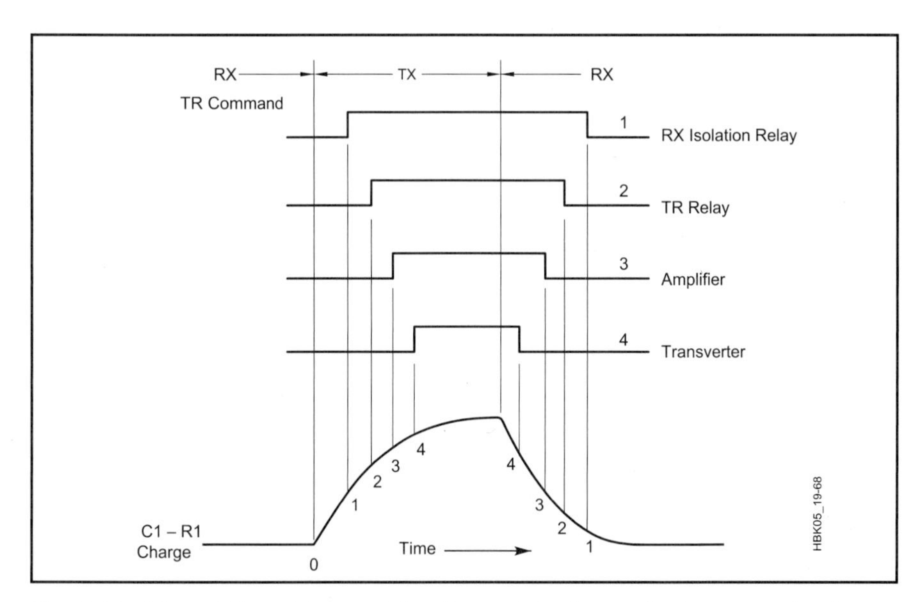

Fig 19.68—The relative states and durations of the four output commands when enabled. This diagram shows the sequence of events when going from receive to transmit and back to receive. The TR delay generator allows about 30 to 50 ms for each relay to close before activating the next one in line.

when mounted to an aluminum plate 12 inches square and 1/4 inch thick. Thermal paste will give better heat transfer to the plate. For long-winded operators, it is a good idea to heat sink the relays even when they are energized only in transmit.

Fig 19.69 shows typical HF power amplifier interconnections. In this application, amplifier in/out and sequencing are all provided. The amplifier will always have an antenna connected to its output before drive is applied.

Many TR changeover schemes are possible depending on system requirements. Most are easily satisfied with this TR delay generator.

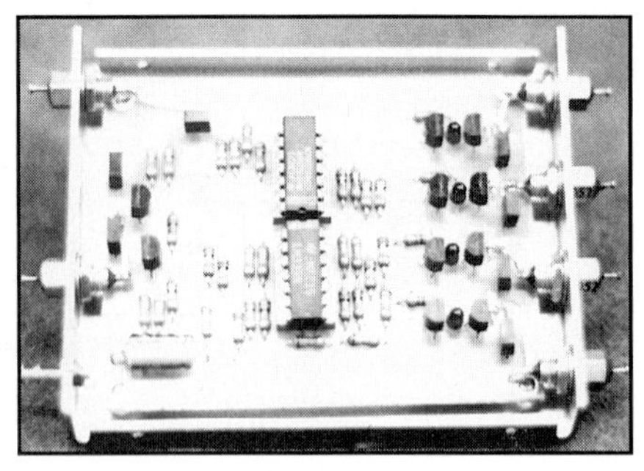

Fig 19.70—The completed time-delay generator fits in a small aluminum box.

The TR delay generator is built on a 2 1/2×3 1/4-inch PC board.[1] See **Fig 19.70**. Connections to the rest of the system are made through feedthrough capacitors. Do not use feedthrough capacitors larger than 2000 pF because peak current through the output switching transistors may be excessive.

[1]From FAR Circuits. Use *TIS Find* program for address information. Template is on CD-ROM.

A SWITCHED ATTENUATOR

How many times has a signal been too strong for the experiment you wish to carry out? It could be from an oscillator on the bench or from signals from an antenna overpowering a mixer. This attenuator will solve those problems and is presented from *Practical Projects*, courtesy of the RSGB.

WHERE CAN I USE IT?

Applications other than those given above are if you are interested in DF and need to attenuate the signal when getting close to the transmitter, or if you have a problem with a TV signal being too strong and causing ghosting on other signals (cross-modulation).

ATTENUATORS

These problems can be averted by using a switched attenuator (pad). The times that we need attenuators occur far more often than first realized. When designing the attenuators, account must be taken of the distinct possibility of poor screening. There is hardly any point in designing a 20-dB attenuator when the leakage around the circuit is approaching this value. It is also important to decide on the accuracy required. If it is intended to do very accurate measurements the construction has to be impeccable, but for comparisons between signals it would be possible to accept attenuation values to a smaller degree of accuracy.

The most useful attenuator is a switched unit where a range from zero to over 60-dB in 1-dB steps can be covered. This is not as difficult as it first seems because, by summing different attenuators, we can obtain the value we need. It takes only seven switches to cover 65-dB. The seven values of attenuation are 1 dB, 2 dB, 4 dB, 8 dB, 10 dB, and two at 20 dB; these can be switched in or out at will. As an example, if 47 dB were needed, switch on the two 20-dB pads plus the 4, 2 and 1-dB pads. [A 'pad' is the name given to a group of components with a known attenuation. — *Ed*]

Construction

The prototype attenuator is shown in **Fig 19.71**. It was constructed 20 years ago and it is still in regular use. It is housed in a box made from epoxy PCB material. The top and sides are cut to size and soldered into a box. It is easier to cut the switch holes prior to making the box. After the box has been constructed, screens made from thin brass shim should be cut and soldered between the switch holes. Next, the switches are fitted and the unit wired up. When this is done, the unit is checked and a back cover, securely earthed to the box, is fitted.

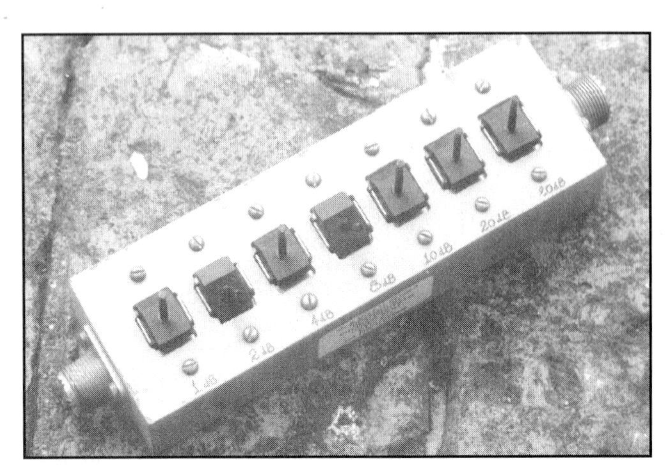

Fig 19.71—The completed attenuator.

Components

The switches must have low capacitance between the contacts and simple slide switches are the best selection. The Maplin DPDT miniature switch (FH36P) would be suitable and Maplin also supplies 1% resistors. Connectors to the unit must be coaxial but can be left to personal preference. (Equivalent components are available from RadioShack.—*Ed.*)

The resistor values shown in **Fig 19.72** determine the attenuator's accuracy at around 5%. This is done for practical reasons. For example, if we wanted to make the attenuation value of the 4-dB cell *exactly* 4 dB, the resistor values would have to be 220.97 and 23.85 Ω. You will see from Fig 19.72 that the values used are 220 and 24 Ω, giving an attenuation value of 4.02 dB.

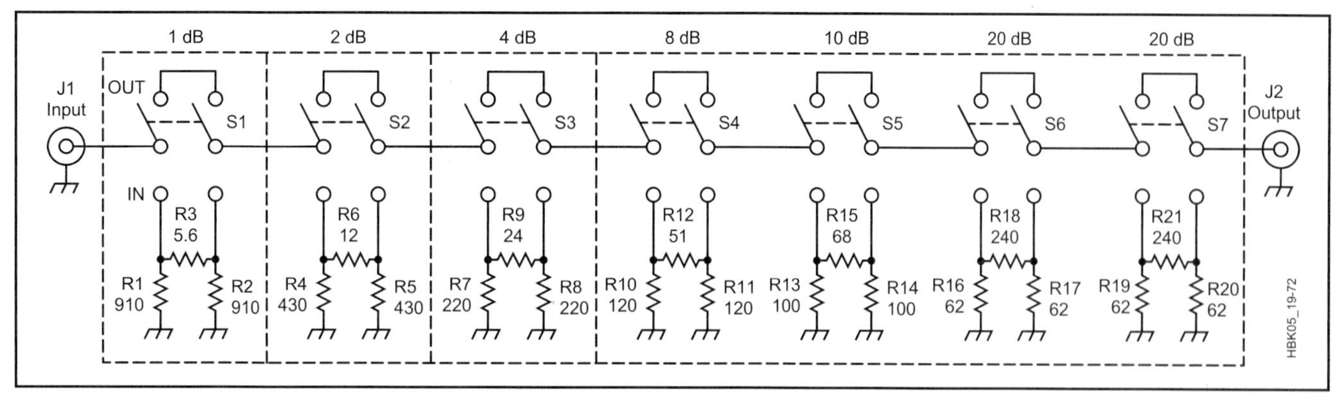

Fig 19.72—The attenuator consists of seven pi network sections, so-called because each pad (eg, R1, R3 and R2) resembles the Greek letter pi (π). Input and output impedances are 50 Ω.

SIMPLE QRP TRANSMIT-RECEIVE CHANGEOVER

When you build a transmitter, you need a method of switching the antenna from the receiver to the transmitter during the transmit period. It is also useful to coordinate the transmitter power supply switching with the antenna changeover switching. The transmitter should not have its power connected during receive periods; if you were to touch the key by accident, the transmitter will operate without any load. In many cases this will cause the PA transistor to fail and may harm your receiver. This project is presented from *Practical Projects*, courtesy of the RSGB.

Fig 19.73—The completed unit.

DOUBLE-POLE SWITCH

The simplest changeover is a double-pole switch. This can change the antenna with one pole and the power with the other. The disadvantage, however, is that you must remember to flick the switch each time you want to change from receive to transmit. With a small QRP transmitter it is possible to flick the little rig right off the bench!

A primitive method of sharing the antenna between the transmitter and the receiver is to have the antenna permanently connected to the transmitter. The receiver is then connected to the antenna via a small capacitor. Two back-to-back diodes across the receiver input are used to prevent large RF voltages from damaging the receiver. The diodes can be 1N4148 or similar and the capacitor can be 50 to 100 pF.

The problems with this arrangement are that some transmitter power is dissipated in the diodes, and a lot of RF still gets into the receiver front-end; on transmit loud

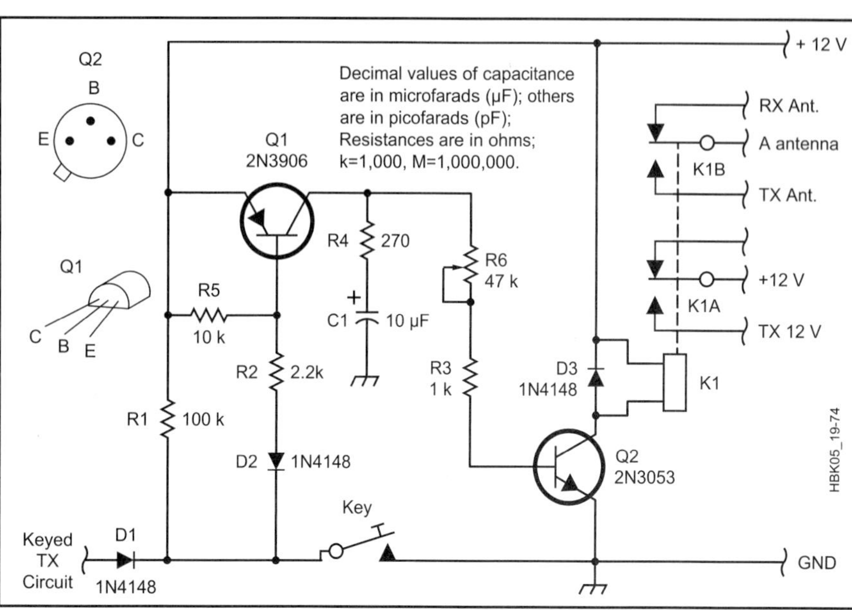

Fig 19.74—Circuit diagram of the electronic transmit/receive antenna changeover switch.

thumps are heard from the receiver. The volume control on the receiver can be turned down but this can become a tiresome chore when using this method.

Electronic Antenna Changeover

A much better method is shown in **Figs 19.73** and **19.74**. The circuit causes a relay to change over when the key is pressed. The circuit is simple and uses a pnp transistor as a switch and an npn transistor as a relay driver. The relay can be any small one with a 12-V coil. Take care to check that the relay does not already have a diode across the coil. If it does, you can leave it in and omit D3, but take care to fit the relay the correct way round or it will not work. The circuit is built on plain perforated board, as shown in **Fig 19.75**, with the components pushed through and solder connections made beneath the board. The solder tags (lugs, in American English—*Ed.*) act as earth contacts to brass standoffs to the case when the board is bolted to them. The changeover board can be housed in the same case as the transmitter.

When complete, check the wiring and connect the 12-V supply. The relay should change over when the key is pressed. The delay is set by R6. The relay should drop out when you pause between words, but not between letters. Diode D1 is included to prevent interaction with the circuits to be keyed on the transmitter.

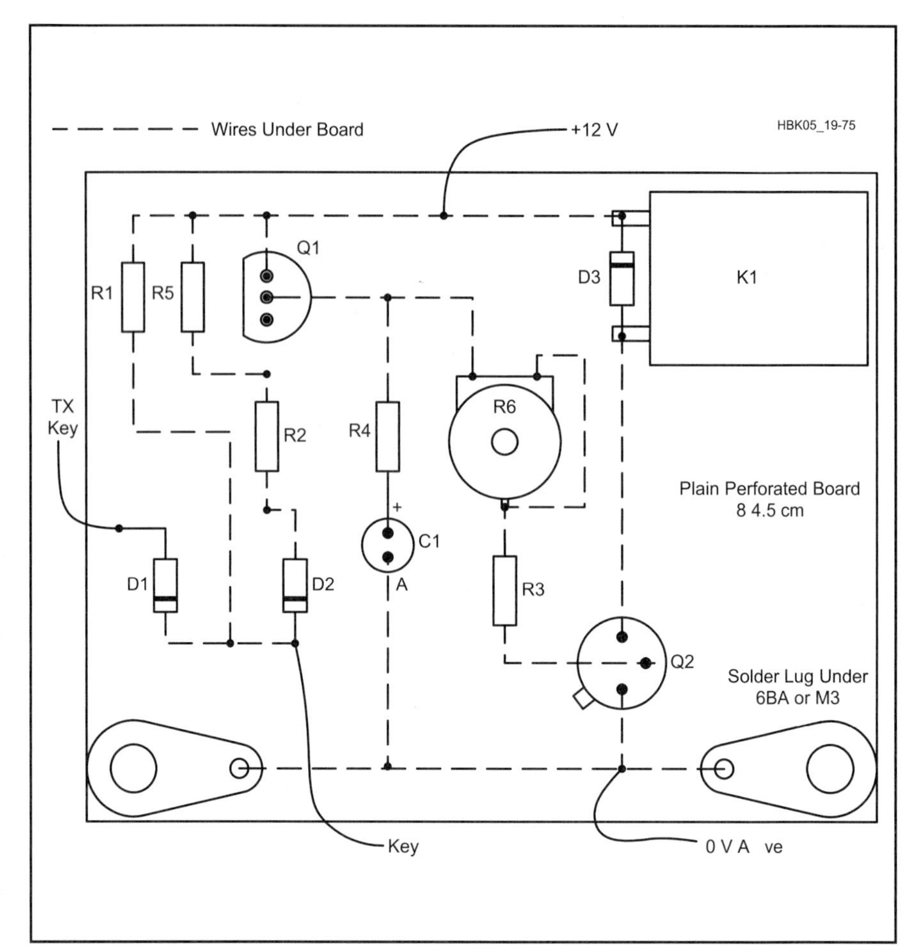

Fig 19.75 — Component layout.

A QRP L-MATCH ATU

The author had good results with this simple L-match tuner feeding a single-wire antenna against quarter-wave counterpoises (elevated 'ground wires'). He used only two counterpoises, 20 m and 5 m long, to cover all bands from 80 m to 10 m. The simple tuner project shown in **Fig 19.76** is presented from *Practical Projects*, courtesy of RSGB.

CONSTRUCTION

The L-match employs just two main components, a coil and a capacitor, as shown in **Fig 19.77**. The coil is wound on a plastic 35-mm film container. It has a total of 50 turns of #24 SWG enameled copper wire, wound tightly and without spaces between the turns, as shown in **Fig 19.78**. It is tapped at each turn up to 10, then at 15, 20, 25, 30, 35, 40 and 45 turns. The use of crocodile clips permits any number of turns to be selected.

Fig 19.76—Front view of the simple L-match ATU, showing crocodile-clip connection to the coil. The knob must be of the insulated type, as both sides of the variable capacitor are potentially live to RF.

The taps on the coil are formed by bending the wire back on itself and twisting a loop. The loops are then scraped and tinned with solder. Some enameled copper wire is self-fluxing and only requires the application of heat from a hot soldering iron bit loaded with fresh solder for a few seconds before it will tin. Make a good job of tinning the loops to ensure the crocodile clips make good contact. The capacitor is a polyvaricon from an old radio. The author used a 2 to 200 pF component with both gangs in parallel, but other values will work.

The ATU was made up on a wooden base with scrap PCB for the front and rear panels, as in **Fig 19.79**. You will need a socket to go to the transmitter or receiver and one each for the wire antenna and the counterpoise or earth. You can use what-

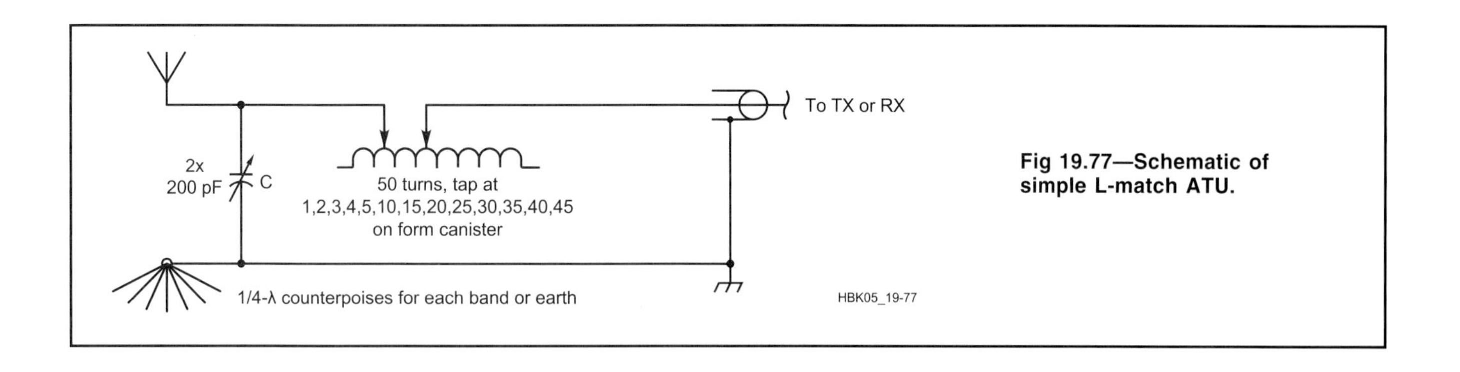

Fig 19.77—Schematic of simple L-match ATU.

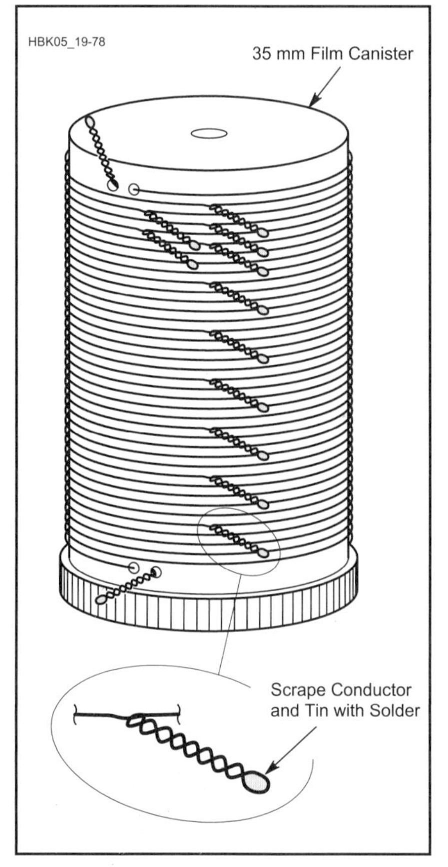

Fig 19.78—The coil is constructed on a 35-mm film container.

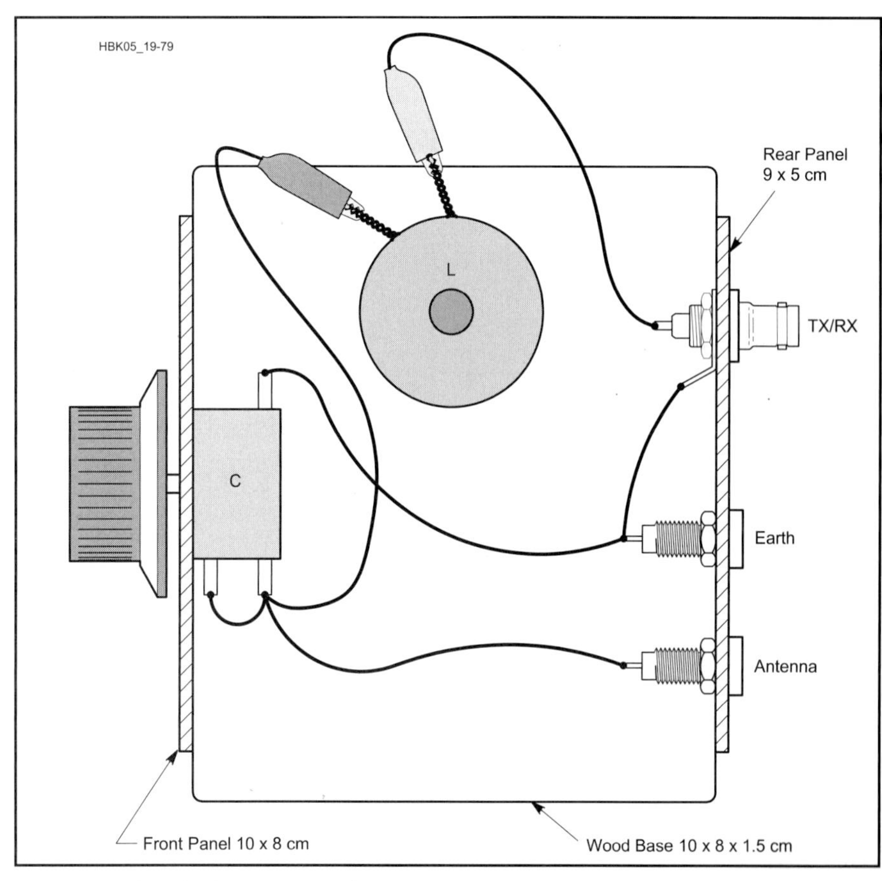

Fig 19.79—Physical layout. The front and rear panels are fixed to the base with wood screws.

Fig 19.80—Rear view of the ATU, showing sockets and connections to the capacitor.

ever matches your existing equipment. The prototype used a phono socket for the transmitter and two 4-mm sockets for the antenna and earth. The coil is mounted by screwing the lid of the film container down on the base, then snapping the completed coil assembly into it. He glued the capacitor to the front panel. Wire up point-to-point, as shown in **Fig 19.80**.

OPERATION

When completed, you can check operation with a receiver or a low-power transmitter (less than 5 W). With a transmitter, use an SWR meter to find the coil tap which gives the lowest SWR, then adjust the capacitor to tune to minimum SWR. Incidentally, remember not to touch any exposed metal within the ATU while on transmit! An insulated knob and a scale help you to note the position of the capacitor for future reference. To set it up using a receiver only, find the best position for the tap and the capacitor by listening to a weak signal and adjusting for the loudest signal (with the help of an S meter, if you have one). As a guide, try 20 turns on 160 m, 10 turns on 40 m, 4 turns on 20 m, and 2 turns on 10 m and 15 m.

A QRP T-MATCH ATU

An antenna tuning unit (ATU) is very useful if you are a licensed radio amateur or a short-wave listener. The purpose of an ATU is to adjust the antenna feed impedance so that it is very close to the 50-Ω impedance of the receiver or transmitter, a process known as *matching*.

When used with a receiver, an ATU can dramatically improve the signal-to-noise ratio of the received signal. On transmit, the antenna must be matched to the transmitter so that the power amplifier operates efficiently. The QRP tuner shown in **Fig 19.81** is also presented from *Practical Projects*, courtesy of the RSGB.

DESIGN

The basic ATU design is called a *T-Match*; you can see the basic shape of the letter 'T' reflected in the layout formed by the components C1, C2 and L1 in **Fig 19.82**. The circuit will match the coaxial output of the transceiver to an end-fed antenna or to a coaxial cable feed to the antenna. This design also uses a BALUN (BALanced to *un*balanced) transformer for use with antennas using twin-wire balanced feeder. This unit will handle up to 5 W and operates over the frequency range of 1.8 to 30 MHz.

CONSTRUCTION

Inductor L1 is wound on a T-130-2 powdered-iron toroid. The inductor is tapped and fixed to the tags (terminals, in American English—*Ed.*) of a 12-way rotary switch. Taps are formed by making a loop about 1 cm long in the wire and twisting it tightly. The loops are scraped clean of enamel and tinned with solder ready to be soldered onto the tags of the 12-way rotary switch. If the loops are about 1 cm long, it is just possible to bend them to fit the tabs of the switch without having to extend them with short wires.

The balun is wound with two wires twisted together; known as BIFILAR

Table 19.9
Components list

Capacitors
C1, 2 200 - 200 pF

Inductors

L1 T-130-2 powdered iron toroid with total 36 turns of #22 SWG enameled copper wire, tapped at 10, 12, 15, 17, 20, 23, 26, 29, 31, 33 and 35 turns from the earth end

T1 12 bifilar turns of #26 SWG enameled copper on FT-50-43 toroid

Additional items
S1 1-pole 12-way rotary switch
S2 SPST toggle switch
RF connectors, SO239 sockets or similar (see text)
Aluminium or die-cast metal box
Three plastic knobs for capacitors and switch
Two 4mm sockets for balanced antenna connection

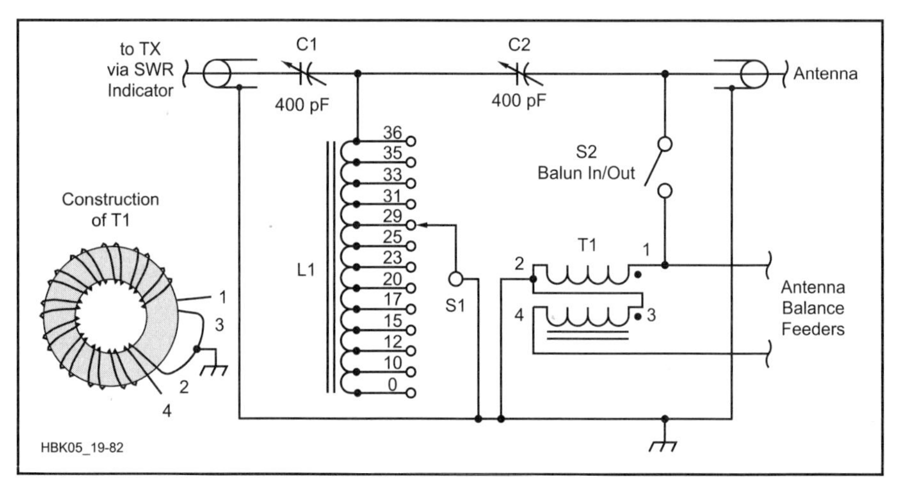

Fig 19.82—Basic circuit of the T-Match ATU.

Fig 19.81—The completed T-Match ATU.

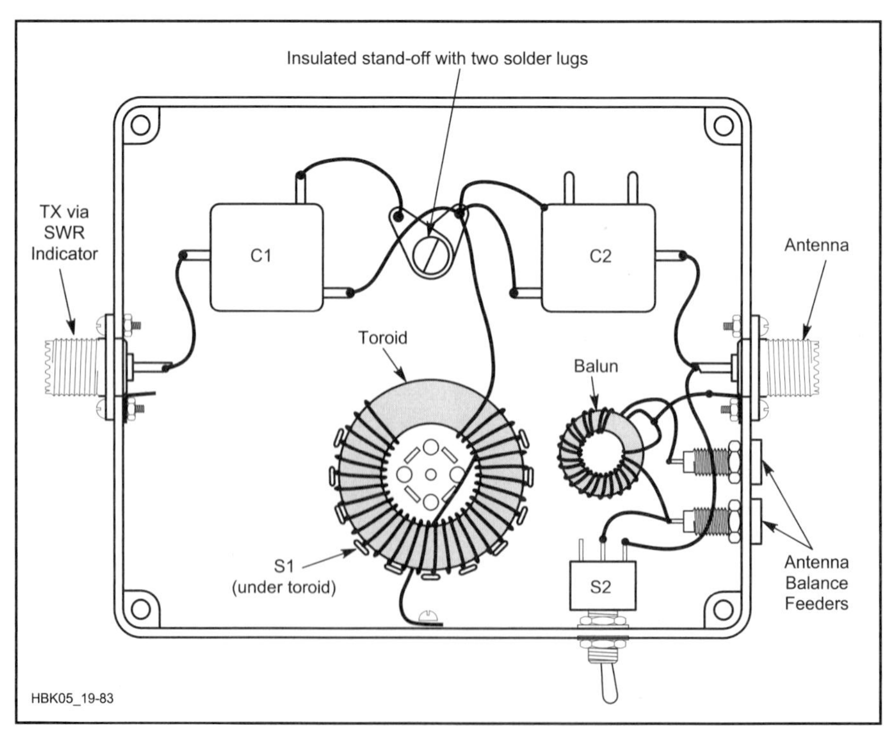

Fig 19.83—Internal view of the ATU. A small metal box makes a suitable housing.

WINDING. These two wires can be twisted together (before winding on to the toroid) by fixing one pair of ends in a vice, the other ends in a small hand drill. The drill is then slowly rotated so that the two wires are twisted together neatly. Identify the start and finish of each winding with a buzzer and battery or an ohmmeter. The finish of the first winding is joined to the start of the second, as shown in Fig 19.82.

The two capacitors are twin 200-pF polyvaricon capacitors with both gangs connected in parallel to give 400 pF max. You have to drill holes in the box for the control shafts of the capacitors and the switch and the RF sockets. The switch is fixed using a nut on the control shaft and the two capacitors are fixed using adhesive (hot melt glue is preferred). Take care not to let any tags from the capacitors touch the box as both sides of both capacitors are not earthed.

An appropriate RF socket, such as a SO-239, BNC or phono socket may be used—the choice is yours and should suit your existing equipment. 2-4 mm sockets may be used for the balanced output. See **Fig 19.83** for layout. It is a good idea to make a graduated dial for each of the three control knobs. An alternative is to use calibrated knobs. You can then calibrate the settings of the three controls so that they can rapidly be reset when you change frequency bands.

OPERATION

The best indication of optimum matching can be achieved using an SWR bridge; the ATU controls are adjusted sequentially and several times for minimum SWR. If used for receive only, the best antenna-to-receiver match can be achieved by adjusting the controls for maximum signal.

AN "UGLY TRANSFORMER" FOR HEAVY-LOAD STATIONS

There is definitely a place for less-than-pretty construction methods and components in Amateur Radio! The phrase *Ugly Construction*, pertaining to circuit boards, is attributed to Wes Hayward, W7ZOI, while the *Ugly Amplifier* was made famous by Rich Measures, AG6K. Transforming away inefficient 120-V ac supply or "copper losses" en route to the shack is the perfect application for the *Ugly Transformer* project described here! For example, vintage radio enthusiasts—especially those with a thirst for kW power levels—will find that many of yesteryear's RF power amplifiers use a 110-120-V plate transformer. The Hallicrafters HT-33B, shown in **Fig 19.84**, is typical of such equipment. AC power line requirements of the HT-33B are specified at 117 Volts and 2350 Watts! Assuming a power factor of 0.95 (nearly unity), this still correlates to greater than 21 A of line current. Even a dedicated branch circuit would have quite a job transferring that current over an appreciable length from the service box without a considerable voltage drop, and accompanying power loss, across the service conductors. Also, not every shack (or the larger structure/dwelling it is part of) lends itself to easily accessible electrical service modifications like installing such a high-current, dedicated supply line. The project described below was created to meet the needs of a high-load, 120 V ac ham station without the need for invasive modifications to existing 120-V service wiring. This project was designed and built by ARRL *Handbook* Editor Dana G. Reed, W1LC.

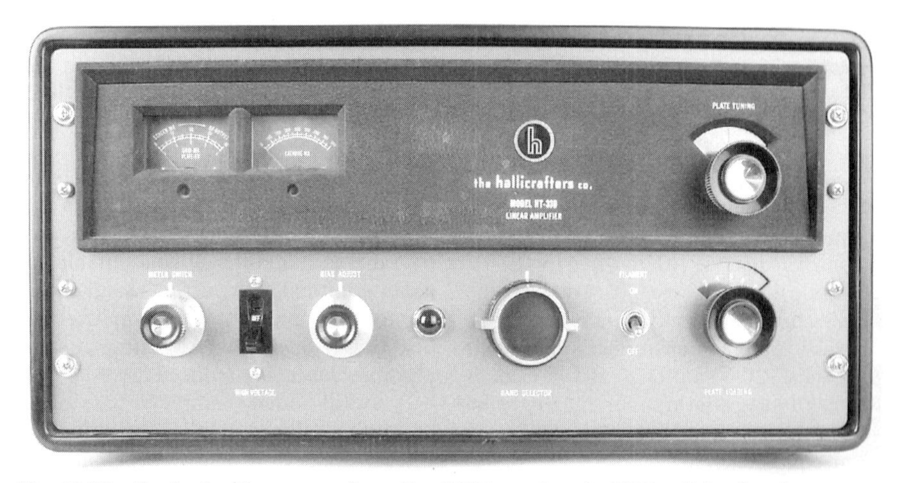

Fig 19.84—Typical of ham gear from the 1950s and early 1960s, this classic Hallicrafters HT-33B Linear Amplifier features a 117 V ac plate transformer. The magnetic circuit breaker on the front panel trips at a line current of 22 Amps! Many existing 110 to 120 V ac branch circuits would be unsuitable for such a load.

The heart of this project is a surplus (or new) 240 to 120 V ac control transformer. The transformer cost $20 on the surplus market. Its nameplate indicates Type SZO, Model D46192 made by HEVI-DUTY Electric Company in Milwaukee, WI, rated for 50 or 60 Cycles, and a 55°C rise. The transformer primary features two separate windings that can be connected in parallel or in series for either 240 or 480 V ac, respectively. The secondary delivers 120 V ac. As you'll see shortly, it was conservatively nameplate-rated at 1000 Volt-Amperes, or "1 kVA"—the *apparent power* rating, equal to the product of the voltage and the load current on either the primary or secondary side.[1] As shown in **Fig 19.85**, the project is contained in a heavy-gauge-steel surplus ammo box along with a small whisper-quiet axial ac fan to keep the transformer cool during extended periods of operation. The air inlet diameter is sized to the fan blades and the outlet diameter is approximately 30% smaller to allow a slight pressurization of the enclosure while still encouraging a premium air transfer through it. The transformer was mounted inside using 1/4-inch metallic standoffs under all four mounting brackets to facilitate better air movement beneath the iron-core and windings in order to remove unwanted heat.

The project uses 240 V ac available at a proximity outlet (perhaps the very source used to power a more modern RF amplifier at the operating position) and the control (step-down) transformer's turns ratio to reduce potentially high, distance-driven, 120-V ac line losses. It accomplishes this by employing a practice used by utility companies and high-voltage transmission line system operators—moving the power over long distances at higher voltage and lower current to minimize power (I^2R) losses along the way. A transformer is key to accomplishing this task.[2] This transformer cuts the load current by one half between the service box and the load. Thus the "Ugly Transformer" looks electrically

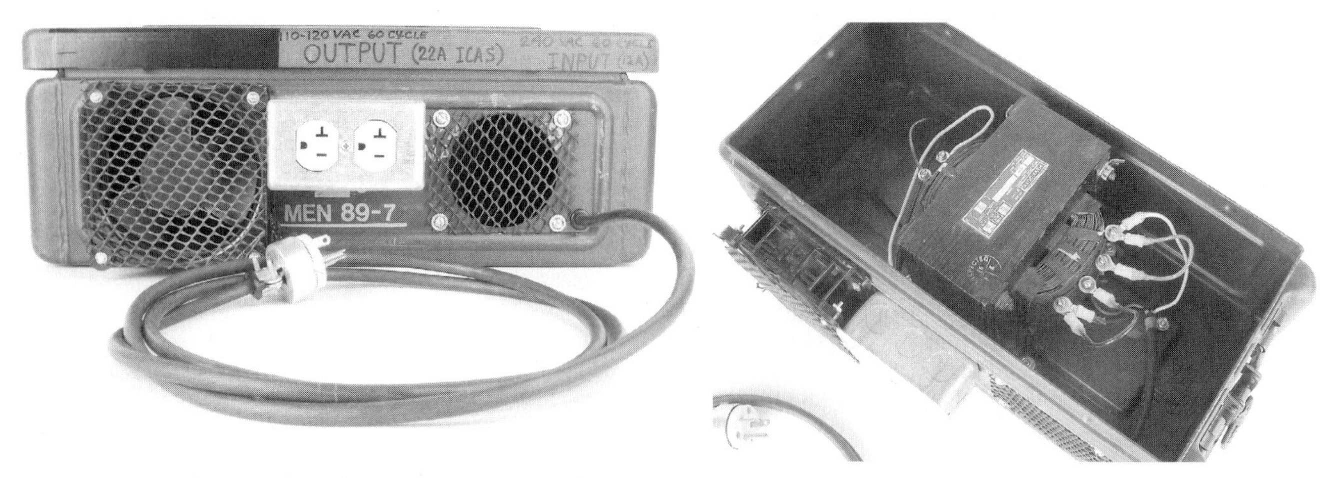

Fig 19.85—W1LC's Ugly Transformer project is contained in a surplus heavy-gauge-steel military ammo box. In the left photo, the small axial cooling fan, 20-A, 120-V ac outlet and fan exhaust vent can be seen, respectively, from left to right. The protective screening was from an old roll of "gutter guard" and is used to prevent accidental contact with components inside. Screening can be of either metallic or non-conductive construction. The photo at the right shows the control transformer and associated wiring inside with the cover off. Case grounding for electric mains purposes is accomplished by connecting the 240-V cord ground to a terminal securely fastened under one of the transformer's mounting brackets. A stranded, insulated ground-wire jumper must also be connected between this grounding point and the cover to ensure a complete case ground. The use of a nylon cable clamp on the cord inside keeps it from being accidentally pulled out of the surplus enclosure.

to the 240-V line just like a more modern RF power amplifier. It draws less than 11 A on the primary side (when wired for 240-V ac) while the 120-V secondary delivers the heavy current, but only over a very short distance from the transformer's secondary terminals/120-V outlet to the device or devices drawing the high current. **Fig 19.86** shows the schematic of the Ugly Transformer project itself.

Before this project was assembled and hailed a success, Reed had to prove its worthiness by conducting a load test on the transformer itself—a recommended exercise on any surplus unit you may procure (consulting the specifications sheet for a new transformer may suffice if overload parameters are listed, or if a unit is selected with a *greater* kVA rating than the load requires.) Retrieving two older electric quartz heaters on hand—each rated at 1500 W—both were connected across the secondary of the transformer from a 117 V ac, no-load setting obtained with a variable-ratio transformer connected to the primary side. The quartz-heater load represented up to three times the rated nameplate load for the transformer. What registered on the voltmeter under overload conditions was nothing short of amazing—a very modest 3-V drop to 114 V ac at 3 kW of resistive heater load! The transformer-under-test was on the bench at this point and had *no external cooling* applied. The transformer was carefully monitored for approximately five minutes under these aggressive and continuous-load conditions. At the end of the test, the taped winding was only slightly warm to the touch with no evidence of dangerous short-term overheating. It was therefore judged suitable for the intermittent high-current task required by my own ancient linear amplifier for SSB or CW operation (AM rating of same is somewhat less at reduced drive levels.)

A useful added feature, in case of an "event" on the system, is an industrial "stop-start station" shown schematically in **Fig 19.87A**. This circuit is typically used for industrial motor control, and as adapted here, *prevents* instantaneous and automatic reactuation of the primary circuit if there is a momentary loss of power. Such surges can be damaging, especially to older equipment where replacement parts (like a blown plate transformer) can be quite a challenge to find, and expensive. The stop-start station of

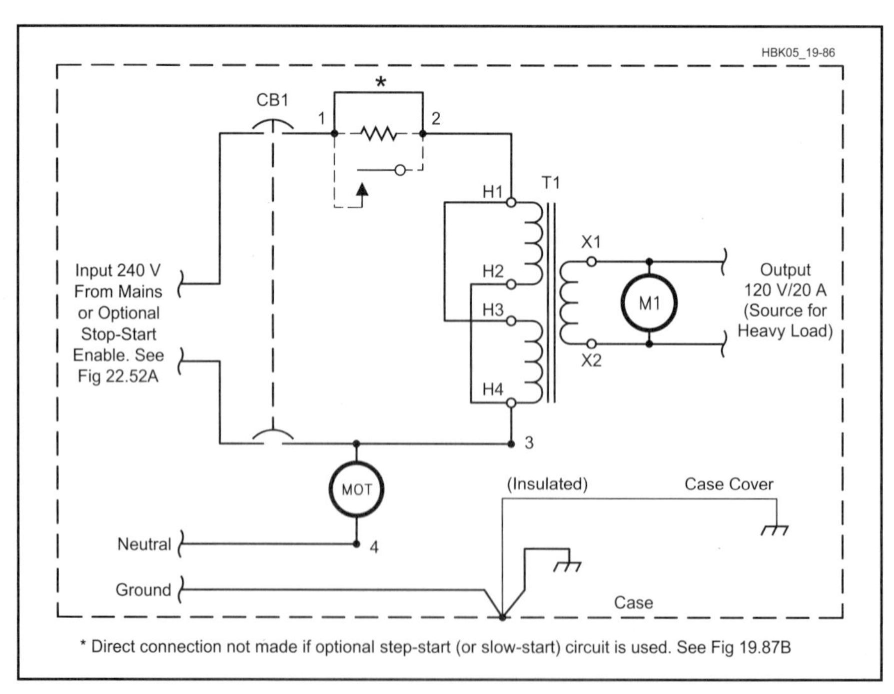

Fig 19.86—Schematic diagram of the Ugly Transformer project.

CB1—15-A, 240-V circuit breaker.
MOT—Small axial cooling fan (see text.)
M1—0-150 Vac/60-Hz Voltmeter (optional.)
T1—Control (stepdown) transformer, 240-V to 120-V, 2-3 kVA.

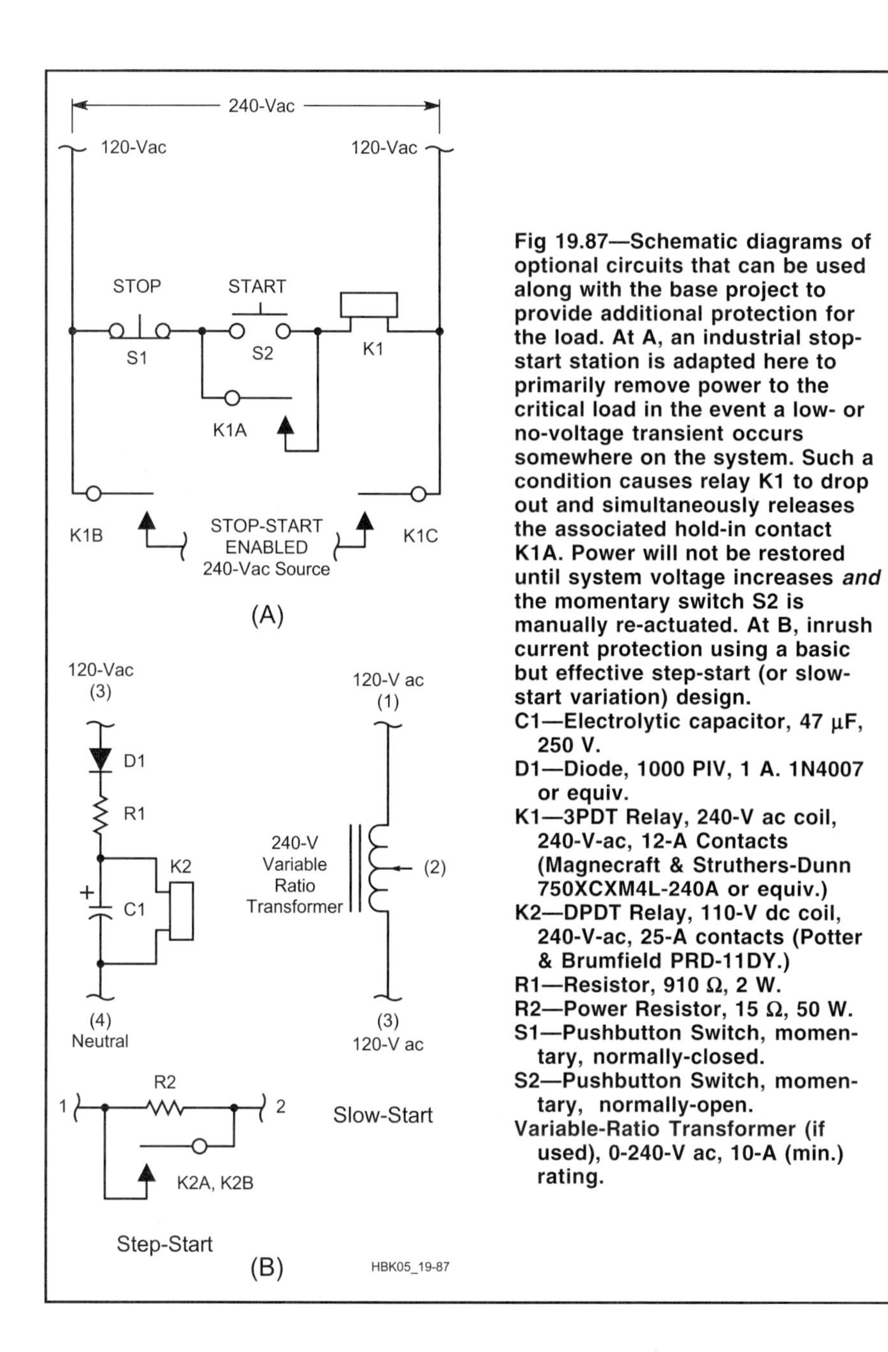

Fig 19.87—Schematic diagrams of optional circuits that can be used along with the base project to provide additional protection for the load. At A, an industrial stop-start station is adapted here to primarily remove power to the critical load in the event a low- or no-voltage transient occurs somewhere on the system. Such a condition causes relay K1 to drop out and simultaneously releases the associated hold-in contact K1A. Power will not be restored until system voltage increases *and* the momentary switch S2 is manually re-actuated. At B, inrush current protection using a basic but effective step-start (or slow-start variation) design.

C1—Electrolytic capacitor, 47 µF, 250 V.
D1—Diode, 1000 PIV, 1 A. 1N4007 or equiv.
K1—3PDT Relay, 240-V ac coil, 240-V-ac, 12-A Contacts (Magnecraft & Struthers-Dunn 750XCXM4L-240A or equiv.)
K2—DPDT Relay, 110-V dc coil, 240-V-ac, 25-A contacts (Potter & Brumfield PRD-11DY.)
R1—Resistor, 910 Ω, 2 W.
R2—Power Resistor, 15 Ω, 50 W.
S1—Pushbutton Switch, momentary, normally-closed.
S2—Pushbutton Switch, momentary, normally-open.
Variable-Ratio Transformer (if used), 0-240-V ac, 10-A (min.) rating.

Fig 19.87A allows the operator to manually start the operation again, when desired, and with a re-sequencing of the surge-inhibiting circuitry if included. An optional, variable-ratio transformer (Variac, Powerstat, etc.), or a step-start circuit as shown in Fig 19.87B, can also be added to the primary or secondary side, or in tandem with a specific 120 V ac outlet(s) mounted on the project case to slow-start vintage equipment, thereby limiting high inrush current.[3] A vintage RF power amplifier is indeed but one example of such a heavy load. Some shacks use a large number of smaller-load devices that can approximate, or even exceed that of a vintage station.

The entire project is fairly simple to construct, even if the optional circuitry discussed above is added. Of course, both ends of the transformer's primary should be fused or otherwise protected by a circuit breaker sized for the transformer rating and the load. If a circuit breaker is used, it can also serve as the means of disconnect otherwise provided by a DPDT switch. A power indicator light is optional and recommended, but not necessary. A built-in voltmeter on the secondary side would be a nice addition, especially when a variable-ratio transformer or step-start circuit is present. The particular enclosure selected should be sized according to how much of the optional circuitry you plan to include, and the physical size of the specific components themselves. Parts used are not critical in terms of manufacturer or model numbers, but rather should be electrically sized for the job at hand. Remember that the objective here is to substantially reduce voltage drops and power loss, chiefly proportional to all series resistances in a service branch circuit. The selection of such items as fuse holders, switches, power cords and connectors, and outlets should all be on the high-current side. That is, do not use underrated components! The only possible exception as noted above is a step-down transformer of stout construction, verifiable under test or by review of the manufacturer's spec sheet.

Judicious selection of conservatively rated components used in the design, as well as a thoughtful approach to project layout, can afford the user a measure of margin above and beyond nameplate ratings as shown here. This is especially true when they're used in an intermittent commercial and amateur service (ICAS) environment. However, not all components are equal, even with identical ratings, and *nothing* can really take the place of component testing prior to usage in a critical and specific application such as this project warrants.

Notes

[1]Apparent power is the trigonometric sum of resistive and reactive power components in the ac-operated device such as the power or "control" transformer described here. It is also the product of the voltage and current as would be read on ordinary metering devices connected into the circuit at such a point. For a review of resistive and reactive components, and impedance in ac circuits, please refer to the **Electrical Fundamentals** chapter of this *Handbook* and review the section on "Impedance."

[2]A power transformer is typically a highly efficient electromagnetic device that transfers power based on a specific turns ratio between one or more windings. For a review of the relationship between voltage, current, and impedance in a power transformer, please refer to the section on "Transformers" in the **Electrical Fundamentals** chapter of this *Handbook*.

[3]Use caution in deciding which item(s) in your shack should or should not be surge protected using the method described here. For example, some vintage transmitters and amplifiers also used mercury vapor rectifier tubes to produce high voltage. These types of rectifier tubes require several minutes of warm-up time *before* high voltage is applied. Solid-state rectifiers can usually be retrofitted in place of the tubes thus allowing step-start inrush current protection and immediate application of high voltage. Conversely, newer solid-state, microprocessor-equipped transceivers *are not* designed to be externally slow or step-started, but rather should have full specified voltage applied via the power switch of the radio.

Chapter 20

Propagation of RF Signals

Radio waves, like light waves and all other forms of electromagnetic radiation, normally travel in straight lines. Obviously this does not happen all the time, because long-distance communication depends on radio waves traveling beyond the horizon. How radio waves propagate in other than straight-line paths is a complicated subject, but one that need not be a mystery. This chapter, by Emil Pocock, W3EP, provides basic understanding of the principles of electromagnetic radiation, the structure of the Earth's atmosphere and solar-terrestrial interactions necessary for a working knowledge of radio propagation. More detailed discussions and the underlying mathematics of radio propagation physics can be found in the references listed at the end of this chapter.

FUNDAMENTALS OF RADIO WAVES

Radio belongs to a family of electromagnetic radiation that includes infrared (radiation heat), visible light, ultraviolet, X-rays and the even shorter-wavelength gamma and cosmic rays. Radio has the longest wavelength and thus the lowest frequency of this group. See **Table 20.1**. Electromagnetic waves result from the interaction of an electric and a magnetic field. An oscillating electric charge in a piece of wire, for example, creates an electric field and a corresponding magnetic field. The magnetic field in turn creates an electric field, which creates another magnetic field, and so on.

These two fields sustain themselves as a composite *electromagnetic wave*, which propagates itself into space. The electric and magnetic components are oriented at right angles to each other and 90° to the direction of travel. The polarization of a radio wave is usually designated the same as its electric field. This relationship can be visualized in **Fig 20.1**. Unlike sound waves or ocean waves, electromagnetic waves need no propagating medium, such as air or water. This property enables electromagnetic waves to travel through the vacuum of space.

Velocity

Radio waves, like all other electromagnetic radiation, travel nearly 300,000 km (186,400 mi) per second in a vacuum. Radio waves travel more slowly through any other medium. The decrease in speed through the atmosphere is so slight that it is usually ignored, but sometimes even this small difference is significant. The speed of a radio wave in a piece of wire, by contrast, is about 95% that of free space, and the speed can be even slower in other media.

The speed of a radio wave is always the product of wavelength and frequency,

Table 20.1
The Electromagnetic Spectrum

Radiation	*Frequency*	*Wavelength*
X-ray	3×10^5 THz and higher	10 Å and shorter
Ultraviolet	800 THz - 3×10^5 THz	4000 - 10 Å
Visible light	400 THz - 800 THz	8000 - 4000 Å
Infrared	300 GHz - 400 THz	1 mm - 0.0008 mm
Radio	10 kHz - 300 GHz	30,000 km - 1 mm

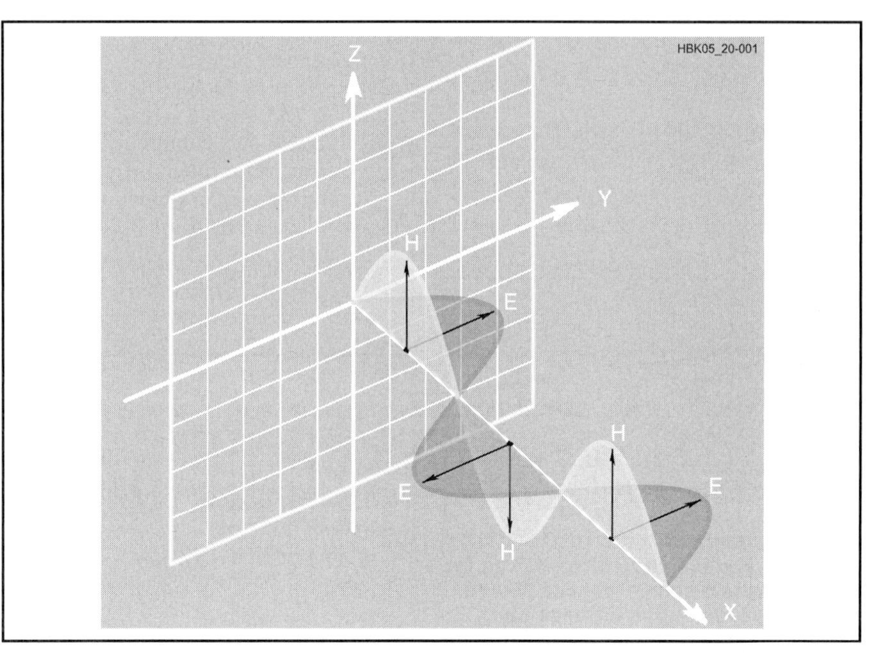

Fig 20.1—Electric and magnetic field components of the electromagnetic wave. The polarization of a radio wave is the same direction as the plane of its electric field.

whatever the medium. That relationship can be stated simply as:

$c = f\lambda$

where

c = speed in m/s
f = frequency in hertz
λ = wavelength in m

The *wavelength* (λ) of any radio frequency can be determined from this simple formula. In free space, where the speed is 3×10^8 m/s, the wavelength of a 30-MHz radio signal is thus 10 m. Wavelength decreases in other media because the propagating speed is slower. In a piece of wire, the wavelength of a 30-MHz signal shortens to about 9.5 m. This factor must be taken into consideration in antenna designs and other applications.

Wave Attenuation and Absorption

Radio waves weaken as they travel, whether in the near vacuum of cosmic space or within the Earth's atmosphere. *Free-space attenuation* results from the dispersal of radio energy from its source. See **Fig 20.2**. Attenuation grows rapidly with distance because signals weaken with the square of the distance traveled. If the distance between transmitter and receiver is increased from 1 km to 10 km (0.6 to 6 mi), the signal will be only one-hundredth as strong. Free-space attenuation is a major factor governing signal strength, but radio signals undergo a variety of other losses as well.

Energy is lost to *absorption* when radio waves travel through media other than a vacuum. Radio waves propagate through the atmosphere or solid material (like a wire) by exciting electrons, which then reradiate energy at the same frequency. This process is not perfectly efficient, so some radio energy is transformed into heat and retained by the medium. The amount of radio energy lost in this way depends on the characteristics of the medium and on the frequency. Attenuation in the atmosphere is minor from 10 MHz to 3 GHz, but at higher frequencies, absorption due to water vapor and oxygen can be high.

Radio energy is also lost during refraction, diffraction and reflection — the very phenomena that allow long-distance propagation. Indeed, any form of useful propagation is accompanied by attenuation. This may vary from the slight losses encountered by refraction from sporadic-E clouds near the maximum usable frequency, to the more considerable losses involved with tropospheric forward scatter or D-Layer absorption in the lower HF bands. These topics will be covered later. In many circumstances, total losses can become so great that radio signals become too weak for communication.

Refraction

Electromagnetic waves travel in straight lines until they are deflected by something. Radio waves are *refracted*, or bent, slightly when traveling from one medium to another. Radio waves behave no differently from other familiar forms of electromagnetic radiation in this regard. The apparent bending of a pencil partially immersed in a glass of water demonstrates this principle quite dramatically.

Refraction is caused by a change in the velocity of a wave when it crosses the boundary between one propagating medium and another. If this transition is made at an angle, one portion of the wavefront slows down (or speeds up) before the other, thus bending the wave slightly. This is shown schematically in **Fig 20.3**.

The amount of bending increases with the ratio of the *refractive indices* of the two media. Refractive index is simply the velocity of a radio wave in free space divided by its velocity in the medium. Radio waves are commonly refracted when they travel through different layers of the atmosphere, whether the highly charged ionospheric layers 100 km (60 mi) and higher, or the weather-sensitive area near the Earth's surface. When the ratio of the refractive indices of two media is great enough, radio waves can be reflected, just like light waves striking a mirror. The Earth is a rather lossy reflector, but a metal surface works well if it is several wavelengths in diameter.

Scattering

The direction of radio waves can also be altered through *scattering*. The effect seen by a beam of light attempting to penetrate fog is a good example of light-wave scattering. Even on a clear night, a highly directional searchlight is visible due to a small amount of atmospheric scattering perpendicular to the beam. Radio waves are similarly scattered when they encounter randomly arranged objects of wavelength size or smaller, such as masses of electrons or water droplets. When the density of scattering objects becomes great enough, they behave more like a propagating medium with a characteristic refractive index.

If the scattering objects are arranged in some alignment or order, scattering takes place only at certain angles. A rainbow provides a good analogy for *field-aligned scattering* of light waves. The arc of a rainbow can be seen only at a precise angle away from the sun, while the colors result from the variance in scattering across the light-wave frequency range. Ionospheric electrons can be field-aligned by magnetic forces in auroras and under other unusual circumstances. Scattering in such cases is best perpendicular to the Earth's magnetic field lines.

Reflection

At amateur frequencies above 30 MHz, reflections from a variety of large objects, such as water towers, buildings, airplanes, mountains and the like can provide a useful means of extending over-the-horizon paths several hundred km. Two stations need only beam toward a common reflector, whether stationary or moving. Contrary to common sense notions, the best position for a reflector is not midway between two stations. Signal strength increases as the reflector approaches one

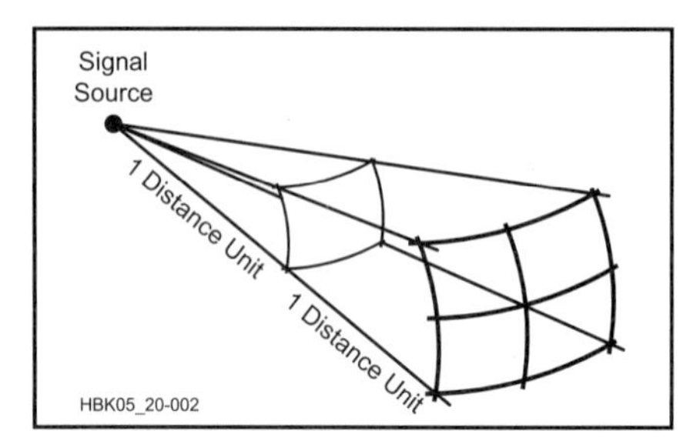

Fig 20.2—Radio energy disperses as the square of the distance from its source. For the change of one distance unit shown the signal is only one quarter as strong. Each spherical section has the same surface area.

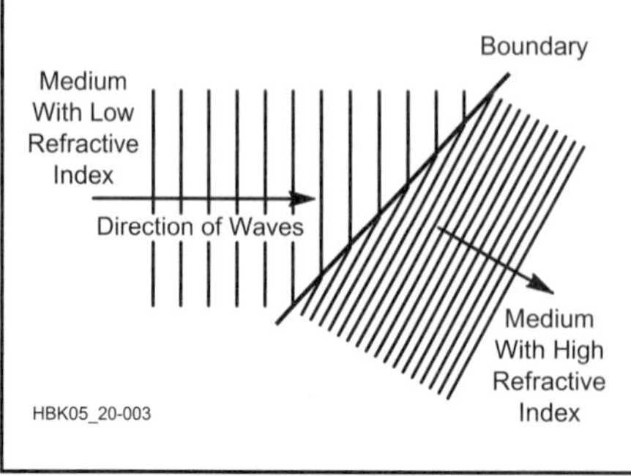

Fig 20.3—Radio waves are refracted as they pass at an angle between dissimilar media. The lines represent the crests of a moving wave front and the distance between them is the wavelength. The direction of the wave changes because one end of the wave slows down before the other as it crosses the boundary between the two media. The wavelength is simultaneously shortened, but the wave frequency (number of crests that pass a certain point in a given unit of time) remains constant.

end of the path, so the most effective reflectors are those closest to one station or the other.

Maximum range is limited by the radio line-of-sight distance of both stations to the reflector and by reflector size and shape. The reflectors must be many wavelengths in size and ideally have flat surfaces. Large airplanes make fair reflectors and may provide the best opportunity for long-distance contacts. The calculated limit for airplane reflections is 900 km (560 mi), assuming the largest jets fly no higher than 12,000 m (40,000 ft), but actual airplane reflection contacts are likely to be considerably shorter.

Knife-Edge Diffraction

Radio waves can also pass behind solid objects with sharp upper edges, such as a mountain range, by *knife-edge diffraction*. This is a common natural phenomenon that affects light, sound, radio and other coherent waves, but it is difficult to comprehend. **Fig 20.4** depicts radio signals approaching an idealized knife-edge. The portion of the radio waves that strike the base of the knife-edge is entirely blocked, while that portion passing several wavelengths above the edge travel on relatively unaffected. It might seem at first glance that a knife-edge as large as a mountain, for example, would completely prevent radio signals from appearing on the other side but that is not quite true. Something quite unexpected happens to radio signals that pass just over a knife-edge.

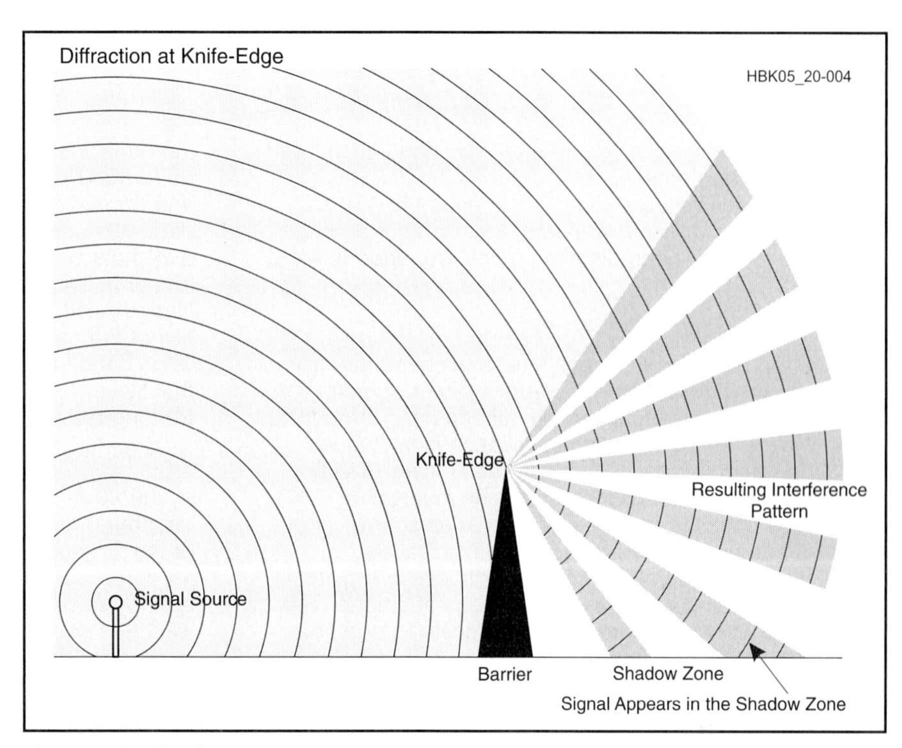

Fig 20.4—Radio, light and other waves are diffracted around the sharp edge of a solid object that is large in terms of wavelengths. Diffraction results from interference between waves right at the knife-edge and those that are passing above it. Some signals appear behind the knife-edge as a consequence of the interference pattern. Hills or mountains can serve as natural knife-edges at radio frequencies.

Normally, radio signals along a wave front interfere with each other continuously as they propagate through unobstructed space, but the overall result is a uniformly expanding wave. When a portion of the wave front is blocked by a knife-edge, the resulting interference pattern is no longer uniform. This can be understood by visualizing the radio signals right at the knife-edge as if they constituted a new and separate transmitting point, but in-phase with the source wave at that point. The signals adjacent to the knife-edge still interact with signals passing above the edge, but they cannot interact with signals that have been obstructed below the edge. The resulting *interference pattern* no longer creates a uniformly expanding wave front, but rather appears as a pattern of alternating strong and weak bands of waves that spread in a nearly 180° arc behind the knife-edge.

The crest of a range of hills or mountains 50 to 100 wavelengths long can serve as a reasonable knife-edge diffractor at radio frequencies. Hillcrests that are clearly defined and free of trees, buildings and other clutter make the best edges, but even rounded hills may serve as a diffracting edge. Alternating bands of strong and weak signals, corresponding to the interference pattern, will appear on the surface of the Earth behind the mountain, known as the *shadow zone*. The phenomenon is generally reciprocal, so that two-way communication can be established under optimal conditions. Knife-edge diffraction can make it possible to complete paths of 100 km or more that might otherwise be entirely obstructed by mountains or seemingly impossible terrain.

Ground Waves

A *ground wave* is the result of a special form of diffraction that primarily affects longer-wavelength vertically polarized radio waves. It is most apparent in the 80- and 160-m amateur bands, where practical ground-wave distances may extend beyond 200 km (120 mi). The term ground wave is often mistakenly applied to any short-distance communication, but the actual mechanism is unique to the longer-wave bands.

Radio waves are bent slightly as they pass over a sharp edge, but the effect extends to edges that are considerably rounded. At medium and long wavelengths, the curvature of the Earth looks like a rounded edge. Bending results when the lower part of the wave front loses energy due to currents induced in the ground. This slows down the lower part of the wave, causing the entire wave to tilt forward slightly. This tilting follows the curvature of the Earth, thus allowing low- and medium-wave radio signals to propagate over distances well beyond line of sight.

Ground wave is most useful during the day at 1.8 and 3.5 MHz, when D-layer absorption makes skywave propagation more difficult. Vertically polarized antennas with excellent ground systems provide the best results. Ground-wave losses are reduced considerably over saltwater and are worst over dry and rocky land.

SKY-WAVE PROPAGATION AND THE SUN

The Earth's atmosphere is composed primarily of nitrogen (78%), oxygen (21%) and argon (1%), with smaller amounts of a dozen other gases. Water vapor can account for as much as 5% of the atmosphere under certain conditions. This ratio of gases is maintained until an altitude of about 80 km (50 mi), when the mix begins to change. At the highest levels, helium and hydrogen predominate.

Solar radiation acts directly or indirectly on all levels of the atmosphere. Adjacent to the surface of the Earth, solar

Propagation Summary, by Band

Medium Frequencies (300 kHz-3 MHz)

The only amateur medium-frequency band is situated just above the domestic AM broadcast band. Ground wave provides reliable communication out to 150 km (90 mi) during the day, when no other form of propagation is available. Long-distance paths are made at night via the F_2 layer.

1.8-2.0 MHz (160 m)

The top band, as it is sometimes called, suffers from extreme daytime D-layer absorption. Even at high radiation angles, virtually no signal can pass through to the F layer, so daytime communication is limited to ground-wave coverage. At night, the D layer quickly disappears and worldwide 160-m communication becomes possible via F_2-layer skip. Atmospheric and man-made noise limit propagation. Tropical and midlatitude thunderstorms cause high levels of static in summer, making winter evenings the best time to work DX at 1.8 MHz. A proper choice of receiving antenna can often significantly reduce the amount of received noise while enhancing desired signals.

High Frequencies (3-30 MHz)

A wide variety of propagation modes are useful on the HF bands. The lowest two bands in this range share many daytime characteristics with 160 m. The transition between bands primarily useful at night or during the day appears around 10 MHz. Most long-distance contacts are made via F_2-layer skip. Above 21 MHz, more exotic propagation, including TE, sporadic E, aurora and meteor scatter, begin to be practical.

3.5-4.0 MHz (80 m)

The lowest HF band is similar to 160 m in many respects. Daytime absorption is significant, but not quite as extreme as at 1.8 MHz. High-angle signals may penetrate to the E and F layers. Daytime communication range is typically limited to 400 km (250 mi) by ground-wave and skywave propagation. At night, signals are often propagated halfway around the world. As at 1.8 MHz, atmospheric noise is a nuisance, making winter the most attractive season for the 80-m DXer.

7.0-7.3 MHz (40 m)

The popular 40-m band has a clearly defined skip zone during the day. D-layer absorption is not as severe as on the lower bands, so short-distance skip via the E and F layers is possible. During the day, a typical station can cover a radius of approximately 800 km (500 mi). Ground-wave propagation is not important. At night, reliable worldwide communication via F_2 is common on the 40-m band.

Atmospheric noise is less troublesome than on 160 and 80 m, and 40-m DX signals are often of sufficient strength to override even high-level summer static. For these reasons, 40 m is the lowest-frequency amateur band considered reliable for DX communication in all seasons. Even during the lowest point in the solar cycle, 40 m may be open for worldwide DX throughout the night.

10.1-10.15 MHz (30 m)

The 30-m band is unique because it shares characteristics of both daytime and nighttime bands. D-layer absorption is not a significant factor. Communication up to 3000 km (1900 mi) is typical during the daytime, and this extends halfway around the world via all-darkness paths. The band is generally open via F_2 on a 24-hour basis, but during a solar minimum, the MUF on some DX paths may drop below 10 MHz at night. Under these conditions, 30 m adopts the characteristics of the daytime bands at 14 MHz and higher. The 30-m band shows the least variation in conditions over the 11-year solar cycle, thus making it generally useful for long-distance communication anytime.

14.0-14.35 MHz (20 m)

The 20-m band is traditionally regarded as the amateurs' primary long-haul DX favorite. Regardless of the 11-year solar cycle, 20 m can be depended on for at least a few hours of worldwide F_2 propagation during the day. During solar-maximum periods, 20 m will often stay open to distant locations throughout the night. Skip distance is usually appreciable and is always present to some degree. Daytime E-layer propagation may be detected along very short paths. Atmospheric noise is not a serious consideration, even in the summer. Because of its popularity, 20 m tends to be very congested during the daylight hours.

18.068-18.168 MHz (17 m)

The 17-m band is similar to the 20-m band in many respects, but the effects of fluctuating solar activity on F_2 propagation are more pronounced. During the years of high solar activity, 17 m is reliable for daytime and early-evening long-range communication, often lasting well after sunset. During moderate years, the band may open only during sunlight hours and close shortly after sunset. At solar minimum, 17 m will open to middle and equatorial latitudes, but only for short periods during midday on north-south paths.

21.0-21.45 MHz (15 m)

The 15-m band has long been considered a prime DX band during solar cycle maxima, but it is sensitive to changing solar activity. During peak years, 15 m is reliable for daytime F_2-layer DXing and will often stay open well into the night. During periods of moderate solar activity, 15 m is basically a daytime-only band, closing shortly after sunset. During solar minimum periods, 15 m may not open at all except for infrequent north-south transequatorial circuits. Sporadic E is observed occasionally in early summer and mid-winter, although this is not common and the effects are not as pronounced as on the higher frequencies.

24.89-24.99 MHz (12 m)

This band offers propagation that combines the best of the 10- and 15-m bands. Although 12 m is primarily a daytime band during low and moderate sunspot years, it may stay open well after sunset during the solar maximum. During years of moderate solar activity, 12 m opens to the low and middle latitudes during the daytime hours, but it seldom remains open after sunset. Periods of low solar activity seldom cause this band to go completely dead, except at higher latitudes. Occasional daytime openings, especially in the lower latitudes, are likely over north-south paths. The main sporadic-E season on 24 MHz lasts from late spring through summer and short openings may be observed in mid-winter.

28.0-29.7 MHz (10 m)

The 10-m band is well known for extreme variations in characteristics and variety of propagation modes. During solar maxima, long-distance F_2 propagation is so efficient that very low power can produce loud signals halfway around the globe. DX is abundant with modest equipment. Under these conditions, the band is usually open from sunrise to a few hours past sunset. During periods of moderate solar activity, 10 m usually opens only to low and transequatorial latitudes around noon. During the solar minimum, there may be no F_2 propagation at any time during the day or night.

Sporadic E is fairly common on 10 m, especially May through August, although it may appear at any time. Short skip, as sporadic E is sometimes called on the HF bands, has little relation to the solar cycle and occurs regardless of F-layer conditions. It provides single-hop communication from 300 to 2300 km (190 to 1400 mi) and multiple-hop opportunities of 4500 km (2800 mi) and farther.

Ten meters is a transitional band in that it also shares some of the propagation modes more characteristic of VHF. Meteor scatter, aurora, auroral E and transequatorial spread-F provide the means of making contacts out to 2300 km (1400 mi) and farther, but these modes often go unnoticed at 28 MHz. Techniques similar to those used at VHF can be very effective on 10 m, as signals are usually stronger and more persistent. These exotic modes can be more fully exploited, especially during the solar minimum when F_2 DXing has waned.

Very High Frequencies (30-300 MHz)

A wide variety of propagation modes are useful in the VHF range. F-layer skip appears on 50 MHz during solar cycle peaks. Sporadic E and several other E-layer phenomena are most effective in the VHF range. Still other forms of VHF ionospheric propagation, such as field-aligned irregularities (FAI) and transequatorial spread F (TE), are rarely observed at HF. Tropospheric propagation, which is not a factor at HF, becomes increasingly important above 50 MHz.

50-54 MHz (6 m)

The lowest amateur VHF band shares many of the characteristics of both lower and higher frequencies. In the absence of any favorable ionospheric propagation conditions, well-equipped 50-MHz stations work regularly over a radius of 300 km (190 mi) via tropospheric scatter, depending on terrain, power, receiver capabilities and antenna. Weak-signal troposcatter allows the best stations to make 500-km (310-mi) contacts nearly any time. Weather effects may extend the normal range by a few hundred km, especially during the summer months, but true tropospheric ducting is rare.

During the peak of the 11-year sunspot cycle, worldwide 50-MHz DX is possible via the F_2 layer during daylight hours. F_2 backscatter provides an additional propagation mode for contacts as far as 4000 km (2500 mi) when the MUF is just below 50 MHz. TE paths as long as 8000 km (5000 mi) across the magnetic equator are common around the spring and fall equinoxes of peak solar cycle years.

Sporadic E is probably the most common and certainly the most popular form of propagation on the 6-m band. Single-hop E-skip openings may last many hours for contacts from 600 to 2300-km (370 to 1400 mi), primarily during the spring and early summer. Multiple-hop E_s provides transcontinental contacts several times a year, and contacts between the US and South America, Europe and Japan via multiple-hop E-skip occur nearly every summer.

Other types of E-layer ionospheric propagation make 6 m an exciting band. Maximum distances of about 2300 km (1400 mi) are typical for all types of E-layer modes. Propagation via FAI often provides additional hours of contacts immediately following sporadic E events. Auroral propagation often makes its appearance in late afternoon when the geomagnetic field is disturbed. Closely related auroral-E propagation may extend the 6-m range to 4000 km (2500 mi) and sometimes farther across the northern states and Canada, usually after midnight. Meteor scatter provides brief contacts during the early morning hours, especially during one of the dozen or so prominent annual meteor showers.

144-148 MHz (2 m)

Ionospheric effects are significantly reduced at 144 MHz, but they are far from absent. F-layer propagation is unknown except for TE, which is responsible for the current 144-MHz terrestrial DX record of nearly 8000 km (5000 mi). Sporadic E occurs as high as 144 MHz less than a tenth as often as at 50 MHz, but the usual maximum single-hop distance is the same, about 2300 km (1400 mi). Multiple-hop sporadic-E contacts greater than 3000 km (1900 mi) have occurred from time to time across the continental US, as well as across Southern Europe.

Auroral propagation is quite similar to that found at 50 MHz, except that signals are weaker and more Doppler-distorted. Auroral-E contacts are rare. Meteor-scatter contacts are limited primarily to the periods of the great annual meteor showers and require much patience and operating skill. Contacts have been made via FAI on 144 MHz, but its potential has not been fully explored.

Tropospheric effects improve with increasing frequency, and 144 MHz is the lowest VHF band at which weather plays an important propagation role. Weather-induced enhancements may extend the normal 300- to 600-km (190- to 370-mi) range of well-equipped stations to 800 km (500 mi) and more, especially during the summer and early fall. Tropospheric ducting extends this range to 2000 km (1200 mi) and farther over the continent and at least to 4000 km (2500 mi) over some well-known all-water paths, such as that between California and Hawaii.

222-225 MHz (135 cm)

The 135-cm band shares many characteristics with the 2-m band. The normal working range of 222-MHz stations is nearly as far as comparably equipped 144-MHz stations. The 135-cm band is slightly more sensitive to tropospheric effects, but ionospheric modes are more difficult to use. Auroral and meteor-scatter signals are somewhat weaker than at 144 MHz, and sporadic-E contacts on 222 MHz are extremely rare. FAI and TE may also be well within the possibilities of

(continued on next page)

222 MHz, but reports of these modes on the 135-cm band are uncommon. Increased activity on 222-MHz will eventually reveal the extent of the propagation modes on the highest of the amateur VHF bands.

Ultra-High Frequencies (300-3000 MHz) and Higher

Tropospheric propagation dominates the bands at UHF and higher, although some forms of E-layer propagation are still useful at 432 MHz. Above 10 GHz, atmospheric attenuation increasingly becomes the limiting factor over long-distance paths. Reflections from airplanes, mountains and other stationary objects may be useful adjuncts to propagation at 432 MHz and higher.

420-450 MHz (70 cm)

The lowest amateur UHF band marks the highest frequency on which ionospheric propagation is commonly observed. Auroral signals are weaker and more Doppler distorted; the range is usually less than at 144 or 222 MHz. Meteor scatter is much more difficult than on the lower bands, because bursts are significantly weaker and of much shorter duration. Although sporadic E and FAI are unknown as high as 432 MHz and probably impossible, TE may be possible.

Well-equipped 432-MHz stations can expect to work over a radius of at least 300 km (190 mi) in the absence of any propagation enhancement. Tropospheric refraction is more pronounced at 432 MHz and provides the most frequent and useful means of extended-range contacts. Tropospheric ducting supports contacts of 1500 km (930 mi) and farther over land. The current 432-MHz terrestrial DX record of more than 4000 km (2500 mi) was accomplished by ducting over water.

902-928 MHz (33-cm) and Higher

Ionospheric modes of propagation are nearly unknown in the bands above 902 MHz. Auroral scatter may be just within amateur capabilities at 902 MHz, but signal levels will be well below those at 432 MHz. Doppler shift and distortion will be considerable, and the signal bandwidth may be quite wide. No other ionospheric propagation modes are likely, although high-powered research radars have received echoes from auroras and meteors as high as 3 GHz.

Almost all extended-distance work in the UHF and microwave bands is accomplished with the aid of tropospheric enhancement. The frequencies above 902 MHz are very sensitive to changes in the weather. Tropospheric ducting occurs more frequently than in the VHF bands and the potential range is similar. At 1296 MHz, 2000-km (1200-mi) continental paths and 4000-km (2500-mi) paths between California and Hawaii have been spanned many times. Contacts of 1000 km (620 mi) have been made on all bands through 10 GHz in the US and over 1600 km (1000 mi) across the Mediterranean Sea. Well-equipped 903- and 1296-MHz stations can work reliably up to 300 km (190 mi), but normal working ranges generally shorten with increasing frequency.

Other tropospheric effects become evident in the GHz bands. Evaporation inversions, which form over very warm bodies of water, are usable at 3.3 GHz and higher. It is also possible to complete paths by scattering from rain, snow and hail in the lower GHz bands. Above 10 GHz, attenuation caused by atmospheric water vapor and oxygen become the most significant limiting factors in long-distance communication.

warming controls all aspects of the weather, powering wind, rain and other familiar phenomena. *Solar ultraviolet (UV) radiation* creates small concentrations of ozone (O_3) molecules between 10 and 50 km (6 and 30 mi). Most UV radiation is absorbed by this process and never reaches the Earth.

At even higher altitudes, UV and X-ray radiation partially ionize atmospheric gases. Electrons freed from gas atoms eventually recombine with positive ions to recreate neutral gas atoms, but this takes some time. In the low-pressure environment at the highest altitudes, atoms are spaced far apart and the gases may remain ionized for many hours. At lower altitudes, recombination happens rather quickly, and only constant radiation can keep any appreciable portion of the gas ionized.

Structure of the Earth's Atmosphere

The atmosphere, which reaches to more than 600 km (370 mi) altitude, is divided into a number of regions, shown in **Fig 20.5**. The weather-producing *troposphere* lies between the surface and an average altitude of 10 km (6 mi). Between 10 and 50 km (6 and 30 mi) are the *stratosphere* and the imbedded *ozonosphere*, where ultraviolet absorbing ozone reaches its highest concentrations. About 99% of atmospheric gases are contained within these two lowest regions.

Above 50 km to about 600 km (370 mi) is the *ionosphere*, notable for its effects on radio propagation. At these altitudes, atomic oxygen and nitrogen predominate under very low pressure. High-energy solar UV and X-ray radiation ionize these gases, creating a broad region where ions are created in relative abundance. The ionosphere is subdivided into distinctive D, E and F regions.

The *magnetosphere* begins around 600 km (370 mi) and extends as far as 160,000 km (100,000 mi) into space. The predominant component of atmospheric gases gradually shifts from atomic oxygen, to helium and finally to hydrogen at the highest levels. The lighter gases may reach escape velocity or be swept off the atmosphere by the solar wind. At about 3,200 and 16,000 km (2000 and 9900 mi), the Earth's magnetic field traps energetic electrons and protons in two bands, known as the *Van Allen belts*. These have only a minor effect on terrestrial radio propagation.

The Ionosphere

The ionosphere plays a basic role in long-distance communication in all the amateur bands from 1.8 MHz to 30 MHz. Ionospheric effects are less apparent in the very high frequencies (30-300 MHz), but they persist at least through 432 MHz. As early as 1902, Oliver Heaviside and Arthur E. Kennelly independently suggested the existence of a layer in the upper atmosphere that could account for the long-distance radio transmissions made the previous year by Guglielmo Marconi and others. Edward Appleton confirmed the existence of the Kennelly-Heaviside layer during the early 1920s and used the letter E on his diagrams to designate the electric waves that were apparently reflected from it.

In 1924, Appleton discovered two additional layers in the ionosphere, as he and Robert Watson-Watt named this atmospheric region, and noted them with the letters D and F. Appleton was reluctant to alter this arbitrary nomenclature for fear of discovering yet other layers, so it has stuck to the present day. The basic physics of ionospheric propagation was largely worked out by the 1920s, yet both amateur and professional experimenters made further discoveries through the 1930s and 1940s. Sporadic E, aurora, meteor scatter and several types of field-aligned scattering were among additional ionospheric phenomena that required explanation.

Ionospheric Refraction

The refractive index of an ionospheric layer increases with the density of free-moving electrons. In the densest regions of the F layer, that density can reach a trillion electrons per cubic meter (10^{12} e/m^3). Even at this high level, radio waves are refracted gradually over a considerable vertical distance, usually amounting to tens of km. Radio waves become useful for terrestrial propagation only when they are refracted enough to bring them back to Earth. See **Fig 20.6**.

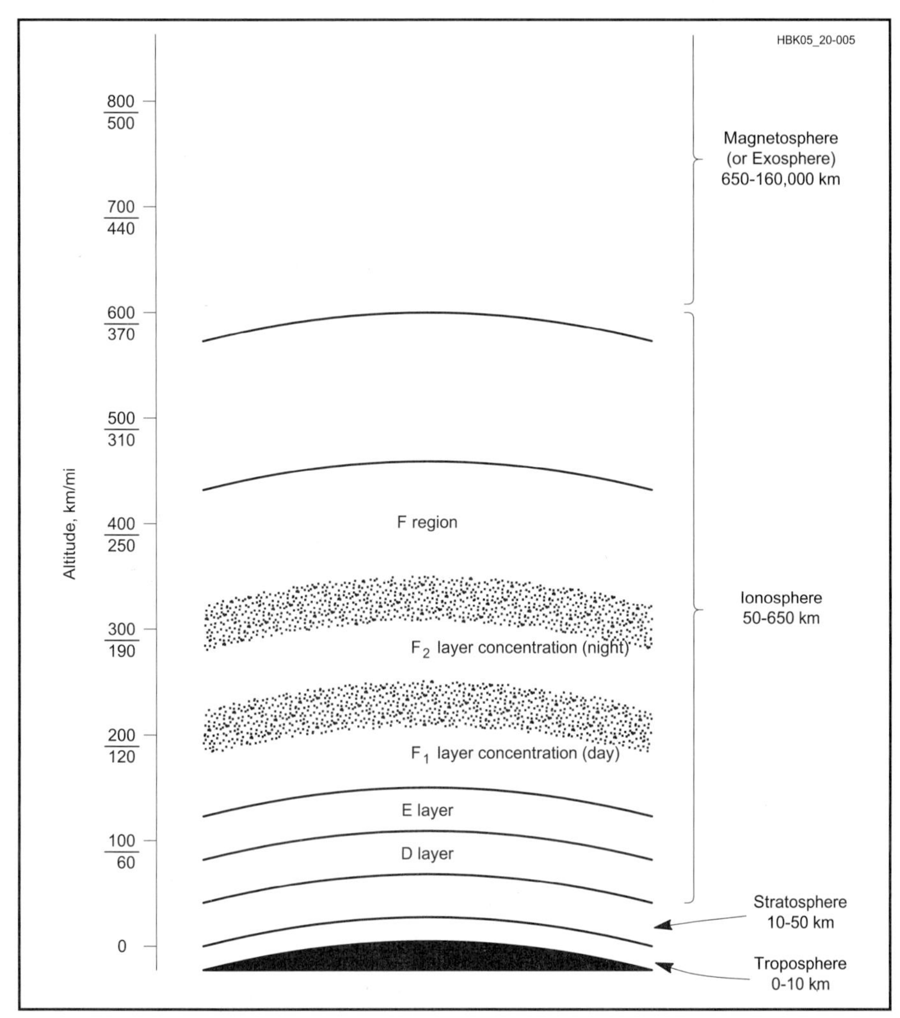

Fig 20.5—Regions of the ionosphere.

Although refraction is the primary mechanism of ionospheric propagation, it is usually more convenient to think of the process as a reflection. The *virtual height* of an ionospheric layer is the equivalent altitude of a reflection that would produce the same effect as the actual refraction. The virtual height of any ionospheric layer can be determined using an ionospheric sounder, or *ionosonde*, a sort of vertically oriented radar. The ionosonde sends pulses that sweep over a wide frequency range, generally from 2 MHz to 6 MHz or higher, straight up into the ionosphere. The frequencies of any echoes are recorded against time and then plotted as distance on an *ionogram*. **Fig 20.7** depicts a simple ionogram.

The highest frequency that returns echoes at vertical incidence is known as the *vertical incidence* or *critical frequency*. The critical frequency is almost totally a function of ion density. The higher the ionization at a particular altitude, the higher becomes the critical frequency. Physicists are more apt to call this the *plasma frequency*, because technically gases in the ionosphere are in a plasma, or partially ionized state. F-layer critical frequencies commonly range from about 1 MHz to as high as 15 MHz.

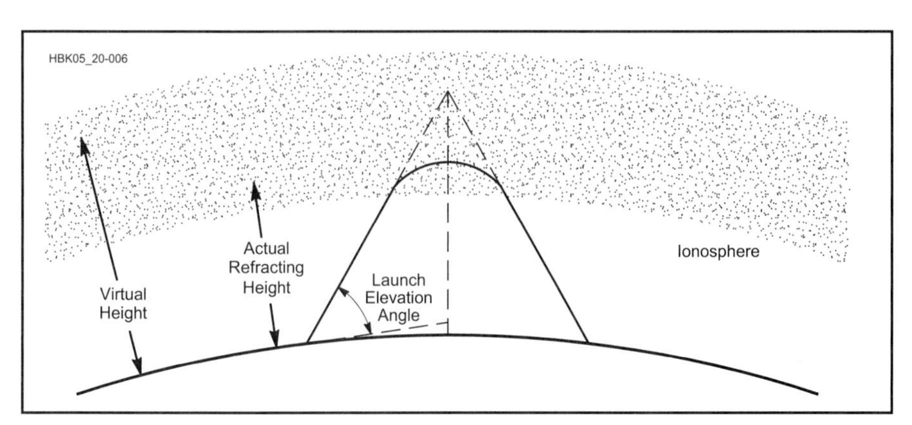

Fig 20.6—Gradual refraction in the ionosphere allows radio signals to be propagated long distances. It is often convenient to imagine the process as a reflection with an imaginary reflection point at some virtual height above the actual refracting region. The other figures in this chapter show ray paths as equivalent reflections, but you should keep in mind that the actual process is a gradual refraction.

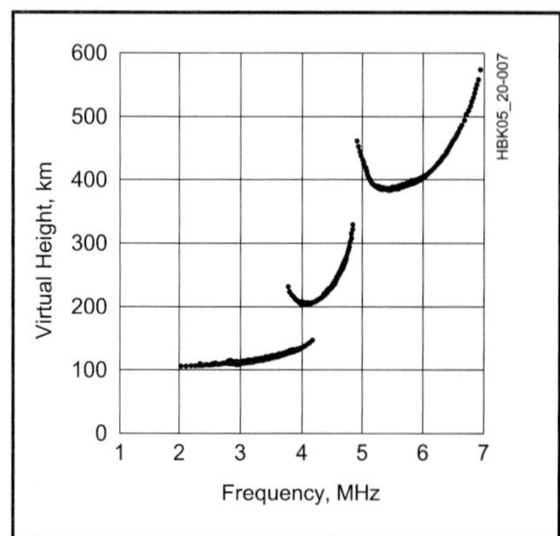

Fig 20.7—Simplified vertical incidence ionogram showing echoes returned from the E, F_1 and F_2 layers. The critical frequencies of each layer (4.1, 4.8 and 6.8 MHz) can be read directly from the ionogram scale.

Maximum and Lowest Usable Frequencies

When the frequency of a vertically incident signal is raised above the critical frequency of an ionospheric layer, that portion of the ionosphere is unable to refract the signal back to Earth. However, a signal above the critical frequency may be returned to Earth if it enters the layer at an *oblique angle*, rather than at vertical incidence. This is fortunate because it permits two widely separated stations to communicate on significantly higher frequencies than the critical frequency. See **Fig 20.8**.

The highest frequency supported by the ionosphere between two stations is the *maximum usable frequency* (MUF) for that path. If the separation between the stations is increased, a still higher frequency can be supported at lower launch angles. The MUF for this longer path is higher than the MUF for the shorter path. When the distance is increased to the maximum one-hop distance, the launch angle of the signals between the two stations is zero (that is, the ray path is tangential to the Earth at the two stations) and the MUF for this path is the highest that can be supported by that layer of the ionosphere at that location. This maximum distance is about 4000 km (2500 mi) for the F_2 layer and about 2300 km (1400 mi) for the E layer. See **Fig 20.9**.

The MUF is a function of path, time of day, season, location, solar UV and X-ray radiation levels and ionospheric disturbances. For vertically incident waves, the MUF is the same as the critical frequency. For path lengths at the limit of one-hop propagation, the MUF can be several times the critical frequency. See **Table 20.2**. The ratio between the MUF and the critical frequency is known as the *maximum usable frequency factor* (MUFF).

The term *skip zone* is closely related to MUF. When two stations are unable to communicate with each other on a particular frequency because the ionosphere is unable to refract the signal from one to the other through the required angle — that is, the frequency is below the MUF — the stations are said to be in the skip zone for that frequency. Stations within the skip zone may be able to work each other at a lower frequency, or by ground wave if they are close enough. There is no skip zone at frequencies below the critical frequency.

The MUF at any time on a particular path is just that — the *maximum* usable frequency. Frequencies below the MUF will also propagate along the path, but ionospheric absorption and noise at the receiving location (perhaps due to thunderstorms, local or distant) may make the received signal-to-noise ratio too low to be usable. In this case, the frequency is said to be below the *lowest usable frequency* (LUF). This occurs most frequently below 10 MHz, where atmospheric and man-made noises are most troublesome. The LUF can be lowered somewhat by the use of high power and directive antennas, or through the use of communications modes that permit reduced receiver bandwidth or are less demanding of SNR — CW instead of SSB, for example. This is not true of the MUF, which is limited by the physics of ionospheric refraction, no matter how high your transmitter power or how narrow your receiver bandwidth. The LUF can be higher than the MUF, in which case there is no frequency that supports communication on the particular path at that time.

Table 20.2
Maximum Usable Frequency Factors (MUFF)

Layer	*Maximum Critical Frequency (MHz)*	*MUFF*	*Useful Operating Frequencies (MHz)*
F_2	15.0	3.3-4.0	1-60
F_1*	5.5	4.0	10-20
E*	4.0	4.8	5-20
Es	30.0	5.3	20-160
D*	Not observed	—	None

* Daylight only

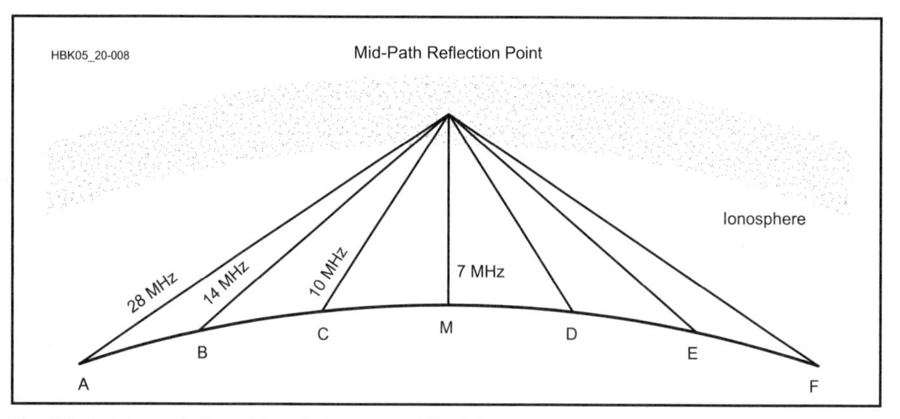

Fig 20.8—The relationships between critical frequency, maximum usable frequency (MUF) and skip zone can be visualized in this simplified, hypothetical case. The critical frequency is 7 MHz, allowing frequencies below this to be used for short-distance ionospheric communication by stations in the vicinity of point M. These stations cannot communicate by the ionosphere at 14 MHz. Stations at points B and E (and beyond) can communicate because signals at this frequency are refracted back to Earth because they encounter the ionosphere at an oblique angle of incidence. At greater distances, higher frequencies can be used because the MUF is higher at the larger angles of incidence (low launch angles). In this figure, the MUF for the path between points A and F, with a small launch angle, is shown to be 28 MHz. Each pair of stations can communicate at frequencies at or below the MUF of the path between them, but not below the LUF—see text.

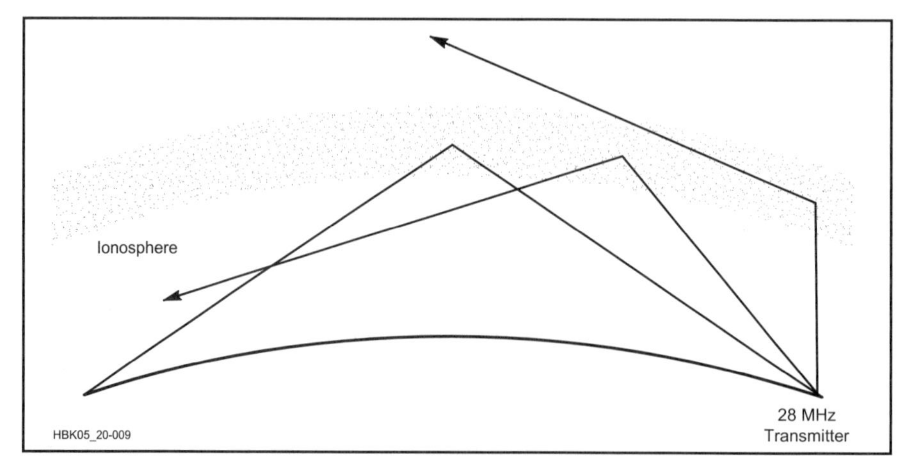

Fig 20.9—Signals at the MUF propagated at a low angle to the horizon provide the longest possible one-hop distances. In this example, 28-MHz signals entering the ionosphere at higher angles are not refracted enough to bring them back to Earth.

Ionospheric Fading

HF signal strengths typically rise and fall over periods of a few seconds to several minutes, and rarely hold at a constant level for very long. Fading is generally caused by the interaction of several radio waves from the same source arriving along different propagation paths. Waves that arrive in-phase combine to produce a stronger signal, while those out-of-phase cause destructive interference and a lower net signal strength. Short-term variations in ionospheric conditions may change individual path lengths or signal strengths enough to cause fading. Even signals that arrive primarily over a single path may vary as the propagating medium changes. Fading may be most notable at sunrise and sunset, especially near the MUF, when the ionosphere undergoes dramatic transformations. Other ionospheric traumas, such as auroras and geomagnetic storms, also produce severe forms of HF fading.

The 11-Year Solar Cycle

The density of ionospheric layers depends on the amount of solar radiation reaching the Earth, but solar radiation is not constant. Variations result from daily and seasonal motions of the Earth, the sun's own 27-day rotation and the 11-year cycle of solar activity. One visual indicator of both the sun's rotation and the solar cycle is the periodic appearance of dark spots on the sun, which have been observed continuously since the mid-18th century. On average, the number of *sunspots* reaches a maximum every 10.7 years, but the period has varied between 7 and 17 years. Cycle 19 peaked in 1958, with an average sunspot number of over 200, the highest recorded to date. **Fig 20.10** shows average monthly sunspot numbers for the past four cycles.

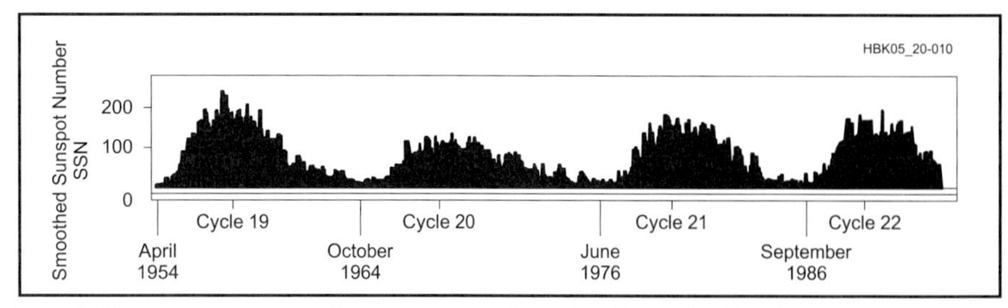

Fig 20.10—Average monthly sunspot numbers for Solar Cycles 19 to 22.

Sunspots are cooler areas on the sun's surface associated with high magnetic activity. Active regions adjacent to sunspot groups, called *plages*, are capable of producing great flares and sustained bursts of radiation in the radio through X-ray spectrum. During the peak of the 11-year solar cycle, average solar radiation increases along with the number of flares and sunspots. The ionosphere becomes more intensely ionized as a consequence, resulting in higher critical frequencies, particularly in the F_2 layer. The possibilities for long-distance communications are considerably improved during solar maxima, especially in the higher-frequency bands.

One key to forecasting F-layer critical frequencies, and thus long-distance propagation, is the intensity of ionizing UV and X-ray radiation. Until the advent of satellites, UV and X-ray radiation could not be measured directly, because they were almost entirely absorbed in the upper atmosphere. The sunspot number provided the most convenient approximation of general solar activity. The sunspot number is not a simple count of the number of visual spots, but rather the result of a complicated formula that takes into consideration size, number and grouping. The sunspot number varies from near zero during the solar-cycle minimum to over 200.

Another method of gauging solar activity is the *solar flux*, which is a measure of the intensity of 2800-MHz (10.7-cm) radio noise coming from the sun. The 2800-MHz radio flux correlates well with the intensity of ionizing UV and X-ray radiation and provides a convenient alternative to sunspot numbers. It commonly varies on a scale of 60-300 and can be related to sunspot numbers, as shown in **Fig 20.11**. The Dominion Radio Astrophysical Observatory, Penticton, British Columbia, measures the 2800-MHz solar flux daily at local noon. (Prior to June 1991, the Algonquin Radio Observatory, Ontario, made the measurements.) Radio station WWV broadcasts the latest solar-flux index at 18 minutes after each hour; WWVH does the same at 45 minutes after the hour. The Penticton solar flux is employed in a wide variety of other applications. Daily, weekly, monthly and even 13-month smoothed average solar flux readings are commonly used in propagation predictions.

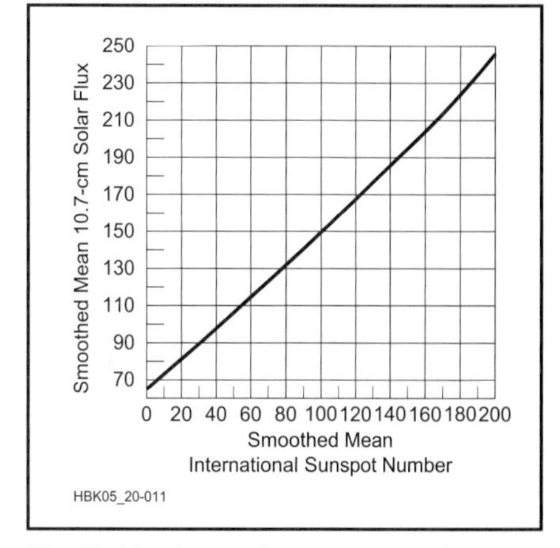

Fig 20.11—Approximate conversion between solar flux and sunspot number.

High flux values generally result in higher MUFs, but the actual procedures for predicting the MUF at any given hour and path are quite complicated. Solar flux is not the sole determinant, as the angle of the sun to the Earth, season, time of day, exact location of the radio path and other factors must all be taken into account. MUF forecasting a few days or months ahead involves additional variables and even more uncertainties.

The Sun's 27-Day Rotation

Sunspot observations also reveal that the sun rotates on its own axis. The sun is composed of extremely hot gases and does not turn uniformly. At the equator, the period is just over 25 days, but it approaches 35 days at the poles. Sunspots that affect the Earth's ionosphere, which appear almost entirely within 35° of the sun's equator, take about 26 days for one rotation. After taking into account the Earth's movement around the sun, the apparent period of solar rotation is about 27 days.

Active regions must face the Earth in the proper orientation to have an impact on the ionosphere. They may face the Earth only once before rotating out of view, but they often persist for several solar rotations. The net effect is that solar activity often appears in 27-day cycles corresponding to the sun's rotation, even though the active regions themselves may last for several solar rotations.

Solar-Ionospheric Disturbances

Like a campfire that occasionally spits out a flaming ember, our sun sometimes erupts spasmodically — but on a much grander scale than a summer campfire here on Earth. After all, any event that violently releases as much as 10 billion tons of solar material traveling up to four and a half million miles per hour has to be considered pretty impressive!

There are two main types of solar eruptions, distinguished partly by where they originate on the sun: solar flares and coronal mass ejections. A *solar flare* erupts from the sun's surface, and its main effect is to launch out into space a wide spectrum of electromagnetic energy, although a big flare can also release matter into space, mainly in the form of energetic protons. Since electromagnetic energy travels at the speed of light, the first indication of a solar flare reaches the Earth in about eight minutes. A large flare shows up as an increase in visible brightness near a sunspot group, accompanied by increases in UV and X-ray radiation and high levels of noise in the VHF radio bands.

A *coronal mass ejection* (CME) originates in the sun's outer atmosphere, its corona. With several sophisticated satellites launched in the mid 1990s, we have gained powerful new tools to monitor the intricacies of solar activity. The reality of how the sun operates is far more complex than initially expected. Using the latest satellite technology (and also some re-engineered earthbound instruments), scientists have observed many CMEs, greatly expanding our knowledge about them. Previously, the only direct observations we had of coronal activity were during solar eclipses — and eclipses don't occur very often.

One surprise has been that a large CME can involve as much as half of the entire solar coronal region. Flares are far more limited spatially — they are launched from the area around active sunspot regions. At one time, scientists believed that flares and CMEs were causally related, but now they recognize that many CMEs occur without an accompanying flare. And while many flares do result in an ejection of some solar material, many do not. It now seems clear that flares don't cause CMEs and vice versa.

While large flares can wreak disastrous effects on HF propagation, discussed further below, CMEs are the main causes of long-lasting magnetic storms here on Earth. Such storms can dramatically affect HF radio propagation — unfortunately, almost always in a negative fashion.

This is not to minimize the effects that a major solar flare can have on ionospheric propagation. After all, NASA rightly calls solar flares "the biggest explosions in the solar system." X-ray radiation from a large flare aimed towards Earth can cause an immediate increase in D- and E-layer ionization known as a *sudden ionospheric disturbance* (SID). Severe D-layer absorption may cause a short-term blackout of all HF communications on the sun-facing side of the Earth. Signals in the 2 to 30-MHz range may completely disappear. In extreme cases, nearly all background noise will be gone as well. SIDs may last up to an hour, after which ionospheric conditions temporarily return to normal.

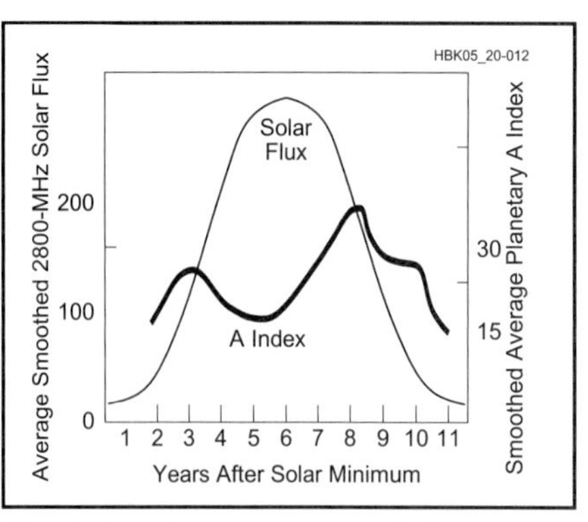

Fig 20.12—Geomagnetic activity (measured as the A-index) also follows an 11-year cycle. Average values over the past few cycles show that geomagnetic activity peaks before and after the peak of solar flux.

Table 20.3
Geomagnetic Storms

Typical Kp	*Description*	*Days per Solar Cycle*
9	Extreme	4
8	Severe	0
7	Strong	130
6	Moderate	360
5	Minor	900

Very energetic protons ejected during a large flare, and arriving in the vicinity of the Earth from several minutes to several hours after the flare, can penetrate deep into the ionosphere at the Earth's poles. This can produce intense ionization and consequent absorption of HF signals known as a *polar cap absorption* (PCA) event. A PCA event may last for days, dramatically affecting transpolar HF propagation.

When a CME occurs, whether or not it accompanies a solar flare, most of the time the electrons and protons ejected from the sun do not reach the Earth. This is because their trajectory takes them in another direction. If they do reach Earth, however, they do so 20 to 40 hours after the CME. As these charged particles sweep past, if their magnetic orientation is just right, they can distort the Earth's geomagnetic field, causing a *geomagnetic storm*. This results in acceleration of the particles to energy levels that permit them to penetrate into the ionosphere at the poles. This tremendous energy influx causes auroral displays at mid-latitudes and can disrupt HF communications for several hours or even much longer. Extraordinary radio noise and interference can accompany geomagnetic storms and associated auroras, especially at HF. Radio emissions from solar flares may be heard as sudden increases in noise on the VHF bands.

Effects on *ionospheric storms* (another name for geomagnetic storms) at HF vary considerably. Communications may be temporarily blacked out during an SID, but ionospheric paths may be generally noisy, weakened or disrupted for several days. Transpolar signals at 14 MHz and higher may be considerably attenuated and take on a hollow multipath sound. The number of geomagnetic storms varies considerably from year to year, with peak geomagnetic activity following the peak of solar activity. See **Fig 20.12**.

Devices known as *magnetometers* monitor geomagnetic activity. These may be as simple as a magnetic compass rigged to record its movements. Small variations in the geomagnetic field are scaled to two measures known as the K and A indexes. The *K index* provides an indication of magnetic activity on a finite scale of 0-9. Very quiet conditions are reported as 0 or 1, while geomagnetic storm levels begin at 4. See **Table 20.3** for the latest NOAA descriptions of geomagnetic storms.

A worldwide network of magnetometers constantly monitors the Earth's magnetic field, because the Earth's magnetic field varies with location. K indices that indicate average planetary conditions are indicated as K_p. Daily geomagnetic conditions are also summarized by the open-ended *A index*, which corresponds roughly to the cumulative K index values. The A index commonly varies between 0 and 30 during quiet to active conditions, and up to 100 and higher during geomagnetic storms.

At 18 minutes past the hour, radio stations WWV and WWVH broadcast the latest solar flux number, the average planetary A-Index and the latest Boulder K-Index. In addition, they broadcast a descriptive account of the condition of the geomagnetic field and a forecast for the next three hours. You should keep in mind that the A-Index is a description of what happened yesterday. Strictly speaking, the K-Index is valid only for Boulder, Colorado. However, the trend of the K-Index is very important for propagation analysis and forecasting. A rising K foretells worsening HF propagation conditions, particularly for transpolar paths. At the same time, a rising K alerts VHF operators to the possibility of enhanced auroral activity, particularly when the K-Index rises above 3.

D-Layer Propagation

The *D layer* is the lowest region of the ionosphere, situated between 55 and 90 km

(30 and 60 mi). See **Fig 20.13**. It is ionized primarily by the strong ultraviolet emission of solar hydrogen and short X-rays, both of which penetrate through the upper atmosphere. The D layer exists only during daylight, because constant radiation is needed to replenish ions that quickly recombine into neutral molecules. The D layer abruptly disappears at night so far as amateur MF and HF signals are concerned. D-layer ionization varies a small amount over the solar cycle. It is unsuitable as a refracting medium for any radio signals.

Daytime D-Layer Absorption

Nevertheless, the D layer plays an important role in HF communications. During daylight hours, radio energy as high as 5 MHz is effectively absorbed by the D layer, severely limiting the range of daytime 1.8- and 3.5-MHz signals. Signals at 7 MHz and 10 MHz pass through the D layer and on to the E and F layers only at relatively high angles. Low-angle waves, which must travel a much longer distance through the D layer, are subject to greater absorption. As the frequency increases above 10 MHz, radio waves pass through the D layer with increasing ease.

Nighttime D Layer

D-layer ionization falls 100-fold as soon as the sun sets and the source of ionizing radiation is removed. Low-band HF signals are then free to pass through to the E layer (also greatly diminished at night) and on to the F layer, where the MUF is almost always high enough to propagate 1.8- and 3.5-MHz signals half way around the world. Long-distance propagation at 7 and 10 MHz generally improves at night as well, because absorption is less and low-angle waves are able to reach the F layer.

D-Layer Ionospheric Forward Scatter

Radio signals in the 25-100 MHz range can be scattered by ionospheric irregularities, turbulence and stratification in the D and lower reaches of the E layers. Signals propagated by ionospheric forward scatter undergo very high losses, so signals are apt to be very weak. Typical scatter distances at 50 MHz are 800-1500 km (500-930 mi). This is not a common mode of propagation, but under certain conditions, ionospheric forward scatter can be very useful.

Ionospheric forward scatter is best during daylight hours from 10 AM to 2 PM local time, when the sun is highest in the sky and D-layer ionization peaks. It is worst at night. Scattering may be marginally more effective during the summer and during the solar cycle maximum due to somewhat higher D-layer ionization. The maximum path length of about 2000 km (1200 mi) is limited by the height of the scattering region, which is centered about 70 km (40 mi). Ionospheric scatter signals are typically weak, fluttery and near the noise level. Ionization from meteors sometimes temporarily raises signals well out of the noise for up to a few seconds at a time.

This mode may find its greatest use when all other forms of propagation are absent, primarily because ionospheric scatter signals are so weak. For best results at 28 and 50 MHz, a 3-element Yagi or larger, several hundred watts of power and a sensitive receiver are required. The paths are direct. CW is preferred, although, under optimal conditions, ionospheric scatter signals may be consistent enough to support SSB communications. Scattering is not efficient below 25 MHz. The very best-equipped pairs of 144-MHz stations may also be able to complete ionospheric scatter contacts.

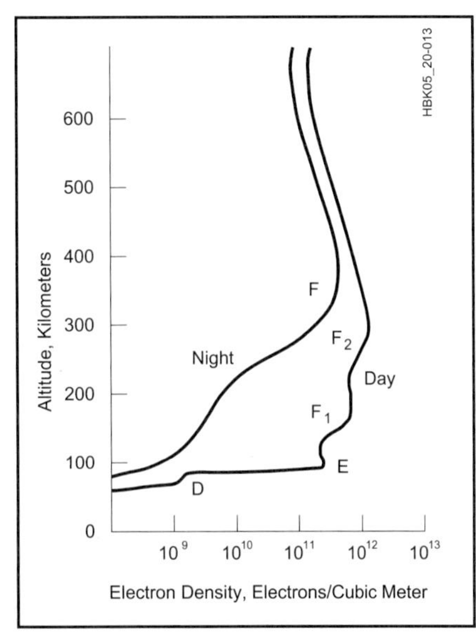

Fig 20.13—Typical electron densities for the various ionospheric regions.

E-Layer Propagation

The *E layer* lies between 90 and 150 km (60 and 90 mi) altitude, but a narrower region centered at 95 to 120 km (60 to 70 mi) is more important for radio propagation. E-layer nitrogen and oxygen atoms are ionized by short UV and long X-ray radiation. The normal E layer exists primarily during daylight hours, because like the D layer, it requires a constant source of ionizing radiation. Recombination is not as fast as in the denser D layer and absorption is much less. The E layer has a daytime critical frequency that varies between 3 and 4 MHz with the solar cycle. At night, the normal E layer all but disappears.

Daytime E Layer

The E layer plays a small role in propagating HF signals but can be a major factor limiting propagation during daytime hours. Its usual critical frequency of 3 to 4 MHz, with a maximum MUF factor of about 4.8, suggests that single-hop E-layer skip might be useful between 5 and 20 MHz at distances up to 2300 km (1400 mi). In practice this is not the case, because the potential for E-layer skip is severely limited by D-layer absorption. Signals radiated at low angles at 7 and 10 MHz, which might be useful for the longest-distance contacts, are largely absorbed by the D layer. Only high-angle signals pass through the D layer at these frequencies, but high-angle E-layer skip is typically limited to 1200 km (750 mi) or so. Signals at 14 MHz penetrate the D layer at lower angles at the cost of some absorption, but the casual operator may not be able to distinguish between signals propagated by the E layer or higher-angle F-layer propagation.

An astonishing variety of other propagation modes finds their home in the E layer, and this perhaps more than makes up for its ordinary limitations. Each of these other modes — sporadic E, field-aligned irregularities, aurora, auroral E and meteor scatter — are aberrant forms of propagation with unique characteristics. They are primarily useful only on the highest HF and lower VHF bands.

Sporadic E

Short skip, long familiar on the 10-m band during the summer months, affects the VHF bands as high as 222 MHz. *Sporadic E* (E_s), as this phenomenon is properly called, commonly propagates 28, 50 and 144-MHz radio signals between 500 and 2300 km (300 and 1400 mi). Signals are apt to be exceedingly strong, allowing even modest stations to make E_s contacts. At 21 MHz, the skip distance may only be a few hundred km. During the most intense E_s events, skip may shorten to less than 200 km (120 mi) on the 10-m band and disappear entirely on 15 m. Unusual multiple-hop E_s has supported contacts up to 10,000 km (6200 mi) on 28 and 50 MHz and more than 3,000 km (1900 mi) on 144 MHz. The first confirmed 220-MHz E_s contact was made in June 1987, but such contacts are likely to remain very rare.

Sporadic E at midlatitudes (roughly 15° to 45°) may occur at any time, but it is most common in the Northern Hemisphere during May, June and July, with a less-intense season at the end of December and early January. Its appearance is independent of the solar cycle. Sporadic E is most likely to occur from 9 AM to noon local time and again early in the evening between 5 PM and 8 PM. Midlatitude E_s events may last only a few minutes to many hours. In contrast, sporadic E is an almost constant feature of the polar regions at night and the equatorial belt during the day.

Efforts to predict midlatitude E_s have not been successful, probably because its causes are complex and not well understood. Studies have demonstrated that thin and unusually dense patches of ionization in the E layer, between 100 and 110 km (60 and 70 mi) altitude and 10 to 100 km (6 to 60 mi) in extent, are responsible for most E_s reflections. Sporadic-E clouds may form suddenly, move quickly from their birthplace, and dissipate within a few hours. Professional studies have recently focused on the role of heavy metal ions, probably of meteoric origin, and wind shears as two key factors in creating the dense patchy regions of E-layer ionization.

Sporadic-E clouds exhibit an MUF that can rise from 28 MHz through the 50-MHz band and higher in just a few minutes. When the skip distance on 28 MHz is as short as 400 or 500 km (250 or 310 mi), it is an indication that the MUF has reached 50 MHz for longer paths at low launch angles. Contacts at the maximum one-hop sporadic-E distance, about 2300 km (1400 mi), should then be possible at 50 MHz. E-skip contacts as short as 700 km (435 mi) on 50 MHz, in turn, may indicate that 144-MHz contacts in the 2300-km (1400 mi) range can be completed. See **Fig 20.14**. Sporadic-E openings occur about a tenth as often at 144 MHz in comparison to 50 MHz and for much shorter periods.

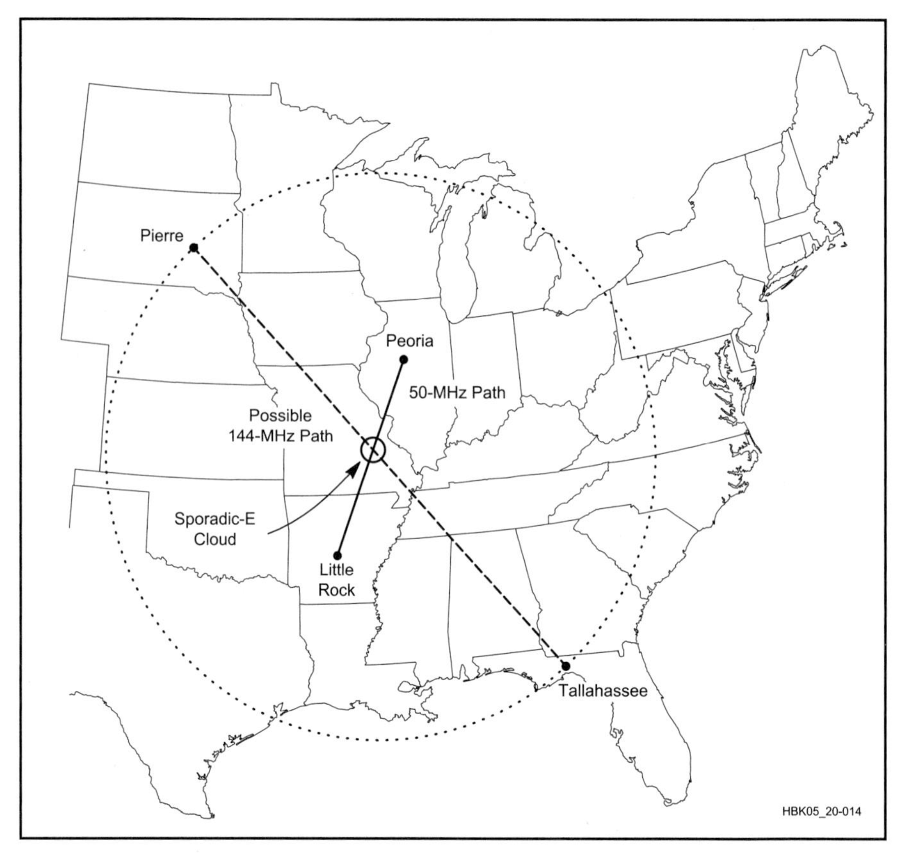

Fig 20.14—50 MHz sporadic-E contacts of 700 km (435 mi) or shorter (such as between Peoria and Little Rock) indicate that the MUF on longer paths is above 144 MHz. Using the same sporadic-E region reflecting point, 144-MHz contacts of 2200 km (1400 mi), such as between Pierre and Tallahassee, should be possible.

Sporadic E can also have a detrimental effect on HF propagation by masking the F_2 layer from below. HF signals may be prevented from reaching the higher levels of the ionosphere and the possibilities of long F_2 skip. Reflections from the tops of sporadic-E clouds can also have a masking effect, but they may also lengthen the F_2 propagation path with a top-side intermediate hop that never reaches the Earth.

E-Layer Field-Aligned Irregularities

Amateurs have experimented with a little-known scattering mode known as *field-aligned irregularities* (FAI) at 50 and 144 MHz since 1978. FAI commonly appear directly after sporadic-E events and may persist for several hours. Oblique-angle scattering becomes possible when electrons are compressed together due to the action of high-velocity ionospheric acoustic (sound) waves. The resulting irregularities in the distribution of free electrons are aligned parallel to the Earth's magnetic field, in something like moving vertical rods. A similar process of electron field-alignment takes place during radio aurora, making the two phenomena quite similar.

Most reports suggest that 8 PM to midnight may be the most productive time for FAI. Stations attempting FAI contacts point their antennas toward a common scattering region that corresponds to an active or recent E_s reflection point. The best direction must be probed experimentally, for the result is rarely along the great-circle path. Stations in south Florida, for example, have completed 144-MHz FAI contacts with north Texas when participating stations were beamed toward a common scattering region over northern Alabama.

FAI-propagated signals are weak and fluttery, reminiscent of aurora signals. Doppler shifts of as much as 3 kHz have been observed in some tests. Stations running as little as 100 W and a single Yagi should be able to complete FAI contacts during the most favorable times, but higher power and larger antennas may yield better results. Contacts have been made on 50 and 144 MHz and 222-MHz FAI seems probable as well. Expected maximum distances should be similar to other forms of E-layer propagation, or about 2300 km (1400 mi).

Aurora

Radar signals as high as 3000 MHz have been scattered by the *aurora borealis* or northern lights (*aurora australis* in the Southern Hemisphere), but amateur aurora contacts are common only from 28 through 432 MHz. By pointing directional antennas generally north toward the center of aurora activity, oblique paths between stations up to 2300 km (1400 mi) apart can be completed. See **Fig 20.15**. High power and large antennas are not necessary. Stations with small Yagis and as little as 10 W output have used auroras on frequencies as high as 432 MHz, but contacts at 902 MHz and higher are exceedingly rare. Auroral propagation works just as well in the Southern Hemisphere, in which case antennas must be pointed south.

The appearance of auroras is closely linked to solar activity. During massive geomagnetic storms, high-energy particles flow into the ionosphere near the polar regions, where they ionize the gases of the E layer and higher. This unusual ionization produces spectacular visual auroral displays, which often spread southward into the midlatitudes. Auroral ionization in the E layer scatters radio signals in the VHF and UHF ranges.

In addition to scattering radio signals, auroras have other effects on worldwide radio propagation. Communication below 20 MHz is disrupted in high latitudes, primarily by absorption, and is especially noticeable over polar and near-polar paths. Signals on the AM broadcast band through the 40-m band late in the afternoon may become weak and watery. The 20-m band may close down altogether. Satellite operators have also noticed that 144-MHz downlink signals are often weak and distorted when satellites pass near the polar regions. At the same time, the MUF in equatorial regions may temporarily rise dramatically, providing transequatorial paths at frequencies as high as 50 MHz.

Auroras occur most often around the spring and fall equinoxes (March-April and September-October), but auroras may appear in any month. Aurora activity generally peaks about two years before and after solar cycle maximum. Radio aurora activity is usually heard first in late afternoon and may reappear later in the evening. Auroras may be anticipated by following the A- and K-index reports on WWV. A K index of five or greater and an A index of at least 30 are indications that a geomagnetic storm is in progress and an aurora likely. The probability, intensity and southerly extent of auroras increase as the two index numbers rise. Stations north of 42° latitude in North America experience many auroral openings each year, while those in the Gulf Coast states may hear auroral signals no more than once a year, if that often.

Aurora-scattered signals are easy to identify. On 28- and 50-MHz SSB, signals sound very distorted and somewhat wider than normal; at 144 MHz and above, the distortion may be so severe that only CW is useful. Auroral CW signals have a distinctive note variously described as a buzz, hiss or mushy sound. This characteristic auroral signal is due to Doppler broadening, caused by the movement of electrons within the aurora. An additional Doppler shift of 1 kHz or more may be evident at 144 MHz and several kilohertz at 432 MHz. This second Doppler shift is the result of massive electrical currents that sweep electrons toward the sun side of the Earth during magnetic storms. Doppler shift and distortion increase with higher frequencies, while signal strength dramatically decreases.

It is not necessary to see an aurora to make auroral contacts. Useful auroras may be 500-1000 km (310-620 mi) away and below the visual horizon. Antennas should be pointed generally north and then probed east and west to peak signals, because auroral ionization is field aligned. This means that for any pair of stations, there is an optimal direction for aurora scatter. Offsets from north are usually greatest when the aurora is closest and often provide the longest contacts. There may be some advantage to antennas that can be elevated, especially when auroras are high in the sky.

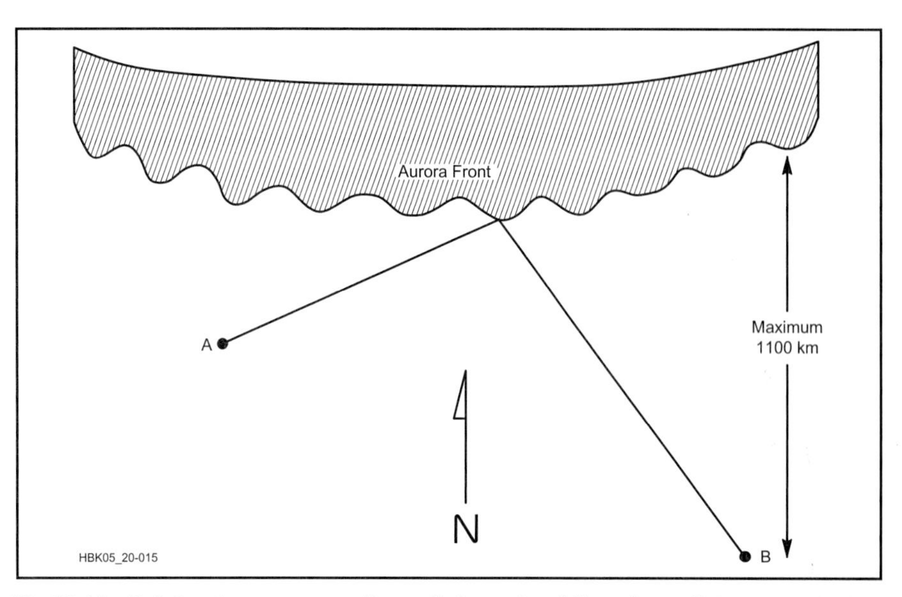

Fig 20.15—Point antennas generally north to make oblique long-distance contacts on 28 through 432 MHz via aurora scattering. Optimal antenna headings may shift considerably to the east or west depending on the location of the aurora.

Auroral E

Radio auroras may evolve into a propagation mode known as *auroral E* at 28, 50 and rarely 144 MHz. Doppler distortion disappears and signals take on the characteristics of sporadic E. The most effective antenna headings shift dramatically away from oblique aurora paths to direct great-circle bearings. The usual maximum distance is 2300 km (1400 mi), typical for E-layer modes, but 28- and 50-MHz auroral-E contacts of 5000 km (3100 mi) are sometimes made across Canada and the northern US, apparently using two hops. Contacts at 50 MHz between Alaska and the east coasts of Canada and the northern US have been completed this way. Transatlantic 50-MHz auroral-E paths are also likely, although only one such contact has been reported.

Typically, 28- and 50-MHz auroral E appears across the northern third of the US and southern Canada when aurora activity is diminishing. This usually happens after midnight on the eastern end of the path. Auroral-E signals sometimes have a slightly hollow sound to them and build slowly in strength over an hour or two, but otherwise they are indistinguishable from sporadic E. Auroral-E paths are almost always east-west oriented, perhaps because there are few stations at very northern latitudes to take advantage of this propagation.

Auroral E may also appear while especially intense auroras are still in progress, as happened during the great aurora of March 1989. On that occasion, 50-MHz propagation shifted from Doppler-distorted aurora paths to clear-sounding auroral E over a period of a few minutes. Many 6-m operators as far south as Florida and Southern California made single- and double-hop auroral-E contacts across the country. At about the same time, the MUF reached 144 MHz for stations west of the Great Lakes to the Northeast, the first time auroral E had been reported so high in frequency. At least two other rare instances of 2-m auroral E have been reported.

Meteor Scatter

Contacts between 800 and 2300 km (500 and 1400 mi) can be made at 28 through 432 MHz via reflections from the ionized trails left by meteors as they travel through the ionosphere. The kinetic energy of meteors no larger than grains of rice are sufficient to ionize a column of air 20 km (12 mi) long in the E layer. The particle itself evaporates and never reaches the ground, but the ionized column may persist for a few seconds to a minute or more before it dissipates. This is enough time to make very brief contacts by reflections from the ionized trails. Millions of meteors enter the Earth's atmosphere every day, but few have the required size, speed and orientation to the Earth to make them useful for meteor-scatter propagation.

Radio signals in the 30- to 100-MHz range are reflected best by meteor trails, making the 50-MHz band prime for meteor-scatter work. The early morning hours around dawn are usually the most productive, because the morning side of the Earth faces in the direction of the planet's orbit around the Sun. The relative velocity of meteors that head toward the Earth's morning side are thus increased by up to 30 km/sec, the average rotational speed of the Earth in orbit. See **Fig 20.16**. The maximum velocity of meteors in orbit around the Sun is 42 km/sec. Thus when the relative velocity of the Earth is considered, most meteors must enter the Earth's atmosphere somewhere between 12 and 72 km/sec.

Meteor contacts ranging from a second or two to more than a minute can be made nearly any morning at 28 or 50 MHz. Meteor-scatter contacts at 144 MHz and higher are more difficult because reflected signal strength and duration drop sharply with increasing frequency. A meteor trail that provides 30 seconds of communication at 50 MHz will last only a few seconds at 144 MHz, and less than a second at 432 MHz.

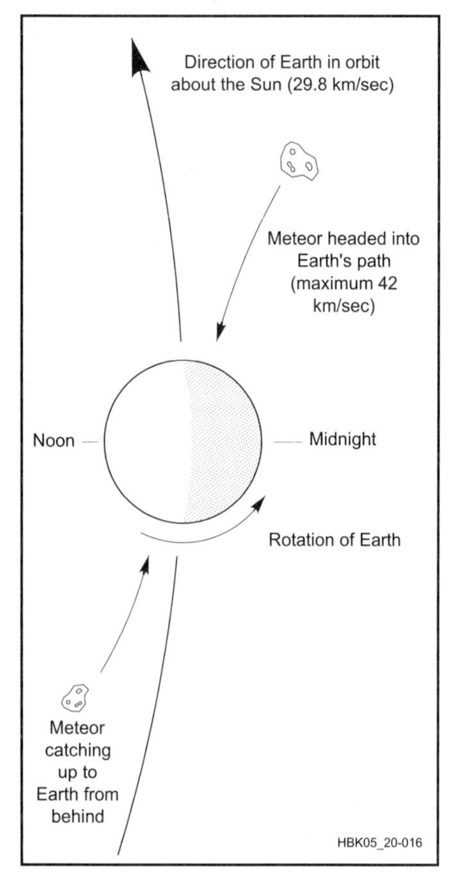

Fig 20.16—The relative velocity of meteors that meet the Earth head-on is increased by the rotational velocity of the Earth in orbit. Fast meteors strike the morning side of the Earth because their velocity adds to the Earth's rotational velocity, while the relative velocity of meteors that "catch up from behind" is reduced.

Meteor scatter opportunities are somewhat better during July and August because the average number of meteors entering the Earth's atmosphere peaks during those months. The best times are during one of the great annual *meteor showers*, when the number of useful meteors may increase tenfold over the normal rate of five to ten per hour. See **Table 20.4**. A meteor shower occurs when the Earth passes through a relatively dense stream of particles, thought to be the remnants of a comet, that are also in orbit around the sun. The most-productive showers are relatively consistent from year to year, although several can produce great storms periodically.

Because meteors provide only fleeting moments of communication even during one of the great meteor showers, special operating techniques are often used to increase the chances of completing a contact. Prearranged schedules between two stations establish times, frequencies and precise operating standards. Usually, each station transmits on alternate 15-second periods until enough information is pieced together a bit at a time to confirm contact. High-speed Morse code of several hundred words per minute, generated and slowed down by special computer programs, can make effective use of very short meteor bursts. Nonscheduled random meteor contacts are common on 50 MHz and 144 MHz, but short transmissions and alert operating habits are required.

It is helpful to run several hundred watts to a single Yagi, but meteor-scatter can be used by modest stations under optimal conditions. During the best showers, a few watts and a small directional antenna are sufficient at 28 or 50 MHz. At 144 MHz, at least 100 W output and a long Yagi are needed for consistent results. Proportionately higher power is required for 222 and 432 MHz even under the best conditions.

F-Layer Propagation

The region of the *F layers*, from 150 km (90 mi) to over 400 km (250 mi) altitude, is by far the most important for long-distance HF communications. F-region oxygen atoms are ionized primarily by ultraviolet radiation. During the day, ionization reaches maxima in two distinct layers. The F_1 layer forms between 150 and 250 km (90 and 160 mi) and disappears at night. The F_2 layer extends above 250 km (160 mi), with a peak of ionization around 300 km (190 mi). At night, F-region ionization collapses into one broad layer at 300-400 km (190-250 mi) altitude. Ions recombine very slowly at these altitudes, because molecular density is relatively low. Maximum ionization levels change significantly with time of day, season and year of the solar cycle.

Table 20.4
Major Annual Meteor Showers

Name	*Peak Dates*	*Approximate Rate (meteors/hour)*
Quadrantids	Jan 3	50
Arietids	Jun 7-8	60
Perseids	Aug 11-13	80
Orionids	Oct 20-22	20
Geminids	Dec 12-13	60

F_1 Layer

The daytime F_1 layer is not important to HF communication. It exists only during daylight hours and is largely absent in winter. Radio signals below 10 MHz are not likely to reach the F_1 layer, because they are either absorbed by the D layer or refracted by the E layer. Signals higher than 20 MHz that pass through both of the lower ionospheric regions are likely to pass through the F_1 layer as well, because the F_1 MUF rarely rises above 20 MHz. Absorption diminishes the strength of any signals that continue through to the F_2 layer during the day. Some useful F_1-layer refraction may take place between 10 and 20 MHz during summer days, yielding paths as long as 3000 km (1900 mi), but these would be practically indistinguishable from F_2 skip.

F_2 and Nighttime F Layers

The F_2 layer forms between 250 and 400 km (160 and 250 mi) during the daytime and persists throughout the night as a single consolidated F region 50 km (30 mi) higher in altitude. Typical ion densities are the highest of any ionospheric layer, with the possible exception of some unusual E-layer phenomenon. In contrast to the other ionospheric layers, F_2 ionization varies considerably with time of day, season and position in the solar cycle, but it is never altogether absent. These two characteristics make the F_2 layer the most important for long-distance HF communications.

The F_2-layer MUF is nearly a direct function of UV solar radiation, which in turn follows closely the solar cycle. During the lowest years of the cycle, the daytime MUF may climb above 14 MHz for only a few hours a day. In contrast, the MUF may rise beyond 50 MHz during peak years and stay above 14 MHz throughout the night. The virtual height of F_2 averages 330 km (210 mi), but varies between 200 and 400 km (120 and 250 mi). Maximum one-hop distance is about 4000 km (2500 mi).

Near-vertical incidence skywave propagation just below the critical frequency provides reliable coverage out to 200-300 km (120-190 mi) with no skip zone. It is most often observed on 7 MHz during the day.

The extraordinary high-angle *Pedersen Ray* can create effective single-hop paths of 5,000 to 12,000 km under certain conditions, but most operators will not be able to distinguish Pedersen-Ray paths from normal F-layer propagation. Pedersen-Ray paths are most evident over high-latitude east-west paths at frequencies near the MUF. They appear most often about noon local time at mid-path when the geomagnetic field is very quiet. Pedersen-Ray propagation may be responsible for 50 MHz paths between the US Northeast and Western Europe, for example, when ordinary MUF analysis could not explain the 5,000-km contacts. See **Fig 20.17E**.

In general, both F_2-layer ionization and MUF build rapidly at sunrise, usually reach a maximum in the afternoon, and then decrease to a minimum prior to sunrise. Depending on the season, the MUF is generally highest within 20° of the equator and lower toward the poles. For this reason, transequatorial paths may be open at a particular frequency when all other paths are closed.

In contrast to all the other ionospheric layers, daytime ionization in the winter F_2 layer averages four times the level of the summer at the same period in the solar cycle, doubling the MUF. This so-called *winter anomaly* is caused by the Earth moving closer to the Sun and tilting. Wintertime F_2 conditions are much superior to those in summer, because the MUF is much higher.

Multihop F-Layer Propagation

Most HF communication beyond 4000 km (2500 mi) takes place via multiple ionospheric hops. Radio signals are reflected from the Earth back toward space for additional ionospheric refractions. A series of ionospheric refractions and terrestrial reflections commonly create paths halfway around the Earth. Each hop involves additional attenuation and absorption, so the longest-distance signals tend to be the weakest. Even so, it is possible for signals to be propagated completely around the world and arrive back at their originating point. Multiple reflections within the F layer may bypass ground reflections altogether, creating what are known as *chordal hops*, with lower total attenuation. It takes a radio signal about 0.15 second to make a round-the-world trip.

Multihop paths can take on many different configurations, as shown in the examples of Fig 20.17. E-layer (especially sporadic E) and F-layer hops may be mixed. In practice, multihop signals arrive via many different paths, which often increases the problems of fading. Analyzing multihop paths is complicated by the effects of D- and E-layer absorption, possible reflections from the tops of sporadic-E layers, disruptions in the auroral zone and other phenomena.

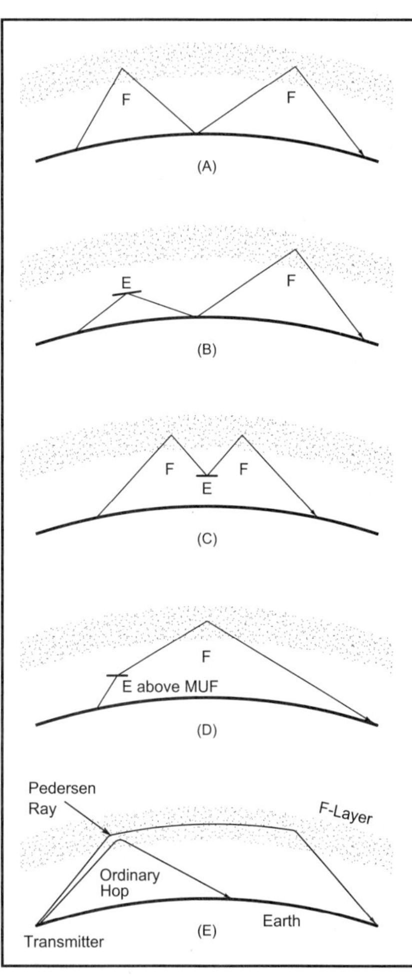

Fig 20.17—Multihop paths can take many different configurations, including a mixture of E- and F-layer hops. (A) Two F-layer hops. Five or more consecutive F-layer hops are possible. (B) An E-layer hookup to the F layer. (C) A top-side E-layer reflection can shorten the distance of two F-layer hops. (D) Refraction in the E layer above the MUF is insufficient to return the signal to Earth, but it can go on to be refracted in the F layer. (E) The Pedersen Ray, which originates from a signal launched at a relatively high angle above the horizon into the E or F region, may result in a single-hop path, 5000 km (3100 mi) or more. This is considerably further than the normal 4000-km (2500 mi) maximum F-region single-hop distance, where the signal is launched at a very low takeoff angle. The Pedersen Ray can easily be disrupted by any sort of ionospheric gradient.

F-Layer Long Path

Most HF communication takes place along the shortest great-circle path between two stations. Short-path propagation is always less than 20,000 km (12,000 mi) — halfway around the Earth. Nevertheless, it may be possible at times to make the same contact in exactly the opposite direction via the *long path*. The long-path distance will be 40,000 km (25,000 mi) minus the short-path length. Signal strength via the long path is usually considerably less than the more direct short-path. When both paths are open simultaneously, there may be a distinctive sort of echo on received signals. The time interval of the echo represents the difference between the short-path and long-path distances.

Sometimes there is a great advantage to using the long path when it is open, because signals can be stronger and fading less troublesome. There are times when the short path may be closed or disrupted by E-layer blanketing, D-layer absorption or F-layer gaps, especially when operating just below the MUF. Long paths that predominantly cross the night side of the Earth, for example, are sometimes useful because they generally avoid blanketing and absorption problems. Daylight-side long paths may take advantage of higher F-layer MUFs that occur over the sunlit portions of the Earth.

F-Layer Gray-Line

Gray-line paths can be considered a special form of long-path propagation that take into account the unusual ionospheric configuration along the twilight region between night and day. The gray line, as the twilight region is sometimes called, extends completely around the world. It is not precisely a line, for the distinction between daylight and darkness is a gradual transition due to atmospheric scattering. On one side, the gray line heralds sunrise and the beginning of a new day; on the opposite side, it marks the end of the day and sunset.

The ionosphere undergoes a significant transformation between night and day. As day begins, the highly absorbent D and E layers are recreated, while the F-layer MUF rises from its pre-dawn minimum. At the end of the day, the D and E layers quickly disappear, while the F-layer MUF continues its slow decline from late afternoon. For a brief period just along the gray-line transition, the D and E layers are not well formed, yet the F_2 MUF usually remains higher than 5 MHz. This provides a special opportunity for stations at 1.8 and 3.5 MHz.

Normally, long-distance communication on the lowest two amateur bands can take place only via all-darkness paths because of daytime D-layer absorption. The gray-line propagation path, in contrast, extends completely around the world. See **Fig 20.18**. This unusual situation lasts less than an hour at sunrise and sunset when the D-layer is largely absent, and may support contacts that are difficult or impossible at other times.

The gray line generally runs north-south, but it varies by 23° either side of true north as measured at the equator over the course of the year. This variation is caused by the tilt in the Earth's axis. The gray line is exactly north-south through the poles at the equinoxes (March 21 and September 21) and is at its 23° extremes on June 21 and December 20. Over a one-year period, the gray line crosses a 46° sector of the Earth north and south of the equator, providing optimum paths to slightly different parts of the world each day. Many commonly available computer programs plot the gray line on a flat map or globe. *The ARRL Operating Manual* provides sunrise and sunset times over the entire year for several hundred worldwide locations. The position of the gray line on any date can also be plotted manually on a globe from these data.

F-Layer Backscatter and Sidescatter

Special forms of F-layer scattering can create unusual paths within the skip zone. *Backscatter* and *sidescatter* signals are usually observed just below the MUF for the direct path and allow communications not normally possible by other means. Stations using backscatter point their antennas toward a common scattering region at the one-hop distance, rather than toward each other. Backscattered signals are generally weak and have a characteristic hollow sound. Useful communication distances range from 100 km (60 mi) to the normal one-hop distance of 4000 km (2500 mi).

Backscatter and sidescatter are closely related and the terminology does not precisely distinguish between the two. Backscatter usually refers to single-hop signals that have been scattered by the Earth or the ocean at some distant point back toward the transmitting station. Two stations spaced a few hundred km apart can often communicate via a backscatter path near the MUF. See **Fig 20.19**.

Sidescatter usually refers to a circuit that is oblique to the normal great-circle path. Two stations can make use of a common side-scattering region well off the direct path, often toward the south. European and North American stations sometimes complete 28-MHz contacts via a scattering region over Africa. US and Finnish 50-MHz operators observed a similar effect early one morning in November 1989 when they made contact by beaming off the coast of West Africa.

When backscattered signals cross an area where there is a sharp gradient in ionospheric density, such as between night and day, the path may take on a different geometry, as shown in **Fig 20.20**. In this case, stations can communicate because backscattered signals return via the day side ionosphere on a shorter hop than the night side. This is possible because the dayside MUF is higher and thus the skip distance shorter. The net effect is to create a backscatter path between two stations within the normal skip zone.

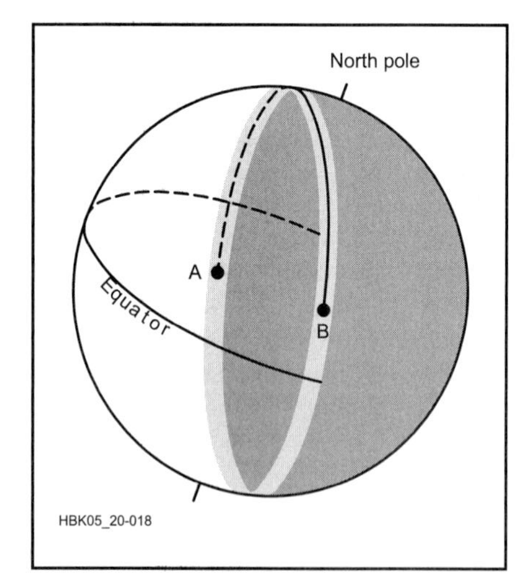

Fig 20.18—The gray line encircles the Earth, but the tilt at the equator to the poles varies over 46° with the seasons. Long-distance contacts can often be made halfway around the Earth along the gray line, even as low as 1.8 and 3.5 MHz. The strength of the signals, characteristic of gray-line propagation, indicates that multiple Earth-ionosphere hops are not the only mode of propagation, since losses in many such hops would be very great. Chordal hops, where the signals are confined to the ionosphere for at least part of the journey, are involved.

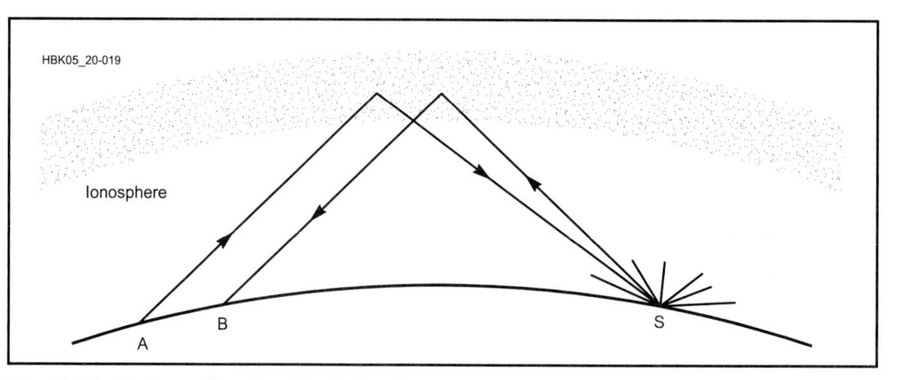

Fig 20.19—Schematic of a simple backscatter path. Stations A and B are too close to make contact via normal F-layer ionospheric refraction. Signals scattered back from a distant point on the Earth's surface (S), often the ocean, may be accessible to both and create a backscatter circuit.

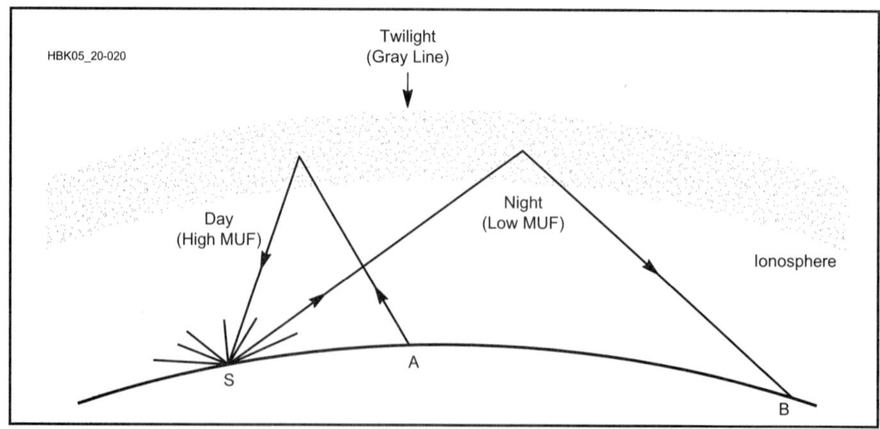

Fig 20.20—Backscatter path across the gray line. Stations A and B are too close to make contact via normal ionospheric refraction, but may hear each other's signals scattered from point S. Station A makes use of a high-angle refraction on the day side of the gray line, where the MUF is high. Station B makes use of a night-time refraction, with a lower MUF and lower angle of propagation. Note that station A points away from B to complete the circuit.

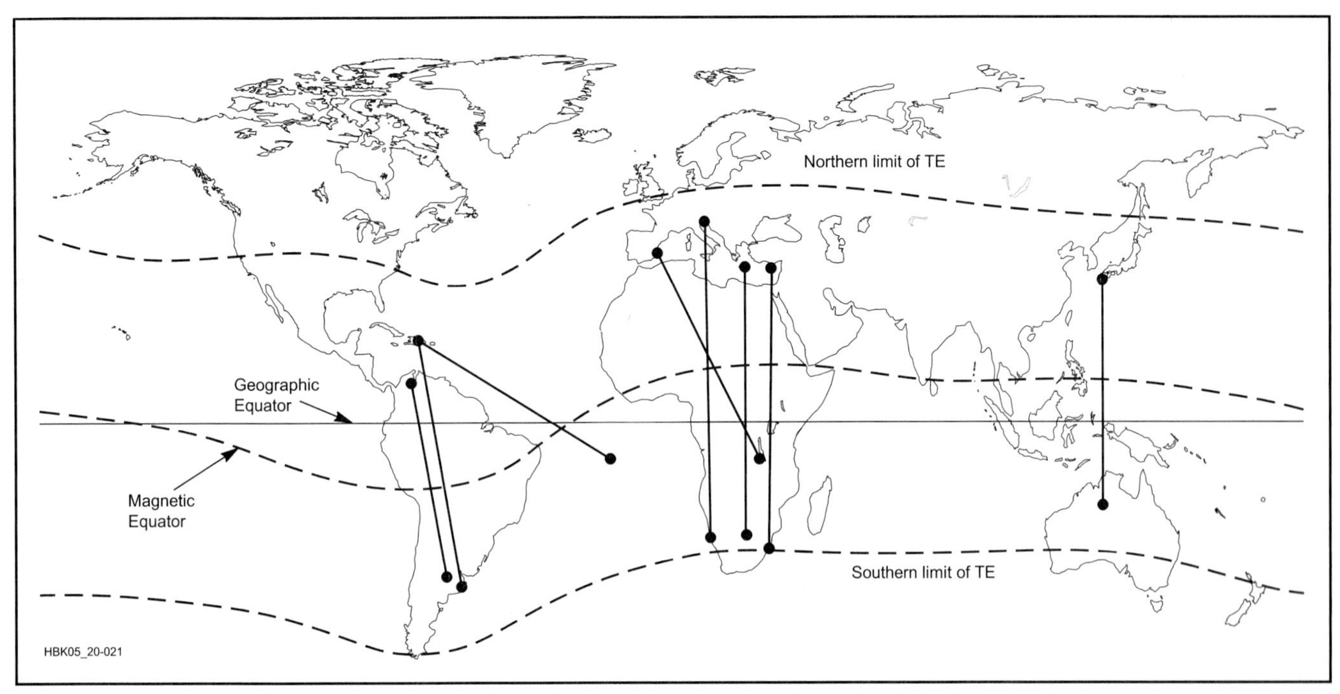

Fig 20.21—Transequatorial spread-F propagation takes place between stations equidistant across the geomagnetic equator. Distances up to 8000 km (5000 mi) are possible on 28 through 432 MHz. Note the geomagnetic equator is considerably south of the geographic equator in the Western Hemisphere.

Transequatorial Spread-F

Discovered in 1947, *transequatorial spread-F* (TE) supports propagation between 5000 and 8000 km (3100 and 5000 mi) across the equator from 28 MHz to as high as 432 MHz. Stations attempting TE contacts must be nearly equidistant from the geomagnetic equator. Many contacts have been made at 50 and 144 MHz between Europe and South Africa, Japan and Australia and the Caribbean region and South America. Fewer contacts have been made on the 222-MHz band. TE signals have been heard at 432 MHz, but so far, no two-way contacts have resulted.

Unfortunately for most continental US stations, the *geomagnetic equator* dips south of the geographic equator in the Western Hemisphere, as shown in **Fig 20.21**, making only the most southerly portions of Florida and Texas within TE range. TE contacts from the southeastern part of the country may be possible with Argentina, Chile and even South Africa.

Transequatorial spread-F peaks between 5 PM and 10 PM during the spring and fall equinoxes, especially during the peak years of the solar cycle. The lowest probability is during the summer. Quiet geomagnetic conditions are required for TE to form. Signals have a rough aurora-like note, sometimes termed *flutter fading*. High power and large antennas are not required to work TE, as VHF stations with 100 W and single long Yagis have been successful.

The best explanation of TE propagation suggests that the F_2 layer near the equator bulges and intensifies slightly, particularly during solar maxima. Irregular field-aligned ionization forms shortly after sunset in an area 100-200 km (60-120 mi) north and south of the geomagnetic equator and 500-3000 km (310-1900 mi) wide. For this reason, the mode is sometimes called *transequatorial field-aligned irregularities*. It moves west with the setting sun. The MUF may increase to twice its normal level 15° either side of the geomagnetic equator.

Field alignment of ionospheric irregularities favors refraction along magnetic field lines, that is north-south. VHF and UHF signals are refracted twice over the geomagnetic equator at angles that normally would be insufficient to bring the signals back toward Earth. See **Fig 20.22**. The geometry is such that two shallow reflections in the F_2 layer can create north-south terrestrial paths up to 8000 km (5000 mi).

Spread-F propagation also occurs over the polar regions, but because of low population densities, amateurs have rarely reported making use of it. Near the northern magnetic pole (located in extreme northeastern Canada), spread-F is a nearly permanent feature of winter. During summer, it appears most summer nights and at least half the time during the day. There is a greater probability of polar spread-F appearing during the equinox periods and during the solar cycle maximum. Field-alignment in the polar regions suggests that some form of backscatter signals, similar to aurora, would be most likely.

MUF PREDICTION

F-layer MUF prediction is key to forecasting HF communications paths at particular frequencies, dates and times, but

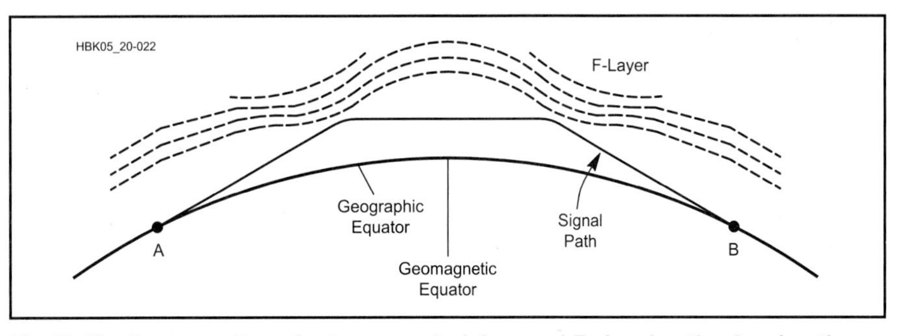

Fig 20.22—Cross-section of a transequatorial spread-F signal path, showing the effects of ionospheric bulging and a double refraction above the normal MUF.

forecasting is complicated by several variables. Solar radiation varies over the course of the day, season, year and solar cycle. These regular intervals provide the main basis for prediction, yet recurrence is far from reliable. In addition, forecasts are predicated on a quiet geomagnetic field, but the condition of the Earth's magnetic field is most difficult to predict weeks or months ahead. For professional users of HF communications, uncertainty is a nuisance for maintaining reliable communications paths, while for many amateurs it provides an aura of mystery and chance that adds to the fun of DXing. Nevertheless, many amateurs want to know what to expect on the HF bands to make best use of available on-the-air time, plan contest strategy, ensure successful net operations or engage in other activities.

MUF Forecasts

Long-range forecasts several months ahead, such as those formerly published in *QST* and other journals, provide only the most general form of prediction. A series of 48 charts on the members-only ARRLWeb site (**www.arrl.org/qst/propcharts/**), similar to **Fig 20.23**, forecast average propagation for a one-month period over specific paths. The charts assume a single average solar flux value for the entire month and they assume that the geomagnetic field is undisturbed.

The uppermost curve in Fig 20.23 shows the highest frequency that will be propagated on at least 10% of the days in the month. The given values might be exceeded considerably on a few rare days. On at least half the days, propagation should be possible on frequencies as high as the middle curve. Propagation will exceed the lowest curve on at least 90% of the days. The exact MUF on any particular day cannot be determined from these statistical charts, but the calculated times when a band will open and close is reliable. You would use a long-range forecast to determine when you should start monitoring a band to see if propagation actually does occur that day, particularly at frequencies above 30 MHz.

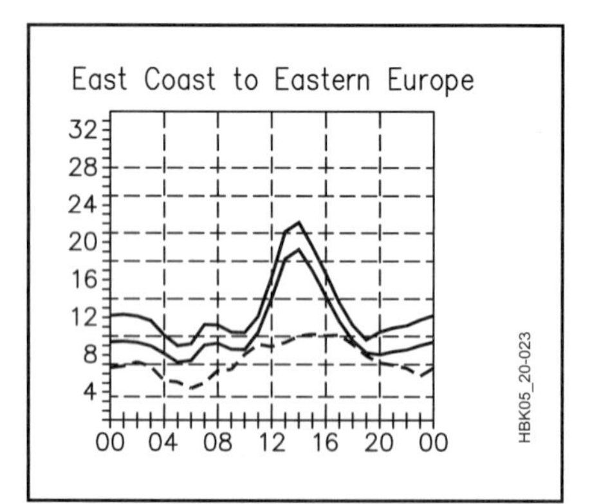

Fig 20.23—Propagation prediction chart for West Coast to Western Europe from the ARRLWeb members-only site for April 2001. An average 2800-MHz (10.7-cm) solar flux of 159 was assumed for the month. On 10% of the days, the highest frequency propagated is predicted to be at least as high as the uppermost curve (the Highest Possible Frequency, or HPF, approximately 33 MHz), and for 50% of the days as high as the middle curve, the MUF. The lowest curve shows the Lowest Usable Frequency (LUF) for a 1500-W CW transmitter.

Short-range forecasts of a few days ahead are marginally more reliable than long-range forecasts, because underlying solar indices and geomagnetic conditions can be anticipated with greater confidence. The tendency for solar disturbances to recur at 27-day intervals also enhance short-term forecasts. Daily forecasts are even more reliable, because they are based on current solar and geophysical data, as well as warnings provided by observations of the sun in the visual to X-ray range.

The CD-ROM bundled with the 20th Edition of *The ARRL Antenna Book* contains even more detailed propagation-prediction tables from 150+ QTHs around the world for six levels of solar activity, for the 12 months of the year. Again, keep in mind that these long-range forecasts assume quiet geomagnetic conditions. Real-time MUF forecasts are also available in a variety of text and graphical forms on the WWW. Forecasts can also be made at home using one of several popular programs for personal computers, including *ASAPS*, *CAPMan*, *VOACAP*, *W6ELProp* and *WinCAP Wizard 2*.

Direct Observation

Propagation conditions can be determined directly by listening to the HF bands. The simplest method is to tune higher in frequency until no more long-distance stations are heard. This point is roughly just above the MUF to anywhere in the world at that moment. The highest usable amateur band would be the next lowest one. If HF stations seem to disappear around 23 MHz, for example, the 15-m band at 21 MHz might make a good choice for DXing. By carefully noting station locations as well, the MUF in various directions can also be determined quickly.

The shortwave broadcast bands (see **Table 20.5**) are most convenient for MUF browsing, because there are many high-powered stations on regular schedules. Take care to ensure that programming is actually transmitted from the originating country. A Radio Moscow or BBC program, for example, may be relayed to a transmitter outside Russia or England for retransmission. An excellent guide to shortwave broadcast stations is the *World Radio TV Handbook*, available through the ARRL.

Table 20.5
Shortwave Broadcasting Bands

Frequency (MHz)	*Band (m)*
2.300-2.495	120
3.200-3.400	90
3.900-4.000	75
4.750-5.060	60
5.959-6.200	49
7.100-7.300	41
9.500-9.900	31
11.650-12.050	25
13.600-13.800	22
15.100-15.600	19
17.550-17.900	16
21.450-21.850	13
25.600-26.100	11

WWV and WWVH

The standard time stations WWV (Ft Collins, Colorado) and WWVH (Kauai, Hawaii), which transmit on 2.5, 5, 10, 15 and 20 MHz, are also popular for propagation monitoring. They transmit 24 hours a day. Daily monitoring of these stations for signal strength and quality can quickly provide a good basic indication of propagation conditions. In addition, each hour they broadcast the geomagnetic A and K indices, the 2800-MHz (10.7-cm) solar flux, and a short forecast of conditions for the next day. These are heard on WWV at 18 minutes past each hour and on WWVH at 45 minutes after the hour. The same information is also available by telephoning the recorded message at 303-497-3235 or various Web sites, such as **dx.qsl.net/propagation/index.html**. The K index is updated every three hours, while the A index and solar flux are updated after 2100 UTC. These data are useful for making predictions on home computers, especially when averaged over several days of solar flux observations.

Beacons

Automated *beacons* in the higher amateur bands can also be useful adjuncts to propagation watching. Beacons are ideal for this purpose because most are designed to transmit 24 hours a day. One of the best organized beacon systems is designed by the Northern California DX Foundation, operating at 14.100, 18.110, 21.150, 24.930 and 28.200 MHz. Eleven beacons on five continents transmit in eighteen successive

Table 20.6
Popular Beacon Frequencies

Frequencies (MHz)	*Comments*
14.100, 18.110, 21.150, 24.930, 28.200	Northern California DX Foundation beacons
28.2-28.3	Several dozen beacons worldwide
50.0-50.1	Most US beacons are within 50.06-50.08 MHz
70.03-70.13	Beacons in England, Ireland, Gibraltar and Cyprus

one-minute intervals. More on this system, along with a longer list of HF, VHF and UHF beacons, can be found in *The ARRL Operating Manual*. Other interested groups publish updated lists of beacons with call sign, frequency, location, transmitter mode, power, and antenna. Beacons often include location as part of their automated message, and many can be located from their call sign. Thus, even casual scanning of beacon subbands can be useful. **Table 20.6** provides the frequencies where beacons useful to HF propagation are most commonly placed.

PROPAGATION IN THE TROPOSPHERE

All radio communication involves propagation through the troposphere for at least part of the signal path. Radio waves traveling through the lowest part of the atmosphere are subject to refraction, scattering and other phenomena, much like ionospheric effects. Tropospheric conditions are rarely significant below 30 MHz, but they are very important at 50 MHz and higher. Much of the long-distance work on the VHF, UHF and microwave bands depends on some form of tropospheric propagation. Instead of watching solar activity and geomagnetic indices, those who use tropospheric propagation are much more concerned about the weather.

Line of Sight

At one time it was thought that communications in the VHF range and higher would be restricted to line-of-sight paths. Although this has not proven to be the case even in the microwave region, the concept of line of sight is still useful in understanding tropospheric propagation. In the vacuum of space or in a completely homogeneous medium, radio waves do travel essentially in straight lines, but these conditions are almost never met in terrestrial propagation.

Radio waves traveling through the troposphere are ordinarily refracted slightly earthward. The normal drop in temperature, pressure and water-vapor content with increasing altitude change the index of refraction of the atmosphere enough to cause refraction. Under average conditions, radio waves are refracted toward Earth enough to make the horizon appear 1.15 times farther away than the visual horizon. Under unusual conditions, tropospheric refraction may extend this range significantly.

A simple formula can be used to estimate the distance to the radio horizon under average conditions:

$$d = \sqrt{2h}$$

where

d = distance to the radio horizon, miles
h = height above average terrain, ft

$$d = \sqrt{17h}$$

where

d = distance to the radio horizon, km
h = height above average terrain, m

The distance to the radio horizon for an antenna 30 m (98 ft) above average terrain is thus 22.6 km (14 mi), a station on top of a 1000-m (3280-ft) mountain has a radio horizon of 130 km (80 mi).

Atmospheric Absorption

Atmospheric gases, most notably oxygen and water vapor, absorb radio signals, but neither is a significant factor below 10 GHz. Attenuation from rain becomes important at 3.3 GHz, where signals passing through 20 km (12 mi) of heavy showers incur an additional 0.2 dB loss. That same rain would impose 12 dB additional loss at 10 GHz and losses continue to increase with frequency. Heavy fog is similarly a problem only at 5.6 GHz and above. More detailed information about atmospheric absorption in the microwave bands can be found in the *ARRL UHF/Microwave Experimenter's Manual*.

Tropospheric Scatter

Contacts beyond the radio horizon out to a working distance of 100 to 500 km (60 to 310 mi), depending on frequency, equipment and local geography, are made every day without the aid of obvious propagation enhancement. At 1.8 and 3.5 MHz, local communication is due mostly to ground wave. At higher frequencies, especially in the VHF range and above, the primary mechanism is scattering in the troposphere, or *troposcatter*.

Most amateurs are unaware that they use troposcatter even though it plays an essential role in most local communication. Radio signals through the VHF range are scattered primarily by wave-length sized gradients in the index of refraction of the

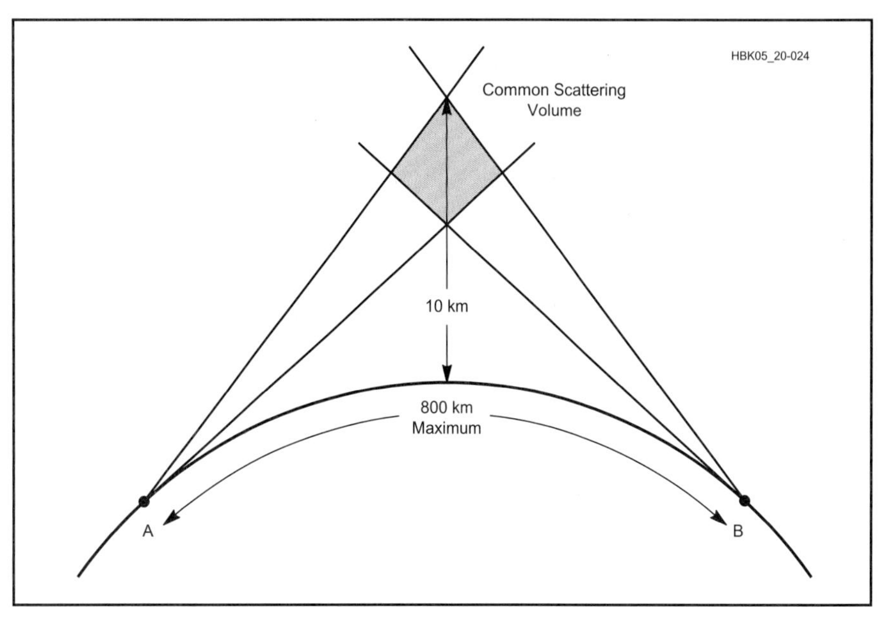

Fig 20.24—Tropospheric-scatter path geometry. The lower boundary of the common scattering volume is limited by the take-off angle of both stations. The upper boundary of 10 km (6 mi) altitude is the limit of efficient scattering in the troposphere. Signal strength increases with the scattering volume.

lower atmosphere due to turbulence, along with changes in temperature. Radio signals in the microwave region can also be scattered by rain, snow, fog, clouds and dust. That tiny part that is scattered forward and toward the Earth creates the over-the-horizon paths. Troposcatter path losses are considerable and increase with frequency.

The maximum distance that can be linked via troposcatter is limited by the height of a scattering volume common to two stations, shown schematically in **Fig 20.24**. The highest altitude for which scattering is efficient at amateur power levels is about 10 km (6 mi). An application of the distance-to-the-horizon formula yields 800 km (500 mi) as the limit for troposcatter paths, but typical maxima are more like half that. Tropospheric scatter varies little with season or time of day, but it is difficult to assess the effect of weather on troposcatter alone. Variations in tropospheric refraction, which is very sensitive to the weather, probably account for most of the observed day-to-day differences in troposcatter signal strength.

Troposcatter does not require special operating techniques or equipment, as it is used unwittingly all the time. In the absence of all other forms of propagation, especially at VHF and above, the usual working range is essentially the maximum troposcatter distance. Ordinary working range increases most dramatically with antenna height, because that lowers the take-off angle to the horizon. Working range increases less quickly with antenna gain and transmitter power. For this reason, a mountaintop is the choice location for extending ordinary troposcatter working distances.

Rain Scatter in the Troposphere

Scatter from raindrops is a special case of troposcatter practical in the 3.3- to 24-GHz range. Stations simply point their antennas toward a common area of rain. A certain portion of radio energy is scattered by the raindrops, making possible over-the-horizon or obstructed-path contacts, even with low power. The theoretical

MUF Prediction on the Home Computer

Like predicting the weather, predicting HF propagation — even with the best computer software available — is not an exact science. The processes occurring as a signal is propagated from one point on the Earth to another are enormously complicated and subject to an incredible number of variables. Experience and a knowledge of propagation conditions (as related to solar activity, especially unusual solar activity, such as flares or Coronal Mass Ejections) are needed when you actually get on the air to check out the bands. Keep in mind, too, that ordinary computer programs are written mainly to calculate propagation for great-circle paths via the F layer. Scatter, skew-path, auroral and other such propagation modes may provide contacts when computer predictions indicate no contacts are possible.

It used to be possible to classify propagation-prediction programs by whether they were used primarily for heavy-duty, long-term forecasting — for planning a high-power shortwave broadcast station, for example — or for making a short-term forecast, perhaps to check out whether a band might be open today for a particular DXpedition. But with the increasing amount of computing power available nowadays, that distinction has blurred. What follows is some brief information about commercially available propagation-prediction programs for the IBM PC and compatible computers. See **Table 20.A**.

ASAPS Version 5

An agency of the Australian government has developed the *ASAPS* program, which stands for Advanced Stand-Alone Prediction System. It rivals *IONCAP* (see below) in its analysis capability and in its prediction accuracy. It is a Windows program that interacts reasonably well with the user, once you become accustomed to the acronyms used. If you change transmit power levels, antennas and other parameters, you can see the new results almost instantly without further menu entries. Available from IPS Radio and Space Services. See: **www.ips.gov.au/index.php**.

IONCAP, CAPMan and VOACAP

IONCAP, short for Ionospheric Communications Analysis and Prediction, was written by an agency of the US government and has been under development for about 30 years in one form or another. The *IONCAP* program has a well-deserved reputation for being difficult to use, since it came from the world of Fortran punch cards and mainframe computers.

CAPMan is a DOS-based version of *IONCAP* that is considerably more "user friendly" than the core program. *CAPMan* produces excellent graphs, some calibrated in S units if the user wishes. It incorporates amateur call signs to specify locations, making it comfortable for amateurs to use. *CAPMan* also allows the user to specify multiple antenna types for both transmitting and receiving. See: **www.taborsoft.com/**.

VOACAP is another version of *IONCAP*, but this one includes a sophisticated Windows interface. The Voice of America (VOA) started work on *VOACAP* in the early 1990s and continued for several years before funding ran out. The program is now maintained by a single, dedicated computer scientist, Greg Hand, at NTIA/ITS (Institute for Telecommunication Sciences), an agency of the US Department of Commerce in Boulder, CO. Although *VOACAP* is not specifically designed for amateurs (and thus doesn't include some features that amateurs are fond of, such as entry of locations by ham-radio call signs and multiple receiving antennas), it is available for free by downloading from: **elbert.its.bldrdoc.gov/hf.html**.

W6ELProp, Version 1.0

In 2001, W6EL ported his well known DOS-based *MINIPROP PLUS* program into the Windows world. It uses the same Fricker-based computation engine as its predecessor. *W6ELProp* has a highly intuitive, ham-friendly user interface. It produces the same detailed output tables as its DOS counterpart, along with a number of useful charts and maps, including the unique and useful "frequency map," which shows the global MUFs from a given transmitting location for a particular month/day/time and solar-activity level. *W6ELProp* is available for free by downloading from: **www.qsl.net/w6elprop**.

WinCAP Wizard 2

Kangaroo Tabor Software sells the *CAPMan* program and is also the creator of the *Active Beacon Wizard* program included with the 19th and 20th Editions of *The ARRL Antenna Book*. They also sell a Windows-based "mini" version of *CAPMan*, called *WinCAP Wizard 2*. This

range for rain scatter is as great as 600 km (370 mi), but the experience of amateurs in the microwave bands suggests that expected distances are less than 200 km (120 mi). Snow and hail make less efficient scattering media unless the ice particles are partially melted. Smoke and dust particles are too small for extraordinary scattering, even in the microwave bands.

Refraction and Ducting in the Troposphere

Radio waves are refracted by natural gradients in the index of refraction of air with altitude, due to changes in temperature, humidity and pressure. Refraction under standard atmospheric conditions extends the radio horizon somewhat beyond the visual line of sight. Favorable weather conditions further enhance normal tropospheric refraction, lengthening the useful VHF and UHF range by several hundred kilometers and increasing signal strength. Higher frequencies are more sensitive to refraction, so its effects may be observed in the microwave bands before they are apparent at lower frequencies.

Ducting takes place when refraction is so great that radio waves are bent back to the surface of the Earth. When tropospheric ducting conditions exist over a wide geographic area, signals may remain very strong over distances of 1500 km (930 mi) or more. Ducting results from the gradient created by a sharp increase in temperature with altitude, quite the opposite of normal atmospheric conditions. A simultaneous drop in humidity contributes to increased refractivity. Useful temperature inversions form between 250 and 2000 m (800-6500 ft) above ground. The elevated inversion and the Earth's surface act something like the boundaries of a natural open-ended waveguide. Radio waves of the right frequency range caught inside the duct will be propagated for long distances with relatively low losses.

uses the *CAPMan* computing engine but limits the number of input parameters to those most commonly used by amateurs. The outputs are customizable and include dynamic summary tables, sunrise/sunset tables and propagation maps.

PropLab Pro, Version 2

PropLab Pro by Solar Terrestrial Dispatch represents the high end of propagation-prediction programs. It is the only commercial program presently available that can do complete 3D ray tracing through the ionosphere, even taking complex geomagnetic effects into account. The number of computations is huge, especially in the full-blown 3D mode and operation can be slow and tedious. The user interface is also very complex and demanding, with a steep user-learning curve. However, it is fascinating to see exactly how a signal can bend off-azimuth or how it can split into the ordinary and extraordinary waves. See: **www.spacew.com/www/proplab.html**.

Table 20.A
Features and Attributes of Propagation Prediction Programs

	ASAPS V. 5	*VOACAP Windows*	*W6ELProp V. 1.00*	*CAPMan*	*WinCAP Wizard 2*	*PropLab Pro*
User Friendliness	Good	Good	Good	Good	Good	Poor
Operating System	Windows	Windows	Windows	DOS	Windows	DOS
Uses k index	No	No	Yes	Yes	Yes	Yes
User library of QTHs	Yes/Map	Yes	Yes	Yes	Yes	No
Bearings, distances	Yes	Yes	Yes	Yes	Yes	Yes
MUF calculation	Yes	Yes	Yes	Yes	Yes	Yes
LUF calculation	Yes	Yes	No	Yes	Yes	Yes
Wave angle calculation	Yes	Yes	Yes	Yes	Yes	Yes
Vary minimum wave angle	Yes	Yes	Yes	Yes	Yes	Yes
Path regions and hops	Yes	Yes	Yes	Yes	Yes	Yes
Multipath effects	Yes	Yes	No	Yes	Yes	Yes
Path probability	Yes	Yes	Yes	Yes	Yes	Yes
Signal strengths	Yes	Yes	Yes	Yes	Yes	Yes
S/N ratios	Yes	Yes	Yes	Yes	Yes	Yes
Long path calculation	Yes	Yes	Yes	Yes	Yes	Yes
Antenna selection	Yes	Yes	Indirectly	Yes	Isotropic	Yes
Vary antenna height	Yes	Yes	Indirectly	Yes	No	Yes
Vary ground characteristics	Yes	Yes	No	Yes	No	No
Vary transmit power	Yes	Yes	Indirectly	Yes	Yes	Yes
Graphic displays	Yes	Yes	Yes	Yes	Yes	2D/3D
UT-day graphs	Yes	Yes	Yes	Yes	Yes	Yes
Area Mapping	Yes	Yes	Yes	Yes	No	Yes
Documentation	Yes	On-line	Yes	Yes	Yes	Yes
Price class	$275 [1]	free [2]	free [3]	$89 [4]	$29.95 [4]	$150 [5]

Prices are for early 2004 and are subject to change.

[1]ASAPS: shipping and handling extra. See: **www.ips.gov.au/index.php**
[2]VOACAP available at: **elbert.its.bldrdoc.gov/hf.html**
[3]W6EL Prop, see: **www.qsl.net/w6elprop**
[4]CAPMan and WinCAP Wizard 2, see: **www.taborsoft.com/**
[5]PropLab Pro, see: **www.spacew.com/www/proplab.html**

Several common weather conditions can create temperature inversions.

Radiation Inversions in the Troposphere

Radiation inversions are probably the most common and widespread of the various weather conditions that affect propagation. Radiation inversions form only over land after sunset as a result of progressive cooling of the air near the Earth's surface. As the Earth cools by radiating heat into space, the air just above the ground is cooled in turn. At higher altitudes, the air remains relatively warmer, thus creating the inversion. A typical radiation-inversion temperature profile is shown in **Fig 20.25**.

The cooling process may continue through the evening and predawn hours, creating inversions that extend as high as 500 m (1500 ft). Deep radiation inversions are most common during clear, calm, summer evenings. They are more distinct in dry climates, in valleys and over open ground. Their formation is inhibited by wind, wet ground and cloud cover. Although radiation inversions are common and widespread, they are rarely strong enough to cause true ducting. The enhanced conditions so often observed after sunset during the summer are usually a result of this mild kind of inversion.

High-Pressure Weather Systems

Large, sluggish, high-pressure systems (or *anticyclones*) create the most dramatic and widespread tropospheric ducts due to *subsidence*. Subsidence inversions in high-pressure systems are created by air that is sinking. As air descends, it is compressed and heated. Layers of warmer air — temperature inversions — often form between 500 and 3000 m (1500-10,000 ft) altitude, as shown in **Fig 20.26**. Ducts usually intensify during the evening and early morning hours, when surface temperatures drop and suppress the tendency for daytime ground-warmed air to rise. In the Northern Hemisphere, the longest and strongest radio paths usually lie to the south of high-pressure centers. See **Fig 20.27**.

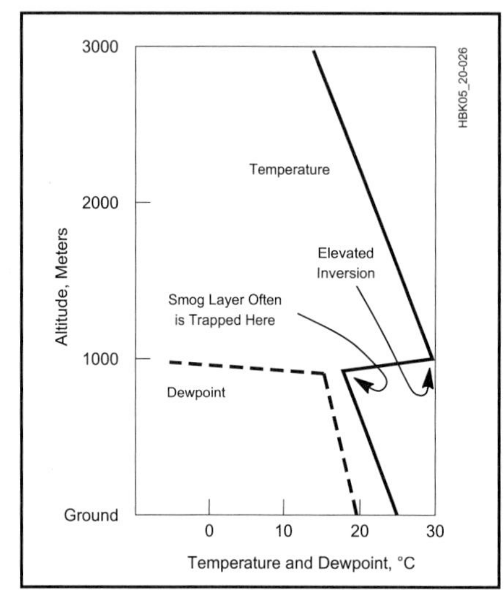

Fig 20.26—Temperature and humidity profile across an elevated duct at 1000-m altitude. Such inversions typically form in summertime high-pressure systems. Note the air is very dry in the inversion.

Sluggish high-pressure systems likely to contain strong temperature inversions are common in late summer over the eastern half of the US. They generally move southeastward out of Canada and linger for days over the Midwest, providing many hours of extended propagation. The southeastern part of the country and the lower Midwest experience the most high-pressure openings; the upper Midwest and East Coast somewhat less frequently; the western mountain regions rarely.

Semipermanent high-pressure systems, which are nearly constant climatic features in certain parts of the world, sustain the longest and most exciting ducting paths. The Eastern Pacific High, which migrates northward off the coast of California during the summer, has been responsible for the longest ducting paths reported to date. Countless contacts in the 4000-km (2500 mi) range have been made from 144 MHz through 5.6 GHz between California and Hawaii. The *Bermuda High* is a nearly permanent feature of the Caribbean area, but during the summer it moves north and often covers the southeastern US. It has supported contacts in excess of 2800 km (1700 mi) from Florida and the Carolinas to the West Indies, but its full potential has not been exploited. Other semipermanent highs lie in the Indian Ocean, the western Pacific and off the coast of western Africa.

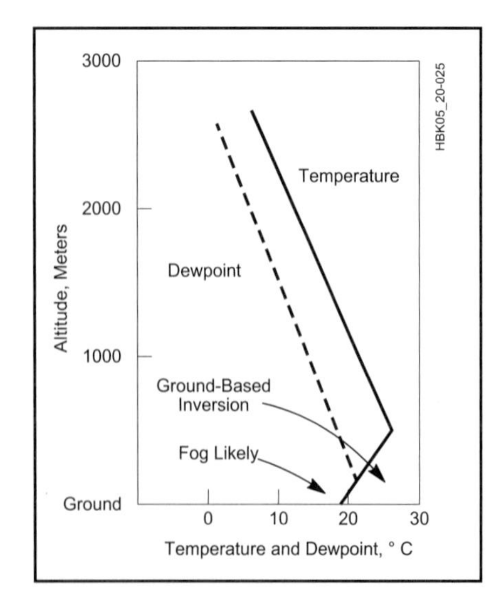

Fig 20.25—Temperature and dewpoint profile of an early-morning radiation inversion. Fog may form near the ground. The midday surface temperature would be at least 30°C.

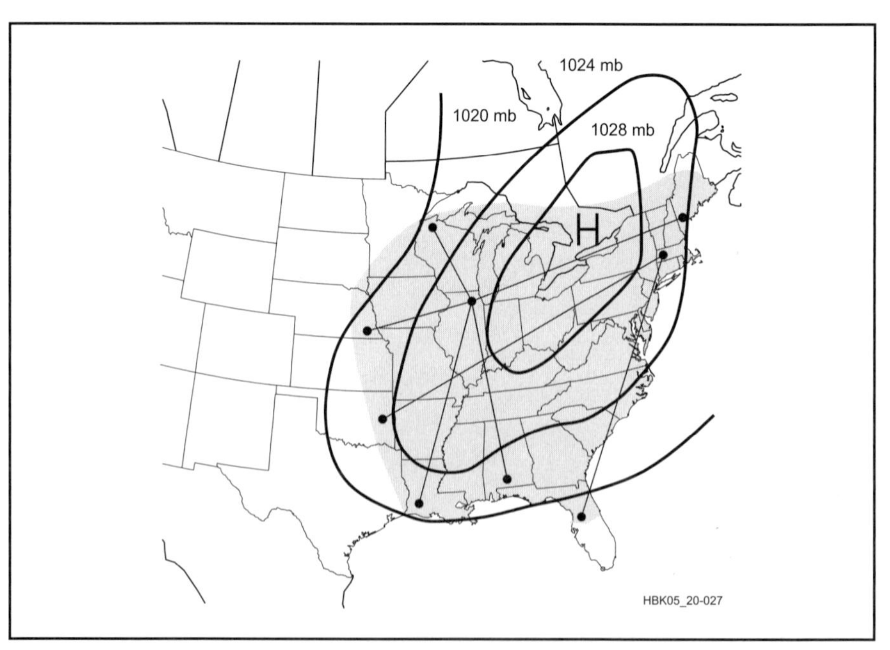

Fig 20.27—Surface weather map for September 13, 1993, shows that the eastern US was dominated by a sprawling high-pressure system. The shaded portion shows the area in which ducting conditions existed on 144 through 1286 MHz and higher.

Wave Cyclone

The *wave cyclone* is a more dynamic weather system that usually appears during the spring over the middle part of the American continent. The wave begins as a disturbance along a boundary between cooler northern and warmer southern air masses. Southwest of the disturbance, a cold front forms and moves rapidly eastward, while a warm front moves slowly northward on the eastward side. When the wave is in its open position, as shown in **Fig 20.28**, north-south radio paths 1500 km (930 mi) and longer may be possible in the area to the east of the cold front and south of the warm front, known as the warm sector. East-west paths nearly as long may also open in the southerly parts of the warm sector.

Wave cyclones are rarely productive for more than a day in any given place, because the eastward-moving cold front eventually closes off the warm sector. Wave-cyclone temperature inversions are created by a southwesterly flow of warm, dry air above 1000 m (3200 ft) that covers relatively cooler and moister gulf air flowing northward near the Earth's surface. Successive waves spaced two or three days apart may form along the same frontal boundary.

Warm Fronts and Cold Fronts

Warm fronts and cold fronts sometimes bring enhanced tropospheric conditions, but rarely true ducting. A warm front marks the surface boundary between a mass of warm air flowing over an area of relatively cooler and more stationary air. Inversion conditions may be stable enough several hundred kilometers ahead of the warm front to create extraordinary paths.

A cold front marks the surface boundary between a mass of cool air that is wedging itself under more stationary warm air. The warmer air is pushed aloft in a narrow band behind the cold front, creating a strong but highly unstable temperature inversion. The best chance for enhancement occurs parallel to and behind the passing cold front.

Other Conditions Associated With Ducts

Certain kinds of wind may also create useful inversions. The *Chinook* wind that blows off the eastern slopes of the Rockies can flood the Great Plains with warm and very dry air, primarily in the springtime. If the ground is cool or snow-covered, a strong inversion can extend as far as Canada to Texas and east to the Mississippi River. Similar kinds of *foehn* winds, as these mountain breezes are called, can be found in the Alps, Caucasus Mountains and other places.

The *land breeze* is a light, steady, cool wind that commonly blows up to 50 km (30 mi) inland from the oceans, although the distance may be greater in some circumstances. Land breezes develop after sunset on clear summer evenings. The land cools more quickly than the adjacent ocean. Air cooled over the land flows near the surface of the Earth toward the ocean to displace relatively warmer air that is rising. See **Fig 20.29**. The warmer ocean air, in turn, travels at 200-300 m (600-1000 ft) altitude to replace the cool surface air. The land-sea circulation of cool air near the ground and warm air aloft creates a mild inversion that may remain for hours. Land-breeze inversions often bring enhanced conditions and occasionally allow contacts in excess of 800 km (500 mi) along coastal areas.

In southern Europe, a hot, dry wind known as the *sirocco* sometimes blows northward from the Sahara Desert over relatively cooler and moister Mediterranean air. Sirocco inversions can be very strong and extend from Israel and Lebanon westward past the Straits of Gibraltar. Sirocco-type inversions are probably responsible for record-breaking microwave contacts in excess of 1500 km (930 mi) across the Mediterranean.

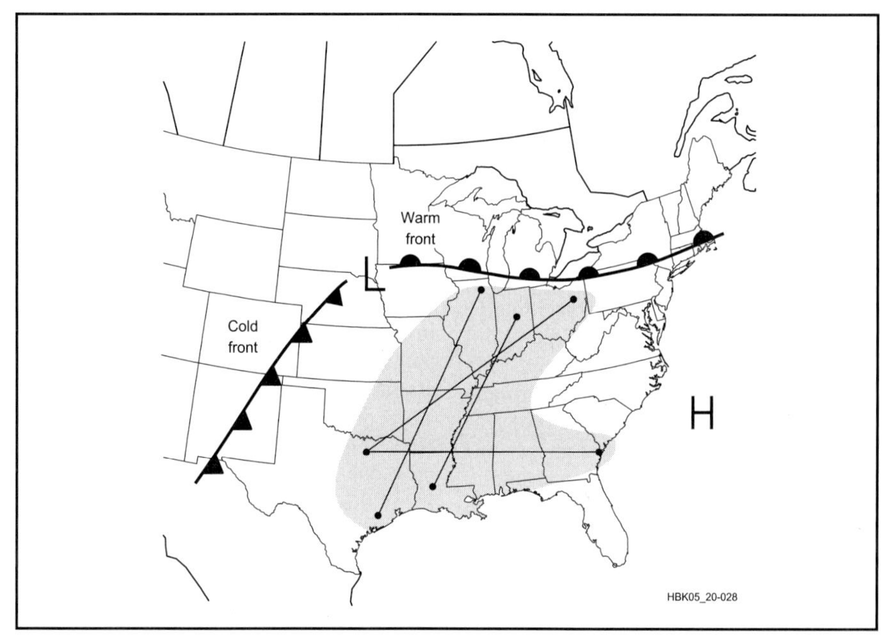

Fig 20.28—Surface weather map for June 2, 1980, with a typical spring wave cyclone over the southeastern quarter of the US. The shaded portion shows where ducting conditions existed.

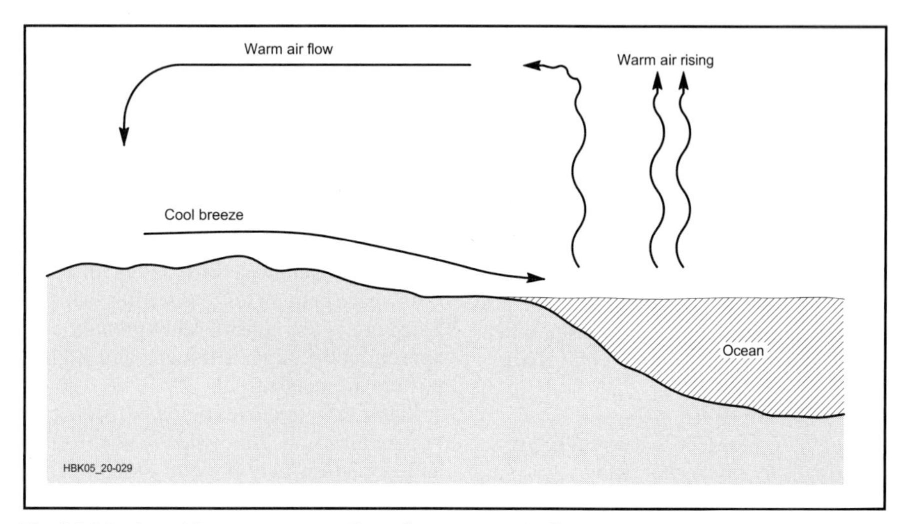

Fig 20.29—Land-breeze convection along a coast after sunset creates a temperature inversion over the land.

Marine Boundary Layer Effects

Over warm water, such as the Caribbean and other tropical seas, *evaporation inversions* may create ducts that are useful in the microwave region between 3.3 and 24 GHz. This inversion depends on a sharp drop in water-vapor content rather than on an increase in temperature to create ducting conditions. Air just above the surface of water at least 30°C is saturated because of evaporation. Humidity drops significantly within 3 to 10 m (10 to 30 ft) altitude, creating a very shallow but stable duct. Losses due to water vapor absorption may be intolerable at the highest ducting frequencies, but breezes may raise the effective height of the inversion and open the duct to longer wavelengths. Stations must be set up right on the beaches to ensure being inside an evaporation inversion.

Tropospheric Fading

Tropospheric turbulence and small changes in the weather are responsible for most fading at VHF and higher. Local weather conditions, such as precipitation, warm air rising over cities and the effects of lakes and rivers, can all contribute to tropospheric instabilities that affect radio propagation. *Fast-flutter fading* at 28 MHz and above is often the result of an airplane that temporarily creates a second propagation path. Flutter results as the phase relationship between the ordinary tropospheric signal and that reflected by the airplane change with the airplane's movement.

EXTRATERRESTRIAL PROPAGATION

Communication of all sorts into space has become increasingly important. Amateurs confront extraterrestrial propagation when accessing satellite repeaters or using the moon as a reflector. Special propagation problems arise from signals that travel from the Earth through the ionosphere (or a substantial portion of it) and back again. Tropospheric and ionospheric phenomena, so useful for terrestrial paths, are unwanted and serve only as a nuisance for space communication. A phenomenon known as *Faraday rotation* may change the polarization of radio waves traveling through the ionosphere, presenting special problems to receiving weak signals. Cosmic noise also becomes an important factor when antennas are intentionally pointed into space.

Faraday Rotation

Magnetic and electrical forces rotate the polarization of radio waves passing through the ionosphere. For example, signals that leave the Earth as horizontally polarized, and return after a reflection from the moon may not arrive with the same polarization. An additional 20 dB of path loss is incurred when polarization is shifted by 90°, an intolerable amount when signals are marginal.

Faraday rotation is difficult to predict and its effects change over time and with operating frequency. At 144 MHz, the polarization of space waves may shift back into alignment with the antenna within a few minutes, so often just waiting can solve the Faraday problem. At 432 MHz, it may take half an hour or longer for the polarization to become realigned. Use of circular polarization completely eliminates this problem, but creates a new one for EME paths. The sense of circularly polarized signals is reversed with reflection, so two complete antenna systems are normally required, one with left-hand and one with right-hand polarization.

Earth-Moon-Earth

Amateurs have used the moon as a reflector on the VHF and UHF bands since 1960. Maximum allowable power and large antennas, along with the best receivers, are normally required to overcome the extreme free-space and reflection losses involved in Earth-Moon-Earth (EME) paths. More modest stations make EME contacts by scheduling operating times when the Moon is at perigee on the horizon. The Moon, which presents a target only one-half degree wide, reflects only 7% of the radio signals that reach it. Techniques have to be designed to cope with Faraday rotation, cosmic noise, Doppler shift (due to the Moon's movements) and other difficulties. In spite of the problems involved, hundreds of stations have made contacts via the Moon on all bands from 50 MHz to 10 GHz. The techniques of EME communication are discussed in the chapter on **Space Communications**.

Satellites

Accessing amateur satellites generally does not involve huge investments in antennas and equipment, yet station design does have to take into account special challenges of space propagation. Free-space loss is a primary consideration, but it is manageable when satellites are only a few hundred kilometers distant. Free-space path losses to satellites in high-Earth orbits are considerably greater, and appropriately larger antennas and higher powers are needed.

Satellite frequencies below 30 MHz can be troublesome. Ionospheric absorption and refraction may prevent signals from reaching space, especially to satellites at very low elevations. In addition, man-made and natural sources of noise are high. VHF and especially UHF are largely immune from these effects, but free-space path losses are greater. Problems related to polarization, including Faraday rotation, intentional or accidental satellite tumbling and the orientation of a satellite's antenna in relation to terrestrial antennas, are largely overcome by using circularly polarized antennas. More on using satellites can be found in the chapter on **Space Communications**.

NOISE AND PROPAGATION

Noise simply consists of unwanted radio signals that interfere with desired communications. In some instances, noise imposes the practical limit on the lowest usable frequencies. Noise may be classified by its sources: man-made, terrestrial and cosmic. Interference from other transmitting stations on adjacent frequencies is not usually considered noise and may be controlled, to a some degree anyway, by careful station design.

Man-Made Noise

Many unintentional radio emissions result from man-made sources. Broadband radio signals are produced whenever there is a spark, such as in contact switches, electric motors, gasoline engine spark plugs and faulty electrical connections. Household appliances, such as fluorescent lamps, microwave ovens, lamp dimmers and anything containing an electric motor may all produce undesirable broadband radio energy. Devices of all sorts, especially computers and anything controlled by microprocessors, television receivers and many other electronics also emit radio signals that may be perceived as noise well into the UHF range. In many cases, these sources are local and can be controlled with proper measures. See the **EMI/DFing** chapter.

High-voltage transmission lines and associated equipment, including transformers, switches and lightning arresters, can generate high-level radio signals over a wide area, especially if they are corroded or improperly maintained. Transmission lines may act as efficient antennas at some frequencies, adding to the noise problem. Certain kinds of street lighting, neon signs and industrial equipment also contribute their share of noise.

Lightning

Static is a common term given to the ear-splitting crashes of noise commonly heard on nearly all radio frequencies, although it is most severe on the lowest frequency bands. Atmospheric static is primarily caused by lightning and other natural electrical discharges. Static may result from close-by thunderstorms, but most static originates with tropical storms. Like any radio signals, lightning-produced static may be propagated over long distances by the ionosphere. Thus static is generally higher during the summer, when there are more nearby thunderstorms, and at night, when radio propagation generally improves. Static is often the limiting factor on 1.8 and 3.5 MHz, making winter a more favorable time for using these frequencies.

Precipitation Static and Corona Discharge

Precipitation static is an almost continuous hash-type noise that often accompanies various kinds of precipitation, including snowfall. Precipitation static is caused by raindrops, snowflakes or even wind-blown dust, transferring a small electrical charge on contact with an antenna. Electrical fields under thunderstorms are sufficient to place many objects such as trees, hair and antennas, into corona discharge. *Corona noise* may sound like a harsh crackling in the radio — building in intensity, abruptly ending, and then building again, in cycles of a few seconds to as long as a minute. A corona charge on an antenna may build to some critical level and then discharge in the atmosphere with an audible pop before recharging. Precipitation static and corona discharge can be a nuisance from LF to well into the VHF range.

Cosmic Sources

The sun, distant stars, galaxies and other cosmic features all contribute radio noise well into the gigahertz range. These *cosmic sources* are perceived primarily as a more-or-less constant background noise at HF. In the VHF range and higher, specific sources of cosmic noise can be identified and may be a limiting factor in terrestrial and space communications. The sun is by far the greatest source of radio noise, but its effects are largely absent at night. The center of our own galaxy is nearly as noisy as the sun. Galactic noise is especially noticeable when high-gain VHF and UHF antennas, such as may be used for satellite or EME communications, are pointed toward the center of the Milky Way. Other star clusters and galaxies are also radio hot-spots in the sky. Finally, there is a much lower cosmic background noise that seems to cover the entire sky.

FURTHER READING

J. E. Anderson, "*MINIMUF* for the Ham and the IBM Personal Computer," *QEX*, Nov 1983, pp 7-14.

B. R. Bean and E. J. Dutton, *Radio Meteorology* (New York: Dover, 1968).

K. Davies, *Ionospheric Radio* (London: Peter Peregrinus, 1989). Excellent, though highly technical text on propagation.

G. Grayer, "VHF/UHF Propagation," Ch 2 of *The VHF/UHF DX Book* (Buckingham, England: DIR, 1992).

G. Jacobs, T. Cohen, R. Rose, *The NEW Shortwave Propagation Handbook*, *CQ* Communications, Inc. (Hicksville, NY: 1995).

L. F. McNamara, *Radio Amateur's Guide to the Ionosphere* (Malabar, Florida: Krieger Publishing Company, 1994). Excellent, quite-readable text on HF propagation.

C. Newton, *Radio Auroras* (Potters Bar, England: Radio Society of Great Britain, 1991).

E. Pocock, "UHF and Microwave Propagation," Ch 3 of *The ARRL UHF/ Microwave Experimenter's Manual* (Newington, Connecticut: ARRL, 1990).

E. Pocock, Ed., *Beyond Line of Sight: A History of VHF Propagation from the Pages of QST* (Newington, Connecticut: ARRL, 1992).

Chapter 21

Transmission Lines

RF power is rarely generated right where it will be used. A transmitter and the antenna it feeds are a good example. To radiate effectively, the antenna should be high above the ground and should be kept clear of trees, buildings and other objects that might absorb energy.

The transmitter, however, is most conveniently installed indoors, where it is out of the weather and is readily accessible. A *transmission line* is used to convey RF energy from the transmitter to the antenna. A transmission line should transport the RF from the source to its destination with as little loss as possible. This chapter was written by Dean Straw, N6BV.

There are three main types of transmission lines used by radio amateurs: coaxial lines, open-wire lines and waveguides. The most common type is the *coaxial* line, usually called *coax*. See **Fig 21.1A**. Coax is made up of a center conductor, which may be either stranded or solid wire, surrounded by a concentric outer conductor. The outer conductor may be braided shield wire or a metallic sheath. A flexible aluminum foil is employed in some coaxes to improve shielding over that obtainable from a woven shield braid. If the outer conductor is made of solid aluminum or copper, the coax is referred to as *Hardline*.

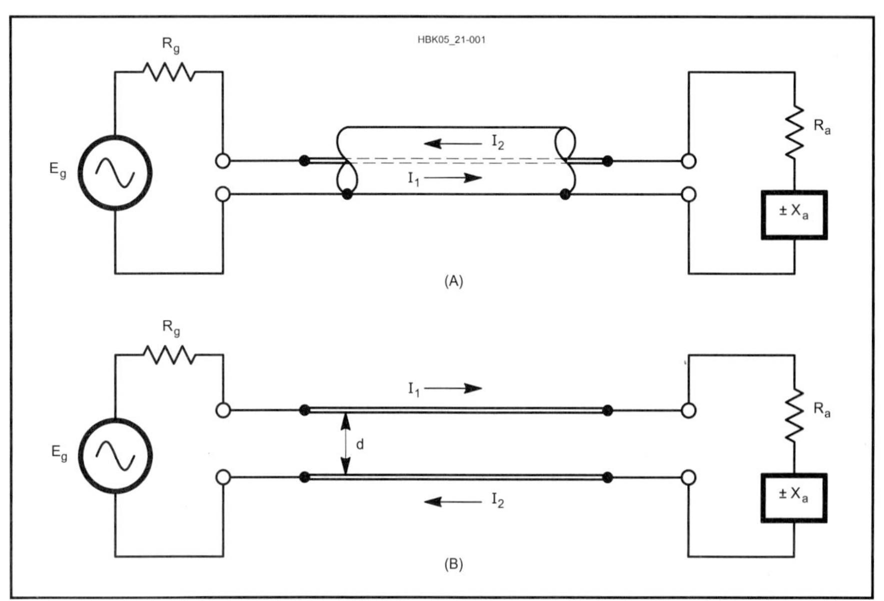

Fig 21.1—In A, coaxial cable transmission line connecting signal generator having source resistance R_g to reactive load $R_a \pm j X_a$, where X_a is either a capacitive (–) or inductive (+) reactance. Velocity factor (VF) and characteristic impedance (Z_0) are properties of the line, as discussed in the text. B shows open-wire balanced transmission line.

The second type of transmission line utilizes parallel conductors side by side, rather than the concentric ones used in coax. Typical examples of such *open-wire* lines are 300-Ω TV ribbon line and 450-Ω ladder line. See Fig 21.1B. Although open-wire lines are enjoying a sort of renaissance in recent years due to their inherently lower losses in simple multiband antenna systems, coaxial cables are far more prevalent, because they are much more convenient to use.

The third major type of transmission line is the *waveguide*. While open-wire and coaxial lines are used from power-line frequencies to well into the microwave region, waveguides are used at microwave frequencies only. Waveguides will be covered at the end of this chapter.

TRANSMISSION LINE BASICS

In either coaxial or open-wire line, currents flowing in each of the two conductors travel in opposite directions. If the physical spacing between the two parallel conductors in an open-wire line is small in terms of wavelength, the phase difference between the currents will be very close to 180°. If the two currents also have equal amplitudes, the field generated by each conductor will cancel that generated by the other, and the line will not radiate energy, even if it is many wavelengths long.

The equality of amplitude and 180° phase difference of the currents in each conductor in an open-wire line determine the degree of radiation cancellation. If the currents are for some reason unequal, or if the phase difference is not 180°, the line will radiate energy. How such imbalances occur and to what degree they can cause problems will be covered in more detail later.

In contrast to an open-wire line, the outer conductor in a coaxial line acts as a shield, confining RF energy within the line. Because of *skin effect* (see the **Real-World Component Characteristics** chapter in this *Handbook*), current flowing in the outer conductor of a coax does so mainly on the inner surface of the outer conductor. The fields generated by the

Table 21.1

Nominal Characteristics of Commonly Used Transmission Lines

RG or Type	*Part Number*	*Nom. Z_0 Ω*	*VF %*	*Cap. pF/ft*	*Cent. Cond. AWG*	*Diel. Type*	*Shield Type*	*Jacket Matl*	*OD inches*	*Max V (RMS)*	*Matched Loss (dB/100') 1 MHz*	*10*	*100*	*1000*
RG-6	Belden 1694A	75	82	16.2	#18 Solid BC	FPE	FC	P1	0.275	600	0.2	.7	1.8	5.9
RG-6	Belden 8215	75	66	20.5	#21 Solid CCS	PE	D	PE	0.332	2700	0.4	0.8	2.7	9.8
RG-8	Belden 7810A	50	86	23.0	#10 Solid BC	FPE	FC	PE	0.405	600	0.1	0.4	1.2	4.0
RG-8	TMS LMR400	50	85	23.9	#10 Solid CCA	FPE	FC	PE	0.405	600	0.1	0.4	1.3	4.1
RG-8	Belden 9913	50	84	24.6	#10 Solid BC	ASPE	FC	P1	0.405	600	0.1	0.4	1.3	4.5
RG-8	CXP1318FX	50	84	24.0	#10 Flex BC	FPE	FC	P2N	0.405	600	0.1	0.4	1.3	4.5
RG-8	Belden 9913F7	50	83	24.6	#11 Flex BC	FPE	FC	P1	0.405	600	0.2	0.6	1.5	4.8
RG-8	Belden 9914	50	82	24.8	#10 Solid BC	FPE	FC	P1	0.405	600	0.2	0.5	1.5	4.8
RG-8	TMS LMR400UF	50	85	23.9	#10 Flex BC	FPE	FC	PE	0.405	600	0.1	0.4	1.4	4.9
RG-8	DRF-BF	50	84	24.5	#9.5 Flex BC	FPE	FC	PE	0.405	600	0.1	0.5	1.6	5.2
RG-8	WM CQ106	50	84	24.5	#9.5 Flex BC	FPE	FC	P2N	0.405	600	0.2	0.6	1.8	5.3
RG-8	CXP008	50	78	26.0	#13 Flex BC	FPE	S	P1	0.405	600	0.1	0.5	1.8	7.1
RG-8	Belden 8237	52	66	29.5	#13 Flex BC	PE	S	P1	0.405	3700	0.2	0.6	1.9	7.4
RG-8X	Belden 7808A	50	86	23.5	#15 Solid BC	FPE	FC	PE	0.240	600	0.2	0.7	2.3	7.4
RG-8X	TMS LMR240	50	84	24.2	#15 Solid BC	FPE	FC	PE	0.242	300	0.2	0.8	2.5	8.0
RG-8X	WM CQ118	50	82	25.0	#16 Flex BC	FPE	FC	P2N	0.242	300	0.3	0.9	2.8	8.4
RG-8X	TMS LMR240UF	50	84	24.2	#15 Flex BC	FPE	FC	PE	0.242	300	0.2	0.8	2.8	9.6
RG-8X	Belden 9258	50	82	24.8	#16 Flex BC	FPE	S	P1	0.242	600	0.3	0.9	3.1	11.2
RG-8X	CXP08XB	50	80	25.3	#16 Flex BC	FPE	S	P1	0.242	300	0.3	0.9	3.1	14.0
RG-9	Belden 8242	51	66	30.0	#13 Flex SPC	PE	SCBC	P2N	0.420	5000	0.2	0.6	2.1	8.2
RG-11	Belden 8213	75	84	16.1	#14 Solid BC	FPE	S	PE	0.405	600	0.2	0.4	1.3	5.2
RG-11	Belden 8238	75	66	20.5	#18 Flex TC	PE	S	P1	0.405	600	0.2	0.7	2.0	7.1
RG-58	Belden 7807A	50	85	23.7	#18 Solid BC	FPE	FC	PE	0.195	300	0.3	1.0	3.0	9.7
RG-58	TMS LMR200	50	83	24.5	#17 Solid BC	FPE	FC	PE	0.195	300	0.3	1.0	3.2	10.5
RG-58	WM CQ124	52	66	28.5	#20 Solid BC	PE	S	PE	0.195	1400	0.4	1.3	4.3	14.3
RG-58	Belden 8240	52	66	28.5	#20 Solid BC	PE	S	P1	0.193	1900	0.3	1.1	3.8	14.5
RG-58A	Belden 8219	53	73	26.5	#20 Flex TC	FPE	S	P1	0.195	300	0.4	1.3	4.5	18.1
RG-58C	Belden 8262	50	66	30.8	#20 Flex TC	PE	S	P2N	0.195	1400	0.4	1.4	4.9	21.5
RG-58A	Belden 8259	50	66	30.8	#20 Flex TC	PE	S	P1	0.192	1900	0.4	1.5	5.4	22.8
RG-59	Belden 1426A	75	83	16.3	#20 Solid BC	FPE	S	P1	0.242	300	0.3	0.9	2.6	8.5
RG-59	CXP 0815	75	82	16.2	#20 Solid BC	FPE	S	P1	0.232	300	0.5	0.9	2.2	9.1
RG-59	Belden 8212	75	78	17.3	#20 Solid CCS	FPE	S	P1	0.242	300	0.6	1.0	3.0	10.9
RG-59	Belden 8241	75	66	20.4	#23 Solid CCS	PE	S	P1	0.242	1700	0.6	1.1	3.4	12.0
RG-62A	Belden 9269	93	84	13.5	#22 Solid CCS	ASPE	S	P1	0.240	750	0.3	0.9	2.7	8.7
RG-62B	Belden 8255	93	84	13.5	#24 Flex CCS	ASPE	S	P2N	0.242	750	0.3	0.9	2.9	11.0
RG-63B	Belden 9857	125	84	9.7	#22 Solid CCS	ASPE	S	P2N	0.405	750	0.2	0.5	1.5	5.8
RG-142	CXP 183242	50	69.5	29.4	#19 Solid SCCS	TFE	D	FEP	0.195	1900	0.3	1.1	3.8	12.8
RG-142B	Belden 83242	50	69.5	29.0	#19 Solid SCCS	TFE	D	TFE	0.195	1400	0.3	1.1	3.9	13.5
RG-174	Belden 7805R	50	73.5	26.2	#25 Solid BC	FPE	FC	P1	0.110	300	0.6	2.0	6.5	21.3
RG-174	Belden 8216	50	66	30.8	#26 Flex CCS	PE	S	P1	0.110	1100	1.9	3.3	8.4	34.0
RG-213	Belden 8267	50	66	30.8	#13 Flex BC	PE	S	P2N	0.405	3700	0.2	0.6	1.9	8.0
RG-213	CXP213	50	66	30.8	#13 Flex BC	PE	S	P2N	0.405	600	0.2	0.6	2.0	8.2
RG-214	Belden 8268	50	66	30.8	#13 Flex SPC	PE	D	P2N	0.425	3700	0.2	0.6	1.9	8.0
RG-216	Belden 9850	75	66	20.5	#18 Flex TC	PE	D	P2N	0.425	3700	0.2	0.7	2.0	7.1
RG-217	WM CQ217F	50	66	30.8	#10 Flex BC	PE	D	PE	0.545	7000	0.1	0.4	1.4	5.2
RG-217	M17/78-RG217	50	66	30.8	#10 Solid BC	PE	D	P2N	0.545	7000	0.1	0.4	1.4	5.2
RG-218	M17/79-RG218	50	66	29.5	#4.5 Solid BC	PE	S	P2N	0.870	11000	0.1	0.2	0.8	3.4
RG-223	Belden 9273	50	66	30.8	#19 Solid SPC	PE	D	P2N	0.212	1400	0.4	1.2	4.1	14.5
RG-303	Belden 84303	50	69.5	29.0	#18 Solid SCCS	TFE	S	TFE	0.170	1400	0.3	1.1	3.9	13.5
RG-316	CXP TJ1316	50	69.5	29.4	#26 Flex BC	TFE	S	FEP	0.098	1200	1.2	2.7	8.0	26.1
RG-316	Belden 84316	50	69.5	29.0	#26 Flex SCCS	TFE	S	FEP	0.096	900	1.2	2.7	8.3	29.0
RG-393	M17/127-RG393	50	69.5	29.4	#12 Flex SPC	TFE	D	FEP	0.390	5000	0.2	0.5	1.7	6.1
RG-400	M17/128-RG400	50	69.5	29.4	#20 Flex SPC	TFE	D	FEP	0.195	1400	0.4	1.1	3.9	13.2
LMR500	TMS LMR500UF	50	85	23.9	#7 Flex BC	FPE	FC	PE	0.500	2500	0.1	0.4	1.2	4.0
LMR500	TMS LMR500	50	85	23.9	#7 Solid CCA	FPE	FC	PE	0.500	2500	0.1	0.3	0.9	3.3
LMR600	TMS LMR600	50	86	23.4	#5.5 Solid CCA	FPE	FC	PE	0.590	4000	0.1	0.2	0.8	2.7
LMR600	TMS LMR600UF	50	86	23.4	#5.5 Flex BC	FPE	FC	PE	0.590	4000	0.1	0.2	0.8	2.7
LMR1200	TMS LMR1200	50	88	23.1	#0 Copper Tube	FPE	FC	PE	1.200	4500	0.04	0.1	0.4	1.3
Hardline														
1/2"	CATV Hardline	50	81	25.0	#5.5 BC	FPE	SM	none	0.500	2500	0.05	0.2	0.8	3.2
1/2"	CATV Hardline	75	81	16.7	#11.5 BC	FPE	SM	none	0.500	2500	0.1	0.2	0.8	3.2
7/8"	CATV Hardline	50	81	25.0	#1 BC	FPE	SM	none	0.875	4000	0.03	0.1	0.6	2.9
7/8"	CATV Hardline	75	81	16.7	#5.5 BC	FPE	SM	none	0.875	4000	0.03	0.1	0.6	2.9
LDF4-50A	Heliax –1/2"	50	88	25.9	#5 Solid BC	FPE	CC	PE	0.630	1400	0.05	0.2	0.6	2.4
LDF5-50A	Heliax –7/8"	50	88	25.9	0.355" BC	FPE	CC	PE	1.090	2100	0.03	0.10	0.4	1.3
LDF6-50A	Heliax – 1¼"	50	88	25.9	0.516" BC	FPE	CC	PE	1.550	3200	0.02	0.08	0.3	1.1
Parallel Lines														
TV Twinlead (Belden 9085)		300	80	4.5	#22 Flex CCS	PE	none	P1	0.400	**	0.1	0.3	1.4	5.9
Twinlead (Belden 8225)		300	80	4.4	#20 Flex BC	PE	none	P1	0.400	8000	0.1	0.2	1.1	4.8
Generic Window Line		405	91	2.5	#18 Solid CCS	PE	none	P1	1.000	10000	0.02	0.08	0.3	1.1
WM CQ 554		420	91	2.7	#14 Flex CCS	PE	none	P1	1.000	10000	0.02	0.08	0.3	1.1
WM CQ 552		440	91	2.5	#16 Flex CCS	PE	none	P1	1.000	10000	0.02	0.08	0.3	1.1
WM CQ 553		450	91	2.5	#18 Flex CCS	PE	none	P1	1.000	10000	0.02	0.08	0.3	1.1
WM CQ 551		450	91	2.5	#18 Solid CCS	PE	none	P1	1.000	10000	0.02	0.08	0.3	1.1
Open-Wire Line		600	92	1.1	#12 BC	none	none	none	**	12000	0.02	0.06	0.2	0.7

Approximate Power Handling Capability (1:1 SWR, 40°C Ambient):

	1.8 MHz	*7*	*14*	*30*	*50*	*150*	*220*	*450*	*1 GHz*
RG-58 Style	1350	700	500	350	250	150	120	100	50
RG-59 Style	2300	1100	800	550	400	250	200	130	90
RG-8X Style	1830	840	560	360	270	145	115	80	50
RG-8/213 Style	5900	3000	2000	1500	1000	600	500	350	250
RG-217 Style	20000	9200	6100	3900	2900	1500	1200	800	500
LDF4-50A	38000	18000	13000	8200	6200	3400	2800	1900	1200
LDF5-50A	67000	32000	22000	14000	11000	5900	4800	3200	2100
LMR500	18000	9200	6500	4400	3400	1900	1600	1100	700
LMR1200	52000	26000	19000	13000	10000	5500	4500	3000	2000

Legend:

**	Not Available or varies	N	Non-Contaminating
ASPE	Air Spaced Polyethylene	P1	PVC, Class 1
BC	Bare Copper	P2	PVC, Class 2
CC	Corrugated Copper	PE	Polyethylene
CCA	Copper Cover Aluminum	S	Single Braided Shield
CCS	Copper Covered Steel	SC	Silver Coated Braid
CXP	Cable X-Perts, Inc.	SCCS	Silver Plated Copper Coated Steel
D	Double Copper Braids	SM	Smooth Aluminum
DRF	Davis RF	SPC	Silver Plated Copper
FC	Foil + Tinned Copper Braid	TC	Tinned Copper
FEP	Teflon ® Type IX	TFE	Teflon®
Flex	Flexible Stranded Wire	TMS	Times Microwave Systems
FPE	Foamed Polyethylene	UF	Ultra Flex
Heliax	Andrew Corp Heliax	WM	Wireman

currents flowing on the outer surface of the inner conductor and on the inner surface of the outer conductor cancel each other out, just as they do in open-wire line.

In a real (non-ideal) transmission line, the energy actually travels somewhat slower than the speed of light (typically from 65 to 97% of light speed), depending primarily on the dielectric properties of the insulating materials used in the construction of the line. The fraction of the speed of propagation in a transmission line compared to the speed of light in free space is called the *velocity factor* (VF) of the line. The velocity factor causes the line's *electrical* wavelength to be shorter than the wavelength in free space. Eq 1 describes the physical length of an electrical wavelength of transmission line.

$$\lambda = \frac{983.6}{f} \times VF \qquad (1)$$

where

λ = wavelength, in ft
f = frequency in MHz
VF = velocity factor.

Each transmission line has a characteristic velocity factor, related to the specific properties of its insulating materials. The velocity factor must be taken into account when cutting a transmission line to a specific electrical length. **Table 21.1** shows various velocity factors for the transmission lines commonly used by amateurs. For example, if RG-8A, which has a velocity factor of 0.66, were used to make a quarter-wavelength line at 3.5 MHz, the length would be (0.66 × 983.6/3.5)/4 = 46.4 ft long, instead of the free-space length of 70.3 ft. Open-wire line has a velocity factor of 0.97, close to unity, because it lacks a substantial amount of solid insulating material. Conversely, molded 300-Ω TV line has a velocity factor of 0.80 to 0.82 because it does use solid insulation between the conductors.

A perfectly lossless transmission line may be represented by a whole series of small inductors and capacitors connected in an infinitely long line, as shown in **Fig 21.2**. (We first consider this special case because we need not consider how the line is terminated at its end, since there is no end.)

Each inductor in Fig 21.2 represents the inductance of a very short section of one wire and each capacitor represents the capacitance between two such short sections. The inductance and capacitance values per unit of line depend on the size of the conductors and the spacing between them. The smaller the spacing between the two conductors and the greater their diameter, the higher the capacitance and the lower the inductance. Each series inductor acts to limit the rate at which current can charge the following shunt capacitor, and in so doing establishes a very important property of a transmission line: its *surge impedance*, more commonly known as its *characteristic impedance*. This is usually abbreviated as Z_0, and is approximately equal to $\sqrt{L/C}$, where L and C are the inductance and capacitance per unit length of line.

The characteristic impedance of an air-insulated parallel-conductor line, neglecting the effect of the insulating spacers, is given by

$$Z_0 = 276 \log_{10} \frac{2S}{d} \qquad (2)$$

where

Z_0 = characteristic impedance
S = center-to-center distance between conductors
d = diameter of conductor (in same units as S).

The characteristic impedance of an air-insulated coaxial line is given by

$$Z_0 = 138 \log_{10} \left(\frac{b}{a} \right) \qquad (3)$$

where

Z_0 = characteristic impedance
b = inside diameter of outer conductors
a = outside diameter of inner conductor (in same units as b).

It does not matter what units are used for S, d, a or b, so long as they are the same units. A line with closely spaced, large conductors will have a low characteristic impedance, while one with widely spaced, small conductors will have a relatively high characteristic impedance. Practical open-wire lines exhibit characteristic impedances ranging from about 200 to 800 Ω, while coax cables have Z_0 values between 25 to 100 Ω.

All practical transmission lines exhibit some power loss. These losses occur in the resistance that is inherent in the conductors that make up the line, and from leakage currents flowing in the dielectric material between the conductors. We'll next consider what happens when a real transmission line, which is not infinitely long, is terminated in real load impedances.

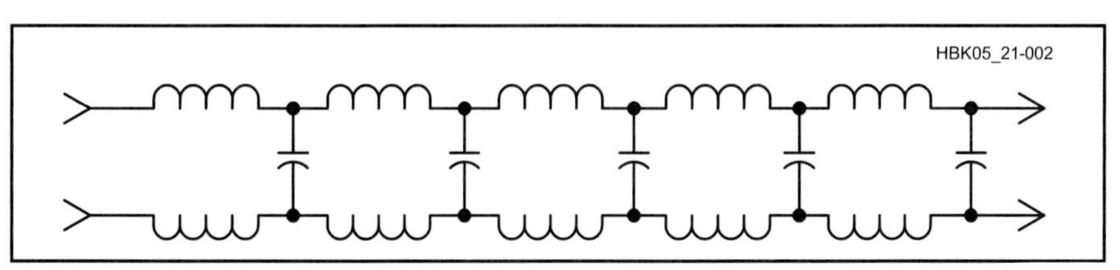

Fig 21.2—Equivalent of an infinitely long lossless transmission line using lumped circuit constants.

Matched Lines

Real transmission lines do not extend to infinity, but have a definite length. In use they are connected to, or *terminate* in, a load, as illustrated in **Fig 21.3A**. If the load is a pure resistance whose value equals the characteristic impedance of the line, the line is said to be *matched*. To current traveling along the line, such a load at the end

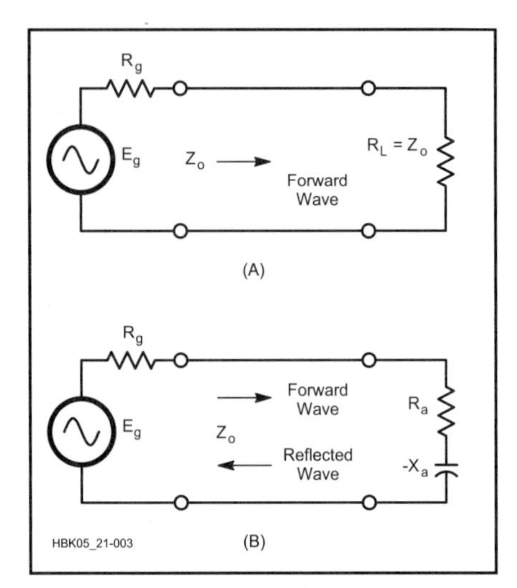

Fig 21.3—At A the coaxial transmission line is terminated with resistance equal to its Z_0. All power is absorbed in the load. At B, coaxial line is shown terminated in an impedance consisting of a resistance and a capacitive reactance. This is a mismatched line, and a reflected wave will be returned back down the line toward the generator. The reflected wave reacts with the forward wave to produce a standing wave on the line. The amount of reflection depends on the difference between the load impedance and the characteristic impedance of the transmission line.

of the line acts as though it were still more transmission line of the same characteristic impedance. In a matched transmission line, energy travels outward along the line from the source until it reaches the load, where it is completely absorbed.

Mismatched Lines

Assume now that the line in Fig 21.3B is terminated in an impedance Z_a which is not equal to Z_0 of the transmission line. The line is now a *mismatched* line. RF energy reaching the end of a mismatched line will not be fully absorbed by the load impedance. Instead, part of the energy will be reflected back toward the source. The amount of reflected versus absorbed energy depends on the degree of mismatch between the characteristic impedance of the line and the load impedance connected to its end.

The reason why energy is reflected at a discontinuity of impedance on a transmission line can best be understood by examining some limiting cases. First, consider the rather extreme case where the line is shorted at its end. Energy flowing to the load will encounter the short at the end, and the voltage at that point will go to zero, while the current will rise to a maximum. Since the current can't develop any power in a dead short, it will all be reflected back toward the source generator.

If the short at the end of the line is replaced with an open circuit, the opposite will happen. Here the voltage will rise to maximum, and the current will by definition go to zero. The phase will reverse, and all energy will be reflected back towards the source. By the way, if this sounds to you like what happens at the end of a half-wave dipole antenna, you are quite correct. However, in the case of an antenna, energy traveling along the antenna is lost by radiation on purpose, whereas a good transmission line will lose little energy to radiation because of field cancellation between the two conductors.

For load impedances falling between the extremes of short- and open-circuit, the phase and amplitude of the reflected wave will vary. The amount of energy reflected and the amount of energy absorbed in the load will depend on the difference between the characteristic impedance of the line and the impedance of the load at its end.

Now, what actually happens to the energy reflected back down the line? This energy will encounter another impedance discontinuity, this time at the generator. Reflected energy flows back and forth between the mismatches at the source and load. After a few such journeys, the reflected wave diminishes to nothing, partly as a result of finite losses in the line, but mainly because of absorption at the load. In fact, if the load is an antenna, such absorption at the load is desirable, since the energy is actually radiated by the antenna.

If a continuous RF voltage is applied to the terminals of a transmission line, the voltage at any point along the line will consist of a vector sum of voltages, the composite of waves traveling toward the load and waves traveling back toward the source generator. The sum of the waves traveling toward the load is called the *forward* or *incident* wave, while the sum of the waves traveling toward the generator is called the *reflected wave*.

Reflection Coefficient and SWR

In a mismatched transmission line, the ratio of the voltage in the reflected wave at any one point on the line to the voltage in the forward wave at that same point is defined as the *voltage reflection coefficient*. This has the same value as the current reflection coefficient. The reflection coefficient is a complex quantity (that is, having both amplitude and phase) and is generally designated by the Greek letter ρ (rho), or sometimes in the professional literature as Γ (Gamma). The relationship

Reflections on the Smith Chart

Although most radio amateurs have seen the Smith Chart, it is often regarded with trepidation. It is supposed to be complicated and subtle. However, the chart is extremely useful in circuit analysis, especially when transmission lines are involved. The Smith Chart is not limited to transmission-line and antenna problems.

The basis for the chart is Eq 4 in the main text relating reflection coefficient to a terminating impedance. Eq 4 is repeated here:

$$\rho = \frac{Z - Z_0}{Z + Z_0} \quad (1)$$

where Z_0 is the characteristic impedance of the chart, and $Z = R + jX$ is a complex terminating impedance. Z might be the feed-point impedance of an antenna connected to a Z_0 transmission line.

It is useful to define a normalized impedance $z = Z/Z_0$. The normalized resistance and reactance become $r = R/Z_0$ and $x = X/Z_0$. Inserting these into Eq 1 yields:

$$\rho = \frac{z - 1}{z + 1} \quad (2)$$

where r and z are both complex, each having a magnitude and a phase when expressed in polar coordinates, or a real and an imaginary part in XY coordinates.

Eq 1 and 2 have some interesting

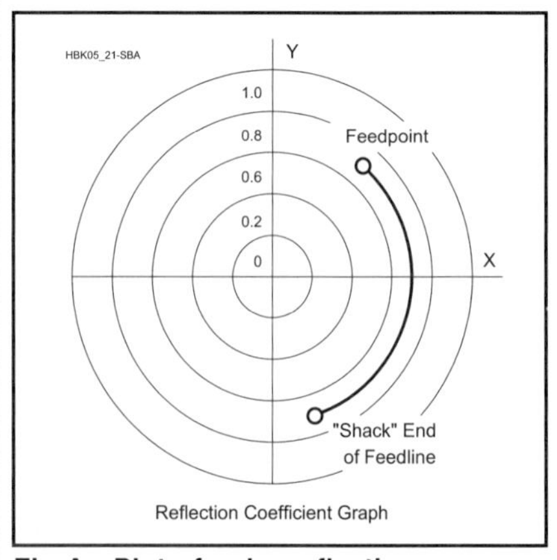

Fig A—Plot of polar reflection coefficient. Circles represent contours of constant ρ. The starting "feed point" value, 0.5 at +45°, represents an antenna impedance of 69.1 + *j* 65.1 Ω with Z_0 = 50 Ω. The arc represents a 15-ft section of 50-Ω, VF 0.66 transmission line at 7 MHz, yielding a shack ρ of 0.5 at –71.3°. The shack z is calculated as 40.3 – *j* 50.9 Ω.

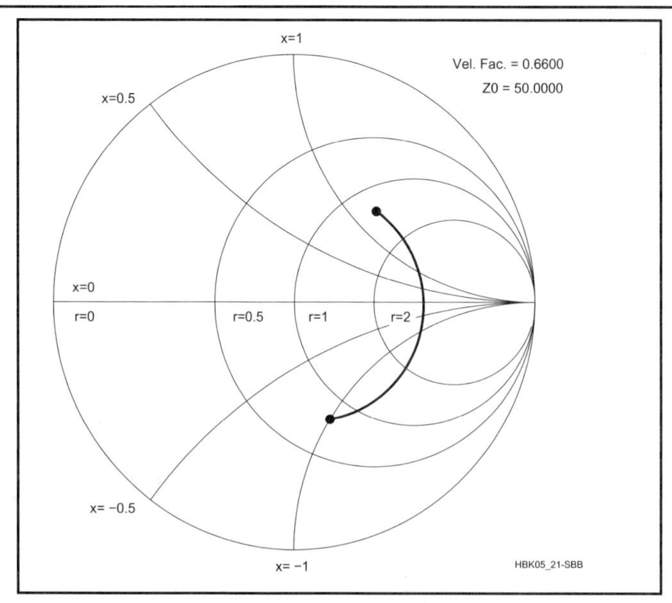

Fig B—This plot shows a Smith Chart. The circles now represent contours of constant normalized resistance or reactance. Note the arc with the markers: This illustrates the same antenna and line used in the previous figure. The plot is the same on the two charts; only the scale details have changed.

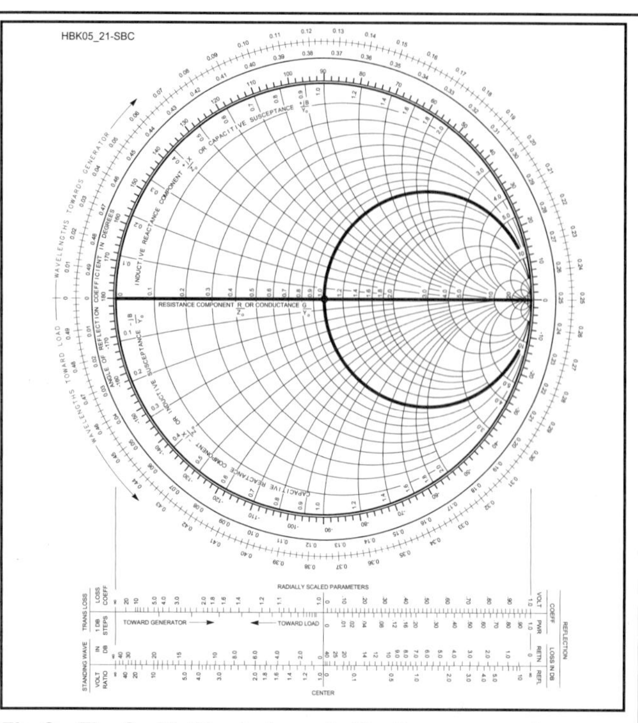

Fig C—The Smith Chart shown in Fig B was computer generated with *winSMITH* 2.0. A much more detailed plot is presented here; this is the chart form used by Smith, suitable for graphic applications. This chart is used with the permission of Analog Instruments.

and useful properties, characteristics that make them physically significant:

- Even though the components of z (and Z) may take on values that are very large, the reflection coefficient ρ, is restricted to always having a magnitude between zero and one if z has a real part, r, that is positive.
- If all possible values for *e* are examined and plotted in polar coordinates, they will lie within a circle with a radius of one. This is termed *the unit circle*. A plot is shown in **Fig A**.
- An impedance that is perfectly matched to Z_0, the characteristic value for the chart, will produce a ρ at the center of the unit circle.
- Real Z values, ones that have no reactance, "map" onto a horizontal line that divides the top from the bottom of the unit circle. By convention, a polar variable with an angle of zero is on the x axis, to the right of the origin.
- Impedances with a reactive part produce ρ values away from the dividing line. Inductive impedances with the imaginary part greater than zero appear in the upper half of the chart, while capacitive impedances appear in the lower half.
- Perhaps the most interesting and exciting property of the reflection coefficient is the way it describes the impedance-transforming properties of a transmission line, presented in closed mathematical form in the main text as Eq 11.

Neglecting loss effects, a transmission line of electrical length θ will transform a normalized impedance represented by ρ to another with the same magnitude and a new angle that differs from the original by –2θ. This rotation is clockwise.

Clearly, the reflection coefficient is more than an intermediate step in a mathematical development. It is a useful, alternative description of complex impedance. However, our interest is still focused on impedance; we want to know, for example, what the final z is after transformation with a transmission line. This is the problem that Phillip Smith solved in creating the Smith Chart. Smith observed that the unit circle, a graph of reflection coefficient, could be labeled with lines representing *normalized impedance*. A Smith Chart is shown in **Fig B**. All of the lines on the chart are complete or partial circles representing a line of constant normalized resistance and reactance.

How might we use the Smith Chart? A classic application relates antenna feed-point impedance to the impedance seen at the end of the "shack" end of the line. Assume that the antenna impedance is known, $Z_a = R_a + j X_a$. This complex value is converted to normalized impedance by dividing R_a and X_a by Z_0 to yield $r_a + j x_a$, and is plotted on the chart. A compass is then used to draw an arc of a circle centered at the origin of the chart. The arc starts at the normalized antenna impedance and proceeds in a clockwise direction for 2θ°, where θ is the electrical degrees, derived from the physical length and velocity factor of the transmission line. The end of the arc represents the normalized impedance at the end of the line in the shack; it is denormalized by multiplying the real and imaginary parts by Z_0.

Antenna feed-point Z can also be inferred from an impedance measurement at the shack end of the line. A similar procedure is followed. The only difference is that rotation is now in a counterclockwise direction. The Smith Chart is much more powerful than depicted in this brief summary. A detailed treatment is given by Phillip H. Smith in his classic book: *Electronic Applications of the Smith Chart* (McGraw-Hill, 1969). I also recommend his article "Transmission Line Calculator" in Jan 1939 *Electronics.* Joseph White presented a wonderful summary of the chart in a short but outstanding paper: "The Smith Chart: An Endangered Species?" Nov 1979 *Microwave Journal. WinSMITH* 2.0 is available from ARRL for $80. The impedance matching tutorial is included.—*Wes Hayward, W7ZOI*

between R_a (the load resistance), X_a (the load reactance), Z_0 (the line characteristic impedance, whose real part is R_0 and whose reactive part is X_0) and the complex reflection coefficient ρ is

$$\rho = \frac{Z_a - Z_0}{Z_a + Z_0} = \frac{(R_l \pm j X_a) - (R_0 \pm j X_a)}{(R_l \pm j X_a) + (R_0 \pm j X_a)} \quad (4)$$

For most transmission lines the characteristic impedance Z_0 is almost completely resistive, meaning that $Z_0 = R_0$ and $X_0 \cong 0$. The magnitude of the complex reflection coefficient in Eq 4 then simplifies to:

$$|\rho| = \sqrt{\frac{(R_a - R_0)^2 + X_a{}^2}{(R_a + R_0)^2 + X_a{}^2}} \quad (5)$$

For example, if the characteristic impedance of a coaxial line is 50 Ω and the load impedance is 120 Ω in series with a capacitive reactance of –90 Ω, the magnitude of the reflection coefficient is

$$|\rho| = \sqrt{\frac{(120-50)^2 + (-90)^2}{(120+50)^2 + (-90)^2}} = 0.593$$

Note that if R_a in Eq 4 is equal to R_0 and X_a is 0, the reflection coefficient, ρ, is 0. This represents a matched condition, where all the energy in the incident wave is transferred to the load. On the other hand, if R_a is 0, meaning that the load has no real resistive part, the reflection coefficient is 1.0, regardless of the value of R_0. This means that all the forward power is reflected since the load is completely reactive. The concept of reflection is often shown in terms of the *return loss*, which is 20 log the reciprocal of the reflection coefficient, in dB. In the example above, the return loss is 4.5 dB.

If there are no reflections from the load, the voltage distribution along the line is constant or *flat*. A line operating under these conditions is called either a *matched* or a *flat* line. If reflections do exist, a voltage *standing-wave* pattern will result from the interaction of the forward and reflected waves along the line. For a lossless transmission line, the ratio of the maximum peak voltage anywhere on the line to the minimum value anywhere on the line (which must be at least 1/4 λ) is defined as the *voltage standing-wave ratio*, or VSWR. Reflections from the load also produce a standing-wave pattern of currents flowing in the line. The ratio of maximum to minimum current, or ISWR, is identical to the VSWR in a given line.

In amateur literature, the abbreviation *SWR* is commonly used for standing-wave ratio, as the results are identical when taken from proper measurements of either current or voltage. Since SWR is a ratio of maximum to minimum, it can never be less than one-to-one. In other words, a perfectly flat line has an SWR of 1:1. The SWR is related to the magnitude of the complex reflection coefficient by

$$SWR = \frac{1 + |\rho|}{1 - |\rho|} \quad (6)$$

and conversely the reflection coefficient magnitude may be defined from a measurement of SWR as

$$|\rho| = \frac{SWR - 1}{SWR + 1} \quad (7)$$

The definitions in Eq 6 and 7 are valid for any line length and for lines which are lossy, not just lossless lines longer than 1/4 λ at the frequency in use. Very often the load impedance is not exactly known, since an antenna usually terminates a transmission line, and the antenna impedance may be influenced by a host of factors, including its height above ground, end effects from insulators, and the effects of nearby conductors. We may also express the reflection coefficient in terms of forward and reflected power, quantities which can be easily measured using a directional RF wattmeter. The reflection coefficient may be computed as

$$|\rho| = \sqrt{\frac{P_r}{P_f}} \quad (8)$$

where

P_r = power in the reflected wave

P_f = power in the forward wave.

If a line is not matched (SWR > 1:1) the difference between the forward and reflected powers measured at any point on the line is the net power going toward the load from that point. The forward power measured with a directional wattmeter (often referred to as a reflected power meter or *reflectometer*) on a mismatched line will thus always appear greater than the forward power measured on a flat line with a 1:1 SWR.

Losses in Transmission Lines

A real transmission line exhibits a certain amount of loss, caused by the resistance of the conductors used in the line and by dielectric losses in the line's insulators. The *matched-line loss* for a particular type and length of transmission line, operated at a particular frequency, is the loss when the line is terminated in a resistance equal to its characteristic impedance. The loss in a line is lowest when it is operated as a matched line.

Line losses increase when SWR is greater than 1:1. Each time energy flows from the generator toward the load, or is reflected at the load and travels back toward the generator, a certain amount will be lost along the line. The net effect of standing waves on a transmission line is to increase the average value of current and voltage, compared to the matched-line case. An increase in current raises I^2R (ohmic) losses in the conductors, and an increase in RF voltage increases E^2/R losses in the dielectric. Line loss rises with frequency, since the conductor resistance is related to skin effect, and also because dielectric losses rise with frequency.

Matched-line loss is stated in decibels per hundred feet at a particular frequency. **Fig 21.4** shows the matched-line loss per hundred feet versus frequency for a number of common types of lines, both coaxial and open-wire balanced types. For example, RG-213 coax cable has a matched-line loss of 2.5 dB/100 ft at 100 MHz. Thus, 45 ft of this cable feeding a 50-Ω load at 100 MHz would have a loss of

$$\text{Matched line loss} = \frac{2.5\ \text{dB}}{100\ \text{ft}} \times 45\ \text{ft}$$

$$= 1.13\ \text{dB}$$

If a line is not matched, standing waves will cause additional loss beyond the inherent matched-line loss for that line. On lines which are inherently lossy, the total line loss (the sum of matched-line loss and additional loss due to SWR) can be surprisingly high for high values of SWR.

Total Mismatched-Line Loss (dB)

$$= 10 \log \left[\frac{a^2 - |\rho|^2}{a(1 - |\rho|^2)} \right] \quad (9)$$

where

$a = 10^{ML/10}$ = matched-line ratio

$$|\rho| = \frac{SWR - 1}{SWR + 1}$$

ML = matched-line loss in dB

SWR = SWR measured at load

Because of losses in a transmission line, the measured SWR at the input of the line is less than the SWR measured at the load end of the line.

$$\text{SWR at input} = \frac{a + |\rho|}{a - |\rho|} \quad (10)$$

For example, RG-8A solid-dielectric coax cable exhibits a matched-line loss per 100 ft at 28 MHz of 1.18 dB. A 250-ft length of this cable has a matched-line loss of 2.95 dB. Assume that we measure the SWR at the load as 6:1.

$$a = 10^{2.95/10} = 1.972$$

$$|\rho| = \frac{6.0-1}{6.0+1} = 0.714$$

$$\text{Total Loss} = 10 \log \frac{1.972^2 - 0.714^2}{1.972^2\,(1-0.714^2)}$$

$$= 5.4 \text{ dB}$$

$$\text{SWR}_{in} = \frac{1.972+0.714}{1.972-0.714} = 2.1$$

The additional loss due to the 6:1 SWR at 28 MHz is 5.4 – 3.0 = 2.4 dB. The SWR at the input of the 250-ft line is only 2.1:1, because line loss has masked the true extent of the SWR (6:1) at the load end of the line.

The losses become larger if coax with a larger matched-line loss is used under the same conditions. For example, RG58A coaxial cable is about one-half the diameter of RG8A, and it has a matched-line loss of 2.5 dB/100 ft at 28 MHz. A 250-ft length of RG58A has a total matched-line loss of 6.3 dB. With a 6:1 SWR at the load, the additional loss due to SWR is 3.0 dB, for a total loss of 9.3 dB. The additional cable loss due to the mismatch reduces the SWR at the input of the line to 1.4:1. An unsuspecting operator measuring the SWR at his transmitter might well believe that everything is just fine, when in truth only about 12% of the transmitter power is getting to the antenna! Be suspicious of very low SWR readings for an antenna fed with a long length of coaxial cable, especially if the SWR remains low across a wide frequency range. Most antennas have narrow SWR bandwidths, and the SWR *should* change across a band.

On the other hand, if expensive $^3/_4$-inch diameter 50-Ω Hardline cable is used at 28 MHz, the matched-line loss is only 0.28 dB/100 ft. For 250 ft of Hardline the matched-line loss is 0.7 dB, and the additional loss due to a 6:1 SWR is 1.1 dB. The total loss is 1.8 dB. See **Table 21.2** for a summary of the losses for 250 ft of the three types of coax as a function of frequency for matched line and 6:1 SWR conditions.

At the upper end of the HF spectrum, when the transmitter and antenna are separated by a long transmission line the use of bargain coax may prove to be a very poor cost-saving strategy. A 7.5 dB linear amplifier, to offset the loss in RG58A compared to Hardline, would cost a great deal more than higher-quality coax. Furthermore, no *transmitter* amplifier can boost *receiver* sensitivity—loss in the line has the same effect as putting an attenuator in front of the receiver.

At the low end of the HF spectrum, say 3.5 MHz, the amount of loss in common coax lines is less of a problem for the range

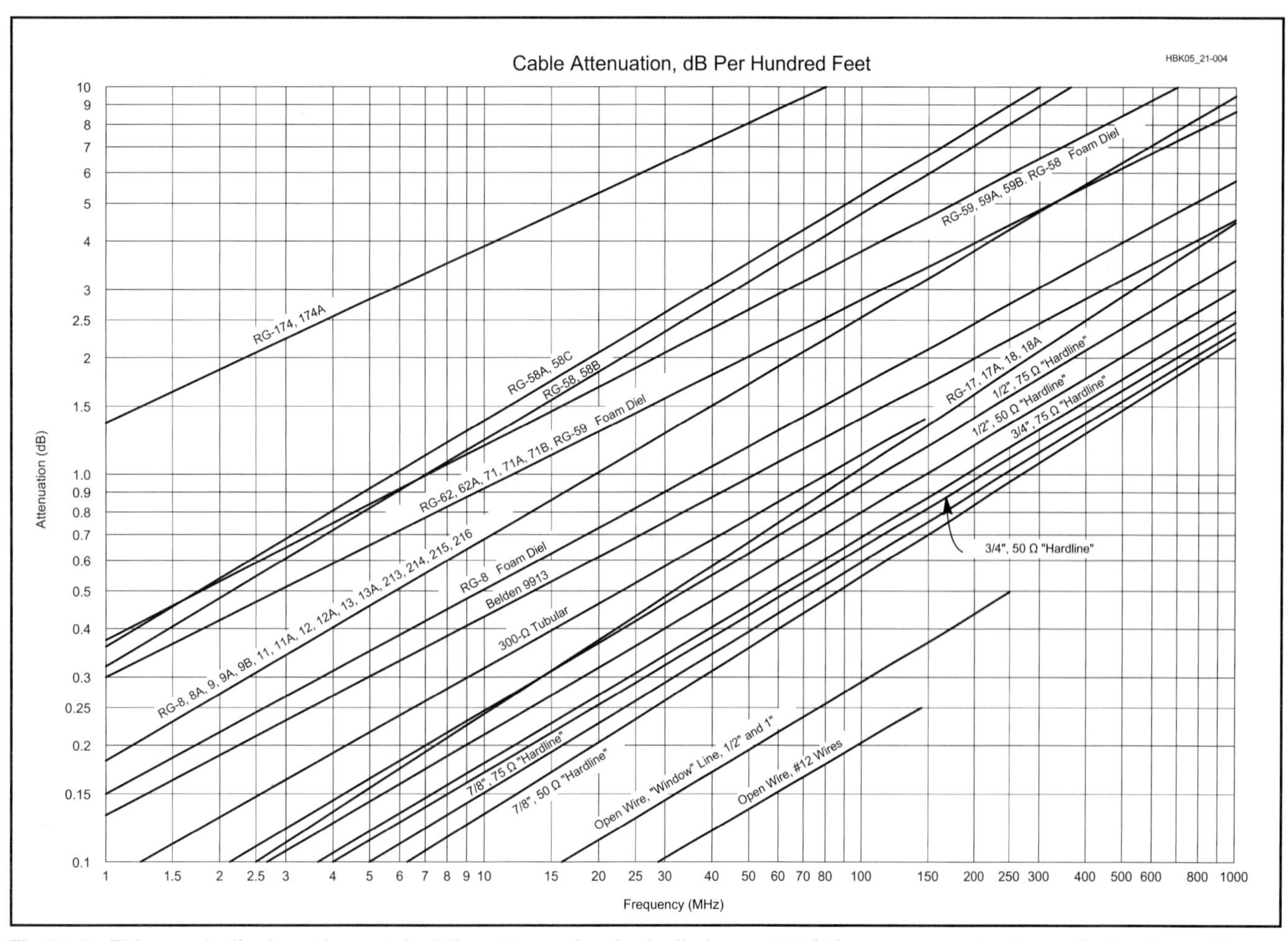

Fig 21.4—This graph displays the matched-line attenuation in decibels per 100 ft for many popular transmission lines. The vertical axis represents attenuation and the horizontal axis frequency. Note that these loss figures are only accurate for properly matched transmission lines.

of SWR values typical on this band. For example, consider an 80-m dipole cut for the middle of the band at 3.75 MHz. It exhibits an SWR of about 6:1 at the 3.5 and 4.0 MHz ends of the band. At 3.5 MHz, 250 ft of RG58A small-diameter coax has an additional loss of 2.1 dB for this SWR, giving a total line loss of 4.0 dB. If larger-diameter RG8A coax is used instead, the additional loss due to SWR is 1.3 dB, for a total loss of 2.2 dB. This is an acceptable level of loss for most 80-m operators.

However, the loss situation gets dramatically worse as the frequency increases into the VHF and UHF regions. At 146 MHz, the total loss in 250 ft of RG-58A with a 6:1 SWR at the load is16.5 dB, 10.8 dB for RG-8A, and 4.2 dB for 3/4-inch 50-Ω Hardline. At VHF and UHF, a low SWR is essential to keep line losses low, even for the best coaxial cable. The length of transmission line must be kept as short as practical at these frequencies.

The effect of SWR on line loss is shown graphically in **Fig 21.5**. The horizontal axis is the attenuation, in decibels, of the line when perfectly matched. The vertical axis gives the additional attenuation due to SWR. If long coaxial-cable transmission lines are necessary, the matched loss of the coax used should be kept as low as possible, meaning that the highest-quality, largest-diameter cable should be used.

Choosing a Transmission Line

It is no accident that coaxial cable became as popular as it has since it was first widely used during World War II. Coax is mechanically much easier to use than open-wire line. Because of the excellent shielding afforded by its outer shield, coax can be run up a metal tower leg, taped together with numerous other cables, with virtually no interaction or crosstalk between the cables. At the top of a tower, coax can be used with a rotatable Yagi or quad antenna without worrying about shorting or twisting the conductors, which might happen with an open-wire line. Class 2 PVC non-contaminating outer jackets are designed for long-life outdoor installations. Class 1 PVC outer jackets are not recommended for outdoor installations. Coax can be buried underground, especially if it is run in plastic piping (with suitable drain holes) so that ground water and soil chemicals cannot easily deteriorate the cable. A cable with an outer jacket of polyethylene (PE) rather than polyvinyl chloride (PVC) is recommended for direct-bury installations.

Open-wire line must be carefully spaced away from nearby conductors, by at least several times the spacing between its conductors, to minimize possible electrical imbalances between the two parallel conductors. Such imbalances lead to line radiation and extra losses. One popular type of open-wire line is called *ladder line* because the insulators used to separate the two parallel, uninsulated conductors of the line resemble the steps of a ladder. Long lengths of ladder line can twist together in the wind and short out if not properly supported.

Despite the mechanical difficulties associated with open-wire line, there are some compelling reasons for its use, especially in simple multiband antenna systems. Every antenna system, no matter what its physical form, exhibits a definite value of impedance at the point where the transmission line is connected. Although the input impedance of an antenna system is seldom known exactly, it is often possible to make a close estimate of its value, especially since sophisticated computer-modeling programs have become available to the radio amateur. As an example, **Table 21.3** lists the computed characteristics versus frequency for a multiband, 100-ft long center-fed dipole, placed 50 ft above average ground having a dielectric constant of 13 and a conductivity of 10 mS/m.

These values were computed using a complex program called *NEC-2* (Numerical Electromagnetic Code), which incorporates a sophisticated Sommerfeld/Norton ground-modeling algorithm for antennas close to real earth. A nonresonant 100-ft length was chosen as an illustration of a practical size that many radio amateurs could fit into their backyards, although nothing in particular recommends this antenna over other forms. It is merely used as an example.

Examine Table 21.3 carefully in the following discussion. Columns three and four show the SWR on a 50-Ω RG-8A coaxial transmission line directly connected to the antenna, followed by the total loss in 100 ft of this cable. The impedance for this nonresonant, 100-ft long antenna varies over a very wide range for the nine operating frequencies. The SWR on a 50-Ω coax connected directly to this antenna would be *extremely* high on some frequencies, particularly at 1.8 MHz,

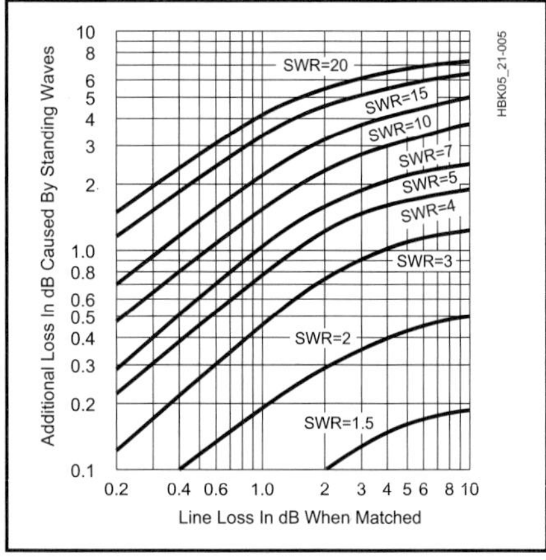

Fig 21.5—Increase in line loss because of standing waves (SWR measured at the load). To determine the total loss in decibels in a line having an SWR greater than 1, first determine the matched-line loss for the particular type of line, length and frequency, on the assumption that the line is perfectly matched (from Fig 21.4). For example, Belden 9913 has a matched-line loss of 0.49 dB/100 ft at 14 MHz. Locate 0.49 dB on the horizontal axis. For an SWR of 5:1, move up to the curve corresponding to this SWR. The increase in loss due to SWR is 0.65 dB beyond the matched line loss.

Table 21.2

Matched-Line Loss for 250 ft of Three Common Coaxial Cables

Comparisons of line losses versus frequency for 250-ft lengths of three different coax cable types: small-diameter RG58A, medium-diameter RG8A, and 3/4-inch OD 50-Ω Hardline. At VHF, the losses for the small-diameter cable are very large, while they are moderate at 3.5 MHz.

Xmsn Line	*3.5 MHz Matched-Line Loss, dB*	*3.5 MHz Loss, 6:1 SWR, dB*	*28 MHz Matched-Line Loss, dB*	*28 MHz Loss, 6:1 SWR, dB*	*146 MHz Matched-Line Loss, dB*	*146 MHz Loss 6:1 SWR, dB*
RG-58A	1.9	4.0	6.3	9.3	16.5	21.6
RG-8A	0.9	2.2	3.0	5.4	7.8	10.8
3/4" 50-Ω Hardline	0.2	0.5	0.7	1.8	2.1	4.2

where the antenna is highly capacitive because it is very short of resonance. The loss for an SWR of 1818:1 in 100 ft of RG-8A at 1.8 MHz is a staggering 25.9 dB.

Contrast this to the loss in 100 ft of 450-Ω open-wire line. Here, the loss at 1.8 MHz is 12.1 dB. While 12.1 dB of loss is not particularly desirable, it is almost 14 dB better than the coax! Note that the RG-8A coax exhibits a good deal of loss on almost all the bands due to mismatch. Only on 14 MHz does the loss drop down to 1.9 dB, where the antenna is just past $^3/_2$-λ resonance. From 3.8 to 28.4 MHz the open-wire line has a maximum loss of only 0.9 dB.

Columns six and seven in Table 21.3 list the maximum RMS voltage for 1500 W of RF power on the 50-Ω coax and on the 450-Ω open-wire line. The maximum RMS voltage for 1500 W on the open-wire line is extremely high, at 7640 V at 1.8 MHz. The voltage for a 100-W transmitter would be reduced by a ratio of $\sqrt{1500/100} = 3.87:1$. This is 1974 V, still high enough to cause arcing in many antenna tuners.

In general, such a nonresonant antenna is a proven, practical multiband radiator when fed with 450-Ω open-wire ladder line connected to an antenna tuner, although a longer antenna would be preferable for more efficient 160-m operation, even with open-wire line. The tuner and the line itself must be capable of handling the high RF voltages and currents involved for high-power operation. On the other hand, if such a multiband antenna is fed directly with coaxial cable, the losses on most frequencies are prohibitive. Coax is most suitable for antennas whose resonant feed-point impedances are close to the characteristic impedance of the feed line.

The Transmission Line as Impedance Transformer

If the complex mechanics of reflections, SWR and line losses are put aside momentarily, a transmission line can very simply be considered as an impedance transformer. A certain value of load impedance, consisting of a resistance and reactance, at the end of the line is transformed into another value of impedance at the input of the line. The amount of transformation is determined by the electrical length of the line, its characteristic impedance, and by the losses inherent in the line. The input impedance of a real, lossy transmission line is computed using the following equation

$$Z_{in} = Z_0 \times \frac{Z_L \cosh(\eta \ell) + Z_0 \sinh(\eta \ell)}{Z_L \sinh(\eta \ell) + Z_0 \cosh(\eta \ell)} \quad (11)$$

where

Z_{in} = complex impedance at input of line = $R_{in} \pm j\,X_{in}$

Z_L = complex load impedance at end of line = $R_a \pm j\,X_a$

Z_0 = characteristic impedance of line = $R_0 \pm j\,X_0$

η = complex loss coefficient = $a + j\,b$

α = matched line loss attenuation constant, in nepers/unit length (1 neper = 8.688 dB; most cables are rated in dB/100 ft)

β = phase constant of line in radians/unit length (related to physical length of line by the fact that 2π radians = 1 wavelength, and by Eq 1)

ℓ = electrical length of line in same units of length measurement as a or above.

Solving this equation manually is tedious, since it incorporates hyperbolic cosines and sines of the complex loss coefficient, but it may be solved using a traditional paper Smith Chart or a computer program. *The ARRL Antenna Book* has a chapter detailing the use of the Smith Chart. *MicroSmith* is a sophisticated graphical Smith Chart program written for the IBM PC, and is available through the ARRL. *TLA* (Transmission Line) is another ARRL program that performs this transformation, but without Smith Chart graphics. *TLA.EXE* is on the CD-ROM accompanying this book.

Lines as Stubs

The impedance-transformation properties of a transmission line are useful in a number of applications. If the terminating resistance is zero (that is, a short) at the end of a low-loss transmission line which is less than $^1/_4$ λ, the input impedance consists of a reactance, which is given by a simplification of Eq 11.

$$X_{in} \cong Z_0 \tan \ell \quad (12)$$

If the line termination is an open circuit, the input reactance is given by

$$X_{in} = Z_0 \cot \ell \quad (13)$$

The input of a short (less than $^1/_4$ λ) length of line with a short circuit as a terminating load appears as an inductance, while an open-circuited line appears as a

Table 21.3

Modeled Data for a 100-ft Flat-Top Antenna

100-ft long, 50-ft high, center-fed dipole over average ground, using coaxial or open-wire transmission lines. Antenna impedance computed using *NEC2* computer program, with ground relative permittivity of 13, ground conductivity of 5 mS/m and Sommerfeld/Norton ground model. Note the extremely reactive impedance levels at many frequencies, but especially at 1.8 MHz. If this antenna is fed directly with RG-8A coax, the losses are unacceptably large on 160 m, and undesirably high on most other bands also. The RF voltage at 3.8 MHz for high-power operation with open-wire line is extremely high also, and would probably result in arcing either on the line itself, or more likely in the Transmatch. Each transmission line is 100 ft long.

Frequency (MHz)	*Antenna Impedance (Ohms)*	*SWR RG-8A Coax*	*Loss 100 ft RG-8A Coax*	*Loss 100 ft 450-Ω Line*	*Max Volt. RG-8A 1500 W*	*Max Volt. 450-Ω Line 1500 W*
1.8 MHz	4.5 – *j* 1673	1818:1	25.9 dB	12.1 dB	1640	7640
3.8 MHz	38.9 – *j* 362	63:1	5.7 dB	0.9 dB	1181	3188
7.1 MHz	481 + *j* 964	49:1	5.8 dB	0.3 dB	981	1964
10.1 MHz	2584 – *j* 3292	134:1	10.4 dB	0.9 dB	967	2869
14.1 MHz	85.3 – *j* 123.3	6.0:1	1.9 dB	0.5 dB	530	1863
18.1 MHz	2097 + *j* 1552	65:1	9.0 dB	0.6 dB	780	2073
21.1 MHz	345 – *j* 1073	73:1	9.8 dB	0.8 dB	757	2306
24.9 MHz	202 + *j* 367	18:1	5.2 dB	0.4 dB	630	1563
28.4 MHz	2493 – *j* 1375	65:1	10.1 dB	0.7 dB	690	2051

capacitance. This is a useful property of a transmission line, since it can be used as a low-loss inductor or capacitor in matching networks. Such lines are often referred to as *stubs*.

A line that is an electrical quarter wavelength is a special kind of a stub. When a quarter-wave line is short circuited at its load end, it presents an open circuit at its input. Conversely, a quarter-wave line with an open circuit at its load end presents a short circuit at its input. Such a line inverts the sense of a short or an open circuit at the frequency for which the line is a quarter-wave long. This is also true for frequencies that are odd multiples of the quarter-wave frequency. However, for frequencies where the length of the line is a half wavelength, or integer multiples thereof, the line will duplicate the termination at its end.

For example, if a shorted line is cut to be a quarter wavelength at 7.1 MHz, the impedance looking into the input of the cable will be an open circuit. The line will have no effect if placed in parallel with a transmitter's output terminal. However, at twice the frequency, 14.2 MHz, that same line is now a half wavelength, and the line looks like a short circuit. The line, often dubbed a *quarter-wave stub* in this application, will act as a trap for not only the second harmonic, but also for higher even-order harmonics, such as the fourth or sixth harmonics.

Quarter-wave stubs made of good-quality coax, such as RG-213, offer a convenient way to lower transmitter harmonic levels. Despite the fact that the exact amount of harmonic attenuation depends on the impedance (often unknown) into which they are working at the harmonic frequency, a quarter-wave stub will typically yield 20 to 25 dB of attenuation of the second harmonic when placed directly at the output of a transmitter feeding common amateur antennas. Because different manufacturing runs of coax will have slightly different velocity factors, a quarter-wave stub is usually cut a little longer than calculated, and then carefully pruned by snipping off short pieces, while monitoring the response at the fundamental frequency, using a grid-dip meter or an SWR indicator. Because the end of the coax is an open circuit while pieces are being snipped away, the input of a quarter-wave line will show a short circuit exactly at the fundamental frequency. Once the coax has been pruned to frequency, a short jumper is soldered across the end, and the response at the second harmonic frequency is measured.

We will examine further applications of quarter-wave transmission lines later in the next section.

Matching the Antenna to the Line

When transmission lines are used with a transmitter, the most common load is an antenna. When a transmission line is connected between an antenna and a receiver, the receiver input circuit is the load, not the antenna, because the power taken from a passing wave is delivered to the receiver.

Whatever the application, the conditions existing at the load, and *only* the load, determine the reflection coefficient, and hence the standing-wave ratio, on the line. If the load is purely resistive and equal to the characteristic impedance of the line, there will be no standing waves. If the load is not purely resistive, or is not equal to the line Z_0, there will be standing waves. No adjustments can be made at the input end of the line to change the SWR at the load. Neither is the SWR affected by changing the line length, except as previously described when the SWR at the input of a lossy line is masked by the attenuation of the line.

Only in a few special cases is the antenna impedance the exact value needed to match a practical transmission line. In all other cases, it is necessary either to operate with a mismatch and accept the SWR that results, or else to bring about a match between the line and the antenna.

Technical literature sometimes uses the term *conjugate match* to describe the condition where the reactance seen looking toward the load from any point on the line is the complex conjugate of the impedance seen looking toward the source. A conjugate match is necessary to achieve the maximum power gain possible from a small-signal amplifier. For example, if a small-signal amplifier at 14.2 MHz has an output impedance of $25.8 + j\,11.0\ \Omega$, then the maximum power possible will be generated from that amplifier when the output load is $25.8 - j\,11.0\ \Omega$. The amplifier and load system is resonant because the ± 11.0-Ω reactances cancel.

Now, assume that 100 ft of 50-Ω RG-213 coax at 14.2 MHz just happens to be terminated in an impedance of $115 - j\,25\ \Omega$. Eq 11 calculates that the impedance looking into the input of the line is $25.8 - j\,11.0\ \Omega$. If this transmission line is connected directly to the small-signal amplifier above, then a conjugate match is created, and the amplifier generates the maximum possible amount of power it can generate.

However, if the impedance at the output of the amplifier is not $25.8 - j\,11.0\ \Omega$, then a matching network is needed between the amplifier and its load for maximum power gain. For example, if 50 ft of RG-213 is terminated in a $72 - j\,34\ \Omega$ antenna impedance, the impedance at the line input becomes $35.9 - j\,21.6\ \Omega$. A matching network is designed to transform $35.9 - j\,21.6\ \Omega$ to $25.8 - j\,11.0\ \Omega$, so that once again a conjugate match is created for the small-signal amplifier.

Now, let us consider what happens with amplifiers where the power level is higher than the milliwatt level of small-signal amplifiers. Most modern transmitters are designed to work into a 50-Ω load. Most will reduce power automatically if the load is not 50 Ω—this protects them against damage and ensures linear operation without distortion.

Many amateurs use an *antenna tuner* between their transmitter and the transmission line feeding the antenna. The antenna tuner's function is to transform the impedance, whatever it is, at the shack-end of the transmission line into the 50 Ω required by their transmitter. Note that the SWR on the transmission line between the antenna and the output of the antenna tuner is rarely exactly 1:1, even though the SWR on the short length of line between the tuner and the transmitter is 1:1.

Therefore, some loss is unavoidable: additional loss due to the SWR on the line, and loss in the antenna tuner itself. However, most amateur antenna installations use antennas that are reasonably close to resonance, making these types of losses small enough to be acceptable.

Despite the inconvenience, if the antenna tuner could be placed at the antenna rather than at the transmitter output, it can transform the $72 - j\,34\ \Omega$ antenna impedance to a nonreactive 50 Ω. Then the line SWR is 1:1.

Impedance matching networks can take a variety of physical forms, depending on the circumstances.

Matching the Antenna to the Line, at the Antenna

This section describes methods by which a network can be installed at the antenna itself to provide matching to a transmission line. Having the matching system up at the antenna rather than down in the shack at the end of a long transmission line does seem intuitively desirable, but it is not always very practical, especially in multiband antennas.

If a highly reactive antenna can be tuned to resonance, even without special efforts to make the resistive portion equal to the line's characteristic impedance, the resulting SWR is often low enough to minimize additional line loss due to SWR. For example, the multiband 100-ft long dipole in Table 21.3 has an antenna impedance of $4.5 - j\,1673\ \Omega$ at 1.8 MHz. Assume that the antenna reactance is tuned out with a network consisting of two symmetrical inductors whose reactance is +836.5 Ω each, with a Q of 200. The inductors are made up of 73.95 µH coils in series with inherent loss resistances of 836.5/200 = 4.2 Ω. The total series resistance is thus 4.5 + 2 × (4.2) = 12.9 Ω, and the antenna reactance and inductor reactance cancel out. See **Fig 21.6**.

If this tuned system is fed with 50-Ω coaxial cable, the SWR is 50/12.9 = 3.88:1, and the loss in 100 ft of RG-8A cable would be 0.47 dB. The radiation efficiency is 4.5/12.9 = 34.9%. Expressed another way, there is 4.57 dB of loss. Adding the 0.47 dB of loss in the line yields an overall system loss of 5.04 dB. Compare this to the loss of 17.1 dB if the RG-8A coax is used to feed the antenna directly, without any matching at the antenna. The use of a moderately high-Q resonator has yielded almost 12 dB of "gain" (that is, less loss) compared to the nonresonator case. The drawback of course is that the antenna is now resonated on only one frequency, but it certainly is a lot more efficient on that one frequency.

The Quarter-Wave Transformer or "Q" Section

The range of impedances presented to the transmission line is usually relatively small on a typical amateur antenna, such as a dipole or a Yagi when it is operated close to resonance. In such antenna systems, the impedance-transforming properties of a quarter-wave section of transmission line are often utilized to match the transmission line at the antenna.

One example of this technique is an array of stacked Yagis on a single tower. Each antenna is resonant and is fed in parallel with the other Yagis, using equal lengths of coax to each antenna. A stacked array is used to produce not only gain, but also a wide vertical elevation pattern, suitable for coverage of a broad geographic area. (See *The ARRL Antenna Book* for details about Yagi stacking.) The feed-point impedance of two 50-Ω Yagis fed with equal lengths of feed line connected in parallel is 25 Ω (50/2 Ω); three in parallel yield 16.7 Ω; four in parallel yield 12.5 Ω. The nominal SWR for a stack of four Yagis is 4:1 (50/12.5). This level of SWR does not cause excessive line loss, provided that low-loss coax feed line is used. However, many station designers want to be able to select, using relays, any individual antenna in the array, without having the load seen by the transmitter change. (Perhaps they might wish to turn one antenna in the stack in a different direction and use it by itself.) If the load changes, the amplifier must be retuned, an inconvenience at best.

See **Fig 21.7**. If the antenna impedance and the characteristic impedance of a feed line to be matched are known, the characteristic impedance needed for a quarter-wave matching section of low-loss cable is expressed by another simplification of Eq 11.

$$Z = \sqrt{Z_1 Z_0} \tag{14}$$

where

Z = characteristic impedance needed for matching section

Z_1 = antenna impedance

Z_0 = characteristic impedance of the line to which it is to be matched.

Example: To match a 50-Ω line to a Yagi stack consisting of two antennas fed in parallel to produce a 25-Ω load, the quarter-wave matching section would require a characteristic impedance of

$$Z = \sqrt{50 \times 25} = 35.4\ \Omega$$

A transmission line with a characteristic impedance of 35 Ω could be closely approximated by connecting two equal lengths of 75-Ω cable (such as RG-11A) in parallel to yield the equivalent of a 37.5-Ω cable. Three Yagis fed in parallel would require a quarter-wave transformer made using a cable having a characteristic impedance of

$$\sqrt{16.7 \times 50} = 28.9\ \Omega$$

This is approximated by using a quarter-wave section of 50-Ω cable in parallel with a quarter-wave section of 75-Ω cable, yielding a net impedance of 30 Ω, quite close enough to the desired 28.9 Ω. Four Yagis fed in parallel would require a quarter-wave transformer made up using cable with a characteristic impedance of 25 Ω, easily created by using two 50-Ω cables in parallel.

T- and Gamma-Match Sections

Many types of antennas exhibit a feed-point impedance lower than the 50-Ω characteristic impedance of commonly available coax cable. Both the so-called *T-Match* and the *Gamma-Match* are used extensively on Yagi and quad beam antennas to increase the antenna feed impedance to 50 Ω.

The method of matching shown in **Fig 21.8** is based on the fact that the impedance between any two points equidistant from the center along a resonant antenna is resistive, and has a value that depends on the spacing between the two points. It is therefore possible to choose a pair of points between which the impedance will have the right value to match a transmission line. In practice, the line cannot be connected directly at these points because the distance between them is much greater than the conductor spacing of a practical transmission line. The T arrangement in Fig 21.8A overcomes this difficulty by using a second conductor paralleling the antenna to form a matching section to which the line may be connected.

The T is particularly well suited to use with parallel-conductor feed line. The

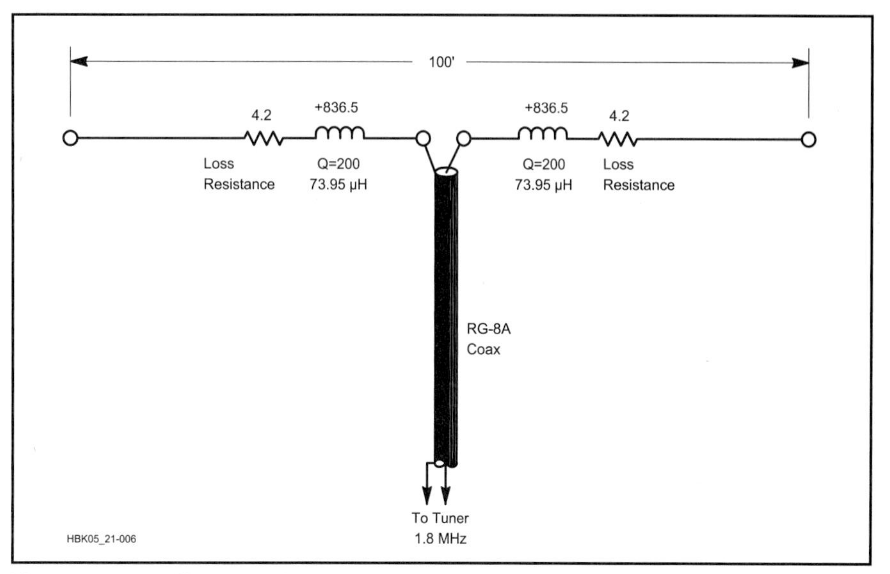

Fig 21.6—The efficiency of the dipole in Table 21.3 can be improved at 1.8 MHz with a pair of inductors inserted symmetrically at the feed point. Each inductor is assumed to have a Q of 200. By resonating the dipole in this fashion the system efficiency, when fed with RG8A coax, is almost 12 dB better than using this same antenna without the resonator. The disadvantage is that the formerly multiband antenna can only be used on a single band.

operation of this system is somewhat complex. Each T conductor (Y in the drawing) forms a short section of transmission line with the antenna conductor opposite it. Each of these transmission-line sections can be considered to be terminated in the impedance that exists at the point of connection to the antenna. Thus, the part of the antenna between the two points carries a transmission-line current in addition to the normal antenna current. The two transmission-line matching sections are in series, as seen by the main transmission line.

If the antenna by itself is resonant at the operating frequency, its impedance will be purely resistive. In this case the matching-section lines are terminated in a resistive load. As transmission-line sections, however, these matching sections are terminated in a short, and are shorter than a quarter wavelength. Thus their input impedance, the impedance seen by the main transmission line looking into the matching-section terminals, will be inductive as well as resistive. The reactive component of the input impedance must be tuned out before a proper match can be obtained.

One way to do this is to detune the antenna just enough, by shortening its length, to cause capacitive reactance to appear at the input terminals of the matching section, thus canceling the reactance introduced. Another method, which is considerably easier to adjust, is to insert a variable capacitor in series with each matching section where it connects to the transmission line, as shown in the chapter on **Antennas**. The capacitors must be protected from the weather.

When the series-capacitor method of reactance compensation is used, the antenna should be the proper length for resonance at the operating frequency. Trial positions of the matching-section taps are then taken, each time adjusting the capacitor for minimum SWR, until the lowest possible SWR has been achieved. The unbalanced (gamma) arrangement in Fig 21.8B is similar in principle to the T, but is adapted for use with single coax line. The method of adjustment is the same.

The Hairpin Match

In beam antennas such as Yagis or quads, which utilize parasitic directors and reflectors to achieve directive gain, the mutual impedance between the parasitic and the driven elements lowers the resistive component of the driven-element impedance, typically to a value between 10 and 30 Ω. If the driven element is purposely cut slightly shorter than its half-wave resonant length, it will exhibit a capacitive reactance at its feed point. A shunt inductor as shown in **Fig 21.9** placed across the feed-point center in-sulator can be used to transform the antenna resistance to match the characteristic impedance of the transmission line, while canceling out the capacitive reactance simultaneously. The antenna's capacitive reactance and the *hairpin*-shaped shunt inductor form an L network.

For mechanical convenience, the shunt inductor is often constructed using heavy-gauge aluminum wire bent in the shape of a hairpin. The center of the hairpin, the end farthest from the driven element, is grounded to the boom, since this point in a balanced feed system is equidistant from the antenna feed terminals. This gives some protection against static buildup and a certain measure of lightning protection.

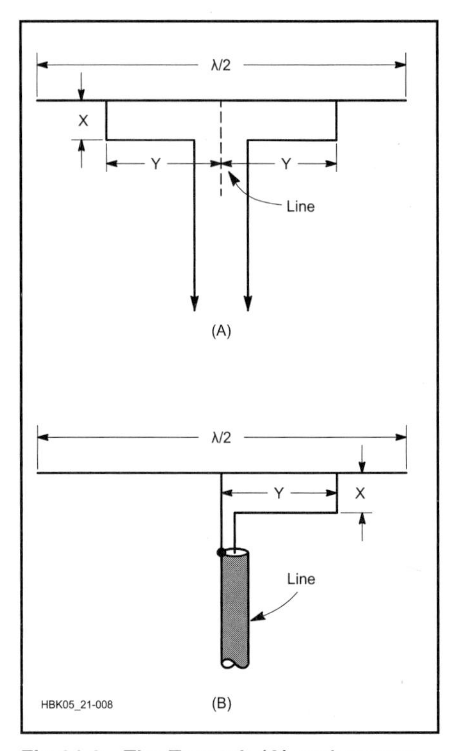

Fig 21.8—The T match (A) and gamma match (B).

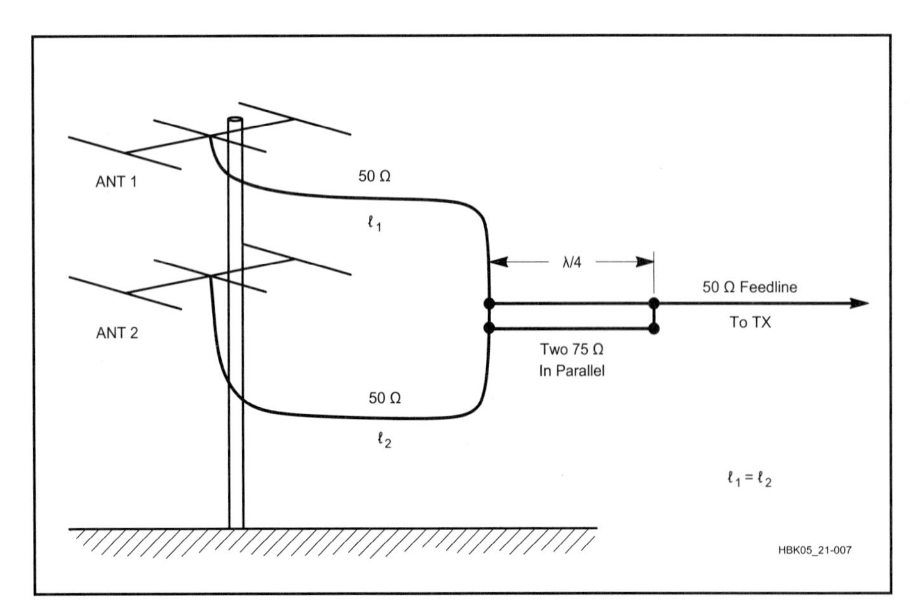

Fig 21.7—Array of two stacked Yagis, illustrating use of quarter-wave matching sections. At the junction of the two equal lengths of 50-Ω feed line the impedance is 25 Ω. This is transformed back to 50 Ω by the two paralleled 75-Ω, quarter-wave lines, which together make a net characteristic impedance of 37.5 Ω. This is close to the 35.4 Ω value computed by the formula.

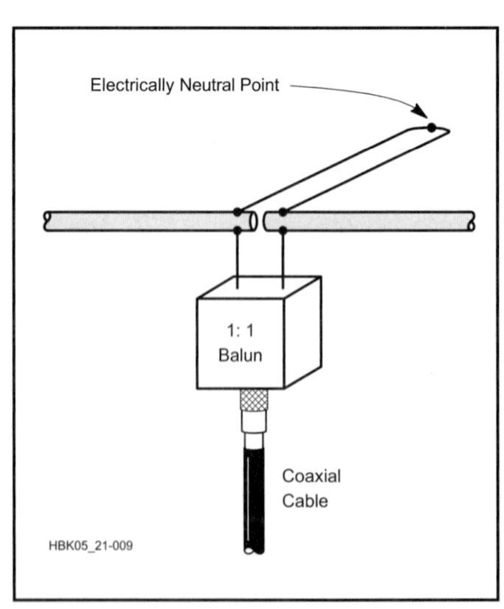

Fig 21.9—Hairpin match, sometimes called the Beta match. The "hairpin" is a shunt inductor, which together with the series capacitive reactance of an electrically short driven element forms an L network. This L network transforms the antenna resistive component to 50 Ω.

The disadvantage of the hairpin match is that it does require that the driven element be split and insulated at its center. The length of the driven element and the value of shunt inductance can be varied in the hairpin match to bring the SWR down to exactly 1:1 at a desired frequency in the band. This can also be achieved with the T or gamma matches previously described.

Matching the Line to the Transmitter

So far we have been concerned mainly with the measures needed to achieve acceptable amounts of loss and a low SWR when real coax lines are connected to real antennas. Not only is feed-line loss minimized when the SWR is kept within reasonable bounds, but also the transmitter is able to deliver its rated output power, at its rated level of distortion, when it sees the load resistance it was designed to feed.

Most modern amateur transmitters use broadband, untuned solid-state final amplifiers designed to work into a 50-Ω load. Such a transmitter very often utilizes built-in protection circuitry, which automatically reduces output power if the SWR rises to more than about 2:1. Protective circuits are needed because many solid-state devices will willingly and almost instantly destroy themselves attempting to deliver power into low-impedance loads. Solid-state devices are a lot less forgiving than vacuum tube amplifiers, which can survive momentary overloads without being destroyed instantly. Pi networks used in vacuum-tube amplifiers typically have the ability to match a surprisingly wide range of impedances on a transmission line. See the **RF Power Amplifiers** chapter in this *Handbook*.

Besides the rather limited option of using only inherently low-SWR antennas to ensure that the transmitter sees the load for which it was designed, we radio amateurs have another alternative. We can use an *antenna tuner*. The function of an antenna tuner is to transform the impedance at the input end of the transmission line, whatever it may be, to the 50 Ω needed to keep the transmitter loaded properly. Do not forget: A tuner does not alter the SWR on the transmission line going to the antenna; it only keeps the transmitter looking into the load for which it was designed. Indeed, some solid-state transmitters incorporate (usually at extra cost) automatically tuned *antenna couplers* (another name for antenna tuner), so that they too can cope with practical antennas and transmission lines that are not perfectly flat. The range of impedances that can be matched is typically rather limited, however, especially at lower frequencies.

Over the years, radio amateurs have derived a number of circuits for use as tuners. At one time, when open-wire transmission line was more widely used, link-coupled tuned circuits were in vogue. See **Fig 21.10**. With the increasing popularity of coaxial cable used as feed lines, other circuits have become more prevalent. The most common form of antenna tuner in recent years is some variation of a T configuration, as shown in **Fig 21.11A**.

The T network can be visualized as being two L networks back to front, where the common element has been conceptually broken down into two inductors in parallel. See Fig 21.11B. The L network connected to the load transforms the output impedance $R_a \pm jX_a$ into its parallel equiva-

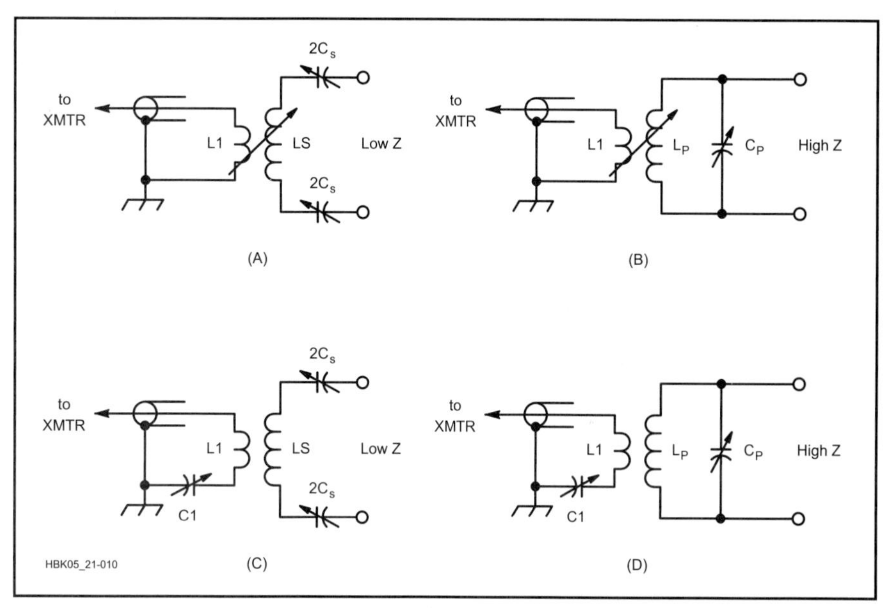

Fig 21.10—Simple antenna tuners for coupling a transmitter to a balanced line presenting a load different from the transmitter's design load impedance, usually 50 Ω. A and B, respectively, are series and parallel tuned circuits using variable inductive coupling between coils. C and D are similar but use fixed inductive coupling and a variable series capacitor, C1. A series tuned circuit works well with a low-impedance load; the parallel circuit is better with high impedance loads (several hundred ohms or more).

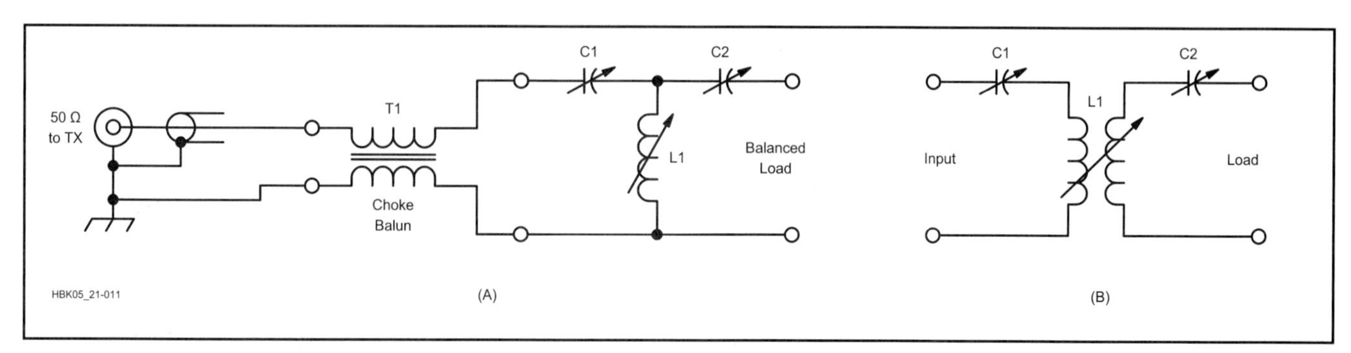

Fig 21.11—Antenna tuner network in T configuration. This network has become popular because it has the capability of matching a wide range of impedances. At A, the balun transformer at the input of the antenna tuner preserves balance when feeding a balanced transmission line. At B, the T configuration is shown as two L networks back to back.

lent by means of the series output capacitor C2. The first L network then transforms the parallel equivalent back into the series equivalent and resonates the reactance with the input series capacitor C1.

Note that the equivalent parallel resistance R_p across the shunt inductor can be a very large value for highly reactive loads, meaning that the voltage developed at this point can be very high. For example, assume that the load impedance at 3.8 MHz presented to the antenna tuner is $Z_a = 20 - j\,1000$. If C2 is 300 pF, then the equivalent parallel resistance across L1 is 66,326 Ω. If 1500 W appears across this parallel resistance, a peak voltage of 14,106 V is produced, a very substantial level indeed. Highly reactive loads can produce very high voltages across components in a tuner.

The ARRL computer program *TLA* calculates and shows graphically the antenna-tuner values for operator selected antenna impedances transformed through lengths of various types of practical transmission lines. The **Station Layout and Accessories** chapter includes antenna tuner projects, and *The ARRL Antenna Book* contains detailed information on tuner design and construction.

Myths About SWR

This is a good point to stop and mention that there are some enduring and quite misleading myths in Amateur Radio concerning SWR.

- Despite some claims to the contrary, a high SWR *does not by itself* cause RFI, or TVI or telephone interference. While it is true that an antenna located close to such devices can cause overload and interference, the SWR on the feed line to that antenna has nothing to do with it, providing of course that the tuner, feed line or connectors are not arcing. The antenna is merely doing its job, which is to radiate. The transmission line is doing its job, which is to convey power from the transmitter to the radiator.
- A second myth, often stated in the same breath as the first one above, is that a high SWR will cause excessive radiation from a transmission line. SWR has nothing to do with excessive radiation from a line. *Imbalances* in open-wire lines cause radiation, but such imbalances are not related to SWR. This subject will be covered more in the section on baluns.
- A third and perhaps even more prevalent myth is that you can't "get out" if the SWR on your transmission line is higher than 1.5:1, or 2:1 or some other such arbitrary figure. On the HF bands, if you use reasonable lengths of good coaxial cable (or even better yet, open-wire line), the truth is that you need not be overly concerned if the SWR at the load is kept below about 6:1. This sounds pretty radical to some amateurs who have heard horror story after horror story about SWR. The fact is that if you can load up your transmitter without any arcing inside, or if you use a tuner to make sure your transmitter is operating into its rated load resistance, you can enjoy a very effective station, using antennas with feed lines having high values of SWR on them. For example, a 450-Ω open-wire line connected to the multiband dipole shown in Table 21.3 would have a 19:1 SWR on it at 3.8 MHz. Yet time and again this antenna has proven to be a great performer at many installations.

Fortunately or unfortunately, SWR is one of the few antenna and transmission-line parameters easily measured by the average radio amateur. Ease of measurement does not mean that a low SWR should become an end in itself! The hours spent pruning an antenna so that the SWR is reduced from 1.5:1 down to 1.3:1 could be used in far more rewarding ways—making contacts, for example, or studying transmission-line theory.

Loads and Balancing Devices

Center-fed dipoles and loops are *balanced*, meaning that they are electrically symmetrical with respect to the feed point. A balanced antenna should be fed by a balanced feeder system to preserve this electrical symmetry with respect to ground, thereby avoiding difficulties with unbalanced currents on the line and undesirable radiation from the transmission line itself. Line radiation can be prevented by a number of devices which detune or *decouple* the line for currents radiated by the antenna back onto the line that feeds it, greatly reducing the amplitude of such so-called *antenna currents*.

Many amateurs use center-fed dipoles or Yagis, fed with unbalanced coaxial line. Some method should be used for connecting the line to the antenna without upsetting the symmetry of the antenna itself. This requires a circuit that will isolate the balanced load from the unbalanced line, while still providing efficient power transfer. Devices for doing this are

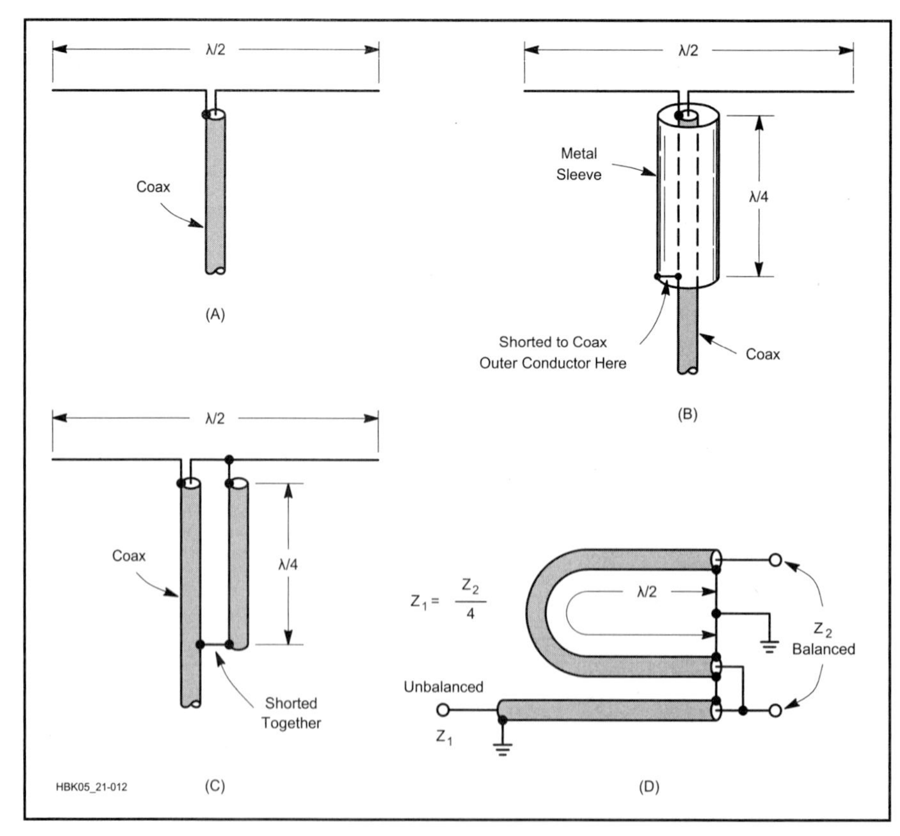

Fig 21.12—Quarter-wave baluns. Radiator with coaxial feed (A) and methods of preventing unbalanced currents from flowing on the outside of the transmission line (B and C). The ½-λ-phasing section shown at D is used for coupling to an unbalanced circuit when a 4:1 impedance ratio is desired or can be accepted.

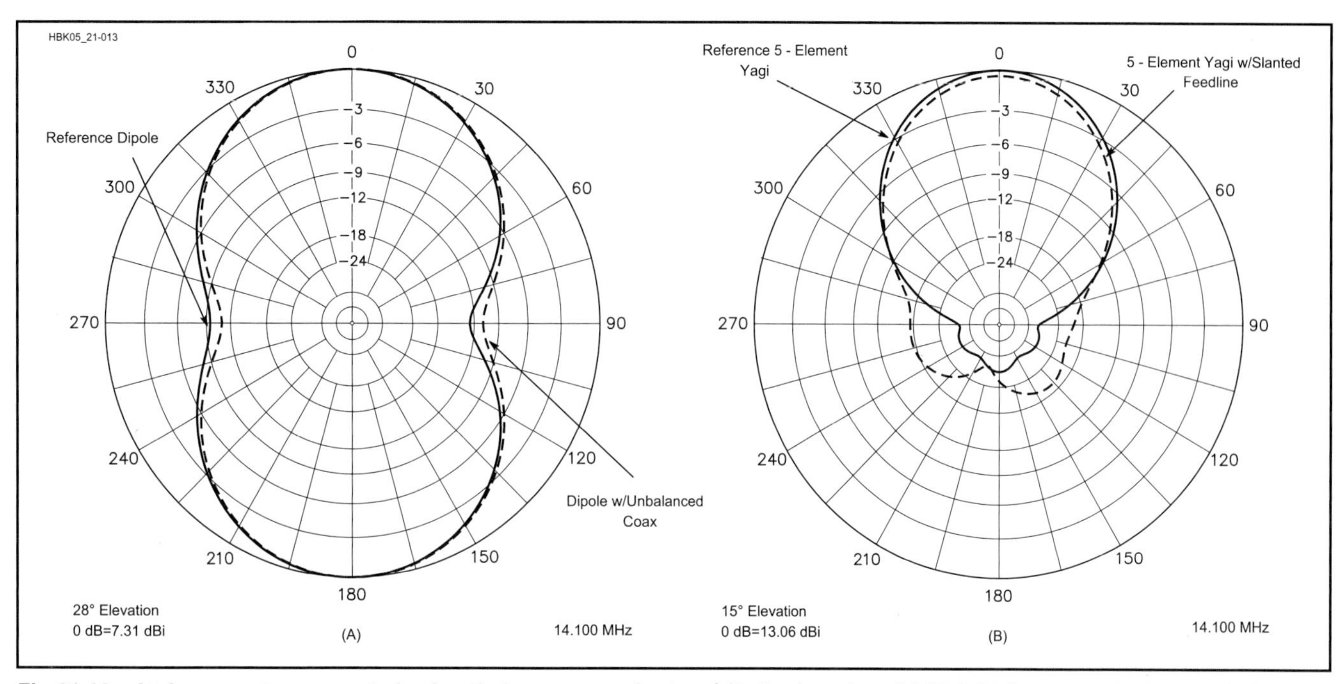

Fig 21.13—At A, computer-generated azimuthal responses for two λ/2 dipoles placed 0.71 λ high over typical ground. The solid line is for a dipole with no feed line. The dashed line is for an antenna with its feed line slanted 45° down to ground. Current induced on the outer braid of the 1-λ-long coax by its asymmetry with respect to the antenna causes the pattern distortion. At B, azimuthal response for two 5-element 20-meter Yagis placed 0.71 λ over average ground. Again, the solid line is for a Yagi without a feed line and the dashed line is for an antenna with a 45° slanted, 1-λ long feed line. The distortion in the radiated pattern is now clearly more serious than for a simple dipole. A balun is needed at the feed point, and most likely 1/4-λ way down the feed line from the feed point, to suppress the common-mode currents and restore the pattern.

called *baluns* (a contraction for "balanced to unbalanced"). A balanced antenna fed with balanced line, such as two-wire ladder line, will maintain its inherent balance, so long as external causes of unbalance are avoided. However, even they will require some sort of balun at the transmitter, since modern transmitters have unbalanced (coax) outputs.

If a balanced antenna is fed at the center through a coaxial line without a balun, as indicated in **Fig 21.12A**, the inherent symmetry and balance is upset because one side of the radiator is connected to the shield while the other is connected to the inner conductor. On the side connected to the shield, current can be diverted from flowing into the antenna, and instead can flow down over the outside of the coaxial shield. The field thus set up cannot be canceled by the field from the inner conductor because the fields inside the cable cannot escape through the shielding of the outer conductor. Hence currents flowing on the outside of the line will be responsible for some radiation from the line.

This is a good point to say that striving for perfect balance in a line and antenna system is not always absolutely mandatory. For example, if a nonresonant center fed dipole is fed with open-wire line and a tuner for multiband operation, the most desirable radiation pattern for general-purpose communication is actually an omnidirectional pattern. A certain amount of feed-line radiation might actually help fill in otherwise undesirable nulls in the azimuthal pattern of the antenna itself. Furthermore, the radiation pattern of a coaxial-fed dipole that is only a few tenths of a wavelength off the ground (50 ft high on the 80-m band, for example) is not very directional anyway, because of its severe interaction with the ground.

Purists may cry out in dismay, but there are many thousands of coaxial-fed dipoles in daily use worldwide that perform very effectively without the benefit of a balun. See **Fig 21.13A** for a worst-case comparison between a dipole with and without a balun at its feed point. This is with a 1-λ long feed line slanted downward 45° under one side of the antenna. *Common-mode currents* are radiated and conducted onto the braid of the feed line, which in turn radiates. The amount of pattern distortion is not particularly severe for a dipole. It is debatable whether the bother and expense of installing a balun for such an antenna is worthwhile.

However, some form of balun should be used to preserve the pattern of an antenna that is purposely designed to be highly directional, such as a Yagi or a quad. Fig 21.13B shows the distortion that can result from common-mode currents conducted and radiated back onto the feed line for a 5-element Yagi. This antenna has purposely been designed for an excellent pattern but the common-mode currents seriously distort the rearward pattern and reduce the forward gain as well. A balun is highly desirable in this case.

Quarter-Wave Baluns

Fig 21.12B shows a balun arrangement known as a *bazooka*, which uses a sleeve over the transmission line. The sleeve, together with the outside portion of the outer coax conductor, forms a shorted quarter-wave line section. The impedance looking into the open end of such a section is very high, so the end of the outer conductor of the coaxial line is effectively isolated from the part of the line below the sleeve. The length is an electrical quarter wave, and because of the velocity factor may be physically shorter if the insulation between the sleeve and the line is not air. The bazooka has no effect on antenna impedance at the frequency where the quarter-wave sleeve is resonant. However, the sleeve adds inductive shunt reactance

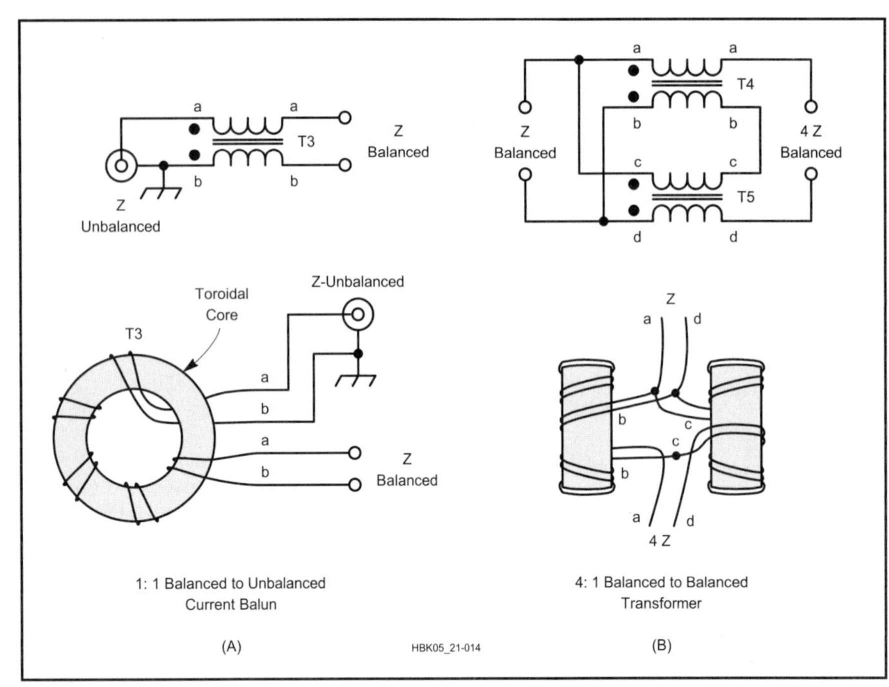

Fig 21.14—Broadband baluns. (A) 1:1 current balun and (B) 4:1 current transformer wound on two cores, which are separated. Use 12 bifilar turns of #14 enameled wire wound on FT240-43 cores for A and B. Distribute bifilar turns evenly around core.

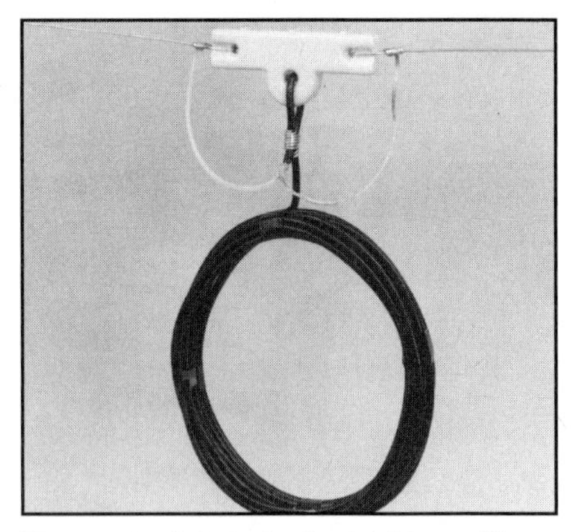

Fig 21.15—RF choke formed by coiling the feed line at the point of connection to the antenna. The inductance of the choke isolates the antenna from the remainder of the feed line.

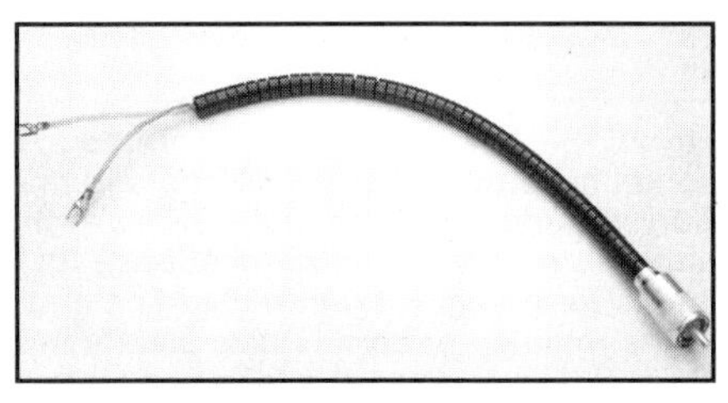

Fig 21.16—W2DU bead balun consisting of 50 FB-73-2401 ferrite beads over a length of RG-303 coax. See text for details.

Table 21.4
Effective Choke (Current Baluns)

Wind the indicated length of coaxial feed line into a coil (like a coil of rope) and secure with electrical tape. The balun is most effective when the coil is near the antenna. Lengths are not critical.

Single Band (Very Effective)

Freq, MHz	*RG-213, RG-8*	*RG-58*
3.5	22 ft, 8 turns	20 ft, 6-8 turns
7	22 ft, 10 turns	15 ft, 6 turns
10	12 ft, 10 turns	10 ft, 7 turns
14	10 ft, 4 turns	8 ft, 8 turns
21	8 ft, 6-8 turns	6 ft, 8 turns
28	6 ft, 6-8 turns	4 ft, 6-8 turns

Multiple Band

Freq, MHz	*RG-8, 58, 59, 8X, 213*
3.5-30	10 ft, 7 turns
3.5-10	18 ft, 9-10 turns
14-30	8 ft, 6-7 turns

at frequencies lower, and capacitive shunt reactance at frequencies higher than the quarter-wave-resonant frequency. The bazooka is mostly used at VHF, where its physical size does not present a major problem. On HF a quarter-wavelength rigid sleeve becomes considerably more challenging to construct, especially for a rotary antenna such as a Yagi.

Another method that gives an equivalent effect is shown at Fig 21.12C. Since the voltages at the antenna terminals are equal and opposite (with reference to ground), equal and opposite currents flow on the surfaces of the line and second conductor. Beyond the shorting point, in the direction of the transmitter, these currents combine to cancel out. The balancing section acts like an open circuit to the antenna, since it is a quarter-wave parallel-conductor line shorted at the far end, and thus has no effect on normal antenna operation. This is not essential to the line balancing function of the device, however, and baluns of this type are sometimes made shorter than a quarter wavelength to provide a shunt inductive reactance required in certain matching systems (such as the hairpin match).

Fig 21.12D shows a third balun, in which equal and opposite voltages, balanced to ground, are taken from the inner conductors of the main transmission line and half-wave phasing section. Since the voltages at the balanced end are in series while the voltages at the unbalanced end are in parallel, there is a 4:1 step-down in impedance from the balanced to the unbalanced side. This arrangement is useful for coupling between a 300-Ω balanced line and a 75-Ω unbalanced coaxial line.

Broadband Baluns

At HF and even at VHF, broadband baluns are generally used nowadays. Examples of broadband baluns are shown in **Fig 21.14**.

Choke or *current baluns* force equal and opposite currents to flow. The result is that currents radiated back onto the transmission line by the antenna are effectively reduced, or "choked off," even if the antenna is not perfectly balanced. If winding inductive reactance becomes marginal at lower frequencies, the balun's ability to eliminate antenna currents is reduced, but (for the 1:1 balun) no winding impedance appears across the line.

If radiated current on the line is a problem, perhaps because the feed line must be run in parallel with the antenna for some portion of its length, additional choke baluns can be placed at approximately $^1/_4$-λ intervals along the line. Choke baluns are particularly useful for feeding asymmetrical antennas with unbalanced coax line.

Broadband Balun Construction

Either type of broadband balun can be constructed using a variety of techniques. Construction of choke (current) baluns is described here. The objective is to obtain a high impedance for currents that tend to flow on the line. Values from a few hundred to over a thousand ohms of inductive reactance are readily achieved. These baluns work best with antennas having resonant feed-point impedances less than 100 Ω or so (400 Ω for 4:1 baluns). This is because the winding inductive reactance must be high relative to the antenna impedance for effective operation. A rule of thumb is that the inductive reactance should be four times higher than the antenna impedance. High impedances are difficult to achieve over a wide frequency range. Any sort of transformer operated at impedances for which it was not designed can fail, sometimes spectacularly.

The simplest construction method for a 1:1 balun for coaxial line is simply to wind a portion of the line into a coil. See **Fig 21.15**. This type of choke balun is simple, cheap and effective. Currents on the outside of the line encounter the coil's impedance, while currents on the inside are unaffected. A flat coil (like a coil of rope) shows a broad resonance that easily covers three octaves, making it reasonably effective over the entire HF range. If particular problems are encountered on a single band, a coil that is resonant at that band may be added. The coils shown in **Table 21.4** were constructed to have a high impedance at the indicated frequencies, as measured with an impedance meter. Many other geometries can also be effective. This construction technique is not effective with open-wire or twin-lead line because of coupling between adjacent turns. A 4:1 choke balun is shown in Fig 21.14B.

Ferrite-core baluns can provide a high impedance over the entire HF range. They may be wound either with two conductors in bifilar fashion, or with a single coaxial cable. Rod or toroidal cores may be used (Fig 21.14A). Current on the shield of a choke balun is the "antenna current" on the line—If the balun is effective, this current is small. Baluns used for high-power operation should be tested by checking for temperature rise before use. If the core overheats, add turns or use a larger or lower-loss core. It also would be wise to investigate the imbalance causing such high line antenna currents.

Type 72, 73 or 77 ferrite gives the greatest impedance over the HF range. Type 43 ferrite has lower loss, but somewhat less impedance. Core saturation is not a problem with these ferrites at HF, since they overheat due to loss at flux levels well below saturation. The loss occurs because there is insufficient inductive reactance, especially at lower frequencies. Ten to twelve turns on a toroidal core or 10 to 15 turns on a rod are typical for the HF range. Winding impedance increases approximately as the square of the number of turns.

Another type of choke balun that is very effective was originated by M. Walter Maxwell, W2DU. A number of ferrite toroids are strung, like beads on a string, directly onto the coax where it is connected to the antenna. The "bead" choke balun in **Fig 21.16** consists of 50 FB-73-2401 ferrite beads slipped over a 1-ft length of Teflon-insulation RG-303 coax. The beads fit nicely over the insulating jacket of the coax and occupy a total length of 9$^1/_2$ inches. Twelve FB-77-1024 or equivalent beads will come close to doing the same job using RG-8A or RG-213 coax. Type-73 material is recommended for 1.8 to 30 MHz use, but type-77 material may be substituted; use type-43 material for 30 to 250 MHz.

The cores present a high impedance to any RF current that would otherwise flow on the outside of the shield. The total impedance is in approximate proportion to the stacked length of the cores. The impedance stays fairly constant over a wide range of frequencies. Again, 70-series ferrites are a good choice for the HF range; use type-43 if heating is a problem or increase the number of beads. Type-43 or -61 is the best choice for the VHF range. Cores of various materials can be used in combination, permitting construction of baluns effective over a very wide frequency range, such as from 2 to 250 MHz.

WAVEGUIDES

A waveguide is a hollow conducting tube, through which microwave energy is transmitted in the form of electromagnetic waves. The tube does not carry a current in the same sense that the wires of a two-conductor line do. Instead, it is a boundary that confines the waves to the enclosed space. Skin effect on the inside walls of the waveguide confines electromagnetic energy inside the guide, in much the same manner that the shield of a coaxial cable confines energy within the coax. Microwave energy is injected at one end (either through capacitive or inductive coupling or by radiation) and is received at the other end. The waveguide merely confines the energy of the fields, which are propagated through it to the receiving end by means of reflections off its inner walls.

Evolution of a Waveguide

Suppose an open-wire line is used to carry UHF energy from a generator to a load. If the line has any appreciable length, it must be well insulated from the supports to avoid high losses. Since high-quality insulators are difficult to make for microwave frequencies, it is logical to support the transmission line with quarter-wave

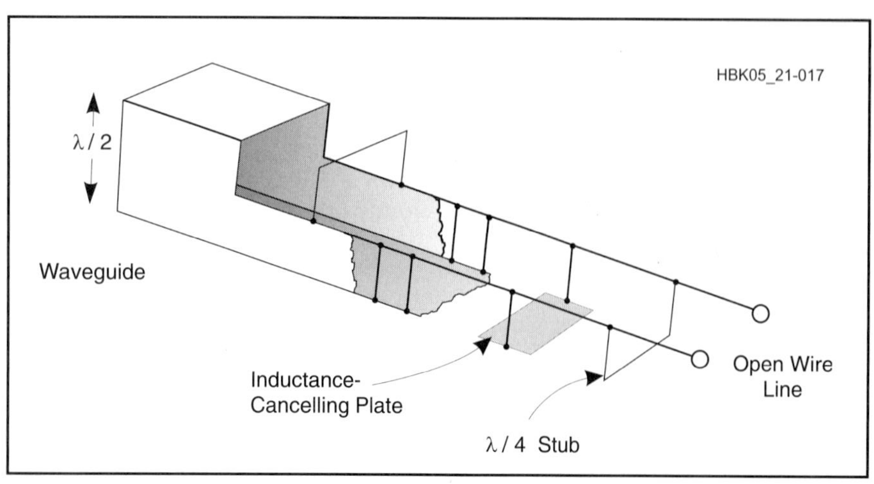

Fig 21.17—At its cutoff frequency a rectangular waveguide can be analyzed as a parallel two-conductor transmission line supported from top and bottom by an infinite number of quarter-wavelength stubs.

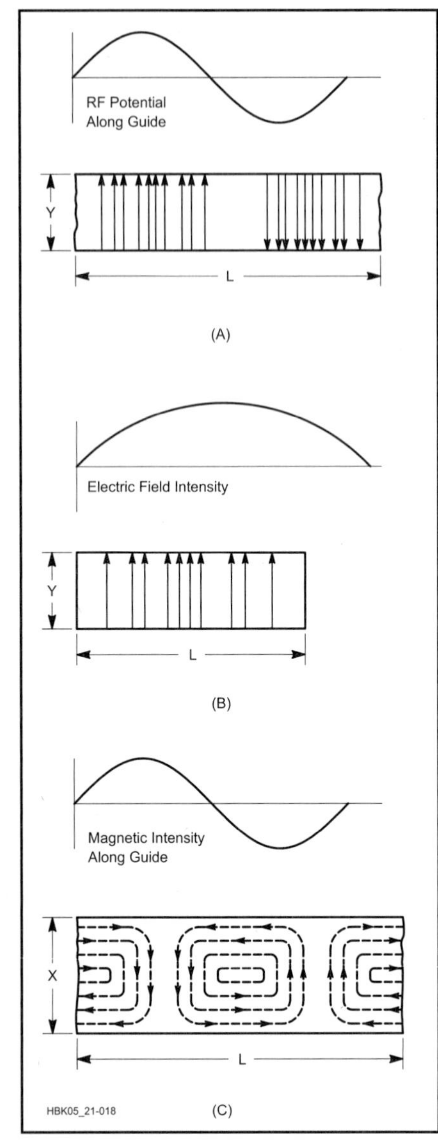

Fig 21.18—Field distribution in a rectangular waveguide. The $TE_{1,0}$ mode of propagation is depicted.

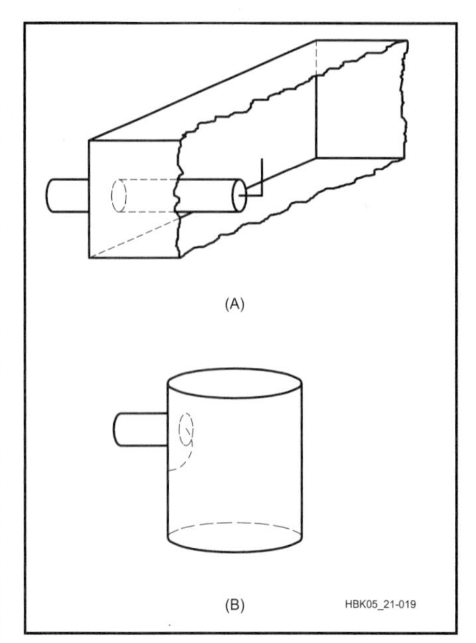

Fig 21.19—Coupling to waveguide and resonators. The probe at A is an extension of the inner conductor of coax line. At B an extension of the coax inner conductor is grounded to the waveguide to form a coupling loop.

length stubs, shorted at the far end. The open end of such a stub presents an infinite impedance to the transmission line, provided that the shorted stub is non-reactive. However, the shorting link has finite length and, therefore, some inductance. This inductance can be nullified by making the RF current flow on the surface of a plate rather than through a thin wire. If the plate is large enough, it will prevent the magnetic lines of force from encircling the RF current.

An infinite number of these quarter-wave stubs may be connected in parallel without affecting the standing waves of voltage and current. The transmission line may be supported from the top as well as the bottom, and when infinitely many supports are added, they form the walls of a waveguide at its cutoff frequency. **Fig 21.17** illustrates how a rectangular waveguide evolves from a two-wire parallel transmission line. This simplified analysis also shows why the cutoff dimension is a half wavelength.

While the operation of waveguides is usually described in terms of fields, current does flow on the inside walls, just as fields exist between the current-carrying conductors of a two-wire transmission line. At the waveguide cutoff frequency, the current is concentrated in the center of the walls, and disperses toward the floor and ceiling as the frequency increases.

Analysis of waveguide operation is based on the assumption that the guide material is a perfect conductor of electricity. Typical distributions of electric and magnetic fields in a rectangular guide are shown in **Fig 21.18**. The intensity of the electric field is greatest (as indicated by closer spacing of the lines of force in Fig 21.18B) at the center along the X dimension and diminishes to zero at the end walls. Zero field intensity is a necessary condition at the end walls, since the existence of any electric field parallel to any wall at the surface would cause an infinite current to flow in a perfect conductor, an impossible situation.

Modes of Propagation

Fig 21.18 represents a relatively simple distribution of the electric and magnetic fields. An infinite number of ways exist in which the fields can arrange themselves in a guide, as long as there is no upper limit to the frequency to be transmitted. Each field configuration is called a *mode*. All modes may be separated into two general groups. One group, designated *TM* (transverse magnetic), has the magnetic field entirely crosswise to the direction of propagation, but has a component of electric field in the propagation direction. The other type, designated *TE* (transverse electric) has the electric field entirely crosswise to the direction of propagation, but has a component of magnetic field in the direction of propagation. TM waves are sometimes called E waves, and TE waves are sometimes called H waves. The TM and TE designations are preferred, however.

The particular mode of transmission is identified by the group letters followed by subscript numbers; for example $TE_{1,1}$, $TM_{1,1}$ and so on. The number of possible modes increases with frequency for a given size of guide. There is only one possible mode (called the *dominant mode*) for the lowest frequency that can be transmitted. The dominant mode is the one normally used in practical work.

Waveguide Dimensions

In rectangular guides the critical dimension (shown as X in Fig 21.18C) must be more than one-half wavelength at the lowest frequency to be transmitted. In practice, the Y dimension is usually made about equal to ½ X to avoid the possibility of operation at other than the dominant mode. Cross-sectional shapes other than

Table 21.5
Wavelength Formulas for Waveguide

	Rectangular	*Circular*
Cut-off wavelength	2X	3.41R
Longest wavelength transmitted with little attenuation	1.6X	3.2R
Shortest wavelength before next mode becomes possible	1.1X	2.8R

rectangles can be used; the most important is the circular pipe.

Table 21.5 gives dominant-mode wavelength formulas for rectangular and circular guides. X is the width of a rectangular guide, and R is the radius of a circular guide.

Coupling to Waveguides

Energy may be introduced into or extracted from a waveguide or resonator by means of either the electric or magnetic field. The energy transfer frequently takes place through a coaxial line. Two methods for coupling are shown in **Fig 21.21.** The probe at A is simply a short extension of the inner conductor of the feed coaxial line, oriented so that it is parallel to the electric lines of force. The loop shown at B is arranged to enclose some of the magnetic lines of force. The point at which maximum coupling will be obtained depends on the particular mode of propagation in the guide or cavity; the coupling will be maximum when the coupling device is in the most intense field.

Coupling can be varied by rotating the probe or loop through 90°. When the probe is perpendicular to the electric lines the coupling will be minimum; similarly, when the plane of the loop is parallel to the magnetic lines, the coupling will be minimum. See *The ARRL Antenna Book* for more information on waveguides.

BIBLIOGRAPHY

J. Devoldere, *ON4UN's Low Band DXing*, 3rd Edition (Newington: ARRL, 1999).

W. Everitt, *Communication Engineering*, 2nd Edition (New York: McGraw-Hill, 1937).

H. Friis, and S. Schelkunoff, *Antennas: Theory and Practice* (New York: John Wiley and Sons, 1952).

M. Maxwell, *Reflections: Transmission Lines and Antennas* (Newington: ARRL, 1990) [out of print].

F. Regier, "Series-Section Transmission Line Impedance Matching," *QST*, Jul 1978, pp 14-16.

A. Roehm, W2OBJ, "Some Additional Aspects of the Balun Problem," *The ARRL Antenna Compendium Vol 2*, p 172.

W. Sabin, WØIYH, "Computer Modeling of Coax Cable Circuits," *QEX*, Aug 1996 pp 3-10.

P. Smith, "Transmission Line Calculator," *Electronics*, Jan 1939.

P. Smith, *Electronic Applications of the Smith Chart* (McGraw-Hill, 1969).

W.D. Stewart, Jr, N7WS, "Balanced Transmission Lines in Current Amateur Practice," *The ARRL Antenna Compendium Vol 6* (Newington: ARRL, 1999).

R. D. Straw, N6BV, Ed., *The ARRL Antenna Book*, 20th Edition (Newington: ARRL, 2003).

F. Witt, "Baluns in the Real (and Complex) World," *The ARRL Antenna Compendium Vol 5* (Newington: ARRL, 1996).

Chapter 22

Antennas

Every ham needs at least one antenna, and most hams have built one. This chapter, by Chuck Hutchinson, K8CH, covers theory and construction of antennas for most radio amateurs. Here you'll find simple verticals and dipoles, as well as quad and Yagi projects and other antennas that you can build and use.

ANTENNA POLARIZATION

Most HF-band antennas are either vertically or horizontally polarized. Although circular polarization is possible, just as it is at VHF and UHF, it is seldom used at HF. *Polarization* is determined by the position of the radiating element or wire with respect to the earth. Thus a radiator that is parallel to the earth radiates horizontally, while a vertical antenna radiates a vertical wave. If a wire antenna is slanted above earth, it radiates waves that have both a vertical and a horizontal component.

For best results in line-of-sight communications, antennas at both ends of the circuit should have the same polarization; cross polarization results in many decibels of signal reduction. However, it is not essential for both stations to use the same antenna polarity for ionospheric propagation (sky wave). This is because the radiated wave is bent and it tumbles considerably during its travel through the ionosphere. At the far end of the communications path the wave may be horizontal, vertical or somewhere in between at any given instant. For that reason, the main consideration for a good DX antenna is a low angle of radiation rather than the polarization.

ANTENNA BANDWIDTH

The *bandwidth* of an antenna refers generally to the range of frequencies over which the antenna can be used to obtain a specified level of performance. The bandwidth is often referenced to some SWR value, such as, "The 2:1 *SWR bandwidth* is 3.5 to 3.8 MHz." Popular amateur usage of the term bandwidth most often refers to the 2:1 SWR bandwidth. Other specific bandwidth terms are also used, such as the *gain bandwidth* and the *front-to-back ratio bandwidth*.

For the most part, the lower the operating frequency of a given antenna design, the narrower is the bandwidth. This follows the rule that the bandwidth of a resonant circuit doubles as the frequency of operation is doubled, assuming the Q is the same for each case. Therefore, it is often difficult to cover all of the 160 or 80-m band for a particular level of SWR with a dipole antenna. It is important to recognize that SWR bandwidth does not always relate directly to gain bandwidth. Depending on the amount of feed-line loss, an 80-m dipole with a relatively narrow 2:1 SWR bandwidth can still radiate a good signal at each end of the band, provided that an antenna tuner is used to allow the transmitter to load properly. Broadbanding techniques, such as fanning the far ends of a dipole to simulate a conical type of dipole, can help broaden the SWR response curve.

CURRENT AND VOLTAGE DISTRIBUTION

When power is fed to an antenna, the current and voltage vary along its length. The current is nearly zero (a current *node*) at the ends. The current does not actually reach zero at the current nodes, because of capacitance at the antenna ends. Insulators, loops at the antenna ends, and support wires all contribute to this capacitance, which is also called the *end effect*. In the case of a half-wave antenna there is a current maximum (a current *loop*) at the center.

The opposite is true of the RF voltage. That is, there is a voltage loop at the ends, and in the case of a half-wave antenna there is a voltage minimum (node) at the center. The voltage is not zero at its node because of the resistance of the antenna, which consists of both the RF resistance of the wire (ohmic loss resistance) and the *radiation resistance*. The radiation resistance is the equivalent resistance that would dissipate the power the antenna radiates, with a current flowing in it equal to the antenna current at a current loop (maximum). The loss resistance of a half-wave antenna is ordinarily small, compared with the radiation resistance, and can usually be neglected for practical purposes.

IMPEDANCE

The *impedance* at a given point in the antenna is determined by the ratio of the voltage to the current at that point. For example, if there were 100 V and 1.4 A of RF current at a specified point in an antenna and if they were in phase, the impedance would be approximately 71 Ω.

Antenna impedance may be either resistive or complex (that is, containing resistance and reactance). This will depend on whether or not the antenna is *resonant* at the operating frequency. You need to know the impedance in order to match the feeder to the feedpoint. Some operators mistakenly believe that a mismatch, however small, is a serious matter. This is not true. The importance of a matched line is described in detail in the **Transmission Lines** chapter of this book. The significance of a perfect match becomes more pronounced only at VHF and higher, where feed-line losses are a major factor.

Some antennas possess a theoretical input impedance at the feedpoint close to that of certain transmission lines. For example, a 0.5-λ (or half-wave) center-fed dipole,

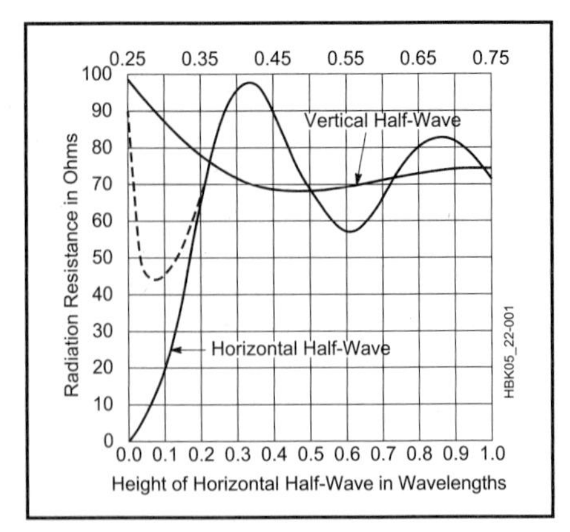

Fig 22.1 — Curves showing the radiation resistance of vertical and horizontal half-wavelength dipoles at various heights above ground. The broken-line portion of the curve for a horizontal dipole shows the resistance over *average* real earth, the solid line for perfectly conducting ground.

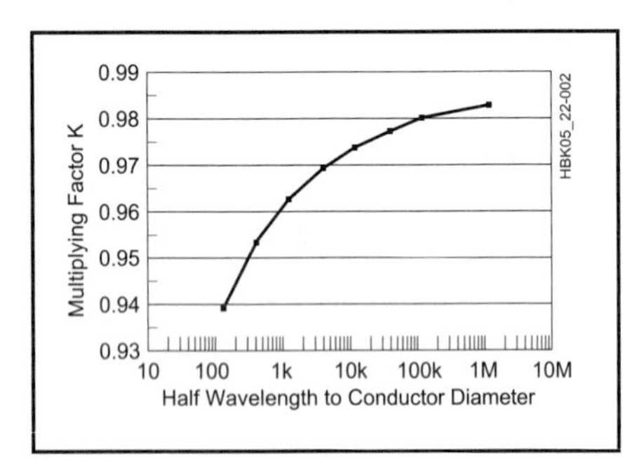

Fig 22.2 — Effect of antenna diameter on length for half-wavelength resonance, shown as a multiplying factor, K, to be applied to the free-space, half-wavelength equation.

Table 22.1
Optimum Elevation Angles to Europe

Band	*Northeast*	*Southeast*	*Upper Midwest*	*Lower Midwest*	*West Coast*
10 m	5°	3°	3°	7°	3°
12 m	5°	6°	4°	6°	5°
15 m	5°	7°	8°	5°	6°
17 m	4°	8°	7°	5°	5°
20 m	11°	9°	8°	5°	6°
30 m	11°	11°	11°	9°	8°
40 m	15°	15°	14°	14°	12°
75 m	20°	15°	15°	11°	11°

Table 22.2
Optimum Elevation Angles to Far East

Band	*Northeast*	*Southeast*	*Upper Midwest*	*Lower Midwest*	*West Coast*
10 m	4°	5°	5°	5°	6°
12 m	4°	8°	5°	12°	6°
15 m	7°	10°	10°	10°	8°
17 m	7°	10°	9°	10°	5°
20 m	4°	10°	9°	10°	9°
30 m	7°	13°	11°	12°	9°
40 m	11°	12°	12°	12°	13°
75 m	12°	14°	14°	12°	15°

Table 22.3
Optimum Elevation Angles to South America

Band	*Northeast*	*Southeast*	*Upper Midwest*	*Lower Midwest*	*West Coast*
10 m	5°	4°	4°	4°	7°
12 m	5°	5°	6°	3°	8°
15 m	5°	5°	7°	4°	8°
17 m	4°	5°	5°	3°	7°
20 m	8°	8°	8°	6°	8°
30 m	8°	11°	9°	9°	9°
40 m	10°	11°	9°	9°	10°
75 m	15°	15°	13°	14°	14°

placed at a correct height above ground, will have a feedpoint impedance of approximately 75 Ω. In such a case it is practical to use a 75-Ω coaxial or balanced line to feed the antenna. But few amateur half-wave dipoles actually exhibit a 75-Ω impedance. This is because at the lower end of the high-frequency spectrum the typical height above ground is rarely more than $^1/_4$ λ. The 75-Ω feed-point impedance is most likely to be realized in a practical installation when the horizontal dipole is approximately $^1/_2$, $^3/_4$ or 1 wavelength above ground. Coax cable having a 50-Ω characteristic impedance is the most common transmission line used in amateur work.

Fig 22.1 shows the difference between the effects of perfect ground and typical earth at low antenna heights. The effect of height on the radiation resistance of a horizontal half-wave antenna is not drastic so long as the height of the antenna is greater than 0.2 λ. Below this height, while decreasing rapidly to zero over perfectly conducting ground, the resistance decreases less rapidly with height over actual ground. At lower heights the resistance stops decreasing at around 0.15 λ, and thereafter increases as height decreases further. The reason for the increasing resistance is that more and more of the induction field of the antenna is absorbed by the earth as the height drops below $^1/_4$ λ.

CONDUCTOR SIZE

The impedance of the antenna also depends on the diameter of the conductor in relation to the wavelength, as indicated in **Fig 22.2**. If the diameter of the conductor is increased, the capacitance per unit length increases and the inductance per unit length decreases. Since the radiation resistance is affected relatively little, the decreased L/C ratio causes the Q of the antenna to decrease so that the resonance curve becomes less sharp with change in frequency. This effect is greater as the diameter is increased, and is a property of some importance at the very high frequencies where the wavelength is small.

DIRECTIVITY AND GAIN

All antennas, even the simplest types, exhibit directive effects in that the intensity of radiation is not the same in all directions from the antenna. This property of radiating more strongly in some directions than in others is called the *directivity* of the antenna.

The *gain* of an antenna is closely related to its directivity. Because directivity is based solely on the shape of the directive pattern, it does not take into account any power losses that may occur in an actual antenna system. Gain takes those losses into account.

Gain is usually expressed in decibels, and is based on a comparison with a *standard* antenna—usually a dipole or an *isotropic radiator*. An isotropic radiator is a theoretical antenna that would, if placed in the center of an imaginary sphere, evenly illuminate that sphere with radiation. The isotropic radiator is an unambiguous standard, and for that reason frequently used as the comparison for gain measurements. When the standard is the isotropic radiator in free space, gain is expressed in dBi. When the standard is a dipole, *also located in free space*, gain is expressed in dBd.

The more the directive pattern is compressed—or focused—the greater the power gain of the antenna. This is a result of power being concentrated in some directions at the expense of others. The directive pattern, and therefore the gain, of an antenna at a given frequency is determined by the size and shape of the antenna, and on its position and orientation relative to the Earth.

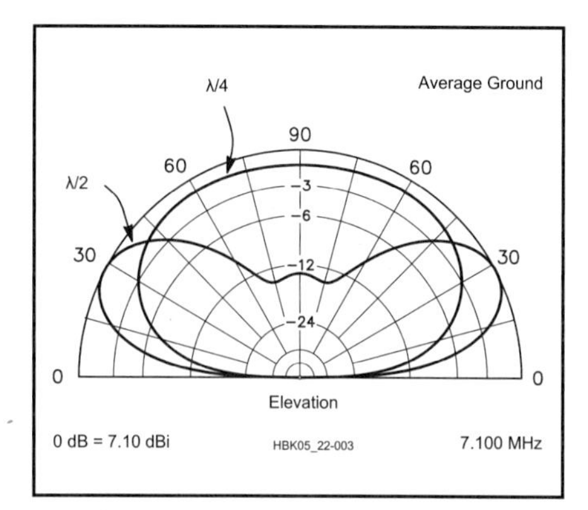

Fig 22.3 — Elevation patterns for two 40-m dipoles over average ground (conductivity of 5 mS/m and dielectric constant of 13) at 1/4 λ (33 ft) and 1/2 λ (66 ft) heights. The higher dipole has a peak gain of 7.1 dBi at an elevation angle of about 26°, while the lower dipole has more response at high elevation angles.

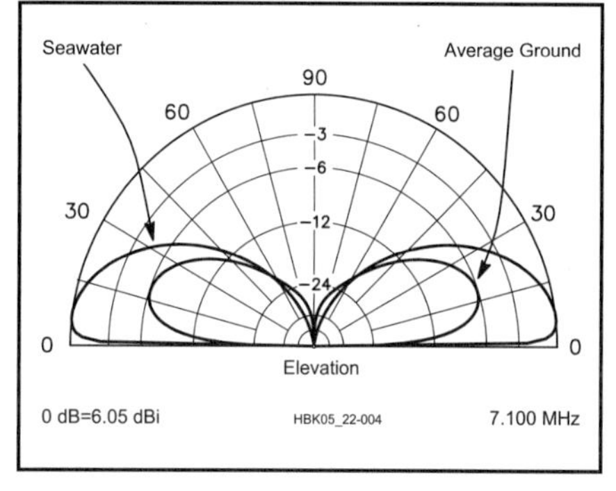

Fig 22.4 — Elevation patterns for a vertical dipole over sea water compared to average ground. In each case the center of the dipole is just over 1/4 λ high. The low-angle response is greatly degraded over average ground compared to sea water, which is virtually a perfect ground.

ELEVATION ANGLE

For long-distance HF communication, the (vertical) *elevation angle* of maximum radiation is of considerable importance. You will want to erect your antenna so that it radiates at desirable angles. **Tables 22.1**, **22.2** and **22.3** show optimum elevation angles from locations in the continental US. These figures are based on statistical averages over all portions of the solar sunspot cycle.

Since low angles usually are most effective for long distance communications, this generally means that horizontal antennas should be high—higher is usually better. Experience shows that satisfactory results can be attained on the bands above 14 MHz with antenna heights between 40 and 70 ft. **Fig 22.3** shows this effect at work in horizontal dipole antennas.

The higher angles can be useful for medium to short-range communications. Dean Straw, N6BV, illustrates this in *The ARRL Antenna Book*. Straw shows that elevation angles between 20 and 65° are useful on the 40 and 80-m bands over the roughly 550-mile path between Cleveland and Boston. Even higher angles may be useful on shorter paths when using these lower HF frequencies. A 75-m dipole between 30 and 70 ft high works well for ranges out to several hundred miles. See the **Propagation of RF Signals** chapter.

IMPERFECT GROUND

Earth conducts, but is far from being a perfect conductor. This influences the radiation pattern of the antennas that we use. The effect is most pronounced at high vertical angles (the ones that we're least interested in for long-distance communications) for horizontal antennas. The consequences for vertical antennas are greatest at low angles, and are quite dramatic as can be clearly seen in **Fig 22.4**, where the elevation pattern for a 40-m vertical half-wave dipole located over average ground is compared to one located over saltwater. At 10° elevation, the saltwater antenna has about 7 dB more gain than its landlocked counterpart.

An HF vertical antenna may work very well for a ham living in the area between Dallas, Texas and Lincoln, Nebraska. This area is pastoral, has low hills, and rich soil. Ground of this type has very good conductivity. By contrast, a ham living in New Hampshire, where the soil is rocky and a poor conductor, may not be satisfied with the performance of a vertical HF antenna.

Dipoles and the Half-Wave Antenna

A fundamental form of antenna is a wire whose length is half the transmitting wavelength. It is the unit from which many more complex forms of antennas are constructed and is known as a *dipole antenna*. The length of a half-wave in free space is

$$\text{Length (ft)} = \frac{492}{\text{f (MHz)}} \quad (1)$$

The actual length of a resonant $^1/_2$-λ antenna will not be exactly equal to the half wavelength in space, but depends on the thickness of the conductor in relation to the wavelength. The relationship is shown in Fig 22.2, where K is a factor that must be multiplied by the half wavelength in free space to obtain the resonant antenna length. An additional shortening effect occurs with wire antennas supported by insulators at the ends because of the capacitance added to the system by the insulators. This shortening is called end effect. The following formula is sufficiently accurate for wire antennas for frequencies up to 30 MHz.

Length of half-wave antenna (ft)

$$= \frac{492 \times 0.95}{\text{f (MHz)}} = \frac{468}{\text{f (MHz)}} \quad (2)$$

Example: A half-wave antenna for 7150 kHz (7.15 MHz) is 468/7.15 = 65.45 ft, or 65 ft 5 inches.

Above 30 MHz use the following formulas, particularly for antennas constructed from rod or tubing. K is taken from Fig 22.2.

$$\text{Length (ft)} = \frac{492 \times K}{\text{f (MHz)}} \quad (3)$$

$$\text{Length (in)} = \frac{5904 \times K}{\text{f (MHz)}} \quad (4)$$

Example: Find the length of a half-wave antenna at 50.1 MHz, if the antenna is made of $^1/_2$-inch-diameter tubing. At 50.1 MHz, a half wavelength in space is

$$\frac{492}{50.1} = 9.82 \text{ ft}$$

The ratio of half wavelength to conductor diameter (changing wavelength to inches) is

$$\frac{(9.82 \text{ ft} \times 12 \text{ in/ft})}{0.5 \text{ in}} = 235.7$$

From Fig 22.2, K = 0.945 for this ratio. The length of the antenna, from equation 3 is

$$\frac{492 \times 0.945}{50.1} = 9.28 \text{ ft}$$

or 9 ft $3^3/_8$ inches. The answer is obtained directly in inches by substitution in equation 4

$$\frac{5904 \times 0.945}{50.1} = 111.4 \text{ in}$$

The length of a half-wave antenna is also affected by the proximity of the dipole ends to nearby conductive and semiconductive objects. In practice, it is often necessary to do some experimental *pruning* of the wire after cutting the antenna to the computed length, lengthening or shortening it in increments to obtain a low SWR. When the lowest SWR is obtained for the desired part of an amateur band, the antenna is resonant at that frequency. The value of the SWR indicates the quality of the match between the antenna and the feed line. If the lowest SWR obtainable is too high for use with solid-state rigs, a Transmatch or line-input matching network may be used, as described in the **Transmission Lines** chapter.

RADIATION CHARACTERISTICS

The radiation pattern of a dipole antenna in free space is strongest at right angles to the wire (**Fig 22.5**). This figure-8 pattern appears in the real world if the dipole is $^1/_2$ λ or greater above earth and is not degraded by nearby conductive objects. This assumption is based also on a symmetrical feed system. In practice, a coaxial feed line may distort this pattern slightly, as shown in Fig 22.5. Minimum horizontal radiation occurs off the ends of the dipole if the antenna is parallel to the earth.

As a horizontal antenna is brought closer to ground, the elevation pattern peaks at a higher elevation angle as shown in Fig 22.3. **Fig 22.6** illustrates what hap-

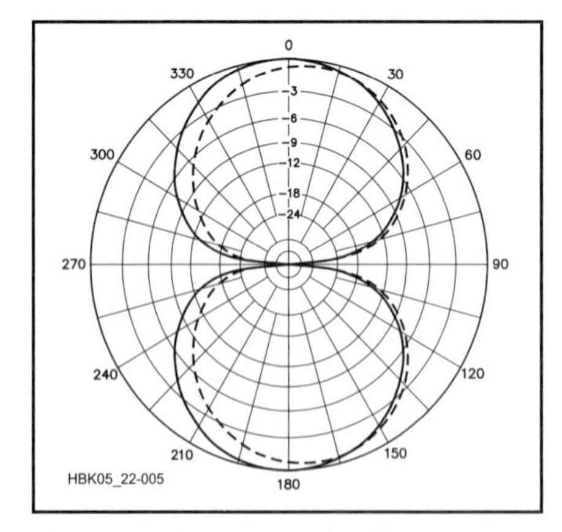

Fig 22.5 — Response of a dipole antenna in free space, where the conductor is along 90° to 270° axis, solid line. If the currents in the halves of the dipole are not in phase, slight distortion of the pattern will occur, broken line. This illustrates the case where a balun is not used on a balanced antenna fed with unbalanced line.

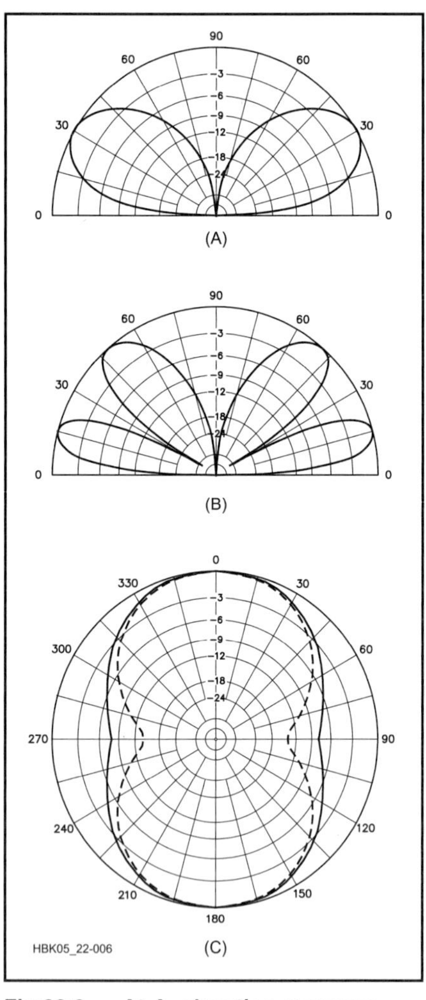

Fig 22.6 — At A, elevation response pattern of a dipole antenna placed $^1/_2$ λ above a perfectly conducting ground. At B, the pattern for the same antenna when raised to one wavelength. For both A and B, the conductor is coming out of the paper at right angle. C shows the azimuth patterns of the dipole for the two heights at the most-favored elevation angle, the solid-line plot for the $^1/_2$-λ height at an elevation angle of 30°, and the broken-line plot for the 1-λ height at an elevation angle of 15°. The conductor in C lies along 90° to 270° axis.

pens to the directional pattern as antenna height changes. Fig 22.6C shows that there is significant radiation off the ends of a low horizontal dipole. For the $1/2$-λ height (solid line), the radiation off the ends is only 7.6 dB lower than that in the broadside direction.

FEED METHODS

Most amateurs use either *coax* or *open-wire* transmission line. Coax is the common choice because it is readily available, its characteristic impedance is close to that of the antenna and it may be easily routed through or along walls and among other cables. The disadvantages of coax are increased RF loss and low working voltage (compared to that of open-wire line). Both disadvantages make coax a poor choice for high-SWR systems.

Take care when choosing coax. Use $1/4$-inch foam-dielectric cables only for low power (25 W or less) HF transmissions. Solid-dielectric $1/4$-inch cables are okay for 300 W if the SWR is low. For high-power installations, use $1/4$-inch or larger cables.

The most common two-wire transmission lines are *ladder line* and *twin lead*. Since the conductors are not shielded, two-wire lines are affected by their environment. Use standoffs and insulators to keep the line several inches from structures or other conductors. Ladder line has very low loss (twin lead has a little more), and it can stand very high voltages (high SWR) as long as the insulators are clean.

Two-wire lines are usually used in balanced systems, so they should have a balun at the transition to an unbalanced transmitter or coax. A Transmatch will be needed to match the line input impedance to the transmitter.

BALUNS

A balun is a device for feeding a balanced load with an unbalanced line, or vice versa (see the Transmission Lines chapter of this book). Because dipoles are balanced (electrically symmetrical about their feedpoints), a balun is often used at the feedpoint when a dipole is fed with coax. When coax feeds a dipole directly (as in **Fig 22.7**), current flows on the outside of the cable shield. The shield can conduct RF onto the transmitter chassis and induce RF onto metal objects near the system. Shield currents can impair the function of instruments connected to the line (such as SWR meters and SWR-protection circuits in the transmitter). The shield current also produces some feed-line radiation, which changes the antenna radiation pattern, and allows objects near the cable to affect the antenna-system performance.

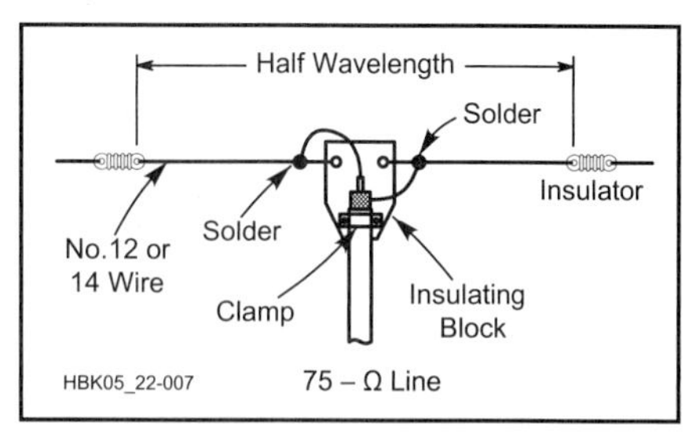

Fig 22.7 — Method of affixing feed line to the center of a dipole antenna. A plastic block is used as a center insulator. The coax is held in place by a clamp. A balun is often used to feed dipoles or other balanced antennas to ensure that the radiation pattern is not distorted. See text for explanation.

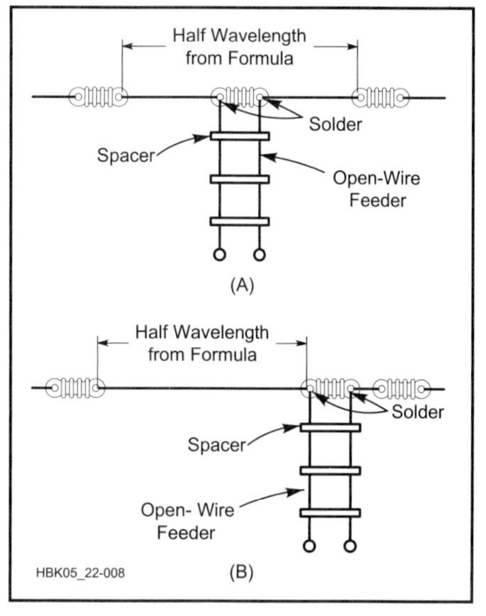

Fig 22.8 — Center-fed multiband *Zepp* antenna at A and an end-fed Zepp at B. See also Fig 22.11 for connection details.

The consequences may be negligible: A slight skewing of the antenna pattern usually goes unnoticed. Or, they may be significant: False SWR readings may cause the transmitter to shut down or destroy the output transistors; radiating coax near a TV feed line may cause strong local interference. Therefore, it is better to eliminate feed-line radiation whenever possible, and a balun should be used at any transition between balanced and unbalanced systems. (The **Transmission Lines** chapter thoroughly describes baluns and their construction.) Even so, balanced or unbalanced systems without a balun often operate with no apparent problems. For temporary or emergency stations, do not let the lack of a balun deter you from operating.

PRACTICAL DIPOLE ANTENNAS

A classic dipole antenna is $1/2$-λ long and fed at the center. The feed-point impedance is low at the resonant frequency, f_0, and odd harmonics thereof. (The impedance is high near even harmonics.) When fed with coax, a classic dipole provides a reasonably low SWR at f_0 and its odd harmonics.

When fed with ladder line (see **Fig 22.8A**) and a Transmatch, the classic dipole should be usable near f_0 and all harmonic frequencies. (With a wide-range Transmatch, it may work on all frequencies.) If there are problems (such as extremely high SWR or evidence of RF on objects at the operating position), change the feed-line length by adding or subtracting $1/8$ λ at the problem frequency. A few such adjustments should yield a workable solution. Such a system is sometimes called a *center-fed Zepp*. A true *Zepp* antenna is an end-fed dipole that is matched by $1/4$ λ of open-wire feed line (see Fig 22.8B). The antenna was originally used on zeppelins, with the dipole trailing from the feeder, which hung from the airship cabin. It is intended for use on a single band, but should be usable near odd harmonics of f_0.

Most dipoles require a little pruning to reach the desired resonant frequency. So, cut the wire 2 to 3% longer than the calculated length and record the length. Next, raise the dipole to the working height and check the SWR at several frequencies. Multiply the frequency of the SWR minimum by the antenna length and divide the result by the desired f_0. The result is the finished length; trim both ends equally to reach that length and you're done.

BUILDING DIPOLE AND OTHER WIRE ANTENNAS

The purpose of this section is to offer information on the actual physical construction of wire antennas. Because the dipole, in one of its configurations, is probably the most common amateur wire antenna, it is used in the following examples. The techniques described here, however, enhance the reliability and safety of all wire antennas.

Wire

Choosing the right type of wire for the project at hand is the key to a successful antenna—the kind that works well and stays up through a winter ice storm or a gusty spring wind storm. What gauge of wire to use is the first question to settle, and the answer depends on strength, ease of handling, cost, availability and visibility. Generally, antennas that are expected to support their own weight, plus the weight of the feed line should be made from #12 wire. Horizontal dipoles, Zepps, some long wires and the like fall into this category. Antennas supported in the center, such as inverted-V dipoles and delta loops, may be made from lighter material, such as #14 wire—the minimum size called for in the National Electrical Code.

The type of wire to be used is the next important decision. The wire specifications table in the **Component Data and References** chapter shows popular wire styles and sizes. The strongest wire suitable for antenna service is *copperclad steel*, also known as *copperweld*. The copper coating is necessary for RF service because steel is a relatively poor conductor. Practically all of the RF current is confined to the copper coating because of *skin effect*. Copper-clad steel is outstanding for permanent installations, but it can be difficult to work with. Kinking, which severely weakens the wire, is a potential problem when handling any solid conductor. Solid-copper wire, either hard drawn or soft drawn, is another popular material. Easier to handle than copper-clad steel, solid copper is available in a wide range of sizes. It is generally more expensive however, because it is all copper. Soft drawn tends to stretch under tension, so periodic pruning of the antenna may be necessary in some cases. Enamel-coated *magnet-wire* is a suitable choice for experimental antennas because it is easy to manage, and the coating protects the wire from the weather. Although it stretches under tension, the wire may be prestretched before final installation and adjustment. A local electric motor rebuilder might be a good source for magnet wire.

Hook-up wire, speaker wire or even ac lamp cord are suitable for temporary installations. Almost any copper wire may be used, as long as it is strong enough for the demands of the installation. Steel wire is a poor conductor at RF; avoid it.

It matters not (in the HF region at least) whether the wire chosen is insulated or bare. If insulated wire is used, a 3 to 5% shortening beyond the standard 468/f length will be required to obtain resonance at the desired frequency. This is caused by the increased distributed capacitance resulting from the dielectric constant of the plastic insulating material. The actual length for resonance must be determined experimentally by pruning and measuring because the dielectric constant of the insulating material varies from wire to wire. Wires that might come into contact with humans or animals should be insulated to reduce the chance of shock or burns.

Insulators

Wire antennas must be insulated at the ends. Commercially available insulators are made from ceramic, glass or plastic. Insulators are available from many Amateur Radio dealers. RadioShack and local hardware stores are other possible sources.

Acceptable homemade insulators may be fashioned from a variety of material including (but not limited to) acrylic sheet or rod, PVC tubing, wood, fiberglass rod or even stiff plastic from a discarded container. **Fig 22.9** shows some homemade insulators. Ceramic or glass insulators will usually outlast the wire, so they are highly recommended for a safe, reliable, permanent installation. Other materials may tear under stress or break down in the presence of sunlight. Many types of plastic do not weather well.

Many wire antennas require an insulator at the feedpoint. Although there are many ways to connect the feed line, there are a few things to keep in mind. If you feed your antenna with coaxial cable, you have two choices. You can install an SO-239 connector on the center insulator and use a PL-259 on the end of your coax, or you can separate the center conductor from the braid and connect the feed line directly to the antenna wire. Although it costs less to connect direct, the use of connectors offers several advantages.

Coaxial cable braid soaks up water. If you do not adequately seal the antenna end of the feed line, water will find its way into the braid. Water in the feed line will lead to contamination, rendering the coax useless long before its normal lifetime is up. It is not uncommon for water to drip from the end of the coax inside the shack after a year or so of service if the antenna connection is not properly waterproofed. Use of a PL-259/SO-239 combination (or other connector of your choice) makes the task of waterproofing connections much easier. Another advantage to using the PL-259/SO-239 combination is that feed line replacement is much easier, should that become necessary or desireable.

Whether you use coaxial cable, ladder line, or twin lead to feed your antenna, an often-overlooked consideration is the mechanical strength of the connection.

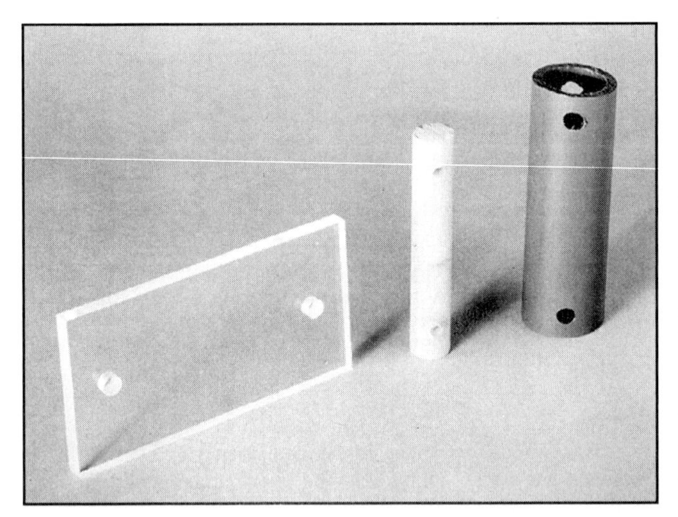

Fig 22.9 — Some ideas for homemade antenna insulators.

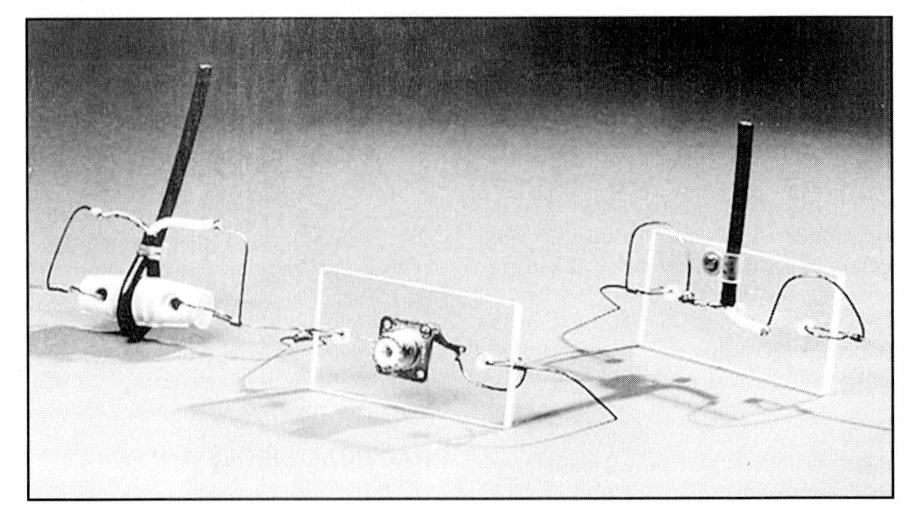

Fig 22.10 — Some homemade dipole center insulators. The one in the center includes a built-in SO-239 connector. Others are designed for direct connection to the feed line. (See the Transmission Lines chapter for details on baluns.)

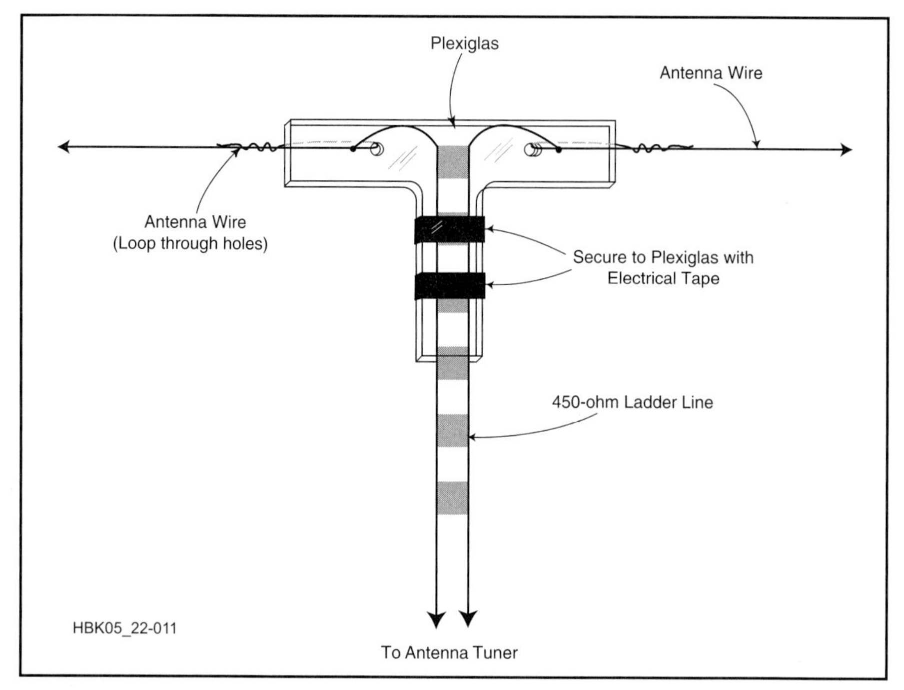

Fig 22.11 — A piece of cut Plexiglas can be used as a center insulator and to support a ladder-line feeder. The Plexiglas acts to reduce the flexing of the wires where they connect to the antenna.

Table 22.4
Dipole Dimensions for Amateur Bands

Freq MHz	*Overall Length*	*Leg Length*
28.4	16' 6"	8' 3"
24.9	18' 9 1/2"	9' 4 3/4"
21.1	22' 2"	11' 1"
18.1	25' 10"	12' 11"
14.1	33' 2"	16' 7"
10.1	46' 4"	23' 2"
7.1	65' 10"	32' 11"
5.37	87' 2"	43' 7"
3.6	130' 0"	65' 0"

Wire antennas and feed lines tend to move a lot in the breeze, and unless the feed line is attached securely, the connection will weaken with time. The resulting failure can range from a frustrating intermittent electrical connection to a complete separation of feed line and antenna. **Fig 22.10** illustrates several different ways of attaching the feed line to the antenna. An idea for supporting ladder line is shown in **Fig 22.11**.

Putting It Together

Fig 22.12 shows details of antenna construction. Although a dipole is used for the examples, the techniques illustrated here apply to any type of wire antenna. **Table 22.4** shows dipole lengths for the amateur HF bands.

How well you put the pieces together is second only to the ultimate strength of the materials used in determining how well your antenna will work over the long term. Even the smallest details, such as how you connect the wire to the insulators (Fig 22.12A), contribute significantly to antenna longevity. By using plenty of wire at the insulator and wrapping it tightly, you will decrease the possibility of the wire pulling loose in the wind. There is no need to solder the wire once it is wrapped. There is no electrical connection here, only mechanical. The high

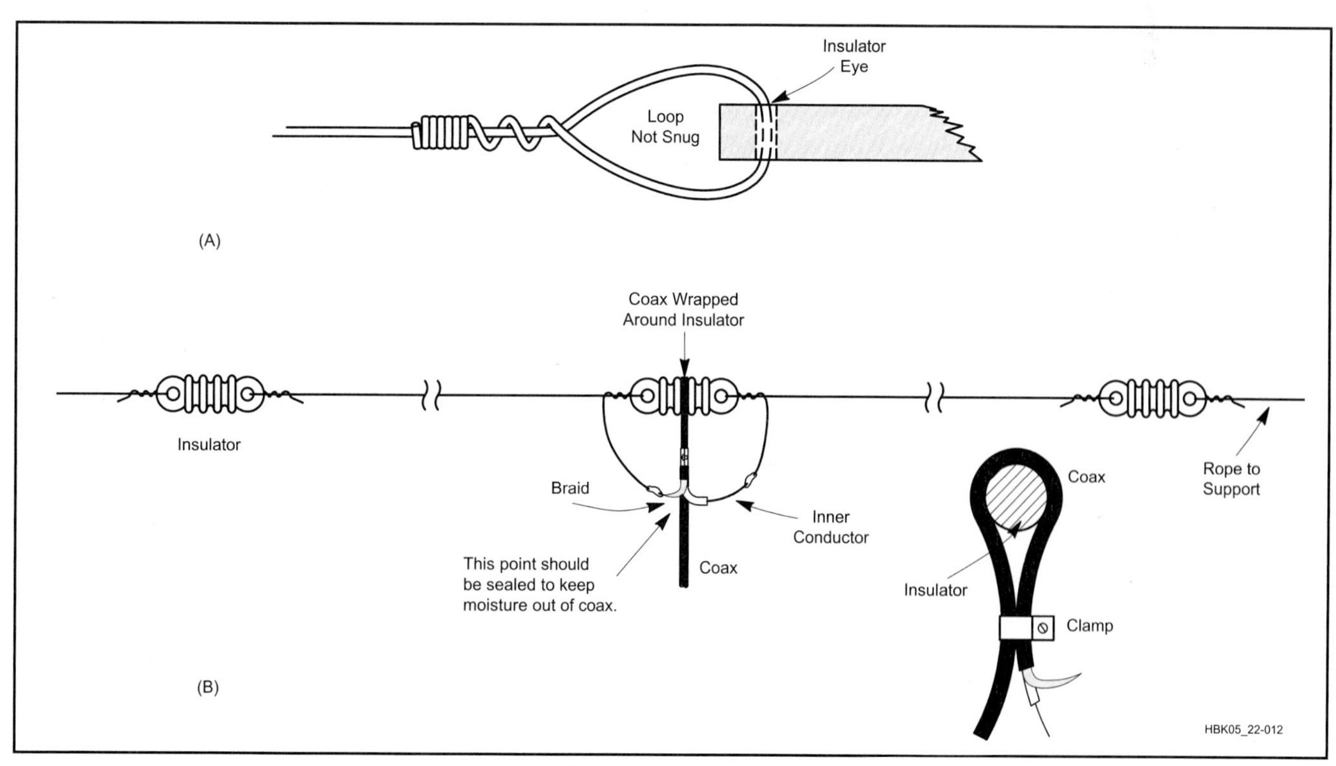

Fig 22.12 — Details of dipole antenna construction. The end insulator connection is shown at A, while B illustrates the completed antenna. This is a balanced antenna and is often fed with a balun.

heat needed for soldering can anneal the wire, significantly weakening it at the solder point.

Similarly, the feed-line connection at the center insulator should be made to the antenna wires after they have been secured to the insulator (Fig 22.12B). This way, you will be assured of a good electrical connection between the antenna and feed line without compromising the mechanical strength. Do a good job of soldering the antenna and feed-line connections. Use a heavy iron or a torch, and be sure to clean the materials thoroughly before starting the job. Proper planning should allow you to solder indoors at a workbench, where the best possible joints may be made. Poorly soldered or unsoldered connections will become headaches as the wire oxidizes and the electrical integrity degrades with time. Besides degrading your antenna performance, poorly made joints can even be a cause of TVI because of rectification. Spray paint the connections with acrylic for waterproofing.

If made from the right materials, the dipole should give a builder years of maintenance-free service—unless of course a tree falls on it. As you build your antenna, keep in mind that if you get it right the first time, you won't have to do it again for a long time.

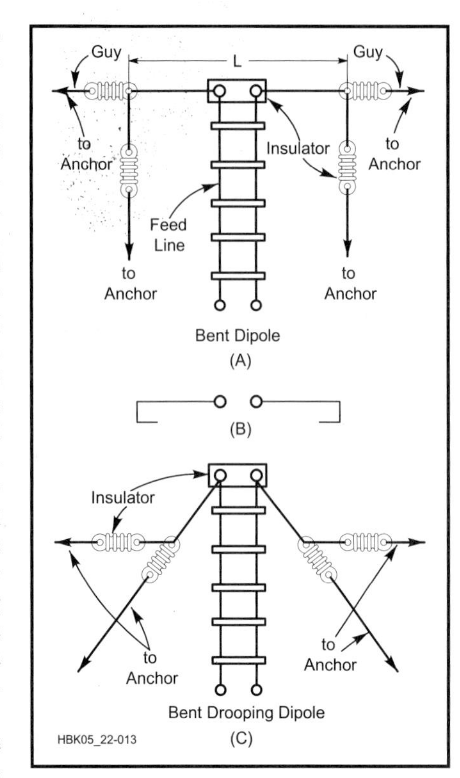

Fig 22.13 — When limited space is available for a dipole antenna, the ends can be bent downward as shown at A, or back on the radiator as shown at B. The inverted V at C can be erected with the ends bent parallel with the ground when the available supporting structure is not high enough.

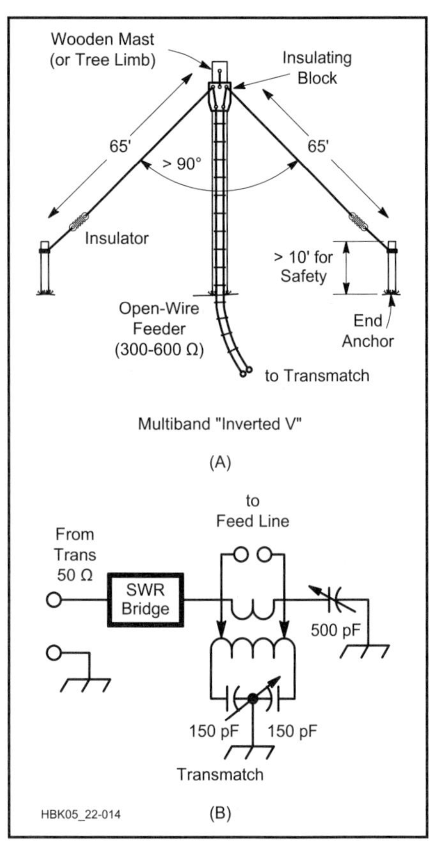

Fig 22.14 — At A, details for an inverted V fed with open-wire line for multiband HF operation. A Transmatch is shown at B, suitable for matching the antenna to the transmitter over a wide frequency range. The included angle between the two legs should be greater than 90° for best performance.

SHORTENED DIPOLES

Inductive loading increases the electrical length of a conductor without increasing its physical length. Therefore, we can build physically short dipole antennas by placing inductors in the antenna. These are called *loaded antennas*, and *The ARRL Antenna Book* shows how to design them. There are some trade-offs involved: Inductively loaded antennas are less efficient and have narrower bandwidths than full-size antennas. Generally they should not be shortened more than 50%.

DIPOLE ORIENTATION

Dipole antennas need not be installed horizontally and in a straight line. They are generally tolerant of bending, sloping or drooping. Bent dipoles may be used where antenna space is at a premium. **Fig 22.13** shows a couple of possibilities; there are many more. Bending distorts the radiation pattern somewhat and may affect the impedance as well, but compromises may be acceptable when the situation demands them. Remember that dipole antennas are RF conductors. For safety's sake, mount all antennas away from conductors (especially power lines), combustibles and well beyond the reach of passersby. When an antenna bends back on itself (as in Fig 22.13B) some of the signal is canceled; avoid this if possible.

DROOPING DIPOLE

A drooping dipole, also known as an *Inverted V dipole*, appears in **Fig 22.14**. While *V* describes the shape of this antenna, this antenna should not be confused with long-wire V antennas, which are highly directive. The radiation pattern and dipole impedance depend on the apex angle, and it is very important that the ends do not come too close to lossy ground. Remember that current produces the radiated signal, and current is maximum at the dipole center. Therefore, performance is best when the central area of the antenna is straight, high and clear of nearby objects.

SLOPING DIPOLE

A sloping dipole is shown in **Fig 22.15**. This antenna is often used to favor one direction (the *forward direction* in the figure). With a nonconducting support and poor earth, signals off the back are weaker than those off the front. With a nonconducting mast and good earth, the response is omnidirectional. There is no gain in any direction with a nonconducting mast.

A conductive support such as a tower

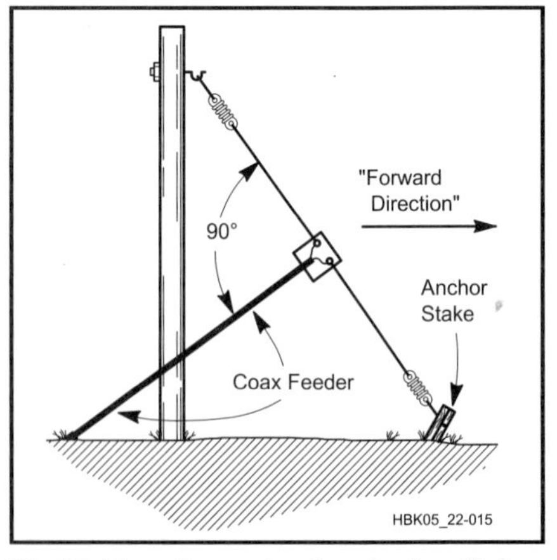

Fig 22.15 — Example of a sloping $^1/_2$-λ dipole, or *full sloper*. On the lower HF bands, maximum radiation over poor to average earth is off the sides and in the *forward direction* as indicated, if a nonconductive support is used. A metal support will alter this pattern by acting as a parasitic element. How it alters the pattern is a complex issue depending on the electrical height of the mast, what other antennas are located on the mast, and on the configuration of any guy wires.

acts as a parasitic element. (So does the coax shield, unless it is routed at 90° from the antenna.) The parasitic effects vary with earth quality, support height and other conductors on the support (such as a beam at the top). With such variables, performance is very difficult to predict.

Losses increase as the antenna ends approach the support or the ground. To prevent feed-line radiation, route the coax away from the feed-point at 90° from the antenna, and continue on that line as far as possible.

HALF-WAVE VERTICAL DIPOLE (HVD)

Unlike its horizontal counterpart, which has a figure-8 pattern, the azimuthal pattern of a vertical dipole is omnidirectional. In other words, it looks like a circle. Look again at Figs 22.3 and 22.4 and note the comparison between horizontal and vertical dipole elevation patterns. These two figures illustrate the fact that performance of a horizontal dipole depends to a great extent on its height above ground. By contrast, *half-wave vertical dipole* (HVD) performance is highly dependent on ground conductivity and dielectric constant.

After looking at these figures, you might easily conclude that there is no advantage to an HVD. Is that really the case? Experiments at K8CH run between 2001 and 2003 showed that the HVD mounted above average ground works well for long-distance (DX) contacts. Two antennas were used in the trials. The first was a 15-m HVD with its base 14 ft above ground (feedpoint at 25 ft). The second (reference) antenna was a 40-m dipole modified to operate with low SWR on 15 m. The reference dipole feedpoint was at 29 feet and the ends drooped slightly to provide a 160° included angle. Signals from outside North America were usually stronger on the HVD. Computer analysis revealed the reasons for this.

Fig 22.16 shows the elevation patterns for the vertical dipole and for the reference dipole at a pattern peak and at a null. The vertical dipole does not look impressive, does it? The large lobe in the HVD pattern at 48° is caused by the antenna being elevated 14 ft above ground. This lobe will shrink at lower heights.

Data compiled by Dean Straw, N6BV, shows that 90% of DX contacts from K8CH should use elevation angles of 10° or less. Further, nearly half of the contacts would use 3° or less. The azimuthal patterns for 10° are shown in **Fig 22.17** and for 3° in **Fig 22.18**. You can clearly see in the patterns the DX potential of an HVD.

Another advantage of the HVD is its radiation resistance at low heights. Look back in Fig 22.1 at the curve for the vertical half-wave antenna. With it's base just above ground, the HVD will have a radiation resistance of over 90 Ω. That can easily be turned to an advantage. Capacitive loading will lower the radiation resistance *and* shorten the antenna. It is possible to make a loaded vertical dipole that is half the height of an HVD and that has a good SWR when fed with 50-Ω coax.

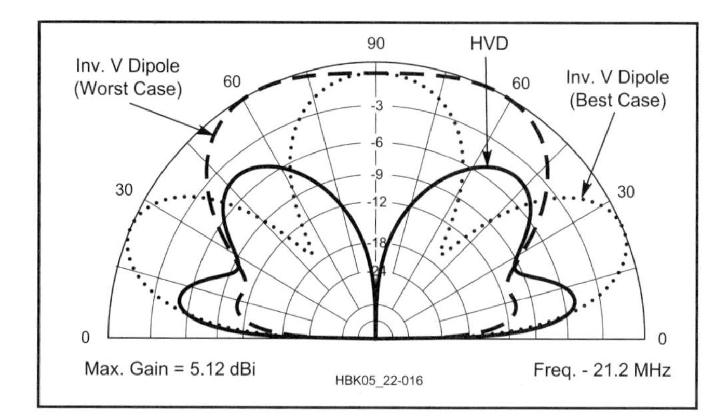

Fig 22.16 — Elevation patterns for the HVD (solid line) and the inverted V comparison antenna in its best case (dashed line) and worst case (dotted line).

MULTIBAND DIPOLES

There are several ways to construct coax-fed multiband dipole systems. These techniques apply to dipoles of all orientations. Each method requires a little more

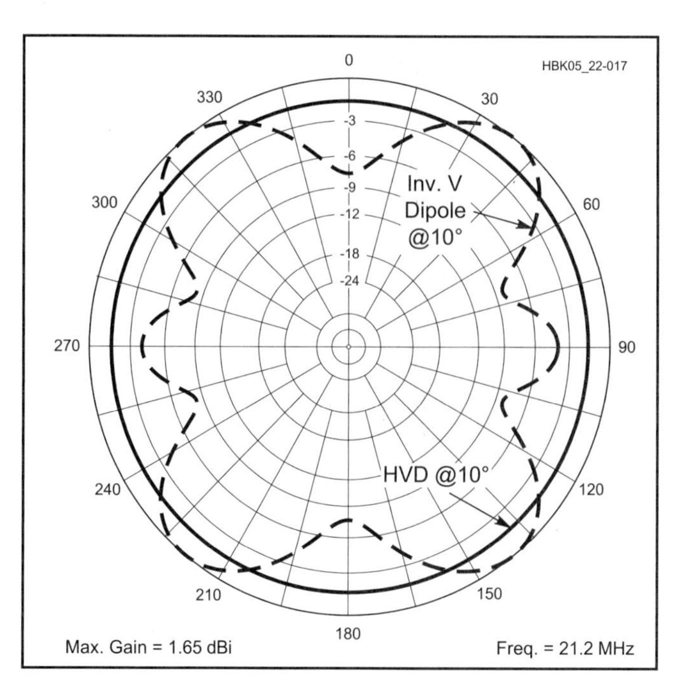

Fig 22.17 — Azimuth patterns at 10° elevation for the HVD (dashed line) and inverted V (solid line).

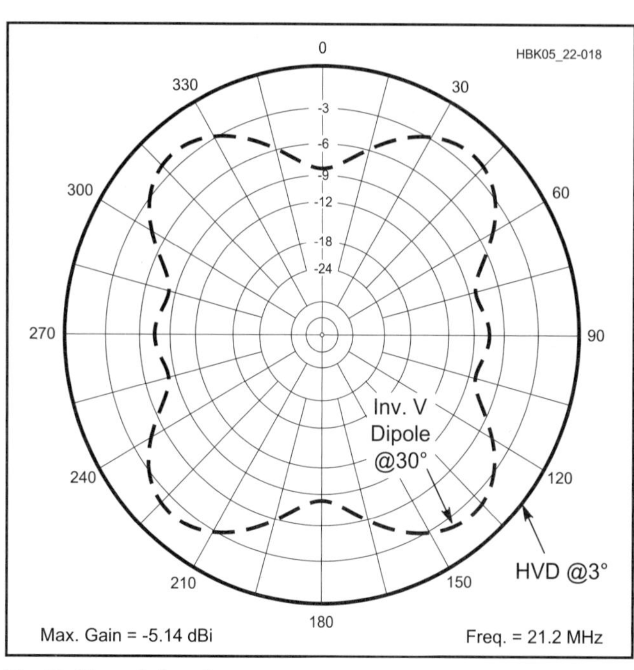

Fig 22.18 — Azimuth patterns at 3° elevation for the HVD (dashed line) and the inverted V (solid line).

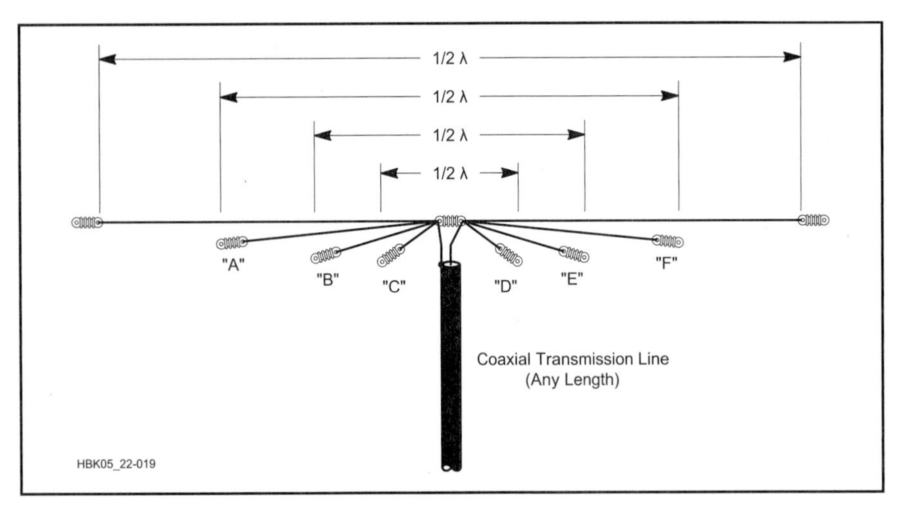

Fig 22.19 — Multiband antenna using paralleled dipoles, all connected to a common 50 or 75-Ω coax line. The half-wave dimensions may be either for the centers of the various bands or selected for favorite frequencies in each band. The length of a half wave in feet is 468/frequency in MHz, but because of interaction among the various elements, some pruning for resonance may be needed on each band. See text.

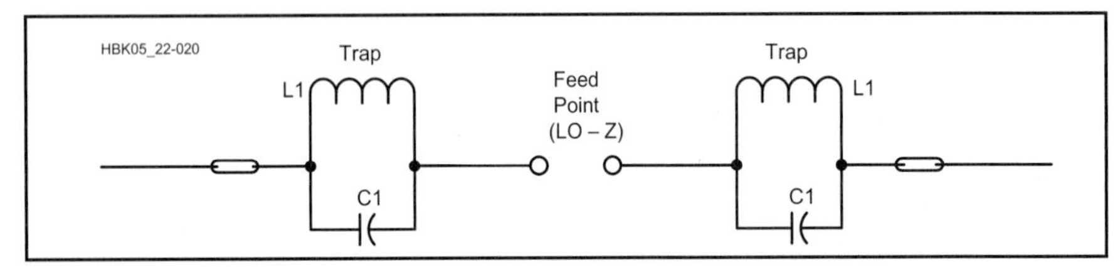

Fig 22.20 — Example of a trap dipole antenna. L1 and C1 can be tuned to the desired frequency by means of a dip meter *before* they are installed in the antenna.

work than a single dipole, but the materials don't cost much.

Parallel dipoles are a simple and convenient answer. See **Fig 22.19**. Center-fed dipoles present low-impedances near f_0, or its odd harmonics, and high impedances elsewhere. This lets us construct simple multiband systems that automatically select the appropriate antenna. Consider a 50-Ω resistor connected in parallel with a 5-kΩ resistor. A generator connected across the two resistors will see 49.5 Ω, and 99% of the current will flow through the 50-Ω resistor. When resonant and nonresonant antennas are parallel connected, the nonresonant antenna takes little power and has little effect on the total feed-point impedance. Thus, we can connect several antennas together at the feedpoint, and power naturally flows to the resonant antenna.

There are some limits, however. Wires in close proximity tend to couple and produce mutual inductance. In parallel dipoles, this means that the resonant length of the shorter dipoles lengthens a few percent. Shorter antennas don't affect longer ones much, so adjust for resonance in order from longest to shortest. Mutual inductance also reduces the bandwidth of shorter dipoles, so a Transmatch may be needed to achieve an acceptable SWR across all bands covered. These effects can be reduced by spreading the ends of the dipoles.

Also, the power-distribution mechanism requires that only one of the parallel dipoles is near resonance on any amateur band. Separate dipoles for 80 and 30 m should not be parallel connected because the higher band is near an odd harmonic of the lower band (80/3 ≈ 30) and center-fed dipoles have low impedance near odd harmonics. (The 40 and 15-m bands have a similar relationship.) This means that you must either accept the lower performance of the low-band antenna operating on a harmonic or erect a separate antenna for those odd-harmonic bands. For example, four parallel-connected dipoles cut for 80, 40, 20 and 10 m (fed by a single Transmatch and coaxial cable) work reasonably on all HF bands from 80 through 10 m.

Trap dipoles provide multiband operation from a coax-fed single-wire dipole. **Fig 22.20** shows a two-band trap antenna. A trap comprises inductance and capacitance in parallel. At resonance that effectively disconnects wire beyond the trap at the resonant frequency. Above resonance, traps provide capacitive loading. Below resonance, they provide inductive loading. Traps may be constructed from coiled sections of coax or from discrete LC components.

Choose capacitors (Cl in the figure) that are rated for high current and voltage. Mica transmitting capacitors are good. Ceramic transmitting capacitors may work, but their values may change with temperature. Use large wire for the inductors to reduce loss. Any reactance (X_L and X_C) above 100 Ω (at f_0) will work, but bandwidth increases with reactance (up to several thousand ohms).

Check trap resonance before installation. This can be done with a dip meter and a receiver. To construct a trap antenna, cut a dipole for the highest frequency and connect the pretuned traps to its ends. It is fairly complicated to calculate the additional wire needed for each band, so just add enough wire to make the antenna 1/2 λ and prune it as necessary. Because the inductance in each trap reduces the physical length needed for resonance, the finished antenna will be shorter than a simple 1/2-λ dipole.

A 135-FT MULTIBAND CENTER-FED DIPOLE

An 80-m dipole fed with ladder line is a versatile antenna. If you add a wide-range matching network, you have a low-cost antenna system that works well across the entire HF spectrum. Countless hams have used one of these in single-antenna stations and for Field Day operations.

For best results place the antenna as high as you can, and keep the antenna and ladder line clear of metal and other conductive objects. Despite significant SWR on some bands, system losses are low. (See the **Transmission Lines** chapter.) You can make the dipole horizontal, or you can install it as an inverted V. ARRL staff analyzed a 135-ft dipole at 50 ft above typical ground and compared that to an inverted V with the center at 50 ft, and the ends at 10 ft. The results show that on the 80-m band, it won't make much difference which configuration you choose. (See **Fig 22.21**.) The inverted V exhibits additional losses because of its proximity to ground.

Fig 22.22 shows a comparison between a 20-m flat-top dipole and the 135-ft flat-top dipole when both are placed at 50 ft above ground. At a 10° elevation angle, the 135-ft dipole has a gain advantage. This advantage comes at the cost of two deep, but narrow, nulls that are broadside to the wire.

Fig 22.23 compares the 135-ft dipole to the inverted-V configuration of the same antenna on 14.1 MHz. Notice that the inverted-V pattern is essentially omnidirectional. That comes at the cost of gain, which is less than that for a horizontal flat-top dipole.

As expected, patterns become more

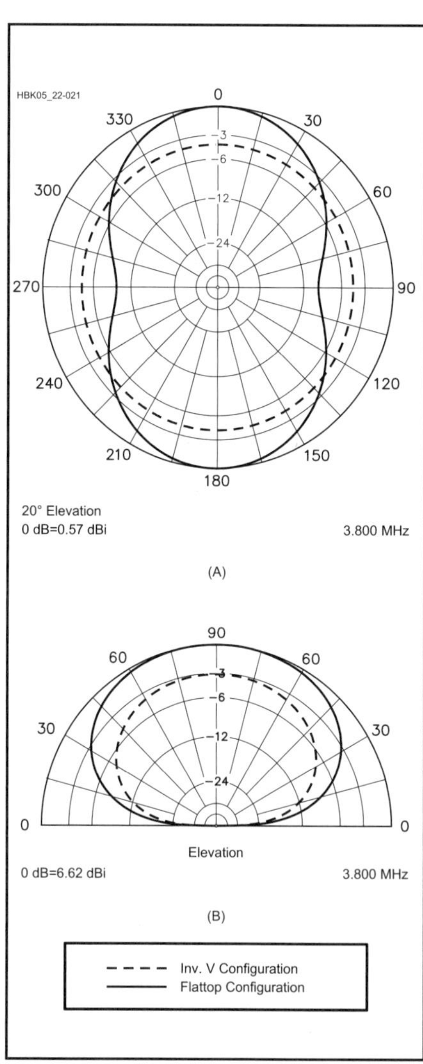

Fig 22.21 — Patterns on 80 m for 135-ft, center-fed dipole erected as a horizontal dipole at 50 ft, and as an inverted V with the center at 50 ft and the ends at 10 ft. The azimuth pattern is shown at A, where the conductor lies in the 90° to 270° plane. The elevation pattern is shown at B, where the conductor comes out of paper at a right angle. At the fundamental frequency the patterns are not markedly different.

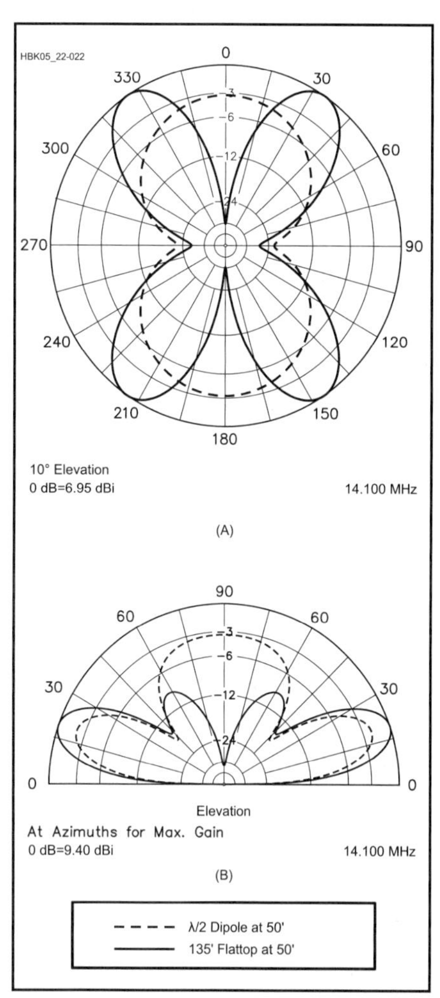

Fig 22.22 — Patterns on 20 m comparing a standard 1/2-λ dipole and a multiband 135-ft dipole. Both are mounted hori-zontally at 50 ft. The azimuth pattern is shown at A, where conductors lie in the 90° to 270° plane. The elevation pattern is shown at B. The longer antenna has four azimuthal lobes, centered at 35°, 145°, 215°, and 325°. Each is about 2 dB stronger than the main lobes of the 1/2-λ dipole. The elevation pattern of the 135-ft dipole is for one of the four maximum-gain azimuth lobes, while the elevation pattern for the 1/2-λ dipole is for the 0° azimuthal point.

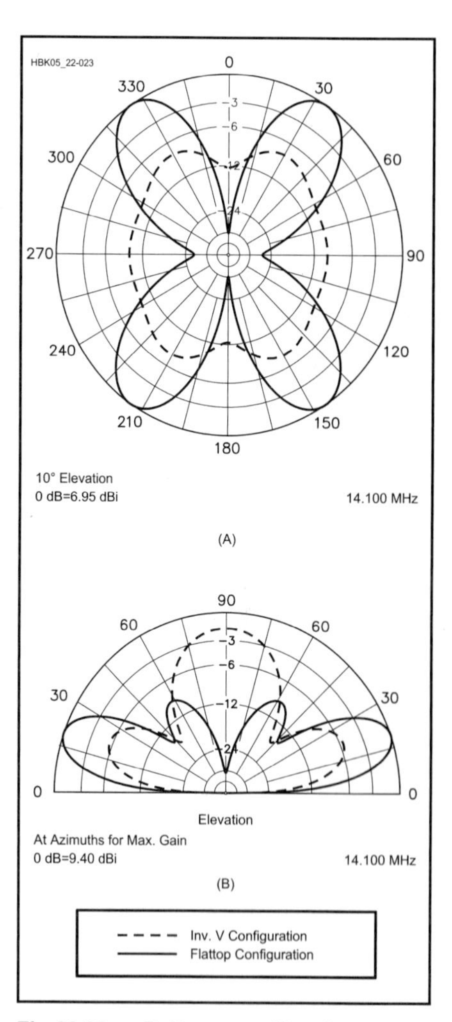

Fig 22.23 — Patterns on 20 m for two 135-ft dipoles. One is mounted hori-zontally as a flat-top and the other as an inverted V with 120° included angle between the two legs. The azimuth pattern is shown at A, and the elevation pattern is shown at B. The inverted V has about 6 dB less gain at the peak azi-muths, but has a more uniform, almost omnidirectional, azimuthal pattern. In the elevation plane, the inverted V has a fat lobe overhead, making it a somewhat better antenna for local communication, but not quite so good for DX contacts at low elevation angles.

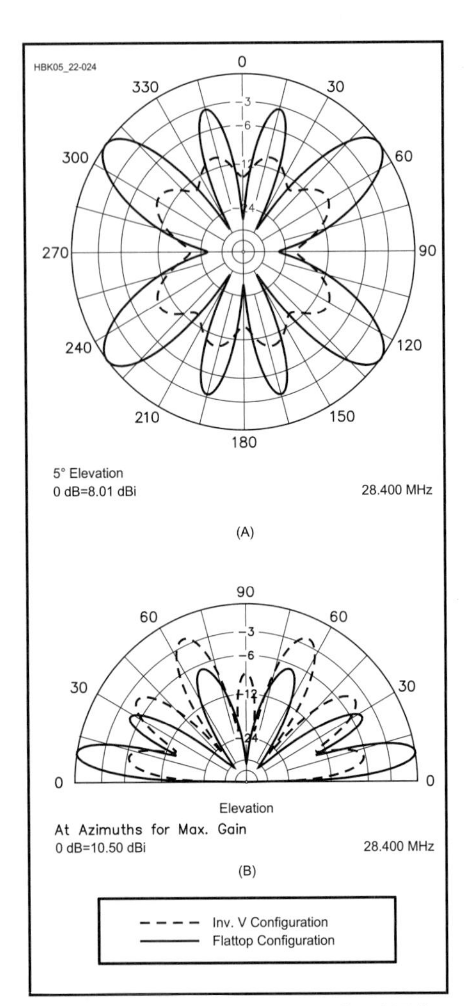

Fig 22.24 — Patterns on 10 m for 135-ft dipole mounted horizontally and as an inverted V, as in Fig 22.23. The azimuth pattern is shown at A, and the elevation pattern is shown at B. Once again, the inverted-V configuration yields a more omnidirectional pattern, but at the expense of almost 8 dB less gain than the flat-top configuration at its strongest lobes.

complicated at 28.4 MHz. As you can see in **Fig 22.24**, the inverted V has the advantage of a pattern with slight nulls, but with reduced gain compared to the flat-top configuration.

Installed horizontally, or as an inverted V, the 135-ft center-fed dipole is a simple antenna that works well from 3.5 to 30 MHz. Bandswitching is handled by a Transmatch that is located near your operating position.

Antenna Modeling by Computer

Modern computer programs have made it a *lot* easier for a ham to evaluate antenna performance. The elevation plots for the 135-ft long center-fed dipole were generated using a sophisticated computer program known as *NEC*, short for "Numerical Electromagnetics Code." *NEC* is a general-purpose antenna modeling program, capable of modeling almost any antenna type, from the simplest dipole to extremely complex antenna designs. Various mainframe versions of *NEC* have been under continuous development by US government researchers for several decades.

But because it is a general-purpose program, *NEC* can be very slow when modeling some antennas—such as long-boom, multi-element Yagis. There are other, specialized programs that work on Yagis much faster than *NEC*. Indeed, *NEC* has developed a reputation for being accurate (if properly applied!), but decidedly difficult to learn and use. A number of commercial software developers have risen to the challenge and created more *user-friendly* versions. Check the ads in *QST*.

NEC uses a *Method of Moments* algorithm. The mathematics behind this algorithm are pretty formidable to most hams, but the basic principle is simple. An antenna is broken down into a set of straight-line wire *segments*. The fields resulting from the current in each segment and from the mutual interaction between segments are vector-summed in the far field to create azimuth and elevation-plane patterns.

The most difficult part of using a *NEC*-type of modeling program is setting up the antenna's geometry—you must condition yourself to think in three-dimensional coordinates. Each end point of a wire is represented by three numbers: an x, y and z coordinate. An example should help sort things out. See **Fig A**, showing a *model* for a 135-foot center-fed dipole, made of #14 wire placed 50 ft above flat ground. This antenna is modeled as a single, straight wire.

For convenience, ground is located at the *origin* of the coordinate system, at (0, 0, 0) feet, directly under the center of the dipole. The dipole runs parallel to, and above, the y-axis. Above the origin, at a height of 50 feet, is the dipole's feedpoint. The *wingspread* of the dipole goes toward the left (that is, in the *negative y* direction) one-half the overall length, or –67.5 ft. Toward the right, it goes +67.5 ft. The *x* dimension of our dipole is zero. The dipole's ends are thus represented by two points, whose coordinates are: (0, –67.5, 50) and (0, 67.5, 50) ft. The thickness of the antenna is the diameter of the wire, #14 gauge.

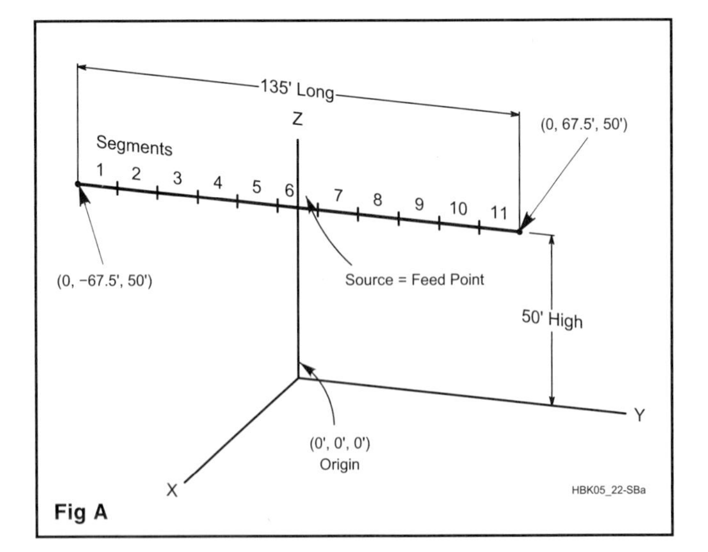

Fig A

To run the program you must specify the

number of segments into which the dipole is divided for the method-of-moments analysis. The guideline for setting the number of segments is to use at least 10 segments per half-wavelength. In Fig A, our dipole has been divided into 11 segments for 80-m operation. The use of 11 segments, an odd rather than an even number such as 10, places the dipole's feedpoint (the *source* in *NEC*-parlance) right at the antenna's center and at the center of segment number six.

Since we intend to use our 135-foot long dipole on all HF amateur bands, the number of segments used actually should vary with frequency. The penalty for using more segments in a program like *NEC* is that the program slows down roughly as the square of the segments—double the number and the speed drops to a fourth. However, using too few segments will introduce inaccuracies, particularly in computing the feed-point impedance. The commercial versions of *NEC* handle such nitty-gritty details automatically.

Let's get a little more complicated and specify the 135-ft dipole, configured as an inverted-V. Here, as shown in **Fig B**, you must specify *two* wires. The two wires join at the top, (0, 0, 50) ft. Now the specification of the source becomes more complicated. The easiest way is to specify two sources, one on each end segment at the junction of the two wires. If you are using the *native* version of *NEC*, you may have to go back to your high-school trigonometry book to figure out how to specify the end points of our droopy dipole, with its 120° included angle. Fig B shows the details, along with the trig equations needed.

So, you see that antenna modeling isn't entirely a cut-and-dried procedure. The commercial programs do their best to hide some of the more unwieldy parts of *NEC*, but there's still some art mixed in with the science. And as always, there are trade-offs to be made—segments versus speed, for example.

However, once you do figure out exactly how to use them, computer models are wonderful tools. They can help you while away a dreary winter's day, designing antennas on-screen—without having to risk life and limb climbing an ice-covered tower. And in a relatively short time a computer model can run hundreds, or even thousands, of simulations as you seek to optimize an antenna for a particular parameter. Doesn't that sound better than trying to optimally tweak an antenna by means of a thousand cut-and-try measurements, all the while hanging precariously from your tower by a climbing belt?!—*R. Dean Straw, N6BV, Senior Assistant Technical Editor*

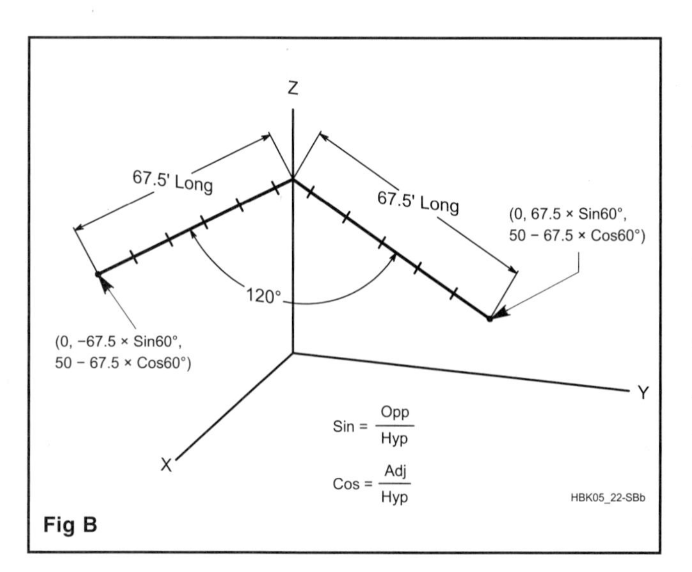

Fig B

A 40-M AND 15-M DUAL-BAND DIPOLE

As mentioned earlier, dipoles have harmonic resonances at odd multiples of their fundamental resonances. Because 21 MHz is the third harmonic of 7 MHz, 7-MHz dipoles are harmonically resonant in the popular ham band at 21 MHz. This is attractive because it allows you to install a 40-m dipole, feed it with coax, and use it without an antenna tuner on both 40 and 15 m.

But there's a catch: The third harmonic resonance is actually higher than three times the fundamental resonant frequency. This is because there is no end effect in the center portion of the antenna.

An easy fix for this, as shown in **Fig 22.25**, is to capacitively load the antenna about a quarter wavelength (at 21.2 MHz) away from the feedpoint in both wires. Known as *capacitance hats*, the simple loading wires shown lower the antenna's resonant frequency on 15 m without substantially affecting resonance on 40 m.

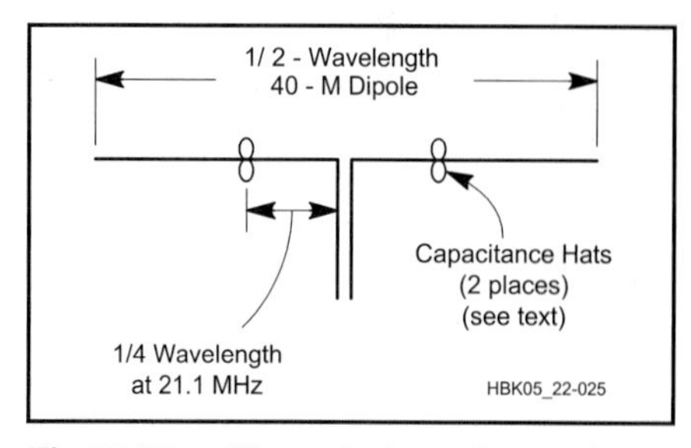

Fig 22.25 — Figure-8-shaped capacitance hats made and placed as described in the text, can make your 40-m dipole resonate anywhere you like in the 15-m band.

To put this scheme to use, first measure, cut and adjust the dipole to resonance at the desired 40-m frequency. Then, cut two 2-ft-long pieces of stiff wire (such as #12 or #14 house wire) and solder the ends of each one together to form two loops. Twist the loops in the middle to form figure-8s, and strip and solder the wires where they cross. Install these capacitance hats on the dipole by stripping the antenna wire (if necessary) and soldering the hats to the dipole about a third of the way out from the feedpoint (placement isn't critical) on each wire. To resonate the antenna on 15 m, adjust the loop shapes (*not while you're transmitting!*) until the SWR is acceptable in the desired segment of the 15-m band.

THE K8SYL 75 AND 10-M DIPOLE

The same idea was adapted by Sylvia Hutchinson, K8SYL, to make a two-band dipole for 75 and 10 m. Her account was published in July 2002 *QST*.

She discovered that a dipole resonant in the General Class portion of the 75-m band is also resonant on 10 m. As in the case of the 40 and 15-m dipole, some additional loading may be required to move the 10-m resonance to the desired portion of the band.

There is another catch. The radiation resistance on 10 m is about 120 Ω. In another words, if you feed this antenna with 50-Ω coax your best SWR will be around 2.4:1. A quarter wavelength (at 10 m) of 75-Ω coax (such as RG-11) will transform that 120-Ω feedpoint impedance to just under 50 Ω. In this case, the SWR will be better than 1.1:1 at resonance.

The length of the matching section is a small fraction of a wavelength at 3.9 MHz. This will tend to narrow the SWR bandwidth, but only slightly. The antenna is shown in **Fig 22.26**.

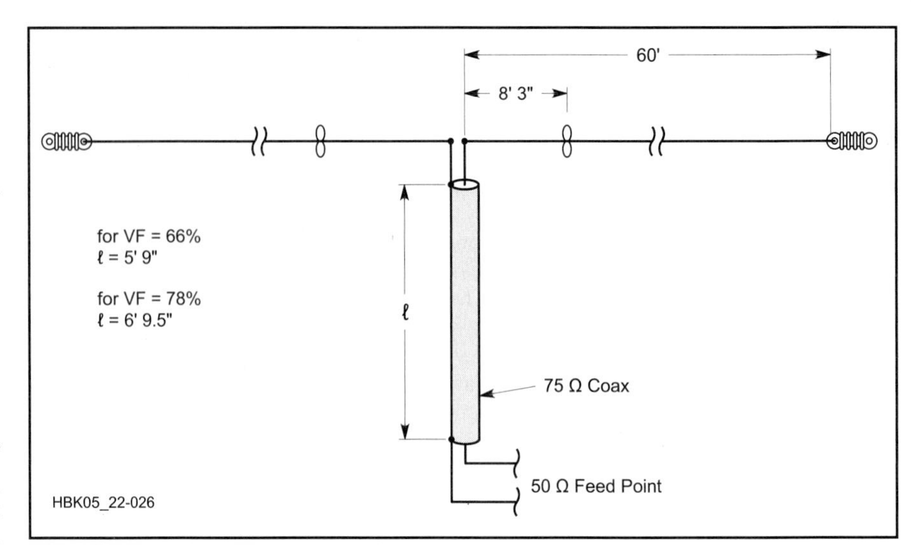

Fig 22.26 — The K8SYL dipole operates on the 75 and 10-m bands. A quarter-wave section of 75-Ω coax transforms the 10-m impedance. See text.

Make each capacitance hat from an 18 to 20-in length of #12 or #14 house wire. Solder the ends together to form a loop, leaving a couple of inches free for attaching to the dipole. Next, twist the loop to form a figure-8. The portion of the loop where the ends meet should be at the center of the figure-8. Strip the crossing wire at this point and the capacitance hat is ready to attach to the dipole.

Tune the antenna on 75 m first. Trim the dipole ends for resonance in your favorite portion of the band. For K8SYL, each leg was 59 ft, 4 inches long. Then check the SWR on 10 m. Adjust the loop shapes (*not while you're transmitting!*) until the SWR is acceptable in the desired segment of the 10-m band.

THE W4RNL INVERTED-U ANTENNA

This simple rotatable dipole was designed and built by L. B. Cebik, W4RNL, for use during the ARRL Field Day. For this and other portable operations we look for three antenna characteristics: simplicity, small size, and light weight. Complex assemblies increase the number of things that can go wrong. Large antennas are difficult to transport and sometimes do not fit the space available. Heavy antennas require heavy support structures, so the overall weight seems to increase exponentially with every added pound of antenna.

Today, a number of light-weight collapsible masts are available. Some will support—when properly guyed—antennas in the 5-10 pound range. Most are suitable for 10-m tubular dipoles and allow the user to hand-rotate the antenna. Extend the range of the antenna to cover 20-10 m, and you put these 20-30-foot masts to even better use. The inverted U meets this need.

THE BASIC IDEA OF THE INVERTED U

A dipole's highest current occurs within the first half of the distance from the feedpoint to the outer tips. Therefore, very little performance is lost if the outer end sections are bent. The W4RNL inverted U starts with a 10-m tubular dipole. You add extensions for 12, 15, 17, or 20 m to cover those bands.

You only need enough space to erect a 10-m rotatable dipole. The extensions hang down. **Fig 22.27** shows the relative proportions of the antenna on all bands from 10 to 20 m. The 20-m extensions are the length of half the 10-m dipole. Therefore, safety dictates an antenna height of at least 20 ft to keep the tips above 10 ft high. At any power level, the ends of a dipole have high RF voltages, and we must keep them out of contact with human body parts.

Not much signal strength is lost by drooping up to half the overall element length straight down. What is lost in bi-directional gain shows up in decreased side-nulls. **Fig 22.28** shows the free-space E-plane (azimuth) patterns of the inverted U with a 10-m horizontal section. There is an undetectable decrease in gain between the 10-m and 15-m versions. The 20-m version shows a little over a half-dB gain decrease and a signal increase off the antenna ends.

The real limitation of an inverted-U is a function of the height of the antenna above ground. With the feedpoint at 20 ft above ground, we obtain the elevation patterns shown in **Fig 22.29**. The 10-m pattern is

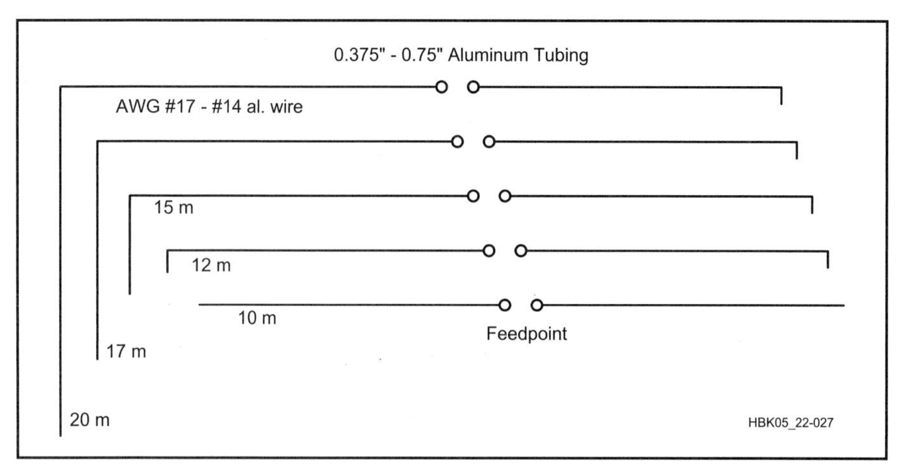

Fig 22.27 — The general outline of the inverted-U field dipole for 20 through 10 m. Note that the vertical end extension wires apply to both ends of the main 10-m dipole, which is constant for all bands.

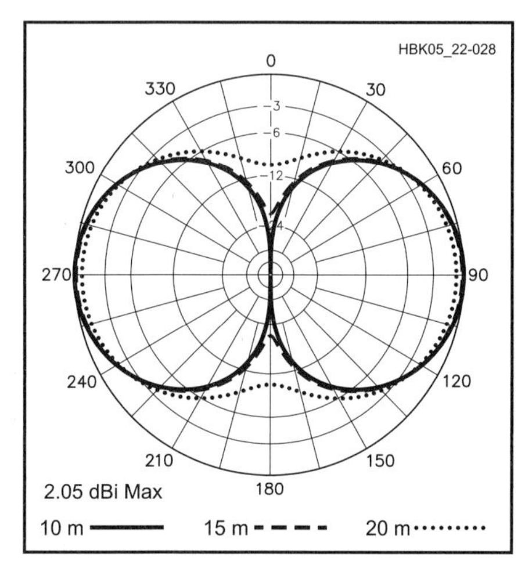

Fig 22.28 — Free-space E-plane (azimuth) patterns of the inverted-U for 10, 15, and 20 m, showing the pattern changes with increasingly longer vertical end sections.

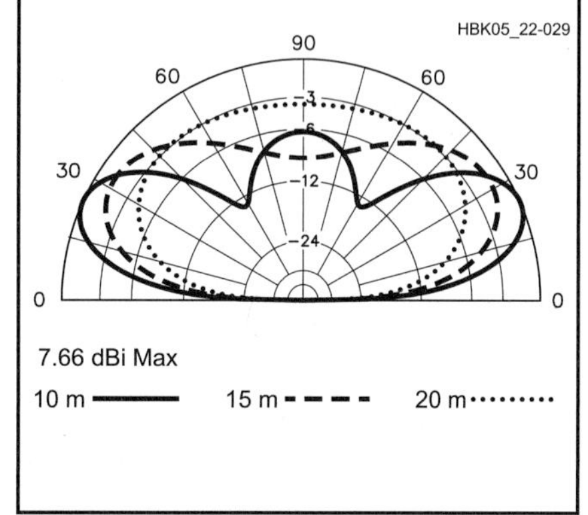

Fig 22.29 — Elevation patterns of the inverted-U for 10, 15, and 20 m, with the antenna feedpoint 20 ft above average ground. Much of the decreased gain and higher elevation angle of the pattern at the lowest frequencies is due to its ever-lower height as a fraction of a wavelength.

Table 22.5
Parts List for the Inverted-U

Amount	*Item*	*Comments*
6'	0.375" OD aluminum tubing	2 - 3' pieces
6'	0.5" OD aluminum tubing	2 - 3' pieces
6'	0.625" OD aluminum tubing	2 - 3' pieces
10"	0.75" OD aluminum tubing	2 - 5" pieces
4"	0.5" nominal (5/8" OD) CPVC	
50'	Aluminum wire AWG #17	
8	Hitch pin clips	Sized to fit tubing junctions.
1	4" by 4" by 1/4" Lexan plate	Other materials suitable.
2	SS U-bolts	Sized to fit support mast
2	Sets SS #8/10 1.5" bolt, nut, washers	SS = stainless steel
2	Sets SS #8 1" bolt, nut, washers	
2	Sets SS #8 .5" bolt, nut, washers	
1	Coax connector bracket, 1/16" aluminum	See text for dimensions and shape
1	Female coax connector	
2	Solder lugs, #8 holes	
2	Short pieces copper wire	From coax connector to solder lugs

Note: 6063-T832 aluminum tubing is preferred and can be obtained from such outlets as Texas Towers (**www.texastowers.com**). Lexan (polycarbonate) is available from such sources as McMasters-Carr (**www.mcmasters.com**), as are the hitch pin clips (if not locally available). Other items should be available from local home centers and radio parts stores.

typical for a dipole that is about 5/8 λ above ground. On 15, the antenna is only 0.45 λ high, with a resulting increase in the overall elevation angle of the signal and a reduction in gain. At 20 m, the angle grows still higher, and the signal strength diminishes as the antenna height drops to under 0.3 λ. Nevertheless, the signal is certainly usable. A full-size dipole at 20 m would show only a little more gain, and the elevation angle would be similar to that of the invert U, despite the difference in antenna shape. If we raise the inverted-U to 40 feet, the 20-m performance would be very similar to that shown by the 10-m elevation plot in Fig 22.29.

The feedpoint impedance of the inverted-U remains well within acceptable limits for virtually all equipment, even at 20 feet above ground. Also, the SWR curves are very broad, reducing the criticalness of finding exact dimensions, even for special field conditions.

BUILDING AN INVERTED-U

Approach the construction of an inverted-U in 3 steps: 1. the tubing arrangement, 2. the center hub and feedpoint assembly, and 3. the drooping extensions. A parts list appears in **Table 22.5**.

The Aluminum Tubing Dipole for 10 meters.

The aluminum tubing dipole consists of three longer sections of tubing and a short section mounted permanently to the feed point plate, as shown in **Fig 22.30**. Let's consider each half of the element separately. Counting from the center of the plate—the feedpoint—the element extends 5 inches using 3/4-inch aluminum tubing. Then we have two 33 inch exposed tubing sections, with an additional 3 inches of tubing overlap per section. These sections are 5/8- and 1/2-inch diameter, respectively. The outer section is 30 inches long exposed (with at least a 3 inches overlap) and consists of 3/8-inch diameter tubing.

Since the 5/8- and 1/2-inch sections are 36 inches long, you can make the outer 3/8-inch section the same overall length and use more overlap, or you can cut the tubing to 33 inches and use the 3 inch overlap. Three inches of overlap is sufficient to ensure a strong junction, and it minimizes excess weight. However, when not in use, the 3 outer tubing sections will nest inside each other for storage, and a 36-inch length for the outer section is a bit more convenient to un-nest for assembly. Keep the end hitch pin on the 3/8 inch tubing as an easy way of pulling it into final position. You may use the readily available 6063-T832 aluminum tubing that nests well and has a long history of antenna service.

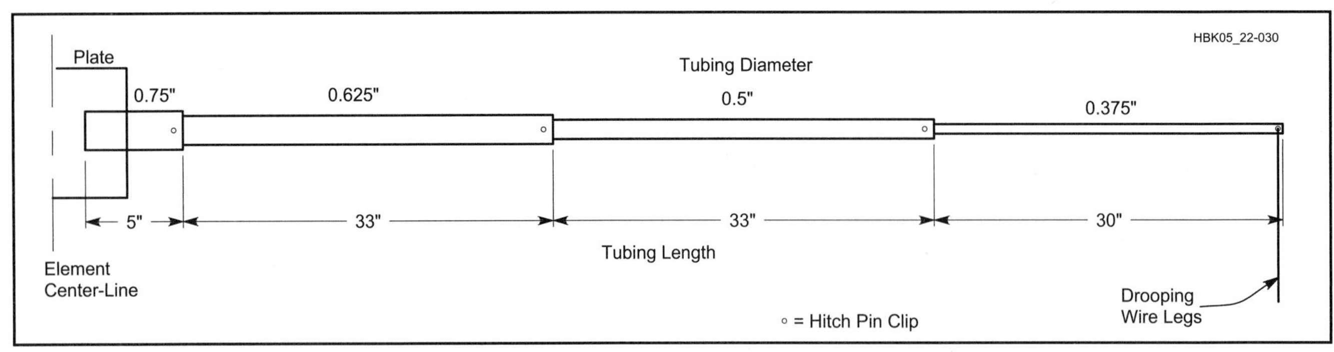

Fig 22.30 — The general tubing layout for the inverted-U for each half element. The opposite side of the dipole is a mirror image of the one shown.

Fig 22.31 — A close-up of the element mounting plate assembly, including the hitch pin clips used to secure the next section of tubing.

The only construction operation that you need to perform on the tubing is to drill a hole at about the center of each junction to pass a hitch pin clip. Obtain hitch pin clips (also called hairpin cotter pin clips in some literature) that fit snugly over the tubing. One size will generally handle about 2 or 3 tubing sizes. In this antenna, I used $^3/_{32}$ (pin diameter) by $2^5/_8$ inch long clips for the $^3/_4$- to $^5/_8$-inch and the $^5/_8$- to $^1/_2$-inch junctions, with $^3/_{32}$ by $1^5/_8$-inch pins for the $^1/_2$- to $^3/_8$-inch junction and for the final hitch pin clip at the outer end of the antenna. Drill the $^1/_8$-inch diameter holes for the clips with the adjacent tubes in position relative to each other. Tape the junction temporarily for the drilling. Carefully deburr the holes so that the tubing slides easily when nested.

The hitch pin clip junctions, shown in **Fig 22.31**, hold the element sections in position. Actual electrical contact between sections is made by the overlapping portions of the tube. Due to the effects of weather, junctions of this type are not suitable for a permanent installation, but are completely satisfactory for short-term use. Good electrical contact requires clean, dry aluminum surfaces, so do not use any type of lubricant to assist the nesting and un-nesting of the tubes. Instead, clean both the inner and outer surfaces of the tubes before and after each use.

Hitch pin clips are fairly large and harder to lose in the grass of a field site than most nuts and bolts. However, you may wish to attach a short colorful ribbon to the loop end of each clip. Spotting the ribbon on the ground is simpler than probing for the clip alone.

Each half element is 101 inches long, for a total 10-m dipole element length of 202 inches (16 ft 10 inches). Length is not critical within about 1 inch, so you may preassemble the dipole using the listed dimensions. However, if you wish a more precisely tuned element, tape the outer section in position and test the dipole on your mast at the height that you will use. Adjust the length of the outer tubing segments equally at both ends for the best SWR curve on the lower 1 MHz of the 10-m band. Even though the impedance will be above 50 Ω throughout the band, you should easily obtain an SWR curve under 2:1 that covers the entire band segment.

The Center Hub: Mounting and Feedpoint Assembly.

Construct the plate for mounting the element and the mast from a 4 × 4 × $^1/_4$-inch-thick scrap of polycarbonate (trade name Lexan), as shown in **Fig 22.32**. You may use other materials so long as they will handle the element weight and stand up to field conditions.

At the top and bottom of the plate are holes for the U-bolts that fit around my mast. Since masts may vary in diameter at the top, size your U-bolts and their holes to suit the mast.

The element center, consisting of 2 5-inch lengths of $^3/_4$-inch aluminum tubing, is just above the centerline of the plate (to allow room for the coax fitting below). $^1/_2$-inch nominal CPVC has an outside diameter of about $^5/_8$ inch and makes a snug fit inside the $^3/_4$-inch tubing. The CPVC aligns the two aluminum tubes in a straight line and allows for a small (about $^1/_2$ inch) gap between them. When centered between the two tubes, the CPVC is the same width as the plate. A pair of 1.5-inch #8 or #10 stainless steel bolts—with washers and a nut—secures the element to the plate.

Note in the sketch that you may insert the $^5/_8$-inch tube as far into the $^3/_4$-inch tube as it will go and be assured of a 3-inch overlap. Drill all hitch pin clip holes perpendicular to the plate. Although this alignment is not critical to the junctions of the tubes, it is important to the outer ends of the tubes when you use the antenna below 10 m.

Mount a single-hole female UHF connector on a bracket made from a scrap of $^1/_{16}$-inch-thick L-stock that is 1 inch on a side. Drill the UHF mounting hole first, before cutting the L-stock to length and trimming part of the mounting side. Then drill two holes for $^1/_2$-inch long #8 stainless steel bolts about 1 inch apart, for a

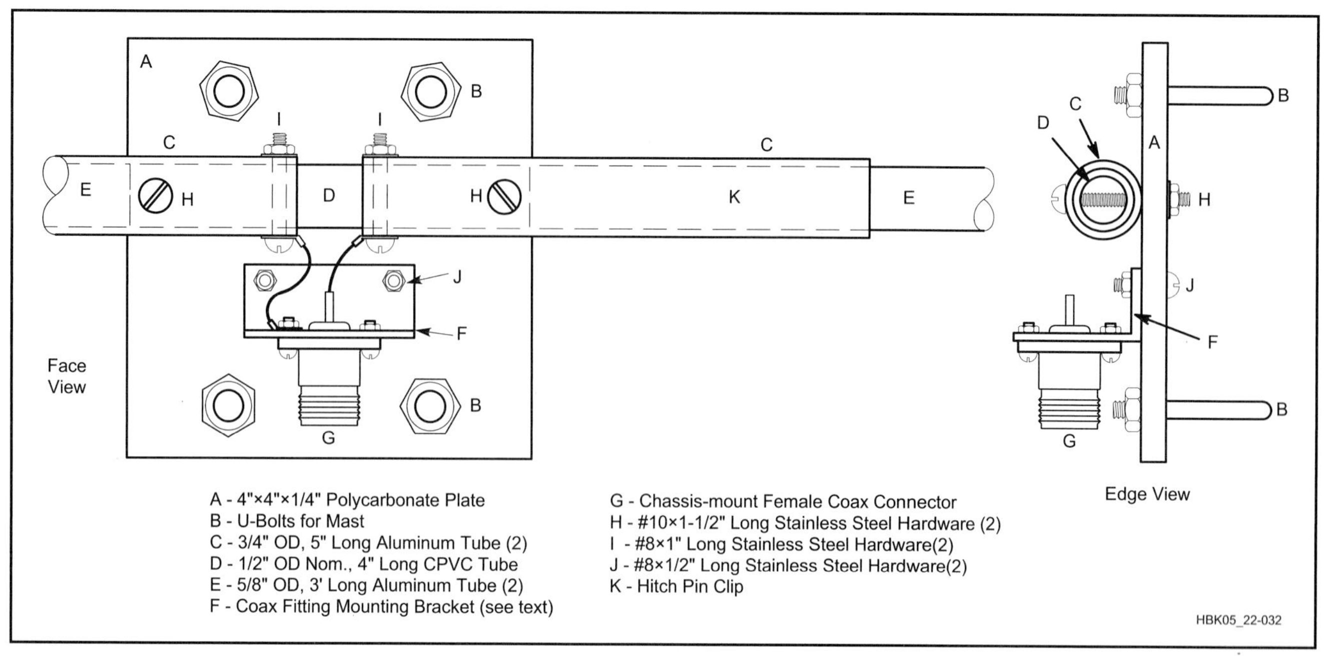

Fig 22.32 — The element and feedpoint mounting plate, with details of the construction used in the prototype.

total length of L-stock of about 1.5 inches. The reason for the wide strip is to place the bolt heads for the bracket outside the area where the mast will meet the plate on the back side. Note in Fig 22.32 that the bracket nuts are on the bracket-side of the main plate, while the heads face the mast. The bracket-to-plate mounting edge of the bracket needs to be only about 3/4 inch wide, so you may trim that side of the L-stock accordingly.

With the element center sections and the bracket in place, drill two holes for 1 inch long #8 stainless steel bolts at right angles to the mounting bolts and as close as feasible to the edges of the tubing at the gap. These bolts have solder lugs attached for short leads to the coax fitting. Solder lugs do not come in stainless steel, so you should check these junctions before and after each use for any corrosion that may require replacement.

With all hardware in place, the hub unit is about 4 × 10 × 1 inch (plus U-bolts). It will remain a single unit from this point onward, so that your only field assembly requirements will be to extend tubing sections and install hitch pin clips. You are now ready to perform the initial 10-m resonance tests on your field mast.

The Drooping Extensions for 12 Through 20 Meters

The drooping end sections consist of aluminum wire. Copper is usable, but aluminum is lighter and quite satisfactory for this application. **Table 22.6** lists the approximate lengths of each extension *below* the element. Add 3 to 5 inches of wire—less for 12 m, more for 20 m—to each length listed.

Common #17 aluminum electric fencing wire works well. Fence wire is stiffer than most wires of similar diameter, and it is cheap. Stiffness is the more important property, since you do not want the lower ends of the wire to wave excessively in the breeze, potentially changing the feedpoint properties of the antenna while it is in use.

When stored, the lengths of wire extensions for 12 and 15 m can be laid out without any bends. However, the longer extensions for 17 and 20 m will require some coiling or folding to fit the same space as the tubing when nested. Fold or coil the wire around any kind of small spindle that has at least a

Table 22.6
Inverted-U Drooping Wire Lengths

Band *m*	*Wire Length* *inches*
10	n/a
12	15.9
15	37.4
17	62.0
20	108.0

Note: The wire length for the drooping ends is measured from the end of the tubular dipole to the tip for AWG #17 wire. Little change in length occurs as a function of the change in wire size. However, a few inches of additional wire length is required for attachment to the element.

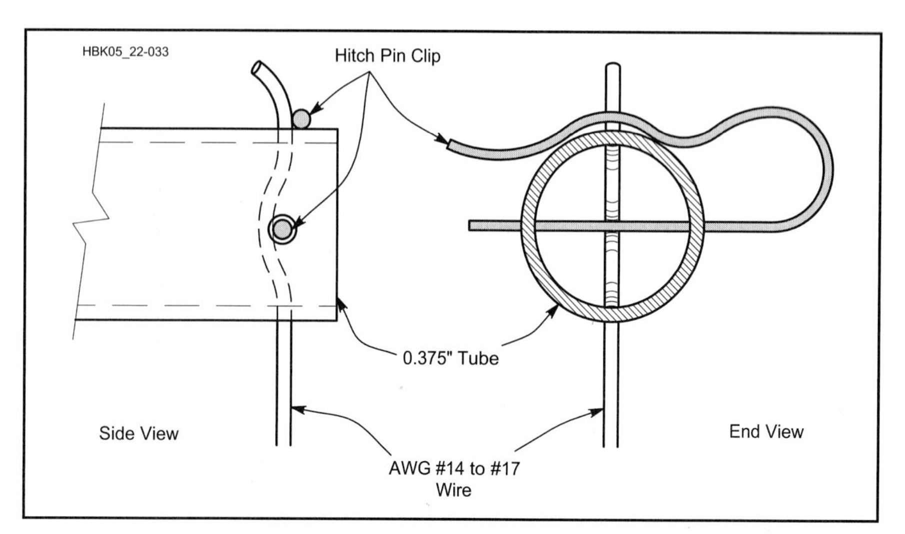

Fig 22.33 — A simple method of clamping the end wires to the 3/8-inch tube end using a hitch pin clip.

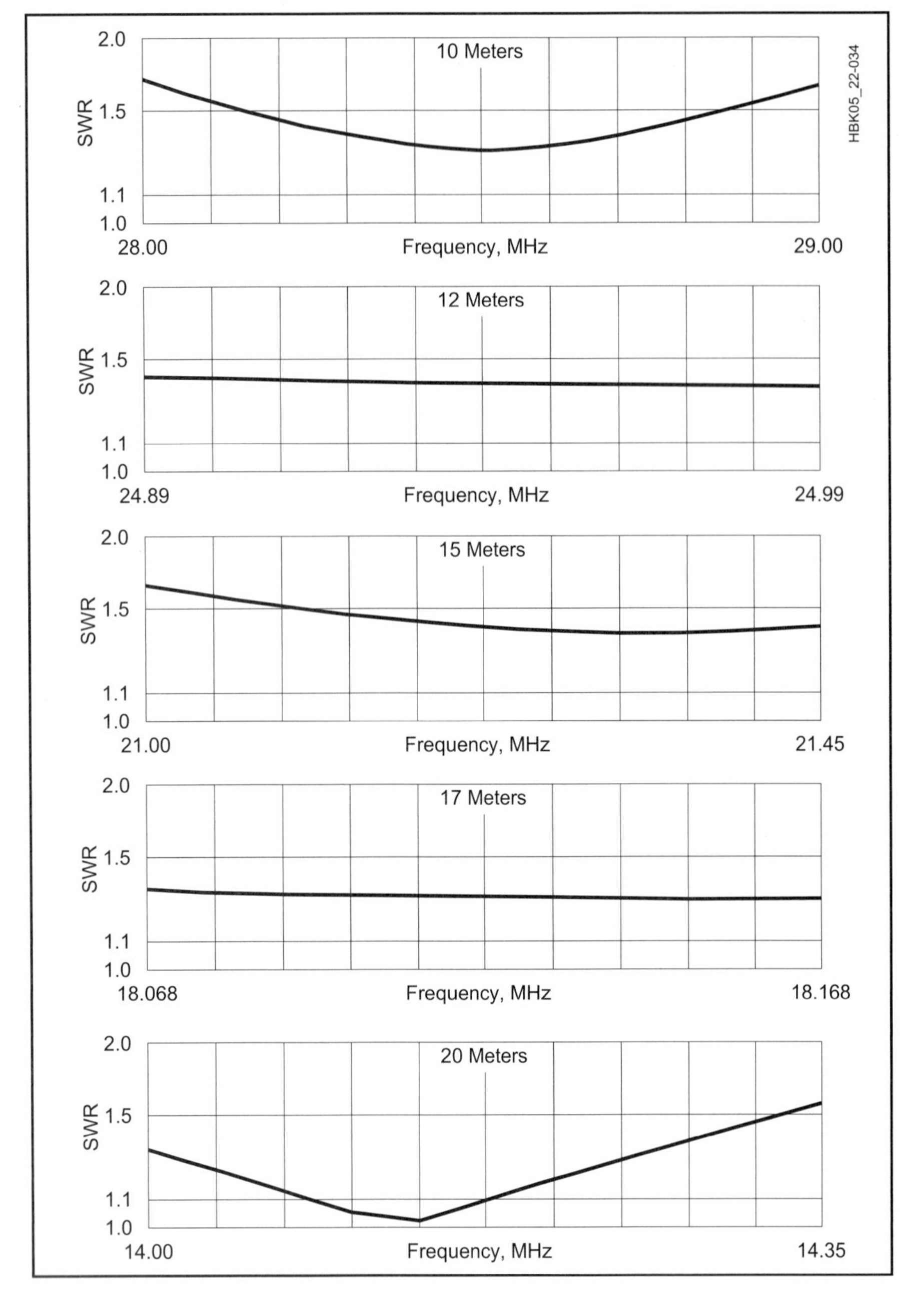

Fig 22.34 — Typical 50-Ω SWR curves for the inverted-U antenna at a feedpoint height of 20 ft.

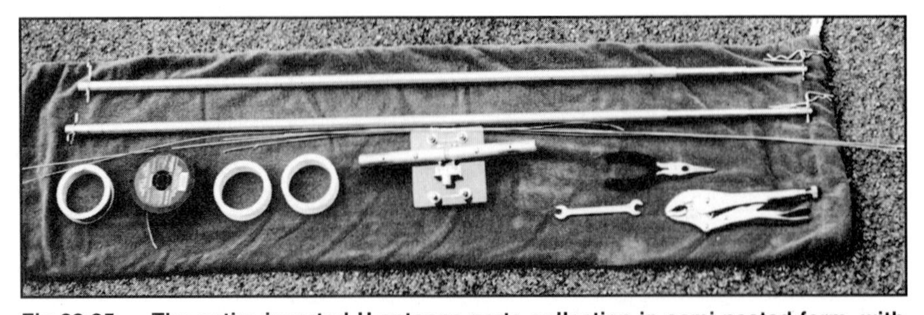

Fig 22.35 — The entire inverted-U antenna parts collection in semi-nested form, with its carrying bag. The tools stored with the antenna include a wrench to tighten the U-bolts for the mast-to-plate mount and a pair of pliers to help remove end wires from the tubing. The pliers have a wire-cutting feature to help replace a broken end wire. A pair of locking pliers makes a good removable handle for turning the mast. The combination of the locking and regular pliers helps to uncoil the wire extensions for any band; give them a couple of sharp tugs to straighten the wire.

2-inch diameter (larger is better). This measure prevents the wire from crimping and eventually breaking. Murphy dictates that a wire will break in the middle of an operating session. So carry some spare wire for replacement ends. All together, the ends require about 50 ft of wire.

Fig 22.33 shows the simple mounting scheme for the end wires. Push the straight wires through a pair of holes aligned vertically to the earth and bend the top portion slightly. To clamp the wire, insert a hitch pin clip though holes parallel to the ground, pushing the wire slightly to one side to reach the far hole in the tube. The double bend holds the wire securely (for a short-term field operation), but allows the wire to be pulled out when the session is over or to change bands.

Add a few inches to the lengths given in Table 22.6 as an initial guide for each band. Test the lengths and prune the wires until you obtain a smooth SWR curve below 2:1 at the ends of each band. Since an inverted-U antenna is full length, the SWR curves will be rather broad and suffer none of the narrow bandwidths associated with inductively loaded elements. **Fig 22.34** shows typical SWR curves for each band to guide your expectations.

You should not require much, if any, adjustment once you have found satisfactory lengths for each band. So you can mark the wire when you finish your initial test adjustments. However, leave enough excess so that you can adjust the lengths in the field.

Do not be too finicky about your SWR curves. An initial test and possibly one adjustment should be all that you need to arrive at an SWR value that is satisfactory for your equipment. Spending half of your operating time adjusting the elements for as near to a 1:1 SWR curve as possible will rob you of valuable contacts without changing your signal strength is any manner that is detectable.

Changing bands is a simple matter. Remove the ends for the band you are using and install the ends for the new band. An SWR check and possibly one more adjustment of the end lengths will put you back on the air.

FINAL NOTES

The inverted-U dipole with interchangeable end pieces provides a compact field antenna. All of the parts fit in a 3-ft long bag. A draw-string bag works very well. **Fig 22.35** shows the parts in their travel form. When assembled and mounted at least 20' up (higher is even better), the antenna will compete with just about any other dipole mounted at the same height. But the inverted-U is lighter than most dipoles at frequencies lower than 10 m. It also rotates easily by hand—assuming that you can rotate the mast by hand. Being able to broadside the dipole to your target station gives the inverted-U a strong advantage over a fixed wire dipole.

With a dipole having drooping ends, safety is very important. Do not use the antenna unless the wire ends for 20 m are higher than any person can touch when the antenna is in use. Even with QRP power levels, the RF voltage on the wire ends can be dangerous. With the antenna at 20 ft at its center, the ends should be at least 10 ft above ground.

Equally important is the maintenance that you give the antenna before and after each use. Be sure that the aluminum tubing is clean—both inside and out—when you nest and un-nest the sections. Grit can freeze the sections together, and dirty tubing can prevent good electrical continuity. Carry a few extra hitch pin clips in the package to be sure you have spares in case you lose one.

TWO W8NX MULTIBAND, COAX-TRAP DIPOLES

Over the last 60 or 70 years, amateurs have used many kinds of multiband antennas to cover the traditional HF bands. The availability of the 30, 17 and 12-m bands has expanded our need for multiband antenna coverage.

Two different antennas are described here. The first covers the traditional 80, 40, 20, 15 and 10-m bands, and the second covers 80, 40, 17 and 12 m. Each uses the same type of W8NX trap—connected for different modes of operation—and a pair of short capacitive stubs to enhance coverage. The W8NX coaxial-cable traps have two different modes: a high- and a low-impedance mode. The inner-conductor windings and shield windings of the traps are connected in series for both modes. However, either the low- or high-impedance point can be used as the trap's output terminal. For low-impedance trap operation, only the center conductor turns of the trap windings are used. For high-impedance operation, all turns are used, in the conventional manner for a trap. The short stubs on each antenna are strategically sized and located to permit more flexibility in adjusting the resonant frequencies of the antenna.

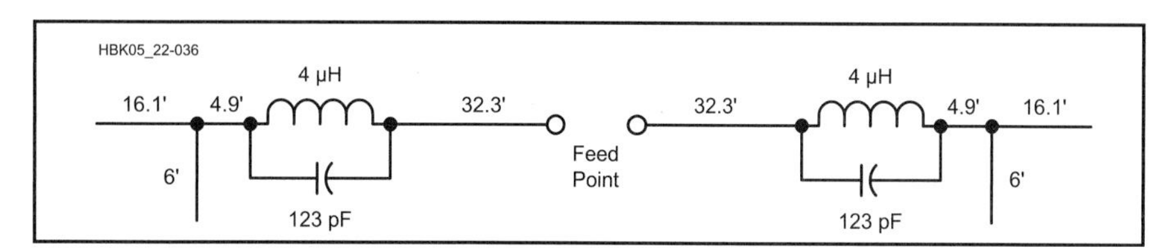

Fig 22.36 — **A W8NX multiband dipole for 80, 40, 20, 15 and 10 m. The values shown (123 pF and 4 μH) for the coaxial-cable traps are for parallel resonance at 7.15 MHz. The low-impedance output of each trap is used for this antenna.**

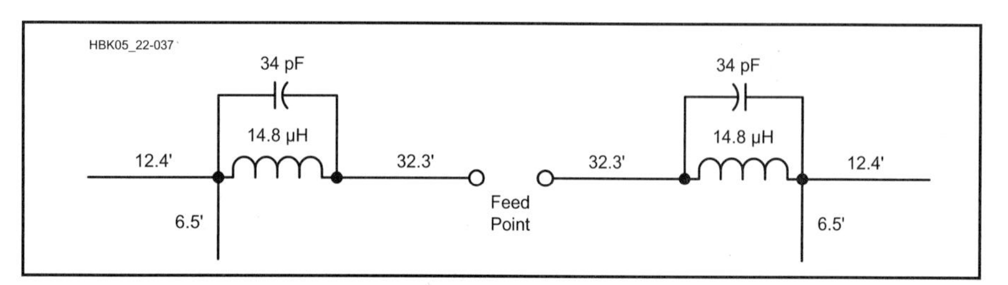

Fig 22.37 — **A W8NX multiband dipole for 80, 40, 17 and 12 m. For this antenna, the high-impedance output is used on each trap. The resonant frequency of the traps is 7.15 MHz.**

80, 40, 20, 15 AND 10-METER DIPOLE

Fig 22.36 shows the configuration of the 80, 40, 20, 15 and 10-m antenna. The radiating elements are made of #14 stranded copper wire. The element lengths are the wire span lengths in feet. These lengths do not include the lengths of the pigtails at the balun, traps and insulators. The 32.3-ft-long inner 40-m segments are measured from the eyelet of the input balun to the tension-relief hole in the trap coil form. The 4.9-ft segment length is measured from the tension-relief hole in the trap to the 6-ft stub. The 16.1-ft outer-segment span is measured from the stub to the eyelet of the end insulator.

The coaxial-cable traps are wound on PVC pipe coil forms and use the low-impedance output connection. The stubs are 6-ft lengths of 1/8-inch stiffened aluminum or copper rod hanging perpendicular to the radiating elements. The first inch of their length is bent 90° to permit attachment to the radiating elements by large-diameter copper crimp connectors. Ordinary #14 wire may be used for the stubs, but it has a tendency to curl up and may tangle unless weighed down at the end. You should feed the antenna with 75-Ω coax cable using a good 1:1 balun.

This antenna may be thought of as a modified W3DZZ antenna due to the addition of the capacitive stubs. The length and location of the stub give the antenna designer two extra degrees of freedom to place the resonant frequencies within the amateur bands. This additional flexibility is particularly helpful to bring the 15 and 10-m resonant frequencies to more desirable locations in these bands. The actual 10-m resonant frequency of the original W3DZZ antenna is somewhat above 30 MHz, pretty remote from the more desirable low frequency end of 10 m.

80, 40, 17 AND 12-METER DIPOLE

Fig 22.37 shows the configuration of the 80, 40, 17 and 12-m antenna. Notice that the capacitive stubs are attached immediately outboard after the traps and are 6.5 ft long, 1/2 ft longer than those used in the other antenna. The traps are the same as those of the other antenna, but are connected for the high-impedance parallel-resonant output mode. Since only four bands are covered by this antenna, it is easier to fine tune it to precisely the desired frequency on all bands. The 12.4-ft tips can he pruned to a particular 17-m frequency with little effect on the 12-m frequency. The stub lengths can be pruned to a particular 12-m frequency with little effect on the 17-m frequency. Both such pruning adjustments slightly alter the 80-m resonant frequency. However, the bandwidths of the antennas are so broad on 17 and 12 m that little need for such pruning exists. The 40-m frequency is nearly independent of adjustments to the capacitive stubs and outer radiating tip elements. Like the first antennas, this dipole is fed with a 75-Ω balun and feed line.

Fig 22.38 shows the schematic diagram of the traps. It illustrates the difference between the low and high-impedance modes of the traps. Notice that the high-impedance terminal is the output configuration used in most conventional trap applications. The low-impedance connection is made across only the inner conductor turns, corresponding to one-half of the total turns of the trap. This mode steps the trap's impedance down to approximately one-fourth of that of the high-impedance level. This is what allows a single trap design to be used for two different multi-band antennas.

Fig 22.39 is a drawing of a cross-section of the coax trap shown through the long axis of the trap. Notice that the traps are conventional coaxial-cable traps, except for the added low-impedance output terminal. The traps are 8¾ close-spaced turns of RG-59 (Belden 8241) on a 2⅜-inch-OD PVC pipe (schedule 40 pipe with a 2-inch ID) coil form. The forms are 4⅛ inches long. Trap resonant frequency is very sensitive to the outer diameter of the coil form, so check it carefully. Unfortunately, not all PVC pipe is made with the same wall thickness. The trap frequencies should be checked with a dip meter and general-coverage receiver and adjusted to within 50 kHz of the 7150 kHz resonant frequency before installation. One inch is left over at each end of the coil forms to allow for the coax feed-through holes and holes for tension-relief attachment of the antenna radiating elements to the traps. Be sure to seal the ends of the trap coax cable with RTV sealant to prevent moisture from entering the coaxial cable.

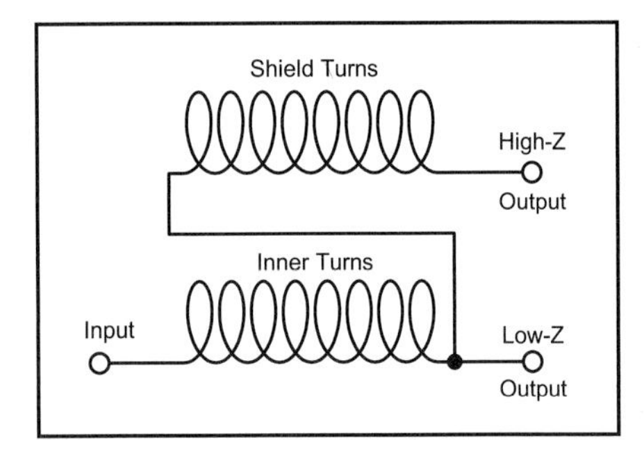

Fig 22.38 — Schematic for the W8NX coaxial-cable trap. RG-59 is wound on a 2⅜-inch OD PVC pipe.

Also, be sure that you connect the 32.3-ft wire element at the start of the inner conductor winding of the trap. This avoids detuning the antenna by the stray capacitance of the coaxial-cable shield. The trap output terminal (which has the shield stray capacitance) should be at the outboard side of the trap. Reversing the input and output terminals of the trap will lower the 40-meter frequency by approximately 50 kHz, but there will be negligible effect on the other bands.

Fig 22.40 shows a coaxial-cable trap. Further details of the trap installation are shown in **Fig 22.41**. This drawing applies specifically to the 80, 40, 20, 15 and 10-m antenna, which uses the low-impedance trap connections. Notice the lengths of the trap pigtails: 3 to 4 inches at each terminal of the trap. If you use a different arrangement, you must modify the span lengths accordingly, All connections can be made using crimp connectors rather than by soldering. Access to the trap's interior is attained more easily with a crimping tool than with a soldering iron.

PERFORMANCE

The performance of both antennas has been very satisfactory. W8NX uses the 80, 40, 17 and 12-m version because it covers 17 and 12 m. (He has a triband Yagi for 20, 15 and 10 m.) The radiation pattern on 17 m is that of 3/2-wave dipole. On 12 m, the pattern is that of a 5/2-wave dipole. At his location in Akron, Ohio, the antenna runs essentially east and west. It is installed as an inverted V, 40 ft high at the center, with a 120° included angle between the legs. Since the stubs are very short, they radiate little power and make only minor contributions to the radiation patterns. In theory, the pattern has four major lobes on 17 m, with maxima to the northeast, southeast, southwest and northwest. These provide low-angle radiation into Europe, Africa, South Pacific, Japan and Alaska. A narrow pair of minor broadside

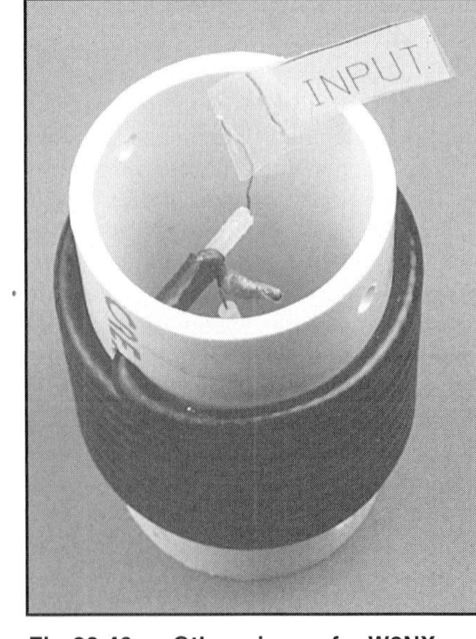

Fig 22.40 — Other views of a W8NX coax-cable trap.

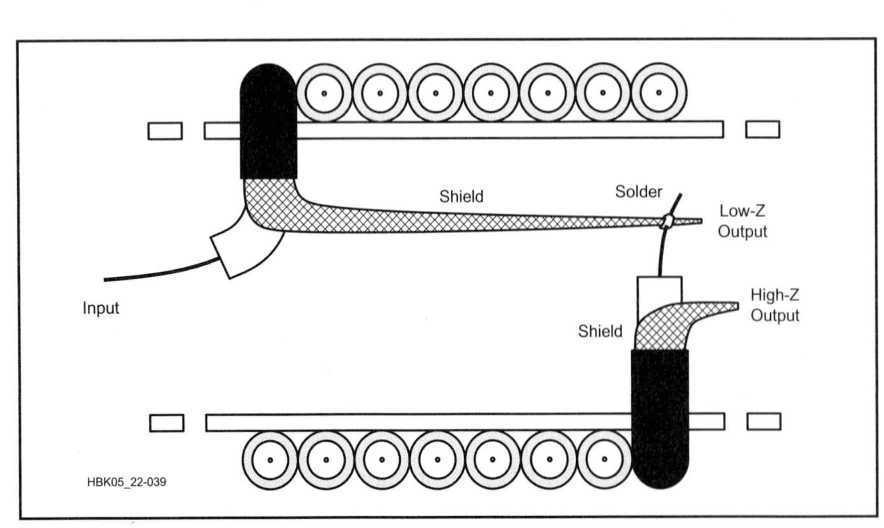

Fig 22.39 — Construction details of the W8NX coaxial-cable trap.

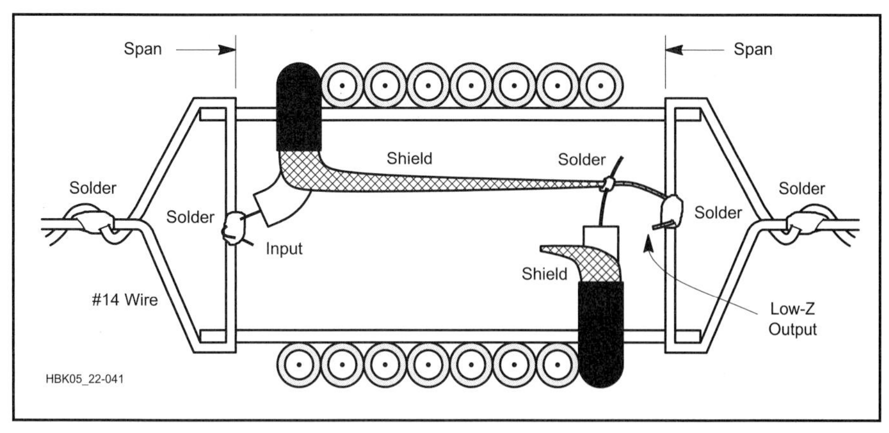

Fig 22.41 — Additional construction details for the W8NX coaxial-cable trap.

lobes provides north and south coverage into Central America, South America and the polar regions.

There are four major lobes on 12 m, giving nearly end-fire radiation and good low-angle east and west coverage. There are also three pairs of very narrow, nearly broadside, minor lobes on 12 m, down about 6 dB from the major end-fire lobes. On 80 and 40 m, the antenna has the usual figure-8 patterns of a half-wave-length dipole.

Both antennas function as electrical half-wave dipoles on 80 and 40 m with a low SWR. They both function as odd-harmonic current-fed dipoles on their other operating frequencies, with higher, but still acceptable, SWR. The presence of the stubs can either raise or lower the input impedance of the antenna from those of the usual third and fifth harmonic dipoles. Again W8NX recommends that 75-Ω, rather than 50-Ω, feed line be used because of the generally higher input impedances at the harmonic operating frequencies of the antennas.

The SWR curves of both antennas were carefully measured using a 75 to 50-Ω transformer from Palomar Engineers inserted at the junction of the 75-Ω coax feed line and a 50-Ω SWR bridge. The transformer is required for accurate SWR measurement if a 50-Ω SWR bridge is used with a 75-Ω line. Most 50-Ω rigs operate satisfactorily with a 75-Ω line, although this requires different tuning and load settings in the final output stage of the rig or antenna tuner. The author uses the 75 to 50-Ω transformer only when making SWR measurements and at low power levels. The transformer is rated for 100 W, and when he runs his 1-kW PEP linear amplifier the transformer is taken out of the line.

Fig 22.42 gives the SWR curves of the 80, 40, 20, 15 and 10-m antenna. Minimum SWR is nearly 1:1 on 80 m, 1.5:1 on 40 m, 1.6:1 on 20 m, and 1.5:1 on 10 m. The minimum SWR is slightly below 3:1 on 15 m. On 15 m, the stub capacitive reactance combines with the inductive reactance of the outer segment of the antenna to produce a resonant rise that raises the antenna input resistance to about 220 Ω, higher than that of the usual $^3/_2$-wave-length dipole. An antenna tuner may be required on this band to keep a solid-state final output stage happy under these load conditions.

Fig 22.43 shows the SWR curves of the 80, 40, 17 and 12-m antenna. Notice the excellent 80-m performance with a nearly unity minimum SWR in the middle of the band. The performance approaches that of a full-size 80-m wire dipole. The short stubs and the low-inductance traps shorten the antenna somewhat on 80 m. Also observe the good 17-m performance, with the SWR being only a little above 2:1 across the band.

But notice the 12-m SWR curve of this antenna, which shows 4:1 SWR across the band. The antenna input resistance approaches 300 Ω on this band because the capacitive reactance of the stubs combines with the inductive reactance of the outer antenna segments to give resonant rises in impedance. These are reflected back to the input terminals. These stub-induced resonant impedance rises are similar to those on the other antenna on 15 meters, but are even more pronounced.

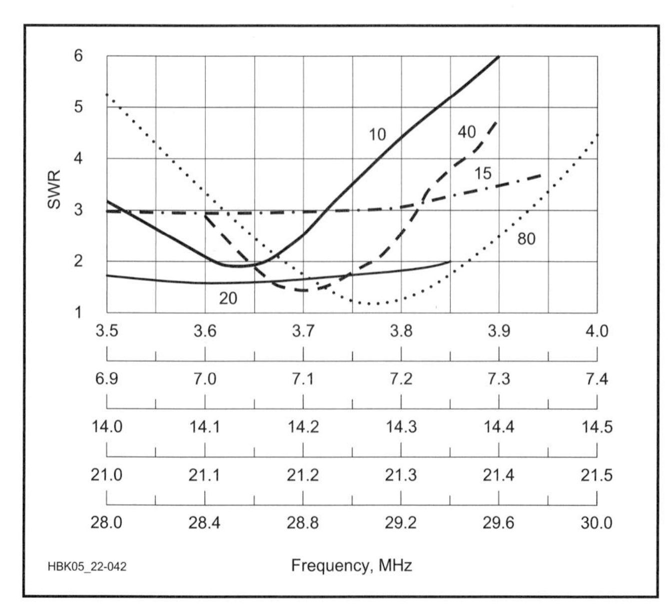

Fig 22.42 — Measured SWR curves for an 80, 40, 20, 15 and 10-meter antenna, installed as an inverted-V with 40-ft apex and 120° included angle between legs.

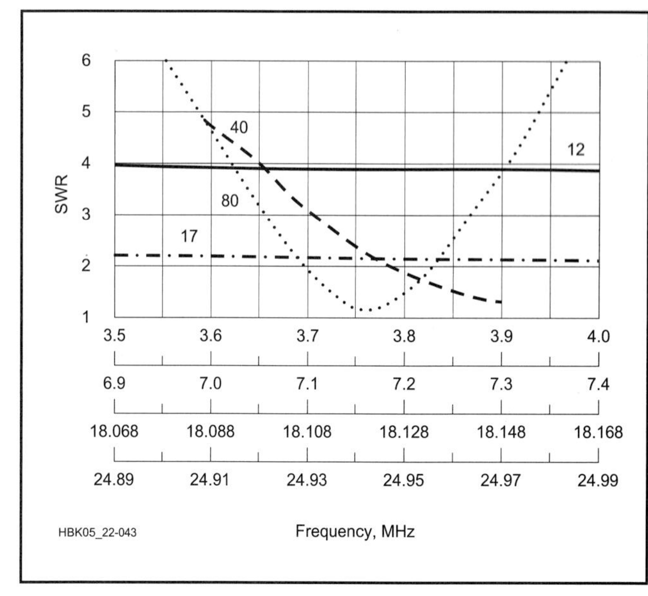

Fig 22.43 — Measured SWR curves for an 80, 40, 17 and 12-meter antenna, installed as an inverted-V with 40-ft apex and 120° included angle between legs.

Too much concern must not be given to SWR on the feed line. Even if the SWR is as high as 9:1 *no destructively high voltages will exist on the transmission line.* Recall that transmission-line voltages increase as the square root of the SWR in the line. Thus, 1 kW of RF power in 75-Ω line corresponds to 274 V line voltage for a 1:1 SWR. Raising the SWR to 9:1 merely triples the maximum voltage that the line must withstand to 822 V. This voltage is well below the 3700-V rating of RG-11, or the 1700-V rating of RG-59, the two most popular 75-Ω coax lines. Voltage breakdown in the traps is also very unlikely. As will be pointed out later, the operating power levels of these antennas are limited by RF power dissipation in the traps, not trap voltage breakdown or feed-line SWR.

TRAP LOSSES AND POWER RATING

Table 22.7 presents the results of trap Q measurements and extrapolation by a two-frequency method to higher frequencies above resonance. W8NX employed an old, but recently calibrated, Boonton Q meter for the measurements. Extrapolation to higher-frequency bands assumes that trap resistance losses rise with skin effect according to the square root of frequency, and that trap dielectric loses rise directly with frequency. Systematic measurement errors are not increased by frequency extrapolation. However, random measurement errors increase in magnitude with upward frequency extrapolation. Results are believed to be accurate within 4% on 80 and 40 m, but only within 10 to 15% at 10 m. Trap Q is shown at both the high- and low-impedance trap terminals. The Q at the low-impedance output terminals is 15 to 20% lower than the Q at the high-impedance output terminals.

W8NX computer-analyzed trap losses for both antennas in free space. Antenna-input resistances at resonance were first calculated, assuming lossless, infinite-Q traps. They were again calculated using the Q values in Table 22.7. The radiation efficiencies were also converted into equivalent trap losses in decibels. **Table 22.8** summarizes the trap-loss analysis for the 80, 40, 20, 15 and 10-m antenna and **Table 22.9** for the 80, 40,17 and 12-m antenna.

The loss analysis shows radiation efficiencies of 90% or more for both antennas on all bands except for the 80, 40, 20, 15 and 10-m antenna when used on 40 m. Here, the radiation efficiency falls to 70.8%. A 1-kW power level at 90% radiation efficiency corresponds to 50-W dissipation per trap. In W8NX's experience, this is the trap's survival limit for extended key-down operation. SSB power levels of 1 kW PEP would dissipate 25 W or less in each trap. This is well within the dissipation capability of the traps.

When the 80, 40, 20, 15 and 10-m antenna is operated on 40 m, the radiation efficiency of 70.8% corresponds to a dissipation of 146 W in each trap when 1 kW is delivered to the antenna. This is sure to burn out the traps—even if sustained for only a short time. Thus, the power should be limited to less than 300 W when this antenna is operated on 40 m under prolonged key-down conditions. A 50% CW duty cycle would correspond to a 600-W power limit for normal 40-m CW operation. Likewise, a 50% duty cycle for 40-m SSB corresponds to a 600-W PEP power limit for the antenna.

The author knows of no analysis where the burnout wattage rating of traps has been rigorously determined. Operating experience seems to be the best way to determine trap burn-out ratings. In his own experience with these antennas, he's had no traps burn out, even though he operated the 80, 40, 20, 15 and 10-m antenna on the critical 40-m band using his AL-80A linear amplifier at 600-W PEP output. He did not make a continuous, key-down, CW operating tests at full power purposely trying to destroy the traps!

Some hams may suggest using a different type of coaxial cable for the traps. The dc resistance of 40.7 Ω per 1000 feet of RG-59 coax seems rather high. However, W8NX has found no coax other than RG-59 that has the necessary inductance-to-capacitance ratio to create the trap characteristic reactance required for the 80, 40, 20, 15 and 10-m antenna. Conventional traps with wide-spaced, open-air inductors and appropriate fixed-value capacitors could be substituted for the coax traps, but the convenience, weatherproof configuration and ease of fabrication of coaxial-cable traps is hard to beat.

Table 22.7

Trap Q

Frequency (MHz)	3.8	7.15	14.18	18.1	21.3	24.9	28.6
High Z out (Ω)	101	124	139	165	73	179	186
Low Z out (Ω)	83	103	125	137	44	149	155

Table 22.8

Trap Loss Analysis: 80, 40, 20, 15, 10-Meter Antenna

Frequency (MHz)	3.8	7.15	14.18	21.3	28.6
Radiation Efficiency (%)	96.4	70.8	99.4	99.9	100.0
Trap Losses (dB)	0.16	1.5	0.02	0.01	0.003

Table 22.9

Trap Loss Analysis: 80, 40, 17, 12-Meter Antenna

Frequency (MHz)	3.8	7.15	18.1	24.9
Radiation Efficiency (%)	89.5	90.5	99.3	99.8
Trap Losses (dB)	0.5	0.4	0.03	0.006

Vertical Antennas

One of the more popular amateur antennas is the *vertical*. It usually refers to a single radiating element placed vertically over the ground. A typical vertical is an electrical 1/4-λ long and is constructed of wire or tubing.

Single vertical antennas are omnidirectional radiators. This can be beneficial or detrimental, depending on the exact situation. On transmission there are no nulls in any direction, unlike most horizontal antennas. However, QRM on receive can't be nulled out from the directions that are not of interest, unless multiple verticals are used in an array.

When compared to horizontal antennas, verticals also suffer more acutely from two main types of losses—*ground return losses* for currents in the near field, and *far-field ground losses*. Ground losses in the near field can be minimized by using many ground radials. This is covered in the sidebar, **Optimum Ground Systems for Vertical Antennas**.

Far-field losses are highly dependent on the conductivity and dielectric constant of the earth around the antenna, extending out as far as 100 λ from the base of the antenna. There is very little that someone can do to change the character of the ground that far away—other than moving to a small island surrounded by saltwater! Far-field losses greatly affect low-angle radiation, causing the radiation patterns of practical vertical antennas to fall far short of theoretical patterns over *perfect ground*, often seen in classical texts. **Fig 22.44** shows the elevation pattern response for two different 40-m quarter-wave verticals. One is placed over a theoretical infinitely large, infinitely conducting ground. The second is placed over an extensive radial system over average soil, having a conductivity of 5 mS/m and a dielectric constant of 13. This sort of soil is typical of heavy clay found in pastoral regions of the US mid-Atlantic states. At a 10° elevation angle, the real antenna losses are almost 6 dB compared to the theoretical one; at 20° the difference is about 3 dB. See *The ARRL Antenna Book* chapter on the effects of the earth for further details.

While real verticals over real ground are not a magic method to achieve low-angle radiation, cost versus performance and ease of installation are incentives that inspire many antenna builders. For use on the lower frequency amateur bands—notably 160 and 80 m—it is not always

Optimum Ground Systems for Vertical Antennas

A frequent question brought up by old-timers and newcomers alike is: "So, how many ground radials do I *really* need for my vertical antenna?" Most hams have heard the old standby tales about radials, such as "if a few are good, more must be better" or "lots of short radials are better than a few long ones."

John Stanley, K4ERO, eloquently summarized a study he did of the professional literature on this subject in his article "Optimum Ground Systems for Vertical Antennas" in December 1976 *QST*. His approach was to present the data in a sort of "cost-benefit" style in Table A, reproduced here. John somewhat wryly created a new figure of merit—the total amount of wire needed for various radial configurations. This is expressed in terms of wavelengths of total radial wire.

Table A

Optimum Ground-System Configurations

Configuration Designation	*A*	*B*	*C*	*D*	*E*	*F*
Number of radials	*16*	*24*	*36*	*60*	*90*	*120*
Length of each radial in wavelengths	0.1	0.125	0.15	0.2	0.25	0.4
Spacing of radials in degrees	22.5	15	10	6	4	3
Total length of radial wire installed, in wavelengths	1.6	3	5.4	12	22.5	48
Power loss in dB at low angles with a quarter-wave radiating element	3	2	1.5	1	0.5	0*
Feed-point impedance in ohms with a quarter-wave radiating element	52	46	43	40	37	35

Note: Configuration designations are indicated only for text reference.

*Reference. The loss of this configuration is negligible compared to a perfectly conducting ground.

The results almost jumping out of this table are:

- If you can only install 16 radials (Case A), they needn't be very long—0.1 λ is sufficient. You'll use 1.6 λ of radial wire in total, which is about 450 feet at 3.5 MHz.
- If you have the luxury of laying down 120 radials (Case F), they should be 0.4 λ long, and you'll gain about 3 dB over the 16-radial case. You'll also use 48 λ of total wire—For 80 meters, that would be about 13,500 feet!
- If you can't put out 120 radials, but can install 36 radials that are 0.15 λ long (Case C), you'll lose only 1.5 dB compared to the optimal Case F. You'll also use 5.4 λ of total wire, or 1,500 feet at 3.5 MHz.
- A 50-Ω SWR of 1:1 isn't necessary a good thing—the worst-case ground system in Case A has the lowest SWR.

Table A represents the case for "Average" quality soil, and it is valid for radial wires either laid on the ground or buried several inches in the ground. Note that such ground-mounted radials are detuned because of their proximity to that ground and hence don't have to be a classical quarter-wave length that they need to be were they in "free space."

In his article John also made the point that ground-radial losses would only be significant on transmit, since the atmospheric noise on the amateur bands below 30 MHz is attenuated by ground losses, just like actual signals would be. This limits the ultimate signal-to-noise ratio in receiving.

So, there you have the tradeoffs—the loss in transmitted signal compared to the cost (and effort) needed to install more radial wires. You take your pick.

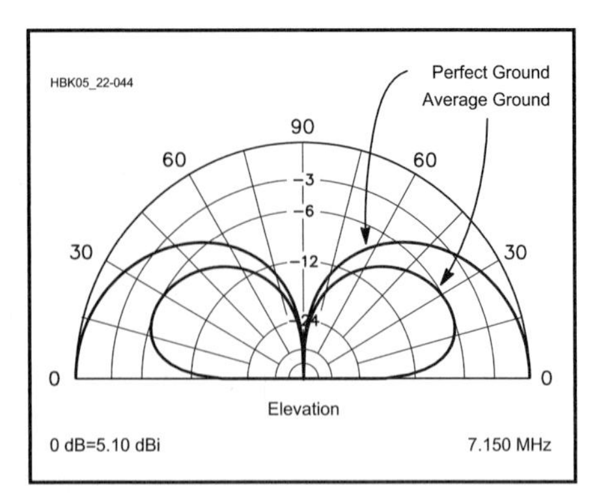

Fig 22.44 — Elevation patterns for two quarter-wave vertical antennas over different ground. One vertical is placed over *perfect* ground, and the other is placed over average ground. The far-field response at low elevation angles is greatly affected by the quality of the ground — as far as 100 λ away from the vertical antenna.

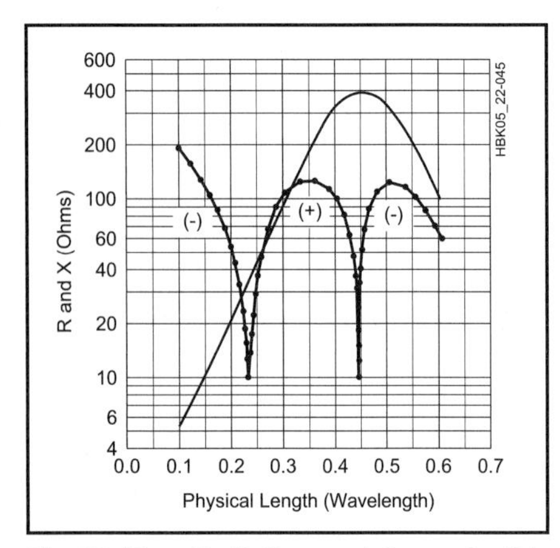

Fig 22.45 — Radiation resistance (solid curve) and reactance (dotted curve) of vertical antennas as a function of their physical height.

practical to erect a full-size vertical. At 1.8 MHz, a full-sized quarter-wave vertical is 130 ft high. In such instances it is often necessary to accept a shorter radiating element and use some form of *loading*.

Fig 22.45 provides curves for the physical height of verticals in wavelength versus radiation resistance and reactance. Although the plots are based on perfectly conducting ground, they show general trends for installations where many radials have been laid out to make a ground screen. As the radiator is made shorter, the radiation resistance decreases—with 6 Ω being typical for a 0.1-λ high antenna. The lower the radiation resistance, the more the antenna efficiency depends on ground conductivity and the effectiveness of the ground screen. Also, the bandwidth decreases markedly as the length is reduced toward the left of the scale in Fig 22.45. It can be difficult to develop suitable matching networks when radiation resistance is very low.

GROUND SYSTEMS

Generally a large number of shorter radials offers a better ground system than a few longer ones. For example, 8 radials of $^1/_4$ λ are preferred over 4 radials of $^1/_4$ λ. Optimum radial lengths are described in the sidebar.

The conductor size of the radials is not especially significant. Wire gauges from #4 to #20 have been used successfully by amateurs. Copper wire is preferred, but where soil is low in acid (or alkali), aluminum wire can be used. The wires may be bare or insulated, and they can be laid on the earth's surface or buried a few inches below ground. Insulated wires will have greater longevity by virtue of reduced corrosion and dissolution from soil chemicals.

When property dimensions do not allow a classic installation of equally spaced radial wires, they can be placed on the ground as space permits. They may run away from the antenna in only one or two compass directions. They may be bent to fit on your property.

A single ground rod, or group of them bonded together, is seldom as effective as a collection of random-length radial wires.

All radial wires should be connected together at the base of the vertical antenna. The electrical bond needs to be of low resistance. Best results will be obtained when the wires are soldered together at the junction point. When a grounded vertical is used, the ground wires should be affixed securely to the base of the driven element.

Ground return losses are lower when vertical antennas and their radials are elevated above ground, a point that is well-known by those using *ground plane* antennas on their roofs. Even on 160 or 80 m, effective vertical antenna systems can be made with as few as four quarter-wave long radials elevated 10 to 20 ft off the ground.

FULL-SIZE VERTICAL ANTENNAS

When it is practical to erect a full-size $^1/_4$-λ vertical antenna, the forms shown in **Fig 22.46** are worthy of consideration. The example at A is the well-known *vertical ground plane*. The ground system consists of four above-ground radial wires. The length of the driven element and $^1/_4$-λ radials is derived from the standard equation

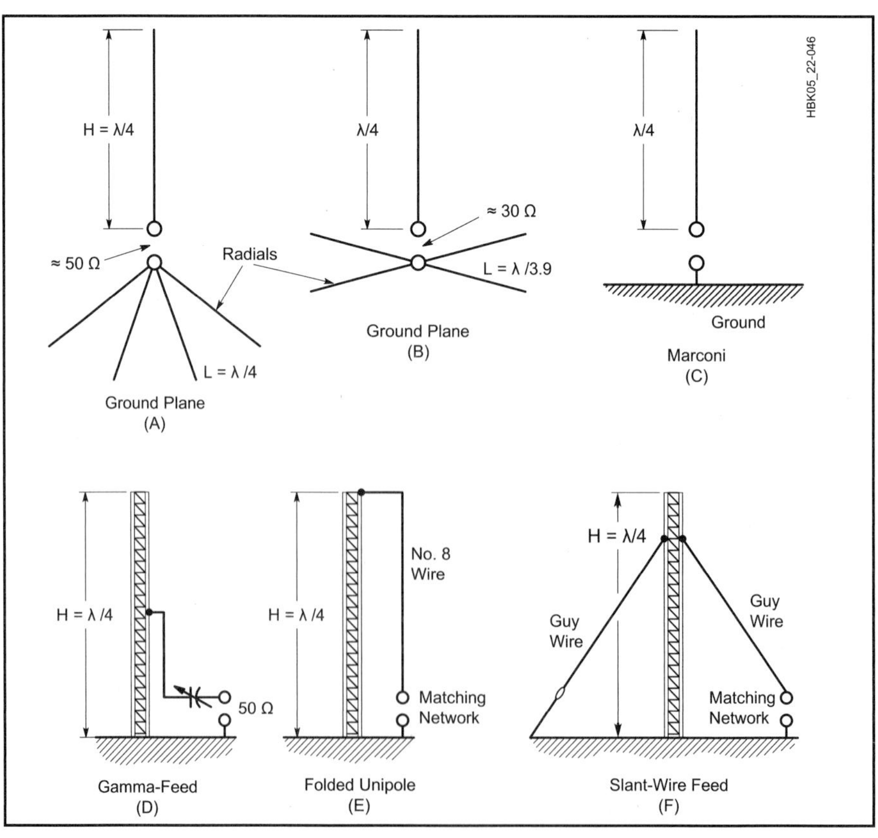

Fig 22.46 — Various types of vertical antennas.

$$L\,(ft) = \frac{234}{f\,(MHz)} \quad (6)$$

With four equidistant radial wires drooped at approximately 30° (Fig 22.46A), the feed-point impedance is roughly 50 Ω. When the radials are at right angles to the radiator (Fig 22.46B) the impedance approaches 36 Ω. Besides minimizing ground return losses, another major advantage in this type of vertical antenna over a ground-mounted type is that the system can be elevated well above nearby conductive objects (power lines, trees, buildings and so on). When drooping radials are used, they can also serve as guy wires for the mast that supports the antenna. The coax shield braid is connected to the radials, and the center conductor to the driven element.

The *Marconi* vertical antenna shown in Fig 22.46C is the classic form taken by a ground-mounted vertical. It can be grounded at the base and shunt fed, or it can be isolated from ground, as shown, and series fed. As always, this vertical antenna depends on an effective ground system for efficient performance. If a perfect ground were located below the antenna, the feed impedance would be near 36 Ω. In a practical case, owing to imperfect ground, the impedance is more apt to be in the vicinity of 50 Ω.

A gamma feed system for a grounded $^{1}/_{4}$-λ vertical is presented in Fig 22.46D. Some rules of thumb for arriving at workable gamma-arm and capacitor dimensions are to make the rod length 0.04 to 0.05 λ, its diameter $^{1}/_{3}$ to $^{1}/_{2}$ that of the driven element and the center-to-center spacing between the gamma arm and the driven element roughly 0.007 λ. The capacitance of C1 at a 50-Ω matched condition will be about 7 pF per meter of wavelength. The absolute value of C1 will depend on whether the vertical is resonant and on the precise value of the radiation resistance. For best results, make the radiator approximately 3% shorter than the resonant length.

Amateur antenna towers lend themselves to use as shunt-fed verticals, even though an HF-band beam antenna is usually mounted on the tower. The overall system should be close to resonance at the desired operating frequency if a gamma feed is used. The HF-band beam will contribute somewhat to *top loading* of the tower. The natural resonance of such a system can be checked by dropping a #12 or #14 wire from the top of the tower (connecting it to the tower top) to form a folded unipole (Fig 22.46E). A four- or five-turn link can be inserted between the lower end of the drop wire and the ground system. A dip meter is then inserted in the link to determine the resonant frequency. If the tower is equipped with guy wires, they should be broken up with strain insulators to prevent unwanted loading of the vertical. In such cases where the tower and beam antennas are not able to provide $^{1}/_{4}$-λ resonance, portions of the top guy wires can be used as top-loading capacitance. Experiment with the guy-wire lengths (using the dip-meter technique) while determining the proper dimensions.

A folded-unipole is depicted at E of Fig 22.46. This system has the advantage of increased feed-point impedance. Furthermore, a Transmatch can be connected between the bottom of the drop wire and the ground system to permit operation on more than one band. For example, if the tower is resonant on 80 m, it can be used as shown on 160 and 40 m with reasonable results, even though it is not electrically long enough on 160. The drop wire need not be a specific distance from the tower, but you might try spacings between 12 and 30 inches.

The method of feed shown at Fig 22.46F is commonly referred to as *slant-wire feed*. The guy wires and the tower combine to provide quarter-wave resonance. A matching network is placed between the lower end of one guy wire and ground and adjusted for an SWR of 1:1. It does not matter at which level on the tower the guy wires are connected, assuming that the Transmatch is capable of effecting a match to 50 Ω.

PHYSICALLY SHORT VERTICALS

A group of short vertical radiators is presented in **Fig 22.47**. Illustrations A and B are for top and center loading. A capacitance hat is shown in each example. The hat should be as large as practical to increase the radiation resistance of the antenna and improve the bandwidth. The wire in the loading coil is chosen for the largest gauge consistent with ease of winding and coil-form size. The larger wire diameters will reduce the resistive (I^2R) losses in the system. The coil-form material should have a medium or high dielectric constant. Phenolic or fiberglass tubing is entirely adequate.

A base-loaded vertical is shown at C of Fig 22.47. The primary limitation is that the high current portion of the vertical exists in the coil rather than the driven element. With center loading, the portion of the antenna below the coil carries high

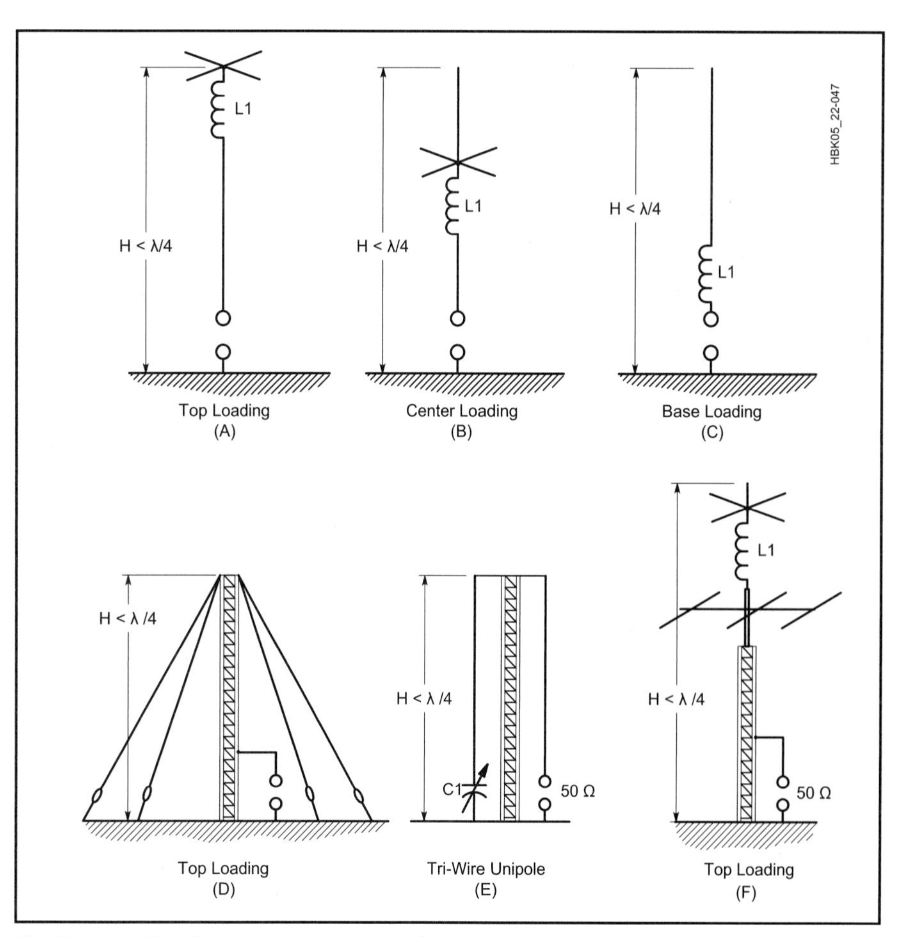

Fig 22.47 — Vertical antennas that are less than one-quarter wavelength in height.

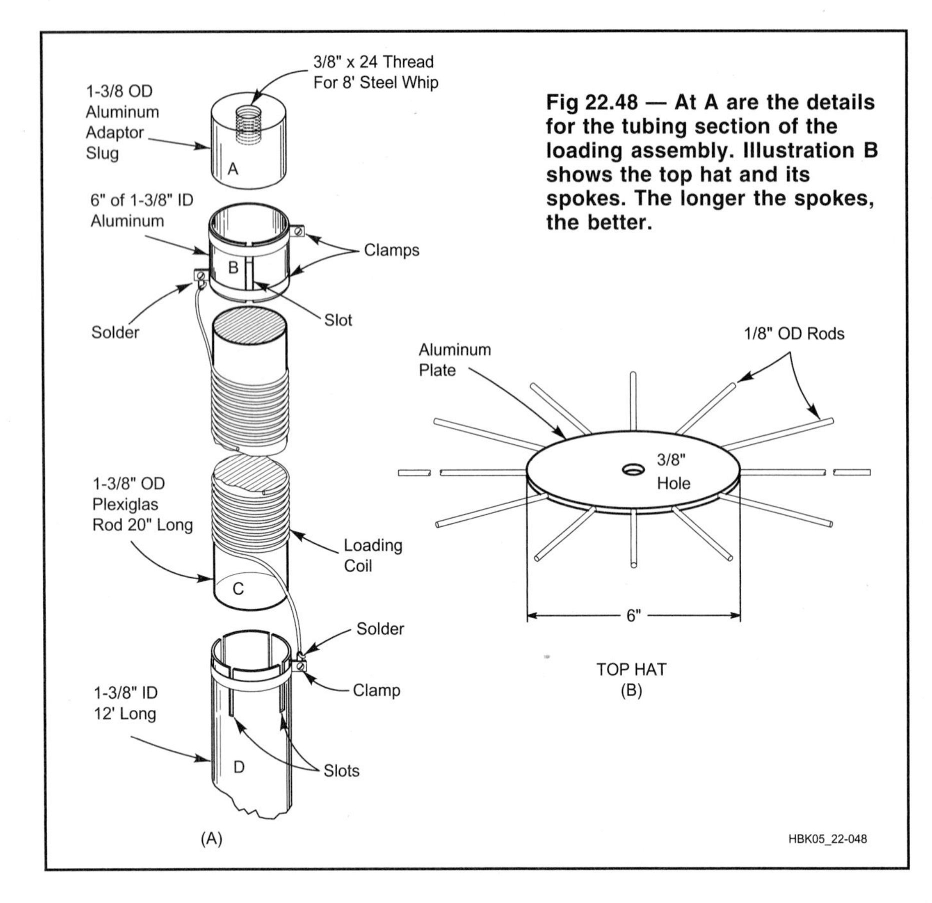

Fig 22.48 — At A are the details for the tubing section of the loading assembly. Illustration B shows the top hat and its spokes. The longer the spokes, the better.

current, and in the top-loaded version the entire vertical element carries high current. Since the high-current part of the antenna is responsible for most of the radiating, base loading is the least effective of the three methods. The radiation resistance of the coil-loaded antennas shown is usually less than 16 Ω.

A method for using guy wires to top load a short vertical is illustrated in Fig 22.47D. This system works well with gamma feed. The loading wires are trimmed to provide an electrical quarter wavelength for the overall system. This method of loading will result in a higher radiation resistance and greater bandwidth than the systems shown at A through C. If an HF or VHF array is at the top the tower, it will simply contribute to the top loading.

A three-wire unipole is shown at E. Two #8 drop wires are connected to the top of the tower and brought to ground level. The wires can be spaced any convenient distance from the tower—normally 12 to 30 inches from one side. C1 is adjusted for best SWR. This type of vertical has a fairly narrow bandwidth, but because C1 can be motor driven and controlled from the operating position, frequency changes can be accomplished easily. This technique will not be suitable for matching to 50-Ω line unless the tower is less than an electrical quarter wavelength high.

A different method for top loading is shown at F. Barry Boothe, W9UCW, described this method in December 1974 *QST*. An extension is used at the top of the tower to effect an electrical quarter-wavelength vertical. L1 is a loading coil with sufficient inductance to provide antenna resonance. This type of antenna lends itself to operation on 160 m.

A method for constructing the top-loading shown in Fig 22.47F is illustrated in **Fig 22.48**. Pipe section D is mated with the mast above the HF-band beam antenna. A loading coil is wound on solid Plexiglas rod or phenolic rod (item C), then clamped inside the collet (B). An aluminum slug (part A) is clamped inside item B. The top part of A is bored and tapped for a 3/8 × 24 stud. This permits a standard 8-ft stainless-steel mobile whip to be threaded into item A above the loading coil. The capacitance hat (Fig 22.48B) can be made from a 1/4-inch-thick brass or aluminum plate. It may be round or square. Lengths of 1/8-inch brazing rod can be threaded and screwed into the edge of the aluminum plate. The plate contains a row of holes along its perimeter, each having been tapped for a 6-32 thread. The capacitance hat is affixed to item A by means of the 8-ft whip antenna. The whip will increase the effective height of the vertical antenna.

CABLES AND CONTROL WIRES ON TOWERS

Most vertical antennas of the type shown in Fig 22.46 consist of towers, usually with HF or VHF beam antennas at the top. The rotator control wires and the coaxial feeders to the top of the tower will not affect antenna performance adversely. In fact, they become a part of the composite antenna. To prevent unwanted RF currents from following the wires into the shack, simply dress them close to the tower legs and bring them to ground level. This decouples the wires at RF. The wires should then be routed along the earth surface (or buried underground) to the operating position. It is not necessary to use bypass capacitors or RF chokes in the rotator control leads if this is done, even when maximum legal power is employed.

TRAP VERTICALS

The 2-band trap vertical antenna of **Fig 22.49** operates in much the same manner as a trap dipole or trap Yagi. The notable

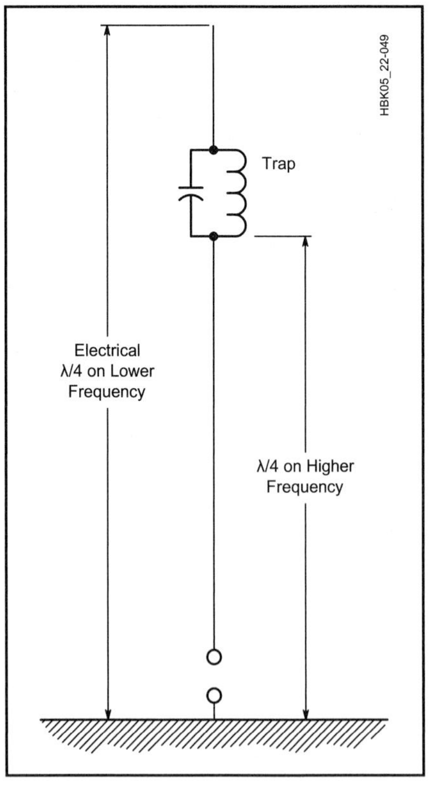

Fig 22.49 — A two-band trap vertical antenna. The trap should be resonated by itself as a parallel resonant circuit at the center of the operating range for the higher frequency band. The reactance of either the inductor or the capacitor range from 100 to 300 Ω. At the lower frequency the trap will act as a loading inductor, adding electrical length to the total antenna.

difference is that the vertical is one half of a dipole. The radial system (in-ground or above-ground) functions as a ground plane for the antenna, and represents the missing half of the dipole. Once again, the more effective the ground system, the better will be the antenna performance.

Trap verticals usually are adjusted as $^1/_4$-λ radiators. The portion of the antenna below the trap is adjusted as a $^1/_4$-λ radiator at the higher proposed operating frequency. That is, a 20/15-m trap vertical would be a resonant quarter wavelength at 15 m from the feedpoint to the bottom of the trap. The trap and that portion of the antenna above the trap (plus the 15-m section below the trap) constitute the complete antenna during 20-m operation. But because the trap is in the circuit, the overall physical length of the vertical antenna will be slightly less than that of a single-band, full-size 20-m vertical.

TRAPS

The trap functions as the name implies: It traps the 15-m energy and confines it to the part of the antenna below the trap. During 20-m operation it allows the RF energy to reach all of the antenna. The trap in this example is tuned as a parallel resonant circuit to 21 MHz. At this frequency it divorces the top section of the vertical from the lower section because it presents a high impedance (barrier) at 21 MHz. Generally, the trap inductor and capacitor have a reactance of 100 to 300 Ω. Within that range it is not critical.

The trap is built and adjusted separately from the antenna. It should be resonated at the center of the portion of the band to be operated. Thus, if one's favorite part of the 15-m band is between 21.0 and 21.1 MHz, the trap should be tuned to 21.05 MHz.

Resonance is checked by using a dip meter and detecting the dipper signal in a calibrated receiver. Once the trap is adjusted it can be installed in the antenna, and no further adjustment will be required. It is easy, however, to be misled after the system is assembled: Attempts to check the trap with a dip meter will suggest that the trap has moved much lower in frequency (approximately 5 MHz lower in a 20/15-m vertical). This is because the trap is part of the overall antenna, and the resultant resonance is that of the total antenna. Measure the trap separate from the rest of the antenna.

Multiband operation is quite practical by using the appropriate number of traps and tubing sections. The construction and adjustment procedure is the same, regardless of the number of bands covered. The highest frequency trap is always closest to the feed end of the antenna, and the lowest frequency trap is always the farthest from the feedpoint. As the operating frequency is progressively lowered, more traps and more tubing sections become a functional part of the antenna.

Traps should be weatherproofed to prevent moisture from detuning them. Several coatings of high dielectric compound, such as Polystyrene Q Dope, are effective. Alternatively, a protective sleeve of heat-shrink tubing can be applied to the coil after completion. The coil form for the trap should be of high dielectric quality and be rugged enough to sustain stress during periods of wind.

DUAL-BAND VERTICALS FOR 17/40 OR 12/30 M

Thanks to the harmonic relationships between the HF ham bands, many antennas can be made to do double duty. The simple verticals described here cover two bands at once. Here's how to turn a 30-m $^1/_4$-λ vertical into a 0.625-λ vertical for the 12-m band, and a 40-m $^1/_4$-λ vertical into a 0.625-λ vertical for the 17-m band. These verticals were designed and constructed by John J. Reh, K7KGP. The write-up first appeared in April 1989 *QST*.

CONSTRUCTION DETAILS

For the 30 and 12-m vertical, an old aluminum multiband vertical was cut to a length of 25 ft, 3 inches. This corresponds to a design frequency of 24.95 MHz. The length-to-diameter ratio is approximately 460. The input impedance of a vertical that is substantially longer than a $^1/_4$-λ (in this case 0.625 λ) is particularly sensitive to the λ/D ratio of the radiating element. If this antenna is duplicated with materials having a significantly different λ/D ratio, the results may be different.

After installing a good ground system, the input impedance was measured and found to have a resistance of about 50 Ω, and a capacitance of about –155 Ω (at 24.95 MHz). At 10.125 MHz, the input impedance was just under 50 Ω, and purely resistive. To tune out the reactance at 24.95 MHz, a series inductor is installed (see **Fig 22.50**) and tapped to resonance at the design frequency. The easiest way to find resonance is by measuring the antenna SWR. Use a good-quality coil for the series inductor. The recommended coil has a diameter of $2^1/_2$ inches, and has 6 turns per inch (B&W stock no. 3029). Resonance on 12 m was established with $3^1/_4$ turns. The SWR on 12 m is 1.1:1, and on 30 m, 1:1. To change bands from 12 to 30 m, move the coil tap to the end of the coil closest to the vertical element. Alternatively, a single-pole switch or remotely operated relay can be installed at the base of the vertical for bandswitching. Later, you'll see how to build the antenna with automatic bandswitching.

THE GROUND SYSTEM

Maximum RF current density—and therefore maximum ground losses—for $^1/_4$-λ verticals occurs in the immediate area of the base of the antenna. Maximum return current ground loss for a 0.625-λ vertical occurs about $^1/_2$ λ away from the base of the antenna. It's important to have the lowest possible losses in the immediate area for both types of verticals. In addition to a ground radial system, 6×6-ft

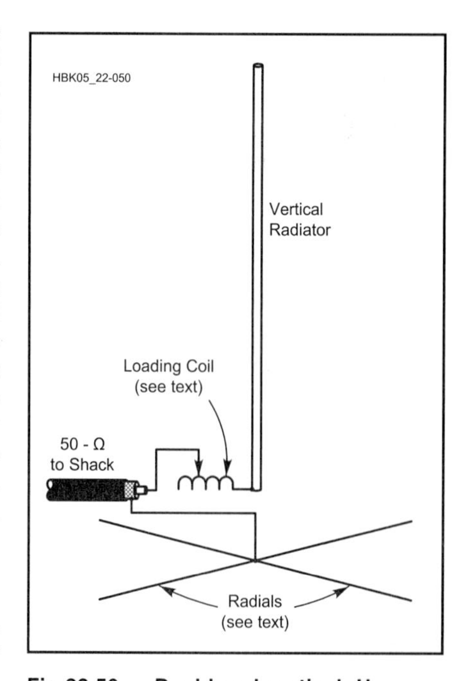

Fig 22.50 — Dual-band vertical. Use a switch or relay to remove the loading coil from the circuit for lower frequency operation. Adjust the coil tap for best SWR on the higher-frequency band. The radial system should be as extensive as possible. See *The ARRL Antenna Book* for more information on ground systems for vertical antennas.

aluminum ground screen is used at the base of the antenna. The screen makes a good tie point for the radials and conducts ground currents efficiently. Seventeen wire radials, each about 33 ft long, are spaced evenly around the antenna. More radials would probably work better. Each radial is bolted to the screen using corrosion-resistant #10-24 hardware. (Do not attempt to connect copper directly to aluminum. The electrical connection between the two metals will quickly deteriorate.) The radials can be made of bare or insulated wire. Make sure the ground screen is bolted to the ground side of the antenna with heavy-gauge wire. Current flow is fairly heavy at this point.

Table 22.10 gives specifications for the dual-band vertical. If your existing 40-m vertical is a few inches longer than 32 ft, 3 inches, try using it anyway—a few inches isn't too critical to performance on 17 m.

Table 22.10
Specifications for Dual-Band Verticals

Bands	*Height*	*Required Matching Inductance (μH)*
12 m & 30 m	23' 5"	0.99
17 m & 40 m	32' 3"	1.36

AUTOMATIC BANDSWITCHING

In October 1989 *QST*, James Johnson, W8EUI, presented this scheme for automatic bandswitching of the 40/17-m vertical. Johnson shortened his 40-m vertical approximately 12 inches and found an inductance that gave him 40 and 17-m band operation with an SWR of less than 1.4:1 across each band. He used an inductor made from B&W air-wound coil stock (no. 3033). This coil is 3 inches in diameter, and has 3 1/8 turns of #12 wire wound at 6 turns per inch, providing an inductance of about 2.8 μH. Johnson experimentally determined the correct tap position.

For the 30/12-m version, start with the vertical radiator 9 inches shorter than the value given in the table. In both cases, radiator height and inductance should be adjusted for optimum match on the two bands covered.

Inverted L and Sloper Antennas

This section covers variations on the vertical antenna. **Fig 22.51A** shows a flat-top T vertical. Dimension H should be as tall as possible for best results. The horizontal section, L, is adjusted to a length that provides resonance. Maximum radiation is polarized vertically despite the horizontal top-loading wire. A variation of the T antenna is depicted at B of Fig 22.51. This antenna is commonly referred to as an *inverted L*. Vertical member H should be as long as possible. L is added to provide an electrical quarter wavelength overall.

THE HALF-SLOPER ANTENNA

Many hams have had excellent results with *half-sloper* antennas, while others have not had such luck. Investigations by ARRL Technical Advisor John S. Belrose, VE2CV, have brought some insight to the situation through computer modeling with *ELNEC* and antenna-range tests. The following is taken from VE2CV's Technical Correspondence in Feb 1991 *QST*, pp 39 and 40. Essentially, the half sloper is a top-fed vertical antenna worked against a ground plane (such as a grounded Yagi antenna) at the top of the tower. The tower acts as a reflector.

For half slopers, the input impedance, the resonant length of the sloping wire and the antenna pattern all depend on the tower height, the angle (between the sloper and tower) the type of Yagi and the Yagi orientation. Here are several configurations extracted from VE2CV's work:

At 160 m—use a 40-m beam on top of a 95-ft tower with a 55° sloper apex angle. The radiation pattern varies little with Yagi type. The pattern is slightly cardioid with about 8 dB front-to-back ratio at a 25° takeoff angle (see Fig 22.51D and E). Input impedance is about 50 Ω.

At 80 m—use a 20-m beam on top of a 50-ft tower with a 55° sloper apex angle. The radiation pattern and input impedance are similar to those of the 160-m half sloper.

At 40 m—use a 20-m beam on top of a 50-ft tower with a 55° sloper apex angle. The radiation pattern and impedance depend strongly on the azimuth orientation of the Yagi. Impedance varies from 76 to 127 Ω depending on Yagi direction.

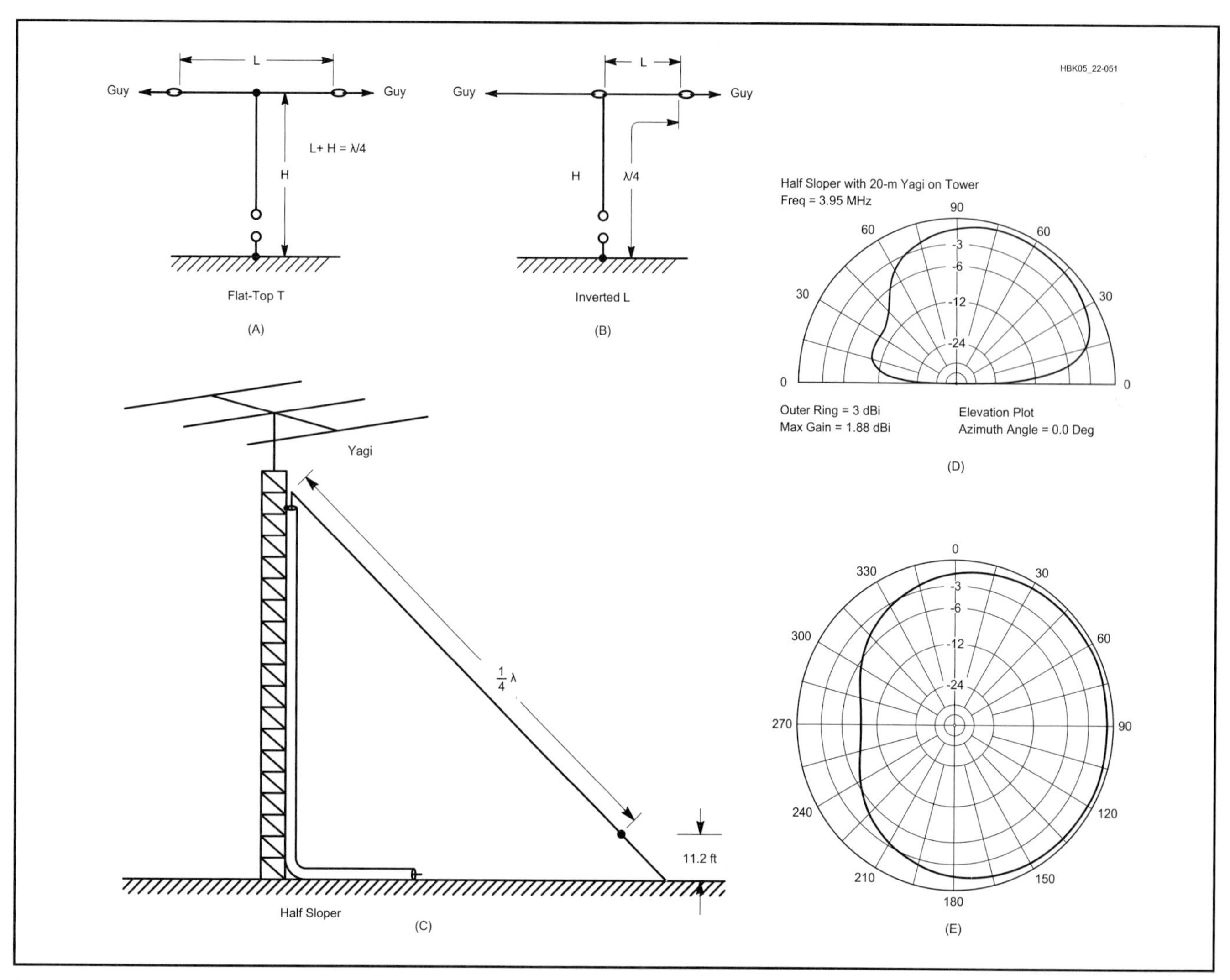

Fig 22.51 — Some variations in vertical antennas. D is the vertical radiation pattern in the plane of a half sloper, with the sloper to the right. E is the azimuthal pattern of the half sloper (90° azimuth is the direction of the sloping wire). Both patterns apply to 160- and 80-m antennas described in the text.

1.8-MHz INVERTED L

The antenna shown in **Fig 22.52** is simple and easy to construct. It is a good antenna for the beginner or the experienced 1.8 MHz DXer. Because the overall electrical length is greater than $^1/_4$ λ, the feed-point resistance is on the order of 50 Ω, with an inductive reactance. That reactance is canceled by a series capacitor, which for power levels up to the legal limit can be an air-variable capacitor with a voltage rating of 1500 V. Adjust antenna length and vari-

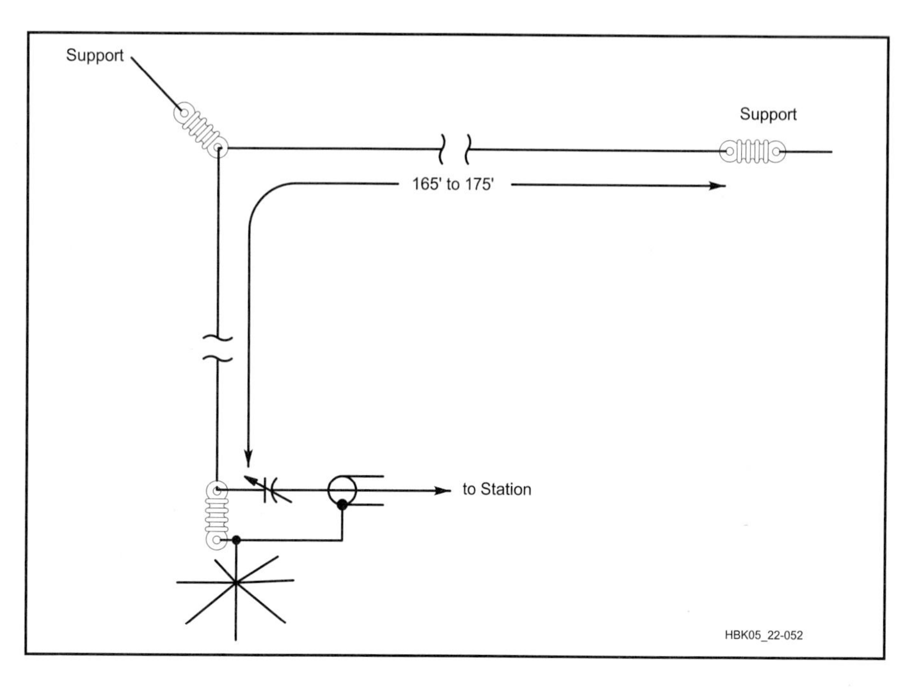

Fig 22.52 — The 1.8-MHz inverted L. Overall wire length is 165 to 175 ft. The variable capacitor has a maximum capacitance of 500 to 800 pF.

able capacitor for lowest SWR.

A yardarm or a length of line attached to a tower can be used to support the vertical section of the antenna. (Keep the inverted L as far from the tower as is practical. Certain combinations of tower height and Yagi top loading can interact severely with the Inverted-L antenna—a 70-ft tower and a 5-element Yagi, for example.) For best results the vertical section should be as long as possible. A good ground system is necessary for good results.

THE HALF-WAVE VERTICAL DIPOLE (HVD)

Chuck Hutchinson, K8CH, describes a 15-m vertical dipole (HVD) that he built in the ARRL book, *Simple and Fun Antennas for Hams*. The performance of this antenna, with its base at 14 ft, compares favorably with a horizontal dipole at 30 ft when making intercontinental QSOs.

CONSTRUCTION OF A 15-M HVD

The 15-meter HVD consists of four 6-ft lengths of 0.875-inch aluminum tube with 0.058 wall thickness. In addition there are two 1-ft lengths of 0.75-inch tubing for splices, and two one-foot lengths of 0.75-inch fiberglass rod for insulators. See **Table 22.11** for dimensions.

Start by cutting off 1 foot from a 6-ft length of 0.875-inch tubing. Next, insert six inches of one of the 1-foot-long 0.75-inch tubes into the machine-cut end of your tubing and fasten the tubes together. Now, slide an end of a 6-ft length of 0.875 tube over the protruding end of the 0.75 tube and fasten them together. Repeat this procedure with the remaining 0.875-inch tubing.

Table 22.11
HVD Dimensions

Length using 0.875-inch aluminum tubing

MHz	*Feet*	*Inches*
18.11	33	11
21.2	22	0
24.94	18	9
28.4	16	5

These lengths should divided by two to determine the length of the dipole legs.

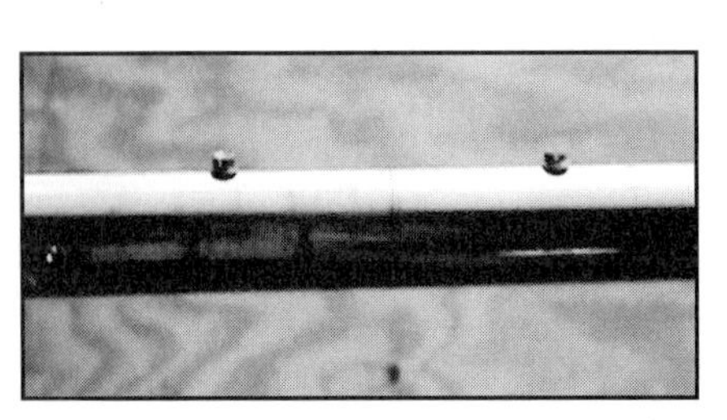

Fig 22.53 — Element splice uses a 1-ft length of 0.75-inch tubing inserted into the 0.875-inch sections to join them together. Self-tapping sheet-metal screws are used in this photo, but aluminum pop rivets or machine screws with washers and nuts can be used.

You should now have two 11-ft-long elements. As you can see in **Fig 22.53**, K8CH was temporarily out of aluminum pop rivets, so he used sheet metal screws. Either will work fine, but pop rivets can easily be drilled out and the antenna disassembled if you ever want to make changes.

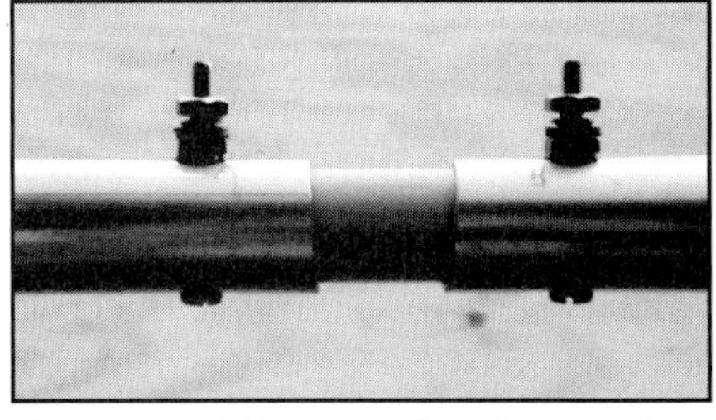

Fig 22.54 — The center insulator of the 15-m HVD is a 1-ft length of 0.75-inch fiberglass rod. Insulator and elements have been drilled to accept #8 hardware.

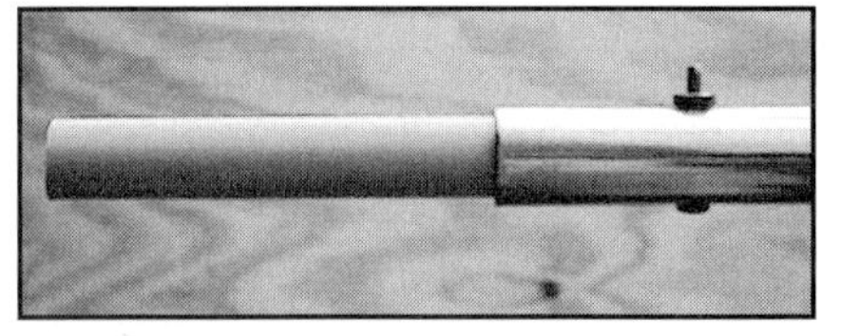

Fig 22.55 — The HVD base insulator is a 1-ft length of 0.75-inch fiberglass rod.

Fig 22.56 — Guys are made of Dacron line that is attached to the HVD by a stainless-steel worm-screw-type hose clamp. A self-tapping sheet-metal screw (not visible in the photo) prevents the clamp from sliding down the antenna.

Because hand-made cuts are not perfectly square, put those element ends at the center of the antenna. Slip these cut ends over the ends of a 1-ft length of 0.75-inch fiberglass rod. This rod serves as the center insulator. Leave about a 1-inch gap at the center. Drill aluminum and fiberglass for #8 hardware as shown in **Fig 22.54**.

Now, slip half of the remaining 1-ft length of 0.75-inch fiberglass rod into one end the dipole. (This end will be the bottom end or base.) Drill and secure with #8 hardware. See **Fig 22.55**.

The final step is to secure the guy wires to your vertical. You can see how K8CH did that in **Fig 22.56**. Start by drilling a pilot hole and then drive a sheet metal screw into the antenna about a foot above the center. The purpose of that screw is to prevent the clamp and guys from sliding down the antenna.

The guys are clean lengths of $^{3}/_{16}$-inch Dacron line. (The Dacron serves a dual purpose: it supports the antenna vertically, and it acts as an insulator.) Tie secure knots into the guy ends and secure these knotted ends to the antenna with a stainless-steel worm-screw-type hose clamp. Take care to not over tighten the clamps. You don't want the clamp to slip (the knots and the sheet-metal screw will help), but you especially don't want to cut your guy lines. Your

Fig 22.57 — At K8CH, the HVD base insulator sits in this saddle-shaped wooden fixture. This was photo was taken before the fixture was painted—a necessary step to protect against the weather.

Fig 22.58 — The HVD installed at K8CH. An eye screw that is used for securing one of the guy lines is visible in the foreground. You can also see the two choke baluns that are used in the feed system (see text).

antenna is ready for installation.

INSTALLATION

Installation requires two things. First, a place to sit or mount the base insulator. Second, you need anchors for the support guys.

K8CH used a piece of 2 × 6 lumber to make a socket to hold the HVD base securely in place. He drilled a ³⁄₄-inch-deep hole with a ³⁄₄-inch spade bit. A couple of pieces of 2 × 2 lumber at the ends of the base form a saddle which nicely straddles the ridge at the peak of his garage roof. You can see how I did this in **Fig 22.57**. The dimensions are not critical, but you should paint your base to protect it from the weather.

BALUN

This antenna needs a common-mode choke to ensure that stray RF doesn't flow on the shield of the coax. This device is also known as a choke balun. Unlike a horizontal dipole, don't consider it an option to omit the common-mode choke when building and installing an HVD.

You can use 8 ft of the RG-213 feed line wound into 7 turns for a balun. Secure the turns together with electrical tape so that each turn lies parallel with the next turn, forming a solenoid coil. Secure the feed line and balun to one of the guy lines with UV-resistant cable ties.

Because the feed line slants away from the antenna, you'll want to do *all* that you can to eliminate common-mode currents from the feed line. For that reason, make another balun about 11.5 ft from the first one. This balun also consists of 8 ft of the RG-213 feed line wound into 7 turns. See **Fig 22.58**.

THE COMPACT VERTICAL DIPOLE (CVD)

An HVD for 20 m will be about 33 ft tall, and for 30 m, it will be around 46 ft tall. Even the 20-m version can prove to be a mechanical challenge. The compact vertical dipole (CVD), designed by Chuck Hutchinson, K8CH, uses capacitance loading to shorten the antenna. Starting with the 15-m HVD described in the previous project, Chuck added capacitance loading wires to lower the resonance to 30 m. Later, he shortened the wires to move resonance to the 20-m band. This project describes those two CVDs.

PERFORMANCE ISSUES

Shortened antennas frequently suffer reduced performance caused by the shortening. A dipole that is less than a half wave in length is a compromise antenna. The question becomes how much is lost in the compromise. In this case there are two areas of primary interest, radiation efficiency and SWR bandwidth.

Radiation Efficiency

Capacitance loading at the dipole ends is the most efficient method of shortening the antenna. Current distribution in the high-current center of the antenna remains virtually unchanged. Since radiation is related directly to current, this is the most desireable form of loading. Computer modeling shows that radiation from a 30-m CVD is only 0.66 dB less than that from a full-size 30-m HVD when both have their bases 8 ft above ground. The angle of maximum radiation shifts up a bit for the CVD. Not a bad compromise when you consider that the CVD is 22-ft long compared to the approximately 46-ft length of the HVD.

SWR and SWR Bandwidth

Shortened antennas usually have lower radiation resistance and less SWR bandwidth than the full-size versions. The amount of change in the radiation resistance is related to the amount and type of loading (shortening), being lower with shorter the antennas. This can be a benefit in the case of a shortened vertical dipole. In Fig 22.1 you can see that vertical dipoles have a fairly high radiation resistance. With the dipole's lower end ¹⁄₈ λ above ground, the radiation resistance is roughly 80 Ω. In this case, a shorter antenna can have a better SWR when fed with 50-Ω coax.

SWR bandwidth tends to be wide for vertical dipoles in general. A properly designed CVD for 7-MHz or higher should give you good SWR (1.5:1 or better) across the entire band!

As you can see, in theory the CVD provides excellent performance in a compact package. Experience confirms the theory.

CONSTRUCTION

To convert the K8CH 15-m HVD to 20 or 30 m, you'll need to add four loading wires at the top and four more at the bottom of the HVD. The lengths are shown in **Table 22.12**. The upper wires droop at a 45° angle and the lower wires run horizontally. The antenna is supported by 4 guy lines. See **Fig 22.59**. You can connect the wires to the vertical portion with #8 hardware. Crimp and solder terminals on the wire ends to make connections easier. The technique is illustrated in **Fig 22.60**.

The upper loading wires can be extended with insulated line and used for additional guying. The lower wires are extended with insulated line and fasten to the guy lines so that the lower wires run horizontally.

Prune the lower wires for best SWR across the band of interest. The K8CH CVD has its base at 14 ft. This antenna has an SWR of less than 1.2:1 on 30 m and less than 1.3:1 across the entire 20-m band.

EXPERIENCE

The 30-m CVD was compared to a ground-mounted quarter-wave vertical and a horizontal dipole at 30 ft. In tests, the CVD was always the superior antenna.

Table 22.12
CVD Loading Wires

Length using #14 insulated copper wire

Band	Feet	Inches	
30 m	6	0	Top & Bottom
20 m	4	2¼	Top
20 m	3	½	Bottom

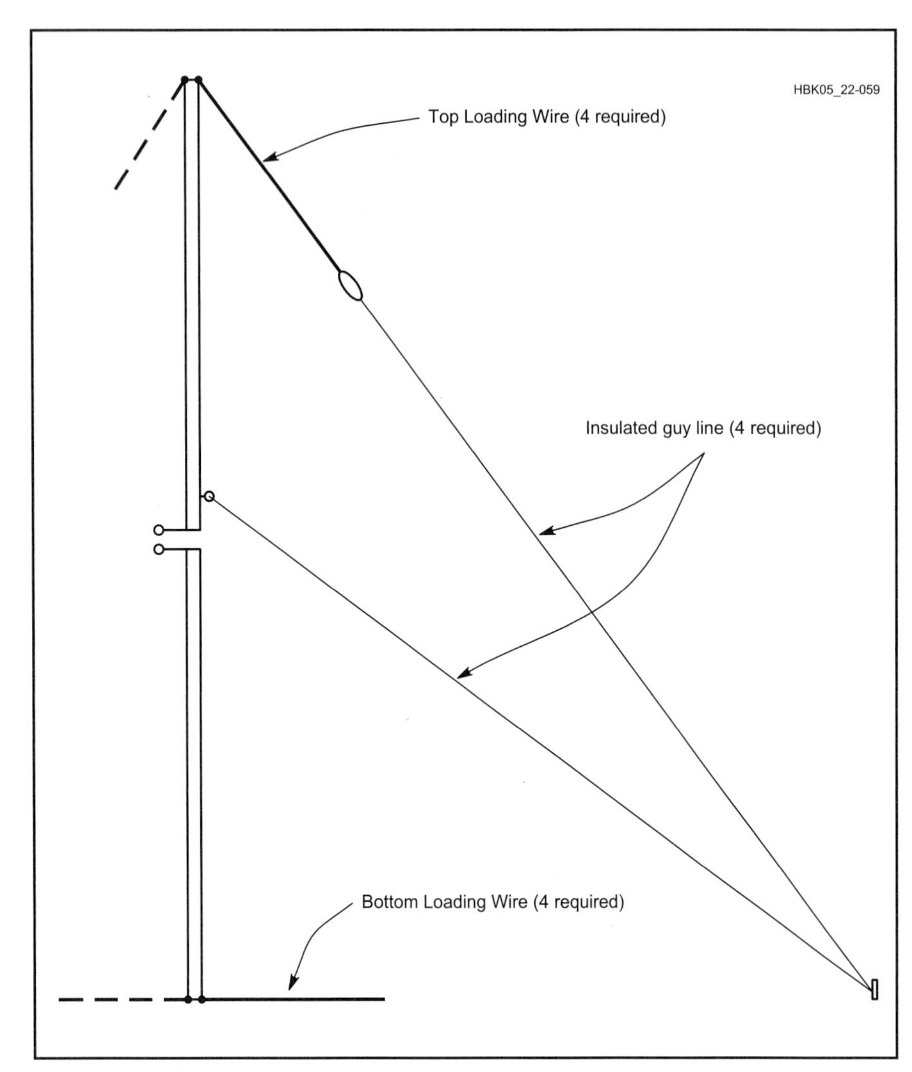

Fig 22.59 — The CVD consists of a vertical dipole and loading wires. Only one set of the four loading wires and only one guy line is shown in this drawing. See text for details.

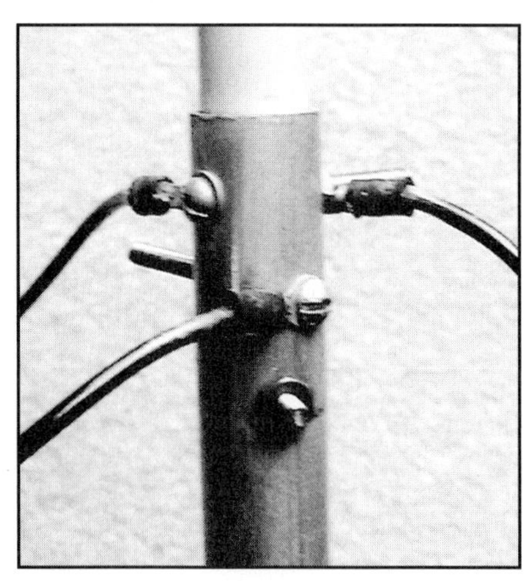
Fig 22.60 — CVD loading wires can be attached using #8 hardware. Crimp and solder terminals on the wire ends to make connections easier.

Every DX station that was called by K8CH responded with one or two calls. What more could you ask for?

Later, the CVD loading wires were shortened for operation on 20 m. Once again the results were very encouraging. Many contest QSOs were entered in the log using this antenna.

Finally, a late winter ice storm deposited about ³/₄ inch of radial ice on the antenna, loading wires and guys. The antenna would probably have survived had it not been for the sustained 45 mph winds that followed. The upper loading wires and their guy lines were not heavy enough to support the load and the antenna bent and broke. This combination of ice and wind is very unusual.

ALL WIRE 30-M CVD

If you have a tree or other support that will support the upper end of a CVD at 32 ft above the ground, you might want to consider an all-wire version of the 30-m CVD. The vertical is 24 ft long and it will have an SWR of less than 1.1:1 across the band. The four loading wires at top and bottom are each 5 ft, 2 inches long.

The configuration is shown in **Fig 22.61**. As with any vertical dipole, you'll need to use a balun between the feedline and the antenna.

Alternatively you can use two loading wires at the top and two at the bottom. In this case each of the loading wires is 8 ft 7¹/₂ inches long.

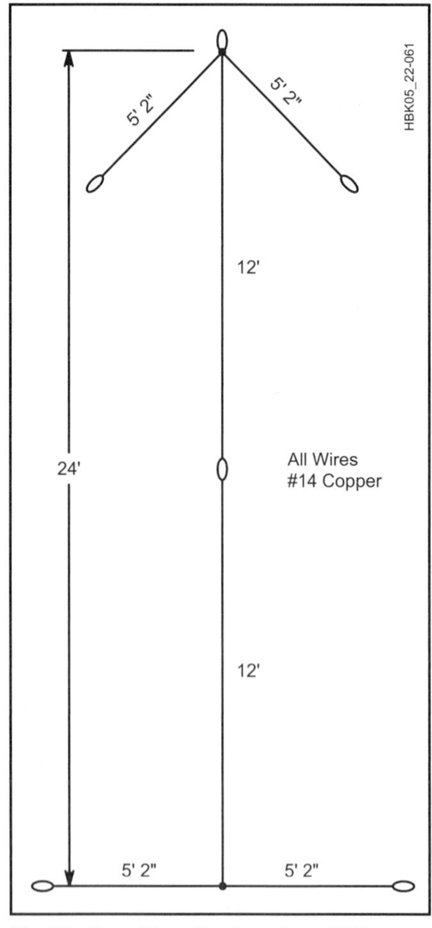

Fig 22.61 — The all-wire 30-m CVD consists of a vertical dipole and loading wires. It can be made entirely with #14 wire. Support lines have been omitted for simplicity. See text for details.

Yagi and Quad Directive Antennas

Most antennas described earlier in this chapter have unity gain compared to a dipole, or just slightly more. For the purpose of obtaining gain and directivity it is convenient to use a Yagi-Uda or cubical quad beam antenna. The former is commonly called a *Yagi*, and the latter is usually referred to as a *quad*.

Most operators prefer to erect these antennas for horizontal polarization, but they can be used as vertically polarized arrays merely by rotating the elements by 90°. In effect, the beam antenna is turned on its side for vertical polarity. The number of elements used will depend on the gain desired and the limits of the supporting structure. Many amateurs obtain satisfactory results with only two elements in a beam antenna, while others have four or five elements operating on a single amateur band.

Regardless of the number of elements used, the height-above-ground considerations discussed earlier for dipole antennas remain valid with respect to the angle of radiation. This is demonstrated in **Fig 22.62** at A and B where a comparison of radiation characteristics is given for a 3-element Yagi at one-half and one wavelength above average ground. It can be seen that the higher antenna (Fig 22.62B) has a main lobe that is more favorable for DX work (roughly 15°) than the lobe of the lower antenna in Fig 22.62A (approximately 30°). The pattern at B shows that some useful high-angle radiation exists also, and the higher lobe is suitable for short-skip contacts when propagation conditions dictate the need.

The azimuth pattern for the same antenna is provided in **Fig 22.63**. Most of the power is concentrated in the *main lobe* at 0° azimuth. The lobe directly behind the main lobe at 180° is often called the *backlobe*. Note that there are small *sidelobes* at approximately 110° and 260° in azimuth. The peak power difference, in decibels, between the *nose* of the main lobe at 0° and the strongest rearward lobe is called the *front-to-rear ratio (F/R)*. In this case the worst-case rearward lobe is at 180°, and the F/R is 12 dB. It is infrequent that two 3-element Yagis with different element spacings and tuning will yield the same lobe patterns. The pattern of Fig 22.63 is shown only for illustrative purposes.

PARASITIC EXCITATION

In most of these arrangements the additional elements receive power by induction or radiation from the driven element and reradiate it in the proper phase relationship to give the desired effect. These elements are called *parasitic elements*, as contrasted to *driven elements*, which receive power directly from the transmitter through the transmission line.

The parasitic element is called a *director* when it reinforces radiation on a line pointing to it from the driven element, and a *reflector* when the reverse is the case. Whether the parasitic element is a director or reflector depends on the parasitic element tuning, which is usually adjusted by changing its length.

GAIN, FRONT-TO-REAR RATIO AND SWR

The gain of an antenna with parasitic elements varies with the spacing and tuning of the elements. Element tuning is a function of length, diameter and taper schedule if the element is constructed with telescoping tubing. For any given spacing, there is a tuning condition that will give maximum gain at this spacing. However, the maximum front-to-rear ratio seldom, if ever, occurs at the same condition that gives maximum forward gain. The impedance of the driven element in a parasitic array, and thus the SWR, also varies with the tuning and spacing.

It is important to remember that all these parameters change as the operating frequency is varied. For example, if you operate both the CW and phone portions

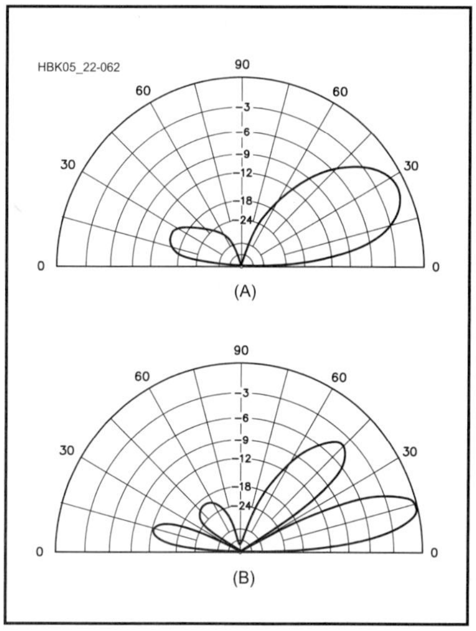

Fig 22.62 — Elevation-plane response of a 3-element Yagi placed ½ λ above perfect ground at A and the same antenna spaced 1 λ above ground at B.

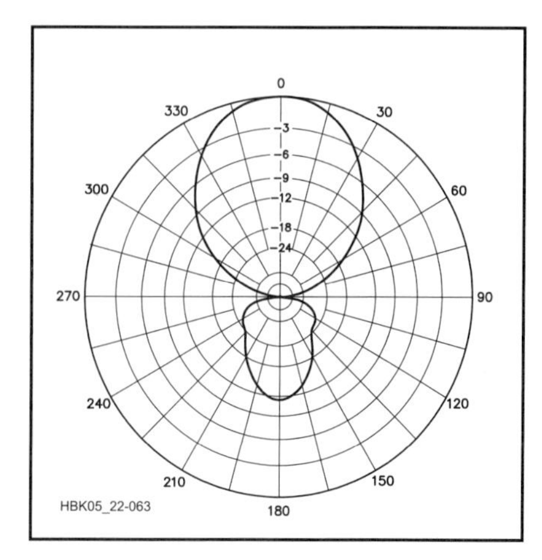

Fig 22.63 — Azimuth-plane pattern of a typical three-element Yagi in free space. The Yagi's boom is along the 0° to 180° axis.

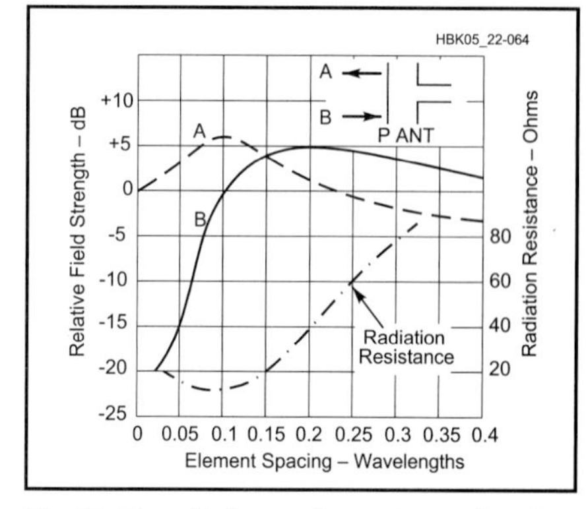

Fig 22.64 — Gain vs element spacing for a 2-element Yagi, having one driven and one parasitic element. The reference point, 0 dB, is the field strength from a half-wave antenna alone. The greatest gain is in the direction A at spacings of less than 0.14 λ, and in direction B at greater spacings. The front-to-rear ratio is the difference in decibels between curves A and B. Variation in radiation resistance of the driven element is also shown. These curves are for the special case of a self-resonant parasitic element, but are representative of how a 2-element Yagi works. At most spacings the gain as a reflector can be increased by slight lengthening of the parasitic element; the gain as a director can be increased by shortening. This also improves the front-to-rear ratio.

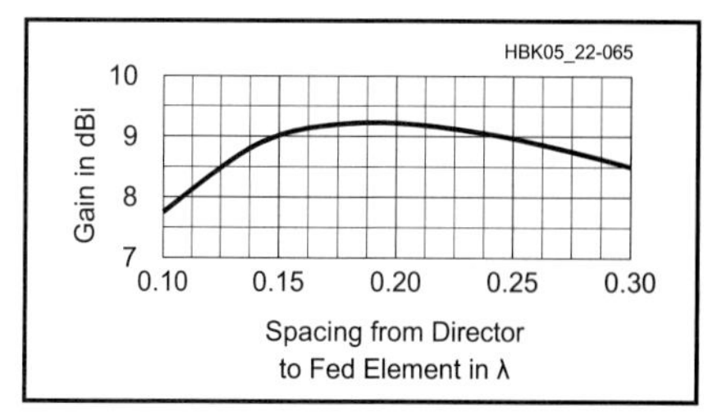

Fig 22.65 — General relationship of gain of 3-element Yagi vs director spacing, the reflector being fixed at 0.2 λ. This antenna is tuned for maximum forward gain.

Table 22.13
10-m Optimized Yagi Designs

	Spacing Between Elements (in.)	Seg 1 Length (in.)	Seg 2 Length (in.)	Seg 3 Length (in.)	Midband Gain F/R
310-08					
Refl	0	24	18	66.750	7.2 dBi
DE	36	24	18	57.625	22.9 dB
Dir 1	54	24	18	53.125	
410-14					
Refl	0	24	18	64.875	8.4 dBi
DE	36	24	18	58.625	30.9 dB
Dir 1	36	24	18	57.000	
Dir 2	90	24	18	47.750	
510-24					
Refl	0	24	18	65.625	10.3 dBi
DE	36	24	18	58.000	25.9 dB
Dir 1	36	24	18	57.125	
Dir 2	99	24	18	55.000	
Dir 3	111	24	18	50.750	

Note: For all antennas, the tube diameters are: Seg 1=0.750 inch, Seg 2=0.625 inch, Seg 3=0.500 inch.

Table 22.14
12-m Optimized Yagi Designs

	Spacing Between Elements (in.)	Seg 1 Length (in.)	Seg 2 Length (in.)	Seg 3 Length (in.)	Midband Gain F/R
312-10					
Refl	0	36	18	69.000	7.5 dBi
DE	40	36	18	59.125	24.8 dB
Dir 1	74	36	18	54.000	
412-15					
Refl	0	36	18	66.875	8.5 dBi
DE	46	36	18	60.625	27.8 dB
Dir 1	46	36	18	58.625	
Dir 2	82	36	18	50.875	
512-20					
Refl	0	36	18	69.750	9.5 dBi
DE	46	36	18	61.750	24.9 dB
Dir 1	46	36	18	60.500	
Dir 2	48	36	18	55.500	
Dir 3	94	36	18	54.625	

Note: For all antennas, the tube diameters are: Seg 1 = 0.750 inch, Seg 2 = 0.625 inch, Seg 3 = 0.500 inch.

of the 20-m band with a Yagi or quad antenna, you probably will want an antenna that *spreads out* the performance over most of the band. Such designs typically must sacrifice a little gain in order to achieve good F/R and SWR performance across the band. The longer the boom of a Yagi or a quad, and the more elements that are placed on that boom, the better will be the overall performance over a given amateur band. For the lower HF bands, the size of the antenna quickly becomes impractical for truly *optimal* designs, and compromise is necessary.

TWO-ELEMENT BEAMS

A 2-element beam is useful—especially where space or other considerations prevent the use of a three element, or larger, beam. The general practice is to tune the parasitic element as a reflector and space it about 0.15 λ from the driven element, although some successful antennas have been built with 0.1-λ spacing and director tuning. Gain vs element spacing for a 2-element antenna is given in **Fig 22.64** for the special case where the parasitic element is resonant. It is indicative of the performance to be expected under maximum-gain tuning conditions. Changing the tuning of the driven element in a Yagi or quad will not materially affect the gain or F/R. Thus, only the spacing and the tuning of the single parasitic element have any effect on the performance of a 2-element Yagi or quad. Most 2-element Yagi designs achieve a compromise F/R of about 10 dB, together with acceptable SWR and gain across a frequency band with a percentage bandwidth less than about 4%. A 2-element quad can achieve better F/R, gain and SWR across a band, at the expense of greater mechanical complexity compared to a Yagi.

THREE-ELEMENT BEAMS

A theoretical investigation of the 3-element case (director, driven element and reflector) has indicated a maximum gain of about 9.7 dBi. A number of experimental investigations have shown that the spacing between the driven element and reflector for maximum gain is in the region of 0.15 to .25 λ. With 0.2-λ reflector spacing, **Fig 22.65** shows that the gain variation with director spacing is not especially critical. Also, the overall length of the array (boom length in the case of a rotatable antenna) can be anywhere between 0.35 and 0.45 λ with no appreciable difference in the maximum gain obtainable.

If maximum gain is desired, wide spacing

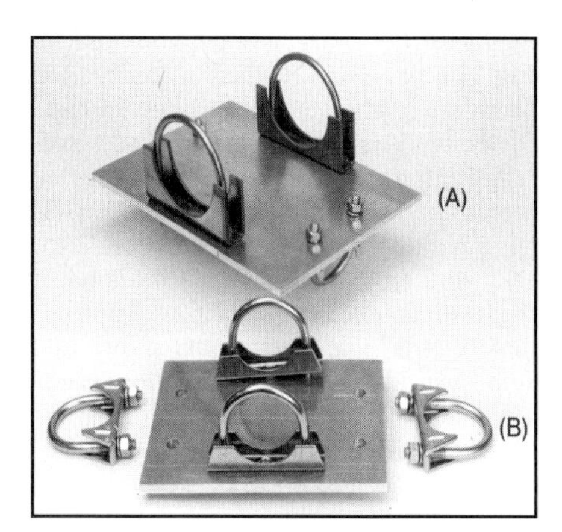

Fig 22.67 — The boom-to-element plate at A uses muffler-clamp-type U-bolts and saddles to secure the round tubing to the flat plate. The boom-to-mast plate at B is similar to the boom-to-element plate. The main difference is the size of materials used.

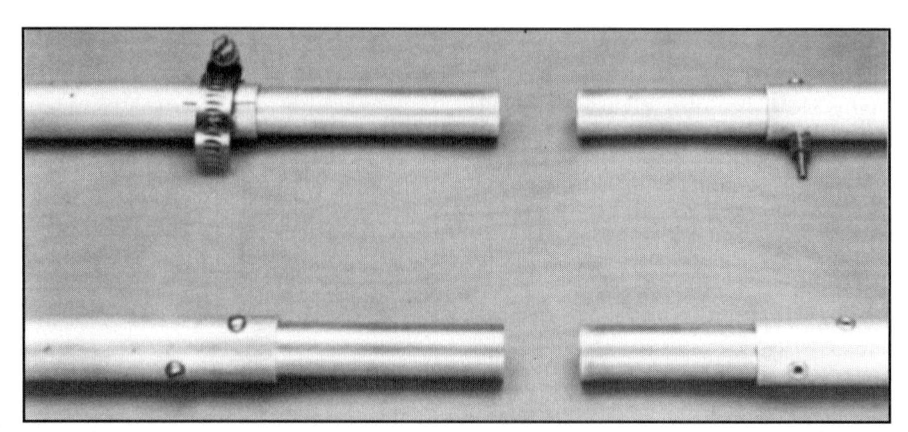

Fig 22.66 — Some methods of connecting telescoping tubing sections to build beam elements. See text for a discussion of each method.

of both elements is beneficial because adjustment of tuning or element length is less critical and the input resistance of the driven element is generally higher than with close spacing. A higher input resistance improves the efficiency of the antenna and makes a greater bandwidth possible. However, a total antenna length, director to reflector, of more than 0.3 λ at frequencies of the order of 14 MHz introduces difficulty from a construction standpoint. Lengths of 0.25 to 0.3 λ are therefore used frequently for this band, even though they are less than opti-

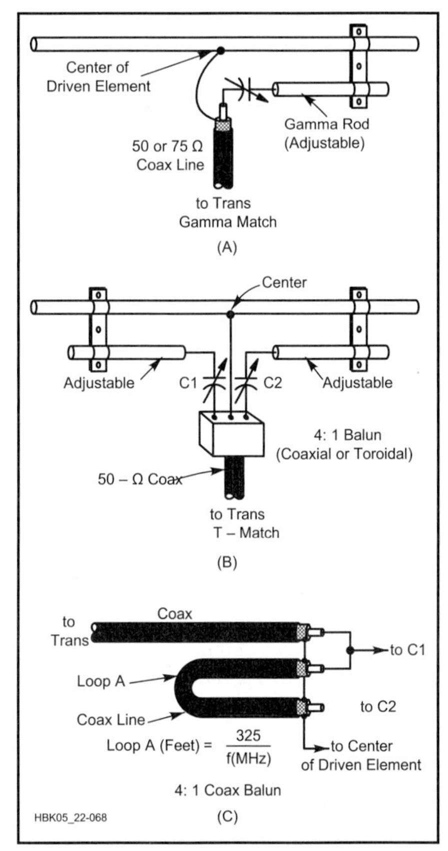

Fig 22.68 — Illustrations of gamma and T matching systems. At A, the gamma rod is adjusted along with C until the lowest SWR is obtained. A T match is shown at B. It is the same as two gamma-match rods. The rods and C1 and C2 are adjusted alternately for a best SWR. A coaxial 4:1 balun transformer is shown at C. A toroidal balun can be used in place of the coax model shown. The toroidal version has a broader frequency range than the coaxial one. The T match is adjusted for 200 Ω and the balun steps this balanced value down to 50 Ω, unbalanced. Or the T match can be set for 300 Ω, and the balun used to step this down to 75 Ω unbalanced. Dimensions for the gamma and T match rods will depend on the tubing size used, and the spacing of the parasitic elements of the beam. Capacitors C, C1 and C2 can be 140 pF for 14-MHz beams. Somewhat less capacitance will be needed at 21 and 28 MHz.

Table 22.15
15-m Optimized Yagi Designs

	Spacing Between Elements (in.)	*Seg 1 Length (in.)*	*Seg 2 Length (in.)*	*Seg 3 Length (in.)*	*Seg 4 Length (in.)*	*Midband Gain F/R*
315-12						
Refl	0	30	36	18	61.375	7.6 dBi
DE	48	30	36	18	49.625	25.5 dB
Dir 1	92	30	36	18	43.500	
415-18						
Refl	0	30	36	18	59.750	8.3 dBi
DE	56	30	36	18	50.875	31.2 dB
Dir 1	56	30	36	18	48.000	
Dir 2	98	30	36	18	36.625	
515-24						
Refl	0	30	36	18	62.000	9.4 dBi
DE	48	30	36	18	52.375	25.8 dB
Dir 1	48	30	36	18	47.875	
Dir 2	52	30	36	18	47.000	
Dir 3	134	30	36	18	41.000	

Note: For all antennas, the tube diameters (in inches) are:
Seg 1 = 0.875, Seg 2 = 0.750, Seg 3 = 0.625, Seg 4 = 0.500.

Table 22.16
17-m Optimized Yagi Designs

	Spacing Between Elements (in.)	*Seg 1 Length (in.)*	*Seg 2 Length (in.)*	*Seg 3 Length (in.)*	*Seg 4 Length (in.)*	*Seg 5 Length (in.)*	*Midband Gain F/R*
317-14							
Refl	0	24	24	36	24	60.125	8.1 dBi
DE	65	24	24	36	24	52.625	24.3 dB
Dir 1	97	24	24	36	24	48.500	
417-20							
Refl	0	24	24	36	24	61.500	8.5 dBi
DE	48	24	24	36	24	54.250	27.7 dB
Dir 1	48	24	24	36	24	52.625	
Dir 2	138	24	24	36	24	40.500	

Note: For all antennas, tube diameters (inches) are: Seg 1=1.000, Seg 2=0.875, Seg 3=0.750, Seg 4=0.625, Seg 5=0.500.

Table 22.17
20-m Optimized Yagi Designs

	Spacing Between Elements (in.)	*Seg 1 Length (in.)*	*Seg 2 Length (in.)*	*Seg 3 Length (in.)*	*Seg 4 Length (in.)*	*Seg 5 Length (in.)*	*Seg 6 Length (in.)*	*Midband Gain F/R*
320-16								
Refl	0	48	24	20	42	20	69.625	7.3 dBi
DE	80	48	24	20	42	20	51.250	23.4 dB
Dir 1	106	48	24	20	42	20	42.625	
420-26								
Refl	0	48	24	20	42	20	65.625	8.6 dBi
DE	72	48	24	20	42	20	53.375	23.4 dB
Dir 1	60	48	24	20	42	20	51.750	
Dir 2	174	48	24	20	42	20	38.625	

Note: For all antennas, tube diameters (inches) are: Seg 1=1.000, Seg 2=0.875, Seg 3=0.750, Seg 4=0.625, Seg 5=0.500.

mum from the viewpoint of maximum gain.

In general, Yagi antenna gain drops off less rapidly when the reflector length is increased beyond the optimum value than it does for a corresponding decrease below the optimum value. The opposite is true of a director. It is therefore advisable to err, if necessary, on the long side for a reflector and on the short side for a director. This also tends to make the antenna performance less dependent on the exact frequency at which it is operated: An increase above the design frequency has the same effect as increasing the length of both parasitic elements, while a decrease in frequency has the same effect as shortening both elements. By making the director slightly short and the reflector slightly long, there will be a greater spread between the upper and lower frequencies at which the gain starts to show a rapid decrease.

We recommend *plumbers delight* construction, where all elements are mounted directly on, and grounded to, the boom. This puts the entire array at dc ground potential, affording better lightning protection. A gamma- or T-match section can be used for matching the feed line to the array.

COMPUTER-OPTIMIZED YAGIS

Yagi designers are now able to take advantage of powerful personal computers and software to optimize their designs for the parameters of gain, F/R and SWR across frequency bands. ARRL Senior Assistant Technical Editor Dean Straw, N6BV, has designed a family of Yagis for HF bands. These can be found in **Tables 22.13**, **22.14**, **22.15**, **22.16** and **22.17**, for the 10, 12, 15, 17 and 20-m amateur bands.

For 12 through 20 m, each design has been optimized for better than 20 dB F/R, and an SWR of less than 2:1 across the entire amateur frequency band. For the 10-m band, the designs were optimized for the lower 800 kHz of the band, from 28.0 to 28.8 MHz. Each Yagi element is made of telescoping 6061-T6 aluminum tubing, with 0.058 inch thick walls. This type of element can be telescoped easily, using techniques shown in **Fig 22.66**. Measuring each element to an accuracy of 1/8 inch results in performance remarkably consistent with the computations, without any need for *tweaking* or fine-tuning when the Yagi is on the tower.

Each element is mounted above the boom with a heavy rectangular aluminum plate, by means of galvanized U-bolts with saddles, as shown in **Fig 22.67**. This method of element mounting is rugged and stable, and because the element is mounted away from the boom, the amount of element detuning due to the presence of the boom is minimal. The element dimensions given in each table already take into account any element detuning due to the boom-to-element mounting plate. The element-to-boom mounting plate for all the 10-m Yagis is a 0.250-inch thick flat aluminum plate, 4 inches wide by 4 inches long. For the 12 and 15-m Yagis, a 0.375-inch thick flat aluminum plate, 5 inches wide by 6 inches long is used, and for the 17 and 20-m Yagis, a 0.375-inch thick flat aluminum plate, 6 inches wide by 8 inches long is used. Where the plate is rectangular, the long dimension is in line with the element.

Each design table shows the dimensions for *one-half* of each element, mounted on one side of the boom. The other half of each element is the same, mounted on the other side of the boom. Use a tubing sleeve inside the center portion of the element so that the element is not crushed by the mounting U-bolts. Each telescoping section is inserted 3 inches into the next size of tubing. For example, in the 310-08.YAG design (3 elements on an 8-ft boom), the reflector tip, made out of 1/2-inch OD tubing, sticks out 66.75 inches from the 5/8-inch OD tubing. For each 10-m element, the overall length of each 5/8-inch OD piece of tubing is 21 inches, before insertion into the 3/4-inch piece. Since the 3/4-inch OD tubing is 24 inches long on each side of the boom, the center portion of each element is actually 48 inches of uncut 3/4-inch OD tubing.

The boom for all these antennas should be constructed with at least 2-inch-OD tubing, with 0.065-inch wall thickness. Because each boom has 3 inches extra space at each end, the reflector is actually placed 3 inches from one end of the boom. For the 310-08.YAG, the driven element is placed 36 inches ahead of the reflector, and the director is placed 54 inches ahead of the driven element.

Each antenna is designed with a driven element length appropriate for a gamma or T matching network, as shown in **Fig 22.68**. The variable gamma or T capacitors can be housed in small plastic enclosures for weatherproofing; receiving-type variable capacitors with close plate spacing can be used at powers up to a few hundred watts. Maximum capacitance required is usually 140 pF at 14 MHz and proportionally less at the higher frequencies.

The driven-element's length may require slight readjustment for best match, particularly if a different matching network is used. *Do not change either the lengths or the telescoping tubing schedule of the parasitic elements*—they have been optimized for best performance and will not be affected by tuning of the driven element.

TUNING ADJUSTMENTS

Preliminary matching adjustments can be done on the ground. The beam should be set up so the reflector element rests on the earth, with the beam pointing upward. The matching system is then adjusted for best SWR. When the antenna is raised to its operating height, only slight touch-up of the matching network may be required.

CONSTRUCTION OF YAGIS

Most beams and verticals are made from sections of aluminum tubing. Compromise beams have been fashioned from less-expensive materials such as electrical conduit (steel) or bamboo poles wrapped with conductive tape or aluminum foil. The steel conduit is heavy, is a poor conductor and is subject to rust. Similarly, bamboo with conducting material attached to it may deteriorate rapidly in the weather. The dimensions shown for the Yagis in the preceding section are designed for specific telescoping aluminum elements, but the elements may be scaled to different sizes by using the information about tapering and scaling in Chapter 2 of *The ARRL Antenna Book*, although with a likelihood of deterioration in performance over the whole frequency band.

For reference, **Table 22.18** details the standard sizes of aluminum tubing, available in many metropolitan areas. Dealers may be found in the Yellow Pages under *Aluminum*. Tubing usually comes in 12-ft lengths, although 20-ft lengths are available in some sizes. Your aluminum dealer will probably also sell aluminum plate in various thicknesses needed for boom-to-mast and boom-to-element connections.

Aluminum is rated according to its hardness. The most common material used in antenna construction is grade 6061-T6. This material is relatively strong and has good workability. In addition, it will bend without taking a *set*, an advantage in antenna applications where the pieces are constantly flexing in the wind. The softer grades (5051, 3003 and so on) will bend much more easily, while harder grades (7075 and so on) are more brittle.

Wall thickness is of primary concern when selecting tubing. It is of utmost importance that the tubing fits snugly where the element sections join. Sloppy joints will make a mechanically unstable antenna. The magic wall thickness is 0.058 inch. For example (from Table 22.18), 1-inch outside diameter (OD) tubing with a 0.058-inch wall has an inside diameter (ID) of 0.884 inch. The next smaller size of tubing, 7/8 inch, has an OD of 0.875 inch. The 0.009-inch difference provides just the right amount of clearance for a snug fit.

Table 22.18
Standard Sizes of Aluminum Tubing

6061-T6 (61S-T6) Round Aluminum Tube in 12-ft Lengths

OD (in.)	Wall Thickness (in.)	stubs ga	ID (in.)	Approx Weight (lb) per ft	per length
3/16	0.035	no. 20	0.117	0.019	0.228
	0.049	no. 18	0.089	0.025	0.330
1/4	0.035	no. 20	0.180	0.027	0.324
	0.049	no. 18	0.152	0.036	0.432
	0.058	no. 17	0.134	0.041	0.492
5/16	0.035	no. 20	0.242	0.036	0.432
	0.049	no. 18	0.214	0.047	0.564
	0.058	no. 17	0.196	0.055	0.660
3/8	0.035	no. 20	0.305	0.043	0.516
	0.049	no. 18	0.277	0.060	0.720
	0.058	no. 17	0.259	0.068	0.816
	0.065	no. 16	0.245	0.074	0.888
7/16	0.035	no. 20	0.367	0.051	0.612
	0.049	no. 18	0.339	0.070	0.840
	0.065	no. 16	0.307	0.089	1.068
1/2	0.028	no. 22	0.444	0.049	0.588
	0.035	no. 20	0.430	0.059	0.708
	0.049	no. 18	0.402	0.082	0.948
	0.058	no. 17	0.384	0.095	1.040
	0.065	no. 16	0.370	0.107	1.284
5/8	0.028	no. 22	0.569	0.061	0.732
	0.035	no. 20	0.555	0.075	0.900
	0.049	no. 18	0.527	0.106	1.272
	0.058	no. 17	0.509	0.121	1.452
	0.065	no. 16	0.495	0.137	1.644
3/4	0.035	no. 20	0.680	0.091	1.092
	0.049	no. 18	0.652	0.125	1.500
	0.058	no. 17	0.634	0.148	1.776
	0.065	no. 16	0.620	0.160	1.920
	0.083	no. 14	0.584	0.204	2.448
7/8	0.035	no. 20	0.805	0.108	1.308
	0.049	no. 18	0.777	0.151	1.810
	0.058	no. 17	0.759	0.175	2.100
	0.065	no. 16	0.745	0.199	2.399
1	0.035	no. 20	0.930	0.123	1.467
	0.049	no. 18	0.902	0.170	2.040
	0.058	no. 17	0.884	0.202	2.424
	0.065	no. 16	0.870	0.220	2.640
	0.083	no. 14	0.834	0.281	3.372
1 1/8	0.035	no. 20	1.055	0.139	1.668
	0.058	no. 17	1.009	0.228	2.736
1 1/4	0.035	no. 20	1.180	0.155	1.860
	0.049	no. 18	1.152	0.210	2.520
	0.058	no. 17	1.134	0.256	3.072
	0.065	no. 16	1.120	0.284	3.408
	0.083	no. 14	1.084	0.357	4.284
1 3/8	0.035	no. 20	1.305	0.173	2.076
	0.058	no. 17	1.259	0.282	3.384
1 1/2	0.035	no. 20	1.430	0.180	2.160
	0.049	no. 18	1.402	0.260	3.120
	0.058	no. 17	1.384	0.309	3.708
	0.065	no. 16	1.370	0.344	4.128
	0.083	no. 14	1.334	0.434	5.208
	*0.125	1/8"	1.250	0.630	7.416
	*0.250	1/4"	1.000	1.150	14.823
1 5/8	0.035	no. 20	1.555	0.206	2.472
	0.058	no. 17	1.509	0.336	4.032
1 3/4	0.058	no. 17	1.634	0.363	4.356
	0.083	no. 14	1.584	0.510	6.120
1 7/8	0.508	no. 17	1.759	0.389	4.668
2	0.049	no. 18	1.902	0.350	4.200
	0.065	no. 16	1.870	0.450	5.400
	0.083	no. 14	1.834	0.590	7.080
	*0.125	1/8"	1.750	0.870	9.960
	*0.250	1/4"	1.500	1.620	19.920
2 1/4	0.049	no. 18	2.152	0.398	4.776
	0.065	no. 16	2.120	0.520	6.240
	0.083	no. 14	2.084	0.660	7.920
2 1/2	0.065	no. 16	2.370	0.587	7.044
	0.083	no. 14	2.334	0.740	8.880
	*0.125	1/8"	2.250	1.100	12.720
	*0.250	1/4"	2.000	2.080	25.440
3	0.065	no. 16	2.870	0.710	8.520
	*0.125	1/8"	2.700	1.330	15.600
	*0.250	1/4"	2.500	2.540	31.200

*These sizes are extruded; all other sizes are drawn tubes.
Shown here are standard sizes of aluminum tubing that are stocked by most aluminum suppliers or distributors in the United States and Canada.

Fig 22.66 shows several methods of fastening antenna element sections together. The slot and hose clamp method shown in Fig 22.66A is probably the best for joints where adjustments are needed. Generally, one adjustable joint per element half is sufficient to tune the antenna—usually the tips at each end of an element are made adjustable. Stainless steel hose clamps (beware—some "stainless steel" models do not have a stainless screw and will rust) are recommended for longest antenna life.

Fig 22.66B, C and D show possible fastening methods for joints that are not adjustable. At B, machine screws and nuts hold the elements in place. At C, sheet metal screws are used. At D, rivets secure the tubing. If the antenna is to be assembled permanently, rivets are the best choice. Once in place, they are permanent. They will never work free, regardless of vibration or wind. If aluminum rivets with aluminum mandrels are employed, they will never rust. Also, being aluminum, there is no danger of corrosion from interaction between dissimilar metals. If the antenna is to be disassembled and moved periodically, either B or C will work. If machine screws are used, however, take precautions to keep the nuts from vibrating free. Use of lock washers, lock nuts and flexible adhesive such as silicone bathtub sealant will keep the hardware in place.

Use of a conductive grease at the element joints is essential for long life. Left untreated, the aluminum surfaces will oxidize in the weather, resulting in a poor connection. Some trade names for this conductive grease are Penetrox, Noalox and Dow Corning Molykote 41. Many electrical supply houses carry these products.

BOOM MATERIAL

The boom size for a rotatable Yagi or quad should be selected to provide stability to the entire system. The best diameter for the boom depends on several factors, but mostly the element weight, number of elements and overall length. Two-inch-diameter booms should not be made any longer than 24 ft unless additional support is given to reduce both vertical and horizontal bending forces. Suitable reinforcement for a long 2-inch boom can consist of a truss or a truss and lateral support, as shown in **Fig 22.69**.

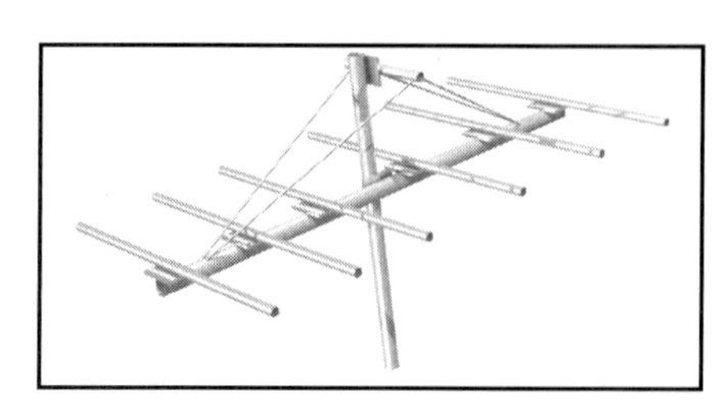

Fig 22.69 — A long boom needs both vertical and horizontal support. The crossbar mounted above the boom can support a double truss, which will help keep the antenna in position.

A boom length of 24 ft is about the point where a 3-inch diameter begins to be very worthwhile. This dimension provides a considerable amount of improvement in overall mechanical stability as well as increased clamping surface area for element hardware. The latter is extremely important to prevent rotation of elements around the boom if heavy icing is commonplace. Pinning an element to the boom with a large bolt helps in this regard. On smaller diameter booms, however, the elements sometimes work loose and tend to elongate the pinning holes in both the element and the boom. After some time the elements shift their positions slightly (sometimes from day to day) and give a ragged appearance to the system, even though this may not harm the electrical performance.

A 3-inch-diameter boom with a wall thickness of 0.065 inch is very satisfactory for antennas up to about a 5-element, 20-m array that is spaced on a 40-ft boom. A truss is recommended for any boom longer than 24 ft. One possible source for large boom material is irrigation tubing sold at farm supply houses.

PUTTING IT TOGETHER

Once you assemble the boom and elements, the next step is to fasten the elements to the boom securely and then fasten the boom to the mast or supporting structure. Be sure to leave plenty of material on either side of the U-bolt holes on the element-to-boom mounting plates. The U-bolts selected should be a snug fit for the tubing. If possible, buy muffler-clamp U-bolts that come with saddles.

The boom-to-mast plate shown in Fig 22.67B is similar to the boom-to-element plate. The size of the plate and number of U-bolts used will depend on the size of the antenna. Generally, antennas for the bands up through 20 m require only two U-bolts each for the mast and boom. Longer antennas for 15 and 20 m (35-ft booms and up) and most 40-m beams should have four U-bolts each for the boom and mast because of the torque that the long booms and elements exert as the antennas move in the wind. When tightening the U-bolts, be careful not to crush the tubing. Once the wall begins to collapse, the connection begins to weaken. Many aluminum suppliers sell 1/4-inch or 3/8-inch plates just right for this application. Often they will shear pieces to the correct size on request. As with tubing, the relatively hard 6061-T6 grade is a good choice for mounting plates.

The antenna should be put together with good-quality hardware. Stainless steel is best for long life. Rust will attack plated steel hardware after a short while, making nuts difficult, if not impossible, to remove. If stainless muffler clamps are not available, the next best thing is to have them plated. If you can't get them plated, then at least paint them with a good zinc-chromate primer and a finish coat or two. Good-quality hardware is more expensive initially, but if you do it right the first time, you won't have to take the antenna down after a few years and replace the hardware. Also, when repairing or modifying an installation, nothing is more frustrating than fighting rusty hardware at the top of a tower.

Quad Antennas

One of the more effective DX arrays is called a *quad* antenna. It consists of two or more loops of wire, each supported by a bamboo or fiberglass cross-arm assembly. The loops are a quarter wavelength per side (full wavelength overall). One loop is driven and the other serves as a parasitic element—usually a reflector. A variation of the quad is called the *delta loop*. The electrical properties of both antennas are the same. Both antennas are shown in **Fig 22.70**. They differ mainly in their physical properties, one being of plumber's delight construction, while the other uses insulating support members. One or more directors can be added to either antenna if additional gain and directivity are desired, though most operators use the 2-element arrangement.

It is possible to interlace quads or deltas for two or more bands, but if this is done the formulas given in Fig 22.70 may have to be changed slightly to compensate for the proximity effect of the second antenna. For quads the length of the full-wave loop can be computed from

$$\text{Full-wave loop} = \frac{1005}{f\,(\text{MHz})}\ \text{ft} \qquad (7)$$

If multiple arrays are used, each antenna should be tuned separately for maximum forward gain, or best front-to-rear ratio, as observed on a field-strength meter. The reflector stub on the quad should be adjusted for this condition. The gamma match should be adjusted for best SWR. The resonance of the antenna can be found by checking the frequency at which the lowest SWR occurs. By lengthening or shortening it, the driven element length can be adjusted for resonance in the most-used portion of the band.

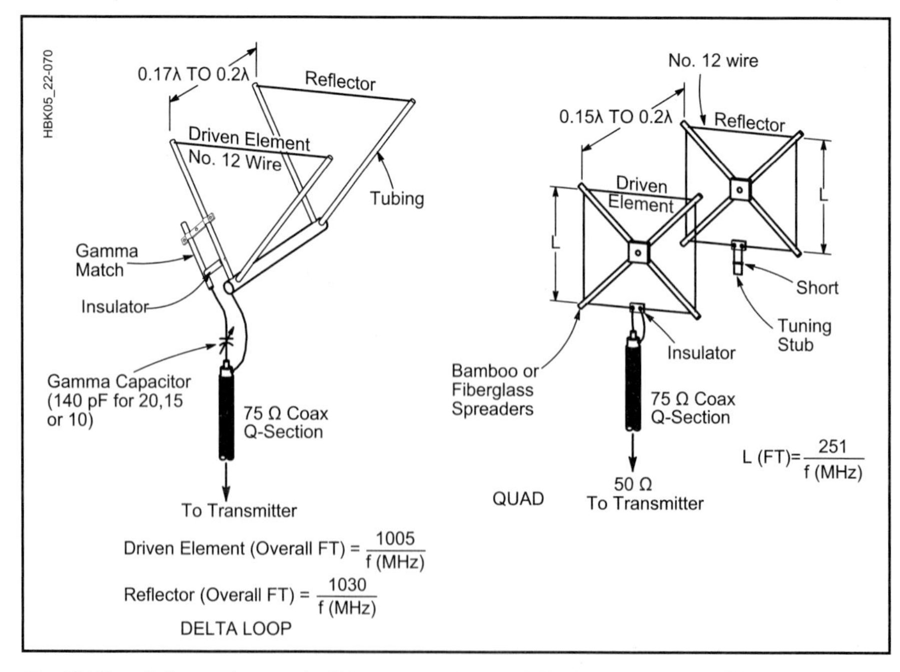

Fig 22.70 — Information on building a quad or a delta-loop antenna. The antennas are electrically similar, but the delta-loop uses plumber's delight construction. The λ/4 length of 75-Ω coax acts as a Q-section transformer from approximate 100-Ω feedpoint impedance of quad to 50-Ω feed line coax.

A FIVE-BAND, TWO-ELEMENT HF QUAD

Two quad designs are described in this article, both nearly identical. One was constructed by KC6T from scratch, and the other was built by Al Doig, W6NBH, using modified commercial triband quad hardware. The principles of construction and adjustment are the same for both models, and the performance results are also essentially identical. One of the main advantages of this design is the ease of (relatively) independent performance adjustments for each of the five bands. These quads were described by William A. Stein, KC6T, in *QST* for April 1992. Both models use 8-ft-long, 2-inch diameter booms, and conventional X-shaped spreaders (with two sides of each quad loop parallel to the ground).

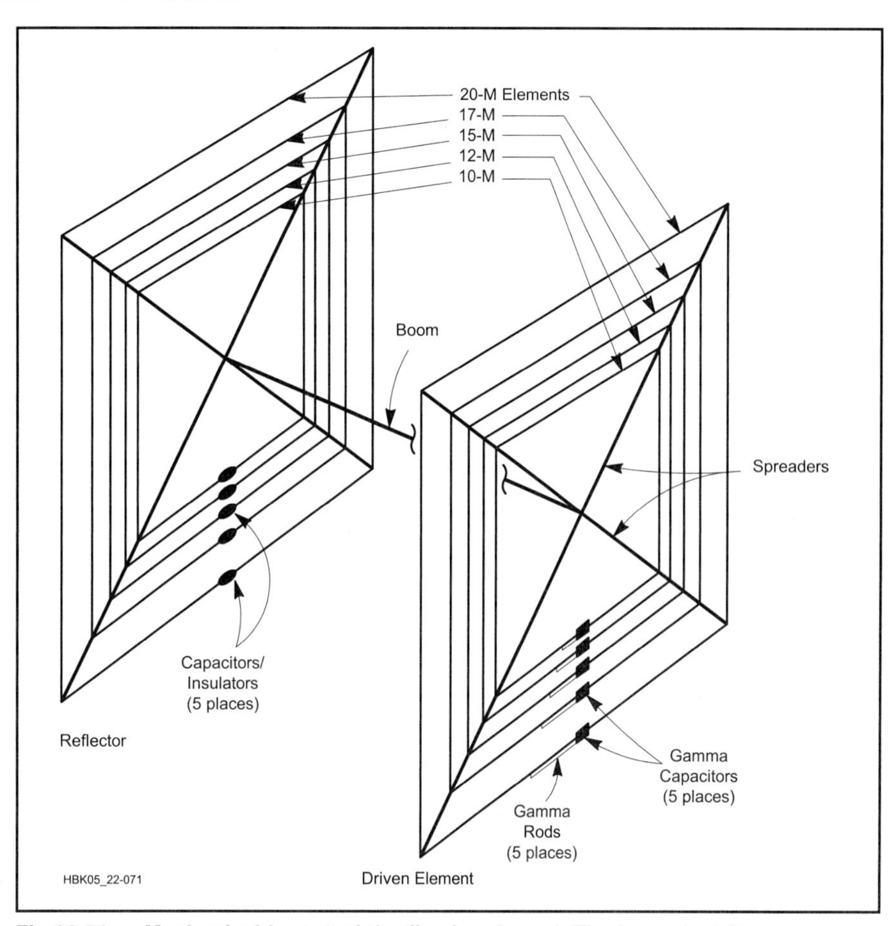

Fig 22.71 — Mechanical layout of the five-band quad. The boom is 8 ft long; see Table 22.19 for all other dimensions.

THE FIVE-BAND QUAD AS A SYSTEM

Unless you are extraordinarily lucky, you should remember one general rule: Any quad must be adjusted for maximum performance after assembly. Simple quad designs can be tuned by pruning and restringing the elements to control front-to-rear ratio and SWR at the desired operating frequency. Since each element of this quad contains five concentric loops, this adjustment method could lead to a nervous breakdown!

Fig 22.71 shows that the reflectors and driven elements are each independently adjustable. After assembly, adjustment is simple, and although gamma-match components on the driven element and capacitors on the reflectors add to the antenna's parts count, physical construction is not difficult. The reflector elements are purposely cut slightly long (except for the 10-m reflector), and electrically shortened by a tuning capacitor. The driven-element gamma matches set the lowest SWR at the desired operating frequency.

As with most multiband directive antennas, the designer can optimize any two of the following three attributes at the expense of the third: forward gain, front-to-rear ratio and bandwidth (where the SWR is less than 2:1). These three characteristics are related, and changing one changes the other two. The basic idea behind this quad design is to permit (without resorting to trimming loop lengths, spacing or other gross mechanical adjustments):

- The forward gain, bandwidth and front-to-rear ratio may be set by a simple adjustment after assembly. The adjustments can be made on a band-by-band basis, with little or no effect on previously made adjustments on the other bands.
- Setting the minimum SWR in any portion of each band, with no interaction with previously made front-to-back or SWR adjustments.

The first of the two antennas described, the KC6T model, uses aluminum spreaders with PVC insulators at the element attachment points. (The author elected not to use fiberglass spreaders because of their high cost.) The second antenna, the W6NBH model, provides dimensions and adjustment values for the same antenna, but using standard triband-quad fiberglass spreaders and hardware. If you have a triband quad, you can easily adapt it to this design. When W6NBH built his antenna, he had to shorten the 20-m reflector because the KC6T model uses a larger 20-m reflector than W6NBH's fiberglass spreaders would allow. Performance is essentially identical for both models.

MECHANICAL CONSIDERATIONS

Even the best electrical design has no value if its mechanical construction is lacking. Here are some of the things that contribute to mechanical strength: The gamma-match capacitor KC6T used was a small, air-variable, chassis-mount capacitor mounted in a plastic box (see **Fig 22.72**). A male UHF connector was mounted to the box, along with a screw terminal for connection to the gamma rod.

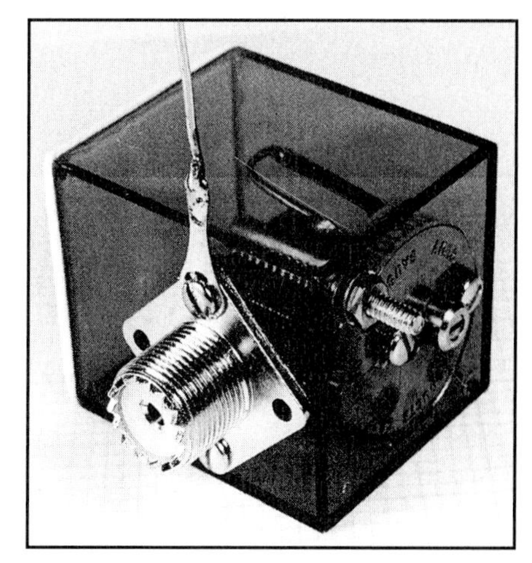

Fig 22.72 — Photo of one of the feed-point gamma-match capacitors.

The terminal lug and wire are for later connection to the driven element. The box came from a local hobby shop, and the box lid was replaced with a piece of 1/32-inch ABS plastic, glued in place after the capacitor, connector and wiring had been installed. The capacitor can be adjusted with a screwdriver through an access hole. Small vent (drain) holes were drilled near corresponding corners of each end.

Enclose the gamma-match capacitor in such a manner that you can tape unwanted openings closed so that moisture can't be directly blown in during wind and rainstorms. Also, smaller boxes and sturdy mounts to the driven element ensure that you won't pick up gamma capacitor assemblies along with the leaves after a wind storm.

Plastic gamma-rod insulators/standoffs were made from 1/32-inch ABS, cut 1/2-inch wide with a hole at each end. Use a knife to cut from the hole to the side of each insulator so that one end can be slipped over the driven element and the other over the gamma rod. Use about four such insulators for each gamma rod, and mount the first insulator as close to the capacitor box as possible. Apply five-minute epoxy to the element and gamma rod at the insulator hole to keep the insulators from sliding. If you intend to experiment with gamma-rod length, perform this gluing operation after you have made the final gamma-rod adjustments.

ELEMENT INSULATORS

As shown in Fig 22.71, the quad uses insulators in the reflectors for each band to break the loop electrically, and to allow reflector adjustments. Similar insulators were used to break up each driven element so that element impedance measurements could be made with a noise bridge. After the impedance measurements, the driven-element loops are closed again. The insulators are made from 1/4 × 2 × 3/4-inch phenolic stock. The holes are 1/2-inch apart. Two terminal lugs (shorted together at the center hole) are used in each driven element. They offer a convenient way to open the loops by removing one screw. **Fig 22.73** shows these insulators and the gamma-match construction schematically. **Table 22.19** lists the component values, element lengths and gamma-match dimensions.

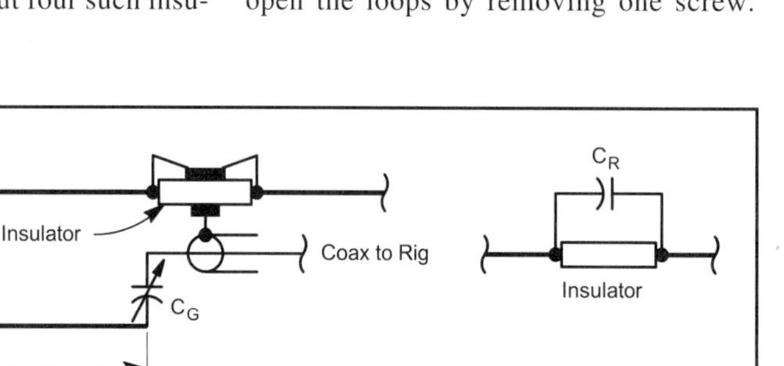

Fig 22.73 — Gamma-match construction details at A and reflector-tuning capacitor (C_R) attachment schematic at B. The gamma matches consist of matching wires (one per band) with series capacitors (C_g). See Table 22.19 for lengths and component specifications.

ELEMENT-TO-SPREADER ATTACHMENT

Probably the most common problem with quad antennas is wire breakage at the element-to-spreader attachment points. There are a number of functional attachment methods; **Fig 22.74** shows one of them. The attachment method with both KC6T and W6NBH spreaders is the same, even though the spreader constructions differ. The KC6T model uses #14 AWG, 7-strand copper wire; W6NBH used #18, 7-strand wire. At the point of element attachment (see **Fig 22.75**), drill a hole through both walls of the spreader using a #44 (0.086-inch) drill. Feed a 24-inch-long piece of antenna wire through the hole and center it for use as an attachment wire.

After fabricating the spider/spreader assembly, lay the completed assembly on a flat surface and cut the element to be installed to the correct length, starting with the 10-m element. Attach the element ends to the insulators to form a closed loop before attaching the elements to the spreaders. Center the insulator between the spreaders on what will become the bottom side of the quad loop, then carefully measure and mark the element-mounting-points with fingernail polish (or a similar substance). Do *not* depend on the at-rest position of the spreaders to guarantee that the mounting points will all be correct.

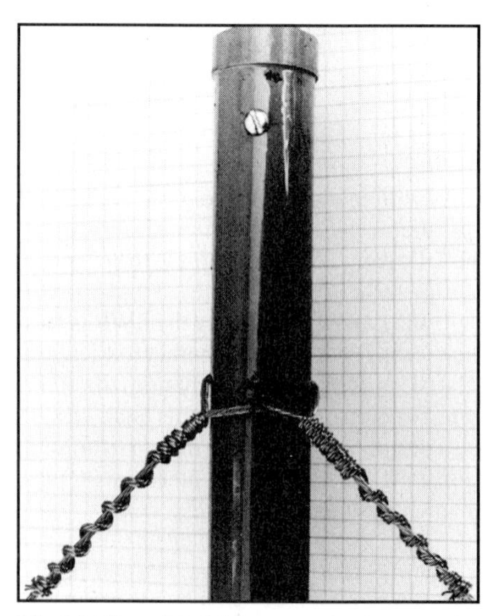

Fig 22.74 — Attaching quad wires to the spreaders must minimize stress on the wires for best reliability. This method (described in the text) cuts the chances of wind-induced wire breakage by distributing stress.

Table 22.19

Element Lengths and Gamma-Match Specifications of the KC6T and W6NBH Five-Band Quads

KC6T Model

Band (MHz)	Driven Element	Length (in.)	Gamma Match Spacing	Gamma Match C_g(pF)	Reflector Length (in.)	C_R (pF)
14	851.2	33	2	125	902.4	68
18	665.6	24	2	110	705.6	47
21	568	24	1.5	90	604.8	43
24.9	483.2	29.75	1	56	514.4	33
28	421.6	26.5	1	52	448.8	(jumper)

W6NBH Model

Band (MHz)	Driven Element	Length (in.)	Gamma Match Spacing	Gamma Match C_g(pF)	Reflector Length (in.)	C_R (pF)
14	851.2	31	2	117	890.4	120
18	665.6	21	2	114	705.6	56
21	568	26	1.5	69	604.8	58
24.9	483.2	15	1	75.5	514.4	54
28	421.6	18	1	41	448.8	(jumper)

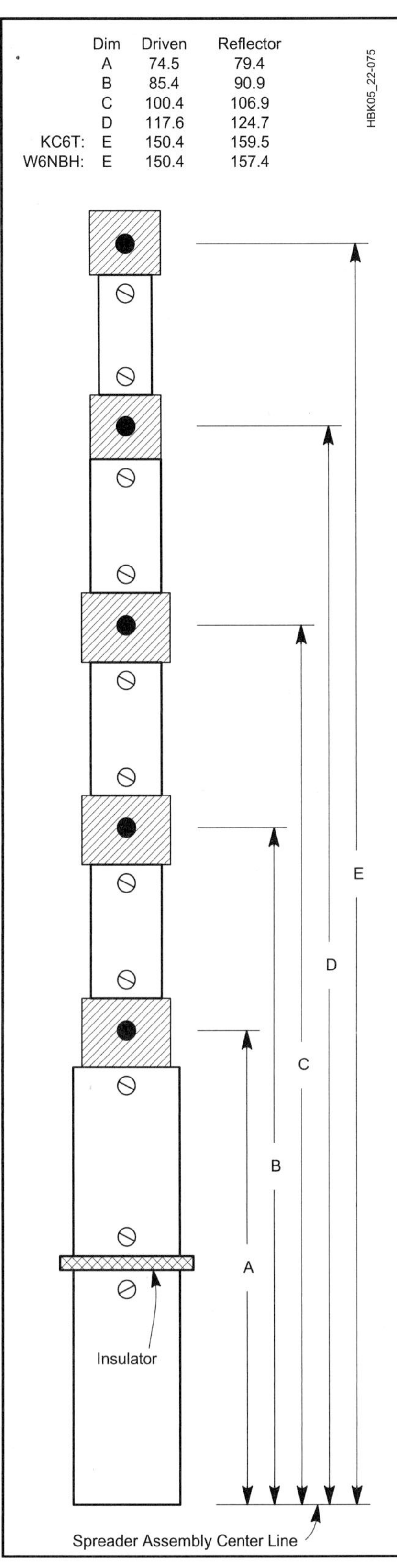

Fig 22.75 — Spreader-drilling diagram and dimensions (in.) for the five-band quad. These dimensions apply to both spreader designs described in the text, except that most commercial spreaders are only a bit over 13 ft (156 inches) long. This requires compensation for the W6NBH model's shorter 20-m reflector as described in the text.

Holding the mark at the centerline of the spreader, tightly loop the attachment wire around the element and then gradually space out the attachment-wire turns as shown. The attachment wire need not be soldered to the element. The graduated turn spacing minimizes the likelihood that the element wire will flex in the same place with each gust of wind, thus reducing fatigue-induced wire breakage.

FEEDING THE DRIVEN ELEMENTS

Each driven element is fed separately, but feeding five separate feed lines down the tower and into the shack would be costly and mechanically difficult. The ends of each of these coax lines also require support other than the tension (or lack of thereof) provided by the driven element at the feed-point. It is best to use a remote coax switch on the boom approximately 1 ft from the driven-element spider-assembly attachment point.

At installation, the cables connecting the gamma-match capacitors and the coax switch help support the driven elements and gamma capacitors. The support can be improved by taping the cables together in several places. A single coaxial feed line (and a control cable from the remote coax switch, if yours requires one) is the only required cabling from the antenna to the shack.

THE KC6T MODEL'S COMPOSITE SPREADERS

If you live in an area with little or no wind, spreaders made from wood or PVC are practical but, if you live where winds can reach 60 to 80 mi/h, strong, lightweight spreaders are a must. Spreaders constructed with electrical conductors (in this case, aluminum tubing) can cause a myriad of problems with unwanted resonances, and the problem gets worse as the number of bands increases.

To avoid these problems, this version uses composite spreaders made from machined PVC insulators at the element-attachment points. Aluminum tubing is inserted into (or over) the insulators 2 inches on each end. This spreader is designed to withstand 80 mi/h winds. The overall insulator length is designed to provide a 3-inch center insulator clear of the aluminum tubing. The aluminum tubing used for the 10-m section (inside dimension "A" in Fig 22.75) is 1 1/8-inch diameter × 0.058-inch wall. The next three sections are 3/4-inch diameter × 0.035-inch wall, and the outer length is made from 1/2-inch diameter × 0.035-inch wall. The dimensions shown in Fig 22.75 are *attachment point* dimensions only.

Attach the insulators to the aluminum using #6 sheet metal screws. Mechanical strength is provided by Devcon no. S 220 Plastic Welder Glue (or equivalent) applied liberally as the aluminum and plastic parts are joined. Paint the PVC insulators before mounting the elements to them. Paint protects the PVC from the harmful effects of solar radiation. As you can see from Fig 22.75, an additional spreader insulator located about halfway up the 10-m section (inside dimension "A") removes one of the structure's electrical resonances not eliminated by the attachment-point insulators. Because it mounts at a relatively high-stress point in the spreader, this insulator is fabricated from a length of heavy-wall fiberglass tubing.

Composite spreaders work as well as fiberglass spreaders, but require access to a well-equipped shop, including a lathe. The main objective of presenting the composite spreader is to show that fiberglass spreaders aren't a basic requirement—there are many other ways to construct usable spreaders. If you can lay your hands on a used multiband quad, even one that's damaged, you can probably obtain enough spreaders to reduce construction costs considerably.

GAMMA ROD

The gamma rod is made from a length of #12 solid copper wire (W6NBH used #18, 7-strand wire). Dimensions and spacings are shown in Table 22.19. If you intend to experiment with gamma-rod lengths and capacitor settings, cut the gamma-rod lengths about 12 inches longer than the length listed in the table. Fabricate a sliding short by soldering two small alligator clips back-to-back such that they can be clipped to the rod and the antenna element and easily moved along the driven element. Note that gamma-rod spacing varies from one band to another. When you find a suitable shorting-clip position, mark the gamma rod, remove the clip, bend the gamma rod at the mark and solder the end to the element.

THE W6NBH MODEL

As previously mentioned, this model uses standard 13-ft fiberglass spreaders, which aren't quite long enough to support the larger 20-m reflector specified for the KC6T model. The 20-m W6NBH reflector loop is cut to the dimensions shown in Table 22.19, 12 inches shorter than that for the KC6T model. To tune the shorter reflector, a 6-inch-long stub of antenna wire (spaced 2 inches) hangs from the reflector insulator, and the reflector tuning capacitor mounts on another insulator at the end of this stub.

GAMMA-MATCH AND REFLECTOR-TUNING CAPACITOR

Use an air-variable capacitor of your choice for each gamma match. Approximately 300 V can appear across this capacitor (at 1500 W), so choose plate spacing appropriately. If you want to adjust the capacitor for best match and then replace it with a fixed capacitance, remember that several amperes of RF will flow through the capacitance. If you choose disc-ceramic capacitors, use a parallel combination of at least four 1-kV units of equal value. Any temperature coefficient is acceptable. NP0 units are not required. Use similar components to tune the reflector elements.

ADJUSTMENTS

Well, here you are with about 605 ft of wire. Your antenna will weigh about 45 pounds (the W6NBH version is slightly lighter) and have about 9 square ft of wind area. If you chose to, you can use the dimensions and capacitance values given, and performance should be excellent. If you adjust the antenna for minimum SWR at the band centers, it should cover all of the lower four bands and 28 to 29 MHz with SWRs under 2:1; front-to-rear ratios are given in **Table 22.20**.

Table 22.20
Measured Front-to-Rear Ratios

Band	*KC6T Model*	*W6NBH Model*
14	25 dB	16 dB
18	15 dB	10 dB
21	25 dB	>20 dB
24.9	20 dB	>20 dB
28	20 dB	>20 dB

Instead of building the quad to the dimensions listed and hoping for the best, you can adjust your antenna to account for most of the electrical environment variables of your installation. The adjustments are conceptually simple: First adjust the reflector's electrical length for maximum front-to-rear ratio (if you desire good gain, and are willing to settle for a narrower than maximum SWR bandwidth), or accept some compromise in front-to-rear ratio that results in the widest SWR bandwidth. You can make this adjustment by placing an air-variable capacitor (about 100-pF maximum) across the open reflector loop ends, one band at a time, and adjusting the capacitor for the desired front-to-rear ratio. The means of doing this will be discussed later.

During these reflector adjustments, the driven-element gamma-match capacitors may be set to any value and the gamma rods may be any convenient length (but the sliding-short alligator clips should be installed somewhere near the lengths specified in Table 22.19). After completing the front-to-rear adjustments, the gamma capacitors and rods are adjusted for minimum SWR at the desired frequency.

ADJUSTMENT SPECIFICS

Adjust each band by feeding it separately. You can make a calibrated variable capacitor (with a hand-drawn scale and wire pointer). Calibrate the capacitor using your receiver, a known-value inductor and a dip meter (plus a little calculation).

To adjust front-to-rear ratio, simply clip the (calibrated) air-variable capacitor across the open ends of the desired reflector loop. Connect the antenna to a portable receiver with an S meter. Point the back of the quad at a signal source, and slowly adjust the capacitor for a dip in the S-meter reading.

After completing the front-to-rear adjustments, replace the variable capacitor with an appropriate fixed capacitor sealed against the weather. Then move to the driven-element adjustments. Connect the coax through the SWR bridge to the 10-m gamma-match capacitor box. Use an SWR bridge that requires only a watt or two (not more than 10 W) for full-scale deflection in the calibrate position on 10 m. Using the minimum necessary power, measure the SWR. Go back to receive and adjust the capacitor until (after a number of transmit/receive cycles) you find the minimum SWR. If it is too high, lengthen or shorten the gamma rod by means of the sliding alligator-clip short and make the measurements again.

Stand away from the antenna when making transmitter-on measurements. The adjustments have minimal effect on the previously made front-to-rear settings, and may be made in any band order. After making all the adjustments and sealing the gamma capacitors, reconnect the coax harness to the remote coax switch.

A SIMPLE QUAD FOR 40 METERS

Many amateurs yearn for a 40-m antenna with more gain than a simple dipole. While two-element rotary 40-m beams are available commercially, they are costly and require fairly hefty rotators to turn them. This low-cost, single-direction quad is simple enough for a quick Field Day installation, but will also make a home station very competitive on the 40-m band.

This quad uses a 2-inch outside diameter, 18-ft boom, which should be mounted no less than 60 ft high, preferably higher. (Performance tradeoffs with height above ground will be discussed later.) The basic design is derived from the N6BV 75/80-meter quad described in *The ARRL Antenna Compendium, Vol 5*. However, since this simplified 40-m version is unidirectional and since it covers only one portion of the band (CW or Phone, but not both), all the relay-switched components used in the larger design have been eliminated.

The layout of the simple 40-m quad at a boom height of 70 ft is shown in **Fig 22.76**. The wires for each element are pulled out sideways from the boom with black 1/8-inch Dacron rope designed specifically to withstand both abrasion and UV radiation. The use of the proper type of rope is very important—using a cheap substitute is not a good idea. You will not enjoy trying to retrieve wires that have become, like Charlie Brown's kite, hopelessly entangled in nearby trees, all because a cheap rope broke during a windstorm! At a boom height of 70 ft, the quad requires a *wingspread* of 140 ft for the side ropes. This is the same wingspread needed by an inverted-V dipole at the same apex height with a 90° included angle between the two legs.

The shape of each loop is rather unusual, since the bottom ends of each element are brought back close to the supporting tower. (These element ends are insulated from the tower and from each other). Having the elements near the tower makes fine-tuning adjustments much easier—after all, the ends of the loop wires are not 9 feet out, on the ends of the boom! The feed-point resistance with this loop configuration is close to 50 Ω, meaning that no matching network is necessary. By contrast, a more conventional diamond or square quad-loop configuration exhibits about a 100-Ω resistance.

Another bonus to this loop configuration is that the average height above ground is higher, leading to a slightly lower angle of radiation for the array and less loss because the bottom of each element is raised higher above lossy ground. The drawback to this unusual layout is that

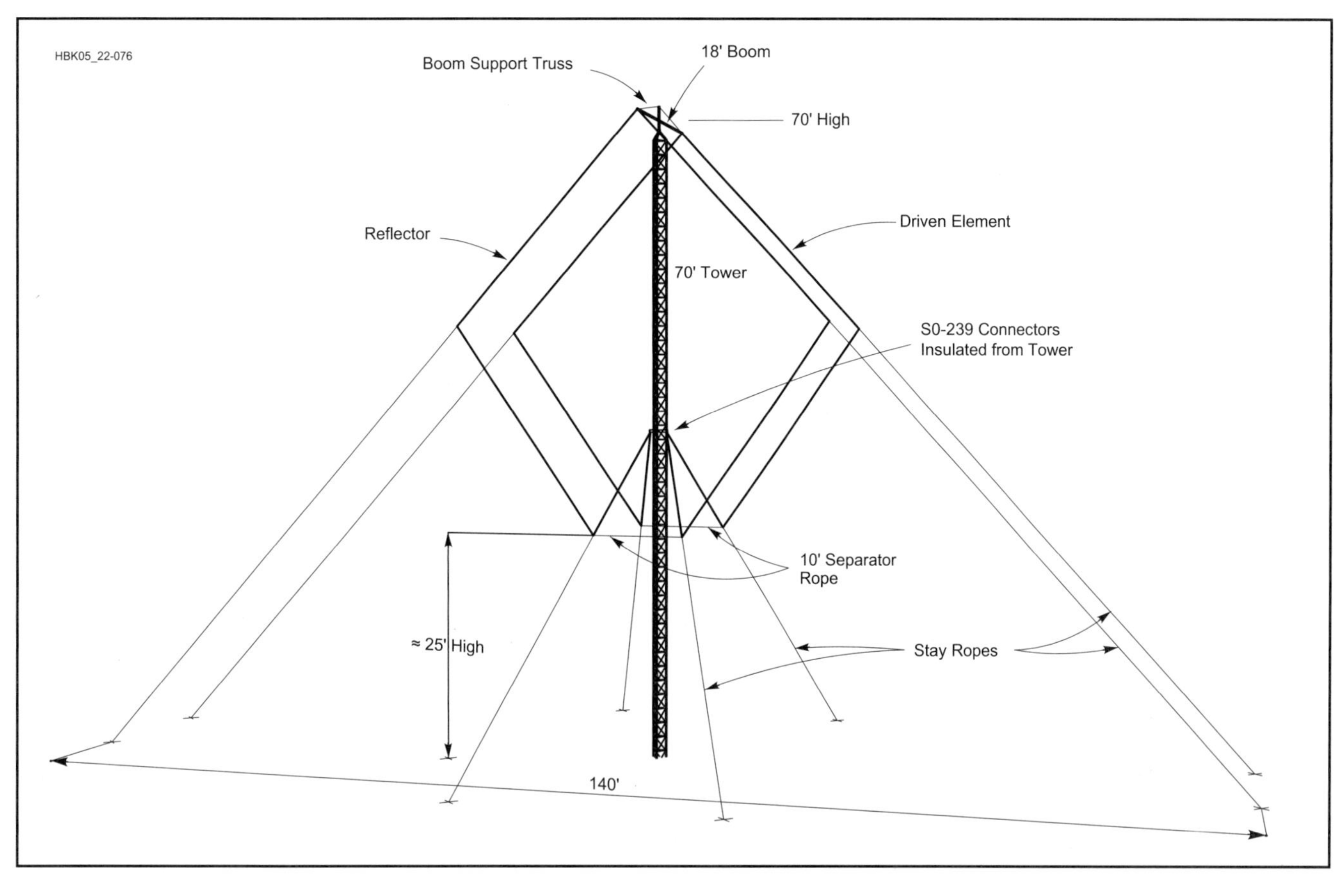

Fig 22.76 — Layout of 40-meter quad with a boom height of 70 feet. The four stay ropes on each loop pull out each loop into the desired shape. Note the 10-foot separator rope at the bottom of each loop, which helps it hold its shape. The feed line is attached to the driven element through a choke balun, consisting of 10 turns of coax in a 1-foot diameter loop. You could also use large ferrite beads over the feed-line coax, as explained in Chapter 21. Both the driven element and reflector loops are terminated in SO-239 connectors tied back to (but insulated from) the tower. The reflector SO-239 has a shorted PL-259 normally installed in it. This is removed during fine-tuning of the quad, as explained in the text.

four more *tag-line* stay ropes are necessary to pull the elements out sideways at the bottom, pulling against the 10-foot separator ropes shown in Fig 22.76.

CONSTRUCTION

You must decide before construction whether you want coverage on CW (centered on 7050 kHz) or on Phone (centered on 7225 kHz), with roughly 120 kHz of coverage between the 2:1 SWR points. If the quad is cut for the CW portion of the band, it will have less than about a 3.5:1 SWR at 7300 kHz, as shown in **Fig 22.77**. The pattern will deteriorate to about a 7 dB F/B at 7300 kHz, with a reduction in gain of almost 3 dB from its peak in the CW band. It is possible to use a quad tuned for CW in the phone band if you use an antenna tuner to reduce the SWR and if you can take the reduction in performance. To put things in perspective, a quad tuned for CW but operated in the phone band will still work about as well as a dipole.

Next, you must decide where you want to point the quad. A DXer or contester in the USA might want to point this single-direction design to cover Europe and North Africa. For Field Day, a group operating on the East Coast would simply point it west, while their counterparts on the West Coast would point theirs east.

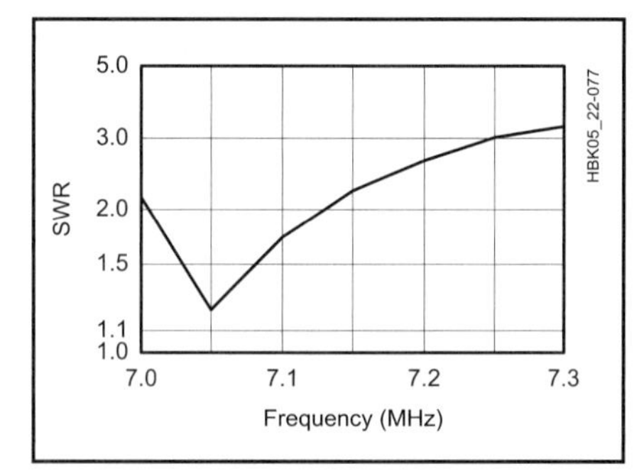

Fig 22.77 — Plot of SWR versus frequency for a quad tuned for CW operation.

The mechanical requirements for the boom are not severe, especially since a top truss support is used to relieve stress on the boom due to the wires pulling on it from below. The boom is 18 ft long, made of 2-inch diameter aluminum tubing. You can probably find a suitable boom from a scrapped triband or monoband Yagi. You will need a suitable set of U-bolts and a mounting plate to secure the boom to the face of a tower. Or perhaps you might use lag screws to mount the boom temporarily to a suitable tree on Field Day! On a 70-ft high tower, the loop wires are brought back to the tower at the 37.5-ft level and tied there using insulators and rope. The lowest points of the loops are located about 25 ft above ground for a 70-ft tower. **Fig 22.78** gives dimensions for the driven element and reflector for both the CW and the Phone portions of the 40-m band.

GUY WIRES

Anyone who has worked with quads knows they are definitely three-dimensional objects! You should plan your installation carefully, particularly if the supporting tower has guy wires, as most do. Depending on where the guys are located on the tower and the layout of the quad with reference to those guys, you will probably have to string the quad loops over

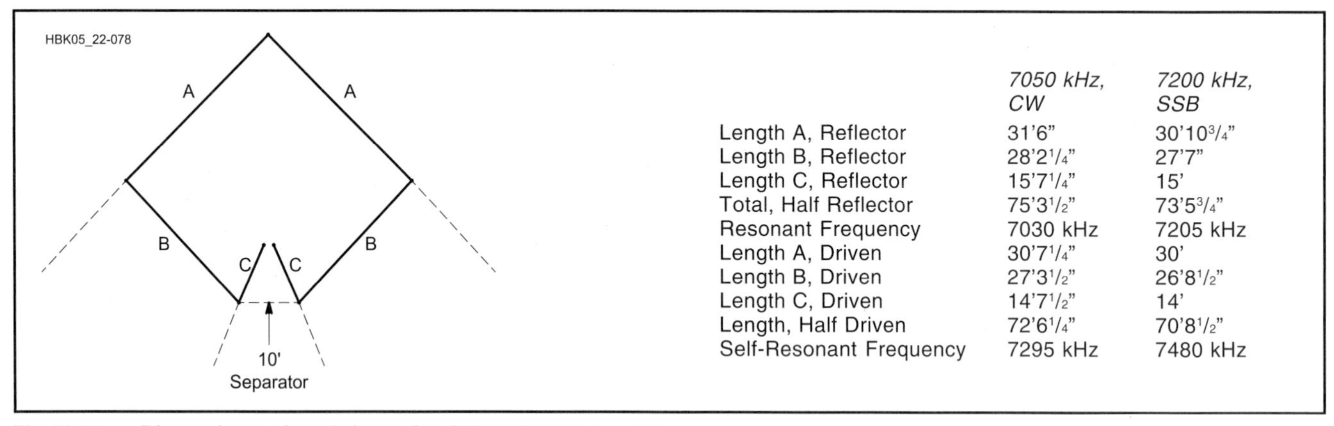

	7050 kHz, CW	7200 kHz, SSB
Length A, Reflector	31'6"	30'10 3/4"
Length B, Reflector	28'2 1/4"	27'7"
Length C, Reflector	15'7 1/4"	15'
Total, Half Reflector	75'3 1/2"	73'5 3/4"
Resonant Frequency	7030 kHz	7205 kHz
Length A, Driven	30'7 1/4"	30'
Length B, Driven	27'3 1/2"	26'8 1/2"
Length C, Driven	14'7 1/2"	14'
Length, Half Driven	72'6 1/4"	70'8 1/2"
Self-Resonant Frequency	7295 kHz	7480 kHz

Fig 22.78 — Dimensions of each loop, for CW or Phone operation.

certain guys (probably at the top of the tower) and under other guys lower down.

It is very useful to view the placement of guy wires using the VIEW ANTENNA function in the *EZNEC* modeling program. This allows you to visualize the 3-D layout of an antenna. You can ROTATE yourself around the tower to view various aspects of the layout. *EZNEC* will complain about grounding wires directly but will still allow you to use the View Antenna function. Note also that it is best to insulate guy wires to prevent interaction between them and the antennas on a tower, but this may not be necessary for all installations.

FINE TUNING, IF NEEDED

We specify stranded #14 hard-drawn copper wire for the elements. During the course of installation, however, the loop wires could possibly be stretched a small amount as you pull and yank on them, trying to clear various obstacles. This may shift the frequency response and the performance slightly, so it is useful to have a tuning procedure for the quad when it is finally up in the air.

The easiest way to fine-tune the quad while on the tower is to use a portable, battery-operated SWR indicator (such as the Autek RF-1 or the MFJ-259) to adjust the reflector and the driven element lengths for specific resonant frequencies. You can eliminate the influence of mutual coupling to the other element by open-circuiting the other element.

For convenience, each quad loop should be connected to an SO-239 UHF female connector that is insulated from but tied close to the tower. You measure the driven element's resonant frequency by first removing the shorted PL-259 normally inserted into the reflector connector. Similarly, the reflector's resonant frequency can be determined by removing the feed line normally connected to the driven element's feed point.

Obviously, it's easiest if you start out with extra wire for each loop, perhaps 6 inches extra on each side of the SO-239. You can then cut off wire in 1/2-inch segments equally on each side of the connector. This procedure is easier than trying to splice extra wire while up on the tower. Alligator clips are useful during this procedure, but just don't lose your hold on the wires! You should tie safety strings from each wire back to the tower. Prune the wire lengths to yield the resonant frequencies (±5 kHz) shown in Fig 22.78 and then solder things securely. Don't forget to reinsert the shorted PL-259 into the reflector SO-239 connector to turn it back into a reflector.

HIGHER IS BETTER

This quad was designed to operate with the boom at least 60 ft high. However, it will work considerably better for DX work if you can put the boom up even higher. **Fig 22.79** shows the elevation patterns for four antennas: a reference inverted-V dipole at 70 ft (with a 90° included angle between the two legs), and three quads, with boom height of 70, 90 and 100 ft respectively. At an elevation angle of 20°, typical for DX work on 40 m, the quad at 100 ft has about a 5 dB advantage over an inverted-V dipole at 70 ft, and about a 3 dB advantage over a quad with a boom height of 70 ft.

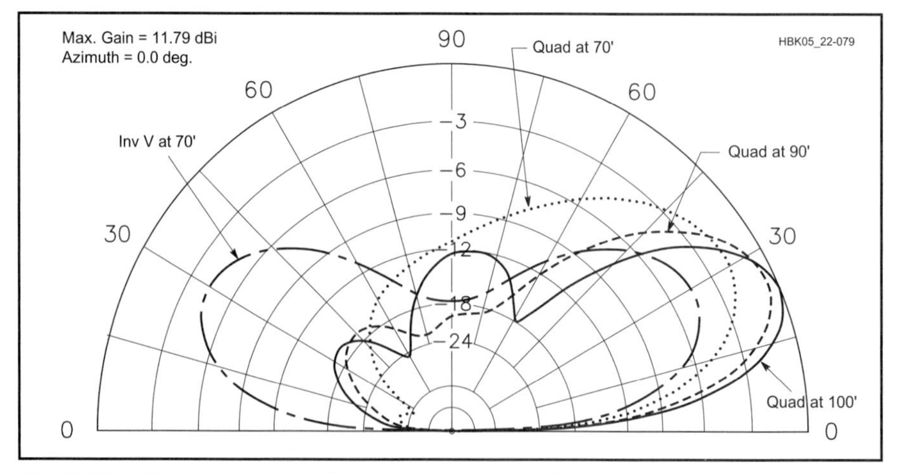

Fig 22.79 — Comparisons of the elevation patterns for quads at boom heights of 70, 90 and 100 ft, referenced to an inverted-V dipole at 70 ft.

A SIMPLE LOOP ANTENNA FOR 28 MHZ

With the large number of operators and wide availability of inexpensive, single-band radios, the 10-m band could well become the hangout for local ragchewers that it was before the advent of 2-m FM, even at a low point in the solar cycle.

This simple antenna provides gain over a dipole or inverted V. It is a resonant loop with a particular shape. It provides 2.1 dB gain over a dipole at low radiation angles when mounted well above ground. The antenna is simple to feed—no matching network is necessary. When fed with 50-Ω coax, the SWR is close to 1:1 at the design frequency, and is less than 2:1 from 28.0-28.8 MHz for an antenna resonant at 28.4 MHz.

The antenna is made from #12 AWG wire (see **Fig 22.80**) and is fed at the center of the bottom wire. Coil the coax into a few turns near the feedpoint to provide a simple balun. A coil diameter of about a foot will work fine. You can support the antenna on a mast with spreaders made of bamboo, fiberglass, wood, PVC or other nonconducting material. You can also use aluminum tubing both for support and conductors, but you may have to readjust the antenna dimensions for resonance.

This rectangular loop has two advantages over a resonant square loop. First, a square loop has just 1.1 dB gain over a dipole. This is a power increase of only 29%. Second, the input impedance of a square loop is about 125 Ω. You must use a matching network to feed a square loop with 50-Ω coax. The rectangular loop achieves gain by compressing its radiation pattern in the elevation plane. The azimuth plane pattern is slightly wider than that of a dipole (it's about the same as that of an inverted V). A broad pattern is an advantage for a general-purpose, fixed antenna. The rectangular loop provides a bidirectional gain over a broad azimuth region.

Mount the loop as high as possible. To provide 1.7 dB gain at low angles over an inverted V, the top wire must be at least 30 ft high. The loop will work at lower heights, but its gain advantage disappears. For example, at 20 ft the loop provides the same gain at low angles as an inverted V.

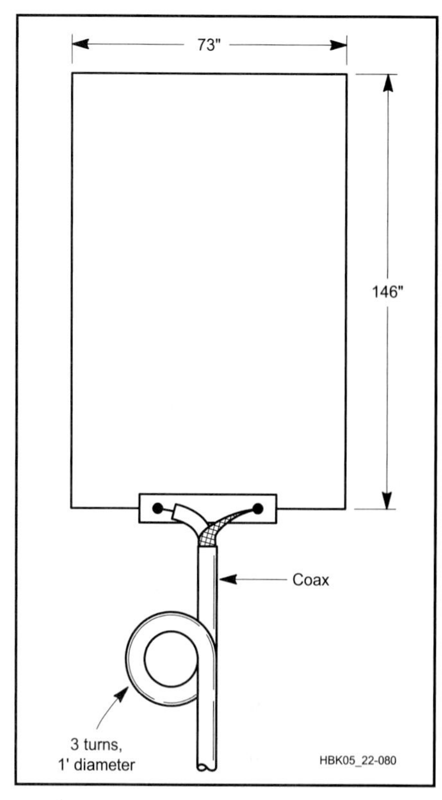

Fig 22.80 — Construction details of the 10-m rectangular loop antenna.

HF Mobile Antennas

This section is by Jack Kuecken, KE2QJ. Jack is an antenna engineer who has written a number of articles for ARRL publications.

An ideal HF mobile antenna is:

1. Sturdy. Stays upright at highway speeds.
2. Mechanically stable. Sudden stops or sharp turns do not cause it to whip about, endangering other vehicles.
3. Flexibly mounted. Permits springing around branches and obstacles at slow speeds.
4. Weatherproof. Handles the impact of wind, rain, snow and ice at high speed.
5. Tunable to all of the HF bands without stopping the vehicle.
6. Mountable without altering the vehicle in ways which lower the resale value.
7. Efficient as possible.
8. Easily removed for sending the car through a car wash, etc.

For HF mobile operation, the ham must use an electrically small antenna. The possibility that the antenna might strike a fixed object places a limitation on its height. On Interstate highways, an antenna tip at 11.5 feet above the pavement is usually no problem. However, on other roads you may encounter clearances of 9.5 or 10 feet. You should be able to easily *tie down* the antenna for a maximum height of about 7 feet to permit passage through low-clearance areas. The antenna should be usable while in the tied-down position.

If the base of an antenna is 1 ft above the pavement and the tip is at 11.5 ft, the length is 10.5 ft which is 0.1 λ at 9.37 MHz, and 0.25 λ at 23.4 MHz. That means that the antenna will require a matching network for all of the HF bands except 10 and 12 m.

The power radiated by the antenna is equal to the radiation resistance times the square of the antenna current. The radiation resistance of an electrically small antenna is given by:

$$Rr = 395 \times (h/\lambda)^2$$

where

h = radiator height in meters

λ = wavelength in meters = 300/Freq in MHz

The capacitance in pF of an electrically small antenna is given approximately by:

$$C = \frac{55.78 \times h}{((\text{den1}) \times (\text{den2}))}$$

where

(den1) = (ln(h/r)–1)

(den2) = $(1 - (f \times h/75)^2)$

Table 22.21

Characteristics of a 10.5-foot whip antenna

F (MHz)	*C (pF)*	*Rr*	*Impedance*	*Efficiency*	*L (μH)*
1.8	30.1	0.146	13.72 –*j*2716	0.01064	240
3.5	30.6	0.55	7.43 –*j*1375	0.074	62.5
7	32.8	2.2	7.04 –*j*644	0.312	14.6
10	36.5	4.5	6.5 –*j*408	0.692	6.49
14	46.5	8.8	10 –*j*232	0.88	2.64

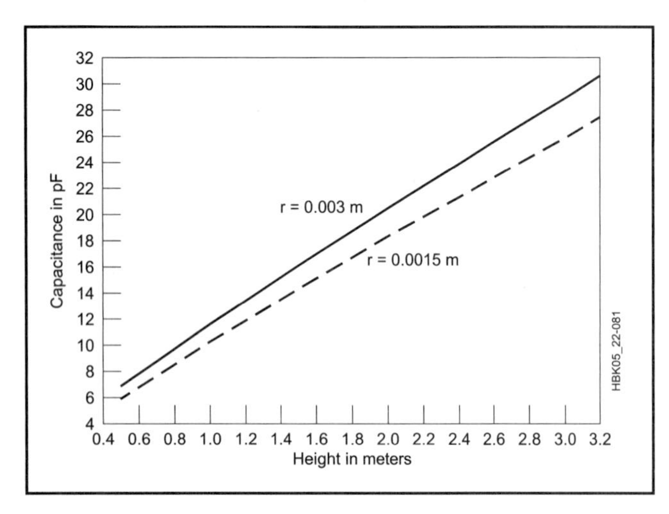

Fig 22.81 — Relationship at 3.5 MHz between vertical radiator length and capacitance. The two curves show that the capacitance is not very sensitive to radiator diameter.

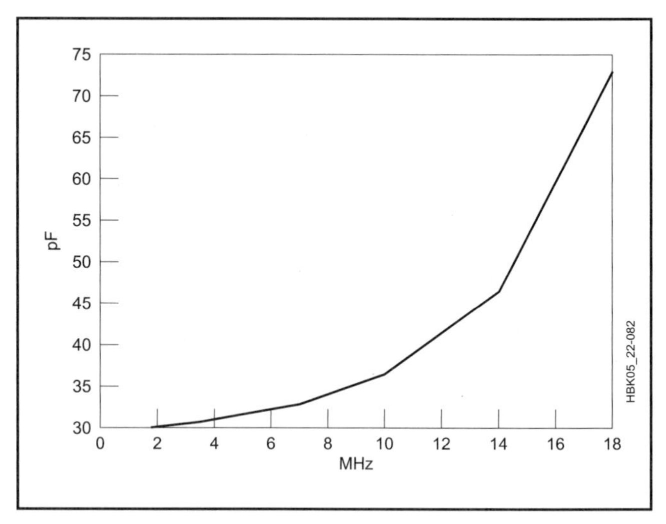

Fig 22.82 — Relationship between frequency and capacitance for a 3.2-meter vertical whip.

ln = natural logarithm
r = conductor radius in meters
f = frequency in MHz

Characteristics of a 10.5-ft (3.2 m) whip with a 0.003 m radius and, assuming a base loading coil with a Q of 200 and coil stray capacitance of 2 pF, are given in **Table 22.21**.

Radiation resistance rises in a nonlinear fashion and the capacitance drops just as dramatically with increase in the ratio h/λ. **Fig 22.81** shows the relationship of capacitance to height. This can be used for estimating antenna capacitance for other heights.

Fig 22.82 shows that capacitance is not very sensitive to frequency for h/λ less than 0.075, 8 MHz in this case. However, the sensitivity increases rapidly thereafter.

Table 22.21 shows that at 3.5 MHz an inductance of 62.5 µH will cancel the capacitive reactance. This results in an impedance of 7.43 Ω which means that additional matching is required. In this case the radiation efficiency of the system is only 0.074 or 7.4%. In other words, nearly 93% of energy at the terminals is wasted in heating the matching coil.

System Q is controlled by the Q of the coil. The bandwidth between 2:1 SWR points of the system = 0.36 × f/Q. In this case, bandwidth = 0.36 × 3.5/200 = 6.3 kHz

If we could double the Q of the coil, the efficiency would double and the bandwidth would be halved. The converse is also true. In the interest of efficiency, the highest possible Q should be used!

Another significant factor arises from the high Q. Let's assume that we deliver 100 watts to the 7.43 Ω at the antenna terminals. The current is 3.67 A and flows through the 1375-Ω reactance of the coil giving rise to 1375 × 3.67= 5046 VRMS (7137 Vpeak) across the coil.

With only 30.6 pF of antenna capacitance, the presence of significant stray capacitance at the antenna base shunts currents away from the antenna. RG-58 has about 21 pF/foot. A 1.5-foot length would halve the radiation efficiency of our example antenna. For cases like the whip at 3.5 MHz, the matching network has to be right at the antenna!

BASE, CENTER OR DISTRIBUTED LOADING

There is no clear-cut advantage in terms of radiation performance for either base or center-loaded antennas for HF mobile. Antennas with distributed (or continuous) loading have appeared in recent use. How do they compare?

Base Loading

In the design procedure, one estimates the capacitance, capacitive reactance and radiation resistance as shown previously. One then calculates the expected loss resistance of the loading coil required to resonate the antenna. There is generally additional resistance amounting to about half of the coil loss which must be added in. As a practical matter, it is usually not possible to achieve a coil Q in excess of 200 for such applications.

Using the radiation resistance plus 1.5 times the coil loss and the power rating desired for the antenna, one may select the wire size. For high efficiency coils, a current density of 1000 A/inch^2 is a good compromise. For the 3.67 A of the example we need a wire 0.068-inch diameter, which roughly corresponds to #14 AWG. Higher current densities can lead to a melted coil.

Design the coil with a pitch equal to twice the wire diameter and the coil diameter approximately equal to the coil length. These proportions lead to the highest Q in air core coils.

The circuit of **Fig 22.83** will match essentially all practical HF antennas on a car or truck. The circuit actually matches the antenna to 12.5 Ω and the transformer boosts it up to 50 Ω. Actual losses alter the required values of both the shunt inductor and the series capacitor. At a frequency of 3.5 MHz with an antenna impedance of 0.55 –*j*1375 Ω and a base capacitance of 2 pF results in the values shown in **Table 22.22.** Inductor and capacitor values are highly sensitive to coil Q. Furthermore, the inductor values are considerably below the 62.5 µH required to resonate the antenna.

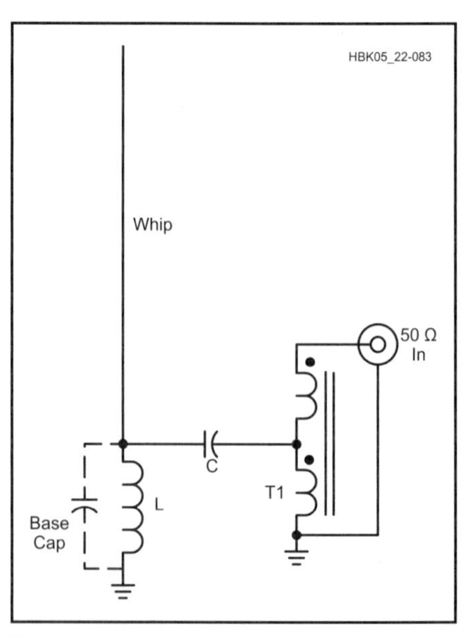

Fig 22.83 — The base-matched mobile whip antenna.

Table 22.22
Values of L and C for the Circuit of Fig 22.83 on 3.5 MHz

Coil Q	*L (μH)*	*C (pF)*	*System Efficiency*
300	44	11.9	0.083
200	29.14	35	0.0372
100	22.2	58.1	0.014

This circuit has the advantage that the tuning elements are all at the base of the antenna. The whip radiator itself has minimal mass and wind resistance. In addition, the rig is protected by the fact that there is a dc ground on the radiator so any accidental discharge or electrical contact is kept out of the cable and rig. Variable tuning elements allow the antenna to be tuned to other frequencies.

Connect the antenna, L and C. Start with less inductor than required to resonate the antenna. Tune the capacitor to minimum SWR. Increase the inductance and tune for minimum SWR. When the values of L and C are right, the SWR will be 1:1.

For remote or automatic tuning the drive motors for the coil and capacitor and the limit switches can be operated at RF ground potential. Mechanical connections to the RF components should be through insulated couplings.

Center Loading

Center loading increases the current in the lower half of the whip as shown in **Fig 22.84**. One can start by calculating the capacitance for the section above the coil just as done for the base loaded antenna. This permits the calculation of the loading inductance. The center loaded antenna is often operated without any base matching in which case the resistive component can be assumed to be 50 Ω for purposes of calculating the current rating and selecting wire size for the inductor.

The reduced size top section results in reduced capacitance which requires a much larger loading inductor. Center loading requires twice as much inductive reactance as base loading. For equal coil Qs, loss resistance is twice as great for center loading. If the coil is above the center, the inductance must be even larger, and the loss resistance increases accordingly. These factors tend to negate the advantage of the improved current distribution.

Because of the high value of inductance required, optimum Q coils are very large. One manufacturer of this type coil does not recommend their use in rain or inclement weather. The large wind resistance necessitates a very sturdy mount for operation at highway speed. Owing to the Q of these large coils the use of a base matching element in the form of either a tapped inductor or a shunt capacitor is usually needed to match to 50 Ω.

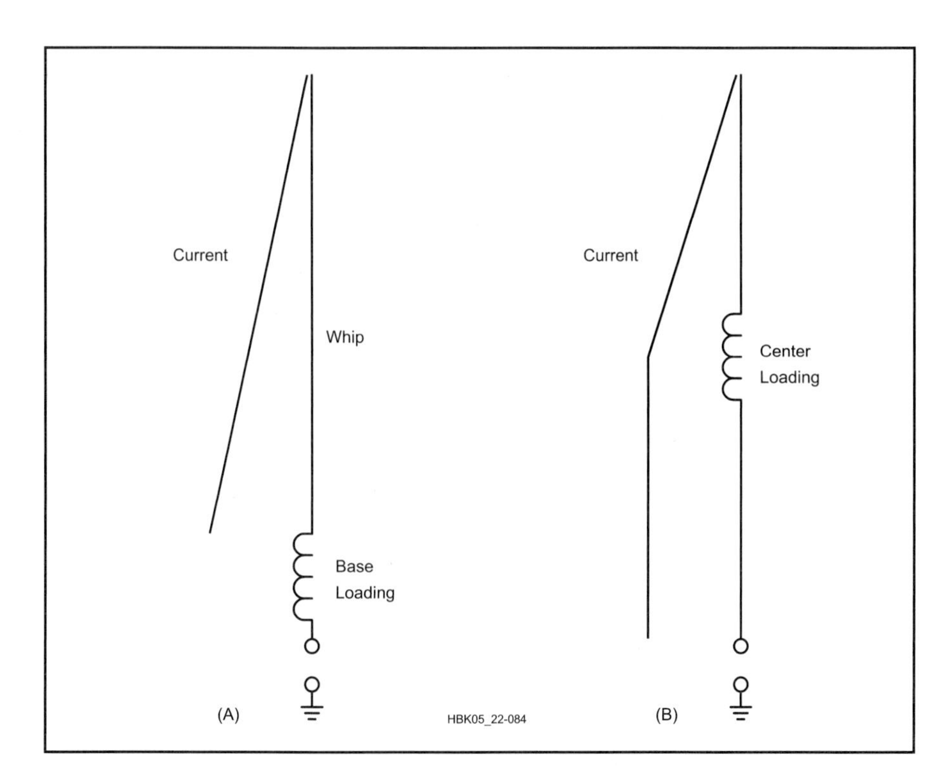

Fig 22.84 — Relative current distribution on a base-loaded antenna is shown at A and for a center-loaded antenna at B.

Another manufacturer places the coil above the center and uses a small extendable *wand* for tuning. To minimize wind resistance, the coil lengths are several times their diameters. These antenna coils are usually close wound with enameled wire. The coils are covered with a heat-shrink sleeve. If used in heavy rain or snow for extended periods water may get under the sleeving and seriously detune and lower the Q of the coils. These antennas usually do not require a base matching element. The resistance seems to come out close enough to 50 Ω.

It is possible to make a center-loaded antenna that is remotely tunable across the HF bands; however, this requires a certain amount of mechanical sophistication. The drive motor, limit switches and position sensor can be located in a box at the antenna base and drive the coil tuning mechanism through an electrically isolated shaft. Alternatively, the equipment could be placed adjacent to the loading coil requiring all of the electrical leads to be choked off to permit RF feeding of the base. The latter choice is probably the most difficult to realize.

Continuously Loaded Antennas

Antennas consisting of a fiberglass sleeve with the radiator wound in a continuous spiral to shorten a CB antenna from 8.65 feet to 5 or 6 feet have been on the market for many years. This modest shortening has little impact on the efficiency but does narrow the bandwidth.

One line of mobile antennas uses periodic loading on a relatively small diameter tube. A series of taps along the length are used to select among the HF bands. An adjustable tip allows one to move about a single band. Because the length to diameter ratio is so large the loading coil Q is relatively low. The antenna is most effective above 20 meters.

THE SCREWDRIVER ANTENNA

The screwdriver antenna consists of a top whip attached to a long slender coil about 1.5 inches in diameter. The coil screws itself out of a base tube which has a set of contact fingers at the top. For lower frequencies more of the coil is screwed out of the base tube and at maximum frequency the coil is entirely *swallowed* by the base tube.

The antenna is tunable over a wide range of frequencies by remote control. It has the advantage that the drive mechanism is operated at ground potential with RF isolation in the mechanical drive shaft. On the other hand, the antenna is not easily extended to 10.5 foot length for maximum efficiency on 80 and 40 meters. Because of its shape, coil Q will not be very high.

DIGITAL VERSUS ANALOG COUPLERS

Digital HF antenna couplers were first used by the military about 1960 for radios with Automatic Link Establishment. In this mode, the military radio has a list of frequencies ranging 2 to 30 MHz. It will try these in some sequence and will *lock* on the frequency giving the best reception. During the search, frequencies change much too fast to permit the use of conventional roller coils and motor driven vacuum capacitors. By comparison the digital coupler can jump from one memory setting to another in milliseconds.

For matching a mobile whip, the circuit shown in **Fig 22.85** will suffice. The inductor and capacitor can each be made up of about 8 binary sequenced steps. For example, at 3.5 MHz, the 10.5-foot antenna has an impedance of about 0.55 $-j1684$ Ω. From Table 22.22 we see that we could use a series inductance sequence of 20, 10, 5, 2.5, 1.25, 0.625, 0.32 and 0.16 µH. We can use a relay to short unwanted elements. In this way we could theoretically produce any value of inductance between 0 and 39.84 µH in steps of 0.16 µH. In reality you will never reach a zero inductance. With all of the relays shorted, the wiring inductance and contact inductance of 8 relays appear in series. Also, each of the coils will have the open circuit capacitance of a relay contact across it in addition to the normal stray capacitance.

With most relays it does not make sense to switch less than 2 pF. For that reason, the capacitance chain would consist of 2, 4, 8, 16, 32, 64, 128 and possibly 256 pF. This would give a maximum of 510 pF and a step size of 2 pF. Each relay has an open circuit capacitance of about 1 pF, and that gives a minimum capacitance of 8 to 9 pF. As a practical matter, there is also the stray capacitance between the relay contacts and the coil windings.

In a high Q matching circuit that handles 100 W, the individual relays must handle 4 or 5 kV with the contacts open and several amperes of RF with the contacts closed. If we can unkey the transmitter so that the coupler will not have to switch under power, we'll still need some sizeable relays. If the inductors have lower Q, the voltages and currents will be correspondingly lower. Some military couplers use Jennings vacuum latching relays. This is expensive, as each of the 16 or 17 relays costs more than $100.

If coil and antenna Qs are kept or forced low, the voltages and currents become more reasonable. However, if the antenna size is restricted this reduction comes only at the cost of decreased efficiency. A commercially available ham/marine digital coupler employs RF reed relays rated for 5 kV and 1 A, and restricts the power at low frequencies if the antenna is small. Another ham/marine unit uses small relays in series where voltage requirements are great and in parallel where current requirements are great—not good engineering practice. A third offering is not too specific about the power rating with very high Q loads.

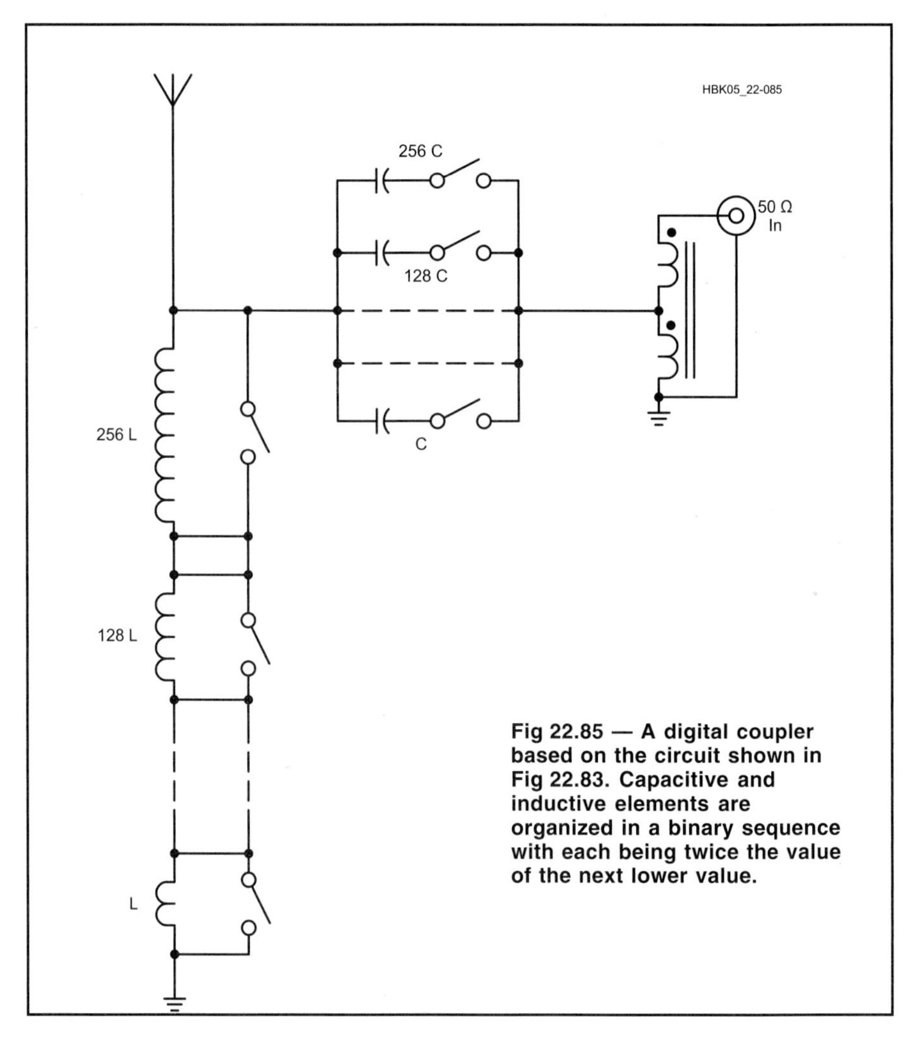

Fig 22.85 — A digital coupler based on the circuit shown in Fig 22.83. Capacitive and inductive elements are organized in a binary sequence with each being twice the value of the next lower value.

There are no successful examples of 100 W plus couplers that use PIN diode switching. Their use is highly problematic given the high-Q loads they would handle.

A REMOTELY TUNED ANALOG ANTENNA COUPLER

KE2QJ built an antenna coupler designed for 100-W continuous-duty operation that will tune an antenna 10 feet or longer to any frequency from 3.5 to 30 MHz. With longer antennas, the power rating is higher and the lowest frequency is lower. The design requires only hand tools to build; however, access to a drill press and a lathe could save labor.

The roller coils and air variable capacitors to be used are not widely manufactured these days. Tube-type linear amplifiers still use air variable capacitors but these are generally built on order for the manufacturer and are not readily available to consumers in small quantity.

Until the 1970s, E. F. Johnson manufactured roller coils and air variable capacitors that were suitable for kilowatt amplifier finals and high power antenna couplers. On occasion one or more of these may be found in the original box, but they tend to be expensive. Ten Tec and MFJ both manufacture antenna couplers and offer some components in small quantities.

The following data refer to generic motors, capacitors and inductors. The descriptions are intended to aid the builder in selecting items from surplus, hamfest flea market offerings or salvage of old equipment.

THE MOTORS

Two motors are required, one to drive the inductor and one to drive the capacitor. The design employs permanent magnet dc gearhead motors with a nominal 12-V rating. A permanent magnet (PM) motor can be reversed by simply reversing the polarity of the drive voltage, and its speed can be controlled over a wide range by pulsing the power on and off with a variable duty cycle. The motor should have an output shaft speed on the order of 60 to 180 r/min (1 to 3 r/s) although this is not critical. New, such motors, can cost as much as $65 to $150 in small quantities. However, they can be found surplus and in repair shops for a few dollars.

The motor you are looking for is 1 to 1.5 inches in diameter and perhaps 2.5 inches long. It might be rated 12 or 24 V and have a 1/4-inch diameter output shaft. At 12 V it should have enough torque to make it hard to stop the shaft with your fingers. Tape recorders, fax machines, film projectors, windshield wipers and copiers often use this type of motor.

LIMIT SWITCHES

On a remotely operated unit it is usually necessary to have limit switches to prevent the device from *crashing* into the ends. On an external roller coil these can be microswitches with paddles mounted on each end of the coil. As the coil is wound to one end, the roller operates the paddle and opens the limit switch which stops the motor.

Fig 22.86A illustrates a simple motor control circuit. Relay K1 is arranged as a DPDT polarity reversing switch. If switch CCW is pressed, the motor rotates CCW and the steering diode D2 prevents the relay from operating. If CW is pressed, relay K1 operates reversing the polarity at the motor. The motor is energized through the steering diode.

Fig 22.86B shows how to add limit switches. The diodes across the switches are called anti-jam diodes. When a switch opens, the diode permits current to flow in the reverse direction and the motor to move the roller away from the open switch.

The photograph of **Fig 22.87** shows the mounting of the switches on the coil. The diode should be a power rectifier type rated for several times the motor current and at least 60 V.

POSITION READER

While not necessary, it is worthwhile to

Fig 22.86 — At A, motor control circuit used by KE2QJ. This circuit uses pulse modulation for speed control with good starting torque. Direction of rotation is controlled by the relay. At B, how to add limit switches to the circuit. See text.

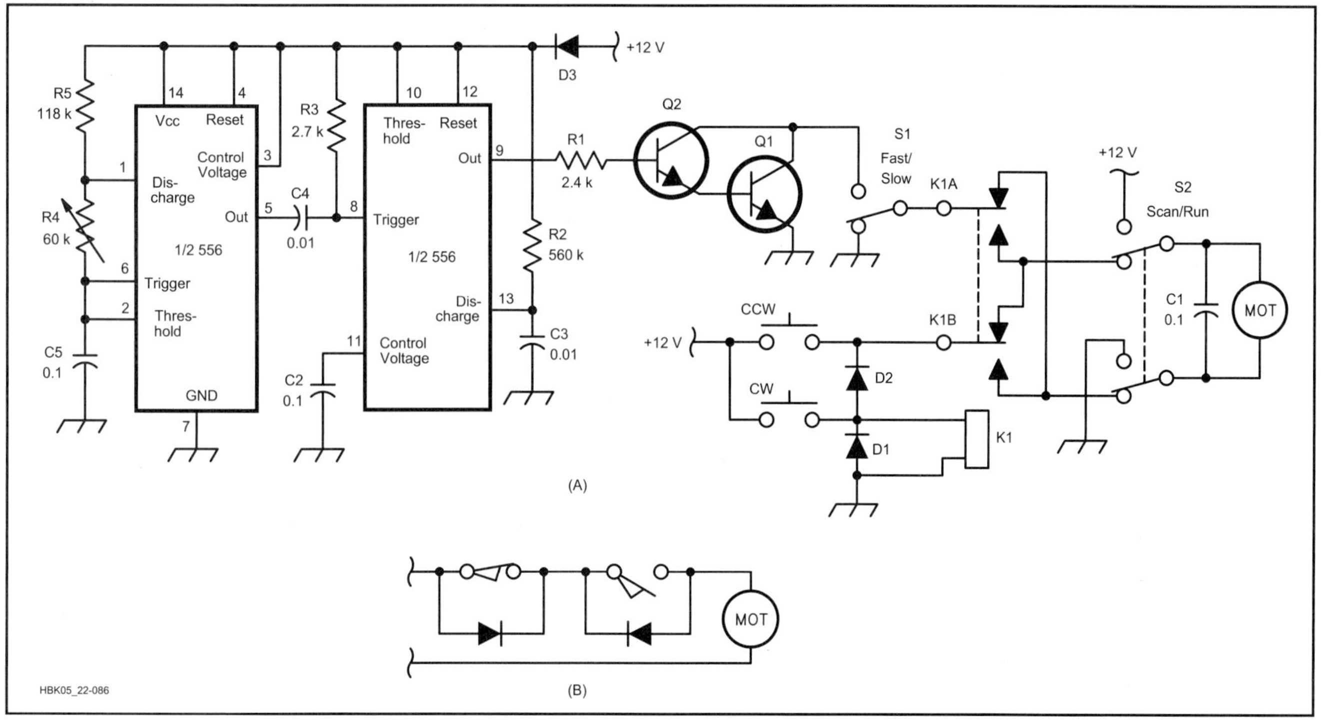

Fig 22.87 — Photo of the inductor drive assembly from KE2QJ's antenna coupler.

have a way to determine inductor position. An easy way to do this is to couple a 10-turn potentiometer to the coil shaft or drive gears. Make sure that the potentiometer turns less than 10 turns between limits. Don't try to make it come out exact.

Because the potentiometer is a light mechanical load, a belt drive reduction works well and won't slip if properly tensioned. Fig 22.87 shows the potentiometer and the gear drive.

You may be able to find suitable gears. However, a belt drive requires less precise shaft positioning than fine tooth gears. With a lathe, pulleys can be made in almost any ratio. Vacuum cleaner belts and O rings make handy belts.

COUPLINGS

In this coupler circuit, both ends of the capacitor are *hot* with RF although the end adjacent to the transformer is at relatively low voltage. Nevertheless, the capacitor shaft must be insulated from the motor shaft. The coil can be driven from the grounded end. Insulation is not necessary, but use a coupler between the motor shaft and the coil to compensate for any misalignment. Universal joints and insulating couplings are available from most electronics supply houses. You can make a coupling from a length of flexible plastic tubing which fits snugly over the shafts. Clamp the tubing to the shafts to avoid slippage.

THE CAPACITOR

The easiest capacitor to use is an air variable. It should have a range of approximately 10 to 250 pF. The plate spacing should be 2 mm (1/16 inch) or more, and the plate edges should be smooth and rounded. The capacitor should be capable of continuous 360° rotation, and it would be nice if it had ball bearings. The straight-line capacitance design is best for this application. Several capacitors of this type are available in military surplus ARC-5 series transmitters. These are approximately 2 × 2 × 3 inches.

The capacitor must be mounted on stand-off insulators although high voltage will not be present on the frame. A cam that briefly operates a microswitch when the capacitor goes through minimum can be used to flash an LED on the remote control panel. This provides an indication that the capacitor is turning.

THE INDUCTOR

As calculated earlier, and assuming an inductor Q a bit under 300 is attainable, the roller inductor for this coupler should have a maximum inductance on the order of 40 μH. The wire should be at least #14 AWG wound about 8 t/inch.

Table 22.23
Data for 40 μH Coils

Diameter	*Length*	*Turns*
2.3 inch (58 mm)	5.625 inch (143 mm)	45
2.8 inch (71 mm)	4.25 inch (108 mm)	34
3.3 inch (84 mm)	3.375 inch (86 mm)	27

You can use **Table 22.23** as a guide to buy a roller coil at a hamfest. The seller may not know the inductance of the coil. The antenna loading coil from an ARC-5 transmitter will work, but the wire is a bit small.

You could make the loading coil by threading 2, 2.5 or 3-inch diameter white, thick-wall PVC pipe with 8 t/inch. If the pipe is threaded in a lathe, the wire can be wound into the threads under considerable tension. This helps to prevent the wires from coming loose with wear or temperature.

THE TRANSFORMER

The transformer consists of a bifilar winding on an Amidon FT-114-61 core. Start with two 2-foot lengths of #18 insulated wire; Teflon insulation is preferable. Twist the wire with a hand drill until there are about 5 t/inch (not critical). Wind 12 turns onto the core. This should about fill it up. Attach the starting end of one wire to the finish end of the other. This is the 12.5-Ω tap. One of the free ends is grounded and the other is the 50-Ω tap. Mount the coil on a plastic or wooden post through the center of the coil. A metal screw can be used as long as it does not make a complete turn around the core.

CONSTRUCTION

For ease of service, mount the inductor, its drive motor, position sensing poten-

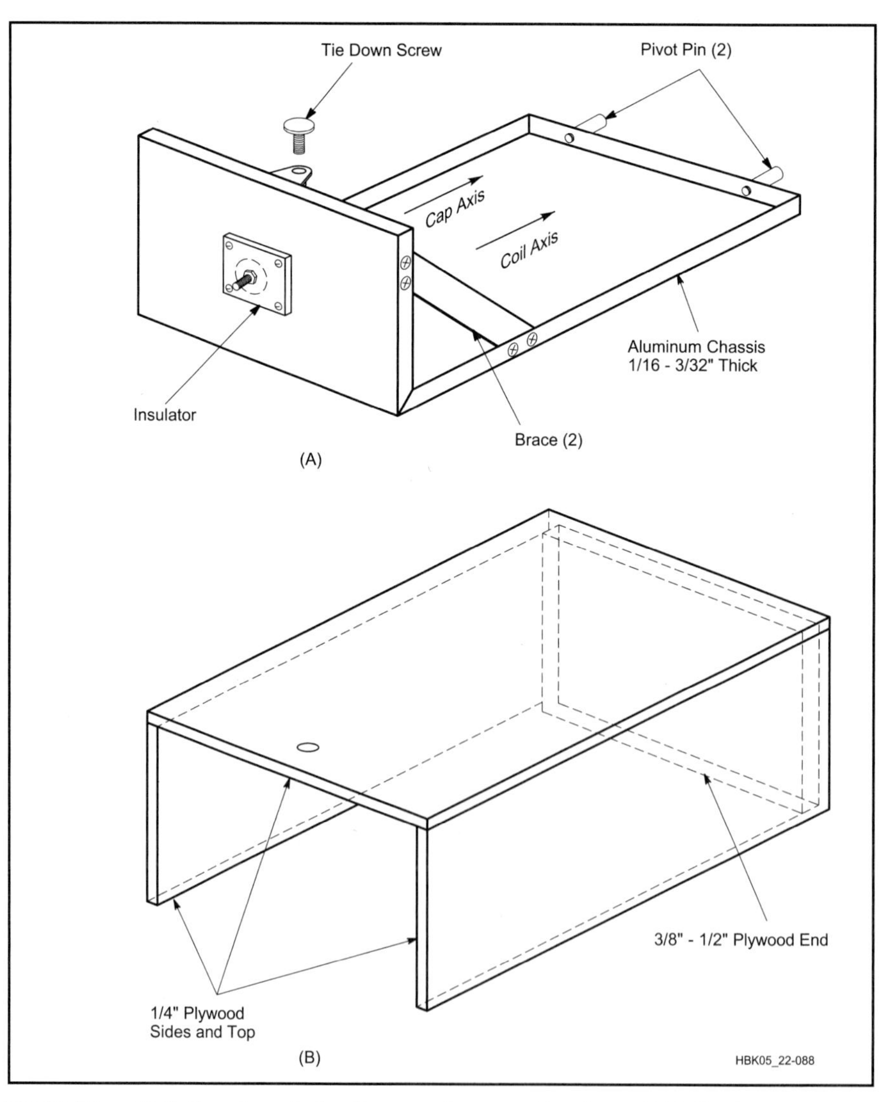

Fig 22.88 — At A, the chassis for the coupler mounting box. At B, the box cover.

tiometer and limit switch assembly on a single aluminum plate. A plug and socket assembly permits rapid disconnection and removal. Make a similar assembly for the capacitor, its drive motor, transformer and the interrupter. Both assemblies should be made on 1/16 to 1/4-inch thick aluminum. These individual assemblies make it easier to fix problems.

The chassis shown in **Fig 22.88A** is made of a single piece 1/16 to 3/32-inch aluminum bent in an L shape. Two chassis-stiffening braces are riveted in place. Alternatively, the chassis can be made of flat sheets with aluminum angles riveted around the edge.

Mount the coil and capacitor assemblies parallel to the long leg of the L. Punch a 1-inch hole in the center of the short end of the L. Cover the hole with an insulator made of PVC, Teflon or other suitable material.

The rest of the case is a 4-sided wooden assembly as shown in Fig 22.88B. The back wall of the box is drilled to accept the two pivot pins. The box is slid over the chassis and the pivot pins engaged. The tie-down screw secures the box. For service, remove the tie-down screw and slide off the cover. The works of the coupler are very easy to get at!

The box is made of 1/4-inch exterior grade plywood except for the back plate, which is 3/8 or 1/2-inch plywood. The sides, top and back should overlap the flanges on the chassis by 1/2 inch. The inside corner seams of the box should be reinforced with 1/2 or 3/4-inch square strips. Assembly can be with any water resistant glue.

Finish the box, inside and out, with several coats of clear urethane varnish, sanding lightly between coats. This leaves a smooth plastic finish. This can be sprayed with an exterior paint that matches your car's color.

If the sides of the box fit closely over the flanges, no fastening beside the tie-down screw is required. A nearly perfect seal will leak out hot air when the sun shines on it and will draw in cold damp air in the evening, trapping moisture inside. A moderate fit will keep rain and snow out and permit the box to *breathe* freely, thereby keeping the inside dry.

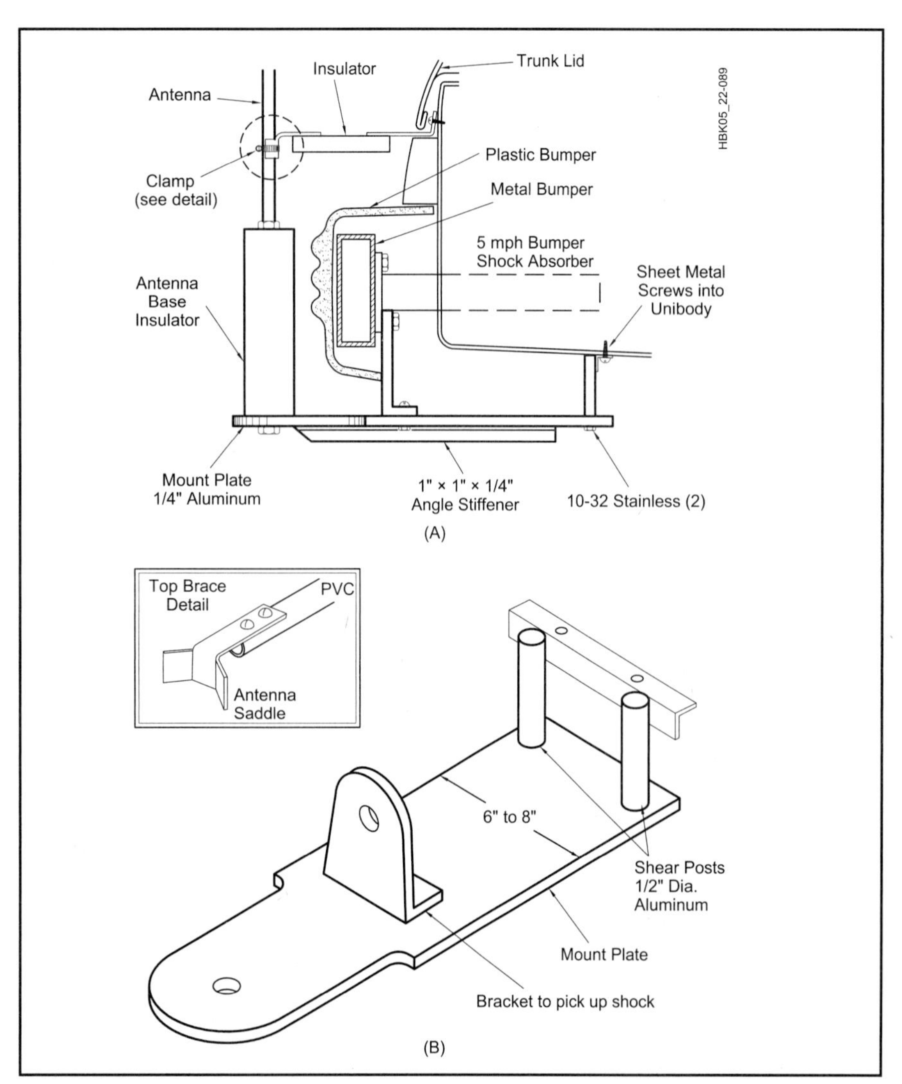

Fig 22.89 — Antenna mounting detail. At A, the overall plan. At B, detail of the mount plate.

MOUNTING THE WHIP AND THE BOX

Plastics in bumpers and bodies makes the mounting of a mobile whip antenna problematic. Modern bumpers are covered with plastic and the bumper is attached to the car unibody through a *5-MPH* shock absorber. The latter item is an unreliable ground.

The arrangement of **Fig 22.89** solves many of these problems. It uses a 1/4-inch aluminum plate 6 to 8 inches wide and long enough to fit between a reasonably strong place on the unibody and the place behind the bumper where the antenna wants to be. This plate is fitted with an angle bracket for the lower bolt on the shock/bumper mounting. This plate is stiffened with a length of 1 × 1 × 1/4 inch aluminum angle bolted in several places.

Near the forward edge of the plate, two 1/2-inch diameter aluminum shear posts are fitted. The bottom of each is tapped 10-32 and bolted through the mount plate with a stainless 10-32 screw. At the top of these shear posts another piece of 1 × 1 × 1/2-inch angle is attached which is screwed to the unibody with three or four #10 stainless sheet metal screws. A bracket attaches the mount plate to the bumper's shock absorber. The angle bracket may either be welded to the plate or bolted with angle stock. In the event that the car is hit from behind or backs into an obstacle, the two 10-32 screws will shear off, thereby preventing the mount from defeating the 5 MPH crushable shock absorber. The part protruding behind the bumper may be cut down in width to 3 inches and rounded for appearance and safety.

Any type of base insulator may be used, but try to bring the base of the antenna to the height of the coupler output terminal. You can make a good base insulator from thick-wall white PVC 1 1/2-inch pipe. Reinforce each end. Start with a 1 1/2-inch length of pipe. Remove a 5/8-inch wide strip so the remaining portion can be rolled and pressed into the open end of the insulator. Apply PVC pipe glue just before pressing in the piece; this gives the insulator a double wall thickness at each end. Aluminum plugs can be turned for a snug fit and tapped for 3/8-24 hardware. The plugs can be held in place with 8-32 stainless screws.

The upper antenna brace has an aluminum plate at one end that goes under the trunk lid (see Fig 22.89). A length of ½-inch diameter heavy-wall white PVC pipe, which serves as an insulator, is screwed to this. At the other end of the insulator, another aluminum piece is bent to form a saddle for the antenna which is clamped to the saddle. This clamp should be as high as convenient above the mount plate, preferably not less than a foot. The mount plate should be sturdy enough for you to stand on and with the brace will easily hold a whip upright at 70 MPH or more.

The coupler box mounting is shown in **Fig 22.90**. Brackets can be made of ⅛ × 2-inch aluminum with a brace going perhaps 2 inches from the corner. The brackets bolt or rivet to the chassis. The bracket reaches through the gap between the trunk lid and the plastic top of the bumper. For reinforcement, a pair of reinforcement plates 1.5 × 2 × ¼-inch thick are bolted to the plastic on the under side of the bumper. Ground the reinforcement plates to the unibody with some ¾ or 1 inch ground braid.

Two 10-32 stainless screws hold each reinforcement plate to the plastic bumper and a central ¼-20 tapped hole holds down the box bracket. One need only remove two screws to get the box off the car for car wash, etc. You have to open the trunk to remove the antenna coupler, and this provides a measure of security.

THE SPRING AND WHIP

A section of 1-inch diameter aluminum tubing extends from the top of the insulator to the base of the spring. It's usually best to have the spring about 4 ft above the pavement. The type used for CB whips works well. A 7-ft whip brings the top to about 11.5 ft above the pavement. The 7-ft whip can be a cut down CB unit. Don't use the type with helical winding. When the antenna is tied down, the bow of the whip should be about 7 ft above the pavement.

TUNING

It is best to initially tune the antenna using low power. A power attenuator just after the transceiver will limit SWR, but your SWR indicator must be on the antenna side of the attenuator.

For a first tune-up, set the capacitor control to SCAN and slowly advance the inductor from minimum inductance toward maximum. As the inductor approaches the correct value the SWR will start to kick down. At this point take the capacitor off of SCAN and JOG it to a best tune. Next, JOG the inductor and repeat; the SWR should go down. Continue until a 1:1 SWR is obtained. Record the potentiometer setting. The next time you want to use this frequency run the coil directly to the logged setting.

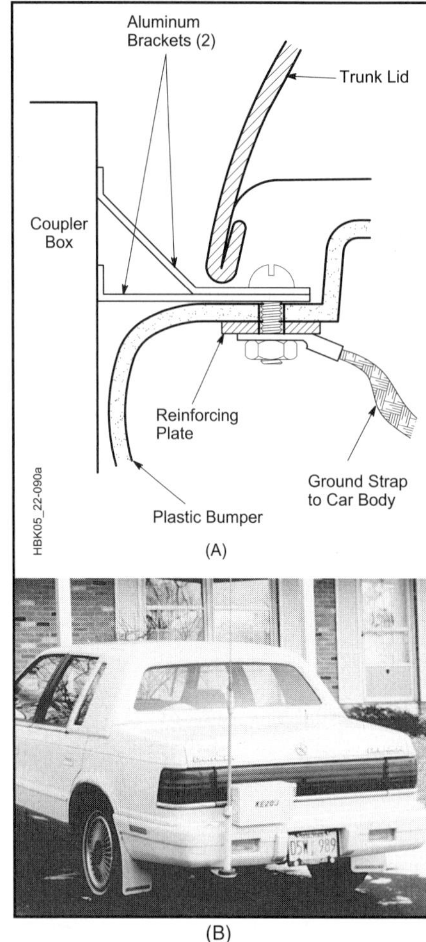

Fig 22.90 — Box mounting detail. At A, mounting-bracket design. At B, photo of KE2QJ's installation.

In the SCAN position the capacitor motor runs at full voltage. When you JOG the capacitor for low SWR you will find the speed far too fast for sharp tuning. The slow speed tuning is provided by using duty-factor modulation of the motor current. The circuit of Fig 22.86A supplies fixed width pulses with variable timing. At the slowest speed, the unit will supply about one pulse per second and the motor shaft will rotate one degree or so per pulse. The full voltage pulse provides good starting torque.

If the SWR cannot be brought to 1:1, examine the coil and capacitor to see whether either is at maximum or minimum. At high frequencies above 24 MHz it may be necessary to place a capacitor between the coupler and the antenna base.

RADIATION PATTERNS

At the lower frequencies the pattern tends to be essentially round in azimuth. At 20 meters the pattern tends to become more and more directive. The patterns in

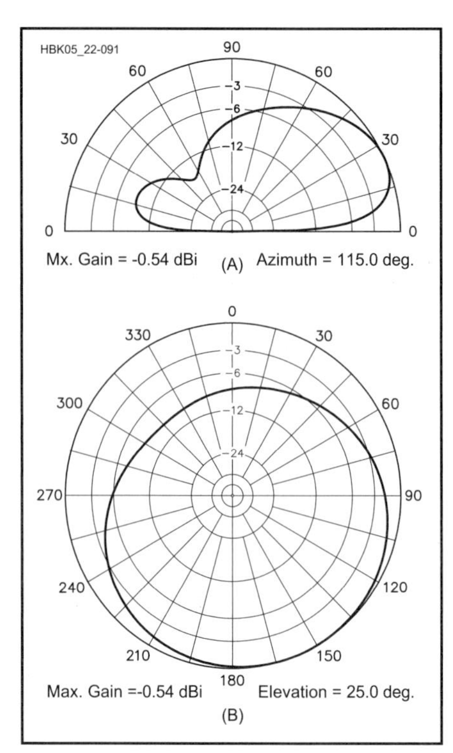

Fig 22.91 — At A, elevation pattern of the KE2QJ mobile antenna. The pattern is in the plane that runs diagonal through the car. At B, azimuth pattern at 25° elevation for the same antenna. The operating frequency is 18.130 MHz.

Fig 22.91 were calculated using *EZNEC*. The frequency is 18.13 MHz and the antenna is mounted at the left rear corner of a mid-size sedan. It may be seen that the pattern has more than 10 dB maximum-to-minimum ratio with the broad maximum along the diagonal of the vehicle occupied by the antenna. If the antenna were mounted in the center of the vehicle, the omnidirectional characteristics would be improved. However, the antenna would have to be much shorter to stay under 11.5 feet. The shorter antenna would likely be weaker in its best direction than the taller antenna is in its worst.

REFERENCES

J. S. Belrose, VE3BLW, "Short Antennas for Mobile Operation," *QST*, Sept 1953, pp 30-35, 109.

J. Kuecken, KE2QJ, *Antennas and Transmission Lines*, MFJ Publishing, Ch 25.

J. Kuecken, KE2QJ, "A High Efficiency Mobile Antenna Coupler," *The ARRL Antenna Compendium, Vol 5*, pp 182-188.

J. Kuecken, KE2QJ, "Easy Homebrew Remote Controls," *The ARRL Antenna Compendium, Vol 5*, pp 189-193.

J. Kuecken, KE2QJ, "A Remote Tunable Center Loaded Antenna," *The ARRL Antenna Compendium, Vol 6*.

VHF/UHF Antennas

Improving an antenna system is one of the most productive moves open to the VHF enthusiast. It can increase transmitting range, improve reception, reduce interference problems and bring other practical benefits. The work itself is by no means the least attractive part of the job. Even with high-gain antennas, experimentation is greatly simplified at VHF and UHF because an array is a workable size, and much can be learned about the nature and adjustment of antennas. No large investment in test equipment is necessary.

Whether we buy or build our antennas, we soon find that there is no one *best* design for all purposes. Selecting the antenna best suited to our needs involves much more than scanning gain figures and prices in a manufacturer's catalog. The first step should be to establish priorities.

GAIN

As has been discussed previously, shaping the pattern of an antenna to concentrate radiated energy, or received signal pickup, in some directions at the expense of others is the only possible way to develop gain. Radiation patterns can be controlled in various ways. One is to use two or more driven elements, fed in phase. Such arrays provide gain without markedly sharpening the frequency response, compared to that of a single element. More gain per element, but with some sacrifice in frequency coverage, is obtained by placing parasitic elements into a Yagi array.

RADIATION PATTERN

Antenna radiation can be made omnidirectional, bidirectional, practically unidirectional, or anything between these conditions. A VHF net operator may find an omnidirectional system almost a necessity but it may be a poor choice otherwise. Noise pickup and other interference problems tend to be greater with omnidirectional antennas. Maximum gain and low radiation angle are usually prime interests of the weak-signal DX aspirant. A clean pattern, with lowest possible pickup and radiation off the sides and back, may be important in high-activity areas, where the noise level is high, or when challenging modes like EME (Earth-Moon-Earth) are employed.

HEIGHT GAIN

In general, the higher a VHF antenna is installed, the better will be the results. If raising the antenna clears its view over nearby obstructions, it may make dramatic improvements in coverage. Within reason, greater height is almost always worth its cost, but height gain must be balanced against increased transmission-line loss. Line losses can be considerable at VHF, and they increase with frequency. The best available line may be none too good, if the run is long in terms of wavelength. Consider line losses in any antenna planning.

PHYSICAL SIZE

A given antenna design for 432 MHz, say a 5-element Yagi on a 1-λ boom, will have the same gain as one for 144 MHz, but being only one-third the size it will intercept only one-ninth as much energy in receiving. Thus, to be equal in communication effectiveness, the 432-MHz array should be at least equal in physical size to the 144-MHz one, requiring roughly three times the number of elements. With all the extra difficulties involved in going higher in frequency, it is well to be on the big side in building an antenna for the UHF bands.

DESIGN FACTORS

Having sorted out objectives in a general way, we face decisions on specifics, such as polarization, type of transmission line, matching methods and mechanical design.

POLARIZATION

Whether to position the antenna elements vertically or horizontally has been a question since early VHF pioneering. Tests show little evidence on which to set up a uniform polarization policy. On long paths there is no consistent advantage, either way. Shorter paths tend to yield higher signal levels with horizontal in some kinds of terrain. Man-made noise, especially ignition interference, tends to be lower with horizontal. Verticals, however, are markedly simpler to use in omnidirectional systems and in mobile work.

Early VHF communication was largely vertical, but horizontal gained favor when directional arrays became widely used. The major trend to FM and repeaters, particularly in the 144-MHz band, has tipped the balance in favor of verticals in mobile work and for repeaters. Horizontal predominates in other communication on 50 MHz and higher frequencies. It is well to check in advance in any new area in which you expect to operate, however, as some localities may use vertical polarization. A circuit loss of 20 dB or more can be expected with cross-polarization.

TRANSMISSION LINES

There are two main categories of transmission lines used at HF through UHF: balanced and unbalanced. Balanced lines include *open-wire lines* separated by insulating spreaders, and *twin-lead*, in which the wires are embedded in solid or foamed insulation. Unbalanced lines are represented by the family of coaxial cables, commonly called *coax*. Line losses in either types of line result from ohmic resistance, radiation from the line and deficiencies in the insulation.

Large conductors, closely spaced in terms of wavelength, and using a minimum of insulation, make the best balanced lines. Characteristic impedances are between 300 to 500 Ω. Balanced lines work best in straight runs, but if bends are unavoidable, the angles should be as gentle as possible. Care should also be taken to prevent one wire from coming closer to metal objects than the other.

Properly built open-wire line can operate with very low loss in VHF and even UHF installations. A total line loss under 2 dB per hundred ft at 432 MHz is readily obtained. A line made of #12 wire, spaced ¾ inch or less with Teflon spreaders, and running essentially straight from antenna to station, can be better than anything but the most expensive *Hardline* coax, at a fraction of the cost. This assumes the use of high-quality baluns to match into and out of the balanced line, with a short length of low-loss coax for the rotating section from the top of the tower to the antenna. A similar 144-MHz setup could have a line loss under 1 dB.

Small coax such as RG-58 or RG-59 should never be used in VHF work if the run is more than a few feet. Half-inch lines (RG-8 or RG-11) work fairly well at 50 MHz, and are acceptable for 144-MHz runs of 50 ft or less. If these lines have foam rather than solid insulation they are about 30% better. Aluminum-jacket *Hardline* coaxial cables with large inner conductors and foam insulation are well worth their cost. Hardline can sometimes even be obtained for free from local Cable TV operators as *end runs*—pieces at the end of a roll. The most common CATV variety is ½-inch OD 75-Ω Hardline. Waterproof commercial connectors for Hardline are fairly expensive, but enterprising amateurs have *home-brewed* low-cost connectors. If they are properly waterproofed, connectors and Hardline can last almost indefinitely. Of course, a disadvantage implied by their name is that Hardline must not be bent too sharply, because it will kink. See *The ARRL Antenna Book* for details on Hardline connectors.

Effects of weather on transmission lines

should not be ignored. A well-constructed open-wire line works well in nearly any weather, and it stands up well. TV type twin-lead is almost useless in heavy rain, wet snow or icing. The best grades of coax are impervious to weather. They can be run underground, fastened to metal towers without insulation, or bent into almost any convenient position, with no adverse effects on performance. However, beware of *bargain* coax. Lost transmitter power can be made up to some extent by increasing power, but once lost, a weak signal can never be recovered in the receiver.

IMPEDANCE MATCHING

Theory and practice in impedance matching are given in detail in the **Transmission Lines** chapter, and in theory, at least, is the same for frequencies above 50 MHz. Practice may be similar, but physical size can be a major modifying factor in choice of methods.

Delta Match

Probably the first impedance match was made when the ends of an open line were fanned out and tapped onto a half-wave antenna at the point of most efficient power transfer, as in **Fig 22.92A**. Both the side length and the points of connection either side of the center of the element must be adjusted for minimum reflected power in the line, but the impedances need not be known. The delta makes no provision for tuning out reactance, so the length of the dipole is pruned for best SWR.

Once thought to be inferior for VHF applications because of its tendency to radiate if adjusted improperly, the delta has come back to favor now that we have good methods for measuring the effects of matching. It is very handy for phasing multiple-bay arrays with low-loss open lines, and its dimensions in this use are not particularly critical.

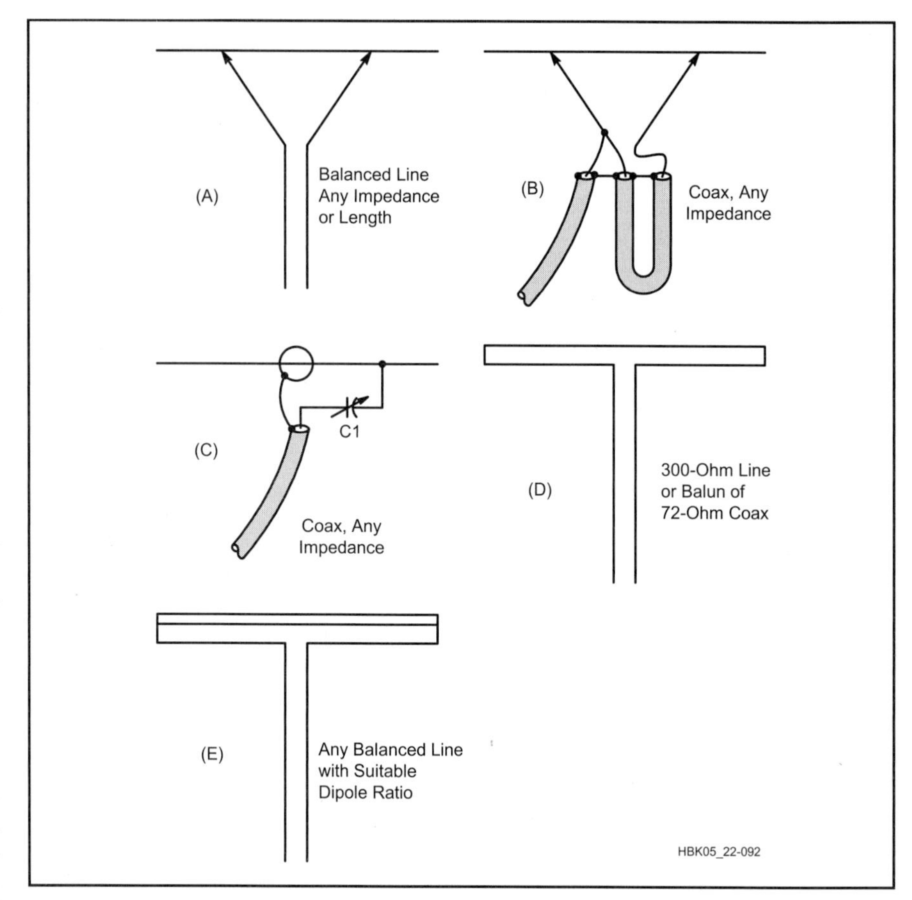

Fig 22.92 — Matching methods commonly used in VHF antennas. In the delta match, A and B, the line is fanned out to tap on the dipole at the point of best impedance match. The gamma match, C, is for direct connection of coax. C1 tunes out inductance in the arm. Folded dipole of uniform conductor size, D, steps up antenna impedance by a factor of four. Using a larger conductor in the unbroken portion of the folded dipole, E, gives higher orders of impedance transformation.

Gamma and T Matches

The gamma match is shown in Fig 22.92C, and the T match is shown in Fig 22.92D. These matches are covered in more detail in the **Transmission Lines** chapter. There being no RF voltage at the center of a half-wave dipole, the outer conductor of the coax is connected to the element at this point, which may also be the junction with a metallic or wooden boom. The inner conductor, carrying the RF current, is tapped out on the element at the matching point. Inductance of the arm is canceled by means of C1. Both the point of contact with the element and the setting of the capacitor are adjusted for zero reflected power, with a bridge connected in the coaxial line.

The capacitor can be made variable temporarily, then replaced with a suitable fixed unit when the required capacitance value is found, or C1 can be mounted in a waterproof box. Maximum capacitance should be about 100 pF for 50 MHz and 35 to 50 pF for 144 MHz. The capacitor and arm can be combined with the arm connecting to the driven element by means of a sliding clamp, and the inner end of the arm sliding inside a sleeve connected to the inner conductor of the coax. It can be constructed from concentric pieces of tubing, insulated by plastic sleeving or shrink tubing. RF voltage across the capacitor is low, once the match is adjusted properly, so with a good dielectric, insulation presents no great problem, if the initial adjustment is made with low power. A clean, permanent, high-conductivity bond between arm and element is important, as the RF current is high at this point.

Because it is inherently somewhat unbalanced, the gamma match can sometimes introduce pattern distortion, particularly on long-boom, highly directive Yagi arrays. The T-match, essentially two gamma matches in series creating a balanced feed system, has become popular for this reason. A coaxial balun like that shown in Fig 22.92B is used from the balanced T-match to the unbalanced coaxial line going to the transmitter.

Folded Dipole

The impedance of a half-wave antenna broken at its center is 72 Ω. If a single conductor of uniform size is folded to make a half-wave dipole, as shown in Fig 22.92D, the impedance is stepped up four times. Such a folded dipole can thus be fed directly with 300-Ω line with no appreciable mismatch. Coaxial line of 70 to 75 Ω impedance may also be used if a 4:1 balun is added. Higher impedance step-up can be obtained if the unbroken portion is made larger in cross-section than the fed portion, as in Fig 22.92E.

BALUNS

Conversion from balanced loads to unbalanced lines, or vice versa, can be per-

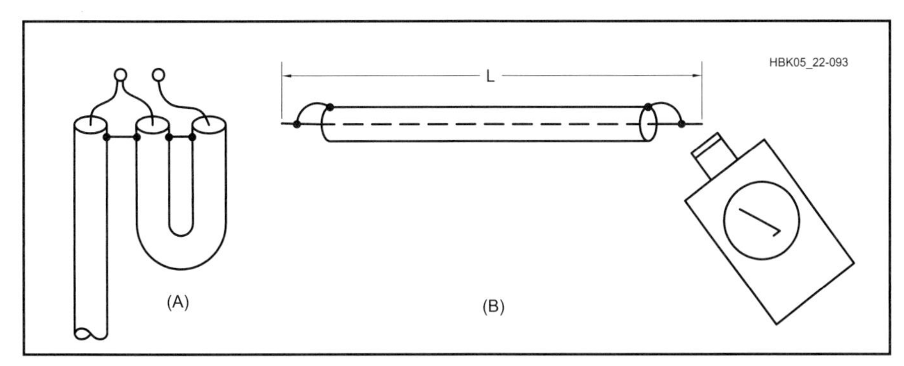

Fig 22.93 — Conversion from unbalanced coax to a balanced load can be done with a half-wave coaxial balun, A. Electrical length of the looped section should be checked with a dip meter, with ends shorted, B. The half-wave balun gives a 4:1 impedance step up.

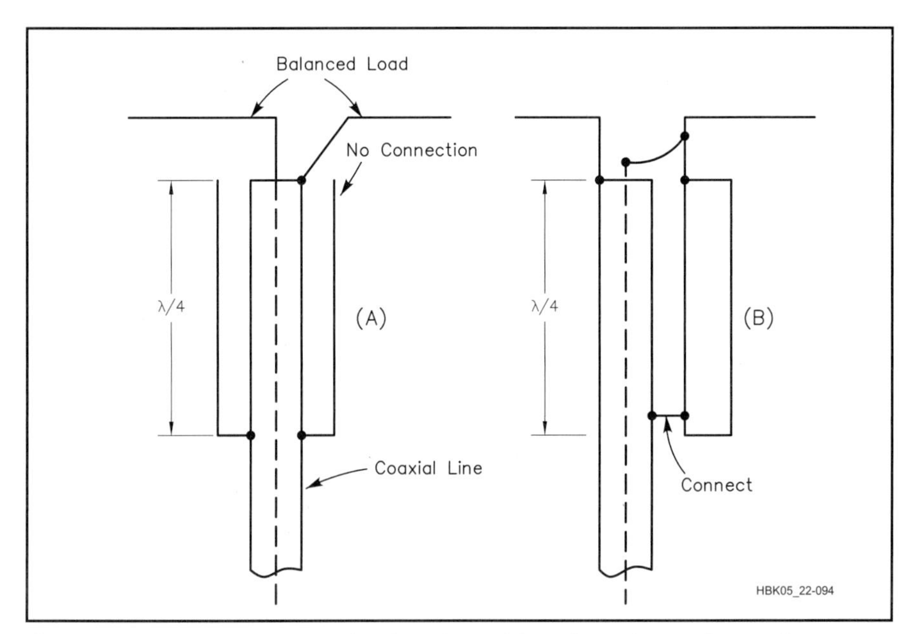

Fig 22.94 — The balun conversion function, with no impedance change, is accomplished with quarter-wave lines, open at the top and connected to the coax outer conductor at the bottom. The coaxial sleeve shown at A is preferred.

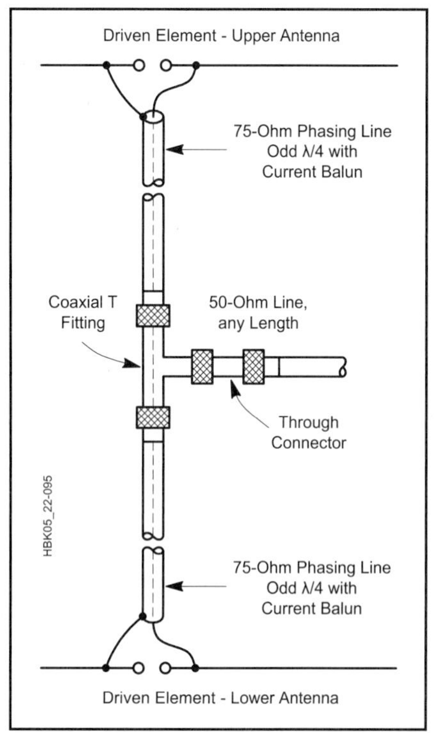

Fig 22.95 — A method for feeding a stacked Yagi array. Note that baluns at each antenna are not specifically shown. Modern-day practice is to use current (choke) baluns made up of ferrite beads slipped over the outside of the coax and taped to prevent movement. See the Transmission Lines chapter for details.

formed with electrical circuits, or their equivalents made of coaxial line. A balun made from flexible coax is shown in **Fig 22.93A**. The looped portion is an electrical half-wave. The physical length depends on the propagation factor of the line used, so it is well to check its resonant frequency, as shown at B. The two ends are shorted, and the loop at one end is coupled to a dip-meter coil. This type of balun gives an impedance step-up of 4:1, 50 to 200 Ω, or 75 to 300 Ω typically.

Coaxial baluns giving a 1:1 impedance transfer are shown in **Fig 22.94**. The coaxial sleeve, open at the top and connected to the outer conductor of the line at the lower end (A) is the preferred type. A conductor of approximately the same size as the line is used with the outer conductor to form a quarter-wave stub, in B. Another piece of coax, using only the outer conductor, will serve this purpose. Both baluns are intended to present an infinite impedance to any RF current that might otherwise tend to flow on the outer conductor of the coax.

STACKING YAGIS

Where suitable provision can be made for supporting them, two Yagis mounted one above the other and fed in phase may be preferable to one long Yagi having the same theoretical or measured gain. The pair will require a much smaller turning space for the same gain, and their lower radiation angle can provide interesting results. On long ionospheric paths a stacked pair occasionally may show an apparent gain much greater than the 2 to 3 dB that can be measured locally as the gain from stacking.

Optimum spacing for Yagis with booms longer than 1 λ is one wavelength, but this may be too much for many builders of 50-MHz antennas to handle. Worthwhile results are possible with as little as $^1/_2$ λ (10 ft), but $^5/_8$ λ (12 ft) is markedly better. The difference between 12 and 20 ft may not be worth the added structural problems involved in the wider spacing, at 50 MHz at least.

The closer spacings give lowered measured gain, but the antenna patterns are cleaner (less power in the high-angle elevation lobes) than with one-wavelength spacing. Extra gain with wider spacings is usually the objective on 144 MHz and higher bands, where the structural problems are not quite as severe as on 50 MHz.

One method for feeding two 50-Ω antennas, as might be used in a stacked Yagi array, is shown in **Fig 22.95**. The transmission lines from each antenna, with a balun feeding each antenna (not shown in the drawing for simplicity), to the common feedpoint must be equal in length and an odd multiple of a quarter wavelength. This line acts as an quarter-wave (Q-section) impedance transformer and raises the feed impedance of each antenna to 100 Ω. When the coaxes are connected in parallel at the coaxial T fitting, the resulting impedance is close to 50 Ω.

CIRCULAR POLARIZATION

Polarization is described as *horizontal* or *vertical*, but these terms have no meaning once the reference of the Earth's surface is lost. Many propagation factors can cause polarization change—reflection or refraction and passage through magnetic fields (Faraday rotation), for example. Polarization of VHF waves is often random, so an antenna capable of accepting any polarization is useful. Circular polarization, generated with helical antennas or with crossed elements fed 90° out of phase, will respond to any linear polarization.

The circularly polarized wave in effect threads its way through space, and it can be left- or right-hand polarized. These polarization senses are mutually exclusive, but either will respond to any plane (horizontal or vertical) polarization. A wave generated with right-hand polarization, when reflected from the moon, comes back with left-hand plarization, a fact to be borne in mind in setting up EME circuits. Stations communicating on direct paths should have the same polarization sense.

Both senses can be generated with crossed dipoles, with the aid of a switchable phasing harness. With helical arrays, both senses are provided with two antennas wound in opposite directions.

SIMPLE, PORTABLE GROUNDPLANE ANTENNA

This utility antenna is built on a coaxial connector. UHF connectors work well, but you may prefer to use type N or BNC connectors. With only two radials, it is essentially two dimensional, which makes it easier to store when not in use.

If the antenna is sheltered from weather, copper wire is sufficiently rigid for the radiating element and radials. Antennas exposed to the wind and weather can be made from brazing rod, which is available at welding supply stores. Alternatively, #12 or #14 copper-clad steel wire could be used to construct this antenna.

The ground-plane antenna is shown in **Fig 22.96** and uses a female chassis-mount connector to support the element and two radials. To eliminate sharp ends, it's a good idea to bend the element and radial ends into a circle or to terminate them with a solder lug. See **Fig 22.97**. The solder lug approach is easier with stiff wire. Crimp and then solder the lug to the wire. Make the overall length of the element and radials the same as shown in Fig 22.96, measuring to the outer tip of the loop or lug.

Radials may be attached directly to the mounting holes of the coaxial connector. Bend a hook at one end of each radial for insertion through the connector. Solder the radials to the connector using a large soldering iron or propane torch.

Solder the element to the center pin of the connector. If the element does not fit inside the solder cup, use a short section of brass tubing as a coupler (a slotted 1/8-inch-ID tube will fit over an SO-239 or N-receptacle center pin).

Tune the antenna by adjusting the radial droop angle for best SWR. If necessary, you can also prune the element length.

One mounting method for fixed-station antennas appears in Fig 22.96. The feed line and connector are inside the mast, and a hose clamp squeezes the slotted mast end to tightly grip the plug body. Once the antenna is mounted and tested, thoroughly seal the open side of the coaxial connector with RTV sealant, and weatherproof the connections with rust-preventative paint.

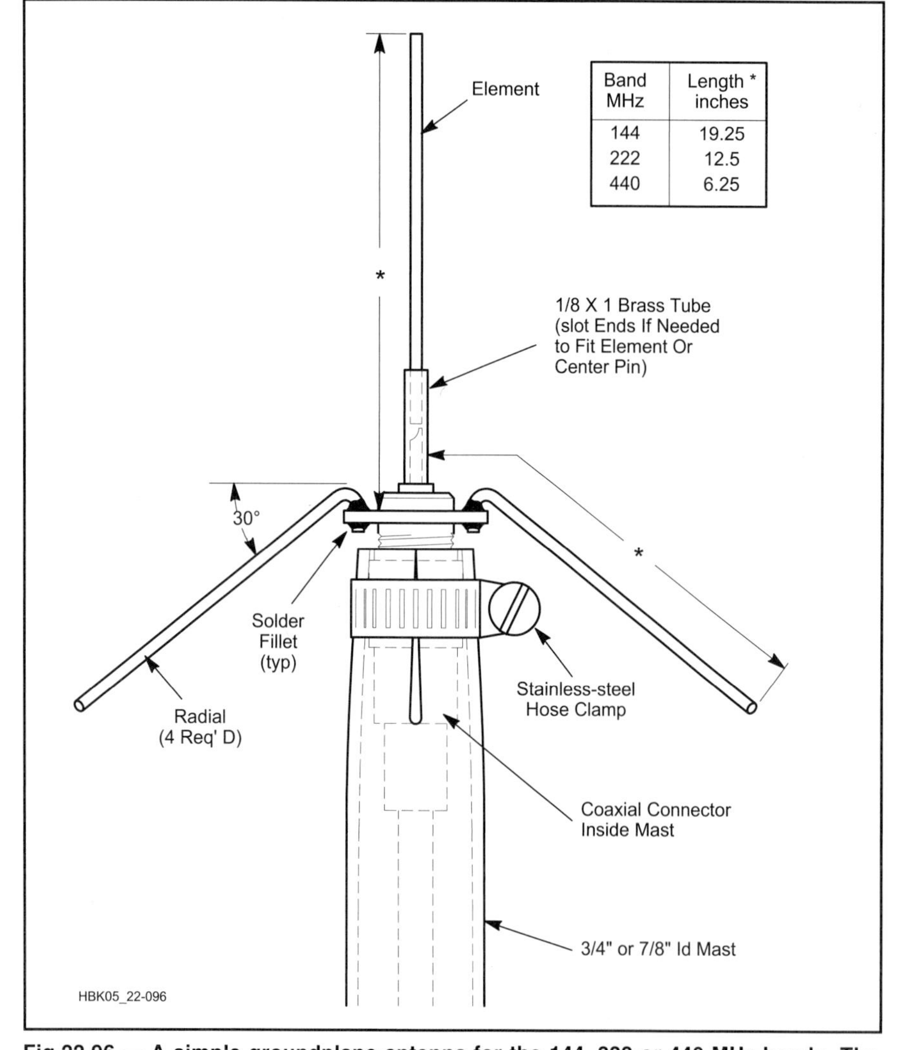

Fig 22.96 — A simple groundplane antenna for the 144, 222 or 440-MHz bands. The feed line and connector are inside the mast, and a hose clamp squeezes the slotted mast end to tightly grip the plug body. Element and radial dimensions given in the drawing are good for the entire band.

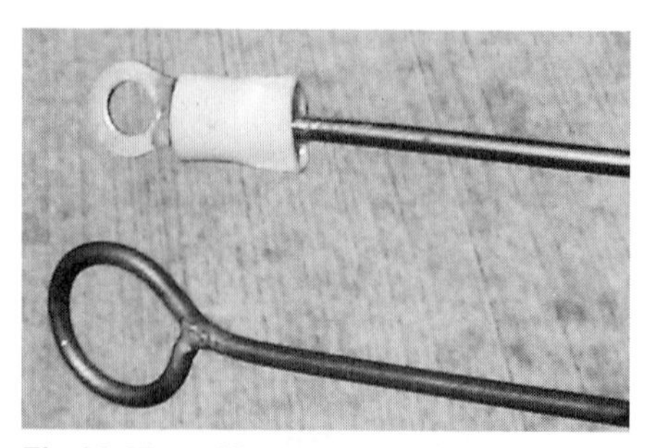

Fig 22.97 — Alternate methods for terminating element and radial tips on the simple groundplane antenna. See text. [Photo by K8CH]

DUAL-BAND ANTENNA FOR 146/446 MHZ

This nifty project by Wayde Bartholomew, K3MF (ex-WA3WMG), first appeared in *The ARRL Antenna Compendium, Volume 5*. This mobile whip antenna won't take long to build, works well and only requires one feed line for the two-band coverage.

Wayde used a commercial NMO-style base and magnetic mount. For the radiator and decoupling stub, He used brazing rod, which he coated with a rust inhibitor after all the tuning was done. You can start with a 2-m radiator that's 20.5 inches long. This is an inch longer than normal so that it may be pruned for best SWR.

Next tack on the 70-cm decoupling stub, which is 6.5 inches long. Trim the length of the 2-m radiator for best SWR at 146 MHz and then tune the 70-cm stub on 446 MHz, moving it up and down for best SWR. There should be no significant interaction between the adjustments for either frequency.

Final dimensions are shown in **Fig 22.98**. The SWR in the repeater portions of both bands is less than 2:1.

ADAPTING WA3WMG'S MOBILE ANTENNA FOR FIXED-STATION USE

You can use the WA3WMG dual-band mobile whip as the radiating element for the groundplane antenna in Fig 22.96. Don't change the 2-m radials. Instead, add two 70-cm radials at right angles to the 2-m set. See **Fig 22.99**. The antenna is no longer two dimensional, but you do have two bands with one feed line *and* automatic band switching.

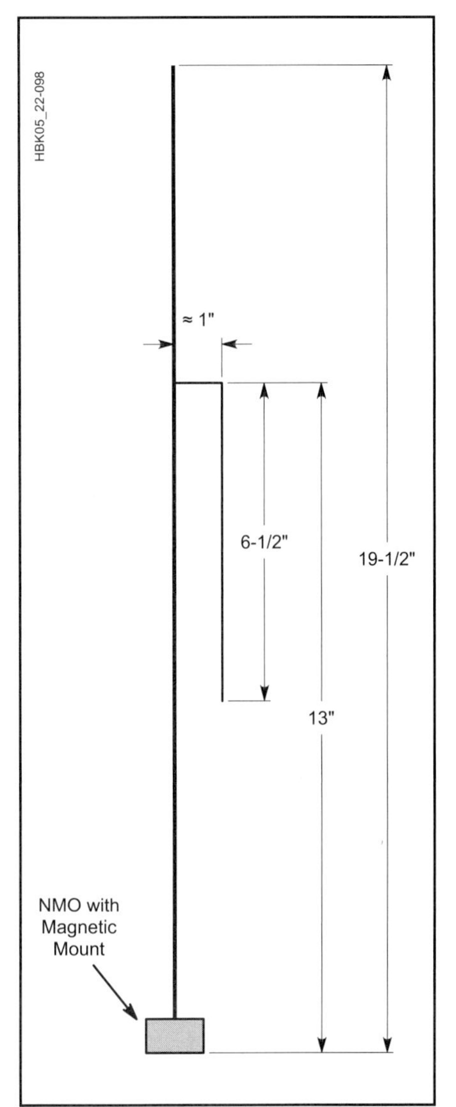

Fig 22.98 — Diagram of WA3WMG's dual-band 146/446-MHz mobile whip. Brazing rod is used for the 2-m radiator and for the 70-cm decoupling stub.

Fig 22.99 — WA3WMG's whip can be used to make a dual-band groundplane antenna. Separate radials for 2-m and 70-cm simplifies tuning. [Photo by K8CH]

A QUICK ANTENNA FOR 223 MHZ

William Bruce Cameron, WA4UZM, built the antenna for 223 MHz shown in **Fig 22.100**. It took less than an hour to build. To make one, you'll need 9 feet of #10 copper wire, 6 inches of small-diameter copper tubing, and a 10-foot length of PVC pipe or some other physical support.

Bend the antenna from one piece of wire. Slide the copper tubing over the top end of the antenna, and adjust how far it extends beyond the wire to get the lowest SWR. (Don't handle the antenna while transmitting—make adjustments only while receiving.) For more precision, you can move the coaxial feed line taps on the antenna's matching stub (the 12-inch section at the bottom) about an eighth of an inch at a time. The antenna shows an SWR of 1.2 at 223 MHz.

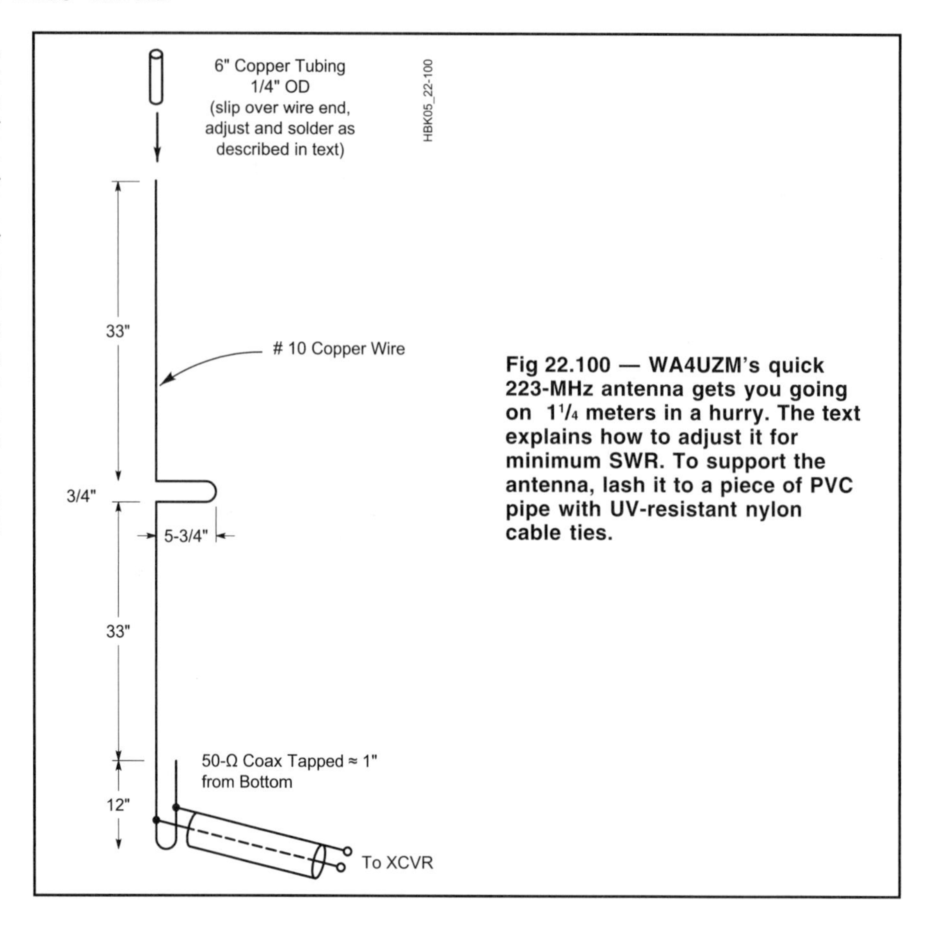

Fig 22.100 — WA4UZM's quick 223-MHz antenna gets you going on 1¼ meters in a hurry. The text explains how to adjust it for minimum SWR. To support the antenna, lash it to a piece of PVC pipe with UV-resistant nylon cable ties.

AN ALL-COPPER 2-M J-POLE

Rigid copper tubing, fittings and assorted hardware can be used to make a really rugged J-pole antenna for 2 m. When copper tubing is used, the entire assembly can be soldered together, ensuring electrical integrity, and making the whole antenna weatherproof. This material came from an article by Michael Hood, KD8JB, in *The ARRL Antenna Compendium, Vol. 4*.

No special hardware or machined parts are used in this antenna, nor are insulating materials needed, since the antenna is always at dc ground. Best of all, even if the parts aren't on sale, the antenna can be built for less than $15. If you only build one antenna, you'll have enough tubing left over to make most of a second antenna.

CONSTRUCTION

Copper and brass is used exclusively in this antenna. These metals get along together, so dissimilar metal corrosion is eliminated. Both metals solder well, too. See **Fig 22.101**. Cut the copper tubing to the lengths indicated. Item 9 is a 1¼-inch nipple cut from the 20-inch length of ½-inch tubing. This leaves 18¾ inches for the λ/4-matching stub. Item 10 is a 3¼-inch long nipple cut from the 60-inch length of ¾-inch tubing. The ¾-wave element should measure 56¾ inches long. Remove burrs from the ends of the tubing after cutting, and clean the mating surfaces with sandpaper, steel wool, or emery cloth.

After cleaning, apply a very thin coat of flux to the mating elements and assemble the tubing, elbow, tee, endcaps and stubs. Solder the assembled parts with a propane torch and rosin-core solder. Wipe off excess solder with a damp cloth, being careful not to burn yourself. The copper tubing will hold heat for a long time after you've finished soldering. After soldering, set the assembly aside to cool.

Flatten one each of the ½-inch and ¾-inch pipe clamps. Drill a hole in the flattened clamp as shown in Fig 22.101B. Assemble the clamps and cut off the excess metal from the flattened clamp using the unmodified clamp as a template. Disassemble the clamps.

Assemble the ½-inch clamp around the ¼-wave element and secure with two of the screws, washers, and nuts as shown in Fig 22.101B. Do the same with the ¾-inch clamp around the ¾-wave element. Set the clamps initially to a spot about 4 inches above the bottom of the *J* on their respective elements. Tighten the clamps only finger tight, since you'll need to move them when tuning.

TUNING

The J-Pole can be fed directly from 50 Ω coax through a choke balun (3 turns of the feed coax rolled into a coil about 8 inches in diameter and held together with electrical tape). Before tuning, mount the antenna vertically, about 5 to 10 ft above the ground. A short TV mast on a tripod works well for this purpose. When tuning VHF antennas, keep in mind that they are sensitive to nearby objects—such as your body. Attach the feed line to the clamps on the antenna, and make sure all the nuts and screws are at least finger tight. It really doesn't matter to which element ($^3/_4$-wave element or stub) you attach the coaxial center lead. Tune the antenna by moving the two feed-point clamps equal distances a small amount each time until the SWR is minimum at the desired frequency. The SWR will be close to 1:1.

FINAL ASSEMBLY

The final assembly of the antenna will determine its long-term survivability. Perform the following steps with care. After adjusting the clamps for minimum SWR, mark the clamp positions with a pencil and then remove the feed line and clamps. Apply a very thin coating of flux to the inside of the clamp and the corresponding surface of the antenna element where the clamp attaches. Install the clamps and tighten the clamp screws.

Solder the feed line clamps where they are attached to the antenna elements. Now, apply a small amount of solder around the screw heads and nuts where they contact the clamps. Don't get solder on the screw

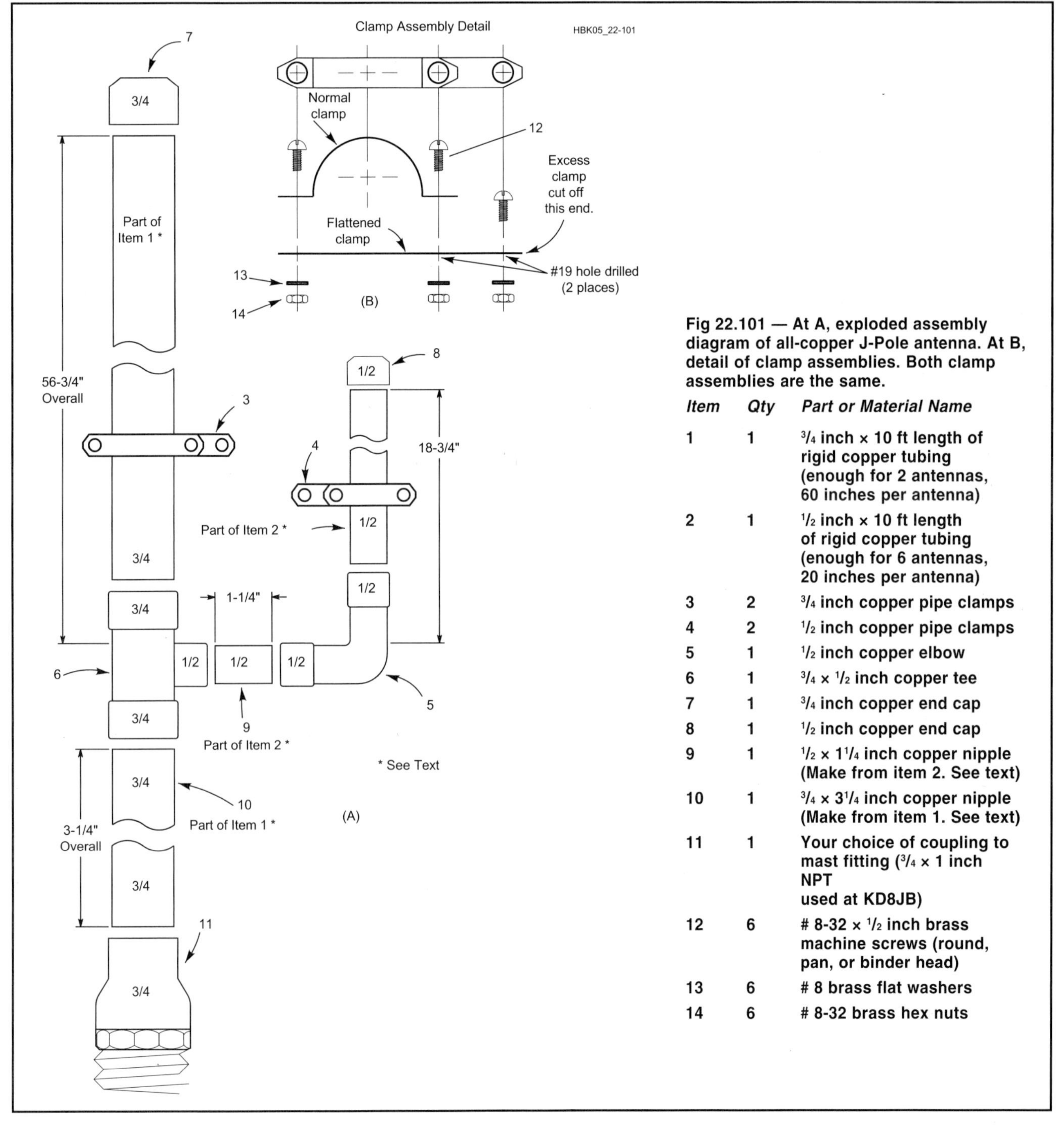

Fig 22.101 — At A, exploded assembly diagram of all-copper J-Pole antenna. At B, detail of clamp assemblies. Both clamp assemblies are the same.

Item	*Qty*	*Part or Material Name*
1	1	$^3/_4$ inch × 10 ft length of rigid copper tubing (enough for 2 antennas, 60 inches per antenna)
2	1	$^1/_2$ inch × 10 ft length of rigid copper tubing (enough for 6 antennas, 20 inches per antenna)
3	2	$^3/_4$ inch copper pipe clamps
4	2	$^1/_2$ inch copper pipe clamps
5	1	$^1/_2$ inch copper elbow
6	1	$^3/_4$ × $^1/_2$ inch copper tee
7	1	$^3/_4$ inch copper end cap
8	1	$^1/_2$ inch copper end cap
9	1	$^1/_2$ × 1$^1/_4$ inch copper nipple (Make from item 2. See text)
10	1	$^3/_4$ × 3$^1/_4$ inch copper nipple (Make from item 1. See text)
11	1	Your choice of coupling to mast fitting ($^3/_4$ × 1 inch NPT used at KD8JB)
12	6	# 8-32 × $^1/_2$ inch brass machine screws (round, pan, or binder head)
13	6	# 8 brass flat washers
14	6	# 8-32 brass hex nuts

threads! Clean away excess flux with a non-corrosive solvent.

After final assembly and erecting/mounting the antenna in the desired location, attach the feed line and secure with the remaining washer and nut. Weather-seal this joint with RTV. Otherwise, you may find yourself repairing the feed line after a couple years.

ON-AIR PERFORMANCE

The author had no problem working various repeaters around town with a $^1/_4$-wave antenna, but simplex operation left a lot to be desired. The J-Pole performs just as well as a Ringo Ranger, and significantly better than the $^1/_4$-wave ground-plane vertical.

VHF/UHF Yagis

Without doubt, the Yagi is king of home-station antennas these days. Today's best designs are computer optimized. For years amateurs as well as professionals designed Yagi arrays experimentally. Now we have powerful (and inexpensive) personal computers and sophisticated software for antenna modeling. These have brought us antennas with improved performance, with little or no element pruning required.

A more complete discussion of Yagi design can be found earlier in this chapter. For more coverage on this topic and on stacking Yagis, see the most recent edition of *The ARRL Antenna Book*.

3 AND 5-ELEMENT YAGIS FOR 6 M

Boom length often proves to be the deciding factor when one selects a Yagi design. ARRL Senior Assistant Technical Editor Dean Straw, N6BV, created the designs shown in **Table 22.24**. Straw generated the designs in the table for convenient boom lengths (6 and 12 ft). The 3-element design has about 8 dBi gain, and the 5-element version has about 10 dBi gain. Both antennas exhibit better than 22 dB front-to-rear ratio, and both cover 50 to 51 MHz with better than 1.6:1 SWR.

Element lengths and spacings are given in the table. Elements can be mounted to the boom as shown in **Fig 22.102**. Two muffler clamps hold each aluminum plate to the boom, and two U bolts fasten each element to the plate, which is 0.25 inches thick and 4·4 inches square. Stainless steel is the best choice for hardware, however, galvanized hardware can be substituted. Automotive muffler clamps do not work well in this application, because they are not galvanized and quickly rust once exposed to the weather.

The driven element is mounted to the boom on a Bakelite plate of similar dimension to the other mounting plates. A 12-inch piece of Plexiglas rod is inserted into the driven element halves. The Plexiglas allows the use of a single clamp on each side of the element and also seals the center of the elements against moisture. Self-tapping screws are used for electrical connection to the driven element.

Refer to **Fig 22.103** for driven element and Hairpin match details. A bracket made from a piece of aluminum is used to mount the three SO-239 connectors to the driven element plate. A 4:1 transmission-line balun connects the two element halves, transforming the 200-Ω resistance at the Hairpin match to 50 Ω at the center connector. Note that the electrical length of the balun is λ/2, but the physical length

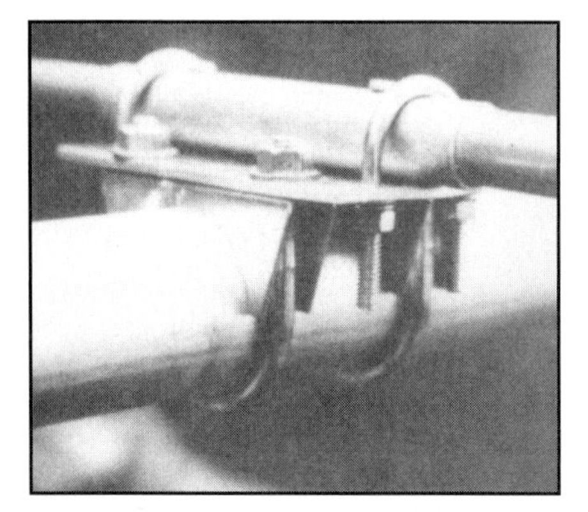

Fig 22.102 — The element-to-boom clamp. Galvanized U bolts are used to hold the element to the plate, and 2-inch galvanized muffler clamps hold the plates to the boom.

Table 22.24
Optimized 6-m Yagi Designs

	Spacing From Reflector (in.)	*Seg 1 Length (in.)*	*Seg 2 Length (in.)*	*Midband Gain F/R*
306-06				
Refl	0	36	22.500	8.1 dBi
DE	24	36	16.000	28.3 dB
Dir 1	66	36	15.500	
506-12				
OD		0.750	0.625	
Refl	0	36	23.625	10.0 dBi
DE	24	36	17.125	26.8 dB
Dir 1	36	36	19.375	
Dir 2	80	36	18.250	
Dir 3	138	36	15.375	

Note: For all antennas, telescoping tube diameters (in inches) are: Seg1=0.750, Seg2=0.625. See figure 22.66 for element details.

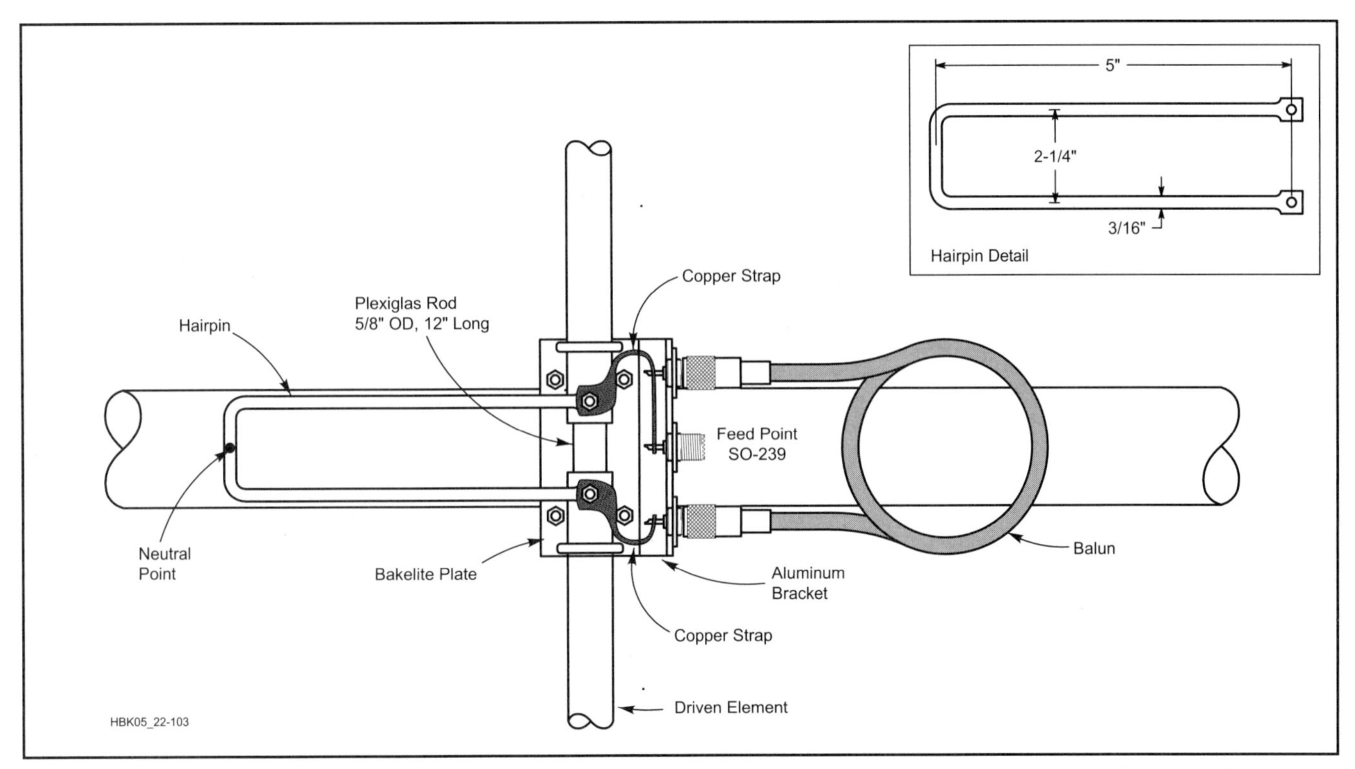

Fig 22.103 — Detailed drawing of the feed system used with the 50-MHz Yagi. Balun lengths: For cable with 0.80 velocity factor — 7 ft, 10$^3/_8$ in. For cable with 0.66 velocity factor — 6 ft, 5$^3/_4$ in.

will be shorter due to the velocity factor of the particular coaxial cable used. The Hairpin is connected directly across the element halves. The exact center of the hairpin is electrically neutral and should be fastened to the boom. This has the advantage of placing the driven element at dc ground potential.

The Hairpin match requires no adjustment as such. However, you may have to change the length of the driven element slightly to obtain the best match in your preferred portion of the band. Changing the driven-element length will not adversely affect antenna performance. *Do not adjust the lengths or spacings of the other elements—they are optimized already.* If you decide to use a gamma match, add 3 inches to each side of the driven element lengths given in the table for both antennas.

A MEDIUM GAIN 2-M YAGI

This project was designed and built by L. B. Cebik, W4RNL. Practical Yagis for 2 meters abound. What makes this one a bit different is the selection of materials. The elements, of course, are high-grade aluminum. However, the boom is PVC and there are only two #6 nut-bolt sets and two #8 sheet metal screws in the entire antenna. The remaining fasteners are all hitch-pin clips. The result is a very durable 6-element Yagi that you can disassemble with fair ease for transport.

THE BASIC ANTENNA DESIGN

The 6-element Yagi presented here is a derivative of the *optimized wide-band antenna* (OWA) designs developed for HF use by NW3Z and WA4FET. **Fig 22.104** shows the general outline. The reflector and first director largely set the impedance. The next 2 directors contribute to setting the operating bandwidth. The final director (Dir. 4) sets the gain. This account is over-simplified, since every element plays a role in every facet of Yagi performance. However, the notes give some idea of which elements are most sensitive in adjusting the performance figures.

Designed on *NEC-4*, the antenna uses 6 elements on a 56 inch boom. **Table 22.25** gives the specific dimensions for the version described in these notes. The parasitic elements are all $^3/_{16}$-inch aluminum rods. For ease of construction, the driver is $^1/_2$-inch aluminum tubing. Do not alter the element diameters without referring to a source, such as RSGB's *The VHF/UHF DX Book*, edited by Ian White, G3SEK, (Chapter 7), for information on how to recalculate element lengths.

The driver is the simplest element to readjust. Table 22.25 shows an alternative driver using $^3/_{16}$-inch diameter material. Of all the elements, the driver is perhaps the only one for which you may extrapolate reasonable lengths for other diameters from the two lengths and diameters shown. However, the parasitic elements may require more work than merely substituting one diameter and length for another. The lower portion of the table shows the design adjusted for $^1/_8$-inch elements throughout. Not all element lengths change by the same amount using any single formula.

The OWA design provides about 10.2 dBi of free-space gain with better than 20 dB front-to-back (or front-to-rear) ratio across the entire 2-m band. Azimuth (or E-plane) patterns show solid performance across the entire band. This applies not only to forward gain but rejection from the rear.

One significant feature of the OWA design is its direct 50-Ω feedpoint impedance that requires no matching network. Of course, a choke balun to suppress any currents on the feedline is desirable, and a simple bead-choke of W2DU design works well in this application. The SWR, shown in **Fig 22.105**, is very flat across the band and never reaches 1.3:1. The SWR and the pattern consistency together create a very useful utility antenna for 2 m, whether installed vertically or horizontally. The only remaining question is how to effectively build the beam in the average home shop.

THE BEAM MATERIALS

The boom is Schedule 40, 1/2-inch nominal PVC. Insulated booms are good for test antennas, since they do not require recalculating the element lengths due to the effects of a metal boom.

White PVC stands up for a decade of exposure in Tennessee, but apparently does not do as well in every part of the US. You may wish to use the gray electrical conduit version. If you use any other material for your boom, be sure that it is UV-protected. You'll find a parts list in **Table 22.26**. Sources for the parts are given in the table. However, you are encouraged to develop your own sources for antenna materials.

Fig 22.106 shows the element layout

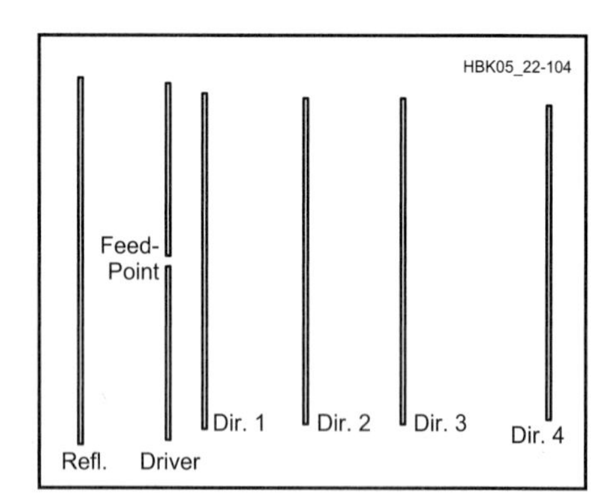

Fig 22.104 — The general outline of the 2-m 6-element OWA Yagi. Dimensions are given in Table 22.25.

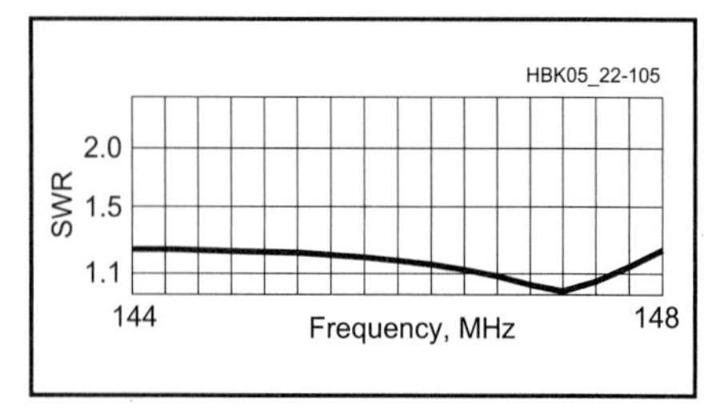

Fig 22.105 — SWR curve as modeled on *NEC-4* for the 2-m 6-element OWA Yagi.

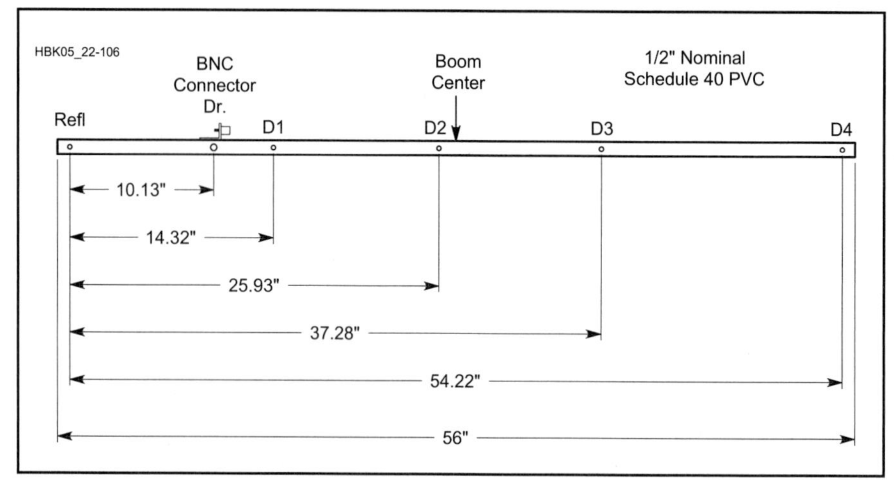

Fig 22.106 — Layout of elements along the PVC boom for the 2-m 6-element OWA Yagi, showing placement of the BNC connector and the boom center.

Table 22.25
2-m OWA Yagi Dimensions

Element	*Element Length in Inches*	*Spacing from Reflector in Inches*	*Element Diameter in Inches*
Version described here:			
Refl.	40.52	—	0.1875
Driver	39.70	10.13	0.5
(Alt. Driver	*39.96*	*10.13*	*0.1875)*
Dir. 1	37.36	14.32	0.1875
Dir. 2	36.32	25.93	0.1875
Dir. 3	36.32	37.28	0.1875
Dir. 4	34.96	54.22	0.1875
Version using 1/8-inch diameter elements throughout:			
Refl.	40.80	—	0.125
Driver	40.10	10.20	0.125
Dir. 1	37.63	14.27	0.125
Dir. 2	36.56	25.95	0.125
Dir. 3	36.56	37.39	0.125
Dir. 4	35.20	54.44	0.125

Table 22.26
Parts List for the 2-Meter OWA Yagi

Qty	*Item*
17'	0.1875" (3/16") 6061-T6 aluminum rod (Source: Texas Towers)
3.5'	0.5" (1/2") 6063-T832 aluminum tubing (Source: Texas Towers)
7'	Schedule 40, 1/2" PVC pipe (Source: local hardware depot)
3	Schedule 40, 1/2" PVC Tee connectors (Source: local hardware depot)
2	Schedule 40, 1/2" PVC L connectors (Source: local hardware depot) depot)
—	Miscellaneous male/female threaded pipe diameter transition fittings (Source: local hardware depot)
1	Support mast
10	Stainless steel hitch-pin clips (hairpin cotter pins), 3/16" to 1/4" shaft range, 0.04" "wire" diameter (McMasters-Carr part number 9239A024)
2	Stainless steel #6 nut/bolt/lock-washer sets, bolt length 1" (Source: local hardware depot)
2	Stainless steel #8 sheet metal screws (Source: local hardware depot)
1	BNC connector (Source: local electronics outlet)
2"	1/16" thick aluminum L-stock, 1" per side (Source: local hardware depot)
1	VHF bead-balun choke (Source: Wireman, Inc.)

along the 56-inch boom. Centering the first element hole 1 inch from the rear end of the boom results in a succession of holes for the $^{3}/_{16}$-inch pass-through parasitic elements. Only the driver requires special treatment. We shall use a $^{3}/_{8}$-inch hole to carry a short length of fiberglass rod that will support the two sides of the driver element. Note that I used a BNC connector, mounted on a small plate that we shall meet along the way.

The boom is actually a more complex structure than initially meets the eye. You need a support for the elements, and a means of connecting the boom to the mast. If you break the boom in the middle to install a Tee connector for the mast junction, you come very close to the 2nd director. **Fig 22.107** shows how to avoid the predicament.

Before drilling the boom, assemble it from common Schedule 40 $^{1}/_{2}$-inch fittings and insert the lengths of PVC pipe. Fig 22.107 shows the dimensions for the center section of the boom assembly. However, PVC dimensions are always *nominal*, that is, meeting certain minimum size standards. So you may have to adjust the lengths of the linking pieces slightly to come up with a straight and true boom assembly.

Use scrap lumber to help keep everything aligned while cementing the pieces together. A 1×4 and a 1×6 nailed together along the edges produces a very good platform with a right-angle. Start with the two upper Tees and the Ls below each one. Dry-fit scrap PVC into the openings except for the short link that joins the fitting. Cement these in place and align them using the dry-fit pieces as guides to keep everything parallel. Next, cement the two short (2$^{3}/_{4}$-inch) links into the third Tee. Then, cement one link into its L, using the dry-fit tube in the upper Tee as an alignment guide.

Before proceeding further, carefully measure the required length of PVC for the boom section between Tees. How well you measure here will determine whether the boom will be straight or whether it will bow up or down. Now, cement both the L and the Tee at the same time, pressing the cemented sections into the 2-board jig to assure alignment.

The final step in the process is to add the 23-inch boom end pieces to the open ends of the upper Tees. For the brief period in which the PVC cement is wet, it is possible to misalign the tubing. Dry-fit end caps on the boom ends and do the cement work using the 2-board jig. By pressing the assembly into the right angle of the boards, you can assure that you have a very true boom. When you've put the PVC cement back onto its shelf, your boom should be ready to drill.

Consider the boom-to-mast connection. The lower Tee in Fig 22.107 receives a short length of $^{1}/_{2}$-inch nominal Schedule 40 PVC. This material has an outside diameter of about $^{7}/_{8}$-inch, not a useful size for joining to a mast. However, PVC fittings have a handy series of threaded couplers that allow you to screw-fit a series of ever-larger sizes until you reach a more useful size. As **Fig 22.108B** shows, enough of these fittings will finish off with a 1$^{1}/_{4}$-inch threaded female side and a 1$^{1}/_{4}$-inch cement-coupling side. To this fitting, cement a length of 1$^{1}/_{4}$-inch tubing that slides over a length of common TV mast. For a tight fit, wrap the TV mast with several layers of electrical tape in two places—one near the upper end of the PVC pipe section and the other close to where the PVC pipe ends. You may then use stainless steel through-bolts or set-screws to prevent the PVC assembly from turning.

BOOM AND ELEMENTS

Before installing the elements, you need to drill the holes in the boom. The two-board jig comes in handy once more. The key goals in the drilling process are to: A) precisely position the holes; B) create holes that are a fairly tight fit for the rod elements; and C) keep the elements aligned in a flat plane. For this purpose, a drill press is almost a necessity for all but those with the truest eyes.

Use the jig and a couple of clamps to

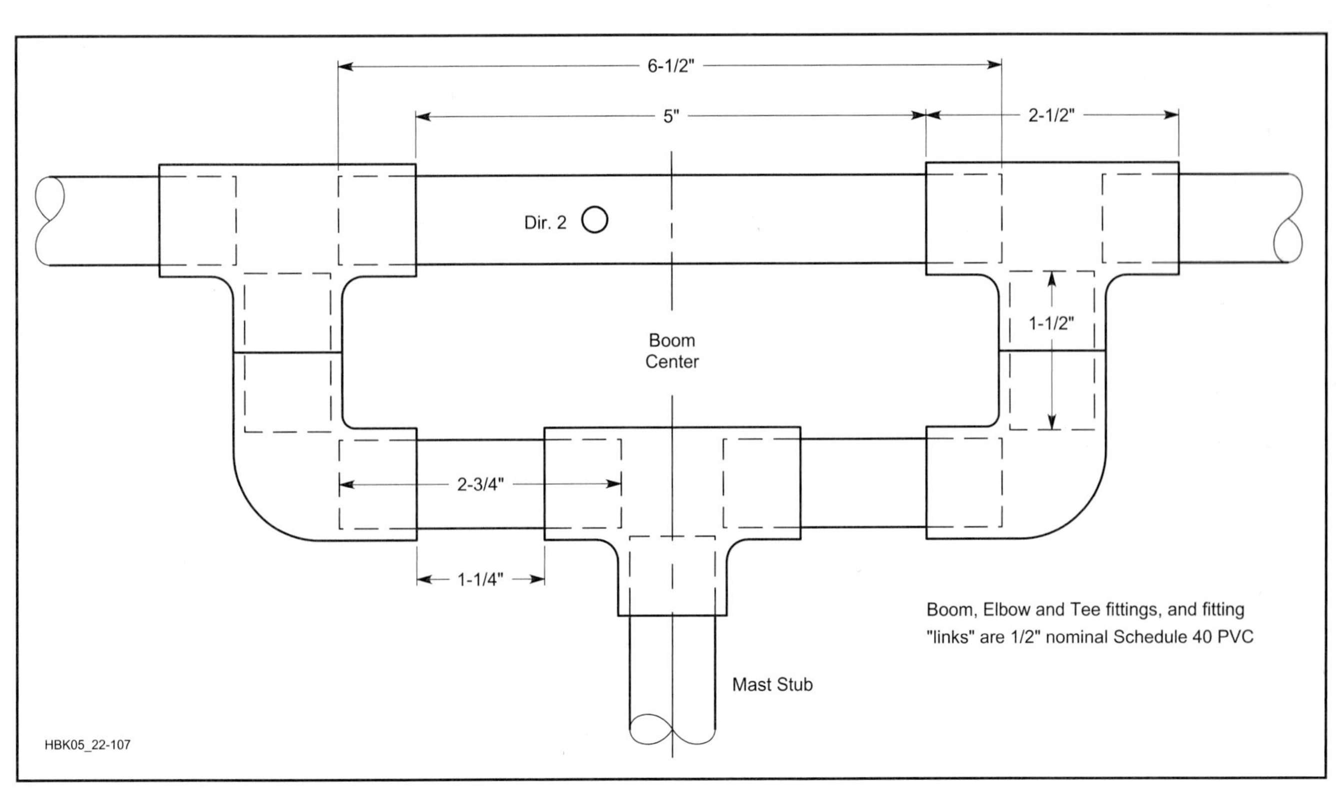

Fig 22.107 — Details of a parallel PVC pipe structure for the Yagi boom and mount.

hold the boom assembly in place. Because the assembly has two parallel sections, laying it flat will present the drill press with the correct angle for drilling through the PVC in one stroke. Drill the holes at premarked positions, remembering that the driver hole is $^3/_8$ inch while all the others are $^3/_{16}$ inch. Clean the holes, but do not enlarge them in the process.

(A)

(B)

Fig 22.108 — The completed Yagi is shown at A. A close-up view of the parallel PVC boom and mount, the sequence of threaded fittings, and the hitch-pin clips used to secure parasitic elements is shown at B.

By now you should have the rod and tube stock in hand. For antenna elements, don't rely on questionable materials that are designed for other applications. Rather, obtain 6063-T832 tubing and 6061-T6 rods from mail order sources, such as Texas Towers, McMasters-Carr, and others. These materials are often not available at local hardware depots.

Cut the parasitic elements to length and smooth their ends with a fine file or sandpaper. Find the center of each element and carefully mark a position about $^1/_{16}$" outside where the element will emerge from each side of the boom. You'll drill small holes in these locations. You may wish to very lightly file a flatted area where the hole is to go to prevent the drill bit from slipping as you start the hole.

Drill $^1/_{16}$-inch holes at each marked location all the way through the rod. De-burr the exit ends so that the rod will pass through the boom hole. These holes are the locations for hitch-pin clips. **Fig 22.109** shows the outline of a typical hitch-pin clip, which is also called a hairpin cotter pin in some catalogs and stores. Obtain stainless steel pins whose bodies just fit tightly over the rod when installed. Initially, install 1 pin per parasitic element. Slide the element through the correct boom hole and install the second pin. Although the upper part of the drawing shows a bit of room between the boom and pin, this space is for clarity. Install the pins as close to each side of the boom as you can.

Pins designed for a $^3/_{16}$-inch rod are small enough that they add nothing significant to the element, and antenna tests showed that they did not move the performance curve of the antenna. Yet, they have held securely through a series of shock tests given to the prototype. These pins—in various sizes—offer the home builder a handy fastener that is applicable to many types of portable or field antennas. Although you may wish to use better fasteners when making permanent metal-to-metal connections, for joining sections of Field Day and similar antennas, the hitch-pin clips perform the mechanical function, while clean tubing sections themselves provide adequate electrical contact for a limited period of use.

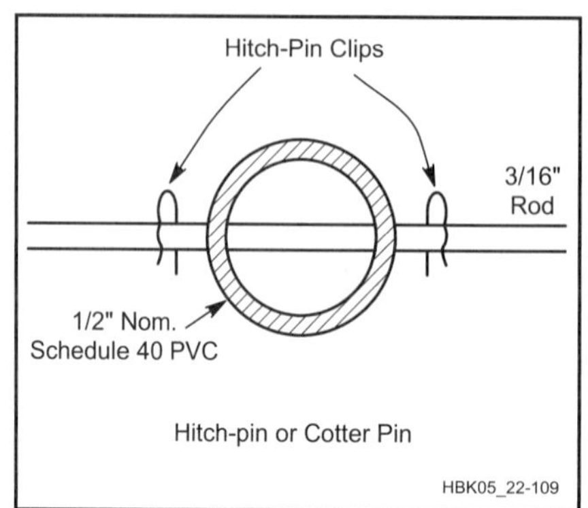

Fig 22.109 — The parasitic element mounting system, showing the placement of the hitch-pin clips and the shape of the clips.

THE DRIVER AND FEEDLINE CONNECTOR

The final construction step is perhaps the one requiring the most attention to detail, as shown in **Fig 22.110**. The driver and feedpoint assembly consists of a 4- to 6-inch length of $^3/_8$-inch fiberglass or other non-conductive rod, two sections of the driver element made from $^1/_2$-inch aluminum tubing, a BNC connector, a homemade mounting plate, two sets of stainless steel #6 nuts, bolts, and lock-washers, and two stainless steel #8 sheet metal screws. Consult both the upper and lower portions of the figure, since some detail has been omitted from each one to show other detail more clearly.

First, trial fit the driver tubing and the fiberglass rod, marking where the rod exits the boom. Now pre-drill $^9/_{64}$-inch holes through the tubing and the fiberglass rod. Do not use larger hardware, since the resulting hole will weaken the rod, possibly to the breaking point. If you use an alternative plastic material, observe the same caution and be certain that the rod remains strong after drilling. Do not use wooden dowels for this application, since they do not have sufficient strength. Position the holes about $^1/_4$ to $^3/_8$ inch from the tubing end where it presses against the boom. One hole will receive a solder lug and the other will connect to an extension of the BNC mounting plate.

Second, install the fiberglass rod through the boom. You can leave it loose, since the elements will press against boom and hold it in place. Alternatively, you may glue it in place with a 2-part epoxy. Slide the driver element tubes over the rod and test the holes for alignment by placing the #6 bolts in them.

Next, cut and shape the BNC mounting plate from $^1/_{16}$-inch thick aluminum. I made my fitting from a scrap of L-stock 1 inch on a side. Before cutting the stock, I drilled the $^3/_8$-inch hole needed for the BNC connector. Then I cut the vertical portion. The horizontal portion requires a curved tab that reaches the bolt on one side of the boom. I used bench vise to bend the tab in a curve and then flatten it for the bolt-hole. It takes several tries to get the shape and tab exact, so be patient. When the squared-edge piece found its perfect

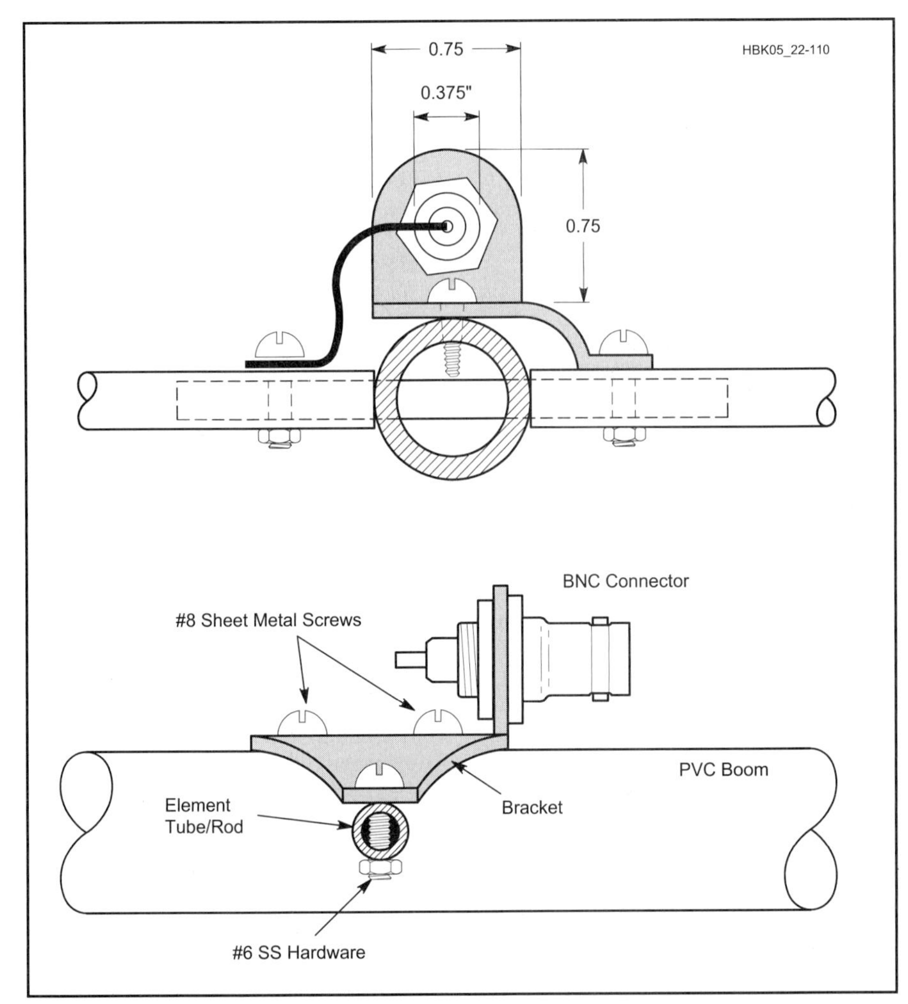

Fig 22.110 — Details of the feedpoint of the Yagi, showing the BNC connector, mounting plate, and connections to the ½-inch driver element halves placed over a central ⅜-inch fiberglass rod.

shape, I took it to a disk sander and rounded the vertical piece to follow the connector shape. I also tapered the tap edges to minimize excess material. The last step is to drill the mounting holes that receive the #8 sheet metal screws.

Mounting the assembly involves loosely attaching both the #6 and #8 hardware and alternatively tightening up all pieces. Be certain that the side of the BNC connector that receives the coax points toward the mast. Next, mount the BNC connector. The shield side is already connected to one side of the driver. Mount the other side of the driver, placing a solder lug under the bolt head. Connect a short wire as directly as possible from the solder lug to the center pin of the BNC connector. After initial testing, you may coat all exposed connections with Plasti-Dip for weather protection.

TUNE-UP

Testing and tuning the antenna is a simple process if you build carefully. The only significant test that you can perform is to ensure that the SWR curve comes close to the one shown in Fig 22.105. If the SWR is high at 148 MHz but very low at 144 MHz, then you will need to shorten the driver ends by a small amount—no more than ⅛ inch per end at a time. I found that shaving the ends with a disk sander was most effective.

Using the antenna with vertical polarization will require good spacing from any support structure with metal vertical portions. One of the easiest ways to devise such a mounting is to create a PVC structure that turns the entire boom by 90 degrees. If you feel the need for added support, you can create an angular brace by placing 45-degree connectors in both the vertical and horizontal supports and running a length of PVC between them.

As an alternative, you can let the rear part of the boom be slightly long. To this end you can cement PVC fixtures—including the screw-thread series to enlarge the support pipe size. Create a smooth junction that you attach with a through-bolt instead of cement. By drilling one side of the connection with two sets of holes, 90-degrees apart, you can change the antenna from horizontal polarization to vertical and back in short order.

The 6-element OWA Yagi for 2 meters performs well. It serves as a good utility antenna with more gain and directivity than the usual 3-element general-use Yagi. When vertically polarized, the added gain confirms the wisdom of using a longer boom and more elements. With a length under 5 feet, the antenna is still compact. The ability to disassemble the parts simplifies moving the antenna to various portable sites.

BIBLIOGRAPHY

J. S. Belrose, "Short Antennas for Mobile Operation," *QST*, Sep 1953.

G. H. Brown, "The Phase and Magnitude of Earth Currents Near Radio Transmitting Antennas," *Proc IRE*, Feb 1935.

G. H. Brown, R. F. Lewis and J. Epstein, "Ground Systems as a Factor in Antenna Efficiency," *Proc IRE*, Jun 1937, pp 753-787.

G. H. Brown and O. M. Woodward, Jr, "Experimentally Determined Impedance Characteristics of Cylindrical Antennas," *Proc IRE*, April 1945.

A. Christman, "Elevated Vertical Antenna Systems," *QST*, Aug 1988, pp 35-42.

R. B. Dome, "Increased Radiating Efficiency for Short Antennas," *QST*, Sep 1934, pp 9-12.

A. C. Doty, Jr, J. A. Frey and H. J. Mills, "Characteristics of the Counterpoise and Elevated Ground Screen," Professional Program, Session 9, Southcon '83 (IEEE), Atlanta, GA, Jan 1983.

A. C. Doty, Jr, J. A. Frey and H. J. Mills, "Efficient Ground Systems for Vertical Antennas," *QST*, Feb 1983, pp 20-25.

A. C. Doty, Jr, technical paper presentation, "Capacitive Bottom Loading and Other Aspects of Vertical Antennas," Technical Symposium, Radio Club of America, New York City, Nov 20, 1987.

A. C. Doty, Jr, J. A. Frey and H. J. Mills, "Vertical Antennas: New Design and Construction Data," *The ARRL Antenna Compendium, Volume 2* (Newington: ARRL, 1989), pp 2-9.

R. Fosberg, "Some Notes on Ground Systems for 160 Meters," *QST*, Apr 1965, pp 65-67.

G. Grammer, "More on the Directivity of Horizontal Antennas; Harmonic Operation—Effects of Tilting," *QST*, Mar 1937, pp 38-40, 92, 94, 98.

H. E. Green, "Design Data for Short and Medium Length Yagi-Uda Arrays,"

Trans IE Australia, Vol EE-2, No. 1, Mar 1966.

H. J. Mills, technical paper presentation, "Impedance Transformation Provided by Folded Monopole Antennas," Technical Symposium, Radio Club of America, New York City, Nov 20, 1987.

B. Myers, "The W2PV Four-Element Yagi," *QST*, Oct 1986, pp 15-19.

L. Richard, "Parallel Dipoles of 300-Ohm Ribbon," *QST*, Mar 1957.

J. H. Richmond, "Monopole Antenna on Circular Disc," *IEEE Trans on Antennas and Propagation*, Vol. AP-32, No. 12, Dec 1984.

W. Schulz, "Designing a Vertical Antenna," *QST*, Sep 1978, pp 19- 21.

J. Sevick, "The Ground-Image Vertical Antenna," *QST*, Jul 1971, pp 16-17, 22.

J. Sevick, "The W2FMI 20-Meter Vertical Beam," *QST*, Jun 1972, pp 14-18.

J. Sevick, "The W2FMI Ground-Mounted Short Vertical," *QST*, Mar 1973, pp 13-18, 41.

J. Sevick, "A High Performance 20-, 40- and 80-Meter Vertical System," *QST*, Dec 1973.

J. Sevick, "Short Ground-Radial Systems for Short Verticals," *QST*, Apr 1978, pp 30-33.

J. Sevick, *Transmission Line Transformers*, 3rd ed. (Tucker, GA: Noble Publishing, 1996).

C. E. Smith and E. M. Johnson, "Performance of Short Antennas," *Proc IRE*, Oct 1947.

J. Stanley, "Optimum Ground Systems for Vertical Antennas," *QST*, Dec 1976, pp 13-15.

R. E. Stephens, "Admittance Matching the Ground-Plane Antenna to Coaxial Transmission Line," Technical Correspondence, *QST*, Apr 1973, pp 55-57.

D. Sumner, "Cushcraft 32-19 'Boomer' and 324-QK Stacking Kit," Product Review, *QST*, Nov 1980, pp 48-49.

W. van B. Roberts, "Input Impedance of a Folded Dipole," *RCA Review*, Jun 1947.

E. M. Williams, "Radiating Characteristics of Short-Wave Loop Aerials," *Proc IRE*, Oct 1940.

TEXTBOOKS ON ANTENNAS

C. A. Balanis, *Antenna Theory, Analysis and Design* (New York: Harper & Row, 1982).

D. S. Bond, *Radio Direction Finders*, 1st ed. (New York: McGraw-Hill Book Co).

W. N. Caron, *Antenna Impedance Matching* (Newington: ARRL, 1989).

L. B. Cebik, *ARRL Antenna Modeling Course* (Newington: ARRL, 2002).

K. Davies, *Ionospheric Radio Propagation*—National Bureau of Standards Monograph 80 2(Washington, DC: U.S. Government Printing Office, Apr 1, 1965).

R. S. Elliott, *Antenna Theory and Design* (Englewood Cliffs, NJ: Prentice Hall, 1981).

A. E. Harper, *Rhombic Antenna Design* (New York: D. Van Nostrand Co, Inc, 1941).

K. Henney, *Principles of Radio* (New York: John Wiley and Sons, 1938), p 462.

C. Hutchinson and R. D. Straw, *Simple and Fun Antennas for Hams* (Newington: ARRL, 2002).

H. Jasik, *Antenna Engineering Handbook*, 1st ed. (New York: McGraw-Hill, 1961).

W. C. Johnson, *Transmission Lines and Networks*, 1st ed. (New York: McGraw-Hill Book Co, 1950).

R. C. Johnson and H. Jasik, *Antenna Engineering Handbook*, 2nd ed. (New York: McGraw-Hill, 1984).

R. C. Johnson, *Antenna Engineering Handbook*, 3rd ed. (New York: McGraw-Hill, 1993).

E. C. Jordan and K. G. Balmain, *Electromagnetic Waves and Radiating Systems*, 2nd ed. (Englewood Cliffs, NJ: Prentice-Hall, Inc, 1968).

R. Keen, *Wireless Direction Finding*, 3rd ed. (London: Wireless World).

R. W. P. King, *Theory of Linear Antennas* (Cambridge, MA: Harvard Univ. Press, 1956).

R. W. P. King, H. R. Mimno and A. H. Wing, *Transmission Lines, Antennas and Waveguides* (New York: Dover Publications, Inc, 1965).

King, Mack and Sandler, *Arrays of Cylindrical Dipoles* (London: Cambridge Univ Press, 1968).

M. G. Knitter, Ed., *Loop Antennas—Design and Theory* (Cambridge, WI: National Radio Club, 1983).

M. G. Knitter, Ed., *Beverage and Long Wire Antennas*—Design and Theory (Cambridge, WI: National Radio Club, 1983).

J. D. Kraus, *Electromagnetics* (New York: McGraw-Hill Book Co).

J. D. Kraus, *Antennas*, 2nd ed. (New York: McGraw-Hill Book Co, 1988).

E. A. Laport, *Radio Antenna Engineering* (New York: McGraw-Hill Book Co, 1952).

J. L. Lawson, *Yagi-Antenna Design*, 1st ed. (Newington: ARRL, 1986).

P. H. Lee, *The Amateur Radio Vertical Antenna Handbook*, 2nd ed. (Port Washington, NY: Cowen Publishing Co., 1984).

D. B. Leeson, *Physical Design of Yagi Antennas* (Newington: ARRL, 1992).

A. W. Lowe, *Reflector Antennas* (New York: IEEE Press, 1978).

M. W. Maxwell, *Reflections—Transmission Lines and Antennas* (Newington: ARRL, 1990). Out of print.

M. W. Maxwell, *Reflections II—Transmission Lines and Antennas* (Sacramento: Worldradio Books, 2001).

G. M. Miller, *Modern Electronic Communication* (Englewood Cliffs, NJ: Prentice Hall, 1983).

V. A. Misek, *The Beverage Antenna Handbook* (Hudson, NH: V. A. Misek, 1977).

T. Moreno, *Microwave Transmission Design Data* (New York: McGraw-Hill, 1948).

L. A. Moxon, *HF Antennas for All Locations* (Potters Bar, Herts: Radio Society of Great Britain, 1982), pp 109-111.

Ramo and Whinnery, *Fields and Waves in Modern Radio* (New York: John Wiley & Sons).

V. H. Rumsey, *Frequency Independent Antennas* (New York: Academic Press, 1966).

P. N. Saveskie, *Radio Propagation Handbook* (Blue Ridge Summit, PA: Tab Books, Inc, 1980).

S. A. Schelkunoff, *Advanced Antenna Theory* (New York: John Wiley & Sons, Inc, 1952).

S. A. Schelkunoff and H. T. Friis, *Antennas Theory and Practice* (New York: John Wiley & Sons, Inc, 1952).

J. Sevick, *Transmission Line Transformers* (Atlanta: Noble Publishing, 1996).

H. H. Skilling, *Electric Transmission Lines* (New York: McGraw-Hill Book Co, Inc, 1951).

M. Slurzburg and W. Osterheld, *Electrical Essentials of Radio* (New York: McGraw-Hill Book Co, Inc, 1944).

G. Southworth, *Principles and Applications of Waveguide Transmission* (New York: D. Van Nostrand Co, 1950).

R. D. Straw, Ed., *The ARRL Antenna Book*, 20th ed. (Newington: ARRL, 2003).

F. E. Terman, *Radio Engineers' Handbook*, 1st ed. (New York, London: McGraw-Hill Book Co, 1943).

F. E. Terman, *Radio Engineering*, 3rd ed. (New York: McGraw-Hill, 1947).

S. Uda and Y. Mushiake, *Yagi-Uda Antenna* (Sendai, Japan: Sasaki Publishing Co, 1954). [Published in English—Ed.]

P. P. Viezbicke, "*Yagi Antenna Design*," NBS Technical Note 688 (U. S. Dept of Commerce/National Bureau of Standards, Boulder, CO), Dec 1976.

G. B. Welch, *Wave Propagation and Antennas* (New York: D. Van Nostrand Co, 1958), pp 180-182.

The GIANT Book of Amateur Radio Antennas (Blue Ridge Summit, PA: Tab Books, 1979), pp 55-85.

IEEE Standard Dictionary of Electrical and Electronics Terms, 3rd ed. (New York: IEEE, 1984).

Radio Broadcast Ground Systems, available from Smith Electronics, Inc, 8200 Snowville Rd, Cleveland, OH 44141.

Radio Communication Handbook, 5th ed. (London: RSGB, 1976).

Radio Direction Finding, published by the Happy Flyer, 1811 Hillman Ave, Belmont, CA 94002.

Chapter 23

Space Communications

An Amateur Satellite Primer

Most amateurs are familiar with repeater stations that retransmit signals to provide wider coverage. Repeaters achieve this by listening for signals on one frequency and immediately retransmitting whatever they hear on another frequency. Thanks to repeaters, small, low-power radios can communicate over thousands of square kilometers. Unfortunately, many amateurs are *not* familiar with the best repeaters that have ever existed. These are the amateur satellites that hams have been using for 40 years. (See the sidebar "Tired of the Same Old QSOs?")

This is essentially the function of an amateur satellite as well. Of course, while a repeater antenna may be up to a few hundred meters above the surrounding terrain, the satellite is hundreds or thousands of *kilometers* above the surface of the Earth. The area of the Earth that the satellite's signals can reach is therefore much larger than the coverage area of even the best Earth-bound repeaters. It is this characteristic of satellites that makes them attractive for communication. Most amateur satellites act as analog repeaters, retransmitting CW and voice signals exactly as they are received, as packet store-and-forward systems that receive whole messages from ground stations for later relay, or as specialized Earth-looking camera systems that can provide some spectacular views. See **Fig 23.2**, an image of a town in the western US.

Amateur satellites have a long history of performing worldwide communications services for amateurs. See the sidebar "Amateur Satellite History."

N1JEZ

Fig 23.1—N1JEZ's portable microwave satellite station.

LINEAR TRANSPONDERS AND THE PROBLEM OF POWER

Most analog satellites are equipped with *linear transponders*. These are devices that retransmit signals within a band of frequencies, usually 50 to 250 kHz wide, known as the *passband*. Since the linear transponder retransmits the entire band, a number of signals may be retransmitted simultaneously. For example, if three SSB signals (each separated by as little

Tired of the Same Old QSOs? Break out of Orbit and Set your Course for the "Final Frontier"

Satellite-active hams comprise a relatively small segment of our hobby, primarily because of an unfortunate fiction that has been circulating for many years—the myth that operating through amateur satellites is overly difficult and expensive.

Like any other facet of Amateur Radio, satellite hamming is as expensive as you allow it to become. If you want to equip your home with a satellite communication station that would make a NASA engineer blush, it will be expensive. If you want to simply communicate with a few low-Earth-orbiting birds using less-than-state-of-the-art gear, a satellite station is no more expensive than a typical HF or VHF setup. In many cases you can communicate with satellites using your present station equipment—no additional purchases are necessary.

Does satellite hamming impose a steep learning curve? Not really. You have to do a bit of work and invest some brain power to be successful, but the same can be said of DXing, contesting, traffic handling, digital operating or any other specialized endeavor. You are, after all, communicating with a *spacecraft!*

The rewards for your efforts are substantial, making satellite operating one of the most exciting pursuits in Amateur Radio. There is nothing like the thrill of hearing someone responding to your call from a thousand miles away and knowing that he heard you through a satellite. (The same goes for the spooky, spellbinding effect of hearing your own voice echoing through a spacecraft as it streaks through the blackness of space.) Satellite hamming will pump the life back into your radio experience and give you new goals to conquer.

Fig 23.2—UO-36 captured this image of a well-known city in the western US. Need a hint? Think "Caesar's Palace."

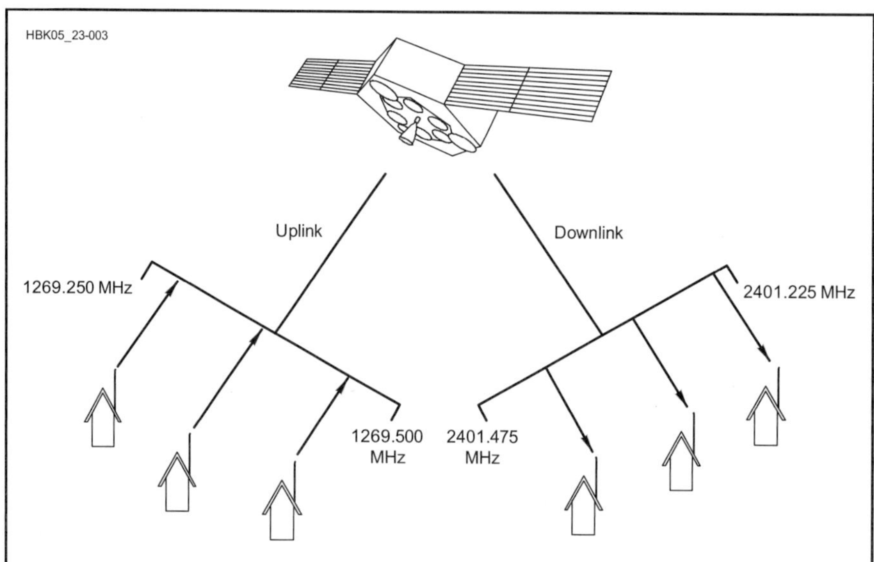

Fig 23.3—A linear transponder acts much like a repeater, except that it relays an entire group of signals, not just one signal at a time. In this example the satellite is receiving three signals on its 23-cm uplink passband and retransmitting them on its 13-cm downlink passband.

as 5 kHz) were transmitted to the satellite, the satellite would retransmit all three signals—still separated by 5 kHz each (see **Fig 23.3**). Just like a terrestrial repeater, the retransmissions take place on frequencies that are different from the ones on which the signals were originally received.

Some linear transponders invert the uplink signals. In other words, if you transmit to the satellite at the *bottom* of the uplink passband, your signal will appear at the *top* of the downlink passband. In addition, if you transmit in lower sideband (LSB), your downlink signal will be in upper sideband (USB). See **Fig 23.4**. Satellite passbands are usually operated according to the courtesy band plan, as shown. Transceivers designed for satellite use usually include features that cope with this confusing flip-flop.

Over the years, the number of amateur bands available on satellites has increased. To help in easily identifying these bands, a system of "Modes" has been created. In the early years reference to these Modes was by a single letter (Mode A, Mode B, etc), but with the launch of more satellites the opportunities greatly increased and it was necessary to show both the uplink and downlink bands. See **Table 23.1,** Satellite Operating Modes.

Linear transponders can repeat any type of signal, but those used by amateur satellites are primarily designed for SSB and CW. The reason for the SSB and CW preference has a lot to do with the hassle of generating power in space. Amateur satellites are powered by batteries, which are recharged by solar cells. "Space rated" solar arrays and batteries are very expensive. They are also heavy and tend to take up a substantial amount of space. Thanks to meager funding, hams don't have the luxury of launching satellites with large power systems such as those used by commercial birds. We have to do the best we

Table 23.1
Satellite Operating Modes

Frequency Band	*Letter Designation*	*New Designation* (Transmit, First Letter; Receive, Second Letter)	*Old Designation*
HF, 21-30 MHz	H	Mode U/V	Mode B
VHF, 144-148 MHz	V	Mode V/U	Mode J
UHF, 435-438 MHz	U	Mode U/S	Mode S
1.26-1.27 GHz	L	Mode L/U	Mode L
2.40-2.45 GHz	S	Mode V/H	Mode A
5.6 GHz	C	Mode H/S	
10.4 GHz	X	Mode L/S	
24 GHz	K	Mode L/X	
		Mode C/X	

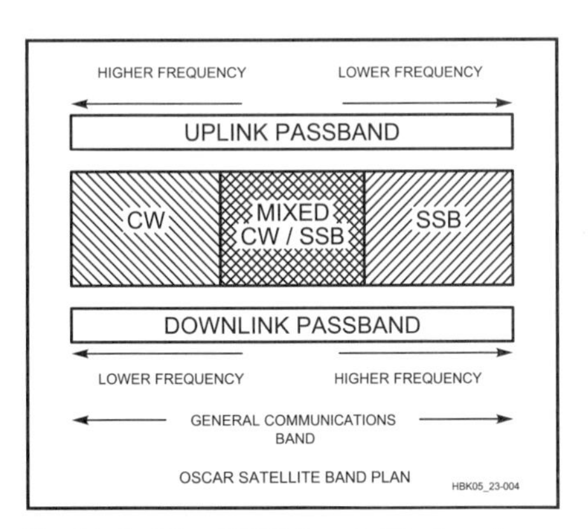

Fig 23.4—The OSCAR satellite band plan allows for CW-only, mixed CW/SSB, and SSB-only operation. Courteous operators observe this voluntary band plan at all times.

can within a much more limited "power budget."

So what does this have to do with SSB or any other mode? Think *duty cycle*—the amount of time a transmitter operates at full output. With SSB and CW the duty cycle is quite low. A linear satellite transponder can retransmit many SSB and CW signals while still operating within the power generating limitations of an amateur satellite. It hardly breaks a sweat.

Now consider FM. An FM transmitter operates at a 100% duty cycle, which means it is generating its full output with every transmission. Imagine how much power a linear transponder would need to retransmit, say, a dozen FM signals—all demanding 100% output!

Having said all that, there *are* a few, very popular FM repeater satellites. However, they do not use linear transponders. They retransmit only one signal at a time.

FINDING A SATELLITE

Before you can communicate through a satellite, you have to know what satellites are available and when they are available. (See sidebar "Current Amateur Satellites.") This isn't quite as straightforward as it seems.

Amateur satellites do not travel in geostationary orbits like many commercial and military spacecraft. Satellites in geostationary orbits cruise above the Earth's equator at an altitude of about 35,000 kilometers. From this vantage point the satellites can "see" almost half of our planet. Their speed in orbit matches the rotational speed of the Earth itself, so the satellites appear to be "parked" at fixed positions in the sky. They are available to send and receive signals 24 hours a day over an enormous area.

Of course, amateur satellites *could* be placed in geostationary orbits. The problem isn't one of physics; it's money and politics. Placing a satellite in geostationary orbit and keeping it on station costs a great deal of money—more than any one amateur satellite organization can afford. An amateur satellite group could ask similar groups in other areas of the world to contribute money tc a geostationary satellite project, but why should they? Would you contribute large sums of money to a satellite that may never "see" your part of the world? Unless you are blessed with phenomenal generosity, it would seem unlikely!

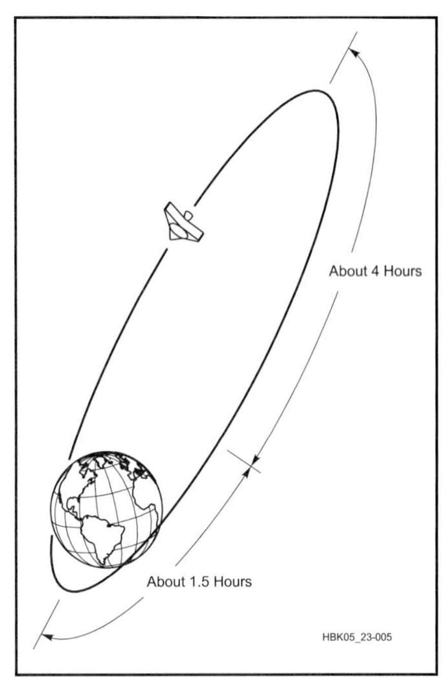

Fig 23.5—An example of a satellite in a high, elliptical orbit.

Instead, all amateur satellites are either low-Earth orbiters (LEO), or they travel in very high, elongated orbits. See **Fig 23.5**. Either way, they are not in fixed positions in the sky. Their positions relative to your station change constantly as the satellites zip around the Earth. This means that you need to predict when satellites will appear in your area, and what paths they'll take as they move across your local sky.

You'll be pleased to know that there is software available that handles this prediction task very nicely. A bare-bones program will provide a schedule for the satellite you choose. A very simple schedule might look something like **Fig 23.6**, showing the antenna pointing angles for each minute of a pass for AO-16.

The time is usually expressed in UTC.

ORBIT NO. 20988 EPOCH: 30.219444444384

UTC	Az	El	D/L Dplr	U/L Dplr	Range	Height	SSP Lat	SSP Long	MA
05:16:00	164	0	9761	-3259	3222	795	1	90	85
05:17:00	164	4	9757	-3258	2821	796	4	91	87
05:18:00	164	9	9694	-3237	2422	796	8	92	90
05:19:00	164	14	9543	-3187	2029	797	12	92	93
05:20:00	163	22	9235	-3083	1649	797	15	93	95
05:21:00	162	33	8601	-2872	1295	797	19	94	98
05:22:00	160	50	7232	-2415	997	798	22	95	100
05:23:00	145	76	4274	-1427	821	798	26	96	103
05:24:00	0	70	-634	211	847	798	29	97	105
05:25:00	352	46	-5179	1728	1061	798	33	98	108
05:26:00	350	31	-7660	2557	1376	799	36	99	110
05:27:00	349	20	-8791	2934	1738	799	40	100	113
05:28:00	348	13	-9319	3110	2121	799	43	101	115
05:29:00	348	8	-9578	3197	2515	799	47	102	118
05:30:00	348	3	-9700	3237	2915	799	50	104	121
05:31:00	348	0	-9746	3253	3316	799	54	106	123

Fig 23.6—Tabular output from an orbit prediction program showing time and position information for AO-16.

Amateur Satellite History

The Amateur Radio satellite program began with the design, construction and launch of OSCAR I in 1961 under the auspices of the Project OSCAR Association in California. The acronym "OSCAR," which has been attached to almost all Amateur Radio satellite designations on a worldwide basis, stands for ***O**rbiting **S**atellite **C**arrying **A**mateur **R**adio.* Project OSCAR was instrumental in organizing the construction of the next three Amateur Radio satellites—OSCARs II, III and IV. *The Radio Amateur's Satellite Handbook*, published by ARRL has details of the early days of the amateur space program.

In 1969, the Radio Amateur Satellite Corporation (AMSAT) was formed in Washington, DC. AMSAT has participated in the vast majority of amateur satellite projects, both in the United States and internationally, beginning with the launch of OSCAR 5. Now, many countries have their own AMSAT organizations, such as AMSAT-UK in England, AMSAT-DL in Germany, BRAMSAT in Brazil and AMSAT-LU in Argentina. All of these organizations operate independently but may cooperate on large satellite projects and other items of interest to the worldwide Amateur Radio satellite community. Because of the many AMSAT organizations now in existence, the US AMSAT organization is frequently designated AMSAT-NA.

Beginning with OSCAR 6, amateurs started to enjoy the use of satellites with lifetimes measured in years as opposed to weeks or months. The operational lives of OSCARs 6, 7, 8 and 9, for example, ranged between four and eight years. All of these satellites were low Earth orbiting (LEO) with altitudes approximately 800-1200 km. LEO Amateur Radio satellites have also been launched by other groups not associated with any AMSAT organization such as the Radio Sputniks 1-8 and the ISKRA 2 and 3 satellites launched by the former Soviet Union.

The short-lifetime LEO satellites (OSCARs I through IV and 5) are sometimes designated the *Phase I* satellites, while the long-lifetime LEO satellites are sometimes called the *Phase II* satellites. There are other conventions in satellite naming that are useful to know. First, it is common practice to have one designation for a satellite before launch and another after it is successfully launched. Thus, OSCAR 40 (discussed later) was known as Phase 3D before launch. Next, the AMSAT designator may be added to the name, for example, AMSAT-OSCAR 40, or just AO-40 for short. Finally, some other designator may replace the AMSAT designator such as the case with Japanese-built Fuji-OSCAR 29 (FO-29).

In order to provide wider coverage areas for longer time periods, the high-altitude Phase 3 series was initiated. Phase 3 satellites often provide 8-12 hours of communications for a large part of the Northern Hemisphere. After losing the first satellite of the Phase 3 series to a launch vehicle failure in 1980, AO-10 was successfully launched and became operational in 1983. AO-13, the follow-up to the AO-10 mission, was launched in 1988 and re-entered the atmosphere in 1996. The successor to AO-13, AO-40 was launched on November 16, 2000 from Kourou, French Guiana.

Satellites providing store-and-forward communication services using packet radio techniques are generically called *PACSATs.* Files stored in a PACSAT message system can be anything from plain ASCII text to digitized pictures and voice.

The first satellite with a digital store-and-forward feature was UoSAT-OSCAR 11. UO-11's Digital Communications Experiment (DCE) was not open to the general Amateur Radio community although it was utilized by designated "gateway" stations. The first satellite with store-and-forward capability open to all amateurs was the Japanese Fuji-OSCAR 12 satellite, launched in 1986. FO-12 was succeeded by FO-20, launched in 1990, and FO-29, launched in 1996.

By far the most popular store-and-forward satellites are the *PACSATs* utilizing the PACSAT Broadcast Protocol. These PACSATs fall into two general categories — the *Microsats,* based on technology developed by AMSAT-NA, and the *UOSATs,* based on technology developed by the University of Surrey in the UK. While both types are physically small spacecraft, the Microsats represent a truly innovative design in terms of size and capability. A typical Microsat is a cube measuring 23 cm (9 in) on a side and weighing about 10 kg (22 lb). The satellite will contain an onboard computer, enough RAM for the message storage, two to three transmitters, a multichannel receiver, telemetry system, batteries and the battery charging/power conditioning system.

Amateur Radio satellites have evolved to provide three primary types of communication services — analog transponders for real-time CW and SSB communication, digital store-and-forward for non real-time communication, and direct "bent-pipe" single-channel FM repeaters. Which of these types interest you the most will probably depend on your current Amateur Radio operating habits. Whatever your preference, this section should provide the information to help you make a successful entry into the specialty of amateur satellite communications.

AO-16 will appear above your horizon beginning at 0516 UTC on January 30. The bird will "rise" at an azimuth of 164°, or approximately south-southeast of your station. The elevation refers to the satellite's position above your horizon in degrees—the higher the better. A zero-degree elevation is right on the horizon; 90° is directly overhead.

By looking at this schedule you can see that the satellite will appear in your south-southeastern sky at 0516 UTC and will rise quickly to an elevation of 70° by 0524. The satellite's path will curve further to the east and then directly to the north as it rises. Notice how the azimuth shifts from 164° at 0516 UTC to 0° at 0524. This is nearly a direct overhead pass of AO-16 and it sets in the north-northwest at 348°.

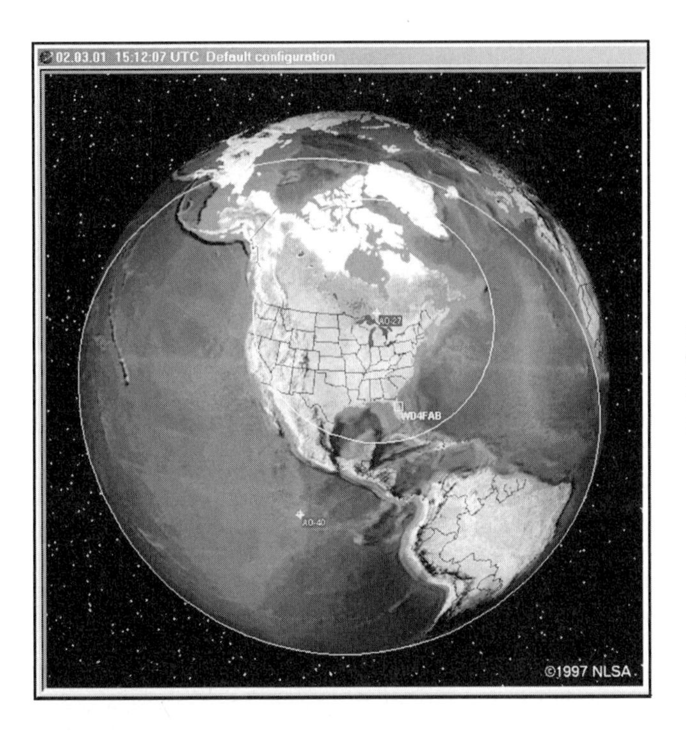

Fig 23.7—The communications range circles, or "footprints" over North America.

The more sophisticated the software, the more information it usually provides in the schedule table. The software may also display the satellite's position graphically as a moving object superimposed on a map of the world. Some of the displays used by satellite prediction software are visually stunning! This view, **Fig 23.7,** is provided by *Nova* for *Windows*, from Northern Lights Software Associates.

Satellite prediction software is widely available on the Web. Some of the simpler programs are freeware. The AMSAT-NA Web site has the largest collection of satellite software for just about any computer you can imagine. Most AMSAT software isn't free, but the cost is reasonable and the funds support amateur satellite programs.

Whichever software you choose, there are two key pieces of information you must provide before you can use the programs:

(1) **Your position**. The software must have your latitude and longitude before it can crank out predictions for your sta-

Current Operational Amateur Satellites

OSCAR 7, AO-7, was launched November 15, 1974 by a Delta 2310 from Vandenberg, CA. AO-7's operating status is semi-operational in sunlight only. After being declared dead in mid 1981 due to battery failure, AO-7 has miraculously sprung back to life. It will only be on when in sunlight and off in eclipse. AO-7 will reset each orbit and may not turn on each time.

OSCAR 11, UO-11, a scientific/educational low-orbit satellite, was built at the University of Surrey in England and launched on March 1, 1984. This UoSat spacecraft has also demonstrated the feasibility of store-and-forward packet digital communications and is operational with telemetry downlinks only.

OSCAR 16, AO-16, also known as PACSAT, was launched in January 1990. A digital store-and-forward packet radio file server, it has an experimental S-band beacon at 2401.143 MHz. AO-16 is only semi-operational with the 1200-baud digipeater for APRS service.

OSCAR 26, IO-26, was launched on September 26, 1993 is semi-operational and now serves as a 1200-baud digipeater for APRS service.

RS 15, launched in December 1994, is a Mode V/H spacecraft; its uplink is on the 2m band, and its downlink is on 10m.

OSCAR 27, AO-27, was launched in September 1993 along with OSCAR 26. It features a mode V/U analog FM repeater. Because of the need to conserve power, OSCAR 27 is usually only available during daylight passes.

OSCAR 29, FO-29, launched from Japan in 1996, in a low earth orbit. It operates with a mode V/U analog transponder.

OSCAR 44, NO-44, also known as PCSAT was launched on September 30, 2001 from Kodiak, Alaska. PCSAT is a 1200-baud APRS digipeater designed for use by stations using hand-held or mobile transceivers. The operational status of PCSAT is uncertain and subject to change due to power availability.

OSCAR 50, SO-50 also known as SAUDISAT-1C, was launched December 20, 2002 aboard a converted Soviet ballistic missile from Baikonur Cosmodrome. SO-50 carries several experiments, including a mode U/V FM amateur repeater. The repeater is available to amateurs as power permits, using a 67.0 Hz uplink tone for on-demand activation.

OSCAR 51, also known as *Echo*, was launched on June 29, 2004 from the Baikonur Cosmodrome. Echo carries a FM repeater capable of 144 MHz and/or 1.2 GHz uplink with a 435 MHz and/or 2.4GHz downlink. The satellite also includes an AX.25 digital PACSAT BBS and a PSK31 uplink on 28 MHz.

VUSat-OSCAR 52 was launched on May 5, 2005. Within 24 hours its Mode U/V transponder was open for business and hams were reporting excellent signals. The first Indian Amateur Radio satellite carries two 1-W linear transponders for SSB and CW communication, although only one transponder is operational at a time. OSCAR 52 travels in a polar sun-synchronous orbit at an altitude of 632 × 621 km with an inclination of 97.8 deg with respect to the equator.

tion. The good news is that your position information doesn't need to be extremely accurate. Just find out the latitude and longitude of your city or town (the public library would have this data, as would any nearby airport) and plug it into the program.

(2) **Orbital elements**. This is the information that describes the orbits of the satellites. You can find orbital elements (often referred to as *Keplerian elements*) at the AMSAT Web site, and through many other sources on the Internet. You need to update the elements every few months. Many satellite programs will automatically read in the elements if they are provided as ASCII text files. The less sophisticated programs will require you to enter them by hand. Automatically updated software is highly recommended; it's too easy to make a mistake with manual entries.

GETTING STARTED WITH THE FM BIRDS

Do you like elevated FM repeaters with wide coverage areas? Then check out the AO-27 and AO-51 FM repeater satellites. From their low-Earth orbits these satellites can hear stations within a radius of 2000 miles in all directions.

You can operate the FM satellites with a basic dual band VHF/UHF FM transceiver and even a good FM HT, as some amateurs have managed. Assuming that the transceiver is reasonably sensitive, you can use a good "rubber duck" antenna as in **Fig 23.8**. Some amateurs have even managed to work the FM birds with HTs coupled to multi-element directional antennas such as the popular Arrow Antenna, **Fig 23.9**. Of course, this means they must aim their antennas at the satellites as they cross overhead.

Fig 23.8—W2RS shows another popular LEO FM antenna for hand-held operations.

High quality omnidirectional antennas for LEO service come in quite a number of forms and shapes. M^2 Enterprises has their EB-144 and EB-432 Eggbeater antennas, **Fig 23.10** and **Fig 23.11**, which have proven to be very useful and do not require any rotators for control. The Quadrifilar omnidirectional antenna, **Fig 23.12**, has been around for a long time, as has the turnstile-over-reflector antenna, **Fig 23.13**.

For even better performance, at the modest cost of a simple TV antenna rotator, check out the fixed elevation Texas Potato Masher antenna by K5OE, **Fig 23.14**. This antenna provides a dual band solution for medium gain directional antennas for LEO satellite operations. This is a considerable improvement over omnidirectional antennas and does not require an elevation control for good performance.

Start by booting your satellite tracking software. Check for a pass with a peak elevation of 30° or higher. As with all satellites, the higher the elevation, the better. If you plan to operate outdoors or away from home, either print the schedule to a printer or jot down the times on a piece of scrap paper that you can keep with you.

When the satellite comes into range, you'll be receiving its signal about 5 kHz higher than the published downlink frequency (see **Table 23.2**, Active Amateur Satellites: Frequencies and Modes) thanks to *Doppler shifting* (see the sidebar, "Down with Doppler"). So, begin listening on the higher frequency. If you suddenly hear the noise level dropping, chances are you are picking up the satellite's signal. At about the midpoint of the pass you'll need to shift your receiver down to the published frequency, and as the satellite is heading away you may wind up stepping down another 5 kHz. Some operators program these frequency steps into memory channels so that they can compensate for Doppler shift at the push of a button.

Once again, these FM satellites behave just like terrestrial FM repeaters. Only one person at a time can talk. If two or more people transmit simultaneously, the result is garbled audio or a squealing sound on the output. The trick is to take turns and keep the conversations short. Even the best passes will only give you about 15 minutes to use the satellite. If you strike up a conversation, don't forget that there are others waiting to use the bird.

The FM repeater satellites are a good way to get started. Once you get your feet

Fig 23.9—The hand-held "Arrow" gain antenna is popular for LEO FM operations. (Photo courtesy *The AMSAT Journal*, Sep/Oct 1998.)

Fig 23.10—Eggbeater antennas are popular for base station LEO satellite operation. This EB144 eggbeater is for 2 m.

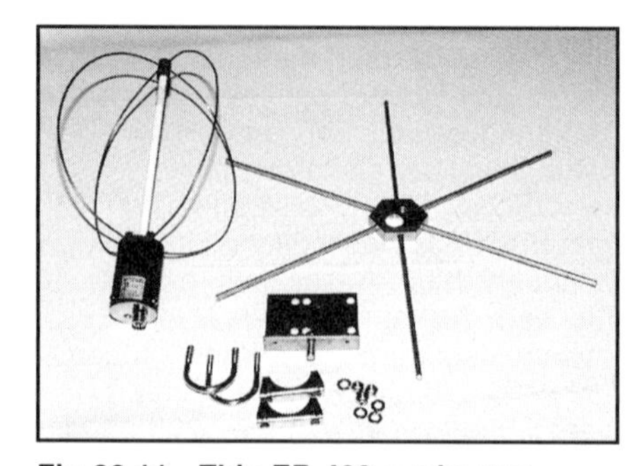

Fig 23.11—This EB-432 eggbeater antenna for 70-cm operation is small enough to put in an attic. Antenna gain pattern is helped with the radials placed below the antenna.

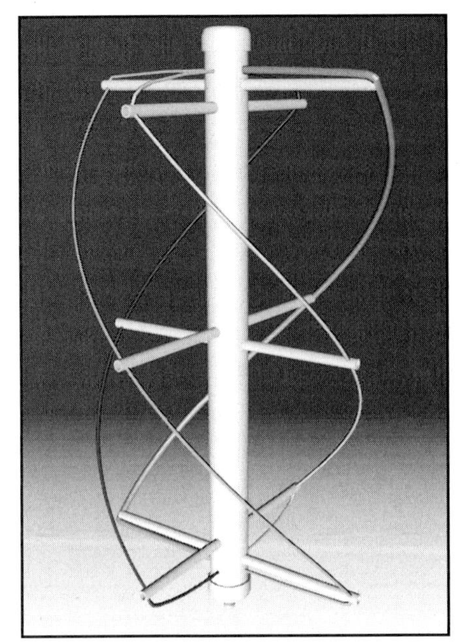

Fig 23.12—W3KH suggests that quadrifilar antennas can serve well for omnidirectional satellite station antenna service.

Table 23.2 – Active Amateur Satellites: Frequencies and Modes

Satellite (SSB/CW)	*Uplink (MHz)*	*Downlink (MHz)*
FO-29	145.900-146.000	435.800-435.900 435.795 (CW beacon)
RS-15	145.858-145.898	29.354-29.394
VO-52	435.225-435.275	145.875-145.925
Packet—1200 bit/s		
AO-16	145.90, .92, .94, .96	437.051
NO-44	145.827	144.390
Packet—9600 bit/s		
GO-32	890/145.850	435.225
FM Voice Repeaters		
AO-51	145.920	435.300
AO-27	145.850	436.795
(Daylight passes *only*)]		

wet, you'll probably wish you could access a satellite that wasn't so crowded, where you could chat for as long as the bird was in range. You may find that a very good assist will be given to you from the book *Working the Easy Sats*, by WA4YMZ and N1JEZ. This is available from AMSAT headquarters.[3] But now it is time to move up!

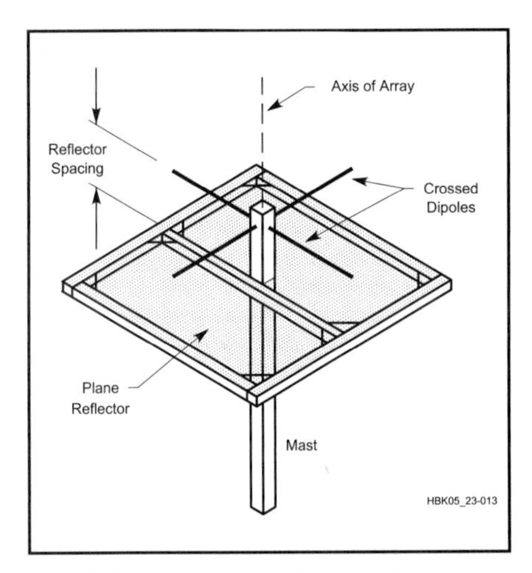

Fig 23.13—The Turnstile Over Reflector antenna has served well for LEO satellite service for a number of years.

MOVING UP TO OSCARS 29 AND 52

VUSat-OSCAR 52 carries a *Mode U/V* transponder, which means that it receives signals on the 70-cm band and retransmits on the 2-m band. Operating this satellite is done with the equipment and antenna setup shown in **Fig 23.15**.

Once you've determined when the satellite is due to rise above the horizon at your location, listen for the satellite's CW telemetry beacon. This signal is transmitted constantly by the satellite and carries information about the state of the satellite's systems, such as its battery voltage, solar-panel currents, temperatures and so on. You should hear it just as the satellite rises above the horizon. As soon as you can hear the beacon, start tuning across the downlink passband.

On an active day you should pick up several signals. They will sound like normal amateur SSB and CW conversations. Nothing unusual about them at all—except that the signals will be slowly drifting downward in frequency. That's the effect of Doppler shift.

Fig 23.14—Jerry Brown, K5OE, uses his Texas Potato Masher antennas for LEO satellite operations.

Now tune your transmitter's frequency to the satellite's uplink passband. OSCAR 52 uses inverting transponders. If you transmit at the low end of the uplink passband, you can expect to hear your signal at the high end of the downlink passband. Generally speaking, CW operators occupy the lower half of the transponder passband while SSB enthusiasts use the upper half, as was shown in Fig 23.4.

Assuming that you cannot hear your own signal from the satellite on a separate receiver, the best thing to do is make your best guess as to where your signal will appear on the downlink and set your receive frequency accordingly. Send several brief CQs ("CQ OSCAR 52, CQ OSCAR 52..."), tuning "generously" around your guesstimated receive frequency after each

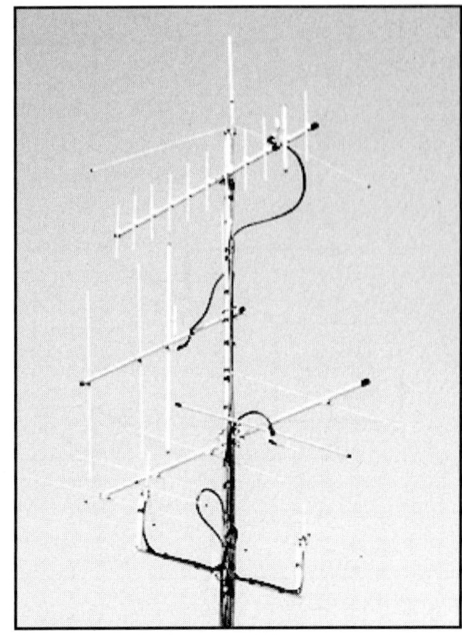

Fig 23.15—Simple ground plane and Yagi antennas can be used for low-Earth-orbit (LEO) satellite contacts.

one. The station that is answering your call will also be making his or her best guess about where you are listening.

The Japanese Fuji OSCAR FO-29, is also a linear transponder bird that functions much like OSCAR 52. The main difference is that it listens on 2 m and retransmits on 70 cm, Mode V/U. (FO-29 also uses inverting transponders.) Few amateurs own receivers that can listen for 70 cm CW and SSB, so this satellite is not very active. During weekend passes, however, you should be able to hear several conversations taking place.

Station Requirements for the OSCAR 29 and 52 Satellites

To work OSCAR 52 you'll need, at minimum, a multiband VHF/UHF SSB or CW transceiver. You *do not* need an amplifier; 50 W is more than enough power for the uplink. In fact, even 50 W may be too much in many instances. The rule of thumb is that your signal on the downlink should never be stronger than the satellite's own telemetry beacon.

The ability to hear yourself simultaneously on the downlink is a tremendous asset for working any satellite. It allows you to operate full duplex as you listen to the Doppler shifting of your own signal, giving you the opportunity to immediately tweak your transmit frequency to compensate (rather than fishing for contacts using the haphazard half-duplex procedure described earlier).

To work OSCAR 52 in Mode U/V, you'll need a 2-m multimode transceiver that can operate in CW or SSB. Remember that OSCAR 52 is listening for signals on 70 cm and retransmitting on 2 m. This means that also need a 70 cm transceiver. Choose your radios carefully. A number of modern HF transceivers also include 2 m and even 70 cm. The problem, however, is that some

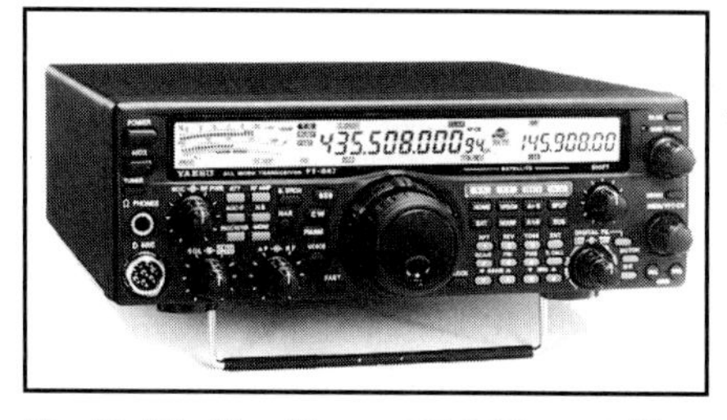

Fig 23.17—The Yaesu FT-847's satellite mode provides the user with a full featured satellite transceiver. As a bonus, it also covers all the HF bands!

Fig 23.16—VUSat-OSCAR 52 heads for orbit!

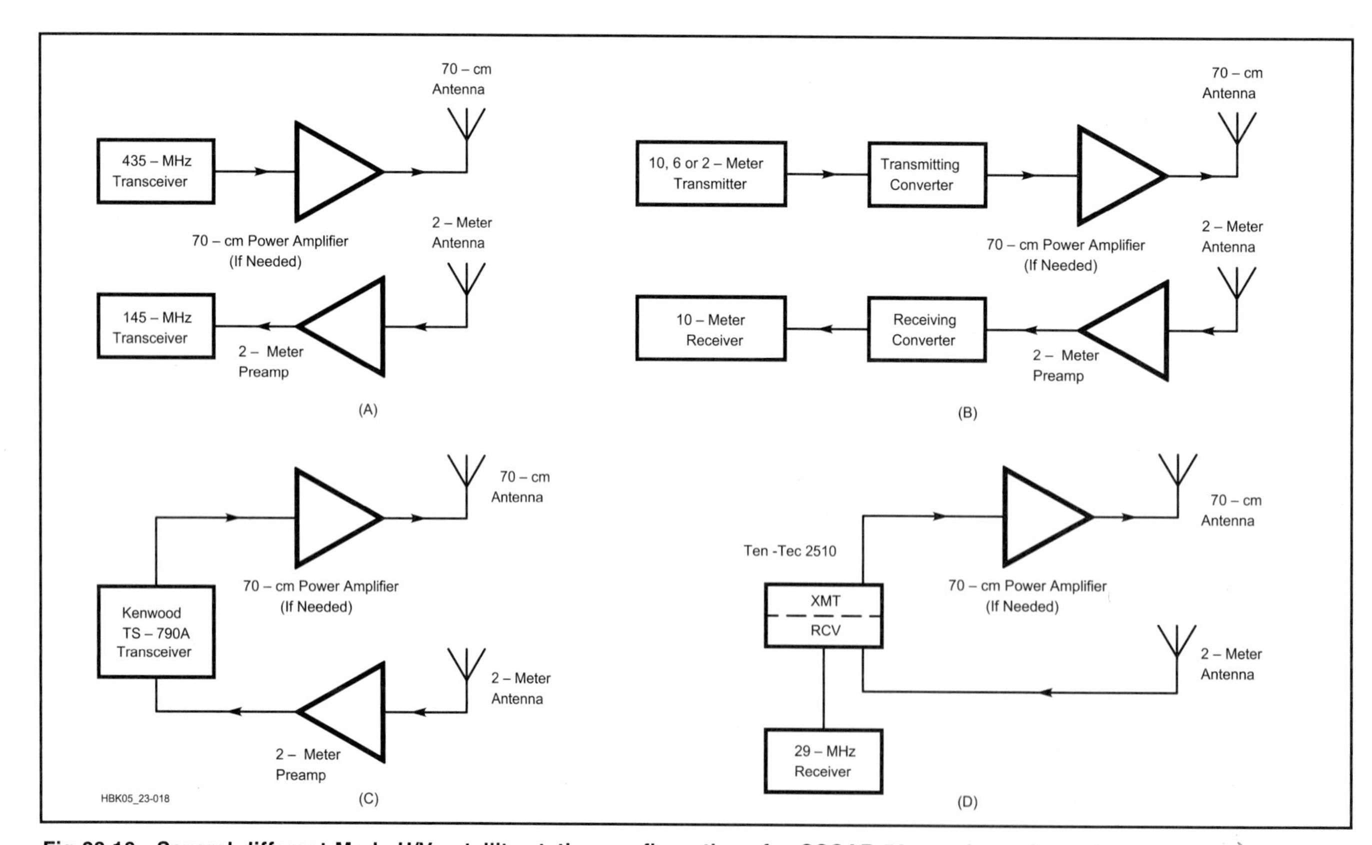

Fig 23.18—Several different Mode-U/V satellite-station configurations for OSCAR 52 are shown here. At A, separate VHF/UHF multimode transceivers are used for transmitting and receiving. The configuration shown at B uses transmitting and receiving converters or transverters with HF equipment. At C, a multimode, multiband transceiver can perform both transmitting and receiving function, full duplex, in one package. The Ten-Tec 2510 shown at D contains a 435-MHz transmitter and a 2-m to 10-m receiving converter.

of these radios do not allow *crossband splits* between VHF and HF. That is, they won't allow you to transmit on 70 cm and receive on 2 m. At the very least they won't allow you to do this simultaneously. One solution that will provide full duplex service for Mode U/V is the Yaesu FT-847 transceiver, **Fig 23.17**, all in one neat package.

Omnidirectional antennas for 2 m are sufficient for listening to OSCAR 52. A beam on 2 m would be even better, but then you incur the cost of an antenna rotator that can move the antenna up and down as well as side to side—the so-called *azimuth/elevation rotator*, or Az-El rotators.

For FO-29 the ability to transmit and receive simultaneously is a must. The Doppler effect is pronounced on the 70 cm downlink. You need to listen to your own signal continuously, making small adjustments to your 2-m uplink so your voice or CW note does not slide rapidly downward in frequency. To achieve this you will need separate 2-m and 70-cm transceivers (such as a couple of used rigs), or a dual band transceiver specifically designed for satellite use. **Fig 23.18** and **Fig 23.19** show several possible combinations of equipment for satellite operations on 2 m and 70 cm. Kenwood, ICOM and Yaesu have such radios in their product lines. These wondrous rigs make satellite operating a breeze, although their price tags may give you a bit of sticker shock (about $1600). They feature full crossband duplex, meaning that you can transmit on 2 m at the same time you are listening on 70 cm. They even have the ability to work with inverting transponders automatically. That is, as you move your receive frequency down, the transmit VFO will automatically move up (and vice versa)!

Although beam antennas and azimuth/elevation rotators are not strictly necessary to work FO-29, they vastly improve the quality of your signal. A good compromise, medium-gain antenna for 2 m and 70 cm is shown in Fig 23.14. If you decide

Down with Doppler

The relative motion between you and the satellite causes *Doppler shifting* of signals. As the satellite moves toward you, the frequency of the downlink signals will increase as the velocity of the satellite adds to the velocity of the transmitted signal. As the satellite passes overhead and starts to move away from you, the frequency will drop, much the same way as the tone of a car horn or a train whistle drops as the vehicle moves past the observer.

The Doppler effect is different for stations located at different distances from the satellite because the relative velocity of the satellite with respect to the observer is dependent on the observer's distance from the satellite. The result is that signals passing through the satellite transponder shift slowly around the published downlink frequency. Your job is to tune your uplink transmitter—*not your receiver*—to compensate for Doppler shifting and keep your frequency relatively stable on the downlink. That's why it is helpful to hear your own signal coming through the satellite. If you and the station you're talking to both compensate correctly, your conversation will stay at one frequency on the downlink throughout the pass. If you don't compensate, your signals will drift through the downlink passband as you attempt to "follow" each other. This is highly annoying to others using the satellite because your drifting signals may drift into their conversations.

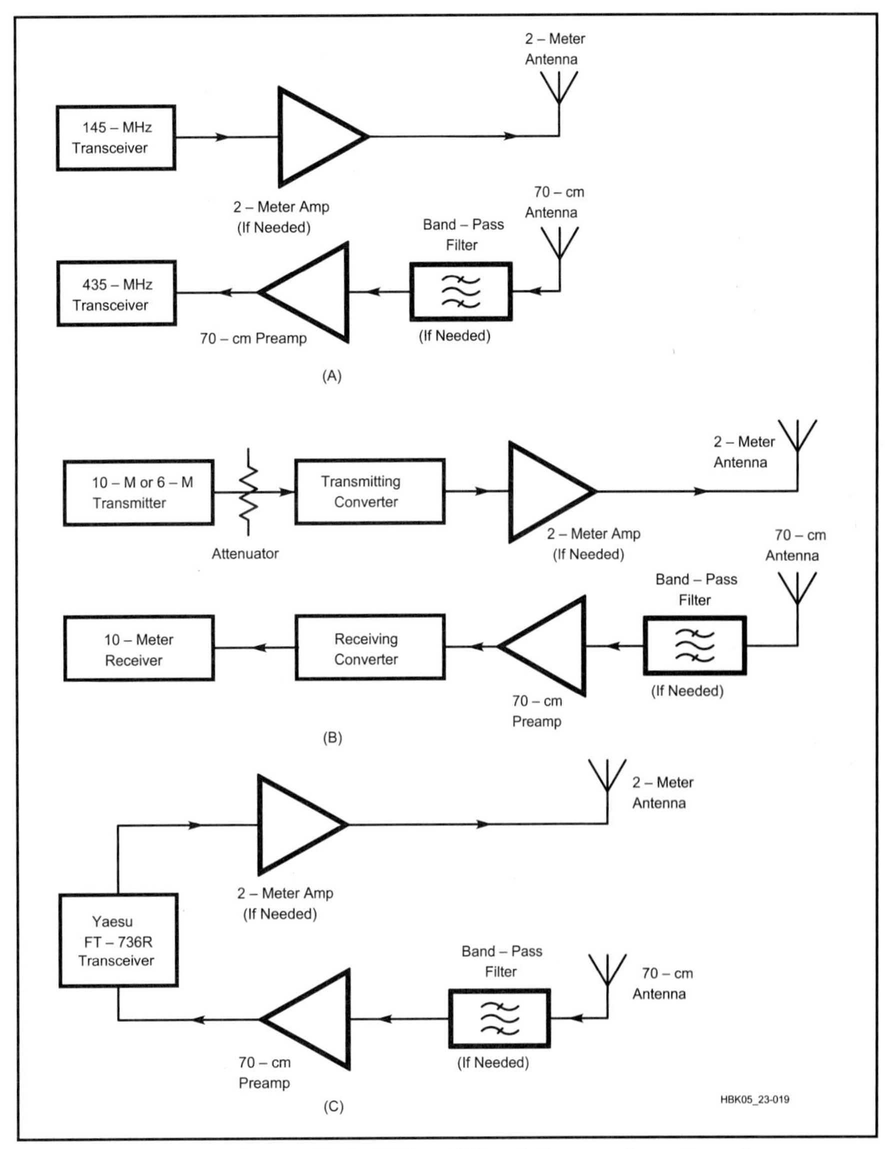

Fig 23.19—Several different Mode-V/U satellite-station configurations for OSCAR 29 are shown here. At A, separate VHF/UHF multimode transceivers are used for transmitting and receiving. The configuration shown at B uses transmitting and receiving converters or transverters with HF equipment. At C, a multimode, multiband transceiver can perform both transmitting and receiving functions, full duplex, in one package.

to go the omnidirectional route, you'll need to add a 70-cm receive preamp at the antenna to boost the downlink signal.

PHASE 3E—THE NEXT GENERATION

The next Phase 3-series satellite planned is Phase 3E or P3E. It is built on a P3C (AO-13) platform. As this *Handbook* went to press, P3E is expected to launch within the next two years. Currently, the planned transmit RF operations on P3E are R Band (47 GHz) very low power — 0.5 W PEP; K Band (24 GHz) moderate power — 5 W PEP; X Band (10.5 GHz) very low, medium and high power — up to 10 W; S Band (2.4 GHz) high power — 25 W PEP, and V Band (145 MHz) high power. Generally, *high power* here means RF output from the satellite in the 5- to 25-W PEP range using mostly linear transmitters that incorporate HELAPS technology. Very low power on X Band is to simulate the signal strength that Earth stations will see from a P5A (Mars mission) transmitter near the red planet. There are other modules on board — such as a star sensor — that are prototypes for the upcoming P5A mission. Receive frequencies on P3E will be in C Band (5.6 GHz); S Band (2.45 GHz); L Band (1.2 GHz), and U Band (435 MHz). Updates and other useful information on P3E can be found on the German AMSAT website at: **www.amsat-dl.org/p3e**.

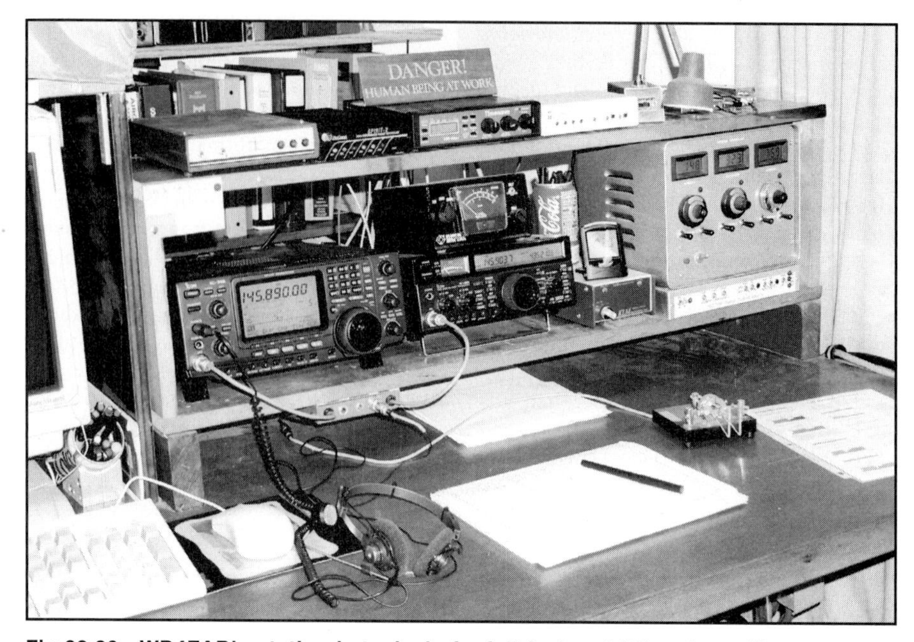

Fig 23.20—WD4FAB's station is typical of a full-featured HF and satellite operation. On the bottom row, L-R are: IC-746 HF to 2 meter transceiver, IC-821H VHF-UHF multimode transceiver topped by a home-brew power meter, KLM 1200GU 1.2-GHz transverter, home-brew multi-antenna rotator controller on top of the station accessory control box. On the top row, L-R are: G3RUH AO-13 Telemetry Demodulator, PacComm SPIRIT2 high performance packet controller, Timewave DSP-599zx audio DSP unit for the VHF-microwave operations through the IC-821H and the TAPR EasyTrak antenna command unit for interfacing the computer tracking to the antenna controller.

SATELLITE GROUND STATIONS

The keys to your enjoyable satellite communications are in the details of your station. One such station is illustrated in **Fig 23.20** and **Fig 23.21**, showing only some of the ways that you can achieve this enjoyable ham radio operation. The following sections will describe that station and the options that are open for you to construct your own satellite station, including discussions of equipment and techniques for transmitting, receiving, antennas, and station accessories, such as audio processing and antenna pointing control.

Transmitting

The 23-cm (L) band is the highest band for which commercial amateur transceivers are readily available. Kits and converters are also available. Commercial equipment, such as the ICOM IC-910H, **Fig 23.22**, with a 23-cm module, the popular Yaesu FT-736R, **Fig 23.23**, with a 23-cm module, and the new Kenwood TS-2000, **Fig 23.24**, with a 23-cm module, all provide about 10 watts of RF output power. Of course multimode transceivers, such as the ICOM IC-821H, **Fig 23.25**, with an outboard 23-cm up-converter will also serve the satellite station very well. See **Table 23.3** for a listing of suppliers of equipment of interest to satellite operators.

Stations that do not have one of the newer transceivers, with their built-in 1.2 GHz band modules, can use a separate transverter or transmitting up-converter to achieve the L-band uplink. Commonly these up-converters employ a 2-m IF for their drive. This generally means that a separate 2-m receiver is needed along with the VHF/UHF transceiver, as often the S-band down-converter also has a 2-m IF. The station shown in Fig 23.20 is of this type, using the 2-m features of the base HF transceiver to provide the IF for the S-band down-converter. Transmitting functions use the VHF/UHF transceiver for uplinks on U band and L band through an up-converter. There are a few L-band-transmitting up-converters on the market just for satellite service. Among these are units from Down East Microwave, with its 1268-144TX; and from SSB-USA, with its MKU130TX; and from Parabolic AB. When you assemble your station, be sure to take the necessary steps to set the RF power drive level needed for your up-converter. This may mean the need for a power attenuator inserted between the transceiver and up-converter.

Operating experience with L-band uplinks has shown that the power and antenna requirements for communications at altitudes up to 40,000 km can be satisfied with 10 W of power *delivered* to a 12-turn helix antenna. This experience has also shown that L-band power levels of 40 W-PEP, or higher, *delivered to the antenna*, along with antenna gains of greater than 20 dBi (4000-5600 W-PEP EIRP) are very workable for operations at the highest altitudes, depending upon the satellite squint angle. A pretty compact L-band antenna arrangement with two 22-element antennas in a stacked array is shown in Fig 23.21. These antennas have a combined gain of about 21.5 dBi. Some other experiences with L-band dish antennas have shown that a 1.2-m offset-fed dish with a helix feed and 10-W of RF power can also provide a superb uplink. This dish antenna will have a practical gain of about 21 dBic, giving an uplink of only 1400 W-PEP EIRP but with RHCP. Circular polarization for L band makes a real uplink difference. These uplinks will provide the user a downlink that is 10-15 dB above the transponder noise floor. In more practical terms this is an S7 to S8 signal over a S3 transponder noise floor in the illustrated station, a very comfortable copy for the capable station.

Fig 23.21—Station antennas at WD4FAB. Mounted are T-B: M2 436-CP30, two stacked M[2] 23CM22EZ, modified PrimeStar dish with homebrew seven-turn Helix feed antenna. The M[2] 2M-CP22 antenna is off the bottom of the photo. See Fig 23.41 for details of the center section.

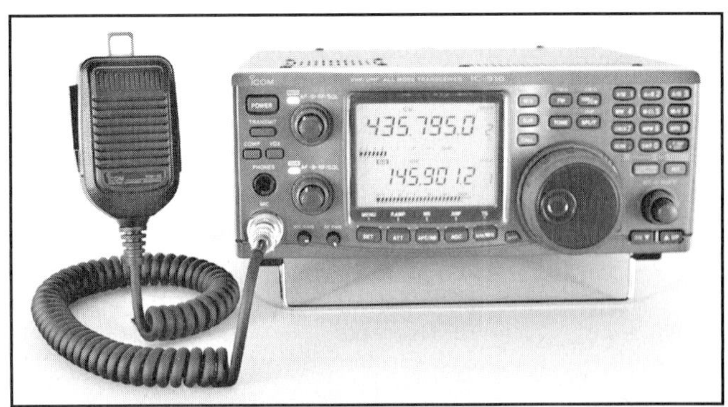

Fig 23.22—ICOM's IC-910H, a recent entry into the multi band VHF/UHF transceiver world, has an available 23-cm module.

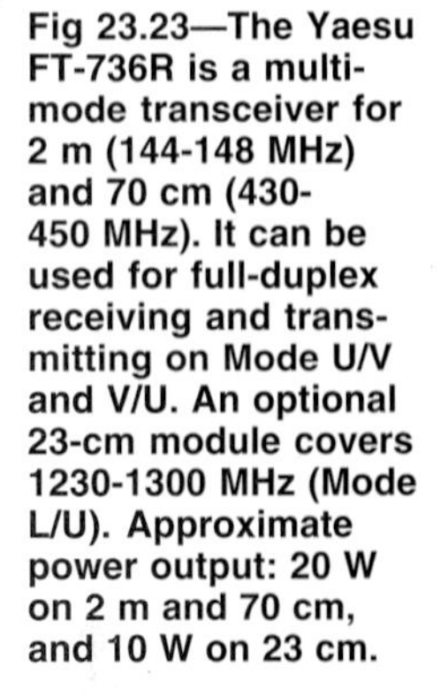

Fig 23.23—The Yaesu FT-736R is a multi-mode transceiver for 2 m (144-148 MHz) and 70 cm (430-450 MHz). It can be used for full-duplex receiving and transmitting on Mode U/V and V/U. An optional 23-cm module covers 1230-1300 MHz (Mode L/U). Approximate power output: 20 W on 2 m and 70 cm, and 10 W on 23 cm.

Fig 23.24—Kenwood's TS-2000 is an all-band, multi-mode transceiver that covers HF, 6 m, 2 m, and 70 cm. A 23-cm (1240-1300 MHz) module is optional. Typical power output is 35-45 W on 144 MHz, 30-40 W on 430 MHz and 10 W on 1240 MHz.

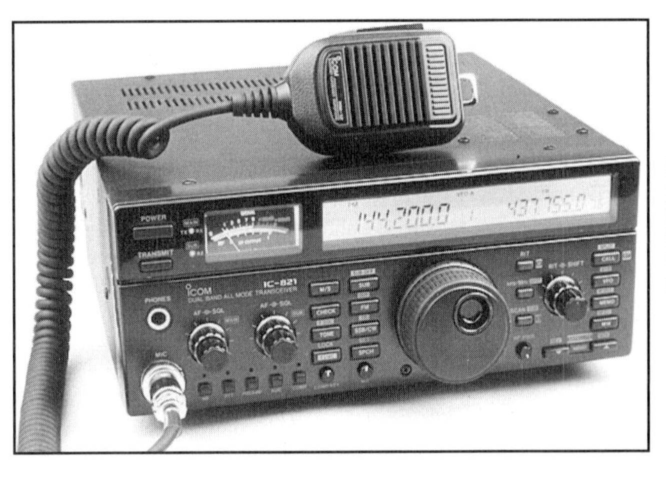

Fig 23.25—ICOM's IC-821H is another full-featured transceiver designed for satellite use.

For the illustrated station, the remotely mounted 23-cm amplifier (see "Double Brick Amplifier Construction") is contained in a tower-mounted-box. The regulated dc power requirements for this amplifier are of too high a current at 13.8 V-dc to bring it up from the ground, as the voltage loss of any reasonable cable would be excessive. Instead, 24 V dc unregulated power is taken from the station power supply before its regulator. This power is brought up the tower and regulated to 13.8 V where it is needed. This power regulator is borrowed from Chapter 17, "28-V, High-Current Power Supply", using only the regulator circuitry, and adjusting the output voltage down to 13.8 V dc.

Much of your station's uplink performance strongly depends upon the satellite's antenna off-pointing, or *squint* angle. L-band and U-band receiving antennas are usually high-gain arrays, with measured gains of 16 dBic and 15 dBic, respectively. This means that the antenna half-power-beamwidth (HPBW) of these arrays are relatively narrow, calculated to be HPBW = 28.5° and 32.0°, respectively. When a satellite's main axis (+Z axis) is off pointed from your ground station that off-pointing angle is known as the squint angle. If the squint angle is less than half of the HPBW, the ground station will be within the spacecraft antenna nominal beam width. It is one of the information outputs from most of the newer computer satellite tracking programs. With the weaker signals from birds at higher altitudes, the effects of the squint angle are more pronounced; making the HPBW values sometimes seem even smaller.

SOME L-BAND PROBLEMS AND SOLUTIONS

An examination of an efficient operating system must include all of the various parts of that system, and how they best fit together. The trick is in turning negatives into positives, as shown by K9EK. For L-band transmitting, the problems are more physical than electronic. **Fig 23.26** illustrates this approach.

First is the fact that transmitters and

Table 23.3 – Suppliers of Equipment of Interest to Satellite Operators

Multimode VHF and UHF Transceivers and Specialty Equipment

ICOM America
Kenwood Communications
Yaesu USA

Converters, Transverters and Preamplifiers

Advanced Receiver Research
Angle Linear
Down East Microwave
Hamtronics
Henry Radio
The PX Shack
Parabolic AB
Radio Kit
RF Concepts
Spectrum International
SSB Electronic

Power Amplifiers

Alinco Electronics
Communications Concepts
Down East Microwave
Encomm
Falcon Communications
Mirage Communications
Parabolic AB
RF Concepts
SSB Electronics
TE Systems

Antennas

Cushcraft Corp
Down East Microwave
KLM Electronics
M2 Antenna Systems
Parabolic AB
Telex Communications

Rotators

Alliance
Daiwa
Electronic Equipment Bank
Kenpro
M2 Antenna Systems
Telex
Yaesu USA

Other Suppliers

AEA
ATV Research
Down East Microwave
Electronic Equipment Bank
Grove Enterprises
M2 Antenna Systems
Microwave Components of Michigan
PacComm
SHF Microwave Parts
Tucson Amateur Packet Radio (TAPR)

Note: This is a partial list. The ARRL does not endorse specific products.

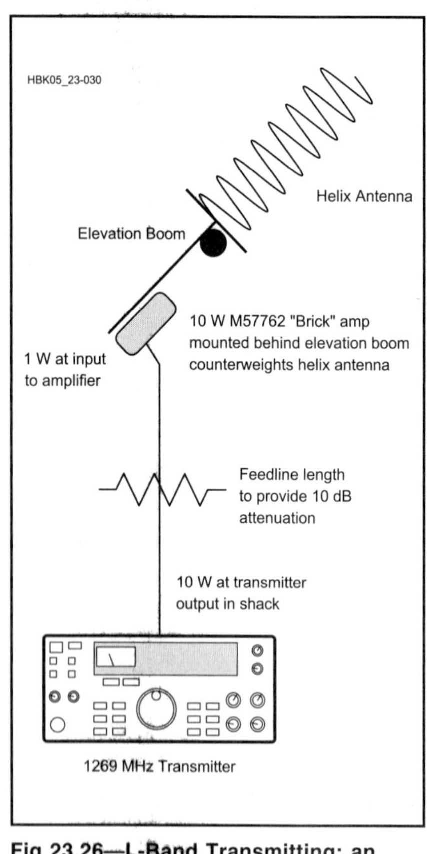

Fig 23.26—L-Band Transmitting: an integrated approach.

antennas are usually in separate places with a long length of coaxial cable connecting the transmitter to the antenna. Coaxial cable is an imperfect RF conductor and with losses increasing with frequency and length of cable. Common (and reasonably priced) coaxial cables have unwanted losses at 1270 MHz. The old standby, RG-8U, loses almost 13 dB per 100 feet at 1270 MHz. A 100-foot feedline would only have 0.5 W coming out from 10 W going in. Of course, we could increase transmitter power by 20 times to compensate, but those are quite expensive watts to be used for heating coaxial cable! Coaxial cable "hardline" is a good but an expensive solution, and even it has significant loss. Low-loss L-band waveguide is much too large and expensive to be considered.

What is needed are methods of getting all of the transmitter power to the antenna. You may not want to remotely mount your entire L-band transmitter at the antenna, although some manufacturers are offering transmitting converters for that service. A much better solution would be to split the transmit power generation into two stages, and integrate the antenna and final amplifier stage by mounting a final amplifier very near the antenna feedpoint. This will greatly reduce the transmitter to antenna feed-line loss by eliminating most of the feed line. Fortunately, an elegant solution exists for the "final amplifier stage" of this integration. This is the M57762 hybrid linear amplifier module. This 1-W input, 10-18 W output "brick" has 50-Ω input and output impedance and requires only the addition of connectors, a heat sink and simple dc power circuitry. A second design of amplifier is also offered, using two of these M57762 modules for higher output power.

There is still the problem of the feedline loss between the shack transmitter and the final amplifier stage, but there are ways to use this to an advantage. Most L-band transmitters use the M57762 brick as part of their output stage, and are rated for 10 W (+40 dBm) RF output. The remotely mounted final amplifier stage requires only 1 W (+30 dBm) RF input, so some method of reducing the transmitter output to match the amplifier input. A very satisfactory way to do this is to use the loss qualities of a length of small coaxial cable as an attenuator. For this feed, a 10-dB loss between the transmitter and amplifier is needed. Referring to cable manufacturer's data, a length of coaxial cable can be determined that will give the required 10-dB attenuation, then use it to connect the transmitter and amplifier.

SINGLE BRICK L-BAND AMPLIFIER

The amplifier of **Fig 23.27** was constructed by K9EK by mounting the M57762 module directly to a heat sink (no insulator required; but thermal compound is highly recommended), then using an etched circuit board, slipped under the leads of the module, to provide both RF and dc connections. The schematic for this amplifier is shown in **Fig 23.28**. The board is made from 0.062-inch thick, G10 board. Keep the input and output lines 0.10-inch wide to maintain the 50-Ω input and output impedance. The type N connectors should be mounted on the end of the heat sink in such a manner that the center conductors lie directly on the board traces. Ensure that the circuit board is well grounded to the heat sink by drilling and tapping several holes through the circuit board as shown.

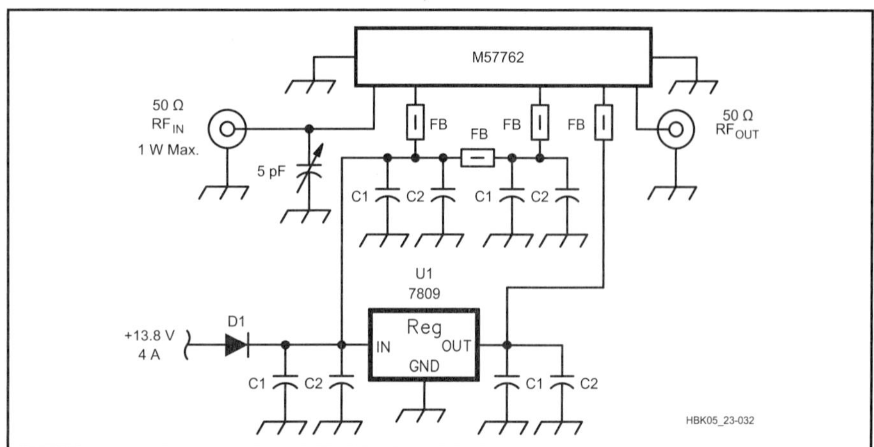

Fig 23.28—Just a handful of parts are needed to connect the brick amplifier module. All capacitor pairs are 10 µF/35 V chip or tantalum units in parallel with 1 nF chip capacitors. D1 is a 4-A (minimum) 50-V power rectifier such as Digi-Key GI820CT-ND. It prevents damage due to reverse connection of the power leads. U1 is a 7809 voltage regulator (9-V, 1-A). Check RF Parts and Down East Microwave for pricing and availability of the amplifier module.

Fig 23.27—The matching brick amplifier can be mounted on the counterbalance side of the Helix antenna boom.

This amplifier operates in class AB linear, and therefore draws some (about 400 mA) dc current when not transmitting. It was decided that the additional circuit complexity to cut the amplifier off completely was not warranted. You just turn off the dc supply during long standby periods. The full load current is approximately 4A.

A template, with additional construction details and a PC board layout, is available from the ARRL. See the *Handbook* CD-ROM **Templates** section for details.

DOUBLE BRICK L-BAND AMPLIFIER

This second L-band amplifier is more complex, offering considerably more output power (≈40 W), and was originally constructed for terrestrial 1296-MHz service by WD4FAB. This amplifier also operates in class AB linear. As the amplifier is pretty broad-banded, it can serve the satellite service as well as for the original terrestrial service plans; all is the matter of not pointing your antenna into the sky for that 1296-MHz contact! The inspirational credit for using the M57762 module and getting this unit started is given to the North Texas Microwave Society. NTMS has a lot of good ideas flowing through their newsletters.

Fig 23.29 shows this completed assembly, as it was just removed from the tower box, for the purpose of the picture taking. This assembly is composed of two flanged extruded heat sinks mounted face-to-face, with the brushless dc blower moving cooling air through the finned chamber of the heat sinks. The identical M57762 module "brick" amplifiers and their bias circuitry PCB assemblies are mounted on the troughs of each heat sink on the opposite sides of the assembly.

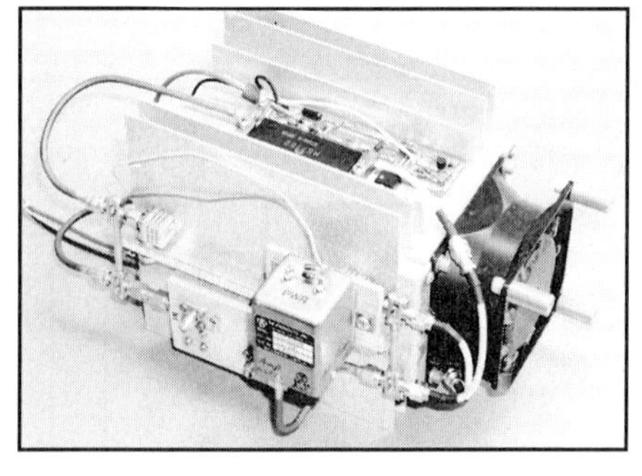

Fig 23.29—This is an alternate, more powerful, design of the L-band brick amplifier. This is a pair of heat sinks mounted face to face with a dc motor cooling fan; the electronics are mounted on each side with the combiner and relay circuitry on the near side. From left to right: output 3-dB hybrid coupler, directional coupler for power measurement, the four-port relay, and the input 3-dB hybrid coupler.

Fig 23.30 and **Fig 23.31** provide the schematic information for this amplifier assembly. The following notes will give the constructor details that may not be obvious from the photographs and schematics. Considerable care is needed in making the RF connections between these module amplifiers and their respective input and output 3-dB hybrid couplers. Slight differences in getting really equal length lines here can make a large difference in the overall performance. Notice in Fig 23.29 that the input coaxial lines, on the right, are of flexible Teflon 50-Ω cable, while the output lines, on the left, are of the semi-rigid UT141 copper jacketed co-

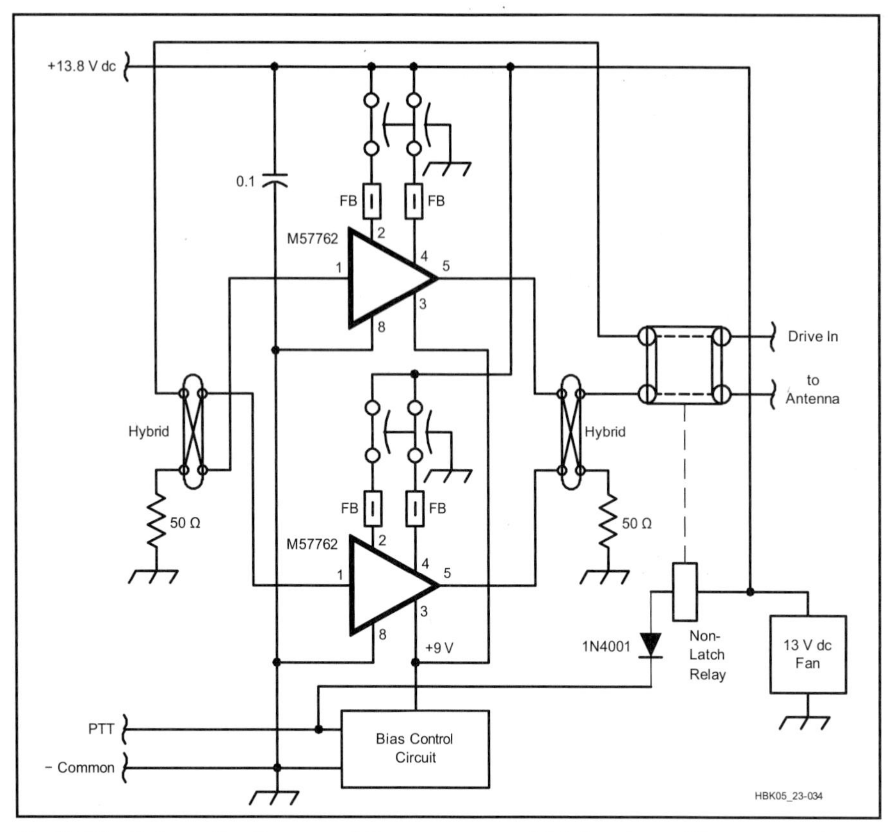

Fig 23.30—Schematic diagram of the more powerful L-band amplifier assembly.

axial cable, selected because of the higher power level at this point. Each end of the UT141 cables is terminated in an SMA connector, for ease of handling and uniformity of cable length.

Also shown along the facing side of the heat sink assembly is a collection of RF parts, which from left to right are output hybrid coupler, directional coupler for power measurement, the four-port relay and the input hybrid coupler. Having a four-port relay of this type surely makes life easier, as it handles both input and output switching. The output from the directional coupler is converted to a dc signal by a diode and capacitor right at the SMA connector (not shown) that mounts to the coupled port. That dc signal is sent back to the station to give a very useful indication of the output power. Note also the 50-Ω load terminations that are SMA-connector-mounted to the couplers. These loads absorb any unbalanced power from the amplifiers through the hybrid couplers. While the input coupler did not need such a large load, it was easier to buy two identical loads.

For the M57762 module assemblies, **Fig 23.32**, there are a number of details that are not so obvious, some mechanical and some electrical. First is to note that measures are taken to insure that the module is very thoroughly clamped to the heat sink, through the use of a pair of 6.4-mm (0.25-inch) square brass bars. Using screws alone would provide insufficient clamping forces. The module must have heat sink compound used between it and the heat sink and that interface must be smooth and flat for reasonably decent heat transfer. These bars are elongated to serve additional purposes as the bar at the output end of the module (at the right) has a two-hole SMA connector embedded in it so as to get a very close termination to the wire lead of the RF output of the module. This bar also serves as a low-resistance dc connection for the power ground lead to the module, thereby not depending upon a casual connection through mounting screws and the heat sink. For the input end, the coaxial cable is fed through a hole in the bar (at the left) to also get the close connection to the module RF input. Note the short, exposed RF leads at these points.

There are three other wire leads to the module, two for +13.8 V and one for the +9 V bias power. These leads are individually filtered with FB43101 ferrite beads, and with the leads running across the top of a 1nF trapezoidal filter capacitor soldered to the PCB. The PCB is a double-sided G10

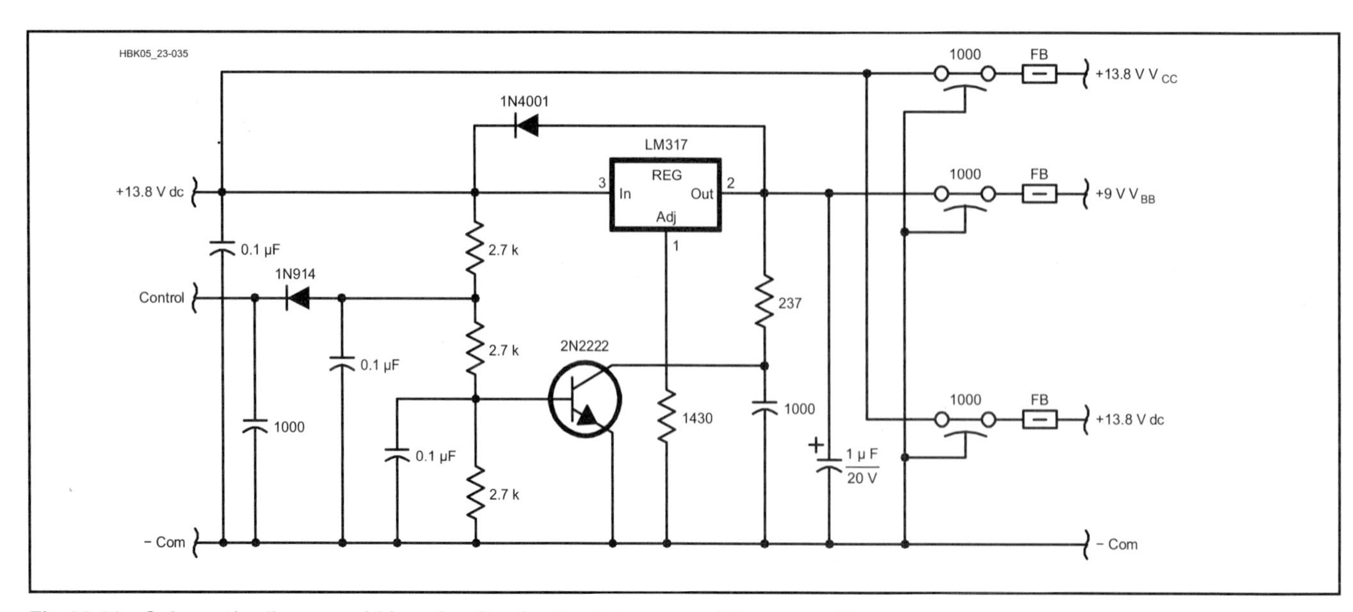

Fig 23.31—Schematic diagram of bias circuitry for the L-band amplifier assembly.

Fig 23.32—One side of the amplifier assembly, showing the brick amplifier and the one-sided PC board with bias regulator and filter circuitry. Note that the input and output coaxial cables are soldered to the brass clamp bars so that the center conductor is closely "presented" to the input and output terminations of the brick amplifier.

board, 0.062-inches thick, with only the topside formed into circuit patterns. The reverse side of the PCB is only used as the ground plane for the power circuits. The two sides of the PCB are connected with copper or brass foil wrapped around the edges at selected locations.

This bias circuitry operates to shut off the bias to the module when not transmitting, through a conventional grounding-on-PTT control line that comes up from the station. This same line, which is controlled by a sequencer in the station, also operates the four-port T/R relay. It was determined necessary to turn off the module to prohibit it from having any possible emissions while the station is in its most sensitive receiving condition, although this is unproven.

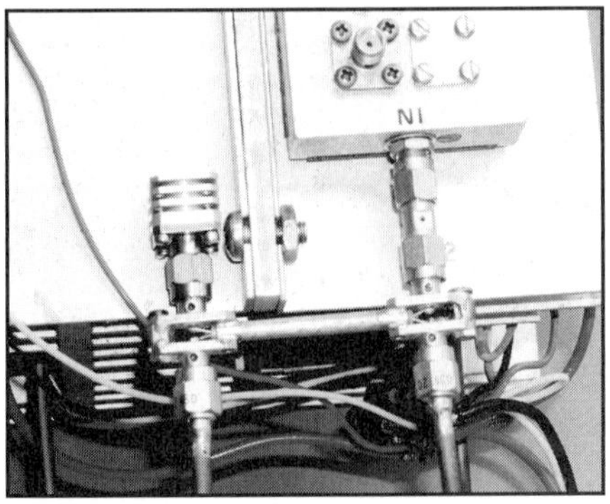

Operation of the bias circuitry is through the diode isolated control line shutting off the bias to a saturated 2N2222, which, in turn, unclamps the LM317 regulator reference lead. It is interesting what tricks can be done with these adjustable regulators. All of the additional capacitive filtering needed for the regulator is added.

One of the two identical 3-dB hybrid coupler assemblies is shown in **Fig 23.33**, with drawing details shown in **Fig 23.34**. The neat tricks of constructing these couplers using the Sage Wireline and SMA connectors was learned from KE3D, who probably "borrowed" it from S57MV. Proper wiring of the Wireline to the SMA connectors is very important here, as is having equal lengths of equal-properties coaxial lines from the amplifiers to the couplers. Careful attention to small details here pay great dividends in performance. Also note the "reversal" of the input lines from the coupler to the amplifiers, needed to compensate for 90° phase shifts of the hybrids. Without this input crossover, there would be no RF output! The first assembly of this unit had ignored that phase shift and the amplifier input lines had to be "patched" with extension cables to correct for that bit of forgetfulness.

Templates with additional construction details and PC board layouts are available for each of these two amplifiers on the CD included with this book.

RECEIVING, PREAMPLIFIERS AND RECEIVE CONVERTERS

Receiving Phase-3E downlinks will require equipment for the microwave bands. Operating experience with AO-40

Fig 23.33—A close-up view of one of the 3-dB hybrid coupler assemblies (the two are identical) showing its construction using Sage Wireline®, four 2-hole SMA connectors, spacers and screw hardware. Note the 50-Ω terminated port.

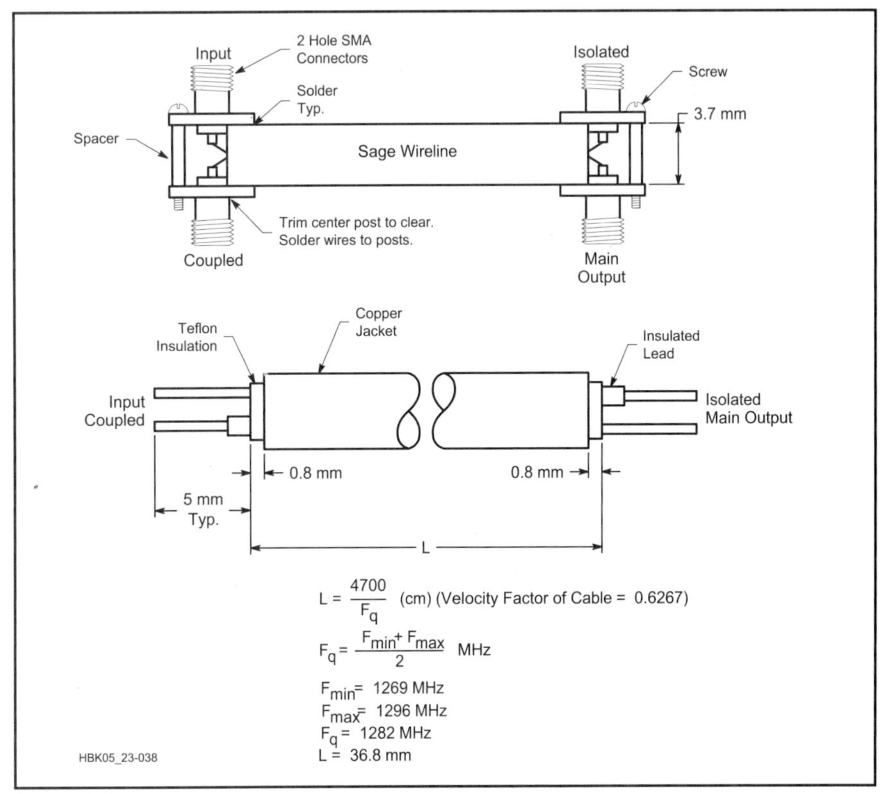

Fig 23.34—Details of the construction of the 3-dB hybrid coupler assemblies.

has shown that receiving antenna gains need to be 22 dBic, minimum. On S band this has clearly been shown to require the use of a dish antenna with at least a diameter of 600 mm. As there are no commercial base-station S band receivers, outboard S band down-converters need to be used and have become available in recent years, with a lot of interest in soaking up the surplus market units provided by Drake and others, **Fig 23.35**. Commercial Amateur S band receive converters have been coming on the market from SSB-USA, with their UEK-3000SAT; from Down East Microwave (DEM), with their 2400-144RX and 2400-432RX (see **Fig 23.36**); from Parabolic AB, with their "Mode S Down-converter"; and from other sources. Amateur microwave operation is not a "strange" activity relegated just to the specialists anymore, as satellite enthusiasts have been exploring these S band downlinks for some time and manufacturers have responded by increasing the supply of useable equipment for S band.

No discussion of satellite receiving systems would be complete without mentioning preamplifiers and their location. The mast mounting of sensitive electronic equipment has been a fact of life for the serious VHF/UHF operator for years, although it may seem to be a strange or difficult technology for HF operators. While a preamplifier can be added at the receiver in the station, it will do no good there. Vastly better results will be obtained if the preamp is mounted *directly* to the antenna. To get the most out of your VHF/UHF satellite station, you'll need to mount a low-noise preamplifier or receive converter on the tower or mast near the antenna, see **Fig 23.37**, so that feed-line losses do not degrade low-noise performance. Feed-line losses ahead of the preamplifier or converter add directly to receiver noise figure. Antenna mounting the preamplifier or down-converter will overcome these noise figure problems. For S band, placing either a preamp or the receiving down-converter *on* your antenna feed point is essential. Not even mast mounting the preamplifier will suffice here.

Fig 23.36—This SSB Electronic unit (left) is one of several solutions for mast-mounted S-band receive converters. Other manufacturers, including Down East Microwave (top), have similar offerings.

Low-noise front ends for down-converters or preamps are essential for receiving weak satellite signals, especially as the operating frequency goes higher. Multimode rigs and most down-converters will hear much better with the addition of a preamplifier employing GaAs FET or PHEMT technology ahead of the front end, albeit at the expense of the reduction in the third-order intercept point of the receiver. DEM provides a weather-resistant assembly in their 13ULNA unit. See photos of S-band dish antennas for views of this DEM product. Other very qualified preamplifier units are available from SSB-USA.

Table 23.3 lists other sources of commercially built preamplifiers and receive converters for most all VHF, UHF and microwave bands. These are available in several configurations. Many models are designed for mounting in a receive-only line for use with a receiving converter or transverter. Others, designed with multimode transceivers in mind, have built-in relays and circuitry that automatically switch the preamplifier or converter out of the antenna line during transmit. Most of these models are housed, with relays, in weatherproof enclosures that mount right at the antenna. For the equipment builder, several suitable designs appear in *The Radio Amateur's Satellite Handbook*.

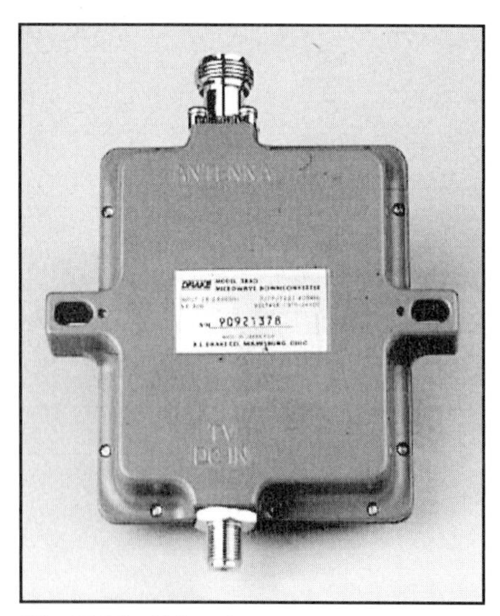

Fig 23.35—S-band converter that was available from Drake, and easily modified for satellite service.

Fig 23.37—Details of WD4FAB's tower cluster of satellite antennas, including a home-brew elevation rotator. Top to bottom: M² 436-CP30, a CP U-band antenna; two each M² 23CM22EZA antennas shown in a CP array for L band; "FABStar" dish antenna with helix feed for S band; M² 2M-CP22, a CP V-band antenna (only partially shown.) To left of dish antenna is a NEMA4 equipment box with an internal 40W L-band amplifier, and also hosts externally-mounted preamplifiers.

ACCESSORIES

Receiver audio output noise on these higher satellite bands is often quite disturbing to many operators, taking away from the "listenablity", or the "arm chair" quality of the QSO audio on these bands. Some of this disruption can also be from the in-band use of such equipment as microwave ovens and the modern spread-spectrum 2.4-GHz cordless telephones that are becoming common in our households. Modern receivers and accessory equipment comes to the rescue here, and is the reason for the use of the HF transceiver (an IC-746) in receiving the S-band signals of the station shown in Fig 23.20. Newer model HF transceivers provide IF DSP and other features that provide very useable "Noise Reduction" and "Noise Blanker" functions. The noise blanker, in particular, removed all the observed noise from a 2.4-GHz Spread-Spectrum cordless telephone near the illustrated station, even with the very high sensitivity of the S-band receiving equipment. In the illustrated station an outboard audio DSP unit (a Timewave DSP599zx) is very helpful to remove the audio "Random Noise" that is not removed by the IF DSP. The combination of the equipment shown does provide "arm-chair" quality comfortable listening. Some of the other most recently available Amateur transceivers may incorporate both the audio and IF DSP noted above, possibly eliminating the need for an outboard audio DSP unit.

Great caution by the operator must be exhibited when using a transceiver for receiving the S-band signals, as even the slightest inadvertent moment of transmitter operation into the "rear end" of a down-converter can make it an instant hot dog without a roll! SSBUSA offers a protection unit "The Protector" with a SSB power rating of 50 W. In the illustrated station with its IC-746 transceiver, the 2-m receive line was internally isolated from the transmitting circuits and brought out to a SMA connector at a convenient point on the rear panel. This separated receive capability on 2 m has saved the down-converter on several occasions.

One very helpful accessory for the station is the remote measurement of RF power. Such power meters are commercially available or easily constructed by the handy operator. This type of measurement is very meaningful for stations with long feed lines between the transmitter and the antennas, making the determination of EIRP more realistic. The illustrated station uses a moving-coil power meter of uncertain origin coupled with surplus or home made directional couplers for power sampling near the antennas in the equipment box on the tower.

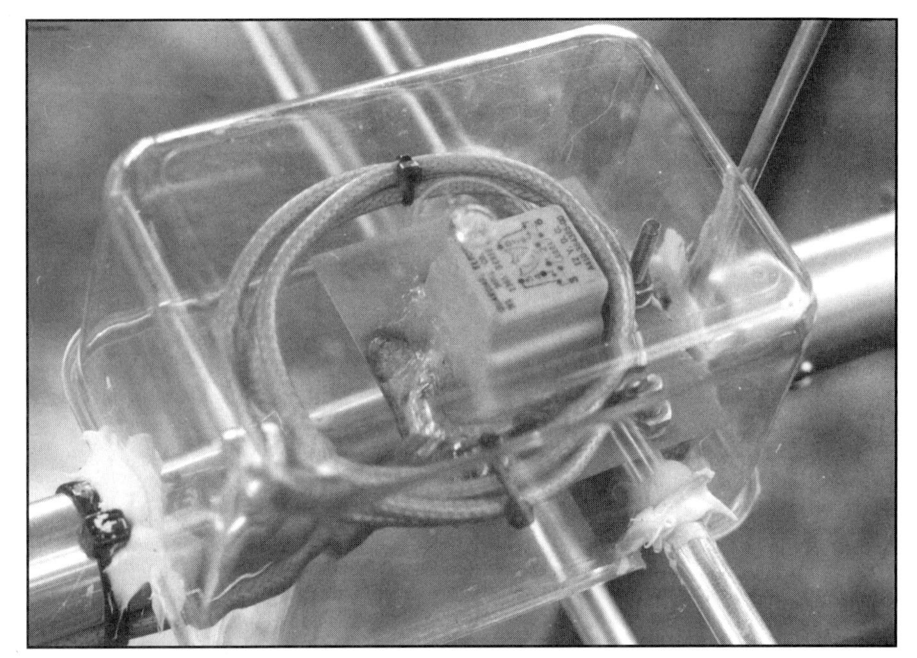

Fig 23.38—KLM 2M-22C antenna CP switching relay with relocated balun. Protective cover is needed for rain protection; see text.

Operators unfamiliar with microwave S-band operation may gain a greater confidence of their receivers with the addition of a 2101-MHz signal source. Such signal sources have been in the literature in recent years, and Down East Microwave offers a kit. An S-band signal source can also provide the user a convenient check for the station antennas and also as a calibrator for the S-band receiver S meter.

For stations using "crossed" Yagi antennas for CP operation, one feature that has been quite helpful for communicating through most of the LEO satellites, has been the ability to switch polarization from RHCP to LHCP. In some satellite operation this switchable CP ability has been essential.

The station shown has an all up capability for analog and digital satellite operations. Not shown just to the left of the photo is the station computer, which provides the satellite tracking information from the Northern Lights Software Association, NLSA, *Nova* for *Windows* tracking program. This provides output for the TAPR EasyTrak antenna controller, as well as providing the interface to the digital communications operations, using a PacComm SPIRIT-2 Packet Controller. Please note the warning sign in Fig 23.20, as it is right and proper for the unwary at that station.

EXPOSED ANTENNA RELAYS AND PREAMPLIFIERS

Experience with the exposed circularity switching relays and preamplifiers mounted on antennas have shown that they are prone to failure caused by an elusive mechanism known as "diurnal pumping." Often these relays are covered with a plastic case, and the seam between the case and PC board is sealed with a silicone sealant. Preamps may also have a gasket seal for the cover, while the connectors can easily leak air. None of these methods create a true hermetic seal and as a result the day/night temperature swings pump air and moisture in and out of the relay or preamp case. Under the right conditions of temperature and moisture content, moisture from the air will condense inside the case when the outside air cools down. Condensed water builds up inside the case, promoting extensive corrosion and unwanted electrical conduction, seriously degrading component performance in a short time.

A solution for those antennas with "sealed" plastic relays, such as the KLM CX series: you can avoid problems by making the modifications shown in **Fig 23.38**. Relocate the 4:1 balun as shown and place a clear polystyrene plastic refrigerator container over the relay. Notch the container edges for the driven element and the boom so the container will sit down over the relay, sheltering it from the elements. Bond the container in place with a few dabs of RTV adhesive sealant. Position the antenna in an "X" orientation, so neither set of elements is parallel to the ground. The switcher board should now be canted at an angle, and one side of the relay case should be lower than the other. An example for the protective cover for an S-band preamp can be seen in the discus-

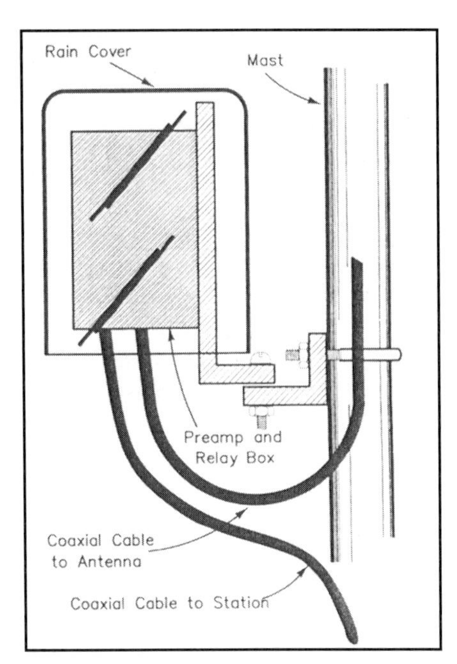

Fig 23.39—Protection for tower-mounted equipment need not be elaborate. Be sure to dress the cables as shown so that water drips off the cable jacket before it reaches the enclosure.

sion on feeds for parabolic antennas.

For both the relay and preamp cases, carefully drill a 3/32-inch hole through the low side of the case to provide the needed vent. The added cover keeps rain water off the relay and preamp, and the holes will prevent any build-up of condensation inside the relay case. Relays and preamplifiers so treated have remained clean and operational over periods of years without problems.

Another example for the protection of remotely, tower mounted equipment is shown in Fig 23.37, illustrating the equipment box and "mast mounted" preamplifiers at the top of WD4FAB's tower. The commercial NEMA4 rated equipment box is used to protect the 23-cm power amplifier and its power supply, as well as a multitude of electrical connections. This steel box is *very* weather resistant, with an exceptionally good epoxy finish, but it is not sealed and so it will *not* trap moisture to be condensed with temperature changes. Be sure to use a box with at least a NEMA3 rating for rainwater and dust protection. The NEMA4 rating is just a little better protection than the NEMA3 rating. Using a well-rated equipment box is very well worth the expense of the box. The box also provides some pretty good flanges to mount the "mast-mounted" preamplifiers for three bands. This box is an elegant solution for the simple need of rain shelter for your equipment, as illustrated in **Fig 23.39.**

ELEVATION CONTROL

Satellite antennas need to have elevation control to point up to the sky, the "El" part of the needed Az-El control of satellite antennas. Generally, elevation booms for CP satellite antennas need to be nonconducting so that the boom does not affect the radiation pattern of the antenna. In the example shown, the elevation boom center section is a piece of extra heavy wall 1 1/2-inch pipe (for greater strength) coupled with a tubular fiberglass-epoxy boom extension on the 70-cm end and a home-brew long extension (not shown in the photo) on the 2-m end, using large PVC pipe reinforced with four braces of Phillystran non-metallic guy cable. (PVC pipe is notoriously flexible, but the Phillystran cables make a quite stiff and strong boom of the PVC pipe.) For smaller installations, a continuous piece of fiberglass-epoxy boom can be placed directly through the elevation rotator.

Elevation boom motion needs to be powered, and one solution, **Fig 23.40**, uses a surplus jackscrew drive mechanism. One operator, VE5FP, found a solution for his Az-El needs by using two low-cost, lightweight TV rotators.[1] See **Fig 23.41**.

Operators through the years have employed many methods for the control of their antenna positions, ranging from true "arm-strong" manual positioning, to manual operation of the powered antenna azimuth and elevation rotators, and to fully automated computer control of the rotators. While computer control of the rotators is not essential, life is greatly assisted with the use of your computer. Fully automated control of your rotators is possible for such tasks as digital message uploading and downloading. For many years, one of the keystone control units for rotators has been the Kansas City Tracker (KCT) board installed in your computer. Most satellite tracking programs can connect to the KCT with ease. There are other options, these days, to replace the KCT unit.

A recent trend for amateur antenna control has been evolving, however, in the form of a standalone controller that translates computer antenna position information into controller commands with an understanding of antenna position limits. These boxes, represented by the new EasyTrak unit, **Fig 23.42**, from the Tucson Amateur Packet Radio (TAPR) group, have made this capability readily available for many amateurs. One of the EasyTrak units is also shown in Fig 23.20. The station computer, to the left and out of the photo, is used to upload antenna positions to EasyTrak. The computer can also control the operation of your station transceiver through the radio interface provided in EasyTrak; you will not need any other radio interface.

Fig 23.40—WD4FAB's home-brew elevation rotator drive using a surplus store drive screw mechanism.

Fig 23.41—VE5FP has a solution for his Az-El rotators by bolting two of them together in his "An Inexpensive Az-El Rotator System," ***QST*****, December 1998.**

ANTENNAS

Antennas have always been favorite subjects for amateur operators. Everyone

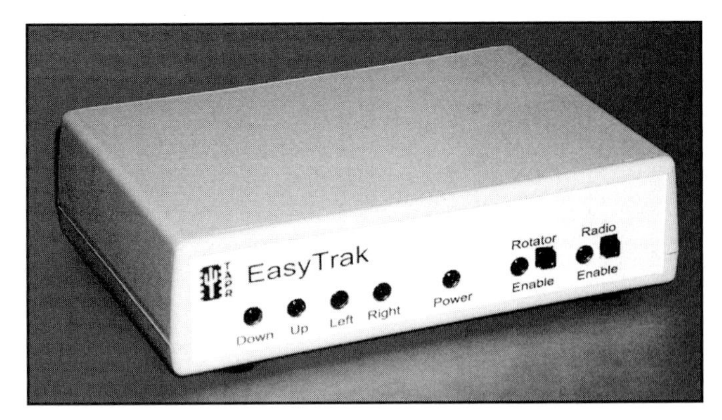

Fig 23.42—The EasyTrak automated antenna rotator and radio controller by TAPR.

Fig 23.43—This L-band helix antenna was designed to be home-brewed without any special tools or shop equipment. It is mounted on a 4-foot-long fiberglass tube.

is an expert! Fig 23.21 shows the example of the antennas for one station. The Yagi antennas are used for the U and L-band uplinks, while the S-band dish antenna is for the downlink. These satellite antennas are tower mounted at 63 feet to avoid pointing into the many nearby trees and suffering from the "green attenuation" provided by those trees. Satellite antennas otherwise do not need to be mounted so high. If the satellite antennas are mounted lower down, the reduced feed-line length and losses are a great benefit.

Linearly polarized antennas are "horizontal" or "vertical" in terms of the antenna's position relative to the surface of the Earth, a reference that loses its meaning in space. The need to use circularly polarized (CP) antennas for space communications is well established, as in space there is no "horizontal" or "vertical," but there is right-hand and left-hand circular polarization, CP. With CP the wavefront describes a rotational path about its central axis, either clockwise (right-hand or RHCP) or counter clockwise (left-hand or LHCP). If spacecraft antennas used linear polarization, ground stations would not be able to maintain polarization alignment with the spacecraft because of changing orientation. Ground stations using CP antennas are not (generally) sensitive to the polarization motions of the spacecraft antenna, and therefore will maintain a better communications link.

UPLINK ANTENNAS

Experience with AO-40 has clearly shown the advantages of using RHCP antennas for both the uplink and downlink communications. The antennas shown in Fig 23.21 are a single-boom RHCP Yagi antenna for U band, a pair of closely spaced Yagi antennas phased for RHCP for L band, and a helix-fed dish antenna for S band. The antenna gain requirements for U band can easily be met with a 30-element "crossed Yagi," two 15 element Yagi antennas mounted on a single boom with one placed a $^1/_4$ wavelength forward of the other. This arrangement allows for an uncomplicated phasing harness to couple the two antennas together to a single feedpoint. Antennas of this size have boom lengths of 4 to $4^1/_2$ wavelengths. The enterprising constructor can build a Yagi antenna from one of several references, such as *The ARRL Antenna Book*, or he might construct a 16 to 20 turn helix antenna for this service. Most of us might prefer, however, to purchase such a well-tested antenna from such commercial sources as M^2 or HyGain. In the past, KLM (now out of business) had offered a 40 element CP Yagi for U band satellite service, and many of these are still in satisfactory use today.

L-band uplink antennas become even more manageable, as their size for a given gain is only one-third of those for U band. Alternatively, higher gains can be obtained for the same boom lengths. With this band there is a narrower difference between using a dish antenna and a Yagi or helix, as a 21 dBic dish antenna would have a 1.2-m (4 foot) diameter. Some of us cannot afford such "real estate" on our elevation rotators and seek the lower wind-loading solutions offered by Yagi antennas. Long-boom rod-element Yagi, or loop-Yagi antennas, are commercially offered by M^2 and DEM, although this band is about the highest for practical Yagi antennas. The example shown in Fig 23.21 is a pair of rod-element Yagi antennas from M^2 in a robust arrangement. These antennas provide an overall gain of about 21.5 dBi. If greater L-band power is available than the 40 W shown in the illustrated station, the constructor may want to consider the helix antenna that follows.

L-BAND HELIX ANTENNA CONSTRUCTION

Fig 23.43 shows an easy-to-build antenna and support structure provided by K9EK. Fiberglass tubing was used for this structure as it has better outdoor longevity. Mounting of a helix of this nature requires that it be placed out "in front" of the elevation boom, so that there are no unwanted structures in the RF field of the antenna. This places a need for the counter weighting of the antenna to remove the unbalanced loading and strain on the elevation rotator. The constructor should take these provisions, and should consider the possible placement of the output amplifier as part of that counterweight, thus also making for a very short, low-loss feed line to the antenna.

Probably the easiest way to construct a helix antenna is to mark the wire, then close wind the spiral on a form, then stretch the winding out to match the required diameter and turns spacing. First, lay out the required length of #6 AWG bare copper wire (available as ground wire wherever house wiring is sold.) To straighten it, clamp one end in a vise, then squeeze the wire between two wood blocks and pull the blocks along the wire. Once the wire is straight and even, clearly mark the wire at each "cumulative wire length" dimension as shown in **Fig 23.44**. These marks show identical length per turn and they should line up along the completed helix. Allow 1-2 feet to remain on either end of the wire for handling, which will be cut off after construction. Cumulative length is used to prevent the compounding of measurement errors.

Then close wind the wire smoothly and tightly around a piece of $2^1/_2$-inch pipe (measures ϕ2.875 inch), with the length marks outward. The wire will spring back some after winding, so using two people to help pull to keep the winding tight gener-

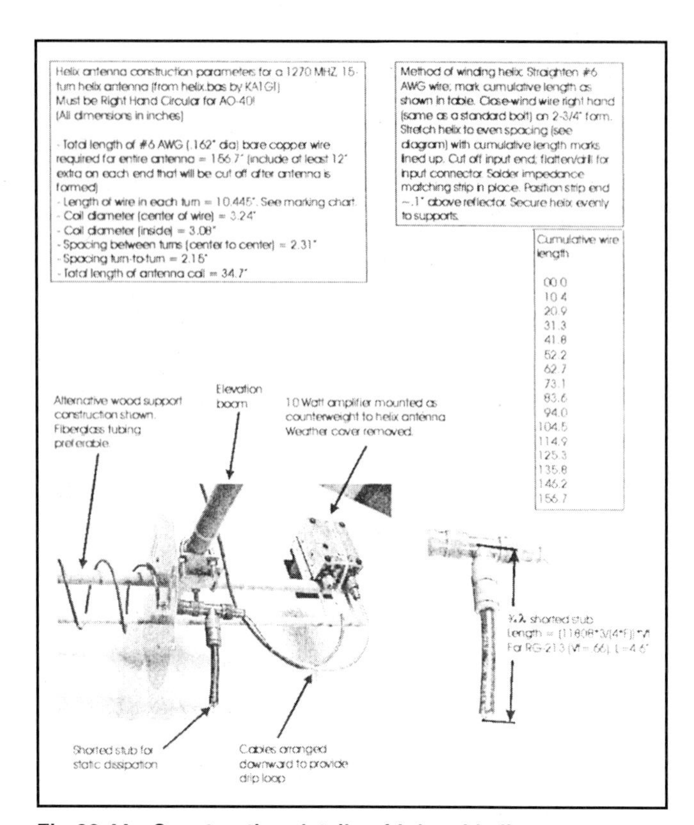

Fig 23.44—Construction details of L-band helix antenna.

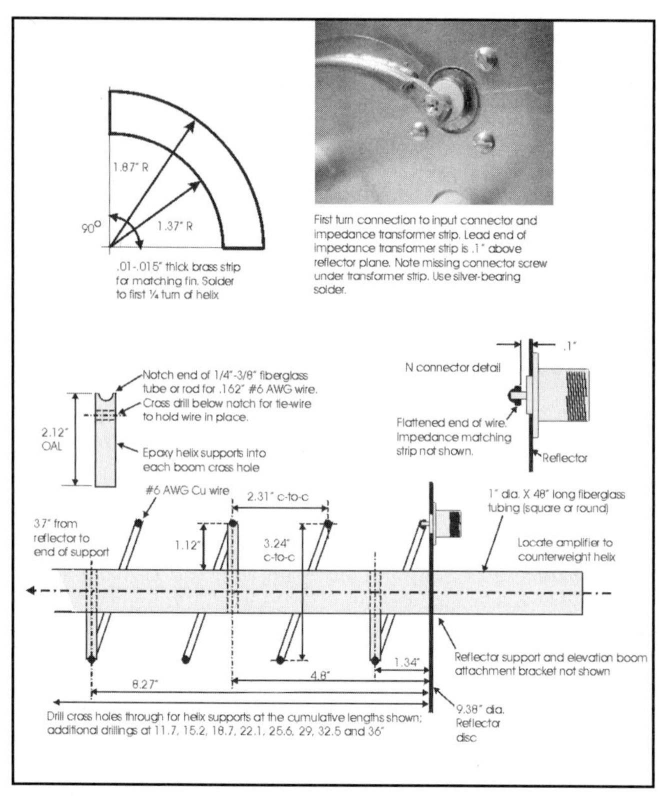

Fig 23.45—More construction details of L-band helix antenna.

ally works better. You must wind the helical element in a right-hand direction (clockwise); look at a standard machine screw or bolt and have the turns proceed in the same direction. Then hang the wire coil over a horizontal tube or form (the same PVC pipe used for winding will do well) that has been marked every 2.31 inches. Do this cumulatively to prevent errors. Holding the end turns (which will eventually be cut off) of the wire coil, gently stretch the helical element to the appropriate turn-to-turn spacing. Ensure that the turn-marks on the wire remain lined up; you may have to gently wind or unwind the helical element as you go. You can make a 2.15-inch (less your wire size) spacer to drop between turns to aid in getting the spacing correct.

Finally, cut off the excess wire on the input end of the helical element. Leave the opposite end uncut so if you goof on the input end, you can simply cut the error off and try again. After the antenna is completed, go back and cut off the excess from the far end. To connect the input end of the wire to the input connector, flatten the last $^1/_4$-inch of the input end of the wire with locking grip pliers or a hammer and steel block. Once it is flat and even, drill a hole in the end of the wire to slip over the center pin on an N connector.

The theoretical impedance of a helix antenna fed from the periphery is 140 Ω. A clever method of changing this to 50 Ω is by the addition of a λ/4 impedance transformer to the first quarter turn of the helical element wire itself. It has been found empirically that a 0.50-inch-wide strip of 0.010-inch-thick sheet brass makes a satisfactory impedance matching device. Cut the brass strip as shown in the figure. The strip is soldered to the first quarter turn of the helical element wire, starting 0.25 inch from the connector drilling. The underside of the strip must be flat relative to the reflector disc. Silver-bearing, lead-free plumbing solder is better than standard lead-tin solder and is highly recommended.

The method of helix antenna construction involves a single central length of 1.0 inch diameter fiberglass tubing with the helical element supported on 0.375-inch diameter fiberglass standoff rods or tubes. **Fig 23.45** shows the dimensions and construction for this helix antenna. Carefully attach the helix support rods to the central boom as shown. Mount a 7 inch to 9$^1/_4$-inch diameter (0.75λ to 1λ) reflector plate to the frame with homemade angle brackets. The helical element is slipped over the boom and stretched onto the support rods and held to the rods with wire ties. A homemade clamp plate and U-bolts are used to connect central rod to the antenna elevation boom.

After attaching the helical element onto the support frame, solder the hole in the end of the first turn to the input connector, an N connector mounted from the back of the reflector plate. Position the impedance matching strip at 0.10-inch (2.5mm) spacing from the reflector plate and solder it in place. Then secure the helical element in place. On two samples of the antenna shown, the input return loss was measured in excess of –16 dB (SWR=1.38:1). If proper equipment is available, lower return loss and SWR values can be achieved with this tuning system.

A practical shortcoming of the axial mode helix antenna is not obvious. The entire helical element itself is a single, long piece of copper wire, connected to the input connector. Since this wire is physically isolated from dc ground, atmospheric static electricity may build up on the element until it damages the attached solid-state device. So, some method is needed to dc ground the helical element itself, without seriously interfering with the RF characteristics of the antenna. One method of preventing static buildup is to add a shorted λ/4 stub to the antenna feed. A shorted quarter-wave stub presents extremely high RF impedance at its non-shorted end, making it virtually invisible to RF energy at its design frequency. At the same time, it provides a dc short to ground for the helical element itself, effectively draining off static electricity before it can build up. This shorted stub can be made from a T connector and coaxial cable. The length of the stub (measured from the center of the main cable to the short itself) must be calculated and must include the VF (velocity factor) of the cable being used. Note that an actual λ/4 stub is too short to be physically realizable; and a λ/2 must be added to the cable to create a 3λ/4 stub. With RG-213 cable (VF = 0.66), the total length of the 3λ/4 stub is 4.60 inches. This stub has been shown to have almost no effect on antenna performance at the design frequency.

PARABOLIC REFLECTOR ANTENNAS FOR S BAND

The satellite S-band downlinks have become very popular for a variety of reasons. Among the reasons are: good performance can be realized with a physically small downlink antenna and good quality down-converters and preamps are available at reasonable prices, as previously discussed. Increased operation on S band has long been advocated by a number of people including Bill McCaa, KØRZ, who led the team that designed and built the AO-13 S-band transponder[2] and James Miller, G3RUH, who operated one of the AO-40 command stations.[3] Ed Krome, K9EK, and James Miller have published many articles detailing the construction of preamps, down-converters, and antennas for S band.[4,5,6,7,8]

WØLMD notes that like a bulb in a flashlight, the parabolic reflector, or dish antenna must have a feed source looking into the surface of the dish. Some dishes are designed so that the feed source is mounted directly in front of the dish. This is referred to as a center-fed dish. Other dishes are designed so that the feed source is off to one side, referred to as an off-center fed dish, or just offset-fed dish. The offset-fed dish may be considered a section of a center-fed dish. The center-fed dish experiences some signal degradation due to blockage of the feed system, but this is usually an insignificantly small amount. The offset-fed dish is initially more difficult to aim, as the direction of reception is not the center axis like the center-fed dishes. The attitude of the offset dish is more likely to be vertical, making it less susceptible to loss from snow accumulation. Offset-fed dishes may have difficulty pointing horizontally for terrestrial communications. One enterprising ham, W4WSR, mounted his offset-fed PrimeStar dish upside down, which matched terrestrial pointing better. Older dishes tend to be center fed; newer dishes are offset fed.

The dish's parabola can be designed so the focus point is closer to the surface of the dish, referred to a "short focal length" dish, or further away from the dish's surface, referred to as a "long focal length" dish. To get the exact focal length, measure the diameter of the dish and the depth of the dish. The diameter squared divided by 16 times the depth is the focal length. The focal length divided by the diameter of the dish gives the *focal ratio*, commonly shown as f/D. Center-fed dishes are usually short focal ratios in 0.3 to 0.35 ranges. Offset-fed dishes are usually longer focal length such as 0.45 to 0.8. If you attach a couple of small mirrors (a couple of the YF's eye shadow mirrors when she isn't looking) to the outer front surface of a dish and then point the dish at the Sun, you will easily find the focus point of the dish. Put the reflector of the patch or helix just beyond this point of focus.

An alternate method for finding a dish's focal length is suggested by W1GHZ, and he provides that calculation on his Web site at: **www.w1ghz.cx/10g/10g_home.htm**. This method literally measures a solid-surface dish by the dimensions of the bowl of water that it will form when properly positioned.

While many of us enjoy building our own antennas, some of these microwave antennas are so small that they don't require a truck to transport them. Surplus market availability of these small dish antennas makes their construction unproductive. Many hams use the practices of AO-13 operators in using a surplus MMDS linear screen parabolic reflector antenna, **Fig 23.46**. These grid-dish antennas are often called "barbeque dishes". K5OE, **Fig 23.47,** and K5GNA, **Fig 23.48**, have been showing us how to greatly improve these linearly polarized reflectors by adapting them for the CP service. See the K5OE site: **members.aol.com/k5oe/**. The work shown us by these operators illustrate that simple methods can be used to "circularize" a linear dish and to

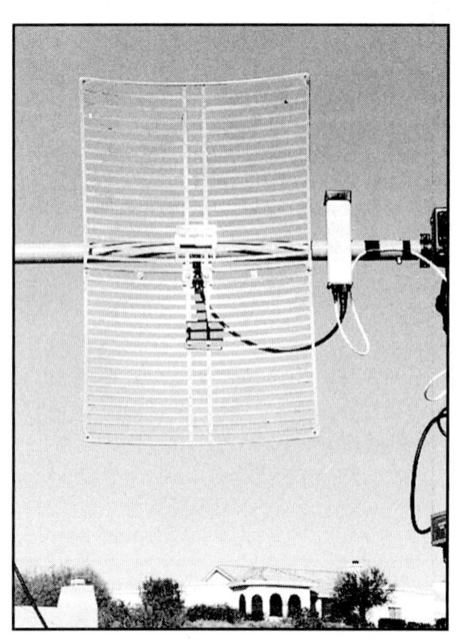

Fig 23.46—Another commercially available solution for S-band reception uses this TVRO antenna.

Fig 23.47—K5OE's mesh modification of a MMDS dish antenna with helix CP feed and DEM preamp.

Fig 23.48—K5GNA's "circularized" mesh modification of an MMDS dish antenna with helix CP feed and DEM preamp.

Fig 23.49—G3RUH's 60-cm spun aluminum dish with CP patch feed, available as a kit.

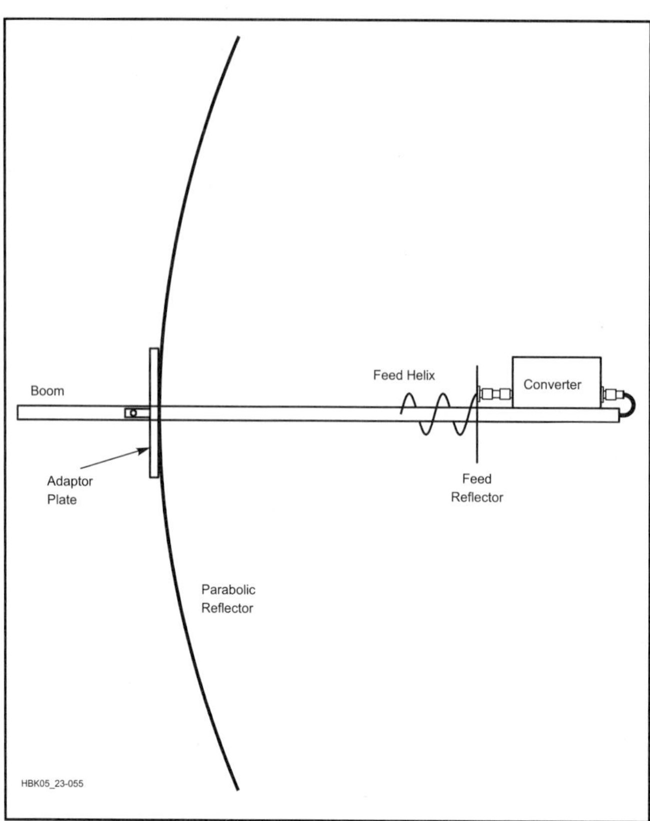

Fig 23.51—Detail of 60-cm S-band dish antenna with feed.

Fig 23.50—PrimeStar offset-fed dish with WD4FAB's helix-feed antenna. NØNSV was so pleased with the modification that he renamed the dish "FABStar," and made a new label!

further add to its gain by using simple methods to increase the dish area and feed efficiency.

Some smaller dishes are constructed from solid aluminum or steel. **Fig 23.49** shows one such 60-cm spun aluminum dish that was designed by G3RUH and ON6UG. This kit, complete with a CP patch feed is available from SSB-USA and has a gain of 21 dBic and provides a 2.5-dB Sun noise signal. Sometimes fiberglass construction is used with a conductive coating, as shown in **Fig 23.50**. Some operators are using their large, 10-foot TVRO dishes for Amateur satellite service, although these types of antennas cannot easily be mounted on a tower to clear the nearby trees. These larger dishes are generally made from perforated steel or aluminum. Dishes with perforations slightly reduce the wind loading. If the perforations are made smaller than λ/10, the dish is equivalent to solid in terms of signal reflection. At 2401 MHz (λ = 125 mm), this is about 0.50 inch, making standard 0.25-inch fence mesh a good option for a dish surface with low wind loading.

In the USA, there are large numbers of dishes that can be obtained either free or at low cost. But in some parts of the world dishes are not so plentiful, so hams make their own. G3RUH shows us his example of creating a dish antenna, see **Fig 23.51**. There are three parts to the dish antenna—the parabolic reflector, the boom, and the feed. There are as many ways to accomplish the construction, as there are constructors. It is not necessary to slavishly replicate every nuance of the design. The only critical dimensions occur in the feed system. When the construction is complete, you will have a 60-cm diameter S-band dish antenna with a gain of about 20 dBi with RHCP and a 3-dB beamwidth of 18°. Coupled with the proper down-converter, performance will be more than adequate for S-band downlink.

DISH FEEDS

On **www.ultimatecharger.com/**, WØLMD describes that the feeding of a dish has two major factors that determine the efficiency. Like a flashlight bulb, the dish feed source should evenly illuminate the entire dish, and none of the feed energy should spillover outside the dish's reflecting surface. No feed system is perfect in illuminating a dish. Losses affect the gain from either under-illuminating or over-illuminating the dish (spillover losses). Typical dish efficiency is 50 percent. That's 3 dB of lost gain. A great feed system for one dish can be a real lemon on another dish. A patch feed system is very wide angle, but a helix feed system is narrow angle. A short focal ratio center-fed dish requires a wide angle feed system to fully illuminate the dish, making the CP patch the preferred feed system. When used with an offset-fed dish, a patch-type feed system will result in a considerable spillover, or over-illumination loss, with an increased sensitivity to off-axis QRM, due to the higher f/D of this dish. Offset-fed dishes do much better when fed with a longer helix antenna.

A helix feed is simplicity personified. Mount a type N connector on a flat reflector plate and solder a couple of turns wire to the inner terminal. Designs are anywhere from two to six turns. The two-turn helices are used for very short focal length dishes in the f/D = 0.3 region, and the six-turn helices are used with longer focal length (f/D ~ 0.6) dishes, typically the offset-fed dishes. Since AO-40 is right circular and the dish reflection will reverse the polarity, the helix should be wound left circular, looking forward from the connector. Helix feeds work poorly on the short focal length dishes but really perform well on the longer focal length offset-fed dishes.

A patch feed is almost as simple. It is typically a type N connector on a flat reflector plate with a tuned flat metal plate soldered to the inner terminal. Sometimes the flat plate is square; sometimes it is rectangular, sometimes round. It could have two feed points, 90° out of phase for circular polarization, as was used in the construction of the AO-40 U-band antennas. Some patches are made rectangular with clipped corners to add circularity. On 2401 MHz, the plate is about 2¼-inches square and spaced from 1/16 inch to ¼-inch away from the reflector. The point of attachment is about halfway between the center and the edge. A round patch for 2401 MHz is $\phi 2^5/_8$ inches. Patches work poorly on the longer focal length offset dishes, but do very well on the shorter focal length center-fed MMDS and TVRO dishes. A well studied CP patch feed for these short f/D dishes is shown to us by G3RUH in Fig 23.53 and **Fig 23.52**.

For additional information on constructing antennas for use at microwave frequencies, see *The ARRL UHF/Microwave Experimenter's Manual*.

Fig 23.52—Details of CP patch feed for short f/D dish antennas by G3RUH and ON6UG.

PARABOLIC DISH ANTENNA CONSTRUCTION

The following is a condensation of several articles written by G3RUH describing a dish antenna that can be easily built and used for reception of an S-band downlink. The dish itself was previously described, and the article continues:

The parabolic reflector used for the original antenna was intended to be a lampshade. Several of these aluminum reflectors were located in department store surplus. The dish is 585 mm in diameter and 110 mm deep corresponding to an f/D ratio of 585/110/16 = 0.33 and a focal length of 0.33 × 585 = 194 mm. The f/D of 0.33 is a bit too concave for a simple feed to give optimal performance but the price was right, and the under-illumination keeps ground noise pickup to a minimum. The reflector already had a 40-mm hole in the center with three 4-mm holes around it in a 25-mm radius circle.

The boom passes through the center of the reflector and is made from 12.7-mm square aluminum tube. The boom must be long enough to provide for mounting to the rotator boom on the backside of the dish. The part of the boom extending through to the front of the dish must be long enough to mount the feed at the focus. If you choose to mount the down-converter or a preamp near the feed, some additional length will be necessary. Carefully check the requirements for your particular equipment.

A 3-mm thick piece of aluminum, 65 mm in diameter is used to support the boom at the center of the reflector. Once the center mounting plate is installed, the center boom is attached using four small angle brackets — two on each side of the reflector. See Fig 23.51 for details of reflector and boom assembly.

A small helix is used for the S-band antenna feed. The reflector for the helix is made from a 125-mm square piece of 1.6-mm thick aluminum. The center of the reflector has a 13-mm hole to accommodate the square center boom described above. The type N connector is mounted to the reflector about 21.25 mm from the middle. This distance from the middle is, of course, the radius of a helical antenna for S band. Mount the N connector with spacers so that the back of the connector is flush with the reflector surface. The helix feed assembly is shown in **Fig 23.53.**

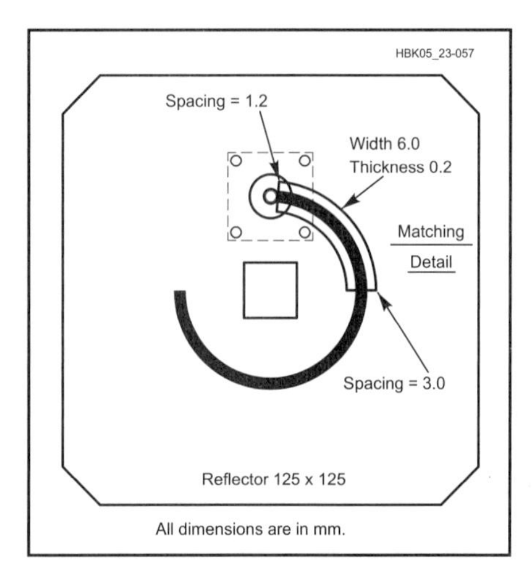

Fig 23.53—Details of helix feed for S-band dish antennas. The Type N connector is fixed with three screws and is mounted on a 1.6-mm spacer to bring the PTFE molding flush with the reflector. An easier mounting can be using a smaller TNC connector. Reflectors should be 95-100 mm, circular. Dimensions are in mm, 1 inch = 25.4 mm.

Copper wire about 3.2 mm in diameter is used to wind the helix. Wind four turns around a 40-mm diameter form. The turns are wound counterclockwise. This is because the polarization sense is reversed from RHCP when reflected from the dish surface. The wire helix will spring out slightly when winding is complete.

Once the helix is wound, carefully stretch it so that the turns are spaced 28 mm (±1 mm). Make sure the finished spacing of the turns is nice and even. Cut off the first half turn. Carefully bend the first quarter turn about 10° so it will be parallel to the reflector surface once the helix is attached to the N connector. This quarter turn will form part of the matching section.

Cut a strip of brass 0.2-mm thick and 6-mm wide matching the curvature of the first quarter turn of the helix by using a paper pattern. Be careful to get this pattern and subsequent brass cutting done exactly right. Using a large soldering iron and working on a heatproof surface, solder the brass strip to the first ¼ turn of the helix. Unless you are experienced at this type of soldering, getting the strip attached just right will require some practice. If it doesn't turn out right, just dismantle, wipe clean and try again.

After tack soldering the end of the helix to the type N connector, the first ¼ turn, with its brass strip in place, should be 1.2 mm above the reflector at its start (at the N connector) and 3.0 mm at its end. Be sure to line up the helix so its axis is perpendicular to the reflector. Cut off any extra turns to make the finished helix have 2¼ turns total. Once you are satisfied, apply a generous amount of solder at the point the helix attaches to the N connector. Remember this is all that supports the helix.

Once the feed assembly is completed, pass the boom through the middle hole and complete the mounting by any suitable method. The middle of the helix should be at the geometric focus of the dish. In the figures shown here, the feed is connected directly to the down-converter and then the down-converter is attached to the boom. You may require a slightly different configuration depending on whether you are attaching a down-converter, preamp, or just a cable with connector. Angle brackets may be used to secure the feed to the boom in a manner similar to the boom-to-reflector mounting. Be sure to use some method of waterproofing if needed for your preamp and/or down-converter.

HELIX FEED FOR AN OFFSET DISH ANTENNA

The surplus PrimeStar offset fed dish antenna with its seven-turn helix feed antenna, Fig 23.50, is described in this section. When the feed antenna is directly coupled with a preamp/down-converter system, this antenna provides superb reception of S-band signals with the satellite transponder noise floor often being the noise-limiting factor in the downlink. This performance is as had been predicted by the W3PM spreadsheet analysis, **Fig 23.54**, and actual operating experience. Operating experience also demonstrates that this antenna can receive the Sun noise 5 dB above the sky noise. Don't try to receive the Sun noise with the antenna looking near the horizon, as terrestrial noise will be greater than 5 dB, at least in a big-city environment. The operator, NØNSV, who provided this dish was rewarded for his effort with a second feed antenna, and he in turn provided new labels for the dish, titling it "FABStar".

The reflector of this dish is a bit out of the ordinary, with a horizontal ellipse shape. It is still a single paraboloid that was illuminated with an unusual feedhorn. At 2401 MHz we must be satisfied with a more conventional feed arrangement. A choice must be made to under-illuminate the sides of the dish while properly feeding the central section, or over-illuminating the center while properly feeding the sides. For the application shown here, the former choice was made. The W1GHZ water-bowl measurements showed this to be a dish with a focal point of f = 500.6 mm and requiring a feed for an f/D = 0.79. The total illumination angle of the feed is 69.8° in the vertical direction and a feedhorn with a 3-dB beamwidth of 40.3°. At 50 percent efficiency this antenna was calculated to provide a gain of 21.9 dBi. A seven-turn helix feed antenna was estimated to provide the needed characteristics for this dish and is shown in **Fig 23.55**.

The helix is basically constructed as described for the G3RUH parabolic dish, noted above. A matching section for the first λ/4 turn of the helix is spaced from the reflector at 2 mm at the start and 8 mm at the end of that fractional turn. Modifications of the design include the use of a cup reflector. For the reflector, a 2-mm thick circular plate is cut for a ϕ94 mm (0.75λ) with a thin aluminum sheet metal cup, formed with a depth of 47 mm. Employment of the cup enhances the performance of the reflector for a dish feed.

The important information for this seven-turn helix antenna is: Boom, 12.7 mm square tube or "C" channel; Element, ϕ1/8-inch copper wire or tubing; Close wind element on a ϕ1.50-inch tube or rod; Finished winding is ϕ40 mm spaced to a helical angle of 12.3°, or 28-mm spacing. These dimensions work out to have the element centerline to be of a cylindrical circumference of 1.0λ.

When WD4FAB tackled this antenna, he felt that the small amount of helical element support that James Miller used was inadequate, in view of the real life

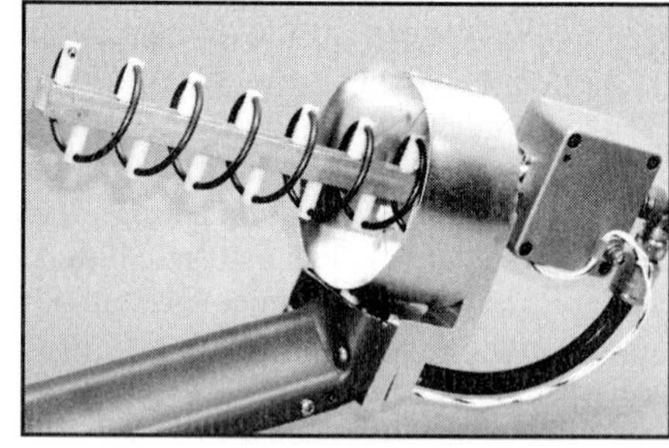

Fig 23.55—Seven-turn LHCP helix dish feed antenna with DEM preamp.

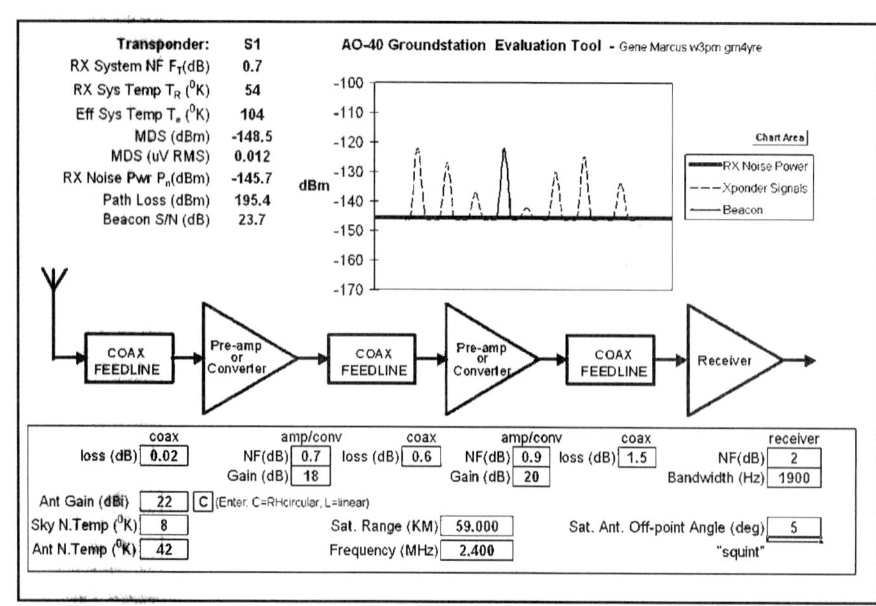

Fig 23.54—Screen display of W3PM's spreadsheet evaluation of ground station operation with AO-40.

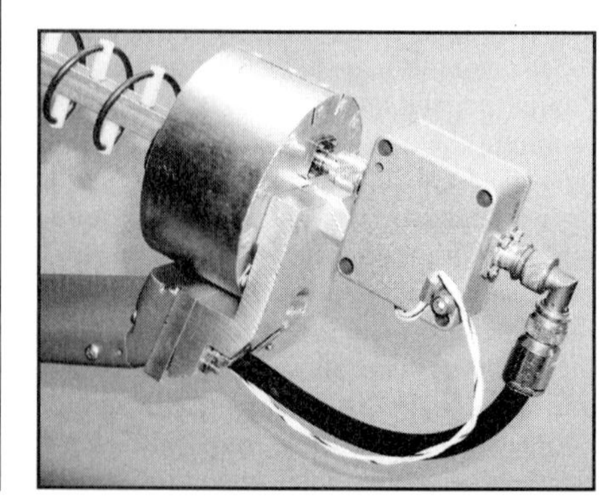

Fig 23.56—Mounting details of seven-turn helix and preamp.

bird traffic on the antennas at his QTH. He chose to use PTFE (Teflon") support posts at every ½ turn. This closer spacing of posts also permitted a careful control of the helix winding diameter and spacing making this antenna *very* robust. A fixture was set up on the drill press to uniformly predrill the holes for the element spacers and boom. Attachment of the reflector is through three very small aluminum angle brackets on the element side of the boom.

Mounting of the helix to the dish requires modification of the dish's receiver mounting boom. **Fig 23.56** shows these modifications using a machined mount. NM2A has constructed one of these antennas and shown that a machine shop is not needed for this construction. He has made a "Z" shaped mount from aluminum angle plate and then used a spacer from a block of acrylic sheet. The key here is to get the dish focal point at the 1.5-turn point of the feed antenna, which is also at about the lip of the reflector cup.

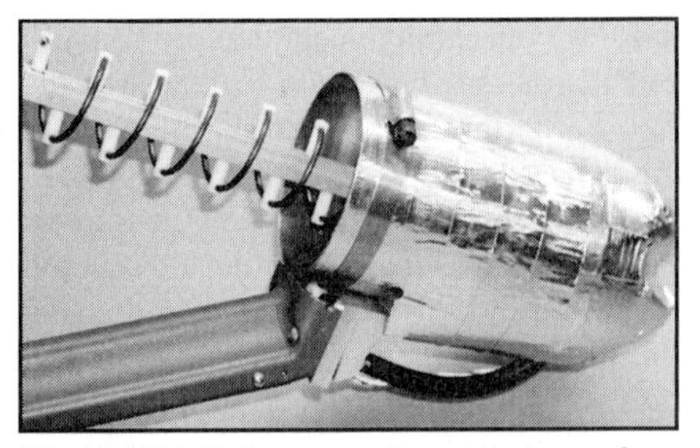

Fig 23.57—Rain cover for preamp using a two-liter soft-drink bottle with aluminum foil tape for protection from sun damage.

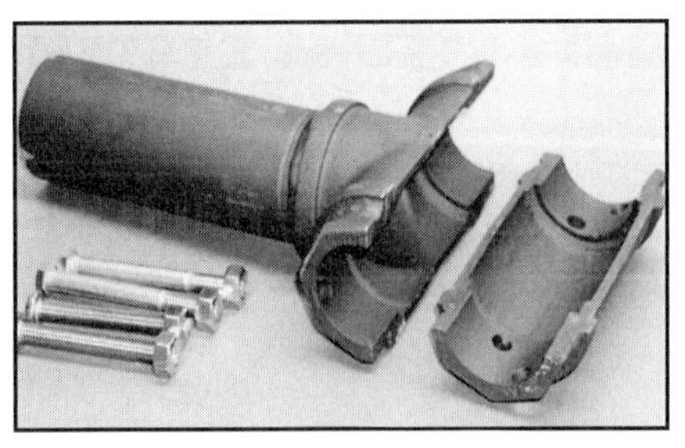

Fig 23.58—Welded pipe fitting mount bracket for FABStar dish antenna.

The W1GHZ data for this focal point is 500.6 mm from the bottom edge of the dish and 744.4 mm from the top edge. A two-string measurement of this point can confirm the focal point, all as shown by Wade in his writings. When mounting this feed antenna the constructor must be cautious to aim the feed at the *beam-center* of the dish, and not the geometric center, as the original microwave horn antenna were constructed. Taking the illumination angle information noted above, the helix feed antenna should be aimed 5.5° down from the geometric center of the dish.

As illustrated in Fig 23.56, a DEM preamp was directly mounted to the feed helix, using TNC connector that had been chosen for this case, as an N connector is quite large for the S-band helix. A male chassis mount connector should be mounted on either the preamp or the antenna so that the preamp can be directly connected to the antenna without any adaptors. This photo also illustrates how the reflector cup walls were riveted to the reflector plate. Exposed connectors must be protected from rainwater. Commonly, materials such as messy Vinyl Mastic Pads (3M 2200) or Hand Moldable Plastic (Coax Seal) are used. Since this is a tight location for such mastic applications, a rain cover was made instead from a two-liter soft-drink bottle as shown in **Fig 23.57**. Properly cutting off the top of the bottle allows it to be slid over the helix reflector cup and secured with a large hose clamp. Sun-damage protection of the plastic bottle must be provided and that was done with a wrapping of aluminum foil pressure-sensitive-adhesive tape.

There are many methods for mounting this dish antenna to your elevation boom. Constructors must give consideration to the placement of the dish to reduce the wind loading and off-balance to the rotator system by this mounting. In the illustrated installation, the off-balance issue was not a major factor, but the dish was placed near the center of the elevation boom, between the pillow block bearing supports. As there is already sizeable aluminum plate for these bearings, the dish was located to "cover" part of that plate, so as to not add measurably to the existing wind-loading area of the overall assembly. A mounting bracket provided with the stock dish clamps to the end of a standard 2-inch pipe (actual measure: ϕ2.38 inch) stanchion. This bracket was turned around on the dish and clamped to the leg of a welded pipe Tee assembly, see **Fig 23.58**. Pipe reducing fittings were machined and fitted in the Tee top bar, which was cut in half for clamping over the 1½-inch pipe used for the elevation boom. Bolts were installed through drilled hole and used to clamp this assembly.

AN INTEGRATED DUAL-BAND ANTENNA SYSTEM

An effective system can be built starting with off-the-shelf components—the Teksharp 1.2-meter dish and AIDC-3731AA downconverter. These systems include not only antennas, but also high quality receiving converters, both of which are covered here.

Operators are finding an increased availability of components for their stations, providing them with a broad selection of the needed equipment.[9] The products described here allow a single dish antenna to provide for the required dual-band operation, with an S-band (2.3 GHz) downlink and L-band (1.2 GHz) uplink.

A new company has surfaced on the amateur equipment horizon, Teksharp. Teksharp is providing kits for both 1.2- and 1.8-meter-diameter parabolic dish antennas. Additionally, S-band receiving converters are available from a number of different suppliers, including Teksharp. Here, Dick Jansson, WD4FAB, selected a well-proven converter, the modified AIDC-3731AA, from Bob Seydler, K5GNA. Bob takes commercial converters and modifies them for S-band amateur service, including an upgrade of the front-end band-pass filtering. This down-converter was selected for its superior and proven passband filtering, allowing an L-band transmitting uplink on the same antenna as the S-band downlink.

The antenna system described here includes the dish antenna, its dual-band feed system and the receiving converter. These three elements are critical for high quality communications. This antenna system was combined with a proven 40-W, tower mounted L-band amplifier, also described in this chapter, resulting in a fully functioning station.

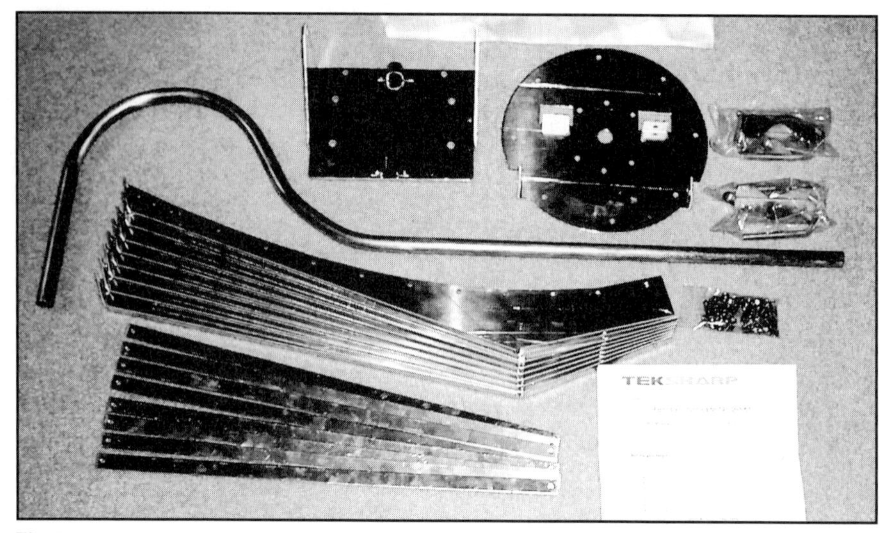
Fig 23.59—The antenna parts as received from Teksharp.

THE TEKSHARP 1.2 METER DISH

The assembly methods are expected to be the same for the 1.8-meter dish. That larger antenna provides additional performance and flexibility, if desired. The gains for the 1.2-meter dish are 21.0 dBi at L-band, and 26.6 dBi at S-band.

For this project, the antenna arrived in a compact, 36×12×4 inch box of parts. **Fig 23.59** shows these parts laid out for assembly. A set of well machined and formed aluminum ribs is the key to any parabolic dish design and these are well executed. WD4FAB measured the dimensions of these ribs to understand the basic dimensional features of the dish. These figures are 1204 mm, 245 mm and 370 mm for diameter, depth and focus, respectively. Therefore, the resulting focus to diameter ratio (F/D) is 0.307.

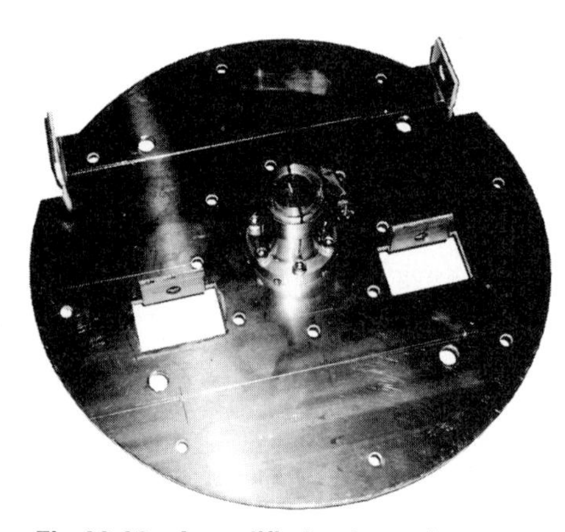
Fig 23.60—A modified antenna base plate assembly. See the text for a description.

Separate plastic packages include the antenna assembly screws and the antenna mounting bolts. Teksharp also sent some well illustrated assembly instructions and a plastic template for the petals (or gores) for the hardware cloth covering the dish.

The included hex-head screws and nuts for the antenna assembly were high quality stainless steel. The U bolts and mast-mounting clamp hardware provided were also high quality, but the bolts and clamps were conventionally plated. It is useful to note that parts fabricated and finished in these materials can quickly corrode and become unusable in a humid climate. The machined parts for these antennas were very well made and finished, and they were a pleasure to work with.

Teksharp provides a well-suited bracket plate that supports the feed boom while also providing for the mounting of the dish to the elevation boom. In the installation shown in **Fig 23.60** at WD4FAB, a modified mounting from a previous dish described elsewhere in this chapter was used. If your arrangement is more conventional, you may be able to make use of the Teksharp mounting hardware without modification.

Before starting the assembly, a copy of a rib shape was made in order to develop some cardboard templates representing the dish surface. A 602-mm circular radius template was also fabricated to aid the author in shaping the perimeter bands to dimension rather than by guess. The dish-surface template was used to check and maintain the parabolic shape of the dish surface as the mesh cloth petals were assembled to the ribs. These petals do not always follow the rib shape in the space between the ribs and the use of the template helped correct those surface inaccuracies while still under construction. The **Antennas** chapter of this *Handbook* and the *ARRL Antenna Book*[10] provide excellent information on dish antenna construction methods.

Following the well-written Teksharp instructions, the ribs were assembled to the circular hub plate and the structure then closed with the formed perimeter bands. In the author's shop, he fashioned the ability to hold the framework in a shop vise, making the assembly task much easier. On one of the rib ends at the top of the dish, a nut was replaced with a riveted-in-place blind anchor nut to allow the coax cables to be clamped to the dish rim without having to simultaneously handle too much loose hardware while performing that step.

It is now time to create a parabolic dish

Fig 23.61—Dish antenna during assembly.

out of this strong and lightweight framework. The constructor must provide his own fabric cloth for this part of the assembly. The kit recommendations are to use ¼-inch galvanized cloth, a low cost item available at most hardware stores. The author had some eight-wires-per-inch aluminum cloth left over from a previous attempt to build a dish antenna. This was ideal, as galvanized material has a habit of eventually rusting away in a humid climate. **Fig 23.61** shows the partially assembled dish in the author's shop.

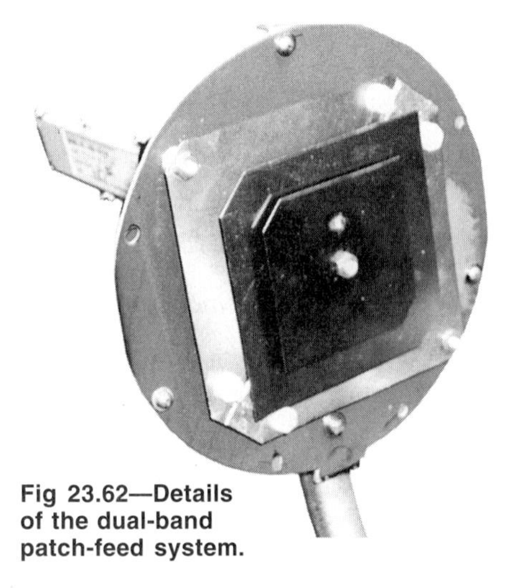

Fig 23.62—Details of the dual-band patch-feed system.

ANTENNA FEED

Early dish antennas for these ranges used helical antennas as feeds. The experiences of other operators in the testing of AO-40 antennas has shown that low F/D parabolic antennas, such as this one, are most effectively fed using patch rather than helical antennas. (See the several articles by Jerry Brown, K5OE, at **members.aol.com/k5oe/**. Also see articles by Robert Suding, WØLMD, at **www.ultimatecharger.com/dish.html**.)

WØLMD provides some useful multiband patch antennas that can be purchased for this service, as does Teksharp. The author decided to follow K5OE's guidance as recently published (see **Fig 23.62**).[11] The sharp-eyed reader may observe that WD4FAB made some modifications to the original patch design, through the addition of some nylon screws on the L-band patch. Please be advised to build this antenna exactly as in the original article unless you have some quality laboratory test equipment! Fortunately, WD4FAB had the use of a good network analyzer at the AMSAT Laboratory.

AIDC-3731AA S BAND DOWNCONVERTER

The next element of this system is the AIDC-3731AA downconverter from K5GNA. This converter offers one of the lower-cost approaches to receiving AO-40's 2401 MHz downlink signal. **Fig 23.63** shows the converter mounted directly to the male Type N connector on the S-band patch antenna. The output port provides a 2-meter IF output through the F connector on the side.

This converter is now available with a low overhead internal voltage regulator, allowing it to be powered by a 13.8 V dc supply source and still obtain proper regulation of the 12 V dc needed for the internal circuits. There is a considerable advantage in using this regulation scheme as it keeps the total power dissipation low, thus holding down the temperature rise from the 2.5 W of power dissipation. There is interest in keeping this power dissipation low to minimize frequency shift due to temperature rise. Converters exposed to the sun will have noticeable frequency shifts with clouds shading the sun. This 13.8 V dc power is supplied to the converter through a bias T in the shack and RG-6 coaxial cable to the IF connector of the converter.

FEED ASSEMBLY

Following the directions in K5OE's article for the dual-band patch antenna, WD4FAB provided a mounting angle on the patch assembly using a piece of steel strapping angle. The author also machined an aluminum sleeve adapter for mounting between the J-shaped feed boom and this steel angle, as can be seen in Fig 23.63.

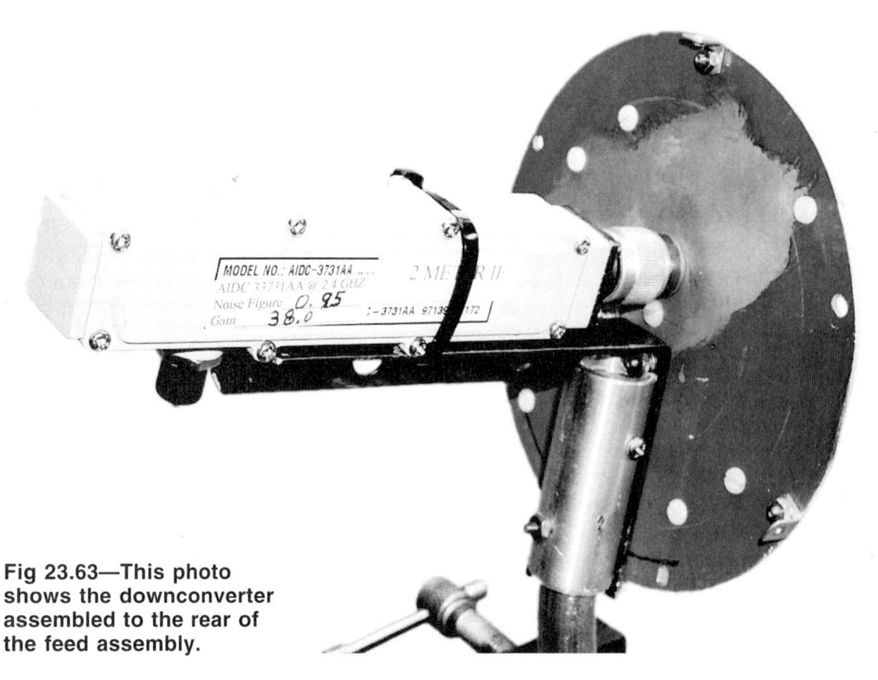

Fig 23.63—This photo shows the downconverter assembled to the rear of the feed assembly.

Glossary of Satellite Terminology

AMSAT—A registered trademark of the Radio Amateur Satellite Corporation, a nonprofit scientific/educational organization located in Washington, DC. It builds and operates Amateur Radio satellites and has sponsored the OSCAR program since the launch of OSCAR 5. (AMSAT, PO Box 27, Washington, DC 20044.)

Anomalistic period—The elapsed time between two successive perigees of a satellite.

AO-#—The designator used for AMSAT OSCAR spacecraft in flight, by sequence number.

AOS—Acquisition of signal. The time at which radio signals are first heard from a satellite, usually just after it rises above the horizon.

Apogee—The point in a satellite's orbit where it is farthest from Earth.

Area coordinators—An AMSAT corps of volunteers who organize and coordinate amateur satellite user activity in their particular state, municipality, region or country. This is the AMSAT grassroots organization set up to assist all current and prospective OSCAR users.

Argument of perigee—The polar angle that locates the perigee point of a satellite in the orbital plane; drawn between the ascending node, geocenter, and perigee; and measured from the ascending node in the direction of satellite motion.

Ascending node—The point on the ground track of the satellite orbit where the sub-satellite point (SSP) crosses the equator from the Southern Hemisphere into the Northern Hemisphere.

Az-el mount—An antenna mount that allows antenna positioning in both the azimuth and elevation planes.

Azimuth—Direction (side-to-side in the horizontal plane) from a given point on Earth, usually expressed in degrees. North = 0° or 360°; East = 90°; South = 180°; West = 270°.

Circular polarization (CP) — A special case radio energy emission where the electric and magnetic field vectors rotate about the central axis of radiation. As viewed along the radiation path, the rotation directions are considered to be right-hand (RHCP) if the rotation is clockwise, and left-hand (LHCP) if the rotation is counterclockwise.

Descending node — The point on the ground track of the satellite orbit where the sub-satellite point (SSP) crosses the equator from the Northern Hemisphere into the Southern Hemisphere.

Desense — A problem characteristic of many radio receivers in which a strong RF signal overloads the receiver, reducing sensitivity.

Doppler effect — An apparent shift in frequency caused by satellite movement toward or away from your location.

Downlink — The frequency on which radio signals originate from a satellite for reception by stations on Earth.

Earth station — A radio station, on or near the surface of the Earth, designed to transmit or receive to/from a spacecraft.

Eccentricity — The orbital parameter used to describe the geometric shape of an elliptical orbit; eccentricity values vary from e = 0 to e = 1, where e = 0 describes a circle and e = 1 describes a straight line.

EIRP — Effective isotropic radiated power. Same as ERP except the antenna reference is an isotropic radiator.

Elliptical orbit — Those orbits in which the satellite path describes an ellipse with the Earth at one focus.

Elevation — Angle above the local horizontal plane, usually specified in degrees. (0° = plane of the Earth's surface at your location; 90° = straight up, perpendicular to the plane of the Earth).

Epoch — The reference time at which a particular set of parameters describing satellite motion (***Keplerian elements***) are defined.

EQX — The reference equator crossing of the ascending node of a satellite orbit, usually specified in UTC and degrees of longitude of the crossing.

ERP — Effective radiated power. System power output after transmission-line losses and antenna gain (referenced to a dipole) are considered.

ESA — European Space Agency. A consortium of European governmental groups pooling resources for space exploration and development.

FO-# — The designator used for Japanese amateur satellites, by sequence number. Fuji-OSCAR 12 and Fuji-OSCAR 20 were the first two such spacecraft.

Geocenter — The center of the Earth.

Geostationary orbit — A satellite orbit at such an altitude (approximately 22,300 miles) over the equator that the satellite appears to be fixed above a given point.

Groundtrack — The imaginary line traced on the surface of the Earth by the subsatellite point (SSP).

Inclination — The angle between the orbital plane of a satellite and the equatorial plane of the Earth.

Increment — The change in longitude of ascending node between two successive passes of a specified satellite, measured in degrees West per orbit.

Iskra — Soviet low-orbit satellites launched manually by cosmonauts aboard Salyut missions. Iskra means "spark" in Russian.

JAMSAT — Japan AMSAT organization.

Keplerian Elements — The classical set of six orbital element numbers used to define and compute satellite orbital motions. The set is comprised of inclination, Right Ascension of Ascending Node (RAAN), eccentricity, argument of perigee, mean anomaly and mean motion, all specified at a particular epoch or reference year, day and time. Additionally, a decay rate or drag factor is usually included to refine the computation.

LEO — Low Earth Orbit satellite such as the Phase 1 and Phase 2 OSCARs.

LHCP — Left-hand circular polarization.

LOS — Loss of signal — The time when a satellite passes out of range and signals from it can no longer

This angle also provided an added support for the AIDC-3731AA converter, shown strapped down with a plastic cable tie. As assembled, this is a robust feed assembly.

Mounting of the patch assembly was done with the patches aligned parallel to the plane of the dish and on the dish center line. Place the S-band reflector at the focal distance, 370 mm (14.57 inches), from the dish surface at the center. This adjustment is made by moving the feed boom along its mounting clamps until the proper measurement is achieved. That is all you have to do. You can play with mirrors, as in the instructions, to focus the sun on the patches, but doing this adjustment by measurement is easier and it is all that really needs to be done.

K5OE cautions that rain, bird droppings and bugs can mess up the tuning and operation of the patch feed antennas. He advises the use of some kind of protection for antennas. In addition, the author has an aversion to having water in hard-to-reach be heard. This usually occurs just after the satellite goes below the horizon.

Mean anomaly (MA) — An angle that increases uniformly with time, starting at perigee, used to indicate where a satellite is located along its orbit. MA is usually specified at the reference epoch time where the Keplerian elements are defined. For AO-10 the orbital time is divided into 256 parts, rather than degrees of a circle, and MA (sometimes called phase) is specified from 0 to 255. Perigee is therefore at MA = 0 with apogee at MA = 128.

Mean motion — The Keplerian element to indicate the complete number of orbits a satellite makes in a day.

Microsat — Collective name given to a series of small amateur satellites having store-and-forward capability (OSCARs 14-19, for example).

Molniya — Type of elliptical orbit, first used in the Russian Molniya series, that features a ground track that more or less repeats on a daily basis.

NASA — National Aeronautics and Space Administration, the US space agency.

Nodal period — The amount of time between two successive ascending nodes of satellite orbit.

Orbital elements — See ***Keplerian Elements***.

Orbital plane — An imaginary plane, extending throughout space, that contains the satellite orbit.

OSCAR — Orbiting Satellite Carrying Amateur Radio.

PACSAT — Packet radio satellite (see ***Microsat*** and ***UoSAT-OSCAR***).

Pass — An orbit of a satellite.

Passband — The range of frequencies handled by a satellite translator or transponder.

Perigee — The point in a satellite's orbit where it is closest to Earth.

Period — The time required for a satellite to make one complete revolution about the Earth. See ***Anomalistic period*** and ***Nodal period***.

Phase I — The term given to the earliest, short-lived, low Earth orbit (LEO) OSCAR satellites that were not equipped with solar cells. When their batteries were depleted, they ceased operating.

Phase 2 — LEO OSCAR satellites. Equipped with solar panels that powered the spacecraft systems and recharged their batteries, these satellites have been shown to be capable of lasting up to five years (OSCARs 6, 7 and 8, for example).

Phase 3 — Extended-range, high-elliptical-orbit OSCAR satellites with very long-lived solar power systems (OSCARs 10 and 40, for example).

Phase 4 — Proposed OSCAR satellites in geostationary orbits.

Polar Orbit — A low, circular orbit inclined so that it passes over the Earth's poles.

Precession — An effect that is characteristic of AO-10 and AO-40 orbits. The satellite apogee SSP will gradually change over time.

Project OSCAR — The California-based group, among the first to recognize the potential of space for Amateur Radio; responsible for OSCARs I through IV.

QRP days — Special orbits set aside for very low power uplink operating through the satellites.

RAAN — Right Ascension of Ascending Node. The Keplerian element specifying the angular distance, measured eastward along the celestial equator, between the vernal equinox and the hour circle of the ascending node of a spacecraft. This can be simplified to mean roughly the longitude of the ascending node.

Radio Sputnik — Russian Amateur Radio satellites (see ***RS #***).

Reference orbit — The orbit of Phase II satellites beginning with the first ascending node during that UTC day.

RHCP — Right-hand circular polarization.

RS # — The designator used for most Russian Amateur Radio satellites (RS-1 through RS-15, for example).

Satellite pass — Segment of orbit during which the satellite "passes" nearby and in range of a particular ground station.

Sidereal day — The amount of time required for the Earth to rotate exactly 360° about its axis with respect to the "fixed" stars. The sidereal day contains 1436.07 minutes (see ***Solar day***).

Solar day — The solar day, by definition, contains exactly 24 hours (1440 minutes). During the solar day the Earth rotates slightly more than 360° about its axis with respect to "fixed" stars (see ***Sidereal day***).

Spin modulation — Periodic amplitude fade-and-peak resulting from the rotation of a satellite's antennas about its spin axis, rotating the antenna peaks and nulls.

SSP — Subsatellite point. Point on the surface of the Earth directly between the satellite and the geocenter.

Telemetry — Radio signals, originating at a satellite, that convey information on the performance or status of onboard subsystems. Also refers to the information itself.

Transponder — A device onboard a satellite that receives radio signals in one segment of the spectrum, amplifies them, translates (shifts) their frequency to another segment of the spectrum and retransmits them. Also called linear translator.

UoSAT-OSCAR (UO #) — Amateur Radio satellites built under the coordination of radio amateurs and educators at the University of Surrey, England.

Uplink — The frequency at which signals are transmitted from ground stations to a satellite.

Visibility Circle — The range of area on the Earth that are "seen" by a satellite. This is also called the "footprint" for that satellite.

Window — Overlap region between acquisition circles of two ground stations referenced to a specific satellite. Communication between two stations is possible when the subsatellite point is within the window.

coaxial connectors. Water in these connectors does not ensure a good satellite signal! To solve both of these problems, WD4FAB went shopping for an antenna cover at his favorite "antenna parts store," K-Mart. He found help from Martha Stewart, no less, in the form of a nice, clear styrene plastic "3.3 quart airtight canister."

One of these canisters was modified to fit over the patch assembly and the feed boom. This canister was also shortened to allow it to fit into the cover lip of an unmodified canister mounted on the "rear" of the feed. The modified canister was mounted to the patch reflector with three small angle brackets, as seen in Fig 23.63. Some of the cutoff plastic cylinder should be saved. This can then be glued to the end of the cutoff container, to shim that OD to fit into the inside diameter of the cover lip of the rear canister. The author also epoxy-bonded a set of blind nuts to the inside of the canister so that the screws through the cover lip of the rear-mounted canister will

Fig 23.64—The 3.3-quart plastic canister converted for use as a radome at the feedpoint.

keep it together. That arrangement is shown in **Fig 23.64**. The slot in the front canister that passes the feed boom has space for the coaxial cables for L and S band. A further protection measure was to cover the exposed plastic with aluminum foil tape, as seen in Fig 23.64. This was not done to area in front of the patch antennas. This tape is there to protect the plastic from sun damage, although this type of clear plastic has a good record in avoiding UV harm.

INSTALLATION

With all preparations made, completion of the installation of the dish antenna and converter was a breeze. **Fig 23.65** shows the dish assembly bolted to the author's old PrimeStar antenna mount. Just remember to use the built-in screw adjustment provided in the PrimeStar mount to aim the dish centerline along the centerline of your antenna pointing system. The beamwidths of L- and S-band operations in this assembly are fairly narrow, so your pointing precision will need to be pretty good. Many of the details of the WD4FAB elevation boom system are shown in Fig 23.65 as well.

Before finally closing the protective plastic containers over the feed antennas, be sure that all of the coaxial cable connections are tight and the cables are dressed properly. This final closure will need to be done at this step in the assembly. Be sure to arrange the coaxial cables so that they don't drag on the feed boom. The cable clamps at the top of the dish are there to dress the cables and support the feed boom. This is shown in **Fig 23.66**.

Fig 23.66 also shows the dish system in operating position along with the U-band (70 cm) cross-polarized (CP) Yagi antenna on the far left, and the two smaller M2 23CM22EZA antennas that were tested previously. The author has kept these two L-band antennas in service for use on 1296 MHz. He placed a ground-commanded relay to switch between the two antenna systems in the amplifier box on the top of the tower. This is also convenient to use to compare the two L-band antenna systems.

ON THE AIR

This configuration was tested with the AO-40 satellite (which is no longer operational at the time of this writing). Tests showed that AO-40 downlink signals on S band, 2401 MHz, were really solid with this system. The pointing requirements were narrow, and had to be good to within 3° to 4°, based on the specified beam width at these frequencies. The noise floor of AO-40 was clearly detectable, about 2 dB above cold sky noise.

For the uplink, WD4FAB had a usable return signal with 40 W of RF on L band when the satellite "squint" was 20° or lower, as reported in note 10. When the squint is low, good downlink signals were realized with only about 5-10 W PEP of L-band power. When the author compared the Teksharp antenna on L band to the M2 23CM22EZA stacked antennas, as shown in Fig 23.66, he saw better than an S-unit (>3-4 dB on his Kenwood TS-2000 transceiver) improvement in downlink signal. This improved performance was seen in the wider ranges of squint angles that can be used with the Teksharp antenna on the L-band uplink. When the AO-40 squint angles were greater than 20°, he still had to use his U-band uplink transmitter with the CP Yagi antenna.

Measurements of Sun noise have been attempted with this antenna system. These have been disappointing, with values of around 2 dB seen. This is compared to the solid 5 dB of Sun noise measured with the WD4FAB previous PrimeStar dish. One of the probable reasons for the lowered performance can be attributed to the front-end band-pass filtering provided in the receive converter. The operating experience with this system has been absolutely solid, however.

IN SUMMARY

It is important to consider all of the interrelated characteristics of an antenna system. It is insufficient to just whip up "any" antenna, marry "any" receive converter to that antenna and then use "any" transmitting scheme to operate with full-duplex signals from high-performance satellites. We've had the ability to employ computer analysis, as shown to us by Gene Marcus, W3PM, with his spreadsheet analysis: **www.amsat.org/amsat/ftp/software/spreadsheet/w3pm-ao40-v2.1.zip**.

Manufacturers: Antenna: Teksharp (Rick Fletcher, KG6IAL), **www.teksharp.com**, 5770 McKellar Dr, San Jose, CA 95129; **inquiries@teksharp.com**; Price: 1.2 meter dish, $165; feed boom, $40; dual-band (S/L) patch feed, $200. Downconverter Bob Seydler, K5GNA, **members.aol.com/k5gna/myhomepage/**, 8522 Rebawood, Humble, TX 77346-1789; 281-852-0252, **bob@k5gna.com**. Price: $100, setup for 12 V operation add $20.

JUST THE BEGINNING

This section barely nicks the surface of satellite operating. There is much more to learn and enjoy. It is suggested that you spend some time at the AMSAT Web site. You'll pick up a wealth of information there. Speaking of "picking up," grab a copy of the *ARRL Radio Amateur's Satellite Handbook* and the *ARRL Antenna Book* (see your favorite dealer, or buy it on *ARRLWeb*). Between these two resources you'll be able to tap just about all the amateur satellite knowledge you're likely to need.

In the meantime, see you in orbit!

NOTES

[1] *QST*, December 1998, "An Inexpensive Az-El Rotator System", Jim Koehler, VE5FP.

[2] McCaa, William D., "Hints on Using the AMSAT-OSCAR 13 Mode S Transponder," *The AMSAT Journal*, Vol 13, no. 1, Mar 1990, pp 21-22.

[3] Miller, James, "Mode S — Tomorrow's Downlink?" *The AMSAT Journal*, Vol 15, no. 4, Sep/Oct 1992, pp 14-15.

[4] Krome, Ed, "S band Reception: Building the DEM Converter and Preamp Kits," *The AMSAT Journal*, Vol 16, no. 2, Mar/Apr 1993, pp 4-6.

Fig 23.65—Mounting details of the antenna system.

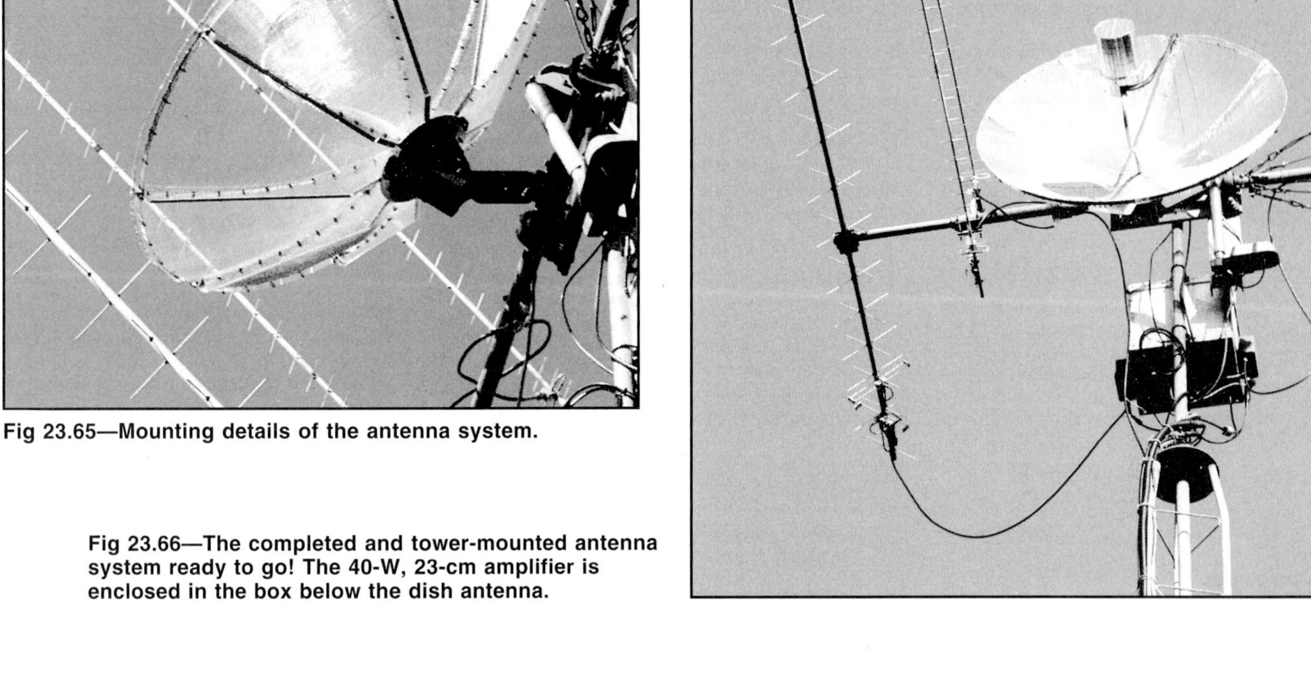

Fig 23.66—The completed and tower-mounted antenna system ready to go! The 40-W, 23-cm amplifier is enclosed in the box below the dish antenna.

[5]Krome, Ed, "Development of a Portable Mode S Ground Station." *The AMSAT Journal*, Vol 16, no. 6, Nov/Dec 1993, pp 25-28.

[6]Krome, Ed, "Mode S: Plug and Play!" *The AMSAT Journal*, Vol 14, no. 1, Jan 1991, pp 21-23, 25.

[7]Miller, James, "A 60-cm S-Band Dish Antenna," *The AMSAT Journal*, Vol 16 no. 2, Mar/Apr 1993, pp 7-9.

[8]Miller, James, "Small is Best," *The AMSAT Journal*, Vol 16, no. 4, Jul/Aug 1993, p 12.

[9]R. Jansson, WD4FAB, "Product Review-M2 23CM22EZA 1.2 GHz Antenna," *QST*, Sep 2002, pp 59-61.

[10]*The ARRL Antenna Book*, 20th Edition, Chapter 19. Available from your local dealer or The ARRL Bookstore. ARRL order no. 9043. Telephone toll-free in the US 888-277-5289, or 860-594-0355, fax 860-594-0303; **www.arrl.org/shop/**; **pubsales@arrl.org**.

[11]G. Brown, K5OE, "Build This No-Tune Dual-Band Feed for Mode L/S," *The AMSAT Journal*, Vol 26, no.1, Jan/Feb 2003, pp 12-16.

SELECTED SATELLITE REFERENCES

Equipment Selection

D. DeMaw, "Trio-Kenwood TS-700S 2-Meter Transceiver," *QST*, Feb 1978, pp 31-32.

C. Hutchinson, "Ten-Tec 2510 Mode B Satellite Station," *QST*, Oct 1985, pp 41-43.

D. Ingram, "The Ten-Tec 2510 OSCAR Satellite Station/Converter," *CQ Magazine*, Feb 1985, pp 44-46.

J. Kleinman, "ICOM IC-290H All-Mode 2-Meter Transceiver," *QST*, May 1983, pp 36-37.

J. Lindholm, "ICOM IC-471A 70-cm Transceiver," *QST*, Aug 1985, pp 38-39.

R. Schetgen, Ed., *The ARRL Radio Buyer's Sourcebook* and *The ARRL Radio Buyer's Sourcebook, Vol 2* (Newington: ARRL, 1991 and 1993). A compilation of Product Reviews from *QST*.

R. Roznoy, Ed. *The ARRL VHF/UHF Radio Buyer's Sourcebook* (Newington: ARRL, 1997). A compilation of VHF and UHF Product Reviews from *QST*.

M. Wilson, "ICOM IC-271A 2-Meter Multimode Transceiver," *QST*, May 1985, pp 40-41.

M. Wilson, "Yaesu Electronics Corp. FT-726R VHF/UHF Transceiver," *QST*, May 1984, pp 40-42.

M. Wilson, "Yaesu FT-480R 2-Meter Multimode Transceiver," *QST*, Oct 1981, pp 46-47.

H. Winard and R. Soderman, "A Survey of OSCAR Station Equipment," *Orbit*, no. 16, Nov-Dec 1983, pp 13-16 and no. 18, Mar-Apr 1984, pp 12-16.

S. Ford, "PacComm PSK-1T Satellite Modem and TNC," *QST*, Jul 1993, p 46.

R. Healy, "Down East Microwave 432PA 432-MHz Amplifier Kit," *QST*, Mar 1993, p 66.

R. Healy, "Down East Microwave DEM432 No-Tune 432-MHz Transverter," *QST*, Mar 1993, p 64.

R. Jansson, "SSB Electronic SP-70 Mast-Mount Preamplifier," *QST*, Mar 1993, p 63.

S. Ford, "QST Compares: SSB Electronic UEK-2000S and Down East Microwave SHF-2400 2.4-GHz Satellite Down-converters," *QST*, Feb 1994, p 69.

S. Ford, "ICOM IC-821H VHF/UHF Multimode Transceiver," *QST*, Mar 1997, pp 70-73.

S. Ford, "Yaesu FT-847 HF/VHF/UHF Transceiver," *QST*, Aug 1998, pp 64-69.

Antennas

M. Davidoff, "Off-Axis Circular Polarization of Two Orthogonal Linearly Polarized Antennas," *Orbit*, no. 15, Sep-Oct 1983, pp 14-15.

J.L. DuBois, "A Simple Dish for Mode-L," *Orbit*, no. 13, Mar-Apr 1983, pp 4-6.

B. Glassmeyer, "Circular Polarization and OSCAR Communications," *QST*, May 1980, pp 11-15.

R.D. Straw, Ed., *The ARRL Antenna Book*

(Newington: ARRL, 2003). Available from your local radio store or from ARRL.
R. Jansson, "Helical Antenna Construction for 146 MHz," *Orbit*, May-Jun 1981, pp 12-15.
R. Jansson,"KLM 2M-22C and KLM 435-40CX Yagi Antennas," *QST*, Oct 1985, pp 43-44.
R. Jansson, "70-Cm Satellite Antenna Techniques," *Orbit*, no. 1, Mar 1980, pp 24-26.
R. Messano, "An Indoor Loop for Satellite Work," *QST*, Jul 1999, p 55.
C. Richards, "The Chopstick Helical," *Orbit*, no. 5, Jan-Feb 1981, pp 8-9.
V. Riportella, "Amateur Satellite Communications," *QST*, May 1985, pp 70-71.
G. Schrick, "Antenna Polarization," *The ARRL Antenna Compendium Vol. 1* (Newington: ARRL, 1985), pp 152-156.
A. Zoller, "Tilt Rather Than Twist," *Orbit*, no. 15, Sep-Oct 1983, pp 7-8.
R. Jansson, "M2 Enterprises 2M-CP22 and 436-CP30 Satellite Yagi Antennas," *QST*, Nov 1992, p 69.
S. Ford, "M2 Enterprises EB-144 Eggbeater Antenna," *QST*, Sep 1993, p 75.

Microsats

Mills, S.E., "Step Up to the 38,400 Bps Digital Satellites," *QST*, Apr 2000, pp 42-45.
Loughmiller, D and B. McGwier, "Microsat: The Next Generation of OSCAR Satellites," *Part 1*: *QST*, May 1989, pp 37-40. *Part 2*: *QST*, Jun 1989, pp 53-54, 70.
Loughmiller, D., "Successful OSCAR Launch Ushers in the '90s," *QST*, May 1990, pp 40-41.
Loughmiller, D., "WEBERSAT-OSCAR 18: Amateur Radio's Newest Eye in the Sky," *QST*, Jun 1990, pp 50-51.
Ford, S., "The Road Less Traveled," *QST*, Apr 1996, p 58.
Ford, S., "KITSAT-OSCAR 23 Reaches Orbit," *QST*, Oct 1992, p 93.
Ford, S., "Satellite on a String: SEDSAT-1," *QST*, Nov 1992, p 113.
Ford, S., "WEBERSAT—Step by Step," *QST*, Dec 1992, p 63.
Ford, S., "AMRAD-OSCAR 27," *QST*, Dec 1993, p 107.
Ford, S., "Two More PACSATs!," *QST*, Oct 1993, p 98.
Soifer, R., "AO-27: An FM Repeater in the Sky," *QST*, Jan 1998, pp 64-65.
Ford, S., "Meet the Multifaceted SUNSAT," *QST*, Mar 1999, p 90.
Ford, S., "Flash! Three New Satellites in Orbit!," *QST*, Mar 2000, p 92.
Soifer, R., "UO-14: A User-Friendly FM Repeater in the Sky," *QST*, Aug 2000, p 64.

OSCAR 40

Coggins, B., "A Box for Phase 3D," *QST*, Jun 1997, p 94.
Ford, S. and Z. Lau, "Get Ready for Phase 3D!" *Part 1*: *QST*, Jan 1997, p 28; *Part 2*: *QST*, Feb 1997, p 50; *Part 3*: *QST*, Mar 1997, p 42; *Part 4*: *QST*, Apr 1997, p 45; *Part 5*: *QST*, May 1997, p 28.
Ford, S., "Phase 3D: The Ultimate EasySat," *QST*, May 1995, p 21.
Ford, S., "Phase 3D Update," Amateur Satellite Communication, *QST*, Dec 1994, p 105.
Krome, E., "Getting Started with AMSAT-OSCAR 40," *QST*, July 2001, p 42.
Krome, E., *Mode S: The Book, 2001 AO-40/P3D Update*. Complete guide to operating satellite S-band. Available from AMSAT-NA.
Tynan, B. and R. Jansson, "Phase 3D—A Satellite for All," Part 1: *QST*, May 1993, p 49; Part 2: *QST*, Jun 1993, p 49.
"Putting the Finishing Touches on Phase 3D," *QST*, Oct 1988, p 20.

Fuji-OSCARs

Ford, S., "K1CE's Secret OSCAR-20 Station," *QST*, Mar 1993, p 47.
Ford, S., "Fuji-OSCAR 20 Mode JA," *QST*, Sep 1993, p 104.

Software

Ford, S., "Short Takes—Nova for Windows 32," *QST*, Apr 2000, p 65.
Ford, S., "Will O' the WiSP," *QST*, Feb 1995, p 90.
Ford, S., "Satellite-Tracking Software," *QST*, Dec 1993, p 89.

General Texts and Articles

M. Davidoff, *The Radio Amateur's Satellite Handbook* (Newington: ARRL, 2000).
S. Ford, "An Amateur Satellite Primer," *QST*, Apr 2000, p 36.
G. McElroy, "Keeping Track of OSCAR: A Short History," *QST*, Nov 1999, p 65.
The ARRL Satellite Anthology (Newington: ARRL, 1999).
The ARRL Operating Manual (Newington: ARRL, 2000).
The AMSAT-Phase III Satellite Operations Manual, prepared by Radio Amateur Satellite Corp and Project OSCAR, Inc. 1985. Available from AMSAT.

Earth-Moon-Earth (EME)

EME communication, also known as "moonbounce," has become a popular form of space communication. The concept is simple: The moon is used as a passive reflector for VHF and UHF signals. With a total path length of nearly 500,000 miles, EME is the ultimate DX. EME is a natural and passive propagation phenomenon, and EME QSOs count toward the WAS, DXCC and VUCC awards. EME opens up the VHF and UHF bands to a new universe of worldwide DX.

The first demonstration of EME capability was done by the US Army Signal Corps just after WW II. In the 1950s, using 400 MW of effective radiated power, the US Navy established a moon relay link between Washington, DC, and Hawaii that could handle four multiplexed Teletype (RTTY) channels. The first successful amateur reception of EME signals occurred in 1953 by W4AO and W3GKP.

It took until 1960 for two-way amateur communications to take place. Using surplus parabolic dish antennas and high-power klystron amplifiers, the Eimac Radio Club, W6HB, and the Rhododendron Swamp VHF Society, W1BU, accomplished this milestone in July 1960 on 1296 MHz. In the 1960s, the first wave of amateur EME enthusiasts established amateur-to-amateur contacts on 144 MHz and 432 MHz. In April 1964, W6DNG and OH1NL made the first 144-MHz EME QSO. 432-MHz EME experimentation was delayed by the 50-W power limit (removed January 2, 1963). Only one month after the first 144-MHz QSO was made, the 1000-ft-diameter dish at Arecibo, Puerto Rico, was used to demonstrate the viability of 432-MHz EME, when a contact was made between KP4BPZ and W1BU. The first amateur-to-amateur 432-MHz EME QSO occurred in July 1964 between W1BU and KH6UK.

The widespread availability of reliable low-noise semiconductor devices along with significant improvements in Yagi arrays ushered in the second wave of amateur activity in the 1970s. Contacts between stations entirely built by amateurs became the norm instead of the exception. In 1970, the first 220- and 2304-MHz EME QSOs were made, followed by the first 50-MHz EME QSO in 1972.

Fig 23.67—Tommy Henderson, WD5AGO, pursues 144-MHz EME from his Tulsa, Oklahoma, QTH with this array. Local electronics students helped with construction.

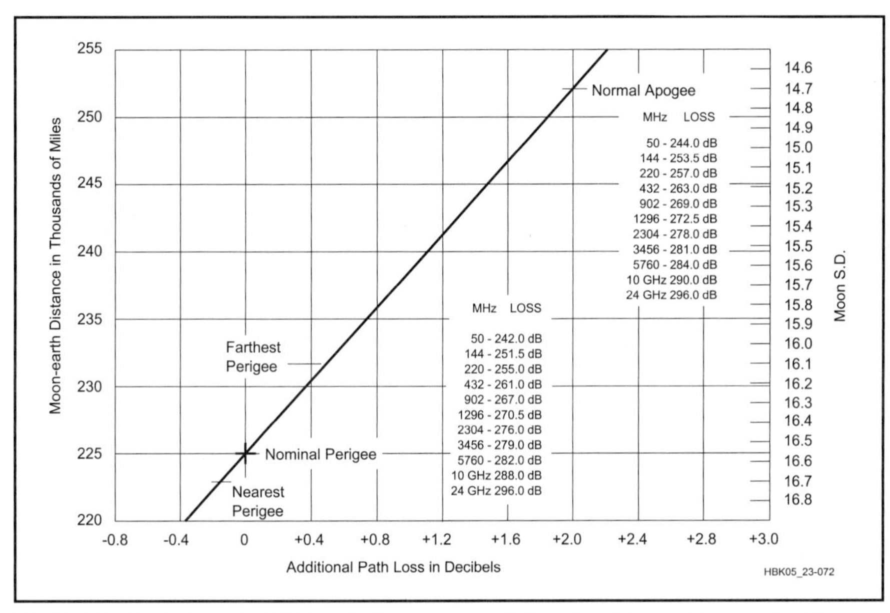

Fig 23.68—Variations in EME path loss can be determined from this graph. SD refers to semi-diameter of the moon, which is indicated for each day of the year in *The Nautical Almanac*.

1970s activity was still concentrated on 144 and 432 MHz, although 1296-MHz activity grew.

As the 1980s approached, another quantum leap in receive performance occurred with the use of GaAsFET preamplifiers. This, and improvements in Yagi performance (led by DL6WU's log-taper design work), and the new US amateur power output limit of 1500 W have put EME in the grasp of most serious VHF and UHF operators. The 1980s saw 144- and 432-MHz WAS and WAC become a reality for a great number of operators. The 1980s also witnessed the first EME QSOs on 3456 MHz and 5760 MHz (1987), followed by EME QSOs on 902 MHz and 10 GHz (1988).

EME is still primarily a CW mode. As stations have improved, SSB is now more popular. Regardless of the transmission mode, successful EME operating requires:

1) As close to the legal power output as possible.

2) A fairly large array (compared to OSCAR antennas).

3) Accurate azimuth and elevation rotation.

4) Minimal transmission-line losses.

5) A low system noise figure, preferably with the preamplifier mounted at the array.

CHOOSING AN EME BAND

Making EME QSOs is a natural progression for many weak-signal terrestrial operators. Looking at EME path loss vs frequency (**Fig 23.68**), it may seem as if the lowest frequency is best, because of reduced path loss. This is not entirely true. The path-loss graph does not account for the effects of cosmic and man-made noise, nor does it relate the effects of ionospheric scattering and absorption. Both short- and long-term fading effects also must be overcome.

50-MHz EME is quite a challenge, as the required arrays are very large. In addition, sky noise limits receiver sensitivity at this frequency. Because of power and licensing restrictions, it is not likely that many foreign countries will be able to get on 50-MHz EME.

144 MHz is probably the easiest EME band to start on. It supports the largest number of EME operators. Commercial equipment is widely available; a 144-MHz EME station can almost be completely assembled from off-the-shelf equipment. 222 MHz is a good frequency for EME, but there are only a handful of active stations, and 222 MHz is available only in ITU Region 2.

432 MHz is the most active EME band after 144 MHz. Libration fading is more of a problem than at 144 MHz, but sky noise is more than an order of magnitude less than on 144 MHz. The improved receive signal-to-noise ratio may more than make up for the more rapid fading. However, 432-MHz activity is most concentrated into the one or two weekends a month when conditions are expected to be best.

902 MHz and above should be considered if you primarily enjoy experimenting and building equipment. If you plan to operate at these frequencies, an unobstructed moon window is a must. The antenna used is almost certain to be a dish. 902 MHz has the same problem that 222 MHz has — it's not an international band. Equipment and activity are expected to be limited for many years.

1296 MHz currently has a good amount of activity from all over the world. Recent equipment improvements indicate 1296 MHz should experience a significant growth in activity over the next few years. 2300 MHz has received renewed interest. It suffers from nonaligned international band assignments and restrictions in different parts of the world.

ANTENNA REQUIREMENTS

The tremendous path loss incurred over the EME circuit requires a high-power transmitter, a low-noise receiver and a high-performance antenna array. Although single-Yagi QSOs are possible, most new EME operators will rapidly become frustrated unless they are able to work many different stations on a regular basis. Because of libration fading and the nature of weak signals, a 1- or 2-dB increase in array gain will often be perceived as being much greater. An important antenna parameter in EME communications is the antenna noise temperature. This refers to the amount of noise received by the array. The noise comes from cosmic noise (noise generated by stars other than the sun), Earth noise (thermal noise radiated by the Earth), and noise generated by man-made sources such as power-line leaks and other broadband RF sources.

Yagi antennas are almost universally used on 144 MHz. Although dish antennas as small as 24 ft in diameter have been successfully used, they offer poor gain-to size trade-offs at 144 MHz. The minimum array gain for reliable operation is about 18 dBd (20.1 dBi). The minimum array gain should also allow a station to hear its own echoes on a regular basis. This is possible by using four 2.2-λ Yagis. The 12-element 2.5-λ Yagi described in the **Antennas** chapter is an excellent choice. When considering a Yagi design, you should avoid old-technology Yagis, that is, designs that use either constant-width spacings, constant-length directors or a combination of both. These old-design Yagis will have significantly poorer side lobes, a narrower gain bandwidth and a sharper SWR bandwidth than modern log-taper designs. Modern wideband designs will behave much more predictably when stacked in arrays, and, unlike many of the older designs, will deliver close to 3 dB of stacking gain.

222-MHz requirements are similar to those of 144 MHz. Although dish antennas are somewhat more practical, Yagis still predominate. The 16-element 3.8-λ Yagi described in the **Antennas** chapter is a good building block for 222-MHz EME. Four of these Yagis are adequate for a minimal 222-MHz EME station, but six or eight will provide a much more substantial signal.

At 432 MHz, parabolic-dish antennas become viable. The minimum gain for reliable 432-MHz EME operation is 24 dBi.

Yagis are also used on 432 MHz. The 22-element Yagi described in the **Antennas** chapter is an ideal 432-MHz design. Four of the 22-element Yagis meet the 24-dBi-gain criteria, and have been used successfully on EME. If you are going to use a fixed polarization Yagi array, you should plan on building an array with substantially more than 24-dBi gain if you desire reliable contacts with small stations. This extra gain is needed to overcome polarization misalignment.

At 902 MHz and above, the only antenna worthy of consideration is a parabolic dish. While it has been proven that Yagi antennas are capable of making EME QSOs at 1296 MHz, Yagi antennas, whether they use rod or loop elements, are simply not practical.

EME QSOs have been made at 1296 MHz with dishes as small as 6 ft in diameter. For reliable EME operation with similarly equipped stations, a 12-ft diameter dish (31 dBi gain at 1296 MHz) is a practical minimum. TVRO dishes, which are designed to operate at 3 GHz make excellent antennas, provided they have an accurate surface area. The one drawback of TVRO dishes is that they usually have an undesirable F/d ratio. More information on dish construction and feeds can be found in *The ARRL Antenna Book* and *The ARRL UHF/Microwave Experimenter's Manual*.

POLARIZATION EFFECTS

All of the close attention paid to operating at the best time, such as nighttime perigee, with high moon declination and low sky temperatures is of little use if signals are not aligned in polarization between the two stations attempting to make contact. There are two basic polarization effects. The first is called spatial polarization. Simply stated, two stations (using az-el mounts and fixed linear polarization) that are located far apart, will usually not have their arrays aligned in polarization as seen by the moon. Spatial polarization can easily be predicted, given the location of both stations and the position of the moon.

The second effect is Faraday rotation. This is an actual rotation of the radio waves in space, and is caused by the charge level of the Earth's ionosphere. At 1296 MHz and above, Faraday rotation is virtually nonexistent. At 432 MHz, it is believed that up to a 360° rotation is common. At 144 MHz, it is believed that the wavefront can actually rotate seven or more complete 360° revolutions. When Faraday rotation is combined with spatial polarization, there are four possible results:

1) Both stations hear each other and can QSO.

2) Station A hears station B, station B does not hear station A.

3) Station B hears station A, station A does not hear station B.

4) Neither station A nor station B hear each other.

At 144 MHz, there are so many revolutions of the signal, and the amount of Faraday rotation changes so fast that, generally, hour-long schedules are arranged. At 432 MHz, Faraday rotation can take hours to change. Because of this, half-hour schedules are used. During the daytime, you can count on 90 to 180° of rotation. If both stations are operating during hours of darkness, there will be little Faraday rotation, and the amount of spatial polarization determines if a schedule should be attempted.

At 1296 MHz and above, circular polarization is standard. The predominant array is a parabolic reflector, which makes circular polarization easy to obtain. Although the use of circular polarization would make one expect signals to be constant, except for the effect of the moon's distance, long-term fading of 6 to 9 dB is frequently observed.

With improved long-Yagi designs, for years the solution to overcoming polarization misalignment has been to make the array larger. Making your station's system gain 5 or 6 dB greater than required for minimal EME QSOs will allow you to work more stations, simply by moving you farther down the polarization loss curve. After about 60° of misalignment, however, making your station large enough to overcome the added losses quickly becomes a lifetime project! See **Fig 23.69**.

At 432 MHz and lower, Yagis are widely used, making the linear polarization standard. Although circular polarization may seem like a simple solution to polarization problems, when signals are reflected off the moon, the polarization sense of circularly polarized radio wave is reversed, requiring two arrays of opposite polarization sense be used. Initially, crossed Yagis with switchable polarization may also look attractive. Unfortunately, 432-MHz Yagis are physically small enough that the extra feed lines and switching devices become complicated, and usually adversely affect array performance. Keep in mind that even at 144 MHz, Yagis cannot tolerate metal mounting masts and frames in line with the Yagi elements.

When starting out on EME, keep in mind that it is best to use a simple system. You will still be able to work many of the larger fixed-polarization stations and those who have polarization adjustment (only one station needs to have polarization control). Once you gain understanding and confidence in your simple array, a more complex array such as one with polarization rotation can be attempted.

RECEIVER REQUIREMENTS

A low-noise receiving setup is essential for successful EME work. Many EME signals are barely, but not always, out of the noise. To determine actual receiver performance, any phasing line and feed line losses, along with the noise generated in the receiver, must be added to the array noise reception. When all losses are considered, a system noise figure of 0.5 dB (35 K) will deliver about all the performance that can be used at 144 MHz, even when low-loss phasing lines and a quiet array are used.

The sky noise at 432 MHz and above is low enough (cold sky is <15 K (kelvins) at 432 MHz, and 5 K at 1296 MHz) so the lowest possible noise figure is desired. Current high-performance arrays will have array temperatures near 30 K when unwanted noise pickup is added in. Phasing line losses must also be included, along with any relay losses. Even at 432 MHz, it is impossible to make receiver noise insignificant without the use of a liquid-cooled preamplifier. Current technology gives a minimum obtainable GaAsFET preamplifier noise figure, at room temperature, of about 0.35 dB (24 K).

GaAsFET preamps have also been standard on 1296 MHz and above for several years. Noise figures range from about 0.4 dB at 1296 MHz (30 K) to about 2 dB (170 K at 10 GHz). HEMT devices are now available to amateurs, but are of little use below 902 MHz because of 1/f noise. At higher frequencies, HEMT devices have already shown impressively low noise figures. Current HEMT devices are capable of noise figures close to 1.2 dB at 10 GHz (93 K) without liquid cooling.

At 1296 MHz, a new noise-limiting factor appears. The physical temperature of the moon is 210 K. This means that just like the Earth, it is a black-body radiator. The additional noise source is the reflection of sun noise off the moon. Just as a full moon reflects sunlight to Earth, the rest of the electromagnetic spectrum is also reflected. On 144 and 432 MHz, the beamwidth of a typical array is wide enough (15° is typical for 144 MHz, 7° for 432 MHz) that the moon, which subtends a 0.5° area is small enough to be insignificant in the array's pattern. At 1296 MHz, beamwidths approach 2°, and moon-noise figures of up to 5 dB are typical at full moon. Stations operating at 2300 MHz and above have such narrow array patterns that many operators actually use moon noise to assure that their arrays are pointed at the moon!

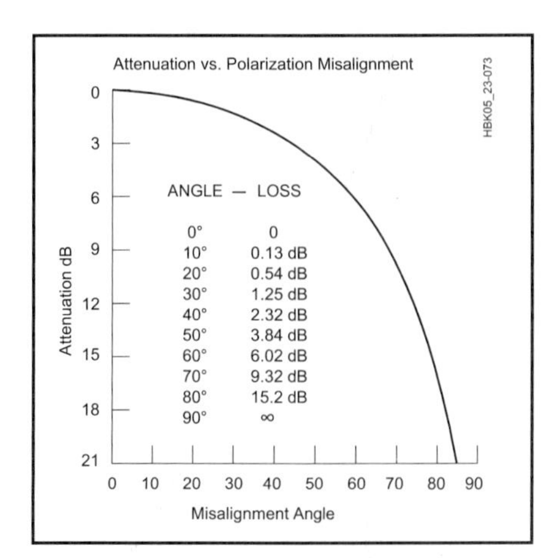

Fig 23.69—The graph shows how quickly loss because of polarization misalignment increases after 45°. The curve repeats through 360°, showing no loss at 0° and maximum loss at 90° and 270°.

A new weak-signal operator is encouraged to experiment with receivers and filters. A radio with passband tuning or IF-shift capability is desired. These features are used to center the passband and the pitch of the CW signal to the frequency at which the operator's ears perform best. Some operators also use audio filtering. Audio filtering is effective in eliminating high-frequency noise generated in the radio's audio or IF stages. This noise can be very fatiguing during extended weak signal operation. The switched-capacitor audio filter has become popular with many operators.

TRANSMITTER REQUIREMENTS

Although the maximum legal power (1500 W out) is desirable, the actual power required can be considerably less, depending on the frequency of operation and size of the array. Given the minimum array gain requirements previously discussed, the power levels recommended for reasonable success are shown in **Table 23.4**

Table 23.4
Transmitter Power Required for EME Success

Power at the array	
50 MHz	1500 W
144 MHz	1000 W
222 MHz	750 W
432 MHz	500 W
902 MHz	200 W
1296 MHz	200 W
2300 MHz and above	100 W

The amplifier and power supply should be constructed with adequate cooling and safety margins to allow extended slow-speed CW operation without failure. The transmitter must also be free from drift and chirp. The CW note must be pure and properly shaped. Signals that drift and chirp are harder to copy. They are especially annoying to operators who use narrow CW filters. A stable, clean signal will improve your EME success rate.

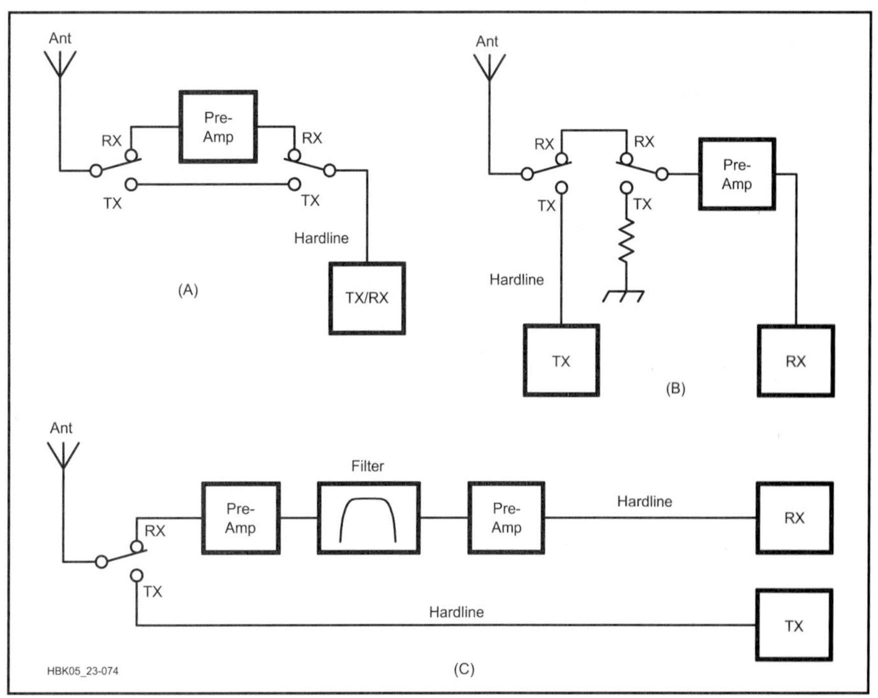

Fig 23.70—Two systems for switching a preamplifier in and out of the receive line. At A, a single length of cable is used for both the transmit and receive line. At B is a slightly more sophisticated system that uses two separate transmission lines. At C, a high-isolation relay is used for TR switching. The energized position is normally used on receive.

CALCULATING EME CAPABILITIES

Once all station parameters are known, the expected strength of the moon echoes can be calculated given the path loss for the band in use (see Fig 23.72). The formula for the received signal-to-noise ratio is:

$$S / N = P_o - L_t + G_t - P_l + G_r - P_n \quad (1)$$

where

P_o = transmitter output power (dBW)
L_t = transmitter feed-line loss (dB)
G_t = transmitting antenna gain (dBi)
P_l = total path loss (dB)
G_r = receiving antenna gain (dBi)
P_n = receiver noise power (dBW).

Receiver noise power, P_n, is determined by the following:

$$P_n = 10 \log_{10} KBT_s \quad (2)$$

where

$K = 1.38 \times 10^{-23}$ (Boltzmann's constant)
B = bandwidth (Hz)
T_s = receiving system noise temperature (K).

Receiving system noise temperature, T_s, can be found from:

$$T_s = T_a + (L_r - 1) T_l + L_r T_r \quad (3)$$

where

T_a = antenna temperature (K)
L_r = receiving feed-line loss (ratio)
T_l = physical temperature of feed line (normally 290 K)
T_r = receiver noise temperature (K).

An example calculation for a typical 432-MHz EME link is:

P_o = +30 dBW (1000 W)
L_t = 1.0 dB
G_t = 26.4 dBi (8 × 6.1-λ 22-el Yagis)
P_l = 262 dB
G_r = 23.5 dBi (15 ft parabolic)
T_a = 60 K
L_r = 1.02 (0.1-dB preamp at antenna)
T_l = 290 K
T_r = 35.4 K (NT = 0.5 dB)
T_s = 101.9 K
P_n = –188.5 dB
S/N = + 5.4 dB

It is obvious that EME is no place for a compromise station. Even relatively sophisticated equipment provides less-than-optimum results.

Fig 23.71 gives parabolic dish gain for a perfect dish. The best Yagi antennas will not exceed the gain curve shown in the **Antennas** chapter. If you are using modern, log-taper Yagis, properly spaced, figure about 2.8 to 2.9 dB of stacking gain. For old-technology Yagis, 2.5 dB may be closer to reality. Any phasing line and power divider losses must also be subtracted from the array gain.

LOCATING THE MOON

The moon orbits the Earth once in approximately 28 days, a lunar month. Because the plane of the moon's orbit is tilted from the Earth's equatorial plane by approximately 23.5°, the moon swings in a sine-wave pattern both north and south of the equator. The angle of departure of the moon's position at a given time from the equatorial plane is termed declination (abbreviated decl). Declination angles of the moon, which are continually changing (a few degrees a day), indicate the latitude on the Earth's surface where the moon will be at zenith. For this presentation, positive declination angles are used when the moon is north of the equator, and negative angles when south.

The longitude on the Earth's surface where the moon will be at zenith is related to the moon's Greenwich Hour Angle, abbreviated G.H.A. or GHA. "Hour angle" is defined as the angle in degrees to the west of the meridian. If the GHA of the moon were 0°, it would be directly over the Greenwich meridian. If the moon's GHA were 15°, the moon would be directly over the meridian designated as 15° W longitude on a globe. As one can readily understand, the GHA of the moon is continually changing, too, because of both the orbital velocity of the moon and the Earth's rotation inside the moon's orbit. The moon's GHA changes at the rate of approximately 347° per day.

GHA and declination are terms that may be applied to any celestial body. *The Astronomical Almanac* (available from the Superintendent of Documents, US Government Printing Office) and other publications list the GHA and decl of the sun and moon (as well as for other celestial bodies that may be used for navigation) for every hour of the year. This information may be used to point an antenna when the moon is not visible. *Almanac* tables for the sun may be useful for calibrating remote-readout systems.

Using the Almanac

The Astronomical Almanac and other almanacs show the GHA and declination of the sun or moon at hourly intervals for every day of the period covered by the book. Instructions are included in such books for interpolating the positions of the sun or moon for any time on a given date. The orbital velocity of the moon is not constant, and therefore precise interpolations are not linear.

Fortunately, linear interpolations from one hour to the next, or even from one day to the next, will result in data that is entirely adequate for Amateur Radio purposes. If linear interpolations are made from 0000 UTC on one day to 0000 UTC on the next, worse-case conditions exist when apogee or perigee occurs near midday on the next date in question. Under such conditions, the total angular error in the position of the moon may be as much as a sixth of a degree. Because it takes a full year for the Earth to orbit the sun, the similar error for determining the position of the sun will be no more than a few hundredths of a degree.

If a polar mount (a system having one axis parallel to the Earth's axis) is used, information from the *Almanac* may be used directly to point the antenna array. The local hour angle (LHA) is simply the GHA plus or minus the observer's longitude (plus if east longitude, minus if west). The LHA is the angle west of the observer's meridian at which the celestial body is located. LHA and declination information may be translated to an EME window by taking local obstructions and any other constraints into account.

Azimuth and Elevation

An antenna system that is positioned in azimuth (compass direction) and elevation (angle above the horizon) is called an *az-el* system. For such a system, some additional work will be necessary to convert the almanac data into useful information. The GHA and decl information may be converted into azimuth and elevation angles with the mathematical equations that follow. A calculator or computer that treats trigonometric functions may be used. *CAUTION:* Most almanacs list data in degrees, minutes, and either decimal minutes or seconds. Generally, computer programs have typically required this information in degrees and decimal fractions, so a conversion may be necessary before the almanac data is entered.

Determining az-el data from equations follows a procedure similar to calculating great-circle bearings and distances for two points on the Earth's surface. There is one additional factor, however. Visualize two observers on opposite sides of the Earth who are pointing their antennas at the moon. Imaginary lines representing the boresights of the two antennas will converge at the moon at an angle of approximately 2°. Now assume both observers aim their antennas at some distant star. The boresight lines now may be considered to be parallel, each observer having raised his antenna in elevation by approximately 1°. The reason for the necessary change in elevation is that the Earth's diameter in comparison to its distance from the moon is significant. The same is not true for distant stars, or for the sun.

Equations for az-el calculations are:

$$\sin E = \sin L \sin D + \cos L \cos D \cos LHA \quad (4)$$

$$\tan F = \frac{\sin E - K}{\cos E} \quad (5)$$

$$\cos C = \frac{\sin D - \sin E \sin L}{\cos E \cos L} \quad (6)$$

where

E = elevation angle for the sun

L = your latitude (negative if south)

D = declination of the celestial body

LHA = local hour angle = GHA plus or minus your longitude (plus if east longitude, minus if west longitude)

F = elevation angle for the moon

K = 0.01657, a constant (see text that follows)

C = true azimuth from north if sin LHA is negative; if sin LHA is positive, then the azimuth = 360 – C.

Assume our location is 50° N latitude, 100° W longitude. Further assume that the GHA of the moon is 140° and its declination is 10°. To determine the az-el information we first find the LHA, which is 140 minus 100 or 40°. Then we solve equation 4:

$$\sin E = \sin 50 \sin 10 + \cos 50 \cos 10 \cos 40$$
$$\sin E = 0.61795 \text{ and } E = 38.2^\circ$$

Solving equation 5 for F, we proceed. (The value for sin E has already been determined in equation 4.)

$$\tan F = \frac{0.61795 - 0.06175}{\cos 38.2} = 0.76489$$

From this, F, the moon's elevation angle, is 37.4°.

We continue by solving equation 6 for C. (The value of sin E has already been determined.)

$$\cos C = \frac{\sin 10 - 0.61795 \sin 50}{\cos 38.2 \cos 50}$$

C therefore equals 126.4°. To determine if C is the actual azimuth, we find the polarity for sin LHA, which is sin 40° and has a positive value. The actual azimuth then is 360 – C = 233.6°.

If az-el data is being determined for the sun, omit equation 5; equation 5 takes into account the nearness of the moon. The solar elevation angle may be determined from equation 4 alone. In the above example, this angle is 38.2°.

The mathematical procedure is the same for any location on the Earth's surface. Remember to use negative values for southerly latitudes. If solving equation 4 or 5 yields a negative value for E or F, this indicates the celestial body below the horizon.

These equations may also be used to determine az-el data for man-made satellites, but a different value for the constant, K, must be used. K is defined as the ratio of the Earth's radius to the distance from the Earth's center to the satellite.

The value for K as given above, 0.01657 is based on an average Earth-moon distance of 239,000 miles. The actual Earth-moon distance varies from approximately 225,000 to 253,000 mi. When this change in distance is taken into account, it yields a change in elevation angle of approximately 0.1° when the moon is near the horizon. For greater precision in determining the correct elevation angle for the moon, the moon's distance from the Earth may be taken as:

$$D = -15{,}074.5 \times SD + 474{,}332$$

where

D = moon's distance in miles

SD = moon's semi-diameter, from the almanac.

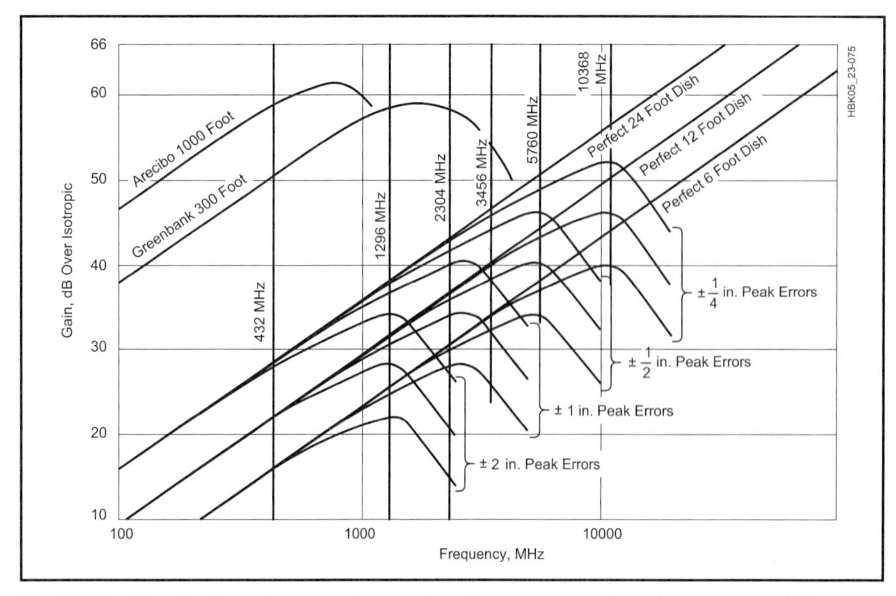

Fig 23.71—Parabolic-antenna gain vs size, frequency and surface errors. All curves assumed 60% aperture efficiency and 10-dB power taper. Reference: J. Ruze, British IEEE.

Computer Programs

Digital modes are also used for EME communications. The *WSJT* software for *Windows* includes a mode known as JT44. This software enables moonbounce contacts using sound-card-equipped PCs, single Yagi antennas and under 200 watts of RF—an accomplishment that seemed virtually impossible in previous years. It now appears possible that this mode, and other digital modes that may appear in the future, will place EME communication within reach of hams with both modest antenna space and budgets. Complete details and operating instructions can be viewed by downloading the *WSJT User Guide* at **pulsar.princeton.edu/~joe/K1JT/WSJT300.PDF**.

As has been mentioned, a computer may be used in solving the equations for azimuth and elevation. For EME work, it is convenient to calculate az-el data at 30-minute intervals or so, and to keep the results of all calculations handy during the EME window. Necessary antenna-position corrections can then be made periodically.

RealTrak prints out antenna azimuth and elevation headings for nearly any celestial object. It can be used with the Kansas City Tracker program described in the satellite section to track celestial objects automatically. *VHF PAK* provides real-time moon and celestial object position information. Two other real-time tracking programs are *EME Tracker* and the *VK3UM EME Planner*.

Libration Fading of EME Signals

One of the most troublesome aspects of

receiving a moonbounce signal, besides the enormous path loss and Faraday rotation fading, is libration fading. This section will deal with libration (pronounced *lie-brayshun*) fading, its cause and effects, and possible measures to minimize it.

Libration fading of an EME signal is characterized in general as fluttery, rapid, irregular fading not unlike that observed in tropospheric scatter propagation. Fading can be very deep, 20 dB or more, and the maximum fading will depend on the operating frequency. At 1296 MHz the maximum fading rate is about 10 Hz, and scales directly with frequency.

On a weak CW EME signal, libration fading gives the impression of a randomly keyed signal. In fact on very slow CW telegraphy the effect is as though the keying is being done at a much faster speed. On very weak signals only the peaks of libration fading are heard in the form of occasional short bursts or "pings."

Fig 23.72 shows samples of a typical EME echo signal at 1296 MHz. These recordings, made at W2NFA, show the wild fading characteristics with sufficient S/N ratio to record the deep fades. Circular polarization was used to eliminate Faraday fading; thus these recordings are of libration fading only. The recording bandwidth was limited to about 40 Hz to minimize the higher sideband-frequency components of libration fading that exist but are much smaller in amplitude. For those who would like a better statistical description, libration fading is Raleigh distributed. In the recordings shown in Fig 23.72, the average signal-return level computed from path loss and mean reflection coefficient of the moon is at about the +15 dB S/N level.

It is clear that enhancement of echoes far in excess of this average level is observed. This point should be kept clearly in mind when attempting to obtain echoes or receive EME signals with marginal equipment. The probability of hearing an occasional peak is quite good since random enhancement as much as 10 dB is possible. Under these conditions, however, the amount of useful information that can be copied will be near zero. Enthusiastic newcomers to EME communications will be stymied by this effect since they know they can hear the signal strong enough on peaks to copy but can't make any sense out of what they try to copy.

What causes libration fading? Very simply, multipath scattering of the radio waves from the very large (2000-mile diameter) and rough moon surface combined with the relative motion between Earth and moon called librations. To understand these effects, assume first that the Earth and moon are stationary (no libration) and that a plane wave front arrives at the moon from your Earthbound station as shown in **Fig 23.73A**.

The reflected wave shown n Fig 23.73B consists of many scattered contributions from the rough moon surface. It is perhaps easier to visualize the process as if the scattering were from many small individual flat mirrors on the moon that reflect small portions (amplitudes) of the incident wave energy in different directions (paths) and with different path lengths (phase). Those paths directed toward the moon arrive at your antenna as a collection of small wave fronts (field vectors) of various amplitudes and phases. The vector summation of all these coherent (same frequency) returned waves (and there is a near-infinite array of them) takes place at the feed point of your antenna (the collecting point in your antenna system). The level of the final summation as measured by a receiver can, of course, have any value from zero to some maximum. Remember that we assumed the Earth and moon were stationary, which means that the final summation of these multipath signal returns from the moon will be one fixed value. The condition of zero relative motion between Earth and moon is a rare event that will be discussed later in this section.

Consider now that the Earth and moon are moving relative to each other (as they are in nature), so the incident radio wave "sees" a slightly different surface of the moon from moment to moment. Since the lunar surface is very irregular, the reflected wave will be equally irregular, changing in amplitude and phase from moment to moment. The resultant continuous summation of the varying multipath signals at your antenna feed-point produces the effect called libration fading of the moon-reflected signal.

The term *libration* is used to describe small perturbations in the movement of celestial bodies. Each libration consists mainly of its diurnal rotation; moon libration consists mainly of its 28-day rotation which appears as a very slight rocking motion with respect to an observer on Earth. This rocking motion can be visualized as follows: Place a marker on the surface of the moon at the center of the moon disc, which is the point closest to the observer, as shown in **Fig 23.74**. Over time, we will observe that this marker wanders around within a small area. This means the surface of the moon as seen from the Earth is not quite fixed but changes slightly as different areas of the periphery are exposed because of this rocking motion. Moon libration is very slow (on the order of 10^{-7} radians per second) and can be determined with some difficulty from pub-

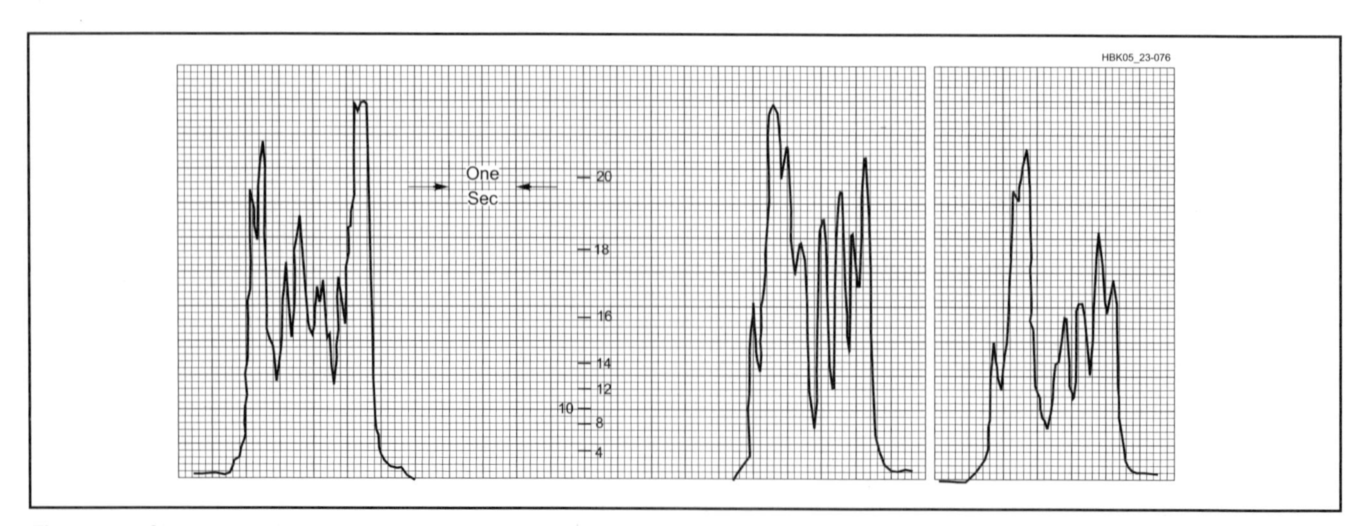

Fig 23.72—Chart recording of moon echoes received at W2NFA on July 26, 1973, at 1630 UTC. Antenna gain 44 dBi, transmitting power 400 W and system temperature 400 K.

lished moon ephemeris tables.

Although the libration motions are very small and slow, the larger surface area of the moon has nearly an infinite number of scattering points (small area). This means that even slight geometric movements can alter the total summation of the returned multipath echo by a significant amount. Since the librations of the Earth and moon are calculable, it is only logical to ask if there ever occurs a time when the total libration is zero or near zero. The answer is yes, and it has been observed and verified experimentally on radar echoes that minimum fading rate (not depth of fade) is coincident with minimum total libration. Calculation of minimum total libration is at best tedious and can only be done successfully by means of a computer. It is a problem in extrapolation of rates of change in coordinate motion and in small differences of large numbers.

EME OPERATING TECHNIQUES

Many EME signals are near the threshold of readability, a condition caused by a combination of path loss, Faraday rotation and libration fading. This weakness and unpredictability of the signals has led to the development of techniques for the exchange of EME information that differ from those used for normal terrestrial work. The fading of EME signals chops dashes into pieces and renders strings of dots incomplete. This led to the use of the "T M O R" reporting system. Different, but similar, systems are used on the low bands (50 and 144 MHz) and the high bands (432 MHz and above). **Tables 23.5** and **23.6** summarize the differences between the two systems.

As equipment and techniques have improved, the use of normal RST signal reports has become more common. It is now quite common for two stations working for the first time to go straight to RST reports if signals are strong enough. These normal reports let stations compare signals from one night to the next. EME QSOs are often made during the ARRL VHF contests. These contacts require the exchange of 4-digit grid locators. On 432 MHz and above, the sending of GGGG has come to mean "Please send me your grid square," or conversely, "I am now going to send my grid square."

The length of transmit and receive periods is also different between the bands. On 50 and 144 MHz, 2-minute sequences are used. That is, stations transmit for two full minutes, and then receive for two full minutes. One-hour schedules are used, with the eastern-most station (referenced to the international date line) transmitting first. **Table 23.7** gives the 2-minute sequence

Table 23.5

Signal Reports Used on 144-MHz EME

T — Signal just detectable
M — Portions of call copied
O — Complete call set has been received
R — Both "O" report and call sets have been received
SK — End of contact

Table 23.6

Signal Reports Used on 432-MHz EME

T — Portions of call copied
M — Complete calls copied
O — Good signal—solid copy (possibly enough for SSB work)
R — Calls and reports copied
SK — End of contact

procedure. On 222 MHz, both the 144 and 432-MHz systems are used.

On 432 MHz and above, $2^1/_2$-minute sequences are standard.

The longer period is used to let stations with variable polarization have adequate time to peak the signal. The last 30 seconds is reserved for signal reports only. **Table 23.8** provides more information on the 432-MHz EME QSO sequence. The western-most station usually transmits first. However, if one of the stations has variable polarization, it may elect to transmit second, to take the opportunity to use the first sequence to peak the signal. If both stations have variable polarization, the station that transmits first should leave its polarization fixed on transmit, to avoid "polarization chasing."

CW sending speed is usually in the 10 to 13-wpm range. It is often best to use greater-than-normal spacing between individual dits and dahs, as well as between complete letters. This helps to overcome libration fading effects. The libration fading rate will be different from one band to another. This makes the optimum CW speed for one band different from another. Keep in mind that characters sent too slowly will be chopped up by typical EME fading. Morse code sent too fast will simply be jumbled. Pay attention to the sending practices of the more successful stations, and try to emulate them.

Doppler shift must also be understood. As the moon rises or sets it is moving toward or away from objects on Earth. This leads to a frequency shift in the moon echoes. The amount of Doppler shift is directly proportional to frequency. At 144

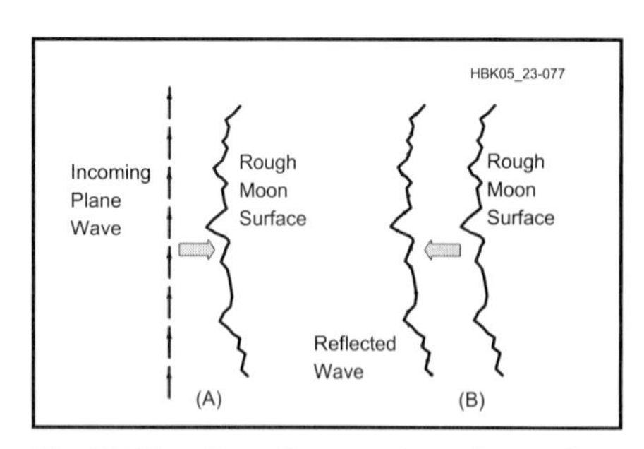

Fig 23.73— How the rough surface of the moon reflects a plane wave as one having many field vectors.

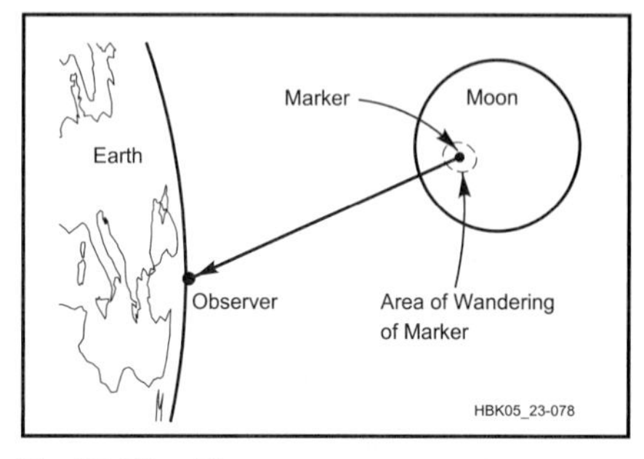

Fig 23.74— The moon appears to "wander" in its orbit about the Earth. Thus a fixed marker on the moon's surface will appear to move about in a circular area.

MHz, about 500 Hz is the maximum shift. On 432 MHz, the maximum shift is 1.5 kHz. The shift is upward on moonrise and downward on moonset. When the moon is due south, your own echoes will have no Doppler shift, but stations located far away will still be affected. For scheduling, the accepted practice is to transmit zero beat on the schedule frequency, and tune to compensate for the Doppler shift. Be careful—most transmitters and transceivers have a built-in CW offset. Some radios read this offset when transmitting, and others don't. Find out how your transmitter operates and compensate as required.

Random operation has become popular in recent years. In the ARRL EME contest, many of the big guns will not even accept schedules during the contest periods, because they can slow down the pace of their contest contacts.

EME Operating Times

Obviously, the first requirement for EME operation is to have the moon visible by both EME stations. This requirement not only consists of times when the moon is above the horizon, but when it is actually clear of obstructions such as trees and buildings. It helps to know your exact EME operating window, specified in the form of beginning and ending GHAs (Greenwich

Table 23.7
144-MHz Procedure — 2-Minute Sequence

Period	1½ minutes	30 seconds
1	Calls (W6XXX DE W1XXX)	
2	W1XXX DE WE6XXX	TTTT
3	W6XXX DE W1XXX	OOOO
4	RO RO RO RO	DE W1XXX K
5	R R R R R R	DE W6XXX K
6	QRZ? EME	DE W1XXX K

Table 23.8
432-MHz Procedure — 2½-Minute Sequence

Period	2 minutes	30 seconds
1	VE7BBG DE K2UYH	
2	K2UYH DE VE7BBG	
3	VE7BBG DE K2UYH	TTT
4	K2UYH DE VE7BBG	MMM
5	RM RM RM RM	DE K2UYH K
6	R R R R R	DE VE7BBG SK

Hour Angle) for different moon declinations. This information allows two different stations to quickly determine if they can simultaneously see the moon.

Once your moon window is determined, the next step is to decide on the best times during that window to schedule or operate. Operating at perigee is preferable because of the reduced path loss. Fig 23.68 shows that not all perigees are equal. There is about a 0.6-dB difference between the closest and farthest perigee points. The next concern is operating when the moon is in a quiet spot of the sky. Usually, northern declinations are preferred, as the sky is quietest at high declinations. If the moon is too close to the sun, your array will pick up sun noise and reduce the sensitivity of your receiver. Finally, choosing days with minimal libration fading is also desirable.

Perigee and apogee days can be determined from the *Astronomical Almanac* by inspecting the tables headed "S.D." (semi-diameter of the moon in minutes of arc). These semi-diameter numbers can be compared to Fig 23.68 to obtain the approximate moon distance. Many computer programs for locating the moon now give the moon's distance. The expected best weekends to operate on 432 MHz and the higher bands are normally printed well in advance in various EME newsletters.

When the moon passes through the galactic plane, sky temperature is at its maximum. Even on the higher bands this is one of the least desirable times to operate. The areas of the sky to avoid are the constellations of Orion and Gemini (during northern declinations), and Sagittarius and Scorpios (during southern declinations). The position of the moon relative to these constellations can be checked with information supplied in the *Astronomical Almanac* or *Sky and Telescope* magazine.

Frequencies and Scheduling

According to the ARRL-sponsored band plan, the lower edge of most bands is reserved for EME operation. On 144 MHz, EME frequencies are primarily between 144.000 and 144.080 MHz for CW, and 144.100 and 144.120 MHz for SSB. Random CW activity is usually between 144.000 and 144.020 MHz. In the US, 144.000 to 144.100 MHz is a CW subband, so SSB QSOs often take place by QSYing up 100 kHz after a CW contact has been established. Because of the large number of active 144-MHz stations, coordinating schedules in the small EME window is not simple. The more active stations usually have assigned frequencies for their schedules.

On 432 MHz, the international EME CW calling frequency is 432.010 MHz. Random SSB calling is done on 432.015 MHz. Random activity primarily takes place between 432.000 and 432.020 MHz. The greater Doppler shift on 432 MHz requires greater separation between schedule frequencies than on 144 MHz. Normally 432.000 MHz, 432.020 MHz and each 5-kHz increment up to 432.070 MHz are used for schedules.

Activity on 1296 MHz is centered between 1296.000 and 1296.040 MHz. The random calling frequency is 1296.010 MHz. Operation on the other bands requires more specific coordination. Activity on 33 cm is split between 902 and 903 MHz. Activity on 2300 MHz has to accommodate split-band procedures because of the different band assignments around the world.

EME Net Information

An EME net meets on 14.345 MHz on weekends for the purpose of arranging schedules and exchanging EME information. The net meets at 1600 UTC. OSCAR satellites are becoming more popular for EME information exchange. When Mode B is available, a downlink frequency of 145.950 MHz is where the EME group gathers. On Mode L and Mode JL, the downlink frequency is 435.975 MHz.

Other Modes

Most EME contacts are still made on CW, although SSB has gained in popularity and it is now common to hear SSB QSOs on any activity weekend. The ability to work SSB can easily be calculated from Eq 1. The proper receiver bandwidth (2.3 kHz) is substituted. SSB usually requires a +3-dB signal-to-noise ratio, whereas slow-speed CW contacts can be made with a 0-dB signal-to-noise ratio. Slow-scan television and packet communication has been attempted between some of the larger stations. Success has been limited because of the greater signal-to-noise ratios required for these modes, and severe signal distortion from libration fading.

Chapter 24

Web, Wi-Fi, Wireless and PC Technology

With the Internet and the Personal Computer (PC) more commonplace in homes and ham shacks around the world, some fear it may actually destroy Amateur Radio! Their reasoning is that increasingly powerful PCs, coupled with access to the Internet or worldwide web (www), will primarily divert youth from the lure, mystique and magic of radio. Could an advancement in technology with the corresponding economies-of-scale actually deprive Amateur Radio of an influx of fresh, young operators and enthusiasts?

The PC and its Internet link to the world around us have surely changed many lifestyles — young and old and those in between. But far from destroying Amateur Radio as we know it, these technologies have greatly enhanced and improved the ham radio experience of countless operators! Many hams with an interest or passion in homebrewing, DXing, contesting, and the wide variety of other related areas of interest now take for granted nearly instantaneous access to information and interest-related Web sites. Computers and modems are sometimes used for the operation of various modes, and can also be a primary part of the actual ham radio equipment (e.g., a *transceiver* or *receiver*).

In this chapter, in addition to covering many of these exciting areas described above, we will also review the personal/commercial *wireless* world around us that is extraneous to ham frequencies, equipment or operation. This chapter was written by Donald R. Greenbaum, N1DG, and Dana G. Reed, W1LC. It also includes contributions from John J. Champa, K8OCL, Reed E. Fisher, W2CQH, Howard S. Huntington, K9KM, and Ronnie P. Milione, KB2UAN.

The World Wide Web (www) — The Internet

SEARCH ENGINES

The Internet encompasses millions and millions of pages of information. There is simply too much information on the subject of ham radio to find without properly using search engines. There are many related ham sites, but the most popular are listed in the sidebar *A Ham's Guide to Useful Internet Sites*. In any event, the most popular search engines are ***Google***, ***Yahoo***, ***Ask Jeeves***, ***Alta Vista***, ***Dogpile***, ***Look Smart***, ***Overture***, ***Teoma*** and ***Find What***.

The secret to a successful search is to be specific by limiting the results or *hits*. Simply typing "ham radio" as the topic of interest to you is not sufficient. A query on *ham* yielded links to Web sites mentioning ham radio, ham (the food) and other hits totaling over 7 million pages! A more specific search of *ham radio* reduced the page count to 1.7 million pages, still too numerous to be helpful. Narrowing and refining your search is as simple as adding

Search Engines — Electronic Parts and Cross Referencing

A QRP list member was trying to search, or cross reference, a semiconductor in Google, but without success. He had probably only entered the exact part number. The part was described as possibly of one family of semiconductor, but apparently in Google he used the part number "IRF331A" which indeed comes up blank.

However, a knowledge of manufacturing prefixes or numbering systems used for semiconductors or other components can be very valuable. Originally, IRF was the prefix for International Semiconductor Parts. Entering that in Google.com, I came up with a Web page: **www.irf.com**. From there, I came to a search engine on the page that eventually led to their cross reference guide — and the fact that the IRF331A can possibly be replaced (depending on the application) by the IRF330. However, an upgraded part is the IRF440, or the 2N6760.

Knowing a general supplier of replacement semiconductors, my next stop — had this not worked out — would have been **www.NTE.com**. The main point here is to study distributors catalogs, for knowledge of part-numbering conventions for whatever electronics you might need. Get to know company logos that are applied to parts as these will shorten your searches if the company still exists, using that logo. For the totally unfamiliar part, a logo can be very helpful to focus your search.—*Stuart Rohre, K5KVH*

A HAM'S GUIDE TO USEFUL INTERNET SITES

Since website URLs change so often, any attempt to present a comprehensive list here would be out of date by the time of publication. Accordingly, we have instead listed general ham-related sites that contain current links to many of the specific radio sites on the Web:

ARRL: **www.arrl.org** Where else would you go for ham radio news?

AC6V: **www.ac6v.com/** Rod has links to over 6,000 other radio related sites covering all topics imaginable.

AA1V: **www.goldtel.net/aa1v/** Don has the usual DX links plus some pretty good NASA links not found on other sites.

DX Zone: **www.dxzone.com/** A good commercially-run site that is well laid out providing 4500 links to other commercial and private ham run sites.

425 DX News: **www.425dxn.org/** Good Italian non-commercial site for DX News and links to DX-related sites.

K4UTE: **www.nfdxa.com/K4UTE/K4UTE.HTML** One of the best sources of QSL routes and links to other DX-related activities.

NG3K: **www.ng3k.com/Misc/adxo.html** Want to know who's going where and the link to their Web site? Since 1996, Bill has been keeping track of these things on this Web site and has everything archived for searching.

QSL.NET: **www.qsl.net/master.htm** Al, K3TKJ, provides free hosting to thousands of hams for radio-related Web pages.

K1BV Awards Directory: **www.dxawards.com/** Are you into wallpaper? Ted has one of the best sites for award information and links to those offering them.

MODS: **www.mods.dk/** Modifying your radio? Find hundreds of links compiled over the last eight years by Erik, OZ2AEP.

Linux Ham Radio Software Directory: **radio.linux.org.au/** Do you prefer *Linux* to *Windows*? This site lists Perl scripts ranging from modeling antennas to European Microwave Beacon propagation forecasting.

Satellite Tracking Software: **www.david-taylor.pwp.blueyonder.co.uk/software/wxtrack.htm**

QSL Museum: **www.hamgallery.com/** In addition to one of the most complete online QSL collections, Tom (K8PX) has one of the more complete ham radio link sites on the web.

more words. This is called using keywords, and choosing them wisely will reduce the page count to manageable levels. *Ham radio software* reduces our search to 427,000 pages while *ham radio dsp software* reduces it to 18,600 pages; you can keep refining your search by adding keywords, and there is no need to include the word *and* between the words.

Another way to reduce hits on your search is to exclude categories. For instance, if you want to search for ham you can exclude the food ham by *ham –food.* If you want to search for an item but only want to search for it on a specific Web site, you can do that too. For instance, to look on the ARRLWeb you can do *license renewal site: www.arrl.org.*

Searching for specific part numbers for your do-it-yourself projects is similarly easy. In the *PC Technology* section of this chapter we discuss isolation transformers. A search for *273-1374* brings you right into the RadioShack catalog, as well as pages on how to build other projects with it. There is a fine line for being too specific in searches. Sometimes the part number can be too exact, and adding either a manufacturer or looking up the description of *isolation transformers* are a broader alternative that will help if the specific part search comes up negative. See also the sidebar *Search Engines — Electronic Parts and Cross Referencing*.

VOICE OVER INTERNET PROTOCOL — VoIP

Like our airwaves, the Internet can carry our transmissions over great distances. Unlike ham radio, the Internet is only digital. If we are use voice or video over that medium those modes must be converted into that digital format before sending and converted back to analog upon receipt. Two software packages have emerged to allow hams to converse over the Internet in what is known as Voice over Internet Protocol, or VoIP. ARRL offers a new publication titled *VoIP: Internet Linking for Radio Amateurs*. The book discusses in great detail the brief outline of the topic that follows.[1]

Echolink

Recognizing the growing use of the Internet by hams, K1RFD wrote a software program called Echolink. Now even antenna-restricted hams can enjoy the fun with a computer, a microphone and an Internet connection. You can connect via Echolink to other licensed hams on computers, mobiles transmitting through repeaters linked to the Internet via connected VHF repeaters and even nodes connected via HF. There are more than 120,000 registered users in 147 countries worldwide. There are also conference nodes, some specific to other hobbies such as aviation. It is the ultimate party line encompassing computers, radio and the Internet. Like radio, it is non-duplex; both stations cannot talk at once. The software is free (you must be licensed) and is downloadable from **www.echolink.org**. The software is very user friendly, and all control features are click-of-a-mouse enabled. The only requirement of your PC is a sound card and microphone.

IRLP

Unlike Echolink, which allows two hams to communicate without having a radio, IRLP takes a different approach to VoIP. David Cameron, VE7LTD, designed the Internet Radio Linking Project (IRLP). As he says on his Web site, he designed the software to provide a simple and easy system to link radio systems together using the Internet as the communications backbone.

IRLP users do not need to own computers. They simply access the network nodes using VHF or UHF FM transceivers.

The IRLP network is *Linux* based and requires a *Linux*-based Pentium PC at each node running *Speak Freely* software, an interface card (purchased inexpensively from VE7LTD), your radio and a sound card. The sound card takes the analog audio from the radio, converts it to compressed digital packets and sends short streams over the

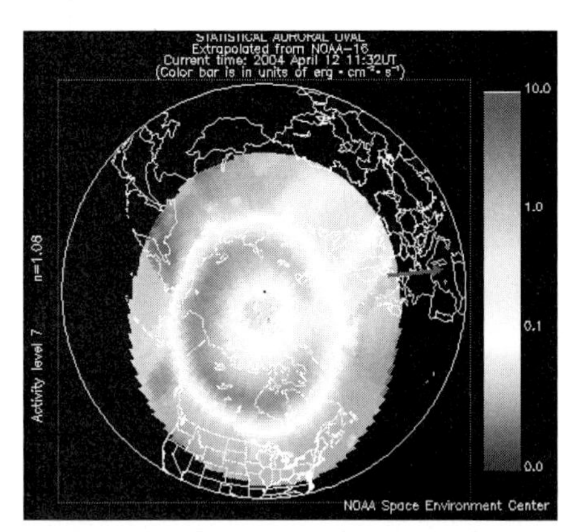

Fig 24.1 — NOAA North Pole map image of recent, real-time auroral activity.

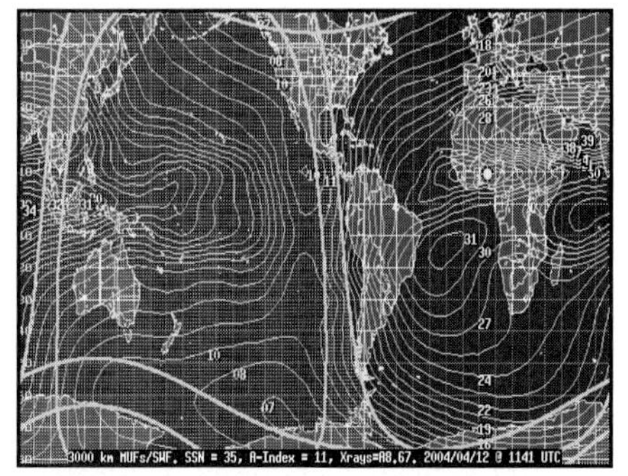

Fig 24.2 — Solar Terrestrial Dispatch graphic depicting frequency usability in real time.

Internet to the connected nodes via the interface card. The control software controls the packets along with continuous tone coded sub audible squelch signals (CTCSS) to start and stop the streams. CTCSS is also used by the transmitting user to pass commands to the nodes for the information as to which node a user wants to connect to, start and stop the connection and a host of other settable parameters to control the process. Streams can be sent to one or many nodes.

The software, interface cards and all manuals are available from the IRLP website at **www.irlp.net**

PROPAGATION FORCASTING

One of the most valuable pieces of information available on the Web is the various real time propagation statistics and tools. Current gray-line paths are vital to the LF operator. MUF values are helpful to HF operators. Current observed sunspot activity, solar storm indexes, etc. are all posted on various ham and government Web sites giving the Internet-connected ham useful tools to know where the best paths to the DX are. Current solar flux numbers are published by the Canadian Space Agency as monitored at its Penticton Observatory: **www.drao-ofr.hia-iha.nrc-cnrc.gc.ca/icarus/www/current_flux.shtml**

Since Auroral activity is harmful to HF transmissions over the globe, NOAA provides both North and South Pole real time auroral maps: **sec.noaa.gov/pmap/gif/pmapN.gif**. A recent North Pole image is shown in **Fig 24.1**.

Knowing which frequencies are usable in real time is also as handy a propagation tool as a ham could want. As shown in **Fig 24.2**, a real-time graphic can be found at the Solar Terrestrial Dispatch Web site: **www.spacew.com/www/realtime.html**.

VIRUS PROTECTION AND FIREWALLS

Lastly, no Internet user should be connecting to the Web without up-to-date virus protection. Norton Antivirus and McAfee are the two most popular commercial products. There are also some shareware products available. However, virus protection alone does not offer full protection from intruders. Hams increase the risk of being infected by opening ports to allow IRLP, Telnet, Instant Messenger, VoIP, or FTP applications. The solution is installing one of two types of firewalls. The first is an inexpensive software-based firewall from Norton or McAfee for under $50. They detect both internal and external attempts to compromise your personal computer (PC). The more expensive but more capable solution is a hardware-based device commonly known as a Firewall Appliance. In addition to the simple intruder detection found in firewall software, they contain an intruder *prevention* system, the ability to connect multiple PCs (as a router or switch), monitor ability on all traffic in and out of the network, open secure ports (VPN) and automatically obtain updates to firmware.

Notes

[1]See *ARRLWeb* at **www.arrl.org/shop**; Order no. 9264.

Wi-Fi Glossary

Access Point (AP) — A wireless bridging device that connects 802.11 stations to shared resources and a wired network such as the Internet.

Ad Hoc — In wireless LAN (WLAN) networks this is a direct wireless connection between two laptop computers without the use of an AP.

AP — Access point.

Carrier Sense Multiple Access / Collision Avoidance (CSMA/CA) — The wireless method that tries to avoid simultaneous access or collisions by not transmitting, if another signal is detected on the same frequency channel.

Direct-Sequence Spread Spectrum (DSSS) — The type of modulation used in 802.11b that is capable of maximum half-duplex data speeds of 11 Mbps.

Frequency Hopping Spread Spectrum (FHSP) — A type of modulation used in early 802.11 devices that uses a time-varied narrow signal to spread the signal over a wide band. Maximum half-duplex data rate is 2 Mbps.

Institute of Electrical and Electronics Engineers (IEEE) — The professional standards setting organization for data networking devices.

Orthogonal Frequency Division Multiplexing (OFDM) — A type of modulation that splits a wide frequency band into many narrow frequency bands. Both 802.11a and 802.11g use OFDM.

Service Set Identity (SSID) — The identification for an AP. It is transmitted continuously in the form of a beacon.

Wired Equivalent Privacy (WEP) — A standard for providing minimal privacy of wireless LAN communication by encrypting individual data frames.

Wireless Fidelity (Wi-Fi) — The Wireless Ethernet Compatibility Alliance certification program to ensure that equipment claiming to be in compliance with 802.11 standards is truly interoperable. The term Wi-Fi5 is sometimes applied to 802.11a equipment that operates on the 5-GHz band.

Wireless Fidelity (Wi-Fi)

COMPUTER CONNECTIONS BY RADIO

Wireless local area networks (WLAN) using spread spectrum transmit in the 902-MHz range (802.11), in the 2.4-GHz frequency range (802.11b and 802.11g), and in the 5-GHz frequency range (802.11a). Combining spread spectrum transmission's characteristics with a low power output (30 to 100-mW range) means it is highly unlikely that one spread spectrum network user will interfere with another. Spread spectrum transmissions distribute or "spread" a radio signal over a broad frequency range. There are various techniques for doing this spreading.

The older WLAN systems operate in the 902-MHz range based on the IEEE 802.11 standard and use frequency hopping spread spectrum modulation, or FHSS. This modulation technique uses what is called a predetermined pseudo-random sequence to transmit data. This pseudo-random sequence is actually a predetermined digital signal pattern that places data on a combination of frequencies across the entire spread spectrum channel. The receiving station must know the specific signal pattern used by the transmitting station to decode the data. These early 802.11 systems generally operated with slightly higher power (250 mW) but at much slower data rates of typically 2 Mbps.

New-generation WLAN

The first of the new generation of WLAN systems is based on the IEEE 802.11b standard and use a modulation technique known as direct-sequence spread spectrum, or DSSS. This modulation technique achieves higher data rates by using a different pseudo-random code known as a Complimentary Sequence. The 8-bit Complimentary Sequence can encode 2 bits of data for the 5.5 Mbps data rate or 6 bits of data for the 11 Mbps data rate. This is known as Complimentary Code Keying (CCK).

The latest generation of WLAN devices are based the same type of modulation, Orthogonal Frequency Division Multiplexing (OFDM) but operate on different frequency bands. OFDM provides its spreading function by transmitting the data simultaneously on multiple carriers. 802.11g operating in the 2.4-GHz range and 802.11a operating in the 5-GHz range both specify 20-MHz wide channels with 52 carriers spaced every 312.5 kHz. OFDM radios can be used to transmit data rates of 6, 9, 12, 18, 24, 36, 48 and 54 Mbps.

The WLAN Advantage

The main advantage of WLAN systems is that the laptop PC users are not tied to an RJ-45 type of wall outlet. At home they can roam between the home office and the patio. At work, they can move with ease between an office and a conference room, for example. There is a significant economic advantage for businesses re-locating office areas. With the use of WLAN technology, it is not necessary to spend the time and money to rewire an entire floor or building. These small radio devices with their small antennas and low power can readily transmit through several layers of drywall. If more than one access point, or AP, is required to cover the business area, several AP devices can all be linked together by putting them in the same virtual LAN in the wired network. Consequently, rather than having to wire each individual office and conference room, only the various AP nodes need to be wired together. This can produce a huge savings in time and expense.

WLAN Security

Although the majority of WLAN operate in the open mode allowing anyone in the area convenient access the network, often a minimum level of information security is desirable. This can be provided by using a WLAN encryption protocol built into the equipment called Wired Equivalent Privacy, or WEP. For those situations requiring a higher level of security, most commercial firms simply re-employ their virtual private network (VPN) or tunneling strategy commonly used to allow secure network remote access for mobile workers and teleworkers. However, in the case of WLAN use, extra steps need to be taken to ensure that the VPN user ID and password are not being wirelessly transmitted in the clear without some form of encryption. A new WLAN security standard called IEEE 802.1x attempts to address this need but is currently plagued with interoperability issues.

Wi-Fi systems being used in Amateur Radio applications are typically referred to as High Speed Multimedia (HSMM), and these techniques are discussed in detail in the HSMM section of the **Modes and Modulation Sources** chapter of this *Handbook*.

Wireless Technology Glossary

AMPS (Advanced Mobile Phone Service) — First standardized cellular service in the world, released in 1983. Uses the 800-900 MHz frequency band.

Analog — A signal that can vary continuously between a maximum and minimum value. For example, the voice voltage waveform from the output of a microphone is analog. RF voltage waveforms (as those from AM, FM and SSB transmitters) are also analog.

Cap Code — A specific address encoded into both a data transmission and the intended receiving equipment so the receiving equipment can discriminate against unintended or unwanted messages.

CDMA (Code Division Multiple Access) — A digital radio system that separates users by digital codes.

Cellular — Characteristic of or pertaining to a system of wireless communication made up of many individual cell units. The term itself is derived from the typical geographic honeycomb shape of the areas into which a coverage region is divided.

CELP (Codebook Excited Linear Predictive coding) — A type of low-bit-rate voice coder that emulates a single human voice tract. Details can be found in Ref. 3 at the end of this chapter.

CSMA/CD (Carrier Sense Multiple Access / Collision Detection) — A set of rules that determine how network devices respond when two devices attempt to use a data channel simultaneously (called a *collision*). After detecting a collision, a device waits a random delay time and then attempts to re-transmit the message. If the device detects a collision again, it now waits twice as long to try to re-transmit the message.

DHCP (Dynamic Host Configuration Protocol) — An external assignment mechanism that provides a "care-of address" to a mobile client (see also *Foreign Agent*).

Digital — A signal that has only discrete values, usually two (logic 1 and logic 0), that changes at predetermined intervals. The value (e.g., voltage) present in a single time period is called a bit. The number of bits transferred per second is called the bit rate that has units of bits per second (bit/s), or kilobits per second (kbit/s), etc.

E-mail — Electronic mail sent and received via computers with modems. Transmission media can be existing telephone or other communication lines, wireless, or not uncommonly—both.

Encode — The process whereby a transmission contains additional data or code added to facilitate proper routing of the transmission to the desired point or points.

Encryption — Technology used to form a secure channel between a wireless client and the server to support user authentication, data integrity, and data privacy.

ESN (Electronic Serial Number) — A manufacturer-assigned identity contained in a data transmission from a call placed to verify that the hardware used belongs to a valid cellular account.

Ethernet — A local-area network (LAN) protocol. Ethernet uses a bus or star topology and supports data transfer rates of 10 Mbit/s, 100 Mbit/s, 1 Gbit/s and 10 Gbit/s. Ethernet uses the CSMA/CD access method to handle simultaneous demands, and is one of the most widely implemented LAN standards.

FDMA (Frequency Division Multiple Access) — A radio system that separates user channels by frequency. Amateur Radio equipment presently uses FDMA.

Footprint — The coverage area of an individual cell.

Foreign Agent — A special "node" which is present on a foreign network and provides mobility services to visiting mobile nodes.

GPS (Global Positioning System) — A Dept. of Defense-developed, worldwide, satellite-based radio navigation system.

Handoff — Process whereby a mobile telephone network automatically transfers a call from cell to cell—possibly to another channel—as a mobile crosses adjacent cells.

Home Agent — A host on a mobile's home network responsible for trapping its packets, and forwarding them to the mobile's present location.

LAN (Local Area Network) — A computer network that spans a relatively small area. Most LANs are confined to a single building or group of buildings. However, one LAN can be connected to other LANs over any distance via telephone lines and radio waves. A system of LANs connected in this way is called a *wide-area network (WAN)*. Most LANs connect workstations and personal computers. Each *node* (individual computer) in a LAN has its own central processing unit (CPU) with which it executes programs, but it is also able to access data and devices anywhere on the LAN. This means that many users can share expensive devices, such as laser printers, as well as data. Users can also use the LAN to communicate with each other, by sending e-mail or engaging in chat sessions.

Mobile Host — Also known as a "mobile node," this addressed entity in the Mobile IP protocol roams between its home network and foreign networks.

Mobile IP — This mobile industry standard enhances the IP protocol to remedy problems associated with using the standard TCP/IP with a mobile entity. It allows for transparent routing of IP datagrams to mobile hosts (nodes) on the Internet.

Modem — A hardware device, either internally or externally connected to a computer that provides a connection from the computer and some of its programs to a landline (phone or communications line).

Network Independence — The ability to roam among networks (e.g., BellSouth Wireless Data Network, CDPD, Wireless LAN, Ethernet), although traditionally accomplished using the same access media such as SLIP, PPP, etc.

Node — A unique host on a network such as a printer, computer device, handheld Personal Digital Assistant (PDA), or a mainframe.

PCMCIA (Personal Computer Memory Card International Association) — An organization consisting of some 500 companies that has developed a standard for small, credit card-sized devices called *PC Cards*, originally designed for adding memory to portable computers.

PDA (Personal Digital Assistant) — A handheld device that functions as a personal organizer. Many PDAs began as pen-based, i.e., using a writing stylus rather than a keyboard for input, thus utilizing handwriting recognition features. Some PDAs feature voice recognition technologies. At present, most PDAs offer either a stylus or keyboard version.

POS (Point of Service) — A generation of narrowband digital, two-way, low-powered wireless services in the 800-900 MHz bands that will support confirmed delivery of message, full two-way data transfer, voice messaging and connectivity via the Internet.

PPP (Point-to-Point Protocol) — A method of connecting a computer to the Internet. PPP is more stable than the older SLIP protocol and provides error-checking features.

Remote Presence — The ability to establish remote network connections and still appear to be connected to the home network.

Security — The ability to create secure channels for user authentication, data integrity, and data privacy.

SLIP (Serial Line Internet Protocol) — An older method of connecting a computer to the Internet. A more commonly used method is PPP. SLIP is an older and simpler protocol, but from a practical perspective, there's not much difference between connecting to the Internet via SLIP or PPP. In general, service providers only offer one protocol, although some support both protocols.

TCP/IP (Transmission Control Protocol/Internet Protocol) — The suite of communications protocols used to connect hosts on the Internet. TCP/IP uses several protocols, the two main ones being TCP and IP. TCP/IP is built into the UNIX operating system and is used by the Internet, making it the *de facto* standard for transmitting data over networks. Even network operating systems that have their own protocols, such as *Netware*, also support TCP/IP.

TDMA (Time Division Multiple Access) — A digital radio system that separates users by time.

Third Party Mobile IP — An Internet technology solution that provides both wireless and wireline IP network and media roaming/communications to both Intranet and Internet services.

Throughput — The amount of data processed, or transferred from one place to another in a specified amount of time. Data transfer rates for disk drives and networks are measured in terms of throughput. Typically, throughput is measured in kbit/s, Mbit/s, and Gbit/s.

TIA (Telecommunications Industry Association) — Telecommunications Industry Association, 2500 Wilson Blvd, Arlington, VA 22001. On the web: **www.tiaonline.org**.

T1P1 — The wireless group in Committee T1. Alliance for Telecommunications Industry Solutions, 122 G St. NW, Suite 500, Washington, DC 20005. On the web: **www.atis.org**.

Virtual Private Network — Network created when a mobile user connects a data terminal to a foreign network and establishes a presence equivalent to a direct connection to the home network.

WAN (Wide Area Network) — A computer network that spans a relatively large geographical area. Typically, a WAN consists of two or more local-area networks (LANs). Computers connected to a wide-area network are often connected through public networks such as the telephone system. They can also be connected through leased lines or satellites. The largest WAN in existence is the Internet.

***Windows* OS** — Microsoft *Windows* Operating System.

Wireless Data — Information or "intelligence," sent or received by wireless transmission/reception without the direct aid of a landline.

WLAN (Wireless Local Area Network) — A local-area network that uses high frequency radio waves rather than wires to communicate between nodes.

Wireless Technology

Most hams have an interest in gadgets. In recent years, the public's use of radio waves to communicate on the telephone and via mobile computer networks has increased tremendously. This section of the chapter is written by Ron Milione, KB2UAN, and explains how these wireless systems work. Reed E. Fisher, W2CQH, also contributed to the cellular subsection.

Radio communications have been with us for a long time, with analog voice as the principal application. Today, tens of millions of people in the United States are using two-way radio for point-to-point or point-to-multipoint voice communications.

The past 25 years have seen an explosion in wireless communications and computer technology. The past eight years have seen the explosion of the Internet. Standing at the center of this convergence is wireless data technology.

Wireless data gives you the freedom to work from almost anywhere and gives you access to personal information when you are on the go. Whether the wireless system is accessing e-mail from an airport or receiving dispatch instructions in a taxi, maintaining a data connection with a remote network from almost anywhere can be realized.

Wire or fiber-based data communications span a wide range of *throughput* and distances—56 kbps over a modem connection; 10/100 Mbps over an *Ethernet* segment; and gigabit speeds over fiber. Similarly, wireless connections span a wide range. The world of wireless data includes fixed microwave links, wireless Local Area Networks (LANs), data over cellular networks, wireless Wide Area Networks (WANs), satellite links, digital dispatch networks, one-way and two-way paging networks, diffuse infrared, laser-based communications, keyless car entry, the Global Positioning System (GPS) and more.

The benefits of wireless include communications when and where no other communication links are possible, connections at lower cost in many scenarios, faster connections, backups to landlines, networks that are much faster to install and data connections for mobile users.

FUNDAMENTAL CONCEPTS IN CELLULAR TECHNOLOGY

In a cellular radiotelephone system the desired radio coverage area is divided up into a number of smaller geographical areas called cells. **Fig 24.3** shows a typical seven-cell (n = 7) cluster. Each cell has a radius r, which may be up to eight miles in

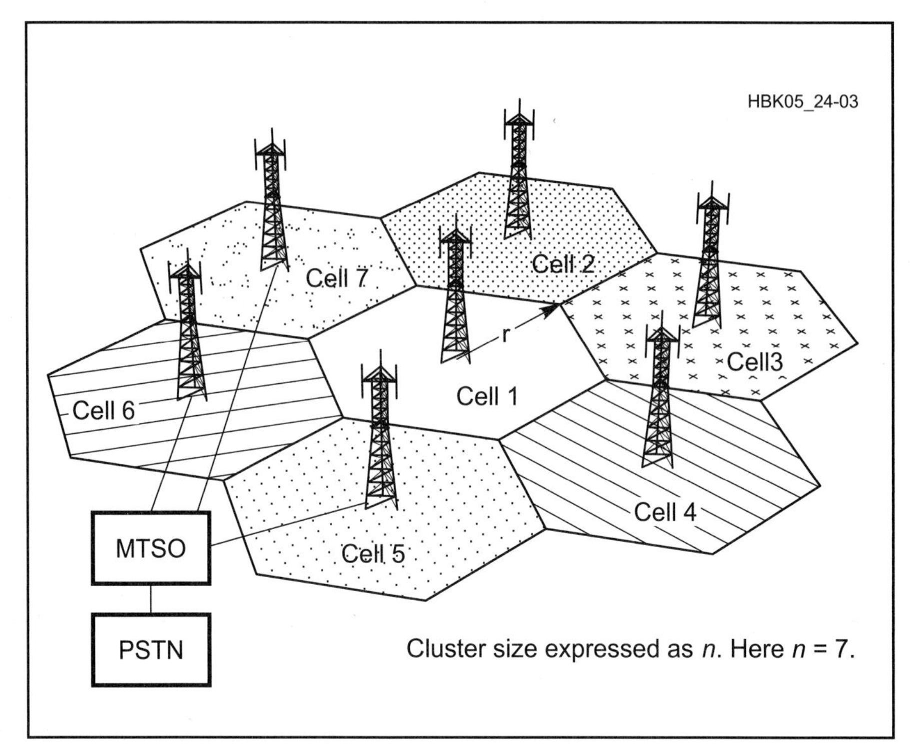

Fig 24.3 — A seven-cell cluster.

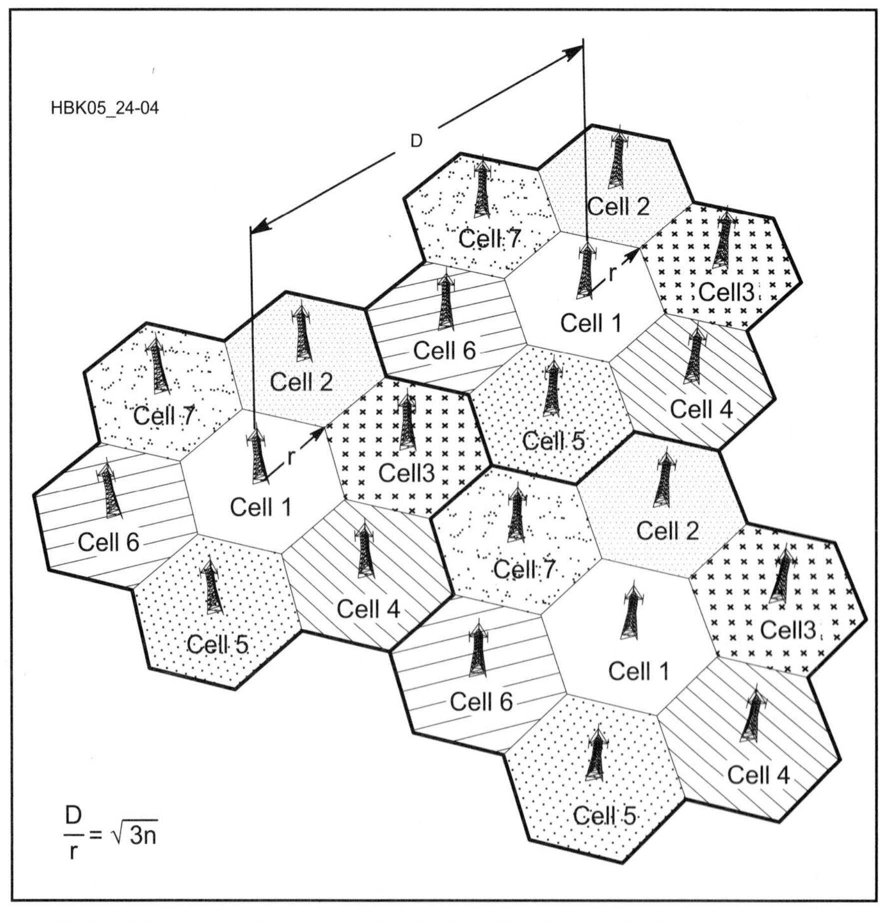

Fig 24.4 — Frequency reuse over an area to conserve spectrum.

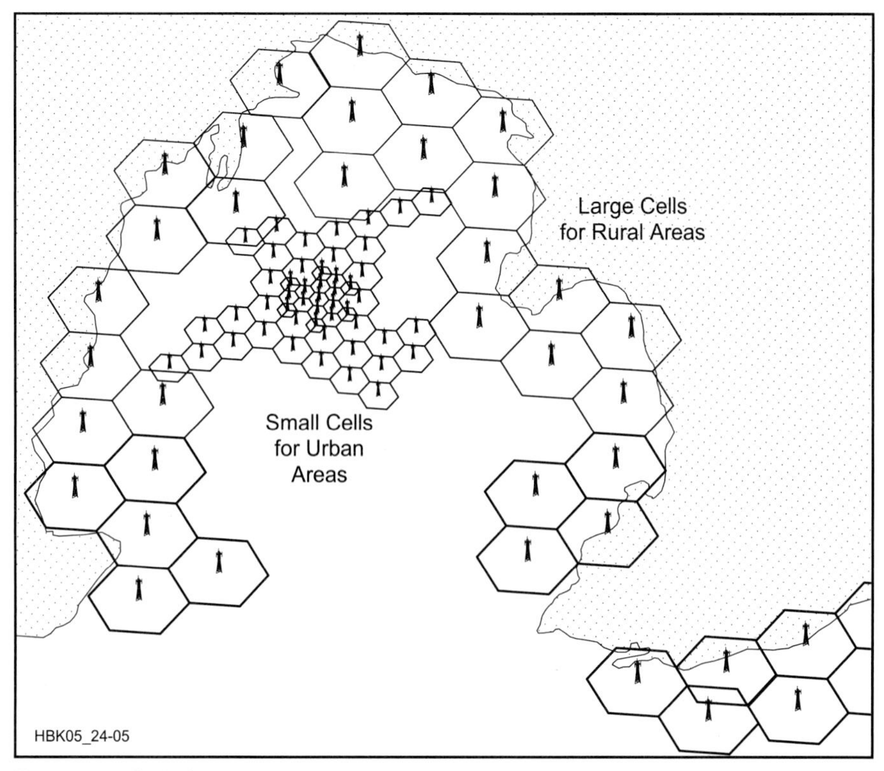

Fig 24.5 — Cell size corresponding to density of population and use.

a startup system. Each cell uses a unique group of radio channels (channel set) that is different from the others. Thus, there is no interference between radio users within a cell cluster. At the center of each cell is located a base station (cell-site) with associated radio tower and antenna system. Mobile stations, located within the cell, communicate with the base station via two-way UHF radio links. The base stations are linked to an MTSO (Mobile Telephone Switching Office) by landline (usually wireline) transmission lines. The MTSO is connected to the PSTN (Public Switched Telephone Network) by additional links. Thus, radio mobile station users can communicate with wireline PSTN users.

Mobile station users can also communicate with other mobile station users, but only through the MTSO and PSTN. The MTSO has other important functions including location and handoff (handover). When a larger serving area is required, cell clusters are placed together as shown in **Fig 24.4**. In this configuration, all cells with the same number use the same channel set. This condition is called frequency reuse, a prime feature of cellular systems. Co-channel interference now becomes important, for a mobile station in cell #1 may interfere with other mobile stations in the surrounding #1 cells. In Fig 24.4, the distance between same-numbered cells is called D. The co-channel interference ratio is a function of D/r. In a hexagonal grid $D/r = \sqrt{3n}$. For n = 7, $D/r = \sqrt{21} = 4.58$. For ground-to-ground UHF radio propagation, the received carrier power (C) falls off with distance raised to the 4th power. The worst case C/I (carrier-to-interference ratio) is: C/I = 40 log (D/r). For n = 7, C/I = 40 log 4.58 = 26.4 dB. A properly designed AMPS cellular system requires C/I > 18 dB in 95% of the coverage area and this value provides satisfactory voice transmission.

As the mobile station moves away from its serving base station, the mobile station's received carrier power (C) becomes smaller and eventually approaches the system noise floor (N). The cell radius r must be kept small enough to provide a satisfactory carrier-to-noise ratio (C/N). An AMPS system with r = 8 miles, serving mobile stations with roof-mounted antennas, is designed to provide C/N > 18 dB in 95% of the coverage area.

Fig 24.5 shows the concept of cell splitting, which is another feature of cellular systems. For example, in densely populated regions such as the center of a city, the cell radius can be reduced from eight miles to four miles. Since the C/I is a function of D/r, it is not reduced when D and r are both made small.

Cellular and PCS Frequency Bands

The 850 MHz cellular frequency band was allocated by the FCC in 1974. Cellular radios operate in the FDD (Frequency Division Duplex) mode that means that both transmitter and receiver are active at the same time. The transmitter of the base station transmits on a forward (downlink) channel, while at the same time the transmitter of the mobile station transmits on a reverse (uplink) channel 45 MHz lower. The cellular uplink band is 824-849 MHz and the downlink band is 869-894 MHz. The bands are divided into two blocks for the two cellular service providers licensed in each coverage area. In 1995, the FCC allocated a new band for a new all-digital, cellular-like service called PCS (Personal Communications Service). The PCS uplink band is 1850-1910 MHz, and the downlink band is 1930-1990 MHz. PCS also uses FDD and the duplex frequency spacing is 80 MHz. The bands are divided into six blocks for the six PCS service providers licensed in each coverage area. There are currently more than 100 million cellular and PCS subscribers in the United States (US).

Cellular Air Interfaces

Air interfaces are the protocol standards used by phone companies to provide cellular voice and data service. The base station, mobile station and MTSO hardware and software are usually implemented with proprietary equipment procured from manufacturers chosen by the cellular/PCS service providers. In contrast, the cellular air interfaces are documented in open (public) standards complying with FCC regulations.

Cellular/PCS standards in the US are generated by the TIA and T1P1, then finally approved by the American National Standards Institute (ANSI). Four standards provide most of the cellular service in the US. Three of the four standards operate in the 850-MHz cellular band. Three different standards operate in the 1.9-GHz PCS band. An obvious disadvantage of multiple air interface standards, and two frequency bands, is the difficulty encountered by "roaming" mobile stations when they enter non-home systems. This problem has necessitated the design of higher cost "dual-band, multi-mode" mobile stations. The four standards are:

• *AMPS (Advanced Mobile Phone Service)* — AMPS was the first cellular system in the US. Service commenced in October 1983 in Chicago. AMPS was standardized by the TIA as IS-553. AMPS is an FDMA (Frequency Division Multiple Access) system that means that user channels are separated by frequency. (Amateur Radio currently uses FDMA.) The channel spacing is 30 kHz, which allows 832 FDD channels in each 25-MHz band near 850 MHz. AMPS, is an analog/digital system. Voice uses analog PM (Phase Modulation) with a peak frequency deviation of ±12 kHz. Digital system control, e.g. call setup and handoff, uses 10 kbit/s binary FM (Frequency Modulation) with a peak frequency deviation of ±7 kHz. There is no digital service provided to the users. A user-provided modem must be attached to the mobile station. At this time, most US cellular customers still use AMPS.. (See Ref. 1 at the end of this chapter for a description of a typical AMPS base station.) By 1986, there was already worry that AMPS service might soon become saturated in city centers. There was a competitive effort to design new cellular air interfaces with increased spectral efficiency — more user channels per MHz. It was decided that all future cellular systems must be digital.

There are many advantages to digital transmission above the obvious benefit to users requiring data transmission services. Unfortunately, there are a few disadvantages. Digital has a sharp threshold, so there are no "fringe" coverage areas. The received signal is either excellent, or absent. Digital requires the use of low-bit-rate voice coders and their associated artifacts. Wireline digital transmission systems encode voice using 64 kbit/s PCM (Pulse Code Modulation) waveform coders that provide nearly distortion-free coding. Since 64 kbit/s PCM cannot fit into the narrowband radio channels, a number of low rate, e.g. 8 kbit/s, coders have been devised. Most use the CELP principle that emulates a single human voice tract. Strange sounds may appear at the receiver when the CELP coder is driven by multiple voices, music or background noises. See Ref. 3 at the end of this chapter for a description of CELP and other voice coders.

• *TDMA (Time Division Multiple Access)* — TDMA was the first all-digital US cellular system. TDMA service started in the late 1980s. It was standardized by the TIA as IS-54 and provided three times the traffic capacity of an AMPS system. It was designed to gradually replace AMPS. The channel spacing was kept at 30 kHz so that TDMA channels could be mixed with AMPS channels. Within each 30 kHz frequency channel are placed three time-divided digital channels. Thus, "TDMA" is actually a hybrid TDMA/FDMA system. QPSK (Quadrature Phase Shift Keying) modulation is used which provides a "raw" channel bit rate of 48.6 kbit/s. The raw user bit rate is 48.3/3 = 16.1 kbit/s. After bit error control techniques are applied, each user is provided an almost error free 8 kbit/s data service or 8 kbit/s VSELP (a version of CELP) voice coder service. An upbanded version of TDMA, with enhanced features, operates in the 1.9 GHz PCS band. It is standardized as IS-136.

• *CDMA (Code Division Multiple Access)* — By 1986, some in the cellular industry believed that CDMA techniques would provide more capacity than TDMA. In the US, CDMA was standardized by the TIA as IS-95. CDMA uses the same up and down bands as AMPS and TDMA, but channel spacing is 1.25 MHz. CDMA separates the user bit streams by digital codes.

More than 40 users can occupy the same 1.25 MHz frequency channel at the same time. The "raw" bit rate provided to each user is 28.8 kbit/s. After bit error control, each user is given a 9.6 kbit/s data channel or 13 kbit/s QCELP (a version of CELP) voice service. In this direct-sequence type of CDMA transmitter, the incoming user-generated bit stream (either digital voice or data) is multiplied (exclusive-OR) with a 1.2288 Mbit/s pseudo-random digital spreading code. Each user is given a unique code. The 1.2288 Mbit/s multiplier output bit stream enters a QPSK modulator, and is then amplified and transmitted with a spreaded RF bandwidth near 1.25 MHz.

In the CDMA receiver, the incoming mixed signal — received from all of the active users—is multiplied by the desired user's unique code that is synchronized with the transmitter code (a challenging task). Only the desired user's de-spreaded 28.8 kbit/s bit stream will emerge from the multiplier. The other codes emerge as wideband noise that is eliminated by a low-pass filter.

Tight mobile-station, transmit-power control must be used for the CDMA system uplink to achieve its expected capacity. All mobile station transmissions received at the base station should be within a few dB of each other. To achieve this severe requirement, mobile station power output is updated 800 times per second by a base station-generated 800 bit/s power control data stream interspersed into the downlink 28.8 kbit/s data stream.

It is difficult to calculate the user capacity of a CDMA system. The per-channel spectral efficiency of CDMA (40 bit streams in 1.25 MHz) is comparable with AMPS (40 channels in 1.20 MHz). However, CDMA is so resistant to co-channel interference that all CDMA radio channels can be used in all cells (n = 1). Therefore, there is a traffic capacity increase of at least seven times that of a standard AMPS system. An upbanded version of CDMA, with enhanced features, operates in the 1.9-GHz PCS band. It is standardized as IS-95A.

• *GSM/PCS-1900* — GSM (Global System for Mobile Communications) was imported from Europe in the mid 1990s. In the US it operates only in the 1.9 GHz US PCS band. It was standardized by the T1P1 as J-STD-007. It is a TDMA/FDMA system. Frequency channel spacing is 200 kHz and the channel bit rate is 270 kbit/s. GMSK (Gaussian Mean Shift Keying), a type of narrowband binary FM, is the modulation method. There are eight time slots in each RF channel, so the user raw bit rate is 270/8 = 33.75 kbit/s. After bit error control, each user is provided a 9.6 kbit/s data channel or a 13 kbit/s RPE-LTP (Regular Pulse Excited-Long Term Prediction) voice coder for voice. The spectral efficiency of GSM (8 bit streams in 200 kHz) is only slightly better than AMPS (8 channels in 240 kHz). This weakness is mitigated by using n = 4, a layout tolerated by the more interference-resistant GSM digital transmission. Thus, GSM can handle about twice as much traffic capacity as AMPS.

Radio Link Operating Protocols

The AMPS system uses three protocols for system communication between the base station and the mobile station:

• *PAGING* — At least one channel in each cell is used for paging mobiles with incoming calls from the MTSO. A downlink paging bit stream, the same in each cell, is continuously monitored by power-ON, non-active mobile stations. When a mobile station recognizes its user number, it responds on the uplink-paging channel. The paging stream then hands off the mobile station to a setup channel. Conversely, the mobile station originates a call by sending its request on the uplink-paging channel.

• *SETUP* — At least three channels are used in each cell for setup; the mobile station-base station link sets up and tears down calls. After the two-way setup procedure is finished, the mobile station is commanded to tune to its assigned voice channel. Mobile authentication is performed during setup. Each mobile station has a NAM (Numeric Address Module) and an ESN (Electronic Serial Number). The NAM contains the user's telephone number. The manufacturer-assigned ESN is unique. At setup, the MTSO verifies that the NAM and ESN numbers are valid.

• *LOCATION AND HANDOFF* — While an active mobile station is linked to its serving base station, its radio signal strength is also being periodically monitored by the six surrounding base stations. When the averaged signal strength at a neighboring base station exceeds that measured at the serving station, the MTSO commands a handoff to the neighbor. This handoff procedure allows un-interrupted trans-mission as the mobile station moves through the cell cluster. Handoff is accomplished by a "blank-and-burst" procedure. The down-link voice channel is interrupted, and a brief 10 kbit/s digital message commands the mobile to move to its newly assigned frequency channel. The mobile station acknowledges, via blank-and-burst on the uplink voice channel, then tunes to the newly assigned channel. If handoff fails, the call is terminated.

The TDMA, CDMA and GSM digital systems use similar procedures. DSP (digital signal processing) techniques provide better authentication and user privacy. Details are found in Ref. 2 at the end of this section.

Virtual Private Network

A Virtual Private Network is created when a mobile user connects a data terminal to a foreign network, either via dial-in or public networks, and establishes a presence equivalent to a direct connection to the home network.

The Wireless Mobile IP solution is intended to enable the creation of Virtual Private Networks by using the Internet as the communications backbone to connect mobile users. The following features characterize Virtual Private Networks:

Remote presence — the ability to establish remote network connections and still appear to be connected to the home network.

Network independence — the ability to roam among networks (e.g., BellSouth Wireless Data Network, CDPD, Wireless LAN, Ethernet). Traditionally, IP network independence (roaming) is done over the same media access (e.g., SLIP, PPP, Ethernet). Virtual Private Network implementation offers the ability to roam across not only single media IP networks, but across multiple wireless and wireline media *without user intervention.*

Security — the ability to help create secure channels for authentication, data integrity, and data privacy.

Third Party Mobile IP — this solution is unique in that it provides wireless and wireline IP network and media roaming/communications to both Intranet and Internet services.

The ability for mobile users to roam seamlessly and without intervention among radio frequency (RF) networks and wireline networks allows the system to operate at maximum system efficiency.

MOBILE COMPUTING

With the rapid growth and availability of wireless data networks, wireless communications tools and Internet standards, mobile workers are finding new ways to do business in today's competitive environment. The need for the mobile worker to access mission-critical information requires access to corporate databases and Internet/Intranet applications. In addition, convenient and reliable file transfer, integrated messaging, and personalized information delivery allow the mobile employee to work at peak productivity levels.

Successful communications between mobile workers and their corporate environment requires the right combination of technologies. From a business standpoint, these technologies must be cost-effective and easy to use. For long-term viability, they should be based on open system architectures and industry standard interfaces.

Virtual Private Networks have emerged to provide networking solutions to a growing mobile workforce. These networks allow businesses to provide their mobile employees with access to corporate information and applications by connecting them to the enterprise via the Internet. A Virtual Private Network provides a low cost extension to the enterprise while offering secure access to an open networking environment.

Mobile IP Security

Mobile IP is an Internet industry standard that enhances the IP protocol to remedy these existing problems and allow the transparent routing of IP datagrams to mobile nodes on the Internet. Security is an integral part of building a Virtual Private Network solution. The Wireless Mobile IP Network Configuration utilizes Mobile IP encryption to form a secure channel between the Wireless client and server to support user authentication, data integrity, and data privacy in mobile environments.

Using the Wireless Mobile IP encryption, a secure channel is formed which allows various foreign networks to become extensions of the home network.

Mobile IP Benefits

The Wireless Mobile IP solution is intended to enable enterprises to create their own Virtual Private Networks, thus providing:

(1) Low initial costs
(2) Low operating costs
(3) Solution flexibility
(4) Significant productivity gains

By providing seamless network roaming and communications capabilities, Mobile IP provides a networking solution to bring enterprises into the 21st century.

Many of the spread spectrum devices on the market today are listed as FCC Part 15 devices. There are three frequency bands allocated to this service:

902-928 MHz (26-MHz bandwidth)
2400-2483.5 MHz (83.5-MHz bandwidth)
5725-5850 MHz (125-MHz bandwidth)

Regulations

In 1999, the FCC greatly liberalized amateur spread spectrum rules, allowing spreading techniques not previously permitted. Amateurs are no longer limited to frequency hopping and direct sequence; any documented spreading code may now be utilized. Amateurs may now freely use SS devices designed under Part 15 of the Commission's rules for amateur applications as well, keeping in mind the identification requirements of the Amateur Service. The maximum power allowed for an amateur SS emission is now 100 watts, up from the previous limit of a single watt! However, stations that transmit more than one watt must utilize automatic power control to limit power output to the minimum necessary to communicate. For information on the rules change, see *ARRLWeb* at: **www.arrl.org/news/stories/1999/09/08/2/**. For the text of the rules, see Section 97.311 in *The ARRL FCC Rule Book*.

KEEPING CURRENT WITH WIRELESS

Changes to the wireless communications industry have been dramatic since the first mobile phone systems were introduced. The high-tech, expanding nature of this fascinating area will continue to drive new wireless communication developments. The following are some related Web sites where interested readers can keep up-to-date on this fast-paced technology:

Cellular Communications and Internet Association (CTIA): **www.wow-com.com/**
Protocols.com Reference page: **www.protocols.com/pbook/cellular.htm**
FCC Cellular Services: **wireless.fcc.gov/services/cellular/**
International Engineering Consortium: **www.iec.org/online/tutorials/cell_comm/**
Wireless Application Protocol (WAP): **www.cellular.co.za/wap.htm**
Waveguide, A brief history of Cellular: **www.wave-guide.org/archives/waveguide_3/cellular-history.html**
Mobile Cellular Technology Newsletter: **www.mobilecomms-technology.com/**
An overview of cellular technologies: **www.ee.washington.edu/class/498/sp98/final/marsha/final.html**

References

1. "Advanced Mobile Phone Service", *The Bell System Technical Journal*, Vol 58, No. 1, Jan. 1979.

2. V.K. Garg and J.E. Wilkes, *Wireless and Personal Communications Systems*, Prentice Hall, 1996, ISBN 0-13-234626-5.

3. J. Bellamy, *Digital Telephony*, John Wiley, 1991. ISBN 0-471-62056-4.

PC Technology (Personal Computers in the Shack)

The Personal Computer debuted in 1980 with offerings by Commodore (PET), Tandy (TRS-80) and Apple (Apple II). Almost immediately, hams found them to be useful additions to the shack. These first computers were little more than a 4-MHz processor with a little RAM (4kB) an audio tape drive and an imbedded *BASIC* operating system. One of the first applications available for hams was an RTTY decoder. The age of digital modes for hams was about to expand exponentially.

As the photo in **Fig 24.6** indicates, today the PC is an important part of any radio station. In most cases, it is the control center of the various pieces of equipment we use. In an extreme example, you can use it to download from the web the test questions to take your exam, learn the code, design your antennas, and even search ebay or QRZ.com to buy a radio. It can control the radio, provide various digital modes through the use of the sound card, connect to the Internet for solar flux numbers to compute which bands have propagation, use TELNET for packet spots to help find a DX station, provide software to log the QSO and keep track of award progress, and finally assist the operator in finding the QSL route to complete the process (or in the case of the new Logbook of The World (LoTW) upload the QSO to the ARRL site and perhaps have that confirmation in hours). What could be a more important tool to the operator?

ERGONOMICS

Since the computer will connect to most of your equipment, it is usually found in the center of the operating arena. A tower unit on the floor will give it some distance from the RF generators and keep the desktop free for equipment that you need to handle. A well-designed station will have the keyboard, mouse and keyer (assuming you are a CW op!) all within a 45-degree arc on either side of your operating chair. Use of an LCD monitor will preserve more space on your desk as well as produce less RF in the shack.

HARDWARE

The computer itself should have a minimum of 256 MB of memory (more is better), hard drive with 40 GB minimum space, a rewritable CD, two serial ports, and a sound card. Serial ports are becoming a thing of the past; newer PCs have replaced them with USB ports. However, there are now many commercial USB-to-Serial Converter devices that allow you to still connect older DB9-serial devices to the newer PCs. If you are connecting your station to the Internet (recommended), you also need to consider a modem or Ethernet port depending on your Internet service provider. To save space on the operating table, locate your PC on the floor if possible (tower) and use an LCD display to preserve maximum desk space — and be less of a heat source.

One of the first things to consider is interference from the radio to the PC or from the PC to the radio. If your radio is properly grounded, your antennas at a proper distance from the shack, and you use well-shielded cables, the chance of RF getting into your PC is remote. It is also good practice to try to keep your radios and computer equipment on separate power circuits.

Receiving computer-generated RF in your radios is often the result of poorly designed monitors or leakage from the monitor cables. Whenever adding new computer equipment into the shack it is good to immediately scan the bands to find any new birdies or new noise. Should there be any, shutting off the monitor and then the computer will isolate which of the two pieces of equipment are at fault. In many cases, the installation of toroids on the monitor cable and all cables (power, speaker, CAT, Keyboard or Microphone) that are a source of radiating will eliminate the problem.

Most modern radios have a CAT (Computer Access Terminal) port that is a serial 9 pin (MINIDIN9). Some older radios required a computer interface device that was usually RS232 based. Yet other radios use DIN6 or CI-V interface devices. Complicating the interface issue is the trend to USB ports instead of serial ports. Fortunately, the commercial ham market has kept up with the lack of standards with a host of good USB to Radio Port products. MicroHam is one such source (**www.microham.com**) covering just about all the USB wiring possibilities. The basic commands to control your radio are available from the radio manufacturer's Web sites. Third-party software applications all support these functions and are discussed in detail below.

Your computer sound card can act as a digital modem enabling you to operate FSK, PSK, SSTV, FAX and other digital modes. However, you also will find the need for a direct-controlled interface to key the transmitter in response to signals over the serial port's RTS or DTR pins. The interface should also include an attenuator to simplify the connection between the computer sound card and the transceiver microphone jack. This type of interface may also be used to connect your

Fig 24.6 — Don, N1DG, and his computer-enriched operating station.

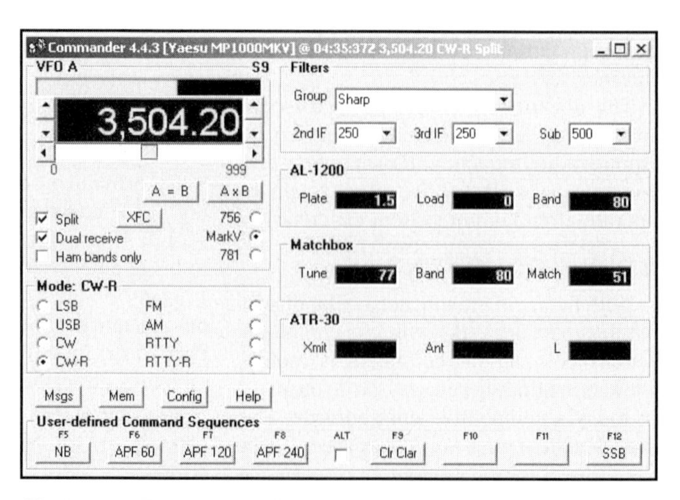

Fig 24.7 — The DXLAB *Commander* Control Panel.

computer to your radio for Echolink or IRLP operation. The RIGblaster from West Mountain Radio is one of several commercial products for this.

Lastly, most logging and contest software generates Morse code. Connecting the output of the radio to the input of the sound card is simple enough with a stereo cable. However, be sure to isolate the two devices with an isolation transformer in the cable (such as a RadioShack 273-1374) to prevent 60-Hz hum from arriving with the signal. The output from the PC to the radio can come either from a parallel or serial port. There are several commercial products available, the best for parallel ports being from Jack Shuster, W1WEF. Two suppliers of interfaces are N3JT (**n3jt@arrl.net**) and W1WEF (**w1wef@arrl.net**).

The recent introduction of *Windows XP* further complicates the software output to serial or parallel ports. *WinXP* no longer allows programs direct access to them. However, there are two shareware programs — *Directio* and *Userport* — that will install drivers on your operating system to allow most ham logging and modem software to reach the ports. Both are free downloads at **www.embeddedtronics.com/design&ideas.html**.

SOFTWARE

Radio Control Software

One of the most popular free programs is *Commander*, (**www.qsl.net/civ_commander/**) by DXLab Software. It is one of many shareware modules by DXLabs and this specific program interfaces with the recent radios from Ten-Tec, ICOM, Kenwood and Yaesu.

Through its graphical interface shown in **Fig 24.7**, a click of the mouse allows you to set filters, frequency ranges pursuant to your license class, memories for favorite frequencies and customization for the way you want your radio setups. All the logging programs listed below have rig control as part of their logging features. However, most of the general logging programs do not have the ability to set filters and menu items as *Commander* does (see Fig 24.7). They are designed for other things.

Logging Software

In 1986, two logging programs appeared for *MS-DOS*-based PCs that would inaugurate an easier way of keeping track of operating awards and contesting scoring. The first, the WB2DND *Log Database*, was built around *DBASE III* and became popular among DXers and DXpeditioners. At the same time, K1EA introduced *CT*. Computerized logging enabled faster contacts, more accurate recording of those contacts, easy scoring and simple QSL management. Today's logging software not only logs, but it controls the radios, uses the sound card to send and decode digital mode transmissions, and even interfaces to packet-spotting networks via VHF packet or Internet nodes.

In 1996, the CEØZ and VKØHI expeditions took computer logging to the next level by posting their logs on their Web sites during the DXpeditions so hams could check their progress while the DXpeditions were still in progress. In order to standardize the format among the many logging programs that appeared over the last 15 years, two standards emerged to allow for the easy use of the information being collected. General logging programs now allow you to export your data in a format known as ADIF (Amateur Data Interface Format), which allows you to use multiple programs and move data among them and share log data with friends or QSL Managers. You never need to worry about obsolescence of your data. If your publisher goes out of business, or no longer supports new platforms, you can choose from a large selection of other programs that will load your logs without retyping everything.

The second format has been written and is used by contesters to submit their scores to the various organizations sponsoring the contests. This format is known as Cabrillo and is available on all contest software today. By submitting all contest logs in a standard format, adjudicators can link the logs with a common database and quickly see errors for scoring adjustments.

It is worth noting that both current general logging programs and contest software also handle both of these standards to make transferring the contest logs into the general logging programs for award tracking.

LoTW

No discussion of general logging and the award tracking can be complete without a discussion of the latest development in this area. In 2003, the ARRL launched a web-based depository of QSO data called the *Logbook of The World* (*LoTW*). The purpose is to electronically verify QSLs and provide those certified QSLs for its award programs (and eventually other ham-organization award programs). *LoTW* has its own website (**www.arrl.org/lotw/default**) that is password protected. Membership is free but digital certificates and passwords are issued only after licensing, and user details are verified by mail to the address of the licensee. This assures that the data submitted is actually done so by the operator.

The free software to obtain your certificate (called *TQSL*), and instructions, are downloadable from **www.arrl.org/lotw/#download**. Once receiving a digital certificate, the Cabrillo or ADIF logs can be *signed* and uploaded over the web to the database. The software automatically matches your data to the database and shows your QSLs in a variety of formats. In the first six months since the web site went live, 7,000 users have uploaded over 40 million QSO records resulting in QSL matches of over 1.2 million records.

Logging/Contesting Software

Some of today's popular logging/contesting software packages include:

CT (ConTest, by K1EA) Web site: **k1ea.com**
Contest logger for DOS/Win85/98/2000 Freeware

Dxbase (by Scientific Solutions). Web site: **dxbase.com/proddxbw.htm**
General logging software for Win95/98 and Windows/NT/2000 $90

DX4WIN (by KK4HD). Web site: **dx4win.com/**
General logging software for Win95/98/ME/XP/2000/NT $99.

EasyLog (by Microware Software). Web site: **easylog.com/eng/index.htm**
Windows full-featured logging software $89.54.

HC Log (Buckmaster Publishing). Web site: **http://hamcall.net**
Logging software integrated to work with *HamCall* CD-ROM.

Logic 7 (Personal Database Applications, Inc.). Web site: **hosenose.com**
Windows full-featured logging software $129.

MiLOG (SA Engineering). Web site: **logwindow.com**
Contesting software for Windows $69.95.

NA (by K8CC). Web site: **datomonline.com**
Contest logger for DOS $60.

N1MM Logger (by N1MM, AB5K, PA3CEF, G4UJS, N7ZFI, N2AMG).
Web site: **http://pages.cthome.net/n1mm/**
Contest logger for 95/98/ME/NT/2000/XP Freeware.

N3FJP Software. Web site: **n3fjp.com**
Windows-based logging/contesting software. Various prices by contest/version.

ProLog2K (by W5VP) Web site: **prolog2k.com/**
General logging software for Win95/98/ME/2000/NT/XP $50.

SWISSLOG (by HB9BJS) Web site: **informatix.li/english/Frame_EN.htm**
General logging software for Win95/98/ME/NT4/2000/XP 70 Euros.

TR Log (by N6TR) Web site: **qth.com/tr/**
Contest logger for DOS $75.

VHF log (by W3KM) Web site: **http://www.qsl.net/w3km/features.htm**
VHF contest logger With CW/PTT/DVK for Win98, ME, 2000 or XP. Freeware.

WinEQF (by N3EQF) Web site: **eqf-software.com**
General logging Program for Win95/98, ME, NT, XP, or Win2K $59.95.

WriteLog (by W5XD) Web site: **writelog.com/**
Contest logging software for Win95/98, NT, Win2K $75.

This list is by no means complete but it will give you a start to explore the products available based on your interests.

PACKETCLUSTER AND AR CLUSTER

During the late 1980s, Dick Newell, AK1A, developed PacketCluster software to link PCs together via VHF radios to DX-related information. Now 15 years later, the software is still in use, but is slowly being replaced by Internet-based Telnet software such as AR Cluster, DX Spider, CLX, DxNet, Clusse, and WinCluster.

These software servers enable users to announce and receive DX spots, general information announcements, and send personal talk messages and email. Almost all logging software today has Telnet software built in to interface to the Internet nodes and retrieve Cluster spots.

AR Cluster (developed by AB5K) allows the user to tailor the information he receives. Since it is an SQL Database, the user can query that database in an almost unlimited way. SH/DX commands can be by call, band, mode, DX Country, State or Country of the spotter, the node the spot came from, DTS (Date/Time stamp), ITU or CQ Zone, or even from a comment associated with the spot. Further, the inquiry can be any one or all of the above parameters. Looking for the last 50 spots for a DXpedition is as simple as sh/dx/50 CALLSIGN. The user can also set specific filters for his use so that spots sent to him are set to the specific parameters he has set for his call. A VHF-only user does not need to see HF spots. Spots outside of licensed frequencies can be ignored, etc. The AR-Cluster user manual can be read at **www.ab5k.net/ArcDocs/UserManual/ArcUserManual.htm**.

One of the best sources of Telnet Ham radio sites is maintained by K6PBT on his website **http://telnet.dxcluster.info/**.

MODE-SPECIFIC SOFTWARE

The ability of using the sound card as a modem has enabled hams to operate many digital modes without addition equipment. Software has emerged that generate and decode these modes and in some cases even log the contacts made. While some of the logging software listed earlier includes mode decoding routines, the most feature-rich software is mode specific. Some popular programs in this software category:

Multi-Modes:

RadioCom 5 (by Computer International, Bonito line). Web site: **computer-init.com**

Skysweep (Skysweep Technologies). Web site: **skysweep.com**.

RTTY:

MMTTY (by JE3HHT) Web site: **www.qsl.net/mmhamsoft**
Win9x Freeware.

SSTV:

DIGTRX (by PY4ZBZ & KB9VAK) Web site: **http://planeta.terra.com.br/lazer/py4zbz/hdsstv/teste1.html#digtrx**
HDSSTV For WIN9X Freeware.

MMSSTV (JE3HHT Makoto Mori) Website: **www.qsl.net/mmhamsoft/mmsstv/**
SSTV Win9x/WinNT/Win2k Freeware.

Mscan (by PA3GPY) Web site: **http://whatasite.com/mscan/products.html**
SSTV Win9x $43.00.

PSK31:

DXPSK (by F6GQK) Web site: **dxfile.free.fr/dxpsk.htm**
Win9x Freeware.

Digipan. Web site: **digipan.net**

WinPSKse Website: **www.winpskse.com**
Win9x Freeware.

WinPSK Website: **www.qsl.net/ae4jy**
Win9x Freeware.

W1SQLpsk > (by W1SQL) Website: **www.faria.net/w1sql/psk31.htm**
Win9x/WinNT Freeware.

Multimode Website: **www.blackcatsystems.com/software/multimode.html**
Mac PowerPC $89.

Meteor Scatter:

WSJT by (K1JT) Website: **http://pulsar.princeton.edu/~joe/K1JT**
JT44 mode for weak signals Win9X/ME/XP Freeware.

WinMSDSP2000 (by 9A4GL) Website: **www.qsl.net/w8wn/hscw/msdsp.html**
Win9X Shareware.

Hellschreiber:

Hellschreiber (by IZ8BLY) Website: **http://iz8bly.sysonline.it/Hell/index.htm**
Feldhell Win9x Freeware.

Feldhell (by G3PPT) Website: **http://members.xoom.com/ZL1BPU/software.html**
Feld-Hell/MTHELL/SMTHEL8 MS-DOS Freeware.

Chapter 25

Test Procedures

This chapter, written by ARRL Technical Advisor Doug Millar, K6JEY, covers the test equipment and measurement techniques common to Amateur Radio. With the increasing complexity of amateur equipment and the availability of sophisticated test equipment, measurement and test procedures have also become more complex. There was a time when a simple bakelite cased volt-ohm meter (VOM) could solve most problems. With the advent of modern circuits that use advanced digital techniques, precise readouts and higher frequencies, test requirements and equipment have changed. In addition to the test procedures in this chapter, other test procedures appear in Chapters 10 and 11.

TEST AND MEASUREMENT BASICS

The process of testing requires a knowledge of what must be measured and what accuracy is required. If battery voltage is measured and the meter reads 1.52 V, what does this number mean? Does the meter always read accurately or do its readings change over time? What influences a meter reading? What accuracy do we need for a meaningful test of the battery voltage?

A Short History of Standards and Traceability

Since early times, people who measured things have worked to establish a system of consistency between measurements and measurers. Such consistency ensures that a measurement taken by one person could be duplicated by others — that measurements are reproducible. This allows discussion where everyone can be assured that their measurements of the same quantity would have the same result. In most cases, and until recently, consistent measurements involved an artifact: a physical object. If a merchant or scientist wanted to know what his pound weighed, he sent it to a laboratory where it was compared to the official pound. This system worked well for a long time, until the handling of the standard pound removed enough molecules so that its weight changed and measurements that compared in the past no longer did so.

Of course, many such measurements depended on an accurate value for the force of gravity. This grew more difficult with time because the outside environment— such things as a truck going by in the street — could throw the whole procedure off. As a result, scientists switched to physical constants for the determination of values. As an example, a meter was defined as a stated fraction of the circumference of the Earth over the poles.

Generally, each country has an office that is in charge of maintaining the integrity of the standards of measurement and is responsible for helping to get those standards into the field. In the United States that office is the National Institute of Standards and Testing (NIST), formerly the National Bureau of Standards. The NIST decides what the volt and other basic units should be and coordinates those units with other countries. For a modest fee, NIST will compare its volt against a submitted sample and report the accuracy of the sample. In fact, special batteries arrive there each day to be certified and returned so laboratories and industry can verify that their test equipment really does mean 1.527 V when it says so.

Basic Units: Frequency and Time

Frequency and time are the most basic units for many purposes and the ones known to the best accuracy. The formula for converting one to the other is to divide the known value into 1. Thus the time to complete a single cycle at 1 MHz = 0.000001 s.

The history of the accuracy of time keeping, of course, begins with the clock. Wooden clocks, water clocks and mechanical clocks were ancestors to our current standard: the electronic clock based on frequency. In the 1920s, quartz crystal controlled clocks were developed in the laboratory and used as a standard. With the

Table 25.1
Standard Frequency Stations

(Note: In recent years, frequent changes in these schedules have been common.)

Call Sign	*Location*	*Frequency (MHz)*
BSF	Taiwan	5, 15
CHU	Ottawa, Canada	3.330, 7.335, 14.670
FFH	France	2.500
IAM/IBF	Italy	5.000
JJY	Japan	2.5, 5, 8, 10, 15
LOL	Argentina	5, 10
RID	Irkutsk	5.004, 10.004, 15.004
RWM	Moscow	5, 4.996, 9.996, 14.996
WWV/WWVH	USA	2.5, 5, 10, 15, 20
VNG	Australia	2.5, 5
ZSC	South Africa	4.291, 8.461, 12.724 (part time)

advent of radio communication time intervals could be transmitted by radio, and a very fundamental standard of time and frequency could be used locally with little effort. Today transmitters in several countries broadcast time signals on standard calibrated frequencies. **Table 25.1** contains the locations and frequencies of some of these stations.

In the 1960s, Hewlett-Packard began selling self-contained time and frequency standards called cesium clocks. In a cesium clock a crystal frequency is generated and multiplied to microwave frequencies. That energy is passed through a chamber filled with cesium gas. The gas acts as a very narrow band-pass filter. The output signal is detected and the crystal oscillator frequency is adjusted automatically so that a maximum of energy is detected. The output of the crystal is thus linked to the stability of the cesium gas and is usually accurate to several parts in 10^{-12}. This is much superior to a crystal oscillator alone; but at close to $40,000 each, cesium frequency standards are a bit extravagant for amateur use.

A rubidium frequency standard is an alternative to the cesium clock. They are not quite as accurate as the cesium, but they are much less expensive, relatively quick to warm up and can be quite small. Older models occasionally appear surplus. As with any precision instrument, it should be checked over and calibrated before use.

Most hams do not have access to cesium or rubidium standards—or need them. Instead we use crystal oscillators. Crystal oscillators provide three levels of stability. The least accurate is a single crystal mounted on a circuit board. The crystal frequency is affected by the temperature environment of the equipment, to the extent of a few parts per million (ppm) per degree Celsius. For example, the frequency of a 10-MHz crystal with temperature stability rated at 3 ppm might vary 60 Hz when temperature of the crystal changes by 2°C. If the crystal oscillator is followed by a frequency multiplier, any variation in the crystal frequency is also multiplied. Even so, the accuracy of a simple crystal oscillator is sufficient for most of our needs and most amateur equipment relies on this technique. For a discussion of crystal oscillators and temperature compensation, look in the **Oscillators and Synthesizers** chapter of this *Handbook*.

The second level of accuracy is achieved when the temperature around the crystal is stabilized, either by an "oven" or other nearby components. Crystals are usually designed to stabilize at temperatures far above any reached in normal operating environments. These oscillators are commonly good to 0.1 ppm per day and are widely used in the commercial two-way radio industry.

The third accuracy level uses a double oven with proportional heating. The two ovens compensate for each other automatically and provide excellent temperature stability. The ovens must be left on continuously, however, and warm-up requires several days to two weeks.

Crystal *aging* also affects frequency stability. Some crystals change frequency over time (age) so the circuit containing the crystal must contain components to compensate for this change. Other crystals become more stable over time and become excellent frequency standards. Many commercial laboratories go to the expense of buying and testing several examples of the same oscillator and select the best one for use. As a result, many surplus oscillators are surplus for a reason. Nevertheless, a good stable crystal oscillator can be accurate to 1×10^{-9} per day and very appropriate for amateur applications.

Time and Frequency Calibration

Many hams have digital frequency counters, which range from surplus lab equipment to new highly integrated instruments with nearly everything on one chip. Almost all of these are very precise and display nine or more digits. Many are even quite stable. Nonetheless, a 10-MHz oscillator accurate to 1 ppm per month can vary ±10 Hz in one month. This drift rate may be acceptable for many applications, but the question remains: How accurate is it?

This question can be answered by calibrating the oscillator. There are several ways to perform this calibration. The most accurate method compares the unit in question by leaving the oscillator operating, transporting it to an oscillator of known frequency and then making a comparison. A commonly used comparison method connects the output of the calibrated oscillator into the horizontal input of a high frequency oscilloscope, and the oscillator to be measured to the vertical input. It helps, but they need not be on the same frequency. By noting how long it takes the sine wave to travel one division at a given sweep speed, one can calculate the resulting drift in parts per million per minute (ppm/min).

Another technique of oscillator calibration uses a VLF phase comparator. This is a special direct-conversion receiver that picks up the signal from WWVB on 60 kHz. Phase comparison is used to compare WWVB with the divided frequency of the oscillator being tested. Many commercial units have a small strip chart printer attached and switches to determine the receiver frequency. Since these 60-kHz VLF Comparator receivers have been largely replaced by units that use Loran signals or rubidium standards, they can be found at very reasonable prices. A very effective 60-kHz antenna can be made by attaching an audio transformer with the low-impedance winding connected to the receiver antenna terminals by way of a series dc blocking capacitor. The high-impedance winding is then connected between ground and a random length of wire. A typical VLF Comparator can track an oscillator well into a few parts in 10^{-10}. This technique directly compares the oscillator with an NIST standard and can even characterize oscillator drift characteristics in ppm per day or week.

Another fairly direct method compares an oscillator with one of the WWV HF signals. The received signal is not immensely accurate, but if the oscillator of a modern HF transceiver is carefully compared, it will be accurate enough for all but the most demanding work.

The last and least accurate way to calibrate an oscillator is to compare it with another oscillator or counter owned by you or another local ham. Unless the calibration of the other oscillator or counter is known, this comparison could be very misleading. True accuracy is not determined by the label of a famous company or impressive looks. Metrologists (people who calibrate and measure equipment) spend more time calibrating oscillators than any other piece of equipment.

DC Instruments and Circuits

This section discusses the basics of analog and digital dc meters. It covers the design of range extenders for current, voltage and resistance; construction of a simple meter; functions of a digital voltmeter (DVM) and procedures for accurate measurements.

Basic Meters

In measuring instruments and test equipment suitable for amateur purposes, the ultimate readout is generally based on a measurement of direct current. There are two basic styles of meters: analog meters that use a moving needle display, and digital meters that display the measured values in digital form. The analog meter for measuring dc current and voltage uses a magnet and a coil to move a pointer over a calibrated scale in proportion to the current flowing through the meter.

The most common dc analog meter is the D'Arsonval type, consisting of a coil of wire to which the pointer is attached so that the coil moves (rotates) between the poles of a permanent magnet. When current flows through the coil, it sets up a magnetic field that interacts with the field of the magnet to cause the coil to turn. The design of the instrument normally makes the pointer move in direct proportion to the current.

Digital Multimeters

In recent years there has been a flood of inexpensive digital multimeters (DMMs) ranging from those built into probes to others housed in large enclosures. They are more commonly referred to as digital voltmeters (DVMs) even though they are multimeters; they usually measure voltage, current and resistance. After some years of refining circuits such as the "successive approximation" and "dual slope" methods, most meters now use the dual-slope method to convert analog voltages to a digital reading. DVMs have basically three main sections as shown in **Fig 25.1**.

The first section scales the voltage or current to be measured. It has four main circuits:

- a chain of multiplier resistors that reduce the input voltage to 0-1 V,
- a converter that changes 0-1 V ac to dc,
- an amplifier that raises signals in the 0-100 mV range to 0-1 V and
- a current driver that provides a constant current to the multiplier chain for resistance measurements.

The second section is an integrator. It is usually based on an operational amplifier that is switched by a timing signal. The timing signal initially shorts the input of the integrator to provide a zero reference. Next a reference voltage is connected to charge the capacitor for a determined amount of time. Finally the last part of the timing cycle allows the capacitor to discharge. The time it takes the capacitor to discharge is proportional to either the input voltage (V_{in}, after it was scaled into the range of 0 to 1 V) or 1 minus V_{in}, depending on the meter design. This discharge time is measured by the next section of the DVM, which is actually a frequency counter. Finally, the output of the frequency counter is scaled to the selected range of voltage or current and sent to the final section of the DVM — the digital display.

Since the timing is quite fast and the capacitor is not used long enough to drift much in value, the components that most determine accuracy are the reference voltage source and the range multiplier resistors. With the availability of integrated resistor networks that are deposited or diffused onto the same substrate, drift is

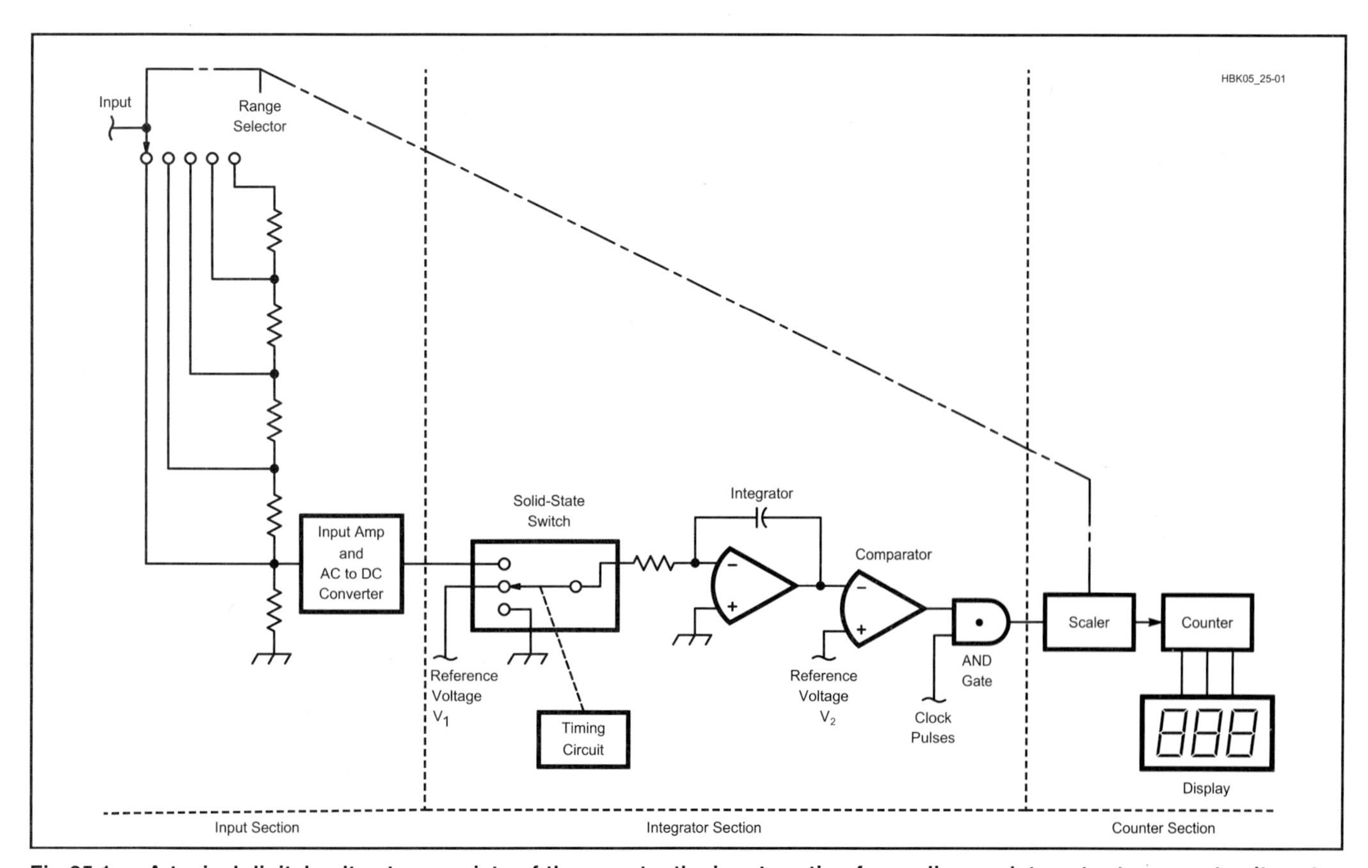

Fig 25.1 — A typical digital voltmeter consists of three parts: the input section for scaling, an integrator to convert voltage to pulse count, and a counter to display the pulse count representing the measured quantity.

automatically compensated because all branches of a divider drift in the same direction simultaneously. The voltage sources are generally Zener diodes on substrates with accompanying series resistors. Often the resistor and Zener have opposite temperature characteristics that cancel each other. In more complex DVMs, extensive digital circuitry can insert values to compensate for changes in the circuit and can even be automatically calibrated remotely in a few moments.

Liquid crystal displays (LCD) are commonly used for commercial DVMs. As a practical matter they draw little current and are best for portable and battery-operated use. The usual alternative, light emitting diode (LED) displays, draw much more current but are better in low-light environments. Some older surplus units use gas plasma displays (orange-colored digits). You may have seen plasma displays on gas-station pumps. They are not as bright as LEDs, but are easier to read. On the down side, plasma displays require high-voltage power supplies, draw considerable current and often fail after 10 years or so.

The advantages of DVMs are high input resistance (10 MΩ on most ranges), accurate and precise readings, portability, a wide variety of ranges and low price. There is one disadvantage, however: Digital displays update rather slowly, often only one to two times per second. This makes it very difficult to adjust a circuit for a peak (maximum) or null (minimum) response using only a digital display. The changing digits do not give any clue of the measurement trend and it is easy to tune through the peak or null between display updates. In answer, many new DVMs are built with an auxiliary bar-graph display that is updated constantly, thus providing instantaneous readings of relative value and direction of changes.

Current Ranges

The sensitivity of an analog meter is usually expressed in terms of the current required for full-scale deflection of the pointer. Although a very wide variety of ranges is available, the meters of interest in amateur work give maximum deflection with currents measured in microamperes or milliamperes. They are called microammeters and milliammeters, respectively.

Thanks to the relationships between current, voltage and resistance expressed by Ohm's Law, it is possible to use a single low-range instrument (for example, 1 mA or less for full-scale pointer deflection) for a variety of direct-current measurements. Through its ability to measure current, the instrument can also be used indirectly to measure voltage. In the same way, a measurement of both current and voltage will obviously yield a value of resistance. These measurement functions are often combined in a single instrument: the volt-ohm-milliammeter or VOM, a multirange meter that is one of the most useful pieces of test equipment an amateur can possess.

Accuracy

The accuracy of a D'Arsonval-movement dc meter is specified by the manufacturer. A common specification is ±2% of full scale, meaning that a 0-100 µA meter, for example, will be correct to within 2 µA at any part of the scale. There are very few cases in amateur work where accuracy greater than this is needed. When the instrument is part of a more complex measuring circuit, however, the design and components can each cause error that accumulates to reduce the overall accuracy.

Extending Current Range

Because of the way current divides between two resistances in parallel, it is possible to increase the range (more specifically to decrease the sensitivity) of a dc current meter. The meter itself has an inherent resistance (its internal resistance) which determines the full-scale current passing through it when its rated voltage is applied. (This rated voltage is on the order of a few millivolts.) When an external resistance is connected in parallel with the meter, the current will divide between the two and the meter will respond only to that part of the current that flows through its movement. Thus, it reads only part of the total current; the effect makes more total current necessary for a full-scale meter reading. The added resistance is called a "shunt."

We must know the meter's internal resistance before we can calculate the value for a shunt resistor. Internal resistance may vary from a fraction of an ohm to a few thousand ohms, with greater resistance values associated with greater sensitivity. When this resistance is known, it can be used in the formula below to determine the required shunt for a given multiplication:

$$R = \frac{R_m}{n-1} \quad (1)$$

where

R = shunt resistance, ohms

R_m = meter internal resistance, ohms

n = the factor by which the original meter scale is to be multiplied.

Often the internal resistance of a particular meter is unknown (when the meter is purchased at a flea market or is taken from a commercial piece of equipment, for example). Unfortunately, the internal resistance of a meter cannot be measured directly with an ohmmeter without risk of damage to the meter movement.

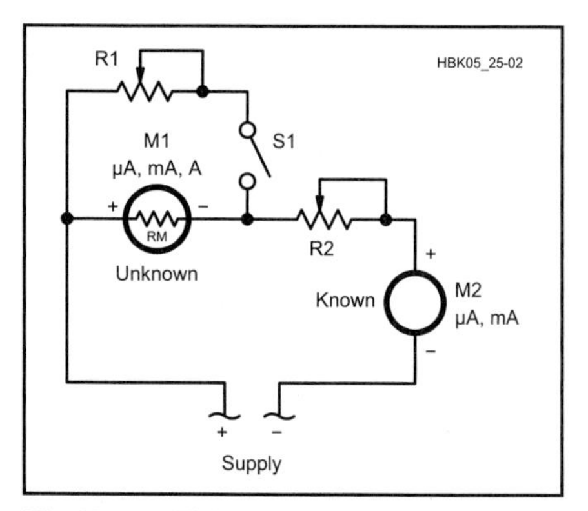

Fig 25.2 — This test setup allows safe measurement of a meter's internal resistance. See text for the procedure and part values.

Fig 25.2 shows a method to safely measure the internal resistance of a linearly calibrated meter. It requires a calibrated meter that can measure the same current as the unknown meter. The system works as follows: S1 is switched off and R2 is set for maximum resistance. A supply of constant voltage is connected to the supply terminals (a battery will work fine) and R2 is adjusted so that the unknown meter reads exactly full scale. Note the current shown on M2. Close S1 and alternately adjust R1 and R2 so that the unknown meter (M1) reads exactly half scale and the known meter (M2) reads the same value as in the step above. At this point, half of the current in the circuit flows through M1 and half through R1. To determine the internal resistance of the meter, simply open S1 and read the resistance of R1 with an ohm-meter.

The values of R1 and R2 will depend on the meter sensitivity and the supply voltage. The maximum resistance value for R1 should be approximately twice the expected internal resistance of the meter. For highly sensitive meters (10 µA and less), 1 kΩ should be adequate. For less sensitive meters, 100 Ω should suffice. Use no more supply voltage than necessary.

The value for minimum resistance at R2 can be calculated using Ohm's Law. For example, if the meter reads 0 to 1 mA and the supply is a 1.5-V battery, the minimum resistance required at R2 will be:

$$R2 = \frac{1.5}{0.001}$$

R2 (min) = 1500 Ω

In practice a 2- or 2.5-kΩ potentiometer would be used.

Making Shunts

Homemade shunts can be constructed from several kinds of resistance wire or from ordinary copper wire if no resistance wire is available. The copper wire table in the **Component Data and References** chapter of this *Handbook* gives the resistance per 1000 ft for various sizes of copper wire. After computing the resistance required, determine the smallest wire size that will carry the full-scale current, again from the wire table. Measure off enough wire to give the required resistance. A high-resistance 1- or 2-W carbon-composition resistor makes an excellent form on which to wind the wire, as the high resistance does not affect the value of the shunt. If the shunt gets too hot, go to a larger diameter wire of a greater length.

VOLTMETERS

If a large resistance is connected in series with a meter that measures current, as shown in **Fig 25.3**, the current multiplied by the resistance will be the voltage drop across the resistance. This is known as a multiplier. An instrument used in this way is calibrated in terms of the voltage drop across the multiplier resistor and is called a voltmeter.

Sensitivity

Voltmeter sensitivity is usually expressed in ohms per volt (Ω/V), meaning that the meter full-scale reading multiplied by the sensitivity will give the total resistance of the voltmeter. For example, the resistance of a 1 kΩ/V voltmeter is 1000 times the full-scale calibration voltage. Then by Ohm's Law the current required for full-scale deflection is 1 milliampere. A sensitivity of 20 kΩ/V, a commonly used value, means that the instrument is a 50-µA meter.

As voltmeter sensitivity (resistance) increases, so does accuracy. Greater meter resistance means that less current is drawn from the circuit and thus the circuit under test is less affected by connection of the meter. Although a 1000-Ω/V meter can be used for some applications, most good meters are 20 kΩ/V or more. Vacuum-tube voltmeters (VTVMs) and their modern equivalent FET voltmeters (FETVOMs) are usually 10-100 MΩ/V and DVMs can go even higher.

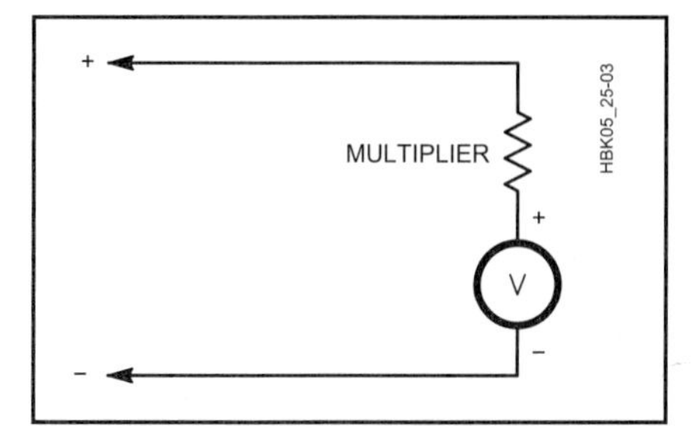

Fig 25.3 — A voltmeter is constructed by placing a current-indicating instrument in series with a high resistance, the "multiplier."

Multipliers

The required multiplier resistance is found by dividing the desired full-scale voltage by the current, in amperes, required for full-scale deflection of the meter alone. To be mathematically correct, the internal resistance of the meter should be subtracted from the calculated value. This is seldom necessary (except perhaps for very low ranges) because the meter resistance is usually very low compared with the multiplier resistance. When the instrument is already a voltmeter with an internal multiplier, however, the meter resistance is significant. The resistance required to extend the range is then:

$$R = R_m (n - 1) \quad (2)$$

where

R_m = total resistance of the instrument
n = factor by which the scale is to be multiplied

For example, if a 1-kΩ/V voltmeter having a calibrated range of 0 to 10 V is to be extended to 1000 V, R_m is 1000 × 10 = 10 kΩ, n is 1000/10 = 100 and R = 10,000 × (100 – 1) = 990 kΩ.

When extending the range of a voltmeter or converting a low-range meter into a voltmeter, the rated accuracy of the instrument is retained only when the multiplier resistance is precise. High-precision, hand-made and aged wire-wound resistors are used as multipliers of high-quality instruments. These are relatively expensive, but the home constructor can do well with 1% tolerance metal-film resistors. They should be derated when used for this purpose. That is, the actual power dissipated in the resistor should not be more than 1/10 to 1/4 the rated dissipation. Also, use care to avoid overheating the resistor body when soldering. These precautions will help prevent permanent change in the resistance of the unit.

Many DVMs use special resistor groups that have been etched on quartz or sapphire and laser trimmed to value. These resistors are very stable and often quite accurate. They can be bought new from various suppliers. It is also possible to "rescue" the divider/multiplier resistors from an older DVM that no longer functions and use them as multipliers. Look for a series of four or five resistors that add up to 10 MΩ: 0.9,9,90,900,9,000,90,000 and 900,000 Ω. There is usually another 1-MΩ resistor in series to isolate the meter from the circuit under test. A few of these high-accuracy resistors in "odd" values can help calibrate less-expensive instruments.

DC Voltage Standards

For a long time NIST has statistically compared a bank of special Weston Cell or cadmium sulfate batteries to arrive at the standard volt. By using a special tapped resistor, a 1.08-V battery can be compared to other voltages and instruments compared. However, these are very high-impedance batteries that deliver almost no current and are relatively temperature sensitive. They are made up of a solution of cadmium and mercury in opposite legs of an "H" shaped glass container. You can read much more about them in *Calibration—Philosophy and Practice*, published by the John Fluke Co of Mount Lake Terrace, Washington.

Hams often use an ordinary flashlight battery as a convenient voltage reference. A fresh *D cell* usually provides 1.56 V under no load, as would be measured by a DVM. The Heath Company, which supplied thousands of kits to the ham community for many years, used such batteries as the calibration references for many of their kits.

Recently, NIST has been able to use a microwave to voltage converter called a "Josephson Junction" to determine the value of the volt. The converter transfers the accuracy of a frequency standard to the accuracy of the voltage that comes out of it. The converter generates only 5 mV, however, which then must be scaled to the standard 1-V level. One problem with high-accuracy measurements is stray noise (low-level voltages) that creates a floor below which measurements are meaningless. For that reason, meters with five or more digits must be very quiet and any comparisons must be made at a voltage high enough to be above the noise.

DC MEASUREMENT CIRCUITS

Current Measurement with a Voltmeter

A current-measuring instrument should have very low resistance compared with the resistance of the circuit being measured; otherwise, inserting the instrument will alter the current from its value when the instrument is removed. The resistance of many circuits in radio equipment is high and the circuit operation is affected little, if at all, by adding as much as a few hun-

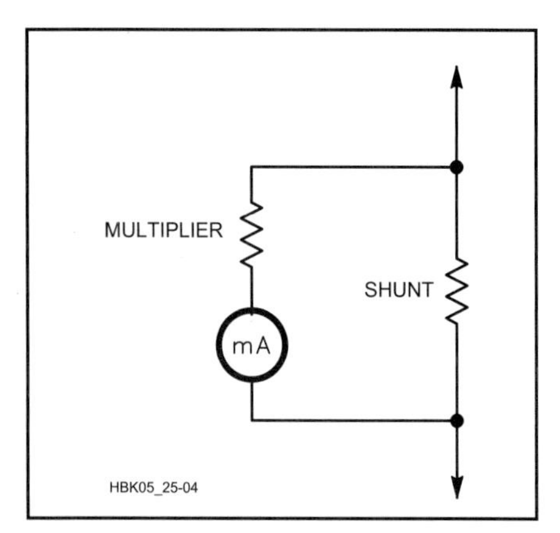

Fig 25.4 — A voltmeter can be used to measure current as shown. For reasonable accuracy, the shunt should be 5% of the circuit impedance or less, and the meter resistance should be 20 times the circuit impedance or more.

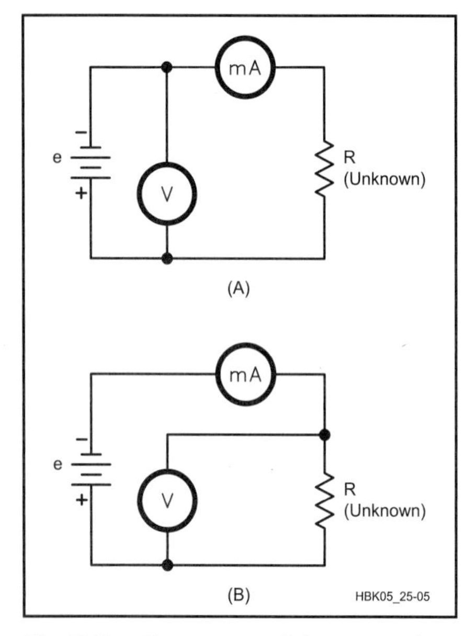

Fig 25.5 — Power or resistance can be calculated from voltage and current measurements. At A, error introduced by the ammeter is dominant. At B, error introduced by the voltmeter is dominant. The text gives an example.

dred ohms in series. [Even better, use a resistor that is part of the working circuit if one exists. Unsolder one end of the resistor, measure its resistance, reinstall it and then make the measurement.—*Ed.*] In such cases the voltmeter method of measuring current in place of an ammeter, shown in **Fig 25.4**, is frequently convenient. A voltmeter (or low-range milliammeter provided with a multiplier and operating as a voltmeter) having a full-scale voltage range of a few volts is used to measure the voltage drop across a suitable value of resistance acting as a shunt.

The value of shunt resistance must be calculated from the known or estimated maximum current expected in the circuit (allowing a safe margin) and the voltage required for full-scale movement of the meter with its multiplier. For example, to measure a current estimated at 15 A on the 2-V range of a DVM, we need to solve Ohm's Law for the value of R:

$$R_{SHUNT} = \frac{2\,V}{15\,A} = 0.133\,\Omega$$

This resistor would dissipate $15^2 \times 0.133 = 29.92$ W. For a short-duration measurement, 30 1.0-Ω, 1-W resistors could be parallel connected in two groups of 15 (0.067 Ω per group) that are series connected to yield 0.133 Ω. For long-duration measurements, 2- to 5-W resistors would be better.

Power

Power in direct-current circuits is usually determined by measuring the current and voltage. When these are known, the power can be calculated by multiplying voltage, in volts, by the current, in amperes. If the current is measured with a milliammeter, the reading of the instrument must be divided by 1000 to convert it to amperes.

The setup for measuring power is shown in **Fig 25.5A**, where R is any dc load, not necessarily an actual resistor. In this measurement it is always best to use the lowest voltmeter or ammeter scale that allows reading the measured quantity. This results in the percentage error being less than if the meter was reading in the very lowest part of the selected scale.

Resistance

If both voltage and current are measured in a circuit such as that in Fig 25.5, the value of resistance R (in case it is unknown) can be calculated from Ohm's Law. For reasonable results, two conditions should be met:

1. The internal resistance of the current meter should be less than 5% of the circuit resistance.

2. The input impedance of the voltmeter should be greater than 20 times the circuit resistance.

These conditions are important because both meters tend to load the circuit under test. The current meter resistance adds to the unknown resistance, while the voltmeter resistance decreases the unknown resistance as a result of their parallel connection.

Ohmmeters

Although Fig 25.5B suffices for occasional resistance measurements, it is inconvenient when we need to make frequent measurements over a wide range of resistance. The device generally used for this purpose is the ohmmeter. Its simplest form is a voltmeter (or milliammeter, depending on the circuit used) and a small battery. The meter is calibrated so that the value of an unknown resistance can be read directly from the scale. **Fig 25.6** shows some typical ohmmeter circuits. In the simplest circuit, Fig 25.6A, the meter and battery are connected in series with the unknown resistance. If a given movement of the meter's needle is obtained with terminals A-B shorted, inserting the resistance to be measured will cause the meter reading to decrease. When the resistance of the voltmeter is known, the following

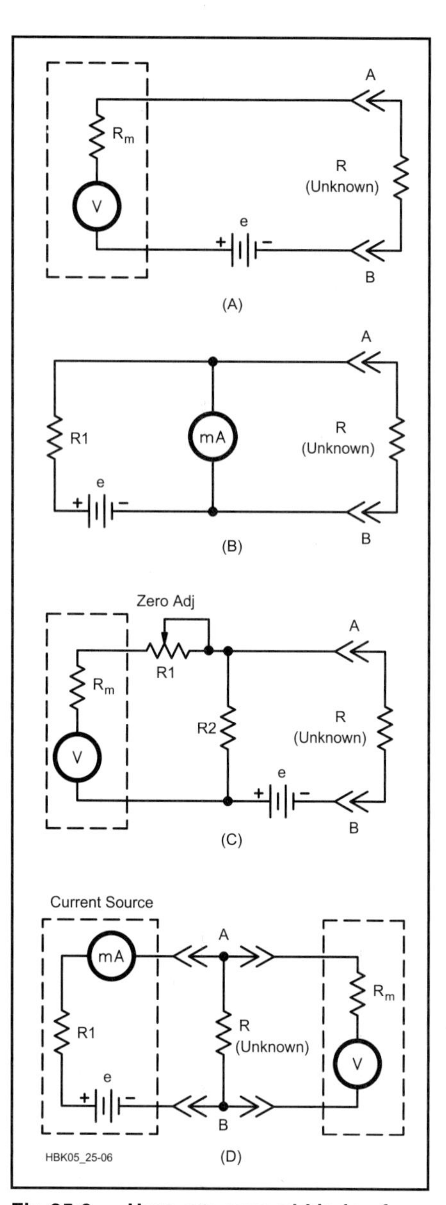

Fig 25.6 — Here are several kinds of ohmmeters. Each is explained in the text.

formula can be applied.

$$R = \frac{eR_m}{E} - R_m \quad (3)$$

where

R = unknown resistance, ohms
e = voltage applied (A-B shorted)
E = voltmeter reading with R connected
R_m = resistance of the voltmeter.

The circuit of Fig 25.6A is not suited to measuring low values of resistance (less than 100 Ω or so) with a high-resistance voltmeter. For such measurements the circuit of Fig 25.6B is better. The unknown resistance is

$$R = \frac{I_2 R_m}{I_1 - I_2} \quad (4)$$

where

R = unknown resistance, ohms
R_m = the internal resistance of the milliammeter, ohms
I_1 = current with R disconnected from terminals A-B, amps
I_2 = current with R connected, amps.

This formula is based on the assumption that the current in the complete circuit will be essentially constant whether or not the unknown terminals are short circuited. This requires that R1 be much greater than R_m. For example, 3000 Ω for a 1-mA meter with an internal resistance of perhaps 50 Ω. In this case, a 3-V battery would be necessary in order to obtain a full-scale deflection with the unknown terminals open. R1 can be an adjustable resistor, to permit setting the open-terminal current to exact full scale.

A third circuit for measuring resistance is shown in Fig 25.6C. In this case a high-resistance voltmeter is used to measure the voltage drop across a reference resistor, R2, when the unknown resistor is connected so that current flows through it, R2 and the battery in series. With suitable R2s (low values for low-resistance, high values for high-resistance unknowns), this circuit gives equally good results for resistance values in the range from one ohm to several megohms. The voltmeter resistance, R_m, must be much greater (50 times or more) than that of R2. A 20-kΩ/V instrument (50-µA movement) is generally used. If the current through the voltmeter is negligible compared with the current through R2, the formula for the unknown is

$$R = \frac{eR2}{E} - R2 \quad (5)$$

where

R and R2 are in ohms
e = voltmeter reading with R removed and A shorted to B.
E = voltmeter reading with R connected.

R1 sets the voltmeter reading exactly to full scale when the meter is calibrated in ohms. A 10-kΩ pot is suitable with a 20-kΩ/V meter. The battery voltage is usually 3 V for ranges to 100 kΩ and 6 V for higher ranges.

Four-Wire Resistance Measurements

In situations where a very low resistance, like a 50-Ω dummy load, is to be measured, the resistance of the test leads can be significant. The average lead resistance is about 0.9 Ω through both leads, which would make a 50.5-Ω dummy load appear to be 51.4 Ω. To compensate for lead resistance, some meters allow for four-wire measurements. Briefly, two wires from the current source and two wires from the measuring circuit exit the meter case separately and connect directly to the unknown resistance (see Fig 25.6D). This eliminates the voltage drop in the current-source leads from the measurement. In practice, four-wire systems use special test clips that are similar to alligator clips, except that the jaws are insulated from each other and a meter lead is attached to each jaw. In some meters, an additional control allows the operator to short the test leads together and adjust the meter for a zero reading before making low-resistance measurements.

Bridge Circuits

Bridges are an important class of measurement circuits. They perform measurement by comparison with some known component or quantity, rather than by direct reading. VOMs, DVMs and other meters are convenient, but their accuracy is limited. The accuracy of manufactured analog meters is determined at the factory, while digital meters are accurate only to some percentage ±1 in the least-significant digit. The accuracy of comparison measurements, however, is determined only by the comparison standard and bridge sensitivity.

Bridge circuits are useful across most of the frequency spectrum. Most amateur applications are at RF, as shown later in this chapter. The principles of bridge operation are easier to understand at dc, however, where bridge operation is simple.

The Wheatstone Bridge

A simple resistance bridge, known as the Wheatstone bridge, is shown in **Fig 25.7**. All other bridge circuits are based on this design. The four resistors, R1, R2, R3 and R4 in Fig 25.7A, are known as the bridge arms. For the voltmeter reading to be zero (null) the voltage dividers consisting of the pairs R1-R3 and R2-R4 must be equal. This means

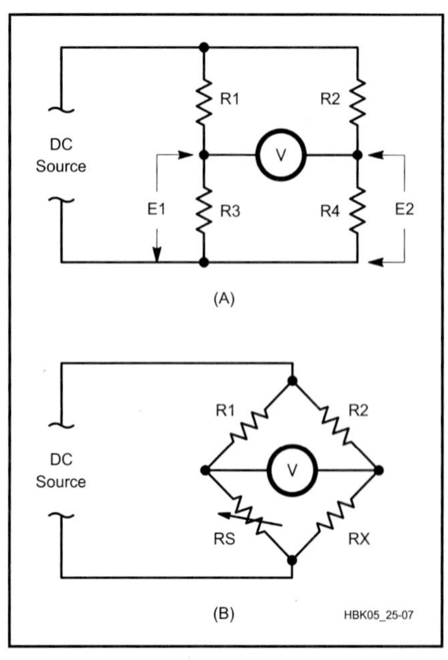

Fig 25.7 — A Wheatstone bridge circuit. A bridge circuit is actually a pair of voltage dividers (A). B shows how bridges are normally drawn.

$$\frac{R1}{R3} = \frac{R2}{R4} \quad (6)$$

When this occurs the bridge is said to be *balanced*.

The circuit is usually drawn as shown at Fig 25.7B when used for resistance measurement. Equation 6 can be rewritten

$$RX = RS\left(\frac{R2}{R1}\right) \quad (7)$$

RX is the unknown resistor. R1 and R2 are usually made equal; then the calibrated adjustable resistance (the standard), RS, will have the same value as RX when RS is set to show a null on the voltmeter.

Note that the resistance ratios, rather than the actual resistance values, determine the voltage balance. The values do have important practical effects on the sensitivity and power consumption, however. Bridge sensitivity is the ability of the meter to respond to slight unbalance near the null point; a sharper null means a more accurate setting of RS at balance. Bridge circuits with narrower bandwidths tend to exhibit a sharper, more precise null, and therefore potentially greater measurement accuracy in contrast to a bridge with a wider bandwidth and a poorly defined or broader null.

The Wheatstone bridge is rarely used by amateurs for resistance measurement, since it is easier to measure resistances with VOMs and DVMs. Nonetheless, it is worthwhile to understand its operation as the basis of more complex bridges.

ELECTRONIC VOLTMETERS

We have seen that the resistance of a simple voltmeter (as in Fig 25.3) must be extremely high in order to avoid "loading" errors caused by the current that necessarily flows through the meter. The use of high-resistance meters tends to cause difficulty in measuring relatively low voltages because multiplier resistance progressively lessens as the voltage range is lowered.

Voltmeter resistance can be made independent of voltage range by using vacuum tubes, FETs or op amps as dc amplifiers between the circuit under test (CUT) and the indicator, which may be a conventional meter movement or a digital display. Because the input resistance of the electronic devices mentioned is extremely high (hundreds of megohms) they have negligible loading effect on the CUT. They do, however, require a closed dc path in their input circuits (although this path can have very high resistance). They are also limited in the voltage level that their input circuits can handle. Because of this, the device actually measures a small voltage across a portion of a high-resistance voltage divider connected to the CUT. Various voltage ranges are obtained from appropriate taps on the voltage divider.

In the design of electronic voltmeters it has become standard practice to use a voltage divider with a total resistance of 10 MΩ, tapped as required, in series with a 1-MΩ resistor incorporated in the meter. The total voltmeter resistance, including probe, is therefore 11 MΩ. The 1-MΩ resistor serves to isolate the voltmeter circuit from the CUT.

AC Instruments and Circuits

Most ac measurements differ from dc measurements in that the accuracy of the measurement depends on the purity of the sine wave. It is fairly easy to measure an ac voltage to between 1% and 5%, but getting down to 0.01% is difficult. Measurements to less than 0.01% must be left to precision laboratories. In general, amateurs measure ac voltages in household circuits, audio stages and RF power measurements, and 1% to 5% accuracy is usually close enough.

This section covers basic measurements, the nature of sine waves and meters. There are four common ways to measure ac voltage:

- Use a rectifier to change the ac to dc and then measure the dc.
- Heat a resistor in a Wheatstone bridge with the ac and measure the bridge unbalance.
- Heat a resistor surrounded by oil and measure the temperature rise.
- Use electronic circuits (such as multipliers and logarithmic amplifiers) with mathematical ac-to-dc conversion formulas. This method is not common, but it's interesting.

Calorimetric Meters

In a calorimetric meter, power is applied to a resistor that is immersed in the flow path of a special oil. This oil transfers the heat to another resistor that is part of a bridge. As the resistor heats, its resistance changes and the bridge becomes unbalanced. An attached meter registers the unbalance of the bridge as ac power. This type of meter is accurate for both dc and ac. They frequently operate from dc well into the GHz range. For calibration, an accurate dc voltage is applied and the reading is noted; a similar ac voltage is applied and the readings are compared. Some calorimetric meters are complicated, but others are simple.

Thermocouple Meters and RF Ammeters

In a thermocouple meter, alternating current flows through a low-resistance heating element. The power lost in the resistance generates heat that warms a thermocouple, which consists of a pair of junctions of two different metals. When one junction is heated a small dc voltage is generated in response to the difference in temperature of the two junctions. This voltage is applied to a dc milliammeter that is calibrated in suitable ac units. The heater-thermocouple/dc-meter combination is usually housed in a regular meter case.

Thermocouple meters are available in ranges from about 100 mA to many amperes. Their useful upper frequency limit is in the neighborhood of 100 MHz. Amateurs use these meters mostly to measure current through a known load resistance and calculate the RF power delivered to the load.

RECTIFIER INSTRUMENTS

The response of a rectifier RF ammeter is proportional (depending on the design) to either the peak or average value of the rectified ac wave, but never directly to the RMS value. These meters cannot be calibrated in RMS without knowing the relationship that exists between the real reading and the RMS value. This relationship may not be known for the circuit under test.

Average-reading ac meters work best with pure sine waves and RMS meters work best with complicated wave forms. Since many practical measurements involve nonsinusoidal forms, it is necessary to know what your instrument is actually reading, in order to make measurements intelligently. Most VOMs and VTVMs use averaging techniques, while DVMs may use either one. In all cases, check the meter instruction manual to be sure what it reads.

Peak and Average with Sine-Wave Rectification

Peak, average and RMS values of ac waveforms are discussed in the **Electrical Fundamentals** chapter. Because the positive and negative half cycles of the sine wave have the same shape, half-wave rectification of either the positive half or the negative half gives exactly the same result.

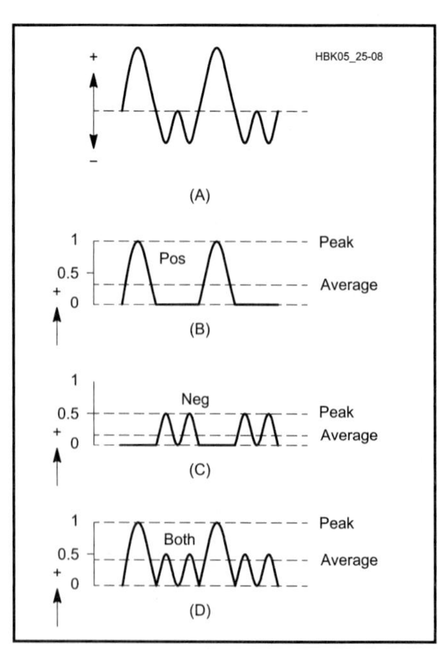

Fig 25.8 — Peak vs average ac values for an asymmetrical wave form. Note that the peak values are different with positive or negative half-cycle rectification.

With full-wave rectification, the peak reading is the same, but the average reading is doubled, because there are twice as many half cycles per unit of time.

Asymmetrical Wave Forms

A nonsinusoidal waveform is shown in **Fig 25.8A**. When the positive half cycles of this wave are rectified, the peak and average values are shown at B. If the polarity is reversed and the negative half cycles are rectified, the result is shown in Fig 25.8C. Full-wave rectification of such a lopsided wave changes the average value, but the peak reading is always the same as that of the half cycle that produces the highest peak in half-wave rectification.

Effective-Value Calibration

The actual scale calibration of commercially made rectifier voltmeters is very often (almost always, in fact) in terms of RMS values. For sine waves, this is satisfactory and useful because RMS is the standard measurement at power-line frequencies. It is also useful for many RF applications when the waveform is close to sinusoidal. In other cases, particularly in the AF range, the error may be considerable when the waveform is not pure.

Turn-Over

From Fig 25.8 it is apparent that the calibration of an average-reading meter will be the same whether the positive or negative sides are rectified. A half-wave peak-reading instrument, however, will indicate different values when its connections to the circuit are reversed (turn-over effect). Very often readings are taken both ways, in which case the sum of the two is the peak-to-peak (P-P) value, a useful figure in much audio and video work.

Average- vs Peak-Reading Circuits

For traditional analog displays, the basic difference between average- and peak-reading rectifier circuits is that the output is not filtered for averaged readings, while a filter capacitor is charged to the peak value of the output voltage in order to measure peaks. **Fig 25.9A** and **B** show typical average-reading circuits, one half-wave and the other full-wave. In the absence of dc filtering, the meter responds to wave forms such as those shown at B, C and D in Fig 25.8; and since the inertia of the pointer system makes it unable to follow the rapid variations in current, it averages them out mechanically.

In Fig 25.9A, D1 actuates the meter; D2 provides a low-resistance dc return in the meter circuit on the negative half cycles. R1 is the voltmeter multiplier resistance. R2 forms a voltage divider with R1 (through D1) that prevents more than a few ac volts from appearing across the rectifier-meter combination. A corresponding resistor can be used across the full-wave bridge circuit.

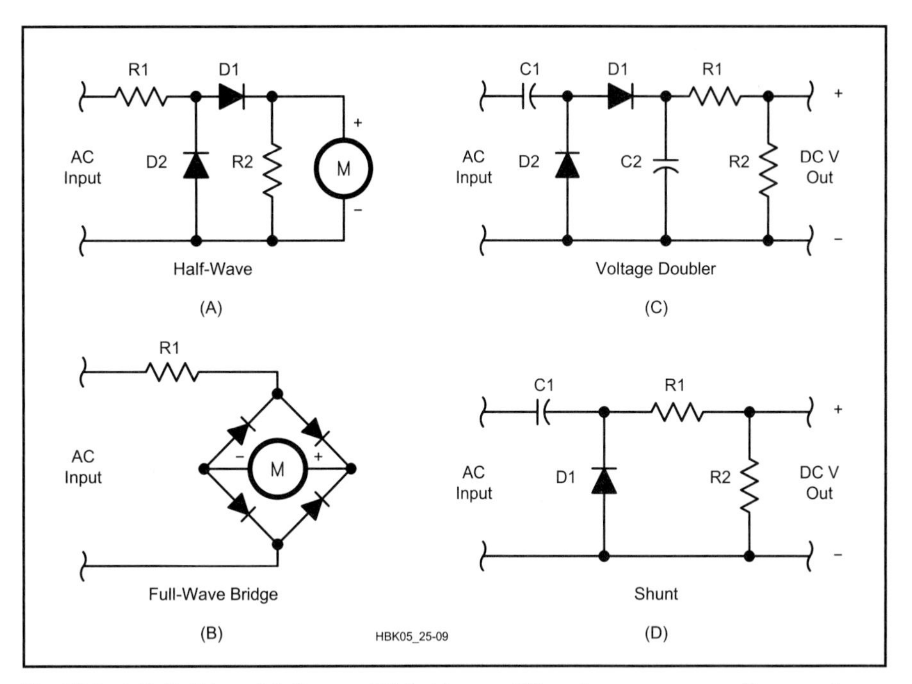

Fig 25.9 — Half (A) and full-wave (B) bridge rectifiers for average-reading, analog-display meters. Peak-reading circuits: a voltage doubler (C) and a shunt circuit (D). All circuits are discussed in the text.

In these two circuits there is no provision to isolate the meter from any dc voltage in the circuit under test. The resulting errors can be avoided by connecting a large nonpolarized capacitor in series with the hot lead. The reactance must be low compared with the meter impedance (at the lowest frequency of interest, more on this later) in order for the full ac voltage to be applied to the meter circuit. Some meters may require as much as 1 µF at line (60 Hz) frequencies. Such capacitors are usually not included in VOMs.

Voltage doubler and shunt peak-reading circuits are shown in Fig 25.9C and D. In both circuits, C1 isolates the rectifier from dc voltage in the circuit under test. In the voltage-doubler circuit, the time constant of the C2-R1-R2 combination must be very large compared with the period of the lowest ac frequency to be measured; similarly with C1-R1-R2 in the shunt circuit. This is so because the capacitor is charged to the peak value (V_{P-P} in C, V_P in D) when the ac wave reaches its maximum and then must hold the charge (so it can register on a dc meter) until the next maximum of the same polarity. If the time constant is 20 times the ac period, the charge will have decreased by about 5% when the next charge occurs. The average voltage drop will be smaller, so the error is appreciably less. The error will decrease rapidly with increasing frequency (if there is no change in the circuit values), but it will increase at lower frequencies.

In Fig 25.9C and D, R1 and R2 form a voltage divider that reduces the voltage to some desired value. For example, if R1 is 0 Ω in the voltage doubler, the voltage across R2 is approximately V_{P-P}; if R1 = R2, the output is approximately V_P (as long as the waveform is symmetrical).

The most common application of the shunt circuit is an RF probe to read V_{RMS}. In that case, R2 is the input impedance of a VTVM or DVM: 11 MΩ. R1 is chosen so that 71% of the peak value appears across R2. This converts the peak reading to RMS for sine-wave ac. R1 is therefore approximately 4.7 MΩ, making the total resistance nearly 16 MΩ. A capacitance of 0.05 µF is sufficient for low audio frequencies under these conditions. Much smaller values of capacitance may be used at RF.

Voltmeter Impedance

The impedance of a voltmeter at the frequency being measured may have an effect on the accuracy similar to that caused by the resistance of a dc voltmeter, as discussed earlier. The ac meter is a resistance in parallel with a capacitance. Since the capacitive reactance decreases with increasing frequency, the impedance also decreases with frequency. The resistance does change with voltage level, particu-

Sources for RF Ammeters

When it comes to getting your own RF ammeter, there's good news and bad news. First, the bad news. New RF ammeters are expensive, and even surplus pricing can vary widely between $10 and $100 in today's market. AM radio stations are the main users of new units. The FCC defines the output power of AM stations based on the RF current in the antenna, so new RF ammeters are made mainly for that market. They are quite accurate, and their prices reflect that!

The good news is that used RF ammeters are often available. For example, Fair Radio Sales in Lima, Ohio has been a consistent RF-ammeter source. Ham flea markets are also worth trying. Some grubbing around in your nearest surplus store or some older ham's junk box may provide just the RF ammeter you need.

RF Ammeter Substitutes

Don't despair if you can't find a used RF ammeter. It's possible to construct your own. Both hot-wire and thermocouple units can be homemade.

Pilot lamps in series with antenna wires, or coupled to them in various ways, can indicate antenna current* or even forward and reflected power.†

Another approach is to use a small low-voltage lamp as the heat/light element and use a photo detector driving a meter as an indicator. (Your eyes and judgment can serve as the indicating part of the instrument.) A feed-line balance checker could be as simple as a couple of lamps with the right current rating and the lowest voltage rating available. You should be able to tell fairly well by eye which bulb is brighter or if they are about equal. You can calibrate a lamp-based RF ammeter with 60-Hz or dc power.

As another alternative, you can build an RF ammeter that uses a dc meter to indicate rectified RF from a current transformer that you clamp over a transmission line wire.††

Copper-Top Battery Testers as RF Ammeters

Finally, there are the *free* RF ammeters that come as the testers with Duracell batteries! For 1.5-V cells, these are actually 3 to 5-Ω resistors with built-in liquid-crystal displays. The resistor heats the liquid-crystal strip; the length of the "lighted" portion (heat turns the strip clear, exposing the fluorescent ink beneath) indicates the magnitude of the current.

Despite their "+" and "–" markings, these indicators are not polarized. Their resistance is low enough to have relatively little effect on a 50-Ω system. (For example, putting one in series with a 50-Ω dummy load would increase the system SWR from 1 to 1.1:1. These testers can measure about 200 to 400 mA. (You can achieve higher ranges by means of a shunt.) Best of all, if you burn out one of these "meters" during your tests, you can replace it at any drugstore, hardware store or supermarket for a few dollars, with some batteries thrown in free. — *John Stanley, K4ERO*

*F. Sutter, "What, No Meters?," *QST*, Oct 1938, p 49.
†C. Wright, "The Twin-Lamp," *QST*, Oct 1947, pp 22-23, 110 and 112.
††Z. Lau, "A Relative RF Ammeter for Open-Wire Lines," *QST*, Oct 1988, pp 15-17.

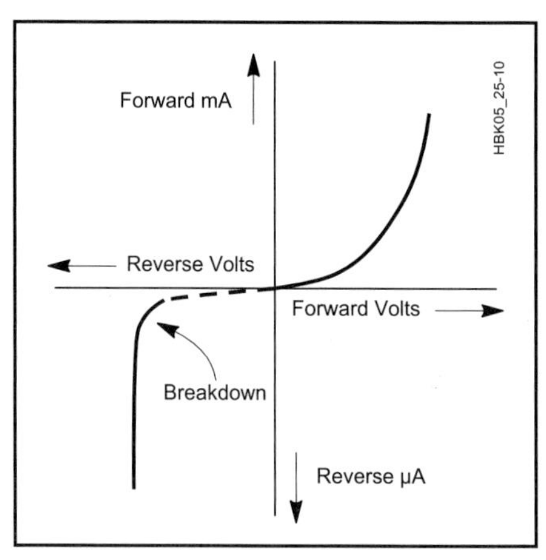

Fig 25.10 — Voltage vs current characteristics for a typical semiconductor diode. Actual values vary with different part numbers, but the forward current will always be increasing steeply with 1 V applied. Note that the forward current scale (mA) is 1000 times larger than that for reverse leakage current (µA). Breakdown voltage varies from 15 V to several hundred volts.

larly at very low voltages (10 V or less) depending on the sensitivity of the meter and the kind of rectifier used.

The ac load resistance represented by a diode rectifier is about one-half of its dc-load resistance. In Fig 25.9A the dc load is essentially the meter resistance, which is generally quite low compared with the multiplier resistance R1. Hence, the total resistance will be about the same as the multiplier resistance. The capacitance depends on the components and construction, test-lead length and location, and other such factors. In general, the capacitance has little or no effect at lower line and audio frequencies, but ordinary VOMs lose accuracy at high audio frequencies and are of little use at RF. Rectifiers with very low inherent capacitance are used at RF and they are usually located at the probe tip to reduce losses.

Similar limitations apply to peak-reading circuits. In the shunt circuit, the resistive part of the impedance is smaller than in the voltage-doubler circuit because the dc load resistance, R1/R2, is directly across the circuit under test and in parallel with the diode ac load resistance. In both peak-reading circuits the effective capacitance may range from 1 or 2 to a few hundred pF, with 100 pF typical in most instruments.

Scale Linearity

Fig 25.10 shows a typical current/voltage chart for a small semiconductor diode, which shows that the forward dynamic resistance of the diode is not constant, but rapidly decreases as the forward voltage increases from zero. The change from high to low resistance happens at much less than 1 V, but is in the range of voltage needed for a dc meter. With an average-reading circuit the current tends to be proportional to the square of the applied voltage. This makes the readings at the low end of the meter scale very crowded. For most measurement purposes, however, it is far more desirable for the output to be linear (that is, for the reading to be directly proportional to the applied voltage), which means that the markings on the meter are more evenly spaced.

To obtain that kind of linearity it is necessary to use a relatively large load resistance for the diode: Large enough that this resistance, rather than the diode resistance, will determine how much current flows. With this technique you can have a linear reading meter, but at the expense of sensitivity. The resistance needed depends on the type of diode; 5 kΩ to 50 kΩ is usually enough for a germanium rectifier, depending on the dc meter sensitivity, but several times as much may be needed for silicon diodes. Higher resistances require greater meter sensitivity; that is, the basic meter must be a microammeter rather than a low-range milliammeter.

Reverse Current

When semiconductor diodes are reverse biased, a small leakage current flows. This reverse current flows during the half cycle when the diode should appear open, and the current causes an error in the dc meter reading. The quantity of reverse current is indicated by a diode's *back resistance* specification. This back resistance is so high that reverse current is negligible with silicon diodes, but back resistance may be less than 100 kΩ for germanium diodes.

The practical effect of semiconductor back resistance is to limit the amount of resistance that can be used in the dc load. This in turn affects the linearity of the meter scale. For practical purposes, the back resistance of vacuum-tube diodes is infinite.

RF Voltage

Special precautions must be taken to minimize the capacitive component of the voltmeter impedance at RF. If possible, the rectifier circuit should be installed permanently at the point where the RF voltage is to be measured, using the shortest possible RF connections. The dc meter can be remotely located, however.

For general RF measurements an RF probe is used in conjunction with a 10 MΩ electronic voltmeter. The circuit of **Fig 25.11**, which is basically the shunt peak-reading circuit of Fig 25.9D, is generally used. The series resistor, which is installed in the probe close to the rectifier, prevents RF from being fed through the probe cable to the electronic voltmeter. (In addition, the capacitance of the coaxial cable serves as a bypass for any RF on the lead.) This resistor, in conjunction with the 10-MΩ divider resistance of the electronic voltmeter, reduces the peak rectified voltage to a dc value equivalent to the RMS of the RF signal. Therefore, the RF readings are consistent with the regular dc calibration.

Of the diodes readily available to amateurs, the germanium point-contact or Schottky diode is preferred for RF-probe applications. It has low capacitance (on the order of 1 pF) and in high-back-resistance Schottky diodes, the reverse current is not serious. The principal limitation is that its safe reverse voltage is only about 50 to 75 V, which limits the applied voltage to 15 or 20 V RMS. Diodes can be series connected to raise the overall rating. At RF, however, it is more common to use capacitors or resistors as voltage dividers and apply the divider output to a single diode.

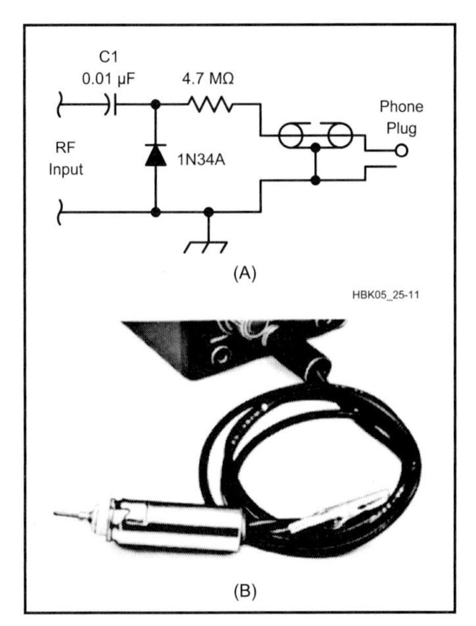

Fig 25.11 — At A is the schematic, at B a photo, of an RF probe for electronic voltmeters. The case of this probe is a seven-pin ceramic tube socket and a 2¼-inch tube shield. A grommet protects the cable where it leaves the tube shield, and an alligator clip on the cable braid connects the probe to the ground of the circuit under test.

RF Power

RF power can be measured by means of an accurately calibrated RF voltmeter connected across a dummy load in which the power is dissipated. If the load is a known pure resistance, the power, by Ohm's Law, is equal to E^2/R, where E is the RMS voltage.

The Hewlett-Packard 410B/C VTVM

The Hewlett-Packard 410B and 410C VTVMs have been standards of bench measurement for industry, and they are now available as industrial surplus. These units are not only excellent VTVMs, but they are also good wide-range RF power meters. Both models use a vacuum-tube detector mounted in a low-loss probe for ac measurements. With an adapter that allows the probe to contact the center conductor of a transmission line, it will give very good RF voltage measurements from 50 mV to 300 V and from 20 Hz to beyond 500 MHz. Very few other measuring instruments provide this range in a single sensor/meter. In addition to the 410B or C, you will also need a probe adapter HP model #11042A.

Do not take the probe apart for inspection because that can change the calibration. You can quickly check the probe by feeling it after it has been warmed up for about 15 minutes. If the body of it feels warm it is probably working. Inside, the 410B is quite different from the C model, with the B being simpler. The meter scales are also different; the 410C offers better resolution and perhaps better accuracy.

To make RF power measurements, remove the ac probe tip and twist lock the probe into the 11042A probe adapter. Attach the output of the adapter to a dummy load and use the formula

$$P = \frac{E^2}{R} \quad (8)$$

where

P = power, in watts

E = value given by the meter, in volts

R = resistance of the dummy load, in ohms.

The resistance of the dummy load should be accurately known to at least ±0.1 Ω, preferably measured with a four-wire arrangement as described in the section on ohmmeters. For frequent measurements, make a chart of voltage vs power at your most often used wattages.

THE MICROWATTER

This simple, easy to build terminating microwattmeter by Denton Bramwell, W7DB appeared in June 1997 *QST*.[1] It can measure power levels from below –50 dBm (10 nW) to -20 dBm (10 µW) with an accuracy of within 1 dB at frequencies from below the broadcast band up to 2 meters. Beyond that, it's still capable of making *relative* power measurements. The range of the meter can be extended by using an attenuator or by adding a broadband RF amplifier.

There are many uses for this meter. Combined with a signal generator, you can use it to characterize crystal and LC filters, for direct, on-air checks of field strength and measuring antenna patterns. If you are fortunate enough to have a 50-Ω oscilloscope probe (or a high-impedance probe), you can also use the Microwatter as an RF voltmeter for circuit testing.

Design

The circuit concept is quite simple (see

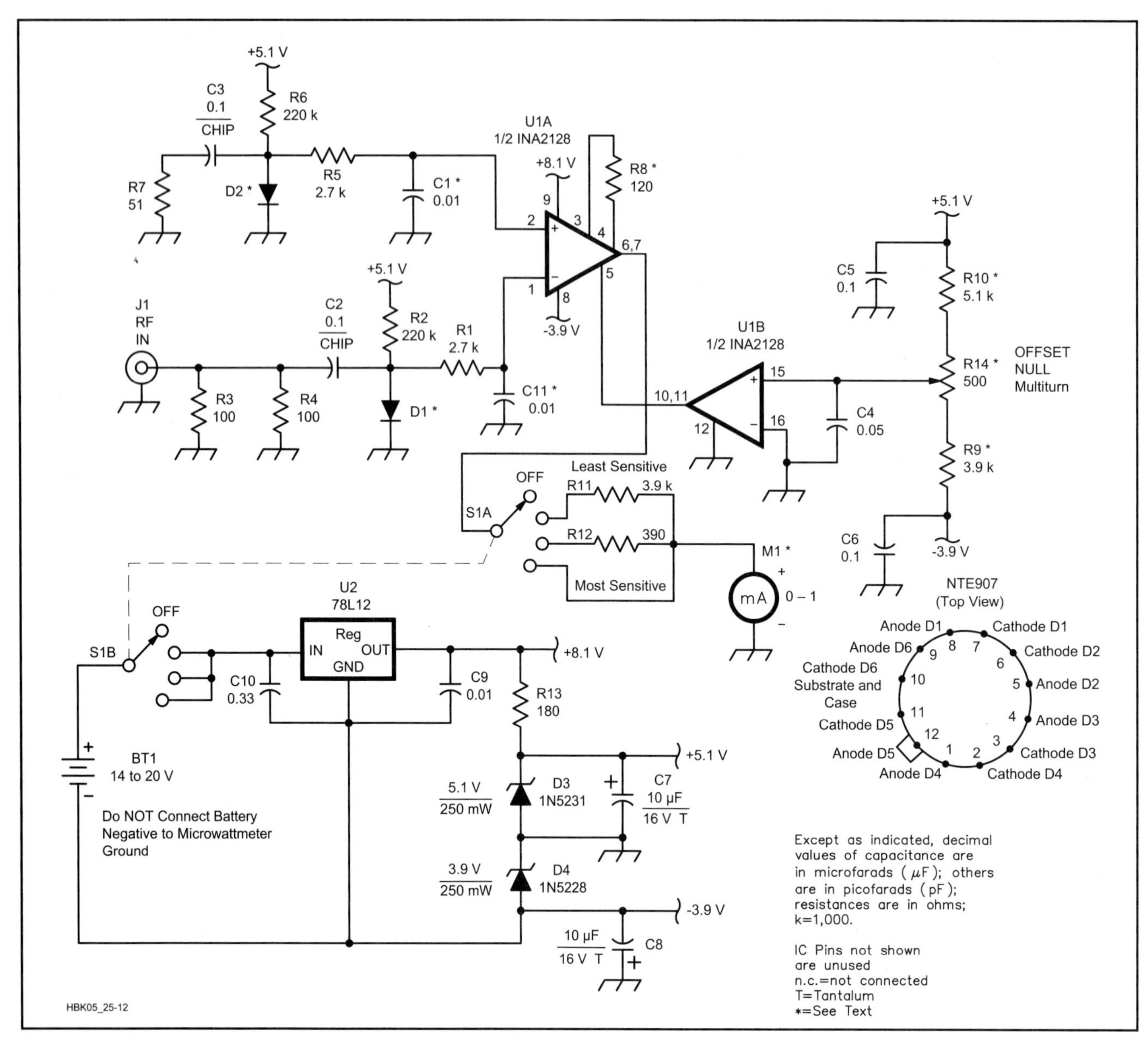

Fig 25.12 — Schematic of the Microwatter circuit. Equivalent parts can be substituted. Unless otherwise specified, resistors are 1/4-W, 5%-tolerance carbon-composition or film units.

D1, D2 — Part of CA3039 or NTE907 diode array; see text and Notes 1 and 2. The NTE907 is available from Mouser Electronics.
M1 — 1-mA meter movement, 50-Ω internal resistance; see text.
R14 — 500-Ω, 10-turn (or more) potentiometer; see text.
S1 — 2-pole, 4-position rotary switch.
U1 — Burr Brown amplifier INA2128P; Digi-Key INA2128P; available from Digi-Key Corp.
U2 — 78L12, positive 12-V, 100-mA voltage regulator; Digi-Key LM78L12ACZ.

Fig 25.12). Two nearly identical diodes (D1 and D2) are biased on by a small dc current. RF energy is coupled to one of the diodes, which then acts as a square-law detector. The difference between the voltages present on the two diodes is amplified by a differential amplifier (U1) and applied to an analog voltmeter equipped with hand-calibrated ranges. Getting adequate stability requires careful selection of parts.

D1 and D2 must have a low junction capacitance and a fast response in order to provide the desired bandwidth. Their temperature coefficients must also be extremely well matched. The solution is to use a very old part: the CA3039 diode array.[2] The diodes in this device are very fast silicon types, with exceptional thermal matching.[3] Any two diodes of the array—except the one tied to the substrate—can be used for D1 and D2. Unused IC pins can be cut off or bent out of the way. Similarly rated arrays of hot-carrier diodes may provide a better frequency response.

The Burr-Brown INA2128 provides two differential amplifiers with exceptional specifications: low thermal drift, superb common-mode rejection and low noise. In this circuit, one half of the INA2128 provides dc amplification of the detected voltage; the other half serves as a low-impedance offset-voltage source used for error nulling.

The RF-input terminating resistor is composed of two 100-Ω resistors (R3 and R4), providing a cleaner 50-Ω termination than a single resistor does. If you have them, use chip resistors. If not, clip the leads off two standard 100-Ω resistors and carefully scrape away any paint at the ends. After installing the resistors, verify the input resistance with an ohmmeter.

A stable power source is required. Operating the circuit from an unregulated battery supply allows slight relative changes in the local ground as the terminal voltage drops, creating drift in the reading. The Zener diodes (D3 and D4) following regulator U2 provide a simple, dynamic means of splitting the 12-V supply into the three voltages required to power the circuit.

C1 and C11 should be as close as physically possible to U1 pins 1 and 2. In combination with R1 and R5, these capacitors provide good immunity to unwanted RF. R14, the OFFSET NULL potentiometer, allows nulling the dc amplifier offset and any offset from slight mismatches in the CA3039 diodes. Use a multiturn potentiometer for R14—10 turns or more. Whether you use a panel-mount pot, or a PC-board-mounted trimmer as I did, R14 must be accessible. On the most-sensitive range, the instrument must be zeroed before each use.

The forward voltage drop difference between any two diodes in the CA3039 is specified as "typically less than 0.5 mV." If your device is typical, the nominal values of 5.1 kΩ for R10 and 3.9 kΩ for R9 will do just fine. The manufacturer does not reject parts until the forward voltage drop difference is *5 mV*—so the 0.5 mV figure isn't totally dependable. You can always rotate the diode array so that pins 1, 2, 3 and 4 occupy the spots designated for pins 5, 6, 7 and 8. That gives you a second set of diodes from which to choose. The simplest general solution for a homebrew project is to use nonidentical values for R9 and R10.

If your diode array isn't typical, your Microwatter won't zero properly. The solution is simple. Remove R14 from the circuit. Temporarily connect one end of a 10-kΩ potentiometer to the +8.1 V supply and attach the other end to the –3.9 V supply. Connect the pot arm to the PC-board point of R14's arm. With this arrangement, you'll be able to zero your Microwatter, although the setting will be sensitive and might not produce an exact zero. Once you have the zero point, turn off the Microwatter and remove the 10-kΩ pot carefully without disturbing its setting. Measure the resistance between the pot arm and end that was connected to the +8.1 V point—this is the value of R10. R9's value is that measured between the pot arm and the end previously connected to the –3.9 V point. The combined value of these two resistors and that of R14 provides a total resistance of about 10.5 kΩ, and the resistance ratios will allow easy meter zeroing.

The Meter Movement

The specified meter movement has a 50-Ω internal resistance. Hence, on the most sensitive scale, a 50 mV output from U1 provides full-scale deflection. The middle and upper scales require 440 mV and 3.95 V, respectively, for full-scale deflection. Here's how you can use a meter that has a different internal resistance or current sensitivity.

Take, for example, a 25-µA meter movement with an internal resistance of 1910 Ω. A signal level of 47.8 mV (25 µA × 1910 Ω) provides full-scale deflection, so this meter can be substituted directly for the 1-mA, 50-Ω meter on the most sensitive scale. On the middle and top scales, the values of R12 and R11 should be 15.69 kΩ and 156.1 kΩ, or values close to those.

If the most sensitive meter you can find requires 80 to 100 mV to drive it to full scale deflection, don't worry. The Microwatter's most-sensitive scale will be compressed, but quite usable. The middle and top scales will still run full scale, probably not even requiring a change in the values of R11 and R12, if you use a 1-mA movement.

Construction and Calibration

For the prototype, components are mounted on double-sided PC board (see Note 1) with one side acting as a groundplane. The cabinet is a 5 × 7 × 3-inch aluminum box, primed and painted gray (see **Fig 25.13**). Power is supplied by a set of NiCd batteries glued to the inside rear of the cabinet. The PC board is small enough to be mounted on the back of the input BNC connector without other support. Once you have completed PC-board assembly, remove the solder flux from the board.

To calibrate the Microwatter, you'll need a 50-Ω RF source with a known output level. A suitable signal source can be made from a crystal oscillator and an attenuator, using at least 6 dB of attenuation between the oscillator and the Microwatter at all times. The output of such a system can be calibrated with an oscilloscope, or with a simple RF meter. Any frequency in the mid-HF range is suitable for calibration—W7DB used 10 MHz.

Remove the meter-face cover and prepare the face for new markings. You can paint it white and use India ink to make the meter scale arcs and incremental marks. Before calibrating the Microwatter, let it stabilize at room temperature for an hour, and turn it on at least 15 minutes before use. (At these low power levels, even the heat generated by the small dc bias on the diodes needs time to stabilize.) Set the Microwatter to its most sensitive scale. Use R14 to set the meter needle to your chosen meter zero point and mark that point with a pencil. Then apply a –40 dBm signal and mark the top end of the scale. Decrease the input signal level in 1-dB steps, marking each step. Recheck your meter zeroing, switch the Microwatter to the middle scale, apply a –30 dBm signal and mark this point. Again, decrease drive in 1-dB steps, marking each. Repeat with the least-sensitive scale, starting with –20 dBm. Once this is done, replace the meter-face cover and your Microwatter is ready to use.

Fig 25.13 — Photo of the Microwatter.

DIRECTIONAL WATTMETERS

Directional wattmeters of varying quality are commonly used by the amateur community. The high quality standard is made by the Bird Electronic Corporation, who call their proprietary line THRULINE. The units are based on a sampling system built into a short piece of 50-Ω transmission line with plug-in elements for various power and frequency ranges.

AC BRIDGES

In its simplest form, the ac bridge is exactly the same as the Wheatstone bridge discussed earlier in the dc measurement section of this chapter. However, complex impedances can be substituted for resistances, as suggested by **Fig 25.14A**. The same bridge equation holds if Z (complex impedance) is substituted for R in each arm. For the equation to be true, however, both phase angles and magnitudes of the impedances must balance; otherwise, a true null voltage is impossible to obtain. This means that a bridge with all "pure" arms (pure resistance or reactance) cannot measure complex impedances; a combination of R and X must be present in at least one arm aside from the unknown.

The actual circuits of ac bridges take

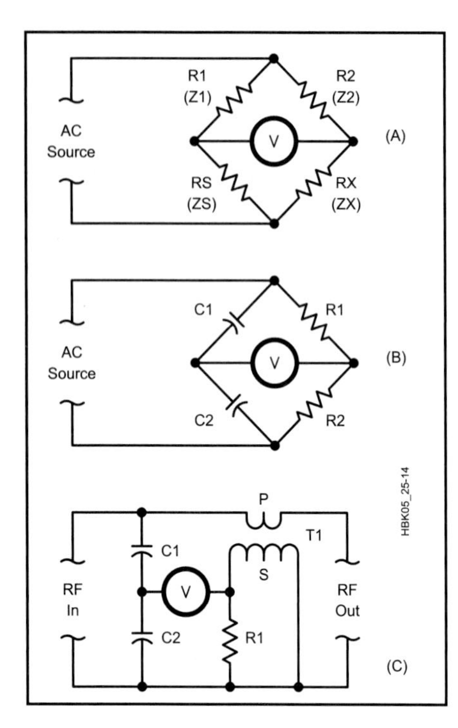

Fig 25.14 — A shows a bridge circuit generalized for ac or dc use. B is a form of ac bridge for RF applications. C is an SWR bridge for use in transmission lines.

many forms, depending on the intended measurement and the frequency range to be covered. As the frequency increases, stray effects (unwanted capacitances and inductances) become more pronounced. At RF, it takes special attention to minimize them.

Most amateur built bridges are used for RF measurements, especially SWR measurements on transmission lines. The circuits at Fig 25.14B and C are favorites for this purpose.

Fig 25.14B is useful for measuring both transmission lines and lumped constant components. Combinations of resistance and capacitance are often used in one or more arms; this may be required for eliminating the effects of stray capacitance. The bridge shown in Fig 25.14C is used only on transmission lines and only on those lines having the characteristic impedance for which the bridge is designed.

SWR Measurement

The theory behind SWR measurement is covered in the **Transmission Lines** chapter and more fully in *The ARRL Antenna Book*. Projects to measure SWR appear in the **Station Layout and Accessories** chapter of this *Handbook*.

Notes

[1]A PC board and some components for this project is available from FAR Circuits. A PC-board template package is not available.

[2]The NTE907 replacement is available from Mouser Electronics.

[3]The absolute value of the difference in forward drop between any two diodes is 1 mV/°C.

Frequency Measurement

The FCC Rules for Amateur Radio require that transmitted signals stay inside the frequency limits of bands consistent with the operator's license privileges. The exact frequency need not be known, as long as it is within the limits. On these limits there are no tolerances: Individual amateurs must be sure that their signal stays safely inside. The current limits for each license class can be found in Chapter 1 of this *Handbook* and in the current edition of *The FCC Rule Book*, published by the ARRL.

Staying within these limits is not difficult; many modern transceivers do so automatically, within limits. If your radio uses a PLL synthesized frequency source, just tune in WWV or another frequency standard occasionally.

Checks on older equipment require some simple equipment and careful adjusting. The equipment commonly used is the frequency marker generator and the method involves use of the station receiver, as shown in **Fig 25.15**.

FREQUENCY MARKER GENERATORS

A marker generator, in its simplest form, is a high-stability oscillator that generates a series of harmonic signals. When an appropriate fundamental is chosen, harmonics fall near the edges of the amateur frequency allocations.

Most US amateur band and subband limits are exact multiples of 25 kHz. A 25-kHz fundamental frequency will therefore produce the right marker signals if its harmonics are strong enough. But since harmonics appear at 25-kHz intervals throughout the spectrum, there is still a problem of identifying particular markers. This is easily solved if the receiver has reasonably good calibration. If not, most marker circuits provide a choice of fundamental outputs, say 100 and 50 kHz as well as 25 kHz. Then the receiver can be first set to a 100-kHz interval. From there, the desired 25-kHz (or 50-kHz) points can be counted. Greater frequency intervals are rarely required. Instead, tune in a signal from a station of known frequency and count off the 100-kHz points from there.

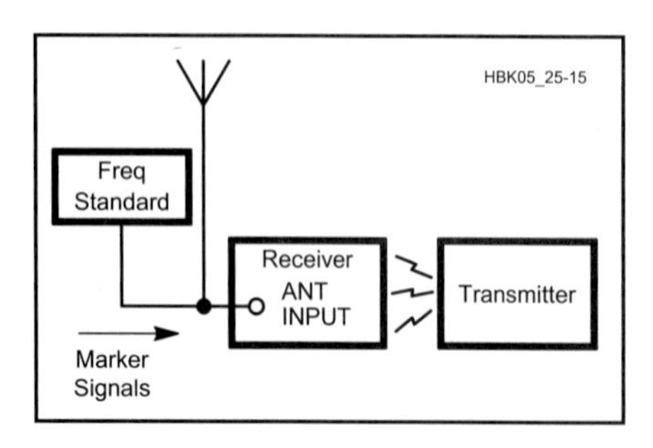

Fig 25.15 — A setup for checking transmitter frequency. Use care to ensure that the transmitter does not overload the receiver. False signals would result; see text.

Transmitter Checking

To check transmitter frequency, tune in the transmitter signal on a calibrated receiver and note the dial setting at which it is heard. To start, reduce the transmitter to its lowest possible level to avoid receiver overload. Also, place a direct short across the receiver antenna terminals, reduce the RF gain to minimum and switch in any available receiver attenuators to help prevent receiver IMD, overload and possible false readings.

Place the transmitter on standby (not transmitting) and use the marker generator as a signal source. Tune in and identify the nearest marker frequencies above and below the transmitter signal. The transmitter frequency is between these two known frequencies. If the marker frequencies are accurate, this is all you need to know, except that the transmitter frequency must not be so close to a band (or subband) edge that sidebands extend past the edge.

If the transmitter signal is inside a marker at the edge of an assignment, to the extent that there is an audible beat note with the receiver BFO turned off, normal CW sidebands are safely inside the edge. (So long as there are no abnormal sidebands such as those caused by clicks and chirps.) For phone the safety allowance is usually taken to be about 3 kHz, the usual width of one sideband. A frequency difference of this much can be estimated by noting the receiver dial settings for the two 25-kHz markers that are either side of the signal and dividing 25 by the number of dial divisions between them. This will give the number of kilohertz per dial division. It is a prudent practice to allow an extra kHz margin when setting the transmitter close to a band or subband edge (5-kHz is a safe HF margin for most modes on modern transmitters).

Transceivers

The method described above is good when the receiver and transmitter are separate pieces of equipment. When a transceiver is used and the transmitting frequency is automatically the same as that to which the receiver is tuned, setting the tuning dial to a spot between two known marker frequencies is all that is required. The receiver incremental tuning control (RIT) must be turned off.

The proper dial settings for the markers are those at which, with the BFO on, the signal is tuned to zero beat (the spot where the beat note disappears as tuning makes its pitch progressively lower). Exact zero beat can be determined by a very slow rise and fall of background noise, caused by a beat of a cycle or less per second. In receivers with high selectivity it may not be possible to detect an exact zero beat, because low audio frequencies from beat notes may be prevented from reaching the speaker or headphones.

Most commercial equipment has some way to match either the equipment's internal oscillator or marker generator with the signal received from WWV on one of its short-wave frequencies. It is a good idea to do this check on a new piece of gear. A recheck about a month later will show if anything has changed. Normal commercial equipment drifts less than 1 kHz after warm up.

Also check the dial linearity of equipment that has an analog dial or subdial. Often analog dials do not track frequency accurately across an entire band. Such radios usually provide for pointer adjustment so that dial error can be minimized at the most often used part of a band.

Frequency-Marker Circuits

The frequency in most amateur frequency markers is determined by a 100-kHz or 1-MHz crystal. Although the marker generator should produce harmonics every 25 kHz and 50 kHz, crystals (or other high-stability resonators) for frequencies lower than 100 kHz are expensive and rare. There is really no need for them, however, since it is easy to divide the basic frequency down to the desired frequency; 50- and 25-kHz steps require only two successive divisions by two (from 100 kHz). In the division process, the harmonics of the generator are strengthened so they are useful up to the VHF range. Even so, as frequency increases the harmonics weaken.

Current marker generators are based on readily available crystals. A 1-MHz basic oscillator would first be divided by 10 to produce 100 kHz and then followed by two successive divide-by-two stages to produce 50 kHz and 25 kHz.

A MARKER GENERATOR WITH SELECTABLE OUTPUT

Fig 25.16 shows a marker generator with selectable output for 100, 50 or 25-kHz intervals. It provides marker signals well up into the 2-m band. The project was first built by Bruce Hale, KB1MW, in the ARRL Lab. A more detailed presentation appeared in the **Station Accessories** chapter of *Handbooks* from 1987 through 1994. An etching pattern and parts-placement diagram are available for this project. See the **Project Templates** section on the *Handbook* CD-ROM.

A 1, 2 or 4-MHz computer-surplus crystal is suitable for Y1. Several prototypes were built with such crystals and all could be tuned within 50 Hz at 100 kHz. The marker division ratio must be chosen once the crystal frequency is selected. This is accomplished by means of several jumpers that disable part or all of U2A, the dual flip-flop IC. When Y1 is a 1-MHz crystal, U2 may be omitted entirely. **Table 25.2** gives the jumper placement for each crystal frequency.

The prototypes were built with TTL logic ICs. Unused TTL gate inputs should always be connected either to ground or to V_{CC} through a pull-up resistor. If low-power Schottky (they have "LS" as part of their numerical designator) parts are available for U1 through U4, use them. They draw much less current than plain TTL, and will greatly extend battery life.

Table 25.2
Marker Generator Jumper Placement

Crystal Frequency	*Jumper Placement*
1 MHz	A to F (U2 not used)
2 MHz	A to B C to F D to +5 V, via 1-kΩ resistor
4 MHz	A to B C to D E to F

DIP METERS

This device is often called a grid-dip meter or oscillator (from vacuum-tube days) or a transistor dip meter. Most dip meters can also serve as absorption frequency meters (in this mode measurements are read at the current peak, rather than the dip). Further, some dip meters have a connection for headphones. The operator can usually hear signals that do not register on the meter. Because the dip meter is an oscillator, it can be used as a signal generator in certain cases where high accuracy or stability are not required.

A dip meter may be coupled to a circuit either inductively or capacitively. Inductive coupling results from the magnetic field generated by current flow. Therefore, inductive coupling should be used when a conductor with relatively high current is convenient. Maximum inductive coupling results when the axis of the pick-up coil is placed perpendicular to a nearby current path (see **Fig 25.17**).

Capacitive coupling is required when current paths are magnetically confined or shielded. (Toroidal inductors and coaxial cables are common examples of magnetic self shielding.) Capacitive coupling depends on the electric field produced by voltage. Use capacitive coupling when a

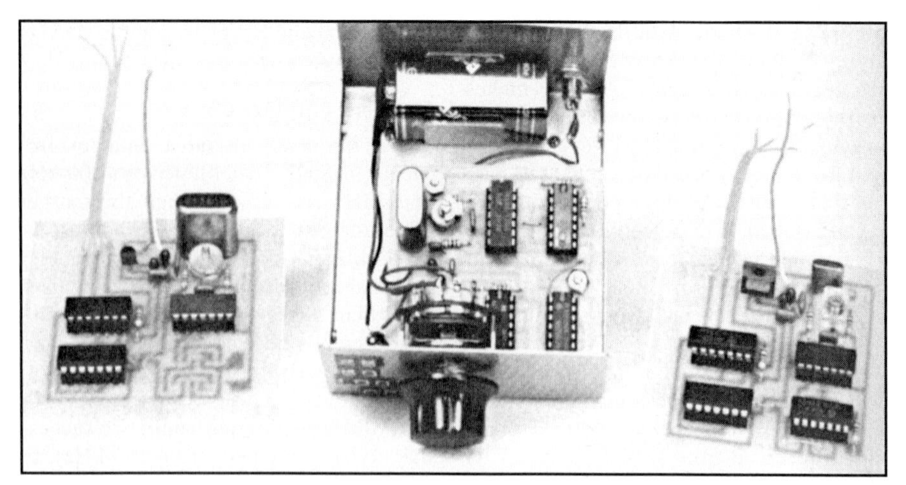

Fig 25.16A—The generator enclosure and views of the circuit board.

point of relatively high voltage is convenient. (An example might be the output of a 12-V powered RF amplifier. *Do not* attempt dip-meter measurements on true high-voltage equipment such as vacuum-tube amplifiers or switching power supplies while they are energized.) Capacitive coupling is maximum when the end of the pick-up coil is near a point of high voltage (see Fig 25.17). In either case, the circuit under test is affected by the presence of the dip meter. Always use the minimum coupling that yields a noticeable indication.

Use the following procedure to make reliable measurements. First, bring the dip meter gradually closer to the circuit while slowly varying the dip-meter frequency.

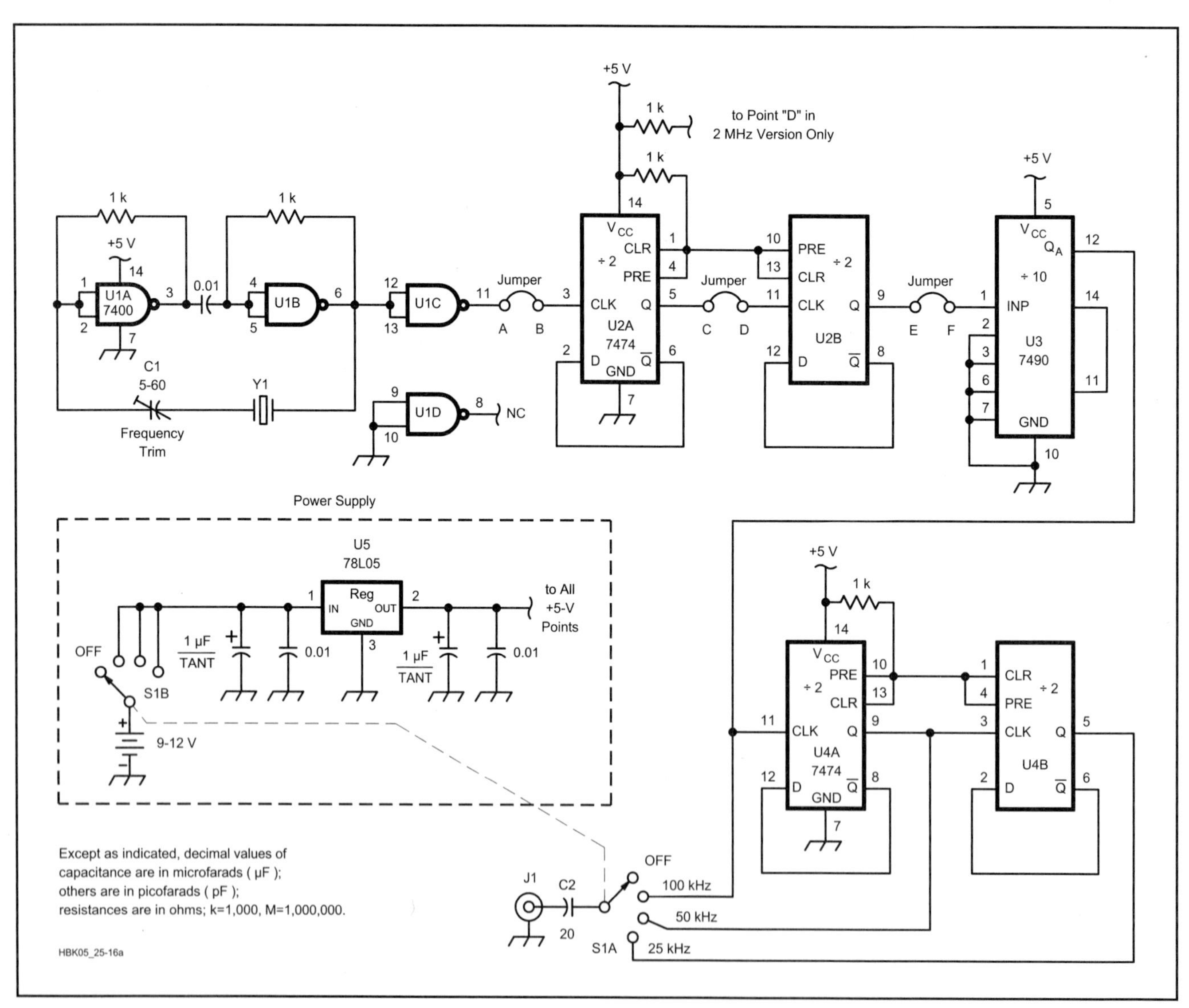

Fig 25.16B — Schematic diagram of the marker generator.

C1 — 5-60 pF miniature trimmer capacitor.
C2 — 20-pF disc ceramic.
U1 — 7400 or 74LS00 quad NAND gate.
U2, U4 — 7474 or 74LS74 dual D flip-flop.
U3 — 7490 or 74LS90 decade counter.
U5 — 78L05, 7805 or LM340-T5 5-V voltage regulator.
Y1 — 1, 2 or 4-MHz crystal in HC-25, HC-33 or HC-6 holder.

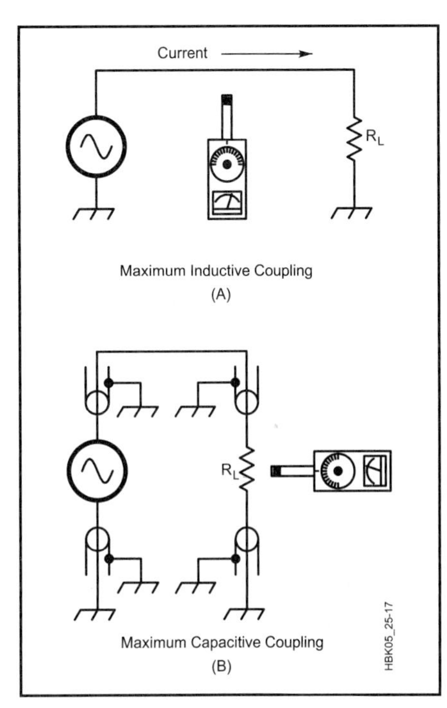

Fig 25.17 — Dip meter coupling. (A) uses inductive coupling and (B) uses capacitive coupling.

When a current dip occurs, hold the meter steady and tune for minimum current. Once the dip is found, move the meter away from the circuit and confirm that the dip comes from the circuit under test (the current reading should increase with distance from the circuit until the dip is gone). Finally, move the meter back toward the circuit until the dip is just noticed. Retune the meter for minimum current and read the dip-meter frequency with a calibrated receiver or frequency meter.

The current dip of a good measurement is smooth and symmetrical. An asymmetrical dip indicates that the dip-meter oscillator frequency is being significantly influenced by the test circuit. Such conditions do not yield usable readings.

A measurement of effective unloaded inductor Q can be made with a dip meter and an RF voltmeter (or a dc voltmeter with an RF probe). Make a parallel resonant circuit using the inductor and a capacitance equal to that of the application circuit. Connect the RF voltmeter across this parallel combination and measure the resonant frequency. Adjust the dip-meter/circuit coupling for a convenient reading on the voltmeter, then maintain this dip-meter/circuit relationship for the remainder of the test. Vary the dip meter frequency until the voltmeter reading drops to 0.707 times that at resonance. Note the frequency of the dip meter and repeat the process, this time varying the frequency on the opposite side of resonance. The difference between the two dip meter readings is the test-circuit bandwidth. This can be used to calculate the circuit Q:

$$Q = \frac{f_0}{BW} \qquad (9)$$

where

f_0 = operating frequency,

BW = measured bandwidth in the same units as the operating frequency.

When purchasing a dip meter, look for one that is mechanically and electrically stable. The coils should be in good condition. A headphone connection is helpful. Battery-operated models are convenient for antenna measurements.

A DIP METER WITH DIGITAL DISPLAY

An up-to-date dip meter was described by Larry Cicchinelli in the October 1993 issue of *QEX*. It consists of a dip meter with a three-digit frequency display. The analog portion of the circuit consists of an FET oscillator, voltage-doubler detector, dc-offset circuit and amplifier. The digital portion of the circuit consists of a high-impedance buffer, prescaler, counter, display driver, LED display and control circuit.

Circuit Description

The dip meter shown in **Fig 25.18** has four distinct functional blocks. The RF oscillator is a standard Colpitts using a common junction FET, Q1, as the active element. Its range is about 1.7 to 45 MHz. The 200-pF tuning capacitor gives a 2:1 tuning range. A 2:1 frequency range requires a capacitor with a 4:1 range. The sum of the minimum capacitance of the variable, the capacitors across the inductor and the strays must therefore be in the order of 70 pF. The values of L1 were determined experimentally by winding the coils and observing the lower and upper frequency values. [**Table 25.3** shows winding data calculated from the author's schematic. — *Ed.*] The coil sizes were experimentally determined, and the coils are constructed on 1 1/4-inch-diameter plug-in coil forms with #20 enameled wire. They are close wound. The number of turns shown are only a starting point; you may need to change them slightly in order to cover the desired frequency range.

The tapped capacitors are mounted inside the coil forms so that their values could be different for each band if required. The frequency spread of the lowest band is less than 2:1 because the tapped capacitor values are larger than those for the other bands. The frequency spread of the highest band is greater than 2:1 because its capacitors are smaller.

The analog display circuit begins with a voltage-doubler detector in order to get higher sensitivity. It drives a dc-offset circuit, U1A. R1 inserts a variable offset that is subtracted from the detector voltage. This allows the variable gain stage, U1B, to be more sensitive to variations in the detector output voltage. Q9 follows U1C to get an output gating voltage closer to ground. The resistor in series with the meter is chosen to limit the meter current to a safe value. For example, if a 1-mA meter is used, the resistor should be 8.2 kΩ.

The prescaler begins with a high-impedance buffer and amplifier, Q2 and Q3. If you are going to use the meter for the entire frequency range described, take care in the layout of both the oscillator and buffer/ amplifier circuits. The digital portion of the prescaler is a divide-by-100 circuit consisting of two divide-by-10 devices, U2 and U3. The devices used were selected because they were available. Any similar devices may be used as long as the reset circuit is compatible. Q5 is a level translator that shifts the 5-V signal to 9 V.

The first part of the digital display block is the oscillator circuit of U1C, which creates the gate time for the frequency counter. R3 adjusts the oscillator to a frequency of 500 Hz, yielding a 1-ms gate. The best way to set this frequency is to

Table 25.3
Calculated Coil Data for the Dip Meter

Frequency (MHz)	*L (µH)*	*Turns*
1.7 to 3.1	48.6	52
2.8 to 5.9	16.3	23
5.6 to 11.9	4.0	9
9.7 to 20.7	1.3	5
19.0 to 45.0	0.3	2

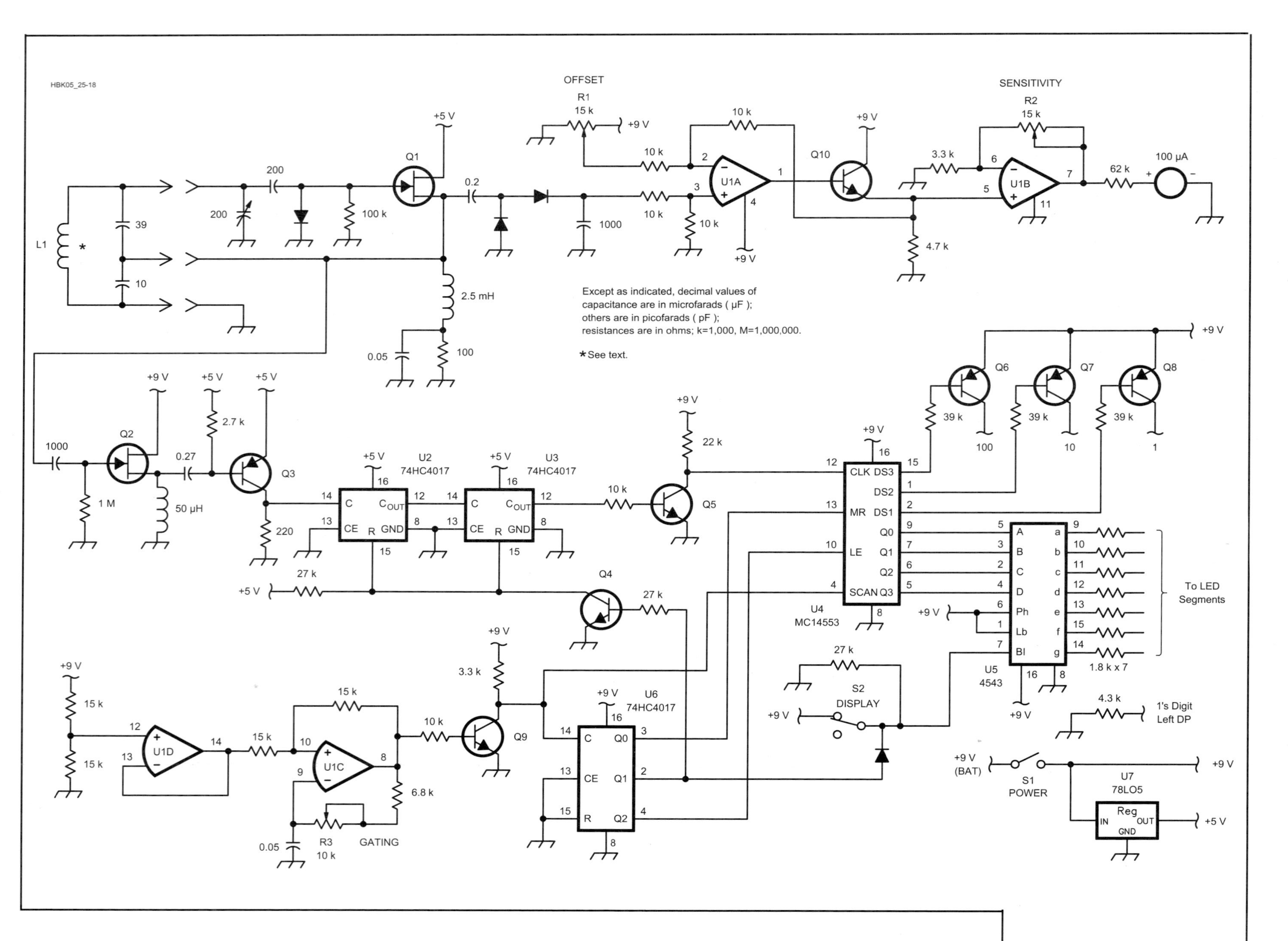

HBK05_25-18
OFFSET
R1
15 k
SENSITIVITY
R2
15 k
100 µA
U1A
U1B
U1C
U1D
Q1
Q2
Q3
Q4
Q5
Q6
Q7
Q8
Q9
Q10
L1
2.5 mH
50 µH
Except as indicated, decimal values of capacitance are in microfarads (µF); others are in picofarads (pF); resistances are in ohms; k=1,000, M=1,000,000.
* See text.
U2
74HC4017
U3
74HC4017
U4
MC14553
U5
4543
U6
74HC4017
To LED Segments
1.8 k x 7
S2
DISPLAY
1's Digit Left DP
+9 V (BAT)
S1
POWER
U7
78LO5
Reg
R3
10 k
GATING

Fig 25.18 — A schematic diagram of the dip meter. All diodes are 1N914 or similar. All resistors are 1/4 W, 5%.
Q1, Q2 — MPF102 JFET transistor.
Q3, Q6, Q7, Q8 — 2N3906 PNP transistor.
Q4, Q5, Q9, Q10 — 2N3563 (or any general purpose NPN).
U1 — LM324.
U2, U3 — 74HC4017.
U4 — MC14553.
U5 — 4543.
U6 — 4017.
U7 — 78L05.

listen for the dip-meter output on a communications receiver and adjust R3 until the display agrees with the receiver. Once this calibration has been made for one of the bands, all the bands are calibrated. U1D gives a low-impedance voltage reference for U1C. Q9 was added to the output of the oscillator to remove a small glitch, which can cause the counter to trigger incorrectly. This type of oscillator has the advantage of simplicity. This circuit is fairly stable, easy to adjust and has a low parts count.

The digital system controller, U6, is a divide-by-10 counter. It has 10 decoded outputs, each of which goes high for one period of the input clock. The Q0 output is used to reset the frequency counter, U4. The Q1 output is used to enable the prescaler and disable the display and the Q2 output latches the count value into the frequency counter. Since the prescaler can only count while Q1 is high, it will be enabled for only 1 ms. Normally a 1-ms gate will yield 1-kHz resolution. Since the circuit uses a divide-by-100 prescaler, the resolution becomes 100 kHz.

The frequency counter, U4, is a three-digit counter with multiplexed BCD outputs. The clock input is driven from the prescaler, hence it is the RF oscillator frequency divided by 100. This signal is present for only 1 ms out of every 10 ms. The digit scanning is controlled by the 500-Hz oscillator of U1C.

U5 is a BCD-to-seven-segment decoder/driver. Its outputs are connected to each of the three common-anode, seven-segment displays in parallel. Only the currently active digit will be turned on by the digit strobe outputs of U5, via Q6, Q7 and Q8. The diode connected to the blanking input of U5 disables the display while U4 is counting. U7 is a 5-V regulator that allows the use of a single 9-V battery for both the circuit and the LEDs. S2 turns on the displays once the unit has been adjusted for a dip.

The circuit draws about 20 mA with the LEDs off and up to 35 mA with the LEDs on.

Many of the resistor values are not critical, and those used were chosen based upon availability; the op-amp circuits depend primarily on resistance ratios. The resistor at the collector of Q3 is critical and should not be varied. Use 0.27-µF monolithic capacitors. They have the required good high-frequency characteristics over the range of the meter.

Most of the parts can be purchased from Digi-Key. They did not have the '4543 IC, which was purchased from Hosfelt Electronics in Steubenville, OH. The 74HC4017 may be substituted with a 74HCT4017. The circuit is built on a 4-inch-square perf board (with places for up to 12 ICs) and is housed in a 7 × 5 × 3-inch minibox.

Operation

To use the unit, set the gain, R2, fully clockwise for maximum sensitivity. With this setting, the output of the offset circuit (Q10 emitter) is at ground. As R1 is rotated, the voltage on the arm approaches and then becomes less than, the detector output. At this point the meter will start to deflect upward. Adjust R1 so that the meter reads about center scale. (Manual adjustment allows for variations in the output level of the RF oscillator.) As L1 is brought closer to the circuit under test, the meter will deflect downward as energy is absorbed by the circuit. For best results use the minimum possible coupling to the circuit being tested. If the dip meter is overcoupled to the test circuit, the oscillator frequency will be pulled.

FREQUENCY COUNTERS

One of the most accurate means of measuring frequency is a frequency counter. This instrument is capable of numerically displaying the frequency of the signal supplied to its input. For example, if an oscillator operating at 8.244 MHz is connected to a counter input, 8.244 would be displayed. At present, counters are usable well up into the gigahertz range. Most counters that are used at high frequencies make use of a prescaler ahead of a basic low-frequency counter. A prescaler divides the high-frequency signal by 10, 100, 1000 or some other fixed amount so that a low-frequency counter can display the operating frequency.

The accuracy of the counter depends on its internal crystal reference. A more accurate crystal reference yields more accurate readings. Crystals for frequency counters are manufactured to close tolerances. Most counters have a trimmer capacitor so that the crystal can be set exactly on frequency. Crystal frequencies of 1 MHz, 5 MHz or 10 MHz have become more or less standard. For calibration, harmonics of the crystal can be compared to a known reference station, such as those shown in Table 25.1, or other frequency standard and adjusted for zero beat.

Many frequency counters offer options to increase the accuracy of the counter timebase; this directly increases the counter accuracy. These options usually employ temperature-compensated crystal oscillators (TCXOs) or crystals mounted in constant temperature ovens that keep the crystal from being affected by changes in ambient (room) temperature. Counters with these options may be accurate to 0.1 ppm (part per million) or better. For example, a counter with a timebase accuracy of 5.0 ppm and a second counter with a TCXO accurate to 0.1 ppm are available to check a 436-MHz CW transmitter for satellite use. The counter with the 5-ppm timebase could have a frequency error of as much as 2.18 kHz, while the possible error of the counter with the 0.1 ppm timebase is only 0.0436 kHz.

Other Instruments and Measurements

This section covers a variety of test equipment that is useful in receiver and transmitter testing. It includes RF and audio generators, an inductance meter, a capacitance meter, oscilloscopes, spectrum analyzers, a calibrated noise source, and hybrid combiners. A number of applications of this equipment for basic transmitter and receiver testing is also included.

RF OSCILLATORS FOR CIRCUIT ALIGNMENT

Receiver testing and alignment uses equipment common to ordinary radio service work. Inexpensive RF signal generators are available, both complete and in kit form. However, any source of signal that is weak enough to avoid overloading the receiver usually will serve for alignment work. The frequency marker generator is a satisfactory signal source. In addition, its frequencies, although not continuously adjustable, are far more precise, since the usual signal-generator calibration is not highly accurate. When buying a used or inexpensive signal generator, look for these attributes: output level is calibrated, the output doesn't "ring" too badly when tapped, and doesn't drift too badly when warmed up. Many military surplus units are available that can work quite well. Commercial units such as the HP608 are big and stable, and they may be inexpensive.

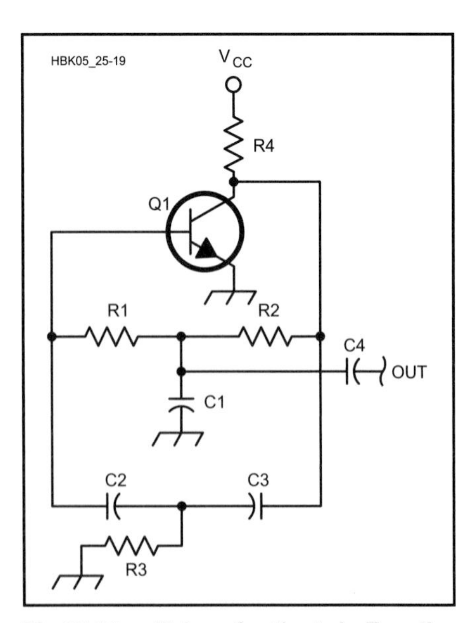

Fig 25.19 — Values for the twin-T audio oscillator circuit range from 18 kΩ for R1-R2 and 0.05 µF for C1 (750 Hz) to 15 kΩ and 0.02 µF for 1800 Hz. For the same frequency range, R3 and C2-C3 vary from 1800 Ω and 0.02 µF to 1500 Ω and 0.01 µF. R4 is 3300 Ω and C4, the output coupling capacitor, can be 0.05 µF for high-impedance loads.

AUDIO-FREQUENCY OSCILLATORS

An audio signal generator should provide a reasonably pure sine wave. The best oscillator circuits for this use are RC coupled, operating as close to a class-A amplifier as possible. Variable frequencies covering the entire audio range are needed for determining frequency response of audio amplifiers.

An oscillator generating one or two frequencies with good waveform is sufficient for most phone-transmitter testing and simple troubleshooting in AF amplifiers. A two-tone (dual) oscillator is very useful for testing and adjusting sideband transmitters.

A circuit of a simple RC oscillator that is useful for general testing is given in **Fig 25.19**. This Twin-T arrangement gives a waveform that is satisfactory for most purposes. The oscillator can be operated at any frequency in the audio range by varying the component values. R1, R2 and C1 form a low-pass network, while C2, C3 and R3 form a high-pass network. As the phase shifts are opposite, there is only one frequency at which the total phase shift from collector to base is 180°: Oscillation will occur at this frequency. When C1 is about twice the capacitance of C2 or C3 the best operation results. R3 should have a resistance about 0.1 that of R1 or R2 (C2 = C3 and R1 = R2). Output is taken across C1, where the harmonic distortion is least. Use a relatively high impedance load — 100 kΩ or more.

Most small-signal AF transistors can be used for Q1. Either NPN or PNP types are satisfactory if the supply polarity is set correctly. R4, the collector load resistor may be changed a little to adjust the oscillator for best output waveform.

A WIDE-RANGE AUDIO OSCILLATOR

A wide-range audio oscillator that will provide a moderate output level can be built from a single 741 operational amplifier (**Fig 25.20**). Power is supplied by two 9-V batteries from which the circuit draws 4 mA. The frequency range is selectable from 8 Hz to 150 kHz. Distortion is approximately 1%. The output level under a light load (10 kΩ) is 4 to 5 V. This can be increased by using higher battery volt-

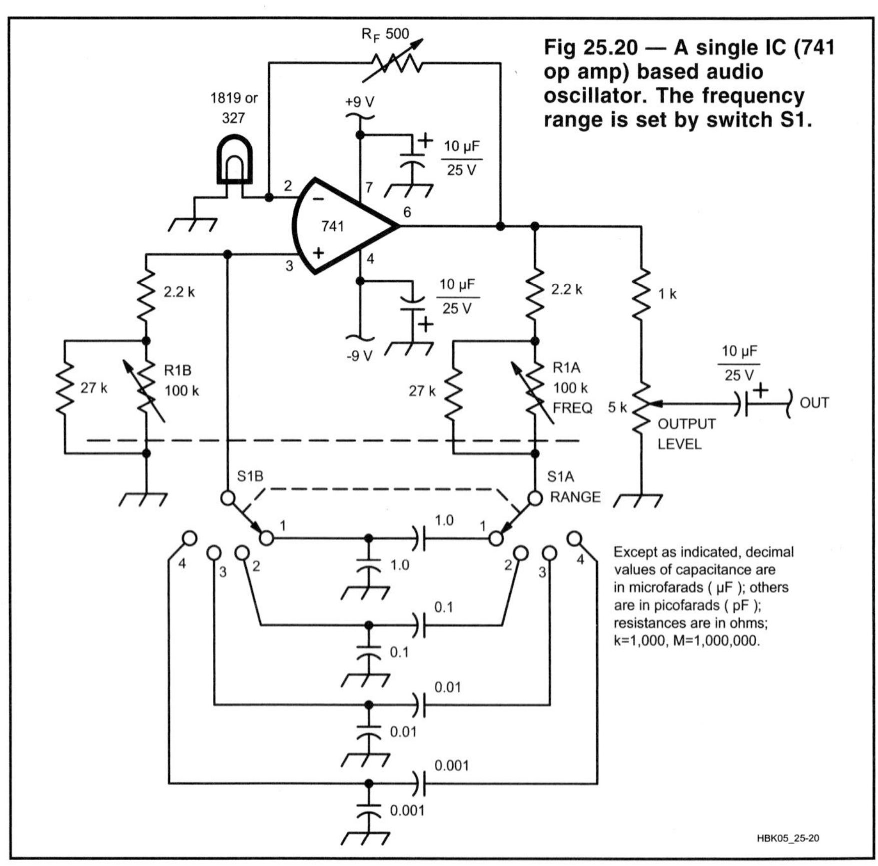

Fig 25.20 — A single IC (741 op amp) based audio oscillator. The frequency range is set by switch S1.

ages, up to a maximum of plus and minus 18 V, with a corresponding adjustment of R_F.

Pin connections shown are for the TO-5 case and the eight-pin DIP package. Variable resistor R_F is trimmed for an output level of about 5% below clipping as seen on an oscilloscope. This should be done for the temperature at which the oscillator will normally operate, as the lamp is sensitive to ambient temperature. This unit was originally described by Shultz in November 1974 *QST*; it was later modified by Neben as reported in June 1983 *QST*.

MEASURE INDUCTANCE AND CAPACITANCE WITH A DVM

Many of us have a DVM (Digital voltmeter) or VOM (volt-ohm meter) in the shack, but few of us own an inductance or capacitance meter. If you have ever looked into your junk box and wanted to know the value of the unmarked parts, these simple circuits will give you the answer. They may be built in one evening (**Fig 25.21** and **Fig 25.22**), and will adapt your DVM or VOM to measure inductance or capacitance. The units are calibrated against a known part. Therefore, the overall accuracy depends only on the calibration values and not on the components used to build the circuits. If it is carefully calibrated, an overall accuracy of 10% may be expected if used with a DVM and slightly less with a VOM.

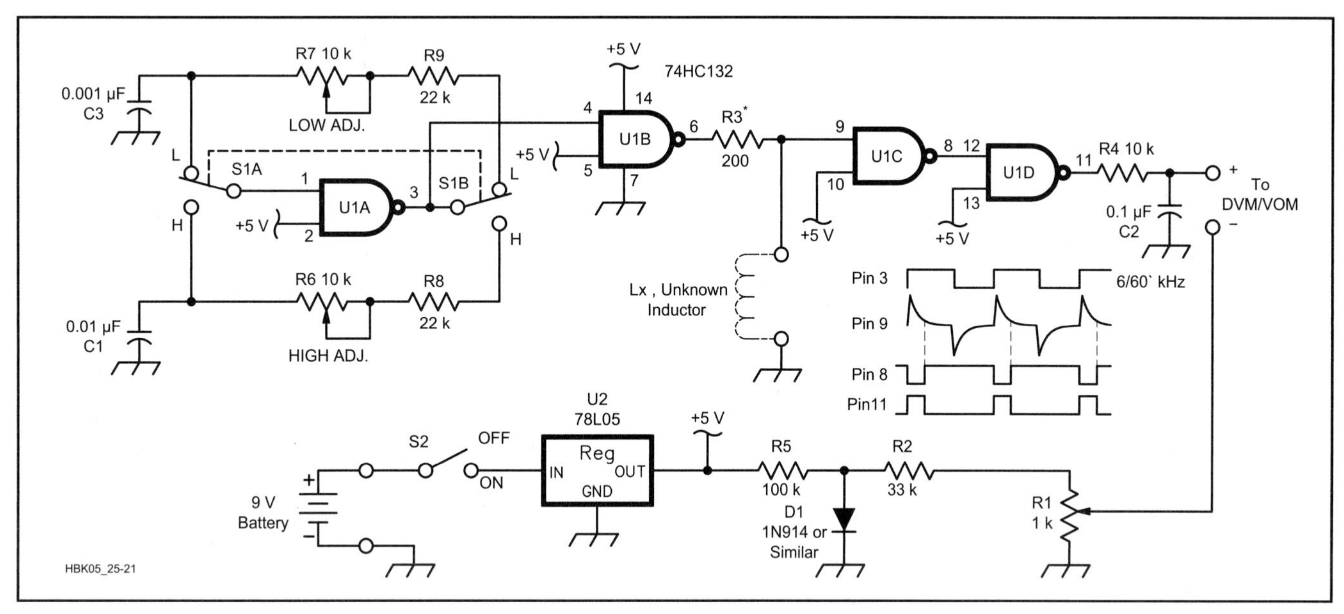

Fig 25.21 — All components are 10% tolerance. 1N4148 or equivalent may be substituted for D1. A LM7805 may be substituted for the 78L05. All fixed resistors are 1/4-W carbon composition. Capacitors are in µF. R3 value may need to be increased or decreased slightly if calibration cannot be achieved as described in text.

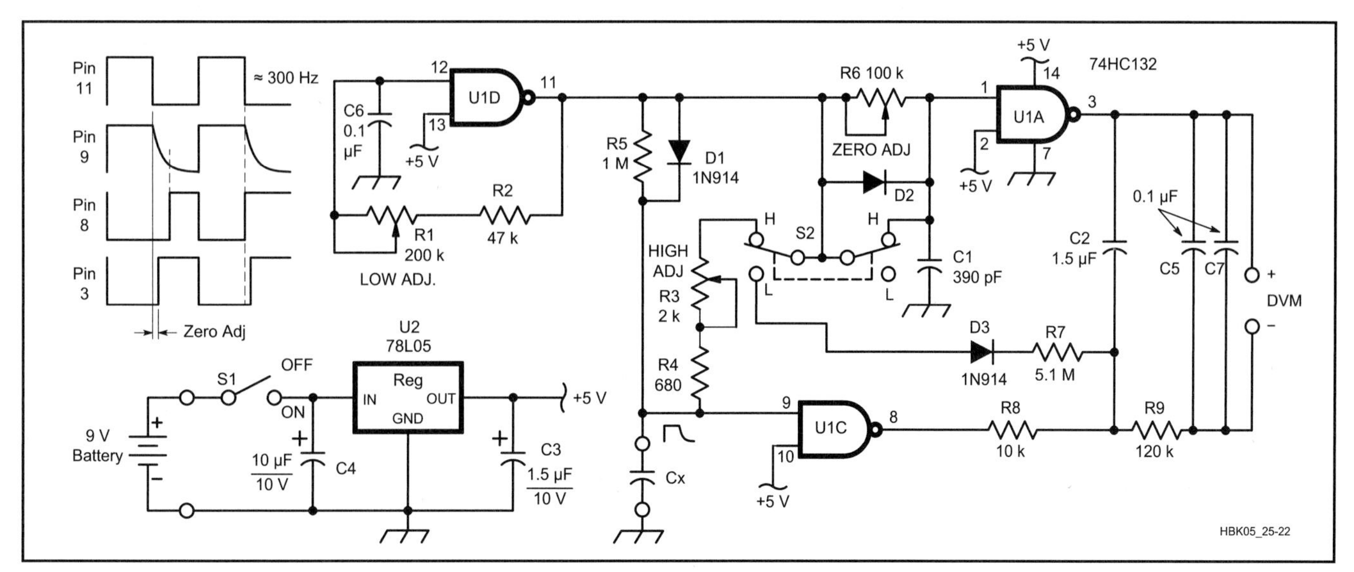

Fig 25.22 — All components are 10% tolerance. An LM7805 may be substituted for the 78L05. All fixed resistors are 1/4-W carbon composition. Capacitors are in µF unless otherwise indicated.

CONSTRUCTION

The circuits may be constructed on a small perf board (RadioShack dual mini board, #276-168), or if you prefer, on a PC board. A template and parts layout may be obtained from the ARRL.[1] Layout is non-critical—almost any construction technique will suffice. Wire-wrapping or point-to-point soldering may be used.

The circuits are available in kit form from Electronic Rainbow Inc., Indianapolis, IN. The IA inductance adapter kit is sold separately, but a cabinet is also available. The PC board is 1.75 × 2.5 inches. The CA-1 capacitance adapter kit comes with a 1.80 × 2.0-inch PC board.

INDUCTANCE ADAPTER FOR DVM/VOM

Description

The circuit shown in Fig 25.21 converts an unknown inductance into a voltage that can be displayed on a DVM or VOM. Values between 3 and 500 µH are measured on the low range and from 100 µH to 7 mH on the high range. NAND gate U1A is a two frequency RC square-wave oscillator. The output frequency (pin 3) is approximately 60 kHz in the low range and 6 kHz in the high range. The square-wave output is buffered by U1B and applied to a differentiator formed by R3 and the unknown inductor, LX. The stream of spikes produced at pin 9 decay at a rate proportional to the time constant of R3-LX. Because R3 is a constant, the decay time is directly proportional to the value of LX. U1C squares up the positive going spikes, producing a stream of negative going pulses at pin 8 whose width is proportional to the value of LX.

They are inverted by U1D (pin 11) and integrated by R4-C2 to produce a steady dc voltage at the + output terminal. The resulting dc voltage is proportional to LX and the repetition rate of the oscillator. R6 and R7 are used to calibrate the unit by setting a repetition rate that produces a dc voltage corresponding to the unknown inductance. D1 provides a 0.7 volt constant voltage source that is scaled by R1 to produce a small offset reference voltage for zeroing the meter on the low inductance range.

When S1 is low, *mV* corresponds to *µH* and when high, *mV* corresponds to *mH*. A sensitive VOM may be substituted for the DVM with a sacrifice in resolution.

Test and Calibration

Short the LX terminals with a piece of wire and connect a DVM set to the 200-mV range to the output. Adjust R1 for a zero reading. Remove the short and substitute a known inductor of approximately 400 µH. Set S1 to the low (in) position and adjust R7 for a reading equal to the known inductance. Switch S1 to the high position and connect a known inductor of about 5 mH. Adjust R6 for the corresponding value. For instance, if the actual value of the calibration inductor is 4.76 mH, adjust R7 so the DVM reads 476 mV.

CAPACITANCE ADAPTER FOR DVM/VOM

Description

The circuit shown in Fig 25.22 measures capacitance from 2.2 to 1000 pF in the low range, from 1000 pF to 2.2 µF in the high range. U1D of the 74HC132 (pin 11) produces a 300 Hz square-wave clock. On the rising edge CX rapidly charges through D1. On the falling edge CX slowly discharges through R5 on the low range and through R3-R4 on the high range. This produces an asymmetrical waveform at pin 8 of U1C with a duty cycle proportional to the unknown capacitance, CX. This signal is integrated by R8-R9-C2 producing a dc voltage at the negative meter terminal proportional to the unknown capacitance. A constant reference voltage is produced at the positive meter terminal by integrating the square-wave at U1A, pin 3. R6 alters the symmetry of this square-wave producing a small change in the reference voltage at the positive meter terminal. This feature provides a zero adjustment on the low range. The DVM measures the difference between the positive and negative meter terminals. This difference is proportional to the unknown capacitance.

Test and Calibration

Without a capacitor connected to the input terminals, set SW2 to the low range (out) and attach a DVM to the output terminals. Set the DVM to the 2-volt range and adjust R6 for a zero meter reading. Now connect a 1000 pF calibration capacitor to the input and adjust R1 for a reading of 1.00 volt. Switch to the high range and connect a 1.00 µF calibration capacitor to the input. Adjust R3 for a meter reading of 1.00 volts. The calibration capacitors do not have to be exactly 1000 pF or 1.00 µF, as long as you know their exact value. For instance, if the calibration capacitor is known to be 0.940 µF, adjust the output for a reading of 940 mV.

Note

[1]See the **Project Templates** section of the *Handbook* CD-ROM.

OSCILLOSCOPES

Most engineers and technicians will tell you that the most useful single piece of test and design equipment is the triggered-sweep oscilloscope (commonly called just a "scope"). This section was written by Dom Mallozzi, N1DM.

Oscilloscopes can measure and display voltage relative to time, showing the waveforms seen in electronics textbooks. Scopes are broken down into two major classifications: analog and digital. This does not refer to the signals they measure, but rather to the methods used inside the scope to process signals for display.

ANALOG OSCILLOSCOPES

Fig 25.23 shows a simplified diagram

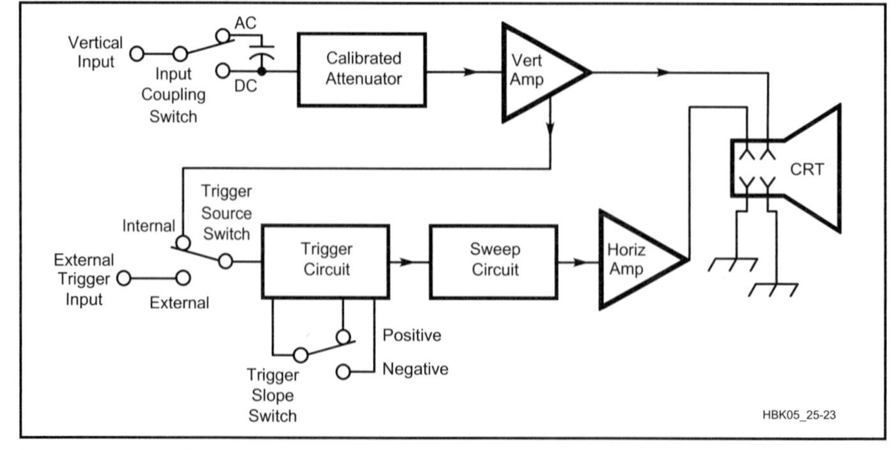

Fig 25.23 — Typical block diagram of a simple triggered-sweep oscilloscope.

of a triggered-sweep oscilloscope. At the heart of nearly all scopes is a cathode-ray tube (CRT) display. The CRT allows the visual display of an electronic signal by taking two electric signals and using them to move (deflect) a beam of electrons that strikes the screen. Unlike a television CRT, an oscilloscope uses electrostatic deflection rather than magnetic deflection. Wherever the beam strikes the phosphorescent screen of the CRT it causes a small spot to glow. The exact location of the spot is a result of the voltage applied to the vertical and horizontal inputs.

All of the other circuits in the scope are used to take the real-world signal and convert it to a form usable by the CRT. To trace how a signal travels through the oscilloscope circuitry start by assuming that the trigger select switch is in the INTERNAL position.

The input signal is connected to the input COUPLING switch. The switch allows selection of either the ac part of an ac/dc signal or the total signal. If you wanted to measure, for example, the RF swing at the collector of an output stage (referenced to the dc level), you would use the dc-coupling mode. In the ac position, dc is blocked from reaching the vertical amplifier chain so that you can measure a small ac signal superimposed on a much larger dc level. For example, you might want to measure a 25 mV 120-Hz ripple on a 13 Vdc power supply. Note that you should not use ac coupling at frequencies below 30 Hz, because the value of the blocking capacitor represents a considerable series impedance to very low-frequency signals.

After the coupling switch, the signal is connected to a calibrated attenuator. This is used to reduce the signal to a level that can be tolerated by the scope's vertical amplifier. The vertical amplifier boosts the signal to a level that can drive the CRT and also adds a bias component to locate the waveform on the screen.

A small sample of the signal from the vertical amplifier is sent to the trigger circuitry. The trigger circuit feeds a start pulse to the sweep generator when the input signal reaches a certain level. The sweep generator gives a precisely timed signal that looks like a triangle (see **Fig 25.24**). This triangular signal causes the scope trace to sweep from left to right, with the zero-voltage point representing the left side of the screen and the maximum voltage representing the right side of the screen.

The sweep circuit feeds the horizontal amplifier that, in turn, drives the CRT. It is also possible to trigger the sweep system from an external source (such as the system clock in a digital system). This is done by using an external input jack with

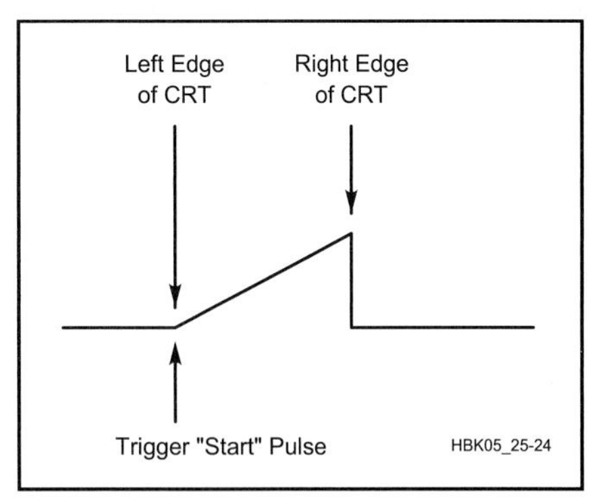

Fig 25.24 — The sweep trigger starts the ramp waveform that sweeps the CRT electron beam from side to side.

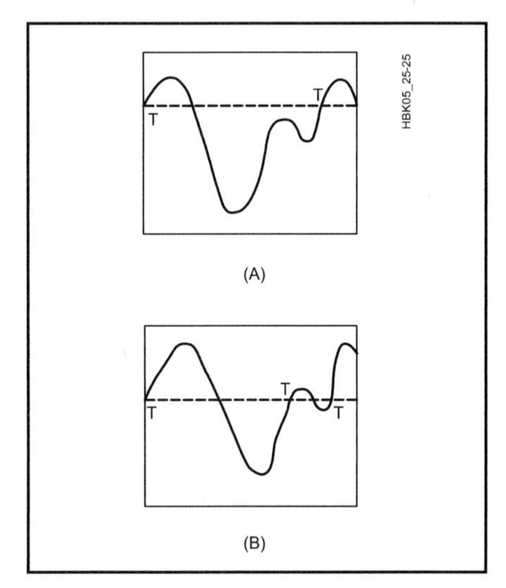

Fig 25.25 — In order to produce a stable display the selection of the trigger point is very important. Selecting the trigger point in A produces a stable display, but the trigger shown at B will produce a display that "jitters" from side to side.

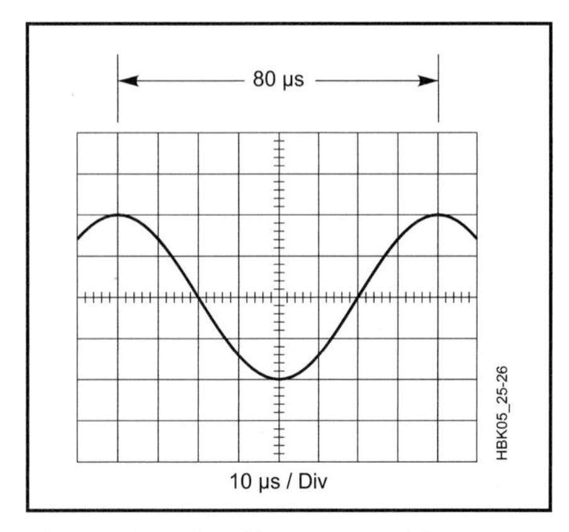

Fig 25.26 — Oscilloscopes with a calibrated sweep rate can be used to measure frequency. Here the waveform shown has a period of 80 microseconds (8 divisions × 10 µs per division) and therefore a period of 1/80 µs or 12.5 kHz.

the trigger select switch in the EXTERNAL position.

The trigger system controls the horizontal sweep. It looks at the trigger source (internal or external) to find out if it is positive- or negative-going and to see if the signal has passed a particular level. **Fig 25.25A** shows a typical signal and the dotted line on the figure represents the trigger level. It is important to note that once a trigger circuit is "fired" it cannot fire again until the sweep has moved all the way across the screen from left to right. In normal operation. the TRIGGER LEVEL control is manually adjusted until a stable display is seen. Some scopes have an AUTOMATIC position that chooses a level to lock the display in place without manual adjustment.

Fig 25.25B shows what happens when the level has not been properly selected. Because there are two points during a single cycle of the waveform that meet the triggering requirements, the trigger circuit will have a tendency to jump from one trigger point to another. This will make the waveform jitter from left to right. Adjustment of the TRIGGER control will fix this problem.

The horizontal travel of the trace is calibrated in units of time. If the time of one cycle is known, we can calculate the frequency of the waveform. In **Fig 25.26**, for example, if the SWEEP speed selector is set at 10 µs/division and we count the number of divisions (vertical bars) between peaks of the waveform (or any similar well defined points that occur once per cycle) we can find the period of one cycle. In this case it is 80 µs. This means that the frequency of the waveform is 12,500 Hz (1/80 µs). The accuracy of the measured frequency depends on the accuracy of the scope's sweep oscillator (usually approximately 5%) and the linearity of the ramp generator. This accuracy cannot compete with even the least-expensive frequency counter, but the scope can still be used to determine whether a circuit is functioning properly.

Dual-Trace Oscilloscopes

Dual-trace oscilloscopes can display two waveforms at once. This type of scope has two vertical input channels that can be displayed either alone, together or one after the other. **Fig 25.27** shows a simplified block diagram of a dual-trace oscilloscope. The only differences between this scope and the previous example are the additional vertical amplifier and the "channel switching circuit." This block determines whether we display channel A, channel B or both (simultaneously). The dual display is not a true dual display

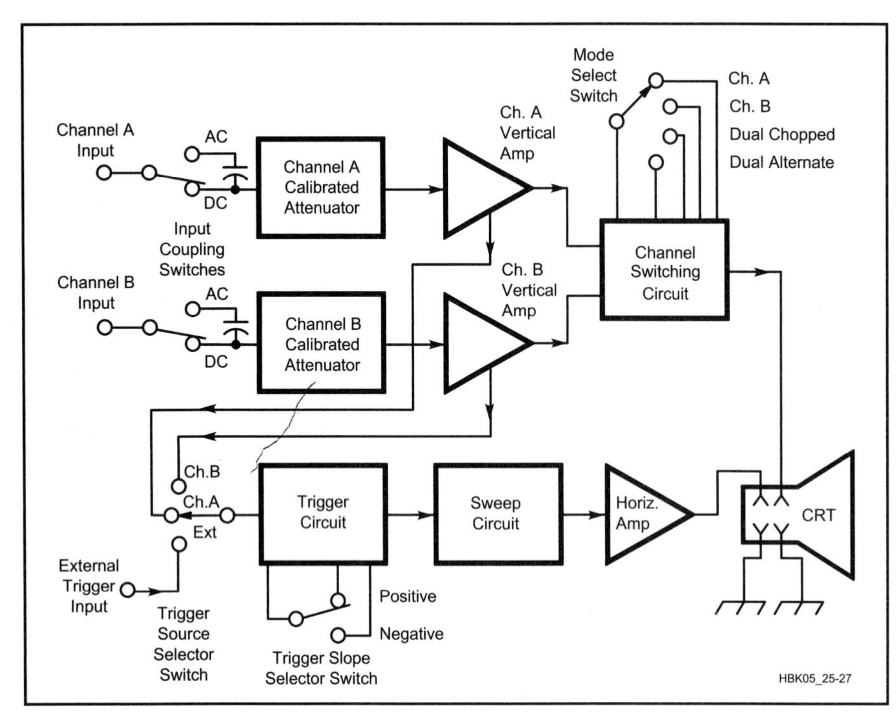

Fig 25.27 — Simplified Dual-trace oscilloscope block diagram. Note the two identical input channels and amplifiers.

(there is only one electron gun in the CRT) but the dual traces are synthesized in the scope.

There are two methods of synthesizing a dual-trace display from a single-beam scope. These two methods are referred to as *chopped mode* and *alternate mode*. In the chopped mode a small portion of the channel A waveform is written to the CRT, then a corresponding portion of the channel B waveform is written to the CRT. This procedure is continued until both waveforms are completely written on the CRT. The chopped mode is especially useful where an actual measure of the phase difference between the two waveforms is required. The chopped mode is usually most useful on slow sweep speeds (times greater than a few microseconds per division).

In the alternate mode, the complete channel A waveform is written to the CRT followed immediately by the complete channel B waveform. This happens so quickly that it appears that the waveforms are displayed at the same time. This mode of operation is not useful at very slow sweep speeds, but is good at most other sweep speeds.

Most dual-trace oscilloscopes also have a feature called "X-Y" operation. This feature allows one channel to drive the horizontal amplifier of the scope (called the X channel) while the other channel (called Y in this mode of operation) drives the vertical amplifier. Some oscilloscopes also have an external Y input. X-Y operation allows the scope to display Lissajous patterns for frequency and phase comparison and to use specialized test adapters such as curve tracers or spectrum analyzer front ends. Because of frequency limitations of most scope horizontal amplifiers the X channel is usually limited to a 5 or 10-MHz bandwidth.

DIGITAL OSCILLOSCOPES

The classic analog oscilloscope just discussed has existed for over 50 years. In the last 15 years, the digital oscilloscope has advanced from a specialized laboratory device to a very useful general-purpose tool, with a price attractive to an active experimenter. It uses digital circuitry and microprocessors to enhance the processing and display of signals. These result in dramatically improved accuracy for both amplitude and time measurements. When configured as a digital storage oscilloscope (DSO) it can read a stored waveform for as long as you wish without time limitations incurred by an analog type of storage scope.

Examine the simplified block diagram shown in **Fig 25.28**. After the signal goes through the vertical input attenuators and amplifiers, it arrives at the analog-to-digital converter (ADC). The ADC assigns a digital value to the level of the analog input signal and puts this in a memory similar to computer RAM. This value is stored with an assigned time, determined by the trigger circuits and the crystal timebase. The digital oscilloscope takes discrete amplitude samples at regular time intervals. If you were to take this data directly from memory, it would put a series

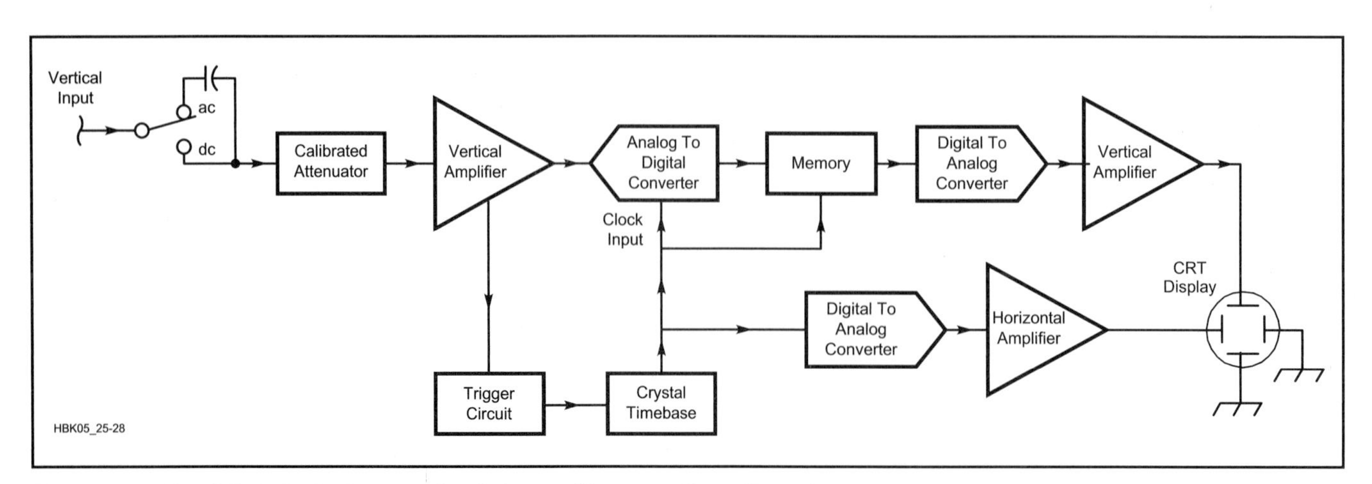

Fig 25.28 — Simplified block diagram of a digital oscilloscope. Note: The microprocessors are not shown for clarity. From "ABC's of Oscilloscopes," copyright Fluke Corporation (reproduced with permission).

of dots on the screen. You would then have to connect the dots to reconstruct the original waveform. The digital scope's microprocessor does this for you by mathematically processing the signal while reading it back from the memory and driving a digital-to-analog converter (DAC), which then drives the vertical deflection amplifier. A DAC also takes the digital stored time data and uses it to drive the horizontal deflection amplifier.

For the vertical signals you will see manufacturers refer to "8-bit digitizing," or perhaps "10-bit resolution." This is a measure of how many digital levels that are shown along the vertical (voltage) axis. More bits give you better resolution and accuracy of measurement. An 8-bit vertical resolution means each vertical screen has 2^8 (or 256) discrete values; similarly, 10 bits resolution yields 2^{10} (or 1024) discrete values.

It is important to understand some of the limitations resulting from sampling the signal rather than taking a continuous, analog measurement. When you try to reconstruct a signal from individual discrete samples, you must take samples at least twice as fast as the highest frequency signal being measured. If you digitize a 100-MHz sine wave, you should take samples at a rate of 200 million samples a second (referred to as 200 Megasamples/second). Actually, you really would like to take samples even more often, usually at a rate at least five times higher than the input signal.

If the sample rate is not high enough, very fast signal changes between sampling points will not appear on the display. For example, **Fig 25.29** shows one signal measured using both analog and digital scopes. The large spikes seen in the analog-scope display are not visible on the digital scope. The sampling frequency of the digital scope is not fast enough to store the higher frequency components of the waveform. If you take samples at a rate less than twice the input frequency, the reconstructed signal has a wrong apparent frequency; this is referred to as *aliasing*. In Fig 25.29 you can see that there is about one sample taken per cycle of the input waveform. This does not meet the 2:1 criteria established above. The result is that the scope reconstructs a waveform with a different apparent frequency.

Many older digital scopes had potential problems with aliasing. Newer scopes use advanced techniques to check themselves. A simple manual check for aliasing is to use the highest practical sweep speed (shortest time per division) and then to change to other sweep speeds to verify that the apparent frequency doesn't change.

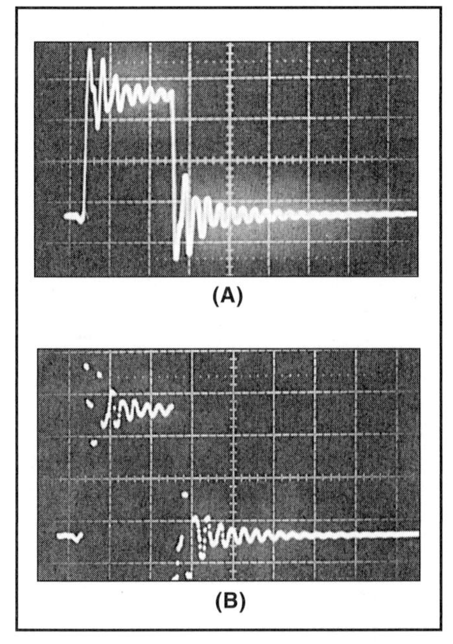

Fig 25.29 — Comparison of an analog scope waveform (A) and that produced by a digital oscilloscope (B). Notice that the digital samples in B are not continuous, which may leave the actual shape of the waveform in doubt for the fastest rise time displays the scope is capable of producing.

LIMITATIONS

Oscilloscopes have fundamental limits, primarily in frequency of operation and range of input voltages. For most purposes the voltage range of a scope can be expanded by the use of appropriate probes. The frequency response (also called the bandwidth) of a scope is usually the most important limiting factor. At the specified maximum response frequency, the response will be down 3 dB (0.707 voltage). For example, a 100-MHz 1-V sine wave fed into a 100-MHz bandwidth scope will read approximately 0.707 V on the scope display. The same scope at frequencies below 30 MHz (down to dc) should be accurate to about 5%.

A parameter called *rise time* is directly related to bandwidth. This term describes a scope's ability to accurately display voltages that rise very quickly. For example, a very sharp and square waveform may appear to take some time in order to reach a specified fraction of the input voltage level. The rise time is usually defined as the time required for the display to show a change from the 10% to 90% points of the input waveform, as shown in **Fig 25.30**. The mathematical definition of rise time is given by:

$$t_r = \frac{0.35}{BW} \quad (10)$$

where
t_r = rise time, µs
BW = bandwidth, MHz

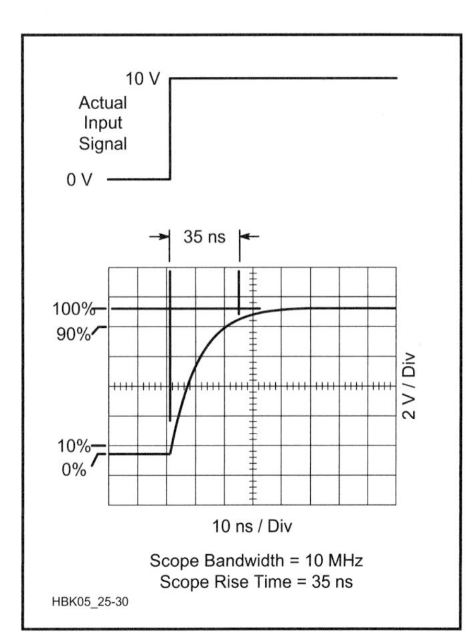

Fig 25.30 — The bandwidth of the oscilloscope vertical channel limits the rise time of the signals displayed on the scope.

It is also important to note that all but the most modern (and expensive) scopes are not designed for precise measurement of either time or frequency. At best, they will not have better than 5% accuracy in these applications. This does not change the usefulness of even a moderately priced oscilloscope, however. The most important value of an oscilloscope is that it presents an image of what is going on in a circuit and quickly shows which component or stage is at fault. It can show modulation levels, relative gain between stages and oscillator output.

Oscilloscope Probes

Oscilloscopes are usually connected to a circuit under test with a short length of shielded cable and a probe. At low frequencies, a piece of small-diameter coax cable and some sort of insulated test probe might do. Unfortunately, at higher frequencies the capacitance of the cable would produce a capacitive reactance much less than the one-megohm input impedance of the oscilloscope. In addition each scope has a certain built-in capacitance at its input terminals (usually between 5 and 35 pF). These two capacitances cause problems when probing an RF circuit with a relatively high impedance.

The simplest method of connecting a signal to a scope is to use a specially de-

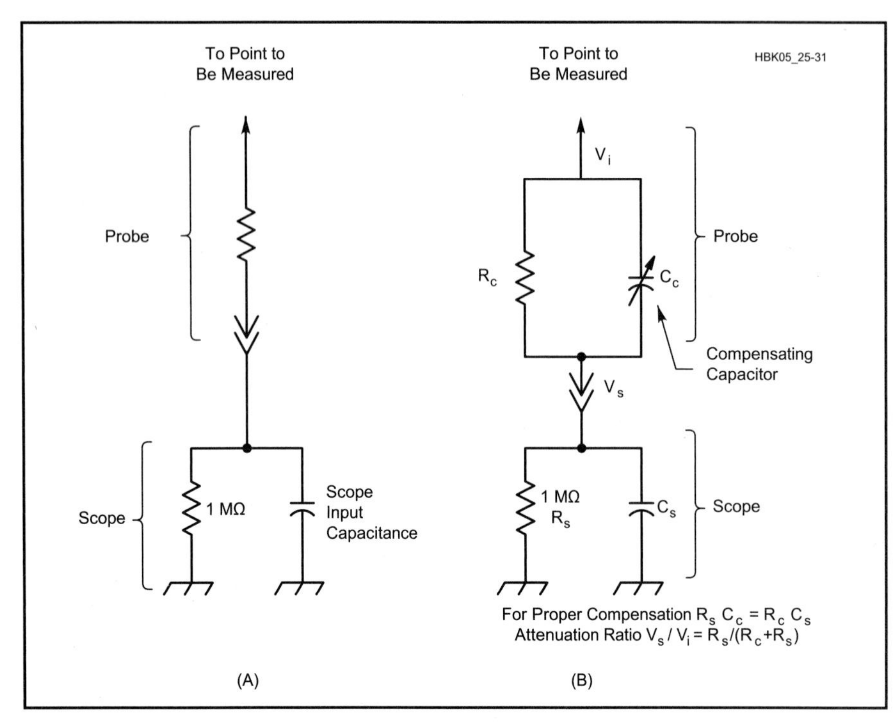

Fig 25.31 — Uncompensated probes such as the one at A are sufficient for low-frequency and slow-rise-time measurements. However, for accurate display of fast rise times with high-frequency components the compensated probe at B must be used. The variable capacitor is adjusted for proper compensation (see text for details).

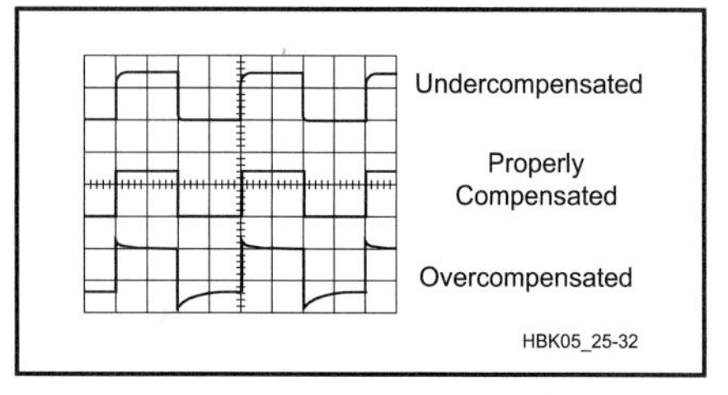

Fig 25.32 — Displays of a square-wave input illustrating undercompensated, properly compensated and overcompensated probes.

signed probe. The most common scope probe is a ×*10 probe* (called a times ten probe). This probe forms a 10:1 voltage divider using the built-in resistance of the probe and the input resistance of the scope. When using a ×10 probe, all voltage readings must be multiplied by 10. For example, if the scope is on the 1 V/division range and a ×10 probe was in use, the signals would be displayed on the scope face at 10 V/division.

Unfortunately a resistor alone in series with the scope input seriously degrades the scope's rise-time performance and therefore its bandwidth. Since the scope input looks like a parallel RC circuit, the series resistor feeding it causes a significant reduction in available charging current from the source. This may be corrected by using a compensating capacitor in parallel with the series resistor. Thus two dividers are formed: one resistive voltage divider and one capacitive voltage divider. With these two dividers connected in parallel and the RC relationships shown in **Fig 25.31**, the probe and scope should have a flat response curve through the whole bandwidth of the scope.

To account for manufacturing tolerances in the scope and probe the compensating capacitor is made variable. Most scopes provide a "calibrator" output that produces a known-frequency square wave for the purpose of adjusting the compensating capacitor in a probe. **Fig 25.32** shows possible responses when the probe is connected to the oscilloscope's calibrator jack.

If a probe cable is too short, do not attempt to extend the length of the cable by adding a piece of common coaxial cable. The cable usually used for probes is much different than common 50 or 75-Ω coax. In addition the compensating capacitor in the probe is chosen to compensate for the provided length of cable. It usually will not have enough range to compensate for extra lengths.

The shortest ground lead possible should be used from the probe to the circuit ground. Long ground leads are inductors at high frequencies. In these circuits they cause ringing and other undesirable effects.

THE MODERN SCOPE

For many years a scope (even a so-called portable) was big and heavy. Computers and modern ICs have reduced the size and weight. Modern scopes can take other forms than the traditional large cabinet with built-in CRT. Some modern scopes use an LCD display for true portability. Some scopes take the form of a card plugged into a PC, where they use the PC and monitor for display. Even if they don't use a PC for a display, many scopes can attach to a PC and download their data for storage and analysis using advanced mathematical techniques. Many high-end scopes now incorporate nontraditional functions, such as Fast Fourier Transforms (FFT). This allows limited spectrum analysis or other advanced mathematical techniques to be applied to the displayed waveform.

BUYING A USED SCOPE

Many hams will end up buying a used scope due to price. If you buy a scope and intend to service it yourself, be aware all scopes that use tubes or a CRT contain lethal voltages. Treat an oscilloscope with the same care you would use with a tube-type high-power amplifier. The CRT should be handled carefully because if dropped it will crack and implode, resulting in pieces of glass and other materials being sprayed everywhere in the immediate vicinity. You should wear a full-face safety shield and other appropriate safety equipment to protect yourself.

Another concern when servicing an older scope is the availability of parts. Many scopes made since about 1985 have used special ICs, LCDs and microprocessors. Some of these may not be available or may be prohibitive in cost. You should buy a used scope from a reputable vendor— even better yet, try it out before you buy it. Make sure you get the operators manual also.

AN HF ADAPTER FOR NARROW-BANDWIDTH OSCILLOSCOPES

Fig 25.33 shows the circuit of a simple piece of test equipment that will allow you to display signals that are beyond the normal bandwidth of an inexpensive oscilloscope. This circuit was built to monitor modulation of a 10-m signal on a scope that has a 5-MHz upper-frequency limit. This design features a Mini-Circuits Laboratory SRA-1 mixer. Any stable oscillator or VFO with an output of 10 dBm can be used for the local oscillator (LO), which mixes with the HF signal to produce an IF in the bandwidth of the oscilloscope.

The mixer can handle RF signal levels up to –3 dBm without clipping, so this was set as an upper limit for the RF input. A toroidal transformer coupler is constructed by winding a 31-turn secondary of #28-AWG wire on an FT-37-75 core for RG-174 or an FT-50-75 core for use with slightly larger coax such as RG-58. The primary is the piece of coaxial cable passed through the core center. The coupler gives 30 dB of attenuation and has a flat response from 0.5 to 100 MHz. An additional 20-dB of attenuation was added for a total of 50 dB before the mixer. One-watt resistors will do fine for the attenuator. The completed adapter should be built into a shielded box.

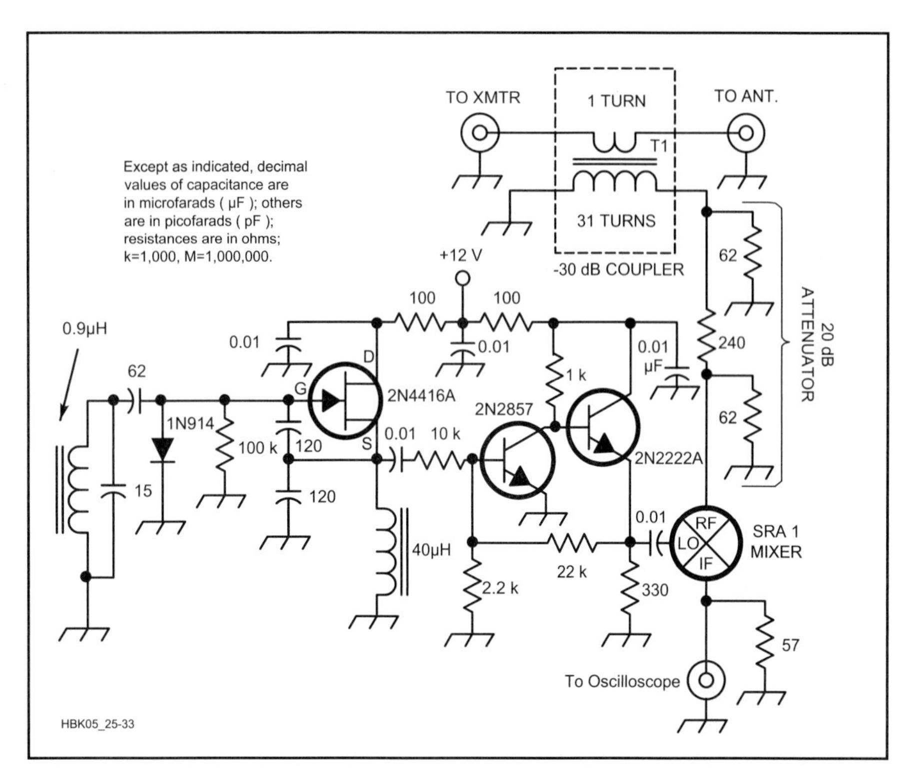

Fig 25.33 — This adapter displays HF signals on a narrow-bandwidth oscilloscope. It uses a 10-dBm 25-MHz LO, –30-dB coupler, 20-dB attenuator and diode-ring mixer. See text for further information.

This circuit, with a 25-MHz LO frequency, is useful on frequencies in the 20 to 30-MHz range with transmitters of up to 50-W power output. By changing the frequency of the LO, any frequency in the range of the coupler can be displayed on a 5-MHz-bandwidth oscilloscope. The frequency displayed will be the difference between the LO and the input signal. As an example, a 28.1-MHz input and a 25-MHz LO will be seen as a 3.1-MHz signal on the oscilloscope.

More attenuation will be required for higher-power transmitters. This circuit was described by Kenneth Stringham Jr, AE1X, in the Hints and Kinks column of February 1982 *QST*.

A CALIBRATED NOISE SOURCE

NOISE FIGURE MEASUREMENT

One of the most important measurements in communications is the noise figure of a receiving setup. Relative measurements are often easy, while accurate ones are more difficult and expensive. One EME (moon bounce) station checks noise and system performance by measuring the noise of the sun reflected off the moon. While the measurement source (use of the sun and moon) is not expensive, the measuring equipment on 2 m consists of 48 antennas (each over 30 ft long). This measurement equipment is not for everyone!

The rest of us use more conventional noise sources and measuring techniques. Coverage of noise figure and its measurement appear in the **Receivers and Transmitters** chapter of this *Handbook.*

Most calibrated and stable noise sources are expensive, but not this unit developed by Bill Sabin, WØIYH. It first appeared in May 1994 *QST*. When hams use a noise source, it is usually included in an RF bridge used to measure impedances and adjust antenna tuners. A somewhat different device (an *accurately calibrated and stable* noise source) is also useful. Combining a broadband RF noise source of known power output and a known output impedance with a true-RMS voltmeter, results in an excellent instrument for making interesting and revealing measurements on a variety of circuits hams commonly use. (Later on, some examples will be described.) The true-RMS voltmeter can be an RF voltmeter, a spectrum analyzer or an AF voltmeter at the output of a linear receiver.[1]

Calibrated noise generators and noise-figure meters are available at medium to astronomical prices. Here is a low-cost approach which can be used with reasonable confidence for many amateur applications where accuracy to tenths of a decibel is not needed, but where precision (repeatability) and comparative measurements are much more important. PC boards are available for this project.[2]

Semiconductor Noise Diodes

Any Zener diode can be used as a source of noise. If, however, the source is to be calibrated and used for reliable mea-

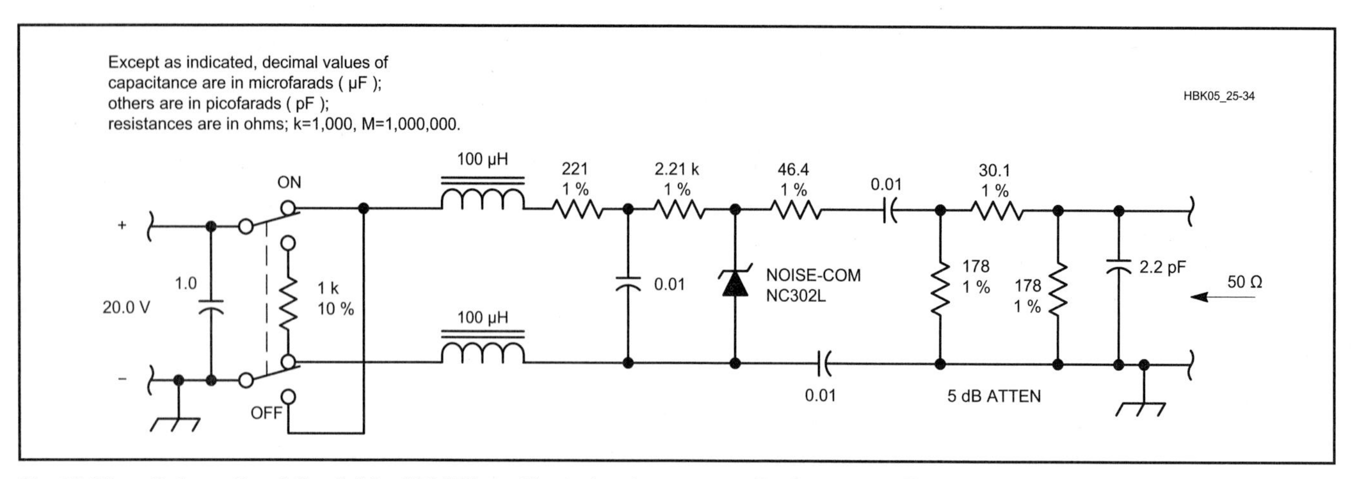

Fig 25.34 — Schematic of the 0.5 to 500-MHz calibrated noise source. Resistors are 1/8-W, 1%-tolerance metal-film units. 1% resistors are available from Digi-Key.

surements, avalanche diodes specially designed for this purpose are preferable by far.[3] A good noise diode generates its noise through a carefully controlled *bulk avalanche* mechanism which exists *throughout* the PN junction, not merely at the junction surfaces where unstable and unreliable surface effects predominate due to local breakdown and impurity.[4] A true noise diode has a very low *flicker noise* (1/f) effect and tends to create a uniform level of truly *Gaussian noise* over a wide band of frequencies.[5] In order to maximize its bandwidth, the diode also has very low junction capacitance and lead inductance.

This project uses the NOISE/COM NC302L diode. It consists of a glass, axial-lead DO-35 package and is rated for use from 10 Hz to 3 GHz, if appropriate construction methods are followed. Prior to sale, the diodes are factory aged for 168 hours and are well stabilized. NOISE/COM has kindly agreed to make these diodes available to amateur experimenters for the special price of $10 each; the usual low-quantity price is about $25.[6]

Noise Source Design

The noise source presents two kinds of available output power. One is the thermal noise (–174 dBm/Hz at room temperature) when the diode is turned off. This is called N_{OFF}. The other is the sum of this same thermal noise and an "excess" noise, N_E, which is created by the diode when turned on, called N_{ON} (equivalent to $N_{OFF} + N_E$). For accurate measurements, the output impedance of the test apparatus must be the same (on or off) so that the device under test (DUT) always sees the same generator impedance. In Amateur Radio work, this impedance is usually 50 Ω, resistive. The circuit design must guarantee this condition.

For maximum frequency coverage, a PC-board layout and coax connector suitable for use at microwaves are needed. For lower frequency usage, a less stringent approach can be employed. Two noise sources are presented here. One is for the 0.5 to 500-MHz region and uses conventional components that many amateurs already have. The other is for the 1-MHz to 2.5-GHz range; it uses chip components and an SMA connector.

Circuit Diagram and Construction

Figs 25.34 and **25.35** show the simple schematics of the two noise sources. In series with the diode is a 46.4-Ω resistor that combines with the dynamic resistance of the diode in the avalanche, noise generator mode (about 4 Ω) to total about 50 Ω. When the applied voltage polarity is reversed, the diode is forward conducting and its dynamic resistance is still about 4 Ω, but the avalanche noise is now turned off. As a result, the noise source output impedance

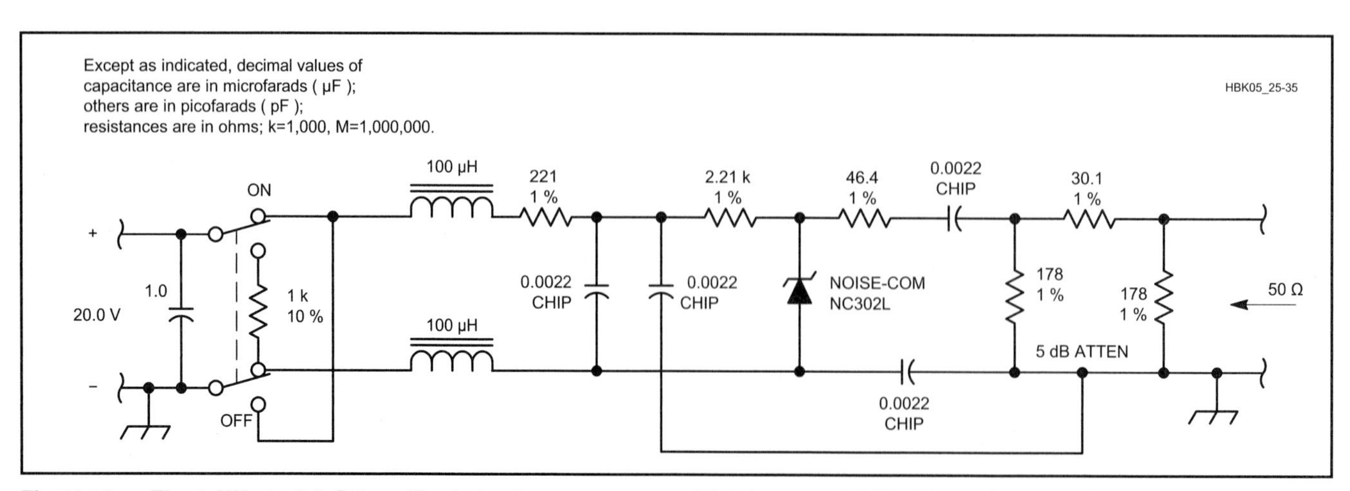

Fig 25.35 — The 1-MHz to 2.5-GHz calibrated noise source uses 1%-tolerance, 0.1-W chip resistors and chip capacitors. 1% resistors are available from Digi-Key.

is always about 50 Ω. The 5-dB pad reduces the effect of any small impedance differences, so that the output impedance is nearly constant from the *on* to the *off* condition, and the SWR is less than 2:1.

Consider the noise situation of the noise diode when it is forward conducting. The resistance of the forward biased PN junction is a *dynamic* resistance. This dynamic resistance is *not* a source of thermal noise, since it is not an actual physical resistance such as in a resistor or lossy network. However, the 0.6-V forward drop across the PN junction does produce a shot noise effect. The mathematics of this shot noise shows that the noise power associated with this effect is only about 50% of the thermal noise power that would be available from a physical resistor having the same value as the dynamic resistance. Therefore, the forward biased junction does *not* add excess noise to the system.[7] There is an 1/f noise effect associated with this shot noise in the diode, but its corner frequency is at about 100 kHz and of no importance at higher frequencies. Also, the small amount of bulk resistance contributes a little thermal noise.

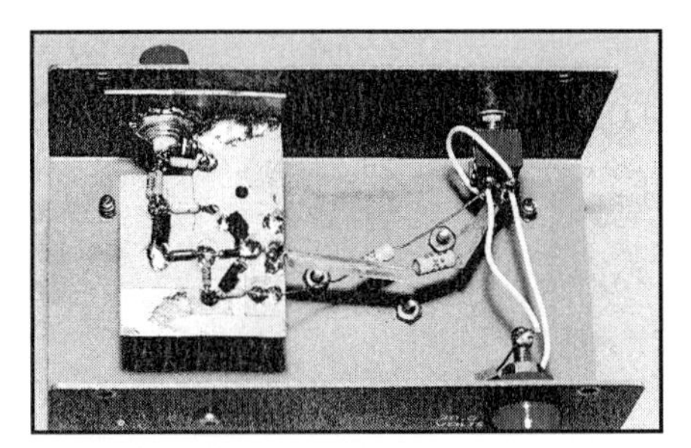

Fig 25.36 — An inside view of the 0.5 to 500-MHz noise source.

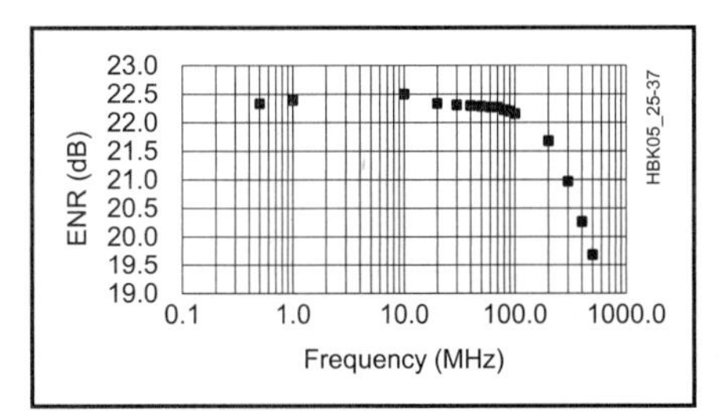

Fig 25.37 — Sample calibration chart for the 0.5 to 500-MHz noise source.

Fig 25.38 — A view inside of the 1-MHz to 2.5-GHz noise source.

In order to maximize the unit's flatness and frequency response bandwidth, noise-source construction methods should aim for RF circuit lead lengths as close to zero as possible as well as minimum inductance in the ground path and the coupling capacitors. The power-supply voltage must be clean, well bypassed and set accurately. **Fig 25.36** shows a 0.5 to 500-MHz unit. This construction method satisfies quite well the electrical requirements wanted for this model. At 500 MHz, the return loss with respect to 50 Ω at the output jack decreased to 10 dB. A calibration chart (**Fig 25.37**) is attached to the unit's top for easy reference. **Fig 25.38** shows the inside of the 1-MHz to 2.5-GHz noise source.

Calibrating the Noise Source

If the construction is solid, the calibration should last for a long time. There are two ways to calibrate the noise source. If the unit *has been carefully constructed and its correct operation verified,* NOISE/COM will calibrate home-built units over the desired frequency range for $25 plus return shipping charges. Note that one factory calibrated unit can be used as a reference for many home calibrated units. **Fig 25.39** shows the NOISE/COM calibration data for both models of prototype noise sources, including SWR data. The noise data is strictly valid only at room temperature, so it's necessary to avoid extreme temperature environments.

The second calibration method requires a signal generator with known output levels at the various desired calibration frequencies. One approach is to build a tunable weak-signal oscillator that can be compared to some accessible high-quality signal generator, using a sensitive receiver as a detector.[8] The level of the signal source in dBm is needed.

Access to a multistage attenuator is also desirable. Build the attenuator using the nearest 1% values of metal-film resistors, so that systematic errors are minimized. A total attenuation of 25 dB in 0.1-dB steps is desirable. Attenuator construction must be appropriate for use at the intended frequency range. In some cases, a high-frequency correction chart may be needed.

With the calibrated signal source and the attenuator feeding the receiver in an SSB or CW mode the techniques discussed in the reference of Note 1 should be used to determine the excess noise (N_E) of the noise source and the noise bandwidth (B_N) of the receiver.

Excess Noise Ratio

A few words about excess noise ratio (ENR) are needed. It is defined as the ratio of excess noise to thermal noise. That is,

$$ENR = \frac{N_{ON} - N_{OFF}}{N_{OFF}} = \frac{N_E}{N_{OFF}} \qquad (12)$$

When the noise source is turned on, its

Frequency (MHz)	*0.5 to 500 MHz Unit*		*1.0 to 2500 MHz Unit*	
	ENR (dB)	*SWR*	*NR (dB)*	*SWR*
0.5	22.33	1.03		
1	22.38	1.03	21.38	1.03
10	22.45	1.04	21.46	1.03
20	22.35	1.06		
30	22.32	1.06		
40	22.32	1.09		
50	22.30	1.11		
60	22.29	1.12		
70	22.25	1.15		
80	22.22	1.17		
90	22.20	1.20		
100	22.15	1.23	21.80	1.07
200	21.65	1.42		
300	20.96	1.62		
400	20.25	1.70		
500	19.60	1.90	20.71	1.44
1000			20.12	1.86
1500			20.00	2.06
2000			20.70	2.14
2500			21.51	1.88

Fig 25.39 — NOISE/COM calibration data for both prototype noise sources. The data is not universal; it varies from unit to unit.

output is $N_{OFF} + N_E$. The ratio of N_{ON} to N_{OFF} is then

$$\frac{N_{ON}}{N_{OFF}} = \frac{N_{OFF} + N_E}{N_{OFF}} = 1 + \frac{N_E}{N_{OFF}} = 1 + ENR \quad (13)$$

Therefore, ENR is a measure of how much the noise increases and the noise generator can be calibrated in terms of its ENR.

Normalizing ENR to a 1-Hz bandwidth and converting to decibels, this is

$$ENR\,(dB) = 174\,(dBm/Hz) + \frac{N_E\,(dBm)}{B_N\,(Hz)} \quad (14)$$

Prepare a calibration chart and attach it to the top of the unit (see Fig 25.37). If the unit is to be factory calibrated, first perform the calibration procedure to ensure everything is working properly. Remember, a factory calibrated unit can be used as a reference for other home calibrated units, once the calibration-transfer procedures have been worked out. This requires some careful thinking and proper techniques. Generally speaking, a NOISE/COM calibration is the best choice.

Noise-Figure Measurement

The thermal noise power available from the attenuator remains constant for any value of attenuator setting. But the excess noise and therefore the ENR (in dB) due to the noise diode is equal to the calibration point of the source minus the setting (in dB) of the attenuator.

The noise-figure measurement of a device under test (DUT) uses the Y method and the setup in **Fig 25.40**. If the DUT has a noise-generator input and a true-RMS noise-measuring instrument at the output, then the total output noise (including the contribution of the measuring instrument) with the noise generator turned off is

$$N_{OFF\,(TOT)} = kTB_N\, F_{TOT}\, G_{DUT}\, G_{NMI} \quad (15)$$

where

kTB_N = thermal noise,

G_{DUT} = gain of the DUT,

G_{NMI} = gain of the noise-measuring instrument, and

F_{TOT} = noise factor of the combination of the DUT and the noise-measuring instrument.

When the noise generator is turned on, the output noise is

$$N_{ON\,(TOT)} = kTB_N\, F_{TOT}\, G_{DUT}\, G_{NMI} + (ENR)kTB_N\, G_{DUT}\, G_{NMI} \quad (16)$$

Where the last term is the contribution of excess noise by the noise generator. Note that none of these values is in dB or dBm.

If we divide equation 16 by equation 15 and say that the ratio

$$\frac{N_{ON\,(TOT)}}{N_{OFF\,(TOT)}} = Y \quad (17)$$

then,

$$\frac{F_{TOT} + ENR}{F_{TOT}} = 1 + \frac{ENR}{F_{TOT}}$$

Note that kTB_N, G_{DUT} and G_{NMI} disappear, so that these quantities need not be known to measure noise factor. If we solve equation 17 for F_{TOT}, we get the noise factor

$$F_{TOT} = \frac{ENR}{Y - 1} \quad (18)$$

If the noise output doubles (increases by 3 dB) when we turn on the noise source, then Y = 2 and the noise factor is numerically equal to the excess noise ratio (ENR). If the attenuator steps are not fine enough or if the attenuator is not reliable over the entire frequency range, use equation 18 to get a better answer. (It's much simpler to use a good fine-step attenuator.) The value of F_{TOT} is that of the DUT in cascade with the noise-measuring instrument. To find F_{DUT}, we must know the noise factor F_{NMI} of the noise-measuring instrument and G_{DUT} and then use the Friis formula, unless G_{DUT} is very large (as it would be if the DUT were a high-gain receiver (see Note 1).

$$F_{DUT} = F_{TOT} - \frac{F_{NMI} - 1}{G_{DUT}} \quad (19)$$

The validity of equation 19 (if we need to use it) requires that the noise bandwidth of the noise-measuring instrument be less than the noise bandwidth of the DUT (see the reference of Note 1). Verify this before proceeding.

There's another advantage to using the power-doubling method. If the 3-dB attenuator of Fig 25.40 is used to maintain a constant noise level into the following stages and the RMS meter, this means that the noise factor, using the calibration scale and the input attenuator (without using equation 18), is

$$F_{DUT} = ENR + \frac{1}{G_{DUT}} \quad (20)$$

If G_{DUT} is large, then the last term can be neglected. If G_{DUT} is small, we need to know its value. However, we do not need to know the noise factor F_{NMI} of the circuitry after the DUT, as we did in the previous discussion.

The 3-dB attenuator method also removes all restrictions regarding the type of noise measuring instrument, since the meter reading is now used only as a reference point. This last statement applies only when two noise (or two signal generator) inputs are being compared.

Frequency Response Measurements

The noise generator, in conjunction with a spectrum analyzer, is an excellent tool for measuring the frequency response of a DUT, if the noise source is much stronger than the internal noise of the DUT and that of the spectrum analyzer. Many spectrum analyzers are not equipped with tracking generators, which can be quite expensive for an amateur's budget.

The spectrum analyzer needs to be calibrated for a noise input, if accurate amplitude measurements are needed, because it responds differently to noise signals than to sine-wave signals. The envelope detection of noise, combined with the logarithmic amplification of the spectrum analyzer, creates an error of about 2.5 dB for a noise signal (the noise is that much greater than the instrument indicates). Also, the noise bandwidth of the IF filter is different from its resolution bandwidth. Some modern spectrum analyzers have internal DSP algorithms that make the corrections so that external noise sources and

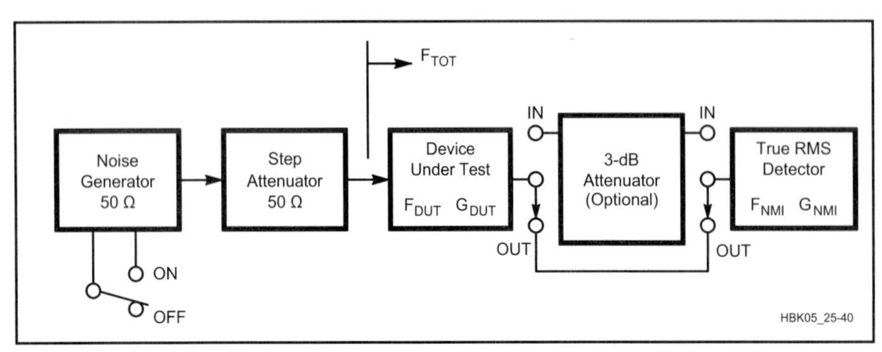

Fig 25.40 — Setup for measuring noise figure of a device under test (DUT).

also carrier-to-noise ratios, normalized to some noise bandwidth like 1.0 Hz, can be measured with fair accuracy if the input noise is a few decibels above thermal. One example is the Tektronix Model 2712. If only relative response readings are needed, then these corrections are not needed.

Also, the noise source itself can be used to establish an accurate reference level (in dBm) on the screen. An accurate, absolute measurement with the DUT in place will then be this reference level (in dBm), plus the increment in decibels produced by the DUT.

The noise-generator output can be viewed as a collection of sine waves separated by, say, 1 Hz. Each separated f requency "bin" has its own Gaussian amplitude and random phase with respect to all the others. So the DUT is simultaneously looking at a collection or "ensemble," of input signals. As the spectrum analyzer frequency sweeps, it looks simultaneously at all of the DUT frequencies that fall within the spectrum analyzer's IF noise bandwidth. The spectrum display is thus the "convolution" of the IF filter frequency response and the DUT frequency response. If the DUT is a narrow filter, a very narrow resolution and a slow sweep are needed in the spectrum analyzer. In addition, the analyzer's video, or post-detection, filter has a narrow bandwidth and also requires some settling time to get an accurate reading. So, some experience and judgment are required to use a spectrum analyzer this way.

Using Your Station Receiver

Your station receiver can also be used as a spectrum analyzer. Place a variable attenuator between the DUT and the receiver. As you tune your receiver, in a narrow CW mode, adjust the attenuator for a constant reference level receiver output. The attenuator values are inversely related to the frequency response.

A calibrated noise source with an adjustable attenuator that can be easily switched into a receiver antenna jack is an excellent tool for measuring antenna noise level or incoming weak signal level (in dBm) or for establishing correct receiver operation.

The noise source can also be combined with a locally generated data-mode waveform of a known dBm value to get an approximate check on modem performance or to make adjustments that might assure correct operation of the system. The rigorous evaluation of system performance requires special equipment and techniques that may be unavailable at most amateur stations. Or, you could evaluate the intelligibility improvement of your SSB transmitter's speech processor in a noise background.

Summary

The calibrated, flat-spectrum noise generator described in this article is quite a useful instrument for amateur experimenters. Its simplicity and low cost make it especially attractive. Getting a good calibration is the main challenge, but once it is achieved, the calibration lasts a long time, if the right diode is used. The ENR of the units described here is in the range of 20 dB. Use of a high-quality, external, 10-dB attenuator barrel will get into the range of 10-dB ENR. If the unit is sent to NOISE/COM the attenuator should also be sent, with the request that it be included in the calibration. That attenuator then "belongs" to the noise source and should be so tagged. If the attenuator is of high quality, the output SWR will also be improved. NOISE/COM suggests periodic recalibration, at your discretion.

Notes

[1]W. Sabin, "Measuring SSB/CW Receiver Sensitivity," *QST*, Oct 1992, pp 30-34. See also Technical Correspondence, *QST*, Apr 1993, pp 73-75.

[2]PC boards are available from FAR Circuits, Dundee, IL. A PC board template for the Sabin noise source is available at: **www.arrl.org/notes**.

[3]The term Zener diode is commonly used to denote a diode that takes advantage of avalanche effect, even though the Zener effect and the avalanche effect are not exactly the same thing at the device-physics level.

[4]The term bulk avalanche refers to the avalanche multiplication effect in a PN junction. A carrier (electron or hole) with sufficient energy collides with atoms and causes more carriers to be knocked loose. This effect "avalanches" and it occurs throughout the volume of the PN junction. This mechanism is responsible for the high-quality noise generation in a true noise diode.

[5]Gaussian noise refers to the instantaneous values of a noise voltage. These values conform to the Gaussian probability density function of statistics.

[6]NOISE/COM Co, Paramus, NJ.

[7]Motchenbacher and Fitchen, Low Noise Electronic Design (New York: Wiley & Sons, 1973), p 22.

[8]W. Hayward and D. DeMaw, *Solid State Design for the Radio Amateur* (Newington: ARRL, 1986).

A SIGNAL GENERATOR FOR RECEIVER TESTING

The oscillator shown in **Fig 25.41** and **Fig 25.42** was designed for testing high-performance receivers. Parts cost for the oscillator has been kept to a minimum by careful design. While the stability is slightly less than that of a well-designed crystal oscillator, the stability of the unit should be good enough to measure most amateur receivers. In addition, the ability to shift frequency is important when dealing with receivers that have spurious responses. More importantly, LC oscillators with high-Q components often have much better phase noise performance than crystal oscillators, because of power limitations in the crystal oscillators (crystals are easily damaged by excessive power).

The circuit is a Hartley oscillator followed by a class-A buffer amplifier. A 5-V regulator is used to keep the power supply output stable. The amplifier is cleaned up by a seven-element Chebyshev low-pass filter, which is terminated by a 6-dB attenuator. The attenuator keeps the filter working properly, even with a receiver that has an input impedance other than 50 Ω. A receiver designed to work with a 50-Ω system may not have a 50-Ω input impedance. The +4 dBm output is strong enough for most receiver measurements. It may even be too strong for some receivers. Note that sensitive components like crystal filters may require a step attenuator to lower the output level.

Construction

This unit is built in a box made of double-sided circuit board. Its inside dimensions are 1 × 2.2 × 5 inches (HWD). The copper

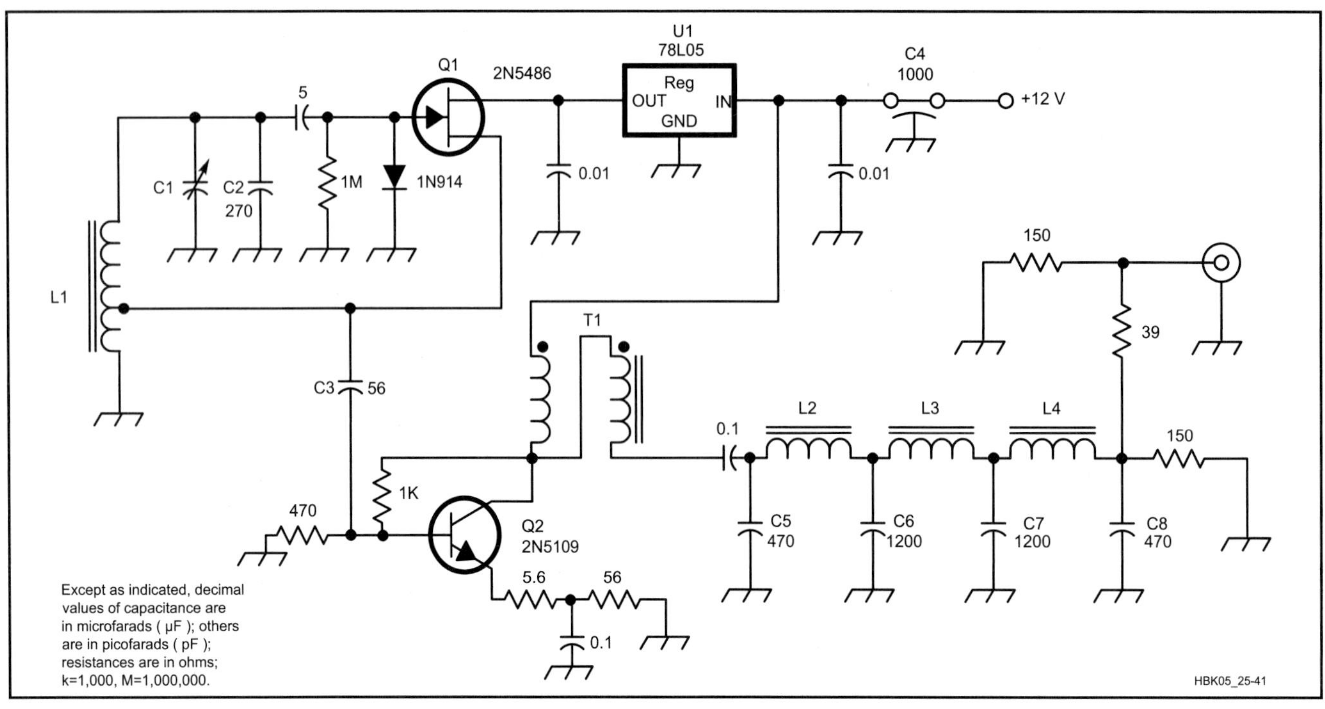

Fig 25.41 — Schematic diagram of the LC oscillator operating at 3.7 MHz. All resistors are 1/4 W, 5% units.

C1 — 1.4 to 9.2-pF air trimmer (value and type not critical).
C2 — 270-pF silver-mica or NPO capacitor. Value may be changed slightly to compensate for variations in L1.
C3 — 56-pF silver mica or NPO capacitor. Value may be changed to adjust output power.
C4 — 1000 pF solder-in feedthrough capacitor. Available from Microwave Components of Michigan.
C5-C8 — Silver-mica, NPO disc or polystyrene capacitor.
D1 — 1N914, 1N4148.
L1 — 31t #18 enameled wire on T-94-6 core. Tap 8 turns from ground end (7.5 µH).
L2, L4 — 21t #22 enameled wire on a T-50-2 core (2.5 µH, 2.43 µH ideal).
L3 — 23t #22 enameled wire on a T-50-2 core. (2.9 µH, 3.01 µH ideal).
T1 — 7t #22 enameled wire bifilar wound on an FT-37-43 core.
Q1 — 2N5486 JFET. MPF102 may give reduced output.
Q2 — 2N5109.
U1 — 78LO5 low-current 5-V regulator.

Return Loss Bridges

Return loss is a measure of how closely one impedance matches a reference impedance in phase angle and magnitude. If the reference impedance equals the measured impedance level with a 0° phase difference it has a return loss of infinity. **Fig A** shows basic return-loss measurement setups. Return-loss bridges are good for measuring filter response because return loss measurements are a more sensitive measure of passband response than insertion-loss measurements.

A 100 Hz to 100-kHz Return-Loss Bridge

Ed Wetherhold, W3NQN, has developed a low-frequency return-loss bridge (RLB) that can be adapted to different impedance levels. (See **Fig B**.) This bridge is used primarily for testing passive-LC filters that have been designed to work at a certain impedance level. Return-loss measurements require that the signal generator and RLB match the specific filter impedance level.

The characteristic impedance of this RLB is set by the values of four resistors (R1 = R2 = R3 = R4 = characteristic impedance). Ed mounted the four resistors on a plug-

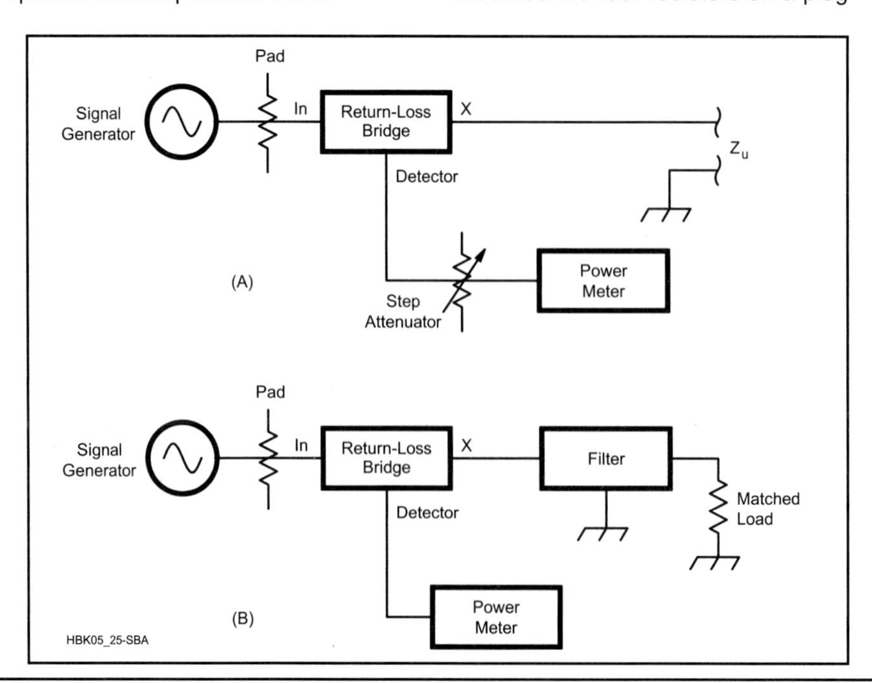

Fig A — A shows a setup to measure an unknown impedance with a return loss bridge. B is for measuring filter response.

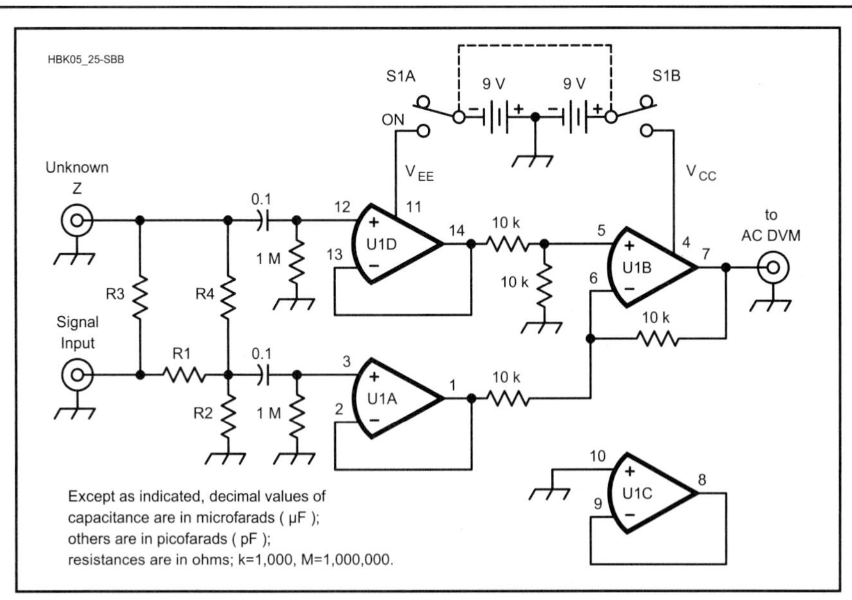

Fig B — An active low-frequency RLB. U1 is an LM324 quad op amp. R1, R2, R3 and R4 are 1/4-W, 1% resistors with the same value as the output impedance of the signal generator and the input impedance of the test circuit.

Table A
Performance and Test Data for the 100-Hz to 100-kHz RLB

Bridge and signal-generator impedance = 500 Ω

Bridge Directivity

Frequency	*Return Loss (dB, Unknown = 500 Ω*
100 Hz to 10 kHz	> 45
10 kHz to 80 kHz	40
80 kHz to 100 kHz	30

Return Loss of Known Resistive Loads

$R_{LOAD} = LF \times Z_{BRIDGE}$

where
R_{LOAD} = Load resistance, ohms
LF = Load factor
Z_{BRIDGE} = characteristic bridge impedance, ohms.

LF	*Return Loss (dB)*
5.848	3
3.009	6
1.925	10
1.222	20
1.065	30

in module and placed their interconnections on the socket. Additional modules may be built for any impedance.

Table A provides computed values of return loss for several known loads and a 500-Ω RLB. If you build this RLB, use the values in the table to check its operation. For this frequency range, use an ac voltmeter in place of the power meter shown in Fig A. Choose one with good ac response well above 100 kHz.

Ed originally described this bridge circuit in an article published in 1993. That article provides complete construction details. Reprints are available for $3 from Ed Wetherhold (W3NQN), 1426 Catlyn Place, Annapolis, MD 21401-4208. The RLB shown is useful from 100 Hz to 100 kHz.

An RF Return-Loss Bridge

At HF and higher frequencies, return-loss bridges are used as shown in Fig A for making measurements in RF circuits. The schematic of a simple bridge is shown in **Fig C**. (Notice that the circuit is identical to that of a hybrid combiner.) It is built in a small box with short leads to the coax connectors. Either 49.9-Ω 1% metal-film or 51-Ω 1/4-W carbon resistors may be used. The transformer is wound with 10 bifilar turns of #30 enameled wire on a high permeability ferrite core such as an FT-23-43 or similar.

Apply the output of the signal generator to the RF INPUT port of the RLB. It may be necessary to attenuate the generator output to avoid overloading the amplifier under test. Connect the bridge DETECTOR port to a power meter through a step attenuator and leave the UNKNOWN port of the bridge open circuited. Set the step attenuator for a relatively high level of attenuation and note the power meter indication.

Now connect the unknown impedance, Z_u, to the bridge. The power meter reading will decrease. Adjust the step attenuator to produce the same reading obtained when the UNKNOWN port was open circuited. The difference between the two settings of the attenuator is the return loss, measured in dB.

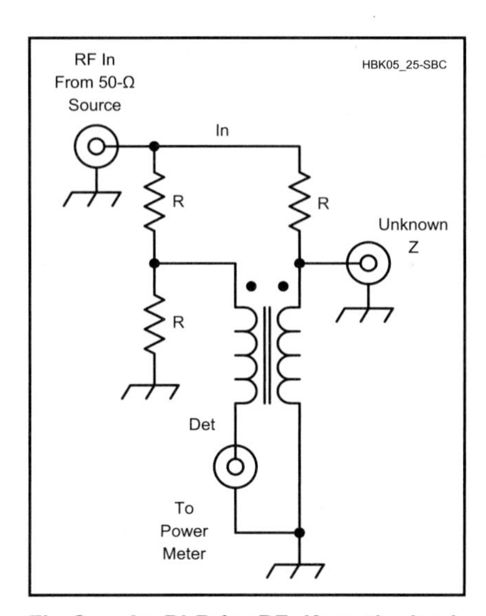

Fig C — An RLB for RF. Keep the lead lengths short. Wind the transformer on a high-permeability ferrite core. Use either 51-Ω carbon or 49.9-Ω 1% metal-film resistors.

The unknown impedance measured by this technique is not limited to amplifier inputs. Coax cable attached to an antenna, a filter, or any other fixed impedance device can be characterized by return loss. Return loss is measured in dB, and it is related to a quantity known as the voltage reflection coefficient, ρ:

$$RL = -20 \log |\rho|$$

$$|\rho| = 10^{\frac{-RL}{20}}$$

where
RL = return loss, dB
ρ = voltage reflection coefficient.

The relationship of return loss to SWR is:

$$SWR = \frac{1+10^x}{1-10^x}$$

where $x = \frac{-RL}{20}$

For example, if RL = 20 dB then

$$SWR = \frac{(1+.1)}{(1-.1)} = \frac{1.1}{0.9} = 1.22$$

Fig 25.42 — A low-cost LC oscillator for receiver measurements. Toroidal cores are used for all of the inductances.

foil of the circuit board makes an excellent shield, while the fiberglass helps temperature stability. Capacitor C2 should be soldered directly across L1 to ensure high Q. Since this is an RF circuit, leads should be kept short. While silver mica capacitors have slightly better Q, NP0 capacitors may offer better stability. Mounting the three inductors orthogonal (axis of the inductors 90° from each other) reduced the second-order harmonic by 2 dB when it was compared to the first unit that was made.

Alignment and Testing

The output of the regulator should be +5 V. The output of the oscillator should be +4 dBm (2.5 mW) into a 50-Ω load. Increasing the value of C3 will increase the power output to a maximum of about 10 mW. The frequency should be around 3.7 MHz. Additional capacitance across L1 (in parallel with C2) will lower the frequency if desired, while the trimmer capacitor (C1) specified will allow adjustment to a specific frequency. The drift of one of the first units made was 5 Hz over 25 minutes after a few minutes of warm up. If the warm-up drift is large, changing C2 may improve the situation somewhat. For most receivers, a drift of 100 Hz while you are doing measurements is not bad.

HYBRID COMBINERS FOR SIGNAL GENERATORS

Many receiver performance measurements require two signal generators to be attached to a receiver simultaneously. This, in turn, requires a combiner that isolates the two signal generators (to keep one generator from being frequency or phase modulated by the other). Commercially made hybrid combiners are available from Mini-Circuits Labs, Brooklyn, NY.

Alternatively, a hybrid combiner is not difficult to construct. The combiners described here (see **Fig 25.43**) provide 40 to 50 dB of isolation between ports (connections) while attenuating the desired signal paths (each input to output) by 6 dB. The 50-Ω impedance of the system is kept constant (very important if accurate measurements are to be made).

The combiners are constructed in small boxes made from double-sided circuit-board material. Each piece is soldered to the next one along the entire length of the seam. This makes a good RF-tight enclosure. BNC coaxial fittings are used on the units shown. However, any type of coaxial connector can be used. Leads must be kept as short as possible and precision resistors (or matched units from the junk box) should be used. The circuit diagram for the combiners is shown in **Fig 25.44**.

Fig 25.43 — The hybrid combiner on the left is designed to cover the 1 to 50-MHz range; the one on the right 50 to 500 MHz.

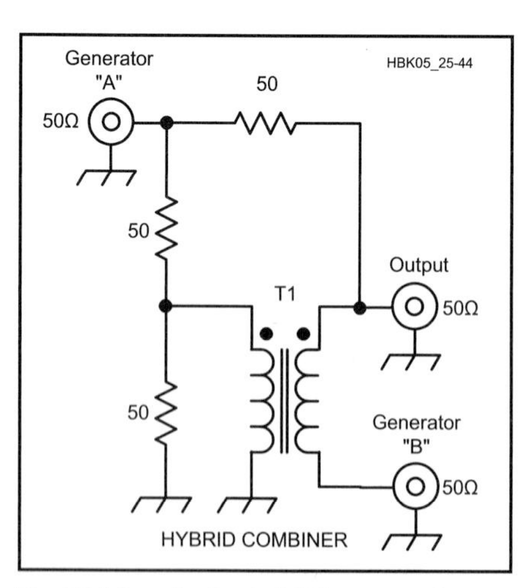

Fig 25.44 — A single bifilar wound transformer is used to make a hybrid combiner. For the 1 to 50-MHz model, T1 is 10 turns of #30 enameled wire bifilar wound on an FT-23-77 ferrite core. For the 50 to 500-MHz model, T1 consists of 10 turns of #30 enameled wire bifilar wound on an FT-23-67 ferrite core. Keep all leads as short as possible when constructing these units.

A COMPENSATED, MODULAR RF VOLTMETER

This versatile and yet simple voltmeter measures levels from 100 mV to 300 V, from 30 MHz to audio frequencies, and is accurate to ± 0.5 dB. The design allows for either a built-in analog meter or an externally connected DVM—or both! This portable, simple, and inexpensive project was originally featured in the March/April 2001 issue of *QEX*. Like the original article, the project described below includes several refinements implemented by its author, Sid Cooper, K2QHE that evolved from a suggestion by *QEX* Managing Editor, Robert Schetgen, KU7G, to increase the voltage range of the instrument.

The requirements for this RF voltmeter (RFVM, **Fig 25.46**) are many:

- Provide accurate, stable measurements
- Measure voltage at frequencies from audio through HF
- Measure voltage levels from QRP to QRO
- Measure voltages inside equipment or in coax
- Operate portably or from ac lines
- Be flexible: work with digital voltmeters already in the shack, or independently
- Be inexpensive

DESIGN APPROACH

To achieve stability, the RFVM uses op amps and any drift in the probe's diode detector is compensated by a matched diode in the RF op amp. This stretches the sensitivity to QRP levels. The dynamic range extends linearly from 0.1 V to 300 V (RMS) by using a series of compensated voltage multipliers.

The frequency response was flattened by using Schottky diodes, which easily reach from 60 Hz to 30 MHz. Both the basic probe and the multipliers very simply adapt to function as a probe, clip to various measurement points on chassis or to screw onto UHF connectors. The active devices are only two op amps. They draw 800 μA of resting current and a maximum of 4.0 mA during operation from a 9-V battery, which is disconnected when a 9-V wall unit is plugged into a jack on the back panel. As designed, the RFVM has a small sloping panel cabinet with a microammeter display. A pair of front-panel banana OUTPUT jacks can be used with a DVM; this eliminates one op amp, the meter and a multiposition switch. Since the meter and switch are the most expensive parts (when they are not bought at a hamfest), how low-cost can this RF voltmeter get?

The Probe

The low-voltage probe is a detector circuit that uses a Schottky diode and a high-impedance filter circuit (**Fig 25.45**). The diode is matched with the one in the feedback circuit of the CA3160 shown in **Fig 25.64**. This match reduces the diode's threshold voltage from about 0.34 V to less than 0.1 V, making the voltage drop comparable to that of a germanium diode. Since I think 0.1 V adequately covers QRP requirements, I made no further tests with germanium diodes to determine how much lower in voltage we could go.[1]

Use the low voltage probe and the output of the CA3160 in the meter unit (Fig 25.64) to find a matched diode pair. Build the CA3160 circuit first for this purpose. Select a diode and place it in the feedback loop of the CA3160, then test each of the remaining diodes in the probe with the three pots set about midrange. Test each diode, first at probe inputs of 100 mV, then at 3.00 V, at 400 Hz. Record the dc output from the CA3160, using the OUTPUT terminals. A bag of 20 diodes from Mouser Electronics[2] contained seven matched pairs, with identical readings at both low and high voltages. My tests used the ac scale (good to 500 Hz) of a Heath 2372 DVM to read the input voltage, and its dc scale to read the output. The number of matched pairs is surprising, but the diodes in the bag may all be from the same production run.

RF Probe Assembly

After selecting matched diodes, all components of the RF probe are mounted

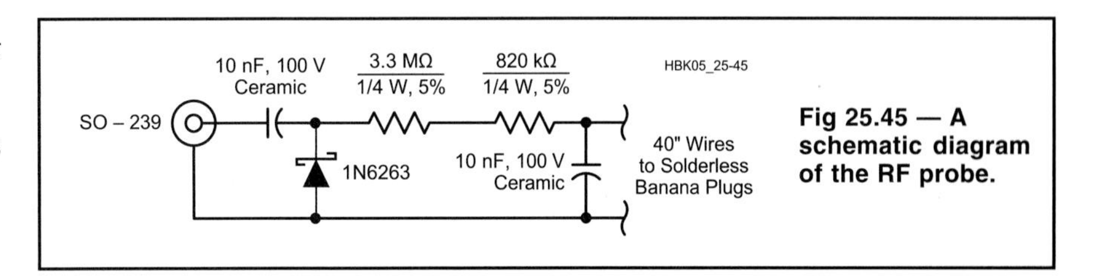

Fig 25.45 — A schematic diagram of the RF probe.

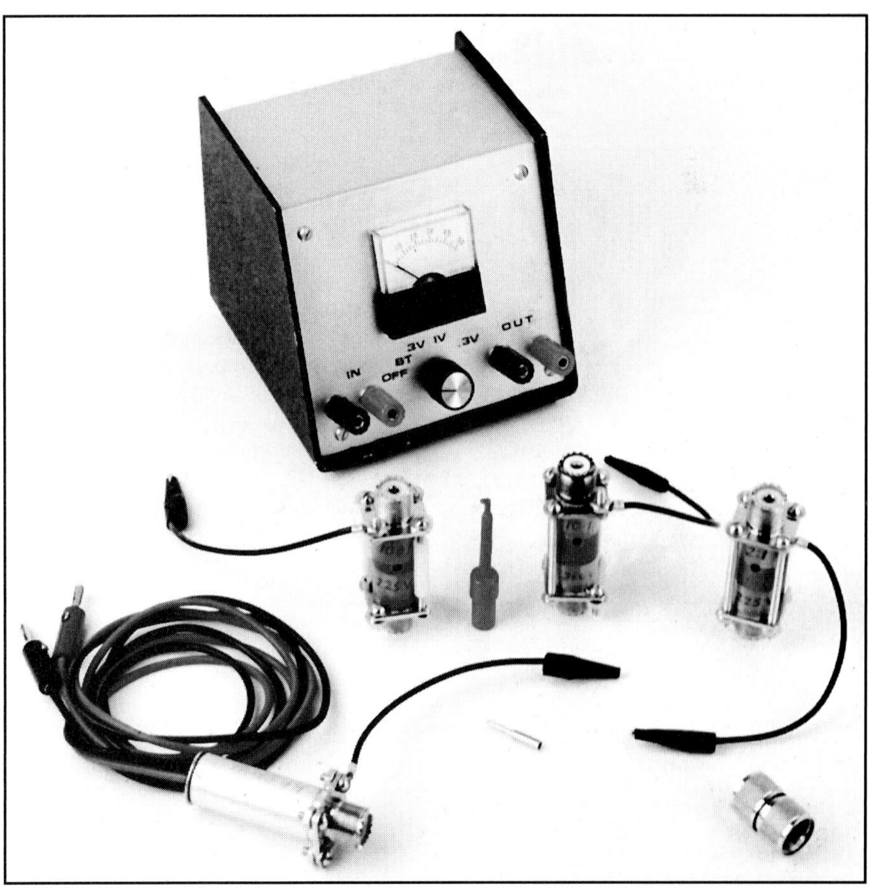

Fig 25.46 — The RF voltmeter with meter unit, RF probe and voltage multipliers.

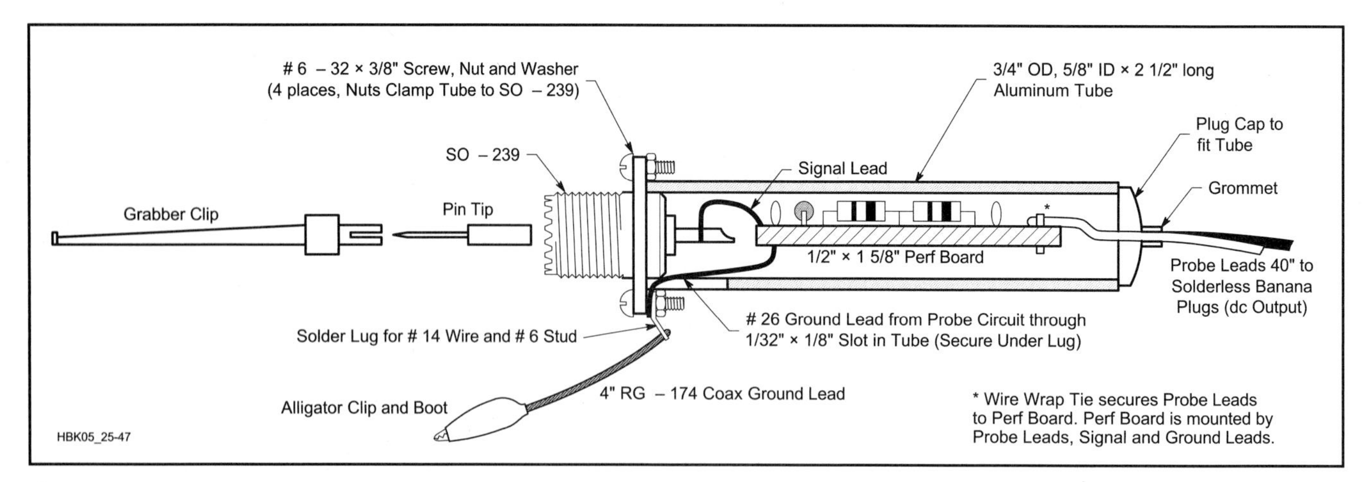

Fig 25.47 — RF probe construction showing the grabber clip (RS 270-334), pin tip (Mouser 534-1600), lug (Mouser 571-34120), for #14-#16 wire and a #6 stud), alligator clip (RS 270-1545), plug cap (Mouser 534-7604), grommet (Mouser 5167-208) and four-inch ground lead made from shield of RG-174.

on a piece of perf board that is inserted in an aluminum tube (**Fig 25.47**). Since the circuit board is very light, it is supported on each end only by its input and output wires. The input RF wire is soldered to the center pin of the SO-239 connector. The two dc-output wires are held to the board by a small tie wrap, consisting of a piece of the #26 insulated wire, after having been passed through the plug cap and grommet. Cut a slit in the aluminum tube, perpendicular to its end. Make it at least 1/8-inch long and 1/32-inch wide to allow a bare #26 ground wire to pass through it.

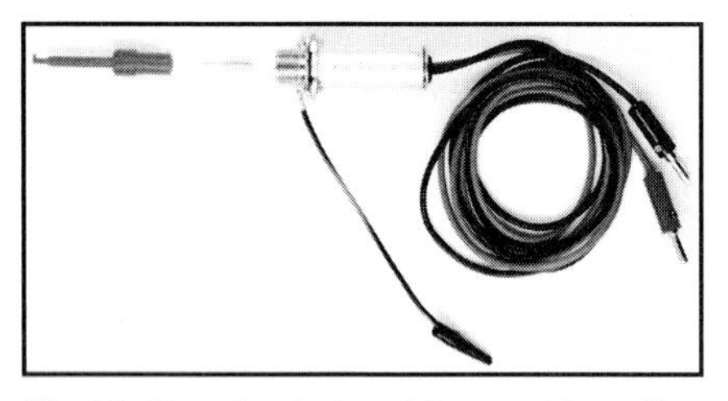

Fig 25.48 — A photo of the grabber clip, pin tip and RF probe.

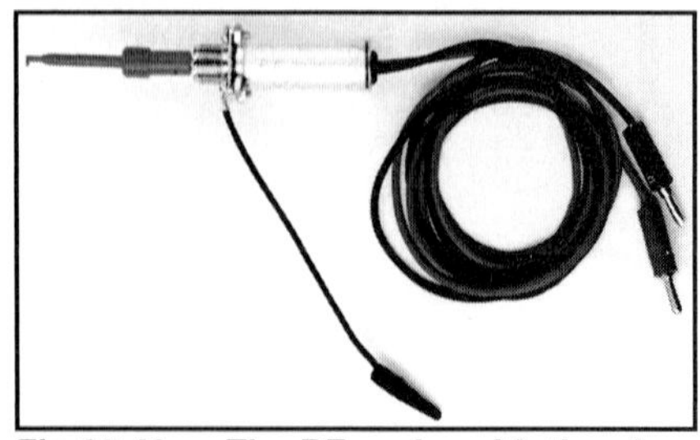

Fig 25.49 — The RF probe with the pin tip and grabber clip in place. The pin tip is not visible, but it provides the physical bridge and electrical connection between the clip and SO-239 of the RF probe.

Let's look at how the probe is assembled. The aluminum tube just fits around the back end of the SO-239 connector. When the four #6-32 screws, nuts and washers are installed in the four holes of the SO-239 connector, the flat face of the nut bears down strongly on the aluminum tube and rigidly holds it in place. Most standard SO-239 connector holes easily accept #6-32 screws; if yours do not, enlarge them with a #27 bit. The slit cut in the aluminum tube allows the bare ground wire from the perf board to pass from inside the tube to the outside and around the screw that secures the ground lead of the probe. As can be seen in Fig 25.47, the threaded part of the SO-239 connector faces away from the aluminum tube.

This probe allows easy voltage measurements in a coaxial cable when the probe SO-239 is secured to a mating T connector. To make a probe reading at any point on a chassis or PC board with this connector, simply insert the larger end of a nicely mating pin tip into the SO-239. The pin tip's diameter is 0.08 inches at the small end and—fittingly—0.14 inches at the mating end. I made this purely fortuitous discovery while rummaging through a very ancient junk box. Further, to make measurements with the probe clipped to a part or test point on a PC board or chassis, a clip or grabber can be mated to the pin tip that protrudes from the SO-239. This is shown in Figs 25.47 and **25.48**. All this may have been known to people in the connector industry, but it appears not to have been known or used elsewhere. **Fig 25.49** shows the grabber connected to the RF probe using the pin tip to join the two.

PROBE MEASUREMENTS

The probe can accept RF signals up to 20 V (RMS) when there is no dc voltage or a combination of dc voltage and peak ac of about 28 V without exceeding the reverse-voltage rating of the diode. I made linearity measurements of the RF probe from 100 mV to 8.0 V at 400 Hz. (The range was limited by my available signal generator.) **Fig 25.50** is a plot of the error at the probe output versus the input amplitude. This was measured using a digital voltmeter with a 10-MΩ input resistance. This input resistance and the 4.1-MΩ resistor in the probe converts the peak voltage of a sine wave signal to the RMS reading of the DVM. The error is –44% at 100 mV, then 10% (0.9 dB) at 1.0 V and finally 2.9% at 8.0 V. The RF probe and DVM are obviously intended for higher voltage readings where the diode voltage drop doesn't affect the accuracy significantly.

The frequency response of the probe was measured at the midpoint of each ham band from 1.9 to 30 MHz with 650 mV input (see **Fig 25.51**). This time, because the input voltage was low and the absolute error would have been high, the probe was connected to the meter unit, which will be described later. The compensating diode loop of the meter unit reduces the error that would otherwise be read by the DVM from 90 mV (14%) to 4 mV (0.6%). The frequency response, however, is controlled by the probe. Use of the meter unit—which operates entirely on the dc signal—does not affect the frequency response, but it does improve the sensitivity of the readings. The measured response shown goes from –3.6% (–0.32 dB) to +1.2% (+0.1 dB).

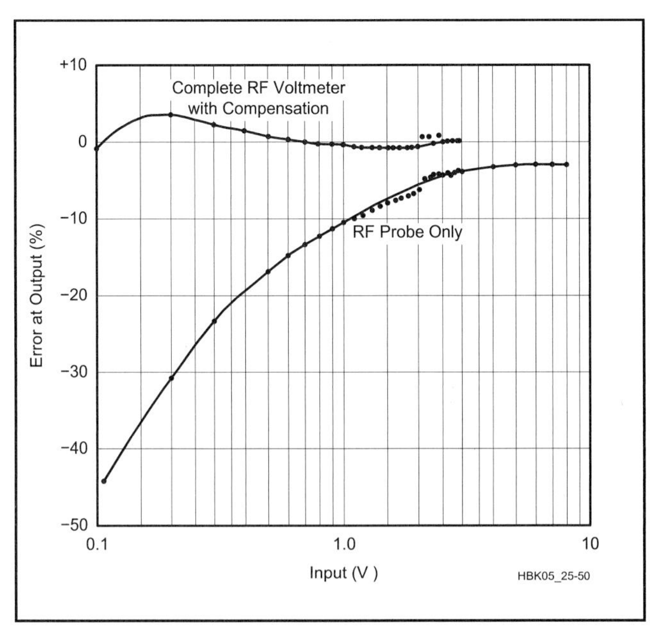

Fig 25.50 — Amplitude linearity of the RF probe alone and in combination with the RF voltmeter.

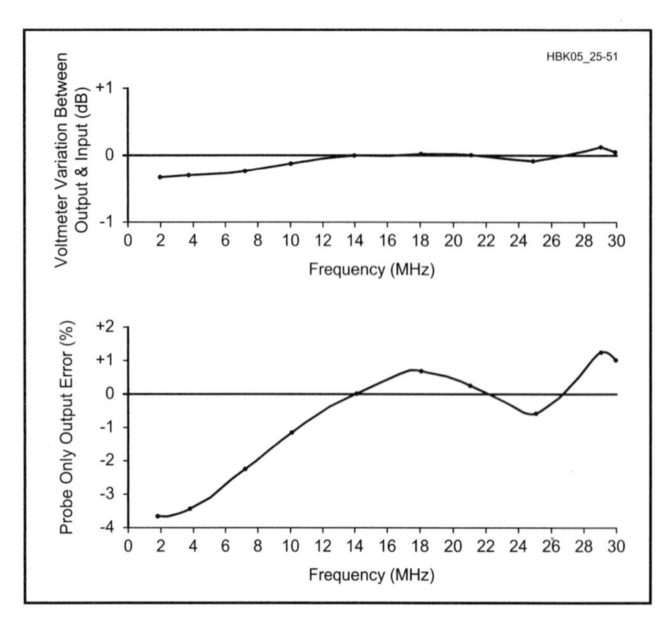

Fig 25.51 — A frequency-response plot for the RF voltmeter. The response of the probe determines the overall system response.

MULTIPLIERS

To extend the voltage range of the RF probe, use a 10× multiplier, which is a compensated divider, shown in **Fig 25.52**. The electrical design is straightforward and includes a small trimmer capacitor to adjust for a flat frequency response. These components are very lightweight, so they require no perf board and are easily supported by their leads, which are anchored at the two SO-239 connectors. To make measurements, the input connector can be mated to connectors in a coaxial cable, a pin tip can be inserted for probing PC boards or a grabber can be added for connections to components on a chassis. The multiplier can connect to the RF probe either through a male-to-male UHF fitting or by a piece of coax with a UHF male connector at each end.

The typical way to adjust the trimmer requires a square-wave input and an oscilloscope on the output. The trimmer is then set to produce a flat square-wave output with no overshoots on the leading edges and no droop across the top. An RF signal generator that covers 1.9 MHz to 30 MHz can be used instead, if a square wave generator is not available. Since this is a 10× multiplier, it is reasonable to expect the input to handle up to 200 V (RMS), but the limit is 150 V because the trimmer capacitor is rated at 225 V dc. You may be able to find small trimmers with higher voltage ratings that will take the multiplier to 200 V and still fit inside the tube housing.

Multiplier construction follows the same methods used for the RF probe but with a few differences. My first model used an aluminum tube like that in the RF probe with #6-32 screws to hold it all together, but the frequency response began to drop off at 21 MHz. By replacing the aluminum tube with a phenolic tube[3] having the same dimensions, the response is adequate to 30 MHz.

Fig 25.53 shows construction methods. Two slits at the ends of the tube allow the ground wires at the input and output ends

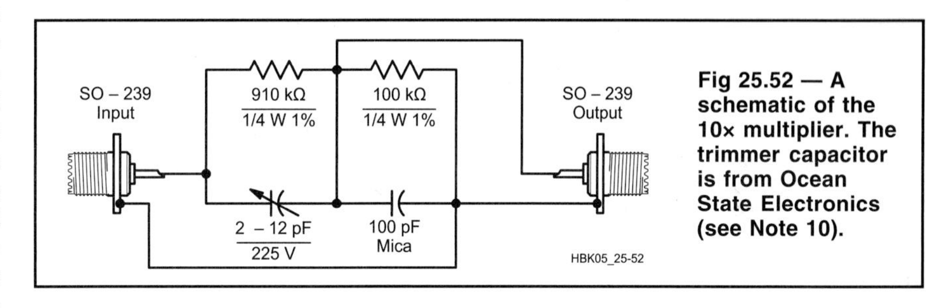

Fig 25.52 — A schematic of the 10× multiplier. The trimmer capacitor is from Ocean State Electronics (see Note 10).

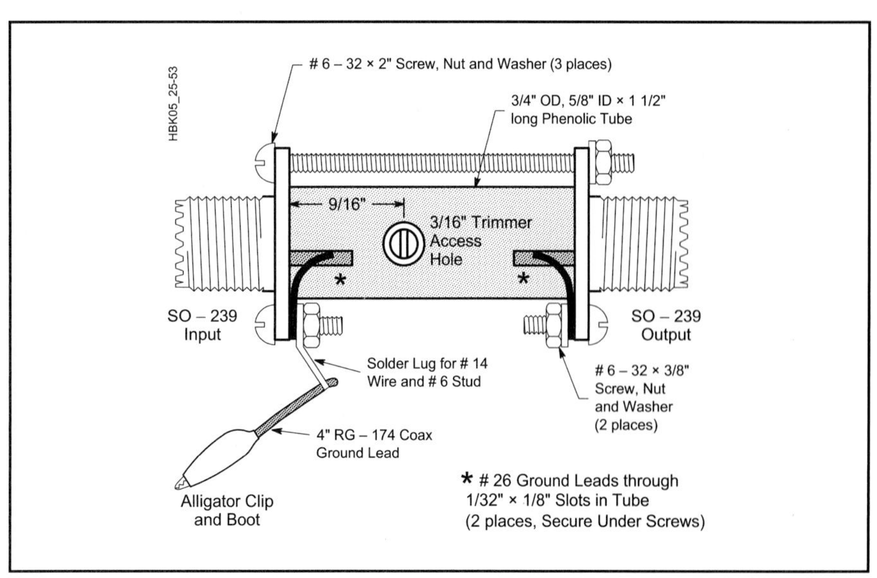

Fig 25.53 — Construction of the 10x multiplier. The phenolic tube is US Plastics #47081, see Note 3. The lug is Mouser 571-34120.

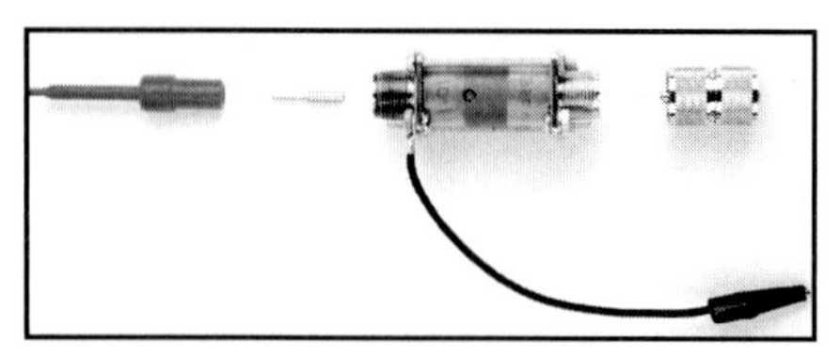

Fig 25.54 — A view showing the 10× multiplier with grabber, pin tip and a male UHF coupler that connects the multiplier to the RF probe.

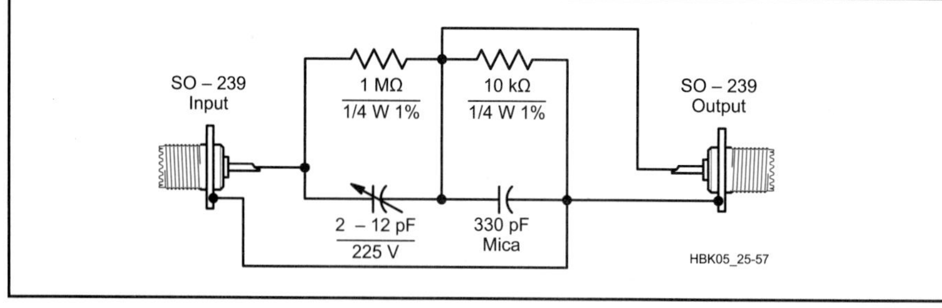

Fig 25.57 — Schematic of the 100x multiplier (maximum allowable input is 150 V RMS). The trimmer capacitor is from Ocean State Electronics (see Note 10).

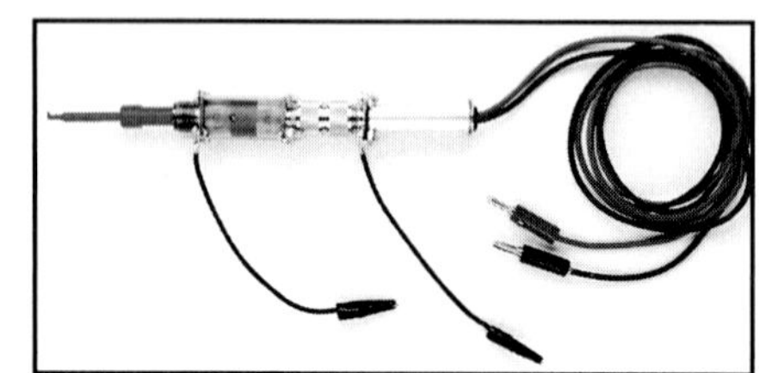

Fig 25.55 — The complete RF probe and 10× multiplier assembly.

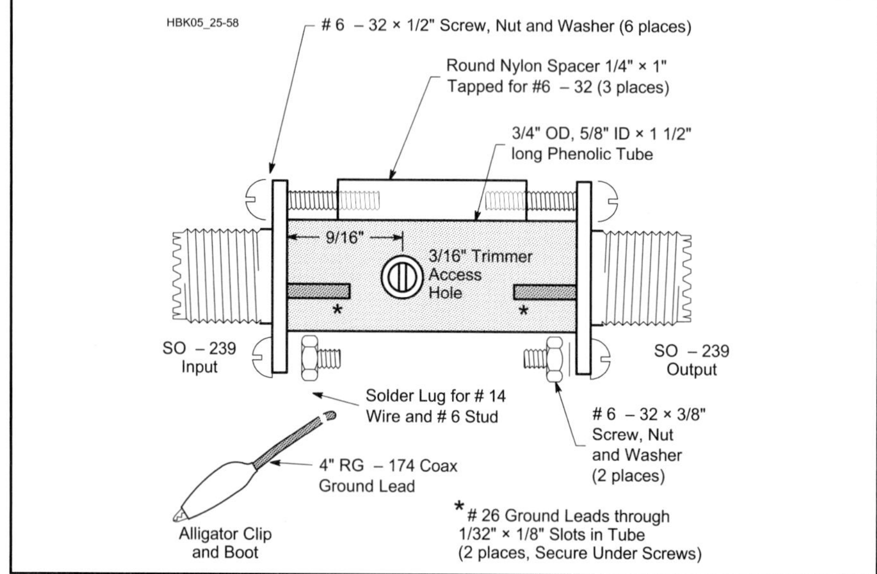

Fig 25.58 — 100x multiplier assembly showing the round Nylon spacer (Mouser 561-TSP10), phenolic tube (US Plastics #47081, see Note 3). The lug is Mouser 571-34120.

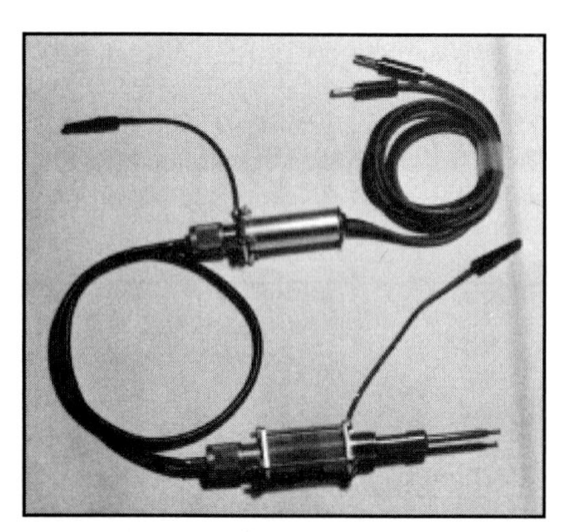

Fig 25.56 — The RF probe and multiplier can be separated by a piece of coax less than 12 inches long if it makes measurements more convenient.

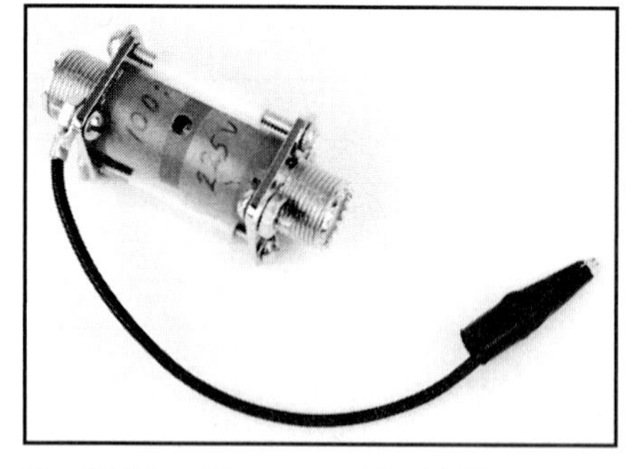

Fig 25.59 — The assembled 100× multiplier.

to exit and wrap around the screws of each connector. A hole in the phenolic tube allows adjustment of the trimmer capacitor.

Fig 25.54 shows the exploded view of the pin tip and grabbers at the input end of the multiplier and a UHF male-to-male connector on the output side to mate with the RF probe. The assembled multiplier and probe are shown in **Fig 25.55**. The whole assembly is only 3½ inches long and is comfortable in the hand. If more flexibility is desired, the two pieces may be separated by as much as one foot of coax (**Fig 25.56**) without doing too much damage to the accuracy of the measurement.

The 100× multiplier follows the same design concepts as the 10× multiplier including the obligatory trimmer capacitor, as in **Fig 25.57**. Here too, the 225 V dc rating of the trimmer limits the maximum RF to 150 V RMS. So, this is not useful should it be used only with the probe and your DVM because the 10× multiplier already covers this range. Later, we will see how it is useful with the RF probe and the meter unit.

The 100× unit also uses a phenolic-tube housing with a hole to adjust the trimmer and a slit at each end to bring out the ground wires, which then wrap around screws in the connectors. In order to reach at least 30 MHz, however, the two-inch steel screws between the two SO-239 connectors must be insulated. Using plastic washers at the connector holes could do this job but would have required #4-40× 2-inch-long screws. After a long search at the biggest hardware stores and catalogs, I found nothing longer than 1½ inches. That's the best way to assemble this multiplier, if you can find the screws. Otherwise, use nylon spacers and shorter screws as shown in **Figs 25.58** and **25.59**. The pin tip and grabber at the input and UHF male-to-male connector or coaxial cable at the output are used as with the 10× multiplier.

The only purpose of the 2× multiplier is to create a 200× multiplier when used with the 100× multiplier in series, so as to extend the voltage range from 150 V to 300 V (1800 W at 50 Ω only under matched conditions[4]). The trimmer in each multiplier now divides in half the input voltage.

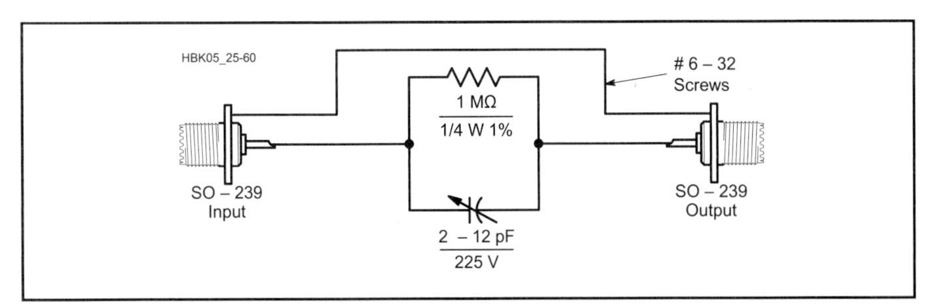

Fig 25.60 — A schematic diagram of the 2x multiplier. Its output connects to the input of the 100x multiplier for 200x measurements up to 300 V RMS. This multiplier is used *only* with the 100x multiplier.

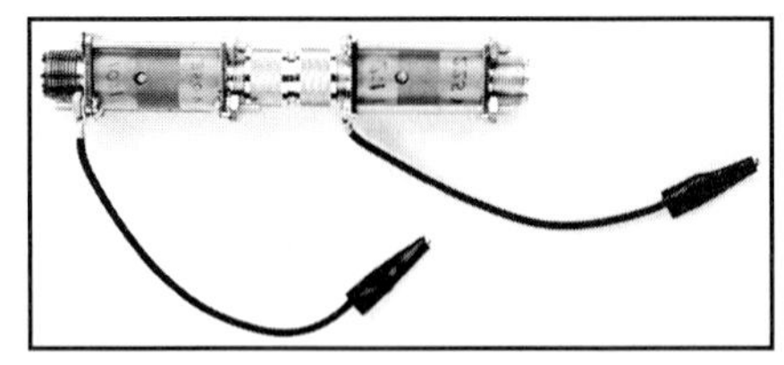

Fig 25.62 — The 2x and 100x multipliers are joined by a male UHF coupler for readings up to 300 V (RMS, 1800 W, see Note 4).

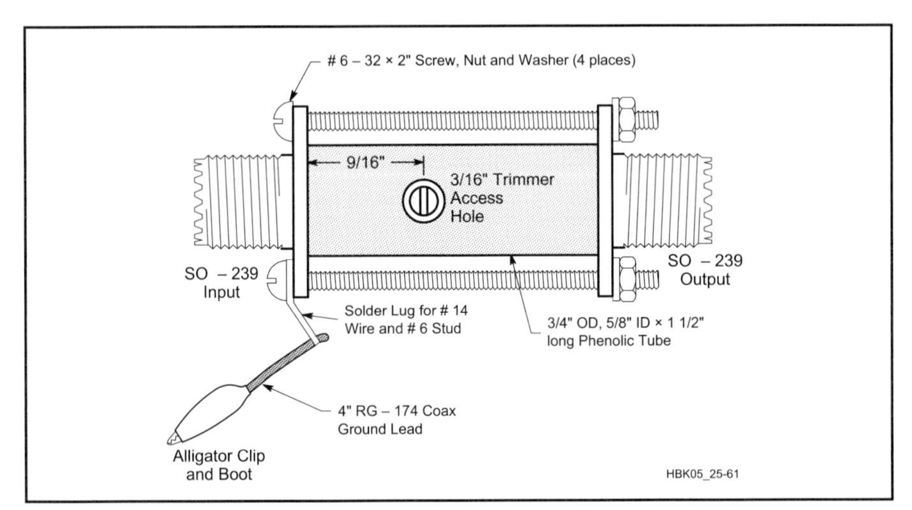

Fig 25.61 — 2x multiplier assembly showing the phenolic tube (US Plastics #47081, see Note 3). The lug is 16-14 #6 stud (Mouser 571-34120).

Fig 25.60 shows the 2× multiplier with only one RC section since it relies on the RC sections in the 100× multiplier to complete the division (multiplication).

The construction, **Fig 25.61**, returns to the tubular format as before, but this time there are no slots to pass the ground wires; the four screws are able to do the job. The two multipliers are shown in series in **Fig 25.62**. The input capacitance for each multiplier is shown in **Table 25.4**.

When measurements are made on a load resistance of 50 Ω, the input capacity of the multipliers has no affect on readings below 30 MHz. When the impedance of the load resistance is larger, consider any error it introduces. To get a sense of the affects of the load resistance, use the equation below. It shows the relationship between the resistance, capacitance and frequency when the measured voltage is 3 dB down from what it would be if it were read by a meter with no input capacity:

$$f = \frac{1}{2\pi R_L C_P} \qquad (21)$$

where
f = 3-dB loss frequency
R_L=load resistance
C_P=probe capacitance

For example, at a load of 1 kΩ and a probe capacity of 13.3 pF, the 3-dB down frequency is about 12 MHz, where the error is 30%. If an attenuator were used instead of the multiplier, it would not have this problem because it works with either a fixed load of 50 Ω at high frequencies or 600 Ω at low frequencies. Probes cannot select their frequency or load impedance and are thereby more flexible in use, so measurements must be made with consideration, but these multipliers would not have a problem at 50 Ω or 600 Ω either.

Table 25.5 summarizes the voltage ranges using only the multiplier with the RF probe and a DVM that has a 10-MΩ input resistance that's available in the shack.

The multipliers are intended for use at high voltages where safety precautions are a primary consideration to avoid personal injury. See the **Safety** chapter of this *Handbook*. In Tektronics' *ABC's of Probes,*[5] an entire section thoroughly covers the hazards and necessary precautions when making measurements with probes. It is worth the little effort to get a copy of it and also the Pomona Catalog,[6] which has some good information on probe use.

The multipliers have panel connectors at both ends and the RF probe has one at one end. Their grounds are connected and brought to the meter unit. This unit is grounded only when the ac-powered 9-V power supply that may be used with it, is grounded. When it is battery operated, the meter unit relies on the ground-clip connection to the ground of the equipment under test. This is satisfactory if there is no unknown break in the chain of ground connections that would make the panel

Table 25.4
Multiplier Range versus Capacitance

Multiplier	10×	100×	200×
Input Cap (pF)	13.3	10.9	13.5

Table 25.5
Multiplier versus Voltage Range

Probe Configuration	*Range (V)*	*Maximum Voltage*
Probe Alone*	1.0-20 V ac	20 V ac
With 10× Multiplier*	10-150 V ac	225 V (ac + dc)
With 100× and 2× Multiplier†	150-200 V ac	450 V (ac + dc)
With 100× and 2× Multiplier††	200-300 V ac	450 V (ac + dc)

*error less than –10% or –0.83 dB
†with error of –45% to –10% or less than 3.2 dB. A trimmer capacitor with a rating of 300 V dc increases the 10× range to 200 V ac and reduces the error from 45% to 10%. The RF probe and meter unit reduces the error to less than –3.5% or less than –0.3 dB.
††with error less than –6% or –0.54 dB

HBK05_25-63

$$e_s - e_{d1} = e_p$$

$$e_o - e_{d2} = e_n$$

$$e_i = e_p - e_n = e_s - e_{d1} - e_o + e_{d2}$$

$$e_i = \frac{e_o}{G} = 0$$

$$e_s - e_o = e_{d1} - e_{d2}$$

e_s represents the input of the RF probe as a dc signal.
e_o represents the output from the high-gain CA3160 op amp.
e_{d1} is the voltage drop across the diode in the RF probe.
e_{d2} is the voltage drop across the diode in the feedback loop.
G=320,000 for the CA3160.
The last equation shows that any difference between the signal and output results from unequal diode drops when the diodes are not matched.

Fig 25.63 — Basics of the diode-compensation method to improve measurement accuracy at low input voltages. It also provides a measure of temperature compensation and drift reduction. See Fig 25.50.

connector hot. Furthermore, as insurance, it is well to wrap these connectors with vinyl tape or to cover the multiplier with a plastic boot to prevent contact with either hands or equipment under test. This is not shown in any photographs because it would have obscured the appearance and construction of the probe and multipliers.

The frequency responses of the multipliers were determined using an RF signal generator set at the midpoint of each ham band from 160 to 10 meters, including 30 MHz. An error of 5.42% (0.46 dB) was measured from 160 to 17 meters, which then decreased to 4.13% (0.35 dB) at 15 meters and then to zero through 30 MHz. The constant error from 160 to 17 meters is probably due to the inherent errors in the test equipment. The oscilloscope has a 60 MHz bandwidth, an input impedance of 1 MΩ and 30 pF, which introduces a load effect on the multiplier outputs. It also has a reading accuracy of ±3%. When the multiplier-components accuracy of 1% is included, it is not surprising to find the overall inaccuracy to be at most 5.42%.[7] The error of the multiplier itself could be inherently less than this. Before the overall frequency response was measured, the trimmer capacitor was adjusted with the input frequency set mid-frequency at 15 MHz. If your interest in RF probes and multipliers has been raised and you have more questions, see the references in Notes 5 and 6.

THE METER UNIT

The meter unit serves several purposes when used with the RF probe. It increases the accuracy of the probe and DVM from –45% to –3.5% at a voltage of 100 mV RF. At higher voltages, it maintains a minimum advantage of 5:1 in reducing the error, when the probe is used with a DVM. This is an increase in both sensitivity and accuracy. The usual non-linearity caused by the diode in the probe is reduced when used with the meter unit. The technique also incidentally provides temperature compensation for diode drift. The meter unit has a ±5% panel meter to display measurements, but it also contains a pair of OUTPUT terminals for a DVM. Add a DVM when more accuracy is desired at the low end of the range or when you want a bit more resolution. Finally, since the meter unit is fully portable, low-level field-strength measurements are possible

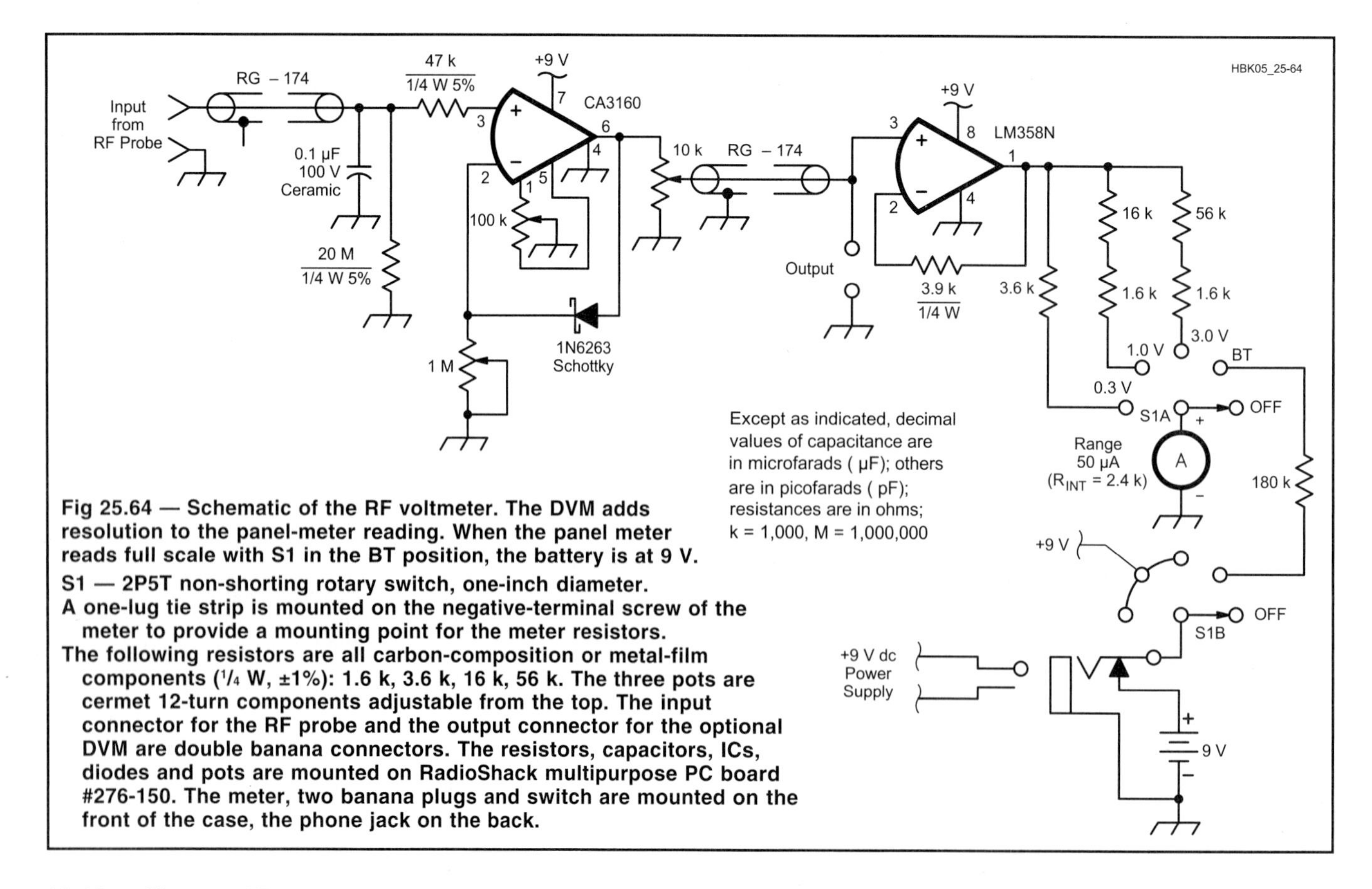

Fig 25.64 — Schematic of the RF voltmeter. The DVM adds resolution to the panel-meter reading. When the panel meter reads full scale with S1 in the BT position, the battery is at 9 V.

S1 — 2P5T non-shorting rotary switch, one-inch diameter.

A one-lug tie strip is mounted on the negative-terminal screw of the meter to provide a mounting point for the meter resistors.

The following resistors are all carbon-composition or metal-film components (1/4 W, ±1%): 1.6 k, 3.6 k, 16 k, 56 k. The three pots are cermet 12-turn components adjustable from the top. The input connector for the RF probe and the output connector for the optional DVM are double banana connectors. The resistors, capacitors, ICs, diodes and pots are mounted on RadioShack multipurpose PC board #276-150. The meter, two banana plugs and switch are mounted on the front of the case, the phone jack on the back.

with a whip antenna at the probe input connector. Due to its sensitivity and accuracy, the probe can be adapted for use in many places around the shack.

The non-linear response of the probe diode is compensated (improved) by a circuit in the meter unit. The feedback loop of a CA3160 op amp contains a diode matched to the one in the probe (see Note 1). **Fig 25.63** and its sequence of equations present a very simple sketch of how matched diodes do this when dc is applied through a diode. The final equation shows that any difference between the op amp input and output is due to a difference in voltage drops across the two diodes. When the diodes are matched, the error disappears. When RF is applied, the average currents through the diodes must be equal to keep the voltage drops equal. Articles by Kuzdrall (Note 1) Grebenkemper[8] and Lewallen[9] are first-class descriptions of the principles used in this RF voltmeter.

Fig 25.64 shows three pots for calibrating the meter unit. This should be performed at 400 Hz to avoid any effects due to RF. The 100-kΩ pot is typically used to null the offset of the CA3160, but is used here to initially set the offset to about 0.5 mV to 1.0 mV with 100 mV input. Then the 1-MΩ pot sets the output to 100 mV with 100 mV input, and the 10 kΩ pot sets the output to 3.0 V when the input is 3.0 V. Finally, the three pots are alternately adjusted until the 100 mV and 3.0 V set points occur together. The 100-kΩ pot is helpful in fine tuning the 100-mV point. A DVM was used at the optional OUTPUT sockets during calibration.

A CA3160 was selected for the input op amp because of its high input impedance of 1.5 TΩ and high gain of 320,000. Although it has diodes that provide protection, I think it could be sensitive to electrostatic discharge, so handle it carefully. Use IC sockets to permit easy replacement of the ICs in case of damage. An LM358N IC follows the CA3160 (see Fig 25.64), primarily to drive the panel meter. If you use a DVM as the display, you can omit the LM358N circuit, meter and multi-position switch. The only functions lost are the battery-voltage check and power on/off switch. Add an on/off switch to the circuit when the multi-position switch is omitted. The CA3160 op amp circuit easily spans the range from 100 mV to 3.0 V without the need for a range switch.

CONSTRUCTION

A 5×5×4½-inch sloping-front instrument case was used to house the components of the meter unit. The meter on the front panel has a 2×2-inch face and requires a 1½-inch hole in the panel. Although a 50-µA meter from the junk box is used here, a 1-mA movement will work as well, provided the series resistors are changed accordingly.

The rotary switch has two poles and five positions for changing the meter range, testing the battery condition and switching the power off. Two sets of double banana binding posts are used, the INPUT pair accepts the dc signal from the RF probe. The OUTPUT pair provides a voltage for a DVM display, whether the panel meter is used or not. On the rear of the case, a miniature phone jack accepts 9-V power from either a battery or a 9-V dc supply. The ICs and other parts are mounted on a RadioShack multipurpose PC board that has very convenient holes and traces. The board is bolted to the back of the case via standoff insulators and wired to the front panel components (see **Fig 25.65**).

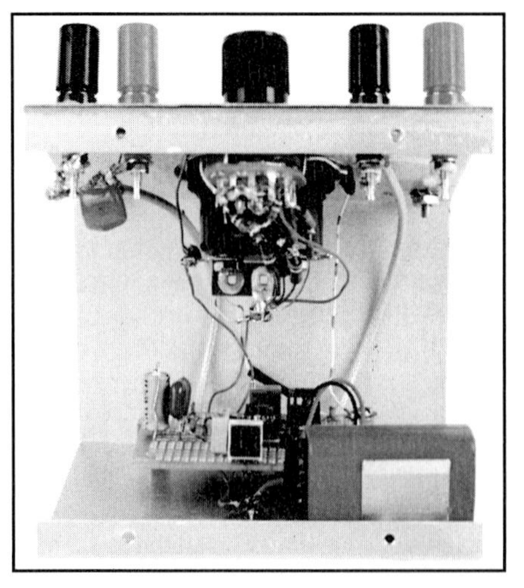

Fig 25.65 — Inside view of the meter unit. RG-174 connects the INPUT and OUTPUT banana binding posts to the circuit board. The rotary switch, one-lug tie strip and battery are also visible.

Notes

[1]J. A. Kuzdrall, "Linearized RF Detector Spans 50-to-1 Range," Analog Applications Issue, *Electronic Design*, June 27, 1994.

[2]Mouser Electronics, 2401 Hwy 287 N, Mansfield, TX 76063; tel 800-346-6873, fax 817-483-0931; E-mail **sales@mouser.com**; **www.mouser.com**.

[3]United States Plastics Corp, 1390 Neubrecht Road, Lima, OH 45801-3196, tel 419-228-2242, fax 419-228-5034.

[4]If the power remains constant, load mismatch multiplies the voltage by the SWR. A 2:1 SWR would produce 600 V, 3:1 900 V and so on.—*Ed*

[5]ABCs of Probes, Tektronix Inc, Literature number 60W-6053-7, July 1998. Tektronix, Inc. Export Sales, PO Box 500 M/S 50-255 Beaverton, OR 97707-0001; 503-627-6877. Johnny Parham, "How to Select the Proper Probe," Electronic Products, July 1997.

[6]Pomona Test and Measurement Accessories catalog. ITT Pomona Electronics, 1500 E Ninth St, Pomona, CA 91766-3835.

[7]A. Frost, "Are You Measuring Your Circuit or Your Scope Probe?" *EDN*, July 22, 1999. E. Feign, "High-Frequency Probes Drive 50-Ω Measurements," *RF Design*, Oct 1998.

[8]J. Grebenkemper, KI6WX, "The Tandem Match—An Accurate Directional Wattmeter," *QST*, Jan 1987, pp 18-26; and "Tandem Match Corrections," *QST*, Jan 1988, p 49.

[9]R. Lewallen, W7EL, "A Simplified and Accurate QRP Directional Wattmeter," *QST*, Feb 1990, pp 19-23, 36.

[10]Ocean State Electronics, 6 Industrial Dr, PO Box 1458, Westerly, RI 02891; tel 800-866-6626, fax 401-596-3590.

Receiver Performance Tests

Comparing the performance of one receiver to another is difficult at best. The features of one receiver may outweigh a second, even though its performance under some conditions is not as good as it could be. Although the final decision on which receiver to purchase will more than likely be based on personal preference and cost, there are ways to compare receiver performance characteristics. Some of the more important parameters are sensitivity, blocking dynamic range and two-tone IMD dynamic range.

Instruments for measuring receiver performance should be of suitable quality and calibration. Always remember that accuracy can never be better than the tools used to make the measurements. Common instruments used for receiver testing include:

- Signal generators
- Hybrid combiner
- Audio ac voltmeter
- Distortion meter (FM measurements only)
- Noise figure meter (only required for noise figure measurements)
- Step attenuators (10 dB and 1 dB steps are useful)

Signal generators must be calibrated accurately in dBm or mV. The generators should have extremely low leakage. That is, when the output of the generator is switched off, no signal should be detected at the operating frequency with a sensitive receiver. Ideally, at least one of the signal generators should be capable of amplitude modulation. A suitable lab-quality piece would be the HP-8640B, no longer manufactured, but a good item to scout on the surplus market.

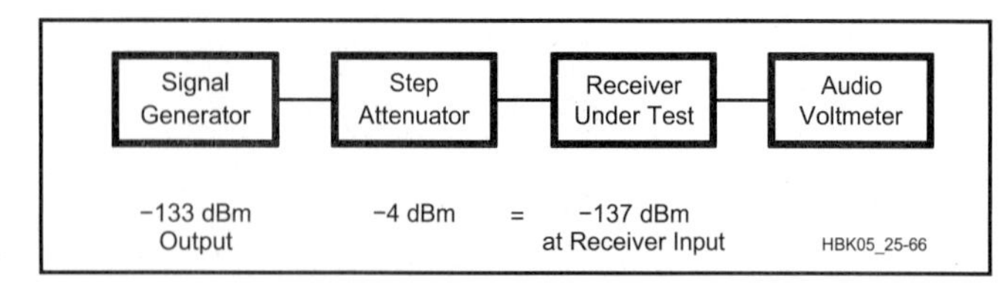

Fig 25.66 — A general test setup for measuring receiver MDS, or noise floor. Signal levels shown are for an example discussed in the text.

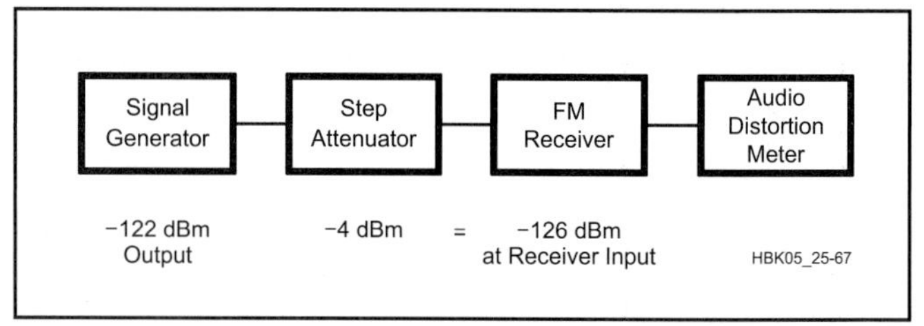

Fig 25.67 — FM SINAD test setup.

While most signal generators are calibrated in terms of microvolts, the real concern is not with the voltage from the generator but with the power available. The unit that is used for most low-level RF work is the milliwatt, and power is often specified in decibels with respect to 1 mW (dBm). Hence, 0 dBm would be 1 mW. The dBm level, in a 50-Ω load, can be calculated with the aid of the following equation:

$$\mathrm{dBm} = 10 \log_{10} \left[20 (V)^2 \right] \tag{22}$$

where

dBm = power with respect to 1 mW

V = RMS voltage available at the output of the signal generator

The convenience of a logarithmic power unit such as the dBm becomes apparent when signals are amplified or attenuated. For example, a –107 dBm signal that is applied to an amplifier with a gain of 20 dB will result in an output increased by 20 dB. Therefore in this example (–107 dBm + 20 dB) = 87 dBm. Similarly, a –107 dBm signal applied to an attenuator with a loss of 10 dB will result in an output of (–107 dBm – 10 dB) or –117 dBm.

A hybrid combiner is a three-port device used to combine the signals from a pair of generators for all dynamic range measurements. It has the characteristic that signals applied at ports 1 or 2 appear at port 3 and are attenuated by 3 dB.However, a signal from port 1 is attenuated 30 or 40 dB when sampled at port 2. Similarly, signals applied at port 2 areisolated from port 1 some 30 to 40 dB. The isolating properties of the box prevent one signal generator from being frequency or phase modulated by the other. A second feature of a hybrid combiner is that a 50-Ω impedance level is maintained throughout the system.

Audio voltmeters should be calibrated in dB as well as volts. This facilitates easy measurements and eliminates the need for cumbersome calculations. Be sure that the step attenuators are in good working order and suitable for the frequencies involved. A distortion meter, such as the Hewlett-Packard 339A, is required for FM sensitivity measurements and a noise figure meter, such as the Hewlett-Packard 8970A, is excellent for certain kinds of sensitivity measurements.

Receiver Sensitivity

Several methods are used to determine receiver sensitivity. The mode under consideration often determines the best choice. One of the most common sensitivity measurements is minimum discernible signal (MDS) or noise floor. It is suitable for CW and SSB receivers.

This measurement indicates the minimum discernible signal that can be detected with the receiver. This level is defined as that which will produce the same audio-output power as the internally generated receiver noise. Hence, the term "noise floor."

To measure MDS, use a signal generator tuned to the same frequency as the receiver (see **Fig 25.66**). With the generator output at 0 or with maximum attenuation of its output note the voltmeter reading. Next increase the generator output level until the ac voltmeter at the receiveraudio-output jack shows a 3-dB increase. The signal input at this point is the MDS. Be certain that the receiver is peaked on the generator signal. The filter bandwidth can affect the MDS. Always compare MDS readings taken with identical filter bandwidths. (A narrow bandwidth tends to improve MDS performance.) MDS can be expressed in µV or dBm.

In the hypothetical example of Fig 25.66, the output of the signal generator is –133 dBm and the step attenuator is set to 4 dB. Here is the calculation:

$$\text{Noise floor} = -133\ \text{dBm} - 4\ \text{dB} = -137\ \text{dBm} \tag{23}$$

where the noise floor is the power available at the receiver antenna terminal and 4 dB is the loss through the attenuator.

Receiver sensitivity is also often expressed as 10 dB S+N/N (a 10-dB ratio of signal + noise to noise) or 10 dB S/N (signal to noise). The procedure and measurement are identical to MDS, except that the input signal is increased until the receiver output increases by 10 dB for 10 dB S+N/N and 9.5 dB for 10 dB S/N (often called "10 dB signal to noise ratio"). AM receiver sensitivity is usually expressed in this manner with a 30% modulated, 1-kHz test signal. (The modulation in this case is keyed on and off and the signal level is adjusted for the desired increase in the audio output.)

SINAD is a common sensitivity measurement normally associated with FM receivers. It is an acronym for "*s*ignal plus *n*oise *a*nd *d*istortion." SINAD is a measure of signal quality:

$$\text{SINAD} = \frac{\text{signal} + \text{noise} + \text{distortion}}{\text{noise} + \text{distortion}} \tag{24}$$

where SINAD is expressed in dB. In this example, all quantities to the right of the equal sign are expressed in volts, and the ratio is converted to dB by multiplying the log of the fraction by 20.

$$\text{SINAD(dB)} = 20 \log \left(\frac{\text{Signal(V)} + \text{Noise(V)} + \text{Distortion(V)}}{\text{Noise(V)} + \text{Distortion(V)}} \right)$$

Let's look at this more closely. We can consider distortion to be a part of the receiver noise because distortion, like noise, is an unwanted signal added to the desired signal by the receiving system. Then, if we assume that the desired signal is much stronger than the noise, SINAD closely approximates the signal to noise ratio. The common 12-dB SINAD specification therefore corresponds to a 4:1 S/N ratio (noise + distortion = 0.25 × signal).

The basic test setup for measuring SINAD is shown in **Fig 25.67**. The level of input signal is adjusted to provide 25% distortion (12 dB SINAD). Narrow-band FM signals, typical for amateur communications, usually have 3-kHz peak deviation when modulated at 1000 Hz.

Noise figure is another measure of receiver sensitivity. It provides a sensitivity evaluation that is independent of the system bandwidth. Noise figure is discussed further in the **Receivers and Transmitters** chapter.

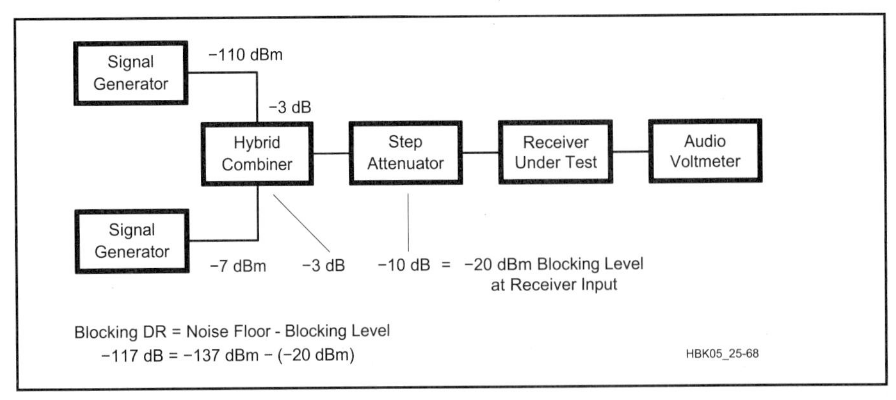

Fig 25.68 — Receiver Blocking DR is measured with this equipment and arrangement. Measurements shown are for the example discussed in the text.

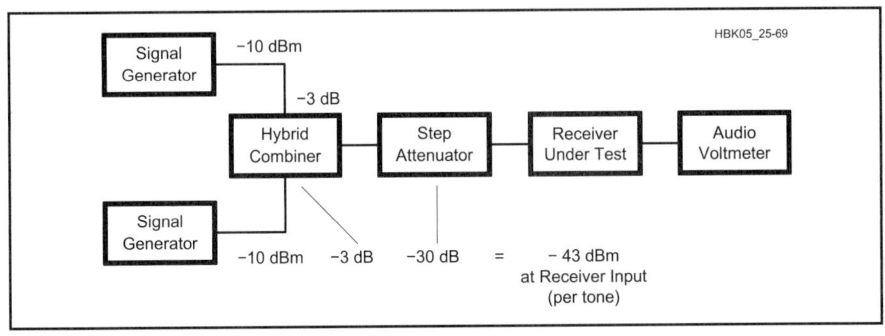

Fig 25.69 — Receiver IMD DR test setup. Signal levels shown are for the example discussed in the text.

Dynamic Range

Dynamic range is the ability of the receiver to tolerate strong signals outside of its band-pass range. Two kinds will be considered:

Blocking dynamic range (blocking DR) is the difference, in dB, between the noise floor and a signal that causes 1 dB of gain compression in the receiver. It indicates the signal level, above the noise floor, that begins to cause desensitization.

IMD dynamic range (IMD DR) measures the impact of two-tone IMD on a receiver. IMD is the production of spurious responses that results when two or more signals mix. IMD occurs in any receiver when signals of sufficient magnitude are present. IMD DR is the difference, in dB, between the noise floor and the strength of two equal incoming signals that produce a third-order product 3 dB above the noise floor.

What do these measurements mean? When the IMD DR is exceeded, false signals begin to appear along with the desired signal. When the blocking DR is exceeded, the receiver begins losing its ability to amplify weak signals. Typically, the IMD DR is 20 dB or more below the blocking DR, so false signals appear well before sensitivity is significantly decreased. IMD DR is one of the most significant parameters that can be specified for a receiver. It is generally a conservative evaluation for other effects, such as blocking, which will occur only for signals well outside the IMD dynamic range of the receiver.

Both dynamic range tests require two signal generators and a hybrid combiner. When testing blocking DR (see **Fig 25.68**), one generator is set for a weak signal of roughly –110 dBm. The receiver is tuned to this frequency and peaked for maximum response. (ARRL Lab procedures require this level to be about 10 dB below the 1-dB compression point, if the AGC can be disabled. Otherwise, the level is set to 20 dB above the MDS.)

The second generator is set to a frequency 20 kHz away from the first and its level is increased until the receiver output drops by 1 dB, as measured with the ac voltmeter.

In the example shown, the output of the generator is –7 dBm, the loss through the combiner is fixed at 3 dB and the step attenuator is set to 10 dB. The 1-dB compression level is calculated as follows:

$$\text{Blocking level} = -7\text{ dBm} - 3\text{ dBm} - 10\text{ dB} = -20\text{ dBm} \quad (25)$$

To express this as a dynamic range, the blocking level is referenced to the receiver noise floor (calculated earlier). Calculate it as follows:

$$\text{Block DR} = \text{noise floor} - \text{blocking level} = -137\text{ dBm} - (-20\text{ dBm}) = -117\text{ dB} \quad (26)$$

This value is usually expressed as an absolute value: 117 dB.

Two-Tone IMD Test

The setup for measuring IMD DR is shown in **Fig 25.69**. Two signals of equal level, spaced 20-kHz apart are injected into the receiver input. When we call these requencies f1 and f2, the so-called third-order IMD products will appear at frequencies of (2f1 – f2) and (2f2 – f1). If the two input frequencies are 14.040 and 14.060 MHz, the third-order products will be at 14.020 and 14.080 MHz. Let's talk through a measurement with these frequencies.

First, set the generators for f1 and f2. Adjust each of them for an output of –10 dBm. Tune the receiver to either of the third-order IMD products. Adjust the step attenuator until the IMD product produces an output 3 dB above the noise level as read on the ac voltmeter.

For an example, say the output of the generator is –10 dBm, the loss through the combiner is 3 dB and the amount of attenuation used is 30 dB. The signal level at the receiver antenna terminal that just begins to cause IMD problems is calculated as:

$$\text{IMD level} = -10\text{ dBm} - 3\text{ dB} - 30\text{ db} = -43\text{ dBm} \quad (27)$$

To express this as a dynamic range the IMD level is referenced to the noise floor as follows:

$$\text{IMD DR} = \text{noise floor} - \text{IMD level} = -137\text{ dBm} - (-43\text{ dBm}) = -94\text{ dB} \quad (28)$$

Therefore, the IMD dynamic range of this receiver would be 94 dB.

Third-Order Intercept

Another parameter used to quantifyreceiver performance is the third-order input intercept (IP^3). This is the point at which the desired response and the third-order IMD response intersect, if extended beyond their linear regions (see **Fig 25.70**). Greater IP^3 indicates better receiver performance. Calculate IP^3 like this:

$$IP^3 = 1.5\ (\text{IMD dynamic range in dB}) + (\text{MDS in dBm}) \quad (29)$$

For our example receiver:

$$IP^3 = 1.5\ (94\text{ dB}) + (-137\text{ dBm}) = +4\text{ dBm}$$

The preferred method for the third-order input intercept, however, is to use S5 signal levels instead of MDS levels. This is due to the tendency for a receiver to exhibit non-linearity at the noise floor level (i.e., near the level where MDS is typically defined).

Thus, Third Order Intercept = (3 × (S5 IMD Level) – (S5 Reference))/2

The results of this method should show close correlation with the intercept point determined by the MDS test. If not, the test engineer determines (from further investigation) which method provides a more accurate result.

In the same way (using the S5 method),

Second Order Interrupt = 2 × (S5 IMD Level) – (S5 Reference)

The example receiver we have discussed here is purely imaginary. Nonetheless, its performance is typical of contemporary communications receivers.

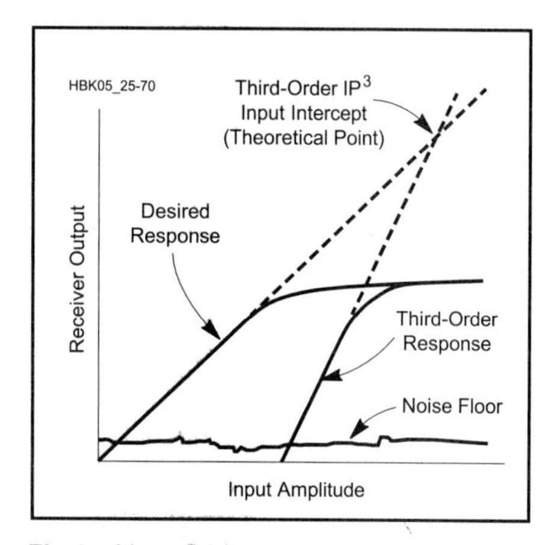

Fig 25.70 — A plot of the receiver characteristics that determine third-order input intercept, a measure of receiver performance.

Evaluating the Data

Thus far, a fair amount of data has been gathered with no mention of what the numbers really mean. It is somewhat easier to understand exactly what is happening by arranging the data as shown in **Fig 25.71**. The base line represents power levels with a very small level at the left and a higher level (0 dBm) at the right.

The noise floor of our hypothetical receiver is at –137 dBm, the IMD level (the level at which signals will begin to create spurious responses) at –43 dBm and the blocking level (the level at which signals will begin to desensitize the receiver) at –20 dBm. The IMD dynamic range is some 23 dB smaller than the blocking dynamic range. This means IMD products will be heard long before the receiver begins to desensitize, some 23 dB sooner.

SPECTRUM ANALYZERS

A spectrum analyzer is similar to an oscilloscope. Both visually present an electrical signal through graphic representation. The oscilloscope is used to observe electrical signals in the time domain (amplitude as a function of time). The time domain, however, gives little information about the frequencies that make up complex signals. Amplifiers, mixers, oscillators, detectors, modulators and filters are best characterized in terms of their frequency response. This information is obtained by viewing electrical signals in the *frequency domain* (amplitude as a function of frequency). One instrument that can display the frequency domain is the spectrum analyzer.

Time and Frequency Domain

To better understand the concepts of time and frequency domain, see **Fig 25.72**. The three-dimensional coordinates show time (as the line sloping toward the bottom right), frequency (as the line rising toward the top right) and amplitude (as the vertical axis). The two discrete frequencies shown are harmonically related, so we'll refer to them as f1 and 2f1.

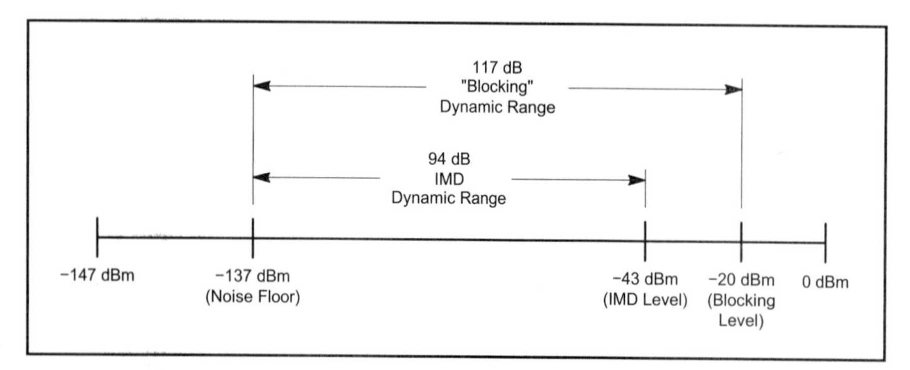

Fig 25.71 — Performance plot of the receiver discussed in the text. This is a good way to visualize the interaction of receiver-performance measurements. Note that –147 dBm is considered the lowest possible noise level in a 500 Hz receiver bandwidth.

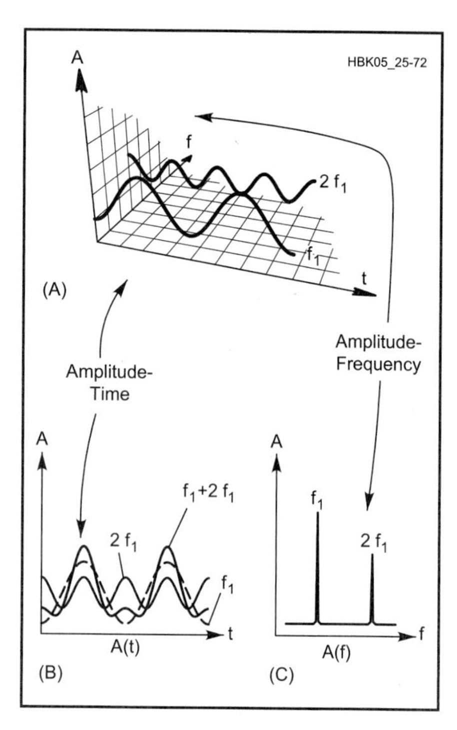

Fig 25.72 — A complex signal in the time and frequency domains. A is a three-dimensional display of amplitude, time and frequency. B is an oscilloscope display of time vs amplitude. C is spectrum analyzer display of the frequency domain and shows frequency vs amplitude.

In the representation of time domain at B, all frequency components of a signal are summed together. In fact, if the two discrete frequencies shown were applied to the input of an oscilloscope, we would see the solid line (which corresponds to f1 + 2f1) on the display.

In the frequency domain, complex signals (signals composed of more than one frequency) are separated into their individual frequency components. A spectrum analyzer measures and displays the power level at each discrete frequency; this display is shown at C.

The frequency domain contains information not apparent in the time domain and therefore the spectrum analyzer offers advantages over the oscilloscope for certain measurements. As might be expected, some measurements are best made in the time domain. In these cases, the oscilloscope is a valuable instrument.

Spectrum Analyzer Basics

There are several different types of spectrum analyzers, but by far the most

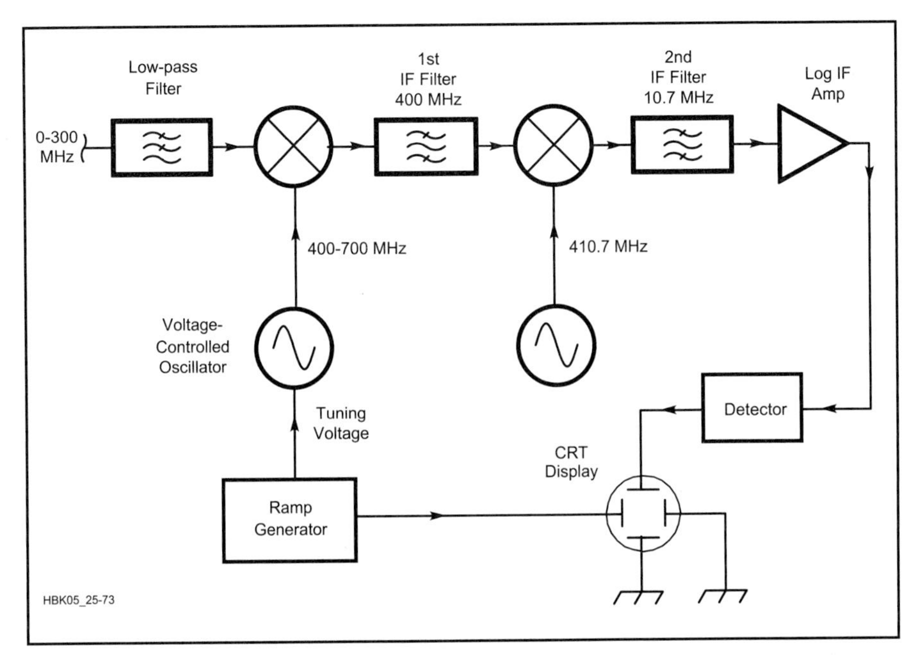

Fig 25.73 — A block diagram of a superheterodyne spectrum analyzer. Input frequencies of up to 300 MHz are up converted by the local oscillator and mixer to a fixed frequency of 400 MHz.

common is nothing more than an electronically tuned superheterodyne receiver. The receiver is tuned by means of a ramp voltage. This ramp voltage performs two functions: First, it sweeps the frequency of the analyzer local oscillator; second, it deflects a beam across the horizontal axis of a CRT display, as shown in **Fig 25.73**. The vertical axis deflection of the CRT beam is determined by the strength of the received signal. In this way, the CRT displays frequency on the horizontal axis and signal strength on the vertical axis.

Most spectrum analyzers use an up-converting technique so that a fixed tuned input filter can remove the image. Only the first local oscillator need be tuned to tune the receiver. In the up-conversion design, a wide-band input is converted to an IF higher than the highest input frequency. As with most up-converting communications receivers, it is not easy to achieve the desired ultimate selectivity at the first IF, because of the high frequency. For this reason, multiple conversions are used to generate an IF low enough so that the desired selectivity is practical. In the example shown, dual conversion is used: The first IF is at 400 MHz; the second at 10.7 MHz.

In the example spectrum analyzer, the first local oscillator is swept from 400 MHz to 700 MHz; this converts the input (from nearly 0 MHz to 300 MHz) to the first IF of 400 MHz. The usual rule of thumb for varactor tuned oscillators is that the maximum practical tuning ratio (the ratio of the highest frequency to the lowest frequency) is an octave, a 2:1 ratio. In our example spectrum analyzer, the tuning ratio of the first local oscillator is 1.75:1, which meets this specification.

The image frequency spans 800 MHz to 1100 MHz and is easily eliminated using a low-pass filter with a cut-off frequency around 300 MHz. The 400-MHz first IF is converted to 10.7 MHz where the ultimate selectivity of the analyzer is obtained. The image of the second conversion, (421.4 MHz), is eliminated by the first IF filter. The attenuation of the image should be great, on the order of 60 to 80 dB. This requires a first IF filter with a high Q; this is achieved by using helical resonators, SAW resonators or cavity filters. Another method of eliminating the image problem is to use triple conversion; converting first to an intermediate IF such as 50 MHz and then to 10.7 MHz. As with any receiver, an additional frequency conversion requires added circuitry and adds potential spurious responses.

Most of the signal amplification takes place at the lowest IF; in the case of the example analyzer this is 10.7 MHz. Here the communications receiver and the spectrum analyzer differ. A communications receiver demodulates the incoming signal so that the modulation can be heard or further demodulated for RTTY or packet or other mode of operation. In the spectrum analyzer, only the signal strength is needed.

In order for the spectrum analyzer to be most useful, it should display signals of widely different levels. As an example, signals differing by 60 dB, which is a thousand to one difference in voltage or a million to one in power, would be difficult to display. This would mean that if power were displayed, one signal would be one million times larger than the other (in the case of voltage one signal would be a thousand times larger). In either case it would be difficult to display both signals on a CRT. The solution to this problem is to use a logarithmic display that shows the relative signal levels in decibels. Using this technique, a 1000:1 ratio of voltage reduces to a 60-dB difference.

The conversion of the signal to a logarithm is usually performed in the IF amplifier or detector, resulting in an output voltage proportional to the logarithm of the input RF level. This output voltage is then used to drive the CRT display.

Spectrum Analyzer Performance Specifications

The performance parameters of a spectrum analyzer are specified in terms similar to those used for radio receivers, in spite of the fact that there are many differences between a receiver and a spectrum analyzer.

The sensitivity of a receiver is often specified as the minimum discernible signal, which means the smallest signal that can be heard. In the case of the spectrum analyzer, it is not the smallest signal that can be heard, but the smallest signal that can be seen. The dynamic range of the spectrum analyzer determines the largest and smallest signals that can be simultaneously viewed on the analyzer. As with a receiver, there are several factors that can affect dynamic range, such as IMD, second- and third-order distortion and blocking. IMD dynamic range is the maximum difference in signal level between the minimum detectable signal and the level of two signals of equal strength that generate an IMD product equal to the minimum detectable signal.

Although the communications receiver is an excellent example to introduce the spectrum analyzer, there are several differences such as the previously explained lack of a demodulator. Unlike the communications receiver, the spectrum analyzer is not a sensitive radio receiver. To preserve a wide dynamic range, the spectrum analyzer often uses passive mixers for the first and second mixers. Therefore, referring to Fig 25.73, the noise figure of the analyzer is no better than the losses of the input low-pass filter plus the first mixer, the first IF filter, the second mixer and the loss of the second IF filter. This often results in a combined noise figure of more than 20 dB. With that kind of noise figure the spectrum analyzer is obviously not a communications receiver for extracting very weak signals from the noise but a measuring instrument for the analysis of frequency spectrum.

The selectivity of the analyzer is called the resolution bandwidth. This term refers to the minimum frequency separation of two signals of equal level that can be resolved so there is a 3-dB dip between the two. The IF filters used in a spectrum analyzer differ from a communications receiver in that the filters in a spectrum analyzer have very gentle skirts and rounded passbands, rather than the flat passband and very steep skirts used on an IF filter in a high-quality communications receiver. This rounded passband is necessary because the signals pass into the filter passband as the spectrum analyzer scans the desired frequency range. If the signals suddenly pop into the passband (as they would if the filter had steep skirts), the filter tends to ring; a filter with gentle skirts is less likely to ring. This ringing, called scan loss, distorts the display and requires that the analyzer not sweep frequency too quickly. All this means that the scan rate must be checked periodically to be certain the signal amplitude is not affected by fast tuning.

Spectrum Analyzer Applications

Spectrum analyzers are used in situations where the signals to be analyzed are very complex and an oscilloscope display would be an indecipherable jumble. The spectrum analyzer is also used when the frequency of the signals to be analyzed is very high. Although high-performance oscilloscopes are capable of operation into the UHF region, moderately priced spectrum analyzers can be used well into the gigahertz region.

A spectrum analyzer can also be used to view very low-level signals. For an oscilloscope to display a VHF waveform, the bandwidth of the oscilloscope must extend from zero to the frequency of the waveform. If harmonic distortion and other higher-frequency distortions are to be seen the bandwidth of the oscilloscope must exceed the fundamental frequency of the waveform. This broad bandwidth can also admit a lot of noise power. The spectrum analyzer, on the other hand, analyzes the waveform using a narrow bandwidth; thus it is capable of reducing the noise power admitted.

Probably the most common application of the spectrum analyzer is the measurement of the harmonic content and other spurious signals in the output of a radio transmitter. **Fig 25.74** shows two ways to connect the transmitter and spectrum analyzer. The method shown at A should not be used for wide-band measurements since most line-sampling devices do not exhibit a constant-amplitude output over a broad frequency range. Using a line sampler is fine for narrow-band measurements, however. The method shown at B is used in the ARRL Lab. The attenuator must be capable of dissipating the transmitter power. It must also have sufficient attenuation to protect the spectrum analyzer input. Many spectrum analyzer mixers can be damaged by only a few milliwatts, so most analyzers have an adjustable input attenuator that will provide a reasonable amount of attenuation to protect the sensitive input mixer from damage. The power limitation of the attenuator itself is usually on the order of a watt or so, however. This means that 20 dB of additional attenuation is required for a 100-W transmitter, 30 dB for a 1000-W transmitter and so on, to limit the input to the spectrum analyzer to 1 W. There are specialized attenuators that are made for transmitter testing; these attenuators provide the necessary power dissipation and attenuation in the 20 to 30-dB range.

When using a spectrum analyzer it is very important that the maximum amount of attenuation be applied before a measurement is made. In addition, it is a good practice to start with maximum attenuation and view the entire spectrum of a signal before the attenuator is adjusted. The signal being viewed could appear to be at a safe level, but another spectral component, which is not visible, could be above the damage limit. It is also very important to limit the input power to the analyzer when pulse power is being measured. The average power may be small enough so the input attenuator is not damaged, but the peak pulse power, which may not be readily visible on the analyzer display, can destroy a mixer, literally in microseconds.

When using a spectrum analyzer it is necessary to ensure that the analyzer does not generate additional spurious signals that are then attributed to the system under test. Some of the spurious signals that can be generated by a spectrum analyzer are harmonics and IMD. If it is desired to measure the harmonic levels of a transmitter at a level below the spurious level of the analyzer itself, a notch filter can be inserted between the attenuator and the

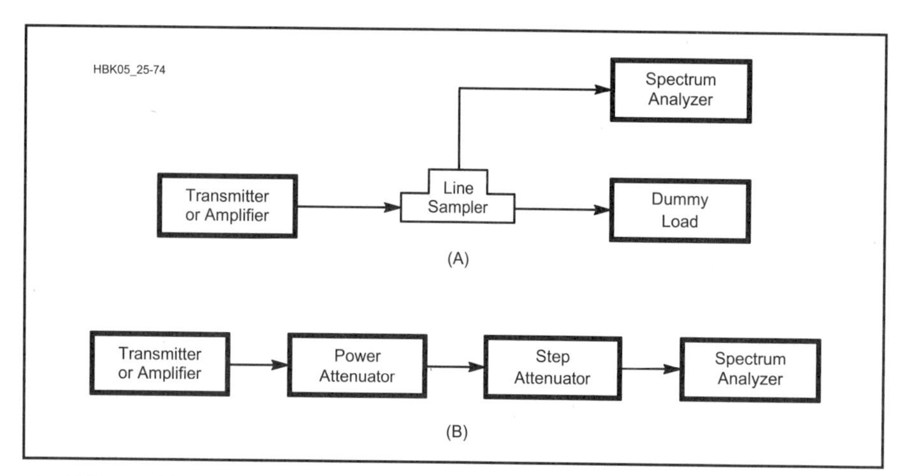

Fig 25.74 — Alternate bench setups for viewing the output of a high power transmitter or oscillator on a spectrum analyzer. A uses a line sampler to pick off a small amount of the transmitter or amplifier power. In B, most of the transmitter power is dissipated in the power attenuator.

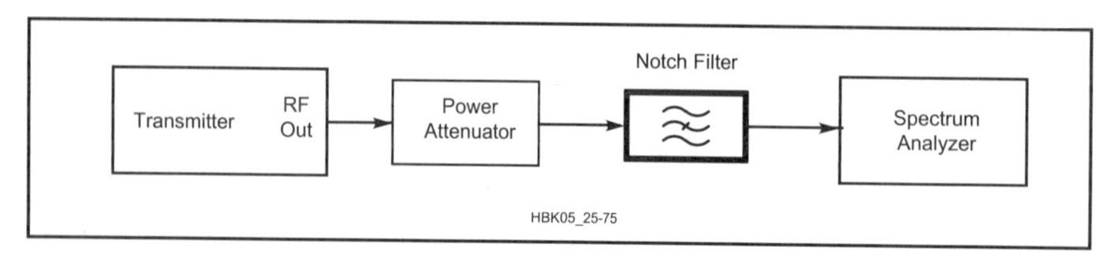

Fig 25.75 — A notch filter is another way to reduce the level of a transmitter's fundamental signal so that the fundamental does not generate harmonics within the analyzer. However, in order to know the amplitude relationship between the fundamental and the transmitter's actual harmonics and spurs, the attenuation of the fundamental in the notch filter must be known.

spectrum analyzer as shown in **Fig 25.75**. This reduces the level of the fundamental signal and prevents that signal from generating harmonics within the analyzer, while still allowing the harmonics from the transmitter to pass through to the analyzer without attenuation. Use caution with this technique; detuning the notch filter or inadvertently changing the transmitter frequency will allow potentially high levels of power to enter the analyzer. In addition, use care when choosing filters; some filters (such as cavity filters) respond not only to the fundamental but notch out odd harmonics as well.

It is good practice to check for the generation of spurious signals within the spectrum analyzer. When a spurious signal is generated by a spectrum analyzer, adding attenuation at the analyzer input will cause the internally generated spurious signals to decrease by an amount greater than the added attenuation. If attenuation added ahead of the analyzer causes all of the visible signals to decrease by the same amount, this indicates a spurious-free display.

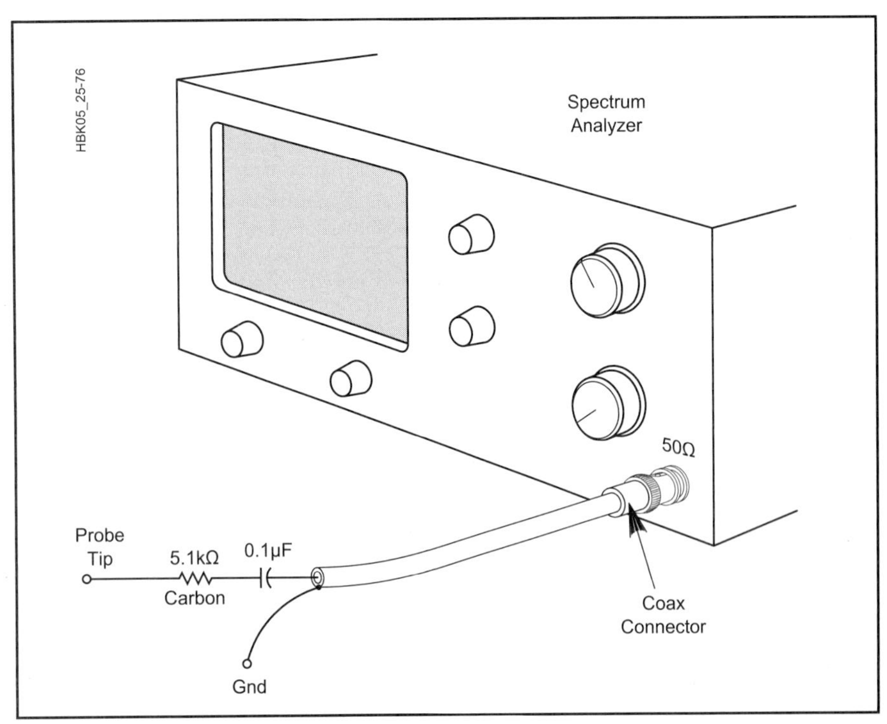

Fig 25.76 — A schematic representation of a voltage probe designed for use with a spectrum analyzer. Keep the probe tip (resistor and capacitor) and ground leads as short as possible.

The input impedance for most RF spectrum analyzers is 50 Ω; not all circuits have convenient 50-Ω connections that can be accessed for testing purposes, however. Using a probe such as the one shown in **Fig 25.76** allows the analyzer to be used as a troubleshooting tool. The probe can be used to track down signals within a transmitter or receiver, much like an oscilloscope is used. The probe shown offers a 100:1 voltage reduction and loads the circuit with 5000 Ω. A different type of probe is shown in **Fig 25.77**. This inductive pickup coil (sometimes called a "sniffer") is very handy for troubleshooting. The coil is used to couple signals from the radiated magnetic field of a circuit into the analyzer. A short length of miniature coax is wound into a pick-up loop and soldered to a larger piece of coax. The use of the coax shields the loop from coupling energy from the electric field component. The dimensions of the loop are not critical, but smaller loop dimensions make the loop more accurate in locating the source of radiated RF. The shield of the coax provides a complete electrostatic shield without introducing a shorted turn.

The sniffer allows the spectrum analyzer to sense RF energy without contacting the circuit being analyzed. If the loop is brought near an oscillator coil, the oscillator can be tuned without directly contacting (and thus disturbing) the circuit. The oscillator can then be checked for reliable starting and the generation of spurious sidebands. With the coil brought near the tuned circuits of amplifiers or frequency multipliers, those stages can be tuned using a similar technique.

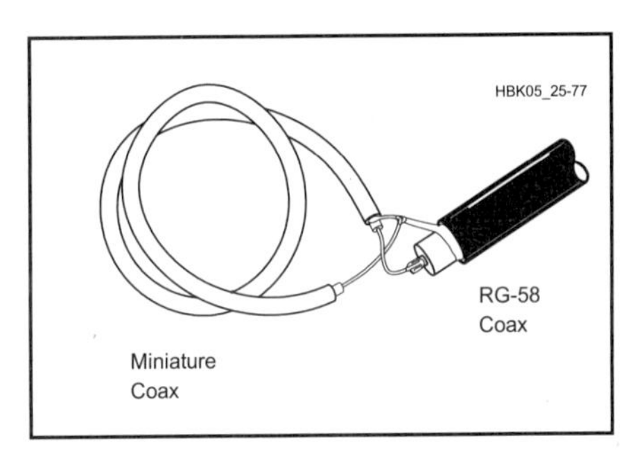

Fig 25.77 — A "sniffer" probe consisting of an inductive pick-up. It has an advantage of not loading the circuit under test. See text for details.

Even though the sniffer does not contact the circuit being evaluated, it does extract some energy from the circuit. For this reason, the loop should be placed as far from the tuned circuit as is practical. If the loop is placed too far from the circuit, the signal will be too weak or the pick-up loop will pick up energy from other parts of the circuit and not give an accurate indication of the circuit under test.

The sniffer is very handy to locate sources of RF leakage. By probing the shields and cabinets of RF generating equipment (such as transmitters) egress and ingress points of RF energy can be identified by increased indications on the analyzer display.

One very powerful characteristic of the spectrum analyzer is the instrument's capability to measure very low- level signals. This characteristic is very advantageous when very high levels of attenuation are measured. **Fig 25.78** shows the setup for tuning the notch and passband of a VHF duplexer. The spectrum analyzer, being capable of viewing signals well into the low microvolt region, is capable of measuring the insertion loss of the notch cavity more than 100 dB below the signal generator output. Making a measurement of this sort requires care in the interconnection of the equipment and a well designed spectrum analyzer and signal generator. RF energy leaking from the signal generator cabinet, line cord or even the coax itself, can get into the spectrum analyzer through similar paths and corrupt the measurement. This leakage can make the measurement look either better or worse than the actual attenuation, depending on the phase relationship of the leaked signal.

Extensions of Spectral Analysis

What if a signal generator is connected to a spectrum analyzer so that the signal generator output frequency is exactly the same as the receiving frequency of the spectrum analyzer? It would certainly ap-

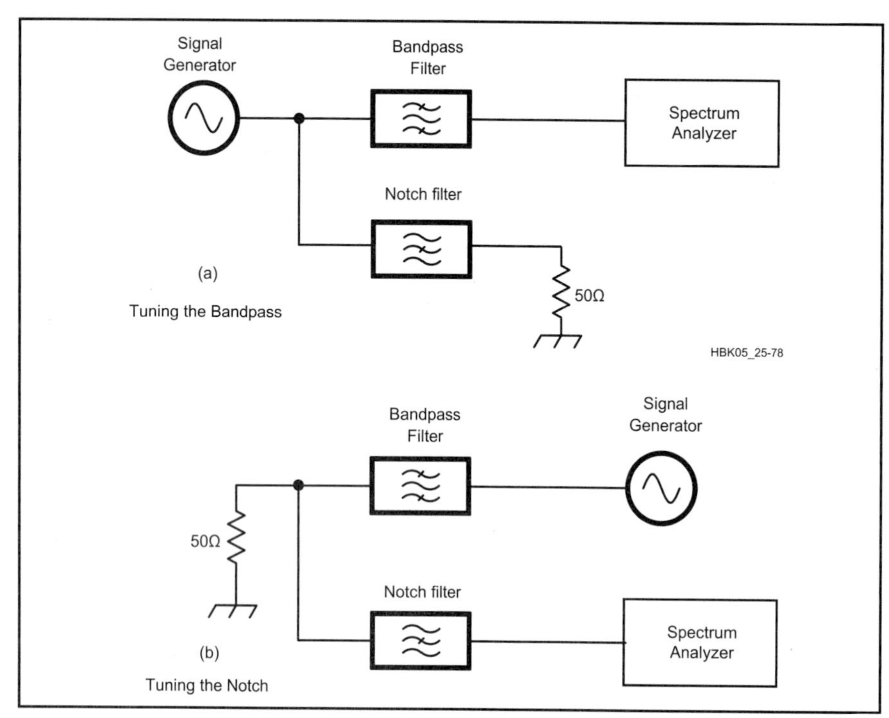

Fig 25.78 — Block diagram of a spectrum analyzer and signal generator being used to tune the band-pass and notch filters of a duplexer. All ports of the duplexer must be properly terminated and good quality coax with intact shielding used to reduce leakage.

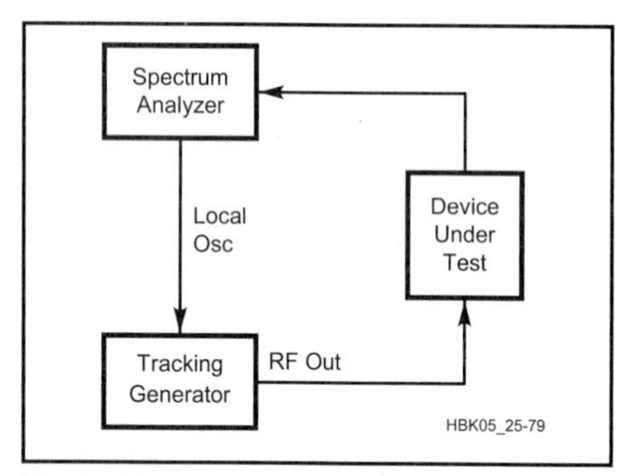

Fig 25.79 — A signal generator (shown in the figure as the "Tracking Generator") locked to the local oscillator of a spectrum analyzer can be used to determine filter response over a range of frequencies.

Fig 25.80 — A network analyzer is usually found in commercial communications development labs. It can measure both the phase and magnitude of the filter input and output signals. See text for details.

pear to be a real convenience not to have to continually reset the signal generator to the desired frequency. It is, however, more than a convenience. A signal generator connected in this way is called a tracking generator because the output frequency tracks the spectrum analyzer input frequency. The tracking generator makes it possible to make swept frequency measurements of the attenuation characteristics of circuits, even when the attenuation involved is large.

Fig 25.79 shows the connection of a tracking generator to a circuit under test. In order for the tracking generator to create an output frequency exactly equal to the input frequency of the spectrum analyzer, the internal local oscillator frequencies of the spectrum analyzer must be known. This is the reason for the interconnections between the tracking generator and the spectrum analyzer. The test setup shown will measure the gain or loss of the circuit under test. Only the magnitude of the gain or loss is available; in some cases, the phase angle between the input and output would also be an important and necessary parameter.

The spectrum analyzer is not sensitive to the phase angle of the tracking generator output. In the process of generating the tracking generator output, there are no guarantees that the phase of the tracking generator will be either known or constant. This is especially true of VHF spectrum analyzers/tracking generators where a few inches of coaxial cable represents a significant phase shift.

One effective way of measuring the phase angle between the input and output of a device under test is to sample the phase of the input and output of device under test and apply the samples to a phase detector. **Fig 25.80** shows a block diagram of this technique. An instrument that can measure both the magnitude and phase of a signal is called a vector network analyzer or simply a network analyzer. The magnitude and phase can be displayed either separately or together. When the magnitude and phase are displayed together the two can be presented as two separate traces, similar to the two traces on a dual-trace oscilloscope. A much more useful method of display is to present the magnitude and phase as a polar plot where the locus of the points of a vector having the length of the magnitude and the angle of the phase are displayed. Very sophisticated network analyzers can display all of the S parameters of a circuit in either a polar format or a Smith Chart format.

Transmitter Performance Tests

The test setup used in the ARRL Laboratory for measuring an HF transmitter or amplifier is shown in **Fig 25.81**. As can be seen, different power levels dictate different amounts of attenuation between the transmitter or amplifier and the spectrum analyzer.

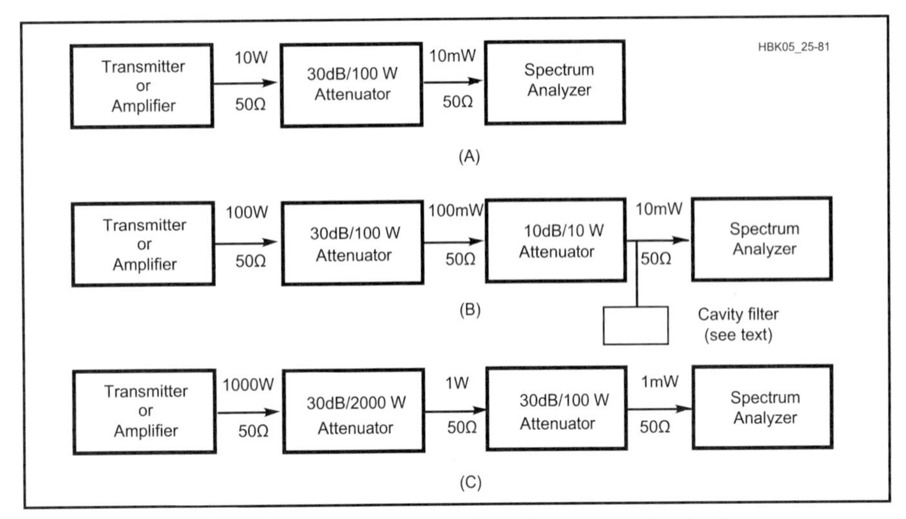

Fig 25.81 — These setups are used in the ARRL Laboratory for testing transmitters or amplifiers with several different power levels.

Spurious Emissions

Fig 25.82 shows the broadband spectrum of a transmitter, showing the harmonics in the output. The horizontal (frequency) scale is 5 MHz per division; the main output of the transmitter at 7 MHz can be seen about 1.5 major divisions from the left of the trace. Although not shown, a very large apparent signal is often seen at the extreme left of the trace. This occurs at what would be zero frequency and it is caused by the first local oscillator frequency being exactly the first IF. All up-converting superheterody ne spectrum analyzers have this IF feed-through; in addition, this signal is occasionally accompanied by a smaller spurious signal, generated within the analyzer. To determine what part of the displayed signal is a spurious response caused by IF feed-through and what is an actual input signal, simply remove the input signal and observe the trace. It is not necessary or desirable that the transmitter be modulated for this broadband test.

Other transmitter tests that can be performed with a spectrum analyzer include measurement of two-tone IMD and SSB carrier and unwanted sideband suppression.

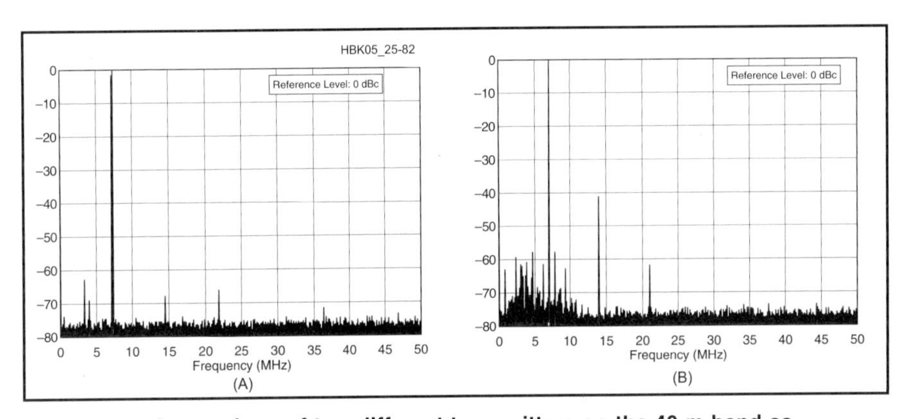

Fig 25.82 — Comparison of two different transmitters on the 40-m band as typically seen on a spectrum analyzer display. The display at the left shows a relatively clean transmitted signal but the transmitter at the right shows more spurious signal content. Horizontal scale is 5 MHz per division; vertical is 10 dB per division. According to current FCC spectral purity requirements both transmitters are acceptable.

Two-Tone IMD

Investigating the sidebands from a modulated transmitter requires a narrow-band spectrum analysis and produces displays similar to that shown in **Fig 25.83**. In this example, a two-tone test signal is used to modulate the transmitter. The display shows the two test tones plus some of the IMD produced by the SSB transmitter. The test setup used to produce this display is shown in **Fig 25.84**.

In this example, a two-tone test signal with frequencies of 700 and 1900 Hz is used to modulate the transmitter. Set the transmitter output and audio input to the manufacturer's specifications. Each desired tone is adjusted to be equal in amplitude and centered on the display. The step attenuators and analyzer controls are then adjusted to set the two desired signals 6 dB below the 0-dB reference (top) line. The IMD products can then be read directly from the display in terms of "dB below Peak Envelope Power (PEP)." (In the example shown, the third-order products are 30 dB below PEP, the fifth-order products are 37 dB down, the seventh-order products are down 40 dB.)

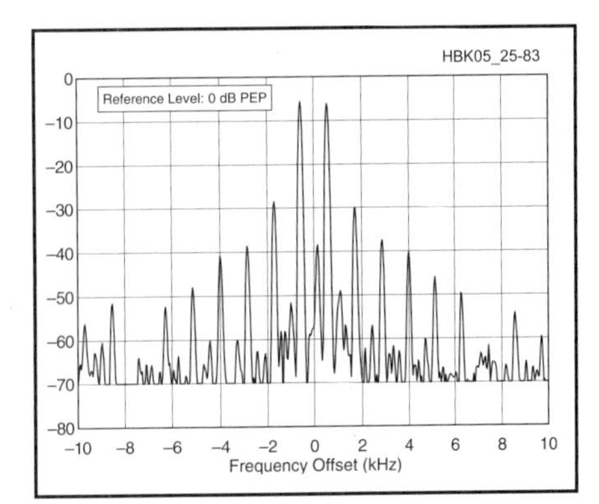

Fig 25.83 — An SSB transmitter two-tone test as seen on a spectrum analyzer. Each horizontal division represents 2 kHz and each vertical division is 10 dB. The third-order products are 30 dB below the PEP (top line), the fifth-order products are down 37 dB and seventh-order products are down 40 dB. This represents acceptable (but not ideal) performance.

Carrier and Unwanted Sideband Suppression

Single-tone audio input signals can be used with the same setup to measure unwanted sideband and carrier suppression of SSB signals. In this case, set the single tone to the 0-dB reference line. (Once the level is set, the audio can be disabled for carrier suppression measurements inorder to eliminate IMD and other effects.)

Phase Noise

Phase/composite noise is also measured with spectrum analyzers in the ARRL Lab. This test requires specialized equipment and is included here for information purposes only.

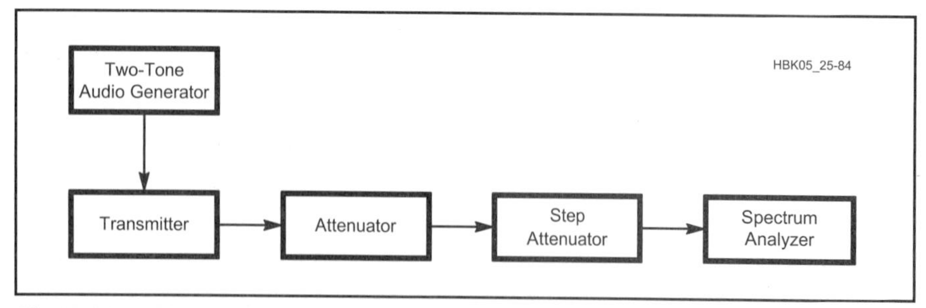

Fig 25.84 — The test setup used in the ARRL Laboratory to measure the IMD performance of transmitters and amplifiers.

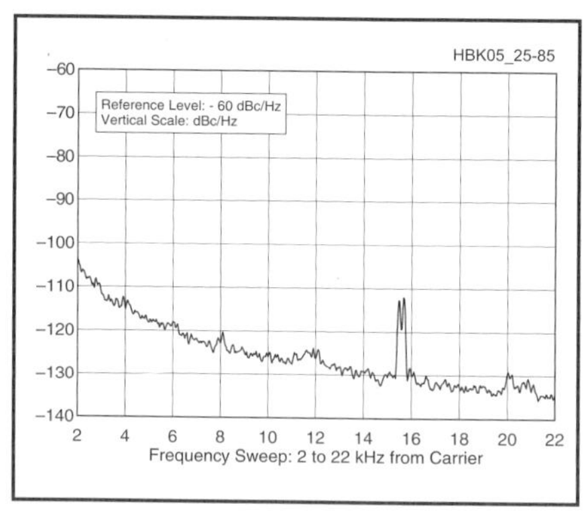

Fig 25.85 — The spectral-display results of a composite-noise test in the ARRL Lab. Power output is 100 W at 14 MHz. Vertical divisions are 10 dB; horizontal divisions are 2 kHz. The log reference level (the top horizontal line on the scale) represents –60 dBc/Hz and the baseline is –140 dBc/Hz. The carrier, off the left edge of the plot, is not shown. This plot shows composite transmitted noise 2 to 22 kHz from the carrier.

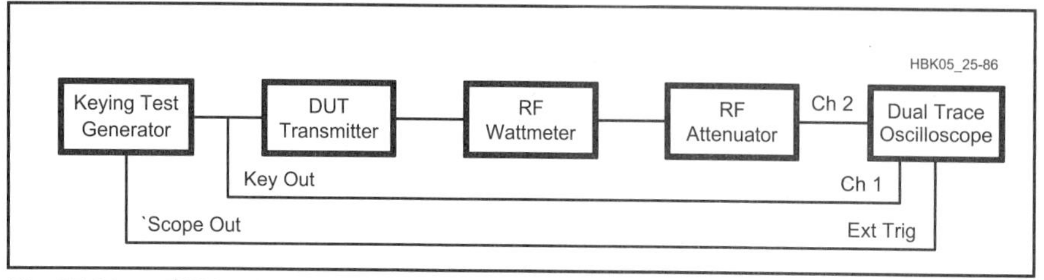

Fig 25.86 — CW keying waveform test setup.

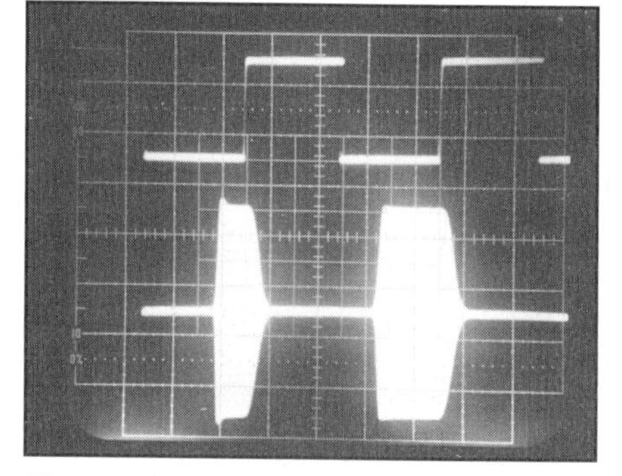

Fig 25.87 — Typical CW keying waveform test results. This display is for the ICOM IC-707 (semi-break-in mode) reviewed in April 94 *QST*. The upper trace is the actual key closure; the lower trace is the RF envelope. Horizontal divisions are 10 ms. The transceiver was being operated at 100 W output at 14 MHz.

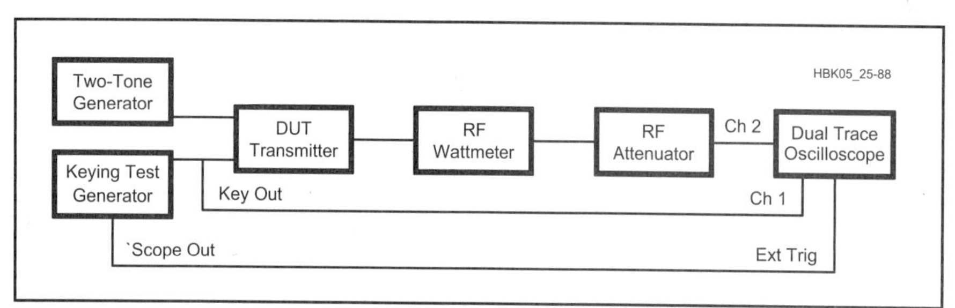

Fig 25.88 — PTT-to-RF-output test setup for voice-mode transmitters.

The purpose of the Composite-Noise test is to observe and measure the phase and amplitude noise, as well as any close-in spurious signals generated by a transmitter. Since phase noise is the primary noise component in any well-designed transmitter, almost all of the noise observed during this test is phase noise.

This measurement is accomplished in the lab by converting the transmitter output down to a frequency band about 10 or 20 kHz above baseband. A mixer and a signal generator (used as a local oscillator) are used to perform this conversion. Filters remove the 0-Hz component as well as any unwanted heterodyne components. A spectrum analyzer (see **Fig 25.85**) displays the remaining noise and spurious signals from 2 to 22 kHz from the carrier frequency (in the CW mode).

Tests in the Time Domain

Oscilloscopes are used for transmitter testing in the time domain. Dual-trace instruments are best in most cases, providing easy to read time-delay measurements between keying input and RF- or audio-output signals. Common transmitter measurements performed with 'scopes include CW keying wave shape and time delay and SSB/FM transmit-to-audio turnaround tests (important for many digital modes).

A typical setup for measuring CW keying waveform and time delay is shown in **Fig 25.86**. A keying test generator is used to repeatedly key the transmitter at a controlled rate. The generator can be set to any reasonable speed, but ARRL tests are usually conducted at 20-ms on and 20-ms off (25 Hz, 50% duty cycle). **Fig 25.87** shows a typical display. The rise and fall times of the RF output pulse are measured between the 10% and 90% points on the leading and trailing edges, respectively. The delay times are measured between the 50% points of the keying and RF output waveforms. Look at the **Receivers and Transmitters** chapter for further discussion of CW keying issues.

For voice modes (SSB/FM), a PTT-to-RF output test is similar to CW keying tests. It measures rise and fall times, as well as the on- and off-delay times just as in the CW test. See **Fig 25.88** for the test setup.

"Turnaround time" is the time it takes for a transceiver to switch from the 50% fall time of a keying pulse to 50% rise of audio output. The test setup is shown in **Fig 25.89**. Turn-around time measurements require extreme care with respect to transmitter output power, attenuation, signal-generator output and the maximum input signal that can be tolerated by the

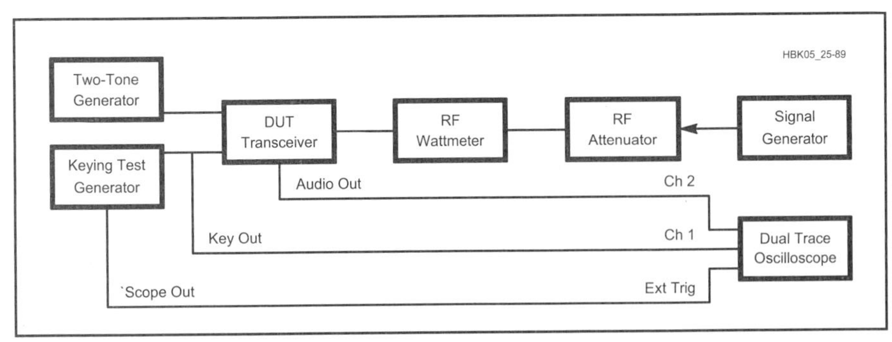

Fig 25.89 — Transmit-receive turn-around time test setup.

generator. The generator's specifications must not be exceeded and the input to the receiver must be at the required level, usually S9. Receiver AGC is usually off for this test, but experimentation with AGC and signal input level can reveal surprising variations. The keying rate must be considerably slower than the turn-around time; rates of 200-ms on/200-ms off or faster, have been used with success in Product Review tests at the ARRL Lab.

Turn-around time is an important consideration with some digital modes. AMTOR, for example, requires a turn-around time of 35 ms or less.

OSCILLOSCOPE BIBLIOGRAPHY

R. vanErk, Oscilloscopes, Functional Operation and Measuring Examples, McGraw-Hill Book Co, New York, 1978.

V. Bunze, Probing in Perspective—Application Note 152, Hewlett-Packard Co, Colorado Springs, CO, 1972 (Pub No. 5952-1892).

The XYZs of Using a Scope, Tektronix, Inc, Portland, OR, 1981 (Pub No. 41AX-4758).

Basic Techniques of Waveform Measurement (Parts 1 and 2), Hewlett-Packard Co, Colorado Springs, CO, 1980 (Pub No. 5953-3873).

J. Millman, and H. Taub, Pulse Digital and Switching Waveforms, McGraw-Hill Book Co, New York, 1965, pp 50-54.

V. Martin, ABCs of DMMs, Fluke Corp, PO Box 9090, Everett, WA 98206.

GLOSSARY

Alternating current (ac) — The polarity constantly reverses, as contrasted to dc (direct current) where polarity is fixed.

Analog — Signals which have a full set of values. If the signal varies between 0 and 10 V all values in this range can be found. Compare this to a *digital* system.

Attenuator — A device which reduces the amplitude of a signal.

Average value — Obtained by recording or measuring *N* samples of a signal, adding up all of these values, and dividing this sum by *N*.

Bandwidth — A measure of how wide a signal is in frequency. If a signal covers 14,200 to 14,205 kHz its bandwidth is said to be 5 kHz.

BNC — A small bayonet-type connector used with coax cable.

Bridge circuit — Four passive elements, such as resistors, inductors, connected as a pair of voltage dividers with a meter or other measuring device across two opposite junctions. Used to indicate the relative values of the four passive elements. See the chapter discussion of Wheatstone bridges.

CMOS — A family of digital logic elements usually selected for their low power drain. See the **Electrical Signals and Components** chapter of this *Handbook*.

Coaxial cable (coax) — A cable formed of two conductors that share the same axis. The center conductor may be a single wire or a stranded cable. The outer conductor is called the shield. The shield may be flexible braid, foil, semirigid or rigid metal. For more information, look in the **Transmission Lines** chapter.

Combiners — See ***Hybrid.***

D'Arsonval meter — A common mechanical meter consisting of a permanent magnet and a moving coil (with pointer attached).

Direct Current (dc) — The polarity is fixed for all time, as contrasted to ac (alternating current) where polarity constantly reverses.

Digital — A system that allows signals to assume a finite range of states. Binary logic is the most common example. Only two values are permitted in a binary system: one value is defined as a logical *1* and the other value as a logical *0*. See the *Handbook* chapter on **Electrical Signals and Components**.

Divider — A network of components that produce an output signal that is a fraction of the input signal. The ratio of the output to the input is the division factor. An analog divider divides voltage (a string of series connected resistors) or current (parallel connected resistors). Digital dividers divide pulse trains or frequency.

DMM (digital multimeter) — A test instrument that usually measures at least: voltage, current and resistance, and displays the result on a numeric digit display, rather than an analog meter.

Dummy antenna or dummy load — A resistor or set of resistors used in place of an antenna to test a transmitter without radiating any electromagnetic energy into the air.

DVM (digital volt meter) — See ***DMM***.

FET voltmeter — See also ***VTVM***. An updated version of a VTVM using field effect transistors (FETs) in place of vacuum tubes.

Flip-flop — A digital circuit that has two stable states. See the chapter on **Electrical Signals and Components**.

Frequency marker — Test signals generated at selected intervals (such as 25 kHz, 50 kHz, 100 kHz) for calibrating the dials of receivers and transmitters.

Fundamental — The first signal or frequency in a series of harmonically related signals. This term is often used to describe an oscillator or transmitter's desired signal.

Harmonic — A signal occurring at some integral multiple (such as two, three, four) of a *fundamental* frequency.

Hybrid (hybrid combiners) — A device used to connect two signal generators to one receiver for test purposes, without the two generators affecting each other.

IC (integrated circuit) — A complete circuit built into a single electronic component.

LCD (liquid crystal display) — A low-power display device utilizing the physics of liquid crystals. They usually need either ambient light or backlighting to be seen.

LED (light emitting diode) — A diode that emits light when an appropriate voltage (usually 1.5 V at about 20 mA) is connected. They are used either as tiny pilot lights or in bar shapes to display letters and numbers.

Loran — A navigation system using very-low-frequency transmitters.

Marker — See ***Frequency marker.***

Multiplier — A circuit that purposely creates some desired *harmonic* of its input signal. For example, a frequency multiplier that takes energy from a 3.5-MHz exciter and puts out RF at 7 MHz is a two times multiplier, usually called a frequency doubler.

N — A type of coaxial cable connector common at UHF and higher frequencies.

NAND — A digital element that performs the *not-and* function. See the **Electrical Signals and Components** chapter.

Noise (noise figure) — Noise is generated in all electrical circuits. It is particularly critical in those stages of a receiver that are closest to the antenna (RF amplifier and mixer), because noise generated in these stages can mask a weak signal. The noise figure is a measure of this noise generation. Lower noise figures mean that less noise is generated and weaker signals can be heard.

NOR — A digital element that performs the *not-or* function. See the **Electrical Signals and Components** chapter.

Null (nulling) — The process of adjusting a circuit for a minimum reading on a test meter or instrument. At a perfect null there is null, or no, energy to be seen.

Ohmmeter — A meter that measures the value of resistors. Usually part of a multimeter. See ***VOM*** and ***DMM***.

Peak value — The highest value of a signal during the measuring time. If a measured voltage varies in value from 1 to 10 V over a measuring period, the peak value would be the highest measured, 10 V.

PL-259 — A connector used for coaxial cable, usually at HF. It is also known as a male UHF connector. It is an inexpensive and common connector, but it is not weatherproof, nor is its impedance constant over frequency.

Prescaler — A circuit used ahead of a counter to extend the counter range to higher frequencies. A counter capable of operating up to 50 MHz can count up to 500 MHz when used with a divide-by-10 prescaler.

Q — The ratio of the reactance to the resistance of a component or circuit. It provides a measure of bandwidth. Lower resistive losses make for a higher Q, and a narrower bandwidth.

RMS (root mean square) — A measure of the value of a voltage or current obtained by taking values from successive small time slices over a complete cycle of the waveform, squaring those values, taking the mean of the squares, and then the square root of the mean. Very significant when working with good ac sine waves, where the RMS of the sine wave is 0.707 of the peak value.

Scope — Slang for oscilloscope. See the Oscilloscopes section of this chapter.

Shunt — Elements connected in parallel.

Sinusoidal (sine wave) — The nominal waveform for unmodulated RF energy and many other ac voltages.

Spectrum — Used to describe a range of frequencies or wavelengths. The RF spectrum starts at perhaps 10 kHz and extends up to several hundred gigahertz. The light spectrum goes from infrared to ultraviolet.

Spurious emissions, or spurs — Unwanted energy generated by a transmitter or other circuit. These emissions include, but are not limited to, *harmonics*.

Thermocouple — A device made up of two different metals joined at two places. If one joint is hot and the other cold a voltage may be developed, which is a measure of the temperature difference.

Time domain — A measurement technique where the results are plotted or shown against a scale of time. In contrast to the frequency domain, where the results are plotted against a scale of frequency.

TTL (Transistor-transistor-logic) — A logic IC family commonly used with 5 V supplies. See the chapter on **Electrical Signals and Components**.

Vernier dial or vernier drive — A mechanical system of tuning dials, frequently used in older equipment, where the knob might turn 10 times for each single rotation of the control shaft.

VOM (volt-ohm-meter) — A multimeter whose design predates digital multimeters (see **DMM**).

VTVM (vacuum tube voltmeter) — A meter that was developed to provide a high input resistance and therefore low current drain (loading) from the circuit being tested. Now replaced by the FET meter.

Wheatstone bridge — See ***Bridge circuit***.

Chapter 26

Troubleshooting and Repair

Traditionally, the radio amateur has maintained a working knowledge of electronic equipment. This knowledge, and the ability to make repairs with whatever resources are available, keeps amateur stations operating when all other communications fail. This troubleshooting ability is not only a tradition; it is fundamental to the existence of the service.

This chapter, by Ed Hare, W1RFI, tells you what to do when you are faced with equipment failure or a circuit that doesn't work. It will help you ask and answer the right questions: "Should I fix it or send it back to the dealer for repair? What do I need to know to be able to fix it myself? Where do I start? What kind of test equipment do I need?" The best answers to these questions will depend on the type of test equipment you have available, the availability of a schematic or service manual and the depth of your own electronic and troubleshooting experience.

Not everyone is an electronics wizard; your set may end up at the repair shop in spite of your best efforts. The theory you learned for the FCC examinations and the information in this *Handbook* can help you decide if you can fix it yourself. If the problem is something simple (and most are), why not avoid the effort of shipping the radio to the manufacturer? It is gratifying to save time and money, but, even better, the experience and confidence you gain by fixing it yourself may prove even more valuable.

Although some say troubleshooting is as much art as it is science, the repair of electronic gear is not magic. It is more like detective work. A knowledge of complex math is not required. However, you must have, or develop, the ability to read a schematic diagram and to visualize signal flow through the circuit.

SAFETY FIRST

Always! Death is permanent. A review of safety must be the first thing discussed in a troubleshooting chapter. Some of the voltages found in amateur equipment can be fatal! Only 50 mA flowing through the body is painful; 100 to 500 mA is usually fatal. Under certain conditions, as little as 24 V can kill.

Make sure you are 100% familiar with all safety rules and the dangerous conditions that might exist in the equipment you are servicing. Remember, if the equipment is not working properly, dangerous conditions may exist where you don't expect them. Treat every component as potentially "live."

Some older equipment uses "ac/dc" circuitry. In this circuit, one side of the chassis is connected directly to the ac line. This is an electric shock waiting to happen.

A list of safety rules can be found in **Table 26.1**. You should also read the **Safety** chapter of this *Handbook* before you proceed.

GETTING HELP

Other hams may be able to help you with your troubleshooting and repair problems, either with a manual or technical help. Check with your local club or repeater group. You may get lucky and find a troubleshooting "wizard." (On the other hand, you may get some advice that is downright dangerous, so be selective.) You can also place a classified ad in one of the ham magazines, perhaps when you are looking for a rare manual.

Your fellow hams in the ARRL Field organization may also help. Technical Coordinators (TCs) and Technical Specialists (TSs) are volunteers who are willing to help hams with technical questions. For the name and address of a local TC or TS, contact your Section Manager (listed in the front of any recent issue of *QST*).

THEORY

To fix electronic equipment, you need to understand the system and circuits you are troubleshooting. A working knowledge of electronic theory, circuitry and components is an important part of the process. If necessary, review the electronic and circuit theory explained in the other chapters of this book. When you are troubleshooting, you are looking for the unexpected. Knowing how circuits are supposed to work will help you to look for things that are out of place.

TEST EQUIPMENT

Many of the steps involved in troubleshooting efficiently require the use of test equipment. We cannot see electrons flow. However, electrons do affect various devices in our equipment, with results we can measure.

Some people think they need expensive test instruments to repair their own equipment. This is not so! In fact, you probably already own the most important instruments. Some others may be purchased inexpensively, rented, borrowed or built at home. The test equipment available to you may limit the kind of repairs you can do, but you will be surprised at the kinds of repair work you can do with simple test equipment.

Senses

Although they are not "test equipment"

Table 26.1
Safety Rules

1. Keep one hand in your pocket when working on live circuits or checking to see that capacitors are discharged.
2. Include a conveniently located ground-fault current interrupter (GFCI) circuit breaker in the workbench wiring.
3. Use only grounded plugs and receptacles.
4. Use a GFCI protected circuit when working outdoors, on a concrete or dirt floor, in wet areas, or near fixtures or appliances connected to water lines, or within six feet of any exposed grounded building feature.
5. Use a fused, power limiting isolation transformer when working on ac/dc devices.
6. Switch off the power, *disconnect equipment from the power source, ground the output of the internal dc power supply*, and discharge capacitors when making circuit changes.
7. Do not subject electrolytic capacitors to excessive voltage, ac voltage or reverse voltage.
8. Test leads should be well insulated.
9. Do not work alone!
10. Wear safety glasses for protection against sparks and metal fragments.
11. Always use a safety harness when working above ground level.
12. Wear shoes with nonslip soles that will support your feet when climbing.
13. Wear rubber-sole shoes or use a rubber mat when standing on the ground or on a concrete floor.
14. Wear a hard hat when someone is working above you.
15. Be careful with tools that may cause short circuits.
16. Replace fuses only with those having proper ratings.

in the classic sense, your own senses will tell you as much about the equipment you are trying to fix as the most-expensive spectrum analyzer. We each have some of these natural "test instruments."

Eyes — Use them constantly. Look for evidence of heat and arcing, burned components, broken connections or wires, poor solder joints or other obvious visual problems.

Ears — Severe audio distortion can be detected by ear. The "snaps" and "pops" of arcing or the sizzling of a burning component may help you track down circuit faults. An experienced troubleshooter can diagnose some circuit problems by the sound they make. For example, a bad audio-output IC sounds slightly different than a defective speaker.

Nose — Your nose can tell you a lot. With experience, the smells of ozone, an overheating transformer and a burned carbon-composition resistor each become unique and distinctive.

Finger — Carefully use your fingers to measure low heat levels in components. Small-signal transistors can be fairly warm to the touch; anything hotter can indicate a circuit problem. (Be careful; some high-power devices or resistors can get downright hot during normal operation.)

Brain — More troubleshooting problems have been solved with a VOM and a brain than with the most expensive spectrum analyzer. You must use your brain to analyze data collected by other instruments.

"Internal" Equipment

Some "test equipment" is included in the equipment you repair. Nearly all receivers include a speaker. An S meter is usually connected ahead of the audio chain. If the S meter shows signals, it indicates that the RF and IF circuitry is probably functioning. Analyze what the unit is doing and see if it gives you a clue.

Some older receivers include a crystal frequency calibrator. The calibrator signal, which is rich in harmonics, is injected in the RF chain close to the antenna jack and may be used for signal tracing and alignment.

Bench Equipment

Here is a summary of test instruments and their applications. Some items serve several purposes and may substitute for others on the list. The list does not cover all equipment available, only the most common and useful instruments. The theory and operation of much of this test equipment is discussed in more detail in the **Test Procedures** chapter.

Multimeters — The multimeter is the most often used piece of test equipment. This group includes vacuum-tube voltmeters (VTVMs), volt-ohm-milliammeters (VOMs), field-effect transistor VOMs (FETVOMs) and digital multimeters (DMMs). Multimeters are used to read bias voltages, circuit resistance and signal level (with an appropriate probe). They can test resistors, capacitors (within certain limitations), diodes and transistors.

DMMs have become quite inexpensive. Their high input impedance, accuracy and flexibility are well worth the cost. Many of them contain other test equipment as well, such as capacitance meters, frequency counters, transistor testers and even digital thermometers. Some DMMs are affected by RF, so most technicians keep an analog-display VOM on hand for use near RF equipment.

When buying an analog meter, look for one with an input impedance of 20 kΩ/V or better. Reasonably priced models are available with 30 kΩ/V ($35) and 50 kΩ/V ($40). The 10 MΩ or better input impedance of DMMs, FETVOMs, VTVMs and other electronic voltmeters makes them the preferred instruments for voltage measurements.

Test leads — Keep an assortment of wires with insulated, soldered alligator clips. Commercially made leads have a high failure rate because they use small wire that is not soldered to the clips; it is best to make your own.

Open wire leads (**Fig 26.1A**) are good for dc measurements, but they can pick up unwanted RF energy. This problem is reduced somewhat if the leads are twisted together (Fig 26.1B). A coaxial cable lead is much better, but its inherent capacitance can affect RF measurements.

The most common probe is the low-capacitance (×10) probe shown in Fig 26.1C. This probe isolates the oscilloscope from the circuit under test, preventing the 'scope's input and test-probe capacitance from affecting the circuit and changing the reading. A network in the probe serves as a 10:1 divider and compensates for frequency distortion in the cable and test instrument.

Demodulator probes (see the **Test Procedures** chapter and the schematic shown in Fig 26.1D) are used to demodulate or detect RF signals, converting modulated RF signals to audio that can be heard in a signal tracer or seen on a low-bandwidth 'scope.

You can make a probe for inductive coupling as shown in Fig 26.1E. Connect a two- or three-turn loop across the center conductor and shield before sealing the end. The inductive pick up is useful for coupling to high-current points.

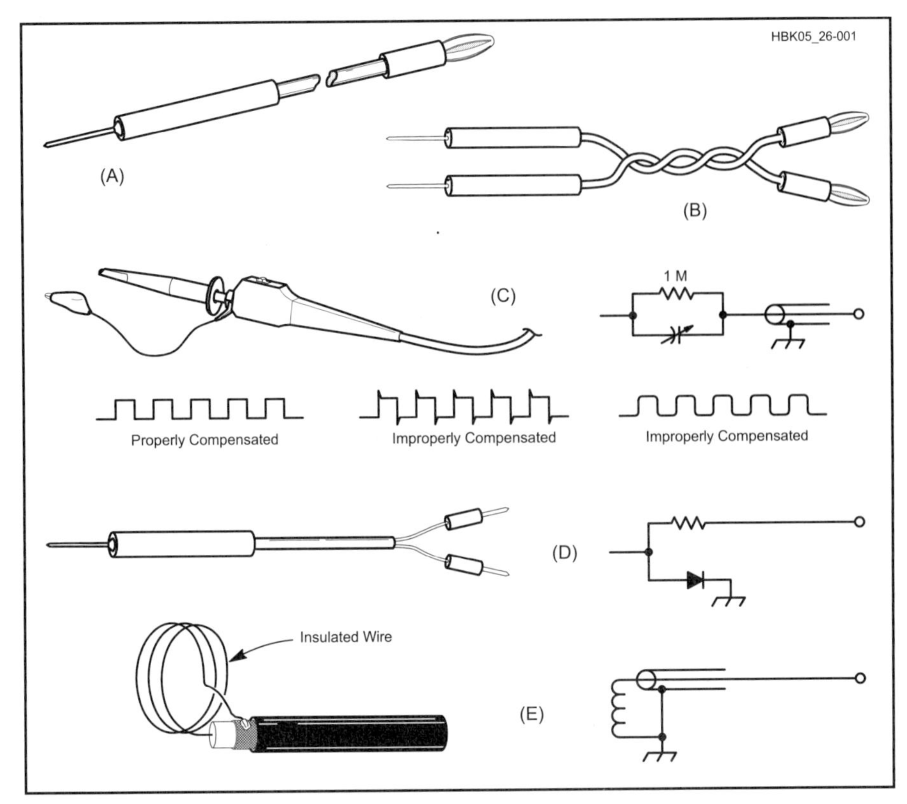

Fig 26.1—An array of test probes for use with various test instruments.

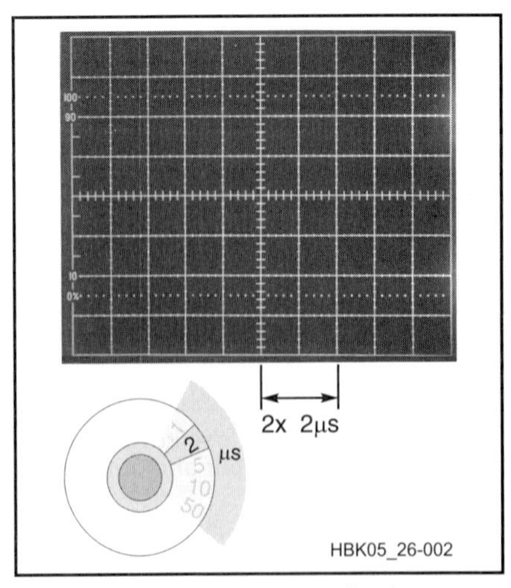

Fig 26.2—An oscilloscope display showing the relationship between time-base setting and graticule lines.

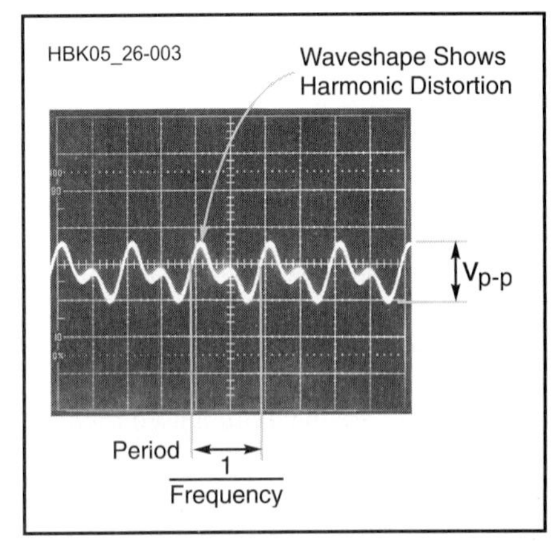

Fig 26.3—Information available from a typical oscilloscope display of a waveform.

RF power and SWR meters — Every shack should have one. It is used to measure forward and reflected RF power. A standing-wave ratio (SWR) meter can be the first indicator of antenna trouble. It can also be used between an exciter and power amplifier to spot an impedance mismatch.

Simple meters indicate relative power SWR and are fine for Transmatch adjustment and line monitoring. However, if you want to make accurate measurements, a calibrated wattmeter with a directional coupler is required.

Dummy load — A "dummy" or "phantom" load is a necessity in any shack. Do not put a signal on the air while repairing equipment. Defective equipment can generate signals that interfere with other hams or other radio services. A dummy load also provides a known, matched load (usually 50 Ω) for use during adjustments.

When buying a dummy load, avoid used, oil-cooled dummy loads unless you can be sure that the oil does not contain PCBs. This biologically hazardous compound was common in transformer oil until a few years ago.

Dummy loads in the shack are often required to dissipate a transmitter's or linear amplifier's full power output, i.e., up to 1500 W, at least for a short time. They are also used in lower power applications—typically smaller sized and rated, and can also be used as impedance standards or as terminations for low-power computer or communications lines.

Dip meter — This device is often called a transistor dip meter or a grid-dip oscillator from vacuum-tube days.

Most dip meters can also serve as an absorption frequency meter. In this mode, measurements are read at the current peak, rather than the dip. Some meters have a connection for headphones. The operator can usually hear signals that do not register on the meter. Because the dip meter is an oscillator, it can be used as a signal generator in certain cases where high accuracy or stability are not required.

When purchasing a dip meter, look for one that is mechanically and electrically stable. The coils should be in good condition. A headphone connection is helpful. Battery operated models are easier to use for antenna measurements. Dip meters are not nearly as common as they once were.[1]

Oscilloscope — The oscilloscope, or 'scope, is the second most often used piece of test equipment, although a lot of repairs can be accomplished without one. The trace of a 'scope can give us a lot of information about a signal at a glance.

The simplest way to display a waveform is to connect the vertical amplifier of the 'scope to a point in the circuit through a simple test lead. When viewing RF, use a low-capacitance probe that has been adjusted to match the 'scope. Select the vertical gain and time-base (horizontal scale, **Fig 26.2**) for the most useful displayed waveform.

A 'scope waveform shows voltage (if calibrated), approximate period (frequency is the reciprocal of the period) and a rough idea of signal purity (see **Fig 26.3**). If the 'scope has dual-trace capability (meaning it can display two signals at once), a second waveform may be displayed and compared to the first. When the two signals are taken from the input and output of a stage, stage linearity and phase shift can be checked (see **Fig 26.4**).

An important specification of an oscilloscope is its amplifier bandwidth. This tells us the frequency at which amplifier response has dropped 3 dB. The instrument will display higher frequencies, but its accuracy at higher frequencies is not known. Even well below its rated bandwidth a 'scope is not capable of much more than about 5% accuracy. This is adequate

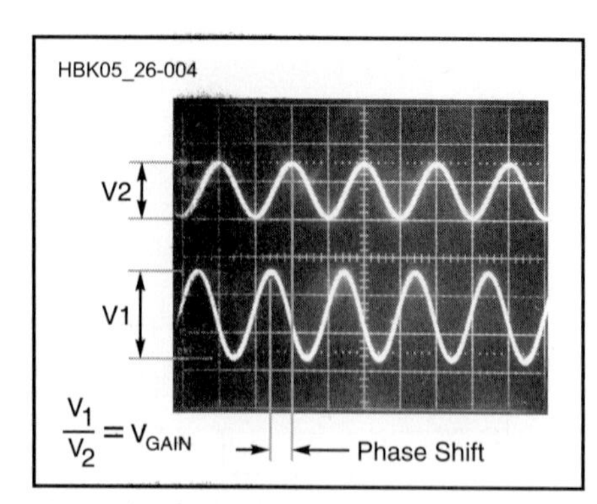

Fig 26.4—A dual-trace oscilloscope display of amplifier input and output waveforms.

for most amateur applications.

An oscilloscope will show gross distortions of audio and RF waveforms, but it cannot be used to verify that a transmitter meets FCC regulations for harmonics and spurious emissions. Harmonics that are down only 20 dB from the fundamental would be illegal in most cases, but they would not change the oscilloscope waveform enough to be seen.

When buying a 'scope, get the greatest bandwidth you can afford. Old Hewlett-Packard or Tektronix 'scopes are usually quite good for amateur use.

Signal generator — Although signal generators have many uses, in troubleshooting they are most often used for signal injection (more about this later) and alignment.

An AF/RF signal-injector schematic is shown in **Fig 26.5**. If frequency accuracy is needed, the crystal-controlled signal source of **Fig 26.6** can be used. The AF/RF circuit provides usable harmonics up to 30 MHz, while the crystal controlled oscillator will function with crystals from 1 to 15 MHz. These two projects are not meant to compete with standard signal generators, but they are adequate for signal injection.[2] A better generator is required for receiver alignment or for receiver quality testing.

When buying a generator, look for one that can generate a sine wave signal. A good signal generator is double or triple shielded against leakage. Fixed-frequency audio should be available for modulation of the RF signal and for injection into audio stages. The most versatile generators can generate amplitude and frequency modulated signals.

Good generators have stable frequency controls with no backlash. They also have multiposition switches to control signal level. A switch marked in dBm is a good indication that you have located a high-quality test instrument. The output jack should be a coaxial connector (usually a BNC or N), not the kind used for microphone connections.

Some older, high-quality units are common. Look for World War II surplus units of the URM series, Boonton, GenRad, Hewlett-Packard, Tektronix, Measurements Inc or other well-known brand names. Some home-built signal generators may be quite good, but make sure to check construction techniques, level control and shielding quality.

Signal tracer — Signals can be traced with a voltmeter and an RF probe, a dip meter with headphones or an oscilloscope, but there are some devices made especially for signal tracing. A signal tracer is primarily a high-gain audio amplifier. It may have a built-in RF detector, or rely on an external RF probe. Most convert the traced signal to audio through a speaker.

The tracer must function as a receiver and detector for each frequency range in the test circuit. A high-impedance tracer input is necessary to prevent circuit loading.

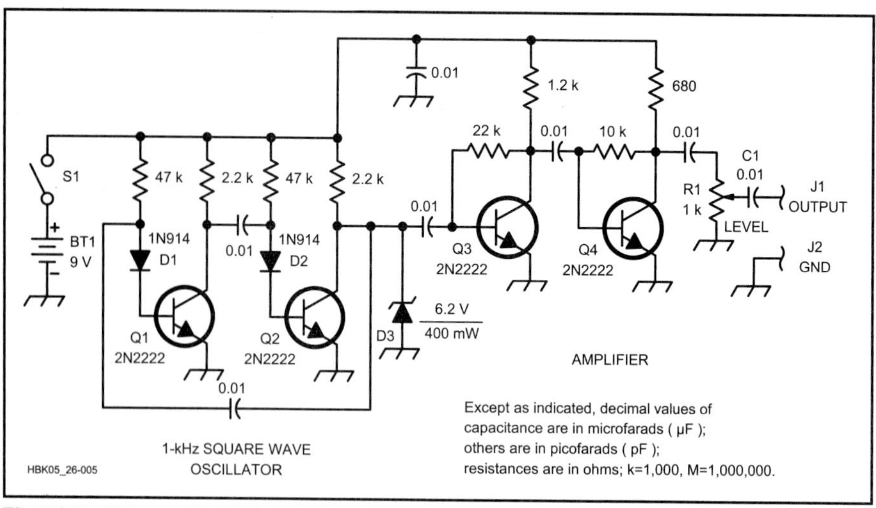

Fig 26.5—Schematic of the AF/RF signal injector. All resistors are ¼-W, 5% carbon units, and all capacitors are disc ceramic.

BT1 — 9-V battery.
D1, D2 — Silicon switching diode, 1N914 or equiv.
D3 — 6.2-V, 400-mW Zener diode.
J1, J2 — Banana jack.
Q1-Q4 — General-purpose silicon NPN transistors, 2N2222 or similar.
R1 — 1-kΩ panel-mount control.
S1 — SPST toggle switch.

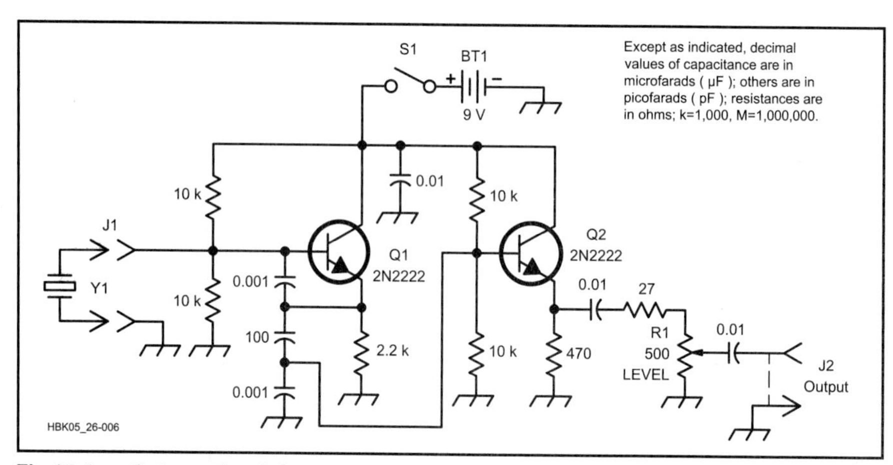

Fig 26.6 — Schematic of the crystal-controlled signal source. All resistors are ¼-W, 5% carbon units, and all capacitors are disc ceramic. A full-size etching pattern and parts-placement diagram can be found on the Templates Section of the *Handbook* CD-ROM.

BT1 — 9-V transistor radio battery.
J1 — Crystal socket to match the crystal type used.
J2 — RCA phono jack or equivalent.
Q1, Q2 — General-purpose silicon NPN transistors, 2N2222 or similar.
R1 — 500-Ω panel-mount control.
S1 — SPST toggle switch.
Y1 — 1 to 15-MHz crystal.

A general-coverage receiver can be used to trace RF or IF signals, if the receiver covers the necessary frequency range. Most receivers, however, have a low-impedance input that severely loads the test circuit. To minimize loading, use a capacitive probe or loop pickup. When the probe is held near the circuit, signals will be picked up and carried to the receiver. It may also pick up stray RF, so make sure you are listening to the correct signal by switching the circuit under test on and off while listening.

Tube tester — Vacuum-tube testers used to be found in nearly every drug or department store. They are scarce now because tubes are no longer used in modern consumer or (most) amateur equipment. Older tube gear is found in many ham shacks or flea markets, though. There are many aficionados of vintage gear who enjoy working with old vacuum-tube equipment.

Most simple tube testers measure the cathode emission of a vacuum tube. Each grid is shorted to the plate through a switch and the current is observed while the tube operates as a diode. By opening the switches from each grid to the plate (one at a time), we can check for opens and shorts. If the plate current does not drop slightly as a switch is opened, the element connected to that switch is either open or shorted to another element. (We cannot tell an open from a short with this test.) The emission tester does not necessarily indicate the ability of a tube to amplify.

Other tube testers measure tube gain (transconductance). Some transconductance testers read plate current with a fixed bias network. Others use an ac signal to drive the tube while measuring plate current.

Most tube testers also check interelement leakage. Contamination inside the tube envelope may result in current leakage between elements. The paths can have high resistance, and may be caused by gas or deposits inside the tube. Tube testers use a moderate voltage to check for leakage. Leakage can also be checked with an ohmmeter using the ×1M range, depending on the actual spacing of tube elements.

Transistor tester — Transistor testers are similar to transconductance tube testers. Device current is measured while the device is conducting or while an ac signal is applied at the control terminal. Commercial surplus units are often seen at ham flea markets. Some DMMs being sold today also include a built-in, simple transistor tester.

Most transistor failures appear as either an open or shorted junction. Opens and shorts can be found easily with an ohmmeter; a special tester is not required.

Transistor gain characteristics vary widely, however, even between units with the same device number. Testers can be used to measure the gain of a transistor. A tester that uses dc signals measures only transistor dc alpha and beta. Testers that apply an ac signal show the ac alpha or beta. Better testers also test for leakage.

In addition to telling you whether a transistor is good or bad, a transistor tester can help you decide if a particular transistor has sufficient gain for use as a replacement. It may also help when matched transistors are required. The final test is the repair circuit.

Frequency meter — Most frequency counters are digital units, often able to show frequency to a 1-Hz resolution. Some older "analog" counters are sometimes found surplus, but a low-cost digital counter will out-perform even the best of these old "classics."

Power supplies — A well-equipped test bench should include a means of varying the ac-line voltage, a variable-voltage regulated dc supply and an isolation transformer.

AC-line voltage varies slightly with load. An autotransformer with a movable tap lets you boost or reduce the line voltage slightly. This is helpful to test circuit functions with supply-voltage variations.

As mentioned earlier, ac/dc radios must be isolated from the ac line during testing and repair. Keep an isolation transformer handy if you want to work on table-model broadcast radios or television sets (check for other ac/dc equipment, too. Even some old phonographs or Amateur Radio transceivers used this dangerous circuit design).

A good multivoltage supply will help with nearly any analog or digital troubleshooting project. Many electronics distributors stock bench power supplies. A variable-voltage dc supply may be used to power various small items under repair or provide a variable bias supply for testing active devices. Construction details for a laboratory power supply appear in the **Power Supplies** chapter.

If you want to work on vacuum-tube gear, the maximum voltage available from the dc supply should be high enough to serve as a plate or a bias supply for common tubes (about 300 to 400 V ought to do it).

Accessories — There are a few small items that may be used in troubleshooting. You may want to keep them handy.

Many circuit problems are sensitive to temperature. A piece of equipment may work well when first turned on (cold) but fail as it warms up. In this case, a cold source will help you find the intermittent connection. When you cool the bad component, the circuit will suddenly start working again (or stop working). Cooling sprays are available from most parts suppliers.

A heat source helps locate components that fail only when hot. A small incandescent lamp can be mounted in a large piece of sleeve insulation to produce localized heat for test purposes.

A heat source is usually used in conjunction with a cold source. If you have a circuit that stops working when it warms up, heat the circuit until it fails, then cool the components one by one. When the circuit starts working again, the last component sprayed was the bad one.

A stethoscope (with the pickup removed — see **Fig 26.7**) or a long piece of sleeve insulation can be used to listen for arcing or sizzling in a circuit.

WHERE TO BEGIN

New Construction

In most repair work, the technician is aided by the knowledge that the circuit once worked. It is only necessary to find the faulty part(s) and replace it. This is not so with newly constructed equipment. Repair of equipment with no working history is a special, and difficult, case. You

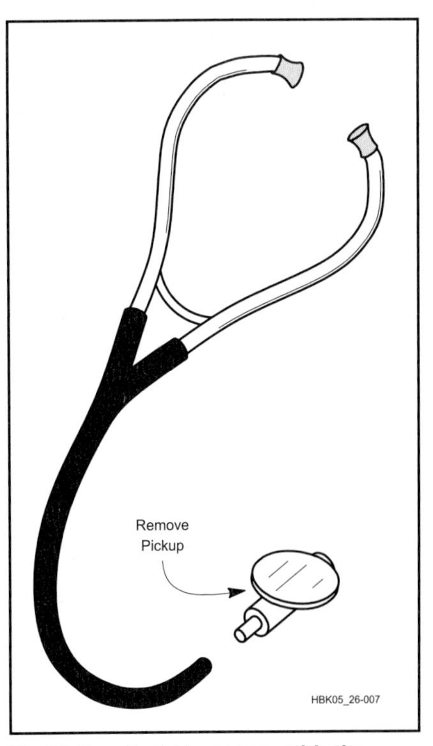

Fig 26.7 — A stethoscope, with the pickup removed, is used to listen for arcing in crowded circuits.

may be dealing with a defective component, construction error or even a faulty design. Carefully checking for these defects can save you hours.

All Equipment

Check the Obvious

Try the easy things first. If you are able to solve the problem by replacing a fuse or reconnecting a loose cable, you might be able to avoid a lot of effort. Many experienced technicians have spent hours troubleshooting a piece of equipment only to learn the hard way that the on/off switch was "off" or that they were not using the equipment properly.

Read the manual! Your equipment may be working as designed. Many electronic "problems" are caused by a switch that is set in the wrong position, or a unit that is being asked to do something it was not designed to do. Before you open up your equipment for major surgery, make sure you are using it correctly.

Next, make sure the equipment is plugged in, that the ac outlet does indeed have power, that the equipment is switched "on" and that all of the fuses are good. If the equipment uses batteries or an external power supply, make sure these are working.

Check that all wires, cables and accessories are working and plugged in to the right connectors or jacks. In a "system," it is often difficult to be sure which component or subsystem is bad. Your transmitter may not work on SSB because the transmitter is bad, but it could also be a bad microphone.

Connector faults are more common than component troubles. Consider poor connections as prime suspects in your troubleshooting detective work. Do a thorough inspection of the connections. Is the antenna connected? How about the speaker, fuses and TR switch? Are transistors and ICs firmly seated in their sockets? Are all interconnection cables sound and securely connected? Many of these problems are obvious to the eye, so look around carefully.

Simplify the Problem

If the broken equipment is part of a system, you need to find out exactly which part of the system is bad. For example, if your amateur station is not putting out any RF, you need to determine if it is a microphone problem, a transmitter problem, an amplifier problem or a problem somewhere in your station wiring. If you are trying to diagnose a bad channel on your home modular stereo system, it could be anything from a bad cable to a bad amplifier to a bad speaker.

Simplify the system as much as possible. To troubleshoot the "no-RF" problem, temporarily eliminate the amplifier from the station configuration. To diagnose the stereo system, start troubleshooting by checking just the amplifier with a set of known good headphones. Simplifying the problem will often isolate the bad component quickly.

Documentation

Once you have determined that a piece of equipment is indeed broken, you need to do some preparation before you diagnose and fix it. First, locate a schematic diagram and service manual. It is possible to troubleshoot without a service manual, but a schematic is almost indispensable.

The original equipment manufacturer is the best source of a manual or schematic. However, many old manufacturers have gone out of business. Several sources of equipment manuals can be located by a web search.

If all else fails, you can sometimes reverse engineer a simple circuit by tracing wiring paths and identifying components to draw your own schematic. If you have access to the databooks for the active devices used in the circuit, the pin-out diagrams and applications notes will sometimes be enough to help you understand and troubleshoot the circuit.

Define Problems

To begin troubleshooting, define the problem accurately. Ask yourself these questions:

1. What functions of the equipment do not work as they should; what does not work at all?

2. What kind of performance can you realistically expect?

3. Has the trouble occurred in the past? (Keep a record of troubles and maintenance in the owner's manual or log book.)

Write the answers to the questions. The information will help with your work, and may help service personnel if their advice or professional service is required.

Take It Apart

All of the preparation work has been done. It is time to really dig in. You usually will have to start by taking the equipment apart. This is the part that can trap the unwary technician. Most experienced service technicians can tell you the tale of the equipment they took apart and were unable to easily put back together. Don't let it happen to you.

Take lots of notes about the way you take it apart. Take notes about each component you remove. Write down the order in which you do things, color codes, part placements, cable routings, hardware notes and anything else you think you might need to be able to reassemble the equipment weeks from now when the back-ordered part comes in.

Put all of the screws in one place. A plastic jar with a lid works well; if you drop it the plastic is not apt to break and the lid will keep all the parts from flying around the work area (you will never find them all). It may pay to have a separate labeled container for each subsystem.

Look Around

Many service problems are visible, if you look for them carefully. Many a technician has spent hours tracking down a failure, only to find a bad solder joint or burned component that would have been spotted in careful inspection of the printed-circuit board. Start troubleshooting by carefully inspecting the equipment.

It is time consuming, but you really need to look at every connector, every wire, every solder joint and every component. A connector may have loosened, resulting in an open circuit. You may spot broken wires or see a bad solder joint. Flexing the printed-circuit board or tugging on components a bit while looking at their solder joints will often locate a defective solder job. Look for scorched components.

Make sure all of the screws securing the printed-circuit board are tight and making good electrical contact. (Do not tighten the adjusting screws, however! You will ruin the alignment.) See if you can find evidence of previous repair jobs; these may not have been done properly. Make sure that each IC is firmly seated in its socket. Look for pins folded underneath the IC rather than making contact with the socket. If you are troubleshooting a newly constructed circuit, make sure each part is of the correct value or type number and is installed correctly.

If your careful inspection doesn't reveal anything, it is time to apply power to the unit under test and continue the process. Observe all safety precautions while troubleshooting equipment. There are voltages inside some equipment that can kill you. If you are not qualified to work safely with the voltages and conditions inside of the equipment, do not proceed. See Table 26.1 and the **Safety** chapter.

Other Senses

With power applied to the unit, listen for arcs and look and smell for smoke. If no problems are apparent, you will have to start testing the various parts of the circuit.

VARIOUS APPROACHES

There are two fundamental approaches to troubleshooting: the systematic approach and the instinctive approach. The systematic approach uses a defined process to analyze and isolate the problem. An instinctive approach relies on troubleshooting experience to guide you in selecting which circuits to test and which tests to perform. The systematic approach is usually chosen by beginning troubleshooters.

At the Block Level

The block diagram is a road map. It shows the signal paths for each circuit function. These paths may run together, cross occasionally or not at all. Those blocks that are not in the paths of faulty functions can be eliminated as suspects. Sometimes the symptoms point to a single block, and no further search is necessary.

In cases where more than one block is suspect, several approaches may be used. Each requires testing a block or stage. Signal injection, signal tracing, instinct or combination of all techniques may be used to diagnose and test electronic equipment.

Systematic Approaches

The instinctive approach works well for those with years of troubleshooting experience. Those of us who are new to this game need some guidance. A systematic approach is a disciplined procedure that allows us to tackle problems in unfamiliar equipment with a reasonable hope of success.

There are two common systematic approaches to troubleshooting at the block level. The first is signal tracing; the second is signal injection. The two techniques are very similar. Differences in test equipment and the circuit under test determine which method is best in a given situation. They can often be combined.

Power Supplies

You may be able to save quite a bit of time if you test the power supply first. All of the other circuits may be dead if the power supply is not working. Power supply diagnosis is discussed in detail later in this chapter.

Signal Tracing

In signal tracing, start at the beginning of a circuit or system and follow the signal through to the end. When you find the signal at the input to a specific stage, but not at the output, you have located the defective stage. You can then measure voltages and perform other tests on that stage to locate the specific failure. This is much faster than testing every component in the unit to determine which is bad.

It is sometimes possible to use over-the-air signals in signal tracing, in a receiver for example. However, if a good signal generator is available, it is best to use it as the signal source. A modulated signal source is best.

Signal tracing is suitable for most types of troubleshooting of receivers and analog amplifiers. Signal tracing is the best way to check transmitters because all of the necessary signals are present in the transmitter by design. Most signal generators cannot supply the wide range of signal levels required to test a transmitter.

Equipment

A voltmeter, with an RF probe, is the most common instrument used for signal tracing. Low-level signals cannot be measured accurately with this instrument. Signals that do not exceed the junction drop of the diode in the probe will not register at all, but the presence, or absence, of larger signals can be observed.

A dedicated signal tracer can also be used. It is essentially an audio amplifier. An experienced technician can usually judge the level and distortion of the signal by ear. You cannot use a dedicated signal tracer to follow a signal that is not amplitude modulated (single sideband is a form of AM). Signal tracing is not suitable for tracing CW signals, FM signals or oscillators. To trace these, you will have to use a voltmeter and RF probe or an oscilloscope.

An oscilloscope is the most versatile signal tracer. It offers high input impedance, variable sensitivity, and a constant display of the traced waveform. If the oscilloscope has sufficient bandwidth, RF signals can be observed directly. Alternatively, a demodulator probe can be used to show demodulated RF signals on a low-bandwidth 'scope. Dual-trace scopes can simultaneously display the waveforms, including their phase relationship, present at the input and output of a circuit.

Procedure

First, make sure that the circuit under test and test instruments are isolated from the ac line by transformers. Set the signal source to an appropriate level and frequency for the unit you are testing. For a receiver, a signal of about 100 μV should be plenty. For other circuits, use the schematic, an analysis of circuit function and your own good judgment to set the signal level.

In signal tracing, start at the beginning and work toward the end of the signal path. Switch on power to the test circuit and connect the signal-source output to the test-circuit input. Place the tracer probe at the circuit input and ensure that you can hear the test signal. Observe the characteristics of the signal if you are using a 'scope (see **Fig 26.8**). Compare the detected signal to the source signal during tracing.

Move the tracer probe to the output of the next stage and observe the signal. Signal level should increase in amplifier stages and may decrease slightly in other stages. The signal will not be present at the output of a "dead" stage.

Low-impedance test points may not provide sufficient signal to drive a high-impedance signal tracer, so tracer sensitivity is important. Also, in some circuits the output level appears low where there is an impedance change from input to output of a stage (see **Fig 26.9**). For example, the circuit in Fig 26.9 is a common-collector current amplifier with a high input impedance and low output impedance. The voltages at TP1 and TP2 are approximately equal and in phase.

There are two signals — the test signal and the local oscillator signal — present in a mixer stage. Loss of either one will result in no output from the mixer stage. Switch the signal source on and off repeatedly to make sure that the tracer reading varies (it need not disappear) with source switching.

Signal Injection

Like signal tracing, signal injection is particularly suited to some situations. Signal injection is a good choice for receiver troubleshooting because the receiver already has a detector as part of the design. It is suitable for either high- or low-impedance circuits and can be used with vacuum tubes, transistors or ICs.

Equipment

If you are testing equipment that does not include a suitable detector as part of the circuit, some form of signal detector is required. Any of the instruments used for signal tracing are adequate.

Most of the time, your signal injector will be a signal generator. There are other injectors available, some of which are square-wave audio oscillators rich in RF harmonics (see Fig 26.5). These are usually built into a pen-sized case with a test probe at the end. These "pocket" injectors do have their limits because you can't vary their output level or determine their frequency. They are still useful, though, because most circuit failures are caused by a stage that is completely dead.

Consider the signal level at the test point when choosing an instrument. The signal source used for injection must be able to supply appropriate frequencies and levels for each stage to be tested. For example, a typical superheterodyne receiver requires AF, IF and RF signals that vary from 6 V

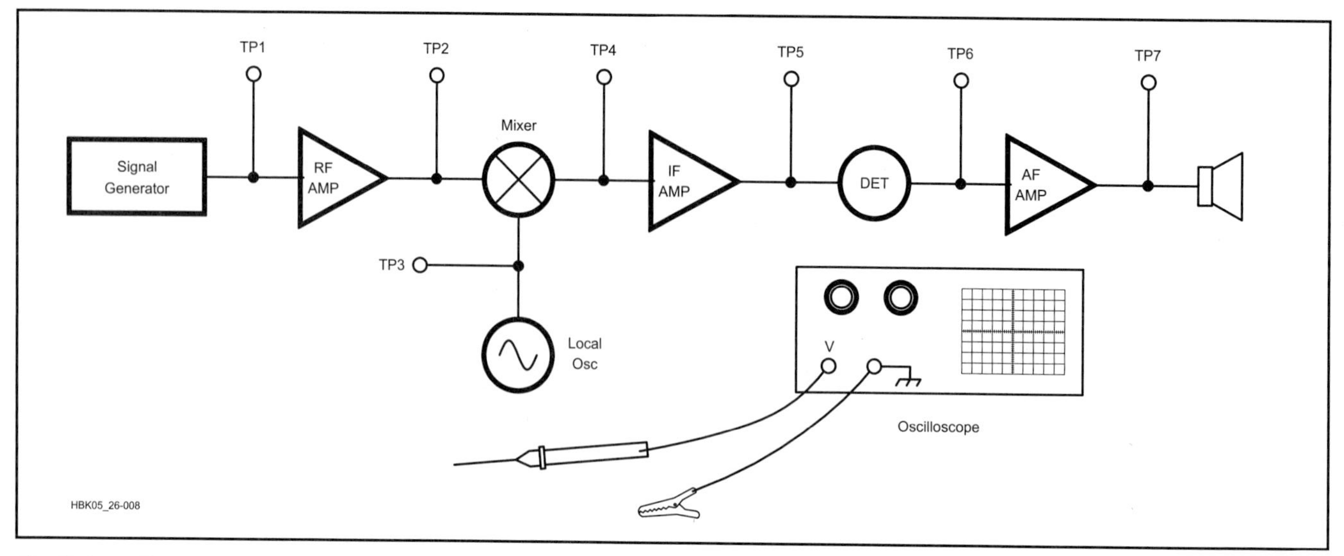

Fig 26.8 — Signal tracing in a simple receiver.

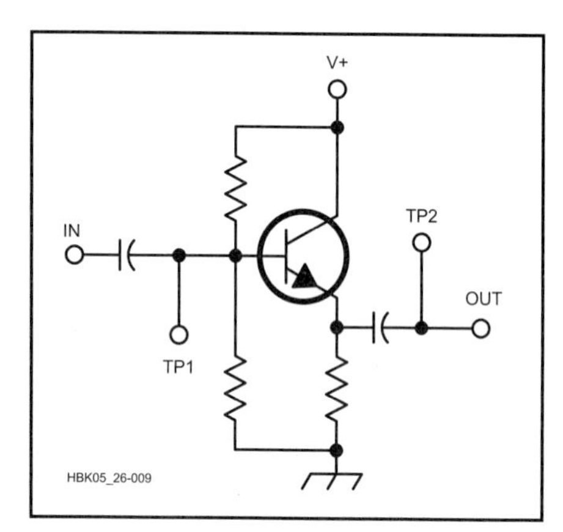

Fig 26.9 — The effect of circuit impedance on an oscilloscope display. Although the circuit functions as a current amplifier, the change in impedance from TP1 to TP2 results in the traces described. This is a common-collector amplifier.

at AF, to 0.2 mV at RF. Each conversion stage used in a receiver requires another IF from the signal source.

Procedure

If an external detector is required, set it to the proper level and connect it to the test circuit. Set the signal source for AF, and inject a signal directly into the signal detector to test operation of the injector and detector. Move the signal source to the input of the preceding stage, and observe the signal. Continue moving the signal source to the inputs of successive stages.

When you inject the signal source to the input of the defective stage, there will be no output. Prevent stage overload by reducing the level of the injected signal as testing progresses through the circuit. Use suitable frequencies for each tested stage.

Make a rough check of stage gain by injecting a signal at the input and output of an amplifier stage. You can then compare how much louder the signal is when injected at the input. This test may mislead you if there is a radical difference in impedance from stage input to output. Understand the circuit operation before testing.

Mixer stages present a special problem because they have two inputs, rather than one. A lack of output signal from a mixer can be caused by either a faulty mixer or a faulty local oscillator (LO). Check oscillator operation with a 'scope or absorption wavemeter, or by listening on another receiver. If none of these instruments are available, inject the frequency of the LO at the LO output. If a dead oscillator is the only problem, this should restore operation.

If the oscillator is operating, but off frequency, a multitude of spurious responses will appear. A simple signal injector that produces many frequencies simultaneously is not suitable for this test. Use a well-shielded signal generator set to an appropriate level at the LO frequency.

Divide and Conquer

Under certain conditions, the block search may be speeded by testing at the middle of successively smaller circuit sections. Each test limits the fault to one half of the remaining circuit (see **Fig 26.10**). Let's say the receiver has 14 stages and the fault is in stage 12. This approach requires only four tests to locate the faulty stage, a substantial saving of time.

This "divide and conquer" tactic cannot be used in equipment that splits the signal path between the input and the output. Test readings taken inside feedback loops are misleading unless you understand the circuit and the waveform to be expected at each point in the test circuit. It is best to consider all stages within a feedback loop as a single block during the block search.

Both signal tracing and signal injection procedures may be speeded by taking some diagnostic short cuts. Rather than check each stage sequentially, check a point halfway through the system. As an example:

An HF receiver is not working. There is absolutely no response from the speaker. First, substitute a suitable speaker — still no sound. Next, check the power supply — no problem there. No clues indicate any particular stage. Signal tracing or injection must provide the answer.

Get out the signal generator and switch it on. Set the generator for a low-level RF signal, switch the signal off and connect the output to the receiver. Switch the signal on again and place a high-impedance signal-tracer probe at the antenna connection. Instantly, the tracer emits a strong audio note. Good; the test equipment is functioning.

Move the probe to the input of the receiver detector. As the tracer probe touches the circuit the familiar note sounds. Next, set the tracer for audio and place the probe halfway through the audio chain. It is silent! Move the probe halfway back to the detector, and the note appears once again. Yet, no signal is present at the output of the stage. You now know that the defect is somewhere between the two

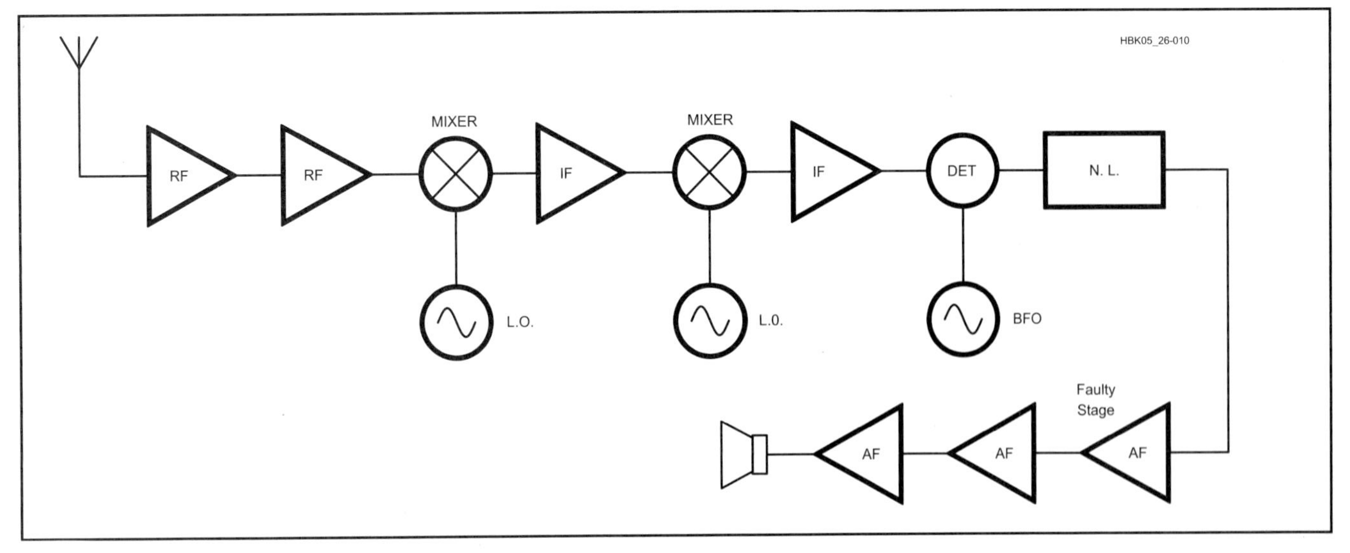

Fig 26.10 — The 14-stage receiver diagnosed by the "divide and conquer" technique.

points tested. In this case, the third audio stage is faulty.

The Instinctive Approach

In an "instinctive" approach to troubleshooting, you rely on your judgment and experience to decide where to start testing, what and how to test. When you immediately check power supply voltages, or the ac fuse on a unit that is completely nonfunctional, that is an example of an instinctive approach. If you are faced with a receiver that has distorted audio and immediately start testing the speaker and audio output stage, or if you immediately start checking the filter and bypass capacitors in an audio stage that is oscillating or "motorboating" you are troubleshooting on instinct.

Most of our discussion on the instinctive approach is really a collection of tips and guidelines. Read them to build your troubleshooting skills.

The check for connector problems mentioned at the beginning of this section is a good idea. Experience has shown connector faults to be so common that they should be checked even before a systematic approach begins.

When instinct is based on experience, searching by instinct may be the fastest procedure. If your instinct is correct, repair time and effort may be reduced substantially. As experience and confidence grow, the merits of the instinctive approach grow with them. However, inexperienced technicians who choose this approach are at the mercy of chance.

TESTING WITHIN A STAGE

Once you have followed all of the troubleshooting procedures and have isolated your problem to a single defective stage or circuit, a few simple measurements and tests will usually pinpoint one or more specific components that need adjustment or replacement.

First, check the parts in the circuit against the schematic diagram to be sure that they are reasonably close to the design values, especially in a newly built circuit. Even in a commercial piece of equipment, someone may have incorrectly changed them during attempted repairs. A wrong-value part is quite likely in new construction, such as a home-brew project.

Voltage Levels

Check the circuit voltages. If the voltage levels are printed on the schematic, this is easy. If not, analyze the circuit and make some calculations to see what the circuit voltages should be. Remember, however, that the printed or calculated voltages are nominal; measured voltages may vary from the calculations.

When making measurements, remember the following points:

- Make measurements at device leads, not at circuit-board traces or socket lugs.
- Use small test probes to prevent accidental shorts.
- Never connect or disconnect power to solid-state circuits with the switch on. Consider the effect of the meter on measured voltages. A 20-kΩ/V meter may load down a high-impedance circuit and change the voltage.

Voltages may give you a clue to what is wrong with the circuit. If not, check the active device. If you can check the active device in the circuit, do so. If not, remove it and test it, or substitute a known good device. After connections, most circuit failures are caused directly or indirectly by a bad active device. The experienced troubleshooter usually tests or substitutes these first.

Analyze the other components and determine the best way to test each. There is additional information about electronic components in the electronic-theory chapters and in the **Component Data and References** chapter.

There are two voltage levels in most circuits (V+ and ground, for example). Most component failures (opens and shorts) will shift dc voltages near one of these levels.

Typical failures that show up as incorrect dc voltages include: open coupling transformers; shorted capacitors; open, shorted or overheated resistors and open or shorted semiconductors.

Noise

A slight hiss is normal in all electronic circuits. This noise is produced whenever current flows through a conductor that is warmer than absolute zero. Noise is compounded and amplified by succeeding stages. Repair is necessary only when noise threatens to obscure normally clear signals.

Semiconductors can produce hiss in two ways. The first is normal — an even white noise that is much quieter than the desired signal. Faulty devices frequently produce excessive noise. The noise from a faulty device is usually erratic, with pops and crashes that are sometimes louder than the desired signal. In an analog circuit, the end result of noise is usually sound. In a control or digital circuit, noise causes erratic operation: unexpected switching and so on.

Noise problems usually increase with temperature, so localized heat may help you find the source. Noise from any component may be sensitive to mechanical vibration. Tapping various components with an insulated screwdriver may quickly isolate a bad part. Noise can also be traced with an oscilloscope or signal tracer.

Nearly any component or connection can be a source of noise. Defective components are the most common cause of crackling noises. Defective connections are a common cause of loud, popping noises.

Check connections at cables, sockets and switches. Look for dirty variable-capacitor wipers and potentiometers. Mica trimmer capacitors often sound like lightning when arcing occurs. Test them by installing a series 0.01-µF capacitor. If the noise disappears, replace the trimmer.

Potentiometers are particularly prone to noise problems when used in dc circuits. Clean them with spray cleaner and rotate the shaft several times.

Rotary switches may be tested by jumpering the contacts with a clip lead. Loose contacts may sometimes be repaired, either by cleaning, carefully rebending the switch contacts or gluing loose switch parts to the switch deck. Operate variable components through their range while observing the noise level at the circuit output.

Oscillations

Oscillations occur whenever there is sufficient positive feedback in a circuit that has gain. (This can even include digital devices.) Oscillation may occur at any frequency from a low-frequency audio buzz (often called "motorboating") well up into the RF region.

Unwanted oscillations are usually the result of changes in the active device (increased junction or interelectrode capacitance), failure of an oscillation suppressing component (open decoupling or bypass capacitors or neutralizing components) or new feedback paths (improper lead dress or dirt on the chassis or components). It can also be caused by improper design, especially in home-brew circuits. A shift in bias or drive levels may aggravate oscillation problems.

Oscillations that occur in audio stages do not change as the radio is tuned because the operating frequency, and therefore the component impedances, do not change. However, RF and IF oscillations usually vary in amplitude as operating frequency is changed.

Oscillation stops when the positive feedback is removed. Locating and replacing the defective (or missing) bypass capacitor may effect an improvement. The defective oscillating stage can be found more reliably with a signal tracer or oscilloscope.

Amplitude Distortion

Amplitude distortion is the product of nonlinear operation. The resultant waveform contains not only the input signal, but new signals at other frequencies as well. All of the frequencies combine to produce the distorted waveform. Distortion in a transmitter gives rise to splatter, harmonics and interference.

Fig 26.11 shows some typical cases of distortion. Clipping (also called flat-topping) is the consequence of excessive drive. The corners on the waveform show that harmonics are present. (A square wave contains the fundamental and all odd harmonics.) These odd harmonics would be heard well away from the operating frequency, possibly outside of amateur bands. Key clicks are similar to clipping.

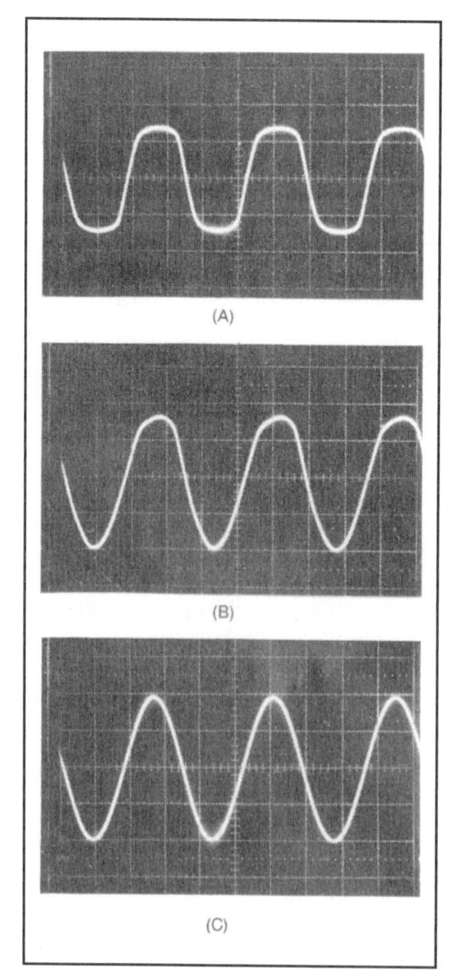

Fig 26.11 — Examples of distorted waveforms. The result of clipping is shown in A. Nonlinear amplification is shown in B. A pure sine wave is shown in C for comparison.

Harmonic distortion produces radiation at frequencies far removed from the fundamental; it is a major cause of electromagnetic interference (EMI). Harmonics are generated in nearly every amplifier. When they occur in a transmitter, they are usually caused by insufficient transmitter filtering (either by design, or because of filter component failure).

Incorrect bias brings about unequal amplification of the positive and negative wave sections. The resultant waveform is rich in harmonics.

Frequency Distortion

If a "broadband" amplifier, such as an audio amplifier, doesn't amplify all frequencies equally, there is frequency distortion. In many cases, this "frequency distortion" is deliberate, as in a transmitter microphone amplifier that has been designed to pass only frequencies from 200 to 2000 Hz. In most cases, the amateur's ability to detect and measure distortion is limited by available test equipment.

Distortion Measurement

A distortion meter is used to measure distortion of AF signals. A spectrum analyzer is the best piece of test gear to measure distortion of RF signals. If a distortion meter is not available, an estimation of AF distortion can sometimes be made with a function generator (sine and square waves) and an oscilloscope.

To estimate the amount of frequency distortion in an audio amplifier, set the generator for a square wave and look at it on the 'scope. (Use a low-capacitance probe.) The wave should show square corners and a flat top. Next, inject a square wave at the amplifier input and again look at the input wave on the 'scope. Any new distortion is a result of the test circuit loading the generator output. (If the wave shape is severely distorted, the test is not valid.) Now, move the test probe to the test circuit output and look at the waveform. Refer to **Fig 26.12** to evaluate square-wave distortion and its cause.

The above applies only to audio amplifiers without frequency tailoring. In RF gear, the transmitter may have a very narrow audio passband, so inserting a square wave into the microphone input may result in an output that is difficult to interpret. The frequency of the square wave will have a significant effect.

Anything that changes the proper bias of an amplifier can cause distortion. This includes failures in the bias components, leaky transistors or vacuum tubes with interelectrode shorts. These conditions

Look for the Obvious

The best example of how looking for the obvious can save a lot of repair time comes from my days as the manager of an electronics service shop. We had hired a young engineering graduate to work for us part time. He was the proud holder of a First-Class FCC Radiotelephone license (the predecessor to today's General Radiotelephone license). He was a likable sort, but, well . . . the chip on his young shoulder was a bit hard to take sometimes.

One day, I had asked him to repair a "tube-type" FM tuner. He had been poking around without success: hooking up a voltmeter, oscilloscope and signal generator, pretty much in that order. Finally, in total exasperation, he pronounced that the unit was beyond economical repair and suggested that I return it to the customer unfixed. The particular customer was a "regular," so I wanted to be sure of the diagnosis before I sent the tuner back. I told the tech I wanted to take a look at it before we wrote it off.

He started to expound loudly that there was no way that I, a lowly technician (even though I was also his boss) could find a problem that he, an engineering graduate and holder of a First Class . . . you get the idea. I did remind him gently that I was the boss, and he, realizing that I had him there, stepped aside, mumbling something about my suiting myself. He stepped back to gloat when I couldn't find it either.

I began by giving the tuner a thorough visual inspection. I looked it over carefully from stem to stern, while listening to our young apprentice proclaiming with certainty that one cannot fix electronic equipment by merely looking at it. I didn't see anything obviously wrong, so I decided to move wires and components around, looking for a bad solder joint or broken component. Of course, I had to listen to him telling me that one cannot possibly find bad components by touch. Unfortunately for our loud friend, he couldn't have been more wrong.

I grabbed hold of a ceramic bypass capacitor to give it a little wiggle, and much to my surprise it was hot enough to cause some real pain. I kept my composure; it was an opportunity for a good learning experience. Ceramic capacitors don't get very hot unless they are either shorted or very leaky. I kept silent and never let on that my finger "probe" had indeed located the bad part. I set the tuner down, sighed a bit, and then looked him right in the eye when I pointed to the capacitor and said "Change that part!"

They probably heard his bellowing in the next county! He went on and on about how there was just no way in the world that I could tell a good part from a bad part by just looking and touching things. He alternated between accusing me of pulling his leg and guessing, then back to just plain bellowing again. After letting this "source of great noise" run his course, I offered the ultimate shop challenge — I bet him a can of soda pop.

The traditional shop challenge did the trick. He smugly grabbed a replacement part from the bin and got out his soldering iron. In a matter of seconds (a new shop record, I believe) the capacitor was installed. He hooked the tuner up to a test amplifier and turned them on. After a couple of seconds, he smugly turned to me and started an "I told ya' so!" Just then, the last tube warmed up and the sounds of our local rock station blasted out of the speaker. He stopped in mid "told-ya" and stared at the tuner in disbelief. The tempo and pitch of his voice jumped by an order of magnitude as he asked me how I managed to fix the tuner without using even an ohmmeter to test a fuse. It was weeks before I told him —the soda pop tasted especially good.

The moral of the story is clear; sophisticated test equipment and procedures are useful in troubleshooting, but they are no substitute for the experience of a veteran troubleshooter. — *Ed Hare, W1RFI, ARRL Laboratory Supervisor*

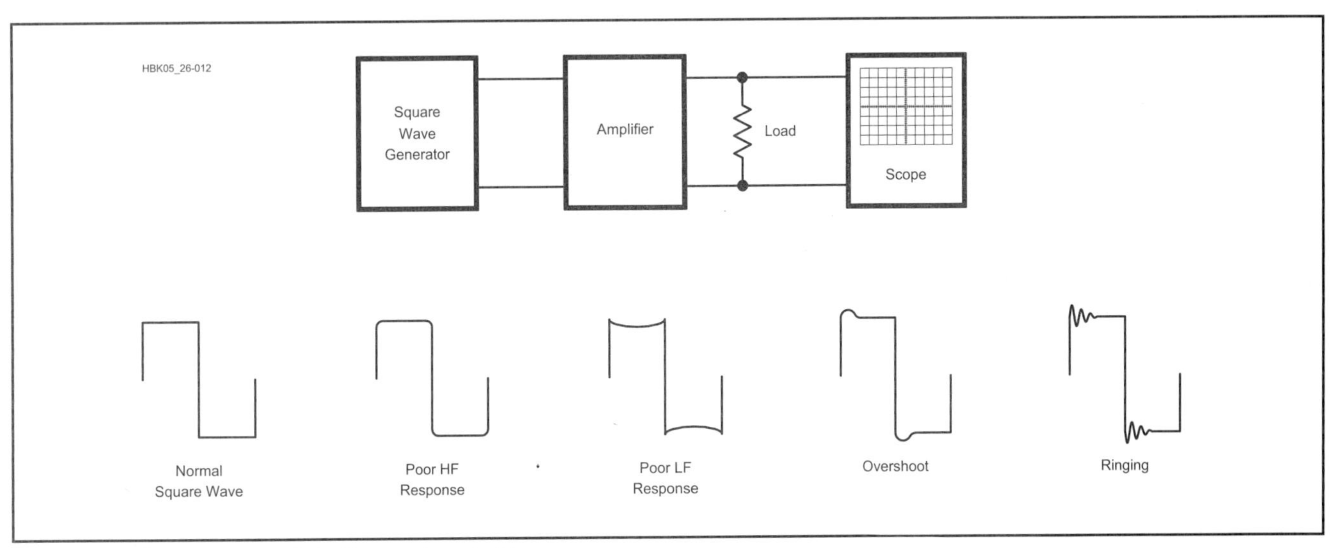

Fig 26.12 — Square-wave distortion and probable causes.

may mimic AGC trouble. Improper bias often results from an overheated or open resistor. Heat can cause resistor values to permanently increase. Leaky, or shorted capacitors and RF feedback can also produce distortion by disturbing bias levels. Distortion is also caused by circuit imbalance in Class AB or B amplifiers.

Oscillations in an IF amplifier may produce distortion. They cause constant, full AGC action, or generate spurious signals that mix with the desired signal. IF oscillations are usually evident on the S meter, which will show a strong signal even with the antenna disconnected.

Alignment

Alignment is rarely the cause of an electronics problem. As an example, suppose an AM receiver suddenly begins producing weak and distorted audio. An inexperienced person frequently suspects poor alignment as a common problem. Even though the manufacturer's instructions and the proper equipment are not available, our "friend" (this would never be one of US!) begins "adjusting" the transformer cores. Before long, the set is hopelessly misaligned. Now our misguided ham must send the radio to a shop for an alignment that was not needed before repairs were attempted.

Alignment does not shift suddenly. A normal signal tracing procedure would have shown that the signal was good up to the audio-output IC, but badly distorted after that. The defective IC that caused the problem would have been easily found and quickly replaced.

Contamination

Contamination is another common service problem. Cold soda pop spilled into a hot piece of electronics is an extreme example (but one that does actually happen).

Conductive contaminants range from water to metal filings. Most can be removed by a thorough cleaning. Any of the residue-free cleaners can be used, but remember that the cleaner may also be conductive. Do not apply power to the circuit until the area is completely dry.

Keep cleaners away from variable-capacitor plates, transformers and parts that may be harmed by the chemical. The most common conductive contaminant is solder, either from a printed-circuit board "solder bridge" or a loose piece of solder deciding to surface at the most inconvenient time.

Solder "Bridges"

In a typical PC-board solder bridge, the solder that is used to solder one component has formed a short circuit to another PC-board trace or component. Unfortunately, they are common in both new construction and repair work. Look carefully for them after you have completed any soldering, especially on a PC-board. It is even possible that a solder bridge may exist in equipment you have owned for a long time, unnoticed until it suddenly decided to become a short circuit.

Related items are loose solder blobs, loose hardware or small pieces of component leads that can show up in the most awkward and troublesome places.

Arcing

Arcing is a serious sign of trouble. It may also be a real fire hazard. Arc sites are usually easy to find because an arc that generates visible light or noticeable sound also pits and discolors conductors.

Arcing is caused by component failure, dampness, dirt or lead dress. If the dampness is temporary, dry the area thoroughly and resume operation. Dirt may be cleaned from the chassis with a residue-free cleaner. Arrange leads so high-voltage conductors are isolated. Keep them away from sharp corners and screw points.

Arcing occurs in capacitors when the working voltage is exceeded. Air-dielectric variable capacitors can sustain occasional arcs without damage, but arcing indicates operation beyond circuit limits. Transmatches working beyond their ability may suffer from arcing. A failure or high SWR in an antenna circuit may also cause transmitter arcing.

Replacing Parts

If you have located a defective component within a stage, you need to replace it. When replacing socket mounted components, be sure to align the replacement part correctly. Make sure that the pins of the device are properly inserted into the socket.

Some special tools can make it easier to remove soldered parts. A chisel-shaped soldering tip helps pry leads from printed-circuit boards or terminals. A desoldering iron or bulb forms a suction to remove excess solder, making it easier to remove the component. Spring-loaded desoldering pumps are more convenient than bulbs. Desoldering wick draws solder away from a joint when pressed against the joint with a hot soldering iron.

In all cases, remember that soldering tools and melted solder can be hot and dangerous! Wear protective goggles and clothing when soldering. A full course in first aid is beyond the scope of this chapter, but if you burn your fingers, run the burn immediately under cold water and seek first aid or medical attention. Always seek medical attention if you burn your eyes; even a small burn can develop into serious trouble.

TYPICAL SYMPTOMS AND FAULTS

Power Supplies

Many equipment failures are caused by power-supply trouble. Fortunately, most power-supply problems are easy to find and repair (see **Fig 26.13**). First, use a voltmeter to measure output. Loss of output voltage is usually caused by an open circuit. (A short circuit draws excessive current that opens the fuse, thus becoming an open circuit.)

Most fuse failures are caused by a shorted diode in the power supply or a shorted power device (RF or AF) in the failed equipment. More rarely, one of the filter capacitors can short. If the fuse has opened, turn off the power, replace the fuse and measure the load-circuit dc resistance. The measured resistance should be consistent with the power-supply ratings. A short or open load circuit indicates a problem.

If the measured resistance is too low, check the load circuit with an ohmmeter to locate the trouble. (Nominal circuit resistances are included in some equipment manuals.) If the load circuit resistance is normal, suspect a defective regulator IC or problem in the rest of the unit. Electrolytic capacitors fail with long (two years) disuse; the electrolytic layer may be reformed as explained later in this chapter.

IC regulators can oscillate, sometimes causing failure. The small-value capacitors on the input, output or adjustment pins of the regulator prevent oscillations. Check or replace these capacitors whenever a regulator has failed.

AC ripple (hum) is usually caused by low-value filter capacitors in the power supply. Less likely, hum can also be caused by excessive load, a regulation problem or RF feedback in the power supply. Look for a defective filter capacitor (usually open or low-value), defective regulator or shorted filter choke. In older equipment, the defective filter capacitor will often have visible leaking electrolyte: Look for corrosion residue at the capacitor leads. In new construction projects make sure RF energy is not getting into the power supply.

Here's an easy filter-capacitor test: Temporarily connect a replacement capacitor (about the same value and working voltage) across the suspect capacitor. If the hum goes away, replace the bad component permanently.

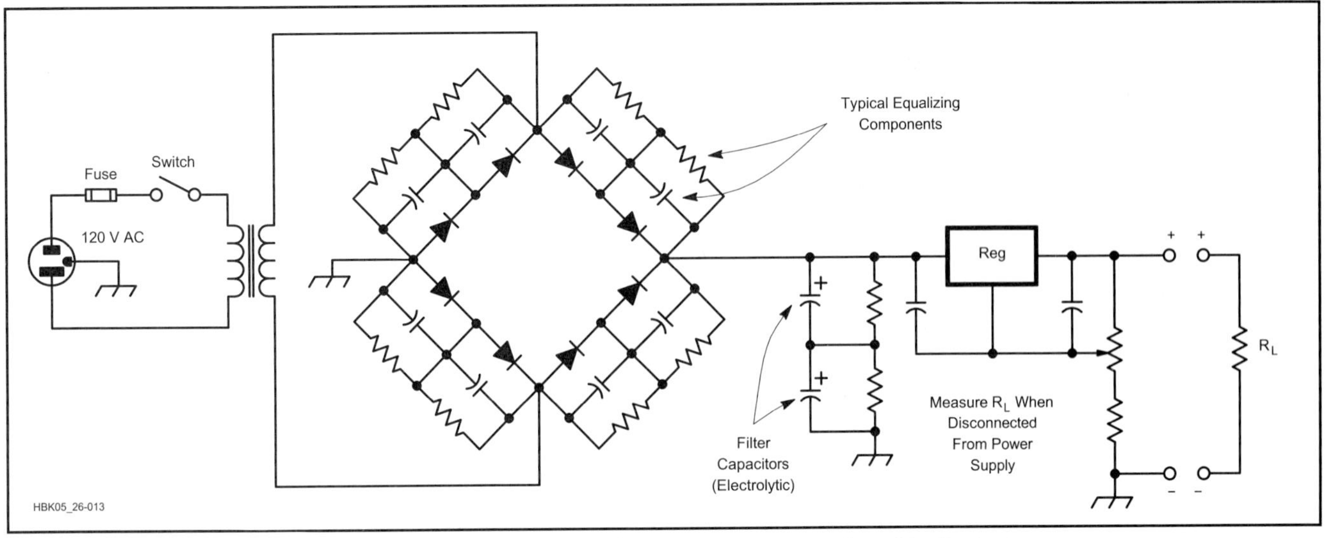

Fig 26.13 — Schematic of a typical power supply showing the components mentioned in the text.

Once the faulty component is found, inspect the surrounding circuit and consider what may have caused the problem. Sometimes one bad component can cause another to fail. For example, a shorted filter capacitor increases current flow and burns out a rectifier diode. While the defective diode is easy to find, the capacitor may show no visible damage.

Switching Power Supplies

Switching power supplies are quite different than conventional supplies. In a "switcher," a switching transistor is used to change dc voltage levels. They usually have AF oscillators and complex feedback paths. Any component failure in the rectifiers, switch, feedback path or load usually results in a completely dead supply. Every part is suspect. While active device failure is still the number one suspect, it pays to carefully test all components if a diagnosis cannot be made with traditional techniques.

Some equipment, notably TVs and monitors, derive some of the power-supply voltages from the proper operation of other parts of the circuit. In the case of a TV or monitor, voltages are often derived by adding secondary low-voltage windings to the flyback transformer and rectifying the resultant ac voltage (usually about 15 kHz). These voltages will be missing if there is any problem with the circuit from which they are derived.

Amplifiers

Amplifiers are the most common circuits in electronics. The output of an ideal amplifier would match the input signal in every respect except magnitude: No distortion or noise would be added. Real amplifiers always add noise and distortion.

Gain

Gain is the measure of amplification. Gain is usually expressed in decibels (dB) over a specified frequency range, known as the bandwidth or passband of the amplifier. When an amplifier is used to provide a stable load for the preceding stage, or as an impedance transformer, there may be little or no voltage gain.

Amplifier failure usually results in a loss of gain or excessive distortion at the amplifier output. In either case, check external connections first. Is there power to the stage? Has the fuse opened? Check the speaker and leads in audio output stages, the microphone and push-to-talk (PTT) line in transmitter audio sections. Excess voltage, excess current or thermal runaway can cause sudden failure of semiconductors. The failure may appear as either a short, or open, circuit of one or more PN junctions.

Thermal runaway occurs most often in bipolar transistor circuits. If degenerative feedback (the emitter resistor reduces base-emitter voltage as conduction increases) is insufficient, thermal runaway will allow excessive current flow and device failure. Check transistors by substitution, if possible.

Faulty coupling components can reduce amplifier output. Look for component failures that would increase series, or decrease shunt, impedance in the coupling network. Coupling faults can be located by signal tracing or parts substitution. Other passive component defects reduce amplifier output by shifting bias or causing active-device failure. These failures are evident when the dc operating voltages are measured.

In a receiver, a fault in the AGC loop may force a transistor into cutoff or saturation. Open the AGC line to the device and substitute a variable voltage for the AGC signal. If amplifier action varies with voltage, suspect the AGC-circuit components; otherwise, suspect the amplifier.

In an operating amplifier, check carefully for oscillations or noise. Oscillations are most likely to start with maximum gain and the amplifier input shorted. Any noise that is induced by 60-Hz sources can be heard, or seen with a 'scope synchronized to the ac line.

Unwanted amplifier RF oscillations should be cured with changes of lead dress or circuit components. Separate input leads from output leads; use coaxial cable to carry RF between stages; neutralize inter-element or junction capacitance. Ferrite beads on the control element of the active device often stop unwanted oscillations.

Low-frequency oscillations ("motorboating") indicate poor stage isolation or inadequate power-supply filtering. Try a better lead-dress arrangement and/or check the capacitance of the decoupling network (see **Fig 26.14**). Use larger capacitors at the power-supply leads; increase the number of capacitors or use separate decoupling capacitors at each stage. Coupling capacitors that are too low in value can also cause poor low-frequency response. Poor response to high frequencies is usually caused by circuit design.

Amplifiers vs Switches

To help you hone your skills, let's analyze a few simple circuits. There is often a big difference in the performance of similar-looking circuits. Consider the differences between a common-emitter amplifier and a common-emitter switch circuit.

Common-Emitter Amplifier

Fig 26.15 is a schematic of a common-emitter transistor amplifier. The emitter, base and collector leads are labeled e, b and c, respectively. Important dc voltages are measured at these points and designated V_e, V_b and V_c. Similarly, the important currents are I_e, I_b and I_c. V+ indicates the supply voltage.

First, analyze the voltages and signal levels in this circuit. The "junction drop," is the potential measured across a semiconductor junction that is conducting. It is typically 0.6 V for silicon and 0.2 V for germanium transistors.

This is a Class-A linear circuit. In Class-A circuits, the transistor is always conducting some current. R1 and R2 form a voltage divider that supplies dc bias (V_b) for the transistor. Normally, V_e is equal to V_b less the emitter-base junction drop. R4 provides degenerative dc bias, while C3 provides a low-impedance path for the signal. From this information, normal operating voltages can be estimated.

The bias and voltages will be set up so that the transistor collector voltage, V_c, is somewhere between V+ and ground potential. A good rule of thumb is that V_c should be about one-half of V+, although this can vary quite a bit, depending on component tolerances. The emitter voltage is usually a small percentage of V_c, say about 10%.

Any circuit failure that changes I_c (ranging from a shorted transistor or a failure in the bias circuit) changes V_c and V_e as well. An increase of I_c lowers V_c and raises V_e. If the transistor shorts from collector to emitter, V_c drops to about 1.2 V, as determined by the voltage divider formed by R3 and R4.

You would see nearly the same effect if the transistor were biased into saturation by collector-to-base leakage, a reduction in R1's value or an increase in R2's value. All of these circuit failures have the same effect. In some cases, a short in C1 or C2 could cause the same symptoms.

To properly diagnose the specific cause of low V_c, consider and test all of these parts. It is even more complex; an increase in R3's value would also decrease V_c. There would be one valuable clue, however; if R3 increased in value, I_c would not increase; V_e would also be low.

Anything that decreases I_c increases V_c. If the transistor failed "open," R1 increased in value, R2 were shorted to ground or R4 opened, then V_c would be high.

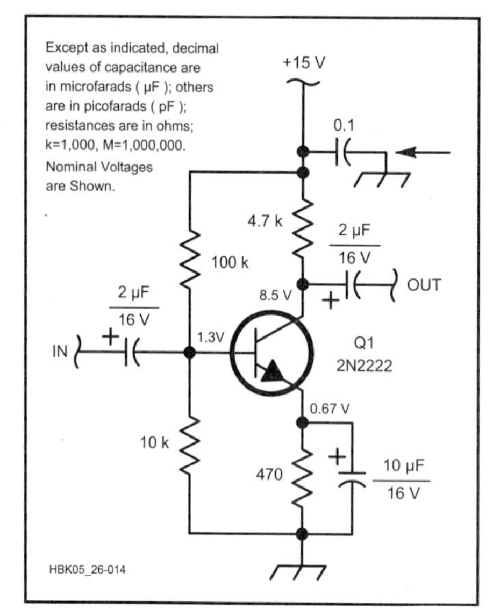

Fig 26.14 — The decoupling capacitor in this circuit is designated with an arrow.

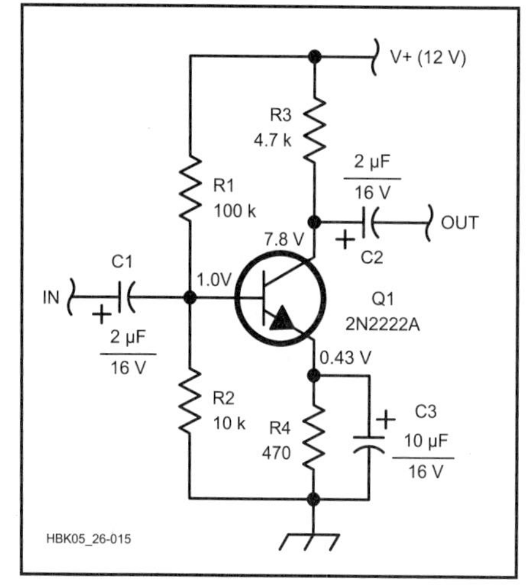

Fig 26.15 — A typical common-emitter audio amplifier.

Common-Emitter Switch

A common-emitter transistor switching circuit is shown in **Fig 26.16**. This circuit functions differently than the circuit shown in Fig 26.15. A linear amplifier is designed so that the output signal is a faithful reproduction of the input signal. Its input and output may have any value from V+ to ground.

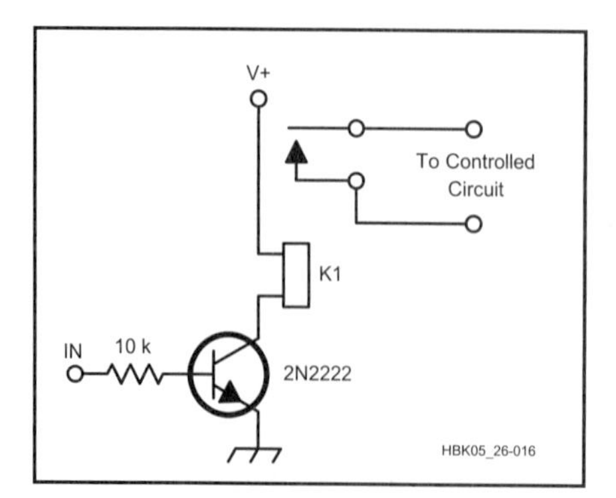

Fig 26.16 — A typical common-emitter switching amplifier.

The switching circuit of Fig 26.16, however, is similar to a "digital" circuit. The active device is either on or off, 1 or 0, just like digital logic. Its input signal level should either be 0 V or positive enough to switch the transistor on fully (saturate). Its output state should be either full off (with no current flowing through the relay), or full on (with the relay energized). A voltmeter placed on the collector will show either approximately +12 V or 0 V, depending on the input.

Understanding this difference in operation is crucial to troubleshooting the two circuits. If V_c were +12 V in the circuit in Fig 26.15, it would indicate a circuit failure. A V_c of +12 V in the switching circuit, is normal when V_b is 0 V. (If Vb measured 0.8 V or higher, Vc should be low and the relay energized.)

DC Coupled Amplifiers

In dc coupled amplifiers, the transistors are directly connected together without coupling capacitors. They comprise a unique troubleshooting case. Most often, when one device fails, it destroys one or more other semiconductors in the circuit. If you don't find all of the bad parts, the remaining defective parts can cause the installed replacements to fail immediately. To reliably troubleshoot a dc coupled circuit, you must test every semiconductor in the circuit and replace them all at once.

Oscillators

In many circuits, a failure of the oscillator will result in complete circuit failure. A transmitter will not transmit, and a superheterodyne receiver will not receive if you have an internal oscillator failure. (These symptoms do not always mean oscillator failure, however.)

Whenever there is weakening or complete loss of signal from a radio, check oscillator operation and frequency. There are several methods:

- Use a receiver with a coaxial probe to listen for the oscillator signal.
- A dip meter can be used to check oscillators. In the absorptive mode, tune the dip meter to within ±15 kHz of the oscillator, couple it to the circuit, and listen for a beat note in the dip-meter headphones.
- Look at the oscillator waveform on a 'scope. The operating frequency can't

be determined with great accuracy, but you can see if the oscillator is working at all. Use a low capacitance (10×) probe for oscillator observations.

- Tube oscillators usually have negative grid bias when oscillating. Use a high-impedance voltmeter to measure grid bias. The bias also changes slightly with frequency.
- Emitter current varies slightly with frequency in transistor oscillators. Use a sensitive, high-impedance voltmeter across the emitter resistor to observe the current level. (You can use Ohm's Law to calculate the current value.)

Many modern oscillators are phase-locked loops (PLLs). A PLL is a marriage of an analog oscillator and digital control circuitry. Read the Digital Circuitry section in this chapter and the **Oscillators and Synthesizers** chapter of this book in order to learn PLL repair techniques.

To test for a failed oscillator tuned with inductors and capacitors, use a dip meter in the active mode. Set the dip meter to the oscillator frequency and couple it to the oscillator output circuit. If the oscillator is dead, the dip-meter signal will take its place and temporarily restore some semblance of normal operation. Tune the dip meter very slowly, or you may pass stations so quickly that they sound like "birdies."

Stability

We are spoiled; modern amateur equipment is very stable. Drift of several kilohertz per hour was once normal. You may want to modify old equipment for more stability, but drift that is consistent with the equipment design is not a defect. (This applies to new equipment as well as old.) It is normal for some digital displays to flash back and forth between two values for the least-significant digit.

Drift is caused by variations in the oscillator. Poor voltage regulation and heat are the most common culprits. Check regulation with a voltmeter (use one that is not affected by RF). Voltage regulators are usually part of the oscillator circuit. Check them by substitution.

Chirp is a form of rapid drift that is usually caused by excessive oscillator loading or poor power-supply regulation. The most common cause of chirp is poor design. If chirp appears suddenly in a working circuit, look for component or design defects in the oscillator or its buffer amplifiers. (For example, a shorted coupling capacitor increases loading drastically.) Also check lead dress, tubes and switches for new feedback paths (feedback defeats buffer action).

Frequency instability may also result from defects in feedback components. Too much feedback may produce spurious signals, while too little makes oscillator start-up unreliable.

Sudden frequency changes are frequently the result of physical variations. Loose components or connections are probable causes. Check for arcing or dirt on printed-circuit boards, trimmers and variable capacitors, loose switch contacts, bad solder joints or loose connectors.

Frequency Accuracy

Dial tracking errors may be associated with oscillator operation. Misadjustments in the frequency-determining components make dial accuracy worse at the ends of the dial. Tracking errors that are constant everywhere in the passband can be caused by misalignment or by slippage in the dial drive mechanism or indicator. This is usually cured by calibration of a simple mechanical adjustment.

In LC oscillators, tracking at the high-frequency end of the dial is controlled by trimmer capacitors. A trimmer is a variable capacitor connected in parallel with the main tuning capacitor (see **Fig 26.17**). The trimmer represents a higher percentage of the total capacitance at the high end of the tuning range. It has relatively little effect on tuning characteristics at the low-frequency end of the dial.

Low-end tracking is adjusted by a padder capacitor. A padder is a variable capacitor that is connected in series with the main tuning capacitor. Padder capacitance has a greater effect at the low-frequency end of the dial. The padder capacitor is often eliminated to save money. In that case, the low-frequency tracking is adjusted by the main tuning coil.

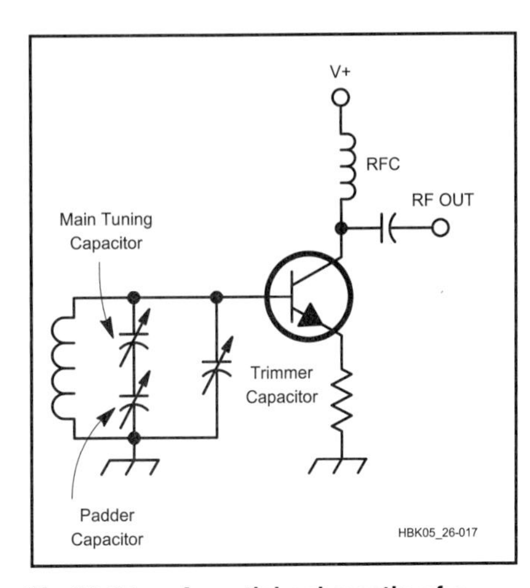

Fig 26.17 — A partial schematic of a simple oscillator showing the locations of the trimmer and padder capacitors.

Control Circuitry

Semiconductors have made it practical to use diodes for switching, running only a dc lead to the switching point. This eliminates problems caused by long analog leads in the circuit. Semiconductor switching usually reduces the cost and complexity of switching components. Switching speed is increased; contact corrosion and breakage are eliminated. In exchange, troubleshooting is complicated by additional components such as voltage regulators and decoupling capacitors (see **Fig 26.18**). The technician must consider many more components and symptoms when working with diode and transistor switched circuits.

Mechanical switches are relatively rugged. They can withstand substantial voltage and current surges. The environment does not drastically affect them, and there is usually visible damage when they fail. Semiconductor switching offers inexpensive, high-speed operation. When subjected to excess voltage or current, however, most transistors and diodes silently expire. Occasionally, if the troubleshooter is lucky, one sends up a smoke signal to mark its passing.

Temperature changes semiconductor characteristics. A normally adequate control signal may not be effective when transistor beta is lowered by a cold environment. Heat may cause a control voltage regulator to produce an improper control signal.

A control signal is actually a bias for the semiconductor switch. Forward biased diodes and transistors act as closed switches; reverse biased components simulate open switches. If the control (bias) signal is not strong enough to completely saturate the semiconductor, conduction may not continue through a full ac cycle. Severe distortion can be the result.

When dc control leads provide unwanted feedback paths, switching transistors may become modulators or mixers. Additionally, any reverse biased semiconductor junction is a potential source of white noise.

Microprocessor Control

Nearly every new transceiver is controlled by a miniature computer. Entire books have been written about microprocessor (μP) control. Many of the techniques are discussed in the Digital Circuitry section. Many microprocessor related problems end up back at the factory for service; however, the surface mounted components are just too difficult for most hams to replace. For successful repair of microprocessor controlled circuits, you should have the knowledge and

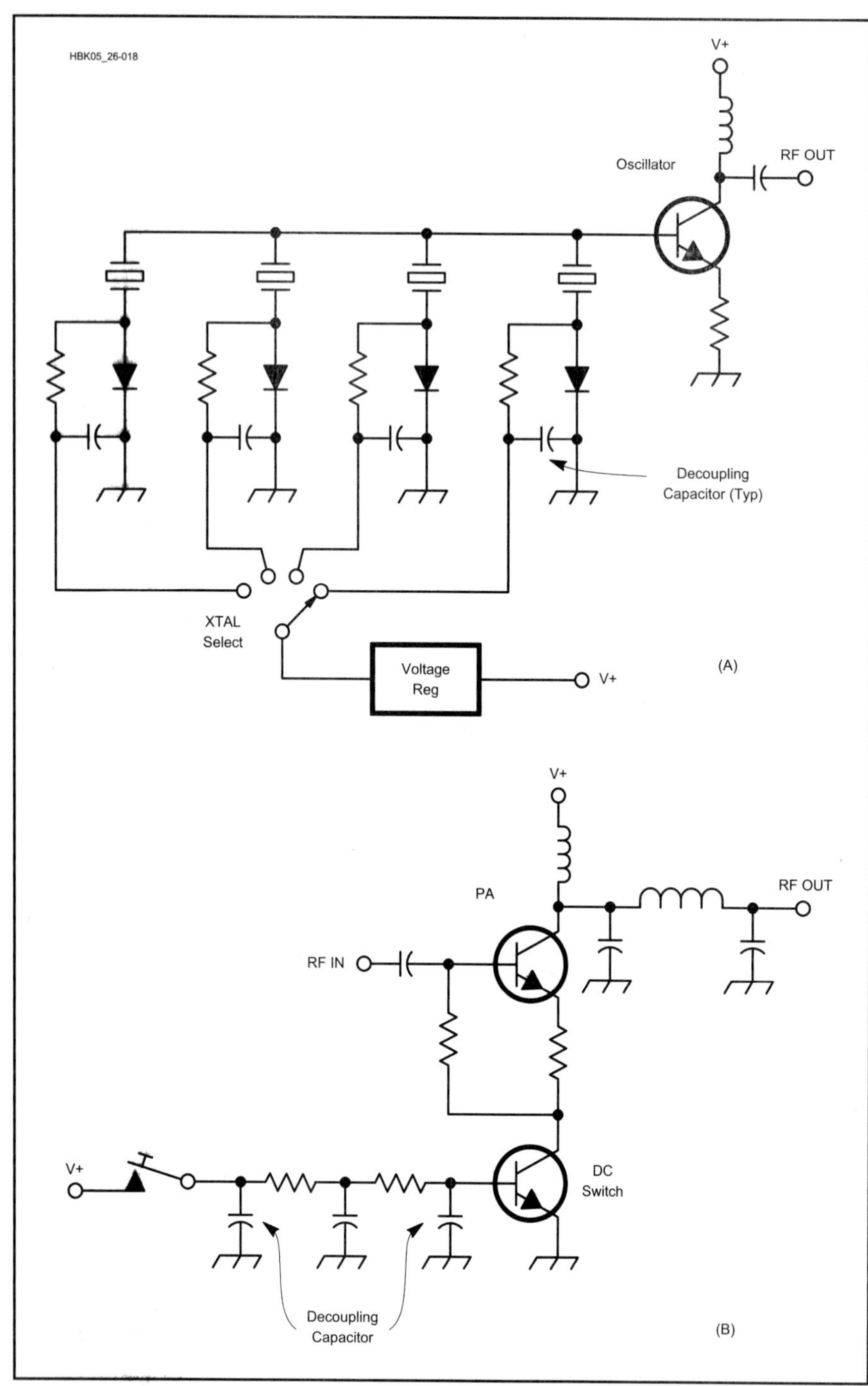

Fig 26.18 — Diode switching selects oscillator crystals at A. A transistor switch is used to key a power amplifier at B.

test equipment necessary for computer repair. Familiarity with machine-language programming may also be desirable.

Digital Circuitry

The digital revolution has hit most ham shacks and amateur equipment. Microprocessors have brought automation to everything from desk clocks to ham transceivers and computer controlled EME antenna arrays. Although every aspect of their operation may be resolved to a simple 1 or 0, or tristate (an infinite impedance or open circuit), the symptoms of their failure are far more complicated. As with other equipment:

- Observe the operating characteristics.
- Study the block diagram and the schematic.
- Test.
- Replace defective parts.

Problems in digital circuits have two elementary causes. First, the circuit may give false counts because of electrical noise at the input. Second, the gates may lock in one state.

False counts from noise are especially likely in a ham shack. (A 15- to 20-µs voltage spike can trigger a TTL flip-flop.) Amateur Radio equipment often switches heavy loads; the attendant transients can follow the ac line or radiate directly to nearby digital equipment. Oscillation in the digital circuit can also produce false counts.

How these false counts affect a circuit is dependent on the design. A station clock may run fast, but a microprocessor controlled transceiver may "decide" that it is only a receiver. It might even be difficult to determine that there is a problem without a logic analyzer or a multitrace oscilloscope and a thorough understanding of circuit operation.

Begin by removing the suspect equipment from RF fields. If the symptoms stop when there is no RF energy around, you need to shield the equipment from RF.

In the mid '90s, microprocessors in general use ran clock speeds up to a few hundred megahertz. (They are increasing all the time.) It may be impossible to filter RF signals from the lines when the RF is near the clock frequency. In these cases, the best approach is to shield the digital circuit and all lines running to it.

If digital circuitry interferes with other nearby equipment, it may be radiating spurious signals. These signals can interfere with your Amateur Radio operation or other services. Digital circuitry can also be subject to interference from strong RF fields. Erratic operation or a complete "lock up" is often the result. *The ARRL RFI Book* has a chapter on computer and digital interference. That chapter discusses interference to and from digital devices and circuits.

Logic Levels

To troubleshoot a digital circuit, check for the correct voltages at the pins of each chip. The correct voltages may not always be known, but you should be able to identify the power pins (V_{cc} and ground).

The voltages on the other pins should be either a logic high, a logic low, or tristate (more on this later). In most working digital circuitry the logic levels are constantly changing, often at RF rates. A dc voltmeter may not give reliable readings. An oscilloscope or logic analyzer is usually needed to troubleshoot digital circuitry.

Most digital circuit failures are caused by a failed logic IC. In clocked circuits, listen for the clock signal with a coax probe and a suitable receiver. If the signal is found at the clock chip, trace it to each of the other ICs to be sure that the clock system is intact. Some digital circuits use VHF clock speeds; an oscilloscope must have a bandwidth of at least twice the clock speed to be useful. If you have a suitable scope, check the pulse timing and duration against circuit specifications.

As in most circuits, failures are catastrophic. It is unlikely that an AND gate will suddenly start functioning like an OR gate. It is more likely that the gate will have a signal at its input, and no signal at the output. In a failed device, the output pin will have a steady voltage. In some cases, the voltage is steady because one of the input signals is missing. Look carefully at what is going into a digital IC to determine what should be coming out. Keep manufacturers' data books handy. These data books describe the proper functioning of most digital devices.

Tristate Devices

Many digital devices are designed with a third logic state, commonly called tristate. In this state, the output of the device acts as if it weren't there at all. Many such devices can be connected to a common "bus," with the devices that are active at any given time selected by software or hardware control signals. A computer's data and address busses are good examples of this. If any one device on the bus fails by locking itself on in a 0 or 1 logic state, the entire bus becomes nonfunctional. These tristate devices can be locked "on" by inherent failure or a failure of the signal that controls them.

Simple Gate Tests

Logic gates, flip-flops and counters can be tested (see **Fig 26.19**) by triggering them manually, with a power supply (4 to 5 V is a safe level). Diodes may be checked with an ohmmeter. Testing of more complicated ICs requires the use of a logic analyzer, multitrace scope or a dedicated IC tester.

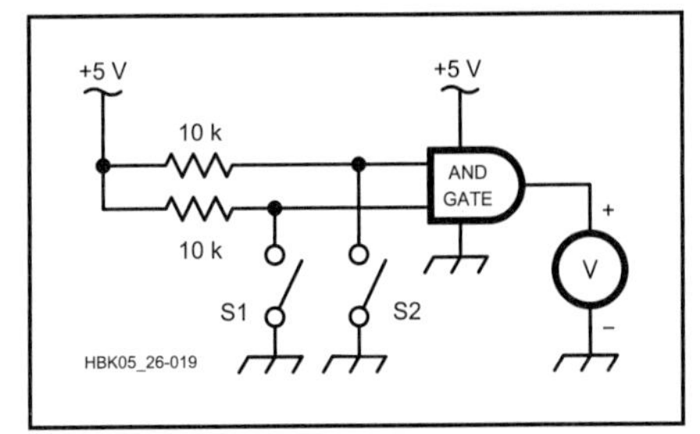

Fig 26.19 — This simple digital circuit can be tested with a few components. In this case, an AND gate is tested. Open and close S1 and S2 while comparing the voltmeter reading with a truth table for the device.

TROUBLESHOOTING HINTS

Receivers

A receiver can be diagnosed using any of the methods described earlier, but if there is not even a faint sound from the speaker, signal injection is not a good technique. If you lack troubleshooting experience, avoid following instinctive hunches. That leaves signal tracing as the best method.

The important characteristics of a receiver are selectivity, sensitivity, stability and fidelity. Receiver malfunctions ordinarily affect one or more of these areas.

Selectivity

Tuned transformers or the components used in filter circuits may develop a shorted turn, capacitors can fail and alignment is required occasionally. Such defects are accompanied by a loss of sensitivity. Except in cases of catastrophic failure (where either the filter passes all signals, or none), it is difficult to spot a loss of selectivity. Bandwidth and insertion-loss measurements are necessary to judge filter performance.

Sensitivity

A gradual loss of sensitivity results from gradual degradation of an active device or long-term changes in component values. Sudden partial sensitivity changes are usually the result of a component failure, usually in the RF or IF stages. Complete and sudden loss of sensitivity is caused by an open circuit anywhere in the signal path or by a "dead" oscillator.

Receiver Stability

The stability of a receiver depends on its oscillators. See the Oscillators section elsewhere in this chapter.

Distortion

Receiver distortion may be the effect of poor connections or faulty components in the signal path. AGC circuits produce many receiver defects that appear as distortion or insensitivity.

AGC

AGC failure usually causes distortion that affects only strong signals. All stages operate at maximum gain when the AGC influence is removed. An S meter can help diagnose AGC failure because it is operated by the AGC loop.

An open AGC bypass capacitor causes feedback through the loop. This often results in a receiver "squeal" (oscillation). Changes in the loop time constant affect tuning. If stations consistently blast, or are too weak for a brief time when first tuned in, the time constant is too fast. An excessively slow time constant makes tuning difficult, and stations fade after tuning. If the AGC is functioning, but the "timing" seems wrong, check the large-value capacitors found in the AGC circuit — they usually set the AGC time constants. If the AGC is not functioning, check the AGC-detector circuit. There is often an AGC voltage that is used to control several stages. A failure in any one stage could affect the entire loop.

Detector Problems

Detector trouble usually appears as complete loss or distortion of the received signal. AM, SSB and CW signals may be weak and unintelligible. FM signals will sound distorted. Look for an open circuit in the detector near the detector diodes. If tests of the detector parts indicate no trouble, look for a poor connection in the power-supply or ground lead. A BFO that is "dead" or off frequency prevents SSB and CW reception. In modern rigs, the BFO frequency is either crystal controlled, or derived from the PLL.

Receiver Alignment

Unfortunately, IF transformers are as enticing to the neophyte technician as a carburetor is to a shade-tree mechanic. In truth, radio alignment (and for that matter, carburetor repair) is seldom required. Circuit alignment may be justified under the following conditions:

- The set is very old and has not been adjusted in many years.
- The circuit has been subject to abusive treatment or environment.
- There is obvious misalignment from a previous repair.
- Tuned-circuit components or crystals have been replaced.
- An inexperienced technician attempted alignment without proper equipment. ("But all the screws in those little metal cans were loose!")
- There is a malfunction, but all other circuit conditions are normal. (Faulty transformers can be located because they will not tune.)

Even if one of the above conditions is met, do not attempt alignment unless you have the proper equipment. Receiver alignment should progress from the detector to the antenna terminals. When working on an FM receiver, align the detector first, then the IF and limiter stages and finally the RF amplifier and local oscillator stages. For an AM receiver, align the IF stages first, then the RF amplifier and oscillator stages.

Both AM and FM receivers can be aligned in much the same manner. Always follow the manufacturer's recommended alignment procedure. If one is not available, follow these guidelines:

1. Set the receiver RF gain to maximum, BFO control to zero or center (if applicable to your receiver) and tune to the high end of the receiver passband.
2. Disable the AGC.
3. Set the signal source to the center of the IF passband, with no modulation and minimum signal level.
4. Connect the signal source to the input of the IF section.
5. Connect a voltmeter to the IF output as shown in **Fig 26.20**.
6. Adjust the signal-source level for a slight indication on the voltmeter.
7. Peak each IF transformer in order, from the meter to the signal source. The adjustments interact; repeat steps 6 and 7 until adjustment brings no noticeable improvement.
8. Remove the signal source from the IF-section input, reduce the level to minimum, set the frequency to that shown on the receiver dial and connect the source to the antenna terminals. If necessary, tune around for the signal — if the local oscillator is not tracking, it may be off.
9. Adjust the signal level to give a slight reading on the voltmeter.
10. Adjust the trimmer capacitor of the RF amplifier for a peak reading of the test signal. (Verify that you are reading the correct signal by switching the source on and off.)
11. Reset the signal source and the receiver tuning for the low end of the passband.
12. Adjust the local-oscillator padder for peak reading.
13. Steps 8 through 11 interact, so repeat them until the results are as good as you can get them.

Transmitters

Many potential transmitter faults are discussed in several different places in this chapter. There are, however, a few techniques used to ensure stable operation of RF amplifiers in transmitters that are not covered elsewhere.

High-power RF amplifiers often use parasitic chokes to prevent instability. Older parasitic chokes usually consist of a 51- to 100-Ω noninductive resistor with a coil wound around the body and connected to the leads. It is used to prevent VHF and UHF oscillations in a vacuum-tube amplifier. The suppressor is placed in the plate lead, close to the plate connection.

In recent years, problems with this style of suppressor have been discovered. Look at the **RF Power Amplifiers** chapter for information about suppressing parasitics.

Parasitic chokes often fail from excessive current flow. In these cases, the resistor is charred. Occasionally, physical shock or corrosion produces an open circuit in the coil. Test for continuity with an ohmmeter.

Transistor amplifiers are protected against parasitic oscillations by low-value resistors or ferrite beads in the base or collector leads. Resistors are used only at low power levels (about 0.5 W), and both methods work best when applied to the base lead. Negative feedback is used to prevent oscillations at lower frequencies. An open component in the feedback loop may cause low-frequency oscillation, especially in broadband amplifiers.

Keying

The simplest form of modulation is on/off keying. Although it may seem that there cannot be much trouble with such an elementary form of modulation, two very important transmitter faults are the result of keying problems.

Key clicks are produced by fast rise and decay times of the keying waveform. Most transmitters include components in the keying circuitry to prevent clicks. When clicks are experienced, check the keying filter components first, then the succeeding stages. An improperly biased power amplifier, or a Class C amplifier that is not keyed, may produce key clicks even though the keying waveform earlier in the circuit is correct. Clicks caused by a linear amplifier may be a sign of low-frequency parasitic oscillations. If they occur in an amplifier, suspect insufficient power-supply decoupling. Check the power-supply filter capacitors and all bypass capacitors.

The other modulation problem associated with on/off keying is called back-

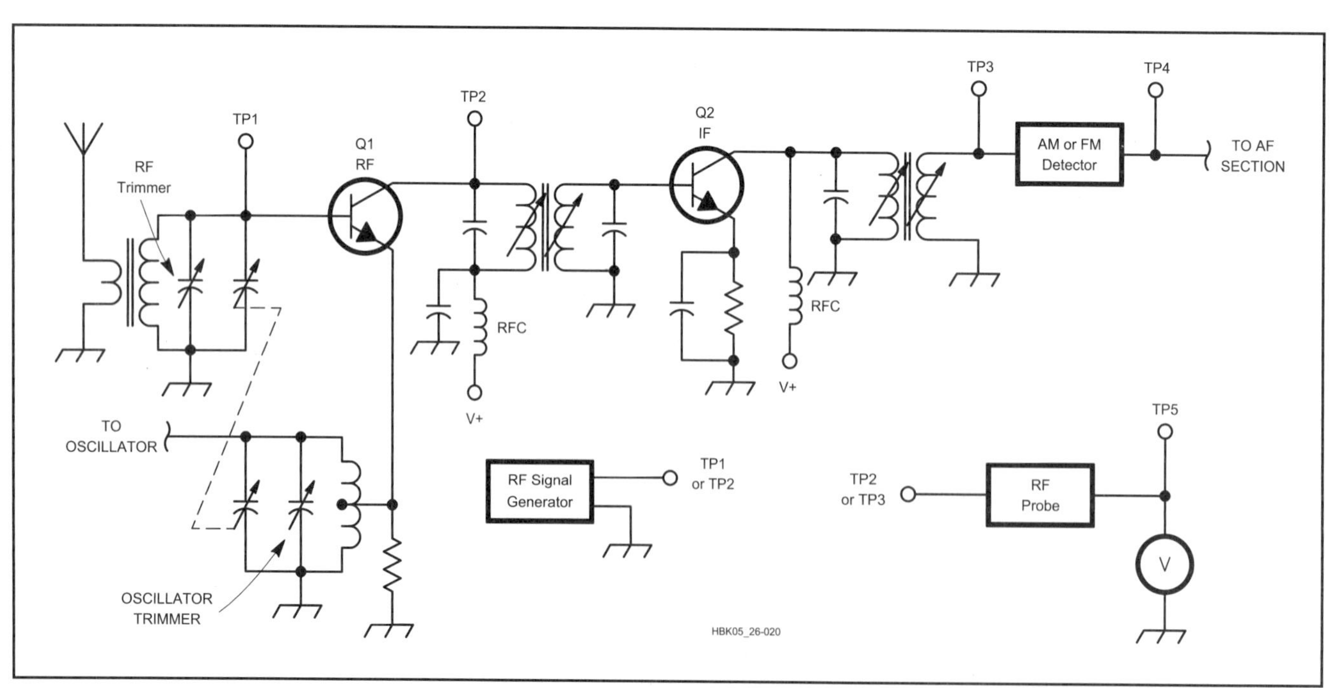

Fig 26.20 — Typical receiver alignment test points. To align the entire radio, connect a dc voltmeter at TP4. Inject an IF signal at TP2 and adjust the IF transformers. Move the signal generator to TP1 and inject an RF signal for alignment of the RF amplifier and oscillator stages. To align a single stage, place the generator at the input and an RF voltmeter (or demodulator probe and dc voltmeter) at the output: TP1/TP2 for RF, TP2/TP3 for IF.

wave. Backwave is a condition in which the signal is heard, at a reduced level, even when the key is up. This occurs when the oscillator signal feeds through a keyed amplifier. This usually indicates a design flaw, although in some cases a component failure or improper keyed-stage neutralization may be to blame.

Low Output Power

Some transmitters automatically reduce power in the TUNE mode. Check the owner's manual to see if the condition is normal. Check the control settings. Transmitters that use broadband amplifiers require so little effort from the operator that control settings are seldom noticed. The CARRIER (or DRIVE) control may have been bumped. Remember to adjust tuned amplifiers after a significant change in operating frequency (usually 50 to 100 kHz). Most modern transmitters are designed to reduce power if there is high (say 2:1) SWR. Check these obvious external problems before you tear apart your rig.

Power transistors may fail if the SWR protection circuit malfunctions. Such failures occur at the "weak link" in the amplifier chain: It is possible for the drivers to fail without damaging the finals. An open circuit in the "reflected" side of the sensing circuit leaves the transistors unprotected, a short "shuts them down."

Low power output in a transmitter may also spring from a misadjusted carrier oscillator or a defective SWR protection circuit. If the carrier oscillator is set to a frequency well outside the transmitter passband, there may be no measurable output. Output power will increase steadily as the frequency is moved into the passband.

Transceivers

Switching

Elaborate switching schemes are used in transceivers for signal control. Many transceiver malfunctions can be attributed to relay or switching problems. Suspect the switching controls when:

- The S meter is inoperative, but the unit otherwise functions. (This could also be a bad S meter.)
- There is arcing in the tank circuit. (This could also be caused by a bad antenna system.)
- Plate current is high during reception.
- There is excessive broadband PA noise in the receiver.

Since transceiver circuits are shared, stage defects frequently affect both the transmit and receive modes, although the symptoms may change with mode. Oscillator problems usually affect both transmit and receive modes, but different oscillators, or frequencies, may be used for different emissions. Check the block diagram.

For example, one particular transceiver uses a single carrier oscillator with three different crystals (see **Fig 26.21**). One crystal sets the carrier frequency for CW, AM and FSK transmit. Another sets USB transmit and USB/CW receive, and a third sets LSB transmit and LSB/FSK receive. This radio showed a strange symptom. After several hours of CW operation, the receiver produced only a light hiss on USB and CW. Reception was good in other modes, and the power meter showed full output during CW transmission. An examination of the block diagram and schematic showed that only one of the crystals (and seven support components) was capable of causing the problem.

VOX

Voice operated transmit (VOX) controls are another potential trouble area. If there is difficulty in switching to transmit in the VOX mode, check the VOX-SENSITIVITY and ANTI-VOX control settings. Next, see if the PTT and manual (MOX) transmitter controls work. If the PTT and MOX controls function, examine the VOX control diodes and amplifiers. Test the switches, control lines and control voltage if the transmitter does not respond to other TR controls.

VOX SENSITIVITY and ANTI-VOX settings should also be checked if the transmitter switches on in response to received audio. Suspect the ANTI-VOX circuitry next. Unacceptable VOX timing results from a poor VOX-delay adjustment, or a bad resistor or capacitor in the timing circuit or VOX amplifiers.

Alignment

The mixing scheme of the modern SSB transceiver is complicated. The signal passes through many mixers, oscillators and filters. Satisfactory SSB communication requires accurate adjustment of each stage. Do not attempt any alignment without a copy of the manufacturer's instructions and the necessary test equipment.

Troubleshooting Charts

Tables 26.2, **26.3**, **26.4** and **26.5** list some common problems and possible cures. These tables are not all-inclusive. They are a collection of hints and shortcuts that may save you some troubleshooting time. If you don't find your problem listed, continue with systematic troubleshooting.

COMPONENTS

Once you locate a defective part, it is time to select a replacement. This is not always an easy task. Each electronic component has a function. This section acquaints you with the functions, failure modes and test procedures of resistors, capacitors, inductors and other components. Test the components implicated by symptoms and stage-level testing. In most cases, a particular faulty component will be located by these tests. If a faulty component is not indicated, check the circuit adjustments. As a last resort, use a shotgun approach — replace all parts in the problem area with components that are known to be good.

Check the Circuit

Before you install a replacement component of any type, you should be sure that another circuit defect didn't cause the failure. Check the circuit voltages carefully before installing any new component. Check the potential on each trace to the bad component. The old part may have "died" as a result of a lethal voltage. Measure twice — repair once! (With apologies to the old carpenter.) Of course, circuit performance is the final test of any substitution.

Fuses

Most of the time, when a fuse fails, it is for a reason — usually a short circuit in the load. A fuse that has failed because of a short circuit usually shows the evidence of high current: a blackened interior with little blobs of fuse element everywhere. Fuses can also fail by fracturing the element at either end. This kind of failure is not visible by looking at the fuse. Check even "good" fuses with an ohmmeter. You may save hours of troubleshooting.

For safety reasons, always use *exact* replacement fuses. Check the current and voltage ratings. The fuse timing (fast, normal or slow blow) must be the same as the original. Never attempt to force a fuse that is not the right size into a fuse holder. The substitution of a bar, wire or penny for a fuse invites a "smoke party."

Wires

Wires seldom fail unless abused. Short circuits can be caused by physical damage to insulation or by conductive contamination. Damaged insulation is usually apparent during a close visual inspection of the conductor or connector. Look carefully where conductors come close to corners or sharp objects. Repair worn insulation by replacing the wire or securing an insulating sleeve (spaghetti) or heat-shrink tubing over the worn area.

When wires fail, the failure is usually caused by stress and flexing. Nearly everyone has broken a wire by bending it back

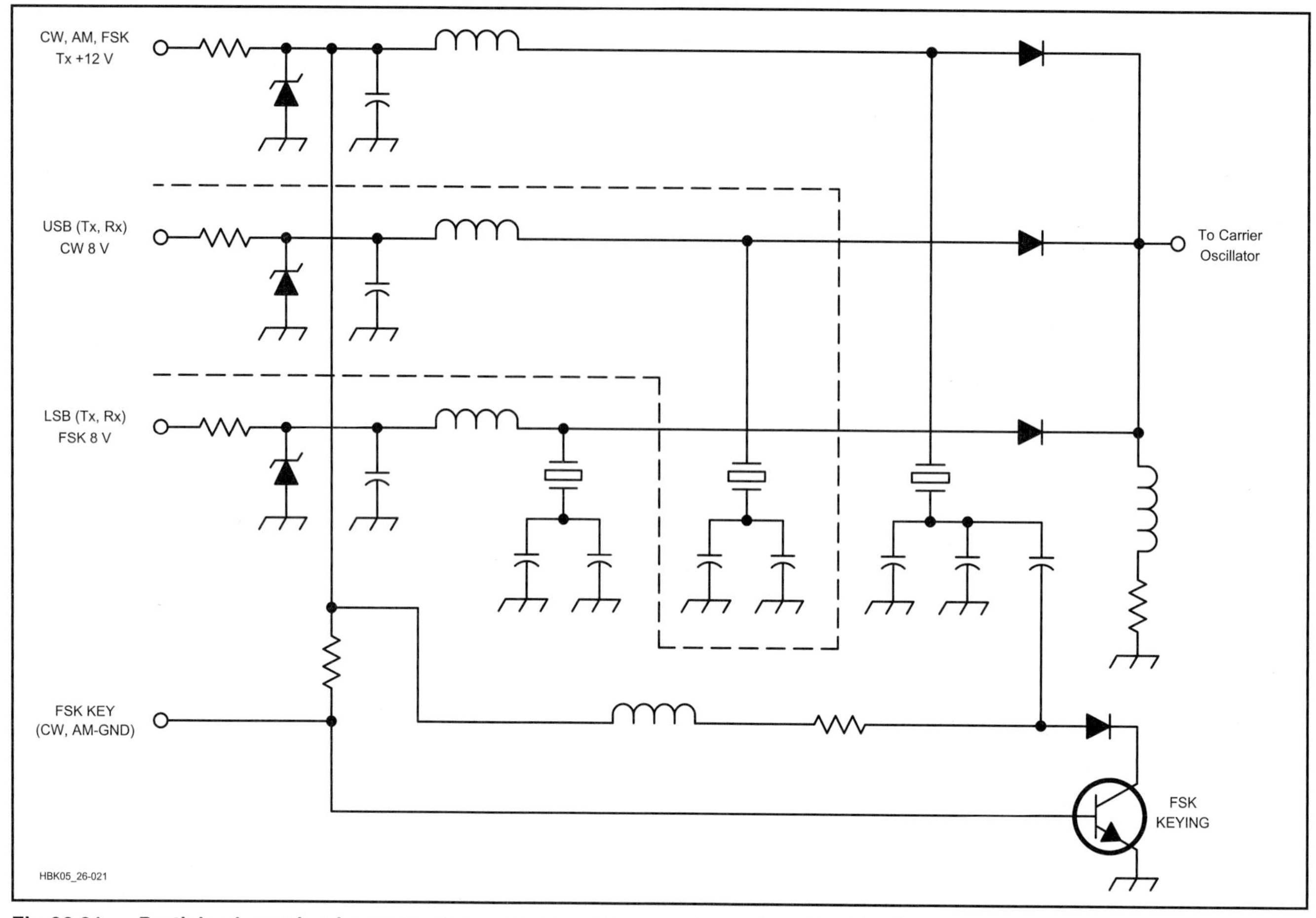

Fig 26.21 — Partial schematic of a transceiver oscillator. The symptoms described in the text are caused by one or more components inside the dashed lines or a faulty USB/CW control signal.

and forth, and broken wires are usually easy to detect. Look for sharp bends or bulges in the insulation.

When replacing conductors, use the same material and size, if possible. Substitute only wire of greater cross-sectional area (smaller gauge number) or material of greater conductivity. Insulated wire should be rated at the same, or higher, temperature and voltage as the wire it replaces.

Connectors

Connection faults are one of the most common failures in electronic equipment. This can range from something as simple as the ac-line cord coming out of the wall, to a connector having been put on the wrong socket, to a defective IC socket. Connectors that are plugged and unplugged frequently can wear out, becoming intermittent or noisy. Check connectors carefully when troubleshooting.

Connector failure can be hard to detect. Most connectors maintain contact as a result of spring tension that forces two conductors together. As the parts age, they become brittle and lose tension. Any connection may deteriorate because of non-conductive corrosion at the contacts. Solder helps prevent this problem but even soldered joints suffer from corrosion when exposed to weather.

The dissipated power in a defective connector usually increases. Signs of excess heat are sometimes seen near poor connections in circuits that carry moderate current. Check for short and open circuits with an ohmmeter or continuity tester. Clean those connections that fail as a result of contamination.

Occasionally, corroded connectors may be repaired by cleaning, but replacement of the conductor/connector is usually required. Solder all connections that may be subject to harsh environments and protect them with acrylic enamel, RTV compound or a similar coating.

Choose replacement connectors with consideration of voltage and current ratings. Use connectors with symmetrical pin arrangements only where correct insertion will not result in a safety hazard or circuit damage.

Resistors

Resistors usually fail by becoming an open circuit. More rarely they change value. This is usually caused by excess heat. Such heat may come from external sources or from power dissipated within the resistor. Sufficient heat burns the resistor until it becomes an open circuit.

Resistors can also fracture and become an open circuit as a result of physical shock. Contamination of a high-value resistor (100 kΩ or more) can cause a change in value through leakage. This contamination can occur on the resistor body, mounts or printed-circuit board. Resistors that have changed value should be replaced. Leakage is cured by cleaning the resistor body and surrounding area.

In addition to the problems of fixed-value resistors, potentiometers and rheostats can develop noise problems,

Table 26.2
Symptoms and Their Causes for All Electronic Equipment

Symptom	*Cause*
Power Supplies	
No output voltage	Open circuit (usually a fuse or transformer winding)
Hum or ripple	Faulty regulator, capacitor or rectifier, low-frequency oscillation
Amplifiers	
Low gain	Transistor, coupling capacitors, emitter-bypass capacitor, AGC component, alignment
Noise	Transistors, coupling capacitors, resistors
Oscillations	Dirt on variable capacitor or chassis, shorted op-amp input
Untuned (oscillations do not change with frequency)	Audio stages
Tuned	RF, IF and mixer stages
Squeal	Open AGC-bypass capacitor
Static-like crashes	Arcing trimmer capacitors, poor connections
Static in FM receiver	Faulty limiter stage, open capacitor in ratio detector, weak RF stage, weak incoming signal
Intermittent noise	All components and connections, band-switch contacts, potentiometers (especially in dc circuits), trimmer capacitors, poor antenna connections
Distortion (constant)	Oscillation, overload, faulty AGC, leaky transistor, open lead in tab-mount transistor, dirty potentiometer, leaky coupling capacitor, open bypass capacitors, imbalance in tuned FM detector, IF oscillations, RF feedback (cables)
Distortion (strong signals only)	Open AGC line, open AGC diode
Frequency change	Physical or electrical variations, dirty or faulty variable capacitor, broken switch, loose compartment parts, poor voltage regulation, oscillator tuning (trouble when switching bands)
No Signals	
All bands	Dead VFO or heterodyne oscillator, PLL won't lock
One band only	Defective crystal, oscillator out of tune, band switch
No function control	Faulty switch, poor connection, defective switching diode or circuit
Improper Dial Tracking	
Constant error across dial	Dial drive
Error grows worse along dial	Circuit adjustment

Table 26.3
Transmitter Problems

Symptom	*Cause*
Key clicks	Keying filter, distortion in stages after keying
Modulation Problems	
Loss of modulation	Broken cable (microphone, PTT, power), open circuit in audio chain, defective modulator
Distortion on transmit	Defective microphone, RF feedback from lead dress, modulator imbalance, bypass capacitor, improper bias, excessive drive
Arcing	Dampness, dirt, improper lead dress
Low output	Incorrect control settings, improper carrier shift (CW signal outside of passband) audio oscillator failure, transistor or tube failure, SWR protection circuit
Antenna Problems	
Poor SWR	Damaged antenna element, matching network, feed line, balun failure (see below), resonant conductor near antenna, poor connection at antenna
Balun failure	Excessive SWR, weather or cold-flow damage in coil choke, broken wire
RFI	Arcing or poor connections anywhere in antenna system or nearby conductors

especially in dc circuits. Dirt often causes intermittent contact between the wiper and resistive element. To cure the problem, spray electronic contact cleaner into the control, through holes in the case, and rotate the shaft a few times.

The resistive element in wire-wound potentiometers eventually wears and breaks from the sliding action of the wiper. In this case, the control needs to be replaced.

Replacement resistors should be of the same value, tolerance, type and power rating as the original. The value should stay within tolerance. Replacement resistors may be of a different type than the original, if the characteristics of the replacement are consistent with circuit requirements.

Substitute resistors can usually have a greater power rating than the original, except in high-power emitter circuits where the resistor also acts as a fuse or in cases where the larger size presents a problem.

Table 26.4
Receiver Problems

Symptom	*Cause*
Low sensitivity	Semiconductor contamination, weak tube, alignment
Signals and calibrator heard weakly	
(low S-meter readings)	RF chain
(strong S-meter readings)	AF chain, detector
No signals or calibrator heard, only hissing	RF oscillators
Distortion	
On strong signals only	AGC fault
AGC fault	Active device cut off or saturated
Difficult tuning	AGC fault
Inability to receive	Detector fault
AM weak and distorted	Poor detector, power or ground connection
CW/SSB unintelligible	BFO off frequency or dead
FM distorted	Open detector diode

Table 26.5
Transceiver Problems

Symptom	*Cause*
Inoperative S meter	Faulty relay
PA noise in receiver	
Excessive current on receive	
Arcing in PA	
Reduced signal strength on transmit and receive	IF failure
Poor VOX operation	VOX amplifiers and diodes
Poor VOX timing	Adjustment, component failure in VOX timing circuits or amplifiers
VOX consistently tripped by receiver audio	AntiVOX circuits or adjustment

Variable resistors should be replaced with the same kind (carbon or wire wound) and taper (linear, log, reverse log and so on) as the original. Keep the same, or better tolerance and pay attention to the power rating.

In all cases, mount high-temperature resistors away from heat-sensitive components. Keep carbon resistors away from heat sources. This will extend their life and ensure minimum resistance variations.

Capacitors

Capacitors usually fail by shorting, opening or becoming electrically (or physically) leaky. They rarely change value. Capacitor failure is usually caused by excess current, voltage, temperature or age. Leakage can be external to the capacitor (contamination on the capacitor body or circuit) or internal to the capacitor.

Tests

The easiest way to test capacitors is out of circuit with an ohmmeter. In this test, the resistance of the meter forms a timing circuit with the capacitor to be checked. Capacitors from 0.01 µF to a few hundred µF can be tested with common ohmmeters. Set the meter to its highest range and connect the test leads across the discharged capacitor. When the leads are connected, current begins to flow. The capacitor passes current easily when discharged, but less easily as the charge builds. This shows on the meter as a low resistance that builds, over time, toward infinity.

The speed of the resistance build-up corresponds to capacitance. Small capacitance values approach infinite resistance almost instantly. A 0.01-µF capacitor checked with an 11-MΩ FETVOM would increase from zero to a two-thirds scale reading in 0.11 s, while a 1-µF unit would require 11 s to reach the same reading. If the tested capacitor does not reach infinity within five times the period taken to reach the two-thirds point, it has excess leakage. If the meter reads infinite resistance immediately, the capacitor is open. (Aluminum electrolytics normally exhibit high-leakage readings.)

Fig 26.22 shows a circuit that may be used to test capacitors. To use this circuit, make sure that the power supply is off, set S1 to CHARGE and S2 to TEST, then connect the capacitor to the circuit. Switch on the power supply and allow the capacitor to charge until the voltmeter reading stabilizes. Next, switch S1 to TEST and watch the meter for a few seconds. If the capacitor is good, the meter will show no potential. Any appreciable voltage indicates excess leakage. After testing, set S1 to CHARGE, switch off the power supply, and press the DISCHARGE button until the meter shows 0 V, then remove the capacitor from the test circuit.

Capacitance can also be measured with a capacitance meter, an RX bridge or a dip meter. Some DMMs (digital multimeters) measure capacitance. Capacitance measurements made with DMMs and dedicated capacitance meters are much more accurate than those made with RX bridges or dip meters. To determine capacitance with a dip meter, a parallel-resonant circuit should be constructed using the capacitor of unknown value and an inductor of known value. The formula for resonance is discussed in the **Electrical Fundamentals** chapter of this book.

It is best to keep a collection of known components that have been measured on accurate L or C meters. Alternatively, a "standard" value can be obtained by ordering 1 or 2% components from an electronics supplier. A 10%-tolerance component can be used as a standard; however, the results will only be known to within 10%. The accuracy of tests made with any of these alternatives depends on the accuracy of the "standard" value component. Further information on this technique appears in Bartlett's article, "Calculating Component Values," in Nov 1978 *QST*.

Cleaning

The only variety of common capacitor that can be repaired is the air-dielectric variable capacitor. Electrical connection to the moving plates is made through a spring-wiper arrangement (see **Fig 26.23**). Dirt normally builds on the contact area, and they need occasional cleaning. Before cleaning the wiper/contact, use gentle air pressure and a soft brush to remove all dust and dirt from the capacitor plates. Apply some electronic contact cleaning fluid. Rotate the shaft quickly several times to work in the fluid and establish contact. Use the cleaning fluid sparingly, and keep it off the plates except at the contact point.

Replacements

Replacement capacitors should match the original in value, tolerance, dielectric, working voltage and temperature coeffi-

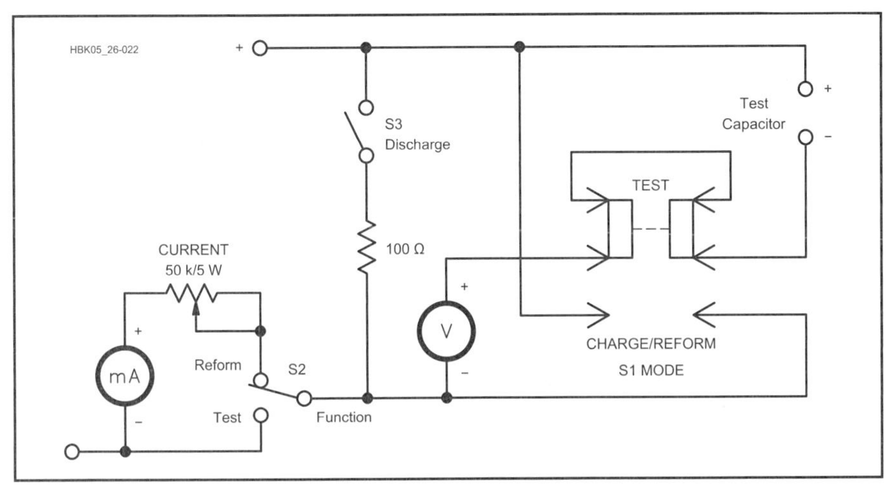

Fig 26.22 — A fixture for testing capacitors and reforming the dielectric of electrolytic capacitors. Use 12 V for testing the capacitor. Use the capacitor working voltage for dielectric reformation.

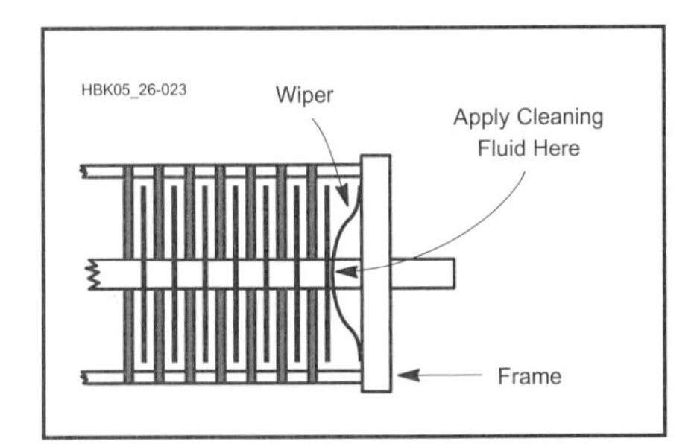

Fig 26.23 — Partial view of an air-dielectric variable capacitor. If the capacitor is noisy or erratic in operation, apply electronic cleaning fluid where the wiper contacts the rotor plates.

cient. Use only ac-rated capacitors for line service. If exact replacements are not available, substitutes may vary from the original part in the following respects: Bypass capacitors may vary from one to three times the capacitance of the original. Coupling capacitors may vary from one half to twice the value of the original. Capacitance values in tuned circuits (especially filters) must be exact. (Even then, any replacement will probably require circuit realignment.)

If the same kind of capacitor is not available, use one with better dielectric characteristics. Do not substitute polarized capacitors for nonpolarized parts. Capacitors with a higher working voltage may be used, although the capacitance of an electrolytic capacitor used significantly below its working voltage will usually increase with time.

The characteristics of each type of capacitor are discussed in the **Real-World Component Characteristics** chapter. Consider these characteristics if you're not using an exact replacement capacitor.

Inductors and Transformers

The most common inductor or transformer failure is a broken conductor. More rarely, a short circuit can occur across one or more turns of a coil. In an inductor, this changes the value. In a transformer, the turns ratio and resultant output voltage changes. In high-power circuits, excessive inductor current can generate enough heat to melt plastics used as coil forms.

Inductors may be checked for open circuit failure with an ohmmeter. In a good inductor, dc resistance rarely exceeds a few ohms. Shorted turn and other changes in inductance show only during alignment or inductance measurement.

The procedure for measurement of inductance with a dip meter is the same as that given for capacitance measurement, except that a capacitor of known value is used in the resonant circuit.

Replacement inductors must have the same inductance as the original, but that is only the first requirement. They must also carry the same current, withstand the same voltage and present nearly the same Q as the original part. Given the original as a pattern, the amateur can duplicate these qualities for many inductors. Note that inductors with ferrite or iron-powder cores are frequency sensitive, so the replacement must have the same core material.

If the coil is of simple construction, with the form and core undamaged, carefully count and write down the number of turns and their placement on the form. Also note how the coil leads are arranged and connected to the circuit. Then determine the wire size and insulation used. Wire diameter, insulation and turn spacing are critical to the current and voltage ratings of an inductor. (There is little hope of matching coil characteristics unless the wire is duplicated exactly in the new part.) Next, remove the old winding—be careful not to damage the form—and apply a new winding in its place. Be sure to dress all coil leads and connections in exactly the same manner as the original. Apply Q dope to hold the finished winding in place.

Follow the same procedure in cases where the form or core is damaged, except that a suitable replacement form or core (same dimensions and permeability) must be found.

Ready-made inductors may be used as replacements if the characteristics of the original and the replacement are known and compatible. Unfortunately, many inductors are poorly marked. If so, some comparisons, measurements and circuit analysis are usually necessary.

When selecting a replacement inductor, you can usually eliminate parts that bear no physical resemblance to the original part. This may seem odd, but the Q of an inductor depends on its physical dimensions and the permeability of the core material. Inductors of the same value, but of vastly different size or shape, will likely have a great difference in Q. The Q of the new inductor can be checked by installing it in the circuit, aligning the stage and performing the manufacturer's passband tests. Although this practice is all right in a pinch, it does not yield an accurate Q measurement. Methods to measure Q appear in the **Test Procedures** chapter.

Once the replacement inductor is found, install it in the circuit. Duplicate the placement, orientation and wiring of the original. Ground-lead length and arrangement should not be changed. Isolation and magnetic shielding can be improved by replacing solenoid inductors with toroids. If you do, however, it is likely that many circuit adjustments will be needed to compensate for reduced coupling and mutual inductance. Alignment is usually required whenever a tuned-circuit component is replaced.

A transformer consists of two inductors that are magnetically coupled. Transformers are used to change voltage and current levels (this changes impedance also). Failure usually occurs as an open circuit or short circuit of one or more windings.

Amateur testing of power transformers is limited to ohmmeter tests for open circuits and voltmeter checks of secondary voltage. Make sure that the power-line voltage is correct, then check the secondary voltage against that specified. There should be less than 10% difference between open-circuit and full-load secondary voltage.

Replacement transformers must match the original in voltage, volt-ampere (VA), duty cycle and operating-frequency ratings. They must also be compatible in size. (All transformer windings should be insulated for the full power-supply voltage.)

Relays

Although relays have been replaced by semiconductor switching in low-power circuits, they are still used extensively in high-power Amateur Radio equipment. Relay action may become sluggish. AC relays can buzz (with adjustment becoming impossible). A binding armature or weak springs can cause intermittent switching. Excessive use or hot switching ruins contacts and shortens relay life.

You can test relays with a voltmeter by jumpering across contacts with a test lead (power on, in circuit) or with an ohmmeter (out of circuit). Look for erratic readings across the contacts, open or short circuits at contacts or an open circuit at the coil.

Most failures of simple relays can be repaired by a thorough cleaning. Clean the contacts and mechanical parts with a residue-free cleaner. Keep it away from the coil and plastic parts that may be damaged. Dry the contacts with lint-free paper, such as a business card; then burnish them with a smooth steel blade. Do not use a file to clean contacts.

Replacement relays should match or exceed the original specifications for voltage, current, switching time and stray impedance (impedance is significant in RF circuits only). Many relays used in transceivers are specially made for the manufacturer. Substitutes may not be available from any other source.

Before replacing a multicontact relay, make a drawing of the relay, its position, the leads and their routings through the surrounding parts. This drawing allows you to complete the installation properly, even if you are distracted in the middle of the operation.

Semiconductors

Diodes

The primary function of a diode is to pass current in one direction only. They can be easily tested with an ohmmeter.

Signal or switching diodes — The most common diode in electronics equipment, they are used to convert ac to dc, to detect RF signals or to take the place of relays to switch ac or dc signals within a circuit. Signal diodes usually fail open, although shorted diodes are not rare. They can easily be tested with an ohmmeter.

Power-rectifier diodes — Most equipment contains a power supply, so power-rectifier diodes are the second-most common diodes in electronic circuitry. They usually fail shorted, blowing the power-supply fuse.

Other diodes — Zener diodes are made with a predictable reverse-breakdown voltage and used as voltage regulators. Varactor diodes are specially made for use as voltage controlled variable capacitors. (Any semiconductor diode may be used as a voltage-variable capacitance, but the value will not be as predictable as that of a varactor.) A Diac is a special-purpose diode that passes only pulses of current in each direction.

Diode tests — There are several basic tests for most diodes. First, is it a diode? Does it conduct in one direction and block current flow in the other? An ohmmeter is suitable for this test in most cases. An ohmmeter will read high resistance in one direction, low resistance in the other. Make sure the meter uses a voltage of more than 0.7 V and less than 1.5 V to measure resistance. Use a good diode to determine the meter polarity.

Diodes should be tested out of circuit. Disconnect one lead of the diode from the circuit, then measure the forward and reverse resistance. Diode quality is shown by the ratio of reverse to forward resistance. A ratio of 100:1 or greater is common for signal diodes. The ratio may go as low as 10:1 for old power diodes.

The first test is a forward-resistance test. Set the meter to read ×100 and connect the test probes across the diode. When the negative terminal of the ohmmeter battery is connected to the cathode, the meter will typically show about 200 to 300 Ω (forward resistance) for a good silicon diode, 200 to 400 Ω for a good germanium diode. The exact value varies quite a bit from one meter to the next.

Next, test the reverse resistance. Reverse the lead polarity and set the meter to ×1M (times one million, or the highest scale available on the meter) to measure diode reverse resistance. Good diodes should show 100 to 1000 MΩ for silicon and 100 kΩ to 1 MΩ for germanium. When you are done, mark the meter lead polarity for future reference.

This procedure measures the junction resistances at low voltage. It is not useful to test Zener diodes. A good Zener diode will not conduct in the reverse direction at voltages below its rating.

We can also test diodes by measuring the voltage drop across the diode junction while the diode is conducting. (A test circuit is shown in **Fig 26.24**.) To test, connect the diode, adjust the supply voltage until the current through the diode matches the manufacturer's specification and compare the junction drop to that specified. Silicon junctions usually show about 0.6 V, while germanium is typically 0.2 V. Junction voltage-drop increases with current flow. This test can be used to match diodes with respect to forward resistance at a given current level.

A final simple diode test measures leakage current. Place the diode in the circuit described above, but with reverse polarity. Set the specified reverse voltage and read the leakage current on a milliammeter. (The currents and voltages measured in the junction voltage-drop and leakage tests vary by several orders of magnitude.)

The most important specification of a Zener diode is the Zener (or avalanche) voltage. The Zener-voltage test also uses the circuit of Fig 26.24. Connect the diode in reverse. Set the voltage to minimum, then gradually increase it. You should read low current in the reverse mode, until the Zener point is reached. Once the device begins to conduct in the reverse direction, the current should increase dramatically. The voltage shown on the voltmeter is the Zener point of the diode. If a Zener diode has become leaky, it might show in the leakage-current measurement, but substitution is the only dependable test.

Replacement diodes — When a diode fails, check associated components as well. Replacement rectifier diodes should have the same current and peak inverse voltage (PIV) as the original. Series diode combinations are often used in high-voltage rectifiers, with resistor and capacitor networks to distribute the voltage equally among the diodes.

Switching diodes may be replaced with diodes that have equal or greater current ratings and a PIV greater than twice the peak-to-peak voltage encountered in the circuit. Switching time requirements are not critical except in RF, logic and some keying circuits. Logic circuits may require exact replacements to assure compatible switching speeds and load characteristics. RF switching diodes used near resonant circuits must have exact replacements as the diode resistance and capacitance will affect the tuned circuit.

Voltage, current and capacitance characteristics must be considered when replacing varactor diodes. Once again, exact replacements are best. Zener diodes should be replaced with parts having the same Zener voltage and equal or better current, power, impedance and tolerance specifications. Check the associated current-limiting resistor when replacing a Zener diode.

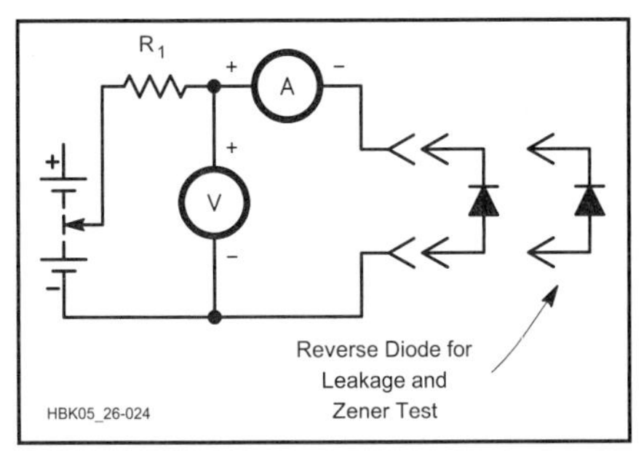

Fig 26.24 — A diode conduction, leakage and Zener-point test fixture. The ammeter should read mA for conduction and Zener point, µA for leakage tests.

Bipolar Transistors

Transistors are primarily used to switch or amplify signals. Transistor failures occur as an open junction, a shorted junction, excess leakage or a change in amplification performance.

Most transistor failure is catastrophic. A transistor that has no leakage and amplifies at dc or audio frequencies will usually perform well over its design range. For this reason, transistor tests need not be performed at the planned operating frequency. Tests are made at dc or a low frequency (usually 1000 Hz). The circuit under repair is the best test of a potential replacement part. Swapping in a replacement transistor in a failed circuit will often result in a cure.

A simple and reliable bipolar-transistor test can be performed with the transistor in a circuit and the power on. It requires a test lead, a 10-kΩ resistor and a voltmeter. Connect the voltmeter across the emitter/collector leads and read the voltage. Then use the test lead to connect the base and emitter (**Fig 26.25A**). Under these con-ditions, conduction of a good transistor will be cut off and the meter should show nearly the entire supply voltage across the emitter/collector leads. Next, remove the clip lead and connect the 10-kΩ re-sistor from the base to the collector. This should bias the transistor into conduction and the emitter/collector voltage should drop (Fig 26.25B). (This test in-dicates transistor response to changes in bias voltage.)

Transistors can be tested (out of circuit) with an ohmmeter in the same manner as diodes. Look up the device characteristics before testing and consider the consequences of the ohmmeter-transistor circuit. Limit junction current to 1 to 5 mA for small-signal transistors. Transistor destruction or inaccurate measurements may result from careless testing.

Use the ×100 Ω and ×1000-Ω ranges for small-signal transistors. For high-power transistors use the ×1 Ω and ×10-Ω ranges. The reverse-to-forward resistance ratio for good transistors may vary from 30:1 to better than 1000:1.

Germanium transistors sometimes show high leakage when tested with an ohmmeter. Bipolar transistor leakage may be specified from the collector to the base, emitter to base or emitter to collector (with the junction reverse biased in all cases). The specification may be identified as I_{cbo}, I_{bo}, collector cutoff current or collector leakage for the base-collector junction, I_{ebo}, and so on for other junctions.[3] Leakage current increases with junction temperature.

A suitable test fixture for base-collector leakage measurements is shown in **Fig 26.26**. Make the required connections and set the voltage as stated in the transistor specifications and compare the measured leakage current with that specified. Small-signal germanium transistors exhibit Icbo and I_{ebo} leakage currents of about 15 µA. Leakage increases to 90 µA or more in high-power components. Leakage currents for silicon transistors are seldom more than 1 µA. Leakage current tends to double for every 10°C increase above 25°C.

Breakdown-voltage tests actually measure leakage at a specified voltage, rather than true breakdown voltage. Breakdown voltage is known as BV_{cbo}, BV_{ces} or BV_{ceo}. Use the same test fixture shown for leakage tests, adjust the power supply until the specified leakage current flows, and compare the junction voltage against that specified.

A circuit to measure dc current gain is shown in **Fig 26.27**. Transistor gain can range from 10 to over 1000 because it is not usually well controlled during manufacture. Gain of the active device is not critical in a well-designed transistor circuit.

The test conditions for transistor testing are specified by the manufacturer. When testing, do not exceed the voltage, current (especially in the base circuit) or dissipated-power rating of the transistor. Make sure that the load resistor is capable of dissipating the power generated in the test.

While these simple test circuits will identify most transistor problems, RF devices should be tested at RF. Most component manufacturers include a test-circuit schematic on the data sheet. The test circuit is usually an RF amplifier that operates near the high end of the device frequency range.

Semiconductor failure is sometimes the result of environmental conditions. Open junctions, excess leakage (except with

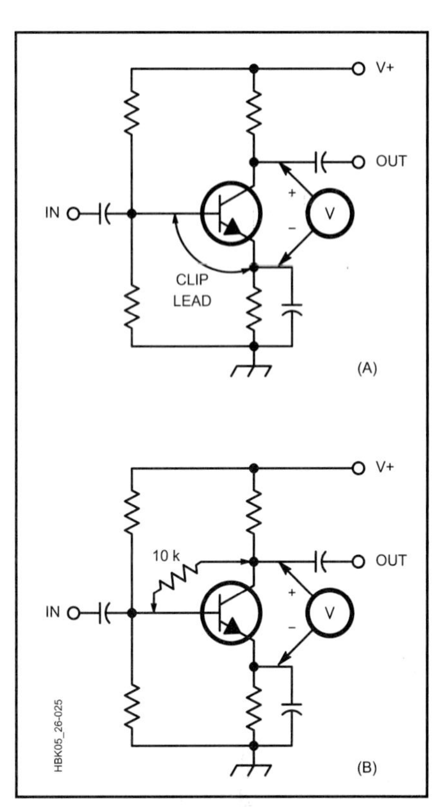

Fig 26.25 — An in-circuit semiconductor test with a clip lead, resistor and voltmeter. The meter should read V+ at (A). During test (B) the meter should show a decrease in voltage, ranging from a slight variation down to a few millivolts. It will typically cut the voltage to about half of its initial value.

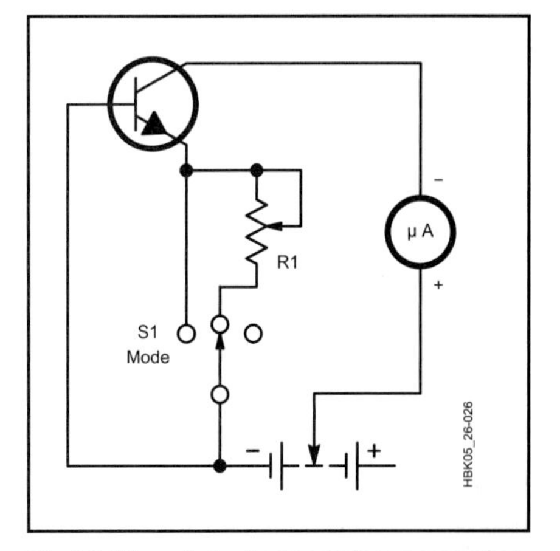

Fig 26.26 — A test circuit for measuring collector-base leakage with the emitter shorted to ground, open or connected to ground through a variable resistance, depending on the setting of S1. See the transistor manufacturer's instructions for test conditions and the setting of R1 (if used). Reverse battery polarity for PNP transistors.

germanium transistors) and changes in amplification performance result from overload or excessive current. Electrostatic discharge can destroy a semiconductor in microseconds. Shorted junctions are caused by voltage spikes. Check surrounding parts for the cause of the transistor's demise, and correct the problem before installing a replacement.

JFETs

Junction FETs can be tested with an ohmmeter in much the same way as bipolar transistors (see text and **Fig 26.28**). Reverse leakage should be several megohms or more. Forward resistance should be 500 to 1000 Ω.

MOSFETs

MOS (metal-oxide semiconductor) layers are extremely fragile. Normal body static is enough to damage them. Even "gate protected" (a diode is placed across the MOS layer to clamp voltage) MOSFETs may be destroyed by a few volts of static electricity.

Make sure the power is off, capacitors discharged and the leads of a MOSFET are shorted together before installing or removing it from a circuit. Use a voltmeter to be sure the chassis is near ground potential, then touch the chassis before and during MOSFET installation and removal. This assures that there is no difference of potential between your body, the chassis and the MOSFET leads. Ground the soldering-iron tip with a clip lead when soldering MOS devices. The FET source should be the first lead connected to and the last disconnected from a circuit. The insulating layers in MOSFETs prevent testing with an ohmmeter. Substitution is the only practical means for amateur testing of MOSFETs.

FET Considerations

Replacement FETs should be of the same kind as the original part: JFET or MOSFET, P-channel or N-channel, enhancement or depletion. Consider the breakdown voltage required by the circuit. The breakdown voltage should be at least two to four times the power-supply and signal voltages in amplifiers. Allow for transients of ten times the line voltage in power supplies. Breakdown voltages are usually specified as $V_{(BR)GSS}$ or $V_{(BR)GDO}$.

The gate-voltage specification gives the gate voltage required to cut off or initiate channel current (depending on the mode of operation). Gate voltages are usually listed as $V_{GS(OFF)}$, V_p(pinch off), V_{TH} (threshold) or $I_{D(ON)}$ or I_{TH}.

Dual-gate MOSFET characteristics are more complicated because of the interaction of the two gates. Cutoff voltage, breakdown voltage and gate leakage are the important traits of each gate.

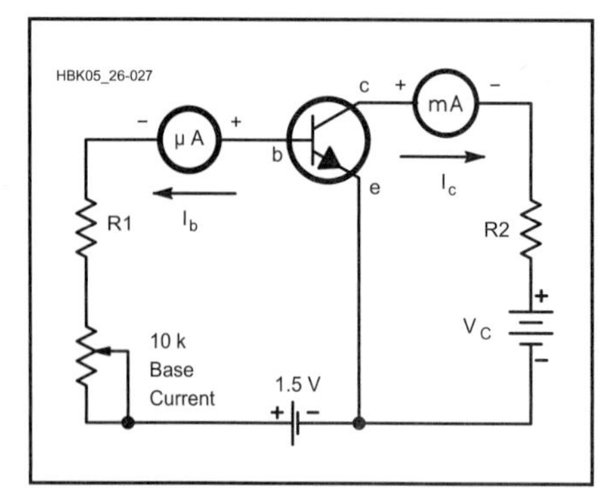

Fig 26.27 — A test circuit for measuring transistor beta. Values for R1 and R2 are dependent on the current range of the transistor tested. Reverse the battery polarity for PNP transistors.

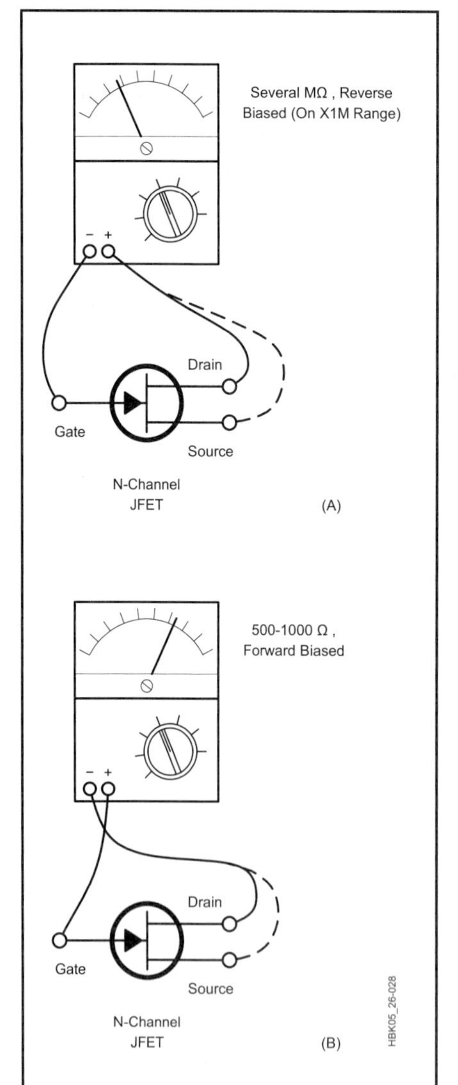

Fig 26.28 — Ohmmeter tests of a JFET. The junction is reverse biased at A and forward biased at B.

Integrated Circuits

The basics of integrated circuits are covered in earlier chapters of this book. Amateurs seldom have the sophisticated equipment required to test ICs. Even a multitrace 'scope can view only their simplest functions. We must be content to check every other possible cause, and only then assume that the problem lies with an IC. Experienced troubleshooters will tell you that — most of the time anyway — if a defective circuit uses an IC, it is the IC that is bad.

Linear ICs — There are two major classes of ICs: linear and digital. Linear ICs are best replaced with identical units. Original equipment manufacturers are the best source of a replacement; they are the only source with a reason to stockpile obsolete or custom-made items. If substitution of an IC is unavoidable, first try the cross-reference guides published by several distributors. You can also look in manufacturers' databooks and compare pinouts and other specifications.

Digital ICs — It is usually not a good idea to substitute digital devices. While it may be okay to substitute an AB74LS00YZ from manufacturer "A" with a CD74LS00WX from a different manufacturer, you will usually not be able to replace an LS (low-power Schottky) device with an S (Schottky), C (CMOS) or any of a number of other families. The different families all have different speed, current-consumption, input and output characteristics. You would have to analyze the circuit to determine if you could substitute one type for another.

Semiconductor Substitution

In all cases try to obtain exact replacement semiconductors. Specifications vary slightly from one manufacturer to the next. Cross-reference equivalents are useful, but not infallible. Before using an equivalent, check the specifications against those for the original part. When choosing a replacement, consider:

- Is it silicon or germanium?
- Is it a PNP or an NPN?
- What are the operating frequency and input/output capacitance?
- How much power does it dissipate (often less than $V_{max} \times I_{max}$)?
- Will it fit the original mount?
- Are there unusual circuit demands (low noise and so on)?
- What is the frequency of operation?

Remember that cross-reference equivalents are not guaranteed to work in every application. There may be cases where two dissimilar devices have the same part number, so it pays to compare the listed replacement specifications with the intended use. If "the book" says to use a diode in place of an RF transistor, it isn't going to work! Derate power specifications, as recommended by the manufacturer, for high-temperature operation.

Tubes

The most common tube failures in amateur service are caused by cathode depletion and gas contamination. Whenever a tube is operated, the coating on the cathode loses some of its ability to produce electrons. It is time to replace the tube when electron production (cathode current, I_c) falls to 50 - 60% of that exhibited by a new tube.

Gas contamination in a tube can often be identified easily because there may be a greenish or whitish-purple glow between the elements during operation. (A faint deep-purple glow is normal in most tubes.) The gas reduces tube resistance and leads to runaway plate current evidenced by a red glow from the anode, interelectrode arcing or a blown power-supply fuse. Less common tube failures include an open filament, broken envelope and interelectrode shorts.

The best test of a tube is to substitute a new one. Another alternative is a tube tester; these are now rare. You can also do some limited tests with an ohmmeter. Tube tests should be made out of circuit so circuit resistance does not confuse the results.

Use an ohmmeter to check for an open filament (remove the tube from the circuit first). A broken envelope is visually ob-vious, although a cracked envelope may appear as a gassy tube. Interelectrode shorts are evident during voltage checks on the operating stage. Any two elements that show the same voltage are probably shorted. (Remember that some interelectrode shorts, such as the cathode-suppressor grid, are normal.)

Generally, a tube may be replaced with another that has the same type number. Compare the data sheets of similar tubes to assess their compatibility. Consider the base configuration and pinout, inter-electrode capacitances (a small variation is okay except for tubes in oscillator service), dissipated power ratings of the plate and screen grid and current limitations (both peak and average). For example, the 6146A may be replaced with a 6146B (heavy duty), but not vice versa.

In some cases, minor type-number differences signify differences in filament voltages, or even base styles, so check all specifications before making a replacement. (Even tubes of the same model number, prefix and suffix vary slightly, in some respects, from one supplier to the next.)

AFTER THE REPAIRS

Once you have completed your troubleshooting and repairs, it is time to put the equipment back together. Take a little extra time to make sure you have done everything correctly.

All Units

Give the entire unit a complete visual inspection. Look for any loose ends left over from your troubleshooting procedures — you may have left a few components temporarily soldered in place or overlooked some other repair error. Look for cold solder joints and signs of damage incurred during the repair. Double check the position, leads and polarity of components that were removed or replaced.

Make sure that all ICs are properly oriented in their sockets and all of the pins are properly inserted in the IC socket or printed-circuit board holes. Test fuse continuity with an ohmmeter and verify that the current rating matches the circuit specification.

Look at the position of all of the wires and components. Make sure that wires and cables will be clear of hot components, screw points and other sharp edges. Make certain that the wires and components will not be in the way when covers are installed and the unit is put back together.

Separate the leads that carry dc, RF, input and output as much as possible. Plug-in circuit boards should be firmly seated with screws tightened and lock washers installed if so specified. Shields and ground straps should be installed just as they were on the original.

For Transmitters Only

Since the signal produced by an HF transmitter can be heard the world over, a thorough check is necessary after any service has been performed. Do not exceed the transmitter duty cycle while testing. Limit transmissions to 10 to 20 seconds unless otherwise specified by the owner's manual.

1. Set all controls as specified in the operation manual, or at midscale.

2. Connect a dummy load and a power meter to the transmitter output.

3. Set the drive or carrier control for low output.

4. Switch the power on.

5. Transmit and quickly set the final-amplifier bias to specifications.

6. In narrowband equipment, slowly tune the output network through resonance. The current dip should be smooth and repeatable. It should occur simultaneously with the maximum power output. Any sudden jumps or wiggles of the current meter indicate that the amplifier is unstable. Adjust the neutralization circuit (according to the manufacturer's instructions) if one is present or check for oscillation. An amplifier usually requires neutralization whenever active devices, components or lead dress (that affect the output/input capacitance) are changed.

7. Check to see that the output power is consistent with the amplifier class used in the PA (efficiency should be about 25% for Class A, 50 to 60% for Class AB or B, and 70 to 75% for Class C).

8. Repeat steps 4 through 6 for each band of operation from lowest to highest frequency.

9. Check the carrier balance (in SSB transmitters only) and adjust for minimum power output with maximum RF drive and no microphone gain.

10. Adjust the VOX controls.

11. Measure the passband and distortion levels if equipment (wideband 'scope or spectrum analyzer) is available.

Other Repaired Circuits

After the preliminary checks, set the circuit controls per the manufacturer's specifications (or to midrange if specifications are not available) and switch the power on. Watch and smell for smoke, and listen for odd sounds such as arcing or hum. Operate the circuit for a few minutes, consistent with allowable duty cycle. Verify that all operating controls function properly.

Check for intermittent connections by subjecting the circuit to heat, cold and slight flexure. Also, tap or jiggle the chassis lightly with an alignment tool or other insulator.

If the equipment is meant for mobile or portable service, operate it through an appropriate temperature range. Many mobile radios do not work on cold mornings, or on hot afternoons, because a temperature-dependent intermittent was not found during repairs.

Button It Up

After you are convinced that you have repaired the circuit properly, put it all back together. If you followed the advice in this book, you have all the screws and assorted doodads in a secure container. Look at the notes you took while taking it apart; put it back together in the reverse order. Don't forget to reconnect all internal connections, such as ac-power, speaker or antenna leads.

Once the case is closed, and all appears

well, don't neglect the final, important step — make sure it still works. Many an experienced technician has forgotten this important step, only to discover that some minor error, such as a forgotten antenna connector, has left the equipment non-functional.

PROFESSIONAL REPAIRS

This chapter does not tell you how to perform all repairs. Repairs that deal with very complex and temperamental circuits, or that require sophisticated test equipment, should be passed on to a professional.

The factory authorized service personnel have a lot of experience. What seems like a servicing nightmare to you is old hat to them. There is no one better qualified to service your equipment than the factory.

If the manufacturer is no longer in business, check with your local dealer or look in the classified ads in electronics and Amateur Radio magazines. You can usually find one or more companies that service "all makes and models." Your local TV shop might be willing to tackle a repair, especially if you have located a schematic.

If you are going to ship your equipment somewhere for repair, notify the repair center first. Get authorization for shipping and an identification name or number for the package.

Packing It Up

You can always blame shipping damage on the shipper, but it is a lot easier for all concerned if you package your equipment properly for shipping in the first place. Firmly secure all heavy components, either by tying them down or blocking them off with shipping foam. Large vacuum tubes should be wrapped in packing material or shipped separately. Make sure that all circuit boards and parts are firmly attached.

Use a box within a box for shipping. (See **Fig 26.29**.) Place the equipment and some packing material inside a box and seal it with tape. Place that box inside another that is at least six inches larger in each dimension. Fill the gap with packing material, seal, address and mark the outer box. Choose a good freight carrier and insure the package.

Don't forget to enclose a statement of the trouble, a short history of operation and any test results that may help the service technician. Include a good description of the things you have tried. Be honest! At current repair rates you want to tell the technician everything to help ensure an efficient repair.

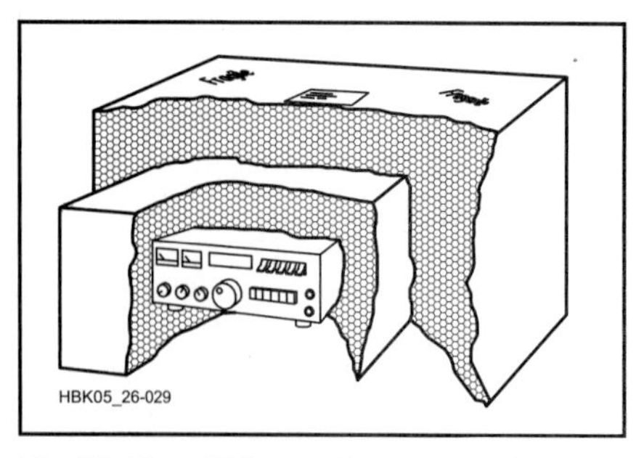

Fig 26.29 — Ship equipment packed securely in a box within a box.

Even if you ended up sending it back to the factory, you can feel good about your experience. You learned a lot by trying, and you have sent it back knowing that it really did require the services of a "pro." Each time you troubleshoot and repair a piece of electronic circuitry, you learn something new. The down side is that you may develop a reputation as a real electronics whiz. You may find yourself spending a lot of time at club meetings offering advice, or getting invited over to a lot of shacks for a late-evening pizza snack. There are worse fates.

References

J. Bartlett, "Calculating Component Values," *QST*, Nov 1978.

J. Carr, *How to Troubleshoot and Repair Amateur Radio Equipment*, Blue Ridge Summit, PA: TAB Books Inc, 1980.

D. DeMaw, "Understanding Coils and Measuring their Inductance," *QST*, Oct 1983.

H. Gibson, *Test Equipment for the Radio Amateur*, London, England: Radio Society of Great Britain, 1974.

C. Gilmore, *Understanding and Using Modern Electronic Servicing Test Equipment*, TAB Books, Inc, 1976.

F. Glass, *Owner Repair of Amateur Radio Equipment*, Los Gatos, CA: RQ Service Center, 1978.

R. Goodman, *Practical Troubleshooting with the Modern Oscilloscope*, TAB Books, Inc, 1979.

A. Haas, *Oscilloscope Techniques*, New York: Gernsback Library, Inc, 1958.

C. Hallmark, *Understanding and Using the Oscilloscope*, TAB Books, Inc, 1973.

A. Helfrick, *Amateur Radio Equipment Fundamentals*, Englewood Cliffs, NJ: Prentice-Hall Inc, 1982.

K. Henney, and C. Walsh, *Electronic Components Handbook*, New York: McGraw-Hill Book Company, 1957.

L. Klein, and K. Gilmore, *It's Easy to Use Electronic Test Equipment*, New York: John R. Rider Publisher, Inc (A division of Hayden Publishing Company), 1962.

J. Lenk, *Handbook of Electronic Test Procedures*, Prentice-Hall Inc, 1982.

G. Loveday, and A. Seidman, *Troubleshooting Solid-State Circuits*, New York: John Wiley and Sons, 1981.

A. Margolis, *Modern Radio Repair Techniques*, TAB Books Inc, 1971.

H. Neben, "An Ohmmeter with a Linear Scale," *QST*, Nov 1982.

H. Neben, "A Simple Capacitance Meter You Can Build," *QST*, Jan 1983.

F. Noble, "A Simple LC Meter," *QST*, Feb 1983.

J. Priedigkeit, "Measuring Inductance and Capacitance with a Reflection-Coefficient Bridge," *QST*, May 1982.

H. Sartori, "Solid Tubes — A New Life for Old Designs," *QST*, Apr 1977; "Questions on Solid Tubes Answered," Technical Correspondence, *QST*, Sep 1977.

B. Wedlock, and J. Roberge, *Electronic Components and Measurements*, Prentice-Hall Inc, 1969.

"Some Basics of Equipment Servicing," series, *QST*, Dec 1981-Mar 1982; Feedback May 1982.

Notes

[1]The ARRL has prepared a list of dip-meter sources. These are available on the **CD-ROM** included with this *Handbook*).

[2]More information about the signal injector and signal sources appears in "Some Basics of Equipment Servicing," February 1982 *QST* (Feedback, May 1982).

[3]The term "I_{cbo}" means "Current from collector to base with emitter open." The subscript notation indicates the status of the three device terminals. The terminals measured are listed first, with the remaining terminal listed as "s" (shorted) or "o" (open).

Advertisers Index

Advertising Department Staff
Janet Rocco, Business Services Manager
Joe Bottiglieri, AA1GW, Accounts Manager
Lisa Tardette, Accounts Manager
Diane Szlachetka, Advertising Graphic Designer

800-243-7768

Direct Line: 860-594-0207 ■ Fax: 860-594-4285 ■ e-mail: ads@arrl.org ■ Web: www.arrl.org/ads

If your company provides products or services of interest to our readers,
please contact the ARRL Advertising Department today for information on building your business.

IC-7800

The Ultimate HF

200 Watt Output, Full Duty Cycle | Four 32 Bit IF-DSPs + 24 Bit AD/DA Converters | 2 Independent Receivers | +40 dBm 3rd Order Intercept Point | Selectable IF Filter Shapes

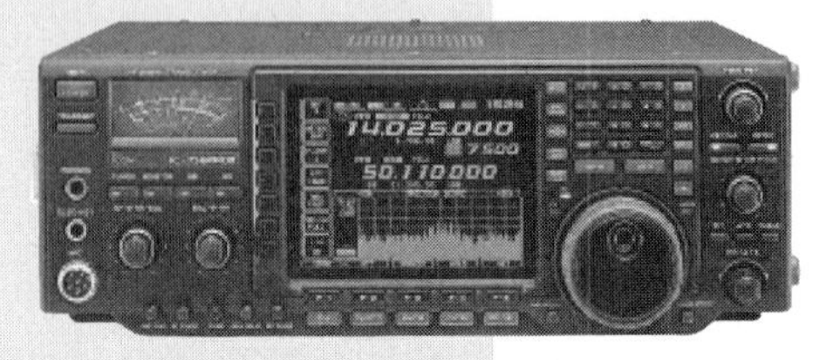

IC-756PROIII

Heard it. Worked it. Logged it.

100 Watt Output, Full Duty Cycle | 32 Bit IF-DSP + 24 Bit AD/DA Converter | +30 dBm 3rd Order Intercept Point | Low-distortion BPF Switching | Real-time Spectrum Scope with Mini Scope

IC-746PRO

HF Hot Rod

100 Watt Output, Full Duty Cycle | 32 Bit IF-DSP + 24 Bit AD/DA Converter | All Mode, including RTTY | Enhanced Receiver Performance | Selectable, "Build Your Own" IF Filter Shapes

IC-718

The "Get into HF" Rig

100 Watt Output, Full Duty Cycle | One-touch Band Switching | RF Gain Control | Built-in CW Keyer | Built-in VOX

IC-910H

Satellite Rig

100 Watt VHF/75 Watt UHF Variable Output | AM, FM, SSB, CW, & Satellite | Work Two Bands at Once | Main & Sub Band Functions for IF Shift, Sweep, NB, & RF Attenuator

IC-7000

New HF to 70CM Multibander

100 Watt HF+6M, 50 Watt 2M, 35 Watt 70CM | DSP² - Dual DSP Processors | Digital IF Filters | Twin Pass Band Tuning | Multiple AGC Loops | MNF² - Dual Manual Notch Filters

IC-706MKIIG

Proven Multiband Performance

100 Watt HF* + 6M, 50 Watt 2M, 20 Watt 70CM | All Mode, Full Duty Cycle | 99 Alphanumeric Memories | CTCSS Encode/Decode with Tone Scan | Built-in DSP

Shack, Car, or Field...

For the love of

DIGITAL

IC-2200H

Digital or Analog 2M Mobile

65W Output | 207 Alphanumeric Memories | Optional Digital Voice & Data | Optional Callsign Squelch | CTCSS & DTCS Encode/Decode w/Tone Scan | Weather Alert

DIGITAL

ID-800

2M/70CM Digital Mobile

55 Watt VHF/50 Watt Output | Wide RX: 118-173, 230-549, 810-999 MHz (Cellular Blocked) | Analog/Digital Voice & Data | Callsign Squelch | CTCSS & DTCS Encode/Decode w/Tone Scan

DIGITAL

ID-1

Advanced 1.2 GHz Digital Mobile

10 Watt Output | High Speed Digital Data, Digital Voice, Analog Voice (FM) | Wireless Internet/Network Access Capable | PC Control via USB port | Digital Callsign & Digital Code Squelch

Introducing the... AntennaSmith™!

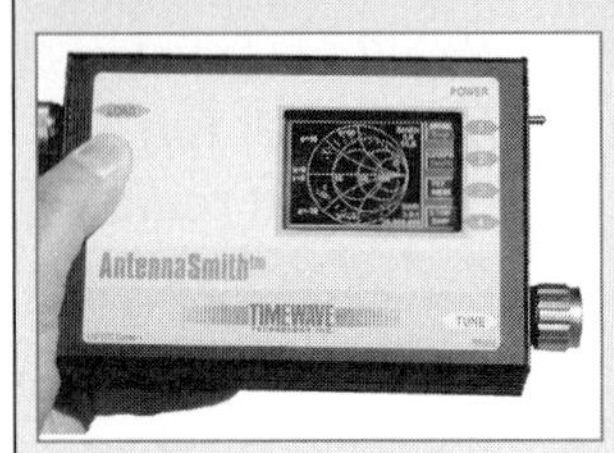

TZ-900 Antenna Impedance Analyzer

Featuring 2 Sec Sweeps, Sweep Memories, 1 Hz steps, manual & computer control w/software, low power. Rugged Aluminum Housing - Take it up the tower!

Full Color TFT LCD Graphic Display - Sunlight Visible!

- 0.5 - 60 MHz
- SWR
- Impedance (Z)
- Reactance (r+jx)
- Reflection coefficient (ρ,θ)
- Smith Chart

The ANC-4 and the DSP-599zx - the perfect complement!

DSP-599zx V5 Signal Processor

The #1 signal processor for reducing noise, eliminating heterodynes and slicing away QRM!

ANC-4 Antenna Noise Eliminator

Reduces BPL & Line Noise up to 40 dB. Wipe out S-9 line noise before it hits your receiver!

PK-232/PSK Multimode Data Controller

Do all the modes - old & new! The most popular data controller with a built-in Sound Card Interface and DSP!

PK-96/100 Packet Radio TNC

Proven value for 1200/9600 bps packet, GPS tracking!

651-489-5080 sales@timewave.com www.timewave.com
1025 Selby Ave., Suite 101 St. Paul, MN 55104 USA

New! USB Interfaces
Free your PC serial ports. USB to CW/PTT & USB to VK-64 versions.

VK-64 Combo Voice/CW Keyer
Voice keyer and full feature CW memory keyer in a single package. Front panel operation or control through your laptop or PC.

BCD-10 Band Decoder
Use band port signals from selected Yaesu® rigs or PC printer port for automatic antenna switching as you change bands.

XT-4 CW Memory Keyer
Battery powered and small size for VHF rover, FD, DXpeditions and vacations. 4 memories.

XT-4BEACON - CW Beacon IDer
Easy to program IDer for VHF beacons. Low power. Selectable speeds 5-25 WPM.

Visit our web site for new products.

Unified Microsystems
PO Box 133
Slinger, WI 53086 262-644-9036
www.unifiedmicro.com

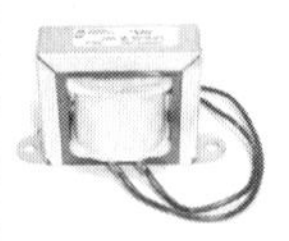

RADIO DAZE — VINTAGE RADIO & ELECTRONICS

• Books • Capacitors Of All Kinds • Chassis-Aluminum/Steel • Chokes • Custom Cloth-Covered Solid/Stranded Wire • Dial Belts & Cord • Dial Lamps & Sockets • Diodes • Enclosures • Fuses & Fuseholders • Grillecloth • Hardware • Kits • Knobs • Potentiometers • Power Cord & Plugs • Refinishing Supplies • Reproduction Glass Dials For Vintage Amateur, Audio & Radio • Resistors • Service Supplies • Sockets • Technical Data • Tools • Transformers - Classic Audio, Power, Filament, Plate, etc. • Vacuum Tubes - One of the largest inventories of NOS tubes ..... and much, much more!

Contact Us For Our Free Full Line Catalog!!

Radio Daze, LLC · 7620 Omnitech Place · Victor, New York 14564
(585) 742-2020 · Fax: (585) 742-2099 · Toll Free Fax: (800) 456-6494
e-mail: info@radiodaze.com · Web Site: www.radiodaze.com

ADVANCED SPECIALTIES INC.

New Jersey's Communications Store

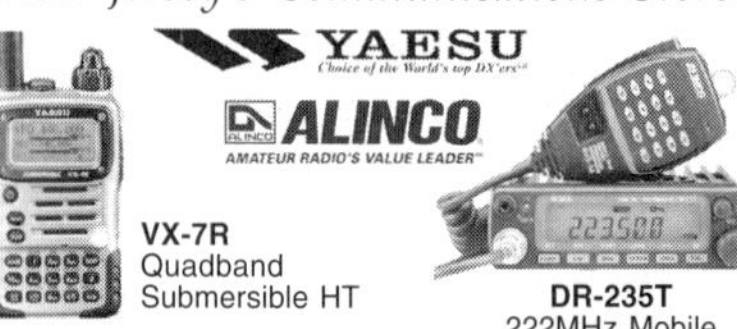

VX-7R
Quadband
Submersible HT

DR-235T
222MHz Mobile
Alphanumeric Display

FT-2800M
65W 2M Mobile

FT-857D
HF, 6M, 2M, 70 CM

Authorized Dealer
* ALINCO * LARSEN
* COMET * MALDOL * MFJ
* UNIDEN * LDG * MAHA* ANLI
* RANGER * YAESU * PRYME

FT-8800R
2M/70 CM
FM Mobile

VX-6R
2M, 222MHz, 70 CM
FM Handheld

AMATEUR RADIO – SCANNERS – BOOKS
ANTENNAS – MOUNTS – FILTERS
ACCESSORIES & MORE

Closed Sunday & Monday
Orders/Quotes 1-800-926-9HAM
(201)-VHF-2067
114 Essex Street,
Lodi, NJ 07644

Big Online Catalog at
www.advancedspecialties.net

Quality Radio Equipment Since 1942

AMATEUR TRANSCEIVERS

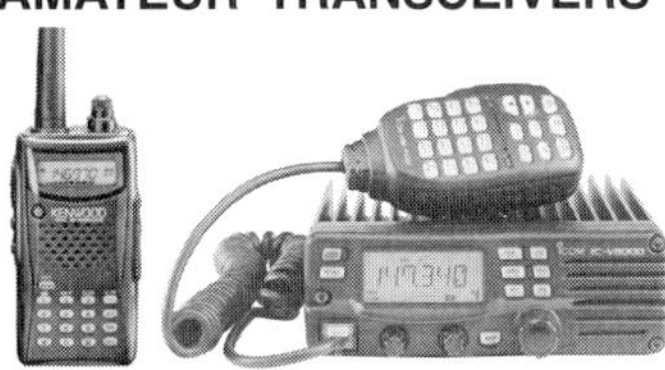

Universal Radio is your authorized dealer for all major amateur radio equipment lines including: Alinco, Icom, Yaesu, Kenwood and SGC.

SHORTWAVE RECEIVERS

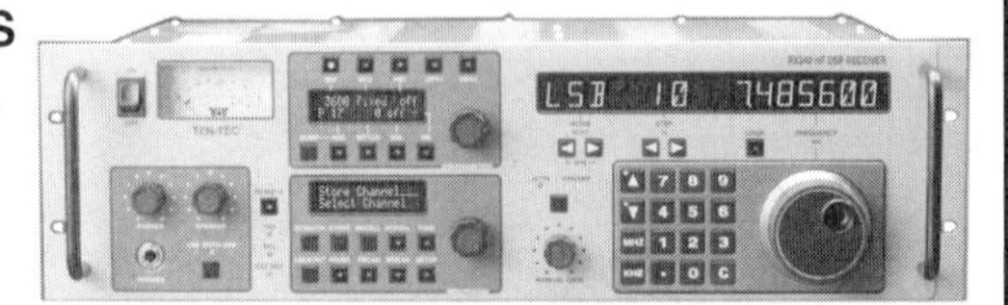

Whether your budget is $50, $500 or $5000, Universal can recommend a shortwave receiver to meet your needs. Universal is the *only* North American authorized dealer for *all* major shortwave lines including: Elad, Icom, Yaesu, Japan Radio Company, Ten-Tec, Grundig, etón, Sony, Sangean, Palstar, RFspace and Kaito.

WIDEBAND RECEIVERS

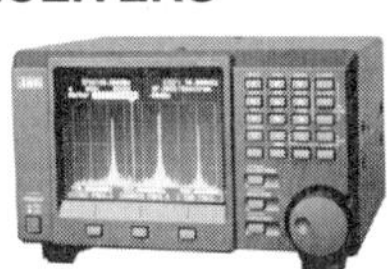

Universal Radio carries a broad selection of wideband and scanner receivers. Alinco, AOR, Icom, Yaesu and Uniden models are offered at competitive prices. Portable and tabletop models are available.

ANTENNAS AND ACCESSORIES

Your antenna is just as important as your radio. Universal carries antennas for every need, space requirement and budget. Please visit our website or request our catalog to learn more about antennas.

BOOKS

Whether your interest is amateur, shortwave or collecting; Universal has the very best selection of radio books and study materials.

■ USED EQUIPMENT

Universal carries an extensive selection of used amateur and shortwave equipment. All items have been tested by our service department and carry a 60 day warranty unless stated otherwise. Request our monthly printed used list or visit our website:

www.universal-radio.com

◆ VISIT OUR WEBSITE

Guaranteed lowest prices on the web? Not always. But we ***do*** guarantee that you will find the Universal website to be the **most** informative.

www.universal-radio.com

◆ VISIT OUR SHOWROOM

FREE 108 PAGE CATALOG

Our informative print catalog covers everything for the amateur, shortwave and scanner enthusiast. All items are also viewable at:

www.universal-radio.com

Universal Radio, Inc.
6830 Americana Pkwy.
Reynoldsburg, OH 43068
➤ 800 431-3939 Orders
➤ 614 866-4267 Information
➤ 614 866-2339 Fax
➤ dx@universal-radio.com

universal radio inc.

✓ Established in 1942.
✓ We ship worldwide.
✓ Visa, Master and Discover cards.
✓ Used equipment list available.
✓ Returns subject to 15% restocking fee.

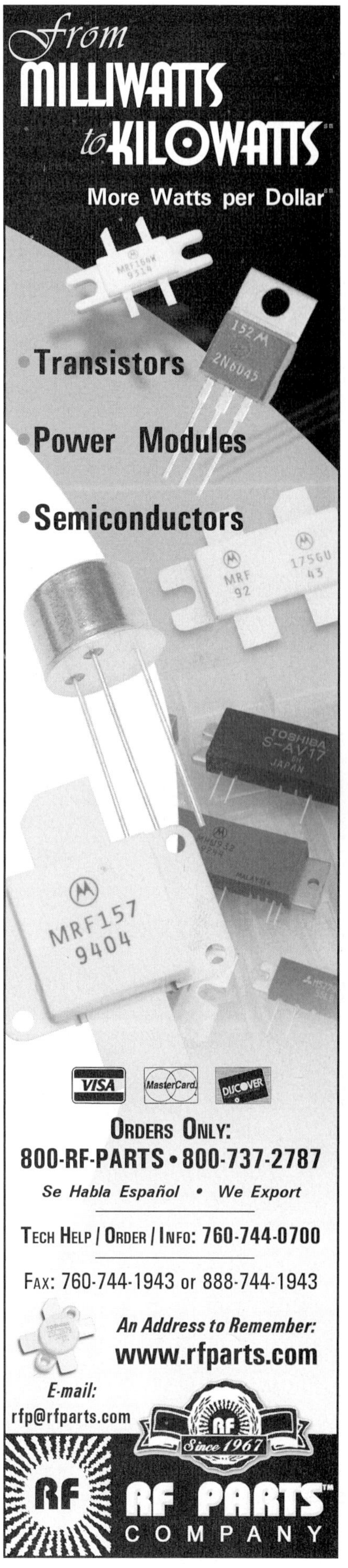

Index

Editor's Note: Except for commonly used phrases and abbreviations, topics are indexed by their noun names. Many topics are also cross-indexed, especially when noun modifiers appear (such as "Voltage, back" and "Back voltage"). Numerous terms and abbreviations pertaining to Amateur Radio but not contained in the index may be found in the glossaries indexed herein. The letters "ff" after a page number indicate coverage of the indexed topic on succeeding pages.

A

B

C

E

F

G

H

I

J

K

L

M

N

O

P

Q

R

S

T

U

V

W

X

Y

Z

FEEDBACK

Please use this form to give us your comments on this book and what you'd like to see in future editions, or e-mail us at **pubsfdbk@arrl.org** (publications feedback). If you use e-mail, please include your name, call, e-mail address and the book title, edition and printing in the body of your message. Also indicate whether or not you are an ARRL member.

Where did you purchase this book?

☐ From ARRL directly ☐ From an ARRL dealer

Is there a dealer who carries ARRL publications within:

☐ 5 miles ☐ 15 miles ☐ 30 miles of your location? ☐ Not sure.

License class:

☐ Novice ☐ Technician ☐ Technician with code ☐ General ☐ Advanced ☐ Amateur Extra

---

Name ______________________ ARRL member? ☐ Yes ☐ No

Call Sign ______________________

Daytime Phone ( ) ______________________ Age ______________________

Address ______________________

City, State/Province, ZIP/Postal Code ______________________

If licensed, how long? ______________________ e-mail address: ______________________

Other hobbies ______________________

____

Occupation ______________________

For ARRL use only	2006 HBK
Edition	83 84 85 86 87 88 89
Printing	1 2 3 4 5 6 7 8 9 10 11 12

From ______________________________

____

____

Please affix postage. Post Office will not deliver without postage.

EDITOR, ARRL HANDBOOK
ARRL—THE NATIONAL ASSOCIATION FOR AMATEUR RADIO
225 MAIN STREET
NEWINGTON CT 06111-1494

— — — — — — — — — — — — — — — — please fold and tape — — — — — — — — — — — — — — — —

About the ARRL

The national association for Amateur Radio

The seed for Amateur Radio was planted in the 1890s, when Guglielmo Marconi began his experiments in wireless telegraphy. Soon he was joined by dozens, then hundreds, of others who were enthusiastic about sending and receiving messages through the air—some with a commercial interest, but others solely out of a love for this new communications medium. The United States government began licensing Amateur Radio operators in 1912.

By 1914, there were thousands of Amateur Radio operators—hams—in the United States. Hiram Percy Maxim, a leading Hartford, Connecticut inventor and industrialist, saw the need for an organization to band together this fledgling group of radio experimenters. In May 1914 he founded the American Radio Relay League (ARRL) to meet that need.

Today ARRL, with approximately 170,000 members, is the largest organization of radio amateurs in the United States. The ARRL is a not-for-profit organization that:

- promotes interest in Amateur Radio communications and experimentation
- represents US radio amateurs in legislative matters, and
- maintains fraternalism and a high standard of conduct among Amateur Radio operators.

At ARRL headquarters in the Hartford suburb of Newington, the staff helps serve the needs of members. ARRL is also International Secretariat for the International Amateur Radio Union, which is made up of similar societies in 150 countries around the world.

ARRL publishes the monthly journal *QST*, as well as newsletters and many publications covering all aspects of Amateur Radio. Its headquarters station, W1AW, transmits bulletins of interest to radio amateurs and Morse code practice sessions. The ARRL also coordinates an extensive field organization, which includes volunteers who provide technical information and other support services for radio amateurs as well as communications for public-service activities. In addition, ARRL represents US amateurs with the Federal Communications Commission and other government agencies in the US and abroad.

Membership in ARRL means much more than receiving *QST* each month. In addition to the services already described, ARRL offers membership services on a personal level, such as the ARRL Volunteer Examiner Coordinator Program and a QSL bureau.

Full ARRL membership (available only to licensed radio amateurs) gives you a voice in how the affairs of the organization are governed. ARRL policy is set by a Board of Directors (one from each of 15 Divisions). Each year, one-third of the ARRL Board of Directors stands for election by the full members they represent. The day-to-day operation of ARRL HQ is managed by an Executive Vice President and his staff.

No matter what aspect of Amateur Radio attracts you, ARRL membership is relevant and important. There would be no Amateur Radio as we know it today were it not for the ARRL. We would be happy to welcome you as a member! (An Amateur Radio license is not required for Associate Membership.) For more information about ARRL and answers to any questions you may have about Amateur Radio, write or call:

ARRL—The national association for Amateur Radio
225 Main Street
Newington CT 06111-1494
Voice: 860-594-0200
Fax: 860-594-0259
E-mail: **hq@arrl.org**
Internet: **www.arrl.org/**

Prospective new amateurs call (toll-free):
800-32-NEW HAM (800-326-3942)
You can also contact us via e-mail at **newham@arrl.org**
or check out *ARRLWeb* at **http://www.arrl.org/**

About the Included CD-ROM

On the included CD-ROM you'll find this entire *Handbook*, including text, drawings, tables, illustrations and photographs—many in color. Using the industry-standard Adobe *Reader* (included), you can view and print the text of the book, zoom in and out on pages, and copy selected parts of pages to the clipboard. A powerful search engine—Acrobat Search—helps you find topics of interest. Also included are Template Packages for many of the projects in this book, the companion software mentioned throughout, the short videos *Amateur Radio Today* and *ARRL goes to Washington*, and the free Adobe *Reader* 7.0 to view the *Handbook* files.

INSTALLING YOUR ARRL HANDBOOK CD

Windows

1. Close any open applications and insert the CD-ROM into your CD-ROM drive. (Note: If your system supports "CD-ROM Insert Notification," the *ARRL Handbook CD* may automatically run the installation program when inserted.)
2. Select **Run** from the *Windows* **Start** menu.
3. Type **d:\setup** (where d: is the drive letter of your CD-ROM drive; if the CDROM is a different drive on your system, type the appropriate letter) and press **Enter**.
4. Click "Install the ARRL Handbook CD" and follow the instructions that appear on your screen.

Macintosh

No installation procedure is needed for the Macintosh. To install so the CD is not required, drag the "ARRL Handbook CD 10.0" icon onto your hard drive.

USING YOUR ARRL HANDBOOK CD

Windows

Note: If your systems supports "CD-ROM Insert Notification," the *ARRL Handbook CD* may run automatically when inserted.

1. To run the program after installation, start *Windows* and place the disc in your CD-ROM drive. (Not required if a "complete" installation was chosen during setup.)
2. From the **Start** menu select **Programs**, then select **ARRL Software** and the **2006 ARRL Handbook CD 10.0** item.

Macintosh

To run the program after installation, place the disc in your CD-ROM drive. Adobe *Reader* should launch automatically to view the book.